Lecture Notes in Civil Engineering

Volume 237

Lecture Notes in Civil Engineering (LNCE) publishes the latest developments in Civil Engineering—quickly, informally and in top quality. Though original research reported in proceedings and post-proceedings represents the core of LNCE, edited volumes of exceptionally high quality and interest may also be considered for publication. Volumes published in LNCE embrace all aspects and subfields of, as well as new challenges in, Civil Engineering. Topics in the series include:

- Construction and Structural Mechanics
- Building Materials
- Concrete, Steel and Timber Structures
- Geotechnical Engineering
- Earthquake Engineering
- Coastal Engineering
- Ocean and Offshore Engineering; Ships and Floating Structures
- Hydraulics, Hydrology and Water Resources Engineering
- Environmental Engineering and Sustainability
- Structural Health and Monitoring
- Surveying and Geographical Information Systems
- Indoor Environments
- Transportation and Traffic
- Risk Analysis
- Safety and Security

To submit a proposal or request further information, please contact the appropriate Springer Editor:

– Pierpaolo Riva at pierpaolo.riva@springer.com (Europe and Americas);
– Swati Meherishi at swati.meherishi@springer.com (Asia—except China, Australia, and New Zealand);
– Wayne Hu at wayne.hu@springer.com (China).

All books in the series now indexed by Scopus and EI Compendex database!

Mahdi Kioumarsi • Behrouz Shafei
Editors

The 1st International Conference on Net-Zero Built Environment

Innovations in Materials, Structures, and Management Practices

 Springer

Editors
Mahdi Kioumarsi
Department of Built Environment
Oslo Metropolitan University
Oslo, Norway

Behrouz Shafei
Department of Civil, Construction,
and Environmental Engineering
Iowa State University
Ames, IA, USA

The 1st International Conference on Net-Zero Built Environment: Innovations in Materials, Structures, and Management Practices (NETZ), netz, netz 2024, NETZ, 1, Oslo, Norway (2024) 6 19 (2024) 6 21. https://netzfuture.com/conference/

ISSN 2366-2557   ISSN 2366-2565   (electronic)
Lecture Notes in Civil Engineering
ISBN 978-3-031-69625-1   ISBN 978-3-031-69626-8   (eBook)
https://doi.org/10.1007/978-3-031-69626-8

This Springer imprint is published by the registered company Springer Nature Switzerland AG
The registered company address is: Gewerbestrasse 11, 6330 Cham, Switzerland

If disposing of this product, please recycle the paper.

Preface

The *1st International Conference on Net-Zero Built Environment: Innovations in Materials, Structures, and Management Practices* was successfully held on June 19–21, 2024, in Oslo, Norway. This conference was part of the *Net-Zero Future* project sponsored by the Research Council of Norway and the Norwegian Directorate for Higher Education and Skills. Through the referenced project, a unique international alliance was formed among Norway, the United States, Germany, South Africa, and India to conduct collaborative research and educational activities toward achieving a net-zero built environment, capitalizing on mutual interests and the diversity of practices. This alliance is in line with sustainable development goals, such as quality education, industry and innovation, sustainable cities, climate action, and partnerships.

The pressing global challenges associated with the degradation of the built environment due to mounting stressors and, on the other hand, the environmental footprint of growth in the built environment serving communities have necessitated a profound transformation in how we design and develop our infrastructure. Civil engineering is the primary field responsible for infrastructural development and is at the forefront of this transformation. The conference series on Net-Zero Built Environment: Innovations in Materials, Structures, and Management Practices has been dedicated to exploring the concept of net-zero within the realm of civil engineering and a variety of relevant domains, emphasizing the critical importance of achieving a net-zero carbon footprint in our built environment.

Net-zero in the built environment refers to the knowledge and practice of delivering, maintaining, and managing civil infrastructures that produce zero net carbon emissions throughout their lifecycle. This ambitious goal requires a holistic effort, encompassing methodological approaches to design, construction, operation, and eventual decommissioning. This motivated us to establish three main themes for this conference series, capturing innovations in materials, structures, and management practices. The first conference of this conference series attracted a large group of participants from the academic, industry, and public sectors, presenting the latest developments in each of the identified main themes.

After a rigorous peer-review process performed by the conference's international scientific committee members and other expert reviewers, 158 full-text papers have been selected to be included in this book. The selected papers represent a diverse group of authors and research groups from around the globe, providing original perspectives and insights into how we can pave the way toward a net-zero built environment. Converging on this ultimate goal, the selected papers offer the latest advances in (i) new materials and manufacturing processes for zero carbon footprint, (ii) robotic construction technologies for minimum formwork and on-site activities, (iii) novel structural designs and details for optimal performance with the least materials, (iv) advanced condition assessment and health monitoring strategies, and (v) innovative life-cycle analyses and civil infrastructure management strategies.

We recognize that achieving net zero is not merely a development goal but a moral imperative to ensure a sustainable future for all. Reducing carbon emissions can help stabilize temperature patterns, decrease the frequency and severity of extreme weather events, and protect vulnerable ecosystems. All who contributed to this book share the same passion to positively impact the communities around the world. Through the dissemination of the latest findings and innovations, this book's main themes and individual chapters directly contribute to the net-zero domain. By equipping scientists, engineers, policymakers, and the general public with relevant knowledge and expertise, we hope to collectively drive a global transition toward a net-zero future.

Oslo, Norway   Mahdi Kioumarsi
Ames, IA, USA   Behrouz Shafei

Contents of Volume I

Contents of Volume II

An Experimental Study of Consistency and Strength Variation of 3D-Printed Concrete Mixes

Dmitry Vysochinskiy ⓘ, Gunnar Madsen, and Ingrid Lande ⓘ

Contents

1 Introduction

Concrete three-dimensional (3D) printing is an innovative technology that allows for building three-dimensional objects layer by layer. Although the process is traditionally called concrete 3D printing, most of the mixes used have a maximum particle size of 2–3 mm, and, thus, calling the process mortar 3D printing would be more technically correct. In this chapter, the terms "3D-printed mortar" and "3D-printed concrete" are used interchangeably. Typical concrete 3D printing involves extrusion of a mortar filament through a nozzle with a diameter ranging from 6 to 50 mm [1]. The printing head is mounted on a gantry or robotic arm that positions the material to form the printed structure layer by layer. The advantages of concrete 3D printing include the absence of the need for formwork and shape flexibility, whereas the challenges include difficulties in applying reinforcement to 3D-printed structures and the lack of standards for the structural use of 3D-printed parts. A substantial

D. Vysochinskiy (✉) · G. Madsen · I. Lande
University of Agder, Grimstad, Norway
e-mail: dmitry.vysochinskiy@uia.no

© The Author(s) 2025
M. Kioumarsi, B. Shafei (eds.), *The 1st International Conference on Net-Zero Built Environment*, Lecture Notes in Civil Engineering 237,
https://doi.org/10.1007/978-3-031-69626-8_1

portion of the research on 3D-printed concrete is dedicated to the properties of the fresh concrete and mix design, ensuring the pumpability, extrudability and buildability of the mix. Interested readers are referred to reviews by Zhang et al. [2] and Chen et al. [3] for more details on mix design and characterization methods. When it comes to the strength and durability of hardened 3D-printed concrete, researchers are mainly concerned with the anisotropy and influence of the layer-interface properties on the overall performance of the structure as reported by e.g. Nerella et al. [4] and Rahul et al. [5]. When a mortar mix is tested in a laboratory, the water-to-binder-ratio (w/b) is usually known and fixed, whereas in the case of an industrial printing process, the concrete printer operator specifies not w/b but the amount of water in litres per hour and routinely adjusts this amount based on a feel of good consistency. Thus, the behaviour of the same mix in the laboratory and at the printing site might vary due to a fluctuating w/b. In this chapter, we present an approach that allows quantifying this variation, which has not yet received enough attention in the literature.

2   Materials and Methods

2.1   3D Printers and Materials

The main piece of equipment is a robot arm-based 1K system concrete printer from Hyperion Robotics™, as shown in Fig. 1. This printer consists of a silo with dry mix, a mortar mixing pump and a robot arm that performs the printing. The printing

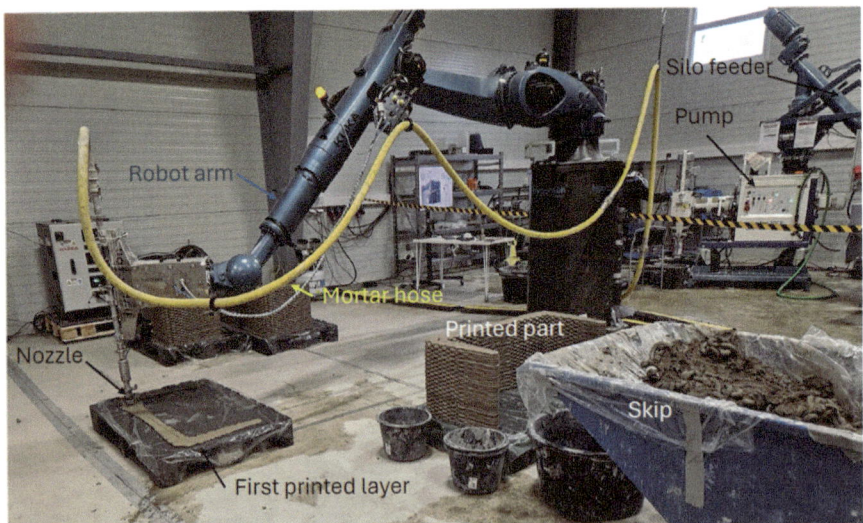

Fig. 1 A view of the 3D printing process

process is shown in Fig. 1. This type of concrete 3D printer uses dry mortar mixes with a maximum particle size of 2 mm. The dry mortar is fed into the pump where it is mixed with water and then pumped through the hose. At another side of the hose is the printing head with the printing nozzle. The robot arm controls the motion of the printing nozzle and creates 3D-printed parts layer by layer. The usual printing velocity is 245 mm/s, and nozzles of 18 mm and 20 mm are used.

Two commercially available dry 3D mortar mixes were tested in this chapter. The mixes are anonymized and denoted simply as mix A and mix B in order not to accidentally disclose any proprietary information or benchmark the mixes against each other. At the time of the demonstrations, the silo was filled with dry mix B, whereas mix A was tested only in the laboratory.

2.2   Testing Equipment and Procedure

The consistency of the mortar needs to be firm enough to support its own weight yet wet enough to flow well, have sufficient bonding between layers and prevent the formation of extrusion cracks. The "right" consistency is determined subjectively by the operator through visual observation of the flow. We supplement the subjective test of the consistency with the quantifiable drop flow table measurements performed in accordance with BS EN 1015-3:1999 [6], as illustrated in Fig. 2. One fills a conical form with mortar, measures the initial diameter of the cone when the form is removed, then performs 15 drops of the flow table and measures the increased cone diameter. The laboratory flow table was transported to the 3D printing site for field measurements i.e. the same flow table was used in the field and in the laboratory.

The exact dry mix design is proprietary information, and the printing operator specifies the amount of water added in the form of litres per hour, which means that the *w/b* of the printed mortar is not directly specified. The water-to-binder ratio (*w/b*)

Fig. 2 Subjective vs. objective control of 3D printing mortar consistency

surpassed the more traditional water-to-cement ratio (w/c) as other binders such as silica fume, fly ash, metallurgical slag and natural pozzolans become commonly added to cement. w/b is calculated as

$$w/b = \frac{w}{c + \sum(k_i \cdot p)} \approx \frac{w}{c + \sum p} \tag{1}$$

where w is the amount of water, c is the amount of cement and $k \cdot p$ is the amount of each additional binder multiplied by a corresponding coefficient [7].

The particle matrix model of concrete [8] interprets all particles smaller than 0.125 mm in diameter as part of the concrete matrix. The mass of the particles included in the concrete matrix can be calculated as

$$m_{0.125} = c + \sum p + m_{filler} \leq c + \sum p \tag{2}$$

where m_{filler} is the mass of particles <0.125 mm coming for the aggregates. Assuming m_{filler} to be small allows us to replace w/b with an approximate water-to-fine particles ratio (w/f)

$$w/f = \frac{w}{m_{0.125}} = \frac{\%wt_{water}}{\%wt_{0.125}} \tag{3}$$

w/f describes the amount of water in the concrete matrix in the same manner as w/c describes the amount of water in the cement paste. For a mix that contains no other binders or fine particles than cement, the matrix would consist of pure cement paste, making $w/f = w/b = w/c$. The advantage of using w/f instead of w/b is that to calculate w/f, one only requires to know the weight percentage $\%wt_{0.125}$ of fine particles (<0.125 mm). For any dry mix, this parameter can be determined by simple sieving analyses performed using a set of sieves with cell diameters of 4.0 mm, 2.0 mm, 1.0 mm, 0.500 mm, 0.250 mm, 0.125 mm and 0.063 mm. Before sieving, the mix was dried in an oven at a temperature of 110 °C for 24 h.

The strength of the hardened mortar was tested in accordance with the mortar strength testing standard [9]. Prismatic samples with dimensions of 40 mm × 40 mm × 160 mm were cast using steel forms, removed from the forms the next day and stored under water at room temperature up until testing at the ages of 1 day, 7 days and 28 days. Figure 3 illustrates the bending and compression testing of the prisms. A total of 6 prisms were tested at each age for each w/f; thus, a minimum of 18 prisms were cast for each w/f tested in the laboratory. For the field tests at the 3D printing site, a total of 24 40 mm × 40 mm × 160 mm prisms [9] and 6100 mm × 100 mm × 100 mm cubes [10] were cast. The cubes were stored under water and tested at the age of 7 days.

To summarize, the following experimental procedure was followed:

• Sieve analyses were performed to determine the content of fine particles (<0.125 mm) in each mix.

Fig. 3 Bending and compression testing of 40 mm × 40 mm × 160 mm prisms

- Field measurements of consistency combined with casting of samples were performed to determine the consistency and strength of the printed mortar. Note that w/f of the printed mortar is not known beforehand.
- Laboratory measurements of consistency and strength were performed for mixes with a known w/f.
- Field measurements and measurements of laboratory mixes were compared.

3 Results

The sieving analysis showed that dry mix A contained about 40.0% weight of particles <0.125 mm and dry mix B contained approximately 43.8% weight of particles <0.125 mm; both numbers provided are an average of more than three measurements. These high % weight of binders are typical for 3D printing mortar mixes, as shown in Zhang et al. [2].

Once the weight percentage of the fine particles was determined, we mixed and tested the four different w/f values in the laboratory for each mix, as shown in Table 1. In Table 1, "b" stands for "bending" and denotes the average flexural strength, whereas "c" stands for "compression" and denotes the average compression strength; the reported flow included in Table 1 is dynamic flow i.e. the diameter of the cone after 15 jolts of the flow table.

In addition to the laboratory mixes, two field measurements were performed to test the consistency and strength of the 3D-printed material. The quality of the first field casts performed on 25th September was deemed insufficient and was not tested at full strength; instead, the strength at the age of 3 days was tested (not included in Table 1). Another test not included in Table 1 is the test of 100 mm × 100 mm × 100 mm cubes at the age of 7 days for the second field cast. Those cubes displayed the strength of 47.7 MPa with the standard deviation of 3.2 MPa as opposed to the

Table 1 An overview of the performed tests

			Prism strength					
			1 day		7 days		28 days	
Casting date	w/f	Flow	b	c	b	c	b	c
dd mm yyyy	[−]	[cm]	[MPa]	[MPa]	[MPa]	[MPa]	[MPa]	[MPa]
Mix A								
Lab mixer								
28 10 2023	0.325	13.8	2.9	14.3	6.9	46.7	8.7	67.4
25 10 2023	0.350	14.9	2.1	11.6	5.4	34.7	7.6	49.8
26 10 2023	0.375	16.2	2.0	9.2	5.7	34.5	8.0	52.1
27 10 2023	0.400	17.3	1.4	6.5	4.7	27.6	7.6	47.7
Mix B								
3D printer								
25 09 2023	Field	12.3	1.5	18.0	2.7	42.2	n/t	n/t
10 11 2023	Field	14.9	2.3	14.7	4.0	49.3	8.0	76.0
Lab mixer								
29 10 2023	0.325	11.1	3.1	16.3	7.3	54.5	9.1	86.0
17 10 2023	0.350	11.3	3.7	17.6	6.7	50.2	9.3	79.1
18 10 2023	0.375	12.8	2.9	15.2	4.4	45.9	7.6	73.2
24 10 2023	0.400	14.9	3.0	14.5	5.7	39.5	7.5	65.5

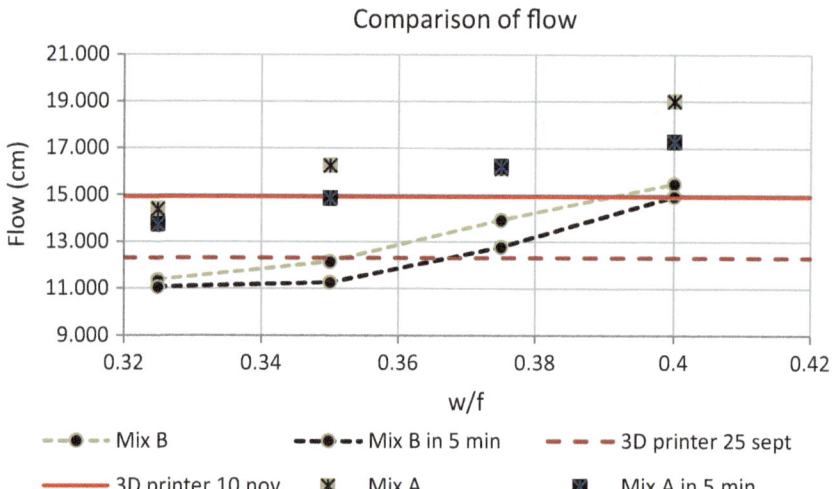

Fig. 4 Comparison of the mortar flow measurements in the laboratory and at the 3D printing site

compressive strength of 49.3 MPa (the value is underscored in Table 1) with the standard deviation of 2.8 MPa displayed by the prisms tested at the same age.

Figure 4 illustrates a comparison of the table flow measurements of the 3D-printed mortar at the site and in the laboratory; only dynamic flow i.e. values after the 15 table jolts are plotted. We can see that the printable consistency window

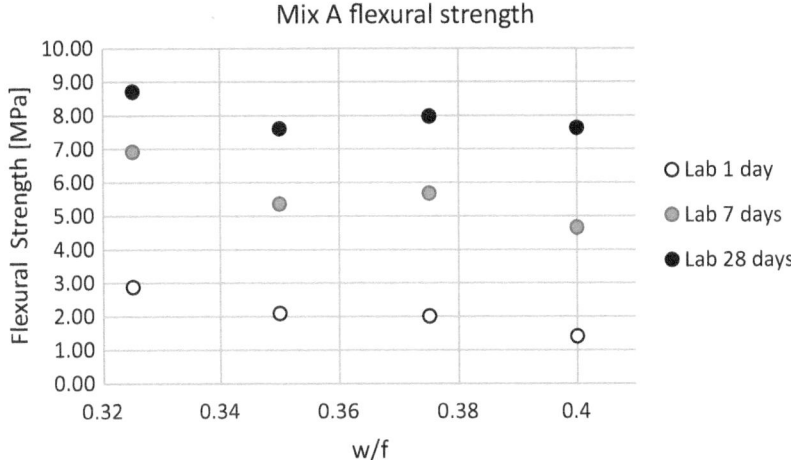

Fig. 5 Flexural strength of the prismatic samples for mix A

for dry mix B is approximately between a *w/f* of 0.35 and 0.40. Mortar made from dry mix A is more flowable at the same *w/f* in comparison with dry mix B, and its printable consistency window is likely to be at lower *w/f* values i.e. around 0.3 *w/f*. The conclusion that mix A's printable consistency lays around a *w/f* of 0.3 could be drawn from the flow measurements in Fig. 4 alone, but it was later confirmed by acquiring a technical data sheet from the producer in which the recommended water content is specified to be 12–13% of the dry mortar mass, which corresponds to *w/f* = 12%/40% = 0.3.

Figures 5 and 6 illustrate the evolution of the flexural strength and compressive strength of the mortar prepared from mix A, respectively, whereas Figs. 7 and 8 provide the same illustration for the dry mix B. The values of strength for the 3D-printed mix are from the second field measurements i.e. from 10 November 2023. Expectedly, one can see a decrease in the material strength with an increase of *w/f*. An interesting observation in Figs. 7 and 8 is that the 1-day strength of the printed mortar is lower than the strength of all mortar mixes tested in the laboratory, whereas for 28 days, the strength of the field samples falls between the strength of the laboratory samples of 0.35 and 0.375 *w/f*. A likely explanation is that the initial hardening of the mortar mixed in the laboratory (at ambient temperature of about 20 °C) is faster than the hardening of the mortar mixed in the printer (the ambient temperature in the hall was around 10 °C on the day of the test).

Figure 4 indicates that the *w/f* of the mortar printed on 10th November should be around 0.4 based on the flow measurements, whereas comparisons of the strength with the laboratory-mixed sample measurements in Figs. 7 and 8 indicate a *w/f* of 0.35 and 0.375, respectively. For the strength measurements for the field test of 25th September, the situation is the opposite; the levels of strength in Table 1 are slightly

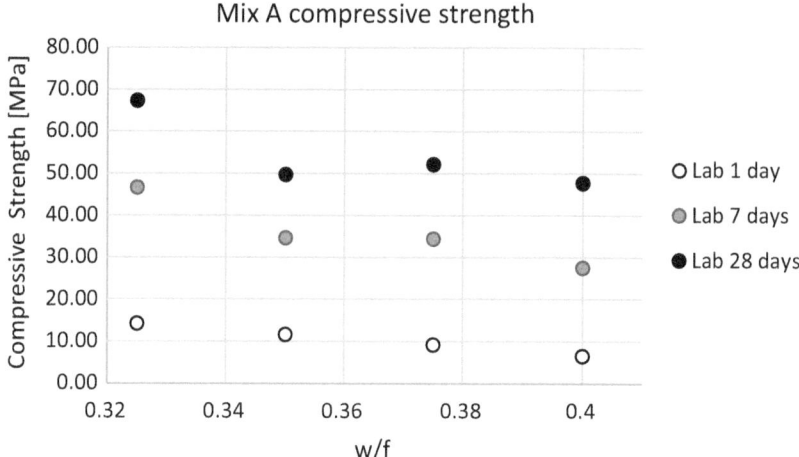

Fig. 6 Compressive strength of the prismatic samples for mix A

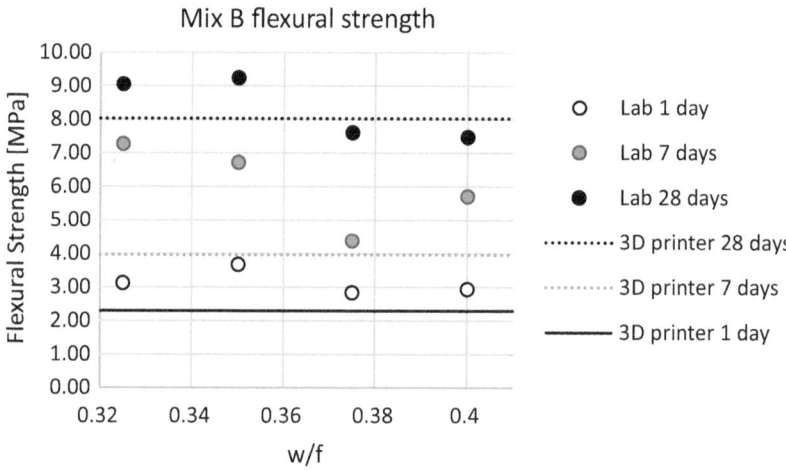

Fig. 7 Flexural strength of the prismatic samples for mix B

lower than one would suggest based on the *w/f* determined by comparing the flow values measured in the laboratory and at the 3D printing site. Thus, the flow measurement comparison serves as a viable but relatively crude estimate of *w/f* with an error of about ±0.05 *w/f*.

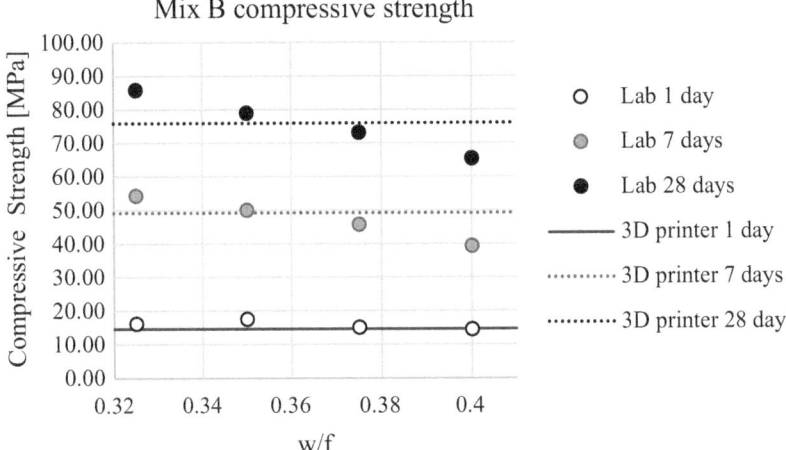

Fig. 8 Compressive strength of the prismatic samples for mix B

4 Conclusions

The aim of this study was to quantify the amount of strength variation one might expect from 3D-printed mortar of unspecified composition due to routine adjustment of the water content by the printer operator. The following conclusions were drawn:

- Replacing *w/b* with *w/f* is a viable way to handle mortar mixes of unspecified composition.
- A comparison of the flow measurements of the laboratory mixes of a known *w/f* and field mix provides an approximate crude estimate of the field mix *w/f*. A comparison with the strength measurements indicates an error of about ±0.05 *w/f*.
- The printable consistency window for mix B was approximately between 0.35 and 0.4 *w/f*, which corresponds to 79.1 MPa and 65.5 MPa of the mean compressive strength. Thus, about 20% of strength variation could be expected within the printable window of consistency.
- The difference in the mean strength between the two field measurements was about 15%, thus reinforcing the previous conclusion.
- The difference between the 7-day mean compressive strength tested on prismatic samples after the mortar standard [6] and the compressive strength of cubes tested after the concrete standard [10] was less than one standard deviation. Thus, use of prismatic samples provides a reliable strength estimate.
- The laboratory cast mixes noticeably overestimate the 1-day flexural strength of the mix compared to the field casts.

5   Discussion and Further Work

The strength measurements in this study were limited to mould cast mortar, whereas for a 3D-printed concrete structure, the bonding strength between the layers also plays an important role. The compressive strength of 3D-printed concrete would be lower than that of the corresponding cast mix due to the influence of the bonding strength between the layers [2, 11, 12]. The results found in the literature show examples of a 12–22% reduction of compressive strength [5] of printed mortar in comparison with the cast mix. Our study has shown about 20% of strength variation in cast mix due to routine adjustment of water content during printing. Combining these factors would suggest that, in the worst case scenario, the strength of a 3D-printed structure might be as low as 60–70% of that of a cast mix used for printing. However, the lowest interlayer bond strength is unlikely to be combined with the highest water content, making the overall reduction of strength less severe. A systematic study of the influence of water content variation in 3D-printed mortar on both the bond strength and mould cast strength is planned as further research work in order to produce more detailed recommendations on the appropriate safety factors on the strength and connection between the strength of mould cast mortar and that of a 3D-printed structure.

References

1. Buswell, R.A., Leal de Silva, W.R., Jones, S.Z., Dirrenberger, J.: 3D printing using concrete extrusion: a roadmap for research. Cem. Concr. Res. **112**, 37–49 (2018). https://doi.org/10.1016/J.CEMCONRES.2018.05.006
2. Zhang, C., et al.: Mix design concepts for 3D printable concrete: a review. Cem. Concr. Compos. **122**, 104155 (2021). https://doi.org/10.1016/J.CEMCONCOMP.2021.104155
3. Chen, Y., He, S., Gan, Y., Çopuroğlu, O., Veer, F., Schlangen, E.: A review of printing strategies, sustainable cementitious materials and characterization methods in the context of extrusion-based 3D concrete printing. J. Build. Eng. **45**, 103599 (2022). https://doi.org/10.1016/J.JOBE.2021.103599
4. Nerella, V.N., Hempel, S., Mechtcherine, V.: Effects of layer-interface properties on mechanical performance of concrete elements produced by extrusion-based 3D-printing. Constr. Build. Mater. **205**, 586–601 (2019). https://doi.org/10.1016/J.CONBUILDMAT.2019.01.235
5. Rahul, A.V., Santhanam, M., Meena, H., Ghani, Z.: Mechanical characterization of 3D printable concrete. Constr. Build. Mater. **227**, 116710 (2019). https://doi.org/10.1016/J.CONBUILDMAT.2019.116710
6. BS EN 1015-3:1999 – Methods of test for mortar for masonry - Determination of consistence of fresh mortar (by flow table)
7. NS-EN 206:2013+A2:2021+NA:2022 Concrete – specification, performance, production and conformity
8. Maage, M.: Betong: regelverk, teknologi og utførelse. Byggenæringens forl, Oslo (2015)
9. NS-EN 196-1:2016 Methods of testing cement. Part 1 Determination of strength
10. ISO 1920-4:2020 – testing of concrete—Part 4: strength of hardened concrete

11. Hou, S., Duan, Z., Xiao, J., Ye, J.: A review of 3D printed concrete: performance requirements, testing measurements and mix design. Constr. Build. Mater. **273**, 121745 (2021). https://doi.org/10.1016/J.CONBUILDMAT.2020.121745
12. Rehman, A.U., Kim, J.H.: 3D concrete printing: a systematic review of rheology, mix designs, mechanical, microstructural, and durability characteristics. Materials. **14**(14), 3800. https://doi.org/10.3390/MA14143800

Three-Dimensional Printing of Fly Ash-Based Geopolymer Materials

Shaurav Alam (ID)**, Stephen Gordon, Blake Bassett, Kevin Cobb, Brandon Jefferson, Nnamdi Nwoha, Tanvir Manzur, Kelly Crittenden, and John Matthews**

Contents

1 Introduction

Innovative materials and technologies in construction have spurred interest in fly ash-based geopolymer materials for three-dimensional (3D) printing. This study aims to design a 3D printer tailored for efficiently constructing multi-layered geopolymer walls. The geopolymer mixture, including fly ash, sand, sodium silicate, sodium hydroxide, and water, undergoes rapid curing via ohmic heating induced by alternating current (AC) voltage. After analyzing existing 3D printer designs, a

S. Alam (✉) · S. Gordon · J. Matthews
Department of CE and CET, LA Tech University, Ruston, LA, USA
e-mail: shaurav@latech.edu

B. Bassett · K. Cobb · B. Jefferson · N. Nwoha · K. Crittenden
Department of ME, LA Tech University, Ruston, LA, USA

T. Manzur
Department of CE, BUET, Dhaka, Bangladesh

M. Kioumarsi, B. Shafei (eds.), *The 1st International Conference on Net-Zero Built Environment*, Lecture Notes in Civil Engineering 237,
https://doi.org/10.1007/978-3-031-69626-8_2

bespoke printer was developed with a rotating auger assembly for controlled extrusion and parallel copper electrodes for AC heating during printing. The resulting printer successfully demonstrates the feasibility of electrically heated geopolymer wall construction. This research investigates ohmic heating's role in expediting the curing process of Class F fly ash-based geopolymer cementitious (GPC) materials and outlines the prototype development of the 3D printer. This advancement presents eco-friendly alternatives in construction, potentially replacing traditional methods of using cement concrete.

Coal-fired power plants are vital for global electricity production, contributing about 20% in the United States and over 36% worldwide [1]. However, coal combustion produces hazardous by-products like fly ash, with the United States alone generating 40.8 million tons and global production of 500 million tons annually [2, 3]. While fly ash is useful in various industries, its main application is in Portland cement concrete, enhancing properties such as hydration and workability. Unfortunately, more than half of the produced fly ash ends up in landfills, posing environmental risks due to potential contamination of soil and groundwater by heavy metals and toxins [4]. There are two main types of fly ash, Class C and Class F, each with distinct properties. While both classes of fly ash are useful for concrete production, the higher calcium content in class C fly ash allows it to be used in higher concentrations than class F fly ash [5]. Geopolymer, a relatively recent development, shows promise as an alternative to traditional concrete, especially in smaller-scale projects. However, further research is needed to make it a viable alternative to conventional cement concrete.

Several studies have been conducted to evaluate the mechanical properties and feasibility of rapid-curing geopolymer concrete (RCGPC) compared to traditional OPC and oven-cured geopolymer concrete (OCGPC) [6, 7]. Additionally, the application of electrically conductive concrete (ECON) in heated-pavement systems (HPS) [8] shows promise in melting ice and snow, although widespread implementation remains a challenge. Ionic conductive mortar presents another innovative application for indoor heating systems, demonstrating stable electrical conductivity and resistance to repetitive heating cycles. Ionic conductive mortar has potential applications in indoor radiant heating floor tiles and partition walls [9]. Experimental findings indicate that the resistivity of this mortar increases with time but stabilizes after 28 days. When subjected to electric voltage, the mortar exhibits temperature increases, with significant drops in resistivity observed under both AC and DC voltages. Repetitive electric-heating cycles do not affect the mortar's hydration products or internal structures, maintaining stable electrical conductivity over time [10]. An increase in reactive recyclable material content beyond 20% led to decreased strength and workability of the material. Longer curing times, between 7 and 28 days, improved compressive strength but reduced flexural strength. Test samples showed varying strengths between layers, indicating a compromise between strength values and recyclable material content in geopolymer development [11].

The development of 3D printing technologies offers new avenues for both geopolymer and conventional concrete applications. Geopolymer's rapid curing property and high compressive strength make it particularly suitable for structural

3D printing applications. This technology allows for the creation of structural members or unique geometries on construction sites without the need for special ordering of materials or longer curing times associated with Ordinary Portland Cement (OPC) construction.

The primary focus of this study is to analyze the printing parameter prerequisites for sustainable 3D printable geopolymer materials. This involves developing a prototype through the creation of engineering drawings, fabrication, and procurement of lab-scale components for the printer, followed by their assembly. Eventually, the study aims to utilize fly ash to produce rapid-curing geopolymer, offering a low-carbon alternative to ordinary Portland cement. Unlike traditional concrete, geopolymers can be rapidly cured with added heat, setting within minutes instead of hours or days. A properly formulated geopolymer mix achieves full strength within 24 hours, making fly ash a potential replacement for concrete rather than just an additive.

2 Design Consideration for a 3D Printer for Geopolymer Specifications

In the realm of 3D printing, meticulous attention to detail in both material selection and system design is paramount to the functionality and longevity of the printing apparatus. This section delves into the critical considerations surrounding the materials chosen for the printer's components, the microcontroller and control system governing its operations, the motion control methodology employed, and the current implementation of axes in the printer's design.

Materials Selection Components of the 3D printer that come into direct contact with geopolymers must exhibit high chemical resistance and inertness to withstand exposure to concentrated base solutions. This requirement ensures that the printer components remain unaffected by the corrosive nature of the geopolymer material over time, thus maintaining the printer's structural integrity and longevity.

Microcontroller and Control System The Prototype 3D Printer utilizes a programmable microcontroller to govern its operations. Specifically, an off-the-shelf Arduino Mega serves as the central processing unit, complemented by a CNC shield. This setup enables precise control over stepper motors, essential for guiding the printer's movements during the printing process. The system also runs the Marlin Firmware, facilitating user interaction through an LCD screen, and providing real-time monitoring and control capabilities.

Motion Control via G-Code Movement along the X and Z axes of the 3D printer is programmed using G-code instructions. Generated as an output of slicer software, such as Cura, G-code allows for seamless integration with a wide range of Computer-Aided Design (CAD) software and slicer applications. This compatibility ensures flexibility in choosing design software and simplifies the workflow for users, enhancing accessibility and usability.

Dual-Axis Implementation The current design of the 3D printer incorporates motion control along only two axes, namely the X and Z directions. While this configuration simplifies the system's complexity and reduces manufacturing costs, it limits the printer's capabilities to printing objects with predominantly planar geometries. However, future iterations of the printer may explore the incorporation of additional axes to enable more complex 3D printing capabilities, further expanding its application potential.

3   Present Design Constraints

Achieving high-quality printed samples in geopolymer materials 3D printing demands meticulous adherence to layer precisely. This section examines essential considerations such as temperature limits, shaft alignment, layer dimensions, and printer frame specifications, all crucial for achieving optimal printing outcomes.

Temperature Limitation It's crucial to ensure that the temperature of printed geopolymer samples does not exceed 80 °C. This limitation is based on current research findings, which have provided performance data up to this temperature threshold. Exceeding this temperature limit could lead to undesirable consequences such as material degradation, warping, or loss of structural integrity. Therefore, maintaining strict temperature control during the printing process is essential to ensure the quality and functionality of the printed samples.

Shaft Alignment The X-gantry shafts, which play a critical role in guiding the movement of the printer's print head along the X-axis, must not exceed 2° of misalignment or deflection. Misalignment beyond this threshold can result in inaccuracies in the printed objects, leading to dimensional errors or inconsistencies. Precise alignment of the gantry shafts is essential for achieving precise and reliable printing results, particularly when printing intricate or detailed designs.

Layer Dimensions Each layer of printed geopolymer must adhere to specific dimensional requirements, with dimensions of 12 inches in length, 1 inch in width, and ½ inch in thickness. Consistency in layer dimensions is essential for ensuring uniformity and structural integrity in the printed objects. Deviations from these prescribed dimensions could compromise the overall quality and functionality of the printed samples, making adherence to these specifications critical for successful printing outcomes.

Printer Frame Specifications The 3D printer frame must be designed to accommodate a build area of 1 meter by 1 meter with a height of 1 meter. A spacious build area provides ample room for printing larger-scale objects and enables versatility in printing various-sized samples. Additionally, a sturdy frame construction is necessary to support the weight of the printing components and maintain stability during the printing process. The specified frame dimensions ensure compatibility with the desired printing requirements and facilitate the production of larger and more complex printed objects.

Overall, precise adherence to these specified requirements is essential for achieving optimal printing outcomes and ensuring the quality, accuracy, and functionality of the printed geopolymer samples. Each requirement addresses specific aspects of the printing process, ranging from temperature control and mechanical precision to dimensional accuracy and structural support, all of which are crucial for successful 3D printing with geopolymer materials.

4 Design Steps and Final Design

A total of four prototypes were developed during this investigative study. The main components in the prototype 3D printer (Fig. 1) include—Base plate, X and Z direction frame, Steppers motors, ACME lead screw, Belt Drive, Hopper to store and feed geopolymer, Extruder, Nozzle, and Electrical curing system. It was found that the viscosity of the geopolymer mix played a significant role, requiring the redesign of the auger inside the hopper. The principal component that required continuous improvement was the auger and based on different redesigned models of augers (see Fig. 2), four different versions of the prototype printer (MK1, MK2, MK3, and MK4) were made.

The first auger design, MK1, produced too much friction between the extruder body and auger blades, preventing the consistent flow of geopolymer out of the nozzle. The MK2 design attempted to reduce the amount of friction by reducing the number of helical revolutions of the auger blades. This increased space allowed

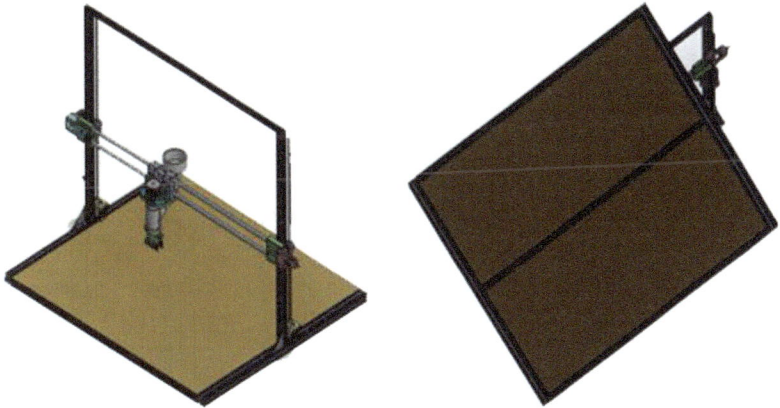

Fig. 1 CAD rendering of prototype form isometric view (left) and bottom view (right)

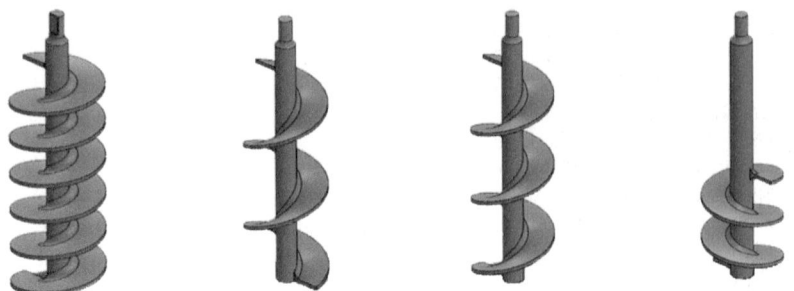

Fig. 2 Auger designs—from left to right—MK1, MK2, MK3, and MK4

Table 1 Different design parameters for Augers

Design	Shaft diameter (in.) mm	Helix length (in.) mm	Helix RPM	Diametral clearance (in.) mm
MK 1	(3/8) 9.53	(6) 152.4	6.0	(0.01) 0.254
MK 2	(3/8) 9.53	(6) 152.4	2.5	(0.01) 0.254
MK 3	(3/8) 9.53	(6) 152.4	3.0	(0.01) 0.254
MK 4	(7/16)11.11	(3) 76.2	2.5	(0.05) 1.27

the geopolymer to leak out of the nozzle. The MK3 design aimed to improve MK2 by increasing the number of revolutions from 2.5 to 3. This helped reduce the leaking but did not eliminate the issue, and ultimately, the amount of geopolymer leaving the nozzle was uncontrollable. Finally, the MK4 design addressed the issues the three previous designs faced, reducing the helical blade length from 6″ to 3″, while using only 2.5 helical revolutions of the blade. In this design, the diametral clearance was also increased from 0.01″ to 0.05″. These changes reduced friction while restricting the flow of geopolymer, allowing for controlled extrusion. The MK4 design was implemented in the final version of the 3D printer. Table 1 shows the dimensions of four different augers tried during the development of this prototype.

The final design of the 3D printer represents an improvement over previous iterations in several key areas. Firstly, the design has been streamlined by simplifying both the auger mechanism within the hopper and the nozzle, enhancing ease of assembly and operation. Additionally, ¼″ ACME lead screws have been replaced with M10 metric lead screws, contributing to smoother and more precise movement during printing. To bolster structural integrity, off-the-shelf corner brackets have been incorporated for the Z-rail brace, ensuring robust support. Furthermore, the addition of three 20 × 20mm extruded aluminum rails at the base of the printer enhances stability and rigidity. Finally, two sheets of 10 mm plywood have been introduced to serve as the build surface, providing a reliable foundation for printing operations. These refinements collectively enhance the functionality and performance of the 3D printer, paving the way for more efficient and reliable printing outcomes. Figures 3, 4, and 5 show some SolidWorks renderings of the final design.

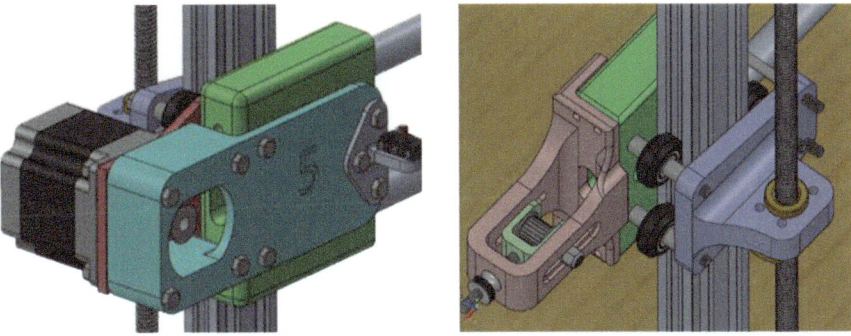

Fig. 3 X-axis motor mount (left) and isometric view of the x-axis belt tensioner

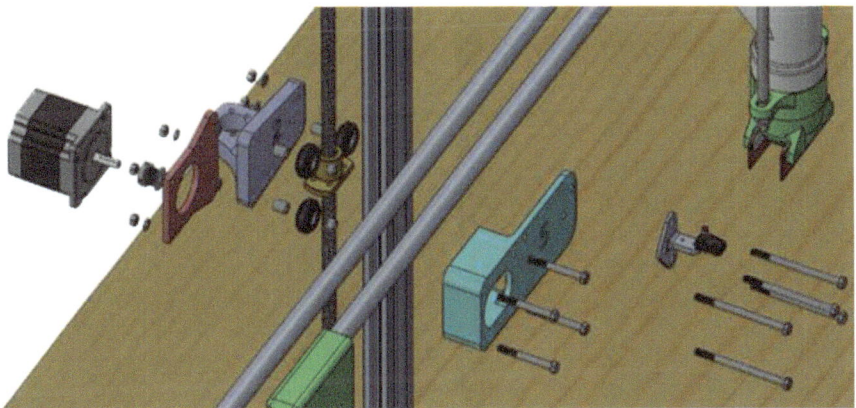

Fig. 4 X-axis motor mount exploded view

Fig. 5 Extruder section view (left) and designed nozzle outlet (right)

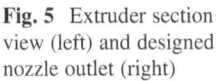

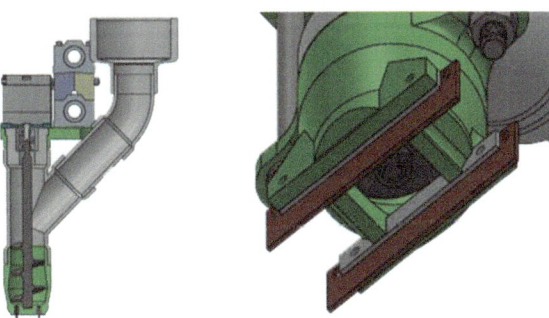

The 3D printer developed for this work deviates significantly from conventional 3D printers, primarily due to the utilization of geopolymer as a printing material. However, inspiration was drawn from established fused deposition modeling (FDM) printers to leverage existing technology and avoid unnecessary duplication of effort.

Unlike commercial 3D printers that employ a hot end to extrude melted thermoplastics through a nozzle, the printer constructed for this project employs a stepper motor-driven auger to propel uncured geopolymer through the nozzle. Upon extrusion, the geopolymer encounters a pair of current-conducting electrodes, triggering rapid setting and curing. Given the unique nature of this extrusion and curing process, each design incorporates a similar extruder setup. Moreover, all considered designs incorporate a hopper, serving as a reservoir to channel geopolymer into the extruder body, where the auger facilitates extrusion through the nozzle. For the construction of a proof-of-concept prototype, off-the-shelf PVC piping and connectors are utilized to fabricate the hopper and extruder body. PVC was selected for its chemical resistance to the sodium hydroxide present in the geopolymer, ensuring the durability and longevity of the components.

5   Results

The comprehensive evaluation of the final design encompassed scrutinizing its motion control capabilities, extrusion and curing functionalities, and print accuracy, all pivotal aspects for successful geopolymer printing. Ensuring the accuracy and reliability of the motion control system is paramount for proper geopolymer printing and the motion control system exhibited reliability and simplicity in calibration, with motors delivering more than sufficient power for the task at hand. The decision to implement metric M10 lead screws was validated by both the printer's performance and subsequent engineering analysis, owing to their low raising torque and minimal rotational backlash, ensuring precise motion and minimizing overworking of the stepper motors. Engineering analysis confirmed the adequacy of the two 16 mm linear motion shafts in preventing deflections exceeding 2°, with no detectable deflection observed upon assembly. Prior to print testing, a thorough analysis of the system's motion in the X and Z directions revealed no issues. Figure 6 below shows the fully assembled extruder and a printed first layer. However, despite these successes, challenges persisted in the extrusion and curing capabilities. Extensive testing of samples informed by static curing analysis revealed optimal feed rates for print moves, with favorable results observed in the printing of initial wall layers. Yet, consistent challenges such as nozzle clogging (Figs. 7 and 8) and runaway heating during subsequent layers were encountered, resulting in print discontinuations. Notably, defects such as wave-like patterns in double-layered prints were observed (see Fig. 9), indicating the need for further improvements to ensure consistent print quality and reliability in future iterations.

Fig. 6 Fully assembled 3D printer extruder and frame

Fig. 7 Clogged nozzle
view from bottom

6 Discussion

The exploration of the rapid curing process in geopolymer reveals its vast potential for diverse applications, particularly as a 3D-printed cementitious material. However, achieving high-quality 3D-printed geopolymer presents several challenges that demand attention. Variations in the prime mover's operation often led to inconsistent extrusion, resulting in either under or over extrusion of the geopolymer. Under extrusion causes premature stutter moves, hindering proper curing, while over extrusion leads to the excessive deposition of geopolymer, surpassing the boundaries of the nozzle and impeding effective curing. Moreover, nozzle clogging occurs due to heat generated by the copper plates, leading to premature curing within the nozzle.

Fig. 8 Failed prints due to
clogged nozzle

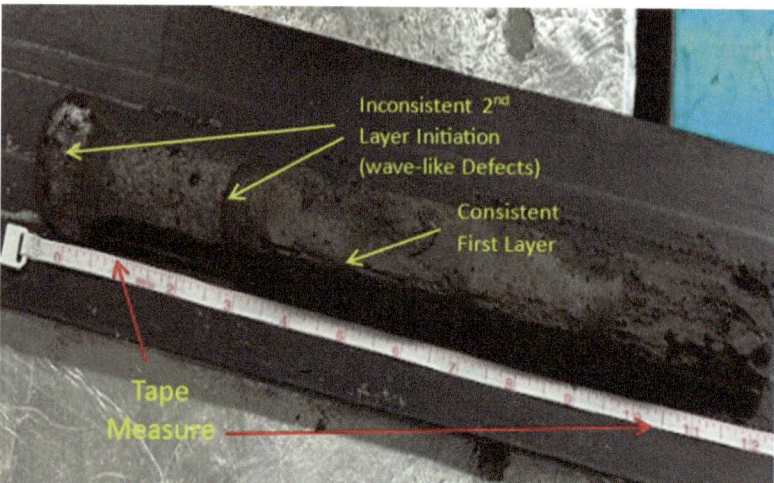

Fig. 9 Example of a two-layer wall with defects

Furthermore, the viscosity of geopolymer exhibits dynamic changes over time, initially presenting a high viscosity akin to the dough, which gradually decreases with continuous mixing, allowing for easier flow. However, prolonged mixing causes an increase in viscosity, posing challenges during 3D printing. These fluidic properties render geopolymer challenging to handle as a 3D printing material. Consequently, substantial research into material properties is warranted. Enhanced mixing procedures aimed at achieving consistency, coupled with further advancements in auger and nozzle designs, hold promise for enabling this machine to produce high-quality 3D-printed geopolymer in the future.

7 Conclusion

The exploration of the rapid curing process of geopolymer unveils its multifaceted potential applications, including its promising role as a 3D printing material, as evidenced by the outcomes of this project. A significant stride in this endeavor was the implementation of static rapid curing tests, which not only enhanced the research team's understanding of the geopolymer mixing and curing processes but also paved the way for practical implementation. The invaluable data gleaned from these tests not only informed the preparation for full-scale printing experiments but also served as a cornerstone for refining and optimizing the printer's performance. The robustness of the printer's frame and components was underscored by their resilience during numerous repeated tests. Engineering analysis conducted on critical components such as lead screws and linear motion shafts yielded a robust motion system, ensuring precise and reliable operation. Leveraging open-source Marlin firmware and widely accessible microcontrollers like the Arduino Mega and RAMPS 1.4 CNC board streamlined the electronic control aspects of the project, offering flexibility for future adaptations and enhancements. As the project neared its conclusion, forward-thinking considerations were made regarding the next steps for 3D-printed geopolymers. Notably, attention was drawn to the sensitivity of the mix design to water content, which directly influenced the viscosity of the mixture. This observation underscores the importance of fine-tuning mix compositions to optimize print quality and performance, laying a solid foundation for future advancements in 3D-printed geopolymer technology.

Acknowledgments This work is supported by the U.S. National Science Foundation under grant number OIA-1946231 and the Louisiana Board of Regents for the Louisiana Materials Design Alliance (LAMDA). A student who worked on this project was funded by an NSF-SURE grant.

References

1. Coal's importance to the world - Society for Mining, Metallurgy & Exploration (n.d.). Retrieved from https://www.smenet.org/What-We-Do/Technical-Briefings/Coal-s-Importance-in-the-US-and-Global-Energy-Supp
2. U.S. electric power industry produces less and recycles more combustion by-product - U.-S. Energy Information Administration (EIA) (n.d.). Retrieved from https://www.eia.gov/todayinenergy/detail.php?id=47336
3. Mathapati, M., Amate, K., Prasad, C.D., Jayavardhana, M., Raju, T.H.: A review on fly ash utilization. Mater. Today Proc. **50**, 1535–1540 (2022). https://doi.org/10.1016/j.matpr.2021.09.106
4. Petrović, M., Fiket, Ž.: Environmental damage caused by coal combustion residue disposal: a critical review of risk assessment methodologies. Chemosphere. **299**, 134410 (2022). https://doi.org/10.1016/j.chemosphere.2022.134410
5. Coal Fly Ash - User guideline - Portland cement concrete - user guidelines for waste and byproduct materials in pavement construction - FHWA-RD-97-148 (n.d.). Retrieved from https://www.fhwa.dot.gov/publications/research/infrastructure/structures/97148/cfa53.cfm

6. Cai, J., Li, X., Tan, J., Vandevyvere, B.: Fly ash-based geopolymer with self-heating capacity for accelerated curing. J. Clean. Prod. **261**, 121119 (2020). https://doi.org/10.1016/j.jclepro.2020.121119

7. Graytee, A., Sanjayan, J.: Development of a high strength fly ash-based geopolymer in short time by using microwave curing. Ceram. Int. **44**(7), 8216–8222 (2018). https://doi.org/10.1016/j.ceramint.2018.02.001

8. Rahman, M.L., Malakooti, A., Ceylan, H., Kim, S., Taylor, P.C.: A review of electrically conductive concrete heated pavement system technology: from the laboratory to the full-scale implementation. Constr. Build. Mater. **329**. Elsevier Ltd (2022). https://doi.org/10.1016/j.conbuildmat.2022.127139

9. Zhao, R., Tuan, C.Y., Xu, A., Fan, D.: Conductivity of ionically-conductive mortar under repetitive electrical heating. Construct. Build. Mater. **173**, 730–739 (2018). https://doi.org/10.1016/j.conbuildmat.2018.04.074

10. Zhao, R.H., Tuan, C.Y., Xu, A., Fan, D.B.: Conductivity of ionically-conductive mortar under repetitive electrical heating. Constr. Build. Mater. **173**, 730–739 (2018). https://doi.org/10.1016/j.conbuildmat.2018.04.074

11. Munir, Q., Peltonen, R., Kärki, T.: Printing parameter requirements for 3d printable geopolymer materials prepared from industrial side streams. Materials. **14**(16) (2021). https://doi.org/10.3390/ma14164758

Material Stories: Assessing Sustainability of Digital Fabrication with Bio-Based Materials Through LCA

Giuliano Galluccio ⓘ, Martin Tamke ⓘ, Paul Nicholas ⓘ, Tom Svilans ⓘ, Nadja Gaudillière-Jami ⓘ, and Mette Ramsgaard Thomsen ⓘ

Contents

1 Introduction

Life cycle assessment (LCA) is an objective procedure for assessing the energy and environmental loads related to a process or activity, carried out through the identification of energy and materials used and waste released into the environment. The assessment shall cover the entire life cycle of the process or activity, including

G. Galluccio (✉)
Department of Architecture, University of Naples Federico II, Naples, Italy
e-mail: giuliano.galluccio@unina.it

M. Tamke · P. Nicholas · T. Svilans · M. R. Thomsen
Center for Information Technology in Architecture – CITA, Royal Danish Academy, Copenhagen, Denmark

N. Gaudillière-Jami
Center for Information Technology in Architecture – CITA, Royal Danish Academy, Copenhagen, Denmark

Department of Architecture, Technische Universität, Darmstadt, Germany

© The Author(s) 2025
M. Kioumarsi, B. Shafei (eds.), *The 1st International Conference on Net-Zero Built Environment*, Lecture Notes in Civil Engineering 237,
https://doi.org/10.1007/978-3-031-69626-8_3

extraction and processing of raw materials, manufacture, transport, distribution, use, reuse, recycling, and final disposal [1].

In its Communication on Integrated Product Policy [2], the European Commission concluded that LCA provides the best framework for assessing the potential environmental impacts in the construction industry. Following the Level (s) guidelines [3] and the European Green Deal [4], all EU countries started embracing the adoption of LCA methods. In Denmark, for instance, LCA is mandatory since 2023 for all new constructions over 1000 m^2 [5]. To support designers, engineers, and policy-makers, LCA protocols and case studies were developed to outline how to perform environmental assessments on buildings and industrial products. Nevertheless, a relatively limited number of studies have investigated the application of LCA to novel technologies and materials that could represent a suitable ecological alternative to current practices in the construction industry. In these cases, LCA could be pivotal to target research and communicate results [6]. Data scarcity, lack of previous research and precedents, and difficulties in applying standard protocols and software are recognized as the main barriers in the application of LCA to digital fabrication (DF) processes and bio-based materials [7, 8]. A methodological development of LCA is necessary in order to use this instrument to evaluate and steer research and innovation projects.

This study aims to design and verify a LCA-oriented methodology for the environmental assessment of DF with bio-based materials, highlighting opportunities and limits. The methodology is framed on the ISO 14040-44 standard [9, 10] and is tested on two case studies within the ERC funded Eco-Metabolistic Architecture project, carried out at the Center for Information Technology in Architecture (CITA) at the Royal Danish Academy: Radicant and Rawlam [11–13] (Fig. 1).

The study investigates how LCA can have a significant impact on the decision-making processes within a research laboratory and why it should be integrated as an early-stage design step. Considering the growing popularity of DF with bio-based materials, the implementation of the here illustrated methodology could be used for

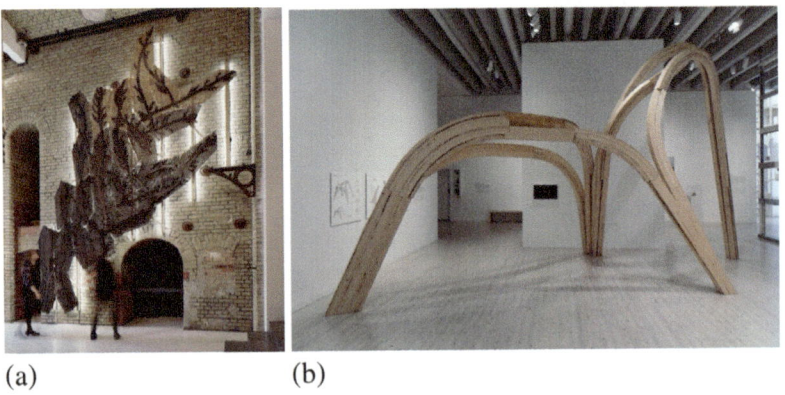

(a)   (b)

Fig. 1 (**a**) The Radicant project by CITA, Aedes Gallery, Berlin, 2022. (**b**) The Rawlam prototype by CITA, Bildmuseet, Umeå, 2021 [11–13]. (Courtesy of CITA)

the realization of environmental guidelines [18, 19] and the creation of digital databases, i.e. a "material library," with dedicated software applications, e.g., a digital platform [20] and plug-in for computational design tools [21].

2 Background

To analyze the current trends in literature (up to 2024) about LCA application to DF with bio-based materials, a scoping review is performed via Scopus web database. The query is prompted by the following combination of keywords, joined by the Boolean operators "AND" and "OR": "life cycle assessment" OR "LCA" OR "life-cycle assessment" AND "bio-based materials" OR "biomaterials" OR "bio-materials" AND "digital fabrication" OR "digital manufacturing" OR "advanced manufacturing" OR "additive manufacturing" OR "robotic fabrication" OR "robotic manufacturing". The query is limited to the fields of title, abstract and keywords.

The results show how this topic is still under-investigated, as "Product Sustainability Assessment" [22] is the only paper available. In this work, authors present an overview of the 3D printing technique and the materials used for the production, discussing how LCA application at every step of the production can help in achieving sustainable development along with the additive manufacturing techniques.

This lack of literature is examined by Kohtala and Hyysal [23], who state how reports on the sustainability of personal fabrication are emerging as the phenomenon spreads, often appearing as grey literature [24]. The few empirical studies that exist mainly focus on additive manufacturing, relevant to some digital fabrication equipment used in our research, such as studies on energy consumption and life cycle analyses [25, 26]. When compared to mass production processes, digital manufacturing has the potential to reduce material, waste, and energy, at least for small batches [27], and may mitigate negative impacts connected to supply chains [28]. However, toxicity of especially additive manufacturing materials remains a concern [29, 30], as well as the high energy consumption of digital fabrication.

3 Materials and Methods

3.1 Research Methodology and Assessment Method

The LCA methodology is organized into 4 steps, namely, flows, data, knowledge, and uncertainty.

In the "Flows" step, goal and scope of the assessment are defined and the life-cycle inventory (LCI) is performed. In this phase, the objective of the whole analysis is identified, the system boundaries are taken into account, and the functional unit of reference is expressed. In the LCI, all materials included in the analysis are listed and tracked down to assess impacts, e.g. CO_2 emissions and energy consumption (modules A1–A3). For each material and each process unit, all the inputs (energy, water, and resources) and outputs (waste, pollution, and by-products) are examined and diagramed.

In the "Data" step, information about the materials and the processes is collected and calculated. Data are collected from different sources:

- The ÖKOBAUDAT generic database, a German-based platform included into the official LCA Danish software LCAbyg [14, 31].
- EPD Denmark [15].
- Other EPDs or studies in literature reporting LCA according to their technological or time representativeness, since they might be referred to different geographical contexts.

The calculations performed in this study are guided by specific impacts-related questions and are performed through a spreadsheet by converting the gathered data into the functional unit and adding impacts from transportation (depending on the different vehicle data from ÖKOBAUDAT) and energy use for manufacturing and construction (referring to the energy mix 2018). Modules A1–A3 (production stage) and A4–A5 (construction process stage) are considered. For transportation, the extraction and processing of the fuel are included. The production of the vehicle is not included in the balancing.

The "Knowledge" step corresponds to the life-cycle impact assessment—LCIA phase, in which the results are graphically explicated.

In the "uncertainty" step, a sensitivity analysis is performed to deal with possible inaccuracy of calculations, e.g. testing different materials or products [16, 17].

3.2   Limitations of the Study

To address the goal of the study, several limits in the application of standard LCA in our research have been identified, and workarounds—mainly based on assumptions—were established.

Only a few data are available about bio-based materials and biopolymers in terms of LCA, e.g. ingredients and machineries. Indeed, DF processes are not standard, as manufacturing technologies are often customized depending on use, materials, scale, etc. This means that data must always be adapted to the specific application. Most LCA software does not allow for custom-made database integration, or this operation is extremely labor intensive. Nonetheless, the use of spreadsheets more easily leads to miscalculations and errors.

3.3   Radicant (Single-Product Assessment)

Radicant is made of 24 custom 3D-printed panels based on a bone glue, water, glycerol recipe, with added fiber fillers (wood flour, wood bark, seagrass, and cotton bark). All panels have different amount of base mixture and fibers. Every panel consists of 2 layers: the "base" layer uses seagrass or wood flour as filler, in percentage around 15% per liter; the "leaf" layer uses cotton bark and wood bark as filler, in percentage around 9% per liter. The production process is divided into three phases:

1. 24 h cooking of a mix of bone glue, glycerol, and water
2. Base mixture mixing with filler fibers (20 min for 1 liter)
3. 3D printing with a custom-fixed extruder mounted on a universal cobot (Fig. 2)

Optional phases might involve the use of refrigerator to store the mixture and the bain-marie to keep parts at temperature before the mixing process.

The LCA focuses on the panels production, transportation to the site, and the whole assembly of the installation (A1–A5).

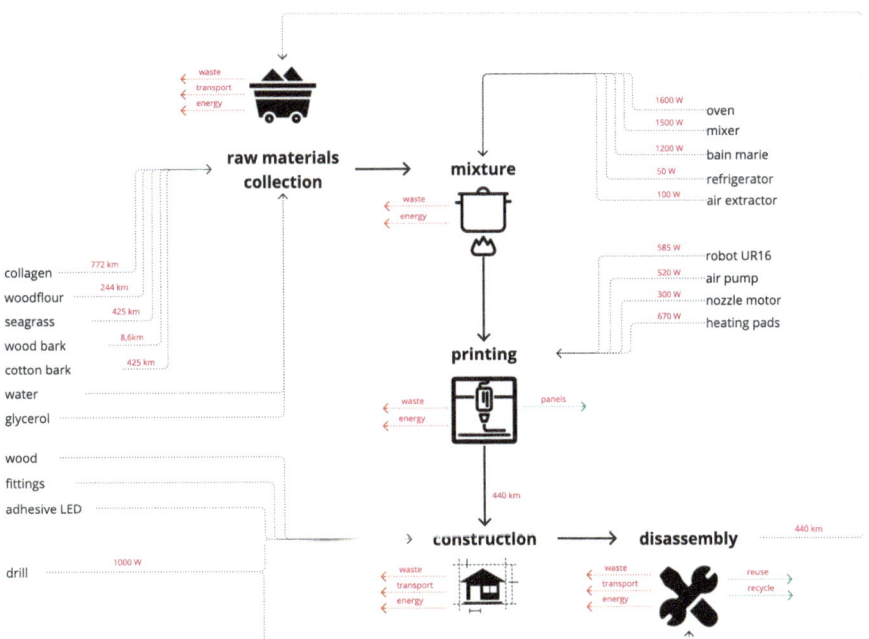

Fig. 2 Radicant material input and output flows map

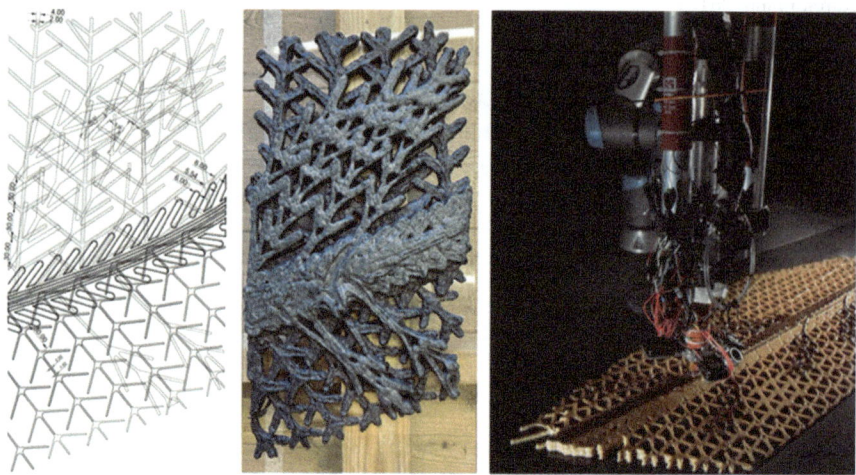

Fig. 3 Radicant panel digital fabrication process. Courtesy of CITA

Table 1 Ingredients and materials data related to their manufacturing phase (module A1)

Ingredient	Original functional unit	Source	Geographical context
Bone glue	1 kg	[20]	N/A
Wood bark	1 m³	[32]	Serbia
Wood flour	1 lt	[33]	USA
Seagrass	1 kg	[15]	Denmark
Cotton bark	1 m²	[15]	Denmark
Glycerol	1 kg	[34]	USA
Wood sticks	1 kg	[14]	Germany
Metal screws	1 kg	[14]	Germany
Adhesive LED stripes	20 million lumen h	[35]	USA

The LCIA is guided by specific impact-related questions, namely:

- Which of the materials from the recipe has the worst environmental score in terms of global warming potential (GWP)
- Which of the 24 panels of the Radicant exhibit has the worst environmental score in terms of GWP
- Which process phase has the worst environmental impact in terms of total use of non-renewable energy as primary resource (PENRT) (Fig. 3)

In A1–A3 LCA modules, all ingredients are mapped and restructured into their production phases (A1). Impacts are collected differently according to the data available for every ingredient. Water and construction materials data (timber sticks, metal fittings, and adhesive led stripes) are taken from the ÖKOBAUDAT database (Table 1). All values are converted to the functional unit of 1 kg. In A3, the manufacturing process is analyzed for each panel, specifically for every "base" and "leaf" layers.

3.4 Rawlam (Comparative Assessment)

The methodology is tested for the comparative analysis between Glulam data from the ÖKOBAUDAT database and the Rawlam prototype.

The Rawlam manufacturing process is divided into four phases:

1. CT scanning of spruce logs to detect wood knots, failures, etc.
2. Logs cutting into lamellas, plaining and thicknessing
3. Glulam pressing and replaining
4. Boards assembly

The LCA focuses on the timber beams manufacturing, transportation to the site, and the whole assembly of the installation (A1–A5). Since the aim of Rawlam is to optimize timber use in structural components manufacturing processes, LCIA was guided by specific impact-related questions, namely, the total amount of nonhazardous waste disposed (HWD), components for reuse (CRU), and materials for recycling (MFR) among phases A2–A3; the total amount of CRU in kg in phase A3 between Glulam and Rawlam; and the total amount of MFR in kg in phase A3 between Glulam and Rawlam.

The functional unit for the comparison is 1 kg. According to the assessment's purposes, A1 phase is considered not relevant for the analysis, as wood source for Glulam and Rawlam might be the same.

4 Results and Discussion

4.1 Radicant

The analysis of materials impact in Radicant helps to spot possible points to improve for environmental impacts reduction. In terms of GWP, animal glue and cotton stand as potential targets for enhancing the sustainability of the recipe.

The comparison between panels in terms of GWP highlights no significant differences in recipes impacts. Nonetheless, panels with wood fibers in both "base" and "leaf" layer as filler fibers demonstrate to have a better environmental score. A1 phase is the most impacting in terms of PENRT. Both transportation phases A2 and A4 have no significant weight on the overall evaluation. Most of energy consumption is thus embodied in material manufacturing, while the exhibition itself has low impact scores (Fig. 4).

To possible guide future printing projects, GWP data for 1 kg are converted to 1 meter of filament. Taking Panel 0 recipe as an example, the "base" layer (8 lt of materials) corresponds to 1.15E-02 kg CO_2 eq. per m. The "leaf" layer (3 lt of materials) corresponds to 2.54E-02 kg CO_2 eq. per m.

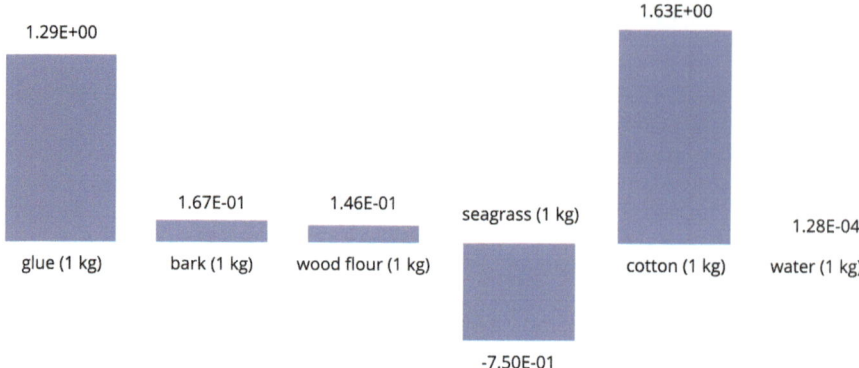

Fig. 4 GWP (Kg CO$_2$ eq.) analysis of Radicant panels ingredients considering A1–A3 LCA modules

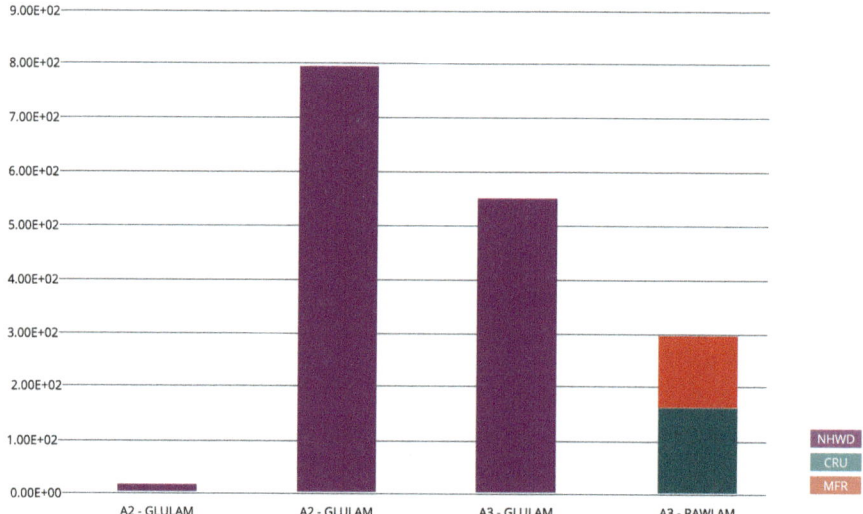

Fig. 5 Rawlam vs Glulam: LCA analysis for A2–A3 modules in terms of NHWD, CRU, and MFR (all expressed in kg)

4.2 Rawlam

Results from comparing A2–A3 phases between Glulam and Rawlam highlight significant differences in terms of NHWD from transportation in the latter, where logs are imported from Trieste (Italy). This suggests that is best to source wood locally when possible. In A3, Glulam produces more than 5.00E+02 kg of NHWD, while in Rawlam this is hugely reduced in favor of CRU and MFR. In particular, the Glulam EPD from ÖKOBAUDAT does not report any CRU or MFR amounts (Fig. 5).

4.3 Sensitivity Analysis

To deal with the data uncertainty and provide for reliable results, for both Radicant and Rawlam assessment, a sentitivity analysis is performed. This allows to understand the impact variation due to possible errors or ambiguity across individual components of the structures (i.e. panels for Radicant, boards for RawLam).

In the first case, the analysis focuses first on the GWP scores of the various panels, as shown in Fig. 6. While panels 1 and 16 display higher emissions due to their larger size (twice the dimensions of the other panels), panel 21 displays lower emissions due to the higher proportion of wood flour in the recipe (12.7%).

Other panels display similar GWP scores (2.56E+01 min, 2.77E+01 max., considering same-size panels). The 24 panels use 20 different recipes, and the study of their GWP scores therefore also provide data for a sensitivity analysis of the recipe's materials quantities. Figure 7 displays the highest and lowest percentages of material

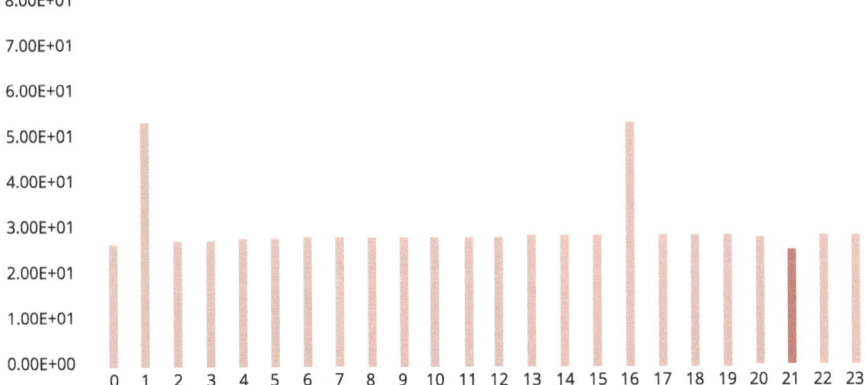

Fig. 6 GWP variation across panels

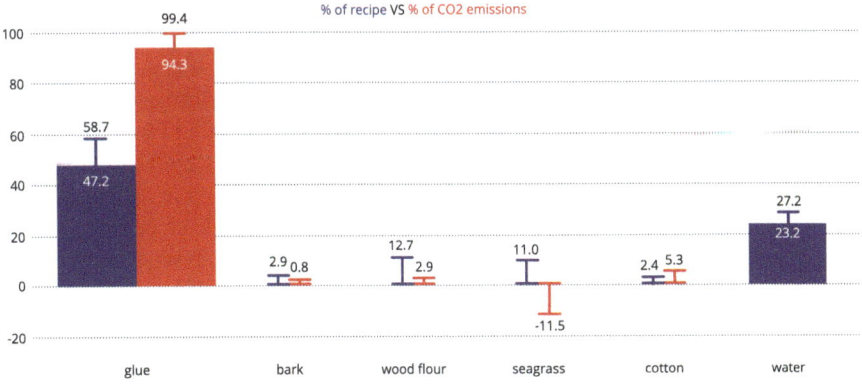

Fig. 7 Recipe variation sensitivity analysis. % of recipe (blue) vs % of CO_2 emissions (red)

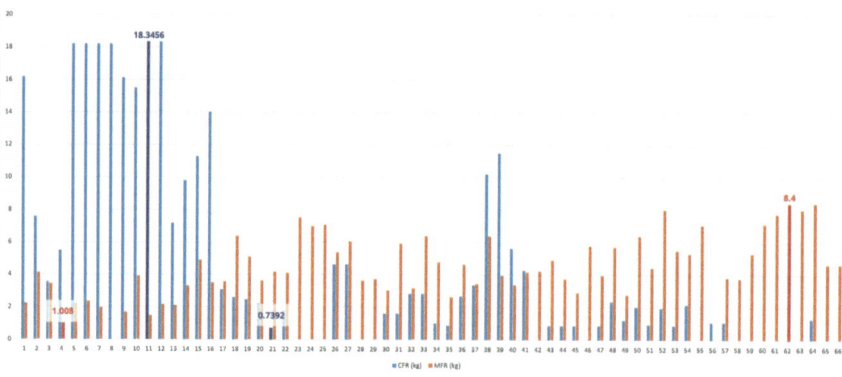

Fig. 8 Impact variation across Rawlam boards

quantities and of CO_2 emissions across all panels and recipes. It highlights which material has the largest influence over GWP variations when modifying the recipes. Bark and wood flour have the lowest impacts in comparison to material amounts, which informs further the low GWP score of panel 21. Cotton and glue display the highest impacts compared to material amount, and in particular the latter would be an ingredient to focus on in further material research.

In the second case, the sensitivity analysis focused on CRU and MFR scores of the different boards, as shown in Fig. 8. NHWD was not considered in the sensitivity analysis as the waste in this category originates in the fabrication process, not the boards themselves. This allows for a best- and worst-case scenario analysis, in which the amount of reusable/recyclable materials vary depending on the quality of the logs.

5   Conclusions

In this paper, the challenge of a LCA analysis of environmental sustainability of digital fabrication with bio-based materials in research stage is addressed by designing and testing a LCA-oriented methodology. This study is performed as an ex-post analysis on two different prototyped realized and exhibited by CITA, i.e. Radicant and Rawlam. In both cases, opportunities and limitations of the application of LCA to experimental practices for ecological alternatives in construction are exposed.

The limitations we encountered are mostly in terms of lack of data and uncertainty. To address these problems, we perform a sensitivity analysis, examining the range of errors and impact scenarios in both assessments, so to improve the reliability of results to inform a decision-making process.

Further developments of this study will focus on direct acquisition of data from manufactures. Also, a better connection between the spreadsheet and the design software (i.e. Rhinoceros/Grasshopper) might indeed help LCA to become a proper decision support system. Eventually, a comparison between ex-post and during-development data might highlight significative changes in the results.

Finally, LCA allows for deeper reflections on sustainability of DF and questions the upscale and replication of experimental manufacturing research projects in different contexts. If applied at the very early stage of the process, LCA can serve as a design and research decision support system in the laboratory.

Acknowledgments The works described in this chapter are part of the Eco-metabolistic Architecture project that has received funding from the European Research Council (ERC) under the European Union's Horizon 2020 research and innovation program (grant agreement No 101019693). The collaboration between CITA and the Department of Architecture, Federico II was carried out since March 2023 within the ERASMUS PLUS KA1_"University for Innovation"—Project Ref. n. 2021-1-IT02 KA131-IIED-000011202 mobility project promoted by the SEND Mobility Consortium, Italy. The activity was developed as part of the FoRWARD—Furniture Waste for circular design (MICS—NextGenEU, PE11, Spoke 4) research project, coordinated by Prof. Massimo Perriccioli and Prof. Marina Rigillo (Department of Architecture, University of Naples "Federico II"), and promoted by the Department of Architecture, the Department of Social Sciences and the Department of Law, University of Naples "Federico II" (IT), the Department of Engineering of the University of Palermo (IT) and the Department of Agricultural, Food, Environmental and Forestry Science and Technology of the University of Florence.

Authors Attributions All authors contributed to the study conception and design. Material preparation, data collection, and analysis are performed by Giuliano Galluccio. The first draft of the manuscript is written by Giuliano Galluccio, Martin Tamke, and Nadja Gaudillière-Jami, and all authors commented on previous versions of the manuscript. Sensitivity analysis is performed by Giuliano Galluccio, Nadja Gaudillière-Jami, and Paul Nicholas. All authors read and approved the final manuscript.

References

1. Hauschild, M.Z., Rosenbaum, R.K., Olsen, S.I.: Life Cycle Assessment. Theory and Practice. Springer, Heidelberg (2018)
2. Commission of the European Communities: COM(2003) 302 final. Integrated Product Policy. Building on Environmental Life-Cycle Thinking. EC, Brussels (2003)
3. Level(s) Europen framework for sustainable buildings Homepage. https://environment.ec. europa.eu/topics/circular-economy/levels_en
4. European Commission: COM(2019) 640 final. The European Green Deal. EC, Brussels (2019)
5. Danish Building Regulation BR18. https://bygningsreglementet.dk/~/media/Br/BR-English/ BR18_Executive_order_on_building_regulations_2018.pdf
6. Albrecht, S., Fischer, M., Leistner, P., Schebek, L.: Progress in Life Cycle Assessment 2019. Springer, Heidelberg (2021)
7. The European Commission's Knowledge Center for Bioeconomy: Prospective LCA for Novel and Emerging Technologies for Bio-based Products. EC, Brussels (2021)
8. Ke Yin Lee, J., Gholami, H., Medini, K., Salameh, A.A.: Hierarchical analysis of barriers in additive manufacturing implementation with environmental considerations under uncertainty. J. Clean. Prod. **408**(137221) (2023)
9. International Organization for Standardization: Environmental Management Life Cycle Assessment Principles and Framework, ISO14040 (2006)

10. International Organization for Standardization: Environmental Management Life Cycle Assessment Requirements and Guidelines, ISO14044 (2006)
11. Thomsen, M.R., Tamke, M.: Towards a transformational eco-metabolistic bio-based design framework in architecture. Bioinspir. Biomim. **17**(4) (2022)
12. Nicholas, P., Lharchi, A., Tamke, M., Goudarzi, H.V., Eppinger, C., Sonne, K., Rossi, G., Ramsgaard Thomsen, M.: A design modeling framework for multi-material biopolymer 3D printing. In: Dorfler, K., Knippers, J., Menges, A., Parascho, S., Pottmann, E., Wortmann, T. (eds.) Advances in Architectural Geometry 2023, pp. 193–206. De Gruyter, Berlin (2023)
13. Tamke, M., Svilans, T., Gatz, S., Ramsgaard Thomsen, M.: Timber elements with graded performances through digital forest to timber workflow. In: Behnejad, S.A., Parke, G.A.R., Samavati, O.A. (eds.) IASS 2020/21 SURREY 7 (2021)
14. ÖKOBAUDAT Platform Homepage. https://www.oekobaudat.de/en.html
15. EPD Denmark Homepage. https://www.epddanmark.dk
16. Wei, W., Larrey-Lassalle, P., Faure, T., Dumoulin, N., Roux, P., Mathìas, J.: How to conduct a proper sensitivity analysis in life cycle assessment: taking into account correlations within LCI data and interactions within the LCA calculation model. Environ. Sci. Technol. **49**, 377–385 (2015)
17. Hanak, D.: How do you compare and benchmark different circular economy strategies using sensitivity analysis in LCA? https://www.linkedin.com/advice/1/how-do-you-compare-benchmark-different-circular (2023)
18. Augustì-Juan, I., Habert, G.: An environmental perspective on digital fabrication in architecture and construction. In: Chien, S., Choo, S., Schnabel, M.A., Nakapan, W., Kim, M.J., Roudavski, S. (eds.) Living Systems and Micro-Utopias: Towards Continuous Designing, Proceedings of the 21st International Conference of the Association for Computer-Aided Architectural Design Research in Asia CAADRIA 2016, pp. 797–806 (2016)
19. Augustì-Juan, I., Habert, G.: Environmental guidelines for digital fabrication. J. Clean. Prod. **142**, 2780–2791 (2017)
20. STICH Sustainability Tools in Cultural Heritage. https://stich.culturalheritage.org
21. One Click LCA. https://academy.oneclicklca.com
22. Singh, H., Bhattacharjee, S.: Product sustainability assessment. In: Kamalpreet, S., Karupppasamy, S., Vivek, S., Neeta Raj, S. (eds.) Application of 3D Printing in Biomedical Engineering, pp. 173–197 (2021)
23. Kohtala, C., Hyysalo, S.: Anticipated environmental sustainability of personal fabrication. J. Clean. Prod. **99**, 333–344 (2015)
24. De Decker, K.: How sustainable is digital fabrication? http://www.lowtechmagazine.com/2014/03/how-sustainable-is-digital-fabrication.html (2015)
25. Baumers, M., Tuck, C., Wildman, R., Ashcroft, I., Rosamond, E., Hague, R.: Transparency built-in. energy consumption and cost estimation for additive manufacturing. J. Ind. Ecology. **17**(3), 418–431 (2013)
26. Faludi, J., Bayley, C., Bhogal, S., Iribarne, M.: Comparing environmental impacts of additive manufacturing vs traditional machining via life-cycle assessment. Rapid Prototyp. J. **21**, 14–33 (2015)
27. ATKINS Project: Manufacturing a Low Carbon Footprint: Zero Emission Enterprise Feasibility Study (Project No: N0012J). Loughborough University. http://www.atkins-project.com/pdf/ATKINSfeasibilitystudy.pdf
28. Huang, S.H., Liu, P., Mokasdar, A., Hou, L.: Additive manufacturing and its societal impact: a literature review. Int. J. Adv. Manuf. Technol. **67**, 1191–1203 (2013)
29. Drizo, A., Pegna, J.: Environmental impacts of rapid prototyping: an overview of research to date. Rapid Prototyp. J. **12**, 64–71 (2006)
30. Short, D.B., Sirinterlikci, A., Badger, P., Artieri, B.: Environmental, health, and safety issues in rapid prototyping. Rapid Prototyp. J. **21**, 105–110 (2015)
31. LCAbyg Homepage. https://www.lcabyg.dk/en/

32. Peric, M., Antonijevic, D., Komatina, M., Bugarski, B., Rakin, M.: Life cycle assessment of wood chips supply chain in Serbia. Renew. Energy. **155**, 1302–1311 (2020)
33. Pokhel, G., Gu, H., Gardner, D.J., O'Neill, S.: Life cycle assessment (LCA) of wood flour and pellets for manufacturing wood-plastic composites (WPCs). Recent Prog. Mater. **4**(1) (2022)
34. United States Environmental Protection Agency, TRACI. https://www.epa.gov/chemical-research/tool-reduction-and-assessment-chemicals-and-other-environmental-impacts-traci
35. U.S. Department of Energy, Life-Cycle Assessment of Energy and Environmental Impacts of LED Lighting Products. https://www1.eere.energy.gov/buildings/publications/pdfs/ssl/lca_factsheet_apr2013.pdf

Increasing the Incorporation of CO_2-Sequestering Materials in Concrete

Alexander Mezhov, Bright Asante, and Wolfram Schmidt

Contents

1 Introduction

Concrete is the most widely used construction material accounting for approximately half of all human production throughout history [1]. The production of one ton of cement emits around 600 kg of CO_2 [2]. With a global cement production estimate of 4.2 billion tons, this results in about 2.5 billion tons of CO_2 emissions [3]. This accounts for roughly 7.2% of the total global carbon emissions in 2021 [4]. Strategies to decrease carbon emissions in the cement and concrete industry include carbon capture and storage (CCS), reduced binders like ordinary Portland cement (OPC) clinker and optimizing material use through improved processes and structural design [5]. This also means changing architectural vision [6], avoiding concrete where its structural performance is not required and use alternative materials instead [7], and optimizing the communication among the entire production value chain [6, 8].

A. Mezhov (✉) · B. Asante · W. Schmidt
Bundesanstalt für Materialforschung und -prüfung (BAM), Berlin, Germany
e-mail: alexander.mezhov@bam.de; bright.asante@bam.de; wolfram.schmidt@bam.de

© The Author(s) 2025
M. Kioumarsi, B. Shafei (eds.), *The 1st International Conference on Net-Zero Built Environment*, Lecture Notes in Civil Engineering 237,
https://doi.org/10.1007/978-3-031-69626-8_4

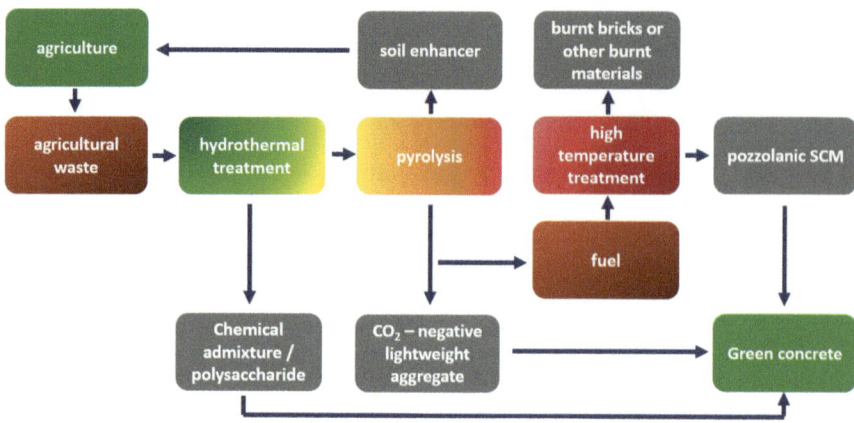

Fig. 1 Possibilities to convert agricultural waste to different materials useful to produce eco-friendly concrete. (Adopted from Schmidt et al. [16])

While CCS technologies are not yet market-ready [5, 6], policies for materials saving and circular technologies also require time to implement. The most promising technologies currently are carbon-reduced concrete types that incorporate supplementary cementitious materials (SCM) as clinker replacement. Since traditional SCMs like fly ash and ground granulated blast furnace slags become scarce and their sustainability questioned, new materials like limestone calcined clay compounds [9] and bio-based reactive SCMs [10–13] are being explored as alternatives. While the feasibility of bio-based cementitious materials depends on regional specifications, supply chains, and infrastructures, it seems to be more promising in the global South and particularly Africa and South America as they are also more suited to reducing the carbon footprint during concrete production rather than cement production [12, 14, 15]. By utilizing agricultural waste materials through processes like hydrothermal treatment, pyrolysis, and ash processing, low carbon or even carbon negative SCMs can be derived, while in parallel generating water reducing agents, energy, and chemical pre-cursors [11] (Fig. 1).

Biochar, which is the product of pyrolyzing process under limited oxygen supply, has shown promise in effectively capturing CO_2 from the atmosphere [17]. This, however, requires the use of bio-based waste materials and not fresh biomass. The carbon sequestration capability of biochar is particularly critical in the context of global efforts to combat climate change and reduce atmospheric CO_2 levels. Potential biomass for clean char production includes agricultural residues, wood waste, agri-food waste, manure, and municipal/industrial sludge. The generation of biomass waste, particularly from agriculture and forestry, is substantial, estimated at approximately 140 gigatonnes per year globally. This significant amount of biomass waste not only contributes to greenhouse gas emissions if left untreated but also represents a considerable environmental pollution challenge [18]. However, when converted into biochar, which is considered as relatively inert, this biomass waste can serve as a valuable resource for CO_2 sequestration.

Research has indicated that converting agricultural waste into biochar could have a notable impact on reducing CO_2 levels. The production of 373 million tons per year of biochar from agricultural waste could sequester around 500 million tons of CO_2 annually. This amount of sequestration equates to roughly 1.5% of the global annual CO_2 emissions, demonstrating the potential of biochar in contributing to global carbon mitigation efforts [19, 20]. The potential benefits of biochar extend beyond carbon sequestration. Utilizing different types of biomass waste for biochar production not only contributes to further reductions in carbon emissions but also supports the advancement of a circular economy. This is achieved through the development and application of biochar in new areas, such as the creation of innovative biochar-infused construction materials. These materials not only help in valorizing waste but also reduce the carbon footprint of the construction sector [21].

Additionally, integrating biochar into the construction industry could yield economic advantages, particularly through mechanisms like carbon trading, thereby encouraging the construction sector to lower its CO_2 emissions [22, 23]. According to the European Biochar Certificate, the conversion ratio of carbon to CO_2 (where 1 ton of biomass can capture up to 3.67 tons of CO_2) illustrates the potential effectiveness of biochar in carbon sequestration. During pyrolysis, the percentage of carbon retained in biochar can vary widely, from a few to approximately 80%, dependent on the biomass's chemical composition and the parameters of the pyrolysis process [24, 25]. This variability underlines the importance of optimizing pyrolysis conditions to maximize the carbon sequestration capability of biochar, further contributing to the broader efforts of decarbonization and environmental sustainability. This paper presents a comprehensive evaluation of the optimal method for incorporating significant amounts of biochar into concrete mixtures, and examines its impact on CO_2 sequestration, thereby contributing to the achievement of carbon neutrality in concrete production.

2　Ways of Incorporating Biochar into Concrete

2.1　Biochar as a Filler

Integrating biochar as a filler in cement (i.e., during cement manufacturing) can lead to the development of a composite cement variant that diminishes the dependency on traditional OPC. This approach will contribute to a reduction in carbon emissions associated with the production processes of cement. The rationale behind this is: biochar, a carbon-rich product derived from the thermal decomposition of organic materials in an oxygen-limited environment, necessitates significantly lower energy inputs for its production compared to the energy-intensive processes required for manufacturing conventional OPC. In this scenario, biochar would play a role as a filler like limestone. Yet, without real contribution to the mechanical properties, the presence of biochar due to the initially embodied carbon would benefit in overall improvement of concrete sustainability.

2.2   Biochar as Supplementary Cementitious Material (SCM)

The utilization of biochar as a SCM for cement in concrete represents a significant advancement in sustainable construction methods. Biochar, which is derived from the pyrolysis of biomass, not only offers an environmentally friendly alternative to traditional cement but also acts as an effective carbon sink. As scientists continue to refine the composition of biochar and its distribution within the concrete mixture, the engineering community is actively exploring the optimal balance between biochar and other components of concrete. Yet, depending on the biochar properties, the maximum cement replacement by biochar without compromising mechanical properties is 5–10 vol.%. Yet, in this scenario, biochar could not be recognized as SCM, since it does not positively contribute to hydration process.

2.3   Biochar as Concrete Aggregates

Incorporating biochar as a substitute for traditional aggregates like sand and stones in concrete mixes is a viable option. The porous structure of biochar aids in lowering the density of concrete, resulting in a lighter and potentially better insulating material. In the world of concrete, where aggregates (fine and coarse aggregates) make up most of the concrete (i.e., about 70–85% of the components), biochar's innovative integration as an aggregate substitute/replacement or lightweight aggregate is a game-changer for sustainable construction. Replacing a part of aggregates with biochar can result in lower density and improved thermal-insulating properties. The lower density and higher porosity are advantageous for lightweight concrete applications. By replacing all normal lightweight aggregates with biochar aggregates, a combination with high strength cement paste can create lightweight high-performance concretes.

It should be noted that the particle size of biochar can potentially impact the overall substitution potential of biochar in concrete. Since it affects surface area and porosity, changing workability and mechanical properties. Fine particles have a larger surface area and possibly can improve bonding with cement. However, they may increase water demand and be challenging to handle. Coarse particles can affect mix homogeneity and reduce certain properties. Optimizing biochar particle size is crucial for balancing carbon sequestration with structural performance and workability. The subsequent section demonstrates and discusses the impact of biochar particle size on CO_2 sequestration.

3 Biochar Particle Size and Its Effect on CO_2 Sequestration

Due to its porous and weak structure, biochar does not contribute to strength; however, it has been shown that small amounts of biochar can increase the strength properties slightly [26]. It was found that the presence of biochar can have a micro-filler effect [27]. Yet, the presence of biochar was observed to contribute to a refinement and better dispersion of the porous network of the cementitious matrix that leads to significantly reduced unwanted large diameter compaction pores [28–30].

A 15% substitution of cement with biochar can lead to a CO_2 reduction similar to Limestone Calcined Clay Cement [31]. However, achieving a true net-zero status requires incorporating larger amounts of biochar, which includes the replacement of portions of sand and aggregates. To improve biochar utilization without significantly affecting concrete performance, a detailed examination of the interactions among cement paste, liquid phase, and biochar particles is necessary. Recent research has highlighted a significant interaction between porous biochar particles and cement paste. The degree of biochar particle pore filling, whether partial or complete, is influenced by factors such as the presence of superplasticizers, age, and biochar type [26]. This interaction is crucial for understanding the impact on the structural behavior of concrete, as filled pore channels exhibit different properties compared to unfilled ones, with filling degree dependent on biochar particle size.

The filling of pore channels, coupled with potential changes in the interfacial transition zone (ITZ) and cement paste chemistry, has a notable impact on the structural behavior of concrete. Factors like pore diameter, influenced by biochar processing, and the rheology of the paste and aqueous phase determine whether biochar channels are filled with hydrates. The aqueous phase of cement paste, particularly in the presence of polymers, plays a key role in rheology [13, 26] and directly affects whether pores are filled with hydrates or left empty. Understanding these intricate relationships is essential for optimizing the incorporation of biochar into concrete systems.

Depending on the biochar particle size, it can be added to concrete as a cement replacement, as fine aggregate replacement and as a coarse aggregate replacement (Fig. 2). It is believed that depending on the size of the biochar particles, the pore filling effect varies. The cement paste can easily access small biochar particles due to their huge surface area. However, as the size of the biochar particles increases, the filling effect decreases. If biochar with particle sizes similar to coarse concrete aggregate is used, the biochar content in the cement-based material would significantly increase. In construction areas where high strength is not a requirement, biochar can compensate for reduced strength properties by providing additional values such as lightness, ion absorption, or enhanced physical properties.

To demonstrate the replacement effect according to the biochar particle size, the negative and positive climate effects of the mixture constituents are projected. The CO_2 impact is calculated according to embodied carbon (EC) criterions calculated in kgCO_2/kg according to [32] (in Tables 1, 2, and 3). In an in-depth analysis

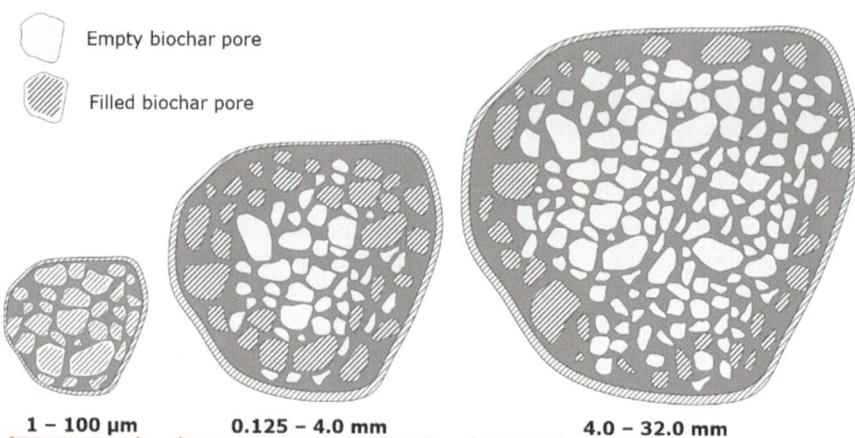

Fig. 2 Schematic representation of biochar particles of different sizes, and pore-filling effect depending on size; (**a**) biochar as cement replacement; (**b**) biochar as fine aggregate replacement; (**c**) biochar as coarse aggregate replacement

Table 1 Scenario A—embodied carbon of standard concrete mix design, no biochar is used

Component	GWP	Density	kg/m^3	l/m^3	kgCO$_2$/kg
Cement	0.83	3.1	350	113	290.5
Biochar	−2.5	1			0.0
Water	0.13	1	200	200	26.0
Aggregates	0.005	2.65	1821	687	9.1
				1000	325.6

Table 2 Scenario B—embodied carbon of standard concrete mix design with 10 vol.% replacement of cement by biochar with a specific gravity of 3.1 and 1.0, respectively

Component	GWP	Density	kg/m^3	l/m^3	kgCO$_2$/kg
Cement	0.83	3.1	315	102	261.5
Biochar	−2.5	1	11	11	−28.2
Water	0.13	1	200	200	26.0
Aggregates	0.005	2.65	1821	687	9.1
				1000	268.3

Table 3 Scenario C—embodied carbon of standard concrete mix design with 20 vol.% replacement of aggregates by biochar

Component	GWP	Density	kg/m^3	l/m^3	kgCO$_2$/kg
Cement	0.83	3.1	350	113	290.5
Biochar	−2.5	1	137	137	−343.5
Water	0.13	1	200	200	26.0
Aggregates	0.005	2.65	1457	550	7.3
				1000	−19.8

contrasting scenarios A and B, as delineated in Tables 1 and 2, an intriguing environmental benefit emerges when 10% of the cement volume is substituted with biochar. This substitution results in quantitatively 18% reduction in the carbon footprint, equating to 57.3 kg of CO_2 per kg of concrete. This finding underscores the potential for biochar as an eco-friendly alternative in concrete production, contributing to a substantial decrease in the concrete's overall carbon footprint. Despite the notable decrease in carbon emissions achieved through the initial substitution process, Scenario C, as detailed in Table 3 (i.e., substituting approx. 20% of the total aggregates with biochar), results in this scenario are particularly compelling, demonstrating that such a replacement strategy even make concrete not only carbon neutral but rather carbon negative. This denotes that the concrete's manufacturing process, from production to usage, has a net carbon emission of zero, effectively balancing the amount of carbon released with the amount offset or sequestered. Consequently, this advancement would be a significant step forward in the quest for sustainable building materials, offering a tangible solution to the construction industry's carbon emission challenges.

At a cement volume, replacement of approximately 60% real net-zero concrete is possible, but it comes at the price of reduced mechanical strength. Utilizing biochar as a replacement for aggregates rather than directly substituting cement could potentially allow for greater biochar content without significantly impacting concrete's binding properties. Nevertheless, an increased biochar amount necessitates a careful balance to maintain concrete's strength and durability. Optimal concrete performance hinges on a meticulous mix design, considering aggregate types and size distribution, water–cement ratio, and other additives. Compliance with industry standards and conducting proper testing are imperative to ensure the concrete fulfills the required structural and safety specifications. The selection of an appropriate method for incorporating biochar depends on the specifications of the specific application, desired properties, and the use of the final product.

4 Conclusions

In summary, biochar can be incorporated in concrete right at cement production (i.e., as additive) or to production of the concrete materials as replacement for binder or powder components, or aggregates. The comparative analysis of the scenarios offers insights into the role of biochar in reducing the carbon footprint of concrete. By systematically replacing a portion of the cement and aggregates with biochar, the research reveals a scalable approach to achieving more sustainable construction practices. The embodiment of higher amount of biochar in concrete allows to make cement-based materials carbon neutral. Utilizing biochar as a replacement for aggregates rather than directly substituting cement can make concrete not only carbon neutral but even carbon negative, depending upon the replacement rate. These findings highlight the potential for considerable environmental benefits and pave the way for future studies on the application of biochar in the construction industry, emphasizing the need for innovative materials that contribute to the mitigation of global climate change.

Acknowledgments Parts of the presented results were derived within the context of research collaboration with the German Development Cooperation (GIZ).

References

1. Venditti, B., Belan, M.: Visualizing the accumulation of human-made mass on earth. https://www.visualcapitalist.com/visualizing-the-accumulation-of-human-made-mass-on-earth/ (2021)
2. Scrivener, K.L., John, V.M., Gartner, E.M., UN Environment: Eco-efficient cements: potential economically viable solutions for a low-CO_2 cement-based materials industry. Cem. Concr. Res. **114**, 2–26 (2018)
3. Andrew, R.M.: Global CO_2 emissions from cement production, 1928-2018. Earth Syst. Sci. Data. **11**(4), 1675–1710 (2019)
4. USDA: Global Carbon Atlas. https://www.climatehubs.usda.gov/hubs/northern-forests/tools/global-carbon-atlas (2023)
5. Scrivener, K.L., John, V.M., Gartner, E.M.: Eco-efficient cements: potential economically viable solutions for a low-CO2 cement-based materials industry. Cem. Concr. Res. **114**(5), 2–26 (2018)
6. Arnold, W., Faber, H.M., Schmidt, W., Scrivener, K.: Decarbonising global construction. In: JCGC (ed.) A policy note within the framework of GLOBE. The Global Consensus on Sustainability in the Built Environment (2022)
7. Schmidt, W., Otieno, M., Olonade, K.A., Radebe, N.W., van Damme, H., Tunji-Olayeni, P., Kenai, S., Tawiah, A.T., Manful, K., Akinwale, A., Mbugua, R.N., Rogge, A.: Specific materials challenges and innovation potentials for minerally bound construction materials in Africa. RILEM Tech. Lett. **4**, 63–74 (2020)
8. Marangu, J.M., Marsh, A.T.M., Panesar, D.K., Radebe, N.W., Puyalto, A.R., Schmidt, W., Valentini, L.: Five recommendations to accelerate sustainable solutions in cement and concrete through partnership. RILEM Tech. Lett. **8**, 1–11 (2023)
9. Zunino, F., Martirena, J.F., Scrivener, K.: Limestone calcined clay cements (LC3). ACI Mater. J. **118**(3), 49–60 (2021)
10. Schmidt, W., Msinjili, N.S., Kühne, H.-C.: Materials and technology solutions to tackle the challenges in daily concrete construction for housing and infrastructure in sub-Saharan Africa. Afr. J. Sci. Technol. Innov. Dev. **11**, 401–415 (2019)
11. Schmidt, W., Olonade, K.A.: Optimised processes for the production of performance concrete constituents based on agricultural wastes. IOP Conf. Ser. Earth Environ. Sci. **1195**, 012001 (2023)
12. Schmidt, W., Commeh, M., Olonade, K., et al.: Sustainable circular value chains: from rural waste to feasible urban construction materials solutions. Dev. Built Environ. **6**, 100047 (2021)
13. Thiedeitz, M., Schmidt, W., Härder, M., Kränkel, T.: Performance of rice husk ash as supplementary cementitious material after production in the field and in the lab. Materials. **13**(19), 4319 (2020)
14. Schmidt, W., Olonade, K.O., Akuffo-Ensaw, A.N., Tawiah, A.T., Asante, B., Ofosu, S.A., Fordjour, A.: Sustainability potentials of the precast industry in West Africa – industry review, part 1. Concr. Plant Int., 122–127 (2023)
15. Schmidt, W., Kanjee, J., Motukwa, G., Olonade, K., Dodoo, A.: A snapshot review of future-oriented standards for cement, admixtures, and concrete: how Africa can spearhead the implementation of green urban construction materials. MRS Adv. **8**, 557–565 (2023)

16. Schmidt, W., et al.: Sustainability potentials of the precast industry in West Africa – industry review, part 2. Concr. Plant Int. **7**, 114 (2024)
17. Dissanayake, P.D., Choi, S.W., Igalavithana, A.D., Yang, X., Tsang, D.C.W., Wang, C.-H., Kua, H.W., Lee, K.B., Ok, Y.S.: Sustainable gasification biochar as a high efficiency adsorbent for CO2 capture: a facile method to designer biochar fabrication. Renew. Sust. Energ. Rev. **124**, 109785 (2020)
18. Tripathi, N., Hills, C.D., Singh, R.S., Atkinson, C.J.: Biomass waste utilisation in low-carbon products: harnessing a major potential resource. NPJ Clim. Atmos. Sci. **2**, 35 (2019). https://doi.org/10.1038/s41612-019-0093-5
19. Windeatt, J.H., Ross, A.B., Williams, P.T., Forster, M.P., Nahil, M.A., Singh, S.: Characteristics of biochars from crop residues: potential for carbon sequestration and soil amendment. J. Environ. Manag. **146**, 189–197 (2014)
20. Yang, J., Tang, L.S., Bai, L., Bao, R.-Y., Liu, Z.-Y., Xie, B.-H., Yang, M.-B., Yang, W.: High-performance composite phase change materials for energy conversion based on macroscopically three-dimensional structural materials. Mater. Horiz. **6**, 250–273 (2019)
21. Zhang, Y., He, M., Wang, L., Yan, J., Ma, B., Zhu, X., Ok, Y.S., Mechtcherine, V., Tsang, D.C.W.: Biochar as construction materials for achieving carbon neutrality. Biochar. **4**(59), 1–25 (2022)
22. Wang, F., Harindintwali, J.D., Yuan, Z., et al.: Technologies and perspectives for achieving carbon neutrality. Innovation. **2**(4), 100180 (2021)
23. Wang, L., Chen, L., Poon, C.S., Wang, C.-H., Ok, Y.S., Mechtcherine, V., Tsang, D.C.W.: Roles of biochar and CO_2 curing in sustainable magnesia cement-based composites. ACS Sustain. Chem. Eng. **9**(25), 8603–8610 (2021)
24. Cha, J.S., Park, S.H., Jung, S.-C., Ryu, C., Jeon, J.-K., Shin, M.-C., Park, Y.-K.: Production and utilization of biochar: a review. J. Ind. Eng. Chem. **40**, 1–15 (2016)
25. Senadheera, S.S., Gupta, S., Kua, H.W., Hou, D., Kim, S., Tsang, D.C.W., Ok, Y.S.: Application of biochar in concrete – a review. Cem. Concr. Compos. **143**, 105204 (2023)
26. Schmidt, W., Midroit, L., Cunningham, P.R., Miller, S.A., Amziane, S.: The influence of biochar on the flow properties, early hydration, and strength evolution of paste. In: Amziane, S., Merta, I., Page, J. (eds.) Bio-Based Building Materials. ICBBM 2023. RILEM Bookseries, vol. 45, pp. 829–838. Springer, Cham (2023)
27. Gupta, S., Kua, H.W., Low, C.Y.: Use of biochar as carbon sequestering additive in cement mortar. Cem. Concr. Compos. **87**, 110–129 (2018)
28. Lorenzoni, R., Cunningham, P., Fritsch, T., Schmidt, W., Kruschwitz, S.: Microstructure of biochar-based concrete: MIP, gas sorption, NMR, and μ-CT analysis. In: 5th International Conference on Bio-Based Building Materials Vienna, Austria (2023)
29. Lorenzoni, R., Cunningham, P., Fritsch, T., Schmidt, W., Kruschwitz, S., Bruno, G.: Microstructure analysis of cement-biochar composites. Mater. Struct. **57**, 175 (2023)
30. Lorenzoni, R., Mezhov, A., Fritsch, T., Schmidt, W., Kruschwitz, S., Giovanni, B.: Influence of microstructural changes caused by biochar on mechanical responses. (in press)
31. Müller, J.: Technologische und klimarelevante Verbesserungspotentiale durch den Einsatz bio-basierter Zementersatzstoffe im Betonbau am Beispiel von Pflanzenkohle, p. 91. Bundesanstalt für Materialforschung und -prüfung (BAM) (2022)
32. Hammond, G.P., Jones, C.I.: Embodied energy and carbon in construction materials. Proc. Inst. Civ. Eng. Energy. **161**, 87–98 (2008)

Developing Eco-friendly Ultra-High-Performance Concrete by Utilizing Recycled Alternatives

Jennifer Wivast, Anette Nyland, Saeed Bozorgmehr Nia, Mahdi Kioumarsi, and Behrouz Shafei

Contents

1 Introduction

Ultra-high-performance concrete (UHPC) represents a groundbreaking advancement in modern construction materials, celebrated for its superior mechanical attributes, including exceptional compressive strength, durability, and resilience against environmental stressors. This innovative material has catalyzed a paradigm shift in construction practices, setting new standards for efficiency and performance. However, the quest for sustainable construction materials has illuminated the environmental footprint of UHPC, particularly its carbon emissions stemming from high Portland cement content—a critical issue given the construction industry's significant contribution to global carbon emissions. UHPC has gained significant attention in recent years as a promising alternative for conventional concrete to improve construction practices [1]. The push for sustainability within the construction sector

J. Wivast · A. Nyland · M. Kioumarsi
Department of Built Environment, Oslo Metropolitan University, Oslo, Norway

S. B. Nia · B. Shafei (✉)
Department of Civil Engineering, Iowa State University, Ames, IA, USA
e-mail: shafei@iastate.edu

© The Author(s) 2025
M. Kioumarsi, B. Shafei (eds.), *The 1st International Conference on Net-Zero Built Environment*, Lecture Notes in Civil Engineering 237,
https://doi.org/10.1007/978-3-031-69626-8_5

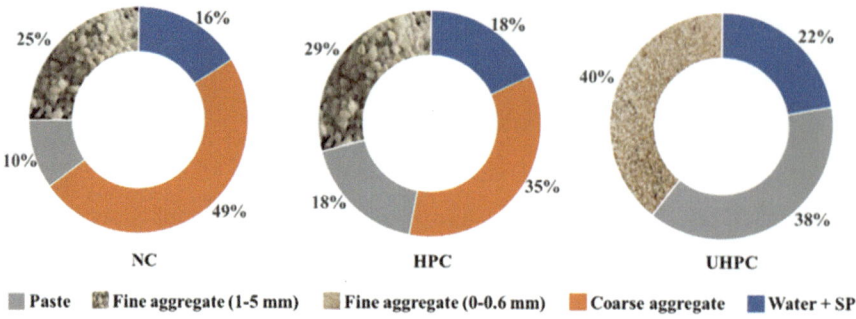

Fig. 1 Components of different types of concrete [6]

has led to significant interest in making UHPC more eco-friendly by optimizing its ingredient mix. Given the cement industry's substantial contribution to global carbon emissions—about 7%—there is a critical focus on reducing the carbon footprint of UHPC, traditionally high in Portland cement content ($800–1000$ kg/m^3) (Fig. 1). This high cement content, while contributing to UHPC's strength, also poses environmental and economic concerns due to the energy-intensive cement production and the resultant unhydrated cement in the low water-to-binder ratio mix [1–5]. These factors challenge the wider adoption of UHPC despite its superior performance. The solution lies in developing a new UHPC mix design with reduced environmental impact, starting with binder system optimization to lower cement use without sacrificing performance [6–9].

Portland limestone cement (PLC) has emerged as a viable, eco-friendlier alternative to ordinary Portland cement (OPC) for UHPC production, as it incorporates limestone during the clinker grinding process, aligning with ASTM C595 [10] and AASHTO M 240 [11] standards for up to 15% limestone content. This substitution not only significantly reduces CO_2 emissions but also maintains or enhances the concrete's performance. Additionally, the use of supplementary cementitious materials (SCMs) like silica fume and glass powder—industrial byproducts with pozzolanic properties— can further decrease cement use in UHPC by reacting with calcium hydroxide to form additional cementitious compounds, thereby improving strength, durability, and reducing permeability. In response to the high costs and environmental impact of proprietary UHPC mixes, this research proposes a nonproprietary mix utilizing PLC to cut CO_2 emissions and glass powder as a pozzolanic replacement for silica fume, aiming to develop an accessible, cost-effective, and environmentally sustainable UHPC mix.

2 Experimental Program

2.1 Materials

The materials used for UHPC binder system consist of OPC Type II, PLC, silica fume, fine aggregate, glass powder, high range water reducer, and water. Two types of cement were used in this research: OPC meets the requirement for ASTM C150-18 [12], and PLC meets ASTM C595 [10] requirements as a Type IL cement. The main physical properties of OPC and PLC are presented in Table 1.

In this study, a dry densified silica fume possessing a purity of over 95% SiO_2 was utilized. This material, which complies with ASTM C 1240 and features a specific gravity of 2.2, was supplied by Master Builders Solutions. In addition, the research also made use of a glass powder derived from waste fiberglass, which has a median particle size of 6 μm and a specific gravity of 2.6.

Fine sand which was a combination of natural sand and crushed sand with specific gravity of 2.68 and maximum size of 2.38 mm was employed.

2.2 Mixture Proportion

After a series of experimental trials, the ratio of sand to binder materials—comprising cement and either silica fume or glass powder—was adjusted. This adjustment was crucial due to its significant impact on the strength development of UHP binders. A comparison between selected materials used in the sustainable and conventional mixtures is presented in Table 2. For all mixtures, the water-to-cement (w/c) ratio was consistently maintained at 0.22, and a high-range water-reducing agent (HRWRA) was incrementally added to achieve the desired fluidity, measured as a flow range from 200 mm to 250 mm.

2.3 Test Plan

Flowability
The flowability tests for the UHPC mortars were carried out in accordance with the ASTM C1437 standard [13], adhering to the criteria outlined in ASTM C1856

Table 1 Physical data for OPC and PLC

	Blaine fineness [cm²/g]	Specific gravity	Initial setting time [min]	Final setting time [min]
OPC	4030	3.15	86	187
PLC	4426	3.08	95	155

Table 2 Selected materials in ordinary vs. sustainable UHPC mixtures

NP-UHPC with ordinary materials	NP-UHPC with sustainable materials
Ordinary Portland cement (OPC)	Portland limestone cement (PLC)
Silica fume	Waste glass powder
Natural sand (MSA: 2.36 mm)	Natural sand (MSA: 2.36 mm)
HRWR	HRWR
Water	Water

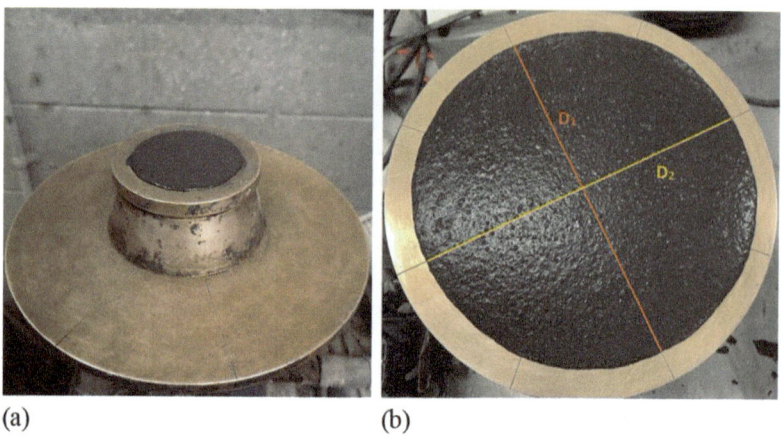

(a)   (b)

Fig. 2 (**a**) Slump flow cone with UHPC mixture and (**b**) average flow measurements

[14]. The cone container, which is shown in Fig. 2a, was filled once with the mortar mixture without tapping. After lifting the cone, the spread of the UHPC mortar was measured at two perpendicular points, D_1 and D_2, after a 2-minute interval, as illustrated in Fig. 2b. The average of these measurements determined the flow, which per standards should range between 200 and 250 mm for optimal flowability performance.

Compressive Strength

Compressive strength is a fundamental parameter in the mix design of UHPC, highlighting the material's ability to withstand loads without failure and indicating its load-bearing capabilities. This research adheres to ASTM C109 [15] standards for assessing the compressive strength of UHPC mortar specimens, utilizing $50 \times 50 \times 50$ mm^3 cubes and subjecting them to a loading rate of 300 lbs./s (1334 N/s). For fiber-reinforced samples, ASTM C1856 standards prevail, requiring cylindrical specimens of 3 by 6 inches (75 mm by 150 mm). Each mixture is represented by three samples to determine its average compressive strength, with tests conducted at four critical ages: 3, 7, 28, 90 days.

Shrinkage

Autogenous shrinkage in concrete, particularly evident in mixes with low water-to-cement ratios, is attributed to volume reduction during the hydration of cementitious materials. This process involves the formation of hydrates as water reacts with cement, leading to a demand for water and its consumption, which in turn causes shrinkage. To accurately measure autogenous shrinkage without the interference of water evaporation, the bulk strain of a sealed mortar specimen is assessed, ensuring that any water movement is solely due to hydration. This method, conducted in accordance with ASTM C1698 [16], utilizes molds made from corrugated plastic tubes, which are filled with mortar. These tubes, detailed in Fig. 3a, measure 420 ± 5 mm in length and have an outer diameter of 29 ± 0.5 mm. A dilatometer

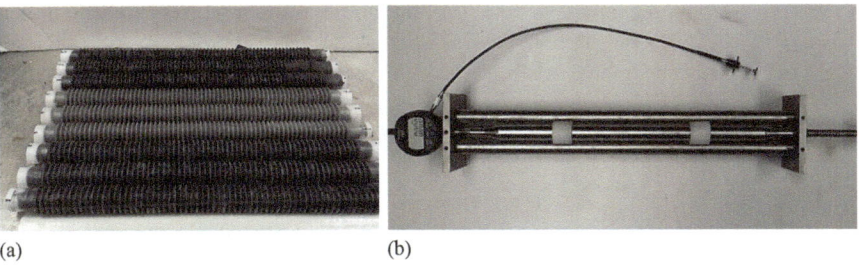

(a)   (b)

Fig. 3 (**a**) Corrugated molds filled with UHPC mixture and (**b**) test set-up for autogenous shrinkage

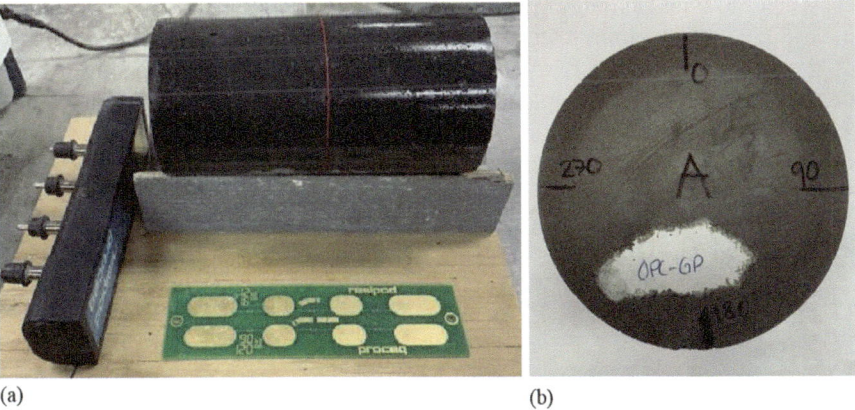

(a)   (b)

Fig. 4 (**a**) Test setup for surface resistivity and (**b**) sample marking

is employed to measure the shrinkage, comparing the length of the mortar-filled tubes to a reference bar as shown in Fig. 3b. These measurements are taken daily over a period of at least 28 days to monitor the progression of autogenous shrinkage.

Surface Resistivity

In this study, electrical surface resistivity (measured in kΩ-cm) serves as a critical metric for assessing the durability, permeability, and potential for chloride penetration, directly influencing the corrosion risk of embedded reinforcement. Utilizing a 4-point Wenner probe, as depicted in Fig. 4a, the electrical resistivity of UHPC mixtures was evaluated. This procedure aligns with AASHTO T 358 [17] standards and involves testing three cylindrical specimens, each 100 mm in diameter and 200 mm in height, after a curing period of 28 days. For minimizing inaccuracies arising from electrical field disruptions caused by steel fiber integration, only the binder samples was utilized for this test. To ensure consistency, the surface of all specimens was moistened before measurements were taken. Each specimen underwent two resistivity measurements at four different angles—0°, 90°, 180°, and 270°—as shown in Fig. 4b. The average resistivity for each specimen was calculated, followed by an overall set average derived from the three samples, providing a comprehensive assessment of the UHPC mixtures' electrical resistivity properties.

3   Results and Discussions

3.1   Flowability

The data presented in Table 3 demonstrate that all examined UHPC mixtures achieved a flow exceeding 200 mm, meeting the ASTM C1856 requirements. A consistent flow range from 230 mm to 250 mm was observed across all mixtures, indicating their self-consolidating properties. These results highlight the capability of the UHPC mixtures to be efficiently cast without the need for mechanical vibration. Further examination of Table 3 indicates that UHPC mixtures containing Type IL and a higher proportion of glass powder demonstrated superior flowability while requiring a reduced quantity of high-range water-reducing agent (HRWRA) compared to mixtures incorporating OPC and silica fume at same water-to-binder ratio. This suggests that the synergy between glass powder and Type IL PLC can significantly enhance the workability of UHPC formulations. Such an improvement points toward a potentially cost-effective strategy for optimizing UHPC mix designs.

3.2   Shrinkage

Table 4 details the autogenous shrinkage-induced deformation in different mixtures after a 28-day period, highlighting the variance in deformation across the mixtures. The OPC mixed with Silica Fume (SF) reported the highest deformation, underscoring significant autogenous shrinkage in mixtures containing silica fume. This is attributed to silica fume's high surface area and its substantial water demand during the hydration process compared to glass powder. Conversely, mixtures incorporating glass powder showcased markedly less deformation than those with silica fume alone, indicating the beneficial impact of glass powder in reducing shrinkage. When comparing OPC- and PLC-based mixtures, OPC mixtures were found to exhibit greater deformation, suggesting a higher susceptibility to long-term

Table 3 Results of flow table test

Mix ID	Flow (mm)	HRWR (%)
OPC-SF	231	4.0
OPC SF-GP	238	3.9
OPC-GP	245	3.6
PLC-SF	241	3.5
PLC-SF-GP	248	3.3
PLC-GP	250	3.2

Table 4 Autogenous shrinkage results

Mix ID	Length deformation (10^{-6} m/m)
OPC-SF	−311.90
OPC-SF-GP	−271.44
OPC-GP	−223.59
PLC-SF	−213.18
PLC-SF-GP	−167.52
PLC-GP	−126.10

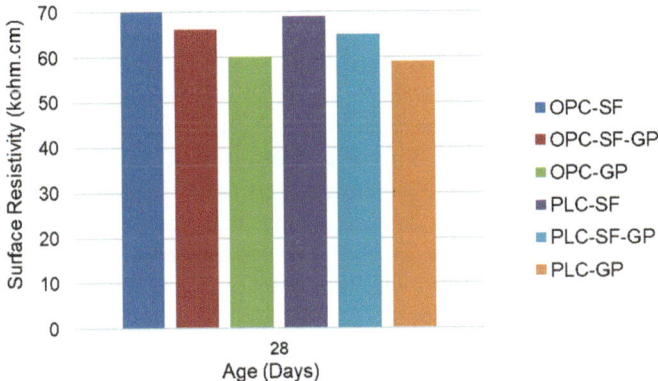

Fig. 5 Electrical resistivity results for tested UHPC binders

autogenous shrinkage. This could be due to OPC's higher clinker content, which increases water consumption for hydration. The inclusion of limestone in PLC mixtures appears to mitigate autogenous shrinkage, possibly by enhancing the pore structure within the concrete.

3.3 Surface Resistivity

The analysis of surface resistivity across all UHPC specimens revealed high electrical resistivity, attributable to the use of pozzolanic materials, a low water-to-cement ratio, and the minimization of voids through fine and well-graded material distribution within the matrix. Specifically, glass powder exhibited excellent performance in enhancing electrical resistivity, signaling its potential to significantly improve UHPC durability. Furthermore, the introduction of silica fume across all cement bases markedly improved resistivity (Fig. 5). This improvement is largely due to silica fume's reaction with calcium hydroxide, leading to the formation of additional calcium silicate hydrate, which effectively densifies the microstructure and impedes ion flow. Silica fume's ultrafine particles also play a critical role in disrupting the connectivity within the concrete's pore network, further enhancing resistivity. Overall, the results suggest that replacing OPC and silica fume with PLC and glass powder does not compromise the concrete's electrical resistivity, thereby maintaining or potentially enhancing its durability.

4   Conclusions

The use of PLC and waste glass not only enhanced the workability of concrete but also reduced the reliance on superplasticizers. This adjustment in the concrete mix design led to optimized material utilization and environmental benefits by minimizing chemical additive usage. PLC-based concrete exhibited lower shrinkage rates, a desirable property for structural durability. Conversely, the addition of silica fume to both OPC and PLC mixes increased shrinkage, attributed to its greater water demand and expansive surface area. In terms of surface resistivity and durability, incorporating silica fume into concrete significantly enhanced surface resistivity in both OPC and PLC mixes after 28 days, signaling improved durability. Notably, all binder combinations surpass the acceptable range for surface resistivity, affirming no concerns regarding durability with the chosen materials. By adopting PLC and waste glass in UHPC formulations, material costs can be significantly reduced, potentially halving expenses while still achieving properties sufficient for a variety of structural applications. This approach not only economizes construction but also advances sustainability in the building and bridge industry.

Acknowledgments   This work is a part of the TRANSFORM project, which has received funding from the Norwegian Directorate for Higher Education and Competence (HK-dir), project number UTF-2020/10107. Additionally, we acknowledge the support from the Net-Zero Future project, an INTPART project funded by the Norwegian Directorate for Higher Education and Competence (HK-dir), and the Research Council of Norway, project number 337262.

References

1. Kazemian, M., Shafei, B.: Investigation of type, size, and dosage effects of superabsorbent polymers on the hydration development of high-performance cementitious materials. Constr. Build. Mater. **422**, 135801 (2024)
2. Karim, R., Shafei, B.: Ultra-high performance concrete under direct tension: investigation of a hybrid of steel and synthetic fibers. Struct. Concr. **25**, 423 (2024)
3. Kazemian, M., Shafei, B.: Effects of supplementary cementitious materials on the hydration of ultrahigh-performance concrete. J. Mater. Civ. Eng. **35**(11), 04023388 (2023)
4. Kazemian, M., Shafei, B.: Carbon sequestration and storage in concrete: a state-of-the-art review of compositions, methods, and developments. J. CO2 Utiliz. **70**, 102443, 1–102443,15 (2023)
5. Kazemian, M., Shafei, B.: Internal curing capabilities of natural zeolite to improve the hydration of ultra-high performance concrete. J. Constr. Build. Mater. **340**(127452), 1–12 (2022)
6. Wang, X., et al.: Design of sustainable ultra-high performance concrete: a review. Constr. Build. Mater. **307**, 124643 (2021)

7. Karim, R., Shafei, B.: Investigation of five synthetic fibers as potential replacements of steel fibers in ultra-high performance concrete, ASCE. J. Mater. Civ. Eng. **34**(7), 04022126, 1–04022126,14 (2022)
8. Karim, R., Shafei, B.: Flexural response characteristics of ultra-high performance concrete made with steel microfibers and macrofibers. J. Struct. Concr. **22**(6), 3476–3490 (2021)
9. Karim, R., Najimi, M., Shafei, B.: Assessment of transport properties, volume stability, and frost resistance of non-proprietary ultra-high performance concrete. J. Constr. Build. Mater. **227**, 117031, 1–117031,10 (2019)
10. ASTM C 595: International Standard Specification for Portland Cement (2022)
11. American Association of State Highway and Transportation Officials, AASHTO M240: Standard Specification for Blended Hydraulic Cement (2015)
12. ASTM C150: International Standard Specification for Blended Hydraulic Cements (2022)
13. ASTM C1437: International Standard Test Method for Flow of Hydraulic Cement Mortar (2020)
14. ASTM C1856: International Standard Practice for Fabricating and Testing Specimens of Ultra-High Performance Concrete (2017)
15. ASTM C109: International Standard Test Method for Compressive Strength of Hydraulic Cement Mortars (Using 2-in. or [50 mm] Cube Specimens) (2021)
16. ASTM C1698: International Standard Test Method for Autogenous Strain of Cement Paste and Mortar (2019)

Reducing the CO$_2$ Footprint of UHPC Through Portland Cement Substitution

Ingrid Lande, Andrej A. Sørensen, Martin Hagen, and Rein Terje Thorstensen

Contents

1 Introduction

Ultra-high-performance concrete (UHPC) might be a competitive alternative to conventional concrete for some purposes. One important argument might be the potential for reducing the CO$_2$ footprint through a reduction in material consumption. This has been shown in previous life cycle assessments (LCA), e.g., [1–3]. A substantial reduction is possible for some structural elements, as the superior strength opens for reducing the material consumption. Durability also favors UHPC.

However, the CO$_2$ footprint is far higher for UHPC, compared to equal volumes of conventional concrete [4]. The main reason is the high amount of cement and high-strength steel fibers [4, 5]. Reducing the content of Portland clinker, which is the main binder in most modern cement types for structural use, will contribute to reducing the CO$_2$ footprint of UHPC. Many studies focus on experimental research on partial cement substitution with various by-products, waste, or natural ores that are locally available. This includes materials such as glass powder [6], quarry-stone

I. Lande (✉) · A. A. Sørensen · M. Hagen · R. T. Thorstensen
University of Agder, Grimstad, Norway
e-mail: Ingrid.lande@uia.no

© The Author(s) 2025
M. Kioumarsi, B. Shafei (eds.), *The 1st International Conference on Net-Zero Built Environment*, Lecture Notes in Civil Engineering 237,
https://doi.org/10.1007/978-3-031-69626-8_6

powders [7], red mud [8], and lead-zinc tailings [9]. There are two restrictions to the industrial relevance of these approaches, at least within a modest time frame. One is lacking compliance with existing standards for most of these materials. The time frame needed for revising standards based on limited research projects is vast. The second limitation is that industrial application normally requires a stable and rich supply of materials. This is often a problem when utilizing locally available by-products from other production. Despite the superb quality and the high relevance of this research, in the long run, industrial utilization of the topics that are researched is often distant. Other researchers have focused on substituting cement with more conventional replacement materials like limestone powder [10], fly ash, and GGBS [11–13]. This is closely related to our topic, as these materials are the dominant cement-substituting materials that are utilized in the pre-accepted blended cement. Only a few researchers have investigated the effect of using pre-accepted blended cement types [14, 15].

In the study presented in this paper, the potential for producing UHPC with alternative pre-accepted cement types is investigated. The contribution of this work is to strengthen the documentation on the use of pre-accepted cement types and to investigate how the use of these binders in UHPC is influenced by using locally available materials in Norway. This includes cement types CEM II and CEM III with 30 and 70% GGBS, respectively, while using the pure ordinary Portland cement (OPC) cement class CEM I as a reference. All classes are stated according to the CEN standard *EN 197-1*. A laboratory investigation was executed, measuring the strength performance of these alternatives. In addition, a measurement of fresh state consistency and a comparison of CO_2 footprint was conducted.

2   Materials and Methods

2.1   Material and Mix-Proportions

Three types of pre-accepted cement were applied: CEM I, CEM II, and CEM III in accordance with the European cement standard *EN 197-1*. The CEM I applied was an ordinary cement with 95% Portland cement clinker. The CEM II was a Portland-composite cement with 30% ground-granulated blast-furnace slag (GGBS), while the CEM III was a Blast furnace slag cement with 70% GGBS. The different types of cement were applied in the same UHPC mix, based on locally available constituents in Norway. Table 1 shows the different materials with specifications according to technical data sheets.

Table 2 shows the compressive strength properties and CO_2-eq of the different cement types. The data is found in the technical data sheets and Environmental Product Declaration (EPD) by the producer.

The UHPC mixes are shown in Table 3. The w/b-ratio and steel fiber content was kept constant for all mixes, 0.23 and 2 vol.%, respectively. The small differences in recipes are due to differences in density between the cement types.

Table 1 Materials

Materials	Specifications
Cement	CEM I 52.5 R *EN 197-1*. Density 3.15 g/cm^3
	CEM II/B-S 52.5 *EN 197-1*. Density 3.05 g/cm^3
	CEM III / B 42.5 L-LH/SR *EN 197-1*, density 2.94 g/cm^3
Microsilica	Undensified microsilica, SiO2 > 90%, bulk density of 200–350 kg/m^3
Filler	Limestone filler for concrete production, density: 2720 kg/m^3
Sand	Sand from local producer, size 0–6 mm, density: 2660 kg/m^3.
Superplasticizer	Specially composed for UHPC production, 40% active content, density: 1080 kg/m^3
Steel fibers	Cord steel wire, brass colored, length (l_f): 12.5 mm, diameter (d_f): 0.175 mm, tensile strength: min. 2800 N/mm^2

Table 2 Properties of cement according to data sheets

		CEM I	CEM II	CEM III
Compressive strength [MPa]	1 day	32	18	–
	2 days	45	29	12
	28 days	72	62	53
kg CO$_2$-eq/1000 kg cement	A1-A4	766.7	539.5	252.1

Table 3 UHPC mixes

Materials	UHPC CEM I kg/m^3	UHPC CEM II kg/m^3	UHPC CEM III kg/m^3
Cement CEM I	782	–	–
Cement CEM II	–	776	–
Cement CEM III	–	–	768
Microsilica	155	153	152
Filler	166	165	163
Sand 0/6 mm	1017	1008	999
Water	214	212	210
Superplasticizer	13	13	13
Steel fibers	157	157	157

2.2 Mixing and Casting

The UHPC mixes were produced in a small planetary mixer for mortars at standard laboratory conditions (*ASTM C305*). Three liters were mixed for each mix. The procedure of mixing was applied according to previous experience [16]:

1. Mixing the dry materials for 30 seconds at low speed (107 rpm)
2. Adding water and superplasticizer
3. Mixing for 30 seconds at low speed (107 rpm)
4. Pausing the mixer for 90 seconds

5. Mixing at medium speed (198 rpm) for 120 seconds
6. Mixing at high speed (361 rpm) for 15–30 seconds

After mixing, the consistency of the fresh mortar was measured by a flow table for cement mortar in accordance with *NS-EN 413-2*, without jolting. For each mix, six test prisms of 40 mm × 40 mm × 160 mm were cast and compacted using a jolting apparatus, according to *NS-EN 196-1*. The test specimens were covered with a plastic film to avoid drying and de-molded after approximately 24 hours.

2.3   Test Program

The prisms were tested according to the European standard *NS-EN 196-1* (Fig. 1). Both compressive strength and flexural strength were tested after 28 days of curing. First, the prisms were tested in bending and then both halves of each prism were tested in compression. The bending test only includes testing until cracking, not the post-cracking behavior. The effect of two different curing regimes was investigated:

- Heat curing: curing at elevated temperature for 72 hours (including gradual increase and decrease in temperature). The specimens were submerged in water in a heat chamber at 90 °C. After the heat curing, the specimens were stored in water tanks at 20 °C until testing.
- Normal curing: The specimens were cured submerged in water at 20 °C until testing.

Table 4 shows an overview of the conducted tests and curing regimes.

3   Results and Discussions

The influence of using cement with a lower content of Portland clinker was studied on the strength of test specimens. In addition, a comparison of CO_2 footprint was conducted. The results are shown in the following.

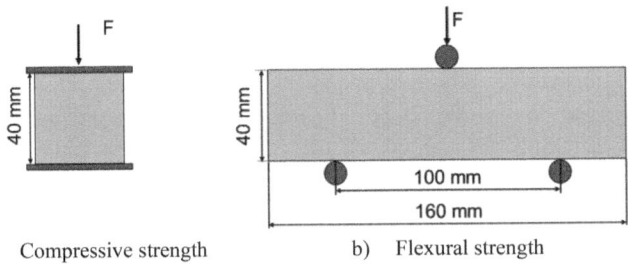

a)   Compressive strength   b)   Flexural strength

Fig. 1 Test setup of strength properties. (**a**) Compressive strength. (**b**) Flexural strength

Table 4 Test program

Mix	Test	Specimens tested	Curing	Test age
UHPC CEM I	Compressive strength	6	90 °C 72 h, 20 °C	28
	Flexural strength	3	90 °C 72 h, 20 °C	28
UHPC CEM II	Compressive strength	6	90 °C 72 h, 20 °C	28
	Flexural strength	3	90 °C 72 h, 20 °C	28
UHPC CEM III	Compressive strength	6	90 °C 72 h, 20 °C	28
	Flexural strength	3	90 °C 72 h, 20 °C	28

Fig. 2 Flow in mm

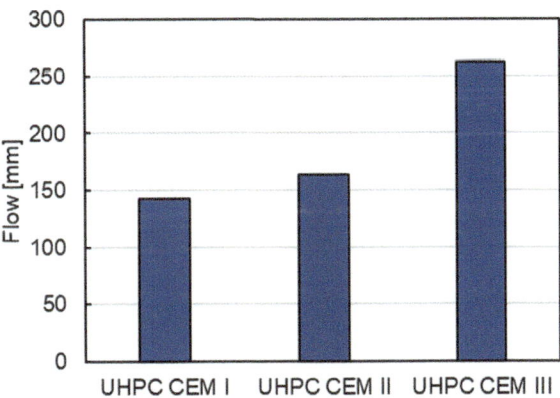

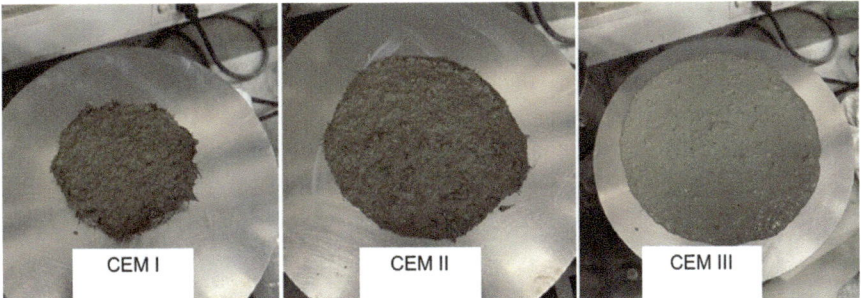

Fig. 3 Flow measurement of the three different UHPC mixes

3.1 Workability

The workability of the mix was measured through flow. The results are presented in Figs. 2 and 3. As can be seen, the water demand decreases with the addition of GGBS, giving higher flow values. Shi et al. (2019) [17] and Wu et al. (2017) [13] also experienced an increase in flow with an increase in GGBS content. This might allow for further reduction in w/b-ratio, which might reveal a potential for further improving the strength properties of UHPC with a high content of GGBS.

3.2 Strength Properties

Figure 4 shows the average 28-day compressive strength for the different mixes, using normal curing and heat curing. Error bars showing the standard deviation are included for all series, to visualize the span in results.

After curing at 20 °C for 28 days, the mixes with CEM I and CEM II obtained compressive strength higher than for the mix with CEM III. There is a slight difference between the two, with the CEM II mix showing the highest compressive strength. However, considering the variation in results for these series, demonstrated with error bars for standard deviation overlapping, it is not claimed that the difference in average values is significant. The compressive strength obtained for the mix with CEM III, which is 10% lower and outside the standard deviation of each, might be significant. The lower level of compressive strength achieved by the CEM III mix after 28 days of curing at 20 °C might be a delay caused by the hardening mechanisms of the GGBS binder being slower than that of the Portland cement clinkers.

Considering that the CEM III is classified 42.5 MPa according to *NS-EN 197* standard while both the CEM I and CEM II are classified 52.5 MPa, this might be a possible explanation for the difference in average compressive strength at 20 °C. However, considering the results after curing a period at elevated temperature seems to overrule this as an explanation for heat curing.

After curing at 90 °C for 72 hours before submerging in water at 20 °C for the remaining time up to 28 days, similar results are obtained for all three cement types. It has been claimed in earlier research that curing cement materials with a high content of pozzolanic binders at elevated temperatures has the potential to create high-capacity C-S-H components that would not be created at lower temperatures, independent of curing duration. Whether this is also relevant for latent hydraulic binders like GGBS, is not known to the authors. Regardless of this, the result of this experiment, is that when curing at 90 °C for 72 hours as the first part of the curing regime, any of the binders CEM I, CEM II, and CEM III applied yield corresponding

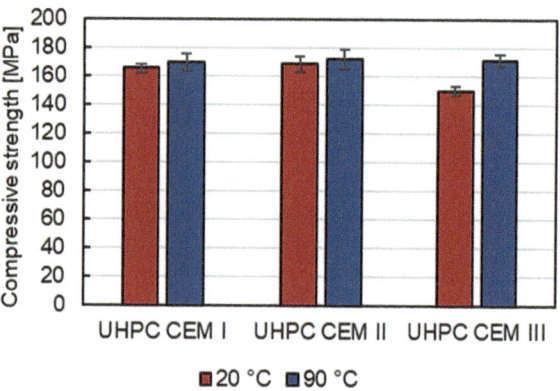

Fig. 4 Compressive strength after 28 days, with error bars showing the standard deviation

Fig. 5 Flexural strength after 28 days. Error bars show the standard deviation

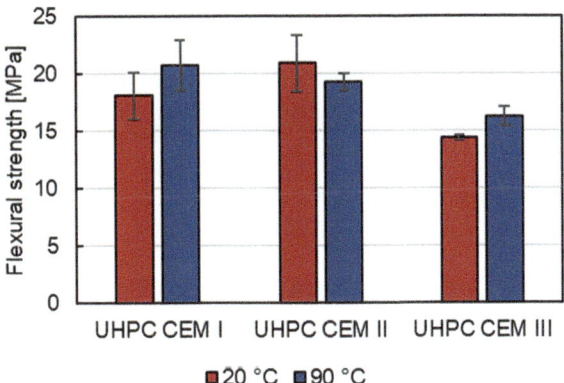

compressive strength of the UHPC mixes as achieved when applying CEM I or CEM II and curing for 28 days given standard curing at 20 °C.

Figure 5 shows the 28-day flexural strength for the different mixes after normal curing and heat curing.

As for the compressive strength, the results for the mixes containing CEM I and CEM II are corresponding. The difference between these two seems to be substantial. However, noting that also the variability for these is higher than for the compressive strength and that the error bars are strongly overlapping, this difference might turn out not to be significant. Calculating significance based on only three results seems not to be meaningful and is consequently not executed. It is well known that the variability of tensile-related strength properties in fiber-reinforced concrete is high. Thus, these results are considered to be "normal."

Despite the variability, the results of the CEM III containing mix might seem to be lower than the CEM I and CEM II mixes. As shown in Fig. 2, the consistency of the mixes is strongly varying, with flow increasing correspondingly to the content of GGBS in the binder. The flow of the CEM III mix is so high that it was suspected that the fibers might have sunk to the bottom as a result of gravity during the jolting treatment, which was part of the casting procedure of all prisms. To reveal whether this had happened, the CEM III prisms where cut into halves, and the cross sections examined. It was confirmed that agglomerations of fibers appeared at the bottom of these prisms during casting. The corresponding was not observed for the prisms made from the CEM I and CEM II mixes, when exposed to the same examination.

To secure the plane and parallel surfaces in the testing machine during bending, all prisms were rotated 90 degrees compared to the casting orientation. Thus, an unproportional part of the fibers was situated at one-half the CEM III mix prisms during testing. This is anticipated as at least a part of the reason for the lower flexural strength of the prisms made from the CEM III mix. Whether this is the explanation for all of the differences, cannot be evaluated.

The presented results are only covering strength-based properties. However, durability properties are crucial for practical application. It is expected that the durability properties of UHPC made with CEM II and CEM III classes of cement will be satisfactory, due to the pre-acceptance and experiences with CEM III in conventional concrete. However, for practical applications, this assumption needs to be confirmed as even the use of well-known cement types in "novel" materials such as UHPC while also utilizing local aggregates, might have some unexpected consequences. The durability properties will constitute the next step in our research following what is presented in this paper.

3.3 Comparison of CO_2 Footprint

A comparison of the CO_2 footprint is executed by using values from the EPDs supplied by the producers (Table 2). The calculations include life phases A1 to A4, meaning the production process (A1-A3), in addition to the transportation of these materials to market (A4).

Figure 6 shows the emission of CO_2-eq/m^3 for each of the three UHPC mixes when only evaluating the emissions stemming from the cement of the UHPC mixes. This simplification is reasonable. The majority (>95%) of the CO_2-eq. emissions from UHPC materials stem from the consumption of cement and micro-steel fibers [4, 18].

Comparing emissions from UHPC mixes with fibers excluded as shown in Fig. 6 is relevant even if the use of UHPC materials without fibers is rare due to brittleness. This is because the actual level of fiber content is adjusted to accommodate the purpose of each single use. A reduction of 67% kg CO_2-eq. was obtained, when substituting CEM I with CEM III.

Figure 7 shows the results of emissions from the UHPC mixes when a volume of 2% micro-steel fibers is included. The emissions from the steel fibers are 2.67 kg CO_2-eq/kg [5], which reduces the effect of substituting one binder with another in the UHPC mix. Still, a potential of 40% reduction in kg CO_2-eq is demonstrated when using CEM III as a binder, compared to using CEM I.

Fig. 6 kg CO_2-eq/m^3 of UHPC only including emissions from cement

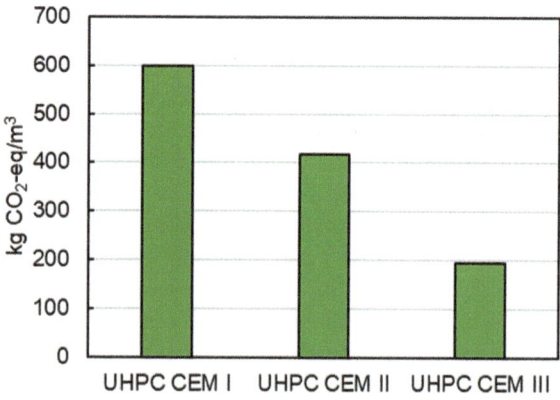

Fig. 7 kg CO$_2$-eq/m^3 of UHPC including emissions from cement and micro steel fibers

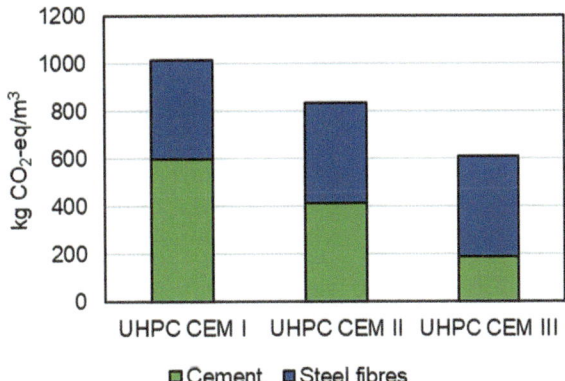

4 Conclusions

In this study, the impact of producing UHPC with alternative pre-accepted cement types was investigated through laboratory experiments. The intention was to investigate the potential for reducing the CO$_2$ footprint related to production of UHPC, based on materials that are accepted for industrial use through the existing standard (at least in Europe):

- After curing at 20 °C, the 28-day compressive strength was similar for the CEM I (OPC cement) and CEM II (30% GGBS) UHPC mixes, and lower for the CEM III (70% GGBS) mix. With curing at elevated temperature (90 °C the first 72 hours), the difference was eliminated.
- The flexural strength was similar for the UHPC mixes made with CEM I and CEM II, but lower for CEM III. However, it was revealed that in the prisms made from the CEM III mix, there was an uneven distribution of fibers, resulting in less amount of fibers to contribute to flexural strength. The prisms made from the mixes containing CEM I and CEM II did not have the same uneven fiber distribution.
- The consistency of the three mixes was varying, with flow increasing correspondingly with the content of GGBS in the binder. The high flow of the CEM III mix combined with the jolting procedure used for compaction during casting (and gravity), is anticipated to explain the uneven distribution of fibers.
- A simple comparison of emission of CO$_2$-eq/m^3 was performed to compare mixes made with CEM I, CEM II, and CEM III, respectively. A considerable reduction in emissions was demonstrated: up to 67% for the mortars excluding fibers, and 40% when including two volume percent of micro fibers made from high-strength steel.
- Considering the difference in flow when using the three different cement types, it seems more reasonable to aim at harmonizing the consistency rather than the w/b ratio. Two arguments for this procedure would be that:

1. This would probably remove the problem with uneven fiber distribution, as the act of gravity during jolting would not influence one UHPC mix more than the other.
2. Reduction of water would be the preferred way to reduce the flow of the mixes with higher GGBS content. In addition to changing the flow of fresh mortar, this would reduce the w/b ratio of the mixes with higher GGBS content. Thus, it would probably also relieve a potential for improving the strength properties that might be advantageous for these materials.

References

1. Habert, G., Denarié, E., Šajna, A., Rossi, P.: Lowering the global warming impact of bridge rehabilitations by using Ultra High Performance Fibre Reinforced Concretes. Cem. Concr. Compos. **38**, 1–11 (2013). https://doi.org/10.1016/j.cemconcomp.2012.11.008
2. Hajiesmaeili, A., Pittau, F., Denarié, E., Habert, G.: Life cycle analysis of strengthening existing RC structures with R-PE-UHPFRC. Sustainability (Switzerland). **11**(24), Art no. 6923 (2019). https://doi.org/10.3390/su11246923
3. Sameer, H., Weber, V., Mostert, C., Bringezu, S., Fehling, E., Wetzel, A.: Environmental assessment of ultra-high-performance concrete using carbon, material, and water footprint. Materials. **12**(6), Art no. 851 (2019). https://doi.org/10.3390/ma12060851
4. Lande, I., Terje Thorstensen, R.: Comprehensive sustainability strategy for the emerging ultra-high-performance concrete (UHPC) industry. Clean. Mater. **8**, 100183 (2023). https://doi.org/10.1016/j.clema.2023.100183
5. Stengel, T., Schießl, P.: 22 - Life cycle assessment (LCA) of ultra high performance concrete (UHPC) structures. In: Pacheco-Torgal, F., Cabeza, L.F., Labrincha, J., de Magalhães, A. (eds.) Eco-efficient Construction and Building Materials, pp. 528–564. Woodhead Publishing (2014)
6. Soliman, N.A., Tagnit-Hamou, A.: Development of ultra-high-performance concrete using glass powder—towards ecofriendly concrete (in English). Constr. Build. Mater. **125**, 600–612 (2016). https://doi.org/10.1016/j.conbuildmat.2016.08.073
7. Yang, R., et al.: Environmental and economical friendly ultra-high performance-concrete incorporating appropriate quarry-stone powders (in English). J. Clean. Prod. **260**, Art no. 121112 (2020). https://doi.org/10.1016/j.jclepro.2020.121112
8. Hou, D., et al.: Sustainable use of red mud in ultra-high performance concrete (UHPC): design and performance evaluation. Cem. Concr. Compos. **115**, 103862 (2021). https://doi.org/10.1016/j.cemconcomp.2020.103862
9. Wang, X., et al.: Development of a novel cleaner construction product: ultra-high performance concrete incorporating lead-zinc tailings (in English). J. Clean. Prod. **196**, 172–182 (2018). https://doi.org/10.1016/j.jclepro.2018.06.058
10. Huang, W., Kazemi-Kamyab, H., Sun, W., Scrivener, K.: Effect of cement substitution by limestone on the hydration and microstructural development of ultra-high performance concrete (UHPC) (in English). Cem. Concr. Compos. **77**, 86–101 (2017). https://doi.org/10.1016/j.cemconcomp.2016.12.009
11. Randl, N., Steiner, T., Ofner, S., Baumgartner, E., Mészöly, T.: Development of UHPC mixtures from an ecological point of view. Constr. Build. Mater. **67**, 373–378 (2014). https://doi.org/10.1016/j.conbuildmat.2013.12.102
12. Yu, R., Spiesz, P., Brouwers, H.J.H.: Development of an eco-friendly Ultra-High Performance Concrete (UHPC) with efficient cement and mineral admixtures uses. Cem. Concr. Compos. **55**, 383–394 (2015). https://doi.org/10.1016/j.cemconcomp.2014.09.024

13. Wu, Z., Shi, C., He, W.: Comparative study on flexural properties of ultra-high performance concrete with supplementary cementitious materials under different curing regimes (in English). Constr. Build. Mater. **136**, 307–313 (2017). https://doi.org/10.1016/j.conbuildmat.2017.01.052

14. Tahwia, A.M., Elgendy, G.M., Amin, M.: Durability and microstructure of eco-efficient ultra-high-performance concrete (in English). Constr. Build. Mater. **303**, Art no. 124491 (2021). https://doi.org/10.1016/j.conbuildmat.2021.124491

15. Tahwia, A.M., Elgendy, G.M., Amin, M.: Mechanical properties of affordable and sustainable ultra-high-performance concrete. Case Stud. Constr. Mater. **16**, e01069 (2022). https://doi.org/10.1016/j.cscm.2022.e01069

16. Lande, I., Thorstensen, R.T.: Towards efficient use of cement in ultra high performance concrete. Nordic Concrete Res. **65**(2), 81–105 (2021). https://doi.org/10.2478/ncr-2021-0017

17. Shi, Y., Long, G., Ma, C., Xie, Y., He, J.: Design and preparation of ultra-high performance concrete with low environmental impact. J. Clean. Prod. **214**, 633–643 (2019). https://doi.org/10.1016/j.jclepro.2018.12.318

18. Stengel, T., Schießl, P.: Sustainable construction with UHPC–from life cycle inventory data collection to environmental impact assessment. In: Second International Symposium on Ultra High Performance Concrete, Kassel, Germany, March 2008, pp. 461–468. Kassel University Press (2008)

Investigation of Lightweight Ultra-High-Performance Concrete for Net-Zero Solutions

Amir Ramezani and Behrouz Shafei ⓘ

Contents

1 Introduction

In response to the urgent need for sustainable development, the construction industry is undergoing a paradigm shift toward more environment-friendly and resource-efficient practices. This transformation is primarily driven by the recognition of the construction sector's environmental impact, including its substantial energy consumption, high carbon dioxide (CO_2) emissions, and extensive resource depletion. Consequently, there is a growing emphasis on the adoption of innovative materials and technologies that can contribute to the realization of net-zero building solutions, aiming to minimize the environmental impact of construction activities while maximizing energy efficiency and occupant comfort.

A notable focus within the field of sustainable construction is the advancement and application of lightweight ultra-high-performance concrete (LUHPC). LUHPC stands as a pivotal development in concrete technology. In comparison to regular

A. Ramezani · B. Shafei (✉)
Department of Civil, Construction, and Environmental Engineering, Iowa State University, Ames, IA, USA
e-mail: shafei@iastate.edu

© The Author(s) 2025
M. Kioumarsi, B. Shafei (eds.), *The 1st International Conference on Net-Zero Built Environment*, Lecture Notes in Civil Engineering 237,
https://doi.org/10.1007/978-3-031-69626-8_7

concrete, this type of material has lower self-weight, reduced environmental impact, and more sustainable materials while providing exceptional mechanical properties and durability, similar to ultra-high-performance concrete (UHPC) [1–5]. The crucial features of LUHPC are its reduced density and relatively high volume of voids, which can improve thermal performance and energy conservation in buildings, hence reducing CO_2 emissions [6]. As a result, LUHPC is an attractive choice for a wide range of construction applications, including net-zero buildings and infrastructure projects.

The adoption of LUHPC in construction projects holds the potential to revolutionize traditional building practices by enabling the construction of lighter, stronger, and more sustainable structures [5, 6]. However, the effective implementation of LUHPC relies on its fresh properties, notably its workability and autogenous shrinkage. Adequate workability is essential for ensuring the quality and mechanical performance of the LUHPC, while autogenous shrinkage is a critical factor in mitigating cracking and other durability concerns [7, 8]. Moreover, conducting a thorough carbon footprint assessment of LUHPC is imperative for its integration into sustainable construction processes.

To address these challenges and capitalize on the potential of LUHPC for sustainable construction, a comprehensive understanding of its fresh properties and carbon footprint is essential. This necessitates an effective analysis employing a multidisciplinary approach that integrates existing literature with a meticulous analysis of experimental data. Through the utilization of the combined knowledge and skills of researchers, engineers, and industry experts, valuable insights can be obtained into the behavior of LUHPC, thereby identifying opportunities for optimization and enhancement.

In this context, the objective of this chapter is to examine the fresh properties and carbon footprint of LUHPC by evaluating the influence of various factors. Through a review of the existing literature and analysis of experimental data, this chapter will contribute to the body of knowledge on LUHPC technology and provide valuable guidance for researchers, practitioners, and policymakers interested in promoting sustainable construction practices. By highlighting the potential benefits associated with the utilization of LUHPC in net-zero solutions, this study aims to facilitate informed decision-making and innovation in the construction industry.

2   Materials

The production of LUHPC involves meticulous selection and processing of raw materials. The key ingredients of LUHPC typically consist of Portland cement, supplementary cementitious materials (SCMs), standard fine aggregates, lightweight aggregates (LWAs), water, and chemical admixtures. In numerous studies, Class I 52.5N Portland cement (CEM 1 52.5N) was used in LUHPC mixtures due to its higher compressive strength compared to other types of Portland cement. SCMs can be replaced with Portland cement to produce cementitious materials. Various SCMs, such as silica fume, flash ash, and natural zeolites, are commonly employed to enhance particle packing density, hydration degree, internal curing, and reactivity [9–12]. In addition, superabsorbent polymers (SAPs) were reported to provide internal curing for high cementitious content mixtures [13].

LWAs are an essential component and a key factor in the manufacturing of LUHPC [14]. According to Eurocode 2 [15], LWAs are characterized by a particle density below 2000 kg/m^3. Previous studies have shown that the inclusion of LWAs can significantly contribute to producing a concrete mixture with a compressive strength exceeding 120 MPa at 28 days, while maintaining a density below 2100 kg/m^3, surpassing the 50 MPa.m^3/t 28-day specific strength threshold suggested by Meng et al. [16] for defining LUHPC characteristics [17]. The addition of large amounts of air voids, which can be supplied by stable foams or chemical gases (commonly known as air-entraining agents), is another effective method for LUHPC production [18]. Notably, specific chemical admixtures such as phase change materials (PCMs) have been utilized to produce LUHPC with promising results [19].

3 Results and Discussion

3.1 Workability

Workability, a fundamental characteristic of LUHPC, is essential for ensuring proper mixing, handling, and placing during construction. The slump flow test is commonly employed to measure the workability of LUHPC. Several factors influence the workability of LUHPC, including the incorporation of LWAs, chemical admixtures, and the water-to-binder (w/b) ratio.

The addition of LWAs significantly impacts the workability of LUHPC due to the physical properties of the aggregates. LWAs with quasi-spherical particle shapes have been reported to enhance workability by reducing water demand [16]. Additionally, LWAs with smooth and spherical shapes provide ball bearing and lubricant effects, further improving workability [20]. Conversely, LWAs with rough particle shapes, such as shale ceramsite, can reduce workability due to increased friction with the cement paste [21]. Importantly, LWAs possess lower density and higher porosity compared to normal-weight aggregates (NWAs), leading to reduced LUHPC density and increased cement paste lubrication and workability [14].

Studies have demonstrated that an increase in LWA content correlates with enhanced workability and higher slump flow [5, 14]. In this regard, Abadel et al. [14] reported that 30% LWA content increased the slump flow of LUHPC by 21%. However, an optimal LWA content exists for workability, as excessive LWA content can lead to a decrease in slump flow due to increased particle packing density and interparticle friction [5]. The water absorption characteristics of LWAs also influence LUHPC workability. Pre-saturation treatments have been shown to enhance workability by providing additional water to the mixture during mixing [7, 22–25].

Table 1 Summarized information on the workability of LUHPC

References	LWA content (by volume) (%)	HRWR range (binder mass %)	w/b ratio	Slump flow range (mm)
Abadel et al. [14]	0–30	2.68	0.210	189–230
Alanazi et al. [27]	0–10	2.67	0.200	184–210
Huang et al. [24]	0–50	0.42	0.200	270–290
Li et al. [28]	0–100	–	0.270	220–250
Liu et al. [29]	0–30	2.83	0.183	125–350
Lu et al. [6]	0–100	3.00	0.160	174–214
Meng et al. [7]	0–75	1.00	0.200	275–290
Meng et al. [16]	0–100	2.00	0.180	80–260
Meng et al. [26]	0–25	0.60–1.00	0.170–0.230	270–290
Shen et al. [25]	0–80	1.00	0.180–0.243	240–300
Teng et al. [23]	0–25	0.65–0.91	0.200	231–244
Valipour et al. [22]	0–60	1.05–2.02	0.200	260–280
Xie et al. [30]	0–100	1.25–2.26	0.200	180–200
Zhang et al. [21]	0–100	2.50	0.183	250–287

For instance, Huang et al. [24] observed a notable enhancement in slump flow by replacing 50% pre-saturated LWAs, resulting in an increase in the slump flow from 270 to 290 mm. Similarly, Shen et al. [25] reported that using 80% pre-saturated LWAs increased the slump flow of LUHPC by 25%.

Chemical admixtures, including air-entraining agents, can also affect LUHPC workability. The use of aerogel as an air-entraining agent has been shown to improve workability by facilitating particle movement [18]. The w/b ratio significantly affects LUHPC workability by determining the amount of water available for lubricating flow between particles [26]. In general, while a higher w/b ratio increases workability, achieving superior mechanical characteristics in LUHPC necessitates a w/b ratio in the range of approximately 0.18–0.23. Consequently, an appropriate dosage of superplasticizers, particularly high-range water reducers (HRWR), plays a critical role in achieving particle dispersion and facilitating the accurate casting of LUHPC [20].

Overall, optimizing LWA content, chemical admixtures, and the w/b ratio is essential for achieving maximum LUHPC workability without segregation. Table 1 provides a summary of the workability characteristics of LUHPC based on the previous studies, including LWA content, HRWR range, w/b ratio, and corresponding slump flow range.

3.2 Autogenous Shrinkage

Autogenous shrinkage, a significant concern in ultra-high-performance concrete (UHPC), stems from self-desiccation during cement hydration, particularly due to its low water-to-binder (w/b) ratio and high cement content [8]. Ineffective control of this shrinkage might result in structural issues like cracking, increased permeability, and loss of durability [31, 32]. Numerous studies have investigated the effectiveness of incorporating pre-saturated LWAs in LUHPC to mitigate autogenous shrinkage by providing internal curing [7, 11, 14, 17, 22, 23, 29]. LWAs possess the ability to release the absorbed water progressively during the cement hydration process. This release of water facilitates internal curing and helps maintain a higher internal relative humidity, effectively mitigating autogenous shrinkage [14]. Moreover, water released from LWAs can contribute to the formation of calcium-silicate-hydrate (CSH) phases, increasing solid volume surrounding aggregates and counteracting shrinkage [11].

In terms of mitigating autogenous shrinkage in LUHPC, Guo et al. [17] reported a 47% reduction in autogenous shrinkage with 100% replacement of saturated LWAs, indicating effective mitigation of self-desiccation by providing internal curing. Additionally, Valipour et al. [22] observed significant reductions in autogenous shrinkage when saturated LWAs were utilized. Specifically, a LUHPC mixture containing 60% saturated LWAs exhibited early expansions followed by 295 μm/ m shrinkage [22].

Besides LWAs, air-entraining agents and SCMs have been employed to reduce autogenous shrinkage in LUHPC. The porous structure of air-entraining agents act as an internal water reservoir, while SCMs lower water demand, capillary porosity, and matrix density. Encouraging results have also been achieved in controlling autogenous shrinkage in LUHPC with the use of shrinkage-reducing admixtures (SRAs) [23]. Teng et al. [23] found that saturated LWAs and SRA reduced the 28-day autogenous shrinkage of LUHPC by 20% due to lower capillary pore pressure and LWA internal curing. However, the inclusion of PCMs in LUHPC exhibited negative results in controlling autogenous shrinkage [19]. Ren et al. [19] observed that the inclusion of microencapsulated PCMs led to an increase in autogenous shrinkage due to the creation of additional pores, which accelerated the propagation of cracks and deformation.

In general, the inclusion of saturated LWAs, SCMs, and SRA in LWUHPC mixtures has been successful in reducing autogenous shrinkage. By utilizing LWAs for internal curing alongside SCMs and SRA, it is possible to decrease autogenous shrinkage significantly. Table 2 presents a comprehensive summary of autogenous shrinkage data for LUHPC, providing further insights. It is important to

Table 2 Summarized information on the autogenous shrinkage of LUHPC

References	Adopted material in LWUHPC production	Material replacement dosage	Autogenous shrinkage (μm/m)
Abadel et al. [14]	LWAs	0–30% by volume	35–100
Guo et al. [17]	LWAs	0–100% by volume	101–164
Huang et al. [24]	LWAs	0–50% by volume	60–170
Meng et al. [7]	LWAs	0–75% by volume	36–115
Meng et al. [26]	LWAs	0–25% by volume	23–502
Teng et al. [23]	LWAs	0–25% by volume	85–320
Shohan et al. [18]	Air-entraining agent	0–25% by fine aggregates weight	20–85[a]
Ren et al. [19]	PCMs	5% and 10% by binder weight	95–315[b]

[a]Just presented as shrinkage
[b]Autogenous shrinkage after 3 days

note that Table 2 displays autogenous shrinkage data after 24 h, as this is the period during which the majority of autogenous shrinkage occurs.

3.3 Carbon Footprint

Currently, the escalating levels of CO_2 emissions present a pressing environmental challenge. Notably, Portland cement manufacturing stands out as a prominent sector that significantly contributes to the escalation of CO_2 emissions [33]. Improving the sustainability performance of LUHPC involves the utilization of SCMs, waste, and recycled materials. Zaid et al. [5] demonstrated that incorporating palm oil fuel ash (POFA) as SCMs and Lytag as LWAs significantly reduced CO_2 emissions by decreasing the Portland cement content and substituting normal aggregates with sustainable alternatives. Integrating waste tire steel fibers into the LUHPC has emerged as an effective method for utilizing waste materials, resulting in decreased CO_2 emissions and landfill waste. These fibers not only enhance the durability and mechanical properties of concrete mixtures but also extend the lifespan of structures, making a significant contribution to the sustainability of the building sector [5].

Moreover, the incorporation of glass microspheres and polyethylene fibers (PE) into LUHPC mixtures led to a reduction in CO_2 emissions and energy usage when compared to the reference mixtures. Specifically, the use of glass microspheres resulted in a 26% reduction in CO_2 emissions, while the substitution of steel fibers with PE fibers led to a 15% decrease. Additionally, the integration of glass microspheres reduced energy consumption by 22%, while the substitution of steel fibers with PE fibers led to a 24% decrease. The aforementioned reductions in CO_2 emissions and energy consumption underscore the environmental benefits associated with incorporating sustainable materials in the manufacturing of LUHPC [17].

4 Conclusions

This study offers a comprehensive examination of LUHPC with a specific focus on its fresh properties and carbon footprint, illustrating its potential for sustainable construction practices. Employing a comprehensive methodology that incorporates a thorough examination of existing literature and meticulous analysis of experimental data, several significant findings about LUHPC emerged.

The assessment of LUHPC's carbon footprint underscores the potential of substantial environmental advantages through the inclusion of sustainable materials and additives. CO_2 emissions and energy consumption of LUHPC can substantially be reduced by substituting Portland cement with SCMs and incorporating waste or recycled materials. Additionally, the utilization of sustainable materials, such as glass microspheres and PE fibers, offers further opportunities for enhancing sustainability while maintaining or improving mechanical performance. These findings underscore the importance of considering the environmental implications of construction materials and emphasize the necessity of adopting comprehensive strategies for sustainable design and construction methods.

References

1. Karim, R., Najimi, M., Shafei, B.: Assessment of transport properties, volume stability, and frost resistance of non-proprietary ultra-high performance concrete. Constr. Build. Mater. **227**, 117031 (2019)
2. Karim, R., Shafei, B.: Investigation of five synthetic fibers as potential replacements of steel fibers in ultrahigh-performance concrete. J. Mater. Civ. Eng. **34**(7), 04022126 (2022)
3. Karim, R., Shafei, B.: Ultra-high performance concrete under direct tension: investigation of a hybrid of steel and synthetic fibers. Struct. Concr. **25**, 423–439 (2024)
4. Karim, R., Shafei, B.: Flexural response characteristics of ultra-high performance concrete made with steel microfibers and macrofibers. Struct. Concr. **22**(6), 3476–3490 (2021)
5. Zaid, O., Alsharari, F., Althoey, F., Elhag, A.B., Hadidi, H.M., Abuhussain, M.A.: Assessing the performance of palm oil fuel ash and Lytag on the development of ultra-high-performance self-compacting lightweight concrete with waste tire steel fibers. J. Build. Eng. **76**, 107112 (2023)
6. Lu, J.-X., et al.: A novel high-performance lightweight concrete prepared with glass-UHPC and lightweight microspheres: towards energy conservation in buildings. Compos. Part B. **247**, 110295 (2022)
7. Meng, W., Khayat, K.: Effects of saturated lightweight sand content on key characteristics of ultra-high-performance concrete. Cem. Concr. Res. **101**, 46–54 (2017)

8. Liu, J., Shi, C., Wu, Z.: Hardening, microstructure, and shrinkage development of UHPC: a review. J. Asian Concr. Fed. **5**(2), 1–19 (2019)
9. Kazemian, M., Shafei, B.: Effects of supplementary cementitious materials on the hydration of ultrahigh-performance concrete. J. Mater. Civ. Eng. **35**(11), 04023388 (2023)
10. Kazemian, M., Shafei, B.: Internal curing capabilities of natural zeolite to improve the hydration of ultra-high performance concrete. Constr. Build. Mater. **340**, 127452 (2022)
11. Lu, J.-X., Shen, P., Ali, H.A., Poon, C.S.: Mix design and performance of lightweight ultra high-performance concrete. Mater. Des. **216**, 110553 (2022)
12. Shekarchi, M., Ahmadi, B., Azarhomayun, F., Shafei, B., Kioumarsi, M.: Natural zeolite as a supplementary cementitious material–a holistic review of main properties and applications. Constr. Build. Mater. **409**, 133766 (2023)
13. Kazemian, M., Shafei, B.: Investigation of type, size, and dosage effects of superabsorbent polymers on the hydration development of high-performance cementitious materials. Constr. Build. Mater. **422**, 135801 (2024)
14. Abadel, A.A.: Physical, mechanical, and microstructure characteristics of ultra-high-performance concrete containing lightweight aggregates. Materials. **16**(13), 4883 (2023)
15. Code, P.: Eurocode 2: Design of Concrete Structures-Part 1–1: General Rules and Rules for Buildings, vol. 668, pp. 659–668. British Standard Institution, London (2005)
16. Meng, L., et al.: Mechanical properties and microstructure of ultra-high strength concrete with lightweight aggregate. Case Stud. Constr. Mater. **18**, e01745 (2023)
17. Guo, P., Meng, W., Du, J., Stevenson, L., Han, B., Bao, Y.: Lightweight ultra-high-performance concrete (UHPC) with expanded glass aggregate: development, characterization, and life-cycle assessment. Constr. Build. Mater. **371**, 130441 (2023)
18. Shohan, A.A.A., Zaid, O., Arbili, M.M., Alsulamy, S.H., Ibrahim, W.M.: Development of novel ultra-high-performance lightweight concrete modified with dehydrated cement powder and aerogel. J. Sustain. Cem.-Based Mater. **13**, 351–374 (2023)
19. Ren, M., Wen, X., Gao, X., Liu, Y.: Thermal and mechanical properties of ultra-high performance concrete incorporated with microencapsulated phase change material. Constr. Build. Mater. **273**, 121714 (2021)
20. Guo, K., Ding, Q.: Effect of shale powder on the performance of lightweight ultra-high-performance concrete. Materials. **15**(20), 7225 (2022)
21. Zhang, G., Chen, H., Yang, J., Ding, Q., Li, Y., Wang, Y.: Relationship between chloride ion permeation resistance of ultra-high performance concrete and lightweight aggregate ratio. J. Build. Eng. **76**, 107360 (2023)
22. Valipour, M., Khayat, K.H.: Coupled effect of shrinkage-mitigating admixtures and saturated lightweight sand on shrinkage of UHPC for overlay applications. Constr. Build. Mater. **184**, 320–329 (2018)
23. Teng, L., Addai-Nimoh, A., Khayat, K.H.: Effect of lightweight sand and shrinkage reducing admixture on structural build-up and mechanical performance of UHPC. J. Build. Eng. **68**, 106144 (2023)
24. Huang, H., Teng, L., Gao, X., Khayat, K.H., Wang, F., Liu, Z.: Use of saturated lightweight sand to improve the mechanical and microstructural properties of UHPC with fiber alignment. Cem. Concr. Compos. **129**, 104513 (2022)
25. Shen, P., et al.: Water desorption characteristics of saturated lightweight fine aggregate in ultra-high performance concrete. Cem. Concr. Compos. **106**, 103456 (2020)
26. Meng, W., Samaranayake, V., Khayat, K.H.: Factorial design and optimization of ultra-high-performance concrete with lightweight sand. ACI Mater. J. **115**(1), 129–139 (2018)
27. Alanazi, H., et al.: Mechanical and microstructural properties of ultra-high performance concrete with lightweight aggregates. Buildings. **12**(11), 1783 (2022)
28. Li, Y., Zhang, G., Yang, J., Ding, Y., Ding, Q., Wang, Y.: Chloride ion transport properties in lightweight ultra-high-performance concrete with different lightweight aggregate particle sizes. Materials. **15**(19), 6626 (2022)

29. Liu, K., et al.: Mechanisms of autogenous shrinkage for ultra-high performance concrete (UHPC) prepared with pre-wet porous fine aggregate (PFA). J. Build. Eng. **54**, 104622 (2022)
30. Xie, Y., Zhou, Q., Long, G., Chaktrimongkol, P., Shi, Y., Umar, H.A.: Experimental investigation on mechanical property and microstructure of ultra-high-performance concrete with ceramsite sand. Struct. Concr. **23**(4), 2391–2404 (2022)
31. Shi, W., Najimi, M., Shafei, B.: Reinforcement corrosion and transport of water and chloride ions in shrinkage-compensating cement concretes. Cem. Concr. Res. **135**, 106121 (2020)
32. Shi, W., Najimi, M., Shafei, B.: Chloride penetration in shrinkage-compensating cement concretes. Cem. Concr. Compos. **113**, 103656 (2020)
33. Kazemian, M., Shafei, B.: Carbon sequestration and storage in concrete: a state-of-the-art review of compositions, methods, and developments. J. CO2 Util. **70**, 102443 (2023)

Carbonation of Hydrated Blast Furnace Slag Cement Powder: Characterization and Application as a Cement Substitute

Hamideh Mehdizadeh ⓘ, Mohammad Hajmohammadian Baghban ⓘ, and Tung-Chai Ling ⓘ

Contents

1 Introduction

In the global effort to mitigate greenhouse gas emissions, the cement and concrete industry has dedicated to reducing CO_2 emissions and increasing CO_2 sequestration through the carbonation process. Research indicates that the manufacture of one ton of Portland cement emits approximately 0.87 tons of CO_2 [1], contributing to 8–10% of global CO_2 emissions. In past years, the production of blended Portland cement

H. Mehdizadeh (✉)
Department of Manufacturing and Civil Engineering, Norwegian University of Science and Technology (NTNU), Gjøvik, Norway

College of Civil Engineering, Hunan University, Changsha, China
e-mail: hamideh.mehdizadeh@ntnu.no

M. H. Baghban
Department of Manufacturing and Civil Engineering, Norwegian University of Science and Technology (NTNU), Gjøvik, Norway

T.-C. Ling
College of Civil Engineering, Hunan University, Changsha, China
e-mail: tcling@hnu.edu.cn

© The Author(s) 2025
M. Kioumarsi, B. Shafei (eds.), *The 1st International Conference on Net-Zero Built Environment*, Lecture Notes in Civil Engineering 237,
https://doi.org/10.1007/978-3-031-69626-8_8

has been considered an effective method for reducing CO_2 emissions by decreasing the use of cement clinker [2, 3]. It is well-known that industrial slags have significant potential for partially replacing cement clinker [4–6].

In recent years, global focus on the use of hydrated cement powder, derived from recycling concrete waste, as an alternative to cement in concrete products has increased significantly [7, 8]. Studies have demonstrated that the type and content of waste hydrated cement influence the fresh properties and compressive strength of the resulting concrete products. Hence, it is crucial to enhance the properties of waste hydrated cement through effective treatment methods; so that a higher value-added material and improved performance could be realized.

Mineral carbonation, also known as accelerated carbonation, has been recognized as a promising method for improving the mechanical properties of cementitious materials, including blended cement waste, as well as for capturing CO_2 [9–11]. In the carbonation process, CO_2 can react with alkaline silicate minerals to form calcium carbonate ($CaCO_3$) and amorphous gel, as described in Eqs. (1) to (4).

$$Ca(OH)_2 + CO_2 \rightarrow CaCO_3 + H_2O \tag{1}$$

$$3CaO.2SiO_2.3H_2O + 3CO_2 \rightarrow 3CaCO_3 + 2SiO_2.3H_2O \tag{2}$$

$$2CaO.SiO_2 + (2 - x)CO_2 + yH_2O \rightarrow xCaO.SiO_2.yH_2O + (2 - x)CaCO_3 \tag{3}$$

$$3CaO.SiO_2 + (3 - x)CO_2 + yH_2O \rightarrow xCaO.SiO_2.yH_2O + (3 - x)CaCO_3 \tag{4}$$

In fact, the natural carbonation process of cementitious binders is an extremely slow reaction. However, the reaction time can be accelerated by increasing the CO_2 concentration in the reaction environment. Previous studies have also revealed that the carbonation reaction rate depends on operating parameters such as relative humidity [12], carbonation time [13, 14], temperature [15], pressure [16], and CO_2 concentration [17].

For decades, ground granulated blast furnace slag (GGBS) has been extensively used as a partial substitute for cement in concrete production, in line with efforts to mitigate CO_2 emissions in the construction industry. Consequently, it is expected that a significant portion of waste hydrated cement, discarded at the end of concrete's life, contains a specific proportion of GGBS. The objective of this study is, therefore, to enhance the cementitious properties of hydrated blast furnace slag cement powder (BSCP) via moisture carbonation method for re-use in new cement paste. To achieve this, hydrated BSCP containing 70% Portland cement and 30% blast furnace slag, cured under water for 90 days, was prepared, and subjected to carbonation curing for 28 days. The microstructure of uncarbonated and carbonated BSCP was studied using thermal analysis (TG-DTG), X-ray diffraction (XRD), and Fourier-transform infrared spectroscopy (FTIR) techniques. Furthermore, the feasibility of using carbonated BSCP as a partial replacement of cement (at 0%, 15%, and 30% by mass) and its effect on flowability and compressive strength of new blended cement paste were investigated.

2 Materials and Experiments

2.1 Raw Materials

An Ordinary Portland cement (OPC; CEM I 42.5 N) and ground granulated blast furnace slag (GGBS) were used in this work to prepare the "model hydrated blast furnace slag cement." The chemical and mineralogical composition of raw materials was analyzed by X-ray fluorescence (XRF), as given in Table 1.

2.2 Preparation and Carbonation of Hydrated Blast Furnace Slag Cement Powder

In this study, the manufacturing process of uncarbonated (U) and carbonated (C) hydrated blast furnace slag cement powder (BSCP) comprises three main steps. First, a blended cement paste containing 70% OPC and 30% GGBS with a water-to-cement ratio (w/c) of 0.3 was mixed for 3 min, cast into 40 mm-cubic molds, and subsequently covered with a plastic sheet for 24 h at a room temperature of 20 ± 1 °C. Following demolding, the blended paste samples were cured in a water tank at 20 ± 1 °C for 3 months. In the next step, the 90-day hydrated samples were pulverized using a ball mill for 20 min to produce the BSCP samples (sized <75 μm). Finally, to examine the impact of carbonation on the microstructure of hydrated BSCP, powders were subjected to carbonation curing (with a CO_2 concentration of 20%, relative humidity of 65 ± 5%, and temperature of 20 ± 1 °C) (referred to as "CBSCP") for 28 days, while natural air curing (temperature of 20 ± 1 °C and relative humidity of 65 ± 5%) (referred to as "UBSCP") was used for comparison.

2.3 Use of Carbonated BSCP as a Cement Substitute

To examine the effect of carbonation on the cementitious properties of BSCP, particularly on the compressive strength and flowability of cement paste, various rations of CBSCP (0%, 15%, and 30% by mass of binder) were incorporated as partial replacements for cement in newly prepared cement paste. In this investigation, different mix proportions of blended cement paste were prepared with a constant water-to-binder ratio of 0.3. Subsequently, the freshly prepared blended

Table 1 Chemical composition (wt.%) of main raw materials determined by XRF

Component	CaO	SiO_2	Al_2O_3	Fe_2O_3	MgO	SO_3	Ignition loss
OPC	64.31	22.01	4.47	3.45	2.45	2.45	1.27
GGBS	38.24	32.65	14.14	9.74	0.75	0.09	–

cement pastes containing CBSCP were poured into 40-mm cubic molds and maintained at room temperature (20 ± 1 °C) for 1 day. Afterward, the samples were removed from molds and cured in water (20 ± 1 °C) for 3, 7, and 28 days.

2.4 Test Methods

Thermal Analysis Both thermogravimetric analysis (TG) and differential thermogravimetric (DTG) tests were used to determine the contents of calcium hydroxide ($Ca(OH)_2$) and calcium carbonate ($CaCO_3$) of UBSCP and CBSCP specimens. For the characterization tests, first, the hydration of the samples was halted by immersing crushed UBSCP and CBSCP in isopropanol solution for 4 days, followed by transfer to a vacuum oven for drying (40 °C) for an additional 5 days. Finally, the dried crushed samples were ground to prepare a powder with particle size below 5 μm for examination. The thermal analysis was performed using a Thermo Plus EV02 TG with a heating rate of 10 °C/min and under a flow of 30 ml/min of N_2 gas. About 15 ± 1 mg of powdered sample was subjected to heating from 30 °C to 1000 °C.

Microstructure Analysis To evaluate the effect of carbonation on the microstructure of BSCP, the phase composition of carbonated and uncarbonated BSCP samples was studied using X-ray diffraction (XRD) and Infrared spectrum (FTIR) techniques. The powder samples were characterized using a an X'Pert MPD PRO diffractometer (D8 Advance Bruker, Germany), featuring a CUKα X-ray radiation source (with a scanning range of 10°-70° 2θ and a step size of 0.02°). For the FTIR test, powder samples were inserted into a Thermo Scientific IS10 infrared spectrometer, and the spectra were recorded for wavenumbers ranging from 400 cm^{-1} to 4000 cm^{-1}.

Flowability The flowability of cement pastes made with various replacement ratios of carbonated BSCP (0%, 15%, and 30% by mass) was measured in accordance with the GB/T 2419–2005 standard. In this measuring method, the freshly prepared blended cement paste was placed into a slump cone mold. The paste sample was then dropped 25 times on a flop table and the average spreading diameter in four directions was reported as the flowability diameter.

Mechanical Properties The compressive strength of blended cement paste containing CBSCP was determined at 3, 7, and 28 days of hydration based on Chinese Standard GB/T 17671–1999, using a Fengzhi YAW-300 compression machine with a loading rate of 2.4 kN/s. The paste strength was determined by recording the average compressive strength values from three specimens. Additionally, the strength activity index (SAI) of the CBSCP-cement paste samples at 3, 7 and 28 days of hydration (t) was calculated according to Eq. 5:

$$SAI\ (\%) = \frac{(\text{compressive strength of cement paste containing CBSCP})_t}{(\text{compressive strength of control sample})_t} \times 100$$

$$(5)$$

3 Results and Discussion

3.1 Characterization of BSCP upon Moisture Carbonation

To evaluate the impact of carbonation on the microstructure of hydrated blast furnace slag cement powder, thermal, XRD, and FTIR analysis of BSCP cured under CO_2 and natural air environment were investigated and compared, as depicted in Figs. 1, 2, and 3, respectively.

Thermal Analysis TG-DTG testing was employed to estimate the contents of hydration and carbonation products in the BSCP sample before and after carbonation. In this study, the contents of hydrated bound water (BW), $Ca(OH)_2$, and $CaCO_3$ were calculated based on the weight loss of BSCP specimens within the temperature ranges of 70–540 °C, 400–500 °C, and 540–950 °C, and they are given in Eqs. (6)–(8), respectively [3]:

$$\%BW = \frac{(W_{70\,°C} - W_{540\,°C})}{W_{950\,°C}} \times 100 \tag{6}$$

$$\%Ca(OH)_2 = \frac{(W_{400\,°C} - W_{500\,°C})}{W_{950\,°C}} \times \frac{74.09}{18.01} \times 100 \tag{7}$$

$$\%CaCO_3 = \frac{(W_{540\,°C} - W_{950\,°C})}{W_{950\,°C}} \times \frac{100.09}{44.01} \times 100 \tag{8}$$

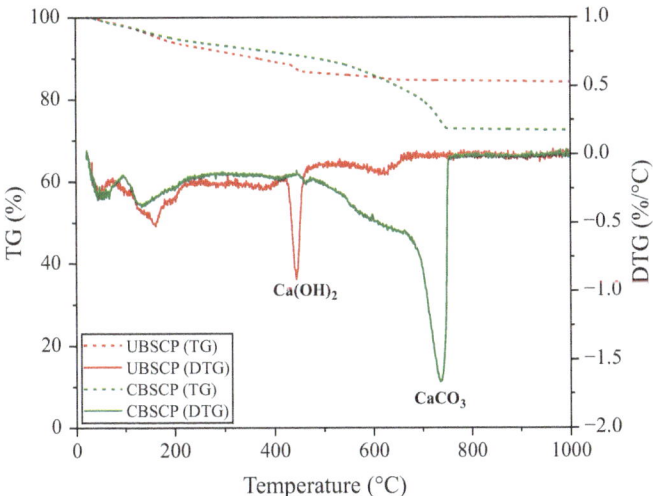

Fig. 1 TG-DTG curves of BSCP samples after 28 days of natural air (UBSCP) and carbonation (CBSCP) curing

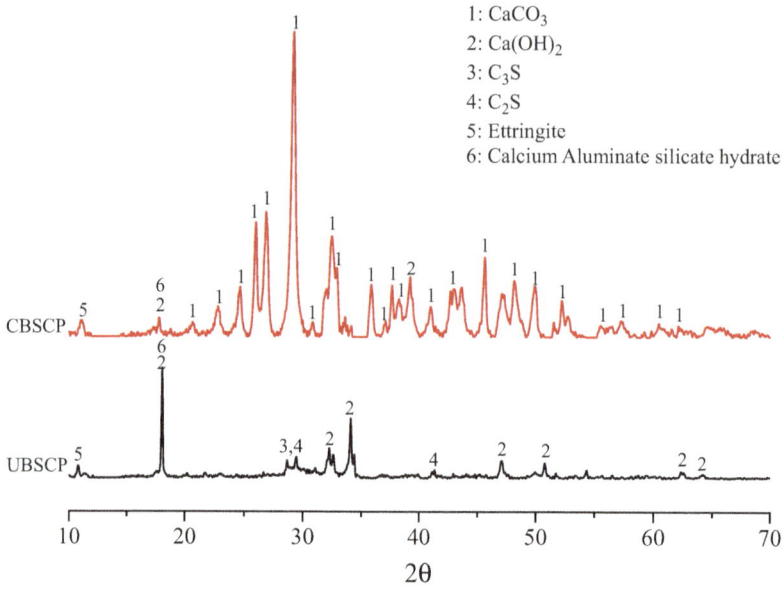

Fig. 2 X-ray diffraction patterns of BSCP samples after 28 days of natural air (UBSCP) and carbonation (CBSCP) curing

As seen in Fig. 1, BSCP cured under natural air demonstrates a peak within 100–200 °C, which can be related to dehydration of the hydrated calcium silicate phase (C-S-H gel) and ettringite (AFt), followed by a larger peak (weight loss of 0.33%) in the temperature range of 400 °C to 500 °C due to decomposition of Ca $(OH)_2$. In comparison to the UBSCP sample, the thermal analysis results for the BSCP sample cured under CO_2 rich condition (CBSCP) indicated a gradual decrease in weight of the sample upon heating up to 680 °C until a significant weight loss of 1.67% appeared from 700 to 800 °C, corresponding to decarbonation of $CaCO_3$.

According to the thermal results given in Table 2 (based on Eqs. (6)–(8)), the total amount of precipitated $CaCO_3$ in the structure of UBSCP and CBSCP samples was 5.49% and 50.91% by weight, indicating a significant amount of $CaCO_3$ has formed in CBSCP sample. This can be associated with an increase in the diffusion rate of CO_2 into the structure of CBSCP during the accelerated carbonation process compared to the natural air carbonation process, which led to heightened CO_2 reaction with calcium-based phases in the samples.

XRD Analysis Microstructure analysis, employing both XRD and FTIR techniques, was performed to further explore the impact of moisture carbonation on the precipitation of $CaCO_3$ in the BSCP sample. As marked in the XRD analysis (Fig. 2), the UBSCP sample consists of $Ca(OH)_2$, a small amount of ettringite (AFt), and unreacted calcium silicate-based phases of cement (C_2S and C_3S), whereas calcium carbonate ($CaCO_3$, calcite) predominates as the main carbonate phase in

Fig. 3 FTIR spectra of BSCP samples after 28 days of natural air (UBSCP) and carbonation (CBSCP) curing

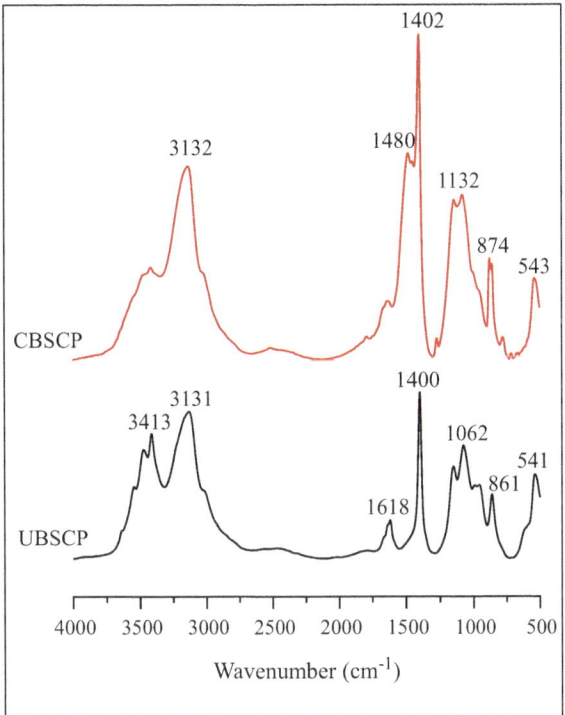

the structure of CBSCP sample. In other words, during the carbonation process, the distinctive peaks of C_2S, C_3S, and $Ca(OH)_2$ in the structure of BSCP specimens were significantly reduced due to the reaction between CO_2 and the mentioned components. This interaction led to the formation of $CaCO_3$ polymorphs, consistent with the TG-DTG curves shown in Fig. 1.

FTIR Analysis The infrared spectroscopy technique was also used to study the amorphous gel formed within the structure of BSCP during the carbonation process. The FTIR spectra of carbonated and uncarbonated BSCP are presented in Fig. 3. It can be observed that vibration bonds appeared at 1618 cm^{-1} and 3413 cm^{-1} in the UBSCP sample, corresponding to the O-H stretching bond in hydrated water molecules of hydration products and portlandite ($Ca(OH)_2$), respectively. However, this bond disappeared after the carbonation process, as evident in the FTIR spectra of the CBSCP sample. Conversely, new vibration peaks of CO_3^{2-} were formed at 874 cm^{-1} and 1480 cm^{-1} after the carbonation of BSCP, indicating the formation of $CaCO_3$ in the structure of the carbonated BSCP sample. Moreover, it is observed that the asymmetric stretching vibration of Si-O bond in the chain structure of C-S-H gel shifted to the higher wave number (from 1062 cm^{-1} to 1132 cm^{-1}) following the carbonation of the BSCP sample. This result is attributed to the decalcification of the C-S-H gel phase and an increase in the degree of polymerization. So, it can be

Table 2 The contents of hydration and carbonation products in uncarbonated and carbonated BSCP (% by weight)

Component	UBSCP	CBSCP
BW	15.19	13.88
Ca(OH)$_2$	13.71	2.60
CaCO$_3$	5.49	50.91

Table 3 Flowability of blended cement paste samples containing carbonated BSCP

Sample	Binder (wt.%)		Flowability (mm)
	OPC	CBSCP	
CBSCP-0 (control sample)	100	0	175
CBSCP-15	85	15	142.5
CBSCP-30	70	30	105

concluded that the carbonation of hydrated blast furnace slag cement powder leads to an increase in silica gel amorphous content in the structure of BSCP. The bond at 3131 cm^{-1} belonging to (Si,Al)-OH bond of water molecule remained unchanged after carbonation of BSCP [11]. The FTIR spectra results are in alignment with the XRD findings (Fig. 2) and thermal results (Fig. 1 and Table 2).

3.2 Performance of Carbonated BSCP as a Cement Substitute

To assess the impact of carbonation on the cementitious reactivity of BSCP, carbonated BSCP with a particle size below 75 μm was used to replace the Portland cement at 0%, 15%, and 30% by mass of binder in the preparation of cement paste.

Flowability (Water Requirement) Table 3 presents the flowability of fresh blended cement paste incorporating different CBSCP contents. As shown in this table, regardless of the cement replacement ratio, the flowability of paste samples containing carbonated BSCP was notably lower compared to that of pure cement paste (CBSCP-0). The reduced flowability of cement pastes containing CBSCP could be mainly attributed to the differences in the porosity and specific surface area of CBSCP samples compared to cement particles (caused by CaCO$_3$ formation), which led to increased water requirement and a subsequent decrease in workability. Moreover, the flowability of freshly blended cement pastes is found to decrease as the CBSCP content in the mix proportion increases. This result can be mainly due to the higher specific surface area of CBSCP compared to cement particles, resulting in greater water demand for CBSCP and contributing to a further reduction in the flowability of the cement paste.

Compressive Strength The effect of CBSCP content on the compressive strength of blended cement paste was investigated. As depicted in Fig. 4 and supported by strength activity index results presented in Table 4, replacing 15% of cement with

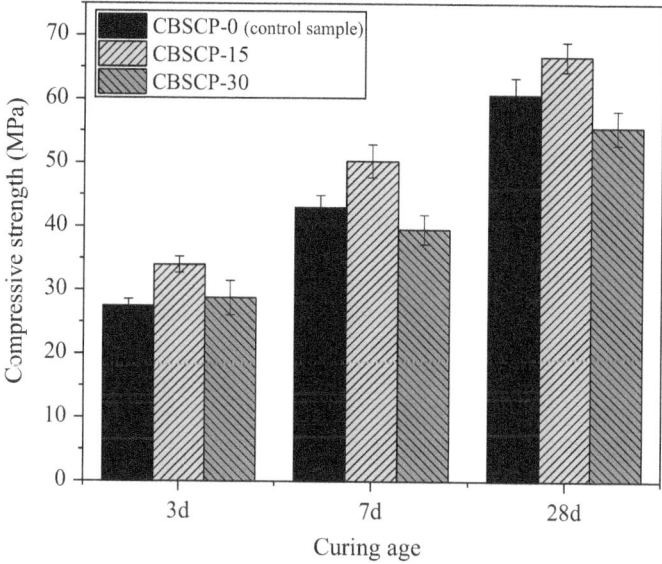

Fig. 4 Compressive strength of blended cement paste containing carbonated BSCP at different curing ages

Table 4 Strength activity index (SAI) of blended cement paste samples

Sample	SAI (%)		
	3 days	7 days	28 days
CBSCP-0 (control sample)	100	100	100
CBSCP-15	123.6	117.0	109.9
CBSCP-30	104.7	92.1	91.6

CBSCP resulted in a notable increase in the compressive strength of paste by 23.6%, 17.0%, and 9.9% for 3, 7, and 28 days of hydration, respectively. The enhanced compressive strength of blended cement pastes containing CBSCP could be mainly attributed to the increase in heterogeneous nucleation sites for cement hydration provided by $CaCO_3$ in carbonated BSCP, along with the stabilization of the ettringite phase [3]. However, the beneficial effect of CBSCP on the strength enhancement of the blended paste became less pronounced when the volume of cement replaced with CBSCP was increased to 30%. This might be attributed to the high water requirement of the paste sample prepared with 30% CBSCP compared to the control sample, resulting in a decrease in hydrated water and a delay in cement hydration, which led to a reduction in the compressive strength of the blended cement paste.

4   Conclusions and Future Perspective

The study investigated the effect of carbonation on the microstructure alteration of hydrated blast furnace slag cement powder (BSCP) as a pozzolanic waste material. To understand the feasibility of re-using BSCP in cementitious products, the study also examined the flowability and compressive strength of blended cement pastes containing 0%, 15%, and 30% of carbonated BSCP (CBSCP) as a cement replacement. The microstructural changes (analyzed by XRD and FTIR tests) and thermal analysis (TG-DTG) revealed that calcium hydroxide is the main crystalline phase of uncarbonated BSCP, while CBSCP samples consist mainly of calcium carbonate ($CaCO_3$) with approximately 50.91% by weight. Moreover, the carbonation increased the silica gel amorphous content in the structure of BSCP samples. The finding also indicated that replacing cement with carbonated BSCP decreased the flowability of cement paste due to the high water absorption of the CBSCP sample compared to cement particles. Furthermore, an increase in the replacement ratio (from 15% to 30% by mass) of cement with CBSCP caused a significant decrease in the flowability of the cement paste. Despite the adverse effect of CBSCP on the flowability of cement paste, substituting 15% of cement with CBSCP enhanced the compressive strength of paste samples at 3, 7, and 28 days. Therefore, it can be concluded that carbonated BSCP can be used as a cement substitute (up to 30% replacement) in the production of cement-based materials, without a significant compromise to mechanical performance.

As a new emerging binder, BSCP can be used as a CO_2 sink and to produce low-carbon cement-based materials with acceptable mechanical performance. The proposed treatment method (moisture carbonation) for improving the pozzolanic reactivity of waste hydrated cement powders demonstrates promising potential for industrial applications. However, scaling up the carbonation process from laboratory scale to industrial scale remains a challenge due to the long carbonation time (28 days). Hence, optimizing moisture carbonation parameters, such as CO_2 concentration, humidity, BSCP particle size, and carbonation temperature, is necessary to reduce carbonation time and enhance the efficiency of the carbonation process for real industrial applications. Additionally, investigating the long-term durability of construction materials incorporating CBSCP is critical to ensure their reliability over time. Conducting a comprehensive life cycle assessment of the carbonation of CBSP and the utilization of carbonated CBSCP as a cement substitute compared to Ordinary Portland cement production methods is also required to assessing the potential environmental benefits of the proposed method and its impact on reducing CO_2 emissions in the construction industry.

Acknowledgments The research funding support from the National Natural Science Foundation of China (NSFC) (grant No. 51950410584) is gratefully acknowledged. Dr. Hamideh Mehdizadeh would also like to acknowledge funding from the European Union's Horizon Europe research and innovation program under the Marie Skłodowska-Curie grant agreement No. 101066240.

References

1. Damtoft, J.S., Lukasik, J., Herfort, D., Sorrentino, D., Gartner, E.M.: Sustainable development and climate change initiatives. Cem. Concr. Res. **38**, 115–127 (2008)
2. Higuchi, A.M.D., dos Santos Marques, M.G., Ribas, L.F., de Vasconcelos, R.P.: Use of glass powder residue as an eco-efficient supplementary cementitious material. Constr. Build. Mater. **304**, 124640 (2021)
3. Mehdizadeh, H., Shao, X., Mo, K.H., Ling, T.C.: Enhancement of early age cementitious properties of yellow phosphorus slag via CO_2 aqueous carbonation. Cem. Concr. Compos. **133**, 104702 (2022)
4. Yang, K.H., Jung, Y.B., Cho, M.S., Tae, S.H.: Effect of supplementary cementitious materials on reduction of CO_2 emissions from concrete. J. Clean. Prod. **103**, 774–783 (2015)
5. Li, X., Mehdizadeh, H., Ling, T.C.: Environmental, economic and engineering performances of aqueous carbonated steel slag powder as alternative material in cement pastes: influence of particle size. Sci. Total Environ. **903**, 166210 (2023)
6. Li, Y., Mehdizadeh, H., Mo, K.H., Ling, T.C.: Co-utilization of aqueous carbonated basic oxygen furnace slag (BOFS) and carbonated filtrate pastes considering reaction duration effect. Cem. Concr. Compos. **138**, 104988 (2023)
7. Martin, C., Manu, E., Hou, P., Adu-Amankwah, S.: Circular economy, data analytics, and low carbon concreting: a case for managing recycled powder from end-of-life concrete. Resour. Conserv. Recycl. **198**, 107197 (2023)
8. Aquino Rocha, J.H., Toledo Filho, R.D.: The utilization of recycled concrete powder as supplementary cementitious material in cement-based materials: a systematic literature review. J. Buil. Eng. **76**, 107319 (2023)
9. Mehdizadeh, H., Wu, Y., Mo, K.H., Ling, T.C.: Evaluation of carbonation conversion of recycled concrete fines using high-temperature CO_2: reaction kinetics and statistical method for parameters optimization. J. Environ. Chem. Eng. **11**(3), 109796 (2023)
10. Li, L., Liu, Q., Huang, T., Peng, W.: Mineralization and utilization of CO_2 in construction and demolition wastes recycling for building materials: a systematic review of recycled concrete aggregate and recycled hardened cement powder. Sep. Purif. Technol. **298**, 121512 (2022)
11. Mehdizadeh, H., Cheng, X., Mo, K.H., Ling, T.C.: Upcycling of waste hydrated cement paste containing high-volume supplementary cementitious materials via CO_2 pre-treatment. J. Build. Mater. **52**, 104396 (2022)
12. Rostami, V., Boyd, A.J.: Durability of concrete pips subjected to combined steam and carbonation curing. Constr. Build. Mater. **25**(8), 3345–3355 (2011)
13. Panesar, D.K., Mo, L.: Properties of binary and ternary reactive MgO mortar blends subjected to CO_2 curing. Cem. Concr. Compos. **38**, 40–49 (2013)
14. Liang, C., Li, B., Guo, M.Z., Hou, S., Wang, S., Gao, Y., Wang, X.: Effects of early-age carbonation curing on the properties of cement-based materials: a review. J. Build. Eng. **84**, 108495 (2024)
15. Zhu, C., Fang, Y., Wei, H.: Carbonation-cementation of recycled hardened cement paste powder. Constr. Build. Mater. **192**, 224–232 (2018)
16. Ye, X., Chen, T., Chen, J.: Carbonation of cement paste under different pressures. Constr. Build. Mater. **370**, 130511 (2023)
17. Sanjuan, M.A., Estevez, E., Argiz, C., del Barrio, D.: Effects of curing time on granulated blast-furnace slag cement mortars carbonation. Cem. Concr. Compos. **90**, 257–265 (2018)

Performance of Metakaolin-Based Alkali-Activated Mortar for Underwater Placement

Joud Hwalla (iD), Mariane Saba (iD), Joseph Assaad (iD), and Hilal El-Hassan (iD)

Contents

1 Introduction

The use of Portland cement in the construction and building industry led to serious environmental challenges, such as the depletion of natural resources, the release of greenhouse gases, and the generation of waste materials [1]. Researchers and scientists are actively exploring alternatives to cement-based materials either partially or fully using wastes or abundant resources [1, 2].

Metakaolin (MK), originating from the calcination of kaolin clay, stands out as a highly reactive pozzolanic material extensively utilized in the production of high-performance concrete and mortar [3]. Alkali-activated or geopolymer (GP) is a cement-free construction product manufactured by activating alumina or silica-rich binders such as metakaolin, fly ash, slag, or perlite, with a sodium or potassium-based alkaline solution [4, 5]. The poly-condensation of silica and alumina

J. Hwalla (✉) · H. El-Hassan
United Arab Emirates University, Al Ain, United Arab Emirates
e-mail: 202090500@uaeu.ac.ae

M. Saba · J. Assaad
University of Balamand, El Koura, Lebanon

© The Author(s) 2025
M. Kioumarsi, B. Shafei (eds.), *The 1st International Conference on Net-Zero Built Environment*, Lecture Notes in Civil Engineering 237,
https://doi.org/10.1007/978-3-031-69626-8_9

oligomers creates dense crystalline and amorphous polymer structures that harden at ambient temperature with durability and mechanical properties comparable to alternative cement-based materials [6]. GP is successfully used in a variety of niche applications such as retrofitting of damaged structures, repair and strengthening works requiring superior bond and attrition resistance, soil injection to reduce seepage, insulating infill used in lightweight sandwich panels, and refractory materials possessing improved resistance to fire and elevated temperature [7]. Recent applications showed that GP could be successfully used as masonry plasters exhibiting shorter settings and higher bonding without the need for moist curing normally required than cement-based plasters [5].

The construction of marine and underwater structures such as dams, bridges, and tunnels requires viscous materials to mitigate washout loss and prevent water penetration into the matrix during placement and curing [8]. As a result, geopolymers have proven to be excellent alternatives to cement-based counterparts due to their higher plastic viscosity and yield stress rheological properties [9]. Favier et al. [10] presented evidence that the increased plastic viscosity of MK-based GP is linked to the high viscosity of alkaline sodium hydroxide and sodium silicate solutions that surpass water by 10–100 times. The negligible colloidal interactions among MK particles suggest that hydrodynamic dissipation of the highly viscous alkaline solution predominantly influences the rheological behavior. Regarding durability, several studies highlighted the superior resistance of GP to degradation by aggressive ions present in marine and wastewater environments such as sulfates, chlorides, and acids [11, 12]. The GP maintained a dense microstructure even after a 30-day immersion in a 5% acid solution with a limited weight loss of 7%, while traditional cement-based concrete experienced about 35% loss under similar conditions [13].

The primary aim of this chapter is to produce MK-based GP mortars intended for underwater casting. Two flowability levels of 16 and 22 ± 1 cm along with three different binder-to-sand ratios of 1:3, 1:1.85, and 1:0.8, are investigated. The sodium-based alkaline solution is adjusted to attain the specified flowability. The washout loss and residual 28-day compressive strength of mortars were determined under dry and submerged casting conditions. The investigation delves into the effects of varying alkaline solution-to-binder and binder-to-sand ratios on the overall performance of GP mortars.

2   Experimental Work

2.1   Materials

Artificial or manufactured MK derived from the calcination of kaolinite clay was utilized to produce the GP. The alumina-silica-rich binder possessed a specific gravity, median particle size (d_{50}), and BET surface area of 2.2, 6.1 μm, and 19,000 m^2/kg, respectively. Figure 1 presents the particle size distribution of MK

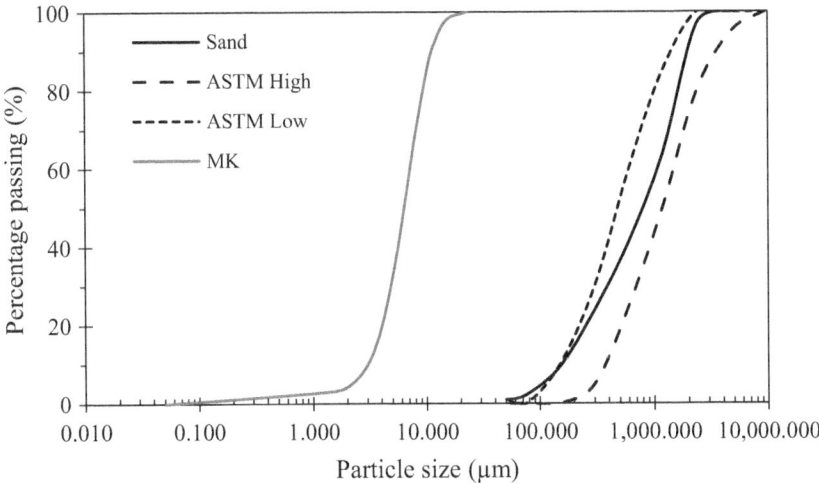

Fig. 1 Particle size distribution of MK and sand

ranging between 0.05 and 22.80 μm. Its chemical composition comprised 55% SiO_2, 39% Al_2O_3, 1.8% Fe_2O_3, 0.35% CaO, 0.25% MgO, and 1.5% TiO_2, with a 1% loss on ignition. Sodium hydroxide pellets (SH) and sodium silicate (SS) with SiO_2-to-Na_2O ratio of 3.25, both supplied by Sigma-Aldrich, were employed to activate the MK.

A well-graded siliceous sand was used; it had a water absorption of 0.91%, fineness modulus of 2.40, and bulk specific gravity of 2.62. Additional water was introduced into all mixes to offset the water absorption of the sand. The particle size gradation of the sand (Fig. 1) conformed to ASTM C33 [14] standard defined by the lowest and highest limits for aggregate gradation.

Tap water was utilized in the production of MK-based mortar. Sea water was not used as it resulted in a slight decrease in the compressive strength of MK-based geopolymer mortar, as zeolite formation was inhibited with increasing sea water concentration [15].

2.2 Mix Design

Six mortars were prepared to assess the impact of flowability and sand content (i.e., binder content) on GP performance. Two different flowabilities (i.e., 16 and 22 ± 1 cm) and three binder-to-sand ratios (i.e., 1:3, 1:1.85, and 1:0.8) were tested. Two flow values, 16 and 22 ± 1 cm, were chosen to evaluate the influence of flowability and alkaline solution content on the performance of MK-based geopolymer mortar in underwater applications. This selection was made to assess a range of flowability conditions and alkaline solution concentrations, which are

Table 1 Mix design of MK-based GP mortar

Mix designation	Binder-to-sand	SH molarity	SS-to-SH	Flow (cm)	AAS-to-binder
B:S(1:3) F(16)	1:3	10	2	16 ± 1	1.37
B:S(1:1.85) F(16)	1:1.85	10	2	16 ± 1	1.24
B:S(1:0.8) F(16)	1:0.8	10	2	16 ± 1	0.78
B:S(1:3) F(22)	1:3	10	2	22 ± 1	1.83
B:S(1:1.85) F(22)	1:1.85	10	2	22 ± 1	1.31
B:S(1:0.8) F(22)	1:0.8	10	2	22 ± 1	0.83

critical factors affecting the behavior of mortar underwater. The volume of sand was systematically increased until a maximum loss of strength between the dry and submerged samples reached 50%.

The mix design for all formulations is provided in Table 1; each mix is designated as $B:S(X)\ F(Y)$, where X represents the binder-to-sand ratio values and Y represents the flow in centimeters.

2.3   Sample Production

The alkali-activated solution (AAS) was prepared 1 day before mortar production to ensure the complete dissipation of heat generated during the preparation of SH and by mixing SH with SS. Initially, tap water was blended with SH pellets to achieve the desired molarity of 10; after that, the SS and SH were mixed at SS-to-SH ratio of 2. This specific molarity and ratio were chosen to attain the desired setting time for underwater placing [9].

The mortar production involved mixing the MK and siliceous sand for 2–3 minutes. Subsequently, the AAS and additional water required to account for sand water absorption were gradually added, and the mixture was blended for another 2 minutes. After a 30-second rest period, mixing was resumed for an additional 3 minutes.

2.4   Testing Procedure

The fresh mortar properties were assessed based on the initial flow, setting time, plastic viscosity, and washout loss. The flowability and setting time of mortars were measured using the mini-slump cone and the Vicat needle in accordance with the ASTM C1437 and ASTM C191, respectively [16, 17]. The average of four diametrical readings was used for determining the flow of mortar without vibration. The plastic viscosity (μ) was determined using a four-bladed slotted vane connected to Anton Paar rheometer [9]. The washout loss ($W3$) of investigated mortars was determined using an adapted version of the CRD C61 test, as described in the

(a)　　　　　　　　　　　(b)

Fig. 2 Casting methods for placing Geopolymer (GP) mortars underwater: (**a**) on concrete slabs and (**b**) inside molds

referenced research study [9]. This consists of measuring the mass of the freshly mixed material placed in a perforated basket before and after dropping it in a 1-m tube filled with water following three drops.

For strength testing, the fresh mortar was poured into prismatic molds measuring $4 \times 4 \times 16$ cm, both in dry and submerged water conditions. In the dry condition, the mortar was poured without vibration and cured in a plastic bag. In the submerged condition (Fig. 2a, b), the molds were positioned in a large container filled with water to a depth of 30 cm. A March funnel cone with an outer diameter of 11 mm, positioned approximately 1 cm above the surface of the molds, was utilized to pour the mortar without any vibration. The specimens remained submerged in water at 21 ± 3 °C until the testing time. The ambient temperature during both mixing and placing varied around 21 ± 3 °C. The compressive strengths of dry $f'c$ (dry) and submerged $f'c$ (submerged) mortars were characterized based on the 28-day compressive strength values. This was measured following the EN 196/1 Test Method, with the average of four readings taken into consideration [18]. Each measurement was obtained from two halves of prisms, resulting in a total of four samples tested for compressive strength.

3　Results and Discussion

3.1　Slump Flow and Alkaline Solution Content

Table 1 summarizes the AAS-to-binder ratio employed in each mix to attain the targeted flowability of 16 and 22 ± 1 cm. Evidently, the rise in the binder-to-sand ratio (i.e., higher binder content) led to a reduced amount of alkaline solution to achieve the desired flow. The AAS-to-binder ratio for mixes with a B:S(1:3) composition increased significantly by 33.5%, rising from 1.37 to 1.83. In contrast, the ratio for mixes produced with B:S ratio of 1:0.8 experienced a modest increase of 6.41%, increasing from 0.78 to 0.83. These adjustments were made to attain flow measurements of 16 and 22 ± 1 cm, respectively. This can be attributed to the higher inter-particles links and internal friction between the fine aggregates and mortar paste [19].

As expected, a lower demand for liquid solution was required for mixes having a flowability of 16 ± 1 cm compared to counterparts having 22 ± 1 cm flow. In the former, the AAS-to-binder ratio increased only from 0.78 to 1.37; in the latter, it increased from 0.83 to 1.83. This phenomenon is directly related to the thickness of the water film surrounding the fine aggregates [20]. The increase in AAS content results in an increase in water film thickness followed by an increase in particles cohesion and a reduction in yield stress that hinders the ease of flow and deformation of fresh mortar [20].

3.2 Setting Time

The final setting times are presented in Table 2. Generally, the setting of different GP mortars hovered between 270 and 470 minutes, which is about half than cement-based countertypes reported in [9]. This suggests that the use of MK-type GP can be relevant to speed up underwater construction, simultaneously enhancing the resistance toward scaling and erosion during the initial phases of casting [21]. It is worth noting that the setting times increased with a decrease in binder content and mixtures possessing higher flow values. Figure 3 shows the relationship between setting and the AAS-to-binder ratio, having a strong correlation coefficient (R^2) of 0.86. Similar findings were observed elsewhere, indicating that the increase in AAS-to-binder content prolongs the setting time of fly ash-slag-based screed mortar [22].

3.3 Plastic Viscosity

Figure 4 illustrates the plastic viscosity of tested mortars, which ranged between 15.6 and 26.3 Pa.s. Comparing these values with values of cement-based counterparts, MK-based mortar possessed higher plastic viscosity, mainly attributed to the higher

Table 2 Fresh and hardened properties of MK-based GP mortar

Mix designation	Setting (min)	⬚ (Pa.s)	$W3$ (%)	$f'c$ (dry) (MPa)	$f'c$ (submerged) (MPa)
B:S(1:3) F(16)	355	24.3	3.27	50.8	28.3
B:S(1:1.85) F(16)	320	25.7	2.78	51.1	28.3
B:S(1:0.8) F(16)	270	26.3	1.93	44.8	22.6
B:S(1:3) F(22)	470	15.6	7.52	23.4	17.7
B:S(1:1.85) F(22)	400	19.5	7.02	39.3	24.6
B:S(1:0.8) F(22)	310	24.6	2.78	47.2	25.3

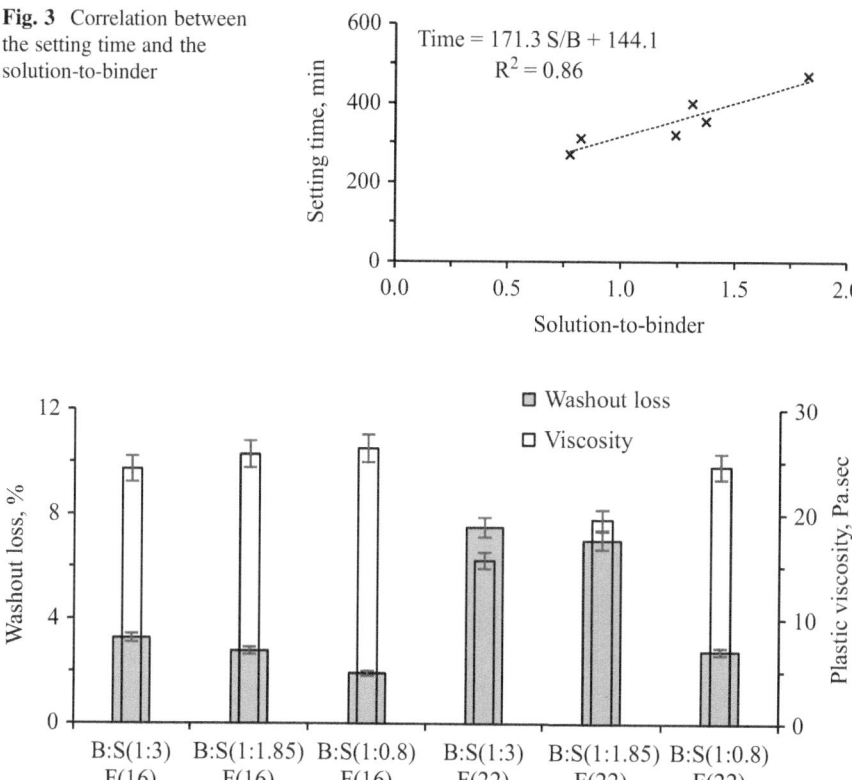

Fig. 3 Correlation between the setting time and the solution-to-binder

$$Time = 171.3 \, S/B + 144.1$$
$$R^2 = 0.86$$

Fig. 4 Plastic viscosity and washout loss values of tested mortars

viscosity of AAS compared to tap water [9]. The plastic viscosity was observed to increase with increased binder content (i.e., lower sand content) and with lower flow values. The rise in stickiness with an increase in plastic viscosity was felt during mortar production and is mainly owed to the drop in AAS content. Similar findings were reported by Alnahhal et al. [23], where the plastic viscosity of GP paste decreased with an increase in the AAS solution, which was attributed to the reduction in the volume fraction of solids in the suspended fluid.

3.4 Washout Loss

The washout loss of tested mortars is presented in Table 2. Compared to cement-based mixtures that normally exhibit between 5% and 20% washout loss [9], all MK-based mortars demonstrated higher resistance against washout with percentages ranging between 1.93% and 7.52%, thus reflecting the suitability of geopolymers for

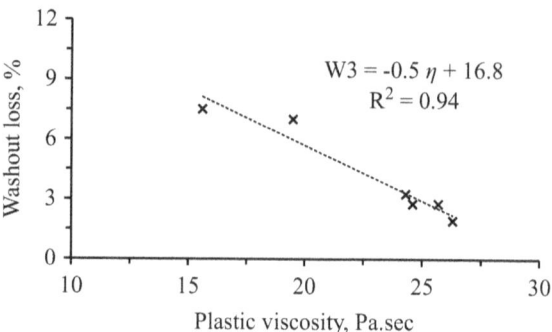

Fig. 5 Correlation between the plastic viscosity and washout loss

underwater applications. This is evidenced in Fig. 2b where the mold remained visible even after pouring the mortar into the water.

Regardless of the flow level, mortars prepared with higher binder-to-sand ratios exhibited higher washout loss, which can be related to the lower plastic viscosity of samples with lower binder content. For example, the washout loss varied from 1.93% to 3.27% when the binder-to-sand ratio changed from 1:0.8 to 1:3, respectively, in mixtures possessing 16 ± 1 cm flow. Such variations ranged from 2.78% to 7.52% for mixtures having 22 ± 1 cm flow. In fact, washout loss was observed to be directly influenced by the plastic viscosity of the produced mortar. The linear correlation of Fig. 5 demonstrates that the increase in viscosity is directly associated to higher resistance against washout loss, which would decrease the material proneness toward dilution upon underwater casting [9].

3.5 Compressive Strength

The 28-day compressive strengths of GP mortars cast in different conditions are illustrated in Fig. 6. Clearly, the effect of underwater casting led to reduced mechanical properties, which can be directly attributed to the water infiltration inside the material coupled with the washout of the MK-based paste [9]. Hence, the 28-day compressive strengths varied from 23.4 to 50.8 MPa and from 17.7 to 28.3 MPa when casting was made in dry or underwater conditions, respectively.

The impact of binder content varied for mixtures produced with different flowability and/or amount of alkaline solution used. Hence, the 28-day compressive strength values of mortars having a flow of 16 ± 1 cm dropped from 50.8 to 44.8 MPa with the increase in binder-to-sand ratio from 1:3 to 1:0.8. Meanwhile, in the other group of mixes having a flow of 21 ± 1 cm, the strength increased from 23.4 to 47.2 MPa. In general, it is common to observe a strength increase with higher paste volume due to a denser paste matrix and improved interfacial transition zone between the paste and aggregates [24, 25]. Similarly, a decrease in the binder-to-sand ratio from 1:3 to 1:7 resulted in a substantial drop in strength, approximately 87%, for fly ash and slag-based geopolymer screed produced with crushed limestone

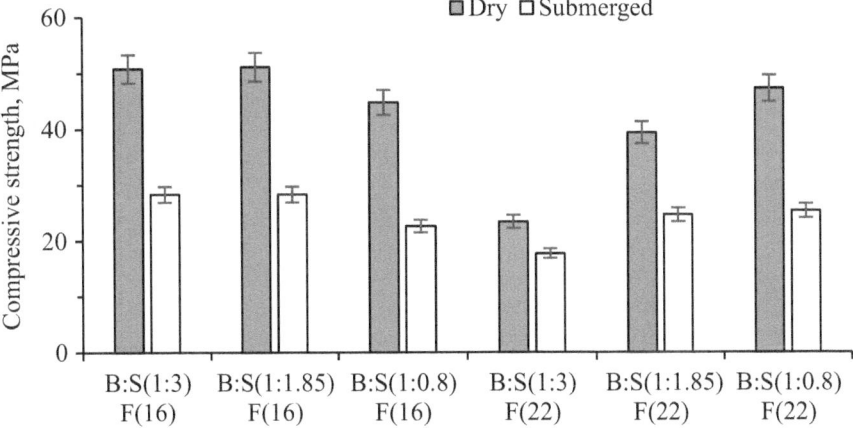

Fig. 6 28-day compressive strength values

Fig. 7 Correlation between the solution-to-binder ratio and residual strength

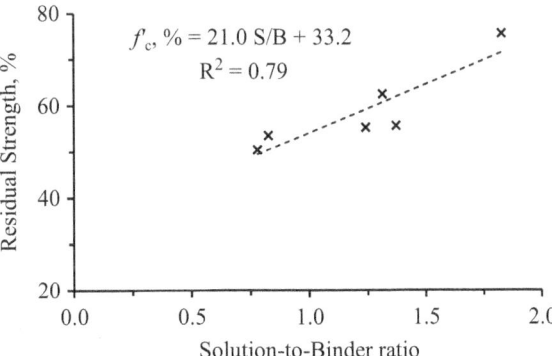

sand as fine aggregates [22]. However, it appears that the amount of AAS added to the first group of mixes having 16 ± 1 cm flow was insufficient for the activation of MK particles, which resulted in a drop in strength. This becomes apparent when comparing the strength values of B:S(1:0.8) F(16) and B:S(1:0.8) F(22), where the strength increased from 44.8 to 47.2 MPa with an increase in AAS-to-binder ratio from 0.78 to 0.83 (Table 1). Similar findings were found elsewhere [22]; the maximum strength values of fly ash and slag-based geopolymer mortar produced with dune sand were achieved with an alkaline solution-to-binder ratio of 0.60.

Figure 7 presents the relationship between the residual compressive strength and alkaline solution-to-binder ratio, with an acceptable R^2 of 0.79. Hence, the higher the solution content, the better the residual compressive strength of GP mortars. This phenomenon is attributed to the high viscosity of the solution, which may act as a barrier preventing water penetration and dilution of the mortar matrix at early age casting.

4   Conclusions

This research focused on the formulation of MK-based alkaline-activated GP mortars intended for underwater applications. The impact of different flowability, alkaline solution content, and binder-to-sand ratios on setting time, plastic viscosity, washout loss, and compressive strength was assessed. From the foregoing, it can be concluded that the setting times increased with a decrease in binder content and mixtures possessing higher flow values. The washout loss was well correlated to the plastic viscosity of the mortar, wherein the increase in sand content and decrease in alkaline solution content (i.e., lower flowability) resulted in greater resistance against washout loss. The difference in compressive strengths of mortars placed in dry and submerged conditions decreases with the increase in the alkaline solution, which worked as a barrier for water intrusion inside the mortar matrix. Such data are quite promising for advancing the adoption of GP materials in underwater construction practices, ultimately contributing to a more sustainable and eco-friendly built environment.

Acknowledgment The authors acknowledge the financial support provided by ACTS, Lebanon, and UAEU under grant 21N235. The support of Balamand University staff is appreciated.

References

1. Bawab, J., El Hassan, H., El-Dieb, A., Khatib, J.: Effect of mix design parameters on the properties of cementitious composites incorporating volcanic ash and dune sand. Dev. Built Environ., 100258 (2023). https://doi.org/10.1016/j.dibe.2023.100258
2. Najm, O., El-Hassan, H., El-Dieb, A.: Optimization of alkali-activated ladle slag composites mix design using Taguchi-based TOPSIS method. Constr. Build. Mater. **327**, 126946 (2022). https://doi.org/10.1016/j.conbuildmat.2022.126946
3. Saba, M., Assaad, J.J.: Effect of recycled fine aggregates on performance of geopolymer masonry mortars. Constr. Build. Mater. **279**, 122461 (2021). https://doi.org/10.1016/j.conbuildmat.2021.122461
4. Hwalla, J.; El-Mir, A.; El-Hassan, H.; El-Dieb, A. Taguchi method for optimizing alkali-activated mortar mixtures using waste perlite powder and granulated blast furnace slag. In Proceedings of the International RILEM Conference on Synergising Expertise Towards Sustainability and Robustness of Cement-based Materials and Concrete Structures; Jędrzejewska, A., Kanavaris, F., Azenha, M., Benboudjema, F., Schlicke, D., Eds.; Springer Nature Switzerland: Cham, 2023; pp. 362–373
5. Assaad, J.J., Saba, M.: Suitability of metakaolin-based geopolymers for masonry plastering. ACI Mater. J. **117** (2020). https://doi.org/10.14359/51725991
6. Davidovits, J.: Geopolymers inorganic polymerie new materials. J. Therm. Anal. Calorim. **37**, 1633–1656 (1991). https://doi.org/10.1007/bf01912193
7. Hwalla, J., Bawab, J., El-Hassan, H., Abu Obaida, F., El-Maaddawy, T.: Scientometric analysis of global research on the utilization of geopolymer composites in construction applications. Sustainability. **15**, 11340 (2023). https://doi.org/10.3390/su151411340

8. Ahmad Zaidi, F.H., Ahmad, R., Al Bakri Abdullah, M.M., Abd Rahim, S.Z., Yahya, Z., Li, L. Y., Ediati, R.: Geopolymer as underwater concreting material: a review. Constr. Build. Mater. **291**, 123276 (2021). https://doi.org/10.1016/j.conbuildmat.2021.123276

9. Hwalla, J., Saba, M., Assaad, J.J.: Suitability of metakaolin-based geopolymers for underwater applications. Mater. Struct. **53**, 119 (2020). https://doi.org/10.1617/s11527-020-01546-0

10. Favier, A., Hot, J., Habert, G., Roussel, N.: Flow Properties of MK-based geopolymer pastes. A comparative study with standard Portland cement pastes. Soft Matter. **10**, 1134–1141 (2014). https://doi.org/10.1039/c3sm51889b

11. Kuri, J.C., Sarker, P.K., Shaikh, F.U.A.: Sulphuric acid resistance of ground ferronickel slag blended fly ash geopolymer mortar. Constr. Build. Mater. **313**, 125505 (2021). https://doi.org/10.1016/j.conbuildmat.2021.125505

12. Hwalla, J., El-Hassan, H., Assaad, J.J., El-Maaddawy, T.: Durability assessment of geopolymeric and cementitious composites for screed applications. J. Build. Eng. **87**, 109037 (2024). https://doi.org/10.1016/j.jobe.2024.109037

13. Zivica, V., Bajza, A.: Acidic attack of cement based materials — a review: Part 1. Principle of acidic attack. Constr. Build. Mater. **15**, 331–340 (2001). https://doi.org/10.1016/S0950-0618 (01)00012-5

14. ASTM C33-18: Standard Specification for Concrete Aggregates. ASTM International (2018)

15. Li, Z., Xiong, Z., Zhang, B., Huang, D., Huang, J., Yan, L., Li, L.: Seawater used to metakaolinite-based geopolymer preparation. Constr. Build. Mater. **392** (2023). https://doi.org/10.1016/j.conbuildmat.2023.131816

16. ASTM C230-20: Standard Specification for Flow Table for Use in Tests of Hydraulic Cement. ASTM International (2020)

17. ASTM C191-19: Standard test methods for time of setting of hydraulic cement by vicat needle. ASTM International (2019)

18. de Normalisation, C.E.: EN 196-1: Methods of Testing Cement—Part 1: Determination of Strength. CEN, Brussels (2016)

19. Hwalla, J., El-Hassan, H., Assaad, J., El Maaddawy, T., Bawab, J.: Effect of type of sand on the flowability and compressive strength of slag-fly ash blended geopolymer mortar, 6 June 2023

20. Ghasemi, Y., Emborg, M., Cwirzen, A.: Effect of water film thickness on the flow in conventional mortars and concrete. Mater. Struct. **52**, 62 (2019). https://doi.org/10.1617/s11527-019-1362-9

21. Assaad, J., Daou, Y., Khayat, K.: Simulation of water pressure on washout on underwater concrete repair. ACI Mater. J. **106**(6), 529–536 (2009). https://doi.org/10.14359/51663336

22. Hwalla, J., El-Hassan, H., Assaad, J.J., El-Maaddawy, T.: Performance of cementitious and slag-fly ash blended geopolymer screed composites: a comparative study. Case Stud. Constr. Mater., e02037 (2023). https://doi.org/10.1016/j.cscm.2023.e02037

23. Alnahhal, M.F., Kim, T., Hajimohammadi, A.: Evolution of flow properties, plastic viscosity, and yield stress of alkali-activated fly ash/slag pastes. RILEM Tech. Lett. **5**, 141–149 (2020). https://doi.org/10.21809/rilemtechlett.2020.123

24. Gugulothu, V., Gunneswara Rao, T.D.: Effect of binder content and solution/binder ratio on alkali-activated slag concrete activated with neutral grade water glass. Arab. J. Sci. Eng. **45**, 8187–8197 (2020). https://doi.org/10.1007/s13369-020-04666-5

25. Assaad, J.J., Mardani, A.: Limestone replacements by fine crushed concrete and ceramic wastes during the production of Portland cement. J. Sustain. Cement-Based Mater. **12**(11), 1447–1459 (2023). https://doi.org/10.1080/21650373.2023.2225189

Production of CSA Clinker Using Municipal Solid Waste Incineration Fly Ash as a Main Raw Material

Carlos Andrés Bedoya-Henao, Oscar Jaime Restrepo-Baena, and Jorge Iván Tobón

Contents

1 Introduction

The pursuit of sustainable practices in the construction industry has led to the exploration of alternative materials that can reduce environmental impact while maintaining or even enhancing performance. Calcium sulphoaluminate (CSA) cements have emerged as a promising substitute for ordinary Portland cement (OPC) due to their inherent advantages, including reduced CO_2 emissions and lower energy requirements during production [1, 2]. These attributes align closely with the objectives of sustainable development, driving efforts to integrate eco-friendly materials into construction practices.

Around 2.01 billion tons of municipal solid waste (MSW) are generated globally each year [3]. With urbanization on the rise, leading to a surge in MSW production, urgent steps must be taken for sustainable waste management. A prevalent solution involves the incineration of these MSW where some by-products are generated such

C. A. Bedoya-Henao (✉) · O. J. Restrepo-Baena · J. I. Tobón
Grupo del Cemento y Materiales de Construcción (CEMATCO), Universidad Nacional de
Colombia, Facultad de Minas, Medellín, Colombia
e-mail: cabedoyah@unal.edu.co; ojrestre@unal.edu.co; jitobon@unal.edu.co

© The Author(s) 2025
M. Kioumarsi, B. Shafei (eds.), *The 1st International Conference on Net-Zero Built
Environment*, Lecture Notes in Civil Engineering 237,
https://doi.org/10.1007/978-3-031-69626-8_10

as fly ashes. In way with circular economy and sustainability, different types of industrial wastes and alternative materials have been extensively studied as raw materials in the production of clinkers [4–7], and the municipal solid waste inciner-ation fly ashes (MSWI FA) have garnered attention as a potential resource within the construction sector. Their incorporation into building materials offers an avenue to improve performance and minimize dependence on traditional raw materials so they have been evaluated as a supplementary cementitious material [8], as a precursor material in geopolymers [9], in asphalt mixtures [10], and as a raw material to produce ordinary portland clinker [11] and CSA cements [12].

Several studies have demonstrated the viability of producing CSA clinker utiliz-ing MSWI FA through analyses of various factors such as clinkering temperature [12], kiln residence time, and different formulations of raw meal modules [13], microstructure [14] and durability [14, 15]. However, MSWI FA have a distinctive condition in their chemical and mineralogical compositions such as the presence of some chlorides and trace metals. The potential impact of these constituents on the formation of cementitious phases during clinkering and consequently their influence on the hydration process of the resulting products have not yet been studied in detail.

This research aims to explore the feasibility and effects of using MSWI FA as a raw material in the production of CSA clinkers and evaluate its effect on the development of cementitious phases and hydration products. This study contributes to a broader discourse on waste management practices and the development of innovative building materials that align with sustainability and circularity goals in the construction industry.

2   Materials and Methods

2.1   Materials

The raw meal used in the production of CSA clinker consists mainly of a calcium source such as calcium carbonate ($CaCO_3$), gypsum ($CaSO_4 \cdot 2H_2O$) as a source of sulfur and calcium, fly ash as a source of silicon and iron, and a source of aluminum. The chemical composition of the raw materials was determined by X-ray fluores-cence (XRF) and is shown in Table 1. The MSWI FA is sourced from the municipal solid waste incineration plant located in San Andres Island, Colombia. The MSWI FA shows as a source of calcium, sulfur, and silicon, with the presence of chlorine and some trace elements shown in Table 2.

2.2   Mix Designs and Samples Preparation

The mix designs for preparing the raw meal were standardized across all formula-tions, ensuring consistency in key parameters. The substitution levels of MSWI FA varied at 10%, 20%, 30%, and 40% in the raw meal composition and each

Table 1 Chemical composition of raw materials (%wt)

Composition	MSWI FA	Limestone	Gypsum	Fly ash	Aluminium oxide
CaO	25.29	54.83	30.83	1.22	
Al_2O_3	4.23	0.26	1.00	22.36	99.9
SO_3	15.30	0.10	42.63	0.11	
SiO_2	6.79	0.84	3.37	55.04	
Na_2O	17.84	0.13	0.15	0.60	
Fe_2O_3	0.91	0.20	0.39	7.20	
MgO	4.72	0.34	0.19	0.75	
Cl-	22.64	–	–	–	
K_2O	0.67	0.04	0.11	2.13	
TiO_2	0.95	-	0.05	1.04	
LOI	-	43.16	21.19	8.9	

Table 2 Trace elements in MSWI FA (%wt)

P_2O_5	ZnO	SrO	Br	BaO	PbO	MnO
0.201	0.190	0.104	0.032	0.029	0.024	0.023
CuO	Cr_2O_3	Sb_2O_3	Rh	NiO	WO_3	Co_3O_4
0.018	0.017	0.017	0.008	0.006	0.006	0.005

Table 3 Trace elements in MSWI FA (%wt)

Sample	MSWI FA	Limestone	Gypsum	Aluminium oxide	Fly ash
CS0	0	48.00	19.52	20.90	11.58
CS1	10.00	44.11	18.85	16.77	10.27
CS2	20.00	40.51	18.05	12.51	8.93
CS3	30.00	36,51	17.40	8.43	7.67
CS4	40.00	32.68	16.69	4.27	6.37

formulation maintained an alkalinity modulus (Cm) of 1.05, an alumina–sulfur ratio (P) of 3.0, and an alumina–silica ratio (N) of 2.5 [13]. Table 3 shows the raw mix composition of each formula.

All raw meals underwent the same burning conditions. The clinkering process involved heating from room temperature to 800 °C at a rate of 15 °C/min holding the temperature at 800 °C for 30 minutes to ensure decarbonation, followed by heating from 800 °C to 1250 °C at a rate of 7 °C/min. The temperature was then held at 1250 °C for 45 minutes. Subsequently, the samples underwent thermal shock to prevent phase retrogradation post-burning.

2.3 Test Methods

2.3.1 X-Ray Diffraction (XRD)

The copper anode was used for the test, operated at 45 kV/40 mA, collecting data from 5° to 69.9° (2θ) with steps of 0.017 °2θ and an accumulation time of 59.7

seconds in continuous mode at 25 °C. For qualitative identification of the crystalline phases and quantitative analysis by Rietveld, refinement was carried out using the software X'Pert-HighScore Plus, ensuring an Rwp (Weighted R Profile) value close to or below 20 and a goodness of fit close to 1.

2.3.2 Thermal Analysis (TGA–DSC)

Thermogravimetric analysis and differential scanning calorimetry (TGA–DSC) was carried out on a TA Instruments brand SDT 650 simultaneous thermogravimetric equipment under an inert nitrogen atmosphere from room temperature to 1250 °C with a heating ramp of 15 °C/min up to 800 °C where the temperature was constant during 30 min, then the heating ramp was adjusted at 7 °C/min until reaching 1250 ° C. These conditions were established in order to coincide with the clinkering conditions.

2.3.3 Isothermal Microcalorimetry

Tam Air Thermostat 90/3116-1 isothermal conduction microcalorimetry equipped with admix ampoules was used to study the samples. 1.43 g of raw meal as a solid part was mixed with 1.00 g of water to guarantee a w/c = 0.70 was tested at 25 °C. The mixing was carried out at a speed of 55 rpm for 150 seconds.

3   Results and Discussions

3.1 Mineralogic Development

The XRD patterns of CSA clinkers are show in the Fig. 1 and the mineralogical quantification of clinkers by Rietveld refinement in the Table 4. The proposed formulation for CS0 clinker (blank), with an alumina–silica ratio (N) of 2.5, aimed to achieve a ye'elemite to gehlenite ratio close to 1:1 [16] allowing for the assessment of MSWI FA's impact on phase development. Rietveld quantification revealed 19.3% ye'elemite and 19.9% gehlenite in the sample. Additionally, the blank sample exhibited the highest anhydrite content at 11.9% and showed the presence of calcium dealuminate, suggesting incomplete ye'elemite formation and development.

CSA1 clinker is characterized as the sample with the lowest percentage of anhydrite (3.9%) and belite (27.2%), alongside the presence of ternesite, a phase that can sometimes be associated with the ye'elemitic–belitic system [9, 10]. Ternesite typically forms in systems characterized by low temperature and short residence time, where the reaction between belite and anhydrite is facilitated [17], hereby accounting for the low amounts of these phases in the clinker. A total of 10% MSWI (CS1) shows to encourage ye'elemite development and anhydrite

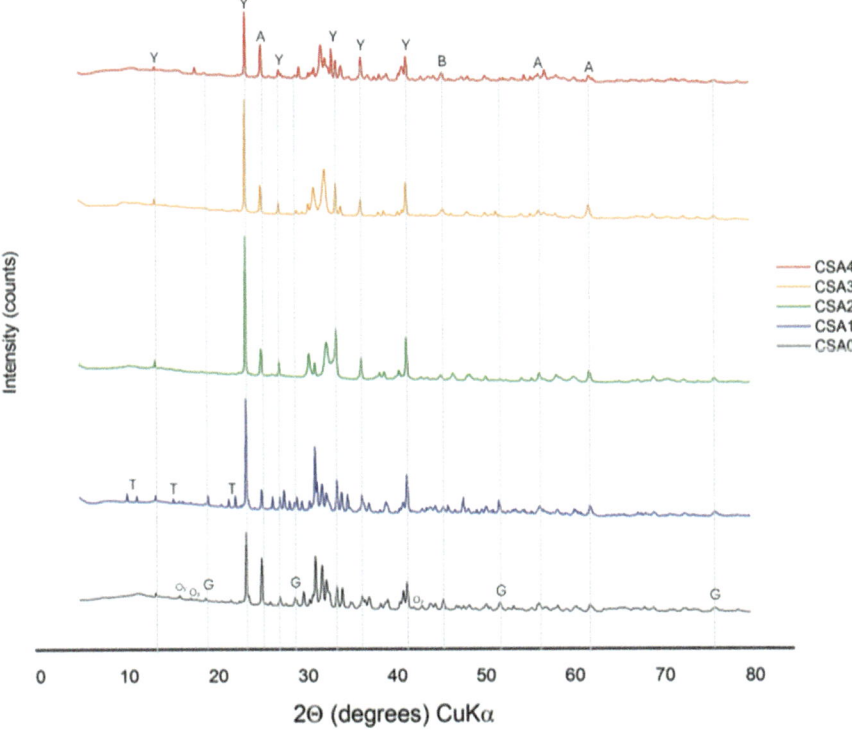

Fig. 1 XRD pattern of CSA clinker. Y: Ye'elimite, Oy: Orthorhombic Ye'elemite, A: Anhydrite, G: Gehlenite, B: Belite, T: Ternisite

Table 4 Mineralogical quantification of clinkers

		CSA0	CSA1	CSA2	CSA3	CSA4
Ye'elemite	$C_4A_3\check{S}$	19.3	27.0	38.7	30.9	18.3
Anhydrite	$C\check{S}$	11.9	3.9	9.8	6.7	11.6
Belite	C_2S	41.5	27.2	51.4	56.9	55.8
Gehlenite	C_2AS	19.9	14.2	0.2	–	–
Calcium dialuminate	CA_2	7.4	–	–	–	–
Ternesite	$C_5S_2\check{S}$	–	27.8	–	–	–
Merwinite	C_3MS	–	–	–	–	–
Alite	C_3S	–	–	–	5.5	–
Wadalite	$C_x(_y)S_z \cdot Cl$	–	–	–	–	14.3

consumption compared to the blank sample; however, ternesite formation evidences a low phase development process.

The main phases identified in the CSA2 (20% MSWI) clinker are ye'elemite with 38.7%, belite with 51.4%, and to a lesser amount anhydrite with 9.8% and gehlenite with 0.2%. This composition makes CSA2 the clinker with the best phase development. CSA3 (30% MSWI) composed of 56.9% belite, 30.9% ye'elemite, 6.7% anhydrite, and 5.5% alite, which has three important characteristics: it boasts the highest belite content among all the clinkers analyzed, the second highest ye'elemite content and the only one where alite was identified. This means that in contrast to 10% MSWI FA, with 20% and 30% of MSWI FA, the ash acts as a mineralizer, promoting the reaction kinetics during clinkerization and also suggests a stabilizing effect possibly due to the contribution of trace metals.

Wadalite was identified in CSA4 (40% MSWI) clinker, a calcium chloride silicate phase. It is important to identify this phase because it shows the possibility of stabilizing chlorine in the CSA cement structure. The stabilization of this phase is due to the fact that by increasing the amount of MSWI FA in the raw meal formulation, so does the amount of magnesium, which is necessary to distort the structure of the calcium silicates in formation and stabilize the chlorine in it [18]. The high MSWI FA content leads to the decomposition of ye'elemite releasing SO_3 forming anhydrite, while the available calcium is stabilized in the form of belite.

Thereby, it is possible to observe a notable variation in the compositions of the main phases across the different CSA clinkers. CSA2 exhibits the highest proportion of ye'elemite at 38.7% followed by CSA3 at 30.9%. All samples show a higher ye'elimite development compared to CSA0, whereas CSA4 having the lowest content with 18.3%, only 1% lower than the blank (19.3%). CSA1 stabilizes ternesite, CSA3 alite with a high belite content, and CSA4 wadalite.

The presence of alite in CSA3 and the effects evidenced in the clinkers suggest that MSWI FA acts as a flux and mineralizer which promotes the development of phases at low temperatures, and that in adequate proportions it acts as a stabilizer of phases of interest such as alite or wadalite. Furthermore, employing MSWI FA obviates the need to augment the alumina–silica ratio (N) in the raw meal formulation to prevent gehlenite formation.

3.2 Thermal Analysis of Raw Meal

Figure 2 shows the TG, DTG, and DSC curves of raw meals CSA0, CSA2, and CSA3 treated up to 1250 °C. Up to 800 °C all the raw meals show a similar behavior as the temperature increases. The first endothermic peak associated to dehydration of gypsum takes place at ~120 °C. Near to ~350 °C, an exothermic effect appears corresponding to the phase transition of γ-$CaSO_4$ into β-$CaSO_4$ form [19]. For CSA2 and CSA3, the decarbonation occurs ~35 °C earlier than the CSA0 blank likely due to the flux effect of the MSWI FA over the system, where the low melting point of chlorides (NaCl + KCl) around 650 °C [20] and high ionic mobility decrease the activation energy required for the decarbonation reaction.

After the full decarbonation of the calcium carbonates, a second endothermic peak related to liquid phase formation appears at 890 °C for the CSA3 raw meal, at 905 °C for the CSA2, and at 980 °C for the blank [21]. This shows again a fluxing effect of the MSWI FA decreasing the formation temperature of the liquid phase as

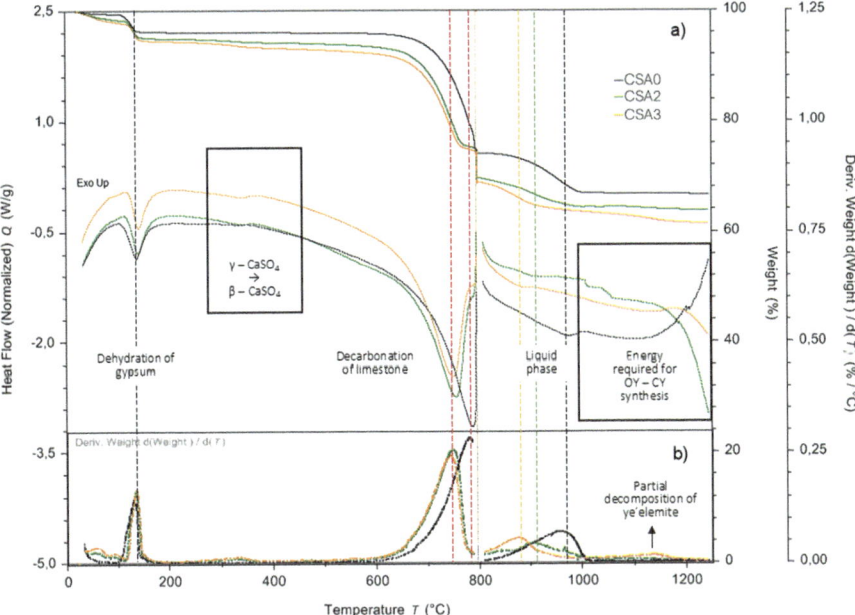

Fig. 2 Thermal analysis of CSA0, CSA2, and CSA3 raw meal. (**a**) TGA-DSC curves (**b**) Derived weight loss curves

the amount of MSWI FA increases. Above 1000 °C, the phase transformations associated with the formation of ye'elemite take place. The energy needed for orthorhombic ye'elemite (OY) synthesis is higher than for cubic ye'elemite (CY), evidenced by a more pronounced endothermic transition in OY, requiring a higher energy flux during the transition to become ye'elemite [22]. This suggests that in the CSA0 clinker, orthorhombic ye'elemite (OY) synthesis is favored, while the samples utilizing MSWI FA promote the formation of cubic ye'elemite (CY). Figure 1 allows to discern this distinction via the XRD patterns between the two polymorphs of ye'elemite, orthorhombic for the CSA0 sample and cubic for CSA2 and CSA3 [23]. This phenomenon is likely attributed to the trace metals within MSWI FA, which act as mineralizers, facilitating the formation of phases with lower energy requirements. During this process, CSA3 exhibits a weight loss near 1150 °C, possibly associated with a minor decomposition of the ye'elemite already formed [22].

3.3 Early Hydration of Cement Paste

The early hydration at 3 days of the clinkers was studied by isothermal microcalorimetry at a water–cement ratio of 0.70 (see Fig. 3). In the curves, four peaks are

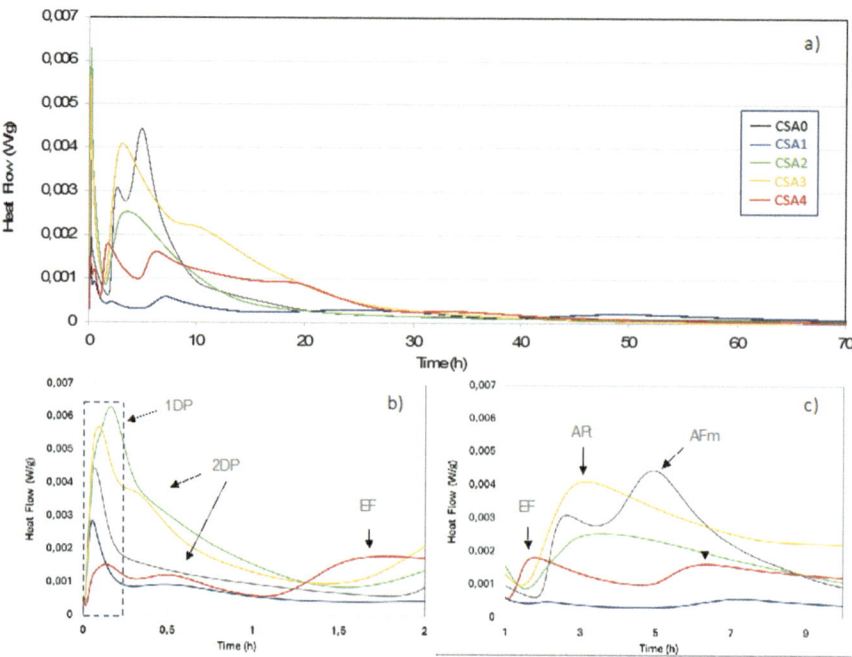

Fig. 3 Heat flow development of clinkers with w/c= 0.70 to: a) 0–70 h, b) 0–2 h, and c) 1–10 h. 1DP (First dilution peak), 2DP (Second dilution peak), EF (early formation)

mainly identified. The first of them occurs in the first 15 minutes of hydration and corresponds to the induction stage, where the dissolution of ye'elemite and C-S-H gel formation occurs. A secondary peak, discernible as a shoulder in the curves, corresponds to the dissolution of anhydrite [24]. These initial two peaks occur before the first hour of hydration, giving way to the dormant period. CSA4 represents the first clinker to break the dormant period, which, in its case, is nearly nonexistent. The peak manifests around 1.5 hours into hydration and correlates with the formation of Friedel's salts stemming from the hydration of calcium chloride aluminosilicate [17, 25]. For the remaining phases, the third peak, indicative of the hydration and precipitation of ettringite and AFt phases [26], typically emerges around the 3-hour mark, while the fourth peak, associated with AFm phases, occurs between 5 and 6 hours. In the case of the blank (CSA0), there's a preference for the formation of AFm phases over AFt attributable to sulfate depletion, as indicated by the absence of the second dissolution peak of anhydrite (2DP) [27]. However, due to the ortho-rhombic polymorph ye'elimite, a notable formation of AFt phases, surpassing even that of the CSA2 clinker.

CSA2 and CSA3 exhibit the highest heat flux released during the first and second dissolution peaks. Both clinkers share an approximate dormant period of one hour before entering the acceleration stage. Then, only a peak is observed within the AFt range, indicating an efficient sulfate consumption for the ettringite formation. Once sulfate is depleted, the formation of AFm phases ensues. The sluggish and low reaction kinetics observed in CSA1 clinker can be attributed to the significant presence of ternesite in its composition. Ternesite is noteworthy due to its intermediate reactivity, serving as a link between the rapid reaction of ye'elemite and the delayed reaction of belite [28]. However, ternesite exhibits reactivity solely in the presence of aluminum phases like $Al(OH)_3$, $C_{12}A_7$, or C_3A, facilitating the formation of AFm-type phases [17]. Additionally, the low anhydrite content contributes to decreased hydration kinetics of the clinker. The third outstanding peak in the CSA4 at about 6 h aligns with the formation of AFm phases. This occurrence is possibly promoted by the fact that being the clinker produced with the highest percentage of MSWI FA, there is an inhibition in the dissolution of some phases such as anhydrite due to the impurities present, causing a sulfate depletion during hydration and promoting the formation of AFm phases [6].

Figure 4 shows the DTG plots of all hydrated clinkers at w/c = 0.70 after 3 days. The first set of peaks are identified approximately at 100–120 °C, mainly due to the dehydration of C-S-H and the release of water molecules from ettringite [19, 20]. Second set of peaks between 200 and 250 °C correspond to the dehydration of AFm-type phases [29] and at ~270 °C the dehydroxilation of aluminum hydroxide gel $Al(OH)_3$. The non-presence of portlandite decomposition at ~450 °C may suggest that after 24 h in the absence of calcium sulfate and ye'elemite [30, 31], portlandite reacts with aluminum hydroxide to produce more AFm phases [32]. CSA2 shows to be the clinker with greater C-S-H and AFt as hydration products, which is consistent with the mineralogy identified by XRD, while the CSA0 clinker shows a greater decomposition of AFm-type phases and aluminum hydroxide. The dehydration of Friedel's salt is identified around 280 °C [33, 34].

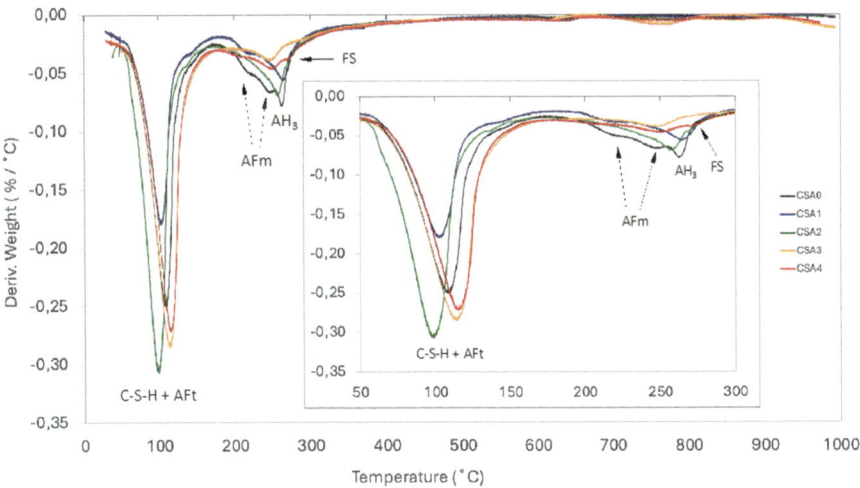

Fig. 4 Thermogravimetric curves of clinkers with w/c = 0.70 to 3 days

4 Conclusions

Utilizing MWSI FA as a raw material for CSA cement production enhances the formation of key phases such as ye'elemite and belite, while also stabilizing alite, owing to its fluxing, mineralizing, and stabilizing attributes derived from its chemical and mineral composition. In this study, incorporating 20 wt% MWSI FA into the raw meal formulation resulted in an improved clinker composition. However, increasing the usage to 30% demonstrates even more favorable effects on phase development.

Employing MSWI FA avoids the need to augment the alumina–silica ratio (N) in the raw meal formulation to prevent gehlenite formation. This is significant, particularly considering one of the drawbacks of CSA cements compared to OPC lies in the availability of aluminum, a resource in limited supply and extensively sought after in other industrial sectors.

The utilization of MSWI FA as a raw material triggers reaction kinetics at lower temperatures and fosters the formation of phases requiring less energy, which occurs in the decarbonation or the liquid phase formation. This implies a potential melting and mineralizing effect exerted by the incineration ash.

In future research, it is essential to assess the distinct impact of the components within MSWI FA, particularly focusing on chlorinated compounds and trace elements.

Acknowledgments This research project received support under grant No. BPIN 2021000100040

Declaration of Competing Interest The authors declared no potential conflicts of interest with respect to the research, authorship, and/or publication of this article.

References

1. Carmen Martín-Sedeño, M., et al.: Aluminum-rich belite sulfoaluminate cements: Clinkering and early age hydration, (2009). https://doi.org/10.1016/j.cemconres.2009.11.003
2. Singh, M., Kapur, P.C., Pradip: Preparation of alinite based cement from incinerator ash. Waste Manag. **28**(8), 1310–1316 (2008). https://doi.org/10.1016/j.wasman.2007.08.025
3. Sondh, S., Upadhyay, D.S., Patel, S., Patel, R.N.: A strategic review on Municipal Solid Waste (living solid waste) management system focusing on policies, selection criteria and techniques for waste-to-value. J Clean Prod. **356**, 131908 (2022). https://doi.org/10.1016/j.jclepro.2022.131908
4. Iacobescu, R.I., Pontikes, Y., Koumpouri, D., Angelopoulos, G.N.: Synthesis, characterization and properties of calcium ferroaluminate belite cements produced with electric arc furnace steel slag as raw material. Cem Concr Compos. **44**, 1–8 (2013). https://doi.org/10.1016/j.cemconcomp.2013.08.002
5. Negrão, L.B.A., da Costa, M.L., Pöllmann, H.: Waste clay from bauxite beneficiation to produce calcium sulphoaluminate eco-cements. Constr Build Mater. **340**, 127703 (2022). https://doi.org/10.1016/j.conbuildmat.2022.127703

6. El Khessaimi, Y., Taha, Y., El Mahdi Safhi, A., Hakkou, R., Benzaazoua, M.: Synthesis of MgO-Belite calcium sulfoaluminate cement from phosphate mine waste rock and phosphogypsum. Mater Today Proc. **58**, 1081–1090 (2022). https://doi.org/10.1016/j.matpr.2022.01.136

7. Tao, Y., Rahul, A.V., Mohan, M.K., De Schutter, G., Van Tittelboom, K.: Recent progress and technical challenges in using calcium sulfoaluminate (CSA) cement. Cem Concr Compos. **137**, 104908 (2023). https://doi.org/10.1016/j.cemconcomp.2022.104908

8. Rémond, S., Pimienta, P., Bentz, D.P.: Effects of the incorporation of Municipal Solid Waste Incineration fly ash in cement pastes and mortars: I. Experimental study. Cem Concr Res. **32**(2), 303–311 (2002). https://doi.org/10.1016/S0008-8846(01)00674-3

9. Niu, M., Zhang, P., Guo, J., Wang, J.: Effect of municipal solid waste incineration fly ash on the mechanical properties and microstructure of geopolymer concrete. Gels. **8**(6), 341 (2022). https://doi.org/10.3390/GELS8060341

10. Romeo, E., Mantovani, L., Tribaudino, M., Montepara, A.: Reuse of stabilized municipal solid waste incinerator fly ash in asphalt mixtures. J Materials Civil Eng. **30** (2018). https://doi.org/10.1061/(ASCE)MT.1943-5533.0002347

11. Saikia, N., Kato, S., Kojima, T.: Production of cement clinkers from municipal solid waste incineration (MSWI) fly ash. Waste Manag. **27**(9), 1178–1189 (2007). https://doi.org/10.1016/J.WASMAN.2006.06.004

12. Shi, H.S., Deng, K., Yuan, F., Wu, K.: Preparation of the saving-energy sulphoaluminate cement using MSWI fly ash. J Hazard Mater. **169**(1–3), 551–555 (2009). https://doi.org/10.1016/j.jhazmat.2009.03.134

13. Wu, K., Shi, H., Guo, X.: Utilization of municipal solid waste incineration fly ash for sulfoaluminate cement clinker production. Waste Manag. **31**(9–10), 2001–2008 (2011). https://doi.org/10.1016/J.WASMAN.2011.04.022

14. Guo, X., Shi, H., Hu, W., Wu, K.: Durability and microstructure of CSA cement-based materials from MSWI fly ash (2013). https://doi.org/10.1016/j.cemconcomp.2013.10.015

15. Guo, X.L., Shi, H.S., Hu, W.P., Wu, K.: Durability of Calcium Sulphoaluminate (CSA) composite cement-based materials made from Municipal Solid Waste Incineration (MSWI) Fly Ash. Appl Mech Materials. **719–720**, 214–217 (2015). https://doi.org/10.4028/WWW.SCIENTIFIC.NET/AMM.719-720.214

16. Berrio, A., Rodriguez, C., Tobón, J.I.: Effect of Al2O3/SiO2 ratio on ye'elimite production on CSA cement. Constr Build Mater. **168**, 512–521 (2018). https://doi.org/10.1016/j.conbuildmat.2018.02.153

17. Montes, M., Pato, E., Carmona-Quiroga, P.M., Blanco-Varela, M.T.: Can calcium aluminates activate ternesite hydration? Cem Concr Res. **103**, 204–215 (2018). https://doi.org/10.1016/j.cemconres.2017.10.017

18. Lampe, F.V., Hilmer, W., and Jost, K.H.: SYr,Whesis, structure and thermal decompos!tio~ OF ALINiTE, 1986

19. El Hazzat, M., Sifou, A., Arsalane, S., El Hamidi, A.: Novel approach to thermal degradation kinetics of gypsum: application of peak deconvolution and Model-Free isoconversional method. J Therm Anal Calorim. **140**(2), 657–671 (2020). https://doi.org/10.1007/s10973-019-08885-3

20. Broström, M., Enestam, S., Backman, R., Mäkelä, K.: Condensation in the KCl–NaCl system. Fuel Proc Tech. **105**, 142–148 (2011). https://doi.org/10.1016/j.fuproc.2011.08.006

21. Bullerjahn, F., Zajac, M., Ben Haha, M.: CSA raw mix design: effect on clinker formation and reactivity. Mater Struct. **48**(12), 3895–3911 (2015). https://doi.org/10.1617/s11527-014-0451-z

22. Berrio, A., Tobón, J.I., De la Torre, A.G.: Kinetic model for ye'elimite polymorphs formation during clinkering production of CSA cement. Constr Build Mater. **321** (2022). https://doi.org/10.1016/j.conbuildmat.2022.126336

23. Zhao, J., Huang, J., Yu, C., Cui, C., Chang, J.: Phosphorus substitution preference in Ye'elimite: experiments and density functional theory simulations. Materials (Basel). **14** (2021). https://doi.org/10.3390/ma14195874

24. Londoño Zuluaga, D.: Tesis doctoral eco-cements containing Belite, Alite and Ye'elimite. Hydration and mechanical properties, (2018). [Online]. Available: http://orcid.org/0000-0002-6842-8754
25. Pliego-Cuervo, Y.B., Glasser, F.P.: The role of sulphates in cement clinkering: subsolidus phase relations in the system CaO-Aℓ2O3-SiO2-SO3. Cem Concr Res. **9**(1), 51–55 (1979). https://doi.org/10.1016/0008-8846(79)90094-2
26. Londono-Zuluaga, D., Tobón, J.I., Aranda, M.A.G., Santacruz, I., De la Torre, A.G.: Influence of fly ash blending on hydration and physical behavior of belite–alite–ye'elimite cements. Mater Struct. **51**(5), 128 (2018). https://doi.org/10.1617/s11527-018-1246-4
27. Bolaños, I., Trauchessec, R., Tobón, J., Lecomte, A.: Influence of the ye'elimite/anhydrite ratio on PC-CSA hybrid cements. Mater Today Commun. **22**, 100778 (2019). https://doi.org/10.1016/j.mtcomm.2019.100778
28. Li, W., Ji, D., Shi, F., Huang, X., Ji, X., Ma, S.: Study on the synthesis of belite-ye'elimite-ternesite clinker. Constr Build Mater. **319**, 126022 (2022). https://doi.org/10.1016/j.conbuildmat.2021.126022
29. Winnefeld, F., Martin, L.H.J., Müller, C.J., Lothenbach, B.: Using gypsum to control hydration kinetics of CSA cements. Constr Build Mater. **155**, 154–163 (2017). https://doi.org/10.1016/j.conbuildmat.2017.07.217
30. Winnefeld, F., Barlag, S.: Calorimetric and thermogravimetric study on the influence of calcium sulfate on the hydration of ye'elimite. J Therm Anal Calorim. **101**(3), 949–957 (2010). https://doi.org/10.1007/s10973-009-0582-6
31. Tobón, J.I., Paya, J., Borrachero, M.V., Soriano, L., Restrepo, O.J.: Determination of the optimum parameters in the high resolution thermogravimetric analysis (HRTG) for cementitious materials. J Therm Anal Calorim. **107**(1), 233–239 (2012). https://doi.org/10.1007/s10973-010-0997-0
32. Londono-Zuluaga, D., Tobón, J.I., Aranda, M.A.G., Santacruz, I., De la Torre, A.G.: Clinkering and hydration of belite-alite-ye´elimite cement. Cem Concr Compos. **80**, 333–341 (2017). https://doi.org/10.1016/j.cemconcomp.2017.04.002
33. Wang, Q., et al.: Characterization of the mechanical properties and microcosmic mechanism of Portland cement prepared with soda residue. Constr Build Mater. **241**, 117994 (2020). https://doi.org/10.1016/j.conbuildmat.2019.117994
34. Uçal, G.O., Mahyar, M., Tokyay, M.: Hydration of alinite cement produced from soda waste sludge. Constr Build Mater. **164**, 178–184 (2018). https://doi.org/10.1016/j.conbuildmat.2017.12.196

Feasibility Study of Precoated Binder-Type Electric Arc Furnace Oxidizing Slags as Aggregates for Cement Mortar

Wei-Ting Lin, Andīna Sprince, Marek Hebda, Gábor Mucsi, An Cheng, and Huang-Hsing Pan

Contents

1 Introduction

Cement and civil engineering industry is a critical infrastructure industry. With the goal of net-zero carbon emissions, research issues such as using aggregates and carbon emissions from cement are complex and urgent industry transformation

W.-T. Lin (✉) · A. Cheng
Department of Civil Engineering, National Ilan University, Yilan, Taiwan
e-mail: wtlin@niu.edu.tw

A. Sprince
Institute of Civil Engineering, Faculty of Civil and Mechanical Engineering, Riga Technical University, Riga, Latvia

M. Hebda
Department of Materials Engineering, Faculty of Materials Engineering and Physics, Cracow University of Technology, Kraków, Poland

G. Mucsi
Institute of Raw Materials Preparation and Environmental Technologies, University of Miskolc, Miskolc, Hungary

H.-H. Pan
Department of Civil Engineering, National Kaohsiung University of Science and Technology, Kaohsiung, Taiwan

© The Author(s) 2025
M. Kioumarsi, B. Shafei (eds.), *The 1st International Conference on Net-Zero Built Environment*, Lecture Notes in Civil Engineering 237,
https://doi.org/10.1007/978-3-031-69626-8_11

challenges. The cement and concrete industry is now the world's largest producer of carbon emissions. To achieve the goal of net zero carbon emissions by 2050, researchers, engineers, and the industry have invested heavily in research and development of materials and equipment and other strategies to reduce carbon emissions and indirectly reduce the payments of carbon tax [1, 2]. As the global steel industry has continued to flourish in recent years, steel has become an indispensable and significant industrial material. In particular, the increased demand in engineering fields, such as high-rise buildings and bridges, has strengthened the utilization of steel materials. Recycled steel waste has increased as reinforced steel and steel materials have been produced at higher rates [3, 4]. Over the past few years, electric arc furnace oxidizing slag (EAFOS) has become the primary recycling material due to the recycling industry's requirements. For a long time, EAFOS reuse has troubled the relevant engineering units. The industry strongly advocates that resources should be recycled and reused, and how to use EAFOS is the research focus at this stage [5, 6]. The EAFOS waste was a typical solid waste with high iron content and hardness, making it unsuitable for grinding into powder for reuse. Reprocessing EAFOS also led to high production costs. This treatment option contributed significantly to alleviating the shortage of natural aggregates by using crushed EAFOS as a direct substitute for aggregates [7, 8]. However, using EAFOS instead of natural aggregation has led to problems with expansion and cracks on the surface of the concrete because of the effect of free calcium oxide (f-CaO) or magnesium oxide (f-MgO) [9]. Therefore, the development of stabilization technology to replace natural aggregates with EAFOS is an essential task at this stage.

The use of EAFOS as aggregates was subject to the requirement of volume stability. In the past, alkali-activated technology was used as the test procedure for the stabilization of EAFOS [10, 11]. Although using such alkali-activation technology to reuse EAFOS was feasible, practical application required the addition of chemical agents, which also increased the cost. It was found in the literature that EAFOS should be able to be successfully used in cement mortars to replace fine aggregates in combination with other industrial by-products such as copper slag and coal bottom ash [12, 13]. Past studies have shown that recycled aggregates have the potential to be precoated to enhance recyclability and adaptability [14, 15]. The same precoated technology can be applied directly to EAFOS, verifying its feasibility and achieving EAFOS stabilization. In this study, EAFOS was selected as an industrial by-product to replace the natural aggregates and to improve the recycling efficiency of the waste. The paste mixing process provided EAFOS with a precoated layer of cement, ground granulated blast furnace slag (GGBS), or fly ash. It verified the most suitable precoated material and the best aggregate replacement ratio for precoated EAFOS. The volumetric stability of the precoated EAFOS mortar has also been continuously verified to verify the stabilizing effect of this solution on EAFOS particles.

2 Test Programs

2.1 *Materials*

The precoated materials used in this study consisted of Portland Type I cement, GGBS, and fly ash. The specific gravities of the three precoated materials were 3.15, 2.88, and 2.72, respectively. The fineness of the three precoated materials was 3690 cm^2/g, 5850 cm^2/g, and 10,500 cm^2/g, respectively. EAFOS was sourced from Tung Ho Steel Enterprise Corporation in Taiwan. EAFOS conducted tests in the range of fine aggregate gradations to replace the natural aggregates in the mortar specimens. EAFOS had a modulus of fineness of 3.90, water absorption of 3.80%, and a specific gravity of 3.60. Results of the EAFOS sieve analysis tests are shown in Fig. 1. It was found that EAFOS aggregates were significantly coarser than fine aggregates, and their grading curves deviated from ASTM requirements. Subsequent tests were performed to improve the particle size distribution of EAFOS using paste-precoated technology.

2.2 *Paste-Precoated Method*

Three types of binders (cement, GGBS, and fly ash) were used as paste-precoated materials in this test. Different water-to-binder ratios (w/b) were used as variables. The test procedure was to mix the paste-precoated materials into a paste specimen at a specified w/b. And the next procedure was to mix the paste with EAFOS using a mixer for 5 min to perform the subsequent procedures. The precoated EAFOS was then allowed to cure in the air for 7 days and tested by sieve analysis. The test procedure is to set the No. 4 (4.75 mm) sieve passing rate to 85% of the paste-precoated group, which was regarded as the test group that met the requirements. The test variables included two parameters, namely the amount of paste-precoated (20%, 25%, 30%, 35%, and 40%) and w/b (0.20, 0.25, 0.30, 0.35, 0.40, 0.45, and 0.50).

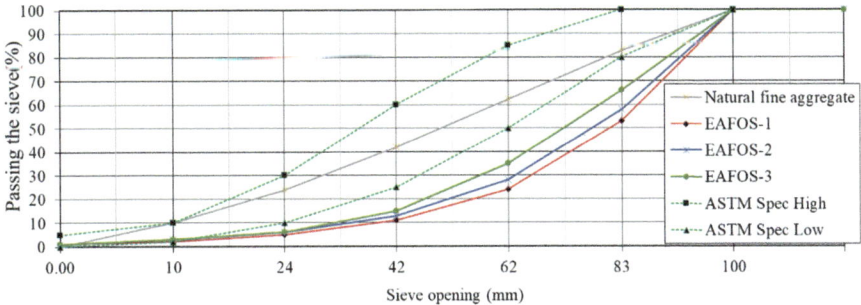

Fig. 1 Results of the EAFOS sieve analysis tests

Table 1 Mix design of control group (kg/m³)	Water	Cement	Fine aggregates
	288.1	523.8	1440.5

2.3 Mix Design

The mix design for the control group is shown in Table 1, and the w/b was fixed at 0.55. EAFOS was used to replace 10–50% of fine aggregates in 10% increments. The ratio of cement to aggregates was 1:2.75.

2.4 Testing Methods

The sieve analysis of natural fine aggregates and EAFOS particles was carried out according to ASTM C136 specification. The specific gravity and water absorption of natural fine aggregates and EAFOS particles were tested with reference to the ASTM C128 specification. The compressive strength tests of mortar specimens were conducted by referring to ASTM C109. The specimen size for compressive strength was 50 × 50 × 50 mm. The dry shrinkage tests of mortar were conducted with reference to ASTM C596, and the size of the specimen was 25 × 25 × 285 mm. The measurement of length change was mainly based on the observed value on the first day as the zero point (reference point), and the measurement values for each age were the observed values. Scanning electron microscope (SEM) observations were conducted in accordance with the recommended procedures of the ASTM C1723 specification. The specimens were selected after compressive strength tests from broken specimen fragments (size less than 3 × 3 × 10 mm).

3 Results and Discussion

3.1 Precoated Properties of EAFOS Particles

EAFOS was mixed with three types of binder using the paste precoated technique and cured for 7 days, and the test groups that passed 85% through a No. 4 sieve (4.75 mm) are listed below. For the cement group, the passing percentages were 83%, 89%, and 86% for the precoated EAFOS with a paste amount of 30% and w/b of 0.35, 0.30, and 0.25, respectively. Aggregates with 35% and 40% paste content showed a less than 15% passing percentage. This means that the EAFOS particles in the fine grades have jumped to the coarse grades due to excessive precoated pastes. For the GGBS group, the passing percentages were 86% and 87% for the precoated EAFOS with a paste amount of 30% and w/b of 0.35 and 0.30, respectively. The same trend was observed in the GGBS group. When the amount of precoated was

Table 2 Specific gravity and water absorption of different types of precoated EAFOS

	Precoated EAFOS					
Indicators	–	Cement	GGBS	GGBS	Fly ash	Fly ash
Parameters for w/b and precoated	–	0.35	0.35	0.30	0.40	0.35
amount	–	30%	30%	30%	40%	40%
Specific gravity	3.60	3.30	3.20	3.28	3.29	3.25
Absorption (%)	3.80	8.80	7.53	4.82	3.39	3.01

20% and 25%, although its passing percentage met the requirement of more than 85%. It can be observed from the external observation of the precoated EAFOS particles that there is insufficient precoated paste. For the fly ash group, the passing percentages were 92% and 92% for the precoated EAFOS with a paste amount of 40% and w/b of 0.40 and 0.45, respectively. Insufficient precoated pastes were also observed in groups with less than 35% precoated content. From the relevant studies, it can be found that these surface coatings were beneficial to forming surface hydration layers on the EAFOS particles to achieve the effect of particle stabilization [16, 17].

Table 2 summarizes the specific gravity and water absorption of EAFOS particles, cement-precoated, fly ash-precoated, and GGBS-precoated EAFOS (curing at room temperature for 7 days). The lower specific gravity of the coated particles compared to the EAFOS particles demonstrates the lower weight for the same volume. The GGBS and fly ash precoated group had a lower specific gravity than the cement group (cement had a higher specific gravity of 3.15 compared to the two binders). In addition, the water absorption of the coated particles was higher than that of the original EAFOS particles. The cement group had the highest water absorption. On the contrary, the GGBS and fly ash groups showed significantly lower water absorption than the cement group. It was evident that the pozzolanic reaction led to an increase in compactness and a decrease in water absorption. Such a trend is consistent with the findings of previous studies [18, 19].

3.2 Compressive Strength

Development curves of EAFOS utilization against compressive strengths for non-precoated group are shown in Fig. 2. The compressive strength was negatively affected by EAFOS doping without paste precoated, and the strength began to decrease significantly at 40% replacement. In addition to being coarser than natural fine aggregates, EAFOS aggregates are distributed in approximate section grades. Large amounts of EAFOS resulted in grain size grades deviating from ASTM specifications. This significantly reduced the strength of the mortar. For the cement-precoated group, a precoated amount of 30% with a w/b of 0.35 was used as a parameter. The development of compressive strength with the corresponding

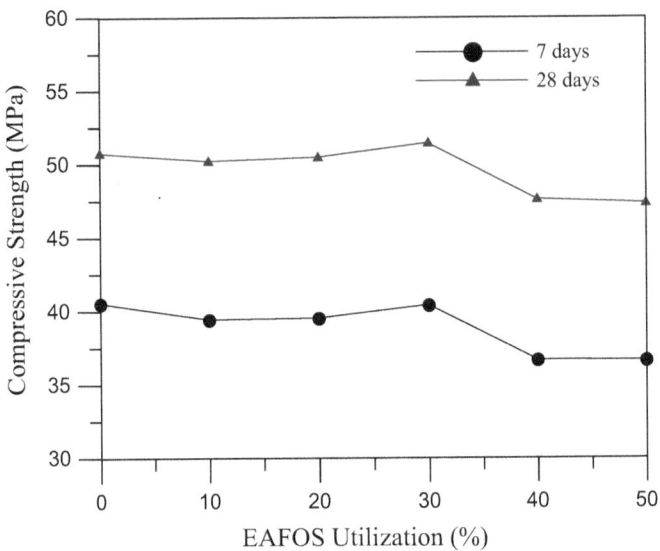

Fig. 2 Development of EAFOS utilization against compressive strengths (non-precoated)

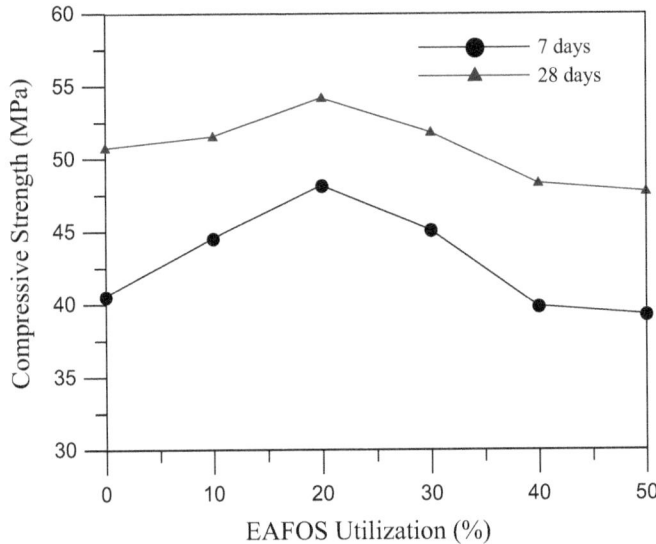

Fig. 3 Development of EAFOS utilization against compressive strengths (cement-precoated)

EAFOS replacement is shown in Fig. 3. Cement-precoated specimens increased their compressive strength, with the maximum compression strength of 20% EAFOS being substituted. Compressive strength has also increased significantly with the

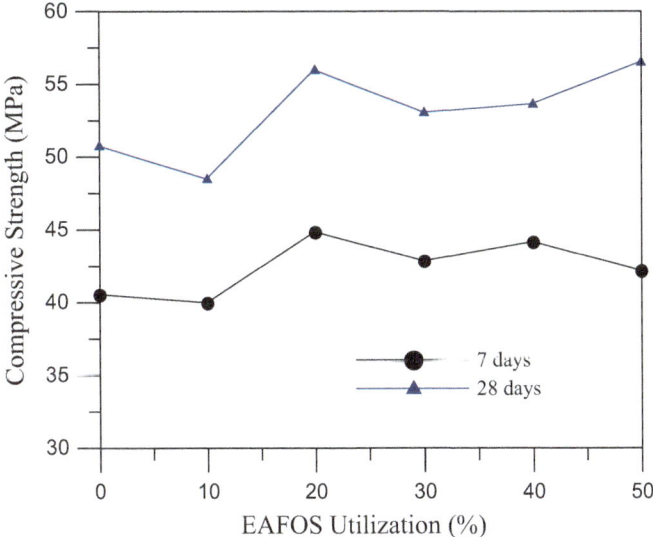

Fig. 4 Development of EAFOS utilization against compressive strengths (GGBS-precoated)

increase in EAFOS replacement. It is clear that the paste-precoated technology contributed to the improved suitability of EAFOS aggregates. Prior to the production of mortar specimens, the paste precoated EAFOS aggregates were allowed to cure for 7 days. To further increase the strength of the specimens, it is recommended to increase the curing time to 28 days.

For the GGBS-precoated group, a precoated amount of 30% with a w/b of 0.30 was used as a parameter. The development of compressive strength with the corresponding EAFOS replacement is shown in Fig. 4. GGBS was slightly more potent than the cement group, and the optimum substitution was also 20%. The maximum strength of the GGBS group (20% EAFOS) was up to 55.92 MPa. Under the same curing conditions, GGBS had a better pozzolanic reaction and a better interfacial transition zone between aggregates and pastes. There was an improvement in the compactness of the paste and an increase in strength [16]. For the fly ash-precoated group, a precoated amount of 40% with a w/b of 0.35 was used as a parameter. The development of compressive strength with the corresponding EAFOS replacement is shown in Fig. 5. Because of the high fluidity of fly ash, a precoated amount of 40% was required. The fly ash-coated specimens were not significantly stronger due to the long curing times required for fly ash to develop the reactive properties required for pozzolanic materials. The maximum strength (47.97 MPa) of the fly ash-precoated specimens was achieved with 30% EAFOS replacement. In conclusion, coated EAFOS particles helped to increase the compressive strength of the mortar. GGBS as paste-precoated materials provided superior compressive strength. Compared to previous findings [8, 20], such paste-

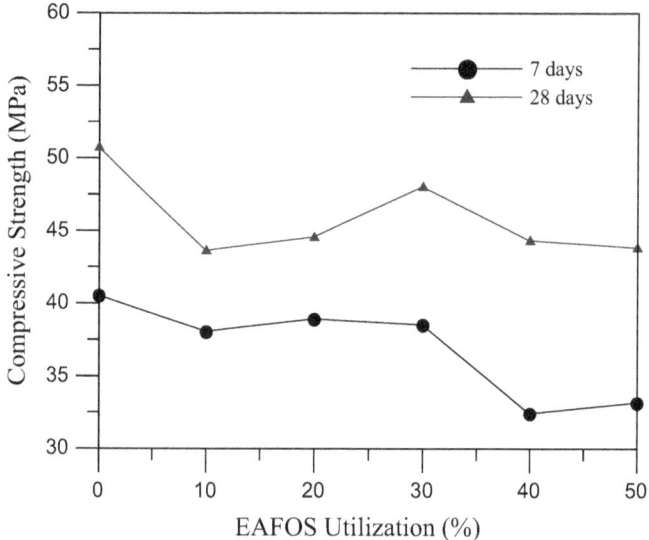

Fig. 5 Development of EAFOS utilization against compressive strengths (fly ash-precoated)

precoated technology provided an effective solution for replacing aggregates with EAFOS and offered superior compressive strength and a cost-effective treatment process.

3.3 Length Variations

Length variations of mortar specimens with different precoated methods are illustrated in Fig. 6. The w/b of cement, GGBS, and fly ash paste-precoated materials were 0.35, 0.30, and 0.35, respectively, for 20% EAFOS replacement (based on better strength considerations). Observing the length variations within 28 days, it was found that the length variations of the paste-precoated materials were similar to those of the un-precoated specimens. The precoated method contributed to the control of the expansion behavior of the specimens. Compared with the results of compressive strength, the paste-coating method is favorable for increasing strength and utilization. This treatment solution contributed to the high-volume reuse of EAFOS and achieved industrial waste reuse efficiency. The appearance of the specimens immersed in 90-degree water for 28 days after demolding is shown in Fig. 7. From the appearance of the specimens, it can be found that there were no abnormalities such as cracking and swelling in the various proportions.

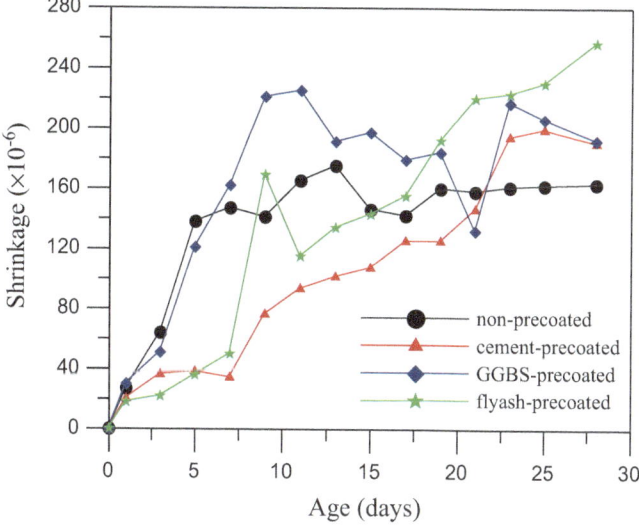

Fig. 6 Length variations of mortar specimens with different precoated methods

Fig. 7 Appearance of specimens immersed in water at 90° for 28 days

3.4 SEM Observations

SEM photographs of the three paste-precoated materials are shown in Figs. 8, 9, and 10. The w/b of cement paste-precoated materials was 0.35, GGBS paste-precoated materials was 0.30, and fly ash paste-precoated materials was 0.35. The replacement rate of EAFOS was fixed at 30%. Significant inter-crystalline gaps were found in the microstructure of cement paste-precoated materials. AFt and calcium hydroxide were also observed, and hydrates filled the remaining voids. The microstructure of

Fig. 8 SEM photo for
cement-precoated specimen

Fig. 9 SEM photo for
GGBS-precoated specimen

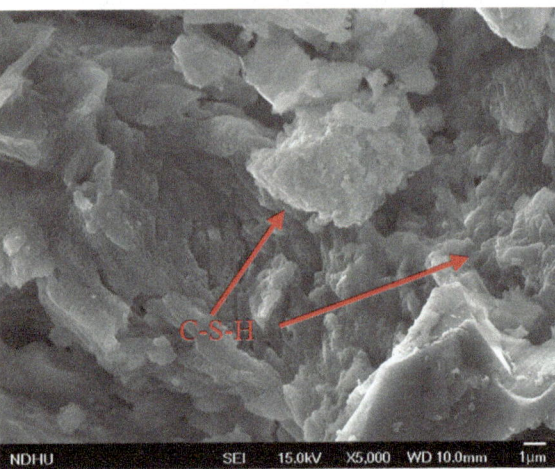

Fig. 10 SEM photo for fly
ash-precoated specimen

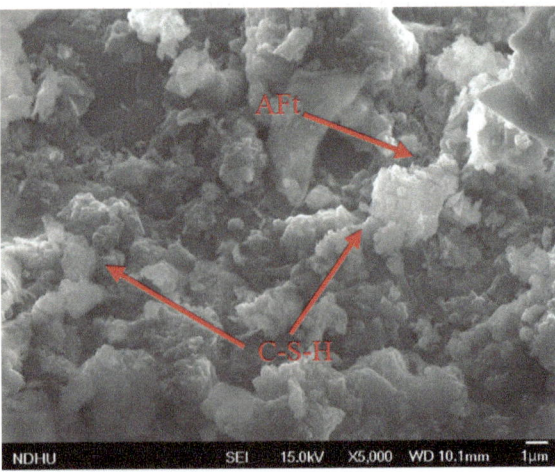

GGBS paste-precoated materials had fewer pores than cement paste-precoated materials, the surface was significantly smoother and denser, and the amount of AFt and sodium hydroxide was reduced considerably. This is the main reason that GGBS paste-precoated materials possess better compressive strength than other compositions. The microstructure of the fly ash paste-precoated materials showed noticeable unreacted fly ash particles, AFt and calcium hydroxide, with smaller gaps between the crystals and a smoother surface. The surface had fine hydrates but required continuous curing to facilitate hydration reaction.

4 Conclusions

The results are summarized as follows:

1. The compressive strength of natural aggregates replaced by uncoated EAFOS in mortar specimens did not increase significantly with increasing amounts of aggregate. Strength began to deteriorate at 30% replacement.
2. The paste-precoated method benefited the compressive strength of EAFOS mortars, and the optimum replacement was between 20% and 30%. The precoated specimens also maintained excellent volume stability.
3. Among the three paste-precoated materials, GGBS-precoated showed the highest compressive strength (55.92 MPa at 20% EAFOS replacement).
4. The paste-precoated method should pay attention to the amount of precoated and the w/b of the binders; the optimal amount of precoated was 30–40%, and the w/b was 0.30 ~ 0.35.
5. This paste-precoated technology provided an effective solution for replacing aggregates with EAFOS and offered superior compressive strength and a cost-effective treatment process.

References

1. Nehdi, M.L., Marani, A., Zhang, L.: Is net-zero feasible: systematic review of cement and concrete decarbonization technologies. Renew. Sust. Energ. Rev. **191**, 114169 (2024)
2. Supriy, Chaudhury, R., Sharma, U., Thapliyal, P.C., Singh, L.P.: Low-CO_2 emission strategies to achieve net zero target in cement sector. J. Clean. Prod. **417**, 137466 (2023)
3. Sabrin, R., Shahjalal, M., Bachu, H.A.E., Habib, M.M.L., Jerin, T., Billah, A.M.: Recycling of different industrial wastes as supplement of cement for sustainable production of mortar. J. Build. Eng. **86**, 108765 (2024)
4. Zhan, P., Xu, J., Wang, J., Zuo, J., He, Z.: Structural supercapacitor electrolytes based on cementitious composites containing recycled steel slag and waste glass powders. Cem. Concr. Compos. **137**, 104924 (2023)

5. You, I., Park, J.J., Lee, N., Ryu, G.S., Kwark, J.W.: Use of electric arc furnace oxidizing slag (EOS) and electric arc furnace reducing slag (ERS) powders in cement pastes for CO_2 sequestrations. J. Build. Eng. **84**, 108631 (2024)
6. Gómez-Casero, M.A., Bueno-Rodríguez, S., Castro, E., Quesada, D.E.: Alkaline activated cements obtained from ferrous and non-ferrous slags. Electric arc furnace slag, ladle furnace slag, copper slag and silico-manganese slag. Cem. Concr. Compos. **147**, 105427 (2024)
7. Pan, S.Y., Adhikari, R., Chen, Y.H., Li, P., Chiang, P.C.: Integrated and innovative steel slag utilization for iron reclamation, green material production and CO_2 fixation via accelerated carbonation. J. Clean. Prod. **137**, 617–631 (2016)
8. Sobhani, J., Komijani, S., Shekarchi, M., Ghazban, F.: Durability of concrete mixtures containing Iranian electric arc furnace slag (EAFS) aggregates and lightweight expanded clay aggregates (LECA). Constr. Build. Mater. **400**, 132597 (2023)
9. Wang, W.C., Mao, Y.J.: Using autoclave pulverization technology to evaluate the expansion potentiality of electric arc furnace oxidizing slag. Case Stud. Constr. Mater. **18**, e01901 (2023)
10. Xu, B., Yi, Y.: Use of ladle furnace slag containing heavy metals as a binding material in civil engineering. Sci. Total Environ. **705**, 135854 (2020)
11. Hafez, H., Kassim, D., Kurda, R., Silva, R.V., de Brito, J.: Assessing the sustainability potential of alkali-activated concrete from electric arc furnace slag using the ECO_2 framework. Constr. Build. Mater. **281**, 122559 (2021)
12. Oyejobi, D.O., Adewuyi, A.P., Yusuf, S.O., Oyebanji, Y.O., Suleiman, I., Hassan, I.A.: Performance of blended cement mortar modified with fly ash and copper slag. Mater. Today Proc. **86**, 104–110 (2023)
13. Bhoi, A.M., Patil, Y.D., Waysal, S.M.: Influence of coal bottom ash and copper slag as a fine aggregate substitute on the characteristics of mortar. Resour. Conserv. Recycl. **93**, 522–529 (2023)
14. Ho, H.L., Huang, R., Lin, W.T., Cheng, A.: Pore-structures and durability of concrete containing pre-coated fine recycled mixed aggregates using pozzolan and polyvinyl alcohol materials. Constr. Build. Mater. **160**, 278–292 (2018)
15. Wenzel, B., Bustamante, M., Muñoz, P., Ortega, J.M., Loyola, E., Letelier, V.: Physical and mechanical behavior of concrete specimens using recycled aggregate coated using recycled cement paste. Constr. Build. Mater. **393**, 132015 (2023)
16. Shaban, W.M., Elbaz, K., Yang, J., Thomas, B.S., Shen, X., Li, L., Du, Y., Xie, J., Li, L.: Effect of pozzolan slurries on recycled aggregate concrete: mechanical and durability performance. Constr. Build. Mater. **276**, 121940 (2021)
17. Hu, H.B., He, Z.H., Fan, K.J., Shibro, T., Liu, B.J., Shi, J.Y.: Properties enhancement of recycled coarse aggregates by pre-coating/pre-soaking with zeolite powder/calcium hydroxide. Constr. Build. Mater. **286**, 122888 (2021)
18. Li, L., Xuan, D., Sojobi, A., Liu, S., Poon, C.S.: Efficiencies of carbonation and nano silica treatment methods in enhancing the performance of recycled aggregate concrete. Constr. Build. Mater. **308**, 125080 (2021)
19. Puente de Andrade, G., de Castro Polisseni, G., Pepe, M., Toledo Filho, R.D.: Design of structural concrete mixtures containing fine recycled concrete aggregate using packing model. Constr. Build. Mater. **252**, 119091 (2020)
20. Ferber, N.L., Al Naimi, K.M., Hoffmann, J., Al-Ali, K., Calvet, N.: Development of an electric arc furnace steel slag-based ceramic material for high temperature thermal energy storage applications. J. Energy Storage. **51**, 104408 (2022)

Early-Day Effects of Graphene on NA$_2$CO$_3$: Activated GGBS Concrete

Yongcong Zhao, Meini Su, and Yong Wang

Contents

1 Introduction

The rapid growth of industrialisation and urbanisation has made the construction industry greater than ever. The global construction volume is expecting a 3.2% yearly increase in the next decade [1]. However, the production process of Ordinary Portland Cement (OPC) is energy intensive. It is reported that the construction sector is responsible for ~39% of the Global greenhouse gas emission [2]. In order to achieve net-zero, alkali-activated material (AAM) has gained increasing attention due to its comparable mechanical properties to OPC and environmentally friendly nature [3]. AAM is produced via the activation of the solid aluminosilicate precursors by the activator solution, where polycondensation takes place and the main hydration product C-(A)-S-H is produced. The most commonly used base materials are ground granulated blast furnace slag (GGBS), Class-F fly ash and metakaolin, while the alkaline activator often consists of one or more chemicals such as sodium

Y. Zhao (✉) · M. Su · Y. Wang
University of Manchester, Manchester, UK
e-mail: yongcong.zhao@manchester.ac.uk; Meini.su@manchester.ac.uk;
yong.wang@manchester.ac.uk

© The Author(s) 2025
M. Kioumarsi, B. Shafei (eds.), *The 1st International Conference on Net-Zero Built Environment*, Lecture Notes in Civil Engineering 237,
https://doi.org/10.1007/978-3-031-69626-8_12

hydroxide (NaOH), sodium silicate ($Na_2O \cdot 2SiO_2$) or sodium carbonate (Na_2CO_3) aqueous solution [4]. Depending on the geographic location, the availability of base material and the alkali activator may vary. Depending on the raw materials used, the cost and energy of manufacturing alkali-activated binder can save up to 45% and 75%, respectively, in comparison to OPC [5].

Properly designed AAM mixes can have superior engineering properties. Literature [6] suggests that when using 8% sodium silicate by the weight of slags, it has an amazing compressive strength of 80.9 MPa at 28 days. When similar amounts of Na_2CO_3, NaOH, and $Na_2O \cdot 2SiO_2$ are used as a sole activator for slag cement, the 28-day compressive strengths are 33.0 MPa, 55.9 MPa and 64.7 MPa, respectively [7]. Despite the outstanding mechanical performance of AAM, it still has many shortcomings, such as hazardous chemicals, flash setting and high drying shrinkage, which are mainly linked to the activator material [8]. At the current stage, the most widely used activator is NaOH, which is very environmentally and user-unfriendly due to its caustic nature. Besides, high temperature ranging 1100–1200 °C is required during the NaOH production process [9]. All of the mentioned problems impede the wide application of AAM. Recently, it has been found that using Na_2CO_3 as a sole activator can solve most of the short comings. However, longer setting times and lower mechanical strengths were seen due to its low alkalinity nature. These two problems limit Na_2CO_3-activated GGBS concrete as a main construction material [5].

Therefore, efforts have been made to enhance the early-day mechanical properties of Na_2Co_3-activated GGBS concrete. For example, the use of MgO and nucleation seeding have been proven effective, which could triple the 28-day compressive strength [8]. So far, no study has considered to utilise graphene as a seeding material in AAM, although graphene has demonstrated outstanding performances on traditional cementitious material due to the nano-filler and nucleation effect [9]. The large surface area of graphene will provide additional nucleation sites between the cement particles. Hence, a denser structure can be achieved.

This study is focusing on the role of graphene on setting time, early-day mechanical strength and working mechanism at the microstructural level of Na_2CO_3-activated GGBS concrete. The aim of this study is to find out the key effects of graphene in AAM systems by quantitative measurement of the changes. This study first tested the 3-day compressive strength performances. Afterwards, the scanning electron microscope (SEM) was employed to observe the microstructure changes. Finally, X-ray diffraction (XRD) was used to analyse the formation of hydration products.

2   Materials and Methodology

2.1   Materials

The graphene powder used in the experiments was PureGRAPH 50 supplied by First Graphene (UK) with an average particle diameter of 50 μm and tapped density of

0.395 g/cm^3. GGBS of this research (density of 2.4–3 g/cm^3) was obtained from Conserv (UK), in line with BS EN 15167-1:2006. The sodium carbonate (Na$_2$CO$_3$) was obtained from Fisher scientific (UK). The activator is in a powdered form and the purity is ≥99.5%. Ordinary river sand and coarse aggregates (up to 10 mm) were used in the concrete samples. Note that mechanical test samples were prepared with aggregates while samples for setting time tests and microstructure imaging were prepared in paste form without aggregates to avoid contamination. The composition of GGBS concrete used was Na$_2$CO$_3$/GGBS = 0.08 and water/binder ratio = 0.375.

2.2 Sample Preparation

Graphene solutions, by the process of ultrasonification, were prepared in five different concentrations (0%, 0.005%, 0.0075%, 0.01% and 0.02% by weight of GGBS), denoted as 0G, 0.005G, 0.0075G, 0.01G and 0.02G, respectively. The Na$_2$CO$_3$ alkali activator was added into the graphene solution (graphene powder and water), dispersed using Cole-Parmer Ultrasonic homogeniser for 1 h around 20 ± 3 °C (20 s dispersion with 3 s break at amplitude of 35%). The graphene/Na$_2$CO$_3$ solution was then added into the cement drum mixer first, followed by GGBS powder. Sand was added until the GGBS paste reached a smooth consistency (the operation took about 2 min). Coarse aggregates were then added and the mixing process lasted 3 min. Finally, the concrete mix was casted into moulds, placed on vibration table for 60 s and cured at 20 ± 3 °C exposed to air until the day of testing.

2.3 Characterisation

Compressive strength measurement was carried out using Controls C56Z00 compression frames, with specimen size of 50 × 50 × 50 mm. The tests were carried out in line with EN12390-3 [10] standard. Five repeat tests were performed for each mix. Setting time measurements, on the paste form of GGBS only, were obtained using Vicatronic E044N, according to EN 480-2 [11].

Scanning electron microscope (SEM) images were obtained by TESCAN MIRA3 FEG-SEM, the accelerating voltage used was 10 kV. Both secondary and backscattered modes were used to observe surface changes and the distribution of unreacted GGBS.

XRD was recorded on Bruker Autochanger using Cu K_α radiation with 40 kV and 30 mA. Calcium fluoride was used as an internal standard. The scanning rate is 0.04° 2Θ/step from 5° to 70° 2Θ. This was used to identify the composition and crystallinity of different phases.

3 Results

3.1 Setting Time

The initial and final setting times of Na_2CO_3– activated GGBS pastes are shown in Fig. 1. The number on the y-axis indicates the percentage of graphene by the weight of GGBS. The control sample (0 wt.%) has a relatively long setting period with an initial and final setting times of 2.93 and 6.67 h. This long setting time is related to the low alkalinity of the pore solutions. The setting time has an inverse relationship with the dosage of graphene. When 0.02 wt.% graphene is used, the initial and final settings are reduced by 1 and 0.7 h, respectively. The reduction on both initial and final setting times shows that graphene influences both the dissolution and hardening stages of GGBS formation. This finding is in line with other graphene-modified traditional cementitious materials. The shortened setting time is related to the nucleation effect of graphene.

3.2 Compressive Strengths

The compressive strengths of specimens with various amounts of graphene are presented in Fig. 2. Clearly, the presence of graphene has an overall positive effect on compressive strength. This finding is consistent with other studies conducted on traditional cementitious materials [9], due to graphene acting as a nucleation site for hydration products to grow on. However, the strength gained is not always proportional to the graphene added. In other words, there is an optimal level of graphene content. As shown in Fig. 2, when 0.005 wt.% graphene is added into the mix, the

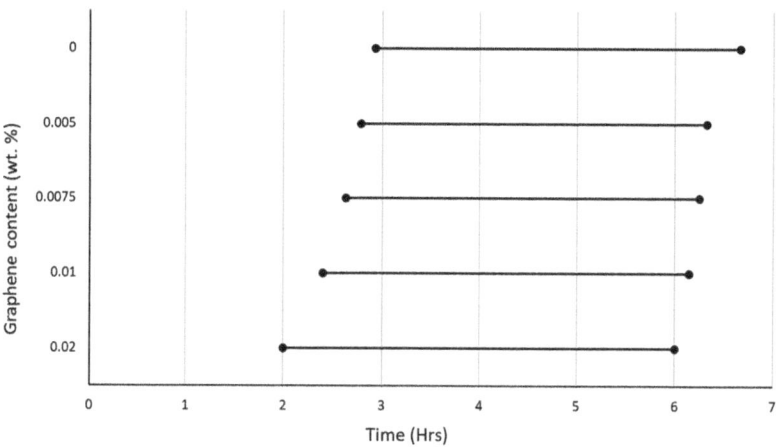

Fig. 1 Initial and final setting times of the samples

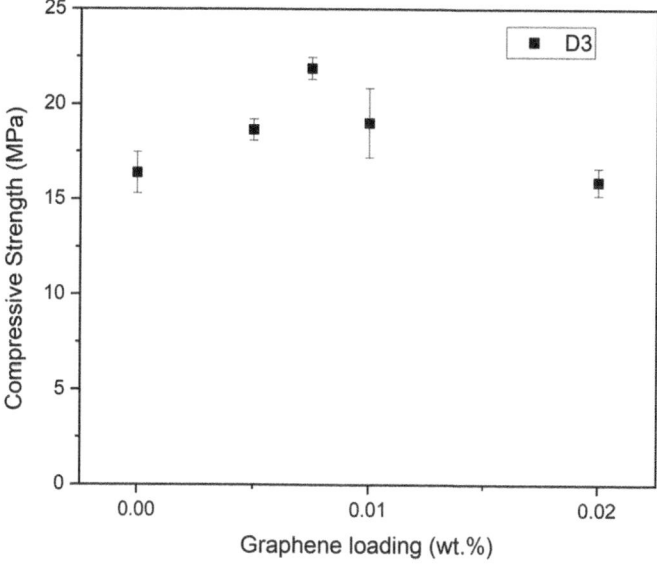

Fig. 2 3-day compressive strength with various graphene loadings

strength increased by 13.8%. When 0.0075 wt.% graphene is added, the increase of compressive strength is around 31%. After reaching the optimum levels, the increases in compressive strength of concrete are much less. Based on these sets of results, the optimum level of graphene content is about 0.0075 wt.% for 3-day compressive strength.

In fact, adding more graphene may lead to a reduction in concrete compressive strength compared with control sample. The observation is that there is an optimal level of graphene content. This suggests that in addition to the beneficial effect of graphene nucleation, incorporating graphene introduces a competing mechanism. Due to the nature of graphene, agglomeration of graphene will often occur at higher dosage, thus reducing the number of nucleation sites, leading to reduced increase in compressive strength. Furthermore, the smaller number of larger nucleation sites may introduce some weakness in concrete, this is attributed to the uneven hydration products formation through the matrix. This can lead to a reduction in concrete strength compared to the specimen without graphene [9, 12].

3.3 SEM Analysis

Figure 3 presents SEM images of undamaged concrete mixes after 3-day curing for different graphene contents. Obviously, the addition of graphene can cause significant microstructure changes. For the control sample, unreacted GGBS particles are often surrounded by large gaps or cracks, due to the 'slow reaction' nature of

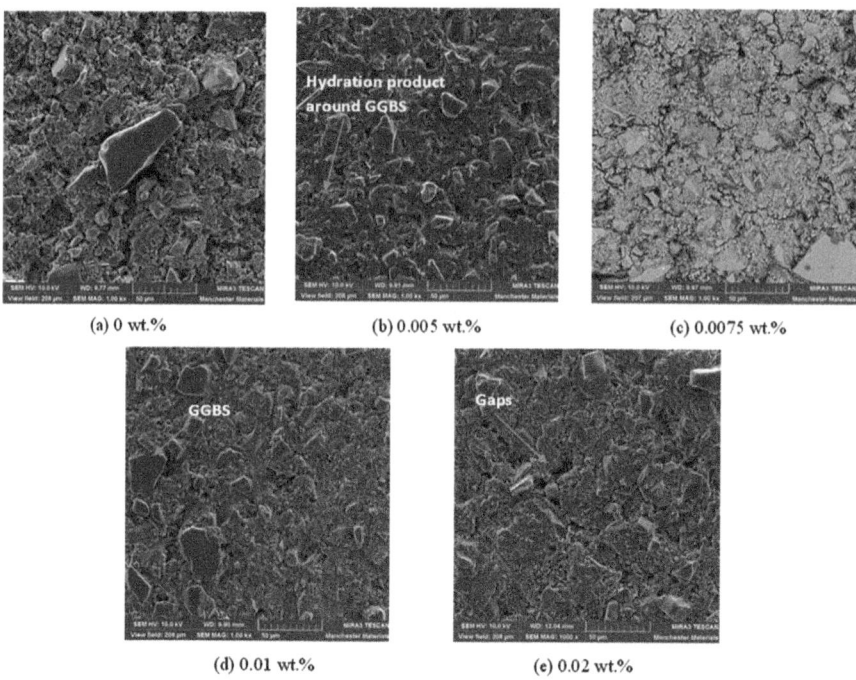

(a) 0 wt.%   (b) 0.005 wt.%   (c) 0.0075 wt.%

(d) 0.01 wt.%   (e) 0.02 wt.%

Fig. 3 SEM image for various 3-day samples (**a**) 0 wt.% (**b**) 0.005 wt.% (**c**) 0.0075 wt.% (**d**) 0.01 wt.% (**e**) 0.02 wt.%

Na_2CO_3 with GGBS as shown in Fig. 3a. When graphene content increases from 0 to 0.005 wt.%, the microstructure of the concrete changes from the control sample having a very porous microstructure with uneven distribution of unreacted GGBS particles (large grey shapes), to much more compact and dense microstructures. The same effect appears in other samples with graphene. With the addition of graphene, there are much less unreacted GGBS particles owing to more nucleation sites enabling more reactions. These can be seen by hydration products covering GGBS particles. Furthermore, the big pores observed in the sample without graphene are spilt into many small micro pores, which is attributed to the nano-filler effect of graphene.

However, when the graphene content increased from 0.01 to 0.02 wt.%, the gaps around the GGBS particles became larger and the overall structure became more porous. The reason might be that higher dosages of graphene lead to agglomeration, resulting in less nucleation site as mentioned in previous paragraph [13]. Moreover, the larger nucleation sites will likely introduce some weakness in concrete. Hence, identifying an optimum graphene level for GGBS concrete design is crucial. Based on the SEM observation, the optimum level remains around 0.005 wt.%, which is in line with the 3-day compressive test results.

An additional beneficial effect of graphene in cementitious material is that graphene makes any crack development in concrete become more tortuous (as shown in the red highlighted lines) [14]. For example, Fig. 4 compares cracks

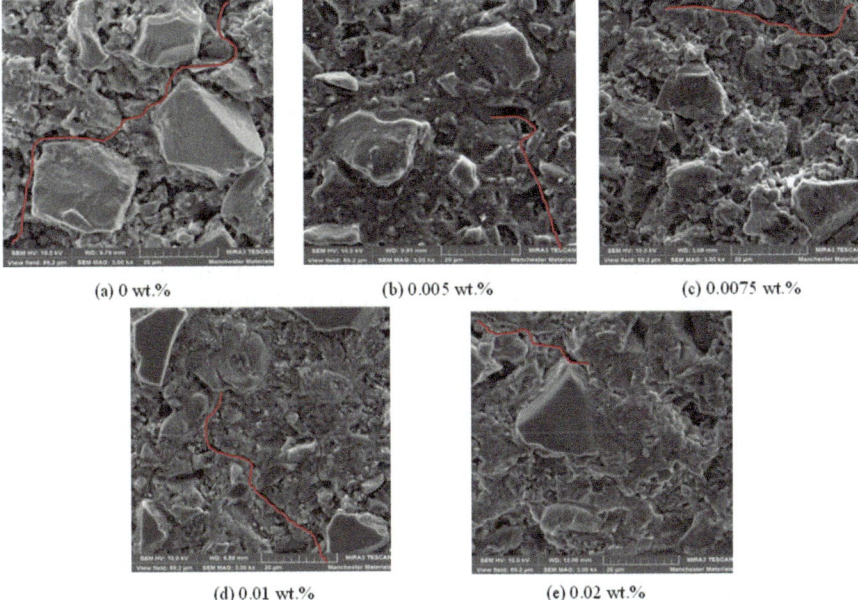

(a) 0 wt.%   (b) 0.005 wt.%   (c) 0.0075 wt.%

(d) 0.01 wt.%   (e) 0.02 wt.%

Fig. 4 Cracks propagation in the 3-day samples (**a**) 0 wt.% (**b**) 0.005 wt.% (**c**) 0.0075 wt.% (**d**) 0.01 wt.% (**e**) 0.02 wt.%

in concrete samples without graphene and with 0.005% graphene after mechanical testing at 3 days. The cracks in the control sample without graphene are almost straight, whereas the graphene-modified sample has several finer and discontinued cracks with occasional branching off. When graphene content increased to higher dosages, the propagation of cracks is stilled being control without competing effect of 'agglomeration'. This can be directly observed from the mechanical strength test. The mechanical strength has an inversely proportional relationship with the size of the cracks. As the graphene dosages increase, the cracks have been mitigated. Hence, the overall compactness has been increased, which leads to an increase in the overall strength. Once the optimum dosage has been reached, the effects of graphene on the cracks decreased. Therefore, the overall compactness is slightly weakened compared to the optimum level. Which explains the decreased strength gains.

3.4 XRD Analysis and Associated Chemical Reactions

The proposed 3-day chemical reaction is listed as follows. In Eq. (1), when the moment activator is added into the dry precursor. The CaO from slag reacts with water leads to an increase in the alkali environment. The reaction product (Ca^{2+}) is a key component in the reaction process, as Ca^{2+} initialises the following reactions (i.e., Eqs. (2) and (3)). In Eq. (2), the activator itself contains Na+ and CO$_3^{2-}$, the Ca^{2+} is from the reaction in Eq. (1), which is the reaction in between CaO from slag

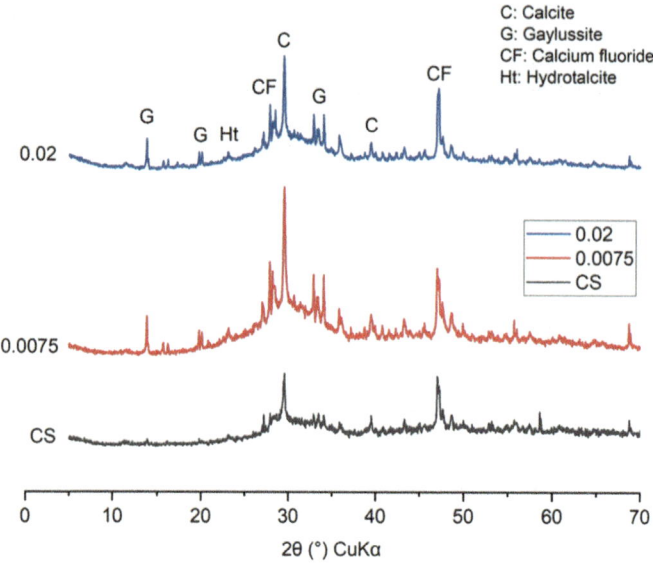

Fig. 5 XRD phases comparison of the 3-day samples

and the water added. Due to the weak bonding in between O and Ca. The O-Ca bond breakage can be carried out in a low alkalinity environment without any additional environment requirements (such as evaluated temperature) [8]. Therefore, the following Eq. (2) will take place to form gaylussite. Afterwards, gaylussite will quickly free ions to form calcite in Eq. (3).

$$CaO + H_2O \rightarrow Ca^{2+} + 2OH^- \tag{1}$$

$$5H_2O + 2Na^+ + Ca^{2+} + 2CO_3^{2-} \rightarrow Na_2Ca(CO_3)_2 \cdot 5H_2O \ (gaylussite) \tag{2}$$

$$Ca^{2+} + CO_3^{2-} \rightarrow CaCO_3 \tag{3}$$

Figure 5 shows the combination view of the XRD patterns of CS, 0.0075 wt.% and 0.02 wt.% graphene samples, it suggests that when graphene is added into the sample, the peaks are constantly sharper and higher than the control sample. Calcium fluoride was included as the internal standard in this test. The main hydration phases are gaylussite ($Na_2Ca(CO_3)_2 \cdot 5H_2O$; PDF #01-072-8410) and calcite ($CaCO_3$; PDF #01-071-3699), which identified via using Highscore. It suggests that the formation of calcite phase increase by 22% when 0.0075 wt.% graphene added. The optimum level has been reached at this point, the reason being that the formation of calcite reduced to 15% increase when 0.02 wt.% graphene added. This finding is closely linked to the mechanical strength test. Figure 2 shows the compressive strength test results, it indicates that the addition of 0.0075G leads to greater than 30%

compressive strength performance at 3 days curing. The 30% increase is related to the amount and crystallinity of phases formed in the sample. The reason being that gaylussite (Na$_2$Ca(CO$_3$)$_2$·5H$_2$O; PDF #01-072-8410) and calcite (CaCO3; PDF #01-071-3699) are the main strength providers during the 3-day reactions. Hence, the mechanical performance is directly linked to the amount of these two hydration phases formed.

With the addition of graphene added into the system, the peaks do not change, the only difference is the intensities of all phases are constantly higher than the plain sample. As shown in Fig. 5, the shapes of all peaks in each sample are the same. When 2θ at 13°, 20°, 30°, 33°, 35° and 40°. The identical peaks suggest that no new crystalline chemical phases were found, and graphene only takes account of the physical reactions. This finding is similar to the findings from others in traditional cementitious material. Which the addition of graphene only will change the shapes of the humps and peaks. In other words, graphene will only have an impact on the crystallinity of the phases. The sharper the peak is, the higher the crystallinity will be.

Overall, a higher degree of formation of all phases are observed in samples with addition of graphene in the system. The intensity increases in calcite and other crystalline phases is due to the nucleation effect of graphene. Due to the nano-size nature, graphene will fill in the macro pores and split the big pores into small micropores. Hence, a much-compacted structure was seen, leading to increased compressive strength. Also, promotes nucleation leading to increased reactions.

4   Conclusions

This paper has presented the results of a research study to investigate the effects of graphene on the 3-day compressive strength and setting time of Na$_2$CO$_3$-activated GGBS concrete. Various microstructure analyses were obtained to explain the fundamental mechanisms of the effects of graphene. Graphene can be used as an effective additive to increase early-day compressive strengths of Na2CO3-activated GGBS concrete, increasing the 3-day result by up to 33% at the optimal level of graphene dosage of 0.0075% in weight of GGBS. The improvement on setting time due to the incorporation of graphene is modest, achieving about 20-min reduction in setting time at the optimal graphene dosage (i.e. 0.005% by weight of GGBS). The effects of graphene on Na$_2$CO$_3$-activated GGBS concrete are controlled by two main mechanisms: nucleation effect (which promotes nucleation leading to increased reactions) and nano-filler effect (which fills in the macro pores and split the big pores into small micropores). Hence, a much-compacted structure was seen, increased compressive strength. Moreover, graphene will not change the chemical reactions during the hydration process, it will affect the amount of hydration products formed in the process. It is found that the inclusion of 0.0075% graphene will lead to 22% and 11% increase in the formation of calcite and gaylussite, respectively. This finding is in line with the compressive strength test results.

References

1. UN Environmental Programme: Building Sector Emissions Hit Record High, But Low-Carbon Pandemic Recovery Can Help Transform Sector – UN Report (2021)
2. Peng, J., Huang, L., Zhao, Y., Chen, P., Zeng, L., Zheng, W.: Modeling of carbon dioxide measurement on cement plants. Adv. Mater. Res. **610–613**, 2120–2128 (2013)
3. Ababneh, A., Matalkah, F., Aqel, R.: Synthesis of kaolin-based alkali-activated cement: carbon footprint, cost and energy assessment. J. Mater. Res. Technol. **9**(4), 8367–8378 (2020)
4. Provis, J., Deventer, J.: Geopolymers and other alkali-activated materials. In: Lea's Chemistry of Cement and Concrete (2019). https://doi.org/10.1016/B978-0-08-100773-0.00016-2
5. Duran Atiş, C., Bilim, C., Çelik, Ö., Karahan, O.: Influence of activator on the strength and drying shrinkage of alkali-activated slag mortar. Constr. Build. Mater. **23**(1), 548–555 (2009)
6. Adesina, A.: Alkali Activated Materials: Review of Current Problems and Possible Solutions. SynerCrete'18 Int. Conf. Interdiscip. Approaches Cem. Mater. Struct. Concr., no. October, 2018, [Online]. Available: https://spectrum.library.concordia.ca/984667/1/Mainfile -ALKALI ACTIVATED MATERIALS REVIEW OF CURRENT PROBLEMS AND POSSIBLE SOLUTIONS.pdf. (2018)
7. Provis, J.L., Bernal, S.A.: Geopolymers and related alkali-activated materials. Annu. Rev. Mater. Res. **44**, 299–327 (2014)
8. Dung, N., Hooper, T., Unluer, C.: Accelerating the reaction kinetics and improving the performance of Na2CO3-activated GGBS mixes. Cem. Concr. Res. **126**, 105927 (2019)
9. Dimov, D., Amit, I., Gorrie, O., Barnes, M., Townsend, N., Neves, A., Withers, F., Russo, S., Craciun, M.: Ultrahigh performance nanoengineered graphene-concrete composites for multifunctional applications. Adv. Funct. Mater. **28**(23), 1705183 (2018)
10. British Standards Institution: BS EN 12390-3: 2019. Testing Hardened Concrete. Compressive Strength of Test Specimens. British Standards Institution, London. https://shop.bsigroup.com/ProductDetail?pid=000000000030360097. (2019)
11. British Standards Institution: BS EN 480-2:2006 Admixtures for concrete, mortar and grout. Test methods – Determination of setting time. British Standards Institution, London. https://knowledge.bsigroup.com/products/admixtures-for-concrete-mortar-and-grout-test-methods-determination-of-setting-time?version=standard. (2019)
12. Wu, Y., Que, L., Cui, Z., Lambert, P.: Physical properties of concrete containing graphene oxide nanosheets. Materials. **12**(10), 1707 (2019)
13. Pan, Z., He, L., Qiu, L., Korayem, A., Li, G., Zhu, J., Collins, F., Li, D., Duan, W., Wang, M.: Mechanical properties and microstructure of a graphene oxide–cement composite. Cem. Concr. Compos. **58**, 140–147 (2015)
14. Wang, B., Shuang, D.: Effect of graphene nanoplatelets on the properties, pore structure and microstructure of cement composites. Mater. Express. **8**(5), 407–416 (2018)

Numerical Modeling of the pH Effect on the Calcium Carbonate Precipitation by *Sporosarcina pasteurii*

Shiva Khoshtinat [iD] and Claudia Marano [iD]

Contents

1 Introduction

With the development of bioinspired green solutions for sustainable construction, Microbially Induced Calcium carbonate Precipitation (MICP), which uses microbial metabolic processes for biocementation to improve the durability of concrete and masonry materials, has received a lot of attention in different sectors [1–5]. This approach has several advantages over traditional methods that use Portland cement: MICP's embodied energy is 43–95% less than that of conventional cement as it occurs at ambient temperature [6]; the carbon footprint from the MICP process is about 18–49.6% lower than traditional cement [6, 7]; due to the relatively low viscosity of the cementation solution and bacterial suspension, they can flow easily through the pores of the concrete, leading to better permeability [8]; and bacteria's small dimensions (<10 μm) compared to cement particles (<40 μm) lead to higher efficacy of MICP for holes as small as 6 mm [9]. Each of the mentioned benefits

S. Khoshtinat (✉) · C. Marano
Department of Chemistry, Materials and Chemical Engineering "Giulio Natta", Milan, Italy
e-mail: shiva.khoshtinat@polimi.it

© The Author(s) 2025
M. Kioumarsi, B. Shafei (eds.), *The 1st International Conference on Net-Zero Built Environment*, Lecture Notes in Civil Engineering 237,
https://doi.org/10.1007/978-3-031-69626-8_13

represents a significant advancement in the development of a low-carbon construction material and contributes to the achievement of net-zero construction.

In this context, the literature extensively discusses the mechanism of biocementation for a large spectrum of microorganisms, from bacteria and cyanobacteria to fungi and microalgae and even the enzymes isolated from them [1–5]. The latter is not discussed here for the sake of brevity. Bio-cementation is an intricate process that is influenced by various factors. The quantity and quality of $CaCO_3$ precipitation are influenced by various factors, including the type and conditions of microorganisms (such as live bacteria, bacterial fractions, or isolated enzymes), the method of application (e.g., mixing, spraying, injection, and percolation), and environmental conditions like pH, temperature, and nutrition media, as well as the availability of nucleation sites for bacterial growth [10–12].

While experimental characterization of the bio-cementation has advanced significantly at the laboratory scale, the feasibility of scaling up this technology to the building site and field scale has yet to be thoroughly investigated. This considerable limitation is primarily due to the complicated nature of the process, which occurs on several temporal and dimensional scales and involves interrelated biological, chemical, hydraulic, and mechanical processes [12]. In this respect, numerical models that predict ideal environmental conditions and application processes, while taking into account interrelated elements, might pave the way for optimization and scalability of the technology. However, compared to the experimental characterization of bio-cementation, computational modeling of this phenomenon has not made as much progress [13]. Only a few studies have been devoted to the computational modeling of bio-cementation, which mainly focused on the application process parameters rather than simulating the effects of the environmental conditions [14, 15]. Sharma et al. [14] developed a simplified model to simulate the kinetics of the bacterial ureolysis, the dynamic balance between the liquid–gas interface and the supersaturation of ions, as well as the kinetics of formation of calcite. Qin et al. [15] utilized PHREEQC software to simulate the development of biofilm, the introduction of nutrition media, and the formation of calcium carbonate under various injection flow rates at pore scales ranging from 18 to 400 μm. Undoubtedly, Paassen [16] has conducted one of the most comprehensive experimental and analytical investigations in the literature, providing physical and chemical models for various influential factors that impact bio-cementation.

The first step in optimizing bio-cementation involves establishing the optimal environmental conditions that promote the highest level of urease activity in microorganisms. In this regard, a computational model able to predict the outcome of bio-cementation influenced by environmental conditions can shed light on maximizing the effectiveness of bio-cementation exploitation. Among the environmental variables (e.g., pH, nutrition supply, temperature, etc.), pH has the most significant impact on microorganisms' metabolic processes, as the by-products of $CaCO_3$ precipitation induce a fluctuation in the pH of the environment, which in turn continuously affects the urease activity in microorganisms [11, 16, 17]. This results in a loop between the urease activity rate and pH variation.

This chapter focuses on the impact of environmental conditions (urea and calcium concentrations) on the urease activity of the most studied bacterium, *Sporosarcina pasteurii*, with a particular emphasis on the effect of pH level variation in the environment over time. First, the theoretical foundation of the bio-cementation phenomena, including environmental variables influencing urease activity and controlling biochemical reactions, is described. Next, the computational model structure in COMSOL Multiphysics® is systematically elucidated. Subsequently, the simulation's prediction for CaCO$_3$ concentration, pH fluctuation over time, and electrical conductivity under certain experimental conditions are compared with the experimental outcomes to assess the model's accuracy. Last but not least, a parametric analysis of the effect of the initial pH of the environment on the final pH is presented.

2 Theoretical Background

2.1 Influential Parameters

The rate of urea hydrolysis is mostly influenced by the enzyme type and population, the environmental conditions, storage, hydrolysis, and precipitation [10, 16, 18]. Among these parameters, environmental factors, including urea and calcium content in the system, pH, and temperature, substantially impact on urease activity [12, 16, 18]. The ideal environmental conditions for bio-cementation differ depending on the microorganism or isolated enzyme type, as each microorganism or enzyme has a distinct optimal pH and temperature for its enzymatic activity [12]. Furthermore, these ideal conditions vary once again depending on whether the microorganism or isolated enzyme is used [12]. As this study specifically examines the impact of pH and considers the hydrolysis of urea at a constant temperature (of 25 °C), the effect of temperature variation is not taken into account.

The effect of urea concentration on the hydrolysis rate of urea (R) is known to follow Michaelis–Menten kinetics (Eq. 1) [14–16], where R_{max}, K_{m_ur}, and C_{ur} are the maximum hydrolysis rate, urea concentration at which the reaction rate is half of R_{max}, and urea concentration, respectively.

$$R = R_{max} \times \frac{C_{ur}}{K_{m\_ur} + C_{ur}} \tag{1}$$

On the other hand, calcium concentration affects the urease activity following the exponential equation shown in Eq. (2), where the coefficient K_{iCa} is the concentration at which the urease activity is reduced to 37% of its original value [16, 18].

$$R = R_{max} \times e^{-C_{Ca}/K_{iCa}} \tag{2}$$

The impact of ambient pH, however, is deeply convoluted. As previously stated, each microorganism or enzyme has distinct ideal environmental conditions for its growth. Similarly, these organisms also experience particular environmental conditions that permanently hinder their enzymatic activity. This impediment, also known as denaturation, is the result of the enzyme undergoing structural modifications and losing its functionality. Therefore, urease activity declines within a specific pH range, both below and above the optimal pH range, generating a bell-shaped inhibition curve that can be represented using the expression (Eq. 3) proposed by Batstone [19], where pH_{LL} and pH_{UL} represent the minimum (lower limit) and maximum (upper limit) pH values, respectively, at which the hydrolysis rate is decreased by 50%.

$$R = R_{max} \times \frac{1 + 2 \times 10^{0.5(pH_{LL} - pH_{UL})}}{1 + 10^{(pH - pH_{UL})} + 10^{(pH_{LL} - pH)}} \qquad (3)$$

It is crucial to emphasize that the values of parameters R_{max}, K_{m_ur}, K_{iCa}, pH_{LL}, and pH_{UL} vary not only between different microorganisms but also between a microorganism and the enzyme extracted from it [16, 19]. In addition, the parameter K_{iCa} is also influenced by the type of calcium source (e.g., calcium chloride, calcium nitrate, or soluble calcium) utilized as a nutritional medium [16, 19]. The K_{m_ur}, pH_{LL}, and pH_{UL} values for the *Sporosarcina pasteurii* bacteria were determined to be 10 (mM), 5, and 9.7, respectively, at a constant temperature of 25 °C [16, 18]. The value of K_{iCa}, on the other hand, which depends on the calcium source, was considered 0.6 mol/l for calcium chloride [16, 18].

Figure 1 depicts the normalized hydrolysis rate (R/R_{max}) of *Sporosarcina pasteurii*, described in Eqs. (1)–(3), as functions of urea and calcium concentrations

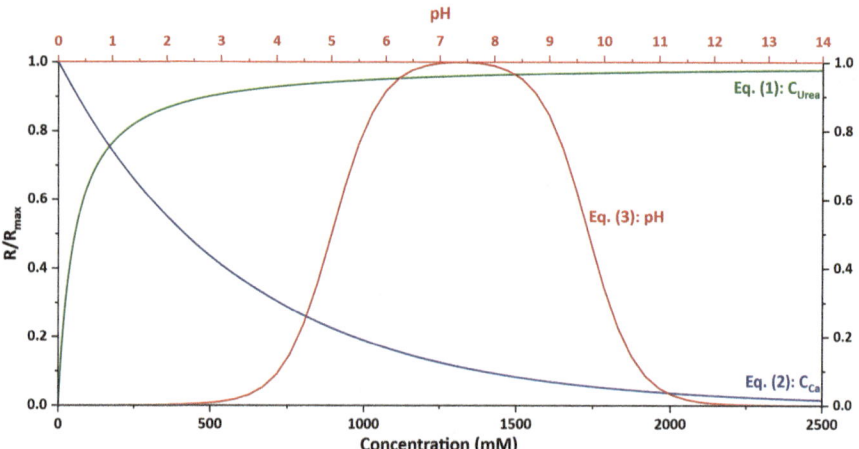

Fig. 1 The effect of urea concentration (green), calcium concentration from calcium chloride source (blue), and environmental pH (red) on *Sporosarcina pasteurii* hydrolysis rate

(from calcium chloride source) and the pH of the environment. As seen in this figure, the hydrolysis rate exhibits a substantial rise as the urea content in the solution increases up to about 1 M; beyond this concentration, the rate stabilizes at a nearly constant level (green line). In contrast, an increase in calcium content up to about 2 M results in a decrease in the rate of hydrolysis, while at higher concentration it does not cause any significant change (blue line). Regarding the impact of pH, the enzyme responsible for bio-cementation in *Sporosarcina pasteurii* denatures at pH levels below 3.5 and at pH above 12 [12]. Although it has been reported that the bacterium can remain alive up to a pH of 13.6, the enzyme essential for bio-cementation is denatured at a pH above 12; hence, no precipitation occurs under these conditions [12]. The bacteria exhibit optimal precipitation performance within a pH range of 6–8.75, where the hydrolysis rate, as influenced by the environmental pH, reaches 90% of its maximum rate (red line).

2.2 Chemical Reactions in Microbially Induced Calcium Carbonate Precipitation

Bio-cementation is an intricate process involving an array of biochemical reactions in which microorganisms, through the activity of the enzyme urease, catalyze the hydrolysis of urea, leading to the precipitation of calcium carbonate. First, urease catalyzes the decomposition of urea into ammonium (NH_4^+) and carbonate (CO_3^{2-}) through hydrolysis, which is an irreversible reaction (Eq. 4).

$$CO(NH_2)_2 + 2H_2O \rightarrow 2NH_4^+ + CO_3^{2-} \tag{4}$$

Then, the presence of carbonate and ammonium in the aqueous solution triggers a series of reversible acid–base reactions within the medium until the system reaches equilibrium (Eqs. (5)–(13)). The equilibrium of each of these subsequent acid–base reactions is expressed by the pK_a value, which represents the acid dissociation constant on a logarithmic scale [20].

In more detail, under neutral pH conditions, bicarbonate (HCO_3^-) is the primary species in the system rather than the carbonate ion (CO_3^{2-}). This leads to an increase in pH in order to maintain a balanced charge; consequently, ammonium undergoes dissociation into ammonia (NH_3) until a state of equilibrium is achieved between NH_4^+/NH_3 and HCO_3^-/CO_3^{2-} (Eqs. (5)–(10)).

$$H_2O \rightleftharpoons OH^- + H^+ \ (pK_w = 14) \tag{5}$$

$$CO_2 + H_2O \rightleftharpoons H_2CO_3 \ (pK_a = 6.1) \tag{6}$$

$$H_2CO_3 \rightleftharpoons HCO_3^- + H^+ \ (pK_a = 6.35) \tag{7}$$

$$HCO_3^- \rightleftharpoons CO_3^{2-} + H^+ \ (pK_a = 10.33) \tag{8}$$

$$CO_2 + OH^- \rightleftharpoons HCO_3^- \ (pK_a = 7.64) \tag{9}$$

$$NH_4^+ + OH^- \rightleftharpoons NH_3 + H_2O \ (pK_a = 9.25) \tag{10}$$

On the other hand, calcium ions (Ca^{2+}) engage in interactions with other compatible species in the system, including carbonate ions that yield calcium carbonate, which is soluble in water (Eqs. (11) and (13)) and eventually precipitates as calcium carbonate crystals (Eq. 14). The equilibrium between the solid and aqueous calcium carbonate and its ions in the system is presented by the solubility product (K_{sp}).

$$Ca^{2+} + H_2O \rightleftharpoons CaOH^+ + H^+ \ (pK_a = 12.78) \tag{11}$$

$$Ca^{2+} + CO_3^{2-} + H^+ \rightleftharpoons CaHCO_3^+ \ (pK_a = 3.22) \tag{12}$$

$$Ca^{2+} + CO_3^{2-} \rightleftharpoons CaCO_3 \ (pK_a = 11.44) \tag{13}$$

$$CaCO_{3(solid)} \rightleftharpoons CaCO_3 \ \left(K_{sp} = 3.36 \times 10^{-9}\right) \tag{14}$$

3   Structure of the Numerical Model

COMSOL Multiphysics® 6.1 was used for the numerical modeling of the system. A one-space dimension with a time-dependent analysis was considered for this model. Figure 2 shows a diagram outlining the arrangement of inputs and workflow of the numerical model in relation to the described theory. The model was defined using two physics: reaction engineering, which accounts for chemical reactions in the system, and electrophoretic transport, which considers the balance of electrical charges of the medium's species in acid–base equilibria. The "thermodynamics" of a "vapor–liquid system" with an ideal solution and gas was added as a subnode to

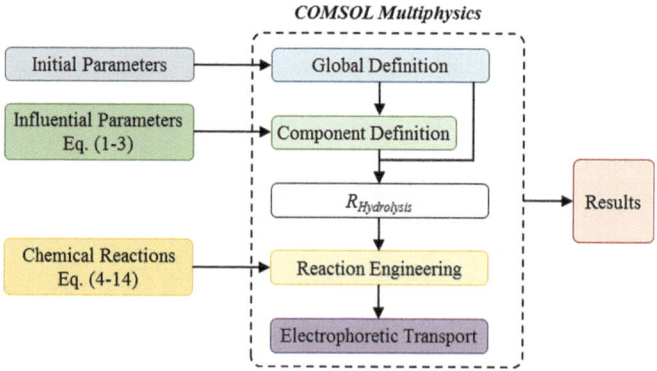

Fig. 2 Schematic representation of the inputs and workflow within *COMSOL Multiphysics*®

"Global Definition," recalling three species: ammonia (NH_3), carbon dioxide (CO_2), and water (H_2O).

The value for R_{max} was considered 40 (mM/h) at 25 °C [16, 18]. The effect of urea and calcium concentration, as well as the pH of the environment, on the activity of the enzyme urease was defined as three separate variables as described by Eqs. (1)–(3) under the "component definition" subnode. The urea concentration (C_{ur}) for Eq. (1) was defined as the variation of urea over reaction time. The pH value for Eq. (3) was defined as the logarithmic function of hydrogen ion concentration over time ($pH(t) = - \log [H^+(t)]$).

A new function ($R_{Hydrolysis}$) has been created that incorporates the rate dependency on urea and calcium concentrations, as well as pH (Eqs. (1)–(3)) with conditional statements to obtain a time-dependent reaction rate that changes as the parameters that affect urease activity change over time. Under the assumption that each function is independent of the other two parameters, $R_{Hydrolysis}$ alters the hydrolysis rate by integrating the values from all three functions in the following manner:

1. If the pH of the environment causes denaturation of the enzyme (pH <3.5 or pH >12), no precipitation occurs, and hence the hydrolysis rate is null.
2. If the pH falls within the permissible range of 3.5–12 and the normalized hydrolysis rate (R/R_{max}) from the effect of pH is above 0.5 and the one from the effect of calcium concentration is above 0.67, the hydrolysis rate is mainly affected by the urea concentration, which triggers the reactions.
3. If one of these conditions is not satisfied, an average value of normalized hydrolysis rate R/R_{max} from the three functions will be taken into consideration.
4. When the system reaches charge equilibrium for all species and the concentration of urea is significantly low to trigger ureolysis, the process comes to a conclusion.

A total of 11 chemical reactions were defined in the reaction engineering node: one irreversible reaction for urea hydrolysis (Eq. 4), nine reversible reactions for acid–base equilibrium in the system, considering the acid dissociation constant (pK_a) for each pertinent reaction (Eqs. (5)–(13)), and one reversible reaction for the solubility of calcium carbonate in an aqueous solution (Eq. 14).

The reaction rate for urea hydrolysis (Eq. 4) has been redefined as denoted function $R_{Hydrolysis}$. Initial concentrations of urea (C_{ur}), calcium (C_{Ca}), water (C_{H2O}), hydrogen ions (C_{H^+}), and hydroxide (C_{OH^-}) have been considered according to the experimental conditions presented in the literature in the "initial values" subnode.

The electrophoretic transport module utilized the predetermined solvent, water. The initial potential was set to zero. Urea was defined as a protein subnode with a neutral average charge (zero). CO_2, NH_3, and $CaCO_3$ were defined as uncharged species. NH_4^+ was specified as a weak acid with monoprotic donation and the pK_a from Eq. (10). H_2CO_3, on the other hand, was defined as a weak acid with polyprotic donation and two dissociation stages, with the first and second ionization constants reported in Eqs. (7) and (8), respectively. Both $CaOH^+$ and $CaHCO_3^+$ were defined as weak bases with monoprotic donation and relevant pK_a from Eqs. (11) and (12).

OH^-, H^+, Ca^{2+}, and CO_3^{2-} were defined as fully dissociated species with the relevant electrical charge. The concentration of all species in the electrophoretic node was determined by analyzing the variation in time of the relevant species resulting from reaction engineering. A time increment of 1 min was used to simulate the process over a period of 20 h. The data obtained from the simulation's global outputs comprised the evolution of species concentration in the system as determined by the reaction engineering module and the electrical conductivity derived from the electrophoretic module over time.

4   Results and Discussion

The first step of validating the numerical methodology outlined in this work involves comparing the simulation results with existing experimental data found in the literature. Figure 3 illustrates the comparison between the simulated concentrations of the main reactants (urea and calcium) and the resulting product (precipitated calcium carbonate) and by-product (ammonium) for a 0.5 mol/l concentration of urea and calcium and an initial pH of 6.67 obtained from calcium chloride and the experimental findings presented by Paassen [16].

The model accurately predicts both the final concentrations of reactants and products, as well as the evolution of these species' concentrations over time. Since the initial pH (6.67) falls within the ideal pH range of *Sporosarcina pasteurii's* urease during the early stage of the reaction, the rate of hydrolysis is maximized, resulting in a rapid occurrence of the reaction. Simultaneous consumption of equal quantities of urea and calcium results in the precipitation of an equivalent amount (1: 1) of calcium carbonate. The ratio of ammonium production to reactant consumption is twice that of calcium carbonate precipitation, as expected. The reactions continue

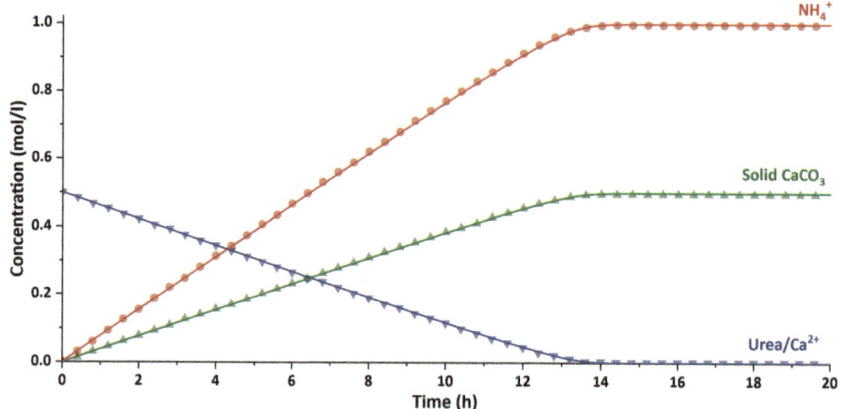

Fig. 3 Main species concentration evolution with time: comparison between this work numerical modeling (continuous lines) and data from Paassen [16] (symbols)

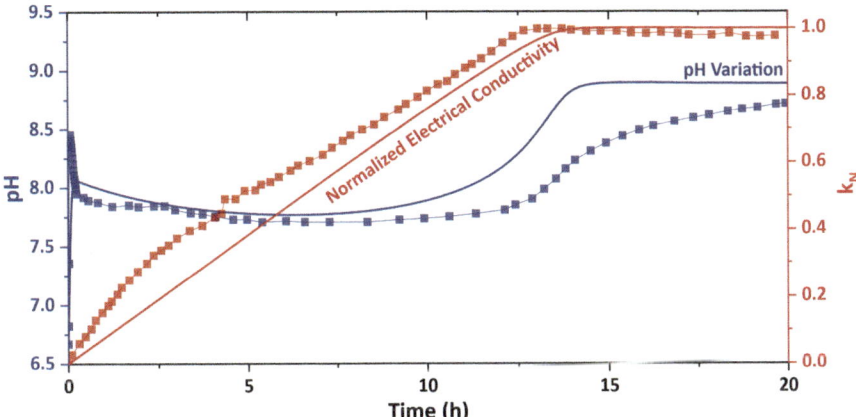

Fig. 4 pH (blue) and normalized electrical conductivity (red) variation during the precipitation process: a comparison between this work's numerical modeling (continuous lines) and data from Paassen [16] (symbols)

at a steady rate for approximately 13.5 h, during which time almost all of the urea and calcium in the system are consumed. After 14 h, the pace began to change and eventually leveled off.

Figure 4 presents the predicted pH variation and the normalized electrical conductivity ($k_N = \left(\frac{k - k_{min}}{k_{max} - k_{min}}\right)$) of the species in the system, compared to the experimental results reported by Paassen [16] for the same experimental setting described in Fig. 3. The numerical model and experimental data show substantial agreement for both the pH fluctuation and normalized electrical conductivity.

In relation to pH variation, during the initial phases of reactions (the first 5 min), the pH increases rapidly (from 6.67 to 8.75) upon hydrolysis of a small amount of urea, when the amount of carbonate produced exceeds the solubility product, leading to the precipitation of calcium carbonate. Then, the pH rapidly decreases, primarily due to the balance between unbound calcium and the calcium bicarbonate complex, which is the most prevalent dissolved inorganic carbon compound during the early stages of the process, when calcium levels are still elevated. When almost 50% of the calcium and urea initially present have been used (approximately 7 h (Fig. 3)) and the pH is still within the optimal range (R/R_{max} from Eq. (3) >0.9), the pH starts to steadily increase due to the acceleration of the hydrolysis rate, which is influenced by the calcium concentration (see Eq. (2) and Fig. 1). Eventually, after 14 h, when nearly all of the calcium and urea have been consumed and equilibrium between the reactants, products, and by-products has been reached, the reaction ends and the pH remains stable. The discrepancy between the computational model and experimental results for pH after 10 h of hydrolysis time is attributable to overlooked influencing factors, such as the decrease of urease activity caused by bacterial cell lysis [16].

As for the normalized electrical conductivity (curve and symbol in red), it is worth noting that both model and experimental data demonstrate an anticipated increase over time. Even though there is a disparity, particularly at the beginning of the process, the model is able to predict the overall trend adequately. Given that electrical conductivity in this scenario is affected by various factors, such as the presence of additional ions in the system, which can lead to a different pattern, as reported in the literature [14], any discussion on this subject necessitates a more comprehensive experimental analysis.

Once confirming the ability of the numerical model to well predict the evolution of calcium carbonate concentration and pH over time, the model was exploited to obtain information regarding the optimal environmental conditions for improving the calcium carbonate precipitation process. For instance, Fig. 5 depicts a parametric analysis of pH evolution based on the starting pH value for the same amounts of urea and calcium considered in Figs. 3 and 4.

When the initial environmental pH values are below 3.5 or above 12 (the threshold of denaturation of the enzyme responsible for bio-cementation), the hydrolysis rate is zero; hence, a horizontal line with no change in pH may be anticipated. At these pH levels, because the enzyme does not initiate the hydrolysis process, all of the subsequent chemical reactions ranging from Eq. (5) to Eq. (14), with the exception of Eq. (11), are not triggered. At these specified pH ranges, the only by-product that may be generated is $CaOH^+$ (Eq. 11), which is formed as a consequence of the charge balance between the different calcium (calcium chloride in this case) species in the solution. Therefore, in an acidic environment (pH <3.5), there is not any change in pH. Conversely, in a basic environment (pH >12), the system attempts to achieve charge balance between calcium species within the first 1 or 2 min, leading to an instant fall of about 0.9 in pH. This aspect may be confirmed by an additional parametric analysis of the concentration of by-product species, which is not included here for brevity.

For an initial pH ranging from 4 to 7, during the early phase of urea hydrolysis (first 10–15 min), there is a significant and rapid increase in pH, as already

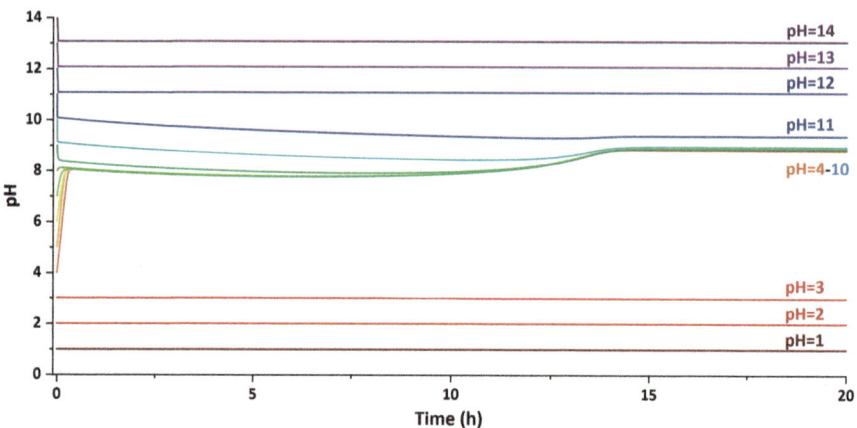

Fig. 5 Parametric analysis of pH evolution estimation in time as a function of the initial pH

commented above (Fig. 4). This phenomenon aids processes that begin at lower pH levels, which may not be ideal for bacterial enzymatic activity, to rapidly rise to higher levels where the normalized hydrolysis rate as a function of pH is greater than 0.9. This prediction aligns with the experimental study performed by Lai et al. [17] on the pH evolution over time influenced by initial environmental pH values ranging from 4 to 8.75, which is not reported here for brevity.

For initial pH levels ranging from 8.75 to 11, as the precipitation of $CaCO_3$ proceeds the pH increase causes a decrease of normalized hydrolysis rate (Fig. 1). The pH drop at the early stage is the result of superposition of decrease of pH by reactions' by-products and increase of pH due to urea hydrolysis. Thus, the overall variation of pH for this range of initial pH is little.

5   Conclusion

The present research provided a concise overview of the theoretical foundation of influencing factors in biochemical processes related to microbially induced calcium carbonate precipitation. A step-by-step procedure has been described for using COMSOL Multiphysics® to develop a computational model for prediction of calcium carbonate precipitation and the variations in environmental conditions during the bio-cementation process. A function has been introduced to define bacterial urease activity based on the real-time influence of three independent environmental parameters: urea and calcium concentrations and the initial pH of the environment. This function controls the urease hydrolysis rate during the process, depending on the current conditions. The comparison between experimental data available in the literature and the numerical model prediction presented in this study resulted in the following conclusions:

- The defined model accurately predicts the trend and amount of consumption of the primary reactants (urea and calcium), as well as the generation of the subsequent products (precipitated calcium carbonate) and by-products (ammonium). It can also forecast the quantity of all other minor by-products, which are not mentioned here for brevity.
- The model can estimate the system electrical conductivity evolution and how the environmental pH fluctuates over time as a result of the bio-cementation process.
- The parametric analysis results suggest the most favorable initial pH range for the precipitation of calcium carbonate and pH variation over time, using *Sporosarcina pasteurii* for bio-cementation, is between 4 and 10. This applies to the specific concentrations of urea and calcium derived from calcium chloride used in this investigation.

However, it is important to note that the model can predict the outcome and serve as a useful tool for optimizing ideal bio-cementation under varying concentrations of primary reactants, different calcium sources, or changes in the type of bacterium or enzyme isolated from a bacterium or plant, as long as the intrinsic parameters for the process mentioned in Eqs. (1)–(3) are known. For future developments, it would be beneficial to use a systematic coding program (e.g., MATLAB) to integrate the introduced hydrolysis rate function ($R_{\text{Hydrolysis}}$) with other influential parameters simultaneously. This will result in a more sophisticated and comprehensive computational model.

References

1. Xu, X., Guo, H., Li, M., Deng, X.: Bio-cementation improvement via $CaCO_3$ cementation pattern and crystal polymorph: a review. Constr. Build. Mater. **297**, 123478 (2021). https://doi.org/10.1016/j.conbuildmat.2021.123478
2. Omoregie, A.I., Palombo, E.A., Nissom, P.M.: Bioprecipitation of calcium carbonate mediated by ureolysis: a review. Environ. Eng. Res. **26**, 200379 (2020). https://doi.org/10.4491/eer.2020.379
3. Mutitu, K.D., Munyao, M.O., Wachira, M.J., Mwirichia, R., Thiong'o, K.J., Marangu, M.J.: Effects of biocementation on some properties of cement-based materials incorporating *Bacillus Species* bacteria – a review. J. Sustain. Cem.-Based Mater. **8**, 309–325 (2019). https://doi.org/10.1080/21650373.2019.1640141
4. Castro-Alonso, M.J., Montañez-Hernandez, L.E., Sanchez-Muñoz, M.A., Macias Franco, M. R., Narayanasamy, R., Balagurusamy, N.: Microbially induced calcium carbonate precipitation (MICP) and its potential in bioconcrete: microbiological and molecular concepts. Front. Mater. **6**, 126 (2019). https://doi.org/10.3389/fmats.2019.00126
5. Mujah, D., Shahin, M.A., Cheng, L.: State-of-the-art review of biocementation by microbially induced calcite precipitation (MICP) for soil stabilization. Geomicrobiol J. **34**, 524–537 (2017). https://doi.org/10.1080/01490451.2016.1225866
6. Porter, H., Mukherjee, A., Tuladhar, R., Dhami, N.K.: Life cycle assessment of biocement: an emerging sustainable solution? Sustain. For. **13**, 13878 (2021). https://doi.org/10.3390/su132413878
7. Bandyopadhyay, A., Saha, A., Ghosh, D., Dam, B., Samanta, A.K., Dutta, S.: Microbial repairing of concrete & its role in CO_2 sequestration: a critical review. Beni-Suef Univ. J. Basic Appl. Sci. **12**, 7 (2023). https://doi.org/10.1186/s43088-023-00344-1
8. Wu, C., Chu, J., Wu, S., Hong, Y.: 3D characterization of microbially induced carbonate precipitation in rock fracture and the resulted permeability reduction. Eng. Geol. **249**, 23–30 (2019). https://doi.org/10.1016/j.enggeo.2018.12.017
9. Wang, Y., Soga, K., DeJong, J.T., Kabla, A.J.: Microscale visualization of microbial-induced calcium carbonate precipitation processes. J. Geotech. Geoenviron. Eng. **145**, 04019045 (2019). https://doi.org/10.1061/(ASCE)GT.1943-5606.0002079
10. Erdmann, N., Strieth, D.: Influencing factors on ureolytic microbiologically induced calcium carbonate precipitation for biocementation. World J. Microbiol. Biotechnol. **39**, 61 (2023). https://doi.org/10.1007/s11274-022-03499-8
11. Hammes, F., Verstraete*, W.: Key roles of pH and calcium metabolism in microbial carbonate precipitation. Rev. Environ. Sci. Biotechnol. **1**, 3–7 (2002). https://doi.org/10.1023/A:1015135629155
12. Khoshtinat, S.: Advancements in exploiting *Sporosarcina pasteurii* as sustainable construction material: a review. Sustain. For. **15**, 13869 (2023). https://doi.org/10.3390/su151813869
13. Bagga, M., Hamley-Bennett, C., Alex, A., Freeman, B.L., Justo-Reinoso, I., Mihai, I.C., Gebhard, S., Paine, K., Jefferson, A.D., Masoero, E., Ofiteru, I.D.: Advancements in bacteria based self-healing concrete and the promise of modelling. Constr. Build. Mater. **358**, 129412 (2022). https://doi.org/10.1016/j.conbuildmat.2022.129412

14. Sharma, M., Satyam, N., Tiwari, N., Sahu, S., Reddy, K.R.: Simplified biogeochemical numerical model to predict pore fluid chemistry and calcite precipitation during biocementation of soil. Arab. J. Geosci. **14**, 807 (2021). https://doi.org/10.1007/s12517-021-07151-x

15. Qin, C.-Z., Hassanizadeh, S.M., Ebigbo, A.: Pore-scale network modeling of microbially induced calcium carbonate precipitation: insight into scale dependence of biogeochemical reaction rates. Water Resour. Res. **52**, 8794–8810 (2016). https://doi.org/10.1002/2016WR019128

16. Van Paassen, L.A.: Biogrout, ground improvement by microbial induced carbonate precipitation (2009)

17. Lai, H.-J., Cui, M.-J., Chu, J.: Effect of pH on soil improvement using one-phase-low-pH MICP or EICP biocementation method. Acta Geotech. **18**, 3259 (2022). https://doi.org/10.1007/s11440-022-01759-3

18. Whiffin, V.S.: Microbial CaCO$_3$ precipitation for the production of biocement – Murdoch University. https://researchportal.murdoch.edu.au/esploro/outputs/doctoral/Microbial CaCO3-precipitation-for-the-production/991005540291407891 (2004)

19. IWA Task Group for Mathematical Modelling of Anaerobic Digestion Processes: Anaerobic Digestion Model No. 1 (ADM1). IWA Publishing. https://doi.org/10.2166/9781780403052 (2005)

20. Equilibrium Constants-Chemistry LibreTexts: https://chem.libretexts.org/Ancillary_Materials/Reference/Reference_Tables/Equilibrium_Constants. Last accessed 10 Feb 2024

Exploring the Potential Utilization of Silicon Manganese Slag as a Supplementary Cementitious Material for Cement Replacement in Developing Low-Carbon Composite Binders

Dileepa Hettiarachchi, S. M. Samindi M. K. Samarakoon, Kjell Tore Fosså, Kidane F. Gebremariam, and Mahmoud Khalifeh

Contents

1 Introduction

The world is becoming more vulnerable to climate change and its associated impacts than ever before, and many attempts are being made toward mitigation initiatives. CO_2 is one of the key factors highlighted in here. Actions are being taken to achieve net-zero CO_2 emissions globally by 2050 to avoid the worst impacts of climate changes [1]. Moreover, the majority of CO_2 emissions associated with the

D. Hettiarachchi (✉) · S. M. S. M. K. Samarakoon · K. T. Fosså
Department of Structural and Mechanical Engineering and Material science,
University of Stavanger, Stavanger, Norway
e-mail: dileepa.c.hettiarachchi@uis.no; samindi.samarakoon@uis.no; kjell.t.fossa@uis.no

K. F. Gebremariam
Museum of Archaeology, University of Stavanger, Stavanger, Norway
e-mail: kidane.f.gebremariam@uis.no

M. Khalifeh
Department of Energy and Petroleum Engineering, University of Stavanger, Stavanger, Norway
e-mail: mahmoud.khalifeh@uis.no

© The Author(s) 2025
M. Kioumarsi, B. Shafei (eds.), *The 1st International Conference on Net-Zero Built Environment*, Lecture Notes in Civil Engineering 237,
https://doi.org/10.1007/978-3-031-69626-8_14

construction industry result from the combination effect of cement production and transportation that involves concrete. This eventually ascribes up to approximately 6–8% of greenhouse gas emissions globally per annum [2], which directly contributes to the effects related to global warming and climate changes [3]. The manufacturing of low-carbon concrete holds significant potential for reducing CO_2 emissions associated with the construction industry. This involves various options such as utilizing alternative SCMs, optimizing mix designs, employing different types of carbon capture techniques, developing new materials, conducting life cycle assessments, and improving energy efficiency. In this context, the exploration of sustainable alternatives to partially replace traditional cement-based materials, and the development of low-carbon construction materials are becoming attractive hot topics in the construction industry nowadays. Achieving sustainability within the construction industry is one of the most challenging goals as the demand for new infrastructures continues growing unabated worldwide.

SCMs are materials that contribute in enhancing the properties of hardened concrete through hydraulic or pozzolanic activity or both [4]. Pozzolans and slags are generally categorized as SCMs or mineral admixtures or additives. SCMs such as ground granulated blast furnace slag (GGBFS), various forms of fly ash (FA), silica fume (SF), metakaolin (MK), limestone (LS), and natural pozzolans including calcined shale and metakaolin are well-known SCMs in this field [3]. In addition to these, there are a number of other SCMs presently being investigated to discover enhanced performances. Many of their effects on concrete are well understood, and most recent research is being focused on areas like exploring new SCM materials, increasing the replacement amount of additives to improve properties, etc. [4].

From various studies conducted about the suitable percentage of replacement of cement, 10% is acceptable for all types of SCMs [4]. Generally, the replacement levels of cement by weight with SCMs like GGBFS, FA, SF, MK, and LS range from 40% to 80% [3], 15% to 35% (some instants up to 50%) [5], 3% to 10%, 5% to 10%, and 5% to 20%, respectively [3, 5]. Moreover, SCMs show a range of performances in different replacement percentages based on various factors like the water demand, fines and shape of particles, and specific density [5]. Most of the research works are mainly directed toward using a single type of SCM and forming a binary mixture. However, current studies are targeting the utilization of a combination of two or three SCMs instead. The main goal of such interest is to use high replacement percentages of cement and take advantage of the synergy to overcome the shortfalls of binary binders. There are some negative effects observed like high demand for water and on the other hand poor mixing with water, hence making the optimization of such binder formulations tedious. However, having a good comparative perspective to estimate the behavior of different SCMs when used in several combinations is an interesting area to study [6]. Moreover, recent research focuses on exploring new SCM materials and optimizing their performance, including investigating increased replacement percentages and using different combinations [7].

One such material of interest is SiMn slag, a by-product of SiMn alloy production [8]. Worldwide production of SiMn alloy is increasing relentlessly and 1.2–1.4 tons of SiMn slag generated during the production of 1.0 ton of alloy [9]. Norway shares the fourth place in the world in this industry [10]. Despite its abundance, SiMn slag remains underutilized, primarily relegated to landfilling. However, there is a growing potential to incorporate SiMn slag into concrete production as a high-volume cement replacement SCM [11]. Hence, it is more important to aim at exploring the feasibility of utilizing SiMn slag in various proportions to develop low-carbon binders with high cement replacement rates.

The objective of the study is to develop a low-carbon composite binder by incorporating SiMn slag in various combinations attempting to replace the maximum amount of cement possible. Previous researchers have identified the high feasibility of the usage of SiMn slag in granulated glassy or even in crystalline hard stone form within the field of SCMs and geopolymer concrete. Accordingly, there is a potential to modify the properties of SiMn slag resembling granulated blast furnace slag (GGBFS) in order to utilize it more effectively. GGBFS is used as a promising early strength gain SCM to verify the possibility of developing a ternary binder. Initially, to develop high-volume replacement binary binder, 30%, 50%, 70%, and up to 90% of cement was replaced with SiMn slag. For ternary binder, 50%, 70%, and up to 90% of cement were replaced with a combination of varied SiMn slag with 20% of GGBFS.

2 Experimental Procedure

2.1 Materials

The experiment utilized INDUSTRISEMENT, CEM I 52,5 R conforming to NS-EN 197-1:2011 standards as the primary cement, sourced from a cement producer in Norway. Various proportions of this cement were replaced with SiMn slag supplied from ERAMET-Norway and GGBFS obtained by Merox AB-Sweden as provided, which served as SCMs in the experiment.

Prior analyses, including particle size distribution and BET-surface area analysis, were conducted on the same materials to assess their characteristics. Accordingly, SiMn slag used in this experiment has a mean particle size of D50 as 24.5 μm in dry analysis and 23.0 μm in wet analysis. The related values of D50 for GGBFS are 13.7 μm in dry analysis and 14.4 μm in wet analysis, respectively. Potable tap water was used in the mixture, and an admixture based on modified acrylic polymers (Dynamon SX-N by MAPEI), was added to ensure consistency throughout the experiment.

The X-ray fluorescence (XRF) analysis data are presented in Table 1. SiO_2, CaO, and Al_2O_3 are the main components of the two types of slags comprising together approximately 85% of the total composition, similar to cement. They have the potential of improving the binding performances. SiMn slag and GGBFS have MgO content of 3.38% and 7.75%, respectively. While SiMn slag contains MnO at a higher percentage of 8.72% of total composition when compared with 0.8% of

Table 1 XRF results of cement and SCMs used in the experimental setup

Material (weight %)	SiO_2	Al_2O_3	Fe_2O_3	CaO	MgO	Na_2O	K_2O	MnO	TiO_2	SO_3	BaO	P_2O_5	SrO_2
Cement	23.59	5.05	3.84	59.14	0.0	0.0	2.01	0.10	0.25	5.51	0.06	0.35	0.09
SiMn slag	48.58	10.83	0.46	24.38	3.38	0.0	1.04	8.72	0.03	0.65	1.00	0.11	0.56
GGBFS	41.74	9.26	0.20	34.45	7.75	1.92	1.20	0.80	1.90	2.44	0.07	0.11	0.07

GGBFS. However, the composition of CaO in SiMn slag is 24.38%, which is lower than that of cement and GGBFS.

2.2 Mixture Proportions and Preparation of Specimens

To explore the feasibility of developing high-volume cement replacement composite binders using SiMn slag and GGBFS as SCMs, a comprehensive experimental setup was designed. This included a control series, consisting of four types of binary binders and three types of ternary binders. These binders are aimed to partially replace cement by weight basis. The replacement percentages of cement were systematically varied: 30%, 50%, 70%, and 90% by weight of the total binder content to assess the maximum achievable cement replacement. Both binary and ternary binder series were investigated. Within the ternary binders, GGBFS accounted for 20% of the total binder weight, with the percentages of SiMn slag being fine-tuned to match the partial replacement levels found in the binary mixes. Notably, no ternary binder was proposed with 30% replacement utilizing GGBFS, as the objective was to evaluate the high cement replacement potential of SiMn slag alongside the use of GGBFS to enhance the strength gain properties of the mixture.

In line with the findings from preliminary trials, a fixed water-to-binder ratio of 0.35 was maintained throughout the study for targeting a high level of compressive strength after the 28-day period. Additionally, a superplasticizer was added at 1% of the binder weight to ensure consistent workability. The superplasticizer dosage was determined based on observations of mini-slump flow, aiming to maintain a minimum flow table diameter of 200 mm through several trials. Control specimens (C), containing only cement without any replacements, were also prepared for comparative analysis. Specimens were labeled according to the type of binder (control, binary, or ternary) and the percentage of replacement (ranging from 30% to 90%). For example, the designation "B1" referred to binary set 1, replacing 30% of the cement, while "T2" indicated ternary set 2, replacing 50% of the cement with a combination of 30% SiMn slag and 20% GGBFS. The specific mixture proportions are shown in Table 2.

2.3 Casting, Curing and Testing of Specimens

The procedures for casting, curing, and testing were carefully carried out to ensure an accurate assessment of the specimens' properties and are detailed below.

Mixing Process All cement binders were mixed using a desktop Hobart mixer with a bowl volume of 4.73 liters, operating at a speed setting of 139 RPM. In order to obtain a homogeneous mixture, dry materials were first gently hand-mixed in the bowl, followed by a 2-min dry mixing process at low speed. Potable tap water was

Table 2 Mixture Quantities for 5 liters

Binder ID	% Replaced	Material requirement/(g)				
		Cement	SiMn slag	GGBFS	Water	Admixture
Control—(C)	0	2610	–	–	913.5	26.1
Binary-1 (B1)	30	1825	785	–		
Binary-2 (B2)	50	1305	1305	–		
Binary-3 (B3)	70	785	1825	–		
Binary-4 (B4)	90	260	2350	–		
Ternary −1 (T2)	50	1305	785	520		
Ternary −2 (T3)	70	785	1305	520		
Ternary −3 (T4)	90	260	1830	520		

Fig. 1 (**a**) Mini flow table test, (**b**) Curing of specimens, and (**c**) Compressive strength test

then added to the mixing vessel, and wet mixing continued for another 2-min to obtain a uniform paste. While the mixer was running, the admixture was slowly added and mixing continued for an additional 2-min [12].

Slump Test and Casting After mixing, consistency tests were performed using a mini-flow table. The mini-slump cone was placed on a flat, nonabsorbent mini-flow board. After filling the binder to the top of the cone, it was carefully lifted vertically to allow the paste to spread out and settle as shown in Fig. 1a. The average diameter of the slump was measured. Subsequently, the paste was poured into oiled three-gang cube molds measuring 50 × 50 × 50 mm in size in two lifts, with 32 hand compressions applied to each lift using a steel tamping rod to ensure complete filling without voids. The top surface of the paste was leveled, and the specimens were placed under ambient conditions while covered with polythene to prevent moisture loss.

Curing and Compressive Strength Testing After 24 h of initial curing under ambient room conditions, the molds were carefully de-molded, and the specimens were stored fully submerged in water to ensure 100% relative humidity [13] inside a curing chamber as shown in Fig. 1b at a constant temperature of 23 °C until subjected to testing. Subsequently, the specimens were removed and allowed to dry at room temperature for 1 h before testing. The average compressive strength was calculated from the results of three samples each at intervals of 3, 7, 14, and 28 days using an automatic compressive strength machine, (UCS test machine, Toni Technik, 300 kN) with a loading rate of 0.5 MPa/s, as shown in Fig. 1c. Here, the compressive strengths of all 3-day specimens were measured using a UCS test machine (Zwick-Roell, 50 kN). The NS-EN 12390-3:2019 is the test standard followed in here.

Microstructural Analysis Microstructural analysis was conducted on samples collected after the 28 days compressive strength test using scanning electron microscope (SEM) (Zeiss Supra 35VP field emission gun scanning electron microscope coupled with an EDS detector of EDAX Octane Elite 25) at University of Stavanger. Small representative samples from the selected test specimens were used for SEM analysis. The specimens selected were coated with sputtering using palladium to get rid of charging effects. Thereafter, samples were crushed and ground into powder for XRD analysis using a Bruker-AXS micro-diffractometer D8 ADVANCE equipped with a CuKα radiation source [8]. Measurement range used was $2\theta = 3$ degrees–70 degrees.

Isothermal Hydration Tests Monitoring of the progress of the hydration process after mixing the binder materials with water was carried out using an isothermal calorimeter [12] to identify the involvement and effect of SCMs in the hydration process compared to the control. The tests were performed at 23 °C to examine the heat generation rate and degree of reaction of SCMs, especially at early ages, using separately mixed small samples of paste (approximately 10 g). The temperature of 23 °C was selected to represent the curing conditions adopted during the test. The rate of heat evolution during the reaction was measured using an eight-test channel isothermal conduction calorimeter (TAM Air, thermometric AB).

3   Results and Discussion

3.1   Effects of SCMs on Compressive Strength

Figure 2 displays the average compressive strength of the binder samples at different ages, including the control specimens consisting of industrial cement without any replacement with SCMs. The compressive strength of all specimens exhibited an increasing trend with time. According to Fig. 2, binary specimens except the B4 category achieved beyond a strength of 60 MPa at 28 days. Out of them, B1 and B3 categories developed higher strengths of 9.4% and 13.1%, respectively, higher than the control. The B3 category achieved a strength of 9.1% lower than the control specimen's strength. The B4 category showed a comparatively lower strength development during the test period and ended up with 61.9% reduced strength compared to the control specimens.

Similarly, a trend of increasing strength with time is observed in ternary binders as shown in Fig. 2. The T2 category exhibits a slightly greater strength development

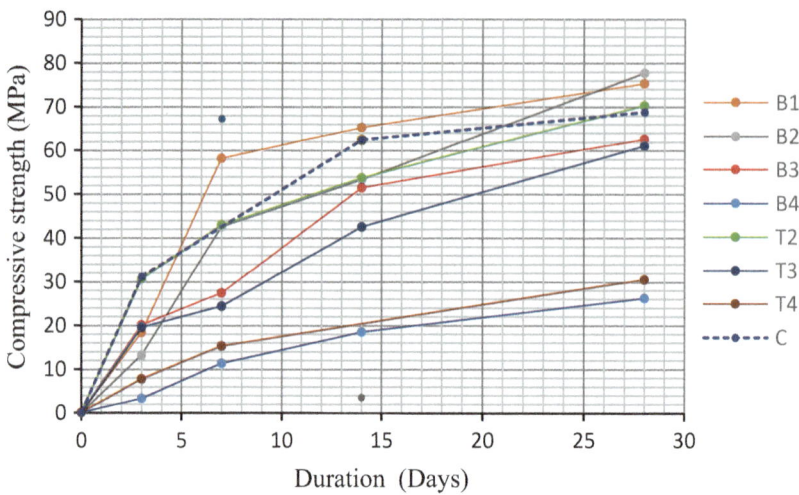

Fig. 2 Compressive strength test results of control, binary, and ternary specimens

when compared to the control, which is 2.2%. Except for T2, the rest of the two categories' strength developments fall at lower value ranges compared to the controls by 11.2% and 55.6%, respectively. However, T2 and T3 have developed strengths over 60 MPa, indicating the capability of developing high strength similar to the binary binders in the same experiment. These binders exhibited the potential for further modification to attain the development of even better high-strength composite binders.

Out of all the binder types investigated, T2 showed a similar strength development pattern at initial and later stages resembling that of the control specimens. This is in agreement with the previous research findings demonstrating the achievement of increased early and later-stage strengths of the binder paste by adding GGBFS in right proportion at about 20% [14]. This result implies a good indication of the capability of utilizing the properties of SCMs to develop new binders with enhanced strength and other desired properties effectively.

The relative compressive strength of all the proposed binders based on average compressive strength results together with respect to those of control specimens are presented in Fig. 3. It demonstrates the gradual development of strengths of the proposed binders from an early stage (3 days) up to later stage (28 days) in terms of the percentage deviations to the control specimens. The common tendency of these proposed binders is starting with low early strength (except T2), gradual development of strength less than the control and achieving a higher later strength (except B4 and T4). On the other hand, these results can be attributed to the low heat of hydration generation at the initial stage which improves durability aspects associated with the heat of hydration process in general practice, especially with high-strength concrete works.

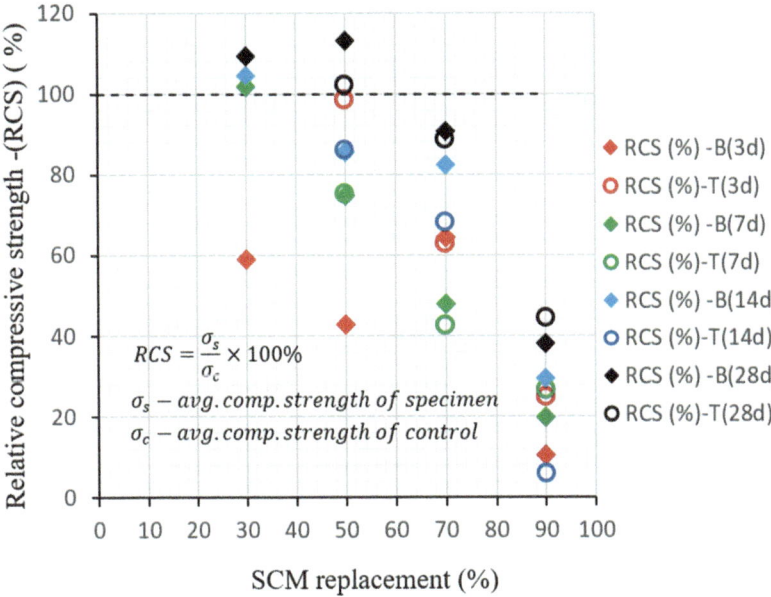

Fig. 3 Variation of relative compressive strength with replacing percentage of SCM

3.2   *Isothermal Conduction Calory Meter Studies*

The heat evolution curves have been obtained by the isothermal conduction calory meter of the samples at 23 °C for 72 h as shown in Fig. 4. The first peak at time starts from 0, representing where it starts the collection of data and is being assigned to the preinduction stage occurring due to wetting and initial dissolution of particles [15]. This occurred due to the procedure adopted, as the mixing was done outside the instrument and the paste was then introduced to the instrument soon after 2–3 min of mixing. Then the gelation process has taken place, and the second peaks were observed. The next peak(s) can be attributed to the precipitation of C-S-H gel with time. This result is correlated with the compressive strength development of the binders along with the timeline [16]. The resulting curves belonging to binary and ternary testes imply low heat evolution directly correlated to the strength developments. They are clearly matched with the strength development rates of binders compared with the control. Especially, the low peak in the case of B4 after approximately 36 h indicates the very low heat of hydration that took place in that specimen type and less progress of the hydration process particularly at high volume replacement. According to the calculated area under each curve, heat evolution during the reaction periods decreased in the following order: C, B1, T2, B2, B3, T3, B4, and T4. This is correlated to the strength characteristics of the specimens noted in the same order and explains well that the rate of hydration process has taken place with an increasing percentage of replacement with SCMs used.

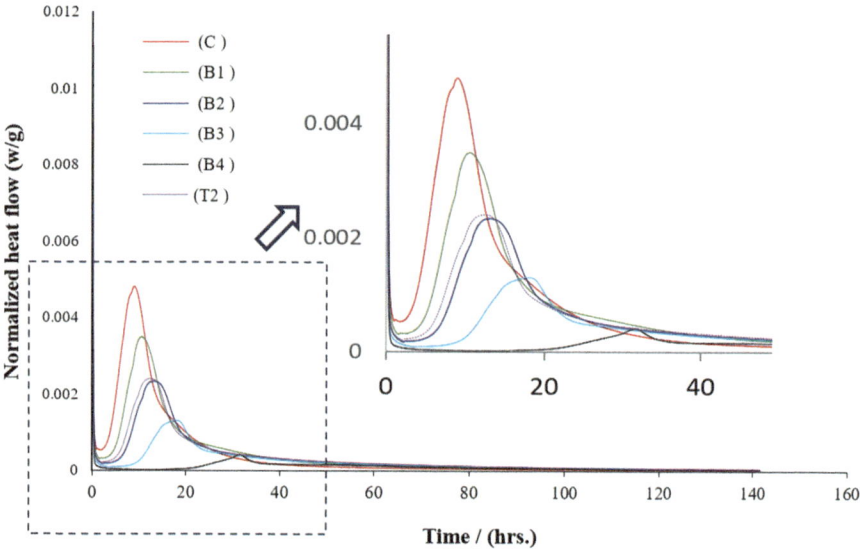

Fig. 4 Time evolution heat production—cement & other binder types

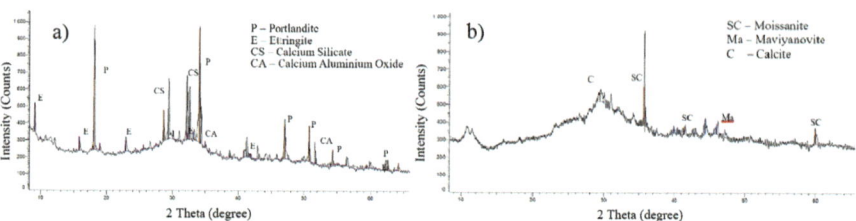

Fig. 5 XRD pattern—(**a**) control specimen (only cement) and (**b**) B4 specimen (90% SiMn slag)

3.3 X-Ray Diffraction Technique (XRD)

XRD was used for the purpose of the identification of crystalline and amorphous phases in raw SiMn slag and cementitious paste after hardening. XRD results of SiMn slag showed Mavlyanovite (Mn_5Si_3), Moissanite (SiC), and Quartz (SiO_2) as the main crystalline components along with amorphous-rich phases. Generally, the amorphous phase is more reactive than the crystalline phase in SCMs. On the other hand, the observations were used to correlate the strength development with time within the binder. When the XRD diffraction pattern of the control specimen at 28 days is compared with that of the B4 specimen at the same duration, the existence of hydration products is clearly identifiable in the control specimen as shown in Fig. 5a. Based on Fig. 5b, the presence of lesser amounts of hydration products and remains of unreacted particles are clearly visible. This evidently displays the difference in the extent of production of cement hydration products with the addition of

SCMs, which on the other hand is also associated with the strength development of the binder. Amorphous components could have contributed to the enhanced strength observed. Diffraction patterns from minor components and trace components cannot be easily determined using XRD due to its limited sensitivity.

3.4 Scanning Electron Microscopy (SEM) Observation

Since XRD results verified the presence of both amorphous and crystalline phases in SiMn slag, a good hydration process followed by strong reaction products associated with cement hydration products can be expected in binary specimens. The comparative examinations of the microstructures of the control specimens, binaries (B1, B2, B3, and B4), and ternary (T2) were conducted after 28 days. An overview of the backscatter electron (BSE) images at similar magnification for selected samples is displayed in Fig. 6. B1, B2, and T2 showed a densified microstructure similar to the control while B3 particularly B4, consists of significantly very loose and unreacted phases within their respective microstructures. B1, B2, and T2, in contrast, showed the formation of ettringite, portlandite, and C-S-H hydrated gel and consequently appear well bonded, supporting the relatively high compressive strengths measured. Partially reacted SiMn slag grains, apparently at an early stage of bond formation on their surfaces, are notable in the microstructure of B4 compared to the control, B1, and B2 at higher magnifications (Fig. 7). Furthermore, in B3 and B4 samples, formation of ettringite looks at a comparatively very low state. The microstructure of B4 is very poor when it comes to the extent of bonding between the different components and at higher magnification levels, it showed the existence of partially unreacted finer grains. In general, the microstructural observations are in good agreement with the 28-days compressive strength results which are increasing with time and decreasing with the addition of progressively higher proportions of SCMs as shown in Fig. 2.

The bond formation mechanism has been studied and compared, and the comparative densifications in the microstructures of C, B1, B2, and B4 as shown in

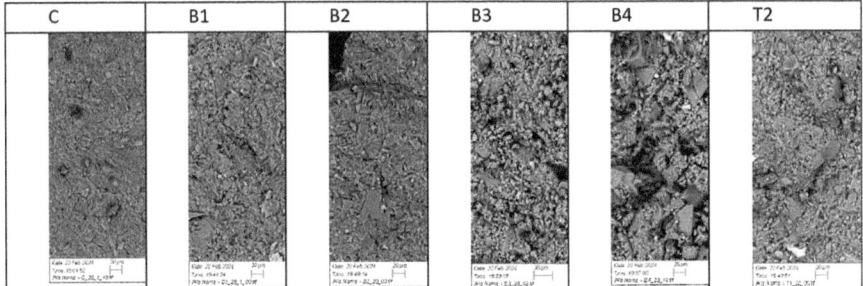

Fig. 6 SEM analysis of microstructures at 500× magnification level

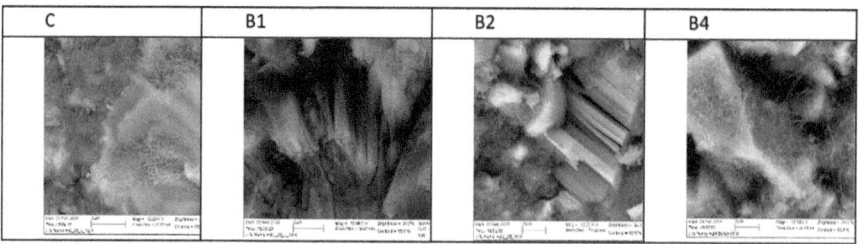

Fig. 7 SEM analysis of microstructures at 15,000× magnification level

Fig. 7. SEM results are also consistent with those acquired from isothermal calorimeter measurements. Mainly, in the B4 specimen, even after 28 days, the smaller grains have not fully entered into the reaction process while other binders have a well densified, good bonding microstructural formations as visible in Fig. 7. Portlandite was presented in every sample at 28 days, which implies further hydration process will take place along the timeline with the possibility of further enhancement in compressive strength with time.

4   Conclusions

The objective of the study is to develop a low-carbon composite binder by incorporating SiMn slag, as well as combinations of SiMn slag and GGBFS, to replace the maximum amount of cement feasible. Based on the experimental investigation, it was found that binary binders (B1 and B2) with SiMn slag replaced cement by 30% and 50%, respectively, and the ternary binder T2 comprised of 30% of SiMn slag and 20% GGBFS, replaced cement by 50%, achieved compressive strengths higher than the control after 28 days. Moreover, utilizing even higher replacement proportions like 70% binary and ternary binders resulted in achieving strengths exceeding 60 MPa after 28 days.

Low strength development has been observed for up to 7 days with the incorporation of SiMn slag, resulting in a reduction in the initial strength of the resultant binder. However, in later stages, this might be outweighted by the pozzolanic characteristics due to the presence of SiO_2 and Al_2O_3, which led to increases in strength. Significant amounts of nonhydrated and isolated particles are observed, and they increased with progressively greater replacement percentages of SCMs. This phenomenon may explain the observed strength loss with increasing SiMn slag replacement percentages.

Studying the characteristics of the binder phase in the initial stage of developing the low-carbon concrete is essential. Based on the results obtained in this experiment, it opens the prospect of exploring and investigating more about the next level of optimal mixing of cement with aggregates in the binder. This exploration aims to obtain higher strength and other desirable performance characteristics. Based on the strength development observations made, several strength classes can be identified along with different replacement percentage of cement with SiMn slag and GGBFS and their varied combinations in future research.

Acknowledgments The author would like to express gratitude to Leif Hunsbedt and Kjorkleiv Oyvind Karstein of ERAMET, Norway for providing the ground SiMn slag necessary to conduct this study.

References

1. I.E.A.: Net Zero by 2050, A Road Map for the Global Energy Sector. OECD Publishing (2021)
2. Zhang, Y., Zhang, J., Luo, W., Wang, J., Shi, J., Zhuang, H., Wang, Y.: Effect of compressive strength and chloride diffusion on life cycle CO2 assessment of concrete containing supplementary cementitious materials. J. Clean. Prod. **218**, 450–458 (2019)
3. Kumar, R., Goyal, S., Srivastava, A.: A comprehensive study on the influence of supplementary cementitious materials on physico-mechanical, microstructural and durability properties of low carbon cement composites. Powder Technol. **394**, 645–668 (2021)
4. Juenger, M.C.G., Siddique, R.: Recent advances in understanding the role of supplementary cementitious materials in concrete. Cem. Concr. Res. **78**, 71–80 (2015)
5. Sankar, B., Ramadoss, P.: Mechanical and durability properties of high strength concrete incorporating different combinations of supplementary cementitious materials a review. In: Proceedings of Fourth International Conference on Inventive Material Science Applications, pp. 543–557 (2022)
6. Ramzi, S., Moradi, M.J., Hajiloo, H.: The study of the effects of supplementary cementitious materials (SCMs) on concrete compressive strength at high temperatures using artificial neural network model. Buildings. **13**(5), 1337 (2023)
7. Juenger, M., Provis, J.L., Elsen, J., Matthes, W., Hooton, R.D., Duchesne, J., Courard, L., He, H., Michel, F., Snellings, R., Belie, N.D.: Supplementary cementitious materials for concrete characterization needs. MRS Proc. **1488**, imrc12-1488 (2012)
8. Kumar, S., García-Triñanes, P., Teixeira-Pinto, A., Bao, M.: Development of alkali activated cement from mechanically activated silico-manganese (SiMn) slag. Cem. Concr. Compos. **40**, 7–13 (2013)
9. Marsh, A.T.M., Yang, T., Adu-Amankwah, S., Bernal, S.A.: 11 – Utilization of metallurgical wastes as raw materials for manufacturing alkali-activated cements. In: de Brito, J., Thomas, C., Medina, C., Agrela, F. (eds.) Waste and Byproducts in Cement-Based Materials, pp. 335–383. Woodhead Publishing (2021)
10. Nath, S.K., Randhawa, N.S., Kumar, S.: A review on characteristics of silico-manganese slag and its utilization into construction materials. Resour. Conserv. Recycl. **176**, 105946 (2022)
11. Jiang, Y., Ling, T.-C., Shi, C., Pan, S.-Y.: Characteristics of steel slags and their use in cement and concrete—a review. Resour. Conserv. Recycl. **136**, 187–197 (2018)
12. Al-Duais, I.N.A., Ahmad, S., Al-Osta, M., Maslehuddin, M., Saleh, T., Al-Dulaijan, S.: Optimization of alkali-activated binders using natural minerals and industrial waste materials as precursor materials. J. Build. Eng. **69**, 106230 (2023)
13. Oh, T., Kim, M.-J., Banthia, N., Yoo, D.-Y.: Influence of curing conditions on mechanical and microstructural properties of ultra-high-performance strain-hardening cementitious composites with strain capacity up to 17.3%. Dev. Built. Environ. **14**, 100150 (2023)
14. Boháč, M., Palou, M., Novotný, R., Másilko, J., Všianský, D., Staněk, T.: Investigation on early hydration of ternary Portland cement-blast-furnace slag–metakaolin blends. Constr. Build. Mater. **64**, 333–341 (2014)

15. Yang, K.-H., Moon, G.D., Jeon, Y.-S.: Implementing ternary supplementary cementing binder for reduction of the heat of hydration of concrete. J. Clean. Prod. **112**, 845–852 (2016)
16. Sikora, P., Cendrowski, K., Abd Elrahman, M., Chung, S.-Y., Mijowska, E., Stephan, D.: The effects of seawater on the hydration, microstructure and strength development of Portland cement pastes incorporating colloidal silica. Appl. Nanosci. **10**, 2627–2638 (2019)

Understanding Carbon-Negative Potential of Hempcrete Using a Life Cycle Assessment Approach

Sejal Sanjay Shanbhag and Manish Kumar Dixit

Contents

1 Introduction

The construction industry is currently grappling with significant environmental challenges, primarily due to its substantial carbon footprint and reliance on non-renewable resources [1]. As one of the main sources of global carbon emissions (39%), the sector is under increasing pressure to adopt more sustainable practices [2]. This includes reducing greenhouse gas emissions using greener materials and energy-efficient construction techniques. Additionally, the industry's heavy consumption of non-renewable resources, such as gravel, sand, and various minerals, raises concerns about resource depletion and environmental degradation [3]. Efforts are being made to address these issues through innovations in material science, recycling, and the adoption of circular economy principles, but the pace of change is a critical factor in mitigating environmental impacts [4]. Bio-based materials offer a

S. S. Shanbhag · M. K. Dixit (✉)
Texas A&M University, College Station, TX, USA
e-mail: mdixit@tamu.edu

© The Author(s) 2025
M. Kioumarsi, B. Shafei (eds.), *The 1st International Conference on Net-Zero Built Environment*, Lecture Notes in Civil Engineering 237,
https://doi.org/10.1007/978-3-031-69626-8_15

promising solution to the environmental challenges previously mentioned. These materials derived from renewable resources are significant for their sustainability, contributing to a reduced carbon footprint and lower environmental impact compared to traditional construction materials [5]. Bio-based materials, such as bamboo, timber, and hempcrete, not only support the reduction of greenhouse gas emissions but also promote the conservation of non-renewable resources. Their renewable nature and biodegradability position them as a cornerstone in the transition toward more sustainable and eco-friendly construction practices, embodying the principles of circular economy and environmental stewardship [6].

Among the myriad of sustainable materials under investigation, hempcrete has become a prominent choice among the many sustainable materials being researched because of its distinctive properties and potential environmental benefits [7]. Hempcrete, a composite material formed from the mixture of hemp shiv (the woody core of hemp plant), lime-based binder, and water, offers a compelling case for sustainable construction through its low density, high thermal insulation, and inherent carbon sequestration capabilities [8]. Notably, hempcrete's lifecycle embodies a carbon-negative profile, primarily due to the hemp plant's rapid CO_2 absorption rate during its growth phase, a characteristic that distinguishes it from traditional construction materials [7].

This paper presents a rigorous analysis of hempcrete's Global Warming Potential (GWP) through a detailed Life Cycle Assessment (LCA) focused on cradle-to-gate stages. Our study quantitatively evaluates the GWP of hempcrete mixes incorporating varying percentages of hemp (5%, 10%, 20%, 30%, 40%, and 50% by weight), aiming to establish a potential correlation between hemp content in hempcrete and environmental impact. By integrating process-based LCA data for industrial hemp production in the United States with comprehensive reviews of literature for other key components such as hydrated lime, and admixtures, this research offers an in-depth understanding of hempcrete's carbon footprint.

Given the construction industry's growing interest in 3D printing and additive manufacturing for its precision, efficiency, and waste reduction capabilities [9], the potential for integrating hempcrete into this technology represents a significant stride toward sustainable construction methodologies [10, 11]. In conducting this analysis, the paper aims to contribute quantitatively substantiated insights into the sustainability discourse, providing a robust foundation for advocating hempcrete's use in reducing the construction industry's environmental impact. Through this endeavor, we seek not only to advance the scientific understanding of bio-based construction materials but also to influence policy and practice toward embracing more ecologically responsible building solutions.

2 Literature Review

2.1 Need for Bio-Based Construction Materials

The construction sector's environmental footprint is notably marked by its use of traditional construction materials, with Portland cement emerging as a primary concern due to its high carbon intensity [12]. Studies quantifying the environmental impact of construction practices have shed light on the substantial CO_2 emissions attributed to Portland cement production, emphasizing its role in exacerbating global warming [13]. Specifically, the cement industry is responsible for approximately 8% of worldwide CO_2 emissions [13], highlighting the critical need for a paradigm shift toward sustainable construction methodologies [14]. This necessitates exploring innovative materials and practices that can mitigate the industry's environmental impact, emphasizing the urgency of integrating sustainability into the core of construction activities. Bio-based construction materials represent a transformative shift in the building industry [5], offering a path to sustainability that diverges from traditional practices known for their environmental toll [6]. These materials are derived from biological sources such as plants, trees, and other organic matter and have gained prominence for their potential to substantially reduce the construction industry's carbon footprint [6]. The emergence of bio-based materials as sustainable alternatives is driven by their renewable nature, reduced energy consumption in production, and their ability to sequester carbon, thereby contributing to a lower overall carbon footprint of construction projects [5].

A critical aspect of bio-based materials is their potential to be carbon-negative. Carbon-negative materials absorb more carbon during their growth and production processes than is emitted throughout their lifecycle [15]. This characteristic is pivotal for mitigating climate change, as it contributes to the reduction of atmospheric CO_2 levels. Materials such as bamboo, hempcrete, and cross-laminated timber not only offer sustainable building solutions but also help combat climate change by sequestering carbon [16]. The role of bio-based materials in construction is not just an alternative approach but a necessary evolution toward sustainable development and climate resilience.

2.2 Hempcrete as a Sustainable Alternative

Hempcrete is a material that exemplifies the shift toward environmentally conscious construction practices. Composed of hemp fibers mixed with lime and water, hempcrete offers a lightweight, durable, and natural alternative to traditional construction materials [17]. Its distinct composition endows it with excellent thermal insulation properties, moisture regulation, and a high degree of breathability, making it an ideal material for a variety of construction applications [18]. Unlike conventional materials such as concrete, hempcrete's production and application processes are associated with lower carbon emissions, contributing to its recognition as a carbon-negative material due to its capacity to sequester carbon over its lifespan [19].

The application of hempcrete in construction has been the subject of various research studies, focusing on its thermal performance, mechanical properties, and compatibility with different construction techniques [7, 8, 17, 19]. These studies emphasize hempcrete's potential to significantly improve energy efficiency in buildings, reduce heating and cooling costs, and enhance indoor air quality.

Life Cycle Assessment (LCA) plays a pivotal role in evaluating the environmental impact of building materials, including bio-based options like hempcrete [20]. LCA methodology assesses materials from production to disposal, providing a thorough understanding of their environmental footprint [7]. Previous Life Cycle Assessment (LCA) studies on hempcrete have emphasized its environmental benefits, particularly highlighting its low Global Warming Potential (GWP) and its efficacy in carbon sequestration [7, 19, 20]. For instance, research conducted by Shareef et al. [21] indicated that hemp, a plant with a long history of use, can be effectively employed in load-bearing wall structures and offers a cost advantage over many traditional insulation materials. Research by Salvatore et al. [22] showed that walls made of hempcrete surpassed both hemp-lime blocks and conventional perforated brick blocks with external polystyrene insulation in performance. The study by Yadav et al. [10] emphasized the role of hempcrete in reducing waste, conserving natural resources, and minimizing energy consumption within the building and construction sector. These studies reinforce the position of hempcrete as a material that not only meets the structural and aesthetic needs of modern construction but also advances the industry toward a more sustainable and environmentally responsible future.

3   Methodology

This section outlines the approach applied to evaluate the global warming potential (GWP) of hempcrete mixes and discusses their suitability for 3D printing applications in construction. Our methodology integrates process-based hybrid life cycle assessment (LCA) data of industrial hemp production with literature review data for other ingredients. The objective is to quantify the environmental impact of hempcrete with varying hemp ratios by weight and assess its carbon negativity potential.

3.1 Hemp LCA Data Calculation Using Process-Based Hybrid Approach

To assess the GWP of hemp used in hempcrete, we employed a life cycle assessment approach. This involved collecting primary data on the cultivation, harvesting, and industrial processing of hemp in the United States. The data encompassed inputs such as energy consumption, transportation, and agricultural material usage, alongside outputs like yield and waste. Specifically, inputs such as seeds, fertilizer, lime, and fuel were collected from the US Department of Agriculture (USDA). Since this data was older, these inputs were adjusted for inflation and advancement in energy efficiency based on the reported Energy Use Intensity data for different sectors of the US economy. Transportation from hemp cultivation site to industrial hemp processing plant was calculated and added to the environmental impacts. Electricity used by the industrial hemp processing plant to produce hemp hurds and fibers was gathered from the plant and added to the energy calculations. Finally, the GWP of hemp was calculated in kilograms of carbon dioxide equivalent per kilogram of hemp ($kgCO_{2e}$/kg). Following the calculation, we undertook a comparative analysis against existing data from the literature to validate our findings. This comparison served not only to benchmark our results but also to ensure the reliability and accuracy of the GWP values determined for hemp. Utilizing the validated GWP value for hemp, we proceeded to calculate the overall GWP of various hempcrete mix designs. This critical step allowed us to quantify the environmental impact of hempcrete, considering the specific contributions of hemp alongside other constituents within the mix.

3.2 Data Collection for Hempcrete Components

We conducted a comprehensive literature review to gather existing GWP values for the other components of hempcrete: lime along with admixtures such as plasticizer. This step was crucial for understanding the individual and collective environmental impacts of the hempcrete mix components.

A range of GWP values for hydrated lime was extracted from literature sources. This range was essential to accommodate the variability in production processes and regional differences affecting the environmental impact of hydrated lime. The literature formed a basis for understanding the broad environmental implications of this material within the hempcrete mix. We also sourced data for plasticizers and superplasticizers that offered a standardized, verified summary of the environmental impacts, including their carbon footprint. By relying on these data for these components, we ensured the reliability of our environmental impact data.

3.3   Hempcrete Mix Design Analysis

Our analysis focused on hempcrete mix designs incorporating 5%, 10%, 20%, 30%, 40%, and 50% hemp by weight per cubic foot of hempcrete. The mix designs were based on the studies conducted by Danché et al. [23] and Tronet et al. [24]. The plasticizer in the mix is considered as 1% lime quantity by weight [25]. These ratios were chosen to explore the impact of varying hemp content on the net GWP of a cradle-to-gate LCA of hempcrete (Fig. 1).

Using the gathered data, we calculated the net GWP for each hempcrete mix design. This calculation included the summation of GWPs for all components, adjusted for their respective proportions in the mix as shown in the equation below.

$$\sum_{c=1}^{4} W_c \times GWP_c \tag{1}$$

In the Eq. (1), "W" denotes the ratio by weight per cubic feet of hempcrete for the four components "c" in the hempcrete mix namely, hemp, hydrated lime, and plasticizer. "GWP_c" is the corresponding GWP value of each of the components.

4   Results

In this section, we present a comprehensive analysis of the Global Warming Potential (GWP) values for hempcrete, focusing on its three main components: hemp, hydrated lime, and plasticizer. The GWP results are derived from a multifaceted approach, incorporating data from process-based calculations and existing literature. This analysis provides a solid foundation for understanding the environmental impact of hempcrete mix designs, with a particular emphasis on their potential contribution to or mitigation of global warming. Following the comparative

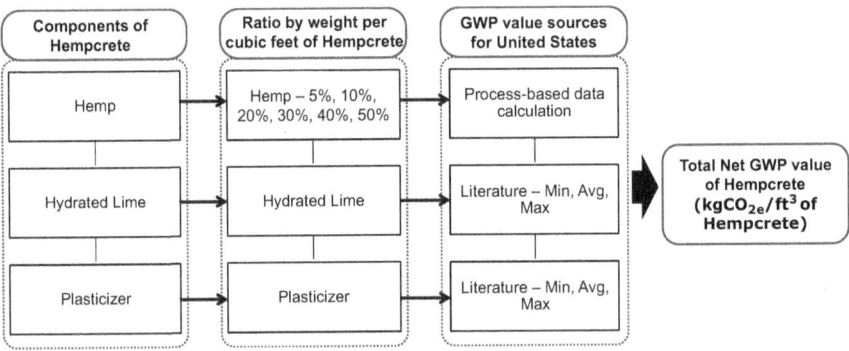

Fig. 1 Total net global warming potential calculation of hempcrete

analysis, we delve into the specific GWP impacts based on varying hempcrete mix designs, culminating with the total net GWP associated with these compositions.

4.1 Global Warming Potential (GWP) of Hempcrete Components

The data collected on the GWP of hempcrete components as shown in Table 1 reveal a significant variation in emissions, uptake, and net GWP values across hydrated lime, hemp, and plasticizer. Hydrated Lime exhibits a range of emissions from 0.823 [26] to 1.085 $kgCO_{2e}$/kg, with an average uptake that interestingly surpasses its min. emissions value based on min value of 0.571 [27] and 1.19 $kgCO_{2e}$/kg [28], indicating a potential for net carbon sequestration under certain conditions. The average GWP for hydrated lime is notably low at 0.252 $kgCO_{2e}$/kg, suggesting a lower environmental impact relative to its counterparts. Hemp, on the other hand, demonstrates a striking characteristic with its emissions and uptake values resulting in a negative average GWP of -1.72 $kgCO_{2e}$/kg, highlighting its strong potential for carbon negativity in hempcrete mixtures. Plasticizers GWP values ranged from 0.0052 to 1.88 $kgCO_{2e}$/kg [29], respectively, with no carbon uptake it positions them as more carbon-intensive components of hempcrete. The absence of carbon uptake for these materials suggests a straightforward contribution to GWP, unlike hydrated lime and hemp, which exhibit a complex interplay between emissions and carbon sequestration capabilities.

4.2 Environmental Benefits of Hempcrete Mix Design Alternatives

As the percentage of hemp increases in the mix, there is a notable trend toward a lower GWP, suggesting a significant environmental advantage as shown in Fig. 2. Specifically, for the 20% hemp mix, the GWP values show that while hydrated lime contributes positively to the GWP, hemp has a large negative impact, which leads to a potential for a net negative GWP in some instances, indicating carbon sequestration. As the hemp content decreases to 10% and 5%, the GWP values for hemp become less negative, yet still maintain a substantial offset against the positive contributions from plasticizer, and hydrated lime.

For instance, with a 20% hemp mix, the GWP for hemp component from a significant uptake of -7.35 $kgCO_{2e}$/ft^3 yields a net negative GWP impact, offsetting the emissions from the other components. This trend continues as the hemp content decreases, with each percentage decrease in hemp content resulting in a less negative GWP for hemp, and consequently, a smaller offset against the GWP contributions

Table 1 Global warming potential values for hemp, hydrated lime, and plasticizer

Material	Emissions ($kgCO_{2e}$/kg)			Uptake ($kgCO_{2e}$/kg)			Global warming potential ($kgCO_{2e}$/kg)		
	Min	Max	Avg	Min	Max	Avg	Min	Max	Avg
Hydrated lime	0.8230	1.0850	0.9540	0.5710	1.1900	0.8805	0.2520	−0.1050	0.0735
Hemp	0.12			1.84			−1.72		
Plasticizer	0.0052	1.8800	0.8485	–			0.0052	1.8800	0.8485

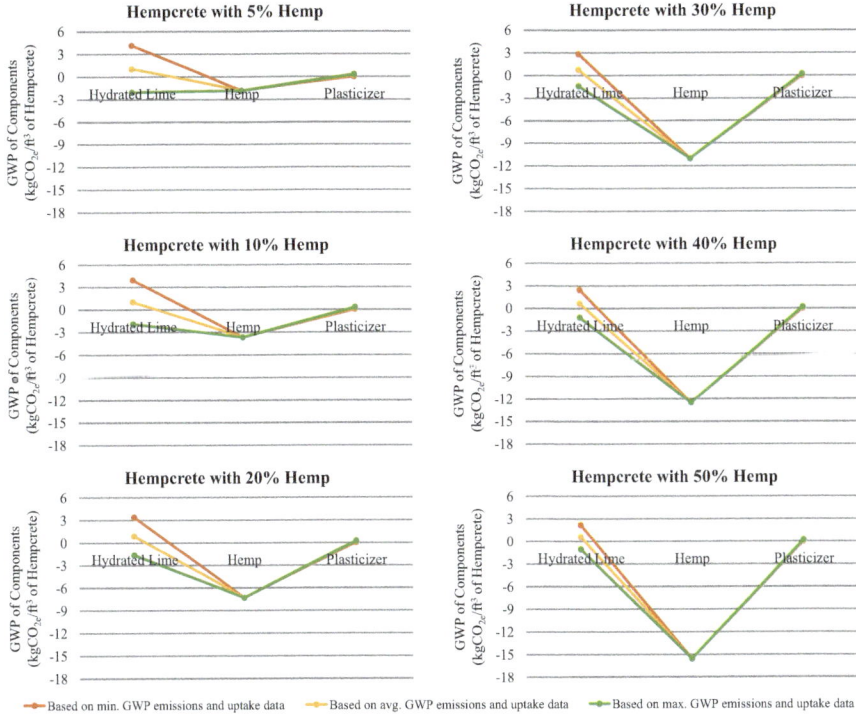

Fig. 2 Global warming potential (GWP) of hempcrete mix design

from the other components. Plasticizers show consistent positive GWP values across all mix designs, indicating a steady contribution to the GWP.

The data reveal that incorporating more hemp into the hempcrete mix design could be a critical factor in reducing the overall GWP of the material, highlighting the environmental benefits of hemp in construction materials. The GWP values also reflect that the other components, while contributing to GWP, do not significantly vary with the change in hemp content, suggesting that the role of hemp is pivotal in the mix designs for achieving lower GWP values. When applying these GWP results to practical scenarios, it is important to consider the specific application of the hempcrete, whether it be for innovative 3D printing construction techniques or for traditional building methods, as the mix design optimized for one may not be as effective or environmentally beneficial for the other.

Table 1 shows the net Global Warming Potential (kg CO_{2e}/ft^3 of hempcrete) for different hemp contents in the mix, based on minimum, average, and maximum GWP emissions and uptake data for lime as mentioned in Table 1 (Fig. 3).

It is evident that as the hemp content increases from 5% to 20%, there is a significant shift from positive to negative GWP values. Specifically, when the minimum lime GWP data are considered, the net GWP shows a descending trend from 2.3 kgCO_{2e}/ft^3 at 5% hemp to -3.98 kgCO_{2e}/ft^3 at 20% hemp, crossing the

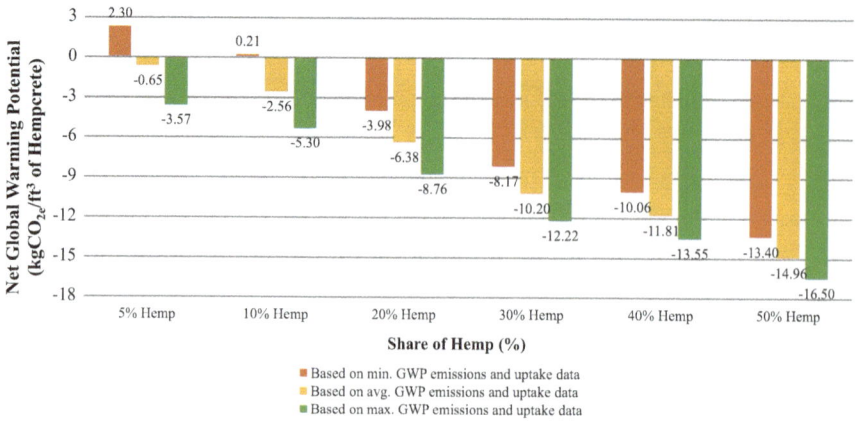

Fig. 3 Total net global warming potential of hempcrete mix designs

threshold to negative values, which implies carbon sequestration, at the highest hemp content. This trend accelerates to 50% hemp content at -13.4 kgCO$_{2e}$/ft^3. The trend becomes more pronounced when average lime GWP data are used, starting at -0.65 kgCO$_{2e}$/ft^3 and moving to 14.96 kgCO$_{2e}$/ft^3. Most strikingly, the maximum lime GWP data present negative net GWP across all hemp contents, with -3.57 kgCO$_{2e}$/ft^3 at 5% hemp, deepening to -16.5 kgCO$_{2e}$/ft^3 at 50% hemp, suggesting that the higher the hemp content, the greater the potential for the hempcrete mix to function as a carbon sink. These results emphasize the critical role of hemp's proportion in the mix design in reducing the overall GWP of hempcrete and enhancing its environmental performance.

5 Discussion

The outcomes presented in the previous section indicate the significant role of hemp in mitigating the GWP of hempcrete. Notably, the inverse relationship between hemp content and GWP indicates that hemp's inclusion is not merely beneficial but pivotal for the environmental performance of hempcrete. The data compellingly suggest that higher hemp proportions substantially enhance carbon sequestration capabilities, with potential for net negative GWP at higher concentrations, which is a promising prospect for sustainable construction practices. Furthermore, the consistency of GWP contributions from admixtures across all mixes indicates that the optimization of hempcrete for specific applications, whether in 3D printing or traditional construction, must prioritize the adjustment of hemp content to leverage its carbon-negative potential. These insights are crucial for advancing the development of environmentally beneficial building materials and for informing industry standards and sustainability benchmarks. Consequently, it is crucial to acknowledge the variability in data quality, which arises due to multiple factors. Utilizing SimaPro

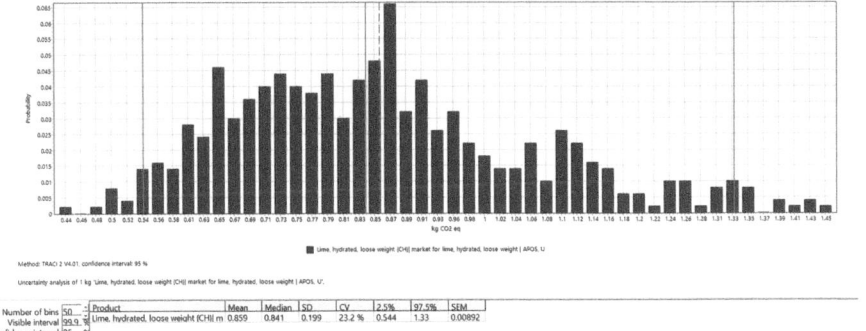

Fig. 4 Uncertainty analysis of hydrated lime GWP using Monte Carlo simulation

for uncertainty analysis, we evaluated the sensitivity of hydrated lime's GWP, with median and mean values at 0.841 and 0.859 $kgCO_{2e}$ respectively, indicating a balanced distribution close to the mean with a standard deviation of 0.199 as shown in Fig. 4. This reflects the complexity in pinpointing an exact GWP value, yet the histogram confirms a 97.5% confidence interval ranging from 0.544 to 1.33 $kgCO_{2e}$. This understanding is vital for stakeholders to refine processes and sustainability strategies.

The environmental impact of hempcrete is profoundly influenced by the characteristics of its primary components. The analysis conducted using SimaPro highlights the uncertainty associated with the GWP of hydrated lime, factors that are critical to understanding the overall environmental footprint of hempcrete.

For hydrated lime, a key ingredient in hempcrete, the variability in GWP data primarily arises from the energy consumption and emissions during its production process. The calcination of limestone, a process that emits a significant amount of CO_2, along with the energy source used (e.g., fossil fuels or renewable energy), greatly affects the GWP figs [30]. The uncertainty analysis from SimaPro highlights these variations, emphasizing the importance of sourcing and production methods in the environmental assessment of hempcrete.

Similarly, the GWP data for hemp present its own set of challenges. The cultivation of hemp, transportation to processing sites, and the processing itself contribute to the variability in the embodied energy and carbon data. Factors such as agricultural practices, the use of fertilizers, and the efficiency of transportation and processing technologies play a crucial role. The biogenic carbon uptake of hemp, a pivotal aspect of its environmental profile, introduces further uncertainty. The value of 1.84 ton CO_2 sequestered per ton of hemp, derived from literature, encompasses a range of conditions and methodologies, emphasizing the difficulty in applying a universal figure for carbon sequestration. Note that the compressive strength of hempcrete may not be comparable to conventional Portland cement concrete and its application in building assemblies may only be compared with non-load bearing components. Results indicate that increasing hemp quantity in a hempcrete mix may not only improve thermal properties but also reduce its GWP. This, however, may adversely impact its compressive strength as well as printability through an extrusion-based 3D printing process, which must be analyzed to fully understand hempcrete's industry potential.

6 Conclusions

This paper conducts a Life Cycle Assessment to assess the Global Warming Potential of hempcrete from production to pre-use stages, analyzing mixes with 5–50% hemp by weight. We merged process-based LCA data for US industrial hemp with literature for components like hydrated lime and plasticizer, offering insights into hempcrete's carbon footprint. A thorough Life Cycle Assessment of hempcrete's potential to cause global warming has shed light on the material's environmental effects, especially regarding its ability to sequester carbon. The study confirms that hempcrete's potential for carbon negativity is directly correlated with the percentage of hemp within the mix, highlighting the importance of mix design in achieving sustainability objectives. With at least 20% hemp content by weight, hempcrete transitions to a carbon-negative material provided the carbon absorption capacity of the lime binder is factored into the calculation. Our analysis has demonstrated that while hempcrete presents a promising pathway toward reducing the construction industry's carbon footprint, the variability in the quality of data for key components like hydrated lime and hemp must be taken into consideration. The uncertainty highlighted by the variability in GWP values underscores the need for careful selection of sources and production methods to ensure the environmental benefits of hempcrete are fully realized. Future research must focus on developing probability distributions of uncertainties associated with not just the GWP impacts of lime but also the carbon sequestration values of both hemp and lime and running a sensitivity analysis to conduct a more robust LCA of hempcrete. Future research must also consider comparing hempcrete's environmental performance based on its compressive strength as compared to other traditional wall options such as wood and steel framed walls. Future discussions should also explore the implications of this research on policy formulation, industry practices, and the goal of greenhouse gas emissions reduction in the construction sector.

References

1. Sizirici, B., Fseha, Y., Cho, C.S., Yildiz, I., Byon, Y.J.: A review of carbon footprint reduction in construction industry, from design to operation. Materials. **14**(20), 6094 (2021). https://doi.org/10.3390/ma14206094. PMID: 34683687; PMCID: PMC8540435
2. IEA: Global Status Report for Buildings and Construction 2019. IEA, Paris (2019). https://www.iea.org/reports/global-status-report-for-buildings-and-construction-2019. Licence: CC BY 4.0

3. Habert, G., Bouzidi, Y., Chen, C., Jullien, A.: Development of a depletion indicator for natural resources used in concrete. Resour. Conserv. Recycl. **54**(6), 364–376 (2010). ISSN 0921-3449, https://doi.org/10.1016/j.resconrec.2009.09.002

4. Thirunavukkarasu, A., Nithya, R., Sivashankar, R., Sathya, A.B.: Bio-based building materials for a green and sustainable environment. In: Bioprocess Engineering for a Green Environment, pp. 47–65. CRC Press (2018)

5. Boros, A., Tőzsér, D.: The emerging role of plant-based building materials in the construction industry—a bibliometric analysis. Resources. **12**(10), 124 (2023). https://doi.org/10.3390/resources12100124

6. Yadav, M., Agarwal, M.: Biobased building materials for sustainable future: an overview. Mater. Today Proc. **43**(5), 2895–2902 (2021). ISSN 2214-7853, https://doi.org/10.1016/j.matpr.2021.01.165

7. Rivas-Aybar, D., John, M., Biswas, W.: Environmental life cycle assessment of a novel hemp-based building material. Materials. **16**(22), 7208 (2023). https://doi.org/10.3390/ma16227208

8. Asghari, N., Memari, A.M.: State of the art review of attributes and mechanical properties of hempcrete. Biomass. **4**(1), 65–91 (2024). https://doi.org/10.3390/biomass4010004

9. Vijayan, D.S., Devarajan, P., Sivasuriyan, A., Stefańska, A., Koda, E., Jakimiuk, A., Vaverková, M.D., Winkler, J., Duarte, C.C., Corticos, N.D.: A state of review on instigating resources and technological sustainable approaches in green construction. Sustainability. **15**(8), 6751 (2023). https://doi.org/10.3390/su15086751

10. Yadav, M., Saini, A.: Opportunities & challenges of hempcrete as a building material for construction: an overview. Mater. Today Proc. **65**, 2021–2028 (2022)

11. Ahmed, A.T.M.F., Islam, M.Z., Mahmud, M.S., Sarker, M.E., Islam, M.R.: Hemp as a potential raw material toward a sustainable world: a review. Heliyon. **8**(1), e08753 (2022). https://doi.org/10.1016/j.heliyon.2022.e08753. PMID: 35146149; PMCID: PMC8819531

12. Adesina, A.: Recent advances in the concrete industry to reduce its carbon dioxide emissions. Environ. Chall. **1**, 100004 (2020). ISSN 2667-0100, https://doi.org/10.1016/j.envc.2020.100004

13. Andrew, R.M.: Global CO_2 emissions from cement production. Earth Syst. Sci. Data. **10**, 195–217 (2018). https://doi.org/10.5194/essd-10-195-2018

14. Meyer, C.: The greening of the concrete industry. Cem. Concr. Compos. **31**(8), 601–605 (2009)

15. Göswein, V., Arehart, J., Huy, P., Pomponi, F., Habert, G.: Barriers and opportunities of fast-growing biobased material use in buildings. Build. Cities. **3**, 745–755 (2022). https://doi.org/10.5334/bc.254

16. Arehart, J.H., Hart, J., Pomponi, F., D'Amico, B.: Carbon sequestration and storage in the built environment. Sustain. Prod. Consum. **27**, 1047–1063 (2021). ISSN 2352-5509, https://doi.org/10.1016/j.spc.2021.02.028

17. Barbhuiya, S., Das, B.B.: A comprehensive review on the use of hemp in concrete. Constr. Build. Mater. **341**, 127857 (2022). ISSN 0950-0618, https://doi.org/10.1016/j.conbuildmat.2022.127857

18. Zuo, J.: Hemp in construction (2021). https://repository.tudelft.nl/islandora/object/uuid:a10845b1-594f-4e1a-b7c1-1be659ccc88e/datastream/OBJ4/download

19. Maalouf, C., Ingrao, C., Scrucca, F., Moussa, T., Bourdot, A., Tricase, C., Presciutti, A., Asdrubali, F.: An energy and carbon footprint assessment upon the usage of hemp-lime concrete and recycled-PET façades for office facilities in France and Italy. J. Clean. Prod. **170**, 1640–1653 (2018). ISSN 0959-6526, https://doi.org/10.1016/j.jclepro.2016.10.111

20. Arrigoni, A., Pelosato, R., Melià, P., Ruggieri, G., Sabbadini, S., Dotelli, G.: Life cycle assessment of natural building materials: the role of carbonation, mixture components and transport in the environmental impacts of hempcrete blocks. J. Clean. Prod. **149**, 1051–1061 (2017). ISSN 0959-6526, https://doi.org/10.1016/j.jclepro.2017.02.161

21. Shareef, S.S., Latif Rauf, H.: Using hemp for walls as a sustainable building material. J. Stud. Sci. Eng. **2**(4), 17–24 (2022)

22. Di Capua, S.E., Paolotti, L., Moretti, E., Rocchi, L., Boggia, A.: Evaluation of the environmental sustainability of hemp as a building material, through life cycle assessment. Environ. Clim. Technol. **25**(1), 1215–1228 (2021)
23. Danché, V., Pierre, A., Ndiaye, K., Ngo, T.T.: Particle bed technique for hempcrete. In: RILEM International Conference on Concrete and Digital Fabrication, pp. 277–282. Springer International Publishing, Cham (2022)
24. Tronet, P., Lecompte, T., Picandet, V., Baley, C.: Study of lime hemp concrete (LHC)–mix design, casting process and mechanical behaviour. Cem. Concr. Compos. **67**, 60–72 (2016)
25. Nagrockiene, D., Pundienė, I., Kicaite, A.: The effect of cement type and plasticizer addition on concrete properties. Constr. Build. Mater. **45**, 324–331 (2013)
26. Randle, D.: Building with hemp and lime. Retrieved from https://www.votehemp.com/wp-content/uploads/2018/09/building_with_hemp_and_lime.pdf
27. Ip, K., Miller, A.: Life cycle greenhouse gas emissions of hemp–lime wall constructions in the UK. Resour. Conserv. Recycl. **69**, 1–9 (2012). ISSN 0921-3449, https://doi.org/10.1016/j.resconrec.2012.09.001. (https://www.sciencedirect.com/science/article/pii/S0921344912001620)
28. Pittau, F., Krause, F., Lumia, G., Habert, G.: Fast-growing bio-based materials as an opportunity for storing carbon in exterior walls. Build. Environ. **129**, 117–129 (2018). https://doi.org/10.1016/j.buildenv.2017.12.006
29. Saade, M.R.M., Passer, A., Mittermayr, F.: (Sprayed) concrete production in life cycle assessments: a systematic literature review. Int. J. Life Cycle Assess. **25**, 188–207 (2020)
30. Laveglia, A., Sambataro, L., Ukrainczyk, N., De Belie, N., Koenders, E.: Hydrated lime life-cycle assessment: current and future scenarios in four EU countries. J. Clean. Prod. **369**, 133224 (2022)

Use of the Fine Fraction from High-Quality Concrete Recycling as an Alternative Cement Substitute

Laurena De Brabandere ⓘ, Vadim Grigorjev ⓘ, Philip Van den Heede ⓘ, Hannah Nachtergaele, Krist Degezelle, and Nele De Belie ⓘ

Contents

1   Introduction

Due to the structural advantages and the relatively low cost, concrete is one of the most used building materials in the world. However, the many benefits are beginning to be overshadowed by the high environmental impact of the production process. It is already well established that the production of cement, one of the main components of concrete, contributes for up to 8% to the global anthropogenic CO_2 emissions [1]. In order to meet the new demand for sustainability and decrease carbon emissions, concrete production needs to be adapted. Recycling of concrete could play a key role in producing more sustainable concrete; however, there is still a lack of knowledge, guidelines and regulations on how fine recycled aggregates can

L. De Brabandere (✉) · V. Grigorjev · P. Van den Heede · H. Nachtergaele · N. De Belie
Magnel-Vandepitte Laboratory, Department of Structural Engineering and Building Materials,
Faculty of Engineering and Architecture, Ghent University, Ghent, Belgium
e-mail: laurena.debrabandere@ugent.be

K. Degezelle
Devagro, Waregem, Belgium

© The Author(s) 2025
M. Kioumarsi, B. Shafei (eds.), *The 1st International Conference on Net-Zero Built Environment*, Lecture Notes in Civil Engineering 237,
https://doi.org/10.1007/978-3-031-69626-8_16

be used to produce high-quality concrete. The quality of the fine recycled aggregates depends heavily on the type of waste concrete and on the applied production and storage method. For fine recycled aggregates, the amount of residual cement paste is higher in comparison to coarse recycled aggregates, which causes an even higher porosity in comparison to natural aggregates [2, 3]. This high porosity causes a large water absorption and therefore inferior performance in the fresh and hardened state [2]. Due to these downsides and a lack of knowledge, the fine fraction of recycled aggregates is now mostly landfilled or used for geotechnical and non-structural applications [2, 4]. An alternative for the use of the finer fraction is to grind it in a ball mill to obtain recycled fines (<63 μm), which can be used as a supplementary cementitious material (SCM) or inert filler. Previous research has shown that the reactivity of recycled fines is quite low; however, replacement rates of maximum 30% have been found to be feasible [5–7]. In this research, recycled concrete fines from the concrete recycling of Devagro (Waregem, Belgium) were investigated as a partial replacement for cement. First of all, the physical and chemical properties of the recycled fines are studied followed by an investigation of the reactivity and strength development using isothermal calorimetry and flexural and compressive strength tests. Finally, two concrete mixes are made: one reference mix and one mix with partial cement replacement by the fines and partial replacement of natural aggregates by recycled aggregates. The durability of these mixes is tested by compressive strength tests, shrinkage, porosity measurements and freeze–thaw tests.

2   Materials

In this research, the fine fraction of recycled concrete is studied as a partial cement replacement. These recycled fines are obtained from the waste of prefabricated concrete pavers, namely pavers with form imperfections and waste concrete from the production process. The binder used in this concrete mixture consists of 70% CEM I 52.5 R and 30% blast furnace slag. When recycled, the concrete is up to 1 year old. The production process of the recycled aggregates contains the following steps: (1) breaking down the concrete pavers into smaller pieces followed by a pre-screening where soil and organic matter are removed from the debris, (2) crushing of the material and washing of the granulates and (3) dividing the recycled aggregates in different fractions. To obtain the recycled concrete fines, the fraction with size 0/6 mm is then sieved until a fraction 0/2 mm and subsequently, this fraction is ground in a ball mill in order to obtain a desired maximal grain size of 63 μm.

To test the influence of the fines on durability, two concrete mixes were made: one reference mix and one mix where 10% of cement is replaced by recycled fines (0.10FC). The mix composition is shown in Table 1. In the mix 0.10FC, 27.8% of the sand and 100% of the coarse aggregates were replaced by recycled sand and recycled coarse aggregates, which have a water absorption rate of 7% and 5.4%, respectively. To increase the workability of the mixtures, two different PCE-based

Table 1 Mix compositions of concrete

Materials	Reference	0.10FC
Sea sand 0/2 [kg/m^3]	345	0
Sand 0/4 [kg/m^3]	515	0
River sand 0/7 [kg/m^3]	0	660
Crushing sand 0/6 [kg/m^3]	0	262
Limestone 4/20 [kg/m^3]	1030	0
Concrete aggregate 4/20 [kg/m^3]	0	802
CEM I 52.5 N [kg/m^3]	310	280
Recycled fines [kg/m^3]	0	30
Water [kg/m^3]	179	224
TechniFlow 92 (superplasticizer) [kg/m^3]	1.5	0
PowerFlow EVO503 (superplasticizer) [kg/m^3]	0	7
Total W/C	**0.546**	**0.540**

(polycarboxylates-ether) superplasticizers were used, TechniFlow 92 and PowerFlow EVO503. To keep the casting conditions as realistic as possible, the mixes were made in the concrete plant of Devagro and cast in a non-controlled environment. For each mix, 15 cubes (150 × 150 × 150 mm^3), 6 prisms (100 × 100 × 400 mm^3) and 3 cylinders (ø 100 mm, height 200 mm) were cast. The specimens were compacted using a vibrating table and covered with a plastic foil after casting to prevent moisture evaporation. All the specimens were demoulded after 24 h.

3 Methods

3.1 Characterization of the Recycled Concrete Fines

Particle Size Distribution and Particle Shape The particle size distribution of both the recycled fines and the CEM I 52.5 N was measured using a laser diffractometer with a dry dispersion unit.

The particle shape was quantified using the Occhio Flowcell FC200M-HR in combination with the Callisto 3D software.

Particle Density and Water Absorption The particle density and the water absorption were determined on the fraction 0/2 (before milling) according to the pycnometer method described in NBN EN 1097-6.

Specific Surface Area The specific surface area of the recycled fines was determined according to the standard NBN EN 196-6, using the air permeability method (Blaine method). First of all, the apparatus was calibrated using the reference material SN3c Portland Cement (CEM I 52.5 N). With this calibration, the apparatus constant K was determined. Finally, the specific surface area of the fines was determined.

XRD and XRF A phase analysis of the fines was performed by X-ray diffraction (XRD) analysis in combination with Rietveld refinement using the Profex XRD software [3]. As an internal standard, 10 wt% ZnO was used.

X-ray fluorescence (XRF) was used to obtain the elemental composition of the fines. In order to do this, three pellets were made by mixing 18 g of fines with approximately 3–4 g of the binding agent $C_{42}H_{83}ON$ from SpectroBlend 44 µm powder.

Isothermal Calorimetry The reactivity of the fines was investigated using an isothermal calorimetry test at a constant temperature of 20 °C. Seven different cement replacement levels were tested, namely 0%, 5%, 10%, 15%, 20%, 25% and 30%. For all samples, 14 g of paste with a w/b of 0.5 was made. The paste was mixed outside the calorimeter and then put in a glass vial to be tested. The heat evolution in all samples was registered for three consecutive days.

Flexural and Compressive Strength of Mortar To investigate the influence of the recycled fines on the strength, 3 mortar mixes with different amounts of cement replacements, 0%, 10% and 25%, were made and tested. Each mixture had a water-to-binder ratio (w/b) of 0.5, and the ratio of sand to cement was 3:1. For these mixes, CEN sand with a fraction of 0/2 and CEM I 52.5 N were used. According to the standard NBN EN 196-1, 3 mortar prisms ($40 \times 40 \times 160$ mm^3) were tested at the ages of 7, 28 and 90 days.

3.2   Strength and Durability Tests on Concrete

Compressive Strength of Concrete The compressive strength of the two concrete mixes was tested on cubes ($150 \times 150 \times 150$ mm^3) according to the standard NBN EN 12390-3. For each mixture, 3 concrete cubes were tested after 2, 7, 28, 56 and 90 days of curing.

Shrinkage Autogenous and drying shrinkage were measured on concrete prisms ($100 \times 100 \times 400$ mm^3) according to the standard NBN EN 12390-16. For each concrete mix, 6 prisms were cast, demoulded 24 h after casting and wrapped in plastic foil to prevent moisture exchange with the environment during transportation to the laboratory. In the laboratory (20 °C, 60% RH), the prisms to measure autogenous shrinkage were wrapped in aluminium tape to prevent drying. Subsequently, two DEMEC-points were placed on the 3 sides of the prisms that were in contact with the formwork during casting, with a spacing of 200 mm between them. Finally, the initial distance between the two points was accurately measured using a demountable mechanical strain gauge (DEMEC). The distance was measured again after 1, 7, 14, 28, 42, 56 and 70 days after the initial measurement.

Freeze–Thaw Resistance According to the standard CEN/TS 12390-9, the freeze–thaw resistance of both concrete mixes was determined. In order to do this, cylinders with a height of 200 mm and a diameter of 100 mm were cast, demoulded after 24 h and transported to the laboratory. Subsequently, the cylinders were immersed in water until the age of 7 days was reached and then they were stored in a climate-controlled room at 20 °C and 60% RH until the age of 25 days. At the age of 21 days, 3 discs with a height of 50 mm were sawn from each cylinder. At 25 days, the discs were glued in a PVC tube with a height of 70 mm and an internal diameter of 104 mm using epoxy resin. After 28 days, the specimens were saturated for 3 days with a layer of demineralized water on top of the samples. After 3 days of saturation, the demineralized water was replaced with a 3 mm layer of 3% NaCl solution and the specimens were placed in a freeze–thaw chamber for 56 cycles of 24 h with a temperature ranging from −20 to 20 °C. After 7, 14, 28, 42 and 56 cycles, the samples were taken out of the chamber and the material which had scaled from the specimen surface was collected, dried at 110 °C and weighed when a constant mass was reached.

Porosity The porosity of both concrete mixes was determined based on the method described in NBN B 24-213. Cylinders with a height of 200 mm and a diameter of 100 mm were cast, demoulded after 24 h and transported to the laboratory where they were cured in a climate-controlled room at 20 °C and 95% RH. At the age of 115 days, 3 discs with a height of 50 mm were sawn from each cylinder. Subsequently, the discs were placed in a vacuum tank where a vacuum was created and maintained for 2 h. After 2 h, the vacuum tank was slowly filled with water while maintaining the vacuum, submerging the specimens. After another hour, the pressure in the tank was restored again until an atmospheric pressure was reached. The specimens were kept underwater until the mass gain was less than 0.1 m% over a 24 h period. When a constant mass was reached, the saturated specimens were weighed in air (m_{sat}) and underwater (m_w). Subsequently, the specimens were placed in an oven at 40 °C and weighed every 24 h until a constant mass ($m_{dry,40}$) was reached (<0.1 m% over 24 h). Then, the specimens were placed in an oven at 105 °C until a constant mass ($m_{dry,105}$) was reached. The capillary and open porosity can be determined using the following equations:

$$CP = \frac{m_{sat} - m_{dry,40}}{m_{sat} - m_w} \times 100 \tag{1}$$

$$OP = \frac{m_{sat} - m_{dry,105}}{m_{sat} - m_w} \times 100 \tag{2}$$

4   Results and Discussion

4.1   *Characterization of the Recycled Concrete Fines*

Particle Size Distribution and Particle Shape   The particle size distribution of the recycled fines and the cement determined with the dry dispersion unit is shown in Fig. 1. The recycled fines have a less uniform size in comparison to the CEM I particles. The high value for d90 can be attributed to a non-optimized milling process. Earlier it was mentioned that the target size for the fines is <63 μm. According to these results, only 73% of the particles can actually be classified as fines, while for CEM I, 97% of the particles have a size smaller than 63 μm. This problem could be resolved by additional sieving and milling since the bigger particles are probably cushioned by the smaller ones [2].

In the recycled fines, oval, rectangular and irregular shaped particles are present. The values of the circularity and roundness in 2D are 0.564 and 0.707, respectively. These values correspond to more elongated particles without sharp edges and corners [8].

Particle Density and Water Absorption   Table 2 shows the particle density of the fine fraction (0/2 mm). The measured values are similar to the values for recycled

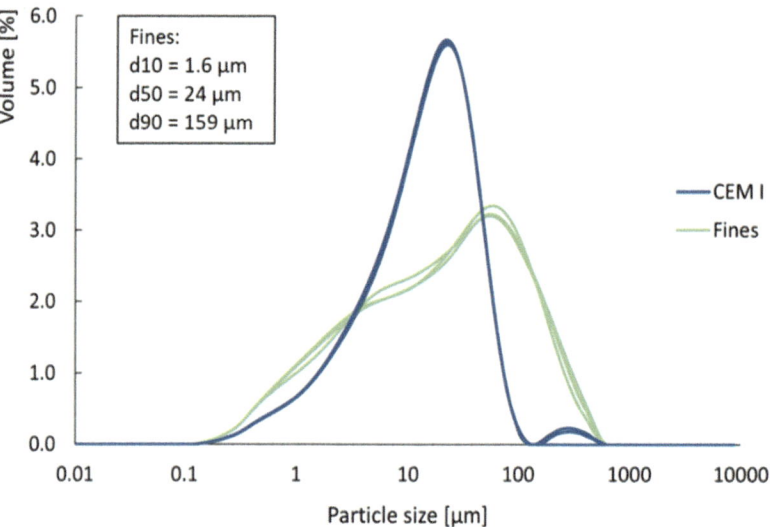

Fig. 1 Particle size distribution of recycled fines and CEM I

Table 2 Particle density of the recycled concrete fine fraction 0/2

Apparent particle density	2710 kg/m^3
Oven-dried particle density	2240 kg/m^3
Saturated and surface-dried particle density	2410 kg/m^3

Table 3 Results of XRD and XRF analysis on the recycled concrete fines

XRD		XRF	
Phase	Quantity [%]	Compound	Quantity [%]
Quartz (SiO_2)	35.86	Na_2O	1.35 ± 0.14
Calcite ($CaCO_3$)	13.61	MgO	2.17 ± 0.05
Belite (C_2S)	2.28	Al_2O_3	5.91 ± 0.11
Gypsum ($CaSO_4 * 2H_2O$)	3.71	SiO_2	36.30 ± 1.37
Anhydrite ($CaSO_4$)	2.78	P_2O_5	0.02 ± 0.01
Portlandite ($Ca(OH)_2$)	0.20	S	2.01 ± 0.09
Albite	4.27	K_2O	0.50 ± 0.06
Muscovite 2M1	6.19	CaO	46.22 ± 1.10
Amorphous	31.10	TiO_2	2.21 ± 0.13
		Mn_2O_3	0.06 ± 0.00
		Fe_2O_3	3.25 ± 0.07

concrete fines in the literature [9, 10]. In comparison to the CEM I 52.5 N, which has a density of 3150 kg/m³, the density is lower.

Regarding the water absorption, a value of 7.7% was obtained for the recycled fine fraction after immersion for 24 h. According to [11], the most restrictive standard (KS F2573) demands a maximum water absorption of 3%, while the least restrictive standard (DIN 4226-100) demands a maximum water absorption of 13%. Hence, the fine fraction complies with the least restrictive standard, but not with the most restrictive one.

Specific Surface Area The calculated specific surface area of the recycled concrete fines is 8690.31 cm²/g. The fines have a higher specific surface area in comparison to the CEM I 52.5 N, which has a specific surface area of 3145 cm²/g. Due to the rather high specific surface area of the fines, an increased reactivity can be expected. This increased reactivity can be explained by the large area and large amount of active atoms that are easily accessible for the reaction [12].

XRD and XRF The results of the XRD and XRF analysis are shown in Table 3. The results of the XRD show that quartz and calcite are the main phases present in the fines. The large amount of quartz originates from the natural aggregates present in the concrete pavers, which were most likely siliceous in nature, such as siliceous river or marine sand [3]. The large quantity of amorphous phases originates from amorphous cement hydration products (e.g. C-S-H), unreacted amorphous binder particles and amorphous $CaCO_3$ due to carbonation of the fines [3]. The presence of belite, gypsum and anhydrite can be explained by the young recycling age of the concrete pavers and therefore the presence of unreacted particles [13].

The results of the XRF analysis correspond with the results obtained with the XRD analysis. The high amount of SiO2 can possibly be explained by the siliceous sand present in the concrete pavers. The high quantity of CaO is inherent to cementitious materials.

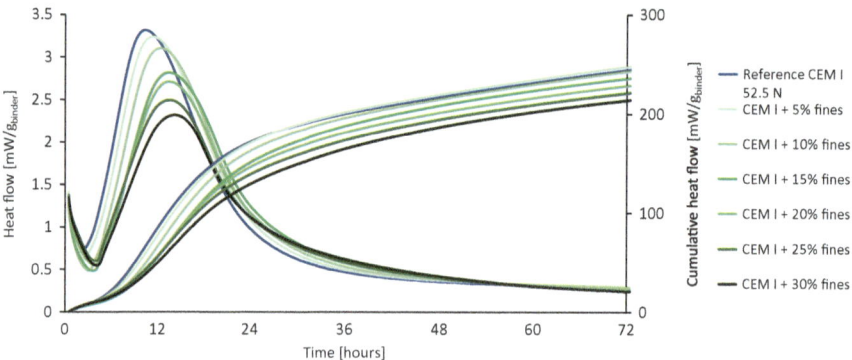

Fig. 2 Heat flow and cumulative heat flow normalized to the total binder mass

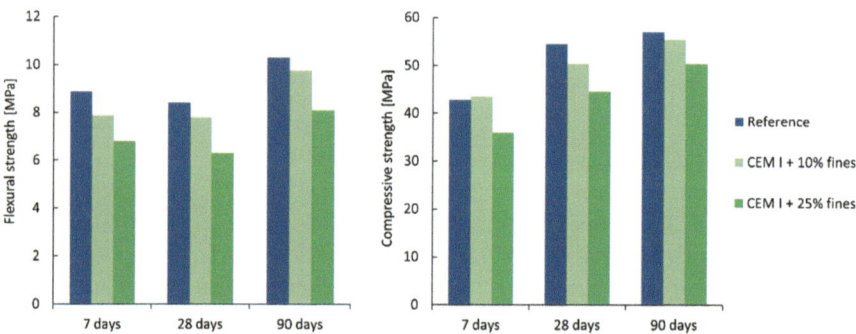

Fig. 3 Flexural and compressive strength of mortar containing 0%, 10% and 25% cement replacement by fines after 7, 28 and 90 days

Isothermal Calorimetry Figure 2 shows the heat flow and cumulative heat flow normalized to the total binder mass for mixes with 0 up to 30% cement replacement by recycled fines. As can be seen in Fig. 2, an increase in cement replacement delayed and decreased the hydration peak. However, the cumulative heat flow shows that 5% and 10% cement replacement resulted in a similar total heat production as CEM I 52.5 N after 72 h, which suggests that the hydration reaction is not negatively affected at this age by these replacement rates.

Flexural and Compressive Strength of Mortar Figure 3 shows the flexural and compressive strength of mortars containing 0%, 10% and 25% recycled fines as cement replacement. It is clear that an increased cement replacement leads to a decreased flexural and compressive strength. The compressive strength after 7 days for mortars with 10% fines is slightly higher in comparison to the reference, which indicates that there is a good early-age strength development. After 90 days, the decrease in flexural and compressive strength is limited to 5.2% and 2.8%, respectively, for mortars containing 10% recycled fines.

4.2 Strength and Durability Tests on Concrete

Compressive Strength Figure 4 shows the compressive strength of REF and 0.10FC. At all tested ages, 0.10FC has a slightly lower compressive strength in comparison to the reference. However, the maximum strength reduction was only 4.2% which occurred at the age of 56 days. Furthermore, a T-test indicates that there is no significant statistical difference between REF and 0.10FC at all tested ages (level of significance $= 5\%$, $p > 6.7\%$).

Porosity The capillary and open porosity of both REF and 0.10FC are shown in Table 4. The incorporation of recycled fines, sand and coarse aggregates in the concrete increased the capillary and open porosity by 45.5% and 47.5%, respectively. This increase in porosity can probably be attributed to the lower reactivity of the binder and the use of recycled aggregates in the 0.10FC in comparison to natural aggregates in the reference.

Shrinkage The autogenous (dashed-dotted line), drying (dashed line) and total (solid line) shrinkage of REF and 0.10FC are shown in Fig. 5. It can be seen that at the age of 70 days, a higher total shrinkage was observed for the 0.10FC mix in comparison to the reference. However, when looking at the autogenous shrinkage, the concrete containing 10% recycled fines expands instead of shrinks. This is probably due to the higher porosity and high water retention capacity of recycled aggregates causing internal curing and preventing autogenous shrinkage [14]. On the

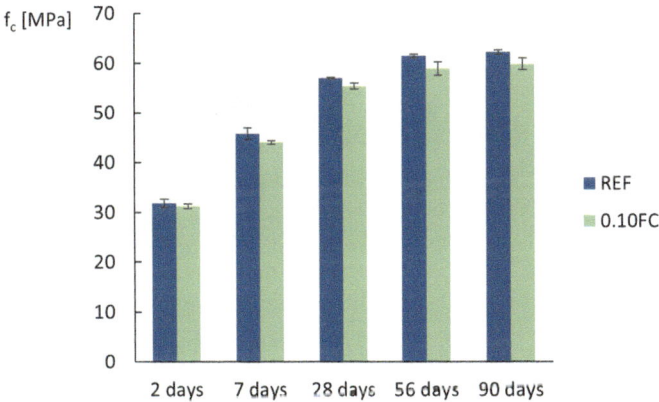

Fig. 4 Compressive strength of REF and 0.10FC after 2, 7, 28, 56 and 90 days. Error bars represent the standard error

Table 4 Porosity of concrete. The standard error is added in the table

	REF	0.10FC
Capillary porosity [%]	6.6 ± 0.3	9.6 ± 0.2
Open porosity [%]	12.0 ± 0.4	17.7 ± 0.2

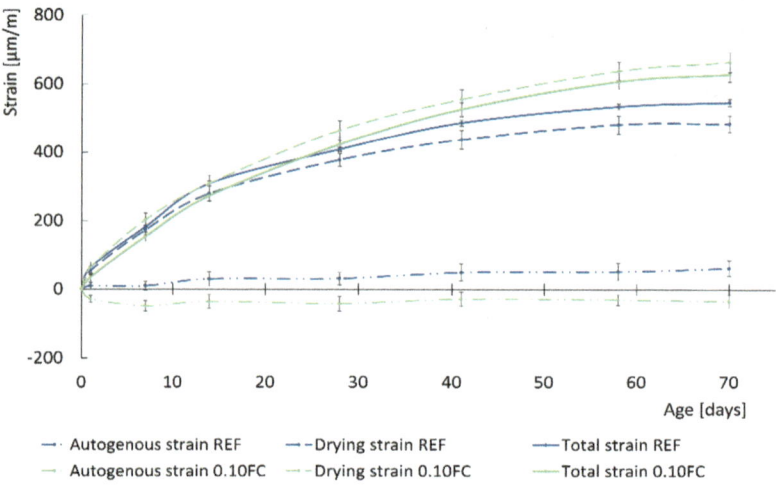

Fig. 5 Autogenous, drying and total shrinkage strain of REF and 0.10 FC

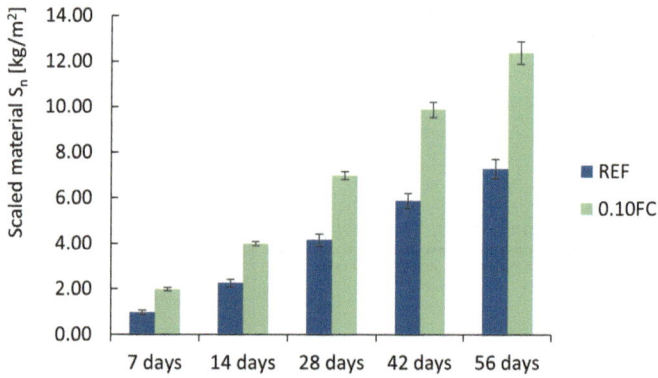

Fig. 6 Scaled material during freeze–thaw with de-icing salts of REF and 0.10FC

other hand, 0.10FC has an increased drying shrinkage, which is probably caused by the evaporation of the higher water content in the recycled aggregates.

Freeze–Thaw Resistance The mass of scaled material during the freeze–thaw test with de-icing salts after 7, 14, 28, 42 and 56 days is shown in Fig. 6. According to EN 12390-9, the limit value for scaling after 56 cycles is 1 kg/m^2, which suggests that both of the materials have a poor freeze–thaw resistance [15]. This was expected since the mix designs belong to an exposure class XF1. On average, the addition of recycled fines, sand and coarse aggregates into the concrete increases the scaling by 77.8% in comparison to the reference mix. This high decrease in freeze–thaw resistance was also found in previous literature [16]. The decrease can possibly be explained by the reduced content of Ca(OH)$_2$ in the hydration products, which can cause fast superficial carbonation and lead to more susceptibility to freeze–thaw scaling [17].

Environmental Impact A preliminary LCA-analysis showed that the global warming potential (GWP) decreased from 313 kg CO_2 eq for the reference concrete to 288 kg CO_2 eq for the concrete with 10% recycled fines and recycled sand and coarse aggregates, which corresponds with a decrease of 8%. The savings are mostly found in the raw material supply, due to a lower cement consumption, and in transport and manufacturing. A disadvantage of the green concrete is the lower recycling potential in comparison to the reference mix.

5 Conclusions

In this research, the use of recycled fines was investigated as a partial cement replacement. This included a characterization of the fines and their reactivity and an investigation of the durability of concrete with 10% cement replacement. It could be concluded that the fines are less uniform in size in comparison to cement which is probably attributable to a non-optimized milling process. The heat production during hydration after 72 h was similar to a CEM I for replacement rates up to 10% and the reduction of strength of mortars containing up to 10% fines is limited. A 'green' concrete containing 10% fines and partially recycled aggregates has a beneficial effect on the autogenous shrinkage in comparison to a reference concrete without fines and with natural aggregates. This can be attributed to the water retention ability of the recycled aggregates which serve as water reservoirs during hydration. The reduction in compressive strength was only 4.2% after 56 days. The freeze–thaw resistance of both the reference mix and the green concrete was insufficient, and the green concrete increased the scaling by 77.8% after 56 days of freeze–thaw cycles. A preliminary LCA analysis showed that the global warming potential is reduced by 8% if the green concrete is used.

References

1. Shah, I.H., Miller, S.A., Jiang, D., Myers, R.J.: Cement substitution with secondary materials can reduce annual global CO2 emissions by up to 1.3 gigatons. Nat. Commun. **13**(1), 5758 (2022). https://doi.org/10.1038/s41467-022-33289-7
2. Villagrán-Zaccardi, Y.A., Marsh, A.T.M., Sosa, M.E., Zega, C.J., De Belie, N., Bernal, S.A.: Complete re-utilization of waste concretes–valorisation pathways and research needs. Resour. Conserv. Recycl. **177**, 105955 (2022). https://doi.org/10.1016/j.resconrec.2021.105955
3. Nedeljković, M., Visser, J., Nijland, T.G., Valcke, S., Schlangen, E.: Physical, chemical and mineralogical characterization of Dutch fine recycled concrete aggregates: a comparative study. Constr. Build. Mater. **270**, 121475 (2021). https://doi.org/10.1016/j.conbuildmat.2020.121475

4. Rodríguez, C., Miñano, I., Aguilar, M.Á., Ortega, J.M., Parra, C., Sánchez, I.: Properties of concrete paving blocks and hollow tiles with recycled aggregate from construction and demolition wastes. Materials. **10**(12). https://doi.org/10.3390/ma10121374

5. Topič, J., Prošek, Z.: Properties and microstructure of cement paste including recycled concrete powder. Acta Polytech. **57**(1), 49–57 (2017). https://doi.org/10.14311/AP.2017.57.0049

6. Topič, J., Prošek, Z., Plachý, T.: Influence of increasing amount of recycled concrete powder on mechanical properties of cement paste. IOP: Mater. Sci. Eng. **236**(1), 012094 (2017). https://doi.org/10.1088/1757-899X/236/1/012094

7. Kwon, E., Ahn, J., Cho, B., Park, D.: A study on development of recycled cement made from waste cementitious powder. Constr. Build. Mater. **83**, 174–180 (2015). https://doi.org/10.1016/j.conbuildmat.2015.02.086

8. Cruz-Matías, I., Ayala, D., Hiller, D., Gutsch, S., Zacharias, M., Estradé, S., Peiró, F.: Sphericity and roundness computation for particles using the extreme vertices model. J. Comput. Sci. **30**, 28–40 (2019)

9. Lotfi, S., Rem, P.: Recycling of end of life concrete fines into hardened cement and clean sand. J. Environ. Prot. **7**(6), 934–950 (2016)

10. Florea, M., Brouwers, H.: Properties of various size fractions of crushed concrete related to process conditions and re-use. Cem. Concr. Res. **52**, 11–21 (2013)

11. Martín-Morales, M., Zamorano, M., Valverde-Palacios, I., Cuenca-Moyano, G., Sánchez-Roldán, Z.: Quality control of recycled aggregates (RAs) from construction and demolition waste (CDW). In: Handbook of Recycled Concrete and Demolition Waste, pp. 270–303. Elsevier (2013)

12. Tang, Q., Ma, Z., Wu, H., Wang, W.: The utilization of eco-friendly recycled powder from concrete and brick waste in new concrete: a critical review. Cem. Concr. Compos. **114**, 103807 (2020)

13. Kulisch, D., Katz, A., Zhutovsky, S.: Quantification of residual unhydrated cement content in cement pastes as a potential for recovery. Sustainability. **15**, 263 (2022)

14. Revilla-Cuesta, V., Evangelista, L., de Brito, J., Skaf, M., Ortega-López, V.: Mechanical performance and autogenous and drying shrinkage of MgO-based recycled aggregate high-performance concrete. Constr. Build. Mater. **314**, 125726 (2022). https://doi.org/10.1016/j.conbuildmat.2021.125726

15. Testing Hardened Concrete – Part 9: Freeze-Thaw Resistance with De-icing Salts – Scaling, EN 12390-9, B. f. Standardisation (2016)

16. Ma, Z., Li, W., Wu, H., Cao, C.: Chloride permeability of concrete mixed with activity recycled powder obtained from C&D waste. Constr. Build. Mater. **199**, 652–663 (2019)

17. Gruyaert, E., De Belie, N.: Effect van hoogovenslak als cementvervanging op de hydratatie, microstructuur, sterkte en duurzaamheid van beton. Ghent University, Ghent (2011)

Discussion of Physical Performance of Hydraulic Lime and Oyster Shell-Based Mortars

Poliana Bellei, Inês Flores-Colen, Isabel Torres, and Manuel F. C. Pereira

Contents

1 Introduction

Most of Portugal's architectural heritage is characterized by walls built with natural stone or solid brick. In addition to these substrates, the most common mortars to be used in rehabilitating of these buildings are produced based on cement or lime (hydraulic or aerial) [1, 2].

P. Bellei (✉)
CERIS, DECivil, IST – University of Lisbon, Lisbon, Portugal

CERENA, IST – University of Lisbon, Lisbon, Portugal
e-mail: poliana.bellei@tecnico.ulisboa.pt

I. Flores-Colen
CERIS, DECivil, IST – University of Lisbon, Lisbon, Portugal

I. Torres
CERIS, DECivil, University of Coimbra, Coimbra, Portugal

Itecons, Coimbra, Portugal

M. F. C. Pereira
CERENA, IST – University of Lisbon, Lisbon, Portugal

© The Author(s) 2025
M. Kioumarsi, B. Shafei (eds.), *The 1st International Conference on Net-Zero Built Environment*, Lecture Notes in Civil Engineering 237,
https://doi.org/10.1007/978-3-031-69626-8_17

According to the manufacturer of Natural Hydraulic Lime NHL 3.5, this material is a natural hydraulic binder suitable for the production of coating mortars. This mortar can be applied to the rehabilitation and conservation of old and historic architectural structures, or new construction work. This product integrates, in a single substance, the aerial setting and hydraulic setting processes, without the need for additional additions. With these characteristics, it can be applied internally and externally, and on the substrates of new masonry in prefabricated blocks or ceramic brick and stone masonry [3].

However, research on mortars generally takes place in a laboratory environment, where various properties are evaluated according to the relevant standards. These standards recommend the manufacture of standardized test specimens, molded into metallic shapes and subjected to specific curing conditions. In practical reality, the application of mortar occurs on substrates made of different materials and under different conditions, which often differ from the controlled conditions found in a laboratory.

Some studies have already been conducted to detect the influences that substrates can have on mortars. The results showed changes when these mortars were compared with mortars molded in standardized molds [4–6].

This research aims to analyze the physical durability of oyster shells in hydraulic lime-based mortars. In addition to the work trying to predict the in-service behavior of hydraulic lime mortars applied to solid brick substrates, one of the mortar mixtures incorporates oyster shell to recover old construction techniques. Mortars based on lime from oyster shells were used in ancient buildings in coastal regions. In the Algarve (Southern region of Portugal), for example, can find buildings with an insulation layer made up of a mixture of medium-grained sand with trampled and compacted cockle shells [7].

If it is feasible to predict the behavior of the mortar on the substrate based on the characteristics determined in the laboratory, the decision to initiate a rehabilitation action will be based on and informed more solidly and consciously.

2   Materials and Methods

The materials for this research were chosen from the perspective of rescuing old construction techniques from coastal regions and rehabilitating these old buildings that still exist. Natural hydraulic lime was chosen as a binder to produce the mortars and apply them to solid ceramic brick substrates ($21.6 \times 10.2 \times 5$ cm). The sand was composed of four types of calibrated sand. The oyster shell was donated by a cooperative in the southern region of Portugal, and was used to replace the sand aggregate. With this, two different types of mortars were produced: a control group and a group with a replacement of 30% of the sand aggregate by oyster shell aggregate (30%-OS).

When the oyster shell arrived at the university laboratory, its preparation included washing with water, drying in an oven until constant mass, and grinding in a cutting

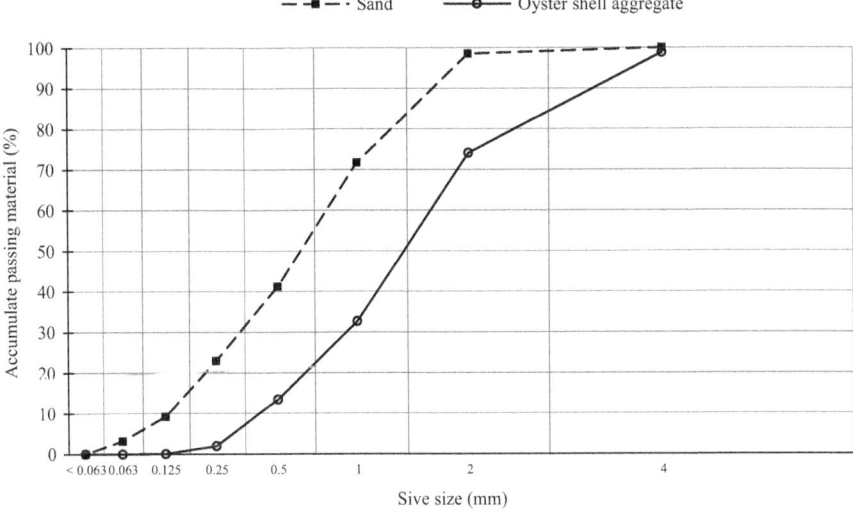

Fig. 1 Particle size distribution (sand and oyster shell as aggregate)

Table 1 Composition of the mortars produced

Mortars	Binder (%)	Sand (%)	Oyster shell aggregate (%)	Water/binder ratio
Control	100	100	0	1.07
30%-OS	100	70	30	1.18

mill (to transform the shell as a substitute for sand). Particle size distribution of sand and oyster shell aggregate were found according to the NP EN 933-1 standard [8] and are shown in Fig. 1.

The mortars were produced with of a mixer and a plastic pail. The mortar ratio corresponded to 1:3, and the sand was replaced by 30% by shell, in volume. The composition of the mortars can be seen in Table 1.

The quantity of water added to the mixtures was adjusted prior to the production of all specimens under study. The quantity of water was determined by a consistency test, carried out in the fresh state of the mortars according to the EN 1015–3 standard [9].

After mixing, the specimens were first molded in standard molds (prisms: 40 × 40 × 160 mm and circles 15 mm thick and 100 mm in diameter). Then, a layer of the same mortar was applied to one of the surfaces of the solid brick substrates. To guarantee the uniformity and thickness of 20 mm of the coating at the time of application, a wooden mold was used. Using a 40 × 40 mm metal mold and a 100 mm diameter polyvinyl chloride (PVC) mold, markings were made on the layer of fresh mortar applied. Immediately before application, the surface of the bricks was moistened by spraying with water, using a spray bottle, to simulate real, normal application conditions. A 5 × 5 mm fiberglass mesh was incorporated to facilitate subsequent detachment of the applied mortar layer. The process of applying

Fig. 2 Specimens preparation under in-service condition: (**a**) molding the mortar on the solid brick substrates; (**b**) marking the specimens on the fresh mortar layer; (**c**) accelerated aging curing; (**d**) detaching of mortar from the substrates

mortar to the solid brick can be seen in Fig. 2a. The marking of the specimens in the still-fresh mortar layer is shown in Fig. 2b.

The curing process of the specimens molded in standardized molds and the mortars applied to the substrates followed the specifications of the EN 1015–11 standard [10]. All tested material was kept in a controlled laboratory environment at 20 °C and 95% relative humidity for 7 days, followed by subsequent days at 20 °C and 65% relative humidity until the test date. In addition to curing in the laboratory for 28 days, the solid brick and mortar models were subjected to accelerated aging cure [11] (Fig. 1c). Basically, the models were subjected to two stages of four cycles, with the first step of the second series being adapted for technical reasons:

- First stage—four cycles: 60 ± 2 °C for 8 h $\pm$ 15 min; 20 ± 2 °C and $65 \pm 5\%$ relative humidity for 30 ± 2 min; -15 ± 1 °C for 15 ± 15 min; 20 ± 2 °C and $65 \pm 5\%$ relative humidity for 30 ± 2 min.
- Second stage—four cycles: The sets were moistened with 300 ml of water each and were conditioned in a climatic chamber at 95% relative humidity and 20 ± 1 °C for 8 h $\pm$ 15 min; 20 ± 2 °C/$65 \pm 5\%$ relative humidity for 30 ± 2 min; -15 ± 1 °C for 15 min $\pm$ 15 min; 20 ± 2 °C and $65 \pm 5\%$ relative humidity for 30 ± 2 min.

After that, the coating layer was detached from the substrates, and specimens with dimensions of $40 \times 40 \times 15$ mm and circles of 15×100 cm were easily obtained (Fig. 1d). The prismatic specimens from the standardized molds were cut into 15 mm thick slices, to have the same dimensions as the cured specimens detached from the substrates. This allowed the comparison of performance between different molding conditions and mortar composition.

The identification of the mortars is as follows:

- Control: control mortar molded in standard molds;
- 30%-OS: mortar with 30% replacement of sand aggregate with oyster shell aggregate molded in standard molds;
- Control/SB: control mortar applied to solid ceramic brick and with accelerated aging curing (representing in-service conditions);

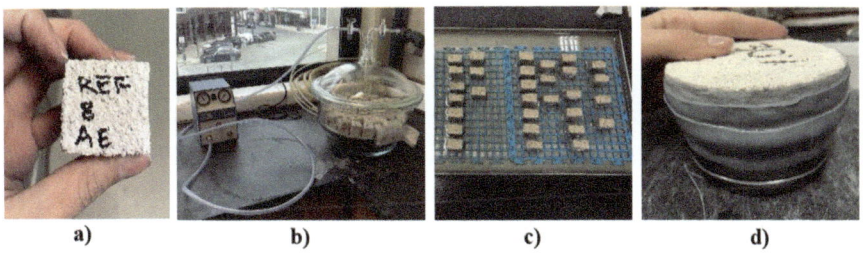

a)   b)   c)   d)

Fig. 3 Tests on mortars in the hardened state: (**a**) bulk density; (**b**) open porosity; (**c**) water absorption by capillarity; (**d**) water vapor permeability

- 30%-OS/SB: mortar with 30% replacement of sand aggregate with oyster shell aggregate applied to solid ceramic brick and with accelerated aging curing (representing in-service conditions).

The results of the characterization tests in the fresh and hardened state of the mortars are presented below in both laboratory and real-world conditions after curing by accelerated aging, application, and detachment from the substrate. Although the specimens did not have the dimensions recommended by current standards, the tests were based on their guidelines. The hardened state tests analyzed were: bulk density and open porosity [12] (Fig. 3a and Fig. 3b), water absorption coefficient by capillarity [13] (Fig. 3c), and water vapor permeability [14, 15] (Fig. 4d).

The bulk density and open porosity began the test with specimens dried in an oven at 70 °C (dry weight). After that, the specimens were transferred to a container connected to a vacuum pump (24 h). Then, water was introduced until the specimens were submerged, and the vacuum pump remained on for another 24 h. A hydrostatic balance was used to record the mass of the specimen immersed in water, as well as the mass of the saturated specimen.

The specimens were waterproofed, with plastic film, on their side faces, to guarantee a unidirectional flow, before starting the water absorption test by capillarity. During the test, the specimens remained in a layer of water with a height of 5–10 mm. The specimens were weighed 5, 10, 30, 60, 90, 180, 300, 480, 1440, 2880, 4320 minutes after the start of the test.

In the case of the water vapor permeability test, the specimens (20 mm thick and 100 mm in diameter, approximately) were tested using the wet trough method, so that vapor diffusion was unidirectional and exclusively through the specimen. After preparing the specimens, they were placed in a climate chamber with constant temperature and relative humidity (20 °C and 50%, respectively). An initial and periodic weighing was carried out until a constant mass variation per unit of time was obtained.

The results of all tests in the hardened state were entered into the Statistic® 7.0 software. An analysis was performed using one-way ANOVA followed by the Tukey test, which assessed the existence of significant differences at the 5% significance level ($p < 0.05$).

3   Results and Discussions

3.1   Fresh State

The slump flow result obtained for the two types of mortar produced was 170 mm. This index corresponds to the suggestion in EN1015–2 [16] that the slump flow measurement be 175 ± 10 mm for mortar with a bulk density greater than 1200 kg/m^3. In the same way, the quantity of water used in the mortar was sufficient to allow it to be applied to solid brick.

3.2   Hardened State

Table 2 presents the density (D) and open porosity (OP) results.

The density result of the mortar produced with oyster shell (30%-OS and 30%-OS/SB) was lower than the density presented by the control mortar (Control and Control/SB), both for the mortars molded in the standard molds and mortars in-service conditions. In the case of open porosity, mortars with oyster shells demonstrated greater results compared to the control mortar. The density of the aggregates explains the results found in the mortars, as the oyster shell aggregate is heavier than the calibrated sand aggregate. This implies lower densities and greater porosities. When studying cement mortars with marine waste, Ez-zaki et al. [17] attribute a decrease in apparent density and consequently an increase in apparent porosity due to the greater water demand these mortars needed to reach an adequate consistency.

After applying the mortars to the solid brick and curing through accelerated aging, the densities of the Control/SB (1820 kg/m^3) and 30%-OS/SB (1679 kg/m^3) mortars increased in relation to the Control (1775 kg/m^3). m^3) and 30%-OS (1619 kg/m^3) mortars, respectively. In the case of open porosity, the mortars reduced their results after in-service conditions. Travincas et al. [6] analyzed industrial cement mortar applied to different substrates (concrete plate; concrete block; lightweight concrete block; and hollow ceramic brick). The authors also concluded that

Table 2 Density and open porosity tests of mortar

Mortars	D (kg/m^3)				OP (%)			
	A	SD	CV (%)	NS	A	SD	CV (%)	NS
Control	1775[a]	34.7	2	3	26[a]	0.03	0.1	3
30%-OS	1619[b]	14.5	1	3	30[b]	0.12	0.4	4
Control/SB	1820[c]	10.7	1	6	24[a]	0.65	2.7	6
30%-OS/SB	1679[d]	23.14	1	8	27[c]	0.78	2.9	9

A average, *SD* standard deviation, *CV* coefficient of variation, *NS* number of specimens
Equal lowercase letters represent that there is no significant difference between lines, while different lowercase letters represent that there is a significant difference between lines

mortars produced in standard molds showed lower density and greater open porosity results, compared to applied and detached mortars, regardless of the type of substrate.

The analyzes prove that the density results are significantly different in terms of mortar composition but that the curing and molding conditions also cause changes in the results. The same happens in the case of open porosity. However, the control mortar submitted in-service situations did not show significant differences in its results. The oyster shell mortar showed a significant difference in behavior when subjected to different conditions than the laboratory standard.

The water absorption coefficients by capillarity are shown in Table 3.

The water absorption coefficients decreased for the Control/SB and 30%-OS/SB mortars, compared to the Control and 30%-OS mortars, considering the same groups of mortars. The highest water absorption coefficients were for mortars that had oyster shell in the composition.

Although the absorption coefficients of the 30%-OS and 30%-OS/SB mortars are greater than the coefficients of the Control and Control/SB mortars, the statistical analysis demonstrated that these differences are not significant in relation to the Control mortar. The Control/SB mortar presented the lowest water absorption coefficient, however the conditions in service did not change this coefficient, to the point that the result was significantly different in relation to Control.

Through the water vapor permeability test it was possible to obtain Table 4.

The application of the control mortar and the mortar with 30% oyster shell on the solid brick, as well as curing by accelerated aging, contributed to the reduction of water vapor permeability results for both mortars. Through the statistical analysis carried out, no results were found with significant differences among any of the mortars studied. The Control/SB and 30%-OS/SB mortars showed smaller differences in behavior (in the order of 8%), while the Control and 30%-OS mortars showed differences in results of approximately 19%.

Table 3 Water absorption coefficient by capillarity of mortar

Mortars	A $(kg/(m^2.s^{0.5}))$	SD $(kg/(m^2.s^{0.5}))$	CV (%)	NS
Control	$0.386^{a,b}$	0.015	4	4
30%-OS	0.415^a	0.017	4	3
Control/SB	0.363^b	0.011	3	6
30% OS/SB	0.387^a	0.014	4	7

Table 4 Water vapor permeability test of mortar

Mortars	A (kg/(m.s.Pa))	SD (kg/(m.s.Pa))	CV (%)	NS
Control	$1.80E-11^a$	6.20E-12	34.51	3
30%-OS	$1.46E-11^a$	7.21E-13	4.94	3
Control/SB	$1.21E-11^a$	1.18E-12	9.74	3
30%-OS/SB	$1.32E-11^a$	5.00E-13	3.79	3

The results of these hydraulic lime mortars differ from the results found by Travincas et al. [6]. The authors concluded that mortars produced in standard molds showed lower water vapor permeability results, compared to applied and detached mortars, regardless of the type of substrate. This trend may have occurred due to the pore distribution checked by MIP.

Although the 30%-OS and 30%-OS/SB mortars presented the highest water vapor permeability values (when analyzing the groups), the standard deviation and coefficient of variation of these mortars were lower.

4 Conclusion

This research aimed to analyze the physical durability of oyster shells in mortars based on hydraulic lime. In this case, a control mortar and a mortar with 30% replacement of sand aggregate were replaced by oyster shell aggregate.

With the same consistency for the mortars, specimens were molded under standard laboratory conditions, and other specimens simulated conditions considered in-service (application of a layer of mortar on a solid ceramic brick support and curing by accelerated aging).

In most results, mortars produced with oyster shell showed lower densities, higher porosities, water absorption coefficient and vapor permeability when compared to control mortar. However, these results showed no significant differences using the Tukey test.

Furthermore, the results presented confirm changes in the performance of the coating layer applied to the substrate surface. In some cases, both the control mortar and the shell mortar showed significant differences in their properties after application to the solid brick support and curing by accelerated aging, compared to laboratory conditions.

Therefore, mortars with oyster shell have demonstrated that their results do not differ from standard mortar, as well as improving their performance under service conditions. A substrate characterization analysis would improve understanding of the effects caused to mortar coatings.

When studying hydraulic lime mortars containing oyster shell applied to a solid ceramic brick support, at least two advantages can be cited: ancient cultural aspects of construction present mainly in coastal locations are valued, as well as important evidence for cases of rehabilitation of old buildings. With this work it can be said that changes occurred in the properties of the mortars being influenced by the support, as well as the oyster shell has potential for use in construction. Future studies with oyster shell could explore whether this bioproduct has potential as a thermal material.

Acknowledgments The study presented was carried out within the scope of the SHELLTER project (FBR_OC2_30) financed by the EEA Grants bilateral fund (Portugal and Norway). The authors would like to thank CERIS (UIDB/04625/2020; https://doi.org/10.54499/UIDB/04625/

2020) and CERENA (UID/ECI/04028/2019), IST's research unit, and FCT (Foundation for Science and Technology) for the financial support with the reference scholarship UI/BD/151151/2021 (https://doi.org/10.54499/UI/BD/151151/2021).

References

1. INE, Buildings Tables—2.03—buildings, according to construction period, by main materials used in construction, in: Portuguese) 2011 Census: XV General Population Census; V General Housing Census, National Institute of Statistics (INE), Lisbon, 2011
2. Flores-Colen, I., Brito, J.: Renders. In: Gonçalves, M., Margarido, F. (eds.) Materials for Construction and Civil Engineering, pp. 53–122. Springer, Cham (2015)
3. SECIL Homepage. https://www.secil.pt/pt/produtos/cimento/cal-hidraulica-natural/cal-hidraulica-natural-nhl-3-5. Last Accessed 20 Feb 2024
4. Silveira, D., Gonçalves, A., Flores-Colen, I., Veiga, M.R., Torres, I., Travincas, R.: Evaluation of in-service performance factors of renders based on in-situ testing techniques. J Building Eng. **34**, 101806 (2020)
5. Bellei, P., Arromba, J., Flores-Colen, I., Veiga, R., Torres, I.: Influence of brick and concrete substrates on the performance of renders using in-situ testing techniques. J Building Eng. **43**, 102871 (2021)
6. Travincas, R., Torres, I., Flores-Colen, I., Francisco, M., Bellei, P.: The influence of the substrate type on the performance of an industrial cement mortar for general use. J Building Eng. **73**, 106784 (2023)
7. Pacheco, M. B. P.: A Evolução Urbana e Arquitectónica da Fuseta. Dissertação de Mestre em Arquitectura. Departamento de Engenharia Civil e Arquitectura, Instituto Superior Técnico, Lisboa, 310 p
8. NP EN 933-1:2014. Ensaios das Propriedades Geométricas dos Agregados, Parte 1: Análise Granulométrica – Método de Peneiração. IPQ: Lisboa, Portugal, 2013
9. EN 1015-3: Methods of Test for Mortar for Masonry—Part 3: Determination of Consistence of Fresh Mortar (by Flow Table). CEN, Brussels (1999)
10. EN 1015-11: Methods of Test for Mortar for Masonry—Part 11: Determination of Flexural and Compressive Strength of Hardened Mortar. CEN, Brussels (2019)
11. EN 1015-21: Methods of Test for Mortar for Masonry—Part 21: Determination of the Compatibility of One-Coat Rendering Mortars with Substrates. CEN, Brussels (2002)
12. NP EN 1936. Métodos de ensaio para pedra natural. Determinação das massas volúmicas real e aparente e das porosidades total e aberta. IPQ, Caparica, 2008
13. BS EN 15148. Hygrothermal performance of building materials and products—Determination of water absorption coefficient by partial immersion. ISO, Brussels, 2002
14. EN 1015-19: Methods of Test for Mortar for Masonry—Part 19: Determination of Water Vapour Permeability of Hardened Rendering and Plastering Mortars. IPQ, Brussels (1998)
15. ISO 12571: Hygrothermal Performance of Building Materials and Products—Determination of Water Vapor Transmission Properties. ISO, Brussels (2013)
16. EN 1015-2: Methods of Test for Mortar for Masonry—Part 2: Bulk Sampling of Mortars and Preparation of Test Mortars. CEN, Brussels (1998)
17. Ez-Zaki, H., Diouri, A., Kamali-Bernard, S., Sassi, O.: Composite cement mortars based on marine sediments and oyster Shell powder. Mater. Constr. **66**, e080 (2016)

CO$_2$ Capturing of Aggregates Extracted from Alkali-Activated GGBS Concrete

Syamak Tavasoli and Wolfgang Breit

Contents

1 Introduction

Carbon dioxide emission reduction has become a prominent global concern in recent years. The construction sector stands out with a substantial 40% contribution to the overall carbon footprint, making it a major player in CO$_2$-emission [1, 2]. The cement industry is considered as the largest CO$_2$-contributor in this sector responsible for about 8% of global emissions [3, 4]. To address this issue, exploring alternative binders, e.g., alkali-activated binders (AAB), has emerged as a promising strategy to mitigate carbon emissions. By only substitution of traditional binders with alternative binders, about 80% of CO$_2$-emission reduction could be possible [5–7]. However, achieving carbon-neutral conditions in the cement industry requires a multifaceted approach, combining various methods aimed at reducing CO$_2$-production along with the adoption of alternative binders. Life cycle assessment (LCA) studies on alternative binders highlight AAB as particularly promising,

S. Tavasoli (✉) · W. Breit
Institute of Construction Material Technology, Rheinland-Pfälzische Technische Universität Kaiserslautern-Landau (RPTU), Kaiserslautern, Germany
e-mail: syamak.tavasoli@rptu.de

© The Author(s) 2025
M. Kioumarsi, B. Shafei (eds.), *The 1st International Conference on Net-Zero Built Environment*, Lecture Notes in Civil Engineering 237,
https://doi.org/10.1007/978-3-031-69626-8_18

demonstrating the lowest CO_2-emissions among the other alternatives. Using carbon-capturing techniques in AAB can lead to a reduction of more than 90%.

Two components of the materials used in concrete have a CO_2-absorption potential. Binders containing alkaline or alkaline earth elements inherently possess the ability to react with CO_2, particularly in humid conditions [8]. Additionally, recycled aggregates containing old cement paste are capable of absorbing CO_2 due to their high calcium content. The reaction of calcium hydroxide with CO_2 is considered as the dominant reaction in ordinary Portland cement (OPC) matrix [9]. The CO_2 diffuses in the pore water of the unsaturated matrix, followed by its reaction with carbonatable solutes, i.e., $Ca(OH)_2$ (calcium hydroxide) solute [8, 9]. Similarly, aggregates extracted from alkali-activated concrete (AAC) can absorb CO_2 through their reaction with alkaline or alkaline earth elements. The rate of carbonation in alkali-activated slag mortar has been reported considerably higher than that of OPC, leading to the decalcification of calcium silicate hydrate (CSH) [10, 8]. Depending on the type of activator used, the carbonation rate of AAB can vary [10]. AAB activated with hydroxide activator shows more rapid carbonation rate compared to the one activated with waterglass [11].

The use of recycled aggregates in concrete production has increased sharply in recent years [12, 13]. This trend can be attributed to the dual objectives of addressing the scarcity of natural resources and mitigating the release of construction and demolition waste (CDW) into the environment [14, 15]. CDW constitutes the largest global waste stream by around 3 billion tons annually [16]. This inclination toward sustainability also extends to concrete manufactured with alkali-activated binders (AAB). Consequently, it becomes imperative to assess the feasibility of recycling such concrete. Furthermore, the potential to absorb carbon dioxide during its service life, as well as after recycling is a key factor for the calculation of CO_2-uptake. Leemann et al. [17] indicated that recycled concrete aggregate (RCA) from OPC absorb 9–21 wt.% CO_2 . The CO_2-uptake of concrete structures was estimated about 17% of the total emission produced during the cement manufacturing in Sweden in 2011 throughout their service life [18].

The reduction of porosity and water absorption of concrete, as well as RCA is associated with the aggregate carbonation [19]. Subsequently, the mechanical behavior of RCA will be improved, followed by the improvement of corresponding behavior of concrete made with them [20]. Due to the current and future tendency for using low-carbon footprint concrete, the recycling of it and the CO_2-uptake potential of recycled materials should be clarified. In addition, the recognition of these recycled materials behavior in treated and non-treated state will be a useful parameter in reusing them in concrete.

Regarding the lower CO_2-emission of AAC in comparison with than of OPC, it is foreseen that this type of concrete will be partly replaced with ordinary concrete in the future. GGBS is considered as one of the main precursors used in this system [7]. However, GGBS production is limited to around 408 million tons in 2022 worldwide [21]. In contrary to the announcement about the reduction of GGBS production, it is reported that its production will reach around 497 million tons by 2035 [21]. In addition, thanks to the replacement of GGBS with supplementary precursors, the GGBS demand could be reduced. Therefore, it is necessary to evaluate the feasibility of recycling this type of concrete after the end of service life of the structures made with them. Regarding the importance of CO$_2$-capturing in civil structures, the sequestration potential of the aggregates extracted from them could also be considered for their utilization in the future.

2 Materials and Testing Procedures

2.1 Materials

In this research, aggregates, for primary mixtures included natural and recycled aggregates type 1 according to EN 12620 and DIN 4226-101. The properties of aggregates in primary mixtures had no influence on the CO$_2$-uptake of recycled aggregates extracted from the concrete made with them. Therefore, they were neglected to be described in this study. Mixes based on alkali-activated ground granulated blast furnace slag (GGBS) using these aggregates were produced. Subsequently, aggregates extracted from these two concrete mixtures containing natural and recycled aggregates were sieved and characterized. Then, the aggregates properties were tested and exposed to carbonation.

In alkali-activated concrete mixtures, GGBS was the only precursor employed in this study. The chemical composition of GGBS is shown in Table 1. The GGBS had a Blaine value of 4050 cm^2/g and a density of 2.87 g/cm^3. It was activated with 8% activators, including sodium hydroxide (NaOH, 16 mol), sodium silicate (Na$_2$O. nSiO$_2$, n = 2.1), and potassium silicate (K$_2$O.nSiO$_2$, n = 1.7).

The description of aggregates extracted from AAC mixes in this study before and after carbonation is depicted in Table 2.

2.2 Mix proportions and Production

For the generation of recycled aggregates from alkali-activated concrete, two reinforced concrete slabs were produced based on the mix proportions M1 and M2, presented in Table 3. Both M1 and M2 shared identical binder compositions, differing solely in the types of aggregates used. M1 was made with natural aggregates. In M2 the coarse aggregates (2/8 mm and 8/16 mm) were recycled concrete aggregate type 1. These slabs were cured in a plastic foil for 28 days. The compressive strength of the specimens corresponding to the slabs was tested to ensure that the targeted strength of slabs was achieved at 28 days. Subsequently, the slabs were crushed to produce recycled materials.

Additionally, mix M3 was produced, which consists of 100% coarse aggregates extracted from the concrete slab made with aggregates A(M1) (recycled from mix M1), see Table 3. CM3 is another mixture with a similar mix composition to M3 in

Table 1 Chemical compositions of GGBS determined by XRF (X-ray fluorescence)

CaO (%)	SiO$_2$ (%)	Al$_2$O$_3$ (%)	MgO (%)	SO$_3$ (%)	K$_2$O (%)	Fe$_2$O$_3$ (%)	MnO (%)	Na$_2$O (%)	Mn (%)	TiO$_2$ (%)	Cr$_2$O$_3$ (%)
42.13	36.62	11.34	6.48	1.42	0.53	0.24	0.16	0.3	0.12	0.91	42.13

Table 2 Recycled AAC aggregate abbreviation description

Aggregate index	Description
A(M1)	Aggregate extracted from mix M1
AC(M1)	Carbonated aggregate extracted from mix M1
A(M2)	Aggregate extracted from mix M2
AC(M2)	Carbonated aggregate extracted from mix M2

Table 3 Mix proportions

Mix	Water	GGBS	Gravel (8/16 mm)	Gravel (2/8 mm)	Sand (0/2 mm)	NaOH	$K_2O.$ $nSiO_2$	$Na_2O.$ $nSiO_2$
	kg/m³							
M1	138.7	400	613.6	613.6	526.0	26.9	26.7	17.8
M2	178.5	400	616.1	616.1	528.1	26.9	26.7	17.8
M3	195.2	400	613.6	613.6	526.0	26.9	26.7	17.8
CM3	177.2	400	611.2	611.2	523.9	26.9	26.7	17.8

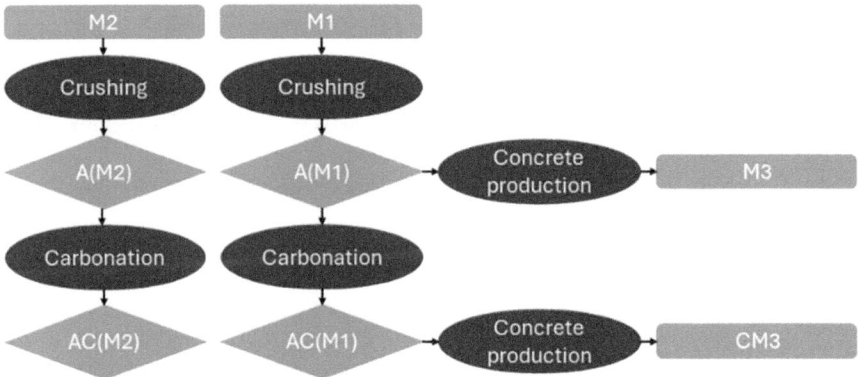

Fig. 1 The visual procedure of material preparation and concrete production

which the extracted aggregates from M1 (A(M1)) were carbonated and then used in concrete production. All the concrete mixes in this study had an equal water to binder ratio (w/b) of 0.38. The difference between the water content of mixes in Table 3 is due to the difference of water absorption in aggregates. Furthermore, the water content of alkaline solutions was also considered in the total water content of mixtures. A visual sketch of material preparation and concrete production is demonstrated in Fig. 1.

The aggregates were dried before concrete production. To prevent a significant workability drop after the production, caused by water suction in dried aggregates, they were initially mixed with the required water for saturation in a concrete mixer for 30 seconds and remained in it for 10 minutes to achieve about 90% of the total saturation. Subsequently, the aggregates were mixed with other ingredients of the mixture for 5 minutes.

2.3   Testing Procedures

The water absorption and densities of the primary aggregates, as well as produced aggregates were determined according to EN 1097-6. For this test, three specimens from each sort of aggregates were tested and the average values were subsequently reported. For CO_2-absorption testing, five specimens from each sort of aggregates, each weighing 70–120 g, were selected. The non-carbonated aggregates were dried in $105 \pm 5°C$ until reaching constant mass state, followed by a cooling procedure to room temperature under vacuum conditions. The non-carbonated aggregates in dried condition were weighed. Subsequently, they were placed in a CO_2 chamber having 1% CO_2 concentration at $20 \pm 2°C$ temperature and a relative humidity (RH) of $57 \pm 5\%$. The aggregates were weighed again after 7 and 14 days to justify their CO_2 absorption. At the end of each interval and before weighing the specimens were dried through the mentioned procedure above.

Fresh concrete properties were tested using slump and flow table tests according to EN 12350-2 and EN 12350-5, respectively. The air content of concrete mixtures was determined based on EN 12350-7.

This study investigated the hardened concrete properties, specifically focusing on compressive strength, tensile splitting strength, and elastic modulus, as outlined in EN 12390-3, EN 12390-6, and EN 12390-13, respectively. The compressive strength tests of concrete mixes were carried out on $150 \times 150 \times 150$ mm cubes. Tensile splitting strength tests were performed on a prism having $100 \times 100 \times 600$ mm dimension. A cylindrical specimen with 150×300 mm dimension was produced for the elastic modulus test. All the three above-mentioned tests were carried out at the ages of 2, 7, 28, and 91 days.

3   Results and Discussions

3.1   Aggregates

Figures 2 and 3 demonstrate the carbon dioxide absorption capacity of the aggregates A(M1) and A(M2) (extracted from recycled AAC mixtures (M1) and (M2)), respectively. It is observed that the maximum carbonation level of recycled aggregates based on AAC is reached after 7 days, requiring no further exposure. Notably, fine recycled aggregates showed a higher carbonation capacity than coarse aggregates. The 8/16 mm recycled AAC aggregate shows the lowest CO_2-absorption potential by around 2 wt.%. The reason could be the higher porosity of the fine recycled aggregates compared with the coarse recycled aggregates, which expedites the CO_2 penetration. Moreover, the higher available specific surface in the fine aggregates and the higher amount of old mortar content in them lead to a superior absorption capacity in fine recycled aggregates than the corresponding coarse recycled aggregates. Indeed, the natural aggregate parts of recycled aggregates can

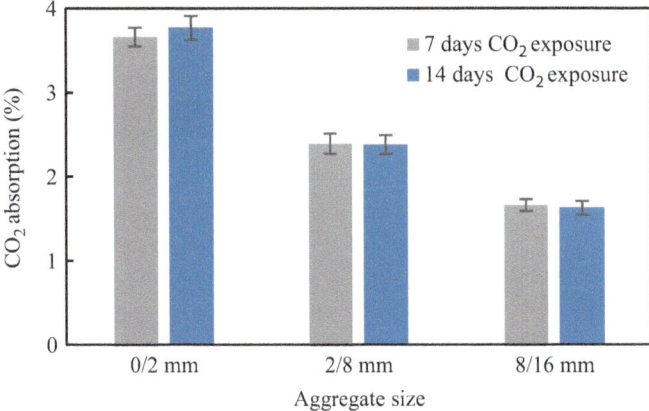

Fig. 2 CO_2-absorption of aggregates A(M1) (recycled from mix M1, including natural aggregates)

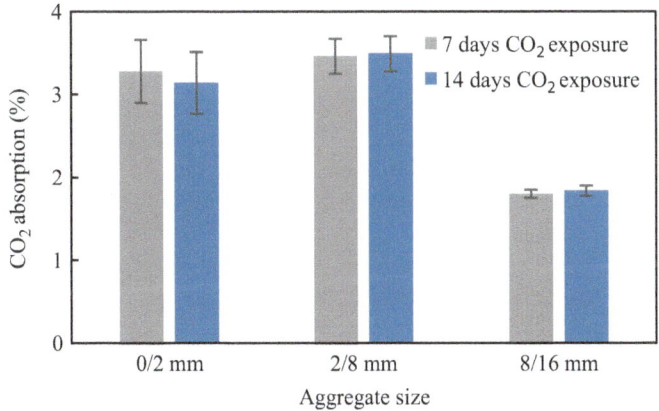

Fig. 3 CO_2-absorption of aggregates A (M2) (recycled from mix M2, including recycled aggregates)

be considered as non-absorbent parts (Fig. 2). Old concrete usually tends to be crushed through their weakest part in the crusher. Old mortar content in recycled aggregate is the dominant weak part compared to the natural stone content [22]. This observation aligns with the differing fracture mechanism of normal-strength and high-strength concrete. In normal-strength concrete, the rupture occurs through the binder, while the high-strength concrete usually ruptures through the aggregate [23]. Given the weaker nature of the old mortar, as well as the interfacial transition zone (ITZ) between the aggregate and mortar in normal-strength concrete, the old mortar content becomes a critical factor [23]. For this reason, the amount of old mortar paste content in the fine recycled aggregates is usually higher than its content in the larger grain fractions. Thus, the finer fractions, which have higher porosity, should have higher CO_2-absorption potential than the others. It is observed that the

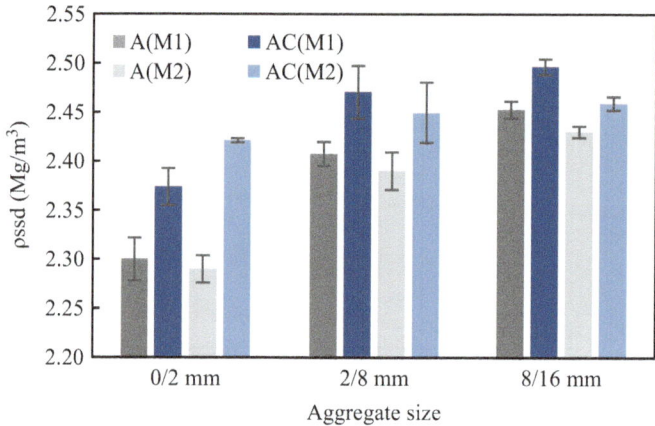

Fig. 4 Density of recycled AAC aggregate in saturated surface dried state

2/8 mm A(M2) absorbed 1% higher amount of CO_2 in comparison with 2/8 mm A (M1). However, the 8/16 mm aggregates from A(M1) and A(M2) showed almost similar CO_2-capturing potential.

The density variations of recycled AAC aggregates before and after carbonation in saturated surface-dried state (SSD) are shown in Fig. 4. It is seen that the fine aggregates with 0/2 mm size from both M1 and M2 mixtures have the lowest densities in comparison to the corresponding 2/8 mm and 8/16 mm aggregates. In contrast, 8/16 mm aggregates showed the highest densities compared to the corresponding aggregates with smaller sizes. This observation confirms the previously mentioned higher porosity in fine aggregates in comparison with coarse aggregates.

Furthermore, the carbonation of recycled AAC aggregate increases the density in the SSD state, regardless of the mix composition of old concrete and the size of the aggregate. The difference between the density of aggregates in the carbonated and non-carbonated state is larger for 0/2 mm than 2/8 mm fraction. Subsequently, that for the 2/8 mm is more significant than that for the 8/16 mm grains. The higher potential of CO_2-absorption in finer fractions, explained before, can be substantiated as the primary reason.

The water absorption of recycled AAC aggregates as one of the crucial parameters for the evaluation of porosity, which subsequently affects their durability properties, is illustrated in Fig. 5. The results highlight the positive influence of carbonation on the water absorption reduction of aggregates. The effect of carbonation on the water absorption variation depends on the size of aggregates. The fine recycled AAC aggregates (0/2 mm) show higher water absorption drop compared with 2/8 mm fraction, as well as 8/16 mm aggregates from the same source. Similarly, the coarse recycled AAC aggregates with 2/8 mm size show a bit larger water absorption than the 8/16 mm aggregates from the corresponding mixture.

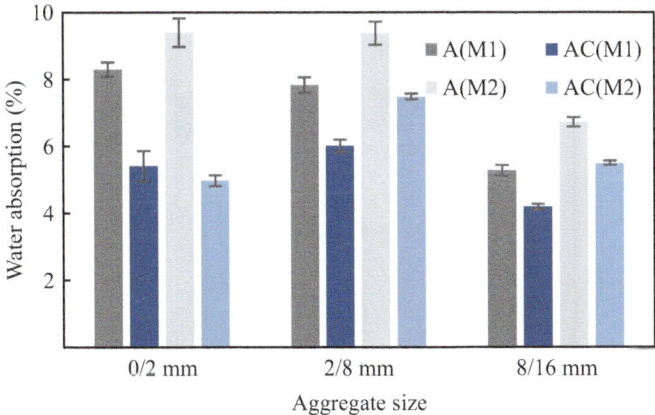

Fig. 5 Water absorption of recycled AAC aggregate

Thus, carbonation is meaningful not only in terms of CO$_2$-footprint decrease but also considering the positive influence on the durability behavior of recycled aggregates. It was also reported that the water absorption and porosity of RCA from OPC decreases through carbonation treatment [19].

3.2 Concrete

Figure 6 depicts the compressive strength of AAC mixture M3 and CM3, produced with recycled aggregates extracted from AAC mixture M1 in both non-carbonated and carbonated states named A(M1) and AC(M1), respectively. The use of carbonated recycled AAC aggregates in AAC concrete shows a higher compressive strength at all ages compared to the concrete made with the corresponding non-carbonated aggregates. The compressive strength development of AAC with non-carbonated aggregates seems to be retarded after 28 days. While its trend in AAC with carbonated aggregates is followed by a moderate increase after 28 days. Higher alkalinity of the aggregate ambient, originated from the presence of alkaline solution in newly made concrete could lead to the reactivation of non-activated precursors in recycled AAC aggregate.

Figure 7 illustrates the tensile splitting strength of AAC made with A(M1) and AC(M1). The trend of strength development is similar to that of compressive strength depicted in Fig. 6. The notable difference between the tensile strength of the mixtures made with the carbonated and non-carbonated aggregates at 91 days could be primarily because of the late reactivity of the recycled AAC aggregates. However, the carbonation of recycled AAC aggregates results in tensile strength enhancement of AAC at all ages.

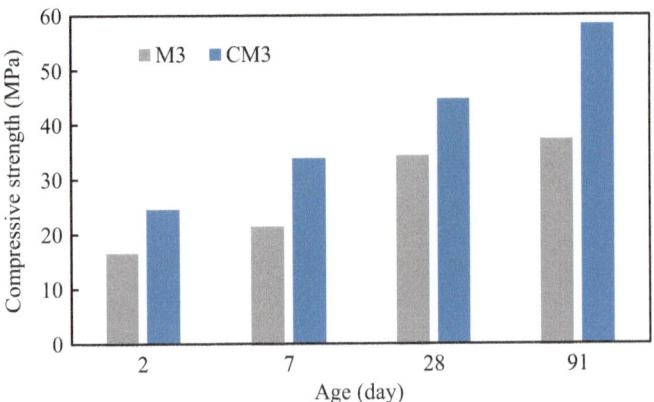

Fig. 6 Compressive strength of AAC with carbonated (CM3) and non-carbonated (M3) recycled AAC aggregate from mix M1

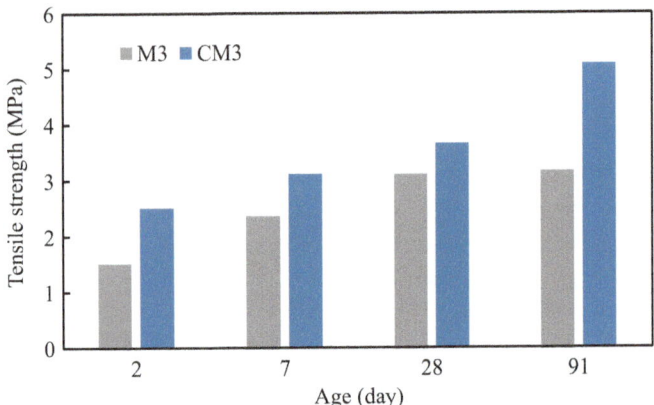

Fig. 7 Tensile splitting strength of AAC mixtures with carbonated (CM3) and non-carbonated (M3) recycled AAC aggregate from mix M1

The elastic modulus of AAC mixtures with A(M1) and AC(M1) aggregates is demonstrated in Fig. 8. The AAC mixture made with non-carbonated aggregates showed higher elastic modulus values than that made with carbonated aggregate, until 28 days age. It is also observed that the difference between the elastic modulus of both mixtures progressively diminishes until 28 days, reaching an equal point after 28 days. In contrary to early ages, the mixture made with carbonated recycled AAC aggregates showed a superior elastic modulus compared to the corresponding value of the mixture made with non-carbonated aggregates at 91 days. Despite the difference between the elastic modulus of both concrete mixes, this is not more than 10% of the total values at 28 or 91 days.

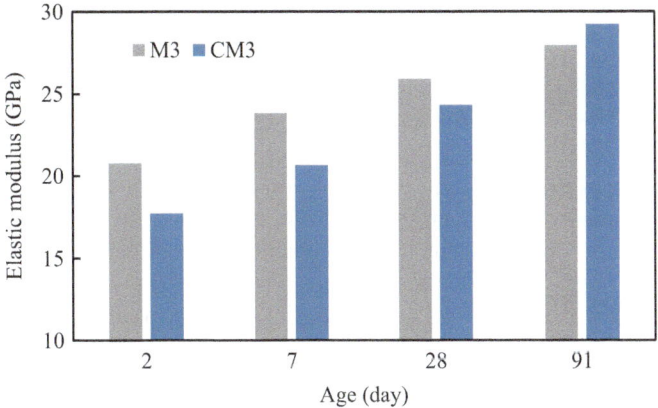

Fig. 8 Elastic modulus of AAC mixtures with carbonated (CM3) and non-carbonated (M3) recycled AAC aggregate from mix M1

4   Conclusion

1. One-week exposure of recycled aggregates to 1% CO$_2$ leads to a complete carbonation in all aggregate sizes (0/2, 2/8, and 8/16) tested.
2. Fine recycled aggregates from ACC can absorb up to 4 wt.% CO$_2$. This value can be considered for the coarse aggregates size having 8/16 mm size 2 wt.%.
3. Carbonation of recycled AAC aggregates increases their SSD densities. In addition, these aggregates can show approximately up to 4% reduction of the water absorption due to the carbonation. Smaller size aggregates are more influenced by carbonation in terms of density and water absorption than larger grains.
4. Compressive and tensile strength of AAC with recycled AAC aggregate improve by using carbonated aggregates. The strength increase increment at 91 days due to the carbonation of aggregates is considerably higher than that at earlier ages.
5. The elastic modulus of AAC with recycled AAC aggregates is negatively influenced by using carbonated recycled AAC aggregates until 28 days. This value shows a slight enhancement by the carbonation of these aggregates.

References

1. Global Alliance for Building and Construction: Global status report for building and construction: towards a zero-emissions, efficient and resilient buildings and construction sector (2019)
2. Global Alliance for Building and Construction: Global status report for building and construction: towards a zero-emissions, efficient and resilient buildings and construction sector (2022)
3. Nehdi, M.L., Yassine, A.: Mitigating Portland cement CO$_2$ emissions using Alkali-activated materials: system dynamics model. Materials. **13**(20) (2020). https://doi.org/10.3390/ma13204685

4. Benhelal, E., Zahedi, G., Shamsaei, E., et al.: Global strategies and potentials to curb CO2 emissions in cement industry. J Cleaner Prod. **51**, 142–161 (2013). https://doi.org/10.1016/j. jclepro.2012.10.049
5. Bernal, S.A., Mejía de Gutiérrez, R., Pedraza, A.L., et al.: Effect of binder content on the performance of alkali-activated slag concretes. Cement Concrete Res. **41**(1), 1–8 (2011). https:// doi.org/10.1016/j.cemconres.2010.08.017
6. Mahmood, A.H., Foster, S.J., Castel, A.: Effects of mixing duration on engineering properties of geopolymer concrete. Const Building Materials. **303**, 124449 (2021). https://doi.org/10. 1016/j.conbuildmat.2021.124449
7. Matthys, S., Ghorbani, S., Krajnovic, I., et al.: The Urban Concrete Innovation. Zenodo (2023)
8. Peter, M.A., Muntean, A., Meier, S.A., et al.: Competition of several carbonation reactions in concrete: a parametric study. Cement Concrete Res. **38**, 1385–1393 (2008)
9. Papadakis, V., Vayenas, C.G., Fardis, M.N.: A reaction engineering approach to the problem of concrete carbonation. AIChE. **10**(35), 1639–1650 (1989)
10. Puertas, F., Palacios, M., Vázquez, T.: Carbonation process of alkali-activated slag mortars. J Materials Sci. **41**(10), 3071–3082 (2006). https://doi.org/10.1007/s10853-005-1821-2
11. Tavasoli S., Breit W.: Investigation of porosity and carbonation depth in alkali-activated GGBS mortar. In: 4th International Rilem Conference on Microstructure Related Durability of Cementitious Composites, pp 842–850. https://repository.tudelft.nl/islandora/object/uuid:4ed97cf1-c6a9-4dac-bd67-adb371ffba04
12. BTB/Bettertimes: Jahresbericht BTB: Große Aufgaben—Gemeinsame Lösungen (2022).
13. Interreg Europe: Collection and recycling of construction and demolition waste: key learnings (2022).
14. Nedeljković, M., Visser, J., Šavija, B., et al.: Use of fine recycled concrete aggregates in concrete: a critical review. J Building Eng. **38**, 102196 (2021). https://doi.org/10.1016/j.jobe. 2021.102196
15. European Parliament, Council of the European Union: Directive 2008/98/EC on waste (98) (November 22/2008).
16. Colangelo, F., Navarro, T.G., Farina, I., et al.: Comparative LCA of concrete with recycled aggregates: a circular economy mindset in Europe. Int J Life Cycle Assess. **25**(9), 1790–1804 (2020). https://doi.org/10.1007/s11367-020-01798-6
17. Leemann, A., Münch, B., Wyrzykowski, M.: CO2 absorption of recycled concrete aggregates in natural conditions. Materials Today Communications. **36**, 106569 (2023). https://doi.org/10. 1016/j.mtcomm.2023.106569
18. Andersson, R., Fridh, K., Stripple, H., et al.: Calculating CO_2 uptake for existing concrete structures during and after service life. Env Sci Technol. **47**, 11625–11633 (2013). https://doi. org/10.1021/es401775w
19. Sereng, M., Djerbi, A., Metalssi, O.O., et al.: Improvement of recycled aggregates properties by means of CO_2 uptake. Appl Sci. **11**(14), 6571 (2021). https://doi.org/10.3390/app11146571
20. Russo, N., Lollini, F.: Effect of carbonated recycled coarse aggregates on the mechanical and durability properties of concrete. J Building Eng. **51**, 104290 (2022). https://doi.org/10.1016/j. jobe.2022.104290
21. Chemanalyst: Granulated Ground Blast Furnace Slag (GGBFS) Market Analysis: Industry Market Size, Plant Capacity, Production, Operating Efficiency, Demand & Supply, End-User Industries, Demand by Type, Sales Channel, Company Share, Foreign Trade, Regional Demand, Manufacturing Process, Policy and Regulatory Landscape, 2015-2035 (2023)
22. Kang, M., Weibin, L., Nicolais, L.: Effect of the aggregate size on strength properties of recycled aggregate concrete. Adv Materials Sci Eng. **2018**, 2428576 (2018). https://doi.org/ 10.1155/2018/2428576
23. Thilakarathna, P., Kristombu Baduge, K.S., Mendis, P., et al.: Understanding fracture mechanism and behaviour of ultra-high strength concrete using mesoscale modelling. Eng Fracture Mech. **234**, 107080 (2020). https://doi.org/10.1016/j.engfracmech.2020.107080

Application of Iron Mine Waste Rock as an Innovative Cement Replacement Material in Mortar

Bruna Figueiredo Cezar, Margareth da Silva Magalhães, and André Rocha Pimenta

Contents

1   Introduction

Crude iron ore reserves are distributed among various countries, with Brazil, Australia, and China collectively contributing to 63% of the world's iron ore production [1]. Brazil, in particular, plays a significant role in global iron ore production, accounting for approximately 19% of the total output [2]. With an annual production of around 430 million tons [3], Brazil holds a prominent position in the global iron ore market, representing about 5% of its GDP [4].

However, the mining sector encounters a significant challenge related to the substantial waste generated during mineral extraction and processing [5]. According to Kalisz et al. (2022), this sector generates an estimated 65 billion tons of waste

B. F. Cezar · M. da Silva Magalhães (✉)
Post-Graduate Program in Civil Engineering, State University of Rio de Janeiro, Rio de Janeiro, Brazil
e-mail: margareth.magalhaes@eng.uerj.br

A. R. Pimenta
Instrumentation and Computational Simulation Laboratory, Federal Institute of Rio de Janeiro, Rio de Janeiro, Brazil

M. Kioumarsi, B. Shafei (eds.), *The 1st International Conference on Net-Zero Built Environment*, Lecture Notes in Civil Engineering 237,
https://doi.org/10.1007/978-3-031-69626-8_19

annually, posing significant storage and environmental management challenges [6]. For instance, North Americans produce over ten times more solid mine waste per capita than household garbage, while China's solid mine waste inventory approaches 70 billion tons [7, 8].

The main wastes of this operation are waste rock and iron ore tailings (IOT). The first represents a layer of soil and rock to be removed to access the ore and, which has no economic value [9], and the second is the material discarded in the ore processing stage [10]. The waste/metallic ore ratio resulting from surface mining is typically between 2 and 5, depending on location-specific conditions [11]. In Brazil, between 2010 and 2019, approximately 3.9 billion tons of waste rock and 1.3 billion tons of IOT were generated [12].

In general, both wastes are deposited in piles and dams [13, 14]. These geotechnical structures require adequate management and constant monitoring to guarantee stability and safety, which represents a high cost to the entrepreneur [12]. There are several records of pile and dam failures that resulted in the loss of human life, loss of equipment, and environmental damage [15–18].

Consequently, utilizing waste materials in cementitious materials represents a more sustainable alternative, than their disposal in piles or dams [12]. While there is ample literature on the use of IOT in civil construction (as summarized in Table 1), studies on the utilization of waste rock remain limited.

Therefore, incorporating waste rock in mortar production not only represents a sustainable solution for mining waste disposal but also offers a readily available means of reducing the amount of cement in mortar mixtures, thus mitigating greenhouse gas emissions associated with cement production, notably carbon dioxide.

Table 1 Use of waste from iron ore extraction in civil construction

Waste	Application	References
IOT	Clinker production	[19, 20]
	Bricks and ceramic tile	[21–24]
	Concrete interlocking blocks	[25]
	Aerated concrete	[26, 27]
	Paving layer	[28–31]
	Mineral admixture in concretes and cement pastes	[13, 32–35]
	Mineral admixture in fiber-reinforced cementitious composites	[36, 37]
	Aggregate in mortars and concretes	[38–43]
	Aggregate in fiber-reinforced cementitious composites	[44]
	Geopolymer	[45]
	Pigment for sustainable paints	[46]
	Dye glaze for ceramics	[47]
Waste rock	Aggregate in concretes	[11, 48, 49]
	Asphalt concretes	[50]
	Mineral admixture	[51]

2 Experimental Program

2.1 Materials

The mortar mixtures comprised Portland cement type V-ARI, conforming to Brazilian standard NBR 16697 [52], waste rock, sand with a maximum diameter of 300 μm and a density of 2.64 g/cm^3, and water. The cement has between 90 and 100% clinker plus calcium sulfate and up to 10% carbonaceous material, density of 3.08 g/cm^3 and a fineness, on a 45 μm sieve, of 0.6%.

The waste rock utilized in this study was supplied by the vale mining company and was proven by the extracted of iron ore in cava conceição, situated in Itabira, quadrilaterro ferrífero region, Minas Gerais, Brazil. Following mineral removal, via detonation, the waste was transported to the disposal site, where it was collected. Prior to using the waste rock, it underwent sieving on a 75 μm sieve.

The waste rock has density of 2.79 g/cm^3, fineness on 45 μm sieve of 27,8%, and moisture content of 0.4%. X-ray fluorescence (XRF) analysis determined the waste rock's oxide composition (refer to Table 2), revealing SiO_2 + Al_2O_3 + Fe_2O_3 content exceeding 70% (93.60%), low calcium oxide (CaO) content (0.59%), and absence of SO_3. Additionally, trace amounts of toxic compounds, including chromium, lead, and copper, were detected, each constituting less than 0.001%. However, the waste rock exhibited a strength activity index with Portland cement at 28 days, as determined by NBR 5752 [53], of 72%, falling below the 75% threshold outlined in ASTM C618 [54].

Figure 1 illustrates the ternary diagram representing the CaO - SiO_2 - (Al_2O_3 + Fe_2O_3) system of waste rock, IOT, and various mineral additions. It is noteworthy that the composition of the waste rock utilized in this study closely mirrors that of fly ash [55–58], metakaolin [59–61], Brazilian IOT [32, 34], and Brazilian waste rock sourced from another source [51].

The waste's morphology was examined via scanning electron microscope (SEM) images, revealing grains with lamellar shapes (refer to Fig. 2). Energy-dispersive X-ray spectroscopy (EDS) analysis showed that the sample is composed of silicon (72.1%), aluminum (13.3%), iron (10.3%), and potassium (4.3%), consistent with its chemical composition.

Mineralogical analysis via X-ray diffraction (XRD) identified crystalline phases in the waste rock sample, mainly composed of quartz (62.8%), kaolinite (27.1%), and biotite (9.0%). Goethite and hematite are present in concentrations below 1%.

Table 2 Major chemical compounds of waste rock

Compound	SiO_2	Al_2O_3	Fe_2O_3	K_2O	TiO_2	P_2O_5	CaO	Other
Waste rock (%)	60.10	21.40	12.10	3.74	1.07	0.59	0.32	< 0.01

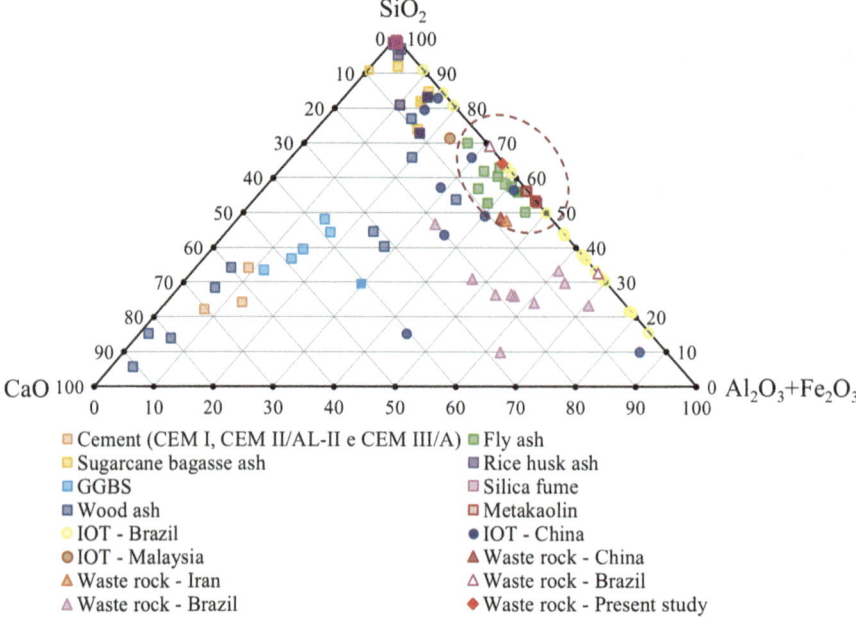

Fig. 1 Ternary diagram of waste rock and other types of mineral additions

Fig. 2 SEM images with approximation of (**a**) 500× and (**b**) 1000×

2.2 Mixtures Design Details

Mortar mixtures were designed with cement partially replaced by 10%, 20% and 30% waste rock (in mass). The water/(cement + waste rock) ratio was maintained at 0.40, and the sand/(cement + waste rock) ratio was maintained at 0.36. Details of the mixture designs are presented in Table 3.

Table 3 Material quantities used in the mortar mixtures

Mix	Waste rock/ (cement + waste rock) (%)	Quantities (kg/m^3)			
		Cement	waste rock	Sand	water
B0	0	1152.90	0	415.03	461.14
B10	10	1026.40	114.04	410.54	456.16
B20	20	902.56	225.64	406.15	451.28
B30	30	781.39	334.88	401.86	446.51

The consistency of mortars was evaluated with spread values determined in accordance with NBR 13276 [62]. Densities of fresh mortars were obtained by weighing the compacted fresh mortar in a vessel of 400 ml capacity. Following molding, mortar specimens were cured in a water tank until the testing day.

2.3 Experimental Tests

The test procedure involved determining the compressive strength, young's modulus, splitting tensile strength, density, water absorption, and total porosity of mortars, by using three cylindrical specimens measuring 50 × 100 mm (diameter × height). Compressive strength was assessed at 7 and 28 days, in line with the NBR 7215 [63] standard. Splitting tensile tests were conducted at 28 days, following the guidelines of NBR 7222 [64], while density, water absorption, and porosity tests, at 28 days, adhered to the NBR 9778 [65] standard. Young's modulus was obtained by fitting a linear trend to the compressive strength versus strain curves up to 30% compressive strength.

3 Results

3.1 Physical Properties

The spread values determined by consistency tests resulted in measurements of 245 mm (B0), 237 mm (B10), 239 mm (B20), and 221 mm (B30). It was noted that the in-corporation of waste rock slightly enhanced the workability of the mixtures. This improvement may potentially enable a reduction in the water/cement ratio while maintaining consistent workability. Densities of fresh mortars presented values of 2.08 g/cm^3 (B0), 2.03 g/cm^3 (B10), 2.02 g/cm^3 (B20), and 2.01 g/cm^3 (B30).

After 28 days of curing, water absorption, porosity, and density tests were conducted on both the reference mortar (B0) and waste rock-mortars (B10, B20 and B30). Figure 3a illustrates a notable augment in water absorption of all waste rock-mortars in comparison to the B0 mortar. This rise in water absorption, observed

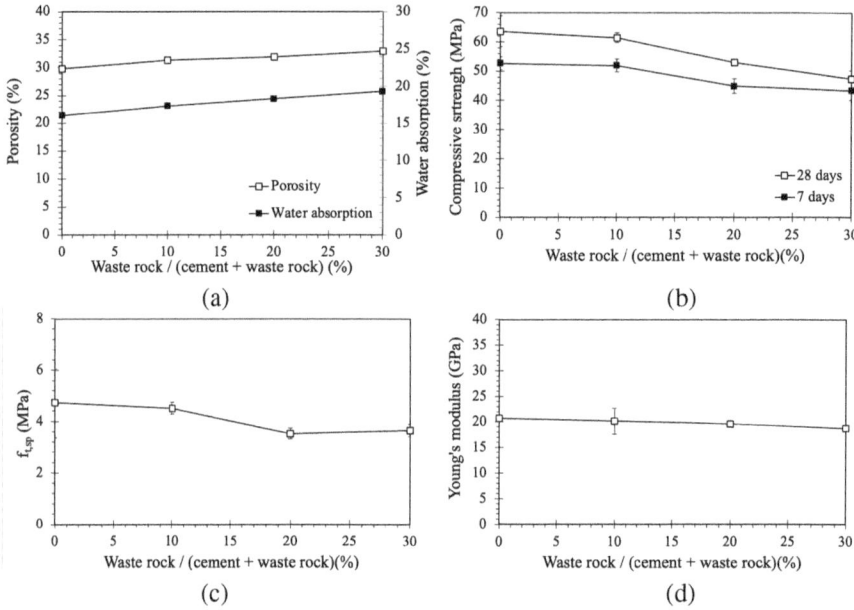

Fig. 3 Influence of waste rock amount on the (**a**) water absorption and porosity, (**b**) compressive strength, (**c**) splitting tensile strength, and (**d**) young's modulus

on waste rock-mortars can be estimated as up to 20%, when used 30% waste rock (B30), and can be attributed to the higher porosity of waste rock-mortars, as depicted in Fig. 3a. About density results, the mortars exhibited densities at 28 days of 1.86 g/cm^3 (B0), 1.81 g/cm^3 (B10), 1.74 g/cm^3 (B20), and 1.70 g/cm^3 (B30). Similar to fresh densities results, the reduced densities were due to lower densities of waste rock compared to cement density.

3.2 Mechanical Properties

In Fig. 3b, the compressive strength of mortars at 7 and 28 days is depicted. At both ages, it is evident that the increased substitution of cement with waste rock led to a reduction in the compressive strength of the mixtures, when compared to the mixture without waste rock (B0). Specifically, with 10% waste rock, the reduction was not significant at 7 days, but was 4%, at 28 days. With 20% waste rock, the compressive strength decreased by 15%, at 7 days, and 17%, at 28 days, while with 30% waste rock, the reductions were 18%, at 7 days, and 26%, at 28 days. This trend primarily stems from the lower cement content in the waste rock-mortars, rather than a slower cement hydration reaction. This is evident when observed the evolution of compressive strength on the time.

The 7-day compressive strength of the reference mortar (B0) equated to approximately 83% of its 28-day compressive strength. Conversely, specimens containing 10%, 20%, and 30% waste rock exhibited a relatively higher rate of compressive strength development (B10: 84%, B20: 85%, and B30: 92%), attributed to the nucleation effect at early ages, when cement was replaced by waste rock, enhancing the hydration reaction of the aluminates [34].

Although the compressive strength of waste rock-mortars has been reduced, an analysis of the eco-efficiency performances of waste rock-mortars at 28 days, by using the binder intensity (bi) index, proposed by Damineli et al. [66], indicates that waste rock-mortar was more efficient than the reference mortar because the bi of reference mortar (18.2 kg/m^3.MPa) was higher than the bi of waste rock-mortars (B10: 17 kg/m^3.MPa, B20: 17.4 kg/m^3.MPa and B30: 17 kg/m^3.MPa). According to Damineli et al. [66], the bi index quantifies the overall amount of binder required to achieve 1 MPa of compressive strength.

Similar to compressive strength results, Fig. 3c indicates that the tensile strength was not significantly changed when 10% cement was replaced by waste rock. However, when cement was replaced by 20% and 30% waste rock, the tensile strength was reduced up to 25%. Young's modulus presented the similar behavior, with a maximum loss of 11% (Fig. 3d), when compared with the reference mortar.

4 Conclusion

The aim of this study was to identify and assess the primary characteristics of mining waste rock produced in Brazilian region. Additionally, the feasibility of using waste rock as a partial substitute for Portland cement in mortar mixture was explored. The investigation yielded the following key findings:

The physical–chemical, mineralogical, and morphological analyses of waste rock revealed it to be an inert material, suitable as an alternative filler for mortar mixtures.

Optimal performance was observed with a 10% replacement of cement with waste rock, as it did not significantly alter mechanical properties. In other way, the use of 20% or 30% waste rock led to reductions in both compressive and tensile strength and young's modulus.

Incorporating waste rock in mortar mixtures resulted in a shortened hydration period, particularly evident in compressive tests conducted at 7 days, attributed to a nucleation effect.

While waste rock had a diminishing effect on the modulus of elasticity and density of the mortars, its impact was less pronounced compared to its influence on compressive strength. Conversely, water absorption and porosity increased in waste rock mortars compared to those without waste rock.

Although the inclusion of waste rock in the mixture led to reduced mechanical strength, it was found that all mixtures containing waste rock were more environmentally efficient compared to those without it. This indicates the feasibility and sustainability of mortars with levels of 10%, 20% and 30% cement replacement by this filler.

Based on the findings, additional strategies could be identified for treating the waste rock prior to its incorporation into cementitious mixtures. To this end, the authors are presently exploring the effects of thermal pretreatment, aiming to enhance the waste's reactivity.

Acknowledgments The funding for this study was provided by Coordenação de Aperfeiçoamento de Pessoal de Nível Superior (CAPES) and Fundação de Amparo à Pesquisa do Estado do Rio de Janeiro (FAPERJ). The authors extend their gratitude to Vale company, for supplying the waste rock used in the research.

References

1. Chemale Jr., Takehara, L., de Ferro, M.: Geologia e Geometarlurgia. Publisher Blucher, São Paulo (2013)
2. United State Geological Survey (USGS): National minerals information center e mineral commodity summaries. 2019. https://www.usgs.gov/centers/nmic/iron-ore-statistics-and-information
3. Brazilian Mining Institute (IBRAM), Mineral economy of Brazil: https://agenciabrasil.ebc.com.br/geral/noticia/2020-02/ibram-producao-de-minerio-em-2019-caiu-mas-faturamento-cresceu
4. Alves, F. E. Brasil Mineral Perspectivas mineração brasileira deve retomar crescimento. 40 anos. 7° Mineração &/X Comunidades. Fórum Brasil Mineral 2022, janeiro/fevereiro, n° 426 (2023)
5. Bian, Z., Lie, S., Chen, S.: The challenges of reusing mining and mineral-processing wastes. Science. **337**, 702–703 (2012)
6. Kalisz, S., Kibort, K., Mioduska, J., Lieder, M., Małachowska, A.: Waste management in the mining industry of metals ores, coal, oil and natural gas - a review. J. Environ. Manag. **304**, 114239 (2022)
7. Jamieson, H.E.: Geochemistry and mineralogy of solid mine waste: essential knowledge for predicting environmental impact. Elements. **7**, 381–386 (2011)
8. Wang, H., Wang, Y., Li, W., Qiao, J.: Report on Conservation and Comprehensive Utilization of Mineral Resources in China. Geological Press, Beijing (2019)
9. Yin, S., Yan, Z., Chen, X., Yan, R., Chen, D., Chen, J.: Mechanical properties of cemented tailings and waste-rock backfill (CTWB) materials: laboratory tests and deep learning modeling. Constr. Build. Mater. **369**, 130610 (2023)
10. BRGM: Management of mining, quarrying, ore processing wastes in the European Union; p. 79, RP-50319-F (2001)
11. Yellishettya, M., Karpe, V., Reddy, E.H., Subhash, K.N., Ranjitha, P.G.: Reuse of iron ore mineral wastes in civil engineering constructions: a case study. Resour. Conserv. Recycl. **52**, 1283–1289 (2008)
12. Agência nacional de mineração – ANM: Análise de Impacto Regulatório—AIR 48051.000384/2020–43, Aproveitamento de estéril e rejeitos (2020)
13. Simonsen, A.M.T., Solismaa, S., Hansen, H.K., Jensen, P.E.: Evaluation of mine tailings' potential as supplementary cementitious materials based on chemical, mineralogical and physical characteristics. Waste Manag. **102**, 710–721 (2020)
14. Edraki, M., Baumgartl, T., Manlapig, E., Bradshaw, D., Franks, D.M., Moran, C.J.: Designing mine tailings for better environmental, social and economic outcomes: a review of alternative approaches. J. Clean. Prod. **84**, 411–420 (2014)

15. Gupta, G., Sharma, S.K., Singh, G.S.P., Kishore, N.: Numerical modelling-based stability analysis of waste dump slope structures in open-pit mines-a review. J. Inst. Eng. India Ser. D. **102**, 589 (2021)
16. World Information Service on Energy (Wise) Uranium: Chronology of major tailings dam failures. WISE Uranium Project. http://www.wiseuranium.org/mdaf.htm (2019)
17. Ke, X., Zhou, X., Wang, X., Wang, T., Hou, H., Zhou, M.: Effect of tailings fineness on the pore structure development of cemented paste backfill. Constr. Build. Mater. **126**, 345–350 (2016)
18. Yang, L., Qiu, J., Jiang, H., Hu, S., Li, H., Li, S.: Use of cemented super-fine unclassified tailings backfill for control of subsidence. Fortschr. Mineral. **7**, 216 (2017)
19. Zheng, Y.C., Liu, Q., Li, Y.J.: Mine tailing as alternative to clay for producing belite cement clinker. Adv. Mater. Res. **726-731**, 2704–2271 (2013)
20. Yang, C., Cuin, C., Qin, J.: Recycling of low-silicon iron tailings in the production of lightweight aggregates. Ceram. Int. **41**, 1213–1221 (2015)
21. Yao, G., Wang, Q., Wang, Z., Wang, J., Lyu, X.: Activation of hydration properties of iron ore tailings and their application as supplementary cementitious materials in cement. Powder Technol. **360**, 863–871 (2020)
22. Yang, C., Cui, C., Qin, J.: Characteristics of the fired bricks with low-silicon iron tailings. Constr. Build. Mater. **70**, 36–42 (2014)
23. Chen, Y., Huang, F., Li, W., Liu, R., Li, G., Wei, J.: Test research on the effects of mechanochemically activated iron tailings on the compressive strength of concrete. Constr. Build. Mater. **118**, 164–170 (2016)
24. Das, S.K., Kumar, S., Ramachandrarao, P.: Exploitation of iron ore tailing for the development of ceramic tiles. Waste Manag. **20**, 725–729 (2000)
25. Sant'ana Filho, J.N.: Estudo de reaproveitamento dos resíduos das barragens de minério de ferro para fabricação de blocos Inter travados de uso em pátios industriais e alto tráfego. Dissertação de Mestrado em Engenharia Civil. CEFET, Belo Horizonte, 130 p (2013)
26. Cai, L., Ma, B., Li, X., Lv, Y., Liu, Z., Jian, S.: Mechanical and hydration characteristics of autoclaved aerated concrete (AAC) containing iron-tailings: effect of content and fineness. Constr. Build. Mater. **128**, 361–372 (2016)
27. Wang, C.L., Ni, W., Zhang, S.Q., Wang, S., Gai, G.S., Wang, W.-K.: Preparation and properties of autoclaved aerated concrete using coal gangue and iron ore tailings. Constr. Build. Mater. **104**, 109–115 (2016)
28. Campanha, A.: Caracterização de rejeitos de minério de ferro para uso em pavimentação. 2011. 106p. Dissertação de Mestrado - Universidade Federal de Viçosa, Programa de Pós-Graduação em Engenharia Civil. Viçosa - MG (2011)
29. Pinto, S.S.S.: Caracterização das propriedades físicas e mecânicas de misturas de diferentes tipos de rejeito para aplicação em pavimentos. Dissertação de Mestrado em Engenharia Civil - Universidade Federal de Viçosa. Viçosa - MG, 100p (2013)
30. Bastos, L.A.C., Silva, G.C., Mendes, J.C., Peixoto, R.A.F.: Using iron ore tailings from tailing dams as road. Mater. J. Mater. Civ. Eng. **28**, 04016102 (2016)
31. Oliveira, T.M., Generoso, F.J., Silva, T.O., Sant'Anna, G.L., Silva, C.H.D.C., Pitanga, H.N.: Geomechanical properties of mixtures of iron ore tailings improved with Portland cement. Acta Sci. Technol. **41**, 38038 (2019)
32. Almada, B.S.: Influência da heterogeneidade de rejeitos de minério de ferro utilizados como adição mineral nas propriedades de microconcretos. Dissertação de mestrado em Engenharia Civil, Universidade Federal de Minas Gerais (2021)
33. Zhao, Y., Qiu, J., Ma, Z.: Temperature-dependent rheological, mechanical and hydration properties of cement paste blended with iron tailings. Powder Technol. **381**, 82–91 (2021)
34. Bezerra, C.G., Rocha, C.A.A., Siqueira, I.S., Filho, R.D.T.: Feasibility of iron-rich ore tailing as supplementary cementitious material in cement pastes. Constr. Build. Mater. **303**, 124496 (2021)
35. Han, F., Li, L., Song, S., Liu, J.: Early-age hydration characteristics of composite binder containing iron tailing powder. Powder Technol. **315**, 322–331 (2017)

36. Pedroso, D.E.: Aproveitamento do rejeito de minério de ferro em compósitos para construção civil. Tese, Doutorado em Engenharia Civil. Universidade Tecnológica Federal do Paraná. Curitiba, 115p (2020)
37. Huang, X., Ranade, R., Li, V.C.: Feasibility study of developing green ECC using Iron Ore Tailings (IOT) powder as cement replacement. ASCE J. Mater. **25**(7), 923–931 (2013)
38. Argane, R., Benzaazoua, M., Bouamrane, A., Hakkou, R.: Cement hydration and durability of low sulfide tailings-based renders: a case study in Moroccan construction. Miner. Eng. **76**, 97–108 (2015)
39. Fontes, W.C., Mendes, J.C., Da Silva, S.N., Peixoto, R.A.F.: Mortars for laying and coating produced with iron ore tailings from tailing dams. Constr. Build. Mater. **112**, 988–995 (2016)
40. Zhu, Q., Yuan, Y.-X., Chen, J.-H., Fan, L., Yang, H.: Research on the high-temperature resistance of recycled aggregate concrete with iron tailing sand. Constr. Build. Mater. **327**, 126889 (2022)
41. Protasio, F.N.M., de Avillez, R.R., Letichevsky, S., de Silva, F.: The use of iron ore tailings obtained from the Germano dam in the production of a sustainable concrete. J. Clean. Prod. **278**, 123929 (2021)
42. Zhao, S., Fan, J., Sun, W.: Utilization of iron ore tailings as fine aggregate in ultra-high performance concrete. Constr. Build. Mater. **50**, 540–548 (2014)
43. Zhang, W., Gu, X., Qiu, J., Liu, J., Zhao, Y., Li, X.: Effects of iron ore tailings on the compressive strength and permeability of ultra-high performance concrete. Constr. Build. Mater. **260**, 119917 (2020)
44. Huang, X., Ranade, R., Ni, W., Li, V.C.: Development of green engineered cementitious composites using iron ore tailings as aggregates. J. Constr. Build. Mater. **44**, 757–764 (2013)
45. Duan, P., Yan, C., Zhou, W., Ren, D.: Development of fly ash and iron ore tailing based porous geopolymer for removal of Cu (II) from wastewater. Ceram. Int. **42**, 13507–13518 (2016)
46. Galvão, J.L.B., Andrade, H.D., Brigolini, G.J., Peixoto, R.A.F., Mendes, J.C.: Reuse of iron ore tailings from tailings dams as pigment for sustainable paints. J. Clean. Prod. **200**, 412–422 (2018)
47. Pereira, O.C., Bernardin, A.M.: Ceramic colorant from untreated iron ore residue. J. Hazard. Mater. **233–234**, 103–111 (2012)
48. Lopes, D.F., Silva, A.C., de Barros, M.R., Silva, E.M.S.: Reuse of mining waste rock as coarse aggregate for the manufacture of concrete. Technol. Metal. Mater. Miner., São Paulo. **17**(1), 30-36 (2020)
49. Gayana, B.C., Chandar, K.R.: A study on suitability of iron ore overburden waste rock for partial replacement of coarse aggregates in concrete pavements. In: 14th International Conference on Concrete Engineering and Technology. IOP Conf. Series: Materials Science and Engineering, vol. 431, p. 102012 (2018)
50. Shamsi, M., Zakerinejad, M.: Production of sustainable hot mix asphalt from the iron ore overburden residues. Transp. Res. D. **123**, 103926 (2023)
51. Seerig, T., Brandão, P.R.G., Gama, E.M.: Characterization of iron mining waste rock for use as raw material for pozzolan production. XXVIII National Meeting on Ore Treatment and Extractive Metallurgy Belo Horizonte-MG, 4th to 8th November (2019)
52. Brazilian Standard NBR 16697: Portland cement—Requirements. Brazilian Association of Technical Standards (ABNT), Rio de Janeiro (2018)
53. Brazilian Standard NBR 5752: Pozzolanic materials — Determination of the performance index with Portland cement at 28 days. Associação Brasileira de Normas Técnicas (ABNT). Rio de Janeiro (2015)
54. ASTM C618-22: Standard Specification for Coal Fly Ash and Raw or Calcined Natural Pozzolan for Use in Concrete. American Society for Testing and Materials (2022)
55. Magalhaes, M.S., Cezar, B.F., Lustosa, P.R.: Influence of Brazilian fly ash fineness on the cementing efficiency factor, compressive strength and Young's modulus of concrete. Dev. Built Environ. **14**, 100147 (2023)

56. Zhang, Y., Ma, Z., Zhi, X., Chen, X., Zhou, J., Wei, L., Liu, Z.: Damage characteristics and constitutive model of phosphogypsum/fly ash/slag recycled aggregate concrete under uniaxial compression. Cem. Concr. Compos. **138**, 104980 (2023)
57. Wang, Q., Yi, Y., Ma, G., Luo, H.: Hybrid effects of steel fibers, basalt fibers and calcium sulfate on mechanical performance of PVA-ECC containing high-volume fly ash. Cem. Concr. Compos. **97**, 357–368 (2019)
58. Li, Y., Li, J., Yang, E.H., Guan, X.: Mechanism study of crack propagation in river sand Engineered Cementitious Composites (ECC). Cem. Concr. Compos. **128**, 104434 (2022)
59. Babaahmadi, A., Machner, A., Kunther, W., Figueira, J., Hemstad, P., Weerdt, K.: Chloride binding in Portland composite cements containing metakaolin and silica fume. Cem. Concr. Res. **161**, 106924 (2022)
60. Da Cruz, T.A.M., Geraldo, R.H., Costa, A.R.D., Maciel, K.R.D., Gonçalves, J.P., Camarini, G.: Microstructural and mineralogical compositions of metakaolin-lime-recycled gypsum plaster ternary systems. J. Build. Eng. **47**, 103770 (2022)
61. Jing, H., Li, M., Zhang, Y., Gao, M.: Hydration kinetics, microstructure and physicochemical performance of metakaolin-blended cementitious composites. Constr. Build. Mater. **408**, 133756 (2023)
62. Brazilian Standard NBR 13276: Argamassa para assentamento e revestimento de paredes e tetos—Determinação do índice de consistência. Associação Brasileira de Normas Técnicas (ABNT), Rio de Janeiro (2016)
63. Brazilian Standard NBR 7215: Cimento Portland - Determinação da resistência à compressão de corpos de prova cilíndricos. Associação Brasileira de Normas Técnicas, Rio de Janeiro (2019)
64. Brazilian Standard NBR 7222: Concreto e argamassa — Determinação da resistência à tração por compressão diametral de corpos de prova cilíndricos. Associação Brasileira de Normas Técnicas (ABNT), Rio de Janeiro (2011)
65. Brazilian Standard NBR 9778: Concreto e argamassa endurecida– Determinação da absorção de água, índice de vazios e massa específica. Brazilian Associação Brasileira de Normas Técnicas (ABNT), Rio de Janeiro (2005)
66. Damineli, B.L., Kemeid, F.M., Aguiar, P.S., John, V.M.: Measuring the eco-efficiency of cement use. Cement Concr. Compos. **32**, 555–562 (2010)

Feasibility of Utilizing Treated Domestic Wastewater (TDW) for the Production of Concrete

Hans Beushausen, Zaid Manuel, Joanitta Ndawula, and Dyllon Randall

Contents

1 Introduction

The global water demand is ever-increasing to such an extent that water scarcity will be one of the biggest problems facing the world in the coming years. Both in South Africa and abroad, water is rapidly becoming a scarce resource. The World Economic Forum ranked both the water crisis and social instability as the third highest threats to conducting business in South Africa in 2021 [1]. The Western Cape in South Africa experienced severe water scarcity from mid-2017 to mid-2018, and more recently in the hydrological year from 2021 to 2022, below-average rainfall resulted in 19% lower dam levels than the previous year [2].

H. Beushausen · Z. Manuel · J. Ndawula (✉) · D. Randall
Department of Civil Engineering, University of Cape Town, Cape Town, South Africa
e-mail: hans.beushausen@uct.ac.za; ndwjoa001@myuct.ac.za

© The Author(s) 2025   231
M. Kioumarsi, B. Shafei (eds.), *The 1st International Conference on Net-Zero Built Environment*, Lecture Notes in Civil Engineering 237,
https://doi.org/10.1007/978-3-031-69626-8_20

An often overlooked factor that may contribute to the alleviation of water scarcity and increase the sustainability of concrete is water management. The United Nations Environmental Program has indicated that over the entire life cycle, the construction industry accounts for as much as 30% of global freshwater use [3]. Water is used throughout the concrete production process, from the generation of power to washing the aggregates, mixing and transportation of fresh concrete and concrete curing. With proper planning and management, the strain on freshwater resources could be reduced globally by using alternative water sources for many of these activities in the concrete construction industry.

Research suggests that non-potable water such as treated domestic wastewater (TDW) could be used to produce concrete of suitable quality for the construction industry, thereby reducing the industry's demand for potable water resources [4]. TDW is domestic wastewater that has been treated to a level that it can safely be discharged into water bodies. TDW is typically 99.9% pure water with the remaining 0.1% being made up of suspended and dissolved solids, and microorganisms [5]. TDW has the potential to substitute the water requirements of the concrete industry, provided it fulfills the requirements for concrete production.

Common substances found in non-potable water which can at certain concentrations be deleterious for concrete include chlorides (may affect the setting time of fresh concrete and promote reinforcement corrosion in hardened concrete), sugar (can result in set retardation), sulfates and acids (chemical attack), and organic matter (may influence cement hydration or cause air entrainment) [6].

While various secondary water resources can be considered for concrete production, each alternative source needs to be investigated with regard to the requirements set in national standards. In South Africa, SANS 51008 [7] (equivalent to ASTM C1602 [8] and BS EN 1008 [9]) stipulates acceptable levels of harmful substances in concrete mixing water, including chlorides, sulfates, alkalis, sugar, phosphates, nitrates, lead, and zinc. According to [7], water recovered from processes in the concrete industry, water from underground resources, natural surface waters, and industrial wastewater can all be considered for use in concrete manufacture, if successfully tested against the limiting values for the harmful substances mentioned above. In cases where the limits set in [7] are not met, the standard stipulates that the water can still be used to produce concrete if certain values of setting time and compressive strength are met.

As of 2018, the total treated effluent capacity of Cape Town's wastewater treatment plants was 164.5 ml per day [10]. The research study presented in this paper was therefore conducted as a pilot project to assess the feasibility of utilizing treated domestic water to produce structurally sound and durable concrete in South Africa. The study was intended to inform local long-term water management strategies and in turn secure water supply for the concrete industry in times of water scarcity.

2 Aims and Scope of the Study

The primary aim of the experimental investigation was to determine if TDW could be used for concrete manufacture according to SANS 51008 [7]. This included chemical analysis of the TDW samples, evaluation of the setting time, slump retention and compressive strength assessment.

3 Methodology

Four samples of treated domestic wastewater were collected from WWTPs around Cape Town at the effluent stage. They were labeled A, B, C, and D for confidentiality reasons. WWTPs A, B, and C employ the activated sludge process (ASP) for wastewater treatment while WWTP D employs membrane bioreactor (MBR) technology for wastewater treatment.

Water was sampled from the respective plants in two batches of 25 l each, the second being collected 4 weeks after the initial samples were collected. The main reason for the double sampling was to check for water quality consistency. It was noted that WWTP B treated both domestic and industrial wastewater with the fractions of each wastewater type varying throughout the day. Maximum industrial wastewater was expected during the afternoon at 13:00 with 40% industrial and 60% domestic wastewater being treated at this time. The sampled water was outsourced to a water treatment lab to assess the chemical impurities present in the waters and was stored in the laboratory prior to concrete mixing. The results of the chemical investigation were compared to the requirements of [7] to check the suitability of the TDW for concrete manufacture and are presented in Sect. 4.1.

The effect of TDW on concrete fresh and hardened properties was investigated for two different w/c ratios, namely, 0.40 and 0.60. The mix designs are summarized in Table 1.

Testing of fresh properties included setting time using a Vicat Apparatus according to [11] and slump testing according to [12]. Setting time tests were conducted 3 days after water collection for all sample waters as well as potable water.

For compressive strength testing, 100-mm cubes using TDW were cast within 2 days of water collection and cured. Concrete cubes cast using potable water were used as controls for all the results obtained from various tests. The cube specimens

Table 1 Mix designs

w/c	CEM II A-M 42.5N (kg/m^3)	Phillipi dune sand (kg/m^3)	Crusher dust (kg/m^3)	19 mm Graywacke (kg/m^3)	Water (l/m^3)	Superplasticizer (l/m^3)
0.4	425	740	0	1050	170	0.667
0.6	283	441	441	1050	170	0

were taken out of their molds 24 h after casting, placed under plastic sheets for an additional 72 h and then stored in an environmental room at 50% relative humidity and a temperature of 23 °C until testing. The curing regime was intended to mimic practical curing procedures and adverse site conditions. Compressive strength was tested at 7 and 28 days following [13].

4 Results and Discussion

4.1 TDW Chemistry

Table 2 summarizes the results of the wastewater chemical analyses and compares these to results from the literature and the requirements given in [7].

As seen in Table 2, the TDW samples passed SANS 51008 [7] requirements. Consequently, the TDW used in this research was expected to produce concrete with acceptable fresh and hardened properties. Table 3 shows the TDW chemical analysis results for the secondary samples and the percentage changes in the measured values between the initial and secondary TDW samples from the four different plants. Although large percentage changes can be noted between the initial and second TDW samples, all the water samples from the second sampling passed the water quality requirements for concrete production outlined in [7].

From Table 3, it was observed that TDW C had an extremely high Chemical Oxygen Demand (COD) value. COD generally indicates the organic content in the water that may lead to retardation in concrete setting time and thus impacts early age strength. Consequently, a high COD value can severely affect concrete properties.

Table 2 Chemical analysis results of the initial samples of TDW and tolerable limits

	Wastewater initial samples				Average from literary sources	Tolerable limits [7]
	A	B	C	D		
pH	7.2	6.9	7.1	6.5	7.7	>4
Sodium (Na)	232.3	179.1	139.7	111.4	29.0	
Chlorine (Cl)	320.0	250.0	149.0	139.0	149.0	1000
Sulfate (SO_4)	108.0	95.0	114.0	97.0	103.9	2000
Phosphorus (P)	6.7	0.3	0.3	0.4		
Ammonium (NH_4)	37.4	20.7	33.9	<0.3	36.0	
Nitrate (NO_3)	<0.4	6.9	5.7	23.9	6.0	500
Chemical oxygen demand (COD)	210.0	52.0	100.0	45.0	61.0	
Phosphates (PO_4)	5.6	0.1	0.1	0.3		100
Total solids (TS)	944.3	602.1	670.6	604.8	515.6	2000
E. coli	>2420	159.0	>2420	16.0		<10

All impurities are measured in mg/l; except acidity and alkalinity (mg/l as $CaCO_3$), E. coli (CFU (colony-forming unit)/100 ml) and pH

Table 3 Chemical analysis results of the secondary TDW samples and the percentage changes in the measured values between the initial and secondary TDW samples

	Wastewater second samples								Limits [7]
	A		B		C		D		
	Sample 2	% Change[a]	Sample 2	% Change[a]	Sample 2	% Change[a]	Sample 2	% Change[a]	
pH	7.6	6	6.9	0	6.6	−7	6.4	−1	>4
Na	233.2	0	206.2	15	131.7	−6	105.7	−5	1000
Cl	327.0	2	262.0	5	136.0	−9	101.3	−27	1000
SO_4	101.0	−6	100.0	5	93.0	−18	72.0	−26	2000
P	3.1	−54	0.1	−50	3.6	1189	0.9	117	
NH_4	42.7	14	21.5	4	0.4	−99	<0.3	0	
NO_3	0.7	100	7.7	11	7.3	27	27.3	14	500
COD	109.0	−48	25.0	−52	773.0	673	26.0	−42	
PO_4	3.0	−48	0.1	−40	3.8	4067	.6	106	100
TS	820.8	−13	706.3	17	780.6	16	460.8	−24	
E. coli	>2420	0	178.0	12	>2420	0	13.0	−19	>2000

All impurities are measured in mg/l; except acidity and alkalinity (mg/l as $CaCO_3$), E. coli (CFU (colony-forming unit)/100 ml) and pH
[a]Percentage change between initial and second sampling

SANS 51008 [7] does not provide any limiting values for COD, nor does any other standard and therefore, it is unknown when COD begins to negatively affect concrete quality.

Additionally, it was noted that Escherichia coli (E. coli) concentrations were significantly high for TDW A and C, being in excess of 2420 CFU/100 ml. This was far from the safe range normally given as <10 CFU/100 ml for drinking water [14]. TDW B had an E. coli concentration of 159–178 CFU/100 ml. While this was lower than TDW A and C, it also fell short of the safe range. It was observed that TDW D was exceptionally close to the safe range with a concentration of 13–16 CFU/100 ml. This was to be expected as WWTP D makes use of MBR technology. This technology employs filters which range from 5 to 240 µm, and thus successfully filters out microorganisms and settleable and suspended solids [15]. WWTPs A, B, and C use the conventional ASP to treat wastewater. This generally produces a lower-quality TDW [15]. From Tables 2 and 3, it was thus evident that further disinfection of TDW A, B, and C would have to be recommended if they are to be used safely in the construction industry.

Cape Town is currently in a phase where wastewater treatment facilities are reaching their end of life as many equipment and facilities need to be replaced. Additionally, many plants are being overwhelmed with the quantity of wastewater being treated and are running over capacity. These factors have significant implications on water quality and consistency thereof. Thus, it can be expected that water quality will vary over time. Therefore, in order to gain a better understanding of how water quality fluctuates as a result of WWTP-related problems, more samples have to be collected over a longer period. If it can be proven from this expanded testing period that the worst sample passes the requirements of SANS 51008 [7], then in essence, TDW in Cape Town should be suitable for use as concrete mixing water. A positive aspect observed, however, was that some of the WWTPs are in the process of expanding which is a good indication for future TDW quality.

4.2   Setting Times

To comply with SANS 51008 [7], initial and final setting times of cement paste samples made with TDW must exceed 60 min and may not exceed 12 h, respectively, and may not differ by more than 25% when compared to paste samples made with potable water. SANS 51008 [7] requires the initial and final setting times of non-potable water used in concrete production to be within 25% of that of potable water. The results of the initial setting time tests using the different TDWs are presented in Fig. 1.

Having recorded an initial setting time of 120 min for the control samples, it was required that the samples made using TDW have an initial setting time between 90 and 150 min. The samples that failed to meet these criteria were those made using TDW B and the sample made using the secondary TDW C. While the delayed initial setting time for the second TDW C may be attributed to its high COD value, TDW B

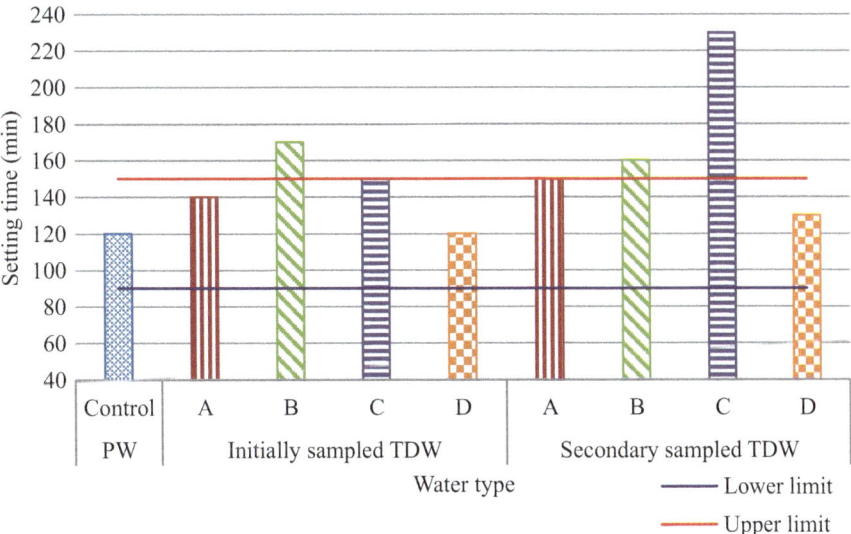

Fig. 1 Initial setting times for all sampled TDW

has significantly lower COD values. The delayed initial setting time for TDW B could, however, be a result of its chemical composition. As discussed previously, WWTP B treats industrial wastewater in addition to domestic wastewater. Industrial wastewater may contain metals such as lead and zinc which are known to retard concrete setting time. It would thus be beneficial to test for these impurities in TDW from this specific WWTP and the other WWTPs that treat industrial wastewater.

The results of the final setting time tests using the different TDWs are presented in Fig. 2. The control sample had a final setting time of 230 min and therefore the samples made using TDW were required to have a final setting time between 173 and 288 min. The initially sampled TDW C sample had a final setting time of 180 min, while the second samples yielded a much higher final setting time of 450 min. This significant increase in the final setting time may be attributed to the significant increase in phosphorus, phosphates, and COD that are known to retard setting time. The color of TDW D was nearly the same as that of potable water. This, in addition to the low COD and low impurities content, resulted in similar setting times for TDW D and the control.

The longer initial setting times of the TDW B samples are at most 50 min longer than the control. Considering that the final setting times of the TDW B samples were all within the required range, TDW B may still be considered suitable for industry application without the use of accelerating admixtures, if the delay in initial setting time is accounted for. On the other hand, the significantly higher initial setting time for the second TDW C sample as compared to the control may require the use of an accelerator to attain more practical initial and final setting times.

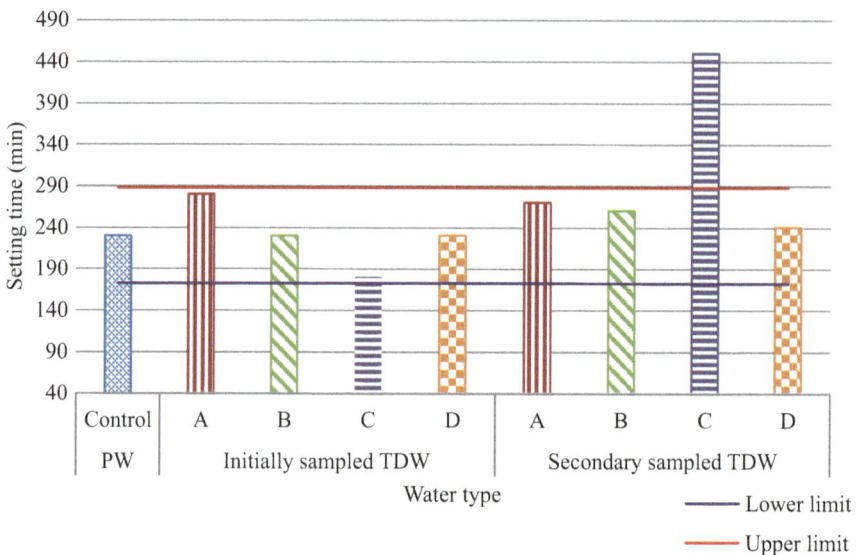

Fig. 2 Final setting times for all sampled TDW

4.3 Workability

The design slump values were 85 ± 25 mm and 65 ± 25 mm for the 0.4 w/c ratio and 0.6 w/c ratio concrete mixes, respectively. Figures 3 and 4 present the slump values for both the initial and second samples of TDW for the 0.4 w/c mix and the 0.6 w/c mix, respectively.

In comparing between the initial and second samples of TDW to assess if any adverse effects arose as a result of water quality inconsistencies, it was observed that the results were similar. It was observed that TDW A and B always had a higher slump than the control and a higher slump than TDW C and D. It was initially thought that this result could be due to the total solids content of the TDW. An increase in total solids increases the sticky content in concrete and thus makes it less workable. However, from the chemical analyses, it is noted that the total solids content is highest in TDW A. Thus, if this were the cause of the observed results, less workability would be expected for TDW A. Additionally, TDW B had similar total solids content to that of TDW C and D. Therefore, the higher workability observed for TDW A and TDW B could not be attributed to the total solids content of the TDW.

While the cause of the observed trend in the slump test values requires further investigation, the test results do show that TDW does not adversely impact concrete workability as all the results were within the tolerance limits of the design slump. This is further supported by the results obtained by [4, 16–18].

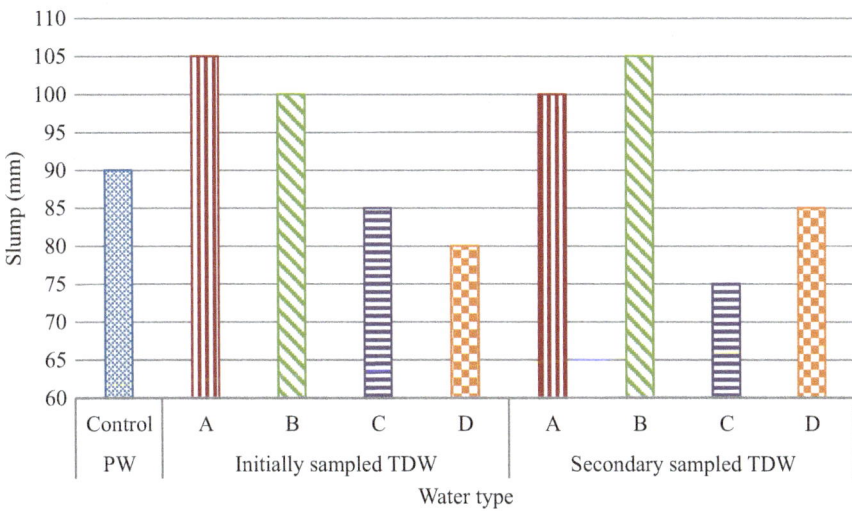

Fig. 3 Slump values for initially and secondary sampled TDW for w/c = 0.4

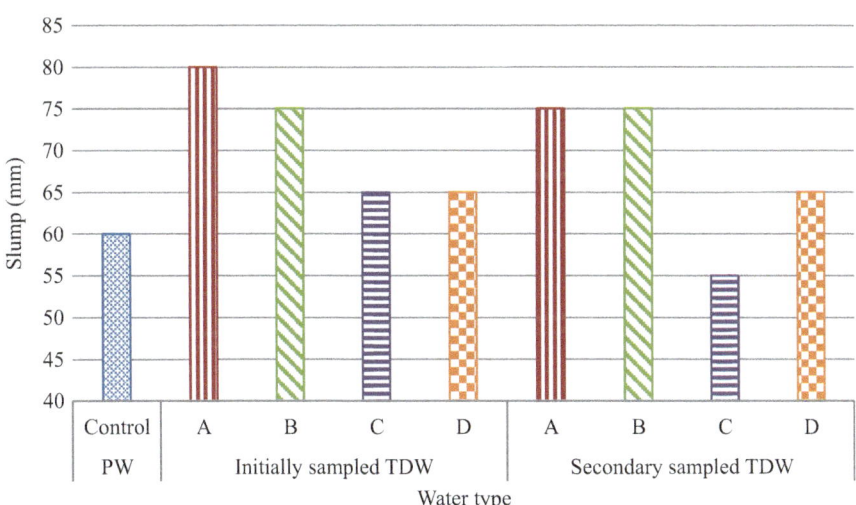

Fig. 4 Slump values for initially and secondary sampled TDW, w/c = 0.6

4.4 Compressive Strength

Concrete compressive strength is used extensively as an index of various concrete properties as well as concrete quality [19]. It is one of the main guidelines used to assess whether non-potable water is suitable for use as mix water in concrete

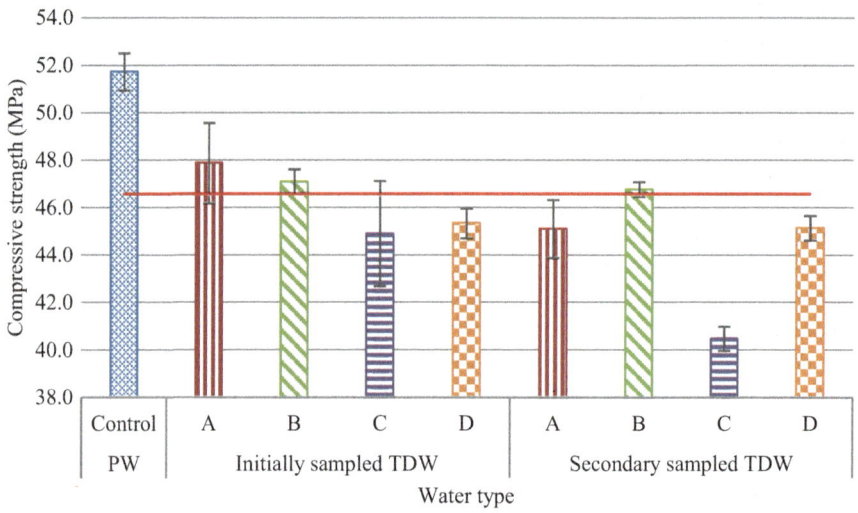

Fig. 5 Seven-day compressive strength results with standard deviations for TDW concrete, w/c = 0.4

manufacture according to SANS 51008 [7]. In this study, 7- and 28-day compressive strengths were assessed as a result of using TDW as mix water.

To pass the 7-day compressive strength check stipulated in SANS 51008 [7] which requires a compressive strength within 90% of the control, a compressive strength of 46.6 MPa was required for the 0.4 w/c ratio concrete mix. The results of the 7-day compressive strength test for the 0.4 w/c mix are presented in Fig. 5.

Only the initial sample of TDW A and the TDW B samples achieved compressive strengths above 46.6 MPa. No clear trend was observed in the 7-day compressive strength results to explain the shortfall for the TDW samples that did not achieve the required compressive strength. However, it was postulated that the much lower compressive strength of the second TDW C concrete samples may be due to the high concentrations of phosphorus, phosphates, and COD which in addition to retarding the setting time also delay the compressive strength development of the concrete. Another observation made was that TDW D yielded compressive strength values of 45.3 MPa and 45.1 MPa for the initial and second samplings, respectively, indicating consistent compressive strength results and corresponding to the consistent setting times recorded.

The results of the 7-day compressive strength test for the 0.6 w/c mix are presented in Fig. 6. A compressive strength of 24.9 MPa was required for samples to meet the 7-day strength criteria stipulated in SANS 51008 [7] for the 0.6 w/c ratio concrete mix. All TDW concrete samples passed this requirement, except for the secondary TDW C with 23.0 MPa. Again, this result may be cautiously attributed to the chemical species present in the TDW causing set retardation and delaying compressive strength gain.

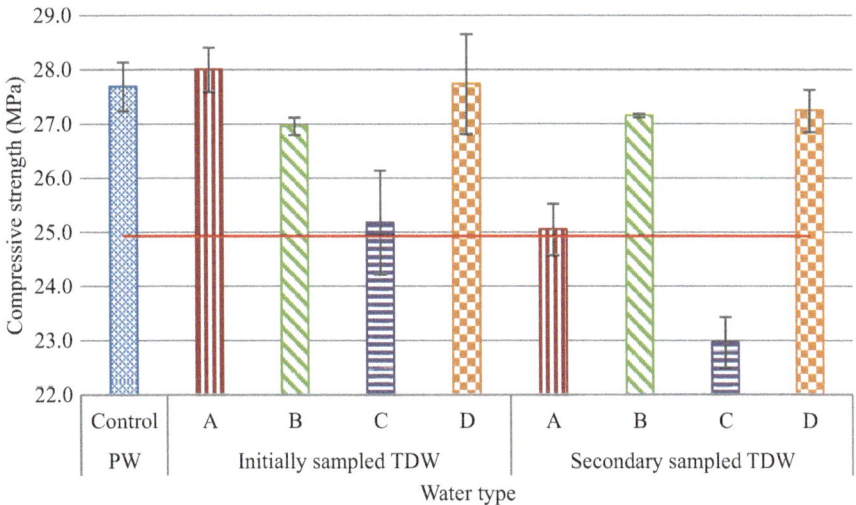

Fig. 6 Seven-day compressive strength results with standard deviations for TDW concrete w/c = 0.6

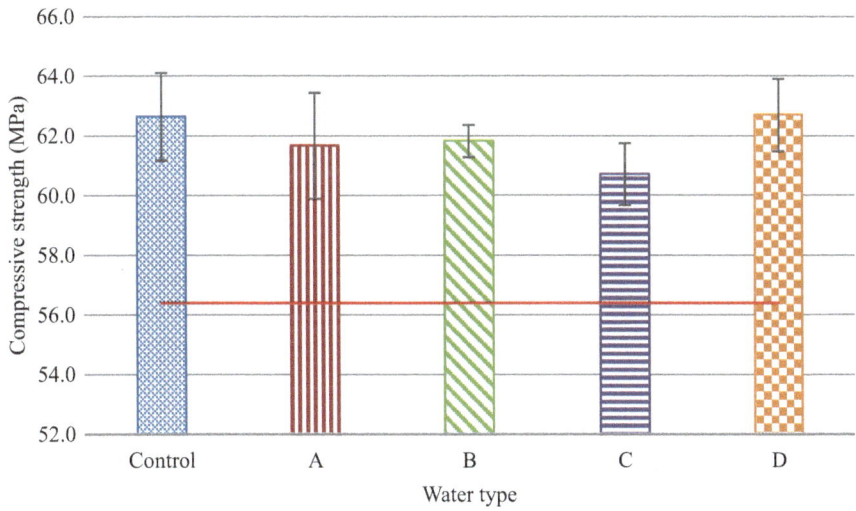

Fig. 7 Twenty-eight-day compressive strength results with standard deviations for TDW concrete, w/c = 0.4

The 28-day compressive strength test was performed on only the samples made using the initial samples of TDW due to the time constraints and is presented in Figs. 7 and 8 for the 0.4 and 0.6 w/c ratio, respectively.

The 0.4 w/c ratio concrete mix required samples to have a compressive strength of 56.4 MPa to pass the 28-day compressive strength requirement outlined in SANS 51008 [7]. At 28 days, all TDW concrete samples passed the requirement with the

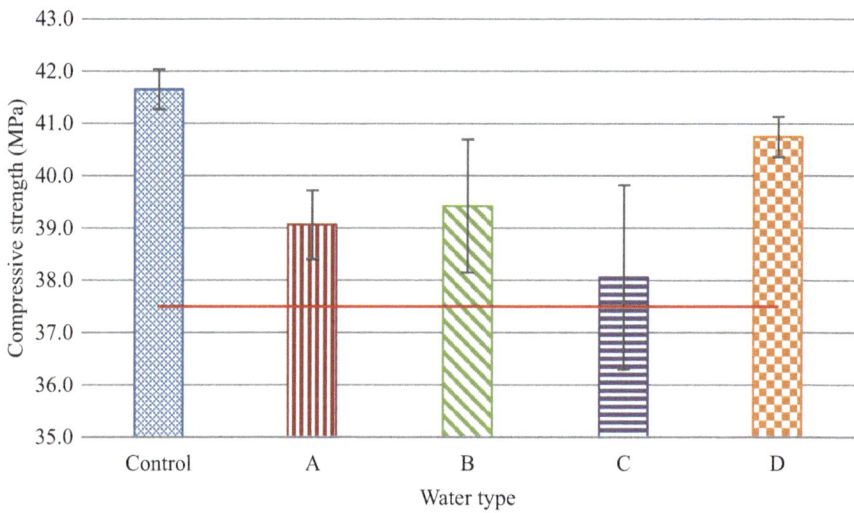

Fig. 8 Twenty-eight-day compressive strength results with standard deviations for TDW concrete, w/c = 0.6

lowest compressive strength recorded being 60.7 MPa for TDW C. For the 0.6 w/c ratio concrete mix, the TDW concrete samples were required to possess a compressive strength of 37.5 MPa to adhere to SANS 51008 [7] requirements. All samples passed this check with the lowest compressive strength being 38.1 MPa for TDW C. Potable water produced the highest compressive strength at 41.7 MPa.

While not all the TDW concrete samples passed the 7-day compressive strength test for both w/c ratios, all the samples passed the 28-day compressive strength requirements stipulated in SANS 51008 [7]. In comparing the 7-day to the 28-day compressive strength test results, it was evident that using TDW as mixing water delays the compressive strength development of the concrete. The variability in the compressive strength test results may be attributed to the chemical composition of each of the TDW samples used and how the chemical species interact in the concrete mix. To fully understand this interaction, more tests than those required in the SANS 51008 [7] and presented in this paper would be necessary.

5   Concluding Remarks and Way Forward

Various tests were conducted to investigate the effects which TDW has on concrete properties when used as mix water. From the initial and final setting time tests, no clear correlation between the setting times and the chemical composition of the TDW could be surmised. While not all the recorded setting times were within 25% of the control samples, the recorded setting times may still be practically acceptable for industry application, especially if accelerating admixtures are used to reduce the setting time. Similarly, a correlation between the measured slump and the chemical composition of the TDW could not be determined. However, all the measured slumps were within the stipulated ranges.

Only the initial sample of TDW A and the TDW B samples achieved the required 7-day compressive strengths for the 0.4 w/c mix, and all TDW concrete samples achieved the required 7-day compressive strengths for the 0.6 w/c mix except for the secondary TDW C. While no clear trend was observed in the results to explain the shortfall for the TDW samples that did not achieve the required 7-day compressive strength, all the samples had gained sufficient compressive strength by the 28th day to pass the requirements stipulated in SANS 51008 [7]. It was thus concluded that using TDW as mixing water delays the compressive strength development of the concrete.

The findings of this study show that TDW produces concrete with sufficient workability, setting time, compressive strength when comparing it to literature and the control. However, quality control of the TDW is of the utmost importance to ensure the consistent production of good quality concrete.

References

1. GreenCape: Water Market Intelligence Report. GreenCape, Cape Town (2022)
2. City of Cape Town: Cape Town Water Outlook, City of Cape Town (2023)
3. Bardhan, S.: Assessment of water resource consumption in building construction in India. WIT Trans. Ecol. Environ. **144**, 93–101 (2011)
4. Asadollahfardi, G., Delnavaz, M., Rashnoiee, V., Ghonabadi, N.: Use of treated domestic wastewater before chlorination to produce and cure concrete. Constr. Build. Mater. **105**, 253–261 (2016)
5. Shekarchi, M., Yazdian, M., Mehrdadi, N.: Use of biologically treated domestic waste water in concrete. Kuwait J. Sci. Eng. **39**(2B), 97–111 (2012)
6. Roxburgh, J.: Mixing water. In: Alexander, M. (ed.) Fulton's Concrete Technology, 10th edn, pp. 149–160. Cement & Concrete SA, Midrand (2021)
7. SANS 51008: Mixing Water for Concrete – Specification for Sampling, Testing and Assessing the Suitability of Water, Including Water Recovered from Processes in the Concrete Industry, as Mixing Water for Concrete. SABS Standards Division, Pretoria (2006)
8. ASTM C1602: Standard Specification for Mixing Water Used in the Production of Hydraulic Cement Concrete. ASTM International, West Conshohocken (2022)
9. BS EN 1008: Mixing Water for Concrete – Specification for Sampling, Testing and Assessing the Suitability of Water, Including Water Recovered from Processes in the Concrete Industry, as Mixing Water for Concrete. British Standards Institution, London (2002)
10. City of Cape Town: Water Services and the Cape Town Urban Water Cycle, City of Cape Town, Cape Town (2018)
11. SANS 50196-3: Methods of Testing Cement. Part 3: Determination of Setting Imes and Soundness. SABS. Standards Division, Pretoria (2006)
12. SANS 5862-1: Concrete Tests – Consistence of Freshly Mixed Concrete – Slump Test. SABS. Standards Division, Pretoria (2006)
13. SANS 5863: Concrete Tests — Compressive Strength of Hardened Concrete. SABS, Pretoria (2006)
14. SANS 241-1: Drinking Water Specification. SABS Standards Division, Pretoria (2015)

15. Iorhemen, O.T., Hamza, R.A.T.J.H.: Membrane bioreactor (MBR) technology for wastewater treatment and reclaimation: membrane fouling. Membranes (Basel). **6**(2), 33 (2016)
16. Al-Ghusain, I., Terro, M.J.: Use of treated wastewater for concrete mixing in Kuwait. Kuwait J. Sci. Eng. **30**(1), 214–228 (2003)
17. Gulve, B., Bothe, R., Dange, P., Gite, P., Jondale, S., Landge, A.S., Nagpure, A.: Effect of treated waste water (TWW) on mechanical properties of concrete. Ijariie. **4**(3), 1390–1396 (2018)
18. Looman, J.: Feasibility of Utilising Treated Domestic Wastewater for the Production of Concrete. University of Cape Town, Cape Town (2018)
19. Beushausen, H., Crosswell, S., Goodman, J., Jooste, P., Perrie, B., Owens, G., Roxburgh, J., Selby, G., Theodosiou, G., Van der Merwe, D.: Fundamentals of Concrete, 3rd edn. The Concrete Institute, Midrand (2013)

Compressive Characteristics of Perforated Re-entrant Auxetic Steel Honeycomb–Mortar Composite

Emmanuel Owoichoechi Momoh ⑩, Mohammad Hajsadeghi ⑩, Amila Jayasinghe ⑩, Raffaele Vinai ⑩, Prakash Kripakaran ⑩, John Orr ⑩, and Ken E. Evans

Contents

1 Introduction

Cementitious materials are the most used construction materials due to their relatively low cost and ease of production, which essentially requires fluid concrete to be poured into moulds and be allowed to set [1, 2]. Although excellent in compressive strength, the inherent brittleness and poor tensile strength of these materials result in low energy absorption and sudden fracture when the material tensile strength is exceeded. Consequently, cementitious materials are unsuitable for high-energy absorption applications such as vibration and impact resistance. Several approaches have been employed to improve the ductility for such brittle materials, such as through embedding of ductile reinforcement materials, e.g. steel rods, steel strands

E. O. Momoh (✉) · M. Hajsadeghi · R. Vinai · P. Kripakaran · K. E. Evans
Department of Engineering, Faculty of Environment Science, and Economy,
University of Exeter, Exeter, UK
e-mail: e.o.momoh@exeter.ac.uk

A. Jayasinghe · J. Orr
Department of Engineering, University of Cambridge, Cambridge, UK

© The Author(s) 2025
M. Kioumarsi, B. Shafei (eds.), *The 1st International Conference on Net-Zero Built Environment*, Lecture Notes in Civil Engineering 237,
https://doi.org/10.1007/978-3-031-69626-8_21

and fibres [3], polymers [4, 5], natural fibres [6, 7] and only recently auxetic materials derived from altering the geometries of conventional materials [2, 8]. While conventional materials expand laterally when compressed axially and contract laterally under axial tension, auxetic materials contract laterally when compressed axially and expand laterally under axial tension. This counter intuitive behaviour of auxetic materials that results in Negative Poisson's Ratio (NPR) can be exploited for enhancing indentation resistance [9, 10], shear resistance [9, 11], strain-hardening, energy absorption, crashworthiness, damping [9, 12, 13]; as well as physical characteristics such as acoustic behaviour [11] and inherent synclasticity under bending moments [13].

Auxetic behaviour has been cheaply achieved in several studies by altering the geometry of conventional materials. The most studied auxetic geometry with potential application in structural engineering is the re-entrant honeycomb structure [14]. Several studies have investigated re-entrant honeycomb structures with concrete infill. The study of Zhong et al. [15] investigated the compressive strength behaviour of fabricated re-entrant honeycomb aluminium structures with 0.4 mm of sheet thickness. The height and width of the re-entrant honeycomb structure were 110 mm and filled with 16 MPa concrete. The elastic modulus and tensile strength of the aluminium sheets were 69 GPa and 274 MPa, respectively. The study reported the enhancement of the composite against compression, buckling, flexure and shear failures. The study investigated the confinement effect of the auxetic honeycomb lattice to the transverse deformation of the concrete matrix. The influence of dimensional parameters such as cell wall thickness, cell angle and cell size on the auxetic response of the composite was also investigated. The method of manufacture of the aluminium re-entrant honeycomb was firstly by folding plain sheets of aluminium, then stacking and gluing the folded sheets. The same type of honeycomb was used in the study of Zhou et al. [16], although with foam concrete as the cementitious matrix. Zhou et al. [16] investigated the in-plane uniaxial quasi-static and dynamic compressive strength of the auxetic mesh-reinforced foam concrete. The sample failure evolved from predominantly compression to shear especially as the density of the foam concrete increased. The study recommended the use of auxetic meshes to enhance load-bearing of porous cementitious structures.

Apart from sheet folding and crimping into re-entrant shapes, another popular method of introducing auxetic behaviour in metal sheets is through perforations. Perforations in the form of orthogonal ellipses have been shown to create nodes of rotation within the perforation grids and hence cause NPR [14, 17]. For example, in the study of Luo et al. [17], 8 mm, 10 mm and 12 mm thick stainless steel sheets were perforated in orthogonal alternating elliptic patterns (using laser cutting technology) and folded into tubes. The tubes were then filled with concrete and tested in compression after 28 days. The elliptical-perforated tubes were compared with circular-perforated tubes with the same area of perforations. The elliptical-perforated tubes provided active restraint to the lateral strain of the concrete, thereby enhancing the axial compressive strength of the composite.

Investigations of auxetic behaviour through perforation of materials are scarce and focused on tubular structures used as external confinement to the cementitious phase. Also in such studies, the auxetic behaviour is limited to the material plane of the confinement. This means that, apart from the confining edge provided by the sheet metal, the greater volume of the cementitious phase is unaffected by the auxetic behaviour of the sheet metal. Furthermore, for structures such as re-entrant honeycombs, there is lack of continuity in the cementitious matrix as each cell is separated by a continuous sheet of metal. In this study, the possibility of introducing auxeticity throughout the composite in re-entrant honeycomb structures through perforations is explored. The perforations are also expected to enhance the bond strength between the metal sheet and the cementitious matrix, as well as additional confinement, thereby improving the mechanical properties of the overall composite.

The effect of the perforations on the compressive capacities of the samples is investigated by comparing the force–displacement curves of the circular-perforated, elliptic-perforated and peanut-perforated samples with the non-perforated samples. Subsequent paragraphs, however, are indented.

2 Materials and Methods

2.1 Re-entrant Honeycomb Structure

Cold-reduced (CR4) steel sheets of 0.9 mm thickness were supplied by The Metal Store, UK, and cut into strips of 1200×70 mm and used for the non-perforated samples. For the perforated samples, a plasma cutter was used to carry out the perforations before the sheet was cut into the 1200×70 mm strips. All the perforated samples have the same volume of materials which is exactly half the volume of the non-perforated samples. Details of the dimensions of the sheets and perforations are shown in Fig. 1.

For both perforated and non-perforated sheets, the samples were folded into re-entrant shapes at 75° angle of inclination as shown in Fig. 2i. The folded metal sheets were stacked upon one another and welded using TIG welding. Non-perforated solid plates of the same thickness were welded on to the top and

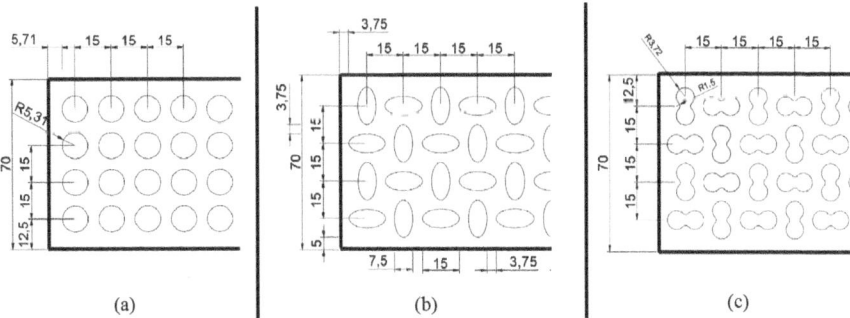

Fig. 1 Dimension details of (**a**) circular-perforated, (**b**) elliptic-perforated and (**c**) peanut-perforated mild steel sheets. All dimensions are in mm

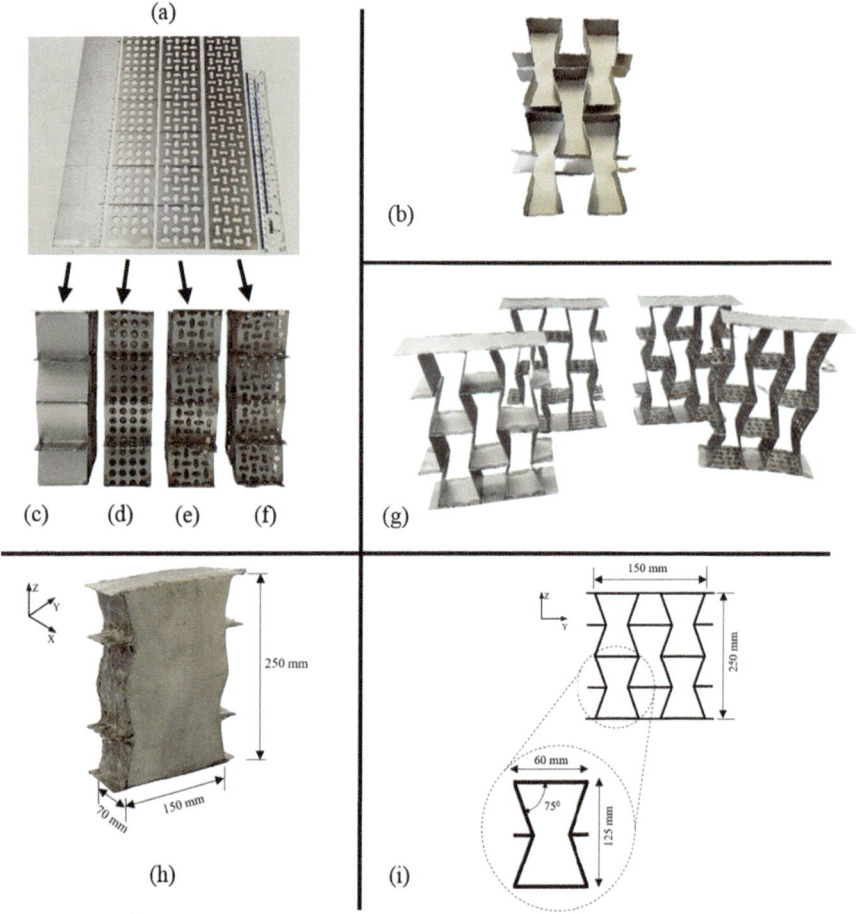

Fig. 2 (a) Non-perforated and perforated mild steel strips; (b) folded strips into re-entrant shapes; side views of welded (c) non-perforated, (d) circular-perforated, (e) elliptic-perforated, (f) peanut-perforated re-entrant honeycomb structures; (g) some welded re-entrant meshes; (h) mesh-mortar composites; (i) dimension details of re-entrant structure

the bottom of each auxetic structure in order to contain the mortar within the structure during casting (see Fig. 2g).

2.2 Mortar

The materials used for the mortar were Hanson general-purpose cement (32.5R) and natural sand that complies with the ASTM C33/C33M as sharp concreting sand. The sand with 4 mm maximum aggregate size together with the cement were all obtained

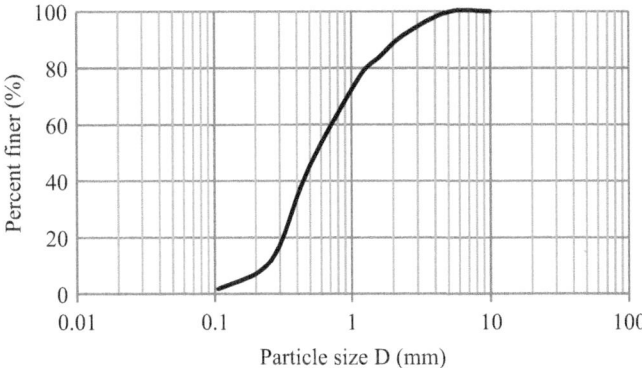

Fig. 3 Particle size distribution of sharp sand used for the cement mortar

from Jewson Exeter, UK. The mortar consisted of the binder and aggregate in 1:5 mix ratio while the water/cement ratio was 0.48 [18]. Sikamix mortar plasticiser was added to the water at a dosage of 2.5 ml plasticiser per kg of cement before mixing with the aggregates. The binder and aggregate were first dry-mixed in a rotating drum after which the measured water containing mortar plasticiser was added. The mixing continued for 5 min, following the requirements of ASTM C33/C33M until a consistent mix was achieved. Figure 3 presents the particle size distribution. The external side of the outer walls of the honeycomb were covered with transparent stick tape (Sellotape) to avoid the concrete seeping through. The re-entrant honey-comb structures were then placed on a flat wooden plate and held onto the plate using steel wires. The mortar was then poured into the plate and vibrated for about 2 min and left undisturbed for 24 h. At the end of the 24 h period, the sticky tapes were removed, and the honeycomb–concrete composites were transferred into a water tank and cured under ambient temperature for 28 days prior to testing.

2.3 Tensile Strength Test

The perforated and non-perforated strips were tested in tension to investigate their tensile capacities and auxetic behaviour. The tensile strength test was carried out in displacement control (at 5 mm/min) in an Instron 34TM-30 universal testing machine, according to ASTM A370 – 11. Figure 4 shows images of some of the tensile test samples before and after testing. The Poisson's ratios of the samples were calculated by obtaining six (6) lateral width measurements along each sample length before and after the tensile test, using a digital vernier caliper and dividing the change in the average final width by the change in the final longitudinal length of each sample.

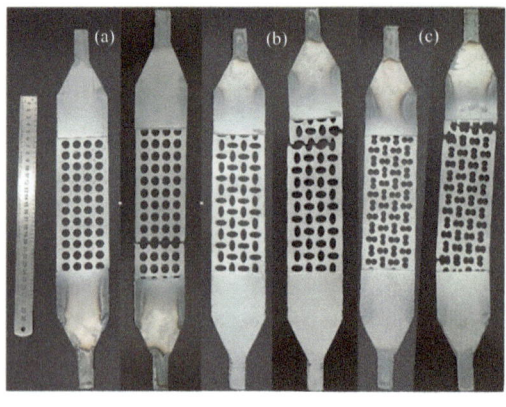
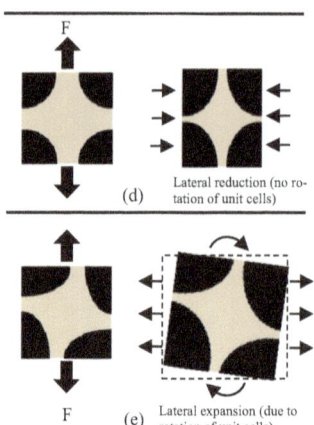

Fig. 4 (**a**) Circular-perforated, (**b**) elliptic-perforated and (**c**) peanut-perforated mild steel samples before and after tensile strength test; (**d**) illustration of positive Poisson's ratio for circular perforated samples; and (**e**) negative Poisson's ratio for elliptic perforated samples

2.4 Compressive Strength Test

The mortar-filled auxetic composites were tested in compression in compliance with ASTM C39 in displacement control at 5 mm/min. The test was stopped after a travel distance of 75 mm.

3 Results and Discussion

3.1 Tensile Strength Test of Perforated Mild Steel Strips

While the circular-perforated samples underwent reduction in the lateral direction, the elliptic- and peanut-perforated samples expanded in the lateral direction, hence indicating NPR (see Fig. 4). The overall behaviour of the samples was governed by the resistance mechanism of each constituting unit cell. For the circular-perforated samples, the axial tension force from the machine was resisted by the material in between the holes with direct axial straining and yielding until failure. Conversely, the orthogonal arrangement of holes in the elliptic- and peanut-perforated sheets, provided alternating material paths for load resistance, hence, causing the rotation of the unit cells (see Fig. 4d, e). The rotations of the cells caused an increase in the lateral dimension until the cells were aligned in the vertical axis of loading.

It was after this alignment, that the cells began to strain and yield in the axial direction of loading subsequently with some strain-hardening until failure. Consequently, the circular-perforated samples possessed relatively higher stiffness and

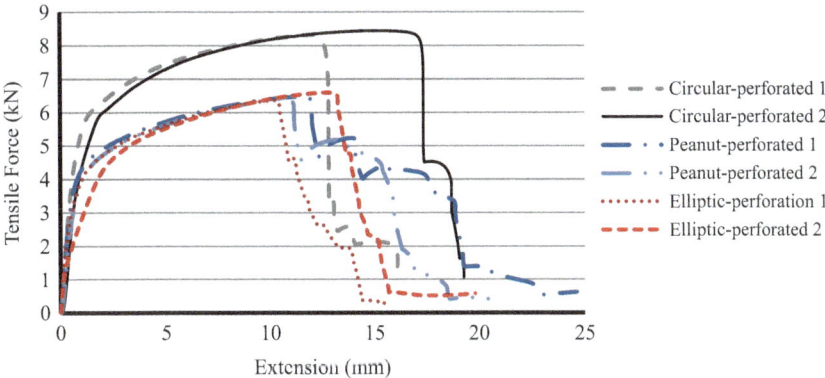

Fig. 5 Tensile strength curves of the perforated steel sheets

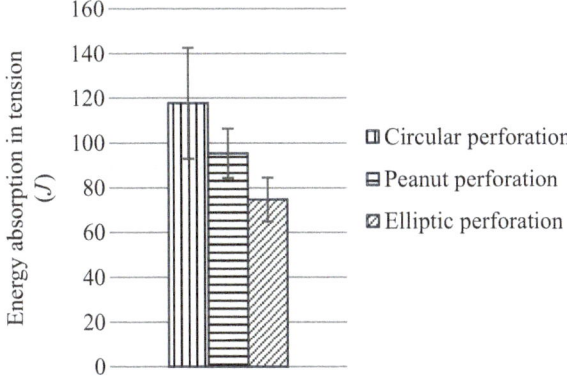

Fig. 6 Energy absorption from tensile strength curves of the perforated steel sheets

tensile strength as shown in Fig. 5. Average tensile capacities and Poisson's ratios of 8.39 kN, 6.47 kN, and 6.44 kN and -0.214, -0.246, and 0.279 were obtained for circular-, elliptic- and peanut-perforated samples, respectively. The variation between individual tensile test strip samples of the same category is due to possible inconsistencies in the perforations as well as in the testing procedure.

A simple uniaxial tensile strength test alone is not sufficient to account for energy dissipation through rotation of the unit cells of the auxetic strips; therefore, the circular-perforated strips seemed to possess more energy absorption (i.e. the area under the load–extension curve). Although the difference in average tensile capacities between the elliptic- and peanut-perforated samples is negligible, the average energy absorption of the latter is about 28% greater due to its relatively higher NPR (see Fig. 6). It is also important to note that only two samples each were tested for tensile strength behaviour for each category of perforated strip. This small sample size is a limitation and will be increased in future studies.

3.2 Compressive Strength Test

Generally, the stress–strain behaviour of re-entrant honeycomb structures is characterised by three major regimes: the linear elastic stage, the softening or plateau stage and the densification/destruction stage [19]. However, in this study, at the start of loading, the compressive load on the auxetic steel-mortar composite, increased linearly with displacement. The slope of the load–displacement curve then reduces at the onset of cracking within the mortar (see Fig. 7). Nevertheless, the capacity of the samples increased until excessive cracking within the mortar resulted in the rapid reduction of the load–displacement curve. This rapid reduction in capacity is followed by a plateau regime, which further evolves into the densification of the composite with subsequent crushing/destruction. The crushing stage succeeded the densification stage with concrete spalling from the cells of the auxetic meshes as depicted in Fig. 8.

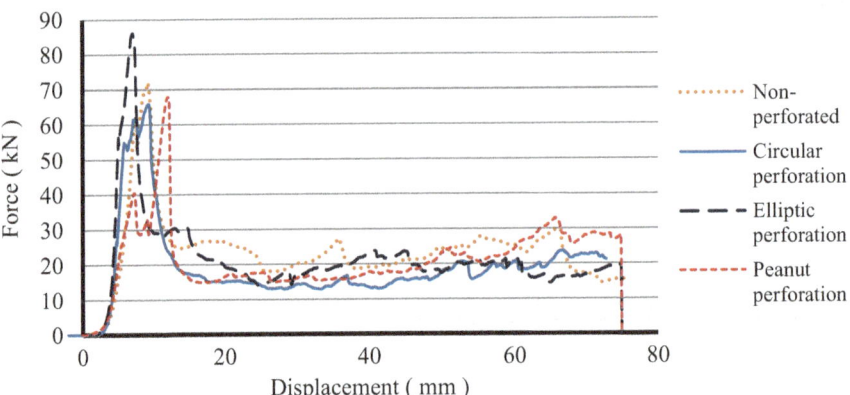

Fig. 7 Compressive behaviour curves of samples with the maximum compressive strengths for non-perforated and perforated re-entrant honeycombs

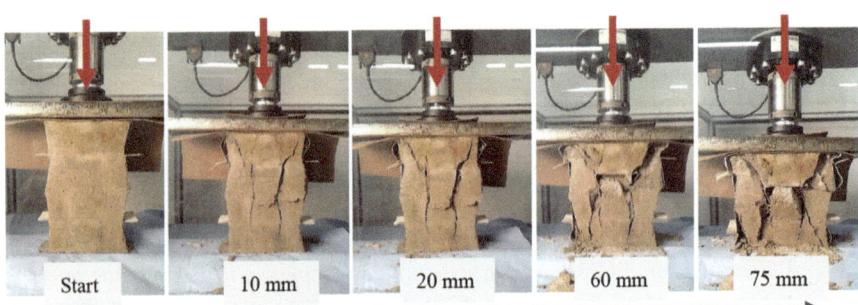

Typical failure pattern for all samples with vertical crosshead movements

Fig. 8 Compressive strength testing of non-perforated and perforated re-entrant honeycomb–mortar composites

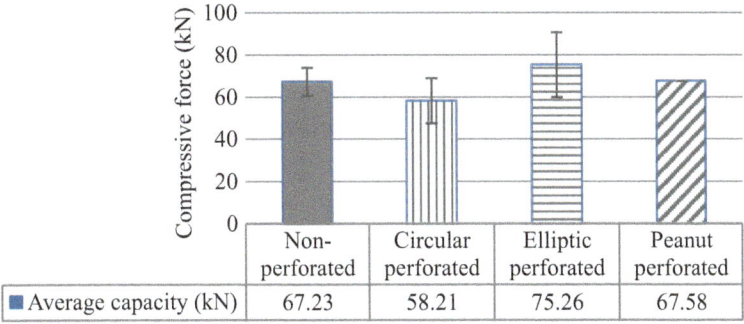

Fig. 9 Average compressive strengths of non perforated and perforated re-entrant honeycombs

Figure 7 is the compressive behaviour curves of selected samples having maximum compressive capacities for non-perforated and perforated re-entrant honeycombs–mortar composites. Figure 9 presents the average compressive capacities (with standard deviations) of all the categories. Due to the enhanced bonding and the auxetic action provided by the perforations, the average compressive capacity of the elliptic-perforated samples increased by about 12% and 19% compared with the unperforated and circular-perforated samples, respectively. The compressive capacity of the peanut-perforated sample was greater than the circular-perforated samples by about 17%. However, there was negligible improvement in the compressive capacity of the peanut-perforated samples compared with the unperforated samples. This could be attributed to the trapping of larger grains of aggregates in between the 'neck' of the peanut perforation, thereby creating weak zones within the composite. The respective energy absorption for the samples was calculated using Eq. (1).

$$\int_0^l Fdl \tag{1}$$

where F is the applied load (in kN) and l is the corresponding displacement (in mm).

The amount of steel used for each perforated steel–mortar composite category is exactly half the mass of the steel used for the non-perforated samples. According to Eq. (2), this corresponds to a 50% mass consumption ratio (TMCR) for the perforated samples [20]. Recent studies show that cellular auxetic structures can be obtained with up to 34% of material in comparison with their homogeneous counterparts [20]. Although the process of perforation could result in extra costs, advancements in manufacturing technologies could offset these costs allowing for innovative engineering load bearing applications.

$$TMCR = \frac{mass\ of\ auxetic\ component}{mass\ of\ homogeneous\ component} \times 100\% \tag{2}$$

The different samples can also be compared using Eq. (3) which specifies a dimensionless design optimisation factor (*Df*) [20]:

$$Df = \frac{k}{\text{displacement} \times \text{TMCR}} \times 100\%, \tag{3}$$

where k is empirically chosen as 1 and having the same unit as the displacement (or strain, deformation, rotation, etc.) [21]. In this study, the highest *Df* resulted from the auxetic reinforcements. These are the samples with the highest compressive capacity and use the least amount of steel reinforcement (i.e. the elliptic- and peanut-perforated samples). The *Df* of the perforated samples are more than that of the non-perforated samples with values of 9.3, 25.5, 19.4 and 16.5 for non-perforated, circular-perforated, elliptic-perforated and peanut-perforated samples respectively. Parameters such as TMCR and *Df* can be used as selection tools for the dimensions of perforation and for the design of other components. For components in flexure, the displacement in Eq. (3) can be replaced with deflection or strain. Although the normalised average energy absorption from the compressive strength curves was highest for the non-perforated samples (Fig. 10), it should be noted that the perforated samples have exactly 50% less steel reinforcement.

The perforated steel re-entrant honeycomb-reinforced mortar composite developed in this study can be used for applications such as lightweight beams, columns, sandwich panels and shear walls. The perforations improve bonding between the reinforcement and matrix through continuity of the matrix as well as through the NPR of the steel sheets. Such composite also does not require additional formwork for making them since the cellular arrangement can keep the wet concrete in place. Further studies are however required to investigate the effects of perforation parameters such as aspect ratio and angular orientation of the holes in order to tailor them for specific load-bearing components. The next stage of this research involves investigations on the mechanical behaviour on a larger sample size in order to enhance the rigor and impact of the results.

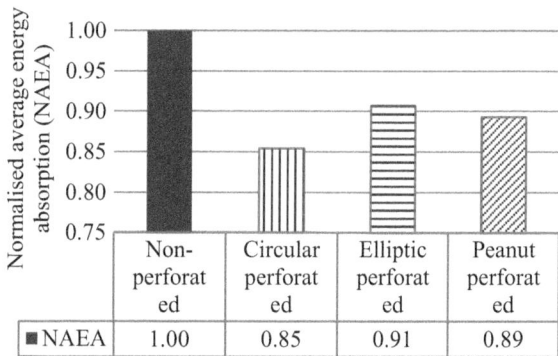

Fig. 10 Normalised average energy absorption for all perforated samples under compression

4 Conclusions

This paper focused on the behaviour of perforated re-entrant auxetic steel honeycomb–mortar composite. The need to improve the bond strength between steel and concrete through auxeticity led to the orthogonal perforation of the steel strips. The tensile strength behaviour of the steel strips was assessed as well as the compressive strengths of the respective composites with circular, elliptic and peanut orthogonal perforations in the re-entrant steel honeycomb. Consequently, the following key findings are highlighted:

- The tensile capacities of circular-perforated sheets are over 29% greater than those of the orthogonal elliptic- and peanut-perforated sheets.
- The difference in average tensile capacities between the elliptic- and peanut-perforated samples is negligible; however, the average energy absorption of the latter is about 28% greater.
- Due to the enhanced bonding and the auxetic action provided by the perforations, the average compressive capacity of the elliptic-perforated samples is about 12% and 19% greater than that of the unperforated and circular-perforated samples.
- Parameters such as the 'design optimisation factor' and 'total mass consumption ratio' can be used to select the optimum dimensions for perforations.
- The perforated steel sheets use only 50% of the material constituting the non-perforated steel sheets, implying that introducing auxeticity through perforations can reduce overall material usage. This can offer benefits in specialised steel-reinforced cementitious load bearing applications such as concrete shear walls.

Acknowledgement and Funding This work was supported by the United Kingdom's Engineering and Physical Sciences Research Council (EPSRC, UK), [EP/W019027/1].

Conflicts of Interest The authors have no conflicts of interest to declare that are relevant to the content of this article.

Data Access Statement Data will be made available on request.

References

1. Xu, Y., Šavija, B.: Architected cementitious cellular materials: peculiarities and opportunities. Heron. **66**(2–3), 239–271 (2021)
2. Rosewitz, J.A., Choshali, H.A., Rahbar, N.: Bioinspired design of architected cement-polymer composites. Cem. Concr. Compos. **96**, 252–265 (2019)
3. Momoh, E.O., Osofero, A.I., Oleksandr, M.: Behaviour of clamp-enhanced palm tendons reinforced concrete. Constr. Build. Mater. **341**, 127824 (2022)

4. Oggeri, C., Ronco, C., Vinai, R.: Validation of numerical D.E.M. modelling of geogrid reinforced embankments for rockfall protection. Geoing. Ambient. Mineraria. **58**(2(3)), 36–45 (2021)
5. Mukhopadhyay, S., Khatana, S.: A review on the use of fibers in reinforced cementitious concrete. J. Ind. Text. **45**(2), 239–264 (2014)
6. Momoh, E.O., Osofero, A.I., Menshykov, O.: Physicomechanical properties of treated oil palm-broom fibers for cementitious composites. J. Mater. Civ. Eng. **32**, 04020300 (2020)
7. Momoh, E.O., Osofero, A.I., Menshykov, O.: Bond behaviour of oil palm broom fibres in concrete for eco-friendly construction. Proc. Inst. Civ. Eng.-Constr. Mater. **174**(1), 47–64 (2021)
8. Luo, C., Han, C.Z., Zhang, X.Y., Zhang, X.G., Ren, X., Xie, Y.M.: Design, manufacturing and applications of auxetic tubular structures: a review. Thin-Walled Struct. **163**, 107682 (2021)
9. Xu, Y., Schlangen, E., Lukovic, M., Savija, B.: Tunable mechanical behavior of auxetic cementitious cellular composites (CCCs): experiments and simulations. Constr. Build. Mater. **266**(B), 121388 (2021)
10. Subramani, P., Rana, S., Oliveira, D.V., Fangueiro, R., Xavier, J.: Development of novel auxetic structures based on braided composites. Mater. Des. **61**, 286–295 (2014)
11. Zahra, T., Dhanasekar, M.: Characterisation of cementitious polymer mortar – auxetic foam composites. Constr. Build. Mater. **147**, 143–159 (2017)
12. Peliński, K., Smardzewski, J.: Static response of synclastic sandwich panel with auxetic wood-based honeycomb cores subject to compression. Thin-Walled Struct. **179**, 109559 (2022)
13. Momoh, E.O., Jayasinghe, A., Hajsadeghi, M., Vinai, R., Evans, K.E., Kripakaran, P., Orr, J.: A state-of-the-art review on the application of auxetic materials in cementitious composites. Thin-Walled Struct. **196**, 111447 (2024)
14. Ren, X., Shen, J., Ghaedizadeh, A., Tian, H., Xie, Y.M.: A simple auxetic tubular structure with tuneable mechanical properties. Smart Mater. Struct. **25**, 065012 (2016)
15. Zhong, R., Ren, X., Zhang, X.Y., Luo, C., Zhang, Y., Xie, Y.M.: Mechanical properties of concrete composites with auxetic single and layered honeycomb structures. Constr. Build. Mater. **322**, 126453 (2022)
16. Zhou, H., Jia, K., Wang, X., Xiong, M.-X., Wang, Y.: Experimental and numerical investigation of low velocity impact response of foam concrete filled auxetic honeycombs. Thin-Walled Struct. **154**, 106898 (2020)
17. Luo, C., Ren, X., Han, D., Zhang, X.G., Zhang, R., Zhang, X.Y., Xie, Y.M.: A novel concrete-filled auxetic tube composite structure: design and compressive characteristic study. Eng. Struct. **268**, 114759 (2022)
18. Bouaich, F.Z., Maherzi, W., Benzerzour, M., Taleb, M., Abriak, N.-E., Rais, Z., Senouci, A.: Mortar mixing using treated wastewater feasibility. Constr. Build. Mater. **352**, 128983 (2022)
19. Alomarah, A., Masood, S.H., Sbarski, I., Faisal, B., Gao, Z., Ruan, D.: Compressive properties of 3D printed auxetic structures: experimental and numerical studies. Virtual Phys. Prototyp. **15**, 1–21 (2020)
20. Menon, H.E., Dutta, S., Krishnan, A., Hariprasad, M.P., Shankar, B.: Proposed auxetic cluster designs for lightweight structural beams with improved load bearing capacity. Eng. Struct. **260**, 114241 (2022)
21. Baumers, M., Dickens, P., Tuck, C., Hague, R.: The cost of additive manufacturing: machine productivity, economies of scale and technology-push. Technol. Forecast. Soc. Chang. **102**, 193–201 (2016)

An Experimental Investigation on Using Seawater in Sustainable Mortar Mixtures

Shekhar Saxena, Harald Justnes,
and Mohammad Hajmohammadian Baghban

Contents

1 Introduction

Water stands as the primary component in concrete mix proportions and is an indispensable resource for sustaining life. Typically, freshwater serves as the primary choice for mixing and curing concrete, yet its availability is finite. A study conducted by the WRI (World Resources Institute Europe) [1] revealed that the demand for water is escalating due to urbanization, population growth and socio-economic advancements. Industries, particularly the construction sector, consume approximately 19% of the global water supply [2]. If current concrete preparation practices persist over the next 35 years, an estimated 590–710 km^3 of water will be expended solely for this purpose [3]. The United Nations' report of 2022 [4]

S. Saxena (✉) · M. H. Baghban
Norwegian University of Science and Technology, Gjøvik, Norway
e-mail: shekhar.saxena@ntnu.no

H. Justnes
Norwegian University of Science and Technology, Trondheim, Norway

© The Author(s) 2025
M. Kioumarsi, B. Shafei (eds.), *The 1st International Conference on Net-Zero Built Environment*, Lecture Notes in Civil Engineering 237,
https://doi.org/10.1007/978-3-031-69626-8_22

highlighted that over 40% of the world's populace faces challenges regarding freshwater availability, exacerbated by the escalating infrastructure development, which intensifies the strain on natural water resources. Hence, the exploration of alternative water sources for concrete production becomes imperative.

Considering that the Earth's surface is predominantly covered by water, with seawater accounting for approximately 96.5% of the total, utilizing seawater for concrete mixing presents a viable solution to conserve freshwater resources and ensure sustainability [5]. Seawater, however, comprises a complex blend of salts, organisms, dissolved gases, suspended sediments and organic matter, which can either positively or negatively impact concrete properties. Despite its controversial reputation, the historical application of seawater in concrete dates back to the Second World War, where it was utilized for constructing coastal structures in California and Florida [6]. Moreover, recent research challenges the prevailing wisdom against using seawater salts in structural concrete. While it is widely believed that these salts can compromise the durability of conventional reinforced concrete, evidence suggests that they do not significantly impact the properties of plain cement concrete and mortar. Nevertheless, the diverse salts found in seawater engage in chemical interactions with the constituents of concrete, resulting in alterations of concrete properties. This study aims to investigate the influence of seawater on the fresh and hardened properties of cement mortar when seawater replaces freshwater for mixing and curing purposes.

2   Materials and Methods

The study utilized Norcem's CEM I 42.5 N – SR3 cement, as per EN 197–1 standard. Table 1 outlines the chemical composition of this cement. Tap water and seawater were used for mixing and curing of mortar mixes. The seawater employed in this study was produced synthetically following the guidelines outlined in ASTM D1141 standards. The seawater used had a pH of 8.4, and its chemical composition is detailed in Table 2. Natural sand with a particle size of 0/8 mm was used as fine aggregate. The specific gravity of the natural sand was measured at 2.65, with a water absorption rate of 0.5% and a particle density of 2.66 Mg/m^3. Mortar specimens were prepared following EN 196–1:2016 guidelines. The proportion of water to cement 0.50 and cement to sand 1:3 were maintained during mortar preparation. This allowed for an examination of how the different water types affected the properties of the cement mortar over time.

Table 1 Major oxides in cement

Oxide	CaO	SiO$_2$	Fe$_2$O$_3$	Al$_2$O$_3$	K$_2$O	SO$_3$	NaO$_2$	MgO
Values (%)	60.66	20.45	4.28	3.36	0.45	2.67	0.28	1.90

Table 2 Elemental content in artificial seawater

Element	Amount (grams per liter)
Sodium chloride (NaCl)	24.53
Sodium sulfate (Na₂SO₄)	4.09
Magnesium chloride (MgCl₂)	5.20
Potassium chloride (KCl)	0.695
Calcium chloride (CaCl₂)	1.16
Sodium bicarbonate (NaHCO₃)	0.201
Boric acid (H₃BO₃)	0.027
Potassium bromide (KBr)	0.101
Sodium fluoride (NaF)	0.003
Strontium chloride (SrCl₂)	0.025

3 Tests on Mortar

Multiple examinations were undertaken to evaluate the characteristics of cement mortar. A heat of hydration test was executed on a paste comprising cement mixed separately with tap water and seawater. An isothermal calorimeter was used for measuring the heat of hydration of cement for 72 h by following the guidelines of EN 196–11 (2018). For testing the mechanical properties of cement mortar, compressive and flexural strengths tests were conducted on all mortar mixes. Mortar test specimens 40 mm × 40 mm × 160 mm prisms were prepared according to EN 196–1 (2016) and tested for flexural and compressive strengths after 1, 3, 28 and 90 days of curing at room temperature. Additionally, the electrical resistance and resistivity of various mortar mixes were assessed according to ASTM C1876–19 at 28 and 90 days of age. The RCON device, utilizing AC impedance technique, developed by Giatec Scientific Inc., was employed for the measurement of electrical properties. For this experimental evaluation, cylindrical mortar samples with a diameter of 100 mm and a length of 200 mm were utilized. Electrode plates were positioned on both end surfaces of the mortar specimens, with water-saturated sponges serving as conductive mediums. A current frequency of 1 kHz was utilized for resistance measurement. By applying an alternating current voltage across the specimen's cross-section between the two electrodes, the current flowing through the pore network within the mortar specimen was quantified. Moreover, the assessment of permeable pore volume, absorption, dry bulk density and apparent density in TWM and SWM mortar specimens was conducted in accordance with the procedure outlined in ASTM C642–13. Four parallel mortar discs from each mortar mix, with 100 mm diameter and 50 mm length were utilized for this examination, each cured for a duration of 90 days.

4   Results and Discussion

4.1   *Heat of Hydration*

The variation of heat flow in cement with tap water and seawater was measured up to 3 days as shown in Fig. 1. It was reported that the peak of cement paste mixed with seawater exhibited greater magnitude and became evident earlier compared to cement paste mixed with tap water. The highest heat flow of cement with tap water was found as 2.45 mW/g at 10.88 h, whereas with seawater, it was found as 3.42 mW/g at 8.89 h. In accordance with existing literature, it has been discovered that the hydration process of cement is expedited by seawater due to its high chloride content [7–9]. Sikora et al. (2020) [10] also found an about 19% increase in the peak of seawater-mixed cement paste than the peak of distilled water cement pastes in heat of hydration test. The possible reasons behind the increased cement hydrations might be:

1. Chloride ions, being smaller than hydroxyl ions, diffuse into the initial hydration products of cement. These products form a protective layer on the surface of cement particles. However, when chloride ions interact with water, they weaken or break down this protective layer. As a result, the breakdown of hydration products such as tri-calcium silicate is enhanced, leading to an increased rate of cement hydration [16].
2. Additionally, a blend of negatively and positively charged ions coexists within cement grains. When these grains encounter water, they tend to flocculate. In contrast, chlorides carry a negative charge. When seawater introduces chlorides into concrete, they are absorbed onto the cement grains surfaces, rendering them to acquire a negative charge. This intriguing process leads to the separation and unique behavior of cement grains, further enhancing the process of hydration of cement [17].

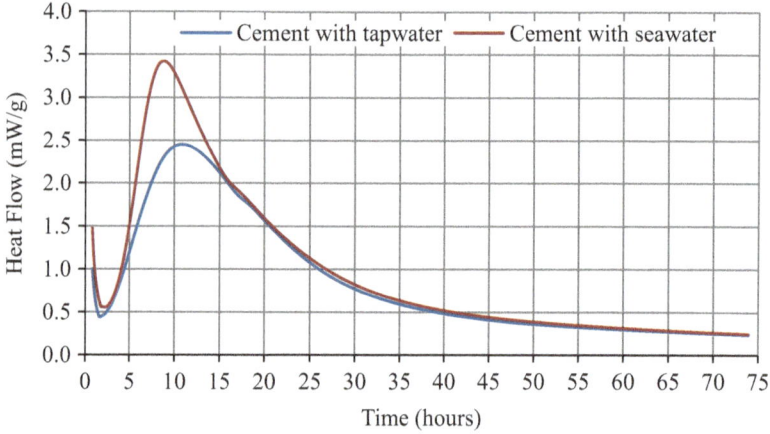

Fig. 1 Heat of hydration of cement with tap water and seawater

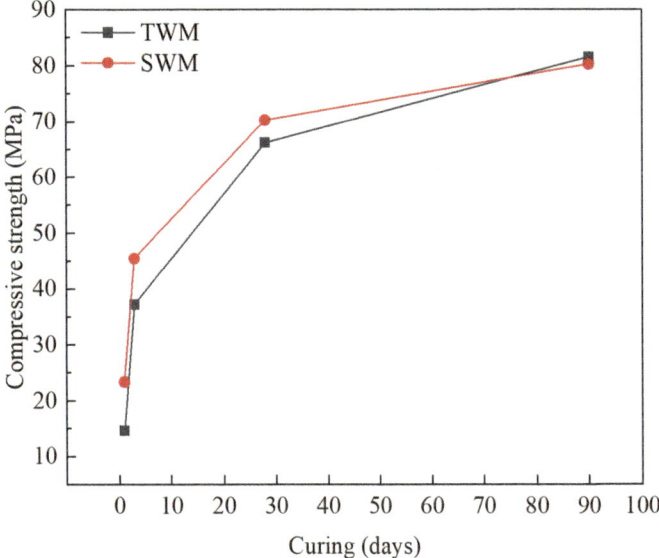

Fig. 2 Compressive strength of TWM and SWM

4.2 Compressive Strength

The compressive strength of the mortar with seawater and tap water after 1, 3, 28 and 90 days of curing is shown in Fig. 2. Seawater mortar gained more compressive strengths at the ages of 1, 3 and 28 days as compared to mortar prepared and cured in tap water whereas little reduction in the later age (90 days) compressive strength of seawater mortar than that of tap water mortar. The increment in compressive strength of seawater mortar was found as 60.1%, 22% and 6% at 1, 3 and 28 days of curing respectively and reduction at 90 days was 1.6% than that of tap water mortar. Previous studies also show that seawater contributes to enhance initial compressive strength in concrete. However, over time, it diminishes the subsequent strength [11, 12]. Justnes et al. (2022) [13] also found the increment in the compressive strength of concrete with seawater as compared with fresh water. In a study by Mori et al. (1981) [14], it was observed that the difference in strength between concrete mixed with seawater and that mixed with freshwater is relatively small after a decade of exposure testing.

4.3 Flexural Strength

Figure 3 shows the flexural strength of cement mortar by using seawater and tap water. Flexural strength of cement mortar with seawater increases by 42.9% at 1 day curing age and then decreases by 12.7%, 21.3% and 27.6% at 3, 28 and 90 days of

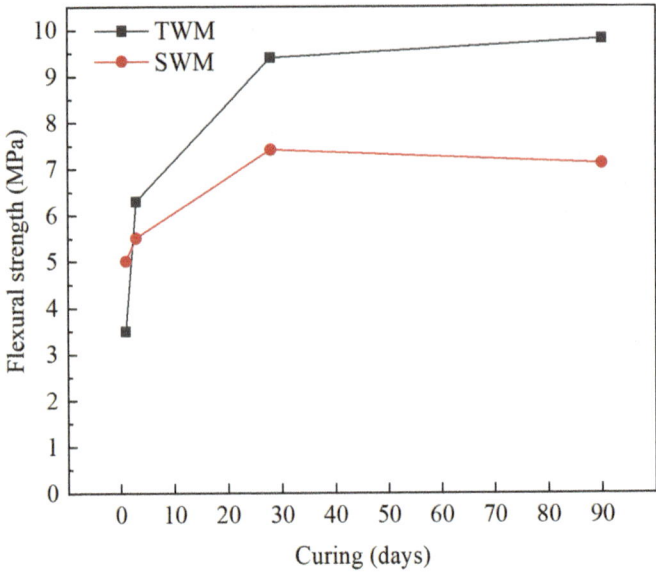

Fig. 3 Flexural strength of TWM and SWM

curing respectively. In Wegian's (2010) [15] study, it was observed that seawater concrete exhibited greater flexural strength during the initial period. However, seawater concrete experienced a reduction in flexural strength by 3.8–14.5% compared to freshwater concrete as it aged. Earlier studies have observed that, concrete mixed with seawater displays enhanced strength in the early phases because of the existence of NaCl. This substance expedites the breakdown of tricalcium silicate, prompting faster hydration reactions [16]. Consequently, additional hydration components are generated, occupying the concrete voids and improving its microscopic arrangement [17].

4.4 Electrical Resistance and Resistivity

The electrical resistance and resistivity of mortar mixes TWM and SWM is presented in Table 3. The use of seawater tends to reduce the resistance and resistivity of mortar, both at 28 and 90 days. This reduction diminishes as the mortar age increases at 28 days, the electrical resistance decreases by 23%, and at 90 days, by 7%. Similarly, resistivity decreases by 19% at 28 days and 4% at 90 days. Li et al. (2021) [8] also observed a decrease in the electrical resistivity of concrete when seawater was used as the mixing water, attributing it to the higher concentration of ions in the pore solution. This suggests that seawater, having more ions than tap water, likely possesses higher conductivity, thus resulting in lower resistivity in

Table 3 Electrical resistance and resistivity of TWM and SWM

Curing days	Electrical resistance (ohm)		Electrical resistivity (ohm-meter)	
	TWM	SWM	TWM	SWM
28	962.0	740.0	37.0	30.0
90	1070.0	995.0	42.5	41.0

Table 4 Water absorption, density and permeable voids of TWM and SWM

Properties	TWM	SWM
Water absorption after immersion (%)	8.08	7.54
Bulk density dry (mg/m^3)	2.16	2.19
Apparent density (mg/m^3)	2.62	2.63
Volume of permeable voids (%)	17.57	16.89

seawater mortar compared to tap water mortar. Jahani et al. (2023) [18] reported an increase in the electrical resistivity of seawater concrete at 7 days but observed a gradual reduction thereafter, at 28, 180, 365 days, and 3 and 4 years. This trend was attributed to the growth in the number and size of expansive crystals in seawater concrete over time, leading to potential cracking and increased porosity due to expansion. Consequently, the electrical resistance of seawater concrete decreases. Saleh et al. (2023) [19] investigated the influence of ion movement through the pore solution on concrete resistivity. The study anticipated that the use of seawater, rich in free chloride ions during mixing and curing, would lower the resistivity of concrete.

4.5   Water Absorption, Density and Permeable Voids

An essential aspect contributing to concrete's durability lies in its ability to withstand the ingress of harmful ions. The porosity of concrete is often assessed through its water absorption behavior, revealing insights into its permeable pore volume and connectivity. The data presented in Table 4 demonstrate that the incorporation of seawater leads to a decrease in water absorption and permeable voids within the mortar. However, the effect is not pronounced. Moreover, both the bulk density and apparent density display similar trends for both TWM and SWM mortar compositions.

The decrease in voids and water absorption highlights the dense microstructure characteristic of SWM. This phenomenon could potentially stem from the filling of concrete pores by salts found in seawater. Additionally, the formation of a common hydration product known as Friedel's salt (FS) occurs through the reaction between seawater chlorides and calcium-aluminate hydrates, along with calcium hydroxide in cement, as illustrated in Eq. 1. The precipitation of FS aids in compacting pore structures, thereby resulting in reduced voids and absorption levels [20].

$$3CaO \cdot Al_2O_3 \cdot 6H_2O + Ca(OH)_2 + 2Cl^- + 4H_2O \rightarrow 3CaO \cdot Al_2O_3 \cdot CaCl_2$$
$$\cdot 10H_2O \text{ (FS)} + 2OH^- \tag{1}$$

5   Limitations and Recommendations for Future Research

The research findings suggest that seawater poses no significant harm to mortar, indicating its potential use in plain concrete construction. Seawater accelerates the hydration rate, leading to enhanced initial strength in mortar. Additionally, it reduces water absorption and diminishes the volume of permeable voids in cement mortar. However, it also results in lower electrical resistivity due to the high concentration of ions in the pore solution. The current study is confined to conducting mechanical tests along with durability for a duration of up to 90 days. However, conducting a long-term study would provide a more comprehensive understanding of concrete performance in seawater environments. Furthermore, exploring additional durability parameters such as microstructure studies and investigating the effects of seawater on concrete performance under various environmental conditions like freezing and thawing, wetting and drying and acid attack could offer valuable insights for future research endeavors.

While seawater holds promise for plain concrete construction, challenges arise in utilizing seawater for reinforced concrete due to the corrosive nature of chlorides which affect steel reinforcement. Alternative solutions, such as non-corrosive reinforcements like fiber-reinforced bars, stainless steel and aluminum, have been employed in the past studies to mitigate corrosion. Yet, the long-term durability of such reinforcements remains uncertain, necessitating further research in this area for more effective seawater utilization. Supplementary cementitious materials (SCMs) offer another avenue for minimizing seawater's impact by binding chloride ions and enhance the durability of seawater concrete.

6   Conclusions

In summary, the present study yields the following key findings:

- Cement mixed with seawater generates more heat and exhibits an earlier peak compared to tap water.
- Seawater mortar shows higher strengths at 1, 3 and 28 days than tap water mortar. There is a slight reduction in 90 days compressive strength for seawater mortar compared to tap water mortar.
- Seawater increases flexural strength by approximately 42.9% at 1 day, but it decreases at 3, 28 and 90 days.
- The use of seawater tends to reduce the electrical resistance and resistivity of mortar at both 28 and 90 days. This reduction diminishes as the mortar ages.
- The incorporation of seawater in mortar results in a reduction in water absorption and permeable voids, indicating a denser microstructure.

Thus, this study presents a comparative analysis of the effects of seawater and tap water on the fresh, mechanical and durability properties of cement mortar. It also provides perspectives on attaining improved durability and sustainability in seawater-based concrete through present research work.

References

1. World Resources Institute Europe. https://www.wri.org/wri-europe. Last accessed 2024/02/23
2. Arosio, V., Arrigoni, A., Dotelli, G.: Reducing water footprint of building sector- concrete with seawater and marine aggregates. IOP Conf Ser. Earth Environ. Sci. **323**(012127), 1–8 (2019). https://doi.org/10.1088/1755-1315/323/1/012127
3. Fry, A.: WBCSD Water- Facts and Trends (2006)
4. UN Water: UN World Water Development Report 2022. United Nation (2022) https://www.unwater.org/publications/un-world-water-development-report-2022. Last accessed 2024/01/26
5. Howard, P., Jack, C., Adam, N.: How Much Water Is There on Earth? (2019). https://www.usgs.gov/special-topics/water-science-school/science/how-much-water-there-earth#
6. Kaushik, S.K., Islam, S.: Suitability of sea water for mixing structural concrete exposed to marine environment. Cem. Concr. Compos. **17**(3), 177–185 (1995)
7. Ebead, U., Lau, D., Lollini, F., Nanni, A., Suraneni, P., Yu, T.: A review of recent advances in the science and technology of seawater-mixed concrete. Cem. Concr. Res. **152** (2022). https://doi.org/10.1016/j.cemconres.2021.106666
8. Li, P., Li, W., Sun, Z., Shen, L., Sheng, D.: Development of sustainable concrete incorporating seawater: a critical review on cement hydration, microstructure and mechanical strength. Cem. Concr. Compos. **121**, 104100 (2021)
9. Saxena, S., Baghban, M.H.: Suitability of seawater for mixing and curing of cementitious materials: a step towards sustainability. In: Proceedings of the International fib Symposium on the Conceptual Design of Concrete Structures held in Oslo, Norway, June 29–July 1, pp. 130–137 (2023)
10. Sikora, P., Lootens, D., Liard, M., Stephan, D.: The effects of seawater and nanosilica on the performance of blended cements and composites. Appl. Nanosci. **10**(12), 5009–5026 (2020). https://doi.org/10.1007/s13204-020-01328-8
11. Younis, A., Ebead, U., Suraneni, P., Nanni, A.: Fresh and hardened properties of seawater-mixed concrete. Constr. Build. Mater. **190**, 276–286 (2018). https://doi.org/10.1016/j.conbuildmat.2018.09.126
12. Saxena, S., Baghban, M.H.: Seawater concrete: a critical review and future prospects. Dev. Built Environ., 100257 (2023)
13. Justnes, H., Danner, T., Sletnes, M.: Aluminium reinforced concrete enabled by calcined clay. In: 6th Fib International Congress at: Oslo, Norway (2022)
14. Mori, Y.: 10 years exposure test of concrete mixed with seawater under marine environment. J. Cem. Assoc. **35**, 341–344 (1981)
15. Wegian, F.M.: Effect of seawater for mixing and curing on structural concrete. IES J. Part A Civil Struct. Eng. **3**(4), 235–243 (2010). https://doi.org/10.1080/19373260.2010.521048
16. Goyal, A., Karade, S.R.: Steel corrosion and control in concrete made with seawater. Innov. Corros. Mater. Sci. **10**(1), 58–67 (2020). https://doi.org/10.2174/2352094909666191121104836
17. Wang, J., Liu, E., Li, L.: Multiscale investigations on hydration mechanisms in seawater OPC paste. Constr. Build. Mater. **191**, 891–903 (2018). https://doi.org/10.1016/j.conbuildmat.2018.10.010

18. Jahani, M., Moradi, S., Shahnoori, S.: 4-year monitoring of degradation mechanisms of seawater sea-sand concrete exposed to tidal conditions: development of chemical composition and micro-performance. Constr. Build. Mater. **409**, 133475 (2023)
19. Saleh, S., Mahmood, A.H., Hamed, E., Zhao, X.L.: The mechanical, transport and chloride binding characteristics of ultra-high-performance concrete utilising seawater, sea sand and SCMs. Constr. Build. Mater. **372**, 130815 (2023)
20. Ting, M.Z.Y., Yi, Y.: Durability of cementitious materials in seawater environment: a review on chemical interactions, hardened-state properties and environmental factors. Constr. Build. Mater. **367** (2023)

Effect of Recycled Aggregate Utilization on Strength Properties of Lightweight Concrete Facade Having a Self-Cleaning Characteristic

Hatice Gizem Şahin ⓘ, Hatice Elif Beytekin ⓘ, and Ali Mardani ⓘ

Contents

1   Introduction

It is known that the resistance to earthquakes increases due to the decrease in the mass of the structure with the use of lightweight concrete articles [1]. Moreover, it has become increasingly popular in the construction industry due to its advantages such as low thermal conductivity, high fire resistance, and low-cost articles [2–4]. Due to these positive effects, lightweight concrete has become more preferred, especially on the building façade [5, 6]. With the green economy approach adopted after the Paris climate agreement, it was understood that various methods have been applied to ensure the sustainability of lightweight concrete facades. These methods include (i) reducing energy consumption in the production process of lightweight concrete [7, 8] and (ii) using recycled materials [9, 10]. It is known that there are

H. G. Şahin · A. Mardani (✉)
Department of Civil Engineering, Faculty of Engineering, Bursa Uludag University, Nilüfer-Bursa, Turkey

H. E. Beytekin
Department of Architecture, Faculty of Architecture, Bursa Uludag University, Nilüfer-Bursa, Turkey

© The Author(s) 2025
M. Kioumarsi, B. Shafei (eds.), *The 1st International Conference on Net-Zero Built Environment*, Lecture Notes in Civil Engineering 237,
https://doi.org/10.1007/978-3-031-69626-8_23

applications such as (iii) reducing the amount of waste [11, 12]. Additionally, it was reported that the use of photocatalytic materials provides self-cleaning properties on the facade and improves indoor air quality [13]. Among many semiconductors, it was determined that nano-TiO_2 (nT) is generally used as a photocatalyst due to its advantages. Among the phases of nT, it was determined that the anatase phase is generally preferred because it shows a faster photocatalytic effect. It was understood that various studies have been carried out on extending the strength, durability, and service life of lightweight concrete facades and increasing the energy efficiency performance [14, 15]. However, it was observed that limited and contradictory research was conducted regarding the use of waste materials in the sustainability and composition of lightweight concrete mixtures. It was reported that construction waste increases rapidly, especially after severe earthquakes [16]. For this reason, encouraging the use of recycled concrete aggregate and effectively disposing of construction-demolition waste is of great importance for the understanding of green economy. In this context, it was understood that new research is necessary on the effect of the use of recycled concrete aggregate on the strength performance of lightweight concrete mixtures. In this study, the effect of changing the recycled concrete aggregate usage rate on the compressive strength and flexural strength performance of self-cleaning lightweight concrete mixtures (SCLWC) was examined. For this purpose, the compressive strength and flexural strength performances of mixtures produced by substituting 25% and 50% recycled concrete aggregate instead of pumice aggregate in an SCLWC mixture containing 1% TiO_2 and 100% pumice aggregate were examined.

2   Material and Method

2.1   Materials

In this study, CEM I 42.5 R type cement complying with TS EN 197-1 Standard was used as the binder. The chemical composition and some physical and mechanical properties of the cements supplied by the manufacturer are shown in Table 1.

Within the scope of the study, some properties of NT used to provide self-cleaning properties to lightweight concrete mixtures are shown in Table 2.

Pumice (P) and recycled concrete aggregate (RCA) with a maximum grain diameter of 2 mm were used in the preparation of photocatalytic self-cleaning lightweight concrete mixtures. The specific gravity and water absorption capacity values of the aggregates determined in accordance with EN 1097-6 are shown in Table 3. A single type of water reducing admixture was used to achieve the target emission value of 240 ± 20 mm. Some properties of the water reducing admixture used, provided by the manufacturer, are shown in Table 3. In order for the change in aggregate grain size not to affect the results, all aggregates were sieved and used in a similar gradation curve.

Table 1 Chemical composition, physical properties, and mechanical properties of cement

Material	(%)	Physical properties		
SiO_2	18.8	Specific gravity		3.15
Al_2O_3	5.71	*Mechanical properties*		
CaO	62.70	Compressive strength (MPa)	1 day	14.7
Fe_2O_3	3.09		2 days	26.80
MgO	1.16		7 days	49.80
$Na_2O + 0.658\ K_2O$	0.92		28 days	58.5
SO_3	2.39	*Fineness*		
Cl^-	0.01	Residual on 0.045 mm sieve (%)		7.6
Loss on ignition (LOI)	3.20	Blaine specific surface (cm^2/g)		3530
Insoluble residue	0.32			
Free CaO	1.26			

Table 2 Some features of NT used in the study

Value	Units	28 nm NT
Purity	%	>99
Size	nm	28
Specific surface area	m^2/g	>60
Loss of weight in drying	%	2 max.
Loss of weight in ignition	%	5 max.
pH	–	5.5–7.0
Color	–	White

Table 3 Water absorption capacities and specific gravity of the aggregates used

Aggregate type	Specific weight	Water absorption capacity (%)	Los Angeles wear loss (%)
Pumice, 0–2 mm	1.1	44	36
Recycled concrete aggregate, 0–2 mm	2.4	5.6	26

Table 4 Some properties of the HRWRA

Type	Intensity (g/cm^3)	Solids content (%)	pH value	Chloride content (%)	Alkaline ratio, Na_2O (%)
HRWRA	1.023–1.063	32	5.8	<0.1	<10

In order to provide the desired slump value in the lightweight concrete mixtures produced, a single type of high-range water reducing admixture (HRWRA) was used. Some technical specifications of the admixture used, provided by the manufacturer, are given in Table 4.

2.1.1 Preparation of Mixtures

Within the scope of the study, experiments were carried out by substituting recycled concrete aggregate in certain proportions (50% and 25% by volume) instead of pumice aggregate into the mixture containing 1% of the cement weight of NT and 100% pumice as aggregate, prepared by taking into account the studies in the literature. A total of 3 self-cleaning lightweight concrete mixtures were prepared (Table 5). Mixture calculation was made on 1 m^3 concrete. Mixtures with a water/cement ratio of 0.46 are designed according to ACI 211-2 Standard. The plasticizer admixture requirements of mixtures with a constant spread value of 220 ± 20 mm have been determined.

The naming of the mixtures was made according to the aggregate type and substitution rate. For example, the nomenclature of the mixture in which 50% recycled concrete aggregate was replaced was 50P-50RCA. The material rates and quantities used in the production of 1 m^3 lightweight concrete are shown in Table 5.

2.2   Method

2.2.1 Determining the Admixture Requirements of Mixtures

Spreading values of mortar mixtures were determined in accordance with ASTM C1437 standard. Admixture requirements for mixtures with a spreading value of 220 ± 20 mm have been determined. The 7-day compressive strength of the mixtures was determined on 50 mm cube samples according to ASTM C109 Standard. The 7-day flexural strength of the mixtures was determined by performing a three-point bending test on 40 × 40 × 160 mm prism samples in accordance with TS EN 196-1 Standard. Each value was calculated by taking the average of 3 measurements. Some images of the experiments carried out within the scope of the study are shown in Fig. 1.

Table 5 Amounts of materials used in the production of 1 m^3 SCLWC (kg/m^3)

Mixture	Cement (kg/m^3)	Aggregate (kg/m^3)		nT (%)	w/b
		Pumice	RCA		
100 P	544.5	605.3	–	1[a]	0.46
75 P-25 RCA	544.5	454.3	373.7		
50 P-50 RCA	544.5	302.9	747.5		

[a]By weight of cement

Fig. 1 Images of the experiments

Table 6 Admixture requirement of mixtures

Mixture	HRWRA requirement (kg/m^3)
100 P	5.45
75 P-25 RCA	5.01
50 P-50 RCA	4.9

3 Experimental Results and Discussion

3.1 Admixture Requirement of Mixtures

The admixture requirements of the mixtures produced within the scope of the study are shown in Table 6. When the admixture requirement of lightweight concrete mixtures was examined, it was observed that the admixture requirement decreased by approximately 8.78% and 10.09% depending on the increase in the use of recycled concrete aggregate. This was attributed to the higher water absorption capacity of pumice compared to recycled concrete aggregate. However, it is thought that the fact that pumice has a more porous and rough structure may also cause this situation. In a study conducted by Karthika et al. [17], the effect of substituting pumice aggregate with 50%, 80%, and 100% coarse aggregate on the spreading performance of concrete mixtures was examined. As a result, it was determined by the researchers that with the increase in the pumice substitution rate, the flow performance of the mixtures decreased by approximately 70%, 75%, and 77%, respectively, compared to the control mixture without pumice. Similarly, in a study conducted by Hossain et al. [18], it was found that the workability of mixtures decreased with the increase in pumice content. It was reported by researchers that this situation is due to the low density and porosity of pumice.

3.2 Compressive Strength of Mixtures

Unit volume weight and compressive strength values of lightweight concrete mixtures are shown in Fig. 2. It was determined that all mixtures produced are classified as lightweight concrete since their unit volume weight values are lower than 2000 kg/m³. As expected, the unit volume weight values of lightweight concrete mixtures increased due to the decrease in the amount of pumice used in the mixtures. This is due to the fact that the specific gravity of pumice is 54% lower than that of recycled concrete aggregate. Similar results were reported by Mardani et al. [15]. It is understood from Fig. 2 that the compressive strength values increased due to the decrease in the amount of pumice usage in the lightweight concrete mixtures produced. Compared to the control mixture using 100% pumice, it was determined that the increase in compressive strength in question was 24% and 53%, respectively, in mixtures with 75% and 50% pumice usage rate. This is thought to be due to the fact that recycled concrete aggregate has a less porous structure than pumice. A similar effect was also expressed by Taşdemir et al. [19] and Remasar et al. [5]. In addition, since the Los Angeles abrasion resistance of recycled concrete aggregate is

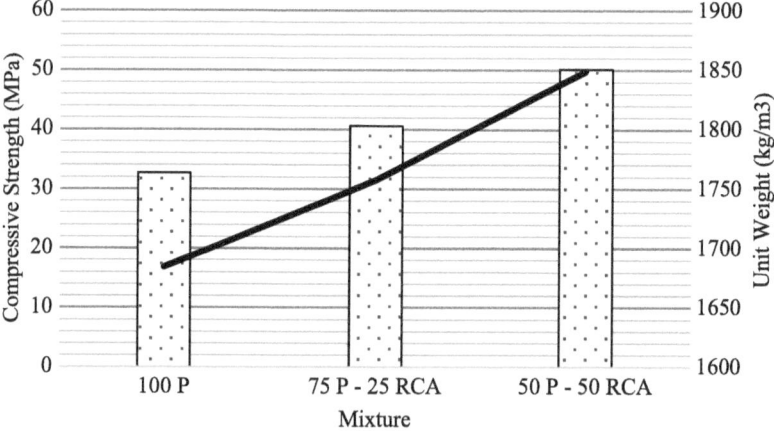

Fig. 2 Compressive strength (column chart) and unit volume weight (line chart) values of mixtures

10% higher than that of pumice aggregate (Table 3), it is thought that the decrease in the use of pumice aggregate in the mixtures affects the compressive strength performance positively. In another study, Lo et al. [20] emphasized that the compressibility performance of aggregates positively affects the compressive strength.

3.3 Flexural Strength of Mixtures

The 7-day flexural strength results of the mixtures produced within the scope of the study are shown in Fig. 3. It was determined that, similar to compressive strength, flexural strength values increase due to the decrease in the pumice usage rate. It was determined that the increase in question was 4% and 7%, respectively, in mixtures containing 75% and 50% pumice, compared to mixtures containing 100% pumice aggregate. It was determined that the loss of flexural strength due to the increase in pumice usage rate is quite low compared to the loss of compressive strength. It was reported by Bideci et al. [21] that in lightweight aggregate concretes, fracture occurs in the aggregate, not at the interlayer transition zone (ITZ), due to the porous and weak structure of the aggregate. Similarly, in another study by Mardani et al. [16], it was stated that the fracture resistance in ITZ decreased due to the lower strength and hardness properties of lightweight aggregates. It is also thought that due to the rough texture of pumice aggregate, it strengthens the ITZ of concrete and reduces the loss of flexural strength. For this reason, with the use of pumice aggregate and nano-TiO_2, in lightweight concrete mixtures that already have a strong ITZ, the flexural strength increased slightly by decreasing the amount of pumice aggregate, which has a very high water absorption capacity and low strength.

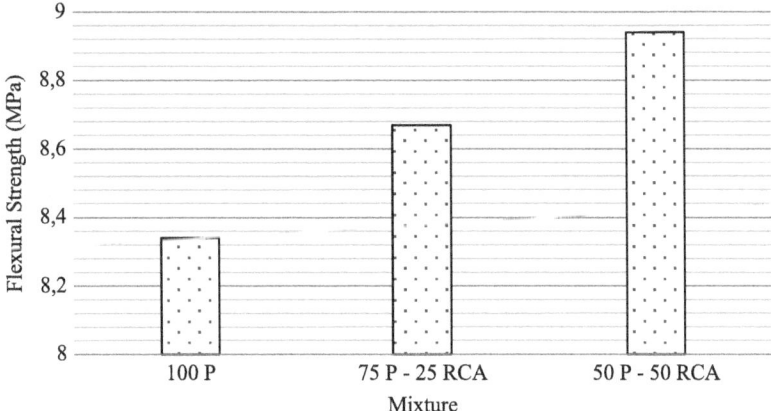

Fig. 3 Flexural strength values of mixtures

4   Conclusion

As a result of the materials used and the experiments carried out within the scope of the study, the following information was obtained:

1. It was determined that the need for admixture decreases due to the increase in the use of recycled concrete aggregate.
2. It was determined that the unit volume weight values of lightweight concrete mixtures increase due to the decrease in the amount of pumice.
3. It was understood that the compressive and flexural strength values increased due to the decrease in the amount of pumice usage. It was measured that the increase in question was lower for flexural strength values.

References

1. Badogiannis, E.G., Christidis, K.I., Tzanetatos, G.E.: Evaluation of the mechanical behavior of pumice lightweight concrete reinforced with steel and polypropylene fibers. Constr. Build. Mater. **196**, 443–456 (2019)
2. Al-Sibahy, A., Edwards, R.: Thermal behaviour of novel lightweight concrete at ambient and elevated temperatures: experimental, modelling and parametric studies. Constr. Build. Mater. **31**, 174–187 (2012)
3. Ustaoglu, A., Kurtoglu, K., Gencel, O., Kocyigit, F.: Impact of a low thermal conductive lightweight concrete in building: energy and fuel performance evaluation for different climate region. J. Environ. Manag. **268**, 110732 (2020)
4. Elshahawi, M., Hückler, A., Schlaich, M.: Shear behavior of infra lightweight concrete (ILC) with stirrups. J. Build. Eng. **72**, 106667 (2023)
5. Remesar, J.C., Vera, S., Lopez, M.: Assessing and understanding the interaction between mechanical and thermal properties in concrete for developing a structural and insulating material. Constr. Build. Mater. **132**, 353–364 (2017)
6. Canım, D.S., Kalfa, S.M.: Development of a new pumice block with phase change material as a building envelope component. J. Energy Storage. **61**, 106706 (2023)
7. Al-Tarbi, S.M., Al-Amoudi, O.S.B., Al-Osta, M.A., Al-Awsh, W.A., Shameem, M., Zami, M. S.: Development of energy-efficient hollow concrete blocks using perlite, vermiculite, volcanic scoria, and expanded polystyrene. Constr. Build. Mater. **371**, 130723 (2023)
8. Shafigh, P., Muda, Z.C., Beddu, S., Almkahal, Z.: Thermo-mechanical efficiency of fibre-reinforced structural lightweight aggregate concrete. J. Build. Eng. **60**, 105111 (2022)
9. Aytekin, B., Mardani-Aghabaglou, A.: Sustainable materials: a review of recycled concrete aggregate utilization as pavement material. Transp. Res. Rec. **2676**(3), 468–491 (2022)
10. Durgun, M.Y., Özen, S., Karakuzu, K., Kobya, V., Bayqra, S.H., Mardani-Aghabaglou, A.: Effect of high temperature on polypropylene fiber-reinforced mortars containing colemanite wastes. Constr. Build. Mater. **316**, 125827 (2022)
11. Sadrolodabaee, P., Hosseini, S.A., Claramunt, J., Ardanuy, M., Haurie, L., Lacasta, A.M., de la Fuente, A.: Experimental characterization of comfort performance parameters and multi-criteria sustainability assessment of recycled textile-reinforced cement facade cladding. J. Clean. Prod. **356**, 131900 (2022)
12. Sanyal, A.P., Mohanty, S., Sarkar, A.: Application of recycled aggregates generated from waste materials towards improvement in acoustical and thermal conductivity of concrete. Mater. Today Proc. (2023). https://doi.org/10.1016/j.matpr.2023.04.079

13. Shen, W., Zhang, C., Li, Q., Zhang, W., Cao, L., Ye, J.: Preparation of titanium dioxide nano particle modified photocatalytic self-cleaning concrete. J. Clean. Prod. **87**, 762–765 (2015)
14. Ducman, V., Mirtič, B.: Water vapour permeability of lightweight concrete prepared with different types of lightweight aggregates. Constr. Build. Mater. **68**, 314–319 (2014)
15. Mardani-Aghabaglou, A., Yoğurtcu, E., Andiç-Çakır, Ö.: Water transport of lightweight concrete with different aggregate saturation levels. ACI Mater. J. **112**(5), 681–692 (2015)
16. Mardani, A., Hatungimana, D., Yazici, Ş., Şahin, H.G., Assaad, J.J.: Use of recycled mortar as fine aggregates in pavement concrete applications. Heliyon. **10**(2), e24264 (2024)
17. Karthika, R.B., Vidyapriya, V., Sri, K.N., Beaula, K.M.G., Harini, R., Sriram, M.: Experimental study on lightweight concrete using pumice aggregate. Mater. Today Proc. **43**, 1606–1613 (2021)
18. Hossain, K.M.A., Ahmed, S., Lachemi, M.: Lightweight concrete incorporating pumice based blended cement and aggregate: mechanical and durability characteristics. Constr. Build. Mater. **25**(3), 1186–1195 (2011)
19. Tasdemir, C., Sengul, O., Tasdemir, M.A.: A comparative study on the thermal conductivities and mechanical properties of lightweight concretes. Energ. Buildings. **151**, 469–475 (2017)
20. Lo, T.Y., Tang, W.C., Cui, H.Z.: The effects of aggregate properties on lightweight concrete. Build. Environ. **42**(8), 3025–3029 (2007)
21. Bideci, A., Bideci, Ö.S., Ashour, A.: Mechanical and thermal properties of lightweight concrete produced with polyester-coated pumice aggregate. Constr. Build. Mater. **394**, 132204 (2023)

The Use of Recycled Construction and Demolition Waste in Low-Strength Concrete Brick and Block Production: A South African Perspective

Jaziitha Simon, Hans Beushausen, and Mark Alexander

Contents

1   Introduction

Over the past decades, South Africa has experienced a steady increase in population, reaching approximately 61.5 million people in 2023, alongside a corresponding rise in urbanization, projected to reach 71% by 2030 and 80% by 2050 [1]. This demographic expansion has fueled increased construction activities, leading to a substantial generation of construction and demolition waste (C&DW). Significant amounts of C&DW are generated from the construction of new buildings and the demolition and maintenance of existing structures. While a portion of this waste is used in low-grade applications such as road base construction and landfill cover materials, most of it ends up in landfills. This waste poses environmental challenges due to its large volume and issues associated with landfill disposal. Notably, the traditional linear approach of using virgin materials prolongs these challenges,

J. Simon (✉) · H. Beushausen · M. Alexander
Department of Civil Engineering, University of Cape Town, Cape Town, South Africa
e-mail: smnjaz001@myuct.ac.za

© The Author(s) 2025
M. Kioumarsi, B. Shafei (eds.), *The 1st International Conference on Net-Zero Built Environment*, Lecture Notes in Civil Engineering 237,
https://doi.org/10.1007/978-3-031-69626-8_24

A linear approach to concrete block production

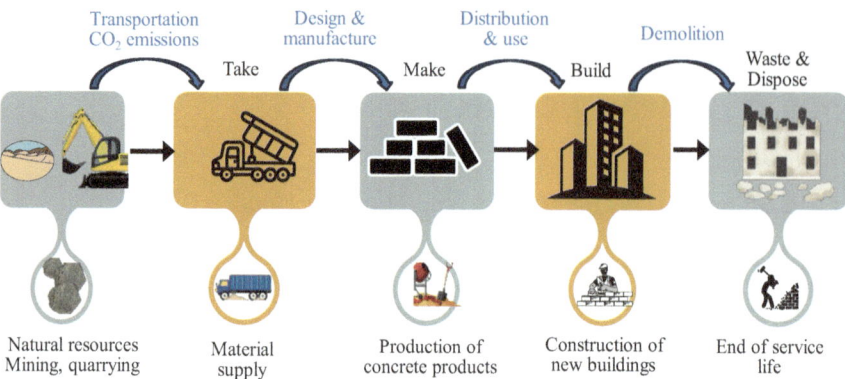

A circular approach to concrete block production

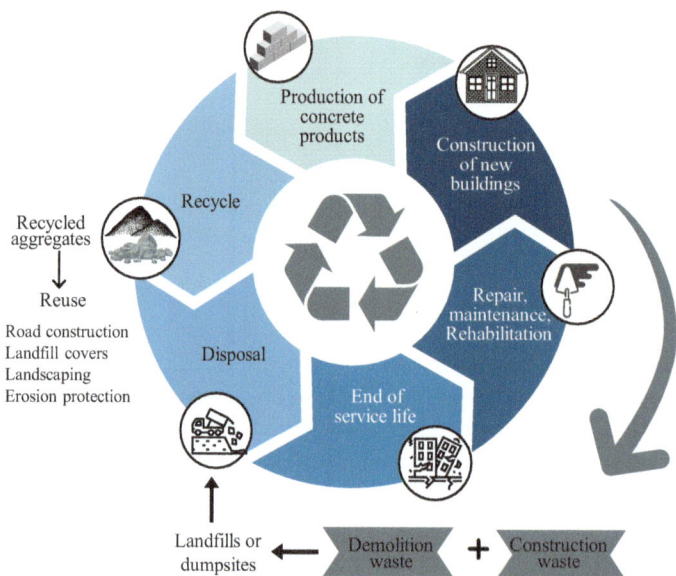

Fig. 1 A traditional linear approach to a circular economy in concrete block production

contributing to environmental issues such as increased energy usage and CO_2 emissions resulting from the mining and transportation of raw materials and the use of cement in concrete production. This approach is economically and environmentally unsustainable. To address these concerns and promote a circular economy, shifting to C&DW reuse and recycling is imperative. A circular economy emphasizes reusing, repairing, maintaining, and recycling materials to maximize societal value over time (see Fig. 1).

In the context of C&DW, this involves exploring alternative materials like recycled aggregates (RAs). The process includes sorting and crushing C&DW and sieving them into fractions suitable for new concrete production. This approach holds significant potential for applications in concrete brick and block production, particularly in the context of housing development in South Africa. Embracing recycling practices is crucial for fostering resource efficiency and reducing the carbon footprint associated with construction activities.

2   C&DW in South Africa

C&DW constitutes a significant portion of waste in South Africa, accounting for up to 30% (7.8 million tons) of the annual 26 million tons of municipal solid waste [2]. This category includes concrete, bricks, mortar, wood, plastics, metals, and soil waste from local cut-and-fill operations [3]. Figure 2 shows an ideal example of C&DW constituents at a specific demolition site in South Africa. Despite its substantial volume, there has been limited focus on circularity and diversion from landfills for C&DW. Concrete from C&DW is commonly reused as aggregates in subbases for road construction and landfill cover. However, hesitations persist regarding using RAs in new concrete due to uncertainties about their properties. On a smaller scale, recovered materials like bricks, blocks, wood, and window or door frames can be used in communities with limited financial means.

Recycling is not a recent development practice in SA. It has only become apparent as the government began to pressure people to use materials consciously,

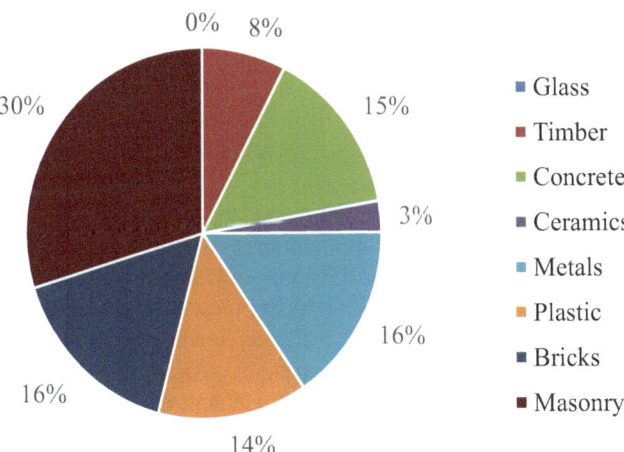

Different type of waste produced

Fig. 2 C&DW makeup. (Data retrieved from Mogodi [4])

as there is an increase in demand for raw materials. However, the country lags behind with a recycling rate of about 20% in 2017 [5], significantly lower than those observed in countries like China, European countries, and the United States (with recycling rates exceeding 70%). Recycling efforts are concentrated in provinces with larger populations, such as the Western Cape and Gauteng. Certain companies have identified and demonstrated opportunities for using RAs from C&DW in the construction industry. These initiatives need replication and support to foster widespread adoption.

South Africa faces challenges related to landfill space shortage due to the closure of older landfills, rising operational costs, and limitations in constructing new ones. The Western Cape Province, specifically Cape Town, anticipates exhausting its landfill capacity by 2032, with escalating disposal costs [6]. This situation drives waste service providers to explore cheaper alternatives, including illegal dumping, which poses environmental, financial, and aesthetic consequences.

Despite robust environmental and waste regulations in South Africa, challenges persist in implementing and enforcing these laws. C&DW is often disposed of as general waste instead of properly categorized for source separation and recycling. Noncompliance issues plague several landfills. Efforts have been made to strengthen waste management regulations by imposing penalties and requiring accreditation for waste service providers and generators in some provinces. Construction projects and demolition companies require an approved integrated waste management plan before project commencement. Additionally, the National Waste Management Strategy (NWMS) has set goals for waste prevention and recycling, but more comprehensive waste management plans and research are needed to manage C&DW in South Africa effectively. The country must address coordination issues between the relevant authorities, allocate resources, and prioritize waste management to achieve the targets set in the Waste Act and promote a circular economy.

3   Recycled Aggregates in Concrete

The construction industry generates a significant amount of C&DW, which can be recycled as RAs to replace the natural aggregates (NAs) in concrete. This recycling initiative has gained prominence globally due to the excessive consumption of approximately 20 billion tons of natural resources annually in fresh concrete production [7]. The overexploitation of natural resources, leading to scarcity, has prompted research and increased use of RAs in some developed countries. However, in South Africa, where NAs were initially low-cost and readily available, the adoption of RAs was limited. Over the years, environmental awareness and the scarcity of NAs have driven the recycling of C&DW to solve the NA shortage. The recycling process involves sorting and crushing C&DW to produce fine and coarse fractions of RAs, which find applications in low- and high-strength concrete production, road construction, and landscaping. Using RAs conserves natural resources and reduces costs, landfill airspace, and cement usage, offering environmental benefits.

The focus of studies on RAs in concrete has been primarily on structural concrete, mainly using concrete waste—recycled concrete aggregates (RCAs). The presence of masonry or clay brick components in C&DW has raised concerns due to their porous microstructures, leading to limitations on masonry to 10% in structural concrete standards [8, 9]. Some studies suggest removing masonry elements from C&DW for structural concrete, whereas others argue for their potential use [10]. The debate centers on the impact of mortar content or masonry on the durability performance of RCAs, requiring further investigation for conclusive evidence. Wickins [11] added that if separation at the source is enforced, then there are potential uses for clay masonry and concrete materials in low- and high-grade structural and nonstructural applications.

An RCA is a two-phase composite material consisting of original NA and adhered mortar from crushed concrete. The adhered mortar, containing both hydrated and unhydrated cement, may be weak and porous, negatively impacting the performance of RCAs in concrete. This can lead to decreased aggregate density, increased water absorption, increased water demand, and subsequent effects on the mechanical and durability properties [12, 13]. Some studies suggest that adhered mortar provides self-cementing properties to RCAs as latent cement particles react with water, producing hydrate compounds. The self-cementing mechanism, primarily influenced by fine fractions of RCAs (<5 mm), is still being understood. High levels of active calcium silicate hydrate (CSH), the primary hydration product of Portland cement, tend to be found in finer fractions. Their intrinsic properties are influenced by the age, grade, and mix proportions of the original concrete [14].

Little research on RCAs has been conducted in South Africa, covering diverse aspects, as detailed in Table 1. The first research conducted by Clayton in 1987 at the University of Cape Town (UCT) examined the feasibility of using RCAs in structural concrete, highlighting the high absorption due to porosity in RAs

Table 1 Research on RCAs in South Africa

Author, year	Aspects covered	Type of RA used	Concrete application
Clayton F., 1987 [15]	Introduce the concept of recycling demolished concrete waste	Coarse RCAs	Structural concrete
Kutegeza B., 2004 [16]	Performance of concrete made with RAs	Coarse and Fine RCAs	Structural concrete
Kearsley and Mostert, 2012 [17]	Parent concrete of different strength classes	Coarse and Fine RCAs	Structural concrete
Paul S.C., 2011 [18]	Volume change and durability of RAC[a]	Coarse RCAs	Structural concrete
Wickins K., 2013 [11]	Influence of on-site recycling and concrete mixing procedures	Coarse RCAs and CMAs[a]	Structural concrete
Immelman D., 2013 [10]	Influence of replacement and concrete properties	Coarse RCAs	Structural concrete
Kahabi N.S., 2022 [19]	Durability of RCAs	Fine RCAs	Structural concrete

[a]*RAC* recycled aggregate concrete, *CMA* concrete masonry aggregates

[15]. Kutegeza's 2004 study at UCT found that fine and coarse recycled aggregates exhibited higher absorption rates by 7 and 10 times, respectively, compared to natural aggregates, which reduced compressive strength when aggregates were combined [16]. Kearsley and Mostert from the University of Pretoria also assessed RCA performance in concrete [17], whereas other studies explored optimal replacement levels for better mechanical and durability properties [10, 11, 18, 19]. As depicted in Table 1, most studies have predominantly focused on structural concrete (with a strength of 25 MPa and upward). The use of RCAs in structural concrete has shown negative impacts depending on factors like adhered mortar quality, replacement level, waste sources, crushing process, water-to-cement ratio (w/c), and moisture state during mixing. The negative impacts are more significant in fine RCAs than in coarse RCAs. It was suggested that good-quality RCAs at 30% for coarse and 25% for fine replacement of NAs might be suitable for structural concrete. Despite these studies attempting to establish expected outcomes, further research is essential to fill gaps and contribute to guideline and specification development.

4   Recycled Aggregates in Concrete Products

The drawbacks associated with using RCAs in structural concrete can be mitigated using concrete mixtures for mechanized molded concrete products like bricks and blocks. The manufacturing process involving molding mixed materials under a combined vibrating and compacting action using mechanized machines reduces the importance of maintaining a workable mix. This process demands only a small amount of water for introduction into the molding machine, easing challenges related to controlling w/c and workability. The production of bricks and blocks requires less cement than that of structural concrete, reducing the carbon footprint.

4.1   The Concrete Product Industry in South Africa

The South African construction industry extensively relies on concrete products, such as bricks and blocks, due to their durability, strength, and cost-effectiveness compared to alternatives like fired clay bricks. The clay brick industry has been noted to have about three times more embodied energy than concrete bricks and blocks. Concrete bricks and blocks are made from Portland cement, fine aggregates, coarse aggregates, water, and admixtures. These concrete products are typically made from virgin materials using a low-strength concrete mix, with a strength range of 7–15 MPa, and they do not require reinforcement. Concrete mixes typically contain 10–15% cement, with a water-to-cement ratio of around 0.3–0.6 and a cement-to-aggregate ratio of around 5–20% by mass. The manufacturing process of these concrete products consists of five basic processes: batching, mixing, molding, curing, and cubing (packaging), as illustrated in Fig. 3 [11]. The process

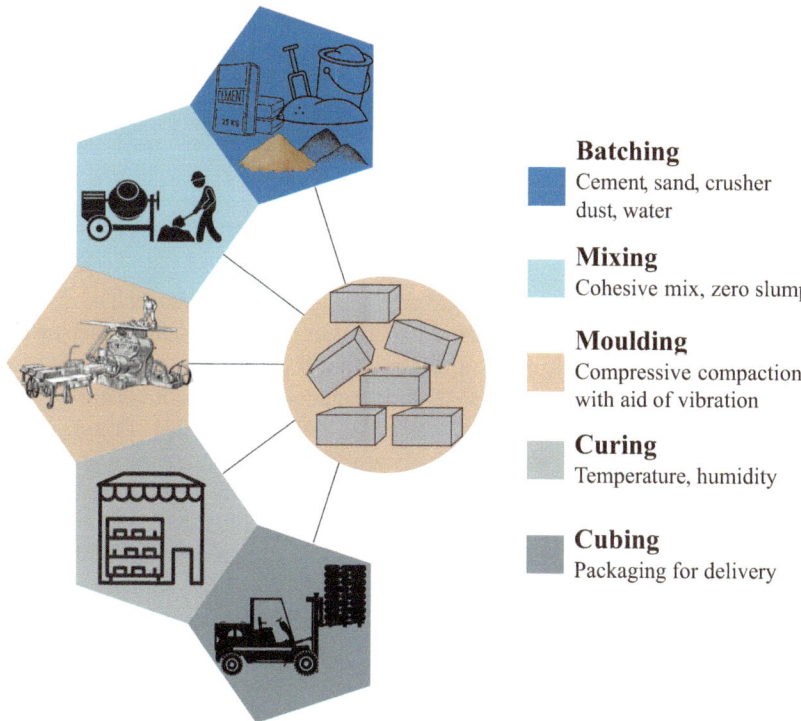

Batching
Cement, sand, crusher
dust, water

Mixing
Cohesive mix, zero slump

Moulding
Compressive compaction
with aid of vibration

Curing
Temperature, humidity

Cubing
Packaging for delivery

Fig. 3 The production process of bricks and blocks

involves mass production in factories using the dry mix pressed method with the aid of vibration forces, allowing for flexibility in shapes, thicknesses, densities, and strengths to meet diverse civil engineering project requirements.

The use of RAs in bricks and blocks is currently practiced in the South African construction industry, specifically in Cape Town. This is done to maximize recycled material usage for environmental benefits while ensuring cost-effectiveness. This shift is driven by the economic advantage of using RA, which costs 10% less than virgin materials. Although about 12 brick and block manufacturing companies are available in Cape Town, only a few incorporate RAs into their products, primarily due to misconceptions surrounding their properties. Quality control measures are lacking across the industry, resulting in varying product quality from one company to another. As a result, each company adopts practices based on their individual preferences, contributing to the inconsistency in product quality within the market. For instance, while some companies advocate for the inclusion of clay brick materials as waste, proclaiming potential enhancements to the final product, others caution against exceeding specific thresholds of 5% or 10% due to concerns about increased water requirements and compromised material density.

In the South African context, the current practice involves using various constituents of C&DW, such as concrete waste, mortar, or clay bricks, without separating them at the source. The application relies on the local availability of materials. Hence, it is crucial to determine the composition and physical properties of these materials before incorporating them into concrete production, as it provides a deeper insight into the materials and their anticipated performance.

4.2 Properties of Recycled Aggregates in Concrete Products

Despite the widespread use of concrete products, there has been a lack of research in South Africa regarding using RAs in low-strength concrete products such as bricks and blocks. However, studies from other countries, particularly China and some European countries, have shown that the mechanical properties of concrete products are significantly influenced by the properties of RAs [20–25]. Factors such as density, water absorption, water content, cement content, and compaction methods play crucial roles in production and performance. Density is the main factor that indicates the packing capacity of the material, which influences the water require-ment. Aggregates with more cement paste on their surface tend to have low densities due to the less dense mortar than NAs. Hence, fine fractions usually have lower densities compared to coarse aggregates. Furthermore, material compaction at the maximum moisture content corresponding to the maximum dry density is critical to achieve the best strength and durability. When compacting materials such as RAs with a certain amount of water, the density tends to increase until the maximum moisture content is reached, where the material obtains the maximum density, after that, it starts to decrease.

Studies have found that RCAs often enhance the strength of concrete products due to their inherent quality and residual cementitious content. Particle size distri-bution, aggregate size, shape and surface texture, density, and water absorption play pivotal roles in influencing workability and compressive strength [14, 23]. However, the presence of contaminants in these aggregates may weaken the bond within the cement matrix, leading to a reduction in the overall concrete strength. Common trends in using masonry in concrete block production include decreased density and compressive strength, along with increased water absorption [22, 23]. Some studies suggest that incorporating crushed masonry as a fine aggregate replacement yields superior results compared to its use as a coarse aggregate replacement due to the fine particles' filler effect [24]. Compared to RCAs, it has also been observed that blocks with masonry materials can experience an increase in strength at a later stage, attributed to further hydration of unhydrated cement phases in the RCA.

While the replacement level of RCAs in structural concrete tends to be limited, concrete bricks and blocks might allow higher replacement of NAs. Studies have shown that replacing fine and coarse aggregates with RAs at 25% and 50% levels, respectively, has little impact on the compressive strength of bricks and blocks [20, 25]. However, a more than 50% replacement level tends to decrease strength. This reduction can be mitigated by increasing cement content, using a combination of fine aggregates, and using proper design considerations and curing techniques.

Despite the perceived negative properties associated with RAs, different beneficiation methods can be used to improve their quality. Most of the methods have been primarily used in structural concrete on RCAs, and only steam and CO_2 curing and the use of admixtures such as plasticizers have been applied in the production of concrete products [26, 27]. Steam and CO_2 curing enhance strength and durability, whereas plasticizers improve the binding properties of the dry mix concrete. Additionally, a two-stage mixing method initially used in structural concrete shows promise for improving properties such as water absorption and strength of the concrete made with RAs [28]. Furthermore, pre-wetting of aggregates is practiced in the industry as it tends to enhance the absorptive capacity of RCAs and reduce dust from crushing.

4.3   Performance Requirements of Concrete Products

Concrete masonry units, encompassing bricks and blocks, are subjected to rigorous codes and standards, such as the South African National Standard (SANS), European Standards (EN), and the American Society for Testing Materials (ASTM). Specific manufacturing, construction, and design directives are laid out in diverse SANS standards like SANS 1215, SANS 10145, and SANS 10164-1, encompassing criteria such as dimensions, compressive strength, and soundness. EN 12620, BS 8500, ASTM C90, and ASTM C129 also cover various performance attributes. However, these codes primarily focus on natural aggregates, with no specifications for using RAs and contamination in concrete products. Therefore, for optimal use of RAs and desired properties in concrete products, careful consideration of replacement levels, mix design, and quality control measures is essential, especially in adherence to the existing standards.

4.4   Potential Impacts of Using Recycled Aggregates

The significance of using recycled C&DW for low-strength concrete production is multifaceted. Primarily, it contributes to environmental benefits by reducing the landfill waste and greenhouse gas emissions associated with transportation and disposal. Moreover, this practice aids in conserving natural resources by reducing the demand for virgin materials such as gravel and sand. Economically, it offers a viable and cost-effective alternative for construction companies. The research by Ohemeng and Ekolu [29] underscores this point, revealing that RAs' production cost is notably lower than that of NAs. For instance, producing one ton of coarse RCA costs approximately 40% less than coarse NAs while offering a significantly higher environmental benefit, approximately 97% greater than NAs [29].

Despite these evident benefits, using RAs has various limitations that need consideration. RAs sourced from demolition sites may exhibit variability in composition, gradation, and contaminants, necessitating rigorous quality control measures to ensure consistent product quality. Persistent concerns revolve around the long-term durability and structural performance of concrete containing RAs, including issues such as weaker interfacial transition zones and reduced bond strength between RAs and cement paste. Furthermore, the absence of standardized testing methodologies and specific regulations governing the use of RAs complicates its widespread adoption, highlighting the need for clear guidelines and compliance measures within the industry.

5   Challenges and Opportunities

Despite the promising potential of RA, integrating these materials into construction applications faces challenges such as perception, inconsistent quality, regulatory hurdles, and the absence of standards. While developed countries have made progress in C&DW recycling, emerging economies like South Africa lag behind due to slow acceptance and recognition of the use of recycled materials, which can be attributed to a lack of awareness and knowledge regarding the benefits and appropriate utilization of these materials.

For decades, research has predominantly focused on RAs in structural concrete, leading to their adoption in some national standards. These codes mandate the recycling of concrete waste, excluding masonry waste. Bricks and blocks can, however, be made from C&DW from various sources like concrete, masonry, clay, and soil. Furthermore, the substitution of NAs with RAs in structural concrete is constrained by limitations in replacement levels. High replacement levels in structural concrete pose challenges to mechanical and durability properties, leading researchers to suggest that RAs might be suitable for low-strength concrete bricks and blocks, especially in the context of housing development in South Africa [11]. Despite the long-standing existence of discussions on recycling, a significant knowledge gap exists regarding low-strength concrete in South Africa, contributing to a need for an understanding of the behavior of recycled materials and their impact on concrete product performance.

The absence of certifications and standards for RA use in concrete, including low-strength products like bricks and blocks, is a hindrance. The current South African standards, SANS 1083:2006, do not provide specifications for recycled materials, necessitating the establishment of guidelines and quality control procedures for consistent and high-quality recycled C&DW concrete products. Additionally, construction companies are not obligated by any laws to use C&DW in their projects. So, to promote the use of C&DW, there is a need for policy changes mandating a minimum percentage of RAs in concrete. This could create market demand and benefit low-cost housing, roads, stormwater management, erosion protection, and dolosse production.

The self-cementing properties of RCAs offer an opportunity to reduce cement consumption in block production, contributing to efforts to reduce the carbon footprint in the construction industry. Investigating the mechanism behind the self-cementing properties of RCAs can provide insights into optimizing concrete product performance.

Existing research advocates for techniques like using water-reducing admixtures such as plasticizers, steam, and CO_2 curing, a two-stage mixing approach, and mix design adjustments to enhance RA properties. Furthermore, pre-wetting of RAs in the industry is also encouraged. However, the practicality of the industry in terms of the sustainability of steam and CO_2 curing is questionable due to cost implications, energy requirements, and environmental impact. The influence of using plasticizers, pre-wetting of aggregates, and the two-stage mixing method on the properties of recycled materials and low-strength concrete needs to be understood.

Challenges in classifying C&DW, which comprises various waste streams, including hazardous waste and contaminants, highlight the importance of source separation. Implementing technologies from other countries, such as European nations, can improve waste separation in the South African context.

Furthermore, the lack of stringent tracking systems for C&DW generation statistics in South Africa, as seen in many countries, poses a challenge to circular economy development. Accurate data on C&DW types and quantities by location are crucial for effective waste management and circular economy initiatives in South Africa.

Lastly, continued research and technological improvements are crucial to for addressing the limitations and enhancing the quality and performance of recycled aggregate-based concrete products. Collaboration among stakeholders is essential to navigate the uncertainties and promote the sustainable integration of RAs in concrete products.

6   Outlook

Concrete production, essential for new construction and maintenance, uses millions of tons of materials annually. Addressing material procurement by specifying RAs in concrete applications could establish a sustainable market, significantly utilizing a large amount of C&DW each year. Converting a percentage of annually landfilled C&DW back into aggregates would lead to substantial benefits, including waste diversion, reduced carbon footprint, cost savings, and improved resource efficiency.

Developing material specifications for using RAs in concrete could attract private sector investment in crushing and sieving operations. However, a potential obstacle arises from the perception that RAs may need to meet standards. Understanding the behavior and influence of recycled materials on product performance is crucial to address this. This understanding can contribute to formulating guidelines and specifications for the construction industry, fostering reliability in using RAs. Comprehensive research in this area is thus essential for promoting the sustainable use of RAs in low-strength concrete products such as bricks and blocks, ensuring environmental benefits and industry acceptance.

References

1. StatsSA: Distribution of Dwellings in South Africa 2021, by Type. Statista Research Department. [Online]. Available: https://www.statista.com/statistics/1116038/distribution-of-dwellings-in-south-africa-by-type/. Accessed 29 June 2023
2. Govender, D., Govender, T., Whyte, C.: 2023 Market Study of the Circular (& Waste) Economy of South Africa, Pretoria, South Africa, August 2023. [Online]. Available: www.lindon.co.za
3. DEA: State of Waste Management Report 2020. [Online]. Available: www.westerncape.gov.za/eadp (2020)
4. Mogodi, M.: Ässessment of Practices and Strategies for Waste Management of Concrete and Cementitious Materials in South Africa. University of the Witwatersrand (2020)
5. Prajapati, R., et al.: Research Monograph: Recycled Concrete Aggregates and Their Influence on Concrete Properties. Scheme for Promotion of Academic and Research Collaboration (SPARC) (2020)
6. GreenCape: 2022 Waste Market Intelligence Report, Cape Town, South Africa (2022)
7. Joseph, H.S., et al.: A comprehensive review on recycling of construction demolition waste in concrete. Sustainability. **15**(4932), 1–27 (2023)
8. EN 12620: Aggregates for Concrete. British Standards (2002)
9. BS 8500-2: Concrete – Part 2: Specification for Constituent Materials and Concrete. British Standards Institution (2016)
10. Immelman, D.W.: The Influence of Percentage Replacement on the Aggregate and Concrete Properties from Commercially Produced Coarse Recycled Concrete Aggregate. University of Stellenbosch (2013)
11. Wickins, K.: The Use of Construction & Demolition Waste in Concrete in Cape Town. Thesis. University of Cape Town (2013)
12. de Andrade Salgado, F., de Andrade Silva, F.: Recycled aggregates from construction and demolition waste towards an application on structural concrete: a review. J. Build. Eng. **52**(March), 1–20 (2022)
13. Vivek Kumar, C., Palanisamy, M., Balakrishna, C., Reddy, S.P.S., Ravi, S.R.: Evaluation of strength characteristics and identifying the optimum dosage with the impact of partial replacement of recycled fine and coarse aggregate from construction and demolition waste. Mater. Today Proc. **66**, 1699–1709 (2022)
14. Poon, C.S., Qiao, X.C., Chan, D.: The cause and influence of self-cementing properties of fine recycled concrete aggregates on the properties of unbound sub-base. Waste Manag. **26**(10), 1166–1172 (2006)
15. Clayton, F.: The Use of Aggregate from Demolition Rubble in the Making of Ordinary and Structural Concretes. University of Cape Town (1987)
16. Kutegeza, B.: The Performance of Concrete Made with Commercially Produced Recycled Coarse and Fine Aggregates in the Western Cape. University of Cape Town (2004)
17. Kearsley, E.P., Mostert, H.F.: The use of recycled building materials as aggregate for concrete. Concr. Beton J. Concr. Soc. S. Afr., 8–12 (2012)
18. Paul, S.C.: Mechanical Behaviour and Durability Performance of Concrete Containing Recycled Concrete Aggregate. Stellenbosch University (2011)
19. Kahabi, N.S.: The Effect on the Durability Properties of Concrete of Partial Replacement of Natural Fine Aggregates with Recycled Concrete Fine Aggregates. University of Cape Town (2022)
20. Poon, C.S., Kou, S.C., Lam, L.: Use of recycled aggregates in moulded concrete bricks and blocks. Constr. Build. Mater. **16**, 281–289 (2002)
21. Jones, N., Millard, S.G., Soutsos, M.N., Bungey, J.H., Tickell, R.G., Gradwell, J.: Developing precast concrete products made with recycled construction and demolition waste. In: Sustainable Waste Management and Recycling: Challenges and Opportunities. Kingston University, London (2004)

22. Poon, C.S., Chan, D.: Paving blocks made with recycled concrete aggregate and crushed clay brick. Constr. Build. Mater. **20**(8), 569–577 (2006)
23. Xiao, Z., Ling, T.C., Kou, S.C., Wang, Q., Poon, C.S.: Use of wastes derived from earthquakes for the production of concrete masonry partition wall blocks. Waste Manag. **31**, 1859–1866 (2011)
24. Meng, Y., Ling, T.C., Mo, K.H.: Recycling of waste for value-added applications in concrete blocks: an overview. Resour. Conserv. Recycl. **138**, 298–312 (2018)
25. Matar, P., El Dalati, R.: Using recycled concrete aggregates in precast concrete hollow blocks. Mater. Sci. Eng. **43**(5), 388–391 (2012)
26. Zhan, B.J., Poon, C.S., Shi, C.J.: Materials characteristics affecting CO_2 curing of concrete blocks containing recycled aggregates. Cem. Concr. Compos. **67**, 50–59 (2016)
27. Martín-Morales, M., Cuenca-Moyano, G.M., Valverde-Espinosa, I., Valverde-Palacios, I.: Effect of recycled aggregate on physical-mechanical properties and durability of vibro-compacted dry-mixed concrete hollow blocks. Constr. Build. Mater. **145**, 303–310 (2017)
28. Wang, R., Yu, N., Li, Y.: Methods for improving the microstructure of recycled concrete aggregate: a review. Constr. Build. Mater. **242**, 118164 (2020)
29. Ohemeng, E.A., Ekolu, S.O.: Comparative analysis on costs and benefits of producing natural and recycled concrete aggregates: a South African case study. Case Stud. Constr. Mater. **13**(e00450), 1–13 (2020)

Scaling Effect on Mechanical Property of Calcium Silicate Hydrate in Cement Using Reactive Molecular Dynamics

Jie Cao, Chao Wang, Jaime Gonzalez-Libreros, Yongming Tu,
Lennart Elfgren, and Gabriel Sas

Contents

1   Introduction

Calcium silicate hydrate (C-S-H) is one of the main cement hydrates in Portland cement. The molecular dynamics method is crucial for studying the properties of cement hydrates at the nanoscale [1, 2]. To simulate the C-S-H gel at the nanoscale is not an easy task due to its long-range disorder. In recent years, the simulation of the structure of C-S-H gel at the nanoscale has been optimized multiple times [3–5], and using the corrected structure of tobermorite 11 Å to simulate C-S-H gel is becoming more and more accurate. However, few researchers have studied the performance difference between the corrected model and the uncorrected C-S-H gel model. Additionally, in terms of mechanical properties, while numerous studies have examined the uniaxial tensile stress–strain curve of models, few researchers have

J. Cao · C. Wang (✉) · J. Gonzalez-Libreros · L. Elfgren · G. Sas
Luleå University of Technology, Luleå, Sweden
e-mail: chao.wang@ltu.se

Y. Tu
Luleå University of Technology, Luleå, Sweden

Southeast University, Nanjing, China

© The Author(s) 2025
M. Kioumarsi, B. Shafei (eds.), *The 1st International Conference on Net-Zero Built Environment*, Lecture Notes in Civil Engineering 237,
https://doi.org/10.1007/978-3-031-69626-8_25

validated its accuracy. This lack of verification hinders comparative studies between simulations and experiments at various scales. Likewise, it is important to explore the impact of scale effects on the mechanical properties of models at the nanoscale for comparison with experimental or simulation results at different scales.

In this work, corrected and uncorrected C-S-H gel models were constructed at different sizes to investigate the influence of model correction and scaling effects on the mechanical properties of C-S-H models.

2 Methodology

2.1 Model Construction

The model construction process was completed in Materials Studio 2023. First, the unit cell of tobermorite 11 Å was extended and orthogonalized to a supercell with a ratio of $5 \times 8 \times 1$. For models 1–3, the silicon chains were broken according to previous research [4, 6]. More details on model construction can be found in reference [6]. In this article, the final calcium-to-silicon ratio in models 1–3 is set to 1.67, and the Q_n distribution is $Q_2/Q_1 = 0.2$. For models 4–6, no corrections were made. Therefore, they can also be regarded as the supercells of tobermorite 11 Å. Based on this, all the supercells were then extended at ratios of $1 \times 1 \times 1$, $2 \times 2 \times 2$, and $3 \times 3 \times 3$. The basic information of the models can be found in Table 1. The initial configuration displays the model dimensions in the x, y, and z directions, with the y direction being the direction in which the silicon chains extend, and the z direction being the layering direction.

2.2 Molecular Dynamics

LAMMPS software was used to perform molecular dynamics simulations, and the reactive force field [7] was adopted in this study. The time step throughout the simulation is set to 0.25 fs. First, the initial model is energy minimized and then

Table 1 Model information

	Silicon chain	Magnification	Initial configuration (Å^3)
Model 1	Breakage	$1 \times 1 \times 1$	$32.95 \times 59.12 \times 22.77$
Model 2	Breakage	$2 \times 2 \times 2$	$65.89 \times 118.24 \times 45.54$
Model 3	Breakage	$3 \times 3 \times 3$	$98.84 \times 177.36 \times 68.31$
Model 4	No breakage	$1 \times 1 \times 1$	$32.94 \times 59.12 \times 22.77$
Model 5	No breakage	$2 \times 2 \times 2$	$65.88 \times 118.24 \times 45.54$
Model 6	No breakage	$3 \times 3 \times 3$	$98.81 \times 177.36 \times 68.31$

relaxed for 100 ps under the NPT ensemble, with the pressure set to 1 atmosphere and the temperature to 300 K. Subsequently, uniaxial tension was applied along the silicon chain (y) direction of each model, with the strain rate set to 0.01 ps^{-1}. Throughout the uniaxial tension process, the strain and pressure values in each direction at each time step were recorded. In addition, atomic trajectories were output every 1 ps, and OVITO was used for data visualization.

3 Results and Discussion

3.1 Stress–Strain Relationship

On a nanoscale, uniaxial tension differs from macroscale tension in that the accuracy of the directly output stress–strain curve is reduced due to pressure oscillations in various directions during the uniaxial tension process [8]. Due to the particularity of analyzing stress–strain curves in molecular dynamics, it is essential to consider the pressures/stresses in directions other than tension when conducting the uniaxial tension tests. This is necessary to avoid excessive pressure oscillations in non-tension directions that could impact the accuracy of the results. After filtering out data points with excessive pressure in the non-tension direction, the stress–strain curves were plotted. The filtered curves of models 1–3 and models 4–6 were presented in Figs. 1 and 2, respectively. It can be seen from the figure that although the pressure oscillation still exists, the pressures in the non-tension direction are much smaller than the pressure in the tension direction. Additionally, from model 1 to model 3, or from model 4 to model 6, it can be seen that the larger the size of the model, the smaller the amplitude of the pressure oscillation. This means that increasing the model size can also reduce the error caused by pressure oscillations. In general, models with larger sizes (or greater total number of atoms) more accurately simulate the uniaxial tension process. Comparing the stress in the tensile direction in Figs. 1 and 2, it can be observed that the tensile strength of the corrected model (models 1–3) drops significantly when compared to the uncorrected model (models 4–6). Because the strength of C-S-H at the microscale is about 100 MPa [9], which is smaller than at the nanoscale, and C-S-H exhibits rapid deterioration in strength as the scale increases, introducing defects in the model can indeed better simulate the amorphous structure of C-S-H. In this simulation, the stress–strain curves of models of various sizes showed minimal changes during the stress increasing stage, whether it was models 1–3 or models 4–6. Considering that the revised model can still be further optimized, the applicability of this inference in a wider range still needs to be further explored.

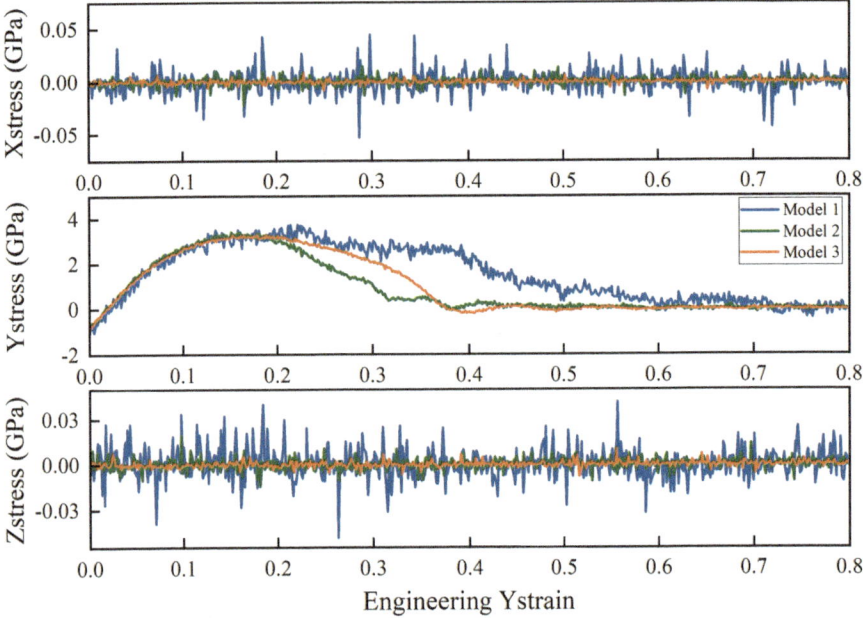

Fig. 1 Changes in stress/pressure in various directions with *y*-direction engineering strains during *y*-direction uniaxial tension in models 1–3

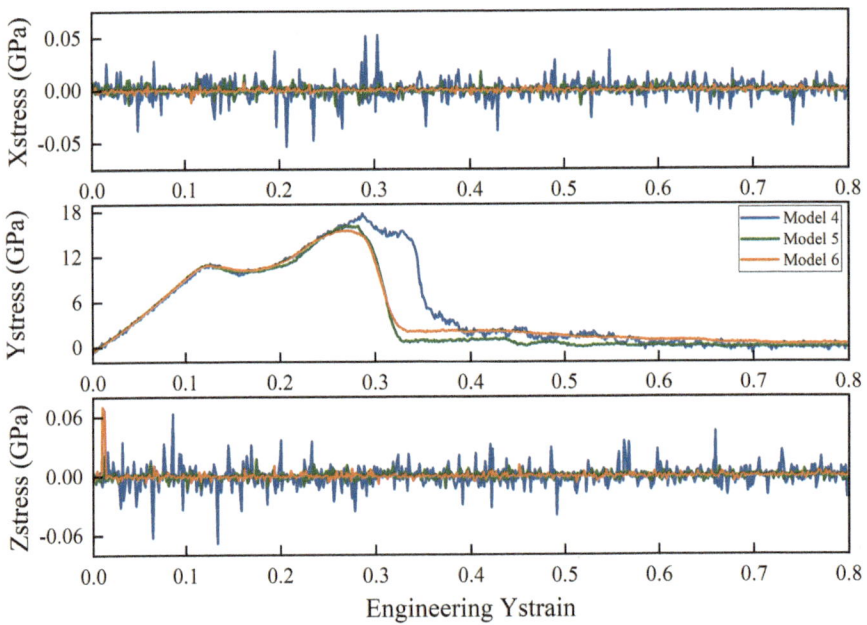

Fig. 2 Changes in stress/pressure in various directions with *y*-direction engineering strains during *y*-direction uniaxial tension in models 4–6

3.2 Mechanical Properties of Models

Observing the stress–strain curves of all models, it can be seen that when the strain value reaches 0.44, except for model 1, the other models have lost their load-bearing capacity and are in a state of fracture. Figure 3 shows the atomic trajectories of models 1–6 when the strain value is 0.44. The left side of Fig. 3 shows models 1–3. It can be observed that the fracture position in model 2 is at the end. This result indicates a failure in the experiment. In macroscopic experiments, uniaxial tensile force is typically applied only at the ends, whereas in nanoscopic simulations, pressure is uniformly distributed across the entire model, which may result in fractures at the ends. The tensile strength of model 2 is not significantly different from that of models 1 and 3; therefore, this tension process is undoubtedly effective. Considering the randomness that may occur during the molecular dynamics simulation process, a second simulation was conducted on the initial configuration of model 2. It was observed that the fracture position after tension was located at the center of the model. Figure 4 shows a snapshot of the atomic trajectory when the strain value is 0.44 during two uniaxial tension tests in model 2. It can be observed that, apart from the varying fracture positions, the crack shapes are nearly identical.

The elastic modulus, ultimate strength, and ultimate strain during the uniaxial tension process of models 1–6 were determined based on the stress–strain curves, and the results are presented in Table 2. Due to the unexpected fracture position of model 2, the initial configuration of model 2 was subjected to uniaxial tension twice, resulting in two sets of data. The underlined data corresponds to the uniaxial tension model in which the crack location occurs at the end. The elastic modulus is determined by linearly fitting stress and strain curves within the 0–0.05 strain range. The ultimate strength is the maximum stress that occurs during uniaxial tension, and the ultimate strain is the strain value when the maximum stress is reached. It can be seen from the two sets of data of model 2 that the fracture location will have a certain impact on the mechanical properties of the models. When the fracture occurs at the end, the elastic modulus is 45.29 GPa, and the ultimate strain is 0.15. When the fracture occurs in the middle, the elastic modulus is 43.58 GPa, and

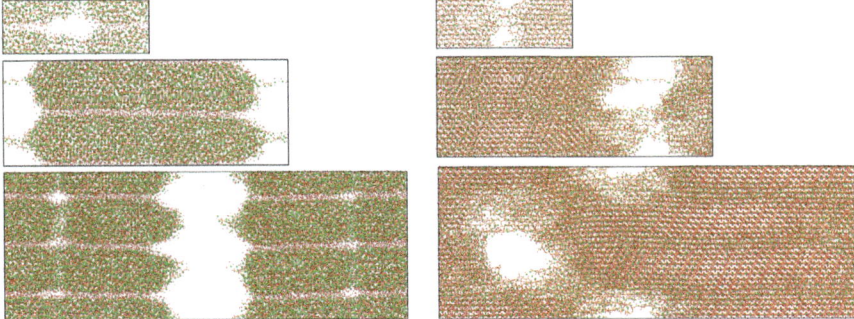

Fig. 3 Snapshots at strain 0.44 in models 1–6

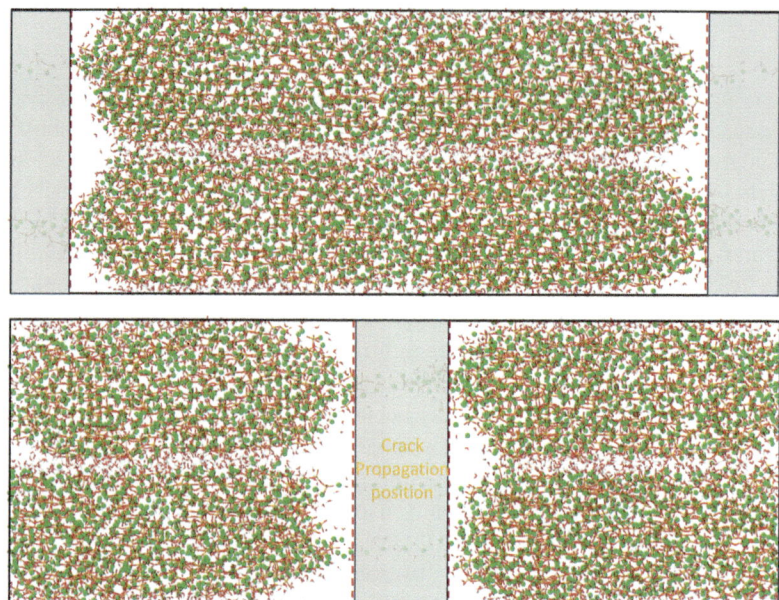

Fig. 4 Comparison of the fracture location after two uniaxial tensions of model 2 (at strain 0.44)

Table 2 Elastic modulus, ultimate strength, and ultimate strain in models 1–6

	Elastic modulus, E (GPa)	E_i/E_{i-1}	Ultimate strength, σ (GPa)	σ_i/σ_{i-1}	Ultimate strain, ε	$\varepsilon_i/\varepsilon_{i-1}$
Model 1	43.64	/	3.75	/	0.22	/
Model 2	45.29	/	3.42	/	0.15	/
	43.58	99.9%	3.42	91.2%	0.19	86.4%
Model 3	43.33	99.4%	3.29	96.2%	0.15	78.9%
Model 4	87.37	/	17.78	/	0.29	/
Model 5	88.77	101.6%	16.11	90.6%	0.27	93.1%
Model 6	89.25	100.5%	15.44	95.8%	0.27	100%

the ultimate strain is 0.19. The elastic modulus and ultimate strain values of the latter are between model 3 and model 1. This result is more consistent with the pattern of mechanical property degradation during scaling up in size. The occurrence of end fractures will not cause the simulation results to fail; however, it does affect the accuracy of the results. When comparing the mechanical properties of models, it is preferable to minimize differences in fracture locations.

According to the values in Table 2, it can be observed that the elastic modulus of the corrected model (models 1–3) is halved compared to the uncorrected model (models 4–6), bringing it closer to the experimental value [10]. The modification of the model significantly reduces the elastic modulus, ultimate strength, and ultimate strain, with the ultimate strength showing the most significant decrease. With the same size, the ultimate strength of the modified model is 21.1%, 21.2%, and 21.3% of that before the modification, respectively. However, the change in elastic modulus caused by increasing the size of the model is not significant. In models 1–3, as the model size increases, these three parameters decrease slightly. However, the magnitude of this decrease is significantly smaller compared to the changes induced by altering the internal structure of the models. When the size increases, mechanical properties degrade. This means that the ultimate strain is more easily affected, while the elastic modulus degrades more slowly.

3.3   Stress–Strain Correction

The strain used in the above analysis is the engineering strain, and its formula is

$$\varepsilon_E = \frac{L_n - L_0}{L_0} \tag{1}$$

where L_0 is the initial length, and the L_n is the current length during tension. This formula overestimates the effect of initial length on strain but is widely used in macro-experiments because it is easy to calculate and the dimensional change is much smaller than the dimension itself, which ensures the high accuracy. On a macro scale, the engineering strain of common concrete structures is usually much less than 4%, and the disparity between the engineering strain and the true strain is minimal. At the nanoscale, the size of the structure is reduced, leading to simultaneous amplification of stress and strain values. In some simulations, the measured value of engineering strain can even reach 1.0 [11], while the actual strain at this time is around 0.69. In other words, if we want to compare stress–strain curves across scales, it will become unreasonable to continue using engineering strains at the nanoscale. In addition, the performance of C-S-H deteriorates rapidly during upscaling [9]. Therefore, the use of engineering strain also impedes drawing correct conclusions when studying scale effects. For the reasons mentioned above, when comparing stress–strain relationships across different scales, it is advisable to use true strains in molecular dynamics simulations instead of engineering strains. The formula for true strain is

$$d\varepsilon_T = \frac{dl}{l} \tag{2}$$

ε_T was obtained by integrating Eq. (2):

$$\varepsilon_T = \int_{L_0}^{L_n} \frac{dl}{l} = ln\left(\frac{L_n}{L_0}\right) \tag{3}$$

The necking phenomenon is not considered here. According to the expressions of engineering strain and true strain, the transformation relationship between the two strains can be derived:

$$\varepsilon_T = ln(\varepsilon_E + 1) \tag{4}$$

Observing the stress–strain curve of the above models, we can find that when the strain is 0 (i.e., when uniaxial tension has not yet begun), the stresses are all negative, which means that there is an undesired pre-compression at the beginning of the simulation process. Considering the definition of true strain (Eq. 2), i.e., the strain value at this moment is the superposition of the instantaneous strains at all previous moments. Therefore, the accumulation of strain can be calculated starting from when the stress is 0. The specific method is to translate the entire curve in the horizontal direction so that it passes the origin, thus eliminating the strain value accumulated in the preloading state. Obviously, the curve obtained in this way is more reasonable than the stress–strain curve obtained by forcing the data to fit through the origin. Taking model 3 with the more reasonable structural configuration and larger size as an example, the strain (including engineering strain, true strain, and corrected true strain) and stress curves are shown in Fig. 5. Comparing the various stress–strain curves, it can be observed that the curves generated using engineering strain and true strain tend to overestimate the deformation capacity and strain of the model during the uniaxial tension process. However, the elastic modulus and ultimate stress remained unchanged before and after the correction.

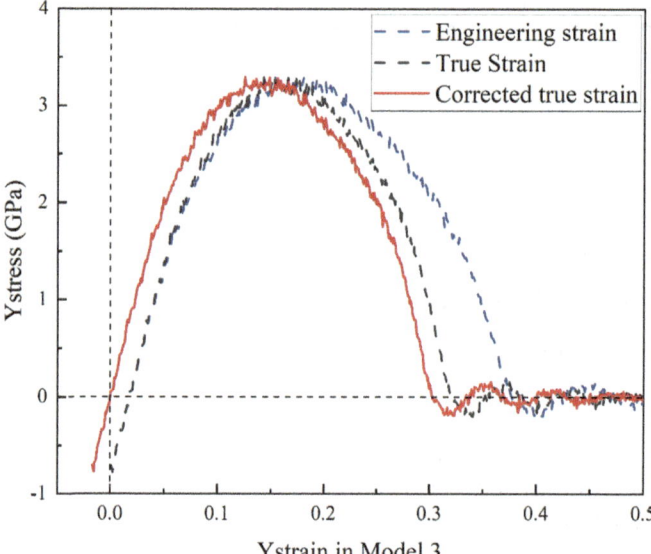

Fig. 5 Stress–strain curve correction diagram

4   Conclusions

In this paper, uncorrected and corrected molecular models with different sizes were built to study the influence of size effect and model structure. Several conclusions can be drawn as follows:

1. Larger-sized models exhibit smaller pressure oscillations in all directions, allowing for a more precise simulation of uniaxial tension processes.
2. The elastic modulus of models of different sizes but of the same type (either corrected or uncorrected) decreases slightly with increasing size. Within the scale range discussed in this paper, differences in stress–strain curves caused by size variations are mainly observed in the stress descent phase.
3. Significantly distinct stress–strain curves exist between the corrected structure and the structure obtained directly from the tobermorite supercell. The ultimate strength of the corrected model is approximately 21.2% of the uncorrected model.
4. Since pressure is uniformly applied to the models in all three directions, fractures occurring at the ends of the models in the tension direction do not invalidate the data. However, different fracture locations in the same models do affect the specific values of the model's elastic modulus and ultimate strain, with no impact on ultimate strength.
5. In uniaxial tension simulations, the strain in the stress–strain curve should represent true strain. After that, shifting the original stress–strain curve horizontally to pass through the origin ensures a more accurate stress–strain curve at the nanoscale.

References

1. Cao, J., Wang, C., Wang, T., Gonzalez-Libreros, J., Tu, Y., Sas, G., Elfgren, L.: Effects of temperature and NaCl concentration on the adsorption of C-S-H gel in cement paste: a multi-fidelity molecular dynamics simulation. In: Llki, A., Cavunt, D., Cavunt, Y.S. (eds.) Building for the Future: Durable, Sustainable, Resilient. Fib Symposium 2023 Lecture Notes in Civil Engineering, vol. 350. Springer, Cham (2023). https://doi.org/10.1007/978-3-031-32511-3_53
2. Cao, J., Kong, L., Guo, T., Shi, P., Wang, C., Tu, Y., Sas, G., Elfgren, L.: Molecular dynamics simulations of ion migration and adsorption on the surfaces of AFm hydrates. Appl. Surf. Sci. **615**, 156390 (2023). https://doi.org/10.1016/j.apsusc.2023.156390
3. Pellenq, J.M., Kushima, A., Shahsavari, R., Vliet, K.J.V., Buehler, M.J., Yip, S., Ulm, F.J.: A realistic molecular model of cement hydrates. Proc. Natl. Acad. Sci. **106**(38), 16102–16107 (2009). https://doi.org/10.1073/pnas.0902180106
4. Kovačević, G., Nicoleau, L., Nonat, A., Veryazov, V.: Revised atomistic models of the crystal structure of C-S-H with high C/S ratio. Zeitschrift Fur Physikalische Chemie. **230**, 1411–1424 (2016). https://doi.org/10.1515/zpch-2015-0718
5. Kunhi Mohamed, A., Parker, S.C., Bowen, P., Galmarini, S.: An atomistic building block description of C-S-H—towards a realistic C-S-H model. Cem. Concr. Res. **107**, 221–235 (2018). https://doi.org/10.1016/j.cemconres.2018.01.007
6. Tu, Y., Cao, J., Wen, R., Shi, P., Yuan, L., Ji, Y., Das, O., Försth, M., Sas, G., Elfgren, L.: Molecular dynamics simulation study of the transport of pairwise coupled ions confined in C-S-

H gel nanopores. Constr. Build. Mater. **318**, 126172 (2022). https://doi.org/10.1016/j.conbuildmat.2021.126172

7. Pitman, M.C., Van Duin, A.C.T.: Dynamics of confined reactive water in smectite clay-zeolite composites. J. Am. Chem. Soc. **134**, 3042–3053 (2012). https://doi.org/10.1021/ja208894m

8. LAMMPS Users Manual, 21 Nov 2023 version. Sandia National Laboratories. https://docs.lammps.org/fix_deform.html

9. Wang, J., Gao, C., Tang, J., Hu, Z., Liu, J.: The multi-scale mechanical properties of calcium-silicate-hydrate. Cem. Concr. Compos. **140**, 105097 (2023). https://doi.org/10.1016/j.cemconcomp.2023.105097

10. Constantinides, G., Ulm, F.J.: The effect of two types of C-S-H on the elasticity of cement-based materials: results from nanoindentation and micromechanical modeling. Cem. Concr. Res. **34**, 67–80 (2004). https://doi.org/10.1016/S0008-8846(03)00230-8

11. Wang, T., Tu, Y., Guo, T., Fang, M., Shi, P., Yuan, L., Wang, C., Sas, G., Elfgren, L.: Molecular dynamics study on structural characteristics and mechanical properties of sodium aluminosilicate hydrate with immobilized radioactive Cs and Sr ions. Appl. Clay Sci. **243**, 107042 (2023). https://doi.org/10.1016/j.clay.2023.107042

Exploring the Impact of Silica-Rich Calcined Clay as Portland Cement Additive to Reduce Carbon Dioxide Emissions

Mohammed Seddik Meddah and Ola Najjar

Contents

1   Introduction

The construction industry stands out as a dynamic and economically robust sector. Yet it endures the dual distinction of being an intensive energy consumer and a prominent exploiter of global natural mineral resources. This duality results in important environmental effects and unsustainable practices. Despite estimable advances toward reducing natural resource depletion and encouraging the integration of several recycled and by-product materials, the dependence on virgin and non-renewable resources remains paramount. Rocks and clays, in their natural form or following manufacturing processes, are one of the pillars of construction materials and continue to play fundamental roles in the construction industry. Ordinary Portland cement (OPC) has been the primary binding material used in the production of pastes, mortar, and concrete [1], annually generating more than 1.6 billion tons

M. S. Meddah (✉) · O. Najjar
College of Engineering, Sultan Qaboos University, Muscat, Oman
e-mail: seddikm@squ.edu.om

© The Author(s) 2025
M. Kioumarsi, B. Shafei (eds.), *The 1st International Conference on Net-Zero Built Environment*, Lecture Notes in Civil Engineering 237,
https://doi.org/10.1007/978-3-031-69626-8_26

and 12 billion tons, respectively [2]. Despite its popularity and wide usage, OPC manufacturing emits substantial greenhouse gases, averaging 0.85 kg of CO_2 per kilogram of cement produced and accounting for around 8% of global CO_2 emissions [3]. Thus, exploring new cementitious and pozzolanic materials to entirely or partially substitute OPC becomes vital in reducing carbon footprints generated by the construction industry.

There are numerous supplementary cementitious materials (SCMs) developed as probable replacements for OPC, including fly ash [4, 5], slag [6], silica fume [7–9], Rice Husk Ash [10, 11], bagasse ash [12], and metakaolin [13]. The limited availability, and sometimes even the scarcity of these conventional SCMs in many regions of the world, especially in many developing countries, causes difficulties in their common and regular utilization in the cement concrete industry [14]. Meanwhile, natural pozzolans like natural clays [15], clay waste [16], and calcined clays [17] are commonly used as alternative pozzolans. However, compared to these alternatives, calcined clay stands out as the most promising pozzolanic material for blended Portland cement [18]. Clay pozzolan was generally prepared through thermal treatment of kaolinitic clay [19]. Other clays, such as illite [20], smectite [21], and bentonite [22], can also be used to prepare clay pozzolans.

Nevertheless, the calcined clay-blended cement's performance depends on various related parameters such as the mineralogical and chemical compositions, degree of amorphousness, dehydroxylation level, and the fineness of the calcined clay, alongside admixture content, water-to-binder (w/b) ratio, and portlandite (CH) content in the cement paste [1].

A previous study [23] showed that pozzolans derived from heat-treated clay can result in concrete with slightly lower mechanical properties than control concrete, in addition to some negative effects on workability. As the clay brick powder replacement rate increases from 5% to 25%, the 28-day compressive strength may be decreased by around 8–25% as [24] reported. Moreover, blended cement with calcined clay exhibits high resistivity to alkali-silica reactions, chloride diffusion, sulfates, and harmful environmental attacks [25].

Using calcined clay alone or in combination with limestone not only enhances mechanical and durability properties but can significantly reduce clinker usage in cement production by 30–50% without negatively impacting the technical properties [1]. Using calcined clay-blended cement and LC3 systems reduces fuel consumption by 650 MJ/ton and 730 MJ/ton, respectively, compared to OPC. Moreover, clay calcination consumes around 50% less energy than clinker production using the same fuel in India [26]. These alternative systems also significantly reduce CO_2 emissions by 26 and 33%, respectively, compared to OPC.

This study's target is to improve the sustainability of Portland cement, which serves as a cornerstone in the construction industry, while simultaneously addressing its status as one of the most environmentally detrimental constituents in cement-based materials.

2 Materials and Testing Procedures

2.1 Materials

Locally available raw and manufactured materials were used as components of mortar mixes studied in this research. Natural well-graded sand with grain size ranging from 600 μm to 150 μm as per ASTM C778-21 [27], a specific gravity of 2.82, and water absorption of 1.5% was used. Ordinary Portland cement (OPC) conforming to ASTM C150-22 [28] with a specific gravity of 3.14, specific surface air permeability of 327 m^2/kg, a normal consistency of 26%, and an initial setting time of 160 minutes was used in all mortar mixes as the main binder.

To keep the flow of all mortar mixes within the target value set, a high-performance superplasticizer (SP) admixture was used in a liquid state and added directly during the mixing. This SP is chloride-free, reddish in appearance with a specific gravity of 1.24, and conforms to BS-EN 934-2 [29] and ASTM C494/C494M-19 [30] types B, D, and G. The mixing water used has a pH of 7.8 and a TDS of 585 ppm. Raw clay from the northeast region of Oman was collected, screened, dried, ground in a ball mill to fine particles, and sieved to retain only the particles smaller than 75 μm.

2.2 Procedures

A series of materials characterization were conducted on both the raw and burnt clay. Raw clay was first characterized using X-ray Fluorescence (XRF) to arbitrate the major oxides and assess its suitability for use as a pozzolana. Thermogravimetric (TG) analysis was conducted to examine the clay's weight loss upon heating up to 1400 °C and to determine the probable suitable calcination temperature(s).

Mortar mixing was performed as per the guidance specified in ASTM C305-20 [31]. Following the mixing, fresh properties, including temperature, density, and flow as per ASTM C1437-20 [32], were measured. The hardened density, flexural, and compressive strengths were all tested after 28 days of curing in water. Compressive strength was tested on 50 mm cubes as per ASTM C109/109 M-21 [33], and flexural strength was tested on prisms of 40 × 40 × 160 mm as per ASTM C348-21 [34]. At 28 days of water curing, the porosity of the plain and modified mortar mixes was determined as per ASTM C642-21 [35]. A rapid chloride permeability test (RCPT) was carried out after 180 days of water curing to assess the mortar's resistance to chloride ions ingress. The test was carried out on mortar discs measuring 100 × 50 mm according to ASTM C1202-22 [36] and AASHTO T277-15 [37].

3   Results and Discussion

3.1   Physical Characterization of Clay-Pozzolan

The clay used had a specific gravity of 2.75, a Brunauer–Emmett–Teller (BET) surface area of 17,900 m^2/kg, and most of the particles (97%) have a size smaller than 75 μm. Scanning electron microscopy (SEM) images show that both OPC and CCF have angular and irregular shapes with rough surfaces. This is one of the factors that led to a reduction in mortar's flow when CCF replaces part of the OPC.

The thermogravimetric analysis of a raw clay sample showed that the dehydroxylation starts at around 450 °C, followed by an acceleration in the mass loss to reach around 12%. The TGA graph indicates that the transition phase, where the crystalline structure and chemical bonds in the clay are likely to occur, is at a temperature varying from 500 to 800 °C. During this stage, the clay structure transforms from crystalline to semi or non-crystalline. Based on this analysis, a calcination temperature of 700 °C was selected.

The crystal structure and mineralogical composition were examined using X-ray diffraction (XRD). The raw clay's XRD patterns show a sharp and intense peak mainly of four major minerals, including quartz, hematite, kaolinite, and calcite. In terms of oxide composition, the CCF clay is rich in silica, alumina, and iron oxide, with a sum of these three oxides exceeding 93%. This amount qualifies the CCF clay as a pozzolana as per ASTM C618-23 [38]. The burning of the clay at a temperature of 700 °C resulted in an XRD pattern with fewer peaks, lower intensity of the calcite peaks, and complete disappearance of kaolinite peaks. This indicates the calcined clay (CCF) transformation to a more amorphous powder material and a reduction in the degree of crystallinity of the heat-treated clay.

3.2   Fresh Properties of Mortar Mixes

The mortar mixes studied were all tested at fresh state for temperature, fresh density, and flow. The temperature of the freshly mixed mortars was immediately measured after the end of the mixing sequence. The temperatures recorded range between 24 and 27 °C. The inclusion of CCF led to a drop in the mortar's fluidity, as depicted in Fig. 1. To keep the mortar's flow within the targeted range (110–140%), an increase in the SP content was needed (Fig. 1). The higher the substitution ratio of CCF, the higher the SP demand. The relationship between blended mortar's flow and SP content is almost linear. In fact, the loss of fluidity exhibited by incorporating different amounts of CCF is attributed to the type of clay used, which is rich with hematite (Fe_2O_3) that accentuates the clay's water absorption. Additionally, the high surface area (17,900 m^2/kg) of the CCF requires a larger amount of water and/or SP to lubricate all the clay's ultrafine particles compared to the moderate fineness of OPC.

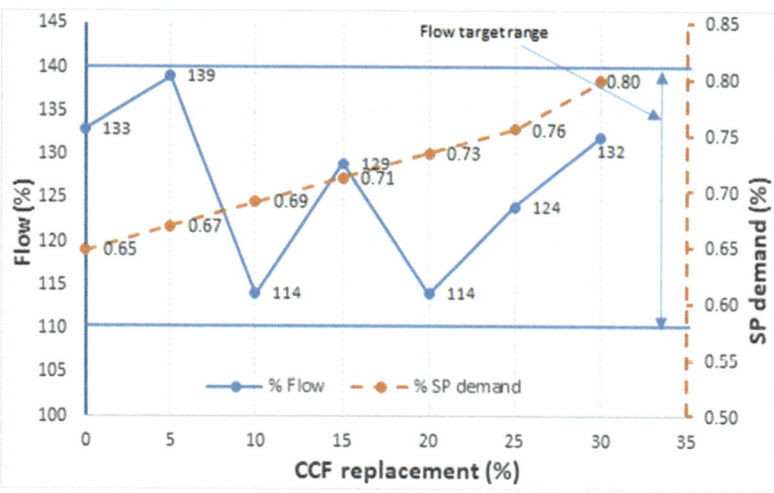

Fig. 1 Mortar's flow and SP demand function the CCF replacement level

The fresh density fluctuated from 2293 kg/m^3 for the control mortar mix to a maximum of 2400 kg/m^3 for the blended mix with 20% CCF. Generally, a slight reduction in the fresh density was found when the substitution ratio of OPC with CCF augmented from 0% to 5%, and 10%. At 15% replacement, the fresh density was equal to the control one, while at 20% replacement, the highest packing was reached with the maximum density. When the replacement level exceeds 20%, the density gradually reduces.

The change in density at a fresh state is likely to be affected by several parameters, including the initial materials' specific gravity, the mix's flow, and the compaction ability. Generally, because CCF has lower specific gravity compared to OPC, this can contribute to the observed fresh density reduction. Also, the lower the flow, the higher the density at the fresh state. The highly fluid mix is loose and, hence, less dense, as seen in Fig. 2.

3.3 Strength and Durability

The results of the 28-day flexural and compressive strengths development of the plain and modified mortar mixes with various percentages of the CCF are illustrated in Fig. 3. The results indicate that incorporating the CCF and increasing its content in the blended mortar mixes had slightly reduced the mortar's 28-day flexural strength from 36.4% to 9%.

However, regardless of the CCF replacement level, incorporating different percentages of the CCF significantly increased the blended mortar's 28-day compressive strength from 68.3% to 9.8%. The highest improvement of 68.3% was achieved

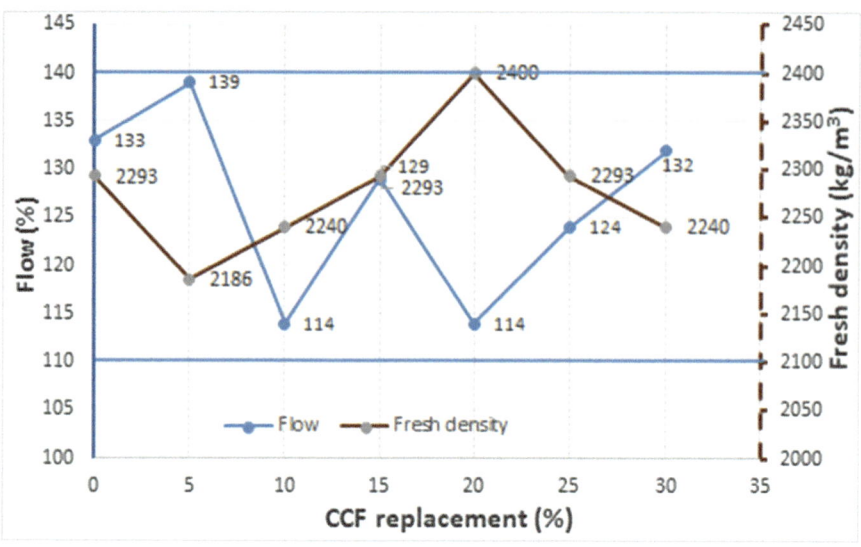

Fig. 2 Mortar flow and fresh density versus the CCF replacement level

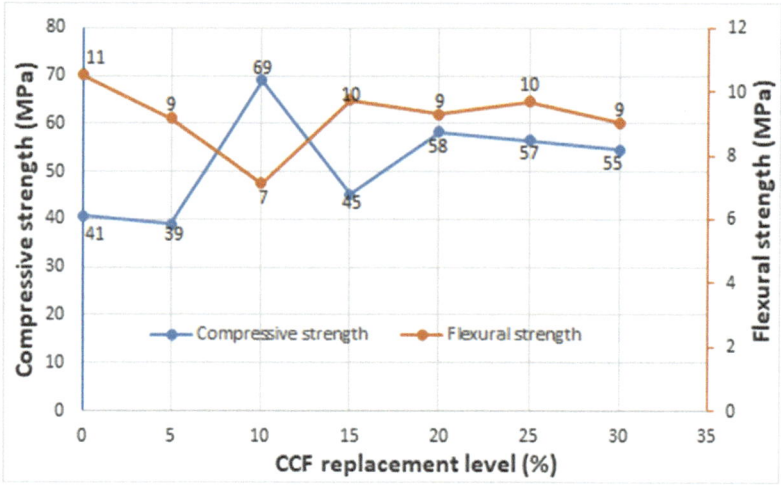

Fig. 3 Mortar's 28-day compressive and flexural strengths versus the CCF substitution level

when 10 wt.% OPC is replaced with CCF. Beyond 10 wt.%, the compressive strength kept higher than the control mortar but reduced gradually compared to the maximum strength reached at 10 wt.%. The strength enhancement of the modified cement mortar is due to the pozzolanic reaction between the amorphous silica-rich CCF and the portlandite of the OPC hydration reaction. The additional CSH formed fills and subdivides the larger capillary pores and, hence, increases mortar compactness and strength.

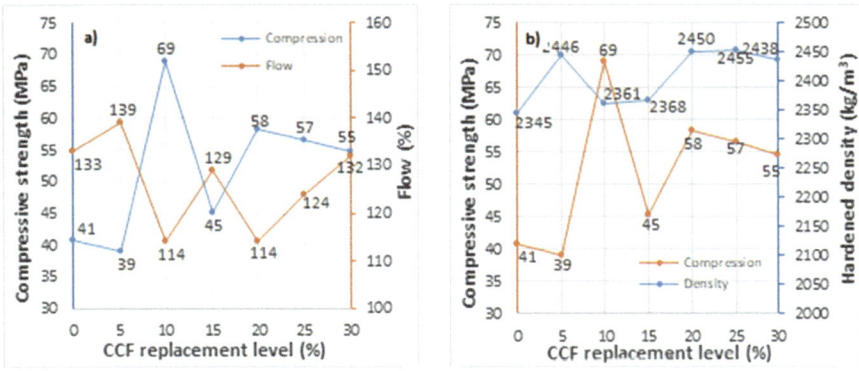

Fig. 4 Relationship between (**a**) the flow-compressive strength and (**b**) the density-compressive strength of mortar designed with various CCF replacement levels

For various replacement levels of OPC with CCF, Fig. 4 depicts the relationship between (a) mortar's flow-compressive strength and (b) hardened density-compressive strength. Generally, the compressive strength was inversely affected by the mortar's flow. The lower the flow, the higher the mortar's compressive strength. However, no apparent trend can be seen relating the mortar's density evolution to its compressive strength when rising the OPC substitution level with the CCF. All blended mortar mixes developed a higher density compared to the plain mortar. At a relatively low replacement ratio (0–15%), the density was moderate, ranging between 2345 kg/m³ for the control mix to 2368 kg/m³ for the blended mortar with a 15% replacement level. At a higher replacement ratio (20–30%), a significant increase in density was recorded. The higher density of blended mortar with CCF is mainly due to three major parameters, including (i) the presence of hematite in the CCF, (ii) the high fineness of CCF increased the packing density, and (iii) the pozzolanic reaction and the resulting CSH that fills the gaps and enhances the mortar's density.

Figure 5 illustrates the relationships between density-compressive strength and density-flexural strength. The results show a linear relationship with a good correlation relating the mortar's density and compressive strength with an R-squared value of 0.965 and a good correlation between the density and flexural strength with an R-squared value of 0.989.

On the other hand, permeable pore volume and chloride ions permeability are two important aspects of the durability of mortar, particularly when exposed to destructive environments such as the presence of chloride or other harmful substances. Figure 6 displays the effect of partially replacing OPC with various CCF content. It can be seen that, generally, both the porosity and permeability to chloride increase when rising the substitution ratio of OPC with CCF. The rate of decrease in permeability is accelerating with the increase in the substitution level, especially beyond 5%. However, porosity reduction is slightly affected by the increase in the CCF content in mortar.

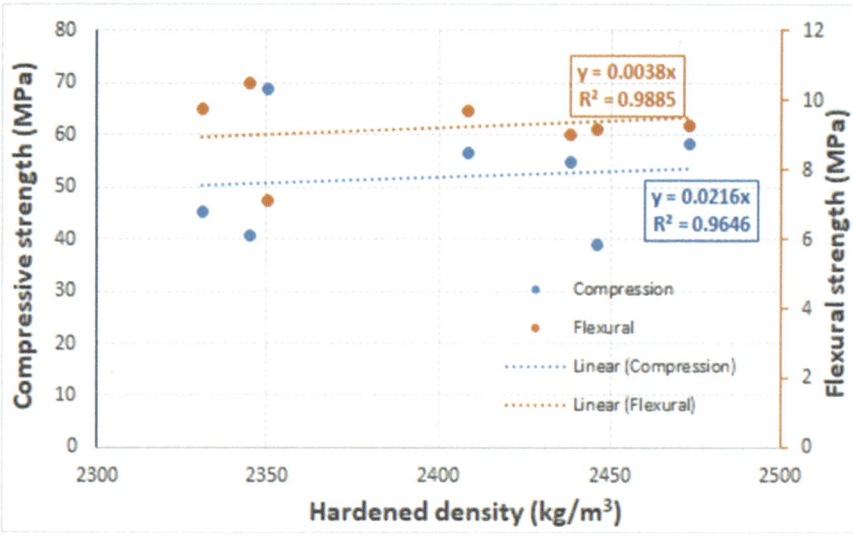

Fig. 5 Relationship between density-compressive strength and density-flexural strength of mortar designed with various CCF replacement levels

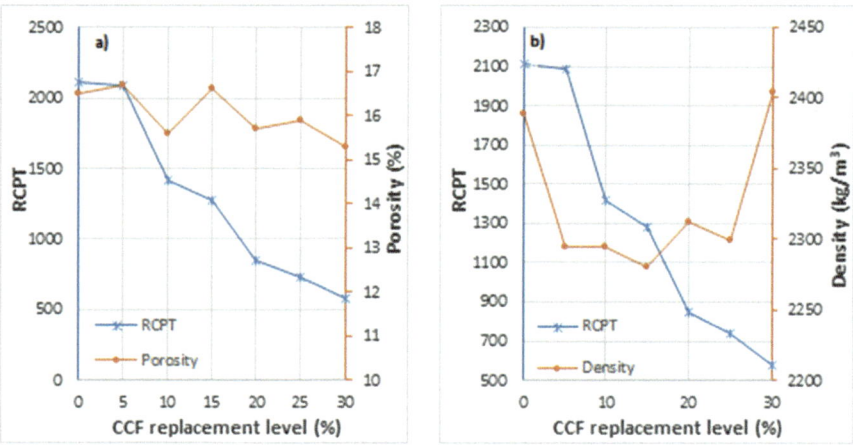

Fig. 6 Relationship between (**a**) RCPT-mortar's porosity and (**b**) RCPT-density of mortar designed with various CCF replacement levels

3.4 Environmental and Economic Assessment

With all its impacts on the quality of life and society and the continuous pressure on national and global economies, climate change has become a contemporary challenge for humanity. Reducing energy consumption, carbon dioxide emissions, raw minerals extraction, wastage generation, and rational usage of available non-renewable natural resources are just some of the main criteria to reduce the environmental impacts of the construction industry. Producing clinker cement is known to generate from 850 to 1 ton of CO_2 per ton of produced cement. This high amount of CO_2 generated is due to the large energy consumption resulting from the calcination temperature of the raw materials used to manufacture clinker, which requires a temperature of around 1600 to 1400 °C. The present study used local clay (less carbon footprint caused by transportation) calcined at only 700 °C. This temperature is precisely half the temperature required for the manufacturing of clinker. Considering the actual clay calcination temperature compared to clinker cement, we can conclude that the energy required, and hence, the CO_2 generated by calcined clay, is half that caused by clinker cement. For each 1 ton of calcined clay, around 425 to 500 kg of CO_2 is generated.

Moreover, the clinker cement cost is mainly due to the energy used during the calcination. Reducing the required energy to half would also reduce the cost to half of the cost of cement, depending on the replacement level. The higher the cement replaced with calcined clay (CCF), the more significant the cost and carbon footprint reduction. Therefore, partially replacing OPC with CCF significantly contributes to environment-friendly concrete/mortar.

4 Conclusions

The findings of this study's experiments allow for the following conclusions to be made:

- The clay used (CCF) in this research work is classified as a pozzolana with a significant content (>93%) of the sum of the three main oxides (SiO_2, Al_2O_3, and Fe_2O_3). Moreover, the XRD analysis indicated that this CCF clay contains mainly four major minerals: quartz, hematite, kaolinite, and calcite.
- The physical properties of the CCF demonstrated a high fineness of 17,900 m^2/kg as a surface area and around 97% of particles passing 75 μm.
- The TG/DT analyses indicated that an optimum calcination temperature of around 800 °C combines enhancement in the mechanical and durability properties with sustainability and low environmental impact.
- The inclusion of various contents of CCF as an OPC partial substitute reduces the mortar's flowability, requiring an SP content increase to keep a satisfactory flow within the target of 110 to 140%.
- Except at a low substitution ratio (5%), the inclusion of CCF has significantly enhanced the compressive strength of mortar at 28 days. The highest strength improvement (68.3%) was obtained with a 10% replacement of OPC with CCF.
- A good correlation was found between the mortar's density and its compressive and flexural strengths, with an R-value of 0.965 and 0.989, respectively.
- The inclusion of CCF slightly reduced the permeable pores and significantly enhanced the mortar's resistance to chloride ions penetration.

- The environmental and economic assessment indicates great potential in reducing the final cost of the concrete/mortar and mitigating CO_2 emissions when partially replacing OPC with the CCF. The higher the replacement levels of OPC with CCF, the higher the reduction in cost and CO_2 emissions.

Acknowledgments His Majesty Trust Fund (HMTF) financially supported the research presented in this paper under the project grant number SR/ENG/CAED/21/01. The authors would like to express their appreciation and thanks for the funds provided.

References

1. Cao, Y., Wang, Y., Zhang, Z., Ma, Y., Wang, H.: Recent progress of utilization of activated kaolinitic clay in cementitious construction materials. Compos. Part B Eng. **211**, 108636 (2021)
2. Rashad, A.M.: A comprehensive overview about the influence of different additives on the properties of alkali-activated slag–A guide for Civil Engineer. Constr. Build. Mater. **47**, 29–55 (2013)
3. Mohammed, S.: Processing, effect and reactivity assessment of artificial pozzolans obtained from clays and clay wastes: a review. Constr. Build. Mater. **140**, 10–19 (2017)
4. Du, S., Zhao, Q., Shi, X.: Quantification of the reaction degree of fly ash in blended cement systems. Cem. Concr. Res. **167**, 107121 (2023)
5. Limbachiya, M.C., Meddah, M.S., Ouchagour, Y.: Use of recycled concrete aggregate in fly-ash concrete. Constr. Build. Mater. **27**(1), 439–449 (2011). https://doi.org/10.1016/j.conbuildmat.2011.07.023
6. Xu, Z., Guo, Z., Zhao, Y., Li, S., Luo, X., Chen, G., Liu, C., Gao, J.: Hydration of blended cement with high-volume slag and nano-silica. J. Build. Eng. **64**, 105657 (2023)
7. Ying, J., Jiang, Z., Xiao, J.: Synergistic effects of three-dimensional graphene and silica fume on mechanical and chloride diffusion properties of hardened cement paste. Constr. Build. Mater. **316**, 125756 (2022)
8. Limbachiya, M.C., Meddah, M.S., Ouchagour, Y.: Performance of Portland/silica fume cement concrete produced with recycled concrete aggregate. ACI Mater. J. **109**(1), 91–100 (2012)
9. Meddah, M.S.: Design of a non-shrinking silica fume high-performance concrete with recycled ceramic tile aggregate. Struct. Concr. **24**(3), 3425–3442 (2023). https://doi.org/10.1002/suco.202200741
10. Meddah, M.S., Praveenkumar, T.R., Vijayalakshmi, M.M., Manigandan, S., Arunachalam, R.: Mechanical and microstructural characterisation of concrete designed with Al_2O_3 nanoparticles and rice husk ash. Constr. Build. Mater. **255**, 119358 (2020). https://doi.org/10.1016/j.conbuildmat.2020.119358
11. Praveenkumara, T.R., Vijayalakshmia, M.M., Meddah, M.S.: Strengths and durability performances of blended cement concrete with TiO_2 nanoparticles and rice husk ash. Constr. Build. Mater. **217**(30), 343–351 (2019). https://doi.org/10.1016/j.conbuildmat.2019.05.045
12. Meddah, M.S., Praveenkumar, T.R., Manigandan, S.: Properties of sugarcane bagasse ash concrete modified with bacterial treatment. ACI Mater. J. **119**(3), 187–196 (2022)
13. Meddah, M.S., Ismail, M.A., El-Gamal, S., Fitriani, H.: Performances evaluation of binary concrete designed with silica fume and metakaolin. Constr. Build. Mater. **166**, 400–412 (2018). https://doi.org/10.1016/j.conbuildmat.2018.01.138
14. Scrivener, K.L.: Options for the future of cement. Indian Concr. J. **88**(7), 11–21 (2014)
15. Ahmad, J., Kontoleon, K.J., Al-Mulali, M.Z., Shaik, S., El Ouni, M.H., El-Shorbagy, M.A.: Partial substitution of binding material by bentonite clay (BC) in concrete: a review. Buildings. **12**(5), 634 (2022)

16. Scheinherrová, L., Doleželová, M., Vimmrová, A., Vejmelková, E., Jerman, M., Pommer, V., Černý, R.: Fired clay brick waste as low cost and eco-friendly pozzolana active filler in gypsum-based binders. J. Clean. Prod. **368**, 133142 (2022)

17. Meddah, M.S., Al Owaisi, M., Abedi, M., Hago, A.W.: Mortar and concrete with lime-rich calcined clay pozzolana: a sustainable approach to enhancing performances and reducing carbon footprint. Constr. Build. Mater. **393**, 132098 (2023). https://doi.org/10.1016/j.conbuildmat.2023.132098

18. Zhao, Y., Zhang, Y.: A review on hydration process and setting time of limestone calcined clay cement (LC3). Solids. **4**(1), 24–38 (2023)

19. Mansour, A.M., Al Biajawi, M.I.: The effect of the addition of metakaolin on the fresh and hardened properties of blended cement products: a review. Mater. Today: Proc. **66**(part 5), 2811–2817 (2022)

20. Irassar, E.F., Bonavetti, V.L., Cordoba, G.P., Rahhal, V.F., Castellano, C.C., Donza, H.A.: Performance of composite Portland cements with calcined illite clay and limestone filler produced by industrial intergrinding. Minerals. **13**(2), 240 (2023)

21. Kaminskas, R., Kubiliute, R., Prialgauskaite, B.: Smectite clay waste as an additive for Portland cement. Cem. Concr. Compos. **113**, 103710 (2020)

22. Ashraf, M., Iqbal, M.F., Rauf, M., Ashraf, M.U., Ulhaq, A., Muhammad, H., Liu, Q.F.: Developing a sustainable concrete incorporating bentonite clay and silica fume: mechanical and durability performance. J. Clean. Prod. **337**, 130315 (2022)

23. Siline, M., Ghorbel, E., Bibi, M.: Valorization of pozzolanicity of Algerian clay: optimization of the heat treatment and mechanical characteristics of the involved cement mortars. Appl. Clay Sci. **132–133**, 711–721 (2016)

24. Aliabdo, A.A., Abd-Elmoaty, A.-E.M., Hassan, H.H.: Utilization of crushed clay brick in concrete industry. Alex. Eng. J. **53**(1), 151–168 (2014)

25. Hossain, M., Karim, M., Hasan, M., Hossain, M., Zain, M.F.M.: Durability of mortar and concrete made up of pozzolans as a partial replacement of cement: a review. Constr. Build. Mater. **116**, 128–140 (2016). https://doi.org/10.1016/j.conbuildmat.2016.04.147

26. Joseph, S., Bishnoi, S., Maity, S.: An economic analysis of the production of limestone calcined clay cement in India. Indian Concr. J. **90**(11), 22–27 (2016)

27. ASTM C778-21: Standard Specification for Standard Sand. ASTM International, West Conshohocken (2021)

28. ASTM C150-22: Standard Specification for Portland Cement. ASTM International, West Conshohocken (2022)

29. BS-EN 934-2: 2009+A1:2012.: Admixtures for Concrete, Mortar and Grout Concrete Admixtures. Definitions, Requirements, Conformity, Marking and Labelling (2012).

30. ASTM C494/C494M-19: Standard Specification for Chemical Admixtures for Concrete. ASTM International, West Conshohocken (2019)

31. ASTM C305-20: Standard Practice for Mechanical Mixing of Hydraulic Cement Pastes and Mortars of Plastic Consistency. ASTM International, West Conshohocken (2020)

32. ASTM C1437-20: Standard Test Method for Flow of Hydraulic Cement Mortar. ASTM International, West Conshohocken (2020)

33. ASTM C109/109M-21: Standard Test Method for Compressive Strength of Hydraulic Cement Mortars (Using 2-in. or [50-mm] Cube Specimens), West Conshohocken (2021)

34. ASTM C348-21: Standard Test Method for Flexural Strength and Modulus of Hydraulic Cement Mortars. Annual Book of ASTM standards, USA (2021)

35. ASTM C642-21: Standard Test Method for Density, Absorption, and Voids in Hardened Concrete, West Conshohocken (2021)

36. ASTM C1202-22: Standard Test Method for Electrical Indication of Concrete's Ability to Resist Chloride Ion Penetration, West Conshohocken (2022)

37. AASHTO T277-15: Electrical Indication of Concrete's Ability to Resist Chloride Ion Penetration. American Association of State and Highway Transportation Officials, Washington, DC (2015)
38. ASTM C618-23e1: Standard Specification for Coal Fly Ash and Raw or Calcined Natural Pozzolan for Use in Concrete. ASTM International, West Conshohocken (2023)

Enhancing Strength and CO_2 Uptake into Mortar Through Supercritical CO_2 Treatment

Gregor Kravanja and Željko Knez

Contents

1 Introduction

The construction industry consumes up to 40% of the global energy, generates about 30–40% of the total municipal solid waste, and emits about 25% of the total global CO_2 emissions [1]. To meet its carbon reduction targets and contribute to the transition to a net-zero built environment by 2050, it is crucial to use energy-efficient technologies and explore new CO_2 sequestration pathways. A promising approach to enhance the properties of cement composite involves accelerated mineral carbonation [2]. In this process, CO_2 reacts with calcium-bearing minerals to

G. Kravanja (✉)
Faculty of Civil Engineering, Transportation and Architecture, University of Maribor, Maribor, Slovenia

Faculty of Chemistry and Chemical Engineering, University of Maribor, Maribor, Slovenia
e-mail: gregor.kravanja@um.si

Ž. Knez
Faculty of Chemistry and Chemical Engineering, University of Maribor, Maribor, Slovenia

Faculty of Medicine, University of Maribor, Maribor, Slovenia

© The Author(s) 2025   315
M. Kioumarsi, B. Shafei (eds.), *The 1st International Conference on Net-Zero Built Environment*, Lecture Notes in Civil Engineering 237,
https://doi.org/10.1007/978-3-031-69626-8_27

form carbonates. This dual action not only improves the mechanical properties of cement composites but also leads to the permanent capture of CO_2 inside the material matrix [3]. However, carbonation at ambient or low-pressure conditions occurs slowly, thereby reducing the practicality of implementing it on an industrial scale. The concept is to explore cement material curing at high pressure, where the carbonation rate can be significantly enhanced through increased pressure.

High-pressure technology involving supercritical fluids is already well established as an energy-efficient technology, with high promise for implementation in the construction industry. Supercritical fluids are unique states of matter that possess properties of both liquids and gases. They are formed when a substance is subjected to high-pressure and temperature conditions, surpassing its critical point (Fig. 1). Their unique solvating power, enhanced mass transfer capabilities, and environmentally friendly nature make them valuable tools in numerous industries [4]. For example, supercritical fluid extraction allows the efficient extraction of valuable compounds from natural sources while preserving their quality [5]. Super-critical fluids are being extensively used for polymer processing including particle formation and encapsulation [6], foaming processes [7], enzymatic reactions [8], supercritical drying of aerogels [9], supercritical-fluid chromatography [10], for jet cutting [11], dry cleaning [12], for sterilization processes, and for powder coatings [13]. Supercritical fluids are frequently found in large-scale operations in petrochemical plants [14], as working fluid in advanced power cycles [15], and carbonation of well cement [16]. A comprehensive understanding of the carbonation process under supercritical conditions is essential for various applications in the construction industry, including CO_2 capture and storage [17], the treatment of recycled concrete [18], and the solidification of toxic waste in building materials [19].

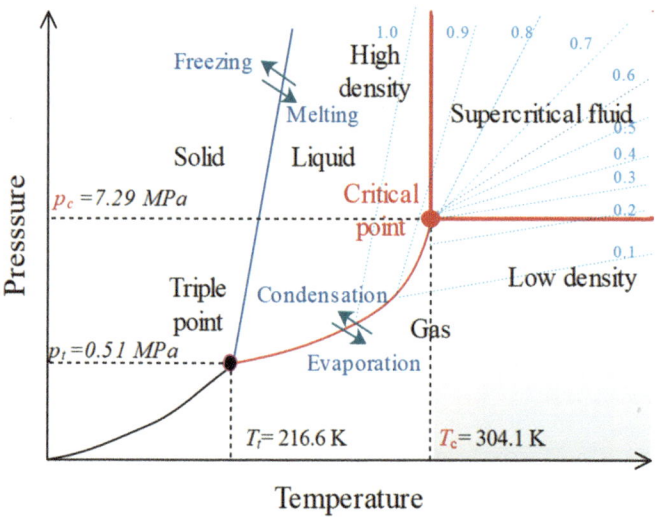

Fig. 1 P-T diagram for CO_2

This study presents the first investigation into the use of high-pressure carbonation of cement mortar mixed with a pozzolanic material, metakaolin. Mortar samples underwent treatment in a high-pressure reactor at 323 K and 15 MPa. The carbonation rate of the carbonation front and the carbonation rate of the cross-section, as well as the compressive and flexural strengths, were measured after 7, 14, and 28 days of exposure to supercritical CO_2. The specific surface area of hydrated mesoporous samples was assessed through N_2 desorption and absorption analysis. Morphological changes were measured with scanning electron microscopy with energy-dispersive spectroscopy (SEM-EDS).

2   Materials and Methods

2.1   Raw Materials

The typical ingredients used in the production of Portland cement mortar (PCM) samples were cement, sand, metakaolin, and water. The Portland cement was CEM II/B-M(LL-V) 42.5 N, known for its high resistance to sulfate attacks due to the clinker composition lacking tricalcium aluminate (C_3A), sourced from Salonit Anhovo d.o.o, Slovenia. A mineral additive, metakaolin (MK), with a mean particle size distribution (d_{50}) of 4.5 μm and density of 890 kg/m^3, supplied by Melanin d.o. o, Slovenia, was employed as a partial replacement for the Portland cement. Standard sand with particle sizes ranging between 0.08 and 2.00 mm was used to prepare the mortar samples (Normensand GmbH, Germany). The phenolphthalein indicator was provided by Sigma Aldrich, Germany. Carbon dioxide gas (3.5) was obtained from Messer d.o.o, Slovenia.

2.2   Samples Preparation and High-Pressure Carbonation Curing

The first set of PCM was prepared with a fixed water-to-cement ratio (w/c) of 0.5 and sand-to-cement ratio (s/c) of 3.0, followed by stirring for 5 minutes. Furthermore, metakaolin (MK) was incorporated as a supplementary cementitious material to create a second type of blended mortar sample, replacing 15% by weight of Portland cement (PCM+MK). The homogenized mortar in its green state was vibrated and poured into molds measuring 40 × 40 × 160 mm, adhering to Standard SIST EN 197-1 specifications. After 24 hours of regular curing, the samples were extracted from the molds and made ready for standard curing in water or high-pressure carbonation curing.

Carbonation was performed in a high-pressure reactor (Sitec AG, Zurich, CH) capable of operating at 150 MPa and 573 K. The CO_2 was pumped inside the reactor by a liquid high-pressure pump (NWA PM 101). The samples were treated at a pressure of 15 MPa and a temperature of 323 K. The required moisture was provided in the reactor as there was water in the lower part of the reactor. Having moisture content is crucial for the carbonation process.

2.3 Characterization Methods

2.3.1 Compression and Flexural Tests

Prism specimens sized 40 mm × 40 mm × 160 mm were prepared for three-point bending flexural (σ_f) and compression (σ_c) strength testing. Compression and flexural strength were tested following Standard SIST EN 197-1. Compression was calculated according to Eq. (1):

$$\sigma_c = F_c/A_0 \tag{1}$$

where Fc represents the instantaneous load applied perpendicular to the specimen cross-section (N), and A_0 is the original cross-sectional area before any load is applied (mm^2). Three-point bending flexural (σ_f) was calculated according to Eq. (2):

$$\sigma_f = \left(1,5 \cdot F_f \cdot l\right)/b^3 \tag{2}$$

where F_f is the load at a given point on the load-deflection curve, (N), l is the distance between the supporting pins (100 ± 0.5 mm) and b is the prism side length (40 ± 0.2 mm).

2.3.2 Carbonation Rate

Carbonated mortar samples were quantified using phenolphthalein indicator. Non-carbonated region with a pH >10, the color changes to purple, while, in the carbonated area with pH <10, the color remains the same. Pictures of the sectioned sample areas were subsequently analyzed using the open-access image processing software, ImageJ [22]. The average carbonation rate and depth was calculated.

2.3.3 Nitrogen Adsorption/Desorption Method

Gas adsorption–desorption analysis was conducted to determine the pore volumes (cm^3/g), specific surface areas (m^2/g), and pore diameters (nm) of fractionated samples. The experiments were performed utilizing an ASAP 2020MP instrument (Micromeritics, Norcross, Georgia, US), with N$_2$ serving as the adsorptive gas.

2.3.4 Microstructural Analysis

The surface morphology and chemical composition of fractured mortar samples were analyzed using a Quanta 200 3D Scanning Electron Microscopy–Secondary Electron Imaging (SEM-SEI) device, manufactured by FEI Company in Hillsboro, OR, USA. The equipment was further equipped with FEI Sirion 400 NC energy-dispersive X-ray spectroscopy (EDS), also from FEI Company.

3 Results and Discussion

3.1 *Compression and Flexural Strength*

The compressive and flexural strength of cement mortar samples exposed to conventional curing and high-pressure carbonation for 7-, 14-, and 28-day test periods are depicted in Figs. 2 and 3. It is evident that the addition of 15 wt% MK significantly enhances the strength of the samples. Under conventional curing conditions, the compression strength increases from 59.34 MPa to 71.4 MPa, and the flexural strength increases from 7.76 MPa to 10.84 MPa at 28 days. The treatment under high pressure further enhances the strength of Portland cement mortars compared to conventional curing methods. For instance, SUP-PCM and SUP-PCM+MK samples carbonized with supercritical CO2 for 28 days show an 18.11% and 13.11% increase in compression strength, and a 31.06% and 10.18% increase in flexural strength, respectively. The increase in strength is attributed to the mineralization of CO_2, which reacts with portlandite ($Ca(OH)_2$) and/or calcium-silicate-hydrate (C-S-H) in the cement matrix, forming carbonates (CaCO3(s)) that enhance the strength at the interface transition zone (ITZ) [20].

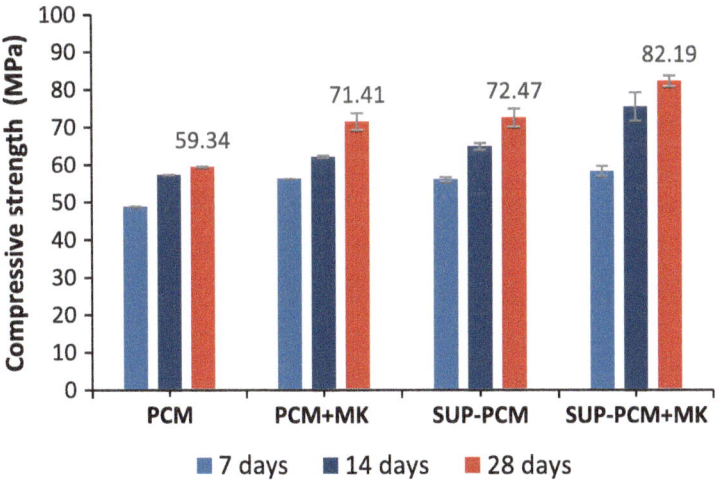

Fig. 2 Compressive strength of cement mortar samples exposed to conventional curing (PCM and PCM+MK) and high-pressure carbonation (SUP-PCM and SUP-PCM + MK)

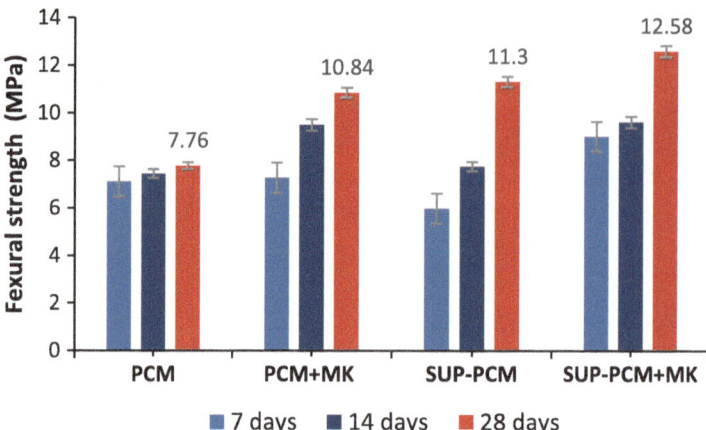

Fig. 3 Flexural strength of cement mortar samples exposed to conventional curing (PCM and PCM +MK) and high-pressure carbonation (SUP-PCM and SUP-PCM + MK)

Table 1 Carbonation rate after 28 days of exposure to supercritical CO_2

PCM	PCM+MK	SUP-PCM	SUP-PCM + MK
Total area: 1560 mm^2 FF colored area: 1560 mm^2	Total area: 1550 mm^2 FF colored area: 1550 mm^2	Total area: 1560 mm^2 FF colored area: 1045 mm^2	Total area: 1595 mm^2 FF colored area: 349 mm^2
Carbonation rate 0%	Carbonation rate 0%	Carbonation rate: 33%	Carbonation rate: 78%

3.2 Carbonation Rate

The carbonation rates at the cross-sections of the mortar samples at 28 days under different curing conditions and compositions are presented in Tables 1, 2 and 3. As expected, the PCM and PCM+MK samples submerged in water and treated at atmospheric pressure exhibited no signs of carbonation. Samples SUP-PCM and

Table 2 BET surface area and BJH pore sizes of mortar samples

Sample	BET surface area (m²/g)	BJH adsorption average pore (nm)	BJH desorption average pore (nm)
PCM	9.0526 m²/g	7.5440	4.9708
PCM-MK	12.1173	10.4334	5.2427
SUP-PCM	5.8763	12.5433	6.8905
SUP-PCM-MK	7.0287	34.6617	12.3251

Table 3 Chemical composition of SUP-PCM and SUP-PCM-MK at the interface

Spectrum	C	O	Mg	Al	Si	S	K	Ca	Fe	Total
SUP-PCM										
Spectrum 1	11.50	54.55	0.66	1.23	8.73	0.56		20.63	2.14	100.00
Spectrum 2	11.47	52.40	0.42	1.32	11.88	0.56		19.47	2.48	100.00
Spectrum 3	9.45	54.28	0.72	1.13	9.23	0.56	0.80	21.84	1.99	100.00
Spectrum 4	8.48	48.34	0.43	1.07	17.99	0.59		20.82	2.27	100.00
Spectrum 5	12.53	55.53	0.41	1.10	7.48	0.38		20.36	2.21	100.00
Spectrum 6	2.02	59.11	0.25	1.49	33.84		0.96	1.57	0.75	100.00
SUP-PCM-MK										
Spectrum 1	12.29	53.71		5.18	9.41	0.75		15.94	2.73	100.00
Spectrum 2	9.46	50.32		5.95	12.08	0.84		16.72	4.62	100.00
Spectrum 3	7.65	48.65		4.96	12.81	0.63		21.52	3.77	100.00
Spectrum 4	11.47	53.86		3.64	9.89	0.31		17.56	3.26	100.00
Spectrum 5	16.82	49.24		1.26	10.89	0.09		19.73	1.96	100.00
Spectrum 6	12.20	47.45		2.96	8.58	0.54		23.54	4.73	100.00

SUP-PCM+MK treated under supercritical CO_2 showed a non-colored area indicating that the carbonation process had occurred. As presented in Fig. 4, the average carbonation rate SUP-PCM and SUP-PCM+MK samples increased with exposure time from 7 days up to 28 days.

Results suggest that with increasing time, more CO_2 was permanently mineralized in the mortar matrix. For example, the carbonation rate and depth increased from 25% and 3.2 mm to 33% and 5.4 mm, respectively. Surprisingly, significantly larger CO_2 uptake was observed in the mortar samples blended with 15 wt% of MK, where the carbonation rate increased from 40% at 7 days to 67% at 14 days and up to 78% at 28 days, suggesting almost complete carbonation.

3.3 Nitrogen Adsorption and Desorption Analysis

The nitrogen adsorption and desorption isotherms were utilized on mortar samples to provide information regarding the pore surface area. According to the IUPAC classifications, there are typically six types of adsorption isotherms [21]. Hysteresis loop can be observed in every sorption isotherm and could be classified as type V

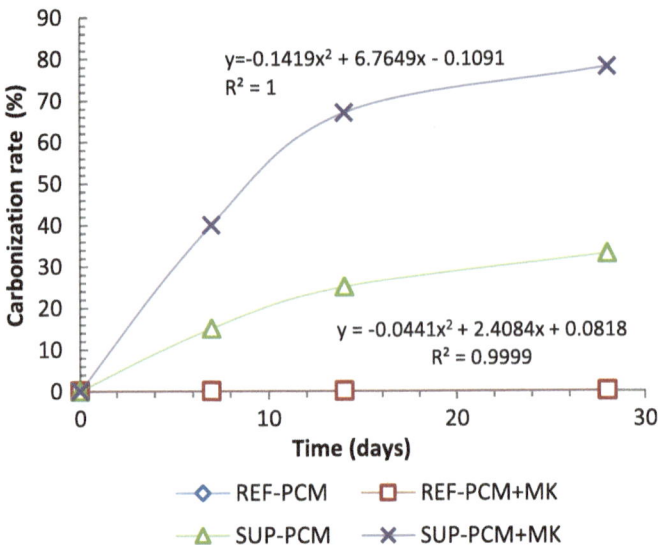

Fig. 4 Increased average high-pressure carbonation rate versus time

isotherms (Fig. 5). These isotherms represent a specific case of adsorption occurring in micro and/or mesopores, characterized by relatively weak interactions between the adsorbent and the adsorptive [22]. The BJH desorption pore volume versus pore size distribution revealed that hydrated mortar samples possess a complex pore system characterized by a significant amount of small gel pores (up to 3 nm) as well as large gel and mesopores ranging from 12 to 50 nm. Macropores are present to a lesser extent (Fig. 6)

Table 2 shows the BET surface area that was used to determine the internal specific area of pores and the average pore sizes of BJH. The specific surface area and quantity of absorbed/desorbed N_2 were slightly higher in the mortar samples containing 15 wt% of MK. Because capillary condensation and surface adsorption contribute to the amount of adsorbed water within the pore structure, the PCM-MK and SUP-PCM-MK samples absorbed more water, leading to higher carbonization rates. The BET surface area decreased upon exposure to supercritical CO_2, suggesting mineralization within the pore system.

3.4   SEM-EDX

SEM-EDX was used to provide a quick nondestructive determination of the elemental composition of mineral additive and carbonated samples.

Figure 7 presents the SEM-EDX analysis of the aluminosilicate mineral additive that was used to produce mortars. The powdery product consists of spillable size particles. Chemical analysis shows a typical representation of elements in MK powder such as silicon (Si) and aluminum (Al).

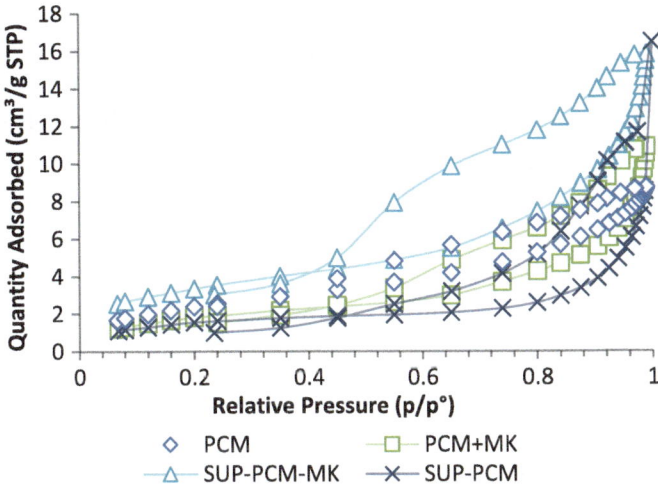

Fig. 5 Adsorption–desorption isotherms for mortar samples treated under atmospheric and supercritical conditions

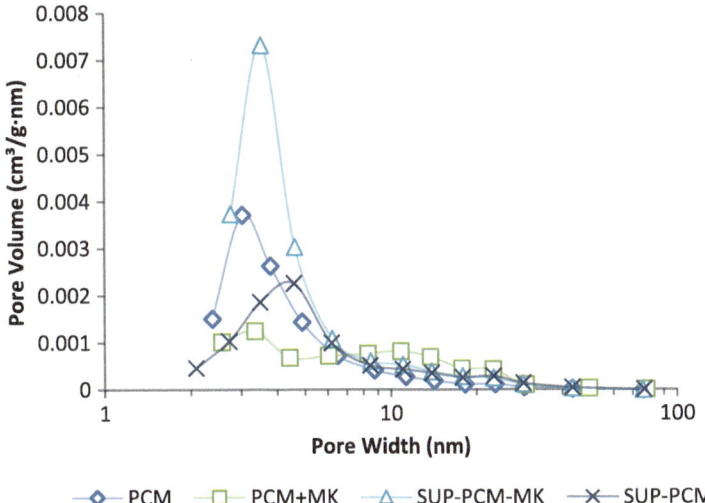

Fig. 6 BJH desorption pore volume for mortar samples treated under atmospheric and supercritical conditions

Figures 8 and 9 depict the SEM-EDX micrographs of mortar samples SUP-PCM and SUP-PCM-MK after exposure to supercritical CO$_2$. A distinct interface delineates the outer carbonized zone from the inner non-carbonated zone, with the carbonated zone exhibiting a rougher surface and crystal formation. Following exposure to supercritical CO$_2$, both samples' outer zones show a decrease in calcium (Ca) atomic percentage and an increase in carbon (C) atomic percentage compared to

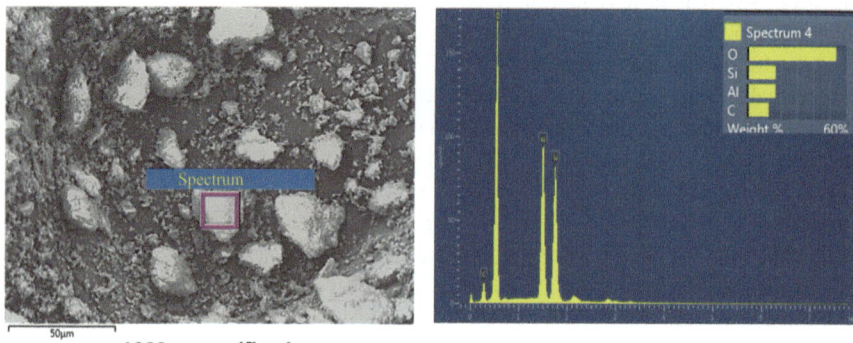

Fig. 7 SEM-EDX analysis of used mineral additive, metakaolin (MK)

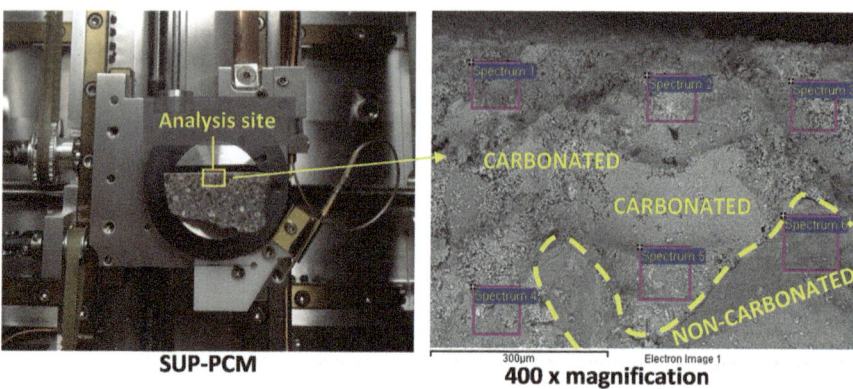

Fig. 8 SEM-EDX analysis of mortar SUP-PCM with a distinguished interface that separates the outer carbonized zone and the inner non-carbonated zone

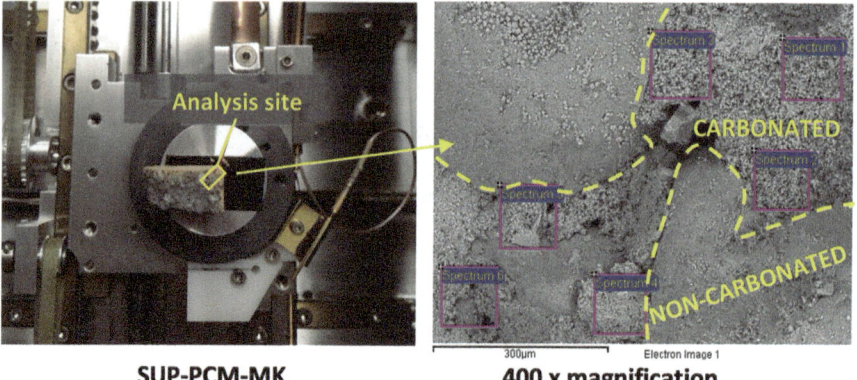

Fig. 9 SEM-EDX analysis of mortar SUP-PCM-MK with a distinguished interface that separates the outer carbonized zone and the inner non-carbonated zone

the non-carbonized inner zone, indicating the formation of $CaCO_3$ polymorphs. Additionally, both samples display a light brown crust on the surface. Table 3 presents the chemical composition of the SUP-PCM and SUP-PCM-MK spectra at 400× magnification.

4 Conclusion

The aim of this study was to investigate the potential of high-pressure carbonation to improve the strength and CO_2 utilization of Portland cement mortar. It was observed that the addition of metakaolin (MK), a supplementary cementitious material with pozzolanic properties, significantly enhanced both strength and CO_2 mineralization under supercritical conditions. These findings suggest promising prospects for the application of high-pressure carbonation in the construction industry, particularly for curing cement composites and recycled concrete where cement mortar is utilized on the surface.

References

1. Oluleye, B.I., Chan, D.W., Saka, A.B., et al.: Circular economy research on building construction and demolition waste: a review of current trends and future research directions. J. Clean. Prod. **357**, 131927 (2022)
2. Jiang, L., Cheng, L., Zhang, Y., et al.: A review on CO_2 sequestration via mineralization of coal fly ash. Energies. **16**, 6241 (2023)
3. Chen, T., Li, L., Gao, X., et al.: New insights into the role of early accelerated carbonation on the calcium leaching behavior of cement paste. Cem. Concr. Compos. **140**, 105103 (2023)
4. Abraham, M.A., Sunol, A.K., Staff, A.C.S.: Supercritical fluids. ACS Publications (1997)
5. Gonenc, Z.S., Akman, U., Sunol, A.K.: Solubility and partial molar volumes of naphthalene, phenanthrene, benzoic acid, and 2-methoxynaphthalene in supercritical carbon dioxide. J. Chem. Eng. Data. **5**(40), 799–804 (1999)
6. Kravanja, G., Hrnčič, M.K., Škerget, M., et al.: Interfacial tension and gas solubility of molten polymer polyethylene glycol in contact with supercritical carbon dioxide and argon. J. Supercrit. Fluids. **108**, 45–55 (2016)
7. Kravanja, G., Primožič, M., Knez, Ž., et al.: Transglutaminase release and activity from novel poly (ε-caprolactone)-based composites prepared by foaming with supercritical CO_2. J. Supercrit. Fluids. **166**, 10503 (2020)
8. Primožič, M., Vasić, K., Kravanja, G., et al.: Immobilized laccase for sustainable technological processes. CET J. Chem. Eng. Trans. **76** (2019)
9. García-González, C.A., Camino-Rey, M., Alnaief, M., et al.: Supercritical drying of aerogels using CO_2: effect of extraction time on the end material textural properties. J. Supercrit. Fluids. **66**, 297–306 (2012)
10. Gonenc, Z.S., Akman, U., Sunol, A.K.: Solubility/retention relationships in supercritical-fluid chromatography. Can. J. Chem. Eng. **73**, 267–271 (1995)
11. Engelmeier, L., Pollak, S., Weidner, E.: Investigation of superheated liquid carbon dioxide jets for cutting applications. J. Supercrit. Fluids. **132**, 33–41 (2018)

12. Lebedev, A.E., Katalevich, A.M., Menshutina, N.V.: Modeling and scale-up of supercritical fluid processes. Part I: Supercritical drying. J. Supercrit. Fluids. **106**, 122–132 (2015)
13. Smith, R.D.: Method of Making Supercritical Fluid Molecular Spray Films, Powder and Fibers: Google Patents (1988)
14. Centi, G., Perathoner, S.: Catalysis and sustainable (green) chemistry. Catal. Today. **77**, 287–297 (2003)
15. Kravanja, G., Zajc, G., Knez, Ž., et al.: Heat transfer performance of CO_2, ethane and their azeotropic mixture under supercritical conditions. Energy. **152**, 190–201 (2018)
16. Kravanja, G., Knez, Ž.: Carbonization of Class G well cement containing metakaolin under supercritical and saturated environments. Constr. Build. Mater. **376**, 131050 (2023)
17. Kravanja, G., Knez, Ž., Hrnčič, M.K.: The effect of argon contamination on interfacial tension, diffusion coefficients and storage capacity in carbon sequestration processes. Int. J. Greenhouse Gas Contr. **71**, 142–154 (2018)
18. Ndiaye, S., Condoret, J.-S., Bourgeois, F., et al.: High-pressure carbonation of mortar as a model for recycled concrete aggregates. J. Supercrit. Fluids. **198**, 105932 (2023)
19. Zha, X., Ning, J., Saafi, M., et al.: Effect of supercritical carbonation on the strength and heavy metal retention of cement-solidified fly ash. Cem. Concr. Res. **120**, 36–45 (2019)
20. Guo, B., Chu, G., Yu, R., et al.: Effects of sufficient carbonation on the strength and microstructure of CO_2-cured concrete. J. Build. Eng. **76**, 107311 (2023)
21. Rahman, M.M., Muttakin, M., Pal, A., et al.: A statistical approach to determine optimal models for IUPAC-classified adsorption isotherms. Energies. **12**, 4565 (2019)
22. Sotomayor, F.J., Cychosz, K.A., Thommes, M.: Characterization of micro/mesoporous materials by physisorption: concepts and case studies. Acc. Mater. Surf. Res. **3**, 34–50 (2018)

Effects of Superabsorbent Polymers and Natural Zeolite on the Properties and Pore Structure of Ultra-High-Performance Concretes

Yuxiang Tan, Weizhuo Shi, Bo Li, and Yung-Tsang Chen

Contents

1   Introduction

In demand for better construction materials, at the end of twentieth century, a novel class of concrete has been developed and referred to as ultra-high-performance concrete (UHPC). It exhibits extremely high strength, ductility, and durability [26]. The typical compressive strength of UHPC lies in the range of 120–200 MPa, its split tensile strength up to 20 MPa, and flexural strength up to 30 MPa [21]. The excellent properties are achieved using ultra-fine silica fume, steel fiber, and low water/binder ratio [9]. Durability of UHPC benefits from the dense matrix formed by carefully graded granular materials which makes UHPC resistant to water penetration, chemical attacks, and carbonation [27].

Y. Tan · W. Shi (✉) · B. Li · Y.-T. Chen
Department of Civil Engineering and New Materials Institute, University of Nottingham Ningbo China, Ningbo, China
e-mail: yuxiang.tan@nottingham.edu.cn; weizhuo.shi@nottingham.edu.cn; bo.li@nottingham.edu.cn; yung.tsang.chen@nottingham.edu.cn

© The Author(s) 2025
M. Kioumarsi, B. Shafei (eds.), *The 1st International Conference on Net-Zero Built Environment*, Lecture Notes in Civil Engineering 237,
https://doi.org/10.1007/978-3-031-69626-8_28

However, low water/binder ratio can lead to autogenous shrinkage up to 2000 to 3000 μm/m due to self-desiccation [17]. To mitigate the autogenous shrinkage, one of the effective methods is internal curing. It provides curing water internally that can participate in early-age chemical reaction in addition to base mixing water [20]. This mechanism can maintain the internal relative humidity (IRH) and thus mitigate the autogenous shrinkage. Superabsorbent polymer (SAP) and natural zeolite are two effective internal curing agents used in many studies [14, 15, 23]. SAP is typically cross-linked poly-electrolyte with acrylamide/acrylic acid or poly-acrylic acid. It can absorb water up to 1000 times of its own mass and gradually release the absorbed water when surrounding IRH is relatively low. The effectiveness of SAP as an internal curing agent is closely related to its chemical characteristics, its dosage and particle size, the water/binder ratio of UHPC [18, 28]. Researchers have divided results regarding the effect of SAP on compressive strength, some indicated an increase in compressive strength [24], and others reported a negative impact [8]. Natural zeolite is a natural pozzolan that is able to act as an internal curing agent similar to SAP particles. Zeolite particles can entrap water molecules in the cavities and channels in their honeycomb-like structure and gradually release it back to the cementitious matrix [14]. Moreover, natural zeolite is considered eco-friendly for several reasons. China currently possesses substantial natural zeolite deposits, estimated at over 3 billion tons. The current annual consumption rate is approximately 3 million tons [10]. To ensure responsible resource management, government oversees the extraction rates, labor safety, environmental impact and operational economic efficiency.

However, there are few efforts spent on studying their effects using the identical setup, that is, same raw materials, mixing method, curing condition and test specifics. Therefore, to facilitate a better visualized comparison of SAP and zeolite, this study aims to investigate the effect of both SAP and natural zeolite on shrinkage reduction, mechanical performance, micropore structure and hydration characteristics of UHPC. To this end, six mixtures were designed, where SAP was added at three different dosages, namely, 0.2%, 0.4%, and 0.6% of cement mass. Silica fume was replaced by zeolite particles at 25% and 50% by mass, respectively. Apart from early-age autogenous shrinkage, drying shrinkage was also recorded to provide a comprehensive understanding of dimensional stability of UHPC. Micropore structure was evaluated by mercury intrusion porosimetry (MIP) to explore its relation with shrinkage and compressive strength. The pore size distribution was supported by the T_2 profiles given by the low-field nuclear magnetic resonance (NMR). The tests and analysis conducted in this study shed a light on how SAP and zeolite influence shrinkage and other properties of UHPC under different dosages. The outcome of this study contributes to making a more sustainable UHPC with better dimensional stability.

2   Experimental Testing Program

2.1   Materials

The UHPC used in this study is a non-proprietary UHPC mixture that consists of ordinary Portland cement (OPC), silica fume (SF), natural zeolite (NZ), SAP, river sand (fine sand), quartz sand (coarse sand), superplasticizer (SP), and water. The OPC, SF and NZ are considered as binders. The median particle sizes of the OPC, SF, SAP, NZ, fine sand are 14.805 μm, 26.368 μm, 29.431 μm, 53.855 μm, and 353.257 μm, respectively. The particle size distributions of materials are shown in Fig. 1. The size range of coarse sand is 600–2000 μm. The OPC used in this study is type P·II52.5R. Steel fibers are not used to avoid the interference of steel fibers on shrinkage restraining. The SP used is provided by SIKA Co. Ltd. in form of solid powder.

2.2   Mix Design and Mixture Proportions

The mix design of this study adopts a non-proprietary mixture proposed by Karim et al. [13]. Minor modifications were made to optimize UHPC performance according to modified Andreasen and Andersen particle packing theory [11], as part of the efforts in adaption to local materials. The mixture proportions are shown in Table 1. For the reference mixture, the binder materials include Portland cement and silica fume, which account for 86.8% and 13.2% of total binder by mass,

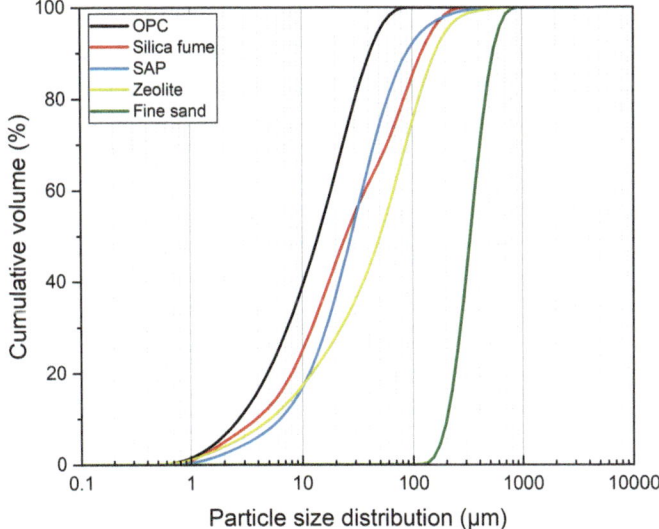

Fig. 1 Particle size distribution of raw materials

Table 1 Proportions of non-proprietary UHPC mixtures

| Mix. ID | Materials (kg/m³) | | | | | | | | | | |
	PC	SF	NZ	SAP	Quartz sand	River sand	Water	Extra water	SP	Eff. w/b	Total w/b
Ref	889.0	135.8	–	–	906.8	349.0	174.2	0.0	14.2	0.170	0.170
S0.2	889.0	135.8	–	1.8	906.8	349.0	174.2	29.2	14.2	0.170	0.198
S0.4	889.0	135.8	–	3.6	906.8	349.0	174.2	58.3	14.2	0.170	0.227
S0.6	889.0	135.8	–	5.3	906.8	349.0	174.2	87.5	14.2	0.170	0.255
Z25	889.0	101.9	33.9	–	906.8	349.0	174.2	5.3	14.2	0.170	0.175
Z50	889.0	67.9	67.9	–	906.8	349.0	174.2	10.6	14.2	0.170	0.180

respectively. The sand (filler) to cement ratio was 1.41:1, and the sand consists of 72.2% of quartz sand and 27.8% of river sand by mass. The water-to-binder ratio was 17.0% for the reference mixture. The amount of superplasticizer is 1.6% by mass of cement across all mixtures. Six mixtures of UHPC were designed in total (including the reference mix). Three of them were mixed with different dosage of SAP, namely, 0.2%, 0.4%, and 0.6% by mass of cement. They were named by the internal curing agent and its dosage, that is, S0.2 denotes the mix with SAP at dosage of 0.2%. Two of them were mixed with different levels of substitution of silica fume for zeolite, namely, 25% and 50% by mass of silica fume. Z25 denotes the mix with zeolite substitution level of 25%.

2.3   Mixing Procedure

The mixing and curing procedures followed ASTM C192 [3] unless specified otherwise below. All mixtures were mixed under 25 °C. First, all dry materials except cement were mixed for five minutes, SAP or zeolite included. Cement was then added to the dry materials and mixed for another five minutes. Water was slowly added for a duration of two minutes while mixing. Finally, the mortar was mixed for another eight minutes.

2.4   Test Methods

2.4.1   Autogenous Shrinkage

To measure the early-age autogenous shrinkage of UHPC mixes, about 185 ml of freshly mixed UHPC mortar was cast in a specialized plastic corrugated tube which can extend or shrink in length depend on its content without creating much resistance. The tube was then sealed and gently put on a dilatometer bench at 23 ± 2 °C in accordance with ASTM C1698 [6]. The length of the tube was recorded by the dilatometer every five minutes for up to 7 days. The result of every mixture is the average of three identical tests.

2.4.2   Drying Shrinkage

The drying shrinkage of UHPC mixtures were evaluated conforming to ASTM C596-23 [5]. For each mixture, three specimens were cast. After being moist cured in molds for 24 hours, specimens were removed from the molds and cured in lime-saturated water for 48 hours. At the age of 72 hours, specimens were removed from water and wiped with a cloth before being placed in air storage until the age of

28 days. Drying shrinkage measurements were recorded by a length comparator at the age of 3, 7, 14, 21, and 28 days. The averaged values of three specimens were used for each mixture.

2.4.3 Compressive Strength

Testing for compressive strength has been employed to evaluate the compressive strength of concrete samples. The test was conducted according to ASTM C109 [2] using a specimen size of 50 ⬚ 50 ⬚ 50 mm. The results were based on the average of three specimens. Immediately after casting, the specimens were covered with a plastic film, until they were removed from the molds at the age of 24 hours. The specimens were then kept in a dedicated moist curing chamber which maintains a temperature at 20 °C and a relative humidity at 95% as per ASTM C511 [4]. The specimens stayed in the chamber until they were tested at the age of 3, 7, and 28 days, respectively.

2.4.4 Mercury Intrusion Porosimetry

To obtain test specimens with dimension of 10 ⬚ 10 ⬚ 10 mm, UHPC cubes were cut with a concrete saw. Then the specimens were submerged in 99.9% isopropanol for 48 hours and were transferred in a vacuum chamber for another 48 hours to dry. The specimens were sealed before they were shipped to a third-party agency for the MIP test, which was conducted with pressure in range of 0.33–227.48 MPa. The MicroActive AutoPore V 9600 program was employed to obtain the pore size distribution.

2.4.5 Low-Field Nuclear Magnetic Resonance (NMR)

A low-field ^{1}H NMR instrument (Lime Echo Co. Ltd., Beijing, China) was used for the NMR experiments. With a constant magnetic field of 0.49 T and a 12 mm coil operating at 2 MHz, the Carr-Purcell-Meiboom-Gill (CPMG) sequence was applied and ^{1}H NMR relaxation data were processed and converted by the Polimar 2021 (Lime Echo Co. Ltd., Beijing, China) application into T_2 relaxation time component. The effective relaxing component of T_2 lies in the range of 0.01 ms to 1000 ms. The longer component of T_2 (100–10,000 ms) corresponds to water with higher degree of freedom, that is, in large pores and surface water. Shorter T_2 components represent gel water (0.01–1 ms) and capillary water (1–100 ms). After repeated trial experiments, the parameters of NMR experiments were optimized to balance the signal/ noise ratio and the running time. The corresponding signal-to-noise ratio for measuring the water in the cement paste can reach 200:1. In this study, NMR tests have been conducted to obtain T_2 signal from UHPC samples at the age of 28 days. The samples were moist-cured per ASTM C511 [4] before test.

3 Results and Discussion

3.1 Shrinkage Remediation

3.1.1 Autogenous Shrinkage

Figure 2 shows the autogenous shrinkage of UHPC mixtures incorporated with SAP and zeolite particles. The reference mixture showed a shrinkage of 1360 μm/m in the first 24 h of hydration. The quick development of autogenous shrinkage of UHPC is related to the use of silica fume and the very low water/binder ratio [16]. The shrinkage of UHPC made with 0.4% of SAP was 1340 μm/m while S0.2 and S0.6 reached 1600 μm/m and 1770 μm/m, respectively. This could be due to that the extra water in S0.2 and S0.4 stimulated an accelerated rate of hydration that could lead to self-desiccation. S0.2 only had a total water/binder ratio of 0.198 which was not sufficient for the increased water consumption. The relatively high total water/binder ratio of S0.6 induced an even faster rate of hydration that quickly consumed the extra water and caused greater self-desiccation and thus, shrinkage. In other words, the total water/binder ratio of S0.4 (0.227) found the balance of increased hydration rate and the amount of extra water that resulted in a reduced early-age autogenous shrinkage. The autogenous shrinkage of S0.2, S0.4 and S0.6 at 7 days maintained the trend at 24 hours, where S0.4 reduced 2.8%, S0.2 and S0.6 increased 15.2% and 20.0%, respectively, when compared to the reference mixture. On the other hand, the UHPC mixtures containing zeolite particles exhibited lower autogenous shrinkage throughout the test duration. But the zeolite replacement level did not have significant effect on shrinkage reduction. At 7 days, the mixtures with 25% and 50%

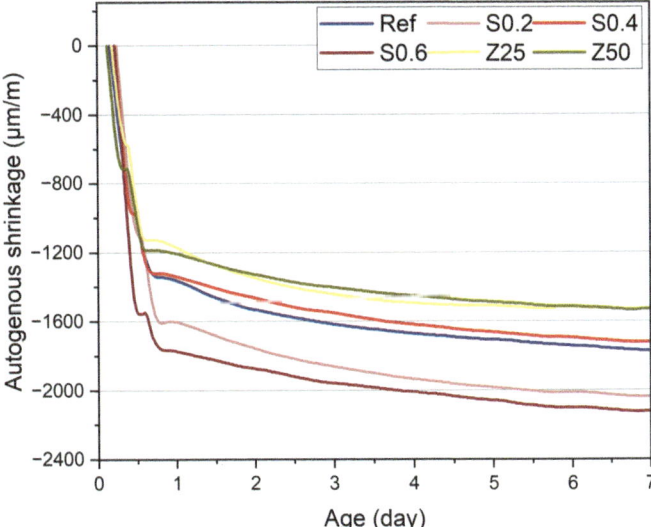

Fig. 2 Early-age autogenous shrinkage of UHPC mixtures

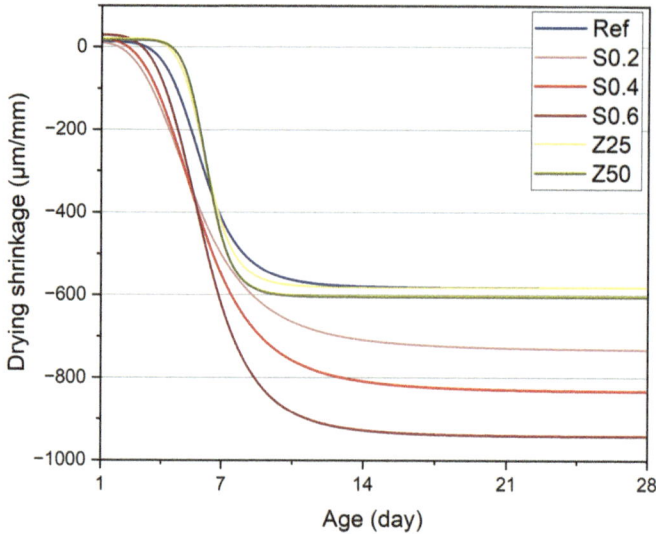

Fig. 3 Drying shrinkage of UHPC mixtures

zeolite replacement experienced 13.4% and 13.5% lower shrinkage than that of the reference, respectively. This could be due to that natural zeolite particles had a larger particle size and smaller surface area than those of silica fume, which make zeolite attract water slower and reduce the self-desiccation.

3.1.2 Drying Shrinkage

The results of drying shrinkage test of UHPC mixtures with SAP and zeolite particles at the different ages are shown in Fig. 3. The specimens were wet-cured under saturated lime water for the second and third days per ASTM C596 [5]. Drying shrinkage of all mixtures developed rapidly for the first seven days then increased slowly thereafter. Noticeably, the amount of shrinkage had a strong positive relation with the total water/binder ratio of the mixtures. The higher the total water/binder ratio, the greater the drying shrinkage, regardless of whether it contained SAP or zeolite particles. This attributes to the fact that more extra water introduces more capillary pores within the UHPC matrix and the rate of water evaporation mainly relates to the size of pore [19]. The increase in capillary pores of UHPC mixtures with high total water/binder ratio can be verified by the results of MIP test where there were more pores within the range of 200–4000 nm in specimens with high total water/binder ratio (Figs. 5 and 6). And this range lies in the range of capillary pores suggested by Aligizaki [1]. Therefore, for thin UHPC sections where drying dominates the total deformation, traditional external curing methods are still necessary to reduce drying shrinkage regardless of the use of internal curing agents, especially when the total water/binder ratio of the UHPC mixture is higher than 0.2.

3.2 Compressive Strength

The results presented in Fig. 4 show decreasing compressive strength of the UHPC
with increased SAP dosage and higher zeolite replacement level. This pattern in
compressive strength of the UHPC mixtures corresponds well with the pattern in
total water/binder ratio as well as the pore size distribution indicated by MIP test
shown in Figs. 5 and 6. The decrease in compressive strength of UHPC with SAP
can be attributed to the porosity introduced by extra water and the void of SAP
particles. Although SAP with small particle size (D_{50} = 29.4 μm) has been used, as
suggested by Justs et al. [12], the SAP voids remained the largest defect in the UHPC
matrix. On the other hand, UHPC mixture with zeolite replacing 25% of silica fume
exhibited comparable compressive strength with the reference. This could be due to
the fact that zeolite particles have similar pozzolanic reactivity to silica fume and can
show internal curing effect without introducing excessive porosity. However, UHPC
with increased level of zeolite replacement at 50% had noticeable decrease in
compressive strength. Ramezanianpour et al. [25] explained that this can be due to
silica fume having higher pozzolanic reactivity than natural zeolite and thus leads to
richer C-S-H gel product and higher compressive strength. Despite having lower
compressive strength than the reference mixture at 28 days, both SAP and zeolite
mixtures exhibited earlier strength development. At seven days, S0.2, S0.4 and S0.6
reached 75.6%, 81.7%, and 83.1% of their 28-day strength, respectively. Z25 and
Z50 reached 82.3% and 84.1%, where the reference mix was only at 65.9%. This
could be due to the internal curing effect of SAP and zeolite particles that helped
maintain a relatively high internal relative humidity (IRH) which provided sufficient
free water to sustain the hydration process at early age.

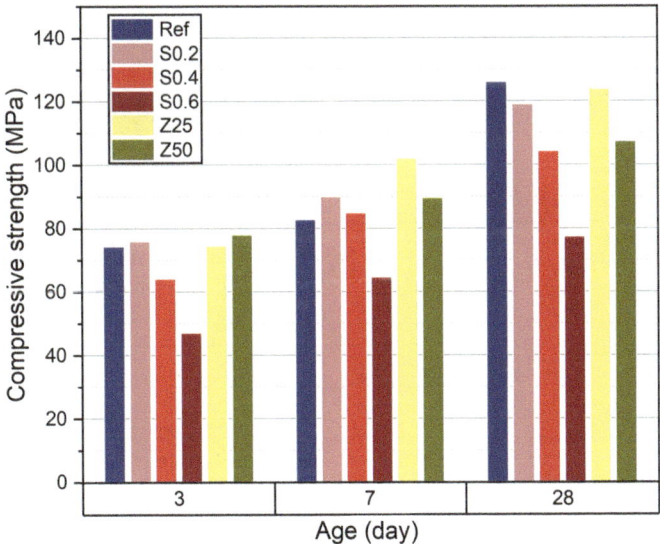

Fig. 4 Compressive strength of UHPC mixtures

3.3   Pore Structure

3.3.1   Mercury Intrusion Porosimetry (MIP)

Figure 5 shows the pore size distribution of UHPC incorporated with SAP at 28 days. As can be seen, the main peak lies within the ranges of 10–40 nm which corresponds to gel and capillary pores [1]. The critical pore size of the SAP mixtures was the same as that of the reference mixture. There were also substantial minor peaks existed in the range of 200–4000 nm which represents the large capillary pores. These were the pores induced by the extra water added for internal curing purpose. Noticeable change in pore structure were found between 200 and 4000 nm where UHPC mixtures with higher SAP content observed significantly greater pore volume, in other words porosity. Figure 6 shows the pore size distribution of UHPC with zeolite replacement. The zeolite mixtures had a generally similar pore structure to the reference mixture where only Z50 had a slightly higher pore volume at both the main peak (critical pore) and the minor peak (capillary pore). This is due to the extra amount of internal curing water added that is proportional to the level of zeolite replacement.

3.3.2   Nuclear Magnetic Resonance (NMR) at 28 Days

Figure 7 shows the transverse relaxation time (T_2) of the UHPC samples that has been moist cured for 28 days then saturated under pressured water. It can be seen that the pore size distribution of all groups presents a bimodal pattern. This pattern agrees

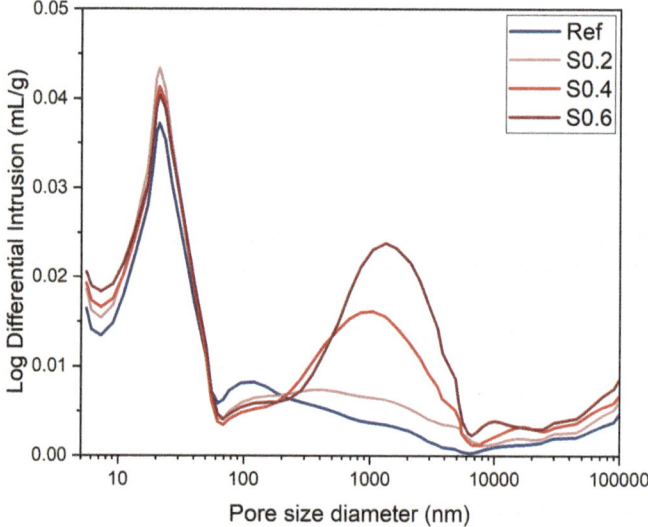

Fig. 5 Pore size distribution of UHPC mixtures with SAP

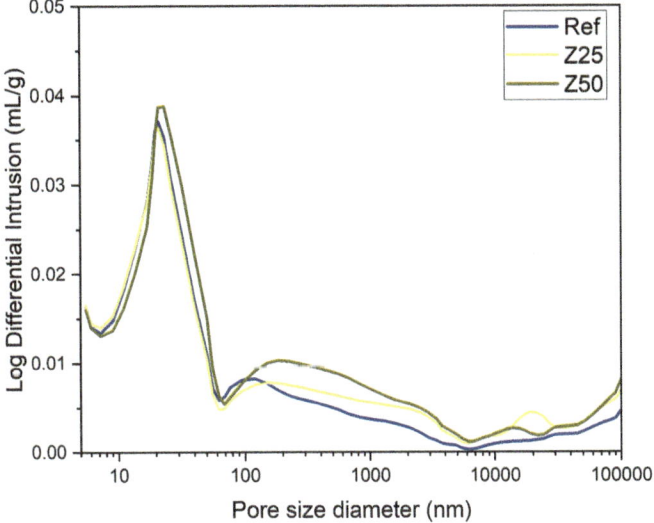

Fig. 6 Pore size distribution of UHPC with natural zeolite

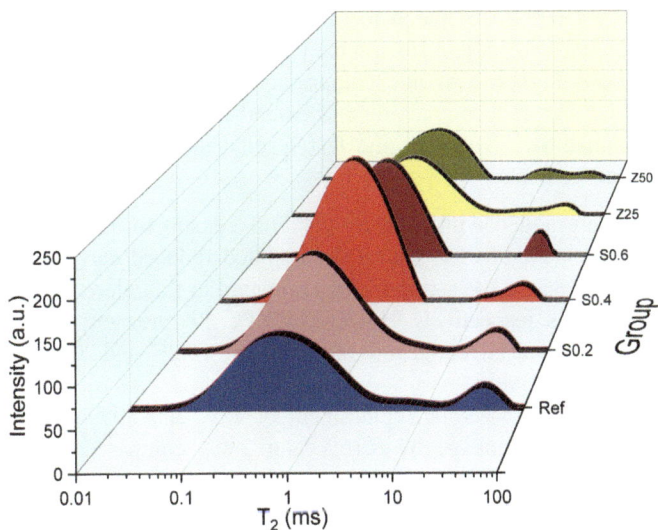

Fig. 7 Transverse relaxation time spectrum of UHPC samples at 28 days

with the findings of Muller et al. [22], where two T_2 distribution peaks were observed in the concrete samples that were first cured for 28 days then dried and saturated under water at a pressure of 300 bars. They reported that these two peaks corresponded to C-S-H gel pores and capillary pores, respectively. For the samples that had only been cured for 28 days and not saturated under pressure, four T_2

distribution peaks had been observed. In this study, with the increase of total water/ binder ratio, the signal strength of the gel pore distribution signal (left peak) also increased in UHPC mixtures with SAP and zeolite replacement. For the capillary water distribution signal (right peak), as the dosage of SAP and extra water increased, the peaks appeared to be narrower and taller. This also held true for UHPC with zeolite replacement, which indicated that excessive extra water could lead to the generation of capillary pores in similar size and reduce the amount of smaller capillary pores. In general, the test results generated by ^{1}H NMR were consistent with the MIP test results in terms of modal pattern and the signal intensity of the gel pore relative to capillary peaks, as well as the relation of signal intensity to the different total water/binder ratio of each UHPC mixture. It can be advised that low-field ^{1}H NMR to be used to verify and complement the results of other pore size distribution evaluation methods.

4 Conclusions

This study evaluates the effect of SAP and zeolite incorporation on shrinkage behavior, mechanical performance and hydration characteristics of a non-proprietary UHPC. Compared to traditional UHPC, the reference UHPC in this study uses less cement and silica fume [7]. Additionally, the SAP/NZ modified UHPC can have higher early-age compressive strength and lower autogenous shrinkage which can reduce cracking and extend the expected lifespan, and thus, promote sustainability. Based on the test results and discussion, the following conclusions can be made:

- UHPC mixture added with SAP by 0.4% of cement mass and mixtures with zeolite replacement of 25% and 50% exhibited reduced early-age autogenous shrinkage by 2.8%, 13.4%, and 13.5% compared to the reference mixture at the age of seven days, respectively. However, UHPC mixtures with SAP at 0.2% and 0.6% dosages increased autogenous shrinkage by 15.2% and 20.0% at the age of seven days, respectively.
- UHPC mixtures with zeolite replacement of 25% and 50% exhibited slightly increased drying shrinkage by 0.16% and 2.4% compared to the reference mixture at the age of 28 days, respectively. On the other hand, all UHPC mixtures added with SAP showed increased drying shrinkage of over 23.4%.
- According to MIP and NMR analysis, increased SAP and zeolite content results in increased overall porosity of UHPC matrix, where SAP incorporation induces higher overall porosity than natural zeolite, especially in the range of capillary pores. But shift of the critical pore size has not been observed in any UHPC mixture in this study.
- Natural zeolite replacement slightly reduced compressive strength of UHPC where Z25 and Z50 reduced 28-day compressive strength by 1.6% and 15.1%, respectively. Notably, Z25 uses 25% less silica fume and effectively reduces

autogenous shrinkage while maintains compressive strength and drying shrink-age, which makes it a more sustainable mixture for further studies.
- By cross-verification of results of low field ^1H NMR test and MIP, hydration heat and TG tests, it is suggested that NMR test can facilitate analysis of hydration characteristics as well as micropore structure of UHPC matrix.

Acknowledgments The authors would like to acknowledge the financial support from the New Materials Institute at the University of Nottingham Ningbo China. The technical assistance provided by Ms. Xinyan Liu in conducting the NMR tests is gratefully acknowledged.

References

1. Aligizaki, K.K.: Pore Structure of Cement-Based Materials: Testing, Interpretation and Requirements. CRC Press (2005)
2. ASTM: C109/C109M-21: Standard Test Method for Compressive Strength of Hydraulic Cement Mortars (Using 2-in. or [50 mm] Cube Specimens), pp. 1–12. American Society for Testing and Materials (2021)
3. ASTM: C192/C192M-19: Standard Practice for Making and Curing Concrete Test Specimens in the Laboratory, pp. 1–8. American Society for Testing and Materials (2019)
4. ASTM: C511-21: Standard Specification for Mixing Rooms, Moist Cabinets, Moist Rooms, and Water Storage Tanks Used in the Testing of Hydraulic Cements and Concretes, pp. 1–3. American Society for Testing and Materials (2021)
5. ASTM: C596-23: Standard Test Method for Drying Shrinkage of Mortar Containing Hydraulic Cement, pp. 1–4. American Society for Testing and Materials (2023)
6. ASTM: C1698-19: Standard Test Method for Autogenous Strain of Cement Paste and Mortar, pp. 1–8. American Society for Testing and Materials (2023)
7. Bajaber, M.A., Hakeem, I.Y.: UHPC evolution, development, and utilization in construction: a review. J. Mater. Res. Technol. **10**, 1058–1074 (2021)
8. Bentz, D.P., Jones, S.Z., Peltz, M.A., Stutzman, P.E.: Influence of Internal Curing on Properties and Performance of Cement-Based Repair Materials. US Department of Commerce, National Institute of Standards and Technology (2015)
9. Farzad, M., Shafieifar, M., Azizinamini, A.: Experimental and numerical study on bond strength between conventional concrete and Ultra High-Performance Concrete (UHPC). Eng. Struct. **186**, 297–305 (2019)
10. Feng, N.Q., Peng, G.F.: Applications of natural zeolite to construction and building materials in China. Constr. Build. Mater. **19**(8), 579–584 (2005)
11. Funk, J.E., Dinger, D.R.: Predictive Process Control of Crowded Particulate Suspensions: Applied to Ceramic Manufacturing. Springer Science & Business Media (2013)
12. Justs, J., Wyrzykowski, M., Bajare, D., Lura, P.: Internal curing by superabsorbent polymers in ultra-high performance concrete. Cem. Concr. Res. **76**, 82–90 (2015)
13. Karim, R., Najimi, M., Shafei, B.: Assessment of transport properties, volume stability, and frost resistance of non-proprietary ultra-high performance concrete. Constr. Build. Mater. **227**, 117031 (2019)
14. Kazemian, M., Shafei, B.: Internal curing capabilities of natural zeolite to improve the hydration of ultra-high performance concrete. Constr. Build. Mater. **340**, 127452 (2022)
15. Liu, J., Shi, C., Ma, X., Khayat, K.H., Zhang, J., Wang, D.: An overview on the effect of internal curing on shrinkage of high performance cement-based materials. Constr. Build. Mater. **146**, 702–712 (2017)

16. Liu, J., Farzadnia, N., Shi, C., Ma, X.: Shrinkage and strength development of UHSC incorporating a hybrid system of SAP and SRA. Cem. Concr. Compos. **97**, 175–189 (2019)
17. Liu, J., Farzadnia, N., Khayat, K.H., Shi, C.: Effects of SAP characteristics on internal curing of UHPC matrix. Constr. Build. Mater. **280**, 122530 (2021)
18. Ma, X., Liu, J., Wu, Z., Shi, C.: Effects of SAP on the properties and pore structure of high performance cement-based materials. Constr. Build. Mater. **131**, 476–484 (2017)
19. Mehta, P.K., Monteiro, P.J.: Concrete: Microstructure, Properties, and Materials. McGraw-Hill Education (2014)
20. Meng, W., Khayat, K.: Effects of saturated lightweight sand content on key characteristics of ultra-high-performance concrete. Cem. Concr. Res. **101**, 46–54 (2017)
21. Mishra, O., Singh, S.P.: An overview of microstructural and material properties of ultra-high-performance concrete. J. Sustain. Cem.-Based Mater. **8**(2), 97–143 (2019)
22. Muller, A.C.A., Scrivener, K.L.: A reassessment of mercury intrusion porosimetry by comparison with 1H NMR relaxometry. Cem. Concr. Res. **100**, 350–360 (2017)
23. Pezeshkian, M., Delnavaz, A., Delnavaz, M.: Effect of natural zeolite on mechanical properties and autogenous shrinkage of ultrahigh-performance concrete. J. Mater. Civ. Eng. **32**(5), 04020093 (2020)
24. Pourjavadi, A., Fakoorpoor, S.M., Hosseini, P., Khaloo, A.: Interactions between superabsorbent polymers and cement-based composites incorporating colloidal silica nanoparticles. Cem. Concr. Compos. **37**, 196–204 (2013)
25. Ramezanianpour, A.M., Esmaeili, K., Ghahari, S.A., Ramezanianpour, A.A.: Influence of initial steam curing and different types of mineral additives on mechanical and durability properties of self-compacting concrete. Constr. Build. Mater. **73**, 187–194 (2014)
26. Richard, P., Cheyrezy, M.: Composition of reactive powder concretes. Cem. Concr. Res. **25**(7), 1501–1511 (1995)
27. Wu, Z., Shi, C., Khayat, K.H.: Multi-scale investigation of microstructure, fiber pullout behavior, and mechanical properties of ultra-high performance concrete with nano-$CaCO_3$ particles. Cem. Concr. Compos. **86**, 255–265 (2018)
28. Zhutovsky, S., Kovler, K.: Influence of water to cement ratio on the efficiency of internal curing of high-performance concrete. Constr. Build. Mater. **144**, 311–316 (2017)

Performance Evaluation of the Quantification of Cement Microphases Using Energy Dispersive X-ray Spectroscopy Imaging

Anuradha Silva ⓘ, Shanaka Baduge, and Priyan Mendis

Contents

1 Introduction

Concrete is a composite material which has superior properties to overcome engineering challenges in the construction industry. It has a heterogeneous microstructure as it is composed of a combination of fine supplementary cementitious material, unreacted cement particles, hydrated products, filler materials, and fibers in various special length scales [1, 2]. In general, ultra-high-performance concrete is known for strengths beyond 120 MPa, higher ductility, high resistance to aggressive chemical environments leading to higher durability, and Young's modulus between 50 and 60 MPa [2]. The performance enhancement of ultra-high-performance concrete at very low water to binder ratios can be achieved by altering its microstructure with partial replacement of cement using low carbon supplementary cementitious materials such as, fly ash, silica fume, limestone, and slag with suitable quantities. For instance, the addition of pozzolanic materials assists the formation of pozzolanic calcium hydrate silicates in addition to the clinker hydration to support the strength while reducing the probability of alkali silica reactions, supporting durability

A. Silva (✉) · S. Baduge · P. Mendis
University of Melbourne, Parkville, VIC, Australia

© The Author(s) 2025
M. Kioumarsi, B. Shafei (eds.), *The 1st International Conference on Net-Zero Built Environment*, Lecture Notes in Civil Engineering 237,
https://doi.org/10.1007/978-3-031-69626-8_29

aspects. Further, by optimizing the particle size distribution of the cement paste, the packing density of the particles is improved and thus reduces the amount of void space in the cement restricting the permeability.

Ultra-high-performance concrete is recognized as reactive powder concrete since finer aggregates and particles provide more compacted strength to the structure. The benefits of finer particles in creating a more denser and compacted structure at micro-nano level with less voids ratio lead researchers to more concentrate on introducing inert nano materials such as graphene oxide [3, 4], graphite nanoplates [5], and carbon nanofibers [6] to improve long-term performance to reduced water sorptivity and permeability, reduced sulfate attacks, freeze thaw effects, etc. The low water to binder ratio makes concrete more prone to alkali-silica reaction causing delayed final setting time, reduced strength, and increased drying shrinkage and permeable voids. This can be controlled in ultra-high-performance concrete by partial replacement of cement with pozzolanic mineral admixtures such as fly ash (10–20%), blast-furnace slag, silica fume (15–30%) by reacting with calcium hydroxide to form calcium hydrate silicate and limiting the availability of alkalis to react with reactive silica [7]. The performance enhancement of ultra-high-performance concrete is greatly contributed by cement, and this leads to long term sustainability issues related to carbon emission. As a solution, carbon-neutralizing methods have drawn the attention of current studies to partially replace cement with phosphorous slag [8], a blend of nano-silica 2% and class F silica fume 10% [9], geopolymer as alkali-activated low carbon binder [10], fine aggregate replacement by mechanically activated pozzolanic gold tailings with a relatively wide particle size distribution (35% of micropowder) [11], and rice husk ash as a reactive filler to increase amorphous silica [12].

The qualitative investigation of microphases in concrete is possible nowadays with a variety of technologies up to atomic level. As discussed in the previous section, ultra-high-performance concrete designs can be altered with low carbon footprint materials and the design can be optimized by conducting a proper characterization at the smallest length scale to investigate the reason behind the macroscale strength improvement. Scanning electron microscopy (SEM) [13] and transmission electron microscopy (TEM) [14] can provide high-resolution images of the micro-structure and capable of revealing the morphology of cement particles [15], the porosity of the cement matrix [16], and the interfacial transition zone [17] between the cement and aggregate. X-ray diffraction (XRD) can be used to precisely identify the crystalline phases present in the sample [18]. Acquiring accurate quantitative information from micro-nano scale is yet an emerging area of study for cementitious materials and essential to overcome engineering challenges in the future. The need for a proper framework to identify the cement microphases is mainly required to understand the underlying cause for any property change in the experiment, so the trial and error-based experiments can be carried out toward a profound direction.

This study exhibits the development of a framework to quantitatively identify the microphases of an ultra-high-performance cement paste consisting of 30% of ground granulated blast furnace slag, silica fume and fly ash altogether with 70% of ordinary Portland cement. The methodology is developed using the information from EDS-based chemical hypermaps, and microphase information is descriptively elaborated in terms of volume fractions of microstructures and distribution of microphases. The image capturing for the chemical characterization was done in such a way that the inhomogeneity of the considered length scale is represented, and the accuracy of the microphase quantification is compared with an independent analytical hydration simulation software, Virtual Cement and Concrete Testing Laboratory (VCCTL).

2 Methodology

2.1 Overview

Chemical phase mapping of cement microstructure using EDS data combined with BSE imaging is a powerful tool when considering the quantification of hydrated and unreacted phases at microlevel. These microphase quantification outcomes were validated with theoretically simulated hydration outcomes from VCCTL and the briefed methodology involved is illustrated in Fig. 1.

2.2 Material Properties and Specimen Preparation

Specimens for the compressive strength test, elastic modulus, and chemical characterization experiments were prepared using the cement mix proportions listed in Table 1; 50 mm × 50 mm × 50 mm cubes were cast for the compressive strength test.

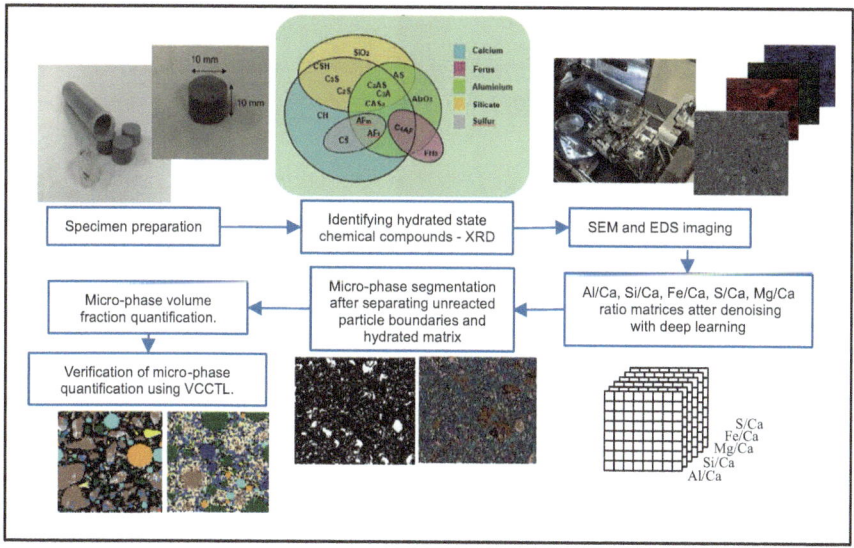

Fig. 1 Methodology of cement microstructure quantification and validation process

Table 1 Mix proportions of the cement paste

Component	Density (kg/m^3)	Weight (kg/m^3)
Ordinary Portland cement (OPC)	3100	1159
Ground granulated blast furnace slag	2860	361
Fly ash—Class F	2290	100
Silica fume	2180	116
Water	1000	385
Superplasticizer 1	1064	16
Superplasticizer 2	1000	3
Retarder	1080	3

The average compressive strengths on 7th and 28th days were 98 MPa and 141 MPa, and elastic modulus was 47 GPa on 28th day.

Specimen dimensions for the chemical characterization were 10 mm in diameter and 10 mm in height (Fig. 1). The hydration was stopped on the 28th day by submerging the specimen in isopropyl alcohol for 7 days to remove free water and afterwards the specimen was exposed to 100 °C temperature for 24 hours in the oven to evaporate the alcohol completely. Then, the observation surface of the specimen was polished and the final mirror like polish was achieved by using 1 μm diamond grit paper. Mirror like polishing of the observation surface was important to remove any surface irregularities that could interfere with chemical analysis. Then the polished surface was coated with carbon up to a thickness of 10 nm and provided a conductive surface to reduce the probability of getting hindered by the positive surface charge accumulation on the observation surface when using the scanning electron microscopy.

2.3 Chemical Composition Identification Using Semi-quantitative XRD

Chemical compound existence of crystalline structures can be accurately found using XRD analysis. Since amorphous compounds exist intermixed with crystalline compounds in hydrated cement, a semi-quantitative XRD analysis was conducted to identify the existence of crystalline compounds. This experimental qualitative identification of chemical compounds from XRD facilitates seeking existing microphases from EDS hypermaps, especially because EDS hypermaps do reveal any information about the potential phases in the mix. In this study, Bruker D8 Advance instrument was used for the XRD analysis, and diffractometer was operated at 40 A and 40 kV. The radiation was CuKα with an angular range of 85° 2θ. Total scan was done for 1 hour with no temperature control. Hydrated cement sample was powdered for this analysis and initial crushing of the sample was done using a mortar and pestle followed by XRD Mill McCrone milling for 5 minutes. Fig. 2 shows the identification of chemical compounds.

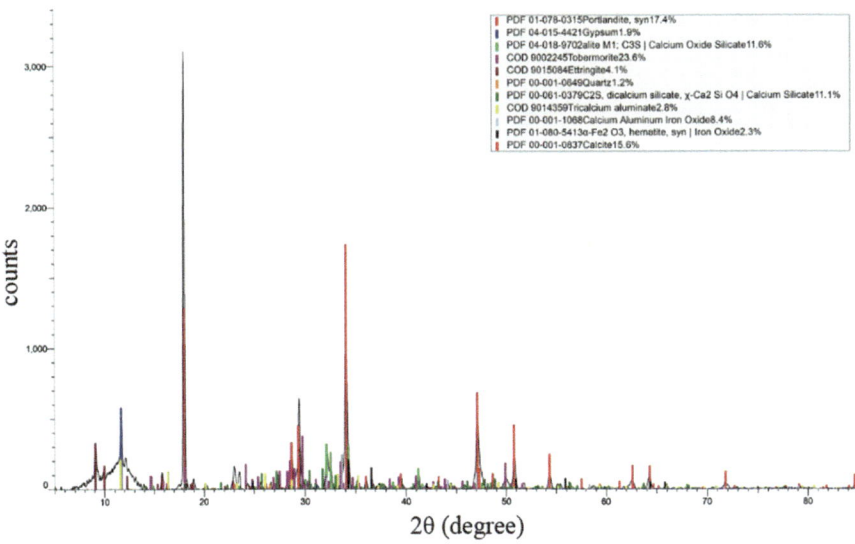

Fig. 2 Phase composition identification using XRD semi-quantitative analysis

2.4 Chemical Data Acquisition by Energy Dispersive Spectroscopy Combined with Scanning Electron Microscopy

Inhomogeneity at nano scale is governed by calcium silicate hydrate, and the Representative Volume Element (RVE) length at nano-level is 1 μm [19]. Considering the scale separation principle in multi-scale modeling, the RVE length at microscale can be taken as 100 μm and the smallest inhomogeneity would be 1 μm. Therefore, in this study the EDS image capturing was done with a resolution of 365 × 319, optimizing the observation of the smallest dimension to 0.74 μm while covering a frame sufficient to represent the inhomogeneity at microlevel.

To quantitatively identify microphase volume fractions along with their topographical distribution, EDS imaging was used. When it comes to EDS map generation, there are two types of maps. EDS maps generated using the characteristic x-ray intensity due to the abundancy of elements are called "intensity maps" and the EDS intensity maps corrected for atomic number, absorption, fluorescence effect are called "quantitative maps." Since the focused interest of this study prefers accurate quantitative chemical compositional outcomes for comparison purposes, quantitative map acquisition with no energy normalization was performed after calibration for both beam and energy measurements.

Generally, elemental concentration maps are noisy, and the noise could be reduced as much as possible using denoising methods such as non-local means (NLM), total variation, wavelet thresholding, etc. With the recent rise of artificial

Table 2 Chemical boundary tolerances to segment microphases

Microphase	Si/Ca	Al/Ca	S/Ca	Mg/Ca	Fe/Ca
Calcium hydrate silicate	$0.5 \pm \xi$	$0.25 \pm \xi$	–	–	–
Portlandite/CaCO$_3$	<=0.20	<=0.20	–	–	–
Ettringite	–	$0.35 \pm \xi$	$0.5 \pm \xi$	–	–
Unreacted cement	$0.3 \pm \xi$	<=0.7	–	–	$0.1 \pm \xi$
Slag	$0.8 \pm \xi$	$0.4 \pm \xi$	–	$0.3 \pm \xi$	–
Silica fume	>50.0	<=3	–	–	–
Fly ash	>1.0	>1.0	–	–	$1.0 \pm \xi$

intelligence applications, data preprocessing functions such as data augmentation, data normalization, feature extraction pretrained models are now available. The denoising Network function in MATLAB creates a deep convolutional neural network for image denoising, which is based on the DnCNN (denoising convolutional neural network) architecture. The network consists of several convolutional layers, followed by batch normalization and ReLU activation, and is trained using a mean squared error loss function to minimize the difference between the denoised and clean images.

To carry out a meaningful microphase segmentation, qualitative phase information from XRD was required to reveal the potential phases of the sample. Considering the possibility of a considered microphase in the mixture being composed of a number of elements depending on the function specific mix-design, it is always advisable to compress the chemical data into data numerical matrices (double type) rather than depending directly on the chemical concentration maps (uint8). Since calcium is the most abundant element in any cement mixture, double type matrices can be developed to represent atomic ratios with respect to Ca using respective EDS maps, for instance Si/Ca, Al/Ca, Fe/Ca, S/Ca, etc. Based on the stoichiometries of the compounds from XRD outcomes (Table 2), the stoichiometric boundaries for microphases can be defined to facilitate the phase segmentation considering the required number of ratio matrices to accurately identify the microphase. This selection can be graphically demonstrated in a 2D ratio plot as shown in Fig. 2, in such a way that a selected point in the 2D ratio plot represents a pixel from the segmented phase map, indicating a particular Si:Al:Ca ratio depending on the microphase stoichiometry.

3 Results and Discussion

This section describes the results of the experimental phase identification analysis carried out in terms of segmented phases, volume fractions, and chemical composition distribution of phases. Further, the hydration outcomes from VCCTL software are used to compare the reliability of the experimentally gathered phase information.

3.1 Phase Segmentation

The expected outcome from the phase segmentation is to visualize the microphase distribution over a two-dimensional plane using EDS images taken from random locations which represent the inhomogeneity of the cement microstructure. The primary task to achieve this outcome from a collected image is to identify the locations of the pixels (pixel IDs) belonging to a particular phase which satisfies the chemical boundary requirements listed in Table 2. A representation of chemical composition distribution of each individual pixel can be illustrated in a ratio plot as shown in Fig. 3.

Once required phases were identified using the stack of chemical ratio matrices, a separate empty RGB map with similar resolution to the BSE image was generated, to store separate RGB colors for each individual phase. Figure 4 shows the segmentation of phase mappings from the 28th day hydration representative areas captured.

It is evidently clear that the slag particle capturing in S28_0 image is comparatively lesser than other specimens. Further, the Portlandite intermixed $CaCO_3$ region seems to be gathered around porous areas and the reason for this might be the reduction of water in these pore areas. Slag particles tend to react slowly over the time and since these specimens are at their 28th day hydration, majority of the slag particles remain as it is. Darker areas represent the porosity intermixed with cracks and these cracks might be due to the polishing effect and dehydration effort to stop the hydration on 28th day using the oven.

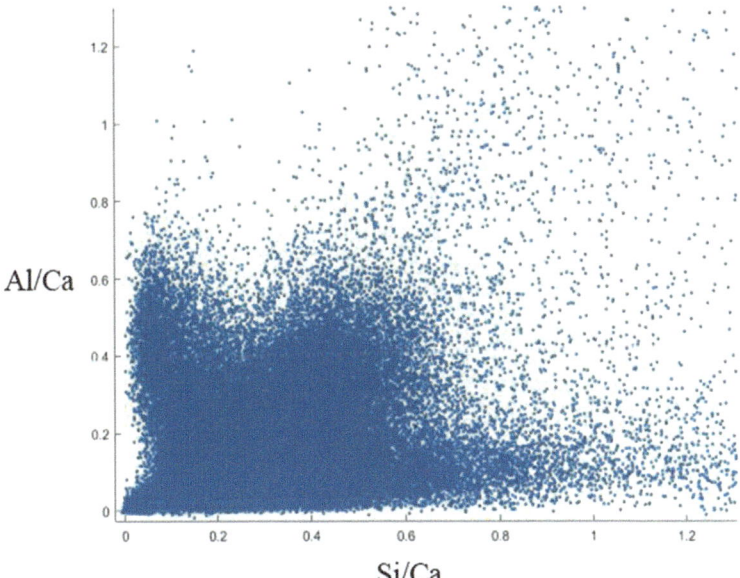

Fig. 3 Si/Ca vs. Al/Ca element distribution in identified phases

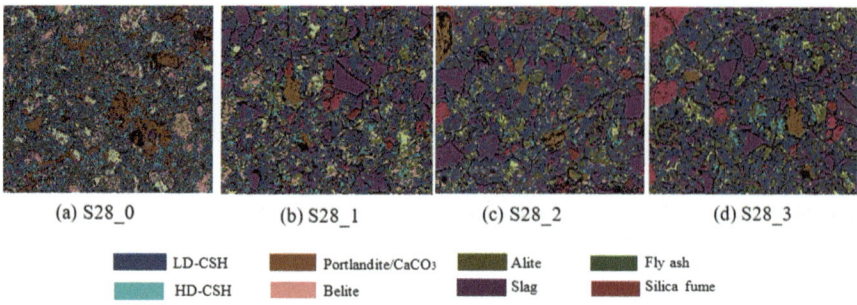

(a) S28_0   (b) S28_1   (c) S28_2   (d) S28_3

▬ LD-CSH	▬ Portlandite/CaCO₃	▬ Alite	▬ Fly ash
▬ HD-CSH	▬ Belite	▬ Slag	▬ Silica fume

Fig. 4 Segmented phase mapping from random locations representing 28th day hydration

3.2 Hydration Simulations Through Virtual Cement and Concrete Testing Laboratory

VCCTL is a cement hydration simulation software which can model the hydration of raw cement constituents with real particle shapes in a 3D space. Further, this software facilitates the hydration simulations with supplementary cementitious materials which paves the way to modeling zero carbon cement mixes revealing theoretical mechanical and transportation properties, upon feeding particle size distributions and chemical composition of raw materials along with dispersion effects due to superplasticizers. Different cement types and other raw cementitious materials in the VCCTL database have been fed by statistical outcomes from a significant amount of SEM/EDS images of 800 magnification over a region of 200 μm × 200 μm [20]. VCCTL has demonstrated precise simulation capabilities for different cement hydrations such as ultra-high-performance concretes [21, 22], blended cements [23], and other normal strength concretes [24].

Simulating the hydration of 0.22 w/b ratio cement paste was done in VCCTL by defining the physical and chemical properties of constituents in the mixture. The cement mixture microstructure and hydrated microstructure are presented in Fig. 5.

It is reported that when preparing databases for different cementitious materials with a significant number of SEM/EDS images for each material, a standard deviation of 3% to 7.5% has been observed [20].

3.3 Microphase Volume Fractions

The microphases from the chemical image analysis precisely reveal the hydration information at microlevel. The unreacted clinker amount is comparatively higher in S28_0 location and the Portlandite/CaCO₃ mixed phase is also comparatively higher. This might be due to the lack of free water for hydration in this area. Further, slag fraction remains low in S28_0 location and this represents the non-homogenous mixing of particles in the cement matrix.

Fig. 5 Cement mixture microstructure (**a**) and hydrated microstructure (**b**) from VCCTL simulation

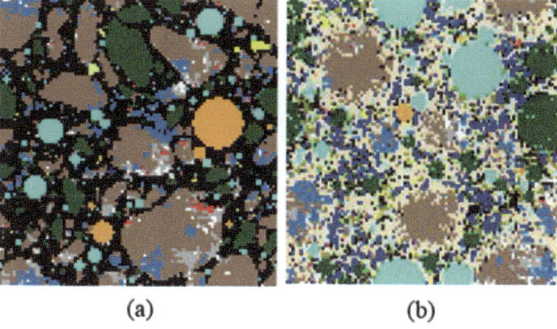

(a)   (b)

Fig. 6 The correlation between experimental vs. VCCTL microphase quantification

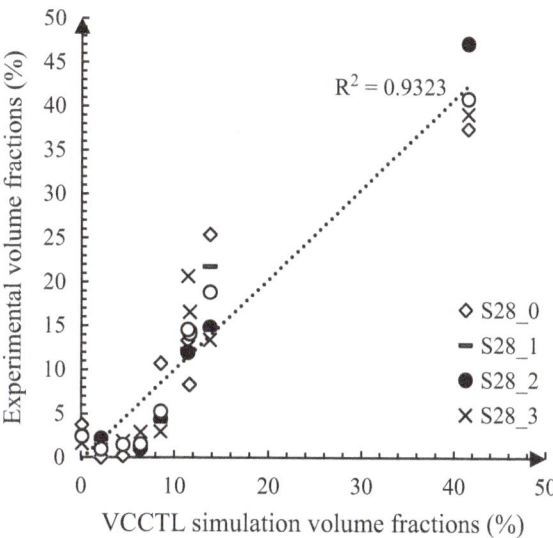

The back scattered electron imaging with higher resolution (1024 × 884) was used to identify the porosity together with cracks, and using an algorithm, the cracks were separated from porosity. The s.d. values from the experiment are within 6% and the linear regression analysis in Fig. 6 shows the mean values from the experimental microphase analysis lies closely near the zero-error line. This result could be more robustly verified with a significant number of sample analysis. Further, larger unreacted particles hold higher s.d. values comparatively to that of smaller particles. This represents the phase heterogeneity sensitiveness within the considered length scale. Reacted phases with high volume fractions also do have a high s.d. compared to less abundant reacted phases (Table 3).

This study shows that VCCTL program can also be used to conduct parametric studies once a control specimen is validated. VCCTL is capable of predicting mechanical and transportation properties from the obtained microphase information. Although most of the studies do confirm that the simulation is correct for generating

Table 3 Volume fractions (%) of hydrated and unreacted cement products

Microphase	S28_0	S28_1	S28_2	S28_3	Mean	Standard deviation (s.d.)	VCCTL simulation
Unreacted cement	25.362	21.757	14.838	13.398	18.839	5.677	13.734
Slag	8.311	17.185	14.359	16.563	14.105	4.048	11.608
Silica fume	0.185	2.282	1.499	1.861	1.457	0.906	4.546
Fly ash	0.020	0.766	2.178	0.869	0.958	0.897	2.172
Calcium hydrate silicate	37.475	39.687	47.169	39.141	40.868	4.305	41.469
Portlandite/CaCO$_3$	10.726	2.951	4.398	2.981	5.264	3.703	8.535
Ettringite	0.830	1.309	1.115	2.907	1.540	0.932	6.413
Porosity	3.748	1.752	2.468	1.618	2.397	0.975	0.100
Others	13.343	12.311	11.976	20.662	14.573	4.101	11.423

phase quantifications but when it comes to predicting mechanical properties, the software underestimates the material properties to a certain extent [24]. If nanoindentation results could be combined with the microphase volume fractions, that would be an alternate method to homogenize the mechanical properties at microscale.

4 Conclusions

The microphase identification of hydrated cement pastes is an essential area of study to overcome many civil engineering challenges. This chapter investigated the feasibility of using EDS chemical hypermaps to develop a MATLAB-based image processing model to identify the existence of microphases in an ultra-high-performance concrete blended with low carbon supplementary cementitious materials in terms of their volume fractions and phase distribution on the 28th day of hydration. A novel algorithm is proposed to develop RGB microphase segmented maps using any desired number of elemental ratio matrices with respect to the most abundant element, Ca. These elemental ratio matrices were developed after denoising with a pretrained deep learning algorithm built in MATLAB. The following are the conclusions from the study.

- The MATLAB inbuilt deep learning denoising algorithm was effective in reducing the noise of chemical maps. Therefore, pixel by pixel Si/Ca, Al/Ca, Mg/Ca, S/Ca, Fe/Ca ratio extraction to segment microphases was accurately done.
- The chemical ratios were stored in numerical matrices (double type) similar to the size of BSE image. Therefore, it facilitated the extraction of any number of required chemical ratios for a given microphase at a considered pixel location with no restriction.

- The phase segmentation volume fractions were in good agreement with the theoretical hydration simulation program—VCCTL outcomes ($R^2 = 0.93$).
- There is commercially available software for phase segmentations with expensive licensing but do require expertise to post process after data acquisition for correct phase segmentation. But this simplified method can be customized/modified for a novel cement paste to get its microphases identified with no post processing time after the feeding quant EDS elemental maps and BSE image.
- It is recommended to improve the robustness of the statistical analysis for the volume fractions by analyzing a larger number of locations from the same specimen. However, the analysis time and cost could impose constraints to this experiment, and they could be detoured if powerful silicon drift detectors can be used EDS instruments to improve the sensitivity, speed, and resolution of the scanning.
- The interfering of heterogeneity within the considered length scale with the volume fractions could be recognized as one of the limitations in this method. The volume fraction of larger unreacted phases was observed to extend the standard deviation up to 6%.

References

1. Hannawi, K., Bian, H., Prince-Agbodjan, W., Raghavan, B.: Effect of different types of fibers on the microstructure and the mechanical behavior of Ultra-High Performance Fiber-Reinforced Concretes. Compos. Part B Eng. **86**, 214–220 (2016). https://doi.org/10.1016/J. COMPOSITESB.2015.09.059
2. Bahmani, H., Mostofinejad, D.: Microstructure of ultra-high-performance concrete (UHPC)—a review study. J. Build. Eng. **50**, 104118 (2022). https://doi.org/10.1016/J.JOBE.2022.104118
3. Lu, Z., Yao, J., Leung, C.: Using graphene oxide to strengthen the bond between PE fiber and matrix to improve the strain hardening behavior of SHCC. Cem. Concr. Res. **126**, 105899 (2019). https://doi.org/10.1016/J.CEMCONRES.2019.105899
4. Wu, Y., Zhang, J., Liu, C., Zheng, Z., Lambert, P.: Effect of graphene oxide nanosheets on physical properties of ultra-high-performance concrete with high volume supplementary cementitious materials. Materials. **13**(8), 1929 (2020). https://doi.org/10.3390/MA13081929
5. Meng, W., Khayat, K.: Effect of graphite nanoplatelets and carbon nanofibers on rheology, hydration, shrinkage, mechanical properties, and microstructure of UHPC. Cem. Concr. Res. **105**, 64–71 (2018). https://doi.org/10.1016/J.CEMCONRES.2018.01.001
6. Wang, H., Gao, X., Liu, J., Ren, M., Lu, A.: Multi-functional properties of carbon nanofiber reinforced reactive powder concrete. Constr. Build. Mater. **187**, 699–707 (2018). https://doi.org/10.1016/J.CONBUILDMAT.2018.07.229
7. Thomas, M.: The effect of supplementary cementing materials on alkali-silica reaction: a review. Cem. Concr. Res. **41**(12), 1224–1231 (2011). https://doi.org/10.1016/J. CEMCONRES.2010.11.003
8. Yang, R.: Low carbon design of an Ultra-High Performance Concrete (UHPC) incorporating phosphorous slag. (2019). https://doi.org/10.1016/j.jclepro.2019.118157
9. Jalal, M.: Influence of class F fly ash and silica nano-micro powder on water permeability and thermal properties of high performance cementitious composites. Sci. Eng. Compos. Mater. **20**(1), 41–46 (2013). https://doi.org/10.1515/SECM-2012-0054/ MACHINEREADABLECITATION/RIS

10. Qaidi, S.: Ultra-high-performance geopolymer concrete: a review. Constr. Build. Mater. **346**, 128495 (2022). https://doi.org/10.1016/J.CONBUILDMAT.2022.128495
11. Wang, J.: A novel design of low carbon footprint Ultra-High Performance Concrete (UHPC) based on full scale recycling of gold tailings. Constr. Build. Mater. **304**, 124664 (2021). https://doi.org/10.1016/J.CONBUILDMAT.2021.124664
12. Kang, S., Hong, S., Moon, J.: The use of rice husk ash as reactive filler in ultra-high performance concrete. Cem. Concr. Res. **115**, 389–400 (2019). https://doi.org/10.1016/J.CEMCONRES.2018.09.004
13. Goldstein, J.: Scanning Electron Microscopy and X-Ray Microanalysis
14. Khanal, S.: Characterization of small-scale surface topography using transmission electron microscopy. Surf. Topogr. **6**(4) (2018). https://doi.org/10.1088/2051-672X/aae5b3
15. Zhang, Z., Scherer, G., Bauer, A.: Morphology of cementitious material during early hydration. Cem. Concr. Res. **107**, 85–100 (2018). https://doi.org/10.1016/J.CEMCONRES.2018.02.004
16. Attari, A., McNally, C., Richardson, M.: A combined SEM–calorimetric approach for assessing hydration and porosity development in GGBS concrete. Cem. Concr. Compos. **68**, 46–56 (2016). https://doi.org/10.1016/J.CEMCONCOMP.2016.02.001
17. Hosan, A., Shaikh, F., Sarker, P., Aslani, F.: Nano- and micro-scale characterisation of interfacial transition zone (ITZ) of high volume slag and slag-fly ash blended concretes containing nano SiO_2 and nano $CaCO_3$. Constr. Build. Mater. **269** (2021). https://doi.org/10.1016/j.conbuildmat.2020.121311
18. de Matos, P., Andrade Neto, J., Sakata, R., Kirchheim, A., Rodríguez, E., Campos, C.: Strategies for XRD quantitative phase analysis of ordinary and blended Portland cements. Cem. Concr. Compos. **131** (2022). https://doi.org/10.1016/j.cemconcomp.2022.104571
19. Thilakarathna, P.: Multiscale Modelling and Homogenization of Ultra-High Strength Concrete (UHSC) from Nano to Macro Scales (2021)
20. Final Report FDOT Project Number: BDK75-977-73 Development of Design Parameters for Virtual Cement and Concrete Testing (2013)
21. de Béjar, L.: Virtual estimation of the Griffith's modulus and cohesive strength of ultra-high performance concrete. Eng. Fract. Mech. **216**, 106488 (2019). https://doi.org/10.1016/J.ENGFRACMECH.2019.106488
22. de Béjar, L., Rushing, T.: Relative compression strength evolution of silica-fume ultrahigh-performance concrete under saturated adiabatic hydration using virtual tests. J. Mater. Civ. Eng. **30**(1), 04017260 (2017). https://doi.org/10.1061/(ASCE)MT.1943-5533.0002117
23. Watts, B., Tao, C., Ferraro, C., Masters, F.: Proficiency analysis of VCCTL results for heat of hydration and mortar cube strength. Constr. Build. Mater. **161**, 606–617 (2018). https://doi.org/10.1016/J.CONBUILDMAT.2017.09.035
24. Watts, B., Ferraro, C.: Prediction of setting for admixture modified mortars using the VCCTL. Cem. Concr. Compos. **78**, 63–72 (2017). https://doi.org/10.1016/j.cemconcomp.2016.11.002

Effect of Discarded Rubber Tire Crumbs and Tile Ceramic Waste Powder on Workability and Strength Properties of Geopolymer Concrete

Iman Faridmehr [ID], Ghasan Fahim Huseien [ID],
Mohammad Hajmohammadian Baghban [ID], and Håvard Lund [ID]

Contents

1 Introduction

Due to the rapid advancements in transportation, the disposal and recycling of waste rubber tires have become a significant environmental concern. As tires become unusable over time, they are discarded as waste, resulting in millions of

I. Faridmehr
Civil Engineering Department, Faculty of Engineering, Girne American University, Istanbul, Turkey

G. F. Huseien
Department of Building, School of Design and Environment, National University of Singapore, Singapore, Singapore

M. H. Baghban (✉)
Department of Manufacturing and Civil Engineering, Norwegian University of Science and Technology (NTNU), Gjøvik, Norway
e-mail: mohammad.baghban@ntnu.no

H. Lund
University College of Vocational Education (HØFY), Gjøvik, Norway

© The Author(s) 2025
M. Kioumarsi, B. Shafei (eds.), *The 1st International Conference on Net-Zero Built Environment*, Lecture Notes in Civil Engineering 237,
https://doi.org/10.1007/978-3-031-69626-8_30

non-recyclable tires being abandoned annually [1]. Improper dumping of these tires in landfills has been shown to cause severe environmental pollution and economic damage, as well as fire-related accidents and the release of pollutant gases. Moreover, the dumped tires create breeding grounds for disease-carrying insects and animals, leading to the spread of infectious diseases [2]. In response to these challenges, considerable efforts have been made to advance waste disposal methods and implement effective waste management strategies [3]. Dumping tires in landfills is now undesirable due to the depletion of available space, leading many countries to enforce strict regulations against such practices. Various strategies have been developed to mitigate tire dumping, including reusing waste tire crumbs in asphaltic concrete, incorporating them into concrete mixtures, incinerating them for steam generation, and utilizing them in plastic and rubberized products. Used tires are repurposed as a fuel source in cement production and for constructing synthetic reefs in ocean habitats [4, 5]. Recycling waste tires in the concrete industry not only helps preserve landfill space but also offers additional benefits such as reducing the demand for natural resources, transportation and production energy costs, carbon dioxide emissions, and increasing the lifespan of concrete structures [6]. Several studies [7–9] have shown that integrating recycled tire materials into concrete mixtures creates more eco-friendly concrete, which has a reduced carbon footprint, is more cost-effective, and consumes less energy than conventional concrete formulations.

The alkali-activation technique (using one or two-part alkaline activator solution) has been devised to create high-performance geopolymer concretes, resulting in lower greenhouse gas emissions than cement-based traditional concrete production [10–14]. Several studies have indicated that calcium in fly ash significantly affects the geopolymer materials' final hardened properties and compressive strength [15, 16]. Calcium oxide reacts with sodium-aluminum-silicate (N-A-S-H) gels to form calcium-silicate-hydrate (C-S-H) gels [17]. These two gels coexist and greatly influence the hybrid compounds of cement and alkali-activated aluminum-silica [18, 19]. Earlier studies have employed synthetic gels to examine the effects of elevated pH levels on distinct gel constituents. Notably, the generation of C-S-H gel was significantly affected by the presence of aluminate in the solution [20]. Furthermore, aqueous calcium has been observed to alter N-A-S-H gels, exchanging sodium ions with calcium ones, thereby leading to the genesis of (N,C)-A-S-H gels [20, 21]. However, the precise mechanisms underlying the synthesis of these gels and further advancements remain fully understood.

The ceramic industry generates significant amounts of waste annually, resulting in a large portion in landfills. However, reusing these wastes in concrete presents a win-win situation. It addresses the ceramic industry's waste problem while promoting more sustainable concrete [22]. Global ceramic tile production exceeds 10 million square meters per year [23]. According to a survey, approximately 15–30% of this production is wasted in the ceramic industry. Currently, this waste is not being reused, but it is accumulating over time. The percentage of waste increases due to ceramic damage during storage, transportation, construction, and house renovations. Additionally, the cost of deposition processes is rising due to restrictions on landfill usage. Consequently, industries need to explore alternative solutions, such as recycling these waste materials into valuable products.

This experiment aims to develop several types of high-performance geopolymer concrete incorporating tile ceramic waste powder (TCWPs) and discarded rubber tire crumbs (DRTCs). The effect of both material's inclusion of geopolymer matrix on workability and strength performance is evaluated. To achieve this aim, various assessments were conducted, including slump tests for fresh concrete and evaluations of compressive, splitting tensile, and flexural strengths. The influence of curing duration on strength development was also examined, with specimen testing occurring at intervals of 3, 7, and 28 days.

2 Methodology

2.1 Materials

To create geopolymer concrete samples, a total of nine materials were utilized, including GBFS, TCWPs, river sand, crushed stone, and fine and coarse DRTC materials, alongside sodium hydroxide, sodium silicate, and water. GBFS and TCWPs were combined to form a dual-component binder, adjusting the proportions of calcium oxide relative to silica and aluminum oxides. Meanwhile, river sand and crushed stone served as the primary fillers (acting as fine and coarse aggregates) and were subsequently substituted with fine and coarse DRTCs. An alkaline activator solution, essential for initiating the geopolymer reaction, was concocted from sodium hydroxide, sodium silicate, and water. To determine the chemical makeup of GBFS and TCWPs, X-ray fluorescence (XRF) analysis was employed, see Table 1. Regarding physical characteristics, TCWPs were characterized by a specific gravity of 2.61, a surface area of 12.2 m^2/g, and a median particle size of 35 μm, while GBFS had a specific gravity of 2.9, a surface area of 13.6 m^2/g, and a median particle size of 12.8 μm.

Natural aggregates with specific properties were selected for this experiment to create geopolymer specimens. The fine aggregates comprised washed river sand, which exhibited a water absorption rate of 2.2%, a specific gravity of 2.65, and a maximal particle diameter of 4.75 mm. Conversely, the coarse aggregates utilized crushed stone, characterized by a water absorption rate of 1.2%, a specific gravity of 2.67, and a maximal particle size of 10 mm. Regarding the DRTC materials, the rubber aggregates were categorized according to the ASTM C136 standard. The details regarding the water absorption rates, specific gravities, and the maximum sizes for both fine and coarse DRTC particles are outlined in Table 2.

Table 1 GBFS and TCWPs chemical composition using XRF spectrometry data (mass%)

Materials	CaO	SiO2	Al2O3	MgO	Na2O	Others	LOI
GBFS	51.8	30.8	10.9	4.6	0.5	1.2	0.2
TCWPs	0.02	72.6	12.2	1.0	13.5	0.6	0.1

Table 2 Water absorption, specific gravity, and maximum size of utilized aggregates

Type of aggregates	Water absorption, %	Specific gravity	Maximum size, mm
Natural river sand	2.2	2.65	4.75
Natural crushed stone	1.2	2.67	10
Fine DRTCs (1–4 mm)	–	1.34	4
Coarse DRTCs (5–8 mm)	–	1.37	8

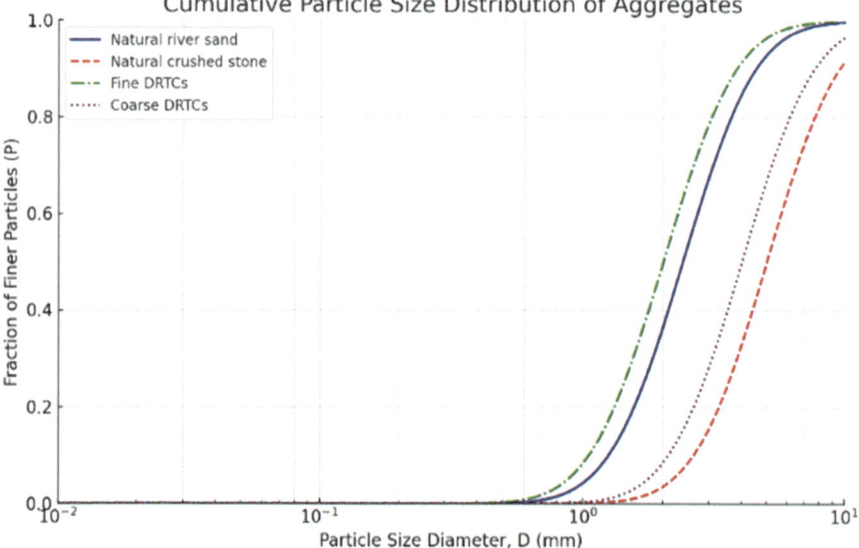

Fig. 1 Cumulative particle size distribution curves of aggregates used in this study

Figure 1 displays the cumulative particle size distribution curves corresponding to the aggregates detailed in Table 2.

2.2 Mix Design of Geopolymer Concrete

Eight distinct concrete mixes were devised to examine the impact of incorporating TCWPs and DRTCs into the geopolymer concrete matrix as substitutes for GBFS and natural aggregates (see Table 3). Initially, five mixes were created to assess the effect of replacing GBFS with TCWPs on the workability and strength characteristics of the envisioned geopolymer concrete. A control sample was established with a geopolymer binder blend consisting of 70% GBFS and 30% TCWPs. Subsequently, geopolymer mixes were prepared with TCWPs replacing GBFS at increments of 40, 50, 60, and 70%. The quantities of river sand and crushed stone were maintained constant across all mixes. Based on the results from this initial phase, the mix

Table 3 Mixes design of geopolymer concrete with various levels of TCWPs and DRTCs

Mix code	Binder, kg/m^3		Natural aggregates, kg/m^3		DRTCs, kg/m^3	
	GBFS	TCWPs	River sand	Crushed stone	Fine	Coarse
GPC$_1$	293.7	125.7	721	995	0	0
GPC$_2$	251.4	167.6	721	995	0	0
GPC$_3$	209.5	209.5	721	995	0	0
GPC$_4$	167.6	251.4	721	995	0	0
GPC$_5$	125.7	293.7	721	995	0	0
GPC$_6$	209.5	209.5	684.9	945.3	18.2	25.4
GPC$_7$	209.5	209.5	648.9	895.5	36.5	50.9
GPC$_8$	209.5	209.5	612.8	845.8	54.7	76.3

demonstrating a compressive strength exceeding 35 MPa at three days was chosen for further experimentation. In other words, the selection criteria for further experimentation were established based on early strength development. Specifically, among the initial mixes, the one that achieved a compressive strength exceeding 35 MPa at the early age of three days was chosen for subsequent detailed study. In the second phase, DRTCs were used to substitute 10, 20, and 30% of the volume of natural aggregates. During the formulation of these geopolymer concrete mixes, the molarity of sodium hydroxide, sodium hydroxide to sodium silicate ratio, and the proportions of alkaline activator solutions were kept constant at 4 M, 0.75, and 0.45, respectively.

2.3 Test Procedure

Following ASTM C579 guidelines, cube molds for the geopolymer concrete, each side measuring 100 mm, were prepared to evaluate compressive strength. Cylindrical molds, 100 mm in diameter and 200 mm in height were utilized for the splitting tensile strength test. Concrete beams of dimensions 100 mm by 100 mm by 500 mm were used to assess flexural strength. Engine oil was applied to the molds prior to casting to ensure the molds could be easily stripped. A dual-component alkaline activator solution, measured by weight, was concocted by mixing sodium hydroxide (4.0 M) and sodium silicate thoroughly at least 24 hours before casting to allow it to cool down to room temperature, reducing the exothermic heat produced during mixing.

The mixing process commenced with blending the fine and coarse aggregates, including river sand, crushed stone, and rubber, in their dry states for 3 minutes to achieve a homogeneous mixture. This was followed by adding the GBFS and TCWPs binary mixture, which was mixed for an additional 5 minutes in a concrete mixer. The alkaline activator solution was then added to kick-start the geopolymerization process, with the mixing continuing for another 5 minutes. The freshly mixed geopolymer concrete was then transferred into the pre-lubricated molds in two layers, compacting each layer with a vibrating table for 30 seconds to remove air pockets.

Post-casting, the geopolymer concrete specimens were left to cure at room temperature (around 27 ± 1.5 °C) and 75% relative humidity for 24 hours. After this initial cure, the specimens were de-molded and kept in the same ambient conditions until their testing dates at 3, 7, and 28 days. Tests for the slump, compressive strength, splitting tensile strength, and flexural strength were performed by ASTM C143, ASTM C109, ASTM C496, and ASTM C78 standards, respectively, with the average values from three specimens being used to determine the geopolymer concrete' s fresh and hardened characteristics.

3   Results and Discussion

3.1   Fresh Properties and Compressive Strength Development

Figure 2a demonstrates the notable effect of replacing GBFS with TCWPs on the workability of the formulated concrete. The results show a progressive increase in slump value with a higher percentage of TCWPs substituting for GBFS in the geopolymer mixture. As the TCWP content escalated from 30% to 40, 50, 60, and 70%, the slump values correspondingly ascended from 85 mm to 92, 110, 122, and 136 mm. The increase in the content of TCWPs and the reduction of GBFS in concrete workability have been attributed to their effects on chemical reaction rate, plasticity, and mixture's workability, as reported in previous studies [24, 25]; this affects the chemical reaction rate and increases the plasticity of the mixture, which improves the concretes' workability.

Figure 2b showcases the influence of substituting natural fine and coarse aggregates with DRTCs on the workability of the engineered concrete. Replacing natural aggregates with 10, 20, and 30% DRTCs led to a reduction in slump values, descending from 110 mm to 102, 90, and 78 mm, respectively. This drop in slump

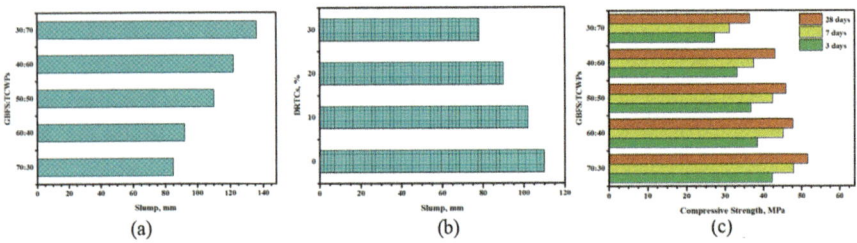

Fig. 2 Workability performance of geopolymer concrete containing TCWPs as GBFS replacement (**a**), Effect of DRTCs content on the workability of proposed geopolymer concrete (**b**), and Effect TCWPs as GBFS replacement on compressive strength development of proposed geopolymer concrete at 3, 7, and 28 days (**c**)

values is linked to the greater water absorption capacity and smaller particle sizes of DRTCs compared to river sand, which negatively impacts the workability of the proposed concrete mixture, thereby diminishing the slump values. A previous study [26] found that the varied morphologies of DRTCs aggregates increased friction among the components of concrete mixes, leading to lower workability when high replacement levels were used. Likewise, concrete mixtures with increased proportions of rubber crumbs exhibited reduced workability, attributable to their more significant specific surface areas and enhanced roughness. Another study [27] observed reduced workability performance with increased DRTCs content as a substitute for river sand.

Before adding DRTCs as aggregate substitutes, the impact of replacing GBFS with TCWPs on geopolymer concrete strength was studied, see Fig. 2c. Results showed that as TCWPs content increased, compressive strength decreased, yet it generally rose with longer curing times. Initially, strength dropped significantly with higher TCWPs levels but all mixtures eventually exceeded 35 MPa in compressive strength after 28 days. The decline in strength is linked to a lower calcium oxide content from TCWPs, altering the chemical balance and slowing down the formation of strength-contributing calcium silicate hydrate (C-S-H) gels [28].

Figure 3 illustrates the effect of substituting natural fine and coarse aggregates with DRTCs on the compressive strength of the designed concrete over curing periods of 3, 7, and 28 days. A consistent pattern of rising compressive strength was noted across all geopolymer concrete samples as the curing time extended from

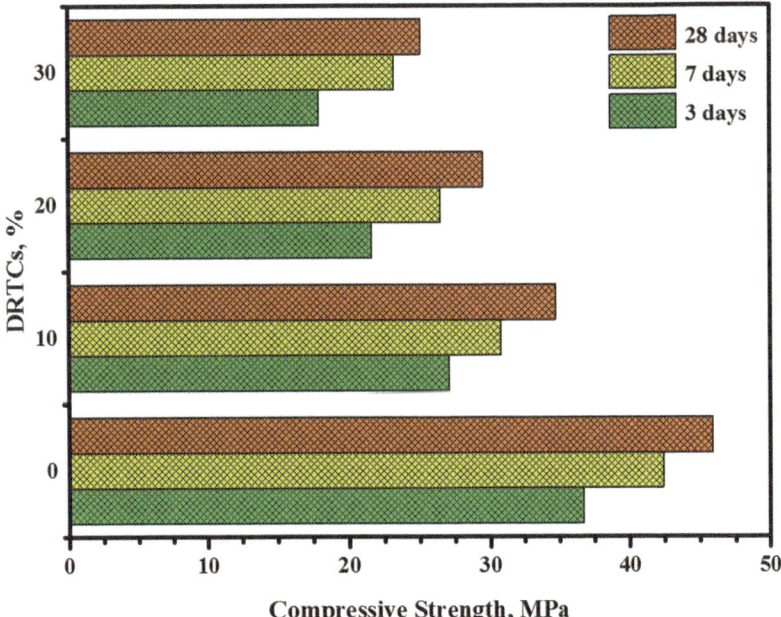

Fig. 3 Influenced compressive strength by DRTCs as fine and coarse aggregate replacement

3 to 28 days. Nonetheless, integrating DRTCs negatively impacted the compressive strength, showing a decrease in values as the DRTC content was augmented from 0% to 10, 20, and 30%.

Specifically, at a curing age of 3 days, replacing natural aggregates with 10, 20, and 30% DRTCs decreased compressive strength from 36.7 MPa to 27.1, 21.6, and 17.9 MPa, respectively. This trend continued at seven days of age, with strength values reducing from 42.4 MPa to 30.8, 26.5, and 23.2 MPa as the DRTCs content increased to 10, 20, and 30%, respectively. By 28 days of curing, the specimens exhibited a lower percentage of strength loss compared to those at 3 and 7 days, with strength values declining from 45.9 MPa to 34.7, 29.5, and 25.1 MPa upon substituting natural aggregates with 10, 20, and 30% DRTCs, respectively.

The decrease in compressive strength can be linked to the smoother texture of rubber aggregates compared to the rougher surface of natural aggregates. This smoothness diminishes the bonding efficacy at the junctions between the cement pastes and rubber particles, decreasing concrete strength as the quantity of rubber in the mix escalates. Other studies [7, 29] have supported these findings, indicating that WRTCs have a negative impact on concrete strength. Similar trends have been reported in several studies [30, 31], demonstrating the negative effect of DRTCs content on strength development.

3.2 Splitting Tensile Strength

Figure 4a highlights the effect of substituting different proportions of TCWPs for GBFS on the splitting tensile strength of geopolymer concrete after a curing period of 28 days. The results demonstrate that the inclusion of TCWPs results in a

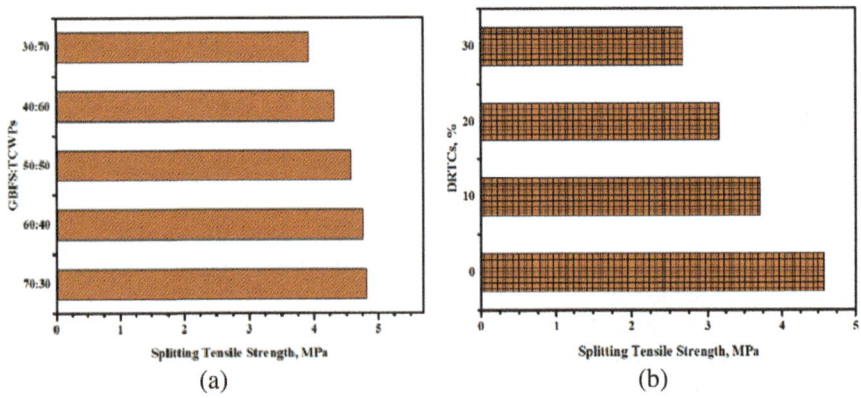

(a)   (b)

Fig. 4 Effect of TCWPs on geopolymer splitting tensile strength after 28 days of curing age (**a**), and Effect of DRTCs on splitting tensile strength of proposed geopolymer concrete after 28 days of curing age (**b**)

diminished splitting tensile strength, with a more significant decline noted with higher replacement levels.

Specifically, substituting GBFS with 40, 50, 60, and 70% of TCWPs decreased tensile strength from an initial value of 4.82 MPa to 4.76, 4.58, 4.31, and 3.92 MPa, respectively. This trend can be attributed to the fact that the strength of geopolymer binders is enhanced by the presence of calcium-based materials, such as calcium oxide, which contributes to the formation of additional C-S-H (Calcium Silicate Hydrate) and C-A-S-H (Calcium Alumino Silicate Hydrate) gels alongside the N-A-S-H (Sodium Alumino Silicate Hydrate) gel. The reduction in tensile strength values with an increase in the level of TCWPs as a replacement for GBFS in the proposed geopolymer matrix is thus explained by the diminished presence of these calcium-enriched materials, which are crucial for strength development [32].

Figure 4b showcases the effects of varying levels of DRTCs as a replacement for natural fine and coarse aggregates on the splitting tensile strength of geopolymer concrete. The results significantly impact tensile strength as the proportion of DRTCs increases. Specifically, substituting natural aggregates with 10, 20, and 30% of DRTCs led to a notable decrease in tensile strength values, dropping from an initial value of 4.58 MPa to 3.71, 3.16, and 2.68 MPa, respectively.

The observed trend underscores a reduction in splitting tensile strength as the incorporation of DRTCs increases. This reduction in tensile strength is linked to inadequate adhesion at the interfaces between the rubber granules and the cement pastes, leading to early failure of the specimens even after a brief curing duration. This phenomenon highlights the difficulties in using DRTCs as replacements for aggregates, particularly concerning preserving the structural integrity and performance of geopolymer concrete [33]. The type and quantity of rubber particles also contribute to the failure, as significant distortion creates a lack of rigidity in the rubberized granules [26, 34].

3.3 Flexural Strength

The impact of TCWPs content as a replacement for GBFS on the flexural strength of geopolymer concrete was evaluated, with the findings presented in Fig. 5a. These findings align with the trends noted in both compressive and splitting tensile strengths at a 28-day curing period, wherein an elevation in the content of TCWPs within the geopolymer matrix corresponds to a modest decline in flexural strength.

The integration of 40, 50, 60, and 70% TCWPs led to a sequential reduction in flexural strength, dropping from an initial value of 1.73 MPa to 1.61, 1.54, 1.42, and 1.29 MPa, respectively. This pattern suggests that the efficiency of the geopolymerization process, heavily dependent on the calcium oxide content, decreases with the rising proportion of TCWPs, culminating in lower strength gains, especially at the early stages of curing. Additionally, the decline in flexural strength is partly due to an increase in the amount of unreacted or partially reacted silicates and a diminished presence of the Calcium Silicate Hydrate (C-S-H) phase, which plays a pivotal role in the strength development of geopolymer concrete [35].

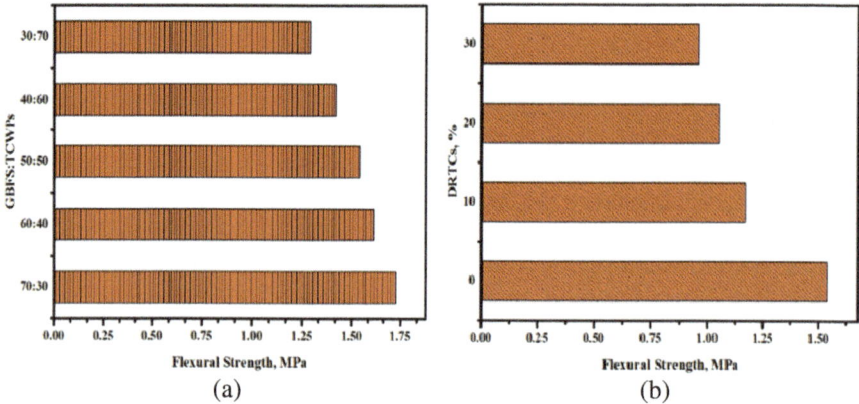

Fig. 5 Flexural strength of geopolymer concrete containing various levels of TCWPs as GBFS replacement (**a**), and flexure strength of geopolymer concrete prepared with various levels of DRTCs as a fine and coarse aggregate replacement (**b**)

Figure 5b illustrates the flexural strength outcomes for geopolymer concrete samples that incorporate DRTCs as replacements for natural fine and coarse aggregates. An inverse correlation was noted between the flexural strength of the geopolymer concrete and the amount of DRTCs used. Specifically, as the replacement level of natural aggregates with DRTCs increased to 10, 20, and 30%, there was a corresponding decrease in flexural strength values, from an initial 1.54 MPa down to 1.17, 1.05, and 0.96 MPa, respectively.

The observed reduction in flexural strength is mainly due to the poor bonding between the rubber particles and the matrix of TCWPs-GBFS pastes in the concrete. Including rubber results in creating voids within the matrix, consequently lowering the concrete's overall density. This effect significantly contributes to the decrease in flexural strength, underscoring the challenges of optimizing performance in geopolymer concrete when replacing traditional aggregates with DRTCs [36]. Additionally, the smooth surface of rubber granules hinders bonding with the geopolymer paste [37, 38].

4 Conclusions

This study explored the development of geopolymer concrete with suitable mechanical properties for various construction applications, utilizing tile ceramic waste powder and discarded rubber tire crumbs as replacements for slag and natural aggregates. The findings of this research are summarized as follows:

(i) *Workability*: The addition of tile ceramic waste powder positively affected the workability of the proposed geopolymer concrete. As the level of slag replacement by tile ceramic waste powder increased, so did the workability, reaching up to 136 mm compared to 85 mm in the control mix. Conversely, incorporating discarded rubber tire crumbs into the geopolymer matrix decreased workability, with a 30% inclusion of rubber aggregates resulting in a loss of strength greater than 25%. This reduction in workability is attributed to the rubber crumbs' excessive fineness and absorption capacity.

(ii) *Compressive Strength*: The inclusion of tile ceramic waste powder and discarded rubber tire crumbs negatively impacted the development of compressive strength in geopolymer concrete at both early and late ages. The trend showed that increasing the content of these materials in the geopolymer matrix led to a decrease in strength.

(iii) *Compressive Strength of Tile Ceramic Powder Mixes*: Specimens prepared with tile ceramic powder as a replacement for slag exhibited acceptable compressive strength after 28 days of curing, with values exceeding 35 MPa. However, increasing the replacement content from 30% to 70% decreased strength from 51.6 MPa to 45.9 MPa, respectively.

(iv) *Impact of Rubber Aggregates*: Replacing natural fine and coarse aggregates with 10, 20, and 30% rubber aggregates resulted in a more than 40% strength loss compared to control specimens. This significant decrease in strength is primarily due to the formation of widened, weaker, and highly porous bond zones, which led to poor adhesion between the rubber particles and the binary binder pastes.

(v) *Tensile and Flexural Strengths*: Similar to the compressive strength results, the inclusion of tile ceramic waste powder and discarded rubber tire crumbs in the geopolymer matrix decreased the splitting tensile and flexural strengths of the tested specimens.

References

1. Eisa, A.S., Elshazli, M.T., Nawar, M.T.: Experimental investigation on the effect of using crumb rubber and steel fibers on the structural behavior of reinforced concrete beams. Constr. Build. Mater. **252**, 119078 (2020)
2. Kordoghli, S., et al.: Managing the environmental hazards of waste tires. J. Eng. Stud. Res. **20**(4), 1–11 (2014)
3. Mhaya, A.M., et al.: Evaluating mechanical properties and impact resistance of modified concrete containing ground Blast Furnace slag and discarded rubber tire crumbs. Constr. Build. Mater. **295**, 123603 (2021)
4. Junqing, X., et al.: High-value utilization of waste tires: a review with focus on modified carbon black from pyrolysis. Sci. Total Environ., 140235 (2020)
5. Youssf, O., ElGawady, M.A., Mills, J.E., Ma, X.: An experimental investigation of crumb rubber concrete confined by fibre reinforced polymer tubes. Constr. Build. Mater. **53**, 522–532 (2014)
6. Jawjit, W., Kroeze, C., Rattanapan, S.: Greenhouse gas emissions from rubber industry in Thailand. J. Clean. Prod. **18**(5), 403–411 (2010)
7. Mhaya, A.M., Huseien, G.F., Abidin, A.R.Z., Ismail, M.: Long-term mechanical and durable properties of waste tires rubber crumbs replaced GBFS modified concretes. Constr. Build. Mater. **256**, 119505 (2020)

8. Dayaratne, S.P., Gunawardana, K.D.: Carbon footprint reduction: a critical study of rubber production in small and medium scale enterprises in Sri Lanka. J. Clean. Prod. **103**, 87–103 (2015)

9. Rahman, R., et al.: The comparison of properties and cost of material use of natural rubber and sand in manufacturing cement mortar for construction sub-base layer. In: IOP Conference Series: Materials Science and Engineering. IOP Publishing (2017)

10. Li, N., et al.: A mixture proportioning method for the development of performance-based alkali-activated slag-based concrete. Cem. Concr. Compos. **93**, 163–174 (2018)

11. Huseien, G.F., et al.: Geopolymer mortars as sustainable repair material: a comprehensive review. Renew. Sust. Energ. Rev. **80**, 54–74 (2017)

12. Kubba, Z., et al.: Impact of curing temperatures and alkaline activators on compressive strength and porosity of ternary blended geopolymer mortars. Case Stud. Constr. Mater. **9**, e00205 (2018)

13. Zhang, Z., Provis, J.L., Reid, A., Wang, H.: Geopolymer foam concrete: an emerging material for sustainable construction. Constr. Build. Mater. **56**, 113–127 (2014)

14. Provis, J.L.: Geopolymers and other alkali activated materials: why, how, and what? Mater. Struct. **47**(1), 11–25 (2014)

15. Phoo-ngernkham, T., et al.: High calcium fly ash geopolymer mortar containing Portland cement for use as repair material. Constr. Build. Mater. **98**, 482–488 (2015)

16. Marinković, S., Dragaš, J., Ignjatović, I., Tošić, N.: Environmental assessment of green concretes for structural use. J. Clean. Prod. (2017)

17. Chindaprasirt, P., et al.: Effect of calcium-rich compounds on setting time and strength development of alkali-activated fly ash cured at ambient temperature. Case Stud. Constr. Mater. **9**, e00198 (2018)

18. Yip, C.K., Lukey, G., Van Deventer, J.: The coexistence of geopolymeric gel and calcium silicate hydrate at the early stage of alkaline activation. Cem. Concr. Res. **35**(9), 1688–1697 (2005)

19. Huseien, G.F., et al.: The effect of sodium hydroxide molarity and other parameters on water absorption of geopolymer mortars. Indian J. Sci. Technol. **9**(48) (2016)

20. Garcia-Lodeiro, I., Palomo, A., Fernández-Jiménez, A., Macphee, D.: Compatibility studies between NASH and CASH gels. Study in the ternary diagram Na_2O–CaO–Al_2O_3–SiO_2–H_2O. Cem. Concr. Res. **41**(9), 923–931 (2011)

21. Palomo, A., et al.: A review on alkaline activation: new analytical perspectives. Mater. Constr. **64**(315), 022 (2014)

22. Senthamarai, R., Manoharan, P.D., Gobinath, D.: Concrete made from ceramic industry waste: durability properties. Constr. Build. Mater. **25**(5), 2413–2419 (2011)

23. Daniyal, M., Ahmad, S.: Application of waste ceramic tile aggregates in concrete. Int. J. Innov. Res. Sci. Eng. Technol. **4**(12), 12808–12815 (2015)

24. Sugama, T., Brothers, L., Van de Putte, T.: Acid-resistant cements for geothermal wells: sodium silicate activated slag/fly ash blends. Adv. Cem. Res. **17**(2), 65–75 (2005)

25. Al-Majidi, M.H., Lampropoulos, A., Cundy, A., Meikle, S.: Development of geopolymer mortar under ambient temperature for in situ applications. Constr. Build. Mater. **120**, 198–211 (2016)

26. Aslani, F., Ma, G., Wan, D.L.Y., Le, V.X.T.: Experimental investigation into rubber granules and their effects on the fresh and hardened properties of self-compacting concrete. J. Clean. Prod. **172**, 1835–1847 (2018)

27. Su, H., et al.: Properties of concrete prepared with waste Tyre rubber particles of uniform and varying sizes. J. Clean. Prod. **91**, 288–296 (2015)

28. Huseien, G.F., Sam, A.R.M., Shah, K.W., Mirza, J.: Effects of ceramic tile powder waste on properties of self-compacted alkali-activated concrete. Constr. Build. Mater. **236**, 117574 (2020)

29. Siddika, A., et al.: Properties and utilizations of waste tire rubber in concrete: a review. Constr. Build. Mater. **224**, 711–731 (2019)

30. Duarte, A., et al.: Tests and design of short steel tubes filled with rubberised concrete. Eng. Struct. **112**, 274–286 (2016)
31. Mhaya, A., Abidin, A., Sarbini, N., Ismail, M.: Role of crumb tyre aggregates in rubberised concrete contained granulated blast-furnace slag. E&ES. **220**(1), 012029 (2019)
32. Huseien, G.F., et al.: Evaluation of alkali-activated mortars containing high volume waste ceramic powder and fly ash replacing GBFS. Constr. Build. Mater. **210**, 78–92 (2019)
33. Akinyele, J.O., Salim, R.W., Kupolati, W.K.: The impact of rubber crumb on the mechanical and chemical properties of concrete. Eng. Struct. Technol. **7**(4), 197–204 (2015)
34. Ganjian, E., Khorami, M., Maghsoudi, A.A.: Scrap-tyre-rubber replacement for aggregate and filler in concrete. Constr. Build. Mater. **23**(5), 1828–1836 (2009)
35. Huseien, G.F., et al.: Properties of ceramic tile waste based alkali-activated mortars incorporating GBFS and fly ash. Constr. Build. Mater. **214**, 355–368 (2019)
36. Aslani, F.: Mechanical properties of waste tire rubber concrete. J. Mater. Civ. Eng. **28**(3), 04015152 (2015)
37. Thomas, B.S., Gupta, R.C.: Properties of high strength concrete containing scrap tire rubber. J. Clean. Prod. **113**, 86–92 (2016)
38. Thomas, B.S., Gupta, R.C., Panicker, V.J.: Recycling of waste tire rubber as aggregate in concrete: durability-related performance. J. Clean. Prod. **112**, 504–513 (2016)

Advancing Toward Net Zero: The Role of Fibers in Sustainable Concrete Construction

Raymond Pepera and Behrouz Shafei (ID)

Contents

1 Introduction

The quest for sustainability within the construction industry has increasingly spotlighted the significance of reducing the carbon footprint associated with concrete mix designs. Concrete, one of the most widely used construction materials globally, plays a pivotal role in the sector's environmental impact, mainly due to its intensive carbon dioxide emissions from cement production and aggregate use. Innovations in concrete mix design, including incorporating various fibers and alternative materials, offer a promising avenue to enhance structural performance while significantly mitigating environmental footprints. By reducing the carbon footprint through thoughtful mix designs, the construction industry can make substantial strides toward sustainability goals, contributing to the global effort to combat climate change, reduce energy consumption, and promote environmental stewardship. This approach aligns with the urgent need for eco-friendly construction practices.

R. Pepera · B. Shafei (✉)
Iowa State University, Ames, IA, USA
e-mail: shafei@iastate.edu

© The Author(s) 2025
M. Kioumarsi, B. Shafei (eds.), *The 1st International Conference on Net-Zero Built Environment*, Lecture Notes in Civil Engineering 237,
https://doi.org/10.1007/978-3-031-69626-8_31

Following the need to reduce carbon footprints in concrete mix designs, integrating fibers into concrete presents a transformative approach to achieving enhanced material properties and environmental benefits. Synthetic fibers such as polyvinyl alcohol (PVA), alkali-resistant Glass (AG), and polypropylene (PP), among others, have been increasingly adopted to improve concrete's strength, durability, and crack resistance properties. These fibers reduce traditional cement content and steel reinforcing bars, indirectly reducing CO_2 emissions associated with cement and steel production. Moreover, using fibers in concrete not only aids in achieving a more sustainable construction material by enhancing its lifespan and reducing the need for repairs but also opens possibilities for incorporating recycled and bio-based fibers, further aligning with eco-friendly construction practices. This broader adoption of fibers signifies a shift toward more resilient and sustainable construction methodologies, where the performance benefits of fiber-reinforced concrete are leveraged to meet the dual goals of structural integrity and environmental responsibility.

Therefore, this study aims to illuminate the focus of enhancing structural benefits while simultaneously addressing the environmental impacts of incorporating fibers in concrete mix design. By carefully balancing the addition of fibers such as PVA, AG, and PP, the study delves into the realm where improved mechanical properties and durability of concrete do not come at the expense of the environment. By evaluating the benefits and impacts of these fibers, the study contributes to the evolving discourse on sustainable construction, advocating for innovations that ensure the longevity and resilience of infrastructure without compromising environmental integrity. This endeavor aims to foster a construction paradigm where environmental and structural considerations are balanced and synergistically advanced, heralding a new era of eco-conscious and structurally sound building practices.

2 Properties of Fibers Used for Concrete Structures

Fiber-reinforced concrete (FRC) is a composite material that has evolved significantly over time, incorporating distinct fibers to bolster concrete's structural characteristics and longevity. These fibers are often classified into two main categories: macrofibers and microfibers, differentiated primarily by size. Microfibers, characterized by their low diameter and high aspect ratio (length to diameter), are typically shorter than 18 mm, whereas macro fibers are more substantial, with a length exceeding 18 mm. This distinction is critical as microfibers and macrofibers are pivotal in mitigating cracking and enhancing the concrete's structural integrity throughout its service life. They achieve this by increasing resistance to plastic and drying shrinkage cracking, significantly reducing crack widths, and improving the concrete's ability to absorb energy and withstand impacts [1, 2].

Including fibers in concrete is not merely for structural enhancement; it fundamentally transforms the mixture's durability and resilience. The spectrum of fibers utilized in FRC includes metallic (notably steel), glass, synthetic, and natural fibers, each bringing unique properties to the concrete mix. The length of these fibers varies widely, ranging from 10 mm (approximately 3/8 inch) to 75 mm (about 3 inches), allowing for a tailored approach to reinforcing concrete based on specific engineering and construction requirements [3].

Carbon steel fibers, widely recognized as the predominant metallic fibers in concrete applications, are available in various forms such as straight, hooked end, crimped, or flat shapes, designed to optimize their mechanical interlock and bonding with the concrete matrix, enhancing tensile strength and crack resistance. In contrast, stainless steel fibers are specifically chosen for environments where corrosion is a significant concern, as their inherent resistance to rust and degradation makes them ideal for ensuring the longevity of concrete structures in harsh conditions [4, 5].

On the other hand, glass fibers, sourced from natural minerals or rocks, are utilized in concrete for their superior mechanical properties; silica and basalt glass fibers are notable examples, with basalt fibers often preferred for their higher elastic modulus and tensile strength, as documented in past studies [3, 6].

In synthetic fibers, materials such as PP, PVA, nylon, polyolefin, carbon, polyethylene, polyester, acrylic, and aramid stand out. These fabricated fibers are essential for controlling concrete shrinkage and temperature-induced stresses, making them suitable for environments prone to chemical exposure and corrosion [7, 8]. However, it is important to note that synthetic fibers enhance durability and resistance to environmental factors. However, they generally do not contribute as significantly to concrete's compressive and flexural strengths as their metallic counterparts. The selection of fiber type for concrete reinforcement is thus a strategic decision influenced by a blend of factors, including mechanical properties, cost considerations, specific demands of the construction project, and prevailing environmental conditions. As pointed out in [1], choosing fiber is crucial for optimizing the performance and sustainability of concrete and cementitious composites in varied applications.

Incorporating synthetic fibers into concrete has recently garnered significant attention in the construction sector, particularly for its potential to fortify the material's durability under adverse environmental conditions. Acknowledged for substantially enhancing concrete's mechanical attributes and longevity, synthetic fibers have become a focal point in the quest for more robust construction materials.

This study thoroughly investigates the design of concrete mixes reinforced with synthetic fibers, focusing on the roles and impacts of PVA, PP, and AG fibers. By thoroughly examining these fibers' contributions, the study aims to elucidate their influence on concrete performance, particularly in enhancing strength, durability, and sustainability. By understanding the specific benefits and functionalities of PVA, PP, and AG fibers within concrete matrices, this analysis seeks to uncover the pathways through which synthetic fiber reinforcement can lead to the development of construction materials that are not only more durable and capable of withstanding harsh environmental challenges but also contribute to the sustainability of the construction industry.

2.1 Polyvinyl Alcohol Fibers

PVA fibers, as shown in Fig. 1b, have consistently demonstrated an ability to augment concrete's durability properties. A study by [9] subjected FRC composites made with varying PVA fiber contents (ranging from 0% to 1.5%) to extreme wet-thermal and chloride salt conditions. It was discerned that a PVA fiber content of 0.3% was optimal for compressive strength, while 0.6% maximized flexural strength. There were noteworthy enhancements, with a 67% increase in maximum flexural strength and a 30% rise in residual compressive strength compared to the non-fiber counterparts. Similarly, it was reported in [10] that blending 1.0 vol.% PVA fibers with 1.0 wt.% nano-SiO_2 in cement mortar can considerably improve its resistance to various deteriorating agents, including chloride-ions. This combination also positively influenced mechanical strength and frost resilience [11], further underscoring the durability of PVA fibers under seawater exposure, observing that even after a decade, the fibers remained robust, offering crack suppression benefits, particularly near rebars. Lee et al. [12] provided additional insights, suggesting that integrating PVA fibers into specific concrete types can considerably mitigate chloride ion and water penetration.

2.2 Polypropylene Fibers

PP fibers, as shown in Fig. 1a, have demonstrated multifaceted benefits and limitations. Wang and Chen [13] indicated that PP fibers could enhance the durability of concrete under freeze-thaw cycles and maintain its residual mechanical properties up to 400 °C. However, it could compromise concrete's strength properties, especially at higher temperatures. Further reinforcing these findings, it was found that concrete with up to 0.15% PP fiber content exhibited minimal water absorption and commendable resistance to chemical attacks [14]. Haq et al. [15] extended the understanding, suggesting that an optimal inclusion of PP fibers in certain concrete mixtures could enhance water absorption, permeability, and resistance to chloride-ion penetration. Nevertheless, Resende et al. [16] highlighted some limitations.

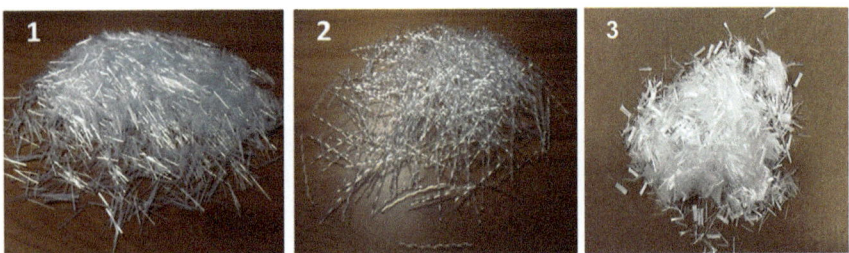

Fig. 1 (a) PP, (b) PVA, and (c) AG fibers

When exposed to a temperature spectrum of 25 to 800 °C, high-strength concrete containing PP fibers consistently revealed greater water absorption than its fiber-free counterpart.

2.3 Glass Fibers

Glass fiber-reinforced concrete (GFRC) presents distinct advantages in construction materials. Research by [17] demonstrated that GFRC initially shows superior carbonation resistance compared to ordinary portland cement (OPC). Despite reducing this resistance over time, GFRC still surpasses OPC in crucial aspects such as reduced porosity and enhanced resistance to chloride-ion penetration. These findings are corroborated by [18], which noted that increasing the glass fiber content in concrete significantly curtails chloride-ion penetration, a benefit primarily due to the robust bond between the fibers and the concrete binders. Experimental studies further highlight the advantages of GFRC, notably its increased resistance to acid attacks compared to steel fiber-reinforced concrete. This resilience is particularly significant in environments where concrete is exposed to corrosive substances. Additionally, even in high volume fractions, glass fibers contribute minimal weight to structural members, a stark contrast to the denser steel fibers, thereby not substantially increasing the self-weight of structural components. Moreover, as shown in Fig. 1c, glass fiber enhances the concrete's durability in terms of fatigue and impact resistance, akin to steel fibers, but with the added benefit of being cost-effective. This makes GFRC not only a financially viable option but also beneficial for improving the flexural tensile strength of concrete, thus offering a comprehensive suite of enhancements to concrete's structural performance. Overall, GFRC stands out for its cost-efficiency, improved mechanical properties, and reduced environmental impact, making it a compelling choice in modern construction practices. Its capacity to enhance the longevity and resilience of concrete structures while mitigating the environmental footprint underscores the importance of further research and adoption in the construction industry.

3 Environmental Impact of Fibers

The environmental impact of the construction industry is a significant concern, primarily due to the extraction, processing, and transportation of raw materials, which contribute to substantial pollution levels and high energy consumption. In the context of this research, using PVA, PP, and AG fibers in concrete necessitates a closer look at their environmental footprint [19].

3.1 Polyvinyl Alcohol Fibers

PVA, denoted chemically as $(C_2H_4O)n$, is a synthetic fiber known for its high strength, initially developed to replace asbestos in asbestos cement products. Despite its advantageous properties, PVA fibers are less prevalent in the construction industry than other concrete fibers, primarily due to their higher cost. Their limited availability in the concrete fiber market means they are sourced from select suppliers. Regarding environmental impact, producing PVA fibers involves processes that can contribute to CO_2 emissions. These emissions stem from the energy-intensive nature of synthesizing PVA, which involves polymerizing vinyl acetate into polyvinyl acetate and then hydrolyzing it to form PVA. This process typically requires significant amounts of energy, often derived from fossil fuels, leading to the generation of CO_2.

Moreover, the raw materials used in the production of PVA, such as ethylene (a petrochemical product), further contribute to its carbon footprint through the emissions associated with petrochemical processing. While PVA fibers can enhance the structural integrity and durability of construction materials, their production's potential CO_2 emissions must be considered in the broader context of sustainable construction practices, assessing whether their performance benefits outweigh their environmental costs [1, 3].

3.2 Polypropylene Fibers

PP fibers, as synthetic materials, undergo a complex and energy-intensive production process. The journey begins with the procurement of raw materials. These materials are primarily derived through the thermal cracking of fuels like naphtha and liquefied petroleum gas, necessitating sophisticated techniques to maintain their liquid state during storage. The domestic supply contributes the remaining 20%, separating propylene from propane at local facilities.

The manufacturing process then advances to the purification stage, where all raw materials are meticulously cleansed of impurities to ensure the stability and quality of the polypropylene production. Post-purification, the material enters polymerization reactors where polypropylene is synthesized. Following polymerization, the product undergoes degassing to expel residual hydrocarbons, leading to the extrusion phase. Here, polypropylene is mixed and melted, combined with additives, and formed into pellets, readying for consumer distribution.

For the fibers to be used in construction, they are further processed through specialized machines that shape and size them appropriately for integration into concrete. Within the concrete matrix, polypropylene fibers contribute to the structural integrity and durability of the material, not posing significant environmental concerns during the structure's service life. However, the environmental implications become pertinent at the structure's end of life, where the disposal or recycling of polypropylene fibers can lead to environmental challenges, emphasizing the need for sustainable end-of-life management practices [19, 20].

3.3 Glass Fibers

Glass fibers, derived from naturally occurring minerals or rocks and primarily consisting of $(SiO_2)n$ monomers, are produced through a process that begins with melting raw materials like silica sand, limestone, and soda ash, along with recycled glass, at temperatures ranging from 1400 to 1600 °C to form molten glass. This molten glass is refined to eliminate bubbles and achieve uniformity before extruding through tiny orifices in a platinum-rhodium alloy bushing, creating continuous filaments. Depending on the production method and intended application, glass strands can be engineered to disperse into microfiber form upon contact with water or remain integral macro fibers. Manufacturing glass fibers is an energy-intensive process that incurs significant CO_2 emissions, primarily due to the high temperatures required for melting the raw materials, often utilizing energy from fossil fuels. The subsequent refining and fiberization processes also add to the carbon footprint, although modern advancements in production technology and energy efficiency are helping to reduce these environmental impacts. In construction, glass fibers are valued for creating robust, lightweight materials, but their environmental footprint includes the CO_2 emissions from their production phase. Utilizing recycled glass in manufacturing can diminish energy consumption and lower CO_2 emissions. Despite the carbon emissions associated with their production, glass fibers play a crucial role in promoting energy-efficient and durable construction, which can lead to a reduction in the overall CO_2 impact throughout the lifecycle of a building. Thus, while glass fiber production produces CO_2 emissions, their application in sustainable construction practices has the potential to balance out these initial environmental costs, contributing to a more sustainable construction industry [1, 3, 21, 22].

3.4 Carbon Footprint Assessment of Mix Designs

In the mix design study focused on concrete beams reinforced with PVA, PP, and AG fibers, an in-depth analysis was conducted to evaluate the environmental impacts, particularly CO_2 emissions associated with these materials. The study methodically integrated these fibers at two distinct concentrations—0.25% and 0.50%—and assessed six beams for each type, dividing equally among the specified percentages. Each fiber type's total usage in the beams amounted to 1.14 kg, allowing for a direct comparison of their respective CO_2 emissions.

Table 1 Synthetic fiber options and their various CO_2 emissions

Fibers	Percentage	Total number of beams	Percentage	Total number of beams	Total weight of fibers used in all beams (kg)	CO_2 emissions per kg (CO_2-kg/kg)
AG	0.25	3	0.50	3	1.14	2.04 [23]
PP	0.25	3	0.50	3	1.14	1.85 [23]
PVA	0.25	3	0.50	3	1.14	2.70 [24]

The mix included AG fibers at the aforementioned percentages was explored for reinforced concrete, with six beams collectively. The total CO_2 emissions for AG, calculated from an emission factor of 2.04 kg CO_2/kg of fiber, were approximately 2.33 kg CO_2 for the entire set of beams. This figure delineates the carbon cost of employing AG fibers in concrete reinforcement. PP reinforced concrete followed a similar experimental setup. Here, PP fibers demonstrated a lower environmental impact, evidenced by a CO_2 emission factor of 1.85 kg CO_2/kg, culminating in total emissions of around 2.11 kg CO_2 for the total fiber quantity. This denotes a comparatively lesser environmental footprint for PP fibers under the same testing conditions.

PVA fibers, utilized likewise in the beam compositions, emerged as the most environmentally taxing, with the highest CO_2 emission factor of 2.70 kg CO_2/kg. The cumulative emissions for PVA fiber reached approximately 3.08 kg CO_2, marking it the most impactful on the environment among the fibers evaluated. The above assessment is also illustrated in Table 1.

This comparative assessment of CO_2 emissions sheds light on the diverse environmental implications of incorporating these fibers in concrete construction. Despite all fibers enhancing concrete's structural attributes, their environmental ramifications vary significantly. PP fibers, in particular, stand out as the most environmentally benign option, emitting the least CO_2. This study's findings are critical in the broader context of sustainable construction practices. While the CO_2 emissions from these fibers are considerably lower than those from major construction sources like cement production and steel reinforcing bars and fibers production, they are nonetheless consequential and warrant careful consideration in the environmental footprint analysis of construction projects. Emphasizing sustainability, the research advocates selecting construction materials that fulfill structural requirements and align with environmental sustainability goals. Reducing CO_2 emissions, even in smaller quantities, is imperative in the construction industry's pursuit of a reduced carbon footprint. The urgency of these reductions becomes apparent when considering the current CO_2 concentration in the Earth's atmosphere, which stands at approximately 415 ppm, or 0.0415%. This level marks a 48% increase from the beginning of the industrial era and a 12% rise since the start of this millennium [25]. Therefore, as demonstrated in this study, the nuanced evaluation of materials' environmental impact is essential for advancing sustainable construction methodologies and achieving long-term ecological balance in the industry.

4   Conclusions

In conclusion, this study has thoroughly examined the integration of synthetic fibers such as PVA, PP, and AG fibers into concrete mixes, highlighting their potential to reduce the construction materials' carbon footprint and contribute to achieving net-zero goals. The comprehensive analysis, encompassing these fibers' production processes, application, and end-of-life phases, underscores their role in sustainable construction practices. The comparative life cycle assessment conducted within this study reveals that while all tested fibers enhance the mechanical properties of concrete, thus improving durability and resilience, PP fibers stand out for their lower CO_2 emission profile. This finding is instrumental in advocating for a more environmentally conscious selection of materials within the construction industry, pointing toward preferable choices of fibers in pursuing sustainability. The mix design and CO_2 emissions analysis results confirm the necessity of integrating energy-efficient manufacturing processes and promoting the use of recycled materials to further mitigate the environmental impact of construction activities.

Acknowledgments This research study was partially sponsored by the American Concrete Institute (ACI) Center of Excellence for Nonmetallic Building Materials. The authors would like to acknowledge and thank the sponsor for this support. Opinions, findings, and conclusions expressed in this chapter are of the authors and do not necessarily reflect those of the sponsor.

References

1. Shafei, B., Kazemian, M., Dopko, M., Najimi, M.: State-of-the-art review of capabilities and limitations of polymer and glass fibers used for fiber-reinforced concrete. Materials. **14**(2), 409 (2021)
2. Karim, R., Shafei, B.: Ultra-high performance concrete under direct tension: investigation of a hybrid of steel and synthetic fibers. Struct. Concr. **25**(1), 423–439 (2024)
3. Dopko, M., Najimi, M., Shafei, B., Wang, X., Taylor, P., Phares, B.M.: Flexural performance evaluation of fiber-reinforced concrete incorporating multiple macro-synthetic fibers. Transp. Res. Rec. **2672**(27), 1–12 (2018)
4. Dopko, M., Najimi, M., Shafei, B., Wang, X., Taylor, P., Phares, B.: Strength and crack resistance of carbon microfiber reinforced concrete. ACI Mater. J. **117**(2), 11–23 (2020)
5. Karim, R., Shafei, B.: Flexural response characteristics of ultra-high performance concrete made with steel microfibers and macrofibers. Struct. Concr. **22**(6), 3476–3490 (2021)
6. Kazemian, M., Shafei, B.: Mechanical properties of hybrid fiber-reinforced concretes made with low dosages of synthetic fibers. Struct. Concr. **24**(1), 1226–1243 (2023)
7. Karim, R., Shafei, B.: Investigation of five synthetic fibers as potential replacements of steel fibers in ultrahigh-performance concrete. J. Mater. Civ. Eng. **34**(7) (2022)
8. Karim, R., Shafei, B.: Performance of fiber-reinforced concrete link slabs with embedded steel and GFRP rebars. Eng. Struct. **229**, 111590 (2021)
9. Zhang, P., Wei, S., Wu, J., Zhang, Y., Zheng, Y.: Investigation of mechanical properties of PVA fiber-reinforced cementitious composites under the coupling effect of wet-thermal and chloride salt environment. Case Stud. Constr. Mater. **17** (2022)
10. Huang, J., Wang, Z., Li, D., Li, G.: Effect of nano-SiO_2/PVA on corrosion behavior of steel rebar embedded in high-volume fly ash mortar under accelerated chloride attack. Materials. **15**(11) (2022)

11. Ito, H., Watanabe, K., Todoroki, S., Suemori, H., Shinjyo, R.: Study on performance of PVA fiber reinforced concrete exposed for 10 years to seawater spray. J. Adv. Concr. Technol. **16**(3), 159–169 (2018) Japan Concrete Institute

12. Lee, S.K., Jeon, M.J., Cha, S.S., Park, C.G.: Mechanical and permeability characteristics of latex-modified fiber-reinforced roller-compacted rapid-hardening-cement concrete. Appl. Sci. **7**(7) (2017)

13. Wang, C., Chen, F.: Durability of polypropylene fiber concrete exposed to freeze-thaw cycles with deicing salts. In: Proceedings of the 2019 International Conference on Electronical, Mechanical and Materials Engineering (ICE2ME 2019). Atlantis Press, Paris (2019)

14. Brundha, S., Vishnuram, B.G.: Fresh and Durability Studies of Polypropylene Fiber Reinforced Self-Compacting Concrete (SCC) [Online]. Available: www.ijstm.com

15. Ul Haq, I., Elahi, A., Nawaz, A., Shah, S.A.Q., Ali, K.: Mechanical and durability performance of concrete mixtures incorporating bentonite, silica fume, and polypropylene fibers. Constr. Build. Mater. **345** (2022)

16. Resende, H.F., et al.: Residual mechanical properties and durability of high-strength concrete with polypropylene Fibers in high temperatures. Materials. **15**(13) (2022)

17. Huang, Q., Shi, X.-S., Wang, Q.-Y., Tang, L., Zhang, H.-E.: The Influence of Fiber on the Resistance to Chloride-Ion Penetration of Concrete under the Environment of Carbonation (2014)

18. Ahmad, J., González-Lezcano, R.A., Majdi, A., Ben Kahla, N., Deifalla, A.F., El-Shorbagy, M. A.: Glass fibers reinforced concrete: overview on mechanical, durability and microstructure analysis. Materials. **15**(15) (2022)

19. Acosta-Calderon, S., Gordillo-Silva, P., García-Troncoso, N., Bompa, D.V., Flores-Rada, J.: Comparative evaluation of sisal and polypropylene fiber reinforced concrete properties. Fibers. **10**(4), 31 (2022)

20. Limbachiya, M.C., Leelawat, T., Dhir, R.K.: Use of recycled concrete aggregate in high-strength concrete. Mater. Struct. **33**(9), 574–580 (2000)

21. Rajak, D.K., Wagh, P.H., Linul, E.: Manufacturing technologies of carbon/glass fiber-reinforced polymer composites and their properties: a review. Polymers. **13**(21), 3721 (2021)

22. Stade, K.H.: The production of glass fiber-reinforced poly(butylene terephthalate) on a continuous kneader. Polym. Eng. Sci. **18**(2), 107–113 (1978)

23. Joshi, S.V., Drzal, L.T., Mohanty, A.K., Arora, S.: Are natural fiber composites environmentally superior to glass fiber reinforced composites? Compos. Part A Appl. Sci. Manuf. **35**(3), 371–376 (2004)

24. Zhang, Z., Ma, H., Qian, S.: Investigation on properties of ECC incorporating crumb rubber of different sizes. J. Adv. Concr. Technol. **13**(5), 241–251 (2015)

25. Kazemian, M., Shafei, B.: Carbon sequestration and storage in concrete: a state-of-the-art review of compositions, methods, and developments. J. CO_2 Util. **70**, 102443 (2023)

Environmental Assessment of Fiber-Reinforced Self-Compacting Concrete Containing Class-F Fly Ash

Behnoosh Khataei, Masoud Ahmadi, and Mahdi Kioumarsi

Contents

1 Introduction

Concrete is a familiar building material that shapes our world. From tall buildings to extensive transportation networks, concrete forms the foundation of modern life. While concrete's strength and long lifespan are undeniable benefits, making

B. Khataei
Department of Civil and Geomechanics Engineering, Faculty of Earth Sciences Engineering, Arak University of Technology, Arak, Iran
e-mail: b.khataei@arakut.ac.ir

M. Ahmadi (✉)
Department of Civil and Geomechanics Engineering, Faculty of Earth Sciences Engineering, Arak University of Technology, Arak, Iran

Department of Civil Engineering, Faculty of Engineering, Ayatollah Boroujerdi University, Boroujerd, Iran
e-mail: ms.ahmadi@arakut.ac.ir; masoud.ahmadi@abru.ac.ir

M. Kioumarsi
Department of Built Environment, OsloMet – Oslo Metropolitan University, Oslo, Norway
e-mail: mahdi.kioumarsi@oslomet.no

© The Author(s) 2025
M. Kioumarsi, B. Shafei (eds.), *The 1st International Conference on Net-Zero Built Environment*, Lecture Notes in Civil Engineering 237,
https://doi.org/10.1007/978-3-031-69626-8_32

traditional concrete comes with a hidden cost to the environment [1]. A key component of concrete, ordinary Portland cement (OPC), is a major culprit behind greenhouse gas emissions, particularly carbon dioxide (CO_2) [2]. Concrete production has critical environmental impacts, essentially in the form of carbon emission (CE) and embodied energy (EE) [3–5]. A method called Life Cycle Assessment (LCA) can be used to track and assess these environmental effects throughout concrete's entire lifespan, from when the raw materials are first collected to when the concrete reaches the end of its useful life [6–8]. The production of concrete may be a highly energy-intensive process that depends on the combustion of fossil fuels, basically in the form of coal and natural gas. This combustion discharges CO_2 into the environment, contributing to worldwide climate change. So, concrete production can produce a significant portion of global carbon emissions [9, 10].

Sources of CEs from concrete production include raw material extraction, cement production, and transportation. One effective strategy to reduce CEs is using alternative cement formulations. Incorporating alternative cementitious materials like fly ash [11–14], slag [15–18], silica fume [19–22], etc., can decrease the carbon footprint of concrete by lowering the amount of clinker needed.

The second factor, EE, refers to the total amount of energy consumed throughout the life cycle of a concrete product, including the energy used in raw material extraction, manufacturing, transportation, and construction. Due to its energy-intensive manufacturing process, concrete retains high energy and contributes significantly to overall energy consumption in the construction industry [23, 24]. Similar to CE, the energy sources involved in concrete include raw material extraction, cement production, and transportation. Therefore, recycling and reuse, local content, and energy efficiency can be considered as mitigation strategies.

Recognizing this environmental impact, researchers and engineers are actively exploring new materials and techniques to make concrete more sustainable. This quest for eco-friendly concrete solutions involves several approaches:

- Reducing reliance on OPC: A crucial part of this effort involves finding replacements or ways to use less OPC in concrete production. Here, we explore some promising strategies:

  - Supplementary cementitious materials (SCMs): When added to concrete mixes, these materials can partially take the place of OPC while still helping the concrete set and harden [25]. A prime example is fly ash, a leftover material from burning coal. Fly ash is particularly attractive because it's plentiful and relatively inexpensive. Fly ash is derived from pulverized coal combustion during power generation and is characterized by its pozzolanic properties, making it a valuable additive in concrete production. The incorporation of fly ash serves several purposes in enhancing the performance and sustainability of concrete. Initially, these materials act as a cementitious binder by reacting with the calcium hydroxide ($Ca(OH)_2$) produced during cement hydration, leading to the formation of additional calcium silicate hydrate (C-S-H) gel [26]. This reaction improves the overall strength and durability of the concrete. Secondly, the fine particles of fly ash fill the voids between cement grains, leading to denser concrete with lower permeability. Because of this denser structure, the concrete becomes more resistant to damage from chemicals, sulfates, and alkali-silica reaction [27]. Moreover, the use of fly ash helps reduce the heat of hydration during concrete curing, which is beneficial for large concrete placements and in mitigating thermal cracking [28].

– Geopolymers: These innovative binders, activated by alkaline solutions, have a smaller carbon footprint than OPC [29]. Geopolymers can be formulated using industrial byproducts like fly ash or even natural materials. While still under development, geopolymers hold significant promise for the future of sustainable concrete production.

– Alternative binders: Researchers are exploring various alternative binders derived from agricultural waste [30] or industrial processes [31]. These binders can potentially offer significant environmental benefits while meeting the performance requirements for concrete applications.

• Enhancing concrete performance: Another key strategy involves developing concrete formulations with improved performance. This can lead to using less concrete overall for a project, indirectly reducing the environmental impact. For example, self-compacting concrete (SCC) offers several advantages over traditional concrete [32]:

– Easier to work with: SCC flows freely under its weight, eliminating the need for vibration during placement. This translates to faster construction times and potentially safer working conditions.

– Reduced noise pollution: Since vibration isn't required, SCC placement generates less noise pollution than traditional concrete.

– Improved quality control: The self-compacting nature of SCC leads to better filling of forms and less segregation of materials, resulting in improved quality control.

Additionally, by incorporating certain fibers into the concrete mix, researchers aim to improve the durability and lifespan of the concrete, potentially leading to less maintenance and longer infrastructure lifespans [33–36].

This investigation explores the environmental impact of SCC reinforced with fibers and incorporating Class-F fly ash. While SCC offers well-documented improvements in workability and construction efficiency, a key question regarding its sustainability remains. Specifically, this study focuses on comparing how SCC production affects the environment throughout its entire lifecycle, compared to regular concrete. This assessment involves evaluating the potential environmental trade-offs linked to higher cement content, which is often necessary for SCC mixtures, and the use of chemical admixtures to achieve self-compacting properties.

The current investigation builds upon a foundational study by one of the authors [36], which focused on how well SCC containing class-F fly ash and reinforced with fibers works and how it responds to stress.

The previous work serves as a solid foundation for the current research, enabling us to investigate the environmental implications of selected fiber reinforcement strategies within the framework of incorporating fly ash as a partial substitute for OPC. By employing a LCA, a well-established method for comprehensively evaluating the environmental footprint of a product or process, this study aims to offer a clear understanding of the environmental impact of SCC with blended fiber reinforcement throughout its entire life cycle, encompassing raw material extraction, production, transportation, construction, use phase, and ultimately, demolition and waste disposal. The findings of this paper will provide valuable knowledge to the ongoing discussion on sustainable construction materials. By measuring the environmental impact of SCC with class-F fly ash and blended fiber reinforcement, this research can inform decision-making processes within the construction industry, guiding the selection of materials and techniques that minimize environmental burdens while maintaining structural integrity and performance.

2 Methodology

This study builds upon a foundational investigation by one of the authors [36] to establish the methodological framework for assessing the environmental impact of fibrous SCC with class-F fly ash. The prior work, which focused on characterizing the workability and mechanical response of SCC, provides a strong foundation for the current research. Key aspects of the methodology, including the utilization of established material properties and the adoption of the mixture design and specimen preparation procedures, are drawn from [36] to ensure consistency and minimize repetition.

2.1 Material Properties

The current investigation utilizes the well-defined material properties meticulously characterized in the previous study by [36]. That study provides comprehensive details on the constituent materials employed in this investigation, including:

- OPC
- Class-F fly ash
- Fine aggregates (limited to a maximum size of 2.36 mm)
- Clean tap water, free from any harmful impurities
- Carboxylate-based high-range water reducer (HRWR)
- Micro steel fibers
- Polyvinyl alcohol (PVA) fibers

Readers are directed to the aforementioned publication for a more in-depth understanding of the individual material properties [36]. Here, the focus is on how these materials are strategically used to assess the environmental impact of the reinforced SCC.

2.2 Mixture Design and Specimen Preparation

The base mixture design adopted here was directly adapted from [36] (as presented in Table 1), ensuring consistency in material composition and fresh concrete properties. This selection allowed for a focused evaluation of the environmental impact introduced by the blended fiber reinforcement. As detailed in [36], the base mixture consisted of OPC, water, and aggregates in a specific ratio of water to cement (0.38). A high-range water reducer was employed to attain the desired workability characteristics for SCC. Varying dosages and combinations of micro steel fibers and PVA fibers were incorporated into the base mixture design to investigate their combined effect on the environmental performance of the SCC. As outlined in [36], three volume fractions (0.5%, 1.0%, and 1.5%) of both fibers were explored in separate mixture series. The blended fiber approach aimed to leverage the synergistic benefits of both fiber types, as discussed in [36]. Micro steel fibers were expected to enhance the post-cracking behavior and tensile capacity, while PVA fibers were anticipated to mitigate plastic cracking and improve spalling resistance. The specific dosages and combinations of fibers were carefully chosen, considering the findings from [36] and aiming for a balance between desired mechanical performance and potential environmental impact.

Table 1 Concrete mix properties [36]

No.	Mix identification	Cement (kg/m^3)	Fine agg. (kg/m^3)	Fly ash (%)	Micro steel fiber (%)	PVA fiber (%)
1	F0S0P0	850	1550	0	0	0
2	F0S0P1.5	850	1550	0	0	1.5
3	F0S0. 5P1	850	1550	0	0.5	1
4	F0S1P0.5	850	1550	0	1	0.5
5	F0S1.5P0	850	1550	0	1.5	0
6	F5S0P0	807.5	1550	5	0	0
7	F5S0P1.5	807.5	1550	5	0	1.5
8	F5S0.5P1	807.5	1550	5	0.5	1
9	F5S1P0.5	807.5	1550	5	1	0.5
10	F5S1.5P0	807.5	1550	5	1.5	0
11	F10S0P0	765	1550	10	0	0
12	F10S0P1.5	765	1550	10	0	1.5
13	F10S0.5P1	765	1550	10	0.5	1
14	F10S1P0.5	765	1550	10	1	0.5
15	F10S1.5P0	765	1550	10	1.5	0

2.3 Life Cycle Assessment

The initial phase of LCA involves establishing its purpose and boundaries, defining the production system based on a functional unit and system scope. In this research, the functional unit was defined as 1 cubic meter of concrete utilizing varying proportions of cement, fly ash, micro steel fibers, and PVA fibers. Additionally, the environmental comparison considered the impact of compressive strength and service life. The subsequent stage of LCA, known as Life Cycle Inventory (LCI), involves gathering necessary data (input and output inventory of materials within the system) aligned with the study's objectives. Following the inventory analysis, the Life Cycle Impact Assessment (LCIA) aids in assessing and interpreting environmental impacts within the established goals and scope.

To assess the environmental impact, we focus on two key factors: Global Warming Potential (GWP) and EE. GWP is designed to quantify the impact on the greenhouse effect caused by human emissions and absorptions. Typically, LCAs utilize GWP over a 100-year timeframe. GWP enables the calculation of a single index, expressed in grams of CO_2 per functional unit of a product, representing the quantity of CO_2 with equivalent global warming potential over 100 years (Eq. 1):

$$GWP = \sum_i m_i \times GWP_i \qquad (1)$$

This calculation considers the mass (in kilograms) of each material used (represented by m_i) and its GWP_i. GWP_i indicates how much a specific material contributes to warming the planet over 100 years, similar to how 1 kg of CO_2 does. The specific GWP_i values for each material are listed in Table 2 [37].

In the same way, Eq. 2 was applied to calculate the amount of EE as MJ per 1 m^3 concrete prepared according to the concrete mix design.

$$EE = \sum_i m_i \times EE_i \qquad (2)$$

The CO_2 emissions and EE due to binder materials (cement, fine aggregate, fly ash, steel fibers, and PVA fibers) were extracted from different studies, as listed in Table 3 [38–40]. As observed, the significant amount of CEs caused by conventional concrete production is directly related to the amount of cement in the composition [41].

Table 2 Global warming potential equivalency factors

Flow (i)	GWP_i (CO_2-equivalent)
Carbon dioxide (CO_2, net)	1
Methane (CH_4)	23
Nitrous oxide (N_2O)	296

Table 3 Carbon emission and embodied energy of concrete components (per 1 kg)

Items	CO_2 (kg)	Energy (MJ)
Cement	0.912	5.5
Water	0	0
Fine agg.	0.0139	0.0048
Fly ash	0.004	0.1
Steel fiber	1.5	1.01
PVA fiber	3.6	106.54

3 Results

In this section, the findings of the compressive capacity of concrete as a mechanical index are presented first. Then, the results related to the carbon emitted and the energy consumed to make 1 m^3 of concrete (as environmental indexes) are presented. In the following, the above indexes are compared.

3.1 Compressive Strength

A previous study by one of the authors [36] investigated the influence of fibers and fly ash on the compressive strength of SCC. The results aligned with existing literature, demonstrating a trade-off between the effects of fibers and fly ash. Figure 1 presents the compressive strength of specimens. While adding fibers can increase CS by limiting crack growth (bridging effect), it can also introduce porosity, potentially reducing CS. This study observed a significant CS increase with micro steel fibers (16.45% for 1.5% content) attributed to their uniform dispersion, rough surface, and high elastic modulus. Conversely, PVA fibers (1.5% content) led to a 9.16% decrease in CS, likely due to their smooth surface and lower stiffness compared to steel fibers. On the other hand, the incorporation of class-F fly ash resulted in a decrease in CS (15% and 8% for 5% and 10% replacement ratios, respectively) due to its slower pozzolanic reaction rate.

3.2 Sustainability Potential

The results of carbon emissions of all specimens are displayed in Fig. 2. In the first sample as a control test, the values of fly ash, steel fibers, and PVA fibers are equal to zero, and the concrete sample consists of 850 kg of cement and 1550 kg of fine aggregates. In this case, the carbon emission is about 796 kg CO_2. In samples 2–5, with cement and fine aggregates remaining constant, two types of fibers were added as a percentage of cement (without fly ash). By using 1.5% PVA fiber, the amount of carbon emission has been significantly increased to 862 kg CO_2. According to

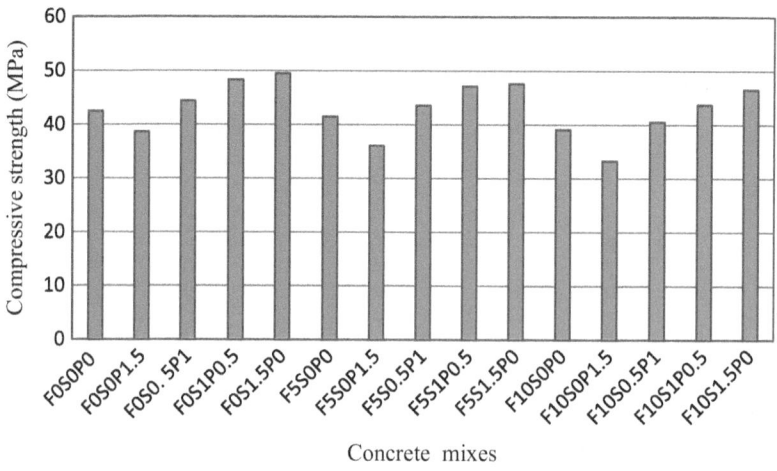

Fig. 1 Compressive strength of specimens

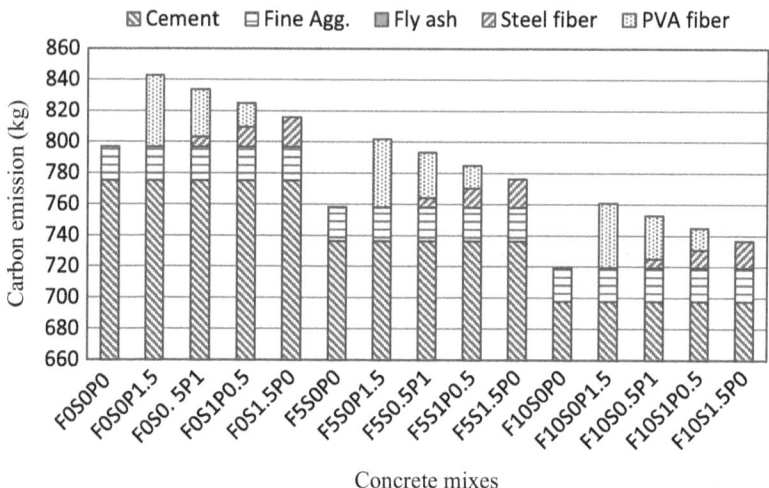

Fig. 2 Carbon emission value of concrete mixes (F: Fly ash; S: Steel fiber; P: PVA fiber)

Table 3, the carbon emission of PVA fiber is higher than other components. However, by reducing the PVA fiber and increasing the percentage of micro steel fiber usage, the amount of carbon emission has decreased. As the carbon emission varied from 842 to 815 kg CO_2.

In samples 6–10, the amount of fine aggregate remained constant, while the amount of cement was considered constant with a smaller amount. In addition, the amount of fly ash was added as a percentage of the used cement and fixed. Different amounts of micro steel and PVA fibers have been investigated in these samples by

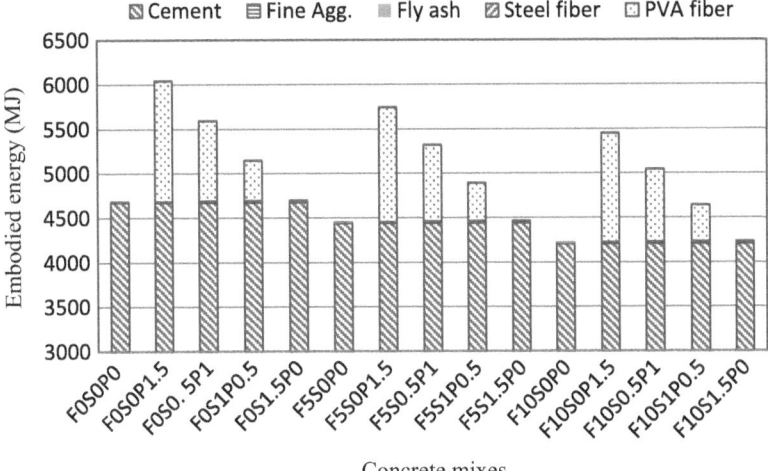

Fig. 3 Embodied energy value of concrete mixes (F: Fly ash; S: Steel fiber; P: PVA fiber)

applying fly ash. Again, it was concluded that PVA fibers have more carbon emission content than micro steel fibers. On the other hand, by reducing cement and replacing it with different types of fibers, carbon emission content has decreased significantly compared to tests 2–5. In tests 10–15, the same trend can be seen well, and the positive effect of replacing cement with PVA and micro steel fibers has been confirmed. So, in the last test with 765 kg of cement, 1550 kg of fine aggregate, 10% fly ash, and 1.5% micro steel fibers, the lowest amount of carbon emission equal to 736 kg as CO_2 was obtained. As the main result, micro steel fibers have an effective reduction in carbon emission compared to PVA fibers.

Figure 3 shows the results related to the energy consumed during the production of 1 m^3 of concrete for all specimens according to the concrete mix design (Table 1). Based on the consumption amount of the materials and the embodied energy values of each material (Table 3), it is observable that first cement and then PVA fibers had the greatest impact on the amount of EE. In different samples, the amount of EE has decreased with the reduction of cement and PVA fibers. On the other hand, considering the high energy value of PVA fibers, it is not logical to replace cement with this type of fiber.

Based on the results of using two types of micro steel and PVA fibers, and their influence on the amount of EE and GWP, it can be concluded that the use of micro steel is more effective. The inclusion of PVA fibers not only failed to reduce EE and carbon emissions but also had a detrimental effect on concrete compressive capacity. As mentioned previously, adding 1.5% PVA fibers decreased compressive capacity by approximately 9%. In contrast, the use of micro steel fibers significantly impacted concrete capacity. Therefore, it can be concluded that incorporating PVA fibers does not positively influence the mechanical or environmental properties of the concrete as intended.

4 Conclusion

The assessment of concrete's life cycle is crucial for comprehending and mitigating its environmental impact during production. By taking into account EE and carbon emissions throughout the concrete life cycle, stakeholders in the construction industry can make informed decisions to advance sustainability and diminish carbon emissions and overall carbon footprint. Implementing sustainable practices such as using alternative materials, adopting energy-efficient production techniques, and reducing carbon emissions will diminish the environmental effect of concrete making and contribute to a more sustainable built environment. In this study, mechanical and environmental indicators for self-compacting concrete were considered. For this purpose, compressive strength, global warming potential (as carbon emission), and EE were investigated for concrete samples reinforced with micro steel and PVA fibers. It should be noted that in the present investigation, fly ash was utilized as a substitute for part of cement. The findings showed that reducing the amount of cement, as expected, resulted in a significant reduction in environmental indicators. Additionally, the use of micro steel fibers improved the mechanical characteristics of SCC. Also, the fact that there is less CO_2 emission and EE compared to PVA fibers has had a significant influence on increasing the mechanical index and reciprocally reducing the environmental indexes on concrete.

References

1. Bahramian, M., Yetilmezsoy, K.: Life cycle assessment of the building industry: an overview of two decades of research (1995–2018). Energ. Buildings. **219**, 109917 (2020)
2. Farahzadi, F., Kioumarsi, M.: Application of machine learning initiatives and intelligent perspectives for CO2 emissions reduction in construction. Clean. Prod. **384**, 135504 (2023)
3. Gursel, A.P., Masanet, E., Horvath, A., Stadel, A.: Life-cycle inventory analysis of concrete production: a critical review. Cem. Concr. Compos. **51**, 38–48 (2014)
4. Habert, G., Miller, S.A., John, V.M., Provis, J.L., Favier, A., Horvath, A., Scrivener, K.L.: Environmental impacts and decarbonization strategies in the cement and concrete industries. Nat. Rev. Earth Environ. **1**(11), 559–573 (2020)
5. Minunno, R., O'Grady, T., Morrison, G.M., Gruner, R.L.: Investigating the embodied energy and carbon of buildings: a systematic literature review and meta-analysis of life cycle assessments. Renew. Sust. Energ. Rev. **143**, 110935 (2021)
6. Vieira, D.R., Calmon, J.L., Coelho, F.Z.: Life cycle assessment (LCA) applied to the manufacturing of common and ecological concrete: a review. Constr. Build. Mater. **124**, 656–666 (2016)
7. Shi, X., Mukhopadhyay, A., Zollinger, D., Grasley, Z.: Economic input-output life cycle assessment of concrete pavement containing recycled concrete aggregate. J. Clean. Prod. **225**, 414–425 (2019)
8. Mesa, J.A., Fúquene-Retamoso, C., Maury-Ramírez, A.: Life cycle assessment on construction and demolition waste: a systematic literature review. Sustainability. **13**(14), 7676 (2021)
9. Hasanbeigi, A., Price, L., Lin, E.: Emerging energy-efficiency and CO2 emission-reduction technologies for cement and concrete production: a technical review. Renew. Sust. Energ. Rev. **16**(8), 6220–6238 (2012)

10. Nie, S., Zhou, J., Yang, F., Lan, M., Li, J., Zhang, Z., Chen, Z., Xu, M., Li, H., Sanjayan, J.G.: Analysis of theoretical carbon dioxide emissions from cement production: methodology and application. J. Clean. Prod. **334**, 130270 (2022)

11. Sahloddin, Y., Dalvand, A., Ahmadi, M., Hatami, H., Houshmand Khaneghahi, M.: Performance evaluation of built-up composite beams fabricated using thin-walled hollow sections and self-compacting concrete. Constr. Build. Mater. **305**, 124645 (2021)

12. Thangaraj, R., Thenmozhi, R.: Industrial and environmental application of high volume fly ash in concrete production. Nat. Environ. Pollut. Technol. **12**(2), 315 (2013)

13. Nath, P., Sarker, P.K., Biswas, W.K.: Effect of fly ash on the service life, carbon footprint and embodied energy of high strength concrete in the marine environment. Energ. Buildings. **158**, 1694–1702 (2018)

14. Lee, J.W., Jang, Y.I., Park, W.S., Yun, H.D., Kim, S.W.: The effect of fly ash and recycled aggregate on the strength and carbon emission impact of FRCCs. Int. J. Concr. Struct. Mater. **14**, 1–13 (2020)

15. Kioumarsi, M., Dabiri, H., Kandiri, A., Farhangi, V.: Compressive strength of concrete containing furnace blast slag; optimized machine learning-based models. Clean. Eng. Technol. **13**, 100604 (2023)

16. Tam, V.W., Le, K.N., Evangelista, A.C.J., Butera, A., Tran, C.N., Teara, A.: Effect of fly ash and slag on concrete: properties and emission analyses. Front. Eng. Manag. **6**, 395–405 (2019)

17. Ahmadi, M., Hakimi, B., Mazaheri, A., Kioumarsi, M.: Potential use of water treatment sludge as partial replacement for clay in eco-friendly fired clay bricks. Sustainability. **15**(12), 9389 (2023)

18. Mohanty, I., Saha, P., Patra, S.R., Jha, S.K.: Waste to valuable resource: application of copper slag and steel slag in concrete with reduced carbon dioxide emissions. Innov. Infrastruct. Solut. **8**(4), 122 (2023)

19. Shekarchi, M., Ahmadi, B., Azarhomayun, F., Shafei, B., Kioumarsi, M.: Natural zeolite as a supplementary cementitious material–a holistic review of main properties and applications. Constr. Build. Mater. **409**, 133766 (2023)

20. Hamad, M.A., Nasr, M., Shubbar, A., Al-Khafaji, Z., Al Masoodi, Z., Al-Hashimi, O., Kot, P., Alkhaddar, R., Hashim, K.: Production of ultra-high-performance concrete with low energy consumption and carbon footprint using supplementary cementitious materials instead of silica fume: a review. Energies. **14**(24), 8291 (2021)

21. Golewski, G.L.: Green concrete based on quaternary binders with significant reduced of CO2 emissions. Energies. **14**(15), 4558 (2021)

22. Kumar, A., Bheel, N., Ahmed, I., Rizvi, S.H., Kumar, R., Jhatial, A.A.: Effect of silica fume and fly ash as cementitious material on hardened properties and embodied carbon of roller compacted concrete. Environ. Sci. Pollut. Res. **29**, 1210–1222 (2022)

23. Reddy, B.V., Jagadish, K.S.: Embodied energy of common and alternative building materials and technologies. Energ. Buildings. **35**(2), 129–137 (2003)

24. Hammond, G.P., Jones, C.I.: Embodied energy and carbon in construction materials. Proc. Inst. Civ. Eng.-Energy. **161**(2), 87–98 (2008)

25. Khataei, B., Nasrollahi, M.: Optimizing the tensile strength of concrete containing coal waste considering the cost. SN Appl. Sci. **2**, 103 (2020)

26. Hu, C.: Microstructure and mechanical properties of fly ash blended cement pastes. Constr. Build. Mater. **73**, 618–625 (2014)

27. Kasaniya, M., Thomas, M.D., Moffatt, E.G.: Efficiency of natural pozzolans, ground glasses and coal bottom ashes in mitigating sulfate attack and alkali-silica reaction. Cem. Concr. Res. **149**, 106551 (2021)

28. Jung, S.H., Choi, Y.C., Choi, S.: Use of ternary blended concrete to mitigate thermal cracking in massive concrete structures—a field feasibility and monitoring case study. Constr. Build. Mater. **137**, 208–215 (2017)

29. Yang, K.H., Song, J.K., Song, K.I.: Assessment of CO2 reduction of alkali-activated concrete. J. Clean. Prod. **39**, 265–272 (2013)

30. Mohamad, N., Lakhiar, M.T., Samad, A.A.A., Mydin, M.A.O., Jhatial, A.A., Sofia, S.A., Goh, W.I., Ali, N.: Innovative and sustainable green concrete–a potential review on utilization of agricultural waste. IOP Conf. Ser.: Mater. Sci. Eng. **601**(1), 012026 (2019). IOP Publishing
31. Kim, T.H., Chae, C.U., Kim, G.H., Jang, H.J.: Analysis of CO2 emission characteristics of concrete used at construction sites. Sustainability. **8**(4), 348 (2016)
32. Dalvand, A., Ahmadi, M.: Impact failure mechanism and mechanical characteristics of steel fiber reinforced self-compacting cementitious composites containing silica fume. Eng. Sci. Technol., Int. J. **24**(3), 736–748 (2021)
33. Paul, S.C., van Zijl, G.P., Šavija, B.: Effect of fibers on durability of concrete: a practical review. Materials. **13**(20), 4562 (2020)
34. Zamani, A.A., Ahmadi, M., Dalvand, A., Aslani, F.: Effect of single and hybrid fibers on mechanical properties of high-strength self-compacting concrete incorporating 100% waste aggregate. J. Mater. Civ. Eng. **35**(1), 04022365 (2023)
35. Blazy, J., Blazy, R., Drobiec, Ł.: Glass fiber reinforced concrete as a durable and enhanced material for structural and architectural elements in smart city—a review. Materials. **15**(8), 2754 (2022)
36. Sattarifard, A.R., Ahmadi, M., Dalvand, A., Sattarifard, A.R.: Fresh and hardened-state properties of hybrid fiber–reinforced high-strength self-compacting cementitious composites. Constr. Build. Mater. **318**, 125874 (2022)
37. Lippiatt, B.C.: Bees 4.0: Building for Environmental and Economic Sustainability. Technical Manual and User Guide (2007)
38. Bheel, N., Mohammed, B.S., Ali, M.O.A., Shafiq, N., Tag-eldin, E.M., Ahmad, M.: Effect of polyvinyl alcohol fiber on the mechanical properties and embodied carbon of engineered cementitious composites. Results Eng. **20**, 101458 (2023)
39. Gholami, E., Afshin, H., Basim, M.C., Sharghi, M.: Ultra-high performance recycled steel fiber reinforced concrete segments under the thrust force of TBM jacks and their environmental potentialities. Structures. **47**, 2465–2484 (2023). Elsevier
40. Lin, J.X., Luo, R.H., Su, J.Y., Guo, Y.C., Chen, W.S.: Coarse synthetic fibers (PP and POM) as a replacement to steel fibers in UHPC: tensile behavior, environmental and economic assessment. Constr. Build. Mater. **412**, 134654 (2024)
41. National Ready Mixed Concrete Association: Concrete CO2 Fact Sheet. NRMCA Publication, (2PCO2) (2012)

Strengthening Brick Masonry Structures with Natural Fiber Elements for Enhancing Earthquake Resistance

Manisha Kushwaha ⓘ, Kusum Saini ⓘ, and Vasant Matsagar ⓘ

Contents

1   Introduction

Every year, earthquakes cause a lot of destruction and damage to civil infrastructures, further leading to economic and life losses. Many regions in countries like India, Japan, Nepal, and the United States have frequent seismic activities. For example, the Gorkha earthquake in 2015 resulted in damage to more than half a million buildings and subsequently life loss in Nepal [1]. Such events show the necessity of advancing seismic design techniques for structures. Especially for commonly used burned clay brick masonry structures in earthquake-prone areas, they are at risk of multiple failure modes. Mercimek [2] and Canditone et al. [3] have shown the common shear failure resulting from cracks along mortar joints under earthquakes. Flexural [4] and diagonal tension failure [5] have also been reported as predominant failures in masonry structures. Moreover, out-of-plane and in-plane failures are major failures in masonry under lateral dynamic loads.

M. Kushwaha · K. Saini · V. Matsagar (✉)
Department of Civil Engineering, Indian Institute of Technology (IIT) Delhi,
Hauz Khas, New Delhi, India
e-mail: matsagar@civil.iitd.ac.in

© The Author(s) 2025
M. Kioumarsi, B. Shafei (eds.), *The 1st International Conference on Net-Zero Built
Environment*, Lecture Notes in Civil Engineering 237,
https://doi.org/10.1007/978-3-031-69626-8_33

Traditional construction practices are unable to provide sufficient resistance to masonry structures under seismic activities. Inadequate material bonding, irregular structural configuration, insufficient reinforcement, heavy constructions, and not implementing seismic standards are some of the main reasons that lead to seismic failure of structures. Therefore, the development of seismic design of civil infrastructures and their strengthening are required, especially in these regions. Various strengthening techniques are adopted in design practices for new construction and existing structures. Strengthening with synthetic fiber-reinforced polymer (FRP) composites (i.e., carbon and glass FRP composites) in different forms, such as fabric, bars, and laminates as embedded and externally bonded reinforcement have been proposed as effective methods [6], to enhance the performance of structures under earthquake loadings. The strengthening provides resistance to the in-plane and out-of-the-plane failures of the masonry under seismic vibrations. However, the production of synthetic composites and their use in construction further lead to an increase in greenhouse gas (GHG) emissions. Moreover, disposal of such materials becomes another challenge to environmental safety.

Therefore, in this study, locally available natural fiber elements are proposed as a sustainable and cost-effective alternative to synthetic FRP composites for the strengthening of buildings. Saini et al. [7] have presented a detailed study on the strengthening of structures with various natural fibers in different forms. The use of sisal, jute, and flax in the form of short fibers in earthen brick, ropes to strengthen block masonry, and fabric to strengthen a masonry wall have been demonstrated [7]. Similarly, the enhancement in seismic resistance of an adobe wall with palm fibers has been obtained [8]. The effectiveness of hemp fiber composite grids in masonry panels and hemp ropes in adobe walls in increasing ductility has been reported by Menna et al. [9] and Abdulla et al. [10], respectively. Bitar et al. [11] also reported a significant increase in flexural strength at a 2.0% reinforcement ratio of flax fabric in masonry. The advantages of flax [12] and coir [13] fiber composites, i.e., high strength and less internal stress in composites, and durability and wear resistance, respectively, make them a suitable option for FRP strengthening.

However, there is no study that shows the seismic behavior of FRP composite strengthened a full-scale building. Therefore, in the present study, the efficacy of strengthening with flax and coir FRP composites on the seismic performance of a masonry building with opening is investigated. The seismic analysis of a structure has different components, which include modal, linear, and nonlinear dynamic analyses to assess the accurate seismic behavior of a structure [14]. Generally, the seismic response of the structure is evaluated in the form of displacement, acceleration, drift [14], and stresses [15, 16] to get a wholesome idea of the dynamic behavior of the structure required in new construction and in retrofitting of the structure against earthquake. Therefore, the main objectives of this study are: (a) to investigate the seismic behavior of a realistic building with openings designed for seismic loads and (b) to assess the effectiveness of flax and coir FRP composites in strengthening the building under earthquake load.

2 Numerical Modeling

A finite element (FE) model of a two-story unreinforced masonry building with openings considering fixed boundary conditions at the base is developed in ETABS® [17]. The concrete columns and beams are modeled as a frame, reinforcement bars in columns and beams are as a uniaxial bar element, and slab and masonry walls are modeled with shell elements. A macro modeling approach, continuum homogeneous modeling, is implemented for the masonry walls. In the study, the mortar is taken as an integrated part of the masonry and not modeled separately [18, 19]. An efficient mesh size of 25 mm is considered after a mesh convergence study. The FRP composite is also modeled with shell elements and considered perfectly bonded to masonry walls. The front elevation of the building has four windows and one door opening, and the back elevation has five window openings, as shown in Figs. 1 and 2.

2.1 Design of Building

The building with plan dimensions, length of 8.68 m, and width of 4.5 m, is provided with 0.23 m thick load-bearing brick masonry walls considered as per NBC guidelines [20], shown in Fig. 1. The size of the square window and door are taken as 1.219 m and 2.134 m long with 0.914 m width. A 0.125 m thick roof slab is provided in the building. Furthermore, lintel and sill seismic bands at heights of 2.134 m and 0.914 m from the base, respectively (see Fig. 2), are included in the design as per the NBC guidelines [20]. The flax and coir composites taken for strengthening the brick masonry walls of the building are designed using ACI 440 guidelines [6]. The FRP composites with 0.125 m width and 0.15 m thickness are applied externally on the masonry wall in a grid configuration (spacing less than 0.8 m), illustrated in Fig. 2.

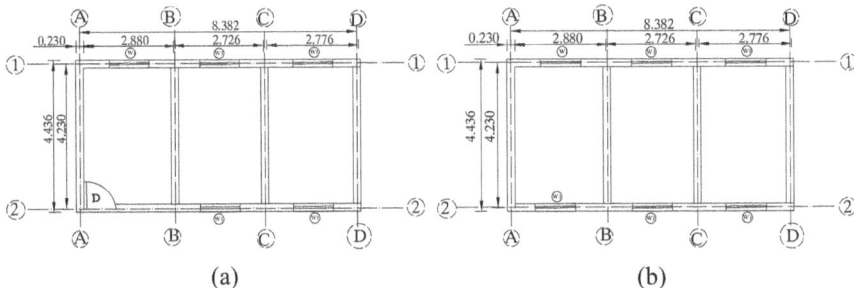

(a)   (b)

Fig. 1 The plan view of (**a**) the ground floor and (**b**) the first floor of the building considered in this study

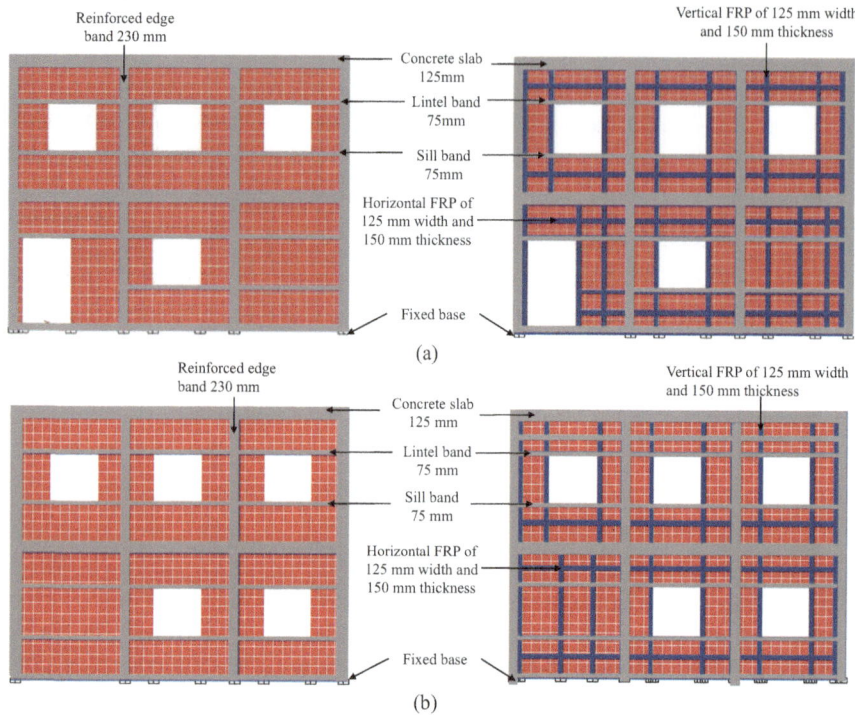

Fig. 2 Masonry buildings considering seismic bands, (**a**) front view and (**b**) back view strengthened without and with FRP composites

2.2 Material Properties

Masonry, concrete, and steel reinforcement are taken as isotropic materials, and FRP composites are considered to be orthotropic materials. The material properties of brick masonry and reinforced concrete used for columns and beams are considered as per NBC guidelines [20], as presented in Table 1. Moreover, the material properties of flax [12] and coir [13, 21] FRP composites are considered from the literature (see Table 2).

2.3 Permissible Stresses in Masonry

The permissible stresses in unreinforced masonry are estimated as per IS 1905 [22]. The design parameters of a masonry unit with 10 MPa are taken as thickness (t), 75 mm and width (w), 110 mm with mortar type M1. The permissible compressive stress ($f_c = f_b \times k_s \times k_a \times k_p$) in masonry depends upon basic compressive stress (f_b), 0.96 MPa, stress reduction factor (k_s), 0.43 (for a slenderness ratio of 27), area

Table 1 Material properties of brick masonry, concrete, and steel reinforcement adopted in the study

S. N.	Properties	Brick masonry	Concrete	Steel reinforcement
1.	Modulus of elasticity (MPa)	2112	22,360.68	200,000
2.	Poisson's ratio	0.3	0.2	–
3.	Density (kN/m^3)	19.2	25.0	76.97
4.	Co-efficient of thermal expansion (1/°C)	3.1×10^{-6}	13×10^{-6}	11.7×10^{-6}
5.	Shear modulus (MPa)	812.31	9316.95	891.25

Table 2 Material properties of flax and coir FRP composites used in the study

S. N.	Properties	Flax composite [12]	Coir composite [13, 21]
1.	Modulus of elasticity (MPa)	22,300	2139
2.	Poisson's ratio	0.44	0.2
3.	Density (kN/m^3)	12.18	7.24
4.	Co-efficient of thermal expansion (1/°C)	1.08×10^{-6}	1.2×10^{-6}
5.	Shear modulus (MPa)	1150	891.25

reduction factor (k_a), and shape modification factor (k_p), 1.0 (for height/width ratio ≤ 0.075). The area reduction factor ($k_a = 0.7 + 1.5A$) depends on the area of the section (m^2), A, and results in 0.71 magnitude. The permissible compressive stress results in 0.29 MPa and the permissible shear stress ($\tau_0 = 0.1 + f_d/6$) depends upon compressive stress due to dead loads (f_d), 0.17 MPa, resulting in approximately 0.13 MPa.

3 Results and Discussion

Modal and linear time history analyses are performed to assess the seismic behavior of masonry buildings without and with flax and coir composites.

3.1 Modal Analysis

A modal analysis of the masonry buildings without and with FRP composite strengthening is performed in this study. The P-Delta effect is insignificant, so it is not considered in the study. The major modes, time periods, and frequencies obtained are listed in Table 3. The mass participation of 92.15% and 91.45% is observed for respective translational Modes 1 and 2, whereas rotation is reported in mode 3. Mode 4 and Mode 5 have showed 7.82% and 8.53% mass participation in translational direction, respectively.

Table 3 Results of the modal analysis of the masonry building without and with flax and coir FRP composites strengthening

Mode	Building without strengthening		Building with flax composite		Building with coir composite	
	Time period (s)	Frequency (Hz)	Time period (s)	Frequency (Hz)	Time period (s)	Frequency (Hz)
1	0.131	7.622	0.099	10.13	0.124	8.052
2	0.118	8.472	0.094	10.655	0.115	8.723
3	0.099	10.063	0.079	12.656	0.096	10.469
4	0.057	17.590	0.042	23.808	0.054	18.613
5	0.046	21.661	0.037	26.807	0.045	22.107

It is observed that the strengthening of buildings with flax FRP composite results in an 8.33% reduction in the fundamental time period and an 8.96 % increase in the frequency of the building, as compared to that of the building without strengthening. Similarly, adding the coir FRP composite leads to a reduction of 2.78 % in the fundamental time period and a increment of 2.54 % in the frequency of the un-strengthened building, respectively. The increase in the fundamental frequency indicates enhancements in the stiffness of the building and subsequent seismic performance. The flax FRP composite has higher stiffness and strength than the coir FRP composite, resulting in higher enhancement in the stiffness of the structure. However, both composites have low density, leading to negligible increment in the mass of the structure. Thus, both lightweight biocomposites are preferred for strengthening structures due to the effective enhanced structural performance of the masonry structure.

3.2 Time History Analysis

Further, to compare the seismic performance of the building without and with FRP composites, a uniaxial linear time history analysis of the building where the seismic load is applied in the direction of X is performed.

The response spectrum is considered per NBC recommendations for zone IV (zone factor 0.24) and soil type 2 ground conditions, where the response reduction factor, 3, the importance factor, 1, and 5% damping are taken. The acceleration time history of ground motion from the 1940 Imperial Valley earthquake ground motion data recorded at El Centro station with a peak ground acceleration of 0.281 g (see Fig. 3), is applied to the structure. The seismic response is measured in terms of peak shear force and story drift, presented in Table 4. Moreover, the peak base shear and acceleration time histories are depicted in Fig. 4. The fundamental frequencies of the building without strengthening and with coir composite are very close, resulting in similar mode shapes and vibrations of the masonry building. However, due to the

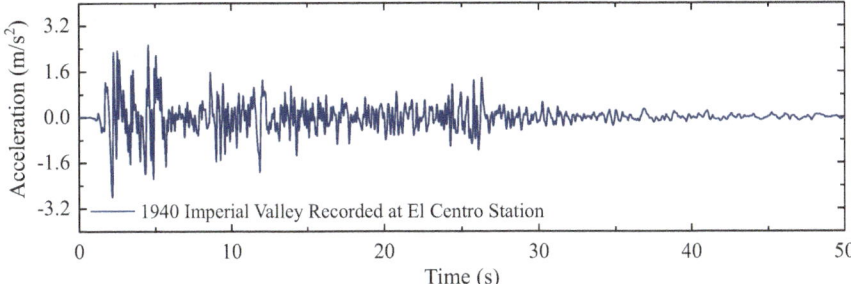

Fig. 3 The ground motion time history recorded at El Centro in the 1940 Imperial Valley Earthquake

Table 4 Response of the building without and with flax and coir FRP composites strengthening subjected to the earthquake ground motion

Response	Building without strengthening	Building with flax composite	Building with coir composite
Shear force (kN)	62.23	34.82	48.60
Peak drift ratio (%)	0.26	0.19	0.24

increase in the stiffness and strength of the building with flax FRP composite, the dynamic behavior of the structure changes compared to the structure without FRP composite.

It is observed that the tensile strength and stiffness enhancement due to strengthening with flax and coir FRP composites lead to 44.1% and 21.9% reduction in peak shear forces, respectively. Moreover, the overall reduction in floor acceleration response at the first story and roof level is reported due to flax and coir composite strengthening, respectively. Similarly, 28.6% and 8.8% reductions in peak deflection of the building due to the addition of flax and coir composites are obtained, respectively. This indicates a notable improvement in energy dissipation and reduction in kinetic energy, resulting in minimizing the impact of seismic forces on the building. The peak drift reduces to 0.19% and 0.24% with flax and coir FRP composites, which also presents the reduction in deformation of the building. These findings showcase the evident impact of flax and coir FRP composites on the seismic resistance of the building.

3.2.1 Stress Distribution

Stress distribution across the two-story buildings reveals a distinct pattern around the openings, where the highest stress concentration at the corners of the openings is obtained, resulting in localized failures in the masonry walls on both the front and back sides of the structure. The discontinuity in stress distribution is caused by openings in the wall, which further leads to stiffness reduction and seismic

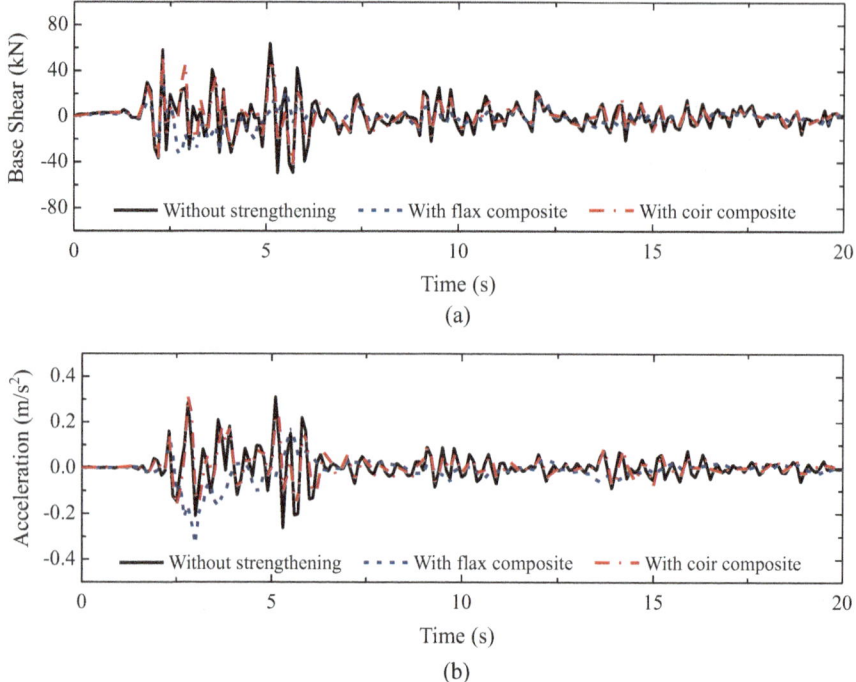

Fig. 4 The peak response time histories of the building without and with flax and coir FRP composites strengthening under the earthquake ground motion, (**a**) base shear and (**b**) acceleration

Table 5 Principal stresses in the masonry buildings without and with flax and coir FRP composites strengthening subjected to the earthquake ground motion

Stress	Building without strengthening (MPa)	Building with flax composite (MPa)	Building with coir composite (MPa)
Normal stress in the longitudinal direction (S11)	0.32	0.27	0.37
Normal stress in transverse direction (S22)	0.75	0.29	0.59
Shear stress in the in-plane direction (S12)	0.3	0.25	0.28

vulnerability of the structure. Therefore, the natural FRP composites are provided around openings as per ACI 440 [6] to reduce the stresses, as shown in Table 5.

It is observed that cracks propagate from corners of the openings diagonally under seismic load. The provided seismic bands also help restrict the stresses to a region (see Figs. 5, 6, and 7). The major stresses are longitudinal (S11) and transverse (S22) normal stresses, and in-plane shear (S12) stress, compared in Figs. 5, 6, and 7, respectively.

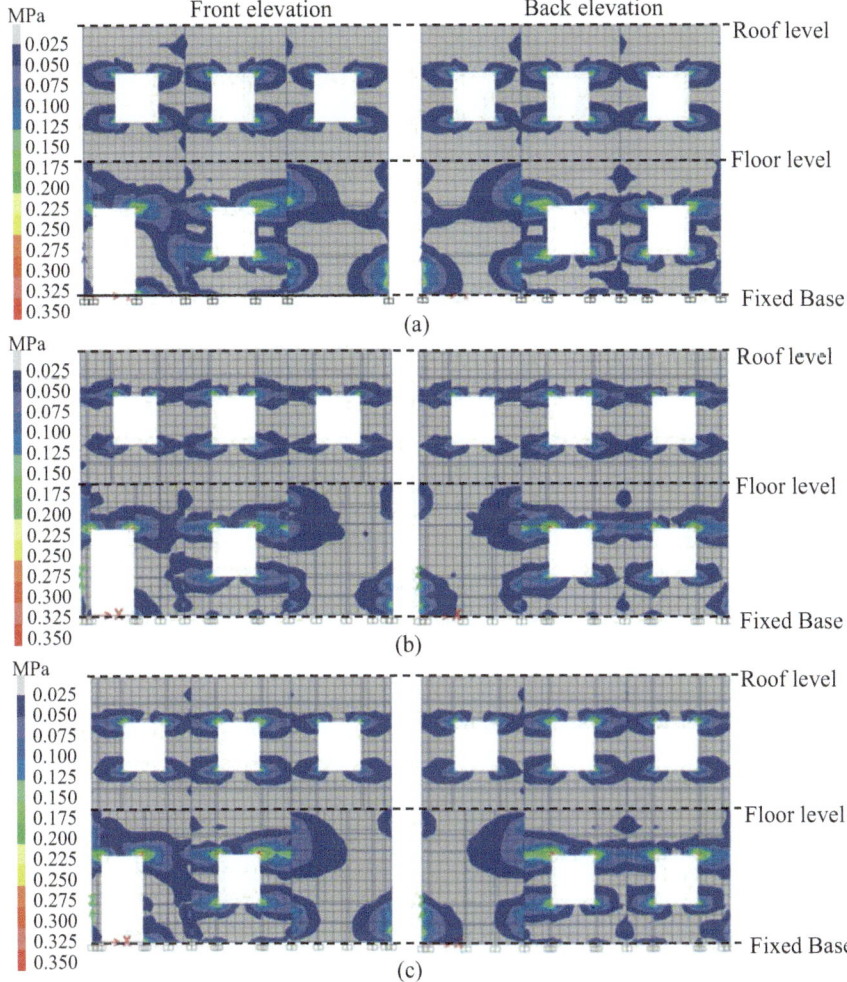

Fig. 5 Longitudinal normal stress (S11) in the building, (**a**) without strengthening, (**b**) with flax FRP composite, and (**c**) with coir FRP composite under the earthquake ground motion

The exceedance of stresses beyond permissible stresses in the case of the building without strengthening is clearly visible. The remaining normal (S33) and shear stresses (S13 and S23) are found insignificant; therefore, not presented here. The highest exceedance is observed for in-plane shear stress (S12) beyond permissible shear stress. This comparative analysis highlights the efficacy of FRP composites in strengthening structural weak points. It illustrates that flax FRP composite stands out as a more effective option for improving the seismic resilience of buildings, particularly by addressing stress concentrations around openings. The strengthening with flax and coir FRP composites results in a reduction of the longitudinal normal stress

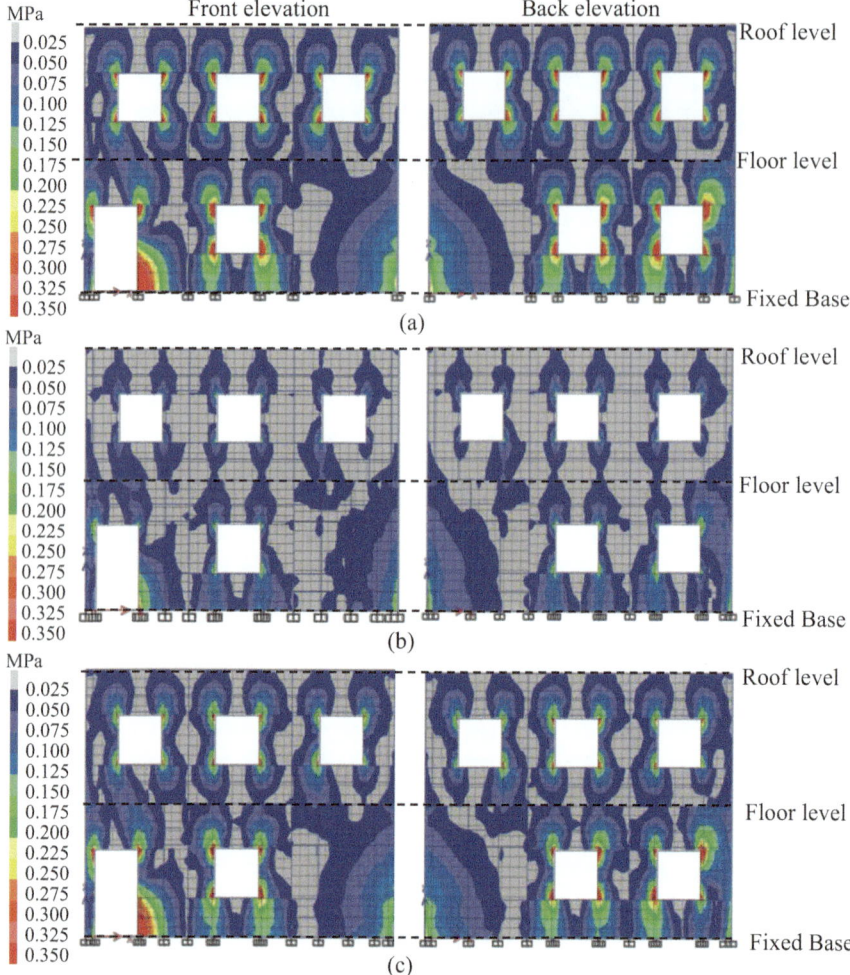

Fig. 6 Transverse normal stress (S22) in the building, (**a**) without strengthening, (**b**) with flax FRP composite, and (**c**) with coir FRP composite under the earthquake ground motion

by 15.6% and approximately insignificant, whereas in transverse normal stress by 61.34% and 21.34%, and in in-plane shear stress by 16.67% and 6.67%, respectively. Both flax and coir FRP composites can reduce the stresses in the masonry wall. However, the current grid configuration with coir FRP composite is not sufficient to reduce stresses and to keep them within the permissible stress limits. On the other hand, the current grid configuration with flax FRP composite sufficiently reduces the normal stresses than the permissible compressive stresses.

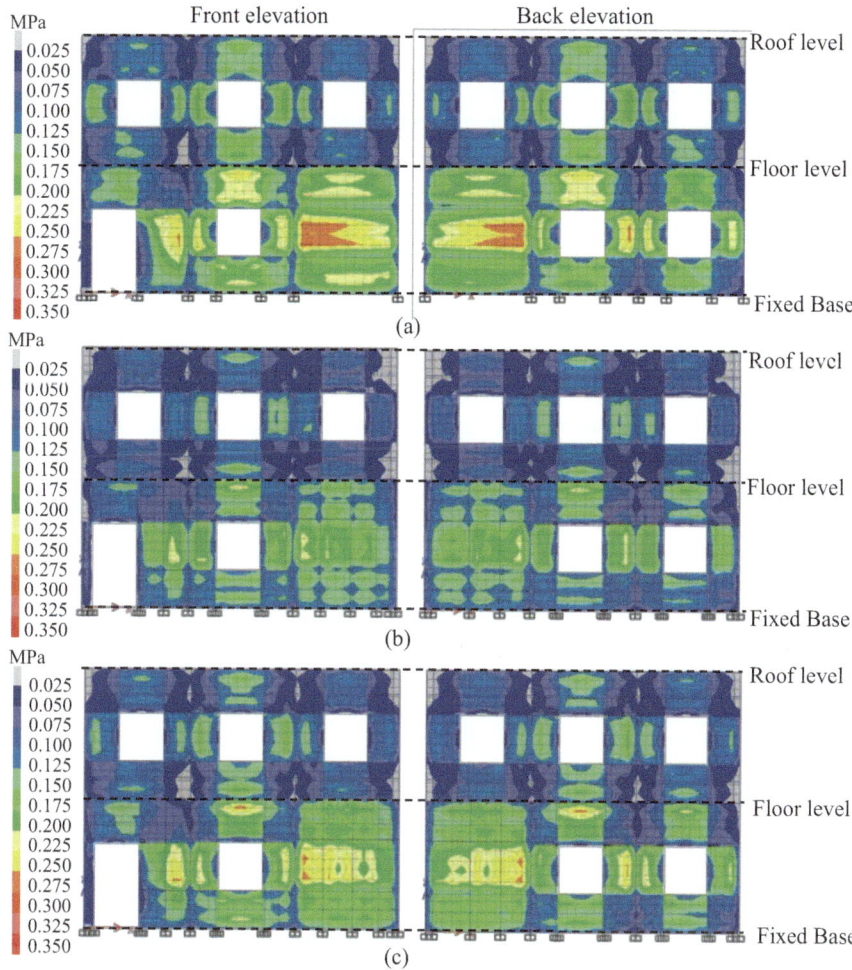

Fig. 7 In-plane shear stress (S12) in the building, (**a**) without strengthening, (**b**) with flax FRP composite, and (**c**) with coir FRP composite under the earthquake ground motion

However, to restrict the shear stress in buildings (to keep within the limit of permissible shear stress), various configurations of strengthening strategies and ways to increase in the stiffness (i.e., design dimensions, treatment of fibers) of composites are further required to be investigated as future scope of study. The design of the composites to strengthen the unreinforced masonry walls is considered as per the ACI 440 [6] recommendations, which are proposed for strengthening with synthetic FRP composites. Therefore, the same design guidelines may not work for strengthening of structures with natural fiber-based composites due to their lesser stiffness and strength than synthetic FRP composites, which reflects the need of new design guidelines for natural fiber-based strengthening techniques.

4 Conclusions

In this study, the seismic performance of two-story unreinforced masonry buildings (designed as per Nepal national building code guidelines)without and with flax and coir fiber-reinforced polymer (FRP) composite strengthening is assessed. A finite element model of the brick masonry building is developed and further, strengthened with the externally bonded FRP composites designed as per ACI 440 recommendations. Modal and time history analyses are conducted to show the efficacy of the natural fiber-based composites in strengthening structures under seismic forces, for advancing their broader adoption in earthquake-prone regions. The major conclusions drawn based on the findings of the study are listed as follows.

1. Strengthening with natural fiber-reinforced polymer composites in the form of externally bonded wrapping in a grid configuration, especially around openings of the unreinforced masonry walls of the building, as per ACI 440 guidelines, can help in reducing the seismic response of the building significantly.
2. The externally bonded wrapping with flax FRP composites for strengthening the masonry building is the most effective in enhancing the seismic resistance of the structure by reducing the base shear, peak acceleration, and stresses under earthquake ground motions.
3. The wrapping configuration in grids with the current design of flax and coir FRP composites for strengthening masonry walls is not sufficient to prevent the exceedance of in-plane shear stress compared to the permissible shear stress for unreinforced masonry under seismic load.
4. The design guidelines for strengthening of masonry structures with externally bonded synthetic FRP composites, i.e., carbon and glass FRP, may not be perfectly valid for the case of strengthening with natural FRP composites. Therefore, there is a need to develop new design standards and guidelines for strengthening masonry structures with natural fiber-based elements.

The study of integrating environmental considerations into structural engineering practices contributes to the development of sustainable strengthening methods for brick masonry structures globally.

References

1. Gautam, D., Chaulagain, H.: Structural performance and associated lessons to be learned from world earthquakes in Nepal after 25 April 2015 (MW 7.8) Gorkha earthquake. Eng. Fail. Anal. **68**, 222–243 (2016). https://doi.org/10.1016/j.engfailanal.2016.06.002
2. Mercimek, Ö.: Seismic failure modes of masonry structures exposed to Kahramanmaraş earthquakes (Mw 7.7 and 7.6) on February 6, 2023. Eng. Fail. Anal. **151** (2023). https://doi.org/10.1016/j.engfailanal.2023.107422
3. Canditone, C., Diana, L., Formisano, A., Rodrigues, H., Vicente, R.: Failure mechanisms and behaviour of adobe masonry buildings: a case study. Eng. Fail. Anal. **150**, 107343 (2023). https://doi.org/10.1016/j.engfailanal.2023.107343

4. Tomaževič, M.: Earthquake-Resistant Design of Masonry Buildings, vol. 1. World Scientific (1999)

5. Paudel, K.: Structural suitability of masonry structure for residential buildings in rural areas of Nepal. Doctoral dissertation, Pulchowk Campus (2021)

6. ACI 440.2R-08: Guide for the Design and Construction of Externally Bonded FRP Systems for Strengthening Concrete Structures. American Concrete Institute (ACI) (2008). https://doi.org/10.14359/51700867

7. Saini, K., Matsagar, V.A., Kodur, V.R.: Recent advances in the use of natural fibers in civil engineering structures. Constr. Build. Mater. **411**, 134364 (2024). https://doi.org/10.1016/j.conbuildmat.2023.134364

8. Meybodian, H., Morshed, R., Eslami, A.: Experimental investigation on the seismic behavior of adobe walls retrofitted with palm meshes. Amirkabir J. Civil Eng. **52**(8), 2129–2142 (2020). https://doi.org/10.22060/CEEJ.2019.15927.6079

9. Menna, C., Asprone, D., Durante, M., Zinno, A., Balsamo, A., Prota, A.: Structural behaviour of masonry panels strengthened with an innovative hemp fibre composite grid. Constr. Build. Mater. **100**, 111–121 (2015). https://doi.org/10.1016/j.conbuildmat.2015.09.051

10. Abdulla, K.F., Cunningham, L.S., Gillie, M.: Out-of-plane strengthening of adobe masonry using hemp fibre ropes: an experimental investigation. Eng. Struct. **245**, 112931 (2021). https://doi.org/10.1016/j.engstruct.2021.112931

11. Bitar, R., Saad, G., Awwad, E., El Khatib, H., Mabsout, M.: Strengthening unreinforced masonry walls using natural hemp fibers. J. Build. Eng. **30**, 101253 (2020). https://doi.org/10.1016/j.jobe.2020.101253

12. Saidane, E.H., Scida, D., Ayad, R.: Thermo-mechanical behavior of flax/green epoxy composites: evaluation of thermal expansion coefficients and application to internal stress calculation. Ind. Crop. Prod. **170**, 113786 (2021). https://doi.org/10.1016/j.indcrop.2021.113786

13. Hasan, K.M.F., Horváth, P.G., Kóczán, Z., Alpár, T.: Thermo-mechanical properties of pretreated coir fiber and fibrous chips reinforced multilayered composites. Sci. Rep. **11**(1), 3618 (2021). https://doi.org/10.1038/s41598-021-83140-0

14. Chopra, A.K.: Dynamics of Structures: Theory and Applications to Earthquake Engineering, 3rd edn. Pearson Education. ISBN: 9788131713297 (2007)

15. Galasco, A., Lagomarsino, S., Penna, A., Resemini, S.: Non-linear seismic analysis of masonry structures. In: 13th World Conference on Earthquake Engineering, vol. 843, 1–6, Vancouver, B. C., Canada (2004)

16. Augenti, N., Parisi, F.: Non-linear static analysis of masonry structures. In: Proceedings of the 13th Italian National Conference on Earthquake Engineering, vol. 1, S4, Bologna, Italy (2009)

17. ETABS®: User's Guide. Computers and Structures, Inc (2016). https://www.csiamerica.com

18. D'Altri, A.M., Sarhosis, V., Milani, G., Rots, J., Cattari, S., Lagomarsino, S., Sacco, E., Tralli, A., Castellazzi, G., de Miranda, S.: Modeling strategies for the computational analysis of unreinforced masonry structures: review and classification. Arch. Comput. Methods Eng. **27**(4), 1153–1185 (2019). https://doi.org/10.1007/s11831-019-09351-x

19. Ferreira, T.M., Mendes, N., Silva, R.: Multiscale seismic vulnerability assessment and retrofit of existing masonry buildings. Buildings. **9**(4), 91 (2019). https://doi.org/10.3390/buildings9040091

20. NBC 202:2015. Guideline On: Load Bearing Masonry. Nepal National Building Code, Department of Urban Development and Building Construction, Ministry of Urban Development, Government of Nepal (2015)

21. Renjith, R., Nair, R.P.: Thermal analysis of natural fiber reinforced composites. Mater. Today: Proc. **72**, 3216–3221 (2023). https://doi.org/10.1016/j.matpr.2022.12.024

22. IS 1905: Code of Practice for Structural Use of Unreinforced Masonry, 3rd edn. Bureau of Indian Standard (BIS) (1987)

Mechanical Reinforcement of Building Materials with Microfibers Produced by Electrospinning

Habtom Daniel, Omar Mohamed Omar, Mahdi Kioumarsi, Shima Pilehvar, and Rafael Borrajo-Pelaez

Contents

1 Introduction

Integrating electrospun micro- and nanofibers into traditional construction materials, such as cement and lime composites, offers a promising avenue for enhancing their mechanical properties and structural integrity. Fibers with diameters in the micron and sub-micron range possess unique attributes, including elevated surface area, flexibility, and strength, making them ideal candidates for reinforcing construction

H. Daniel · O. M. Omar
Department of Built Environment, Faculty of Technology, Art and Design, Oslo Metropolitan University, Oslo, Norway
e-mail: S341705@oslomet.no

M. Kioumarsi
Department of Built Environment, Faculty of Technology, Art and Design, Oslo Metropolitan University, Oslo, Norway

Department of Engineering, Østfold University College, Halden, Norway

© The Author(s) 2025
M. Kioumarsi, B. Shafei (eds.), *The 1st International Conference on Net-Zero Built Environment*, Lecture Notes in Civil Engineering 237,
https://doi.org/10.1007/978-3-031-69626-8_34

materials [1, 2]. Electrospinning, a widely utilized process for generating ultrafine fibers from various materials, is pivotal in producing micro- and nanofibers for construction applications [3].

The electrospinning process typically involves three main components: a high-voltage power supply, a polymer solution reservoir equipped with a syringe and a small diameter needle, and a conductive collector [4]. Electrospinning generates a charged polymer jet from the solution, which solidifies into microfibers upon reaching the collector surface [2]. This process can be adapted to vertical [2, 4] or horizontal setups, each offering distinct advantages in terms of fiber orientation and production efficiency [5]. Vertical setup has gained the attention of many researchers because of its simplicity and flexibility in blending fibers with cementitious materials [1, 2, 4, 6].

In construction materials, producing microfibers involves careful consideration of various parameters, including polymer concentration, solvent selection, and electrospinning parameters such as voltage, needle diameter, tip–collector distance, and flow rate. These parameters significantly influence the morphology and properties of the resulting microfibers. For instance, higher polymer concentrations produce thicker fibers, while lower flow rates result in thinner fibers with greater polarization [5]. Additionally, the distance between the needle tip and collector surface is crucial in facilitating proper fiber formation and evaporation of solvent [5, 7].

The literature shows that researchers have used a variety of polymers to produce micro- and nanofibers and blend them with cement to increase the mechanical properties, for instance, Nylon 66. The addition of Nylon 66 has been credited with an increase of 41% and 33% in tensile strength and compressive strength, respectively [8]. Other studies reported that Nylon 66 could increase tensile and compressive strength by 25%, 75%, and 7% respectively [1].

This study focuses on understanding and optimizing the electrospinning parameters essential for tailoring microfiber-reinforced composites to specific construction applications by systematically investigating production methods and materials for creating electrospun microfibers. Furthermore, different blending processes between microfibers and cementitious materials are examined and compared. Finally, results from preliminary mechanical tests are reported, analyzing the resulting structural and mechanical properties of microfiber-reinforced cement composites.

S. Pilehvar
Department of Engineering, Østfold University College, Halden, Norway

R. Borrajo-Pelaez (✉)
Department of Mechanical, Electrical and Chemical Engineering, Oslo Metropolitan University, Oslo, Norway
e-mail: rafael.borrajo@oslomet.no

2   Materials

2.1   Chemicals and Reagents

The micro-fibers used in this study were obtained by electrospinning using the polymer Cellulose Acetate (CA) (Sigma-Aldrich, from Schnelldorf Distribution density: 1.3 g/ml). Acetone, $(CH_3)_2CO$, was the solvent used in the electrospinning process to obtain cellulose acetate fibers (VWR International, density: 0.792 g/ml).

In addition to CA, the literature indicates that several other polymers are promising candidates for the reinforcement of cement materials. The choice of CA for this study, over other relevant chemicals, is due to safety constraints since the use of alternative polymers required the use of toxic solvents, which was not feasible in the laboratory premises available for this work.

2.2   Solution of CA

The CA solution was prepared by dissolving cellulose acetate into acetone. The CA solution was prepared following the work of [9]. During the electrospinning process, the coagulation of the solution at the emitter tip was observed. Multiple studies have observed coagulation during electrospinning with CA solutions. To mitigate this issue, this study followed the suggestion of increasing the viscosity of the solutions, as found in the literature [5, 9].

2.3   Cementitious Composite

The cement used in this experiment to make a cement-CA electrospun microfiber composite material was an industry-grade binder known as "CEM I 52.5 R" from Heidelberg Material AS (NORCEM). This binder has an early-stage compressive strength of 52.5 MPa and meets NS-EN 197-1 standard.

3   Experimental Apparatus and Methods

3.1   Electrospinning Experiments

Microfibers are created using an electrospinning system manufactured by the company Bioinicia Fluidnatek, model number P 2022623-P1, equipped with an internal power supply. The liquid solution is transferred from a syringe to the electrospinning source, specifically to the needle's tip, through a capillary tube using a pumping device from Kd Scientific, allowing careful flow rate control (Fig. 1).

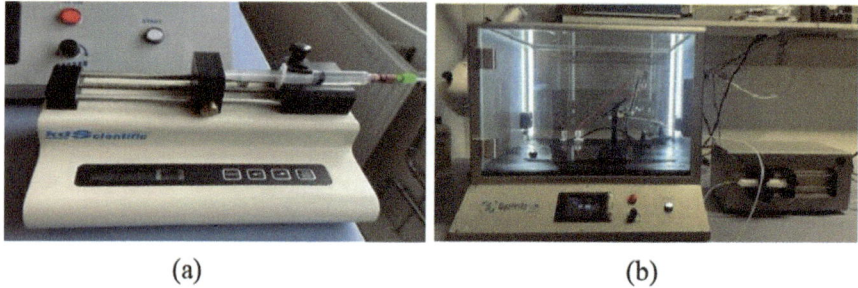

Fig. 1 Electrospinning system: (**a**) syringe pump and (**b**) electrospinning chamber, controlling the flow rate in the experiment

Table 1 The parameters tested with various gauge numbers

Test	List of parameters tested	Gauge number
1	12.5 g CA solution for flow rate effect	G15
2	12.5 g CA solution for flow distance effect	G15
3	12.5 g CA solution for flow voltage effect	G15
4	12.5 g CA solution for flow rate effect	G18
5	12.5 g CA solution for flow distance effect	G18
6	12.5 g CA solution for flow voltage effect	G18
7	12.5 g CA solution for flow rate effect	G21
8	12.5 g CA solution for flow distance effect	G21
9	12.5 g CA solution for flow voltage effect	G21

Standard stainless-steel needles are used as electrospinning emitter tip, with various gauges, including G15, G18, and G21. The solution was prepared by dissolving 12.5 g of CA in 100 ml of acetone. A summary of the tests that were conducted is presented in Table 1.

3.2 Mixing of Fibers and Cement

This study employed two different methods for mixing fibers and cement powder, to check the effect of the mixing approach on the resulting composite. For this purpose, experiments with two distinct collectors for the fiber were conducted.

The first approach entailed the collection of fibers over a flat surface, particularly a flat aluminum foil horizontally located beneath the vertically installed emitter, grounding the foil. Subsequently, a layer of 30 g cement was applied over the aluminum, followed by the spraying of fibers atop the cement layer. Another 30 g layer of cement was then applied over the fibers, which was iteratively repeated. See Fig. 2a.

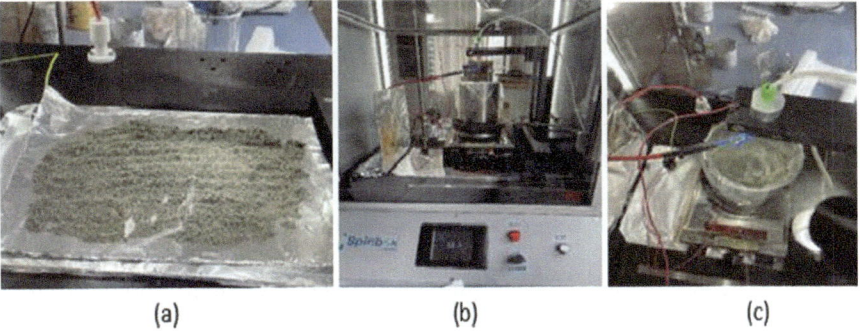

(a)   (b)   (c)

Fig. 2 (**a**) First mixing approach, mixing CEM I and fiber horizontally over aluminum foil; (**b**) and (**c**) second mixing method, involving a metal bowl collector on a magnetic stirrer

Fig. 3 The outcome of combining fiber and cement using a magnetic stirrer

The second fiber collection method involved using a metal bowl collector placed on top of a magnetic stirrer machine. The top of the bowl was covered with plastic to keep the fibers from adhering to its sides. This arrangement allowed for the effective mixing of the fibers with the cement powder. The fibers were collected at the bottom of a negatively charged cylindrical stainless steel bowl, where a 50 mm long magnetic stirrer rotated at 500 rpm [10], as illustrated in Fig. 2b, c.

3.3 Optical Microscope

A compound microscope, specifically the OLYMPUS GX71 model, examines the fibers and measures their diameter following the production process (Fig. 3).

3.4 Mixing and Casting of Fiber–Cement Composite Samples

Mechanical Testing Sample Preparation In compressive testing, specimens are commonly in the form of cubes measuring 50 mm on each side. These size

Fig. 4 Sample preparation for compressive test

specifications align with the Norwegian Standard NS-EN 196-1:2016 guidelines for testing cement. The mold and samples used in this experiment are presented in Fig. 4.

As mentioned, this experiment used two methods to mix the fiber with the cement. Three samples were prepared when mixing fiber and cement powder, and the CA fiber content was 0.1 wt% and 0.5 wt%.

Mixture of Cement and Water The casting and curing process of the cement powder/fiber and water mixture involves several steps outlined in the Norwegian standard NS-EN 196-1:2016. The ingredients are mixed according to the specified experimental ratio and method to ensure consistency and quality, with a water-to-binder ratio of 0.5 chosen to maintain consistency with prior research and facilitate comparison with existing literature [1, 2, 6]. Following mixing, the mixture is poured into a mold and left to cure for 24 hours. After this initial curing period, the mold is removed, and the specimen undergoes immersion in water for 7 days for compressive testing in accordance with standard requirements. This curing process is crucial for the cement mixture to achieve its maximum strength and durability, rendering it suitable for various applications, including construction.

3.5 Testing

Compressive Strength Test The compressive strength of cubic samples of fiber–cement composite was tested using a FORM+TEST Prufsysteme machine with a maximum load capacity of 200 kN. The machine determined the compressive strength of the 50 × 50 mm samples by subjecting them to a maximum load across their cross-sectional area.

The work reported in this chapter contains the preliminary results of a larger ongoing project, where additional mechanical testing is planned to be conducted. This will extend testing to bending test, providing a more holistic understanding of the impact of fiber blending in the cement material.

4 Results and Discussion

4.1 *Effect of Production Parameters in Electrospun Fibers*

Various electrospinning parameters, including polymer solution concentration, emitter voltage, working distance between emitter and collector, and solution flow rate greatly influence polymer fibers' morphology and cross-section diameter. Investigating these parameters is crucial for optimizing the electrospinning process of cellulose acetate and achieving fiber formation without bead defects and with a controlled diameter.

A systematic approach is adopted to assess the thickness of the fibers. Initially, fibers are electrospun directly onto a horizontally positioned aluminum base beneath the emitter (needle). Subsequently, samples are quickly transferred to an optical microscope for immediate evaluation, allowing for observing parameter-induced effects on fiber structure. The graphs below (Figs. 5, 6 and 7) represent the average data collected from each test. We selected ten different locations for each test and measured the fiber diameter. Then, we calculated the average of these measurements to determine the average diameter of the fiber sample.

Furthermore, an aluminum foil is positioned atop an inverted microscope for careful examination. Utilizing a magnification level of 100x, fibers are examined and

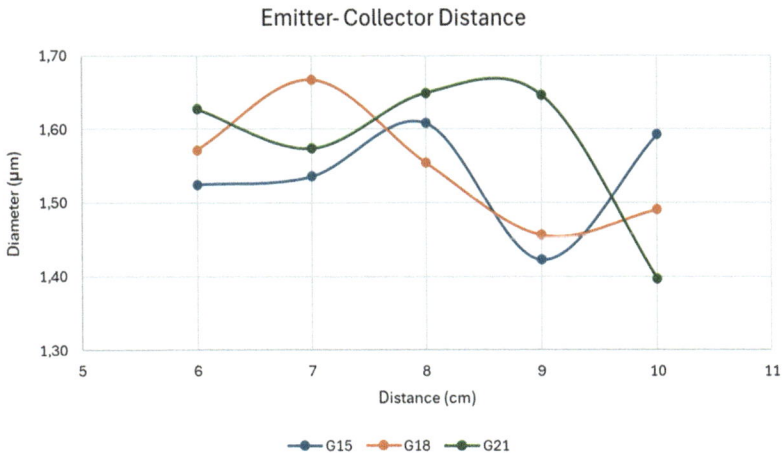

Fig. 5 The effect of the distance between the emitter and the collector on fiber diameter using different gauges

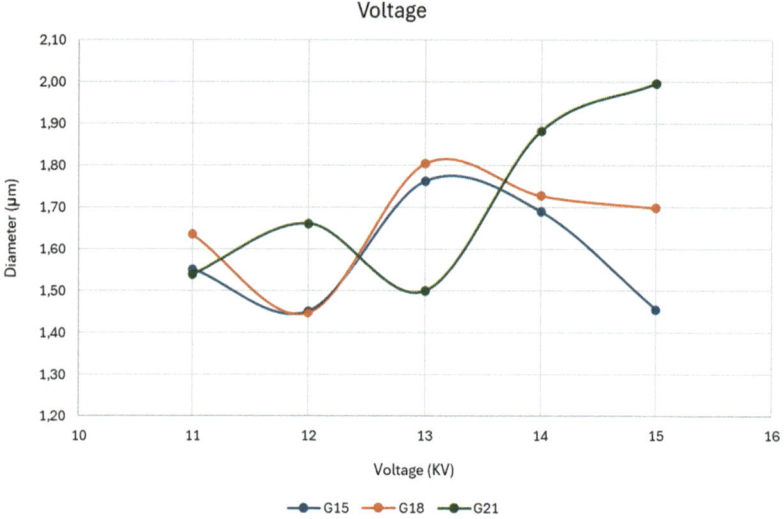

Fig. 6 The effect of the emitter voltage on fiber diameter using different emitter tip gauges

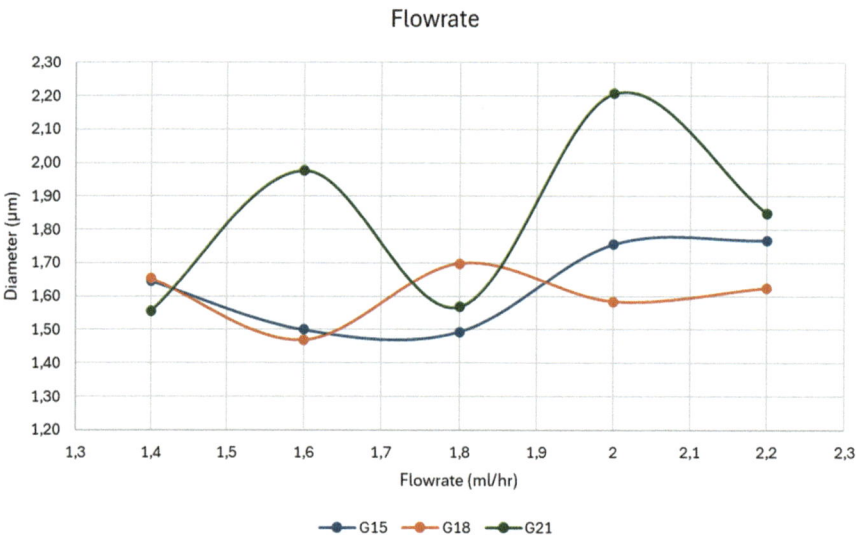

Fig. 7 Flow rate effect on fiber diameter using different gauges

measured with precision. This careful process ensures comprehensive data collection for subsequent analysis, facilitating the identification of optimal electrospinning conditions.

Distance The data analysis reveals insights into the relationship between emitter–collector distance and average fiber diameter across different gauge sizes (G-15, G-18, and G-21). Overall, variations in emitter–collector distance appear to influence average fiber diameter differently depending on the gauge used. For G-15 and G-18, slight fluctuations in average fiber diameter are observed as the emitter–collector distance decreases from 10 cm to 6 cm. These fluctuations suggest a relatively consistent response to emitter–collector distance changes, with both gauges showing comparable trends. In contrast, G-21 exhibits more pronounced changes in average fiber diameter with decreasing emitter–collector distances. The data indicates that larger emitter–collector distances significantly decrease the average fiber diameter for G-21 compared to G-15 and G-18. This suggests that the effect of emitter–collector distance on fiber diameter may vary depending on the specific properties of the polymer solution and the electrospinning process, highlighting the importance of careful parameter optimization for achieving desired fiber characteristics.

Voltage The data provided represents the average fiber diameter for three different gauge sizes (G-15, G-18, and G-21) at varying emitter voltage levels during micro-fiber production. Across all gauge sizes, there is a notable variation in average fiber diameter as the voltage changes.

For G-15, there is a fluctuating trend in average fiber diameter as voltage increases from 11 to 15. While there is a general increase in average fiber diameter from 11 to 13 KV, it decreases at 14 KV and then slightly increases again at 15 KV.

For G-18, there is a similar fluctuation in average fiber diameter with changes in voltage. However, the trend is less pronounced than G-15, with more subtle changes observed.

For G-21, there is a more erratic pattern in average fiber diameter as voltage increases. While there is an initial decrease in average fiber diameter from 11 to 12 KV, it then increases sharply at 13 KV before decreasing again at 14 KV and then increasing once more at 15 KV.

Overall, the data suggests that voltage plays a significant role in determining the average fiber diameter during microfiber production, but the relationship is complex and varies depending on the specific gauge size. Further analysis and experimentation may be necessary to fully understand and optimize the relationship between voltage and fiber diameter for each gauge size.

Flow Rate The data suggests that the flow rate influences the fiber diameter during microfiber production and varies across gauge sizes (G-15, G-18, and G-21).

For G-15 and G-18, a flow rate of 1.6 leads to a smaller average fiber diameter compared to adjacent flow rates of 1.4 and 1.8. However, no clear trend was observed for the average fiber diameter at other flow rates for both G-15 and G-18.

In contrast, for G-21, a flow rate of 1.6 results in a significantly larger average fiber diameter compared to adjacent flow rates of 1.4 and 1.8. Similar to G-15 and G-18, there are fluctuations in the average fiber diameter at other flow rates for G-21.

Overall, the effect of flow rate on average fiber diameter varies depending on the gauge size. Further analysis and experimentation are needed to optimize the flow rate and fiber diameter relationship for each gauge size.

The main objective is to create microfibers using CA and explore how they improve hardened cement paste's mechanical properties. The smaller the diameter of the fiber in micrometers, the closer it is to our target microfiber-sized fiber diameter.

The impact of distance, voltage, and flow rate on fiber diameter was investigated using different gauges. Results indicate that a smaller diameter is achieved at a distance of 10 cm, while a better diameter is attained at 12 kV voltage. The optimal flow rate for achieving better fiber diameter is 1.6 ml/h. When all three parameters are applied, gauge G-18 exhibits the best results for fiber diameter (the smallest possible diameter, close to the nanometer range). Consequently, for further investigation in compressive testing, parameters including a flow rate of 1.6 ml/h, an emitter–collector distance of 10 cm, and a voltage of 12 kV are selected. Moreover, G-18 was chosen for production based on its superior fiber diameter performance.

4.2   Compressive Strength

The compressive strength of cement samples was evaluated after a 7-day curing period, with varying percentages of CA fibers. The reference sample, conformed by cement without fiber reinforcement, exhibited a compressive strength of 14.2 MPa. See Fig. 8.

Incorporating 0.5% CA fibers resulted in a notable increase in compressive strength, reaching 16.3 MPa. This represents a percentage increase of approximately 15.5% compared to the reference sample.

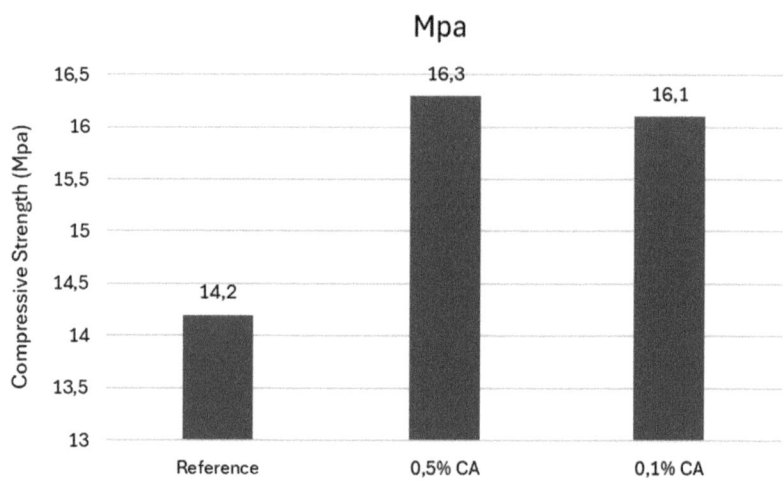

Fig. 8 Compressive strength of specimens after 7 days

Similarly, at a lower fiber concentration of 0.1% CA, the compressive strength remained high at 16.1 MPa. This corresponds to a percentage increase of about 14.6% compared to the reference sample.

These findings show CA fibers' effectiveness in enhancing cementitious materials' compressive strength, even after a relatively short curing period of 7 days. The significant percentage increases in compressive strength indicate the substantial reinforcement CA fibers provide within the cement matrix.

The results suggest that even at lower fiber concentrations, considerable improvements in compressive strength can be achieved within a relatively short time frame. This highlights CA fiber reinforcement's cost-effectiveness and potential applicability in cement-based materials.

5 Conclusion

The addition of CA fibers offers a viable means of enhancing the mechanical properties of cement composites, with notable percentage increases in compressive strength observed at both 0.5% and 0.1% fiber concentrations after 7 days. Further research could explore optimal fiber percentages and their effects on other mechanical properties to fully leverage the benefits of CA fiber reinforcement in cementitious materials.

This study represents an initial exploration, highlighting the need for further research to thoroughly examine the morphological features of CA through scanning electron microscopy (SEM).

References

1. Nguyen, T.N.M., Lee, D.H., Kim, J.J.: Effect of electrospun nanofiber additive on selected mechanical properties of hardened cement paste. Appl. Sci. (Switzerland). **10**(21), 1–12 (2020)
2. Nguyen, T.N.M., et al.: Microstructural and mechanical properties of cement blended with TEOS/PVP nanofibers containing CNTs. Appl. Sci. (Switzerland). **13**(2), 1–2 (2023)
3. Nabeel Zabar Abed, A.-H.: Nanofibers and electrospinning method. In: George, Z.K., Athanasios, C.M. (eds.) Novel Nanomaterials, p. 2. Rijeka, IntechOpen (2018)
4. Nguyen, T.N.M., Moon, J., Kim, J.J.: Microstructure and mechanical properties of hardened cement paste including Nylon 66 nanofibers. Constr. Build. Mater. **232**, 1–3 (2020)
5. Nabeel Zabar Abed, A.-H.: Nanofibers and electrospinning method. In: George, Z.K., Athanasios, C.M. (eds.) Novel Nanomaterials, pp. 193–196. Rijeka, IntechOpen (2018)
6. Nguyen, T.N.M., Yoo, D.Y., Kim, J.J.: Cementitious material reinforced by carbon nanotube-Nylon 66 hybrid nanofibers: mechanical strength and microstructure analysis. Mater. Today Commun. **23**, 2–3 (2020)
7. Rodoplu Solovchuk, D., Mutlu, M.: Effects of electrospinning setup and process parameters on nanofiber morphology intended for the modification of quartz crystal microbalance surfaces. J. Eng. Fiber Fabr. **7**(2), 120–121 (2012)
8. Nguyen, T.N.M., Lee, D.H., Kim, J.J.: Incorporation of silica particles attached to nylon 66 electrospun nanofibers with cement. Materials. **15**(19), 4–9 (2022)

9. Tedros, H.B.M., Teshale, Z.: Mechanical Reinforcement of Cement Pastes with Cellulose Acetate Microfibers Produced by Electrospinning, p. 33, OsloMet - Storbyuniversitetet, Oslo, Norway (2022)
10. Elmaghraby, N.A., et al.: Electrospun composites nanofibers from cellulose acetate/carbon black as efficient adsorbents for heavy and light machine oil from aquatic environment. J. Iran. Chem. Soc. **19**(7), 3013–3027 (2022)

Affordable Phase Change Materials in Lightweight Concrete Walls for Superior Hygrothermal Performance

Saeed B. Nia, Raymond Pepera, and Behrouz Shafei ⓘ

Contents

1 Introduction

Functional building materials and additives in products like adhesives and paints can significantly improve indoor air quality, lower emissions, and manage water absorption. Porous materials, including lightweight aggregates or rigid polyurethane foams and expanded lightweight clay (LECA), are shown to enhance thermal insulation in concrete, potentially increasing thermal efficiency by up to 50%. Studies point to lightweight concrete with LECA or perlite as effective alternatives for improving thermal performance [1–5].

The rising energy demand and increasing CO_2 emissions highlight the urgent need for sustainable building solutions. Innovations such as phase change materials (PCMs) and hygroscopic materials (IIMs), like lightweight concrete—which can absorb and retain more moisture than conventional concrete—are pivotal in wall technologies. These materials enhance thermal storage and regulate humidity,

S. B. Nia (✉) · R. Pepera · B. Shafei
Iowa State University, Ames, IA, USA
e-mail: saeed@iastate.edu

© The Author(s) 2025

M. Kioumarsi, B. Shafei (eds.), *The 1st International Conference on Net-Zero Built Environment*, Lecture Notes in Civil Engineering 237,
https://doi.org/10.1007/978-3-031-69626-8_35

resulting in significant energy savings and a reduction in greenhouse gas emissions. By improving the building's ability to maintain stable indoor temperatures and moisture levels, HMs reduce the reliance on heating, cooling, and ventilation systems, thereby lowering energy consumption and the building's overall carbon footprint [5–8].

Integrating PCMs and HMs into buildings provides a unique temperature and humidity control method, enhancing energy efficiency and performance without relying on mechanical systems. This strategy utilizes the thermal storage and phase change capabilities of PCMs, along with the moisture regulation of HMs, supporting sustainable construction and comfort standards. Techniques like shape-stabilized PCM (SSPCM) and microencapsulated PCM (MPCM) have been developed to address PCM leakage, highlighting the growing focus on hygrothermal performance to manage indoor humidity, health, building longevity, and energy use effectively [9–15].

Concrete deformation caused by heat and moisture can lead to cracking, reducing durability and thermal efficiency. Temperature changes lead to thermal stresses and shrinkage, while moisture induces volume changes, creating internal stresses that can cause cracks. These cracks allow external agents to penetrate, accelerating deterioration and reducing concrete's service life. Furthermore, cracks compromise concrete's effectiveness as a thermal mass in buildings, disrupting its heat storage and release capabilities, thereby diminishing building energy efficiency [15–21].

The importance of analyzing hygrothermal performance in building materials has become critical for ensuring indoor comfort. This study aims to enhance lightweight concrete by adding a PCM layer into walls or ceiling, improving heat insulation, water absorption, and storage. While lightweight concrete has benefits like sound insulation and fire resistance, risks include shrinkage, mold growth, and ventilation issues, affecting health and durability. Investigating hygrothermal performance is essential for a moisture-problem-free and comfortable indoor environment.

This research aims to address the gap in studies on hygrothermal performance at normal temperatures by investigating the effects of applying a PCM layer on walls in comparison to uncoated walls, particularly in response to heat and thermal variations over time. Experimental methods are used to evaluate shrinkage, water absorption, and thermal conductivity under humidity-induced stress and hygrothermal deformation in concrete. This approach enhances the understanding of lightweight concrete behavior and encourages the integrated use of construction materials, including PCMs, to optimize building performance.

2 Materials and Setups

This study investigates glycerin's effectiveness as a PCM in improving the energy efficiency of lightweight concrete walls. Using experimental and analytical methods, this study examines the insulation benefits of a glycerin wax layer across temperature variations, guided by Hundt's parameters. Figure 1 shows the experiments that

(a)	(b)

Fig. 1 (**a**) Fine LECA and (**b**) coarse LECA

Table 1 Chemical composition of Portland limestone cement and LECA

Chemical composition (% by mass)	PLC	Expanded clay (LECA)
SiO_2	18.69	63.60
Al_2O_3	3.78	16.70
Fe_2O_3	2.77	6.45
CaO	64.10	3.05
MgO	1.67	3.85
Na_2O	0.17	1.24
K_2O	0.56	2.85
SO_3	2.80	0.18
LOI	6.50	1.40

employ Portland limestone cement Type 1L (ASTM C 595) and lightweight LECA aggregates (Table 1), aiming to capture glycerin's role in energy conservation for residential settings.

In addition to lightweight aggregates, this study uses ordinary aggregates, natural sand, and crushed pea gravel (with grains passing through a 1/2 inch diameter). Table 2 outlines the properties and gradation of these aggregates. A high-range water reducer (HRWR), with a 35% solid content and a density of 1.1 g/cm^3 at 20 °C, was added to enhance mix flowability.

Glycerin, a highly effective phase change material (PCM), significantly enhances the thermal performance of building walls by utilizing its latent heat during phase transitions. It solidifies at about 17 °C and melts just above this, at 18 °C, making it ideal for stabilizing indoor climates—refer to Fig. 2 for two phases of glycerin. With a melting enthalpy of approximately 200 kJ/kg, glycerin can store and release substantial amounts of heat, thereby maintaining stable temperatures. Its properties include a high viscosity that decreases with temperature and a density of about 1260 kg/m^3. Additionally, its non-toxic nature makes glycerin a safe and efficient choice for thermal regulation in construction. The thermal characteristics of glycerin are summarized in Table 3.

Table 2 Aggregates grading details

Components			Fine aggregates		Coarse aggregates	
			Natural sand	Fine LECA	Crushed gravel	Coarse LECA
		Sieve opening	Cumulative passing (%)			
		3/4 in. (19.0 mm)	–	–	100	100
		1/2 in. (12.5 mm)	–	–	100	100
Sieve		3/8 in. (9.5 mm)	100	100	63.0	90
analysis		No. 4 (4.75 mm)	100	98	4	0
		No. 8 (2.36 mm)	90	11	0	0
		No. 16 (1.18 mm)	74	0	–	0
		No. 30 (600 μm)	45	0	–	0
		No. 50 (300 μm)	22	0	–	0
		No. 100 (150 μm)	7	0	–	0
		No. 200 (75 μm)	0	0	–	0
Maximum nominal size (mm)			4.75	4.75	12.5	9.5
Specific gravity			2.65	0.450	2.70	0.310

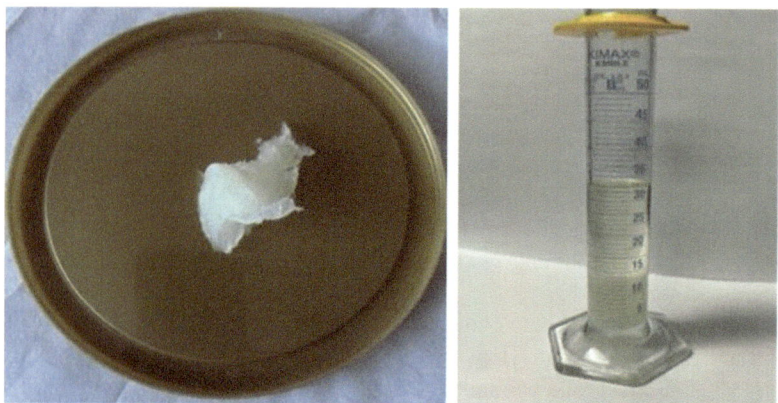

Fig. 2 Glycerin showcased in semi-solid form at 16 °C and liquid form at 30 °C

Table 3 Thermal properties of glycerol

Material	Melting temperature (°C)	Melting enthalpy (kJ/kg)	Thermal conductivity (W/(m·K))	Viscosity (Pa·s)	Density (kg/m³)
Glycerol	18	~200	0.14	1.41	1260

Table 4 Mix design details

| Mix ID/ingredient | Cement type IL (PLC) | Natural sand | Lightweight aggregate | | Water |
			Fine	Coarse	
LWC-PCM	380 kg/m^3	450 kg/m^3	250 kg/m^3	150 kg/m^3	152 kg/m^3

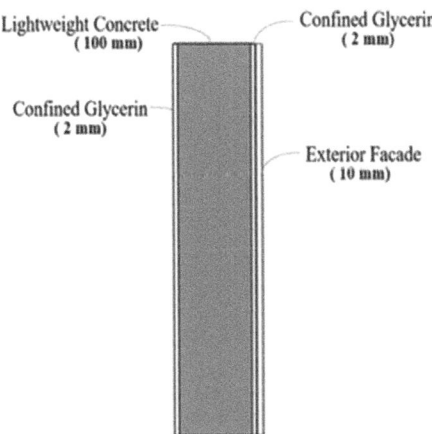

Fig. 3 Cross-section of a lightweight concrete wall with a 2 mm layer of glycerin-based PCM confined on the exterior façade

Glycerin's molecular structure, $C_3H_8O_3$, with three hydroxyl groups, boosts its thermal energy absorption and retention, making it a stable and eco-friendly PCM for enhancing building energy efficiency.

The structural lightweight concrete was produced with a water-to-cement ratio of 0.4, as mix design detailed in Table 4. Samples cured for 28 days at 20 °C demonstrated a compressive strength exceeding 30 MPa, confirming their suitability for both non-structural and structural wall applications. In our experiments, we assessed conductive heat transfer through lightweight concrete, which is influenced by factors such as temperature gradient, material thickness, and thermal conductivity. Additionally, drying shrinkage and water absorption tests were conducted to evaluate moisture loss and the moisture retention capabilities of lightweight walls with and without a PCM layer. Figure 3 illustrates a cross-section of a lightweight concrete wall featuring a 2 mm layer of glycerin-based PCM confined on the exterior façade. The experimental results evaluate the PCM's phase change contribution to both thermal and moisture resistance in the wall. This provides valuable insights for optimizing wall designs to improve thermal efficiency under varying heat loads, ultimately enhancing strategies for better energy efficiency (Fig. 3).

3 Test Methods

In the experimental program, three distinct test methods were employed to evaluate the hygrothermal properties of lightweight concrete samples, which included both samples with and without PCM applied to their surfaces. The hygrothermal properties of the lightweight concrete samples were evaluated using three distinct test methods, examining samples with and without PCM applied to their surfaces. The first method, focusing on drying shrinkage, adhered to the ASTM C596 standard for small prisms measuring $25 \times 25 \times 285$ mm. Comparative readings and calculations were conducted following ASTM C157, with the length change of each specimen determined daily for up to 42 days, using Eq. (1).

$$\varepsilon_{\text{dry}} = \frac{l_{(t)} - l_{(t_o)}}{l_{(t_o)}} \times 10^6 \mu \frac{m}{m} \tag{1}$$

where $l_{(t)}$ is the length of the sample at the measured time t, and $l_{(t_o)}$ is the specimen's initial length at 72 h.

The second method assessed capillary water absorption based on ASTM C1585. Before water contact, the initial weight of the specimen was obtained from Eq. (2):

$$I_c = \frac{m_t - m_o}{A_c} \tag{2}$$

where I_c represents the mean water absorption rate in mm. A_c denotes the water-contact surface area of the specimen in mm^2. ρ_c is the water's density in g/mm^3.

To measure the thermal conductivity of lightweight high-performance concrete (LWHPC), the hot plate method from BS EN 12664 was applied after 35 days. Specimens of $300 \times 300 \times 50$ mm were dried for 24 h at 105 ± 5 °C and then subjected to temperatures of 40 and 18 °C on hot and cold plates, respectively. Temperature was recorded every 10 min daily, and these readings were averaged to calculate thermal conductivity. This method provided detailed insights into LWHPC's thermal efficiency and moisture management.

4 Results and Discussion

4.1 Drying Shrinkage

Applying a 0.5 mm glycerin layer to lightweight concrete specimens significantly reduces drying shrinkage over time, as shown in Fig. 4. This reduction, compared to control samples without glycerin PCM, is due to glycerin's water retention within the concrete, acting as a barrier that decreases evaporation and shrinkage. The

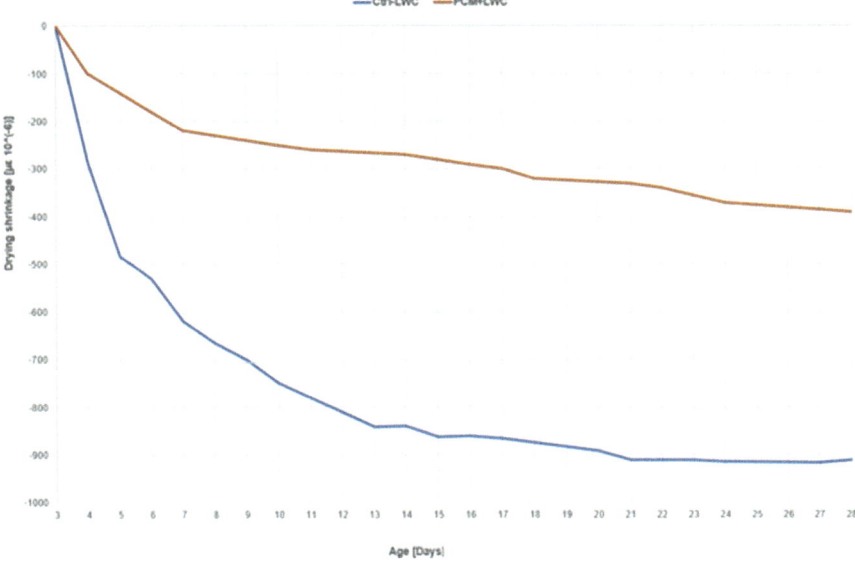

Fig. 4 Comparative drying shrinkage for Ctrl-LWC and PCM+LWC

glycerin layer counters the high porosity of lightweight aggregates, which typically leads to more shrinkage by promoting better moisture retention. This aids the internal curing process, improving hydration and the concrete's overall characteristics. Additionally, glycerin PCM as a curing aid ensures a more even drying rate, essential for maintaining the concrete's durability and preventing defects like cracks.

4.2   Capillary Water Absorption

The data indicate that integrating a PCM into lightweight concrete (LWC) substantially reduces its water absorption at various testing intervals (3, 7, and 28 days) compared to control lightweight concrete without PCM (Ctrl-LWC), as illustrated in Fig. 5. This reduction in water absorption suggests that PCM addition not only impedes water ingress by blocking the pathways within the lightweight aggregate's cavities, but also modifies the concrete's pore structure. These changes enhance the concrete's resistance to water penetration and could potentially improve its durability and thermal properties.

The addition of PCM to lightweight concrete (LWC + PCM) significantly enhances its resistance to moisture by drastically reducing the capillary water absorption rate by approximately 86% compared to control lightweight concrete (Ctrl-LWC), as shown in Table 5. This improvement highlights PCM's effectiveness in minimizing water ingress in porous construction materials.

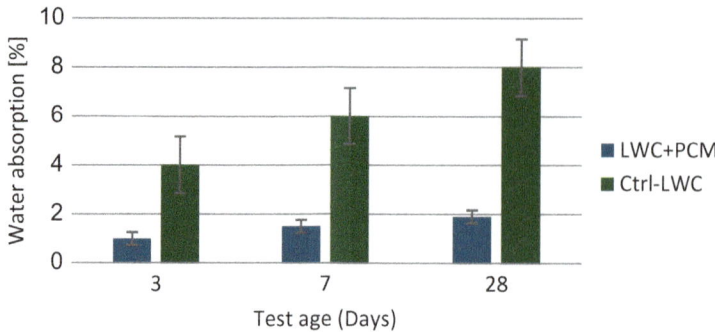

Fig. 5 Comparative percentage of water absorption for Ctrl-LWC and PCM+LWC at different ages

Table 5 Comparative capillary water absorption rates for Ctrl-LWC and PCM+LWC specimens at 28 days

Mix ID	Capillary water absorption rate (m/s$^{1/2}$)
Ctrl-LWC	7.90×10^{-6}
PCM+LWC	1.10×10^{-6}

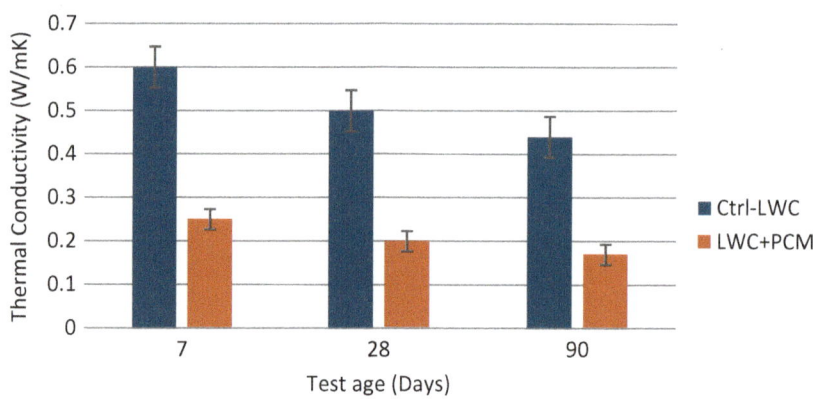

Fig. 6 Comparative thermal conductivity for Ctrl-LWC and PCM+LWC specimens at different ages

4.3 Thermal Conductivity

Figure 6 shows that lightweight concrete with glycerin PCM significantly reduces thermal conductivity, enhancing thermal resistance and insulation, which is especially beneficial in cold climates. Thermal conductivity drops by 71–76% across testing periods, attributed to glycerin's latent heat storage, moderating temperature fluctuations. This improvement supports energy conservation by stabilizing indoor temperatures and minimizing heating and cooling needs.

5 Conclusions

This study has demonstrated that applying a 2 mm layer of glycerin as a phase change material to lightweight concrete significantly enhances its hygrothermal performance and sustainability.

References

1. Ruiz-Herrero, J.L., Nieto, D.V., López-Gil, A., Arranz, A., Fernández, A., Lorenzana, A., Merino, S., De Saja, J.A., Rodríguez-Pérez, M.Á.: Mechanical and thermal performance of concrete and mortar cellular materials containing plastic waste. Constr. Build. Mater. **104**, 298–310 (2016)
2. Yang, S., Yue, X., Liu, X., Tong, Y.: Properties of self-compacting lightweight concrete containing recycled plastic particles. Constr. Build. Mater. **84**, 444–453 (2015)
3. Coppola, B., Courard, L., Michel, F., Incarnato, L., Di Maio, L.: Investigation on the use of foamed plastic waste as natural aggregates replacement in lightweight mortar. Compos. Part B. **99**, 75–83 (2016)
4. Hajilar, S., Shafei, B.: Thermal transport properties at interface of fatty acid esters enhanced with carbon-based nanoadditives. Int. J. Heat Mass Transf. **145**, 118762 (2019)
5. Zahir, M.H., Irshad, K., Shafiullah, M., Ibrahim, N.I., Islam, A.K., Mohaisen, K.O., Sulaiman, F.A.: Challenges of the application of PCMs to achieve zero energy buildings under hot weather conditions: a review. J. Energy Storage. **64**, 107156 (2023)
6. Nia, S.B., Chari, M.N.: Applied development of sustainable-durable high-performance lightweight concrete: toward low carbon footprint, durability, and energy saving. Results Mater. **20**, 100482 (2023)
7. Žmak, I., Hartmann, C.: Current state of the plastic waste recycling system in the European Union and in Germany. Tehnički glasnik. **11**(3), 138–142 (2017)
8. Musiał, M., Lichołai, L., Katunský, D.: Modern thermal energy storage systems dedicated to autonomous buildings. Energies. **16**(11), 4442 (2023)
9. Cai, S., Gou, Z.: Defining the energy role of buildings as flexumers: a review of definitions, technologies, and applications. Energ. Build. **303**, 113821 (2023)
10. Pandey, R.N., Pandey, S.K., Ribeiro, J.W.: Complete and satisfactory solutions of luikov equations of heat and moisture transport in a spherical capillary—porous body. Int. Commun. Heat Mass Transf. **27**(7), 975–984 (2000)
11. Hinojosa, J.F., Moreno, S.F., Maytorena, V.M.: Low-temperature applications of phase change materials for energy storage: a descriptive review. Energies. **16**(7), 3078 (2023)
12. Gérard, B., Marchand, J.: Influence of cracking on the diffusion properties of cement-based materials: part I: influence of continuous cracks on the steady-state regime. Cem. Concr. Res. **30**(1), 37–43 (2000)
13. Ichikawa, Y., England, G.L.: Prediction of moisture migration and pore pressure build-up in concrete at high temperatures. Nucl. Eng. Des. **228**(1–3), 245–259 (2004)
14. Isgor, O.B., Razaqpur, A.G.: Finite element modeling of coupled heat transfer, moisture transport and carbonation processes in concrete structures. Cem. Concr. Compos. **26**(1), 57–73 (2004)
15. Nia, S.B., Salimi, A., Chari, M.N.: Enhancing energy efficiency in lightweight concrete walls: validating the effectiveness of glycerin wax as a phase change material through finite volume method (FVM). PREPRINT (Version 1) (2023). Available at Research Square https://doi.org/10.21203/rs.3.rs-3329090/v1

16. Qin, M., Belarbi, R.: Development of an analytical method for simultaneous heat and moisture transfer in building materials utilizing transfer function method. J. Mater. Civ. Eng. **17**(5), 492–497 (2005)
17. Chen, B., Han, M., Zhang, B., Ouyang, G., Shafei, B., Wang, X., Hu, S.: Efficient solar-to-thermal energy conversion and storage with high-thermal-conductivity and form-stabilized phase change composite based on wood-derived scaffolds. Energies. **12**(7), 1283 (2019)
18. Bentz, D.P., Garboczi, E.J., Quenard, D.A.: Modelling drying shrinkage in reconstructed porous materials: application to porous Vycor glass. Model. Simul. Mater. Sci. Eng. **6**(3), 211 (1998)
19. Grasley, Z.C., Lange, D.A.: Thermal dilation and internal relative humidity of hardened cement paste. Mater. Struct. **40**, 311–317 (2007)
20. Hajilar, S., Shafei, B.: Multiscale investigation of interfacial thermal properties of n-octadecane enhanced with multilayer graphene. J. Phys. Chem. C. **123**(38), 23297–23305 (2019)
21. Chen, D.: Study on Numerical Simulation of Hygro-Thermal Deformation of Concrete Based on Heat and Moisture Transfer in Porous Medium and Its Application. Southeast University, Nanjing (2007)

Effects of Low-Carbon Binders on the Mechanical and Thermal Properties of Biobased Insulation Materials

Houssam Affan, Badreddine El Haddaji, and Fouzia Khadraoui

Contents

1 Introduction

The construction sector currently represents 43% of the annual energy consumption in France and contributes significantly, accounting for 23% of greenhouse gas emissions [20]. This makes it a major player in climate change within the sector. Globally, the construction industry is the second-largest contributor to CO_2 emissions, trailing only the transportation sector [15]. Given the urgent need for an ecological transition, there is a crucial necessity to innovate and decarbonize the entire construction sector. Experts suggest a promising solution: Transitioning construction systems toward the adoption of biobased materials, which are notably less environmentally impactful than traditional materials and are already transforming our construction practices.

H. Affan (✉) · B. El Haddaji · F. Khadraoui
BUILDERS Ecole d'ingénieurs, Unité de Recherche "Builders Lab", ComUE NU, Epron, France
e-mail: houssam.affan@builders-ingenieurs.fr

© The Author(s) 2025
M. Kioumarsi, B. Shafei (eds.), *The 1st International Conference on Net-Zero Built Environment*, Lecture Notes in Civil Engineering 237,
https://doi.org/10.1007/978-3-031-69626-8_36

Biobased materials, derived from biomass and animals such as wood, wool, hemp [6], flax [7], seaweed [2], miscanthus [18], sunflower [1], etc., offer distinct advantages over traditional concrete materials. Unlike traditional concrete, which requires significant amounts of carbon-intensive energy for manufacturing (such as grinding and thermal processing), the production of biomaterials involves minimal energy quantities. Moreover, these materials have the unique ability to serve as carbon sinks, as they can store the CO_2 emitted into the air throughout their life cycle. This means that certain buildings constructed with biobased materials can absorb more CO_2 than they emit, contributing positively to carbon neutrality efforts.

One of the key advantages of biobased materials is their potential for regular and infinite recycling [14]. Unlike traditional materials that often end up as waste in landfills, biobased materials can be reused or reintegrated into natural cycles, such as composting. This not only reduces waste but also minimizes the environmental impact associated with the disposal of construction materials [16]. Additionally, the recyclability of biobased materials supports the principles of circular economy, where resources are continuously reused, ultimately leading to a more sustainable and resource-efficient construction industry.

Furthermore, biobased materials offer versatility in design and application, allowing for innovative and environmentally friendly construction solutions. With advancements in material science and technology, biobased materials can now meet the performance requirements of various construction applications while offering environmental benefits. From insulation and structural components to finishing materials and furniture, biobased materials offer a wide range of options for sustainable building design [9].

Overall, the use of biobased materials in construction presents a promising pathway toward reducing carbon emissions, minimizing waste generation, and promoting sustainable practices in the built environment. By harnessing the benefits of biobased materials, the construction industry can play a significant role in mitigating climate change and transitioning toward a more sustainable future.

In order to achieve a low-carbon footprint material, we are incorporating various biobased materials, such as hemp and flax fibers, in combination with different types of hydraulic binders, namely, natural prompt cement and CL 90-S air lime. This study aims to analyze the influence of these two types of binders and two types of biobased aggregates on the material's variability concerning:

Physical and mechanical properties

- Porosity accessible to water (according with the standard NF EN ISO 5017 [23])
- Compressive strength

Thermal properties

- Thermal conductivity (according with the standard ISO 8301 [21])
- Specific heat capacity (according with the standard NF EN ISO 11357-4 [22])

2 Materials and Methods

2.1 Materials

In the context of this research, natural prompt cement and air lime CL 90-S were used for our mixtures. The physical, mechanical, and thermal properties of biobased aggregates play a crucial role in defining the characteristics of insulating concrete. Hemp and flax shives were chosen with a length between 5 and 25 mm, as illustrated in Fig. 1. The data in Table 1 provide detailed information on the physical properties (bulk density measured according to RILEM TC 236-BBM [3]), hydric properties (water absorption measured according to RILEM TC 236-BBM [3]), thermal properties (thermal conductivity measured according to ISO 8301 [21], and mass-specific heat capacity measured according to NF EN ISO 11357.4 [5]) of the biobased aggregates examined in this study. The obtained results are consistent with the literature [3–5, 12]. Additionally, the accompanying Table 2 presents the properties of the binders used in this study.

The objective of this study was to explore the physical, mechanical, and thermal properties through four distinct formulations: cement/hemp, lime/hemp, cement/flax, and lime/flax. Biobased aggregates are widely recognized for their excellent thermal and hydric properties. The aim of this study was to develop a material with insulation properties, both thermal and hydric. With the aim of achieving these objectives, a substantial amount of plant aggregates was deliberately included to enhance these properties. The chosen percentages align with the compositions

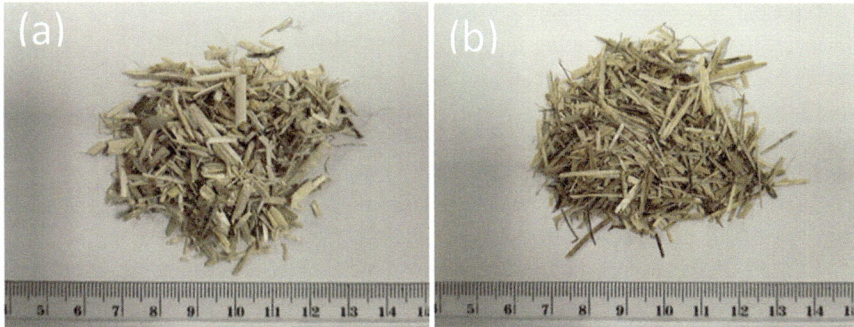

Fig. 1 (a) Hemp shives and (b) flax shives with a diameter between 5 and 25 mm

Table 1 Mechanical, thermal, and hydric properties of hemp and flax shives

Mixture	Bulk density (kg/m³)	Water absorption (%)	Thermal conductivity (W·m⁻¹·K⁻¹)	Specific heat capacity (J/(kg·k))
Hemp shives	129.57 ± 3.62	$363.54 \pm 4.86\%$	0.0529 ± 0.0023	2925.67 ± 111.15
Flax shives	105.52 ± 5.99	$334.37 \pm 17.83\%$	0.0472 ± 0.0016	3149 ± 67.26

Table 2 The physical and mechanical properties of low-carbon binders

Low-carbon binder	Bulk density (kg/m^3)	Blaine specific surface (cm^2/g)	Compressive strength at 24 h (MPa)	Compressive strength at 9 months (MPa)	References
Natural prompt cement	2980	7600	7.8	40	[24]
Air lime CL 90-S	450	13,260	–	2.5	[19]

Table 3 Composition of the mixture

Formulations	Prompt natural cement (kg/m^3)	Air lime CL 90-S (kg/m^3)	Hemp shives (kg/m^3)	Flax shives (kg/m^3)	Water (kg/m^3)
C100H	250	–	116	–	271
L100H	–	250	116		271
C100F	250	–	–	116	271
L100F	–	250	–	116	271

derived from the studies of Benmahiddine et al. [4], and they also adhere to the guidelines recommended by Vicat. Consequently, the formulations will include the addition of 46.5% of the binder mass in the form of plant aggregates while keeping the water quantity constant and maintaining a water-to-cement ratio of 1.08. Specific compositions for each mixture are detailed in Table 3.

Physical assessments (open porosity) were conducted using samples ranging from 50 to 200 cm^3 following the NF EN ISO 5017 standard [23]. Additionally, cubes measuring 150 × 150 × 150 mm^3 were used to evaluate compressive strength, and prismatic panels measuring 200 × 200 × 40 mm^3 were employed for thermal tests, assessing thermal conductivity according to ISO 8301 [21] and mass-specific heat capacity according to the NF EN ISO 11357-4 standard [22].

The mixing process begins by first placing the components (2.15-kg binder and 1-kg biobased aggregates). These elements are manually mixed for 30 seconds. Then, half of the water quantity is added, which is 1.161 kg, and the mixture is manually stirred for an additional 30 seconds. Finally, the remaining water (1.161 kg) is introduced, and the mixing process is repeated for 30 seconds before initiating the mold filling.

The fabrication of cubic and prismatic samples follows an identical procedure. Initially, the mold is filled one-third full to form the initial layer. Subsequently, compaction is carried out 30 times using a wooden stick for both this layer and the two subsequent layers. To achieve a smooth surface, horizontal compaction is applied during the final layer of the pouring process.

2.2 Methods

2.2.1 Physical Properties

Porosity accessible to the water test: This method is based on Archimedes' principle [13], as specified in the NF EN ISO 5017 standard [23]. The process involves vacuum saturation for 24 hours of three samples from each formulation. The mass of the samples saturated with water and in the air are then measured for the calculation of porosity, following Eq. (1).

$$Pa(\%) = \frac{M_{air} - M_{dry}}{M_{air} - M_{water}} \times 100 \qquad (1)$$

The variable "Pa" represents the water-accessible porosity in percentage. Other terms include "M_{air}" for the sample mass after air saturation in kilograms, "M_{water}" for the mass of the sample saturated and immersed in water in kilograms, and finally, "M_{sec}" for the mass of the dry sample, obtained after drying at 60 °C with mass monitoring, also measured in kilograms.

2.2.2 Mechanical Properties

Compressive strength test: In order to study this parameter, three cubic samples measuring $150 \times 150 \times 150$ mm^3 of each formulation were utilized. Compressive strength tests were conducted on three of these samples after a period of 28 days. The tests took place on an IGM press with a constant speed rate of 0.3 mm/minute.

2.2.3 Thermal Properties

Thermal conductivity: The thermal conductivity was investigated using a NETZSCH heat flow meter (model HFM 446 Lambda). The sample, with parallel-epiped dimensions of $200 \times 200 \times 40$ mm^3, is positioned between the cooling plate and the heating plate according to the ISO 8301 standard [21].The thermal conductivity value is calculated using the following equation:

$$\lambda = (q\Delta T)/\Delta X \qquad (2)$$

where

λ: Thermal conductivity of the sample
q: Heat flux
ΔT: Temperature difference across the sample
ΔX: Thickness of the sample

Specific heat mass: Mass-specific heat capacity (Cp) is a crucial parameter indicating a material's ability to store heat. In this study, the specific heat capacity of the 200 × 200 × 40 mm^3 samples is tested within a temperature range of 15–25 °C according to the NF EN ISO 11357-4 standard [22]. The value of the mass-specific heat capacity Cp is calculated using the following equation:

$$Q = M \times C \times \Delta T \tag{3}$$

where

Q: Heat quantity in kilojoules
M: Processed mass in kilograms
C: Specific heat in [kJ/kg °C]

3 Results and Discussion

3.1 Physical and Mechanical Properties

Porosity accessible to water and compressive strength: Concrete incorporating biobased aggregates is distinguished by its high level of porosity, attributed to the significant absorption of these aggregates. The drying process leads to the formation of voids, resulting in a notable increase in porosity [5, 17]. Figure 2 presents the results of porosity accessible to water measurements for various mix samples examined at the age of 28 days.

The cement/hemp mixture is characterized by particularly high porosity, reaching 70.2%. In contrast, the lime/hemp mixture shows a reduction in this percentage, decreasing to 67%. Mixtures incorporating flax shives with cement exhibit a higher

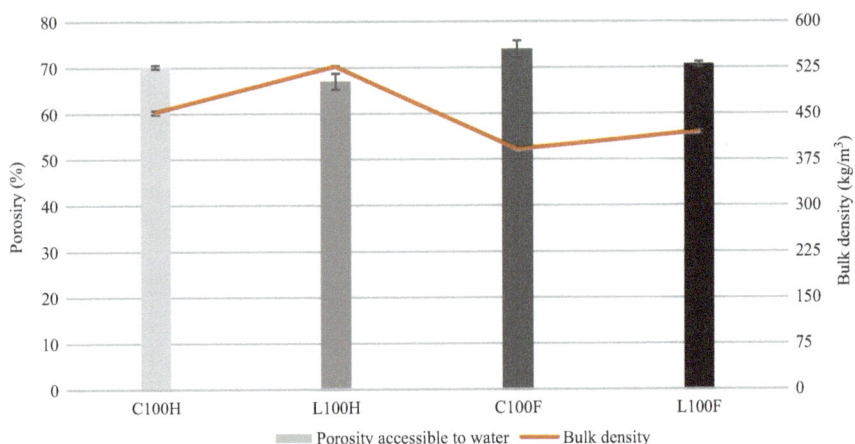

Fig. 2 Porosity accessible to water measured for various formulations at 28 days

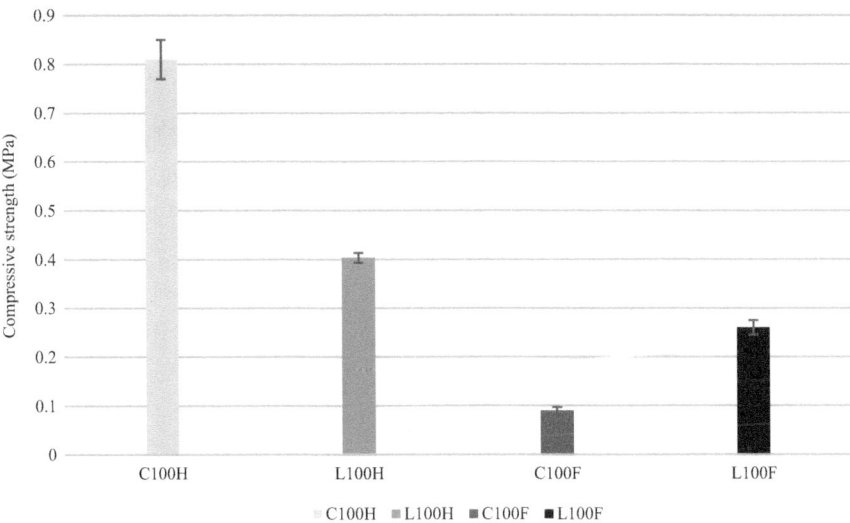

Fig. 3 Compressive strength of various formulations at 28 days

porosity value than those made with lime, with a value of 74%, compared to lime mixtures, which show a value of 70%. The observed variation in porosity between formulations based on cement and those based on lime with both types of biomass can be explained by the significant difference in specific surface area between natural prompt cement (7000 cm^2/g) [24] and air lime CL 90-S (13,260 cm^2/g) [19]. The latter being twice as high means that lime is finer than cement, so its particles fill the voids caused by the biomass. Additionally, this disparity increases the importance of cohesion between the binder and plant aggregates, thus leading to a decrease in the percentage of porosity.

The results from the compressive strength test, depicted in Fig. 3, indicate that the cement/hemp formulation exhibits higher compressive strength compared to the cement/flax formulation. Specifically, the compressive strength value of 0.80 MPa for the C100H formulation aligns with findings in the literature [8, 17], surpassing that of the cement/flax formulation, which measures 0.09 MPa. Similarly, the lime/hemp formulation demonstrates a compressive strength of 0.40 MPa, exceeding that of the lime/flax formulation, which registers at 0.26 MPa. The notable disparities in compressive strength between the cement/hemp and lime/hemp formulations can be attributed to the distinct mechanical behaviors of natural prompt cement and air lime CL 90-S. Notably, technical data reveal a considerably higher compressive strength for cement, reaching 19 MPa at 28 days [24], whereas that of lime is evaluated at 2.5 MPa at 9 months [19]. Similarly, the differences in compressive strength between the cement/hemp and cement/flax formulations may arise from variations in porosity percentage, which are more advantageous for the cement/flax (74.1%) blend than for the cement/hemp (70.2%) one, leading to a lower density in the cement/flax mixture and consequently lower compressive strength.

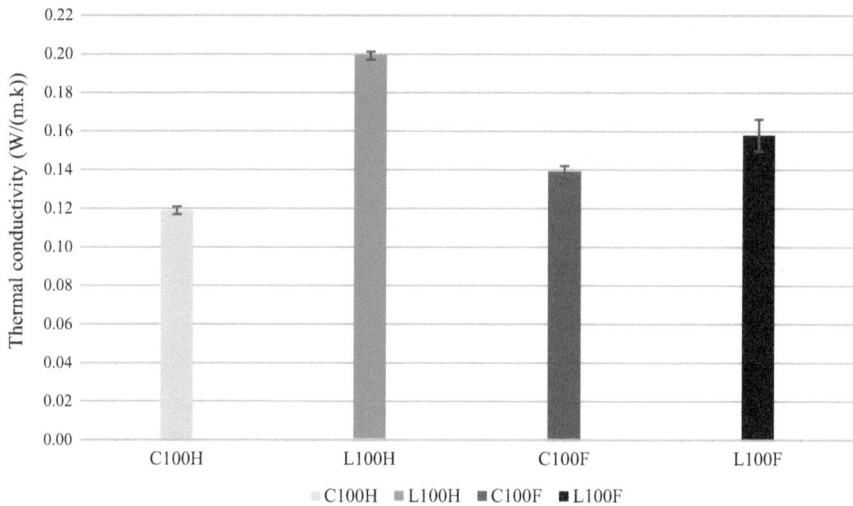

Fig. 4 Thermal conductivity measurement for various formulations at 28 days

3.2 *Thermal Properties*

Thermal conductivity: Figure 4 illustrates the thermal conductivity results for various mixture samples examined at the age of 28 days. The thermal conductivity increases progressively, going from 0.12 (W/(m·K)) for the cement/hemp mixture to 0.139 for the cement/flax mixture. However, this value shows a significant increase from 0.158 (W/(m·K)) for the lime/flax mixture to 0.20 (W/(m·K)) for the lime/hemp mixture.

The low thermal conductivity of the cement/hemp mixture results from several factors, including the relatively low bulk density and a high porosity percentage reaching 70.2%. In contrast, the high thermal conductivity of the lime/hemp mixture is attributable to the significant bulk density of the mixture [10, 11]. However, the slight disparity between the cement/hemp mixture and the cement/flax mixture can be explained by the differences in thermal conductivity values between hemp and flax shives.

3.2.1 Specific Heat Mass

Figure 5 presents the results of the mass-specific heat capacity for the various mixture samples examined after 28 days. The variation in specific heat capacity follows a gradual progression, starting at 1313 J/(kg·K) for the combination of cement and hemp, reaching 1375 for the lime and hemp mixture. Nevertheless, a significant decrease occurs, dropping to 1155 J/(kg·K) for the cement/flax mixture. More notably, this value increases further, reaching 1274 for the lime/flax formulation.

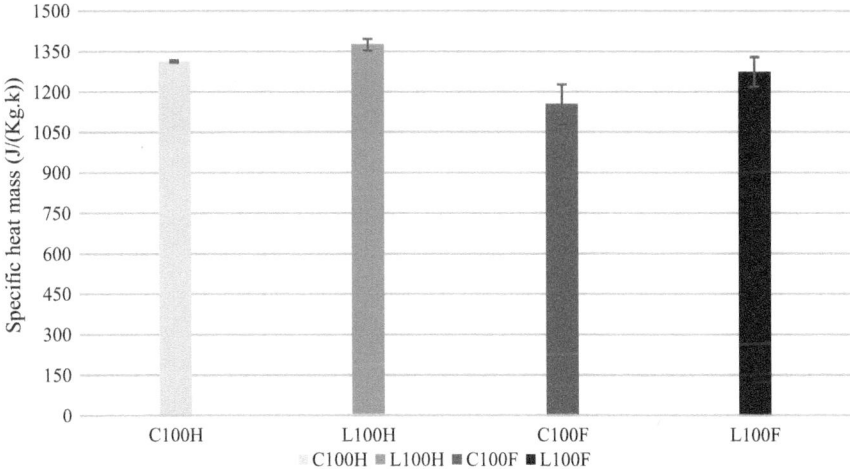

Fig. 5 Specific heat capacity measurement for various formulations at 28 days

The specific heat mass of the mixture is strongly influenced by parameters such as bulk density, porosity, and the heat capacity of the components [11]. The higher specific heat capacity (Cp) observed in the cement/hemp mixture compared to the cement/flax mixture can be attributed to the favorable bulk density of the cement/hemp composite over that of the cement/flax composite. These results are also reflected in the lime/hemp and lime/flax mixtures, where the Cp value for lime/hemp is more favorable than that of lime/flax.

4 Conclusions

Biobased aggregates are recognized for their excellent thermal properties and light-weight nature, imparting exceptional thermal characteristics to vegetable concrete. The main conclusions of this study can be summarized as follows:

- The cement/hemp formulation, characterized by high porosity, imparts excellent thermal properties to the material, resulting in high thermal resistance (thermal conductivity of 0.12 W/(m·K)).
- The mechanical and physical properties of hemp make the cement/hemp formulation a material with good mechanical properties, with a compressive strength of 0.80 MPa.
- The thermal capacity reaches its maximum for the lime/hemp formulation, with a value of 1333 J/(kg·K), attributed to its high bulk density compared to the cement/hemp formulation.

- The cement/hemp formulation exhibits the best mechanical and thermal properties, followed by the lime/hemp formulation from a mechanical perspective, and the cement/flax mixture from a thermal perspective.

Regarding future prospects, the goal is to develop a material with excellent thermal and hydric insulation properties while maintaining robust mechanical characteristics. In this regard, particular attention will be given to evaluating hydric features such as water vapor permeability, sorption, and desorption.

References

1. Affan, H., Arairo, W., Arayro, J.: Mechanical and thermal characterization of bio-sourced mortars made from agricultural and industrial by-products. Case Stud. Constr. Mater. **18**, e01939 (2023)
2. Affan, H., Touati, K., Benzaama, M.-H., Chateigner, D., El Mendili, Y.: Earth-based building incorporating Sargassum muticum seaweed: mechanical and hygrothermal performances. Buildings. **13**(4), 932 (2023)
3. Amziane, S., Collet, F., Lawrence, M., Magniont, C., Picandet, V., Sonebi, M.: Recommendation of the RILEM TC 236-BBM: characterisation testing of hemp shiv to determine the initial water content, water absorption, dry density, particle size distribution and thermal conductivity. Mater. Struct. **50** (2017)
4. Benmahiddine, F., Bennai, F., Cherif, R., Belarbi, R., Tahakourt, A., Abahri, K.: Experimental investigation on the influence of immersion/drying cycles on the hygrothermal and mechanical properties of hemp concrete. J. Build. Eng. **32**, 101758 (2020)
5. Benmahiddine, F., Cherif, R., Bennai, F., Belarbi, R., Tahakourt, A., Abahri, K.: Effect of flax shives content and size on the hygrothermal and mechanical properties of flax concrete. Constr. Build. Mater. **262**, 120077 (2020)
6. Collet, F., Chamoin, J., Pretot, S., Lanos, C.: Comparison of the hygric behaviour of three hemp concretes. Energ. Build. **62**, 294–303 (2013)
7. Garikapati, K.P., Sadeghian, P.: Mechanical behavior of flax-lime concrete blocks made of waste flax shives and lime binder reinforced with jute fabric. J. Build. Eng. **29**, 101187 (2020)
8. Hirst, E., Walker, P., Paine, K., Yates, T.: Characterisation of low density hemp-lime composite building materials under compression loading. In: Second International Conference on Sustainable Construction Materials and Technologies, pp. 28–30 (2010)
9. Khalid, M.Y., Al Rashid, A., Arif, Z.U., Ahmed, W., Arshad, H., Zaidi, A.A.: Natural fiber reinforced composites: sustainable materials for emerging applications. Results Eng. **11**, 100263 (2021)
10. Khushefati, W.H., Demirboğa, R., Farhan, K.Z.: Assessment of factors impacting thermal conductivity of cementitious composites—A review. Clean. Mater. **5**, 100127 (2022)
11. Lagouin, M., Magniont, C., Sénéchal, P., Moonen, P., Aubert, J.-E., Laborel-préneron, A.: Influence of types of binder and plant aggregates on hygrothermal and mechanical properties of vegetal concretes. Constr. Build. Mater. **222**, 852–871 (2019)
12. Mohamad, A.: Semi-Structural Insulating and Waterproof Biobased Concrete. Normandie Université (2021)
13. Montes, F., Haselbach, L.: Measuring hydraulic conductivity in pervious concrete. Environ. Eng. Sci. **23**(6), 960–969 (2006)
14. Pacheco-Torgal, F., Jalali, S.: Vegetable fibre reinforced concrete composites: a review, 1 April 2009
15. Mardiana, A., Riffat, S.B.: Building energy consumption and carbon dioxide emissions: threat to climate change. J Earth Sci. Clim. Change. **s3** (2015)

16. Sáez-Pérez, M.P., Brümmer, M., Durán-Suárez, J.A.: A review of the factors affecting the properties and performance of hemp aggregate concretes. J. Build. Eng. **31**, 101323 (2020)
17. Walker, R., Pavia, S., Mitchell, R.: Mechanical properties and durability of hemp-lime concretes. Constr. Build. Mater. **61**, 340–348 (2014)
18. Wu, F., Yu, Q., Brouwers, H.J.H.: Long-term performance of bio-based miscanthus mortar. Constr. Build. Mater. **324**, 126703 (2022)
19. Chaux de Wasselonne: https://www.chauxdewasselonne.com/cl90 (2016). Accessed 06 Dec 2023
20. Construction et performance environnementale du bâtiment: Ministères Écologie Énergie Territoires. https://www.ecologie.gouv.fr/construction-et-performance-environnementale-du-batiment (2022). Accessed 22 Dec 2023
21. ISO 8301:1991: Afnor EDITIONS. https://www.boutique.afnor.org/fr-fr/norme/iso-83011991/isolation-thermique-determination-de-la-resistance-thermique-et-des-proprie/xs007216/101342 (2010). Accessed 18 Dec 2023
22. NF EN ISO 11357-4: Afnor EDITIONS. https://www.boutique.afnor.org/fr-fr/norme/nf-en-iso-113574/plastiques-analyse-calorimetrique-differentielle-dsc-partie-4-determination/fa199556/238238 (2021). Accessed 04 Dec 2023
23. NF ISO 5017; Dense Shaped Refractory Products—Determination of Bulk Density, Apparent Porosity and True Porosity—Produits Réfractaires Façonnés Denses. AFNOR: Saint-Denis, France, 2013
24. Tout savoir sur le ciment naturel Prompt Vicat | Ciment Prompt Vicat: https://www.ciment-prompt-vicat.fr/produit (2022). Accessed 27 Oct 2023

The Effect of Phase Change Materials (PCM) on the Thermophysical Properties of Cement Mortar

Iman Asadi ⓘ, Guomin Ji ⓘ, Gerald Steiner ⓘ, and Mohammad Hajmohammadian Baghban ⓘ

Contents

1 Introduction

Cement mortar can be used widely for rendering, masonry bedding, and joint glue in buildings. Producing energy-efficient cement-based materials is desired due to the cost of energy. Phase change materials (PCMs) are the potential materials to reduce energy consumption in buildings due to high thermal energy storage capacity [1]. The heat is absorbed and released during the changing phase from solid to liquid and vice versa [2, 3]. Despite their energy-saving advantages, they have

I. Asadi (✉) · G. Steiner
Department of Knowledge and Communication Management, University for Continuing Education Krems, Krems an der Donau, Austria
e-mail: iman.asadi@donau-uni.ac.at; gerald.steiner@donau-uni.ac.at

G. Ji · M. H. Baghban
Department of Manufacturing and Civil Engineering, Norwegian University of Science and Technology, Gjøvik, Norway
e-mail: guomin.ji@ntnu.no; mohammad.baghban@ntnu.no

© The Author(s) 2025
M. Kioumarsi, B. Shafei (eds.), *The 1st International Conference on Net-Zero Built Environment*, Lecture Notes in Civil Engineering 237,
https://doi.org/10.1007/978-3-031-69626-8_37

drawbacks when incorporated into cement-based materials, like leakage and chemical reactions. The current solution is microencapsulating phase change materials (MPCM), consisting of a core as PCM and a polymer shell as a cover [4]. A set of mechanical, physical, and thermal properties should be evaluated to apply the cement mortar in practical projects. Workability, density, compressive strength, water absorption, and thermal conductivity are some of the most critical parameters that should be considered for any new cement mortar and concrete. Previous literature revealed that adding MPCM can affect cement mortar's mechanical and thermal properties [5]. Different studies reported that MPCM reduced the workability, compressive strength, and thermal conductivity of concrete and mortar. The increment in air voids was mentioned as one of the main reasons for the reduction in physical and mechanical properties and thermal conductivity.

X-ray CT scanner has strong advantages for analyzing the internal structure of cementitious materials while providing information about internal defects (i.e., pores) and visualized 3D anatomical images. Image capture and reconstruction are used in a CT scanner to create a 3D image of the pores inside the concrete. In the first step, several 2D projections are acquired, and then the back projection algorithm is used to reconstruct it as a 3D tomographic image. Further analysis of the 3D image provides volumetric information about the solid phase and internal defects based on 3D volumetric elements (voxels). In the tomographic image, a denser phase has a high X-ray attenuation and appears bright, while a low-density phase—consisting of air or gas—will appear quite dark due to low X-ray attenuation [6, 7]. It should be noted that air void content and, consequently, thermophysical properties of mortar can be varied depending on the type of MPCMs, mix proportions, type of cement paste, curing conditions, temperatures, etc.

While numerous researchers have explored various cement-to-sand ratios (C:S) tailored to specific applications, existing literature predominantly focuses on ratios ranging from 1:0.5 to 1:8, with the most common falling between 1:2 and 1:4, as indicated by multiple studies [8–12]. However, an important research gap exists as the impact of incorporating Microencapsulated Phase Change Materials (MPCMs) into cement mortar with different cement content remains largely unexplored.

Therefore, our study's main objective is to address this gap by evaluating the effects of MPCMs on the mechanical and thermophysical properties of cement mortar, specifically focusing on workability, density, compressive strength, water absorption, and thermal conductivity. To achieve our research goals, we have introduced MPCMs into cement mortar with three different grades (C:S ratios of 1:2, 1:3, and 1:4) at 10% and 20% weight of cement. Furthermore, we conducted a microstructural analysis of all mixtures using an X-ray CT scanner and image analysis to visualize potential defects and better understand the material's microstructure.

2 Materials and Methods

2.1 Mix Design

Norcem cement (CEM 152.5R), NorsTone sand (Årdal 0–8 mm), and Micronal-24D were applied as cement, aggregate, and MPCM, respectively. Micronal-24D is a polymer-coated MPCM composite of water (45–65%), acrylic polymer (10–20%), silicon dioxide (10–20%), cellulose derivative (1–5%), and paraffin (1–5%). The appearance is a white powder (mean particle size: 50–300 µm). The heat of fusion and melting point of Micronal-24D are 105 J/g and 24 ± 2 °C, respectively. The lab analysis showed that the Micronal-24D density is 0.4 g/cm^3 and contains 99.5% solid and 0.5% water. All samples were prepared based on NS-EN 196-1. The control mixes (CM2, CM3, and CM4) were prepared with a cement-to-sand ratio of 1:2, 1:3, and 1:4, respectively. Then, MPCM was added to the control mix in 10% (CM210, CM310, and CM410) and 20% (CM220, CM320, and CM420) weight of cement. Adding MPCM reduced the workability of mortar significantly. Gradually, the water was added to the mix to satisfy the required workability (180 ± 10). The fresh workability was measured based on the flow table test conforming EN-EN 1015-3:1999+A1. Table 1 summarizes the mixed proportions of all samples. Samples were demolded after 24 h and cured in water (20 °C) for 28 days.

2.2 Test Procedures

The compressive strength test was carried out on prism samples by the dimension of 40 × 40 × 160 conforming NS-EN 196-1:2016. The dry density of all specimens was measured according to the NS-EN 1015-10. The water absorption was calculated based on the ratio of differences between the sample's weight in dried and saturated conditions and the weight of the wet sample. Thermal conductivity was measured based on the transient plane source (TPS) using TPS 2500-S. Thermal conductivity

Table 1 Mix proportion

ID	Cement (gr)	Sand (gr)	MPCM (gr)	W/C	Flow (mm)
CM2	645	1290.0	0	0.42	190 ± 10
CM210			64.5	0.49	
CM220			129	0.55	
CM3	488.8	1466.6	0	0.50	
CM310			48.4	0.55	
CM320			97.6	0.61	
CM4	389.8	1559.2	0	0.64	
CM410			38.9	0.66	
CM420			77.9	0.68	

was measured when the sample temperature was above 27 °C (solid phase) and below 21 °C (Liquid phase) to evaluate the effect of the PCM phase. It should be noted that all specimens were dried in an oven for 24 h (70 °C for density and 105 °C for compressive strength and thermal conductivity tests). It should be noted three samples were cast for each mixture and the reported results are the average values for each mix proportions. The METROTOM 1500 Scanner was applied to scan the prism samples, with a size of 40 mm × 40 mm × 80 mm (after 28 days of curing and drying in an oven at 105 °C for 24 hours). The VG Studio software was used to analyze the achieved image.

3   Results and Discussion

3.1   Test Results

Table 2 presents the test results for each individual test. The provided values represent the averages of three measurements for each mix proportion. It is observed that the dry density and compressive strength of cement mortar containing Microencapsulated Phase Change Materials (MPCM) were lower compared to the conventional sample. It can be attributed to the less compacted structure, weak interface, and inhomogeneity distribution in the matrix. Conversely, there was a

Table 2 The experimental measurement results. The parentheses numbers are the standard deviation, based on three samples

ID	Density (kg/m^3)	Compressive strength (MPa)	Water absorption (%)	Thermal conductivity	
				Solid phase	Liquid phase
CM2	2164.5 (29.5)	32.2 (0.24)	6.72 (0.28)	1.13 (0.03)	
CM210	1978.2 (33.2)	22.1 (0.28)	7.59 (0.07)	0.83 (0.03)	0.81 (0.01)
CM220	1767.8 (24.3)	17.7 (0.04)	8.16 (0.03)	0.81 (0.19)	0.68 (0.07)
CM3	2172.2 (16.4)	29.4 (1.56)	7.01 (0.06)	1.04 (0.08)	
CM310	2045.4 (21.2)	21.7 (0.56)	7.03 (0.11)	0.98 (0.06)	0.87 (0.20)
CM320	1894.6 (27.4)	15.8 (0.35)	8.86 (0.02)	0.87 (0.05)	0.71 (0.06)
CM4	2189.4 (18.9)	22.0 (0.33)	7.53 (0.25)	1.00 (0.05)	
CM410	2038.6 (31.3)	19.3 (0.18)	7.45 (0.01)	0.96 (0.03)	0.95 (0.02)
CM420	1956.9 (14.2)	15.6 (0.13)	7.55 (0.03)	0.82 (0.08)	0.86 (0.01)

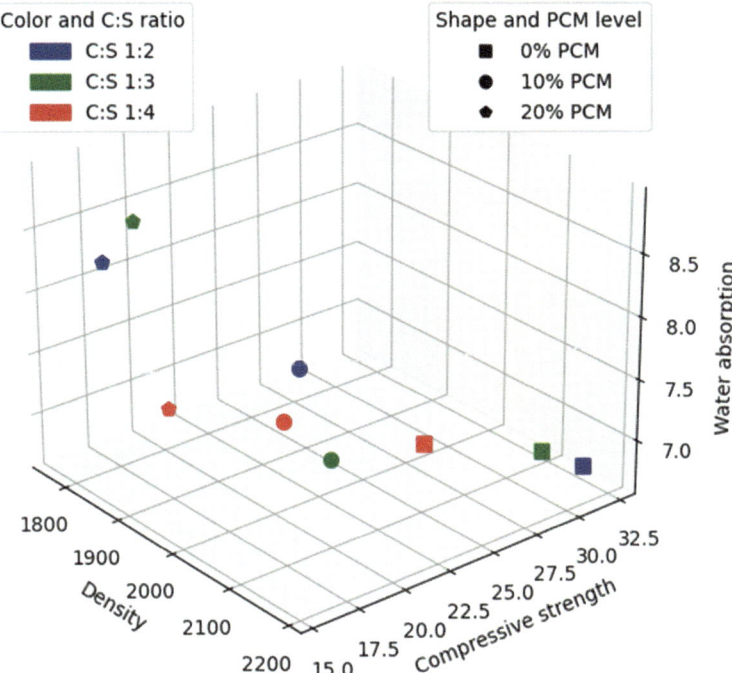

Fig. 1 Dry density, compressive strength, and water absorption

significant increase in water absorption (Fig. 1) due to pore structure, the hydrophi-licity of MPCMs, and incomplete bindings. Concerning thermal conductivity, the addition of MPCM to the mortar results in a decrease in its thermal conductivity (Fig. 2). Typically, mortar containing MPCM exhibits reduced thermal conductivity, particularly in the liquid phase. This observation aligns with findings from existing literature [13, 14], suggesting that the decrease can be attributed to the lower conductivity of paraffin in its liquid phase compared to the solid phase.

Figure 3 illustrates the percentage of change in the sample properties due to adding MPCMs. In general, it can be concluded that using MPCMs reduces the workability, density, compressive strength, and thermal conductivity of cement mortar. The potential reasons for reduction are an increment of W/C ratio to the samples with higher amounts of MPCMs, air bubbles increment due to PCM usage, and poor bonding between MPCMs and cement paste [13]. However, different samples exhibit varying changes in properties compared to their reference samples. Some samples show significant changes, while others show minor changes in certain properties. These variations can be crucial for selecting the appropriate material for specific applications based on desired properties. It was noted that the same amount of PCM incorporation in samples with different C:S consequences has different effects. For instance, Density Change: CM220 shows the highest decrease in density (−18.33%), while CM310 exhibits the lowest decrease at (5.84%). Compressive

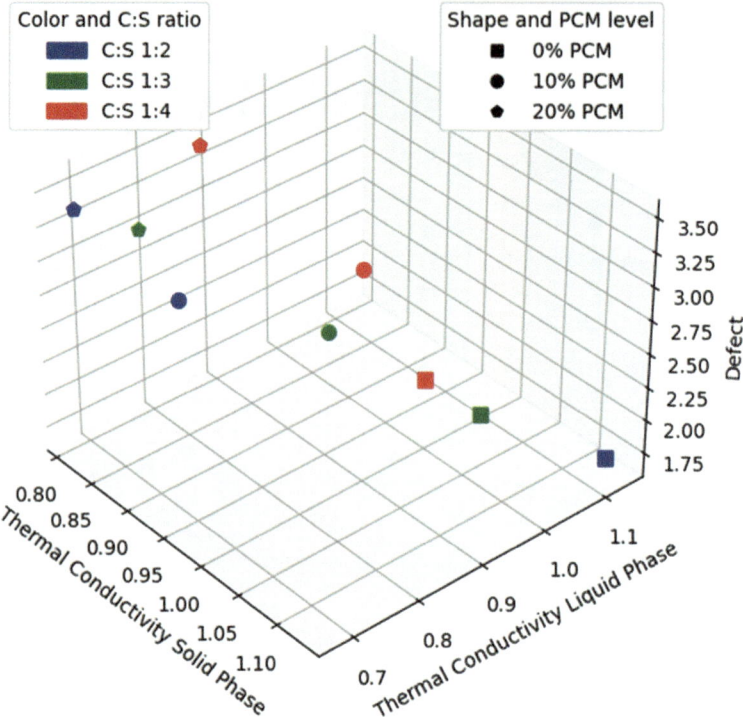

Fig. 2 Thermal conductivity versus defect

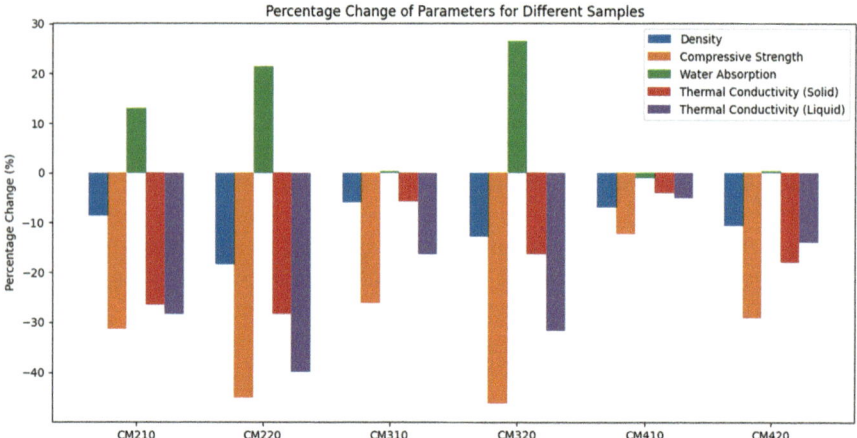

Fig. 3 The effect of MPCMs on the mechanical and thermophysical properties of cement mortar with various grades

Strength Change: CM320 experiences the largest decrease in compressive strength at (46.26%), whereas CM410 shows the least decrease (12.27%). Water Absorption Change: CM320 displays the highest increase in water absorption (26.39%), while CM410 has the smallest increase (−1.06%) (a decrease).Thermal Conductivity (Solid) Change: CM420 demonstrates the most substantial decrease in thermal conductivity (solid) (18.00%), while CM410 shows the least decrease (−4.00%). Thermal Conductivity (Liquid) Change: CM220 exhibits the highest decrease in thermal conductivity (liquid) (−39.82%), while CM410 has the lowest decrease (−5.00%).

A multivariate regression analysis was conducted to establish predictive models for each parameter based on the cement-to-sand ratio of cement mortar and the amount of MPCM, as depicted in Fig. 4. This analytical approach allowed for a comprehensive examination of the interdependencies among the variables, facilitating insights into how changes in cement-to-sand ratio and MPCM content influence the properties of the mortar. Through the regression analysis, equations were derived to estimate the density, compressive strength, water absorption, and thermal conductivity (both solid and liquid phases) of the mortar samples. These equations serve as predictive tools, enabling the assessment of the impact of varying cement-to-sand ratios and MPCM amounts on the performance characteristics of the mortar. The regression models provide valuable insights into the relationships between the independent variables (cement-to-sand ratio and MPCM content) and the dependent variables (density, compressive strength, water absorption, and thermal conductivity). By analyzing the coefficients and significance levels of the regression equations, it is possible to identify the relative importance of each factor and discern their contributions to the observed changes in mortar properties. Furthermore, the predictive capabilities of the regression models offer practical utility in optimizing mortar formulations for specific applications. Engineers and researchers can utilize these equations to tailor mortar compositions to meet desired performance criteria, balancing considerations such as strength, thermal insulation, and moisture resistance.

3.2 X-Ray Tomography

There is a challenge in distinguishing between porosity and MPCM through X-ray scanning. Due to the low density of the air voids and MPCMs, both look like dark colors. Therefore, distinguishing between air voids and MPCM is not possible with a CT scanner. Although some studies used shape and size analysis to visualize the air voids and MPCM in different samples, they can also face a bias. The PCM may not remain in spherical shapes and may be broken during mixing. Also, some small pores are almost spherical. Furthermore, MPCM can be agglomerated and make a much bigger colony than an individual one. Moreover, there are different types of pores in the range of individual MPCM used in this study (50–300 microns). Considering all these, we defined a binary image as follows: 1. Material and

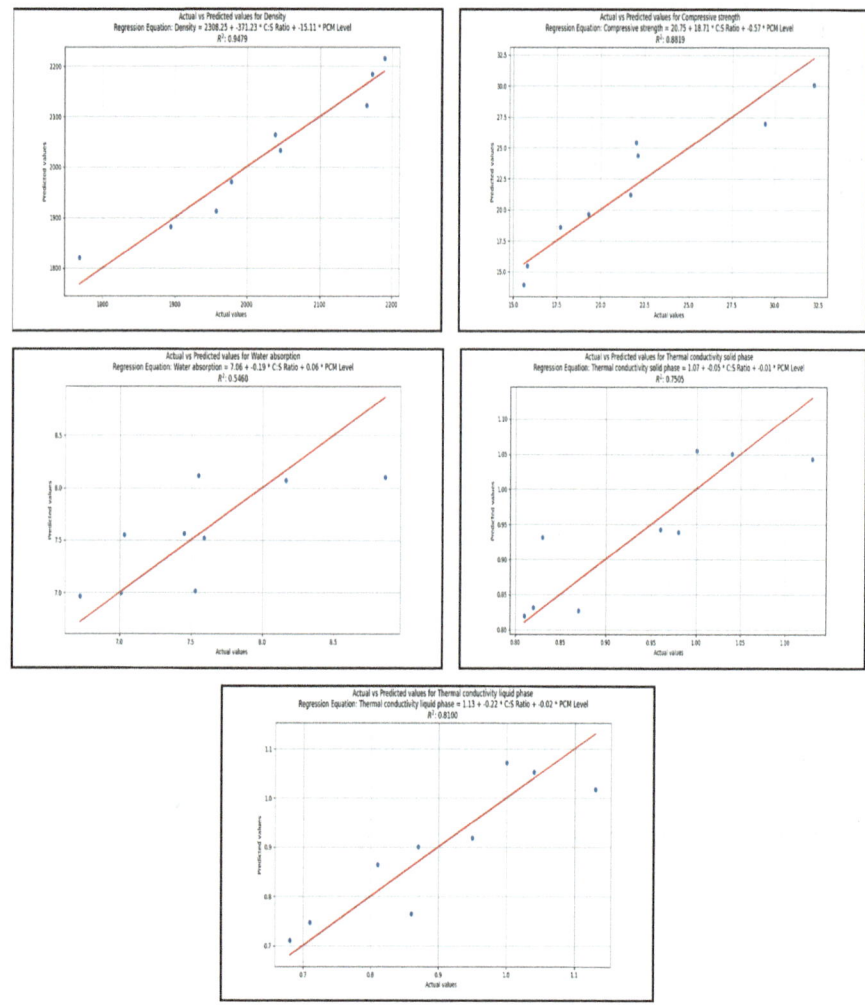

Fig. 4 The relationship between thermomechanical properties with the c:s and MPCM content

2. Defect. Then the defects were analyzed and categorized based on their size and volume (Fig. 5). The material is the white part of the image (sand, hydration products, and unhydrated cement), and the defect is the black part of the image (air voids for CM2, CM3, CM4, and air voids+MPCM for CM210 CM220, CM310, CM320, CM410, CM420). Figure 6 demonstrates the correlation between cement-to-sand ratio, MPCM content, and detected defects.

The equation represents a predictive model for estimating the defect level in a cement mortar based on two key factors: the cement-to-sand ratio and the content of MPCM. The coefficient associated with the cement-to-sand ratio indicates that an

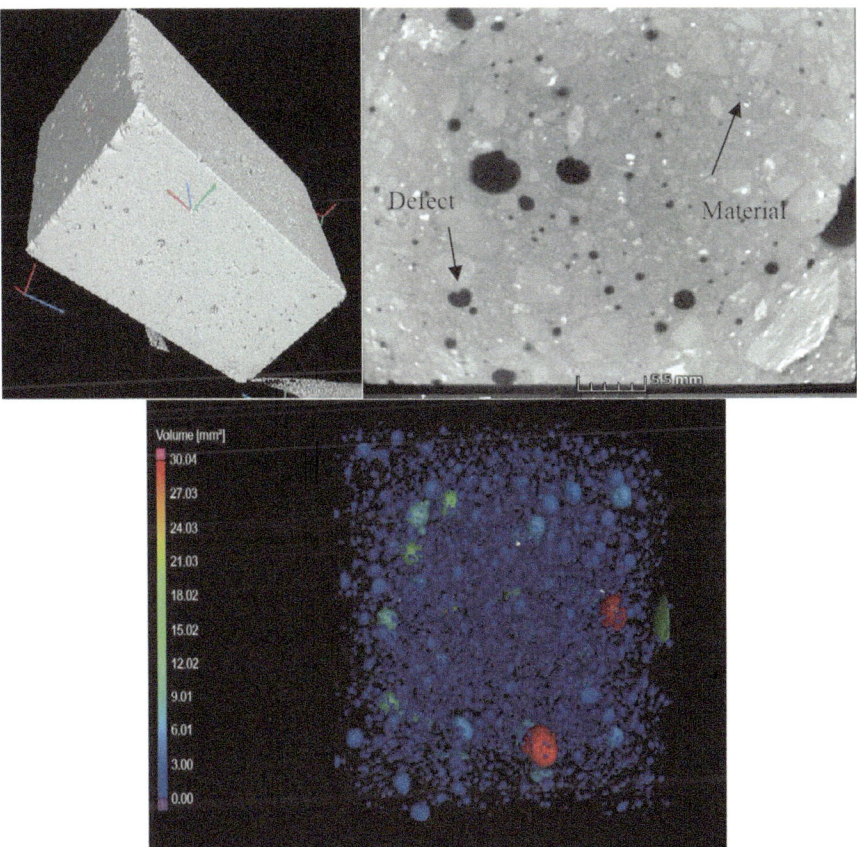

Fig. 5 The defect detection by SC Scanner

increase in this ratio is expected to lead to a decrease in the defect level, with a magnitude of 1.14 units per unit increase in the ratio. Conversely, a decrease in the ratio is anticipated to result in an increase in the defect level by the same magnitude. On the other hand, the coefficient linked to the MPCM content suggests that an increase in the MPCM content would lead to a higher defect level, with each unit increase in MPCM content corresponding to an increase of 0.080.08 units in the defect level. Conversely, a decrease in MPCM content is expected to reduce the defect level by the same amount. This equation provides valuable insights into how changes in these two factors can influence the defect level of the material, aiding in the optimization of material composition and defect mitigation strategies.

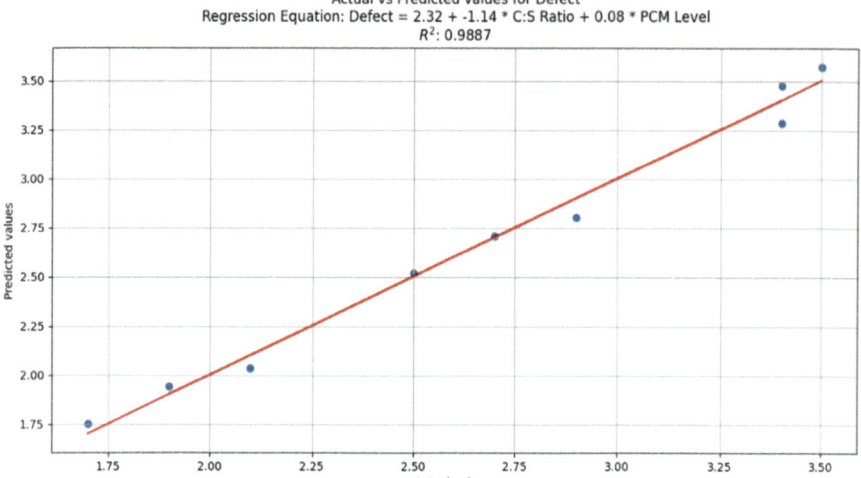

Fig. 6 The relationship between defects with the c:s and MPCM content

4 Conclusion

In conclusion, the experimental findings and further analysis throughout this study shed light on the multifaceted effects of incorporating MPCMs into cement mortar with different grades. The results indicate a notable reduction in dry density and compressive strength of the mortar containing MPCMs compared to conventional samples. This decrease can be attributed to factors such as less compacted structure, weak interface, and heterogeneous distribution within the matrix. Conversely, there is a significant increase in water absorption, attributed to alterations in pore structure, the hydrophilicity of MPCMs, and incomplete bindings. However, the addition of MPCMs leads to a decrease in thermal conductivity, particularly in the liquid phase, aligning with existing literature. The variations observed among samples underscore the importance of considering specific material compositions for tailored applications. Multivariate regression analysis provided predictive models for estimating various properties of mortar based on cement-to-sand ratio and MPCM content. These models offer valuable insights into the interdependencies among variables and facilitate the optimization of mortar formulations for desired performance criteria. Overall, this comprehensive investigation offers insights into the complex interactions between MPCMs and cement mortar properties. By understanding these relationships, researchers can make informed decisions to develop mortar compositions fitted to specific research studies based on their intended applications within the field of mortar technology.

References

1. Pilehvar, S., Cao, V.D., Szczotok, A.M., Valentini, L., Salvioni, D., Magistri, M., Pamies, R., Kjøniksen, A.-L.: Mechanical properties and microscale changes of geopolymer concrete and Portland cement concrete containing micro-encapsulated phase change materials. Cem. Concr. Res. **100**, 341–349 (2017)
2. Ling, T.-C., Poon, C.-S.: Use of phase change materials for thermal energy storage in concrete: an overview. Constr. Build. Mater. **46**, 55–62 (2013)
3. Regin, A.F., Solanki, S., Saini, J.: Heat transfer characteristics of thermal energy storage system using PCM capsules: a review. Renew. Sust. Energ. Rev. **12**(9), 2438–2458 (2008)
4. Pomianowski, M., Heiselberg, P., Jensen, R.L., Cheng, R., Zhang, Y.: A new experimental method to determine specific heat capacity of inhomogeneous concrete material with incorporated microencapsulated-PCM. Cem. Concr. Res. **55**, 22–34 (2014)
5. Djamai, Z.I., Salvatore, F., Larbi, A.S., Cai, G., El Mankibi, M.: Multiphysics analysis of effects of encapsulated phase change materials (PCMs) in cement mortars. Cem. Concr. Res. **119**, 51–63 (2019)
6. Asadi, I., Henry Trussell, N., Stangeland Hårr, M.K., Gaute, P.E., Endrerud, E.G., Jacobsen, S.: Macro pore content in wet-sprayed concrete characterized by CT scanning. In: fib International Congress, Oslo (2022)
7. Asadi, I., Trussell, N., Hårr, M.S., Kjeka, G., Endrerud, P.E., Grøv, E., Jacobsen, S.: Macro pore content in wet-sprayed concrete characterized by CT scanning. In: Fib International Congress, Oslo (2022)
8. Demirboğa, R.: Influence of mineral admixtures on thermal conductivity and compressive strength of mortar. Energ. Buildings. **35**(2), 189–192 (2003)
9. Lertwattanaruk, P., Makul, N., Siripattarapravat, C.: Utilization of ground waste seashells in cement mortars for masonry and plastering. J. Environ. Manag. **111**, 133–141 (2012)
10. Olmeda, J., De Rojas, M.S., Frías, M., Donatello, S., Cheeseman, C.: Effect of petroleum (pet) coke addition on the density and thermal conductivity of cement pastes and mortars. Fuel. **107**, 138–146 (2013)
11. Baite, E., Messan, A., Hannawi, K., Tsobnang, F., Prince, W.: Physical and transfer properties of mortar containing coal bottom ash aggregates from Tefereyre (Niger). Constr. Build. Mater. **125**, 919–926 (2016)
12. Zhang, H.: Building Materials in Civil Engineering. Elsevier (2011)
13. Asadi, I., Baghban, M.H., Hashemi, M., Izadyar, N., Sajadi, B.: Phase change materials incorporated into geopolymer concrete for enhancing energy efficiency and sustainability of buildings: a review. Case Stud. Constr. Mater. **17**, e01162 (2022)
14. Cui, H., Liao, W., Mi, X., Lo, T.Y., Chen, D.: Study on functional and mechanical properties of cement mortar with graphite-modified microencapsulated phase-change materials. Energ. Buildings. **105**, 273–284 (2015)

Thermal Performance Characterization of Recycled Textile-Based Materials for Building Insulation

Alessandro Dama (iD)**, Esra Abdelhalim Mohamed Khalfallaand, Andrea Alongi** (iD)**, and Adriana Angelotti** (iD)

Contents

1 Introduction

The EU generates 12.6 million tons of textile waste per year. Clothing and footwear alone accounts for 5.2 million tons of waste, equivalent to 12 kg of waste per person every year. Currently, only 22% of post-consumer textile waste is collected separately for re-use or recycling, while the remainder is often incinerated or landfilled. Today, the Commission is proposing rules to accelerate the development of the separate collection, sorting, reuse and recycling sector for textiles in the EU, in line with the EU Strategy for Sustainable and Circular Textiles [1]. Common EU Extended Producer Responsibility (EPR) rules will also make it easier for Member States to implement the requirement to collect textiles separately from 2025, in line with current legislation. Therefore a large amount of separated textile waste is

A. Dama (✉) · E. A. M. Khalfallaand · A. Alongi · A. Angelotti
Politecnico di Milano, Milan, Italy
e-mail: alessandro.dama@polimi.it

© The Author(s) 2025
M. Kioumarsi, B. Shafei (eds.), *The 1st International Conference on Net-Zero Built Environment*, Lecture Notes in Civil Engineering 237,
https://doi.org/10.1007/978-3-031-69626-8_38

expected to be available soon and there is a great interest in the possibility to recycle what is not suitable for reuse in the clothing sector, as raw material for other sectors, such as in building applications.

In the last decade, many researches have been carried out with the purpose of exploring solutions and potentialities of recycling waste materials for making insulating panels, and some of them have specifically considered textile waste. Zach et al. [2] tested thermal insulation panels made from sheep wool, with different bulk densities and at different temperatures. The results showed thermal insulation performances comparable with mineral/rock wool. Patnaik et al. [3] studied thermal and sound insulation samples developed from waste wool and recycled polyester fibers (RPET) for building industry applications. Waste wool fibers were mixed with RPET fibers in 50/50 proportions in the form of a two-layer mat. The behavior of the samples was evaluated also under high humidity conditions, but in such case the waste wool components were sprayed with silicon, concluding that an adequate moisture resistance may be achieved preserving insulation and acoustic performances. Drochytka et al. [4] tested samples produced in laboratory from polyester fibers and bi-component fibers, made by combining hot air and pressure, in order to simulate the real producing technology named "Airlay" [5]. Different percentage of binding bi-component, different bulk densities and operating temperatures were considered. The results show thermal insulation performances comparable with mineral/rock wool. In [6], Diassanayake et al., by facing the recycling complexity associated to the increasing use of multi-materials in textile production, and the environmental impact of synthetic textiles, developed, tested and modeled the performances of a novel insulating material made of postindustrial cutting waste, Nylon/Spandex and Polyurethane. Rubino et al. [7] modeled and experimentally investigated the effective thermal conductivity in highly porous fibrous materials like Merino wool panels, realized with a natural binder made by a chitosan solution. Different samples were obtained and studied, by combining the same percentages of binder and fibrous matrix (60% and 40% respectively), with different bulk density values (ranging between 80 and 197 kg/m^3), in order to characterize and compare their thermal behavior. Finally, a recent review on the factors influencing the thermal conductivity of insulating materials was given in [8]. The review considers both conventional and alternative materials, including natural and fibrous ones, and summarizes results obtained from different experimental investigations. One major conclusion is that the moisture content strongly affects the thermal conductivity of both organic and inorganic materials.

In this context originated MATE.RIA (Methods and Actions for the Ecological Treatment of Post-Consumer Textiles and their Innovative Recycling in Architecture), a collaborative and interdisciplinary research project funded by MASE (Italian Ministry of Environment and Energy Security), spanning the years 2022–2024. This study was carried within MATE.RIA with the objective to investigate and optimize the thermal performances of semifinished panel obtained by recycled textile materials.

The experimental analysis concerning the heat transfer properties was carried out on mats obtained from the post-consume textile waste selected and processed with "Airlay" technique by the industrial partners. The analyzed prototypal panels have been produced from three kind of textile fabric selection (mostly polyester, mostly cotton or mixed) and with different density (low, medium and high) and percentage of binding component (10% or 20%).

The heat transfer properties were measured through the dynamic methodology of the hot disc, both in laboratory conditions and in a climatic chamber (where temperature and humidity are controlled). The aim of the study is to assess the thermal insulation performance of the panels, considering the influence of density, temperature and humidity, and to investigate to what extent such performance is influenced by the kind of waste fabric material.

Although in literature different kinds of waste fabric materials are used to create building insulation, they are rarely processed in the same way and tested with the same methodology. In turn, in this work the performances of different combination of materials can actually be compared, as the same production process and the same experimental technique are adopted.

2 Case Study

The prototypes of products for building applications, based on recycled textile materials, were developed in collaboration between University and industrial partners. The production process of the mats for these semi-finished panels includes: selection of the post-consume textile materials, granulation of the fabric, addiction of a binding component and a thermal mechanical process called "Airlay technique" for the mixing and cohesion of the granulates. Different fabric materials and percentage of binding component (10% or 20%) were used for the productions of the mats, resulting in different density and rigidity. Three family of compositions were considered for the recycled material selection, mainly cotton (>90%), mainly polyester (>90%) and a mix (34% polyester, 30% wool, 23% viscous and 13% cotton). For each composition, three levels of mat density were investigated, in a range from 50 to 150 kg/m^3. Table 1 reports the main features of tested panels.

3 Methods

3.1 Hot Disc Method for Thermal Conductivity and Diffusivity Measurements

The heat transfer properties by conduction (thermal conductivity λ, thermal diffusivity α) of the semi-finished panels were obtained through the dynamic measurement technique called Transient Plane Source (TPS) or hot disk method, also

Table 1 Main features of the semi-finished panels

Recycled textile material	Process/ code	% of Binding component (%)	Nominal thickness [cm]	Density [kg/m³]
Polyester	RUN 13	10	3	57
	RUN 12	20	4	82
	RUN 11	20	3.2	124
Mix	RUN 8	10	3	60
	RUN 7	20	3	65
	RUN 10	10	4	91
	RUN 9	10	3.3	100
	RUN 6	20	3	155
Cotton	RUN 4	10	4	46
	RUN 3	10	3.5	57
	RUN 2	20	4	116

Fig. 1 Hot disk sensor (left) and placement of the sensor sandwiched between two samples (right)

described in the ISO technical standard 22007-2:2022 [9]. The measurement methodology is based on the application of a step pulse heat source at a sample surface and the simultaneous measurement of its temperature increase over time. The sensor (hot disk), which is sandwiched between two samples of the same material (Fig. 1), acts both as a heat source and a temperature probe, and is powered by an electric current. It consists of a thin double spiral resistor enclosed in an electrical insulating material (Kapton in the case of the apparatus in use). Moreover, great care has been taken to ensure good thermal contact between the sensor and the two samples, while keeping the surfaces as planar as possible. The variation over time of the sensor's electrical resistance is measured, it is proportional to the temperature increase. The measuring apparatus therefore includes, in addition to the measuring disc, a power supply, a resistor bridge which allows the variation in resistance of the sensor to be measured in terms of an unbalancing potential of the bridge, a digital voltmeter and a control system. The temperature variation over time is then interpreted via a model, thus allowing the thermal conductivity and diffusivity of the material to be derived.

Fig. 2 Samples of semi-finished panels tested with the hot disc method, from left to right: blend, cotton and polyester

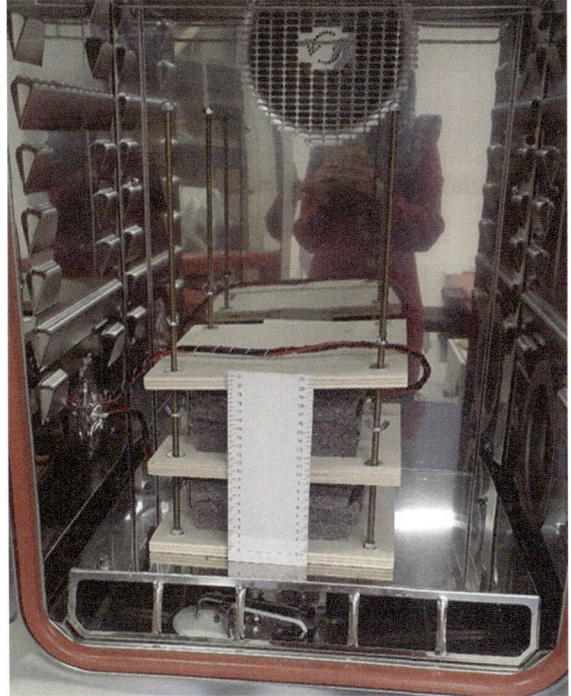

Fig. 3 The sample fixing system inside the climatic chamber of the Textile Hub

In the case of the samples of semi-finished panels in recycled textile (Fig. 2), in order to guarantee good thermal contact between the hot disk and the faces of the samples, a panel fixing system was realized, consisting of two wooden panels 25 cm × 25 cm and a screw system (Fig. 3). At the same time, a graduated scale allows to measure the reduction in thickness of the samples caused by the (modest) pressure applied. It was thus possible to record and limit the thickness reduction in the order of 10% and evaluate the effective density of the samples in the measurement condition.

3.2   TPS Measurements in Climatic Chamber

The thermal transport parameters of the material can generally be influenced by the temperature and, if the material is porous, by the water content, which in turn depends on the degree of humidity in the environment. Therefore, after having carried out a first series of measurements at the Building Physics Laboratory of the Department of Energy, with temperature varying between 18 °C and 24 °C and the ambient relative humidity between 35% and 65%, a second series of measurements was performed, where temperature and humidity conditions were controlled via the climatic chamber available at the Interdepartmental Textile Hub Laboratory.

The operating range of the Votsch VC 7018 climatic chamber (Fig. 3) is equal to 10–95 °C for the air temperature, and 10–100% for the relative humidity, but depending on the temperature the minimum attainable relative humidity changes (being the 10% for temperatures higher than 40 °C and rising up to 65% for temperatures of 10 °C).

In the case of the MATE.RIA project, the TPS measurements in climatic chamber were performed on a small number of samples, i.e., those with the lowest densities, at a temperature of 20 °C and for three relative humidity, i.e., 20%, 50% and 80%. Low density samples were chosen, as from the first characterization in laboratory conditions they were found to have the best insulating performance.

For this second series of measurements, the samples were pre-conditioned in the controlled climatic chamber at the same temperature and relative conditions of the measurement for at least 3, 6 or 12 h before, depending on the starting and target condition. Larger preconditioning times, up to 72 h, were also tested without evidence of significant differences in the measurements results.

4   Results

4.1   Thermal Conductivity and Diffusivity Measurements Versus Density

The results of the first group of TPS measurements are reported in Table 2 and represented in Fig. 4 as a function of the effective density under measurement condition.

As it can be seen in Fig. 4, thermal conductivity has an increasing trend with density. This trend is consistent with the literature relating to cellular materials [10] which predicts that thermal conductivity decreases with density until transmission is dominated by radiative exchange, reaches a minimum and then begins to increase linearly with density again when conduction through the solid matrix and the air becomes prevalent. It is therefore deduced that in the density range analyzed the heat transmission through the semi-finished panels is dominated by the conduction mechanism and it is possible to expect a minimum at even lower densities.

Table 2 TPS Measurements in laboratory environment without sample precondition

Sample	Effective density [kg/m³]	Average temperature [°C]	Average relative humidity	Thermal conductivity [W/(m.K)] Mean ±	Std. dev	Thermal diffusivity [mm²/s] Mean ±	Std. dev	Num. of meas.
Polyester								
Run 13	64	22.8 ± 1	51%	0.0422	0.0005	0.470	0.023	7
Run 12	86	22.7 ± 1	50%	0.0443	0.0011	0.403	0.016	6
Run 11	127	22.6 ± 1	54%	0.0524	0.0006	0.369	0.019	7
Mix								
Run 8	65	21.1 ± 2	52%	0.0460	0.0008	0.514	0.021	9
Run 7	65	21.0 ± 2	46%	0.0452	0.0008	0.568	0.046	6
Run 10	95	19.0 ± 0.5	49%	0.0512	0.0007	0.477	0.011	6
Run 9	106	20.3 ± 1	58%	0.0510	0.0008	0.365	0.029	6
Run 6	159	22.3 ± 1	52%	0.0588	0.0012	0.386	0.015	6
Cotton								
Run 4	50	20.7 ± 1	58%	0.0482	0.0017	0.672	0.069	9
Run 3	61	21.4 ± 2	54%	0.0525	0.0011	0.626	0.041	8
Run 2	120	22.6 ± 0.5	53%	0.0626	0.0012	0.436	0.017	5

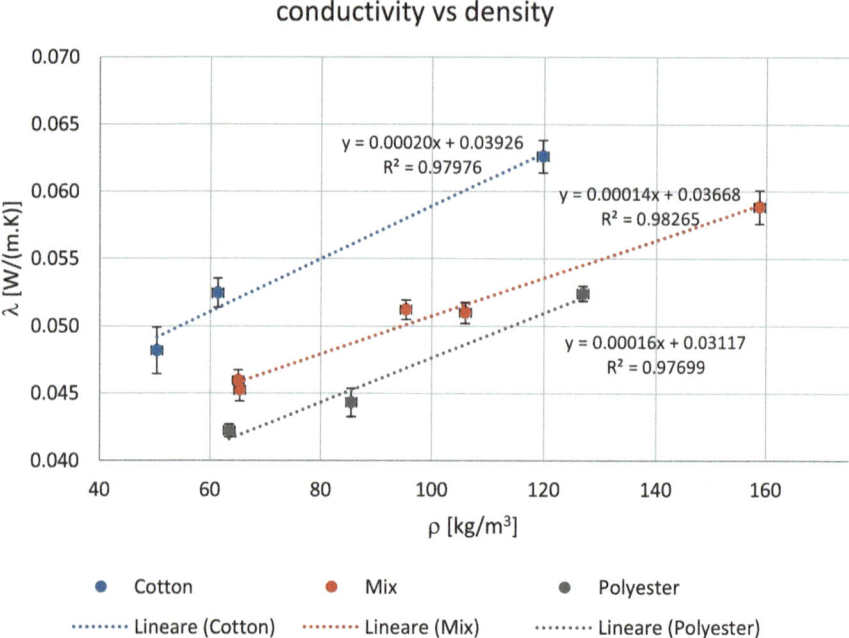

Fig. 4 Thermal conductivity as function of probing density, TPS measurements in laboratory without samples pre-conditioning

The linear trend of conductivity with density is distinct for the three types of materials used. The interpolation lines shown in Fig. 4 allow to compare different materials with the same density: it can thus be observed that polyester-based panels show the lowest conductivities. Specifically, in the density range studied, the cotton-based panels appear to have a higher conductivity than polyester by an average of 26%, while the mixed material panels appear to have a conductivity higher than that of the polyester-based panels by 5–9%. Therefore, it is observed that the insulating performance of mixed material-based panels is intermediate between polyester-based panels and cotton-based panels, at a short distance from the former.

Moreover, Tables 1 and 3 allow to evaluate the possible role of the fraction of the bi-component used as binder. The Run 7 and Run 8 samples, both obtained from mixed textile waste, have a very similar effective density and differ substantially in the percentage of the binding component used, equal to 20% and 10%, respectively. Since the resulting heat transfer properties are the same within the measurement uncertainty, it can be deduced that the different amount of binder does not significantly affect the conductivity.

Table 3 TPS measurements in climatic chamber (CC) with sample preconditioning

Sample	Effective density [kg/m³]	Temperature [°C]	Relative humidity	Thermal conductivity [W/(m.K)] Mean ±	Std.dev	Thermal diffusivity [mm²/s] Mean ±	Std.dev	Num. of meas.
Polyester								
Run 13	63.1 ± 1	20	20%	0.0408	0.0006	0.558	0.076	3
			50%	0.0413	0.0005	0.586	0.050	3
			80%	0.0445	0.0010	0.512	0.034	3
Mix								
Run 8	65.2 ± 2.3	20	20%	0.0421	0.0007	0.488	0.101	5
			50%	0.0446	0.0010	0.539	0.044	3
			80%	0.0482	0.0010	0.502	0.046	4
Run 7	68.1 ± 0.5	20	20%	0.0434	0.0018	0.612	0.110	3
			50%	0.0455	0.0008	0.539	0.064	7
			80%	0.0465	0.0017	0.530	0.069	4
Cotton								
Run 4	49.7 ± 2.2	20	20%	0.0456	0.0008	0.662	0.171	4
			50%	0.0483	0.0007	0.708	0.069	3
			80%	0.0523	0.0015	0.597	0.066	5
Run 3	61.7 ± 1.6	20	20%	0.0477	0.0004	0.662	0.061	6
			50%	0.0513	0.0012	0.623	0.067	4
			80%	0.0551	0.0016	0.559	0.147	6

4.2 Thermal Conductivity and Diffusivity Versus Relative Humidity

The results of the second group of TPS measurements are reported in Table 3 and Fig. 5. The latter shows for each of the five low density samples, belonging to the three family of recycled textile (A: Polyester, B: Mix and C: Cotton), the trend of the thermal conductivity measured in the climatic chamber as a function of the relative humidity. The same graph also shows for each sample the comparison with the average thermal conductivity previously obtained in the laboratory (first group), at the average relative humidity present in free running conditions and without preconditioning.

A good consistency can be observed between the measurements performed in laboratory conditions and those performed with fully controlled hygrothermal conditions in the climatic chamber.

As it is evident from the angular coefficient of the interpolation lines in Fig. 5, the thermal conductivity for the cotton-based samples (Run 3 and 4) is more sensitive to ambient relative humidity than the polyester-based sample (Run 13). The mixed-based samples (Run 7 and 8) also appear to have an intermediate behavior between cotton and polyester. These results suggest that cotton-based samples may have a greater hygroscopicity compared to polyester, as also found for similar panels by [11].

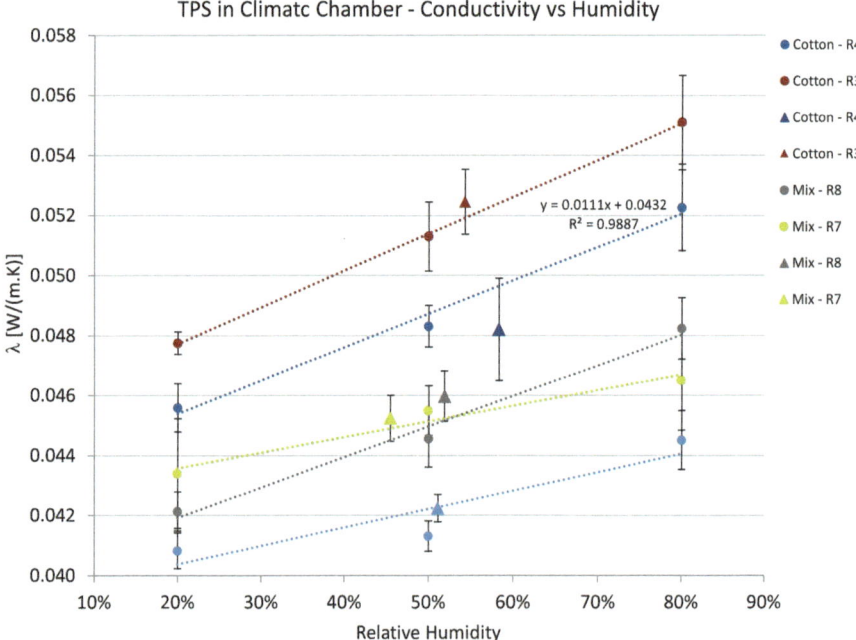

Fig. 5 TPS measurements in climatic chamber (CC). Thermal conductivity as function of effective density (A: Polyester, B: Mix, C: Cotton). Linear trends and comparisons with measurements performed in laboratory without sample preconditioning

5   Conclusions

A good thermal insulation performance was observed for all the panels, so that the general conclusion is that all the panels can be used in building envelope solutions. For a given kind of waste textile, the thermal conductivity was found to increase with density, leading to identify the low-density panels as the most ones. Moreover, thermal conductivity was found to increase with relative humidity, in line with the expected hygroscopic behavior of the porous matrix. For the same density and relative humidity, the thermal conductivity of the cotton-based panels is higher by 26% on average compared to the polyester-based panels. The question then arises whether it is worth the effort of sorting the textile waste into the different fibers. Indeed the thermal conductivity of the mixed waste based panels was found to be only slightly higher than that of the polyester based ones. If this outcome could be confirmed for any kind of mixed textile waste deriving from post-consume clothing, namely collected in other regions and periods of the year, it could be concluded that fibers sorting can be avoided for this kind of building application.

The development of this work will further investigate the hygroscopic properties of the sample panels, their relation with the environmental conditions and their influence on the thermal performances.

Acknowledgments This research has been carried out within the project "MATE.RIA: Methods and actions for ecological treatment of post-consume textiles and innovative recycling in architecture," funded by the Italian Ministry of Ecological Transition (MITE), call D.D. n. 72/2020, project code D33B22000030004.

References

1. EUROPEAN COMMISSION EU Strategy for Sustainable and Circular Textiles Brussels, 30.3.2022 COM (2022) https://environment.ec.europa.eu/strategy/textiles-strategy_en. Author, F., Author, S.: Title of a proceedings paper. In: Editor, F., Editor, S. (eds.) CONFERENCE 2016, LNCS, vol. 9999, pp. 1–13. Springer, Heidelberg (2016)
2. Zach, J., Hroudová, J., Korjenic, A.: Environmentally efficient thermal and acoustic insulation based on natural and waste fibers: environmentally efficient insulations based on natural and waste fibers. J. Chem. Technol. Biotechnol. **91**, 2156–2161 (2012). https://doi.org/10.1002/jctb.4940. Author, F.: Contribution title. In: 9th International Proceedings on Proceedings, pp. 1–2. Publisher, Location (2010)
3. Patnaik, A., Mvubu, M., Muniyasamy, S., Botha, A., Anandjiwala, R.D.: Thermal and sound insulation materials from waste wool and recycled polyester fibers and their biodegradation studies. Energy Build. **92**, 161–169 (2015). https://doi.org/10.1016/j.enbuild.2015.01.056
4. Drochytka, R., Dvorakova, M., Hodna, J.: Performance evaluation and research of alternative thermal insulation based on waste polyester fibers. Proc. Eng. **195**, 236–243 (2017). https://doi.org/10.1016/j.proeng.2017.04.549
5. Zhang, D.: Chapter 6 – nonwovens for consumer and industrial wipes. In: Applications of Nonwovens in Technical Textiles, pp. 103–119. Woodhead Publishing (2010). https://doi.org/10.1533/9781845699741.2.103

6. Dissanayake, D.G.K., Weerasinghe, D.U., Wijesinghe, K.A.P., Kalpage, K.M.D.M.P.: Developing a compression moulded thermal insulation panel using postindustrial textile waste. Waste Manag. **79**, 356–361 (2018)

7. Rubino, C., Bonet-Aracil, M., Liuzzi, S., Martellotta, F., Stefanizzi, P.: Thermal characterization of innovative sustainable building materials from wool textile fibers waste. J. Eng. Sci. **63**(2–4), 277–283 (2019)

8. Hung Anh, L.D., Pasztoryn, Z.: An overview of factors influencing thermal conductivity of building insulation materials. J. Build. Eng. **44** (2021). https://doi.org/10.1016/j.jobe.2021.102604

9. ISO 22007-2:2022 Plastics, Determination of Thermal Conductivity and Thermal Diffusivity Part 2: Transient Plane Heat Source (Hot Disc) Method

10. Gibson, L.J., Ashby, M.F.: Cellular Solids: Structure and Properties, 2nd edn. Cambridge University Press, Cambridge (1997)

11. Dieckmannn, E., Onsiong, R., Nagy, B., Sheldrick, L., Cheeseman, C.: Valorization of waste feathers in the production of new thermal, insulation materials. Waste Biomass Valorization. **12**, 1119–1131 (2021)

Temperature-Dependent Dual-Operation Mode for Energy Tunnel Integrated with Phase-Change Materials in Geothermal Environment

Qiling Wang, Jiaolong Zhang, Peng Xiao, Xi Chen, Eddie Koenders, and Yong Yuan

Contents

Q. Wang
College of Civil Engineering, Tongji University, Shanghai, China

Institute of Construction and Building Materials, Technical University of Darmstadt, Darmstadt, Germany

J. Zhang · Y. Yuan (✉)
College of Civil Engineering, Tongji University, Shanghai, China
e-mail: yuany@tongji.edu.cn

P. Xiao · E. Koenders
Institute of Construction and Building Materials, Technical University of Darmstadt, Darmstadt, Germany

X. Chen
College of Civil Engineering and Architecture, Jiaxing University, Jiaxing, China

© The Author(s) 2025
M. Kioumarsi, B. Shafei (eds.), *The 1st International Conference on Net-Zero Built Environment*, Lecture Notes in Civil Engineering 237,
https://doi.org/10.1007/978-3-031-69626-8_39

1 Introduction

The application of energy tunnels has garnered attention from engineers and researchers in recent years due to its alignment with carbon reduction initiatives [1, 2]. High geothermal tunnels, characterized by a geothermal environment, represent a promising energy underground structure owing to substantial heat energy reservoirs within surrounding rock formations [3]. However, the heat energy stored in these tunnels presents challenges when following traditional construction methods [4, 5]. Limited research has explored the harnessing of geothermal energy from high geothermal tunnels. High-temperature water within these tunnels can be extracted directly to heat surrounding buildings [6]. Previous studies by Li et al. [7] and Xu et al. [8] have examined energy tunnel applications in tropical zones and proposed hazard mitigation systems for utilizing heat energy in mining tunnels.

Phase-change materials (PCMs) offer significant advantages in heat energy storage due to their high latent heat capacities, prompting their exploration for use in energy tunnels [9]. Zhang et al. [10] proposed a solution for heat transportation in geothermal tunnels utilizing PCM energy storage units, while Cao et al. [11] integrated PCM plates with lining ground heat exchangers to extract geothermal energy. Wang et al. [12] proposed PCM-modified support structures to serve as insulation layers in high geothermal tunnels. However, there is limited research addressing the operational modes of PCM-modified energy tunnels.

This study introduces a temperature-dependent dual-operation mode for energy tunnels with PCMs, capitalizing on the advantageous properties of PCMs. The performance of heat reduction and energy extraction are evaluated, taking factors such as PCM mass fraction and thermal properties of PCMs into account.

2 Concept of Temperature-Dependent Dual-Operation Mode

The schematic depiction of the establishment procedure for an energy tunnel incorporating PCMs is presented in Fig. 1. The foundational material for constructing the tunnel structure comprises cement mixed with PCMs, constituting the base material for concrete formulation. The PCMs are homogeneously blended with cement in a small-scale mixer to achieve uniformity, resulting in the formation of cement with PCMs. Subsequently, meticulous mixing of this composite material with water and aggregates yields concrete integrated with PCMs, serving as the lining material installed within the geothermal sections of the tunnels. Thermal sensors are strategically embedded within these distinct linings to facilitate real-time monitoring of the temperature dynamics within the linings. Concurrently, absorber pipes are embedded within the linings, facilitating fluid circulation within them to facilitate

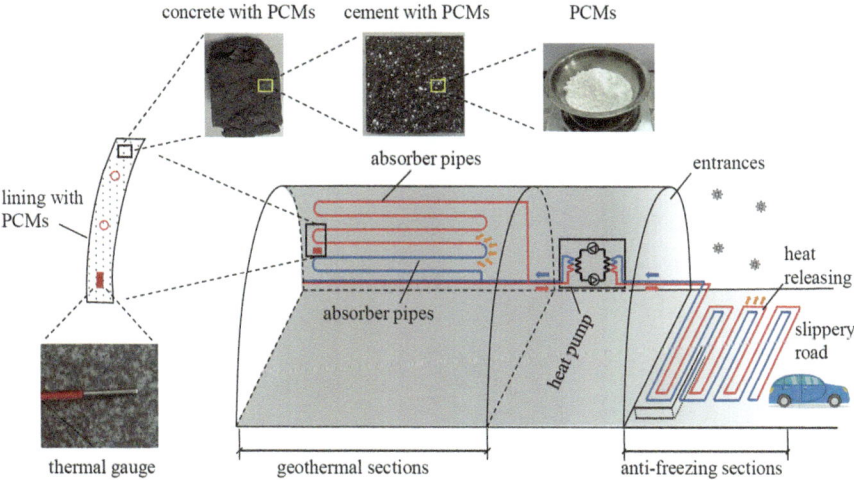

concrete with PCMs   cement with PCMs   PCMs

absorber pipes   entrances

lining with
PCMs

heat
releasing

absorber pipes   slippery
road

heat pump

thermal gauge   geothermal sections   anti-freezing sections

Fig. 1 Establishment procedure of energy tunnel integrated with PCMs

efficient heat exchange processes. A heat pump system is integrated into the infrastructure to provide the necessary power for fluid circulation, enabling the transportation of energy-laden fluid to mitigate against freezing conditions, such as the tunnel entrances susceptible to frost heave, as well as on road surfaces prone to snow-induced slippery conditions.

Based on the system of energy tunnel with PCMs, we develop a temperature-dependent dual-operation mode, consisting of storing mode and cooling mode, see Fig. 2. The temperature of the lining is the indicator of the operation status of the system. The cooling mode is required when the lining temperature T_{lin} increases the temperature at the end of the solid-liquid phase transition T_o. The Ground Source Heat Pump (GSHP) is opened at this time, where the fluid with low temperature enters the absorber pipes and extracts the heat stored in concrete and PCMs, see the blue area in Fig. 2a and b. Both the temperatures of lining and air decrease, when huge heat is extracted out of tunnels for anti-freezing. Meanwhile, the liquid PCMs are converted into solid PCMs with the release of heat. The system enters the storing mode, when the lining temperature T_{lin} decreases the temperature at the start of the solid-liquid phase transition T_c, see the pink area in Fig. 2a and b. The GSHP is closed at this time, no heat exchange between pipes and linings. The heat from high temperature surrounding rock is stored in linings, where the latent heat of PCMs augments the heat-storing capacity of concrete to a large extent through the solid-liquid phase transition. The lining temperature increases until the T_o, where the GSHP is reopened, starting the next cooling-storing cycle.

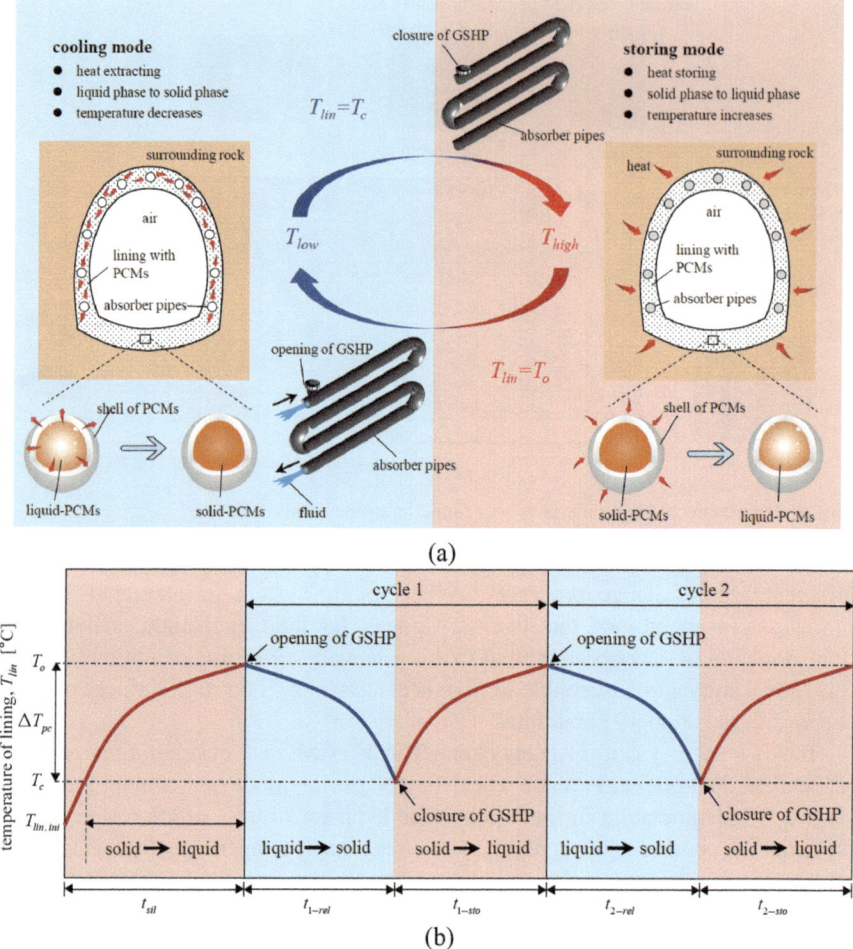

(a)

(b)

Fig. 2 Design principle of energy tunnel with PCMs under the temperature-dependent dual-operation mode. (**a**) Schematic illustration of energy tunnel with PCMs switching between storing mode (in pink area) and cooling mode (in blue area). (**b**) Temperature evolution of lining switching between storing mode (in pink area) and cooling mode (in blue area)

3   Model Setup

3.1   *Governing Equations*

The thermal behavior within the energy tunnel incorporating PCMs is governed by the combined effects of heat conduction and non-isothermal flow fields. In order to achieve reasonable simulation accuracy while minimizing computational efforts, several fundamental assumptions are employed: (1) Homogeneity and isotropy of materials throughout the tunnel structure are assumed. (2) Thermal resistances at interfaces are neglected. (3) The heat exchange process between the fluid and pipes is approximated as quasi-steady heat conduction, considering the relatively short time required to reach a steady state. (4) Temperature gradients along the thickness direction of pipes and ventilation ducts are disregarded due to their small thicknesses. (5) The effects of stress within linings on the thermal properties of PCMs are not accounted for.

The heat conduction for objectives in simulating follow [13, 14].

$$\rho_i c_{p,i} \frac{\partial T_i}{\partial t} = \nabla (\lambda_i \nabla T_i) + \dot{\Phi}_i (i = 1, 2, 3, 4),$$ (1)

where T_i and t denote the temperature (°C) and time (s), respectively. ρ_i, $c_{p, i}$, and λ_i represent the density (kg/m^3), specific heat (J/(kg · ° C)), and thermal conductivity, respectively. $\dot{\Phi}_i$ is the internal heat source (W/m^3). $i = 1, 2, 3, 4$ denote the surrounding rock, linings, pipes, and air, respectively. Meanwhile, quasi-steady heat conduction is imported to describe the convective heat transfer between pipes and fluid [15–18], following

$$\rho_f c_{p,f} u \frac{dT_f}{dz} (z) = h_{pf} (T_p - T_f(z)),$$ (2)

driving

$$T_f(z) = T_p + (T_{in} - T_p) e^{-\frac{h_{pf}}{\rho_f c_{p,f} u} z}$$ (3)

for

$$h_{pf} = \frac{0.023 \, Re^{0.8} Pr^{0.3} \lambda_f}{d_p},$$ (4)

$$Re = \frac{u d_p}{\nu_f},$$ (5)

$$Pr = \frac{\nu_f \rho_f c_{p,f}}{\lambda_f}.$$ (6)

Among them, $T_f(z)$ denotes the fluid temperature at the z position of absorber pipes. u, ν_f, λ_f, and $c_{p, f}$ represent the flow velocity, viscosity coefficient, thermal conductivity, and specific heat of fluid, respectively. T_p and d_p represent the temperature and internal diameter of pipes, and h_{pf} is the convective heat transfer coefficient between pipes and fluid.

3.2 Establishment of Finite Element Model

Based on the principles of heat conduction, a two-dimensional finite element model (FEM) is constructed utilizing the COMSOL software. The model exhibits a square geometry with sides measuring 60 m, encompassing a horseshoe-shaped lining at its center with dimensions of 8.26 m in width and 10.43 m in height. This configuration mirrors the linings of the Sangzhuling tunnel, a representative tunnel with geotherm in the southwestern region of China, as illustrated in Fig. 3. The lining itself is 0.25 m wide, accommodating heat exchanger pipes with internal diameters of 4 cm and a thickness of 3 mm, strategically embedded within. To facilitate a temperature-dependent dual-operation mode, the Implicit Event Component in COMSOL, functioning as a temperature sensors, is introduced into the linings. This component monitors the temperature evolution within the linings and effectively controls the on-and-off status of the GSHP, as depicted in Fig. 3d. This integrated approach ensures a comprehensive simulation of the dynamic thermal behavior of the system.

The basic values of materials used in the simulations refer to Table 1. The description of the phase-change process of PCMs is based on the apparent heat capacity [19], following

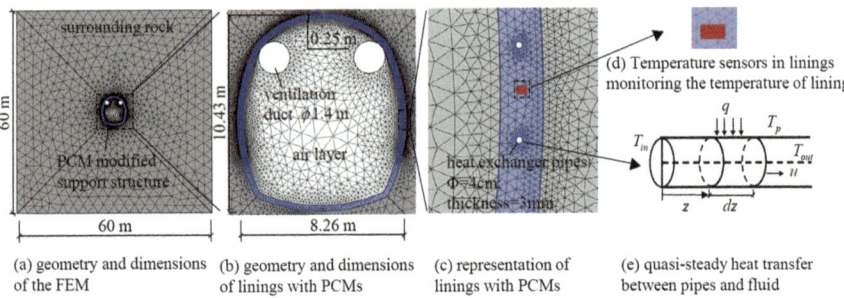

(a) geometry and dimensions of the FEM (b) geometry and dimensions of linings with PCMs (c) representation of linings with PCMs (d) Temperature sensors in linings monitoring the temperature of lining (e) quasi-steady heat transfer between pipes and fluid

Fig. 3 Establishment of the finite element model of energy tunnel with PCMs

Table 1 Values of materials used in simulations

Materials	Thermal conductivity [W/(m · K)]	Specific heat [J/(kg · K)]	Density [kg/m³]	Latent heat [kJ/kg]
Surrounding rock	2.9	850	2600	–
Concrete	1.65	837	2300	–
Air layer	3	1005	1.128	–
Heat exchanger pipes	0.6	400	936	
Circulating fluid	0.599	4183	998.2	
PCM (solid)	0.136	1800	856	325.57
PCM (liquid)	0.127	1762.8	779	325.57

$$k_c = \theta(T)k_s + (1 - \theta(T))k_l, \tag{7}$$

$$\rho_c = \theta(T)\rho_s + (1 - \theta(T))\rho_l. \tag{8}$$

$\theta(T)$ is the mass fraction of solid phase at temperature, taken as a linear function. k and ρ denote the thermal conductivity and density, respectively. The subscript c, s, and l represent the PCMs, solid phase of PCMs, and liquid phase of PCMs, respectively. The calculation of heat capacity c_p of PCMs follows

$$c_{p,c} = \frac{1}{\rho_c} \left(\theta(T)\rho_s c_{p,s} + (1 - \theta(T))\rho_l c_{p,l} \right) + L\frac{\partial \alpha(T)}{\partial T}, \tag{9}$$

where the $\alpha(T)$ is a transition equation, following

$$\alpha(T) = \frac{(1 - \theta(T))\rho_l - \theta(T)\rho_s}{2(\theta(T)\rho_s + (1 - \theta(T))\rho_l)}. \tag{10}$$

The estimation of properties of concrete mixed with PCMs, including density, thermal conductivity, and heat capacity, follow

$$P_{\text{eff}} = (1 - \phi_c)P_{\text{con}} + \phi_c P_c, \tag{11}$$

where the ϕ_c is the mass fraction of PCMs. The subscript eff and con denote the concrete mixed with PCMs and the concrete, respectively.

Based on the on-site experiment conducted at Sangzhuling tunnel, initial temperature values for the surrounding rock, lining, and air layer are determined to be $62\,°\text{C}$, $32\,°\text{C}$, and $30\,°\text{C}$, respectively. The boundary condition for the surrounding rock is set as a constant temperature of $62\,°\text{C}$ to simulate the geothermal environment accurately. Additionally, the temperature of the ventilation ducts is specified as $28\,°\text{C}$. Moreover, parameters including the inlet temperature T_{in}, flow velocity of circulating fluid u, and the total length of heat exchanger pipes Z are defined as $20\,°\text{C}$, 0.6m/s, and 400 m, respectively.

3.3 Performance Merits

In the present study, a comprehensive assessment is conducted to appraise the efficacy of energy tunnels with PCMs under the temperature-dependent dual-operation mode. The average air temperature T_a and the average lining temperature T_{lin}, derived from the simulation results of a 2D numerical model, serve as indicators for assessing thermal load reduction effectiveness. Additionally, we analyze the heat exchange rate of the heat exchanger pipes q and the increment in energy extraction E_{inc} to evaluate the efficiency of energy extraction. The heat exchange rate is estimated by

$$q = c_{p,w}\rho_w\pi\left(\frac{d_{inn}}{2}\right)^2 u(T_{out} - T_{in}), \tag{12}$$

where the temperature of circulating fluid at the outlet is quantified through Eq. (3). The energy extraction increment E_{inc} is defined as the relative difference between the annual energy extraction through the energy tunnel with PCMs Q_L and through the ordinary tunnel $Q_{L, o}$ expressed as,

$$E_{inc} = \frac{Q_L - Q_{L,o}}{Q_{L,o}}, , \tag{13}$$

where the energy extraction Q_L and $Q_{L, o}$ are respectively defined as:

$$Q_L = \pi\left(\frac{d_{inn}}{2}\right)^2 \rho_w u c_{p,w} \int_0^{1a} (T_{out} - T_{in})dt \tag{14}$$

and

$$Q_{L,o} = \pi\left(\frac{d_{inn}}{2}\right)^2 \rho_w u c_{p,w} \int_0^{1a} (T_{out,o} - T_{in})dt. \tag{15}$$

4 Results and Discussions

Based on the temperature-dependent dual-operation mode, the performances of thermal load reduction and energy extraction of energy tunnel integrated with PCMs through one year are investigated, considering the mass fraction of PCMs and thermal properties of PCMs.

4.1 Influence of Mass Fraction of PCMs

The mass fraction of PCMs considered in this simulation group varies from 0% to 20%. The phase-change temperature window is defined within the range of 35 ° C to 40 ° C. Additionally, the specified temperatures for the closure and opening of the GSHP are 35 ° C and 40 ° C, respectively. The simulation condition corresponding to an ordinary lining, denoted as $\phi_c = 0$, serves as the reference state for calculating $Q_{L, o}$ in Eq. (15). The predicted temperature evolution of air T_a, the lining T_{lin}, the heat exchange rate q, and the increment in energy extraction E_{inc} are shown in Fig. 4.

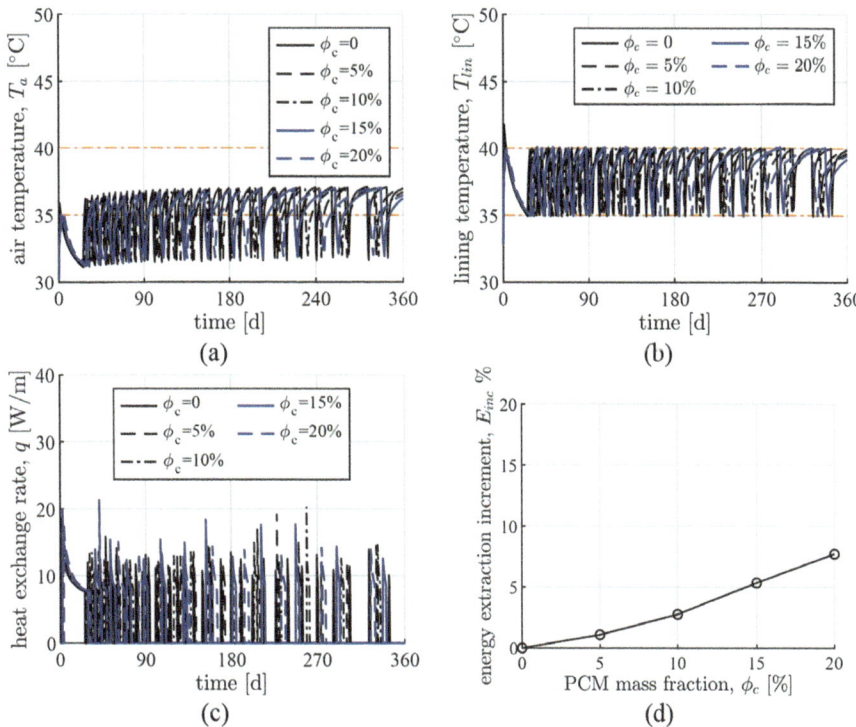

Fig. 4 Performances of energy tunnel with different mass fractions of PCMs based on the temperature-dependent dual-operation mode: (**a**) the air temperature as a function of time; (**b**) the lining temperature as a function of time; (**c**) the heat exchange rate as a function of time; (**d**) energy extraction increment as a function of mass fraction of PCMs

The air temperature changes with the opening and closure of GSHP, mainly ranging from 31.2 ° C to 36.9 ° C, see Fig. 4a. The mass fraction of PCMs exhibits negligible influence on the magnitude of air temperature variations. However, an increase in the mass fraction of PCMs results in a reduction in the frequency of air temperature fluctuations over the course of one year. The temperature of the lining can be controlled in the defined opening-closure temperature, i.e. from 35 ° C to 40 ° C, see Fig. 4b. The evolution of the lining temperature demonstrates reduced fluctuations over the course of one year with increasing mass fraction of PCMs, attributable to the latent heat characteristics of PCMs. The energy tunnel with PCMs extracts the thermal energy stored in the lining integrated with PCMs when the GSHP is opened, see Fig. 4c. The higher the PCM mass fraction, the longer the lasting time of each opening-closure cycle, characterized by the number of opening-closure cycles in one year decreasing from 33 to 11 as the escalating from 5% to 20%. The PCMs mass fraction has less effect on the total opening time of GSHP in one year, ranging from 75.6 d to 79.1 d. Although the number of opening-closure cycles decreases with the increase of PCMs mass fraction, the heat exchanger rate increases. The amount of energy extraction of lining with PCMs is higher than that of ordinary lining, where the energy extraction increment under the five simulating conditions are calculated, see Fig. 4d The energy extraction increment increases from 0% to 7.73% throughout the year as the mass fraction ranging from 0% to 20%.

4.2 Influence of Phase-Change Temperature Window

The performances of energy tunnels with $\phi_c = 15\%$ for ΔT_{pc} ranging from $1\,^\circ$C to $5\,^\circ$C are investigated. The closure and opening temperatures of GSHP keep consistency with the phase-change temperature window. The efficacy in heat reduction and energy extraction is depicted in Fig. 5.

The evolution of the air temperature shows a drop and rise with the opening and closure of GSHP, see Fig. 5a. The amplitude of air temperature exhibits an escalation as the phase-change temperature window widens. Conversely, the frequency of air

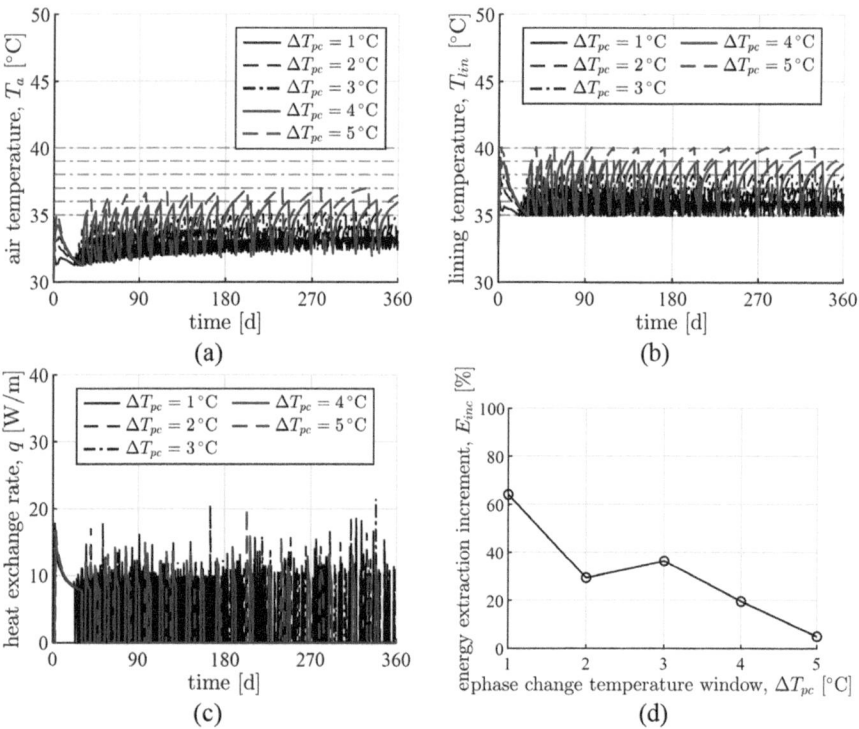

Fig. 5 Performances of energy tunnels with 15% PCM mass fraction for ΔT_{pc} ranging from 1 to 5 $^\circ$ C based on the temperature-dependent dual-operation mode: (**a**) illustrates the air temperature as a function of time; (**b**) illustrates the lining temperature as a function of time; (**c**) illustrates the heat exchange rate as a function of time; (**d**) illustrates energy extraction increment as a function of phase-change temperature window

temperature fluctuations over the span of one year diminishes concomitantly. The lining temperature can be controlled in the corresponding phase-change temperature window effectively under the temperature control operation mode, see Fig. 5b. The maximum heat exchanger rate shows a significant difference as the phase-change temperature window widens, see Fig. 5c. As the phase-change temperature window expands, there is a reduction in the frequency of opening-closure cycles and an extension in the duration of each individual cycle. The total opening time of GSHP in one year shows a decreasing trend with the increase of phase-change temperature window, where the working days are 121.8 d, 99.9 d, 107.3 d, 94.7 d, and 76.9 d for ΔT_{pc} ranging from 1 °C to 5 °C. Compared with the amount of extracted energy from the ordinary lining (the $\phi_c = 0$ simulation condition in Sect. 4.1), the energy extraction increment is calculated, see Fig. 5d. The energy extraction increment decreases from 64.13% to 5.01% as the phase-change temperature window ΔT_{pc} increases from 1 °C to 5 °C. When the phase-change temperature window is 3 °C, a larger energy extraction increment is obtained due to the longer lasting time of opening of GSHP in one year. These indicate that the improvement of the phase-change temperature window of PCMs will be accompanied by less opening-closure cycle and longer lasting time of each opening-closure cycle. Therefore, a suitable phase-change temperature window of PCM should be determined based on the application.

4.3   Influence of Latent Heat

The effects of latent heat L ranging from 100 to 500kJ/kg on the performances of energy tunnel with PCMs are discussed. The mass fraction of PCMs ϕ_c, phase-change temperature T_{pc}, and phase-change temperature window ΔT_{pc} are taken as 15%, 35 °C, and 3 °C, respectively. The performances of thermal load reduction and energy extraction are summed up in Fig. 6.

The evolution of the air temperature shows fluctuations with the working status of GSHP, where the opening of GSHP results in a drop in air temperature, see Fig. 6a. Higher latent heat narrows the frequency of fluctuations of air temperature in one year but does not affect the amplitude of air temperature after the first cycle. The temperature of the lining can be limited in the phase-change temperature window (i.e. from 35 °C to 38 °C) after the first opening closure cycle of GSHP, see Fig. 6b. The latent heat has more effects on the lasting time of the opening-closure cycle rather than the maximum heat exchanger rate, see Fig. 6c, where the larger the latent heat, the less of the number of opining-closure cycles and the longer lasting time of each opening-closure cycle. The total opening time of GSHP in one year decreases with the increase of latent heat, whereas the opening time of GSHP in one year decreases from 115.7 d to 71.9 d as the latent heat escalates from 100 to 500kJ/kg. Compared with the amount of extracted energy from the ordinary lining $Q_{L,\,o}$, the energy extraction increment is calculated, see Fig. 6d. The energy extraction increment decreases from 44.50% to 17.17% as the latent heat L increases from

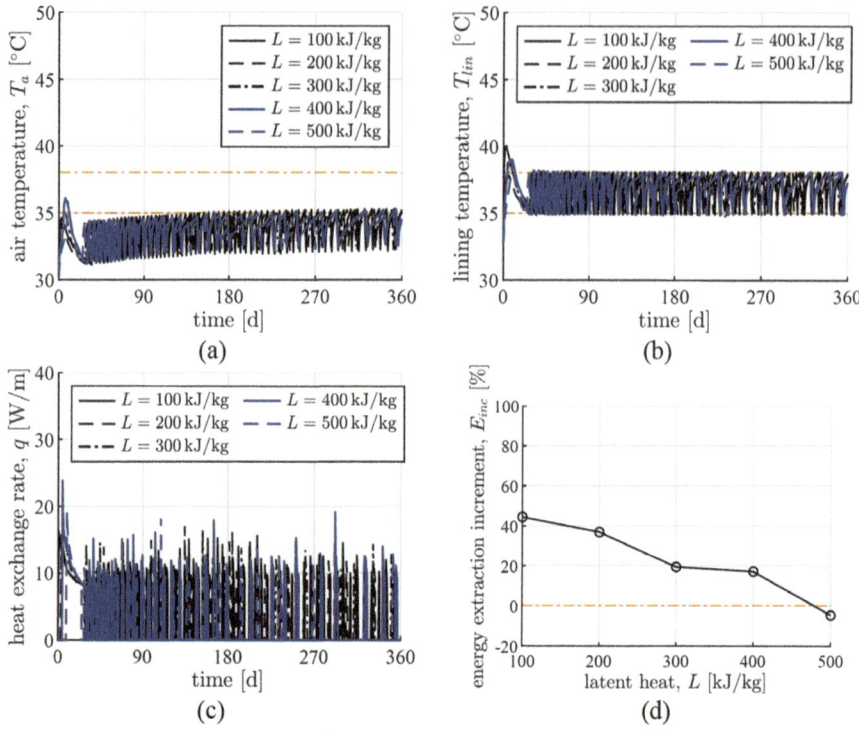

Fig. 6 Performances of energy tunnels with 15% PCM mass fraction for latent heat L ranging from 100 to 500kJ/kg based on the temperature-dependent dual-operation mode: (**a**) illustrates the air temperature as a function of time; (**b**) illustrates the lining temperature as a function of time; (**c**) illustrates the heat exchange rate as a function of time; (**d**) illustrates energy extraction increment as a function of phase-change temperature window

100 to 400kJ/kg. When the latent heat is 500kJ/kg, less heat energy is extracted compared with the reference ordinary energy tunnel. Thus, under this operation mode, the PCMs with low latent heat can also show advantages in energy extraction.

5 Conclusions

In this study, we introduce a temperature-dependent dual-operation mode for energy tunnels with PCMs. Using a finite element model, we assess the feasibility of implementing this mode in a geothermal environment. The performances incorporating heat reduction and energy extraction are evaluated, considering the mass fraction, phase-change temperature window, and latent heat of PCMs. The conclusions are delineated as following

1. The temperature-dependent dual-operation mode maximizes the energy storage potential of PCMs. Compared to conventional energy tunnels, the increment in energy extraction ranges from 0% to 7.73% annually with increasing PCM mass fraction from 0% to 20%.
2. Under this mode, the air temperature can be effectively controlled within the desired range. Energy tunnels with PCMs demonstrate superior heat reduction performance, particularly with smaller phase-change temperature windows and higher latent heat capacities of PCMs.
3. Larger phase-change temperature windows and higher latent heat lead to reduced energy extraction due to less operation time GSHP under this mode. The PCMs with low latent heat show advantages in energy extraction under this operation mode.

Overall, the proposed operation mode for energy tunnels with PCMs in geothermal environments offers notable advantages. However, further investigations regarding the safety of linings incorporating PCMs are warranted. Additionally, the influence of stress within linings on the thermal properties of PCMs remains unidentified, potentially impacting system performance and operational efficacy. Furthermore, future design considerations should entail a comprehensive multi-factor and multi-level analysis to optimize the operation mode and the overall energy tunnel system.

Acknowledgments This research was sponsored by the Sino-German Center for Research Promotion (Grant No. GZ 1574). In addition, the authors would like to thank the support from the Science and Technology Commission of Shanghai Municipality (Grant No. 21DZ1203505), the Department of Transportation of Jiangxi Province (Grant No. 2021C0008), and China Scholarship Council (Grant No. 202306260177).

References

1. Brandl, H.: Energy foundations and other thermo-active ground structures. Geotechnique. **56**(2), 81–122 (2006)
2. Loveridge, F., McCartney, J.S., Narsilio, G.A., Sanchez, M.: Energy geostructures: a review of analysis approaches, in situ testing and model scale experiments. Geomechan. Energy Environ. **22**, 100173 (2020)
3. Jácome-Paz, M.P., Pérez-Zárate, D., Prol-Ledesma, R.M., Rodríguez-Díaz, A.A., Estrada-Murillo, A.M., González-Romo, I.A., Magaña-Torres, E.: Two new geothermal prospects in the Mexican Volcanic Belt: La Escalera and Agua Caliente—Tzitzio geothermal springs. Michoacán, México. Geothermics. **80**, 44–55 (2019)
4. Hu, Y., Wang, Q., Wang, M., Liu, D.: A study on the thermo-mechanical properties of shotcrete structure in a tunnel, excavated in granite at nearly 90 °C temperature. Tunn. Undergr. Space Technol. **110**, 103830 (2021)
5. Wang, Q., Wang, M., Yuan, Y., Hu, Y., Wang, H.: Thermomechanical behavior of tunnel linings in the geothermal environment: field tests and analytical study. Tunn. Undergr. Space Technol. **137**, 105109 (2023)
6. Wilhelm, J., Rybach, L.: The geothermal potential of Swiss Alpine tunnels. Geothermics. **32**, 557–568 (2003)

7. Li, C., Zhang, G., Xiao, S., Xie, Y., Liu, X., Cao, S.: Long-term operation of tunnel-lining ground heat exchangers in tropical zones: energy, environmental, and economic performance evaluation. Renew. Energy. **196**, 1429–1442 (2022)
8. Xu, Y., Li, Z., Chen, Y., Jia, M., Zhang, M., Li, R.: Synergetic mining of geothermal energy in deep mines: an innovative method for heat hazard control. Appl. Therm. Eng. **210**, 118398 (2022)
9. Han, C., Yu, X.: Sensitivity analysis of a vertical geothermal heat pump system. Appl. Energy. **170**, 148–160 (2016)
10. Zhang, G., Cao, Z., Xiao, S., Guo, Y., Li, C.: A promising technology of cold energy storage using phase change materials to cool tunnels with geothermal hazards. Renew. Sust. Energ. Rev. **163**, 112509 (2022)
11. Cao, Z., Zhang, G., Wu, Y., Yang, J., Sui, Y., Zhao, X.: Energy storage potential analysis of phase change material (PCM) energy storage units based on tunnel lining ground heat exchangers. Appl. Therm. Eng. **235**, 121403 (2023)
12. Wang, Q., Wang, H., Yuan, Y.: Numerical study on the temperature field of a high geothermal tunnel with PCM-modified support structure. In: Building for the Future: Durable, Sustainable, Resilient, pp. 450–458. Springer (2023)
13. Wang, M., Hu, Y., Liu, D., Jiang, C., Wang, Q., Wang, Y.: A study on the heat transfer of surrounding rock-supporting structures in high-geothermal tunnels. Appl. Sci. **10**(7), 2307 (2020)
14. Chen, Q., Zhang, H., Zhu, Y., Chen, S., Ran, G.: Study on distributions of airflow velocity and convective heat transfer coefficient characterizing duct ventilation in a construction tunnel. Build. Environ. **188**, 107464 (2021)
15. Zhang, G., Xia, C., Sun, M., Zou, Y., Xiao, S.: A new model and analytical solution for the heat conduction of tunnel lining ground heat exchangers. Cold Reg. Sci. Technol. **88**, 59–66 (2013)
16. Kreith, F., Manglik, R.M.: Principles of Heat Transfer. Cengage Learning (2016)
17. Yang, S., Tao, W.: Heat Transfer (in Chinese). Higher Education Process (2006)
18. Wang, Q., Wang, H., Koenders, E., Zhang, J., Yuan, Y.: An innovative PCM-modified lining system for energy tunnel: from concept to numerical investigations. Appl. Therm. Eng. **248**, 123232 (2024)
19. Han, C., Yu, X.: An innovative energy pile technology to expand the viability of geothermal bridge deck snow melting for different United States regions: computational assisted feasibility analyses. Renew. Energy. **123**, 417–427 (2018)

Effect of Lightweight Masonry on Life Cycle Energy: A Case Study of Residential Buildings in India

Pradip Sarkar (ID) **and Nikhil P. Zade** (ID)

Contents

1 Introduction

The current era demands sustainability in the built environment to address the escalating concern of greenhouse gas (GHG) emissions and their adverse environmental impacts. GHG emissions from the construction sector contribute significantly, accounting for one-third of total emissions and 40% of global energy consumption [1]. Mitigating the environmental effects requires a reduction in carbon emissions from construction activities. Considering the significant natural resource utilization in the construction sector, it is imperative to assess the environmental footprint of emerging building materials. This assessment should consider both GHG emissions and natural resource conservation before recommending their use in construction. Green materials like glass-fiber reinforced gypsum (GFRG), fly-ash-based building materials (cement, bricks, etc.), and vegetal concrete are recognized as environmentally friendly options in the construction industry [1, 2], promoting sustainable development.

P. Sarkar (✉) · N. P. Zade
National Institute of Technology Rourkela, Rourkela, Odisha, India
e-mail: sarkarp@nitrkl.ac.in

© The Author(s) 2025
M. Kioumarsi, B. Shafei (eds.), *The 1st International Conference on Net-Zero Built Environment*, Lecture Notes in Civil Engineering 237,
https://doi.org/10.1007/978-3-031-69626-8_40

Bricks, cement, and steel reinforcement are the three key elements that considerably subsidize the embodied energy of a building [1, 3, 4]. Conventional bricks, which constitute a predominant building material, are manufactured predominantly in Asia, accounting for approximately 87% of global brick production [5]. The production of conventional fired clay bricks depletes the natural agricultural soil along with the emission of significant CO_2 by burning fossil fuels, making it a non-environmentally friendly material. In India, the brick sector is the third largest consumer of coal [5]. Given this context, the construction industry offers various sustainable alternatives to conventional clay brick, including fly ash bricks, GFRG panels, lightweight cellular and aerated blocks, and others that have the potential to reduce the carbon footprint and embodied energy of building construction.

Over the past two decades, there has been a notable surge in the utilization of lightweight infill materials such as autoclaved aerated concrete (AAC) and cellular lightweight concrete (CLC) in urban construction projects, particularly for load-bearing and framed structures. This trend can be attributed to their numerous advantages over conventional fired clay bricks, including reduced weight, superior thermal and acoustic insulation, heightened fire resistance, and consistent geometrical properties. Despite these benefits, the adoption of these infill blocks remains limited in rural areas of developing nations. Factors contributing to this disparity include the scarcity of production facilities, the high initial investment required for establishing such plants, elevated transportation expenses, inadequate public awareness, and the perceived lower structural strength of these materials [6, 7].

The carbon content of AAC blocks and fired clay bricks is approximately 0.23–0.24 kg CO_2/kg, as per the published literature [8]. However, AAC blocks are recognized as environmentally friendly, and certified by several environmental agencies. This may be due to the lightweight nature of AAC blocks, resulting in significantly lower consumption by weight compared to fired clay bricks in identical buildings. However, there is insufficient detailed research available in the area of the comparative advantages of replacing traditional masonry with AAC block masonry concerning embodied energy.

The primary goal of the present study is to quantitatively evaluate the life cycle energy (*LCE*) of a chosen AAC-infilled residential building. Additionally, this study also compared the material flow and initial embodied energy of a selected building infilled with AAC to a comparable hypothetical building infilled with conventional fired clay bricks. The findings of this investigation add to the rising body of information surrounding sustainable construction practices, emphasizing the advantages of AAC as a green material. The research aims to guide decision-makers, architects, and builders in making informed choices that align with environmentally conscious building practices. Ultimately, this work seeks to foster a more sustainable future in the construction industry by promoting the adoption of materials that reduce embodied energy and environmental impact.

2 Methodology

The present research centers on assessing the embodied energy of residential buildings constructed with AAC block masonry infill, utilizing the life cycle assessment (LCA) methodology. The method employed to calculate embodied energy and *LCE* adheres to recognized standards and existing literature [8, 9].

2.1 Embodied Energy and Life Cycle Assessment

LCA is a comprehensive examination of the environmental impacts related to a product, device, or process from its creation to its disposal [9]. In the present study, the LCA was carried out for the chosen residential building infilled with AAC, evaluating its *LCE* from cradle to grave, covering the manufacturing, construction, operation, and demolition phases. Figure 1 illustrates a schematic presentation of the methodology followed in the present study.

Embodied energy in the building sector refers to the overall energy consumed per unit weight or volume of a building, typically quantified in MJ/kg or MJ/m^3. The cumulative energy requirement encompassing the production of raw materials, their transportation to the construction site, the construction process itself, ongoing operation and maintenance throughout the building's lifespan, and eventual

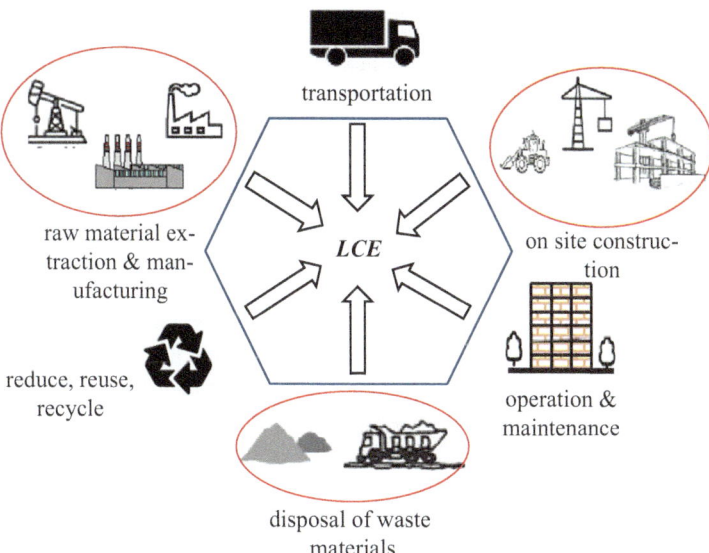

Fig. 1 Schematic presentation of *LCE* of building

demolition is referred to as the Life Cycle Energy (*LCE*) of a building [10], as illustrated in Eq. 1.

$$LCE = IEE + (REE + OE) \times Lifespan + DE \qquad (1)$$

where IEE is the initial embodied energy, the total energy used in the production of raw material and building construction phase, REE is the recurring embodied energy, the energy used for the maintenance and renovation of a structure over its lifetime, OE is the operational energy, the energy used on a daily basis for tasks such as ventilation, lighting, air conditioning, space heating, and cooking, and DE is the demolition energy, the energy consumed during the demolition process, including dismantling the building and transporting the resulting waste to landfills.

The embodied energy linked with the manufacturing of raw materials used for building construction is approximated by considering the quantity of material (θ_m) utilized during construction and the material energy content (EE$_m$) [1, 3], as depicted in Eq. (2). The energy consumption during building construction (E_c) constitutes the cumulative energy expenditure for material transportation (E_{tm}), equipment transportation (E_{te}), labor transportation (E_{tmp}), and on-site construction activities requiring construction equipment (E_{ce}), as outlined in Eq. (3). The IEE denotes the aggregate energy consumed in both raw material production (E_m) and construction activity (E_c), as presented in Eq. (4).

$$E_m = \sum \theta_m \times EE_m \qquad (2)$$

$$E_c = E_{tm} + E_{tmp} + E_{ce} \qquad (3)$$

$$EII = E_m + E_c \qquad (4)$$

2.2   Selected Building Details

The present study focuses on conducting an LCA of a seven-storey RC building situated at the National Institute of Technology in Rourkela, India. This building features an open ground floor designated for parking. Rourkela is situated at latitude 22.12 N and longitude 84.54E, positioned in the extreme northwest of Odisha state within Sundergarh district. The climate in this region is characterized by intense heat during the summer months, with the rainy season spanning from June to September. The average annual rainfall ranges from 750 to 1000 mm, while summer temperatures typically fluctuate between 30 °C and 42 °C.

The selected residential building consists of two identical towers joined by a connecting block, separated by a 50 mm expansion joint. Each tower accommodates six residential units per floor, resulting in a total of 72 units across the entire building. All units are uniform, with each offering a carpet area of 122 m^2. The overall floor

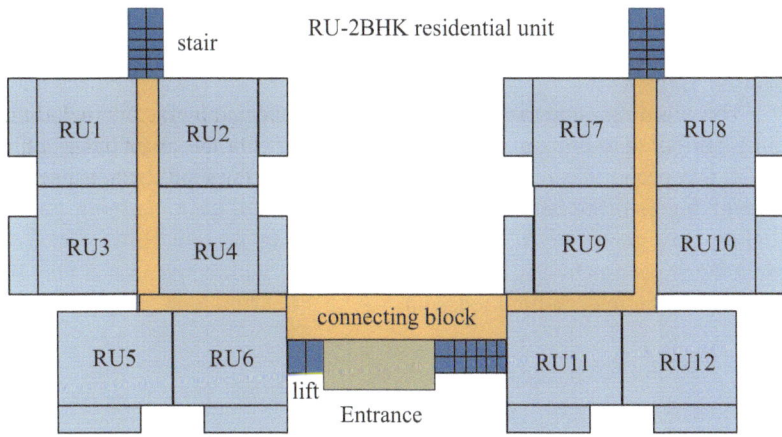

Fig. 2 Schematic presentation of the selected building

Fig. 3 Photograph of the actual structure

area, inclusive of residential units, common spaces like corridors and stairs, the connecting block, and the stilt floor, spans 11,810 m^2. Vertical access across different floors is facilitated by two elevators and three staircases. Figure 2 provides a schematic representation of the typical layout of the selected building, while Fig. 3 offers a visual depiction of the actual structure. The chosen framed building incorporates nonintegral AAC block masonry as infill, with structural design tailored accordingly. The selected building underwent analysis and redesign to conduct a comparative evaluation of embodied energy resulting from the substitution of AAC infill (Case 1) with conventional clay infill (Case 2).

In both scenarios, the dimensions of the partition and main walls stay unchanged, with a thickness of 115 mm for the partitions and 230 mm for the main walls. Analysis of various building elements' contributions to the combined dead and live load of the building superstructure reveals that approximately 35% of the total weight is attributed to brick masonry infill when conventional clay brick is utilized. However, this proportion reduces to 16% when AAC block masonry is employed. Because of the restricted bearing capacity (125 kN/m^2) of the underlying soil, a raft foundation was implemented for the initial building design. However, in Case 2, where the superstructure is assumed to be infilled with conventional clay brick, the raft foundation underwent a redesign.

3 Results and Discussion

As stated earlier, the *LCE* encompasses initial, recurring, operational, and demolition energy components. In this section, each aspect of the *LCE* is calculated and presented for the Case 1 building. However, the *IEE* is obtained for both cases (Case 1 and Case 2), which includes the energy linked with the construction activity and construction materials.

3.1 Initial Embodied Energy and Material Flow

The quantity of materials used in the construction of the building (Case 1) was determined through an analysis of the inventory of materials and structural plans. This data was cross-referenced with the materials utilized in construction, and necessary adjustments were made to ensure accuracy. Comparable estimates for the hypothetical building (Case 2) were derived based on its design specifications. These estimates were then adjusted to accommodate anticipated variations observed during the construction of the Case 1 building.

3.1.1 Embodied Energy for Building Construction

Table 1 offers a comprehensive overview of the material utilized for both selected buildings. The greater RC sections in the Case 2 building, resulting from the increased self-weight of the superstructure, necessitated a greater usage of rebar, cement, fine and coarse aggregate, and water for both the foundations and superstructure. As indicated in Table 1, it is evident that the building in Case 2 requires an additional 6386.8 tons of material compared to the building in Case 1.

The estimation of embodied energy in building materials involves assessing the quantity of material used during construction and its energy content, as outlined in Eq. (2). When specific embodied energy data for construction materials are not

Table 1 Selected building's material flow

Building elements	Constitutive material	Weight (t) Case 1	Case 2
Substructure up to plinth level (RCC, PCC, Filling)	Cement, Sand, Coarse aggregate, Clay brick, Rebar, Quarry dust fill, murum, boulders, Water	26,757.8	27,338.6
Superstructure (RC frame elements, wall, slab, flooring)	Cement, Sand, Coarse aggregate, Water, Rebar' Clay brick, AAC block, Granite, Ceramic/Mosaic tiles, Stone flooring (marble/kota)	14,889.7	20,254.9
Finishing work (plastering, waterproofing, and painting)	Cement, Sand, Aggregate, Water, Primer, Waterproofing, Wall putty, Paint, Turpentine, Lime, Aluminium	1783.0	2223.7
Doors, Windows, Railings, and welding's Miscellaneous	Wood, PVC, Mild Steel, Welding, Stainless steel, Aluminium, Zink' Anti-termite treatment, Glass	98.6	98.6
	Cement, Sand, Aggregate, Water, Porcelain, Primer, Paint, PVC, PTMT, CPVC, MS, Stainless steel, Brass, Iron (CI + GI)	93.6	93.6
Firefighting equipment (heavy-duty pipes, fire hydrant, CO_2 extinguisher, etc.)	Mild steel, Stainless steel, Brass, CI+GI (Iron), EPDM rubber	15.4	15.4
Total weight of the materials used (t)	–	43,638.2	50,025.0

Table 2 The energy content of the selected building materials

Material	Embodied energy coeff. (GJ/t)
Cement	6.85
Sand	0.15
Aggregate	0.4
Water	0.01
Rebar	35.1
Clay brick	1.2–4.05
AAC block	3.54
Boulders/Dust fill	0.15
Mild steel	20.1
Stainless steel	56.7

Note: The embodied energy of materials for various building materials can be found in elsewhere [1, 3, 10–14].

available in the Indian region, the 'carbon and energy inventory database' developed by Hammond and Jones [11] is utilized. Table 2 presents the energy content of various commonly used building materials as obtained from published sources. Given the notable deviation in the embodied energy of the conventional clay bricks in India [12], the mean of the highest and lowest values range is employed to compute the embodied energy associated with the bricks used in the present study.

In cases where embodied energy coefficients for sanitary and hardware items are lacking, such as mild steel (MS) grills, door and window fittings, cast iron and galvanized iron (CI + GI) pipes, MS pipes, china pedestals, kitchen sinks, washbasins, and faucets, energy consumption is estimated based on the embodied energy coefficients of their parent materials.

For buildings in Case 1 and Case 2, the flow material per unit built-up area is calculated to be 3.71 t/m^2 and 4.25 t/m^2, respectively. These values align with existing literature on RC-framed building construction in India [3, 4]. This suggests that the weight of buildings with AAC block masonry infill is reduced by approximately 13% related to assumed clay brick-infilled buildings. Several factors contribute to this weight reduction, including the lightweight nature of AAC blocks, smaller RC section sizes, fewer masonry joints due to larger AAC block sizes, and thinner plaster due to AAC block smooth surfaces. Figure 4 illustrates the comparative embodied energy contribution of different building materials for the chosen buildings.

The embodied energy for Case 1 and Case 2 buildings is calculated at 5.19 GJ/m^2 and 6.32 GJ/m^2, respectively. Published literature [1, 4] reported an embodied energy range of 4.21–10.8 GJ/m^2 for the Indian region. The energy associated with building materials per unit built-up area is observed to be 18% lower for the chosen building infilled with AAC (Case 1) compared to the assumed building (Case 2) infilled with clay brick. Cement, rebar, and clay bricks/AAC blocks emerge as the major contributors to the embodied energy, collectively constituting approximately 66% (Case 1) and 70% (Case 2) of the overall embodied energy of construction materials. The replacement of conventional clay bricks with AAC block is observed to reduce the embodied energy of infill material by approximately 40%.

3.1.2   Construction Embodied Energy

During construction, various activities were carried out, including material and labor transportation, manufacturing of formwork and materials onsite, and the utilization of construction equipment. This study takes into account the energy linked with the equipment employed for tasks such as excavation, material transport, and concreting only. Field data concerning the construction phase were gathered from the project manager and site engineer, enabling the determination of the corresponding embodied energy of this phase.

Table 3 presents an assessment of the *IEE* for the chosen AAC block-infilled building and a theoretical building infilled with clay brick masonry, showing a 17% lower *IEE* per unit area for the AAC-infilled structure. Additionally, the AAC-infilled building shows a 6% transportation energy savings, attributed to the absence of AAC block manufacturing plants nearby, unlike locally available clay bricks. The wide variation in onsite construction energy for residential buildings, ranging from 0.14 to 0.88 GJ/m^2 in the published literature [15, 16], aligns with the obtained value.

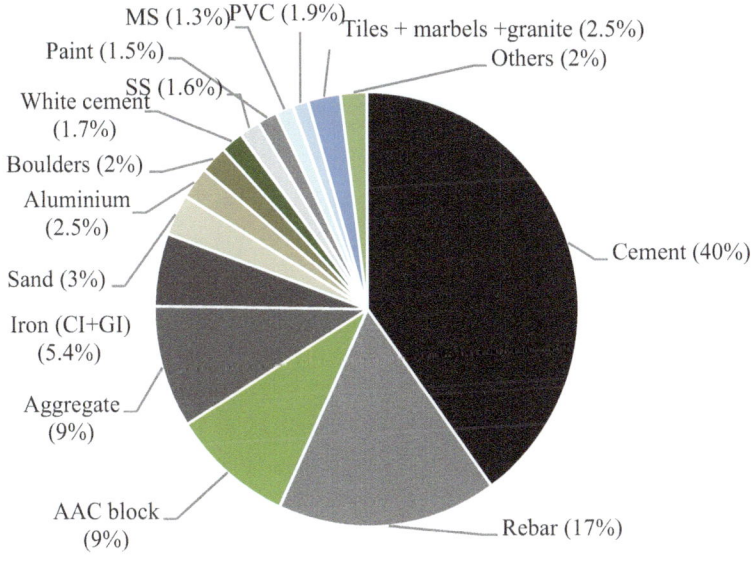

(a) Case 1

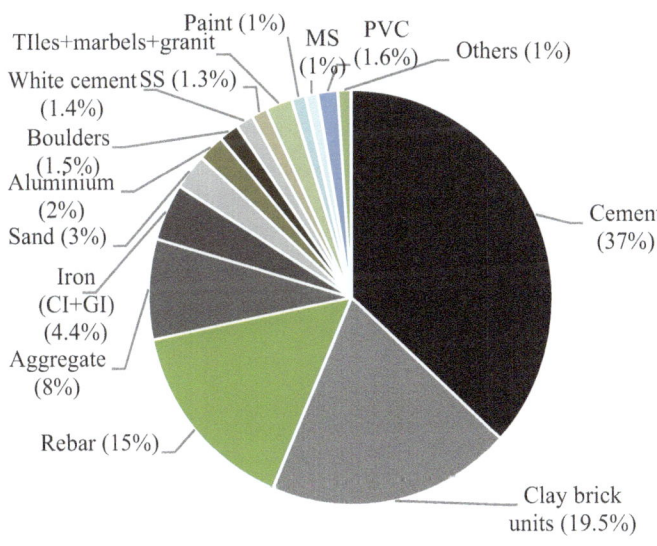

(b) Case 2

Fig. 4 Relative contribution of materials embodied energy

Table 3 *IEE* of the selected buildings

| Infill type | Material EE, E_m (GJ/m^2) | Construction EE, E_c | | Total *IEE* (GJ/m^2) |
		Transportation EE, E_{tm} (GJ/m^2)	Onsite construction EE, E_{eq} (GJ/m^2)	
Case 1 (AAC)	5.19	0.249	0.133	5.57
Case 2 (Clay)	6.32	0.264	0.155	6.74
% saving	17.9	5.7	14.2	17.4

3.2   Operational Energy

The energy required to operate a building is determined by monitoring both the electricity consumption of the residential units and the usage of liquefied petroleum gas (LPG) for cooking purposes. Electricity data were gathered from the electricity meters of all 72 residential units over 12 months. Additionally, records of LPG consumption were obtained through direct communication with residents from 12 selected residential units.

The combined electricity usage of all residential units amounted to 157,515 kWh during the typical year (2020–2021). It was noted that, on average, each residential unit changed its cylinder once every two months. The overall operational energy consumption of the chosen building amounts to 2241.6 GJ/year, encompassing both electricity and LPG usage, equating to 0.19 GJ/m^2/year. Comparative analysis reveals that operational energy ranges from 0.01 to 0.22 GJ/m^2/year for urban housing across different climatic zones in India [17], aligning with the obtained value. However, the observed operational energy is slightly lower than findings from other published literature [18], possibly due to variations in environmental conditions and architectural characteristics. Notably, improved thermal insulation of AAC results in a reduction of approximately 20–25% in energy usage for cooling during summer and heating in winter, as compared to conventional infilled structures [19, 20].

3.3   Life Cycle Energy

The life cycle energy of a building encompasses the initial embodied, recurrent embodied, operational, and demolition energy over its anticipated lifespan [16]. However, due to the relatively low significance of *REE* in *LCE* [1, 21] and a lack of requisite data, *REE* could not be considered in this study. Although the energy required for demolition is not explicitly calculated in this study, it is estimated to be around 3% of the *IEE* based on existing literature [15].

Table 4 *LCE* of the chosen residential building infilled with AAC over a service span of 50 years

Different phase	Energy (GJ)	Relative contribution (%)
Construction (material + construction activity)	65789.9	36.6
Operation	112,079.9	62.3
Demolition	1973.7	1.1
LCE	179,843.5	100.0

The *LCE* of the chosen residential building, which is infilled with AAC block masonry, is detailed in Table 4, for a service period of 50 years. Unfortunately, due to insufficient data, a comparable analysis could not be conducted for the assumed clay-infilled building (Case 2). The relative contributions of *IEE* (material and construction) and *OE* are determined to be 36.6% and 62.3% of the *LCE* calculated for a service period of 50 years, respectively. The derived *LCE* of 15.2 GJ/m^2, which equates to 0.30 GJ/m^2/year, aligns with the range of values (0.12–0.66 GJ/m^2/year) documented in the published literature [22, 23].

4 Conclusions

The sustainability of building materials is crucial for construction decisions. Despite their non-eco-friendly characteristics, conventional fired clay bricks remain popular in developing countries like India due to limited awareness and convenience. This study evaluates the environmental impact of AAC block masonry compared to traditional clay brick masonry, focusing on initial embodied energy. Additionally, a detailed life cycle energy assessment of an AAC masonry-infilled building is conducted. The primary findings are outlined as follows:

i) Replacing traditional clay bricks with AAC blocks can result in a 13% reduction in material usage attributed to several factors, including the lower density of AAC blocks, smaller section sizes of RC elements, reduced number of mortar joints, and decreased plaster thickness.

ii) For clay-infilled buildings (Case 1), the sequence of building materials in terms of increasing embodied energy is observed to be cement > masonry unit > rebar. However, this sequence shifts to cement > rebar > masonry unit when AAC blocks are employed as infill. This alteration results in savings of approximately 17% in overall initial embodied energy for an infilled RC frame.

iii) The life cycle energy estimation for a service span of 50 years of selected AAC-infilled buildings is determined to be 0.30 GJ/m^2/year.

References

1. Cherian, P., Palaniappan, S., Menon, D., Anumolu, M.P.: Comparative study of embodied energy of affordable houses made using GFRG and conventional building technologies in India. Energ. Buildings. **223**, 110138 (2020)
2. Ahmad, M.R., Pan, Y., Chen, B.: Physical and mechanical properties of sustainable vegetal concrete exposed to extreme weather conditions. Constr. Build. Mater. **287**, 123024 (2021)
3. Devi, P.L., Palaniappan, S.: Life cycle energy analysis of a low-cost house in India. Int. J. Constr. Educ. Res. **15**(4), 256–275 (2018)
4. Shukla, A., Tiwari, G.N., Sodha, M.S.: Embodied energy analysis of adobe house. Renew. Energy. **34**(3), 755–761 (2009)
5. CCAC Report.: Mitigating black carbon and other pollutants from brick production. Climate and Clean Air Coalition, CCAC SECRETARIAT hosted by the United Nations Environment Programme, Paris, France. Retrieved March 27, 2022, from https://breathelife2030.org/wp-content/uploads/2016/09/Fact-Sheet-5-Bricks-FINAL-Digital-May2015.pdf (2015)
6. Raj, A.: Strength Enhancement of Autoclaved Aerated Concrete (AAC) Block and Its Masonry. Ph.D. Thesis, Department of Mechanical Engineering, Indian Institute of Technology Guwahati, India (2020). http://hdl.handle.net/10603/443969
7. Bhosale, A., Zade, N.P., Davis, R., Sarkar, P.: Experimental investigation of autoclaved aerated concrete masonry. J. Mater. Civ. Eng. **31**(7), 04019109 (2019)
8. Sen, R., Bhattacharya, S.P., Chattopadhyay, S.: Are low-income mass housing envelops energy efficient and comfortable? A multi-objective evaluation in warm-humid climate. Energ. Buildings. **245**, 111055 (2021)
9. ISO 14044: Environmental management—life cycle assessment—requirements and guidelines. International Organization for Standardization. Geneva, Switzerland (2006)
10. Ramesh, T., Prakash, R., Shukla, K.K.: Life cycle energy analysis of buildings: an overview. Energ. Buildings. **42**(10), 1592–1600 (2010)
11. Hammond, G.P., Jones, C.I.: Embodied energy and carbon in construction materials. Proc. Inst. Civ. Eng. Energy. **161**(2), 87–98 (2008)
12. Praseeda, K.I., Reddy, B.V., Mani, M.: Embodied energy assessment of building materials in India using process and input–output analysis. Energ. Buildings. **86**, 677–686 (2015)
13. Mohammed, T., Greenough, R., Taylor, S., Ozawa-Meida, L., Acquaye, A.: Operational vs. embodied emissions in buildings – a review of current trends. Energ. Buildings. **66**, 232–245 (2013)
14. Barber, A.: New Zealand fuel and electricity total primary energy and life cycle greenhouse gas emission factors 2010. AgriLINK New Zealand Ltd, Kumeu, New Zealand. Source: https://agrilink.co.nz/wp-content/uploads/2016/08/Fuel_LCA_emission_factors_2011.pdf (2011)
15. Cole, R.J., Kernan, P.C.: Life-cycle energy use in office buildings. Build. Environ. **31**(4), 307–317 (1996)
16. Ramesh, T., Prakash, R., Shukla, K.K.: Life cycle approach in evaluating energy performance of residential buildings in Indian context. Energ. Buildings. **54**, 259–265 (2012)
17. Praseeda, K.I., Reddy, B.V., Mani, M.: Embodied and operational energy of urban residential buildings in India. Energ. Buildings. **110**, 211–219 (2016)
18. Pullen, S.F.: Energy used in the construction and operation of houses. Archit. Sci. Rev. **43**(2), 87–94 (2000)
19. Islam, S.: Eco-friendly AAC blocks for construction in smart cities. 3rd Smart Cities Symposium. The Institution of Engineering and Technology, London, United Kingdom (2020)
20. Khalil, A. E.: Impact of autoclaved aerated concrete (AAC) on modern constructions: a case study in the new Egyptian administrative capital, Master's thesis, Construction Engineering Department, The American University in Cairo (AUC), Egypt. Knowledge Fountain. https://fount.aucegypt.edu/etds/804 (2020)

21. Sartori, I., Hestnes, A.G.: Energy use in the life cycle of conventional and low-energy buildings: a review article. Energ. Buildings. **39**(3), 249–257 (2007)
22. Johnstone, I.M.: Energy and mass flows of housing: a model and example. Build. Environ. **36**(1), 27–41 (2001)
23. Buchanan, A.H., Honey, B.G.: Energy and carbon dioxide implications of building construction. Energ. Buildings. **20**(3), 205–217 (1994)

Seasonal Waste Heat Storage in Energy-Efficient Finnish Apartment Buildings

Janne Hirvonen (ID)**, Santeri Sirén** (ID)**, and Piia Sormunen** (ID)

Contents

1 Introduction

In Finland, most urban areas are heated with district heating systems. The heat is mostly generated by combustion of local biomass or fossil fuels like peat, coal and natural gas. Thanks to the low carbon electricity generation, electrification of heating by switching to heat pumps has been found to be a very effective solution for decarbonizing heating [1]. In large heat pump systems, ground-source heating is a

J. Hirvonen (✉)
Tampere University, Tampere, Finland
e-mail: janne.hirvonen@tuni.fi

S. Sirén
Tampere University, Tampere, Finland

Ramboll Finland Oy, Helsinki, Finland

P. Sormunen
Tampere University, Tampere, Finland

Granlund Oy, Helsinki, Finland

© The Author(s) 2025
M. Kioumarsi, B. Shafei (eds.), *The 1st International Conference on Net-Zero Built Environment*, Lecture Notes in Civil Engineering 237,
https://doi.org/10.1007/978-3-031-69626-8_41

common option. However, in urban areas, geothermal systems with high energy coverage may not be feasible due to the space requirement of the borehole field. The boreholes need to be spaced far enough apart to prevent the ground in the borehole field from cooling too much. Otherwise, this might cause premature freezing of the water-grouting in the boreholes and malfunctioning of the heat pump system.

A borehole thermal energy storage system (BTES) is a borehole field that is used not only to extract energy but also to store it [2]. BTES has been considered for seasonal energy storage with solar energy systems in Canadian [3] and Finnish conditions [4]. Industrial waste heat is another potential heat source to store in borehole fields [5]. While industrial sources are not widely available, ambient air is another promising alternative. In [6], a dry-cooler was connected to a two-part BTES system for heat and cold storage. Sewage heat also shows some potential for energy storage. Sewage heat recovery was compared to a solar energy solution and was found to be economical when electricity prices are low and the heat source is plentiful [7].

Ventilation heat recovery (HR) is required in new buildings in Finland, but the standard method only preheats incoming supply air with air-to-air heat exchangers. In warm weather, the potential heat is lost as there is no way to store the heat. The building regulations do not require sewage heat recovery, which is why it generally is not implemented. Using these additional heat sources for borehole regeneration and direct load reduction, a tightly spaced borehole field in a space-limited urban area might be feasible.

This study examines the effect of tight borehole spacing on borehole field ground temperature and heat transfer fluid temperature in the case of an up-to-code apartment building in Finland. The potential of ventilation and sewage heat recovery for borehole field regeneration is shown. The research questions are How is the long-term ground temperature of ground-source heat pump systems impacted by borehole field spacing and size? What is the benefit of thermal regeneration of borehole fields by residential waste heat?

2   Methodology

2.1   Simulation Tools

To estimate the impact of various borehole field and heating configurations, a heating demand profile was generated using the building energy simulation tool IDA-ICE [8]. The borehole field seasonal storage, heat recovery and heat pump systems were modeled in TRNSYS [9], which used the generated demand profile as input data. The TRNSYS model was used to estimate the temperature of the borehole field and the heat transfer fluid.

2.2 Building Description

The simulation study is based on a cold climate case study in Finland, where space heating is needed for 8 months of the year. The examined building was an apartment building that meets the requirements of the current Finnish building code. U-values of the envelope were 0.17, 0.09, 0.16 and 1.0 W/(m²K) for the external walls, roof, floor and windows, respectively. Ventilation was handled by a balanced mechanical ventilation system with a heat recovery efficiency of 65%. The annual space and ventilation heating demand was 110 MWh (22.1 kWh/m²), while the annual domestic hot water (DHW) heating demand was 128 MWh (25.7 kWh/m²). The heated floor area of the building was 4994 m², with 88 occupants. Heating demand was calculated using the Helsinki TRY2020 weather file [10].

The ventilation system was operated constantly, with a total exhaust air flow of 7200 kg/h. The minimum temperature of the exhaust air was 21 °C, according to the indoor temperature setpoint. The air-to-air heat recovery system was allowed to reduce the waste air temperature down to 2 °C. Total water consumption in the building was assumed to be 120 L/occupant/day, of which 35% was hot water (42 L/occupant/day) [11]. The average sewage temperature was assumed to be 26 °C [12]. This is for the total water flow, as even cold water gets heated up while standing in pipes and toilets.

2.3 Waste Heat Recovery System

In the reference system, waste heat was only recovered from ventilation exhaust air to preheat incoming fresh air. The studied heat recovery scheme is shown in Fig. 1. In this system, two additional heat recovery methods were added. An air-to-liquid HR system was included to recover any heat remaining after the reference HR system. This is used to take advantage of energy contained in the air outside the heating season. The air-to-liquid heat exchanger was assumed to operate at a constant 60% efficiency. Sewage heat was recovered using a HR tank with an indirect heat exchanger of 74% efficiency [12]. The tank was filled with brine, allowing direct flow of the heat pump fluid through it.

The same fluid was circulated from system to system. First, cold heat transfer fluid from the heat pump was fed into the borehole field. After initial heating, it was further heated by the ventilation excess heat and then by the recovered sewage heat. After the three stages, the fluid was sent to the heat pumps, which produced heat to the hot water storage tank for DHW and space heating. For simplicity of control, the heat pump was divided into two parts which heat different sections of the same hot water storage tank. The low-temperature heat pump keeps the lower section of the hot water tank at 45 °C, while the high-temperature heat pump keeps the top part of the tank at above 55 °C to ensure adequate DHW temperature level. If the heat transfer fluid temperature remained high after the heat pumps, any excess heat would

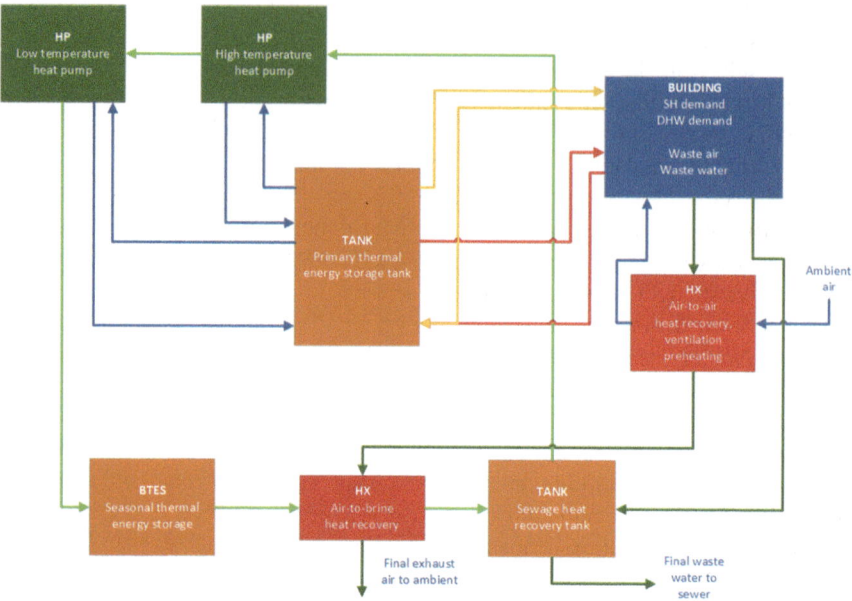

Fig. 1 Heating system schematic

be stored into the borehole field. Thus, the borehole field works in two directions and can be considered to be a borehole thermal energy storage system (BTES). The heat conductivity of the ground was assumed to be 3.5 W/(mK), the heat capacity was 2200 kJ/(m³K), and the starting temperature of the ground was 5.6 °C. The heat collector was a U-pipe of 40 mm diameter, filled with a glycol-water mixture of 3.67 J/(kgK) specific heat capacity. The grout around the borehole was water.

2.4 Studied Scenarios

The reference borehole field was sized according to a simple rule of thumb, where a typical borehole will provide 80–100 kWh/m of annual heating. The various system configurations are shown in Table 1. The 'exhaust air heat recovery' column indicates whether the air-to-liquid HR was utilized in addition to the default air-to-air HR system, which was always included. In cases where seasonal storage charging was active, the heat transfer fluid was kept in constant circulation, even when the heat pumps weren't running. This way, the BTES was injected with additional heat recovered from ventilation and sewage during period of low energy consumption.

Scenario 1 describes the default situation with the heat pumps dimensioned to 80% of peak heating load, using typical borehole spacing. The total borehole length was estimated to cover all of the annual heating needs of the building. Scenario 2 was otherwise the same, but the boreholes were closer together. Scenario 3 included

Table 1 Scenario descriptions

Scenario	# of boreholes	Borehole depth (m)	Borehole spacing (m)	Exhaust air heat recovery	Sewage heat recovery	Seasonal storage charging	BH field horizontal surface area (m^2)
1A	10	300	12	No	No	No	1250
2A	10	300	5	No	No	No	216
3A	10	300	5	Yes	No	No	216
4A	10	300	5	Yes	Yes	No	216
5A	10	300	5	Yes	Yes	Yes	216
1B	100	300	12	No	No	No	12,500
2B	100	300	5	No	No	No	2160
3B	100	300	5	Yes	No	No	2160
4B	100	300	5	Yes	Yes	No	2160
5B	100	300	5	Yes	Yes	Yes	2160

additional heat recovery from ventilation, thus reducing somewhat the energy drain on the borehole field. Scenario 4 added the sewage heat recovery as the third heat source. In Scenario 5, the borehole field was really used as a BTES, with constant charging whenever heat was available.

The A scenarios describe a case with a single apartment building, while the B scenarios show the same cases for a block of ten identical buildings. Instead of ten separate small borehole fields, there was a single large borehole field evenly shared by all the buildings. In both groups, reducing the borehole spacing reduced the surface area requirement of the borehole field by 83% compared to the reference case (Scenario 1).

3 Results

3.1 Ground Temperature

The purpose of the seasonal waste heat storage was to maintain borehole temperatures while reducing the surface area requirement of the borehole field through tighter spacing. To estimate the feasibility of the recovery system, the average ground temperature in the borehole field is presented in Fig. 2. Scenario 1A shows the reference system with heat pumps receiving all heat from the ground in a conventional setup. Over five years of operation, the average ground temperature gradually decreased from 5.6 °C down to 4.2 °C but remained at reasonable levels for continued use. In Scenario 2A, the borehole spacing was switched from 12 m to 5 m, significantly reducing the total energy storage capacity around each borehole. Now, the final ground temperature was reduced to 2.0 °C. In both cases, a decreasing temperature trend was in effect at the end of the simulation, indicating further 0.2–0.3 °C/year temperature reduction with continued operation. In Scenario 3A,

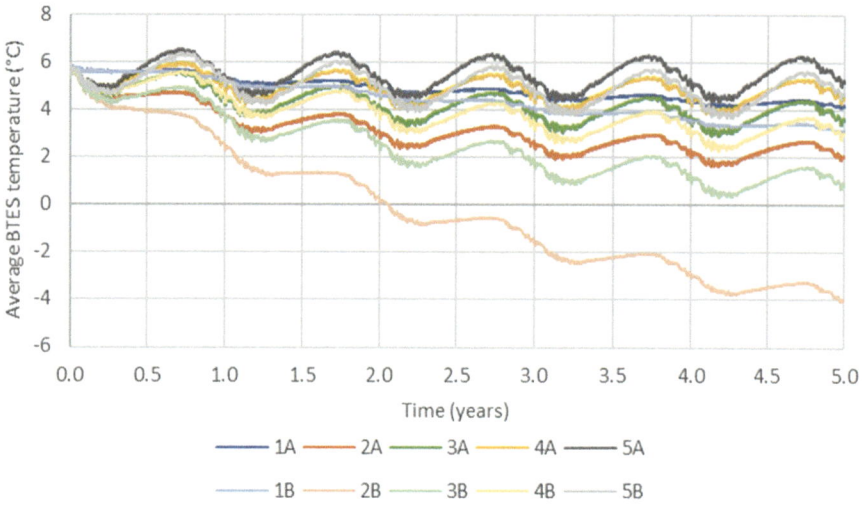

Fig. 2 Development of the average borehole field temperature in different scenarios

however, the additional energy provided by the utilization of ventilation waste heat slowed the cooling of the ground, ending up at 3.6 °C at the end of the fifth year. A clear seasonal variance was now observed compared to the pure GSHP cases, as the ground was allowed to rest during summer and some heat was even charged into the ground. In Scenario 4A, sewage waste heat was also utilized, which shifted the whole temperature curve upwards, ending with 4.5 °C at the end of the fifth year. In both Scenarios 3A and 4A, the BTES temperature was still reducing by 0.1 °C/year at the end of the simulation. However, in Scenario 5A, the heat supply flow into the BTES was increased by constantly injecting recovered heat, even if the heat pumps weren't running. This further raised the temperature levels in the ground and resulted in the final temperature remaining at 5.2 °C in years 3, 4 and 5. This means that the borehole field temperature could be maintained indefinitely without fear of excessive cooling, despite much tighter borehole spacing.

The B scenarios represent cases with larger borehole fields. In these cases, the surface-to-volume ratio of the BTES went down by 76–82% compared to the A scenarios, which means that there was less contact with the surrounding ground. This resulted in lower natural regeneration of the borehole field from the heat contained in the environment. Thus, Scenario 1B resulted in a similar, but faster temperature reduction trend as 1A. The final temperature after the fifth year was only 3.1 °C, with an expected further reduction of 0.5 °C/year. In the tightly spaced system of Scenario 2B, this effect was even more drastic. The final average borehole field temperature was −4.0 °C, which would mean total freezing of the water-filled boreholes. While further reduction would still be expected, in practice the system would stop functioning long before this, preventing the temperatures from actually reaching such levels. In a large GSHP system without additional heat sources, such a tight spacing is clearly infeasible and would drastically reduce the energy coverage. However, with the addition of the waste heat sources in Scenarios 3C and 3D, despite the tight spacing, there was a notable rise in the average ground temperature. The ventilation heat alone was not enough, as further temperature drop of 0.5 °C/year was still to be expected. With sewage heat, the expected reduction at the end of simulation was 0.2 °C/year. Finally, when active charging of the ground was included, the final temperature after the fifth year of operation was 4.6 °C, with an expected drop of only 0.1 °C/year.

3.2   Heat Transfer Fluid Temperature

It was seen that regenerating the boreholes with residential waste heat can sustain the average temperature in the BTES indefinitely, at least in the case of the smaller field. However, low temperatures within the water-filled boreholes themselves may cause local freezing, which may cause malfunctions in the heating system. Freezing may occur if the heat transfer fluid temperature remains below zero for extended periods (e.g. a whole day). Figure 3 shows the temperature of the heat transfer fluid exiting the BTES during the fifth year of operation in each scenario. The local temperature in the borehole may be assumed high enough if the exiting fluid temperature is above 0 °C.

Scenario 2A is the only one of the single building scenarios where the exit temperature is below zero for extended periods of times. However, the reducing trends in average borehole temperatures show this risk remains during long-term operation of the system. Scenario 3A shows elevated temperature levels during the summer, but in winter there is little energy left from the conventional ventilation heat

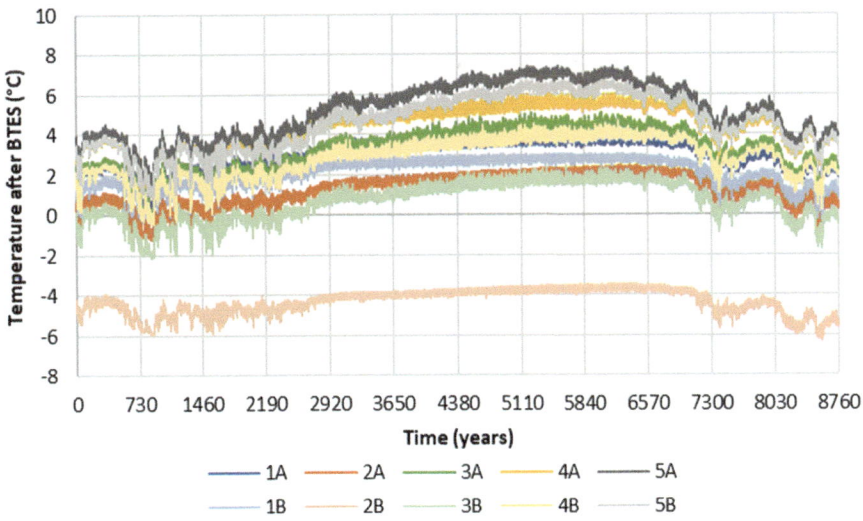

Fig. 3 Temperature of the fluid exiting the BTES during the fifth year of operation

Table 2 Temperature statistics

Scenario	1A	2A	3A	4A	5A	1B	2B	3B	4B	5B
Final BTES temperature (°C)	4.2	2.0	3.6	4.5	5.2	3.1	– 4.0	0.8	3.0	4.6
Total time output below 0 °C (h)	0	537	11	0	0	162	8760	2202	63	0
Longest period output below 0 °C (h)	0	197	5	0	0	58	8760	404	9	0

recovery to provide further benefits to the ground heat collection loop. The addition of sewage heat recovery (4A) notably raises the temperature level of the heat collection loop and seasonal storage further increases the temperature (5A).

Seasonal trends are identified in all of the scenarios. During the coldest parts of the year, the risk of low temperatures in the boreholes was increased. The heat transfer fluid temperature correlated with the BTES temperature. Thus, the fluid temperature was lower in the larger borehole fields in the B scenarios. Tight borehole spacing exacerbated the situation as shown in the extreme case of 2B. Higher temperatures compared to the reference case (1B) were realized in the large BTES with sewage recovery (4B) and regeneration (5B), despite the tighter borehole spacing.

The temperature levels may be better illustrated by duration data, shown in Table 2. It shows the total hours the BTES output is below 0 °C and the longest consecutive period of subzero hours. In the reference Scenario 1A, the heat transfer fluid temperature exiting the BTES remained quite stable and always stayed above 0. However, with the tighter spacing in Scenario 2A, there was several week's worth of subzero hours in the borehole output temperature. Scenarios 1B, 2B and 3B are identified as other problematic cases.

4   Discussion

This study examined borehole field regeneration under a limited amount of scenarios. No optimization of the connections was done, and the reference scenario was designed in a simplified way. A more complete study with a wider range of examined parameters could reveal further information about the potential of regeneration.

Electricity consumption in circulation pumps was ignored, but according to [4, 13], it is negligible compared to the electricity consumption of heat pumps. Heat losses in piping were not included in the calculation model, and the practicalities of connecting the various heat sources were not considered. Horizontal piping from the borehole field to the buildings was similarly ignored. According to [13], heat losses in a low temperature heating grid could be 3–5% of delivered heat. These things should be accounted for in more detailed life cycle calculations, but here the idea was to demonstrate the basic idea and potential of the residential waste heat recovery.

The block of several buildings was modeled by simply multiplying the flow rates from a single building. In a more practical case, there could be individual variations between load profiles of buildings within the same location, which might reduce peak heating loads compared to simple multiplication. This would make it easier to keep the system temperatures within acceptable limits.

5 Conclusions

This simulation study examined the seasonal storage of residential waste heat to reduce the ground space requirements of boreholes in a heat pump-based heating system in a cold climate. Reducing the distance between boreholes can reduce the surface are requirement of a borehole field by over 80%. This improves the feasibility of borehole systems designed in urban areas with very little usable space around buildings. Borehole fields of 10 and 100 boreholes were examined.

The simulation results showed that if borehole spacing was reduced without the addition of borehole regeneration, after five years of operation the borehole temperatures were reduced enough to cause significant risk of freezing and system malfunction due to low heat transfer fluid temperatures for long periods of time. This was especially the case in larger borehole fields, with a very low surface-to-volume ratio and resulting low natural regeneration rate. Keeping the final borehole field temperature above 4 °C improved the chance of keeping outlet temperatures high enough.

Utilizing the waste heat from the building's ventilation and sewage flows helped mitigate the long-term cooling of the boreholes when tight borehole spacing was used. However, to fully stabilize the temperature levels in the long term required active borehole regeneration with the excess heat whenever there was no immediate need for heating. With larger borehole fields, the stabilization process may need more time, alternative control strategies, or additional energy input.

The seasonal storage of residential waste heat shows great potential for reducing the space need of urban ground-source heat pump systems. This is especially true of larger energy systems, where a block of buildings could share the same borehole field or thermal energy storage system. Further studies could examine optimal borehole configurations and heat source connection methods and how to reduce the total borehole field length and cost.

Acknowledgements This study received funding from the research program 'Hiilineutraalit energiaratkaisut ja lämpöpumpputeknologia', which is funded by Paavo V. Suomisen rahasto, Sähkötekniikan ja energiatehokkuuden edistämiskeskus STEK ry, Granlund Oy, Ramboll Finland Oy, Senaatti-kiinteistöt, HUS Tilakeskus and HUS Kiinteistöt Oy.

References

1. Niemelä, T., Kosonen, R., Jokisalo, J.: Energy performance and environmental impact analysis of cost-optimal renovation solutions of large panel apartment building in Finland. Sustain. Cities Soc. **32**, 9–30 (2017)
2. Sadeghi, H., Jalali, R., Sing, R.M.: A review of borehole thermal energy storage and its integration into district heating systems. Renew. Sust. Energ. Rev. **192**, 114236 (2024)
3. Karasu, H., Dincer, I.: Life cycle assessment of integrated thermal energy storage systems in buildings: a case study in Canada. Energ. Buildings. **217**, 109940 (2020)
4. Hirvonen, J., ur Rehman, H., Sirén, K.: Techno-economic optimization and analysis of a high latitude solar district heating system with seasonal storage, considering different community sizes. Sol. Energy. **162**, 472–488 (2018)
5. Nilsson, E., Rohdin, P.: Performance evaluation of an industrial borehole thermal energy storage (BTES) project—experiences from the first seven years of operation. Renew. Energy. **143**, 1022–1034 (2019)
6. Allaerts, K., Coomans, M., Salenbien, R.: Hybrid ground-source heat pump system with active air source regeneration. Energy Convers. Manag. **90**, 230–237 (2015)
7. Pokhrel, S., Amiri, L., Poncet, S., Sasmito, A.P., Groreishi-Madiseh, S.A.: Renewable heating solutions for buildings; a techno-economic comparative study of sewage heat recovery and solar borehole thermal energy storage system. Energ. Buildings. **259**, 111892 (2022)
8. IDA ICE—Simulation Software. https://www.equa.se/en/ida-ice. Last accessed 05 Apr 2024.
9. TRNSYS 18. https://www.trnsys.de/en/trnsys18. Last accessed 05 Apr 2024.
10. Energialaskennan testivuodet TRY2020. https://www.ilmatieteenlaitos.fi/energialaskenta-try2020. Last accessed 16 Jan 2024.
11. Korhonen, A., Kuusela, M., Liski-Markkanen, S., Marjomaa, T.: Kestävä veden käyttö—vedenkäyttöselvitys. Motiva, Rajamäki (2020)
12. Ecowec waste heat recovery calculation example. https://www.ecopal.fi/wp-content/uploads/2019/11/esimerkkilaskelma.pdf. Last accessed 01 2023/12/01
13. Kauko, H., Kvalsvik, K.H., Rohde, D., Nord, N., Utne, Å.: Dynamic modeling of local district heating grids with prosumers: a case study for Norway. Energy. **151**, 261–271 (2018)

Optimization of Nativo Wood Fibre Insulation Board Through LCA Analysis

Angela Daniela La Rosa, Gaute Thomassen, and Petter Erlandsen

Contents

1 Introduction

Hunton Fiber AS (hereafter; Hunton) is a manufacturer of building materials, founded in Gjøvik, Norway, in 1889. They are the only producer of porous wood fiber-based building materials in the Nordic region. Hunton started producing wood fiber-based insulation in 2019. They produce both insulation boards and blown-in insulation based on wood-fibre with a production capacity of 32,000 tons/600000 m^3 of insulation boards. Hunton aims to contribute achieving UN's Sustainable Development Goals, especially SDG9 (Industry, Innovation, and Infrastructure), SDG12 (Responsible Consumption and Production and SDG13 (Climate Action), and for this reason it is active in performing LCA studies and Environmental Product Declarations (EPDs) to its products. Hunton conducted an LCA study [1] on the

A. D. La Rosa (✉)
Department of Manufacturing and Civil Engineering, Norwegian University of Science and Technology (NTNU), Gjøvik, Norway
e-mail: angela.d.l.rosa@ntnu.no

G. Thomassen · P. Erlandsen
Hunton Fiber AS, Gjøvik, Norway

© The Author(s) 2025
M. Kioumarsi, B. Shafei (eds.), *The 1st International Conference on Net-Zero Built Environment*, Lecture Notes in Civil Engineering 237,
https://doi.org/10.1007/978-3-031-69626-8_42

insulation products shortly after they started manufacturing. The study was based on 6 months of production from the first 6 months after initializing manufacturing. This, in addition to marginal internal involvement in the LCA process has led to some uncertainty regarding the data produced and the resulting EPD. There are certain themes included in the existing LCA which contain larger uncertainties than others. The bulk of uncertainties are related to allocation of impacts stemming from the phase of extraction and production of raw materials (A1), impacts from the transportation of raw materials (A2) and impacts of different scenarios of waste processing (C3).

The study at hand has focused on the environmental impacts, in terms of greenhouse gas emissions, from the product life of Huntons Nativo® wood fiber insulation boards.

2 Methodology

A Life Cycle Assessment (LCA) is a comprehensive method for evaluating the environmental impact of a product or system throughout its entire lifecycle. The method is standardized by the ISO14040-44:2006 standards [2, 3]. A LCA consists of the four following phases: 1. Goal & Scope, 2. Life cycle inventory (LCI), 3. Life cycle impact assessment (LCIA) 4. Interpretation of results. In the present paper, the software Simapro 9.3 and the Ecoinvent v3 database were used.

2.1 Goal and Scope

The goal of this study is to address the uncertainties mentioned in the introduction. Several different aspects will be analysed, namely: (i) different scenarios for allocation of impacts from raw materials (wood chips), (ii) the impact of different modes of transportation of raw materials and (iii) few different waste treatment scenarios. This will be studied by a comparative analysis of several different LCA scenarios. The focus of the study will be on the product's impact on climate change indicators.

The product is manufactured for the purpose to provide insulation for residential and professional buildings. The functional unit associated with these types of products is dictated by the Product Category Rules [4]. This document specifies that the functional unit is to be calculated to an R-value of 1 per m^2. In this case, the functional unit will be identical to the one stated in the existing EPD for the product: 1 m^2 wood fibre insulation installed in a thickness of 38 mm and a thermal resistance of $R = 1$ m^2K/W from cradle-to-grave with a reference lifecycle of 60 years.

In pertinence to system boundaries, the compared LCA scenarios will focus on different life-cycle stages. For the themes related to raw materials, the study will focus on the cradle-to-gate part of the life cycle (A1–A3). For the part of the study investigating waste treatment scenarios, the focus is limited to the waste processing

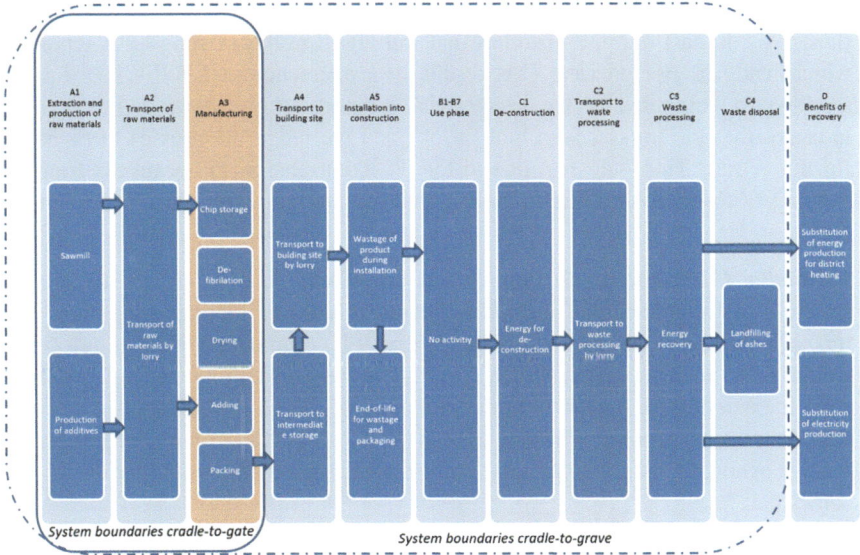

Fig. 1 Flowchart cradle-to-grave for Hunton Nativo insulation boards

stage (C3) (Fig. 1). To present an overview of the complete process, a simple LCA is conducted for the entire process for A1–C4, but with some crude approximations made, in stages not particularly relevant for the main focus themes.

2.2 The Life Cycle Inventory (LCI) Analysis

The LCI is an obligatory part that describes what data and prerequisites have been applied in the LCA performed. In general, it can be viewed as a recipe of sorts for the product at hand.

2.2.1 Data Collection Procedure

Most of the data were collected from the life cycle inventory of product phase (A1–A3) from the existing LCA-report for the insulation [1]. We were not able to find all the known inputs from technosphere (materials/fuels/electricity/heat) based on Ecoinvent, which made us rely on multiple proxies.

Raw Materials A short description of the actual components that go into the production of Hunton Nativo insulation boards.

Woodchips The inventory inputs for woodchips are based on processes found in ecoinvent.

Fire-Retardant The inventory inputs for the fire-retardant are split into two materials. We couldn't find the materials that were used in the existing LCA-report [1] which made us use proxies. This results in a difference in GWP-values for this LCA-report compared to the existing LCA-report. The closeness of the values indicates that the proxies were not far off, and we judge the proxies to be usable for this report. This is in part due to the close correlation to the results from the existing LCA-report and due to the fact that the fire-retardant is not a focus in this report.

Polyolefin Fibre (Bico) Hunton uses a polyolefin fibre to aid in making the structure of the wood fibre insulation boards. These are bicomponent fibres of a mix of polypropylene (PP) and polyethylene (PE) granulates. We couldn't find a similar material in ecoinvent, which also made us rely on a proxy material. We know from the existing LCA-report [1] that the data for the *bico* were labelled as confidential, from the supplier. The data used in the existing LCA-report are therefore not readily available. The proxy used likely underestimates the contribution of *bico* somewhat.

Production Process (A1–A3) The main component in producing the insulation boards is wood fibre, stemming from wood chips originating from local Norwegian sawmills in the vicinity of the production site in Gjøvik. The wood chips are transported by lorry, an average of approximately 50 km from the sawmills. The lumber entering the product at the sawmill is sourced from sustainably managed forestry. The actual wood chips, used by Hunton, to produce insulation is a by-product from a process in which the final product is sawn construction timber.

The wood chips are processed using heat and mechanical manipulation to break it down to wood fibre. The fire-retardant is added to the fibre prior to a drying stage. After drying, the *bico* is added to aid with the structuring of the board. The insulation boards are then formed before a heating process is applied to partly melt the *bico*, solidifying the structure of the final product. After this, the insulation boards are trimmed and cut to the desired measurements before packaging.

The production process is summarized in the illustration in Fig. 2.

Data entering the LCA stems from a data survey conducted by the manufacturer when creating the existing LCA-report.

To investigate the impact of different allocations of impacts from wood chips, separate processes were created for the input of wood chips in the process. Processes were created for allocation based on three different scenarios, in addition to 100% allocation: volume based, market price based (economic) and allocation of all impacts to sawn timber (0% allocation). The estimates for allocation were obtained from Jungmeier et al. [5]. To investigate the impact of transportation of raw materials, we created an alternative scenario where all transportation based on fossil fuel-driven lorries were conducted using lorries conforming to the EURO6 standards instead of EURO4/EURO5 which is the case in the existing documentation. In addition, a scenario was created where the transportation of raw materials inside of Norway was conducted using lorries driven on biodiesel. The scenario for

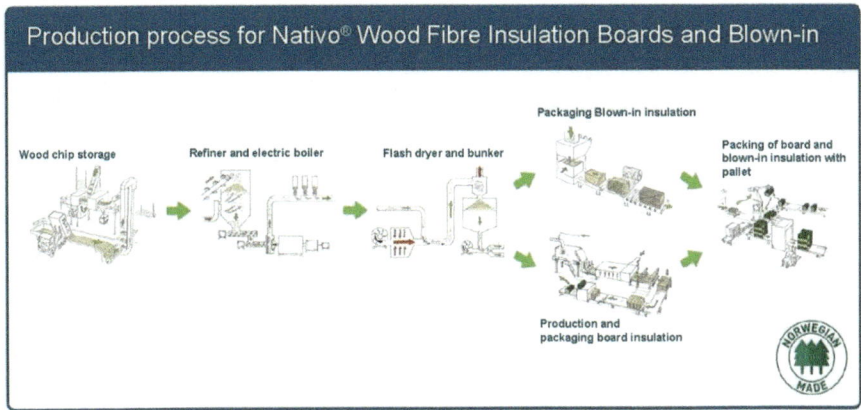

Fig. 2 Flowchart of the production process for Nativo Wood Fibre Insulation

transportation using lorries powered by electricity for the transportation within Norway was not included as there no process for electrically powered lorries in the Ecoinvent database. The possible benefits of such a substitution are discussed in the latter parts of the report.

Transport to Building Site (A4) The transportation from the production site to the building site occurs by lorry. In general, the product is transported by large lorries to warehouses, from which transportation to the actual building site is conducted by medium-sized lorries. In accordance with national product category rules [2], the total transport distance is set to 300 km. This is split into 250 km of transportation by means of large lorries and 50 km of transportation by means of medium lorries. The product has a high volume compared to weight. Therefore, the capacity utilization of the vehicles has been adjusted to volume limitations, in the LCA conducted.

Installation of the Product (A5) The impacts, related to the LCA in the installation phase, is constituted by the use of electricity and waste treatment for 2% scrap and the packaging, which is discarded during this stage.

Use Phase (B1–B7) During the use phase, wood has natural emissions, but no emissions have been found relevant to the LCA.

De-construction (C1) It is assumed that it is necessary to expend a certain amount of electric energy during the de-construction of the structure into which the product is installed. A small part of this is allocated to the product under investigation. The allocated amount is sourced from the existing EPD, which cites Wærp et al. [6] as its source for information.

Transport to Waste Processing (C2) The transport distance is set to 85 km, in accordance with Raadal et al. [7]. The weight of the transported waste corresponds with the weight of the product without packaging, due to the assumption made, that the packaging is discarded during the installation phase (A5).

Waste Processing (C3) The waste processing applied is based on assumptions made in the existing LCA report [1] using scenarios available in Ecoinvent. The practice for waste processing of mixed wood waste in Norway is dominated by incineration with energy recovery, so the scenarios that are applied are based on municipal incineration. Some scenarios are proxies, due to the lacking existence of specific scenarios for all the relevant product components. There were no data available for ammonium polyphosphate, but it is inorganic and therefore ash was used as a proxy. There are no specific for *bico* either, but these are fossil-based organic materials, so plastic packaging was used as a proxy.

To investigate into the potential effect of substituting parts of the existing waste processes with composting, we created waste treatment scenarios where 25, 50 and 75% of the waste processing was handled through composting. The remaining proportion of waste treatment was maintained as incineration with energy recovery in each scenario.

Waste Disposal (C4) Following incineration, there is an amount of residual ash that needs to be deposited in a landfill. In the data entering the LCA for the existing LCA report [7], Ecoinvent processes for municipal incineration, where all flows except landfilling related have been removed, are used in C4. We have not found a way to conduct this in SimaPro software [8]. Therefore, the proxy process for *bico* has been omitted. This is a very crude approximation, but for our purposes, it is considered adequate since C4 is not a stage in which we are conducting detailed analysis.

2.3 The Life Cycle Impact Assessment (LCIA)

In the LCIA, we will introduce the methods used, and the results of our LCA relevant to analyse the selected topics highlighted in goal and scope.

Methods To evaluate the impact on climate change relevant to our selected focus areas, we have analysed the data using the method IPCC 2021 GWP100. The method is based on the final government distribution version of the IPCC report "AR6 Climate Change 2021: The Physical Science Basis". This is a mid-point method developed to yield insight into climate change impact, using relevant units used for comparisons within the building manufacturing industry [9]. It also excludes CO_2 uptake and biogenic CO_2 emissions, which is in line with current practice for climate calculations after Norwegian building regulations.

Results An overview of the analysis, presented as a network flowchart can be seen in Fig. 3. The visualization shows climate change impact represented as GWP-values in the unit kg CO_2-equivalents. From the flowchart, it is apparent that the largest contribution stems from the A1 stage (extraction and production of raw materials). This includes the fire-retardant compound made of inorganic nitrogen fertilizer. Interestingly, it is also the largest contributor to GWP when looking at absolute numbers for amounts, even though it only constitutes approximately 8% of the total

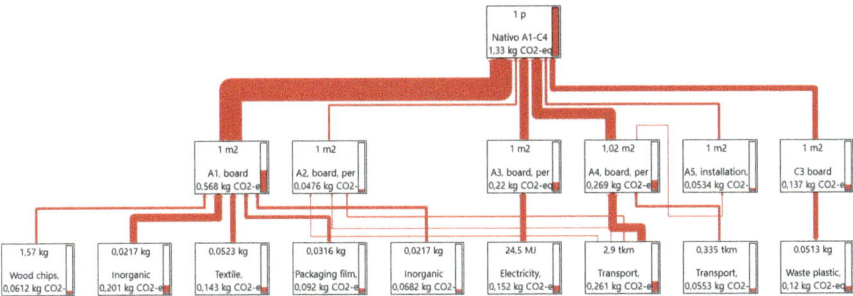

Fig. 3 Network flowchart showing GWP-contributions for the largest contributing processes for Nativo insulation boards

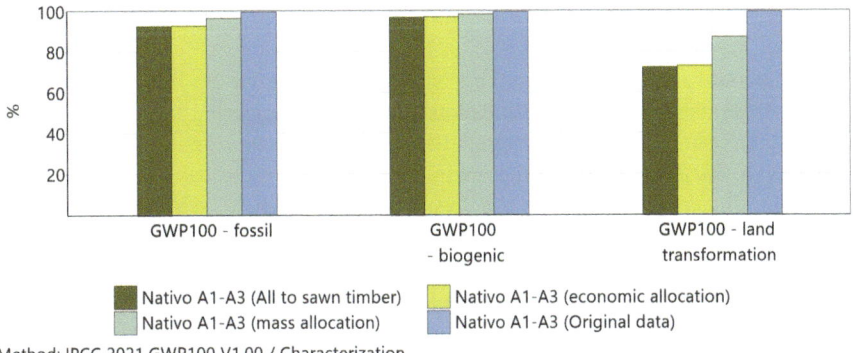

Fig. 4 Illustration showing results of a sensitivity analysis comparing four scenarios for allocation of impacts from wood chips. Method IPCC 2021 GWP100

product mass. Second largest contribution is from the A4 stage (transport to building site). This is interesting to us considering that two of our focus themes concern these two stages. The last focus theme concerns C3 (waste processing) which also has a quite substantial contribution.

3 Allocation and Sensitivity Analysis

3.1 Allocation of Impacts from Wood Chips

In Fig. 4, we can see the results of a sensitivity analysis comparing the four different allocation scenarios for impacts imparted by the raw material wood chips. The analysis is conducted on the processes A1–A3 as this is the value that is routinely compared, when selecting building materials based on climate change impact, in Norway.

Table 1 GWP numbers for the A1 stage for four different allocation scenarios for climate change impact from wood chips

Damage category	Unit	Allocation scenario			
		All to sawn timber	Economic allocation	Mass allocation	Original data
GWP100	Kg CO_2-eq	0.508	0.510	0.541	0.568

Table 2 GWP numbers for the A1–A3 stages for three different scenarios for climate change impact from raw material transport

Damage category	Unit	Transportation scenario		
		EURO6	EURO6 + Biodiesel	Original data
GWP100	Kg CO_2-eq	0.834	0.840	0.836

Compared to the data used in the existing LCA-report, there is a relatively large impact from using alternative allocation models. If considering the A1 stage separately, there is a reduction of GWP of 10.6% if all impacts from the wood chips are allocated to the final product in the sawmill. Corresponding values are 4.8 and 10.3% for mass allocation and economic allocation, respectively. The actual numbers for GWP in A1 for the different scenarios are shown in Table 1. According to Jungmeier et al. [5], the following percentages of CO2 emissions from a sawmill contributed to wood chips in different scenarios: 35.4% for mass allocation, 1.6% for economic allocation and 0% when all emissions are allocated to sawn timber. These percentages are included as a basis for our analysis.

3.2 Different Transportation Modes

The differences between the scenarios are minuscule. The scenario with the highest number for climate impact is the scenario with EURO6 and Biodiesel. The original scenario has a 0.5% smaller impact, while the scenario where all transport is based on EURO6 has a 0.7% smaller impact than the scenario including biodiesel (Table 2).

3.3 Analysis of Waste Treatment Scenarios

When looking at the C3 stage, composting appears to have a large effect on the GWP-numbers (Fig. 5). Compared to no composting, composting of all the waste after use infers a climate change impact reduction of 80.3%. The reduction is linear; therefore, the scenarios of 25% and 75% composting are intermediate at a reduction of 20.1% and 40.2%, respectively.

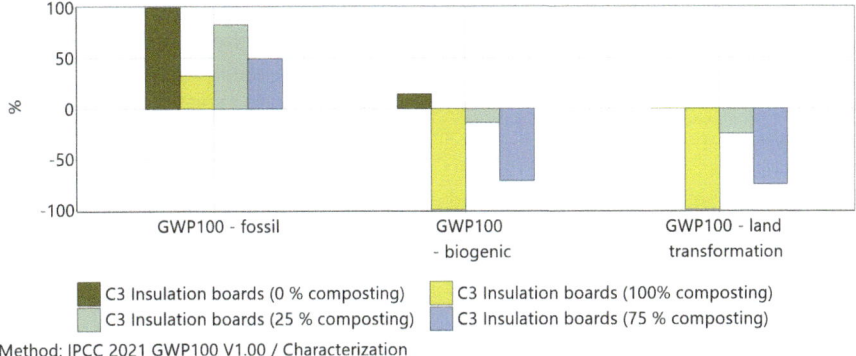

Fig. 5 Illustration showing results of a sensitivity analysis comparing four scenarios for different quantities of waste treated by means of composting

4 Results and Discussion

4.1 Life Cycle Interpretation

From the LCA conducted, it is apparent that the stage mainly contributing to the climate change impact imparted by the Hunton Nativo insulation boards is A1 (extraction and production of raw materials). Other important stages in this respect are A4 (transportation to building site), A3 (manufacturing) and C3 (waste processing).

The analysis of the impact of allocation of the wood chips, which is the largest raw material source in terms of volume, yielded some interesting results. Choosing one of the described allocation scenarios massively alters the climate change indicator GWP for the wood chips in isolation. However, the relative contribution to the impact from A1 to A3 as a whole is quite modest. Considering the distribution of the total climate change impact indicator illustrated in the flowchart in Fig. 3, this is not that surprising. Even though the wood chips make up approximately 77% of the finished product (including packaging), its contribution to the GWP for A1–A3 is less than 11%. Even though other raw materials do have much higher impact on the GWP, it could be worth looking into whether the allocation used in the existing LCA-report is correct and fair.

The results from the analysis of different transportation scenarios for the raw materials were meager. The change in GWP resulting from a scenario consisting of upgrading the lorries used for transportation from EURO4/5 to EURO6 yielded next to no effect on the climate change impact indicator. Introducing biodiesel for the transportation conducted in Norway resulted in a somewhat higher GWP-number for A4. Even though biofuels do have certain characteristics making them a good alternative in some fossil fuel-driven processes, there are large uncertainties considering their negative impacts. These arise from a wide array of sources, such as land use, use of fertilizers and pesticides, refining with coal or natural gas. The result is that biofuels are oftentimes considered to have a greater impact on climate change than fossil fuels, for example, UN Energy (2007) [10]. We also wanted to investigate the effect of substitution some of the fossil fuel-driven transportation with electrically powered lorries. We were, however, unable to locate a process for electrically powered lorries in the Ecoinvent database, so we were not able to carry out such an analysis. Considering the electrical mix in the Norwegian market, we are confident that such an analysis would have yielded positive results.

Introducing composting as a waste treatment substituting incineration with energy recovery appears to have a significant effect on GWP. After A1, A3 and A4, C3 (waste processing) is the largest contributor to the total climate change impact of the complete product. It is uncertain how large a percentage of the waste processing it is realistic to convert to composting, but this appears to be a potentially fruitful issue to explore.

4.2 Conclusions

One conclusion of the study is that the most imperative material input for climate change impact is the fire-retardant compound made of inorganic nitrogen fertilizer (Fig. 3). Therefore, it is highly recommended that a search for an alternative fire-retardant is conducted for further development. If this is not seen as feasible, it is recommended to conduct a critical review of the amount of fire retardant added to the product. Could one obtain the desired fire-retarding properties of the end product with a smaller addition of fire-retardant.

About the different allocation scenarios, it could be useful to investigate whether the allocation applied in the existing LCA-report is correct and fair. Should one uncover that other allocation models are more accurate and fairer, one might get a more accurate and more beneficial GWP value for the insulation boards through a modest effort.

The analysed scenarios for transportation of raw materials yielded such miniscule changes to the GWP value for the product that it is considered not worthy of any effort to employ them. In our view, it might, however, be interesting to investigate how other transportation modes might affect the climate change potential of the product, if it could be beneficial to redirect some of the transport in the Norwegian market to electrically powered lorries. The production of electricity in Norway is based on renewable sources, to such an extent that we believe this will offset other, potentially negative, impacts from other processes concerning electrically powered lorries, to be deemed interesting to investigate.

About waste management, the analysis results suggest that composting produces avoided impact concerning climate change, and it would be worthwhile to exert some effort into applying the process. Hunton is currently involved in a process to obtain an approval from authorities concerning the feasibility of composting these products. Should such approval be granted, we recommend that Hunton work together with Norwegian municipal recycling companies in this issue. Even greater effect could be obtained through employing a take back-system where used products can be reused in Hunton's manufacturing, to make new isolation boards. This would both yield avoidance of impact from waste treatment processes and reduce impact from today's practice of using virgin raw materials. Although this would be a win–win scenario, it is unlikely to be able to take back 100% of the used products. Could a significant amount of the used products, not taken back, be treated through composting, rather than incineration, this appears to result in a decent reduction of the climate impact at the end-of-life stage.

Acknowledgements The authors acknowledge the project "1-2-TRE-STEG: Steps towards circularity in wood-based".

References

1. Tellnes, L.G.: EPD Wood Fibre Insulation, LCA-Report. OR.04.20 Østfoldforskning AS (2020)
2. ISO 14040:2006 Environmental management. Life cycle assessment. Principles and framework
3. ISO 14044:2006 Environmental management. Life cycle assessment. Requirements and guidelines
4. NPCR012 v.2.: Product category rules for thermal insulation products. EPD-Norge (2018)
5. Jungmeier, G., Werner, F., Jarnehammar, A., Hohenthal, C., Richter, K.: Allocation in LCA of wood-based products experiences of cost action E9. Int. J. LCA. **7**, 369–375 (2002)
6. Wærp, S., Flæte, P.O., Svanæs, J.: MIKADO—Miljøegenskaper for treog trebaserte produkter over livsløpet. Et litteraturstudium, SINTEF (2009) 67 pp
7. Raadal, H.L., Modahl, I.S., Lyng, K.-A.: Klimagassregnskap for avfallshåndtering, Fase I og II. OR.18.09 Østfoldforskning AS (2009)
8. Simapro software. https://simapro.com/
9. Masson-Delmotte, V., Zhai, P., Pirani, A., Connors, S.L., Péan, C., Berger, S., Caud, N., Chen, Y., Goldfarb, L., Gomis, M.I., Huang, M., Leitzell, K., Lonnoy, E., Matthews, J.B.R., Maycock, T.K., Waterfield, T., Yelekçi, O., Yu, R., Zhou, B. (eds.): IPCC, 2021: Climate Change 2021: The Physical Science Basis. Contribution of Working Group I to the Sixth Assessment Report of the Intergovernmental Panel on Climate Change. Cambridge University Press, Cambridge and New York., 2391 pp. https://doi.org/10.1017/9781009157896
10. UN Energy.: Sustainable bioenergy: a framework for decision makers (2007). https://www.fao.org/3/a1094e/a1094e00.htm. Accessed 09 Nov 2023.

Dynamic Characterization of Indian Pond Ash Through Cyclic Simple Shear Tests

Ghanta Naga Sireesha and Prishati Raychowdhury

Contents

1 Introduction

Ash is the by-product obtained after combustion of pulverized coal in thermal power plants. In general, two kinds of ashes are produced in a thermal power plant, namely, bottom ash and fly ash. Bottom ash is comparatively heavier and coarser, and collected from the bottom furnace after burning of coal, whereas the fly ash is the comparatively lighter and finer counterpart, which is collected from the electrostatic precipitators. After combining a significant amount of water to the bottom ash and fly ash, and transforming the mixture to a slurry form, it is generally transported to nearby storage ponds. It is evident that a large amount of space is required for this disposal. Not only the material poses a huge space issue, it can also cause environmental conscquence by being airborne due to its weightlessness. Utilization of pond ash as a fill material in various infrastructure projects can control the above-mentioned disposal problem in addition to reducing the cost of the project.

G. N. Sireesha · P. Raychowdhury (✉)
Indian Institute of Technology Kanpur, Kanpur, India
e-mail: gnsiree@iitk.ac.in; prishati@iitk.ac.in

M. Kioumarsi, B. Shafei (eds.), *The 1st International Conference on Net-Zero Built Environment*, Lecture Notes in Civil Engineering 237,
https://doi.org/10.1007/978-3-031-69626-8_43

Numerous studies carried out in the past few decades focused on behaviour of pond ash material and assessing its potential as a fill material. Martin et al. [1] investigated the geotechnical properties of ashes from coal-burning utilities in different cities of USA and found that fly ash can be rapidly compacted with methods similar to those used for granular soils, whereas construction is less sensitive to compaction-moisture content than the fine-grained soils. Gandhi et al. [2] carried out extensive field trials to evaluate the effectiveness of deep blasting for densification of deposited fly ash. Das and Yudhbir [3] performed a series of laboratory experiments to characterize some Indian fly ashes in terms of their geotechnical properties such as grain size, specific gravity, compaction characteristics, and unconfined compression strength in order to evaluate their suitability as embankment material. Jakka et al. [4] performed a detailed experimental study on the strength and other geotechnical characteristics of pond ash samples collected from inflow and outflow points of two ash ponds in India. Mohanty et al. [5] studied the strength and deformation behaviour of Rae-Bareli pond ash under cyclic loading by conducting a series of stress controlled one-way cyclic compressive triaxial tests on reconstituted samples of pond ash. Mohanty and Patra [6] performed strain-controlled cyclic triaxial tests on pond ash samples collected from plants from different seismic zones of India. Vijayasri et al. [7] performed a series of cyclic triaxial tests on pond ash samples obtained from Renusagar pond ash embankment of Northern India and evaluated the effect of inclusion of geotextiles in improving the shear modulus, strength, and liquefaction potential of the material.

Although several studies had focused on evaluating the properties of Indian pond ash through element level triaxial tests, dynamic characterization of the material using cyclic simple shear tests are very limited. As we know that a soil sample in a direct shear test setup resemblance better with an in situ soil deposit subjected to an earthquake induced vibration, compared to other element level tests such as triaxial test. In this context, the present study ventures to present some experimental observations on a pond ash sample collected from Parichha power plant in North-Central India through a series of cyclic simple shear tests.

2 Geotechnical Characterization of the Pond Ash Sample

The pond ash material was collected from the **inflow points of the** ash ponds at Parichha thermal power station near Jhansi town of Uttar Pradesh state in India. Prior to determination of the dynamic properties of the material, basic characteristic tests including specific gravity, gradation, and particle mineralogical tests were performed. Further, standard proctor test and direct shear tests were also conducted to estimate the maximum dry density, optimum moisture content, and the static shear strength properties of the material. The observations from these static tests are described in the following subsections.

2.1 X-Ray Fluorescence (XRF) Test

XRF test was performed on the pond ash sample to determine the chemical composition of the pond ash sample by measuring the fluorescent (or secondary) X-ray emitted from the sample when it is excited by a primary X-ray source. In this case, an amount of 4.5 g of pond ash sample was placed in a beaker and then was mixed with 0.5 g of boric acid to bind the pond ash particles. After that, the pressed powder pellets were prepared at Advanced Centre for Material Sciences (ACMS), Indian Institute of Technology Kanpur (IITK) with the help of stainless steel dye-cast having a diameter of 30 mm. Then the samples were pressed on a hydraulic pre-machine with a pressure of 12–13 ton to make 30 mm diameter pressed powder pellets and heated in oven for about 20 minutes at 80 °C to remove any moisture absorbed by the pellets. Afterwards, the sample was placed in Rigaku ZSX Primus II wavelength dispersive X-ray fluorescence spectrometer to determine the major oxides and some of the trace elements (in parts per million, ppm) present in the pond ash sample. The chemical composition of pond ash from XRF tests are shown in Table 1. It may be observed that Oxides of silica, alumina and iron together constitute the major proportion (of about 95%). According to the American Society for Testing Materials standard (ASTM C618), ashes containing more than 70% of SiO_2 + Al_2O_3 + Fe_2O_3 (by weight) are defined as Class F, while those with a SiO_2 + Al_2O_3 + Fe_2O_3 content between 50% and 70% (by weight) are defined as Class C. The free lime content in the Pond ash samples is low, indicating that the ashes belong to class F category. It is to be noted that Class F category ashes are suitable for use a geotechnical fill material, whereas the chemically active Class C category ashes are not. Hence, it is concluded that the sample used in the present study is in the suitable category.

Table 1 Chemical composition (%) of Parichha thermal power station pond ash

Chemical composition	%
Silica (SiO2)	61.6
Alumina (Al2O3)	22.6
Iron oxide (Fe2O3)	6.84
Titanium (TiO2)	1.81
Potash (K2O)	1.23
Lime (CaO)	0.68
Phosphorus pentoxide (P2O5)	0.29
Magnesia (MgO)	0.11
Soda (Na2O)	0.11
Manganese (MnO)	0.09
Nickel (NiO)	0.006

2.2 Morphological Characteristics

Particle morphology indicating the shape and surface of the particle of the sample was analysed using scanning electron micrographs (SEM) facility available at Indian Institute of Technology, Kanpur. Figure 1 shows the obtained SEM graphs for the collected pond ash sample. It may be observed that the shapes of the ash particles are angular and irregular in shape and has complex pore structure. When a higher magnification has been used, some agglomerations of smaller ash particles are observed. Further, **intra-particle voids were observed in the sample. It was observed to be a more porous material than initially thought, making the particles more crushable.**

2.3 Specific Gravity

The specific gravity of the sample was obtained using a standard Pycnometer test as per IS: 2720: Part-3 [8]. The specific gravity was obtained to be 2.22. This lies in the range of standard Indian pond ashes, which is reported to be 1.64–2.66 by several previous studies including Das and Yudhbir [3].

Fig. 1 Scanning electron micrographs of Parichha pond ash

2.4 Grain Size Distribution

The grain size distribution of the pond ash was obtained following IS: 2720-Part 4. First, the sieve analysis was carried out, followed by a hydrometer analysis for the remaining sample passing through the 75 μm. The grain-size distribution indicated the fraction of gravel, sand, silt, and clay to be 0%, 79.2%, 20.8%, and 0%, respectively, by weight. As per unified soil classification system, thee pond ash considered in this study can be classified as silty sand with the symbol SM. Table 2 provides the fractions of different components in the sample, whereas Fig. 2 shows the grain-size distribution plot obtained from the test. In this plot, apart from the grain-size distribution of the chosen pond ash sample, the ranges for liquefiable soils as proposed by Tsuchida [9] have also been shown. It can be seen that Parichha pond ash definitely fall in the liquefiable range confirming potential for liquefaction.

Table 2 Grain size characteristics of Parichha pond ash

Physical properties	Values
Gravel (> 4.75 mm)	0
Sand (4.75–0.075 mm)	79.2% (by weight)
Silt (0.075–0.002 mm)	20.8% (by weight)
Clay % (<0.002 mm)	0
Coefficient of uniformity Cu	2.57
Coefficient of Curvature Cc	0.681
D_{10}	0.068 mm
D_{30}	0.09 mm
D_{60}	0.175 mm
USCS Group symbol	SM
USCS Group name	Silty Sand

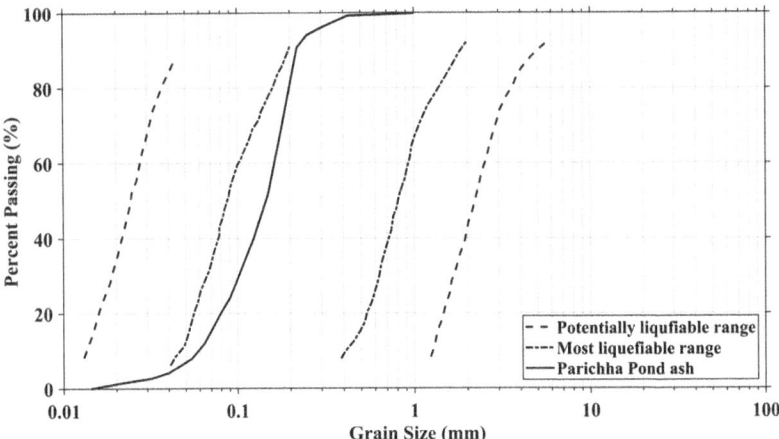

Fig. 2 Grain size distribution curves for Parichha pond ash sample overlaid with a range of liquefiable soil as given in Tsuchida [9]

2.5   Compaction Characteristics

To obtain the compaction characteristics of the pond ash, standard Proctor test was done. The compaction curve obtained from the test is shown in Fig. 3. The maximum dry unit weight and optimum moisture content of the material were estimated as 1090 kg/m3 and 38.5%, respectively.

2.6   Direct Shear Test

In order to estimate the shear strength properties of the pond ash sample, a series of direct shear tests had been carried out as per the guidelines of IS:2720 (Part-13), [10]. The sample was prepared at a density of 65%. Normal stresses of 50 kPa, 100 kPa, and 150 kPa were applied on the specimen. The deformation applied at a rate of 0.6 mm/min. Figure 4 shows the shear stress versus shear strain plots at different normal stress levels. It can be observed from the plot that the sample shows dilative behaviour corresponding to 150 kPa normal stress case. The angle of internal friction was estimated as 33°.

3   Cyclic Simple Shear Test Setup and Sample Preparation

A series of strain-controlled cyclic tests was performed on the Parichha pond ash using the cyclic simple shear test at the Indian Institute of Technology Kanpur. The primary components of equipment include: a loading unit, a pneumatic control panel, an electronic control and a display unit. The photograph of the testing facility is given in Fig. 5.

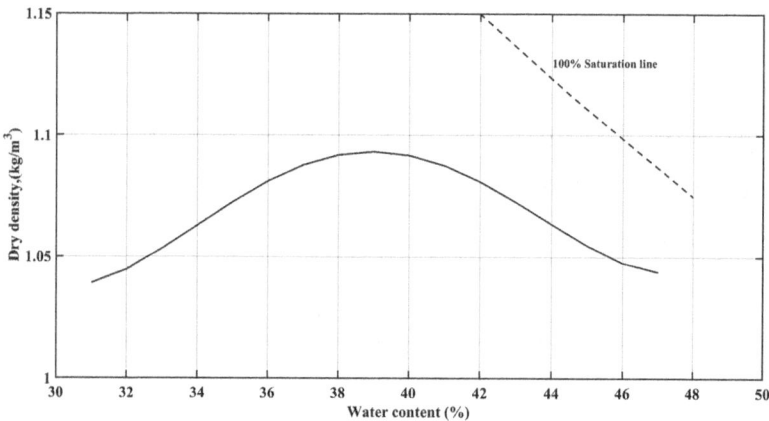

Fig. 3 Compaction curve of Parichha pond ash

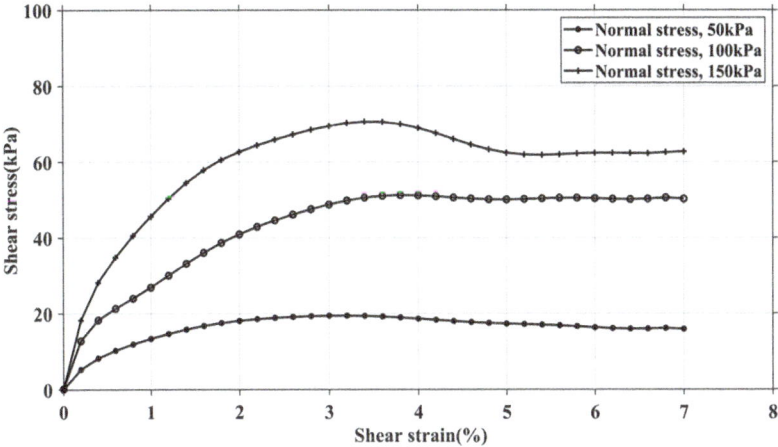

Fig. 4 Shear deformations vs. shear stress curve of Parichha pond ash

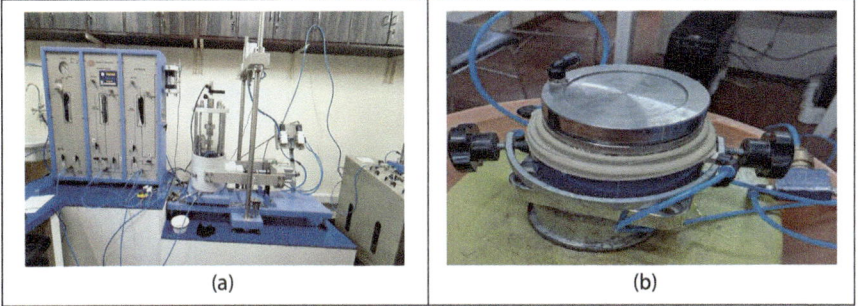

Fig. 5 (**a**) Cyclic simple shear apparatus and (**b**) sample preparation mould

During a test, the required cell pressure and back pressure can be controlled through the computer-based controller unit, which operates on a closed loop control mechanism. In the present study, the pond ash specimen, after compacted into desired relative density of 65%, was confined in a plain membrane. The specimen prepared was of size 10 cm diameter and 3 cm height, making an aspect ratio of 0.3, which is well below the maximum limit 0.4 as per ASTM D8296.

A split mould was used for the sample preparation (Fig. 5b). A simple moist tamping technique is adopted to achieve the desired 65% relative density. The ash sample was placed in the mould, water was added and then the sample was compacted mechanically in three layers. After the sample was placed on the lower plate, a nominal vacuum pressure was applied to keep the sample in vertical position, after which, the mould was removed. After the sample was placed on the mould, the hammer and the normal load cell were placed on top of the sample. The box was then closed and with the upper box and consequently was filled with water. The cell pressure and back pressure were applied in stages by maintaining a variation of 20 kPa. In each stage, Skempton's pore water pressure parameter B was estimated.

The consolidation process had been started with a confining pressure of 100 kPa after Skempton's pore water pressure parameter B reached 0.95. At this point, the soil was considered to be fully saturated. When the volume change approached zero, the back pressure valve was closed to end the consolidation process and to maintain an undrained condition during the shearing stage. After the consolidation phase was complete, a strain-controlled cyclic loading was started to be applied.

A cyclic loading in sinusoidal wave forms having peak strain amplitude (δ_c) 3.83% and 5% with frequency of 0.2 Hz was applied on the consolidated pond ash specimen which is under the confining pressure of 100 kPa and 200 kPa. According to Jakka et al. [4], the frequency of load application is set to provide for enough time to equalize the pore pressure.

4 Results and Discussion

The following observations were made from the cyclic simple shear tests on the Parichha pond ash sample.

4.1 Dynamic Properties

The cyclic stress–strain behaviour of pond ash sample subjected to sinusoidal waves of strain amplitude, γ_c 5% with frequency, f of 0.2 Hz under the confining pressure, σ_c of 100 kPa is shown in Fig. 6. It can be observed from Fig. 6 that as the number of cycles increases, the hysteresis loops progressively become flatter indicating higher

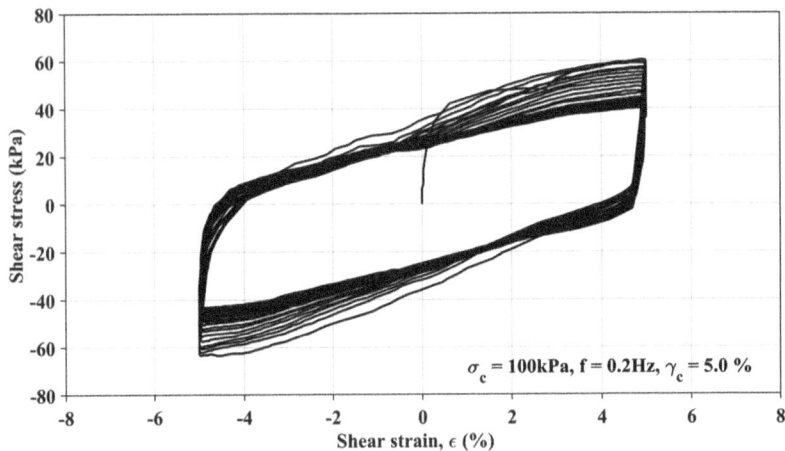

Fig. 6 A typical stress–strain behaviour of Parichha pond ash sample of 65% relative density subjected to 5% cyclic shear strain amplitude under 100 kPa confining pressure

Table 3 Dynamic properties of Parichha pond ash

Shear-strain amplitude (%)	Confining pressure (kPa)	Maximum shear stress (τ_{max}) at the first cycle	Shear modulus (kPa) (G_{sec})	Damping ratio
3.83	100	50	1306	0.2898
3.83	200	67	1747	0.2892
5	100	60	1207	0.3028
5	200	81	1641	0.3019

energy dissipation and gradually increasing inelastic behaviour. Furthermore, each hysteresis loop indicates nonlinear behaviour of the sample with clear indication of strength and stiffness degradation with each loading cycle. The tests were repeated for confining pressure of 200 kPa and shear strain amplitude of 3.83%. The effect of the shear strain and the confining pressure on different dynamic characteristics was examined.

The important dynamic properties including strain-dependent secant shear modulus, damping ratio, and peak shear stresses observed from the tests for different confining pressures (100 kPa and 200 kPa) and strain amplitudes (3.83% and 5%) are provided in Table 3. It can be noted that the dynamic shear modulus increases from 1306 to 1747 kPa, i.e., an increase of about 34% for an increase in the confining pressure from 100 kPa to 200 kPa. Similarly, the peak stress level also shows an increase of about 35%. The damping ratio, on the other hand, is insensitive to the change in confining pressure, showing a change less than 0.2%.

When the effect of strain amplitude is considered, it is observed that for the 100 kPa confining stress, the peak stress increases from 50 kPa to 60 kPa (i.e., an increase of about 20%) for an increase in the shear strain amplitude from 3.83% to 5%. The shear modulus reduces from 1300 MPa to 1200 MPa (i.e., a reduction of 7.7%), whereas the damping ratio increases from 29% to 30% (i.e., an increase of about 3.33%). For the higher confining pressure, i.e., 200 kPa confining pressure case, the influence of strain amplitude is similar, with an increase in peak shear stress of 21%, reduction in shear modulus of about 5.7% and increase in damping ratio as 3.33%. The observations indicate that at higher strain level, the ash sample behaves as softer material with lower stiffness and higher deformation potential, as also expected for any conventional loose to medium cohesionless soil.

4.2 Liquefaction Potential

The pore pressure developed during the undrained loading was using pore pressure transducer attached to the apparatus. From the pore pressure development pattern, the liquefaction potential of the sample was estimated. Figure 7 shows the excess pore pressure ratio versus number of cycles at 5% strain amplitude confined with

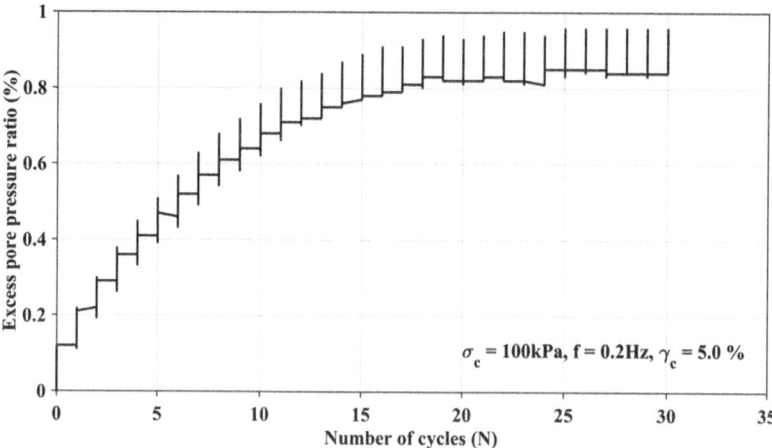

Fig. 7 Excess pore pressure ratio versus number of cycles of Parichha pond ash (65% relative density) subjected to 5% cyclic shear strain amplitude under 100-kPa confining pressure

100 kPa pressure. The gradual increase in pore pressure is observed as normal for any typical cohesionless soil. The maximum pore pressure ratio is observed to reach a value up to 98% at 25th cycle. The number of cycles needed for liquefying the sample is observed to reduce from 29 to 25 when strain amplitude is increased from 3.833% to 5%. This reinforces the strain dependency of the ash sample and increased vulnerability with respect to liquefaction hazard when subjected to large intensity loadings.

4.3 Degradation Index

To quantify the amount of stiffness reduction with increase in strain, a parameter, namely, *Degradation Index*, δ, has been introduced herein. The degradation index, δ, is defined as ratio of the shear modulus of the first cycle and the shear modulus of Nth cycle, as shown below.

$$G_N = \delta \, G_1$$

where G_N and G_1 are the secant shear moduli, after the Nth cycle and the first cycle, respectively.

Figure 8 presents the degradation index versus number of cycles for both 5% and 3.83% strain level cases under 100 kPa confining pressure. It may be observed from Fig. 8 that the dynamic shear modulus constantly decreases with increasing shear strain for both strain amplitude test series. The degradation index is as much as 0.55 (i.e., 45% loss) for 5% strain level and 0.75 (i.e., 25% loss) for 3.83% strain level. Further, when the rate of reduction is considered, it is observed that the rate of degradation is higher for the 5% strain amplitude case compared to the 3.83% strain amplitude case. Moreover, it is also evident from Fig. 8 that during the initial 5–10 cycles, the rate of degradation is most prominent for both strain amplitude tests, whereas after 10 cycles, the curves reach a plateau and degradation index that remains approximately constant afterwards.

4.4 Comparison with Other Studies

In Table 4, a summary of observations on various critical parameters pertaining to dynamic loading for the chosen Parichha pond ash sample is provided. The results

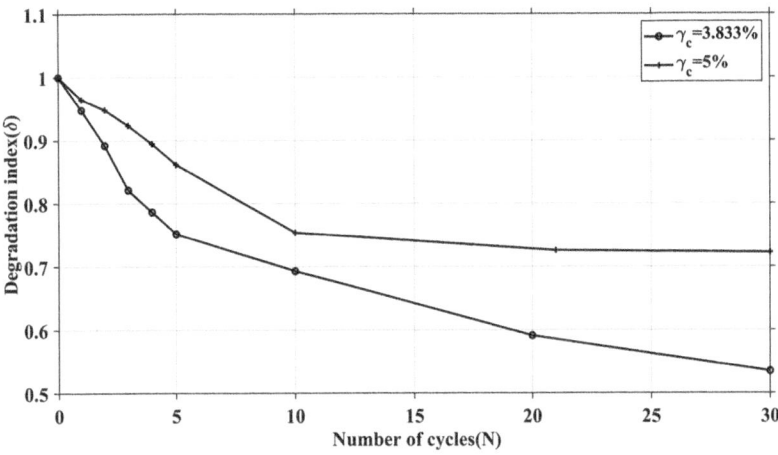

Fig. 8 Degradation index versus number of cycles for Parichha pond ash (for 100 kPa confining pressure case)

Table 4 Comparison of dynamic properties of pond ash and other resources of fly ash

Powerplant name (ash source)	Frequency (Hz)	Shear-strain amplitude (%)	Secant shear modulus (MPa)	Damping ratio (%)	Number of cycles for initiation of liquefaction
Parichha[a]	0.2	3.8	1600	28.9	28
		5.0	1100	30.2	25
Badarpur[b]	1	0.1	9200	7.8	–
		1.0	2500	14.5	15
Talcher[c]	0.3	0.6	10,400	9.0	470
		1.2	8600	11.0	445
Panki[c]	0.3	0.6	8800	11.0	475
		1.2	6500	14.0	420

(continued)

Table 4 (continued)

Powerplant name (ash source)	Frequency (Hz)	Shear-strain amplitude (%)	Secant shear modulus (MPa)	Damping ratio (%)	Number of cycles for initiation of liquefaction
Panipat[c]	0.3	0.6	8100	12.5	481
		1.2	6000	17.0	434
Renusagar[d]	0.5	0.5	4700	11.5	164
		1.05	2900	14.5	89
Gandhinagar[e]	1	0.5	2800	48.0	15

[a]Present study
[b] Jakka et al. [11]
[c]Mohanty and Patra [6]
[d]Vijaysri et al. [7]
[e]Shrivastava et al. [12]

are also compared with reported values for different ash samples taken from other powerplants located in various parts of India. It can be noted that the properties obtained for Parichha pond ash is within the range reported in the previous studies.

5 Conclusions

The present study focuses on static as well as dynamic characterization of pond ash collected from Parichha power plant in North-Central India. For the dynamic characterization, cyclic simple shear tests with varying shear strain amplitude and confining pressure had been performed. The following key observations were obtained:

- The ash falls into the category of silty sand (SM) as per USCS classification system. The shear strength parameters indicated zero cohesion and a friction angle of 33° indicating it as a medium dense cohesionless material. Chemical and mineralogical compositions indicated that the material falls under Class-F category with low potential for adverse environmental impact, and hence suitable for its usage as geotechnical fill material.
- The dynamic shear modulus was observed to be highly sensitive to the change in confining pressure, showing an increase of about 34% when the confining pressure was increased from 100 kPa to 200 kPa. The damping ratio, on the other hand, remained unchanged.
- The shear strain amplitude has shown reasonable effect on all dynamic properties. The shear modulus was observed to reduce about 7.7%, whereas the damping ratio increased about 3.3%, when strain amplitude increased to 5% from 3.83%.
- The excess pore pressure generation and related strength loss indicated the vulnerability of the material for liquefaction-induced failure. The number of

cycles required to initiate liquefaction were observed as 29 and 25, for the strain amplitudes of 3.83% to 5%, respectively, indicating increased vulnerability of the material during higher intensity earthquake events.

- The modulus degradation was observed to be as much as 0.55 and 0.75, for 5% and 3.83% strain level, respectively. Further, the rate of degradation was prominent during the first 5–10 cycles and then attained almost a constant value for both test series.

References

1. Martin, J.P., Collins, R.A., Browning, J.S., Biehl, F.J.: Properties and use of fly ashes for embankments. J. Energy Eng. ASCE. **117**(2), 71–86 (1990)
2. Gandhi, S.R., Dey, A.K., Selvam, S.: Densification of pond ash by blasting. ASCE J. Geotech. Geoenviron. Eng. **125**(10), 889–899 (1999)
3. Das, S.K., Yudhbir: Geotechnical characterization of some Indian Fly ashes. ASCE J. Mater. Civ. Eng. **17**(5), 544–552 (2005)
4. Jakka, R.S., Datta, M., Ramana, G.V.: Liquefaction strength of fly ash reinforced with randomly distributed fibers. Soil Dyn. Earthq. Eng. **30**, 580–590 (2002)
5. Mohanty, B., Patra, N. R., Chandra, S., Cyclic Triaxial Behavior of Pond Ash, GeoFlorida 2010: Advances in Analysis, Modeling & Design, Geotechnical Special Publication-199(2010)
6. Mohanty, S., Patra, N.R.: Cyclic behavior and liquefaction potential of Indian pond ash located in seismic zones III and IV. J. Mater. Civ. Eng. ASCE, ISSN 0899-1561/06014012 (5)(2014)
7. Vijayasri, T., Patra, N.R., Raychowdhury, P.: Cyclic behavior and liquefaction potential of Renusagar pond ash reinforced with geotextiles. J. Mater. Civ. Eng. **28**(11), 1–1 (2016)
8. IS: 2720: Part 3 Methods of Test for Soils—Determination of Specific Gravity. Bureau of Indian Standards, New Delhi (1980)
9. Tsuchida, H.: Prediction and Countermeasure against Liquefaction in Sand Deposits. Port and Harbour Research Institute, pp. 3.1–3.33. Ministry of Transport, Yokosuka (1970)
10. IS: 2720: Part 4 Methods of Test for Soils—Grain Size Analysis. Bureau of Indian Standards, New Delhi (1985)
11. Jakka, R.S., Ramana, G.V., Datta, M.: Seismic slope stability of embankments constructed with pond ash. Geotech. Geol. Eng. **29**, 821–835. https://doi.org/10.1007/s10706-011-9419-8 (2011)
12. Shrivastava, A., Luhar, H., Sachan, A.: An experimental investigation on shear strength and liquefaction response of pond ash for road embankment construction. Transp. Infrastruct. Geotechnol. **11**, 1231–1248. https://doi.org/10.1007/s40515-023-00324-z (2024)

Analysis of the Mechanical Behavior of Rammed Earth Mixes Using Soft Computing Methods with an Emphasis on Sustainability

Aryan Baibordy, Mohammad Yekrangnia, and Fatemeh Khodabakhshian

Contents

1   Introduction

In 2015, the United Nations introduced the 2030 Agenda for Sustainable Development to address the social, economic, and environmental issues of societies by setting 17 sustainable development goals (SDGs) and 169 targets [1, 2]. By introducing this agenda, it was expected to tackle most of the sustainability issues by 2030. However, at the midpoint of this journey, reports regarding the sustainability issues show unsatisfactory and unexpected results. On top of everything, the global mean temperature has increased by about 1.1 °C above preindustrial levels (1850–1900). Moreover, global carbon dioxide (CO_2) emissions from industrial processes grew by 0.9 percent to a new all-time high of 36.8 billion metric tons [3].

A. Baibordy · M. Yekrangnia (✉) · F. Khodabakhshian
Shahid Rajaee Teacher Training University, Tehran, Iran
e-mail: yekrangnia@sru.ac.ir

© The Author(s) 2025
M. Kioumarsi, B. Shafei (eds.), *The 1st International Conference on Net-Zero Built Environment*, Lecture Notes in Civil Engineering 237,
https://doi.org/10.1007/978-3-031-69626-8_44

The construction industry contributes to some of the issues mentioned earlier [4]. However, the UN has recently stated that to build sustainable cities and societies, efforts should concentrate on implementing inclusive, resilient, and sustainable urban development policies and practices. These should prioritize access to basic services, affordable housing, efficient transportation, and green public spaces for all [3]. Despite this recommendation, civil engineers have recently begun to incorporate sustainability approaches into their projects. In terms of sustainable construction, they have integrated green building materials and techniques [5–7]. Among these, earth-based materials have garnered significant attention from engineers and practitioners [8].

Earth-based materials are cost effective and can mitigate environmental issues. These materials can be incorporated into the construction industry using sustainable techniques for constructing both structural and nonstructural elements of buildings, pavements, and other infrastructures. These techniques, inspired by historical construction methods, include adobe, rammed earth, and compressed earth block [9].

Rammed earth, also known as pisé, is a sustainable construction technique where a wet soil mixture is compacted layer by layer at its optimum moisture content and maximum dry density [9]. Furthermore, the mechanical properties of rammed earth can be enhanced by stabilizing it with additives such as cement [10], lime [11], natural additives [12], and geopolymers [13]. Despite the superior mechanical performance, particularly in terms of compressive strength, of stabilized rammed earth compared to unstabilized rammed earth, both have lower tensile strength due to the inherent properties of soil. To overcome this limitation, rammed earth can be reinforced with both natural and artificial fibers [14].

While rammed earth appeals to engineers due to its sustainability benefits and historical context, the information surrounding its trends, implications, and properties remains unclear. To address these gaps, this research study first evaluates public comments and sentiments toward the implementation of rammed earth in the construction industry using an AI-based tool. Subsequently, the trends related to rammed earth are studied and critically compared. In the second section, the gaps in the analysis of the mechanical properties of rammed earth are identified and will be addressed in the subsequent section. In the third section, the gaps in the mechanical properties of rammed earth are analyzed using soft computing and machine learning methods. Within this process, the dataset includes sustainable rammed earth (with a lower carbon footprint) and conventional rammed earth (with a higher carbon footprint), which could be helpful to interpret the correlations between the sustainability factors of rammed earth and its mechanical properties, especially compressive strength. This interpretation, along with the critical analysis, is included in discussion section of this study.

2 Sentiment Analysis of Social Media

Sentiment analysis or opinion mining is primarily a business tool that can be beneficial for evaluating customer satisfaction with a product. By doing so, companies can identify the weaknesses of their products and take steps toward improving and enhancing their quality. Sentiment analysis is so expansive that it can be conducted on anything from a single text to numerous texts and comments across various platforms. Social media can be a great source of data for sentiment analysis as it contains a vast number of datasets and texts where people share their opinions on specific subjects [15]. The most important social media platforms for sentiment analysis are Reddit, Facebook, X (Twitter), and Instagram [16, 17].

Sentiment analysis is conducted using artificial intelligence techniques such as deep learning (DL) and natural language processing (NLP). In this context, data, primarily text, plays a crucial role in facilitating efficient sentiment analysis. For sentiment analysis of social media, data can be gathered using a technique called web scraping. The extracted data is then analyzed and modeled using NLP. As a result of this modeling, sentiments can be interpreted in two categories: positive and negative, or on an n-point scale (good, neutral, and bad) [15].

In this study, due to the restrictions and security policies on the APIs of reputable social media platforms, especially X, sentiment analysis is conducted using an AI-based tool introduced by North Carolina State University, USA. This platform is capable of visually analyzing sentiments based on emotional models, and it can display a heatmap and word cloud of the sentiments [16].

The keyword "rammed earth" was entered into the query section of the platform, which then analyzed the sentiments surrounding this keyword within the Reddit database. It extracted 102 comments regarding "rammed earth" from 2014 to 2023. Figure 1 shows the sentiment visualization, indicating that the public's attitudes toward "rammed earth" are generally acceptable and pleasant. This result was expected, as the sentiment analysis of keywords related to sustainable development has mostly been positive [18].

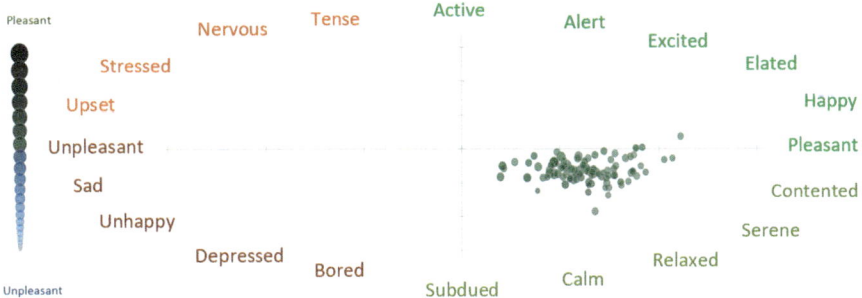

Fig. 1 The visualization of sentiment analysis regarding rammed earth

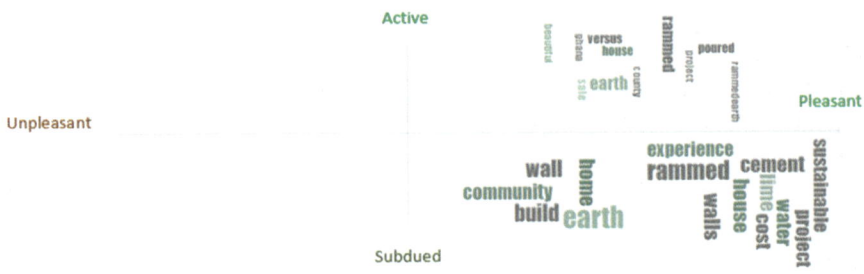

Fig. 2 The word cloud extracted from the sentiment analysis regarding rammed earth

Furthermore, from the extracted data, frequently used words can be highlighted using a word cloud. Figure 2 demonstrates the word cloud of the posts regarding "rammed earth." In the word cloud, the size of each word corresponds to its frequency in the context of rammed earth construction. It can be concluded that most of the posts and comments are shared regarding the sustainability capabilities, especially cost-effectiveness, of cement and lime-stabilized rammed earth elements such as walls. The word "water" has been mentioned as an influential factor on its properties as well.

It is worth mentioning that these results are based on the Reddit database, which is one of the applicable platforms in social media [16]. The results might change if other social media platforms are included in the study.

3   Bibliography of Rammed Earth

This section will provide a bibliography of the rammed earth technique, not only to study the trends in its number of publications but also to identify specific gaps in this material compared to other materials. This bibliometric analysis utilizes the Scopus dataset, focusing on articles published within the last 15 years, from 2008 to 2023.

First and foremost, regarding the search criteria, the keywords "rammed earth," "concrete," "masonry," "compressed earth block," and "timber" have been searched in the Scopus database within the article titles. This includes the field of engineering disciplines and excludes the field of pharmacy and its related fields.

The total number of publications from 2008 to 2023 regarding rammed earth, concrete, masonry, compressed earth block (CEB), and timber are 737, 159,000, 12,392, 181, and 13,556, respectively. Although rammed earth and CEB techniques are highly sustainable and more historical compared to other materials and techniques, concrete, timber, and masonry have gained more attention. The number of publications regarding rammed earth is 0.5% of the number of articles published for concrete. While it is true that the mechanical properties of concrete are superior to those of rammed earth, resulting in better performance against earthquakes and other strong lateral loads, rammed earth has superior sustainability properties. This allows

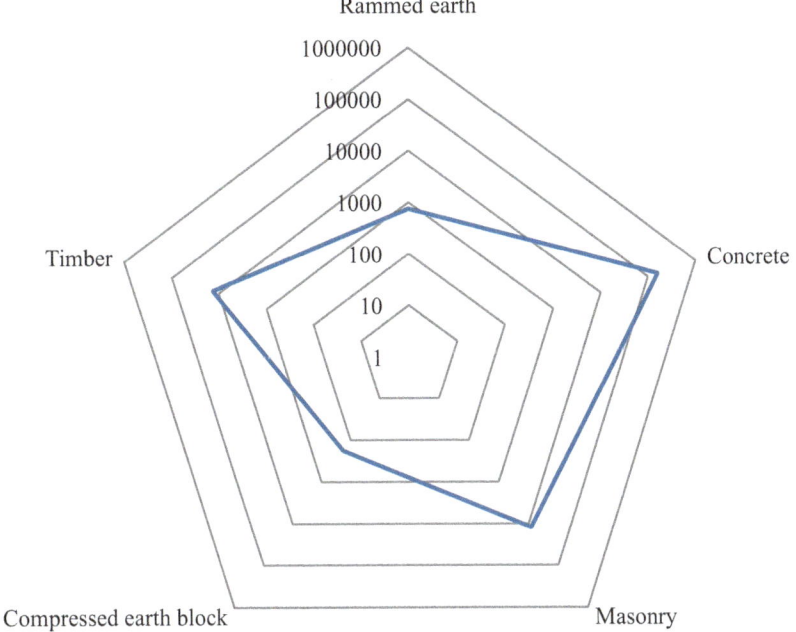

Fig. 3 The total number of publications from 2008 to 2023 on a logarithmic scale

it to contribute to a sustainable and resilient society, which is a pressing need today. Figure 3 demonstrates the total number of publications from 2008 to 2023 on a logarithmic scale.

In addition to the total number of publications regarding these materials, it is crucial to evaluate the trend each year from 2008 to 2023. Figure 4 shows the trend in the number of publications each year from 2008 to 2023 for the mentioned materials. While concrete, masonry, and timber materials have shown an increasing trend up until 2023, rammed earth has experienced fluctuations since 2008.

As a result of the aforementioned statistics, engineers and scholars need to research more on rammed earth and compressed earth blocks in order to identify and address their gaps. However, from the bibliometric study of rammed earth, other gaps could be identified. For instance, as the rammed earth technique requires more energy and time to conduct experiments regarding its properties, the application of soft computing and numerical methods could be beneficial in this regard. According to the Scopus dataset, the number of publications on the application of soft computing methods in rammed earth and compressed earth block research is lower compared to other construction materials such as concrete, masonry, and timber. It is so low that the sum of the number of publications for both rammed earth and compressed earth block is 0.19%, 4.8%, and 5.2% of the number of articles published for concrete, masonry, and timber, respectively. Nevertheless, this result was expected, as based on the results of the bibliometric study in this article, the total number of

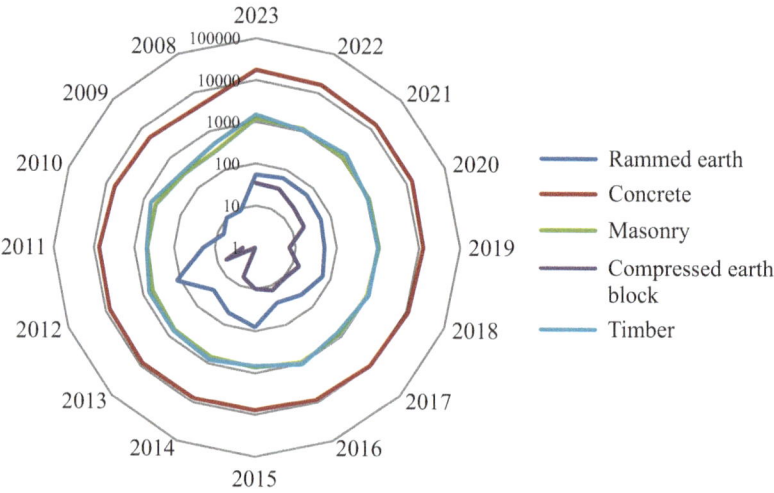

Fig. 4 The trend of the number of publications in each year from 2008 to 2023

publications regarding both rammed earth and compressed earth block is much lower than for other construction building materials. Although in other types of soil-based materials which are more related to geotechnical applications (e.g., soil stabilization), numerous research exist that apply soft computing and numerical methods to their methodology, the focus of this study is just rammed earth in the field of structural and earthquake engineering, or in other words, rammed earth building structures. Furthermore, according to the Scopus dataset, there are only six publications regarding the application of soft computing in rammed earth. Therefore, it can be concluded that it is expected to conduct more research in this field of rammed earth material in order to bridge the gaps.

4   The Analysis of the Mechanical Properties of Rammed Earth Using Data Science Principles and Machine Learning

In Sect. 3, it was concluded that there is a lack of research, especially in the field of soft computing, concerning rammed earth materials compared to other construction building materials. Regarding the use of soft computing in rammed earth and compressed earth block projects, there has been a paucity of research, most of which is discussed in the following paragraphs.

Mustafa et al. [19] used machine learning techniques, namely multi-linear regression (MLR) and artificial neural networks (ANN), to predict the 28-day unconfined compressive strength of both unstabilized and stabilized rammed earth. The dataset for this study consisted of 488 entries with feature vectors including fine, sand, and gravel content, soil liquid limits, plastic index and linear shrinkage, stabilizer type (none, cement, and lime), stabilizer dosage, optimum moisture content (OMC), maximum dry density (MDD), aspect ratio, and testing condition (dry or wet). This study showed that using ANN resulted in better performance compared to MLR. Narloch et al. [20] predicted the compressive strength of cement-stabilized rammed earth by fitting a deep learning technique, namely a convolutional neural network (CNN), into 4284 SEM images. The accuracy of the model was about 84%, which was a satisfactory performance. Turco et al. [21] modeled an ANN on a study with a dataset of 332 entries. In this study, the feature vector comprised clay content, OMC, cement content, fiber content, fiber length, fiber tensile strength, and the specimen's age. Moreover, the target was the compressive and tensile strength of compressed earth blocks. The results demonstrated better performance in predicting compressive strength rather than tensile strength. Anysz and Narloch [22] and Anysz et al. [23] predicted the compressive strength of cement-stabilized rammed earth using ANN, decision tree (DT), and random forest. The feature vectors for both studies were clay, silt, sand, and gravel contents, cement content, OMC, and MDD. Kardani et al. [24] gathered datasets of 83 and 891 entries for the prediction of compressive strength and stress-strain properties, respectively. In this study, a new feature called loading rate was considered. Moreover, the dataset was trained using XGBoost algorithms and techniques. The results revealed an appropriate performance of using XGBoost, especially with smaller datasets.

A review of the background of studies regarding the application of machine learning in rammed earth materials reveals that many gaps still remain and future studies should focus on these gaps. For instance, the combination of lime and cement, information regarding the types of fibers, clay mineralogy, the shape of the specimen, curing conditions, and novel machine learning methods could all be considered to advance the prediction of the compressive strength of rammed earth.

This study will focus on a part of this gap in the following sections. The compressive strength of rammed earth materials will be predicted using a dataset extracted from Scopus, which includes some features that have not yet been investigated. It is worth mentioning that this dataset includes the application of natural-based materials which can improve the sustainability properties of rammed earth. Through data science principles, the dataset will be preprocessed and then modeled using two machine learning methods, namely artificial neural networks (ANN) and decision trees (DT). Finally, the model will be evaluated using the R-squared metric.

4.1 Data Collection and Analysis

A dataset of an acceptable size, 382, has been provided from the Scopus database by searching according to the keywords "rammed earth" and "compressive strength." The dataset, part of which is shown in Table 1, comprises both feature and target values. The feature values are regarded as the factors affecting the target value. Fine x1 (the combination of clay and silt), sand x2, and gravel x3 contents (%), type of

Table 1 Partial data of dataset for compressive strength of rammed earth

x1	x2	x3	x4	x5	x6	x7	x8	x9	x10	x11	x12	x13	x14	x15	x16	x17	y
20	60	20	No	0	No	0	0	0	0	10	1950	CY	3.17	5.5	70	10	0.71
20	60	20	C	3	No	0	0	0	0	10	2219	CY	3.17	5.5	70	10	2.01
16	42	35	L	7	N	0.25	30	3.5	50.5	8.9	2050	CY	1.00	2	28	28	2.58
20	60	20	CL	5	No	0	0	0	0	7.4	1919.41	CY	2.00	5	28	28	4.93
67	33	0	No	0	A	1	30	0.8	600	13.35	1900	CY	2.00	5	49	28	4.012

Table 2 Statistical information of data set for compressive strength of rammed earth

	Count	Mean	Std	Min	25%	50%	75%	Max
x1	382	29.78623	23.85786	4.5	11.94	21	37.5	100
x2	382	56.99136	23.26134	0	40	61.5	71.95	93.5
x3	382	13.54885	14.75998	0	0	10	20	49.76
x5	382	5.829843	5.14576	0	0	5	10	20
x7	382	0.209398	0.72518	0	0	0	0	5
x8	382	5.082461	12.12751	0	0	0	0	50
x9	382	0.220575	0.756744	0	0	0	0	4.5
x10	382	24.55794	94.58026	0	0	0	0	600
x11	382	15.23202	8.675183	5.3	10.1	12.4	18.475	43
x12	382	1864.844	191.2936	1315	1720	1890	1985	2315
x14	382	1.728816	0.989535	1	1	2	2	10
x15	382	3.748691	1.726643	1	3	3	5	15
x16	382	31.18848	20.97095	2	28	28	38	120
x17	382	25.3822	18.03774	2	10	28	28	120
y	382	2.994916	2.306959	0.187	1.3025	2.27	4.21175	12.45

stabilizer x4, stabilizer content x5 (%), type of fiber x6, fiber content x7 (%), fiber properties including length x8 (mm), diameter x9 (mm), and tensile strength x10 (MPa), moisture content x11 (%), dry density x12 (kg/m3), specimen shape (cylinder (CY), cube (CU), prism (PR)) x13, specimen's aspect ratio x14 (height/width), number of layers x15, the age of the specimen until conducting the compressive test x16 (day), and the duration of curing for each specimen x17 (day) are provided as features that can affect the compressive strength y (MPa) of rammed earth in dry condition. It is worth mentioning that both chemical (cement (C), lime (L), the combination of cement and lime (CL), the combination of cement and additives such as super-plasticizers (CA)) and natural (cow dung (CD)) stabilizers are considered in the dataset. Besides, the natural (N) (e.g., straw, coir, date palm, and Curaua) and artificial (A) (e.g., polypropylene and waste tire textile) are the types of fibers used in the dataset.

After gathering and preparing the dataset, it is essential to extract the statistical information from the dataset and then visualize it for better interpretation. Table 2 shows the statistical information of the dataset.

Furthermore, before determining the correlations between the features and targets, it is necessary to ensure that there are no missing data, outliers, or noise. By coding in Python, it was confirmed that there are no missing data, as the sum of is null values is zero. Moreover, to check and evaluate outliers, the boxplot of each feature and target was compared to z-score values. In the existing dataset, no values were identified as outliers.

The next step is to normalize the dataset. However, due to the existence of nominal and qualitative data, data normalization and the subsequent process of modeling may be affected by this discrepancy in the data. To address this problem,

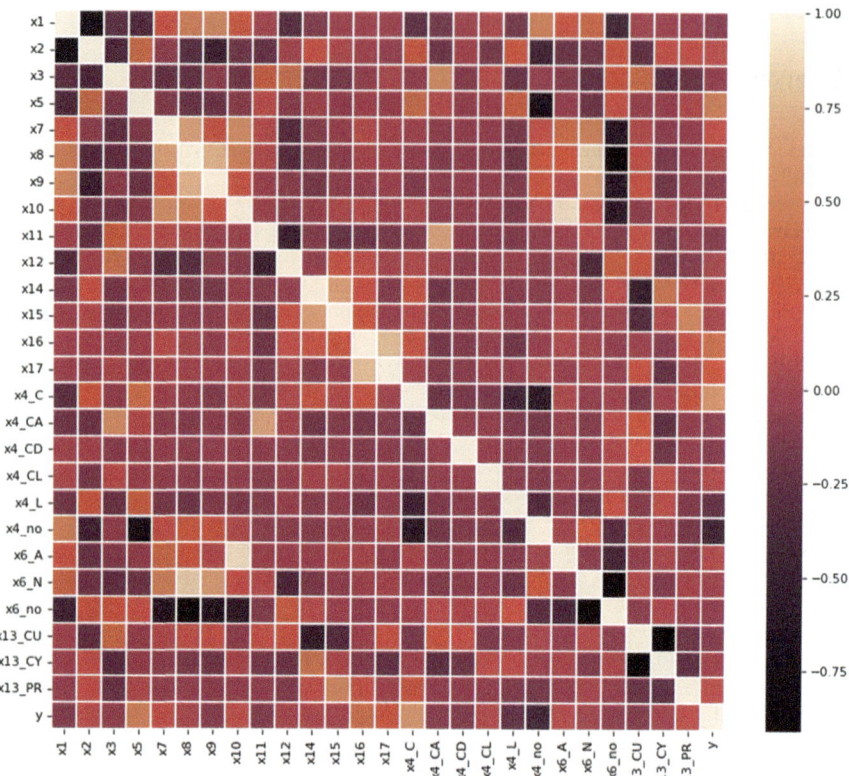

Fig. 5 The correlations between the feature and target values

encoding such data could be a viable solution. Dummy encoding is one of the solutions to appropriately encode the nominal data [25]. Dummy encoding is implemented for the nominal data, after which all the data is normalized using the Min-Max Scaler.

A Pearson correlation was conducted between the normalized features and the target. The Pearson correlation is an effective tool to represent the correlations between the feature and target values. Figure 5 depicts the Pearson correlations of the dataset.

Based on Fig. 5, compressive strength is directly affected by factors including the type and content of stabilizer, as well as curing time. On the other hand, a higher amount of fines can reduce the compressive strength of rammed earth, especially in cases involving cement-stabilized rammed earth.

4.2 Modeling Using ANN and DT

Modeling the scaled dataset is conducted using two machine learning methods, namely artificial neural networks (ANN) and decision trees (DT). The hyperparameters of these two techniques are tuned and optimized using a combination of grid search and cross-validation techniques. Moreover, in ANN, the gradient descent technique is implemented to optimize the rate of change of the weights in the neural networks. After modeling, the performance of the model is evaluated using the R-squared metric. The R-squared values for ANN and DT are 74% and 81%, respectively. The uncertainties in the nature of the soil led to the differences between these two models. Improving the performance of these models could involve employing optimization methods and expanding the dataset. Additionally, integrating additional factors such as soil chemical composition could be feasible.

5 Discussion

Rammed earth, a sustainable material, is less noticed than concrete despite its potential for building resilient societies. With the sustainable development goals (SDGs) unlikely to be met by 2030, focusing on adaptation is crucial for progress. To shed light on this, according to a report from the UN in 2023, approximately 1.1 billion people live in slums or slum-like conditions, and this number is expected to rise to more than 2 billion in the next 30 years. Moreover, air pollution issues are not only confined to urban areas but also affect rural areas [3]. Accordingly, serious policies should be implemented in this regard. Using earth-based materials such as rammed earth and compressed earth blocks (CEBs) could be applicable in these situations as these kinds of buildings are cost-effective and, in some cases, have a negative carbon footprint. Furthermore, with the increase in the world's mean temperature, the majority of wildfires and fire-related disasters will occur [3]. Despite the existence of sustainable building materials such as timber, earth-based materials would be more viable compared to other sustainable building materials in this regard.

Regarding the correlations between the sustainability properties of rammed earth and its mechanical properties, such as compressive strength, some information can be extracted from Fig. 5. This figure suggests that in rammed earth construction, using less cement and more natural materials such as fibers and eco-friendly stabilizers can be sustainable. The type and amount of stabilizer greatly influence compressive strength, with cement being the most effective. However, sustainable alternatives like lime or cow dung also show promise. More research is needed to better understand these natural stabilizers' effects on both sustainability and mechanical strength.

Further research into rammed earth can enhance AI models, making them more reliable. Since rammed earth construction is energy intensive due to compaction, AI can optimize resource use and assist engineers in making efficient decisions in time-sensitive projects.

6 Conclusion

This research article first evaluates the public's opinion regarding the rammed earth technique through sentiment analysis using an NLP-based tool. As a result of the sentiment analysis, the public's attitude toward rammed earth was pleasant and positive. According to the word cloud generated by the analysis, most of the comments are related to the sustainability and stabilization perspective of rammed earth.

Furthermore, through a bibliometric study, the trend of rammed earth research was compared with other construction building materials such as concrete, masonry, compressed earth blocks (CEB), and timber. The results reveal that despite the sustainability capabilities of rammed earth and CEB, their trend is not as high as other materials. Accordingly, an increasing trend in research is needed to make these sustainable and historical materials and techniques applicable. More importantly, this increase in research is expected in the numerical and soft computing perspectives of these materials, as these techniques are required for making sustainable decisions to save time and cost.

Lastly, to address the gaps identified from the bibliometric study, machine learning methods, namely artificial neural networks (ANN) and decision trees (DT), were used, along with data science principles, to predict the compressive strength of rammed earth by considering various features including soil composition (fine, sand, and gravel), type of stabilizer, stabilizer content, type of fiber, fiber content, fiber properties including length, diameter, and tensile strength, moisture content, dry density, specimen shape (cylinder, cube, prism), specimen's aspect ratio (height/width), number of layers, the age of the specimen until conducting the compressive test, and the duration of curing for each specimen. According to the Pearson correlation, the type and content of stabilizer, the curing time, and the age of the specimen highly affect the compressive strength. On the contrary, a higher number of fines in soil composition can negatively affect the compressive strength of rammed earth, especially in cement-stabilized rammed earth. Moreover, the R-squared values of both ANN and DT models were lower than 82%. This performance can be enhanced by utilizing optimization techniques and a larger dataset. Moreover, implementing other features such as soil chemical composition would be viable.

Future studies should focus on both the experimental and numerical perspectives of rammed earth materials. The increasing trend in research can lead to the generation of a large amount of data, thereby resulting in robust AI models.

References

1. Sachs, J.D., et al.: Six transformations to achieve the sustainable development goals. Nat. Sustain. **2**(9), 805–814 (2019)
2. Stafford-Smith, M., et al.: Integration: the key to implementing the sustainable development goals. Sustain. Sci. **12**, 911–919 (2017)
3. Nations, U.: The Sustainable Development Goals Report 2023: Special Edition (2023)
4. Ahmed, N., et al.: Impact of sustainable design in the construction sector on climate change. Ain Shams Eng. J. **12**(2), 1375–1383 (2021)
5. Sharma, M., et al.: Limestone calcined clay cement and concrete: a state-of-the-art review. Cem. Concr. Res. **149**, 106564 (2021)
6. Zandifaez, P., et al.: AI-assisted optimisation of green concrete mixes incorporating recycled concrete aggregates. Constr. Build. Mater. **391**, 131851 (2023)
7. Gaudenzi, E., et al.: The use of lignin for sustainable asphalt pavements: a literature review. Constr. Build. Mater. **362**, 129773 (2023)
8. Gomaa, M., et al.: Automation in rammed earth construction for industry 4.0: precedent work, current progress and future prospect. J. Clean. Prod. **398**, 136569 (2023)
9. Reddy, B.V., Reddy, B.V.: Compressed Earth Block & Rammed Earth Structures. Springer (2022)
10. Tripura, D.D., Singh, K.D.: Behavior of cement-stabilized rammed earth circular column under axial loading. Mater. Struct. **49**, 371–382 (2016)
11. Ávila, F., Puertas, E., Gallego, R.: Mechanical characterization of lime-stabilized rammed earth: lime content and strength development. Constr. Build. Mater. **350**, 128871 (2022)
12. Darshan, H.C., Mamatha, K.H., Dinesh, S.V., Latha, B.M.: Effectiveness of cow dung for rammed earth application. In: Problematic Soils and Geoenvironmental Concerns: Proceedings of IGC 2018, pp. 493–502. Springer, Singapore (2021)
13. Toufigh, V., Karamian, M.H., Ghasemalizadeh, S.: Study of stress–strain and volume change behavior of fly ash-GBFS based geopolymer rammed earth. Bull. Eng. Geol. Environ. **80**, 6749–6767 (2021)
14. Liu, L., et al.: Study on the mechanical properties of modified rammed earth and the correlation of influencing factors. J. Clean. Prod. **374**, 134042 (2022)
15. Wankhade, M., Rao, A.C.S., Kulkarni, C.: A survey on sentiment analysis methods, applications, and challenges. Artif. Intell. Rev. **55**(7), 5731–5780 (2022)
16. North Carolina State University, Social Media Sentiment Visualization. https://www.csc2.ncsu.edu/faculty/healey/social-media-viz/production/. Last accessed 30 Mar 2024
17. Agarwal, A., Xie, B., Vovsha, I., Rambow, O., Passonneau, R.J.: Sentiment analysis of twitter data. In Proceedings of the workshop on language in social media (LSM 2011) (pp. 30–38) (2011)
18. Shen, C.-W., Luong, T.-H., Pham, T.: Exploration of social media opinions on innovation for sustainable development goals by topic modeling and sentiment analysis. In: Research and Innovation Forum 2020: Disruptive Technologies in Times of Change. Springer (2021)
19. Mustafa, Y.M.H., et al.: Analysis of unconfined compressive strength of rammed earth mixes based on artificial neural network and statistical analysis. Materials. **15**(24), 9029 (2022)
20. Narloch, P., et al.: Predicting compressive strength of cement-stabilized rammed earth based on SEM images using computer vision and deep learning. Appl. Sci. **9**(23), 5131 (2019)
21. Turco, C., et al.: Artificial neural networks to predict the mechanical properties of natural fibre-reinforced compressed earth blocks (CEBs). Fibers. **9**(12), 78 (2021)
22. Anysz, H., Narloch, P.: Designing the composition of cement stabilized rammed earth using artificial neural networks. Materials. **12**(9), 1396 (2019)
23. Anysz, H., et al.: Feature importance of stabilised rammed earth components affecting the compressive strength calculated with explainable artificial intelligence tools. Materials. **13**(10), 2317 (2020)

24. Kardani, N., et al.: Experimental study and machine learning aided modelling of the mechanical behaviour of rammed earth. Geotech. Geol. Eng. **40**(10), 5007–5027 (2022)
25. Jolly, S., Gupta, N.: Understanding and implementing machine learning models with dummy variables with low variance. In: International Conference on Innovative Computing and Communications: Proceedings of ICICC 2020, Volume 1, pp. 477–487. Springer, Singapore (2021)

Towards Circular and Sustainable Insulation Solutions: Resolving Uncertainty in the Thermal Conductivity of Mycelium-Based Composites (MBCs)

Joni Wildman 🆔, Andrew Shea 🆔, Daniel Henk 🆔, Martin Naido, and Pete Walker 🆔

Contents

1 Introduction

A key challenge facing the future of the built environment is the urgent need to reduce the energy consumption of buildings, from material extraction to demolition [1]. The United Nations Environment Programme (UNEP) identifies that buildings are responsible for 40% of energy usage, 25% of water consumption and 40% of resource use globally. The energy performance of buildings in their operational stage is only one aspect of the environmental impact of the built environment. Life cycle

---

J. Wildman (✉)
Department of Architecture and Civil Engineering, University of Bath, Bath, UK

Department of Life Sciences, Milner Centre for Evolution, University of Bath, Bath, UK
e-mail: jlw89@bath.ac.uk

A. Shea · M. Naido · P. Walker
Department of Architecture and Civil Engineering, University of Bath, Bath, UK

D. Henk
Department of Life Sciences, Milner Centre for Evolution, University of Bath, Bath, UK

© The Author(s) 2025
M. Kioumarsi, B. Shafei (eds.), *The 1st International Conference on Net-Zero Built Environment*, Lecture Notes in Civil Engineering 237,
https://doi.org/10.1007/978-3-031-69626-8_45

assessment (LCA) is key to a more holistic consideration of building environmental performance. To address whole-life environmental concerns, there has been a shift in recent years toward the development and implementation of bio-based thermal insulators. These nature-based solutions are often plant-based (e.g. hemp, straw, sisal) and biodegradable [2–4]. To address waste streams generated in construction and other sectors the re-purposing of would-be waste products such as straw, wood waste, paper etc. into insulation materials is of importance in the transition to a circular economy. Where such materials can be bound together with natural materials, rather than, for example, plastic-based adhesives, their use and re-use would become better aligned to circular principles. Mycelium has been demonstrated to be a potentially suitable binder for plant-based and other waste materials [5]. Additionally, mycelium-based composites (MBCs) can adapt to regional resource availability which can substantially reduce transportation costs and associated carbon emissions [6, 7]. The thermal and physical properties of these materials are tunable through a selection of species, substrates, growth conditions and manufacturing techniques, thereby enhancing their utility in design optimisation [8, 9].

Mycelium is the vegetative part of a fungus comprised of a mass of branching, thread-like, filaments that form a network structure. As mycelium colonises a substrate it creates a denser, tougher, outer layer due to the exposure to air and moisture evaporation. The result is a lightweight, porous structure with notable thermal insulation properties [10]. Figure 1 presents the stages of the production process and Fig. 2 shows a representative MBC used in this study and its interior and exterior, clearly depicting the distinct core and skin layers.

The reliable assessment of thermal conductivity and the comprehensive characterisation of thermal properties is important for their practical application and wider acceptance. Furthermore, the need to compare the results of thermal measures-.

ments between laboratories and materials databases etc. is self-evident. There are two common approaches to measuring the thermal conductivity of materials, namely, steady-state and non-steady-state [11]. Guarded Hot Plate (GHP) and Heat Flow Meter (HFM) methods are standard steady-state methods [12], widely used in the testing of building materials. Steady-state methods determine thermal

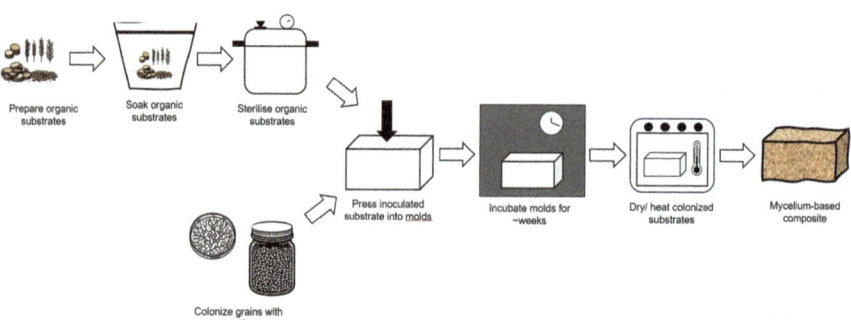

Fig. 1 Mycelium-Based Composite (MBC) manufacturing process

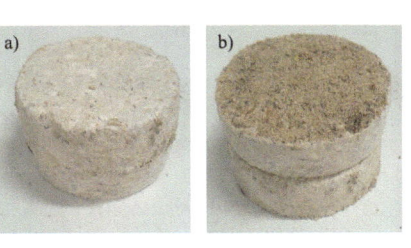
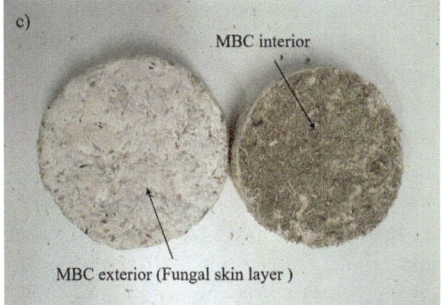

Fig. 2 Two MBC cylinders with an exterior surface removed, as used in this study. In (**a**) their interior surfaces are touching (skin-off configuration), and in (**b**) their exterior surfaces are touching (skin-on configuration). In (**c**) the exterior (fungal skin) and interior are shown. Cylinders are 100 mm in diameter and 30 mm in height

conductivity directly from the proportionality between heat flow and an applied temperature difference under thermal equilibrium. Measurement accuracy requires large samples and a test duration in the order of hours [13]. Due to the requirement for thermal equilibrium, throughout the test specimen, the method is largely insensitive to heterogeneity at the centimetre, or smaller, scale but specimens should be dry as the movement of water vapour will restrict the ability to reach thermal equilibrium [14].

The steady-state thermal conductivity of MBCs has been reported in [15–19]. Schmidt et al. [15] and Wimmers et al. [18] employed the heat flow meter technique, recording thermal conductivity values of MBCs, made using *Fomes fomentarius* and hemp shives, and *Polyporus arcularius, Trametes suaveolens* and *Trametes pubescens* each with birch sawdust, respectively, in the ranges of 0.041–0.046 W/mK and 0.051–to 0.055 W/mK, respectively. Zhang et al. utilised the divided bar method, yielding a measurement of 0.069 W/mK for an MBC based on rye berries with *Pleurotus ostreatus*. Dias et al. and Travaglini et al. both used the Guarded Hot Plate method to measure MBC thermal conductivity of MBCs made using *Ganoderma resinaceum* with Miscanthus and *Ganoderma lucidium* and *Laetiporus sulphureus* respectively with results in the range of 0.088 to 0.104 W/mK and 0.053 to 0.077 W/mK, respectively.

Transient, non-steady, methods measure the feedback response to a short pulse of heat and thus temperature distribution throughout the specimen varies with time. The test duration can be relatively short, as reaching equilibrium is not required. However, computing the heat conduction equation is more complex as it requires resolving time-dependent heat flow Eqs. [20]. Test specimen sizes can be much smaller than those required for GHP or HFM testing, which is a significant advantage in research and development laboratories. The Hot Disk (HD) device, a transient plane source (TPS) technique, is a popular tool for rapid and accurate thermal conductivity measurement [12]. In the Hot Disk device, the sensor performs the function of the heat source and a temperature measuring device. Although single-sided testing is possible, measurements are optimised when the sensor is sandwiched between two pieces of test specimen material. In this arrangement, the two sample specimens sandwiching the sensor must be of sufficient diameter and depth to ensure that the 'thermal wave' does not reach any boundary of the test specimen during the transient recording period. Studies using the Hot Disk have presented systematic errors when applied to low-conductivity insulation materials [21–23]. These errors are attributed to the idealised model assumptions that the Hot Disk employs for data analysis. Comparisons with steady-state techniques have shown reproducibility errors of up to 30% due to inhomogeneity and poor thermal contact [24]. Analysis using an isotropic model can overestimate thermal conductivity by around 25% and Colinart et al. conclude that even with the anisotropic functionality, the Hot Disk method's sensitivity to volumetric heat capacity measurements remains a source of difficulty when seeking to determine thermal conductivity for heterogeneous bio-based materials [24].

Schritt et al. used a dynamic hot-wire method to measure thermal conductivity at the centre and the edge of MBCs [25]. Measurements showed significant species-dependent differences in the fungal skin thermal conductivity. However, the device settings are not presented and the potential influence of the room environment on the probe at the material edge is not discussed. Whilst it is stated that for the edge measurement, the hot wire was placed between two fungal skins it is unclear what thickness of material was present.

The characteristic skin layer of the MBC has properties such as density and porosity that differ from the bulk core of the specimen. Consequently, the size of the specimen alters the ratio of skin to core material, which influences the measured thermal conductivity. Accordingly, measured thermal conductivity may not be representative of the entire composite, which is significant in the context of scaling up MBCs to the size required for insulation in building elements.

In the current body of research, a total of 17 studies reporting thermal conductivity measurements of MBCs are documented; these present a range of λ values from 0.0321 to 0.180 W/mK. These studies include 12 distinct fungal species and 19 substrates. The lowest recorded thermal conductivity value for an MBC was observed in a composite comprising *Pleurotus ostreatus* mycelium with ash wood chips. The authors attribute this low conductivity to the fungus-substrate combination. However, this conclusion ignores the potentially significant role of the measurement technique employed and the experimental arrangement.

In this study, we seek to (i) evaluate the influence of the fungal skin of MBCs and (ii) undertake a direct, same specimen, comparison of Hot Disk (HD) and Heat Flow Meter (HFM) thermal conductivity measurements. The uncertainty in specimen thermal conductivity introduced by the MBC skin layer is evaluated through HD measurements of 12 MBC specimens. Each specimen was made using the same fungal strain derived from the species (*Trametes versicolor*) and a range of cellulose substrates and measured both with and without the skin layer. The samples range in bulk density from 157 to 233 kg/m^3 and the test arrangement employed the HD anisotropic measurement module. The anisotropic measurement setup allows for the assessment of thermal conductivity in both the radial and axial directions, offering a comprehensive analysis of the specific influence of the fungal skin and its influence on the specimen's thermal conductivity. The Heat Flow Meter measurements were taken on a single large specimen, which was then cut to enable measurement with the HD apparatus. The values obtained are compared and an effective heat capacity is obtained which can be used in HD measurements to allow for comparison with steady-state results.

2   Methods

2.1   Sample Preparation

The mycelium samples used in this work were provided by Mykor Ltd. The samples were prepared using the following methodology: Mycelium spawn is produced by removing a sample from a mushroom fruit, placing it in an agar dish, and allowing the mycelium to colonise the agar. Cellulose-based waste products are sterilised in a steam autoclave. Mycelium spawn is then weighed and added to the sterilised cellulose. The cellulose and spawn are churned before being used to fill moulds. The spawn is left to colonise the substrate in the moulds for one week. The composites are then de-moulded and heated to 70 °C to stop hyphae growth [26].

  The moulds used in this study are cylinders of diameter 100 mm and height 30 mm, and a 500 mm by 500 mm by 105 mm panel with a density 172 kg/m^3 of which a 110 mm by 110 mm by 80 mm cuboid is cut from, which is cut in half for HD measurement. Each sample was formed from a cellulose-based substrate, with samples 1–12 varying in composition with different additives with a density in the range of 157–233 kg/m^3. Figure 2 shows a representative cylinder with the fungal skin present and removed. Sample 13 was obtained by cutting a 110 mm × 110 mm × 40 mm section from the larger sample to enable measurement with the HD is shown in Fig. 4a.

2.2   Hot Disk

The 12 cylindrical samples and the 110 mm × 110 mm × 40 mm cuboid sample (sample 13) were subject to thermal testing using the Hot Disk equipment according to ISO 22007-2:2022 [27]. Samples were tested with the skin layer on and again with the skin layer off. For samples 1–12, the skin layer was removed by slicing it off, and for sample 13 the skin layer was absent on one face as it was cut from the larger sample. Five thermal analysis tests were performed on each sample using a Hot Disk TPS 3500 device in a controlled environment at 23 °C. The samples were allowed to

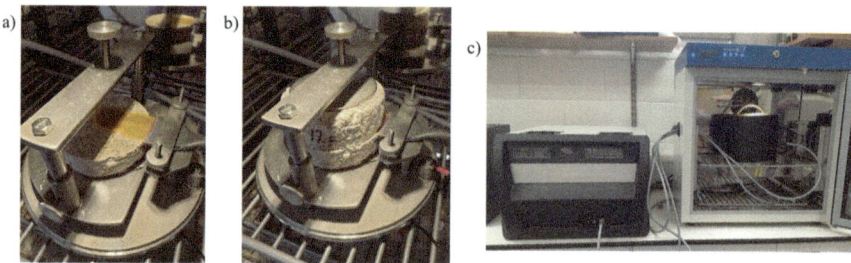

Fig. 3 (**a**) Hot Disk set up for skin-off measurements (an identical sample on placed on top of the sensor). (**b**) Hot Disk set up with sensor sandwiched between identical samples. (**c**) Hot Disk apparatus alongside incubation unit to maintain a stable temperature during measurements

stablise before testing. Between measurements, the samples were allowed to cool for 320 minutes to equilibrate back to 23 °C. The tests were conducted using a 14.6 mm Kapton Sensor. The thermal tests used 40 W with a duration of 640 seconds, ensuring a mean deviation of less than 0.0001 K in all cases. The mean deviation provides an indicator of the quality of the data, where a value of 0.0001 K or less is considered acceptable 'noise'. The measurements were conducted in the anisotropic module of the Hot Disk TPS 3500 which performs measurements in the radial and axial directions. To perform anisotropic thermal conductivity measurements, the volumetric heat capacity (ρC_ρ) was measured using the isotropic thermal conductivity set-up for the HD and used as an input in the anisotropic module to determine the ratio between axial and radial thermal conductivity. Figure 3 presents the setup for the HD measurements conducted in this experiment.

The density of the samples was measured to be in the range of 157 kg/m^3 to 244kgm/^{3}, with a weakly positive correlation between density and isotropic thermal conductivity observed (R^2 = 0.33).

2.3 Heat Flow Meter

A single specimen (sample 13) of 500 × 500 × 105 mm was produced for Heat Flow Meter testing according to ASTM-C518 [28] (Fig. 4b), and subsequently cut to form two smaller samples for Hot Disk testing, as described earlier. The specimen was stored in the laboratory at room conditions (approx. 20 °C and 50% RH) for three days prior to testing. The mass of the sample as tested was 4523 g, giving a density of 172 kg/m^3. The test was conducted at mean temperatures of 10 °C and 20 °C with a temperature difference of 20 K in both cases and the test duration was 7 hours for each set point temperature.

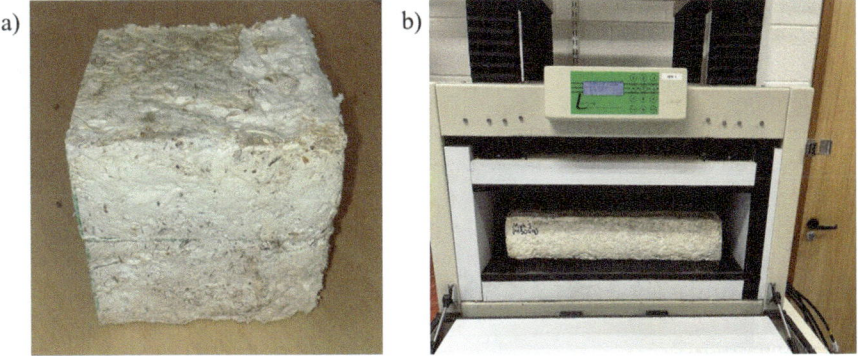

Fig. 4 (**a**) Two 110 × 110 × 40 mm samples cut from larger sample 13. (**b**) Heat Flow Meter testing of sample 13

3 Results and Discussion

3.1 Skin on vs Skin Off

Figure 5 presents the results from the measurements of the anisotropic thermal conductivity measurements of the samples with and without the skin layer in both the axial and radial direction. The difference in λ_{axial} vs λ_{radial} is likely attributable to the material structure, with the difference between the two diminishing at higher densities. In the axial direction, 10 out of 12 samples tested showed a higher thermal conductivity with skin on than with skin off, whereas radially 8 out of 12 samples had a higher measured thermal conductivity with the skin on. This inconsistency suggests that the specific composition of each sample and the characteristics of the fungal skin influence thermal performance in a way that requires further investigation into the material properties of each sample before assumptions can be made about whether the fungal skin layer will lead to an increased or decreased thermal conductivity for any specific MBC. Overall, these results show the presence of a skin layer on the samples introduces additional discrepancies in the measurement of thermal conductivity. This layer, while sometimes removable, presents a challenge in accurately determining the thermal properties of materials, particularly for thin samples. For thicker specimens, the removal of this skin layer may be feasible and can lead to more accurate representations of the material's inherent thermal conductivity, and the thermal conductivity of similar scaled-up samples. However, this is not always possible, especially for thinner samples, where the skin layer constitutes a more significant proportion of the sample's overall structure. Consequently, the thermal conductivity values obtained for samples with the skin layer intact may not accurately reflect the true thermal properties of the material. This limitation underscores the necessity of considering the potential impact of the skin layer when interpreting thermal conductivity measurements, especially when evaluating small prototypes before scaling up dimensions.

Effect of Fungal Skin on Thermal Conductivity Measurement of Mycelium Based Composites

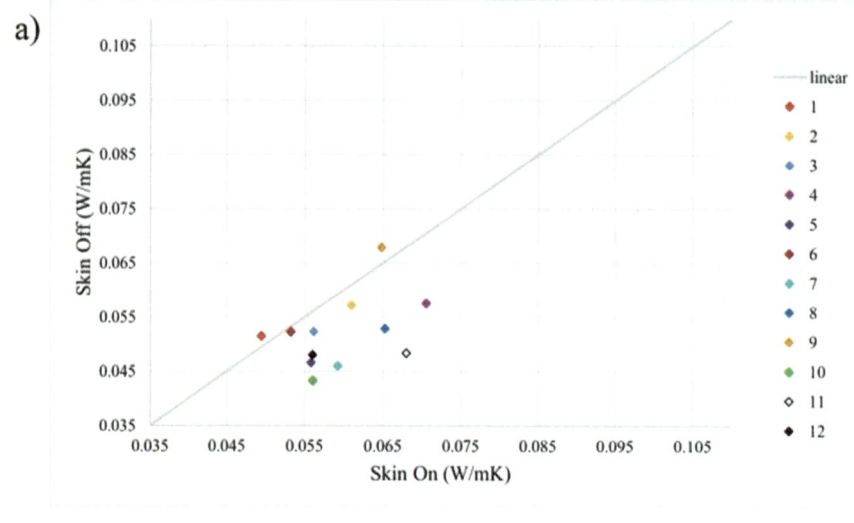

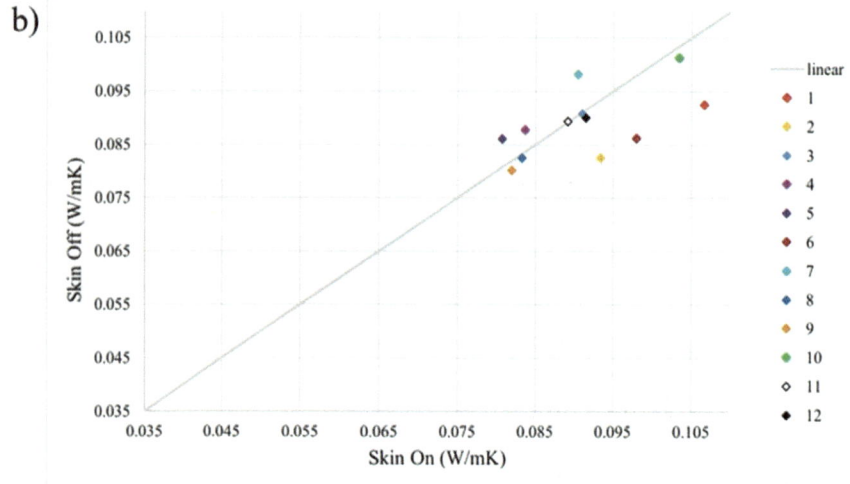

Fig. 5 (**a**) λ_{axial} of MBC samples measured using the Hot Disk Method with and without the fungal skin layer. All errors <0.001 W/mK. (**b**) λ_{radial} of MBC samples measured using the Hot Disk Method with and without the fungal skin layer. All errors <0.003 W/mK

3.2   Heat Flow Meter vs Hot Disk

Measured using the Hot Disk anisotropic module, the λ_{axial} of sample 13 with skin on was (0.0775 ± 0.0003) W/mK and with skin off was (0.0695 ± 0.0002) W/mK. Measured using the HFM λ_{HFM} of sample 13 was (0.05536 ± 0.00003) W/mK at 10 °C, and (0.06260 ± 0.00003) W/mK at 20 °C. The isotropic measurement of thermal conductivity (skin on) was (0.0773 ± 0.0001) W/mK which is within error of λ_{axial}, we assume $\lambda_{axial} = \lambda_{HFM}$. Comparing these values, λ measured using HFM at 10 °C, the Hot Disk overestimates thermal conductivity by 40% for the sample with skin on, and 26% for the sample with skin off. In comparison to the HFM results at 20 °C, the Hot Disk overestimates thermal conductivity by 24% for the sample with skin on, and 11% for the sample with skin off. This observation highlights the inherent challenge in comparing the thermal conductivities of samples when different measurement methodologies are employed. Furthermore, the presence of fungal skin introduces an additional layer of discrepancy, particularly pronounced in the case of small samples.

Volumetric heat capacity (ρC_ρ) is used as an input in the Hot Disk anisotropic thermal conductivity measurement. This value can be measured using the isotropic thermal conductivity set-up. The anisotropic measurement is sensitive to ρC_ρ and therefore by using an effective ρC_ρ the values of thermal conductivity measured using the HFM and the Hot Disk can be compared. A specified value of ρC_ρ can be used in the isotropic set-up of the Hot Disk; however, the isotropic measurement is far less sensitive to ρC_ρ with an increase in ρC_ρ of 1% yielding a decrease in thermal conductivity of 0.39%, as calculated by opting to override the default setting of measuring ρC_ρ during the measurement and using varying ρC_ρ values as inputs. By sampling values of ρC_ρ between 0.3 and − 0.5 J/m³K, the value of ρC_ρ that yielded λ_{HFM10} was approximately 0.42 J/m³K for sample 13 with skin on, and 0.39 J/m³K for sample 13 with skin off, and the value of ρC_ρ that yielded λ_{HFM20} was approximately 0.38 J/m³K for sample 13 with the skin on and 0.34 J/m³K for sample 13 with skin off. Figure 6 presents the results of this calibration. Sensitivity analysis

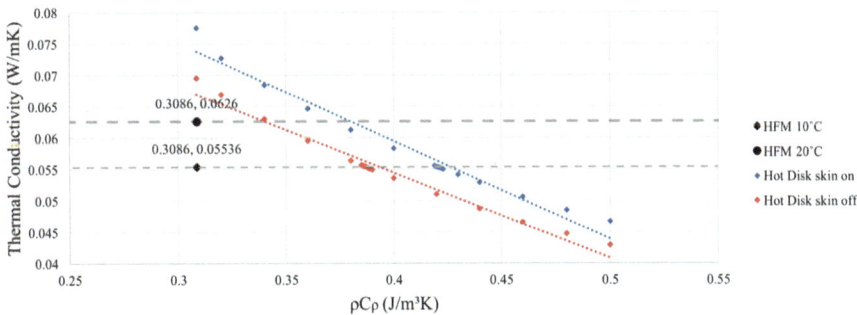

Fig. 6 Axial thermal conductivity of sample 13 (skin on and skin off) as measured using the Hot Disk anisotropic module for various input values of ρC_ρ between 0.3–0.5 J/m³K, and the thermal conductivity measurement using the HFM (assuming a value of ρC_ρ of 0.308 J/m³K as measured using the isotropic Hot Disk set-up)

based on the slope of the linear regression line that describes the relationship between ρC_ρ and λ_{axial}, a 1% increase in ρC_ρ yields a 1.09% increase in λ_{axial} with this relationship being approximately linear over the range of ρC_ρ sampled.

4 Conclusions

In this study, we aimed to resolve some of the uncertainties associated with measuring the thermal conductivity of mycelium-based composites to facilitate their use as sustainable insulation materials. To do this we compared the thermal conductivity measurements of MBC samples with and without the fungal skin layer to assess its influence on thermal conductivity measurement using the Hot Disk apparatus. We then compared the thermal conductivity measured using transient and steady-state techniques by using the heat flow meter to measure the thermal conductivity of a large sample before cutting the sample into smaller dimensions to measure thermal conductivity using the Hot Disk. To calibrate the Hot Disk measurements to the Heat Flow meter we determined an effective volumetric heat capacity for the MBC. The following main conclusions can be drawn:

– The presence of the fungal skin influences thermal conductivity measurements of MBCs. In 10 out of 12 samples measured, the presence of the fungal skin leads to an increase in measured thermal conductivity compared to without the fungal skin. Consideration is therefore needed when measuring the thermal conductivity of small samples with the fungal skin present, as this will influence the measurement and will not scale up to larger samples where the thickness of the skin assumes a different dimension.
– The presence or absence of the fungal skin is not enough to explain the large discrepancy between measurements conducted with the Hot Disk apparatus (transient methodology) and the heat flow meter (steady-state methodology). The choice of measurement technique is the largest source of uncertainty in this instance.
– To allow for comparison between the Hot Disk apparatus results and the heat flow meter results, an effective volumetric heat capacity can be used with the anisotropic Hot Disk set-up measuring axial thermal conductivity. We found that the effective ρC_ρ that allowed for this comparison was 0.42 J/m^{3}K with the fungal skin off, and 0.39 J/m^{3}K with the fungal skin on.

Acknowledgements The authors gratefully acknowledge Mykor Ltd. for their support in material fabrication and development, with special thanks to Fred Robinson, Max Frost, Valentina Dipietro, Olivia Page, and Annie Ferrari. Many thanks to the laboratory and technical staff in both the Faculty of Engineering & Design and the Faculty of Science at the University of Bath. Dr. Shea and Mykor Ltd. gratefully acknowledge the funding support of InnovateUK under grant agreement number TS/X004236/1.

References

1. Asdrubali, F., D'Alessandro, F., Schiavoni, S.: A review of unconventional sustainable building insulation materials. Sustain. Mater. Technol. **4**, 1–17 (2015)
2. J. Jefferson Andrew and H.N. Dhakal. Sustainable biobased composites for advanced applications: recent trends and future opportunities—a critical review. Compos. Part C Open Access, 7: 100220, March 2022.
3. Raja, P., Murugan, V., Ravichandran, S., Behera, L., Mensah, R.A., Mani, S., Kasi, A.K., Balasubramanian, K.B.N., Sas, G., Vahabi, H., Das, O.: A review of sustainable bio-based insulation materials for energy-efficient buildings. Macromol. Mater. Eng. **308**, 2300086 (2023)
4. Bourbia, S., Kazeoui, H., Belarbi, R.: A review on recent research on bio-based building materials and their applications. Mater. Renew. Sustain. Energy. **12**(2), 117–139 (2023)
5. Alemu, D., Tafesse, M., Mondal, A.K.: Mycelium-based composite: the future sustainable biomaterial. Int. J. Biomater., 1–12 (2022)
6. Alves, R.M.E., Alves, M.L., Campos, M.J.: Morphology and Thermal Behaviour of New Mycelium-Based Composites with Different Types of Substrates Lecture Notes in Mechanical Engineering, pp. 189–197. Springer International Publishing, Cham (2020)
7. Bansfield, D., Spilling, K., Mikola, A., Piiparinen, J.: Growth of fungi and yeasts in food production waste streams: a feasibility study. BMC Microbiol. **23**(1), 328 (2023)
8. Elsacker, E., Vandelook, S., Brancart, J., Peeters, E., De Laet, L.: Mechanical, physical and chemical characterisation of myceliumbased composites with different types of lignocellulosic substrates. PLOS ONE. **14**(7), e0213954 (2019)
9. Yang, L., Park, D., Qin, Z.: Material function of myceliumbased bio-composite: a review. Front. Mater. **8**, 737377 (2021)
10. Lai, J., Walczyk, D., Mooney, L., Putney, S.: Manufacturing of mycelium-based biocomposites. In: International SAMPE Technical Conference, pp. 1944–1955 (2013)
11. Speyer, R.F.: Thermal Analysis of Materials. CRC Press, New York (1994)
12. Yüksel, N.: The review of some commonly used methods and techniques to measure the thermal conductivity of insulation materials. In: Insulation Materials in Context of Sustainability. IntechOpen (2016)
13. Kraemer, D., Chen, G.: A simple differential steady-state method to measure the thermal conductivity of solid bulk materials with high accuracy. Rev. Sci. Instrum. **85**, 025108 (2014)
14. Langlais, C., Hyrien, M., Klarsfeld, S.: Moisture migration in fibrous insulating materials under the influence of a thermal gradient and its effect on thermal resistance. In: Moisture Migration in Buildings, ASTM STP 779, pp. 191–206 (1982)
15. Schmidt, B., Freidank-Pohl, C., Zillessen, J., Stelzer, L., Guitar, T.N., Lühr, C., Müller, H., Zhang, F., Hammel, J.U., Briesen, H., Jung, S., Gusovius, H.-J., Meyer, V.: Mechanical, physical and thermal properties of composite materials produced with the basidiomycete fomes fomentarius. Fungal Biol. Biotechnol. **10**(1), 22 (2023)
16. Zhang, X., Jianying, H., Fan, X., Xiong, Y.: Naturally grown mycelium-composite as sustainable building insulation materials. J. Clean. Prod. **342**, 130784 (2022)
17. Dias, P.P., Jayasinghe, L.B., Waldmann, D.: Investigation of mycelium-miscanthus composites as building insulation material. Results Mater. **10**, 100189 (2021)
18. Wimmers, G., Klick, J., Tackaberry, L., Zwiesigk, C., Egger, K., Massicotte, H.: Fundamental studies for designing insulation panels from wood shavings and filamentous fungi. BioResources. **14**(3), 5506–5520 (2019)
19. Sonia Travaglini, C.K., Dharan, H., Ross, P.G.: Thermal properties of mycology materials. Mater. Sci. Environ. Sci. (2015)
20. He, Y.: Rapid thermal conductivity measurement with a hot disk sensor: part 1. Theoretical considerations. Thermochim. Acta. **436**(1–2), 122–129 (2005)
21. Zheng, Q., Kaur, S., Dames, C., Prasher, R.S.: Analysis and improvement of the hot disk transient plane source method for low thermal conductivity materials. Int. J. Heat Mass Transf. **151**, 119331 (2020)

22. Jannot, Y., Degiovanni, A.: Thermal Properties Measurement of Materials. Wiley, Hoboken (2018)
23. Jannot, Y., Degiovanni, A., Félix, V., Bal, H.: Measurement of the thermal conductivity of thin insulating anisotropic material with a stationary hot strip method. Meas. Sci. Technol. **22**(3), 035705 (2011)
24. Colinart, T., Pajeot, M., Vinceslas, T., Hellouin De Menibus, A., Lecompte, T.: Thermal conductivity of biobased insulation building materials measured by hot disk: Possibilities and recommendation. J. Build. Eng. **43**, 102858 (2021)
25. Schritt, H., Vidi, S., Pleissner, D.: Spent mushroom substrate and sawdust to produce mycelium-based thermal insulation composites. J Clean Prod. **313**, 127910 (2021)
26. Mykor. https://www.mykor.co.uk/. Last accessed 30 Nov 2023.
27. International Organization for Standardization. Plastics—determination of thermal conductivity and thermal diffusivity—part 2: Transient plane heat source (hot disc) method. https://www.iso.org/standard/81244.html (2022). ISO 22007-2.
28. ASTM International.: Standard test method for steady-state thermal transmission properties by means of the heat flow meter apparatus (2024).

A Comparative Life Cycle Assessment of Ordinary Portland Cement, Limestone Calcined Clay Cement and Mining Waste Marl Cement

Yasmine Rhaouti, Yassine Taha, and Mostafa Benzaazoua

Contents

1 Introduction

Cement production is responsible for around 8% of global CO_2 emissions [1]. Two-thirds of these emissions come from limestone calcination during clinker production [2]. To address this issue, one of the main sustainable strategies targeted by cement manufacturers is that of clinker substitution by supplementary cementitious materials (SCMs) [3]. SCMs can either be sourced naturally (clay, natural zeolite, limestone) or industrially (fly ash, granulated blast furnace slag, glass powder) [4]. They have proven to be effective, with limestone calcined clay, in particular, yielding positive results: a 45% substitution of clinker enables a 40% reduction of CO_2 emissions, with the advantage of clay being a widely available material [5, 6]. However, clay is also a valuable resource for other industries, such as agriculture [7]. For that reason, reusing clay waste instead of extracting natural clay to substitute clinker could be a balanced solution tackling both the issue of CO_2 emissions and of resource availability.

Y. Rhaouti (✉) · Y. Taha · M. Benzaazoua
Mohammed VI Polytechnic University, Benguerir, Morocco
e-mail: yasmine.rhaouti@um6p.ma; Yassine.taha@um6p.ma

M. Kioumarsi, B. Shafei (eds.), *The 1st International Conference on Net-Zero Built Environment*, Lecture Notes in Civil Engineering 237,
https://doi.org/10.1007/978-3-031-69626-8_46

551

In this light, previous studies have proposed the valorisation of mining waste marl as an SCM in cement [8–10]. Marl, a material rich in clay and lime, presents promising pozzolanic properties when calcined, thus enabling the production of mechanically performant calcined marl cement (CMC). It is particularly abundant in Moroccan phosphate mines due to the country's substantial phosphate mining activities. As Morocco endeavours to reduce its environmental impacts according to a planned sustainable transition [11], the valorisation of phosphate mining waste marl as clinker substitutes in cement will be a favourable way to stimulate the circularisation of the country's economy.

To monitor such efforts and help orientate decisions within this context, a life cycle assessment (LCA) can be used. Regulated by ISO 14040 and ISO 14044 standards, it is applied to evaluate, in a systematic manner, the environmental impacts of products or services throughout their life cycle.

In the available literature, there are many LCAs of various SCM-based cements, from naturally sourced [12–16] to industrially sourced ones [17, 18]. Several contexts have been explored: American, Brazilian, Indonesian, etc. Results of these LCAs have supported decision-making in several countries, such as Cuba, in which blended cement manufacturing has been encouraged by prolific research [19] as well as the establishment of new regulations (e.g. ASTM WK70466). As far as waste-clay cement is concerned, only one study has been conducted so far [15]. It compares the environmental impacts of Ordinary Portland Cement (OPC) and various LC3, including ceramic waste clay cement. A systematic review of the LCA of blended cement has compiled all these studies and analysed them from the points of view of methodology and results [20]. Several issues were highlighted, such as the choice of a functional unit (FU) of 1 ton of cement or cradle-to-gate system boundaries, which hinder the representativeness of the models. Based on this, future LCAs of blended cement were recommended to increase the precision of their FU by considering the mechanical and durability properties of cement on the one hand and extending system boundaries to the grave on the other hand. This, in turn, would require data on the service life (SL) of the cement which would enrich the blended cement literature.

This paper thus aims to address these issues through a comparative LCA of waste marl cement, OPC and LC3 in a Moroccan context. The objectives are defined as follows: (1) Estimate the environmental potential of waste marl cement (2) Compare the environmental impacts of waste marl cement with OPC and natural clay based-LC3 (3) Conduct the LCA for two different FUs: 1 ton of cement and 1 ton of cement per MPa per year (4) Estimate the SL of the cements (5) Conduct a sensitivity analysis of the results to the impact method to test the robustness of the study.

2 Materials and Methods

2.1 Life Cycle Assessment

The methodology adopted for the LCA is defined by ISO 14040-14,044. It consists of a goal and scope definition, life cycle inventory (LCI) analysis, life cycle impact assessment (LCIA) and interpretation.

Table 1 Characteristics of the cement under study (V_{ACC}: measured accelerated carbonation, ρ_0: measured electrical resistivity)

Cement	Compressive strength (MPa)	ρ_0 (Ω. m)	V_{ACC} (mm.d$^{-1/2}$)	CaO content (%)	Source
OPC	40	80	0.59	59	Lab
30CMC	37	155	1.27	52	Lab
20CMC	38	145	0.82	57	Lab
LC3	43	219	0.76	41	[21, 22]

Goal Definition This LCA aims to estimate, in a Moroccan context, the environmental potential of calcined marl cement and benchmark it against the conventionally used OPC and the environmentally high-potential LC3.

Scope Definition *System description.* This study focuses on three cement types, with two distinct formulas for CMC. In total, there will be four scenarios compared: OPC with a 0.95 clinker content; LC3 with a 0.70 clinker, 0.20 calcined clay and 0.05 limestone content; 20CMC with a 0.75 clinker and 0.20 calcined marl content, and 30CMC with a 0.65 clinker and 0.30 calcined marl content. The marl used in CMC is waste from the OCP phosphate mine in Benguerir, Morocco. The physico-chemical characteristics of the four cements under study are provided by lab experiments or taken from previous papers and displayed in Table 1. For the lab experiments, concrete with a water/binder ratio of 0.45 and the remaining volume divided according to a fine/coarse aggregate ratio of 0.55 was used for mechanical and durability performance evaluation.

Functional Unit The FU is defined for the first part of the study as 1 ton of cement to enable a simpler collection and clearer listing of data. Later, for the sake of a fairer comparison, the FU is upscaled to "1 ton of cement per MPa per year" by dividing the impact assessment results by the respective compressive strength and SL of each cement. *Reusing waste: a cut-off approach.* The marl used in CMC is a by-product of phosphate rock extraction. In this study, it is assumed to be retrieved directly from a pile of marl waste, sorted and deposited in the mine to be valorised. In the present reality, waste rocks generated by phosphate mining are deposited in a way that makes it difficult for marl to be recovered separately in enough reusable quantities. In any case, this marl remains unexploited by the mine, making it a "burden-free" waste material. A cut-off approach is thus adopted for the modelling: no impacts are allocated to the processes producing marl. On the other hand, the transport of the marl from the mine to the cement plant and the subsequent processing are accounted for. *System boundaries.* The boundaries of the analysed product systems extend from the raw material extraction stage to the production stage. This therefore is a cradle-to-gate LCA. The extracted raw materials, in addition to the recovered marl, are transported to the cement plant for processing: the marl and clay are ground and calcined to be used as SCMs, whereas limestone, iron ore, bauxite and shale are crushed, ground and calcined to produce clinker. The kiln used for clinker production is different from the one used for marl and clay calcination. The fuel used for the kilns is petroleum coke, which is also extracted, processed then transported to the plant for grinding and burning. For the kiln used to produce clinker, in addition to petroleum coke, used tires are also burned with an 85:15 petroleum coke/tires ratio.

Clinker and gypsum are ground together to produce OPC. Calcined marl or clay is ground with clinker and gypsum to respectively produce CMC or LC3. It must be noted that even though in the second part of the paper, the SL of the cement is considered, the use and end-of-life stages are not modelled as they are assumed to be the same for all four cement.

Life Cycle Inventory Analysis Whenever possible, foreground processes such as marl calcination or clinker production were modelled using primary data—on materials, energy use and emissions—provided by three Moroccan cement plants. Otherwise, secondary data taken from the Ecoinvent database version 3.8 or online literature was used. For marl calcination for instance, CO_2 emissions generated by a 28.8 wt% content of dolomite in marl were estimated using the IPCC emission factor for dolomite calcination equal to 0.47732 t CO_2/t dolomite [23].

Basis for Impact Assessment LCIA calculations were carried out on SimaPro version 9.4.0.3. To have a more complete, detailed and accurate estimation, a midpoint method was adopted: IMPACT World+ midpoint version 1.29. Its impact categories are the following: climate change (CC), fossil and nuclear energy use (FNE), mineral resources use (MRU), photochemical oxidant formation (POF), ozone layer depletion (OLD), freshwater ecotoxicity (FEco), human toxicity cancer (HTc), human toxicity non-cancer (HTnc), freshwater acidification (FA), terrestrial acidification (TA), freshwater eutrophication (FEutro), marine eutrophication (ME), particulate matter formation (PMF), ionising radiation (IR), land transformation (LT), land occupation (LO) and water scarcity (WS).

Basis for Sensitivity Analysis The parameter tested for sensitivity is the impact assessment method. Two scenarios are compared: the base scenario applying IMPACT World+ and an alternative scenario applying ReCiPe 2016 version 1.07.

2.2 Mechanical and Durability Properties

In the second part of the study, the durability of the four cements is accounted for in the LCIA. To that end, the SL of each cement is quantified.

Service Life Estimation Khaldi's model of cement SL estimation, based on reinforced concrete behaviour during carbon induced corrosion, was adopted [24]. It considers SL as the sum of initiation and propagation times of corrosion as shown in eq. (1).

$$SL\ (years) = t_{ini} + t_{prop} = \frac{1}{K_{eff}} \left(\frac{c}{3.03 k_{HR} V_{ACC}} \right)^2 + \rho_0 \frac{11c/d}{11.61 V_0 k_T k_{RH,prop}} \quad (1)$$

With Keff a wetting/drying cycle parameter, c the concrete cover (mm), k_{HR} and $k_{RH,prop}$ humidity parameters, V_{ACC} the measured accelerated carbonation rate (mm. d-1/2), ρ_0 the measured electrical resistivity (Ω.m), d the reinforcement diameter (mm), V_0 a constant (= 168.9 $\mu A\Omega m.cm-2$) and k_T a temperature parameter (T = 296 K).

SL results were obtained for a relative humidity of 55% and a humidification ToW ratio of 0, in accordance with Benguerir's climatic conditions. SL of the four product systems was calculated and compared based on c and d, with c inferior to d because the concrete cover must be larger than the reinforcement diameter, and c superior to 5 mm due to the NF EN 10080 standard. The data generated by this calculation was further consolidated on Microsoft Excel version 2008. The values for which LC3's and CMC's SL was superior to OPC's alone were retained. This allows for a thorough and complete study of the cements' SL. The (c;d) combination maintained for LCIA considering durability was arbitrarily chosen among the combinations generated during the SL calculations.

3 Results and Discussion

3.1 LCIA Results and Interpretation

FU: 1 Ton of Cement (Fig. 1) *Characterised results.* Characterised LCIA results are scaled to the score of the cement with the highest impact in each category (100%), as shown in Fig. 1. It appears that OPC presents the highest impacts for 13 categories. For these 13 categories, environmental impact scores follow the trend OPC > 20CMC > 30CMC > LC3. LC3 presents the highest impacts for the remaining four categories, particularly MRU, for which the score difference with other cements is substantial. Similar results have been found in a previous study comparing OPC with four other LC3, including a natural clay-based LC3, which

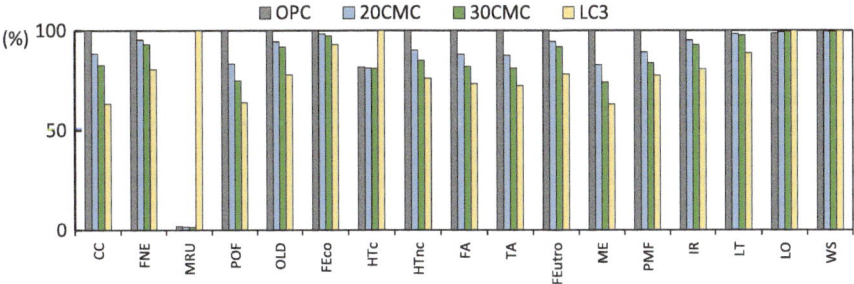

Fig. 1 Life cycle impact assessment normalised results for FU: 1 ton of cement

scored the highest for the abiotic depletion category [15]. *Process contribution analysis.* Apparent gaps between environmental impact scores and trends in the characterised results can be explained with the help of a process contribution analysis. The contribution of each process to each impact category can be seen in Fig. 2. Across all cements, it appears that clinker is mainly responsible for impacts on CC, POF and ME, with a contribution exceeding 75%. This is in line with previous LCAs of various types of cements [13, 17, 18]. On the other hand, fuel is

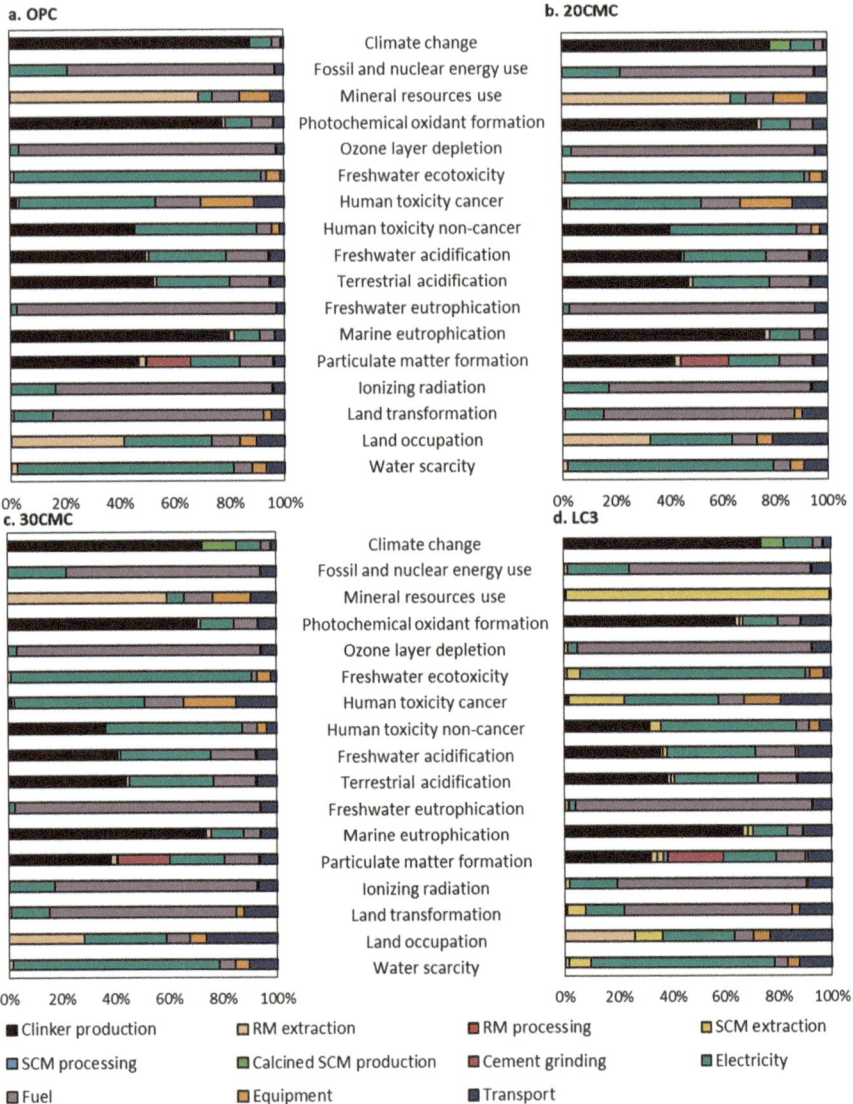

Fig. 2 Contribution analysis results for FU: 1 ton of cement

primarily responsible for impacts on FNE, ODP, FEutro, IR and LT, with a contribution superior to 75%. Moreover, electricity is mainly responsible for impacts on FEco and WS, with a contribution exceeding 70%. Besides, it seems that the impact of SCM calcination on climate change is of less than 10% for the three blended cements under study. Calcined marl's impact on CC was 370 gCO_2eq/kg while calcined clays were 191 gCO_2eq/kg. This is in harmony with what can be found in literature as far as calcined clays are concerned. Malacarne et al. [15] found global warming potentials (GWP) ranging between 66 and 185 gCO_2eq/kg of calcined clay while Pillai et al. [25] found emissions of 127 gCO_2eq/kg of calcined clay. The difference between calcined marl and clay emissions is due to the dolomitic content of phosphate waste marl which carbonates and releases CO_2 to the air. Finally, clay extraction for LC3 appears to be responsible for 99% of the impact on MRU.

Sensitivity Check Sensitivity analysis of the results to the choice of the impact method was conducted to evaluate the robustness of the study. Results for most impact indicators present the same trend for both impact methods. The five exceptions are MRU, WS, ME, LO and FEco which are shown in Fig. 3. *Mineral resources use.* When calculated with IMPACT WORLD+, LCIA results suggest a major problem with the MRU impact indicator for LC3. This poses a challenge for the LC3 low-carbon solution, as it appears to be burdensome for mineral resource consumption.

However, ReCiPe 2016 offers a different perspective, according to which 30CMC helps best with resource consumption, then followed by LC3. The reason behind such striking differences lies in the approach of each impact method to the mineral resource consumption problem. While IMPACT WORLD+ applies the material

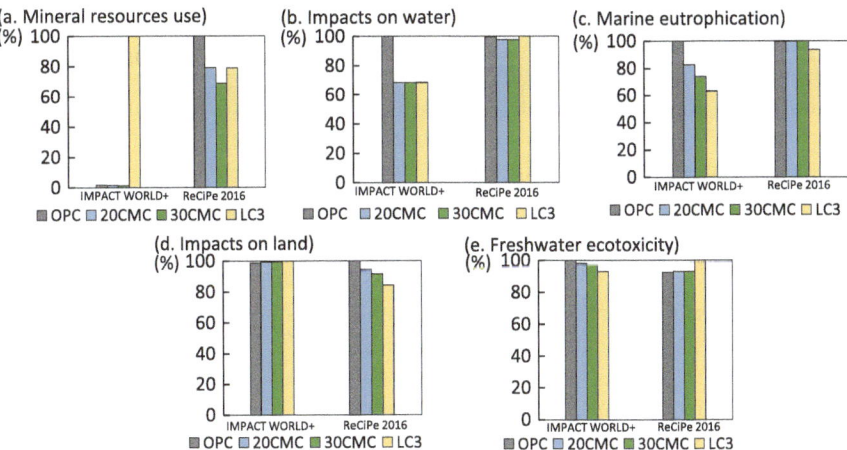

Fig. 3 Sensitivity analysis results for mineral resources use, freshwater ecotoxicity, water consumption, marine eutrophication and land occupation

competition scarcity index (MACSI) as a midpoint indicator for the MRU impact category [26], ReCiPe 2016 uses average surplus ore potential (SOP) as the midpoint indicator [27]. The difference between these two approaches is crucial in understanding the disparity in the results. MACSI looks at resource depletion; it considers the needs of initial users and their ability to meet their needs through the potential substitution of a resource by a depleted one [28]. On another hand, SOP looks at the impact of a mineral's extraction on its future ore production [27]. It is on this basis that characterisation factors (CFs) for clay differ, hence yielding the results in Fig. 3. *Impacts on water.* Results for impacts on water are similar for all four types of cement when the ReCiPe 2016 method is applied. On the contrary, there is a manifest gap between OPC and the three other cements when the IMPACT WORLD+ method is applied. This difference in trend in Fig. 3b lies in the different approaches to water impacts adopted by each method. Indeed, while ReCiPe 2016 considers water consumption as an impact indicator, IMPACT WORLD+ considers water scarcity. Water consumption refers to water that is no longer available in its original watershed [27], whereas water scarcity refers to the water remaining for one user (humans or ecosystems) in a watershed once the needs of another user have been met [29]. *Marine eutrophication.* Results obtained with ReCiPe 2016 show that OPC, 20CMC and 30CMC have a similar score, which is not the case when IMPACT WORLD+ is applied. This difference is due to the fact that unlike ReCiPe 2016, IMPACT WORLD+ considers nitrogen oxides emitted to the air in its ME indicator [25, 26]. *Impacts on land.* IMPACT WORLD+ and ReCiPe 2016 have different impact indicators to assess environmental impacts on land. While IMPACT WORLD+ has two separate indicators——one for land occupation and one for land transformation, ReCiPe 2016 has only one indicator combining the two under the name of land use. This thus explains the difference in trends seen in Fig. 3d. *Freshwater ecotoxicity.* LC3 presents the lowest impact on FEco when IMPACT WORLD+ is applied and the highest impact when ReCiPe 2016 is applied. This is due to aluminium, accounted for by IMPACT WORLD+ (with a CF of 3.2E7 CTUe/kg) but not by ReCiPe 2016 [26]. All in all, it must be underlined that impact assessment methods differ on the basis of their approach to environmental impact mechanisms. This contrast, which results from a difference in philosophies, may lead to major disparities in results at times. However, it also contributes to broader perspectives and richer discussions within the LCA community.

FU: 1 Ton of Cement per MPa per Year *Service life.* The longest SL for OPC, 20CMC, 30CMC and LC3 is respectively equal to 323, 207, 124 and 415 years. In their study, Khaldi et al. [24] have found service lives ranging from 10 to 300 years for (c;d) equal to (25;15) mm. For c ranging from 6 to 29 mm, and d from 5 to 8 mm, LC3 has a longer SL than that of OPC, ranging from 64 to 415 years. In other words, LC3 has a longer SL than OPC for 201 (c;d) combinations. On the contrary, 20CMC and 30CMC respectively have a higher SL than OPC for only 54 and 26 (c;d) combinations, with c varying from 6 to 22 mm for 20CMC; 6 to 16 mm for 30CMC; and d varying from 5 to 8 mm for 20CMC and 30CMC. As can be seen in Table 2,

Table 2 Calculated service life for (c;d) = (10;8) mm

Cement type	t_{ini} (yrs)	t_{prop} (yrs)	Service life (yrs)
30CMC	6	60	65
20CMC	14	56	70
OPC	26	31	57
LC3	16	84	100

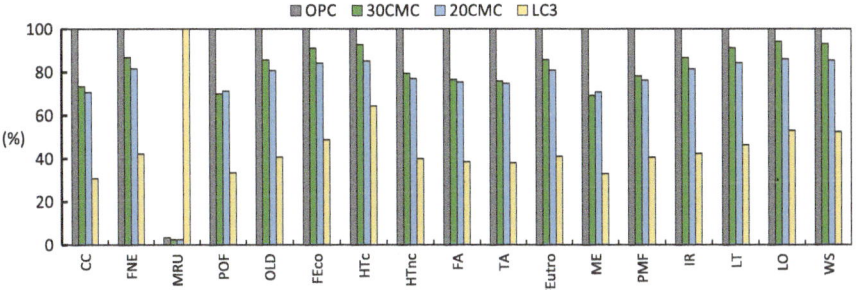

Fig. 4 LCIA impacts for OPC, 30CMC, 20CMC and LC3 for FU: 1 ton of cement/MPa/year

the reason behind this trend is the cements' electrical resistivity which influences the propagation time of corrosion. LC3's electrical resistivity far exceeds that of the other cements, which makes its t_{prop} and thus SL longer. On the other hand, the reason why 20CMC and 30CMC fall behind LC3 and OPC is their reduced resistance to carbonation, which dramatically decreases their t_{ini} and SL. This highlights the strength of LC3 as a low-carbon cement with high durability performance. Moreover, it also stresses the importance of improving CMC's resistance to carbonation since it would allow for a powerful low-carbon and durable solution in the Moroccan context. Lastly, LC3's durability performance can be furthered if its resistance to carbonation is enhanced.

Scaled Environmental Impacts When mechanical and durability performances are considered, OPC appears to be the worst-case scenario for 13 impact categories. It presents the highest impact on CC of 0.200 kgCO$_2$eq/t/MPa/year, whereas LC3 presents the lowest impact of 0.084 kgCO$_2$eq/t/MPa/year, 70% lower than OPC. 20CMC and 30CMC, respectively, present a GWP 29% and 27% lower than OPC. It must be noted that despite having a higher clinker content compared to 30CMC, 20CMC's environmental impacts are lower. That is due to 20CMC's higher resistance to compressive strength and longer SL. Apart from its MRU score, LC3 remains the best scenario even when we modify the FU, with its impact scores lower than 50% for most impact categories, mainly due to its long SL. Overall, it seems that the trends in this figure show that CMC and LC3 are both positive alternatives to OPC, with LC3 being the better of the two (Fig. 4).

4   Conclusion

This study consists of a comparative LCA of waste marl cement, OPC and LC3 in the Moroccan context. It seeks to evaluate the environmental potentials of waste marl cement production, and at the same time addresses several methodological issues pointed out in previous works regarding LCAs of blended cements.

The key takeaways of this paper are the following:

- LC3 proves to be the best low-carbon solution for the cement industry, with a GWP 70% lower when mechanical and durability performances are considered.
- Through the reuse of phosphate waste, CMC enables gains of >95% compared to LC3 and > 20% compared to OPC in terms of MRU.
- SL calculations have shown that LC3 has the best durability performance, followed by OPC then CMC. Taking this parameter into account in the FU leads to different LCIA results, with LC3 presenting the lowest environmental impacts in all cases.
- Reading LCIA results in the light of the chosen impact method is crucial for a fair interpretation. Sensitivity analysis of the impact method offers a broader and more comprehensive view of LCIA results.

The limitations and perspectives highlighted by this study are the following:

- Changing the FU should involve a whole rebuild of the LCI rather than a mere division of LCIA results by new parameters.
- SL results were retained for the case where OPC's SL was the lowest. This limits interpretation as cases where OPC would have a longer SL were not evaluated.
- Further investigation is encouraged regarding waste clay-based LC3 to enable better comparison with CMC, another waste-based blended cement.

References

1. Kajaste, R., Hurme, M.: Cement industry greenhouse gas emissions–management options and abatement cost. J. Clean. Prod. **112**, 4041–4052 (2016)
2. Czigler, T., et al.: Laying the Foundation for Zero-Carbon Cement, vol. 9. McKinsey & Company (2020)
3. Habert, G., et al.: Environmental impacts and decarbonization strategies in the cement and concrete industries. Nat Rev Earth Envir. **1**(11), 559–573 (2020)
4. Snellings, R.: Assessing, understanding and unlocking supplementary cementitious materials. RILEM Tech Lett. **1**, 50–55 (2016)
5. U. N. Environment, et al.: Eco-efficient cements: potential economically viable solutions for a low-CO2 cement-based materials industry. Cem. Concr. Res. **114**, 2–26 (2018)
6. Scrivener, K., et al.: Calcined clay limestone cements (LC3). Cem. Concr. Res. **114**, 49–56 (2018)
7. Newman, A.C.D.: The significance of clays in agriculture and soils, philosophical transactions of the Royal Society of London. Series A Mathemat Phys Sci. **311**(1517), 375–389 (1984)
8. Bahhou, A., et al.: Use of phosphate mine by-products as supplementary cementitious materials. Materials Today Proc. **37**, 3781–3788 (2021)

9. Snellings, R., et al.: Future and emerging supplementary cementitious materials. Cem. Concr. Res. **171**, 107199 (2023)
10. Bahhou, A., et al.: Assessment of hydration, strength, and microstructure of three different grades of calcined marls derived from phosphate by-products. J Building Eng, 108640 (2024)
11. IEA: Energy policies beyond IEA countries: Morocco 2014, IEA, 2014
12. Sánchez-Berriel, S, et al.: Impacts assessment of local and industrial LC3 in cuban context: challenges and opportunities. In: Calcined Clays for sustainable concrete: Proceedings of the 3rd international conference on calcined clays for sustainable concrete, pp. 263–270. Springer, 2020
13. Çankaya, S., Pekey, B.: A comparative life cycle assessment for sustainable cement production in Turkey. J. Environ. Manag. **249**, 109362 (2019)
14. Huntzinger, D.N., Eatmon, T.D.: A life-cycle assessment of Portland cement manufacturing: comparing the traditional process with alternative technologies. J. Clean. Prod. **17**(7), 668–675 (2009)
15. Malacarne, C.S., et al.: Environmental and technical assessment to support sustainable strategies for limestone calcined clay cement production in Brazil. Constr. Build. Mater. **310**, 125261 (2021)
16. Berriel, S.S., et al.: Assessing the environmental and economic potential of limestone calcined clay cement in Cuba. J. Clean. Prod. **124**, 361–369 (2016)
17. Yang, D., et al.: Comparative study of cement manufacturing with different strength grades using the coupled LCA and partial LCC methods—a case study in China. Resour. Conserv. Recycl. **119**, 60–68 (2017)
18. Hossain, M.U., et al.: Comparative LCA on using waste materials in the cement industry: a Hong Kong case study. Resour. Conserv. Recycl. **120**, 199–208 (2017)
19. Vizcaíno, L., et al.: Industrial manufacture of a low-clinker blended cement using low-grade calcined clays and limestone as SCM: the Cuban experience, in calcined clays for Sustainable Concrete: Proceedings of the 1st international conference on calcined clays for Sustainable Concrete, pp. 347–358. Springer, 2015
20. Rhaouti, Y., et al.: Assessment of the environmental performance of blended cements from a life cycle perspective: a systematic review. Sust Prod Consump. **36**, 32–48 (2023)
21. Balestra, C.E.T., et al.: Contribution to low-carbon cement studies: effects of silica fume, fly ash, sugarcane bagasse ash and acai stone ash incorporation in quaternary blended limestone-calcined clay cement concretes. Envir Devel. **45**, 100792 (2023)
22. Martirena, F., et al.: Calcined Clays for Sustainable Concrete: Proceedings of the 2nd International Conference on Calcined Clays for Sustainable Concrete, vol. 16. Springer (2017)
23. Eggleston, H.S., et al.: 2006 IPCC guidelines for national greenhouse gas inventories, 2006
24. Elkhaldi, I., et al.: Towards global indicator of durability performance and carbon footprint of clinker-slag-limestone cement-based concrete exposed to carbonation. J. Clean. Prod. **380**, 134876 (2022)
25. Pillai, R.G., et al.: Service life and life cycle assessment of reinforced concrete systems with limestone calcined clay cement (LC3). Cem. Concr. Res. **118**, 111–119 (2019). https://doi.org/10.1016/j.cemconres.2018.11.019
26. Bulle, C., et al.: IMPACT World+: a globally regionalized life cycle impact assessment method. Int. J. Life Cycle Assess. **24**(9), 1653–1674 (2019). https://doi.org/10.1007/s11367-019-01583-0
27. Huijbregts, M.A.J., et al.: ReCiPe2016: a harmonised life cycle impact assessment method at midpoint and endpoint level. Int. J. Life Cycle Assess. **22**(2), 138–147 (2017). https://doi.org/10.1007/s11367-016-1246-y
28. De Bruille, V.: Impact de l'utilisation des ressources minérales et métalliques dans un contexte cycle de vie: une approche fonctionnelle, 2014
29. Boulay, A.-M., et al.: The WULCA consensus characterization model for water scarcity footprints: assessing impacts of water consumption based on available water remaining (AWARE). Int. J. Life Cycle Assess. **23**, 368–378 (2018)

R³ Reactivity Test on Biochar from Pyrolyzed Green Waste, Wood Waste, and Screen Overflow

Maximilian Mayer, Neven Ukrainczyk, and Eduardus Koenders

Contents

1 Introduction

The anthropogenic greenhouse gas emissions (GHG) are at the highest level in human history [1]. Since the last century, GHG emissions due to human activity have increased from about 5 $GtCO_2$-eq/yr. to about 60 $GtCO_2$-eq/yr., where CO_2 from fossil fuels and industry play a major role [2]. Cement production has a share of about 6% of global fossil CO_2 emissions in 2021, 2022 and 2023 [3]. Therefore, reducing the emissions from cement production is mandatory to reach net zero emissions.

Carbon dioxide removal (CDR) is a promising solution for reducing the CO_2 content in the atmosphere. Intergovernmental Panel on Climate Change (IPCC) mentioned Biochar as an available and scalable solution for CDR [4]. One of the most common methods for producing biochar is pyrolysis. Pyrolysis is mainly used to produce biochar from dry biomass through thermochemical conversion in the absence of oxygen [5]. It can be divided into slow and fast pyrolysis differentiated

M. Mayer (✉) · N. Ukrainczyk · E. Koenders
Technical University Darmstadt, Darmstadt, Germany
e-mail: mayer@wib.tu-darmstadt.de

© The Author(s) 2025
M. Kioumarsi, B. Shafei (eds.), *The 1st International Conference on Net-Zero Built Environment*, Lecture Notes in Civil Engineering 237,
https://doi.org/10.1007/978-3-031-69626-8_47

through the residence time of the biomass in the pyrolization chamber. Various forms of biomass can be used for pyrolysis (e.g., wood waste, screen overflows, rice husk, green waste, food waste, sugar cane, etc.). The chemical and physical properties of the biochar depend on the used biomass, production method, and pyrolysis temperature [6–8].

Screen overflows have a high potential for being used as biochar because in Germany they are mainly used in biomass power plants and waste-fueled power plants. Through pyrolysis, the screen overflow is converted into a higher-level product that can be used in agriculture or building materials, while storing CO_2 and generating power from the pyrolysis gas.

Biochar as SCM changes the behavior of fresh and hydrated mortar and concrete. Zahra Asadi Zeidabadi et al. [9] showed that the addition of biochar enhanced the compressive and tensile strength, whereas the amount added shouldn't exceed more than 5% in relation to the mass of the cement. In addition, Biochar has the ability to increase the durability and hydration of concrete because of the introduction of a large pore volume which enhances the retention capacity and therefore improves the self-curing of the concrete according to Hamid Maljaee et al. [10]. Furthermore, Souradeep Gupta et al. [11] conducted experiments to investigate the flow rate and air content of fresh mortar with biochar. The addition of biochar increased the air content in the mortar and led to a reduction of flowability. These influences of biochar need to be taken into consideration when evaluating its usability in concrete.

The content of aluminosilicates of the respective biochar mainly influences if this biochar can be used as SCM and which amount can be added/substituted for cement. This is due to the pozzolanic reaction of amorphous aluminosilica with calcium hydroxide (CH) to form calcium silicate/aluminate hydrates phases (CSH/CAH) and other reaction products [12]. The amount of hydration products mainly influences the strength, permeability, and durability of the concrete. To determine the suitability of materials for use as SCM, Avet et al. [13] introduced the R^3 test (rapid, relevant, reliable). The R^3 test measures the reactivity of the respective SCM by eliminating the influence of cement [13–15]. The R^3 test is mostly used for calcined clays, blast furnace slag, fly ash, and silica fume with high aluminosilicate contents [16]. The R^3 test simulates the alkaline pore solution of concrete with controlled masses of $Ca(OH)_2$, $CaCO_3$, KOH, and K_2SO_4.

Therefore, this research aims to determine the potential of biochar made from screen overflow as SCM using the R^3 test. Furthermore, the influence of different particle sizes is investigated.

2 Materials and Methods

2.1 Materials

The materials used for the preparation of the R^3-emulsion are calcium hydroxide ($\geq$96%), potassium hydroxide in flakes ($\geq$85%), powdered potassium sulfate ($\geq$99%), and calcite ($\geq$99%) which were obtained from Carl Roth GmbH + Co. KG (Karlsruhe, Germany).

The biochar employed in this study are from two different composting plants, Abfallwirtschaftsbetrieb Alzey Worms (AWB, Alzey, Germany) and Gesellschaft für Abfallentsorgung Lippe (GAL, Lippe, Germany). Each of the composting plants supplies their screen overflow for pyrolysis. The screen overflows are a mixture of wood waste, green waste, paper waste, metal waste, plastic waste, and other fine, partially mineral aggregates <10 mm.

The substrates were pyrolyzed by Carbontechnik Schuster (CTS, Dischingen, Germany) at 750 °C with a residence/dwell time of 210 minutes. After the pyrolysis, the biochar was ground with a mortar and then separated by particle size using a sieve tower. The particle sizes prepared from the two different biochar are <40 μm, 40–125 μm, and 125–250 μm.

The phase composition of the biochar was determined by qualitative X-ray diffraction (XRD) with estimates of the perspective shares of the determined phases and amorphous content using the software Diffrac.Eva Version 6.0.0.7 from Bruker Corporation (Billerica, USA). The corresponding XRD results of the screen overflows are shown in Fig. 1. Both biochar have very similar mineral compositions even though they are sourced from different composting plants. They have an amorphous content of approximately 51.5% for the screen overflow from AWB and 41.7% for the screen overflow from GAL. The major phases are quartz, calcite, albite, and cristobalite as trace mineral. Magnesite occurs only as a trace in the screen overflow from AWB.

The particle size distribution is given in Fig. 2. The distribution shows that even if a sieve with a screen width of 40 μm is used higher particle sizes than 40 μm are measured. This can occur because of the irregular shape and length-to-width ratio of the biochar. While the maximum particle width is 250 μm the respective particle

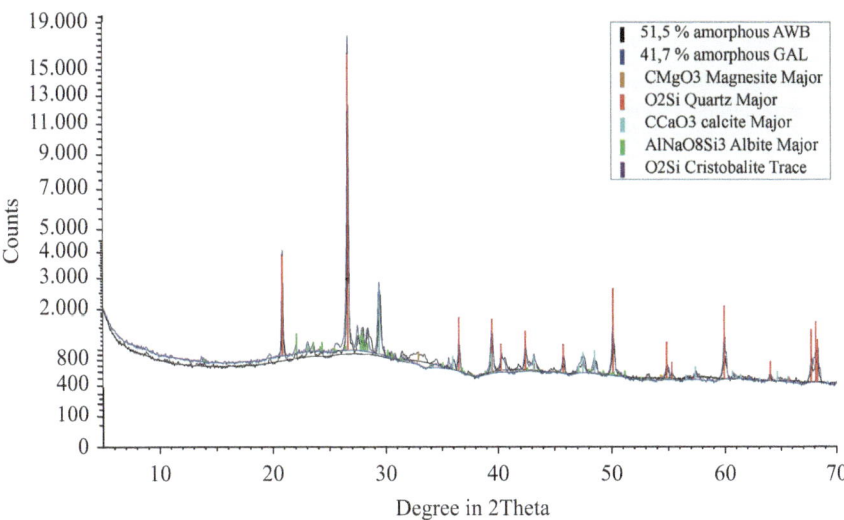

Fig. 1 Powder XRD and qualitative analysis of biochar from screen overflow from both sourced composting plants AWB and GAL

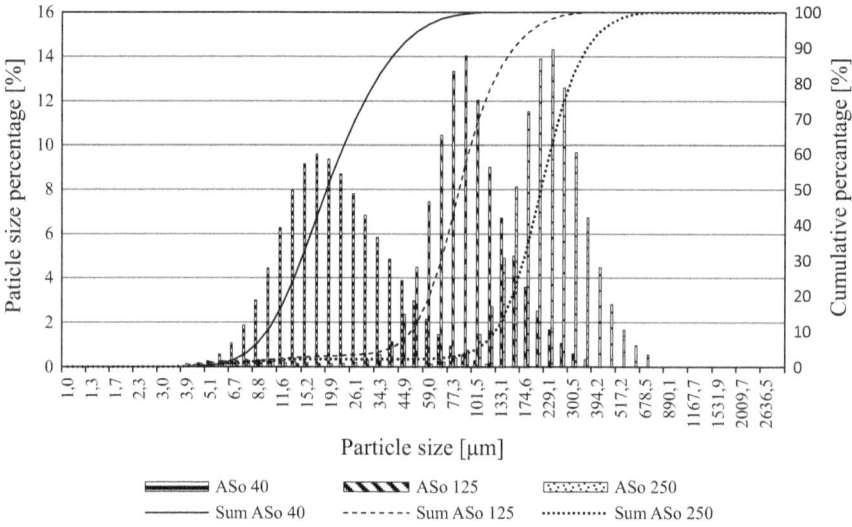

Fig. 2 Particle size distribution of the screen overflow from AWB for the different particle sizes

Table 1 R³ mix design by Li et al. [16]

Biochar	Ca(OH)₂	CaCO₃	K₂SO₄	KOH	H₂0
11.11 g	33.33 g	5.56 g	1.2 g	0.24 g	60 g

length may exceed 250 µm (i.e., strongly disproportionate oblong particles). There-fore, the sieve analysis with the sieve tower can deviate from the actual measured particle sizes. Especially for the particle size group 125–250 µm, the true particle size measured may be up to 650 µm. For the particle size measurement, the laser scattering particle size distribution analyzer "particle LA950V2" from HORIBA (Kyoto, Japan) was used.

The particle size distribution of the screen overflows from GAL is similar to the results from AWB. The same grinding and sieving technic were used for both biochar.

2.2   Sample Preparation

The biochar was mixed with the R³ test mixture as proposed by Avet et al. [13] and Li et al. [16] with the respective mix design for the emulsion shown in Table 1. The R³ emulsion mimics the alkaline pore solution of cement paste and was mixed with the biochar (SCM) according to Weise et al. [15] and Avet et al. [13]. Firstly, the alkaline R³ emulsion was prepared with powdered calcium hydroxide, potassium hydroxide in flakes, powdered potassium sulfate, calcite, and deionized water. Then the R³ emulsion was mixed with the biochar [13, 15, 16].

The mixed solutions for thermogravimetric analysis (TGA) and XRD measurements were then stored in an airtight container in an oven at 40 ° C for 7, 14, 28, and 56 days. After the respective storage times, the reaction process was stopped by treating the samples with isopropanol and diethyl ether according to Scrivener et al. [17] and Weise et al. [15] storing them in an excavator for 24 hours before testing by TGA and XRD.

2.3 Measurement Method

The reaction products and the reaction process of the biochar with the R³ emulsion were tested by TGA and XRD. For TGA, "STA 449 F5 Jupiter" from NETZSCH (Selb, Germany) was used. Alumina crucibles were filled with 40–50 mg of fine powder material after the hydration stoppage. Inert gas (nitrogen) was used to avoid oxidation of the samples during the measurement. The heating program of the TGA measurement started by heating the samples to 40 °C and keeping the temperature constant for 30 minutes. After this initial constant heating phase, the samples were further heated up to 1000 °C at a constant heating rate of 20 K per minute.

For the XRD measurements, the "Bruker D2 Phaser" from Bruker Corporation (Billerica, USA) was used, configured with $CuK\alpha_{1,2}$ radiation (40 kV and 10 mA) and a linear Lynxeye detector (5-degree opening). Powder samples were measured with a step size of 0.02 two-theta in a range of 5–70 two-theta and a measurement time of 2 seconds per step.

3 Results

The pozzolanic reaction of the SCM turns calcium hydroxide and water into hydrated phases. This results in chemically bound water (CBW) incorporated into the reaction products. The formation of calcium carbonate as a result of carbonization during the reaction process and during the drying procedure also consumes part of the calcium hydroxide. The actual amount of new calcium carbonates in the sample is needed to calculate the actual consumption of calcium hydroxide by the pozzolanic reaction of the biochar [18, 19]. The total consumption of calcium hydroxide is an indicator for the pozzolanic reactivity of the samples.

Therefore, the results of the TGA measurements were analyzed according to Weise et al. [19] and Scrivener et al. [17] to obtain the most accurate results possible. These analysis method takes the formation of carbonates out of calcium hydroxide into account. It was assumed that the sample mass remained constant during the whole reaction process.

The mass loss of CH for both biochar after 7 and 56 days is shown in Table 2. The derivate of the thermogravimetric curve (DTG) is shown in Fig. 3 for the screen

Table 2 Mass loss of CH in wt.% of the R^3-samples after 7 and 56 days for all particle sizes measured by TGA

		7 days	56 days	
	Particle size [µm]	CH [wt.%]	CH [wt.%]	ΔCH [wt.%]
AWB	<40	11.57	10.33	1.24
	40–125	11.89	10.74	1.15
	125–250	12.41	11.36	1.05
GAL	<40	12.19	11.13	1.06
	40–125	12.78	12.19	0.59
	125–250	13.96	13.48	0.48

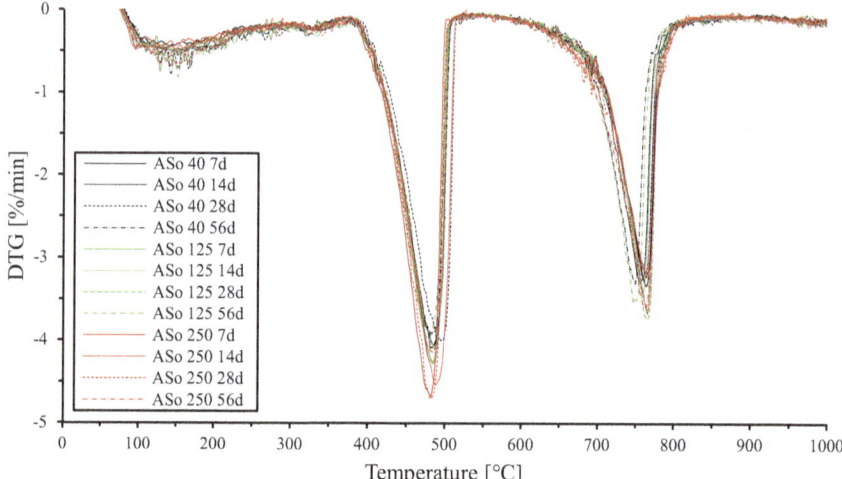

Fig. 3 DTG-R^3 emulsion with biochar (screen overflow AWB) with different particle sizes and reaction days

overflow from AWB. The three different particle sizes (<40 µm labeled as 40, 40–125 µm labeled as 125, and 125–250 µm labeled as 250) are measured after 7, 14, 28, and 56 days.

As shown in Fig. 3 and Table 2, the mass loss of CH (allocated to the mass loss from 400 to 500 °C) is higher with increasing particle size and lower with increasing days of reaction [17]. The consumption of CH increases with decreasing particle size. This shows, that the pozzolanic reactivity increases with decreasing particle sizes [20]. Comparing the mass loss of CH after 7 and 56 days the trend for all the particle size groups shows the same behavior. This indicates that with an increasing amount of time, the amount of reaction products is increasing. An indicator for the higher amount of reaction products (C-S-H phases) is the CBW in these phases [17]. The CBW of the reaction products (mass loss from approximately 100–350 °C) increases with the days of reaction for all samples.

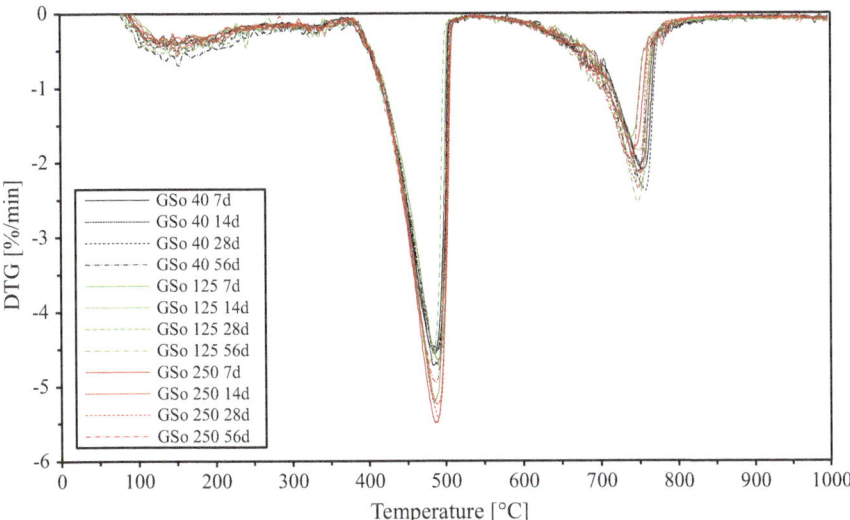

Fig. 4 DTG-R³ emulsion with biochar (screen overflow GAL) with different particle sizes and reaction days

In Fig. 4, the DTG curve for the biochar from GAL is shown. The mass losses listed in Table 2 and illustrated in Fig. 4 indicate that the reactivity increases with smaller particle sizes like the biochar from AWB. Looking at the time-dependent reaction the behavior is also similar to the biochar from AWB. The reaction products increase with increasing reaction time. As a result, the pozzolanic reaction of the biochar is still ongoing until 56 days.

In total, the CH mass losses are higher for the biochar from GAL compared to the samples from AWB. This indicates that the biochar from GAL is less reactive. The CBW increases with the reaction days of the sample but is also lower compared to the biochar from AWB. The lower reactivity can be explained by the estimated lower amorphous content of the biochar from GAL compared to AWB.

The amount of calcium carbonate in the samples from AWB is higher than in the samples from GAL. The content of calcium carbonate (mass loss from 650 to 850 °C) is about 6 wt.% for GAL and 9 wt.% for AWB. This is caused by a higher calcium carbonate content of the biochar from AWB.

In Fig. 5, the consumption of CH per 100 g SCM is plotted over the content of CBW per 100 g SCM. The samples after 7 days and 56 days are taken into consideration.

The consumption of CH and the content of CBW were calculated according to Weise et al. [19] while taking the carbonization (formation of calcium carbonate from calcium hydroxide) of the R³ samples into consideration. This is important to not overestimate the consumption of CH by the biochar and therefore the pozzolanic reactivity.

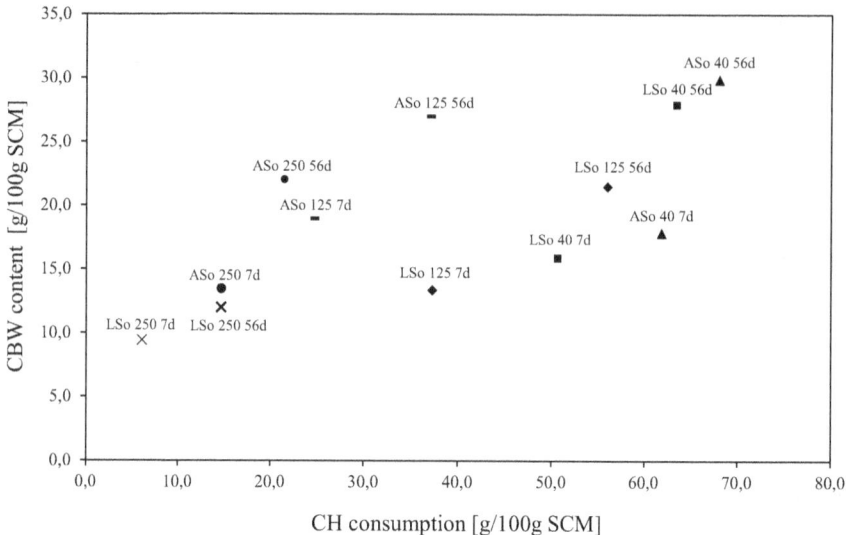

Fig. 5 CBW content over CH consumption in g/100 g SCM after 7 and 56 days

The consumption of CH and the content of CBW per 100 g SCM increase with reaction time. This is also indicated by the results of the DTG curves. The CBW is not consistently increasing with the consumption of CH considering all samples. This could happen because different reaction products are forming for the different particle sizes with different amounts of CBW. Nevertheless, a moderate pozzolanic reactivity is found and the results show that the pozzolanic reactivity increases with smaller particle sizes.

In Figs. 6 and 7, the XRD for the AWB <40 μm and 125–250 μm is shown. The results clearly show the formation of hydration products like ettringite and calcium hemicarboaluminate [21]. Portlandite, calcite, and quartz from the biochar and the R^3 emulsion are also present. The reaction products are similar for both considered particle sizes.

The results from TGA and XRD show the formation of C-S-H and C-A-H phases in small amounts increasing with smaller particle size and hydration days.

4 Conclusion

The results of the R^3 test in combination with TGA and XRD of the biochar show moderate pozzolanic reactivity. Smaller particle sizes have higher reactivity because of their higher surface area. The formation of hydration products still occurs after 56 days of reaction time. Therefore, the longer the reaction time the higher the consumption of CH and formation of hydration products.

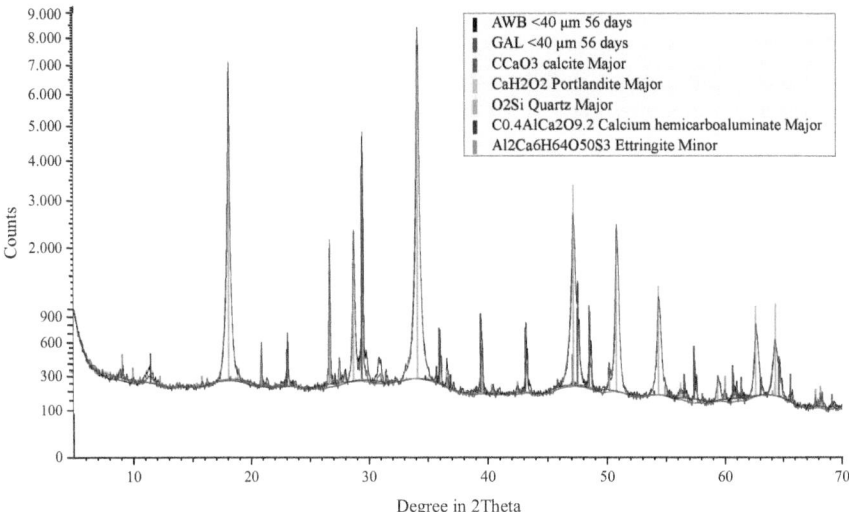

Fig. 6 Powder XRD and analysis for the particle size <40 μm after 56 days of the R³ test from both sourced composting plants AWB and GAL

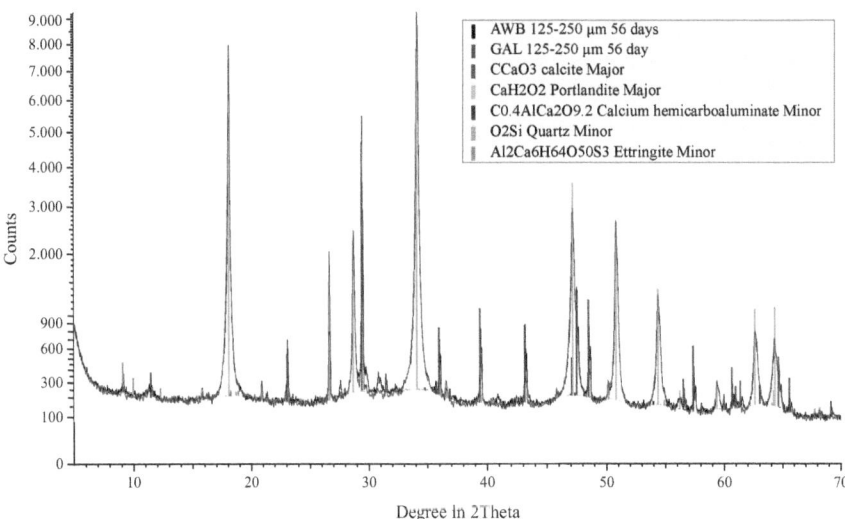

Fig. 7 Powder XRD and analysis for the particle size 125–250 μm after 56 days of the R³ test from both sourced composting plants AWB and GAL

Nevertheless, there is some uncertainty in the results regarding the correlation between the CBW and the consumption of CH, especially regarding particle sizes bigger than 40 μm. Many variables, such as chemical composition, the proportion of amorphous phases, and the actual particle size distribution, influence the results. Therefore, further investigations are needed to fully understand the reaction mechanism to ensure a correct evaluation of the results.

The amorphous content of the original biochar is directly influencing the reactivity. The higher the amorphous content the higher the reactivity.

Biochar from screen overflows shows potential to be used as SCM. Therefore, an optimization of the pyrolysis which leads to an increased amorphous content of the biochar should be investigated. Furthermore, other pozzolanic reactivity tests should be conducted and compared with the results obtained by the R^3 test.

Acknowledgments The research is related to the Innovation Space BioBall and the industry-research project PYROCEM. The innovation space BioBall promotes bioeconomy in the metropolitan region Rhine-Main and is supported by the Bundesministerium für Bildung und Forschung (BMBF) funding measure "Innovation Spaces Bioeconomy" within the framework of the "National Research Strategy Bioeconomy 2030." The PYROCEM project investigates the innovative use of different biomasses for the production of PYROlyzed biochar as an additive for CEMent-based construction materials with integrated CO_2 reduction. The authors would also acknowledge Luca Salpietro, Yvette Schales, and Helga Janning who helped with the experiments and measurements.

References

1. International Energy Agency Homepage. https://www.iea.org/reports/co2-emissions-in-2023. Last Accessed 26 Mar 2024
2. Lee, H., Romero, J., et al.: Climate change 2023: synthesis report: contribution of working groups I, II and III to the sixth assessment report of the Intergovernmental Panel on Climate Change, IPCC, (2023)
3. Friedlingstein, P., O'Sullivan, M., Jones, M.W., et al.: Global Carbon Budget 2023. Earth System Sci Data. **15**(12), 5301–5369 (2023)
4. Shukla, P.R., Skea, J., Slade, R., et al.: Climate change 2022: mitigation of climate change. Contribution of working group III to the sixth assessment report of the Intergovernmental Panel on Climate Change. IPCC, (2022)
5. Goel, C., Mohan, S., Dinesha, P.: CO2 capture by adsorption on biomass-derived activated char: a review. Sci. Total Environ. **798**, 149296 (2021)
6. Aman, A.M.N., Selvarajoo, A., Lau, T.L., Chen, W.: Biochar as cement replacement to enhance concrete composite properties: a review. Energies. **15**(20), 7662 (2022)
7. Senadheera, S.S., Gupta, S., Kua, H.W., et al.: Application of biochar in concrete—a review. Cem. Concr. Compos. **143**, 105204 (2023)
8. Sun, Y., Gao, B., Yao, Y., et al.: Effects of feedstock type, production method, and pyrolysis temperature on biochar and hydrochar properties. Chem. Eng. J. **240**, 574–578 (2014)
9. Zeidabadi, Z.A., Bakhtiari, S., Abbaslou, H., Ghanizadeh, A.R.: Synthesis, characterization and evaluation of biochar from agricultural waste biomass for use in building materials. Constr. Build. Mater. **181**, 301–308 (2018)
10. Maljaee, H., Madadi, R., Paiva, H., et al.: Incorporation of biochar in cementitious materials: a roadmap of biochar selection. Constr. Build. Mater. **283**, 122757 (2021)
11. Gupta, S., Kua, H.W., Koh, H.J.: Application of biochar from food and wood waste as green admixture for cement mortar. Sci. Total Environ. **619–620**, 419–435 (2018)
12. Maljaee, H., Paiva, H., Madadi, R., et al.: Effect of cement partial substitution by waste-based biochar in mortars properties. Constr. Build. Mater. **301**, 124074 (2021)
13. Avet, F., Li, X., Ben Haha, M., et al.: Report of RILEM TC 267-TRM phase 2: optimization and testing of the robustness of the R3 reactivity tests for supplementary cementitious materials. Mater. Struct. **55**, 92 (2022)

14. Londono-Zuluaga, D., Gholizadeh-Vayghan, A., Winnefeld, F., et al.: Report of RILEM TC 267-TRM phase 3: validation of the R3 reactivity test across a wide range of materials. Mater. Struct. **55**(5), 142 (2022)
15. Weise, K., Ukrainczyk, N., Endell, L.M., Koenders, E.: R^3-test for Pozzolanic reactivity: experimental issues and practical recommendations for hydration stoppage with isopropanol. In: Jędrzejewska, A., Kanavaris, F., Azenha, M., Benboudjema, F., Schlicke, D. (eds.) International RILEM Conference on Synergising Expertise towards Sustainability and Robustness of Cement-Based Materials and Concrete Structures SynerCrete 2023, RILEM Bookseries, vol. 44, pp. 55–64. Springer, Cham (2023)
16. Li, X., Snellings, R., Antoni, M., et al.: Reactivity tests for supplementary cementitious materials: RILEM TC 267-TRM phase 1. Mater. Struct. **51**(6), 151 (2018)
17. Scrivener, K., Snellings, R., Lothenbach, B.: A Practical Guide to Microstructural Analysis of Cementitious Materials. CRC Press Taylor & Francis Group, Boca Raton (2015)
18. Kim, T., Olek, J.: Effects of sample preparation and interpretation of thermogravimetric curves on calcium hydroxide in hydrated pastes and mortars. Transp. Res. Rec. **2290**(1), 10–18 (2012)
19. Weise, K., Ukrainczyk, N., Koenders, E.: A mass balance approach for thermogravimetric analysis in Pozzolanic reactivity R3 test and effect of drying methods. Materials. **14**(19), 5859 (2021)
20. Reschke., T., Siebel, E., Thielen, M.: Influence of the granulometry and reactivity of cement and additions on the development of the strength and microstructure of mortar and concrete. beton, (1999)
21. Runčevski, T., Dinnebier, R.E., Magdysyuk, O.V., Pöllmann, H.: Crystal structures of calcium hemicarboaluminate and carbonated calcium hemicarboaluminate from synchrotron powder diffraction data. Acta Crystallogr. B. **68**(5), 493–500 (2012)

Low Carbon-Oriented Concrete Mix Optimization Using Ensemble Learning and NSGA-II

Lin DENG ⑩ and Xueqing Zhang ⑩

Contents

1 Introduction

Concrete, a widely used material in modern construction and infrastructure projects, ranks second only to water in terms of mass consumption. However, several issues need to be addressed. From a technical standpoint, ensuring concrete quality is a complex undertaking. Instances of inadequate compressive strength have resulted in significant rework or, in extreme cases, the complete demolition of structures [1]. Concrete production also accounts for approximately 8–9% of global anthropogenic CO_2 emissions and 2–3% of global energy consumption [2]. Furthermore, the massive consumption of concrete entails significant economic costs.

Concrete mix design refers to determining the composition, proportions, aggregate gradation, water/cement ratio, admixtures, and other factors required to attain desired concrete properties. Traditionally, mix design has primarily focused on

L. DENG · X. Zhang (✉)
Hong Kong University of Science and Technology, Hong Kong, People's Republic of China
e-mail: lncs@springer.com; ldengah@connect.ust.hk

© The Author(s) 2025   575
M. Kioumarsi, B. Shafei (eds.), *The 1st International Conference on Net-Zero Built Environment*, Lecture Notes in Civil Engineering 237,
https://doi.org/10.1007/978-3-031-69626-8_48

technical properties, with compressive strength and slump being crucial indicators. However, the mix design also exerts influence over the cost and carbon emissions throughout the concrete's life cycle. The interdependence between the concrete mixture and its technical, environmental, and economic performance implies that an optimal mix design can achieve the most favorable trade-off for specific concrete applications. The question is how we can effectively integrate these three aspects and achieve their optimal balance?

Traditional methods for concrete strength prediction are empirical models, which are not accurate due to limited data. The data-driven machine learning models have been widely used for concrete property prediction, including support vector machine, k nearest neighbors, tree-based models, and artificial neural network, etc. However, few studies use ensemble learning methods such as stacking and voting for compressive strength prediction. Multi-objectives optimization (MOO) algorithms have been widely used for concrete mix optimization due to their fast convergence speed and excellent robustness. This chapter proposes a hybrid framework of ensemble learning and MOO algorithms to produce technically capable, environmentally friendly, and economically efficient concrete.

2   Data

The concrete strength experimental data of measured by [3] is used in this study. The input features include the content of cement (C), blast furnace slag (BFS), fly ash (FA), water (W), superplasticizer (SU), coarse aggregate (CA), fine aggregate (FA) of concrete per unit volume, and the curing age (AG) of the test specimen. The target variable is the compressive strength of concrete. The data distribution of each input feature and relationships between each factor and compressive strength are shown in Fig. 1. The correlation matrix plot is illustrated in Fig. 2. Cement and water have the largest positive and negative correlation coefficients, respectively.

3   Concrete Strength Prediction

3.1   Ensemble Learning

Ensemble learning is a weighted combination of multiple base learners to solve a particular classification or regression problem [4], including boosting, bagging, stacking, and voting. Boosting integrates a set of weak learners into a strong learner to minimize training errors. Bagging uses the bootstrap sampling method to draw training data and calculate an aggregated value of base learner's predictions as the final prediction. Stacking is a meta-learning technique for learning from learners in multiple learning stages. Voting aggregates the predictions of different base models

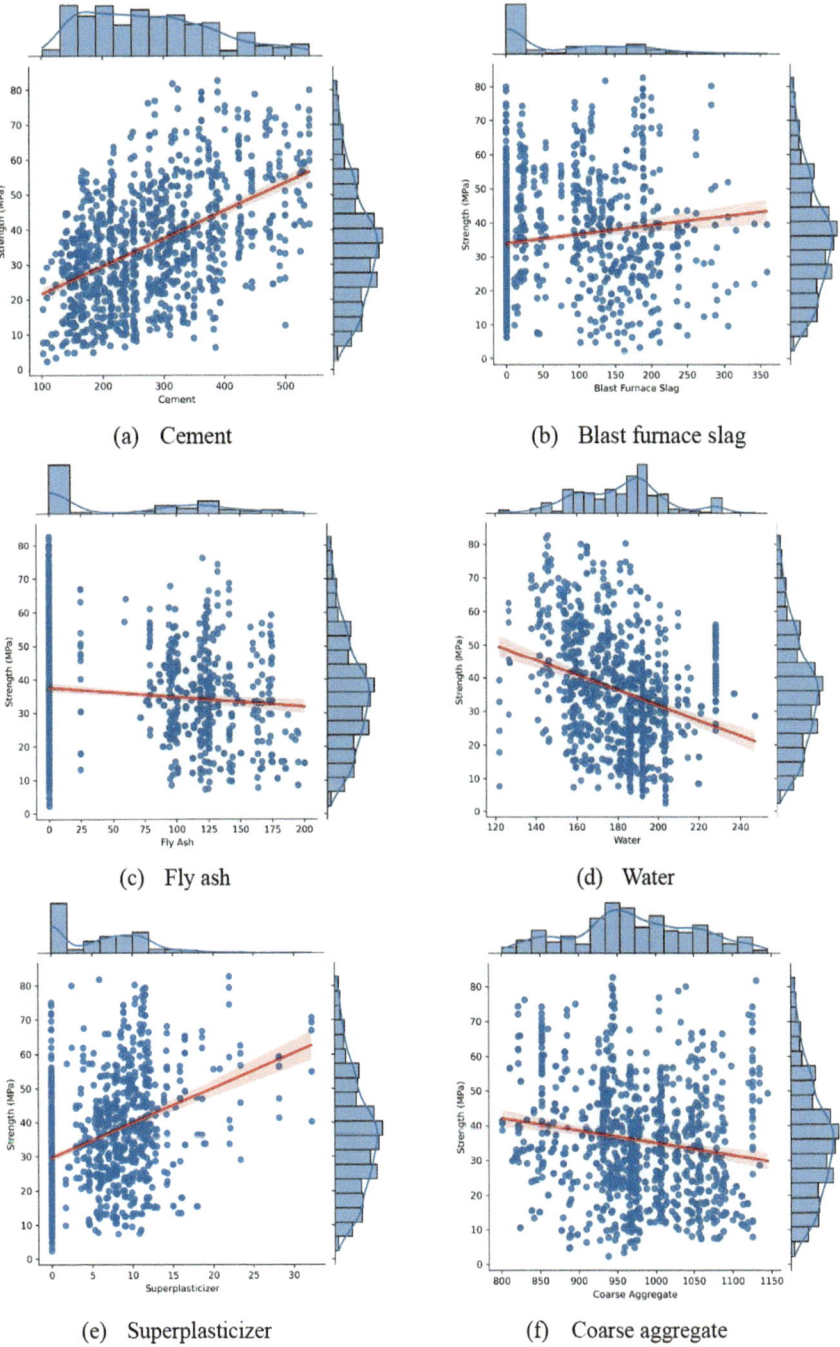

Fig. 1 Marginal histogram plots for the compressive strength versus (**a**) cement, (**b**) blast furnace slag, (**c**) fly ash, (**d**) water, (**e**) superplasticizer, (**f**) coarse aggregate, (**g**) fine aggregate, and (**h**) curing age

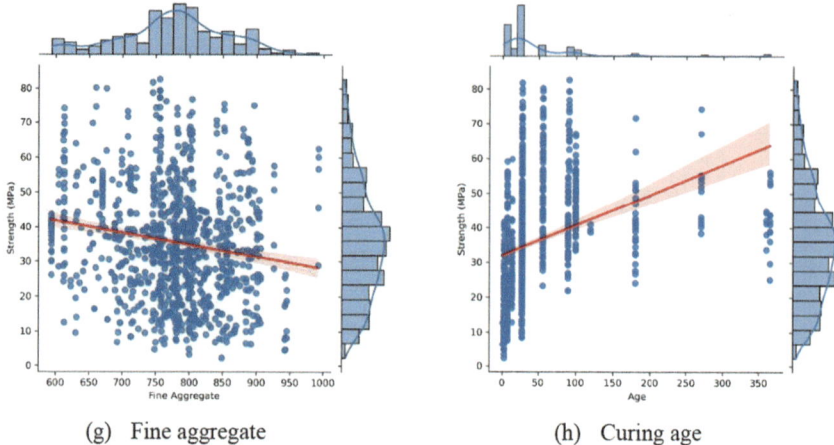

      (g) Fine aggregate   (h) Curing age

Fig. 1 (continued)

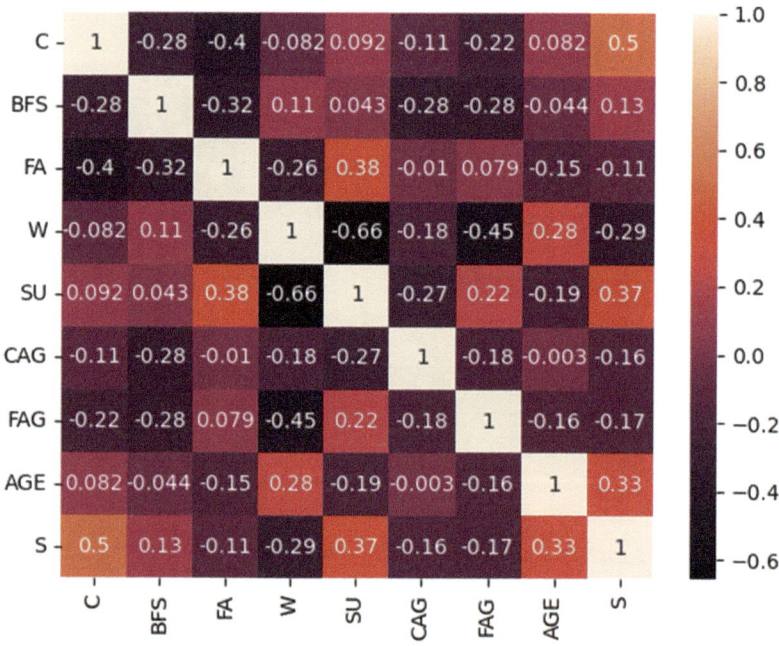

Fig. 2 Correlation matrix plot for input features and compressive strength

to output the final predictions, such as majority voting for classifiers and weighted sum for regressors. In this study, two bagging models (Random Forest, Extra Tree), two boosting models (XGBoost and LightGBM), three ensemble strategies (bagging, two-layer stacking, and weighted voting) were used in this study.

Four Base Leaners Random Forest (RF) new dataset with a replacement (bootstrap sampling) from an existing dataset and train several decision tree models with randomly selected features on the new dataset. The predictions of each decision tree model are aggregated into the final prediction. Extra Trees (ET) splits nodes by choosing cut points fully at random and that it uses the whole learning sample (rather than a bootstrap replica) to grow the trees. XGBoost is a gradient boosting tree (GBDT) based algorithm by constructing the objective function of model deviation and regularization term to prevent over-fitting. LightGBM is also a GBDT base algorithm that proposes two novel techniques, i.e., gradient-based one-side sampling (GOSS) and exclusive feature bundling (EFB), to realize faster training efficiency and lower memory usage.

Three Ensemble Strategies Bagging fits the base learner on random subsets of the original dataset and then aggregates their individual predictions to form a final prediction. Stacking uses the predictions of base learner as new training dataset and use a meta learner to learn the data. Voting fits the base learners on the original dataset and obtains a weighted average of each base learner's predictions.

3.2 Bayesian Optimization

Bayesian optimization (BO) was used in this study to optimize the hyperparameters. BO is an iterative stochastic optimization framework for the combined algorithm selection and hyperparameter (CASH) optimization problem [5]. It first builds a probabilistic surrogate model (Gaussian process or tree-based model) mapping from the hyperparameters to the objective metrics. Then, it defines an acquisition function to decide which hyperparameter configuration to evaluate next, balancing the exploration and exploitation during the search process.

3.3 Three-Level Ensemble Learning Framework

The proposed three-level ensemble learning framework is shown in Fig. 3. We split the data into training, validation, and test datasets. Level 1 includes four base learners: RF, ET, XGBoost, and LightGBM. Bayesian optimization is used to optimize the hyperparameters. The hyperparameter search space of the four base learners is presented in Table 1. Level 2 uses the optimized base learners of Level 1 to build six meta learners: four bagging models and one stacking model with default hyperparameters, and one voting model using the cross-validation loss of four optimized base learners as weights. Level 3 ensembles all base and meta learners using stacking and voting methods. Each trained model is applied for predictions based on the test dataset, and best machine learning model is selected after model performance evaluation for model explainability analysis.

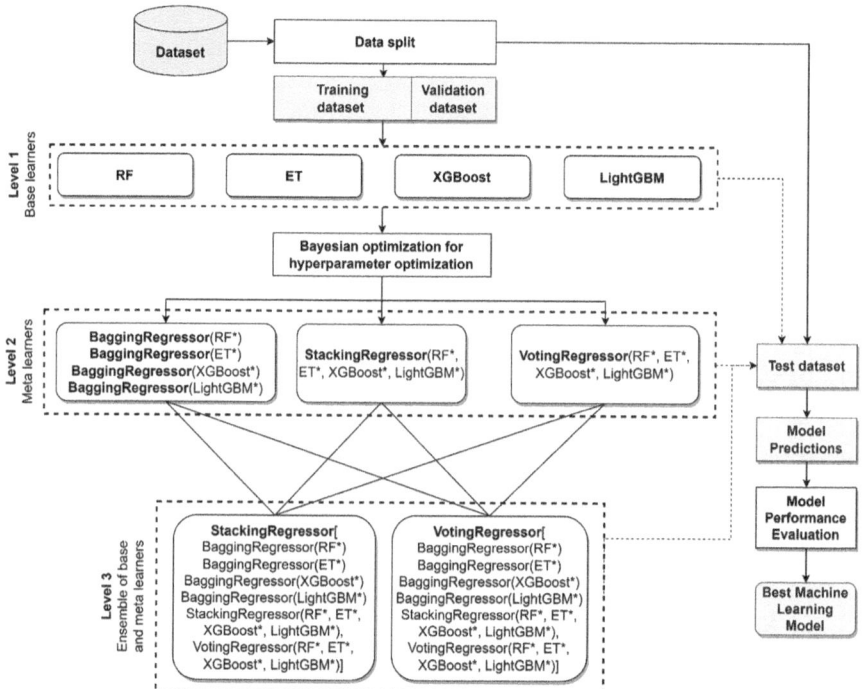

Fig. 3 Three types of ensemble strategies

Table 1 Hyperparameter search space

Model	Selected hyperparameters	Search space
RF	Number of trees in the forest (*n _ estimators*)	Range (50, 150)
	Maximum depth of the tree (*max _ depth*)	Range (10, 100)
ET	Number of trees in the forest (*n _ estimators*)	Range (50, 150)
	The maximum depth of the tree (*max _ depth*)	Range (10, 100)
LightGBM	Maximum number of estimators (*n _ estimators*)	Range (50, 150)
	Maximum tree depth for base learners (*max _ depth*)	Range (10, 100)
	Maximum tree leaves for base learners (*num _ leavess*)	Range (10, 50)
XGBoost	Number of gradient boosted trees (*n _ estimators*)	Range (50, 150)
	Maximum tree depth for base learners (*max _ depth*)	Range (10, 100)

3.4   XAI Methods

SHAP explains the prediction of an instance by computing the contribution of each feature to the prediction by computing Shapley values from coalitional game theory [6]. The SHAP value for feature j of observation x, $\phi_j(x)$, is defined as:

$$\phi_j(x) = \sum_{S \subseteq \{1,2,\ldots,M\} \setminus \{j\}} \frac{|S|!(M-|S|-1)!}{M!} \left[f\left(x_S \bigcup \{j\}\right) - f(x_S) \right] \qquad (1)$$

where S is a subset of the features with feature j excluded. M is the total number of features. f is the model's prediction function. Features with large absolute Shapley values are important, and therefore SHAP feature importance (FI) for feature j is calculated as (Molnar 2020):

$$\text{SHAP}_j = \frac{1}{n} \sum_{i=1}^{n} |\phi_j(i)| \qquad (2)$$

where $|\phi_j(i)|$ is the absolute value of SHAP value for feature j of observation i. PDP captures the relationship between the target variable and one or two feature by deriving the feature's marginal effect on the predictions [7]. The partial dependence function for given value(s) of features S shows the average marginal effect on the prediction and is defined as [8]:

$$\widehat{f}_S(x_S) = E_{X_C}\left[\widehat{f}(x_S, X_C)\right] = \frac{1}{n} \sum_{i=1}^{n} \widehat{f}\left(x_S, x_C^{(i)}\right) \qquad (3)$$

The x_S are the features for which the partial dependence function should be plotted and X_C are the other features used in the machine learning model $\widehat{f}$. The partial function $\widehat{f}_S$ is estimated with Monte Carlo method by calculating averages in the training data. $x_C^{(i)}$ is actual feature values of other features in which we are not interested, and n is the number of instances in the dataset.

4 Concrete Mix Optimization

4.1 Establishment of Objective Functions

The objectives are to minimize carbon emission $E(X)$, production cost $C(X)$, and maximize compressive strength $S(X)$ of 1 m³ of concrete during use phase according to mix design X. The compressive strength $S(X)$ is predicted with the ensemble learning methods. The carbon emission $E(X)$ can be calculated as:

$$E(X) = \sum_{m=1}^{M} e_m q_m \qquad (4)$$

Where e_m and q_m are the carbon emission factor and content of material m in concrete per volume. Production cost $C(X)$ can be calculated as:

$$C(X) = \sum_{m=1}^{M} c_m q_m \tag{5}$$

Where c_m is the unit price and content of material m in concrete per volume.

4.2 Establishment of Constraints

The constraints include the ratio constraints [Eq. (3)], absolute volume constraint [Eq. (4)], and range constraints [Eq. (5)]

$$R_k^l \le R_k \le R_k^u, k = 1, 2, 3 \ldots, K \tag{6}$$

$$\sum_{m=1}^{M} \frac{q_m}{\rho_m} = 1, m = 1, 2, 3 \ldots, M \tag{7}$$

$$q_i^l \le q_m \le q_i^u, m = 1, 2, 3 \ldots, M \tag{8}$$

where R_i^l, R_i^u are the lower and upper bounds of i-th ratio. q_m is the content of material m per cubic meter of concrete. ρ_m is the density of material m. q_m^l, q_m^u are the lower and upper bounds of material m. The parameters of carbon emission factors, unit prices, material density, and constraints are set the same as [9].

4.3 NSGA-II

Non-dominated sorting genetic algorithm (NSGA-II) is an efficient algorithm for solving multi-objective optimization problem. It is a genetic algorithm based on Pareto theory to improve mating and survival selection strategies [10]. The NSGA-II procedures are:

Step 1: Initiate the population.

Step 2: Double the size of initial parent population using mutation and crossover strategies.

Step 3: Sort the generated population into different ranks according to Pareto dominating rule.

Step 4: Calculate the crowding distance between the nearest two solutions in the same Pareto rank and a large crowding distance means a large space for group diversity.

Step 5: Select solutions with higher Pareto rank and crowding distance to form the parent group of the next generation.

Step 6: Repeat from Step 2 to Step 5 until the maximum number of loops is reached.

4.4 TOPSIS-Based Pareto Pruning

A Pareto pruning method applies a set of predefined rules to the resolution process to identify a focused subset of Pareto optimal solutions. In this study, we will use a multi-criteria decision-analysis (MCDA) based pruning method to obtain the optimal solution from Pareto front, namely, technique for Order Preference by Similarity to an Ideal Solution (TOPSIS). It scores the Pareto solution by calculating the distance between the positive and negative points [11]. The smaller the distance from the ideal positive point is, the higher the score is up to one. The ideal positive point and the ideal negative point can be obtained from the following equations:

$$d_{i+} = \sqrt{\sum_{j=1}^{n} \left(F_{ij} - F_j^{\text{ideal}} \right)^2} \tag{9}$$

$$d_{i-} = \sqrt{\sum_{j=1}^{n} \left(F_{ij} - F_j^{\text{non-ideal}} \right)^2} \tag{10}$$

Where d_{i-} and d_{i+} are the distances to the negative and positive ideal points, respectively. n is the number of objectives; i refers to a solution in the Pareto front set. F_j^{ideal} and $F_j^{\text{non-ideal}}$ are the ideal and non-ideal values for the j-th objective, respectively. The closeness coefficient is given by the following equation:

$$c_i = \frac{d_{i-}}{d_{i+} + d_{i-}} \tag{11}$$

The Pareto solution with the highest c_i is the final optimal solution.

5 Results and Discussion

5.1 Model Performance Results

A deep neural network (DNN) model with 8 inputs, 3 hidden layers (each with 20 neurons), and one output (8-20-20-20-1) was developed for comparison and the R squared is 0.8348. Table 2 compares the model performance and stacking method in level 2 (L2-Stacking) has the highest model performance in all performance evaluation criteria. Bagging methods are not always better than individual models except XGBoost, while stacking and voting have better prediction accuracy than individual models, indicating the necessity and advantages of aggregating the prediction results of individual models. A deeper level of stacking and voting does not yield better results, which may be due to the accumulation of model error. The L2-Stacking model is used for the following model interpretation analysis.

Table 2 Model performance comparison

Model	RMSE	%RMSE	MAE	%MAE	MAPE	R^2
DNN	6.4858	18.78%	4.7770	13.83%	0.1597	0.8348
L1-RF	4.5853	13.22%	3.2429	9.35%	0.1208	0.9202
L1-ET	4.3077	12.42%	2.7136	7.82%	0.0962	0.9295
L1-XGBoost	4.6444	13.39%	2.9745	8.58%	0.1076	0.9181
L1-LGBM	4.0461	11.66%	2.74	7.90%	0.0968	0.9378
L2-Bagging (RF)	4.8447	13.97%	3.5166	10.14%	0.1321	0.9109
L2-Bagging (ET)	4.3684	12.59%	2.9441	8.49%	0.1042	0.9275
L2-Bagging (XGBoost)	4.529	13.06%	3.1003	8.94%	0.1109	0.9221
L2-Bagging (LGBM)	4.226	12.18%	2.8888	8.33%	0.1016	0.9322
L2-Stacking	**3.8237**	**11.02%**	**2.5261**	**7.28%**	**0.0885**	**0.9445**
L2-Voting	3.9994	11.53%	2.6263	7.57%	0.0962	0.9393
L3-Meta Stacking	3.8538	11.11%	2.5665	7.40%	0.0919	0.9436
L3-Meta Voting	4.0192	11.59%	2.6885	7.75%	0.0977	0.9387

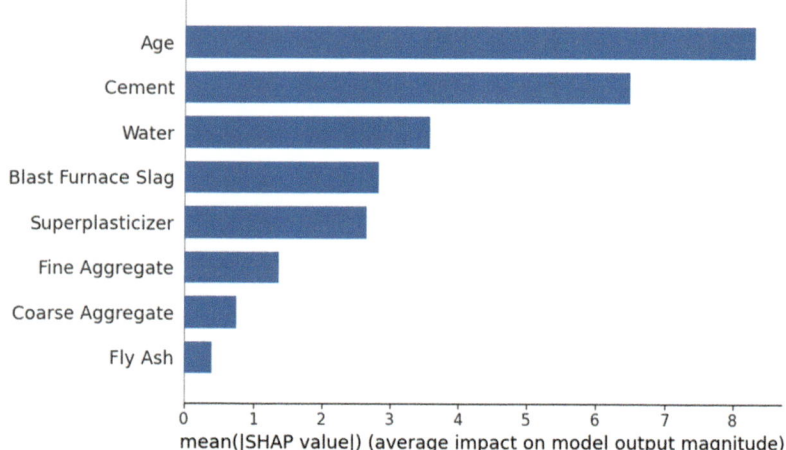

Fig. 4 SHAP summary plot for input features

5.2   Model Interpretation Results

Figure 4 provides the SHAP summary plot for input features and the top three important features are curing age, cement, and water. Fly ash has the least feature importance. To explore the relationships between input features and strength, we provide the PDP results of each input feature in Fig. 5. Most features have nonlinear relationships with the compressive strength, such as fly ash, water, superplasticizer, fine aggregate. It indicates that there might be an optimal content value for these materials to achieve largest strength. The strength increases linearly with more cement and slag. Longer curing time will bring higher compressive strength (Fig. 5).

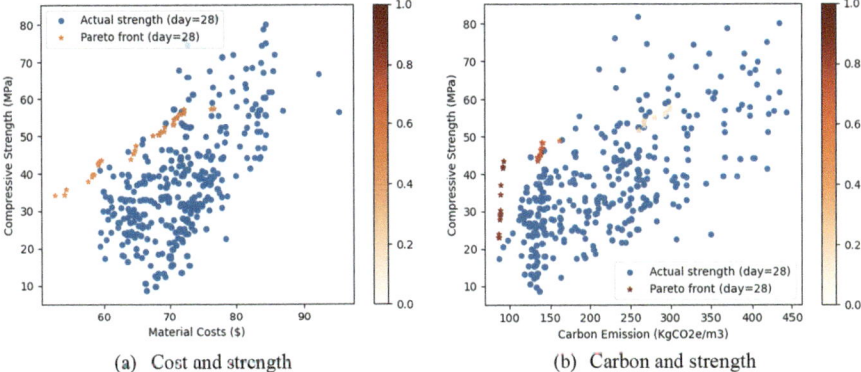

(a) Cost and strength (b) Carbon and strength

Fig. 5 Partial dependence plots for compressive strength and (**a**) cement, (**b**) blast furnace slag, (**c**) fly ash, (**d**) water, (**e**) superplasticizer, (**f**) coarse aggregate, (**g**) fine aggregate, and (**h**) curing age

5.3 MOOP Solutions for Concrete Mix

The curing age of the concrete was set to 28 days before optimization. Figure 6 shows the Pareto front and actual data points by setting objectives as (a) cost and strength, (b) carbon and strength, respectively. Higher compressive strength brings higher material costs and more carbon emissions. The TOPSIS score of all Pareto solutions are calculated and represented with color bar. The solution with highest score is selected as the optimal concrete mix. It should be noted that the NSGA-II fails to obtain an effective solution at strength higher than 50 MPa, which may be due to the limitation of the collected dataset. Therefore, a larger concrete mix dataset is required to increase the robustness of the optimization model.

6 Conclusion

This study firstly proposes a three-level ensemble learning framework for concrete strength prediction and then use NSGA-II for concrete mix optimization with two objectives: cost and strength, carbon emissions and strength, respectively. Ensemble learning methods such as stacking and bagging have better performance than individual models and should be well considered in the field of concrete property prediction. NSGA-II can effectively obtain the Pareto solutions to achieve a better tradeoff between technical properties and economic costs/environmental impacts. However, a larger concrete mix dataset is needed to increase the robustness of optimization results. Future research directions can be incorporation of domain knowledge into machine learning model, a hybrid of MOO algorithms, more interactive Pareto pruning methods based on decision makers' preference, etc.

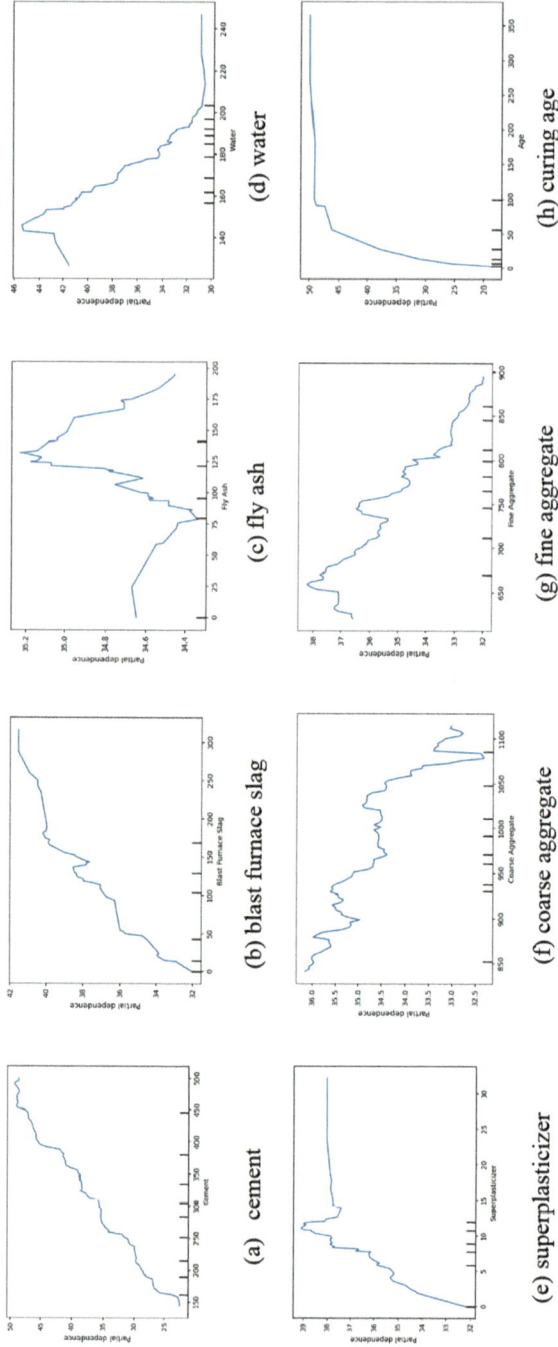

Fig. 6 MOO results of (**a**) cost and strength and (**b**) carbon and strength

References

1. Ministry of Housing and Urban-Rural Development (MHURD): Construction Supervision Penalty (2021), No. 40, Beijing, China (2021)
2. Monteiro, P.J.M., Miller, S.A., Horvath, A.: Towards sustainable concrete. Na Materials. **16**, 698–699 (2017)
3. Yeh, I.C.: Modeling of strength of high-performance concrete using artificial neural networks. Cement Concrete Res. **28**(12), 1797–1808 (1998)
4. Sagi, O., Rokach, L.: Ensemble learning: a survey. Wiley Interdiscip Rev Data Mining Know Dis. **8**(4), e1249 (2018)
5. Bischl, B., Binder, M., Lang, M., Pielok, T., Richter, J., Coors, S., et al.: Hyperparameter optimization: foundations, algorithms, best practices, and open challenges. Wiley Interdiscip Rev Data Mining Know Dis. **13**(2), e1484 (2023)
6. Lundberg, S.M., Lee, S.I.: A unified approach to interpreting model predictions. Adv Neural Infor Proc Sys. **30** (2017)
7. Friedman, J.H.: Greedy function approximation: A gradient boosting machine. Ann Stat, 1189–1232 (2001)
8. Molnar, C.: Interpretable machine learning. Lulu. com. (2020)
9. Jin, L., Zhang, Y., Liu, P., Fan, T., Wu, T., Wu, Q.: Carbon-footprint based concrete proportion design using LSTM and MOPSO algorithms. Materials Today Comm. **38**, 107837 (2024)
10. Deb, K., Pratap, A., Agarwal, S., Meyarivan, T.A.M.T.: A fast and elitist multiobjective genetic algorithm: NSGA-II. IEEE Trans Evol Comp. **6**(2), 182–197 (2002)
11. Hwang, C.L., Yoon, K., Hwang, C.L., & Yoon, K.: Methods for multiple attribute decision making. Multiple attribute decision making: methods and applications a state-of-the-art survey, 58–191 (1981)

Preliminary Environmental Assessment of Ultra-High-Performance Concrete Mixtures

Leila Farahzadi, Saeed Bozorgmehr Nia, Behrouz Shafei (iD),
and Mahdi Kioumarsi

Contents

1   Introduction

Ultra-high-performance concrete (UHPC) has become increasingly popular in the construction industry due to its exceptional mechanical properties and versatility in construction applications [1, 2]. UHPC offers a promising way to meet the rising need of robust infrastructure solutions with its high compressive strength, low permeability, and high durability [3]. However, the conventional production methods and mixture design of UHPC, which heavily rely on ordinary Portland cement (OPC), raise concerns about its environmental sustainability due to the significant carbon footprint associated with cement production [4–6]. Cement pro-

L. Farahzadi (✉) · M. Kioumarsi
Department of Built Environment, OsloMet – Oslo Metropolitan University, Oslo, Norway
e-mail: leilafar@oslomet.no

S. Bozorgmehr Nia · B. Shafei
Department of Civil, Construction and Environmental Engineering, Iowa State University, Ames, IA, USA

© The Author(s) 2025
M. Kioumarsi, B. Shafei (eds.), *The 1st International Conference on Net-Zero Built Environment*, Lecture Notes in Civil Engineering 237,
https://doi.org/10.1007/978-3-031-69626-8_49

duction is one of the considerable contributors to global carbon emissions [7–9], given estimates that cement production contributes about 8% of total CO_2 emissions globally [10].

Traditional UHPC compositions typically use OPC amounting to more than 800–1000 kg/m^3 [11, 12]. This extreme OPC proportion is considered unfavorable because of the immense energy required in its production, thus adding to environmental issues. To reduce or substitute the use of OPC in UHPC with alternative cementitious materials [13], researchers have searched for ways to reduce or replace OPC [14–17]. This includes directly replacing a portion of OPC with SCMs [18, 19] such as ground granulated blast furnace (GGBF) slag, fly ash (FA), and silica fume (SF).

This study aims to assess and compare the environmental impact of six UHPC mix designs incorporating SCMs of different proportions of GGBFS, FA, and SF as partial substitutes for Portland cement. A life cycle assessment (LCA) method, which is a standardized method to evaluate the environmental burdens associated with a product or process throughout its entire life cycle, is applied in this study [20].

The subsequent section details the materials and methods employed, material characteristics, and the production of different UHPC mixtures, ensuring a detailed examination of the research scope. Then, the life cycle inventory, system boundary, and impact assessment methodology are explained to maintain transparency and clarity in the evaluation process. Following this, the results section interprets the LCA findings, emphasizing reductions in global warming potential (GWP) achieved by alternative UHPC mix designs. Finally, the article concludes by highlighting the significant findings.

2   Materials and Methods

The materials and methods section is divided into two subsections: (1) Description of UHPC materials, mix design, and compressive strength, and (3) LCA methodology, as follows:

2.1   *UHPC Materials, Mixture Design, and Compressive Strength*

The formulation of UHPC relies on a unique combination of materials, with Portland cement as the foundation, adhering to ASTM C150-18 standards [21], significantly impacting UHPC strength and durability. SF, acting as a micro-filler from Master Builders Solutions, enhances strength and reduces concrete permeability, while GGBF slag, obtained from Levy Co., contributes to strength development, and offers environmental benefits. FA (Class C) from National Minerals Corporation,

Table 1 Mix design developed for UHPC mixtures, data for 1 m^3 of concrete

MIX ID	Portland cement	Silica fume	GGBF slag	Fly ash	Sand	HRWR	Steel fibers	W/B
UHPC-1	800	200	–	–	1100	37.0	156 (2% vol)	0.2
UHPC-2	600	100	–	300	1100	35.0	156 (2% vol)	0.2
UHPC-3	500	100	400	–	1100	34.5	156 (2% vol)	0.2
UHPC-4	500	100	300	100	1100	35.6	156 (2% vol)	0.2
UHPC-5	400	–	600	–	1100	32.0	156 (2% vol)	0.2
UHPC-6	250	50	500	200	1100	33.2	156 (2% vol)	0.2

known for its pozzolanic activity, is also utilized to improve long-term strength and durability. MasterGlenium 7920, a high-range water-reducing admixture (HRWA), and smooth straight steel fibers are incorporated into the mix.

The UHPC mix design aims to optimize material gradation for high packing density and maintain a low water-to-binder (w/b) ratio. A consistent w/b ratio of 0.2 is applied, following research suggesting optimal cement hydration [22]. Six UHPC mixtures were investigated, with variations in the proportions of GGBF slag, FA, SF, alongside the base components, as detailed in Table 1.

Mechanical properties, including compressive strength, of the UHPC mixes were evaluated, with all mixes falling within the same UHPC class. The results of this evaluation are essential to determine the optimal UHPC mix design, balancing environmental sustainability with desired mechanical properties.

2.2 Life Cycle Assessment Methodology

LCA was conducted to evaluate the environmental impact of different UHPC mix designs. LCA is a standardized framework used to assess the environmental impacts associated with a product or service throughout its entire life cycle [23, 24]. The framework follows four key phases: (1) Goal and Scope Definition, (2) Life Cycle Inventory (LCI) Analysis, (3) Life Cycle Impact Assessment (LCIA), and (4) Interpretation.

For this study, a "cradle-to-gate" LCA approach was employed according to the EN 15804 method [25]. This means the assessment focuses on the environmental impacts of the UHPC mix design, from raw material acquisition to production at the factory gate. Data for the LCA was collected from the Ecoinvent database (version 3.9.1) and analyzed using Simapro software 9.5.

3 Life Cycle Assessment

3.1 Goal and Scope Definition

This study investigates the environmental impact of six UHPC formulations. The main goal is to assess the potential of alternative materials to reduce the GWP associated with UHPC production and inform design strategies that prioritize sustainability. LCI was conducted using a global perspective. The analysis focuses on the cradle-to-gate stages of the UHPC production process, encompassing: (A1) Raw Material Acquisition: This includes the extraction and processing of all raw materials used in the UHPC mix; (A2) Transportation: This considers the environmental burdens associated with transporting raw materials to the production facility; (A3) Processing and Production: This stage covers the energy consumption and emissions generated during UHPC manufacturing. This framework aligns with the "cradle-to-gate" approach outlined in the EN 15804 standard [25].

To ensure a fair comparison, as recommended by ISO 14040 [24], a functional unit of 1 cubic meter (m^3) of UHPC with a specified compressive strength (meeting UHPC class requirements) was established. This functional unit ensures all UHPC mixtures are evaluated based on equivalent performance, facilitating a consistent comparison across production methods, strength classes, and potential applications. Furthermore, for this study, all UHPC mixes were selected for comparable durability characteristics such as surface resistivity, compressive strength, and overall durability. These similarities allow us to isolate the environmental impact of different mix designs while maintaining equal performance capabilities within the same UHPC class.

3.2 Life Cycle Inventory

A conventional UHPC comprises OPC, water and fine aggregates, SF, and reinforcement fibers. Six different UHPC combinations have been developed. All compositions belong to the same compressive strength class. The alternative compositions incorporate different proportions of SCMs as replacements for cement.

The LCI for this study focused on the "cradle-to-gate" stage of the UHPC production process, encompassing all materials, energy consumption, and transportation associated with UHPC mix design. The inventory for the production phase relied on generic data from Ecoinvent 3.9.1, while the concrete production process is modeled based on North American high-strength concrete (50 MPa). However, data for electricity, heat, and raw material markets were adjusted to reflect global or rest-of-the-world averages to ensure global applicability. All major raw materials, energy inputs, and transportation accounted based on Ecoinvent v3.9.1 cut-off datasets, aligning with the data quality and allocation requirements outlined in EN 15804. Transportation of materials to the factory for UHPC production is represented according to market processes in Ecoinvent for each respective raw material.

3.3 Life Cycle Impact Assessment

This study employed the EN 15804 method [25] for the LCIA, specifically the most recent approach, EN 15804 +A2 Method V1.01/EF 3.0 normalization and weighing set. It is categorized as a midpoint method. This midpoint method is recognized as the current industry standard and calculates impact indicators across various environmental aspects.

The chosen impact categories encompass various impact categories, including climate change, ozone depletion, ionizing radiation, photochemical ozone formation, particulate matter formation, human toxicity (both carcinogenic and noncarcinogenic), acidification, eutrophication (freshwater, marine, and terrestrial), freshwater ecotoxicity, land and water use, and resource depletion (fossil fuels and minerals & metals).

This LCIA approach served a dual purpose. Firstly, it enabled the calculation of specific environmental impacts, particularly the GWP. Secondly, it facilitated the proposal of optimized UHPC alternatives based on characterization and normalization results at the midpoint level. This streamlined approach condenses diverse environmental burdens into a single metric, boosting result interpretation and enabling a comparative assessment of the sustainability implications associated with different material usage scenarios within UHPC formulations.

3.4 Results and Interpretation

This section will provide an overview of the results of the life cycle impact assessment. It summarizes the results obtained in the LCIA of 1 m³ UHPC. LCIA results provide a comprehensive understanding of the environmental performance of different UHPC mixtures.

Figure 1 shows the weighted environmental impacts of UHPC alternatives, exhibiting higher values across most impact assessment categories for UHPC1,

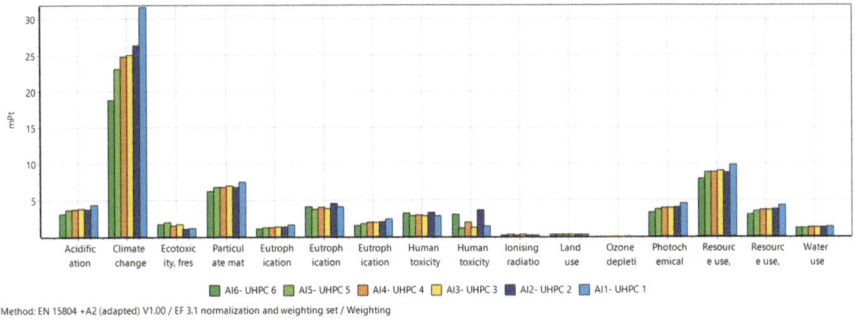

Fig. 1 Impact contribution analysis of UHPC mixture

Global Warming Potential (mPt)

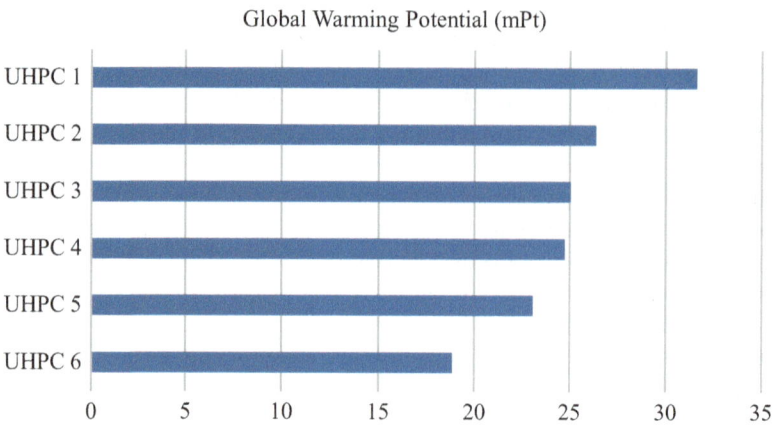

Fig. 2 Global warming potential of UHPC mixtures

except for ionizing radiation and human toxicity. However, the effects of all damage categories including these two mentioned ones, are far less than those of climate change and GWP. This underscores the importance of addressing GWP in UHPC production.

Figure 2 illustrates the GWP of six UHPC mix designs. UHPC1 which contains 800 kg of OPC and 200 kg of SF in 1 m^3 of UHPC has the highest GWP. UHPC 6 including 250 kg OPC and 750 kg SCMs (500 kg GGBF slag, 50 kg SF, and 200 kg FA) has the lowest GWP. UHPC 5 ranks second in terms of having the lowest GWP. This mix design contains 400 kg OPC and 600 kg GGBF slag. As the content of OPC reduces, the impact of GWP and climate change decreases. UHPC 3 and UHPC 4 have almost similar GWPs (same OPC proportion); however, they are a bit lower for UHPC 4 due to 100 kg FA content.

As the amount of superplasticizer is a bit different in each UHPC mixture, and it may affect the results to some extent, in Fig. 3, this effect has been excluded, and only the effect of OPC replacement with SCMs on GWP is considered. By replacing 75% of OPC with SCMs, climate change and GWP can be minimized by almost 40% (UHPC 6).

4   Discussion

This study investigated the environmental impact of incorporating SCMs such as GGBF slag, FA, and SF as partial replacements for OPC in UHPC formulations. A LCA was conducted to assess the environmental impact of six innovative UHPC compositions. The findings demonstrate that significant reductions in GWP, a key indicator of climate change impact, can be achieved by strategically incorporating SCMs at tested amounts in the laboratory.

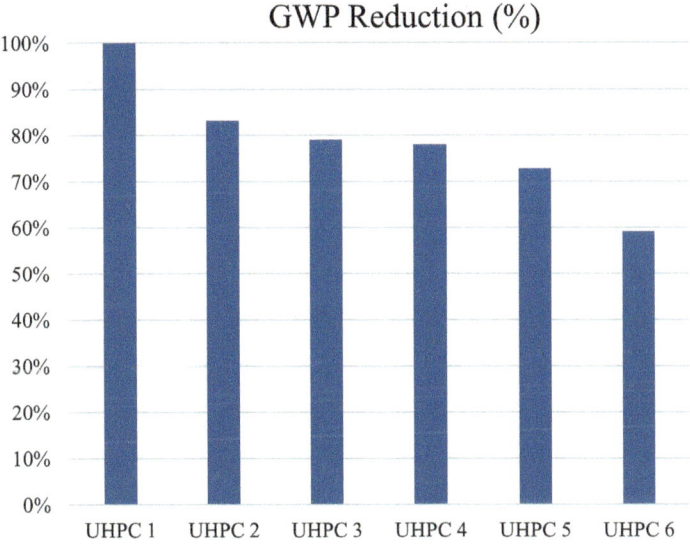

Fig. 3 Global warming potential percentage reduction of UHPC mixtures

16.7 to 40.4 % of reduction in GWP was achieved by replacing OPC with different proportions of GGBF slag, FA, and SF. Replacing three-quarters of the OPC with SCMs, climate change and GWP reduced by up to 40%. This finding is consistent with other research highlighting the potential of SCMs to minimize the environmental impact of concrete production. This finding reinforces the notion that OPC production is a major contributor to the environmental impact of UHPC due to the high energy consumption and CO_2 emissions. Existing research also shows that replacing a significant portion of OPC with SCMs in UHPC can significantly reduce the environmental impact, particularly in terms of CO_2 emissions and GWP [26, 27]. In a study by Miller [28], the author states that the replacement of significant amounts of SCMs does not invariably lead to a consistent reduction in greenhouse gas emissions during the production of UHPC on a per unit strength basis. This indicates complexities and trade-offs are involved in optimizing concrete mixtures for both strength and environmental sustainability. Therefore, mix designs tested in the laboratory and categorized within similar strength classes are crucial. Researchers and engineers can identify the optimal balance between strength and environmental sustainability by evaluating mixtures with comparable strength classes, ensuring that concrete structures meet structural requirements while minimizing their carbon footprint.

A key strength of this study lies in the innovative mix design that incorporates various SCMs in different ratios. These novel UHPC formulations have been experimented in the laboratory to evaluate their mechanical properties, particularly compressive strength, ensuring they meet the specified class requirements. This combination of environmental assessment through LCA and mechanical performance testing strengthens the overall significance of this research.

One limitation of this study is the focus on the "cradle-to-gate" LCA stage. While this approach provides valuable insights into the environmental impact of UHPC production, a more comprehensive assessment could include the "use" and "end-of-life" stages. Future research could explore the potential benefits of incorporating recycled content in UHPC formulations and investigate the environmental implications of the entire UHPC life cycle. Furthermore, exploring regional consideration of material characteristics, investigating the long-term durability of these innovative mixtures, and considering a life cycle cost analysis for a more comprehensive understanding of environmental and economic benefits can be considered.

5 Conclusion

This study successfully demonstrates that incorporating SCMs as partial replacements for OPC in UHPC formulations presents a viable strategy for reducing the environmental impact of UHPC production, particularly concerning GWP. A structured LCA focused on a cradle-to-gate perspective, encompassing raw material extraction, transportation, and production phases. EN 15804 standard and the Simapro software, incorporating data from the Ecoinvent v.3.9.1 and EPDs, were utilized for the assessments. By strategically utilizing SCMs like GGBF slag, fly ash, and silica fume, UHPC production can become more sustainable while maintaining the desired performance characteristics. These findings highlight the importance of exploring alternative cementitious materials and optimizing mix designs to promote sustainable practices in the construction sector. Future research can further explore the potential of SCMs in UHPC formulations, considering incorporation of recycled content, optimal replacement levels, and life cycle cost implications.

Acknowledgments This work is a part of the TRANSFORM project, which has received funding from the Norwegian Directorate for Higher Education and Competence (HK-dir), project number UTF-2020/10107. Additionally, we acknowledge the support from the Net-Zero Future project, an INTPART project funded by the Norwegian Directorate for Higher Education and Competence (HK-dir) and the Research Council of Norway, project number 337262.

References

1. Jonnalagadda, S., Chava, S.: Ultra-High-Performance Concrete (UHPC): a state-of-the-art review of material behavior, structural applications and future. Electron. J. Struct. Eng. **23**, 25–30 (2023)
2. Khayat, K.H., et al.: Rheological properties of ultra-high-performance concrete — an overview. Cem. Concr. Res. **124**, 105828 (2019)
3. Azmee, N.M., Shafiq, N.: Ultra-high performance concrete: from fundamental to applications. Case Stud. Constr. Mater. **9**, e00197 (2018)

4. Rahla, K.M., Mateus, R., Bragança, L.: Comparative sustainability assessment of binary blended concretes using Supplementary Cementitious Materials (SCMs) and Ordinary Portland Cement (OPC). J. Clean. Prod. **220**, 445–459 (2019)
5. Ghafari, E., Costa, H., Júlio, E.: Critical review on eco-efficient ultra high performance concrete enhanced with nano-materials. Constr. Build. Mater. **101**, 201–208 (2015)
6. Farahzadi, L., Kioumarsi, M.: Application of machine learning initiatives and intelligent perspectives for CO2 emissions reduction in construction. J. Clean. Prod. **384**, 135504 (2023)
7. Farahzadi, L., Kioumarsi, M.: Intelligent initiatives to reduce CO2 emissions in construction. In: 8th European Congress on Computational Methods in Applied Sciences and Engineering (ECCOMAS), Oslo, Norway (2022)
8. Farahzadi, L., et al.: Assessment of alternative building materials in the exterior walls for reduction of operational energy and CO2 emissions. Int. J. Eng. Adv. Technol. **5**(5), 183–189 (2016)
9. Hosseini, S.A., Farahzadi, L., Pons, O.: Assessing the sustainability index of different post-disaster temporary housing unit configuration types. J. Build. Eng. **42**, 102806 (2021)
10. Khaiyum, M., Sarker, S., Kabir, G.: Evaluation of carbon emission factors in the cement industry: an emerging economy context. Sustainability. **15**, 15407 (2023)
11. Li, P.P., Yu, Q.L., Brouwers, H.J.H.: Effect of coarse basalt aggregates on the properties of Ultra-High Performance Concrete (UHPC). Constr. Build. Mater. **170**, 649–659 (2018)
12. Shi, Y., et al.: Design and preparation of Ultra-High Performance Concrete with low environmental impact. J. Clean. Prod. **214**, 633 (2019)
13. Farahzadi, L., Kioumarsi, M., Shafei, B.: Life cycle assessment of concrete mixes for global warming mitigation and carbon footprint reduction. In: Global Cleaner Production Conference (JCPC 2023), Shanghai, China (2023)
14. Farahzadi, L., Kioumarsi, M.: Preliminary life cycle assessment and environmental impact evaluation of RC Bridge Deck: a case study in Norway. In: 6th International Conference on Civil Engineering and Architecture (ICCEA 2023), Bali Island, Indonesia (2023)
15. Wang, Y., et al.: Preparation of sustainable ultra-high performance concrete (UHPC) with ultra-fine glass powder as multi-dimensional substitute material. Constr. Build. Mater. **401**, 132857 (2023)
16. Rai, B., Wille, K.: Recycled glass powder as an alternative to fly ash in non-proprietary UHPC: a comparative study of resource-efficient design, mechanical and durability properties. J. Clean. Prod. **451**, 141907 (2024)
17. Zhang, X., et al.: Trends toward lower-carbon ultra-high performance concrete (UHPC) – a review. Constr. Build. Mater. **420**, 135602 (2024)
18. Shekarchi, M., et al.: Natural zeolite as a supplementary cementitious material – a holistic review of main properties and applications. Constr. Build. Mater. **409**, 133766 (2023)
19. Azarhomayun, F., et al.: Combined use of sewage sludge ash and silica fume in concrete. Int. J. Concr. Struct. Mater. **17**(1), 34 (2023)
20. Laurin, L., Overview of LCA—History, Concept, and Methodology, in Encyclopedia of Sustainable Technologies, M.A. Abraham, Editor. Elsevier: Oxford. p. 217–222 (2017)
21. ASTM International, Standard Specification for Portland Cement. ASTM International: West Conshohocken, PA, USA (2022)
22. Zhang, Z., et al.: Mitigating shrinkage in ultra-high performance concrete using MgO expansion agents with different activity levels. Front. Mater. **9**, 1033467 (2022)
23. British Standards Institute Staff: ISO 14044—Environmental Management—Life Cycle Assessment—Requirements and Guidelines, London, UK (2006)
24. British Standards Institute Staff: ISO 14040—Environmental Management—Life Cycle Assessment—Principles and Framework, London, UK (2006)
25. British Standards Institution: EN 15804—Sustainability of Construction Works: Environmental Product Declarations; Core Rules for the Product Category of Construction Products, London, UK (2011)

26. Srivastava, V., et al.: Supplementary cementitious materials in construction – an attempt to reduce CO2 emission. J. Environ. Nanotechnol. **7**, 31–35 (2018)
27. Park, S., et al.: The role of supplementary cementitious materials (SCMs) in ultra high performance concrete (UHPC): a review. Materials. **14** (2021)
28. Miller, S.A.: Supplementary cementitious materials to mitigate greenhouse gas emissions from concrete: can there be too much of a good thing? J. Clean. Prod. **178**, 587–598 (2018)

Compressive Strength Gain of Glass Powder–Portlandite: An Investigation Toward Maximizing the Use of Waste Glass as Cement Replacement in Concrete

Gaurav Chand ⓘ, Mithila Achintha ⓘ, and Yong Wang ⓘ

Contents

1 Introduction

Supplementary cementitious materials (SCMs) such as fly ash and Ground Granulated Blast Furnace slag (GGBS) have proven to be successful partial replacement to cement, but their availability is declining due to shutting down of the coal-based power plants in various parts of the World. Use of glass in buildings has been increased in recent years [1, 2] which eventually contributes to high volumes of waste glass from construction and demolition (C & D). Recycling waste glass as a partial replacement to cement is a useful recycling method for construction sector waste glasses, which are usually not recycled [3]. Although attempts were made to use waste glass coarse aggregate is concrete, the effects of Alkali Silica Reaction (ASR) prevent the successful application of the technology [4].

G. Chand · M. Achintha (✉) · Y. Wang
School of Engineering, The University of Manchester, Manchester, UK
e-mail: Mithila.Achintha@manchester.ac.uk

© The Author(s) 2025
M. Kioumarsi, B. Shafei (eds.), *The 1st International Conference on Net-Zero Built Environment*, Lecture Notes in Civil Engineering 237,
https://doi.org/10.1007/978-3-031-69626-8_50

There have been studies on the use of glass powder as partial replacement to cement (up to 20% of cement replacement with glass powder) in concrete where the results showed comparable compressive strength as the reference concrete [5, 6]. The cementitious characteristic of glass powder was believed to be due to the formation of calcium–silicate–hydrate (C–S–H) as a result of the reactions between the Silica (SiO_2) in glass powder and the Portlandite (by-product from cement hydration) [5, 6]. However, the formation of C–S–H from the reaction between glass powder–Portlandite is not well understood. No research which investigates the chemical reactions between glass powder and Portlandite has been reported in the literature. Understanding of the reaction between glass powder–Portlandite is required to go beyond the current wisdom of 20% cement replacement with glass powder.

The present combined theoretical and experimental study investigated the compressive strength gain and the formation of C–S–H of calcium hydroxide (CH) (representative of Portlandite available from cement hydration in a real concrete mix) and glass powder (GP) mixes. Results from a mole-concept based theoretical analysis of the chemical reaction were used to determine a mix design for CH and GP mix (this mix is denoted by CH–GP in the present paper). Reference mixes, 100% calcium hydroxide with water (denoted as 100CH) and 100% CEM-I (denoted as 100CEM-I), were used for the comparisons. The compressive strength of CH–GP, 100CH and 100CEM-I mixes was investigated and the formation of C–S–H in CH–GP specimens was studied using XRD analysis.

2 Materials and Properties

2.1 Glass Powder

The washed and cleaned waste glass was crushed using a Jaw crusher machine available at the University of Manchester [7] (400 kW power and 200 rpm). The crushed waste glass was then prepared as a powder of particle size less than 75 μm using a vibratory ball milling machine [8] equipped with Zirconium ball media. The glass powder of particle size less than 75 μm was chosen in the present study because the results reported in literatures suggest this particle size reacts in cement/concrete mixes and ensure no adverse effect due to ASR [5, 6]. The glass powder obtained after milling is shown in Fig. 1a. The particle size distribution of the glass powder was investigated using a Mastersizer 3000, Laser Particle size analyzer [9] where it was determined that 50% of the glass powder particle size was <40 μm and 90% was <75 μm.

The oxides of Silicon and Calcium of the used glass were determined using X-ray Fluorescence (XRF) method [10] where the oxides of Silicon and Calcium were 70.4% and 10.1%, respectively shown in Table 1.

(a)   Glass powder   (b)   Calcium Hydroxide powder

Fig. 1 (a) Glass powder (b) Calcium hydroxide powder

Table 1 The main chemical compounds of the glass

Chemical compounds	SiO_2	CaO	Al_2O_3	$Na_2O + K_2O$	Fe_2O_3	SO_3
% content	70.4	10.1	3.7	12.5	<1	1–2

2.2 Calcium Hydroxide Powder

Calcium hydroxide powder with particle size less than 100 μm was used in the present study because it was the nearest particle size to that of the glass powder (75 μm) available to purchase from a commercial supplier [11]. The similar sizes of both materials were usually expected to ensure chemical reactions given their similar surface areas [12]. The specific gravity of the calcium hydroxide powder is 2.24 g/cm^3 at 20 °C with a bulk density of 540 kg/m^3. The used calcium hydroxide powder is shown in Fig. 1b.

2.3 Cement (CEM-I)

The cement used in this study was CEM-I, grade 52.5. The percentage amount of C_3S and C_2S in the CEM-I cement was experimentally determined to be 54 and 20.7, respectively [13]. The particle size distribution of the cement was investigated using a Mastersizer 3000, Laser Particle size analyzer [9] where it was determined that 50% of the cement particles were less than 18 μm in size and 90% of the particles were less than 67 μm.

3 Glass Powder Reaction with Calcium Hydroxide: The Theoretical Analysis

The equation of the chemical reaction between calcium hydroxide and silica from glass powder in the presence of water can be represented as:

$$3\left[Ca(OH)_2\right] + \quad 2SiO_2 + H_2O \rightarrow \qquad (CaO)_3\,(SiO_2)_2(H_2O)_4$$

$$CH \qquad\qquad S \qquad\qquad\qquad C{-}S{-}H$$

$$3\times(40+(16+1)\times2)\;\; 2\times(28+16\times2)\;\; (40+16)\times3+(28+16\times2)\times2+(1\times2+16)\times4$$

$$222 \qquad\qquad 120 \qquad\qquad\qquad 360$$

$$z \qquad\qquad 0.54z \qquad\qquad\qquad 1.62z$$

$$(1)$$

As can be noted from Eq. (1), C–S–H (i.e., the same main chemical compound resulted in from hydration of regular cement) is expected from the chemical reaction between glass powder and CH. The microstructure of C–S–H is complex and not unique. For example, based on the results of research investigations focusing on cement hydration, it is believed Ca/Si ratio of ~1.5 in C–S–H ensures high degree of polymerization of the silicate chains ensuring crystalline/ordered microstructure, thereby resulting in high strength in concrete [12, 14]. Although C–S–H with Ca/Si ratio greater than 1.5 and less than 1.5 could also form, their strength contributing characteristics is expected to be less than that of the C–S–H with Ca/Si ratio 1.5 [12]. Therefore, considering C–S–H with Ca/Si as 1.5 to be the most desirable in terms of strength contribution, the present study aimed to achieve C–S–H with Ca/Si = 1.5. As demonstrated in Eq. (1), the C–S–H with Ca/Si ratio of 1.5 can be expected with the weight ratio between SiO_2 (S) and $Ca(OH)_2$ (CH) at 0.54 (S/CH ratio) on the reactants side (i.e., left hand side) of the equation.

3.1 Mix Design of Calcium Hydroxide and Glass Powder

Given glass powder contains CaO (see Table 1). It was assumed that this CaO would also contribute to $Ca(OH)_2$ in the mix in addition to $Ca(OH)_2$ directly provided to the mix as $Ca(OH)_2$.

Let assume the required amount of direct $Ca(OH)_2$ was y and CaO from glass powder was x (considering the chemical compounds shown on Table 1, the amount of Silica was about 7 times that of CaO, so the amount of SiO_2 was $7x$ and total weight of glass powder was $10x$).

$$
\begin{array}{ccccc}
\text{CaO} & + & \text{H}_2\text{O} & \rightarrow & \text{Ca(OH)}_2 \\
40 + 16 & & 1 \times 2 + 16 & & 40 + (16 + 1) \times 2 \\
56 & & 18 & & 74
\end{array}
\tag{2}
$$

Based on Eqs. (1) and (2),

$$
\left(\frac{74}{56}\right)x + y = 222 \ \ (\text{i.e. total Ca(OH)}_2 \text{ in Eq.}(1))
\tag{3}
$$

$$
7x = 120 \ (\text{i.e., total SiO}_2 \text{ in Eq.}(1))
\tag{4}
$$

On solving Eqs. (3) and (4),

$$
x = 17.14, \ \ y = 200
$$

Therefore, required glass powder $= 10x = 10 \times 17.14 = 172$ and $Ca(OH)_2 = 200$
Therefore, CH–glass powder mix with 46% wt. of glass powder (i.e., 172/372) and 54% (i.e., 200/372) wt. of $Ca(OH)_2$ was required.

4 Experimental Program

4.1 Specimen Preparation

The glass powder was mixed with calcium hydroxide and about 5% of water by weight (this quantity of water ensured the desired consistency of the paste) to prepare CH–GP test specimens. Cubes (2 cm) were cast (see Fig. 2a), and the specimens were taken out of the molds ~24 h after casting and kept in water for final curing. Figure 2b shows a CH–GP specimen obtained after opening the molds. The first category of reference specimens of the same size was made by mixing calcium

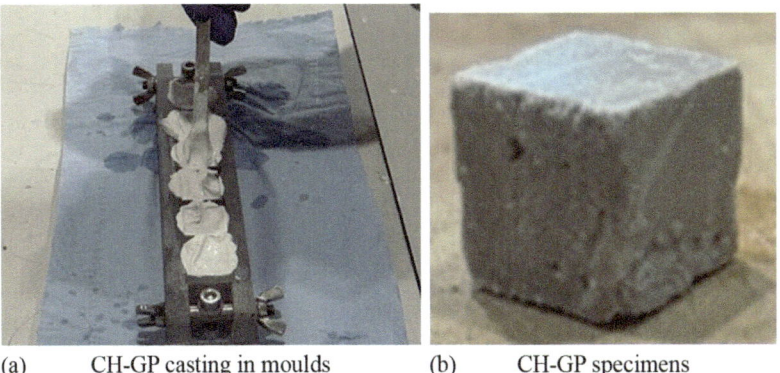

(a)　　　CH-GP casting in moulds　　　(b)　　　CH-GP specimens

Fig. 2 (a) CH–GP casting in molds (b) CH–GP specimens

hydroxide powder with water (denoted by 100CH in the paper). Similarly, the second category of reference specimens with 100% CEM-I (denoted by 100CEM-I) were also prepared. The experimental program focused on two main investigations: (1) compressive strength and (2) formation of C–S–H using XRD analysis.

4.2 Results and Discussion

Compressive strength The compressive strength of the test specimens was determined using a compression test carried out on a displacement controlled at a rate of 0.3 mm/min (a rate as suggested in the literature [12, 15]). The reference 100CH specimens were cracked soon after they were cast (see Fig. 3a) and the specimens were broken into pieces once they were taken out of the molds (see Fig. 3b). Therefore, no compression tests were carried out on 100CH test specimens and it was assumed that the compression strength of the 100CH test specimens was zero.

At least three test specimens of CH–GP and 100CEM-1 categories were tested at 3, 14, 28 and 90 days after curing. Given the reaction between glass powder and calcium hydroxide were reported to be slow compared to that of the cement hydration [5, 6], it was decided to monitor the compressive strength gain and the formation of C–S–H in both CH–GP and the reference 100CEM-1 specimens for longer duration than usual 28-day reference for regular concrete. The present study is still ongoing and only the results up to 90 days after curing are reported in the present paper.

The color of the CH–GP specimen was more whitish compared to the 100CEM-I specimens during the entire curing period. This could be attributed to the less presence of Fe_2O_3 in glass powder (~1%, see Table 1) compared to that in cement

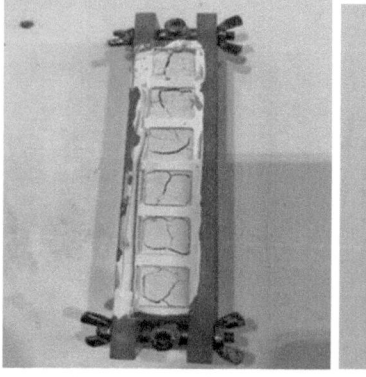 

(a) 100CH specimen in moulds   (b) Broken pieces of 100CH

Fig. 3 (a) 100CH specimen in molds (b) Broken pieces of 100CH

Table 2 Failure characteristics of 100CEM-I and CH–GP specimens before and after compression test

	3 days		14 days	
CH–GP	Before	After	Before	After
100 CEM-I				
	28 days		90 days	
CH–GP	Before	After	Before	After
100 CEM-I				

($\sim$5% of Fe_2O_3 in cement) [12, 13]. The cubes edges of the CH–GP specimens were blunt compared to the sharp edges in 100CEM-I specimen.

Table 2 shows the failure characteristics of the test specimens at 3, 14, 28 and 90 days. All specimens, expect CH–GP specimens at 3 days of curing, failed in brittle manner. The brittle failure of the CH–GP specimens (i.e., at 14 days and later) suggests the microstructure of the specimens was similar to that of the 100CEM-1 specimens. The crumbling nature of failure of the CH–GP specimens at 3 days after casting suggests a different microstructure compared to 100CEM-1 specimens and CH–GP specimens at later days of curing. The authors believe this was due to the ongoing development of the microstructure of the CH–GP test specimens at early days of curing and the more stable microstructure would only fully develop at a later stage of curing

The average compressive strength of the CH–GP specimens at 3, 14, 28, and 90 days were 3.8 (range 3.9–4.1 MPa), 5.0 (range 4.7–5.3 MPa), 7.8 (range 7.6–8.1 MPa) and 12.3 (range 11.8–12.9 MPA), respectively. The average compressive strength of 100CEM-I specimen at 3, 14, 28 and 90 days were 32.1 (range 29.5–34.9 MPa), 41.6 (range 40.2–43.4 MPa), 56.1 (range 53.5–58.6 MPa) and 65.4 (range 63.6–67.4 MPa), respectively. Figure 4a shows the recorded compressive strength data of both CH–GP and 100CEM-1 test specimens. Although the

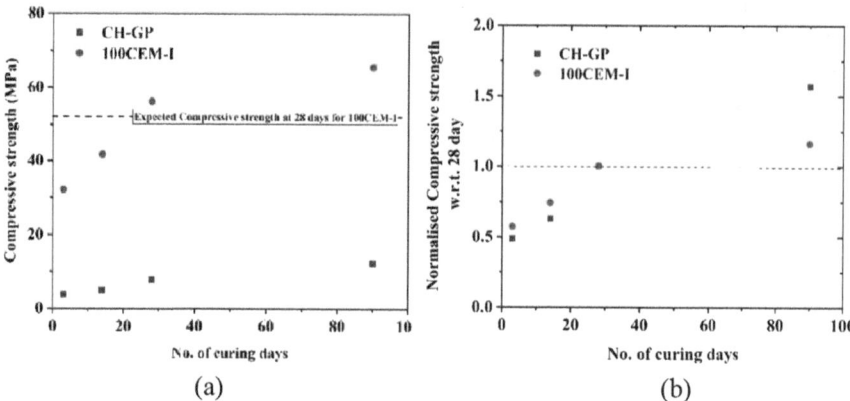

Fig. 4 (a) Average compressive strength of CH–GP and 100CEM-I (b) Normalized compressive strength w.r.t 28 days of CH–GP and 100CEM-I

compressive strength of 100CEM-1 specimens was significantly higher than that of CH–GP specimens on all test days, the results confirm notable compressive strength of CH–GP specimens. Given it was previously observed that 100CH (i.e., hydrated $Ca(OH)_2$ only) test specimens showed almost zero compressive strength, the observed compressive strength in CH–GP specimens was owing to a chemical reaction between calcium hydroxide and glass powder with possible formation of C–S–H. Figure 4b shows that the 90 days compressive strength of CH–GP mixes was 57% higher than its 28-day strength, whereas the 90-day compressive strength of 100CEM-I specimens was 16% higher than its 28-day strength. This result suggests the strength contributing compounds in CH–GP specimens were likely form at later days (after 28 days) of curing.

X-ray Diffraction (XRD) analysis The XRD analysis of CH–GP, 100CEM-I and 100CH were conducted to identify the formed C–S–H and the remaining content of CH in CH–GP and the reference test specimens. The XRD results were obtained using Bruker D8 Autochanger + X-ray diffractometer equipped with graphite-monochromatized Cu Kα radiation, operating at 40 kV and 100 mA available at the University of Manchester [16]. The powdered form of the crushed samples obtained after compressive strength tests were used in the XRD investigation. The results of the XRD analysis of the CH–GP specimens showed in Fig. 5a suggest that XRD peaks of calcium hydroxide (at 2Θ angles namely 29.369, 34.936, 47.448 [12]) decreases and the XRD peaks for C–S–H (at 2Θ angles namely 28.943, 29.004, 30.008, 29.746, 49.537 [12]) increases with the increase of the curing time of the specimen. The decrease in the amount of $Ca(OH)_2$ in the mix suggests a possible reaction in the mix where $Ca(OH)_2$ was consumed for the chemical reaction. On the other hand, the presence of C–S–H peaks suggest the formation of C–S–H in the mix, and similar C–S–H peaks were noticed in 100CEM-I specimen at same 2Θ angles (28.943, 29.004, 30.008, 29.746, 49.537) shown in Fig. 5b. This suggests that the C–S–H formed in CH–GP specimens were of the same type produced in

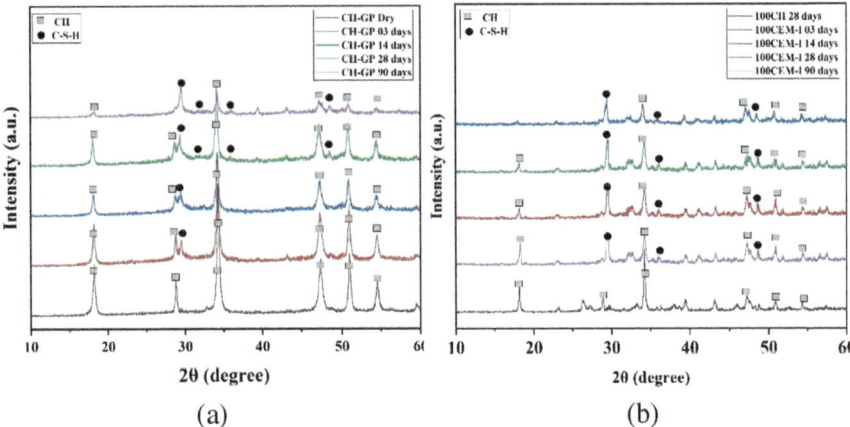

Fig. 5 (**a**) XRD peaks of C–S–H and CH in CH–GP specimen (**b**) XRD peaks of C–S–H and CH in 100CEM-I and 100CH specimen

100CEM-I specimens. Further, the increment in the C–S–H peaks with time (3–90 days) in CH–GP specimen showed increment in the quantity of the C–S–H formation which aligns with the findings noticed in compressive strength results where the strength increased with time. Therefore, it could be concluded that the compressive strength noted in the CH–GP specimens (shown in compressive strength section in Sect. 4.2) were likely due to the C–S–H in the mix. Further, Fig. 5b shows no XRD peaks of C–S–H in 100CH specimens which suggest there was no or negligible C–S–H formation in 100CH specimens.

5 Conclusions

The present experimental study aimed to understand the chemistry of the reaction between calcium hydroxide (CH) and glass powder (GP) to quantify the gain in the compressive strength and to identify the formation of C–S–H using XRD investigation. The results of the CH–GP mixes were compared with 100% CH and 100% CEM-I specimens.

- CH–GP specimens possessed noticeable compressive strength whereas no strength was noticed in 100CH specimens. The results suggest a likely chemical reaction between calcium hydroxide and glass powder which led to the formation of strength contributing compound C–S–H.
- The brittle failure of the CH–GP specimens (except 3-day cured specimens) suggests the microstructure of the specimens was similar to that of the 100CEM-1 specimens. The results suggests there was an ongoing development of the microstructure of the CH–GP test specimens and more stable microstructure developed at a later stage of curing.

- The 90-day compressive strength of CH–GP mixes was 12.3 MPa which was lower than the same 90-day strength of 100CEM-I (65.4 MPa). However, the 90-day strength of CH–GP specimen was 57% higher than its 28-day strength (the observed strength increase of CEM-I specimens within the same period was 16%). This suggests the major strength imparting compounds in CH–GP specimens formed at later stages of curing.
- The XRD analysis confirms no peaks of C–S–H in the 100CH specimen but clear intense peaks of C–S–H in CH–GP specimens. The C–S–H peaks in CH–GP specimens were similar to the C–S–H peaks noticed in 100CEM-I specimens and they were increasing as the curing period increases from 3 to 90 days. This implies that the compressive strength gain in CH–GP specimens could be attributed to the formation of C–S–H in the mix.

Acknowledgments Gaurav Chand wishes to express his gratitude to Government of India for supporting his PhD study (K-11015/46) at the University of Manchester, UK, and SIBELCO Sheffield for providing waste glass and X-ray Diffraction Suite, University of Manchester, for the facilities to carry out the XRF experiment under the technical support of Mr Gary Harrison.

References

1. Achintha, M., Balan, B.: Characterisation of the mechanical behaviour of annealed glass–GFRP hybrid beams. Constr. Build. Mater. **147**, 174–184 (2017)
2. Achintha, M., Balan, B.A.: An experimentally validated contour method/eigenstrains hybrid model to incorporate residual stresses in glass structural designs. J. Strain Anal. Eng. Des. **50**(8), 614–627 (2015)
3. Achintha, M.: Sustainability of glass in construction. In: Sustainability of Construction Materials, pp. 79–104. Woodhead Publishing, Kidlington (2016)
4. Khan, M.N.N., Saha, A.K., Sarker, P.K.: Evaluation of the ASR of waste glass fine aggregate in alkali activated concrete by concrete prism tests. Constr. Build. Mater. **266**, 121121 (2021)
5. Chand, G., Happy, S.K., Ram, S.: Assessment of the properties of sustainable concrete produced from quaternary blend of Portland cement, glass powder, metakaolin and silica fume. Clean. Eng. Technol. **4**, 100179 (2021)
6. Jani, Y., Hogland, W.: Waste glass in the production of cement and concrete–a review. J. Environ. Chem. Eng. **2**(3), 1767–1775 (2014)
7. Mesto jaw crusher machine Nordberg® C200™ jaw crusher – Metso
8. Vibration Mill Machine Vibration Mill – ALPA Powder Equipment
9. Malvern Panalytical Mastersizer 3000 – Particle Size Analyzer I Malvern Panalytical
10. XRF analyzers XRF Analyzers I XRF Spectrometers I Malvern Panalytical
11. Source Chemicals Calcium Hydroxide from Source Chemicals Ltd I Powders I Source Chemicals Ltd
12. Taylor, H.F.: Cement Chemistry, vol. 2, p. 459. Thomas Telford, London (1997)
13. Chand, G., Achintha, M., Wang, Y.: Exploration of waste glass powder as partial replacement of cement in concrete. In: International RILEM Conference on Synergising Expertise Towards Sustainability and Robustness of CBMs and Concrete Structures, pp. 260–270. Cham, Springer Nature (2023)

14. Pelisser, F., Gleize, P.J.P., Mikowski, A.: Effect of the Ca/Si molar ratio on the micro/nanomechanical properties of synthetic CSH measured by nanoindentation. J. Phys. Chem. C. **116**(32), 17219–17227 (2012)
15. Zhan, B.J., Xuan, D.X., Poon, C.S.: The effect of nanoalumina on early hydration and mechanical properties of cement pastes. Constr. Build. Mater. **202**, 169–176 (2019)
16. XRD AUTO-CHANGER Spec Sheet DOC-S88-EXS028 V2 high.pdf (bruker.com)

Impact of High-Strength Low-Alloy Steel in Reducing the Embodied Water of Buildings: A Case Study

Manish Dixit ⓘ, **Pranav Pradeep Kumar** ⓘ, **and Sarbajit Banerjee** ⓘ

Contents

1   Introduction

1.1   Steel as a Construction Material

Steel is considered as one the most important materials in the construction industry second to only cement concrete. With its incredible mechanical and structural properties, steel is widely used in different sectors of construction. Although concrete structures form a massive share (60%) of the buildings across the world, most concrete structures use steel in the form of either reinforcement bars or prestressing steel [1]. In 2023, 1.89 billion tonnes of steel were produced around the world with China alone producing a little over 1 billion tonnes [2]. According to the World Steel Association (WSA), 52% of the total steel produced goes into the building and infrastructure sector [2]. Reinforcement bars make up to 44% of the steel in the

M. Dixit (✉) · S. Banerjee
Texas A&M University, College Station, TX, USA
e-mail: mdixit@tamu.com

P. Pradeep Kumar
Massachusetts Institute of Technology, Cambridge, MA, USA

© The Author(s) 2025   611
M. Kioumarsi, B. Shafei (eds.), *The 1st International Conference on Net-Zero Built Environment*, Lecture Notes in Civil Engineering 237,
https://doi.org/10.1007/978-3-031-69626-8_51

building sector, followed by structural products such as cladding, roofing, etc. that make up to 31% and structural steel sections make up to 25% [3, 4]. In the infrastructure sector, reinforcement bars make up to 60% of this sector with structural steel, rail sections, and steel plates making up the remainder 40% [4]. These numbers clearly show that, from a design perspective, steel is primarily used as reinforcement bars in concrete structures and as structural steel sections.

1.2   Water Footprint of Steel Industry

Choudhury et al. evaluated the optimization of water footprint in steel industries [5]. This study reported that the average intake of water during the manufacture of every tonne of steel is estimated to range from 71 to 706 ft^3, not including the wastewater generation. This demand of water is used essentially for cooling purposes as "*make-up water*" which is used for compensating the water lost in the evaporations and mechanical losses and does not require any treatment. The water utilized for unit operations such as rolling and coking operations generates wastewater with several pollutants. These contaminants and toxic substances vary based on the operation and the point of generation of the wastewater. Treatment of wastewater contaminants generates hazardous sludge which is required to be recycled and managed. The study reports that blast furnace, coke plants and sinter plants have the most water footprint among the different operating units in steel manufacture. Efficient methods such as electrochemical precipitation, adsorption, membrane bioreactor facilities, and oxidation can be used to recover and treat the wastewater. This recovered water can then be utilized as make-up water for cooling purposes. The authors identified the beneficial use of several compounds in sludge such as Fe_2O_3 and CaO that are needed in the steel and cement industries and Nitrogen which is used to produce soil fertilizers. Despite the methods of recycling and recovery, the leachable heavy metals in sludge are a challenging issue.

Optimizing water footprint is a crucial step in the conservation of water. Water is a valuable resource, and NAE highlighted that providing clean water is one of the challenges for engineering [6]. Identifying that the construction industry utilizes water and a definition for water footprint of steel can help in estimating and conserving the resource. The estimation of water footprint is based on various stages of the use of water: the use of water during the manufacture of the product, the water utilized during the use of the product, and the use of water during the manufacture of the raw materials used for the product [7]. Therefore, a cradle-to-grave analysis of the manufacture of steel and its utilization during the function of the building is needed to compute the direct and indirect water use. This includes the water utilized during all the three stages, namely the manufacture of steel, the use of the building which uses the steel in its functionality, and during the manufacture of the raw materials used to produce the steel. The manufacture of coke, which is a part of the process of the manufacture of steel is one of the most water intensive processes [8]. In fact, the water utilized in the production of coke needs to be treated before it can be reused whereas 90% of the water used in beneficiation can be recirculated. The steel industry values its role in water management and strives to improve water conservation. More than 90% of the water used in cooling operations is recycled with the losses arising due to evaporation [9].

1.3 High-Strength Low-Alloy Steel

Given the development in environmental policies, there is a need for lightweight materials that can be beneficial to improving the environment impact and pressure from the respective industry. Branco and Berto explain how high-strength low-alloy (HSLA) steels can offer a better solution for an environmentally friendly structural material [10]. The high strength, formability, hardness, good fatigue behavior, and toughness, in addition to its resistance to corrosion and weldability, make them an efficient choice from utility and environmental standpoint. HSLA steel has mechanical properties that make them comparable to superalloys. The microalloying processes make the production of HSLA steel often more expensive than mild steel; however, the enhanced strength and other mechanical properties help in improved strength-to-weight ratio thereby reducing the quantity of material required, making HSLA cost-effective [3, 11]. Graedel and Miatto highlight that vanadium can be used to improve the performance of metals [12]. Vanadium is used to strengthen steel as an alloying element because of the improved toughness, strength, and better resistance to corrosion [13]. This helps in making the steel sections lighter with less use of material, thus having groundbreaking implications in the construction industry. Studies also show the benefits of microalloying vanadium in steel reinforcement bars and sections and the role of vanadium HSLA steel in decarbonizing the construction industry [3, 14].

2 Research Methodology

The aim of this paper is to study the embodied water impact of using high-strength low-alloy steel instead of mild (carbon) steel in buildings. The research objectives of this study are:

1. Develop an input–output-based hybrid (IOH) model to compute not only the embodied energy (EE) and carbon (EC) but also the embodied water (EW) of steel, concrete and microalloying material.
2. Compute the embodied water at a building level to understand the trade-off in using HSLA steel instead of mild steel from not only EE and EC but also from EW perspective.

An input–output-based hybrid (IOH) model is developed using the most recent US census data [15]. The model uses macroeconomic data between different industry sectors. The economic flows are converted into energy as well as water flows using the energy and water tariffs, respectively. The benchmark input–output data in the form of *Use-* and *Make-* tables from United States Bureau of Economic Analysis (USBEA) is collected [16]. The *Use-* and *Make-* tables present how commodities are used as inputs and are manufactured by each industry, respectively. It is important to account for industry scrap and adjust the *Make-* table to ensure that the Market share is computed accurately. The *Use-* and *Make-* coefficient matrices are calculated using the total values for each industry and are then used to determine the direct requirement matrix that presents the monetary value of a commodity used to manufacture another commodity worth one monetary unit. Using the principles of Leontief's inverse matrix, the direct requirement matrix is used to determine the total requirement matrix. This matrix shows the total monetary value of a commodity to produce a dollar worth of another commodity by an industry which includes both direct as well as indirect components. Detailed description of the development of the IOH model can be found in literature [17–19].

The total requirement matrix is expressed in the form of monetary units and cannot be directly used for consistent estimations because the cost of the commodities does not reflect the energy and water flows within the system. Therefore, the representation of the total requirement matrix quantifying the energy and water commodities utilized by the steel industry needs to be converted from a monetary representation to physical unit representation in MBtu for energy and Gal. for water. This is achieved by using the reported data of the energy and water usage and prices [20–22]. The data converted into the physical units is then used by the IO model. Several research studies have explained the procedure of collecting data of the amount of energy and water consumption [17–19]. The model incorporates information from various reports, articles, and online databases reported by various federal organizations such as the United States Department of Agriculture (USDA) [23], United States Department of Energy (USDOE) [24], United States Census Bureau (USCB) [15], and United States Geological Survey (USGS) [21, 22, 25]. Another advantage of embedding the data in physical units instead of price units is that the output generated is per unit dollar and is simply multiplied with the cost of the materials and items to generate the energy and water use. Wastewater and sewage form the other area that utilizes water, and it needs to be accounted for in the computations of embodied water. Therefore, in this study, the aggregated water commodity is separated into the water use component and the other component encompasses all the other commodities such as wastewater and sewage. The annual revenue for the total aggregate water was divided based on the portion of water utilities for disaggregation purposes. The data was incorporated into the *Use-* table to obtain the results of embodied water intensities that are reported in million gallons per dollar. The total EW is quantified by multiplying the costs of the materials.

After computing the embodied water and embodied energy, Energy Information Administration (USDOE) was referred to for the carbon emission factors (CEFs). These CEFs are used to quantify the EC of the materials. To compute the total EW, EE and EC, a cradle-to-site analysis is conducted that encompasses the extraction of raw materials, transportation of the materials, the manufacturing process, and the delivery of the material to the destination.

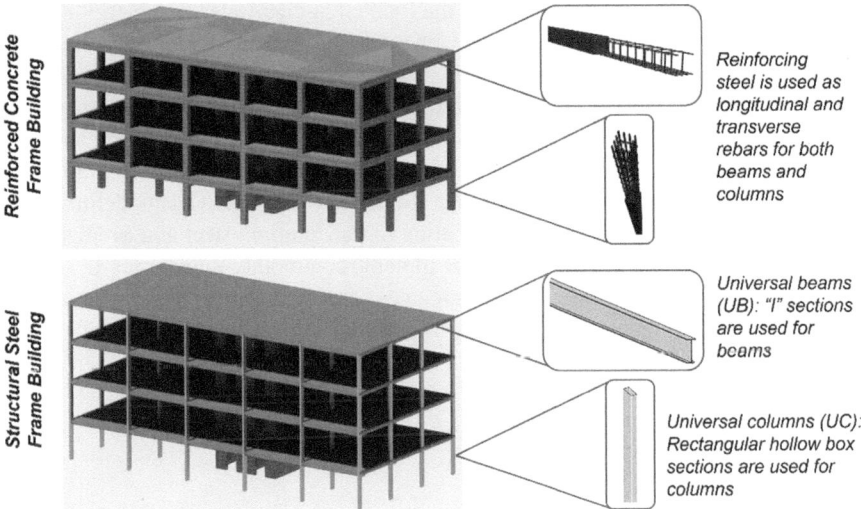

Fig. 1 The two 3-story buildings that are modeled to study the embodied impacts of HSLA steel. Reinforced concrete building (top) and structural steel section (bottom)

The IOH model is also extended to compute the total EW, EE, and EC of a case study, a 3-story building. Figure 1 shows the two building models developed to perform the structural analysis of the system. The analysis helps to calculate the total quantity of steel and concrete required to achieve the same load-carrying capacity. The design conforms to the Eurocode standards [26–29]. The two buildings represent the two commonly used frameworks: reinforced concrete and structural steel section frames. It is important to note that the quantity of materials and thereby the embodied impacts of only the structural frame, i.e., beams and columns are calculated to ensure consistency in comparison between the steel sections and reinforced concrete frameworks.

For the case study building in this paper, vanadium microalloyed steel is considered as HSLA steel. As shown in Fig. 1, the reinforced concrete frames are made of concrete and steel reinforcement bars in both longitudinal and transverse direction. Grade 250 (250 MPa) steel is used as reference mild steel for analysis while Grade 400, 500, and 600 (400, 500, and 600 MPa) steel are used to represent the vanadium microalloyed HSLA steel. Steel sections conforming to BS4 Part 1 1993 are used for the section steel frames with I-sections from universal beams and hollow box sections from universal columns selected for analysis [30]. Grade 235 (235 MPa) steel is used as reference mild steel for analysis while Grade 350 and 450 (350, 450 MPa) steel are used to represent the vanadium microalloyed HSLA steel. A 5 × 3 bay building with 4 m story height is modeled with H-type roof, C1 building category and loading details conforming to EN 1991 [27]. Weight percent of vanadium required to manufacture the respective steel grades were derived from the machine learning model presented in literature [3].

The IOH model is used to compute the total embodied water and energy per unit dollar of commodity. It is important to calculate the EW and EE of concrete, steel, and the microalloying material which is vanadium for the case study building. Revenue data for Iron and steel and ferroalloy manufacturing commodity (NAICS 331110) from USCB [15] is used to disaggregate the commodity into iron and steel mills (331111) and electrometallurgical ferroalloy product manufacturing (331112). To incorporate vanadium microalloying commodity into the IOH model, the total vanadium production, average consumption of vanadium in steel and average price of vanadium are obtained from USGS mineral commodity summaries [31]. This information is used to further disaggregate vanadium from electrometallurgical ferroalloy product manufacturing (331112). Thus, the IOH model is now modified to incorporate alloy steel and vanadium separately.

3 Results and Findings

The modified IOH model is used to obtain the embodied water (EW), energy (EE), and carbon (EC) of the different commodities such as concrete and steel used in the construction of a building as well as vanadium which is used as a microalloying material in steel to produce HSLA steel. Figure 2 presents the EW, EE, and EC of concrete and structural steel per kg of the respective commodity. For each commodity considered, the result obtained from the IOH model is multiplied with the price per unit weight of each commodity, collected from market reports [31–33] to compute the embodied impact per unit weight as presented in Fig. 2. Concrete and steel have EW less than 2 Gal./kg. Similarly, the EE of concrete and steel is 1.23 and 20.53 MJ/kg, respectively.

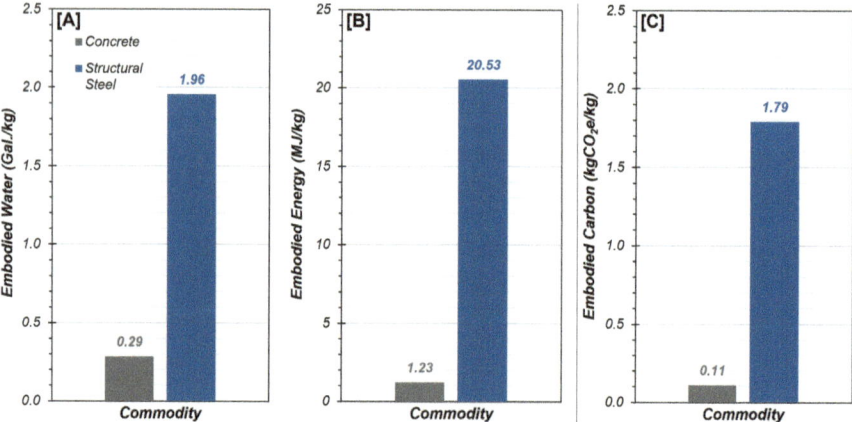

Fig. 2 Embodied impacts of concrete and structural steel commodities per unit weight of the respective commodity obtained from the input–output-based hybrid (IOH) model. (**a**) Embodied water (Gal./kg); (**b**) Embodied energy (MJ/kg); and (**c**) Embodied carbon (kgCO$_2$e/kg)

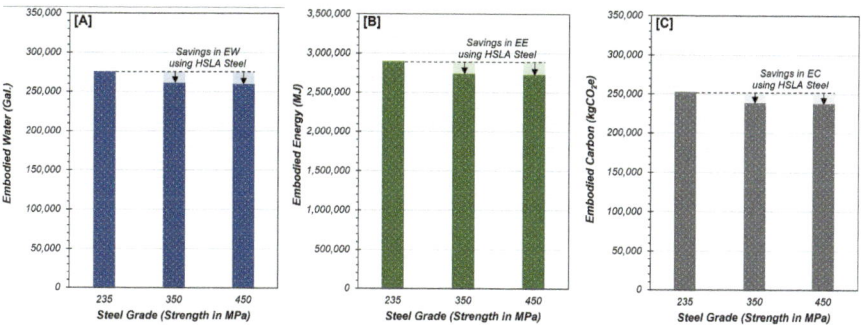

Fig. 3 Embodied impacts in a 3-story building with structural steel section frames analyzed using different grades of section steel (Grade 235 is the reference mild steel structural section grade). (**a**) Embodied water (Gal.); (**b**) Embodied energy (MJ); and (**c**) Embodied carbon (kgCO₂e)

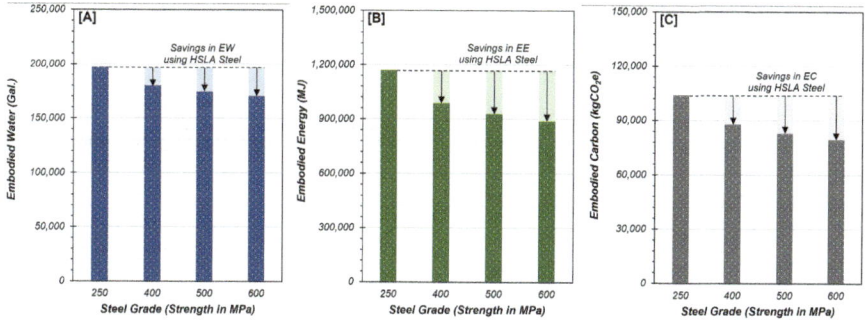

Fig. 4 Embodied impacts in a 3-story building with reinforced concrete frames analyzed using different grades of reinforcement bar steel (Grade 250 is the reference mild steel reinforcement bar grade). (**a**) Embodied water (Gal.); (**b**) Embodied energy (MJ); and (**c**) Embodied carbon (kgCO₂e)

Figure 3 presents the total EW, EE, and EC of the 3-story building with structural steel section frames. It may be observed from the figure that with respect to the reference grade 235 (mild steel) the vanadium microalloyed steel grades 350 and 450 have lower embodied impacts. The percentage vanadium content for the steel grades were taken as 0.035% and 0.11% for grade 350 and 450, respectively, from the machine learning model in literature [3]. There are ca. 15,000–16,500 Gal. savings in EW, ca. 156,700–168,700 MJ savings in EE, and ca. 13,600–14,700-kgCO₂e savings in EC with the use of HSLA steel instead of mild steel sections.

Figure 4 presents the total EW, EE, and EC of the 3-story building with reinforced concrete frames. It may be observed from the figure that with respect to the reference grade 250 (mild steel) the vanadium microalloyed reinforcement bar steel grades 400, 500, and 600 have lower embodied water, energy, and carbon. The percentage vanadium content for the steel grades were taken as 0.013%, 0.095%, and 0.177% for grade 400, 500, and 600, respectively, from the machine learning model presented in literature [3, 14]. There are ca. 17,000–27,000 Gal. savings in EW, ca. 183,000–281,000 MJ savings in EE, and ca. 16,000–25,000 kgCO$_2$e savings in EC with the use of HSLA steel instead of mild steel reinforcement bars.

4 Discussion and Conclusion

The embodied energy and carbon results for vanadium, steel, and concrete obtained from the IOH model are compared with results obtained in literature to understand variability. The EE results of vanadium (412 MJ/kg), steel (20.53 MJ/kg), and concrete (1.23 MJ/kg) are within 20%, 2%, and 50% of the reported values in literature [34, 35]. The EC results obtained from the IOH model are within 10% for vanadium and concrete and within 25% for steel, when compared with results from previous studies [34–36]. The results calculated using the input–output-based hybrid (IOH) model from IOH model are thus within reasonable limits of the values reported in literature except for EE of concrete. The high variability observed in concrete material may be due to the dependency of these values on the strength and mix design of the concrete. The system boundary coverage differences of the IOH methods and the approaches of the referred studies may also add to this variability. Looking at the comparable IOH-based EE and EC values, it can be expected that IOH-based EW values may also show reasonable variability. Since EW of vanadium is not available, we compared the results of EW of steel and concrete with previous studies. The calculated IOH-based EW of steel (1.96 Gal./kg) is more than double of the reported value of 0.7 Gal./kg [37]. This may be due to a much wider system coverage of the IOH model. In the case of concrete, the calculated EW value of 0.29 Gal./kg is comparable to 0.26 Gal./kg value reported in literature [38].

Building level analysis clearly presents the benefits of using HSLA steel in terms of savings in EW, EE, and EC. However, it is important to observe that the savings in EW may not directly reflect the benefits in terms of EE and EC savings. Figure 5 presents the percentage savings in using HSLA steel in lieu of mild steel. From Fig. 5 (left), it may be observed that the use of HSLA steel sections help in savings of

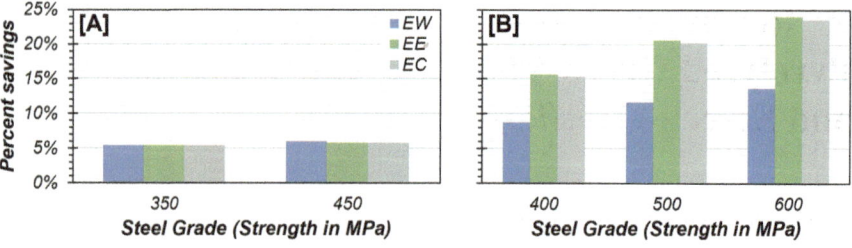

Fig. 5 Percentage savings in embodied water (EW), embodied energy (EE), and embodied carbon (EC) of using HSLA steel instead of mild (carbon) steel. (**a**) Steel section frames (Grade 235 is the reference mild steel structural section grade); and (**b**) Reinforced concrete frames (Grade 250 is the reference mild steel reinforcement bar grade)

ca. 5–6% of EW, EE, and EC. However, from Fig. 5 (right), it is evident that HSLA steel reinforcement bars help in savings of ca. 9–14% of EW and ca. 14–24% of EE and EC. The figure also underscores that the percentage savings in EW and EE are not similar for steel sections and reinforcement bars. The primary difference between the two types is due to the presence of concrete in the latter case. The quantity of concrete used in reinforced concrete structures is massive, and therefore, its water and energy use influences the total EW and EE of the building.

Thus, it is important for design engineers to ensure that the environmental sustainability in design is achieved by considering not only embodied energy and carbon but also the embodied water footprint of structures. Water intensive manufacturing of materials such as steel draws attention to the need for a comprehensive life cycle assessment to account for the water footprint. The robustness of the analysis may be improved by performing a sensitivity study to understand the impact of other key parameters such as material sources and transportation aspects on the tradeoff between EW and EE. The present study shows the benefits of using HSLA steel in buildings from embodied water perspective. Thus, one may conclude that the use of HSLA steel in buildings provides high potential for efficient design with the goal of attaining NetZero structures.

References

1. Aytekin, B., Mardani-Aghabaglou, A.: Sustainable materials: a review of recycled concrete aggregate utilization as pavement material. In: Transportation Research Record, pp. 468–491. SAGE Publications Ltd (2022)
2. WSA: 2023 World Steel in Figures. World Steel Association, Brussels (2023)
3. Pradeep Kumar, P., Santos, D.A., Braham, E.J., Sellers, D.G., Banerjee, S., Dixit, M.K.: Punching above its weight: life cycle energy accounting and environmental assessment of vanadium microalloying in reinforcement bar steel. Environ. Sci. Process. Impacts. **23**, 275–290 (2021). https://doi.org/10.1039/d0em00424c
4. WSA: Steel in Buildings and Infrastructure. World Steel Association, Brussels (2023)
5. Choudhury, A.R., Singh, N., Veeraraghavan, A., Gupta, A., Palani, S.G., Me-hdizadeh, M., Omidi, A., Al-Taey, D.K.A.: Ascertaining and optimizing the water footprint and sludge management practice in steel industries. Water. **15**, 2177 (2023)
6. National Academy of Engineering: NAE Grand Challenges for Engineering. NAE, Washington, DC (2008)
7. Ogaldez, J., Barker, A., Zhao, F., Sutherland, J.W.: Water footprint quantification of machining processes. In: Leveraging Technology for a Sustainable World, pp. 461–466. Springer, Berlin Heidelberg (2012)
8. Kluender, E.J.: Quantification of water footprint: calculating the amount of water needed to produce steel. J. Purdue Undergrad. Res. **3**, 50–57 (2013). https://doi.org/10.5703/jpur.03.1.08
9. WSA: Life Cycle Assessment in the Steel Industry. World Steel Association, Brussels (2020)
10. Branco, R., Berto, F.: High-strength low-alloy steels. Metals. **11**, 1000 (2021)
11. Kljestan, N., McWilliams, B.A., Knezevic, M.: Fatigue strength of an ultra-high strength low alloy steel fabricated via laser powder bed fusion. Mater. Sci. Eng. A. **896**, 146269 (2024). https://doi.org/10.1016/j.msea.2024.146269
12. Graedel, T.E., Miatto, A.: Vanadium: a U.S. perspective on an understudied metal. Environ. Sci. Technol. **57**, 8933–8942 (2023). https://doi.org/10.1021/acs.est.3c01009

13. Yang, Q.Q., Fang, Y.W., Mu, Y.W., Zhang, S.H., Zhao, Y.S., Li, Y.G., Liu, J.T., Wu, Z.J.: Summary of the application of vanadium. Appl. Mech. Mater. **598**, 55–59 (2014). https://doi.org/10.4028/www.scientific.net/AMM.598.55
14. Santos, D.A., Dixit, M.K., Pradeep Kumar, P., Banerjee, S.: Assessing the role of vanadium technologies in decarbonizing hard-to-abate sectors and ena-bling the energy transition. iScience. **24**, 103277 (2021). https://doi.org/10.1016/j.isci.2021.103277
15. USCB: Economic census data. https://www.census.gov/programs-surveys/economic-census/data/tables.2012.List_1822683115.html
16. USBEA: 2012 Input-output accounts data (2012 data files)
17. Dixit, M.K., Singh, S.: Embodied energy analysis of higher education build-ings using an input-output-based hybrid method. Energy Build. **161**, 41–54 (2018). https://doi.org/10.1016/j.enbuild.2017.12.022
18. Pradeep Kumar, P., Venkatraj, V., Dixit, M.K.: Evaluating the temporal repre-sentativeness of embodied energy data: a case study of higher education buildings. Energy Build. **254**, 111596 (2022). https://doi.org/10.1016/j.enbuild.2021.111596
19. Dixit, M.K., Kumar, P.P., Haghighi, O.: Embodied water analysis of higher education buildings using an input-output-based hybrid method. J. Clean. Prod. **365**, 132866 (2022). https://doi.org/10.1016/j.jclepro.2022.132866
20. EIA: Monthly energy review, June 2020 (2020)
21. Diehl, T.H., Harris, M.A.: Withdrawal and Consumption of Water by Thermoelectric Power Plants in the United States 2010. U.S. Geological Survey, Reston (2014)
22. Harris, M.A., Diehl, T.H.: Withdrawal and Consumption of Water by Thermoelectric Power Plants in the United States, 2015: U.S. Geological Survey Scientific Investigations Report 2019–5103. U.S. Geological Survey, Reston (2019)
23. USDA: Quick stats tool. https://quickstats.nass.usda.gov/
24. USDOE: Manufacturing energy consumption survey (MECS). https://www.eia.gov/consumption/manufacturing/
25. Diehl, T.H., Harris, M.A., Murphy, J.C., Hutson, S.S., Ladd, D.E.: Methods for Estimating Water Consumption for Thermoelectric Power Plants in the United States. U.S. Geological Survey, Reston (2013)
26. EN 1990:2002+A1: Eurocode – basis of structural design (2005)
27. EN 1991-1-1: Eurocode 1: actions on structures – Part 1-1: general actions – densities, self-weight, imposed loads for buildings (2002)
28. EN 1992-1-1: Eurocode 2: design of concrete structures – Part 1-1: general rules and rules for buildings (2004)
29. EN 1993-1-1: Eurocode 3: design of steel structures – Part 1-1: general rules and rules for buildings (2005)
30. BS4 Part1: Structural section. In: British Standards, pp. 1–45 (1993)
31. Polyak, D.E.: Vanadium. U.S. Geological Survey, Reston (2016)
32. Ben m'barek, B., Hasanbeigi, A., Gray, M.: Global Steel Production Costs. TransitionZero, London (2022)
33. CFI: Ready mixed concrete volume & price trends. https://concretefinancialinsights.com/us-concrete-industry-data
34. Nuss, P., Eckelman, M.J.: Life cycle assessment of metals: a scientific synthesis. PLoS One. **9**, 1–12 (2014). https://doi.org/10.1371/journal.pone.0101298
35. ICE: Inventory of Carbon and Energy (2011)
36. Weber, S., Peters, J.F., Baumann, M., Weil, M.: Life cycle assessment of a vanadium redox flow battery. Environ. Sci. Technol. **52**, 10864–10873 (2018). https://doi.org/10.1021/acs.est.8b02073
37. Strezov, V., Evans, A., Evans, T.: Defining sustainability indicators of iron and steel production. J. Clean. Prod. **51**, 66–70 (2013)
38. Netz, J., Sundin, J.: Water Footprint of Concrete. KTH Royal University of Technology (2015)

Environmental Impact of Timber Concrete Composites: An Overview

Alemayehu Darge Dalbiso ⓘ **and Mohammad Haj Mohammadian Baghban**

Contents

1 Introduction

Timber concrete composite (TCC) structures provide an innovative alternative to conventional floor slabs by merging the advantages of pure timber and pure concrete into a singular structure. Special shear connectors are used to connect the timber and concrete element making this system suitable for medium to long span (7–15 m) floors [1]. TCC structural elements are commonly employed in horizontal components like building floors or bridge decks, designed to withstand uniaxial bending in one-way spanning elements. Typically, timber is placed in the tension zone, while concrete occupies the compression zone. However, variations exist, including reversed TCC structural elements with concrete positioned in the tension zone [2] and TCC wall systems [3, 4]. In addition, recently [5] has developed and investigated a novel two-way panning slab system.

A. D. Dalbiso (✉) · M. H. M. Baghban
Norwegian University of Science and Technology, Gjøvik, Norway
e-mail: alemayehu.dalbiso@ntnu.no

© The Author(s) 2025
M. Kioumarsi, B. Shafei (eds.), *The 1st International Conference on Net-Zero Built Environment*, Lecture Notes in Civil Engineering 237,
https://doi.org/10.1007/978-3-031-69626-8_52

TCC is first introduced to the construction industry for refurbishment of old timber floors in Europe, where special shear connectors are placed on existing timber beams and the floor is strengthened by adding concrete on top [6]. A TCC system with nails as shear connectors was first patented by Paul Müller in the year 1922 [7]. After years of stagnation, interest in TCC systems has resurged as an appealing option in new building and bridge construction [1, 8].

The growing interest in TCCs can be attributed to their capacity to offer significantly increased stiffness, improved acoustic separation, and higher thermal mass in comparison to timber-only floors [9]. In contrast to reinforced concrete floors, TCCs offer advantages such as reduced weight, resulting in decreased loads on foundations and seismic forces. Additionally, they provide the potential for prefabrication, lower embodied energy, and reduced CO_2 emissions, attributed to timber's carbon-neutral nature [1].

Research in previous years has focused on the performance aspects of TCC systems such as the shear connection methods [10–13], stiffness [14], load-bearing capacity [15], acoustic performance [16], fire resistance [11, 17], and dynamic response [18, 19]. However, limited studies have been conducted on the environmental impacts of TCCs. Furthermore, no comprehensive literature review has been undertaken on the environmental impact of TCCs. The current study aims to review existing research on TCCs, specifically focusing on environmental impact assessment through the utilization of the life cycle assessment method.

2   Method

The literature search is conducted electronically using two online databases: Web of Science and Scopus. Additionally, a search for grey literature was conducted online across various databases. Despite the growing popularity of Timber Concrete Composite (TCC) structures, there is a lack of studies specifically addressing the life cycle assessment (LCA) of TCCs. Consequently, this study includes 11 peer-reviewed articles that specifically focus on the LCA of TCC structures.

The systematic procedure outlined in Fig. 1 was adopted for conducting the literature survey, following the guidelines proposed by [20].

3   Environmental Impact of Building Materials

The building and construction industry alone consumes 50% of global raw material extraction [21] and contributes to 37% global carbon emissions with 23% resulting from operational energy and 14% resulting from embodied carbon related to production, transport, construction as well as the end-of-life process of construction materials [22]. To reduce emissions, the industry must decrease the carbon footprint

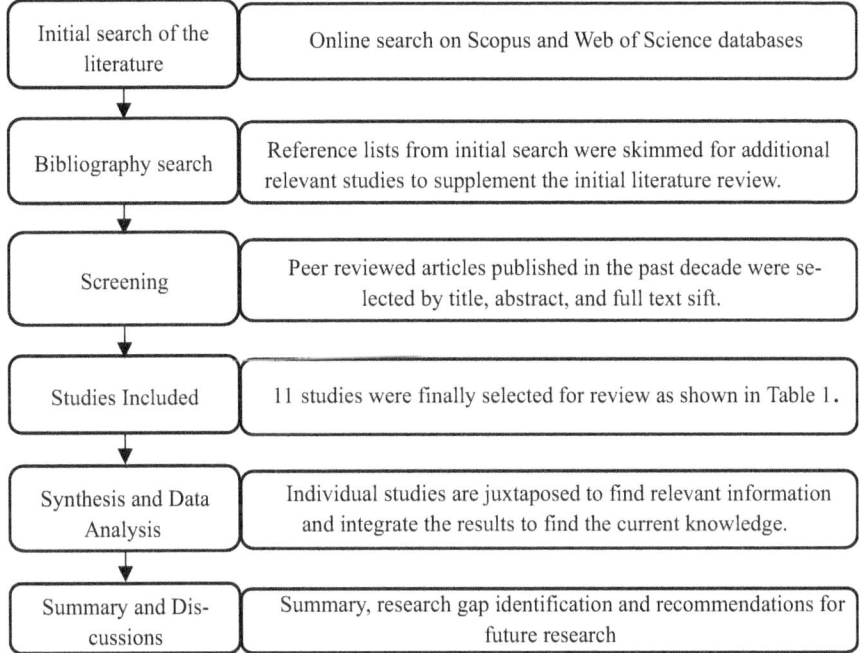

Fig. 1 Literature survey process

of building materials and enhance the energy efficiency of buildings. Considerable attention has been given in the last decades by the architecture, engineering, and construction (AEC) industry in making buildings energy efficient by moving towards renewable energy sources, which has lowered operational carbon emissions [23]. However, the share of embodied carbon emission is projected to increase from 14% to 18.5% (half of total carbon footprint) by 2050 [24].

Embodied carbon in building and construction industry comes predominantly from the energy-intensive production of cement, steel, aluminum, and insulating materials. The strategies to reduce embodied carbon include use of low-carbon materials, deign improvements and building less [25]. In addition, use of circular construction materials (construction materials with recycled, reused and repaired contents) can reduce emissions [21].

In recent years, life-cycle assessment (LCA) method has gained popularity for evaluating a product's environmental effects throughout its life cycle, aiming to increase resource-use efficiency and reduce liabilities. LCA involves identifying and quantifying environmental loads, such as energy and raw materials consumption, emissions, and waste generation, while assessing alternatives for reducing environmental impacts [26].

3.1   Life Cycle Assessment (LCA)

The idea of LCA was conceived in the 1960s in response to concerns in environmental degradation and natural resource depletion of human activities [27]. The methodology and applications have evolved in recent decades, achieving scientific consensus and adhering to established standards. LCA has become an internationally standardized method (ISO 14040:2006) to assess the direct and indirect environmental consequences of resource use along their life cycle stages. ISO defines LCA as a process of compilation and evaluation of the inputs, outputs, and the potential environmental impacts of a product system throughout its life cycle [28]. EN 15978 is an LCA standard specific to the European region, harmonized with ISO 14040, and employed for evaluating the environmental performance and potential impact of buildings and construction products. EN 15978:2011 is utilized alongside EN 15804:2019, which establishes a LCA methodology for construction products, including detailed guidelines for creating Environmental Product Declarations (EPDs).

Analyzing the environmental impacts of processes and products throughout their entire lifecycle, as per EN 15978:2011 [26], involves four stages: production, construction, use, and end of life, which are further divided into 16 substages (refer to Fig. 2). The benefits and loads stage of reuse, recovery, and recycling are separately shown outside the system boundary as an optional fifth stage in module D.

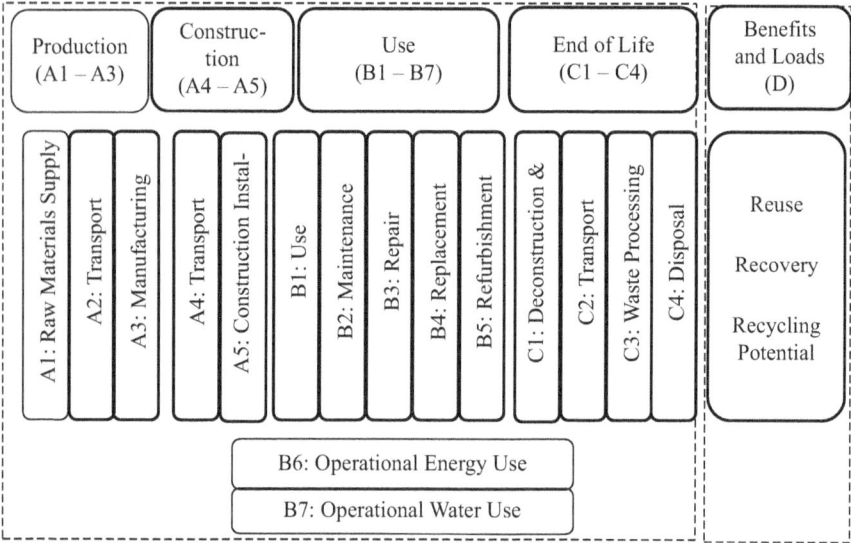

Fig. 2 Life cycle stages as per EN 15978 [26]

4   LCA Studies of TCC

This section presents the review results by addressing each stage of the LCA process, followed by a comparison of the studies. Additionally, it explores end-of-life scenarios and considerations beyond system boundaries. Table 1 outlines the studies included in this review, providing essential characteristics such as year of publication, country, scope, system boundary, and structure type. Approximately half of the studies are conducted in Europe, with the remainder distributed across Asia, North America, and Oceania.

4.1   Goal and Scope Definition

Environmental impact assessment of TCCs using the LCA approach has been conducted for building floor slabs [29–33], bridges [32, 34–36], and walls [3, 4]. The studies compared the TCC system in terms of life cycle carbon foot print with reinforced concrete [29, 30, 35, 37], prestressed concrete [35, 36], ultra-high-performance fibre reinforced cementitious composite (UHPFRC) [34], steel

Table 1 Summary of reviewed paper

Reference	Year	Country	Study scope	System boundary	Structure type
[37]	2023	Korea	CLT-concrete/ RC/ CLT	Cradle-to-gate	Floor
[29]	2014	Australia	GLT-concrete/ CLT/RC	Cradle-to-gate	Long span floor
[36]	2017	Portugal	GLT-concrete/ RC/PC	Cradle-to-grave	Bridge deck
[33]	2022	Malaysia	GLT-concrete/ light weight steel-concrete/precast hollow core concrete/cofradal	Cradle-to-grave	Floor
[31]	2021	Canada	CLT-concrete/GLT-concrete		Floor
[35]	2022	Czech Republic	Transverse prestressed wooden concrete/steel concrete composite	Cradle-to-gate	Bridge deck
[4]	2023		CLT-UHPC/ CLT/RC	Cradle to cradle	Wall
[30]	2023	Luxemburg	CLT-concrete/RC/steel concrete composite	Cradle to cradle	Floor
[34]	2021	Switzerland	Timber-UHPFRC/ RC/UHPFRC	Cradle-to-grave	Bridge deck
[3]	2018	Italy	Concrete glulam framed panel (CGFP)	Cradle to gate	Wall
[39]	2017	Italy	Concrete CLT framed panel (CGFP)	Cradle-to-gate	Wall

concrete composite (SCC) [30, 33], hollow core precast concrete slab [33], cofradal slab [33], and cross-laminated timber [4, 29, 37]. The TCC systems studied included CLT-Concrete, GLT-Concrete and CLT-UHPFRC.

Few studies have conducted a cradle-to-cradle LCA [4, 30], while majority of the studies conducted either a cradle-to-gate or cradle-to-grave LCA. The cradle-to-cradle assessment scope examines total environmental impact of a product from extraction of raw materials to reusing, repurposing or recycling into a new product at the end of its life so that the product never gets wasted and it is kept in the economic loop.

4.2 Life Cycle Inventory

A diverse range of databases was employed by the reviewed literatures, reflecting the regional specificity of certain databases [30, 31] and the absence of environmental profiles for materials such as UHPFRC & CLT [34] in some databases. Most of the cases adopted the generic Ecoinvent database followed by environmental product declaration (EPD), KBOB, GaBi, ÖKOBAUDAT and Athena. According to [31], the carbon intensity of CLT on Athena showed an exaggerated value compared with EPD values from major regional CLT producers in Canada which affected final LCA outcomes. Due to the absence of CLT environmental profile in databases [37], used GLT profile as a substitute for CLT, which may have an influence on the final result. According to [38], the comparison of five common LCI databases showed that the greenhouse gas emission values are different for the same wood product across the databases, which indicates that the use of appropriate database is necessary.

4.3 Life Cycle Impact Assessment

Global warming potential (GWP) and life cycle energy were the most evaluated environmental impact indicators on the reviewed papers. Other indicators such as Ozone layer depletion (ODP), acidification potential (AP) and eutrophication potential (EP), Photochemical oxidation (POCP) and abiotic depletion (ADP) are also identified in some of the studies as shown in Table 2.

4.4 Interpretation

The most comprehensive LCAs of TCCs to date were conducted by [4, 30]. In [4], modules A1–3, B4, C1–4, and D were considered, while [30] focused on A1–3, B1, C1–4, and D, as indicated in Table 3. However, none of the studies encompassed all

Table 2 Environmental impact indicators

Reference	GWP	ODP	AP	EP	POCP	ADP	Others
[37]	✓						
[29]	✓	✓					
[36]	✓		✓	✓	✓	✓	
[33]	✓		✓		✓	✓	
[31]	✓						
[35]	✓	✓				✓	✓
[4]	✓						
[30]	✓						✓
[34]	✓					✓	
[3]	✓					✓	✓
[39]	✓					✓	✓

Table 3 Life cycle modules studied

Reference	A1–A3	A4	A5	B1	B2	B3	B4	B5	B6–B7	C1–C4	D
[37]	✓										
[29]	✓	✓									
[36]	✓	✓	✓	✓	✓					✓	
[33]	✓	✓	✓							✓	✓
[31]	✓	✓	✓								
[35]	✓	✓	✓								
[4]	✓						✓			✓	✓
[30]	✓	✓								✓	✓
[34]	✓	✓									
[3]	✓	✓	✓								
[39]	✓	✓	✓								

modules. The emphasis across the studies was on the product stage (A1–A39), suggesting that more comprehensive data are available for this stage compared to others.

4.5 Comparison of Studies

Comparing research cases is challenging due to differences in structure type, system boundary, database, location, and functional unit. The GWP of selected studies where the studies have a likely similar functional unit, system boundary, and life span are given in Fig. 3. According to [37], TCCs emitted 50–70% less CO_2 for increasing span length compared with RC slabs. The study analyzed the effect of carbon storage in timber and found that TCC slabs can reduce CO_2 emissions by 77% due to the carbon storage potential of timber.

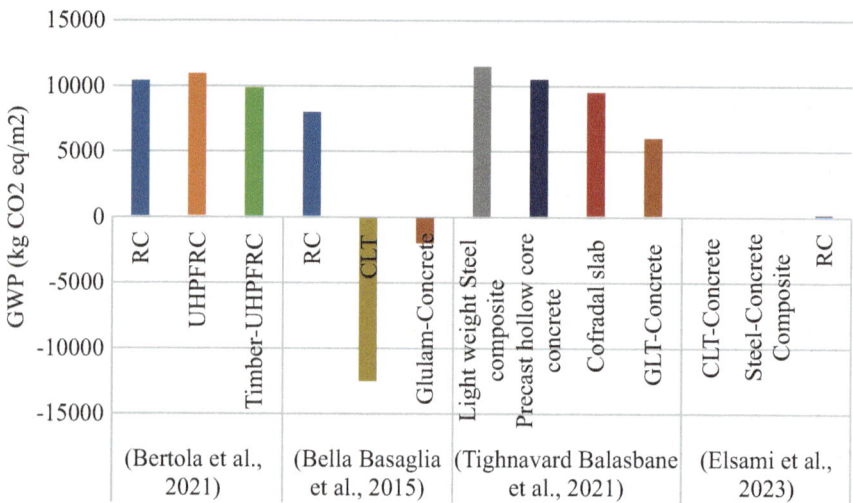

Fig. 3 Global warming potential comparison

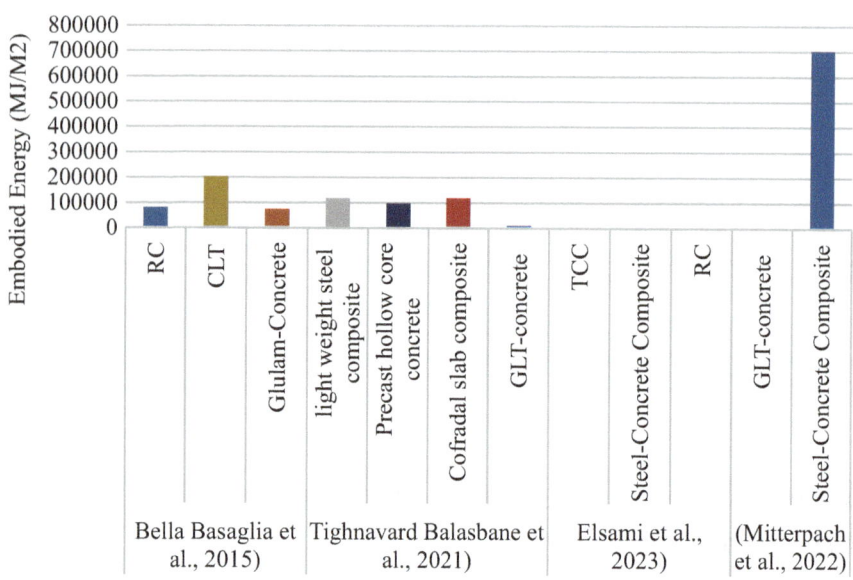

Fig. 4 Embodied energy comparison

Another study [35], compared the environmental impact of a TCC bridge with a steel–concrete bridge. The findings revealed that the TCC bridge exhibited 28% lower GWP and 30% lower nonrenewable energy consumption compared to the steel-concrete bridge. The comparison of embodied energy consumption among the selected studies is provided in Fig. 4.

4.5.1 End of Life Scenarios and Considerations for Benefits and Loads Beyond the System Boundary

Comprehensive cradle-to-grave analysis requires information about material disposal impacts which occurs for several decades. In recent decades, the popularity of TCCs has grown; however, estimating the disposal impacts (C4) has proven unattainable due to insufficient information on the demolition and disposal of TCC structures. Data on the GWP of the disposal process for Ultra-High Performance Fiber-Reinforced Concrete (UHPFRC) were not available in databases, as the material has been used only recently in TCC structures [4, 34]. For CLT members, incineration is considered as end of life scenario in module C3 and the benefit of thermal energy recovered is accounted in module D [4].

The study by [30] analyzed three end-of-life scenarios for CLT in CLT-concrete floor systems. The first scenario involves full energy recovery through wood incineration to recover heat energy. The second scenario, widely adopted in Europe, allocates approximately one-third of wood products for disposal, recycling, and energy production, respectively. The third scenario considers the full reusability of CLT panels after their service life, assuming easily demountable shear connections are used. The cradle-to-grave analysis revealed that the third scenario exhibits approximately 25% and 15% lower GWP compared to scenarios 1 and 2, respectively. Furthermore, an analysis of the benefits and loads (Module D) of the three scenarios demonstrated that the potential GWP benefits of scenario 3 are twice as high as those of scenarios 1 and 2.

5   Summary and Conclusion

This study conducted a literature review on the environmental impacts of TCCs utilizing the LCA methodology. The results unveiled significant environmental advantages of TCCs compared to conventional structural systems. These benefits are particularly notable when accounting for the carbon storage potential of timber in TCCs and further enhanced by implementing an end-of-life scenario allowing for the deconstruction and reconstruction of timber and concrete elements.

Variations in assessment outcomes have been noted in the literature, attributed to differences in LCA methodologies, varying goal and scope definitions, diverse data sources, and regional disparities. This underscores the need for a standardized and globally recognized LCA methodology to accurately evaluate the environmental impacts of TCCs.

Moreover, it has been observed that the selection of end-of-life scenarios significantly influences the outcome of the LCA. Specifically, the environmental impact of TCCs can be substantially reduced when designed for deconstruction and the reuse of timber and concrete elements, highlighting the necessity for further research on easily demountable and reconstruct able TCC shear connection systems.

Future research endeavors should aim to further explore the environmental impacts of TCCs, given the current insufficiency in both the quantity and depth of existing studies. It is recommended to conduct more comprehensive investigations using a cradle-to-grave approach, which includes assessing the carbon sequestration potential of wood and considering the impact of different end-of-life scenarios. This holistic approach will enable decision-makers to access comprehensive data when considering TCCs as an environmentally beneficial structural system.

References

1. Yeoh, D., Fragiacomo, M., De Franceschi, M., Heng Boon, K.: State of the art on timber-concrete composite structures: literature review. J. Struct. Eng. **137**(10), 1085 (2011)
2. Meena, R., Schollmayer, M., Tannert, T.: Experimental and numerical investigations of fire resistance of novel timber-concrete-composite decks. J. Perform. Constr. Facil. **28**(6) (2014)
3. Boscato, G., Mora, T.D., Peron, F., Russo, S., Romagnoni, P.: A new concrete-glulam prefabricated composite wall system: thermal behavior, life cycle assessment and structural response. J. Build. Eng. **19**, 384–401 (2018)
4. Wagner, A., Winter, K., Nestmann, A., Ott, S., Winter, S., Fischer, O.: Comparative life cycle assessment of timber-concrete-composite walls with concrete and CLT-wall elements. In: World Conference on Timber Engineering (WCTE 2023), pp. 967–975 (2023)
5. Kreis, B., Kübler, W., Frangi, A.: Development and investigation of an innovative, light-weight, two-way spanning timber-concrete composite slab. Eng. Struct. **286**, 116087 (2023)
6. Holschemacher, K., Klotz, S., Klug, Y., Weibe, D.: Application of steel fibre reinforced concrete for timber-concrete composite constructions. In: Bontempi, F. (ed.) System-Based Vision for Strategic and Creative Design, vol. 2, pp. 1393–1398. Balkema, Leipzig, Germany (2003)
7. Mueller, P.: Decke aus hochkantig stehenden Holzbohlen oder Holzbrettern und Betondeckschicht, Germany (1922)
8. Wacker, J.P., Hosteng, T.K., Dias, A.: Investigation of early timber–concrete composite bridges in the United States. In: Gustaffson, A., Pousette, A., Hagman, O., Ekevad, M. (eds.) 3rd International Conference on Timber Bridges 2017. Skellefteå, Sweden (2017)
9. Pastori, S., Sergio Mazzucchelli, E., Wallhagen, M.: Hybrid timber-based structures: a state of the art review. Constr. Build. Mater. **359**, 129505 (2022)
10. Frohnmüller, J., Seim, W., Umbach, C., Hummel, J.: Adhesively bonded timber-concrete composite construction method (Atcc) – pilot application in a school building in Germany. In: World Conference on Timber Engineering (WCTE 2023), pp. 4164–4172 (2023)
11. Shi, D., Hu, X., Hong, W., Zhang, J., Du, H.: Review of connections for timber-concrete composite structures under fire. BioResources. **17**(4) (2022)
12. Tao, H., Shi, B., Yang, H., Wang, C., Ling, X., Xu, J.: Experimental and finite element studies of prefabricated timber-concrete composite structures with glued perforated steel plate connections. Eng. Struct. **268**, 114778 (2022)
13. Thai, M., Ménard, S., Elachachi, S., Galimard, P.: Performance of notched connectors for CLT-concrete composite floors. Buildings. **10**(7), 122 (2020)
14. Yang, H., Lu, Y., Ling, X., Tao, H., Shi, B.: Experimental and theoretical investigation on shear performances of glued-in perforated steel plate connections for prefabricated timber–concrete composite beams. Case Stud. Constr. Mater. **18**, e01885 (2023)
15. Martín-Gutiérrez, E., Estévez-Cimadevila, J., Suárez-Riestra, F., Otero-Chans, D.: Flexural behaviour of a new timber-concrete composite structural flooring system. Full scale testing. J. Build. Eng. **64**, 105606 (2023)

16. Martins, C., Santos, P., Almeida, P., Godinho, L., Dias, A.: Acoustic performance of timber and timber-concrete floors. Constr. Build. Mater. **101**, 684–691 (2015)
17. Shephard, A.B., Fischer, E.C., Barbosa, A.R., Sinha, A.: Fundamental behavior of timber concrete-composite floors in fire. J. Struct. Eng. **147**(2) (2021)
18. Wen, B., Tao, H., Shi, B., Yang, H.: Dynamic properties of timber–concrete composite beams with crossed inclined coach screw connections: experimental and theoretical investigations. Buildings. **13**(9) (2023)
19. Movaffaghi, H., Pyykkö, J.: Vibration performance of timber-concrete composite floor section – verification and validation of analytical and numerical results based on experimental data. Civ. Eng. Environ. Syst. **39**(2), 165–184 (2022)
20. Booth, A., Sutton, A., Papaioannou, D.: Systematic Approaches to a Successful Literature Review, 2nd edn. SAGE Publications Ltd (2016)
21. Circle Economy: The Circularity Gap Report 2022. Report No.: 1.0, Amsterdam (2022)
22. UNEP: 2022 Global Status Report for Buildings and Construction, Nairobi (2022)
23. Material Economics: The Circular Economy a Powerful Force for Climate Mitigation Transformative Innovation for Prosperous and Low-Carbon Industry. Material Economics Sverige AB, Sweden (2018)
24. OECD: Global Material Resources Outlook to 2060: Economic Drivers and Environmental Consequences, Paris (2019)
25. PEEB: Embodied carbon – a hidden heavyweight for the climate: how financing and policy can reduce the carbon footprint of building materials and construction (2021)
26. Standardization ECf: Sustainability of Construction Works – Assessment of Environmental Performance of Buildings – Calculation Method, Brussels (2011)
27. Bjørn, A., Owsianiak, M., Molin, C., Hauschild, M.Z.: LCA history. In: Hauschild, M.Z., Rosenbaum, R.K., Olsen, S.I. (eds.) Life Cycle Assessment: Theory and Practice. Springer Nature, Switzerland (2018)
28. ISO: ISO 14040:2006. Environmental Management — Life Cycle Assessment — Principles and Framework, Switzerland (2006)
29. Basaglia, B., Lewis, K., Shrestha, R., Crews, K. (eds.): A comparative life cycle assessment approach of two innovative long span timber floors with its reinforced concrete equivalent in an Australian context. In: 2nd International Conference on Performance-based and Life-cycle Structural Engineering. The University of Queensland, Brisbane, Australia (2015)
30. Eslami, H., Yaghma, A., Bhagya Jayasinghe, L., Waldmann, D.: Influence of different end-of-life cycle scenarios on the environmental impacts of timber-concrete composite floor systems. In: World Conference on Timber Engineering (WCTE 2023), pp. 982–988 (2023)
31. Mirdad, M.A.H., Daneshvar, H., Joyce, T., Chui, Y.H., Mazzotti, C.: Sustainability design considerations for timber-concrete composite floor systems. Adv. Civil Eng. **2021**, 1–11 (2021)
32. Movaffaghi, H., Yitmen, I.: Multi-criteria decision analysis of timber-concrete composite floor systems in multi-storey wooden buildings. Civ. Eng. Environ. Syst. **38**(3), 161–175 (2021)
33. Tighnavard, B.A., Sher, W., Yeoh, D., Koushfar, K.: LCA & LCC analysis of hybrid glued laminated timber–concrete composite floor slab system. J. Build. Eng. **49**, 104005 (2022)
34. Bertola, N., Küpfer, C., Kälin, E., Brühwiler, E.: Assessment of the environmental impacts of bridge designs involving UHPFRC. Sustainability. **13**(22) (2021)
35. Mitterpach, J., Fojtík, R., Machovčáková, E., Kubíncová, L.: Life cycle assessment of a road transverse prestressed wooden–concrete bridge. Forests. **14**(1) (2022)
36. Rodrigues, J.N., Providência, P., Dias, A.M.P.G.: Sustainability and lifecycle assessment of timber-concrete composite bridges. J. Infrastruct. Syst. **23**(1), 04016025 (2017)

37. Oh, J.-W., Park, K.-S., Kim, H.S., Kim, I., Pang, S.-J., Ahn, K.-S., et al.: Comparative CO2 emissions of concrete and timber slabs with equivalent structural performance. Energy Build. **281**, 112768 (2023)
38. Takano, A., Winter, S., Hughes, M., Linkosalmi, L.: Comparison of life cycle assessment databases: a case study on building assessment. Build. Environ. **79**, 20–30 (2014)
39. Fortuna, S., Mora, T.D., Peron, F., Romagnoni, P.: Environmental performances of a timber-concrete prefabricated composite wall system. Energy Procedia. **113**, 90–97 (2017)

Influence of Spatial Coarsening of Pore Space in Porous Asphalt on Simulated Infiltration for Enhanced Stormwater Management

Rebecca Allen [ORCID]

Contents

1 Introduction

Typical roads are designed with a relatively impermeable top layer, for example, stone mastic asphalt. Water that accumulates on top of this type of asphalt, due to snow melt or rain events, either evaporates or runs off from the surface into a collection system. It is possible some amount of water is absorbed into the asphalt and underlying pavement layers, particularly if there is a crack or other such damage in the road.

However, another type of asphalt design exists and has been used in various locations (for example, the Netherlands) that is more porous than typical asphalt with the purpose of allowing water to infiltrate the road. This accomplishes several objectives: one is to handle stormwater directly 'on the spot' and thus minimize the need for construction of a rainfall runoff collection system alongside the road. Another objective is to minimize water splash that occurs when vehicles drive over wet asphalt. A third objective is minimizing noise that is created when vehicle tires

R. Allen (✉)
Oslo Metropolitan University, Oslo, Norway
e-mail: rebeccaa@oslomet.no

© The Author(s) 2025
M. Kioumarsi, B. Shafei (eds.), *The 1st International Conference on Net-Zero Built Environment*, Lecture Notes in Civil Engineering 237,
https://doi.org/10.1007/978-3-031-69626-8_53

635

drive over asphalt; however, this current study focuses on the simulation of water infiltration through porous asphalt and not on the application of a noise model.

Modelling of the physical process of water infiltration through porous asphalt is similar to other porous media applications, and a wealth of knowledge already exists about this domain. Fluid flow through porous media is typically categorized as either saturated or unsaturated, and can also involve more than one phase (i.e. liquid and gas) and more than one component (i.e. hydrocarbon, CO_2 and water in an oil reservoir). While the underlying physics is relatively well understood, computational advances have allowed engineers and researchers to focus on applying multi-physics models and using high-performance computing to simulate fluid behaviour at smaller and smaller scales. That is, the porous media is modelled at a scale where individual aggregates or grains, air voids and other components of the media are resolved locally in space, rather than representing the media as a homogenized material with effective parameters such as void content (or porosity).

In order to perform this 'pore-scale' modelling and simulation, the precise distribution of materials in the media must be resolved spatially. To do this, one can take a digital image of the media with a high-resolution camera, an X-ray computed tomography (CT) scanner, or other similar apparatuses, depending on the resolution desired. This current study focuses on the use of X-ray CT scanning to acquire images of a porous asphalt sample, and the steps that follow after image acquisition, in particular: image processing, construction of a three-dimensional (3D) digital representation of the sample and fluid flow simulation applied to the sample. The objectives are to: (1) demonstrate the image processing workflow and compare it to other similar studies, (2) simulate saturated permeability in the sample and (3) study how coarsening of the digital sample affects both the computational cost of the simulations and the results of the simulated permeability.

The paper is outlined as follows: first, the general physics of saturated fluid flow through porous media is presented, second, the simulator used to solve the general physics is briefly described, third, the porous asphalt sample and the imaging details are described, forth, results of the simulations are presented considering various spatial coarsening levels in the sample, and lastly, the results are discussed in the context of how the workflow of imaging and simulation can provide relevant information about porous asphalt in a way that is computationally practical.

2   Method

2.1   Materials and 3DXRCT Scanning

A 10-cm high, 6-cm diameter, cylindrical porous asphalt sample was created in the lab by the Norwegian Public Road Authority (Statens Vegvesen), as part of a previous research work. Findings from investigations done on this sample, as well as other field-cut samples, are found in [1] and [2]. In [2], details of the sample used in this current study (called sample 'A') were presented, as well as details of the

3DXRCT scanning work that was conducted at Norwegian Geotechnical Institute (NGI) to produce the 7.5 GB dataset of grayscale values representing the physical 3D sample, and the details of an algorithm developed to detect the top-to-bottom connected air voids. This algorithm was written in MATLAB programming language, and is available on an online public repository; see link given in [2].

2.2 Fluid Flow Simulation

The fundamental physics of the fluid flow problem is represented by the Navier-Stokes equations. However, for flow that is steady and sufficiently slow, the NSE reduces to the Stokes equation. In this work, the lattice Boltzmann (LB) equation is used to approximate a solution to the Stokes equation, and the open-source software PALABOS is used to do this [3]. PALABOS stands for Parallel Lattice Boltzmann Solver and is ideal for handling a relatively large dataset because of its parallel computing capabilities. Simulations of this work were run using up to four processors.

The output of the lattice Boltzmann simulations are pore-scale fluid velocities, that is, the velocity of water at each spatially resolved point through the 3D porous asphalt sample. These velocities can be visualized in order to gain a better understanding of the infiltration behaviour; however, an important quantity that can be computed from these velocities is the sample's permeability. This is done by taking an average of the pore-space velocities to get the Darcy velocity, and then by using this Darcy velocity along with the applied pressure gradient in the simulation to calculate the saturated permeability from the Darcy equation.

Permeability results from the LB simulations are given in lattice units. The conversion from lattice units (k_{LB}) to physical units (k_{phy}) is according to the following relationship:

$$k_{phy} = k_{LB} \left(\Delta x_{phy} \right)^2 \tag{1}$$

where Δx_{phy} is the voxel size in meters. The physical permeability calculated by this relationship is in units of meters squared, and this value can be expressed in Darcy, where 1 Darcy is equivalent to approximately 10^{-12} m^2. It can also be expressed as a saturated hydraulic conductivity in meters per second:

$$K = \frac{\rho g}{\mu} k \tag{2}$$

where μ is the fluid viscosity in pascals-seconds, ρ is fluid density in kilograms per meter cubed and g is gravitational acceleration in meters per second squared. The fluid viscosity and density are assumed to be 0.001 Pa-s and 1000 kg/m^3, respectively, in all cases.

Other interesting quantities that can be computed from the pore-scale fluid velocities include tortuosity. However, knowledge of the saturated permeability of the porous asphalt is considered more important, since it plays a role in further fluid flow modelling efforts, particularly for unsaturated flow.

2.3 Spatial Coarsening

The full dataset consisted of $2000^3 = 8 \times 10^9$ voxels, and after removing some of the voxels outside the region of interest (ROI), the dataset consists of $1543 \times 1544 \times 901 = 2.15 \times 10^9$ voxels (including the empty voxels surrounding the cylindrical sample that was imaged as a rectangular volume), and 1.87×10^6 voxels of the cylindrical ROI. And even though the pores or air voids take up 16.14% of the cylindrical ROI (and the connected pores take up 12.30% of the cylindrical ROI), it is still quite computationally costly to run a fluid flow simulation on a lattice grid made up of 1.87×10^6 fluid-solid nodes. The lattice grid is rectangular, so the dataset is further reduced by cropping the sides. This is finally our rectangular ROI of the porous asphalt sample. In order to reduce the computational cost, we investigate the effect of spatial coarsening. That is, we coarsen the rectangular ROI in the x-, y- and z-directions by the same degree of coarsening and run simulations on each coarsened case. We then compare the simulated velocity field and simulated permeability.

3 Results

3.1 Uncoarsened Digital Model

The rectangular ROI is shown in Fig. 1. The vertical distribution of porosity is shown in Fig. 5, the line is labelled as CL = 1, which means the coarsening level of 1, that is, no coarsening applied. In this porosity distribution, the total porosity fluctuates around the average of 13.54%. The top-to-bottom connected porosity fluctuates around an average slightly less than the total porosity, and is always (slightly) less than the total porosity in each XY plane slice.

3.2 Coarsened Cases

The structure of the pore space (both total and top-to-bottom connected) is visualized in Figs. 2, 3 and 4 for a coarsening level of 5, 10 and 25, respectively. Also visualized in these figures are the velocities magnitudes, given in lattice units, to show any dominant infiltration pathways through the sample.

Fig. 1 The uncoarsened dataset. In (**a**), the grey colour represents pores or air voids that are found within the porous asphalt sample. In (**b**), the pores or air voids are represented by the white colour, and both the solids and the space outside the ROI are black

(a) 3D volume of pores (air voids)

(b) Slice-plane view of pores (air voids)

Results of the spatial coarsening and simulations are shown in Table 1. Details of the uncoarsened case are shown in the table for reference; however, a simulation was not conducted on this uncoarsened case due to the high computational cost of simulating on a lattice grid comprised of over 1 billion nodes. It is possible a simulation could be run using a more powerful computer with more than four available cores but this has not been investigated further in this present study.

The difference between these coarsened cases is shown in Fig. 5. Coarsening did not have a significant effect on the total pore space in the sample, and the fluctuation in the vertical direction was relatively similar in all cases. But coarsening had a significant effect on the top-to-bottom connected pore space in the sample, presumably because removing some of the micropores that keep the pore space well connected resulted in more isolated 'pockets' or isolated 'pathways' in the pore space. This is particularly evident in Fig. 4b, where much of the pore space has been removed as it does not connect to the one remaining top-to-bottom pore branch.

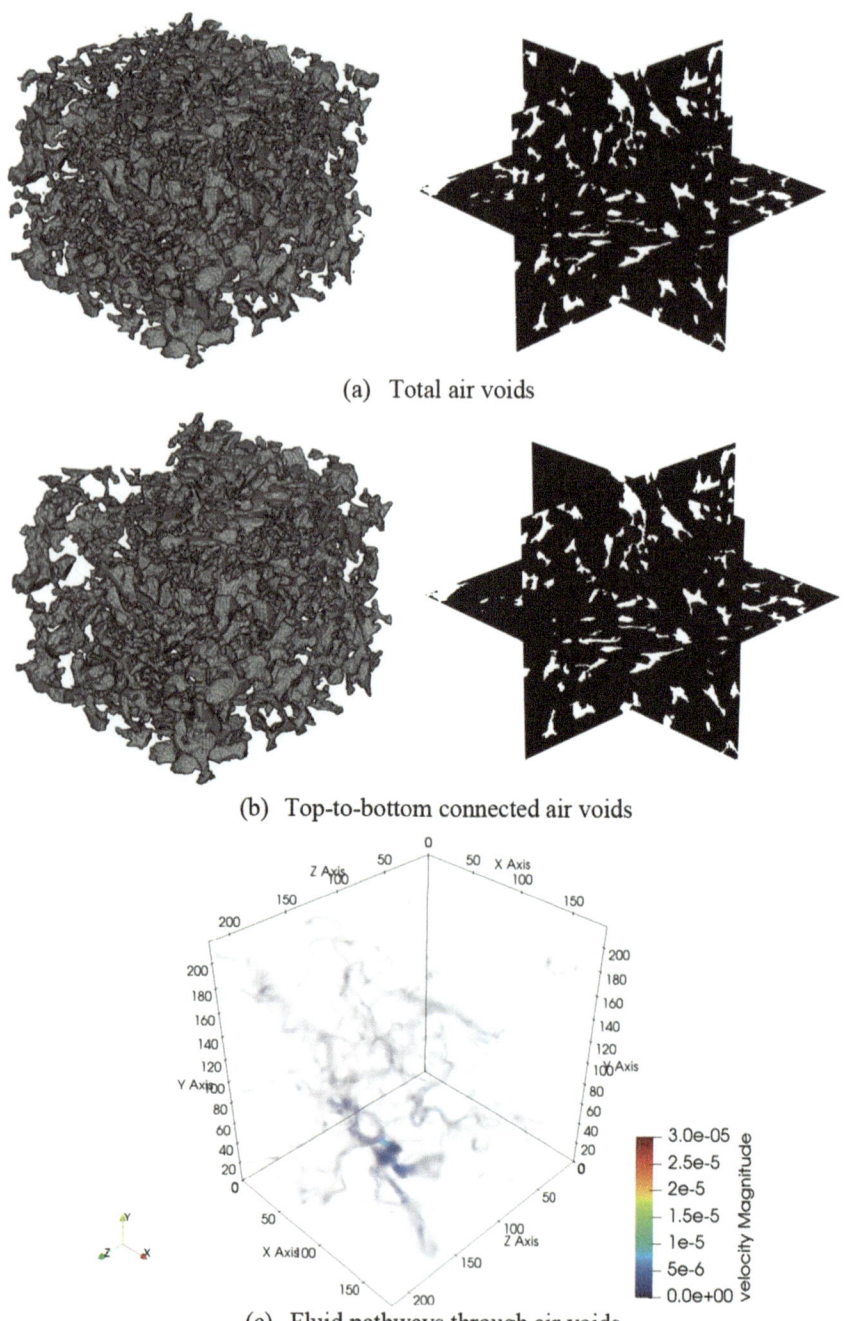

(a) Total air voids

(b) Top-to-bottom connected air voids

(c) Fluid pathways through air voids

Fig. 2 Air voids and fluid pathways in rectangular ROI with a spatial coarsening level of CL = 5

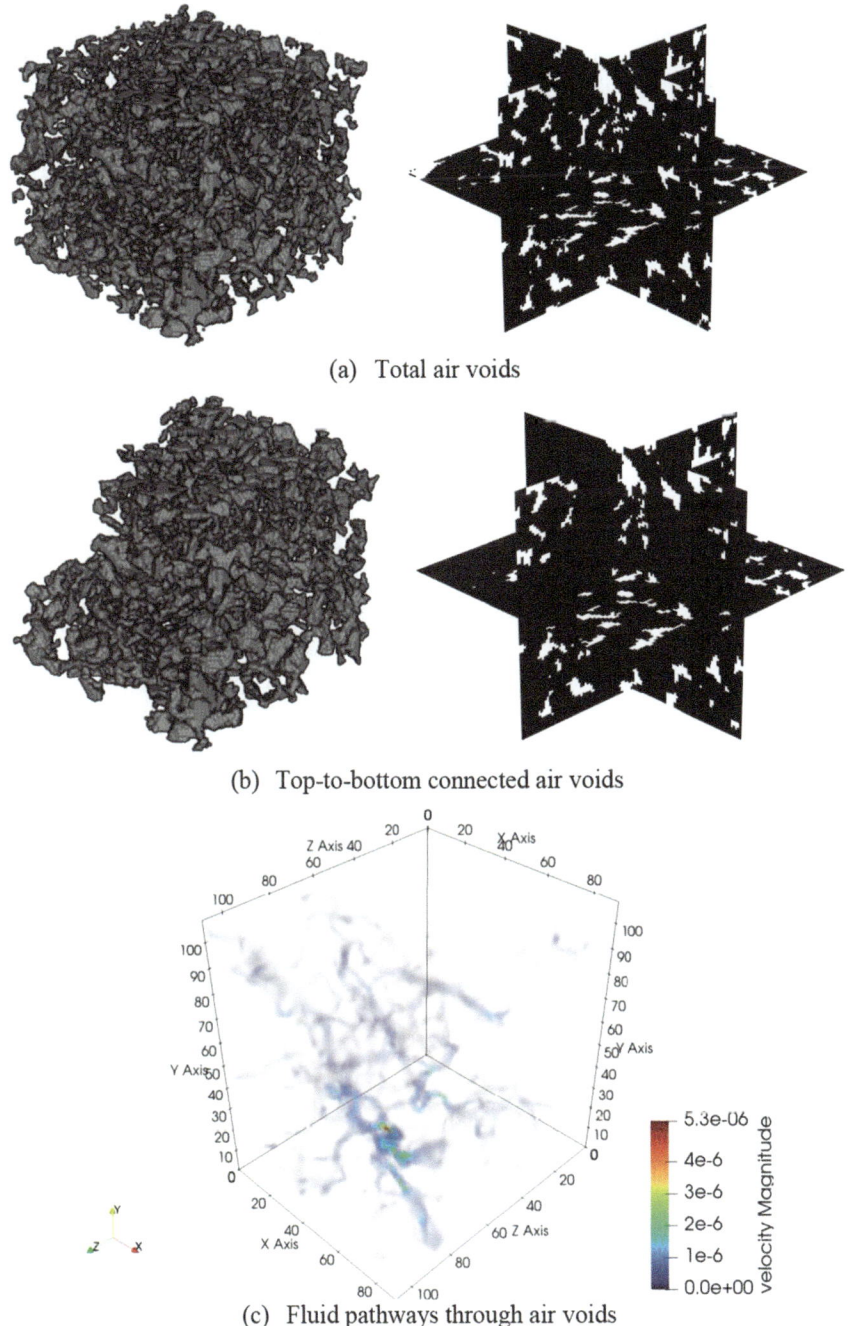

(a) Total air voids

(b) Top-to-bottom connected air voids

(c) Fluid pathways through air voids

Fig. 3 Air voids and fluid pathways in rectangular ROI with a spatial coarsening level of CL = 10

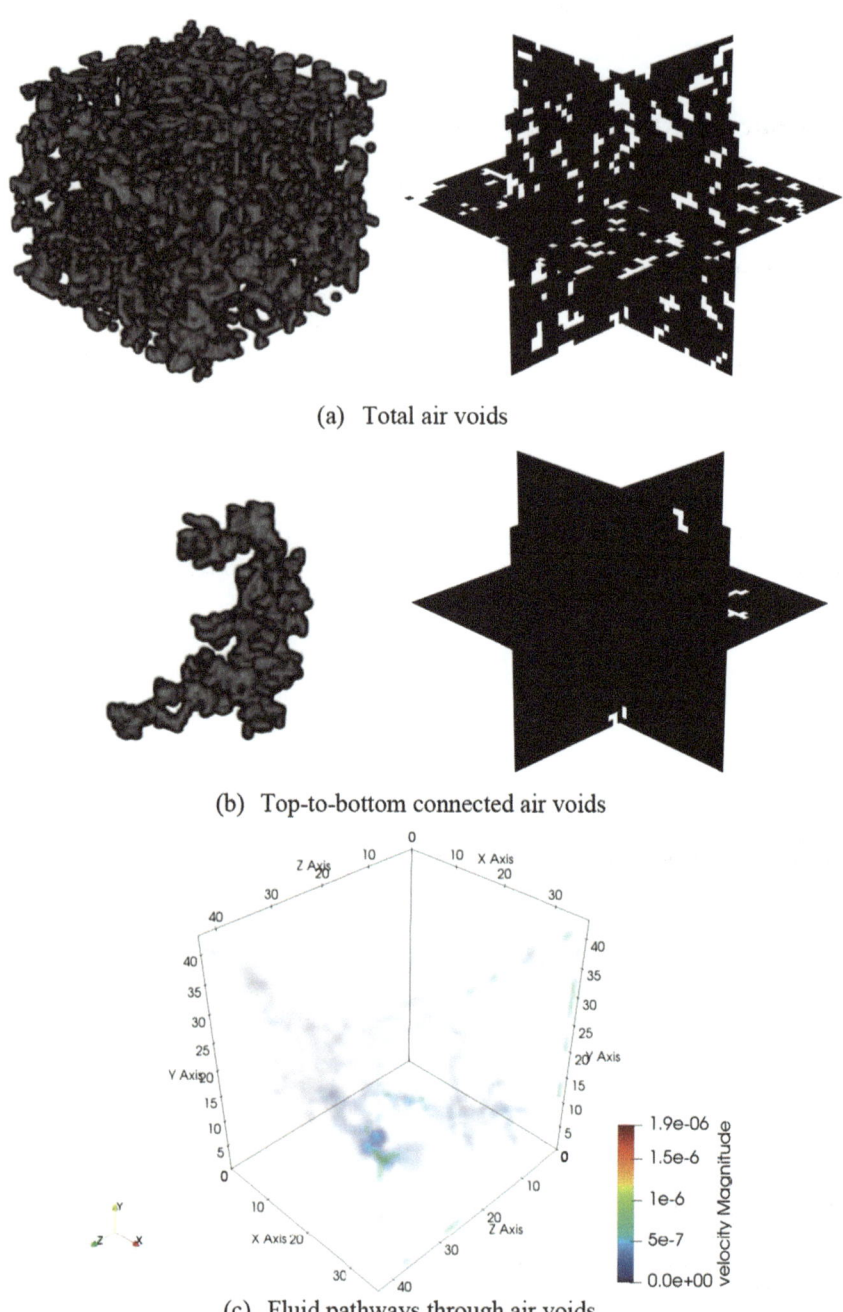

(a) Total air voids

(b) Top-to-bottom connected air voids

(c) Fluid pathways through air voids

Fig. 4 Air voids and fluid pathways in rectangular ROI with a spatial coarsening level of CL = 25

Table 1 Spatial coarsening and corresponding results of a rectangular ROI that measures 6.82 cm × 6.82 cm × 5.64 cm

| CL | Voxel size (cm) | (N_x, N_y, N_z) | # Voxels (× 10^6) | Porosity | | Simulated k | | | Simulated K |
				Total	Connected	LB units	Physical units (m^2)	Physical units (Darcy)	Physical units (m/s)
1	0.0063	(1089, 1089, 901)	1068.5	0.1354		–	–	–	–
5	0.0313	(218, 218, 181)	8.60	0.1356	0.1266	0.00625122	6.1×10^{-10}	621	6.0×10^{-3}
10	0.0626	(109, 109, 91)	1.08	0.1359	0.1113	0.00182735	7.2×10^{-10}	726	7.0×10^{-3}
25	0.1566	(44, 44, 37)	0.072	0.1362	0.0170	0.000199997	4.9×10^{-10}	497	4.8×10^{-3}

"CL" refers to spatial coarsening level, where 1 means uncoarsened

"–" means simulation not conducted

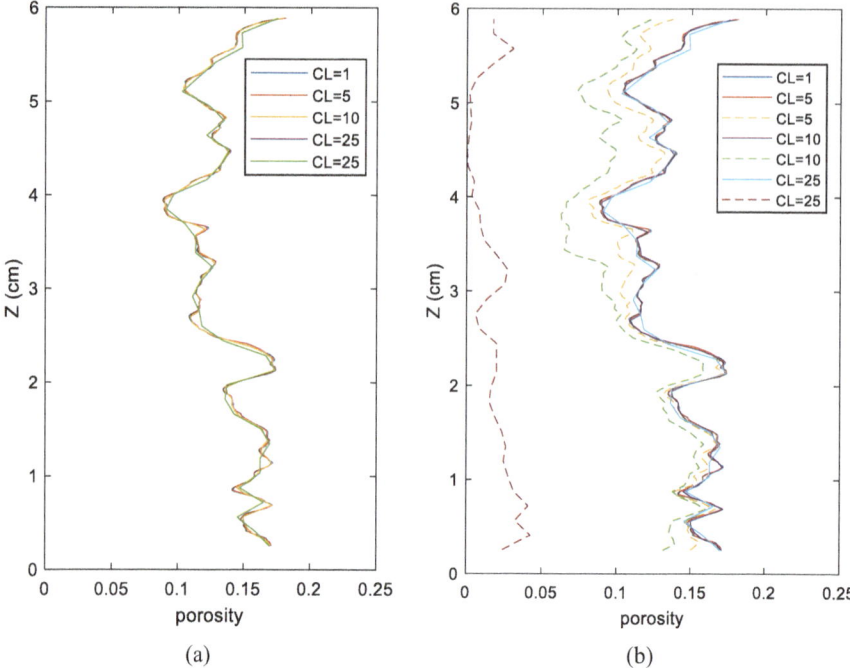

Fig. 5 Distribution of porosity in vertical direction for various cases of spatial coarsening. (**a**) Only shows total porosity. (**b**) Showing both total porosity (solid-coloured lines) and top-to-bottom connected porosity (dashed coloured lines). 'CL' in legend refers to coarsening level

The simulated permeabilities of the coarsened cases are all high (between 500 and 700 Darcy), and it is noted that gravel can have a permeability of up to 10^5 Darcy. It is interesting to note that there is not an observable trend between porosity and permeability between these three coarsened cases. This could be related to an issue with convergence of the simulation, and it was noted during this work that some pore throats present in the geometries were not properly discretized in space with the lattice mesh. This issue can be solved by refining these pore throats with more lattice nodes; however, this increases the computational cost of the simulations. While it is important to achieve convergence in the solution, we leave this issue for a future work and note that results in the present work are intended for visualization of the flow paths rather than accurate representation of the velocity magnitude. Permeability values are expected to change as grid refinement is done to for convergence purposes; however, it is possible the values will not change by many orders of magnitude.

4   Conclusions

This work helps visualize the fluid flow pathways through a sample of porous asphalt, and shows how spatial coarsening can be used to speed up simulation work while still resulting in a permeability of around 400–700 Darcy. The permeability should be viewed as demonstrative since it was noted that simulation results had not converged appropriately due to narrow pore throats present in the geometry, and that refinement of the lattice mesh is required; however, this was left for future work.

Other future work includes studying other coarsening approaches, and studying the influence of the ROI's volume on simulated permeability. That is, the size of the representative element volume (REV) for this porous asphalt sample should be assessed, even though this work assumed a rectangular ROI measuring 6.82 cm × 6.82 cm × 5.64 cm was sufficient.

References

1. Kassem, A., Linnerud, A.H., Martinussen, T.S.: Digitale tvillinger for kvalitetskontroll av asfaltdekker. Oslo Metropolitan University, Oslo (2023)
2. Allen, R., Dongmo-Engeland, B., Al-Batat, S.: Development of a MATLAB-based code for quantification of effective void space in porous pavement. In: Proceedings of the 64th International Conference of Scandinavian Simulation Society, SIMS 2023 Västerås, Sweden, September 25–28, 2023, Västerås, Sweden (2023)
3. Latt, J., Malaspinas, O., Kontaxakis, D., Parmigiani, A., Lagrava, D., Brogi, F., Belgacem, M.B., Thorimbert, Y., Leclaire, S., Li, S., Marson, F., Lemus, J., Kotsalos, C.: Palabos: parallel lattice Boltzmann Solver. Comput. Math. Appl. **81**, 334–350 (2021)

Controllable Generation of Porous Microstructures Through Generative Adversarial Networks

Jakob Torben and Kai Bao

Contents

1   Introduction

The use of computed tomography (CT) imaging has become increasingly popular in studying pavement structures, offering detailed insights into their complex micro-structural properties. The imaging technique allows for a non-destructive examination of asphalt's pore structures, such as porosity, a key factor affecting the material's performance and durability. However, CT imaging also comes with its challenges, including high costs and time-intensive processes, especially when large datasets are required for comprehensive analysis.

Accurate methods for statistical reconstruction of the three-dimensional (3D) pore structures from limited sample images have been explored to tackle these challenges. Generative adversarial networks (GANs) [1] have emerged as a promising alternative, offering a more efficient and flexible approach to augmenting the available dataset for pavement structure studies. It has the potential to synthesise high-fidelity

J. Torben · K. Bao (✉)
SINTEF Digital, Oslo, Norway
e-mail: kai.bao@sintef.no

© The Author(s) 2025
M. Kioumarsi, B. Shafei (eds.), *The 1st International Conference on Net-Zero Built Environment*, Lecture Notes in Civil Engineering 237,
https://doi.org/10.1007/978-3-031-69626-8_54

asphalt microstructures from a limited set of real CT images. This technique not only mitigates the constraints associated with traditional CT imaging but also introduces new potentials based on the development of deep learning techniques.

Porosity distribution is a crucial parameter that significantly influences the mechanical behaviour and longevity of pavement materials. The vertical distribution of the porosity of asphalt materials can be of a certain pattern based on the conditions of the asphalt materials, which are crucial because it influences the material's durability, permeability and overall performance under traffic and environmental conditions.

In this work, we aim to develop a controllable technique that allows us to generate microstructures with user-defined porosity profiles. This can potentially allow the creation of large amount of diverse, realistic asphalt microstructures that can be used for further simulations and analyses. With this technique, we can generate microstructures in a more flexible manner so that a more targeted investigation into how porosity distribution impacts the overall performance of pavement systems.

2   Related Works

Generative Adversarial Networks (GANs) is a class of generative models originally introduced by Goodfellow et al. in 2014 [1]. GANs typically consist of two neural networks, the generator and the discriminator, which are trained simultaneously through adversarial processes. Since its introduction, different variants have been proposed to enhance its capability and performance. Deep convolutional GANs (DCGAN) [2] apply convolutional neural networks within the GAN framework, leading to more stable training and better quality of the generated images. Self-Attention Generative Adversarial Network (SAGAN) [3] incorporates a self-attention mechanism into a GAN architecture, which improves the performance significantly for long-range dependency image generation tasks. Since then, GANs have been used to successfully generate various data, such as images, text, videos, etc.

Mosser et al. [4] applied GAN to reconstruct 3D microstructure of a granular porous structure. SliceGAN [5] resolves the dimensionality incompatibility between two-dimensional (2D) training images and 3D generated volumes so that it can generate 3D volumes from 2D images, which are more accessible and of better resolution and quality. However, SliceGAN holds limited capabilities in handling the heterogeneity of the microstructures.

Conditional GAN (GAN) [6] extends the GAN framework by introducing class labels (categorical conditions), which can be used to condition the discriminator and generator so that the single generator can generate images of various categories. However, this method does not work for continuous scalar conditions, which is the case for many porous structure-related properties. Continuous conditional GAN (CCGAN) [7, 8] addresses challenges in conditioning on continuous scalar labels by reformulating empirical cGAN losses and introducing a novel method for integrating regression labels with the generator and discriminator. This approach effectively overcomes limitations associated with traditional cGANs when dealing with continuous scalar conditions. CCGAN has served as the foundation for our research. In this study, porosity has been explored as the scalar condition, so that we can generate porous structure for specified porosity. Furthermore, like [9], the latent space interpolation technique is used to synthesise 3D microstructure volume based on the trained CCGAN.

3 Methods

3.1 Dataset and Preprocessing

In the study, we use micro-CT images of Porous Asphalt (PA) sample (detailed in [10]) as our training dataset. This lab-crafted sample is designed with sufficiently high porosity to allow surface water to permeate the material, facilitating drainage on roads. The original dataset features a dimension of 2000^3 and a resolution of 62.6 μm. It is of cylindrical core shape. To be used for deep learning training, using the processing technique from [10], a region of interest was extracted, noise filtered, and the pore space segmented. The resulting processed image is shown in Fig. 1. For efficient training samples, we further cropped it to a square within the core sample, yielding a 3D data sample with dimensions of $1000 \times 1000 \times 901$.

To effectively train the CCGAN model used in this study, a substantial number of real samples are needed. To compile the necessary training samples, we randomly selected 32,000 image patches from the dataset. To capture enough reprehensive pore structure, while maintaining a small enough image size that can be used for training the CCGAN, we down-sampled the patches by a factor of four to get a patch of size 128×128. This process is shown in Fig. 1.

To understand the porosity distribution among these samples, we plotted their porosity values in a histogram, as shown in Fig. 2. GAN models aim to approximate the real data's distribution through a high-dimensional probability function, with a particular emphasis on capturing the likelihood of various porosity levels within the dataset in this study. The success of a GAN model is closely tied to how well this distribution is represented in its training set. It is therefore crucial to have a dataset that is both diverse and balanced enough to cover the range of the observed porosity values. As a result, the model is expected to perform well within the porosity value range that is adequately represented. As depicted in Fig. 2, the 5th and 95th percentiles of the training samples' porosity values fall within the [0.07, 0.19] range, which also represents the focus of the experiments conducted in this study.

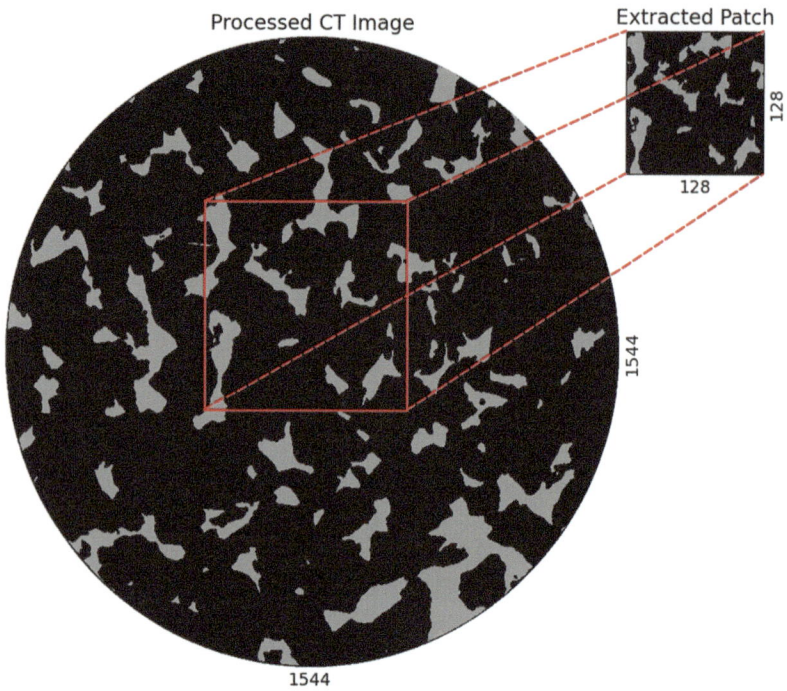

Fig. 1 An example of a slice of asphalt CT image and an extracted patch from the dataset. The patches are down-sampled by a factor of four

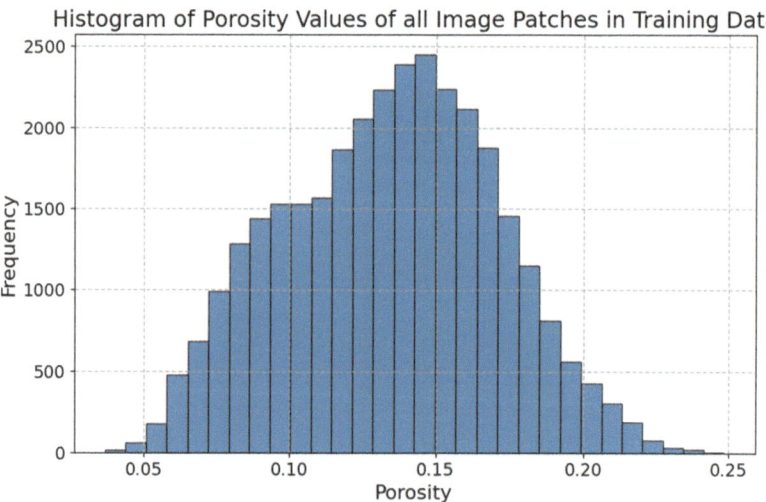

Fig. 2 Distribution of porosity values of the 32,000 image patches used in the training data

3.2 Network Architecture and Training Parameters

Figure 3 shows the overall workflow and architecture of the neural network structure used. In a typical GAN, the network structure consists of a generator and a discriminator. CCGAN extends the structure of a GAN by introducing a regression label to condition the training of both the generator and discriminator. In this work, the porosity of the porous structure is used as the regression label. The generator takes both a random noise vector (latent space) and the porosity label as inputs. The discriminator receives both real and generated data, along with the porosity label, and is trained to determine whether the data is real or fake, considering the input porosity. During training, the generator aims to generate data that the discriminator cannot distinguish from real data, given the input porosities. At the same time, the discriminator learns to better distinguish between real and generated data with given porosities.

Training GANs with high-resolution images enhances the quality and detail of generated images, enabling the model to capture more complex features. However, training GANs with higher resolution images demands more training data, computational resources and memory. It also increases the complexity of the learning process, potentially destabilising the balance between the discriminator and the generator. To manage these challenges, we limit our training images to a resolution of 128 × 128.

In microstructure generation, accurately identifying the phase of each pixel is crucial. In this study, we incorporate image segmentation directly into the neural network model. Our approach involves one-hot encoding of the training images to designate each phase with distinct channels, setting pixels to 1 where a phase exists and 0 elsewhere. By using a SoftMax activation function, we transform the output into a probability distribution for each phase. By using an argmax function we can easily get the segmented image from the output, with clearly defined phases for each pixel.

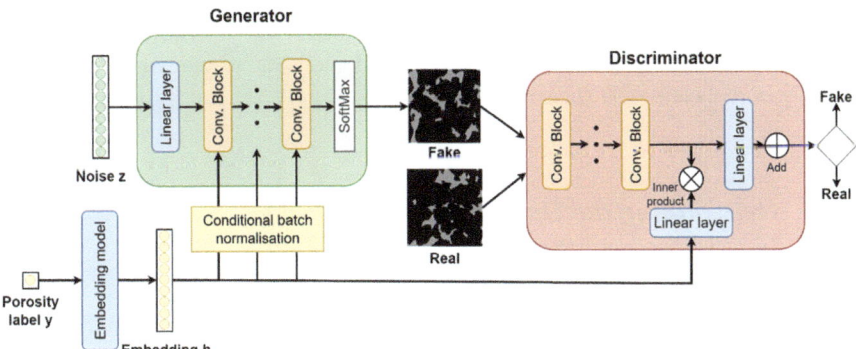

Fig. 3 Overview of deep learning architecture used. The architecture is primarily based on SAGAN architecture but with some modifications for our purpose, such as using a SoftMax activation function in the generator. The porosity is used as the regression label into the generator and discriminator using the improved label input method developed in CCGAN [8]

3.3 Latent Space Interpolation

Once the generator is trained, it can produce 2D microstructures or slices that follow a specified porosity. To use this 2D generator to produce 3D volumes, latent space interpolation [9] is used to generate correlated 2D microstructure slices. These slices can then be stacked to synthesise a cohesive 3D microstructure. For simplicity, we use a linear interpolation, given by the expression

$$f_{z_1, z_2}(\lambda) = (1 - \lambda)z_1 + \lambda z_2 \tag{1}$$

The relative weight of the two latent vectors is thus controlled with the parameter λ. Here we set the parameter λ using the following equation

$$\lambda = \frac{i}{N} \qquad i = 0, 1, 2, \ldots, N - 1 \tag{2}$$

where N is a control over how much we want the microstructure to change between the slices. To generate more images than N, we can perform this procedure in several cycles, where after each N steps we sample a new random latent vector. Using this approach, we can generate an arbitrary number of unique images.

4 Results

4.1 Generating 2D Slices with Controllable Porosity

After training the generator using the real sample images, along with their corresponding porosity labels, it can produce new, unique 2D images with given porosity values. Figure 4 displays a side-by-side comparison of real images against images generated by the model, using the porosity labels from their real counterparts. This qualitative comparison reveals that the structure of the generated images bears a close resemblance to that of the real images, demonstrating the generator's proficiency in replicating 2D slices.

4.2 Determining the Optimal Number of Steps for the Interpolation Method

The latent interpolation technique used in our work is a method to make our generated 2D slices into a cohesive 3D volume. Our goal is to create a 3D volume with material properties that match our dataset. For CT images of microstructures, the individual 2D images are usually highly correlated with their neighbouring slices. The degree of correlation depends on the size of the structures in the material

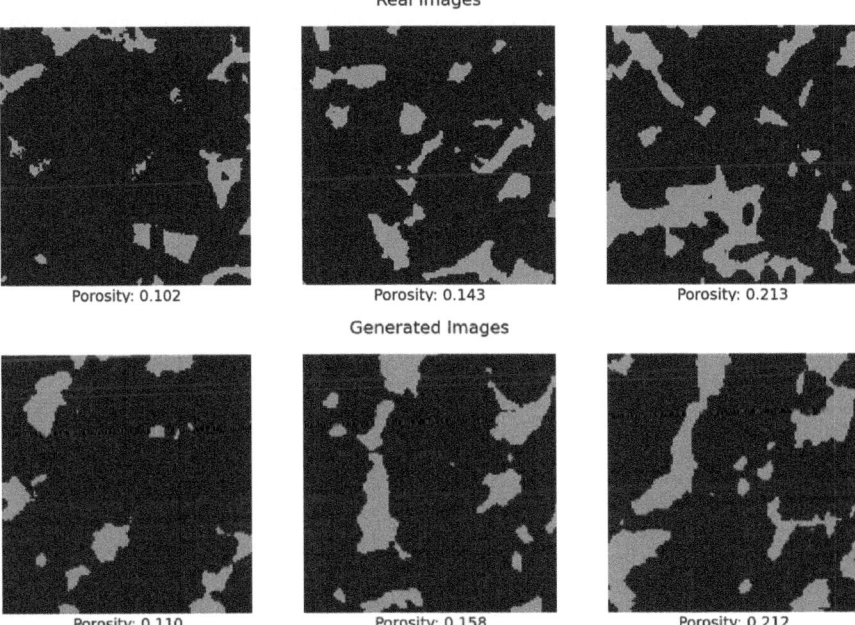

Fig. 4 Comparison between real images from the dataset, and generated images, using the porosity values of the real images as input labels

and the distance between the 2D images. For our generator, we can use the number of interpolation steps N, to control how many images we want to generate before they are independent of each other. This gives us an explicit method to control the correlation properties of our generated 3D volume.

To get the best match between the real dataset and our generated volume, we calculated the mean correlation value between an image and its neighbouring images, shown with the blue line in Fig. 5. This was done for the full volume of size $1000 \times 1000 \times 901$, where neighbours up to 100 images away were considered. Using the measured porosity profile from the real dataset as an input to our generator, we generated corresponding 3D volumes of shape $128 \times 128 \times 901$ for varying values of N. To find the best match to our dataset, the value of N with the minimal root mean square error (RMSE) between the correlation curve of the real dataset and our generated volume was selected, which was found to be 55. Figure 5 shows the correlation curve for varying values of N, where we can see that our optimal value closely matches the correlation behaviour of the real dataset.

4.3   Generating 3D Microstructures

The vertical distribution of porosity in asphalt materials significantly influences their structural integrity, filtration capabilities and overall performance. This study

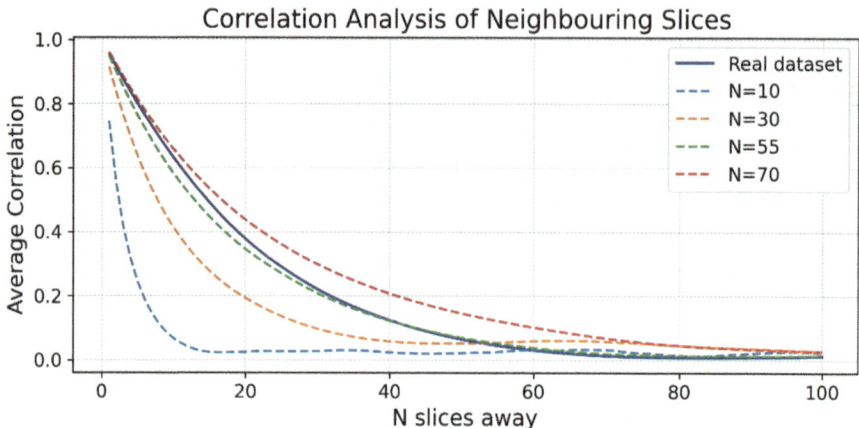

Fig. 5 Analysis of the correlation between neighbouring slices for the dataset. This is used to determine the parameter N for the number of slices in the z direction to sample before they should be independent of each other

focuses on generating asphalt microstructures that accurately capture the specific vertical porosity distribution.

In Fig. 6, the right panel illustrates the porosity profile of the actual dataset in the vertical direction, represented by the blue line. Utilising this profile as a reference and input, we aim to generate 3D microstructures that replicate this specific porosity distribution. The left panel of Fig. 6 displays three slices from the generated 3D microstructures, indicating that our generator indeed generates 2D microstructures to match the input porosity accurately. An orange dashed line in the right panel of Fig. 6 depicts the vertical porosity profile of the generated 3D microstructures. A comparison between the blue and orange dashed lines reveals similar porosity profiles, although deviations are observed.

Utilising the same generator, instead of generating a vertical porosity profile from a real dataset, we generated 3D microstructures based on a user-defined linear porosity profile, shown in Fig. 7. The vertical porosity profile of the resulting 3D microstructure demonstrates a general linear trend, while with noticeable deviations from the specified porosity profile. This observation demonstrates the capability of the current method to generate asphalt microstructures based on custom porosity profiles, which can be challenging with real samples. This approach could potentially set new ways in asphalt material design and optimisation.

A significant advantage of the deep learning-based generation technique is its efficiency: once the generator is trained, it can reconstruct 3D microstructures very quickly. In this study, generating a 3D microstructure with dimensions $128 \times 128 \times 901$ requires only 0.7 s using an NVIDIA 3090 RTX Ti GPU. This efficiency suggests the feasibility of producing numerous realisations of 3D microstructures, each with identical or varied porosity profiles.

Fig. 6 Visualisation of generated 3D volume from real porosity profile: Left: Three selected slices from the volume. Middle: 3D rendering of the generated porous volume. Right: A comparison of the porosity profile extracted from the dataset (blue line), which was used as input to the generator, and the calculated porosity profile from the generated volume (orange dashed line)

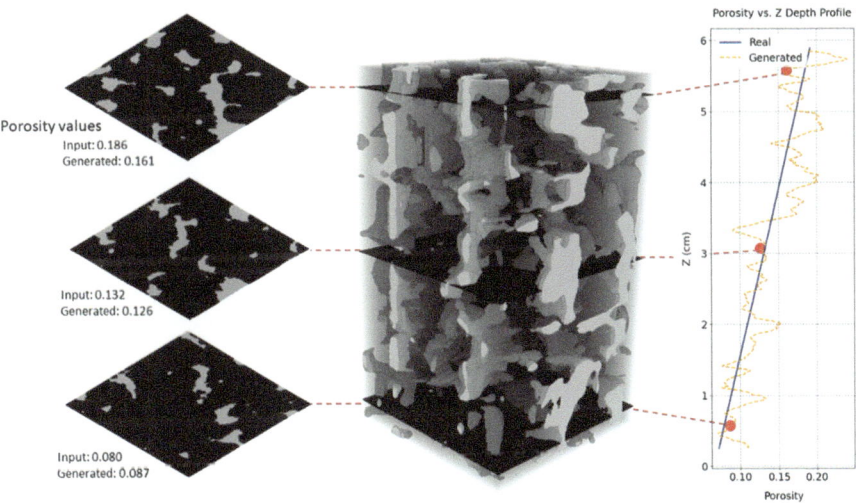

Fig. 7 Visualisation of generated 3D volume from linearly increasing porosity profile: Left: Three selected slices from the volume. Middle: 3D rendering of the generated porous volume. Right: A comparison of the defined linearly increasing porosity profile (blue line), which was used as input to the generator, and the calculated porosity profile from the generated volume (orange dashed line)

4.4   Evaluating the Properties of the Generator

In Figs. 6 and 7, we can see that the generated porosity profile has a similar profile to the real data but has some fluctuations. To evaluate the degree of fluctuations and average behaviour, 50 volumes were generated for a given porosity profile. The porosity of the generated volumes was calculated, and the mean and standard deviation calculated across the 50 volumes. Figure 8 shows the result for a user-defined linearly increasing porosity profile and a porosity profile extracted from the real dataset. The results indicate that the average porosity profile follows the specified input porosity profiles but with a significant variation. For the linear porosity profile, we can see that the generator consistently generates higher values for low porosities and lower values for higher porosities. This regression to the mean is likely a result of low and high porosity values not being well represented in our dataset.

Figure 9 shows a comparison of the material properties—porosity, surface area and effective diffusivity—between the real dataset and generated volumes. The

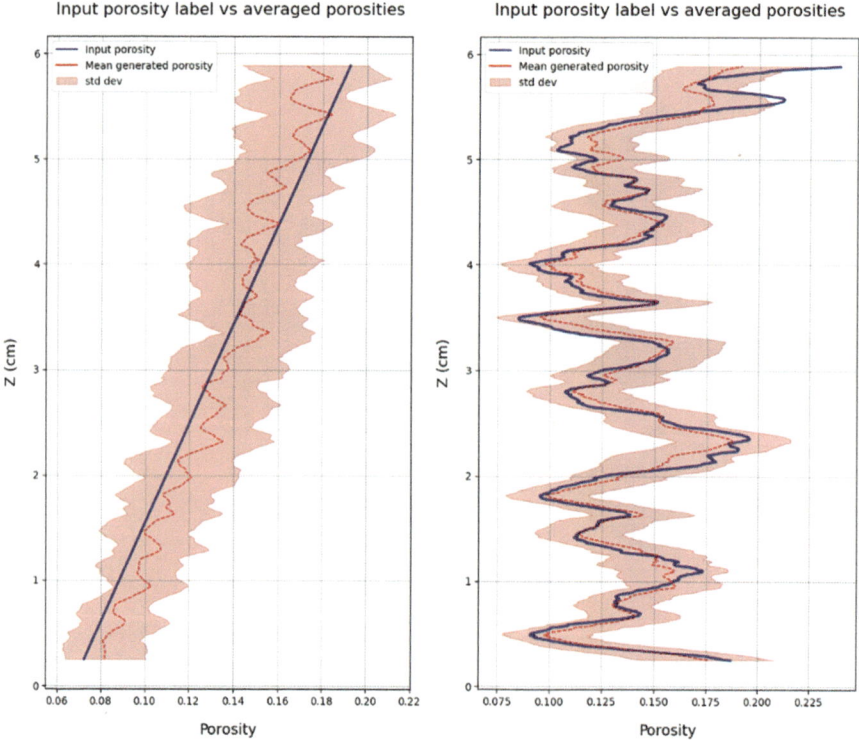

Fig. 8 Comparing the input porosity label to the averaged porosity profiles of 50 different generated microstructures. Left: For linearly increasing porosity values. Right: For a porosity profile extracted from the real dataset

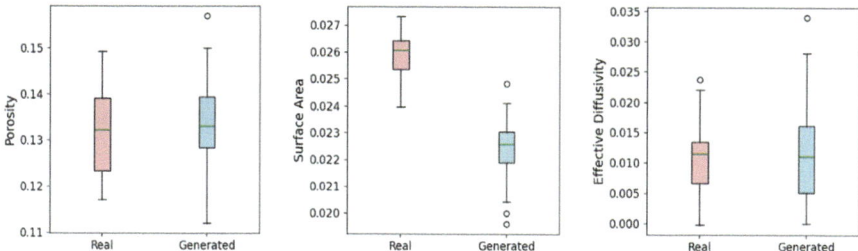

Fig. 9 Comparison of the material properties porosity, surface area and effective diffusivity, between the real dataset and generated volumes. The boxes extend from the first quartile (Q1) to the third quartile (Q3), with the central green line indicating the median. With whiskers extending 1.5 times beyond the interquartile range (IQR), (Q1 − 1.5 IQR below and Q3 + 1.5 IQR above). Outliers are indicated as circles

metrics have been calculated with TauFactor [11]. There is a close match between the median porosity of the real and generated data. However, the generated data have a narrower distribution, except for some outliers. The surface area of the generated volumes has a small consistent bias toward lower values. The effective diffusivity is a metric related to the flow transport of the material, which is determined by the 3D material structure. This is a metric that describes the emergent 3D behaviour of the volume. In the comparison, there is a close match between the median values of the real and generated data, demonstrating that the generator can create representative 3D volumes.

5   Conclusion

This study introduces a framework that incorporates Continuous Conditional Generative Adversarial Networks (CCGAN) and latent space interpolation techniques for the controllable generation of microstructures with specified vertical porosity profiles. The ability to fine-tune the generation of 3D microstructures provides a flexible and controllable approach that could potentially pave a new path for asphalt material design and optimisation. We believe the approach used in this work can potentially be extended to generate microstructures based on other properties.

Within the framework, by using the presented CCGAN model, we can generate 2D microstructures that closely resemble real samples, guided by specified porosity labels. Latent space interpolation technique is shown to be effective in assembling the 2D microstructure slices into cohesive 3D volumes. However, we have observed deviations in the porosity profile of the generated 3D microstructures compared to the input porosity profile. Exploring the possibility of combining our controllable generation technique with a 3D-aware training approach offers a promising direction for future research.

Acknowledgement We acknowledge Rebecca Allen from OsloMet for providing the asphalt CT images (detailed in [10]). These images, derived from an asphalt sample prepared by the Norwegian Public Roads Administration (Statens Vegvesen), were produced using micro-CT by the Norwegian Geotechnical Institute (NGI) to support thesis work at OsloMet. We also thank the authors of [7, 8] and [11] for publishing their developed source code, which are used in this work.

References

1. Goodfellow, I., Pouget-Abadie, J., Mirza, M., Bing, X., Warde-Farley, D., Ozair, S., Courville, A., Bengio, Y.: Generative adversarial nets. Adv. Neural Inf. Proces. Syst. **27**(2), 2672–2680 (2014)
2. Radford, A., Metz, L., Chintala, S.: Unsupervised representation learning with deep convolutional generative adversarial networks. arXiv preprint arXiv:1511.06434 (2015)
3. Zhang, H., Goodfellow, I., Metaxas, D., Odena, A.: Self-attention generative adversarial networks. In: International Conference on Machine Learning, pp. 7354–7363. PMLR (2019)
4. Mosser, L., Dubrule, O., Blunt, M.J.: Reconstruction of three-dimensional porous media using generative adversarial neural networks. Phys. Rev. E. **96**(4), 043309 (2017)
5. Kench, S., Cooper, S.J.: Generating three-dimensional structures from a two-dimensional slice with generative adversarial network-based dimensionality expansion. Nat. Mach. Intell. **3**, 299–305 (2021). https://doi.org/10.1038/s42256-021-00322-1
6. Mirza, M., Osindero, S.: Conditional generative adversarial nets. arXiv preprint arXiv:1411.1784 (2014)
7. Ding, X., Wang, Y., Xu, Z., Welch, W.J., Jane Wang, Z.: Ccgan: continuous conditional generative adversarial networks for image generation. In: International Conference on Learning Representations. OpenReview (2021). https://iclr.cc/Conferences/2021
8. Ding, X., Wang, Y., Xu, Z., Welch, W.J., Wang, Z.J.: Continuous conditional generative adversarial networks: novel empirical losses and label input mechanisms. IEEE Trans. Pattern Anal. Mach. Intell. **45**(7), 8143–8158 (2023)
9. Liu, X., Park, K., So, M., Ishikawa, S., Terao, T., Shinohara, K., Komori, C., Kimura, N., Inoue, G., Tsuge, Y.: 3D generation and reconstruction of the fuel cell catalyst layer using 2D images based on deep learning. J. Power Sources Adv. **14**, 100084 (2022)
10. Allen, R., Dongmo-Engeland, B., Al-Batat, S.: Development of a MATLAB-Based Code for Quantification of Effective Void Space in Porous Pavement, pp. 386–392. Scandinavian Simulation Society (2023)
11. Cooper, S.J., Bertei, A., Shearing, P.R., Kilner, J.A., Brandon, N.P.: TauFactor: an open-source application for calculating tortuosity factors from tomographic data. SoftwareX. **5**, 203–210 (2016)

Marine-Based Photocatalytic Protection of Building Envelopes on Behalf of Climate Change

Jéssica Deise Bersch [ID], Ana Paula Soares Dias [ID], Denise Dal Molin [ID], Angela Borges Masuero [ID], and Inês Flores-Colen [ID]

Contents

1   Introduction

Buildings can act against air pollution through photocatalysis [1], which also attributes self-cleaning and antimicrobial effects to surfaces [2], mainly benefiting façades in urban environments. Photocatalysis involves electron transfers from the valence to the conduction band of semiconductors under irradiation, the reaching of

J. D. Bersch (✉)
CERIS, Instituto Superior Técnico, Universidade de Lisboa, Lisbon, Portugal

NORIE, PPGCI, Universidade Federal do Rio Grande do Sul, Porto Alegre, Brazil
e-mail: jessica.d.bersch@tecnico.ulisboa.pt

A. P. S. Dias
CERENA, Instituto Superior Técnico, Universidade de Lisboa, Lisbon, Portugal

D. Dal Molin · A. B. Masuero
NORIE, PPGCI, Universidade Federal do Rio Grande do Sul, Porto Alegre, Brazil

I. Flores-Colen
CERIS, Instituto Superior Técnico, Universidade de Lisboa, Lisbon, Portugal

© The Author(s) 2025
M. Kioumarsi, B. Shafei (eds.), *The 1st International Conference on Net-Zero Built Environment*, Lecture Notes in Civil Engineering 237,
https://doi.org/10.1007/978-3-031-69626-8_55

electron–hole pairs to the particles' surface, and their reaction with air, forming reactive species to mineralize pollutants [3].

Several photocatalysts exist, like titanium dioxide (TiO_2), zinc oxide (ZnO), and iron (III) oxide (Fe_2O_3) [4]. Their conventional production with physical and chemical methods is environmentally unfriendly, involving toxic chemicals (stabilizing and reducing agents, chemical solvents), being costly and time consuming [5]. The large bandgap energy of well-established photocatalysts, like TiO_2, limits their photoactivation to ultraviolet (UV) radiation [6], and their efficiency may be further harmed by the high recombination between photoinduced electrons and holes [7]. Moreover, there are concerns about nanomaterials' ecotoxicity and long-term impacts on human health [8].

The green synthesis of nanoparticles is valuable for biological and environmental applications [5] since converting waste into functional substances should be encouraged [9] to pursue a circular economy and fight climate change. Blue economy accounts for efforts promoting the use of marine resources in search of sustainability [10]. So, marine algae have been employed in the green synthesis of nanoparticles due to their fast harvesting, renewability, low toxicity, and simple handling [11]. Products like mollusk shells derive from aquaculture, and their large availability, combined with the usual landfill or seawater disposal, pollutes the environment [12], demanding novel usages.

This research dived into the green synthesis of photocatalysts based on sea resources through a comprehensive literature review in several research fields, seeking resilience and climate mitigation. The goals were to answer whether marine bio-photocatalysts are already embraced by the construction sector and identify experimental starting points for their use in building envelopes, fostering preservation and better air quality.

2 Methods and Overview

For the primary question, "How to synthesize (and evaluate) an effective bio-photocatalyst produced with marine resources following a green route?", the proposed query string was: "(bio* OR natur* OR green OR waste OR sustainab* OR organic) AND (synthes* OR produc* OR compos* OR fabric*) AND (fish OR shell OR *algae OR marine OR maritime OR sea*) AND (photocatal* OR photoact*)". The survey was broad to inquire if any result concerned buildings or, if not, to identify potential experimental approaches. The title, abstract, and keywords of journal papers in English from Scopus were searched, including documents from the beginning of 2020 up to February 2024. Filtering rounds began with the titles, followed by abstracts and full-text analysis. The selected full papers explicitly described the extraction of raw marine materials; experiments using purchased chitosan, alginate, or algae powder, for instance, were discarded. Literature reviews were disqualified in the final selection.

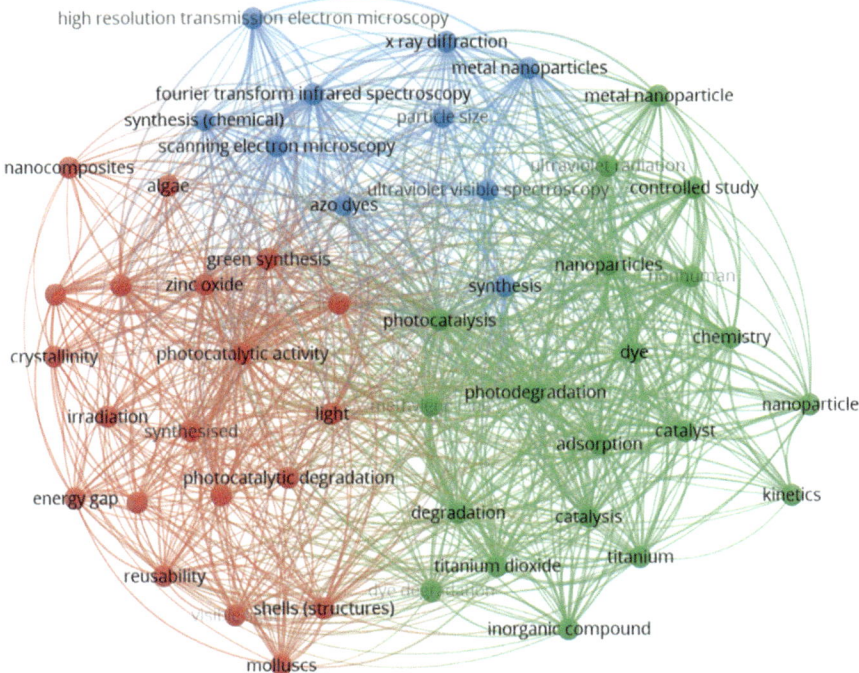

Fig. 1 Network visualization of all keywords from the 64 selected papers (VOSviewer v1.6.18: co-occurrence analysis, full counting method, five minimum keyword occurrences)

In total, 1894 papers were initially retrieved. Title and abstract analysis led to the selection of 186 documents, of which 122 were discarded after full paper reading. So, the final sample had 64 papers, whose keywords network is depicted in Fig. 1, from VOSviewer v1.6.18. Green synthesis is within the three main clusters, and shells, mollusks, and algae connect with photocatalysis. Photoactivity inherent aspects, like energy gap, kinetics, and reusability, as well as characterization techniques, are also shown in Fig. 1.

At least 68.8% of the papers had wastewater or water treatment as a primary goal [4, 7, 9, 11–51]. Meanwhile, 20.3% usually referred to more than one photocatalysis application [52–64] despite typically testing dyes and organic molecule degradation. Around 6.2% of the studies focused on medicine or human health [65–68] and 4.7% on indoor air purification [69–71], the closest topic to buildings but not related to façades.

Algae were used to synthesize photocatalysts in 42.2% of the papers, including seaweed [11, 17, 30, 31, 34, 38, 41, 47, 48, 51–55, 58, 64, 66, 68] and microalgae [15, 28, 29, 36, 37, 43, 45, 49, 57]. Different shell types were studied in 37.5% of the articles, comprising abalones [39, 70, 71], clams [7, 27, 39], crabs [42], crawfishes [20], mussels [12, 18, 26, 32, 39, 50, 56, 69], oysters [13, 19, 21, 23, 39], scallops [4, 39, 61], snails [62], shrimps [40], and seashells in general [9, 14]. The remaining documents addressed other sea species: one for bacteria [44], marine oyster extract [60], sand [59], sponge [24], and tunicates [46]; two for fungi [25, 35]; and three for fish waste (bones [16, 22] or scales [63]) and plants [33, 65, 67]. So, algae and shells were the most used bio-sources.

Three papers referred to the application of bio-photocatalysts, mentioning tetraethylorthosilicate encapsulation [38], thin films of chitosan (from crab shells) with polyvinylidene chloride [42], and cellulose hydrogels with extraction from tunicates [46]. None of the selected documents addressed environmental impacts with life cycle assessments. Safat et al. [60] tested cytotoxicity against fibroblast cells, Sundar et al. [65] and Rajivgandhi et al. [67] against MCF-7 cells, and Lavanya et al. [66] against breast human hepatoma (Huh-7) cells. However, no article presented ecotoxicity assessments.

3   Characterization and Photocatalytic Evaluation Methods

X-ray diffraction (XRD) and UV-Vis or UV-Vis/NIR spectroscopy were the most frequent tests to characterize novel photocatalysts. XRD provided mineralogical compositions and, in addition, crystallographic data, such as crystalline sizes, using the Scherrer formula [29, 35]. Bandgaps were determined from spectroscopy results with Tauc plots [7, 9]. Surface compositions and valence states were measured with X-ray photoelectron spectroscopy (XPS) [56], and Fourier-transform infrared spectroscopy (FTIR) complemented chemical analysis by examining functional groups [4].

Microscopy techniques, like scanning electron microscope (SEM), field emission scanning electron microscopy (FE-SEM), transmission electron microscopy (TEM), and high-resolution transmission electron microscopy (HRTEM), were shown in most papers, combined or not with energy dispersive spectrometer (EDX/EDS). Thermogravimetric analysis (TGA) and particle size/specific surface area also characterized the photocatalysts, including N_2 adsorption–desorption isotherms [20]. Electron–hole separation, charge transfer, and migration [22] were analyzed with photoluminescence spectra (PL), and free radical and hole scavenging experiments were common to investigate photochemical-produced charges [9]. Several papers modeled kinetics and detailed photocatalytic mechanisms. Although not so frequent, other characterization tests, such as Raman measurements [11] and zeta potential [4], were also carried out.

Most studies performed dye degradation to evaluate photoactivity. Dyes such as rhodamine B (RhB) [18, 26, 30], methylene blue (MB) [19, 55, 56], methyl orange [40, 44, 65], and congo red [11, 16, 34] were considered reference pollutants. In some cases, antibiotics were degraded by the photocatalysts in aqueous solution, including amoxicillin [4], enrofloxacin [20], procaine penicillin [21], sulfadiazine [56], and tetracycline [36], further than phenols [33], and herbicides like glyphosate [23]. For testing, the marine-based photocatalysts were mixed in a solution with the compound to be degraded and kept in the dark for variable times (30 min [9, 24, 53,

55, 66], 1 h [26], 2 h [27], etc.), until reaching adsorption–desorption equilibrium [50], and under stirring to achieve good dispersion [55]. Then, the prepared suspensions were exposed to light sources, generally under continuous agitation [20]: mainly visible light [17, 31, 51, 61, 65], sunlight [19, 21, 24, 52, 53], led light [36, 49, 54, 59], xenon [18, 20, 23, 55, 57] and UV lamps [4, 22, 29, 30] were used. At given intervals (3 min [11], 10 min [66], 30 min [53], etc.), aliquots from the solutions were removed for UV-Vis or UV-Vis/NIR spectroscopy. Supernatants were usually collected and centrifuged for sampling [9].

Some studies, like Saikia et al. [11] and Serrà et al. [36], evaluated the total organic carbon (TOC) of the reactions before and after pollutants' degradation. Serrà et al. [36] also performed liquid chromatography-mass spectroscopy (LC–MS) to identify intermediate products during tetracycline photodegradation. Reusability tests were typically achieved by submitting the synthesized photocatalysts to several consecutive degradation runs, like three [7], four [65], five [11], six [4], or ten [36]. Generally, the catalysts were regenerated with filtering [40] or centrifugation, washing, and drying [9, 14, 52].

For indoor air purification, formaldehyde and toluene were studied [69–71]. The photocatalysts were coated on quartz glass, and the tests were run by turning on light sources in reactors with injected gas volumes [69–71] after a period in the dark (1 h [69], 2 h [71]). Qiu et al. [69] measured carbon dioxide (CO_2) generation in time and formaldehyde degradation with a gas detector of photoacoustic spectrometer. Wang et al. [70] and Wang et al. [71] used a photoacoustic infrared multigas monitor.

4 Synthesis and Behavior of Marine-Based Photocatalysts

Three main groups were identified regarding the objectives behind employing marine materials to synthesize photocatalysts. The sea species were mainly used as carriers or supports for well-known photocatalysts, as catalysts without combining additional photoactive materials, or as reducing and capping agents.

Since nano-scaled particles tend to agglomerate, using carriers with adsorption abilities and hydrophilic surfaces to support photocatalysts may reduce their aggregation [72]. Employing sea materials to face environmental pollution can represent the "control of waste by waste," being potentially upgraded by trace metal elements in their composition, such as ZnO and TiO_2, and by their contribution to higher availability of active sites [39] for pollutant degradation and surface area for adsorption [12].

Wang et al. [39] studied powders from seashells calcined at 300 °C for 5 h and 500 °C for 2 h to support TiO_2 with a sol–gel process; Rabie et al. [51] supported nano-ZnO and Co-ZnO in algae of the *Sargassum* species using microwave irradiation; and Cai et al. [18] used a solvothermal method to prepare powders of mussel shells calcined at 900 °C for 3 h to support yttrium-doped Bi_2MoO_6. Wang et al. [39] observed that calcined abalone shell supports increased the concentration of O vacancies, enhancing charge separation on the TiO_2 surface and, therefore, its photoactivity, reaching even better results than commercial P25 nanoparticles in a 23.4% catalyst concentration. Rabie et al. [51] verified around 58% and 75% of enhancement in the photocatalytic performance of ZnO/algae and Co-ZnO/algae composites compared with pure ZnO; in fact, 2.45 eV and 2.32 eV were the calculated bandgaps for the composites, lower than the usual values for ZnO and Co-ZnO. Regarding the photocatalytic mechanism, Cai et al. [18] indicated that O_2^- and h^+ represented the primary active species, with a crucial role of ·OH. On stability and reusability, after four cycles, Cai et al. [18] observed a decrease from 99.7% of RhB degradation to 83.3%, probably due to a loss of catalyst during centrifugation and by the adsorption of intermediates onto the catalyst surface.

For a Bi_2WO_6/calcined mussel shell composite, Li et al. [12] measured 52.3% TOC removal after 150 min of reaction, suggesting a gradual mineralization of RhB. In the presence of a SiO_2-doped photocatalyst derived from a waste mussel shell, Wang et al. [50] got mineralization rates of 33.5% and 16.5% for MB and RhB solutions; so the dyes were decomposed and only partially mineralized under visible light. Zhang et al. [23] studied glyphosate degradation kinetics by cerium (Ce) and nitrogen (N) co-doped TiO_2 nano-photocatalysts loaded on modified oyster shell powders with the Langmuir–Hinshelwood mechanism, reporting an apparent first-order model over 120 min. The glyphosate degradation rate by the composite with 0.5% Ce doping molar ratio (0.5%-CeNT@Oys), with a 2.91 eV bandgap, was 8 times higher than pure TiO_2 (3.20 eV), indicating the synergy of adsorption and photocatalysis under xenon light [23].

Koladia et al. [14] compared the formal quantum efficiency (FQE) of photocatalysts synthesized by combining CaO from seashells and powdered molybdenum disulfide (MoS_2) in different mass proportions (1:3, 2:3, 9:11, 1:1, and 3:1) with pure CaO, MoS_2, and TiO_2. In this case, the FQE illustrated the photonic efficiency concerning industrial effluent treatment, dividing the reaction rate by the photon flux. The photocatalyst 45-CaMo (9:11) had the highest FQE, 0.32%, compared to 0.24% from TiO_2, 0.13% from pure CaO, and 0.02% from pure MoS_2 [14]. Bhole et al. [9] also reported the FQEs for MB molecules treated by incident photons, considering photocatalysts produced from seashell-derived CaO and synthesized Fe_2O_3 powder (rust); the highest FQE was 0.27% for the mixture with 10% Fe_2O_3 and 90% CaO.

Considering the photocatalysts produced without additional doping, Eddy et al. [13, 19, 21] and Qu et al. [27] focused on oyster and clam shells, respectively. Eddy et al. [13, 19, 21] applied the sol–gel method, counting on hydrochloric acid (HCl) and sodium hydroxide (NaOH) solutions and calcination to convert calcium carbonate ($CaCO_3$) into calcium oxide (CaO) nanoparticles; Qu et al. [27] also used HCl and calcination in the synthesis. Compared to pure photolysis of crystal violet dye, using CaO nanoparticles enhanced photodegradation from 3.9% to 98% under sunlight; higher degradation was observed with increasing contact times and catalyst dosages [13]. Eddy et al. [21] got an average bandgap of 4.42 eV for CaO nanoparticles calcined up to 850 °C, while Qu et al. [27] found 2.64 eV for calcination at 1000 °C, which led to 99.7% and 90.8% MB and congo red degradation by adsorption-photodegradation mechanisms, respectively, compared to 25.9% and 70.55% removal rates in the dark.

Campalani et al. [63] produced carbon dots from fish scales with a hydrothermal treatment, while Prosad Moulick [16] obtained hydroxyapatite by calcinating fish bones at 900 °C for 2 h. With 0.08 g fish bone catalyst and a 10.0 ppm congo red solution, nearly 82% of degradation was achieved within 3 h [16]. Carbon dots from fish scales had 3 eV of bandgap, converting 56.6% methyl viologen, with better photoelectron transfer than N-doped carbon dots [63]. Lastly, with dead marine microalgae *Scenedesmus quadricauda*, Wu et al. [37] got estrogen photodegradation under visible light.

Finally, several papers used marine materials (mostly algae) as reducing or capping agents. For instance, Maduraimuthu [53] prepared a seaweed extract from *Ulva lactuca* and added it to silver nitrate ($AgNO_3$) solution to form Ag nanoparticles. Indeed, phytochemicals like proteins, saponins, and quinines may reduce Ag^+ ions, further preventing aggregation and providing stability [53]. Wang et al. [17] produced an aqueous extract with *Solieria tenuis*, added with a chloroauric acid ($HAuCL_4$) solution to synthesize Au nanoparticles; phycoerythrin from the seaweed improved catalytic ability. Lavanya et al. [66], who synthesized silver nanoparticles using *Halimeda macroloba* seaweed extract, observed binding energy peaks at 367.5 eV and 373.8 eV with XPS, showing that Ag (0) was dominant and Ag (I) was completely reduced.

Azeem et al. [34] used 1.0 mmol of radical scavengers to study the reaction mechanism of macroalgae-based Ag nanoparticles under visible light. Butanol and ethylene diamine tetra-acetic acid sodium affected the photoactivity significantly, indicating the decisive roles of ·OH and h^+ on MB degradation, respectively, since hydrogen peroxide (H_2O_2) and superoxide dismutase had low effects, $·O_2^-$ had a minor participation in the results [34]. Saikia et al. [11] applied ammonium oxalate, isopropyl alcohol, and 1,4-benzoquinone as scavengers for h^+, ·OH, and $·O_2^-$, respectively, separately mixing them with congo red solution under sunlight; the radicals' role was $h^+ < ·OH < ·O_2^-$, with quenching of the $·O_2^-$ reducing degradation from 85.4% to 21.7% [11].

Regarding indoor air purification to tackle sick-house pollution [69] and incomplete photocatalytic degradation [70, 71], none of the papers worked with building materials or surface applications. However, concerning the $TiO_2/C/CaCO_3$ heterojunctions produced with an 11.8% mass fraction of mussel shell extract by Qiu et al. [69], formaldehyde was wholly degraded in 80 min under light, while the pure TiO_2 only led to a slight degradation. Wang et al. [70] explained, for the ternary $TiO_2/C/MnO_2$ with abalone shells, that electrons could be transferred from the conduction band of the TiO_2 and MnO_2 to carbon (C) promoting electron–hole separation and, thereby, the lifespan of charge carriers; carbon materials enhanced visible light absorption. In Wang et al. [71], pure TiO_2 had a bandgap energy of 3.3 eV, decreased from 3.1 eV to 2.9 eV by adding 0.2–0.8 g of insoluble matrix proteins from abalone shell waste.

5   Conclusions

This chapter reviewed marine-based photocatalysts inquiring about their presence in construction. Wastewater treatment was the main reason for bio-photocatalyst synthesis, while indoor air pollution was the closest topic concerning buildings. Algae, shells, fish waste, and other sea species have been studied, and titanium, zinc, and silver were the metals most combined with the bio-sources. Photoactivity was usually evaluated through the degradation of contaminants in solution.

Building envelopes coated with photocatalysts benefit from depolluting, antimicrobial, and self-cleaning effects, and green synthesis could further enhance these advantages. The availability of sea waste and underused resources has yet to be explored within construction since no retrieved study has applied marine-based photocatalysts to buildings and their façades. Experimental gaps in bio-photocatalysis should be fulfilled, envisioning buildings' resilience to climate change and their adaptability to requirements, mainly regarding air quality, sustained by toxicity and life cycle assessments.

Acknowledgments The authors are grateful for the Foundation for Science and Technology's support through funding UIDB/04625/2020 from the research unit CERIS (DOI: 10.54499/UIDB/04625/2020) and the grant number 2023.05316.BD (DOI: 10.54499/2023.05316.BD) and for EEA Grants bilateral fund through project financing SHELLTER (FBR_OC2_30). The authors also thank the CERENA research unit from IST, PPGCI from UFRGS, CAPES, and CNPq.

References

1. Sapiña, M., Jimenez-Relinque, E., Castellote, M.: Controlling the levels of airborne pollen: can heterogeneous photocatalysis help? Environ. Sci. Technol. **47**, 11711–11716 (2013)
2. Klienchen de Maria, V.P., Guedes de Paiva, F.F., Tamashiro, J.R., Silva, L.H.P., da Silva Pinho, G., Rubio-Marcos, F., Kinoshita, A.: Advances in ZnO nanoparticles in building material: antimicrobial and photocatalytic applications—systematic literature review. Constr. Build. Mater. **417**, 135337 (2024). https://doi.org/10.1016/j.conbuildmat.2024.135337
3. Hot, J., Dasque, A., Topalov, J., Mazars, V., Ringot, E.: Titanium valorization: from chemical milling baths to air depollution applications. J. Clean. Prod. **249**, 119344 (2020)
4. Alshandoudi, L.M., Al Subhi, A.Y., Al-Isaee, S.A., Shaltout, W.A., Hassan, A.F.: Static adsorption and photocatalytic degradation of amoxicillin using titanium dioxide/hydroxyapatite nanoparticles based on sea scallop shells. Environ. Sci. Pollut. Res. **30**, 88704–88723 (2023)
5. Malik, A.Q., Mir, T.U.G., Kumar, D., Mir, I.A., Rashid, A., Ayoub, M., Shukla, S.: A review on the green synthesis of nanoparticles, their biological applications, and photocatalytic efficiency against environmental toxins. Environ. Sci. Pollut. Res. **30**, 69796–69823 (2023)
6. Gonçalves, S.G., Milanez, P.R., de Souza Pereira, M., Benetti, R.M., Bellettini, G.C., Elyseu, F., Dal-Bó, A.G., Bernardin, A.M.: Sensitization of TiO_2 NPs with curcumin and chlorella and photocatalytic performance for degradation of phenol and organic dyes. Chem. Eng. Sci. **285**, 119598 (2024)
7. Ling, M.F.C., Hui, K.C., Sambudi, N.S.: Modification of TiO_2 with clam-shell powder for photodegradation of methylene blue. J. Solgel Sci. Technol. **102**, 412–421 (2022)

8. Ferreira, M.T., Soldado, E., Borsoi, G., Mendes, M.P., Flores-Colen, I.: Nanomaterials applied in the construction sector: environmental, human health, and economic indicators. Appl. Sci. **13**, 12896 (2023)

9. Bhole, A., Koladia, G.C., Koladia, S.P., Bora, N.V., Bora, L.V.: Treatment of waste using waste-derived materials and free energy: a practical concept of circular-economy. Sustain. Chem. Pharm. **37**, 101341 (2024)

10. Hossain, M.A., Islam, M.N., Fatima, S., Kibria, M.G., Ullah, E., Hossain, M.E.: Pathway toward sustainable blue economy: consideration of greenhouse gas emissions, trade, and economic growth in 25 nations bordering the Indian ocean. J. Clean. Prod. **437**, 140708 (2024)

11. Saikia, P., Borah, P., Borah, D., Gogoi, D., Rout, J., Ghosh, N.N., Bhattacharjee, C.R.: Facile green synthesis of rGO and NiO, and fabrication of a novel ternary nanoheterostructure NiO@g-C_3N_4-rGO as earth abundant superior photocatalyst for dye degradation. Mater. Today Sustain. **24**, 100595 (2023)

12. Li, S., Wang, C., Liu, Y., Xue, B., Chen, J., Wang, H., Liu, Y.: Facile preparation of a novel Bi_2WO_6/calcined mussel shell composite photocatalyst with enhanced photocatalytic performance. Catalysts. **10**, 1–11 (2020)

13. Eddy, N.O., Jibrin, J.I., Ukpe, R.A., Odiongenyi, A., Iqbal, A., Kasiemobi, A.M., Oladele, J.O., Runde, M.: Experimental and theoretical investigations of photolytic and photocatalysed degradations of crystal violet (CVD) in water by oyster shells derived CaO nanoparticles (CaO-NP). J. Hazard. Mater. Adv. **13**, 100413 (2024)

14. Koladia, G.C., Bhole, A., Bora, N.V., Bora, L.V.: Biowaste derived UV–visible-NIR active Z-scheme CaO/MoS_2 photocatalyst as a low-cost, waste-to-resource strategy for rapid wastewater treatment. J. Photochem. Photobiol. A Chem. **446**, 115172 (2024)

15. Zamani, W., Rastgar, S., Hedayati, A.: Capability of TiO_2 and Fe_3O_4 nanoparticles loaded onto Algae (*Scendesmus sp.*) as a novel bio-magnetic photocatalyst to degration of Red195 dye in the sonophotocatalytic treatment process under ultrasonic/UVA irradiation. Sci. Rep. **13**, 18182 (2023)

16. Prosad Moulick, S., Sahadat Hossain, M., Zia Uddin Al Mamun, M., Jahan, F., Farid Ahmed, M., Sathee, R.A., Sujan Hossen, M., Ashraful Alam, M., Sha Alam, M., Islam, F.: Characterization of waste fish bones (*Heteropneustes fossilis* and *Otolithoides pama*) for photocatalytic degradation of Congo red dye. Results Eng. **20**, 101418 (2023)

17. Wang, L., Qiang, X., Song, Y., Wang, X., Gu, W., Niu, J., Sun, Y., Srinuanpan, S., Wang, G.: Green synthesis of gold nanoparticles by phycoerythrin extracted from *Solieria tenuis* as an efficient catalyst for 4-nitrophenol reduction and degradation of dyes in wastewater. Mater. Today Sustain. **23**, 100435 (2023)

18. Cai, L., Zhou, Y., Guo, J., Sun, J., Ji, L.: Mussel shell-supported yttrium-doped Bi_2MoO_6 composite with superior visible-light photocatalytic performance. Water (Switzerland). **15**, 3478 (2023)

19. Eddy, N.O., Ukpe, R.A., Ameh, P., Ogbodo, R., Garg, R., Garg, R.: Theoretical and experimental studies on photocatalytic removal of methylene blue (MetB) from aqueous solution using oyster shell synthesized CaO nanoparticles (CaONP-O). Environ. Sci. Pollut. Res. **30**, 81417–81432 (2023)

20. Xiao, L., Zhang, S., Chen, B., Wu, P., Feng, N., Deng, F., Wang, Z.: Visible-light photocatalysis degradation of enrofloxacin by crawfish shell biochar combined with g-C_3N_4: effects and mechanisms. J. Environ. Chem. Eng. **11**, 109693 (2023)

21. Eddy, N.O., Odiongenyi, A.O., Garg, R., Ukpe, R.A., Garg, R., El Nemr, A., Ngwu, C.M., Okop, I.J.: Quantum and experimental investigation of the application of *Crassostrea gasar* (mangrove oyster) shell–based CaO nanoparticles as adsorbent and photocatalyst for the removal of procaine penicillin from aqueous solution. Environ. Sci. Pollut. Res. **30**, 64036–64057 (2023)

22. Ahamed, A.F., Kalaivasan, N., Thangaraj, R.: Probing the photocatalytic degradation of acid orange 7 dye with chitosan impregnated hydroxyapatite/manganese dioxide composite. J. Inorg. Organomet. Polym. Mater. **33**, 170–184 (2023)

23. Zhang, W., You, Q., Shu, J., Wang, A., Lin, H., Yan, X.: Photocatalytic degradation of glyphosate using Ce/N co-doped TiO_2 with oyster shell powder as carrier under the simulated fluorescent lamp. Front. Environ. Sci. **11**, 1131284 (2023)

24. Bhatti, M.A., Almani, K.F., Shah, A.A., Tahira, A., Chana, I.A., Aftab, U., Ibupoto, M.H., Mirjat, A.N., Aboelmaaref, A., Nafady, A., Vigolo, B., Ibupoto, Z.H.: Renewable and eco-friendly ZnO immobilized onto dead sea sponge floating materials with dual practical aspects for enhanced photocatalysis and disinfection applications. Nanotechnology. **34**, 035602 (2023)

25. Kumar, R.V., Vinoth, S., Baskar, V., Arun, M., Gurusaravanan, P.: Synthesis of zinc oxide nanoparticles mediated by *Dictyota dichotoma* endophytic fungi and its photocatalytic degradation of fast green dye and antibacterial applications. S. Afr. J. Bot. **151**, 337–344 (2022)

26. Mohammad, I., Jeshurun, A., Ponnusamy, P., Reddy, B.M.: Mesoporous graphitic carbon nitride/hydroxyapatite (g-C_3N_4/HAp) nanocomposites for highly efficient photocatalytic degradation of rhodamine B dye. Mater. Today Commun. **33**, 104788 (2022)

27. Qu, T., Yao, X., Owens, G., Gao, L., Zhang, H.: A sustainable natural clam shell derived photocatalyst for the effective adsorption and photodegradation of organic dyes. Sci. Rep. **12**, 2988 (2022)

28. Putri, R.M., Almunadya, N.S., Amri, A.F., Afnan, N.T., Nurachman, Z., Devianto, H., Saputera, W.H.: Structural characterization of polycrystalline titania nanoparticles on *C. striata* biosilica for photocatalytic POME degradation. ACS Omega. **7**, 44047–44056 (2022)

29. Al-Enazi, N.M.: Optimized synthesis of mono and bimetallic nanoparticles mediated by unicellular algal (diatom) and its efficiency to degrade azo dyes for wastewater treatment. Chemosphere. **303**, 135068 (2022)

30. Pachiyappan, J., Nirmala, G., Sivamani, S., Govindasamy, R., Thiruvengadam, M., Derkho, M., Burkov, P., Popovich, A., Gribkova, V.: Biogenic synthesis, characterization, and photocatalytic evaluation of pristine and graphene-loaded $Zn_{50}Mg_{50}O$ nanocomposites for organic dyes removal. Nanomaterials. **12**, 2809 (2022)

31. Fouda, A., Eid, A.M., Abdelkareem, A., Said, H.A., El-Belely, E.F., Alkhalifah, D.H.M., Alshallash, K.S., Hassan, S.E.D.: Phyco-synthesized zinc oxide nanoparticles using marine macroalgae, *Ulva fasciata* Delile, characterization, antibacterial activity, photocatalysis, and tanning wastewater treatment. Catalysts. **12**, 756 (2022)

32. Sheng, W., Zhang, Y., Song, K., Xu, J., Wu, J., Liu, B., Zhao, X.: Mesoporous boron-doped TiO_2/mussel shell for photocatalytic degradation of dye X-3B. Water Air Soil Pollut. **233**, 128 (2022)

33. Morjène, L., Schwarze, M., Seffen, M., Schomäcker, R., Tasbihi, M.: Immobilization of TiO_2 semiconductor nanoparticles onto *Posidonia Oceanica* fibers for photocatalytic phenol degradation. Water (Switzerland). **13**, 2948 (2021)

34. Abdel Azeem, M.N., Hassaballa, S., Ahmed, O.M., Elsayed, K.N.M., Shaban, M.: Photocatalytic activity of revolutionary *Galaxaura elongata*, *Turbinaria ornata*, and *Enteromorpha flexuosa*'s bio-capped silver nanoparticles for industrial wastewater treatment. Nanomaterials. **11**, 3241 (2021)

35. Ameen, F., Dawoud, T., AlNadhari, S.: Ecofriendly and low-cost synthesis of ZnO nanoparticles from *Acremonium potronii* for the photocatalytic degradation of azo dyes. Environ. Res. **202**, 111700 (2021)

36. Serrà, A., Gómez, E., Michler, J., Philippe, L.: Facile cost-effective fabrication of $Cu@Cu_2O@CuO$–microalgae photocatalyst with enhanced visible light degradation of tetracycline. Chem. Eng. J. **413**, 127477 (2021)

37. Wu, P.H., Yeh, H.Y., Chou, P.H., Hsiao, W.W., Yu, C.P.: Algal extracellular organic matter mediated photocatalytic degradation of estrogens. Ecotoxicol. Environ. Saf. **209**, 111818 (2021)

38. Garcia-Bedoya, D., Ramírez-Rodríguez, L.P., Quiroz-Castillo, J.M., Esquer-Miranda, E., Castellanos-Moreno, A.: *Caulerpa sertularioides* extract as a complexing agent in the synthesis of ZnO and $Zn(OH)_2$ nanoparticles and its effect in the azo dye's photocatalysis in water. Bioresources. **16**(1), 1548–1560 (2021)

39. Wang, W., Lin, F., Yan, B., Cheng, Z., Chen, G., Kuang, M., Yang, C., Hou, L.: The role of seashell wastes in TiO_2/seashell composites: photocatalytic degradation of methylene blue dye under sunlight. Environ. Res. **188**, 109831 (2020)
40. Aadnan, I., Zegaoui, O., Daou, I., Esteves da Silva, J.C.G.: Synthesis and physicochemical characterization of a ZnO-Chitosan hybrid-biocomposite used as an environmentally friendly photocatalyst under UV-A and visible light irradiations. J. Environ. Chem. Eng. **8**, 104260 (2020)
41. de Bittencourt, M.A., Novack, A.M., Scherer Filho, J.A., Mazur, L.P., Marinho, B.A., da Silva, A., de Souza, A.A.U., de Souza, S.M.A.G.U.: Application of $FeCl_3$ and TiO_2-coated algae as innovative biophotocatalysts for Cr(VI) removal from aqueous solution: a process intensification strategy. J. Clean. Prod. **268**, 122164 (2020)
42. El-Ella, A.A., Youssef, A.M., Ghannam, H.E., Zedan, A.F., Aboulthana, W.M., Al-Sherbini, A.S.A.: Synthesis of high efficient CS/PVDC/TiO_2-Au nanocomposites for photocatalytic degradation of carcinogenic ethidium bromide in sunlight. Egypt. J. Chem. **63**, 1619–1638 (2020)
43. Serrà, A., Pip, P., Gómez, E., Philippe, L.: Efficient magnetic hybrid ZnO-based photocatalysts for visible-light-driven removal of toxic cyanobacteria blooms and cyanotoxins. Appl Catal B. **268**, 118745 (2020)
44. Zhang, H., Xie, J., Sun, Y., Zheng, A., Hu, X.: A novel green approach for fabricating visible, light sensitive nano-broccoli-like antimony trisulfide by marine Sb(v)-reducing bacteria: revealing potential self-purification in coastal zones. Enzym. Microb. Technol. **136**, 109514 (2020)
45. Serrà, A., Artal, R., García-Amorós, J., Sepúlveda, B., Gómez, E., Nogués, J., Philippe, L.: Hybrid Ni@ZnO@ZnS-microalgae for circular economy: a smart route to the efficient integration of solar photocatalytic water decontamination and bioethanol production. Adv. Sci. **7**, 1902447 (2020)
46. Wang, J., Li, X., Cheng, Q., Lv, F., Chang, C., Zhang, L.: Construction of β-FeOOH@tunicate cellulose nanocomposite hydrogels and their highly efficient photocatalytic properties. Carbohydr. Polym. **229**, 115470 (2020)
47. Dumbrava, A., Matei, C., Diacon, A., Moscalu, F., Berger, D.: Novel ZnO-biochar nanocomposites obtained by hydrothermal method in extracts of *Ulva lactuca* collected from Black Sea. Ceram. Int. **49**, 10003–10013 (2023)
48. Balaraman, P., Balasubramanian, B., Liu, W.C., Kaliannan, D., Durai, M., Kamyab, H., Alwetaishi, M., Maluventhen, V., Ashokkumar, V., Chelliapan, S., Maruthupandian, A.: *Sargassum myriocystum*-mediated TiO_2-nanoparticles and their antimicrobial, larvicidal activities and enhanced photocatalytic degradation of various dyes. Environ. Res. **204**, 112278 (2022)
49. Pinna, M., Binda, G., Altomare, M., Marelli, M., Dossi, C., Monticelli, D., Spanu, D., Recchia, S.: Biochar nanoparticles over TiO_2 nanotube arrays: a green co-catalyst to boost the photocatalytic degradation of organic pollutants. Catalysts. **11**, 1048 (2021)
50. Wang, Z., Xia, L., Chen, J., Ji, L., Zhou, Y., Wang, Y., Cai, L., Guo, J., Song, W.: Fine characterization of natural SiO_2-doped catalyst derived from mussel shell with potential photocatalytic performance for organic dyes. Catalysts. **10**, 1–13 (2020)
51. Rabie, A.M., Abukhadra, M.R., Rady, A.M., Ahmed, S.A., Labena, A., Mohamed, H.S.H., Betiha, M.A., Shim, J.J.: Instantaneous photocatalytic degradation of malachite green dye under visible light using novel green Co–ZnO/algae composites. Res. Chem. Intermed. **46**, 1955–1973 (2020)
52. Borah, D., Saikia, P., Gogoi, D., Das, A., Rout, J., Ghosh, N.N., Pandey, P., Das Gupta, M., Bhattacharjee, C.R.: Marine alga-mediated facile green synthesis, antibacterial and enhanced catalytic activity of highly stable superparamagnetic NiO nanostructure. Inorg. Chem. Commun. **156**, 111182 (2023)
53. Maduraimuthu, V., Ranishree, J.K., Gopalakrishnan, R.M., Ayyadurai, B., Raja, R., Heese, K.: Antioxidant activities of photoinduced phycogenic silver nanoparticles and their potential applications. Antioxidants. **12**, 1298 (2023)

54. López-Miranda, J.L., Mares-Briones, F., Molina, G.A., González-Reyna, M.A., Velázquez-Hernández, I., España-Sánchez, B.L., Silva, R., Esparza, R., Estévez, M.: *Sargassum natans I* algae: an alternative for a greener approach for the synthesis of ZnO nanostructures with biological and environmental applications. Mar. Drugs. **21**, 297 (2023)

55. Amina, M., Al Musayeib, N.M., Alterary, S., El-Tohamy, M.F., Alhwaiti, S.A.: Advanced polymeric metal/metal oxide bionanocomposite using seaweed *Laurencia dendroidea* extract for antiprotozoal, anticancer, and photocatalytic applications. PeerJ. **11**, e15004 (2023)

56. Cao, J., Ju, P., Chen, Z., Dou, K., Li, J., Zhang, P., Zhu, Z., Sun, C.: Trash to treasure: green synthesis of novel Ag_2O/Ag_2CO_3 Z-scheme heterojunctions with highly efficient photocatalytic activities derived from waste mussel shells. Chem. Eng. J. **454**, 140259 (2023)

57. Amereh, F., Jokar, R., Ghasemi, A.H.B., Yazdanbakhsh, A., Rafiee, M., Alast, F.H., Naiem, S. M.: Sunlight-active hierarchical Ag@insulator@ZnO core-shell array based on natural diatoms for environmental remediation. Appl. Mater. Today. **30**, 101698 (2023)

58. Jerlin, G., Ashok, M.: Synthesis and characterization of efficient photocatalyst from seaweed extract under solar irradiation. Funct. Mater. Lett. **15**, 2251047 (2022)

59. Kamali, A.R., Zhu, W., Shi, Z., Wang, D.: Combustion synthesis-aqueous hybridization of nanostructured graphene-coated silicon and its dye removal performance. Mater. Chem. Phys. **277**, 125565 (2022)

60. Safat, S., Buazar, F., Albukhaty, S., Matroodi, S.: Enhanced sunlight photocatalytic activity and biosafety of marine-driven synthesized cerium oxide nanoparticles. Sci. Rep. **11**, 14734 (2021)

61. Manoj, M., Manaf, O., Ismayil, K.M., Sujith, A.: Composites based on poly(ethylene-co-vinyl acetate) and silver-calcined scallop shell powder: mechanical, thermal, photocatalytic, and antibacterial properties. J. Elastom. Plast. **53**, 902–921 (2021)

62. Elemike, E.E., Onwudiwe, D.C., Mbonu, J.I.: Green synthesis, structural characterization and photocatalytic activities of chitosan-ZnO nano-composite. J. Inorg. Organomet. Polym. Mater. **31**, 3356–3367 (2021)

63. Campalani, C., Cattaruzza, E., Zorzi, S., Vomiero, A., You, S., Matthews, L., Capron, M., Mondelli, C., Selva, M., Perosa, A.: Biobased carbon dots: from fish scales to photocatalysis. Nano. **11**, 1–14 (2021)

64. Balaraman, P., Balasubramanian, B., Kaliannan, D., Durai, M., Kamyab, H., Park, S., Chelliapan, S., Lee, C.T., Maluventhen, V., Maruthupandian, A.: Phyco-synthesis of silver nanoparticles mediated from marine algae *Sargassum myriocystum* and its potential biological and environmental applications. Waste Biomass Valorization. **11**, 5255–5271 (2020)

65. Sundar, V., Balasubramanian, B., Sivakumar, M., Chinnaraj, S., Palani, V., Maluventhen, V., Kamyab, H., Chelliapan, S., Arumugam, M., Patricia Zuleta Mediavilla, D.: An eco-friendly synthesis of titanium oxide nanoparticles mediated from *Syringodium isoetifolium* and evaluate its biological activity and photocatalytic dye degradation. Inorg. Chem. Commun. **161**, 112125 (2024)

66. Lavanya, G., Anandaraj, K., Gopu, M., Selvam, K., Selvankumar, T., Govarthanan, M., Kumar, P.: Green chemistry approach for silver nanoparticles synthesis from *Halimeda macroloba* and their potential medical and environmental applications. Appl. Nanosci. (Switzerland). **13**, 5865–5875 (2023)

67. Rajivgandhi, G.N., Ramachandran, G., Kannan, M.R., Velanganni, A.A.J., Siddiqi, M.Z., Alharbi, N.S., Kadaikunnan, S., Li, W.J.: Photocatalytic degradation and anti-cancer activity of biologically synthesized Ag NPs for inhibit the MCF-7 breast cancer cells. J. King Saud. Univ. Sci. **34**, 101725 (2022)

68. Jun, E.S., Kim, Y.J., Kim, H.H., Park, S.Y.: Gold nanoparticles using *Ecklonia stolonifera* protect human dermal fibroblasts from UVA-induced senescence through inhibiting MMP-1 and MMP-3. Mar. Drugs. **18**, 433 (2020)

69. Qiu, S., Wang, W., Yu, J., Tian, X., Li, X., Deng, Z., Lin, F., Zhang, Y.: Enhanced photocatalytic degradation efficiency of formaldehyde by in-situ fabricated $TiO_2/C/CaCO_3$ heterojunction photocatalyst from mussel shell extract. J. Solid State Chem. **311**, 123110 (2022)

70. Wang, W., Lin, F., An, T., Qiu, S., Yu, H., Yan, B., Chen, G., Hou, L.: An: photocatalytic mineralization of indoor VOC mixtures over unique ternary $TiO_2/C/MnO_2$ with high adsorption selectivity. Chem. Eng. J. **425**, 131678 (2021)
71. Wang, W., Yu, H., Li, K., Lin, F., Huang, C., Yan, B., Cheng, Z., Li, X., Chen, G., Hou, L.: Insoluble matrix proteins from shell waste for synthesis of visible-light response photocatalyst to mineralize indoor gaseous formaldehyde. J. Hazard. Mater. **415**, 125649 (2021)
72. Tang, X., Feng, Q., Liu, K., Luo, X., Huang, J., Li, Z.: A simple and innovative route to remarkably enhance the photocatalytic performance of TiO_2: using micro-meso porous silica nanofibers as carrier to support highly-dispersed TiO_2 nanoparticles. Microporous Mesoporous Mater. **258**, 251–261 (2018)

Importance of Optimally Combining Binder Types and Rebar Cover in Reducing Lifetime CO$_2$ Emissions in RC Structures

Sakib Hasnat ⓘ and Tanvir Manzur ⓘ

Contents

1 Introduction

Concrete, the most widely manufactured solid material in the world, contributes significantly to global warming due to the high embodied energy of Ordinary Portland Cement (OPC). The production of OPC contributes approximately 5–7% towards global annual CO$_2$ emissions [1]. Compared to OPC, Supplementary Cementitious Materials (SCMs) such as fly-ash and slag have miniscule levels of embodied energy in terms of carbon dioxide equivalent (CO$_2$-e) [2]. However, replacement of OPC with fly-ash and slag alone is not sufficient for reduction of global CO$_2$ emissions stemming from the construction industry. Reinforced Concrete (RC) structures around the world in coastal environments are susceptible to abrupt and unwanted repair and maintenance due to damage induced by corrosion of

S. Hasnat · T. Manzur (✉)
Department of Civil Engineering, Bangladesh University of Engineering and Technology,
Dhaka, Bangladesh
e-mail: tanvirmanzur@ce.buet.ac.bd

© The Author(s) 2025
M. Kioumarsi, B. Shafei (eds.), *The 1st International Conference on Net-Zero Built Environment*, Lecture Notes in Civil Engineering 237,
https://doi.org/10.1007/978-3-031-69626-8_56

the embedded reinforcement. RC structures present in chloride-laden conditions often approach serviceability and durability limit states, that is, cracking and spalling of the concrete cover in a more rapid manner. In countries such as Bangladesh, the absence of performance-based standards for RC structures exposed to harsh marine conditions often lead to premature patchworks for continued functionality [3]. Such patchworks not only escalate the life-cycle costs of the RC structure in consideration but also exacerbate the structures lifetime CO_2 emissions due to additional use of OPC.

The service lives of RC structures exposed to chloride-induced corrosion can be assumed to consist of two separate phases, these are the corrosion initiation (T_i) and the crack propagation (T_{cr}) phases [4]. The initiation phase, T_i, represents the time required for chloride ions to reach the level of the embedded reinforcement, causing depassivation of the protective layer around the reinforcement, thereby initiating corrosion. The second phase, T_{cr}, represents the time taken for the rust products of corrosion to accumulate and cause the concrete cover to crack and spall, thus reaching the serviceability limit state. After this point, routine patchworks are required to replace the deteriorated original concrete in order for the continued functionality of the structure. Hence, the lifetime CO_2 emissions for an RC structure exposed to chloride-induced corrosion shall correspond to the initial CO_2 emission caused by the construction of the structure and the incremental CO_2 emissions caused by continuous patchwork operations [5]. Suitable corrosion management strategies should be employed in order to delay the time to repair in RC structures and lessen the number of patchworks required within the prescribed service life. One of these strategies is to set back the corrosion initiation time by use of Supplementary Cementitious Materials (fly-ash, slag), subsequently reducing the number of repair cycles [6]. Additionally, concrete cover can be utilized as an energy-efficient strategy in reducing lifetime CO_2 emissions. Previous studies [7] have demonstrated that concrete cover is a simple and cost-effective tool for reducing the number of patchworks required in a structures service life, provided the cover thickness is within a feasible limit.

In this study, the effectiveness of OPC replacement by SCMs such as fly-ash and slag combined with different concrete covers in reducing the lifetime CO_2 emissions of a typical coastal structure has been assessed within a prescribed service life of 50 years. A time-dependent probabilistic analysis using the Monte Carlo Simulation approach has been conducted for both the initiation and the propagation phases, considering the relevant limit states and corresponding failure criteria. Patchwork is applied when the failure probability exceeds a certain threshold value for the serviceability limit state and is continued routinely. The CO_2 emissions over the service life of the structure is determined for the different binder and cover alternatives by considering the emission factor (CO_2-e) for OPC only. The outcomes of the limited study performed can be used as a stepping stone in establishing performance-based local guidelines for extending service lives and reducing the carbon footprint of RC structures, advancing towards a NetZero future.

2   Experimental Program

2.1   Materials Used

The materials used for producing the concrete mixes in this study were Ordinary Portland Cement (OPC) or CEM-I and different replacements of fly-ash and slag as SCMs. Class F fly-ash and blast furnace slag obtained from local sources were used. The specific gravity and unit weight of CEM-I used were 3.15 and 1440 kg/m^3, respectively. The aggregates used in this study were locally available 19 mm downgraded crushed stone chips as coarse aggregates and local "Sylhet Sand" as fine aggregates with fineness modulus of 2.82.

2.2   Concrete Mix Design

The proportions for concrete mix designs were determined using the ACI Mix Design Manual [8]. The target slump was chosen to be 75–100 mm and the w/b ratio used was 0.40, corresponding to a specified design strength of 35 MPa. Three different mixes were produced by replacing CEM-I with fly-ash and slag. One mix with 35% fly-ash replacement, one with 40% slag replacement, and one control mix with CEM-I only and no SCM replacement. Cylindrical samples of 100 × 200 mm dimensions were cast for different tests. A summary of the mix design details is provided in Table 1, where FA and S represent fly-ash and slag, respectively.

3   Conducted Tests

3.1   Compressive Strength Test

Since the compressive strengths of the different concrete mixes were used as a target parameter, it was evaluated as per the specifications of ASTM C39 [9]. After curing

Table 1 Mix design details for the mixes cast in this study (per m^3 of concrete)

Mix ID	Proportion of SCMs	CEM-I (kg)	Fly-ash (kg)	Slag (kg)	Course Aggregate (kg)	Fine Aggregate (kg)
W/B ratio: 0.40						
C1	CEM-I (100%)	513	–	–	1001	630
35F	CEM-I (65%) + FA (35%)	333	180	–	1001	615
40S	CEM-I (60%) + S (40%)	308	–	205	1001	622

the standard cylindrical specimens in lime water, the strengths were measured at 28 and 180 days. The C1 and 40S mixes achieved the design strength of 35 MPa in 28 days, while the 35F mixed achieved it in the later ages.

3.2 Non-steady State Rapid Migration Test

The chloride diffusion coefficient (D_{rcm}) indicates the resistance of the different concrete mixes to degradation by ingress of chloride ions. Additionally, the D_{rcm} serves as an input parameter in modelling the corrosion initiation phase, representing unique concrete mixes. The Rapid Migration Test (RMT) was conducted as per the specifications of Nordtest Method NT BUILD 492 [10]. After submergence of the concrete cylinders in the curing pond for 180 days, they were taken out and sawn into disk specimens of dimensions 100×50 mm. The disks were sawn from the middle portion of the concrete cylinder. After vacuum saturation in a saturated Ca $(OH)_2$ solution, the cylinders were kept submerged in the solution for 24 hours. An electrical potential was then applied to the disk specimens for a certain duration, forcing chloride ions to migrate into the concrete as described in the standard [10]. After the test duration, the specimens were split into two, and a silver nitrate solution was sprayed on the inside of the specimen, resulting in the formation of a white precipitate representing the depth of chloride ion penetration. The chloride diffusion coefficient (D_{rcm}) was obtained by determining the average depth of chloride ion penetration, applied voltage, temperature of the anolyte and the length of the specimen as per the standard [10].

3.3 Electrical Resistivity Test

The surface resistivity values of the concrete mixes were assessed using the Wenner Four-Probe [11] method as per the specifications of AASHTO T 358-15 [12]. The commercially available Proseq Resipod was used for measuring the surface resistivity. This involves the application of an electrical current through two outer probes, while the resulting voltage difference is measured through the inner electrodes. The cylindrical specimens were kept in submergence up until the testing age at 180 days. Multiple correction factors have been applied to the surface resistivity readings, which are factors for specimen geometry, temperature, and degree of saturation. The application of these factors has been discussed in previous studies [13] and not described in details for this chapter.

4 Service Life Prediction

The service life of an RC structure subjected to chloride-induced corrosion can be divided into two phases; these are the initiation (T_i) and the crack propagation (T_{cr}) phases as described previously. The serviceability limit state is assumed to be reached when the concrete cover cracks, after which continued service of the RC structure requires routine maintenance works.

The initiation phase can be defined as the time required for chloride ions to reach the level of the reinforcement, causing the depassivation of the alkaline protective layer. The diffusion of chloride ions through concrete can be approximated using Fick's Second law of diffusion. Corrosion is assumed to initiate when the chloride concentration at the reinforcement level exceeds a critical concentration, known as the threshold chloride content (C_{th}). According to fib Bulletin 34 [14] model code for service life design, the chloride concentration ($C_{x,t}$) at a certain depth 'x_c' at the corrosion initiation time 'T_i' can be expressed by:

$$C(x,t) = C_o + (C_{s,\Delta x} - C_o) \cdot \left[1 - \text{erf} \frac{x_c - \Delta x}{2 \cdot \sqrt{K_e \cdot K_t \cdot D_{rcm(t_o)} \cdot \left(\frac{t_o}{T_i}\right)^m \cdot T_i}} \right] \quad (1)$$

$$K_e = \exp\left[b_e \left(\frac{1}{T_{ref}} - \frac{1}{T_{real}} \right) \right] \quad (2)$$

The parameters associated with Eqs. (1) and (2) for modelling the corrosion initiation period in tidal, splash and spray zones (exposure class XS3) [14] are presented in Table 2. The probabilistic failure curves for the initiation period are based on the parameters described in Table 2 and the experimental D_{rcm} values presented in Table 4. The variable covers considered for the probabilistic modelling are 37.5, 50, and 62.5 mm, which is denoted by 'x_c' in Table 2.

The model used for predicting the time for first corrosion induced crack to occur is based on the assumption of a perfectly perfect plastic behaviour of concrete. The accumulation of rust products on the porous zone surrounding the reinforcement puts pressure on the concrete cover, eventually causing the cover to crack. Thus, the propagation period (T_{cr}) can be expressed as [18]:

$$T_{cr} = \frac{\rho_s z F \left(\frac{2 x_c f_{ct} k}{D} + \delta_o \right)}{M(\alpha_v - 1) i_{corr}} \quad (3)$$

In Eq. (3), the corrosion rate 'i_{corr}' has been predicted using the model suggested by Pour-Ghaz et al. [19], which establishes independent correlations among the corrosion rate of steel, ambient temperature, kinetic parameters, concrete resistivity and limiting current density. The experimental resistivity measurements were

Table 2 Parameters used for modelling the corrosion initiation period

Description	Notation of Parameter	Distribution (Mean (μ), Standard deviation (σ))
Threshold chloride concentration (wt-%/binder) [15]	C_{th}	Beta (0.60, 0.15); l.b: 0.20, u.b: 2.0
Initial chloride concentration (wt-%/binder) [15]	C_o	0
Surface chloride concentration (wt-%/binder) [15]	$C_{s,\Delta x}$	LN (3, 1.35)
Depth of concrete cover (mm) [16]	x_c	LN (x_c, 8)
Depth of convection zone (mm) [15]	Δx	Beta (10, 5); l.b: 0, u.b: 50
A transfer parameter [15]	K_t	1
A reference point of time (years)	t_o	0.4932 (180 days)
Non-steady state migration coefficient of chloride concrete (m²/s) [10]	D_{rcm}	As determined from RMT test following NT BUILD 492 [10]
The ageing exponent [15]	m	OPC: Beta (0.30, 0.12) Fly-ash: Beta (0.60, 0.15) Slag: Beta (0.45, 0.20); l.b: 0.0, u.b: 1.0.
A regression variable (K) [15]	b_e	N (4800, 700)
A reference temperature (K) [15]	T_{ref}	293
Temperature of the structural element or the ambience (K), taken for the coastal region of Bangladesh [17]	T_{real}	N (299, 5)

l.b Lower bound of Beta distribution, *u.b* Upper bound of Beta distribution, *LN* Lognormal distribution, *N* Normal distribution

incorporated into this model for prediction of the corrosion rate 'i_{corr}'. In Eq. (3), 'k' is the hole flexibility and can be expressed as [18]:

$$k = \frac{\left(\frac{1+\nu+(D+2\delta_o)^2}{2x_c(x_c+D+2\delta_o)}\right)(D+2\delta_o)(1+\varphi_c)}{2E_c} \qquad (4)$$

The descriptions of the different parameters used for the prediction of the propagation period using Eqs. (3) and (4) are provided in Table 3.

Cracks form in the concrete cover at the end of the propagation period. In order to resume functional service of the RC structure, routine patchwork is applied by replacing the cracked concrete cover with new concrete. Thus, the total time to cracking can be expressed as:

$$T_L = T_i + T_{cr} \qquad (5)$$

Due to the nature of the models used and the stochastic variables involved, the time to corrosion initiation and cracking for an RC structure is modelled using a probabilistic approach. In this study, the Monte Carlo Simulation was conducted for 100,000 iterations, and the time-dependent failure probability for both the initiation

Table 3 Parameters used for modelling the crack propagation period

Description	Notation of Parameter	Distribution (Mean (μ), Standard deviation (σ))
Elastic modulus of concrete (MPa) [20]	E_c	$N(4733\sqrt{(f_c)}, 0.12\ \mu)$
Tensile strength of concrete (MPa) [20]	f_{ct}	$N\ (0.6227\sqrt{(f_c)}, 0.20\ \mu)$
Poisson's ratio of concrete [18]	ν	0.20
Creep coefficient of concrete [18]	φ_c	2.35
Rebar diameter (mm)	D	10
Mass density of steel (gm/m^3)	ρ_s	7,850,000
Relative volume ratio [21]	α_v	Beta (3.01, 0.8097); l.b: 1.695, u,b: 6.3)
Porous zone thickness (m) [21]	δ_o	U (l.b: 5E-06, u.b: 12E-05, 0.53 μ)

f_c 28-day compressive strength of concrete determined from standard tests, U Uniform distribution

and the propagation phases was modelled by evaluating the failure criteria. The T_i and T_{cr} have been determined by adopting a target failure probability or reliability criteria. In this study, the probability of failure is taken to be 10% and 6.7%, corresponding to reliability index of 1.28 and 1.5, as suggested by fib 34 [14] and DuraCrete [22] for the criteria of corrosion initiation and cracking of concrete cover, respectively. After the application of routine patchwork, it is assumed that crack appears once again after the duration of the crack propagation period. Thus, the number of repair cycles required could be assumed by considering a design service life of 50 years and determining the T_i and T_{cr} from the probabilistic analysis.

5 Estimation of Embodied CO$_2$

For determination of the embodied CO$_2$ and the impact of the different types of binders considered on emissions, the following equation has been used:

$$E = \sum M \times EF \qquad (6)$$

where 'E' is the total embodied greenhouse gas emissions (in tons CO$_2$-e) for all the building materials, considering the lifecycle of the concrete from cradle to grave, 'M' is the amount of building materials (in tons) and 'EF' is the CO$_2$ emission factor for concrete. For this study, the emission factor for CEM-I was considered as 0.913 t CO$_2$/t [23]. In comparison to CEM-I, fly-ash and slag have comparatively low values of the emission factor at 0.027 t CO$_2$/t [3] and 0.022 t CO$_2$/t [23], respectively. Thus, for the blended mixes 35F and 40S, only the emission factors for CEM-I were considered.

For the calculation of the total amount of materials used, a typical bridge section constructed in the coastal region of Bangladesh has been considered (Fig. 1). In the

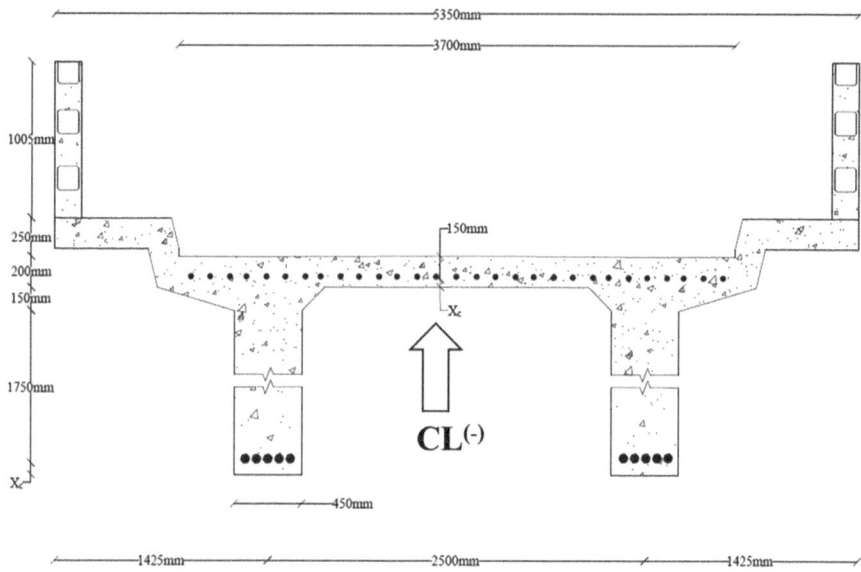

Fig. 1 Typical section of a bridge deck considered as a case study

figure, the parameter 'X_c' represents the variable covers considered in this study for analysis. After calculating the number of patchworks required in the 50-year design service life of the RC structure, the total number of materials used was calculated per metre of the bridge span. The concrete required for construction was determined from the bridge dimensions, and the concrete required for patchworks was calculated by assuming that 10% of the area exposed to chlorides require concrete replacement at each interval. It is assumed that the concrete required for maintenance works is of the same quality as that of the original structure.

6 Results and Discussion

6.1 Probabilistic Service Life Analysis

The principles of the probabilistic analysis conducted have been described previously in this article. The time-dependent failure probabilities have been computed considering the exposure class XS3 (tidal, splash and spray zones) in the coastal region of Cox's Bazar, Bangladesh. The concrete covers considered are 37.5, 50, and 62.5 mm, respectively. The summary of the probabilistic calculus for all the alternatives considered is presented in Table 4, along with the experimental Chloride Diffusion Coefficient (D_{rcm}) and Electrical Resistivity values. Figures 2 and 3 display a sample of the computed probabilistic curves for cover thickness of

Table 4 Summary of the probabilistic analysis for all the alternatives considered

Mix ID	Chloride Diffusion Coefficient ($\times 10^{-12}$ m²/s)	Electrical Resistivity (kΩ·cm)	37.5 mm cover		50 mm cover		62.5 mm cover	
			T_i (years)	T_L (years)[a]	T_i (years)	T_L (years)[a]	T_i (years)	T_L (years)[a]
C1	4.73	11.12	0.3	1.6	1.5	3.8	2.9	6.2
35F	1.79	29.19	2.4	8.2	16.7	24.1	52	62.5
40S	1.97	22.60	1.1	6.4	5.5	12.3	14.3	23.4

[a]$T_L = T_i + T_{cr}$

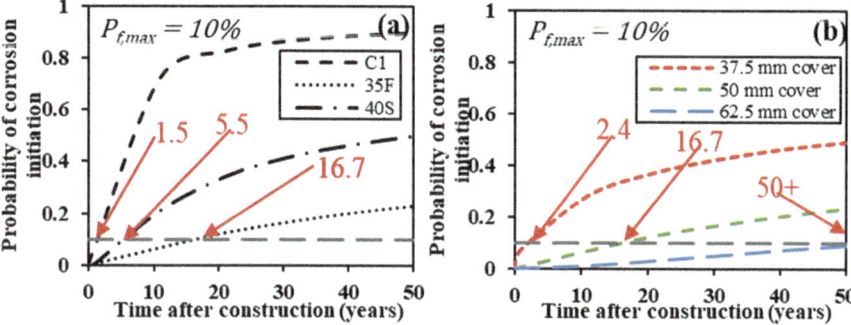

Fig. 2 Probabilities of corrosion initiation with time for (**a**) Different binder types considering 50 mm cover and (**b**) 35F mix considering different cover thicknesses

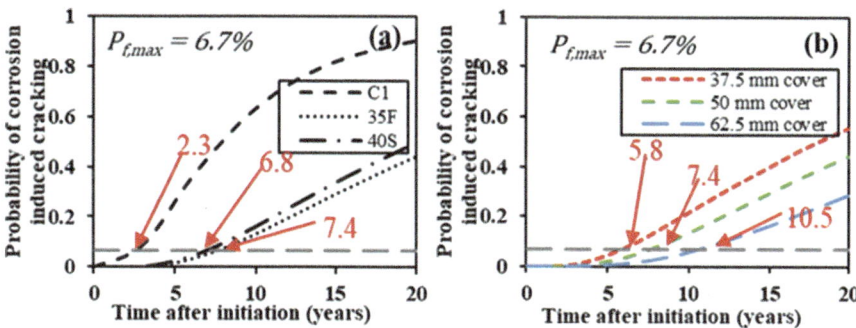

Fig. 3 Probabilities of crack initiation with time for (**a**) Different binder types considering 50 mm cover and (**b**) 35F mix considering different cover thicknesses

50 mm (Figs. 2a and 3a) and the 35F (Figs. 2b and 3b) mix for the initiation and the propagation phases, respectively.

From Figs. 2, 3, and Table 4, it is clear that blended mixes perform substantially better than the C1 mix in terms of both T_i and T_{cr}. As it can be seen from Figs. 2a and 3a, for a cover of 50 mm in case of C1 mix, corrosion initiates only 1.5 years after construction and the first crack appears only 2.3 years after initiation. Whereas, for the 35F mix, the initiation time is greatly extended to 16.7 years (Fig. 2b) and the time to first-cracking is also increased to 7.4 years (Fig. 3b). It is also evident that the slag mix 40S perform substantially better than the C1 mix as well, in terms of both T_i and T_{cr}. The pozzolanic activity of fly-ash causes long-term pore refinement due to the formation of secondary Calcium Silicate Hydrate (C-S-H) gel, resulting in a denser microstructure and thus delaying the onset of corrosion [24]. For the 40S mix, the C-S-H formation occurs earlier, resulting in an inferior late-age pore refinement compared to fly-ash [25]. The differences in the microstructure tortuosity impact the corrosion rate in steel as well since the electrical resistivity is affected. Thus, the crack propagation period is longer for 35F and 40S mixes than that for C1 mix.

Along with the incorporation of SCMs, increasing cover thickness can significantly enhance the T_i and T_{cr} for the mixes considered. From Fig. 2b, it is evident that increasing the cover thickness from 37.5 to 62.5 mm extends T_i from only 2.4 years to 52 years for the 35F mix. This can be observed for all other mixes as well (Table 4). However, for C1 mix, even 62.5 mm cover results in corrosion initiating within 3 years only. The crack propagation time is considerably extended by increasing cover as well. Larger cover will require more force to initiate cracking and hence, requiring accumulation of more corrosion products, thus lengthening the time to first cracking. From Fig. 3b, the crack propagation time increases from almost 6 years to 10.5 years for 35F mix by increasing the cover thickness from 37.5 to 62.5 mm.

6.2   Repair Cycles and Embodied CO_2

The number of patchworks required in the 50-year design service life of the structure is calculated. It is assumed that after the appearance of the first crack in the concrete cover, patchwork will be applied where the cracked cover will be replaced by concrete of original quality. The next patchwork will be required after the duration of the crack propagation period of that particular binder type. The total embodied CO_2 in the 50-year service life of the structure is calculated using Eq. (6) for all the different alternatives considered.

Figure 4 presents the results of this analysis. From the figure, it can be seen that the total CO_2-e emissions for each particular binder type directly corresponds to the number of repair-work required in the structure's service life. The number of patchworks required decreases with increasing cover for all binder types. However, for the C1 mix, even at 62.5 mm cover, the required number of repair intervals is 11, whereas at 62.5 mm cover, the 35F mix requires no patchworks at all in the 50-year period. The 35F mix performs most satisfactorily, requiring the least number of repairs for each cover considered.

The embodied CO_2-e emissions vary mostly with respect to cement replacement. Since, C1 has a high embodied energy, the binder with no replacement results in the highest CO_2 emission of about 2800 kg CO_2-e, whereas the mixes with 35% and 40% binder replacement result in almost 1400 kg CO_2-e emissions for 37.5 mm cover. Additionally, increasing concrete cover results in fewer patchworks and

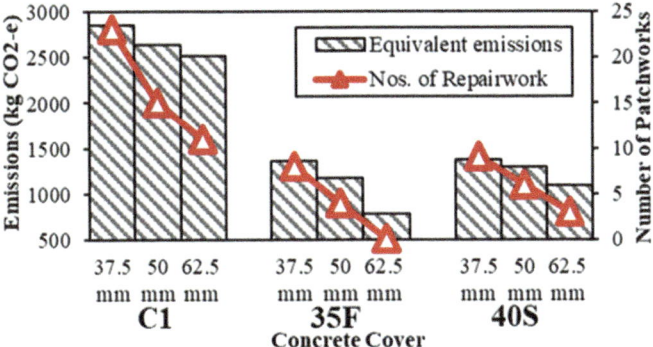

Fig. 4 Total CO_2 emissions per metre length of the bridge and number of repair-works required considering 50-year design service life

consequentially, lower CO_2-e emissions. This is evident from Fig. 4 as well. For all the mixes, increasing concrete cover results in considerable decrease in CO_2-e emissions. The reduction is more prominent for the blended mixes. For C1, increasing concrete cover from 37.5 to 62.5 mm results in a nearly 14% decrease in CO_2-e emissions. The same decrease is around 43% for the 35F mix. Considering a combination of binder replacement and increased cover, the embodied CO_2-e emissions are seen to decrease from 2800 kg CO_2-e (C1, 37.5 mm) to only 790 kg CO_2-e (35F, 62.5 mm). This clearly signifies the importance of optimally combining SCMs as cement replacement with cover design for RC structures under chloride exposure in achieving the NetZero target.

7   Conclusions

In this limited study performed, a probabilistic performance-based approach is utilized for assessing the service life of ordinary and blended concrete mixes and their CO_2-e emission potential in harsh, marine exposures. Based on the time-variant failure probabilities, the total time to cracking for combinations of different binder and cover alternatives could be estimated. The analysis showed that the SCM blended mixes significantly lengthened the time to first cracking for RC structures, and increment in cover thickness reduced the number of repair intervals within a 50-year design service life. The embodied CO_2-e emissions determined from this analysis highlight the significance of incorporating SCMs into blended mixes, combined with sufficient cover thickness for reducing the carbon footprint of the rapidly growing construction industry, particularly in extreme exposures. The framework hereby presented could be used to evaluate the carbon footprint and environmental impact of new, emerging binder types as well.

References

1. Huntzinger, D.N., Eatmo, T.D.: A life-cycle assessment of cement manufacturing:comparing traditional process with alternative technologies. J. Clean. Prod. **17**(7), 668–675 (2009)
2. Turner, L.K., Collins, F.G.: Carbon dioxide equivalent (CO2-e) emissions: a comparison between geopolymer and OPC cement concrete. Constr. Build. Mater. **43**, 125–130 (2013). https://doi.org/10.1016/j.conbuildmat.2013.01.023
3. Manzur, T., Baten, B., Hasan, M.J., Akter, H., Tahsin, A., Hossain, K.M.A.: Corrosion behavior of concrete mixes with masonry chips as coarse aggregate. Constr. Build. Mater. **185**, 20–29 (2018). https://doi.org/10.1016/j.conbuildmat.2018.07.033
4. Tuutti, K.: Service life of structures with regard to corrosion of embedded steel. ACI Symp. Publ. **65** (1980). https://doi.org/10.14359/6355
5. Petcherdchoo, A.: Environmental impacts of combined repairs on marine concrete structures. J Adv. Concr. Technol. **13**, 205–213 (2015). https://doi.org/10.3151/jact.13.205
6. Liu, G., Hua, J., Wang, N., Deng, W., Xue, X.: Material alternatives for concrete structures on remote islands: based on life-cycle-cost analysis. Adv. Civ. Eng. **2022** (2022). https://doi.org/10.1155/2022/7329408
7. Menna Barreto, M.F.F., Timm, J.F.G., Passuello, A., Dal Molin, D.C.C., Masuero, J.R.: Life cycle costs and impacts of massive slabs with varying concrete cover. Clean. Eng. Technol. **5**, 100256 (2021). https://doi.org/10.1016/j.clet.2021.100256
8. ACI 211: Standard Practice for Selecting Proportions for Normal Heavyweight and Mass Concrete. American Concrete Institute, Michigan (1991)
9. ASTM C39 – 14a: Standard Test Method for Compressive Strength of Cylindrical Concrete Specimens. American Society for Testing and Materials (2014)
10. NT Build 492: Concrete, Mortar and Cement-Based Repair Materials: Chloride Migration Coefficient from Non-Steady-State Migration Experiments. NORDTEST, Finland (1999)
11. Wenner, F.: A method of measuring earth resistivity. Bull. Natl. Bur. Std. **12**, 469–478 (1916)
12. AASHTO T 358-15: Standard Method of Test for Surface Resistivity Indication of Concrete's Ability to Resist Chloride Ion Penetration. American Association of State Higheway and Transportation Official, Washington, DC (2015)
13. Baten, B., Manzur, T.: Formation factor concept for non-destructive evaluation of concrete's chloride diffusion coefficients. Cem. Concr. Compos. **128**(December 2021), 104440 (2022). https://doi.org/10.1016/j.cemconcomp.2022.104440
14. Fib Bulletin 34: Model code for service life design. Fib. Lausanne (2006)
15. Fib Bulletin 76: Benchmarking of deemed to satisfy provisions in standards: durability of reinforced concrete structures exposed to chlorides. Fib. Lausanne (2015)
16. Rengaraju, S., Pillai, R.G., Gettu, R.: Input parameters and nomograms for service life-based design of reinforced concrete structures exposed to chlorides. Structures. **56**(July), 104847 (2023)
17. NASA Power Data Access Viewer: https://power.larc.nasa.gov/data-access-viewer/. Last accessed on 15 Dec 2023
18. El Maaddawy, T., Soudki, K.: A model for prediction of time from corrosion initiation to corrosion cracking. Cem. Concr. Compos. **29**(3), 168–175 (2007). https://doi.org/10.1016/j.cemconcomp.2006.11.004
19. Pour-Ghaz, M., Isgor, O.B., Ghods, P.: The effect of temperature on the corrosion of steel in concrete. Part 1: simulated polarization resistance tests and model development. Corros. Sci. **51**(2), 415–425 (2009). https://doi.org/10.1016/j.corsci.2008.10.034
20. ACI: Building Code Requirements for Structural Concrete (ACI 318-05) and Commentary (ACI 318R-05). American Concrete Institute, Michigan (2005)
21. Papakonstantinou, K.G., Shinozuka, M.: Probabilistic model for steel corrosion in reinforced concrete structures of large dimensions considering crack effects. Eng. Struct. **57**, 306–326 (2013). https://doi.org/10.1016/j.engstruct.2013.06.038

22. DuraCreteR17: Final technical report, DuraCrete – probabilistic performance-based durability design of concrete structures. The European Union–Brite EuRam III, 2000 (Document BE95-1347/R17) (2000)
23. Şanal, I.: Discussion on the effectiveness of cement replacement for carbon dioxide (CO2) emission reduction in concrete. Orig. Res. Artic. **8**, 1–13 (2017)
24. Mahmud, S., Manzur, T., Samrose, S., Torsha, T.: Significance of properly proportioned fly ash based blended cement for sustainable concrete structures of tannery industry. Structures. **29**, 1898–1910 (2021)
25. Hooton, R.D.: Canadian use of ground granulated blast-furnace slag as a supplementary cementing material for enhanced performance of concrete. Can. J. Civ. Eng. **27**(4), 754–760 (2000). https://doi.org/10.1139/l00-014

Corrosion Risk Assessment of Bridges in Oslo, Norway, Based on Visual Inspection

Amirhosein Shabani ⓘ, Wahid Amin, and Sven Kirschhausen

Contents

1 Introduction

The economic and social development of many countries hinges significantly on a well-established transportation network. Within this network, bridges emerge as pivotal and complex structures, fostering fast and effective transportation and enhancing connectivity between different regions and nations [1–3]. Nevertheless, a noteworthy percentage of these bridges face diverse forms of deterioration, encompassing fatigue, erosion, acid assaults on concrete components, carbonation, and the corrosive impact of chloride on steel elements [4–6].

The corrosion process affecting structural steel is an electrochemical phenomenon that requires the concurrent presence of both moisture and oxygen [7]. In the context of steel reinforcement, the alkalinity inherent in concrete prompts the embedded reinforcing bar to display characteristics similar to stainless steel. This

A. Shabani (✉) · W. Amin · S. Kirschhausen
Section of Road Operation and Maintenance, Department of Road and Street, The Mobility Division, Oslo Urban Agency (Bymiljøetaten), Oslo Municipality, Oslo, Norway
e-mail: amirhosein.shabani@bym.oslo.kommune.no

© The Author(s) 2025
M. Kioumarsi, B. Shafei (eds.), *The 1st International Conference on Net-Zero Built Environment*, Lecture Notes in Civil Engineering 237,
https://doi.org/10.1007/978-3-031-69626-8_57

corrosion-resistant state endures until chloride ions, sourced from substances like seawater or roadway deicers, accumulate on the surface of the reinforcing bar in significant concentration. Upon reaching this critical level, the steel initiates the corrosion process, giving rise to the expansion of rust that has the potential to compromise the surrounding concrete [8, 9]. Corrosion of steel elements can reduce structural integrity, decrease load-bearing capacity, increase maintenance costs, and reduce the service life of bridges [10]. Evaluating the condition of bridges is crucial because it allows for the timely detection of corrosion-related issues, ensuring proactive measures to address the detrimental effects and safeguard the structural integrity and safety of the infrastructure [11, 12].

The most common way to check a bridge's condition is through visual inspection, where qualified inspectors assess the structure based on what they see [13]. This approach is generally cost-effective, as it typically does not require tests or cause disruptions to traffic. However, its reliance on expert opinions introduces potential limitations due to subjectivity [14]. In contemporary practices, integrating various technologies like structural health monitoring or unmanned aerial vehicles into the inspection process proves beneficial in identifying hidden critical elements and defects [15–17]. Despite their effectiveness, these technologies are not universally available and cannot entirely replace expert assessments. Visual inspection findings remain valuable in supporting the planning of maintenance and rehabilitation efforts [18].

Various computer-aided bridge management systems have been developed to support timely decision-making for scheduling maintenance, repair, and rehabilitation activities [19, 20]. Brutus is an online bridge management system developed by the Norwegian Public Roads Administration (NPRA) for the maintenance of bridges in Norway [21]. Simple inspections and main inspections are planned within the system, and all data from visual inspections are recorded in individual panels provided for each bridge, including information such as construction year, structural system, and material type.

This study aims to assess the corrosion risk of bridges in Oslo, Norway, based on visual inspection data provided in the Brutus system during 2023. An overview of the case studies, including their material type, construction year, length, and structural system type, is provided. The inspection methodology and risk assessment approach provided by the NPRA are presented. Finally, the corrosion risk of the bridges is assessed, and the bridges are categorized into three different damage states.

2   Overview of the Case Studies

This study focuses on the condition assessment of 337 bridges for which the Oslo municipality is responsible for preservation and maintenance. The investigation results reveal that over half of these bridges were constructed during the period from 1950 to 1999, primarily driven by population growth, economic expansion, and

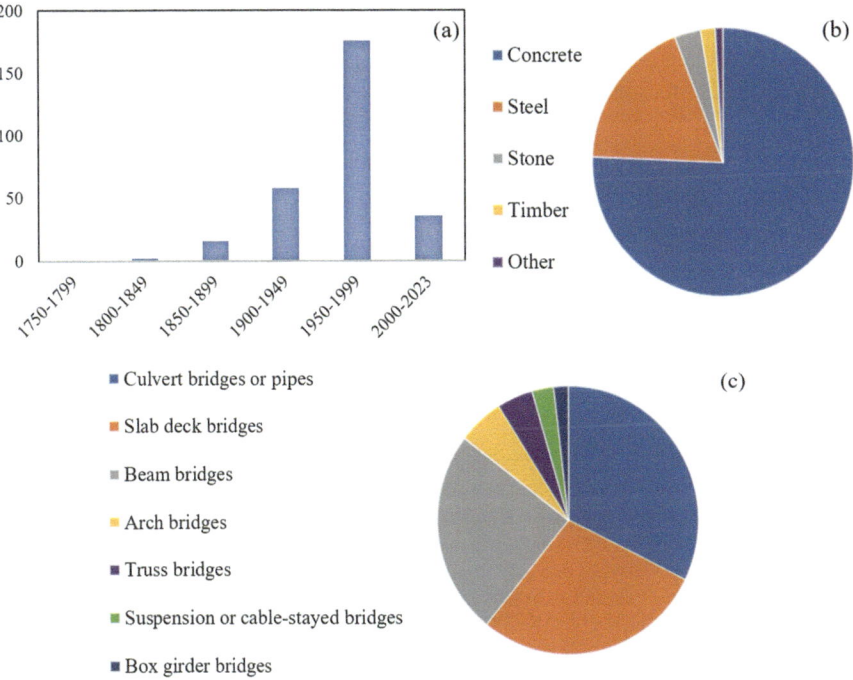

Fig. 1 Categorization of the bridges by (**a**) year of construction, (**b**) construction material, and (**c**) type of structural system

advancements in transportation infrastructure as illustrated in Fig. 1a. Analysis of bridge materials indicates that concrete is the predominant construction material, accounting for more than 75% of the bridges, followed by steel as the second most common material as shown in Fig. 1b. The bridge bearing systems are categorized into seven types: culvert bridges or pipes, slab deck bridges, beam bridges, box girder bridges, arch bridges, truss bridges, and suspension or cable-stayed bridges [21]. The evaluations indicate that culvert bridges or pipes, plate bridges, and beam bridges are the most frequently used structural systems among the bridges under investigation as shown in Fig. 1c.

3   Risk Assessment Methodology

The condition assessment was done based on the Norwegian standard for inspection of bridges [22]. Simple and main visual inspections should be conducted annually and every 5 years, respectively based on [22]. Simple inspection involves visually examining elements above water without using access equipment, often requiring observation from a distance on larger bridges. During the main inspection, the entire bridge is visually inspected to ensure all elements are functioning correctly

Table 1 Degree of R- and NR-corrosion damages and qualitative description. (Adapted from [22])

| DoD | Qualitative description of the damage level | |
	R-corrosion	NR-corrosion
1	Steel surface is generally without rust	Steel surface is generally without rust
2	Steel surface has started to rust and loosen	Steel surface has started to rust and loosen.
3	Steel surface where the scale layer has rusted away or can be scraped off. Some visible rust pits	Steel surface is rusted away or can be scraped off. Visible rust pits are not formed
4	Steel surface where the scale layer has rusted away. Visible rust pits have largely formed	Steel surface is rusted away. Visible rust pits have largely formed

Table 2 FDIs, the corresponding risk levels, and qualitative descriptions. (Adapted from [22])

FDI	Risk level	Qualitative description
1 or 2 or 4	Low	The corrosion risk is negligible however for a longer than expected service life, and serious consideration should be given
6 or 9	Moderate	The corrosion risk is not negligible. Special inspections or tests should be planned and done to avoid propagation of the corrosion and avoid increasing the maintenance costs. Maintenance usually occurs in this phase.
12 or 16	High	The bridge condition has degraded to poor, and the restoration costs may become prohibitive. In some special situation, the structural components should be replaced, if possible, after specialists' evaluation.

[22]. Representative areas of large, uniform surfaces such as steel and concrete are chosen for closer examination to detect expected damage. Four degrees of damage (DoD) were assumed and the corresponding qualitative description for each of them are presented in Table 1. Negligible differences can be found between the description of the R- and NR-corrosion types for all damage degrees. Nevertheless, a more specific description of R-corrosion damage should be presented to distinguish these two damage types. For instance, the loss of bond between reinforcement and concrete is a potential indicator that can be added to the description.

The consequences of the damages were categorized into four main groups which are damages that can affect the load-bearing system, traffic safety, maintenance cost, and environment or aesthetics [22]. Furthermore, the grade of the consequences is rated from 1, when no analysis or measurement is required to 4, when the engineers or the owners should be contacted. In this chapter, the final damage index (FDI) was calculated as the DoD multiplied by the degree of damage consequence. Note that the difference between the DoD and the degree of damage consequence would be less than 2. The possible FDI and the corresponding qualitative description for each risk level are shown in Table 2.

4 Results and Discussion

The findings of the visual inspection of bridges in 2023 were graded and FDI was calculated for different components of each bridge which is subjected to corrosion damage. In total, 193 bridges suffer corrosion damages in such a way that 154 and 87 bridges are subjected to the R-corrosion and NR-corrosion damages. Most corrosion damages lead to increased maintenance costs and the second most repeated consequences would be the environment or aesthetics. Figure 2 shows the risk level of bridge elements that were categorized as low, moderate, and high levels, subjected to corrosion damages.

The majority of bridge elements exhibit low R-corrosion based on Fig. 2a, with only three elements identified as having a high level of corrosion risk, necessitating imminent rehabilitation or replacement. Among these, concrete slabs, which serve as the primary load-bearing system in slab deck bridges, are particularly susceptible to R-corrosion risk. The findings highlight that 39 bridge elements pose a moderate corrosion risk, indicating the need for maintenance to prevent corrosion propagation to larger areas or other elements. Figure 3a and b illustrate a bridge slab with high R-corrosion risk level and Fig. 3c shows a column with moderate R-corrosion risk level. To prevent R-corrosion, it's important to use high-quality concrete mixes with corrosion-inhibiting admixtures, ensure proper concrete placement techniques and effective curing, apply surface coatings or membranes for additional protection, consider implementing cathodic protection systems for existing structures, and conduct regular inspections and proactive maintenance to address corrosion issues promptly [23–26].

Most bridge elements exhibit a low NR-corrosion risk based on Fig. 2b, while five elements have been identified as being at a high level of corrosion risk, necessitating rehabilitation or replacement of components soon. Notably, the railing stands out as the bridge element with the highest NR-corrosion rate among other bridge components. Consequently, stricter regulations regarding the material

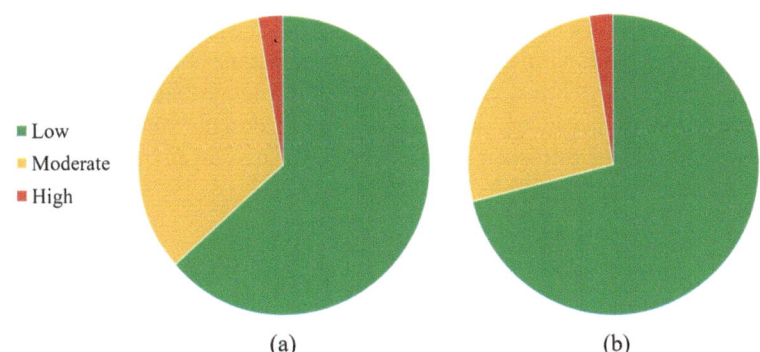

(a)   (b)

Fig. 2 Risk level of investigated bridge elements subjected to the (**a**) R-corrosion and (**b**) NR-corrosion damages

Fig. 3 (**a**) A bridge plate with a high risk of R-corrosion damage, (**b**) a close-up view of corroded reinforcement in the element, highlighting the extent of corrosion, and (**c**) a bridge column with medium risk of R-corrosion damage

selection for this component should be considered to reduce maintenance costs and mitigate the risk of collapses. Additionally, the results indicate that 52 bridge elements are at a moderate corrosion risk and require maintenance. Figure 4a illustrates a beam with a high NR-corrosion risk level and Fig. 4b shows a bridge railing with moderate NR-corrosion risk level. To combat the NR-corrosion in bridges, engineers can employ corrosion-resistant materials, routine maintenance, protective coatings, proper drainage, and moisture-minimizing design features [27–29]. Innovative solutions like electrochemical protection systems and environmental controls further bolster preventive efforts [30, 31].

5   Conclusion

Bridges suffer from various forms of deterioration, including corrosion of steel elements. The assessment of corrosion risk in bridges based on visual inspection data offers valuable insights into maintenance needs and underscores the importance

Fig. 4 (**a**) A steel beam with a high level of NR-corrosion damage and (**b**) railing of a bridge with a medium level of NR-corrosion damage

of proactive measures to preserve infrastructure integrity. The risk of reinforcement corrosion (R-corrosion) and non-reinforcement corrosion (NR-corrosion) in various elements of 337 bridges, for which the Oslo municipality is responsible for maintenance, has been assessed through visual inspection. Visual inspections of bridges led to the grading and calculation of the final damage index (FDI) for various components affected by corrosion. The main findings of this study are as follows:

- The investigation highlights that a significant portion of the bridges were constructed between 1950 and 1999 due to factors such as population growth, economic expansion, and advancements in transportation infrastructure.
- Concrete emerges as the predominant construction material, with over 75% of the bridges, followed by steel.
- Culvert bridges, slab deck bridges, and beam bridges are among the most commonly used structural systems.

- The visual inspection unveiled corrosion damage in 193 of the surveyed bridges, with 154 bridge elements showing signs of R-corrosion and 87 bridge elements affected by NR-corrosion. Moreover, most corrosion damages of the studied bridges lead to increased maintenance costs.
- Most bridge elements show low R-corrosion risk, with only three elements requiring urgent rehabilitation or replacement due to high corrosion levels. Concrete slabs, integral to slab deck bridges, are notably prone to R-corrosion risk.
- Similarly, most bridge elements exhibit low NR-corrosion risk, although five elements require prompt attention for rehabilitation or replacement. Remarkably, among bridge components, railings display the highest NR-corrosion rate.

References

1. Khan, S.A., Kabir, G., Billah, M., Dutta, S.: An integrated framework for bridge infrastructure resilience analysis against seismic hazard. Sustain. Resilient Infrastruct. **8**(sup1), 5–25 (2023). https://doi.org/10.1080/23789689.2022.2126624
2. Lounis, Z., McAllister, T.P.: Risk-based decision making for sustainable and resilient infrastructure systems. J. Struct. Eng. **142**(9), F4016005 (2016)
3. Shabani, A., Kioumarsi, M., Plevris, V.: Performance-based seismic assessment of a historical masonry arch bridge: effect of pulse-like excitations, Frontiers of structural and civil. Engineering. **17**(6), 855–869 (2023). https://doi.org/10.1007/s11709-023-0972-z
4. Srikanth, I., Arockiasamy, M.: Deterioration models for prediction of remaining useful life of timber and concrete bridges: a review. J. Traf. Transport. Eng. (English Edition). **7**(2), 152–173 (2020). https://doi.org/10.1016/j.jtte.2019.09.005
5. Teymouri, M., Behfarnia, K., Shabani, A., Saadatian, A.: The effect of mixture proportion on the performance of alkali-activated slag concrete subjected to sulfuric acid attack. Materials. **15**(19), 6754 (2022) [Online]. Available: https://www.mdpi.com/1996-1944/15/19/6754
6. Shabani, A., Kioumarsi, M.: Seismic assessment and strengthening of a historical masonry bridge considering soil-structure interaction. Eng. Struct. **293**, 116589 (2023). https://doi.org/10.1016/j.engstruct.2023.116589
7. Hunkeler, F.: 1—Corrosion in reinforced concrete: processes and mechanisms. In: Böhni, H. (ed.) Corrosion in Reinforced Concrete Structures, pp. 1–45. Woodhead Publishing (2005)
8. Isgor, O.B., Razaqpur, A.G.: Modelling steel corrosion in concrete structures. Mater. Struct. **39**(3), 291–302 (2006). https://doi.org/10.1007/s11527-005-9022-7
9. Zhou, Y., Gencturk, B., Willam, K., Attar, A.: Carbonation-induced and chloride-induced corrosion in reinforced concrete structures. J. Mater. Civ. Eng. **27** (2014). https://doi.org/10.1061/(ASCE)MT.1943-5533.0001209
10. Mahboubi, S., Kioumarsi, M.: Damage assessment of RC bridges considering joint impact of corrosion and seismic loads: a systematic literature review. Constr. Build. Mater. **295**, 123662 (2021). https://doi.org/10.1016/j.conbuildmat.2021.123662
11. Ngo Le, H.M., Hashimoto, K., Ha, M.T., Kita, S., Fukada, S., Ueno, T.: Health assessment and self-powered corrosion monitoring system for deteriorated bridges. J. Civ. Struct. Heal. Monit. **13**(2), 799–810 (2023). https://doi.org/10.1007/s13349-023-00678-5
12. Simoncelli, M., Aloisio, A., Zucca, M., Venturi, G., Alaggio, R.: Intensity and location of corrosion on the reliability of a steel bridge. J. Constr. Steel Res. **206**, 107937 (2023). https://doi.org/10.1016/j.jcsr.2023.107937

13. Bertola, N.J., Brühwiler, E.: Risk-based methodology to assess bridge condition based on visual inspection. Struct. Infrastruct. Eng. **19**(4), 575–588 (2023). https://doi.org/10.1080/15732479. 2021.1959621

14. Abdallah, A., Atadero, R., Ozbek, M.: A state-of-the-art review of bridge inspection planning: current situation and future needs. J. Bridg. Eng. **27** (2022). https://doi.org/10.1061/(ASCE)BE. 1943-5592.0001812

15. Bah, A.S., Sanchez, T., Zhang, Y., Sasai, K., Conciatori, D., Chouinard, L., Power, G.J., Zufferey, N.: Assessing the condition state of a concrete bridge combining visual inspection and nonlinear deterioration model. Struct. Infrastruct. Eng. **20**(2), 149–164 (2024). https://doi. org/10.1080/15732479.2022.2081987

16. Sun, L., Shang, Z., Xia, Y., Bhowmick, S., Nagarajaiah, S.: Review of bridge structural health monitoring aided by big data and artificial intelligence: from condition assessment to damage detection. J. Struct. Eng. **146**(5), 04020073 (2020)

17. Ngeljaratan, L., Moustafa, M.A.: Structural health monitoring and seismic response assessment of bridge structures using target-tracking digital image correlation. Eng. Struct. **213**, 110551 (2020). https://doi.org/10.1016/j.engstruct.2020.110551

18. Barrufet, A.R., Taddesse, E.: Condition assessment of Norwegian bridge elements using existing damage records. In: Cham, J.P., Liyanage, J.A.-E., Mathew, J. (eds.) Engineering Assets and Public Infrastructures in the Age of Digitalization, pp. 716–722. Springer International Publishing (2020)

19. Darbani, B.M., Hammad, A.: Critical review of new directions in bridge management systems. Comput. Civil Eng. **2007**, 330–337 (2007)

20. Habeenzu, H., McGetrick, P., Hester, D., Taylor, S.: Bridge Management Systems-a Review of the State of the Art and Recommendations for Future Practice, Bridge Maintenance, Safety, Management, Life-Cycle Sustainability and Innovations, pp. 926–933 (2021)

21. N-V440 Bruregistrering, Norwegian Public Roads Administration (NPRA), Statens vegvesen, Norway, (2023)

22. N-V441 Bruinspeksjon, Norwegian Public Roads Administration (NPRA), Statens vegvesen, Norway, (2023)

23. Agboola, O., Kupolati, K.W., Fayomi, O.S.I., Ayeni, A.O., Ayodeji, A., Akinmolayemi, J.J., Olagoke, O., Sadiku, R., Oluwasegun, K.M.: A review on corrosion in concrete structure: inhibiting admixtures and their compatibility in concrete. J. Bio- and Tribo-Corros. **8**(1), 25 (2021). https://doi.org/10.1007/s40735-021-00624-2

24. James, A., Bazarchi, E., Chiniforush, A.A., Panjebashi Aghdam, P., Hosseini, M.R., Akbarnezhad, A., Martek, I., Ghodoosi, F.: Rebar corrosion detection, protection, and rehabilitation of reinforced concrete structures in coastal environments: A review. Constr. Build. Mater. **224**, 1026–1039 (2019). https://doi.org/10.1016/j.conbuildmat.2019.07.250

25. Broomfield, J.P.: Corrosion of Steel in Concrete: Understanding, Investigation and Repair. Crc Press (2023)

26. Su, M.-n., Wei, L., Zhu, J.-H., Ueda, T., Guo, G.-p., Xing, F.: Combined impressed current cathodic protection and FRCM strengthening for corrosion-prone concrete structures. J. Compos. Constr. **23**(4), 04019021 (2019)

27. Nazeer, A.A., Madkour, M.: Potential use of smart coatings for corrosion protection of metals and alloys: a review. J. Mol. Liq. **253**, 11–22 (2018). https://doi.org/10.1016/j.molliq.2018. 01.027

28. Furuya, K., Kitagawa, M., Nakamura, S.-i., Suzumura, K.: Corrosion mechanism and protection methods for suspension bridge cables. Struct. Eng. Int. **10**(3), 189–193 (2018). https://doi.org/ 10.2749/101686600780481518

29. Han, X., Yang, D.Y., Frangopol, D.M.: Optimum maintenance of deteriorated steel bridges using corrosion resistant steel based on system reliability and life-cycle cost. Eng. Struct. **243**, 112633 (2021). https://doi.org/10.1016/j.engstruct.2021.112633

30. Rajahram, S.S., Harvey, T.J., Wood, R.J.K.: Erosion–corrosion resistance of engineering materials in various test conditions. Wear. **267**(1), 244–254 (2009). https://doi.org/10.1016/j.wear.2009.01.052
31. Fernández-Solis, C.D., Vimalanandan, A., Altin, A., Mondragón-Ochoa, J.S., Kreth, K., Keil, P., Erbe, A.: Fundamentals of electrochemistry, corrosion and corrosion protection. Soft Matter at Aqueous Interfaces, 29–70 (2016)

Reducing the Carbon Footprint of New Reinforced Concrete Structures in Aggressive Environments: From Real Experience to Future Applications

Juan Daniel Cassiani Hernandez (iD) and Sylvia Keßler (iD)

Contents

1   Introduction

To reach the reduction goals of CO_2 in new reinforced concrete (RC) structures, their carbon emissions must be considered throughout their entire life cycle. Since RC structures are designed to last for decades, the carbon emissions produced during their operation phase are relevant [1]. This is critical for structures exposed to aggressive environments, such as chloride contamination, given the elevated risk of reinforcement corrosion. Corrosion-induced degradation leads to considerable repairs, increasing the total CO_2 footprint of the structure [2].

The new generation of the Eurocode for reinforced concrete structures introduces the lab-performance concept in the design rules for durability [3]. In the new approach, the dimensioning of the concrete elements is based on the actual resistance of the concrete mix design against aggressive environments. This is achieved through the introduction of the Exposure Resistance Classes (ERC). The ERC

J. D. C. Hernandez (✉) · S. Keßler
Helmut Schmidt University/University of the Federal Armed Forces Hamburg, Hamburg, Germany
e-mail: cassiani@hsu-hh.com

© The Author(s) 2025
M. Kioumarsi, B. Shafei (eds.), *The 1st International Conference on Net-Zero Built Environment*, Lecture Notes in Civil Engineering 237,
https://doi.org/10.1007/978-3-031-69626-8_58

697

classifies the concrete into predefined categories concerning its resistance against corrosion induced by carbonation, chlorides, and damage caused by freeze/thaw attacks. The new durability rules offer advantages by simplifying the evaluation of current (and future) eco-efficient binder systems. Therefore, the lab-performance approach paves the way for durable and more sustainable reinforced concrete structures.

This work proposes a dimensioning concept that integrates lab-performance concept and sustainability assessment. The framework enables the design optimization of reinforced concrete elements in view of durability, environmental footprint, and budget. The framework applicability is demonstrated through a case study.

2 Methods

2.1 Dimensioning Concept for Sustainable and Durable Structures

The framework extends the lab-performance design concept described by the new Eurocode 2 [3]. The design process is divided into five steps.

1. *Definition of the Exposure Classes and Design Service Life*

   The exposure class (EC) is assigned according to the environmental exposure conditions that leads to the element degradation as per [4]. The design service life is 50 years for current buildings and 100 years for civil engineering structures such as bridges or tunnels [3].

2. *Determination of the Exposure Resistance Class (ERC) and Minimum Concrete Cover*

   Concrete is classified in exposure resistance classes (ERC) against deterioration in the form of corrosion induced by carbonation (XRC) and corrosion induced by chlorides (XRDS). The classification into the resistance class is based on the performance test results under predefined conditions described in [3]. The lab-performance defines the ERC and the minimum concrete cover based on the EC and the design service life. Since the ERC and the minimum concrete cover are dependent on each other, two approaches shall be considered to define both parameters. The first approach defines a minimum concrete cover on basis of structural performance that is the preference for the lowest cover possible. The second approach determines a desired ERC considering the materials availability, costs, or other relevant considerations.

3. *Structural Design*

   The Eurocode presents the procedures for designing reinforced concrete elements. The design must consider the concrete mix design's minimum concrete cover and mechanical properties, as defined in the previous step. The design provides the amount of steel reinforcement required for the environmental assessment and cost analysis.

4. *Environmental Assessment and Costs*

The next step is to determine the total costs and environmental impacts associated with the design alternative of the reinforced concrete element over the design life. The environmental impacts are estimated using the life cycle assessment (LCA) as per [5]. The analysis must define the life phases to be considered in the LCA and LCC. The assessment should cover at least the raw material production, installation, and operational phase. The evaluation of costs and environmental impacts are linked to a functional unit covering a specified temporal frame. Some examples of functional units are fractions of elements (m^2), individual elements, or the whole structure with a specific service life.

5. *Selection of the Most Sustainable Alternative*

To select the design with the most balanced performance in relation to the environmental footprint and costs, a multi-criteria decision-making technique is employed. The goal of the multi-criteria technique is to assess the several alternatives through a composite indicator that represents the results of the environmental impacts and costs. A comprehensive selection of state-of-the-art techniques can be found in [6].

The construction of the individual indicators consists of two main steps: Normalisation and weighting. Normalisation is the conversion of the unit of the impact indicator (e.g. kg-CO_2) into a standardised, dimensionless unit that enables a direct comparison between the various impact indicators. Methods for standardising indicators can be found in [7]. Weighting is the process that quantifies the preferences of stakeholders in relation to the various indicators and thus enables aggregation between them.

The final part of the evaluation process comprises the evaluation of uncertainty. The sources of uncertainty are divided into parameter uncertainty, model uncertainty and scenario uncertainty [8].

3   Case Study

The application of the dimensioning concept presented in Sect. 2.1 is demonstrated for a structure exposed to chloride-induced corrosion. The case study compromises a column in an underground parking garage exposed to chloride contamination. Figure 1 shows the column's cross-section.

The case study is inspired by an existing structure. After 14 years of service, the columns were repaired due to chloride contamination. Therefore, the elements did not achieve the designed service life of 50 years as recommended by the current standards for similar structures.

Step one: Definition of the Exposure Classes and Design Service Life.

Mud and snow contaminated with salt are carried by the cars to the underground parking lot, thus contaminating the structure with chlorides. Besides, the concrete

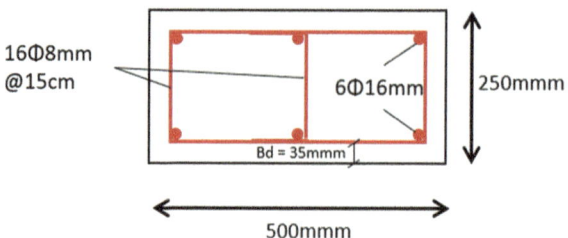

Fig. 1 Column cross-section taken from the original design

Table 1 Considered concrete mix design along with their Exposure Resistance Classes and corresponding minimum concrete cover after [3]

Mix design	Binder	w/b [−]	XRDS	XRC	C_{min}[a] [mm]	C_{min}[a] [mm]	C_{min} [mm]
Concrete A	CEM II/B-S 42,5 R	0.40	XRDS 3	XRC 1	55	10	55
Concrete B	CEM III/A 32,5 N	0.50	XRDS 4	XRC 5	60	30	60
Concrete C	CEM III/A 42,5 N	0.42	XRDS 1.5	XRC 2	40	15	40
Concrete D	CEM III/A 42,5 N	0.50	XRDS 4	XRC 4	60	25	60
Concrete E	CEM III/B 42,5 L-LH/HS/NA	0.40	XRDS 0.5	XRC 3	30	20	30

[a]C_{min} based on the RC for chlorides and carbonation, respectively

bears a risk of carbonation due to the carbon dioxide in the atmosphere and coming from the vehicles. Due to the intermittent moisture brought by the cars, it can be assumed that the element faces cyclic contact with water. Given the environmental conditions, the element risks reinforcement corrosion due to chloride contamination and concrete carbonation. Thus, the exposure classes XD3 and XC4 are assigned. For current buildings, the design service life is 50 years.

Step two: Determination of the Exposure Resistance Class (ERC) and minimum concrete cover

The Exposure Resistance Class (ERC) and minimum concrete cover are defined following the approach's structural performance and material availability. The first approach aims to achieve the best structural performance by choosing the smallest possible concrete cover. Considering a service life of 50 years, the minimum concrete cover for exposure classes XD3 and XC4 are 30 and 10 mm, respectively. Thus, the smallest possible concrete cover is 30 mm. Consequently, the required ERCs are XRDS 0.5 and XRC 5. The concrete mix 'E' in Table 1 meets both criteria.

The second approach defines the Exposure Resistance Classes based on the materials availability. This study considers study 5 concrete mix designs, which were characterized within the framework of the DAUPERF Project. Table 1 contains the considered concrete mixes, along with their Exposure Class Resistance and their corresponding minimum concrete cover. Table 1 shows that the ERC for chlorides (XRDS) dominates the assignment of the minimum concrete covers. In the next steps, the concrete mixes are exclusively referenced based on their XRDS classification.

Table 2 Considered concrete mix design along with their Exposure Resistance Classes and corresponding nominal concrete cover

Mix design	ERC	C_{min} [mm]	C_{nom} [mm]
Beton A	XRDS 3	55	65
Beton B	XRDS 4	70	80
Beton C	XRDS 1.5	45	55
Beton D	XRDS 4	70	80
Beton E	XRDS 0.5	30	40

Step Three: Structural design

The reinforced concrete column is designed following the European Normative for reinforced concrete elements (EN 1992-1). First, the nominal concrete cover C_{nom} must be determined. The nominal concrete cover is defined as the minimum concrete cover C_{min} plus allowance in design to account for deviation ΔC_{dev}. Assuming a tolerance class 1, the allowable deviation ΔC_{dev} is 10 mm [3]. Table 2 presents the C_{nom} in connection with the considered concrete mix designs in step 2 (see Table 1).

The design goal is to conserve the original cross-section while meeting the specified nominal concrete cover. For this purpose, the longitudinal steel is moved towards the element centre, and the reinforcement ratios are adjusted accordingly. However, this is constrained by the minimum clearance between the steel bars stated in the standards [9]. For this reason, the nominal concrete 80 mm is not plausible; thus, alternatives with concretes B and D (XRDS 4) are suppressed in the analysis. Therefore, only the Exposure Resistance Classes with XRDS 3 and below are feasible for the case study. This indicates that the element geometry also limits the selection of the ERC.

Figure 2 plots the resulting cross-sections of the feasible alternatives, along with their corresponding column interaction diagram. The interaction diagram also contains the results of the original design.

Step four: Environmental Assessment and Costs

The system boundaries of the environmental and cost assessment are the raw material production, transport, and installation. The operation phase is not considered since all designs are expected to meet the service life without repair. Additionally, it is assumed that the decommissioning and disposal of the element have the same environmental impacts and costs since the designs have the same volume of reinforced concrete. Thus, the end-of-life phase is not considered in this example.

The economic assessment takes into account the direct and indirect costs. The directs include the materials required for the concrete mixes (binder, aggregates, and admixtures), steel reinforcement, and workforce. The workforce includes the staff for setting the formwork, reinforcing steel placement, concreting, and curing, whose costs are computed after [10]. The indirect costs include external service costs, risk costs, administration, and profit, which are assumed to be the same as the direct costs. Since the alternatives will meet the design service life, no repair actions are considered within the life cycle.

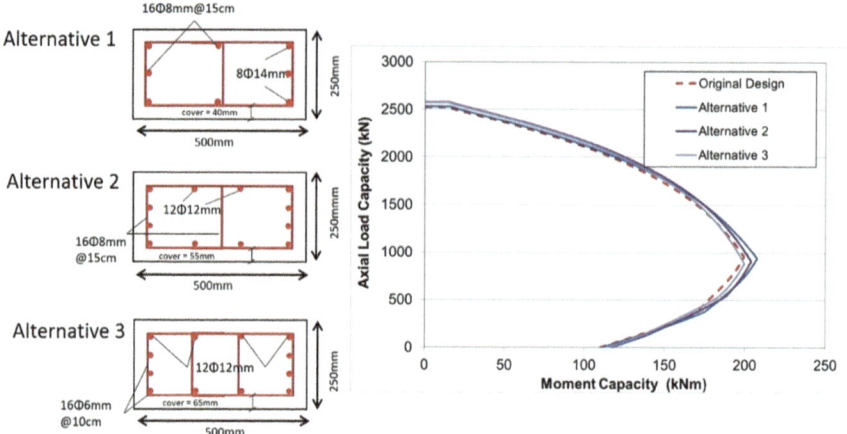

Fig. 2 Reinforcement detailing for the three alternatives (left). Iteration diagram of the columns (right)

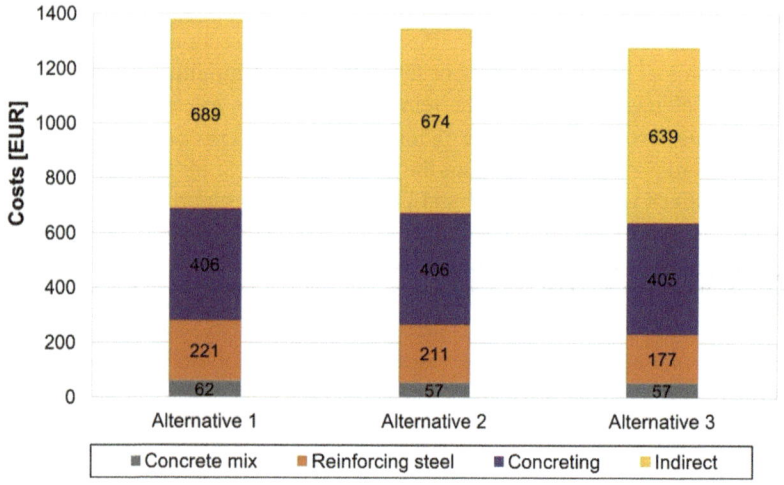

Fig. 3 Distribution of the construction cost for one column

Figure 3 provides the total costs of the column across the considered design alternatives. The results show that the ERC selection influences the elements total cost. The changes in the reinforcement ratios drive the cost differences. The best ERC (i.e. XRDS 0.5) allows the minimal nominal cover, leading to a more efficient reinforcement ratio and thus leading to the lowest total costs. It is worth mentioning that the concrete mix has the lowest contribution to the total costs of the element. The indirect and labour costs bear most of the total costs in the element.

The LCA includes the construction of the new column with dimensions of 250 × 500 × 2500 mm^3. Figure 4 shows the system boundaries. The construction of the elements includes the production of concrete and reinforcing steel. The concrete production includes the transport of cement, aggregates, and admixtures from the production site to the mixing plant. The transport of reinforcing steel to the construction site is also included. All transport distances are assumed to be 50 km. The concrete pumping and compaction processes are also modelled. The environmental data of the materials were taken from the Environmental Product Declarations (EPD) of the IBU [11]. The installation and transport processes and their impacts were taken from the Ecoinvent database [10]. The global warming potential (GWP) in kg-CO_2 eq is used in the project as the impact category for the LCA.

Figure 5 shows the GWP results across the considered design alternatives. Analog to the trend found in the total costs, the selection of the best ERC impacts the Global Warming Potential of the element. The higher ERC (i.e. XRDS 0.5) significantly reduces the CO_2 emissions to −38% compared with the lowest ERC (i.e. XRDS 3). The reduction in CO_2 emissions is attributed to the binder employed in the concrete mix since the binder is the main contributor to the element GWP. The

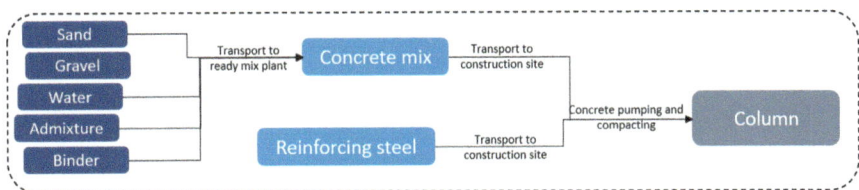

Fig. 4 System boundary of the column construction

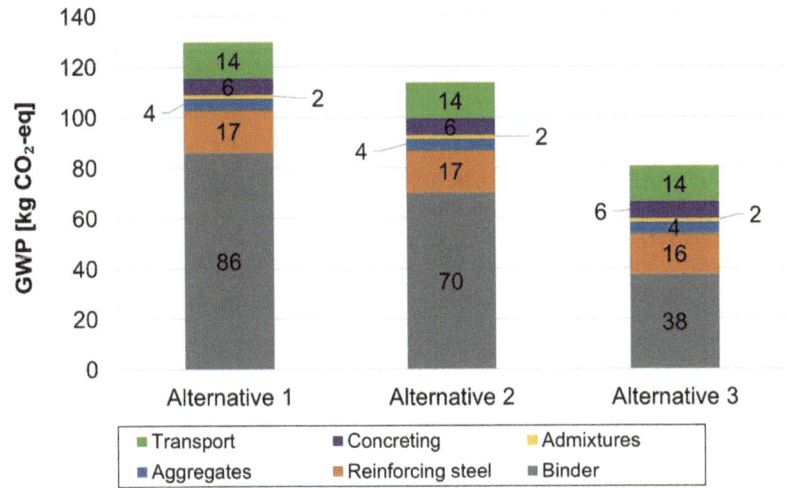

Fig. 5 Global Warming Potential of the column construction

binders account for up to 67% of total CO_2 emissions. The concrete mix design in alternative 3 (XRDS 0.5) contains a low-clinker binder (CEM III/B), whose emissions are −60% less than the binder in alternative 1 (XRDS 3), i.e. CEM II/B-S. Besides the binder, the production of steel reinforcement and transportation of materials significantly affect the GWP emissions.

Step five: Selection of the Most Sustainable Alternative

A multi-criteria assessment is necessary to select the alternative with the most balanced performance regarding total costs and GWP. In the case example, the decision is trivial since alternative 3 has the lowest costs and emissions; it is thus the most sustainable alternative. However, the multi-dimensional assessment is performed to compare the obtained results with reduction goals and planed budget.

The normalization technique is now modified to the "Distance to a reference," as per [7]. The approaches compare the indicator results to a reference value or objective (see Eqs. 1 and 2). This could be a target to be reached in each time frame; for example, the reduction goal of emissions is 65% by 2030 as per [VDZ]. The resulting indicators behave as key performance indicators, thus indicating the relative achievement of the reduction goals. The multi-criteria method for evaluating the results is simple weight aggregation (SAW). The scheme assumes equal weights for the costs and environmental ($W_{costs} = W_{CO2} = 0.5$), see Eq. 3.

$$I_{Costs-i} = \frac{Cost_{ref} - Cost_i}{Cost_{budget}} \tag{1}$$

$$I_{CO2-i} = \frac{GWP_{ref} - GWP_i}{GWP_{reduction}} \tag{2}$$

$$I_{Sustainability-i} = W_{costs} * I_{Costs-i} + W_{CO2} * I_{CO2-i} \tag{3}$$

Based on the actual structure, the reference GWP emissions per column is 152 kg-eq. According to [12], the aimed GWP reduces this value by 65%, thus leading to a GWP reduction of 53 kg-eq per column. Based on the case study information, the reference total cost per column is 2207 EUR, which includes construction and repair. The goal is to suppress the repair costs (830 EUR).

Table 3 presents the results of the sustainability score considering the reduction goals and budget. The sustainability indicator designates alternative 3 as the most sustainable design. The alternative 3 yields 98% of the cost reductions, thus being almost in line with the planned budget. Although this alternative delivers significant reduction, these are still below from the reduction goal. Alternative 3 reaches 72% of the reduction targets for the CO_2 emissions.

Table 3 Sustainability scores of the design variants

Alternative	ERC	I_{Costs}	I_{CO2}	$I_{Sustainability}$
Alternative 1	XRDS 3	0.89	0.22	0.55
Alternative 2	XRDS 1.5	0.92	0.38	0.65
Alternative 3	XRDS 0.5	0.98	0.72	0.85

4 Conclusions and Outlook

This research extended the lab-performance design concept for reinforced concrete structures, consigned in the new generation of EUROCODE, to include the essential aspect of sustainability. The applicability of the proposed dimensioning concept is demonstrated through an application example, a reinforced concrete column exposed to chlorides. The case study is inspired by an actual structure. The following conclusions are drawn from the results of this study.

The dimensioning concept successfully integrates the lab-performance concept, economic costs, and environmental impacts in the design of durable and sustainable reinforced concrete elements. The proposed framework simultaneously considers the environmental footprint and costs by employing and applying multi-criteria decision methods, thus facilitating the selection of the most balanced alternative given the environmental burdens and costs.

The Exposure Resistance Class (ERC) selection leads to more cost-efficient and environmentally friendly results. The optimal ERC (i.e. XRDS 0.5) permits the reduction of the minimal concrete cover, hence achieving a more efficient reinforcement ratio. Given the reduced reinforcing steel amount, the total costs are reduced. Furthermore, the ERC significantly influences the emissions of the element. However, the reduction in emissions is mainly attributed to the binder used in the concrete mix design. Therefore, the selection of the binder system needs to be carefully assessed to obtain the aimed reduction in emissions.

In this study, only two dimensions were considered to assess sustainability: the environment and the economy. Future studies must consider the social aspects by using relevant indicators for the impact on civil infrastructure. Furthermore, the assessment of the environmental dimension only contemplates emissions. Other relevant impacts need to be included for a more comprehensive study, such as energy demand, water consumption, and waste generation. Additionally, the dimensioning concept can extend the system boundaries up to the end of life. Accounting possible impacts and costs linked to refurbishing, recycling, and disposal. In this way, the concept of circular economy can be integrated into the design of reinforced concrete elements. Finally, the multi-criteria decision process must be constructed with a panel of technical experts, academics, and stakeholders. The weighting scheme and aggregation techniques must be selected according to the project objectives and shareholders' preferences.

Funding Note and Acknowledgment This work was supported by the German Committee for Reinforced Concrete (DAfStb) under the grant number V 519. The authors gratefully acknowledge the financial support. Furthermore, the authors would like to thank the Engineering Consultancy Schiessl-Gehlen-Sodeikat[12] for providing the documented case study.

---

[1] DAfStb: Deutscher Ausschuss für Stahlbeton.

[2] Engineering Consultancy Schiessl-Gehlen-Sodeikat GmbH.

References

1. Navarro, I.J., Martí, J.V., Yepes, V.: Reliability-based maintenance optimization of corrosion preventive designs under a life cycle perspective. Environ. Impact Assess. Rev. **74**, 23–34 (2019)
2. Keßler, S.: Comparative life-cycle analysis of two repair measures for chloride contaminated concrete structures. In: XV International Conference on Durability of Building Materials and Components. eBook of Proceedings. URL https://www.scipedia.com/public/p587a (2020)
3. Eurocode 2: Design of Concrete Structures – Part 1-1: General Rules – Rules for Buildings, Bridges and Civil Engineering Structures. German and English Version prEN 1992-1-1:2021. CEN/TC 250, https://www.eurocode-online.de/de/norm-entwurf/din-en-1992-1-1/334621614 (2020)
4. DIN EN 1045-2:2008-08: Tragwerke aus Beton, Stahlbeton und Spannbeton_- Teil_2: Beton_- Festlegung, Eigenschaften, Herstellung und Konformität_- Anwendungsregeln zu DIN_EN_206-1. Beuth Verlag GmbH, Berlin (2008)
5. DIN EN ISO 14040:2021-02: Umweltmanagement_- Ökobilanz_- Grundsätze und Rahmenbedingungen (ISO_14040:2006_+ Amd_1:2020); Deutsche Fassung EN_ISO_14040: 2006_+ A1:2020. Beuth Verlag GmbH, Berlin (2021)
6. Navarro, I.J., Yepes, V., Martí, J.V.: A review of multicriteria assessment techniques applied to sustainable infrastructure design. Adv. Civ. Eng. **2019**, 1–16 (2019)
7. OECD: Handbook on Constructing Composite Indicators. Methodology and User Guide, Paris (2008)
8. Scope, C., Ilg, P., Muench, S., Guenther, E.: Uncertainty in life cycle costing for long-range infrastructure. Part II: guidance and suitability of applied methods to address uncertainty. Int. J. Life Cycle Assess. **21**, 1170–1184 (2016)
9. DIN EN 1992-1-1:2011-01: Eurocode_2: Bemessung und Konstruktion von Stahlbeton- und Spannbetontragwerken_- Teil_1-1: Allgemeine Bemessungsregeln und Regeln für den Hochbau; Deutsche Fassung EN_1992-1-1:2004_+ AC:2010. Beuth Verlag GmbH, Berlin (2013)
10. PRé: Ecoinvent Database v.3.8.: https://ecoinvent.org/the-ecoinvent-database/. Last accessed 27 Dec 2023
11. Institut Bauen und Umwelt e.V.: EPD Programm. https://ibu-epd.com/ibu-data-start/. Last accessed 27 Dec 2023
12. VDZ: Dekarbonisierung von Zement und Beton–Minderungspfade und Handlungsstrategien. Düsseldorf (2020)

Misconceptions Around Strength Requirements for Concrete Repair Materials and Related Sustainability Issues

Nicholas Jarratt and Hans Beushausen

Contents

1 Introduction

Concrete repair and rehabilitation are vital disciplines in structural and civil engineering, given the age of concrete infrastructure and the need to extend its service life. Most concrete repair projects entail removing and replacing the defective concrete with a repair mortar or bonded overlay. Proprietary repair materials are often used for repairs of this kind. Marketed as 'high-performance', these materials tend to have very high compressive strength exceeding 40 MPa and give the illusion that they are more crack-resistant [1].

Compressive strength is one of the main parameters for the structural design of concrete structures. It serves as a quality control and conformity assessment of concrete during construction [2]. However, an undue emphasis has been placed on

N. Jarratt · H. Beushausen (✉)
University of Cape Town, Cape Town, South Africa
e-mail: JRRNIC001@myuct.ac.za; hans.beushausen@uct.ac.za

© The Author(s) 2025
M. Kioumarsi, B. Shafei (eds.), *The 1st International Conference on Net-Zero Built Environment*, Lecture Notes in Civil Engineering 237,
https://doi.org/10.1007/978-3-031-69626-8_59

strength and not enough on durability when it comes to repair materials. When Vaysburd et al. reviewed 120 North American repair projects, not one project prescribed a compressive strength in the region of 20 MPa [1], nor did any prescribe a limitation on shrinkage. It is of no surprise then that when Tilly [3] investigated the performance of 130 patch repairs, only 50% were found to be intact after 5 years of service and that incorrect repair design and material selection were two of the reasons for these failures. A more recent study by [4], who interviewed Clients, Contractors and Engineers, found that 90% of the respondents had listed cracking as the primary damage in concrete structures. While poor quality of work was listed as one of the primary reasons for cracking, the respondents also listed the concrete specs to be higher than required, making the concrete more brittle.

There are standards and guidelines to aid designers and contractors in selecting and specifying material performance requirements for concrete repairs, like the EN 1504 series [5]. However, the values specified in these standards are often viewed as arbitrary, allowing materials to be tailored to meet these requirements but not the repair needs on-site [6]. Furthermore, since these tailored materials are often proprietary, their composition and cement contents are unknown.

The cement and concrete industry is one of the largest contributors to greenhouse gas (GHG) emissions [7]. Various steps have been taken to reduce the GHG emissions associated with the industry, which include optimising cement production processes, carbon capturing, replacing clinker contents with supplementary cementitious materials, and using alternative fuels [8, 9]. Another alternative is to reduce the cement contents in the mortar or concrete mix by increasing the water/cement ratio. Such an approach would inevitably lower the material's compressive strength and influence the structural design of a new concrete structure. Olsson et al. [10] considered compressive strength and the quantity of steel to optimise the structural design of reinforced concrete columns and slabs and found a potential to reduce GHG emissions by 18–25%.

This paper investigates the negative impact of high-strength mortars in concrete repair. It presents recommendations for reducing this negative impact by amending the compressive strength requirement in repair materials. A summary of compressive strength and elastic modulus requirements for concrete repair, prescribed by European and American codes, is first presented. Five proprietary cementitious mortars available in South Africa that comply with the compressive strength requirements of these standards are then presented. Their cement contents are then estimated using an existing compressive strength prediction model for mortar based on Abrams' generalisation law. In addition, two mortars of known composition and strength were considered to validate the model, one of which illustrates how prescribing a lower strength can reduce the cement contents of a mortar. Previous studies on the structural contribution of cementitious materials and the correlation between compressive strength and cracking in repair materials are then discussed. The paper then concludes with recommendations on prescribing material requirements for a more durable repair, which could inherently reduce GHG emissions.

2 Standards and Guidelines on Repair Materials

The ACI 546.3R and EN 1504-3 are the most prominent documents on material selection for concrete repair. The ACI 546.3R provides guidelines for selecting materials for concrete repair, including the material types available and their characteristics and suitability to various repair applications. The EN1504-3 does not give any background on the material types available. Instead, the standard lists the performance characteristics required, how to assess these characteristics and what the requirement is for a particular repair application. While both documents follow different approaches, they both prescribe performance criteria. Table 1 shows the criteria for compressive strength and elastic modulus.

The classes listed under EN 1504-3 refer to the repair method, which is non-structural (Class R1 and R2) and structural (Class R3 and R4). As for ACI 546.3R, the criteria listed only apply to Portland-cement (PC) based materials.

Except for the elastic modulus value prescribed by ACI 546.3R, all the criteria listed in Table 1 are prescribed a minimum value with no upper limit. While this approach seems beneficial, some have criticised these limits as arbitrary, with Wood commenting that this approach by EN 1504 has led to products being tailored to meet the CE requirements and not those on-site [6]. Furthermore, the approach of prescribing minimum performance criteria has unknowingly led to the large-scale production of commercial repair mortars, which may meet the minimum criteria but are over-designed in terms of compressive strength and have excessively high amounts of cement.

3 Predicting the Cement Contents of Repair Mortars in South Africa

A total of five commercially available cementitious repair mortars in South Africa were considered in this study, as well as a repair mortar (RM) designed in the lab and a low-strength mortar taken from [12] (Mix 25). The information available in the commercial repair mortar datasheets is provided in Table 2, and the mixed design of

Table 1 Performance requirements of repair materials for concrete repair

Property	EN 1504-3 [5]	ACI 546.3R [11] (PC based only)
Compressive strength (MPa)	Class R1: ≥ 10 Class R2: ≥ 15 Class R3: ≥ 25 Class R4: ≥ 45	Similar to substrate but ≥ 28
Elastic modulus (GPa)	Class R1 & R2: No requirement Class R3: ≥ 15 Class R4: ≥ 20	E (repair) $\approx E$ (substrate)

Table 2 Datasheet information on commercial cementitious repair mortars in South Africa

Mortar ID	Compressive strength (MPa)	Water consumption (kg)	Yield (L)
1	49	3.825	13.7
2	58	2.7	12.5
3	45	3.05	10
4	55	3.3	12
5	55	3.3	12

Table 3 Mix design of repair mortars

Material	Unit	RM	Mix 25 [12]
CEM II 42.5 A-L	kg/m^3	500	323
Water	kg/m^3	210	297
Dune sand	kg/m^3	730	747
Greywacke crusher sand	kg/m^3	820	843
Compressive strength	MPa	55	15

the two repair mortars is provided in Table 3. It is also worth noting that only three of the five commercial repair mortars provided drying shrinkage data.

The datasheets of these commercial products described these mortars as suitable for structural repair applications, with all the 28-day compressive strength values (σ_{C28}) meeting the EN 1504-3 strength requirements for Class R4 materials in Table 1. Since these repair mortars are proprietary, their constituents are not included, meaning their cement contents are publicly unknown. However, appropriate compressive strength prediction models can help estimate these cement contents. One suitable model is that developed by Abrams [13], which only requires the water/cement (w/c) ratio and is defined as:

$$\sigma_{C28} = A/B^{w/c} \qquad (1)$$

The coefficients A and B are constants for a material tested at a particular age under certain conditions. Although the model was initially developed for concrete mixes, Rao [14] also found Abrams' generalisation law applicable to mortars. Several mortar mixes with differing cement/sand ratios were considered in this study, each of which had incremental water/cement ratios ranging from 0.3 to 0.65. The cement used in this study was 43-grade Portland cement, conforming to IS: 8112-1989.

The study revealed the water/cement ratio as the single largest factor influencing the mortars' compressive strength. These findings were also reported by [15], whose extensive study on various mortar mixes found the water/cement ratios' influence on strength to be unaffected by the type and grading of the fine aggregate. The more generalised form proposed by Rao is provided in Eq. 2 and applicable to low- and high-strength mortars with a water/cement ratio greater than 0.40.

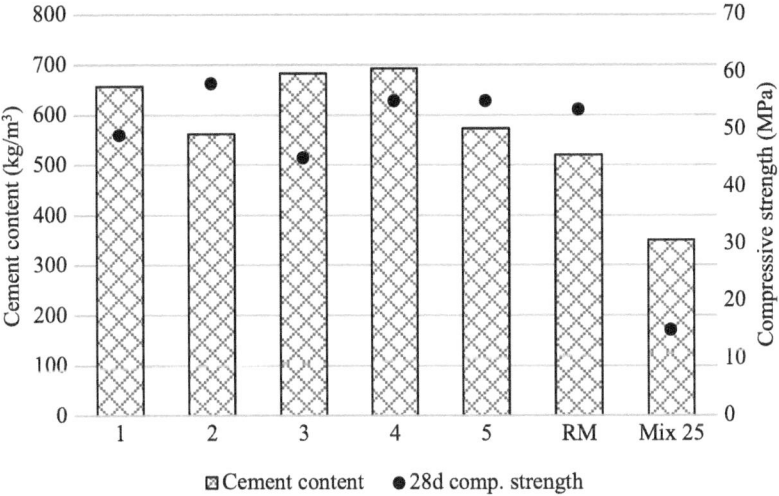

Fig. 1 Cement content results of the repair mortars considered and their compressive strength

$$\sigma_{C28} = 11.3 \, (\text{w/c})^{-1.713} \tag{2}$$

The cement content of each mortar was determined using its compressive strength and water content values in Table 2. Equation 2 and the compressive strength yield the water/cement ratio, allowing the mortars' cement content to be estimated using its water content. The cement contents of repair mortars RM and mix 25 were also estimated and compared to those in Table 3 to validate the model. The estimated cement contents and compressive strength of each mortar considered are presented in Fig. 1. While the cement content results of mortars RM and Mix 25 are over-predicted, the difference is only 4% and 8% of the actual value in Table 3, respectively. Rao [14] reported similar variations, commenting that the model was suitable for practical design purposes and was also deemed suitable for the purposes of this study.

As illustrated in Fig. 1, the model results showed the cement content in all five commercial repair mortars to be above 550 kg/m³ and higher than mortar RM. Mortar 4 had the highest cement content of nearly 700 kg/m³, which was 33% more than mortar RM. The cement content estimated in Mix 25 was below 350 kg/m³ and 1.4–2 times lower than the other mortars considered. While it can be argued that mortar Mix 25 is not a structural repair mortar due to its low compressive strength, the definition of what constitutes a structural repair needs to be considered and if these high-strength repair materials meet this requirement.

4   The Contribution of Cementitious Materials in Structural Repair

The ACI Committee 364 [16] defines a repair as structural when it can carry the applied load. This requirement has led to the misconception that if the cementitious repair material has a high compressive strength, the repair will structurally contribute to the member's load-carrying capacity. While this notion may be partly true in the short term, since the elastic modulus of cementitious materials is closely tied to their compressive strength, it does not consider the long-term effects of shrinkage and creep in the repair.

Naharaghi et al. [17, 18] explored these effects by developing an analytical model for unreinforced repaired axial members loaded in compression. Based on Hooke's law and Euler–Bernoulli beam theory, the model considers the applied load, elastic modulus of the concrete substrate and repair material and shrinkage and creep characteristics of the repair material. Creep and shrinkage in the substrate concrete were ignored as the model assumes the repairs to be done on aged concrete members where such time-dependent deformations have already ensued. The model computes the stresses in the concrete substrate and repair material for a defined period that is discretised into days.

The study presented a square column repaired with a commercially available high-strength cementitious grout bulked with 9 mm Greywacke stone. Figure 2 depicts the column dimension and extent of repair, which was 20% of the column's cross-sectional area. Experimental work was undertaken to determine the material characteristics of the repair grout, which are presented in Table 4. Due to the cementitious grout containing stone aggregate, its cement content could not be predicted using Rao's model. A temperature and relative humidity of 22 ± 1 °C

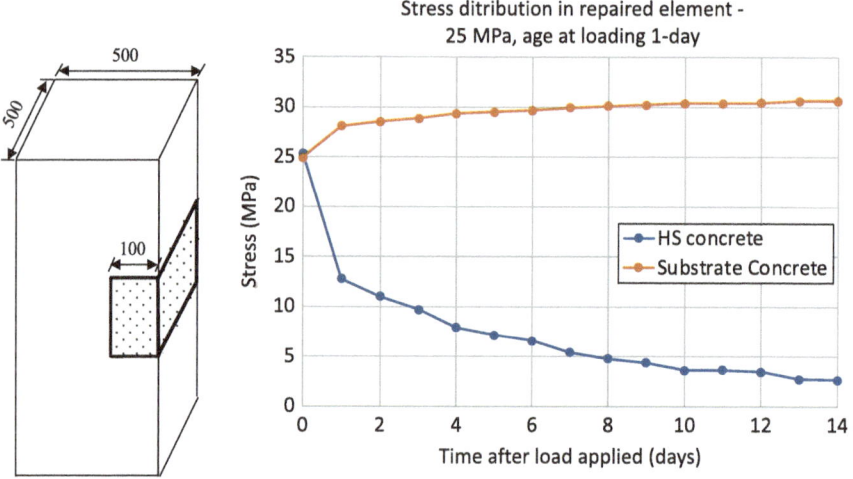

Fig. 2 Repair geometry and stress outputs from analytical model over time [17]

Table 4 Cementitious grout properties from testing and considered in the analytical problem

Age (Day)	Compressive strength (MPa)	Elastic modulus (GPa)
1	42	25.5
28	77	28.5

and $55 \pm 5\%$ were used for shrinkage and creep testing, which started the day after the samples were cast. After 90 days, the shrinkage and creep micro-strains were 340 μ-ε and 120 μ-ε/MPa.

The properties of the substrate concrete were assumed to have a compressive strength and elastic modulus of 50 MPa and 25 GPa, respectively. All loads were presumed to have been removed before the column was repaired, and the repair duration was assumed to be 1 day long. This short period allowed the assumption that only elastic strain was recovered in the substrate. Following the one-day repair period, an axial compressive load of 25 MPa was applied and maintained.

The stress results from the model in Fig. 2 show that despite equivalent load sharing between the repair and substrate on the day of loading, due to the similarities in elastic moduli, the stress in the repair begins to rapidly decline, with approximately 50% of its load contribution being lost after 1 day. This drop in load contribution was owed to the repair materials' shrinkage and creep development. Consequently, this reduction of stress in the repair increased the load contribution of the substrate. The increase in stress in the substrate was not as significant as the repair, though, since the substrate's cross-section was four times larger than the repair. The model showed that stress transfer from the repair to the substrate began to stagnate after 14 days of loading, with the repair only taking 10% of its original load contribution. Thus, despite the high strength of the repair concrete, its ability to structurally contribute became redundant after a short period.

The result above begs the question of whether cementitious materials can carry any significant load in the long term and whether high-strength repair materials are always necessary.

5 Compressive Strength and Cracking Tendency in Repair Materials

A crucial comment raised by the ACI Committee 364 [16] is that, while a structural repair needs to contribute to the load, it must also protect the underlying concrete and reinforcing steel from deterioration and corrosion. Repair material cracking is one of the biggest concerns in repair as this enables easy access of deleterious agents into the repaired structure. From a structural perspective, cracking can also reduce the bond strength between the repair material and substrate, preventing any stress transfer between the substrate and repair. One of the leading causes of cracking is restrained shrinkage of the repair material.

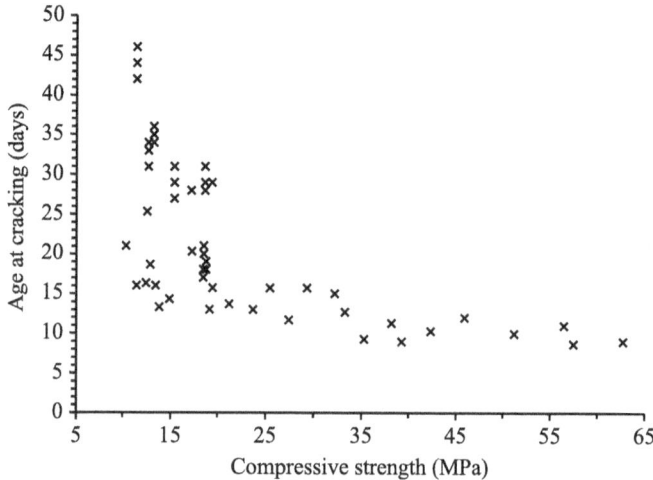

Fig. 3 Age at cracking versus compressive strength at 28 days [19]

A study by Beushausen and Arito [19] found a strong correlation between compressive strength and shrinkage cracking tendency in mortars. This study examined the age of cracking for 40 different repair mortar mixes. These mortar mixes contained various cement types, chemical admixtures, water/binder ratios, water contents and aggregate combinations, with 28-day compressive strengths ranging between 10 and 60 MPa. The cracking age was determined using the ASTM C1581 [20] restrained shrinkage test method, commonly known as the ring test, and enables a relative comparison of the repair materials' susceptibility to crack under restrained shrinkage conditions.

The compressive strength and age of cracking results are illustrated in Fig. 3. As shown, the higher strength mortars cracked at earlier ages than those with lower strength, a trend also found in [21]. Despite the scatter in results, the study suggested an inverse relationship exists between the compressive strength and cracking age in the mortars. It was also noted that increasing the 28-day strength beyond 35 MPa did not significantly change the cracking age in this study, labelling this value as the 'cracking threshold'.

The reason for this occurrence was owed to the fact that an increase in compressive strength is generally complemented by an increase in elastic modulus and a drop in tensile relaxation. These changes would create higher stresses under restrained shrink-age conditions should other factors be held constant, thus increasing the material's susceptibility to crack, a reasoning also noted by [1]. Furthermore, since the elastic modulus of a cementitious material is influenced by the stiffness of its paste phase, which is dependent on its porosity and water/cement ratio [22], it could be posited that increasing the water/cement ratio can improve the materials crack resistance.

6   Recommendations

The current approach of prescribing minimum strength criteria of repair materials for structural repair can be viewed as more detrimental than beneficial. Not only do the studies above suggest that cementitious materials with high compressive strength carry virtually no axial compressive load in the long term, but they can increase the material's susceptibility to crack. Wire meshing and shear dowel reinforcement can be included with the overlay to address the risk of cracking and debonding from the substrate. Still, this approach is not very common in practice and is also viewed as counterintuitive since this would only increase the total amount of embodied carbon in the repair with no added benefit to structural performance. The compressive strength prediction model, discussed in Sect. 3, also illustrates that these high-strength repair mortars can have 40–100% more cement than one with much lower strength. Conversely, a low-strength mortar with better crack resistance can potentially perform the same as one with high strength in a structural repair scenario.

It must be emphasised that this does not mean low-strength materials should be prescribed for all structural repair scenarios and that high-strength materials should never be considered. Instead, the requirements of each repair scenario must be carefully considered, and the repair material must be designed to meet such requirements without automatically specifying compressive strength as the primary or sole performance criterion. Repair standards and guidelines are essential here as they serve as the conduit between research and industry by accepting and applying research findings, like those above, and broadcasting them to engineers and practitioners in concrete repair. Thus, updating existing codes is the most logical step to addressing the issues of premature failure in concrete patch repairs.

One simplistic approach to remedy the concerns above is to revisit the compressive strength requirement and prescribe an upper bound limit. While this suggestion does not guarantee the repair material will have a lower cement content, given that the variability in GHG emissions for producing cementitious materials increases with strength (see [10]), it does ensure a more crack-resistant material and durable repair.

A more sophisticated option would be to include guidelines of what parameters influence the repair materials' susceptibility to crack, an example of which is provided in Table 5. The benefit of this table is that it enables the designer to consider both the material properties and composition when designing a repair. Inspection of this table also reveals that most of the parameters discussed in this study, like compressive strength, elastic modulus and cement content, are significant factors that control the material's sensitivity to crack. It is worth highlighting that the water to cementitious materials ratio in Table 5 has a minor impact on crack sensitivity because this parameter considers cements blended with SCMs, not necessarily cements only.

In addition to including guidelines, mandatory steps toward testing the crack sensitivity of a repair material should also be considered. While the EN 1504-3 does prescribe a restrained shrinkage test method [23] to evaluate such a characteristic, the

Table 5 Parameters influencing the crack sensitivity of cementitious repair materials [24]

Parameter	Effect		
	Major	Moderate	Minor
Drying shrinkage	x		
Modulus of elasticity	x		
Creep		x	
Compressive strength	x		
Early strength	x		
Paste content	x		
Cement content and type	x		
Aggregate content, type and size	x		
Coefficient of thermal expansion			x
Water to cementitious materials ratio			x
Accelerating admixtures	x		
Plasticisers		x	
Silica fume	x		
Fly ash		x	
Slag		x	
Water content	x		
Slump (within typical ranges)			x

test is impractical as it requires a specific reference concrete substrate to be cast 6 months prior to testing and its surface prepared to a specified roughness index, after which the repair material is only then applied and tested. A more straightforward approach would be to consider a free and restrained shrinkage test, like the ring test [20], as these tests do not require a concrete substrate and are easier to conduct.

7 Conclusion

This paper investigated the negative impacts of high-strength proprietary mortars in concrete repair by estimating the cement contents, its inability to contribute towards the load-bearing capacity of a repaired concrete member subjected to axial compression and its susceptibility to cracking when restrained. The compressive strength and elastic modulus requirements for concrete repair in EN 1504-3 and ACI 546.3R were summarised, where it was suggested that the prescription of arbitrary lower bound values has resulted in a surge of proprietary repair mortars that meet these requirements but not the needs of the repair.

Based on Abrams' generalisation law, a compressive strength prediction model for mortars was then presented to estimate the cement contents of five proprietary cementitious repair mortars available in South Africa. These proprietary mortars had a compressive strength of 45–55 MPa and were computed to have cement contents of 550–700 kg/m^3. When comparing these results to one of low strength, it was revealed that lowering the compressive strength of the mortar can reduce the cement contents by as much as 50%.

Previous studies were then reviewed in the context of high-strength materials. The studies illustrated that not only do materials with high compressive strength increase the likelihood of shrinkage cracking but they are also incapable of carrying compressive axial loads in the long term due to shrinkage and creep of the repair material. Based on the model outputs and reviews, an argument was made on how lowering the compressive strength requirement in repair materials can not only reduce its cement content but can also yield a more durable repair.

Various remedies to attain a more durable, which could reduce the cement content in a repair material, were then put forward. The most simplistic was to prescribe an upper bound limit for compressive strength, while a more sophisticated option was to include material composition and property parameter guidelines that influence the repair materials' susceptibility to crack.

References

1. Vaysburd, A., Bissonnette, B., Garbacz, A., Courard, L.: Specifying concrete repair materials. Materiały Budowlane. **1**(3), 44–47 (2016). https://doi.org/10.15199/33.2016.03.13
2. Jarratt, N., Beushausen, H.: Misconceptions around compressive strength of cementitious repair materials for structural repair. In: Young Concrete Researchers, Engineers and Technologist Symposium (YCRETS) Proceedings, pp. 80–87. Cement & Concrete SA, South Africa (2023)
3. Tilly, G.: Durability of concrete repairs. In: Concrete Repair: A Practical Guide, 1st edn, pp. 231–247. Routledge, Oxfordshire (2011). https://doi.org/10.1201/b12433-15
4. Gardner, D., Lark, R., Jefferson, T., Davies, R.: A survey on problems encountered in current concrete construction and the potential benefits of self-healing cementitious materials. Case Stud. Constr. Mater. **8**, 238–247 (2018)
5. British Standard: Products and Systems for the Protection and Repair of Concrete Structures – Definitions, Requirements, Quality Control and Evaluation of Conformity – Part 3: Structural and Non-structural Repair (BS EN 1504-3:2005). BSI (2005)
6. Wood, J.: Structural repair of defects and deterioration to extend service life. In: Concrete Solutions: Proceedings of the International Conference on Concrete Solutions, pp. 329–335, Padua, Italy (2009). https://doi.org/10.1201/9780203864005.ch57
7. Monteiro, P., Miller, S., Horvath, A.: Towards sustainable concrete. Nat. Mater. **16**, 698–699 (2017). https://doi.org/10.1038/nmat4930
8. Huntzinger, D.N., Eatmon, T.D.: A life-cycle assessment of Portland cement manufacturing: comparing the traditional process with alternative technologies. J. Clean. Prod. **17**, 668–675 (2009)
9. Habert, G., Billard, C., Rossi, P., Chen, C., Roussel, N.: Cement production technology improvement compared to factor 4 objectives. Cem. Concr. Res. **40**, 820–826 (2010)
10. Olsson, J.A., Miller, S.A., Alexander, M.G.: Near-term pathways for decarbonizing global concrete production. Nat. Commun. **14**, 4574 (2023). https://doi.org/10.1038/s41467-023-40302-0
11. ACI Committee 546: Guide to Materials for Concrete Repair (ACI 546.3R-14). American Concrete Institute, Farmington Hills (2014)
12. Arito, P.: Influence of Mix Design Parameters on Restrained Shrinkage Cracking in Non-structural Concrete Patch Repair Mortars. University of Cape Town (2018)

13. Abrams, D.: Design of Concrete Mixtures. Structural Materials Research Laboratory. Lewis Institute (1919)
14. Rao, G.A.: Generalization of Abrams' law for cement mortars. Cem. Concr. Res. **31**, 495–502 (2001)
15. Currie, D., Sinha, B.P.: Survey of Scottish sands and their characteristics which affect mortar strength. Chem. Ind. **19**, 631–645 (1981)
16. ACI Committee 364: Importance of Modulus of Elasticity in Surface Repair Materials (ACI 364.5T-10). American Concrete Institute, Farmington Hills (2010)
17. Naraghi, V., Jarratt, N., Beushausen, H.: Investigating the structural contribution of patch repairs to reinforced concrete elements. MATEC Web Conf. **361**, 03002). EDP Sciences (2022)
18. Naraghi, V.: Effectiveness of High Strength Repair Mortars in Structural Patch Repairs of Concrete Members Under Axial Compression. University of Cape Town (2022)
19. Beushausen, H., Arito, P.: The influence of mix composition, w/b ratio and curing on restrained shrinkage cracking of cementitious mortars. Constr. Build. Mater. **174**, 38–46 (2018)
20. ASTM C1581: Standard Test Method for Determining Age at Cracking and Induced Tensile Stress Characteristics of Mortar and Concrete Under Restrained Shrinkage. West Conshohocken, American Society for Testing and Materials (2018)
21. Dittmer, T., Beushausen, H.: The effect of coarse aggregate content and size on the age at cracking of bonded concrete overlays subjected to restrained deformation. Constr. Build. Mater. **69**, 73–82 (2014)
22. Beushausen, H., Arito, P., van Zijl, G., Alexander, M.: Deformation and volume change of hardened concrete. In: Alexander, M. (ed.) Fulton's Concrete Technology, 10th edn, pp. 265–342. Cement & Concrete SA, Midrand (2021)
23. British Standard: Products and Systems for the Protection and Repair of Concrete Structures – Test Methods – Part 4: Determination of Shrinkage and Expansion (BS EN 12617-4:2002). BSI (2002)
24. Vaysburd, A.M., Bissonnette, B., Fay, K.F.: Compatibility Issues in Design and Implementation of Concrete Repairs and Overlays. US Bureau of Reclamation. (Report No. MERL-2014-87) (2015)

The Use of Electrodeposition Technology to Promote Self-Repairing of Cracks in Concrete: A Mini Review

Mohammed H. Alzard ⓘ and Hilal El-Hassan ⓘ

Contents

1   Introduction

Since its initial discovery, concrete has garnered substantial usage in construction endeavors due to its unique characteristics. Concrete's elevated compressive strength, wide availability, durability, harmonious interaction with reinforcement materials, economical cost, ready access to primary components (including aggregates, cement, and water), straightforward fabrication process, and adaptability for molding into diverse configurations render it a paramount choice across diverse construction applications [1–3]. Despite its commendable performance, concrete exhibits inherent weaknesses, notably its comparatively limited tensile strength and ductility, rendering it susceptible to cracking, and the tremendous embodied greenhouse gasses that result from its production [2–5]. Since the world is moving toward establishing large-scale and complicated structures built in a highly aggressive environment, these weaknesses have greatly concerned researchers and practitioners. Different means have been established to overcome some of these

M. H. Alzard (✉) · H. El-Hassan
United Arab Emirates University, Alain City, UAE
e-mail: 201570286@uaeu.ac.ae

© The Author(s) 2025   719
M. Kioumarsi, B. Shafei (eds.), *The 1st International Conference on Net-Zero Built Environment*, Lecture Notes in Civil Engineering 237,
https://doi.org/10.1007/978-3-031-69626-8_60

weaknesses. For example, concrete structures are usually reinforced with steel bars to overcome the low tensile strength of concrete and its ductility. These reinforcement steel bars apparently restrict the crack width by controlling the plastic shrinkage, but they cannot stop cracks from forming. Consequently, the concrete's durability deteriorates [1, 6].

The formation of early age cracks might not jeopardize the concrete's strength, but they expose its interior, making it easier to ingress fluids and gasses [7]. With the presence of moisture, different ions, such as chloride and carbon dioxide, accelerate the carbonation of concrete, and the corrosion of steel reinforcement causes expansion. This poses a severe risk to concrete's lifespan in the long term since these accelerated processes will reduce structural durability, significantly shortening the concrete's life cycle [1, 2]. While these cracks can be reduced once detected by technical means, their formation is still expected. Manual repair methods are typically employed to mitigate the decline in long-term durability and functionality. However, conventionally, these approaches prove neither efficient nor entirely satisfactory, often exhibiting a lifespan of approximately 10–15 years [2, 4]; additionally, they are very costly [3]. Moreover, repairing cracked concrete structures can present challenges and, in some cases, may be exceedingly difficult or even deemed impossible, particularly when considering factors such as crack location, size, and the specific service requirements of the structure [1, 2, 5]. These challenges have fueled researchers for the past few decades to look for a way to make concrete repair, or "heal," without the need for this enormous amount of money, time, and effort. As a result, self-healing concrete was extensively explored.

Self-healing is a phenomenon intrinsic to the material itself, entailing the restoration and consequent performance enhancement following a prior event that had compromised the material's functionality. Self-healing, or self-repairing, concrete can automatically seal small cracks with no exterior analysis or human involvement [3]. Two distinct approaches for self-repair exist within concrete: autogenous self-healing and autonomous self-healing. Microcrack closure in concrete has been observed, wherein cracks are naturally sealed through the rehydration of unhydrated or inadequately hydrated cement particles within the affected areas [8]. This phenomenon is termed "Autogenous self-healing."

Autogenous healing is an unapparent characteristic of cementitious materials where small cracks, usually less than 0.06 mm, can autogenously heal themselves. As shown in Fig. 1, the self-healing processes can be attributed to (i) chemical causes (carbonation of calcium hydroxide and continuous hydration of cement grains), (ii) mechanical causes (cracks blockage due to the loose concrete particles and any other particles in water), and (iii) physical causes (size increase of the hydrated cement in cracks) [3, 4, 9–14].

Although this appears encouraging, the amount of self-healing products generated from the ongoing hydration of the anhydrous cementitious materials is minimal, potentially inadequate for effectively sealing the crack [15]. Despite realizing the concept of autogenous healing in concrete and its capability to seal its cracks without external activation, researchers were swayed to develop "Autonomous self-healing."

Physical cause	**Chemical causes**		**Mechanical causes**	
			Fine particles:	
Swelling	Continued hydration	Calcium carbonate formation	Broken of from fracture surface	Originally in the water

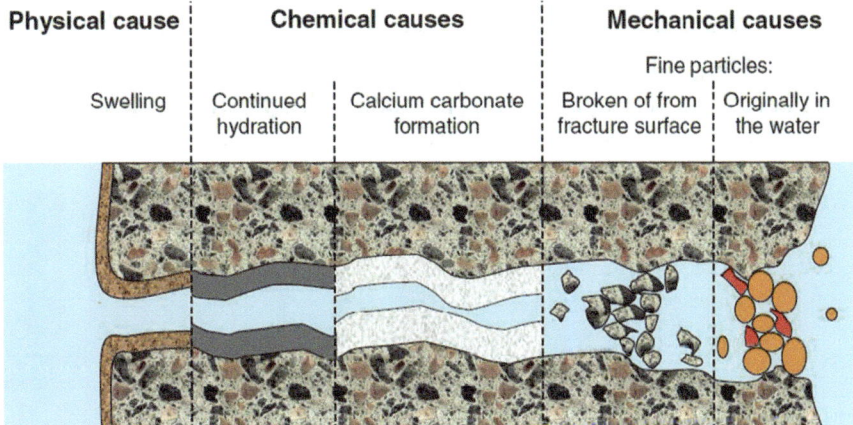

Fig. 1 Different factors contributing to autogenic self-healing in concrete [9]

This approach designs and adds engineered materials that act almost immediately when cracks happen. Researchers aimed to create a more efficient state-of-the-art mechanism than autogenous self-healing.

Unlike autogenous healing, autonomous healing employs engineered additives to accomplish self-healing. Autonomous healing also has the potential to mend larger cracks, and it has been shown to improve concrete's ability to self-heal [16]. Autonomous healing can be divided into passive and active self-healing [16, 17]. Active autonomous self-healing requires human involvement. A sensor monitors cracks until they become larger than a critical width, and then are repaired. In contrast, passive autonomous self-healing does not require any human involvement. This approach embeds functional elements in the concrete matrix, such as hollow pipes or capsules (usually brittle material). These functional elements contain the self-healing agents released once the crack reaches and breaks them [17]. Several autonomous self-healing techniques were proposed and investigated. These include electrodeposition technology, embedding shape memory alloy (SMA), capsules and vascular technologies, and microbially induced calcite precipitation using bacteria [18].

Electrodeposition technology is often used in seaport concrete structures. The electrodeposition method aims to fill cracks with insoluble inorganic compounds, such as zinc oxide (ZnO), magnesium hydroxide ($Mg(OH)_2$), calcium carbonate ($CaCO_3$), etc., precipitated from cations in seawater. These inorganic compounds form a physical barrier around rebars and cover the concrete surface. As a result, resistance against substance penetrability is increased. Electrodeposition technology takes advantage of the reinforced concrete and the surrounding aquatic environment. The healing process initiates with the application of a low-level direct electric current between a reinforcing steel bar (serving as the cathode) embedded within the concrete and an external electrode (acting as the anode) positioned in seawater to serve as an electrolyte solution (Fig. 2). Electrodeposits will occur and fill cracks in concrete and seal its surface, protecting the steel reinforcement and preventing

Fig. 2 An application of
self-healing using
electrodeposition in marine
structure [21]

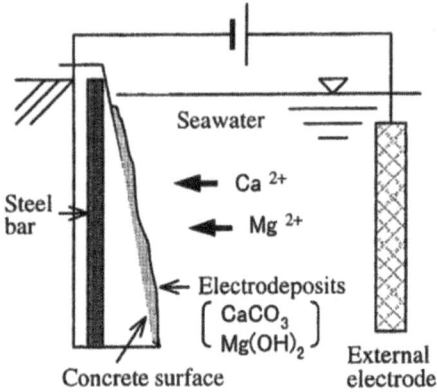

further corrosion. Certain conditions must be met for this technique to give the best results. These include conductive concrete, electric current, and electrolyte solution [19, 20].

Much work has been directed toward studying self-healing concrete, leading to abundant literature on the topic. This literature review aims to collect, summarize, and understand the existing research and debates relevant to electrodeposition autonomous self-healing concrete by comparing and contrasting each relevant source and critically evaluating each. Also, it aims to pinpoint where the research community is up to in self-healing concrete and highlight gaps in the existing research. The novelty of this review work is that it introduces a comprehensive review of literature that covers the self-healing field as a whole and can be used as a basis for future work.

2   Performance of Electrodeposition

Several studies have explored the potential of electrodeposition technology in promoting self-healing of concrete [19, 21–27] in addition to developing models to investigate the damage healing process [28, 29]. One of the earlier studies on electrodeposition by Ryu (2001) investigated the efficiency of the method in sealing drying shrinkage cracks in concrete specimens immersed in two electrolyte solutions (zinc sulfate ($ZnSO_4$) and magnesium chloride ($MgCl_2$)) and applying a 0.25 A/m^2 electrical charge. It was observed that after 14 days, 0.05–0.10 mm wide cracks were almost entirely sealed. The surface coating rate was 70%, irrespective of the immersion solution. The presence of electrodeposits led to achieving a coefficient of permeability less than that of the control specimen after 14 days, improving concrete's water tightness and protection against detrimental materials [24].

It should be noted that the type of the electrodeposits depends on the immersion solution; for example, depositions such as ZnO, $Mg(OH)_2$, $CaCO_3$, Copper(II) oxide (CuO), Silver (Ag), and Copper(II) sulfate ($CuSO_4$) are precipitated when the

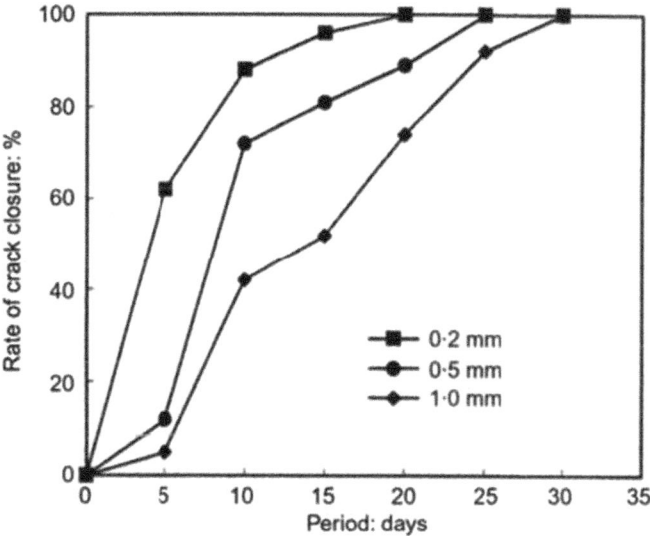

Fig. 3 Crack closure rate versus crack width [30]

immersion solutions contain $ZnSO_4$, $MgCl_2$/Magnesium nitrate $(Mg(NO_3)_2)$, Ca^{2+}, Copper(II) chloride $(CuCl_2)$, Silver nitrate $(AgNO_3)$, and Brochantite $(Cu_4SO_4(OH)_6)$, respectively, are used. Another investigation conducted by Ryu (2003) aimed to explore the potential of the electrodeposition technique for repair purposes and the formation of deposits on the surface of concrete, alongside assessing the impact of parameters such as crack width, concrete cover depth, water–cement ratio, and solution temperature. The immersion solution used in this study contained $ZnSO_4$, and the applied current was 0.5 A/m^2. Similar to his previous study, it was concluded that cracks (0.2–1 mm) were fully closed after immersion for 30 days, and the crack closure rate increased after 10 days. X-ray diffraction (XRD) results revealed that the electrodeposits formed were ZnO. A notable trend emerged, indicating an increase in crack filling depth with the increase in crack width, in contrast to the fluctuation observed in closing speed, as shown in Fig. 3 [30].

Ryu and Otsuki (2002) observed that when $ZnSO_4$ was employed as the submersion solution and a current density of 0.5 A/m^2 was administered, the electrodeposited layer formed on the surface of concrete and its thickness ranged from 0.5 to 2 mm. These electrodeposits penetrated approximately 2 mm and 8 mm deep for cracks measuring 0.2 mm and 0.6 mm in width, respectively, over a treatment duration of 56 days [21]. Conversely, Chu et al. (2014) documented crack filling depths ranging from 4.5 to 13.2 mm for reinforced concrete submerged in a $ZnSO_4$ solution, utilizing a current density four times higher than that employed by Ryu and Otsuki (2002). Consequently, electrodeposition technology enhanced mechanical strength, water tightness and permeability resistance, carbonation and corrosion resistance, and resistivity of reinforced concrete [23]. A comparison of the findings from these two studies underscores the crucial role of current density in the successful implementation of electrodeposition and the extent of deposition achieved.

Yang et al. (2019) tried manifesting a greater current density by incorporating graphene and carbon fibers. Both types of fibers exhibit outstanding mechanical, electrical, and thermal characteristics, contributing to enhancing the physical and durability properties of concrete and augmenting the impact of electrodeposition by decreasing the concrete's resistivity. Consequently, this facilitates a higher current density on the surface of the crack [20, 31, 32]. Cracks with widths of 0.3–0.5 mm were utterly closed when the specimens were immersed in $ZnSO_4$, and a 5-voltage current was applied. Lower current densities tend to take longer, but increasing the current density without caution might increase alkalinity due to OH^- formation.

A study by Zeng et al. (2022) investigated the integration of the surfactant cetyltrimethylammonium bromide (CTAB) into an electrolyte solution to enhance the efficacy of concrete crack healing facilitated by electrodeposition. The impact of varying CTAB concentrations (0.25%, 0.50%, 1.00%, 1.50%, and 2.00%) on the healing process of concrete cracks and the long-term durability of mended mortar samples was examined. The authors noted that immediately upon connecting the circuit, air bubbles started to appear within the apparatus. Subsequently, after 20 days, the cracks of all specimens were nearly filled with white sediments, and a layer of sediment covered the specimen surfaces. XRD results implied that the sediments comprised solely of crystalline ZnO and crystalline $Mg(OH)_2$ when 0.25%, 0.5%, and 2% CTAB were added to the electrodeposition solution. Deposits observed on the sample surfaces appeared loose, irregular, and thin.

Conversely, with CTAB concentrations of 1 and 1.5%, the deposits exhibited uniform and compact distribution, forming a thick layer that filled the cracks. Irrespective of the electrolyte solution used, specimens repaired with 0.25 and 0.5% CTAB demonstrated a crack healing rate (the ratio of length of the closed crack (mm) to the initial length of the crack) of approximately 98%. However, faster healing was observed with CTAB concentrations of 1.0%, 1.5%, and 2.0%, with a crack healing rate reaching 100%. This suggests that a lower CTAB concentration may not effectively adsorb onto the surface of newly formed particles, hindering the optimal combination of Zn^{2+} and OH^- or Mg^{2+} and OH^- ions to form deposits within the cracks, thereby impeding deposit formation. Within the tested concentration range, findings revealed that the optimal CTAB concentration for maximizing the healing of concrete cracks was 1%. However, exceeding this concentration resulted in diminishing returns on the healing effectiveness. Notably, specimens repaired using an electrolyte solution containing 1% CTAB exhibited notably improved steel-mortar adhesive strength, resistance to chloride ion penetration, and corrosion resistance compared to those treated with electrodeposition alone, without CTAB [26].

Wang et al. (2022) took a different route. They examined the effect of changing the anode material on the repair efficiency of a 0.6 mm wide rust-cracked reinforced concrete and its ability to extend the service life effectively. The researchers deduced that the repaired specimens' carbonation resistance and waterproofing performance

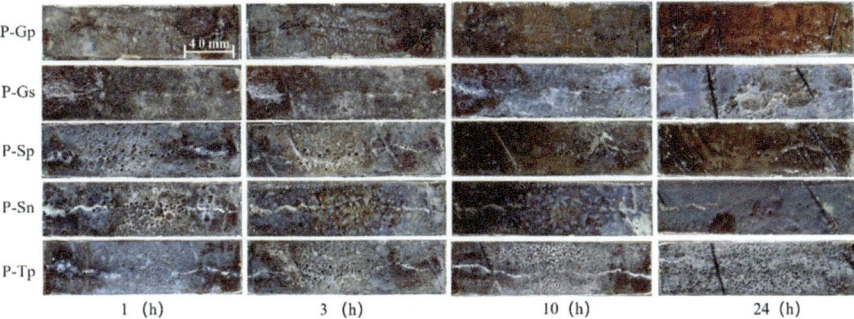

P-Gp P-Gs P-Sp P-Sn P-Tp

1 (h)   3 (h)   10 (h)   24 (h)

Fig. 4 Surface morphology of specimens with different anode material [33]

were enhanced irrespective of the type of material employed as an anode. However, among the various anode materials tested, including graphite plates (Gp), graphite stick (Gs), stainless plate (Sp), stainless net (Sn), and titanium plate (Tp), graphite plates exhibited marginally superior repair efficiency, as depicted in Fig. 4. The slight difference in performance was attributed to the difference in conductivity of these materials and their shape. Various tests were employed to arrive at this conclusion, encompassing mass increase assessments, epoxy resin surface thickness measurements, depth of filling evaluations, ultrasonic pulse velocity tests, water absorption analyses, and accelerated carbonation depth testing. Notably, the authors utilized a water-based cationic epoxy resin with a mole ratio of epoxy resin, diethanolamine, and glacial acetic acid set at 1:1:0.8 [33].

In another study, Wang et al. (2023) investigated the same electrophoretic deposition repair method on 0.5 mm wide rust-cracked reinforced concrete using a high-performance colloid solution containing water-based cationic epoxy resin. The research findings indicated that the approach effectively decreased water absorption and accelerated carbonation depth while improving reinforced concrete's internal compactness, unconfined compressive strength, and split tensile strength [34].

A deep dive into the various studies that investigated electrodeposition shows that plenty of parameters have a significant influence on the performance of this method, such as healing time, electrolyte solution, water/cement (W/C) ratio, current and type density, applied voltage, and anode material, and have been investigated [19, 22, 25, 27, 29, 33]. Among the several immersion solutions, including $MgCl_2$, $ZnSO_4$, $AgNO_3$, $CuCl_2$, $Mg(NO_3)_2$, $CuSO_4$, $Ca(OH)_2$, and Sodium bicarbonate ($NaHCO_3$) solutions, both $MgCl_2$ and $ZnSO_4$ were found to be the most appropriate solutions for the precipitation of deposition products [23, 25, 30]. The increase in the water-to-cement (W/C) ratio also benefits crack closure. Increasing W/C ratio increases the porosity and reduces the resistivity of concrete, which ultimately causes a higher electric current. Research conducted by Chu et al. (2017) demonstrated that pulse current yields superior healing effects compared to direct current. According to the microstructure analysis conducted by the researchers, the porous, honeycomb-like configuration structure of sediments is transformed into a uniform, dense, layered structure while maintaining the same composition when pulse current is employed

[22]. Overall, pulse current tends to result in increased ratios of weight gain, surface coating, crack closure, and greater depth of crack filling. However, further investigation is warranted to understand its effects fully.

3 Challenges and Future Prospects

While electrodeposition for promoting self-healing in concrete is a promising approach, implementing it to promote self-healing in concrete faces several challenges. These include the need to strategically place and distribute electrodes within structures to achieve uniform healing across the entire material, fine-tune electrodeposition parameters (voltage, current density, and duration) to optimize healing agent release without causing any adverse effects, and identify compatible and sustainable healing agents that are compatible with the concrete matrix and ensuring that they do not adversely affect its structural or mechanical properties. Additionally, there is a lack of long-term stability in the self-healing process, high initial cost concerns, and the need for a reliable monitoring system for real-time control. Finally, scaling up the technology for large structures, integrating it into existing construction practices, and ensuring durability under harsh environmental conditions are significant challenges.

Overcoming these obstacles requires collaborative efforts to advance the practical application of electrodeposition for self-healing concrete in the construction industry. Future research in advancing electrodeposition for self-healing concrete should prioritize the optimization of electrodeposition parameters, emphasizing comprehensive studies on voltage, current density, immersion solutions, and duration for efficient and controlled healing agent release without compromising structural integrity. It should also focus on studying the combined effect of electrodeposition and other self-healing methods, i.e., shape memory alloys. It is crucial to identify and develop sustainable healing agents compatible with concrete matrix and focus on their long-term stability. Moreover, advancements in monitoring systems, integrating innovative technologies and sensors, should be explored to provide real-time feedback on concrete conditions and the effectiveness of the healing process. Robust studies on the durability of electrodeposition systems under diverse environmental challenges, scalability for large structures, and seamless integration into existing construction practices are essential. Generally, self-healing concrete aligns with the principles of sustainable development and environmental conservation; however, self-healing technology has numerous shortcomings, spanning from material selection to assessing repair outcomes to practical engineering applications. Hence, life cycle assessments are imperative to evaluate the economic viability and comprehensive environmental implications.

The effectiveness of treatments employing electrodeposition in concrete relies on many variables, as discussed earlier, and they fall under material characteristics, wetting/pore fluid characteristics, and treatment characteristics. Attaining maximum treatment effectiveness necessitates a precise blend of factors concerning the solution's chemistry and the intensity and nature of electrical stimulation. While lower concentrations and lesser magnitudes of electrical stimulation have demonstrated improved efficacy for specific pore/fluid compositions, identifying the optimal treatment parameters for each issue remains complex. Despite significant advancements, research and development in this area are still in their early stages.

Apart from the difficulties mentioned above, two main constraints impede the current comprehension and utilization of electrodeposition in concrete. Firstly, most existing knowledge is empirical, lacking comprehensive theoretical frameworks and simulation tools. This limited theoretical comprehension inhibits prediction accuracy and the optimization of treatments employing electric stimulation. Second, most studies rely on laboratory investigations, with limited full-scale experimental studies demonstrating technical and economic viability at scale. While laboratory experiments provide insights, many phenomena remain unclear, and field studies are scarce. Therefore, large-scale electrodeposition treatments' technical potential and economic feasibility are inadequately assessed. To bridge these knowledge gaps, it is imperative to undertake theoretical inquiries, sophisticated laboratory experiments, and comprehensive field studies. Advanced investigative methodologies like molecular dynamics, computerized tomography, and full-scale induced polarization offer potential solutions to surmount these obstacles, facilitating scientific progress in electrodeposition for concrete restoration.

Electrochemical deposition self-healing offers notable advantages in specific environments, particularly marine settings, and holds promise for reinforced concrete structures, but its applicability is not yet universal. Furthermore, the efficacy of repair is influenced by factors such as the concentration and pH of the external solution, presenting limitations in engineering applications that warrant further exploration. Despite the limited research on electrodeposition technology, with appropriate technological parameters, it is possible to enhance the properties of reinforced concrete to a certain degree, thus paving the way for the broader adoption of this method.

4   Conclusion

The exploration of electrodeposition technology for self-healing concrete presents a promising avenue for enhancing the longevity and durability of concrete structures. Despite its widespread use and advantageous properties, concrete faces weaknesses such as low tensile strength, susceptibility to cracking, and deterioration over time. These vulnerabilities have prompted extensive research efforts to develop effective concrete repair and restoration methods.

Integrating electrodeposition technology into concrete repair processes offers several benefits, including the ability to fill cracks with insoluble inorganic compounds, such as ZnO and $Mg(OH)_2$, thus forming a protective barrier against further deterioration. Studies have shown that electrodeposition can significantly improve concrete's mechanical strength, water tightness, carbonation resistance, and corrosion resistance. However, the effectiveness of electrodeposition is influenced by various factors, including the choice of electrolyte solution, current density, immersion duration, and electrodeposition parameters.

Despite its potential, the widespread adoption of electrodeposition for self-healing concrete faces several challenges. These include the need for strategic placement of electrodes within structures, optimization of electrodeposition parameters, identification of compatible and sustainable healing agents, and development of reliable monitoring systems for real-time control. Additionally, scalability for large structures and integration into existing construction practices are significant considerations.

Moving forward, future research in electrodeposition for self-healing concrete should prioritize the optimization of parameters, comprehensive studies on voltage, current density, and immersion solutions and developing sustainable healing agents compatible with concrete matrices. Advancements in monitoring systems and integration of smart technologies should also be explored to ensure real-time feedback on concrete conditions and healing process effectiveness. Overall, while electrodeposition holds promise for enhancing the sustainability and longevity of concrete structures, further research and development are needed to overcome existing challenges and realize its full potential in the construction industry.

References

1. Seifan, M., Samani, A.K., Berenjian, A.: Bioconcrete: next generation of self-healing concrete. Appl. Microbiol. Biotechnol. **100**, 2591–2602 (2016). https://doi.org/10.1007/s00253-016-7316-z
2. Wang, X.F., Yang, Z.H., Fang, C., Han, N.X., Zhu, G.M., Tang, J.N., Xing, F.: Evaluation of the mechanical performance recovery of self-healing cementitious materials – its methods and future development: a review. Constr. Build. Mater. **212**, 400–421 (2019). https://doi.org/10.1016/j.conbuildmat.2019.03.117
3. Zhang, W., Zheng, Q., Ashour, A., Han, B.: Self-healing cement concrete composites for resilient infrastructures: a review. Compos. Part B Eng. **189**, 107892 (2020). https://doi.org/10.1016/j.compositesb.2020.107892
4. Li, W., Dong, B., Yang, Z., Xu, J., Chen, Q., Li, H., Xing, F., Jiang, Z.: Recent advances in intrinsic self-healing cementitious materials. Adv. Mater. **30**, 1705679 (2018). https://doi.org/10.1002/adma.201705679
5. Qureshi, T., Al-Tabbaa, A.: Self-healing concrete and cementitious materials. Adv. Funct. Mater. (2020). https://doi.org/10.5772/intechopen.92349
6. Xue, C., Li, W., Li, J., Tam, V.W.Y., Ye, G.: A review study on encapsulation-based self-healing for cementitious materials. Struct. Concr. **20**, 198–212 (2019). https://doi.org/10.1002/suco.201800177
7. Sidiq, A., Gravina, R., Giustozzi, F.: Is concrete healing really efficient? A review. Constr. Build. Mater. **205**, 257–273 (2019). https://doi.org/10.1016/j.conbuildmat.2019.02.002
8. Xu, J., Yao, W.: Multiscale mechanical quantification of self-healing concrete incorporating non-ureolytic bacteria-based healing agent. Cem. Concr. Res. **64**, 1–10 (2014). https://doi.org/10.1016/j.cemconres.2014.06.003

9. de Rooij, M., Van Tittelboom, K., De Belie, N.: Self-healing phenomena in cement-based material. In: State-of-the-Art Report of RILEM Technical Committee 221-SHC: Self-Healing Phenomena in Cement-Based Materials. Springer, Dordrecht (2013). https://doi.org/10.1007/978-94-007-6624-2_1

10. Hong, G., Song, C., Choi, S.: Autogenous healing of early-age cracks in cementitious materials by superabsorbent polymers. Materials. **13**, 690 (2020). https://doi.org/10.3390/ma13030690

11. Mahmoodi, S., Sadeghian, P.: Self-healing concrete: a review of recent research developments and existing research gaps. Presented at the June 12 (2019)

12. Mauludin, L.M., Oucif, C.: Modeling of self-healing concrete: a review. J. Appl. Comput. Mech. **5**, 526–539 (2019). https://doi.org/10.22055/jacm.2017.23665.1167

13. Talaiekhozani, A., Keyvanfar, A., Shafaghat, A., Andalib, R., Abd Majid, M.Z., Fulazzaky, M. A., Zin, R.M., Lee, C.T., Hussin, M.W., Hamzah, N., Marwar, N.F., Haidar, H.I.: A review of self-healing concrete research development. J. Environ. Treat. Tech. **2**, 1–11 (2014)

14. Van Tittelboom, K., De Belie, N.: Self healing in cementitious materials—a review. Materials. **6**, 2182–2217 (2013). https://doi.org/10.3390/ma6062182

15. Sonali Sri Durga, C., Ruben, N., Sri Rama Chand, M., Venkatesh, C.: Performance studies on rate of self healing in bio concrete. Mater. Today Proc. **27**, 158–162 (2020). https://doi.org/10.1016/j.matpr.2019.09.151

16. Nasim, M., Dewangan, U.K., Deo, S.V.: Autonomous healing in concrete by crystalline admixture: a review. Mater. Today Proc. **32**, 638–644 (2020). https://doi.org/10.1016/j.matpr.2020.03.116

17. Mihashi, H., Nishiwaki, T.: Development of engineered self-healing and self-repairing concrete-state of the art report. J. Adv. Concr. Technol. **10**, 170–184 (2012). https://doi.org/10.3151/jact.10.170

18. Alzard, M.H., El-Hassan, H., El-Maaddawy, T., Alsalami, M., Abdulrahman, F., Hassan, A.A.: A bibliometric analysis of the studies on self-healing concrete published between 1974 and 2021. Sustain. For. **14**, 11646 (2022). https://doi.org/10.3390/su141811646

19. Jiang, Z., Xing, F., Sun, Z., Wang, P.: Healing effectiveness of cracks rehabilitation in reinforced concrete using electrodeposition method. J. Wuhan Univ. Technol. Mater. Sci. Ed. **23**, 917–922 (2008) http://dx.doi.org.uaeu.idm.oclc.org/10.1007/s11595-007-6917-x

20. Yang, Q., Jinbang, W., Lianwang, Y., Zonghui, Z.: Effect of graphene and carbon fiber on repairing crack of concrete by electrodeposition. Ceram. - Silik. **63**, 403–412 (2019). https://doi.org/10.13168/cs.2019.0037

21. Ryu, J.-S., Otsuki, N.: Crack closure of reinforced concrete by electrodeposition technique. Cem. Concr. Res. **32**, 159–164 (2002). https://doi.org/10.1016/S0008-8846(01)00650-0

22. Chu, H., Jiang, L., Song, Z., Xu, Y., Zhao, S., Xiong, C.: Repair of concrete crack by pulse electro-deposition technique. Constr. Build. Mater. **148**, 241–248 (2017). https://doi.org/10.1016/j.conbuildmat.2017.05.033

23. Chu, H., Jiang, L., Xiong, C., You, L., Xu, N.: Use of electrochemical method for repair of concrete cracks. Constr. Build. Mater. **73**, 58–66 (2014). https://doi.org/10.1016/j.conbuildmat.2014.09.031

24. Ryu, J.-S.: An experimental study on the repair of concrete crack by electrochemical technique. Mater. Struct. Constr. **34**, 433–437 (2001). https://doi.org/10.1617/13666

25. Yodsudjai, W., Suwanvittaya, P.: Experimental study on application of electrode position method for decreasing carbonation and chloride penetration of cracked reinforced concrete. Asian J. Civil Eng. **12**, 197–204 (2011)

26. Zeng, Y., Zuo, Q., Jiang, S., Guo, M.-Z., Wang, T., Chu, H.: Effect of CTAB on the healing of concrete cracks repaired by electrodeposition and the durability of repaired concrete. Constr. Build. Mater. **326**, 126757 (2022). https://doi.org/10.1016/j.conbuildmat.2022.126757

27. Zhang, Q., Yuan, L., Zhou, Z., Wang, J.: Accelerating electrochemical repair rate for cracked cement composites: effect of carbon nanofiber. Constr. Build. Mater. **312**, 125349 (2021). https://doi.org/10.1016/j.conbuildmat.2021.125349

28. Chen, Q., Zhu, H., Ju, J., Li, H., Jiang, Z., Yan, Z.: Stochastic micromechanics-based investigations for the damage healing of unsaturated concrete using electrochemical deposition method. Int. J. Damage Mech. **29**, 1361–1378 (2020). https://doi.org/10.1177/1056789520925868

29. Zhu, H., Chen, Q., Ju, J.W., Yan, Z., Jiang, Z.: Electrochemical deposition induced continuum damage-healing framework for the cementitious composite. Int. J. Damage Mech. **30**, 945–963 (2021). https://doi.org/10.1177/1056789521991871

30. Ryu, J.S.: Influence of crack width, cover depth, water-cement ratio and temperature on the formation of electrodeposits on the concrete surface. Mag. Concr. Res. **55**, 35–40 (2003). https://doi.org/10.1680/macr.2003.55.1.35

31. Lavagna, L., Musso, S., Ferro, G., Pavese, M.: Cement-based composites containing functionalized carbon fibers. Cem. Concr. Compos. **88**, 165–171 (2018). https://doi.org/10.1016/j.cemconcomp.2018.02.007

32. Tong, T., Fan, Z., Liu, Q., Wang, S., Tan, S., Yu, Q.: Investigation of the effects of graphene and graphene oxide nanoplatelets on the micro- and macro-properties of cementitious materials. Constr. Build. Mater. **106**, 102–114 (2016). https://doi.org/10.1016/j.conbuildmat.2015.12.092

33. Wang, Y., Wang, C., Zhou, S., Liu, K.: Influence of anode material on the effect of electrophoretic deposition for the repair of rust-cracked reinforced concrete. Constr. Build. Mater. **335**, 127466 (2022). https://doi.org/10.1016/j.conbuildmat.2022.127466

34. Wang, Y., Wang, C., Zhou, S., Sun, M., Liu, K., Ma, W., Xu, H.: Experimental study of repairing rust-cracked reinforced concrete by electrophoresis deposition method. Cem. Concr. Compos. **143**, 105261 (2023). https://doi.org/10.1016/j.cemconcomp.2023.105261

Predicting Dry Shrinkage Using Machine Learning Methods

Peyman Khodabandeh, Fazel Azarhomayun, Mohammad Shekarchi, and Mahdi Kioumarsi

Contents

1 Introduction

Shrinkage in concrete structures, which can lead to significant deformation and damage, necessitates understanding its time-dependent behaviors for the structures' durability. This shrinkage is categorized into spontaneous and drying shrinkage [1]. Spontaneous shrinkage, occurring without moisture exchange with the environment, is thought to result from self-drying capillary contractions. Some researchers attribute it to internal stress within the concrete's solid framework. On the other hand, drying shrinkage, driven by water loss through evaporation, is linked to capillary actions [2]. Moreover, several shrinkage processes contribute to the overall shrinkage observed in concrete, such as autogenous shrinkage, chemical shrinkage,

P. Khodabandeh (✉) · F. Azarhomayun
School of Civil Engineering, College of Engineering, University of Tehran, Tehran, Iran
e-mail: p.khodabandeh18@ut.ac.ir

M. Shekarchi
Department of Civil Engineering, University of Tehran, Tehran, Iran

M. Kioumarsi
Department of Built Environment, OsloMet – Oslo Metropolitan University, Oslo, Norway

© The Author(s) 2025
M. Kioumarsi, B. Shafei (eds.), *The 1st International Conference on Net-Zero Built Environment*, Lecture Notes in Civil Engineering 237,
https://doi.org/10.1007/978-3-031-69626-8_61

carbonation shrinkage, drying shrinkage, plastic shrinkage, and thermal shrinkage [2–5]. In particular, autogenous shrinkage plays a significant role in the early behavior of high-performance concrete (HPC) and ultra-high performance concrete (UHPC) after they are cast [6–8]. These types of concrete are noted for their enhanced durability and superior mechanical characteristics compared to standard concrete, largely due to their lower water-cement ratio requirements [9]. The scarcity of water available for cement hydration leads to increased self-desiccation within the capillary pores as the relative humidity decreases, causing capillary contraction within the cement matrix [6, 10]. Autogenous shrinkage, occurring under constant temperature conditions, leads to a noticeable reduction in external volume due to these changes, which can result in cracking in the early days or weeks following casting [11–13]. Predictive models are crucial to effectively managing concrete's longevity and performance, accounting for various factors such as environmental conditions, material composition, and loading history. Despite concrete's complex microstructure making accurate modeling challenging, empirical and theoretical models have been developed and used, such as those by the Comite Euro-International du Beton and the ACI 209R-92 model in North America. These models primarily rely on experimental data but are limited in forecasting long-term shrinkage effects [14]. Recent advancements have focused on alkali-activated materials, which exhibit more complex shrinkage behaviors than traditional Portland cement, highlighting the need for innovative predictive methods [15]. The use of non-parametric approaches, particularly artificial intelligence and machine learning techniques like neural networks, has emerged as a promising avenue for accurately predicting concrete's drying shrinkage, especially in long-term scenarios involving alkali-activated materials, where research remains limited.

This research investigates the capability of both traditional and ensemble-based machine learning models to predict drying shrinkage in cementitious materials, including those enhanced with supplementary cementitious components. To facilitate this analysis, a dedicated database was compiled, incorporating data from the NU database along with findings from published studies. The paper elaborates on the theoretical foundations and methodologies employed by the models, with a subsequent evaluation of their predictive outcomes. This study specifically aims to: Assess the predictive accuracy of various machine learning models for drying shrinkage. Identify the most influential factors affecting shrinkage using SHapley Additive exPlanation (SHAP) theory. Quantitatively determine the impact of different features on the accuracy of shrinkage predictions using partial difference plots. The novelty of this research lies in its comprehensive use of SHAP values to decipher the influence of individual features in complex predictive models and its unique approach to integrating multiple types of machine learning strategies to enhance predictive accuracy. These aspects significantly contribute to the field, advancing our understanding of material behavior under drying conditions and improving predictive methodologies

2   Methodology

2.1   Database and Analysis

The required database for predicting self-compacting contractions is collected from published literature and the NU database, totaling 2889 data points extracted from 1 to 28 days. The variables used in predicting self-compacting contractions include 12 inputs comprising the water-to-cement ratio, the water-to-binder ratio, the aggregate-to-cement ratio, height, compressive strength, the volume ratio of chemical composition, the volume-to-surface ratio, temperature, initial measurement time, initial height, cement initial measurement time, water to cement ratio. Therefore, the focus is on creating a prediction model using the abovementioned variables. The data description and the variables' distribution frequency are shown in Table 1.

2.2   Shrinkage Prediction of Typical Machine Learning Model

The Random Forest (RF) technique is a type of ensemble learning used for tasks like classification and regression, leveraging the strengths of bootstrap aggregation (bagging) and the random subspace method. In this process, n bootstrap samples are created by selecting N instances from the training set with replacement, typically with fewer instances and features than the original dataset. Each of these samples forms the basis for a decision tree, where each node within the tree evaluates a specific attribute of the data. The progression through the branches of the tree is determined by the outcomes at each node, leading to a final classification or regression result at the leaves. The unique aspect of RF is how it aggregates the results from all individual trees, with the collective output reflecting the majority vote or average prediction across the forest. This method effectively enhances prediction accuracy and robustness by mitigating overfitting common in single decision trees:

Statistical Measures for Model Evaluation The performance of the predictive model of individual and ensemble learners is evaluated using the mentioned statistical indicators as listed below see Eqs. (1) and (2).

$$\text{MAE} = \frac{1}{n} \sum_{i=1}^{n} |x_i - x| \tag{1}$$

$$\text{RMSE} = \sqrt{\frac{\left(y_{\text{pred}} - y_{\text{ref}}\right)^2}{N}} \tag{2}$$

Table 1 Description of the database utilized in this research

	No	Unit	Min	Q25%	Mean	Median	Q75%	Max	Std	Skw
W/C (water–cement ratio)	1	–	0.15	0.28	0.38	0.35	0.42	1.6	0.16	2.216
a/c (aggregate–cement ratio)	2	–	0.00	1.14	3.08	3.18	4.81	11.56	0.12	0.31
C (Cement content)	3	kg/m^3	167.4	360	472.1	440.0	496	1630	224.4	2.8
AEA (Content of air entraining agent)	4	%Cem	−0.11	0	0.89	0	0	1.5	0.1	11.2
fc28(MPa)	5	MPa	16	45	67	63	85.9	165	29.1	0.7
2VS (volume to surface ratio)	6	mm	25	50	66.79	70	76.2	500	33.5	5.59
H_0 (environmental humidity of specimen preconditioning)	7	%	50	98	94.36	99	101	101	12.53	2.1
t_0 (age at loading)	8	Days	0.42	3	38.21	8	28	625	84.03	8
h (specimen height)	9	mm	10	50	67	65	99	100	24.98	0.18
tt_0 (exposure time for shrinkage)	10	Days	0	14	235.67	65.85	245	8960	586.87	7.23
Time of measurement	11	Days	1	2	9.5	6.8	14	28	9.6	1
Temperature	12	°C	18	20	28.44	20	23	130	18.58	2.74
Shrinkage/swelling		μm	−1988	−260	−193	−123	−31.5	79	247.897	−2.98

$n, N =$ number of data samples
$xi, y_{pred} =$ predicted data sets
$x, y_{ref} =$ experimental or reference data sets

2.3 Random Forest

The term "Random Forest," also known as "Random Decision Forests" and "Randomized Trees," denotes an ensemble approach in machine learning that leverages multiple decision trees to tackle various regression and classification (DT) problems. Moreover, a random forest consists of a series of distinct DTs, all independent of each other. Breiman [16] illustrated that the random forest method boasts strong generalization capability. Random forests offer a flexible framework that enables the selection of task-specific objective functions and encompasses a broad range of separation function classifications and posterior models. The count of trees and their depth in a random forest are significant hyperparameters. With an increase in the number of trees, predictions may become more accurate, leading to a steady decrease in prediction error [17]. Figure 1 shows a prediction of dry shrinkage. Figure 1 demonstrates the forecast of dry shrinkage using the Random Forest method, revealing that this approach yields reasonably accurate outcomes with minimal variance between the observed and predicted values. Furthermore, the effectiveness of the model is assessed through the coefficient of determination (R^2) and statistical evaluation using the mean absolute error (MAE) and the root mean square error (RMSE). The regression analysis presented in Fig. 1 indicates the model's strong performance, as evidenced by an R^2 value of 0.89.

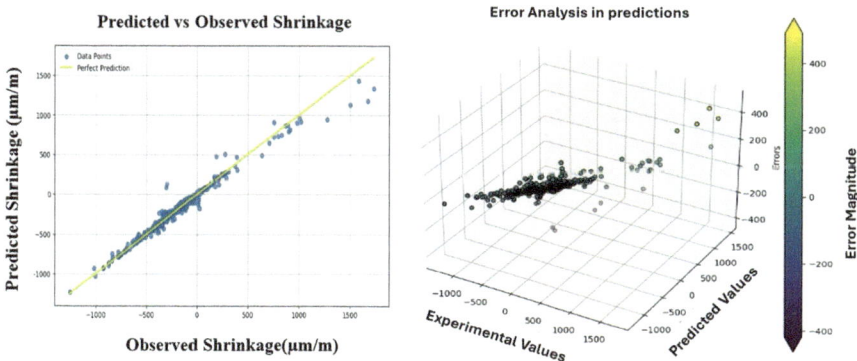

Fig. 1 Random forest; (**a**) regression analysis with the experimental and predicted result; (**b**) errors of experimental and targeted via RF

2.4 Feature Importance Analysis Using Shapley Additive Explanations (SHAP)

In the current study, SHAP value diagrams were utilized to examine the relative importance and impact of various features on the predictions of our advanced model. This diagram, based on concepts derived from game theory and principles of equitable distribution, allows the analyst to understand how each specific feature, individually and in relation to other features, influences the outcome of the model. In this diagram, features are ordered from top to bottom based on their net effect on the model's output. Each feature is represented by a collection of colored points indicating the positive and negative SHAP values. Red points represent higher feature values that positively impact the model's output, while blue points indicate lower feature values that may have a negative or lesser impact. The feature "H" at the top of the diagram is identified as the most impactful, while the feature "c" at the bottom has the least impact. The horizontal axis of the diagram displays the SHAP values, and the distribution of points around the zero line indicates the type of impact. Features with points leaning toward the right of the zero line have had a more positive influence on the model's output, and this influence increases as the points move further from the zero line. Conversely, points to the left of the zero line indicate a negative influence on the predictions. This information can be used to optimize models and develop more accurate decision-making strategies. By carefully analyzing the provided SHAP value diagram, it becomes apparent that features like "H" and "CCEB" are highly strategic and should be given special attention in modeling-related decisions. Ultimately, as shown in Fig. 2, this diagram allows us to move toward creating more transparent and interpretable predictive models with a deeper understanding of the impact of model features.

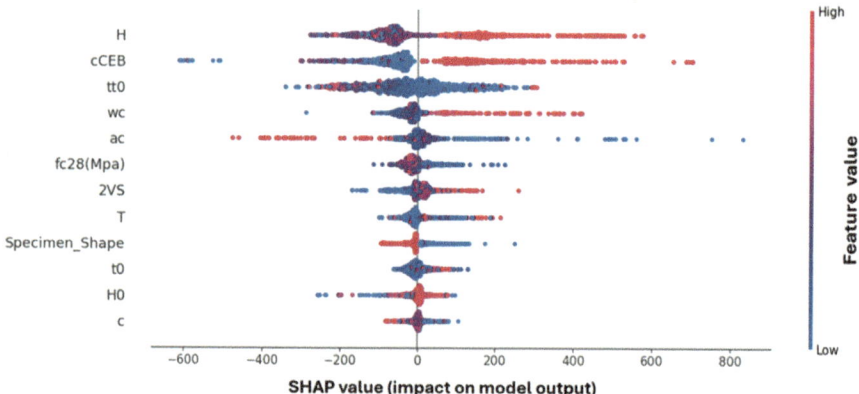

Fig. 2 Feature importance analysis using the Shape library in Python code

3 Conclusion

Machine learning models have increasingly shown their prowess in achieving high accuracy levels, rivaling those of neural networks as reported in existing scholarly literature. This is particularly true in the context of forecasting autogenous or drying shrinkage, where precision in prediction is critical. Among the various models explored, the Random Forest (RF) machine learning model stands out for its superior performance:

Performance Comparison: The RF model not only matched but also surpassed the accuracy of previously studied models, reaffirming the effectiveness of machine learning techniques in this domain.

Accuracy Achievements: In both training and testing phases, the RF model demonstrated high reliability, achieving a coefficient of determination (R^2) of 0.89. This indicates a strong predictive power, with 89% of the variability in shrinkage being explained by the model.

Implications: These results underscore the potential of using advanced machine learning models like the RF model for more accurate predictions of concrete behaviors, which is essential for the engineering and construction industries.

By harnessing sophisticated machine learning approaches such as the RF model, we can enhance our predictive capabilities and improve outcomes in material science applications.

Acknowledgments This work is a part of the TRANSFORM project, which has received funding from the Norwegian Directorate for Higher Education and Competence (HK-dir), project number UTF-2020/10107.

References

1. Azarhomayun, F., Haji, M., Kioumarsi, M., Shekarchi, M.: Effect of calcium stearate and aluminum powder on free and restrained drying shrinkage, crack characteristic and mechanical properties of concrete. J. Cem. Concr. Compos. **125**, 104276 (2022)
2. Kioumarsi, M., Azarhomayun, F., Haji, M., Shekarchi, M.: Effect of shrinkage reducing admixture on drying shrinkage of concrete with different w/c ratios. Materials. **13**(24), 5721 (2020)
3. Khan, K., Jalal, F.E., Khan, M.A., Salami, B.A., Amin, M.N., Alabdullah, A.A., Samiullah, Q., Arab, A.M.A., Faraz, M.I.: Iqbal, prediction models for evaluating resilient modulus of stabilized aggregate bases in wet and dry alternating environments: ANN and GEP approaches. Materials. **15**, 4386 (2022)
4. Ren, G., Yao, B., Ren, M., Gao, X.: Utilization of natural sisal fibers to manufacture eco-friendly ultra-high-performance concrete with low autogenous shrinkage. J. Clean. Prod. **332**, 130105 (2022)
5. Li, S., Mo, L., Deng, M., Cheng, S.: Mitigation on the autogenous shrinkage of ultra-high-performance concrete via using MgO expansive agent. Constr. Build. Mater. **312**, 125422 (2021)

6. Yio, M.H.N., Mac, M.J., Yeow, Y.X., Wong, H.S., Buenfeld, N.R.: Effect of autogenous shrinkage on microcracking and mass transport properties of concrete containing supplementary cementitious materials. Cem. Concr. Res. **150**, 106611 (2021)
7. Sun, Y., Yu, R., Shui, Z., Wang, X., Qian, D., Rao, B., Huang, J., He, Y.: Understanding the porous aggregates carrier effect on reducing autogenous shrinkage of Ultra-High-Performance Concrete (UHPC) based on response surface method. Constr. Build. Mater. **222**, 130–141 (2019)
8. Roberti, F., Cesari, V.F., de Matos, P.R., Pelisser, F., Pilar, R.: High- and ultra-high-performance concrete produced with sulfateresisting cement and steel microfiber: autogenous shrinkage, fresh-state, mechanical properties and microstructure characterization. Constr. Build. Mater. **268**, 121092 (2021)
9. Kheir, J., Klausen, A., Hammer, T.A., De Meyst, L., Hilloulin, B., Van Tittelboom, K., Loukili, A., De Belie, N.: Early age autogenousshrinkage cracking risk of an ultra-high-performance concrete (UHPC) wall: modelling and experimental results. Eng. Fract. Mech. **257**, 108024 (2021)
10. Ashfaq, M., Iqbal, M., Khan, M.A., Jalal, F.E., Alzara, M., Hamad, M., Yosri, A.M.: GEP tree-based computational AI approach to evaluate unconfined compression strength characteristics of Fly ash treated alkali contaminated soils. Case Stud. Constr. Mater. **17**, e01446 (2022)
11. Yang, L., Shi, C., Wu, Z.: Mitigation techniques for autogenous shrinkage of ultra-high-performance concrete—a review. Compos. Part B Eng. **178**, 107456 (2019)
12. Farooq, F., Jin, X., Faisal Javed, M., Akbar, A., Izhar Shah, M., Aslam, F., Alyousef, R.: Geopolymer concrete as sustainable material: a state of the art review. Constr. Build. Mater. **306**, 124762 (2021)
13. Tang, S., Huang, D., He, Z.: A review of autogenous shrinkage models of concrete. J. Build. Eng. **44**, 103412 (2021)
14. Shariati, M., Saeed Mafipour, M., Ghahremani, B., Azarhomayun, F., Ahmadi, M., Thoi Trung, N., Shariati, A.: A novel hybrid extreme learning machine–grey wolf optimizer (ELM-GWO) model to predict compressive strength of concrete with partial replacements for cement. J. Eng. Comput. **38**, 757–779 (2022)
15. Shekarchi, M., Ahmadi, B., Azarhomayun, F., Shafei, B., Kioumarsi, M.: Natural zeolite as a supplementary cementitious material–a holistic review of main properties and applications. J. Constr. Build. Mater. **409**, 133766 (2023)
16. Bierman, L.: Random forests. Mach. Learn. **45**, 5–32 (2001)
17. Aslam, F., Elkotb, M.A., Iqtidar, A., Khan, M.A., Javed, M.F., Usanova, K.I., Khan, M.I., Alamri, S., Musarat, M.A.: Compressive strength prediction of rice husk ash using multiphysics genetic expression programming. Ain Shams Eng. J. **13**, 101593 (2022)

Reducing Carbon Footprint of RC Structure in Saline Exposure: Bangladesh Perspective

Nazmus Sakib Pallab (iD), Mahin Sultana (iD), Saadman Sakib (iD),
Amrita Barua (iD), and Tanvir Manzur (iD)

Contents

1 Introduction

In recent years, the NetZero concept, that is, balancing the equilibrium between the amount of greenhouse gases (GHGs) produced and expelled from the atmosphere, has emerged as the primary framework for climate action, with countries and organizations setting ambitious targets. To limit increases in global temperatures to less than 2 °C, or even 1.5 °C, leaders worldwide committed to balancing greenhouse gas (GHG) emissions from human activities by the second half of the century under the Paris Agreement [1]. Achieving net-zero emissions involves sources and removals of GHGs that are in equilibrium, although the specifics can vary in terms of coverage, timing, and aggregation [2]. From this perspective, it is imperative to decrease CO_2 levels to halt the progression of global warming. The target of reaching NetZero within 2050 to mitigate severe climate effects rests on the

N. S. Pallab · M. Sultana · S. Sakib · A. Barua · T. Manzur (✉)
Department of Civil Engineering, Bangladesh University of Engineering & Technology, Dhaka, Bangladesh
e-mail: tanvirmanzur@ce.buet.ac.bd; 1804113@ce.buet.ac.bd

© The Author(s) 2025
M. Kioumarsi, B. Shafei (eds.), *The 1st International Conference on Net-Zero Built Environment*, Lecture Notes in Civil Engineering 237,
https://doi.org/10.1007/978-3-031-69626-8_62

widespread implementation of technologies that don't contribute to CO_2 emissions. However, the success of this adoption significantly relies on innovating and coordinating efforts within the construction sector to optimize concrete mixtures [3].

A significant durability risk in coastal or marine environments is the corrosion of embedded reinforcement caused by chloride intrusion in reinforced concrete (RC) [4]. The movement of chloride ions within the pore solution substantially diminishes the longevity of susceptible systems [5]. Therefore, comprehending the ionic transport properties of concrete and their impact on service life is of paramount research importance for ensuring durable concrete. Traditional methods for determining chloride diffusion coefficients (D_{rcm}) typically involve destructive core extraction methods, which are costly and time-consuming and can compromise structural integrity. Alternatively, electrical resistivity measurements offer a promising, quick, and non-destructive method to evaluate ion transport in concrete, as both parameters are influenced by the same factor, that is, concrete microstructure [5]. The splash zone is typically the worst place for attacks on concrete, whereas the submerged zone is less vulnerable [6]. Unfortunately, there is a serious risk to the safety of RC structures in coastal locations in Bangladesh due to the absence of locally focused mix design regulations that prioritize primarily strengths. Consequently, restoration of deteriorated RC structures in corrosion-prone areas may demand a notably large financial allowance at later stages of their service life [7, 8]. In this view, to improve the life cycle of marine concrete structures, it is crucial to verify concrete mix designs and assess chloride-induced corrosion risks.

This research has used varying proportions of cement blends, that is, both commercially available (CEM II and CEM III) and customized (mixture of CEM I and class F fly ash/blast furnace slag), to quantitatively assess concrete durability enhancement against chloride ingression in RC structures, leading to increased service life values in saline environments. The longer it takes for corrosion-induced cracks to appear, the fewer repairs will be required during the structure's service life. This eventually results in reduced CO_2 emissions from cement production over time. In addition, the life cycle cost of a typical RC wall made of different blended mixes of the study and exposed to chloride-induced corrosion has been estimated considering 100 years of design life. This research also includes an estimation of CO_2 emissions for different binders used to compare and identify best-performing mix, prioritizing durability and repair frequency. The finding of the limited study performed would assist in recommending policies and guidelines for the local construction supply chain to reduce the carbon footprint of RC construction work exposed to salinity and eventually, contribute towards NetZero achievement.

2   Materials and Methodology

This research has utilized chloride migration coefficients of concrete mixes, determined according to the Nernst-Einstein equation [9], to evaluate the service life and likelihood of RC wall corrosion in the Life-365 software [10] exposed to severe saline exposure. The primary focus has been on binder types, as the presence of pozzolanic characteristics in the binder directly influences the permeability of the resulting concrete, affecting the time for chloride ions to reach the embedded reinforcement. Additionally, an evaluation has been conducted to ascertain the CO_2 emissions associated with binders used in constructing and repairing RC walls.

2.1 Mix Design

This study considers five mixes to represent variation in the mix proportion of supplementary cementitious materials (SCMs) having a water-cement ratio corresponding to particular design strengths. As per the ACI Mix Design manual [11], a slump value of 75–100 mm has been considered to represent the commonly used workability of concrete mixes. The concrete mixtures incorporated single-type fine aggregate variation, which is, in this case, local sand with a Fineness Modulus (FM) of 2.5. Table 1 shows the amount of different ingredients of mixes used, where FA and S denote class F fly ash and slag, respectively. The numerical values, that is, 30FA means 30% fly ash, 20FA means 20% fly ash, and 30S means 30% slag.

The tests were performed on laboratory-cast cylindrical concrete specimens prepared using commonly used and available stone aggregate, with gradation limited to the 19 mm downgraded particle size. The water-binder ratio (w/b) has also been kept constant at 0.39, corresponding to a design strength of 35 MPa. The main types of binders used were CEM I, CEM II-B, and CEM III-A, and two blended mixes having different proportions of class F fly ash and slag mixed with CEM I. The chemical compositions of different binders and SCMs (obtained through XRF analysis) are provided in Table 2. In the case of mixes that cannot achieve slump value, a very small amount of additional water or plasticizing admixture (in case water addition does not increase slump) has been added to achieve the target slump, keeping the water-binder ratio constant.

Table 1 Composition of different concrete mixes considered in the study

Mixture	w/b	Cement (kg/m^3)	Fly ash (kg/m^3)	Slag (kg/m^3)	Coarse aggregate (kg/m^3)	Fine aggregate (kg/m^3)
CEM I	0.39	526	–	–	1048	630
CEM II		526	–	–	1048	612
CEM III		526	–	–	1048	618
CEM I + 30FA		368	158	–	1048	565
CEM I + 20FA + 30S		263	105	158	1048	580

Table 2 Chemical composi-
tions of different binders
and SCMs

Analyte	Mass %				
	CEM I	CEM IIB	CEM III-A	FA	Slag
CaO	68.24	45.40	54.95	1.14	45.86
SiO_2	17.02	26.57	23.82	68.12	31.35
Al_2O_3	2.53	10.40	8.59	18.00	8.52
Fe_2O_3	4.17	3.13	1.97	5.40	1.40
MgO	1.57	5.78	3.58	0.59	8.76
SO_3	4.51	2.48	2.62	0.20	1.76
Na_2O	0.25	–	–	0.08	0.26
K_2O	0.59	2.87	1.13	2.45	0.74

2.2 Estimation of Chloride Diffusion Coefficient (D_{rcm})

Concrete Surface Resistivity A concrete surface resistivity test has been conducted to evaluate the chloride diffusion coefficient (D_{rcm}). The standard techniques to assess surface resistivity are two-point uniaxial and four-point techniques [12]. For this study, the four-point technique has been used. There are four electrodes evenly distributed in a linear arrangement. When the external electrodes apply an alternating current of intensity, I, to the concrete, the two inside electrodes detect electric potential, V [13]. Proceq Resipod has been used to obtain the resistivity values. Correction factors for geometric, temperature, and degree of saturation have been applied to obtain the corrected resistivity from the Proceq Resipod reading. The details of correction factors can be found elsewhere [5].

Chloride Diffusion Coefficient The chloride diffusion coefficient (D_{rcm}) was calculated using the correlation between D_{rcm} and concrete surface resistivity. Andrade et al. [14] suggested a correlation borrowed from the Nernst-Einstein equation.

$$D_{eff} = \frac{12 \times 10^{-5}}{\rho} \tag{1}$$

Where ρ is the resistivity of concrete, and D_{eff} is the effective diffusion coefficient.

2.3 Estimation of Service Life

A life cycle analysis of an exposed reinforced concrete (RC) wall (built of specified concrete mixtures of this study) with a 62.5 mm thick cover against corrosion from chloride ingress has been carried out using the Life-365 [10] environment. The Life-365 uses the migration coefficient of any concrete mix to compute chloride diffusion within the concrete by applying Fick's second law. According to the model's recommendation, time-dependent changes in the mixes' migration coefficients

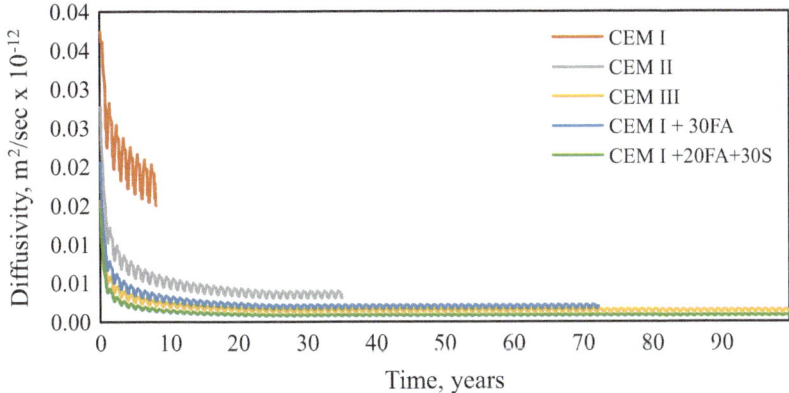

Fig. 1 Diffusivity variation of mixes with time

have been considered using a decay index, m [10]. With a maximum replacement level limit of 70% for slag and 50% for fly ash, the values of m have been determined by the percentage of SCMs in the binder. In order to ensure that the chloride ions reach the implanted rebars after covering the whole cover distance, Life-365 offers time-dependent modifications of migration coefficient values. Figure 1 illustrates the variation in migration coefficients of the mixes with time, as obtained from Life-365, for a 62.5 mm cover. It is evident from Fig. 1 that the incorporation of SCMs with a relatively higher amount of active silica has a considerable impact on the long-time pore refinement within the concrete matrix. As a result, the diffusivity of chloride ions through concrete pores reduces significantly.

As the exposure class, the marine splash zone, namely, XS3 [15], has been considered. The maximum concentration of chloride on the concrete surface and critical chloride content required for corrosion initiation has been assumed to be 1.0 and 0.05 wt. % of concrete, respectively. It has been assumed that maximum surface concentration occurred instantaneously due to the considered exposure class [10]. The concrete exposure temperature was determined by taking the monthly fluctuations in the average temperature of the coastal city of Cox's Bazaar, Bangladesh. The combined duration for both corrosion initiation and propagation is commonly employed to determine the service life of an RC structure affected by chloride-induced corrosion. The propagation phase is assumed to remain constant for a particular rebar type and is set at 5 years in this instance. This study considers the sum of corrosion initiation and propagation period to compare the service life values of different binders.

Following corrosion initiation and propagation, each interval of the corrosion propagation period includes one repair. Mixtures with extended service life are less frequently in need of repairs, thus fostering environmental sustainability. Furthermore, the life cycle cost due to chloride-induced corrosion has also been evaluated in cumulative present dollar (USD) value using Life-365 [10], factoring in inflation and discount rates. It includes the summation of concrete and reinforcement costs, repair costs, and their cycles. The concrete and reinforcement costs have been estimated at USD 165/m^3 and USD 1.05/kg, respectively, and the repair cost for the RC wall has been considered as USD 190/m^2 [16]. Inflation and discount rates were approximated at 5% and 4%, respectively, to assess the impact of binding material selection on marine RC structure [16]. The repair cost for the RC wall was assumed to be constant, with a repair cycle of 5 years and a 10% repair area for each cycle. The life cycle cost was projected for 100 years of service, with a base year of 2022.

2.4   Estimation of Embodied Carbon Dioxide (CO_2) Emission

As mentioned earlier, five different types of binders have been examined for their impact on emissions. Only the CO_2 emissions from the CEM I portion have been calculated, while fly ash and slag within the blended mixes were considered waste by-products with zero CO_2-e contribution from raw materials. The construction mix ratio has been adopted according to the ACI Mix Design manual [11]. An arbitrary RC wall having dimensions of 10.0 m in length, 3.0 m in height, and a total thickness of 250 mm, with a concrete cover of 62.5 mm, has been considered for calculation. During construction, the CO_2 emissions from cement usage have been calculated based on the mix ratio and the quantity required according to the dimensions of the structure. Additionally, it has been assumed that 10% of the total area of the wall would undergo repair after each repair cycle. For repair purposes, non-shrink grout has been chosen as the material of choice. Non-shrink grout is a hydraulic cement grout composed of hydraulic cement, sand, and proprietary admixtures. Following industrial practices and ASTM specifications, this study has employed a grout mixture comprising 80% Ordinary Portland Cement (OPC) by weight. This ratio falls within the typical range of 70 to 90% OPC commonly used. Additionally, a grout-to-water ratio of 50:7 has been considered as per ASTM specifications. Also, the unit weight of non-shrink grout, as per ASTM specifications, typically ranges from 2240 to 2400 kilograms per cubic meter (kg/m^3). For this study, a unit weight of 2300 kg/m^3 for non-shrink grout has been adopted aligning with ASTM specifications.

After calculating the volume of concrete to be used for construction and repair from the dimensions, the weight of cement to be used has been estimated from its unit weight. Equation (2) has been used to calculate the embodied CO_2 emissions of the cement used for construction and repair.

$$E = \sum M \times \text{EF} \tag{2}$$

where E is the total embodied GHG emissions of cement (in tons CO_2-e) (CO_2-e: CO_2 equivalent), M is the amount of particular building material (in tons), and EF is the GHG emission factor for that building material (in ton CO_2-e/ton) [17–19]. The value of EF for CEM I has been used as 0.913 tons CO_2-e/ton without considering emission during transportation [19].

3 Result and Discussion

This section presents and discusses the impact of different binder types on the initiation period of corrosion in typical RC wall construction in saline exposure. Additionally, it compares repair frequency and cumulative CO_2 emission for alternative binder options.

Table 3 presents resistivity values measured using the Proceq Resipod and corresponding diffusion coefficients for different binder types. These data demonstrate that resistivity values rise with the addition of SCMs, while the D_{rcm} values exhibit an inverse correlation as anticipated.

3.1 Corrosion Initiation Period

Figure 2 illustrates the time required for corrosion initiation of various mixes for a typical RC wall construction in extreme saline exposure. Binders incorporating CEM III (having around 45% SCMs) and CEM I with a blend of 30% slag and 20% fly ash experienced significantly prolonged corrosion initiation times as compared to other mixes. This behavior is attributed to the pozzolanic characteristics of active silica that facilitate the conversion of $CaOH_2$ (initial hydration products with less strength potential) into a relatively stable secondary CSH gel through pozzolanic reactions [20, 21]. Such pore refinement due to delayed secondary CHS formation hinders the pore interconnectivity.

3.2 Repair Frequency and Life Cycle Cost

The durability of concrete structures in marine environments is determined by three factors: the time to corrosion initiation (CIT), the time to crack initiation, and the time required for crack propagation [22]. The initial repair is usually scheduled at the end of the service life, followed by subsequent maintenance at particular intervals, that is, every 5 years in this case as discussed above. Table 4 outlines the number of repair cycles initiated for CEM I, CEM II, and CEM I with a blend of 30% fly ash as 18, 12, and 5 times, respectively. Conversely, CEM III (with around 45% SCMs,

Table 3 Resistivity and diffusion coefficient values of different binder types

Design strength	Mixture	w/b	Resistivity(Ω-m)	D_{rcm} (m²/s 10^{-12})
35 MPa	CEM I	0.39	39.1	3.1
	CEM II		52.9	2.3
	CEM III		92.5	1.3
	CI + 30FA		70.5	1.7
	CI + 20FA + 30S		103.3	1.2

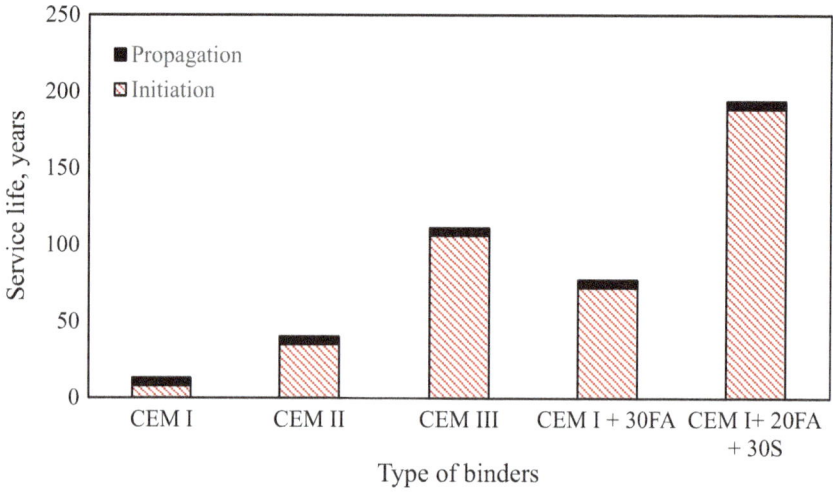

Fig. 2 Corrosion initiation and propagation time (obtained from Life-365) of different types of binders

Table 4 Service life and repair frequency of different types of binder

Types of binder	Initiation (years)	Propagation (years)	Service life (years)	No. of repair cycles required
CEM I	8.6	5	13.3	18
CEM II	37.8	5	42.8	12
CEM III	100.0+	5	100.0+	0
CI + 30FA	72.2	5	77.2	5
CI + 20FA + 30S	100.0+	5	100.0+	0

primarily slag) and CEM I with a blend of 30% slag and 20% fly ash, required no repairs during the 100-year analysis period.

Figure 3 illustrates the ratio of life cycle cost (LCC) to initial construction cost for typical RC wall construction with a 62.5 mm cover, analyzed over 100 years under extreme marine exposure conditions. It is apparent from Fig. 3 that the present value multiplier of life cycle costs is notably lower for CEM III and CEM I with 30% slag and 20% fly ash, compared to other mixes. This difference can be attributed to the considerably more extended initiation period experienced by these mixes.

3.3 Environmental Sustainability

Extending the service life of marine RC elements reduces the need for frequent repairs, thereby lowering the overall CO_2 emissions. Additionally, allocating budgets for periodic maintenance is economically not feasible. Figures 4 and 5 illustrate

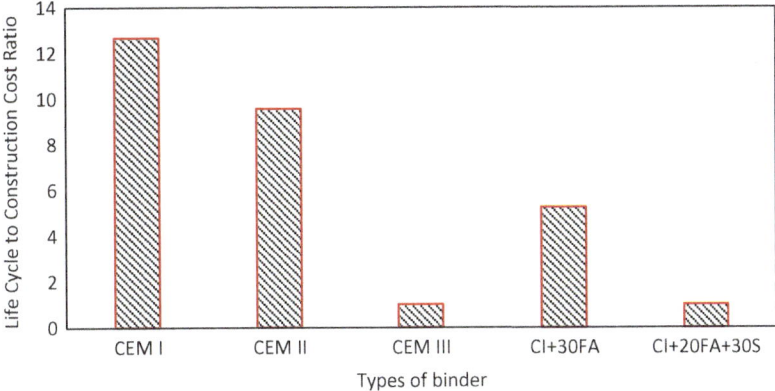

Fig. 3 Life cycle cost to construction cost ratio for different types of binder

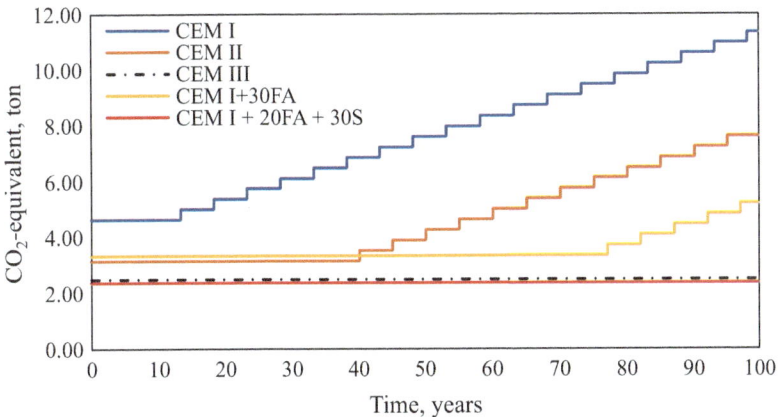

Fig. 4 Cumulative equivalent CO_2 emission of different types of binder

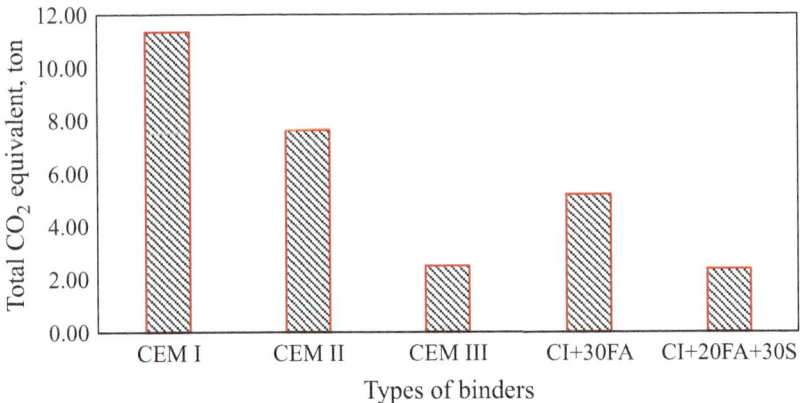

Fig. 5 Total equivalent CO_2 emission of different types of binder

that the total CO_2 emissions remain nearly constant when integrating higher SCMs into binders, aligning closely with constructional requirements. This highlights the potential of SCMs with pozzolanic characteristics as a significant means to work towards the NetZero goal by reducing the carbon footprint of concrete under extreme exposure.

4 Conclusions

The study highlights the significance of utilizing blast furnace slag and class F fly ash as SCMs into blended cement to enhance durability, particularly by ensuring resistance to chloride penetration. This approach ultimately aims to lower the LCC and carbon footprint of any reinforced concrete structures exposed to extreme marine conditions in Bangladesh. For instance, concrete prepared with CEM III and a customized mix having a blend of CEM I and 20% fly ash + 30% slag showed significantly reduced LCC and total CO_2 emissions in comparison with other mixes used in this study. This research, therefore, quantitatively indicates that incorporating a cement blend with higher proportions of fly ash and/or blast furnace slag can decrease CO_2 emissions significantly by extending service life. However, the findings are based on a limited study involving only five concrete mixes. Therefore, future studies with higher numbers of representative local mixes focusing on practical applications would definitely yield substantial benefits, emphasizing the need for further comprehensive research to establish more precise guidelines for Bangladesh.

Acknowledgments The authors extend their sincere appreciation to the diligent staff of the Concrete Laboratory at the Department of Civil Engineering, BUET, for their invaluable assistance in conducting the tests. The financial support for this research has been provided by the Research and Innovation Centre for Science and Engineering (RISE), BUET (Project ID: 2022-01-016).

References

1. Rogelj, J., Geden, O., Cowie, A., Reisinger, A.: Net-zero emissions targets are vague: three ways to fix. Nature. **591**(7850), 365–368 (2021). https://doi.org/10.1038/d41586-021-00662-3
2. Bistline, J.E.T.: Roadmaps to net-zero emissions systems: emerging insights and modeling challenges. In: Joule, vol. 5, Issue 10, pp. 2551–2563. Cell Press (2021). https://doi.org/10.1016/j.joule.2021.09.012
3. Davis, S.J., Lewis, N.S., Shaner, M., Aggarwal, S., Arent, D., Azevedo, I.L., Benson, S.M., Bradley, T., Brouwer, J., Chiang, Y.M., Clack, C.T.M., Cohen, A., Doig, S., Edmonds, J., Fennell, P., Field, C.B., Hannegan, B., Hodge, B.M., Hoffert, M.I., et al.: Net-zero emissions energy systems. In: Science, vol. 360, Issue 6396. American Association for the Advancement of Science (2018). https://doi.org/10.1126/science.aas9793
4. Baten, B., Manzur, T., Ahmed, I.: Combined effect of binder type and target mix-design parameters in delaying corrosion initiation time of concrete. Constr. Build. Mater., Elsevier. **242** (2020). https://doi.org/10.1016/j.conbuildmat.2020.118003

5. Baten, B., Manzur, T.: Formation factor concept for non-destructive evaluation of concrete's chloride diffusion coefficients. Cem. Concr. Compos. **128** (2022). https://doi.org/10.1016/j.cemconcomp.2022.104440
6. ENG-TIPS.COM: Engineering Forum; https://www.eng-tips.com/viewthread.cfm?qid=451608#google_vignette. Last accessed: 23/02/24
7. Magnat, P.S., Ojedokun, O.O., Lambert, P.: Chloride-initiated corrosion in Alkali activated reinforced concrete. Cem. Concr. Compos. **115**, 103823 (2021). https://doi.org/10.1016/j.cemconcomp.2020.103823
8. Vu, K.A.T., Stewart, M.G.: Structural reliability of concrete bridges including improved chloride-induced corrosion models. Struct. Safety, Elsevier. **22**(4), 313–333 (2000). https://doi.org/10.1016/S0167-4730(00)00018-7
9. Yuan, Q.: Fundamental Studies on Test Methods for the Transport of Chloride Ions in Cementitious Materials. Ghent University (2008)
10. Life-365: Service Life Prediction Model and Computer Program for Predicting the Service Life and Life-Cycle Cost of Reinforced Concrete Exposed to Chlorides, Version 2.2.3 (2020)
11. ACI 211: Standard Practice for Selecting Proportions for Normal Heavyweight, and Mass Concrete. American Concrete Institute, Michigan (1991)
12. Layssi, H., Ghods, P., Alizadeh, A.R., Salehi, M.: Electrical resistivity of concrete. Concr. Int. **37**(5), 41–46 (2015)
13. Azarsa, P., Gupta, R.: Electrical resistivity of concrete for durability evaluation: a review. Adv. Mater. Sci. Eng. **2017** (2017). https://doi.org/10.1155/2017/8453095
14. Andrade, C., Alonso, C., Arteaga, A., Tanner, P.: Methodology based on the electrical resistivity for the calculation of reinforcement service life. In: 5th Canmet/ACI Int. Conference on Durability of Concrete–Supplementary Papers volume, pp. 899–915, Barcelona, Spain (2000)
15. McCarter, W.J., Linfoot, B.T., Chrisp, T.M., Starrs, G.: Performance of concrete in XS1, XS2 and XS3 environments. Mag. Concr. Res. **60**(4), 261–270 (2008). https://doi.org/10.1680/macr.2008.60.4.261
16. Manzur, T., Noor, M.A., Torsha, T.: Reducing chloride induced corrosion risk and associated life cycle cost of marine RC structure: Bangladesh perspective. In: Gupta, R., et al. (eds.) Proceedings of the Canadian Society of Civil Engineering Annual Conference 2022. CSCE 2022. Lecture Notes in Civil Engineering, vol. 359. Springer, Cham (2024). https://doi.org/10.1007/978-3-031-34027-7_67
17. Rakib, F.H., Anowar, M.R.S.: Assessment of greenhouse gas emissions in building construction: a case study of SWC building at KUET in Bangladesh. J. Constr. Eng. Manag. Innov. **2**, 215–229 (2019)
18. Carbon dioxide equivalent (CO_2-e) emissions: a comparison between geopolymer and OPC cement concrete. Constr. Build. Mater. **43**, 125–130 (2013). https://doi.org/10.1016/j.conbuildmat.2013.01.023
19. Sanal, I.: Discussion on the effectiveness of cement replacement for carbon dioxide (CO_2) emission reduction in concrete. Origin. Res. Article. **8**, 1–13 (2017)
20. Mahmud, S., Manzur, T., Samrose, S., Torsha, T.: Significance of properly proportioned fly ash based blended cement for sustainable concrete structures of tannery industry. Structures, Elsevier. **29**, 1898–1910 (2021). https://doi.org/10.1016/j.istruc.2020.12.065
21. Baten, B., Manzur, T., Torsha, T., Alam, S.: A parametric study on the graphical approach to assess corrosion vulnerability of concrete mixes in chloride environment. Constr. Build. Mater., Elsevier. **309** (2021). https://doi.org/10.1016/j.conbuildmat.2021.125115
22. Liu, Y., Presuel-Moreno, F.J., Paredes, M.A.: Determination of chloride diffusion coefficients in concrete by electrical resistivity method. ACI Mater. J. **112**(5), 631–640 (2015). https://doi.org/10.14359/51687777

Upcycling of Single-Use Pallet Wood to Cross-Laminated Timber

Niels H. Vonk, Jan Niederwestberg, Ron W. A. Oorschot, and Jan de Jong

Contents

1 Introduction

Cross-laminated timber (CLT) is an engineered wood product panel that is built from a minimum of three wood or wood-based material layers that are glued together [1]. Commonly, within CLT, the fiber direction of neighboring layers alters, usually by 90° between adjacent layers. Since symmetrical build-up is desired and the fiber orientation of neighboring layers alters, most CLT panels have an odd number of layers. As a result of the glued CLT lay-up, the panels show high in-plane rigidity and the ability to transfer loads in both panel directions when loaded out-of-plane. Consequently, CLT can be used as walls and flooring in various construction projects [1]. CLT is a proven, versatile building element that can be used for constructing multi-story houses while yielding a significantly lower environmental impact compared to other building elements based on concrete and steel [2]. Hence, the worldwide demand for CLT is rising [3].

N. H. Vonk (✉) · J. Niederwestberg · R. W. A. Oorschot · J. de Jong
TNO – Building Materials and Structures, Delft, The Netherlands
e-mail: niels.vonk@tno.nl

© The Author(s) 2025
M. Kioumarsi, B. Shafei (eds.), *The 1st International Conference on Net-Zero Built Environment*, Lecture Notes in Civil Engineering 237,
https://doi.org/10.1007/978-3-031-69626-8_63

In the production of CLT, wood is harvested, and planks are sawn and dried before the planks undergo strength grading [4]. Afterwards, planks meeting the required strength and stiffness criteria are finger-jointed together by the use of adhesive, forming a theoretically endless wooden plank. Subsequently, the endless plank is planned to the desired thickness and cut to length. Afterward, an adhesive is applied to the wide face of the planks before they are arranged in the required CLT lay-up and are then pressed. In the end, the CLT undergoes a finishing process in which the panels are cut to size.

During its growing phase, wood captures and binds carbon dioxide (CO_2), which is later released during natural decomposition or incineration (for energy). Thereby, wood that is used in the production of (long-term) goods (e.g., building structures), replacing other materials with a higher CO_2 demand, especially materials that are non-carbon-binding, contributes to the reduction of emissions. Within most European countries, the majority of the "waste" wood is incinerated for the production of energy after the end of the service life of the wooden product [5]. Ideally, wood should be reclaimed and reused after its initial service life, resulting in prolonging the storage of the stored CO_2 within the material.

Reusing reclaimed wooden products poses some key challenges. Structural grading (e.g., based on visual grading [6], X-ray CT scanning [7, 8], and ultrasound techniques [9]) is instrumental for re-use. At the time of writing this work, no grading standard for the regrading of reclaimed wood was in place. Efforts to publish such a standard are ongoing within Europe. Furthermore, reclaimed wooden products often hold metal fasteners at random locations, which must be removed before planing and (re-)shaping. Interestingly, the locations of the nails used to assemble single-use pallets are well-defined [10], thereby simplifying the issue of fastener location detection and, possibly, enabling automated fastener removal. In 2016, approximately 4 billion single-use pallets were in rotation within Europe [11], indicating sufficient supply for possible scaled-up re-use. Therefore, in this work, the feasibility of upcycling reclaimed pallet wood into CLT panels is studied.

An exploratory study revealed that CLT produced from solely reclaimed wood did not comply with the strength requirements in the ANSI standards [12]. To progress on this study, we produced CLT from combinations of virgin and reclaimed wood, in different layups, and their mechanical performance was directly compared with commercial (virgin) CLT. Incorporation of reclaimed materials in the cross-layers should have little influence on the overall bending properties (along the direction of the outer, longitudinal, layers) of these hybrid CLT panels. However, this incorporation may compromise the panel's rolling shear properties. Hence, to study these phenomena, (i) the feasibility of adopting common production processes on the prototype level and (ii) the bending- and shear strength, moduli, and failure modes of the panels, similar to [13–17], are assessed. Since grading regulations for the re-use of reclaimed wood are currently unavailable and the number of test specimens is relatively limited, the results are indicative of evaluating the general feasibility of re-using reclaimed wood for use in CLT. Nevertheless, the results indicate the potential for adequate re-application of reclaimed wood in CLT, thereby extending the carbon storage duration.

In the following, the production process of the CLT panels is described. Then, the considered bending and shear testing methods are given. Afterward, the results are presented and discussed, and, at last, the conclusions are drawn.

2 Materials and Methods

2.1 CLT Panel Lay-Up and Production

Overall, five CLT specimens were manufactured, acquired, and tested, consisting of both new and reclaimed (pallet) wood layers. All CLT panels consist of five layers, where layers 1 and 5 are the top and bottom layers, respectively. The specimen names, layer thicknesses (t_i), associated material, and total panel thickness (*total*) of all panels are given in Table 1. Commercial CLT is labeled "N" (new), while CLT made from pallet wood is labeled "C" (circular). Figure 1 shows a CLT panel with an indication of the layers, their fiber orientation, as well as the global coordinate system (1, 2, 3) that is used within this work.

Table 1 CLT test specimens, with specimen names (N = new, C = circular), layer thicknesses (t_i) with 1 and 5 being the outer layers, layer material type, and total thickness (t_{total})

	Layers 1 and 5		Layers 2 and 4		Layer 3		CLT
	t_1 and t_5	Material	t_2 and t_4	Material	t_3	Material	t_{total}
Name	(mm)	(-)	(mm)	(-)	(mm)	(-)	(mm)
N-CLT-1	28.0	New	21.0	New	22.0	New	120.0
N-CLT-2	28.0	New	21.0	New	22.0	New	120.0
C-CLT-1	23.5	New	16.0	Pallet	25.0	New	104.0
C-CLT-2	15.5	Pallet	15.0	Pallet	22.5	New	83.5
C-CLT-3	15.5	Pallet	15.0	Pallet	22.5	New	83.5

Materials: New, new material; pallet, material reclaimed from pallets

Fig. 1 Lay-up of CLT with the global coordinate system (1, 2, 3), layer numbers, and thicknesses. (Based on an image from [18])

Fig. 2 Production process of C-CLT: (1) pallet acquisition, (2) disassembly and de-nailing, (3) drying, (4) finger-jointing, (5) edge-gluing, (6) face-gluing, and (7) planning and shaping

Two N-CLT panels of 1.2 × 3 m were kindly provided by *Heko Spanten* (*Ede, The Netherlands*), which does not produce the N-CLT themselves. Their supplier used C24 grade timber to create the panels. Unfortunately, some production and material details are unknown; however, it is believed, after thorough inspection, that the N-CLT panels consist of spruce, pine, Douglas fir, or other coniferous species that are commonly used in commercial CLT production. Furthermore, no sign of edge-gluing was found. Finally, a formaldehyde-free polyurethane (structural) adhesive was used to create the panels.

To create (parts of) the C-CLT panels, wood recovered from single-use pallets was used. It was assumed here that the wood is mainly spruce, as this is commonly used for pallets [11]. It is important to note that the exposure history of the pallets is unknown. Pallets may have experienced severe wetting, loading stress levels, chemicals, etc. The new wood used to create some C-CLT panels was spruce, yellow pine, and Douglas fir. Figure 2 displays the C-CLT production process, which is briefly described below:

1. *Pallet acquisition:* Approximately 110 pallets were acquired, all different types and signs of usage.
2. *Disassembly and de-nailing:* Various tools were used to manually disassemble and de-nail the pallets, which was a labor-intensive process. As the locations of the nails on most pallets are well-defined [10], this process can potentially be automated.
3. *Drying:* Since some of the wood showed high moisture content (*MC*) levels, the reclaimed wood was stacked and dried to an *MC* of below 20% in a 20 °C and 65% RH environment, to ensure adequate adhesion in the following gluing steps.
4. *Finger-jointing:* To create six-meter-long laminates, first a tooth-like profile was milled at both ends of the planks, which were, subsequently, glued using a non-structural adhesive. This was done at *Woodjoint* (*Veenendaal, The Netherlands*).

5. *Edge-gluing:* Three-meter long laminates were sawn from the finger-jointed laminates, which were edge-glued and pressed using a melamine urea-formaldehyde (MUF) adhesive, to form single-layer panels. These panels were planned prior to face-gluing. These steps were facilitated by *Boerboom (Bergeijk, The Netherlands)*.
6. *Face-gluing:* Five-layer CLT panels were created by gluing the new and pallet layers according to the lay-up presented in Table 1, using alternating layer directions. A MUF adhesive was used with an applied pressure of 0.03 MPa. This was facilitated by *Boerboom (Bergeijk, The Netherlands)*.
7. *Planning and shaping:* After adhesive curation, the CLT panels were planned and shaped to the desired dimensions. This was facilitated by *Boerboom (Bergeijk, The Netherlands)*.

All panels (N-CLT and C-CLT) were stored at 20 °C and 50–60% RH, before being processed to the desired shape for bending and shear testing. As the above-described panel production process is labor-intensive, the sample size is currently limited.

2.2 Mechanical Testing

Following common definitions as shown in Fig. 1, in the global coordinate system of the panels, axis 1 is along the fiber direction of the outer layers, axis 2 is in the fiber direction of the even layers (2 and 4), and axis 3 is in the thickness direction of the panel. To assess the mechanical performance of the new and circular CLT panels, four-point bending and planar shear testing were employed (Fig. 3) [13–17].

Figure 3a displays a four-point bending test to determine the global bending strength and stiffness of the CLT panels described in EN 408 [19]. The test setup with four support/loading points leads to a shear-free zone between the central two

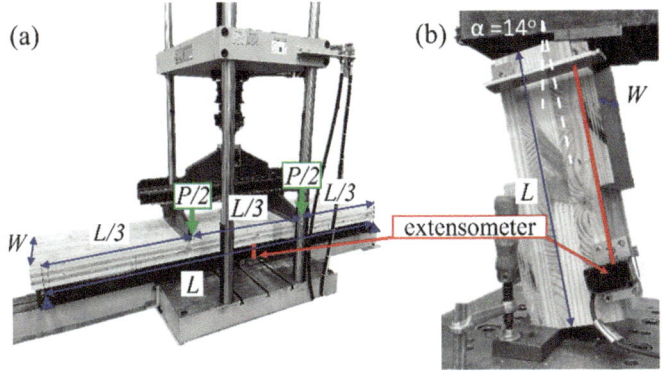

Fig. 3 Setups for mechanical testing, with (**a**) four-point bending and (**b**) planar shear

supports when the load is applied. Consequently, bending failures at the center are unaffected by the influence of shear. The four support/loading points were spaced at one-third of the overall span of the panel (L). Finally, the applied load (P) and the associated displacement (d) of the crosshead and the specimen at the center, using two laser extensometers, were recorded during the test. The sample amounts, dimensions, and displacement rates are given in Table 2. The displacement rates vary due to adjustments to reach failure within a desired time of 6–10 min.

From the load-displacement curves, the global modulus of elasticity of the CLT panel (E_{global}) is determined by,

$$E_{\text{global}} = \frac{23P \cdot L^3}{108\Delta \cdot \left(W \cdot t_{\text{total}}^3\right)},$$ (1)

in which P/Δ is the slope of the initial linear (elastic) part (between 0.1 and 0.4 times the failure load (P_{fail}), as indicated in Fig. 4) of the load-deformation curve and the dimensions are given in Tables 1 and 2. Additionally, the bending strength (f_m) is determined using,

Table 2 Bending and shear test specimen amounts (#), sample length (L) and width (W), and displacement rate (*d-rate*)

Bending					Shear				
Name	No.	L	W	d-rate	Name	No.	L	W	d-rate
		(mm)	(mm)	(mm/s)			(mm)	(mm)	(mm/s)
N-CLT-1	1	2600	735	3.0	N-CLT	3	371–376	98.5–101	0.5
N-CLT-2	1	2600	551	3.0	1 and 2				
C-CLT-1	1	2700	600	10.0	C-CLT-1	4	327–333	99.2–101	0.5
C-CLT-2	1	2400	600	18.0	C-CLT-2	3	278–279	100–101	2.0
C-CLT-3	1	2400	600	10.0	C-CLT-3	5	249–279	99.9–101	1.0

Fig. 4 Load-displacement curves of a bending test

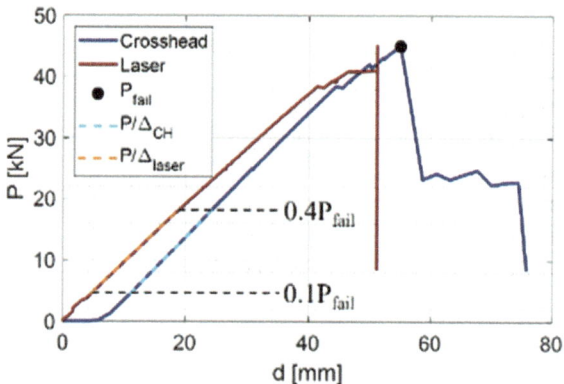

$$f_m = \frac{P_{\text{fail}} \cdot L}{W \cdot t_{\text{total}}^2}. \tag{2}$$

Due to its layered structure, CLT is prone to "so-called" rolling shear issues. This refers to the significantly lower stiffness of the cross-laminations, commonly assumed to be only 10–20% of the shear stiffness of the longitudinal layers [20], and their effect on the global behavior of the panels during out-of-plane loading. The setup used for determining the global rolling shear stiffness and strength is displayed in Fig. 3b; similar test setups can be found in the work by Gong et al. [14] and Niederwestberg [13]. The method used herein was deduced from the shear property evaluation tests described in ASTM D2718-00 [21]. In this work, the specimen is placed in a compression stage under an optimized angle (α) of 14° [14]. Through the inclination and the dimensions of the specimen, it is ensured that the vector of the applied load travels through the center of the specimen.

Due to the lack of available material, samples were cut from the ends of the bending specimens where no failure had occurred. Note that despite no obvious failure occurring in these areas, a pre-loading in shear must be kept in mind. A total of 15 specimens were tested, and their amounts, dimensions, and displacement rates are given in Table 2. Due to the panels' thickness differences (Table 1), the length varies to maintain the 14° inclination and force alignment with the center. The displacement rates vary due to adjustments to reach failure within a desired time of 6–10 min. During testing, the applied load, the crosshead displacement, and the relative displacement between the outer layers were recorded using laser extensometers. Some specimens exhibited crushing failure in the loading regions, this was remedied by gluing additional material in the loading areas (Fig. 3b). This adjustment is deemed acceptable since the deformation within the longitudinal layer is considered to be small due to its high stiffness [21].

From the load-displacement curve, the global shear modulus (G_{13}) is obtained by,

$$G_{13} = \frac{P \cdot (t_2 + t_4)}{\Delta \cdot (L \cdot W)}, \tag{3}$$

in which P/Δ is the slope of the initial linear (elastic) part (between 0.1 and 0.4 P_{fail} as indicated in Fig. 6) of the load-deformation curve and the dimensions are given in Tables 1 and 2.

Due to the inclination, the applied load and the shear plane were not aligned. Hence, the applied load is corrected using a correction factor, k_{angle} ($= \cos(\alpha)$), to find the load in the shear plane. The shear strength ($f_{v,13}$) is then determined using,

$$f_{v,13} = \frac{k_{\text{angle}} \cdot P_{\text{fail}}}{L \cdot W}. \tag{4}$$

Finally, for both bending and shear testing, failure modes were established based on the recognized failure characteristics and the load-displacement behavior.

3   Results and Discussion

3.1   Bending Tests

Figure 4 displays the observed load-displacement curve of the bending tests using the laser extensometers and crosshead displacement. The failure load (P_{fail}) and P/Δ (between 0.1 and 0.4 times P_{fail}), which are used to determine E_{global}, f_m, G_{13}, and $f_{v,13}$, are annotated. The regression range between 0.1 and 0.4 times P_{fail} is chosen to address the initial settlement of the system at the lower boundary and avoid the inclusion of plastic deformation at the upper limit. The laser load-displacement curve is considered for determining the mechanical properties as the P/Δ values in both curves are similar and P_{fail} is identical. A difference in displacement and stiffness between the crosshead and the laser measurement can be seen. This is the result of these measurements reflecting different locations along the beams. While the laser reflects the displacements at the center, the displacement of the crosshead technically reflects the displacements at the loading locations.

Figure 5 displays the determined E-values (E_{global}), failure loads (P_{fail}), and bending strengths (f_m) for all tested CLT panels, including images of failure modes. The E-values for N-CLT are significantly larger than for all C-CLT specimens; the reasons are briefly discussed here. First, and likely most significantly, is the insufficient stiffness and strength of the finger-joints, which lead to a discontinued lamination that cannot transfer the axial tension loads as intended. This practically renders such a layer ineffective within the tension zone and reduces

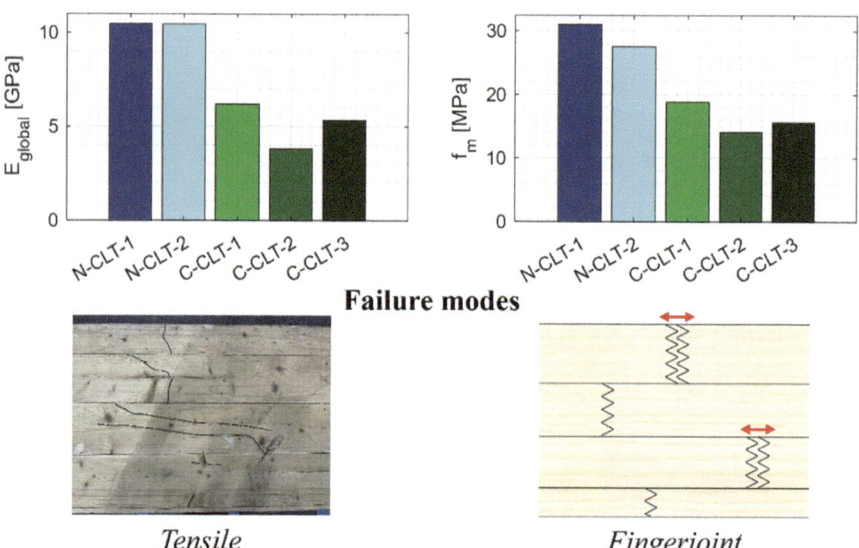

Fig. 5 Mechanical properties of the CLT panels made of new and reclaimed material obtained by the conducted bending tests. Images of tensile and finger-joint failures are added

the statical thickness of the CLT to a thickness where a layer can function as a continuous layer and can transfer the tensile forces. Given that the glue used in the finger-joints is not a structural adhesive, such a failure is reasonable. Second, lower quality of the wooden laminates in C-CLT, especially in the outer layers, compromises the overall panel stiffness. Thirdly, the thickness of the outer layers of the C-CLT specimens constitutes a lower percentage of the total thickness compared to the N-CLT panels (see Table 1), that is, 46.7% and 37.1–45.2% for N-CLT and C-CLT, respectively. Fourthly, lower bonding line stiffness leads to a less rigid overall cross-section, thereby compromising the panel stiffness.

Figure 5 also displays the failure modes from the bending experiments, which vary significantly between the two main specimen groups (N-CLT and C-CLT). While all N CLT panels show the expected tensile laminate failure in the outer layers, all C-CLT specimens showed premature failure within the finger-joints. This strongly indicates that failure in the finger-joints is the main reason for the lower stiffness of the C-CLT panels. In future work, structural glues must be considered for finger-jointing.

Regarding the bending strength (f_m), similarly to E_{global}, the observed finger-joint failure significantly affected the determined bending strength of the C-CLT panels. Due to the prematurely failing finger-joint, the tensile capacity of the outer layers is not utilized, consequently leading to a lower bending strength of the C-CLT panels.

3.2 Shear Tests

Ideally, P/Δ is determined from the load-displacement behavior referencing the relative displacement of the outer layers, which is recorded by the laser extensometer. Unfortunately, technical problems affected the measurements of the laser extensometers, leading to a stepwise load-displacement curve (see Fig. 6), resulting in a faulty determination of P/Δ. Therefore, an additional load-displacement behavior, considering the influence of the angle correction factor, k_{angle}, was calculated.

Fig. 6 Load-displacement curves of a shear test

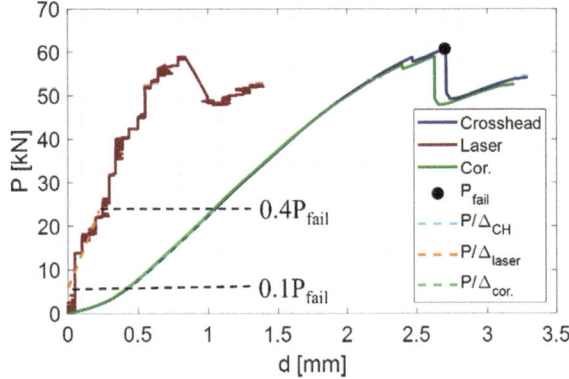

Collectively, Fig. 6 displays three load-displacement curves, that is, based on (i) the applied load and crosshead displacement (Crosshead), (ii) the load corrected by k_{angle} and the laser extensometer displacement (Laser), and (iii) both load and crosshead displacement both corrected by k_{angle} (Cor.). Additionally, the associated P/Δ slopes are displayed. The stepwise curve obtained using the laser extensometer is significantly steeper than those for the other two graphs, which yield similar slopes. While the determined P/Δ for the laser measurement appears faulty, the shear properties obtained using the k_{angle} corrected load and crosshead displacement (Cor. in Fig. 6) still allow for a qualitative comparison between N-CLT and C-CLT. It must be noted that these measurements are affected by the inclined specimen and technically include deformations within the directly loaded (outer) layers. The deformation of these outer layers can be considered small within the load range of the regression for the shear moduli (0.1–$0.4\ P_{\mathrm{fail}}$), as the material for this load range is unaffected by failures.

Figure 7 displays the shear moduli (G_{13}) and bending strength ($f_{v1,3}$) for all CLT panels, including images of the failure modes. Regarding G_{13}, the crosshead-based values (17.34–34.63 MPa) are lower than values that can be found in the literature, while values based on laser measurement (72.06–95.63 MPa, deemed unreliable) are within the range of values from the literature [13]. Furthermore, on average, G_{13} is only slightly larger for N-CLT than for C-CLT.

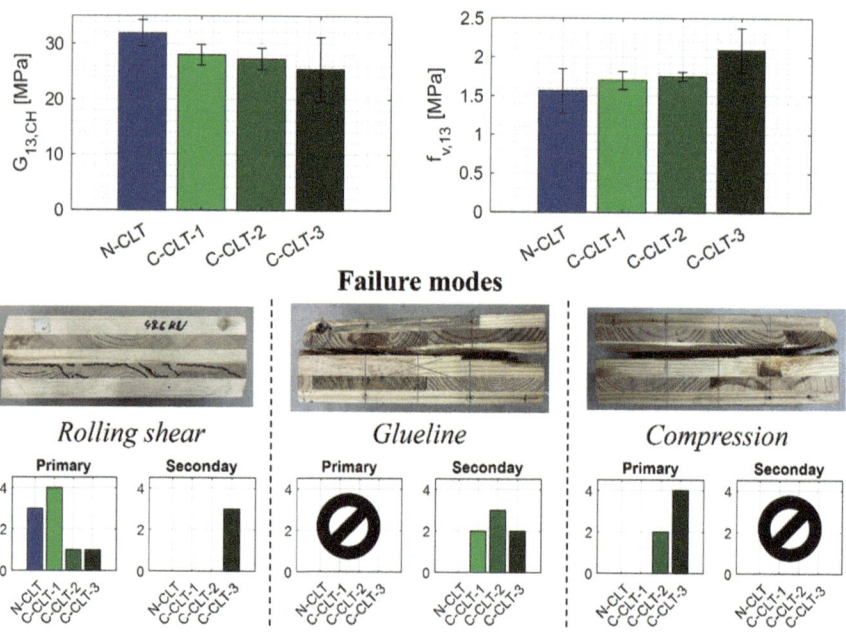

Fig. 7 Mechanical properties of the CLT panels made of new and reclaimed material obtained by the conducted shear tests. Images of all failure modes are added, along with their appearance as primary or secondary failure modes

Regarding the shear strength, $f_{v,13}$, the values for both groups given in Fig. 7 (1.24–2.35 MPa) are within the range of the literature [13]. Furthermore, on average, higher values are found for C-CLT than N-CLT. Lastly, the associated variation in $f_{v,13}$ and G_{13} for C-CLT is comparable and at times even lower than those for N-CLT.

Figure 7 displays three failure modes from the shear experiments, that is, the so-called rolling shear failure due to shear in the cross-layer (which is the expected failure mode as it confirms proper loading), glue line failure due to inadequate laminate bonding, and compression failure in the loading area, commonly due to insufficient contact area or low-quality fibers (as discussed before). The issue related to glue line failure in CLT with yellow pine is a known issue [22]. Furthermore, the graphs display the count of the failure modes occurring first or second. A lot of C-CLT specimens first exhibited compression failure, which is associated with the loading application rather than the desired failure. Nevertheless, since the failure occurred differently from rolling shear failure, the resulting failure loads are here considered to be conservative for the determination of the shear strength. Furthermore, failure in the glue line was also observed. This is promoted by oils within yellow pine bonding surfaces or uneven bonding surfaces [22]. Glue line failure was only observed as a secondary failure and, therefore, has no direct influence on the strength values. Nevertheless, it is advised to ensure higher surface quality in future work. Collectively, most specimens displayed rolling shear failure as primary or secondary (after compression failure). Hence, the obtained $f_{v,13}$ values are deemed appropriate.

Collectively, because the finger-joints are insignificant in the shear tests and the N-CLT and C-CLT specimens exhibit similar shear properties, it is believed that the reclaimed material yields similar potential as virgin material. However, adequate manufacturing considerations are essential. Even though the results of this work indicate that pallet wood can potentially be used to produce CLT, various open questions remain. For example, regarding durability, wood quality, stock variation, contaminants, etc., all require further research. Currently, digital technologies are evaluated to reduce the manual labor involved in the above-described panel production process. As the production process is refined, it is intended to produce additional samples to increase testing efforts and results.

4   Conclusion

The potential of upcycling single-use pallet wood into cross laminated timber (CLT) is explored by means of feasible production processes and bending and shear property characterization. Generally, common CLT production processes can be followed, while additional steps in the preparation of the material are needed, for example, fastener removal. This particular step shows potential for automation. In contrast to the new CLT specimens, all tested pallet CLT specimens exhibited premature failure in the finger-joints of the parallel layers when exposed to bending.

This led to significantly lower determined moduli of elasticity and bending strengths compared to the new CLT. Therefore, the bending test results of the pallet CLT are deemed not representative. It is believed that fully functioning finger-joints within pallet CLT can significantly improve its performance since no other failures were observed. The shear test results suggested that the pallet CLT has similar potential as the new CLT. This is reasonable since the functionality of the finger-joints is not significant to the shear tests. The obtained shear moduli and strength are similar for the pallet and the new CLT. Even though measurement issues occurred, the shear strength of the pallet CLT is comparable to the literature. In summary, this work presents promising results for the potential CLT made from pallets and other secondary sources and highlights key improvement aspects to be considered in a potential follow-up project, namely requiring high-quality finger-joints and more versatile testing methods.

References

1. Karacabeyli, E., Gagnon, S.: Canadian CLT Handbook. fpinnovations.ca, Pointe-Claire (2019)
2. Shin, B., Wi, S., Kim, S.: Assessing the environmental impact of using CLT-hybrid walls as a sustainable alternative in high-rise residential buildings. Energy Build. **294**, 113228 (2023)
3. Polaris Market Research: Cross Laminated Timber Market Share, Size, Trends, Industry Analysis Report, By Type (Adhesive Bonded, Mechanically Fastened); By Industry; By End-Use; By Region; Segment Forecast, 2022–2030 (2021)
4. Swedish Wood: Wood Grades. Swedish Wood. https://www.swedishwood.com/wood-facts/about-wood/wood-grades/. Last accessed 2023/07/13
5. Borzecka, M.: Absorbing the Potential of Wood Waste in EU Regions and Industrial Bio-Based Ecosystems—BioReg. IUNG-PIB (2018)
6. Johansson, C., Brundin, J., Gruber, R.: Stress Grading of Swedish and German Timber. A Comparison of Machine Stress Grading and Three Visual Grading Systems. RISE, Trätek (1992)
7. Huber, J., Broman, O., Ekevad, M., Oja, J., Hansson, L.: A method for generating finite element models of wood boards from X-ray computed tomography scans. Comput. Struct. **260**, 106702 (2022)
8. Huber, J.: Numerical modelling of timber building components to prevent disproportionate collapse. Luleå University of Technology: Doctoral dissertation (2021)
9. Kovryga, A., Khaloian Sarnaghi, A., Van de Kuilen, J.: Strength grading of hardwoods using transversal ultrasound. Eur. J. Wood Wood Prod. **78**, 951–960 (2020)
10. Klaus Timber: Euro pallets. Klaus Timber. https://www.klaustimber.cz/en/euro-pallets#:~:text=Euro%20pallets%20are%20made%20of,machines%20from%20all%20four%20sides. Last accessed 2023/10/23
11. United Nations: Trends and Perspectives for Pallets and Wooden Packaging. Commission for Europe, Geneva (2016)
12. Munandar, W., Purba, R., Christiyanto, A.: Exploratory study on the utilization of recycled wood as raw material for cross laminated timber. IOP Conf. Ser. Mater. Sci. Eng. **669**, 012011 (2019)
13. Niederwestberg, J.: Influence of Laminate Characteristics on Properties of Single-Layer and Cross Laminated Timber Panels. University of New Brunswick (2019)

14. Gong, M., Tu, D., Li, L., Chui, Y.: Planar shear properties of hardwood cross layer in hybrid cross laminated timber. In: 5th International Scientific Conference on Hardwood Processing, Québec City (2015)
15. Wang, T., Yang, Y., Li, Y., Wang, Z., Gong, Y., Zhou, J., Gong, M.: Rolling shear failure damage evolution process of CLT based on AE technology and DIC method. Eur. J. Wood Wood Prod. **80**(3), 719–730 (2022)
16. Navaratnam, S., Ngo, T., Christopher, P., Linforth, S.: The use of digital image correlation for identifying failure characteristics of cross-laminated timber under transverse loading. Measurement. **154**, 107502 (2020)
17. Olsson, A., Schirén, W., Segerholm, K., Bader, T.: Relationships between stiffness of material, lamellas and CLT elements with respect to out of plane bending and rolling shear. Eur. J. Wood Wood Prod. 1–16 (2023)
18. Mayr Melnhof Holz. https://www.mm-holz.com/. Last accessed 2023/07/13
19. CEN European Committee for Standardization: EN 408: Timber Structures—Structural Timber and Glued Laminated Timber - Determination of some Physical and Mechanical Properties. CEN European Committee for Standardization, Brussels (2003)
20. Sebera, V., Muszyński, L., Tippner, J., Noyel, M., Pisaneschi, T., Sundberg, B.: FE analysis of CLT panel subjected to torsion and verified by DIC. Mater. Struct. **48**, 451–459 (2015)
21. ASTM American Society for Testing and Materials: D2718-00: Standard Test Methods for Structural Panel in Planar Shear (Rolling Shear). ASTM American Society for Testing and Materials, West Conshohocken (2011)
22. Lum, W., Norshariza, M., Nordin, M., Ahmad, Z.: Overview on bending and rolling shear properties of cross-laminated timber (CLT) as engineered sustainable construction materials. Green Infrastructure. Springer (2022)

Designing Eco-Friendly Materials Intended for Repairing Reinforced Concrete Structures

Karim Belmokretar ⓘ, Kada Ayed, Nordine Leklou ⓘ,
Djamel El Ddine Kerdal, Mohammed Mouli, and Meftah Allal

Contents

1 Introduction

The imperative to reduce our construction's carbon footprint necessitates embracing natural or ecologically manufactured building materials, as well as those sourced from recycled waste. This adoption is not just an option but a critical step forward. By integrating these materials into our building practices, we can significantly mitigate environmental impact and move toward a more sustainable future. The environmental benefits would be greatly enhanced by choosing this type of concrete for repairing damaged concrete structures [1]. With the continual rise in the number

K. Belmokretar (✉) · D. E. D. Kerdal
University of Science and Technology of Oran Mohammed Boudiaf, Oran, Algeria

K. Ayed · M. Mouli
National Polytechnic School of Oran Maurice Audin, Oran, Algeria

N. Leklou
Nantes Université, Ecole Centrale Nantes, Scientific Research National Center (CNRS), Civil and Mechanical Engineering Research Institute (GeM), UMR 6183, Nantes, France

M. Allal
University of Msila, Msila, Algeria

© The Author(s) 2025
M. Kioumarsi, B. Shafei (eds.), *The 1st International Conference on Net-Zero Built Environment*, Lecture Notes in Civil Engineering 237,
https://doi.org/10.1007/978-3-031-69626-8_64

of deteriorating concrete structures necessitating maintenance, the significance of this approach is magnified.

Sandcrete represents one of the concrete formulations engineered to contribute toward meeting this objective. It was first used by François Coignet in the construction of the "Port Saïd-Egyp" lighthouse. Then widely used by the USSR in the reconstruction of roads, airfields, and buildings after the end of the Second World War [2]. In France, a program called "SABLOCRETE" was launched in 1988 [3]. The aim was to demonstrate the economic and ecological benefits that could be derived from using natural or crushed sand in concrete products, of which it would be the main component.

Several researchers have set out to formulate sand-based concretes using different types of sand. Abidelah et al. have developed a self-compacting sand-based concrete, reinforced by the incorporation of limestone [4]. The concrete formulations developed by these researchers fully comply with the standards set by the French Civil Engineering Association (AFGC) for self-compacting concrete. Li et al. studied the shear strength of concrete beams made from dune sand (DS) from the Gurbantunggut desert in China's province [5]. The results obtained demonstrate the viability of concrete made from dune sand for use in construction. In addition, the shear strength values measured were found to perfectly agree with the requirements specified by current design codes in China. Bédérina et al. undertook a study on the substitution of coarse aggregates by naturally available dune sand (DS) and river sand (RS) in the manufacture of sand concrete [6], Their research led to the creation of a sand concrete offering improved workability, compactness, and strength.

This experimental study aims to develop a sand-based concrete while making it self-compacting, to further enhance the ecological aspect of concrete. Consequently, the concrete was named "Self-Compacting Sand Concrete (SCSC)." The SCSC formulation has been optimized to achieve the characteristics of category 3a self-placing concrete (BAP) [7]. This category is distinguished by several criteria, including maneuverability in an unconfined environment, measured by Abrams cone spread of between 760 and 850 mm (class SF3), a minimum L-box value confined by 3 bar, which must be greater than or equal to 0.8, and a sieve test value not exceeding 15% [7]. Three types of SCSC were produced: SCSC-LF based on limestone fillers, SCSC-PZ based on natural pozzolan, and SCSC-PR based on perlite. The mixes were tested in the fresh state to determine their density and Abrams cone flow following standard EN 12350-8 [8], the filling rate of the L-box following standard EN 12350-10 [9], and finally, the percentage of laitance retained per sieve following standard EN 12350-11 [10]. In the hardened state, the compressive strength of all the specimens was measured following EN 13791, after curing periods of 7, 28, and 90 days. The SCSC developed has been used to repair ordinary concrete substrates. The composites B-SCSC were evaluated in tension by the PULL OFF test according to ASTM 1583 [11].

Finally, SCSCs developed from naturally available or even waste materials, and used without the need for vibration, have the potential to be environment-friendly concretes with a low carbon footprint.

2 Experimental Program

2.1 *Materials*

Cement The cement used is CEM I 52.5R from the LAFARGE Group, located in OGGAZ/MASCARA (north-west Algeria). This cement has a fineness, according to the BLAINE method, of between 4400 and 5400 cm^2/g, and an absolute density equal to 3.09 g/cm^3. The chemical and mineralogical compositions of clinker are given in Tables 1 and 2 respectively.

Aggregates Two sand types were used in this study. The first is a naturally available quartzite sand (SM) with a fraction $\leq$1 mm. The second is sand from quarry waste (SC), fraction 0/4 mm. Limestone gravel with a 3/5 mm fraction was also used. All aggregates were supplied by the SECH Spa quarries of the HASNAOUI Group, Algeria.

The size distribution and physical properties of the aggregates are presented in Table 3 and Fig. 1, respectively.

Mineral Additions

Limestone fillers (LF). The limestone fillers used in this study are manufactured by the Algerian Aggregates Company (ENG). These waste products, commonly called "dust," originate from the aggregate industry's operations within quarries.

Natural Perlite (PR). Perlite was extracted in the form of volcanic rock from the Boughrara area (Maghnia/Tlemcen/Algeria). It was washed, dried for 24 h at 80 °C, crushed and ground. The powder was passed through an 80 μm sieve, with 80% of fractions $\leq$63 μm.

Natural pozzolan (PZ). Pozzolan was extracted as a volcanic rock from the Beni Saf region (Ain Timouchent, Algeria). The same perlite preparation procedure was used to prepare the pozzolan. Seventy-eight percent of its fractions are $\leq$63 μm.

Table 1 Chemical composition of cement [%]

SiO	Al_2O_3	Fe_2O_3	CaO	MgO	K_2O	SO_3	Na_2O	FAP	Cl−
17.40	4.12	2.97	61.15	1.8	0.66	2.8	0.13	8.85	<0.1

Table 2 Mineralogical composition of clinker [%]

C_3S	C_2S	C_3A	C_4AF
55	19.42	9	16.58

Table 3 Physical characteristics of sand and gravel

	Absorption coefficient [%]	Absolute density [g/cm^3]	Fines content [%]	Sand equivalent [%]	Finesse modulus
SM	3.0	2.63	2.4	87	1.4
SC	4.0	2.62	8.1	63	3.31
G3/5	0.5	2.62	0.9	/	/

The chemical compositions physical characteristics and particle size distribution of the three mineral additions are presented in Tables 4 and 5 and Fig. 2, respectively.

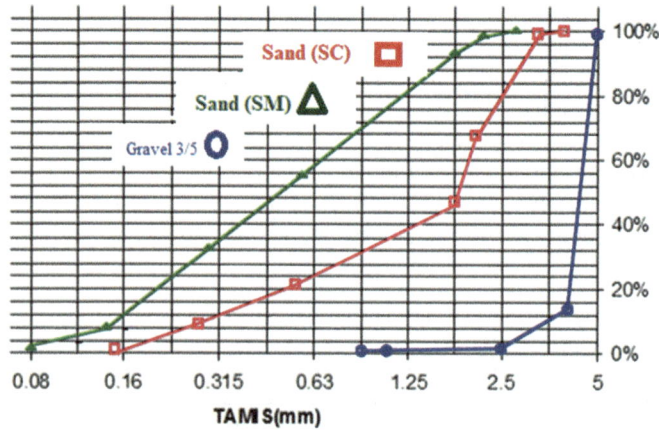

Fig. 1 Grain size distribution curve of aggregates

Table 4 Chemical composition of mineral additions [%]

	SiO	Al₂O₃	Fe₂O₃	CaO	MgO	K₂O	SO₃	Na₂O	FAP	Cl−
LF	17.4	4.12	2.97	61.15	1.8	0.66	2.8	0.13	0.10	17.4
PR	75.3	13.43	2.55	3.22	0.37	4.1	0.01	0.82	0.01	75.3
PZ	43.01	15.98	10.14	9.41	4.20	1.39	2.34	0.05	0.00	43.01

Table 5 Physical characteristics of mineral additions

	Absolute density [g/cm³]	Blaine fineness [cm²/g]
LF	2.63	4350
PR	2.62	4400
PZ	2.62	4390

Fig. 2 Particle size distribution of Lf, PR, and PZ

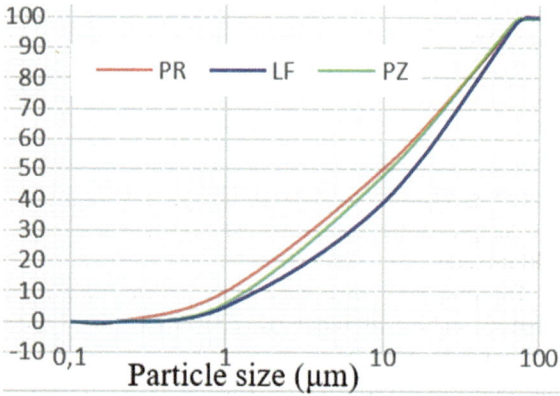

Admixtures The admixture used is MEDAFLOW 30. Manufactured by the Algerian company GRANITEX. It is a superplasticizer with high water-reducing properties. It is based on ether polycarboxylates. Its absolute density is 1.07 g/cm³. Its solids content is 30%.

2.2 Preparation Method and Tests

This research aimed to formulate a self-compacting sand concrete (SCSC), by obtaining the best rheological parameters (slump, pourability, and resistance to segregation), as well as excellent mechanical characteristics, while complying with the recommendations of the French Civil Engineering Association (AFGC), relating to BAP. For this purpose, the most important BAP parameters were used [12]. These are the content of additions, which improves workability and increases paste volume; the equilibrium between the W/C ratio and the content of superplasticizing, water-reducing admixtures, which contributes to reducing the quantity of water used, improving the felicity of the mix and preserving its homogeneity and stability.

Three types of self-compacting sand concrete were produced: SCSC-LF, SCSC-PR, and SCSC-PZ based on limestone fillers, natural perlite, and natural pozzolan, respectively. For each type we modified the W/C ratios (0.45%, 0.5%, and 0.55%), so nine SCSC mixes were prepared (Table 6). To improve the flowability of (SCSC) and thus achieve the 80 cm spreading target, adjustments were made to the proportions of the components. In particular, the quantities of quartzite sand (SM) and limestone sand from waste (SC) have been increased, while coarse aggregates have been reduced and limited to a maximum size of 5 mm. This modification aims to increase the concentration of fine sand while reducing the proportion of coarse aggregate, thereby reducing friction between particles and promoting a smoother

Table 6 Proportion of materials making up the various SCSCs

	SCSC-LF			SCSC-PR			SCSC-PZ		
	1	2	3	1	2	3	1	2	3
CEMI-52.5 N [KG/m³]	420	420	420	420	420	420	420	420	420
Limestone fillers (LF) [KG/m³]	170	170	170	/	/	/	/	/	/
Perlite (PR) [KG/m³]	/	/	/	170	170	170	/	/	/
Pozzolan (PZ) [KG/m³]	/	/	/	/	/	/	170	170	170
Sand (SM) [KG/m³]	619	654	674	610	630	660	619	646	670
Sand (SC) [KG/m³]	480	500	535	464	500	525	480	508	540
Gravel G3/5[KG/m³]	380	380	380	380	380	380	380	380	380
M30 (SP) [KG/m³]	12.6	12.6	12.6	12.6	12.6	12.6	12.6	12.6	12.6
Water (W) [L/m³]	231	210	189	231	210	189	231	210	189
W/C	0.55	0.50	0.45	0.55	0.50	0.45	0.55	0.50	0.45
G/S	0.35	0.33	0.31	0.35	0.34	0.32	0.35	0.33	0.31

Fig. 3 Slump flow test

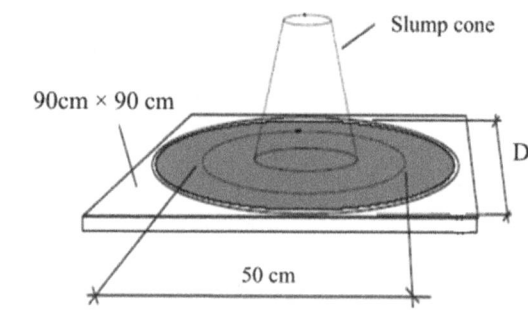

Fig. 4 Passing ability in
L-Box

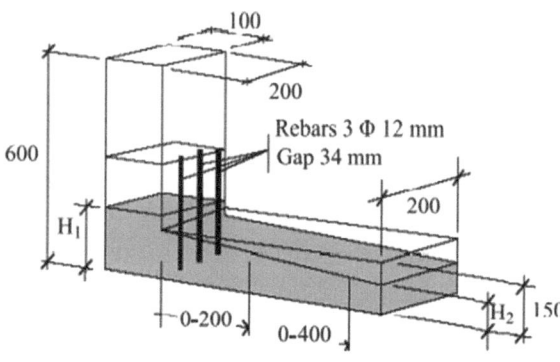

Fig. 5 Segregation
resistance

flow of SCSC, thus avoiding the risk of blockages [13]. The level of MEDAFLOW
30 superplasticizer was set at 3% by weight of cement (approx. 12.60 L/m³).

The tests carried out are slump flow with ABRAMS cone (Fig. 3), passing ability
in L-box (Fig. 4), segregation resistance (Fig. 5), and compressive strength (Fig. 6).

The SCSCs were applied as repair layers on slabs with dimensions (21 × 28 × 7)
cm³. The concrete used to make the slabs was ordinary concrete (OC). The average
compressive strength of OC varies between 40 and 50 MPa. All slabs were subjected
to wet curing for 6 months, after which the slab surfaces were treated with a

Fig. 6 Compressive strength

Fig. 7 Surface preparation by sandblasting

sandblasting technique (Fig. 7) to create the roughness required for bonding between the old and new concrete. For each type of SCSC, three slabs were prepared and smoothed in a wet cure.

The PULL OFF test according to American standard ASTM 1583 [11] was used to assess the bond between SCSC and OC slabs. After a curing period of 28 days, the composites (SCSC-BO) were removed from the cure, and 50 cm diameter cores, spaced 50 cm apart, were drilled to a depth of 2 cm into the OC slabs. Metal disks 50 cm in diameter were bonded with epoxy adhesive. A tensile force was progressively applied by the "PROCEQ dy-216" apparatus until failure (Fig. 8).

2.3   *Results and Discussion*

Slump flow Table 7 shows the slump flow results for SCSC specimens. Mixtures with a W/C ratio of 0.45 gave a spread between 51 and 53 cm for the three SCSCs (Fig. 9). The slump-flow class was SF1 according to the AFGC guide. Mixes with a W/C ratio of 0.50 produced a spread between 63 and 67 cm for the three SCSCs (Fig. 10). The slump-flow class was SF2. In mixes with a water/cement ratio (W/C)

Fig. 8 PULL OFF test

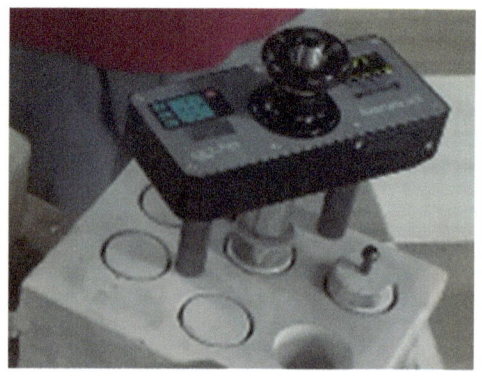

Table 7 Slump flow results of SCSCs

Slump flow [cm]	W/C = 0.45	W/C = 0.5	W/C = 0.55
SCSC-LF	51	63	78
SCSC-PR	53	67	79
SCSC-PZ	50	65	79

Fig. 9 Slump flow of SCSC with W/C = 0.45

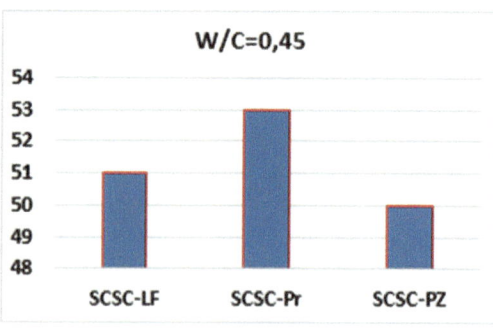

Fig. 10 Slump flow of SCSC with W/C = 0.50

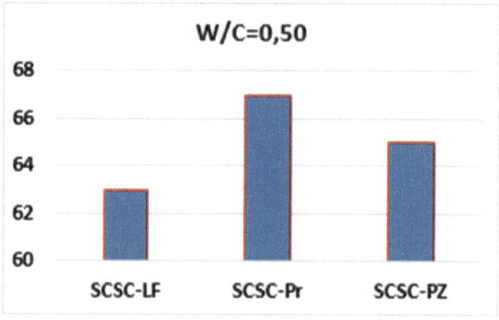

of 0.55, we observed spread values between 78 and 79 cm for all SCSCs (Fig. 11). The slump-flow class was SF3. Additions (LF, PR, and PZ) had no significant influence on slump flow, except for a slight increase for SCSC-PR, perhaps due to

Fig. 11 Slump flow of SCSC with W/C = 0.55

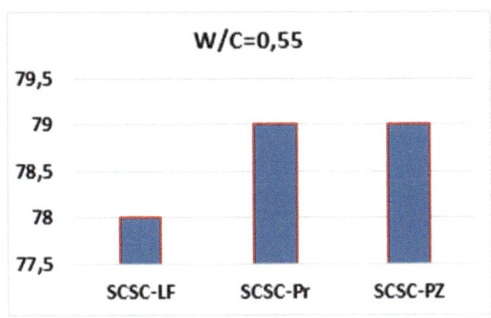

Table 8 Passing ability in L-Box of SCSCs

Passing ability in L-Box	W/C = 0.45	W/C = 0.5	W/C = 0.55
SCSC-LF	0.8	0.9	1
SCSC-PR	0.8	0.9	1
SCSC-PZ	0.8	0.9	1

Fig. 12 L-Box passing ability with W/C = 0.45

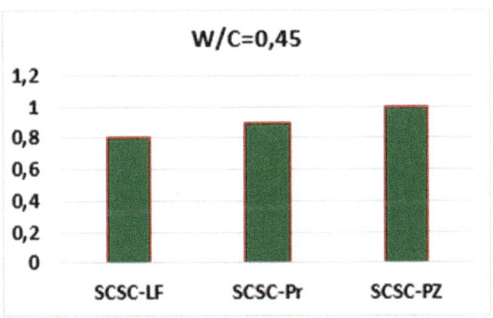

the PR's excellent finesse. Increasing the (W/C) ratio from 0.45 to 0.55 had a significant impact on slump-flow. An improvement of 49% was observed. A W/C ratio of 0.55 and an SP dosage of 3% considerably facilitated the achievement of the desired fluidity (80 cm). No signs of bleeding were observed on the edges of the wafers for any of the formulations tested.

Passing ability in L-Box Table 8 and Figs. 12, 13, and 14 show the L-box passing ability obtained for the SCSCs. All SCSCs gave excellent results. The $h1/h2$ ratios were between 0.8 and 1. The small-sized aggregates used ($D_{max} = 5$ mm) facilitated the flow of mixtures into the L-box, despite the confinement provided by 03 bars with a diameter of 12 mm. Mixtures with a W/C ratio of 0.55 gave total flow ($h1/h2 = 1$). Raising the W/C ratio from 0.45 to 0.55 resulted in a notable enhancement in passing ability, with a recorded increase of 25%. The W/C ratio of 0.55 and the SP dosage of 03% were sufficient to fluidize all the grains and prevent any blockage caused by friction between them. As a result, the flow was completely in the L-box. The type of addition did not significantly influence the passing ability through the L-box, and the results were close for all SCSC types for each W/C ratio.

Fig. 13 L-Box passing
ability with W/C = 0.50

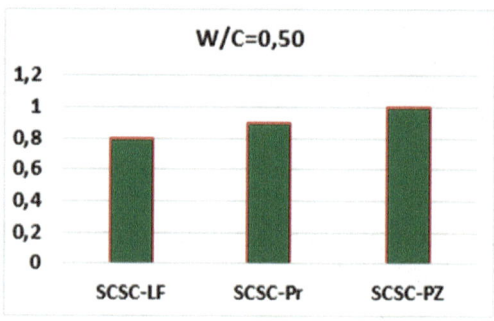

Fig. 14 L-Box passing
ability with W/C = 0.55

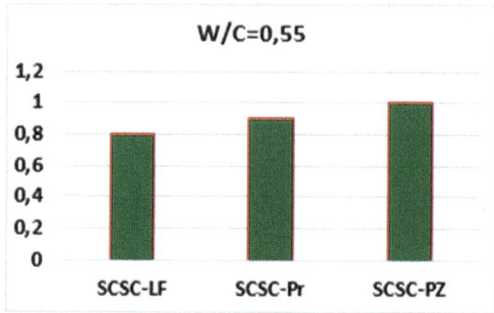

Table 9 Segregation resistance of SCSCs

Segregation resistance	W/C = 0.45	W/C = 0.5	W/C = 0.55
SCSC-LF	0	2	6
SCSC-PR	0	2	7
SCSC-PZ	0	1.7	5

Segregation resistance Table 9 shows the segregation resistance results for SCSCs. All SCSC concretes have proven to be highly stable, with a homogeneous distribution of mixed constituents. The levels of laitance passing through the sieve varied between 0% for W/C = 0.50, 1.7 to 2% for W/C = 0.45, and 5 to 7% for W/C = 0.55 (Fig. 15). A dosage of 3% "MEDAFLOW 30" super plasticizing admixture had a positive influence on the mixes' homogeneity and viscosity. Their influence was highly significant for SCSC formulated with a W/C ratio 0.55. The type of addition did not significantly influence segregation resistance results. They were very close for all SCSC types, depending on the W/C ratio. The linear trend between the increase in the quantity of laitance passed and the increase in the W/C ratio appears more clearly than the influence of changing the type of addition.

Compressive strength Table 10 and Fig. 16 show the compressive strengths recorded. All the SCSCs tested developed acceptable compressive strengths. From an early age, SCSCs develop acceptable compressive strengths for all formulations (between 37 and 40 MPa). At 90 days of age, the SCSCs achieved excellent

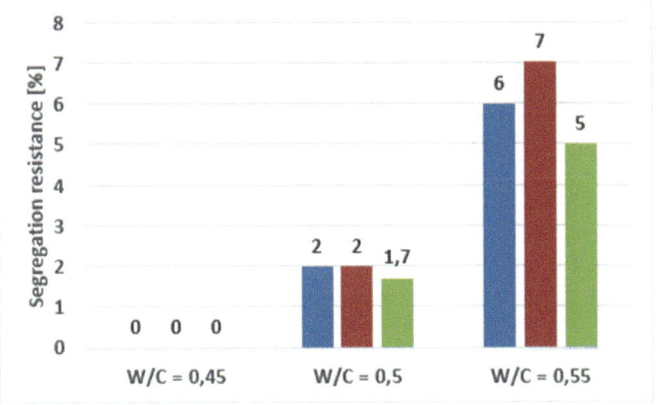

Fig. 15 Segregation resistance of SCSCs

Table 10 Compressive strength of SCSCs

Compressive strength	7D	SD	28 D	SD	90 D	SD
SCSC-LF	40	0.95	60	1.4	62	0.73
SCSC-PR	37	3.47	56	0.75	59	3.09
SCSC-PZ	38	2.81	54	0.9	61	2.26

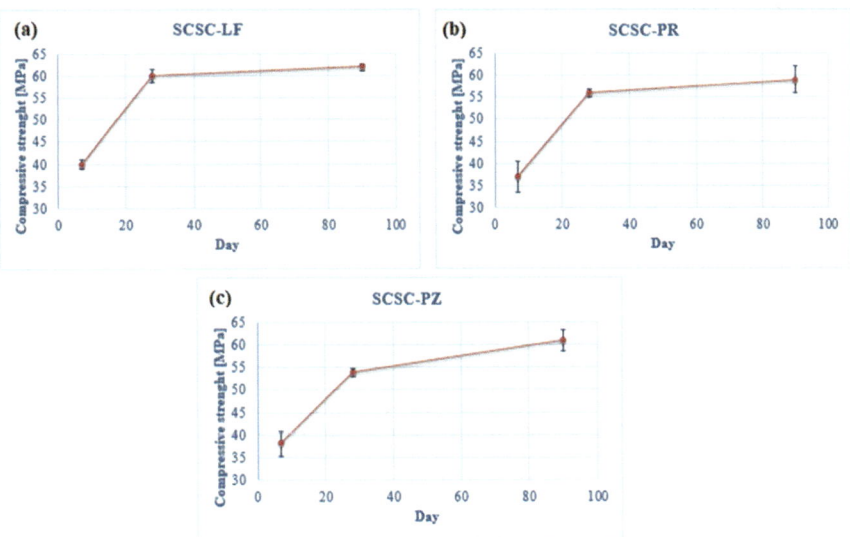

Fig. 16 Compressive strength of SCSCs: (**a**) SCSC-LF, (**b**) SCSC-PR, and (**c**) SCSC-PZ

compressive strengths (59, 61, and 62 MPa for SCSC-PR, SCSC-PZ, and SCSC-LF, respectively). The types of additions did not significantly influence compressive strength, although SCSC-LF gave the best results. Increasing W/C ratios up to 0.55 (57% increase) did not affect compressive strength.

Fig. 17 Bond strength
between SCSCs and OC
substrates

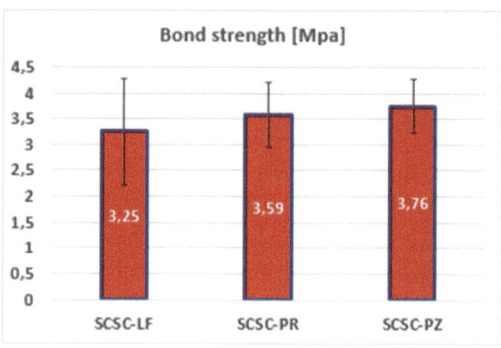

PULL OFF test Figure 17 shows the results obtained by the PULL OFF test at 28 days of age. The forces of adhesion recorded were excellent. The pozzolan-based SCSC gave the best bond strength (3.76 MPs), followed by the perlite-based SCSC (3.59 MPa) and lastly, the SCSC formulated with limestone fillers (3.25 MPa). The bond between the OC substrates and the SCSCs was considerably influenced by the fluidity of the SCSCs, as well as their ability to penetrate effectively into the interface texture created by the sandblasting technique. All specimen failures were planned on the substrates, highlighting the robust bond between the OC substrates and the SCSC repair material. The tests showed that SCSCs have a higher tensile strength than OC since all the breaks were located in the substrates made by OC.

3   Conclusion

This experimental study aimed to design concrete with a low carbon footprint, in this case, self-compacting sand concrete (SCSC), to meet growing environmental requirements while maintaining optimum mechanical properties. At the end of this study, the following conclusions can be drawn:

- By using naturally available quartzite sand, as well as sand from quarry waste, and integrating mineral additions such as limestone filler, also a recovered waste, as well as perlite and pozzolan, recovered natural resources, we have undertaken the development of an environment-friendly concrete.
- The reduction in the carbon footprint of self-placing sand concrete composites (SCSC) is evident in the reduced use of industrial aggregates and the absence of mechanical vibration.
- Self-compacting sand concrete (SCSC) stands out for its exceptional fluidity and high strength, with a compressive strength of up to 62 MPa. This makes it ideal for repair work, with a bond strength of up to 3.76 MPa.

References

1. Karim, B., Kada, A., Djamel-Eddine, K., Nordine, L., Mohamed, M.: New repair material for ordinary concrete substrates: investigating self-compacting sand concrete and its interaction with roughness of the substrate surface. J. Mater. Civ. Eng. **35**, 5023004 (2023). https://doi.org/10.1061/JMCEE7.MTENG-15348
2. Chauvin, J.-J., Grimaldi, G.: Les bétons de sable. Bull. Liaison Des Lab. Des Ponts Chaussées, 9–15 (1988)
3. Press of the National School of Bridges and Roads, France, Sablocrete: Sand Concrete: Characteristics and Use Practices (1994)
4. Abidelah, A., Bouchair, A., Kerdal, D., Ayed, K.: Characterization of a self-compacting sand concrete using the quarry waste. Can. J. Civ. Eng. **36**, 1773–1782 (2009)
5. Li, Z., Ma, R., Li, G.: Experimental study on the shear strength of dune sand concrete beams. Adv. Civ. Eng. **2020**, 8062691 (2020)
6. Bédérina, M., Khenfer, M.M., Dheilly, R.M., Quéneudec, M.: Reuse of local sand: effect of limestone filler proportion on the rheological and mechanical properties of different sand concretes. Cem. Concr. Res. **35**, 1172–1179 (2005)
7. Malier, Y., Guérinet, M.: AFGC: Recommendations for the Use of Self-Composition Concrete, p. 64. French Association of Civil Engineering, AFGC (Association Française Civ. Eng.) (2008). https://www.afgc.asso.fr/publication/recommandations-pour-lemploi-des-betons-auto-placants/
8. B.S. EN, 12350-8: 2010 Testing Fresh Concrete Self-Compacting Concrete. Slump-Flow Test, Br. Stand. Institute, London (2010)
9. B.S. EN, 12350-10: 2010 Testing fresh concrete part 10: self-compacting concrete–L-box test (London: British Standard). IOP Conf. Ser. Mater. Sci. Eng. **271**, 12004 (2017)
10. France and European standards, 12350-11: Essai pour béton frais–Partie 11: béton auto-plaçant–Essai de stabilité au tamis [Test for fresh concrete-Part 11: self-compacting concrete-sieve stability test] (2010). https://www.boutique.afnor.org/fr-fr/norme/nf-en-1235011/essai-pour-beton-frais-partie-11-beton-autoplacant-essai-de-stabilite-au-ta/fa165665/36326#AreasStoreProductsSummaryView
11. ASTM, C1583 – 13: STANDARD test method for tensile strength of concrete surfaces and the bond strength or tensile strength of concrete repair and overlay materials by direct tension (pull-off method). Annu. B. ASTM Stand. **93**, 1–5 (2010)
12. Okamura, H.: Self-compacting high-performance concrete. Concr. Int. **19**, 50–54 (1997)
13. Turcry, P., Loukili, A.: Différentes approches pour la formulation des bétons autoplaçants: influence sur les caractéristiques rhéologiques et mécaniques. Rev. Française Génie Civ. **7**, 425–450 (2003)

Multicriteria Performance Index Analysis of Geopolymeric and Cementitious Screed Flooring Materials

Joud Hwalla (iD), **Hilal El-Hassan** (iD), **Joseph Assaad** (iD),
and **Tamer El-Maaddawy** (iD)

Contents

1 Introduction

Geopolymer (GP) or alkali-activated building materials are cement-free composites produced primarily with alumina-silica-rich waste materials. These materials are activated using sodium, potassium, or calcium-based alkaline solutions [1]. GP composites exhibit superior mechanical and durability performance compared to conventional building materials produced with Portland cement (CM). An early scientometric analysis of the use of geopolymer composites demonstrated their excellent compatibility for various construction applications [2]. GP composites were shown to be used for underwater placement, serve as effective fire and heat protection layers, contribute to repairing, retrofitting, and strengthening old and damaged structures, and function as masonry plaster or block materials [2, 3]. In addition to the aforementioned applications, fly ash and slag-based GP composites

J. Hwalla (✉) · H. El-Hassan · T. El-Maaddawy
United Arab Emirates University, Al Ain, United Arab Emirates
e-mail: 202090500@uaeu.ac.ae

J. Assaad
University of Balamand, El Koura, Lebanon

© The Author(s) 2025
M. Kioumarsi, B. Shafei (eds.), *The 1st International Conference on Net-Zero Built Environment*, Lecture Notes in Civil Engineering 237,
https://doi.org/10.1007/978-3-031-69626-8_65

were developed and compared with CM-based material for screed flooring applications [4]. Screed is a building material used in buildings, warehouses, and bridges [5]. Its primary purpose is to ensure a level surface and act as a protective layer for slabs or bridge decks [4]. The British standard (BS 8204 [6]) categorizes screed flooring materials based on the indentation formed on the surface of the screed. This indentation results from the impact of a drop of a specific mass through a cylindrical rod with a cylinder-to-screed circular contact area of 500 mm^2.

Different studies have focused on analyzing the environmental and economic impact of the production of GP- and CM-based materials [7–9]. According to Kanagaraj [7], the production of fly ash-based self-compacting GP and a blend of fly ash and ground granulated blast furnace slag-based self-compacting GP instead of cement-based counterparts reduced the greenhouse gas emissions by around 9.45% and 5.88%, respectively. In addition, for the same compressive strength values, fly ash-based-GP concrete had lower CO_2 emissions per m^3 than CM-based concrete [8]. Similarly, the production of metakaolin-based concrete decreased the CO_2 emissions by 27–45% [9]. Indeed, despite the high environmental impact of alkaline solution, which contributes the most to increasing the global warming potential (GWP) indices, the production of GP composites reduced greenhouse gas emissions compared to the production of Portland cement-based building materials. Thus, incorporating such materials in construction applications is environmentally beneficial while providing superior mechanical and durability characteristics.

This study presents an ongoing research work on the use of GP- and CM-based screed flooring composites produced with dune sand (DS) and crushed limestone sand (CS) as fine aggregates [4, 10]. Herein, the economic and environmental impact analyses of different types and categories of screed are examined. The environmental impact is measured based on the global warming indices of the total carbon dioxide emitted per m^3 of screed flooring production. Furthermore, the economic impact is analyzed according to different parameters, including the cost of raw materials needed to produce 1 m^3 of screed material. Additionally, the durability performance, including resistance to acids and salt attack and abrasion forces, is considered by evaluating the costs per unit strength before and after exposure to these durability tests. Furthermore, a multicriteria performance index analysis is carried out to determine the optimum mix based on mechanical and durability properties and economic and environmental impacts of various screed mixes.

2   Experimental Work

2.1   Materials

Type I OPC and fly ash (FA) were used to produce the cement-based (CM) screed flooring material at a ratio of 4:1. At the same time, ground granulated blast furnace slag and FA were blended at a 1:1 and activated with a sodium-based alkaline solution to produce the geopolymer (GP) screed composites. The OPC, FA, and BFS were characterized by a specific gravity and Blaine fineness values of 3.16/3350 g/cm^2, 2.32/3680 g/cm^2, and 2.5/4350 g/cm^2, respectively. Moreover, X-ray fluorescence (XRF) analysis showed that the FA consisted mainly of silica and alumina, while calcium oxide and silica were predominant in Portland cement and BFS. Details of the chemical composition of the binding materials can be found elsewhere [4, 10].

The alkaline-activated solution (AAS) was produced by mixing a sodium hydroxide solution (SH) having a molarity of 8 and a sodium silicate solution (SS) with an SH-to-SS ratio of 1.5. The SS solution, with a density of 1506 kg/m^3, was composed of 26.3% SiO_2, 10.3% Na_2O, and 53.4% H_2O. Polycarboxylate superplasticizers, having a specific gravity of 1.2, and additional water were added to CM and GP mixes, respectively, to achieve the minimal targeted flow of 15 cm.

Dune sand (DS) and crushed limestone sand (CS) served as fine aggregates with a maximum particle size of 0.60 mm and 2.36 mm, respectively. Their corresponding d_{50} values were 0.80 mm and 0.26 mm. XRF analysis showed that CS is primarily composed of CaO, as it is made from crushed limestone aggregates, while DS was found to be mainly composed of SiO_2 and CaO. Detailed chemical and physical properties of the aggregates used herein can be found elsewhere [4, 10].

2.2 Mix Design

Tables 1 and 2 present the mix design of CM and GP screeds, respectively. Mixes were designated as X/Y/Z/U/W, where X, Y, Z, U, and W stand for screed category (A or B), type of screed (CM or GP), type of sand (DS or CS), binder-to-sand ratio, and liquid-to-binder ratio (water or AAS), respectively. The binder-to-sand and

Table 1 Mix design of CM screeds

Mix designation	Screed category	Binder content (kg/m^3)	Fly ash-to-cement	Binder-to-sand	Type of sand	Water-to-binder	Superplasticizer (%)
A/CM/D/1: 2/0.50	A	625	1:4	1:2	Dune sand	0.50	–
B/CM/D/1: 2/0.60	B	590	1:4	1:2	Dune sand	0.60	–
B/CM/D/1: 3/0.60	B	475	1:4	1:3	Dune sand	0.60	1.5
A/CM/C/1: 3/0.60	A	488	1:4	1:3	Crushed sand	0.60	–
A/CM/C/1: 4/0.60	A	413	1:4	1:4	Crushed sand	0.60	–
B/CM/C/1: 4/0.65	B	400	1:4	1:4	Crushed sand	0.65	–

Table 2 Mix design of GP screeds

Mix designation	Screed category	Binder content (kg/m^3)	Fly ash-to-slag	Binder-to-sand	Type of sand	AAS-to-binder	Additional water-to-binder
A/GP/D/1:2/0.55	A	520	1:1	1:2	Dune sand	0.55	0.05
B/GP/D/1:3/0.45	B	420	1:1	1:3	Dune sand	0.45	0.25
A/GP/D/1:3/0.55	A	400	1:1	1:3	Dune sand	0.55	0.20
A/GP/C/1:3/0.55	A	440	1:1	1:3	Crushed sand	0.55	0.05
A/GP/C/1:3/0.60	A	430	1:1	1:3	Crushed sand	0.60	0.05
B/GP/C/1:7/0.60	B	230	1:1	1:7	Crushed sand	0.60	0.53

Table 3 GWP indices and cost of raw material

	OPC	FA	Slag	DS	CS	Water	SS	SH	SP
Cost ($/kg)	0.076	0.089	0.076	0.005	0.007	0.002	0.272	0.544	1.928
GWP (kgCO$_2$eq./kg)	0.898	0.027	0.042	0	0.003	0.013	0.424	0.460	1.880
Reference	[11]	[11]	[11]	[11]	[11]	[11]	[12]	[13]	[11]

liquid-to-binder ratios were adjusted to produce Categories A and B screed flooring materials. Category A represents screeds with an indentation less than 3 mm due to BRE drop hammer testing, while Category B includes screed flooring with indentation values ranging between 3 and 4 mm. For example, mix B/CM/D/1:2/0.60 represents a Category B screed mix made with cement and fly ash as the blended binder, dune sand as the fine aggregate, a binder-to-sand ratio of 1:2, and a water-to-binder ratio of 0.60. In GP screed mixes, additional water was added to ensure an adequate flow for placement and compaction (flow >15 cm). A superplasticizer was used for the same reason in CM screed composites.

2.3 Global Warming Indices and Cost of Materials

The global warming potential (GWP) indices were measured by calculating the total mass of CO_2 emitted from the production of 1 m^3 of screed material (kgCO$_2$eq./m^3). To normalize the GWP against strength, the total mass of CO_2 emitted from the production of 1 m^3 of material was calculated per 1 MPa of 28-day compressive strength. The GWP indices and the cost per 1 kg of the raw materials used in this study are presented in Table 3.

The economic impact analysis of different screed samples was conducted based on the total price (in USD, $) of materials needed for the production of 1 m^3 of screed material. Furthermore, the cost of 1 m^3 of screed material per 1 MPa of 28-day compressive strength ($/m^3/MPa) was calculated prior to any durability testing. Subsequently, the cost per mass retained from abrasion ($/% LA) and the cost per strength retained from acetic acid ($/% f_c Acetic), hydrochloric acid ($/% f_c Hydrochloric), and salt attack ($/% f_c Salt) was determined to evaluate the cost of screed mixes as a function of their durability performance. The costs were calculated using Eqs. (1)–(6), where Price_i, Volume_i, 28d f'_c, Final Mass, Initial Mass, 56d f_c After Attack, and 56d f_c represent the price per unit mass of raw material i (in $/kg or $/m^3), volume of raw material i required per m^3 of screed, compressive strength achieved after 28 days of curing, mass of the screed after abrasion testing, initial mass of the material before abrasion testing, compressive strength after exposure ambient cured for 28 days and exposed to salt and acids for another 28 days, and compressive strength achieved after 56 days of curing, respectively:

$$\text{Cost (\$)} = \Sigma(\text{Price\_}i \times \text{Volume\_}i) \tag{1}$$

$$\text{Cost }(\$/\text{m}^3/\text{MPa}) = \text{Cost (\$)}/28\text{d}f'_c \tag{2}$$

$$\text{Cost }(\$/\%\text{LA}) = \text{Cost (\$)}/\text{Mass\_retained} \tag{3}$$

$$\text{Mass\_retained }(\%) = \text{Final mass}/\text{Initial mass} \times 100 \tag{4}$$

$$\text{Cost }(\$/\%f_c \text{ Acetic or hydrochloric or salt}) = \text{Cost (\$)}/\text{Strength\_retained} \tag{5}$$

$$\text{Strength\_retained} = 56\text{d}f_c \text{ After attack}/56\text{d}f_c \times 100 \tag{6}$$

2.4 Multicriteria Performance Index Analysis

Multicriteria Performance Indexing (PI) is a comprehensive tool designed to determine the optimal combination by considering various measures and data, encompassing mechanical characteristics, durability properties, cost, and GWP indices. This management tool effectively combines different performance criteria associated with distinct datasets into a single evaluation measure. The indexing and analysis process involve detailed steps, as shown in other work [14]. In this work, four distinct scenarios were selected to determine the optimal mix, as follows:

- Scenario 1: Mechanical properties including the BRE indentation, 28-day compressive strength (28d f_c), flexural strength (f_r), and pull-off bonding strength (f_b). Testing was carried out in accordance with BS 8204 [6], ASTM C109 [15], ASTM C348 [16], and ASTM D7234 [17], respectively.
- Scenario 2: Durability properties include water absorption (WA), mass retained caused by LA abrasion (LA), and compressive strength retained after acetic acid

attack (56d f_c Acetic), hydrochloric acid attack (56d f_c Hydrochloric), and salt attack (56d f_c Salt). Water absorption and LA abrasion were done following the ASTM C642 [18] and C1747 [19], while acid and salt attack testing was done following the ASTM C267 [20].

• Scenario 3: Cost and environmental impact analyses data.
• Scenario 4: All response criteria.

3 Results and Discussion

3.1 Environmental Impact

Figure 1 highlights the GWP indices for the various screed mixes. The results revealed a significant difference between the GWP and GWP/MPa values of CM and GP screed, ranging from 298.2 to 456.3 kgCO$_2$eq./m^3 and from 11.7 to 22.5 kgCO$_2$eq./m^3/MPa for CM and 57.2 to 105.6 kgCO$_2$eq./m^3 and from 1.7 to 7.9 kgCO$_2$eq./m^3/MPa for GP, respectively. This notably higher GWP in CM-based screeds is attributed to the substantial CO$_2$ emissions generated during the production of OPC, which stands at 0.898 kgCO$_2$eq./kg. While variations in the water content during CM production led to lower GWP values, GP screed experienced higher GWP with an increase in the AAS content. This outcome is primarily due to the elevated GWP values of the SS and SH (Table 3).

When considering the sand-to-binder content ratio, CM and GP screed materials exhibited reduced GWP when a higher volume of sand was used in the mix. Similarly, replacing CS with DS generally decreased the GWP. This reduction is attributed to the lower processing requirements of DS. It should be noted that GWP values for CM mix with similar design ratios, but different sand types showed minimal difference. For example, mix A/CM/D/1:3/0.6 had a GWP of

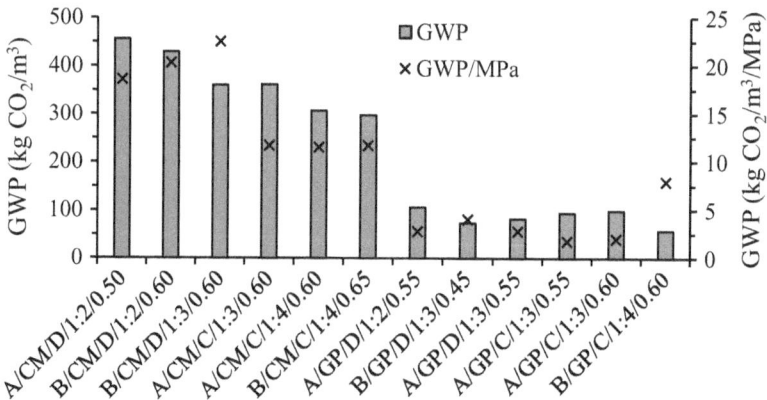

Fig. 1 Global warming potential indices of screed composites

360.7 kgCO$_2$eq./m^3, while mix A/CM/C/1:3/0.6 had a GWP of 361.5 kgCO$_2$eq./m^3. Although DS had a lower footprint than CS, the use of high-range water reducers in DS-based mixes for flowability purposes caused a slight increase in the GWP, leading to a similar overall GWP of the two mixes. Contrarily, for GP screed mixes, replacing CS with DS resulted in a noticeable decrease in the GWP, with values decreasing from 94.0 to 82.0 kgCO$_2$eq./m^3.

3.2 Economical Impact

The economic impact was measured using the cost per 1 m^3 of production ($/m^3), the cost per 1 m^3 of production per 1 MPa of compressive strength ($/m^3/MPa), and the ratio of the cost of 1 m^3 per mass or strength retained under abrasion, acid, and salt attack. For these calculations, the mechanical properties of the screed mixes before and after durability testing are presented in Table 4. Graphs, plots, and analysis of different parameter can be found elsewhere [4, 10].

The cost of screed mixes is shown in Fig. 2a. Meanwhile, the cost per unit strength and the cost as a function of a durability property are shown in the remaining columns of Fig. 2b. Upon analyzing the data, it is evident that the production cost of GP screed material is higher than that of CM-based mixes, with CM mixes ranging from $42.8 to $58.9/m^3 and GP mixes in the range of $60.2–$111.1/m^3. This increase in cost for GP materials can be attributed to the higher production costs associated with AAS. Furthermore, the cost of production tended to decrease with an increase in the sand and water contents, a decrease in AAS and binder contents, and the replacement of CS with DS. This indicates that these factors contribute to lowering the overall production expenses.

While GP mixes seemed to have higher production costs, the economic impact per unit strength led to different conclusions. For example, in mix A/GP/C/1:3/0.55, the cost of 1 m^3 of GP screed was reduced to as low as 1.8 $/MPa. As the mixtures incorporate more sand and water, reduced AAS and binder content, and replacement of CS by DS, the cost per 1 MPa of strength tended to increase. This suggests that while these adjustments may decrease production costs, they may increase the cost-to-strength ratio due to lower strength values. For example, the cost of mixes A/GP/D/1:3/0.55 and A/GP/C/1:3/0.55 were $87.5/m^3 and $98.4/m^3, respectively. However, when assessing the cost-to-strength ratio, these two mixtures displayed costs of $3.0/m^3/MPa and $1.8/m^3/MPa, respectively.

The cost analysis of screed composites was also conducted as a function of durability performance, including mass and compressive strength retention following abrasion, acid, and salt exposure. To obtain the cost per mass retention, the cost (Fig. 2b) was divided by the LA abrasion mass retention (Table 4). As for the cost per strength retained, the ratio of the 56-day strength after durability testing to the 56-day strength before durability testing was first found. Then, the cost was divided by this ratio to obtain the cost per unit strength retained. The cost ranged between $0.4 and $1.8/m^3/% of either mass or strength retained, except for Mix B/CM/D/1:3/

Table 4 Mechanical and durability properties of screed flooring materials

Mix designation	Ind. (mm)	28d f'_c (MPa)	56d f_c (MPa)	f_r (MPa)	f_b (MPa)	WA (%)	LA (%)	56d f_c Acetic (MPa)	56d f_c Hydrochloric (MPa)	56d f_c Salt (MPa)
A/CM/D/1:2/0.50	2.3	24.5	29.4	4.7	1.4	10.9	39.4	16.3	17.0	26.0
B/CM/D/1:2/0.60	3.1	21.1	27.6	4.0	1.1	11.6	28.8	14.6	18.5	25.1
B/CM/D/1:3/0.60	3.3	16.0	20.4	3.6	0.9	15.6	12.7	6.6	8.8	13.7
A/CM/C/1:3/0.60	2.3	30.8	33.5	5.2	1.5	6.9	61.7	16.4	19.9	31.0
A/CM/C/1:4/0.60	2.4	26.6	3.8	5.0	1.7	7.2	57.7	13.7	18.1	31.6
B/CM/C/1:4/0.65	3.2	25.5	29.6	4.9	1.3	8.0	53.6	12.5	19.2	30.5
A/GP/D/1:2/0.55	1.7	37.6	37.3	5.7	1.0	8.2	62.1	32.4	31.7	55.3
B/GP/D/1:3/0.45	3.9	18.3	18.2	3.5	0.5	9.8	45.4	15.1	14.1	23.1
A/GP/D/1:3/0.55	2.0	29.3	29.1	4.3	0.9	9.6	57.7	24.7	20.3	38.0
A/GP/C/1:3/0.55	0.9	53.2	52.3	6.9	1.6	5.9	74.4	44.2	44.4	77.7
A/GP/C/1:3/0.60	0.6	49.2	53.2	6.9	1.4	6.3	72.1	37.1	34.7	66.7
B/GP/C/1:7/0.60	3.3	7.2	7.7	2.3	0.4	8.7	46.5	6.9	5.9	10.1

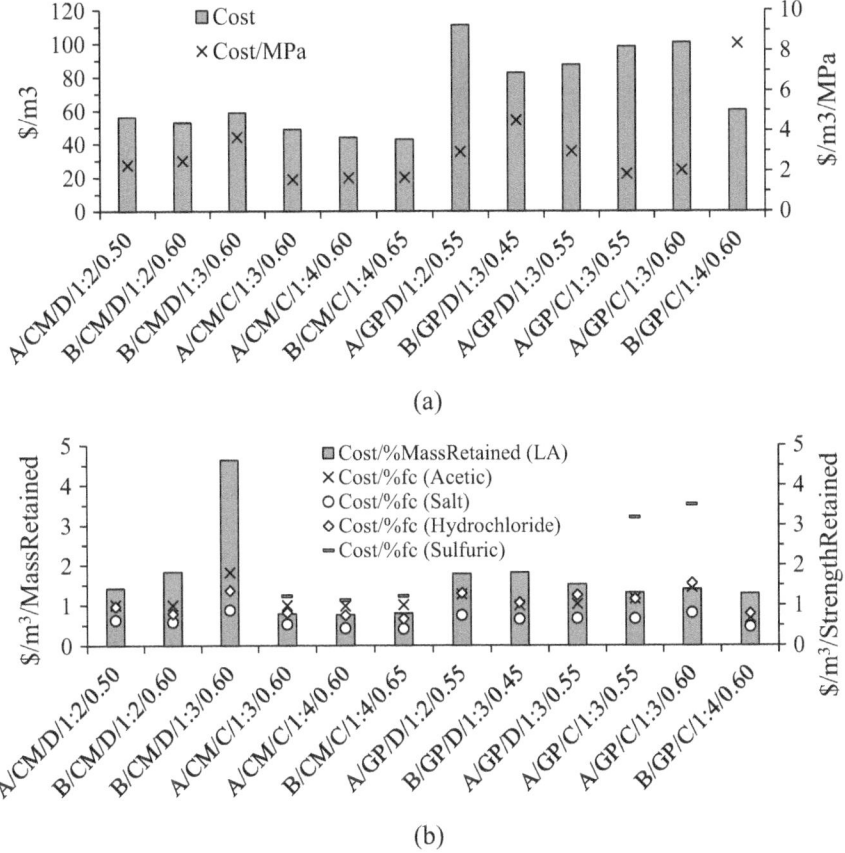

Fig. 2 (**a**) Cost per 1 m³ of screed production, (**b**) cost based on durability performances

0.6, which reached $4.6/m³/% Mass Retained (LA). This shows that the variations in the mix design had a limited effect on the cost of screed mixes as a function of durability performance.

3.3 Performance Index Analysis

Following the four scenarios of Sect. 2.4, the PI results for various screed flooring mixes are summarized in Table 5. The mixes were classified under categories A and B. Among Category A screeds, GP-based mixes incorporating CS as fine aggregates demonstrated superior performance for "Mechanical," "Durability," and "All Results" scenarios. However, when cost and environmental impacts were individually considered, the A/CM/C/1:4/0.6 mix, composed of cement and fly ash as binder

Table 5 Performance index analysis for categories A and B screed mixes

	Mechanical	Durability	Environmental and economic impact	All results
Category A				
A/CM/D/1:2/ 0.50	8.8	1.4	4.9	0.0
A/CM/C/1:3/ 0.60	13.2	4.7	55.2	1.6
A/CM/C/1:4/ 0.60	11.9	3.3	100.0	1.8
A/GP/D/1:2/0.55	15.9	22.4	12.5	1.1
A/GP/D/1:3/0.55	7.2	6.0	34.9	0.4
A/GP/C/1:3/0.55	82.4	100.0	30.1	100.0
A/GP/C/1:3/0.60	100.0	50.7	13.2	24.3
Category B				
B/CM/D/1:2/ 0.60	59.0	33.7	5.7	1.9
B/CM/D/1:3/ 0.60	30.9	1.3	0.5	0.0
B/CM/C/1:4/ 0.65	100.0	100.0	39.0	100.0
B/GP/D/1:3/0.45	16.2	46.0	14.3	1.0
B/GP/C/1:7/0.60	4.0	4.4	100.0	0.1

and CS as fine aggregates, was found to be the most suitable mix, notably because of the high cost and GWP of the AAS solution used in producing GP screeds.

For Category B screeds, CM-based mixes incorporating CS as fine aggregates demonstrated superior performance across all scenarios except for the "Cost and Environmental Impact" scenario, where B/GP/C/1:7/0.60 mix was the most suitable mix, owing to its high dune sand content.

4 Conclusions

This paper evaluates the environmental and economic impacts of various screed composites produced with cementitious (CM) or geopolymeric (GP) binder and dune sand (DS) or crushed limestone sand (CS). A multicriteria performance index analysis was conducted to identify the optimal mixes for different scenarios. The following conclusions can be drawn:

- The environmental impact assessment of screed composites, as determined by the global warming potential, demonstrated that GP-based screeds exhibited lower values ranging from 57.2 to 105.6 $kgCO_2eq./m^3$ compared to their CM counterparts, which ranged from 298.2 to 456.3 $kgCO_2eq./m^3$.

- Economically, GP screed materials had a higher production cost per 1 m^3 than CM-based mixes. However, GP screeds exhibited a reduced difference when considering cost per 1 MPa of compressive strength, indicating competitive strength-related costs. Furthermore, GP samples showed competitive durability-related expenses ranging from $0.4 to $1.8/m^3/% Mass-Strength Retained.
- Among Category A screeds, GP-based mixes with CS showed superior performance in "Mechanical," "Durability," and "All Results" scenarios. In Category B, CM mix with CS excelled in all scenarios except "Cost and Environmental Impact," where the B/GP/C/1:7/0.60 mix was superior, attributed to its high dune sand content.

Acknowledgments The authors acknowledge the financial support provided by UAEU under grant 21N235. The support of UAEU staff is appreciated.

References

1. Hwalla, J., El-Mir, A., El-Hassan, H., El-Dieb, A.: Taguchi method for optimizing alkali-activated mortar mixtures using waste perlite powder and granulated blast furnace slag. In: Jędrzejewska, A., Kanavaris, F., Azenha, M., Benboudjema, F., Schlicke, D. (eds.) Proceedings of the International RILEM Conference on Synergising Expertise towards Sustainability and Robustness of Cement-based Materials and Concrete Structures, pp. 362–373. Springer Nature Switzerland, Cham (2023)

2. Hwalla, J., Bawab, J., El-Hassan, H., Abu Obaida, F., El-Maaddawy, T.: Scientometric analysis of global research on the utilization of geopolymer composites in construction applications. Sustainability. **15**, 11340 (2023). https://doi.org/10.3390/su151411340

3. Hwalla, J., Al-Mazrouei, M., Al-Karbi, K., Al-Hebsi, A., Al-Ameri, M., Al-Hadrami, F., El-Hassan, H.: Performance of alkali-activated slag concrete masonry blocks subjected to accelerated carbonation curing. Sustainability. **15**, 14291 (2023). https://doi.org/10.3390/su151914291

4. Hwalla, J., El-Hassan, H., Assaad, J.J., El-Maaddawy, T.: Performance of cementitious and slag-fly ash blended geopolymer screed composites: a comparative study. Case Stud. Constr. Mater. **18**, e02037 (2023). https://doi.org/10.1016/j.cscm.2023.e02037

5. Georgin, J.F., Ambroise, J., Péra, J., Reynouard, J.M.: Development of self-leveling screed based on calcium sulfoaluminate cement: modelling of curling due to drying. Cem. Concr. Compos. **30**, 769–778 (2008). https://doi.org/10.1016/j.cemconcomp.2008.06.004

6. British Standard Institution: *Screeds, Bases and In Situ Floorings – Concrete Bases and Cement Sand Leveling Screeds to Receive Floorings – Code of Practice*. British Standard (2003)

7. Kanagaraj, B., Anand, N., Johnson Alengaram, U., Samuvel Raj, R.: Engineering properties, sustainability performance and life cycle assessment of high strength self-compacting geopolymer concrete composites. Constr. Build. Mater. **388**, 131613 (2023). https://doi.org/10.1016/j.conbuildmat.2023.131613

8. Shi, X., Zhang, C., Liang, Y., Luo, J., Wang, X., Feng, Y., Li, Y., Wang, Q., Abomohra, A. E.-F.: Life cycle assessment and impact correlation analysis of fly ash geopolymer concrete. Materials. **14**, 7375 (2021). https://doi.org/10.3390/ma14237375

9. Nguyen, L., Moseson, A.J., Farnam, Y., Spatari, S.: Effects of composition and transportation logistics on environmental, energy and cost metrics for the production of alternative cementitious binders. J. Clean. Prod. **185**, 628–645 (2018). https://doi.org/10.1016/j.jclepro.2018.02.247

10. Hwalla, J., El-Hassan, H., Assaad, J.J., El-Maaddawy, T.: Durability assessment of geopolymeric and cementitious composites for screed applications. J. Build. Eng. **87**, 109037 (2024). https://doi.org/10.1016/j.jobe.2024.109037

11. Alzard, M.H., El-Hassan, H., El-Maaddawy, T.: Environmental and economic life cycle assessment of recycled aggregates concrete in the United Arab Emirates. Sustainability. **13**, 10348 (2021). https://doi.org/10.3390/su131810348

12. Davidovits, J.: False Values on CO2 Emission for Geopolymer Cement/Concrete Published in Scientific Papers. Geopolymer Institute Library (2015)

13. Alhassan, M., Alkhawaldeh, A., Betoush, N., Alkhawaldeh, M., Huseien, G.F., Amaireh, L., Elrefae, A.: Life cycle assessment of the sustainability of alkali-activated binders. Biomimetics. **8**, 58 (2023). https://doi.org/10.3390/biomimetics8010058

14. El-Hassan, H., Elkholy, S.: Enhancing the performance of alkali-activated slag-fly ash blended concrete through hybrid steel fiber reinforcement. Constr. Build. Mater. **311**, 125313 (2021). https://doi.org/10.1016/j.conbuildmat.2021.125313

15. ASTM C109-20: Test Method for Compressive Strength of Hydraulic Cement Mortars (Using 2-in. or [50-Mm] Cube Specimens). ASTM International (2020)

16. ASTM C348: Test Method for Flexural Strength of Hydraulic-Cement Mortars. ASTM International (2021)

17. ASTM D7234: Test Method for Pull-Off Adhesion Strength of Coatings on Concrete Using Portable Pull-Off Adhesion Testers. ASTM International (2019)

18. ASTM C642-13: Test Method for Density, Absorption, and Voids in Hardened Concrete. ASTM International (2013)

19. ASTM C1747-13: Test Method for Determining Potential Resistance to Degradation of Pervious Concrete by Impact and Abrasion. ASTM International (2013)

20. ASTM C267-20. Test Methods for Chemical Resistance of Mortars, Grouts, and Monolithic Surfacings and Polymer Concretes. ASTM International (2020)

CO_2 Reductions Utilising Self-Stressing Steel Fibre Reinforced Concrete

Martins Suta, Liga Gaile, and Rolands Cepuritis

Contents

1 Introduction

The introduction of various standards and guidelines aimed at the design of steel fibre reinforced concrete (SFRC) structures has made the implementation of SFRC structures more widespread. For a multitude of years, SS812310 [1], an SFRC standard has been used, and now with the introduction of the new generation of Eurocode 2, which includes the informative Annex L [2] on fibre reinforced concrete

M. Suta (✉)
Primekss SIA, R&D Department, Riga, Latvia

Institute of Structural Engineering, Riga Technical University, Riga, Latvia
e-mail: martins.suta@primekss.com

L. Gaile
Institute of Structural Engineering, Riga Technical University, Riga, Latvia

R. Cepuritis
Department of Structural Engineering, Norwegian University of Science and Technology, Trondheim, Norway

© The Author(s) 2025
M. Kioumarsi, B. Shafei (eds.), *The 1st International Conference on Net-Zero Built Environment*, Lecture Notes in Civil Engineering 237,
https://doi.org/10.1007/978-3-031-69626-8_66

design, the application of SFRC in structures should grow even more. Although SFRC as a material is not new per se, the design methods and safety of the structures [3, 4] among other topics are still disputed and researched to achieve safe yet economically and environmentally viable solutions. One of the structures that are used extensively would be also ground-level slab supported on piles, which is discussed in detail in ACI544.6R-15 [5].

A novelty in the field of SFRC structures is the so-called steel fibre reinforced self-stressing concrete (SFRSSC), which is gaining popularity in the European, North American and other markets. The introduction of self-stressing agent in the randomly distributed fibre concrete matrix allows mimicking the traditional post-tensioned systems that would otherwise utilise steel cables, yet the self-stress would be achieved chemically. The system proves to be feasible for ground-level elevated slabs on piles [6, 7] and for other structures [8]. The chemical self-stressing of the structure acts similarly to traditional tensioned structures, giving the structure enhanced stiffness, reducing deflections, reducing cracking while also reducing the amount of necessary deformation joints. Previously, such a system has been looked at for slabs on ground where it proved to be a viable method of reducing CO_2 emissions [8] that can deliver tangible benefits already today which exceed those of introduction of GGBS or other supplementary materials to reduce the cement content in the concrete matrix.

The current publication is aimed to inform the community of the system being tested in full scale for ground-level elevated slabs on piles, once more proving the systems capability of reducing the CO_2 emissions by providing a stiffer, stronger slab when compared to conventional methods. The study shows the possible thickness reductions and subsequent CO_2 emission reductions and the differences in the behaviour of the SFRSSC in direct comparison to SFRC under the same conditions. The approbation of the system in said uses follows Eurocode 0 Section 5 Structural Analysis and Design Assisted by Testing.

2   Experimental Programme

The conducted full-scale experiments have been introduced to the public recently [9]. Nevertheless, the most crucial data will also be presented here for better context.

The full-scale test was conducted on two slabs, one of SFRC and one of SFRSSC. The tested slabs were directly placed on a grid of columns, mimicking piles, spaced at a grid of 3000 mm by 3000 mm, diameter of the columns being 300 mm. To imitate the internal span behaviour without influence from close edges, additional spans of various distances were introduced as shown in Fig. 1. The slabs were constructed as flat slabs, without drop panels or pile heads, freely supported and unconnected to the columns. The thickness of both slabs was 150 mm, with a span-to-depth ratio being 20 which is in the common range of industrial elevated ground slabs.

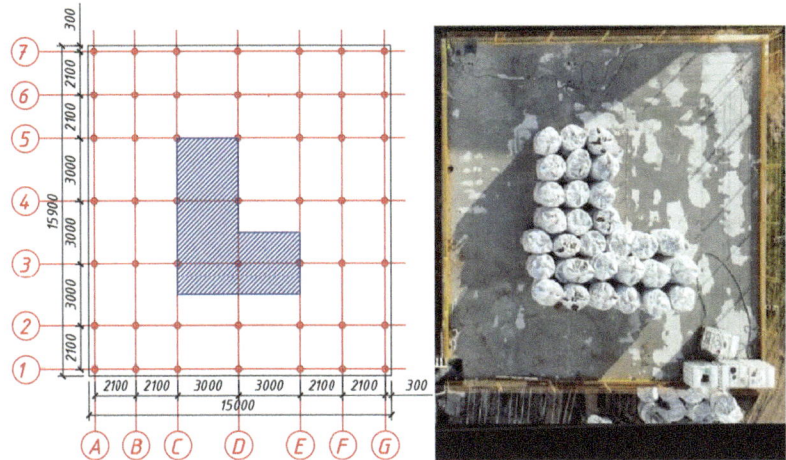

Fig. 1 Load layout: left: scheme and right: actual

Table 1 Material composition for SFRC and SFRSSC

	Amount (kg/m^3)	
Material	SFRC	SFRSSC
CEM II/A-LL 52.5 N	360	360
Water	162	162
w/c ratio	0.45	0.45
0–8 mm sand	952	952
5–16 mm crushed dolomite	914	914
PrīmX flow HRWRA	3.6	3.6
Hooked end steel fibres ($L = 60$ mm, $d = 0.9$ mm)	50	50
PrīmX DC	–	40

2.1 SFRC and SFRSSC Composite Composition

The material composition of the SFRC and SFRSSC is given int Table 1.

The mix designs show the only difference between the two slabs being a self-stressing agent with a brand name PrīmX DC. The reinforcement of the slabs is fibre only, 0.9 mm in diameter, 60 mm in length, and hooked-end steel fibres with 1500 MPa tensile strength. The concrete met the requirements of C30/37 concrete class with an S4 slump.

To reduce the necessary water amount and maintain workability, PrīmX Flow high-range water reducing admixture (HRWRA) was introduced for both mix designs.

2.2   Loading

The full-scale slabs were tested by placing loads on the slab in SLS levels and up to collapse; thus, the test was load-governed. The assumed SLS level for the comparison was set to be approximately 25 kN/m², but the actual SLS levels for the slabs given the criteria of deflections and cracking for such slabs were observed after the test based on measurements of the behaviour of the slabs. The loading layout was chosen following a similar test done for a similar structure [10].

The loading pattern allows achieving the maximum possible positive and negative bending moments and the largest possible deflections, see Fig. 1.

The slab was loaded by placing sandbags in layers, where each layer is regarded as a load level (LL). Seven distinctive points in the loading programme can be detailed, called loading situations (LSs), see Fig. 2.

The loading material chosen was dry, sieved sand, weighed, and secured in big-bags of 1000 kg or 10 kN per bag. This allows for the loading steps to be of uniform value, and the weight of the bags did not change due to rain and water puddling on the slabs. The usage of sand in big-bags allowed to alleviate any possible interlock between the bags, thus eliminating any arching effect which would otherwise be present in stiff materials, *p.e.* bricks, pavement bricks, concrete blocks etc. The assumed SLS load level, LL3, is set to be 25 kN/m².

2.3   Instrumentation on Site

To measure the structural response of the slabs, a multitude of measuring systems was introduced in the experiment, *p.e.* fibre optic sensors for strain measuring, digital image correlation, linear variable differential transformer displacement

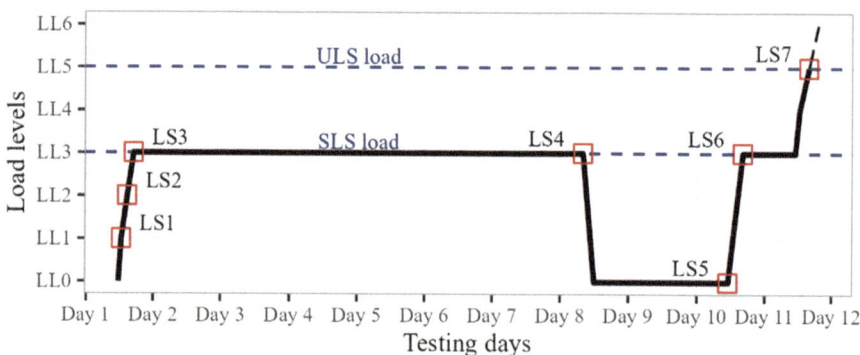

Fig. 2 Loading sequence with characteristic points denoted as loading situations

sensors (LVDTs) for deflection, piezoelectric strain gauges and vibrating wire strain gauges (VWSG), among others. The instrumentation is described in more detail in the following publication [9] and will not be further discussed in this publication.

2.4   Lab Tests

A multitude of samples were prepared during the pouring of the slabs to assess the various properties of the used concrete, *p.e.* concrete compressive strength, residual flexural strength and plastic tensile strength, among others. The samples created ranged from cubes of $100 \times 100 \times 100$ mm^3 according to EN12390-2, $150 \times 150 \times 600$ mm^3 beams according to EN14651 and EN12390-5, slabs $d = 800$ mm and $h = 100$ mm according to ACI544.6R-15, slabs $d - 1500$ mm and $h = 180$ mm according to ACI544.6R-15, up to samples of beams, cores, cylinders and slabs cut out and drilled from the tested structure itself.

The post-cracking residual strength and the plastic tensile strength of the concrete were evaluated for a comparison of a similar testing methodology for a similar composition of SFRSSC continuing a previously done experiment [11, 12] where a comparison between different testing methods and their results was done to evaluate the impact of chosen testing methodology on the necessary thickness of the structure.

3   Equivalent Design

The results of the two slabs revealed that the SFRC slab exhibited larger deflections, larger crack openings and longer cracks lengths under the same loading conditions as SFRSSC, see Fig. 3.

The final failure load of the two slabs was also significantly different where the SFRC collapsed at 45.7 kN/m^2 and the SFRSSC collapsed at 60.5 kN/m^2, thus a 32.3% increase over the performance of the SFRC. At the same loading levels, the SFRSSC exhibited 33% lower deflections at the assumed SLS load, LL3. 6 mm for the SFRSSC and 9 mm for the SFRC. Additionally, the top surface cracking of the SFRSSC was limited to a single crack of 0.04 mm width, while the SFRC exhibited a noticeable cracking fan pattern on the top surface after LL3 loading, see Fig. 4.

Taking the results into account, it can be concluded that the actual values of SLS loading that meets the SLS criteria for the two slabs is 19 and 28 kN/m^2 for SFRC and SFRSSC, respectively, thus a 47% increase in the loading intensity. To achieve the same load bearing capacity, the slab of SFRC with the same material properties should therefore be designed with a thickness of 180 mm versus the 150 mm SFRSSC slab, thus a 16.6% decrease in thickness.

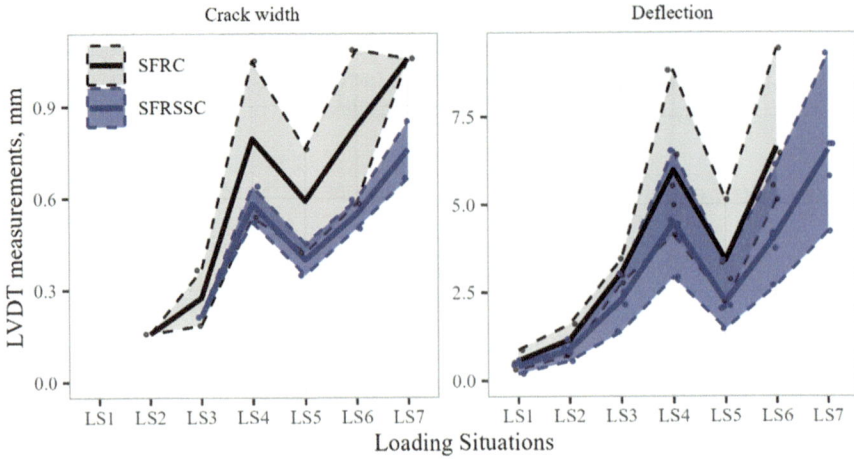

Fig. 3 Comparison of crack width and deflections measured by LVDTs for SFRC and SFRSSC slabs at different loading situations

Fig. 4 Comparison of crack patterns in place of maximums negative bending moment after LL3, left: SFRC and right: SFRSSC

4   CO$_2$ Emission Calculations

Based on the equivalent designs discussed in Sect. 4, a comparison of CO$_2$ emissions for multiple solutions, i.e., SFRSSC, SFRC and welded wire mesh reinforced concrete (WWMRC), can be done, see Table 2.

To develop an envelope of CO$_2$ emissions for the solutions ranging from the worst case to the best, a combination of the materials was used using either the greenest or the least green options. The outcomes and their comparisons are shown in Table 3. The greenest options are denoted with a letter "G." The solutions are ranked from 1 to 6 with 1 being the greenest and 6 the least green solution.

From Table 3, it can be clearly seen that at the same circumstances the SFRSSC solution emits the least amount of CO$_2$, while the WWMRC emits the most. This is true for both the greenest solutions and the less green solutions. Depending on the solutions to be compared, the SFRSSC solutions provide from 11% to 59% reduction while SFRSSC G solutions provide from 8% to 24% reduction when comparing similarly green solutions.

The presented calculation covers only the materials part and does not include other areas of productions, *p.e.* transportation of materials to site etc. Following that there is less material to be transported and poured in situ, the actual CO$_2$ savings are even higher than the values presented here, yet it is not feasible to cover this topic in this publication due to the transportation costs being highly linked with the location of the project and may highly vary.

Table 2 Comparison of CO$_2$ emissions for different components of concrete

Material	Specific emissions	CO$_2$ emissions (kg/m^2)		
		SFRSSC	SFRC	WWMRC
Concrete C30/37 [13]	273 kg CO$_2$eq/m^3	40.95	49.14	49.14
Green concrete C30/37 [13]	230 kg CO$_2$eq/m^3	32.25	38.70	38.70
Reinforcement [14]	1.14 kg CO$_2$eq/kg	–	–	36.94
Green reinforcement [15]	0.30 kg CO$_2$eq/kg	–	–	9.72
Steel fibres [16]	1.24 kg CO$_2$eq/kg	9.26	11.12	–
XCarb steel fibres [16]	0.37 kg CO$_2$eq/kg	2.74	3.29	–
HRWRA [17]	0.30 kg CO$_2$eq/kg	0.16	0.2	0.2
PrimX DC [18]	0.65 kg CO$_2$eq/kg	3.9	–	–

Table 3 Comparison of CO$_2$ emissions for different systems

System	CO$_2$ emissions (kg/m^2)	Rank
SFRSSC	54.28 kg CO$_2$eq/m^2	4
SFRSSC G	39.05 kg CO$_2$eq/m^2	1
SFRC	60.45 kg CO$_2$eq/m^2	5
SFRC G	42.18 kg CO$_2$eq/m^2	2
WWMRC	86.27 kg CO$_2$eq/m^2	6
WWMRC G	48.62 kg CO$_2$eq/m^2	3

5   Results and Discussion

The full-scale tests showed that the SFRSSC slab exhibited smaller deflections, less crack widths and less overall cracking at the same load levels as explained previously. The actual SLS load for SFRC was of 19 kN/m^2, while the SFRSSC reached SLS state at 28 kN/m^2; thus, for the same loading conditions, the SFRSSC slab can be thinner when compared to the SFRC slab due to the self-stressing effect of PrīmX DC.

From the premise that the SFRSSC slab can be thinner than the SFRC slab for the same loading conditions, it is possible to estimate the CO_2 footprint of the systems, which shows significant benefit in reduction of the CO_2 gas emissions by up to 59%. The calculations of the CO_2 emissions also show that at the worst case the SFRSSC solution has less emissions than at any of the best cases of the comparative solutions. In case green concrete and green fibres can be used, the greenhouse gas savings are increased significantly by up to 24%.

6   Conclusion

The full-scale tests of SFRC and SFRSSC along with appropriate calculations shown in this publication reveal the following conclusions:

- At failure, the SFRSSC slab held 33% higher load than SFRC.
- At the same loading conditions, SFRSSC exhibited approximately 33% lower deflections.
- At the same loading conditions, the SFRC exhibited almost no cracking on the top surface.
- For the same loading conditions, the SFRC would render a slab of 180 mm thickness versus 150 mm for SFRSSC.
- The usage of SFRSSC system lowers carbon emissions by 8–55% when compared to regular SFRC and by 10–51% when compared to WWMRC.

Acknowledgement The data presented in the publication are based on the research project executed by Primekss SIA, which has been funded based on an agreement No. 5.1.1.2.i.0/1/22/A/ CFLA/005 between CFLA and Primekss SIA, which has been granted from "Atveseļošanās un noturības mehānisma līdzfinansējums."

References

1. SS812310: Fibre Concrete – Design of Fibre Concrete Structures (2014)
2. LVS EN 1992-1-1: Eurocode 2 – Design of Concrete Structures – Part 1-1: General rules and rules for buildings, pp. 350–363 (2023)
3. Destree, X., Mobasher, B.: Industrial floors with fiber-reinforced concrete: a review of knowledge and experience. Concr. Int. **44**(7), 28–33 (2022)
4. Silfwerbrand, J.: Industrial fiber concrete floors. Concrete International. **43**(5), 33–36 (2021)
5. ACI Committee 544: ACI 544.6R-15 – Report on Design and Construction of Steel Fiber Reinforced Concrete Elevated Slabs. American Concrete Institute (2015)
6. Cepuritis, R., Pease, B.J., Kamars, J., Oslejs, J.: Pile-supported slabs for sites with poor geotechnical conditions. Concr. Int. **40**(11), 38–43 (2018)
7. Destree, X., Cepuritis, R., Fischer, G., Suta, M.: Effect of intrinsically post-tensioned steel fibre reinforced concrete on the structural response and design of slabs on grade and suspended slabs – From design stage to application, FIB 2018 – Proceedings for the 2018 fib Congress: Better, Smarter, Stronger, pp. 3410–3420 (2019)
8. Destree, X., Pease, B.J.: Reducing CO$_2$ Emissions of Concrete Slab Constructions with the PrimeComposite Slab System, vol. 299, pp. 1–12. American Concrete Institute (ACI) Special Publications (2015)
9. Suta, M., et al.: Full-scale tests of steel fibre reinforced concrete and self-stressing steel fibre reinforced concrete – Overview, XI International Symposium on Fiber Reinforced Concrete, BEFIB (unpublished at the time of writing this article) (2024)
10. Suta, M., et al.: Investigation of possibility of using steel fibre reinforced self-stressing concrete (SFRSSC) in watertight concrete structures. In: Engineering for Rural Development, pp. 638–644 (2020)
11. Suta, M., Lukasenoks, A., Cepuritis, R.: Determination of material design values for steel fibre reinforced self-stressing concrete (SFRSSC) structures. In: Engineering for Rural Development, pp. 631–637 (2020)
12. Suta, M., Cepuritis, R., Zegelis, A.: Determination of material design values for steel fibre reinforced self-stressing concrete (SFRSSC) and regular steel fibre reinforced concrete (SFRC) in statically indeterminate round panel tests. Mater. Sci. Forum. **1053**, 297–302 (2022)
13. Betong, S.: Vägledning klimatföbättrad betong – Utgåva 2 Available at: https://www.svenskbetong.se/images/pdf/Vagledning_Utgava_2_Webb. pdf. Accessed 8 Apr 2024 (2022)
14. Environmental Product Declaration, Steel Fibres EAF-Base, Arcelor Mittal
15. XCarb Recycled and renewably produced. https://barsandrods.arcelormittal.com/news/3406/XCarb_for_rebars. Accessed 8 Apr 2024
16. Decarbonised steel fibre reinforcement, XCarb Recycled and renewably produced. https://barsandrods.arcelormittal.com/repository2/Fibres/XCarb%20recycled%20and%20renewably%20produced%20steel%20fibres%20-%20tunnels.pdf. Accessed 8 Apr 2024
17. Nayana, A.Y., Kavitha, S.: Evaluation of CO$_2$ emissions for green concrete with high volume slag, recycled aggregate, recycled water to build eco environment. Int. J. Res. Civ. Eng. Technol. **8**(5), 703–708 (2017)
18. Chaunsali, P., Vaishnav, K.S.: Calcium Sulfoaluminate-Belite cements: opportunities and challenges. Indian Concr. J. **2020**, 18–25 (2020)

Evaluation of the Compressive Strength of Fly Ash- Based Geopolymer Concrete Using Machine Learning

Maryam Bypour, Mohammad Yekrangnia, and Mahdi Kioumarsi

Contents

1   Introduction

The conventional method of producing concrete with Ordinary Portland Cement (OPC) significantly contributes to pollution, primarily through the release of carbon dioxide during cement production and construction activities. Additionally, the disposal of concrete waste further exacerbates environmental issues, as it remains in the environment for extended periods of time. Consequently, researchers have turned their attention to finding sustainable alternatives to OPC. One such alternative gaining significance is geopolymer, which is formed through the reaction of various

M. Bypour (✉)
Department of Civil Engineering, Semnan University, Semnan, Iran
e-mail: mbypour@alum.semnan.ac.ir

M. Yekrangnia
Department of Civil Engineering, Shahid Rajaee Teacher Training University, Tehran, Iran
e-mail: yekrangnia@sru.ac.ir

M. Kioumarsi
Department of Built Environment, OsloMet – Oslo Metropolitan University, Oslo, Norway
e-mail: mahdi.kioumarsi@oslomet.no

© The Author(s) 2025
M. Kioumarsi, B. Shafei (eds.), *The 1st International Conference on Net-Zero Built Environment*, Lecture Notes in Civil Engineering 237,
https://doi.org/10.1007/978-3-031-69626-8_67

materials, including natural pozzolans, industrial by-products like fly ash (FA) and Ground Granulated Blast Furnace Slag (GGBS), and agricultural waste such as rice husk ash and sugarcane bagasse. When activated with alkalis, these materials offer a promising substitute for OPC in concrete production [1, 2].

Geopolymer concrete (GPC) is a viable alternative to conventional concrete. Recent research on this rather new type of concrete has revealed promising outcomes concerning its tensile and compressive strength and durability across different curing conditions [3].

Researchers have examined the performance of GPC using various nanomaterials and additives, with varied percentages of components under various conditions [4]. One extensively studied variant of GPC is fly ash-based GPC, which has been subjected to rigorous experimental studies and comparison with the outcomes obtained from incorporating other additives such as GGBS [2, 5].

This study aimed to assess the impact of various components on the compressive strength of fly ash-based GPC. For that purpose, a comprehensive experimental dataset was gathered. Initially, the accuracy of linear regression to estimate the compressive strength of the specimens was investigated. Subsequently, a machine learning approach, such as XGBoost, was implemented. Finally, the significance of each feature in influencing the compressive strength of fly ash-based GPC was analyzed and evaluated.

2   Dataset Construction

2.1   Experimental Data Collection

In the framework of this study, the effect of various constituents within fly ash-based geopolymer concrete on compressive strength (f'_c) of GPC has been explored. The dataset collected to assess through a machine learning approach is outlined in Table 1. Notably, all examined specimens feature fly ash-based geopolymer concrete, with some incorporating Ground Granulated Blast Furnace Slag (GGBS), Coal Bottom Ash (CBA), and Copper Slag (CS), alongside additional additives like steel fiber and rubber. A comprehensive dataset comprising 157 specimens has been compiled for analysis.

In the selected specimens, GPC with alkaline activator of NaOH and Na_2SiO_3 with concentration ranging from 1.4 to 16 M was studied. To ensure a comparable dataset, the specimens that are cured in ambient condition and tested on the 28th day were selected.

2.2   Data Visualization

The components constituting a standard FA are elaborated in Table 2. It is evident from the data that SiO_2 comprises the largest portion of FA. Following closely is Al_2O_3, with Fe_2O_3 being the third most prevalent component.

Table 1 Experimental dataset

Research	Source Material	Alkaline solution	Coarse aggregates	fine aggregates	Extra additive	Curing condition	Testing age (Day)
Hassan et al. [6]	FA	NaOH and Na$_2$SiO$_3$ (10 M)	✓	✓	–	Ambient	28
Bellum et al. [7]	FA + GGBS	NaOH and Na$_2$SiO$_3$ (8 M)	✓	✓	–	Ambient	28
Xie et al. [8]	FA + GGBS	NaOH and Na$_2$SiO$_3$ (8 M)	✓	✓	–	Ambient	28
Ghafoor et al. [9]	FA	NaOH and Na$_2$SiO$_3$ (8, 10, 12, 14, and 16 M)	✓	✓	–	Ambient	28
Okoye et al. [10]	FA + Silica fume	NaOH (14 M)	✓	✓	-	Ambient	28
Mehta and Siddique [11]	FA	NaOH and Na$_2$SiO$_3$ (10 M)	✓	✓	–	Ambient	28
Farhan et al. [12]	FA + GGBS	NaOH and Na$_2$SiO$_3$ (12 M)	✓	✓	–	Ambient	28
Law et al. [13]	FA	NaOH and Na$_2$SiO$_3$ (10 M)	✓	✓	–	Ambient	28
Haque et al. [14]	FA+ CBA+ CS	NaOH and Na$_2$SiO$_3$ (12 M)	✓	✓	–	Ambient	28
Zheng et al. [15]	FA + GGBS	NaOH and Na$_2$SiO$_3$ (1.4 M)	✓	✓	Steel fiber	Ambient	28
Ojha and Aggarwal [16]	FA	NaOH and Na$_2$SiO$_3$ (14 M)	✓	✓	–	Ambient	28
Park et al. [17]	FA	NaOH and Na$_2$SiO$_3$ (8 and 14 M)	✓	✓	Rubber	Ambient	28
Laxmi et al. [18]	FA + GGBS	NaOH and Na$_2$SiO$_3$ (8 M)	✓	✓	–	Ambient	28
Nath and Sarker [19]	FA + GGBS	NaOH and Na$_2$SiO$_3$ (14 M)	✓	✓	–	Ambient	28
Lee and Lee [20]	FA + Slag	NaOH (4 and 6 M)	✓	✓	–	Ambient	28

FA: Fly ash, *GGBS*: Ground granulated blast furnace slag, *CBA*: Coal bottom ash, *CS*: Copper slag

Table 2 Components of FA particles (%)

Research	SiO_2	Al_2O_3	Fe_2O_3	CaO	P_2O_5	SO_3	K_2O	TiO_2	MgO	Na_2O	MnO	Mn_2O_3
Haque et al. (2023) [14]	60.57	26.92	5.01	1.43	0	0.22	1.31	1.69	0.83	0.12	0	0
Bellum et al. (2020) [7]	58.23	25.08	4.56	2.87	0.2	1.16	0.87	0.83	1.21	0.41	2.94	0
Farhan et al. (2019) [12]	62.2	27.5	3.92	2.27	0.3	0.08	1.24	0.16	1.05	0.52	0	0.09
Law et al. (2015) [13]	49.45	29.61	10.72	3.47	0.53	0.27	0.54	1.76	1.3	0.31	0	0.17

As the target is to evaluate the effectiveness of each component on the compressive strength (f'_c) of GPC, the first step it to compare the effect of all features on f'_c using Probability and Statistics. A total of 16 features are investigated, in which SiO_2, Al_2O_3, Fe_2O_3, CaO, P_2O_5, SO_3, K_2O, TiO_2, MgO, Na_2O, MnO, steel fiber, rubber, coarse aggregates, and fine aggregates are in Kg/m^3 and NaOH is in Mol/Liter.

The correlation matrix offers statistical insights into the relationships between input and output parameters [21, 22], which is shown in Fig. 1a, b. The figure indicates that there is a strong correlation between the amount of CaO and MgO, with values of 0.55 and 0.54, respectively, and f'_c of the selected specimens. These positive values signify that as the quantities of these features increase, it results in an increase in f'_c. Additionally, the weakest correlation is related to SiO_2 and f'_c.

2.3 Linear Regression

Linear regression (LR) can be used to estimate the relationship between a dependent response and independent variables [23, 24]. To mitigate overfitting, where a model excels on the training dataset but fails to perform properly on new data, it is crucial to evaluate the model using a separate test dataset. This evaluation enhances the accuracy of the model. Additionally, employing the R^2 score method, as depicted in Eq. (1), enables the determination of the level of agreement between two parameters, namely the actual values and the predicted targets [25]. In Eq. (1), $y_{test,\,i}$ and $\widehat{y}_{test,i}$ represent the actual and predicted values for this set, respectively. The mean value of the target for this dataset is denoted by $\bar{y}_{test}$, and N represents the number of specimens.

$$R^2 = 1 - \frac{\sum_{i=1}^{N} \left(\widehat{y}_{test,i} - y_{test,i}\right)^2}{\sum_{i=1}^{N} \left(\widehat{y}_{test,i} - \bar{y}_{test}\right)^2} \tag{1}$$

Using LR, as shown in Fig. 2, R^2 value for train and test data was obtained 0.73 and 0.78, respectively. This indicates that LR can predict the f'_c of the gathered specimens with an acceptable accuracy.

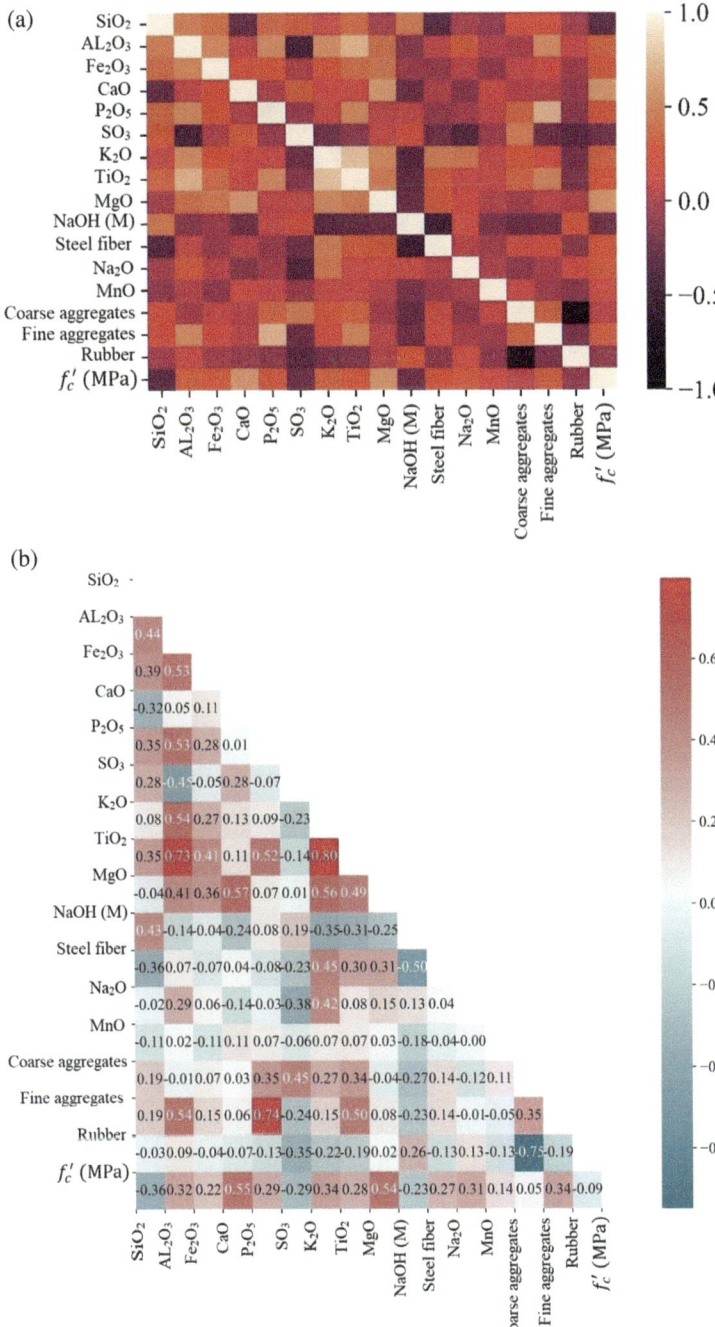

Fig. 1 Correlation matrix between input and output parameters

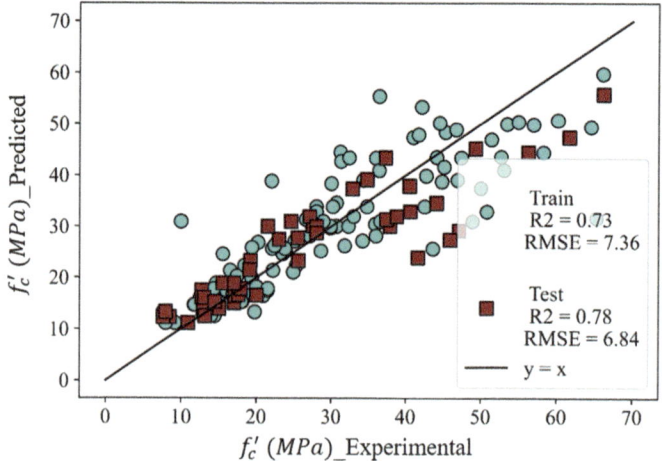

Fig. 2 f'_c prediction using linear regression

2.4   Machine Learning Method

2.4.1   XGBoost Method

Recent research across various domains, particularly in structural engineering, has demonstrated the effectiveness of machine learning techniques in accurately predicting the characteristics and behavior of diverse structural systems and materials [26–30].

In this study, XGBoost method was employed to evaluate the effect of different components of fly ash-based GPC on f'_c.

3   Results and Discussion

3.1   XGBoost Output

Figure 3 compares the experimental data and the predicted f'_c using XGBoost approach. As can be seen, R^2 values of the train and test data are 0.98 and 0.86, respectively. This indicates that this approach can predict the f'_c using the provided dataset with excellent accuracy. In addition, this shows that the method is more accurate than LR method in foreseeing the f'_c of the specimens.

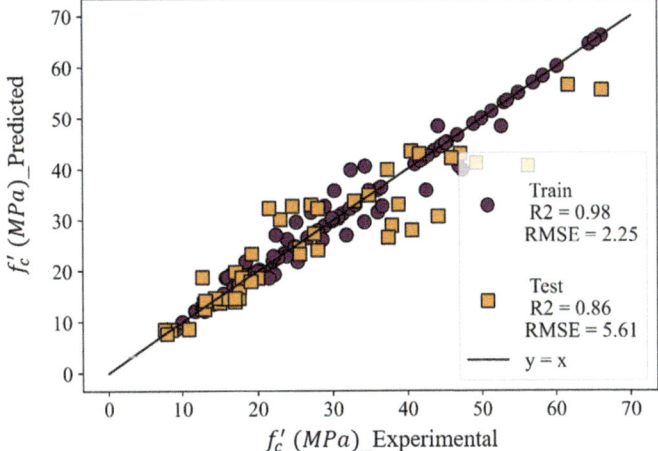

Fig. 3 Prediction of f'_c using XGBoost method

3.2 Shapley Values Technique

Figure 4a, b illustrates the findings of Shapley values technique, which assesses the significance of each feature on the target value (f'_c). It is evident from Fig. 4b that elevated levels of SO_3, SiO_2, and MnO are associated with a negative impact on f'_c, whereas the remaining features demonstrate the opposite trend.

Figure 4a shows that MgO is considered to be the most influential factor affecting the value of f'_c. In addition, as depicted in Fig. 4b, a high value of this feature leads to an increase in the value of f'_c.

Coarse aggregates and SO_3, identified as the second and third features with significant effects on f'_c, demonstrated completely opposite impacts. While an increase in the quantity of coarse aggregates positively influences the target value, SO_3 follows a reverse trend.

Interestingly, despite SiO_2 and Na_2O sharing the same level of significance, as seen in Fig. 4a, their impacts on f'_c are opposite. An increase in the quantity of SiO_2 results in a decline in f'_c, whereas the opposite is true for Na_2O.

Though SiO_2, Al_2O_3, and Fe_2O_3 account for a significant portion of the specimens, as demonstrated in Table 2, the impact of Al_2O_3 and Fe_2O_3 on the target value was found to be less than that of other features with lower values. Furthermore, SiO_2 negatively influences f'_c. This suggests that these features are not as significant when providing concrete mixtures where compressive strength is of utmost importance.

Additionally, the figures highlight the importance of fine aggregates in determining the f'_c of the specimens under study, though its importance is identified to be less than half that of coarse aggregates.

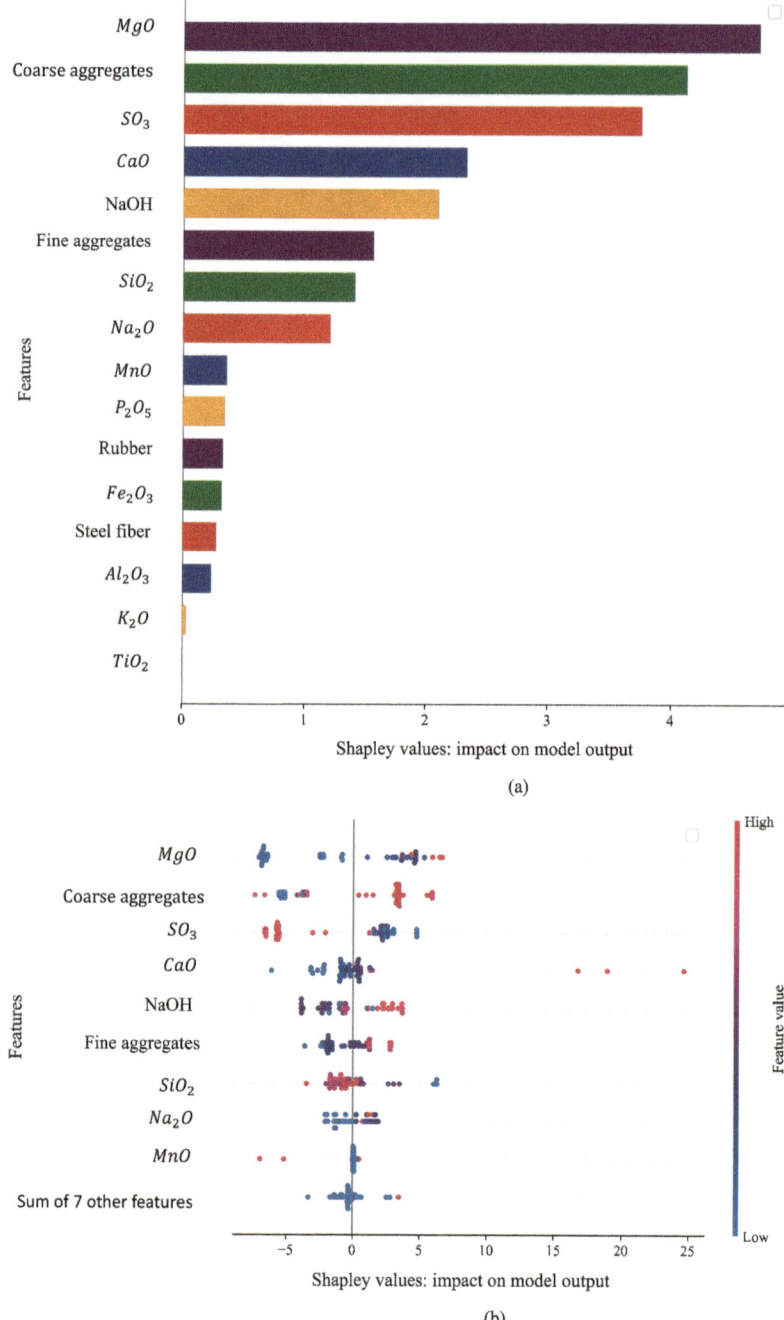

Fig. 4 Utilizing Shapley values technique to evaluate the significance of various features on f'_c

The increase in the molarity of the alkaline activator, NaOH, has a positive impact on the f'_c. Furthermore, according to the findings, TiO_2 is considered the least effective component of FA-based GPC on the target value. In addition, additives such as steel fiber and rubber showed a similar response, indicating an insignificant impact on compressive strength.

To ensure the validation of the obtained results, other machine learning methods, namely Decision Tree, Random Forest, or other methods, which are out of the scope of this research, can be used. In addition, this study focused on the experimental specimens cured in the ambient condition. To obtain a more comprehensive outcome, in future studies, the dataset acquired in varied temperatures can be selected to investigate the role of the curing temperature as a feature.

4 Conclusion

To evaluate the compressive strength (f'_c) of fly ash-based GPC, a machine learning approach was employed. The effect of various features, namely SiO_2, Al_2O_3, Fe_2O_3, CaO, P_2O_5, SO_3, K_2O, TiO_2, MgO, Na_2O, MnO, steel fiber, rubber, coarse aggregates, fine aggregates, and NaOH, was assessed using Linear Regression and XGBoost methods. Additionally, the effect of each feature on the target output (f'_c) was assessed utilizing the Shapley values technique. The findings of this research can be highlighted as follows:

- The linear regression method successfully predicts f'_c of fly ash-based GPC, achieving R^2 values of 0.73 for the training data and 0.78 for the test dataset. However, when utilizing the XGBoost method, the predicted f'_c demonstrates even higher accuracy. The R^2 values for the training and test data are notably improved, reaching 0.98 and 0.86, respectively. This suggests that the XGBoost approach can accurately predict f'_c using the provided dataset. Consequently, the XGBoost method outperforms the linear regression method in forecasting f'_c of the specimens.
- The increased concentrations of SO_3, SiO_2, and MnO are linked to a negative effect on f'_c, while the other features indicated a contrasting pattern.
- MgO is regarded as the most significant factor influencing the value of f'_c. Furthermore, a high level of this feature results in an increase in the f'_c value. Coarse aggregates and SO_3, identified as the second and third features with significant effects on f'_c, exhibited contrasting impacts. An increase in the quantity of coarse aggregates positively affects the target value, whereas SO_3 follows a reverse trend. Similarly, though SiO_2 and Na_2O are equally significant, their impacts on f'_c are contradictory. While an increase in SiO_2 decreases f'_c, the opposite is observed for Na_2O.
- The outcomes underscore the significance of fine aggregates in determining the f'_c of the specimens under study, although their importance is found to be less than

half that of coarse aggregates. Additionally, an increase in the molarity of the alkaline activator, *NaOH*, positively impacts f'_c.

- According to the results, TiO_2 is regarded as the least effective component of FA-based GPC on the target output. Moreover, additives such as steel fiber and rubber exhibited a similar response, both of which were considered insignificant.
- Despite SiO_2, Al_2O_3, and Fe_2O_3 being a significant portion of the specimens, the findings indicated that these features are not as crucial when it comes to providing concrete mixtures where compressive strength is of importance.

References

1. Saini, G., Vattipalli, U.: Assessing properties of alkali activated GGBS based self-compacting geopolymer concrete using nano-silica. Case Stud. Constr. Mater. **12**, e00352 (2020)
2. Chiranjeevi, K., Vijayalakshmi, M., Praveenkumar, T.: Investigation of fly ash and rice husk ash-based geopolymer concrete using nano particles. Appl. Nanosci. **13**(1), 839–846 (2023)
3. Zhang, P., et al.: High-temperature behavior of polyvinyl alcohol fiber-reinforced metakaolin/ fly ash-based geopolymer mortar. Compos. Part B. **244**, 110171 (2022)
4. Zhang, P., et al.: Strengthening mechanism of polyvinyl alcohol fibers on mechanical properties of geopolymer concrete subjected to a wet-hot-salt environment. Polym. Test. **127**, 108199 (2023)
5. Ryu, G.S., et al.: The mechanical properties of fly ash-based geopolymer concrete with alkaline activators. Constr. Build. Mater. **47**, 409–418 (2013)
6. Hassan, A., Arif, M., Shariq, M.: Effect of curing condition on the mechanical properties of fly ash-based geopolymer concrete. SN Appl. Sci. **1**, 1–9 (2019)
7. Bellum, R.R., Muniraj, K., Madduru, S.R.C.: Influence of slag on mechanical and durability properties of fly ash-based geopolymer concrete. J. Korean Ceram. Soc. **57**(5), 530–545 (2020)
8. Xie, J., et al.: Effects of combined usage of GGBS and fly ash on workability and mechanical properties of alkali activated geopolymer concrete with recycled aggregate. Compos. Part B. **164**, 179–190 (2019)
9. Ghafoor, M.T., et al.: Influence of alkaline activators on the mechanical properties of fly ash based geopolymer concrete cured at ambient temperature. Constr. Build. Mater. **273**, 121752 (2021)
10. Okoye, F., Durgaprasad, J., Singh, N.: Effect of silica fume on the mechanical properties of fly ash based-geopolymer concrete. Ceram. Int. **42**(2), 3000–3006 (2016)
11. Mehta, A., Siddique, R.: Properties of low-calcium fly ash based geopolymer concrete incorporating OPC as partial replacement of fly ash. Constr. Build. Mater. **150**, 792–807 (2017)
12. Farhan, N.A., Sheikh, M.N., Hadi, M.N.: Investigation of engineering properties of normal and high strength fly ash based geopolymer and alkali-activated slag concrete compared to ordinary Portland cement concrete. Constr. Build. Mater. **196**, 26–42 (2019)
13. Law, D.W., et al.: Long term durability properties of class F fly ash geopolymer concrete. Mater. Struct. **48**, 721–731 (2015)
14. Haque, M.M., et al.: Carbonation and permeation behaviour of geopolymer concrete containing copper slag and coal ashes. Dev. Built Environ. **16**, 100276 (2023)
15. Zheng, J., et al.: Mechanical properties and compressive constitutive model of steel fiber-reinforced geopolymer concrete. J. Build. Eng. **80**, 108161 (2023)
16. Ojha, A., Aggarwal, P.: Durability performance of low calcium Flyash-based geopolymer concrete. In: Structures, vol. 54, p. 956. Elsevier (2023)
17. Park, Y., et al.: Compressive strength of fly ash-based geopolymer concrete with crumb rubber partially replacing sand. Constr. Build. Mater. **118**, 43–51 (2016)

18. Laxmi, G., et al.: Effect of hooked end steel fibers on strength and durability properties of ambient cured geopolymer concrete. Case Stud. Constr. Mater. **18**, e02122 (2023)

19. Nath, P., Sarker, P.K.: Effect of GGBFS on setting, workability and early strength properties of fly ash geopolymer concrete cured in ambient condition. Constr. Build. Mater. **66**, 163–171 (2014)

20. Lee, N., Lee, H.-K.: Setting and mechanical properties of alkali-activated fly ash/slag concrete manufactured at room temperature. Constr. Build. Mater. **47**, 1201–1209 (2013)

21. Taleshi, M.M., et al.: Prediction of pull-out behavior of timber glued-in glass fiber reinforced polymer and steel rods under various environmental conditions based on ANN and GEP models. Case Stud. Constr. Mater. **20**, e02842 (2024)

22. Mahmoudian, A., et al.: Ensemble machine learning-based approach with genetic algorithm optimization for predicting bond strength and failure mode in concrete-GFRP mat anchorage interface. In: Structures, vol. 57, p. 105173. Elsevier (2023)

23. Bypour, M., Yekrangnia, M., Kioumarsi, M.: Predicting the shear capacity of composite steel plate shear wall with the application of RSM. Eng. Struct. **301**, 117263 (2024)

24. Hamidia, M., et al.: Machine learning-based seismic damage assessment of non-ductile RC beam-column joints using visual damage indices of surface crack patterns. In: Structures, vol. 45, p. 2038. Elsevier (2022)

25. Tajik, N., et al.: Explainable XGBoost machine learning model for prediction of ultimate load and free end slip of GFRP rod glued-in timber joints through a pull-out test under various harsh environmental conditions. Asian J. Civ. Eng. **25**, 1–17 (2023)

26. Hamidia, M., Kaboodkhani, M., Bayesteh, H.: Vision-oriented machine learning-assisted seismic energy dissipation estimation for damaged RC beam-column connections. Eng. Struct. **301**, 117345 (2024)

27. Afzali, M., Hamidia, M., Safi, M.: Data-driven strength-based seismic damage index measurement for RC columns using crack image-derived parameters. Measurement. **218**, 113155 (2023)

28. Ahmadi, M., Kioumarsi, M.: Predicting the elastic modulus of normal and high strength concretes using hybrid ANN-PSO. Mater. Today Proc. (2023). https://doi.org/10.1016/j.matpr.2023.03.178

29. Kandiri, A., Sartipi, F., Kioumarsi, M.: Predicting compressive strength of concrete containing recycled aggregate using modified ANN with different optimization algorithms. Appl. Sci. **11**(2), 485 (2021)

30. Farahzadi, L., Kioumarsi, M.: Application of machine learning initiatives and intelligent perspectives for CO_2 emissions reduction in construction. J. Clean. Prod. **384**, 135504 (2023)

Sustainable Enhancement of Lightweight Concrete: A Comprehensive Investigation into GGBS and Waste Steel Fiber Incorporation for Improved Strength and Durability

Babak Behforouz, Davoud Tavakoli, Behrouz Naderi, and Mohammad Hajmohammadian Baghban

Contents

1   Introduction

Today, concrete is the most widely used construction material in the world [1]. Also, Portland cement is considered to be the main ingredient in all types of concrete. Considering the environmental problems of Portland cement production, it is tried to

B. Behforouz
Water Studies Research Center, Isfahan (Khorasgan) Branch, Islamic Azad University, Isfahan, Iran

D. Tavakoli (✉)
Materials & Construction, Department of Civil Engineering, KU Leuven, Leuven, Belgium
e-mail: davoud.tavakoli@kuleuven.be

B. Naderi
Department of Civil Engineering, Shahid Ashrafi Esfahani University, Isfahan, Iran

M. Hajmohammadian Baghban
Department of Manufacturing and Civil Engineering, Norwegian University of Science and Technology, Gjøvik, Norway

© The Author(s) 2025
M. Kioumarsi, B. Shafei (eds.), *The 1st International Conference on Net-Zero Built Environment*, Lecture Notes in Civil Engineering 237,
https://doi.org/10.1007/978-3-031-69626-8_68

produce concretes with alternative materials. Therefore, many natural and artificial minerals have been tested for this purpose. Among these materials are silica fume, fly ash, Metakaolin, and slag [2]. With the increase in steel production in the world, the amount of blast furnace slag, which is a byproduct of this industry, is also increasing. Since the conducted researches point to the high potential of using steel slag in the concrete industry, both as aggregate and as a cement supplement, currently the most important part of slag consumption in the world is the cement and concrete industry [3]. Of the four types of slag produced in the steel industry, ground granulated blast-furnace slag (GGBS) is the most widely used cement material [3]. GGBS is an industrial waste that is rich in CaO, SiO_2, and Al_2O_3, which makes its use as a substitute for cement valuable. The use of GGBS as a part of the cement can lead to a reduction in cement consumption and thus a reduction in greenhouse gas production [4, 5]. By improving the cement paste structure and strengthening the transition zone, GGBS improves the strength and durability of concrete [6]. Many articles have confirmed the positive effect of GGBS on different properties of concrete [7–9].

On the other hand, in recent years, the problems of using steel, such as heavy weight, corrosion, and its difficult implementation, have led researchers to seek to replace steel with a suitable material. Using fibers is one of the alternative methods. One of the most famous known ways to improve tensile and bending strength is the use of fibers [10]. Many studies have shown the positive effect of steel fibers on concrete properties, especially tensile strength [11–13].

Lightweight concrete has less weight than ordinary concrete, and the reason for this lightness is due to the use of lighter aggregates in its structure. Making lightweight concrete with these aggregates makes its compressive strength and density lower than ordinary concrete, and for this reason, lightweight concrete is used less as structural concrete. On the other hand, in lightweight concrete, the bond between aggregate and matrix is stronger than in normal concrete. Due to the porous nature of the aggregates, the cement paste penetrates it. Therefore, there is no transition zone between the aggregates and the matrix, or if there is, it is very small. This issue is very important in terms of concrete reliability because this area is the weakest area in normal concrete [14].

In this study, the simultaneous effect of waste steel fibers and GGBS in lightweight concrete has been investigated. Considering that the use of GGBS improves the durability and compressive strength of concrete in the long term, and also that steel fiber can improve the tensile strength of concrete, the use of these two materials in lightweight concrete can compensate for the loss of strength and durability caused by the use of lightweight aggregates in concrete. In addition, the use of two waste materials in concrete is an action towards sustainable development.

2 Materials and Methods

The fine aggregate used in this study is river aggregate. Coarse aggregate is also light aggregate obtained from natural mines. The properties of the aggregates are given in Table 1.

Table 1 Properties of aggregate

Properties	Fine aggregate	Lightweight coarse aggregate
Specific gravity (kg/m^3)	1843	466
Maximum size of aggregate (mm)	4.75	12
Water absorption (%)	1.7	11.1
Pore content (%)	2	30
Los Angeles Abrasion (%)	–	23.3

Table 2 Chemical composition of GGBS and cement

	SiO$_2$	Al$_2$O$_3$	Fe$_2$O$_3$	MgO	SO$_3$	Na$_2$O	CaO	LOI
Cement	21.56	4.48	3.2	2.09	2.37	2.6	63.39	1.86
GGBS	37.51	8.82	0.13	9.85	0.001	0.97	39.47	0.02

Fig. 1 Steel Fiber

Used cement is ASTM type II from Nayin factory, Iran, and GGBS is obtained from Isfahan Zobahan Factory, Iran. The chemical composition of GGBS and cement is shown in Table 2.

In this research, waste steel fibers (Fig. 1) have been used, the specifications of which are given in Table 3.

The mix designs are shown in Table 4. In this table, the word S means slag and the word F means steel fiber, and the percentage written in front of each letter means the percentage of its use in the mix design.

To make the samples, in the first stage, cement and fine aggregate were completely mixed together, and then, if used, slag or fibers were added, and the slag or fibers were completely mixed, and then light coarse aggregate was added and mixed. In the last stage, water and superplasticizer are added and mixed for 10 minutes until a homogeneous mixture is obtained. After preparation, the samples were cured in saturated lime water until the test was performed, and then compressive strength (7, 28, and 90 days), tensile strength (28 and 90 days), and water absorption (28 and 90 days) tests were performed on the samples. Compressive

Table 3 Steel fiber properties

Density (g/cm^3)	8.7
Diameter (mm)	1
Length (mm)	30
Young's modulus (GPa)	200
Tensile strength (GPa)	1.7

Table 4 Mix designs (kg/m^3)

Mix Design	Water	Cement	GGBFS	Steel fiber	Fine Aggregate	Light Weight Coarse Aggregate	SP
C	230	510	0	0	565	740	25.5
CS0F1.5	230	510	0	1.5	565	740	25.5
CS0F3	230	510	0	3	565	740	25.5
CS10F0	230	459	51	0	565	740	25.5
CS10F1.5	230	459	51	1.5	565	740	25.5
CS10F3	230	459	51	3	565	740	25.5
CS20F0	230	408	102	0	565	740	25.5
CS20F1.5	230	408	102	1.5	565	740	25.5
CS20F3	230	408	102	3	565	740	25.5
CS30F0	230	357	153	0	565	740	25.5
CS30F1.5	230	357	153	1.5	565	740	25.5
CS30F3	230	357	153	3	565	740	25.5
CS40F0	230	306	204	0	565	740	25.5
CS40F1.5	230	306	204	1.5	565	740	25.5
CS40F3	230	306	204	3	565	740	25.5

strength tests were conducted in accordance with the BS EN 12390 standard, tensile strength tests were performed following the ASTM C496 standard, and water absorption tests were carried out according to ASTM C642.

Each test was performed on three samples and 316 samples were prepared for 15 mix designs.

3   Results and Discussion

The compressive strength test has been performed for 7-, 28-, and 90-day cubic samples ($10 \times 10 \times 10$ cm). The results are shown in Fig. 2.

The compressive strength has not changed much with the use of fiber. Fiber has little positive effect on compressive strength due to the confinement of steel fibers in samples [15] and no negative effect has been observed. In higher amounts of fiber use, the use of slag has the effect of increasing fiber effect and has led to improved strength. One of the reasons for increasing the compressive strength by using fibers is the reduction of microcracks in the cement paste structure [16].

The use of slag has probably had a positive effect. At the age of 7 days, a slight decrease in strength has been observed in some samples, at older ages, especially at

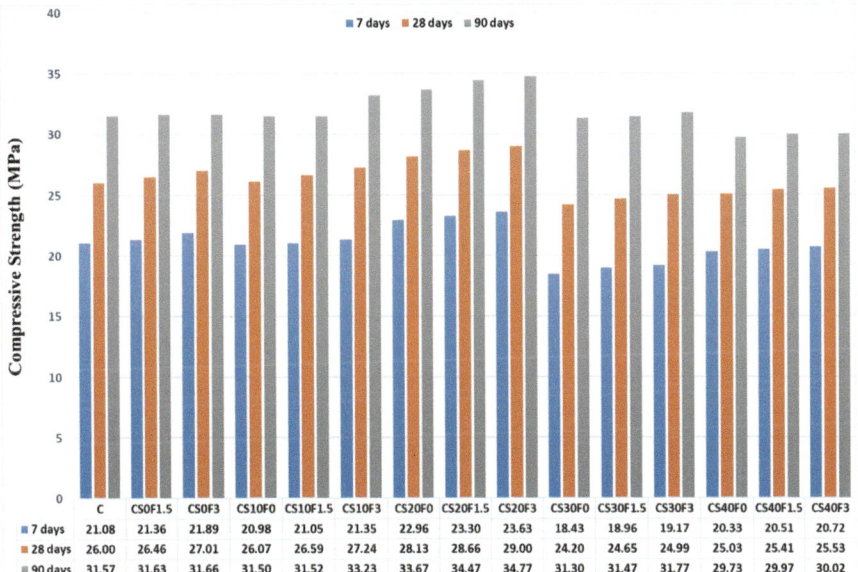

	C	CS0F1.5	CS0F3	CS10F0	CS10F1.5	CS10F3	CS20F0	CS20F1.5	CS20F3	CS30F0	CS30F1.5	CS30F3	CS40F0	CS40F1.5	CS40F3
7 days	21.08	21.36	21.89	20.98	21.05	21.35	22.96	23.30	23.63	18.43	18.96	19.17	20.33	20.51	20.72
28 days	26.00	26.46	27.01	26.07	26.59	27.24	28.13	28.66	29.00	24.20	24.65	24.99	25.03	25.41	25.53
90 days	31.57	31.63	31.66	31.50	31.52	33.23	33.67	34.47	34.77	31.30	31.47	31.77	29.73	29.97	30.02

Fig. 2 Compressive strength

the age of 90 days, the increase in strength with the use of slag is quite evident. The reaction of SiO_2 in GGBS with CH leads to the production of new cement components, which improves the structure of cement paste and improves the strength properties of concrete [10].

The maximum compressive strength obtained was related to the sample containing 20% slag and 3% steel fiber, which had a compressive strength of about 10% more than the control sample. Because of the dilution effect, the use of 30% slag has led to a decrease in strength [10]. Therefore, it can be said that 20% of slag is the optimal amount of use.

The tensile strength test has been performed for 28- and 90-day cylindrical samples (15 × 30). The results are shown in Fig. 3.

The effect of slag on the tensile strength was similar to the effect of this material on the compressive strength of concrete, but steel fiber had a better effect on the tensile strength as expected. Some studies have shown that the use of GGBS has a better effect on the compressive strength compared to the tensile strength of concrete [3, 6].

Steel fiber leads to an increase in tensile strength and this positive effect is increased by increasing the amount of fiber used. The highest strength obtained is related to the samples containing 10 and 20% of slag along with 3% of steel fiber, which is about a 27% increase compared to the control sample. Therefore, the positive effect of steel fibers can be seen in this study. Usually, small cracks are formed in concrete due to various reasons, including thermal stresses [17], which lead to a decrease in tensile strength. Fibers with increased tensile stress lead to a reduction of these cracks and improvement of concrete structure. Improvement of tensile strength using steel fiber has been reported in many studies [10–13, 18].

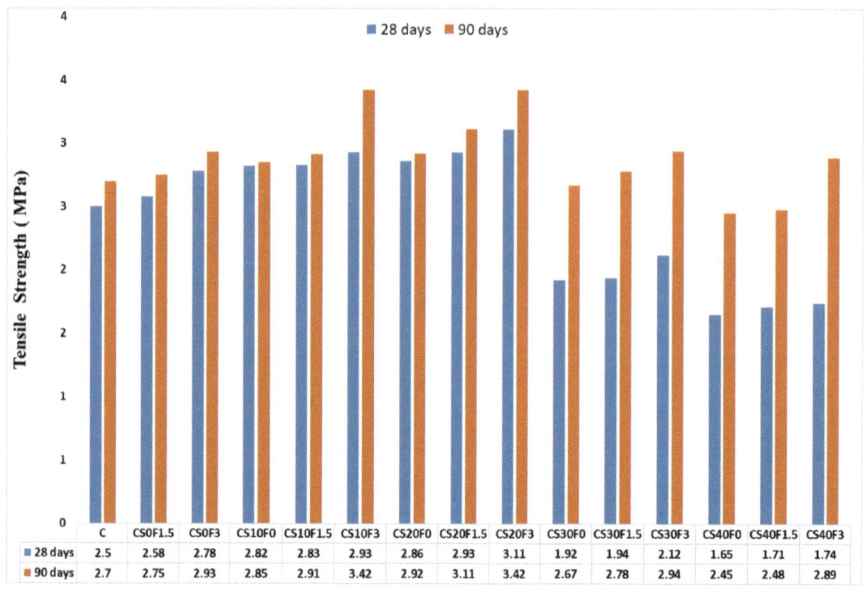

Fig. 3 Tensile strength

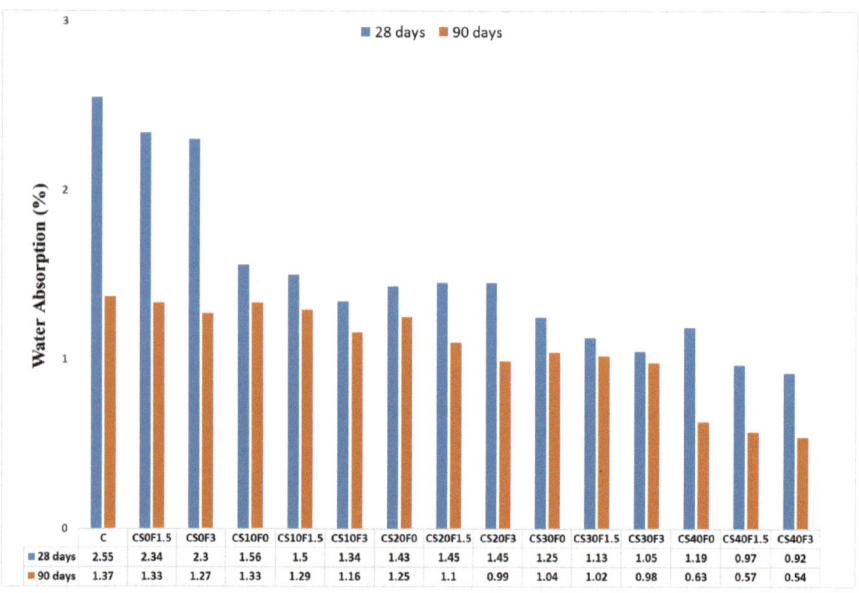

Fig. 4 Water Absorption

The water absorption test has been performed for 28- and 90-day cubic samples (15 × 15 × 15). The results are shown in Fig. 4.

The use of fiber has not had a great effect on water absorption and no specific conclusion can be drawn in this regard, but the use of slag has reduced the water absorption of concrete. This reduction in water absorption has increased with the increase in the use of slag. For the 28-day sample, the maximum reduction in water absorption compared to the control sample was about 60%. The production of a less porous cement paste structure using GGBS has led to a decrease in the permeability of concrete [19].

4   Conclusion

In this study, the simultaneous effect of GGBS and steel fibers in lightweight concrete was investigated. After the tests, the following results were obtained:

- GGBS can compensate for the decrease in strength caused by lightweight concrete particles and improve the compressive strength and durability of concrete in the long term. The optimal percentage of using this material in lightweight concrete is 20%.
- The use of steel fiber can improve the tensile strength of lightweight concrete while not having any negative effect on other properties of concrete.
- The use of GGBS as a waste material as a substitute for part of cement, as well as the use of steel fiber waste, is an action to reduce greenhouse gases and an action in line with sustainable development.

References

1. Tavakoli, D., Gao, P., Tarighat, A., Ye, G.: Multi-scale approach from atomistic to macro for simulation of the elastic properties of cement paste. Iran. J. Sci. Technol. Trans. Civ. Eng. **44**, 861–873 (2020)
2. Behforouz, B., Tavakoli, D., Gharghani, M., Ashour, A.: Bond strength of the interface between concrete substrate and overlay concrete containing fly ash exposed to high temperature. Structures. **49**, 183–197 (2023)
3. El-Hassan, H., Kianmehr, P., Tavakoli, D., El-Mir, A., Dehkordi, R.: Synergic effect of recycled aggregates, waste glass, and slag on the properties of pervious concrete. Dev. Built Environ. **15**, 100189 (2023)
4. Flower, D.J., Sanjayan, J.G.: Green house gas emissions due to concrete manufacture. Int. J. Life Cycle Assess. **12**, 282–288 (2007)
5. Tavakoli, D., Hashempour, M., Heidari, A.: Use of waste materials in concrete: a review. Pertanika J. Sci. Technol. **26**(2), 499–522 (2018)
6. Özbay, E., Erdemir, M., Durmuş, H.İ.: Utilization and efficiency of ground granulated blast furnace slag on concrete properties–a review. Constr. Build. Mater. **105**, 423–434 (2016)
7. Babu, K.G., Kumar, V.S.R.: Efficiency of GGBS in concrete. Cem. Concr. Res. **30**(7), 1031–1036 (2000)
8. Suresh, D., Nagaraju, K.: Ground granulated blast slag (GGBS) in concrete–a review. IOSR J. Mech. Civ. Eng. **12**(4), 76–82 (2015)
9. Ahmad, J., Kontoleon, K.J., Majdi, A., Naqash, M.T., Deifalla, A.F., Ben Kahla, N., et al.: A: comprehensive review on the ground granulated blast furnace slag (GGBS) in concrete production. Sustain. For. **14**(14), 8783 (2022)

10. Ahmad, J., Manan, A., Ali, A., Khan, M.W., Asim, M., Zaid, O.A.: Study on mechanical and durability aspects of concrete modified with steel fibers (SFs). Civ. Eng. Archit. **8**, 814–823 (2020)
11. Olivito, R.S., Zuccarello, F.A.: An experimental study on the tensile strength of steel fiber reinforced concrete. Compos. Part B. **41**(3), 246–255 (2010)
12. Song, P.S., Hwang, S.: Mechanical properties of high-strength steel fiber-reinforced concrete. Constr. Build. Mater. **18**(9), 669–673 (2004)
13. Behbahani, H.P., Nematollahi, B., Farasatpour, M.: Steel fiber reinforced concrete: a review. In: Proceedings of the International Conference on Structural Engineering Construction and Management (ICSECM2011) (2011)
14. Thienel, K.C., Haller, T., Beuntner, N.: Lightweight concrete—from basics to innovations. Materials. **13**(5), 1120 (2020)
15. Ahmad, J., Martínez-García, R., Szelag, M., de Prado-Gil, J., Marzouki, R., Alqurashi, M., Hussein, E.E.: Effects of steel fibers (Sf) and ground granulated blast furnace slag (ggbs) on recycled aggregate concrete. Materials. **14**(24), 7497 (2021)
16. Khaloo, A.R., Kim, N.: Influence of concrete and fiber characteristics on behavior of steel fiber reinforced concrete under direct shear. Mater. J. **94**(6), 592–601 (1997)
17. Al-Jabri, K.S., Hago, A.W., Tavakoli, D., Waris, M.B., Hassan, H.F., Mohamedzein, Y.: Investigating thermal cracking in mass concrete of a bridge abutment: field measurements and numerical modelling. Aust. J. Civ. Eng., 1–17 (2022)
18. Abbass, W., Khan, M.I., Mourad, S.: Evaluation of mechanical properties of steel fiber reinforced concrete with different strengths of concrete. Constr. Build. Mater. **168**, 556–569 (2018)
19. Li, K., Zeng, Q., Luo, M., Pang, X.: Effect of self-desiccation on the pore structure of paste and mortar incorporating 70% GGBS. Constr. Build. Mater. **51**, 329–337 (2014)

Life Cycle Assessment of Geopolymer Concrete Made with Tailings from Ilmenite Mining

Simon Brekke and Reyn O'Born (iD)

Contents

1 Introduction

Cement production represents 5-8% of global CO_2-equivalent (CO_2-eq) emissions and is a major challenge on the path towards a future of net zero emissions while remaining a critical building material without many viable replacements [1, 2]. It is imperative to find low-emissions solutions that can reduce the overall global impact of concrete by finding new mixtures and binders for cement production, which are the main contributing factor for emissions from concrete and the largest contributor to emissions from construction materials [3–5]. Alternative concrete binders that have lower emissions are being developed, and one such binder is called geopolymer

S. Brekke
Norconsult, Grimstad, Norway

University of Agder, Grimstad, Norway

R. O'Born (✉)
University of Agder, Grimstad, Norway
e-mail: reyn.oborn@uia.no

© The Author(s) 2025   821
M. Kioumarsi, B. Shafei (eds.), *The 1st International Conference on Net-Zero Built Environment*, Lecture Notes in Civil Engineering 237,
https://doi.org/10.1007/978-3-031-69626-8_69

cement. Geopolymer cement is made by mixing industrial wastes (typically from mining or incineration processes) with an alkaline solution, such as sodium- (NaOH) or potassium hydroxide (KOH), which can completely replace ordinary Portland cement [6]. Geopolymer concrete can be used in identical applications to standard concrete and can even have improved mechanical strength [7]. Life cycle assessment (LCA) studies of geopolymer concrete have shown a potential CO_2 emissions reduction estimated at 32–70% compared to standard concrete [8–11]. Further studies evaluating KOH show that compressive strength after 28 days is nearly equivalent to standard cement with a tangible reduction in environmental impacts [12, 13], though there are few LCA studies focusing on KOH as an alkaline solution.

In Norway, the company Saferock is developing a new geopolymer cement which utilises mine tailings from ilmenite production combined together with KOH [14]. Saferock utilises feedstock in their production process from ilmenite mine tailings found in the Norwegian area of Sokndal. The Sokndal site has more than 100 million tons of easily accessible norite materials for Saferock to use for producing geopolymer concrete. It is important to understand the environmental impacts of alternative materials and to see if they are beneficial for exploitation or not. Thus, the purpose of this study is to evaluate the environmental impacts of Saferock's geopolymer cement. This is done by using life cycle assessment (LCA) to quantify the emissions that come from producing the geopolymer cement and concrete.

2 Methods

This study uses standard LCA methods as described in ISO 14040, where goal and scope, inventory analysis, impact assessment and interpretation are carried out [15]. LCA is a useful tool for understanding the environmental impacts of systems or processes with a standardised methodology that can be used to quantify impacts throughout the life cycle of a product or system (Fig. 1).

The goal of this study is to understand the environmental impacts of the Saferock concrete mix based on their pilot production facility near Stavanger, Norway. The scope of the study is cradle-to-gate, which covers the emissions of the production of their geopolymer concrete based on real-inputs from the Saferock process. Life cycle inventory (LCI) data was collected for this study from communication and meetings with Saferock, technical documents from the company, and basic calculation from an assumed distance of 80 km for materials for aggregate materials and transport to the Saferock production site. Material mixes for aggregate in the concrete mix were given from Saferock. The assumed energy in production is based on the Norwegian market energy mix.

SimaPro was used to carry out the analysis with the EcoInvent 3.7.1—"-allocation, cut-off by classification" database and the ReCiPe 2016 (midpoint, H) impact assessment method. The functional unit is 1 m^3 of B35 geopolymer concrete with the geographical boundaries of Norway. The study is a cradle to gate study encompassing the production at Saferock in Sokndal, Norway.

Fig. 1 Life cycle
assessment methodology

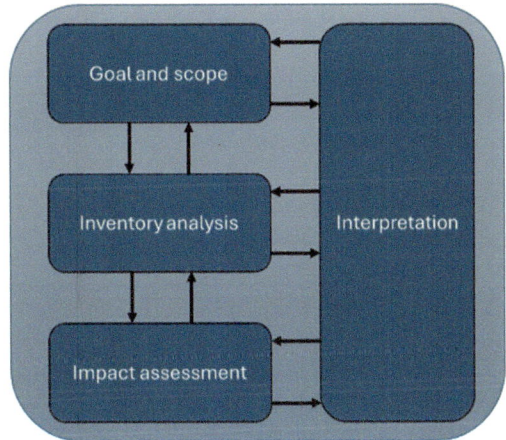

3 Case

The Norwegian company Saferock is in the development phase of its geopolymer concrete. They make the cement and the concrete in a laboratory, but they have received support to build a pilot factory to test production on a larger scale, which is scheduled to be completed in early 2024 and have a production capacity of 30 m^3 geopolymer concrete per day. The next step is to build a factory at the Titania norite mine, where norite waste materials from ilmenite production can ground into a fine powder of less than 63 μm (Fig. 2). This is then sent out, along with activator in powder form, for customers who will mainly be concrete mixing plants. In the mixing plants, this powder is mixed with KOH as an activator, water and aggregates. Saferock's geopolymer concrete has compressive strengths equivalent to B35. It is assumed that the concrete is reinforced in the same way as normal concrete, although this has not been tested when this study was carried out. There are also challenges with the regulations for which cement can be used in constructions. This geopolymer concrete is not approved for use in Eurocode 2. According to the concrete standard NS-EN 206, only cements from three standards can be used: EN 197-1, NS-EN 14647 and NS-EN 15743, where geopolymer concrete is not yet approved [16] (Fig. 3).

The basic system description and system boundaries of this study are shown in Fig. 4.

The system includes all processes from collecting and grinding mine tailings from the Titania site, addition of sand and gravel in the aggregate, KOH and water in activator material, and mixing in the plant. The resource and emission loads in the life cycle inventory are normalised to 1 m^3 of geopolymer concrete. The study does not account for the transport of concrete to construction site, nor the construction, maintenance, and end-of-life processes, which are assumed to be identical to standard B35 concrete.

Fig. 2 The Titania ilmenite mine tailings in Sokndal, Norway [14]

Fig. 3 The Titania ilmenite mine and tailings in Sokndal as seen from above

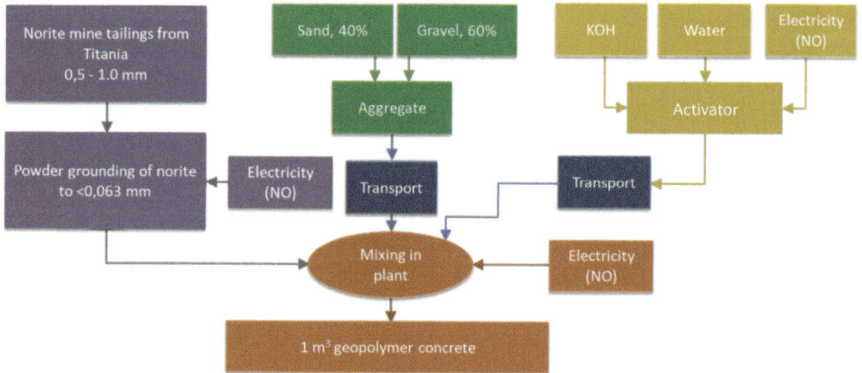

Fig. 4 Saferock B35 geopolymer concrete production LCA system boundaries

4 Results

Table 1 shows the LCI of the production mix of the Saferock concrete. The mix is based on the Saferock's production recipe for normal geopolymer concrete as well as an optimal mix that reduces the amount of norite powder and KOH powder considerably. The optimal mix is what Saferock would like to achieve in the future where water is reduced to under 4.25% of the total mix of the activator materials (KOH and water) is under 5% of the total mass. The optimal mix is a theoretical mix tested in the Saferock laboratory while the Normal geopolymer concrete mix is the mix that Saferock has achieved in their pilot testing.

Figure 5 shows the total impact assessments results for the normal geopolymer concrete mix. The dominant process in all impact categories is the activator mix, almost entirely due to the KOH powder, which is assumed to have a European average production mix. The aggregate, sand and gravel process also has an impact ranging between 10 and 30% in most categories, where the norite powder is

Table 1 Saferock material inventory for 1 m³ B35 geopolymer concrete

Material	Normal geopolymer concrete		optimal mix	
	%	kg	%	kg
Norite powder	35	1120	25	800
KOH-powder	2.25	46	0.75	15
Water	12.75	127	4.25	42
Sand	20	540	28	810
Gravel	30	780	42	1040
Additional materials	0	0	>0.001	2
Total	100	2613	100	2710

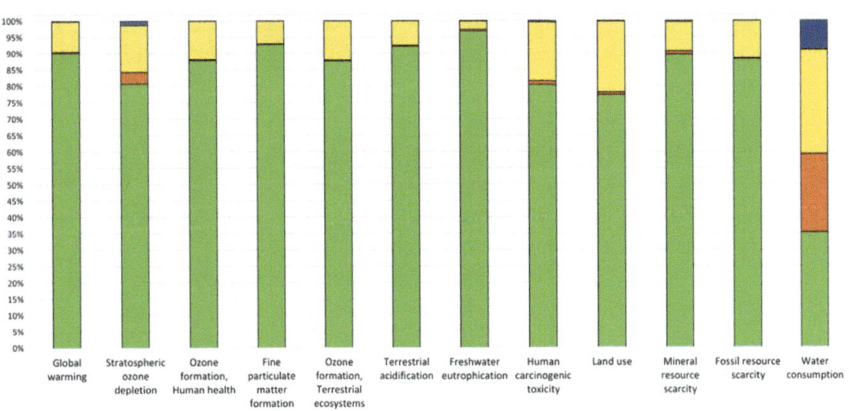

Fig. 5 Impact assessment results for 1 m³ of Saferock normal geopolymer concrete (ReCiPe 2016. Midpoint H)

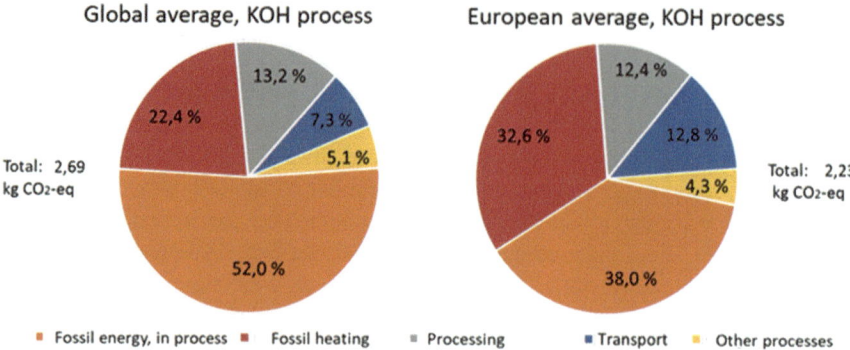

Fig. 6 Kg CO$_2$-eq. emissions per kg KOH produced, global and European averages

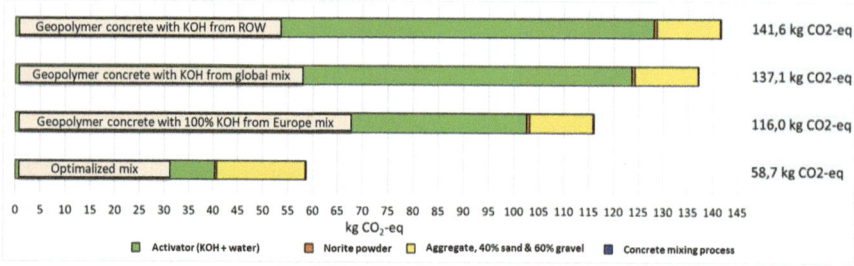

Fig. 7 GWP results of 1 m^3 B35 Saferock concrete based on alternative sources for KOH and optimal mix

insignificant in almost all impact categories. The dominance of the KOH process is due the choice of KOH in the EcoInvent database. At the time of this study, there were no significant or relevant LCA studies on KOH production, so the European average KOH process was used. This process is assumed to use a high level of fossil fuels in the production process.

Figure 6 shows the CO$_2$-eq. emissions profile of the KOH production process for the global and European average processes within EcoInvent 3.7.1 database. Fossil energy is assumed to be used in the production process, for heating, for additional processing, and in transport of the materials. The European average has slightly lower CO$_2$-eq. emissions due to overall higher process efficiency.

Figure 7 shows the GWP of 1 m^3 of Saferock concrete mix using the global average KOH, a generic rest-of-world mix, a European average KOH mix and an optimised concrete mix that reduces the need for KOH by more than 65% while also significantly reducing the norite powder as well. The range for 1 m^3 B35 Saferock concrete is 116 to 141.6 kg of CO$_2$-eq., while the optimised mix is significantly lower at 58.7 kg CO$_2$-eq per 1 m^3.

5   Discussion

The results of the LCA study show that the CO_2 emissions from the geopolymer concrete are less than half of those of a comparable standard concrete. Saferock concrete meets the Norwegian branch standard for "Low Carbon Extreme" concrete when using a generic European KOH mix. The KOH accounts for 90% of the emissions, but the background data on the emissions from this process are insufficient. The lack of data on KOH emissions is a considerable uncertainty in this study. It is recommended that Saferock acquire potassium hydroxide produced using renewable resources to further reduce their emissions and that their KOH supplier should carry out a full life cycle assessment of their production process (Table 2).

Previous LCA studies on sodium hydroxide-based (NaOH) geopolymer concretes had higher emissions than Saferock's geopolymer concrete, but had a GWP reduction of 10 to 45% relative to normal B35 concrete [18, 19]. These studies are not directly comparable to the Saferock process given the advancements in cement production and different admixtures used. The Saferock study results confirm previous study results showing the environmental advantages of geopolymer concrete.

The Saferock concrete is still in the development phase and has not yet been used in constructions at the time this study was carried out. This means that there are unknown uncertainties regarding logistics, use and safety of the final product, although these are assumed to be like normal concrete production.

An additional uncertainty is the use of reinforcing steel. Concrete technology is a complicated field that involves understanding many different chemical processes and how they affect each other. The basic environment in the geopolymer concrete should be a good starting point for the reinforcement, as it is similar to normal concrete. It is nevertheless conceivable that unwanted chemical reactions may occur due to the various ingredients in the concrete. This can have negative consequences for the reinforcement and should be investigated before the concrete can be put into use.

Saferock concrete is currently not approved for structural use in the Norwegian market. As the regulations are today, it is not allowed to use cements other than those

Table 2 Norwegian branch standards for GWP per 1 m³ B35 concrete [17]

Concrete standard	kg CO_2-eq. per m3 B35 concrete	Reduction compared to branch standard (%)
Branch reference standard	270	0
Low Carbon A standard	210	−22
Low Carbon Plus standard	160	−41
Low Carbon Extreme standard	120	−55
Saferock (European KOH mix)	116	−57
Saferock optimal mix	59	−78

described in the concrete standard NS-EN 206 [16]. Saferock will have to go through extensive verification and testing to be able to use the geopolymer concrete in future construction projects. This study has not taken this into account but recognises the potential emissions savings and material savings from using such a concrete. Future studies should investigate more closely the mechanical properties of the Saferock geopolymer concrete through laboratory and pilot testing.

6   Conclusion

The potential for emissions reductions using geopolymer cement is significant and crucial for reducing the overall emissions of the concrete industry. The Saferock concrete process investigated in this study uses mining waste and potassium hydroxide as a replacement for standard Portland cement in concrete mixtures. Although there are uncertainties with the emissions associated with the potassium hydroxide production process used in this this study, the overall potential for reduced emissions for geopolymer concrete is confirmed. Saferock concrete contributes to a broader circular economy in the construction industry by utilising waste materials and drastically reducing emissions of a key construction material. Future studies should on geopolymer concrete should focus on carrying out more comprehensive LCA studies on activator materials and testing of mechanical properties of geopolymer concrete.

Acknowledgements This study was supported by the research project "Green Platform—Sustainable value chain and materials in road construction" as financed by the Norwegian Research Council grant project number 340901.

References

1. Le Quéré, C., et al.: Global Carbon Budget 2016. Earth Syst. Sci. Data. **8**, 605–649 (2016). https://doi.org/10.5194/essd-8-605-2016
2. Andrew, R.M.: Global CO_2 emissions from cement production, 1928–2018. Earth Syst. Sci. Data. **10**(4), 1–20 (2018)
3. Barcelo, L., Kline, J., Walenta, G., Gartner, E.: Cement and carbon emissions. Mater. Struct. **47**(6), 1055–1065 (2013). https://doi.org/10.1617/s11527-013-0114-5
4. Lande, I., Terje Thorstensen, R.: Comprehensive sustainability strategy for the emerging ultra-high-performance concrete (UHPC) industry. Clean. Mater. **8**(March), 100183 (2023). https://doi.org/10.1016/j.clema.2023.100183
5. Habert, G., et al.: Environmental impacts and decarbonization strategies in the cement and concrete industries. Nat. Rev. Earth Environ. **1**(11), 559–573 (2020). https://doi.org/10.1038/s43017-020-0093-3
6. Azad, N.M., Samarakoon, S.M.S.M.K.: Utilization of industrial by-products/waste to manufacture geopolymer cement/concrete. Sustainability. **13**(2), 1–22 (2021). https://doi.org/10.3390/su13020873

7. Akhtar, N., et al.: Ecological footprint and economic assessment of conventional and geopolymer concrete for sustainable construction. J. Clean. Prod. **380**, 134910 (2022). https://doi.org/10.1016/j.jclepro.2022.134910

8. Ouellet-Plamondon, C., Habert, G.: Life cycle Assessment (LCA) of Alkali-Activated Cements and Concretes. Woodhead Publishing Limited (2015). https://doi.org/10.1533/9781782422884.5.663

9. Salas, D.A., Ramirez, A.D., Ulloa, N., Baykara, H., Boero, A.J.: Life cycle assessment of geopolymer concrete. Constr. Build. Mater. **190**, 170–177 (2018). https://doi.org/10.1016/j.conbuildmat.2018.09.123

10. Meshram, R.B., Kumar, S.: Comparative life cycle assessment (LCA) of geopolymer cement manufacturing with Portland cement in Indian context. Int. J. Environ. Sci. Technol. **19**(6), 4791–4802 (2022). https://doi.org/10.1007/s13762-021-03336-9

11. Teh, S.H., Wiedmann, T., Castel, A., de Burgh, J.: Hybrid life cycle assessment of greenhouse gas emissions from cement, concrete and geopolymer concrete in Australia. J. Clean. Prod. **152**, 312–320 (2017). https://doi.org/10.1016/j.jclepro.2017.03.122

12. Stengel, T., Reger, J., Heinz, D.: Life cycle assessment of geopolymer concrete – what is the environmental benefit. In: Proceeding of the 24th Biennial Conference of the Concrete Institute of Australia (2009) [Online]. Available: https://www.researchgate.net/profile/Thorsten-Stengel/publication/373107380_Life_Cycle_Assessment_of_Geopolymer_Concrete_-_What_is_the_Environmental_Benefit/links/64d9e461ad846e288290459e/Life-Cycle-Assessment-of-Geopolymer-Concrete-What-is-the-Environmental-Benefit.pdf

13. Firdous, R., Nikravan, M., Mancke, R., Vöge, M., Stephan, D.: Assessment of environmental, economic and technical performance of geopolymer concrete: a case study. J. Mater. Sci. **57**(40), 18711–18725 (2022). https://doi.org/10.1007/s10853-022-07820-6

14. Saferock: Saferock Homepage – About Technology, 2024. https://saferock.no/#about

15. ISO: ISO14040: Environmental Management-Life Cycle Assessment Principles and Frameworks. British Standards Institution, London (2006)

16. Standard Norway: NS-EN 206:2013+A2:2021+NA:2022 – Betong – Spesifikasjon, Egenskaper, Framstilling og Samsvar (Concrete –Specification, Performance, Production and Conformity). Standard Norway, Oslo (2022)

17. Norwegian Concrete Association: Publication 37 – Low-carbon conrete (Lavkarbonbetong), Oslo (2024)

18. Habert, G., d'Espinose de Lacaillerie, J.B., Roussel, N.: An environmental evaluation of geopolymer based concrete production: reviewing current research trends. J. Clean. Prod. **19**(11), 1229–1238 (2011). https://doi.org/10.1016/j.jclepro.2011.03.012

19. Turner, L.K., Collins, F.G.: Carbon dioxide equivalent (CO2-e) emissions: A comparison between geopolymer and OPC cement concrete. Constr. Build. Mater. **43**, 125–130 (2013). https://doi.org/10.1016/j.conbuildmat.2013.01.023

Performance Evaluation of Recycled Concrete Aggregates as Drainage Material in Combination with Geosynthetics for Landfill Cover Systems

Sayeeda Syed ⓘ and Anumita Mishra ⓘ

Contents

1 Introduction

The development of cover systems for Municipal Solid Waste (MSW) landfills has advanced from simple soil coverings to intricate, multi-component structures designed to effectively regulate infiltration and mitigate landfill gas (LFG) emissions. Cost considerations and resource availability have posed challenges to the use of typical landfill cover materials, such as gravels, sands, and clays, in recent years. The use of alternative materials, such as those obtained from waste, has garnered interest due to their potential to save expenses, conserve natural resources, and improve sustainability by reusing waste streams. Furthermore, this falls in line with international environmental regulations that give priority to minimizing waste, promoting recycling, harnessing energy, and conserving resources. Nevertheless, these materials must satisfy certain requirements for hydraulic, shear strength, and compatibility to be considered as feasible alternative materials. Prior studies on the use of alternate materials in landfill cover systems have mostly

S. Syed · A. Mishra (✉)
Department of Civil Engineering, Indian Institute of Technology, Roorkee, India
e-mail: s_syed@ce.iitr.ac.in; anumita.mishra@ce.iitr.ac.in

© The Author(s) 2025
M. Kioumarsi, B. Shafei (eds.), *The 1st International Conference on Net-Zero Built Environment*, Lecture Notes in Civil Engineering 237,
https://doi.org/10.1007/978-3-031-69626-8_70

focused on the hydraulic efficiency of materials, disregarding a thorough investigation of veneer slope stability. Several researchers have investigated the appropriateness of using materials such as recycled asphalt pavement [1], tire-derived aggregates [2], glass culets [3], recycled concrete aggregates [4], and steel slags [4, 5] as drainage components in landfill cover systems. Significantly, previous research has extensively examined the failure of veneer shear at geosynthetic interfaces in traditional landfill covers [6–13]. However, there is a dearth of studies focussing on the interaction between geosynthetics and alternative waste materials.

In this context, the present study aims at investigation of the feasibility of use of recycled concrete aggregates (RCA) as an alternative material for the drainage layers in landfill final cover systems. The first phase involves the determination of fundamental geotechnical properties of RCA, followed by the assessment of the interface friction angle between RCA and HDPE geomembrane through large-scale direct shear tests. To showcase the practicality of employing RCA within the drainage layer, a numerical model of a municipal solid waste landfill located in Este, Italy, was developed utilizing Finite Element Method (FEM) software, GeoStudio. The validation of this model was achieved by applying the analytical two-wedge theory of veneer shear failure within layered strata, as introduced by Koerner and Soong (2005) [14]. An investigation on the use of recycled asphalt pavement (RAP) aggregates, employing identical methodology, has also been conducted by the authors [15]. While RAP aggregates demonstrated favourable results as a potential alternative material for drainage, it must be noted that most of the RAP produced during road construction and demolition activities is commonly reintegrated on site.

2 Materials and Methods

2.1 Description of Site

The MSW landfill under investigation in this study is owned by *Società Estense Servizi Ambientali* (SESAS p.a.), a limited liability company operating in Italy that specializes in waste collection and treatment. The landfill is situated in the municipality of Este, which is in the southwestern region of Padova Province. The landfill in its entirety encompasses a surface area of 13 ha and a volume of approximately 1.5 million cubic metres [16].

2.2 Sample Procurement and Preparation

The RCA for the present study were obtained from the structural laboratory of the Department of Civil Engineering at the Indian Institute of Technology Roorkee (IITR). Several pre-existing concrete cubes from the laboratory, originally cast for determining the characteristic compressive strength of concrete in various projects,

Table 1 Properties of recycled concrete aggregates

Parameter	Unit	Value	Test Procedure
Specific gravity (G_s)	–	2.55	ASTM D 854
Maximum bulk density (γ_{max})	kPa	12.8	ASTM D 4253
Minimum bulk density (γ_{min})	kPa	11.9	ASTM D 4253
Hydraulic conductivity (k)	cm/s	1.4×10^{-2}	ASTM D 2434
Internal friction angle (ϕ)	Degree	40	–
Cohesion (c)	kPa	0	–

Table 2 Properties of geomembrane

Parameter	Unit	Value	Test Procedure
Material	–	HDPE	–
Thickness	mm	2	ASTM D 5199
Density	kg/m^3	940	ASTM D 792
Tensile Strength	kN/m	57	ASTM D 6693
Puncture resistance	N	675	ASTM D 4833
Tear resistance	N	249	ASTM D 1004

were manually crushed using a hand rammer. The concrete cubes did not conform to a predetermined strength specification. Rather, a varied composition was used to reflect the inherent heterogeneity of RCA typically obtained from construction and demolition (C&D) waste processing facilities. For compatibility with the large-scale direct shear test, particles over 20 mm were excluded from the sample through sieving, and those under 4.75 mm were omitted to prevent material washout. The fundamental geotechnical characteristics of the RCA were determined at the geotechnical laboratory IITR (Table 1). The shear strength parameters of RCA were taken from literature [17].

The HDPE geomembrane (GM) utilized in the studies was obtained from Megaplast India Pvt. Ltd. and its properties (provided by the manufacturer) are presented in Table 2. The GM had a thickness of 2 mm and was supplied in rolls with a width of 1 meter. The GM was divided into sheets of 35 cm × 55 cm to be positioned on the lower shear box, which has an overall size of 35 cm. To secure the GM in position, 10 cm overhangs were maintained on both sides, positioned perpendicular to the shearing direction.

2.3 Experimental Program

To assess the functionality of the drainage layer in combination with the geosynthetic membrane (GM), the shear strength of the interface between RCA and GM was evaluated using the guidelines outlined in ASTM D5321-12. For this purpose, a direct shear apparatus with a shear box of size 30 cm × 30 cm × 30 cm was utilized. The GM interface was set over the top of the lower half of the shear box, which was first filled with standard sand as filler material. This was done by pouring

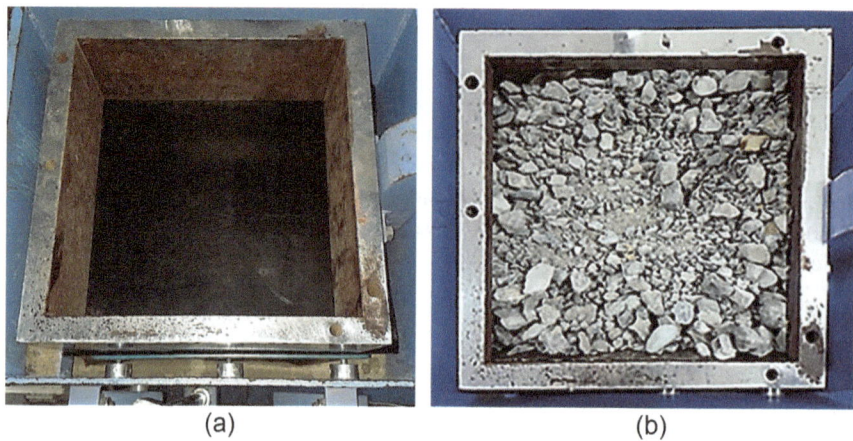

Fig. 1 (**a**) Geomembrane fastened to lower box using a fabricated clamping device (**b**) Upper box filled with RCA

sand in 4 layers and tamping each layer 25 times by a hand tamper. The 35 cm × 55 cm GM sheet was then firmly secured over the top of the lower box using a custom-made clamp designed to hold the GM through reaction force against the outer box as shown in Fig. 1a. The upper box was then placed over the GM and properly aligned with the lower box. The upper box was filled with RCA in four layers and tamping each layer by 25 blows of hand rammer, to achieve a relative density of 95%. After assembly of the direct shear apparatus, the sample was subjected to shearing at standard loads of 50 kPa, 175 kPa, and 275 kPa, with a strain rate of 1 mm/min. The displacements were measured LVDTs, while the associated shear load was recorded using a data logger. During testing, it was observed that the shearing resulted in stripping of cement mortar from the aggregates a shown in Fig. 1b. This suggests a need for selecting RCA with minimal adhered cementitious material to ensure the optimal performance and durability of the drainage layer. Further testing may be necessary to identify suitable aggregates for drainage applications.

2.4 Numerical Program

A numerical analysis of the failure of the cover slope was conducted for the case study of Este landfill, Italy, to illustrate the practicality of utilizing RCA as a drainage medium in conjunction with geosynthetics. The landfill simulation was conducted using the SLOPE/W module of the widely accessible FEM program GeoStudio. The Morgenstern-Price approach was used to analyse the factor of safety (FOS) as it considers both force and moment equilibrium. Below are the specifics of the simulation:

Landfill Geometry and Material Properties The research utilized a 2-D plane strain geometry, focussing specifically on the steepest cross section of the landfill. The analysis was performed on the left half section of the geometry because of the structural symmetry. Figure 2 depicts the landfill geometry, including the slope that was examined for potential collapse. The numerical simulation utilized material characteristics reported by Trivellato (2014) [16], which are outlined in Table 3.

Modelling of Geomembrane The software GeoStudio facilitates the application of geosynthetics as reinforcements with primary purpose of determining their pull-out resistance. However, this approach fails to consider the influence of the interface friction angle on the FOS. Consequently, the GM was constructed using two techniques. The initial approach involved representing the GM as reinforcement, following the conventional protocol. In the second approach, the GM layers were represented as thin layers with thicknesses significantly less than the cover layers (10 mm) using Mohr-Coulomb model, and the interface adhesion and friction angle were taken as the shear strength properties [18].

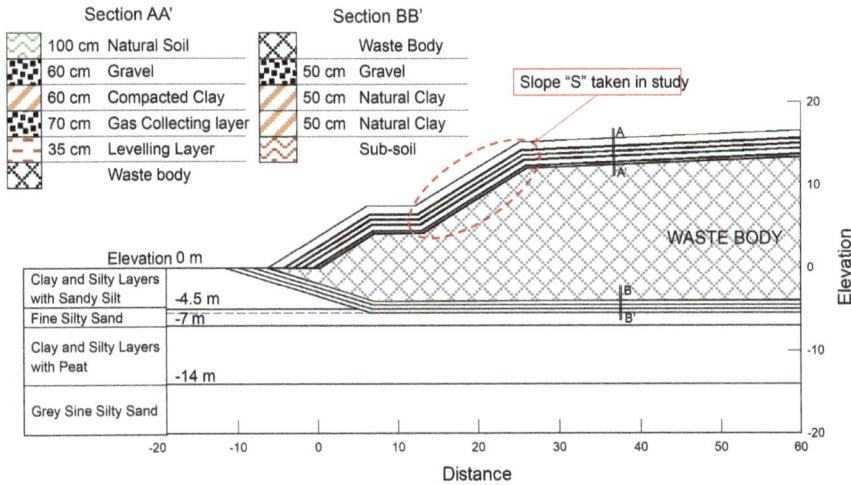

Fig. 2 Geometry of Este MSW landfill taken up in this study [15]

Table 3 Properties of the materials used in numerical model

Material	γ_{dry}	c	ϕ
Natural soil	11	8	19.6
Compacted clay	16	24	29.2
Levelling layer	18.2	0	33.8
Gravel	18	0	28.3
Clay with silty layers	17	4	20.4
Fine silty sand	16	0	24.7

From Trivellato (2014) [16]

γ_{dry}: dry unit weight (kPa); c: cohesion(kPa); ϕ: internal friction angle (°);

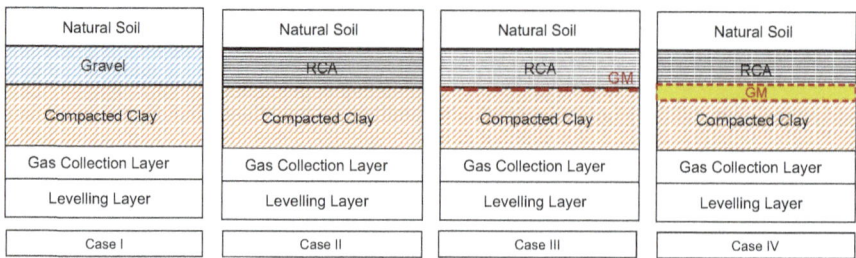

Fig. 3 Cover system profiles for numerical model cases

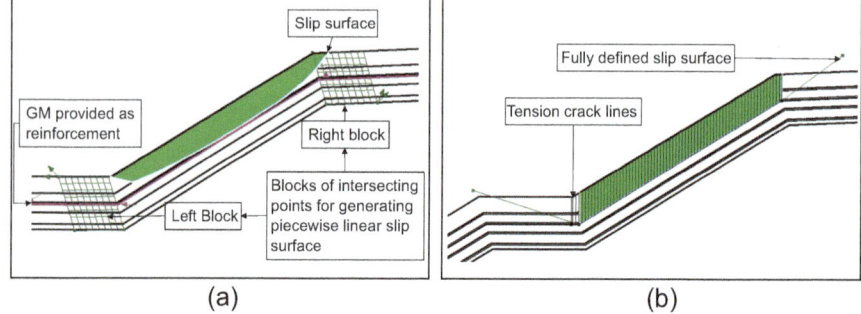

Fig. 4 (**a**) Block specified slip surface method (**b**) Fully specified slip surface method [15]

Modelling of Landfill Cover Cross-Section The landfill cover slope S was analysed in for four cases as depicted in Fig. 3.

Case I: Landfill analysed in its original form.
Case II: Gravel from the drainage layer replaced by RCA.
Cases III: GM was added to the analysis below the drainage layer as reinforcement.
Case IV: GM was added as a thin layer (10 mm) below the drainage layer.

Modelling the Slip Surface The analysis of the landfill cover slope was conducted using two distinct methods (Fig. 4): (i) the block specified slip method, which generates slip surfaces that are parallel to the cover slope, indicative of translational slope failure. This method involved positioning right and left point grids at the slope's crest and toe, respectively, with the software identifying the most critical slope along the cover system. (ii) For calculating the FOS along the actual interface between RCA and geomembrane, a fully specified slip surface parallel to the slope was used, accompanied by a tension crack line based on the approach outlined by Krahn (2004) [19].

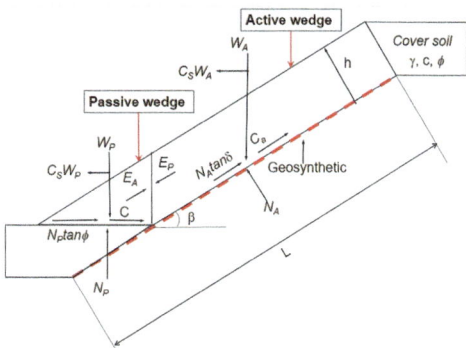

	W_A: Total weight of the active wedge; W_P: Total weight of the passive wedge; N_A: Effective force normal to the failure plane of the active wedge; N_P: Effective force normal to the failure plane of the passive wedge; β: Soil slope angle; δ: Interface friction angle between cover soil and geomembrane; ϕ: Angle of internal friction of the material; c: cohesion; C_a: Adhesive force between cover soil of the active wedge and the geomembrane; L: Length of slope measured along the geomembrane; C: Cohesive force along the failure plane of the passive wedge; h: Thickness of the cover soil;

Fig. 5 Free body diagram of forces on cover soil. (After Koerner & Soong (2005) [14])

2.5 *Analytical Program*

The validation of the numerical model was performed using an analytical approach for veneer failure analysis, as described by Koerner & Soong (2005) [14]. This method considers a linear potential failure surface for veneer cover soils, where the cover soil slides in relation to the interface having the lowest friction angle in the section below. The FOS is determined as the ratio of total resisting forces to destabilizing forces. Considering that the slope is relatively shorter as compared to those typically assumed for infinite slope analysis, the evaluation incorporates the consideration of a passive soil wedge at the toe of the slope. Figure 5 illustrates the forces on the cover slope and the identified failure surface. For comprehensive details, readers are encouraged to consult Koerner and Soong (2005) [14]. To validate the model, failure interfaces within the numerical model were identified for each scenario, followed by the derivation of analytical solutions tailored to these specific failure interfaces.

3 Results and Discussion

The shear strength parameters for RCA were derived from available literature, showing a range of 40–50° [17]. A sensitivity analysis was conducted to establish a specific value for use in the simulation. The internal friction angle of RCA was conservatively varied from 35° to 45°. As can be seen in Fig. 6a, the FOS consistently exceeded 1.5 across the entire range. Based on this, a friction angle of 40° was selected for subsequent analyses. Further, the FOS was found to be insensitive to variations in the interface friction angle of the geosynthetic material when it was specified as a reinforcement in the software. This indicates that the program will produce inaccurate results if the geosynthetic membrane is incorporated in as per conventional method.

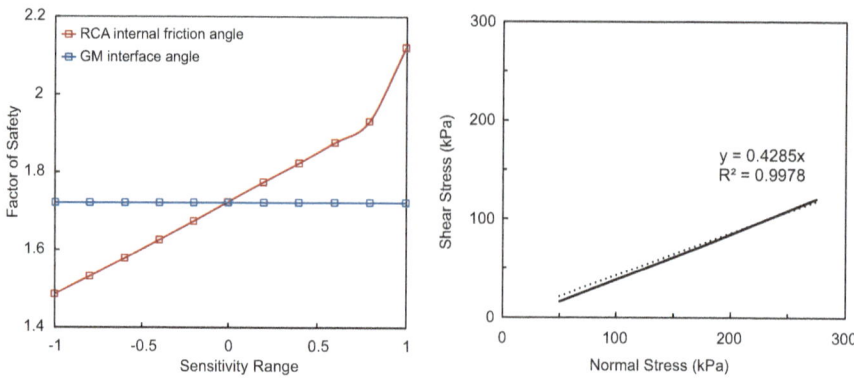

Fig. 6 (**a**) Sensitivity Analysis for internal friction angle of RCA aggregates and interface friction angle of GM (**b**) Results of large-scale direct shear test

Table 4 Results of analytical and numerical analysis

Slip surface specification method	Case	Factor of safety		Error
		Numerical Results	Analytical Results	
Block specified method	Case 1	1.171	1.161	0.8%
	Failure interface	*Gravel-clay*	*Gravel-clay*	
	Case 2	1.72	1.65	4%
	Failure interface	*RCA-clay*	*RCA-clay*	
	Case 3	1.72	1.65	4%
	Failure interface	*GM*	*RCA-clay*	
	Case 4	1.46	1.22	16%
	Failure interface	*GM*	*GM*	
Fully specified slip surface	*RCA-GM interface*	1.39	1.22	16%

The agreement between the outcomes of the analytical and numerical investigations, which are presented in Table 4, provides strong support for the numerical model. It is critical to acknowledge that a steeper landfill cover slope (1:1.6) was utilized in this specific instance, deviating from the recommended range of 1:3 to 1:5. As a result, the study is mostly concerned with comparing several cases as opposed to identifying particular FOS values.

The FOS for Case I, i.e. original cover system with gravel was 1.171 while on replacing the drainage layer material with RCA in Case II, the FOS increased to 1.72. Hence, the RCA layer demonstrated superior shear strength compared to the gravel layer, primarily due to a higher internal friction angle. Additionally, RCA exhibited favourable drainage properties with a hydraulic conductivity of 1.4×10^{-4} m/s. These positive hydraulic characteristics underscore the potential of RCA to contribute to improved slope stability in landfill cover systems.

In order to evaluate the efficacy of alternative cover systems utilizing geosynthetics, GM was integrated into the model for Cases III and IV. The shear strength characteristics of the RCA-GM interface, as determined by large-scale direct shear experiments, are illustrated in Fig. 6b. A negligible intercept of the curve was obtained, which was fixed to zero by linear curve fitting, yielding a Pearson's r value of 0.999 (R-Square $= 0.9994$). The angle of contact friction (δ) was calculated to be 23.2°.

Case III exhibited no sensitivity to FOS when GM was incorporated as reinforcement, while in Case IV, the FOS value decreased from 1.72 to 1.48 due to formation of a weak interface in the cover system. This, in addition to the results of sensitivity analysis highlights that, the software exhibits a limitation in recognizing the effects of a weak interface along the geosynthetic in Case III. The software demonstrates a constraint in identifying the consequences of a weak interface along the geosynthetic in Case III, as evidenced by this and the sensitivity analysis results. Syed and Mishra (2023) [18] explain that in order to address this concern and guarantee precise evaluation of slope stability, it is crucial to incorporate the geomembrane as a thin layer beneath the drainage layer.

It is worth mentioning that the analytical techniques employed in this research were specifically designed to compute the Factor of Safety (FOS) at the interfaces between soil and geosynthetic materials. In contrast, the block failure slip surface method enables the software to independently identify the most critical slip surface along a cover slope. To facilitate a significant comparison, the numerical analysis was used to identify the materials at the top and bottom of the slip surface in each case, which was then used to substitute the properties of the failure interface for each case. The results obtained from this comparative analysis indicated that the analytical model and the numerical model exhibited a significant level of agreement.

4 Conclusion

The study delves into the integration of recycled concrete aggregate (RCA) in landfill cover systems, examining its potential as a sustainable alternative to natural aggregates. Through geotechnical tests and numerical simulations, the performance of RCA in drainage layers and its interaction with geomembranes were assessed. Following are the conclusions drawn from the study:

- The geotechnical tests conducted on recycled concrete aggregates (RCA) have demonstrated their suitability for use as a drainage materials in cover systems. With a hydraulic conductivity of 1.4×10^{-4} m/s, the material possesses the desirable permeability characteristics for effective drainage layers in various applications. However, it is recommended that aggregates with minimum cementitious mortar adhered should be used to ensure the optimal performance and durability of the drainage layer.

- RCA aggregate layer outperforms the gravel layer in terms of shear strength mainly on account of higher value of internal friction angle. There was a 47% increase in the FOS value from of Case I and Case II.
- The interface friction angle between RCA and GM was experimentally determined to be 23.2°. Upon numerical simulation, this resulted in the formation of a weak interface in the cover system.
- Sensitivity analysis and numerical analysis result for Case III revealed that the software is insensitive to changes in interface properties of geomembrane when applied as a reinforcement.
- On the introduction of the GM as a thin layer in the cover system, the FOS experienced a decline from 1.71 to 1.46. This was supported by similar outcomes obtained from the analysis of a fully specified slip surface along the RCA-GM interface.
- Analytical methods for determination of FOS at soil-geosynthetic interfaces were utilized to obtain FOS values for soil-soil interfaces with good agreement.

Acknowledgements The authors are thankful to the management at Megaplast Pvt. Ltd. for providing the geomembrane specimens and their friendly cooperation. Additionally, we acknowledge Mr. Nishant Varshney, an intern under the SPARK Program at IIT Roorkee, for his invaluable contributions to the experiments.

References

1. Rahardjo, H., Satyanaga, A., Leong, E., Wang, J.: Unsaturated properties of recycled concrete aggregate and reclaimed asphalt pavement. Eng. Geol. **161**, 44 (2013)
2. Praveen, V., Sunil, B.M.: Potential use of waste rubber shreds in drainage layer of landfills-an experimental study. Adv. Environ. Res. **5**(3), 201–211 (2016)
3. Cortellazzo, G., Bello, E., Busana, S., Favaretti, M.: Experimental acceptance procedure for using cullet in the gas collection layer of MSW landfill. Indian Geotech. J. **51**, 877–886 (2021)
4. Roque, A.J., da Silva, P.F., de Almeida, M.: Recycling of crushed concrete and steel slag in drainage structures of geotechnical works and road pavements. J. Mater. Cycl. Waste Manag. **24**(6), 2385–2400 (2022)
5. Reddy, K.R., Grubb, D.G., Kumar, G.: Innovative biogeochemical soil cover to mitigate landfill gas emissions. In: International Conference on Protection and Restoration of the Environment XIV (2018)
6. Koerner, R.M., Hwu, B.L.: Stability and tension considerations regarding cover soils on geomembrane lined slopes. Geotext. Geomembr. **10**(4), 335–355 (1991)
7. Sharma, H.D., Reddy, K.R.: Geoenvironmental Engineering: Site Remediation, Waste Containment, and Emerging Waste Management Technologies. Wiley (2004)
8. Giroud, J.P., Bachus, R.C., Bonaparte, R.: Influence of water flow on the stability of geosynthetic-soil layered systems on slopes. Geosynth. Int. **2**(6), 1149–1180 (1995)
9. Koerner, R.M., Soong, T.Y.: Leachate in landfills: the stability issues. Geotext. Geomembr. **18**(5), 293–309 (2000)
10. Soong, T.Y., Koerner, R.M.: Seepage induced slope instability. Geotext. Geomembr. **14**(7–8), 425–445 (1996)
11. Feng, S.J., Gao, L.Y.: Seismic stability analyses for landfill cover systems under different seepage buildup conditions. Environ. Earth Sci. **66**, 381–391 (2012)

12. Khoshand, A., Fathi, A., Zoghi, M., Kamalan, H.: Seismic stability analyses of reinforced tapered landfill cover systems considering seepage forces. Waste Manag. Res. **36**(4), 361–372 (2018)
13. Nadukuru, S., Zhu, M., Gokmen, C., Bonaparte, R.: Combined seepage and slope stability analysis of a landfill cover system. In: Geotechnical Frontiers 2017, pp. 170–179 (2017)
14. Koerner, R.M., Soong, T.Y.: Analysis and design of veneer cover soils. Geosynth. Int. **12**(1), 28–49 (2005)
15. Syed, S., Mishra, A.: Performance evaluation of alternative drainage material for landfill cover systems in combination with geosynthetics. In: Accepted in 5th Pan-American Conference on Geosynthetics, Toronto Canada (2024)
16. Massimo, T.: Geotechnical Slope Stability of the Este MSW Landfill. Universita' Degli Studi di Padova, Padova (2014)
17. Arulrajah, A., Piratheepan, J., Ali, M., Bo, M.: Geotechnical properties of recycled concrete aggregate in pavement sub-base applications. Geotech. Test. J. **35**(5), 743–751. E (2012)
18. Syed, S., Mishra, A.: Stability and sensitivity analyses of an industrial waste landfill with a novel final cover system. In: 9th International Congress on Environmental Geotechnics, Chania, Greece (2023)
19. Krahn, J.: Stability Modeling with SLOPE/W: an Engineering Methodology. GEOSLOPE/W International Ltd., Calgary (2004)

Bio-Based Polymer Composites Used in the Building Industry: A Review

Chinyere O. Nwankwo ⓘ and Jeffrey Mahachi ⓘ

Contents

1   Introduction

The building/construction industry significantly contributes to resource depletion and environmental pollution [1]. Twenty per cent of the environmental impact stems from processes associated with buildings, encompassing the manufacturing and disposal of building materials [2]. The industry contributed 38% of all energy-related carbon dioxide (CO_2) emissions in 2020 [3] and is also one of the largest consumers of virgin materials. In terms of environmental impact, construction and demolition (C&D) wastes make up a large percentage of materials that end up in landfills. For example, 40% of the landfill volume in the USA is from C&D wastes,

C. O. Nwankwo (✉)
Center of Applied Research and Innovation in the Built Environment (CARINBE), Faculty of Engineering and the Built Environment, University of Johannesburg, Johannesburg, South Africa

J. Mahachi
Department of Civil Engineering Technology, University of Johannesburg, Johannesburg, South Africa

© The Author(s) 2025
M. Kioumarsi, B. Shafei (eds.), *The 1st International Conference on Net-Zero Built Environment*, Lecture Notes in Civil Engineering 237,
https://doi.org/10.1007/978-3-031-69626-8_71

mainly wood, drywall, and plastic [4]. A fibre-reinforced polymer (FRP) material is typically a composite of synthetic fibres, particularly carbon, glass, or aramid, and a polymer matrix, typically polyester, vinyl ester, or epoxy resins [5]. Owing to its high strength, corrosion resistance, and lightweight nature, the popularity of this material has surged in the construction industry. Specifically, they have been used to develop structural shapes that could be used in various building and bridge applications, sandwich construction, and as internal and external reinforcement for structural elements [6]. Since FRP composites are commonly made with synthetic materials, rising environmental concerns are associated with their production and disposal. These concerns fuel the development of bio-based composites, which describe composites made from natural sources that can invariably reduce the consumption of virgin materials and are more eco-friendly. These bio-based polymer composites, or biocomposites, are poised to be non-abrasive and reduce CO_2 emissions and dependence on petroleum products [7]. Given the recycling challenges and toxic emissions associated with synthetic FRPs [8], biocomposites present a promising avenue for embracing the principles of the circular economy, particularly if they exhibit biodegradability [9]. A variety of natural fibres sourced from plants and animals have been utilised to create functional biocomposites. These biocomposites, in comparison to synthetic fibres, exhibit lower density, are biodegradable, potentially more accessible, and are less abrasive [10]. Consequently, they present a promising alternative to reinforcement material for FRP composites.

Biocomposites are becoming increasingly attractive, and their manufacture and supply in recent years have been accompanied by political incentives and tax reductions in some countries [11, 12]. Natural fibres are now used instead of synthetic fibres to create biocomposites across the automobile, sports, aerospace, and other industries. The automobile industry has embraced the use of natural fibre biocomposites to achieve weight reductions and improved mechanical properties of automobile components like door panels, interior carpets, boot lining, dashboards, bumpers, and so on [13, 14]. Specific policies like those established by the European Union and Japan that required 85% (95% for Japan) of a vehicle to be either recycled or reused as of 2015 have encouraged the research and development of biocomposites [15]. Cost is also a crucial factor that encourages the use of natural fibres over synthetic fibres. Natural fibres are cheap compared to their synthetic counterparts, increasing their commercial and research potential [12].

Properties of biocomposites, such as their lightweight, high strength-to-weight ratio, corrosion resistance, and thermal insulation properties, make them suitable building materials, and their adoption in the industry offers several benefits. From an environmental perspective, biocomposites reduce reliance on fossil fuels and mitigate greenhouse gas emissions throughout the product lifecycle [16, 17]. Biocomposites are often biodegradable or recyclable, minimising waste generation and contributing to a circular economy. Economically, bio-based composites can reduce material costs and create new opportunities for rural development and job creation in agricultural communities. The potential for biocomposites in the construction industry lies in the current applications of conventional synthetic FRPs. Bakis et al. [6] outlined the specific applications of FRP composites for construction, including bridge decks, sandwich construction, internal reinforcement, and externally bonded reinforcement. Biocomposites have also found their place in the building industry, where they have been developed for non-structural applications such as furniture, insulation boards, and partitions. This study explores the various practical applications of biocomposites in the construction industry as available in existing literature. Biocomposites, though promising, still have their limitations; standardisation, durability, and strength concerns have widely limited their acceptance for their practical usage in the construction industry. This paper also seeks to examine the challenges limiting their use in the building industry.

2 Bio-Based Polymer Composites

An FRP composite comprises a matrix material and a reinforcement material. The matrix defines the shape of the material and guards against chemical and mechanical damage, while the fibres provide the strength and stiffness of the composite [12, 18]. Biocomposites can be a blend of organic and inorganic components [16], i.e. natural fibres with a synthetic polymer, synthetic fibres with a bio-based polymer, a hybrid of natural and synthetic polymers, or a pure biocomposite with all organic components. Natural fibres can be sourced from plant sources like leaves (sisal, pineapple, banana), bast (jute, flax, hemp, kenaf, ramie), seed/fruit (cotton, coir), stalk (rice, wheat), cane (bamboo), wood, animal (wool, silk), and mineral sources (basalt, asbestos) [19]. For the polymer matrix, thermosets or thermoplastics like polyester, vinyl ester, and epoxy, polypropylene, low-density polyethene, high-density polyethene, and nylon serve as the matrix. Typical thermosets and thermoplastics are not biodegradable, so the ideal polymers for the sustainable development of FRP products would be biodegradable bio-based polymers, also known as bioplastics or biopolymers. Bioplastics are crop-derived renewable resources, such as cellulose plastics, polylactides, starch plastic, soy-based plastic, and polyhydroxyalkanoate polymers [20].

3 Biocomposites in the Building Industry

Biocomposites can be used for non-structural (non-load-bearing) and structural (load-bearing) applications in the building industry. These applications are presented in Fig. 1 and will be examined in this section.

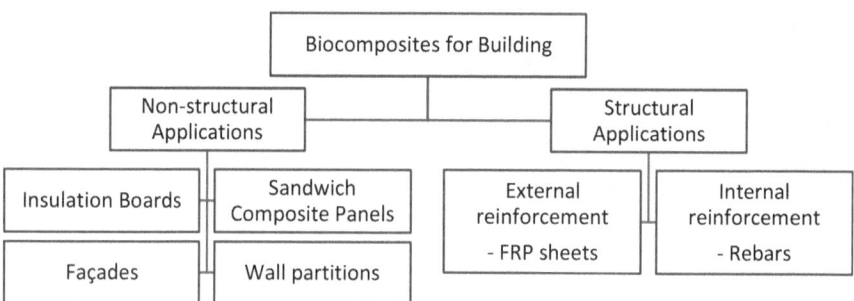

Fig. 1 Classification of biocomposites for building

3.1   Non-Structural Applications

Building Insulation Material Energy consumption in the building industry is responsible for 40% of global CO_2 emissions [21], and this can be reduced with building thermal insulation. The commonly used materials for thermal insulation in buildings are fibreglass, mineral wool, polymers, aerogels, etc. [22]. More eco-friendly materials like primarily sourced plant fibres or agricultural wastes, can be used to develop more sustainable building thermal insulators [23]. These fibres have lower density and heat retardant properties, which make them suitable insulators [24]. Several authors have developed biocomposites for specific use as thermal insulation materials. Madival et al. [22] developed rice straw/furcraea foetida boards with an epoxy matrix for thermal insulation. They evaluated the fabricated composite's thermal stability, conductivity, transmittance, resistance, specific heat capacity, and flammability properties. The authors were able to create a composite that was thermally stable up to 272 °C. Le and Pásztory [24] developed boards with rice straw and reed/phenol formaldehyde, coir fibre/phenol formalde-hyde, and a binder-less coir fibre panel that can be used as thermal insulating material. Though the thermal resistance of the developed material was evaluated and found to be suitable, no comparisons were made with existing commercial thermal insulation boards. Bhuiyan et al. [21] fabricated jute/polypropylene biocomposite boards and established that the porous morphology and low thermal conductivity of cellulose fibres make them good heat insulators. The developed biocomposite's water absorption and mechanical properties were then compared to those of a commercial gypsum/polyvinyl chloride plasterboard and found to be superior. The biocomposite had superior thermal resistance to the commercial board, but the commercial board had higher thermal stability at higher temperatures. Low thermal conductive fillers in the biocomposites were proposed to improve the heat barrier performance.

Acoustic insulation is also an essential aspect of the service life of a building. Insulating materials are sometimes developed to have both thermal and acoustic insulation properties. Vasconcelos et al. [25] developed green high-density polyethylene composites with natural powdered cork to have thermal and acoustic insulation properties. It was found that the acoustic absorption coefficient was dependent on the reinforcement ratio within the composite. Urdanpilleta et al. [26] used the wool fibres of the Latxa sheep breed combined with a soy protein isolate matrix to develop a sound-insulating biocomposite material. It was found that the fibrillary microstructure in natural fibres and the presence of empty cells ensure high porosity and absorption properties, making them have good sound absorption coefficients at medium and high frequencies. Alencar et al. [27] proposed using expanded natural rubber as the polymer matrix reinforced with a wood filler (eucalyptus) in developing biocomposites with acoustic insulation. The developed material had three times the acoustic insulation capacity of polyurethane foam, a commercially available acoustic absorbent.

The use of waste in developing biocomposites is also a significant circular economy consideration. Polyethylene (PE) and polypropylene (PP) are projected to account for over 60% of global plastic production by 2050 and are projected to account for 20% of global petroleum consumption by 2050 [28]. The construction industry is responsible for 19% of the world's consumption of plastics and is the second largest stack globally [25]. Waste plastics can also be repurposed to reduce the environmental pollution associated with their disposal [28].

Sandwich Composite Panels A sandwich composite is a type of material typically consisting of two outer stiff layers called the face sheets bonded to an inner core material sandwiched between them [6]. The face sheets are usually made of strong and stiff materials, while the core material is often considerably thicker and lighter and can use foam [17], honeycomb [29], or wood design. This design provides high strength and stiffness while keeping the structure's overall weight relatively low. The facing typically provides the bending resistance, while the core provides shear resistance and insulation [30, 31]. This type of material has applications in the aerospace, marine, and automotive industries, and in the building industry, it has found its place. Chomachayi et al. developed a sandwich structure from polyhydroxyalkanoate, polylactic acid, and cellulose microfibres to act as a building envelope vapour retarder, reducing mould growth and providing proper building insulation. The developed composite had a water vapour permeance value within the limits set in the national building code of Canada. Fu and Sadeghian [30] developed sandwich composite beams with flax/bio-based epoxy facing and a paper honeycomb core that could be used as building and cladding systems.

Sandwich panels can also be made to be prefabricated building modules that can be assembled off-site to ensure fast assembly and construction. Laraba et al. [32] developed a sandwich panel with an alfa fibre/epoxy core and jute/metallic mesh sheets to be potentially used as building partitions or roof panels. Conventional fabrication techniques such as manual lay-up, vacuum moulding and resin transfer moulding were used. Recently, 3D printing technology has also been used to fabricate sandwich composites with more intricate geometry [29]. Dweib et al. [17] developed a structural sandwich panel for potential residential building applications such as roof, wall, or floor material. The authors made sandwich beams with different face sheets made of flax fibre, chicken feathers, and cellulose from recycled paper with a soybean oil-based resin as the matrix. The sandwich beams were designed to have vertical webs and a structural polyurethane foam core that provided both acoustic and thermal insulation. The manufactured samples, except those made of flax fibre, had comparable strength and flexural rigidity to the available lumber sections. Their further work in [33] saw the successful development of the recycled paper composite into a prototype roof structure.

Façade and Partitions Biocomposites can be developed for their specific use as vertical wall panels. Ks et al. [34] developed a sandwich composite with a flax/epoxy sheet and coir/polyurethane foam core as an alternative to conventional building partitions. The light weight of biocomposites and their insulation properties make them suitable alternatives. Astudillo et al. [35] set out to develop four different biocomposite products: an interior partition system with 40% bio-content, a multi-layered wall system with 70% bio-content, and a curtain wall with 30% bio-content. The products were designed according to the European Building Code and tested for structural performance, fire performance, air quality, thermal insulation, and durability. The developed biocomposites also underwent a monitoring phase where full-scale prototypes were found to have comparable performance to commercially available alternatives. A green vertical system is an innovative solution to introduce green areas directly in urban cities by having vertical gardens either supported by the façade or incorporated directly into the façade [36]. This green wall has been shown to improve air quality and the thermal and acoustic performance of a building. Cork agglomerate boards produced from waste from cork oak agglomerated with natural resins are one such nature-based solution in the building system [37].

Given their natural source, biocomposites have been shown to have less environmental impact when used as partition and façade systems. Morganti et al. [16] examined the global warming potential of three prefabricated façade system modules made with bio-based resin reinforced with basalt fibres and wooden particles. The biocomposite modules, through the product lifecycle from cradle to practical completion, reduced associated CO_2 emissions across all the categories considered compared to a conventional alternative.

3.2 Structural Applications

External Reinforcement Among the different structural retrofitting techniques, using FRP sheets, which are typically bonded to concrete members, is one of the least disruptive. In recent years, natural fibre composites have been developed to be a more sustainable alternative to the popularly used carbon and glass fibre FRP sheets used as external reinforcement to retrofit structural members. A holistic review of their use was done in [38], which saw the strengthening of beams, columns, walls, and slabs across concrete, masonry, and wooden structural elements. Structural elements were strengthened with biocomposites made with different natural fibres, including bamboo [39], hemp [40], kenaf [41], pineapple leaf [42], sisal [43], flax and jute [44], and so on. Bio-based polymers have also been used in this regard to develop completely green biocomposites [40, 45], or to make partial biocomposites when in conjunction with synthetic fibres [46]. In instances where the biocomposite was designed and the thickness increased to compensate for the lower strengths of natural fibres as compared to their synthetic counterparts, the RC beam strengthened with the biocomposite had a comparable load-carrying capacity with those strengthened with carbon FRP [41, 44].

The advantage of biocomposites as an external reinforcing material lies in their eco-friendliness, cost, and lower strength. Since they are developed from bio-based materials, the composite could be biodegradable. Composites could also be cheaper since they are produced from locally available materials. The high strength and stiffness of synthetic FRPs have resulted in more abrupt debonding failures [19], but natural FRPs have lower strengths that are more compatible with concrete.

Internal Reinforcement Due to durability considerations, FRP rods have been developed as a non-corrosive alternative to conventional steel rebars. Its advantage lies in its non-reactive nature, which makes properties unchanged in aggressive environments where steel rebars would corrode [47]. Attai et al. fabricated 12 mm diameter glass FRP and jute FRP reinforcement bars with the hand lay-up technique. Though the glass FRP had much superior performance, the developed jute FRP rebars had tensile strength and elastic modulus up to 178.42 MPa and 12.15 GPa, respectively. Though these strength values are lower than the steel alternative, they can be used in low-loading structural applications. It was also established that the fibre volume fraction greatly influences the strength of the developed composite rebars. Sharkawi et al. [48] used an infusion technique to make reinforcement bars with jute and flax yarn fibres. Though the bars had modest strength, they still enhanced the strength of a normal-weight concrete slab and that of a lightweight concrete slab by 175% and 56%, respectively. Elbehiry et al. [49] made banana fibre/polyester as reinforcement bars in a concrete beam. Though the natural FRP rebars had a much lower strengthening effect than conventional steel bars, they enhanced the beam's flexural strength by up to 25% more than the plain concrete beam. The natural FRP bar can be a low-cost alternative in low-strength applications. Joyklad et al. [50] used sisal fibre rope embedded in epoxy as post-installed shear reinforcement in flat concrete slabs. The natural sisal fibre performed better than the carbon and aramid FRPs, although it had the lowest strength.

4 Challenges and Future Directions

There are associated challenges with adopting biocomposites in the building industry, some of which are discussed in this section.

1. Natural fibre viability: The mechanical properties of natural fibres are not consistent and predictable; they depend on the age of the plant, plant species, plant

origin, and the soil and weather conditions in which the plant was cultivated [51]. This viability can affect the consistent and predictable material performance of biocomposites.

2. Processing technique: Various techniques, such as wet lay-up, compression moulding, resin transfer moulding, and vacuum infusion, have been used to make biocomposites. Sophisticated apparatus is hardly needed to make samples for laboratory testing and prototyping. However, in developing larger sizes and amounts, some of the techniques used become labour-intensive, and the quality of the manufactured composite becomes challenging to ensure as there could be inconsistencies within the produced samples [18]. The specific process used to manufacture biocomposites affects the quality of the composite produced. As long as the manufacturing process remains unstandardised and has no specific methodology, the quality of biocomposites with natural fibres will remain uncertain [52].

3. Water absorption: Polymers are hydrophobic, while lignocellulosic natural fibres absorb moisture. The increase in fibre content in biocomposites directly increases the moisture absorption capabilities of natural fibre composites. When biocomposites absorb moisture, they swell and return to their original state after drying; this causes the fibres to lose their bond with the matrix material [53] and consequently reduces the mechanical performance of the developed biocomposite.

4. Thermal performance: When there is a fire outbreak in a building, elements within should be able to maintain their intensity for particular durations. Natural fibres are organic materials that degrade at high temperatures. The initial degradation of natural fibres, which results in the breakdown of hemicellulose, occurs between 220 °C and 280 °C, while a second and much faster degradation, involving the breakdown of lignin, occurs between 280 °C and 300 °C [54]. If biocomposites are to be used as a building material, their fire performance must be ensured.

5. Degradation behaviour: Degradation behaviour: The biodegradability of biocomposites is essential in advocating for their adoption. Natural fibres are often paired with resins that are not eco-friendly or degradable.

The methodology for material development, from the initiation of an idea to product testing in a laboratory to final integration into a building as a usable product, was outlined in [35]. A semi-industrial-scale sample prototype development should be done after material development and laboratory testing. A life cycle cost analysis and monitoring should then follow the prototype development. As researchers continue to develop biocomposites for use in the building industry, more prototype development should be done, and the abovementioned issues should be addressed further. Standardisation is encouraged within the industry to create more consistent composites. The durability of biocomposites given actual service life loads should be considered, and pure biocomposites with bio-based polymers should be developed.

5 Conclusion

Biocomposites have great potential for creating a more sustainable built environment. Natural fibres sourced directly from plants or indirectly from waste materials and bio-based polymers hold potential as raw materials in creating more eco-friendly building materials that can reduce the buildings industry's associated CO_2 emissions, reduce the non-biodegradable waste that ends up in landfills, reduce the dependence on virgin materials, and reduce building costs by utilising locally available materials while empowering local communities. Researchers have made substantial efforts to develop biocomposites that can be used within the building industry for structural and non-structural applications. The weight, strength, and insulation properties of specific biocomposite modules in the form of panels, sandwich composites, partitions, or building facades make them suitable building materials. This review also saw the development of natural fibre reinforcement rebars for concrete and bio-based structural retrofitting materials. Though the potential for the increased adoption of biocomposites within the building industry exists, associated stakeholders, from researchers to manufacturers should address making the industry more standardised and making more durable biocomposites.

Acknowledgements This research is funded by the Intra-Africa Mobility Scheme of the European Union in partnership with the African Union under the Africa Sustainable Infrastructure Mobility (ASIM) scheme (624204-PANAF-1-2020-1-ZA-PANAF-MOBAF). Opinions and conclusions are those of the authors and are not necessarily attributable to ASIM. The work is supported and part of collaborative research at the Centre of Applied Research and Innovation in the Built Environment (CARINBE).

References

1. Van Erp, G., Rogers, D.: A highly sustainable fibre composite building panel. In: Sustainable Procurement Conference (2008)
2. Galan-Marin, C., Rivera-Gomez, C., Garcia-Martinez, A.: Use of natural-fiber bio-composites in construction versus traditional solutions: operational and embodied energy assessment. Materials. **9**(6), 465 (2016). https://doi.org/10.3390/ma9060465
3. Almpani-Lekka, D., Pfeiffer, S., Schmidts, C., Seo, S.-i.: A review on architecture with fungal biomaterials: the desired and the feasible. Fungal Biol. Biotechnol. **8**(1), 17 (2021). https://doi.org/10.1186/s40694-021-00124-5
4. Christian, S., Billington, S.: Sustainable biocomposites for construction. In: Composites & Polycon (2009)
5. Qureshi, J.: Fibre-reinforced polymer (FRP) in civil engineering. In: Next Generation Fiber-Reinforced Composites – New Insights. IntechOpen (2022). https://doi.org/10.5772/intechopen.107926
6. Bakis, C.E., et al.: Fiber-reinforced polymer composites for construction — state-of-the-art review. J. Compos. Constr. **6**(2), 73–87 (2002). https://doi.org/10.1061/(ASCE)1090-0268(2002)6:2(73)

7. Ekundayo, G., Adejuyigbe, S.: Reviewing the development of natural fiber polymer composite: a case study of sisal and jute. Am. J. Mech. Mater. Eng. **3**(1), 1–10 (2019). https://doi.org/10.11648/j.ajmme.20190301.11

8. Shah, A.U.M., Sultan, M.T.H., Jawaid, M., Cardona, F., Talib, A.R.A.: A review on the tensile properties of bamboo fiber reinforced polymer composites. Bioresources. **11**(4), 10654–10676 (2016). https://doi.org/10.15376/biores.11.4.Shah

9. Wagner, A., Ott, S.: Cascading wood-based construction products – a guideline for product development. In: E3S Web of Conferences (2022). https://doi.org/10.1051/e3sconf/202234904009

10. Júnior, d.O., Nunes, J., Lopes, F.P.D., Simonassi, N.T., Lopera, H.A.C., Monteiro, S.N., Vieira, C.M.F.: Ecofriendly panels for building with eucalyptus sawdust and vegetal polyurethane resin: a mechanical evaluation. Case Stud. Constr. Mater. **18**, e01839 (2023). https://doi.org/10.1016/j.cscm.2023.e01839

11. Väisänen, T., Haapala, A., Lappalainen, R., Tomppo, L.: Utilization of agricultural and forest industry waste and residues in natural fiber-polymer composites: a review. Waste Manag. **54**, 62–73 (2016). https://doi.org/10.1016/j.wasman.2016.04.037

12. Lau, K., Hung, P., Zhu, M.-H., Hui, D.: Properties of natural fibre composites for structural engineering applications. Compos. B Eng. **136**, 222–233 (2018). https://doi.org/10.1016/j.compositesb.2017.10.038

13. Fortea-Verdejo, M., Bumbaris, E., Burgstaller, C., Bismarck, A., Lee, K.Y.: Plant fibre-reinforced polymers: where do we stand in terms of tensile properties? Int. Mater. Rev. **62**(8), 441–464 (2017). https://doi.org/10.1080/09506608.2016.1271089

14. Partanen, A., Carus, M.: Biocomposites, find the real alternative to plastic – an examination of biocomposites in the market. Reinf. Plast. **63**(6), 317–321 (2019). https://doi.org/10.1016/j.repl.2019.04.065

15. Holbery, J., Houston, D.: Natural-fiber-reinforced polymer composites in automotive applications. JOM. **58**(11), 80–86 (2006)

16. Morganti, L., Vandi, L., Larraz, J.A., Garc, J., Muedra, A.N., Pracucci, A.: A1–A5 embodied carbon assessment to evaluate bio-based components in Façade system modules. Sustain. For. **16**, 1190 (2024). https://doi.org/10.3390/su16031190

17. Dweib, M.A., Hu, B., O'Donnell, A., Shenton, H.W., Wool, R.P.: All natural composite sandwich beams for structural applications. Compos. Struct. **63**(2), 147–157 (2004). https://doi.org/10.1016/S0263-8223(03)00143-0

18. Mazumdar, S.K.: Composites Manufacturing: Materials, Product and Process Engineering. CRC Press, Boca Raton (2002)

19. Nwankwo, C.O., Ede, A.N., Olofinnade, O.M., Osofero, A.I.: NFRP strengthening of reinforced concrete beams. IOP Conf. Ser. Mater. Sci. Eng. **640**, 12074 (2019). https://doi.org/10.1088/1757-899X/640/1/012074

20. Ray, D., Rout, J.: Thermoset biocomposites. In: Mohanty, A.K., Misra, M., Drzal, L.T. (eds.) Natural Fibers, Biopolymers, and Biocomposites. Taylor & Francis (2005)

21. Bhuiyan, M.A.R., et al.: Flame resistance and heat barrier performance of sustainable plain-woven jute composite panels for thermal insulation in buildings. Appl. Energy. **345**(October 2022), 121317 (2023). https://doi.org/10.1016/j.apenergy.2023.121317

22. Madival, A.S., Shetty, R., Doreswamy, D., Maddasani, S.: Characterization and optimization of thermal properties of rice straw and Furcraea foetida fiber reinforced polymer composite for thermal insulation application. J. Build. Eng. **78**(July), 107723 (2023). https://doi.org/10.1016/j.jobe.2023.107723

23. Ali, M., et al.: Sunflower and watermelon seeds and their hybrids with pineapple leaf fibers as new novel thermal insulation and sound-absorbing materials. Polymers. **15**, 4422 (2023). https://doi.org/10.3390/polym15224422

24. Le, D.H.A., Pásztory, Z.: Experimental study of thermal resistance values of natural fiber insulating materials under different mean temperatures. SEEFOR. **14**(1), 93–99 (2023). https://doi.org/10.15177/seefor.23-03

25. de Vasconcelos, G.C.M.S., et al.: Thermal and acoustic performance of green polyethylene/cork composite for civil construction applications. Mater. Res. **26**(e20220232), 16–20 (2023). https://doi.org/10.1590/1980-5373-MR-2022-0232

26. Urdanpilleta, M., Leceta, I., Guerrero, P., de la Caba, K.: Sustainable sheep wool/soy protein biocomposites for sound absorption. Polymers. **14**, 5231 (2022). https://doi.org/10.3390/polym14235231

27. de Alencar, L.N., et al.: Natural rubber/wood composite foam: thermal insulation and acoustic isolation materials for construction. Cell. Polym. **42**(2), 55–72 (2023). https://doi.org/10.1177/02624893231151364

28. Bhuiyan, M.A.R., Darda, M.A., Ali, A., Talha, A.R., Hossain, M.F.: Heat insulating jute-reinforced recycled polyethylene and polypropylene bio-composites for energy conservation in buildings. Mater. Today Commun. **37**, 106948 (2023). https://doi.org/10.1016/j.mtcomm.2023.106948

29. Essassi, K., Rebiere, J.L., El Mahi, A., Ben Souf, M.A., Bouguecha, A., Haddar, M.: Dynamic characterization of a bio-based sandwich with auxetic core: experimental and numerical study. Int. J. Appl. Mech. **11**(2), 1950016 (2019). https://doi.org/10.1142/S1758825119500169

30. Fu, Y., Sadeghian, P.: Bio-based sandwich beams made of paper honeycomb cores and flax FRP facings: flexural and shear characteristics. Structures. **54**(August 2022), 446–460 (2023). https://doi.org/10.1016/j.istruc.2023.05.064

31. Cui, Y., Hao, H., Li, J., Chen, W., Zhang, X.: Structural behavior and vibration characteristics of geopolymer composite lightweight sandwich panels for prefabricated buildings. J. Build. Eng. **57**, 104872 (2022). https://doi.org/10.1016/j.jobe.2022.104872

32. Laraba, S.R., et al.: Development of sandwich using low-cost natural fibers: Alfa-Epoxy composite core and jute/metallic mesh-Epoxy hybrid skin composite. Ind. Crop. Prod. **184**-(February), 115093 (2022). https://doi.org/10.1016/j.indcrop.2022.115093

33. Dweib, M.A., Hu, B., Shenton, H.W., Wool, R.P.: Bio-based composite roof structure: manufacturing and processing issues. Compos. Struct. **74**(4), 379–388 (2006). https://doi.org/10.1016/j.compstruct.2005.04.018

34. Ks, M., Prabhakaran, S., Gautham, R., Ramji Gautham, D.R., Sathish Kumar, M., Bala Ganesan, A., Nithish, C.: Investigation of mechanical and thermal behaviour of natural sandwich composite materials for partition walls. Int. J. Res. Rev. **7**(5), 211–216 (2020)

35. Astudillo, J., et al.: New biocomposites for innovative construction façades and interior partitions. J. Facade Des. Eng. **6**(2), 65–83 (2018)

36. Manso, M.: Thermal analysis of a new modular system for green walls. J. Build. Eng. **7**, 53–62 (2016). https://doi.org/10.1016/j.jobe.2016.03.006

37. Cortês, A., Almeida, J., De Brito, J., Tadeu, A.: Water retention and drainage capability of expanded cork agglomerate boards intended for application in green vertical systems. Constr. Build. Mater. **224**, 439–446 (2019). https://doi.org/10.1016/j.conbuildmat.2019.07.030

38. Nwankwo, C.O., Mahachi, J., Olukanni, D.O., Musonda, I.: Natural fibres and biopolymers in FRP composites for strengthening concrete structures: a mixed review. Constr. Build. Mater. **363**, 129661 (2023). https://doi.org/10.1016/j.conbuildmat.2022.129661

39. Chin, S.C., Moh, J.N.S., Doh, S.I., Mat Yahaya, F., Gimbun, J.: Strengthening of reinforced concrete beams using bamboo fiber/epoxy composite plates in flexure. Key Eng. Mater. **821**, 465–471 (2019). https://doi.org/10.4028/www.scientific.net/kem.821.465

40. Cervantes, I., AungYong, L., Chan, K., Ko, Y.-F., Mendez, S.: Flexural retrofitting of reinforced concrete structures using green natural fiber reinforced polymer plates. In: ICSI 2014: Creating Infrastructure for a Sustainable World, pp. 1051–1062. American Society of Civil Engineers (2014)

41. Nwankwo, C.O., Ede, A.N.: Flexural strengthening of reinforced concrete beam using a natural fibre reinforced polymer laminate: an experimental and numerical study. Mater. Struct. Constr. **53**(6), 142 (2020). https://doi.org/10.1617/s11527-020-01573-x

42. Chin, S.C., Tong, F.S., Doh, S.I., Gimbun, J., Ong, H.R., Serigar, J.P.: Strengthening performance of PALF-epoxy composite plate on reinforced concrete beams. IOP Conf. Ser. Mater. Sci. Eng. **318**, 012026 (2018). https://doi.org/10.1088/1757-899X/318/1/012026

43. Sen, T., Reddy, H.N.J.: Flexural strengthening of RC beams using natural sisal and artificial carbon and glass fabric reinforced composite system. Sustain. Cities Soc. **10**, 195–206 (2014). https://doi.org/10.1016/j.scs.2013.09.003

44. Chen, C., et al.: Eco-friendly and mechanically reliable alternative to synthetic FRP in externally bonded strengthening of RC beams: natural FRP. Compos. Struct. **241**, 112081 (2020). https://doi.org/10.1016/j.compstruct.2020.112081

45. Ghorbel, E., Limaiem, M., Wardeh, G.: Mechanical performance of bio-based FRP-confined recycled aggregate concrete under uniaxial compression. Materials. **14**(1778), 1–17 (2021)

46. Machado, M., et al.: Bio-based pultruded CFRP laminates: bond to concrete and structural performance of full-scale strengthened reinforced concrete beams. Materials. **16**, 4974 (2023). https://doi.org/10.3390/ma16144974

47. Attia, M.M., Ahmed, O., Kobesy, O., Malek, A.S.: Behavior of FRP rods under uniaxial tensile strength with multiple materials as an alternative to steel rebar. Case Stud. Constr. Mater. **17**(June), e01241 (2022). https://doi.org/10.1016/j.cscm.2022.e01241

48. Sharkawi, A.M., Mehriz, A.M., Showaib, E.A., Hassanin, A.: Performance of sustainable natural yarn reinforced polymer bars for construction applications. Constr. Build. Mater. **158**, 359–368 (2018). https://doi.org/10.1016/j.conbuildmat.2017.09.182

49. Elbehiry, A., Elnawawy, O., Kassem, M., Zaher, A., Mostafa, M.: FEM evaluation of reinforced concrete beams by hybrid and banana fiber bars (BFB). Case Stud. Constr. Mater. **14**, e00479 (2021). https://doi.org/10.1016/j.cscm.2020.e00479

50. Joyklad, P., Yooprasertchai, E., Wiwatrojanagul, P., Chaiyasarn, K.: Use of natural and synthetic fiber-reinforced composites for punching shear of flat slabs: a comparative study. Polymers. **14**(718), 1–17 (2022). https://doi.org/10.3390/polym14040719

51. Naveen, J., Jawaid, M., Vasanthanathan, A., Chandrasekar, M.: Finite element analysis of natural fiber-reinforced polymer composites. In: Modelling of Damage Processes in Biocomposites, Fibre-Reinforced Composites and Hybrid Composites, pp. 153–170. Elsevier (2019). https://doi.org/10.1016/B978-0-08-102289-4.00009-6

52. Adekomaya, O., Adama, K.: A review on application of natural fibre in structural reinforcement: challenges of properties adaptation. J. Appl. Sci. Environ. Manag. **22**(5), 749–754 (2018)

53. Petroudy, S.R.D.: Physical and mechanical properties of natural fibers. In: Fan, M., Fu, F. (eds.) Advanced High Strength Natural Fibre Composites in Construction, pp. 59–83. Woodhead Publishing (2017)

54. Saheb, D.N., Jog, J.P.: Natural fiber polymer composites: a review. Adv. Polym. Technol. **18**(4), 351–363 (1999)

Decarbonizing Conventional Building Materials for Net-Zero Emissions: A Feasibility Study in Canada

Chi Dara ⓘ

Contents

1 Introduction

According to the World Green Building Council, the building sector contributes about 40% of global emissions, 75% of which comes from building operations and 25% from carbon embodied in building materials [1]. As buildings are becoming more energy-efficient and moving toward reduced operational carbon, embodied carbon emissions will constitute a higher proportion of the whole-life carbon in the future [2]. Yet, efforts to mitigate embodied carbon from the design, manufacturing, and deployment of building materials like cement, steel, and aluminum are significantly lagging behind [3]. The majority of material economies remain linear rather than circular, relying heavily on energy-intensive production of virgin and non-renewable materials, applying the take–make–use–throwaway approach. Consequently, there is a shortage of recyclables in terms of both quantity and quality to

C. Dara (✉)
Mount Royal University, Calgary, AB, Canada
e-mail: cdara@mtroyal.ca

© The Author(s) 2025
M. Kioumarsi, B. Shafei (eds.), *The 1st International Conference on Net-Zero Built Environment*, Lecture Notes in Civil Engineering 237,
https://doi.org/10.1007/978-3-031-69626-8_72

meet the current material demands. Adopting the product circularity concept that aims at minimizing the use of virgin resources, energy, and waste flows has the potential to reduce embodied carbon emissions [4].

Structural systems such as concrete, steel, and masonry contribute the most to the total embodied impact associated with buildings. Adapting building codes and educating construction professionals can result in reduction in associated emissions, up to 25% from cement and concrete. Currently, concrete accounts for 7% of the global carbon emissions, thus emphasizing the need for immediate decarbonization through low-carbon mix and other design innovations [5]. With steel ranking as the second most used material in the architecture, construction, and engineering (ACE) industry and contributing 7.2% to global greenhouse gas (GHG) emissions, prioritizing its reuse and recycling has become crucial as part of the decarbonization effort [8]. This joint effort is necessary to avoid an increase in raw material extraction as the world's population continues to increase, given that manufacturing steel products from scrap can save between 60% and 80% of the embodied energy. However, a widening gap between the demand for reusable steel components and the limited quantity of available steel scraps remains a key challenge to the massive adoption of recycled steel in the industry.

According to reviews in the World Green Building Council report, global cement consumption is expected to rise by up to 23% by 2050 and global steel production is projected to increase by 30% during the same period. It is, however, believed that recycled secondary steel will be a critical component of this increase, outpacing the virgin steel content used in the manufacturing process [1]. Beyond structural materials, the building envelope, particularly components like window systems (e.g., aluminum), and insulating materials also contribute significantly to embodied carbon emissions, accounting for 48–50% of the associated emissions in a typical house [1, 5, 6]. For example, the research findings by Azari and Abbasabadi show that windows designed with aluminum framing with no recycled content have the highest impact from building envelopes [7].

This chapter presents preliminary research on the low-carbon options available for conventional building materials in the Canadian ACE industry. It investigates the performance of structural concrete and steel, as the two key materials having relatively high carbon intensity and are expected to remain the most used construction materials in the future. The chapter summarizes opportunities to reduce embodied carbon emissions using strategies to decrease the material quantities of these high-impact constituents and replace them with low-carbon options and alternatives with higher recycled content. The results from this investigation can perhaps serve in developing innovative design concepts and guidelines for the creation of low-carbon building materials, applicable in the (ACE) industry and the built environment at large.

2 Methodology

Embodied carbon represents the global warming potential (GWP), measured in kilograms of carbon dioxide, CO_2 equivalent ($kgCO_2e$), of a material or product. GWP quantifies the heat absorbed by the atmosphere due to greenhouse gas emissions. For a whole life cycle approach to evaluating the carbon performance of materials, the International Organization for Standardization (ISO) considers embodied carbon life cycle assessment (LCA) as consisting of the: (i) product stage (A1–A3) from raw material extraction to manufacturing; (ii) construction process stage from transportation to installation (A4–A5); (iii) use stage (B2–B5) from maintenance and repair to replacement and refurbishment; and (iv) end-of-life stage from deconstruction and waste transportation to waste disposal (C1–C4) [9]. Embodied carbon emissions are irreversible and primarily stem from upfront emissions, occurring before construction. Geographical location significantly influences the embodied energy and carbon impact from construction due to variations in manufacturing and production practices, energy supply grid systems, construction methods, local energy infrastructures, transportation modes, distance to local or international suppliers, and other region-specific economic factors.

This chapter provides preliminary information on low-carbon structural materials applicable to the construction of high rise commercial buildings in Canada, utilizing mainly concrete and steel frame design. It summarizes alternative strategies for material selection, design specifications, and construction approaches to reduce embodied carbon impact, particularly for concrete and steel. However, specific design specifications information will need to be directed toward individual project details and design team expectations during the design development (DD) stage before actual construction of the project.

This research employs quantitative data analysis by reviewing manufacturers' and suppliers' websites to gather data on the low-carbon material options available in the market. This includes reviewing online resources as third-party information available to the general public, utilizing embodied carbon information available from environmental product declarations (EPDs), identifying the key stages in the life cycle of products at which carbon emissions are most substantial or finding evidence of potential reduction in impacts throughout the life cycle stages. An environmental product declaration (EPD) is built upon a comprehensive product life cycle assessment and is thus a verifiable key research tool for understanding the life cycle embodied impact of products and manufacturers' commitment to product circularity. Further analysis includes telephone interviews and site visits to obtain specific information on manufacturing procedures and industry techniques not made available to the public.

3   Research Findings and Analysis

3.1   Carbon Reduction in Concrete

Concrete mix comprises of crushed stone, sand, water, and Portland cement. Cement in concrete is the primary contributor to embodied impact due to the high carbon intensity during its production. Thus, to attain low carbon in concrete, there is a need to target reducing the quantity of cement content in the mix, as it is primarily responsible for 75% of the entire CO_2 emissions [8]. Mix aggregates such as crushed stone and sand contribute less than 20%, which is often linked to electricity usage and, to a lesser extent, to excavation, hauling, blasting, and transportation. The contemporary approach to low-carbon concrete involves incorporating by-products (e.g. fly ash, slag) into the concrete mix. By minimizing the use of virgin materials and maximizing the utilization of by-products, a substantial reduction in the carbon footprint can be achieved.

This subsection discusses the strategies to reduce the embodied carbon emissions associated with concrete, focusing on ready mix concrete products, manufacturer and suppliers.

Manufacturer: CarbonCure Ready-Mix Solution   The CarbonCure mix solution highlights the effective use of carbon dioxide (CO_2) to mitigate the environmental impact of concrete through carbon removal technology. The process involves injecting a precise dosage of CO_2 into the concrete during mixing, leading to a chemical reaction that converts it into a mineral, specifically calcium carbonate, permanently sequestering CO_2 in a stable form. CarbonCure collaborates with concrete mix plant locations across provinces in Canada, seamlessly integrating the CarbonCure Valve Box and Control Box for precise CO_2 injection during mixing. This sustainable concrete technology significantly enhances performance and compressive strength by reacting with calcium ions, enabling adjustments in cementitious content, reducing carbon footprints, and yielding cost savings. CarbonCure Ready Mix, applicable to various concrete production methods, demonstrates substantial carbon savings per unit, thus fostering environmental sustainability without compromising either quality or performance of the concrete mix. The observed improvements, up to a 10% increase in compressive strength beyond 28 days, highlight the technology's potential for delivering low carbon concrete while maintaining strength in performance [10]. Research findings show that by integrating CarbonCure Ready Mix with CarbonCure Reclaimed Water, the cementitious content and water in the slurry tank can be revitalized, converting them into valuable upcycled resources. This innovative approach results in additional CO_2 savings of 10–25 lb per cubic yard (5–11 kilograms per cubic meter), contributing to an overall carbon reduction of approximately 10% compared to traditional concrete practices [11] (Fig. 1).

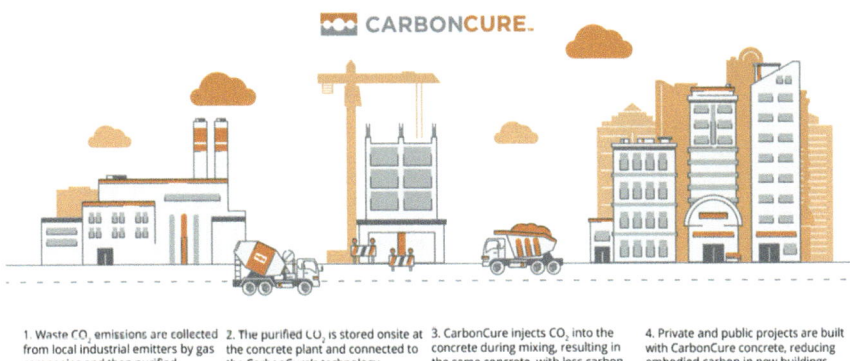

Fig. 1 The CarbonCure Ready Mix process. (Adopted from CarbonCure Technologies Inc, 2024)

Supplier: LafargeHolcim and Lehigh Hanson LafargeHolcim and Lehigh Hanson are the two main local suppliers of low carbon ready-mix concrete in across Canadian provinces. These mixes include supplementary cementitious materials (SCMs) and admixture ingredients. ECOPact by Lafarge offers a significant reduction in embodied carbon emissions (scope A1–A3) compared to standard ready-mix concrete without cement substitution [12]. ECOPact is available at various plants across six provinces in Canada, including British Columbia, Alberta, Manitoba, Saskatchewan, Ontario, and Quebec.

Low-carbon product range:

- ECOPact: 30–50% CO_2 reduction from a baseline mix with 0% SCMs, utilizing blended cement
- ECOPact Prime: 50–70% CO_2 reduction, engineered with higher blends and supplementary materials
- ECOPact Max: 70–90% CO_2 reduction; the lowest carbon range manufactured with a cement alternative technology like alkali activators

When comparing concrete alternatives, Lafarge ECOPact demonstrates an advantage over other conventional designs with a high SCM content, ensuring stability in various weather conditions and enabling year-round pouring without compromising performance. When regulatory conditions allow, ECOPact integrates upcycled construction and demolition materials, thus contributing to a circular design. The ECOPact Prime variant emphasizes specifying a 56-day strength for effective design and construction. The carbon reduction achieved with ECOPact depends on concrete strength, requiring early collaboration with concrete suppliers to determine optimal mixes for different structural elements due to potential performance implications.

Lehigh Hanson, on the other hand, has introduced low-carbon concrete mixes in their EvoBuild Bronze and EvoBuild Silver series. Both series come with verified environmental product declarations (EPDs) that support reductions in global

Table 1 Average material content for 1 metric ton (1000 kg) of the GU and GUL types are converted to percentages.

Material inputs	Type GU (%)	Type GUL (%)
Clinker	92	83
Limestone	3	12
Gypsum (including anhydrites)	5	5
Total	**100**	**100**

Adopted from CAC [14]

warming potential (GWP). The Bronze series generally achieves a reduction of 30–50%, whereas the Silver series attains a more substantial 50–70% reduction compared to the industry average mix. Leigh Hanson's EvoBuild places a priority on carbon optimization without compromising on mix performance, striving to minimize any design and construction implications [13]. It is worth noting that certain Silver mixes may have elevated supplementary cementitious material (SCM) content and necessitate a 56-day design strength, as indicated in the relevant EPDs.

Product Ingredient: General Use Limestone (GUL) Cement General use limestone (GUL) cement can be a viable alternative to general use (GU) Portland cement in various concrete structural constructions, providing approximately 10% reduction in GWP compared to GU cement. However, it is important to recognize that GUL cement is susceptible to "sulfate exposure," posing potential risks in specific elements like footings, piles, and slab-on-grade applications. Some concrete suppliers have developed specialized GUL cement formulations to address these risks, emphasizing the need for verification with the supplier before design specifications [14] (Table 1).

The positive synergy between GUL cement and SCMs has been the focal point of contemporary research studies. This interaction enhances concrete sustainability and reduces embodied carbon impact compared to standard GU cement. Utilizing GUL cement requires a nuanced approach, considering both its environmental benefits and potential challenges in specific structural applications. Further research and development could refine the understanding and application of GUL cement in sustainable construction practices.

Product Ingredient: Supplementary Cementitious Materials (SCMs) Supplementary cementitious materials (SCMs) are natural or industrial by-products that exhibit cementitious properties when mixed with water or other compounds. Examples include fly ash, slag cement, and silica fume from coal power stations of steel and silicon metal production. Natural SCMs, like calcined shale, calcined clay, and metakaolin, undergo controlled heating and purifying processes. These materials are sustainable, promoting recycling and reducing cement in concrete mixes, thereby lowering GWP. Hydraulic SCMs, like ground granulated blast furnace slag (GGBFS), react directly with water, whereas pozzolanic SCMs, such as fly ash and silica fume, chemically react with calcium hydroxide.

SCMs significantly influence the concrete properties affecting water requirements, workability, and setting time during the fresh stage as well as strength, permeability, and other characteristics once hardened. Their positive effects encompass reduced bleeding, enhanced strength, and lower permeability. However, their drawbacks include extended set time and potential construction cost implications. General recommendations include specifying performance criteria over SCM limits, utilizing maximum percentages under freeze–thaw conditions, and close collaboration with stakeholders for optimal low-carbon mix design, considering the potential impacts on construction schedules.

SCMs Availablity in Canada by Provinces SCMs have a long history in Canada. Fly ash and slag are extensively used, with the former being predominantly and widely adopted in most provinces. Currently, 60% of concrete manufacturers in Ontario incorporate GGBFS into cast-in-place concrete production. In the prairies and western Canada (e.g. Alberta), approximately 90% of concrete manufacturers integrate fly ash as cement replacement Fly ash, a coal combustion by-product, is classified based on its calcium oxide (CaO) content, with high-calcium fly ash acting as a sole cementing material, providing moderate strength, and mitigating heat of hydration. Silica fume, on the other hand, mainly serves for specific purposes, primarily used to enhance specific mix properties. Silica fume, derived from silicon metal manufacturing, enhances the durability of structures subjected to corrosion, abrasion, and chemical attacks. Natural pozzolans, like volcanic ashes, have limited applications in Canada.

Regulating Supply Market for SCMs in Canada The supply chain for SCMs, such as slag and fly ash, is not managed by the ready-mix industry, posing a potential risk to interested end users such as contractors and design team. For instance, slag, a by-product of steel manufacturing, is supplied by steel manufacturers in Ontario and sourced from the United States. Lafarge offers products like NewCem and NewCem Plus, blending slag and fly ash and allowing for high replacement rates. Several suppliers, including Carbon Upcycling Technologies, Lehigh Hanson, and Holcim, contribute to the SCM supply chain in different regions of Canada.

3.2 Carbon Reduction in Steel Products

Steel is produced using a blend of raw primary materials (ore) and recycled materials (scrap steel). The majority of the embodied carbon in fabricated structural steel, approximately 90%, occurs during the cradle-to-mill-gate stage. Opting for electric arc furnace (EAF) structural products, boasting up to 97% recycled content, can significantly reduce embodied carbon by up to half (50%) compared to basic oxygen furnace (BOF) steel, which relies on virgin iron. Table 2 presents characteristics of EAF and BOF (Table 2). The choice between EAF and BOF depends on factors such as scale, raw material availability, energy sources, and environmental considerations. Despite EAF's sustainability advantages, BOF remains crucial in the steel

Table 2 Comparison of electric arc furnace (EAF) and basic oxygen furnace (BOF)

Factors	Electric arc furnace (EAF)	Basic oxygen furnace (BOF)
Raw materials	Utilizes scrap steel as the primary raw material, making it more environmentally friendly by recycling existing steel	Relies on iron ore, which requires mining, and metallurgical coal, contributing to a higher carbon footprint
Energy consumption	Consumes less energy, particularly when powered by electricity from renewable sources	Requires significant energy input, mainly from burning fossil fuels during iron ore smelting
Carbon emissions and environmental impact	Tends to have lower carbon emissions, especially when powered by cleaner energy sources. Generally considered more environmentally sustainable due to its use of recycled materials and lower emissions	Emits more carbon dioxide due to the use of coal in the reduction of iron ore. Involves more environmental concerns, including deforestation for iron ore extraction and higher greenhouse gas emissions
Cost	Initially more expensive to set up but can be cost-effective in the long run, especially with the availability of affordable scrap	May have lower initial setup costs but is more susceptible to fluctuations in raw material prices, impacting long-term profitability
Technology and innovation	Allows for easier integration of advanced technologies, including enhanced process control and automation	Traditional technology with limitations on rapid integration of cutting-edge advancements
Flexibility in design	Offers flexibility in adjusting production levels and can quickly switch between different steel grades	Typically designed for continuous production of large batches, making it less flexible for smaller-scale operations

industry, balancing cost efficiency and production scale, as approximately 71% of global steel production relies on the BOF process [15–17]. Scrap metals originate from pre- and post-consumer steel, car scraps, household steel wastes, etc., from recycling centers (Table 2).

Manufacturer: Structural Steel Products Numerous manufacturers of structural steel, steel deck, and cold-formed steel have released EPDs tailored to their products, enabling the direct specification of maximum CO_2 for these elements. The ArcelorMittal XCarb™ initiative, for example, dedicated to achieving carbon-neutral steel, integrates reduced carbon products, steelmaking activities, and low-carbon innovations. XCarb™ structural steel, produced using EAF technology with 97% recycled content and powered by renewable energy, exhibits a 60% lower CO_2 than does conventional steel and a 73% reduction compared to the North American average. This initiative aligns with ArcelorMittal North America's commitment to a 25% CO_2 reduction by 2030, contributing to the broader goal of carbon neutrality by 2050. The use of "recycled and renewably" produced steel, powered by 100% renewable energy in the EAF process, results in an impressively low CO_2 footprint, potentially reaching 300 kg of CO_2 per ton of finished steel with 100% scrap metallics [14].

Product Ingredient: High-Recycled-Content Scrap-based steels offer lower embodied carbon than most virgin steel manufacturing [16, 18]. High-recycled-content steel from EAFs, with more recycled content than BOFs, can effectively reduce embodied carbon in construction projects. However, challenges in the supply chain may impact the desired reductions. It is recommended for design consulting team to engage early with steel manufacturers to identify feasible low-carbon options within project locations. The recycled content in structural steel varies based on factors like scrap availability, product type, chemical composition, and production route. Importing steel to regions without shape mills, such as the west coast of North America, may increase carbon impact.

ArcelorMittal manufactures rolled sections for the North American market in Luxembourg, Europe, using EAF with 51–98% recycled iron content [19]. Given that GHG emissions of overseas transportation are associated with steel manufacturing in Luxembourg, conducting a comprehensive embodied carbon study comparing ArcelorMittal's steel to North American EAF-produced steel is advisable. North American EAFs may have higher manufacturing emissions, especially when compared to potentially less environmentally friendly electrical grids. However, ArcelorMittal's EPD for structural steel from Luxembourg (GWP = 0.84 kg CO_2e/kg for A1–A3) demonstrates a 31% lower GWP than does North America's national average according to the American Institute of Steel Construction (AISC). In Canada, ArcelorMittal has four tubular facilities producing mechanical steel tubing and seamless/welded precision tubes with 30–60% recycled content. ArcelorMittal Contrecoeur (Quebec) is Canada's largest rebar manufacturer, using locally extracted iron from northern Quebec mines [20]. The flat mill in Hamilton, Ontario, employs a blast oxygen furnace. To minimize embodied carbon, rolled sections should be prioritized over hollow structural sections.

Promoting the Use of Locally Produced Steel Specifying domestic steel can mitigate embodied carbon in steel production and transportation, depending on project and source locations. Domestic hot-rolled sections are EAF-produced, whereas structural plate and coils forming high strength steel (HSS) may come from BOF or EAF furnaces. Imported structural products are more likely BOF-produced, but the global BOF to EAF ratio is 70:30. Not all EAF steel is equal; Canadian BOF-produced steel may have lower embodied carbon than foreign EAF steel, depending on the mill's input mix of scrap, direct reduced iron (DRI) or pig iron, and energy grid source. Checking producer-specific EPDs is recommended [21].

Examples of Canadian concrete rebar suppliers include AltaSteel in Edmonton, Alberta, using EAF technology with 100% scrap [22]. ArcelorMittal Montreal in Contrecoeur, Quebec, offers a wide range of long products, including concrete reinforcement bars, with a published EPD showing 1.29 kg CO_2e/kg GWP [23]. Due to complexities in the steel supply chain, reducing carbon emissions in steel involves minimizing tonnage, favoring high-strength steel for efficient designs, smaller foundations, and replacing complicated sections with rolled sections.

However, confirming the carbon emissions of higher-strength steel with the producer is advisable. The greatest challenge remains the lack of publicly available EPDs for steel products [21].

4 Conclusions

This research provides valuable insights into the challenges and opportunities associated with reducing embodied carbon in the construction industry, focusing on key materials such as concrete and steel. Here is a summary of the key points:

1. Structural concrete: There is evidence of low-carbon options for concrete, emphasizing the use of CarbonCure Ready Mix and Lafarge ECOPact. The findings highlight the potential for substantial carbon reduction through innovative technologies and replacing cement content with SCMs as an alternative mix design. Improving strength in concrete is considered a key priority for decarbonization, such as reducing the clinker-to-cement ratio, embracing renewable energy for production, and incorporating carbon capture technologies. Decarbonizing cement requires a shift to alternative binders, prefabrication of reusable circular units, and transitioning to electric kilns powered by a decarbonized electric grid [3].
2. Structural steel and steel reinforcement: In primary steel production, transitioning from blast furnace to direct reduced iron technology, utilizing electric arc furnaces powered by renewable energy, and integrating high recycled content in manufacturing product ingredients show potential for substantial reduction in carbon emissions.

Initiating a prompt and collaborative process during the design development stage is crucial to the overall decarbonization effort. This design iterative approach should be supported by both the "supply and demand sides of stakeholders" [1]. Exploring hybrid alternatives and adopting systemized manufacturing further optimizes construction techniques [24]. Emphasizing carbon reduction strategies from project inception creates an environment conducive to achieving systemic reductions, offering significant opportunities for substantial reductions throughout the life cycle of the project. This chapter emphasizes the importance of transitioning to sustainable practices in material production, construction techniques, and overall project planning. Overall, this study provides a foundation for developing innovative design concepts and guidelines to create low-carbon building materials' specifications, contributing to sustainability in the architecture, construction, and engineering (ACE) industry.

References

1. World Green Building Council: Bringing Embodied Carbon Upfront: Coordinated Action for the Building and Construction Sector to Tackle Embodied Carbon. World Green Building Council, London, United Kingdom office & Toronto-Ontario (2019)
2. International Energy Agency (IEA): Net Zero by 2050: A Roadmap for the Global Energy Sector (2022)
3. United Nations Environment Programme, UNEP: Building Materials and the Climate: Constructing a New Future. United Nations Environment Programme, Nairobi (2023)
4. Luís, B., Meri, C., Rand, A., Viorel, U. (eds.): Creating a Roadmap Towards Circularity in the Built Environment, 1st edn. Springer Nature, Switzerland (2024). https://doi.org/10.1007/978-3-031-45980-1
5. Mithraratne, N., Vale, B.: Life cycle analysis model for New Zealand houses. Build Environ. **39**(4), 483–492 (2004). https://doi.org/10.1016/j.buildenv.2003.09.008
6. Hu, M.: The embodied impact of existing building stock. In: Examining the Environmental Impacts of Materials and Buildings, pp. 1–31. IGI Global (2020)
7. Azari, R., Abbasabadi, N.: Embodied energy of buildings: a review of data, methods, challenges, and research trends. Energy Build. **168**, 225–235 (2018). https://doi.org/10.1016/j.enbuild.2018.03.003
8. Architecture2030: Carbon Smart Materials Palette: Concrete. http://www.materialspalette.org/concrete/. Last accessed 12 May 2023
9. ISO 21930: Sustainability in Buildings and Civil Engineering Works — Core Rules for Environmental Product Declarations of Construction Products and Services. 2017-07 International Organization for Standardization
10. CarbonCure: CarbonCure Ready Mix Durability FAQ. https://www.carboncure.com/wp-content/uploads/2023/05/Brochure_FAQforEngineers_2021.pdf (2021)
11. CarbonCure: CarbonCure Ready Mix: Same Concrete, Less Carbon. https://www.carboncure.com/ready-mix/. Last accessed 20 Feb 2024
12. Lafarge: ECOpact. The Green Concrete: Sustainable Construction Starts Here (2024)
13. Heidelberg Materials: EvoBuild™ Low Carbon Concrete Available in All North American Markets. https://www.heidelbergmaterials.us/home/news/news/2023/12/13/evobuild-low-carbon-concrete-available-in-all-north-american-markets. Last accessed 11 Aug 2023
14. Cement Association of Canada, CAC: Industry Average GU and GUL Cements Environmental Product Declaration (EPD) # 5357-9431 (2016)
15. American Institute of Steel Construction: More than Recycled Content: The Sustainable Characteristics of Structural Steel. A White Paper by the American Institute of Steel Construction May 2017. https://www.aisc.org/globalassets/aisc/publications/white-papers/more-than-recycled-content.pdf (2017)
16. Architecture 2030: Carbon Smart Materials Palette Steel CARBON IMPACTS OF STEEL. https://www.materialspalette.org/steel/. Last accessed 20 Feb 2024
17. Nadoushani, Z., Nezhad, A., Ali.: Effects of structural system on the life cycle carbon footprint of buildings. Energy Build. **102**, 337–346 (2015). https://doi.org/10.1016/j.cnbuild.2015.05.044
18. ASCE: Structural Materials and Global Climate: A Primer on Carbon Emissions for Structural Engineers. https://ascelibrary.org/doi/book/10.1061/9780784414934 (2017)
19. ArcelorMittal: Climate Action Report 2. July 2021 (2021)
20. ArcelorMittal Long Products Canada: https://northamerica.arcelormittal.com/our-operations/arcelormittal-long-products-canada. Last accessed 5 Jan 2024
21. Reinforcing Steel Institute of Canada: Source: Reinforcing Steel Institute of Canada – 2024. https://rebar.org/mill-members/. Last accessed 5 Jan 2024
22. AltaSteel: Rebar (Reinforcing Bar). https://www.altasteel.com/rebar-product. Last accessed 5 Jan 2024

23. ArcelorMittal Long Products Canada: Steel Rebar and Merchant Bar Quality. Environmental Product Declaration (EPD) # 4937-6316 (2019)
24. World Business Council for Sustainable Development (WBCSD): Net-Zero Buildings: Halving Construction Emissions Today. Publication prepared by Arup in collaboration with WBCSD, January 2023. Geneva. Retrieved from https://www.arup.com/perspectives/publications/research/section/net-zero-buildings-halving-construction-emissions-today. CS Homepage. http://www.springer.com/lncs (2023). Last accessed 21 Nov 2016

Greenhouse Gas Emissions and Service Life of Concrete Infrastructures

Gro Markeset ⓘ

Contents

1 Introduction

Due to the increased concern for sustainability and climate change, environmental aspects have now become important in the planning, design, and construction of buildings and infrastructures. To answer and quantify the environmental impacts of construction projects, a method based on Life Cycle Assessment (LCA) of environmental data from raw material extraction, production, construction, operation and maintenance, and disposal, is frequently applied [1]. To meet the need to document environmental impacts for a product or service, standardized environmental declarations have been developed, the so-called Environmental Product Declaration (EPD), based on this LCA method.

Concrete is the world's most widely used building material and is dominant in infrastructure projects. On a worldwide basis, the production of cement and concrete accounts for about 5–8% of the total greenhouse gas emissions (i.e., CO_2 emissions), in which about 90% of the emissions come from the production of ordinary Portland

G. Markeset (✉)
Oslo Metropolitan University, Oslo, Norway
e-mail: gro.markeset@oslomet.no

© The Author(s) 2025
M. Kioumarsi, B. Shafei (eds.), *The 1st International Conference on Net-Zero Built Environment*, Lecture Notes in Civil Engineering 237,
https://doi.org/10.1007/978-3-031-69626-8_73

867

cement (OPC). However, by replacing some of the cement with supplementary cementitious materials (SCMs), like fly ash, silica fume, and slag, the emissions can be reduced. It is considered that the SCMs do not emit CO_2, since they are waste products from other materials production.

Important concrete infrastructures, like bridges, dams, harbors, etc., are designed with a working life (service life) of 100 years or more. In determining the service life of steel-reinforced concrete structures exposed to marine environments or de-icing salts, the ability of the concrete to resist the ingress of chloride ions is crucial. From long-term field studies of concrete exposed to the marine environment, it has been observed a more rapid reduction in the ionic diffusivity through concrete with time for concretes containing fly ash or slag [2–4] compared to plain OPC concrete. This reduced diffusion rate is beneficial in delaying the onset of chloride-induced corrosion, and thus enhancing the service life of the structures. The chloride ingress in concrete may be modelled through Fick's second law of diffusion. In this model, a reduction of diffusion rate with time is included in a parameter describing the observed time dependency of the diffusion coefficient.

This paper deals with the application of low-carbon concretes in concrete infrastructures exposed to chloride-induced corrosion. The main objective is to highlight the importance of including durability (or service life) properties of relevant low-carbon concretes in the assessment of greenhouse gas emissions from concrete infrastructures. Section 2 presents calculations of CO2 emissions from road culvert solutions limited to the production and construction process stage. In Sect. 3, Fick's second law of diffusion is applied to study the effect of various concrete binders on chloride ingress and the service life of concrete structures exposed to the marine environment.

2 Greenhouse Gas Emissions from Road Culvert Solutions

2.1 Background

Assessments of greenhouse gas emissions have been carried out for two alternative road culvert solutions, a concrete box culvert and a steel pipe culvert, respectively. The culvert is a part of a road construction project in Norway and allows pedestrians and light traffic to pass under the roadway. As a part of road infrastructure, the culvert was designed for a service of 100 years.

The concrete box culvert was a cast-in-place construction, whereas the steel pipe culvert was made of corrugated steel elements transported to the construction site. The mass quantities for materials were based on the proposed design of the two culvert solutions.

The calculations of greenhouse gas emissions (Global Warming Potentials) were performed for the production stage, phases A1-A3, and the transport of the products to the construction site, phase A4, as defined in [1]. The Global Warming Potential (GWP) is given as amount of kg CO_2 equivalents (kg CO_2-eq). The calculations of

GWP related to the transport of the products to the construction site are based on generic GWP data from known databases in the software package One Click LCA [5].

2.2 Data for GWP Calculations

Concrete culvert solution A concrete strength class C45 (characteristic compressive strength of 45 MPa at 28 days) was required for the concrete culvert solution. No further measures were set for corrosion protection beyond the requirements given for concrete cover to the reinforcement for achieving a service life of 100 years.

The concrete volume of the box culvert was estimated to be 524 m^3, which included the needed transition plate for the overlying road. The amount of steel reinforcement was set to 180 kg/m^3 of concrete. Two types of low-emission concretes (low-carbon concretes) were included in the study, a low-carbon concrete B with a GWP of 270 kg CO_2-eq/m^3 and a low-carbon concrete A with a GWP of 220 kg CO_2-eq/m^3, given in accordance with the Norwegian Concrete Association Publication No 37 [6]. This publication defines low-carbon concrete as structural concrete produced in accordance with the rules in NS-EN 206 [7] where measures have been taken to limit greenhouse gas emissions. The GWP from the reinforcement was based on an EPD for a relevant steel reinforcement product for concrete.

In the calculations of GWP from the transport of materials to the construction site, phase A4, the transport distance was set to 10 km for the ready-mixed concrete and 30 km for the reinforcement.

Steel culvert solution To achieve a design service life of 100 years of the steel culvert, double-sided hot-dip galvanizing and epoxy painting on the outer parts were used. The mass of the corrugated steel elements is estimated to be 33.5 tons. The GWP for the production stage (phases A1-A3) was based on a product-specific EPD for galvanized corrugated steel plates.

The steel culvert elements were produced in Poland and transported to the construction site; a distance estimated to about 1300 km.

2.3 Results and Discussion

The results from the calculated greenhouse gas emissions are presented in Fig. 1 for both culvert solutions. The total GWPs (phases A1-A4) from the concrete culvert solution with low-carbon concrete B is estimated to be 176 tons CO_2-eq. Using low-carbon concrete A, the total CO_2 emissions are reduced to 149 tons. The total GWP from the steel culvert solution is only 76.5 tons.

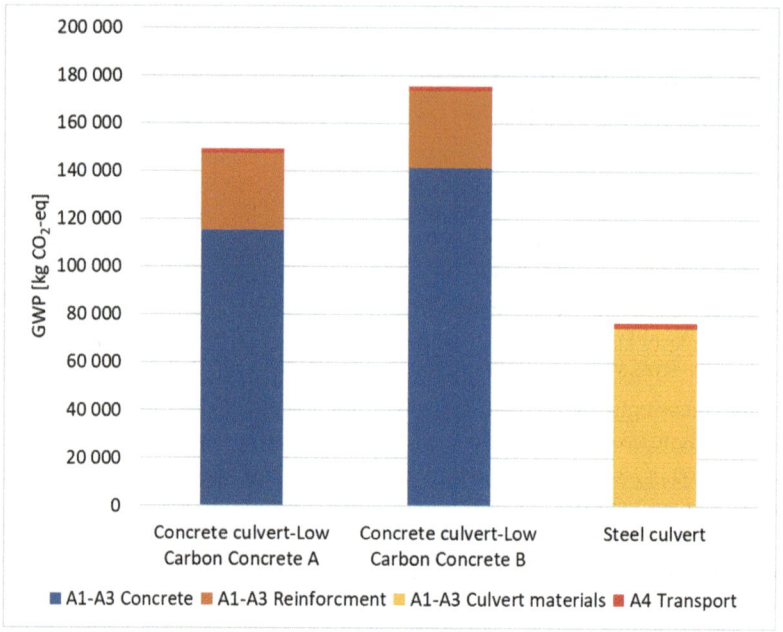

Fig. 1 Calculated Global Warming Potential (GWP) for the concrete and the steel culvert solutions. Results are presented for the construction stage (A1-A3), and transport (A4)

As can be seen, the production stage (A1-A3) is dominating in terms of CO_2 emissions, and the emissions from the transport of the products to the construction site are only 1–3% of the total GWP for the various solutions. The steel culvert has by far the lowest carbon footprint with the emissions of about 50% of the emissions from the low-carbon A culvert. To further reduce the emissions from the culvert with low-carbon concrete A, the quantity of concrete must be reduced by optimization of the structural design.

For the construction project in question, the steel culvert solution was chosen and built as shown in Fig. 2.

3   Impact of Concrete Binders on Chloride Ingress and Service Life

3.1   General

In the construction case presented in Sect. 2, only the greenhouse gas emissions (GWP) related to product and construction process stages (A1-A4) were included in the Life Cycle Assessment (LCA) of the culvert solutions. As the specified design service life is assumed obtained through the provisions given in the codes and

Fig. 2 Steel culvert chosen for a Norwegian road project. (Photo G. Markeset)

standards for the design of concrete structures and the associated materials standards, the GWP linked to the use stage (phases B1-B7) of the LCA is rarely included. However, the provisions for durability and service life within codes and standards are typically based on experiences from old and outdated concrete qualities and types where the focus has been more on strength than on greenhouse gas emissions. In the LCA of concrete infrastructure, the emission related to service life is an important factor, especially for new low-emission concretes with different types and amounts of SCMs.

In the design of concrete structures exposed to chloride-induced corrosion, the end of service life is often defined at the time when the chloride concentration at the surface of the reinforcement has exceeded a critical level resulting in depassivation and corrosion initiation. The critical chloride concentration, or chloride threshold level (C_{crit}) for corrosion initiations, becomes thus a key parameter in the prediction of the design service life. During the "corrosion-free service life" there should be no need for repairs and durability-enhancing measures.

3.2 Modelling of Chloride Ingress in Concrete

The chloride ingress in concrete can be modelled by Fick's second law of diffusion:

$$C(x,t) = C_i + (C_0 - C_i)\,\mathrm{erfc}\left(\frac{x}{2\sqrt{D_a(t)\cdot t}}\right) \tag{1}$$

where $C(x, t)$ = chloride concentration at depth x at time t, C_0 = chloride concentration at the exposed concrete surface, C_i = initial chloride content in the concrete, erfc = error function complement, and $D_a(t)$ = apparent (average) chloride diffusion coefficient at time t.

In Eq. (1) the chloride concentration at the concrete surface C_0 is assumed constant. Values of apparent diffusion coefficient $D_a(t)$ represent an integrated value over the period of exposure. The diffusivity decreases with time caused by the long-term evolution of the micro-structure of the concrete and pore blocking of the exposed concrete surface layer. A commonly used empirical expression for the time-dependent diffusion coefficient in Fick's second law is:

$$D_a(t) = D_0 \left(\frac{t_0}{t} \right)^\alpha \tag{2}$$

where D_0 = apparent chloride diffusion coefficient at the age t_0, and α = ageing exponent.

Several investigations have been conducted worldwide to stipulate the aging exponent α, and many different values have been found in the literature. Most of the values are based on data from marine concrete structures with different concrete compositions and under varying exposure conditions. However, similar systematic investigations for de-icing salt exposure are lacking.

A systematic long-term field investigation suitable for determine reliable data for this aging effect for relevant concrete recipes, with and without SCMs, is presented in [3, 8]. These studies included concrete recipes with CEM I and fly ash contents varying from zero to 35% by weight of cement, and one series of ternary blend concrete with 4% silica fume (SF) and 20% fly ash (FA). The water binder ratio was in the order of 0.40–0.45 for all the concretes. The concrete samples were exposed to seawater in the tidal zone over a period of 9 years, and chloride profiles were measured at different time intervals. Table 1 presents the parameter values for D_0 at 28 days, α and C_0, determined for the concretes with CEM I, 20% FA (FA20), 20% FA and 4% SF (FA20SF4), and 35% FA (FA30), respectively.

Referring to Table 1, the aging exponent α varies from 0.19 for plain CEM I concrete to 0.52 for the FA35 concrete. However, the FA35 concrete has the highest

Table 1 Parameters (means values) determined on concrete samples exposed to seawater in the tidal zone/8/

Type of binder	Aging exponent α	Calculated apparent diffusion coefficient D_0 at 28 days [m²/s]	Surface chloride concentration C_0 [% of concrete weight]
CEM I	0.19	$7.9 \ 10^{-12}$	0.61
CEM I with 20% FA (FA20)	0.40	$8.7 \ 10^{-12}$	0.95
CEM I with 20% FA and 4% SF (FA20SF4)	0.46	$4.1 \ 10^{-12}$	0.87
CEM I with 35% FA (FA35)	0.52	$9.7 \ 10^{-12}$	0.85

apparent diffusion coefficient at 28 days. The chloride content at the concrete surface obtained after 9 years of exposure was lower for the CEM I concrete compared to the concretes with SCMs.

3.3 Chloride Ingress and Service Life for Concretes with Different Binders

Fick's second law (see Eqs. (1) and (2)) is applied to simulate the chloride ingress in the four concretes based on the parameters given in Table 1. Figure 3 shows the calculated chloride profiles, i.e. the relation between chloride concentration and penetration depth, for the different concretes after 50 years of exposure. Comparing the CEM I concrete and the concretes with 20% and 35% FA, respectively, the reduction in diffusivity through concrete with time, modelled through the aging parameter, has a clear effect on the chloride ingress after 50 years of exposure. However, by including 4% silica fume a much lower early-age diffusion coefficient was achieved compared to the FA concretes as illustrated by the FA20SF4 concrete in Table 1, and thus a lower chloride penetration over time. Furthermore, despite the lower chloride concentration at the concrete surface of the CEM I concrete, the chlorides have penetrated much deeper into the concrete compared to the concretes containing SCMs.

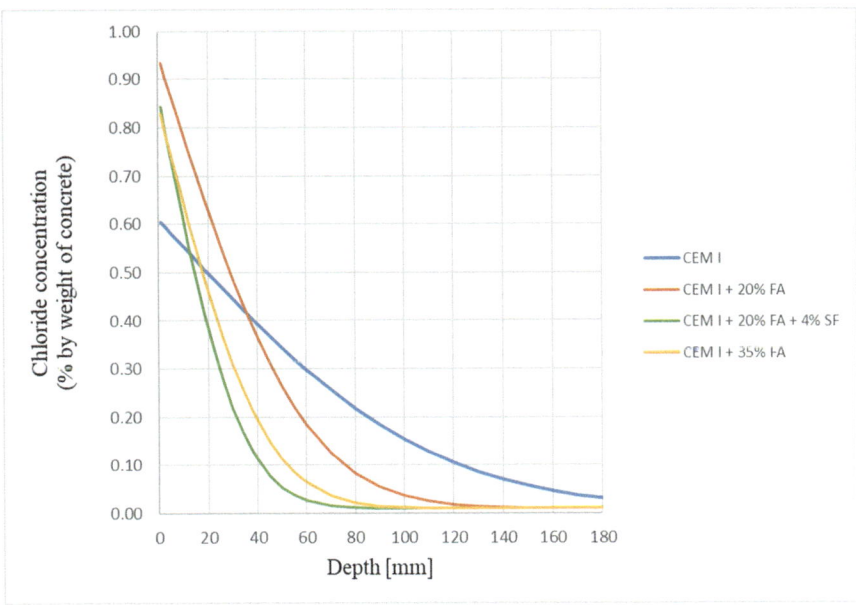

Fig. 3 Predicted chloride ingress profiles for the different concretes after 50 years of exposure

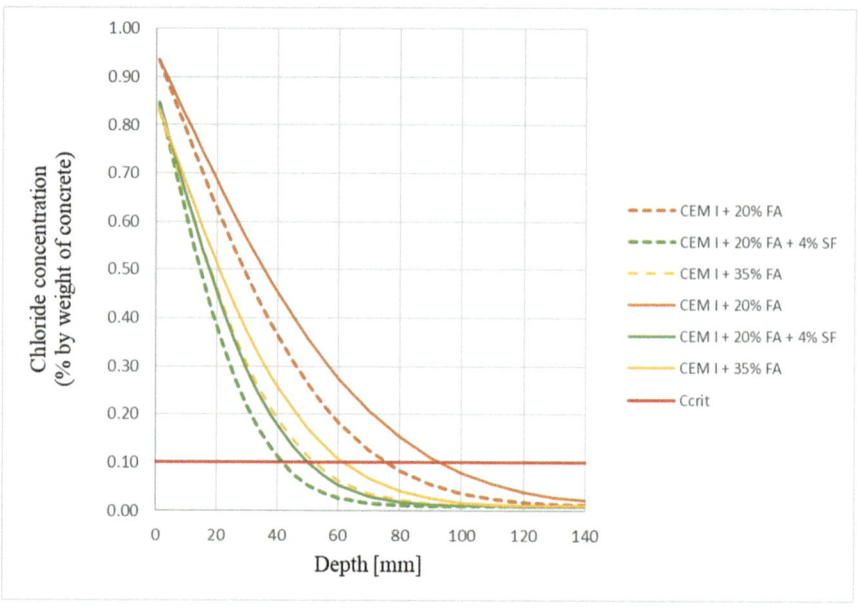

Fig. 4 Chloride profiles predicted for SCM concretes after 50 years (dotted lines) and 100 years (solid lines) of exposure. The horizontal red line is assumed $C_{crit} = 0.1$

Plain OPC concrete is not a suitable concrete for reinforced concrete structures exposed to chloride-induced corrosion (exposure classes XS2, XS3, XD2, XD3), both due to high CO_2 emissions (high cement content) and low chloride resistance (short service life). This is accounted for in NS-EN-1992 National Annex [9] through the requirement that the concrete binder must contain a minimum of 6% silica fume or a minimum of 14% in total of fly ash, silica fume, and slag.

Calculation of the achieved service life for different concrete compositions may be estimated by a probabilistic service life procedure, as presented in e.g [13, 16]. In this paper, a deterministic approach has been chosen to study the effect of different concrete binders on achieved service life. In Fig. 4 the chloride profiles predicted after 50 and 100 years of exposure are presented for the SCM concretes FA20, FA35, and FA20SF4, respectively. As can be observed, the chloride ingress profile of the FA35 concrete after 50 years of exposure coincides with the chloride profile of the FA20SF4 concrete after 100 years, and further, the concrete with only 20% FA will achieve the shortest service life of the studied concretes. The critical chloride concentration C_{crit} needed to initiate corrosion on the surface of the reinforcement is not known for the different SCM concretes. The normal value considered in numerous standards is a C_{crit} of 0.4% by weight of cement (or about 0.07% by weight of concrete). However, there is a huge scatter in reported values of C_{crit} [10–12]. Based on experience from probabilistic service life calculations a reasonable deterministic value of C_{crit} would be 0.1 by % of concrete weight. Using this C_{crit}-value, the needed concrete cover to the reinforcement for achieving a "corrosion-free" service life of 100 years varied from about 50 mm for the FA20SF4 concrete to 95 mm for the FA20 concrete.

As illustrated in Fig. 4, the required cover thickness for the FA20 concrete is about twice the cover needed for the FA20SF4 concrete, which also results in higher CO_2 emissions. Moreover, the service life of a reinforced concrete structure can be twice as long by using FA20SF4 concrete instead of FA35 concrete. If the use stage in the life cycle assessment is neglected, the FA35 concrete will emit less CO_2 than the FA20SF4 concrete. In structural design, the compressive strength of the concrete is an important parameter. For the SCM concretes, the compressive strength at 28 days will be lowest for the FA35 concrete due to high FA content, and highest for the FA20SF4 caused by the addition of 4% SF.

The parameters D_0, C_0, and α applied in the simulations (see Table 1) are based on mean values, i.e. the uncertainties in the model parameters are not included. Further, the predicted service lives are very sensitive to the model parameters α and C_{crit} [3, 13]. The C_{crit} for the different concretes is not known and may depend on the type and amount of SCMs as a reduction of pH in the concrete may reduce the C_{crit}. Further, the apparent diffusion coefficient $D_a(t)$ decreases with exposure time and is found to reach an asymptotic value after 10–20 years of exposure depending on exposure condition and type of binders [2, 3, 8, 14, 15].

4 Discussion and Conclusions

To reduce CO_2 emission from concrete structures, different strategies can be used, e.g. by substituting Portland cement with supplementary cementitious materials (SCMs) and by optimizing the design to reduce concrete volume. In some cases, the optimal low-emission solution may be to apply other materials and structural solutions, as discussed in Sect. 2. Based on the calculations of greenhouse gas emissions (GWPs) from the two culvert solutions, the main source of the emissions comes from the production of the products (ready-mixed concrete, reinforcement, and steel elements). Despite using a low-carbon concrete with a low GWP emission (220 kg CO_2-eq/m^3) in the concrete culvert solution, the amount of concrete was far too high to compete with the carbon footprint of the steel culvert solution.

In the development of different low-emission concretes, the focus has mainly been on the reduction of carbon footprint by using SCMs. However, the resistance against chloride ingress may vary considerably among the different types of SCM concretes, as discussed in Sect. 3. Using Fick's second law of diffusion, the calculated service life for a reinforced concrete structure exposed to seawater in the tidal was found to be twice as long by using concrete with 20% fly ash and 4% silica fume than by using concrete with 35% fly ash. The parameter values used in the calculations are determined from concrete samples exposed to the marine environment in Norwegian coastal areas. For other marine environments or de-icing salt exposures relevant parameter values should be found by performing systematic long-term field studies.

The chloride ingress calculations, and thus the predicted service lives, are very sensitive to the model parameters α (aging exponent) and C_{crit} (critical chloride concentration). The C_{crit} may depend on the type and amount of SCMs, and thus, further investigations are recommended. Care should also be taken by using α-values obtained from short-term exposure tests to extrapolate the chloride ingress to 100 years of exposure and more, as too high α-value will overestimate the service life.

For concrete infrastructures like bridges, the safety and service life requirements may limit the application of very low-emission concretes due to low early strength and reduced durability performance. The study presented in Sect. 3 demonstrates the importance of including durability-related parameters (diffusion rate, service life, cover thickness, etc.) in the assessment of CO_2 emissions from reinforced concrete infrastructures. The GWP from concrete is typically given per volume unit. However, a more appropriate unit should be to include service life and strength parameters in the GWP calculations, and instead present the CO_2 emissions as GWP of the required concrete volume per unit of compressive strength and service life (or other durability-related parameters).

References

1. NS-EN ISO 14040: 2006 Environmental management – Life cycle assessment – Principles and framework, Standard Norge (2006)
2. Michael, D.A., Bamforth, P.B.: Modelling chloride diffusion in concrete - effect of fly ash and slag. Cem. Concr. Res. **29**, 487–495 (1999)
3. Markeset, G., Skjølsvold, O.: Time dependent chloride diffusion coefficient - field studies of concrete exposed to marine environment in Norway. In: Service Life Design for Infrastructure, pp. 83–90. Rilem Publications. ISBN 978-2-35158-096-7 (2011)
4. Pack, S.-W., Jung, M.-S., Song, H.-W., Kim, S.-H., Ann, K.Y.: Prediction of time dependent chloride transport in concrete structures exposed to a marine environment. Cem. Concr. Res. **40**, 302–312 (2010)
5. One Click LCA Homepage, https://www.oneclicklca.com/no/programvare-for-livssyklusanalyser-innen-byggebransjen. Last accessed 2021/09/08
6. Norwegian Concrete Association Publication No 37: Low-carbon concrete, (in Norwegian) (2024)
7. NS-EN 206:2013+A2: 2021+NA:2021 Concrete - Specification, performance, production and conformity, Standard Norge (2021)
8. Skjølsvold, O.: Chloride diffusion into concrete. Evaluation of the ageing effect based on results from field studies, COIN Project Report 11, ISBN 978–82–536-1088-7, (in Norwegian) (2009)
9. NS-EN 1992-1-1:2004+NA:2021, Design of concrete structures (2021)
10. Breit, W.: Critical corrosion inducing chloride content - state of the art and new investigation results. In: Betontechnische Berichte 1998–2000, VDZ, Düsseldorf, Germany (2001)
11. Alonso, M.C., Sanchez, M.: Analysis of the variability of chloride threshold values in the literature. Mater. Corros. **60**(8), 631–637 (2009)
12. Cao, Y., Gehlen, C., Angst, U., Wang, L., Wang, Z.D., Yao, Y.: Critical chloride content in reinforced concrete - an updated review considering Chinese experience. Cem. Concr. Res. **117**, 58–68 (2019)

13. Markeset, G.: Critical chloride content and its influence on service life predictions. Mater. Corros. **60**(8), 593–596 (2009)
14. Li, Q., Li, K., Qhou, X., Zhang, Q., Fan, Z.: Model-based durability design of concrete structures in Hong Kong-Zhuhai-Macau Sea link project. Struct. Saf. **53**, 1–12 (2015)
15. Pang, L., Li, Q.: Service life prediction of RC structures in marine environment using long term chloride ingress data: comparison between exposure trial and real surveys. Constr. Build. Mater. **113**, 979–987 (2016)
16. Larssen, R.M.: Cost optimal design and maintenance of concrete structures in marine environments. In: Krokeborg (ed.) Proc. of Strait Crossings 2001, pp. 156–159, Balkema, Rotterdam (2001)

Review of South African Waste Management Practices and Their Integration into the Construction Industry

Areej Gamieldien, Hans Beushausen, and Mark Alexander

Contents

1   Introduction

South Africa is facing a waste issue in the form of limited landfill airspace availability, with major metropolitan areas like the cities of Cape Town and Johannesburg expected to reach landfill capacity by the next decade. The diversion of construction and demolition waste (CDW) from these landfill sites has high potential not only in terms of slowing the rate that landfills will reach capacity, but also in terms of potential economic gains both in the formal and informal economic sectors [1].

CDW is one of the largest contributors to waste in South Africa by mass and volume. Through the National Waste Information Baseline Report, the South African Department of Environmental Affairs (DEA) [2] reported that 4.7 million tonnes of CDW were produced annually in South Africa as of 2011. This was estimated to be 20% of total general waste produced by mass and, of this value, 16%

A. Gamieldien (✉) · H. Beushausen · M. Alexander
Concrete Materials and Structural Integrity Research Unit (CoMSIRU), Department of Civil Engineering, University of Cape Town, Cape Town, South Africa
e-mail: gmlare001@myuct.ac.za

M. Kioumarsi, B. Shafei (eds.), *The 1st International Conference on Net-Zero Built Environment*, Lecture Notes in Civil Engineering 237,
https://doi.org/10.1007/978-3-031-69626-8_74

was reported to be recovered for reuse or recycling. Annual CDW production as of 2017 was reported by the DEA [1] in the State of Waste Report to have decreased to 4.5 million tonnes. If these values are both taken to be true, the higher rate of CDW production in 2011 may be related to infrastructure development encouraged by the FIFA World Cup which took place in South Africa in 2010. Nineteen per cent of the 4.5 million tonnes of CDW was reported as being recovered for reuse, recycling, or use as backfill material [1]. These values, however, may be an underestimation of CDW production. Values extrapolated using national gross domestic product (GDP), GDP contribution of the Western Cape province, and waste generation statistics from the Western Cape range between 8.7 and 10.8 million tonnes in 2017 [3, 4]. Discrepancies in these values may be due to inaccurate reporting in the formal waste management sector as well as the omission of informal waste management and illegal dumping practices in both the National Waste Information Baseline Report and the State of Waste Report.

While the overall proportion of CDW relative to total waste generated in the Western Cape is consistent with the national value of approximately 20%, between 30% and 40% of the 1.7 million tonnes of CDW produced in the province in 2017 is estimated to have been diverted from landfills to be reused or recycled [4]. The percentage of waste diverted increased to between 61% and 69% in the years 2018 to 2021 [5]. Over the period from 2018 to 2021, between 22% and 29% of total waste in the Western Cape was diverted from landfill. Forty-four per cent of this waste, totalling 1.3 million tonnes, was comprised of CDW [5].

The discrepancy between the recovery rates for CDW observed nationally and in the Western Cape suggests that recovery rates and waste management practices in other provinces are far below that of the Western Cape. This paper presents a review of various factors linked to the existence and enforcement of legislation at a national, provincial, and municipal level with a view to evaluate why the Western Cape is showing better performance and how other provinces can improve performance.

2 Role of Government in Waste Management

Based on the constitutional right of all South Africans to an environment that is not harmful to health and well-being and to have the environment protected, the DEA of the South African national government has developed the National Environmental Waste Act which is supported by the National Waste Management Strategy. Key principles governing this strategy include a waste hierarchy prioritising reduction, reuse, and recycling of waste materials over disposal, and extended producer responsibility where producers of waste are responsible for planning and financing waste management [6]. This, in turn, is linked to financial incentives and penalties throughout the product lifecycle as illustrated in Fig. 1. Incentives include subsidies linked to product recycling and reuse [7]. Penalties include [7]:

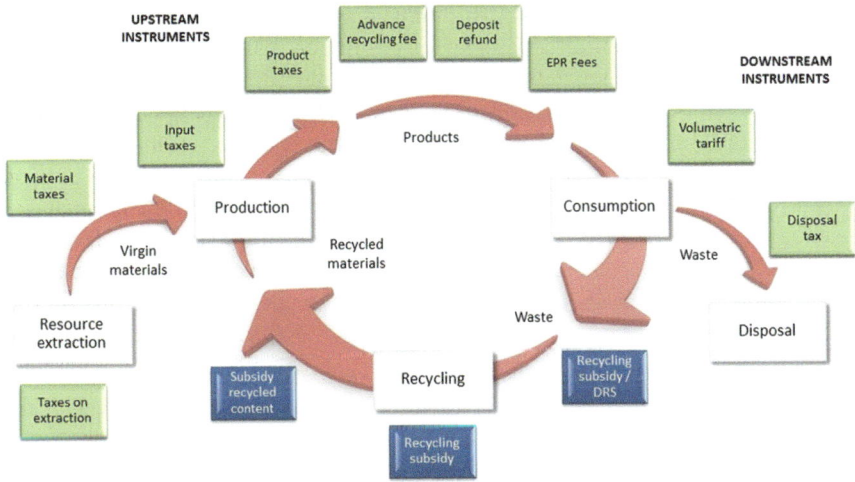

Fig. 1 Financial incentives throughout the lifecycle of a product, according to the South African Department of Environmental Affairs [7]

- Taxes on extraction and materials to minimise resource extraction and encourage either a reduction of materials needed or the use of secondary resources instead of virgin material resources.
- Integrated fees paid upon purchase to cover recycling at the end of the product lifecycle in the form of advanced recycling fees.
- A deposit refund system where the customer gets a rebate when the product is returned, recycled, or disposed of in an environmentally friendly way.
- Extended producer responsibility (EPR) fees where producers pay fees corresponding to the quantity of product being consumed to cover downstream waste management costs.
- A volumetric tariff corresponding to consumption with the goal of reducing consumption.
- A disposal tax to discourage waste disposal to landfill.

The National Waste Management Act mandates that all provinces develop, submit, and implement waste management plans in line with their development goals and waste management needs. Information to be included is the types and quantities of waste, strategies on how waste will be minimised, plans promoting reuse and recycling of waste, and potential methods of disposal that are environmentally friendly and less reliant on landfill usage [6]. Based on the relevant provincial waste management plan, municipalities must then develop, submit, and implement waste management plans and pass by-laws on waste removal, storage, and disposal services. Municipal governments are then responsible for the compliance to and enforcement of the by-laws.

3 Waste Management at the Local Level

Because municipal governments, guided by the relevant waste management plans and by-laws, are responsible for monitoring waste management of local industries and businesses on behalf of the provincial and national governments, the waste management plans and by-laws of municipal governments were compared to evaluate if this was causing the discrepancy in CDW recovery rates observed.

South Africa has a total of 257 municipalities, 8 of which are metropolitan municipalities. Metropolitan municipalities represent 80.4% of public sector revenue and are the centre of rapid and consistent infrastructure development [1]. Consequently, this is where a majority of CDW is generated. Furthermore, the positive link between population size, income, GDP, and waste generation can be leveraged. Therefore, only the four most populous metropolitan municipalities, namely, the City of Cape Town Metropolitan Municipality in the Western Cape, the City of Johannesburg Metropolitan Municipality in Gauteng, eThekwini Metropolitan Municipality in Kwa-Zulu Natal, and Ekurhuleni Metropolitan Municipality in Gauteng, were chosen. Provinces and metropolitan municipalities are shown in Fig. 2.

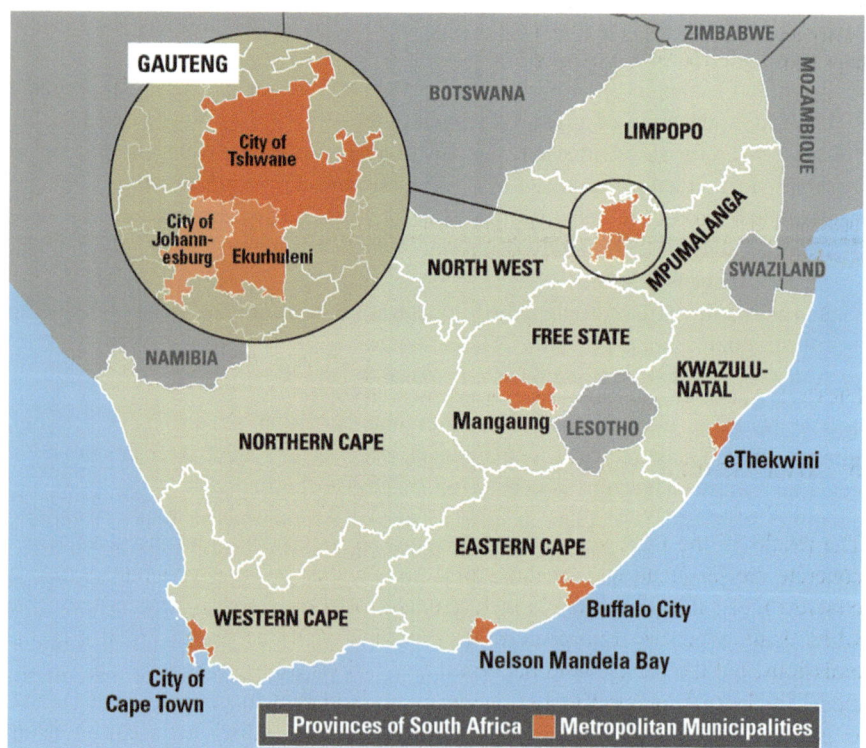

Fig. 2 Map of South Africa showing provinces and metropolitan municipalities [8]

Waste management plans are included in the national waste management strategy as a tool to assist provinces and municipalities in planning their goals with respect to waste management as well as how these goals will be achieved and monitored. However, fewer than 50% of municipalities and provinces have waste management plans in place [7].

Of the four municipalities considered, only eThekwini Metropolitan Municipality is located in a province that does not have a provincial waste management plan. All four have municipal waste management plans and associated by-laws. Although the waste management plans refer to the specific context and development goals of each municipality, all four sets of by-laws require those generating CDW, termed building waste, to develop a project-specific waste management plan which is to be submitted and approved by the local waste management officer before any waste can be generated [9–12]. In addition to the requirements of a waste management plan detailed in Sect. 2, these must include [9–12]:

- an impact assessment of the proposed works on the local environment,
- a projection of resources that will be consumed and may result in waste,
- a list of the services that will be used to store, collect, transport, and dispose of waste generated, with confirmation if an external contractor is being used,
- a plan on how recyclable and non-recyclable waste sources will be defined and separated, and,
- targets for minimisation of resources consumed, minimisation of waste produced, and optimisation of opportunities to reuse, recycle, and recover waste.

The fact that the by-laws governing the behaviour of those who produce and manage CDW in these three municipalities are the same, suggests that the differences in statistics reporting waste management behaviour are due to the enforcement and penalties related to these by-laws rather than the by-laws themselves.

In addition to the completion and submission of a waste management plan, those who generate building waste in the City of Cape Town municipality must complete a standardised form. The form asks for all information included in the waste management plan and also includes information about the site, the name and contact details of the owner, the scope and origin of the building work (partial demolition, full demolition, new building, and alterations to existing building). This allows the application process to be streamlined.

The city of Cape Town has also provided additional centralised CDW recycling facilities that can be used as an alternative to traditional landfills which are located outside of the city borders. The use of these facilities reduces the transport costs associated with dumping CDW and there is no associated dumping fee when sending materials to a recycling facility. In contrast, from 2019, Cape Town landfills have started to charge R20 (approx. €1) per tonne of uncontaminated mineral waste from CDW (including concrete, mortar, masonry components, sand, and stone) [4]. Prior to this, all CDW was accepted for free on the basis that it would be used as cover material in landfill. As shown in Fig. 3, the total volume of mineral CDW received was reduced from approximately 1.24 million tonnes in 2016/17 to 0.41 million tonnes in 2018/19 [4]. The resulting proportion of unused mineral CDW has,

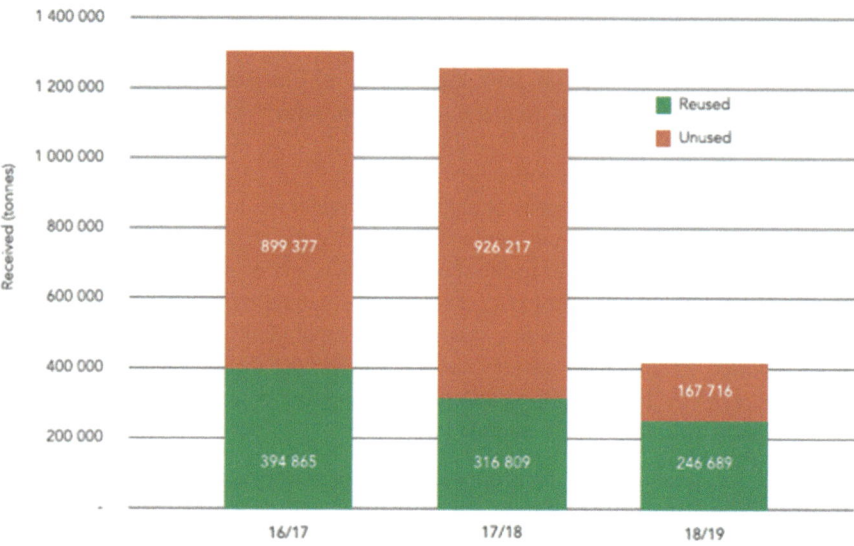

Fig. 3 Mineral fraction of CDW received at landfill sites in the City of Cape Town Municipality from 2016–2017 to 2018–2019 [4]

therefore, been reduced as has excess CDW in landfill not needed as cover material. This demonstrates that in this case the financial penalties detailed in Fig. 1 act in support of the legislation when it comes to changing consumer behaviour.

The mineral waste that is received at landfills and recycling facilities are sorted, cleaned, and processed to be used as bricks and blocks where discrete units can be separated out or as crushed recycled aggregate. The latter is then used as backfill material, landscaping material, aggregate in the sub-base and base of road construction, or aggregates in the production of concrete bricks and blocks and other concrete products in precast yards. Several government-funded projects and private enterprises, both formal and informal, exist in the vicinities of where this waste is being sent and are using it for resale value (e.g. scrap metal and intact clay bricks) or as input materials in the production of new products (e.g. the production of bricks and blocks using recycled aggregate). The presence of these enterprises not only promotes local economic growth and development, but also provides a use for the material that may have been otherwise treated as waste.

The other three municipalities all classify all CDW as zero-rated waste based on the premise that it can be used as landfill cover. The waste management plans and by-laws of all four of the municipalities encourage the recovery of waste material, but don't prohibit sending CDW to landfill. So long as the waste management plan is approved by local government, CDW produced can be dumped. If there is no additional factor encouraging the producers of waste to use alternative disposal methods, taking CDW to landfill may be chosen as the most cost and time efficient option. As in Cape Town before 2019, when tariffs on disposal were introduced, the existence of legislation alone without the support of financial penalties and incentives may not be sufficient to encourage a change in consumer behaviour.

4 Recommendations

While there is a robust legislative framework to guide waste management in South Africa, there is a lack of enforcement of legislation and the relevant by-laws, support to encourage compliance, and consequences where there is non compliance [13]. Furthermore, a lack of oversight and support from local government to those producing waste may limit the prioritisation of reducing, reusing, and recycling waste materials over disposal. Where guidelines are not promoted or enforced and cost-effective and efficient alternatives to disposal not provided, disposal will continue to be the default. As mentioned above, the city of Cape Town Municipality has streamlined the administrative tasks associated with CDW management and provided funds to enterprises using CDW to resell or produce new products. This has encouraged those producing waste to participate in the waste management system and provides an alternative to disposal for waste produced.

An additional use for this waste materials may be found through improved coordination of different government sectors. As outlined in the 2023 Market Study of the Circular (& Waste) Economy of South Africa [14] report commissioned by the Dutch government to identify potential economic opportunities in the South African waste management sector, CDW produced from government infrastructure can be used in new public infrastructure projects such as low-cost housing and roads.

However, due to a lack of funds and underperformance in service delivery, local government may require private sector participation to further increase waste recovery and improve waste management efficiency. This may take the form of public-private partnerships where, in a collaboration between parties from the private sector and government, an institution from the private sector performs a service usually performed by government or uses government-owned assets to perform a service [13]. In such partnerships, the private institution assumes most of the operational, technical, and financial risk. It is believed that through this model, the chances of corruption are reduced, the public has access to better and more economical services, and the provision of business opportunities stimulates the economy. Examples of public-private partnerships in the South African waste sector as shown in the city of Cape Town may include facilities to collect, characterise, process, and stockpile CDW for recycling or facilities using products from CDW for resale or to produce new products. Businesses created may be able to channel any profits generated into improved mechanisation of operations as well as training of employees.

It is estimated that 60,000 jobs were created in the South African waste sector as of 2016 [13]. These jobs are often directed towards unskilled, informal labour and are beneficial in the South African context where there is a large unemployed population with a lack of vocational training. Jobs take the form of waste picking, recovery, and recycling [13]. Informal waste pickers are estimated to have saved South African municipalities between R309 and R749 million (approx. between €15.5 and €37.5 million) [15]. They currently operate at little to no cost to municipalities. Training these workers and equipping them with skills to manage the waste value chain may result in improved space maximisation and management in landfills and waste management sites, ensure environmental compliance, and allow waste streams to be more closely monitored. This may require improved integration of the formal and informal waste management sectors. Integration may include increased regulation and monitoring in the form of registration and personal protective equipment requirements, increased support in assisting waste stream separation through mechanisation or provision of buy-back facilities, the establishment of businesses to employ these workers and pay them a regular wage, or in the full absorption of the informal sector by the formal sector [15]. Implementation of any of these measures would require consideration of the context and needs of the local population.

5 Conclusion

The efficient and appropriate management of CDW is key in overcoming the issue of South African landfill scarcity and environmental protection from waste. Findings suggest that robust legislation alone may not be sufficient in encouraging consumers to minimise the waste produced, optimise recovery, reuse and recycling opportunities, and avoid material disposal. As is shown in the case of the City of Cape Town Metropolitan Municipality, changes to waste producer and material consumer behaviour may have to be encouraged by improving the ease with which environmentally friendly behaviours like recycling waste or using this waste as a secondary resource can be enacted as well as implementing financial penalties and incentives which may negatively or positively impact potential profits.

References

1. Department of Environmental Affairs: South Africa State of Waste Report. Department of Environmental Affairs, Pretoria (2018)
2. Department of Environmental Affairs: National Waste Information Baseline Report. Department of Environmental Affairs, Pretoria (2012)
3. Berge, S., von Blottnitz, H.: An estimate of construction and demolition waste quantities and composition expected in South Africa. S. Afr. J. Sci. **118**, 14–18 (2022)
4. GreenCape: Waste Market Intelligence Report. GreenCape, Cape Town (2020)

5. Department of Environmental Affairs and Development Planning. Western Cape Integrated Waste Management Plan 2023–2027. (2023)
6. Department of Environmental Affairs: National Waste Management Strategy. Department of Environmental Affairs, Pretoria (2020)
7. Department of Environmental Affairs. Status of Waste Management in South Africa. (2022)
8. Africa Research Institute, https://www.africaresearchinstitute.org/newsite/blog/african-national ists-democrats-and-fighters-finding-room-for-compromise-in-sas-metros/. Last accessed 2024.02/28
9. City of Cape Town Municipality. Integrated Waste Management By-law 2009. (2009)
10. City of Johannesburg Metropolitan Municipality. Waste Management By-law 2021. (2021)
11. eThekwini Municipality. eThekwini Municipality Waste Removal By-law 2016. (2016)
12. City of Ekurhuleni Metropolitan Municipality. Integrated Waste Management By-Laws. (2021)
13. Pariatamby, A., Shahul Hamid, F., Bhatti, M.: Sustainable Waste Management Challenges in Developing Countries. IGI Global (2019)
14. Lindon Corporation. 2023 Market Study of the Circular (& Waste) Economy of South Africa. (2023)
15. Godfrey, L., Strydom, W., Phukubye, R.: Integrating the Informal Sector into the South African Waste and Recycling Economy in the Context of Extended Producer Responsibility. (2016)

Advancing Sustainable Construction Materials: Wood and Rubber Geopolymer Masonry Mix Development

Firesenay Zerabruk Gigar (iD), Amar Khennane (iD), Jong-leng Liow (iD),
Safat Al-Deen (iD), Biruk Hailu Tekle (iD), Cooper J. Fitzgerald (iD),
Anthony Basaglia (iD), and Charlie Webster (iD)

Contents

1 Introduction

As the environmental impact of modern society grows, the construction industry is actively seeking environmentally friendly materials to reshape construction methods, maintenance processes, and waste management [1–3]. Recycling emerges as a fundamental strategy, particularly in utilizing industrial wastes, providing a promising avenue for sustainable construction practices [4, 5]. Moreover, the overconsumption of natural resources is another factor the construction sector needs to address. With the building industry responsible for approximately 30% of total natural resource depletion [6], the recycling of end-of-service materials such as

F. Z. Gigar (✉) · A. Khennane · J.-l. Liow · S. Al-Deen · C. J. Fitzgerald · A. Basaglia ·
C. Webster
School of Engineering & Technology, University of New South Wales, Sydney, Australia
e-mail: f.gigar@unsw.edu.au; f.gigar@adfa.edu.au

B. H. Tekle
Institute of Innovation, Science and Sustainability, Federation University, Mount Helen, VIC,
Australia

© The Author(s) 2025
M. Kioumarsi, B. Shafei (eds.), *The 1st International Conference on Net-Zero Built
Environment*, Lecture Notes in Civil Engineering 237,
https://doi.org/10.1007/978-3-031-69626-8_75

wood and rubber could play a critical role in reducing the carbon footprint of the construction sector.

Bio-sourced products emerge as an alternative in construction, offering a versatile and sustainable array of benefits. Not only do these products regulate sound and heat effectively, creating a conducive indoor environment, but they also play a crucial role in carbon sequestration, contributing to environmental sustainability. Furthermore, wood-based materials exhibit excellent energy-dissipating properties, enhancing safety and resilience in various applications. Moreover, their lightweight nature and adaptability for modular construction provide flexibility and efficiency in building processes [7–9].

Within a similar framework, exploring alternative uses for end-of-service material, such as scrap tires, can produce innovative solutions for sustainable construction practices. Scrap tires, designated as hazardous waste, create havens for pests like rats, snakes, and mosquitoes and pose a significant fire risk, potentially releasing harmful substances such as carbon dioxide and dioxins into the environment. In response to these environmental challenges, recycling scrap tires into masonry units emerges as an environmentally friendly solution, offering a dual advantage. By repurposing this waste material, it is possible to simultaneously eliminate shelters for pests and reduce the risk of fires that could result in the release of hazardous substances. Furthermore, integrating recycled tires into masonry units enhances their physical and mechanical properties. Rubberized concrete, a byproduct of recycled tires, exhibits a lower density and higher energy absorption capacity compared to traditional concrete, making it a valuable construction material. In addition to these structural benefits, rubberized concrete also offers advantages in acoustic and thermal properties, exhibiting a higher noise reduction coefficient and lower heat transfer properties [10–13].

By using this by-product material with sustainable alternatives, such as geopolymer, the overall sustainability of these materials can be further improved. Such a combination facilitates rapid strength development and enhances durability while providing increased protection against fire and weathering. This study focused on the innovative reuse of end-of-service wood and crumb rubber to develop environmentally friendly materials that can be used in the production of masonry units. As such, the viability of producing wood geopolymer composites (WGC) with the required workability and strength was explored. The study also explored the integration of crumb rubber to create wood–rubber–geopolymer composites (WRGC).

2 Materials

End-of-service wood chips and crumb rubber were repurposed for this study. Additionally, fly ash and ground granulated blast furnace slag (GGBS), both industrial by-products were used as the base materials for the geopolymer binder. The overall appearance and particle size distribution of the wood chips and crumb rubber are shown in Figs. 1 and 2.

Fig. 1 General ppearance: (**a**) wood chips, (**b**) powder rubber, (**c**) 1–3 mm rubber granules

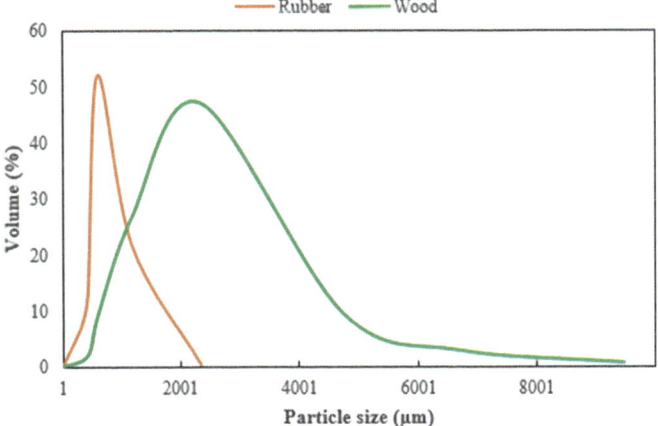

Fig. 2 Particle size distribution

The wood chips used in this study were chipped from recycled untreated pine pallets. These wood chips exhibit an apparent density of 443 kg/m^3 and a moisture content of 11%. The water absorption, determined following the method outlined in [14–16], was found to be approximately 154%. The rubber for this study was supplied by Tyrecycle and comprised a mixture from two different grades, as illustrated in Fig. 1. The rubber used in all mixes was an equal combination of powder rubber and 1–3 mm crumb rubber granules.

Geopolymer cement was used as a binderto further promote sustainability. It was produced by incorporating fly ash and GGBS as the primary constituents. The chemical compositions of these materials were assessed through XRFT (X-ray fluorescence spectroscopy), and the results are presented in Table 1.

A combination of sodium hydroxide and sodium silicate was used as an alkaline solution to activate the base materials. The concentration of the sodium hydroxide was 12M and was prepared from 98% pure flakes. A laboratory grade D sodium silicate solution with a solid ratio in weight % of 44.6 was used. The sodium silicate solution had a SiO$_2$: Na$_2$O ratio of 2.05.

Table 1 Chemical composition of binders

Compound (wt.%)	SiO_2	Al_2O_3	Fe_2O_3	CaO	MgO	Na_2O	K_2O	LOI[a]
Fly ash	67.62	19.94	3.7	1.46	0.53	0.61	2.23	2.02
GGBS	34.38	13.31	0.71	41.25	4.75	0.35	0.28	<0.01

[a]Loss on ignition

3 Experimental Program

The study involved the production and testing of cube samples to establish an optimal mix design by examining various proportions of wood and rubber content through a partial-factorial experimental design. The objective was to develop a wood–rubber–geopolymer composite mix design that not only delivered the required workability and compressive strength but also maximized the utilization of waste aggregates (Wood and crumb rubber). This involved testing of various mix designs through 50 mm × 50 mm × 50 mm cubes to identify the most suitable composition for constructing masonry units. While structural requirements took precedence, factors such as material workability and appearance were also integral considerations. To achieve this, the mix design by Gigar et al. [15, 16], developed for wood-geopolymer composite, served as the foundation, with adaptations made to modify the workability and incorporate rubber. The overarching goal was to devise a mix design for wood–geopolymer composite and wood–rubber–geopolymer composites that could potentially be used in the production of masonry units.

The optimal mix for the wood-geopolymer composite was first established through the examination of seven distinct wood-to-binder ratios (0.1, 0.15, 0.2, 0.25, 0.3, 0.4, and 0.5), designated as Set-A. Following this, the impact of partially replacing wood with rubber was explored at wood chips to binder (WC/B) ratios of 0.25 and 0.4, categorized as Set-B and Set-C, respectively. To assess the effect of partially replacing wood chips with rubber, a 5 × 2 factorial design was implemented, creating 10 different mixes with varying WC/B ratios (2 levels) and five levels of partial replacement. The partial replacements of the wood with rubber were examined at replacement levels of 0%, 25%, 50%, 75%, and 100% by volume. The complete experimental design, encompassing Sets A to C, is outlined in Table 2.

The slag-to-binder ratio and alkaline solid-to-binder ratio were consistently maintained at 0.4 and 0.18, respectively, across all mixes. The water utilized in this investigation was classified into geopolymer water and compensating water, following the methodology proposed by [14–16]. The total water quantity, designated as TW, was calculated using the formula TW = 0.29GPS + 0.5WC, where geopolymer solid (GPS) represents the combined weight of the binder and the solid portion of the alkaline solution, and WC corresponds to the weight of the wood chips.

The samples were created in a controlled laboratory setting using a pan-type concrete mixer. The mixing process involved two steps to produce a consistent mix. Solid components such as binders, dry wood chips, and rubber were mixed for the first 3 min. The alkaline solution and water were then added and mixed well for

Table 2 Mix proportions

Specimen designation	Mixture composition	
	WS/B	V_R/V_{WSR}
Set-A-W-10—R-0-C-0	10	–
Set-A-W-15—R-0-C-0	15	–
Set-A-W-20—R-0-C-0	20	–
Set-A-W-25—R-0-C-0	25	–
Set-A-W-30—R-0-C-0	30	–
Set-A-W-40—R-0-C-0	40	–
Set-A-W-50—R-0-C-0	50	–
Set-B-W-25—R-0-C-0	25	0
Set-B-W-25—R-5-C-0	25	5
Set-B-W-25—R-50-C-0	25	50
Set-B-W-25—R-75-C-0	25	75
Set-B-W-25—R-100-C-0	25	100
Set-C-W-40—R-0-C-0	40	0
Set-C-W-40—R-5-C-0	40	5
Set-C-W-40—R-50-C-0	40	50
Set-C-W-40—R-75-C-0	40	75
Set-C-W-40—R-100-C-0	40	100

another 5 min. Following the mixing process, the freshly mixed material was placed into the mold and subjected to vibration for 30 s to ensure adequate compaction. Following compaction, each sample was moved to the environmental control room, where the relative humidity was kept at 50% and the temperature was kept at 20 ± 3 °C until the testing day. This controlled environment guaranteed uniform conditions for precise evaluation and permitted the composites to cure as intended.

4   Results and Discussion

The compressive strength of the cube samples was tested according to the ASTM D3501 standard [17], and the results are presented in Table 3.

4.1   *Wood–Geopolymer Composites*

The results of the cube tests for WGC shown in Table 3 are plotted in Fig. 3. for comparison. The graph also shows the targeted 5 MPa and 10 MPa strengths for non-load-bearing and load-bearing masonry units, respectively. The compressive strength of the WGC was in the range of 4–25 MPa, while the corresponding density

Table 3 Experimental results

Mix ID	Compressive Strength [MPa]						Density [kg/m³]
	7th day	s.d.	14th day	s.d.	28th day	s.d.	
Set-A-W-10—R-0-C-0	17.65	0.60	23.42	1.15	24.56	1.21	1520
Set-A-W-15—R-0-C-0	13.75	1.06	18.33	0.67	19.34	1.02	1412
Set-A-W-20—R-0-C-0	13.35	1.30	14.72	0.22	16.82	0.21	1320
Set-A-W-25—R-0-C-0	10.98	0.68	15.08	0.61	15.14	0.99	1240
Set-A-W-30—R-0-C-0	8.18	0.51	10.70	0.39	11.60	0.18	1172
Set-A-W-40—R-0-C-0	3.90	0.36	5.57	0.14	6.18	0.16	951
Set-A-W-50—R-0-C-0	2.50	0.41	3.00	0.49	4.00	0.66	807
Set-B-W-25—R-0-C-0	10.98	0.68	15.08	0.61	15.14	0.99	1240
Set-B-W-25—R-5-C-0	9.98	0.44	12.40	0.30	15.38	0.73	1306
Set-B-W-25—R-50-C-0	10.31	0.03	11.83	0.62	12.16	0.82	1368
Set-B-W-25—R-75-C-0	9.87	0.79	12.46	0.88	13.19	0.08	1456
Set-B-W-25—R-100-C-0	10.25	0.24	13.17	0.45	14.15	1.68	1563
Set-C-W-40—R-0-C-0	3.90	0.36	5.57	0.14	6.18	0.16	951
Set-C-W-40—R-5-C-0	4.33	0.19	5.43	0.21	6.55	0.33	1059
Set-C-W-40—R-50-C-0	5.22	0.15	6.35	0.70	6.63	0.36	1172
Set-C-W-40—R-75-C-0	6.55	0.68	8.37	0.36	8.68	0.13	1334
Set-C-W-40—R-100-C-0	7.21	0.54	8.50	0.21	9.84	0.79	1466

Fig. 3 Compressive strength of WGC with vffig4arying wood content

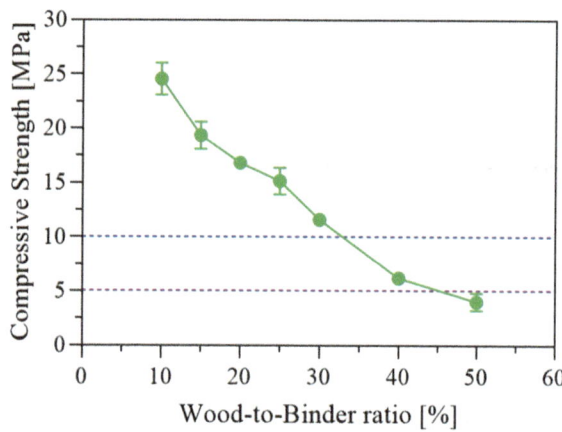

was in the range of 0.807–1.52 g/cm³. The graph highlights that the compressive strength of the WGC decreased as the wood content increased. This is mainly attributed to the increase in porosity and weak interface between the wood aggregate and the geopolymer matrix [15]. A similar trend was also observed for the density.

The WGC with WS/B ratios of 0.1–0.3, with compressive strength higher than 10 MPa, can potentially be used to produce load-bearing masonry units, while WGC with WS/B ratio 0.4, with compressive strength just above 5 MPa could potentially be used for the manufacturing of non-load-bearing masonry units. WGC with the maximum wood content of WS/B ratio of 0.5 fails just short of the minimum compressive strength to produce masonry units. The following stages explored the effect of partial replacement of the wood with rubber. WS/B ratios of 0.25 and 0.4 were selected for this (Fig. 3).

4.2 Wood–Rubber–Geopolymer Composite

The effect of incorporating rubber in the wood-geopolymer mix can be seen in Fig. 4. As can be observed, the strength of the composites is inversely proportional to the total volume of waste aggregate. The 25% waste aggregate cubes achieved between 12 and 15 MPa, while the 40% cubes achieved between 6 and 10 MPa strengths.

The graph also shows that the compressive strength of the wood–rubber–geopolymer composite (WRGC) generally increased with the relative rubber content. The relationship between wood and rubber also impacts the workability of the mix, with an increase in wood causing a drier and stiffer mix, while an increased proportion of rubber caused a much wetter mix. This is mainly attributed to the relatively small water absorption capacity of the crumb rubber.

Since the target is developing sustainable material with a minimum compressive strength of 10 MPa while maximizing the recycling of wood and rubber aggregates, a ratio of half wood to half crumb rubber was selected to provide an appropriate consistency for cinderblock manufacture. Given that the 25% samples exceeded the desired 10 MPa by an average of 4 MPa and the 40% samples were 3.5 MPa lower on average, the ideal waste aggregate content of 30% aggregate mix with half rubber and half wood can potentially be used to produce load-bearing masonry units.

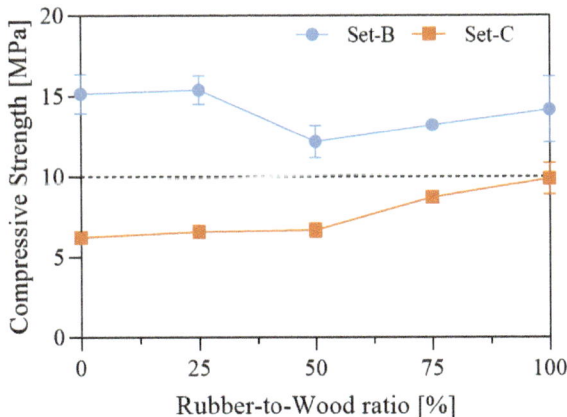

Fig. 4 Compressive strength of WRGC with varying wood and rubber content

5 Conclusion

In summary, this experimental program aimed at formulating an environmentally sustainable construction material through the incorporation of end-of-service materials, such as wood and rubber. It focused on optimizing mix designs with considerations for workability, strength, and the goal of maximizing waste recycling. The key findings of the study are summarized below:

- The compressive strength of the composites decreased with an increase in the wood content.
- The incorporation of rubber affected the appearance, workability, and compressive strength of the composite.
- The wood-binder ratio of 0.2–0.4 showed promising results for the production of load-bearing and non-load-bearing non-traditional masonry units. In addition to their sustainability, these composites are characterized by their lightweight nature and enhanced energy-absorbing capacity, distinguishing them from conventional composites.

Preliminary results showed significant promise in the viability and performance of these innovative composites for the production of masonry units manufactured from recyclable materials. The findings of this study will contribute to the ongoing efforts within the construction industry to adopt sustainable practices and materials, aligning with the imperative to mitigate environmental impacts. Future works will explore the production of masonry units using wood and rubber as base materials, evaluating their physical, mechanical, and durability properties. The performance of this units will be compared with conventional msonry units and assessed against international standards. Additionally, future research will explore the feasibility of integrating other recyclable materials, such as cenosphere, into the manufacturing process.

References

1. United Nations Environment Programme: 2020 Global Status Report for Buildings and Construction: Towards a Zero-Emission. Efficient and Resilient Buildings and Construction Sector, Nairobi (2020)
2. Meyer, C.: The greening of the concrete industry. Cem. Concr. Compos. **31**, 601–605 (2009). https://doi.org/10.1016/j.cemconcomp.2008.12.010
3. Olawumi, T.O., Chan, D.W.M.: A scientometric review of global research on sustainability and sustainable development. J. Clean. Prod. **183**, 231–250 (2018). https://doi.org/10.1016/j.jclepro.2018.02.162
4. Cobîrzan, N., Muntean, R., Thalmaier, G., Felseghi, R.A.: Recycling of mining waste in the production of masonry units. Materials. **15**, 594 (2022). https://doi.org/10.3390/MA15020594
5. Wasim, M., Roychand, R., Barnes, R.T., Talevski, J., Law, D., Li, J., Saberian, M.: Performance of reinforced foam and geopolymer concretes against prolonged exposures to chloride in a normal environment. Materials. **16**, 149 (2023). https://doi.org/10.3390/ma16010149

6. Ahmad, J., Kontoleon, K.J., Majdi, A., Naqash, M.T., Deifalla, A.F., Kahla, N.B., Isleem, H.F., Qaidi, S.M.A.: A comprehensive review on the ground granulated blast furnace slag (GGBS) in concrete production. Sustainability. **14**, 8783 (2022). https://doi.org/10.3390/su14148783

7. Berger, F., Gauvin, F., Brouwers, H.J.H.: The recycling potential of wood waste into wood-wool/cement composite. Constr. Build. Mater. **260**, 119786 (2020). https://doi.org/10.1016/j.conbuildmat.2020.119786

8. Amziane, S., Sonebi, M.: Overview on biobased building material made with plant aggregate. RILEM Tech. Lett. **1**, 31–38 (2016). https://doi.org/10.21809/rilemtechlett.2016.9

9. Karade, S.R.: Cement-bonded composites from lignocellulosic wastes. Constr. Build. Mater. **24**, 1323–1330 (2010). https://doi.org/10.1016/j.conbuildmat.2010.02.003

10. Najim, K.B., Hall, M.R.: A review of the fresh/hardened properties and applications for plain-(PRC) and self-compacting rubberised concrete (SCRC). Constr. Build. Mater. **24**, 2043–2051 (2010). https://doi.org/10.1016/j.conbuildmat.2010.04.056

11. Sukontasukkul, P., Chaikaew, C.: Properties of concrete pedestrian block mixed with crumb rubber. Constr. Build. Mater. **20**, 450–457 (2006). https://doi.org/10.1016/j.conbuildmat.2005.01.040

12. Youssf, O., ElGawady, M.A., Mills, J.E., Ma, X.: An experimental investigation of crumb rubber concrete confined by fibre reinforced polymer tubes. Constr. Build. Mater. **53**, 522–532 (2014). https://doi.org/10.1016/j.conbuildmat.2013.12.007

13. Gheni, A.A., ElGawady, M.A., Myers, J.J.: Mechanical characterization of concrete masonry units manufactured with crumb rubber aggregate. ACI Mater. J. **114**, 65–76 (2017). https://doi.org/10.14359/51689482

14. da Gloria, M.Y.R., Andreola, V.M., dos Santos, D.O.J., Pepe, M., Toledo Filho, R.D.: A comprehensive approach for designing workable bio-based cementitious composites. J. Build. Eng. **34**, 101696 (2021). https://doi.org/10.1016/j.jobe.2020.101696

15. Gigar, F.Z., Khennane, A., Liow, J., Tekle, B.H., Katoozi, E.: Recycling timber waste into geopolymer cement bonded wood composites. Constr. Build. Mater. **400**, 132793 (2023). https://doi.org/10.1016/j.conbuildmat.2023.132793

16. Gigar, F.Z., Khennane, A., Liow, J., Tekle, B.H.: Effect of wood/binder ratio, slag/binder ratio, and alkaline dosage on the compressive strength of wood-geopolymer composites. In: Ilki, A., Çavunt, D., Çavunt, Y.S. (eds.) Building for the Future: Durable, Sustainable, Resilient, pp. 658–667. Springer Nature Switzerland, Cham (2023)

17. ASTM D3501: Standard Test Methods for Wood-Based Structural Panels in Compression. The American Society for Testing and Materials (1994)

Optimizing the Embodied Carbon of Concrete, Timber, and Steel Piles with a Case Study

Kareem Abushama ⓘ, Will Hawkins ⓘ, Loizos Pelecanos ⓘ, and Tim Ibell ⓘ

Contents

1 Introduction

The construction sector is facing heightened scrutiny in response to the urgent need to combat climate change and reduce greenhouse gas emissions [1]. Governments and institutions are implementing immediate measures to achieve a 70% decrease in carbon emissions by 2030, which is a pivotal step towards the overarching objective of achieving net-zero emissions by 2050 [2]. Over recent decades, there has been increasing interest in reducing the negative environmental impacts of buildings to address global challenges and fulfil governmental obligations. On one hand, researchers are exploring ways to reduce the carbon footprint associated with material production and construction processes by investigating alternative materials with lower embodied carbon and refining construction methods to minimize waste and energy use [3–5]. On the other hand, there is a focus on designing buildings and infrastructure with energy efficiency and sustainability at the forefront, incorporating

K. Abushama (✉) · W. Hawkins · L. Pelecanos · T. Ibell
University of Bath, Bath, UK
e-mail: kaa71@bath.ac.uk

© The Author(s) 2025
M. Kioumarsi, B. Shafei (eds.), *The 1st International Conference on Net-Zero Built Environment*, Lecture Notes in Civil Engineering 237,
https://doi.org/10.1007/978-3-031-69626-8_76

passive design strategies and renewable energy systems [6–8]. Furthermore, advancements in digital technologies like Building Information Modelling (BIM) are facilitating more precise simulation and optimization of building performance over its lifecycle, aiding in identifying areas for improvement and reducing energy consumption [9–10]. However, the structural optimization of reinforced concrete foundations has been widely disregarded [11], even though foundations contribute significantly to the structural embodied carbon in various structures like high-rise buildings, fences, lightweight structures and wind turbines [12–13]. This lack of attention is largely attributed to a scarcity of prior data and uncertainties regarding the measured properties of the soil [14]. This chapter aims to determine the optimal pile designs made of three different materials concrete, timber and steel and compare their embodied carbon emissions. The authors have previously introduced a detailed study on optimizing the embodied carbon of reinforced concrete deep foundations [15]; however, the carbon emissions associated with steel and timber piles construction are still not covered in the literature. A genetic algorithm tool is proposed and used to compare the values of the resultant embodied carbon for the different pile materials at different load capacities to assess their environmental impact. Finally, a case study for an existing pile design in London clay is presented, and the proposed algorithm is then used to compare the different piling scenarios and compare their environmental impact to the as-built design to assess the potential carbon savings for future designs.

2   Methodology

A safe pile design should satisfy at least the following three design criteria:

- Structural capacity (N): The applied load should be less than the capacity of the pile as a short concrete column.
- Geotechnical capacity (Q_t): The applied load is less than the factored sum of the base bearing capacity and the pile's skin friction.
- Serviceability: The pile's settlement under the working load is within an acceptable range given the structure's type and sensitivity.

Following are the details of the adopted models to assess the three design criteria for different materials.

2.1   Structural Capacity

The structural capacity of piles is assessed in accordance with Eurocodes [16–17]. The structural capacity of short concrete columns ($N_{con.}$) is calculated as shown in eq. (1). While eq. (2) calculates the structural capacity of timber piles subjected to axial loading (N_{tim}) [18]. Equation (3) is simplified to calculate the structural capacity of stocky steel columns ($N_{stl.}$); the steel pipe's diameter-to-thickness ratio is assumed to be <80 to avoid local buckling of the structural element ($D/t < 80$) [19].

$$N_{con.} = 0.4 f_{ck} A_c + 0.75 f_y A_s \tag{1}$$

$$N_{tim.} = \frac{f_{ctk}}{\gamma_m} . A_t \tag{2}$$

$$N_{stl.} = \frac{f_y}{\gamma_p} . A_s \tag{3}$$

where

- A_c , A_t and A_s = concrete area, timber area and reinforcement steel area
- f_{ck}, f_{ctk} and f_y = compressive strength of concrete, timber and steel
- γ_m = partial factor for timber compressive strength
- γ_p = partial factor for steel compressive strength

2.2 Geotechnical Capacity

The geotechnical capacity of piles is the sum of the base bearing and the total skin friction. Steel pipes are assumed to be fully plugged when driven to the desired depths in clay [20]; therefore, the geotechnical capacity of concrete and steel pipes is calculated using eqs. (4, 5). However, timber piles are usually naturally tapered [21]; therefore, the geotechnical capacity of tapered piles is being assessed using the cavity expansion theory proposed by Khan *et al.* in 2008 [22], authors proposed adding a taper factor (k_t) to account for the effect of taper angles as shown in eqs. (6, 7 and 8).

$$Q_t = Q_b + Q_s \tag{4}$$

$$Q_b = q_{soil}.\pi.r_b^2 \tag{5}$$

$$Q_s = 2\pi \int_0^L L.r_i.\tau_s \, d(L) \tag{6}$$

$$\tau_{st} = k_s.k_t.\sigma_v'.tan\delta \tag{7}$$

$$K_t = \frac{tan\ (\alpha + \varphi).cot\delta}{1 + 2\xi tan\alpha.\, tan(\alpha + \varphi)} + \frac{4Gtan\alpha.\, tan(\alpha + \varphi).cot\varphi}{k_s[1 + 2\xi tan\alpha.\, tan(\alpha + \varphi)]} \frac{\Delta_r}{\sigma_v'}$$
$$+ \left(\frac{c.\sec^2\alpha}{(1 - tan\alpha\ tan\varphi)} / [1 + 2\xi tan\alpha.\, tan(\alpha + \varphi)] \, k_s\ \sigma_v'\ tan\varphi \right) \tag{8}$$

where

- Q_t, Q_i and Q_s = total capacity, base bearing capacity and shaft friction resistance
- q_{soil} = soil bearing capacity
- r_i and r_{ob} = pile radius and pile base radius
- τ_s = pile's surface friction
- α, φ and δ are the taper angle, the angle of internal friction and the concrete–soil friction angle
- σ'_v and φ' = soil effective stress at a given point and effective angle of internal friction

2.3 Settlement Calculation Model

A 2D model is coded to deploy the load–displacement (t-z) method for pile calculation as proposed by Randolph and Wroth in 1978 [23]. Equations (9, 10 and 11) show a summary of the adopted approach, the assigned settlement limit is assumed to be 10% of the pile's diameter, as recommended by Eurocodes [24]:

$$\{F\} = \left[K_p + K_s\right] \cdot \{u\} \tag{9}$$

$$\{u\} = \left[K_p + K_s\right]^{-1} \cdot \{F\} \tag{10}$$

$$k_{base} = \frac{G}{r \cdot \ln\left(2.5\, L\frac{1-\nu}{r}\right)} \tag{11}$$

where

- $\{F\}$ and $\{u\}$ = vectors of axial force and vertical displacements
- $[K_p]$ and $[K_{base}]$ = stiffness matrices of pile and soil end bearing
- G and ν = shear modulus and Poisson's ratio of the soil
- L and r = length and radius of the pile

2.4 Carbon Calculation Model

According to standard and codified life cycle carbon assessment practice [25], the embodied carbon of any building is split into several stages [A1-D]. For this study, only the 'cradle-to-gate' stages [A1–A3] are considered, as calculated in eq. 12. In line with convention, biogenic carbon stored in timber is ignored in this analysis.

$$\sum TEC = (m_x)(ECF_x) \tag{12}$$

where

- $\sum TEC$ = total embodied carbon (kgCO$_2$e)
- m_x = mass of the construction material (kg)
- ECF_x = embodied carbon factor for a given material (kgCO$_2$e/kg)

2.5 The Optimization Algorithm

In this research, the primary solving method for the suggested hybrid optimization tool is the genetic algorithm. Subsequently, a local minimization function in MATLAB (minion) is employed to refine and validate the results obtained from the genetic algorithm. The tool's layout is shown in Fig. 1. The optimization tool is used for fulfilling the analysis detailed in Sect. 2.6.

2.6 Analysis Setup

The algorithm presented in Fig. 1 is run to test the following:

- The optimal pile design corresponding to the lowest embodied carbon for solid and hollow concrete piles constructed in clayey soil with properties summarized in Table 1 at different load capacities. The algorithm is capable of determining the optimal pile geometry; length and diameter (L and D) as well as the optimal concrete grade (f_{ck}). Analysis results are shown in Fig. 2.
- The optimal Douglas fir pile designs for square and circular piles constructed in the same soil at different load capacities; the results are compared in Fig. 3.
- The optimal pile design for steel pipes, including optimal pile diameter, length, and thickness (L, D and t) for different load capacities. Results are presented in Fig. 4.

The input material properties are shown in Table 2, the chosen geometry dimensions ranges are based on recommendations from three different major contractors.

3 Results and Discussion

3.1 Concrete Piles Optimization Results

Figure 2 shows the optimal concrete grades f_{ckc} and slenderness ratios $(L/D)_c$ at different load capacities for both solid and hollow concrete piles. The results show that

- For high-capacity piles, optimal hollow concrete piles provide a significant cut of up to 75% less embodied carbon compared to conventional solid piles despite the construction complexities associated with hollow pile construction.

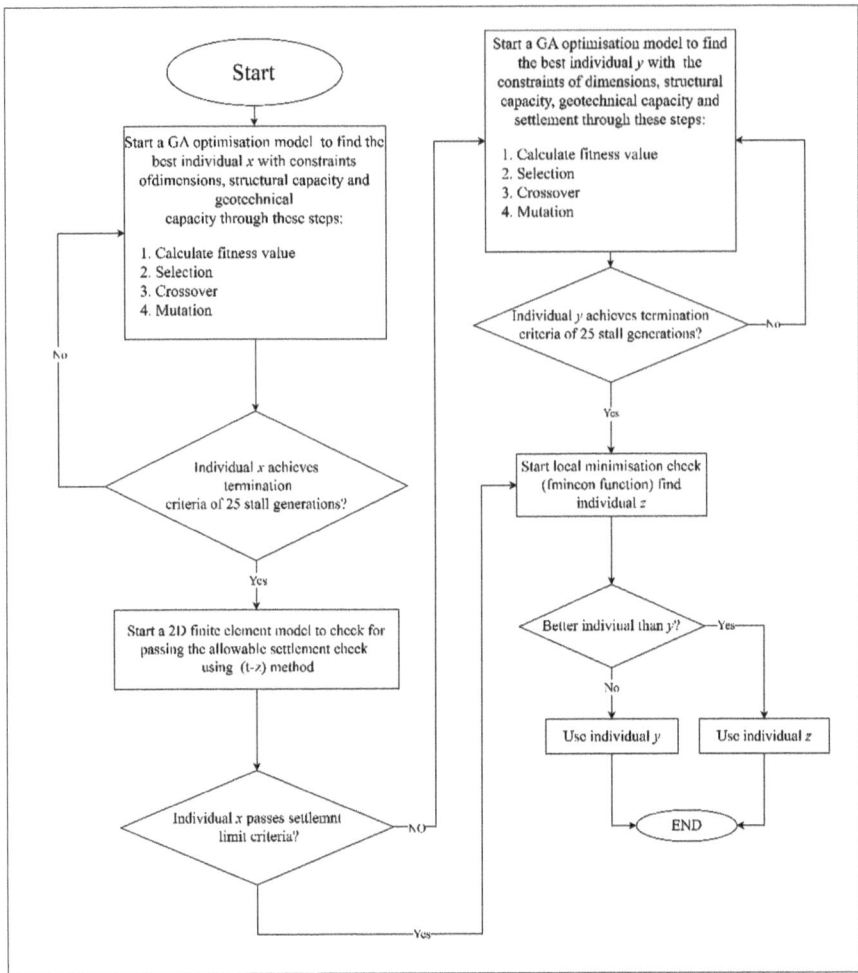

Fig. 1 A schematic of the proposed optimization algorithm

Table 1 Assumed properties of clay soil [23]

Soil property	Symbol	Value
Shape effect factor	N_q	9
Unit weight (kN/m^3)	γ	15
Angle of friction (degrees)	φ	18
Poisson's ratio	v	0.2
Shear modulus (kPa)	G	5000 + 500 z*

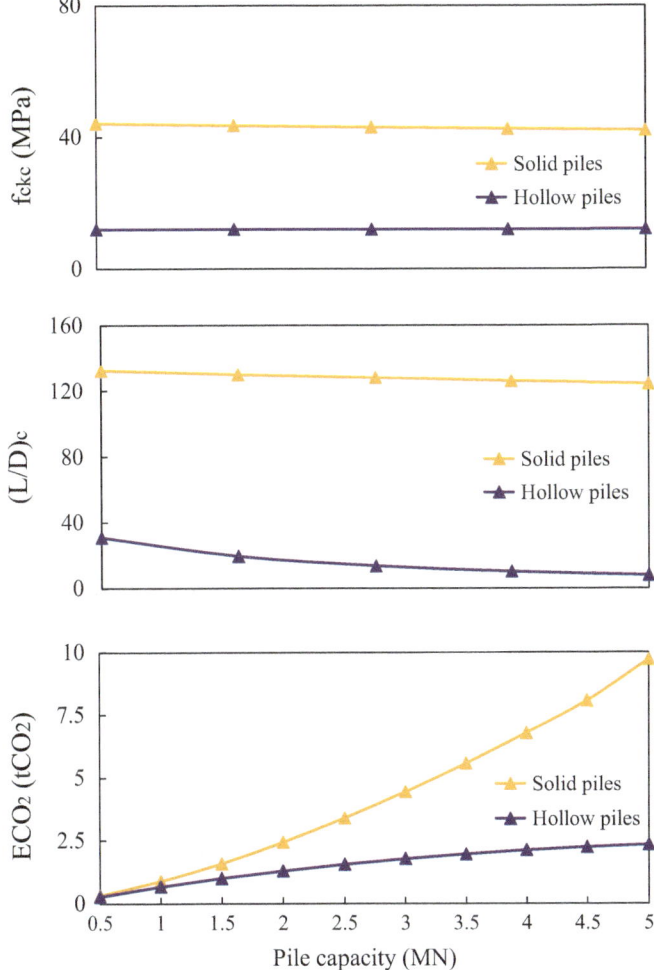

Fig. 2 Optimization results for solid and hollow concrete piles

- Optimal solid piles are very slender (high $(L/D)_c$) and this raises concerns regarding the practicality of constructing optimal solid piles [27], particularly augering.
- Optimal solid piles are usually corresponding to optimal concrete grades of (f_{ckc}) ≥ 40 MPa; however, hollow concrete piles make use of low concrete grades to get wider piles at lower embodied carbon.

3.2 Timber Piles Optimization Results

Figure 3 shows the optimization results for circular and square timber piles with material properties summarized in Table 2; the following are the main observed points:

- Timber pile designs produce significantly lower embodied carbon values compared to concrete piles; however, timber piles failed to provide pile capacities higher than 2.5 MN as the geometric limitations are met.
- Square piles produced slightly lower values of embodied carbon compared to circular piles; however, optimal square timber piles are more slender than optimal circular timber piles. It should also be noted that square piles are created through sawing of an initially cylindrical log and that the excess timber is not included in this study.

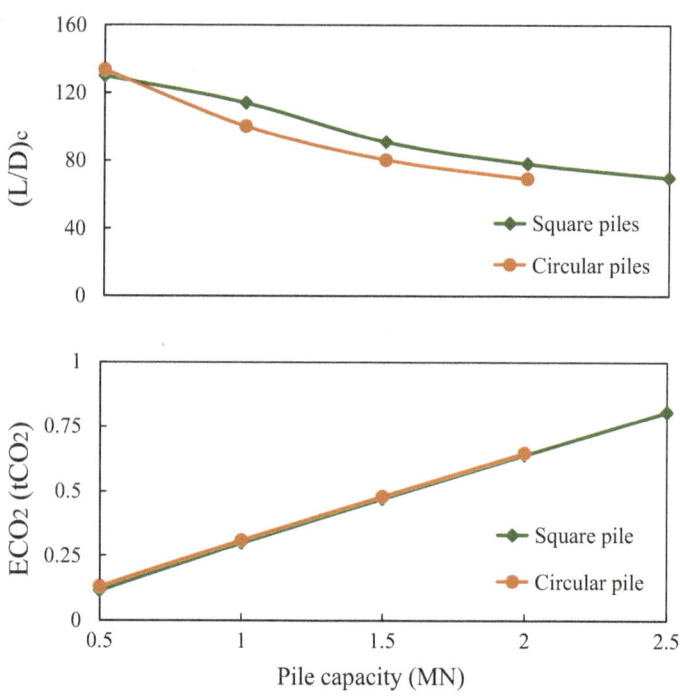

Fig. 3 Optimization results for circular and square Douglas fir timber piles

Table 2 Material properties and dimensions input for the optimization algorithm

Group	Input	Symbol	Value/range
Material strength (MPa)	Concrete grade range [16]	f_{ck}	[12–90]
	Timber compressive strength [18]	f_{ctk}	21
	Steel yield stress [16]	f_y	360
Geometry dimensions (m)	Concrete pile length	l_c	[1–200]
	Concrete pile diameter	D_c	[0–5]
	Minimum concrete cover	t_c	0.05
	Timber pile length	l_t	[1–30]
	Timber pile diameter	D_t	[0–0.45]
	Steel pipe length	L_s	[1–40]
	Steel pipe diameter	D_s	[0–2]
	Inner-to-outer diameter ratio for hollow concrete piles	D_i/D_o	[0.2–0.8]
Partial factors	Partial factor for timber [26]	γ_m	1.3
	Partial factor for steel [17]	γ_p	1.3
ECF (kg of CO_2e/kg)	Embodied carbon factor for concrete, 25% GGBS [15]	$ECF_{conc.}$	0.082 + f_{ck}·0.002
	Embodied carbon factor of reinforcement steel, worldwide steel	$ECF_{str.}$	1.99
	Embodied carbon factor of timber [25]	$ECF_{tim.}$	0.263
	Embodied carbon factor of steel sections, worldwide steel [25]	$ECF_{ste.}$	1.55

3.3 Steel Piles Optimization Results

Figure 4 shows the optimization results for steel piles; the following are the main observed points:

- Steel piles are the most emitting amongst all the different tested materials and designs, this is believed to be due to geometric limitations and local buckling of pipes. However, a significant cut in the embodied carbon can be achieved if stiffeners are used and piles are designed to avoid local buckling effects.
- The optimal values of slenderness ratios $(L/D)_c$ for steel pipes are between 85 and 90 as these values are corresponding to the lowest carbon emissions.

4 Case Study

This section applies the proposed algorithm to an existing case study to test the potential for emission savings.

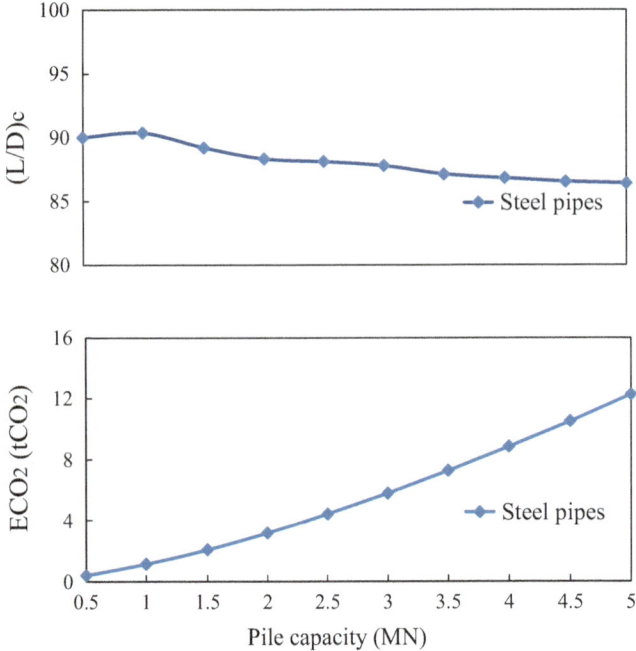

Fig. 4 Optimization results for tubular steel piles

4.1 Definition and Soil Properties

The case study is the design of the foundation system for the Hyde Park Cavalry Barracks, a 31-storey residential tower block located in central London and constructed in 1970 [28]. The main foundation system used is bored concrete piles with various diameters to support a piled raft. The soil encountered is typical of London, with 7 m of terrace sand and gravel [29] over a deep layer of undrained clay soil of unit weight $\gamma = 18$ kN/m^3 and a soil cohesion c_u (kPa) profile defined as shown in eq. 13, where z is the depth.

$$c_u = 6.1z + 60 \tag{13}$$

The soil properties were input to the optimization algorithm and was run to compare the embodied carbon of the different pile designs and materials at different load capacities. Concrete, timber and steel piles were compared given the described soil properties. Moreover, the as-built designs are compared to the different optimal designs to assess potential carbon savings for future projects.

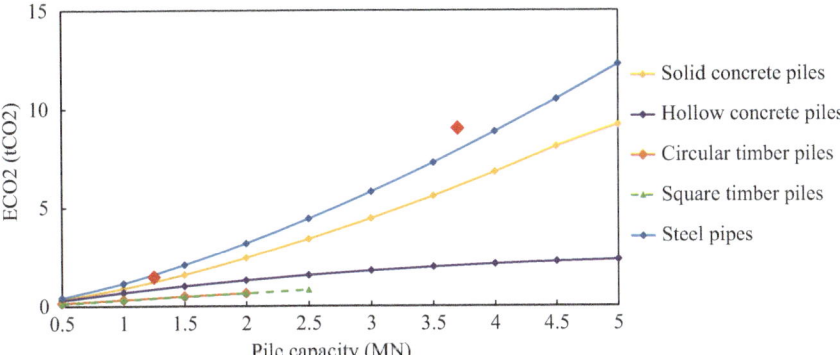

Fig. 5 The optimization results showing as-built designs as red marks

4.2 Optimization Results

Figure 5 shows the optimal embodied carbon values corresponding to the different pile designs resulting from the optimization algorithm. On the graph, the two red marks represent two as-built solid pile designs of capacities 1250 kN and 3760 kN, respectively. The analysis shows significant embodied carbon reduction possibilities if the optimal designs were utilized. For low-capacity pile, utilizing an optimized timber pile could have resulted in 68% less embodied carbon. For high-capacity pile, utilizing an optimized hollow concrete pile could have saved 76% of the embodied carbon. Therefore, the proposed optimization algorithm demonstrated an excellent carbon reduction efficiency and is highly recommended for geotechnical designers.

5 Conclusion

This paper is part of an ongoing research project that aims to optimize the design of deep foundations and build robust carbon-cost optimization models. The paper has introduced a pile optimization algorithm, which is used to compare the embodied carbon of different pile geometries and materials. The following are the main conclusions:

- Despite difficulties associated with construction, hollow concrete piles provide an excellent alternative to common solid piles as they provide the same pile capacity at significantly lower embodied carbon, with a more practical slenderness and concrete strength.
- Timber piles are excellent substitutes for concrete piles at low capacities; however, they can provide a limited threshold of pile capacities, this is mainly due to geometric limitations and the availability of adequate tree sizes.

- The embodied carbon associated with steel piles construction is significantly higher than that of concrete and timber piles.
- The algorithm is applied to a case study of London clay and showed significant potential to cut embodied carbon.
- This study emphasizes the potential for a substantial decrease in the overall embodied carbon linked to substructures. Most common pile designs are usually overdesigned, which leads to an unnecessary increase in values of embodied carbon. Moreover, there is an imminent need to introduce novel pile construction techniques to accommodate the optimal pile designs and geometries [30].
- The findings of this study are well timed, given the growing importance of assessing embodied carbon in construction, especially within the built environment.

References

1. United Nations Environment Programme. 2022 Global status report for buildings and construction: towards a zero-emission, efficient and resilient buildings and construction sector. Nairobi (2022)
2. United Nations Environment Programme. 2021 Global status report for buildings and construction: towards a zero-emission, efficient and resilient buildings and construction sector. Nairobi. (2021)
3. Ellis, L.D., Badel, A.F., Chiang, M.L., Park, R.J.Y., Chiang, Y.M.: Toward electrochemical synthesis of cement—an electrolyzer-based process for decarbonating $CaCO_3$ while producing useful gas streams. Proc Nat Acad Sci. **117**(23), 12584–12591 (2020)
4. Zhang, X., Jiao, K., Zhang, J., Guo, Z.: A review on low carbon emissions projects of the steel industry in the World. J Cleaner Prod. **306**, 127259 (2021)
5. Fan, Z., Friedmann, S.J.: Low-carbon production of iron and steel: technology options, economic assessment, and policy. Joule. Elsevier. **5**(4), 829–862 (2021)
6. Balali, A., Yunusa-Kaltungo, A., Edwards, R.: A systematic review of passive energy consumption optimisation strategy selection for buildings through multiple criteria decision-making techniques. Renewable Sust Energy Rev. **171**, 113013 (2023)
7. Gauch, H.L., Hawkins, W., Ibell, T., Allwood, J.M., Dunant, C.F.: Carbon vs. cost option mapping: a tool for improving early-stage design decisions. Automat Const. Elsevier B.V. **136**(January), 104178 (2022)
8. Kanyilmaz, A., Tichell, P.R.N., Loiacono, D.: A genetic algorithm tool for conceptual structural design with cost and embodied carbon optimization. Eng Appl Artif Intel. Elsevier Ltd. **112**-(January), 104711 (2022)
9. Evins, R.: A review of computational optimisation methods applied to sustainable building design. Renew Sust Energy Rev. **22**, 230–245 (2013)
10. Kiamili, C., Hollberg, A., Habert, G.: Detailed assessment of embodied carbon of HVAC systems for a new office building based on BIM. Sustainability. **12**(8), 3372 (2020)
11. Sandanayake, M., Zhang, G., Setunge, S.: Environmental emissions at the foundation construction stage of buildings—two case studies. Building Envir. Elsevier Ltd. **95**, 189–198 (2016)
12. Poulos, H.G.: Tall building foundations: design methods and applications. Innov. Infrastruct. Solut. **1**, 10 (2016). https://doi.org/10.1007/s41062-016-0010-2
13. Berndt, M.L.: Influence of concrete mix design on CO_2 emissions for large wind turbine foundations. Ren Energy. Elsevier Ltd. **83**, 608–614 (2015)

14. Luo, W., Sandanayake, M., Zhang, G.: Direct and indirect carbon emissions in foundation construction–two case studies of driven precast and cast-in-situ piles. J Cleaner Prod. **211**, 1517–1526 (2019)
15. Abushama K, Hawkins W, Pelecanos L, Ibell T.: Minimising the embodied carbon of reinforced concrete piles using a multi-level modelling tool with a case study. InStructures. **58**, 105476 (2023). Elsevier
16. Standard, B.: Eurocode 2: design of concrete structures. Part. 2004 Dec 23;1(1):230
17. Standard, B.: Eurocode 3—design of steel structures. BS EN. **1**(1), 2005 (1993)
18. Leijten, A.J., Larsen, H.J., Van der Put, T.A.: Structural design for compression strength perpendicular to the grain of timber beams. Const Building Materials. **24**(3), 252–257 (2010)
19. Steel T.: Steel building design: design data. The Steel Construction Institute and The British Constructional Steelwork Association Limited (2011)
20. Knappett, J., Craig, R.F.: Craig's soil mechanics. CRC Press LLC, Florida (2012)
21. Horvath, J.S., Trochalides, T.: A half century of tapered-pile usage at the John F. Kennedy International Airport
22. Khan, M.K., El Naggar, M.H., Elkasabgy, M.: Compression testing and analysis of drilled concrete tapered piles in cohesive-frictional soil. Canadian Geotechn J. **45**(3), 377–392 (2008). https://doi.org/10.1139/T07-107
23. Randolph, M.F., Wroth, C.P.: An analysis of the vertical deformation of pile groups. Geotechnique. **29**(4), 423–439 (1979)
24. British Standards Institution.: Eurocode 7. Part 1, General rules Part 1, General rules. London, British (1995)
25. Gibbons, O.P., Orr, J.J., Archer-Jones, C., Arnold, W., Green, D.: How to calculate embodied carbon. Institution of Structural Engineers (ISTRUCTE) (2022)
26. EN 1995.: Eurocode 5: design of timber structures (2004)
27. Abushama K, Hawkins W, Pelecanos L, Ibell T. Effect of slenderness ratio on the environmental impact of piles bored in clayey soil-case study. In14th fib PhD Symposium in Civil Engineering, 2022 Dec 31 (pp. 673–680). fib. The International Federation for Structural Concrete
28. Letsios, C., Lagaros, N.D., Papadrakakis, M.: Optimum design methodologies for pile foundations in London. Case Stud Struct Eng. **2**, 24–32 (2014)
29. Baxter, D.J., Dixon, N., Fleming, P.R., Cromwell, K.: Refining shear strength characteristic value using experience. Proc Instit Civil Eng Geotechn Eng. **161**(5), 247–257 (2008)
30. Abushama K., Hawkins W., Pelecanos L., Ibell T.: Optimizing the embodied carbon of concrete piles-case study. In The International fib Symposium on the Conceptual Design of Concrete Structures. fib. The International Federation for Structural Concrete 2023 Jun 29 (pp. 53–62)

Suitability of Excavation Clay Wastes for Sustainable Earthen Construction

India Harding ⓘ, Sripriya Rengaraju ⓘ, and Abir Al-Tabbaa ⓘ

Contents

1 Introduction

Cement production is responsible for up to 8% of global CO_2 emissions [1]; whilst a proportion of this is due to processes such as mixing and transportation, emissions affiliated with the calcination of limestone and associated decomposition make up two-thirds of emissions attributed to cement. To tackle such large emissions, cement alternatives such as ground granulated blast furnace slag (GGBS) or pulverized fuel ash (PFA) are often used as a partial, but direct replacement. However, a critical limitation lies in the finite supply of such alternatives. The search for replacement materials with similar pozzolanic properties has led to increasing utilization of calcined clay.

Clay is a globally abundant and low-cost material, however, is unreactive when raw. Therefore, dependent on application, calcination is often a necessity. Clay calcination typically takes place at 700–800 °C, approximately 600 °C lower than that needed for cement. In addition, clay calcination itself does not release CO_2. In

I. Harding (✉) · S. Rengaraju · A. Al-Tabbaa
Department of Engineering, University of Cambridge, Cambridge, UK
e-mail: pih24@cam.ac.uk

© The Author(s) 2025
M. Kioumarsi, B. Shafei (eds.), *The 1st International Conference on Net-Zero Built Environment*, Lecture Notes in Civil Engineering 237,
https://doi.org/10.1007/978-3-031-69626-8_77

recent years, calcined clay has been integrated into cement mixtures, with the most prominent combination called LC3—limestone calcined clay cement–which comprises 50% of clinker being replaced by 30% calcined clay, 15% limestone powder and 5% gypsum. Earth is a traditional building material with minimal associated emissions; however, its inherent weaknesses in terms of strength and durability often require stabilization. While cement is frequently employed for this purpose, the adoption of geopolymers presents an opportunity to reduce emissions further. Although excavation waste often serves as the primary matrix in earthen construction, its potential as a stabilizer is often overlooked. Kaolinite-based geopolymers are often used due to kaolinite's pronounced reactivity; however, literature suggests that despite the prevalence of 2:1 minerals, their application and thus contribution to SCM reactivity remains insufficiently explored [2–5].

Calcined clay within the construction industry is almost exclusively targeted for use within concrete. This chapter takes a more innovative approach focusing on calcined waste excavation clay and investigates the feasibility for use as a geopolymer precursor within earthen construction materials.

1.1 Earthen Construction Stabilization

Earthen construction can take many forms. Throughout literature, compressed earth blocks (CEBs) and rammed earth are the most common. However, in the United Kingdom, earthen construction materials are often disregarded, with more durable materials such as concrete taking precedence. In 2018, 21.7% of UK excavation waste was reused as backfill; however, 50.4% (equivalent to 29.5 Mt) of total generated waste soil ended up in landfill [6]. While small quantities may be deemed hazardous and unsuitable for use, it is evident that there are ample opportunities to capitalize on soil reuse.

Soil stabilization represents a multifaceted concept aimed at enhancing engineering properties to fulfill specific performance requirements. Such criteria will often address strength, durability, shrinkage and workability. For instance, in the context of load-bearing walls, CEBs and rammed earth must have dry compressive strengths of 4 MPa and 2 MPa, respectively [7, 8]. Without stabilization, compressive strength can be significantly lower [5, 9, 10]. Central to earthen stabilization are pozzolanic reactions—a mechanism drawing upon the high reactivity of pozzolans. In the realm of earthen construction, conventional practices involve cement or lime stabilization, however high associated emissions remain [8, 11, 12]. Geopolymers offer an alternative solution, harnessing the reactivity of alkali-activated aluminosilicates to yield significant mechanical improvements, thereby enabling earthen stabilization. The realization of geopolymers as binders has led to lower carbon stabilization methods, in some cases achieving compressive strengths much higher than those obtained with ordinary Portland cement [13, 14].

Preethi and Reddy examined the effects of combining raw soil with an alkali activator, GGBS and PFA on the compressive strength of CEBs [5]. The alkali-activated samples experienced marginal improvement in comparison to

GGBS and PFA, owing to the inherent low reactivity of raw clay minerals. Consequently, significant emphasis should be placed on calcined clay. Several investigations [15–17] have explored the impact of metakaolin-based geopolymers on earthen blocks. However, similar to work by Dhanapani et al. [4], this study will be centered around waste excavation clays as geopolymer precursors, with particular focus on reactivity due to mineral composition. Subsequent investigations will explore the potential of geopolymer formation through alkali activation and its subsequent influence on the durability of earthen structures.

1.2 Reactivity Potential

Geopolymer precursors, characterized by their rich content of alumino-silicate oxides and high reactivity, often utilize pozzolanic materials like GGBS, PFA and metakaolin. Their highly amorphous nature, fine particle size and minimal inert material content render them ideal choices as precursors. Conversely, waste excavation clay can exhibit significant pozzolanic tendencies and appropriate grain sizes. However, a notable limitation lies in the presence of inert materials. The reactivity of clay is intricately linked to its mineralogy; samples with higher kaolinite content demonstrate greater reactivity, while those with high quartz proportions exhibit lower reactivity.

Various methods exist to quantify an SCM's reactivity [18–20] with the R^3 test emerging as one of the most effective options. This modern procedure is designed to emulate the reaction environment observed during cement hydration thereby facilitating a rapid assessment of reactivity [21]. Despite offering two routes for analysis—the cumulative heat release study and the bound water study—both pathways yield highly correlated results, ultimately leading to conclusions comparable to classical methods such as the modified Chapelle test [20, 22, 23].

The R^3 test homes in of the nature of geopolymers as SCMs are mixed with an alkali solution to simulate a reactive environment. Calcined clay R^3 test results, obtained from literature, are provided in Fig. 1 [20–27]. Among the findings, approximately one-third were categorized as metakaolin; however, considering the R^3 values, the majority are presumed metakaolin. Avet et al. [23] reported the lowest value of 225 J/g of SCM, utilizing a 95% kaolinite mixture. It is pertinent to note that this study employed a calcined clay-to-alkali ratio surpassing ASTM C1897 [28] standards by 13%. Given that Avet et al. [23] introduced the R^3 procedure, the recent studies reporting higher values are deemed more accurate.

Clay reactivity is heavily influenced by mineral reactivity, with kaolinite recognised as the most reactive mineral. Consequently, calcined clays containing high proportions of kaolinite demonstrate enhanced performance [23, 30]. The complexity of clay reactivity becomes more apparent when 2:1 minerals are considered, as their reactions tend to be slower and less extensive, often with the influence of individual minerals becoming blurred [4]. Weise et al. reported a result of 300 J/g of SCM [27]; this was the only example disclosed as a mixed clay, comprising 25%

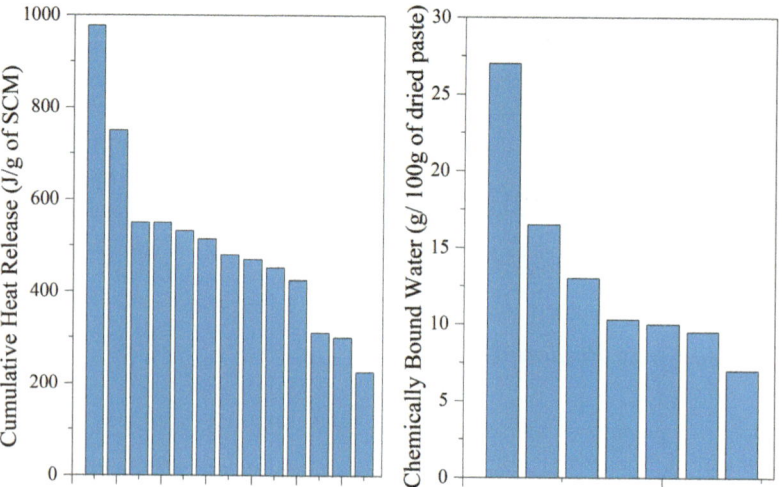

Fig. 1 Comparative analysis of calcined clay R^3 test results obtained from literature [20–27, 29]

kaolinite, 30% muscovite, 11% illite, 6% chlorite and 18% quartz. While numerous studies suggest a 40% kaolinite [3, 4, 31] threshold for SCMs, the above results indicate that comparable results can be achieved using lower grade or mixed clays. This study therefore utilizes waste excavation clay that is likely to comprise moderate to low levels of reactive minerals, to evaluate its suitability as a geopolymer precursor.

2 Methodology

2.1 Materials and Mixtures

The minerology and reactivity of four excavation waste clays will be studied, with the results compared against reagent kaolin and bentonite. The low expandability and high chemical reactivity of kaolin render it a prevalent choice as a geopolymer precursor. Conversely, bentonite is less utilized as a construction material due to its high demand within the geotechnical field. All waste clays were obtained from live UK construction projects; the site locations are illustrated below (Fig. 2).

According to the British Geological Survey [32], the A30 sample originates from the Mylor Slate Formation, the A391 from the St Austell Intrusion—a granite formation recognised as a global kaolin resource, the Copthall sample from the London Clay Formation and the RDP North sample from the lower/middle Pennine Coal Measures formations.

Fig. 2 Site locations of four excavation waste clays to be used for analysis

2.2 Preparation of Samples

All waste clays were air dried upon receival then ground using a mortar and pestle, sieved to a maximum grain size of 75 μm and stored in air-tight containers. Each powdered sample was calcined at 800 °C for a burning time of 1 h, in line with the work carried out by Avet et al. [23] and Abiodun et al. [33]. After calcination, all samples demonstrated significant redish discolouration, indicating the presence of iron.

2.3 Experimental Program

2.3.1 Characterization Using Thermogravimetric Analysis and X-Ray Diffraction

The chemical composition of both raw and calcined samples was characterized by thermogravimetric analysis (TGA) and X-ray diffraction (XRD). The thermographic analyzer used in this study is the PerkinElmer Simultaneous Thermal Analyzer. For each study, circa 15 mg of material was placed into a crucible, and the change in mass was monitored. The TGA began at 30 °C and saw a steady heating rate of 10 °C/min until 950 °C was reached. Upon reaching 950 °C, the samples were held at temperature for 1 min. In addition, derivative thermograms were studied to demonstrate dehydration and dehydroxylation patterns within the samples.

The XRD was carried out independently using a Siemens D500/501 diffractometer. The XRD patterns were generated using Cu-Kα radiation and power parameters 40 kV and 20 mA, from a range of 2–70° 2θ, a step size of 0.02° 2θ at 2 s per step. Further to this, bulk mineralogy was determined using semi-quantitative analysis.

2.3.2 R^3 Test

The pozzolanic reactivity was determined in accordance with R^3 requirements, as advised by ASTM C1897 [28]. The R^3 pastes were prepared using proportions discussed in Table 1. R^3 mixture proportions, with the potassium solution comprising 4 g KOH and 20 g K_2SO_4 dissolved in 1 L of deionized water. The dry materials were combined using the Heidolph RZR 1 mechanical stirrer for 2 min. Thereafter, the alkaline solution was added and mixed for a further 2 min until a homogenous mixture was achieved. The paste was then poured into air-tight plastic molds for curing or isothermal calorimetry containers for immediate testing.

The cumulative heat study was conducted using a Calmetrix I-Cal 8000 HPC isothermal calorimeter set at 40 °C, whereas the bound water study was undertaken using the Carbolite CWF 1200 furnace at 350 °C and a desiccator. The chemically bound water was calculated using Eq. (1), where ω_0 is the total mass of the dried paste and crucible after stabilisation at 40 °C, ω_h is the total mass of the paste and crucible following 2 h at 350 °C and an hour in a desiccator, and ω_c is the mass of the empty crucible.

$$H_2O_{bound} \ (g/100 \ g \ dried \ paste) = \frac{\omega_0 \ - \ \omega_h}{\omega_0 \ - \ \omega_c} \times 100 \tag{1}$$

The cumulative heat release is calculated using Eq. (2), where H is the cumulative heat release from 75 min until 7 days (168 h $\pm$ 10 min) after the start time of mixing by integration of the recorded heat release.

$$H_{SCM} \ (J/(g \ of \ SCM)) = \frac{H}{\left(m_p \times 0.101\right)} \tag{2}$$

Table 1 R^3 mixture proportions

	Clay-based sample	Ca(OH)$_2$	CaCO$_3$	Potassium sol^n
Mass (g)	10.00	30.00	5.00	54.00

3   Results and Discussion

3.1   Chemical Composition

The XRD patterns are given in Fig. 3. The analysis revealed significant proportions of quartz in all waste clay samples, as indicated by significant peaks at 26.6°. Given its inert characteristics, quartz can be detrimental to reactivity. Among the samples examined, A30 and A391 exhibited the highest quartz contents, with semi-quantitative analysis, i.e. the normalized reference intensity ratio (RIR) method, confirming beyond 50% content, notably higher than the average observed in the remaining waste clays. Conversely, Copthall and RDP North each comprise over 20% 1:1 minerals, thereby indicating the likelihood of increased reactivity. Following calcination, peak intensities reduced, indicating a decrease in mixture crystallinity. Similarly, the XRD analysis identified shifts in mixture composition. Despite a reduction in crystalline minerals, 2:1 minerals continued to be identified, thereby indicating an incomplete transformation into amorphous minerals. The trade-off between heightened calcination temperatures and influence on reactivity is an area for potential study.

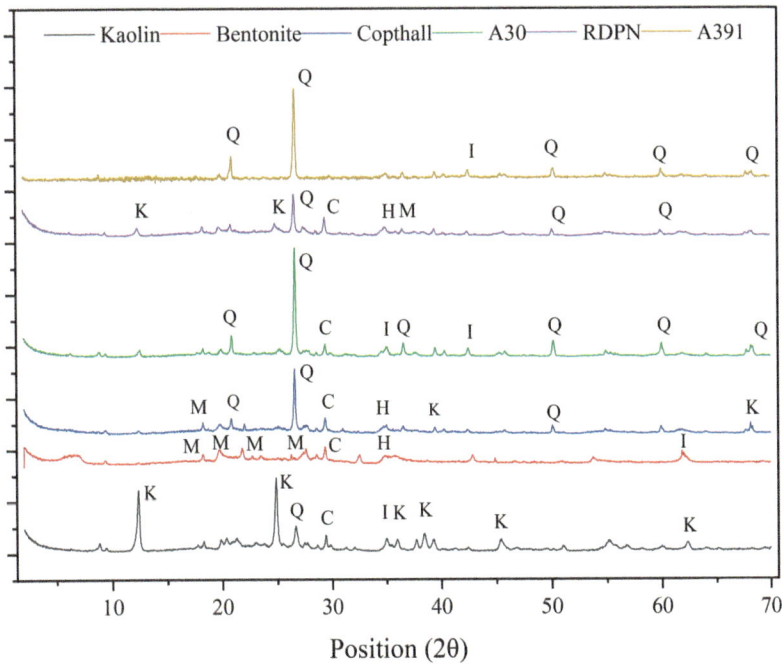

Fig. 3 XRD patterns of raw clays. C: Calcite, I: Illite, H: Halloysite, K: Kaolinite, M: Montmorillonite, Q: Quartz

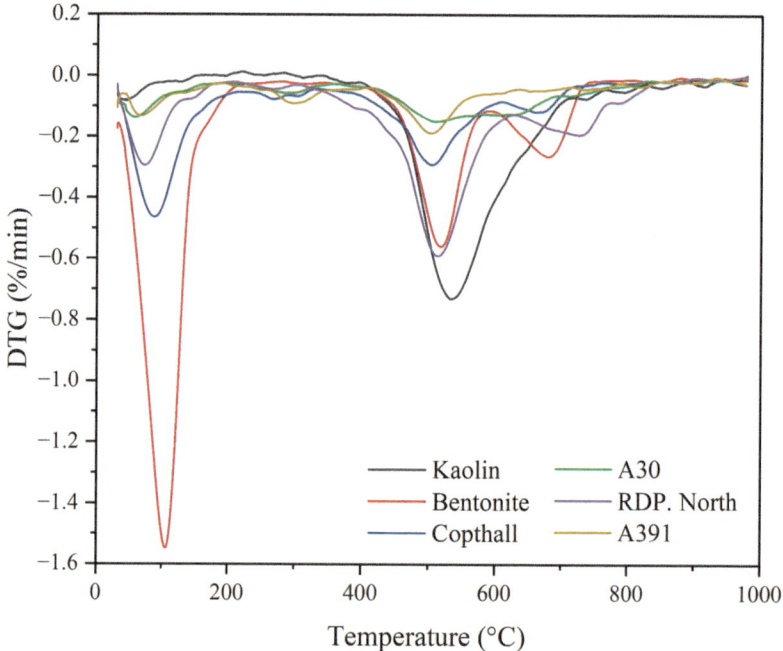

Fig. 4 Derivative thermograms of raw excavation waste clays, kaolin and bentonite

Figure 4 presents the differential thermogravimetric (DTG) curves for all samples. Owing to the mixed mineral composition of waste clays, the DTG troughs overlap. Below 200 °C, the dehydration of 2:1 minerals and evaporation of free water are observed. Above these temperatures, minerals are dehydroxylated. Kaolinite is dehydroxylated between 400 and 600 °C, whereas dehydroxylation of illite and montmorillonite will occur between 450–700 °C and 600–800 °C, respectively [2, 34]. Figure 4 reinforces the XRD findings. In terms of kaolinite content, RDP North is comparable to bentonite. Furthermore, in the analysis of 2:1 minerals, bentonite displays the largest trough owing to its substantial montmorillonite content, with Copthall following closely.

3.2 R^3 Test: Cumulative Heat Release

Figure 5 illustrates the cumulative heat released during the R^3 test, with kaolin exhibiting a notably value compared to other samples, as expected given its renowned reactivity. This value falls within the range reported in the literature, reinforcing the study's accuracy. Among the excavation waste samples, RDP North exhibited promising performance as a geopolymer precursor, aligning closely with the outcome observed with bentonite. While the use of calcined bentonite as an

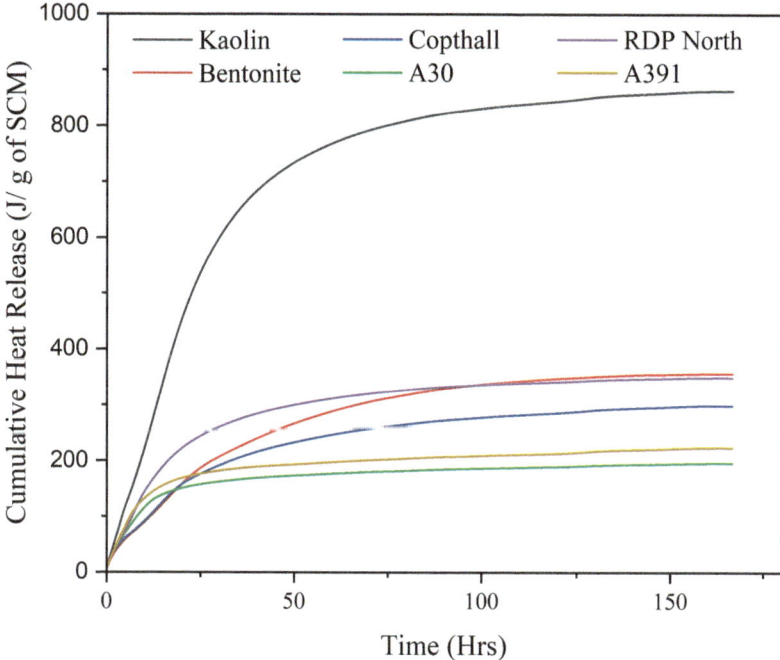

Fig. 5 Reactivity assessment of waste clays against kaolin and bentonite using R^3 cumulative heat release method

SCM is briefly discussed in literature, it acknowledges that pozzolanic activity is comparable with 30–35% kaolinite clays [29]. According to Fig. 5, kaolinite content is comparable, thus yielding similar outcomes. In addition, it has been demonstrated that low-quality kaolinitic clays and high smectitic clays can produce geopolymers with a cumulative heat release beyond 450 J/g of SCM [29]. This finding supports the notion that lesser utilized clays, such as bentonite or more readily available common clays, hold promise as viable geopolymer precursors.

According to literature [20–27], excavation clays Copthall and RDP North exhibit values within the realm of kaolin. However, this study confirms that kaolin presents a higher reactivity, with Copthall and RDP North aligning more closely with bentonite, a pure 2:1 clay. Samples A30 and A391 result in cumulative heat below the range depicted in Fig. 1. The lack of reactive minerals within A30 and A391, as confirmed through XRD and TGA, are consistent with the low R^3 test results.

An intriguing observation emerges from the near-identical heat released from bentonite and Copthall samples within the first 24 h. The identification of kaolinite and montmorillonite via XRD, coupled with similar dehydroxylation patterns at higher temperatures, indicates similar proportioning and thus could explain this similarity in reactivity. This finding corroborates Diaz's [3] and Vallina's work [29], reaffirming that montmorillonite-rich clays can be selected for use as a SCM.

3.3 *R³ Test: Bound Water*

The R^3 values obtained using the chemically bound water study are presented in Fig. 6. The strong correlation observed between R^3 procedures underscores the reliability of the methodology. Notably, kaolin demonstrating significantly higher performance than the waste excavation clays, with those aligning closely with the outcomes observed in the R^3 cumulative heat study.

Consistent with the heat release study, the performance of the kaolin sample corresponds to ranges documented in the existing literature [20–27]. Notably, the excavation waste samples from RDP North and Copthall share similarities with those derived from GGBS and PFA—materials well-known for their rich silicate and aluminate compositions. On the other hand, A30 and A391 samples align more closely with values obtained from natural and blended pozzolans, such as volcanic ash and siliceous PFA. Pure quartz, however, yields values of 1–1.2 g/100 g of dried paste [24, 25]. This reaffirms that despite a reduced reactivity, hydration reactions occur in the high quartz R^3 paste samples, albeit to a lesser degree.

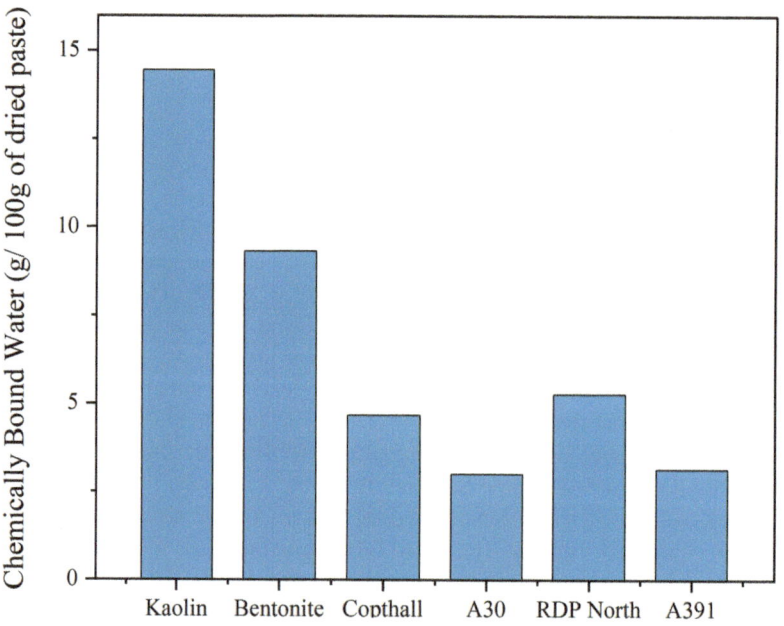

Fig. 6 Reactivity assessment of waste clays against kaolin and bentonite using R^3 bound water method

4   Conclusion and Future Work

In conclusion, this study sheds light on the reuse of excavation waste clays as SCMs. The effectiveness of R^3 tests in simulating cement hydration environments indicates the promise of common clays like Copthall and RDP North, both materials low in quartz but with sufficiently high 1:1 and 2:1 minerals, as geopolymer precursors. Achieving reactivity levels akin to high-quality kaolin proves challenging; however, attaining a reactivity comparable to pure smectitic clays indicates that common clay precursors, with appropriate mineral compositions, can yield satisfactory reactivity levels, enabling the advancement of alternative geopolymer precursors. While the presence of kaolinite significantly influences reactivity, reactivity comparable to traditional SCMs like GGBS of PFA can be achieved using waste, illite- and montmorillonite-rich clays [20, 22, 24, 29]. The advancement of common clay geopolymer precursors will not only mitigate emissions associated with earthen stabilizers but it will also alleviate demand for kaolin, PFA, GGBS while concurrently reducing the volume of waste clay directed towards landfill sites.

Particle size, though often overlooked, is a secondary factor affecting reactivity, especially in hand-processed clays where larger grain sizes may dominate. It is possible that much of the accepted material could be silt or sand-sized grains, thereby impacting pozzolanic reactions. Physical properties of the clays such as particle size distribution and fineness will therefore be studied. Furthermore, while semiquantitative analysis offers estimations, more precise techniques such as XRF and XRD Rietveld refinement are crucial for better quantification of mineralogy. This study not only highlights the potential of excavation waste clays as SCMs but also emphasizes the importance of thorough mineralogical characterization to optimize their utility in earthen applications.

Alongside compressive strength, durability is a key indicator for earthen materials. Future studies will therefore explore the implications of utilizing waste excavation clays on the mechanical and durability properties of earthen construction following their transformation into geopolymers.

References

1. Lehne, J., Preston, F.: Making Concrete Change: Innovation in Low-Carbon Cement and Concrete. Chatham House, London (2018)
2. Hanein, T., et al.: Clay calcination technology: state-of-the-art review by the RILEM TC 282-CCL. Mater. Struct. **55**(1), 3 (2022)
3. Diaz, A.A., et al.: Properties and occurrence of clay resources for use as supplementary cementitious materials: a paper of RILEM TC 282-CCL. Mater. Struct. **55**(5), 139 (2022)
4. Dhandapani, Y., et al.: Suitability of excavated London clay as a supplementary cementitious material: mineralogy and reactivity. Mater. Struct. **56**(10), 174 (2023)
5. Preethi, R.K., Venkatarama Reddy, B.V.: Experimental investigations on geopolymer stabilised compressed earth products. Constr. Build. Mater. **257**, 119563 (2020)

6. Department for Environment, Food & Rural Affairs: UK statistics on waste data. In: ENV 23 – UK Statistics on Waste. Office for National Statistics, London (2022)
7. AFNOR: XP P13-901. In: *Earth Bricks and Earth Blocks for Walls and Partitions*. AFNOR (2022)
8. Walker, P., et al.: *Rammed Earth. Design and Construction Guidelines*. BRE Bookshop, Bracknell (2005)
9. Ávila, F., Puertas, E., Gallego, R.: Characterization of the mechanical and physical properties of stabilized rammed earth: a review. Constr. Build. Mater. **325**, 126693 (2022)
10. Maniatidis, V., Walker, P.: A Review of Rammed Earth Construction. Natural Building Technology Group, Bath (2003)
11. Jaquin, P., Augarde, C.: Earth Building: History, Science and Conservation. IHS BRE Press, Watford (2012)
12. Amendment: Manual of Contract Documents for Highway Works. Series 600 Earthworks (2016)
13. Liu, Z., et al.: Feasibility study of loess stabilization with fly ash–based geopolymer. J. Mater. Civ. Eng. **28**, 04016003 (2016)
14. Meek, A.H., et al.: Alternative stabilised rammed earth materials incorporating recycled waste and industrial by-products: life cycle assessment. Constr. Build. Mater. **267**, 120997 (2021)
15. Djibo, K., et al.: Physico-mechanical performances of compressed earth blocks stabilized with calcined clay-based geopolymer. NanoWorld. **9**, S268–S273 (2023)
16. Idriss, E., et al.: Engineering and structural properties of compressed earth blocks (CEB) stabilized with a calcined clay-based alkali-activated binder. Innov. Infrastruct. Solut. **7**(2), 157 (2022)
17. Sore, S.O., et al.: Stabilization of compressed earth blocks (CEBs) by geopolymer binder based on local materials from Burkina Faso. Constr. Build. Mater. **165**, 333–345 (2018)
18. Pinheiro, V.D., et al.: Methods for evaluating pozzolanic reactivity in calcined clays: a review. Materials. **16**(13), 4778 (2023)
19. Ferraz, E., et al.: Pozzolanic activity of metakaolins by the French standard of the modified Chapelle test: a direct methology. Acta Geodyn. Geomater. **12**, 289–298 (2015)
20. Ramanathan, S., et al.: Linking reactivity test outputs to properties of cementitious pastes made with supplementary cementitious materials. Cem. Concr. Compos. **114**, 103742 (2020)
21. Snellings, R., Londoño-Zuluaga, D., Scrivener, K.: Interlaboratory test program to determine the precision of the R3 test method (ASTM C1897-20) for measuring reactivity of supplementary cementitious materials. Adv. Civ. Eng. Mater. **11**(2), 500–519 (2022)
22. Wang, Y., et al.: Reactivity of unconventional fly ashes, SCMs, and fillers: effects of sulfates, carbonates, and temperature. Adv. Civ. Eng. Mater. **11**(2), 639–657 (2022)
23. Avet, F., et al.: Development of a new rapid, relevant and reliable (R3) test method to evaluate the pozzolanic reactivity of calcined kaolinitic clays. Cem. Concr. Res. **85**, 1–11 (2016)
24. Kasaniya, M., et al.: Exploring the efficacy of emerging reactivity tests in screening pozzolanic materials. Constr. Build. Mater. **325**, 126781 (2022)
25. Li, X., et al.: Reactivity tests for supplementary cementitious materials: RILEM TC 267-TRM phase 1. Mater. Struct. **51**(6), 151 (2018)
26. Londono-Zuluaga, D., et al.: Report of RILEM TC 267-TRM phase 3: validation of the R3 reactivity test across a wide range of materials. Mater. Struct. **55**(5), 142 (2022)
27. Weise, K., Ukrainczyk, N., Koenders, E.: A mass balance approach for thermogravimetric analysis in pozzolanic reactivity R3 test and effect of drying methods. Materials. **14**(19), 5859 (2021)
28. ASTM: *ASTM C1897-20: Standard Test Methods for Measuring the Reactivity of Supplementary Cementitious Materials by Isothermal Calorimetry and Bound Water Measurements*. ASTM International, West Conshohocken (2020)
29. Vallina, D., et al.: Supplementary cementitious material based on calcined montmorillonite standards. Constr. Build. Mater. **426**, 136193 (2024)

30. Maier, M., Beuntner, N., Thienel, K.C.: Mineralogical characterization and reactivity test of common clays suitable as supplementary cementitious material. Appl. Clay Sci. **202**, 11 (2021)
31. Kanavaris, F., et al.: Suitability of excavated London Clay from tunnelling operations as a supplementary cementitious material and expanded clay aggregate. In: Proceedings of the International Conference on Calcined Clays for Sustainable Concrete (CCSC 2022) (2022)
32. British Geological Survey: Bedrock geology 1:625000. In: *BGS Geology 625k* British Geological Survey. UKRI (2023)
33. Abiodun, Y.O., Sadiq, O.M., Adeosun, S.O.: Microstructural, mechanical and pozzolanic characteristics of metakaolin-based geopolymer. Geol. Geophys. Environ. **46**(1), 57 (2020)
34. Snellings, R., Mertens, G., Elsen, J.: Supplementary cementitious materials. Rev. Mineral. Geochem. **74**(1), 211–278 (2012)

Machine Learning Integration in LCA: Addressing Data Deficiencies in Embodied Carbon Assessment

Ming Hu (ID), **Chaoli Wang** (ID), **Siavash Ghorbany** (ID), **Siyuan Yao** (ID), and **Ali Nouri**

Contents

1 Introduction

While buildings stand as a testament to human progress and fulfill our essential need for shelter, they inadvertently contribute to the unsustainable consumption of resources and a significant environmental footprint [1]. The global building sector

M. Hu (✉)
School of Architecture, University of Notre Dame, Notre Dame, United States
e-mail: mhu1@nd.edu

C. Wang · S. Yao
Department of Computer Science, University of Notre Dame, Notre Dame, United States

S. Ghorbany
University of Notre Dame, Notre Dame, IN, USA

Department of Civil and Environmental Engineering and Earth Science, University of Notre Dame, Notre Dame, United States

A. Nouri
Department of Civil and Environmental Engineering and Earth Science, University of Notre Dame, Notre Dame, United States

© The Author(s) 2025
M. Kioumarsi, B. Shafei (eds.), *The 1st International Conference on Net-Zero Built Environment*, Lecture Notes in Civil Engineering 237,
https://doi.org/10.1007/978-3-031-69626-8_78

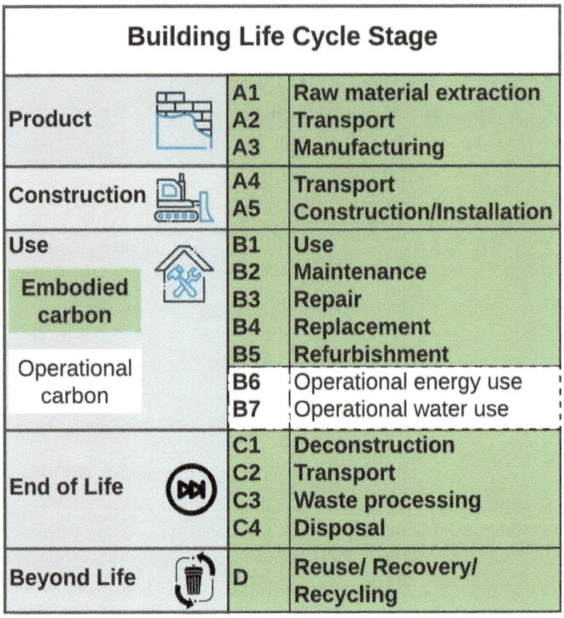

Fig. 1 Building life cycle stage and modules

Building Life Cycle Stage			
Product		A1	Raw material extraction
		A2	Transport
		A3	Manufacturing
Construction		A4	Transport
		A5	Construction/Installation
Use / Embodied carbon		B1	Use
		B2	Maintenance
		B3	Repair
		B4	Replacement
		B5	Refurbishment
Operational carbon		B6	Operational energy use
		B7	Operational water use
End of Life		C1	Deconstruction
		C2	Transport
		C3	Waste processing
		C4	Disposal
Beyond Life		D	Reuse/ Recovery/ Recycling

is responsible for consuming half of the raw materials extracted, one-third of energy use, and more than half of global electricity, leading to 37% of greenhouse gas emissions, of which 9% is attributed to the manufacturing of building materials—termed "embodied carbon" (as illustrated in green color in Fig. 1). Projections indicate a potential doubling of embodied carbon emissions under the current trajectory and a rise of up to 49% of total carbon emissions from the building and construction sector [2]. In light of these findings, it is vital to acknowledge the pivotal role of the global building sector in advancing future sustainability goals, necessitating considerable attention to achieve sustainable development.

On the other hand, such an increase in embodied carbon highlights a critical gap in sustainability practices due to the lack of research efforts [3, 4]. The 2021 Global Status Report for Buildings and Construction by the United Nations Environment Programme underscores a glaring void in research on strategies to curtail embodied carbon [5]. Unlike operational carbon, which has been extensively studied, embodied carbon reduction strategies lack comprehensive long-term evaluation, particularly in manufacturing where small- to medium-sized enterprises struggle to both quantify and mitigate their environmental impact. This gap is not only a barrier to sustainability but also hinders progress toward the ambitious goals set by the United Nation for environmental stewardship. This challenge is particularly acute for small-to medium-sized manufacturers, which may lack the necessary knowledge, resources, and expertise to measure and monitor the environmental impact of their products.

Life Cycle Assessment (LCA) is instrumental in assessing environmental impacts and promoting sustainable practices. The efficacy of LCA depends on the integrity of Life Cycle Inventory (LCI) data, which requires high-quality data. Current LCI databases exhibit significant shortcomings, thereby undermining the precision and reliability of LCA findings [6]. As shown in Fig. 1, life cycle carbon emission comprises operational carbon (grey color) and embodied carbon (green color). Stage A encompasses the product and construction phases and includes A1–A5; they are all embodied carbon [4]. Stage B is the use stage, including substages B1 through B7. B6 and B7 generate operational carbon, and the rest are associated with embodied carbon [4]. Stage C is the end-of-life stage and includes substages C1 through C4. The beyond-life stage is the D stage [6]. Within the embodied carbon, A3 is the major contributor (45%) [7] followed by C stage (40%) and B1–B5 stage (10%) [8].

Environmental Product Declarations (EPDs) are standardized documents providing detailed and quantified information about the environmental impact of products or systems, and they are essential for gauging the environmental impact of products across their life cycle (ISO, 2010). As a result, adopting EPDs information enables stakeholders to determine embodies impacts of materials by utilizing data sources that outline environmental footprint of products or systems over their life cycle. Hence, the utilization of EPDs data offers a useful means to enhance the transparency of sustainable assessment, as well as to compare the impacts of production processes. Also, it is a crucial tool in decision-making in sustainable construction [9, 10]). Consequently, the reliability and completeness of EPDs are of paramount importance in ensuring the credibility and accuracy of environmental footprint assessment.

2   Background and Motivation

The previous work, including the authors', highlights two major barriers to obtaining reliable and precise LCI data [3, 11, 12]. First, data scarcity for life stages B–D is a significant challenge, primarily due to the changes in regulatory frameworks. Compliance with standard EN 15804 + A1 is limited to the product's initial life stages (A1–A3), leaving later stages at the manufacturer's discretion [13]. The newer EN 15804 + A2 standard, effective from July 2022, extends this requirement to include stages C1–C4 and mandates an end-of-life scenario analysis (stage D), introducing more complex regulatory expectations [14]. EPDs largely reflect the earlier standard, thus perpetuating the data gaps in stages B–D. Second, the A3 stage suffers from data inaccuracy issues [15]. Accurate A3 data, crucial to evaluating the environmental impact of a product, necessitates detailed recording of all manufacturing processes. Traditional methods for collecting this data are resource-intensive and cost-prohibitive, leading to limited LCA adoption by manufacturers and affecting the generation of EPDs.

The 2021 Federal Sustainability Plan, via Executive Order 14057, established the Buy Clean initiative to prefer low-carbon construction materials in federal projects [16]. This policy aims to reduce the carbon footprint of construction while boosting the domestic industry and job market. Federal agencies, like the General Services Administration (GSA) and the Department of Defense (DoD), encourage using sustainable materials with EPDs to inform their purchasing decisions, aligning with broader environmental and sustainability objectives [16]. Other government procurement (e.g., Department of Transportation) policies may favor or require products that have demonstrated environmental performance through EPDs or similar documentation. This is part of broader efforts to ensure government purchases support environmental and sustainability goals.

This proposed framework aims to fill the knowledge and data gaps by leveraging machine learning (ML) and collaboration with manufacturers, catalyzing a significant leap toward sustainable manufacturing practices. To this extend, this research proposal aims to leverage ML techniques to address data deficiencies in the LCI stages B through D and to integrate live manufacturing data to optimize operations during the A3 life cycle stage. The goal is to reduce the embodied carbon of construction products. Anticipated outputs include a data-driven framework and tool to help small and medium-sized manufacturers optimize their production processes, leading to more sustainable construction products. This effort is expected to contribute to developing precise and credible LCA results and EPD, thereby enhancing transparency and reliability in the construction industry's environmental assessments. The specific research objectives are:

Objective 1: Employ Machine Learning for Life Cycle Inventory Data Enhancement: Utilize ML to augment LCI data from existing EPDs, targeting life cycle stages B through D.

Objective 2: Develop screening Life Cycle Assessment Tool: Develop and pilot a plug-in tool integrated into existing open access LCA tools (e.g., OpenLCA) powered by Python scripts.

3   Proposed Methodology

Figure 2 outlines the proposed research plan comprised of (a) enhanced B through D stage data imputation, (b) A3 stage live data integration, and (c) tool development. At its core, the strategy begins with establishing a robust LCI database enriched by leveraging EPD and ML techniques. The subsequent phase leverages this data through an embodied carbon assessment, aided by an integrated screening LCA tool, to refine manufacturing processes through collaboration with manufacturers leveraging research trams' long-time collaboration with a wood manufacturer (refer to Sect. 4). In the following section, the research tasks under the two research objectives are explained.

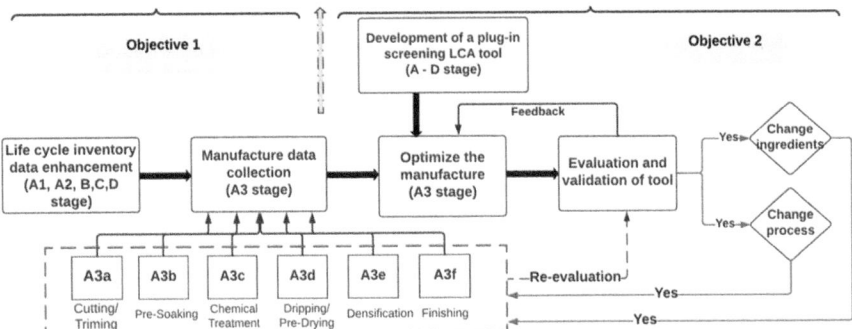

Fig. 2 Conceptual framework of proposed research plan

3.1 Employ Machine Learning for Life Cycle Inventory Data Enhancement

As of December 2023, there are over 130,000 construction product EPDs available globally. A largest number of construction product EPDs are reported to come from the United States, with over 80,000 concrete EPDs (conformed to ISO 21930) listed in the United States. Additionally, there are over 16,000 Verified EPDs (conformed to EN 15804) of construction products registered globally [10, 17]. These figures indicate a significant and growing number of EPDs for construction products, reflecting the increasing emphasis on environmental performance and transparency in the construction industry. We will acquire the existing EPD data via an automated web-scraping tool developed in Python. The available EPD data from multiple sources (e.g., EPD International Library) will be downloaded [18]. Available in an XML format, the downloaded EPDs will undergo parsing, consolidating their contents into a database for future reference. Critical information extracted from the EPDs includes all key input and output of product LCA results, such as product or service names, material classifications, the geographical context of the study, quantitative references, and the life cycle stage impact assessment results concerning specific impact categories (e.g., global warming, acidification potential, ozone depletion potential). This descriptive and categorical information will serve as inputs for the ML algorithm responsible for predicting impact assessment results in the given category.

Natural language processing (NLP) and random forest algorithms are chosen for enhancing Life Cycle Inventory (LCI) data because they provide powerful and complementary capabilities to address the specific challenges associated with LCI data.

NLP is crucial for LCI enhancement because it enables the extraction and structuring of relevant information from large volumes of unstructured text. In the context of LCA, much of the LCI data is embedded in textual product documentation, such as Environmental Product Declarations (EPDs). These documents contain detailed but unstructured data about the environmental impact of products. NLP facilitates the interpretation, categorization, and transformation of this data into a structured format that can be further analyzed and compared. Through steps such as named entity recognition and text classification, NLP can organize and prepare LCI data for quantitative analysis, which is essential for a reliable LCA.

The random forest algorithm, an ensemble tree-based ML method, was chosen due to its robustness and accuracy in dealing with complex datasets that may have nonlinear relationships and interactions between variables. Random forests are particularly effective for prediction and classification tasks, even when datasets are imperfect or incomplete, which is often the case with LCI data. They work by constructing multiple decision trees during training and outputting the mode of the classes for classification tasks. This method is less likely to overfit than individual decision trees and is known for handling large datasets with many input variables— common in LCI datasets.

By employing these methods, the framework can process and enhance the LCI data, particularly addressing the common issues of data scarcity and inaccuracy in stages B to D of the LCI. This approach is expected to improve the quality and reliability of LCI datasets, thus leading to better-informed decisions in the context of sustainable construction and environmental stewardship.

3.1.1 Data Collection and Processing

The intention is to transform the EPD data into a structured LCI database using NLP [19], thus facilitating the analysis and comparison of products based on their environmental impacts. This structured, numeric database could then be utilized for various assessments and evaluations in the field of sustainability and environmental science. As illustrated in Fig. 3, the process comprises six steps.

The initial stage is text preprocessing, where raw text data is cleaned and formatted to ensure uniformity. This step may include converting all text to a standard case, removing special characters, and correcting typos, among other things. It will be followed by the second step, named entity recognition (NER) [20], which involves identifying and classifying key information from the text into predefined categories such as product names, and geographical locations. This is crucial for extracting specific pieces of data from unstructured text. After identifying entities, the third step is text classification, categorizing descriptions into predefined

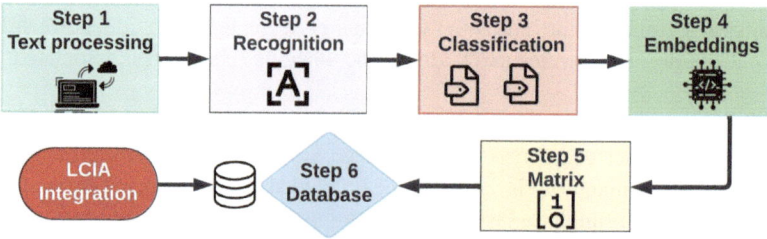

Fig. 3 Flowchart depicting the six-step process of data collection and processing

classes. For example, material descriptions from EPDs will be classified into material types or life stages of the product [21]. This categorization helps in organizing the data for subsequent analysis. Following classification, the fourth step is term frequency-inverse document frequency (TF-IDF) scores and word embeddings. These are techniques to quantify the importance of words or terms within the dataset. TF-IDF highlights words that are more relevant to a document in a corpus, while word embeddings provide a vector representation of words that captures their context and semantics. These methods help us understand the significance of terms and their associations with different categories [22]. After the processed text is classified and highlighted, the text data will then be converted into a numeric matrix in the fifth step, which is expected to be large, with thousands of rows and columns. This matrix quantifies the text data, with each row representing an item (such as an EPD) and each column representing a feature extracted from the text. In the last step, the matrices created from the above steps will be encoded and stored in a database. This structured format is essential for efficient retrieval and analysis. By integrating NLP, the EPD data can be transformed into a structured, numeric LCI database, and stored in the open-access database (repository) along with encoded information for further use in the following.

3.1.2 Quality Assessment and Data Imputation

This step involves examining data integrity, consistency, and reliability. We will employ data profiling techniques and statistical analysis to identify outliers, inconsistencies, and any discrepancies that may compromise data quality. The data profiling techniques will include but are not limited to range and frequency analysis, pattern recognition, and null/missing value analysis [23]. The statistical analysis includes a variety of techniques, such as correlation analysis and anomaly detection. Additionally, we will assess the completeness of the datasets to ensure that all necessary variables are present and properly populated. In parallel, we will evaluate the relevance of the collected datasets to the context of construction product EPDs. This evaluation is crucial to determine whether the data aligns with specific performance goals related to embodied energy and environmental impact categories. By undertaking this task, we aim to provide an informed analysis of the collected data's quality, completeness, and relevance, ensuring that only high-quality and pertinent datasets are used for further analysis.

ML models, like decision trees and clustering algorithms, will be utilized to detect missing data patterns and anomalies in the data. Clustering algorithms will help identify data points that do not conform to established patterns, while dimensionality projection can unveil latent structures and identify inconsistencies. We will also explore outlier detection methods based on ML models to flag and rectify data errors, ensuring the dataset's integrity. Additionally, we will conduct temporal analysis to understand the temporal variations in the LCI data using techniques like time series analysis to identify trends and patterns over time. By the end of this subtask, we will have a comprehensive understanding of the data deficiencies and a strategy for applying ML methods to rectify them.

3.1.3 ML-Driven Data Enhancement

In statistics, imputation refers to the process of replacing missing data with substituted values. We will develop imputation models using both traditional and advanced methods that reflect a strategic approach to address the varying complexities and nuances of missing data in datasets. The traditional methods include k-nearest neighbors (k-NN) and regression analysis. For example, the k-NN algorithm can be used for estimating continuous variables via a weighted average of the k-NN, weighted by the inverse of their distance. For regression imputation, available information for both complete and incomplete data is used to predict the missing value of a specific variable. Fitted values from the regression model are then used to impute the missing data. However, such a model does not supply uncertainty about the data. To mitigate this issue, we can leverage stochastic regression that adds the average regression variance to the regression imputations to introduce error [24]. These models will learn from LCI data patterns and relationships to impute missing values reliably.

Besides the above-mentioned traditional statistical methods, we will employ advanced imputation techniques such as autoencoders and generative adversarial networks (GANs) to enhance the imputation process. An autoencoder consists of two networks: an encoder and a decoder [25]. The encoder encodes an input data sample to a compressed representation in the latent space. The decoder decodes the latent representation back to reconstruct the data sample as close as possible. There are two kinds of autoencoders: regularized autoencoders and variational autoencoders. Variations of regularized autoencoders include sparse, denoising, and contractive autoencoders. Unlike autoencoders, GANs are explicitly set up to optimize for generative tasks [26]. A GAN has two networks: a generator (G) and a discriminator (D), which contend with each other in a zero-sum game. G learns to synthesize a generic sample from an unknown noise distribution or a fake sample with a specific condition or characteristic, while D distinguishes between instances from the true data distribution and candidates produced by G.

To deal with the problem of increased noise due to imputation, we will experiment with multiple imputation that averages the outcomes across multiple imputed datasets. In particular, multiple imputation with denoising autoencoders (MIDAS) [27] uses denoising autoencoders to learn fine-grained latent representations of the observed data, which is more accurate and efficient than traditional multiple imputation methods. Furthermore, we will investigate the use of GAN, such as the generative adversarial imputation net (GAIN) [28] for imputing missing data. GAIN uses a hint vector to ensure that D forces G to learn the desired distribution. The hint reveals partial information D about the missingness of the original sample, which is used by D to focus its attention on the imputation quality of particular components. To impute multivariate time series, we will employ a modified gate recurrent unit for imputation (GRUI) [29] in GAN to model the temporal irregularity of the temporal data by introducing a time decay vector to decrease the memory of the GRUI cell. In addition, we will apply the noised compressing and reconstructing strategy [20, 30] to improve imputation accuracy [27, 28]. We anticipate that the diverse datasets we collect from multiple sources and across different construction products will exhibit various characteristics and distributions. This renders the same network trained on one dataset less effective for the imputation of another dataset. To cope with this issue, we will apply transfer learning where a pre-trained model on one dataset can be used as the starting point for repurposing on another dataset, allowing rapid progress or improved performance when modeling the new dataset. At the end of this task, we will have (i) a set of reliable imputation models tailored for LCI data and (ii) guidelines for the application of transfer learning in the context of LCI data imputation. Hypothesis 1a will be tested; the ML-enhanced LCI dataset will reveal previously unrecognized environmental impacts during the B through D.

3.1.4 Develop a Screening Life Cycle Assessment Tool

To develop this tool, we will use open-access LCA software to integrate real-time production stage data. We will embed the developed ML models into a plug-in interface compatible with one selected open-access LCA software tool. This will include establishing interfaces for data input, user interaction, and output visualization. We will utilize the API to design an intuitive UI/UX for the plug-in that allows users to easily input data, run the ML models, and interpret the results. This plug-in tool aims to streamline the LCA process, making it more user-friendly and action-able for small manufacturers who seek to optimize their processes and reduce their environmental footprint. Figure 4 illustrates the architecture for a plug-in software tool. It delineates the integration of an open-access LCA database with an enhanced

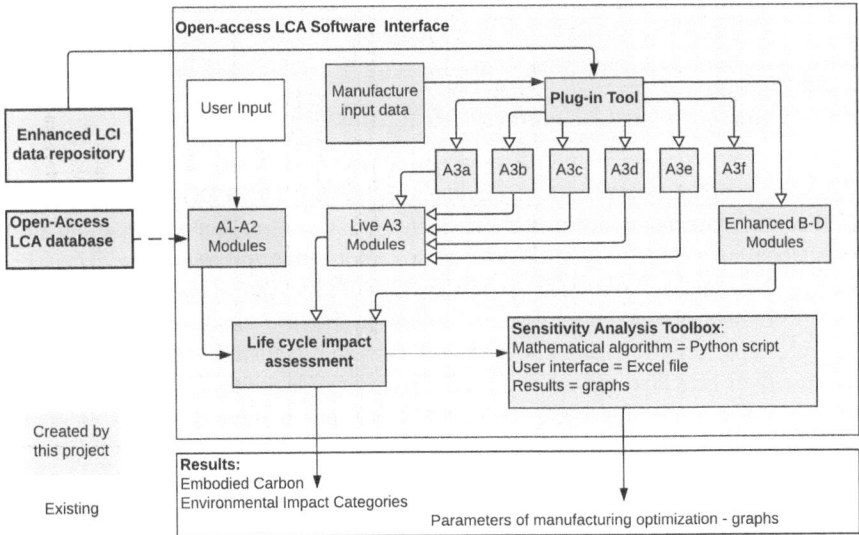

Fig. 4 Architecture of the developed plug-in tool within open-access LCA software

LCI data repository, both of which are foundational elements of the tool. The user interacts with the software through an interface that accepts user input, which informs the A1–A2 modules representing the initial stages of product life. It also interfaces with live A3 modules, which are updated in real time and reflect the manufacturing stage of the product life cycle, denoted as A3a through A3f. These modules feed into a central plug-in tool that processes manufacture input data, combining it with the LCIA to determine the environmental impact. This will focus on embodied carbon outputs along with other environmental impact categories. The tool's sensitivity analysis component, powered by a Python script and accessible through an Excel user interface, allows for the examination of different parameters and their effects on the manufacturing process. The outcome of this analysis is displayed through graphs, providing a visual representation of potential optimization pathways for reducing environmental impact.

After initiating the design of the interface, we will conduct performance tuning of the ML models within the plug-in, optimizing for speed and resource efficiency to handle the computational load of running complex LCA analyses. Such an interface will allow users to input enhanced B–D data and live A3 data collected from the manufacturing process to produce immediate and accurate outputs of embodied carbon and other environmental impacts of the product. The principle of the developed tool is schematized in Fig. 4; the modules stored in the enhanced LCI data repository are parameterized models written in Python™, which calculate chemicals and energy consumption and substance emissions at the level of each production process (e.g., A3a cutting, A3b pre-soaking). Evaluation metrics that will be used to evaluate the effectiveness of the proposed tool are (a) simulation time, (b) ease of use by non-LCA experts, (c) the understandability of the interface, (d) the clarity and visual presentation of results, and (e) the usefulness of information to the decision maker (i.e., manufacture).

3.2 Challenges and Potential Mitigation

The proposed framework's first challenge is related to EPD's availability and quality. A significant concern is the frequent omission of critical information, such as material properties (e.g., density or grammage) of the reference flow. The EPD will be downloaded from different sources in different languages (e.g., EPD Norway, EPD Italy). Even though the EPD follows a similar format, NLP models often struggle to perform well across diverse languages and specialized domains due to the specificity of language and context [31]. The mitigation strategy is to build models that are either designed to understand multiple languages or are trained explicitly on domain-specific data, which can improve performance across different languages and specialized fields [32]. The second challenge is the applicability of developed ML-enabled tools to other categorical construction products (e.g., steel beams). Over-customization can be avoided by providing the functionality in the tool to allow users to revise and define the sub-steps in A3 (e.g., A3a) and group and ungroup some sub-steps. The research group will consult and test the function with the industry collaborators (e.g., the American Institute of Steel Construction). Such flexibility and functionality will ensure the adaptability of the proposed tool to other products with different processes.

4 Preliminary Work

Since 2021, the research team has been working with a startup wood manufacturer to commercialize an innovative wood cladding panel product that has exceptional durability, strength, and toughness that is comparable to steel, yet with much lower carbon intensity. Meanwhile, compared to conventional wood, the innovative product is fire-resistant and dimensionally stable at varying temperatures, hence making it sustainable and resistant to extreme weather. To produce such high-performing products, iterative LCA has been used to optimize the manufacturing process; carbon emission and other environmental impact categories are used as measure units. The research team has conceptualized a Screen LCA framework (using a conventional LCA tool) that enables the integration of real-time A3 stage process data to produce a preliminary analysis for InventWood. Figure 5 illustrates the created framework that divides the A3 stage into sub-stages to ease input variables by InventWood. Those sub-stages are aligned with the manufacturing process

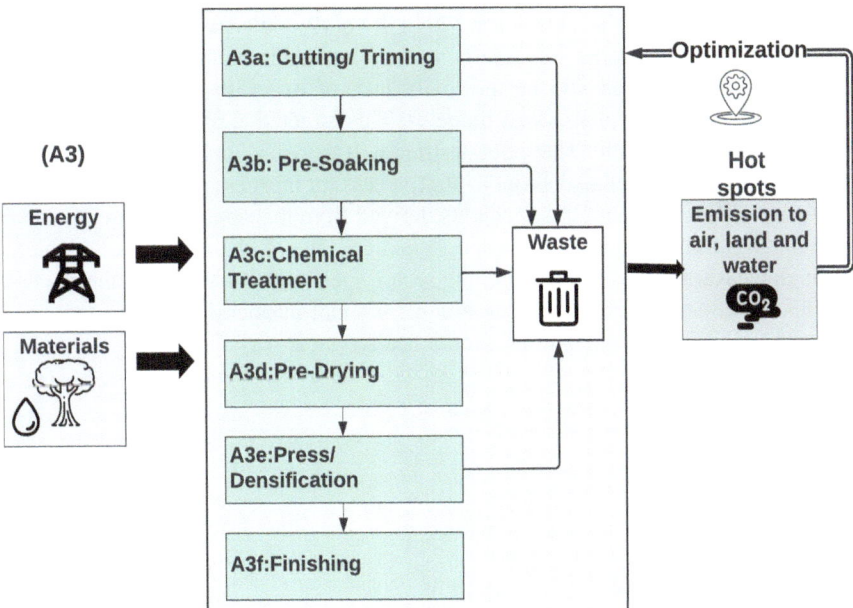

Fig. 5 Framework of integrating live A3 data

and can be easily understood by manufacturers who are not professional LCA professionals, making it easier for manufacturers to collect and input data. With such a framework, the team used the preliminary emission results to hone in on the most carbon-extensive sub-steps for improvement. For example, using the design data, we identified the top three carbon-intensive sub-processes as chemical treatments, press/densification, and pre-drying systems. Based on this, manufacturers have been working on optimizing the drying system to reduce carbon emissions and energy use. The manufacturing optimization process is ongoing, and the target time to commercialize wood panel products is three years.

5 Conclusion

The proposed research represents a fundamental methodology and data processing shift in environmental impact assessment in the building and construction sector. The proposed research is poised to bridge three critical knowledge gaps. First, by infusing ML into the screening phase of LCA, we aim to revolutionize data management, surpassing the efficiency and precision of conventional methodologies. Second, we plan to enhance the current LCI database, leveraging ML algorithms to remedy the prevalent absence of data from life stages B–D. The ML insights are designed to be transferrable, enabling the rectification of similar data voids across various sectors. Third, the research endeavors to conceive and validate a novel ML-enabled tool that seamlessly integrates with existing open-access LCA platforms. Implementing the algorithm within InventWood's manufacturing process offers a strategic showcase of its efficacy, promoting its uptake among small-to-medium manufacturers, who are normally lack of expertise and funding to hire professional life cycle assessment modeler. This targeted application will function as a live demonstration. The pilot testing will be invaluable for accruing user feedback, which will refine the tool's intuitiveness for businesses lacking specialized LCA knowledge. The empirical insights derived from this application are pivotal in formulating a set of guidelines and best practices. By validating the tool's benefits through measurable results and enabling sector-specific customizations, the research will establish a potent, accessible solution for small businesses aspiring to elevate their sustainability measures in an economically viable way.

References

1. Pomponi, F., Moncaster, A.: Embodied carbon mitigation and reduction in the built environment – What does the evidence say? J. Environ. Manag. **181**, 687–700 (2016)
2. AIA. Architecture 2030 Challenge. https://architecture2030.org/2030_challenges/2030-challenge/
3. Hu, M., Esram, N.W.: The status of embodied carbon in building practice and research in the United States: a systematic investigation. Sustainability. **13**, 12961 (2021)

4. Hamilton, D., McKechnie, J., Edgerton, E., Wilson, C.: Immersive virtual reality as a peda-gogical tool in education: a systematic literature review of quantitative learning outcomes and experimental design. J. Comput. Educ. **8**, 1–32 (2021)
5. Hamiltton, I., Rapf, O., Kockat, D. & Zuhaib, D. Global Status Report for Buildings and Construction. https://www.unep.org/resources/report/2021-global-status-report-buildings-and-construction (2021).
6. Cardoso, V.E.M., Sanhudo, L., Silvestre, J.D., Almeida, M., Costa, A.A.: Challenges in the harmonisation and digitalisation of Environmental Product Declarations for construction prod-ucts in the European context. Int. J. Life Cycle Assess. (2024). https://doi.org/10.1007/s11367-024-02279-w
7. Hu, M.: Life-cycle environmental assessment of energy-retrofit strategies on a campus scale. Build. Res. Inf. (2019). https://doi.org/10.1080/09613218.2019.1691486
8. Hu, M.: Balance between energy conservation and environmental impact: Life-cycle energy analysis and life-cycle environmental impact analysis. Energ. Buildings. (2017). https://doi.org/10.1016/j.enbuild.2017.01.076
9. Gelowitz, M. D. C., & McArthur, J. J.: Insights on environmental product declaration use from Canada's first LEED® v4 platinum commercial project. Resour. Conserv. Recycl. **136**, 436–444 (2018)
10. Waldman, B., Huang, M., Simonen, K.: Embodied carbon in construction materials: a frame-work for quantifying data quality in EPDs. Buildings Cities. **1**, 625–636 (2020)
11. Cooper, J., Fava, J., Simonen, K., Boyd, S., Baer, S.: Status of North American life cycle inventory data. J. Ind. Ecol. **16**, 287–289 (2012)
12. Dong, Y., Ng, S.T., Liu, P.: A comprehensive analysis towards benchmarking of life cycle assessment of buildings based on systematic review. Build. Environ. **204**, 108162 (2021)
13. European Standard. EN 15804:2012+A1:2013: Sustainability of construction works - Environ-mental product declarations - Core rules for the product category of construction products. (2013).
14. European Standard. EN 15804:2012+A2:2019: Sustainability of construction works - Environ-mental product declarations - Core rules for the product category of construction products. (2019).
15. Zargar, S., Yao, Y., Tu, Q.: A review of inventory modeling methods for missing data in life cycle assessment. J. Ind. Ecol. **26**, 1676–1689 (2022)
16. Office of the Federal Chief Sustainability Officer. Federal Buy Clean Initiative. https://www.sustainability.gov/buyclean/
17. ECO platform. EPD Facts & Figures. https://www.eco-platform.org/epd-facts-figures.html.
18. EPD. The International EPD System. https://www.environdec.com/about; https://www.environdec.com/library-us/global-house-of-; https://www.environdec.com/home
19. Joseph, S.R., Hlomani, H., Letsholo, K., Kaniwa, F., Sedimo, K.: Natural language processing: A review. Int. J. Res. Eng. Appl. Sci. **6**, 207–210 (2016)
20. Gemmeke, J., Cranen, B.: Missing data imputation using compressive sensing techniques for connected digit recognition. in 1–8 (IEEE, 2009).
21. Kadhim, A.I.: Survey on supervised machine learning techniques for automatic text classifica-tion. Artif. Intell. Rev. **52**, 273–292 (2019)
22. Widaningrum, I., Mustikasari, D., Arifin, R., Tsaqila, S.L., Fatmawati, D.: Algoritma Term Frequency–Inverse Document Frequency (TF-IDF) dan K-Means Clustering Untuk Menentukan Kategori Dokumen. Prosid. SISFOTEK. **6**, 145–149 (2022)
23. Hasan, M.K., et al.: Missing value imputation affects the performance of machine learning: A review and analysis of the literature (2010–2021). Informat. Med. Unlock. **27**, 100799 (2021)
24. Basu, A.P., Ebrahimi, N.: On the reliability of stochastic systems. Statist. Probab. Lett. **1**, 265–267 (1983)
25. Probst, P., Wright, M.N., Boulesteix, A.: Hyperparameters and tuning strategies for random forest. WIREs Data Min & Knowl. **9**, e1301 (2019)

26. Creswell, A., et al.: Generative adversarial networks: An overview. IEEE Signal Process. Mag. **35**, 53–65 (2018)

27. Lu, H., Perrone, G., Unpingco, J.: Multiple imputation with denoising autoencoder using metamorphic truth and imputation feedback. arXiv preprint arXiv:2002.08338. (2020)

28. Yoon, J., Jordon, J. Schaar, M.: Gain: Missing data imputation using generative adversarial nets. in 5689–5698 (PMLR, 2018).

29. Koyamparambath, A., Adibi, N., Szablewski, C., Adibi, S.A., Sonnemann, G.: Implementing artificial intelligence techniques to predict environmental impacts: Case of construction products. Sustain. For. **14**, 3699 (2022)

30. Patruno, L., et al.: A review of computational strategies for denoising and imputation of single-cell transcriptomic data. Brief. Bioinform. **22**, bbaa222 (2021)

31. Min, B., et al.: Recent advances in natural language processing via large pre-trained language models: A survey. ACM Comput. Surv. **56**, 1–40 (2023)

32. Chang, Y., et al.: A survey on evaluation of large language models. ACM Trans. Intell. Syst. Technol. (2023)

A Case Study of Sustainable Resource Management Through Reuse of Building Materials

Sunniva Baarnes and Emma Zheng Liang

Contents

1 Introduction

Nowadays, there is a noticeable shift toward putting more emphasis on reusing building elements rather than constructing new ones. This paper is dedicated to revealing the practical implications of reusing building materials from the demolition phase to construction process, as demonstrated through a case study—Grensen 9b in Oslo, Norway. With its ambitious goal of attaining BREEAM-NOR v6.0 certification at the "Very Good" level, alongside an 93% reduction in CO_2eq-emissions from building materials compared to a new build, Grensen 9b serves as a pilot study of the potential impact of practices on sustainable development within the building industry.

By scrutinizing the steps involved in identifying and preserving building elements, the study seeks to not only address the environmental consequences of reuse but also contribute to a significant reduction in the demand for using new materials, subsequently minimizing the associated carbon footprint.

S. Baarnes (✉) · E. Z. Liang
Rambøll Norge AS, Oslo, Norway
e-mail: Sunniva.Baarnes@ramboll.no; emma.liang@ramboll.no

© The Author(s) 2025
M. Kioumarsi, B. Shafei (eds.), *The 1st International Conference on Net-Zero Built Environment*, Lecture Notes in Civil Engineering 237,
https://doi.org/10.1007/978-3-031-69626-8_79

The study explores design strategies that promote future reuse, aiming to significantly reduce carbon emissions and support climate change mitigation and resource
optimization. It is part of a broader research project that investigates the feasibility
and best practices for reusing building elements. The ultimate goal is to demonstrate
the transformative potential of a sustainable material reuse platform, offering
insights applicable across the building industry.

In Sect. 2, we introduce the methodology: concept of sustainability in reusing.
Section 3 presents the case study and illustrates the findings from the case study:
practical barriers in reusing of building materials providing systematic implications
of mapping reusable building component in case study in Norway, while Sect. 4
concludes this paper.

2 Methodology: Concept of Sustainability in Reusing in Norway

2.1 Definition of Reuse in the Built Environment

2.1.1 Used Materials

Reuse in the built environment is not a new concept; it is rooted in common sense.
Before the industrialization of the building industry, building materials were the
result of hard work and not something to be wasted. The old Norwegian lumber
buildings provide a great example of design for reuse—they could be disassembled,
moved to another location, and reassembled.

As the industrial revolution progressed and an efficient global logistics ecosystem
emerged, building materials became increasingly affordable. Consequently, the
perceived value of materials no longer needed to exceed the calculated payback
time for building projects. This has led to a significant volume of resources with
plenty of technical lifespan left being treated as waste.

At the same time, our planet is undergoing a resource crisis. The extraction and
processing of raw materials have catastrophic impacts on biodiversity, untouched
lands, contributing to widespread air, water, and soil pollution, as well as significant
CO_2 emissions. Far from all environmentally certified supply chains can be trusted.

The contemporary circular economy movement aims to reintroduce common
sense into the value chain of our built environment, this time addressing these
environmental challenges.

To reuse a building material or component is to utilize it after it has already served
a function in a building or other installation. It can be used as it is or can be modified,
repaired, and refinished. It can serve its original purpose or be repurposed entirely.
Material recycling involves melting, shredding, or processing a material so that it can
be used in its shredded state, molded into a new product of the same material, or
combined with other chemicals or components to create a new product.

Fig. 1 The waste reduction pyramid shows the preferred means of reducing waste starting with reduce [1]

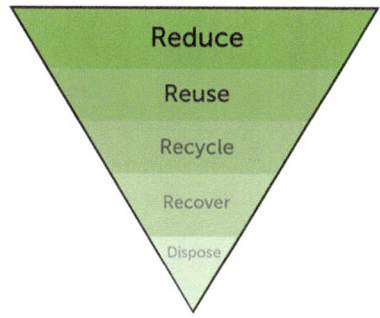

Reuse is ranked second in the waste pyramid (see Fig. 1, below waste reduction and above recycling. However, reusing is an effective way to reduce waste, and it could therefore be argued that it should be prioritized as both the first and second steps in waste and resource management schemes for building projects.

2.1.2　Design for Future Reuse

In the Guide for procuring mapping of reusable building materials [2], a material that is suitable for reuse is described with the following properties, which also describes what properties to design after for future reuse.

- Robust materials: Resistant and homogeneous materials and components that can reused for several generations of buildings.
- Flexible connections: Reversible connections between building components such as simplifies disassembly and reassembly. This can be the use of mechanical couplings, such as bolts or screws rather than gluing or welding.
- Remaining life: Building components with a long remaining life, that is high technical quality, little wear and damage.
- Volume: Large lots can yield a greater return on the effort to dismantle for reuse. Greater dividends can, for example, be in the form of greenhouse gas savings, financial savings, or possibility of turnover.
- Demand: Building components with cultural–historical value, local identity, sentimental value, high economic value or other special characteristics can increase the attractiveness of reuse.
- Environmental effect: Building components that provide large environmental savings when reused, for example, where production or transport causes large emissions or other environmental problems.
- Cost/benefit: Building components that provide cost savings when reused compared to buying a new product, that is, building components with high economic value.
- Associated drawings and documentation: Information on building components among other drawings, user-specific product information, product documentation, installation instructions, and information on building systems and maintenance.

2.2 Legislative and Regulatory Conditions for Reuse in Norway

In Norway, there have recently been made changes in legislation regulating the sale of used building materials. Effective 1 July 2021, it is no longer mandatory to re-document the CE marking of used building materials originally introduced to the market before 1 January 2014. However, this exemption does not apply to products intended for use in the main structural frame of a building. The responsibility for documentation now lies with the buyer and their consultants, who must sign off on the finished project.

The legislation changes regarding the mapping of reusable building materials, effective 1 June 2023, have greatly impacted the supply side of the market. It requires every refurbishment or teardown project over 100 m^2 or generating over 10 metric tons of waste to map any reusable building materials and compile the quantities and relevant mandatory parameters into a report. This is an essential first step toward establishing a viable market for reusable materials. Subsequent steps will involve the publication and promotion of available stock, as well as regulatory changes to set requirements for reuse in refurbishments and new building projects.

Another legislative change that could significantly benefit the reuse market is the removal of mandatory VAT on used goods sold for reuse purposes. Currently, such legislation exists for the B2C market, including typical thrift stores and secondhand furniture markets. With some adjustments, this legislation could also regulate the B2B market for reusable building materials.

2.3 Market Conditions for Reuse in Norway

2.3.1 Internal and Open Markets in the Building Industry

In the early stages of reuse within the building sector, external reuse was hindered by the requirement for redocumentation of the CE marking for many product groups whenever there was a change of ownership. This led to a common understanding that internal reuse within a single property portfolio manager was the most efficient resource management strategy.

As legislation changed without barrier of the documentation requirement, a new challenge emerged: the lack of an open two-sided marketplace to efficiently distribute the substantial volumes of building materials mapped and registered each year. Rehub.no, an online two-sided marketplace, was launched in 2021 to address this need. Although it initially had basic functionalities, its minimum viable product (MVP) format lacked the capacity to handle high volumes. A relaunch is scheduled for later in 2024, on a new platform designed to handle large volumes of data.

To facilitate the large volumes of reuse, it is crucial that each project has access to extensive databases containing materials available within the project's timeframe.

Relying solely on a database representing internal portfolios is insufficient to optimize resources across the entire building industry.

Timing is another critical factor in optimizing the reuse of building materials. Buyers require sufficient time to design with available reusable materials without incurring the added expense of intermediate storage for materials that may not yet be secured for final design. Accessing the databases of materials, a year or two in advance could secure an efficient process for the next project. Keeping internal databases off the open market until all internal use options have been exhausted, limits opportunities for other projects to utilize the materials in the portfolio. Opportunities for reuse can then be lost due to time constraints.

2.3.2 Markets Across Industries

One aspect of reuse is its ability to foster creativity and innovation. As projects seek to optimize resource utilization, reuse is creating markets across industries. Many sectors outside the building industry specialize in trading high-quality materials and have a tradition of waste management at the lower levels of the waste pyramid. The transport, offshore, and industrial sectors all represent such markets with significant potential for symbiosis with the building industry.

An example is the reuse of subway steel tracks that were originally destined for the scrap metal pathway as described in the case study Grensen 9b in Sect. 2 of this paper.

2.4 Impacts of Reuse

2.4.1 Resources

If the supply chain for any given product does not begin with derivatives of other production lines or recycled materials, it starts with the extraction of raw materials. Annually, raw materials amount to nearly 40 billion tons [3]. By 2025, it is estimated that the global volume of waste generated will reach 2.2 billion tons [4]. According to the same source, construction and demolition waste represent approximately 30% of that number.

Applying the estimated potential for reuse, which is 10% according to Norwegian reports [5], the potential for reuse is 66 million tons of waste each year. This equates to removing 66 million tons of waste from the waste stream, along with all the transport and post-consumer processes associated with these materials. Additionally, the corresponding amount of raw material extraction linked to producing 66 million tons of finished products, plus packaging, can be eliminated from the value chain annually. Altogether, this represents a powerful impact on resource conservation and waste reduction.

2.4.2 CO$_2$ Emissions

By reusing building materials, there are significant direct CO$_2$ reductions throughout the entire lifecycle of a material. The primary CO$_2$ reduction arises from the avoidance of using new materials, thus impacting the entire supply chain, theoretically from raw material extraction to delivery to the building site (A1–A4 stages). While a single project may not directly affect the volume of production, the cumulative effects must be attributed to the replacement of individual materials, creating incentive and logic in CO$_2$ accounting.

It is important to note that the new function of the used material may differ from its original use. For example, a used carpet may be reused as a sound-insulating layer under floorboards. In this case, the CO$_2$ reduction is attributed to the emissions avoided from producing a new sound-insulating layer, not the carpet itself.

Another aspect of CO$_2$ emission reduction through reuse is at the end of the material's life. By diverting the used material from the waste chain, CO$_2$ emissions from transport and end-of-life processing, such as recycling or incineration, are avoided. However, there are additional CO$_2$ emissions associated with transport, testing, storage, washing, refinishing, and the potentially shorter remaining lifespan of reused materials, which may result in an extra replacement.

Analyzing the total CO$_2$ reduction is crucial for making informed choices, particularly for significant building components. In Norway, there is a standard consensus that reused materials represent approximately 20% of the A1-A3 CO2 emissions of new materials, according to the methodology outlined in Futurebuilt's guideline [6].

2.4.3 Logistics

The impacts on logistics can be categorized into two main areas: logistics at the project or material level, and logistical impact on society as a whole.

At the project or material level, logistical impacts often necessitate additional planning, coordination, and risk management practices that are not yet standard in the building industry. Materials can be sourced from various places, including the refurbishment of the building itself, a physical marketplace, or a donor source such as another building or installation in a different industry. If the material comes from a donor source, a simplified chain of events might look like this:

1. Seller and buyer agree on the sale terms and responsibilities for the material at each stage. They also agree on a general timeframe for pickup.
2. The seller dismantles and packages the materials.
3. The logistics service and seller coordinate the timing and pickup location for the materials.
4. The logistics service picks up the materials and transports them to intermediate storage facilities.

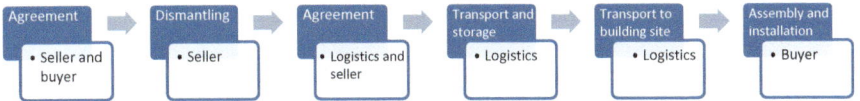

Fig. 2 Diagram of supply chain of logistics for material that comes from a donor source

5. The logistics service transports the materials from storage to the building site.
6. The materials are assembled and installed on the site (Fig. 2).

2.4.4 Waste

Waste poses a significant burden on the planet and society, despite heavy regulation in Western countries aimed at preventing pollution and logistical congestion. Many examples of western waste ending up in unregulated countries have been documented. To address this, as of 1 January 2021, Norway has implemented regulations prohibiting the export of plastics to developing countries. Furthermore, waste logistics companies are now required to document environmentally friendly treatment of all waste fractions [7].

In the building material supply chain, waste is generated during the extraction of raw materials, the production process, installation, and most significantly, at the end of life. After material reduction through design and cofunction, reuse is the single most effective way to reduce waste. The potential for waste reduction by reusing affects the whole material life cycle.

2.4.5 Economy

The transition to a circular economy is still in its infancy, but significant changes have occurred in the past 5 years regarding the actual and perceived economics of reusing building materials. In 2019, there was a prevailing belief, based on actual cases, that reuse automatically increased project costs. During this time of uncertainty, the expenses associated with research, technical testing, risk management, legal inquiries, chaotic logistics, and redesign often exceeded the costs of purchasing new materials. An example of the wide range of costs connected to reuse that could underline these perceptions, the project Kristian August gate 13, an ambitious reuse project finished in 2019, reported that reuse costs ranged from 66% more expensive and 63% saved costs compared to new materials (Excluding project design costs) [8].

With insights garnered from each reuse project over time, a clearer understanding emerges regarding the relevant costs, risks, and benefits associated with each product. Different segments of the building industry have recognized the value of developing cost-efficient processes for their respective product groups.

3   The Case Study: Grensen 9b

3.1   An Introduction to the Project

Hidden in a back alley in the heart of Oslo, Norway, lies a small office building slated for an overhaul to welcome new tenants. Grensen 9b, a 3400 m^2, six-story high refurbishment project, is owned by two pension investment funds: *Oslo pensjonsforsikring* (OPF) and *Pensjonskassen for Helseforetakene i Hovedstadsområdet (PKH)*. They have set ambitious targets for reuse in this project and have enlisted Magna Construction Management to lead the process, with Ramboll Norge AS ensuring the project reaches its ambitious sustainability goals. MAD Architects and SANE Interior Designers are behind the new design, while the rest of the team has been handpicked by the project manager. Grensen 9b aims to achieve BREEAM-NOR v. 6.0 Very Good Certification and has also set a goal to meet the FutureBuilt criteria v. 2.0 for a circular building as part of the BREEAM certification.

3.2   Reflections Around the Impacts of Reuse in This Project

The project's ambitious goal to meet the FutureBuilt criteria v 2.0 for a circular building was just the beginning, as the project has aimed to exceed these goals. As of March 2024, as the project nears its completion, the status is as follows (Fig. 3):

The total saved resources from reuse in the project are almost 50 tons. These 50 tons only represent the reuse of added materials and does not consider the reuse of the existing building which we classify as retention of the existing building.

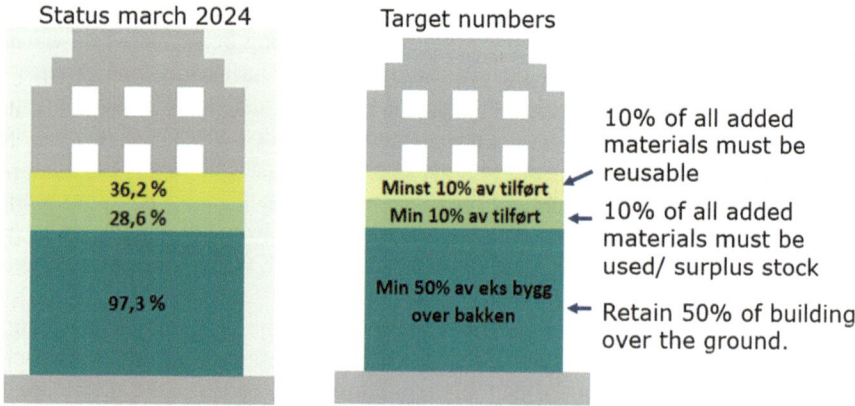

Fig. 3 Future Built goals on Veight % versus stats as of March 2024

3.2.1   CO2 Emissions

Most of the reduction in CO_2 emissions in this project comes from the reusing of the existing load-bearing structure, roof, facade, and floor slabs. This alone surpassed the goal of a 60% reduction in CO_2 emissions from building materials compared to a new office building. Therefore, the evaluation of whether to reuse a particular material or product was not solely based on its contribution to overall CO_2eq emissions but was assessed on a cost/effort versus effect basis. Factors considered included time and cost implications, potential impacts on the project timeline, adherence to Future Built weight percentage goals, and potential effects on CO_2eq emissions and resource consumption beyond this project. Throughout the project, the goal has been to develop use cases that are relevant and transferable to other projects on a larger scale, thereby creating value for the entire building industry.

3.2.2   Logistics

Grensen 9b is located at the end of a back alley in one of the busiest parts of Oslo. It is a small project with limited storage and logistical flexibility. This necessitates a well-organized building site and additional resources for handling the dismantled reusable building materials.

3.2.3   Storage

There are two options for storage: inside Grensen 9b and at the "Sirkulær ressurssentral," a storage facility dedicated to the reuse market on the outskirts of Oslo center. While there are numerous services offering various types of storage options, we chose these two for this project. However, this decision had both advantages and disadvantages:

Inside Grensen 9b:

- Challenges: The limited storage space within the small project area meant that the inventory had to be frequently moved to accommodate ongoing refurbishment activities. This resulted in time-consuming efforts and damage to the materials. Additionally, materials not packed in plastic were exposed to dust from the teardown process. While internal storage reduced costs associated with transport and external storage, the overall cost of moving materials internally exceeded the savings. In hindsight, external storage would have been a better option from the start.
- Benefits: Storage conditions inside Grensen 9b were generally dry and warm. CO_2 emissions from transport to and from intermediate external storage were saved.

External storage facility "Sirkulær ressurssentral":

- Usage: Primarily used for materials sourced externally.
- Description: The facility comprised a large tent repurposed from an outdoor construction project. The climate inside the tent mirrored the outside temperature and air humidity, although the materials were shielded from rain, snow, sun, and wind.
- Additional notes: Later into the storage period, an inflatable tent was introduced to provide partial warmth, but it was considered too late to utilize effectively. In hindsight, this decision may have been incorrect, as the dry and warm conditions of the inflatable tent would have been beneficial, particularly for wooden panels intended for use by carpenters.

3.2.4 Transport

Most of the materials sourced from Grensen 9b itself did not require transportation outside the building, except for those materials that needed cleaning at an external facility. Two examples of this are the metal tubes from the ventilation system and the glass bricks, which were picked up by the cleaning service provider and returned appearing brand new. However, materials sourced from other locations needed to be transported. In most cases, this transportation was organized using personnel and vehicles from the main contractor. For materials requiring off-site processing, such as steel beams and wooden materials, the welder or carpenter would arrange transportation to and from their facility.

3.2.5 Waste

Reusing played a significant role in reducing waste from the teardown phase of Grensen 9b. A total of 31.4 tons of building materials and components were salvaged from the teardown process and repurposed as high-value building materials in the finished product. Additionally, 32.5 tons of building materials and components were diverted from waste streams of external donor projects and reused in Grensen 9b, further contributing to waste reduction and the utilization of high-value building materials and components.

3.2.6 Economy

Although reuse has become a significant aspect of Grensen 9b's identity, the project stands out for achieving this level of reuse without incurring any additional costs. In fact, the cost per square meter is lower than that of other similar refurbishment projects in central Oslo. By reallocating funds from the purchase of new materials to the logistics and processing of used materials, the overall project budget has remained unaffected. Smart decision-making regarding what to reuse and where has been a key factor in this success. For example, the team opted to reuse structural components only in the secondary structural system, such as staircases. Additionally, the project benefited from a highly competent team that understood how to manage the risks associated with the reuse of each item.

3.3 Practical Implications

Reuse is still in its early stage of development in the building industry, and for many participants in Grensen 9b, it marked their first encounter with it. While there are some good examples of best practices and pitfalls in reuse projects, there is still a lack of established infrastructure or procedures to rely upon. Moreover, building projects vary greatly in their contractual organization, which can significantly impact the reuse process. In Grensen 9b, the contractual structure involved client-managed subcontracts, with *Magna Construction Management* hired to represent the developer throughout the project phases. Having a consistent project team from the early stages through to the design phase and even handover proved to be impactful, particularly in considering the long-term implications of decisions regarding reuse made in the early stages.

Creating a consistent procedure for reuse was not initially a priority for this project. Instead, each member of the design team, management, and subcontractors sought opportunities for reuse within their own disciplines and networks. Despite this decentralized approach, driven by individual desires to make reuse work, the project achieved more than anticipated. Moving forward, it is our responsibility to share both successful strategies and lessons learned from Grensen 9b, and to outline general procedures that emerged from the project.

4 Discussion: Analysis and Findings

4.1 Analysis and Findings of the Impacts of Reuse in the Case Study

4.1.1 CO2 Emissions

Grensen 9b is close to, but not yet completed and therefore the final CO_2-calculations have not been completed, preliminary figures indicate total CO_2 emissions of 97,293 kg CO_2eq within the system boundaries set by *Tek17* [9]. These calculations encompass A1–A3 (production of material), A4 (transport to the building site), and B4 (replacement throughout the building's lifetime, set at 60 years). With a building area of 3400 m2, this equates to total CO_2eq emissions from building materials of 0.48 kg CO_2-e/m2/year. Comparatively, the reference number provided by BREEAM-NOR V6.0 is 6.8 CO_2-e/m2/year. As of November 2023, the project has achieved a 93% reduction in CO_2 emissions. These calculations were conducted by Ramboll Norway.

The primary driver of this reduction is the retention of almost all structural elements and the building envelope. While reuse has contributed to CO_2 reductions, it has not been the sole determining factor in achieving the project's goal of a 60% CO_2 reduction for Grensen 9b. For this project, the CO_2 calculation tool OneClick LCA was utilized. This program employs the same methodology as the FutureBuilt methodology for calculating emissions from reused materials, attributing 20% of emissions from A1 to A3 of the replaced new material if it is replaced by the same type of material.

4.1.2   Logistics

One of the key insights gained from this project was the recognition that storing materials within the building being renovated, with the aim of saving costs on external storage, ultimately proved to be more expensive. This was primarily due to the substantial resources required to continually move materials around within the building while the renovation progressed from one area to another.

It is estimated that approximately 400,000 NOK was spent on reallocating all the materials multiple times during the construction phase. Moreover, this internal storage approach resulted in damage to many materials and components, further adding to the project's costs and inefficiencies.

4.1.3   Waste

As the construction phase nears its completion, the project is compliant with Norwegian regulations, which requires that 70% of all waste must be prepared for reuse or material recycling. However, it is important to note that some final reports from the renovation company are still pending, and these reports will provide additional clarity on the project's waste management efforts. The project is scheduled for completion in May 2024 (Fig. 4).

4.1.4   Economy

For this project, the goal was to achieve as much reuse as possible without impacting the project's bottom line. The final calculation for Grensen 9b has not yet been completed as the project is ending in May 2024. So far, reuse has not increased the project's budgeted costs.

Areas where costs associated with reuse could have been reduced include the following:

Storage: Utilizing an external facility for storage could have reduced costs associated with reallocating materials internally multiple times.

Quality control of purchased used materials: Ensuring the quality of purchased used materials, such as carpet tiles, required more quality control measures beyond checking samples from the stock.

Fig. 4 Distribution of waste in Grensen 9b for demolition and construction phase

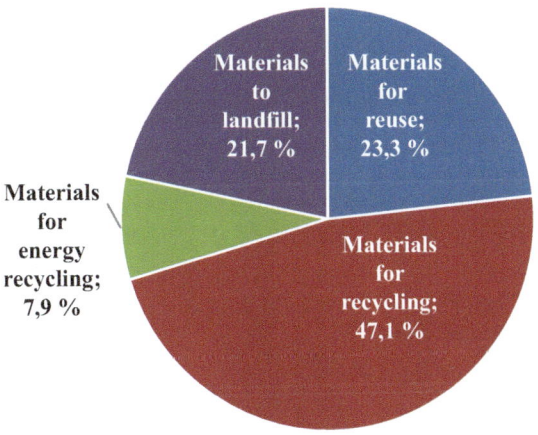

4.2   Conclusion: Lessons of Experience

Through analysis of the reuse in the building industry and case study such as Grensen 9b, it can be concluded that reuse of building materials, although slowly, is headed in the right direction. Grensen 9b provides a much-needed example of reuse being affordable and not impacting the bottom line. There are still many logistical challenges and risks to be managed on a commercial level, but with the right political and legislative incentives the building industry is eager to implement more reuse into our projects. A critical gatekeeper for higher levels of reuse to become plausible in the future is a well-stocked, functioning, two-sided marketplace for national exchange of reusable building materials and local physical and affordable storage across the country.

References

1. https://www.re-tek.co.uk/re-tek-news/importance-of-the-waste-hierarchy/
2. Statsbygg og Grønn Byggallianse. Bestilling av ombrukskartlegging – slik gjør du det v. 2.0 (2023)
3. Luca Valentini. Sustainable sourcing of raw materials for construction: from the earth to the moon and beyond. (2022)
4. Daniel Hoornweg and Perinaz Bhada-Tata. What a waste a global review of solid waste management (2012)
5. Anne Sigrid Nordby. Utredning av barrierer og muligheter for ombruk av byggematerialer og tekniske installasjoner i bygg (dfo.no) (2018)
6. Erik Resch. FutureBuilt Zero—Materialer og Energi Metodebeskrivelse (2021)
7. Miljøverndepartementet. Forskrift om eksport og import av farlig avfall (1990)
8. Anne Sigrid Nordby. KA13 Erfaringsrapport ombruk Rev1. (2020)
9. Direktoratet for byggkvalitet. Byggteknisk forskrift (2017)

Lecture Notes in Civil Engineering

Volume 237

Series Editors

Marco di Prisco, Politecnico di Milano, Milano, Italy

Sheng-Hong Chen, School of Water Resources and Hydropower Engineering, Wuhan University, Wuhan, China

Ioannis Vayas, Institute of Steel Structures, National Technical University of Athens, Athens, Greece

Sanjay Kumar Shukla, School of Engineering, Edith Cowan University, Joondalup, Australia

Anuj Sharma, Iowa State University, Ames, USA

Nagesh Kumar, Department of Civil Engineering, Indian Institute of Science Bangalore, Bengaluru, India

Chien Ming Wang, School of Civil Engineering, The University of Queensland, Brisbane, Australia

Zhen-Dong Cui, China University of Mining and Technology, Xuzhou, China

Xinzheng Lu, Department of Civil Engineering, Tsinghua University, Beijing, China

Lecture Notes in Civil Engineering (LNCE) publishes the latest developments in Civil Engineering—quickly, informally and in top quality. Though original research reported in proceedings and post-proceedings represents the core of LNCE, edited volumes of exceptionally high quality and interest may also be considered for publication. Volumes published in LNCE embrace all aspects and subfields of, as well as new challenges in, Civil Engineering. Topics in the series include:

- Construction and Structural Mechanics
- Building Materials
- Concrete, Steel and Timber Structures
- Geotechnical Engineering
- Earthquake Engineering
- Coastal Engineering
- Ocean and Offshore Engineering; Ships and Floating Structures
- Hydraulics, Hydrology and Water Resources Engineering
- Environmental Engineering and Sustainability
- Structural Health and Monitoring
- Surveying and Geographical Information Systems
- Indoor Environments
- Transportation and Traffic
- Risk Analysis
- Safety and Security

To submit a proposal or request further information, please contact the appropriate Springer Editor:

– Pierpaolo Riva at pierpaolo.riva@springer.com (Europe and Americas);
– Swati Meherishi at swati.meherishi@springer.com (Asia—except China, Australia, and New Zealand);
– Wayne Hu at wayne.hu@springer.com (China).

All books in the series now indexed by Scopus and EI Compendex database!

Mahdi Kioumarsi • Behrouz Shafei

Editors

The 1st International Conference on Net-Zero Built Environment

Innovations in Materials, Structures, and Management Practices

 Springer

Editors
Mahdi Kioumarsi
Department of Built Environment
Oslo Metropolitan University
Oslo, Norway

Behrouz Shafei
Department of Civil, Construction,
and Environmental Engineering
Iowa State University
Ames, IA, USA

The 1st International Conference on Net-Zero Built Environment: Innovations in Materials, Structures, and Management Practices (NETZ), netz, netz 2024, NETZ, 1, Oslo, Norway (2024) 6 19 (2024) 6 21. https://netzfuture.com/conference/

ISSN 2366-2557 ISSN 2366-2565 (electronic)
Lecture Notes in Civil Engineering
ISBN 978-3-031-69625-1 ISBN 978-3-031-69626-8 (eBook)
https://doi.org/10.1007/978-3-031-69626-8

This Springer imprint is published by the registered company Springer Nature Switzerland AG
The registered company address is: Gewerbestrasse 11, 6330 Cham, Switzerland

If disposing of this product, please recycle the paper.

Preface

The *1st International Conference on Net-Zero Built Environment: Innovations in Materials, Structures, and Management Practices* was successfully held on June 19–21, 2024, in Oslo, Norway. This conference was part of the *Net-Zero Future* project sponsored by the Research Council of Norway and the Norwegian Directorate for Higher Education and Skills. Through the referenced project, a unique international alliance was formed among Norway, the United States, Germany, South Africa, and India to conduct collaborative research and educational activities toward achieving a net-zero built environment, capitalizing on mutual interests and the diversity of practices. This alliance is in line with sustainable development goals, such as quality education, industry and innovation, sustainable cities, climate action, and partnerships.

The pressing global challenges associated with the degradation of the built environment due to mounting stressors and, on the other hand, the environmental footprint of growth in the built environment serving communities have necessitated a profound transformation in how we design and develop our infrastructure. Civil engineering is the primary field responsible for infrastructural development and is at the forefront of this transformation. The conference series on Net-Zero Built Environment: Innovations in Materials, Structures, and Management Practices has been dedicated to exploring the concept of net-zero within the realm of civil engineering and a variety of relevant domains, emphasizing the critical importance of achieving a net-zero carbon footprint in our built environment.

Net-zero in the built environment refers to the knowledge and practice of delivering, maintaining, and managing civil infrastructures that produce zero net carbon emissions throughout their lifecycle. This ambitious goal requires a holistic effort, encompassing methodological approaches to design, construction, operation, and eventual decommissioning. This motivated us to establish three main themes for this conference series, capturing innovations in materials, structures, and management practices. The first conference of this conference series attracted a large group of participants from the academic, industry, and public sectors, presenting the latest developments in each of the identified main themes.

After a rigorous peer-review process performed by the conference's international scientific committee members and other expert reviewers, 158 full-text papers have been selected to be included in this book. The selected papers represent a diverse group of authors and research groups from around the globe, providing original perspectives and insights into how we can pave the way toward a net-zero built environment. Converging on this ultimate goal, the selected papers offer the latest advances in (i) new materials and manufacturing processes for zero carbon footprint, (ii) robotic construction technologies for minimum formwork and on-site activities, (iii) novel structural designs and details for optimal performance with the least materials, (iv) advanced condition assessment and health monitoring strategies, and (v) innovative life-cycle analyses and civil infrastructure management strategies.

We recognize that achieving net zero is not merely a development goal but a moral imperative to ensure a sustainable future for all. Reducing carbon emissions can help stabilize temperature patterns, decrease the frequency and severity of extreme weather events, and protect vulnerable ecosystems. All who contributed to this book share the same passion to positively impact the communities around the world. Through the dissemination of the latest findings and innovations, this book's main themes and individual chapters directly contribute to the net-zero domain. By equipping scientists, engineers, policymakers, and the general public with relevant knowledge and expertise, we hope to collectively drive a global transition toward a net-zero future.

Oslo, Norway   Mahdi Kioumarsi
Ames, IA, USA   Behrouz Shafei

Contents of Volume II

Contents of Volume I

Circularity Index and Benchmarks for Buildings: A Novel and Transferable Approach to Evaluating the Circularity Performance Tested in a Norwegian Context

Marianne Kjendseth Wiik ⓘ, Freja Nygaard Rasmussen ⓘ,
Shabnam Homaei ⓘ, Kristin Fjellheim ⓘ, Anne Sigrid Nordby ⓘ,
and Reidun Aasen Vadseth

Contents

1 Introduction

The construction and real estate sector's activities are highly resource intensive. In Europe, these activities account for more than one third of annual raw material demand [1] and generate almost 40% of all waste [2]. The introduction of policies to

M. K. Wiik (✉) · S. Homaei · K. Fjellheim
SINTEF, Oslo, Norway
e-mail: marianne.wiik@sintef.no

F. N. Rasmussen
NTNU, Trondheim, Norway

A. S. Nordby
Gjenbrukbar AS, Hvalstad, Akershus, Norway

R. A. Vadseth
FutureBuilt, Trondheim, Norway

© The Author(s) 2025
M. Kioumarsi, B. Shafei (eds.), *The 1st International Conference on Net-Zero Built Environment*, Lecture Notes in Civil Engineering 237,
https://doi.org/10.1007/978-3-031-69626-8_80

close resource flows within the economy holds the potential of avoiding greenhouse gas (GHG) emissions and reducing other environmental impacts [3]. However, most construction waste is landfilled, energy recovered, or downcycled at a low quality level, much owing to vast amounts of mineral materials being crushed and used for low-value purposes such as backfill [3]. Thus, there is a need to establish systems and practices for monitoring, measuring, and increasing circularity in relation to building and construction.

Several approaches have been developed to quantify the circularity potential in buildings [4]. Some of these are characterized by a broad scope of evaluation covering various circularity aspects in specification, design, and management [4, 5]. Others are strictly material- and component-oriented, based on life cycle assessment methods [6], or allowing for integration into digital design tools [7]. Additionally, a diversity of mass-based assessment systems is available via green building certification systems such as BREEAM and DGNB [8, 9]. However, there is a lack of circularity benchmarking systems to support the evaluation of performance. Reference systems for comparison are missing, meaning that circular buildings are mostly compared with alternatives of their own design or other exemplary buildings.

This article addresses these research gaps with a Norwegian example, by identifying current and future references for circularity performance of buildings and testing this reference in the FutureBuilt Circularity Index system. Our approach fills a gap in the practical planning and design of buildings that meets the need of implementing ambitious circularity policies.

2   Background

FutureBuilt is a Norwegian green building innovation programme. A FutureBuilt project has to fulfil a set of mandatory criteria (urban environment and architecture, social sustainability, innovation, GHG emissions [10], and transport) and elective criteria (plus house, circularity, biodiversity and storm water, landscape, and plastic use). This article focuses on the elective criteria, FutureBuilt's Circularity Index system.

The FutureBuilt Circularity Index system rates a building's circularity performance between 0% and 100% [11]. The indexing system discerns between building materials and fill masses used during construction ('Present—Building' and 'Present—Fill Masses') and building materials available at the end of life of a building ('Future—Building'), see Fig. 1, displaying results of a case study building that is an exemplary rehabilitation project with a high focus on circularity. Each of these phases are broken down into a set of categories: 'Conserved', 'Reused', 'Surplus', 'Recycled', 'New', 'Reusability', 'Recyclability', or 'Waste'.

A normative weighting system is introduced to emphasize the importance of reusing materials at the highest level of functional quality. Here, the present time is weighted above the future since it is a prerogative to reduce environmental impacts

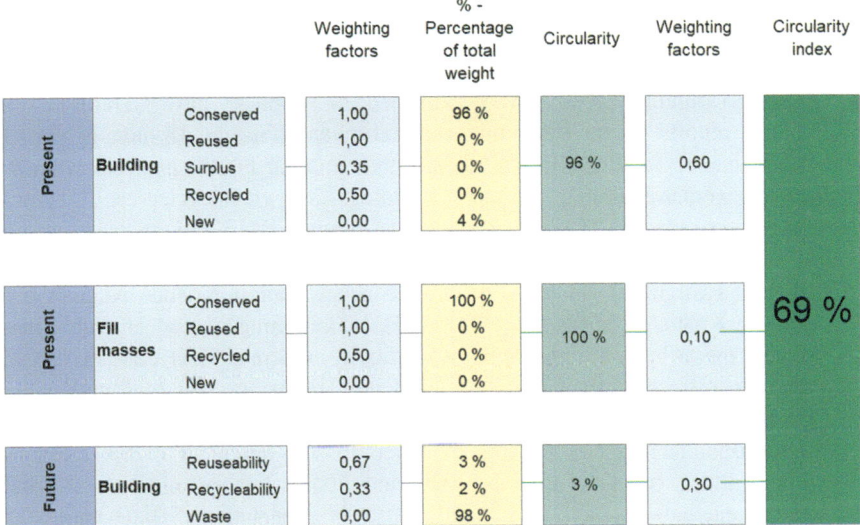

Fig. 1 Calculated circularity index based on Futurebuilt's criteria for the case study building Kristian Augusts gate 23 (KA23)

now and there is a higher uncertainty connected to the future. The waste hierarchy pyramid is used to prioritize measures. Measures connected to the building are prioritized above measures relating to fill masses since measures relating to fill masses are regarded as an easy measure to implement. Therefore, 'Conserved' and 'Reused' have a higher weighting of 1 compared to 'Reusability' at 0.67, 'Recycled' at 0.5, 'Surplus' at 0.35, 'Recyclability' at 0.33, and 'New/Waste' at 0. In addition, the importance of current resource circularity ('Present—Building' and 'Present—Fill Masses') versus circularity at the building's end of life ('Future—Building') is also weighted 0.6, 0.1, and 0.3, respectively. It should be noted that the rating system is subjective to the choice of weighting factors and gives incentives according to timeline and the waste hierarchy pyramid. FutureBuilt's Circularity Index performance requirements start at 50% in 2020 and increase incrementally towards 95% in 2050. However, FutureBuilt wishes to have an empirical reference baseline in which they can compare different building projects to. This paper documents the methods and results used for establishing this reference circularity index baseline.

3 Method

A mixed methods approach is used to establish an empirical reference circularity index baseline for Norwegian buildings based on the FutureBuilt Circularity Index method. Scenarios for future circularity towards 2050 are also developed. Data is gathered from national statistics [12–19], GHG emission accounting reports, building material inventories, pre demolition audit (PDA) reports, environmental product declarations (EPDs) [20], and industry interviews. Data is gathered from industry actors during the period 2020–2023.

In all, 83 building projects were identified, of which 47 have GHG emission accounting reports, 28 have building material inventories, and 18 have PDAs. Of these projects, 43 were detailed sufficiently to be included in the analysis, whereby 16 projects were used to map 'Present—Building—Reused/Recycled/New', 11 projects were used to map 'Present—Building—Conserved', 15 projects were used to map 'Future—Building'. In addition, four building projects have tested the FutureBuilt's Circularity Index as proof of concept, namely Kristian Augusts gate 13 (KA13), Kristian Augusts gate 23 (KA23), Nedre Sem gård, and Tradlab. These building projects have a high focus on circularity, whereby KA13 and KA23 are rehabilitated offices, whilst Nedre Sem gård and Tradlab are new buildings with a high focus on circularity.

In addition, three scenarios are developed for the trajectory of the reference circularity index baseline towards 2030 and 2050, namely, business-as-usual, European and national ambitions, and industry's expectations. The business-as-usual scenario takes the results from the reference circularity index baseline for 2020–2023 and projects results with a 1% increase from year to year towards 2050. This correlates with FutureBuilt ZEROs 1% annual increase in technological developments [10]. The European and national ambitions scenario is based on European and national legislation, goals, and ambitions for circular buildings, and includes the following:

- 2020: EU waste directive with minimum requirements on 70% waste sorting [21]
- 2023: Norwegian building code minimum requirements on 70% waste sorting [22]
- 2023: Norwegian building code requirement to PDA report [22]
- 2024: EU's Circular Economy Action Plan [23]
- 2027: EU taxonomy and Level(s) introduced for all new buildings [24]
- 2030: Political will to introduce restrictions on demolition of buildings [25]
- 2040: Percentage of building stock that is conserved increases to 50% [26]
- 2050: Full circular economy [3]

According to national statistics, Norway has around 4,300,000 existing buildings. Each year there is around a 2% change in this building stock whereby around 13,000 buildings are demolished (ca. 14%), 80,000 are newly built (ca. 84%), and 1400 are refurbished (ca. 2%) in Norway [16]. 'Present—Building—Conserved' is expected to increase from 2% in 2020–2023 to 16% in 2030. 'Present—Building—Reused' is based on an expected 5% increase by 2050, whilst 'Present—Building—Recycled' is based on an expected 35% increase by 2050. There is currently no legislation, goals, or ambitions relating to 'Present —Fill Masses', so this part of the scenario is based on responses from interviews. For 'Future—Building', a full circular economy is expected by 2050; however, there is no agreed upon definition of what a full circular economy is, therefore a moderate approach in line with FutureBuilt has been adopted at 95%. Results for years between 2020, 2030, 2040, and 2050 are evenly distributed to create a trajectory.

The industry's expectations scenario is based on interviews with actors from the construction industry on future expectations relating to the development of circularity in the building industry. The interview objects were asked to base their expectations on their company's strategy and goals, as well as on the trends regarding circularity in their forthcoming construction projects. Based on this they were asked to give a value to each element of the circularity index (Conserved/Reused/Recycled/Reusability/Recyclability).

The scope of data collection was limited to the production phase (A1–A3 according to EN 15978 [27]) and to the following building elements: groundworks and foundations, superstructure, outer walls, inner walls, floors, roofs, balconies, and stairs, since these are the life cycle modules and buildings parts most often reported in GHG emission accounting reports, building material inventories, and PDAs. 'Present—Building—Surplus' is not included in the analysis and is therefore set to zero. The analysis is limited to a mass-based assessment.

In some cases, building material quantities had to be converted from meters, square meter, or cubic meter to kilogrammes. In other cases, it was necessary to ascertain the recycled content of common building materials, for example 90% for steel. Data gathered from 2020 to 2023 is averaged instead of choosing a representative year.

4 Results

According to national statistics, Norway has around 4,300,000 existing buildings. Annually, ca. 13,000 buildings are scheduled for demolition, approximately 80,000 new constructions and roughly 1400 existing buildings undergo conversion. This indicates a 2% annual change in the country's building stock, with approximately 84% being new constructions, 14% being demolished buildings, and 2% being rehabilitated buildings. Considering these national statistics, the FutureBuilt Circularity Index focuses on annual change and the current industry standard for 'Present—Building—Conserved' is set at 2%.

Figure 2 illustrates the outcomes of reviewing material inventory lists extracted from GHG emission reports during the data collection and mapping of building projects. This mapping exercise includes a small sample of buildings (43 projects) that may not be representative of the national building stock and highlights an area for further research. The results indicate the distribution of conserved, reused, recycled, and new building materials used in both new (a) and rehabilitated (b) buildings. For new buildings, approximately 96% comprise new building materials, 4% contain recycled content, with no instances of reuse. In contrast, rehabilitated buildings consist of approximately 78% conserved building materials, 20% new building materials, 2% reused building materials, and 0.4% of building materials with recycled content. The current industry standard for 'Present—Building—

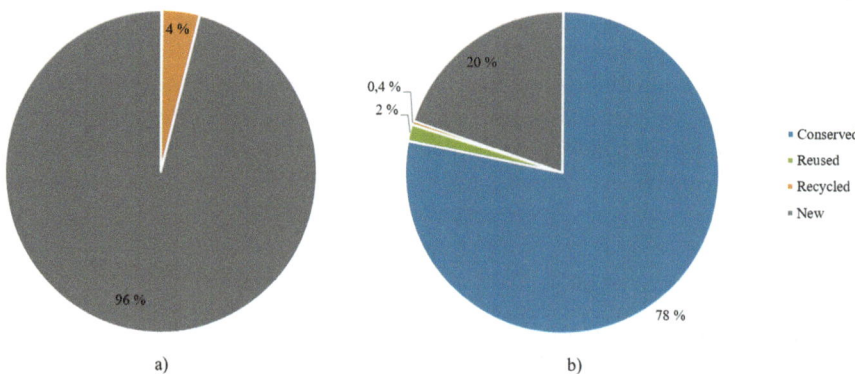

Fig. 2 Share of circularity (Conserved, Reused, Recycled, or New) for materials in new (**a**) and rehabilitated buildings (**b**)

Conserved/Reused/Recycled/New' is therefore weighted 2%/0%/4%/94% accordingly. Interviews with the construction industry revealed that 'Present—Fill Masses' are currently minimally conserved, reused, or recycled. This limited utilization aligns with statistics from Statistics Norway [18] and example projects [28]. For this reason, the current industry standard for 'Present—Fill Masses' is set at 100% new.

For 'Future—Buildings—Reusability', our examination revealed that none of the 15 PDA reports focus on quantifying reusability according to weight. In addition, there is no guarantee that today's materials and components will fulfil the same technical or regulatory requirements in the future. There is a notable variation in the system boundaries used across the different reports. However, it is evident that load-bearing structures, windows, and doors, fixed and loose fixtures, and technical equipment are the most frequently reported building components for reusability, as these are the building components that are typically most suited to reuse. Considering these findings, the current industry standard for 'Future—Buildings—Reusability' has been set at 0%. Statistics show that the average material recycling rate in the construction industry between 2013 and 2021 was 47% (excluding energy recovery). This has been used to calculate the current industry standard for 'Future—Building—Recyclability'. Taking these assumptions into account and utilizing the FutureBuilt Circularity Index system, the reference baseline circularity for Norwegian building projects is around 7% in 2020–2023 (see Fig. 3). The three future scenarios described in the methods section have been used to project a future scenario range from 35% to 80% circularity in 2050. In addition, FutureBuilt Circular (green line) shows the best practice criteria for FutureBuilt projects. Figure 3 also shows the results of four prestigious building projects that have a high focus on circularity and have tested the FutureBuilt Circularity Index. They have building circularity results ranging from 44% to 69%.

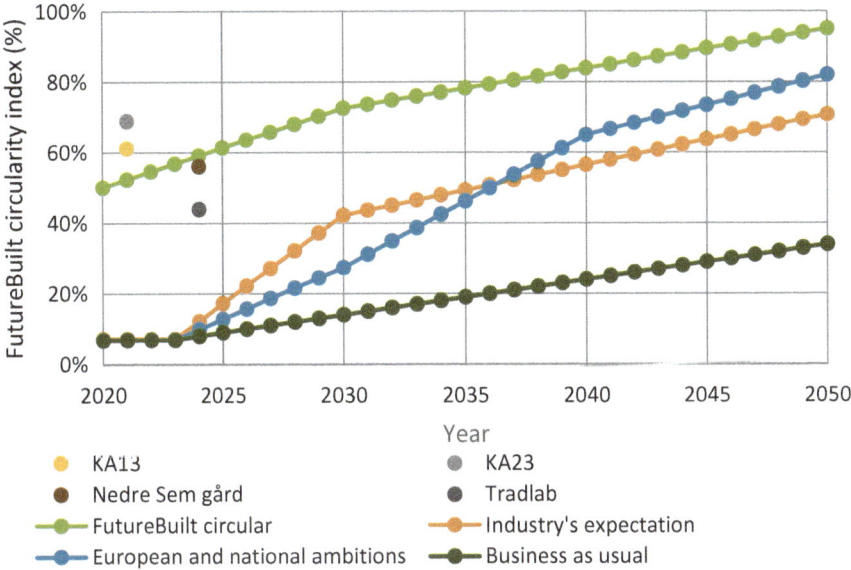

Fig. 3 Circularity projections towards 2050 in Norwegian buildings

5 Discussion

The results show that the reference baseline circularity for Norwegian buildings is around 7% in 2020–2023 when using the FutureBuilt Circularity Index, and that future projects are expected to vary from 35% to 80% circularity in 2050. The four projects that have calculated building circularity according to FutureBuilt's Circularity Index have achieved circularity according to FutureBuilt's definition in 2021 and have come close in 2023. The results show that it may be easier for rehabilitation projects (KA13 and KA23) to achieve a higher level of circularity than new builds (Nedre Sem Gård and Tradlab) since the weighty existing foundations and superstructure can be conserved and conserved/reused components score highly in the circularity index. This result implies that the normative weighting system functions according to intention, as rehabilitation is favoured over new construction. The FutureBuilt Circularity projections will become more ambitious to achieve with each passing year, and the building projects will therefore require further circularity measures to achieve circularity projections in the future. Regardless, all projects have achieved higher scores than those in the reference baseline and business-as-usual scenario.

The proposed current and future references for circularity performance of buildings are based on some assumptions and limitations. Firstly, there is a lack of harmonized definitions and methods. Secondly, the scope is limited to the production phase and main building elements since these parts are most often reported. Thirdly, the data collection period is from 2020 to 2023, and not just a single year, to increase the amount of project data that could be collected. This is because there are few Norwegian projects that document circularity, and recent years have seen large changes due to the pandemic, increased construction activity, increases and decreases in building prices, increased material prices, difficulties with importing building materials due to the war in Ukraine, inflation, as well as increased interest rates. Finally, a large range of varying data sources have been used dependent on the kind of circularity data that was available at the time of study. Therefore, some data may be aggregated, missing, or overlapping. All these factors increase the level of uncertainty in the results.

In addition, the assumed future use of materials, reusability, and recyclability is based on scenarios for the current situation and does not necessarily account for future developments. This can give a lower score on the circularity index than if technological developments are accounted for. On the contrary, when evaluating the reusability of products, the reuse of materials may be technologically possible, but the market and industry may not be organized in such a way to make reuse economically competitive or feasible. Similarly, reusability and recyclability of products needs to be better defined as there is a large disparity in materials sent to recycling, materials that are de-facto recycled, the actual recycled content in new materials, and the amount of recycled content that can potentially be used in new products without affecting the technical performance characteristics of the product. There is also a question as to whether the circularity index should distinguish between internal and external reuse. For example, if building materials are reused within the same building project or if a building material is transferred from one building to another building project. Internal reuse is preferable over external reuse since it will most likely require less processing and transport for reuse, thus reducing GHG emissions. Internal reuse means that less resources are required for storing materials, and it is usually easier to plan for reuse.

PDA reports play an important role in advancing sustainable practices within the construction industry. However, existing PDA reports have significant limitations that hinder their efficient implementation. Consequently, establishing standardized data formats and centralized repositories for building material information could facilitate PDA efforts. The complexity of building materials further intensifies the necessity for standardized reports, including essential details such as material types, quantities, condition, and location within projects. The absence of standardized reporting formats leads to discrepancies in data collection, organization, and presentation, making it difficult for stakeholders to compare, evaluate, and effectively communicate PDA opportunities and challenges.

Future work for the circularity index could focus on harmonizing definitions as well as reporting on the scope, methods, and assumptions employed. For example, what if other weighting systems or units of measure were explored such as volume (m^3) or GHG emissions saved ($kgCO_2e$) instead of weight (kg) of building materials. A key assumption for full circularity relates to the availability of waste materials to cover the demand for construction services. The Circularity Gap Report for Norway 2020 shows how the demand for construction services in the Norwegian economy drives a material consumption of 58 Mt annually. In comparison, only 5 Mt of material from all economic activities are cycled back into the economy, and 10 Mt are lost [28]. Research points to an untapped potential for European countries to recycle more construction and demolition wastes [3]. However, even if the 10 Mt could be cycled back into the economy, the gap between demand and available waste materials points to the need for drastically reducing demand for construction to achieve full circularity.

6 Conclusion

This article explores a mass-based circularity indexing system and identifies current and future references for circularity performance of buildings, for use in FutureBuilt, a Norwegian green building innovation programme. A mixed methods approach using statistics, EPDs, GHG emission reports, PDAs, and interviews is applied to establish a reference baseline for Norwegian buildings (i.e. 7% circularity) and scenarios for future circularity in 2050 (from 35% to 80% circularity). Actors in the Norwegian construction industry can evaluate the level of ambition for circular buildings by using this novel indexing system and benchmarking approach which can also be adopted to other national or regional contexts.

Acknowledgements This article has been written within the Research Centre on Zero Emission Neighbourhoods in Smart Cities (FME ZEN). The authors gratefully acknowledge the support from the ZEN partners and the Research Council of Norway (project no. 257660).

References

1. Eurostat: Material Flow Accounts Statistics – Material Footprints. https://ec.europa.eu/eurostat/statistics-explained/index.php?title=Material_flow_accounts_statistics_-_material_footprints. Last accessed 22 Jan 2024
2. Eurostat: Waste Statistics. https://ec.europa.eu/eurostat/statistics-explained/index.php?title=Waste_statistics. Last accessed 22 Jan 2024
3. Joint Research Centre (European Commission), Cristóbal García, J., Caro, D., Foster, G., Pristerà, G., Gallo, F., Tonini, D.: Techno-economic and Environmental Assessment of Construction and Demolition Waste Management in the European Union: Status Quo and Prospective Potential. Publications Office of the European Union, LU (2023)
4. Amarasinghe, I., Hong, Y., Stewart, R.A.: Development of a material circularity evaluation framework for building construction projects. J. Clean. Prod. **436**, 140562 (2024). https://doi.org/10.1016/j.jclepro.2024.140562
5. European Innovation Council, SMEs Executive Agency (European Commission), Brincat, C., de Graaf, I., León Vargas, C., Mitsios, A., Neubauer, N., Adams, K., Hobbs, G.: Study on Measuring the Application of Circular Approaches in the Construction Industry Ecosystem: Final Study. Publications Office of the European Union, LU (2023)
6. Lam, W.C., Claes, S., Ritzen, M.: Exploring the missing link between life cycle assessment and circularity assessment in the built environment. Buildings. **12** (2022). https://doi.org/10.3390/buildings12122152

7. CircularEcoBIM/CircularityTool4Revit: https://github.com/CircularEcoBIM/CircularityTool4 Revit (2023)
8. Grønn byggallianse: BREEAM-NOR v6.0 for nybygg. Teknisk manual SD5076NOR (2022)
9. DGNB: DGNB System – Sustainable and Green Building. https://www.dgnb-system.de/en/index.php. Last accessed 7 Jan 2022
10. Resch, E., Wiik, M.K., Tellnes, L.G., Andresen, I., Selvig, E., Stoknes, S.: FutureBuilt Zero – a simplified dynamic LCA method with requirements for low carbon emissions from buildings. IOP Conf. Ser.: Earth Environ. Sci. **1078**, 012047 (2022). https://doi.org/10.1088/1755-1315/1078/1/012047
11. Nordby, A.S., Stoknes, S., Vadseth, R.A., Seilskjær, E., Hay, N.H.: FutureBuilt Sirkulær – kriterier for sirkulære bygg, Oslo, Norway (2023)
12. SSB: 10513: Avfallsregnskap for Norge (1 000 tonn), etter behandlingsmåte, materialtype, statistikkvariabel og år. Statistikkbanken. https://www.ssb.no/system/. Last accessed 31 May 2023
13. SSB: 09247: Genererte mengder avfall (tonn), etter statistikkvariabel, aktivitet, materialtype og år. Statistikkbanken. https://www.ssb.no/system/. Last accessed 31 May 2023
14. SSB: 10514: Avfallsregnskap for Norge (1 000 tonn), etter kilde, materialtype, statistikkvariabel og år. Statistikkbanken. https://www.ssb.no/system/. Last accessed 31 May 2023
15. SSB: 09781: Behandling av avfall fra nybygging, rehabilitering og riving (tonn), etter materialtype, behandlingsmåte, statistikkvariabel og år. Statistikkbanken. https://www.ssb.no/system/. Last accessed 31 May 2023
16. SSB: 03158: Eksisterende bygningsmasse. Alle bygg, etter region, bygningstype, år og statistikkvariabel. Statistikkbanken. https://www.ssb.no/system/. Last accessed 31 May 2023
17. SSB: 10785: Byggeareal. Avgang av bygninger, etter statistikkvariabel, region, bygningstype og år. Statistikkbanken. https://www.ssb.no/system/. Last accessed 31 May 2023
18. SSB: 05940: Boligbygg, etter statistikkvariabel, region, bygningstype og år. Statistikkbanken. https://www.ssb.no/system/. Last accessed 31 May 2023
19. SSB: 11358: Byggeareal. Ombygging til boliger, etter region, bygningstype, statistikkvariabel og år. Statistikkbanken. https://www.ssb.no/system/. Last accessed 31 May 2023
20. EPD Norge: Environmental Product Declarations. http://www.epd-norge.no/. Last accessed 1 Jan 2016
21. European Commission: Waste Framework Directive. https://environment.ec.europa.eu/topics/waste-and-recycling/waste-framework-directive_en
22. TEK 17: The Norwegian Building Regulations (Byggteknisk forskrift, TEK 17). https://dibk.no/byggereglene/byggteknisk-forskrift-tek17/9/9-8/ (2017)
23. European Commission: Circular Economy Action Plan. https://environment.ec.europa.eu/strategy/circular-economy-action-plan_en. Last accessed 21 Oct 2022
24. EC: Level(s) Common Framework. https://susproc.jrc.ec.europa.eu/product-bureau/product-groups/412/documents. Last accessed 8 July 2021
25. Miljødirektoratet: Klimakur 2030: Tiltak og virkemidler mot 2030. Miljødirektoratet – Norwegian Environment Agency (2020)
26. Sandberg, N.H., Dokka, T.H., Lien, A.G., Sartori, I., Skeie, K.S., Delgado, B.M., Lassen, N.: Energisparepotensialet i bygg fram mot 2030 og 2050. Hva koster det å halvere energibruken i bygningsmassen? SINTEF Academic Publishing, Oslo (2023)
27. EN 15978: Sustainability of Construction Works – Assessment of Environmental Performance of Buildings – Calculation Method. CEN, Europe (2011)
28. Circle Economy, Circular Norway: The Circularity Gap Report – Norway – Closing the Circularity Gap in Norway. https://www.circularity-gap.world/norway#wf-form-CGR-NOR-Report-Downloads (2020)

Smart Technologies: New Perspectives for the Heritage Environment

Marta Calzolari ⓘ, Pietromaria Davoli ⓘ, and Valentina Frighi ⓘ

Contents

1   Smart Technologies in the Construction Sector

1.1 *Potentiality of the Digital Era*

Today, around 56% of the world's population (approximately 4.4 billion people) live in cities;[1] this trend is expected to grow, reaching about 68% by 2050 [1]. These numbers reveal the inherent potential in tackling climate change, such as the optimization and improved management of available resources; therefore, the construction sector needs to be closely studied, especially existing and historic buildings, which cover approximately 24–35% of the entire European building stock [2]. In this increasingly digital era, technological advancements represent crucial factors in future transitions, thanks to their potential to continuously improve information and communication technologies (ICTs), to spur on remarkable

---

[1] World Bank data. Source: https://www.worldbank.org/en/topic/urbandevelopment/overview#1

M. Calzolari (✉) · P. Davoli · V. Frighi
University of Ferrara, Department of Architecture, Ferrara, Italy
e-mail: marta.calzolari@unife.it

© The Author(s) 2025
M. Kioumarsi, B. Shafei (eds.), *The 1st International Conference on Net-Zero Built Environment*, Lecture Notes in Civil Engineering 237,
https://doi.org/10.1007/978-3-031-69626-8_81

advances in artificial intelligence [3], and to reinforce and spread the Internet of Things (IoT). IoT represents a quite well-known concept, which refers to the technological development path that allows frequently used devices to connect over the Internet and gives each its own identity in the digital world. It is based on the interconnection of "intelligent" objects that exchange the information they have, collect, and/or process, and it takes on its full meaning within the network that interconnects these objects. Digital ecosystems based on IoT, whether tangible or not, comprising both material and immaterial technologies, are generally designed to generate and collect complex data in real time and in a highly structured manner. In the construction sector, IoT conventionally refers to the introduction/adoption of intelligent digital technologies with various purposes, mostly aimed at enhancement and improvement, including the optimization of asset management. This latter aspect is increasingly subject to research to determine the field's experimental guidelines and practical solutions, aiming to play a strategic role in the broader strategy to mitigate climate-altering greenhouse gas emissions (GHGs) and energy resource scarcity while ensuring high standards of indoor comfort for end users.

1.2 Smart Technologies and the Heritage Environment

The impact of this breakthrough on buildings has been significant, exhibiting the fast spread of intelligent devices in the construction sector, which has led to the development of so-called smart buildings. The origins of smartness in buildings dates back to the introduction of the first information technologies and automations through building management systems (BMSs) [4]. However, this initial definition focused only on the strictly technical aspects, though recently, it has been considered also in building interactions and relationships with surrounding environments and occupants. Although most definitions of what constitutes a smart technology focus on the application of advanced technologies, connectivity, and data analytics for optimizing building operations and management [3], agreement on its definition remains elusive in that different researchers emphasize different aspects. The currently available smart building models cover multiple domain areas [5], even if, specifically in the field of cultural heritage (CH), some gaps remain.

The high potential for innovation, which is intrinsic to IoT, seems to only marginally apply to CH; even though it can be used at a strategic level in climate adaptation planning, operationalization through the development and implementation of concrete measures, including monitoring, is lacking [6]. The first explicit reference to smart heritage buildings was made by Thwaites [7], and since then, the literature has been populated with an increasing quantity of references related to smart heritage, conceived as an independent discourse that intertwines autonomous and automatic capabilities and the innovation of smart technologies with the contextual and subjective interpretation of the past [8].

In general terms, the concept of smartness, in this specific field, refers to the implementation of CH features through the adoption of smart technologies, to promote interaction between people and the environment while collecting real-time data.

CH plays a key role in mitigating the causes of climate change in relation to its extension with respect to the built environment, and new smart technologies can augment its performance and improve its energy efficiency toward the decarbonization of the built environment.

2 Literature Review

2.1 Smartness of Cultural Heritage: Main Uses and Approaches

Although most of the existing literature has shown a relationship between smart city strategies (SCSs) and CH, emphasizing that SCSs should be built on the existing assets of the city in a way that incorporates its identity [9], research on the specificities of this relationship between urban smartness and CH remains limited, particularly when referring to strategies for climate mitigation. Many researchers have explored adjacent areas between innovative technology and heritage while referring to digital heritage, augmented heritage, and virtual heritage. For example, Chung et al. [10] investigated how augmented reality influences visitors' intention to come to heritage sites, while Cayla [11] examined new technologies' intersection with geological histories, specifically digital-mapping and data-recording processes. Technologies in use within the CH field can be divided into two main categories: enabling technologies and visualization technologies [12]. Enabling technologies are mostly related to data collection, allocation, storing, and exchange, and they include (i) IoT technology, understood as a huge network of various devices with embedded sensors for transmitting data connected to the Internet; (ii) cloud computing (CC), which complements IoT-enabled devices in that it allows the sharing of information and resources; (iii) wireless sensor networks (WSNs), usually dependent on autonomously distributed devices using sensors to monitor specific parameters, often related to the nearby environmental conditions; (iv) mobile broadband (MB), defined as the wireless Internet accessible through mobile devices [12], including those using Wi-Fi protocols; and (v) short-range wireless (SW) technologies, which allow for communication between various smart devices and objects near each other (also known as near-field communication, or NFC). Visualization technologies, on the other hand, are those that communicate information by encoding it as visual objects [9], making complex data more understandable, accessible, and usable. They could range from simple info-graphic representation to more advanced 3D modeling, and include Virtual Reality (VR), Geo-visualization, and Augmented Reality (AR) winking also at Artificial Intelligence (AI). As the concepts of smart heritage and the Internet of cultural things are becoming more and more widespread in the scientific literature, understanding the main scopes and research fields for the

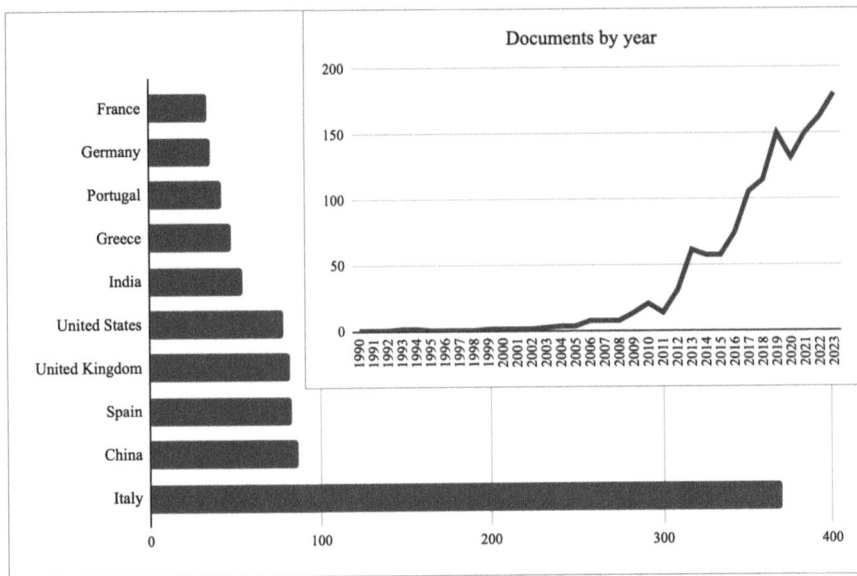

Fig. 1 Distribution of the main published documents sorted according to country (left) and year (right), by relating cultural/historical heritage with the diverse uses of smartness

Authors' elaboration from Scopus and ScienceDirect using a combination of the keywords: "cultural heritage" (and its synonymous/alternative forms) and "IoT/keys/digital technologies/ smartness." The Scopus advanced query (TITLE-ABS-KEY "cultural heritage" OR "historic* building*" OR "heritage" OR "ancient building*" AND TITLE-ABS-KEY "smart*" OR "ket*" OR "key enable technology*") generated more than 1300 documents, which were then filtered according to subject area and the main topic

exploitation of digital technologies for use in cultural assets is now a priority. To best link cultural/historical heritage with the diverse ways of applying smartness when searching in the main scientific online databases (mainly Scopus and ScienceDirect), Italy is the main country for the implementation of this kind of research (Fig. 1). The recent and rapid development of new technologies can also be traced to the time-wise distribution of publications on the topic, which shows a significant increase in papers starting from 2012 to the present (Fig. 1). The two academic disciplines most affected by the research are computer science and the social sciences, followed by engineering (Fig. 2).

In this chapter, a selection of the literature review results is presented.

These documents confirm that the topic is treated differently depending on the perspective. The primary divergence lies in the intended use of technologies to enable or visualize CH. Particularly well-explored areas include new procedures and tools for the valorization of the user's experience in visiting museums, archeological sites, historical archives, religious buildings, and cultural events. In particular, new frontiers feature instruments such as a novel mobile sensory-augmented context-aware multimedia guide [13], applications for assisting users in the exploration of single smart spaces [14], generative pretrained transformers

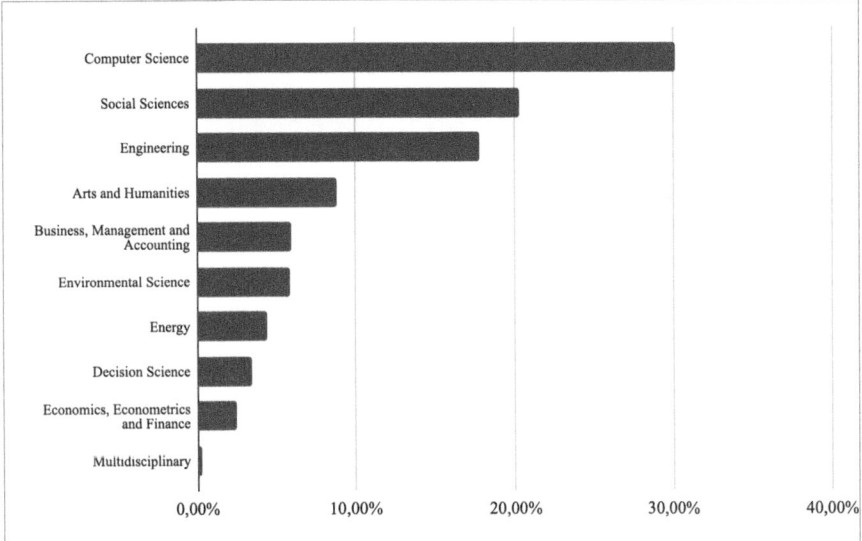

Fig. 2 Distribution of the main published documents by subject area, by relating cultural/historic heritage with the diverse uses of smartness. (Authors' elaboration from Scopus and ScienceDirect, as described in the caption of Fig. 1)

(GPTs) by OpenAI, large language models (LLMs) as digital storytelling machines [15], and the application of augmented reality for classical music concerts [16]. Another possible option is utilizing smart technologies to bolster urban management effectiveness by investigating citizens' needs from smart city advancement and how they align with community opinions in areas rich in cultural and natural characteristics [17]. The literature shows that a significant amount of research has focused on the wider concept of a smart (historical) city and its management. In this context, studies have aimed at listing the fundamental factors of cultural heritage management systems, classified according to the components of a smart city [18–20]. Conservation is the other main reason to implement digital technologies for CH. Innovative applications require the use of sensors and cyber-physical systems or AI for preservation to manage vulnerabilities, functional service provision [21–25], the safe movement of artworks [23], deterioration risks [26, 27], and security [28, 29]. Among the most used technologies related to CH are tools for 3D surveying and visualization [30, 31], useful devices for creating high-resolution digital archives that collect and manage information [32], and so-called heritage building information modeling (HBIM) [33–36]. The use of big data (BD) and statistical predictive models is increasingly developing platforms for data management (e.g., BD, cloud-computing, and edge-computing platforms) [37–39] and particularly for the analysis of energy consumption [40–42] and the external environmental performance of heritage buildings [43, 44]. The above literature shows how much the field of applications for these technologies is developing for the preservation, enhancement, and better management of historical heritage.

main scope	Building's type/use						Specific aim			
	Historical city	Heritage asset	Museum	Archeological sites/Historical archives/Cultural events/Religious sites	Building's materials	Artworks	User's experience enhancement/Tourist guide	Management	Energy consumption	Indoor/outdoor performance
Valorization	[14] [17] [24] [33] [34] [35] [36] [10]	[24] [33] [34] [35] [36] [31] [10] [11]	[13] [14] [58]	[13] [14] [10] [15][16]			[14] [15] [58] [16] [17] [20] [24] [10]	[14] [15] [17] [11]		
Conservation	[24]	[21] [22] [25] [59] [23] [24] [31]	[23]	[23]	[24]	[27]	[59] [23] [24]	[21] [22]		
Security		[29]	[28]		[29]	[28]				
Smart city	[18] [19] [20] [37]						[18] [20]	[18] [19]		
3D survey & visualization		[30] [31] [32]								
Big Data & Data analysis	[37]	[39] [40] [41] [42] [43] [44]					[37] [39]	[37] [38]	[40] [41] [42]	[43] [44]
Climate mitigation	[52] [53] [54] [55] [56] [57]	[60] [45] [46] [47] [48] [49] [50] [51] [60] [62][63] [64][65]	[48] [64] [65]						[60] [45] [46] [47] [48] [49] [61]	[64] [49] [50] [65] [51] [52] [53] [54] [55] [60][61] [66][67]

Fig. 3 Literature review results: applications of digital technologies in CH and how they relate to building types and uses and to the aim of this study

Moreover, understanding how the research world is shifting toward this field is necessary to take advantage of technological developments to control the effects of climate change while proposing intervention strategies. The state-of-the-art is undergirded by a great number of studies that have addressed the procedures for the hygrothermal analysis of building components [45–49] for assessing indoor comfort and air quality [50, 51] as well as outdoor environmental performance [52–57]. For this kind of study, however, the use of digital technologies is often limited to advanced simulation tools and sensor systems for monitoring campaigns, and it does not reach the level of digitalization of other research areas. Figure 3 presents a summarizing matrix of the preliminary results of the outlined literature review.

2.2   Analysis of Best Practices

By focusing on the application of pervasive digital technologies to historical heritage, with the aim of mitigating climate change, a series of best practices can be identified for the exploitation of enabling technologies for redevelopment projects and/or for the recovery of historic buildings. According to the literature review, three best practices were considered particularly relevant to open up new reflections on specific innovative approaches to cultural heritage. Among the many cases in the review, three examples were identified, differing in the type of smart technology used and the scale of intervention. Notable examples include energy-efficient initiatives implemented on a local scale, focusing on delineating priority thermal zones within expansive architectural complexes. Additionally, design strategies for decarbonizing buildings can prioritize the integration of green systems over direct alterations to a historic envelope.

SHCITY—Smart Heritage City

The SHCITY project [68], funded by the Interreg VB South West Europe program, aims to develop an open-source tool for managing historic urban centers, focusing particularly on the walled city of Ávila, Spain. It utilizes data from cloud computing and sensor networks to monitor the risks, energy consumption, and visitor flow at the UNESCO heritage site. By integrating ICTs and interdisciplinary knowledge related to heritage management on an urban scale, the project extends the smart city concept to a smart heritage city, shifting the monitoring focus from individual buildings to the urban complex. This initiative has resulted in the creation of a sensor infrastructure and two open-source applications: SHCity tourist, for spreading information about the preventive conservation of heritage elements, adding social and economic value to the innovative approach, and SHCity manager, for public administrations. These tools improve the management of heritage assets, thereby also improving conservation efforts and decision-making processes. The systematization of these tools allows for the unification of different domains, such as preventive conservation and energy efficiency, enabling tailored configurations for various needs. Overall, this system improves the planning of maintenance interventions, leading to significant savings in management, time, and costs.

Complex of San Pietro in Vincoli, Rome

The sixteenth-century cloister designed by Giuliano da Sangallo in the complex of San Pietro in Vincoli in Rome was the subject of a study [69] aimed at investigating the different aspects related to internal and external thermo-hygrometric comfort by using predictive models and by analyzing its innovative materials and their repercussions on energy consumption. Starting from the analysis on the external spaces, virtual models were created that allowed for the evaluation of the impact of some innovative finishing surfaces on the microclimatic factors of the area. The research developed simulations and in situ monitoring campaigns not only within the cloister but also in the urban canyon of Via delle Carrozze, verifying the impact of strategies aimed at increasing urban greening and/or the use of materials with a high albedo coefficient. The simulated models were then experimentally validated to verify their reliability in reproducing the investigated phenomena.

Palazzo Gulinelli Canonici Mattei, Ferrara

The eco-sustainable restoration project [70] of Palazzo Gulinelli, damaged by an earthquake, aimed to harmonize client requirements with the building's intrinsic purpose by utilizing digital technologies. Emphasizing eco-sustainability, it integrated Leadership in Energy and Environmental Design by the Green Building Council (LEED–GBC) certification mechanisms according to the Historic Building (HB) protocol. The approach relied on an integrated Building Information Modeling process. The restoration considered the details and characteristics of elements and materials, including diagnostics. Laser scanner surveys collected information for each element, which was then modeled and cataloged systematically. This database facilitated timely simulations, predicted processing times and costs, and managed maintenance interventions efficiently.

The notable actions undertaken in this recovery project are as follows:

- Restoring the existing proto-climatization system from the Victorian era, comprising light wells/vertical ventilation ducts that run from the underground space to the upper levels, with cooling provided by fountains in summer for building microventilation
- Establishing a rooftop garden, one of the first in Italy within a protected historic building
- Replacing an incongruous superstructure with a new technical structure made of Cross Laminated Timber (CLT) from certified forests, alongside the utilization of dry technologies and reversible installation systems for both ancient elements and new additions
- Designing and implementing a wireless electrical system to avoid disturbing historic masonry structures, complemented by energy consumption sensors (for Domestic Hot Water (DHW) and air conditioning) to optimize internal comfort through individual room regulation and reduce the overall energy consumption of the building, accounting for the nonsimultaneous use of school spaces

3   Conclusion: Toward the Smart Heritage Environment and New Perspectives for Research

In the era of data technology, exploiting data as a source of knowledge is a recurring and pervasive research topic. A primary area of interest that emerged following this discussion is that aimed defining an effective and innovative relationship of interest between the technological potential inherent in the world of sensors, the production and programming of data on shared platforms and of advanced modeling, and cultural heritage, addressed to develop efficient governance processes for preventive conservation and energy-efficient and environmental retrofit projects for buildings.

To this end, thanks to the conducted analysis, partially unexplored fields of investigation—systematizing concepts that until now pertained to independent domains—could be explored. Indeed, at present, the literature shows widespread applications of enabling technologies and IoT within the construction sector; however, most of these technologies have been poorly applied to historic buildings, for energy efficiency and, more generally, the mitigation of the causes of climate change. In addition, in this specific field, the literature shows a prevalence of research that comes from non-EU countries, with different climates and historic features than the Mediterranean ones, such as the role of the courtyards in the architecture of the past [62–65].

The results of this preliminary literature review are that a new perspective of research could focus on the use of pervasive technologies and digital innovation for a new type of smart management for CH, one that will effectively implement strategies for climate-change mitigation. In this field, the TECHSTART project, specifically the part that concerns the research unit to which the authors belong, has worked to study the so-called liminal spaces (porches, courtyards, cloisters, underground/attic spaces, etc.) of historic buildings to collaborate in the creation of smart environments by using enabling technologies as tools for investigation, knowledge management, and the enhancement of the passive functioning of liminal spaces of ancient cities for climate-change mitigation. The workflow developed by the research team integrates environmental monitoring campaigns, numerical simulations, statistical methods for data analysis, and advanced knowledge management systems, such as HBIM, with the goals of creating a simplified method for analyzing both single buildings and entire urban compounds and of creating a smart tool that automatically optimizes the management of liminal spaces to reach optimal outdoor thermal comfort.

Acknowledgments This chapter has been developed within the TECHSTART–PRIN2017 project (PI: M. Losasso, University of Naples), as part of the research activities conducted by the authors in the task group of the Department of Architecture at the University of Ferrara. This interdisciplinary research group consists of P. Davoli (Head of the task group), F. Conato, M. Calzolari, V. Frighi, V. Modugno, B. Caglioti, D. Vincenzi, G. Mangherini, L. Gabrielli (University of Ferrara), and A.G. Ruggeri (University of Venice). https://in-out-heritage.it/.

Credit Authors' Statement Conceptualization: M.C., P.D., and V.F.; methodology: M.C. and V.F.; writing: M.C. and V.F.; and supervision: P.D.

References

1. World Urbanization Prospects: The 2018 Revision. https://population.un.org/wup/Publications/Files/WUP2018-Report.pdf. Last accessed 28 Feb 2024
2. Blumberga, A., Freimanis, R., Muizniece, I., Spalvins, K., Blumberga, D.: Trilemma of historic buildings: smart district heating systems, bioeconomy and energy efficiency. Energy. **186**, 115741 (2019)
3. Falorca, J.G.: An overview of 'intelligent' and 'smart' approaches in the context of buildings. In: International Conference on Recovery, Maintenance and Rehabilitation of Buildings, pp. 589–600. Springer Nature, Switzerland (2023)

4. Conato, F., Frighi, V.: The smart's era. L'edificio intelligente e le sue declinazioni. L'UFFICIO TECNICO. **9**, 6–14 (2020)
5. Vale, Z., Gomes, L., Ramos, C.: An overview on smart buildings. In: Encyclopedia of Electrical and Electronic Power Engineering, pp. 431–440. Elsevier (2023)
6. Guzman, P., Daly, C.: Cultural heritage in climate planning. In: The HiCLIP Pilot Project for Understanding the Integration of Culture into Climate Action: A Report on the Climate Heritage Network. WG4. ICOMOS (2021)
7. Thwaites, H.: Digital heritage: what happens when we digitize everything? In: Ch'ng, E., Gaffney, H., Chapman, H. (eds.) Visual Heritage in the Digital Age, pp. 327–348. Springer, London (2013)
8. Batchelor, D., Schnabel, M.A., Dudding, M.: Smart heritage: defining the discourse. Heritage. **4**, 1005–1015 (2021)
9. Khalaf, M.: Smart cultural heritage: technologies and applications. In: 2nd Smart Cities Symposium (SCS 2019), pp. 1–6. IET, Bahrain (2019)
10. Chung, N., Han, H., Joun, Y.: Tourists' intention to visit a destination: the role of augmented reality (AR) application for a heritage site. Comput. Hum. Behav. **50**, 588–599 (2015)
11. Cayla, N.: An overview of new technologies applied to the management of geoheritage. Geoheritage. **6**, 91–102 (2014)
12. Borda, A., Bowen, J.P.: Smart cities and cultural heritage – a review of developments and future opportunities. In: Proceedings of EVA London 2017. BCS Learning and Development Ltd., London (2017)
13. Roffia, L., et al.: Requirements on system design to increase understanding and visibility of cultural heritage. In: Styliaras, G., Koukopoulos, D., Lazarinis, F. (eds.) Handbook of Research on Technologies and Cultural Heritage: Applications and Environments, pp. 259–284. IGI Global, Hershey (2011)
14. Colace, F., et al.: A context-aware framework for cultural heritage applications. In: 2014 Tenth International Conference on Signal-Image Technology and Internet-Based Systems, pp. 469–476. IEEE (2014)
15. Trichopoulos, G.: Large language models for cultural heritage. In: Proceedings of the 2nd International Conference of the ACM Greek SIGCHI Chapter, pp. 1–5. Association for Computing Machinery, New York (2023)
16. Kaimaris, D., et al.: The development of a model system for the visualization of information on cultural activities and events. Electronics. **12**(23), 4769 (2023)
17. Khamwachirapithak, P., Khongouan, W.: Smart city development in a tourist city with valuable sites of cultural and natural environment: case study of Amphawa Subdistrict Municipality, Samut Songkhram Province. Kasetsart J. Soc. Sci. **45**(1), 257–268 (2024)
18. Hatami, M., Koramaz, T.K.: Smart solutions for heritage sites: Florence and Yazd. Preserv. Digit. Technol. Cult. **52**(4), 143–155 (2023)
19. Lombardo, L., Saeli, M., Campisi, T.: Smart technological tools for rising damp on monumental buildings for cultural heritage conservation. A proposal for smart villages implementation in the Madonie mountains (Sicily). Sustain. Futures. **6**, 100116 (2023)
20. Geng, S., et al.: Understanding place identity in urban scale Smart Heritage using a cross-case analysis method. Int. J. Tour. Cities. **9**(3), 729–750 (2023)
21. Nota, G., Petraglia, G.: Heritage buildings management: the role of situational awareness and cyber-physical systems. J. Ambient Intell. Humaniz. Comput. **15**, 2227–2239 (2024)
22. Prieto, A.J., et al.: Artificial intelligence applied to the preventive conservation of heritage buildings. In: Ortiz Calderón, P., et al. (eds.) Science and Digital Technology for Cultural Heritage-Interdisciplinary Approach to Diagnosis, Vulnerability, Risk Assessment and Graphic Information Models, pp. 245–249. CRC Press, London (2019)
23. Costantini, S., et al.: DALICA: agent-based ambient intelligence for cultural-heritage scenarios. IEEE Intell. Syst. **23**(2), 34–41 (2008)
24. Liu, Y., Cheng, P., Li, J.: Application interface design of Chongqing intangible cultural heritage based on deep learning. Heliyon. **9**(11), e22242 (2023)

25. Gribaudo, M., Iacono, M., Levis, A.H.: An IoT-based monitoring approach for cultural heritage sites: the Matera case. Concurr. Comput. Pract. Exp. **29**, e4153 (2017)
26. Appolonia, L., et al.: Computer-aided monitoring of buildings of historical importance based on color. J. Cult. Herit. **7**(2), 85–91 (2006)
27. Liu, B., et al.: Environmental monitoring by thin film nanocomposite sensors for cultural heritage preservation. J. Alloys Compd. **504**, S405–S409 (2010)
28. Girolami, M., La Rosa, D., Barsocchi, P.: A CrowdSensing-based approach for proximity detection in indoor museums with Bluetooth tags. Ad Hoc Netw. **154**, 103367 (2024)
29. Ciardelli, L., et al.: Multi-sensor cognitive-based approach to critical infrastructure protection. WIT Trans. Built Environ. **108**, 71–81 (2009)
30. Erenoglu, R.C., Akcay, O., Erenoglu, O.: An UAS-assisted multi-sensor approach for 3D modeling and reconstruction of cultural heritage site. J. Cult. Herit. **26**, 79–90 (2017)
31. Pepe, M., et al.: Data for 3D reconstruction and point cloud classification using machine learning in cultural heritage environment. Data Brief. **42**, 108250 (2022)
32. Pepe, M., et al.: Scan to BIM for the digital management and representation in 3D GIS environment of cultural heritage site. J. Cult. Herit. **50**, 115–125 (2021)
33. Parrinello, S., Sanseverino, A., Fu, H.: HBIM modelling for the architectural valorisation via a maintenance digital eco-system. Int. Arch. Photogramm. Remote Sens. Spat. Inf. Sci. **48**, 1157–1164 (2023)
34. Fiorillo, F., Bolognesi, C.M.: Cultural heritage dissemination: BIM modelling and AR application for a diachronic tale. Int. Arch. Photogramm. Remote Sens. Spat. Inf. Sci. **48**, 563–570 (2023)
35. Martinelli, L., Calcerano, F., Gigliarelli, E.: Methodology for an HBIM workflow focused on the representation of construction systems of built heritage. J. Cult. Herit. **55**, 277–289 (2022)
36. Costa, A.P., Cuperschmid, A.R.M., Neves, L.O.: HBIM and BEM association: systematic literature review. J. Cult. Herit. **66**, 551–561 (2024)
37. Alamgir Hossain, S.K., Anisur Rahman, M., Hossain, M.A.: Edge computing framework for enabling situation awareness in IoT based smart city. J. Parallel Distrib. Comput. **122**, 226–237 (2018)
38. Huang, L., Chiu, Y., Chan, Y.: The design of building management platform based on cloud computing and low-cost devices. In: Proceedings of the 36th ISARC, pp. 407–414, Banff, Canada (2019)
39. Guidara, A., et al.: Towards occupant activity driven smart buildings via WiFi-enabled IoT devices and deep learning. Energy Build. **177**, 12–22 (2018)
40. Ahmad, T., et al.: A comprehensive overview on the data driven and large scale based approaches for forecasting of building energy demand: a review. Energy Build. **165**, 301–320 (2018)
41. Li, W., et al.: Modeling urban building energy use: a review of modeling approaches and procedures. Energy. **141**, 2445–2457 (2017)
42. Catalina, T., Iordache, V., Caracaleanu, B.: Multiple regression model for fast prediction of the heating energy demand. Energy Build. **57**, 302–312 (2013)
43. Ali-Toudert, F., Mayer, H.: Numerical study on the effects of aspect ratio and orientation of an urban street canyon on outdoor thermal comfort in hot and dry climate. Build. Environ. **41**(2), 94–108 (2006)
44. Shah, R., Pandit, R.K., Gaur, M.K.: Urban physics and outdoor thermal comfort for sustainable street canyons using ANN models for composite climate. Alex. Eng. J. **61**(12), 10871–10896 (2022)
45. Muradov, M., et al.: Non-destructive system for in-wall moisture assessment of cultural heritage buildings. Measurement. **203**, 111930 (2022)
46. Rymarczyk, T., et al.: Historical buildings dampness analysis using electrical tomography and machine learning algorithms. Energies. **14**(5), 1307 (2021)

47. Lombardo, L., Saeli, M., Campisi, T.: Smart technological tools for rising damp on monumental buildings for cultural heritage conservation. A proposal for smart villages implementation in the Madonie montains (Sicily). Sustain. Futures. **6**, 100116 (2023)
48. Andreotti, M., et al.: Hygrothermal performance of an internally insulated masonry wall: experimentations without a vapour barrier in a historic Italian Palazzo. Energy Build. **260**, 111896 (2022)
49. Raffler, S., Bichlmair, S., Kilian, R.: Mounting of sensors on surfaces in historic buildings. Energy Build. **95**, 92–97 (2015)
50. Stazi, F., et al.: Design of a smart system for indoor climate control in historic underground built environment. Energy Procedia. **134**, 518–527 (2017)
51. Satiko Nomiso, L., Hideki Tanaka, E., da Costa, D.A.G.: Design of a smart building thermal comfort device. Int. J. Comput. Sci. Inf. Syst. **11**(2), 246–250 (2016)
52. Rodríguez Algeciras, J.A., Gómez Consuegra, L., Matzarakis, A.: Spatial-temporal study on the effects of urban street configurations on human thermal comfort in the world heritage city of Camagüey-Cuba. Build. Environ. **101**, 85–101 (2016)
53. Tablada, A., et al.: On natural ventilation and thermal comfort in compact urban environments – the Old Havana case. Build. Environ. **44**(9), 1943–1958 (2009)
54. Castaldo, V.L., et al.: Microclimate and air quality investigation in historic hilly urban areas: experimental and numerical investigation in central Italy. Sustain. Cities Soc. **33**(04), 27–44 (2017)
55. Ragheb, A.A., El-Darwish, I.I., Ahmed, S.: Microclimate and human comfort considerations in planning a historic urban quarter. Int. J. Sustain. Built Environ. **5**(1), 156–167 (2016)
56. Ambrosini, D., et al.: Evaluating mitigation effects of urban heat islands in a historical small center with the ENVI-Met® climate model. Sustainability. **6**(10), 7013–7029 (2014)
57. Mitro, N., Krommyda, M., Amditis, A.: Smart Tags: IoT sensors for monitoring the microclimate of cultural heritage monuments. Appl. Sci. **12**, 2315 (2022)
58. Fujita, H., Guizzi, G.: Augmented reality coupled with deep convolution neural networks to enhance archaeological sites experience. In: Fujita, H., Watanobe, Y., Azumi, T. (eds.) New Trends in Intelligent Software Methodologies, Tools and Techniques. Proceedings of the 22nd International Conference on New Trends in Intelligent Software Methodologies, Tools and Techniques (SoMeT_23), vol. 371, p. 97. IOS Press (2023)
59. Casillo, M., et al.: An IoT-based system for expert user supporting to monitor, manage and protect cultural heritage buildings. In: Nedjah, N., et al. (eds.) Robotics and AI for Cybersecurity and Critical Infrastructure in Smart Cities, pp. 143–154. Springer International Publishing, Cham (2022)
60. Ismaeel, W.S., Mohamed, A.G.: A structural equation modelling paradigm for eco-rehabilitation and adaptive reuse of cultural heritage buildings. Build. Environ. **242**, 110604 (2023)
61. Silva, H.E., Coelho, G.B.A., Henriques, F.M.A.: Climate monitoring in World Heritage List buildings with low-cost data loggers: the case of the Jerónimos Monastery in Lisbon (Portugal). J. Build. Eng. **28**, 101029 (2020)
62. Zamani, Z., Heidari, S., Hanachi, P.: Reviewing the thermal and microclimatic function of courtyards. Renew. Sustain. Energy Rev. **93**(05), 580–595 (2018)
63. Rodríguez-Algeciras, J., et al.: Influence of aspect ratio and orientation on large courtyard thermal conditions in the historical centre of Camagüey-Cuba. Renew. Energy. **125**, 840–856 (2018)
64. Carlos, R.G., et al.: Tempering potential-based evaluation of the courtyard microclimate as a combined function of aspect ratio and outdoor temperature. Sustain. Cities Soc. **51**(08), 101740 (2019)
65. Soflaei, F., Shokouhian, M., Mofidi Shemirani, S.M.: Traditional Iranian courtyards as microclimate modifiers by considering orientation, dimensions, and proportions. Front. Archit. Res. **5**(2), 225–238 (2016)

66. Gaudenzi Asinelli, M., et al.: The smARTS_Museum_V1: an open hardware device for remote monitoring of Cultural Heritage indoor environments. HardwareX. **4**, e00028 (2018)
67. Dong, C., Zhang, Y., Li, J., Zhang, H.: The integrated design for micro – environment monitoring system of showcase in museum. MATEC Web Conf. **100**, 05009 (2017)
68. Mar, A., et al.: An application to improve smart heritage city experience. In: Ioannides, M., et al. (eds.) Advances in Digital Cultural Heritage, pp. 89–103. Springer International Publishing, New York (2018)
69. Salata, F., et al.: Outdoor thermal comfort in the Mediterranean area. A transversal study in Rome, Italy. Build. Environ. **96**, 46–61 (2016)
70. Ferrari, C.: Palazzo Gulinelli rigenerato in modo sostenibile con il protocollo GBC-Historic Building. Recupero e Conservazione. **158**(03/04), 62–70 (2020)

Energy Efficiency Through Building Renovation: A Study of Challenges and Solutions

Sina Moradi ⓘ, Janne Hirvonen ⓘ, Natalia Lastovets ⓘ, and Piia Sormunen ⓘ

Contents

1 Introduction and Theoretical Background

There is a growing focus on improving building energy efficiency due to the global imperative to combat climate change and reduce greenhouse gas emissions [1]. Buildings account for a substantial portion of total energy consumption and carbon emissions worldwide. In Europe, it is estimated that a significant proportion of the buildings that will still be in use by 2050 already exist today [2]. Therefore, achieving energy efficiency in existing buildings through renovation is a crucial step toward meeting energy efficiency targets and mitigating the environmental impact of the building sector. The Renovation Wave initiative is one way that the European Union is encouraging increased energy retrofitting in the existing building stock.

However, the task of renovating existing buildings for energy efficiency is not straightforward. Unlike existing construction, which can be designed with the latest energy-efficient technologies and standards, existing buildings often present unique

S. Moradi (✉) · J. Hirvonen · N. Lastovets · P. Sormunen
Civil Engineering Unit, Tampere University, Tampere, Finland
e-mail: sina.moradi@tuni.fi

© The Author(s) 2025
M. Kioumarsi, B. Shafei (eds.), *The 1st International Conference on Net-Zero Built Environment*, Lecture Notes in Civil Engineering 237,
https://doi.org/10.1007/978-3-031-69626-8_82

challenges. These challenges arise from factors such as outdated building services systems, architectural constraints, budgetary restrictions, and the need to minimize disruptions to occupants during renovation [3]. Causes of building energy performance gap can be linked to all phases of the building life cycle: design, planning, construction, and commissioning and operation [4]. Energy systems in existing buildings are often operated in suboptimal ways. For example, this can be due to poor energy performance guidelines and documentation or because of too large control zones for heating, ventilation, and air-conditioning (HVAC) equipment [5]. Discrepancies between the design and operation stages can also be caused by lower-than-expected equipment efficiency, differences between simulated and experienced weather, and significantly different than expected occupant behavior [6]. Facility managers could mitigate many of the operational inefficiencies but are often afraid of making necessary adjustments for fear of getting blamed for bad outcomes [7]. A review of green retrofitting has revealed many barriers to effective retrofit implementation [8]. These include financial barriers, such as high investment cost and lack of life cycle cost knowledge as well as lack of energy efficiency information and skills by the designers, contractors, and building owners. Insufficient performance data of green building systems, risks of untested technologies, and poor communication and leadership are the other obstacles to successful energy retrofitting of projects.

Energy-efficient building renovations require effective collaboration among all stakeholders and project monitoring to ensure successful implementation. It involves technological advancements and changes in building management practices, workforce training, and other incentives [9]. Furthermore, building renovation projects involve a range of professionals, including clients, designers, contractors, and maintenance experts. Their insights and expertise are critical for navigating the energy-efficient renovation complexities. Information must also flow between designers and users of buildings, to ensure that the users understand how the building is supposed to be operated [10]. Facility managers should be tasked with continual improvement of building performance, but this mission should have already started during the planning stages of building renovation projects [11]. A multidisciplinary approach is needed so that resolving energy performance issues does not create new problems in other areas [12].

2 Methodology

This study aims to address the challenges and solutions of achieving energy efficiency in building renovation projects. To do so, qualitative data collection and analysis methods were employed based on which 21 semi-structured interviews were conducted in Finland. These interviews were conducted with experts representing client project management (five interviews); contractor project management (five interviews); architectural, structural, and HVAC design (five interviews); and building operation as well as maintenance (six interviews). Half of the

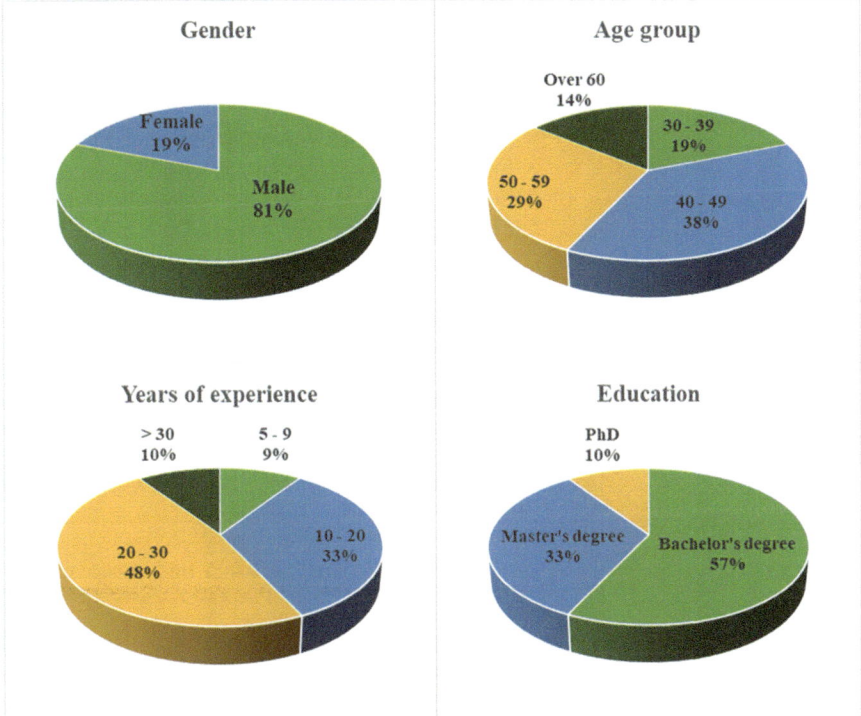

Fig. 1 Demographic information of the interviewees

interviewees' latest project was building construction and the other half building renovation. The demographic information (gender, age, experience, education) about the interviewees is shown in Fig. 1. Following the completion of interviews, they were analyzed through thematic analysis and content analysis methods [13]. Table 1 shows the whole process of the conducted research, including data collection and analysis.

3 Results

Analyzing the interview transcripts resulted in the identification of the challenges and enablers for realizing energy efficiency in building renovation projects. The most frequently mentioned challenges and enablers are shown in Tables 2 and 3, respectively.

Some of the challenges to energy efficiency come from the buildings themselves and are not easily resolved. There might not be enough physical space to install modern systems needed for enhanced operations or the structure might not be strong enough to support heavier equipment. In addition, many historical buildings have a

Table 1 The research process

Step 1	
Formulating the research questions	What are the challenges/barriers to achieving energy efficiency in building renovation projects? How can the abovementioned challenges/barriers be overcome?
Step 2	
Undertaking a literature study	The conducted literature study showed limited research-based knowledge on the topic under study
Step 3	
Selecting the data collection method	A semi-structured interview was selected as the data collection method
Step 4	
Conducting semi- structured interviews	A total of 21 semi-structured interviews were conducted in Finland with project professionals representing clients, design/planning experts, contractors, and building operation/maintenance experts The interviews were audio recorded based on the obtained consent from the interviewees
Step 5	
Transcription and translation of interviews	The conducted interviews were transcribed and translated into the English language by a native Finnish-speaking member of the research team
Step 6	
Coding	This was undertaken by inductively coding the extracted research data as a result of analyzing the interview transcripts. The labels of the codes were data-derived by the researcher
Step 7	
Validating the generated codes	The generated 51 codes in the previous step were reviewed again, and 18 of them were renamed. As a result of this effort, the remaining 34 codes were considered valid for thematic analysis
Step 8	
Thematic analysis	The validated codes were then analyzed through thematic analysis to see what themes the data are suggesting. Nine themes were established as a result of the thematic analysis The themes were established based on the sameness or similarity of the codes in terms of the meaning and/or title
Step 9	
Content analysis	The challenges/barriers and solutions/enablers representing different codes and themes were listed separately and synthesized based on the similarity or sameness of the title and/or meaning

protected status that prevents visible changes to their façade, while general rules on cityscape could, for example, forbid the use of energy-efficient but highly reflective glazing.

Other challenges, however, can be mitigated. Financing issues, for example, can be controlled by focusing more on life cycle costs compared to investment costs, as many energy efficiency measures have not only high initial costs but also significant cost-saving potential. Insufficient monitoring of energy use in different systems and

Table 2 Challenges/barriers to realizing energy efficiency in building renovation projects

Theme	Challenge/barrier	Reference
Existing building's architecture and structure	Insufficient space for new building services systems (e.g., low floor height, narrow shaft, small machine rooms)	(DM) INT 2, (PMClient) INT 2, (DM) INT 3, (PMContractor) INT 3, (PMContractor) INT 4, (DM) INT 5, (PMClient) INT 5, (PMContractor) INT 5, (PropertyMgmt) INT 5, (PropertyMgmt) INT 6
	Incompatibility of the existing buildings' structure, space, and technology with the modern building services systems (i.e., equipment)	(PMClient) INT 1, (DM) INT 2, (PMClient) INT 2, (PropertyMgmt) INT 2, (DM) INT 3, (PMClient) INT 3
Finance	High investment cost of replacing the old energy system with the new one with an uncertain payback period	(DM) INT 3, (PMContractor)INT 4, (DM) INT 5, (PropertyMgmt) INT 5
	Insufficient funding for energy efficiency measures	(PMClient) INT 2, (DM) INT 5
	Cost and time of reinforcing the structure of the existing building for installing modern and heavy equipment	(DM) INT 3, (PMClient) INT 1
Regulatory issues	Building protection issues for façade renovation	(PMClient) INT 1, (PropertyMgmt) INT 2, (DM) INT 3, (PMClient) INT 5
Building's energy system	Inadequate and inaccurate metering of energy consumption in heating	(PropertyMgmt) INT 3, (PropertyMgmt) INT 2
Building operation, maintenance, and optimization	Lack of highly intelligent building automation system for continuous and real-time monitoring and control of building operation	(PropertyMgmt) INT 3, (PMClient) INT 4

Legend for the Reference column:
DM → Design Experts Interviewee Group
PMClient → Client Project Management Interviewee Group
PMContractor → Contractor Project Management Interviewee Group
Property Mgmt → Building Operation and Maintenance Interviewee Group

the lack of data analysis prevent or complicate the identification of energy efficiency issues. These challenges can be resolved by the addition of more detailed monitoring systems and the employment of staff specialized in energy efficiency. Modern energy systems are becoming more complex, and it should not be expected that people responsible for day-to-day building maintenance are also experts on

Table 3 Solutions/enablers of realizing energy efficiency in building renovation projects

Theme	Challenge/barrier	Reference
Competence development	Competent maintenance workforce for handling and optimization of modern and complex building services systems	(PropertyMgmt) INT 3, (PropertyMgmt) INT 6, (PropertyMgmt) INT 5, (PMClient) INT 5
	Providing practice-oriented training and education for maintenance and operation staff	(DM) INT 3, (PropertyMgmt) INT 4, (PropertyMgmt) INT 1, (PMContractor) INT 1
Building operation, maintenance, and optimization	Continuous monitoring and improvement/ optimization of building operation	(PropertyMgmt) INT 4, (PropertyMgmt) INT 3, (PMContractor) INT 3
Project delivery	Adequate investment of time and money for design, planning, and implementation of renovation activities (including energy efficiency measures)	(PMClient) INT 2, (PMContractor) INT 1
	Documenting and using lessons learned in building renovation projects	(PMClient) INT 2, (PMClient) INT 5
Building's energy system	Detailed and purposeful measurement/ metering of energy consumption as the insightful source of information for guiding building occupant behavior	(PMContractor) INT 2, (PropertyMgmt) INT 2
	Improving the resilience of the building energy system using hybrid systems	(PMClient) INT 4, (PropertyMgmt) INT 5

Legend for the Reference column:
DM → Design Experts Interviewee Group
PMClient → Client Project Management Interviewee Group
PMContractor → Contractor Project Management Interviewee Group
Property Mgmt → Building Operation and Maintenance Interviewee Group

interconnected hybrid energy systems. Thus, education of staff for optimal energy system operation and identification of potential improvements in building portfolios are important tasks. An important improvement needed for project delivery is the proper sharing of responsibilities, such that the designers and builders remain invested in the correct operation of the building energy systems even after construction is finished.

4   Discussion

This study provides insights into the challenges and solutions associated with energy efficiency in building renovation projects, particularly in the context of existing Finnish building stock. An earlier study by Häkkinen and Belloni [14] presented issues related to sustainable buildings in Finland, as reported by building professionals. Some of the biggest issues were lack of sustainability assessment methods for clients, short time perspective of developers, lack of client understanding of the benefits and alternatives in sustainable buildings, lack of designer knowledge about modern energy systems, lack of sustainability management team in the process, maintenance services not covering user guidance and building monitoring, overly high priority given to district heating and lack of sustainable refurbishment concepts. Similar issues have also been observed outside of Finland [8]. These include lack of life cycle perspectives, insufficient knowledge of solutions and their benefits by all the stakeholders, and lack of leadership and vision to commit to environmentally friendly buildings.

Many of the issues reported more than 10 years ago still remain. For example, there is still a need for improved monitoring to ensure that the building operates as designed and is maintained at that level. Despite significant developments in the uptake of new technologies like heat pumps, the low investment cost criteria are still common and traditional district heating solutions can be selected due to their ease of use and fear of alternative long-term investments. Maintenance staff still lack the knowledge to ensure optimal operation of buildings. All this implies the need for public knowledge sharing on various options and the need for new ways of operation and marketing for consultants and contractors, to give more weight to sustainable long-term solutions. The clients must be assured that even their complex energy systems will work as intended over their whole life cycle.

Although this study has made valuable contributions, further research and actions are needed to address the challenges and effectively implement the solutions outlined in this study. It is important to note that the study findings are limited to the European and Finnish context and may not be applicable to other regions with different building stock characteristics, regulations, and economic conditions. For this reason, future research should thoroughly examine regional variations to better understand the optimal options for energy-efficient building renovations and to identify useful universal principles that may be utilized in any country. Additionally, combining qualitative and quantitative data can provide a more comprehensive understanding of the economic and environmental impacts of energy-efficient renovations. Thus, long-term studies are needed on energy consumption, cost savings, and carbon emission reductions. This study examined issues that are specifically related to building renovation, but many of the building energy efficiency challenges are shared by new construction as well. Project delivery issues are one such example.

In addition, it is crucial to understand how occupant behavior impacts energy consumption in renovated buildings. Therefore, studying behavior change interventions and occupant engagement strategies can lead to more effective energy-saving measures. Finally, in-depth case studies of successful energy-efficient renovation projects can offer practical insights and lessons learned for industry professionals and policymakers.

5 Conclusions

This study aimed to explore the challenges/barriers to and the solutions/enablers of achieving energy efficiency in building renovation projects. This was accomplished through a qualitative study of semi-structured interviews. The opinions of project professionals representing client, design, contractor, and property management were obtained and analyzed. Accordingly, it is concluded that:

- Sufficient adequacy and accuracy of energy consumption metering and its public availability provide a valuable means for renovation planning and an insightful source of information for evaluating the added value of the completed renovation.
- Using hybrid systems improves the resilience of building energy systems.
- Up-to-date and practice-oriented education and maintenance staff training are necessary for harnessing the potential of modern energy systems.
- Creating more tempting incentives for the renovated buildings with high energy efficiency provides a reasonable justification for more investments.

This study recognizes the significance of the perspectives of project professionals and seeks to understand their views on the challenges and solutions in the field. The research also highlights the need for greater collaboration between all stakeholders. Additionally, this study provides insights into how technology can be used to streamline the building renovation process.

Acknowledgments This study was conducted as part of the research program "Hiilineutraalit energiaratkaisut ja lämpöpumpputeknologia" (HybE). We sincerely thank all the funders of the program: Paavo V. Suomisen Rahasto, Sähkötekniikan ja energiatehokkuuden edistämiskeskus STEK ry, Granlund Oy, Ramboll Finland Oy, HUS Tilakeskus, HUS Kiinteistöt Oy, and Senaatti-kiinteistöt. We also thank all the interviewees for their helpful collaboration in providing the research materials.

References

1. Nejat, P., Jomehzadeh, F., Taheri, M.M., Gohari, M., Majid, M.Z.A.: A global review of energy consumption, CO2 emissions and policy in the residential sector (with an overview of the top ten CO2 emitting countries). Renew. Sust. Energ. Rev. **43**, 843–862 (2015). https://doi.org/10.1016/j.rser.2014.11.066
2. Camarasa, C., Mata, É., Navarro, J.P.J.: A global comparison of building decarbonization scenarios by 2050 towards 1.5–2 C targets. Nat. Commun. **13**(1), 3077 (2022). https://doi.org/10.1038/s41467-022-29890-5
3. Gram-Hanssen, K.: Existing buildings–users, renovations and energy policy. Renew. Energy. **61**, 136–140 (2014). https://doi.org/10.1016/j.renene.2013.05.004
4. Frei, B., Sagerschnig, C., Gyalistras, D.: Performance gaps in Swiss buildings: an analysis of conflicting objectives and mitigation strategies. Energy Procedia. **122**, 421–426 (2017). https://doi.org/10.1016/j.egypro.2017.07.425

5. Borgstein, E.H., Lamberts, R., Hensen, J.L.M.: Mapping failures in energy and environmental performance of buildings. Energy Build. **158**, 476–485 (2018). https://doi.org/10.1016/j.enbuild.2017.10.038
6. Zare, N., Saryazdi, S.M.E., Bahman, A.M., Shafaat, A., Sartipipour, M.: Investigation of heating energy performance gap (EPG) in design and operation stages of residential buildings. Energy Build. **301**, 113747 (2023). https://doi.org/10.1016/j.enbuild.2023.113747
7. Willan, C., Hitchings, R., Ruyssevelt, P., Shipworth, M.: Talking about targets: how construction discourses of theory and reality represent the energy performance gap in the United Kingdom. Energy Res. Soc. Sci. **64**, 101330 (2020). https://doi.org/10.1016/j.erss.2019.101330
8. Jagarajan, R., Asmoni, M.N.A.M., Mohammed, A.H., Jaafar, M.N., Mei, J.L.Y., Baba, M.: Green retrofitting–a review of current status, implementations and challenges. Renew. Sust. Energ. Rev. **67**, 1360–1368 (2017)
9. Zou, P.X., Alam, M.: Closing the building energy performance gap through component level analysis and stakeholder collaborations. Energy Build. **224**, 110276 (2020)
10. Day, J.K., Gunderson, D.E.: Understanding high performance buildings: the link between occupant knowledge of passive design systems, corresponding behaviors, occupant comfort and environmental satisfaction. Build. Environ. **84**, 114–124 (2015). https://doi.org/10.1016/j.buildenv.2014.11.003
11. Boge, K., Salaj, A., Bjørberg, S., Larssen, A.K.: Failing to plan – planning to fail: how early phase planning can improve buildings' lifetime value creation. Facilities. **36**(1/2), 49–75 (2018). https://doi.org/10.1108/F-03-2017-0039
12. Rasmussen, H.L., Jensen, P.A.: A facilities manager's typology of performance gaps in new buildings. J. Facil. Manag. **18**(1), 71–87 (2020). https://doi.org/10.1108/JFM-06-2019-0024
13. Saunders, M.N.K., Lewis, P., Thornhill, A.: Research Methods for Business Students, 8th edn. Harlow, Pearson Education Limited (2019)
14. Häkkinen, T., Belloni, K.: Barriers and drivers for sustain-able building. Build. Res. Inf. **39**(3), 239–255 (2011). https://doi.org/10.1080/09613218.2011.561948

Enhancing Climate Resilience in Mixed-Mode Buildings: A Study of Hybrid Ventilation Strategies in a Cold Climate

Mehrdad Rabani, Guilherme B. A. Coelho, and Arnkell Jonas Petersen

Contents

1 Introduction

In an era where environmental unpredictability and sustainability challenges are at the forefront of architectural design and urban planning, the resilience of the built environment has never been more crucial. The concept of resilience in this context refers to the ability of buildings to adapt and maintain functionality in the face of diverse and changing conditions [1]. Building's resilience is increasingly important

M. Rabani (✉)
Department of Built Environment, Oslo Metropolitan University, Oslo, Norway

Multiconsult Norge AS, Oslo, Norway
e-mail: mehrab@oslomet.no

G. B. A. Coelho
Department of Built Environment, Oslo Metropolitan University, Oslo, Norway

CERIS and Departamento de Engenharia Civil, Faculdade de Ciências e Tecnologia, Universidade NOVA de Lisboa, Lisbon, Portugal

A. J. Petersen
Faculty of Science and Technology, Norwegian University of Life Sciences, Ås, Norway

© The Author(s) 2025
M. Kioumarsi, B. Shafei (eds.), *The 1st International Conference on Net-Zero Built Environment*, Lecture Notes in Civil Engineering 237,
https://doi.org/10.1007/978-3-031-69626-8_83

as the impacts of climate change have to be considered, which is anticipated to influence both building requirements and functionality. Specifically, for the purposes of this paper, it is important to note that climate change is expected to shorten the heating season in Norway to varying degrees based on geographic location, as referenced in [2]. Concurrently, there may be a proportional increase in the cooling season. This shift suggests that climatization strategies that are not designed to handle higher future loads might face operational challenges, and this would be an example of a lack of resilience.

In response to these evolving conditions, mixed-mode climatization concepts, which integrate natural and mechanical ventilation systems, are emerging as a paradigm of sustainable design. These concepts align with the concept of resilience by offering solutions to improve indoor air quality, reduce energy consumption, and ensure occupant comfort conditions [3]. However, as future climate scenarios are assessed, it becomes evident that hybrid ventilation systems in mixed-mode buildings might encounter challenges, particularly in adapting to the increased frequency of extreme weather events and fluctuating temperature patterns. This underscores the necessity for further research focused on analyzing hybrid ventilation in mixed-mode buildings, with an aim to ensure their sustainability, resilience, and efficacy in the face of evolving climatic conditions.

2 Related Studies

Multiple scholarly studies have been conducted to evaluate the resilience and adaptability of various building typologies in response to projected future climate scenarios. Attia et al. [4] performed a scoping review, examining existing definitions of resilience and exploring diverse approaches to achieving resilient cooling of buildings. The aim was to investigate the definition of resilience against overheating and power outage considering the scale of built domain and timeline of resilience. The study emphasizes a holistic approach to resilient cooling design, involving a combination of passive and active measures, renewable energy, and storage capacities, underpinned by the necessity of building operation systems and user collaboration for real-time optimization and adaptation. Farahani et al. [5] conducted an evaluation of the resilience of residential buildings in Finland, focusing on their adaptability to cold climate countries amid changing climatic conditions. Their results revealed that in Finland, residential buildings, both old and new, struggle with overheating during hot summers, with passive strategies like all-day ventilation boost and openable windows proving effective but insufficiently alone. Future building codes and retrofitting measures need to address this by incorporating sustainable technologies for cooling, especially under projected hotter conditions for 2050 and 2080. Chen et al. [6] studied an innovative resilient design framework incorporating climate uncertainties from 2020 to 2090 significantly enhances building life cycle performance, considering factors like infiltration, wall and windows U-values, and thermal mass. The framework's application in a cold region office building case study resulted in 44.0% reduction in life cycle carbon emissions

(LCCE), 8.2% in life cycle cost (LCC), and 4.3% in indoor discomfort hours (IDH), showcasing its efficacy in improving building resilience and sustainability under future climate scenarios. Gremmelspacher et al. [7] carried out a study on retrofitting residential buildings for future climate resilience. Simulations of four case studies in Denmark and Germany revealed a significant decrease in heating energy demand (up to 25% reduction) and a reduction in heating peak loads with retrofits. However, increasing thermal discomfort due to heat stress was also observed.

Furthermore, numerous researchers have concentrated on investigating the impact of passive ventilation strategies on the thermal resilience of buildings, and this chapter highlights some of the most recent studies in this area. Peterson [8] carried out research analyzing the energy efficiency of Danish office buildings in the face of climate change, deploying over 313,000 simulations. The study concludes that existing design methodologies might fall short of guaranteeing resilience. Findings project that by the 2080s, the demand for cooling could surge by more than 200%, whereas the need for heating could decrease to less than 50%. Schünemann et al. [9] carried out research to examine the effect of residents' window opening behaviors on overheating, using simulations in two residential structures in Germany. Their findings showed that effective ventilation, like fully opening windows during cooler hours, can significantly reduce overheating, while tilted windows lead to higher temperatures (up to 35 °C). Tavakoli et al. [10] critically reviewed recent research on thermal resilience in low-energy, non-residential buildings with ventilative cooling. Studies highlighted the effectiveness of passive interventions like solar shading and night ventilation in reducing overheating. However, future climate change poses challenges, with some regions likely to see diminished effectiveness of ventilative cooling strategies.

The literature review highlights a growing need to improve the resilience of climatization solutions, beyond just passive design methods, especially considering expected future climate changes. Particularly, there is a significant research gap in exploring the resilience of climatization strategies in colder climates, leading to uncertainties about their effectiveness in such environments. Our study focuses on testing an automated window opening system that enhances hybrid ventilation in buildings located in colder areas, particularly with an eye toward future climatic conditions. The main goal is to investigate if hybrid ventilation can use less energy than traditional mechanical systems while keeping indoor temperatures comfortable, offering a potentially more sustainable option for future building designs.

3   Methodology

3.1   General Considerations

Figure 1 shows the modeling process for generating climate data scenarios (steps 1–3) and implementing them in the building energy simulation software (BES) for analysis of indoor thermal climate and air quality along with their energy use for different heating, ventilation, and air conditioning (HVAC) scenarios.

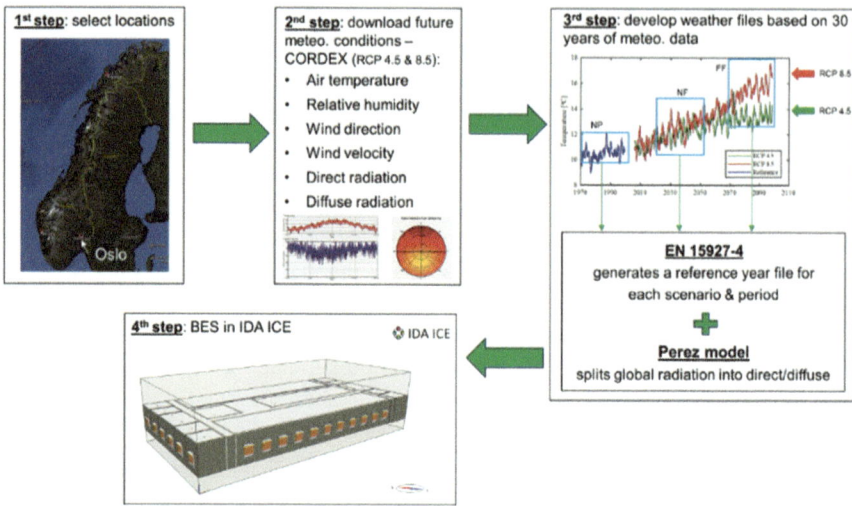

Fig. 1 Scheme of the procedure process, for generating climate data scenarios and running indoor climate and energy simulations, used in this study

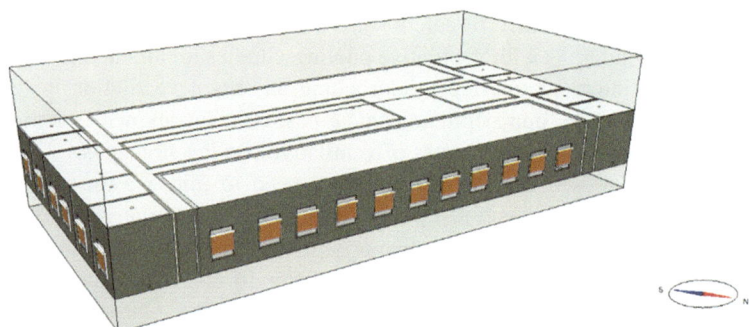

Fig. 2 3D model of office building in IDA-ICE

3.1.1 Case Study

In this study, we examine an office building topology with total floor area of 763 m^2 that serves as a representative example of a floor plan archetype prevalent in Norway, the model floor is depicted in Fig. 2. This topology integrates a variety of spaces, including individual offices, open office landscapes, and meeting rooms. The building envelope and technical properties adhere to the stipulations outlined in the Norwegian Building Code, TEK 17 [11]. Factors contributing to internal gains—namely occupancy, lighting, and equipment—along with their respective utilization profiles, were established in accordance with the Norwegian technical specifications used for evaluating building energy performance against SN-NSPEK 3031 [12].

The window design incorporates a dual-part configuration, with the upper segment being operable, measuring 0.4 meters in height, and the lower segment, non-operable, extending 1.8 meters in height, complemented by an automated external solar shading system. The building properties, internal gains, external shading control, and mechanical ventilation specifications are described in [13].

3.1.2 Ventilation Control and Space Heating Strategies

Our study examines three distinct approaches to ventilation and space heating:

- **Strategy 1**: Employs a full Variable Air Volume (VAV) mechanical ventilation system, enhanced with CO_2 and temperature control. It features water radiators along external walls, particularly beneath windows, for space heating.
- **Strategy 2**: Utilizes a hybrid ventilation system, combining mechanical ventilation with the option to open windows for additional airflow. This setup is activated when mechanical ventilation alone does not achieve desired thermal comfort or CO_2 levels. Space heating is provided through the same method as in Scenario 1.
- **Strategy 3**: Also adopts the hybrid ventilation strategy outlined in Scenario 2 but differs in its heating and cooling approach, utilizing floor heating and cooling systems for temperature control.

In our hybrid solutions, we carefully adjusted the air flow rate to prevent façade openings from operating during the coldest times of the year. This approach, while conservative, reduced the risk of cold air influx compromising indoor comfort. This strategy is grounded in our practical experience with designing robust real-life systems in a Nordic climate. Detailed information on airflow rates applied in different ventilation strategies for two distinct building typologies is presented in Table 1. The simulations were performed in the BES software IDA-ICE 5.0.

Table 1 Airflow rates in different ventilation control and space heating strategies

Ventilation strategies		Strategy 1	Strategy 2	Strategy 3
Cell offices	Max (m^3/h/m^2)	7	3.5	3.5
	Min (m^3/h/m^2)	7	3.5	3.5
Meeting room	Max (m^3/h/m^2)	2.5	13	10
	Min (m^3/h/m^2)	20	13	10
Office landscape	Max (m^3/h/m^2)	2.5	5	4
	Min (m^3/h/m^2)	10	5	4

3.1.3 Weather File Building Methodology

A set of future and contemporary weather files were created, using validated methodologies [14, 15], to evaluate the energy performance of three ventilation strategies under future conditions in relation to contemporary scenarios.

Two distinct emission scenarios were considered (Fig. 3): RCP 4.5 (mid-emission scenario [16]) and RCP 8.5 (high-emission scenario [16]), worst-case scenario. These scenarios account how the climate can change in the future given different scenarios of development and response to climate change, mostly in terms of demographic and socioeconomic development as well as technological evolution and land use change [17]. Three different periods were considered (Fig. 3): the *near future* (NF, 2020–2049), *far future* (FF, 2069–2098), and a *reference period* or *near past* (NP, 1970–1999).

Oslo, classified as a Dfb by Köppen classification [19], that is, humid continental climate, was selected for analysis. Table 2 presents climate assessment for this location for several periods in terms of temperatures, heating degree days (HDD), and heating season (HS). The rise in air temperature across all layers is evident, especially toward the century's close under the RCP 8.5 scenario. Naturally, this warming leads to a reduction in both HDD and HS across all examined periods and scenarios. The HDD and HS were calculated using the methodology described in [2].

Then, the weather files were prepared following the methodology presented in standard EN ISO 15927-4 [20], adhering to a 30-year interval, as recommended by the *World Meteorological Organization* [21]. Each weather file is created based on multi-year records of, at least, temperature, relative humidity, and global radiation (first tier) and wind speed (second tier). The *Finkelstein–Schafer statistic* is calculated for each month of the multi-year period and each meteorological parameter. This will result in a ranking. Then, the three lowest overall ranking months are selected. The final month selection is carried out based on wind speed data.

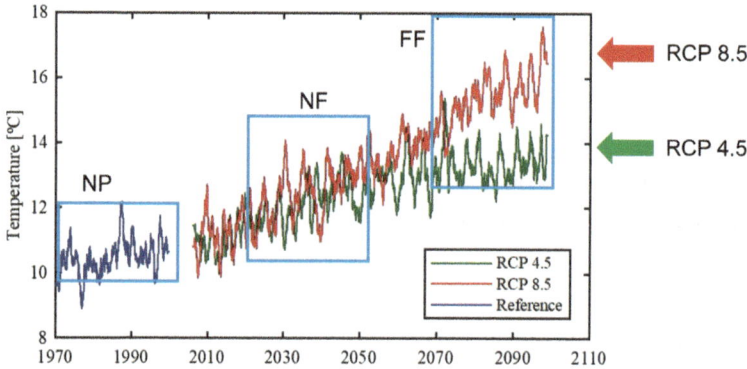

Fig. 3 Annual outdoor air temperature variation from 1970 until 2100 based on data retrieved from CORDEX. (Figure adapted from Ref. [18])

Table 2 Climate conditions comparison—historical climate (Ref.) and near future (NF) and far future (FF) for scenario RCP 4.5 & RCP 8.5—in terms of air temperature, heating degree days, and heating season for Oslo (NO) based on multi-year data (30 years of data each)

Location: OSLO, NO		Temperature (°C)			Heating degree days, HDD (°C)	Heating season, HS (days)
		Percentile 1°	Percentile 50°	Percentile 99°		
Reference		−21.1 ± 3.1	0.9 ± 1.0	18.0 ± 1.8	5779 ± 349	308 ± 17
RCP 4.5	NF	−16.1 ± 2.9	3.4 ± 0.8	18.9 ± 1.5	4890 ± 293	268 ± 13
	FF	−13.4 ± 3.4	4.6 ± 0.8	20.4 ± 1.4	4460 ± 310	251 ± 9
RCP 8.5	NF	−16.4 ± 2.9	3.3 ± 1.0	19.4 ± 1.4	4906 ± 346	268 ± 15
	FF	−10.2 ± 3.7	6.5 ± 1.0	22.4 ± 1.4	3822 ± 361	227 ± 13

Subsequently, the Perez model [22, 23] was employed for splitting the global radiation into its direct and diffuse components and writing to the energy plus weather (EPW) format commonly used by BES, using the authors' developed code [15].

4   Results and Discussion

This section discusses the results from simulations that test energy use and indoor climate under different ventilation strategies in an office building, considering three ventilation systems and five climate scenarios.

Figure 4 displays the energy consumption of the building under three scenarios and five sets of climate data. When looking at the five climate scenarios, there is a noticeable increase in cooling needs, especially under climate scenarios RCP 4.5 and 8.5, for both scenarios in all HVAC cases. The largest increase is seen in Case 1, with fully mechanical ventilation, where cooling needs jumped up to 170% under the RCP 8.5 scenario compared to the standard climate scenario. However, the need for heating energy drops by about 60%. This trend is similar in Cases 2 and 3.

Interestingly, the highest energy savings due to warmer climates are found in Case 3, with savings of about 40% under the RCP 8.5 FF scenario, compared to around 16% in Case 1. This is likely because of longer periods of window opening in Cases 2 and 3, which use hybrid ventilation (as shown in Fig. 5). It highlights the effectiveness of hybrid ventilation as a passive design strategy for reducing energy consumption and improving indoor temperatures in cold climates.

Figure 6 displays the changes in average monthly temperatures in a south-facing office under two conditions (Case 1 and Case 2) and different climate change scenarios. For both cases, the operative hourly temperatures stay within the acceptable range set by Norwegian building codes, that is, a maximum of 50 hours above

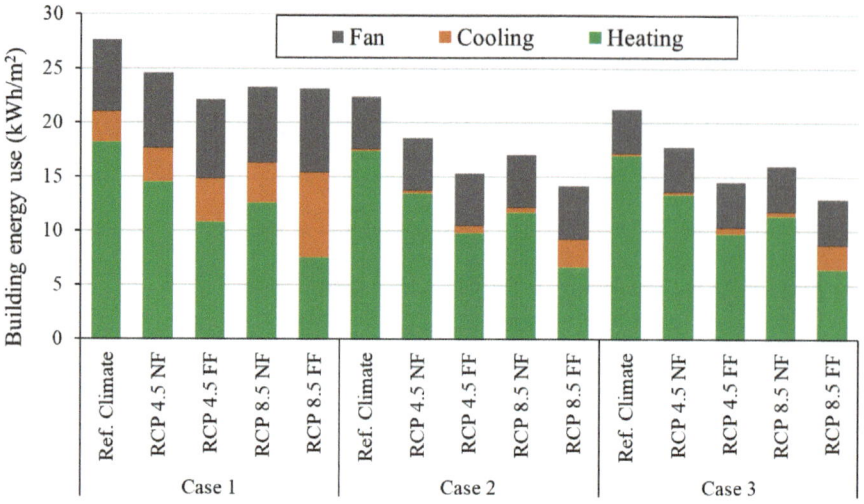

Fig. 4 Energy use of the office building across three different HVAC strategies

Fig. 5 Percentage of window opening in Cases 2 and 3 across various climate change models

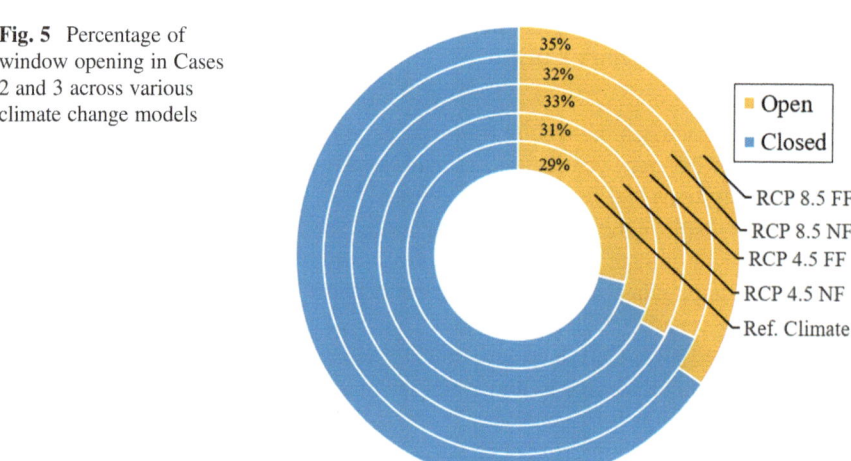

26 °C. However, in the climate change scenarios identified as FF, the average indoor temperature remained higher compared to other climate scenarios.

Figure 7 shows a carpet plot of CO_2 variation during the year in a south-facing office for Case 2, for both the reference climate and the near future climate scenario.

As noted earlier in Figs. 4 and 5, under climate change scenarios, an increase in outdoor temperature facilitates the reduction of indoor CO_2 levels. This reduction is facilitated by prolonged window opening, which increases air exchange rates within the room. Such findings underscore the viability of hybrid ventilation systems, particularly in regions with cold climates and low pollution levels, exemplified by

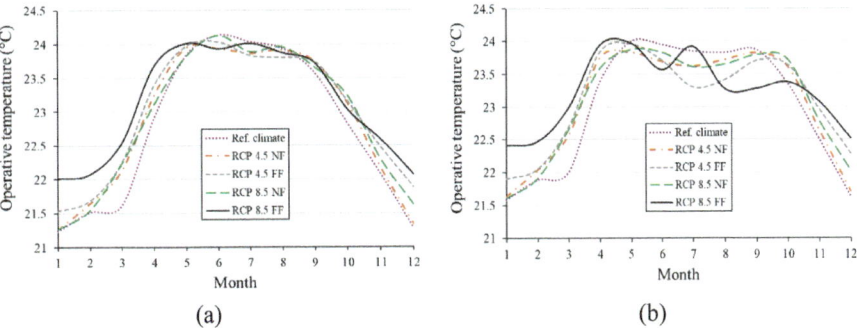

Fig. 6 Average monthly temperature changes in a south-facing office for: (**a**) Case 1 and (**b**) Case 2

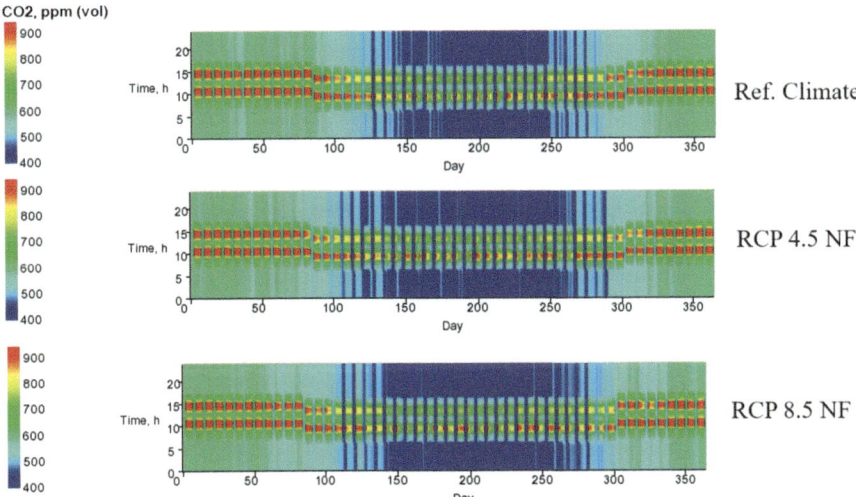

Fig. 7 Carpet plot of yearly CO_2 variation in Case 2 for the reference and near future climate

Norway. This is further supported by the expanded coverage of light and dark blue regions in Fig. 7. This trend is also observed in FF climate scenarios but to a greater degree.

5   Conclusions

This study embarked on an investigation into the resilience of climatization strategies within the context of architectural design and urban planning, focusing primarily on the adaptability of buildings in cold climates amid changing environmental conditions. By exploring the efficacy of hybrid ventilation systems in office buildings, specifically under future climate scenarios, we aimed to bridge the gap identified in the literature regarding the resilience and sustainability of climatization solutions. The results of our simulations, analyzing different HVAC strategies under a range of climate scenarios, reveal significant insights into the performance of hybrid ventilation systems. Particularly, the findings demonstrate that hybrid ventilation, facilitated by automated window opening, presents a promising strategy for reducing energy consumption while maintaining comfortable indoor temperatures. The adaptability of this approach to varying climate conditions—especially considering the projected increase in cooling demand and the corresponding decrease in heating requirements—highlights its potential as a resilient climatization strategy.

Moreover, the extended periods of window opening in scenarios employing hybrid ventilation underscore the system's ability to leverage natural ventilation effectively, thereby reducing reliance on mechanical systems. This not only contributes to energy savings but also ensures a healthier indoor environment by facilitating improved air quality, as evidenced by the observed reduction in CO_2 levels.

In conclusion, our research underlines the importance of integrating passive design strategies, such as hybrid ventilation via window opening, into the architectural design and urban planning discourse in cold climate. Given the projected shifts in climate conditions and the increasing emphasis on sustainable development, such strategies offer a viable pathway to enhancing the resilience and sustainability of the built environment in cold climates.

Acknowledgments This work was based on BES-models supplied by: *Hybridene—Optimal hybrid ventilasjon i fremtidens bygg*, financed by The Research Council of Norway (project n. 327591).

References

1. McAllister, T.: Developing Guidelines and Standards for Disaster Resilience of the Built Environment: A Research Needs Assessment. US Department of Commerce. National Institute of Standards and Technology (2013)
2. Skaugen, T.E., Tveito, O.E.: Present Conditions and Scenario for the Period 2021–2050. Norwegian Meteorological Institute (2002)
3. Peng, Y., Lei, Y., Tekler, Z.D., Antanuri, N., Lau, S.-K., Chong, A.: Hybrid system controls of natural ventilation and HVAC in mixed-mode buildings: a comprehensive review. Energ. Buildings. **276**, 112509 (2022). https://doi.org/10.1016/j.enbuild.2022.112509
4. Attia, S., Levinson, R., Ndongo, E., Holzer, P., Berk Kazanci, O., Homaei, S., Zhang, C., Olesen, B.W., Qi, D., Hamdy, M., Heiselberg, P.: Resilient cooling of buildings to protect against heat waves and power outages: key concepts and definition. Energ. Buildings. **239**, 110869 (2021). https://doi.org/10.1016/j.enbuild.2021.110869
5. Farahani, A.V., Jokisalo, J., Korhonen, N., Jylhä, K., Kosonen, R.: Simulation analysis of Finnish residential buildings' resilience to hot summers under a changing climate. J. Build. Eng. **82**, 108348 (2024). https://doi.org/10.1016/j.jobe.2023.108348
6. Chen, R., Samuelson, H., Zou, Y., Zheng, X., Cao, Y.: Improving building resilience in the face of future climate uncertainty: a comprehensive framework for enhancing building life cycle

performance. Energ. Buildings. **302**, 113761 (2024). https://doi.org/10.1016/j.enbuild.2023.113761

7. Gremmelspacher, J.M., Sivolova, J., Naboni, E., Nik, V.M.: Future Climate Resilience through Informed Decision Making in Retrofitting Projects, in: 2020, pp. 352–364. https://doi.org/10.1007/978-3-030-58808-3_26

8. Petersen, S.: The Effect of Local Climate Data and Climate Change Scenarios on the Thermal Design of Office Buildings in Denmark, in: BuildSim-Nordic. Oslo, Norway (2020)

9. Schünemann, C., Schiela, D., Ortlepp, R.: How window ventilation behaviour affects the heat resilience in multi-residential buildings. Build. Environ. **202**, 107987 (2021). https://doi.org/10.1016/j.buildenv.2021.107987

10. Tavakoli, E., O'Donovan, A., Kolokotroni, M., O'Sullivan, P.D.: Evaluating the indoor thermal resilience of ventilative cooling in non-residential low energy buildings: a review. Build. Environ. **222**, 109376 (2022). https://doi.org/10.1016/j.buildenv.2022.109376

11. TEK17. Technical Regulations for Planning and Execution of Construction Work. §13–4 Thermal indoor climate: Norwegian Ministry of Local Government and Modernisation., (2017)

12. SN-NSPEK 3031:2023, Energy performance of buildings Calculation of energy needs and energy supply (in Norwegian), Standard Norge, (2023)

13. Rabani, M., Petersen, A.J.: Detailed assessment of hybrid ventilation control system in a mixed-mode building in cold climate. J. Phys. Conf. Ser. **2600**, 102006 (2023). https://doi.org/10.1088/1742-6596/2600/10/102006

14. Coelho, G.B.A., Kraniotis, D.: A multistep approach for the hygrothermal assessment of a hybrid timber and aluminium based facade system exposed to different sub-climates in Norway. Energ. Buildings. **296**, 113368 (2023). https://doi.org/10.1016/j.enbuild.2023.113368

15. Coelho, G.B.A., Henriques, F.M.A.: Performance of passive retrofit measures for historic buildings that house artefacts viable for future conditions. Sustain. Cities Soc. **71**, 102982 (2021). https://doi.org/10.1016/j.scs.2021.102982

16. Intergovernmental Panel on Climate Change (IPCC), IPCC, 2014, Climate Change 2014: Synthesis Report. Contribution of Working Groups I, II and III to the Fifth Assessment Report of the Intergovernmental Panel on Climate Change Core Writing Team, R.K. Pachauri and L.A. Meyer (eds.).., (2014) 151. https://doi.org/10.1017/CBO9781107415324

17. K.G. N. Nakicenovic, J. Alcamo, G. Davis, B. de Vries, J. Fenhann, S. Gaffin, W. A. Griibler, T.Y. Jung, T. Kram, E.L. La Rovere, L. Michaelis, S. Mori, T. Morita, M. Pepper, H. Pitcher, L. Price, K. Riahi, A. Roehrl, H.-H. Rogner, A. Sankovski, Z.D. Schlesinger, P. Shukla, S. Smith, R. Swart, S. van Rooijen, N. Victor, Special Report on Emissions Scenarios–A Special Report of Working Group III of the Intergovernmental Panel on Climate Change, 2000

18. Coelho, G.B.A., Rebelo, H.B., De Freitas, V.P., Henriques, F.M.A., Sousa, L.: Current and future geographical distribution of the indoor conditions for high thermal inertia historic buildings across Portugal via hygrothermal simulation. Build. Environ. **245**, 110877 (2023). https://doi.org/10.1016/j.buildenv.2023.110877

19. Kottek, M., Grieser, J., Beck, C., Rudolf, B., Rubel, F.: World map of the Köppen-Geiger climate classification updated. Meteorol. Z. **15**, 259–263 (2006). https://doi.org/10.1127/0941-2948/2006/0130

20. EN ISO 15927-4, Hygrothermal performance of buildings - Calculation and presentation of climatic data - Part 4: Hourly data for assessing the annual energy use for heating and cooling, European Committee for Standardization (CEN), (2005)

21. World Meteorological Organization, Guide to Climatological Practices WMO-No. 100, 2011. https://doi.org/WMO-No. 100

22. Perez, R.R., Ineichen, P., Maxwell, E.L., Seals, R.D., Zelenka, A.: Dynamic global-to-direct irradiance conversion models. ASHRAE Trans. **98**, 354–369 (1992)

23. Perez, R., Ineichen, P., Seals, R., Zelenka, A.: Making full use of the clearness index for parameterizing hourly insolation conditions. Sol. Energy. **45**, 111–114 (1990). https://doi.org/10.1016/0038-092X(90)90036-C

Recent Progress in Net-Zero-Energy Buildings in Tropical Climates: A Review of the Challenges and Opportunities

Mengesha Asefie Mengaw, Wubishet Jekale Mengesha, and Habtamu Bayera Madessa

Contents

1 Introduction

The world shares the same principal mission to reduce energy consumption and pursue a sustainable development path for all. To achieve this goal Energy utilization in buildings is a critical issue right now. Because buildings are a major primary energy consumer in the world energy sector, at about 40% of the total global energy consumption [1–3]. In this way, optimizing building energy utilization is imperative in the network of global energy needs, with considerable energy investment funds needed for viable plans, development, and the operation of buildings [2]. Subsequently, the energy productivity of buildings is broadly recognized as a basic

M. A. Mengaw (✉)
Addis Ababa Science and Technology University, Department of Civil Engineering, Addis Ababa, Ethiopia

W. J. Mengesha
Addis Ababa University, Ethiopian Institute of Architecture, Building Construction and City Development, Addis Ababa, Ethiopia

H. B. Madessa
Oslo Metropolitan University, Department of Built Environment, Oslo, Norway

© The Author(s) 2025
M. Kioumarsi, B. Shafei (eds.), *The 1st International Conference on Net-Zero Built Environment*, Lecture Notes in Civil Engineering 237,
https://doi.org/10.1007/978-3-031-69626-8_84

arrangement for handling energy deficiencies, carbon outflows, and their unfavorable environmental impacts [3]. Various activities have been attempted to present imaginative energy-saving innovations and promote ecofriendly buildings [4]. From those activities, net zero energy building, and net positive energy building are sustainable energy thematic areas.

A net-zero-energy building (NZEB) is characterized as one whose total sum of energy used annually is approximately equal to the total sum of the renewable energy produced on-site for the building's use [5]. Due to advancements in building innovation, renewable energy systems, and academic research, the development of NZEBs is quickly becoming more feasible [7].

An NZEB is seen as a promising arrangement to improve productivity within the worldwide development industry by combining energy productivity with the integration of nearby renewable energy sources [4]. Customizing building frameworks and plan strategies to suit particular territorial climate conditions is fundamental to realizing the vision of NZEBs. As laid out by the Universal Energy Office [5], accomplishing net-zero-energy outflows by 2050, a necessity to meet the 1.5 °C Paris Agreement target, requires expanding the development of renewable energy innovations, increasing energy effectiveness, decreasing methane emissions, and extending net-zero energy goals. According to the International Energy Agency report entitled as World Energy Outlook 2023; "These technologies hold the potential to achieve over 80% of the fundamental emissions cuts by 2030."

A tropical climate is characteristic of regions in the tropics, which lie between the Tropic of Cancer and the Tropic of Capricorn. In most of the literature, tropical climates' monthly average temperature throughout the year is above 18 °C [6]. In tropical climates, renewable energy sources, particularly solar energy, are abundant. Research endeavors into solar energy utilization encompass a broad spectrum of applications, spanning from the capture of low-temperature thermal energy applications [7] to the integration of photovoltaic systems within building structures [8–10].

The design of NZEBs in tropical climates should follow a hierarchical approach that prioritizes passive design features, followed by active energy-efficiency measures and finally by renewable energy technologies. The passive design features of NZEBs aim to reduce the cooling and ventilation loads of buildings by optimizing the building orientation, shape, envelope, shading, natural ventilation, daylighting, and landscaping [11]. Active energy-efficiency measures aim to reduce the energy consumption of mechanical and electrical systems, such as air conditioning, lighting, electric appliances, and controls, by using high-efficiency equipment, sensors, and automation. Renewable energy technologies aim to supply the majority of the energy demand of buildings by using solar, wind, biomass, or other sources, either on-site or off-site [1].

Recently, NZEBs have been growing rapidly in developed countries, though facing challenges and barriers. More than 90% of over 300 NZEB projects listed on the International Energy Agency (IEA) Solar Heating and Cooling (SHC) Program task 40 world map are located in the developed regions of the European Union (EU) and the United States [12]. This chapter endeavors to offer an overall assessment of the most recent progression in achieving net-zero-energy developments in tropical climates while recognizing the essential obstacles and exploring the potential prospects for sustainable development in these regions. By overcoming these obstacles and taking advantage of the openings, significant head-way can be made toward a greener, more energy-efficient future. More specifically, the main objective of this chapter is to provide a compressive systematic literature review on the recent progress, challenges, and opportunities in achieving net-zero-energy buildings in tropical climates.

2   Research Methods

This chapter aimed to assess the recent progress in constructing net-zero-energy buildings in tropical climates, with a particular focus on the challenges and oppor-tunities associated with optimizing building energy. Various research articles were summarized for this purpose. A methodical combination of string-detecting searches in the Scopus database and secondary snowball search techniques were employed. This comprehensive approach was selected to bridge the gap in existing literature. Peer-reviewed articles written in English and published between 2014 and 2023 were considered for this review, using the title, abstract, and keywords fields in the databases. In total, 1261 records were initially extracted, and removing duplications reduced this number to 66 articles. After thorough analysis, 23 articles were chosen for inclusion in the review.

The study relied solely on original research publications and resulted in a detailed overview. Table 1 presents the keywords employed in the article search process, and Fig. 1 illustrates the criteria for inclusion and exclusion at various stages, adhering to the Preferred Reporting Items for Systematic Reviews and Meta-analyses (PRISMA) guidelines.

Table 1 Summary of keywords and their respective definitions

Keyword	Definition
Net-zero emissions	Achieving a balance between the volume of greenhouse gases emitted and the volume removed from the atmosphere
Decarbonization	The process of reducing carbon emissions from energy production, transpor-tation, and industry
Energy efficiency	Improving building design and systems to minimize energy consumption
Renewable energy	Sources like solar, wind, and geothermal that produce clean energy
Passive design	Designing buildings to maximize natural heating, cooling, and lighting
Active energy	The energy consumption of mechanical and electrical systems, such as air conditioning, lighting, electric appliances, and controls, by using high-efficiency equipment, sensors, and automation
Grid integration	Integrating renewable energy sources into existing power grids

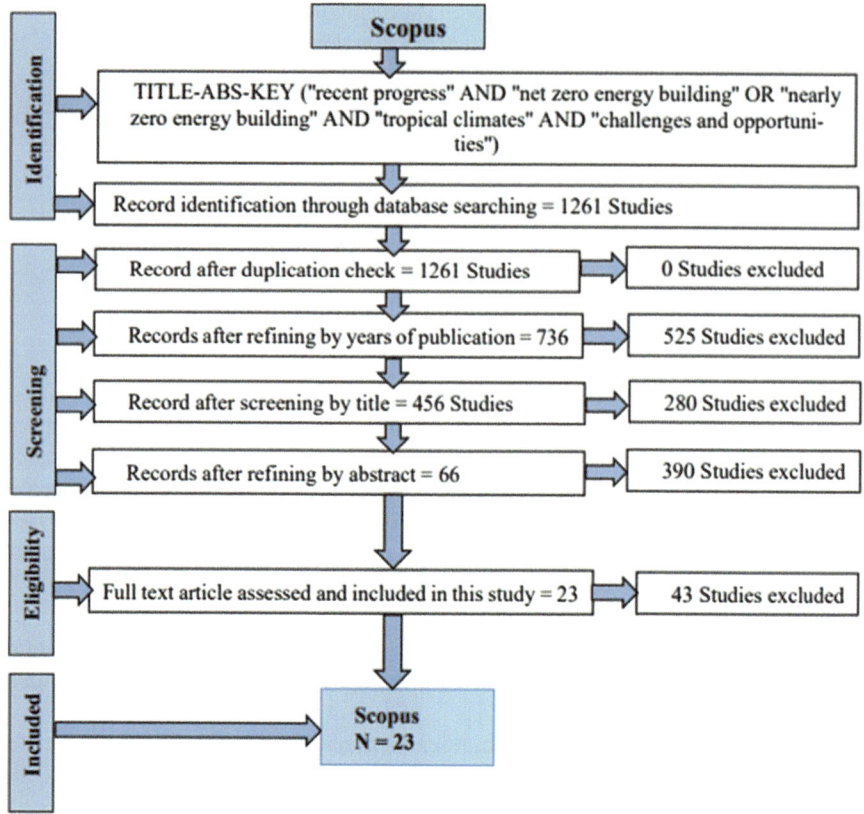

Fig. 1 Flowchart of the article search for this study

3   Results and Discussion

Figure 2 illustrates the annual publication trends of research articles focusing on nearly zero-energy buildings (nZEBs) from 2014 to 2023, in accordance with the inclusion and exclusion criteria outlined in the PRISMA guidelines. Different research articles show that significant progress has been made since 2000 in response to global warming from carbon emissions; researchers have focused on new building technologies and sustainable science and technology. After that, the research areas became a sensitive issue in reducing global warming by introducing the concepts of nearly zero-energy buildings, net-zero-energy buildings, and zero-emission buildings (Table 2).

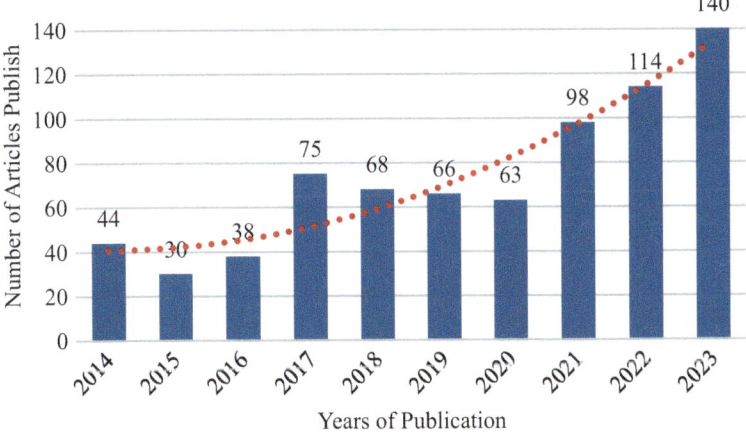

Fig. 2 Net-zero-energy building research trends over time

3.1 Recent Progress in NZEBs

No clear picture of the recent progress of NZEBs in tropical climates is available, but the Net Zero Stocktake 2022 report shows that the share of national emissions categorized by states' and regions' net-zero targets is similar for G7 countries and China, at around 40–50%, where the United States and EU countries' states and regions cover around 30–40%, but other countries' states and regions cover below 20% of net-zero targets [18]. Similarly, among G7 countries, about 70% of people living in large cities are categorized by net-zero targets; the same figure holds for the European Union; the United States covers about 60% of its population with a net-zero target; and China has a low population coverage rate, below 20% [18]. The global net-zero target has nearly the same coverage rate due to the population living in large cities, and the coverage rate for all states and regions is 36–37%.

3.2 Opportunities

Various studies have been carried out on the opportunity to build sufficient energy demand to construct NZEBs. Among them, Uspenskaia et al. [14] and others presented empirical evidence from the smart city project SPARCS on the challenges and barriers to constructing net-zero/positive-energy buildings and districts. The growing demand for energy-efficient buildings, advances in technology and construction materials, and government policies and incentives promoting NZEBs are some of the drivers of constructing NZEBs [19, 20, 23]. Similarly, another study has shown several building-integrated photovoltaic/thermal (BIPVT) applications and

Table 2 Summary of opportunities for and challenges to adopting net-zero-energy buildings

	Opportunities	Challenges	References
Barriers to and driving forces of industrial energy-efficiency improvements in African industries: a case study of Ghana's largest industrial area	Cost reductions from lowered energy use Threats of rising energy prices	Lack of access to funds/ financial constraints Lack of management awareness Weak energy policy frameworks Weak information systems	[19–22]
Challenges for and barriers to net-zero/positive-energy buildings and districts: empirical evidence from the smart city project SPARCS	Growing concerns about climate change and energy security Increasing demand for energy-efficient buildings Advancements in building technologies and materials Government policies and incentives promoting NZEBs	Lack of awareness and knowledge about NZEBs among stakeholders Lack of incentives and regulations to promote NZEBs Limited availability of skilled labor and professionals Technical challenges in integrating renewable energy sources	[2, 19, 23]
Building-integrated photovoltaic/thermal (BIPVT) systems: applications and challenges	Produce electrical and thermal energies Remove cooling and heating loads from buildings Reduce the electricity bills of a building Reduce operational energy costs Decrease the carbon emission of the building Act as thermal insulation Increase building value	The BIPVT systems are required to overcome some barriers that still limit their increased adoption, including the considerations of maintenance and replacement, standards and codes, building loads, cost-efficiency, and social and psychological factors Overheating of the system and building The use of highly expensive materials Feasibility dependent on the national and local supporting policy The electricity cost exceeding than that of grid electricity	[15]
Solar power utility sector in India: challenges and opportunities	Availability of solar energy Availability of wasteland Minimum dependency of solar installations on external costs	Technological barriers, policy and regulatory barriers, financing barriers, transparency, and accountability	[19, 20, 22, 23]

(continued)

Table 2 (continued)

	Opportunities	Challenges	References
Concentrated photovoltaics as light harvesters: outlook, recent progress, challenges, sustainable energy technologies, and assessments	Low cost, high efficiency, and environmental friendliness	Direct normal irradiance, high cell temperature, soiling, optical design, reliability, and durability	[16]
Technological advancements toward net-zero-energy communities: a review of 23 case studies from around the globe	Offers the sharing of needs, costs, and resources among the buildings within the community in a way that is beneficial and cost-effective A microgrid at a community level capable of supplying excess energy to the national energy grid, which reaps additional earnings from the system	Retrofitting or refurbishment is one of the major barriers to the integration of renewable energy tools and technologies within existing buildings The net-zero-energy (NZE) target is more feasible for low-rise residential buildings than high-rise buildings	[2]
Toward zero carbon in a hot, humid subtropical climate	An evaluation of existing NZEBs and net-zero-carbon buildings (NZCBs) shows that several key strategies can achieve net zero: (1) reducing energy demand through the use of low-energy passive design measures; (2) increasing efficiency by using energy-efficient building systems and technologies; and (3) using renewable energy sources to supply the remaining energy demand	Designed target vs. operating target: the disparity between the intended use and the actual use of the facilities at NZCBs Ongoing testing, commissioning and recommissioning Occupant behavior, comfort, and satisfaction Balancing the goal of zero carbon with the provision of a comfortable indoor environment Reliable renewable energy generation: significant overshadowing from surrounding tall buildings limits energy generation by photovoltaics (PVs), particularly during wintertime The legislative and policy framework and energy infrastructure currently limit the wider application of NZEBs The lack of ready infrastructure and incentives for the grid feed-in of electricity generated by renewable sources	[17]

(continued)

Table 2 (continued)

	Opportunities	Challenges	References
A review of hybrid renewable energy systems: solar and wind-powered solutions—challenges, opportunities, and policy implications	Energy reliability: storage units can store excess generated energy, offering a buffer when renewable sources are not generating power Grid stability: in on-grid configurations, energy storage can help in load leveling Optimized use of resources: advanced control systems can intelligently manage energy storage Microgrid capability: in microgrid systems, storage units can enable the system to operate independently of the central grid if needed	Technical challenges: integration complexity, intermittency, infrastructure development, energy storage, and power quality Economic challenges: high initial costs, return on investment, uncertainty, and market maturity Environmental challenges: land use and resource assessment Regulatory and policy challenges: inconsistent policies, grid integration policies, licensing, and standards	[19, 20, 23]
Overview and future challenges of nearly zero-energy building (nZEB) design in Southern Europe		Lack of local governance and a national strategy to create an infrastructure for nZEB implementation Societal and technical barriers to nZEB implementation Lack of policy framework Lack of access to funds	[19, 20, 23]
A review of net-zero-energy buildings in hot, and humid Climate: experience learned from a case study on 34 buildings	NZEBs tend to adopt advanced energy-efficient designs and technologies Natural ventilation and other passive technologies can effectively help NZEBs reduce their cooling energy use Solar energy is the most abundant	The adoption of advanced technologies in NZEBs is strongly influenced by their countries' building energy codes and standards A key barrier is their high upfront cost Lack of personnel for NZEB design, construction, operation, and management	[1]

has identified challenges and opportunities, such as removing cooling and heating loads from buildings, reducing buildings' electricity bills, reducing operating energy costs, and reducing carbon emissions, and increasing the value of buildings [16, 18].

By promoting energy-efficient and low-carbon buildings, many cross-cutting challenges and opportunities can emerge for the construction industry and its stakeholders [20]. They are associated with lower energy consumption, zero carbon emissions, and macroeconomic benefits through indirect cost savings from improving energy security and resilience; creating jobs; improving the health, productivity, and comfort of building occupants; and enhancing assets [21]. These benefits are also well aligned with and contribute to various United Nations sustainable development goals (SDGs).

3.3 Challenges

The challenges identified in this study include, but are not limited to, political, economic, social, and knowledge barriers.

- Policy barriers: Policy design and implementation often require extensive coordination across government agencies and different levels of government, which also includes inconsistent policies, grid integration, licensing, and standards books [22].
- Economic barriers: These barriers include increasing the upfront costs of more-energy-efficient buildings and equipment, a lack of available capital to meet upfront costs, and a poor understanding of the life-cycle costs of maintaining high performance in operations [16, 19–23].
- Social barriers: The behavior and social practices of building occupants significantly influence the gap between a building's predicted and actual energy performance. These factors are complex and thus difficult to address and are strongly influenced by emotional rather than rational decision-making [13, 16]. Studies have shown that whether or not residents are directly responsible for energy bills has a strong influence on their energy-consumption behavior in buildings [24].
- Knowledge barriers: These range from a lack of professional capacity to design and construct high-performance buildings to a lack of capacity or training in policy design and implementation [19, 20, 23].

Overcoming these obstacles by implementing policies that realize benefits will inevitably entail costs. The design and implementation of a carbon-reduction and energy-efficiency policy for buildings, as in any policy, will generate many costs and benefits, which must be taken into account when evaluating the cost-effectiveness of the policy. Therefore, determining the cost-effectiveness of a policy is both a technical and a political calculation [23]. On the contrary, several studies have shown that the preceding points impact both the opportunities and challenges related to the global energy crisis, sustainability strategies, a future-oriented green society, and government policies and regulations.

4 Conclusion

This chapter highlighted the challenges faced in constructing NZEBs in tropical and humid climates, where adoption remains limited when compared to temperate and developed regions. The key challenges identified include the high initial costs of building NZEBs, a lack of design and operational expertise, and difficulties in implementing effective policies. To overcome these barriers, supportive policies and incentives, such as financial support and standards for development, need to be implemented to encourage the wider adoption of NZEBs. Raising awareness among and educating stakeholders, using abundant renewable energy sources, and considering user behavior and comfort are also essential strategies. Efforts to promote NZEBs in tropical climates could help to reduce carbon emissions and ensure reliable energy supplies while addressing rising electricity costs and environmental concerns.

References

1. Feng, W., et al.: A review of net zero energy buildings in hot and humid climates: experience learned from 34 case study buildings. Renew. Sustain. Energy Rev. **114**, 109303 (2019). https://doi.org/10.1016/j.rser.2019.109303
2. Ullah, K.R., Prodanovic, V., Pignatta, G., Deletic, A., Santamouris, M.: Technological advancements towards the net-zero energy communities: a review on 23 case studies around the globe. Sol. Energy. **224**, 1107–1126 (2021). https://doi.org/10.1016/j.solener.2021.06.056
3. Tabatabaei, M., et al.: Reactor technologies for biodiesel production and processing: a review. Prog. Energy Combust. Sci. **74**, 239–303 (2019). https://doi.org/10.1016/j.pecs.2019.06.001
4. Cao, X., Dai, X., Liu, J.: Building energy-consumption status worldwide and the state-of-the-art technologies for zero-energy buildings during the past decade. Energy Build. **128**, 198–213 (2016). https://doi.org/10.1016/j.enbuild.2016.06.089
5. Kumar, R., Chakraborty, S., Elangovan, D., Padmanaban, S.: Concept of net zero energy buildings (NZEB) – a literature review. Clean. Eng. Technol. **11**, 100582 (2022). https://doi.org/10.1016/j.clet.2022.100582
6. Sudhakar, K., Winderla, M., Priya, S.S.: Net-zero building designs in hot and humid climates: a state-of-art. Case Stud. Therm. Eng. **13**, 100400 (2019). https://doi.org/10.1016/j.csite.2019.100400
7. Madessa, H.B., et al.: Investigation of solar absorber for small scale solar concentrating parabolic dish. In: Proceedings of Solar World Congress, Kassel-Germany, 28 August–2 September 2011
8. Yau, Y.H., et al.: A techno-economical study of a solar-assisted under-floor air distribution system for buildings in the tropics. Sustain. Energy Technol. Assess. **51**, 101915 (2022)
9. Yau, Y.H., Lim, K.S.: Energy analysis of green office buildings in the tropics—photovoltaic system. Energy Build. **126**(15), 177–193 (2016)
10. Shabunko, V., et al.: Evaluation of in-situ thermal transmittance of innovative building integrated photovoltaic modules: application to thermal performance assessment for green mark certification in the tropics. Energy. **235**(15), 121316 (2021)
11. Behzadi, A., Arabkoohsar, A.: Feasibility study of a smart building energy system comprising solar PV/T panels and a heat storage unit. Energy. **210**, 118528 (2020). https://doi.org/10.1016/j.energy.2020.118528

12. International Energy Agency: Net Zero by 2050: A Roadmap for the Global Energy Sector. OECD Publishing, Paris (2021)
13. Apeaning, R.W., Thollander, P.: Barriers to and driving forces for industrial energy efficiency improvements in African industries – a case study of Ghana's largest industrial area. J. Clean. Prod. **53**, 204–213 (2013). https://doi.org/10.1016/j.jclepro.2013.04.003
14. Uspenskaia, D., Specht, K., Kondziella, H., Bruckner, T.: Challenges and barriers for net-zero/positive energy buildings and districts—empirical evidence from the smart city project sparcs. Buildings. **11**(2), 1–25 (2021). https://doi.org/10.3390/buildings11020078
15. Maghrabie, H.M., Elsaid, K., Sayed, E.T., Abdelkareem, M.A., Wilberforce, T., Olabi, A.G.: Building-integrated photovoltaic/thermal (BIPVT) systems: applications and challenges. Sustain. Energy Technol. Assess. **45**, 101151 (2021). https://doi.org/10.1016/j.seta.2021.101151
16. Ejaz, A., et al.: Concentrated photovoltaics as light harvesters: outlook, recent progress, and challenges. Sustain. Energy Technol. Assess. **46**, 101199 (2021). https://doi.org/10.1016/j.seta.2021.101199
17. To, C., Li, J., Kam, M.: Towards zero carbon in a hot and humid subtropical climate. Procedia Eng. **180**, 413–422 (2017). https://doi.org/10.1016/j.proeng.2017.04.200
18. Axelsson, K., et al.: Disclaimer Download (2023). [Online]. Available: www.zerotracker.net/analysis/net-zero-stocktake-2023
19. Cielo, D., Subiantoro, A.: Net zero energy buildings in New Zealand: challenges and potentials reviewed against legislative, climatic, technological, and economic factors. J. Build. Eng. **44**, 102970 (2021). https://doi.org/10.1016/j.jobe.2021.102970
20. Attia, S., et al.: Overview and future challenges of nearly zero energy buildings (nZEB) design in Southern Europe. Energy Build. **155**, 439–458 (2017). https://doi.org/10.1016/j.enbuild.2017.09.043
21. Kamal, A., Al-Ghamdi, S.G., Koc, M.: Revaluing the costs and benefits of energy efficiency: a systematic review. Energy Res. Soc. Sci. **54**, 68–84 (2019). https://doi.org/10.1016/j.erss.2019.03.012
22. Badiola, M., Basurko, O.C., Gabiña, G., Mendiola, D.: Integration of energy audits in the Life Cycle Assessment methodology to improve the environmental performance assessment of Recirculating Aquaculture Systems. J. Clean. Prod. **157**, 155–166 (2017). https://doi.org/10.1016/j.jclepro.2017.04.139
23. UNEP: Letter from the Executive UNEP in 2019, pp. 1–16 (2019). [Online]. Available: https://wedocs.unep.org/bitstream/handle/20.500.11822/32374/AR2019.pdf?sequence=1&isAllowed=y
24. Delzendeh, E., Wu, S., Lee, A., Zhou, Y.: The impact of occupants' behaviors on building energy analysis: a research review. Renew. Sustain. Energy Rev. **80**, 1061–1071 (2017). https://doi.org/10.1016/j.rser.2017.05.264

BREEAM Excellent and Zero-Emission Educational Building: A Case Study on Energy Performance and Indoor Climate Evaluation

Emma Zheng Liang, Bente Hellum, Carl Fredrik Melle, Adrian Spoletini, and Hermann Sliper Langeteig

Contents

1 Introduction

Nowadays, sustainability has become a highly influential concept and a decisive factor in economic, social, and environmental decisions of developed countries. Buildings, which serves as an important component in the society, have a prevailing role as a hub for the realization of sustainable development. In order to facilitate and implement sustainability in both the construction and use of buildings, advanced assessment methods such as BREEAM have been developed by the Building Research Establishment (BRE) in the United Kingdom since 1990. Relying on dedicated criteria and specified processes, multidisciplinary teamwork toward sustainable development can be coordinated and realized (Lowe, Watts, Jack, Norman, 2011) [1]. The framework of BREEAM can promote integration of diverse knowledge and know-hows from different experts.

E. Z. Liang (✉)
Rambøll Norge AS, Oslo, Norway
e-mail: emma.liang@ramboll.no

B. Hellum · C. F. Melle · A. Spoletini · H. S. Langeteig
OsloMet-Oslo Metropolitan University, Oslo, Norway

M. Kioumarsi, B. Shafei (eds.), *The 1st International Conference on Net-Zero Built Environment*, Lecture Notes in Civil Engineering 237,
https://doi.org/10.1007/978-3-031-69626-8_85

However, the dilemma of optimizing energy performance while maintaining a comfortable indoor climate is a notable paradox, particularly evident in educational buildings. Achieving energy performance to passive house level can be difficult due to the energy demand of cooling, while maintaining an optimal indoor climate for BREEAM certification.

As mentioned, the optimized project values, such as PPD (Predicted Percentage of Dissatisfied) or PMV (Predicted Mean Vote), are primarily derived from an integrated approach involving various experts. However, these parameters are largely based on subjective perception, underscoring the importance of user-oriented studies to address this dilemma.

This study concentrates on a pilot project, specifically highlighting one of the Norway's most environmentally friendly schools, which has attained BREEAM certification at the Excellent level. The objective is to evaluate the relationship between energy performance and indoor climate by user-oriented interviews and field inspections. The investigation goes beyond theoretical analysis, aiming to uncover practical correlations among energy supply solutions, building envelope, and technical details.

This paper draws from a bachelor's project conducted by three students [2] at Oslo Metropolitan University. Their contribution adds a practical and objective dimension to the research by interviews and field inspections.

In Sect. 1, we introduce the certification systems in Norway-BREEAM NOR. Sect. 2 presents the case study with a detailed and objective exploration of the challenges faced in achieving both energy efficiency and an optimal indoor climate in educational settings. Sect. 3 illustrates the findings and concludes this paper.

2   Methodology: Case Study

The study employs a qualitative research methodology, specifically through case study interviews. The complexity of paradox of energy optimization objectives while maintaining optimal indoor climate conditions for educational buildings underscores the reason of the chosen case study approach. By examining a real pilot case study, employing in-depth interviews and practical fieldwork, this methodological approach offers a robust framework for elucidating the interrelationships among certification processes, indoor environmental quality, and energy performance parameters. The questionnaire is based on and compared to standardized values for school buildings from Örebro model based on a World Health Organization strategy at the early 1980s (Andersson, K. Fagerlund, I. Bodin L, and Ydreborg B, 1988) [3]. The questionnaires of MM060 were chosen based on the background of the project, as outlined in Tables 8 and 12 of the School Environment manual (Andersson, K. Strihd, G. Fagerlund, I. Aslaksen, W) [4–7]. To check whether the indoor climate is good, we look at survey responses related to an Örebro survey. If people are happy with the indoor climate, their answers should match

Örebro's reference values. This study presents preliminary findings from the early stage of the proposed research topic. Future research aims to explore how indoor climate evaluation can be leveraged to optimize technical settings.

2.1 The Case Study

Zero-emission educational building—This school was chosen as the case study for its ambitious environmental goals, which pose a challenge in maintaining a comfortable indoor climate while minimizing energy consumption. The data for the study were collected through interviews conducted using Google Forms (Fig. 1).

The project aims to set a pilot project in environmental and climate-friendly construction in 2022. With ambitious goals set by the municipality to achieve BREEAM-Excellent certification, passive house standards, zero-emission building (ZEB-OM) status, and the extensive use of cross-laminated timber.

The school, designed to accommodate approximately 360 students and 45 staff members within a total gross area of 4500 m^2, a multifaceted team of experts, including external consultants specializing in zero-emission buildings, has been assembled to tackle the project's diverse requirements. In meeting the municipality's requirement for ZEB—O ÷ EQ [8] compliance, a comprehensive renewable energy

Fig. 1 The case study

solution has been devised, primarily leveraging solar energy. For heating, a demand-controlled ventilation system was chosen over traditional methods. The centralized heating system incorporates a highly efficient liquid-to-water heat pump, thermal storage capacity, and geothermal heat pumps for both heating and cooling needs. The choice of extensive use of cross-laminated timber is committed to reducing carbon emissions. The school aims to be a living embodiment of sustainability principles, fostering synergies between education and environmental stewardship.

The bachelor's project focuses on examining critical aspects of the building's energy efficiency and indoor climate quality in relation to the achieved BREEAM certification. This involves conducting field work, interview, and analysis to assess the interrelations between high-energy performance and indoor climate.

2.2 Energy Efficiency and Indoor Climate

Through fieldwork, interviews, and energy analysis, it was found that while the focus on energy efficiency could be achieved, it comes at the expense of user comfort in terms of indoor climate. The survey revealed firstly dissatisfaction among students and teachers regarding indoor temperatures, with classrooms often feeling too warm. Compared to standard guidelines of indoor climate from BREEAM(Hea03) [9], temperatures were consistently higher than recommended, possibly influenced by the survey's timing in April, a period marked by fluctuating outdoor temperatures. For instance, starting at 5°C at 8:00 am, temperatures rose to 14°C by 1:00 pm. Such temperature swings could cause discomfort indoors, exacerbated by the building's effective insulation, and extensive window coverage. Secondly, the survey explored symptoms related to poor indoor air quality, with a significant number of respondents reporting such issues, indicating a need for indoor environment enhancements.

2.2.1 The Perceived PPD/PMV Versus Required PPD/PMV According to BREEAM/Hea03

The results of the in-depth interview on perceived thermal comfort are presented as PPD and PMV values with the graph in Fig. 2. Despite the participants of the study being distributed across four different rooms, the graph shows that one point of the case study that represents the average of all responses regardless of various locations of the room. The perceived PMV values presented here are based on responses from the survey participants (refer to Appendix 1[10]) and not representing a precisely calculated result. Even though the calculated PMV value for a building is more detailed and includes several factors such as room air velocity, operative temperature, metabolism, clothing, and relative humidity than the perceived PMV value. However, this perceived PMV value can be used to provide an overview of how the building's occupants perceive the PMV value.

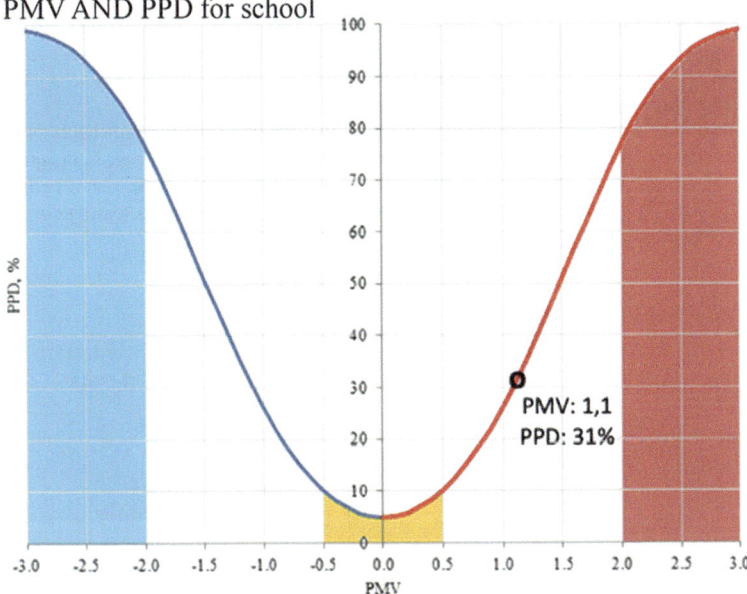

Fig. 2 The PMV/PPD value of the case study (PPD is shown as a function of PMV with the requirement from NS-EN ISO 7730:2005, Table A.1, category B marked in yellow)

In the BREEAM-NOR 2016 manual, Chapter HEA 03—Thermal Comfort, points can only be achieved if the PMV and PPD values in the building's regularly occupied area meet the requirements in category B from Table A.1 in NS-EN ISO 7730:2005 [11]. The requirement is a PMV value between -0.5 and 0.5 with a corresponding PPD value below 10%, as illustrated by the yellow shaded area in Fig. 2.

The survey resulted from in-depth interview in a PMV value of 1.1 and a PPD value of 31%, which is much higher than the values used when the building was certified. This suggests a general dissatisfaction with the thermal indoor climate among students and teachers at the school, with reports of excessively high temperatures in the classrooms than calculated PPD/PMV which is documented according to the requirement from HEA03 for BREEAM. Compared to the standard *Ergonomi i termisk miljø* (see Fig. 3) according to NS-EN-ISO 7730:2005, this result is far above what is expected of modern buildings, as the lowest level (class C) requires a PPD < 15% while the case study has a PPD of 31%. This may be caused by the survey being conducted in mid-April, a period with highly fluctuating outdoor temperatures, which is also reflected in some responses shown in Appendix 1 [10]. The day the survey was conducted is also an example of this, with a temperature of 5°C at school start at 8:00 a.m. increasing to 14°C when the responses were submitted at 1:00 p.m. (climate data according to YR [12]).

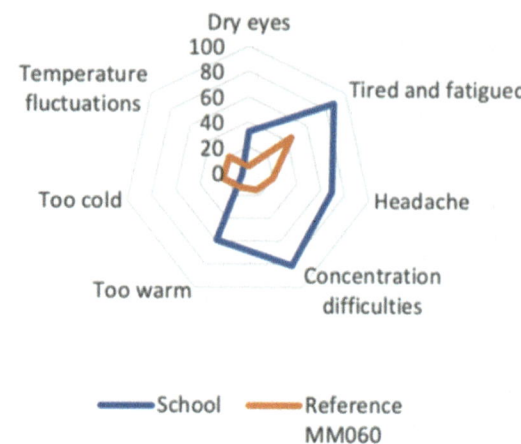

Fig. 3 Frequency of perceived indoor climate-related symptoms of the case study in percentage compared to reference values

The survey also revealed whether participants have experienced one or more symptoms commonly associated with an unsatisfactory indoor climate. The results are presented as the percentage of participants who have experienced the respective symptoms and are illustrated using a rose diagram in Fig. 3.

The results shown in the rose diagram are based on survey answers, so they only give an example to check if results are different from the reference frequency using Table 1 in the School Environment manual according to the Örebro model. The rose diagram shows differences according to School Environment Table 1 for six out of eight symptoms people were asked about. The symptoms people often experience in this survey include feeling too hot, dry eyes, tiredness, headaches, and difficulty concentrating. This shows clearly that people are not happy with how warm it is indoors. To ensure objective results from the survey, interviews were conducted with a total of 74 participants from four different classes. It is improbable that factors like insufficient sleep, allergies, or unengaging teaching influenced the outcome.

The survey results can help to identify which indoor climate factors might be causing the reported symptoms. In this case, it seems that participants feel there's poor air circulation or not enough ventilation in the classrooms. This could be because the need for cooling wasn't properly estimated, leading to insufficient ventilation.

The gray graph in Fig. 4 represents the energy consumption recorded by the SD system from January to April, while the orange graph illustrates the projected energy demand at the design phase.

Figure 4 indicates that the energy consumption in January exceeds projections by over 10,000 kWh, suggesting potential underestimation of heating needs during that period. Subsequent months show a stabilization in consumption, with February slightly higher than expected and lower consumption in both March and April.

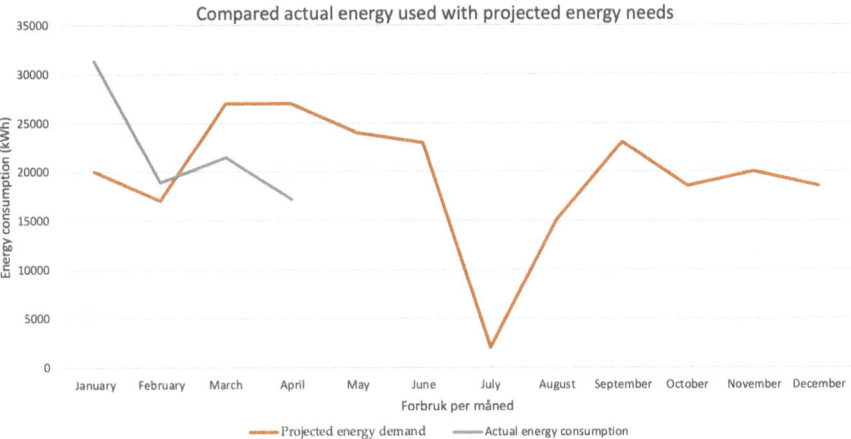

Fig. 4 The comparison of energy consumption between the designed and actual usage from January to April without additional energy consumption

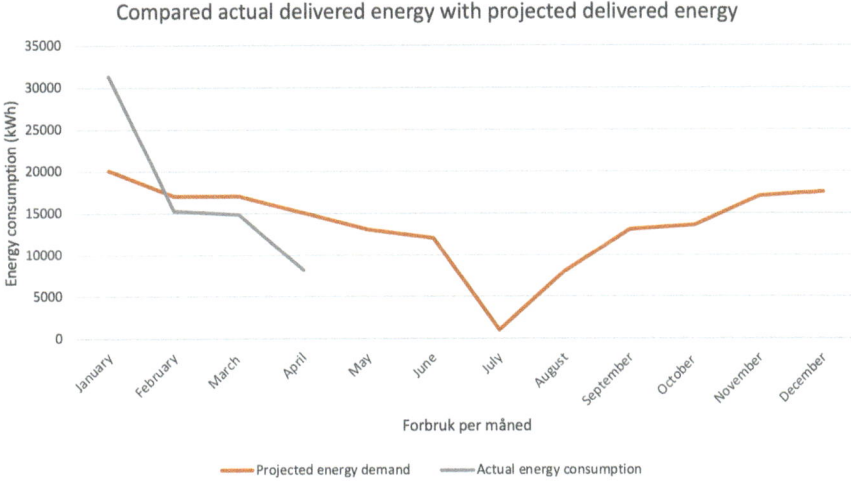

Fig. 5 The actual energy consumption from January to April compared to the projected energy demand with additional energy consumption

Further insights into the source of additional energy consumption are provided in Fig. 5, which displays the case study's delivered energy consumption alongside projected values.

It is evident here that the actual energy consumption in January contrasted similarly to Fig. 5. Thus, the building actual energy consumption was 1.5 times more than projected energy demand in January. For the remaining period, delivered energy consumption was slightly lower than projected, attributed to higher-than-expected electricity production from solar panels and slightly lower total consumption.

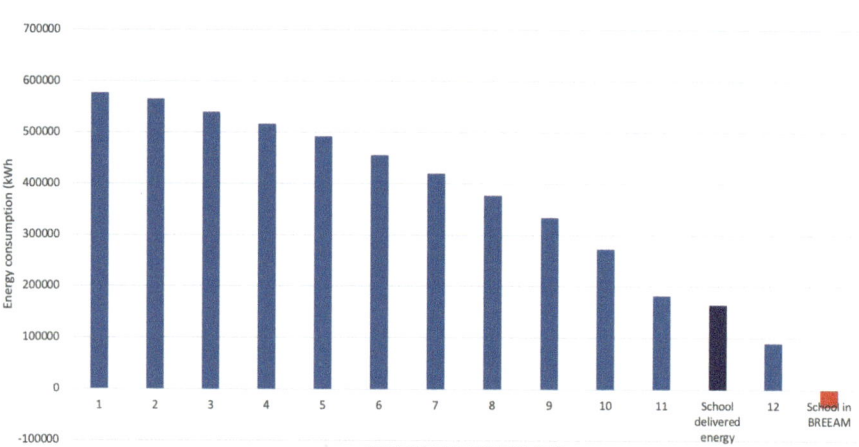

Fig. 6 Comparison of energy performance for BREEAM and energy production

The school is equipped with a 338 kWp solar panel system, contributing to climate-neutral operation and achieving ZEB certification. There is a correlation between energy efficiency and ZEB certification, and the school's potential to improve student comfort through data from the BMS is further examined. The school achieves maximum points for energy efficiency, largely thanks to solar power production.

In the BREEAM-NOR certification of the case study, the building earned 12 out of 12 points in the Ene 01 category by achieving over a 100% improvement from energy grade C. Despite delivered energy comprising 69.7% of the projected annual consumption, the certification overlooks the building's energy load as long as energy export is significant. This is exemplified in Fig. 6, where the blue columns represent the case study's various requirements for maximum annual energy consumption, and the dark blue column denotes the projected consumption of delivered energy used for point allocation for the certification.

Figure 6 illustrates that for the case study to achieve full compliance in this aspect of the BREEAM-NOR manual, there is a requirement for an annual consumption not exceeding 90,000 kWh to achieve 12 points. During the building's certification under Ene 01, a consumption of −34,348 kWh was assumed, reflecting an expected export of self-generated solar energy of 200,000 kWh subtracted from the projected consumption of delivered energy, totaling 164,000 kWh. This adjustment stems from compliance note SN5 in the BREEAM-NOR manual. If only a solar panel system sufficient to cover the remaining 30.3% of the building's energy consumption had been installed, scoring in Ene 01 would have been based solely on the building's delivered energy of 164,000 kWh, resulting in 11 out of 12 possible points. This raises questions about the actual benefits derived from the 1656 m^2 of solar panels installed. It is plausible that all the solar panels were installed to ensure an overproduction of solar energy greater than the building's consumption of delivered energy, thus guaranteeing full compliance with Ene 01 in the BREEAM-NOR manual. Therefore, compliance note SN5 shifts the focus of Ene 01 away from rewarding energy-efficient buildings to favoring those with high-energy production.

2.2.2 Discussions and Findings

One proposed solution involves adjusting the cooling system or reducing morning heating to maintain comfortable temperatures throughout the day. However, careful consideration is required to balance comfort and energy consumption. Alternative approaches, such as improving insulation or implementing shading mechanisms, could also be explored to create a more conducive indoor learning environment.

A more energy-saving solution during periods of greater outdoor temperature variations would be to reduce heating early in the day and accept a slightly lower temperature in the morning hours. These are two solutions that would contribute to an indoor climate where the conditions for a good learning environment are more present.

With such large temperature variations in a relatively short time, clothing that was suitable at the beginning of the school day may become too warm later on during the interview time. This effect could also be exacerbated by the building being highly insulated while having large window areas that allow in a lot of sunlight.

PMV and PPD values that are so far from the recommended range indicate that measures should be taken for improvement. A natural measure in this situation would be to investigate if room cooling system be fine-tuned based on external factors throughout the day, although this would contribute to increased energy consumption during such periods.

In the case study, even though the energy performance standards for BREEAM certification were met, the PPD value is critical and should not be compromised, as it is central to the building's purpose. A suitable indoor climate is essential for effective daily learning.

The frequency of reported symptoms in this survey is suggesting that measurements should be taken to improve the situation. If the cause is seasonal, simple short-term measures might help. The Norwegian Institute of Public Health (*Folkehelseinstituttet*) has suggested some relevant measures, such as ensuring students leave the classroom during breaks to let in fresh air or adjusting the timetable to manage the number of people in the room. Longer term measures could include improving lighting and using external shading device to prevent high temperatures in the classrooms. These measures can help make the school a better learning environment and reduce the frequency of reported symptoms.,

3 Conclusion: Lessons of Experience

The case study,a BREEAM "Excellent" educational building, was investigated regarding its thermal indoor climate in relation to BREEAM themes of energy, materials, and indoor climate. Energy consumption was closely linked to the thermal indoor climate, necessitating a study of energy consumption and indoor climate environment.

An Örebro survey revealed that the majority of students (77 interviewees) experienced the indoor climate as too warm, with temperature fluctuations being either too cold or too hot. This led to inquiries into why the temperature was perceived as too warm and whether the actual energy consumption deviated from the projected [13]. By utilizing the cooling system, which had not previously been used at the school, the PMV value could be reduced, and thermal comfort is improved. However, implementing a cooling system could potentially impact energy efficiency and thus negatively influence BREEAM scoring.

In providing this recommendation, careful consideration was given to whether the building could still achieve full points in the BREEAM energy section, despite increased energy consumption from cooling. Analyses showed that actual consumption for March and April, when high temperatures were reported, was lower than projected. This suggests that the school has the potential to improve thermal comfort while retaining BREEAM points.

The question of whether electricity produced from the PV system returned to the grid should count positively in the energy balance was also considered. While BREEAM's benefits for the external environment are evident, improvement in the indoor environment remains uncertain. The balance between energy efficiency and thermal comfort should be considered in future measures to optimize user experience. This emphasizes once again the importance of post-occupancy evaluation to compare projected values with actual delivered values.

It is crucial to note that it is still in the early stages of the research project. Conducting a more extensive user survey in the future to assess ongoing perceptions of building comfort would yield more accurate results. Establishing a feedback system for continuous monitoring could also be considered.

References

1. Lowe, W., Jack, N.: An evaluation of A BREEAM Case Study Project (PDF). Sheffield Hallam University Built Environment Research Transaction: 42–53 (2011)
2. Bachelor's project conducted by three students: Carl Fredrik Melle, Adrian Spoletini, Hermann Sliper Langeteig
3. Andersson, K., Fagerlund, I., Bodin, L., Ydreborg, B.: Questionnaire as an instrument when evaluating indoor climate. Healthy Buildings'88 Stockholm. **1**, 139–146 (1988)
4. Andersson, K., Stridh, G.: The use of standardized questionnaires in BRI/SBS surveys. In: Levy, F., Maroni, M. (eds.) Pilot study on Indoor Air Quality. NATO/CCMS, Oslo (1991)

5. Andersson, K., Stridh, G., Fagerlund, I., Larsson, B.: The MM-questionnaires—a tool when solving indoor climate problems. Department of Occupational and Environmental Medicine, Örebro University Hospital, Örebro, Sweden. 1993, (17 pages)
6. Andersson K.: Management of the sick building—investigation strategies. Proceedings of the International Symposium on New Epidemics in Occupational Health, Helsinki: 117–120 (1994)
7. Andersson K., Fagerlund I., Larsson, B.: Reference data for questionnaire MM 040 NA—Indoor Climate (work environment). Report M5/90. Department of Occupational and Environmental Medicine. Örebro University Hospital
8. A Norwegian ZEB-definition embodied emission, SINTEFbok
9. Refer to BREEAM manual (page86): byggalliansen.no/wp-content/uploads/2019/06/SD-5075NOR-BREEAM-NOR-2016-New-Construction-v.1.2.pdf
10. Appendix 1: the chart with survey results about how people felt in the building. It will be provided upon request
11. NS-EN-ISO 7730:2005 Ergonomi i termisk miljø—Analytisk bestemmelse og tolkning av termisk velbefinnende ved kalkulering av PMV- og PPD-indeks og lokal termisk comfort
12. Meteorologisk institutt.: Tveten skole. Yr.no. https://www.yr.no/nb/historikk/graf/1-2623713/Norge/Vestfold%20og%20Telemark/Porsgrunn/Tveten%20skole?q=2023-04-17 Yr.no, provided by the Norwegian Meteorological Institute and NRK, uses numerical weather prediction models, observational data, and statistical methods to deliver precise forecasts and climate data worldwide. It is the most widely used weather forecast website in Norway; (2023, 17. april)
13. Fieldwork indicates that the cooling system is not functioning properly for the Building Management System (BMS) or Sentral Driftskontroll

Sustainable Metamodules. Disseminating Sustainable Practices in Design Workflow Via BIM-Based Approaches

Fabio Conato ⓘ, Ilaria Spasari ⓘ, and Habtamu Bayera Madessa ⓘ

Contents

1 Introduction

1.1 Complexity in Construction

The complexity of the construction industry has, over time, attracted the attention of many researchers, who have analyzed its main causes in order to reduce the possible negative effects. The dimension of some projects often leads to delays and cost overruns, which makes the construction industry one of the most dynamic, risky, and challenging business activities in existence [1]. Being aware of this situation facilitates anticipatory planning, both from a technical and a managerial point of view. Complexity is a significant criterion in choosing an appropriate organizational form for the project, influencing its temporal, economic, and qualitative objectives by

F. Conato · I. Spasari (✉)
Department of Architecture, University of Ferrara, Ferrara, Italy
e-mail: cntfba@unife.it; spslri@unife.it

H. B. Madessa
Department of Built Environment, Oslo Metropolitan University, Oslo, Norway
e-mail: habama@oslomet.no

1027

M. Kioumarsi, B. Shafei (eds.), *The 1st International Conference on Net-Zero Built Environment*, Lecture Notes in Civil Engineering 237,
https://doi.org/10.1007/978-3-031-69626-8_86

contributing to the specification of planning, coordination, and control requirements [2]. Among the main factors that make a project complex, the one that has the greatest impact is organizational complexity, which is generated by insufficient utilization of communication channels and limited generation and utilization of information [3]. It is therefore evident that a crucial aspect in managing complexity within the construction sector is the involvement of individuals in the process, who consequently generate, transmit, and exchange information. Additionally, social, cultural, cognitive, and operational complexities further compound the task's intricacies, escalating as the available time diminishes [4]. The emergence of the 4.0 revolution underscores a shift in how buildings are conceived, constructed, and operated. Digital technologies like Building Information Modeling (BIM), artificial intelligence, and the Internet of Things (IoT) are reshaping industry practices, offering efficiencies and innovations. However, they bring forth challenges in data handling, cybersecurity, and workforce preparedness. Additionally, the integration of advanced materials with smart, sensitive, and active functionalities demands a careful design approach to navigate the opportunities and risks associated with their widespread adoption [5]. In this context, the aim of this paper is to codify the meaning and characteristics of smart architectures capable of addressing and managing this complexity, establishing a structured framework to assist designers in making decisions that will be easily digitalizable, and facilitating the dissemination of sustainability via a BIM-based workflow.

1.2 Smart Architecture

In light of the simultaneity and heterogeneity of management of the generative factors of contemporary complexity, the architect, as director of the building process and interpreter of social needs is faced with contemporary challenges that impose paradigm shifts on the way of living and perceiving reality. It turns out to be increasingly valuable to create an architecture that, as enunciated by Mies van der Rohe, is a process of searching for truth and constructive clarity brought to its exact expression [6]. It is inferred that architecture only achieves constructive coherence if it is able to process, integrate and resolve an open matrix, capable of interpreting design instances by intercepting an ever-increasing and changing flow of information, facilitating the task of the architect as an intellectual who must manage, at different scales of detail, the evolution of a multidisciplinary process. Currently, the term "smart feature" within the realm of architecture commonly refers to sophisticated materials and technologies meticulously designed to intelligently adapt to their surroundings. These elements can react intelligently to external influences, altering their properties, structure, or functions accordingly [7]. Accordingly, this notion has led to the conceptualization of smart buildings as systems capable of adapting, responsively, not only to the variability of external conditions but also to the continuously evolving needs of their occupants [8]. Indeed, Casini [9] defines a smart building as the evolution from a nearly zero-energy building, where alongside the utilization of high-performance materials and technologies, there exists the potential to dynamically control the building's environmental parameters, contingent upon changes in external or internal conditions. In order to avoid confusion between the definitions of *smart building* and *smart architecture*, semantic researches were conducted within the Department of Architecture at Ferrara University.[1] This research first investigated the meaning of the adjective *smart*, followed by an examination of the term *architecture*. The outcome of this research led to the definition of smart architecture as "an architecture that, through innovative methodologies and/or tools, responds to the current needs of society" [10].

1.3 Metamodules

Throughout history, architects have frequently utilized modular elements to create architecture that can effectively address social needs and challenges during pivotal moments in society. The module, which can have varying interpretations depending on the context and time period, is typically associated with hierarchy, standardization, and repetition. The principles of modularity suggest the use of consistently logical criteria to generate ordered relationships between parts and an indefinite number of variables.

Modularity in Architecture One of the earliest examples of modularity can be found in vernacular architecture, particularly in the housing typology of the Nuraghe. Their shape is due to the construction system that characterizes them, the "tholos," which, consisting of rings of blocks arranged progressively tighter from the base to the top, has given classic nuraghi their distinctive conical tower shape. Each nuraghe and human settlement constituted an integral part of a complex system, resulting in a vast array of specific contexts in which the Nuragic communities adapted their settlement methods in response to the terrain, resource distribution, demographic and social development, and internal and external competition dynamics. With the demolition of the city walls and the development of extra-moenia cities, a logical flow of information with causal connections between statements became necessary. In the nineteenth century, the construction feasibility of the module was crucial for the development of urban plans in major European cities, such as the Projecte de reforma i Eixample de Barcelona by the engineer Idelfonso Cerdà y Suñer and the redesign of the city of Paris by the then prefect of the department of the Seine, Georges Eugène Haussmann. Although they have different methods and purposes, both urban plans exhibit a modular principle in the construction characteristics of the urban block and in its relationship with the infrastructure network it connects to. In the first half of the twentieth century, Le Corbusier departed from the concept of a regulatory grid and theorized the Modulor as a tool for controlling

---

[1] Semantic research conducted within the International Doctoral Program in Architecture and Urban Planning. Doctoral candidate: Ilaria Spasari, Supervisor: Fabio Conato.

proportion and an open system not consisting of a sequence of measurements but rather a conceptual dimension [11].

The second half of the twentieth century was marked by the advent of globalization, a phenomenon that perceives the world as a single entity comprising new forms of interaction that transcend national borders [12].

Metamodularity in Contemporary Architecture The semantic evolution of the module, from a simple measuring tool linked to specific construction techniques and materials to an expression of a specific idea, prompts us to consider a possible new configuration of modularity.

Considering the complexity of contemporary architecture, both economic and social, it is necessary to codify architecture in order to respond to contemporary needs with innovative methods and tools. Contemporary architecture is characterized by intrinsic complexities that are difficult to visually perceive [13]. It is produced in multiple locations and is embedded in a social context marked by global cultural, social, and technological revolutions [14]. For these reasons, the modular codification of smart architecture tends to transcend material constraints and abstracts to an informative level, becoming a metamodule.

The metamodule consists of a core of smartness indicators which are related and coherent information that creates order among its parts. It systematizes the complexity of contemporary instances with the vastness of innovative design components and solutions that respond to these instances, using a multi-scale approach.

BIM and Metamodularity The informative nature of the metamodule is easily digitizable as it is based on attributing information to specific architectural components or sets of components. This is compatible with BIM software, where three-dimensional modeled objects are associated with both geometric and non-geometric information. The meta-module consists of an information system that can enrich a digital model. For example, a meta-module template can be created within a BIM Authoring software that incorporates all predetermined Smartness Indicators, associating them as non-geometric properties to specific three-dimensional objects. Through exporting the digital model in an open format file or IFC (Information Foundation Classes), the meta-module contributes to the interoperability among the actors of the building process. This allows them to be aware of the consequences and design interactions of their individual intellectual actions.

2   Smartness Indicators

In order to identify the smartness indicators that make up the metamodules of the smart architecture, categories (see Table 1) have been identified that correspond to the main characteristics that the smart architecture must possess to meet the needs of contemporary society. The identification of categories and therefore the needs of contemporary society have been carried out through the analysis of the main regulations issued in the last 50 years, literature reviews on Scopus and Web of

Table 1 The table shows the smartness indicators that characterizing the metamodules of smart architecture when they match the most appropriate design strategies

Smartness indicators		
Categories	Sub-categories	Indicators
Sustainability	Environmental	Environmental design
		Environmental certification
		e-LCA (environmental LCA)
	Economic	LCC
	Social	SLSA
		Users' and workers' satisfaction
Dynamicity	Urban	Versatility
		Multifunctionality
		Adaptability
	Architectural	Versatility
		Multifunctionality
		Adaptability
	Technological	Versatility
		Multifunctionality
		Adaptability
Coherence	External requirements	Context
		Typology
		Environment
		Comfort
	Internal Alterations	Design alteration
		Constructive alteration
Controllability	Architects' competence	Innovation
		Experience
		Educational Aspects

Science, and through the observation of case studies of contemporary architectures that have generated both positive and negative effects in the context in which they were built. The numerous legislations enacted since the 1970s have focused on the primary need of contemporary society: livability in a more sustainable world. For this reason, sustainability is identified as the foremost category of requirements. It has been subdivided into three subcategories based on the three pillars defined by the 1987 Brundtland Report: environmental, economic, and social. The indicators within these subcategories serve as tools and/or characteristics through which contemporary architecture addresses the needs of society. Thus, the environmental subcategory encompasses passive strategies inherent to environmental design, those aimed at obtaining environmental certifications, and those enabling the achievement of a favorable Life Cycle Assessment (LCA) result. Regarding LCA, it has been decided to consider environmental LCA (e-LCA), which is the environmental component of Life Cycle Sustainability Assessment (LCSA) [15]. Similarly, the economic aspect of sustainability is represented by optimal parameters of Life Cycle Costing (LCC), such as labor cost, material cost, annual capital charge, and end-of-life costs. The social aspect of sustainability is expressed through Social Life Cycle Assessment (SLCA) indicators, which encompass workers' categories (child labor, fair salary, working hours, social benefits, job satisfaction), consumers (health and safety, consumer privacy, transparency), the local community (safe and healthy living conditions, community engagement, local employment), society (public commitment to sustainability issues, contribution to economic development, technology development), and the Value Chain (fair competition, respect for intellectual property rights). Another social aspect of sustainability is related to the comfort and satisfaction of users and workers [16], which can be measured through values of daylight, external views, high-frequency lighting, natural ventilation, indoor air quality, thermal comfort, microbial contamination, acoustic performance, presence of color, and vegetation [17]. The second category of smartness indicators is represented by dynamicity, indicating an object's ability to change its initial condition in response to external stimuli. It is expressed through indicators of versatility, multifunctionality, and adaptability, which vary in significance depending on the scale (urban, architectural, and technological). The third category of smart indicators has been identified as coherence, meaning the architecture's ability to remain faithful to both the external objectives imposed by the client (environmental, urban context, comfort, and building typology) and to express its essence at all project detail scales (also through variations in progress both in the design and construction phases). Finally, the last category of smart indicators is the controllability of information flow in the design and construction process, addressed to architects' competences in solving design challenges through innovative strategies, years of experience in the realization of a specific work, and their ability to continually update themselves on new methods and operational tools.

3 Results and Discussion

3.1 Decision Support Matrix's Structure

The categories of indicators selected represent the main demands that contemporary architecture must respond to. If specific design solutions or best practices satisfy these demands, they constitute sets of structural information for smart architecture. Therefore, the four demand categories of sustainability, dynamicity, coherence, and controllability can be assimilated to four generative metamodules of smart architecture. Consequently, the sub-categories identified earlier become sub-metamodules, which, in turn, contain sets of smartness indicators.

Based on these considerations, a decision-support matrix structure (see Table 2) was developed to assist architects in managing the multiple pieces of information during the design process. The matrix consists of metamodules, which are sub-metamodules that contain smartness indicators at various scales of project development. To ensure easy digitization, the project development scales have been assimilated to the LODs (Level of Developments) commonly used in BIM (Building Information Modelling).

Table 2 The decision support matrix's structure, characterized by metamodules declined at each LOD

Decision support matrix structures						
Categories	Sub-categories	LOD 100	LOD 200	LOD 300	LOD 400	LOD 500
Metamodules	**Sub-metamodules**	Urban scale	Arch. scale	Tech. scale	As built	Management
SM	*Environmental*	*BP*	*BP*	*BP*	**BP**	**BP**
	Economic	**BP**	**BP**	**BP**	**BP**	**BP**
	Social	**BP**	**BP**	**BP**	**BP**	**BP**
DM	S/F/T	**BP**	**BP**	**BP**	**BP**	**BP**
CM₁	Coherence	**Y/N**	**Y/N**	**Y/N**	**Y/N**	**Y/N**
CM₂	Controllability	**S/M/T**	**S/M/T**	**S/M/T**	**S/M/T**	**S/M/T**

SM sustainable metamodules, *DM* dynamic metamodules, *CM₁* coherent metamodules, *CM₂* controllable metamodules, *LOD* level of development, *S/F/T* spatial, functional, technological, *BP* best practice, *S/T/M* subjects//methods/tools

3.1.1 Sustainable Metamodules

Environmental Design Best Practice and Their Integration in the BIM Workflow

As an example, this contribution delves into the informative contents of the sustainable metamodule, specifically the "environmental" sub-metamodule, which contains smartness indicators related to environmental design (highlighted in red in Table 2). These indicators were identified through previous research activities carried out by the Department of Architecture at the University of Ferrara and have been adapted to the various project LOD (see Fig. 1).

Integration into the BIM workflow According to ISO 19650 BIM standards, information metamodules can be inserted into BIM authoring software and associated with specific design objects as non-geometrical data at various LODs. The IFC classification allows for customization, enabling the metamodules to be added to the BIM as customized property sets (see Figs. 2, 3, and 4). These can then be extracted through a quantity take-off, which is useful for updating the merged model in a shared environment and improving interoperability with building process stakeholders.

4 Conclusion

The article discusses the pressing problem of complexity in the construction sector and its implications for contemporary architecture. It highlights the need for a proactive approach to managing this complexity, examining the main organizational, social, cultural, and technological issues. The emergence of digital technologies such

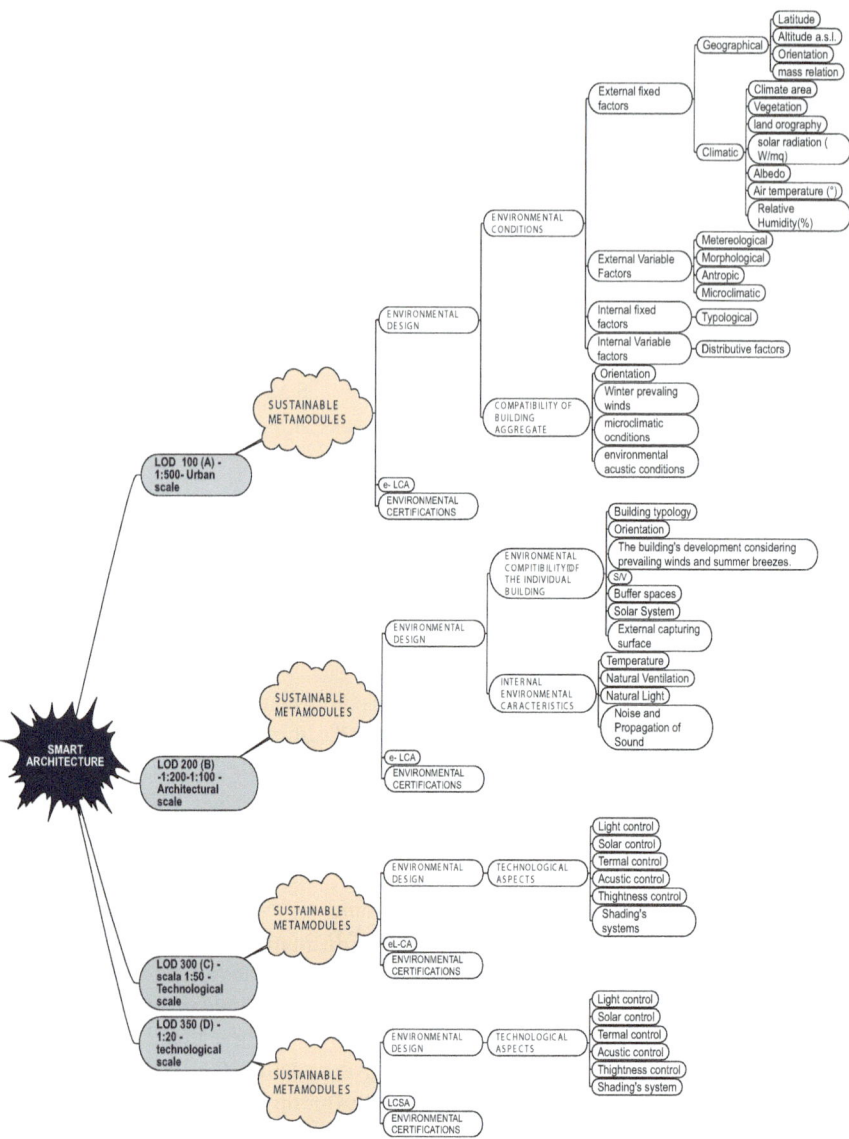

Fig. 1 Development of sustainable metamodules at each LOD: environmental design aspects

as Building Information Modelling (BIM), artificial intelligence, and the Internet of Things (IoT) requires a more conscious design approach that can manage the potential and risks associated with their widespread use.

The concept of smart architecture has emerged as a response to the multifaceted nature of contemporary complexity. It is defined as an architecture that meets current needs through innovative methodologies and tools, integrating materials and

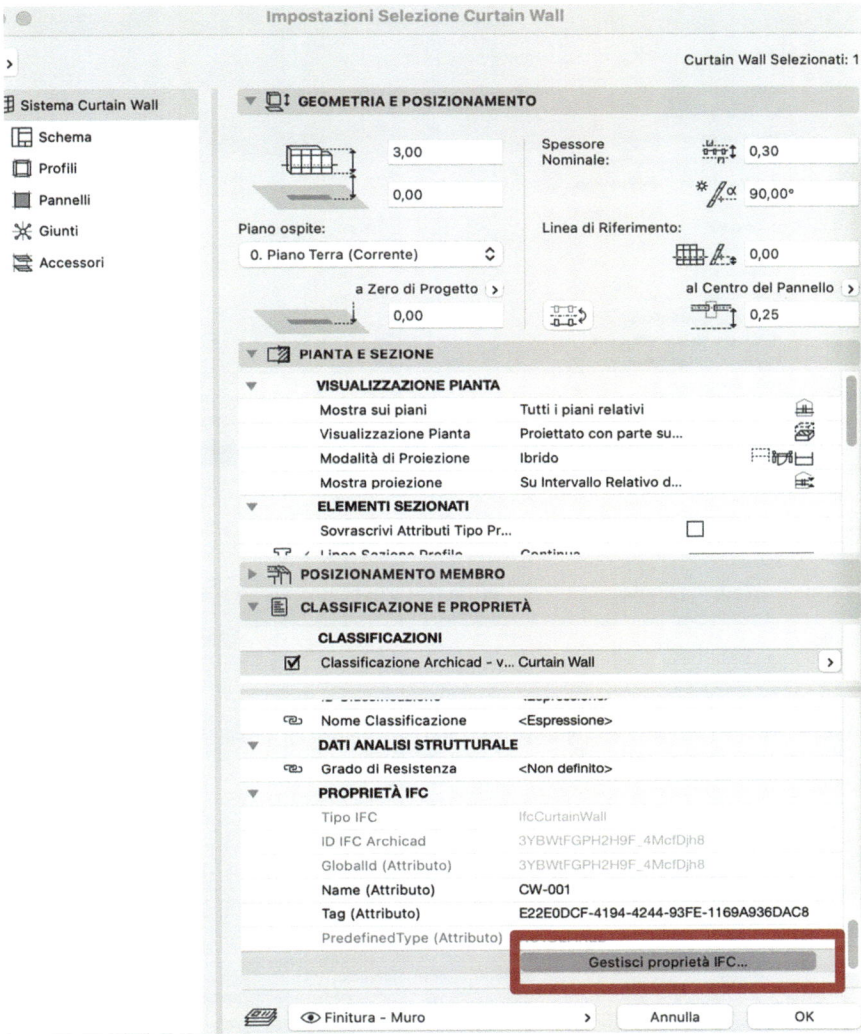

Fig. 2 The image illustrates the integration of sustainable Metamodules in a BIM-based authoring, throughout the insertion of the smartness indicators as property sets, associable as IFC properties to BIM objects. STEP 1: Click on "Manage Properties"

technologies to create buildings that adapt and coexist with the surrounding environment and occupants. The research highlights the importance of semantic research to clarify the distinction between smart buildings and smart architecture and emphasizes the role of the architect as an interpreter of social needs and manager of the construction process.

The concept of metamodule has been stated as a systemic approach to smart architecture that transcends material constraints and reaches the level of knowledge. The metamodules and their sub-metamodules consist of smartness indicators, which correspond to the key characteristics that smart architecture should possess. The

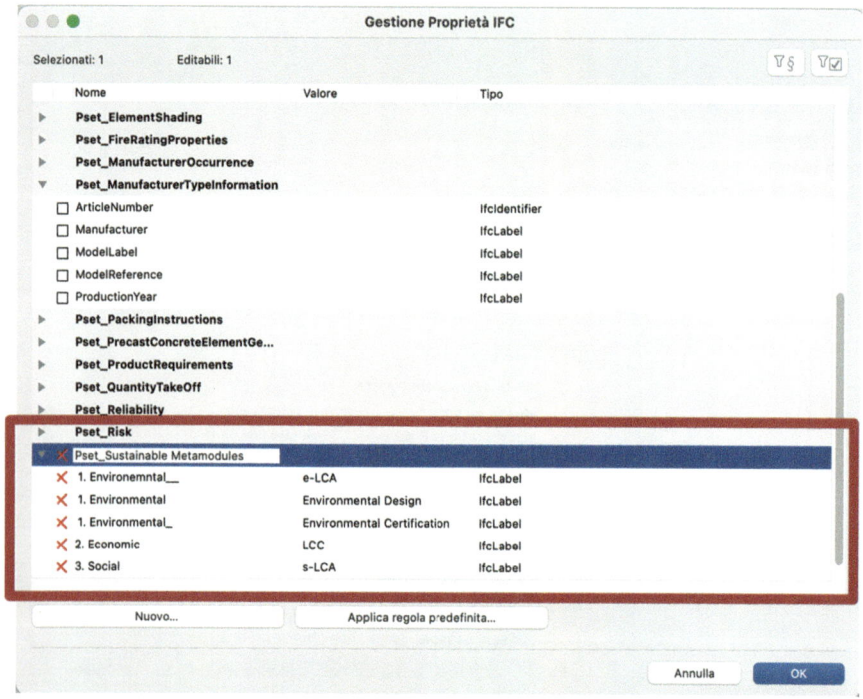

Fig. 3 STEP 2: Create new Properties Set (Pset) according to the predetermined metamodules. In this case, it is shown the Pset_Sustainable metamodules

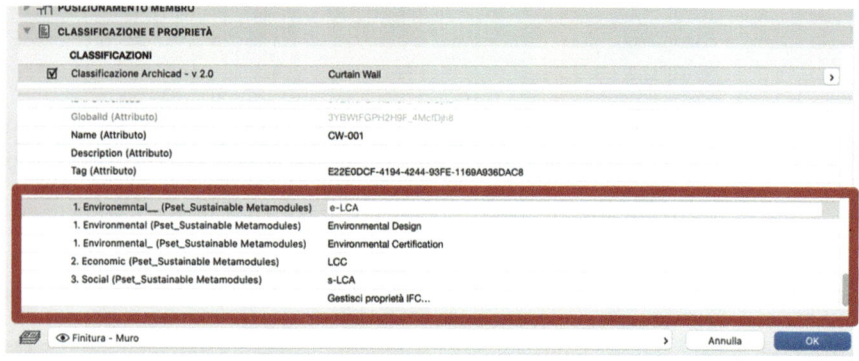

Fig. 4 STEP 3: The Pset metamodules are allowed to be filled as in the IFC classification of the designed object

Building Information Modelling (BIM) system provides a framework for addressing social demands and design challenges, facilitating the creation of a decision support matrix in choosing specific design solutions. BIM also enables easy digitization and integration of metamodules into the design process, facilitating interoperability among the stakeholders of the building process.

References

1. Mills, A.: A systematic approach to risk management for construction. Struct. Surv. **19**, 252–254 (2001)
2. Baccarini, D.: The concept of project complexity: a review. Int. J. Constr. Manag. **14**, 201–204 (1996)
3. Wood, H., Ashton, P.: Factors of complexity in construction projects. In: Proceeding 25th Annual ARCOM Conference, vol. 2, pp. 857–866, Nottingham (2009)
4. Brockmann, C., Kähkönen, K.: Evaluating construction project complexity. In: Niraj, T. (ed.) Management of Construction: Research to Practice (MCRP) Conference Proceeding, pp. 716–727. Joint CIB International Symposium of W055, W065, W089, W118, TG76, TG78, TG81 and TG84, Montreal, Canada (2012)
5. Leone, M.F.: Innovazione Tecnologica e Materiali Avanzati ad alte prestazioni ed eco-efficienza: Nanotecnologie per l'evoluzione dei materiali cementizi. PhD thesis, Prof. Mario Losasso, Tutor (2009)
6. Hilberseimer, L.: Mies Van der Rohe. Città studi Edizioni (1956)
7. Drossel, W.G., et al.: Smart   smart materials for smart applications. In: CIRP – 25th Design Conference for Innovative Product Creation, Haifa, Israel (2015)
8. Frighi, V.: Smart Architecture – A Sustainable Approach for Transparent Building Components Design. Springer Nature (2021)
9. Casini, M.: Smart buildings. Advanced materials and nanotechnology to improve energy-efficiency and environmental performance. Civ. Struct. Eng. **69**, 50–51 (2016)
10. Conato, F., Spasari, I.: Architettura Smart. Definizioni semantiche e modalità di risposta dell'architettura contemporanea alla complessità delle esigenze attuali della società. Ufficio Tecnico. **3**(2023), 5–14 (2023)
11. Moriconi, M.: Le misure di Le Corbusier. Spazio e Società. **76**, 28–37 (1996)
12. Mirzaliyeva, G., Erdoğan, N.: The role of Globalisation in Baku's New City. Period. Polytech. Arch. **53**(3), 186–206 (2022)
13. Moel, K.: Integrated Design in Contemporary Architecture. Princeton Architectural Press, New York (1999)
14. Carrara, G.: Complexity and crisis of design collaboration and knowledge. TECHNE – J. Technol. Arch. Environ. **13**, 50–54 (1999)
15. Nikolić, D., Jovanović, S., Skerlić, J., Šušteršič, V., Radulović, J.: Methodology of life cycle sustainability assessment. In: 13th IQC – International Quality Conference, Turkey (2019)
16. Seungjun, R., Sungho, T., Rakhyun, K., Daniela, M.M.: Analysis of worker category social impacts in different types of concrete plant operations: a case study in South Korea. Sustain. For. **10**, 366 (2018)
17. Smith, A.J., Pitt, M.: Sustainable workplaces and building user comfort and satisfaction. J. Corp. Real Estate. **13**(3), 144–156 (2011)

REVERT Framework: Stakeholder Perspective to Enable Circular Transformation of Construction Industry

Hafize Büşra Bostancı ⓘD, Ali Murat Tanyer ⓘD, and Guillaume Habert ⓘD

Contents

1 Introduction

The circular economy (CE) model is increasingly popular among variable actors [1–3]. The construction industry actors thus seek innovative ways to implement circularity principles for minimizing resource consumption and waste generation [4]. This seeking has revealed many different circularity strategies, approaching different life cycles of a building and construction process. Various frameworks have evolved due

H. B. Bostancı (✉)
Middle East Technical University, Ankara, Türkiye

ETH Zürich, Zürich, Switzerland

Hitit University, Çorum, Türkiye
e-mail: busrab@metu.edu.tr

A. M. Tanyer
Middle East Technical University, Ankara, Türkiye
e-mail: tanyer@metu.edu.tr

G. Habert
ETH Zürich, Zürich, Switzerland
e-mail: habert@ibi.baug.ethz.ch

© The Author(s) 2025   1039
M. Kioumarsi, B. Shafei (eds.), *The 1st International Conference on Net-Zero Built
Environment*, Lecture Notes in Civil Engineering 237,
https://doi.org/10.1007/978-3-031-69626-8_87

to the efforts of many scholars, industrial partners, and governmental institutions to pave the way for circular transformation. The construction stakeholders need more reinforced collaborations to allow for a vital transformation since the industry has a complex organizational structure [5–8]. These collaborations are supported by systematic regulations and clear roadmaps [4]. As there are fundamental managerial challenges in the construction industry [9], the European Commission (EC) published a report about circular building design principles among seven target groups (TG): (i) building users, facility managers, and owners; (ii) design teams; (iii) contractors and builders; (iv) manufacturers of construction products; (v) deconstruction and demolition teams; (vi) investors, developers, and insurance providers; and (vii) Government/Regulators/Local authorities [10]. However, this report only introduced waste management, adaptability, and durability as the fundamental circularity approaches by claiming macro-objectives.

Despite the proliferation of frameworks, the current ones focus on specific aspects such as resources, materials, buildings, digitalization, or city policies. However, the collaboration between stakeholders in all building processes should be strengthened for a robust transformation. But, the concept of stakeholder engagement for a comprehensive circular transformation at three distinct scales is still in its early stages. This study, therefore, presents a unique framework that enables stakeholder collaboration at three scales. By reinforcing stakeholder collaboration in the circular supply chain, this framework offers a novel approach to the circular construction industry to enhance material-building-city integration and foster a regenerative environment.

This study takes a unique approach by investigating the strategies from the perspective of the target groups defined by the EC. It then develops a framework from the stakeholders' perspective at the micro, meso, and macro-scales. The study adopts the micro-scale for materials, meso-scale for buildings, and macro-scale for cities. Methodologically, it conducts a systematic literature review to gather the circular economy strategies from target groups and scale-based approaches. A key innovation of this study is using natural language processing (NLP) to demonstrate the topics of the strategies. This methodology allows for a comprehensive understanding of the strategies in the circular construction industry. The study also defines critical success factors (CSFs) that align with the strategies on a common platform. This is the first study to employ NLP in this context, and its main contribution is the definition of thirty-seven CSFs to enable stakeholder collaboration at multiple scales. Furthermore, the output of the study is the REVERT framework, which underscores the essential synergy between resource, envisagement, validation, entity, regulation, and technology to support actor collaborations at material, building, and city scales. Thus, approaching the gap in the existing frameworks, the REVERT framework reinforces the material-building-city integration by supporting stakeholder collaboration in resource extraction, design and assessment process, organizational capabilities, governmental policies, and digitalization.

2 Methodology

This study aimed to develop a circular transformation framework reinforcing the stakeholder collaboration at the micro, meso, and macro-scales to promote a regenerative environment through the material-building-city trinity for environmental, economic, and social sustainability. The exploratory research method, therefore, was applied to construct insights. It examined the circular economy strategies (CES) from scholars' perspectives and focused on the target groups defined by EC.

This study conducted a systematic literature review to extract the circular economy strategies in the literature. One main keyword query representing the "circular economy and construction industry" was merged with one additional query (see Table 1) representing seven target groups defined by the EC. Seven queries were questioned on the Scopus database, and results were limited to English articles published from 2018 onward as of the data update on December 05, 2022. The scopus database was selected because it contains more results than the other databases. The literature results were deeply analyzed, and the strategies in the articles presented by the scholars were listed on an Excel spreadsheet. Material-related strategies were listed on the micro-scale, building and design-related strategies on the meso-, and city and stakeholder-related strategies on the macro-scale. Additionally, target group numbers, paper IDs, scales, and strategies were applied as the data labels during the listing process.

Table 1 Main query and additional queries defined for systematic literature review

Target groups	Main query	Additional queries
Building users, facility managers, and owners	("circular economy" OR "circular*") AND ("built environment" OR "construction industry" OR "construction sector" OR aec OR aeco)"	AND ("building user" OR "facility manager" OR "owner")
Design teams		AND ("design team" OR "architect" OR "engineer" OR "designer")
Contractors and builders		AND ("contractor" OR "builder" OR "constructor")
Manufacturers of construction products		AND ("manufacturer")
Deconstruction and demolition teams		AND (("demolition" OR "deconstruction") AND ("firm" OR "company" OR "organization" OR "worker" OR "employee"))
Investors, developers, and insurance providers		AND ("investor" OR "developer" OR "insurance provider")
Government/ regulators/ local authorities		AND ("regulator" OR "local authorit*" OR "policymaker" OR "policy maker")

Following that the Excel spreadsheet was processed in the Jupyter Notebook environment. Lemmatization by the 'nltk.stem', stop word removal by the 'nltk. corpus', and tokenization by the 'nltk.tokenize' functions included in the natural language toolkit (NLTK) library were applied to tokenize the circular economy strategies. After the tokens were listed row by row by natural language processing (NLP), the script employed 'wordcloud' function to indicate the frequency of the words in each token at three scales. As a further step, the latent dirichlet allocation (LDA) topic modelling algorithm by the 'gensim.models.LdaMulticore' from the 'gensim' library was applied across three scales. The output of this process enabled the building of critical success factors (CSFs). The critical success factors (CSFs) covered similar lemmatized tokens. The heatmap plots were appealed to demonstrate the density of lemmatized tokens in each CSF at both the target groups and the scales. Hence, circular economy strategies appeared in the most and the least. Consequently, CSFs constructed the new framework to encourage all stakeholders to pave the way for the circular transformation for regenerative development by enabling the material-building-city trinity.

3   Results

3.1   Quantitative Results

The quantitative literature results from the Scopus database screened 112 articles in total for seven target groups: (TG1) building users, facility managers, and owners, (TG2) design teams, (TG3) contractors and builders, (TG4) manufacturers of construction products, (TG5) deconstruction and demolition teams, (TG6) investors, developers, and insurance providers, and (TG7) Government/Regulators/Local authorities. After screening, the articles were deeply analyzed to eliminate the irrelevant and synthesize the results. After synthesizing, some articles were transferred from the existing group to the relevant group, and strategies in 97 articles were critically evaluated for all target groups in the synthesizing process (see Fig. 1). The second group (design teams) demonstrated more results, followed by the seventh (policy makers) and third groups (contractors and builders), respectively.

Micro-scale, meso-scale, and macro-scale strategies listed separately on the Excel spreadsheet, exposed 84, 91, and 88 rows, respectively. The Excel file (.xlsx) processed by Python on the Jupyter Notebook resulted in 804 lemmatized tokens at three scales (206, 337, and 261, respectively), which proved the excessive circular economy strategies. The 'wordcloud' function utilized the lemmatized tokens to demonstrate the word clouds in Fig. 2, indicating the frequency of words in tokens at three scales. Based on these outputs, it was interpreted that materials and resources showed repetition with reuse, recovery, recycling, bio, process, storage, value, generation, value, secondary resources, and loop at the micro-scale. Building and design at the meso-scale highlighted reuse, demolition, maintenance, management,

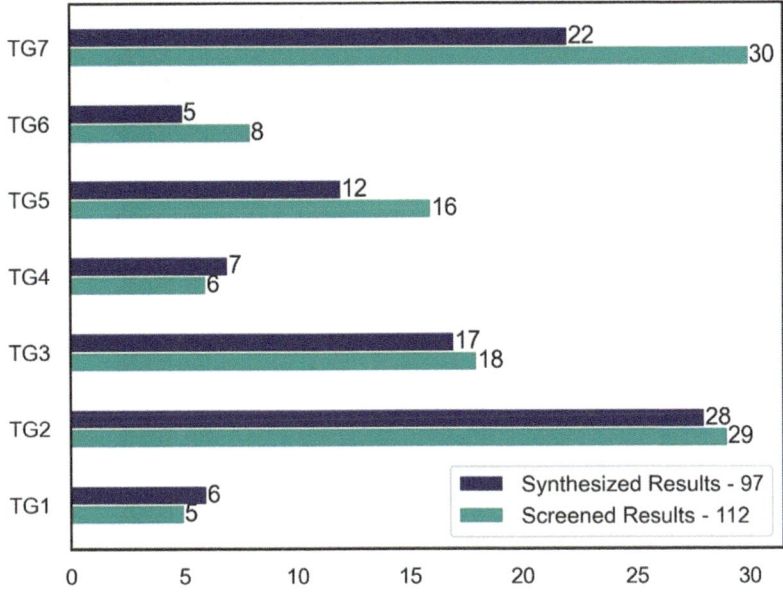

Fig. 1 The number of literature results and snow-balling results based on TGs

Fig. 2 The keywords in the strategies at micro-scale (**a**), meso-scale (**b**), and macro-scale (**c**)

recycling, components, and elements. City and stakeholder-based understanding at the macro-scale revealed the establish, reuse, recycle, market, share, collaboration, information, management, and policy.

The 804 results of the lemmatization process manifested many strategies that could cause chaos due to the unprecedented activities of many stakeholders in accomplishing the projects. Therefore, clearer understanding was required. This purification was significant as the researchers put great efforts into improving circular understanding. Consequently, the authors of this study analyzed the strategies by considering the lemmatized tokens and constructed thirty-seven CSFs to cover similar tokens. These CSFs accommodated similar tokens and optimized the allocation of circular strategies. They focused on all activities during the whole life cycle of a building process, from raw material extraction to end-of-life.

3.2 Qualitative Results

After employing topic modelling, the qualitative result highlighted the most common strategies among seven target groups and three scales. Following this step, generating CSFs was significant in developing a naïve roadmap referring to the needs of all target groups. Analyzing 804 lemmatized tokens and topic models pioneered constructing CSFs. Corresponding CSFs categorized the similar lemmatized tokens, and thus, thirty-seven CSFs were constructed. Figure 3 indicates the quantitative distribution of lemmatized tokens in each CSF by scales and target groups.

During the construction of the CSFs, the most common and unique strategies in the articles were considered to develop a comprehensive framework, including environmental, economic, and social approaches. Hence, a fully sustainable and circular transformation was enabled in the construction industry. Therefore, the heatmaps in Fig. 3 demonstrated the comprehensiveness of CSFs for the most common and unique strategies. Six main dimensions comprehended the thirty-seven CSFs for an understandable and monitorable framework interpreted in Table 2. Resource, envisagement, validation, entity, regulation, and technology dimensions included the CSFs for developing the REVERT framework.

Through this framework, the resource dimension referred the material and resource-oriented factors, including strategies such as maintaining the technical quality of resources and materials, closing, slowing, narrowing, regenerating, and optimizing resource loops [11–13], enabling zero-waste building and construction [14, 15], estimating material stock [16–18], and implementing agricultural waste biomass and low-impact and regenerative materials [19–23].

Fig. 3 Quantitative distribution of the CSFs by scales and target groups

Table 2 Critical success factors (CSFs) for REVERT framework

CSF	Critical success factors (CSFs)
RESOURCE	
CSF1	Adapting narrow/slow/close loop understanding
CSF2	Avoiding any kind of resource waste
CSF3	Encouraging material and resource optimization
CSF4	Managing stocks and flows for resource and waste
CSF5	Promoting use of agricultural waste biomass
CSF6	Promoting use of low-impact, durable, healthy, bio-based, and high-quality materials
CSF7	Promoting use of secondary resources
ENVISAGEMENT	
CSF8	Developing comprehensive disassembling methods
CSF9	Developing design methods to enable reuse and recycling
CSF10	Enabling interchangeability and versatility for building components
CSF11	Encouraging building optimization
CSF12	Promoting adaptability for building design
CSF13	Promoting retrofitting of existing buildings
VALIDATION	
CSF14	Assessing carbon for all life cycle phases and promote carbon accounting
CSF15	Constructing resource chain for circular flow
CSF16	Deploying new facilities to manage resource processes
CSF17	Promoting cost-benefit analysis and encourage low-cost construction
ENTITY	
CSF18	Adjusting policies for circular transition
CSF19	Demarcating understanding for needs and corresponding stakeholders
CSF20	Developing awareness for climate change and sustainability
CSF21	Encouraging circular businesses models
CSF22	Encouraging local businesses to promote new materials
CSF23	Encouraging personal trust and instrument/resource sharing between stakeholders
CSF24	Enhancing collaboration and communication of stakeholders
CSF25	Offering affordable and comparative prices
CSF26	Promoting knowledge for circular transition
CSF27	Promoting supply chain management among all actors
CSF28	Supporting resource allocation and resource management among suppliers and users
REGULATION	
CSF29	Adopting circularity principles for cleaner production
CSF30	Promoting building certification, energy labeling, and legislation
CSF31	Promoting capital investment and competitiveness
CSF32	Promoting R-imperatives
CSF33	Supporting stakeholders activities by legal legislation and increasing commitment to legal framework
TECHNOLOGY	
CSF34	Creating digital passports for materials
CSF35	Integrating industry 4.0 technologies for automation of design-construction-deconstruction process

(continued)

Table 2 (continued)

CSF	Critical success factors (CSFs)
CSF36	Promoting LCA for materials and construction processes (pre-design, design, construction, maintenance and demolition)
CSF37	Promoting scheduling and management/track system

Envisagement dimension referred the design and building-oriented factors including the strategies such as circular design methods, building adaptability, and retrofitting [24–32]. Validation dimension referred the assessment-oriented factors including the strategies such as assessing biogenic carbon [33], resource processes and chain [34–37], and low-cost construction [17, 20, 38]. Entity dimension referred the organizations and supply chain-oriented factors, including strategies such as circular business models [13, 15, 39–41] and local business models [22, 42], knowledge increasing [34, 36, 37, 43]. The regulation dimension referred the legislation-oriented factors, including strategies such as adopting circularity principles and R-imperatives [4, 44], building certification [20, 45], capital investment, and competitiveness [12, 46]. The technology dimension, lastly, referred the digitalization-oriented factors, including strategies such as creating material passports [13, 18, 47], automating processes [12, 18, 48], monitoring, and scheduling systems [18, 34, 46, 47].

4 Conclusion

The circular economy has many strategies that might make it complex for stakeholders to implement. It is essential to evaluate the circular economy from the perspective of all stakeholders to create a shared understanding. Reinforcing stakeholder collaboration in the construction industry is vital as the circular transformation requires intense responsibility. Numerous frameworks have evolved to accelerate the shift towards circularity. On the other hand, the existing frameworks do not support stakeholder engagement, which is an enabler to achieving a circular transformation. Besides, there needs to be more assimilation of the three scales for circularity initiatives. The micro, meso, and macro-scales should collaborate to create material, building, and city trinity and to take bold steps towards achieving a circular economy.

Stakeholder engagement at three scales accelerates the circular transformation process by promoting solid relationships. Therefore, the REVERT framework fills this gap by encouraging stakeholders at three scales to participate in regenerative development. It includes critical success factors to provide a clear roadmap. It demonstrates the fundamental dimensions of a circular transformation: resource, envisagement, validation, entity, regulation, and technology. Many strategies in the literature include the emergence of R-imperatives, the need for low-carbon, bio-based, and regenerative materials, adaptable design principles, and competitive environments for circular and local businesses. It is essential to encourage actors to follow the most common strategies to get more circular and to enable a strong collaboration to implement the strategies at the material, building, and city scales for a regenerative industry. Accelerating the synergies between actors for the material-building-city trinity, regulating legal frameworks, and supporting innovative technologies help stakeholders see the bigger picture and foster a smooth transition towards a circular economy. Thanks to the REVERT framework, three pillars of sustainability are achieved since the framework considers not only the environmental dimension by focusing on low-carbon and regenerative materials and R-imperatives but also the social and economic dimensions by strengthening the collaborative synergies. Consequently, this study is the first since it employs NLP to demonstrate the tokens in the circular economy strategies in the literature, and it is an innovative study as it focuses on the stakeholder perspective to support the collaboration. This provides a dynamic understanding for future studies.

References

1. Kirchherr, J., Yang, N.H.N., Schulze-Spüntrup, F., Heerink, M.J., Hartley, K.: Conceptualizing the circular economy (revisited): an analysis of 221 definitions. Resour. Conserv. Recycl. **194**, 107001 (2023)
2. Kirchherr, J., Reike, D., Hekkert, M.: Conceptualizing the circular economy: an analysis of 114 definitions. Resour. Conserv. Recycl. **127**, 221–232 (2017)
3. Merli, R., Preziosi, M., Acampora, A.: How do scholars approach the circular economy? A systematic literature review. J. Clean. Prod. **178**, 703–722 (2018)
4. Munaro, M.R., Tavares, S.F., Bragança, L.: Towards circular and more sustainable buildings: a systematic literature review on the circular economy in the built environment. J. Clean. Prod. **260**, 121134 (2020)
5. Manavalan, E., Jayakrishna, K.: An analysis on sustainable supply chain for circular economy. Procedia Manuf. **33**, 477–484 (2019)
6. Mangla, S.K., Luthra, S., Mishra, N., Singh, A., Rana, N.P., Dora, M., Dwivedi, Y.: Barriers to effective circular supply chain management in a developing country context. Prod. Plan. Control. **29**, 551–569 (2018)
7. Geissdoerfer, M., Morioka, S.N., de Carvalho, M.M., Evans, S.: Business models and supply chains for the circular economy. J. Clean. Prod. **190**, 712–721 (2018)
8. Farooque, M., Zhang, A., Thürer, M., Qu, T., Huisingh, D.: Circular supply chain management: a definition and structured literature review. J. Clean. Prod. **228**, 882–900 (2019)
9. Eberhardt, L.C.M., Birgisdottir, H., Birkved, M.: Potential of circular economy in sustainable buildings. In: IOP Conference Series: Materials Science and Engineering, vol. 471. IOP Publishing (2019)
10. EC: Circular Economy – Principles for Building Design. https://ec.europa.eu/docsroom/documents/39984
11. Utrilla, P.N.C., Górecki, J., Maqueira, J.M.: Simulation-based management of construction companies under the circular economy concept-case study. Buildings. **10**, 94 (2020)
12. Ghaffar, S.H., Burman, M., Braimah, N.: Pathways to circular construction: an integrated management of construction and demolition waste for resource recovery. J. Clean. Prod. **244**, 118710 (2020)
13. Çetin, S., Gruis, V., Straub, A.: Digitalization for a circular economy in the building industry: multiple-case study of Dutch social housing organizations. Resour. Conserv. Recycl. Adv. **15**, 200110 (2022)

14. Oluleye, B.I., Chan, D.W.M., Olawumi, T.O., Saka, A.B.: Assessment of symmetries and asymmetries on barriers to circular economy adoption in the construction industry towards zero waste: a survey of international experts. Build. Environ. **228**, 109885 (2023)

15. Jahan, I., Zhang, G., Bhuiyan, M., Navaratnam, S., Shi, L.: Experts' perceptions of the management and minimisation of waste in the Australian construction industry. Sustain. For. **14**, 11319 (2022)

16. Sprecher, B., Verhagen, T.J., Sauer, M.L., Baars, M., Heintz, J., Fishman, T.: Material intensity database for the Dutch building stock: towards big data in material stock analysis. J. Ind. Ecol. **26**, 272–280 (2022)

17. Arora, M., Raspall, F., Cheah, L., Silva, A.: Buildings and the circular economy: estimating urban mining, recovery and reuse potential of building components. Resour. Conserv. Recycl. **154**, 104581 (2020)

18. Kovacic, I., Honic, M., Sreckovic, M.: Digital platform for circular economy in aec industry. Eng. Proj. Organ. J. **9**, 1–16 (2020)

19. Bakos, N., Schiano-Phan, R.: Bioclimatic and regenerative design guidelines for a circular university campus in India. Sustain. For. **13**, 8238 (2021)

20. Haselsteiner, E., Rizvanolli, B.V., Villoria Sáez, P., Kontovourkis, O.: Drivers and barriers leading to a successful paradigm shift toward regenerative neighborhoods. Sustain. For. **13**, 5179 (2021)

21. Hahladakis, J.N., Purnell, P., Aljabri, H.M.S.J.: Assessing the role and use of recycled aggregates in the sustainable management of construction and demolition waste via a mini-review and a case study. Waste Manag. Res. **38**, 460–471 (2020)

22. Duque-Acevedo, M., Lancellotti, I., Andreola, F., Barbieri, L., Belmonte-Ureña, L.J., Camacho-Ferre, F.: Management of agricultural waste biomass as raw material for the construction sector: an analysis of sustainable and circular alternatives. Environ. Sci. Eur. **34**, 70 (2022)

23. Abdelshafy, A., Walther, G.: Exploring the effects of energy transition on the industrial value chains and alternative resources: a case study from the German federal state of North Rhine-Westphalia (NRW). Resour. Conserv. Recycl. **177**, 105992 (2022)

24. Bertin, I., Mesnil, R., Jaeger, J.M., Feraille, A., Le Roy, R.: A BIM-based framework and databank for reusing load-bearing structural elements. Sustain. For. **12**, 3147 (2020)

25. Cruz Rios, F., Grau, D., Bilec, M.: Barriers and enablers to circular building design in the US: an empirical study. J. Constr. Eng. Manag. **147**, 04021117 (2021)

26. Nian, S., Pham, T., Haas, C., Ibrahim, N., Yoon, D., Bregman, H.: A functional demonstration of adaptive reuse of waste into modular assemblies for structural applications: the case of bicycle frames. J. Clean. Prod. **348**, 131162 (2022)

27. Roithner, C., Cencic, O., Honic, M., Rechberger, H.: Recyclability assessment at the building design stage based on statistical entropy: a case study on timber and concrete building. Resour. Conserv. Recycl. **184**, 106407 (2022)

28. Gillott, C., Davison, B., Densley Tingley, D.: Drivers, barriers and enablers: construction sector views on vertical extensions. Build. Res. Inf. **50**, 909–923 (2022)

29. van den Berg, M., Voordijk, H., Adriaanse, A.: Recovering building elements for reuse (or not)—ethnographic insights into selective demolition practices. J. Clean. Prod. **256**, 120332 (2020)

30. Passoni, C., Palumbo, E., Pinho, R., Marini, A.: The LCT challenge: defining new design objectives to increase the sustainability of building retrofit interventions. Sustain. For. **14**, 8860 (2022)

31. Rust, R.T., Lemon, K.N., Zeithaml, V.A.: Return on marketing: using customer equity to focus marketing strategy. J. Mark. **68**, 109–127 (2004)

32. Minunno, R., O'Grady, T., Morrison, G.M., Gruner, R.L., Colling, M.: Strategies for applying the circular economy to prefabricated buildings. Buildings. **8**, 125 (2018)

33. Saadé, M., Erradhouani, B., Pawlak, S., Appendino, F., Peuportier, B., Roux, C.: Combining circular and LCA indicators for the early design of urban projects. Int. J. Life Cycle Assess. **27**, 1–19 (2022)

34. Tirado, R., Aublet, A., Laurenceau, S., Habert, G.: Challenges and opportunities for circular economy promotion in the building sector. Sustain. For. **14**, 1569 (2022)
35. Dewagoda, K.G., Ng, S.T., Chen, J.: Driving systematic circular economy implementation in the construction industry: a construction value chain perspective. J. Clean. Prod. **381**, 135197 (2022)
36. Coenen, T.B.J., Visscher, K., Volker, L.: A systemic perspective on transition barriers to a circular infrastructure sector. Constr. Manag. Econ. **41**(1), 22–43 (2023)
37. Knoth, K., Fufa, S.M., Seilskjær, E.: Barriers, success factors, and perspectives for the reuse of construction products in Norway. J. Clean. Prod. **337**, 130494 (2022)
38. Campbell-Johnston, K., ten Cate, J., Elfering-Petrovic, M., Gupta, J.: City level circular transitions: barriers and limits in Amsterdam, Utrecht and The Hague. J. Clean. Prod. **235**, 1232–1239 (2019)
39. Luciano, A., Cutaia, L., Cioffi, F., Sinibaldi, C.: Demolition and construction recycling unified management: the DECORUM platform for improvement of resource efficiency in the construction sector. Environ. Sci. Pollut. Res. **28**, 24558–24569 (2021)
40. Ramos, M., Martinho, G.: Influence of construction company size on the determining factors for construction and demolition waste management. Waste Manag. **136**, 295–302 (2021)
41. Locurcio, M., Tajani, F., Anelli, D., Ranieri, R.: A multi-criteria composite indicator to support sustainable investment choices in the built environment. Valori e Valutazioni. **30**, 85–100 (2022)
42. Dabaieh, M., Maguid, D., El-Mahdy, D.: Circularity in the new gravity—re-thinking vernacular architecture and circularity. Sustain. For. **14**, 328 (2022)
43. Wuni, I.Y.: A systematic review of the critical success factors for implementing circular economy in construction projects. Sustain. Dev. **31**, 1195–1213 (2023)
44. Eberhardt, L.C.M., van Stijn, A., Kristensen Stranddorf, L., Birkved, M., Birgisdottir, H.: Environmental design guidelines for circular building components: the case of the circular building structure. Sustain. For. **13**, 5621 (2021)
45. Sparrevik, M., de Boer, L., Michelsen, O., Skaar, C., Knudson, H., Fet, A.M.: Circular economy in the construction sector: advancing environmental performance through systemic and holistic thinking. Environ. Syst. Decis. **41**, 392–400 (2021)
46. Azcarate-Aguerre, J.F., Conci, M., Zils, M., Hopkinson, P., Klein, T.: Building energy retrofit-as-a-service: a total value of ownership assessment methodology to support whole life-cycle building circularity and decarbonisation. Constr. Manag. Econ. **40**, 676–689 (2022)
47. Cambier, C., Galle, W., De Temmerman, N.: Research and development directions for design support tools for circular building. Buildings. **10**, 142 (2020)
48. Véliz, K.D., Ramírez-Rodríguez, G., Ossio, F.: Willingness to pay for construction and demolition waste from buildings in Chile. Waste Manag. **137**, 222–230 (2022)

Integrated Technical Building Installations for Zero-Emission Building

Emma Zheng Liang and Habtamu Bayera Madessa

Contents

1 Introduction

It is widely acknowledged that the built environment significantly contributes to greenhouse gas emissions. Studies have indicated that integrated technical installations play a crucial role in achieving zero-emission buildings by optimizing energy efficiency and reducing environmental impact. García-Monge et al. [1] demonstrated that regulating heating, ventilation, and air conditioning (HVAC) systems based on CO_2 concentration levels could lead to energy savings ranging from 40% to 70%. This range of energy savings underscores the importance of integrated technical installations in enhancing energy efficiency.

However, there is evidence indicating that buildings are not meeting performance expectations, leading to a significant gap between what is expected and what is happening. Ardehali et al. [2] reported that inadequate commissioning and poor

E. Z. Liang (✉)
Rambøll Norge AS, Oslo, Norway
e-mail: emma.liang@ramboll.no

H. B. Madessa
OsloMet-Oslo Metropolitan University, Department of Built Environment, Oslo, Norway
e-mail: habama@oslomet.no

© The Author(s) 2025
M. Kioumarsi, B. Shafei (eds.), *The 1st International Conference on Net-Zero Built Environment*, Lecture Notes in Civil Engineering 237,
https://doi.org/10.1007/978-3-031-69626-8_88

maintenance practices often lead to increased energy consumption and reduced equipment lifespan. Lawrence and Keime [3] assessed two school buildings and uncovered widespread dissatisfaction among occupants, along with higher energy usage and associated costs than initially thought.

While numerous studies have focused on materials and energy aspects, there is a noticeable gap in research related to the impact of integrated technical building installations (ITB) processes on reducing emission from buildings. The contemporary building complex presents more complicated functional requirements, as well as challenges in managing greenhouse gas emissions. It is therefore crucial to examine the whole ITB process, especially for buildings with ambitious environmental goals. The early involvement of ITB role in the building process is important to allocate the resources for planning, commissioning, handover, and aftercare. The integrated installation is crucial for both Heating, Ventilation, Air Conditioning (HVAC) installations and building automation systems (BAS) from the design phase onward.

The integrated installation is vital to ensure the realization of zero-emission buildings in the sustainable design process through multidisciplinary coordination during the commissioning phase. In this regard, testing schedules and responsibilities should align with local regulations such as NS3935 [4], and multidisciplinary coordination is crucial for forming an interactive basis to address system errors.

During the handover and aftercare phase, integrating Post Occupancy Evaluation (POE) is essential. This involves reviewing design intent, gathering feedback from building users, and optimizing the entire system for zero-emission buildings.

This study focuses on the application of the Integrated Technical Building (ITB) process through a detailed case study of a project aiming for a zero-emission building. The study provides a comprehensive analysis of how ITB is applied, detailing the integration of different technical systems, installation procedures, and reaching ambitious environmental goals: passive house standards, zero-emission building. By examining this case study, the research seeks to demonstrate how ITB contributes to achieving ambitious environmental goals in procedural details and operation.

The paper documents the various stages of implementing the ITB process, from initial design and planning to integration with building automation systems (BAS) and ongoing aftercare. It identifies best practices that lead to significant reductions in carbon emissions and energy consumption, offering insights into how the process facilitates a more sustainable building process.

A key component of this research is the emphasis on the early adoption of integrated systems and continuous monitoring throughout the building's lifecycle. This includes the use of advanced technology, energy management systems, and automation, all working in harmony to minimize environmental impact. Through the case study analysis, the study explores common challenges and proposes solutions that other projects with similar environmental objectives could adopt.

The paper is structured as follows: The Sect. 1 of the paper introduces the ITB process within construction projects in Norway. A case study implementing ITB process aimed at achieving a zero-emission educational building, exploring the practical barriers encountered during the ITB process implementation in Norway, and providing systematic resolutions is presented in the Sect. 2. Finally, the Sect. 3 presents the main findings of the chapter.

2 Integrated Technical Building installations in Norway

Modern buildings in Norway usually contain various technical systems including lighting, heating, cooling, ventilation, sun shading, building automation, fire detection, access control, intrusion alarm, surveillance cameras, energy monitoring meter, elevators, and data/telecommunication. Along with continuously stricter demands for lower energy usage, improved indoor climate, increased security, and enhanced user experiences, the technical installations have become more advanced and integrated.

In Norway, the standard-NS 3935 has defined the roles and responsibilities for stakeholders, consultants, and contractors to ensure proper functioning of technical installation. The standard is designed to establish a clear understanding of the process ensuring successful procurement of integrated technical building installation, and it ensures smooth interaction and coordination between the building's technical systems to satisfy the business, environmental, and safety requirements. The standard covers different phases of a project, emphasizing the importance of an ITB coordinator who maintains focus on the interaction and economy of technical systems from planning to handover.

The role of the ITB coordinator is critical [5] in bridging the gap between different functions and stakeholders. While not accountable for the technical content, this individual serves as the primary point of contact, facilitating communication and coordination across diverse teams. They serve as a control mechanism and resource ensuring technical solutions are addressed throughout all stages of a construction project, from initial planning to handover. The ITB coordinator not only facilitates collaboration but also implements necessary actions.

In some cases, the ITB coordinator may take on additional tasks beyond the standard role, such as providing technical support if the client desires both technical oversight and facilitation throughout the project. This individual has expertise in automation from HVAC or electrical engineering and can provide technical support alongside their ITB responsibilities. The role of the ITB coordinator naturally evolves based on experience and expertise.

A key document/tool for the ITB coordinator is the interface matrix (NS3935, 2019) [4]. This document, reviewed with various contractors/disciplines alongside the ITB coordinator, outlines responsibilities for each deliverable. ITB specialists are required to have good knowledge for each system to ensure compatibility and functionality.

The systematic completion is a term often associated with ITB process. The systematic completion describes the process of ensuring that the building meets its technical and non-technical requirements, which is outlined by BA2015 [6] starting

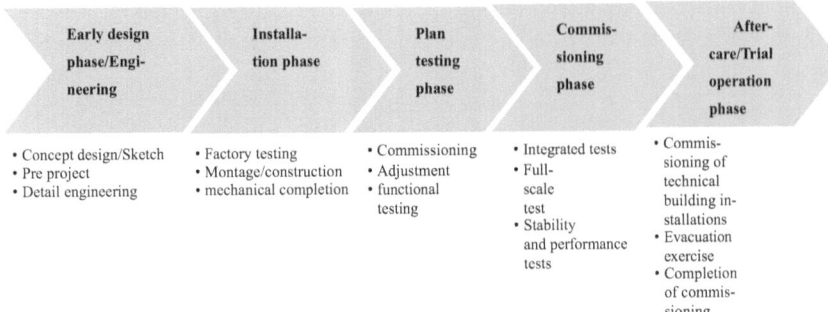

Early design phase/Engineering	Installation phase	Plan testing phase	Commissioning phase	Aftercare/Trial operation phase
• Concept design/Sketch • Pre project • Detail engineering	• Factory testing • Montage/construction • mechanical completion	• Commissioning • Adjustment • functional testing	• Integrated tests • Full-scale test • Stability and performance tests	• Commissioning of technical building installations • Evacuation exercise • Completion of commissioning.

Fig. 1 ITB process according to NS6450: 2016

from early stages of the project until the building is fully operational with functional technical systems.

BA2015, was a nationwide effort to improve the construction industry in Norway. Key players teamed up with SINTEF, NTNU, and Metier to make construction more efficient and sustainable. It aimed to enhance both industry practices and partnerships by adopting and refining processes, systems, and knowledge. The objective was to establish Norway's construction industry as a global leader through the integration of customized international best practices. BA2015 prioritized practical action, focusing on implementing proven methods and knowledge to achieve results.

The processes from the early phase to the delivery are outlined in NS6450 [7] (see Fig. 1 below). Commissioning is a period during which there is an opportunity to uncover errors in equipment, settings, setpoints, etc., before the building is finally handed over to the client. The processes for this are described in NS6450 and could just as well be included in NS3935 for ITB as it is a crucial part of an automation delivery.

Utilizing the standard requires that the client plans for testing, commissioning, and trial operation and describes these processes in the tender documentation. Commissioning takes place after occupancy to test the technical building installations under actual internal load and usage which again have an impact to the justification for setting of various technical systems.

2.1 The Case Study: The ITB Process for Zero-Emission Educational Building

The case study aims to pioneer environmentally friendly construction from 2022, with goals including BREEAM-Excellent certification, passive house standards, zero-emission building (ZEB-OM) [8], and extensive use of cross-laminated timber. Ventilation and heating systems are designed as a combination of demand-controlled

Table 1 The early premise of the interface matrix

System component description	Project planning	Delivery	Assembly	Cabling/connection	Function	Integration
Non-load-bearing external walls	⚒🔒⚲⇨⚡	⚒🔒⚲ ⇨⚡	⚒🔒⚲ ⇨⚡	⚒🔒⚲ ⇨⚡	⚒ 🔒⚲⇨ ⚡	▦🔒 ◊▣
Glass facades	⚒🔒⚲⇨⚡	⚒🔒⚲ ⇨⚡	⚒🔒⚲ ⇨⚡	⚒🔒⚲ ⇨⚡	⚒🔒⚲ ⇨⚡	▦🔒 ◊▣
Windows, doors and gates	⚒🔒⚲⇨⚡	⚒🔒⚲ ⇨⚡	⚒🔒⚲ ⇨⚡	⚒🔒⚲ ⇨⚡	⚒🔒⚲ ⇨⚡	▦🔒 ◊▣
Exterior cladding and surface	⚒🔒⚲⇨⚡	⚒🔒⚲ ⇨⚡	⚒🔒⚲ ⇨⚡	⚒🔒⚲ ⇨⚡	⚒🔒⚲ ⇨⚡	▦🔒 ◊▣
Inner surface	⚒🔒⚲⇨⚡	⚒🔒⚲ ⇨⚡	⚒🔒⚲ ⇨⚡	⚒🔒⚲ ⇨⚡	⚒🔒⚲ ⇨⚡	▦🔒 ◊▣
Solar shading	⚒🔒⚲⇨⚡	⚒🔒⚲ ⇨⚡	⚒🔒⚲ ⇨⚡	⚒🔒⚲ ⇨⚡	⚒🔒⚲ ⇨⚡	▦🔒 ◊▣

ventilation system with sensors, geothermal heat pumps integrated with solar energy systems for both heating and cooling needs. All sensors and energy meters are connected to the Building Management System.

Early Design Phase The ITB coordinator has been engaged from the pre-project phase to establish the concept and standards for planning, commissioning, handover, and aftercare. This early involvement led to the client's bidding documents incorporating the building functional description, as well as the establishment of ITB coordination premises such as the technical room diagram and interface matrix.

The interface matrix, as shown in Table 1, is used to identify and clarify all interfaces in the project, where the interface between the plumbing contractor, electrical contractor, ventilation contractor, and building automation can be managed by the main contractor.

In this table, the symbols represent the following: ⚒ (construction), 🔒 (lock and fittings), ⚲ (plumber), ⇨ (ventilation), ⚡ (electro and automatic) for project planning, delivery, assembly, cabling/connection, and function interfaces. ▦ (BMS), ▣ (top system), 🔒 (access control), ◊ (fire), ▣ (EOS) for the integration interface.

This symbolic approach simplifies the repetition and makes it visually easier to understand the interface matrix. Technical room diagram (see Tables 2 and 3) for climate requirements and equipment delivery is made according to the standard NS6450. Contractors must adhere to these specifications to ensure the HVAC installation meets all building functions and requirements. Outdoor conditions are

Table 2 The early premise of technical room diagram part 1

Room: *trinnarealer*	Room type	Description	Person load	Dimensional indoor temperature (°C)		Air volume (m³/t)	Ventilation type
	Classroom	Teaching room	30	Summer: 19–26	Winter: 21–24	Per person: 26 Per emission: 7.2	VAV
	Group room	Ordinary group room	6	Summer: 19–26	Winter: 21–24	Per person: 26 Per emission: 7.2	VAV
	Group room	Bovis/intro	14	Summer: 19–26	Winter: 21–24	Per person: 26 Per emission: 7.2	VAV
	Toilets	Toilets for students	N/A	Summer: 19–26	Winter: 21–24	–	CAV

Table 3 The early premise of technical room diagram part 2

Room: trinnarealer	Room type	Description	Ventilation control	Heating	Lighting type	Regulation	Local thermostat	Setpoint temperature (°C)
	Classroom	Teaching room	CO2/TEMP/PIR	Supply air	Two zones Presence, dimming, scenario, OFF/ON Shielding/darkening evaluated	Trinnløs	Yes	21
	Group room	Ordinary group room	CO2/TEMP/PIR	Supply air	Presence, dimming, OFF/ON	Trinnløs	Yes	21
	Group room	Bovis/intro	CO2/TEMP/PIR	Supply air	Two zones Presence, dimming, scenario, OFF/ON Shielding/darkening evaluated	Trinnløs	Yes	21
	Toilets	Toilets for students	N/A	Radiator	OFF/ON	N/A	No	22

Table 4 The commissioning plan from detail engineering

Activity	System number		Start	Duration
Assumption				
Mechanical completion—from progress plan			12.10.2022	
Delivery			02.12.2022	
Main systems—into the building				
Trafo and supply to main switchboard				
Fiber				
Builder deliveries—TE coordinates timing and duration with BH				
Network equipment assembly				
Deliveries VA (outdoors)				
Grease separator with pump sump	731	001	12.10.2022	1
Deliveries TE				
Roller grille in front of canteen	= 249	001	12.10.2022	1
Roller grille at the amphitheater	= 249	002	12.10.2022	1
Locks and fittings—door pumps and door automation	= 547	001	12.10.2022	10
Lift	= 621	001	12.10.2022	1

determined by the nearest representative weather station. The highest and lowest 3-day average temperatures from the Meteorological Institute are used for design outdoor temperatures. Climate requirements must be met year-round, even without internal loads. The project's Technical Room Diagram sets indoor climate standards, including ventilation rates according to guidelines from the Labour Inspection Authority and regulations on environmental health for schools [9], in Norway.

Detail Engineering, Installation, and Aftercare Phase The premises have been developed to the commissioning plan according to NS6450 [7], as shown in Table 4, in collaboration with various technical contractors (plumbing, electrical, and ventilation). These contractors are tasked with executing the commissioning and testing of their respective systems. The ITB coordinator is responsible for overseeing technical start-up operations, including monitoring, programming, and preparing for commissioning and testing, as well as recommissioning of building services and control systems.

The specifications outlining the ITB responsibilities and mandate are detailed within the contract agreement forms between the main contractor and the ITB coordinator.

The ITB coordinator also responsible for design reviews and offers advice on suitability for ease of commissioning.

From Commissioning to Aftercare Support In the case study, the ITB coordinator has assumed the responsibility for aftercare support for building occupants. This includes scheduling meetings between the aftercare team or individual and building occupants or management from initial occupation, introducing available aftercare support such as the building user guide and training schedule and content.

Essential information about the building, including design intent, instructions for efficient operation, and details about building systems, is provided and communicated.

Seasonal commissioning, in accordance with standard NS6450:2015 [7], has been conducted as per the contract from the detailed engineering phase. Testing of all building services occurred under full load conditions and periods of extreme occupancy (e.g., heating in midwinter). Interviews with building occupants have been conducted to identify user problems, as illustrated in Table 5.

The project has achieved ambitious environmental goals, including BREEAM-Excellent certification, passive house standards, and zero-emission building (ZEB-OM). In the case study, the ITB systems process enables the integration of various technical systems in an energy-efficient way. This integration has significantly facilitated the optimization of energy consumption during the installation, integration, and operation of complex technical systems. Different sensors are interactively connected to the central system, allowing for energy optimization through automatic control.

Table 6 below demonstrates how this setup aids in identifying potential malfunctions or deactivations in the cooling component of the ventilation system during the trial operation phase, particularly in the heating season.

3   Conclusion

As the complexity of building technical systems evolves, the need for a deeper understanding of both technology and sustainability becomes increasingly important. Neglecting to integrate sustainability knowledge can damage the project progress. This again emphasizes the critical role of multidisciplinary comprehension, particularly in sustainable building practices like integrated technical building (ITB) processes.

This achievement of the case study is related to consistent adherence to predefined standards and transparent working requirements for the ITB process, spanning from project inception to post-project care. The ITB coordinator's closely oversight throughout commissioning and recommissioning phases ensures alignment with specified standards-NS3935, 2019 [4].

The early engagement of ITB experts with clients is also crucial in following precise requirements, ensuring quality assurance, closely monitoring, and providing support at every project stage. Addressing deviations promptly and refraining from assuming control of automation systems until all essential components are in place is highly important. Commissioning emerges as an opportune moment to fine-tune building automation, aligning it seamlessly with tenant needs while optimizing energy efficiency. Effective project planning, robust competence, and capacity

Table 5 The plan of the trial operation phase

Component group	Component	What is checked?	Scope	Inspection method	Note/recommended method	When
General	All	Fire and escape safety	Entire facility	Full-scale test	N/A	After 11 months
	All	Compliance between FDV documentation and what has been delivered	System/facility	Documentation control	N/A	Ongoing
234	Windows, doors, and gates	Lock system, local closer/opener (door pump and local operation such as elbow switch), access control, locking magnet, UPS, lock and fittings	Doors not in daily use	Function test	Simulate night, mains failure, fire signal, check projected function, capacity UPS (time)	After 4 and 8 months
234	Windows, doors, and gates	Smoke hatches	All	Function test	N/A	After 6 months
312	Wiring network	Consumption water meter	All	All reading and compare with corresponding building	N/A	Every month
312	Wiring network	Roof vents, etc.	All	All inspect functionality external drainage conditions and intake	Does the water flow where it should? Consider heating cable, falls, heating mats, etc.	Autumn and winter

Table 6 Data showing the trial operation phase on the cooling battery

System	Energy demand (kW)	Total energy consumption (kWh)
Main system 320.001	33.76	28,911
Zone battery 320.003	19.28	12,557
Floor heating system 320.004	2.49	4958
ICT room cooling 370.002	0	0
Ventilation heating OE501	37.17	1398
Ventilation cooling OE502	0	0
Ventilation cooling OE501	37.45	1023
Ventilation cooling OE502	0	0

form the cornerstone of project success, particularly in energy and sustainability endeavors. The contractual framework delineates various requirements for the commissioning plan, empowering the ITB coordinator to engage the process.

In summary, the research's detailed approach to implementing ITB, including recommended standards, tools, and procedures, serves as a valuable reference for other construction projects targeting high-energy performance and sustainability. It underscores the potential of ITB in optimizing energy efficiency and reducing the carbon footprint of buildings, ultimately contributing to broader sustainable development goals.

Overall, the study provides a roadmap for achieving zero-emission buildings, demonstrating that integrating advanced technical systems and adhering to rigorous environmental standards can lead to successful outcomes. The findings are particularly relevant for architects, engineers, project managers, and policymakers seeking to design and implement green buildings that meet stringent environmental requirements.

References

1. García-Monge, M., Zalba, B., Casas, R., Cano, E., Guillén-Lambea, S., López-Mesa, B., Martínez, I.: Is IoT monitoring key to improve building energy efficiency? Case study of a smart campus in Spain. Energy Build. **285**, 112882 (2023)
2. Ardehali, M.M., Smith, T.F., House, J.M., Klaassen, C.J.: Building energy use and control problems: an assessment of case studies. ASHRAE Trans. **109**, 111–122 (2003)
3. Lawrence, R., Keime, C.: Bridging the gap between energy and comfort: post-occupancy evaluation of two higher-education buildings in Sheffield. Energy Build. **130**, 651–666 (2016)
4. NS3935: Integrated Technical Building Installations (ITB) Design, Implementation and Commissioning. Norge, Oslo (2019)
5. INTEGRA: ITB – Integrated Technical Building Installations: A Profitable Choice for Builders. INTEGRA (2011)
6. Final report BA2015: https://prosjektnorge.no/wp-content/uploads/2017/12/Sluttrapport.pdf

7. NS6450:2016: Commission and Testing of Technical Building Installations. Standard Norge, Oslo (2016)
8. The definition of Zero Emission building in Norway: http://www.zeb.no/index.php/en/about-zeb/zeb-definitions
9. Refer to the website of Labor Inspection Authority and regulations: https://www.arbeidstilsynet.no/en/

Urban Heat Islands in the Urban Built Environment: Quantifying the Spatial Patterns of UHIs Intensity in Oslo, Norway, Using High-Resolution Crowdsourced Weather Observations

Joanna Badach [iD], Guilherme B. A. Coelho [iD], Dimitrios Kraniotis [iD], and Peter Schild [iD]

Contents

J. Badach (✉)
Department of Built Environment, Faculty of Technology, Art and Design, Oslo Metropolitan University, Oslo, Norway

Department of Urban Architecture and Waterscapes, Faculty of Architecture, Gdańsk University of Technology, Gdańsk, Poland
e-mail: joanna.badach@pg.edu.pl

G. B. A. Coelho
Department of Built Environment, Faculty of Technology, Art and Design, Oslo Metropolitan University, Oslo, Norway

CERIS and Departamento de Engenharia Civil, Faculdade de Ciências e Tecnologia, Universidade NOVA de Lisboa, Caparica, Portugal
e-mail: g.coelho@fct.unl.pt

D. Kraniotis · P. Schild
Department of Built Environment, Faculty of Technology, Art and Design, Oslo Metropolitan University, Oslo, Norway
e-mail: dimkra@oslomet.no; petsch@oslomet.no

© The Author(s) 2025
M. Kioumarsi, B. Shafei (eds.), *The 1st International Conference on Net-Zero Built Environment*, Lecture Notes in Civil Engineering 237,
https://doi.org/10.1007/978-3-031-69626-8_89

1 Introduction

Climate change leads to many undesired effects in cities, such as extreme weather episodes and air temperature anomalies, which result in a reduction in safety and life quality of urban inhabitants. It is becoming increasingly important to better understand the local impacts of climate change to develop more suitable adaptation and mitigation strategies. Urban heat islands (UHIs), defined as urban areas with higher temperatures than the surrounding rural areas, are one of the consequences of urban growth. They are the result of heat absorption and re-emission by improperly designed urban infrastructure [1]. Typically associated with low- and mid-latitude cities, they are a worldwide phenomenon, becoming a source of concern in northern settlements likewise [2].

In the past years, UHIs have been studied extensively. Although their main drivers across different climates are generally known, a complete understanding of this phenomenon and its dynamics is still missing, both at the global or regional scale [3–5] and when it comes to their local patterns and various contributing factors [6–8]. Reports show that UHIs are becoming more severe in urban areas with higher energy consumption and where urban planning fails to provide a sufficient amount of open and green spaces for effective ventilation and cooling or surfaces with high albedo. However, these UHIs drivers vary across different urban areas.

Therefore, to mitigate UHIs with a more targeted approach and to account for them in planning strategies and scenarios, it is important to explore the local seasonal and spatial distribution of UHIs and its links with urban morphology. This is because previous studies into the local driver of UHIs indicate that they are significantly impacted by particular features of urban morphology, such as impervious surface ratio or building density. These studies used statistical methods [9–13] or various modeling approaches [14–17]. When investigating the spatial and temporal distribution of UHIs and their links with urban form, historical weather data is necessary for the analysis or model validation. However, this sort of spatial data is not widely available in many urban areas. One possibility to fill this gap is to employ the rapidly evolving low-cost sensing technologies [18].

It is important to note that although there is already a lot of evidence concerning these phenomena, many previous studies focus on urban areas with specific spatial or climatic conditions. As a result, their implications are not easily transferable to other urban areas. Studies addressing northern European cities are still too scarce and this gap needs to be addressed as these cities are also affected by the UHIs effect. For example, it is predicted that in the city of Oslo, the temperatures will rise by over 5 ° C in the warmest month by 2050 [19]. Addressing this gap may be facilitated by the growing popularity of low-cost personal weather sensors in the city. Recently, their number has increased so significantly that the private weather stations are also becoming a source of valuable observations for scientists, including the Norwegian Meteorological Institute [20]. As a result, researchers now have access to high-coverage observations throughout Oslo, which supports the studies exploring the distribution and local drivers of UHIs.

The objective of this study is to verify the potential of weather data crowdsourcing with a high spatial resolution for UHIs intensity estimation. To this end, this chapter focuses on exploring the use of the low-cost observations from Netatmo weather stations in Oslo, where UHIs phenomenon has not yet been fully investigated. It estimates the seasonal temporal distribution of UHIs in the city, providing an updated basis for further investigation. In addition, this study helps to better understand UHIs determinants in urban areas in this climatic region and shows directions for further research into the relationship between urban form characteristics and UHIs effect. In the future, this may bring more guidelines for the integration of UHIs mitigation within the urban planning practice.

2   Materials and Methods

2.1   Study Area

The study focuses on the local climate and built environment in the capital of Norway, Oslo. The municipality (59°54 N, 10°44 E) has a population of over 700,000 inhabitants. It is located at the head of the Oslo fjord, enclosed by the peri-urban forest zone of the Marka Belt (see Fig. 1), which is an area restricted from development [21].

According to Köppen climate classification, its climate is defined as Dfb, i.e., warm-summer humid continental [22]. The occurrence of UHIs was identified in Oslo in previous studies [19, 23], but it is important to further monitor this phenomenon and look further into its links with the spatial structure of the city. In addition, Oslo is an interesting case study to better understand UHIs drivers due to its varied urban morphology and land uses (downtown high-density areas, residential zones with different morphology, large and vegetated areas) and complex topography. Another advantage is the availability of high-resolution long-term weather data from low-cost weather stations, supplementing municipal monitoring systems. Oslo municipality has very ambitious climate goals, i.e., to become zero-emission by 2030, and it is looking into sustainable solutions for climate change adaptation [24]. Hence, the prospects of using low-cost weather data to better evaluate local climate and predict UHIs will be of key importance to design better spatial policies in the city.

2.2   Weather Data Acquisition

The study uses observations from low-cost Netatmo weather stations [25]. They are a passive crowdsourced method and a useful and promising tool to collect local weather data of high spatial resolution for the UHIs effect quantification,

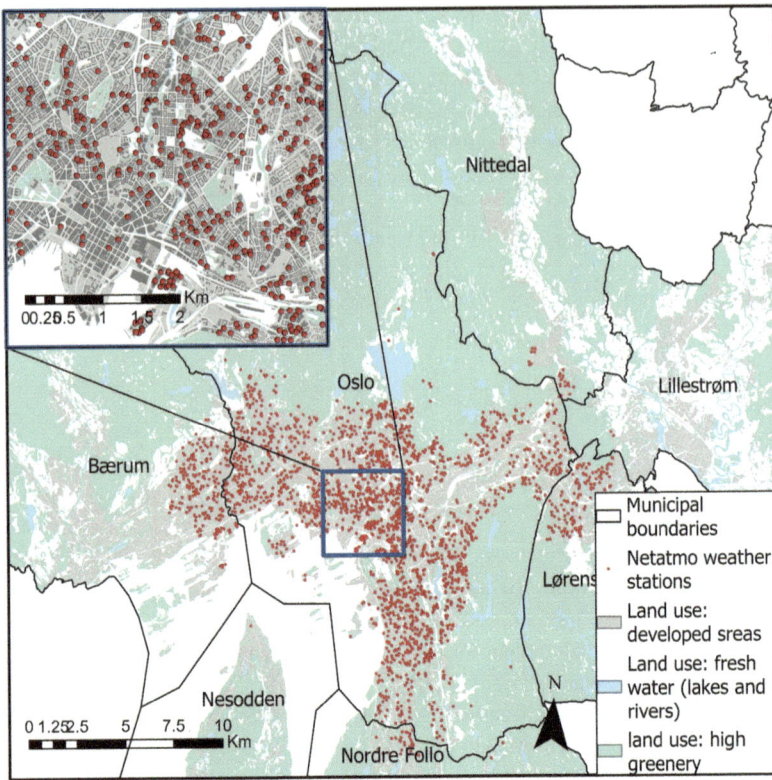

Fig. 1 Oslo municipality and the location of 2826 Netatmo stations (2626 of which were active in 2023)

supplementing greatly the data from municipal weather stations. The Netatmo observations were retrieved using a dedicated macro-enabled Microsoft Excel work-book MeteoDataXL that automatically and independently downloads historical hourly meteorological observations directly from Netatmo's weather API [26].

The whole year of 2023 was selected for the analysis in the study. A list of more than 2800 stations operating in 2023 within the borders of Oslo municipality and the adjacent areas was collected and initially mapped in ArcGIS Pro [27] to check for any errors, such as repeating points and clearly wrong coordinates. This may happen as the Netatmo weather units are calibrated by the users and errors may occur. All such stations were deleted from the dataset. Then the list was used to automatically download the hourly temperature observations with the MeteoDataXL notebook from January 1 to December 31, 2023.

2.3 Data Quality Check

The high coverage with Netatmo stations offers the potential to investigate the relationship between UHIs and the urban morphology characteristics [28]. However, the data needs to be subjected to appropriate validation [29]. Although Netatmo sensors were tested in climate chambers [18, 30] and in field tests [28], the main problem, which is still not fully resolved, is the unshielded design of the units, resulting in warm bias caused either by direct solar radiation or proximity to building walls due to emitted heat [31]. Among the first studies to address the quality control of Netatmo sensors for UHIs quantification were those done for Berlin [28, 30] and for London [31]. Data processing scheme (quality check) was proposed and further developed and tested in subsequent case studies. The scheme is a combination of data processing and statistical methods, which now consists of several steps summarized in Table 1.

A separate quality control scheme was also proposed by other authors, consisting of a "buddy check" (removing observations with a high deviation from the mean neighborhood observation), removing isolated stations (as their values cannot be cross-referenced) and a spatial consistency test (spatially grouping the values and for each region and removing outliers) [35, 36]. However, so far, the quality scheme

Table 1 Data processing scheme

Problems and solutions	Reference
Invalid metadata (latitude, longitude, altitude, time-stamp) *Solution*: Removing values from stations with uncertain (e.g., repeating) latitude and longitude	[28, 30, 32–34]
Obviously erroneous values (e.g., sensors placed indoors or close to a heat source) *Solution*: (1) Comparing temperature data from Netatmo sensors and from municipal weather stations and removing observations when the difference between measured temperature is beyond the range of reference thresholds.	[12, 14]
(2) Using only Netatmo observations and eliminating outliers	[31–34]
Missing values (temporary disturbances in sensor operation, resulting in insufficient observations for a given day or month) *Solution*: Removing hourly values for a given day if there is less than 80% hourly data per day, removing monthly values for a given month if there is less than 80% daily data per month	[28, 30, 32, 35]
No solar radiation shield (e.g., sensors placed in direct sunlight) *Solution*: (1) Comparing temperature data from Netatmo sensors and from municipal weather stations and removing observations when the difference between measured temperature is beyond the range of reference thresholds.	[30]
(2) Comparing temperature data from Netatmo sensors and global radiation and temperature data from municipal stations for selected day-time values, and removing observations when there is a significant link between Netatmo observations and solar radiation (systematic radiative filter error), and when the difference between measured temperature is beyond the range of reference thresholds (single value radiative error)	[28, 33]

proposed by Meier et al. (2017) [28] and Napoly et al. (2018) [32], later modified by Feichtinger et al. (2020) [33], are so far considered the most advanced ones and were used in subsequent case studies (e.g., [37]). In one of the most recent works, Fenner et al. (2021) proposed a modification of the scheme so that it does not require additional data (e.g., temperature or solar radiation from municipal stations) but uses only the Netatmo sensors data [34]. Despite not using solar radiation data, the scheme is very rigorous. It consists of the following steps:

m1—*Metadata Check*,

m2—*Distribution Check*—removing statistical outliers—any observations in the upper and lower part of the hourly distribution (1% and 95% quantile),

m3—*Data Validity*—when too many values are flagged as false due to the procedure in—m2 step (more than 20%) then removing values for a given month for this station,

m4—*Temporal Correlation*—removing values when the Pearson correlation coefficient between the individual station and the median of all stations is lower than 0.9 for a specific month,

m5—*Spatial Buddy Check* (additional outlier detection in the given neighborhood)—mean and standard deviation are calculated for stations within the given radius, and the values in the lower and upper part of the distribution are flagged as false,

Optional steps:

o1—*Temporal Interpolation*—filling in the missing hourly values for a single time step by interpolating from the two closest values from the same station,

o2—*Daily Validity*—removing hourly values for a given day if there is less than 80% hourly data per day,

o3—*Validity in Time Period*—removing monthly values for a given month if there is less than 80% daily data per month,

It was recently updated and published as an R package CrowdQCplus [34], which performs the full data quality check scheme or its selected steps.

In this study, CrowdQCplus was used for the retrieved Netatmo data, only step o1 was omitted. Despite the initial metadata check in ArcGIS Pro, step m1 was also included. Collected observations were used for further analysis if went through the data quality check scheme.

2.4   UHIs Intensity Calculations

The UHIs intensity was selected to calculate the spatial distribution of UHIs. This indicator is calculated as the difference between the 2 m temperature for each grid point (°C) and the 2 m average temperature (°C) measured at all rural stations [38]:

$$\text{UHI}_i = T_i - \overline{T_{\text{rur}}} \tag{1}$$

Since the Netatmo observations were at different heights and the urban thermal gradient is too complex to easily recalculate it to 2 m, the Netatmo observations were used as they are, without accounting for the height difference. The rural stations data were retried using the MeteDataXL workbook, which also downloads data from Frost API [39] and AgroMetBase [40] in Norway. According to the Copernicus definition, the rural area is represented by the rural classes of CORINE land cover covering grassland, cropland, shrubland, woodland, broadleaf forest, and needleleaf forest [38]. The CORINE land cover inventory is updated every 6 years, and the latest update was from 2018 [41]. It was retrieved for Oslo and its adjacent areas directly from the Geonorge portal of the Norwegian Mapping Authority [42]. The CORINE classes and the rural stations selected for UHIs intensity calculations are shown in Fig. 2.

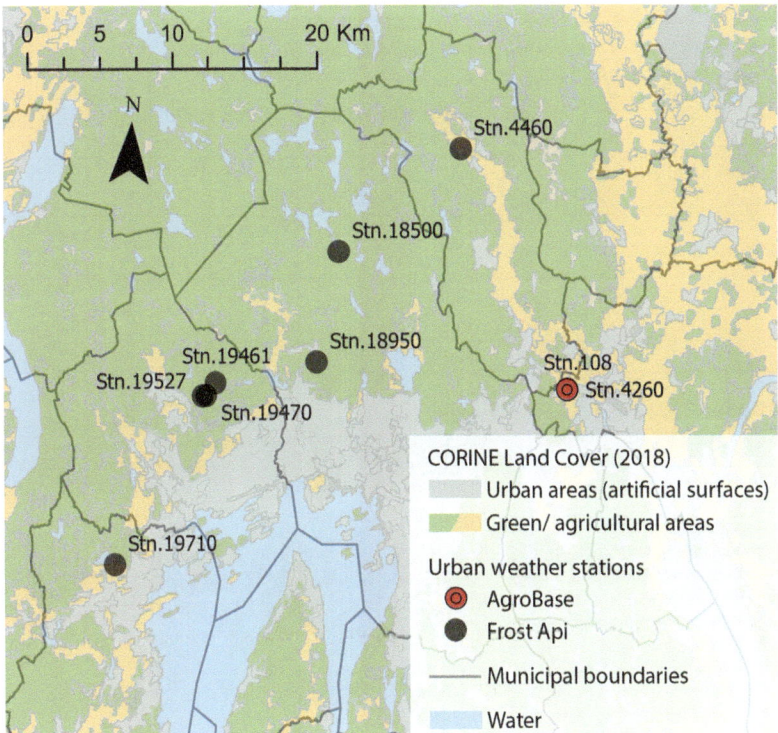

Fig. 2 CORINE land cover inventory for Oslo and the AgroMetBase and FrostApi stations that were used for the UHIs intensity calculations

3   Results

3.1   Data Quality Check

The processing of the weather data with CrowdQCplus resulted in over 60% of data loss. This is consistent with previous case studies, in which the data loss was even over 70% [34]—see the comparison in Table 2. The difference may be explained by the fact that in the Oslo case study the initial metadata check was already performed before running the data quality scheme.

3.2   UHIs Spatial Distribution

UHIs intensity was calculated for selected days and mapped in ArcGIS Pro to show its spatial distribution—see Figs. 3 and 4.

Table 2 Percentage of hourly data availability at each level of the quality check in previous case studies in Amsterdam and Toulouse and in the new case study in Oslo

CrowdQC+ step	Amsterdam (2019) (%)	Toulouse (2020) (%)	Oslo (2023) (%)
Raw data (number of weather stations)	100 (368)	100 (1214)	100 (2626)
m1	69.4	92.0	84.1
m2	59.9	82.5	74.3
m3	59.4	82.2	74.0
m4	58.2	81.4	73.4
m5	47.1	70.6	65.6
o1	47.8	70.2	–
o2	36.9	56.6	52.5
o3	20.7	29.5	33.7

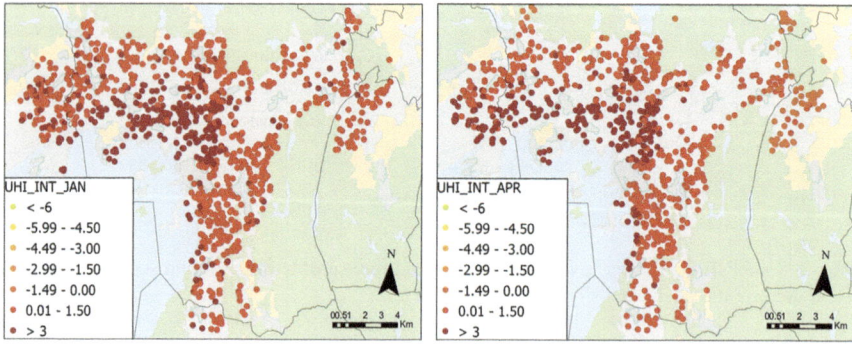

Fig. 3 UHIs intensity in Oslo at 12.00 on 15 January and 15 April 2023

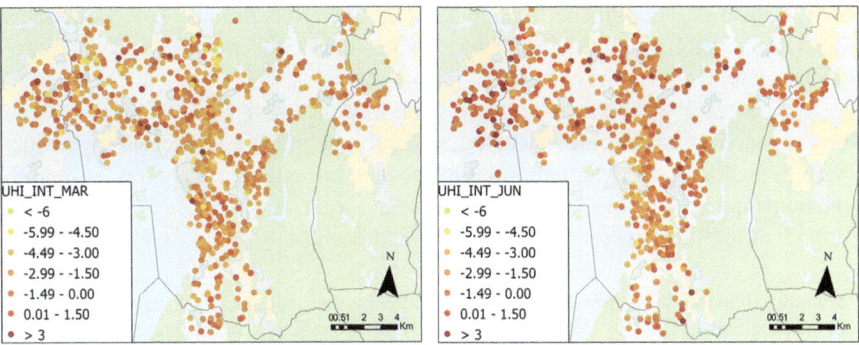

Fig. 4 UHIs intensity in Oslo at 12.00 on 15 March and 15 June 2023

What is interesting when comparing the UHIs intensity maps is that it varies with seasonality. In the first two cases (January and April), it follows a typical gradient with a stronger UHIs effect in downtown areas (even up to 3° higher), while in the other two cases (March and June), the distribution is random or even has an inversed pattern. In some cases, the temperatures captured in downtown areas were lower than at the agricultural monitoring sites (with up to 6.5° of difference). The results are consistent with previous studies, which suggest that UHIs intensity in Oslo is relatively high in selected seasons and marginal in others. However, a previous study identified autumn and winter as seasons with the strongest UHIs effect [23]. Further analysis is still needed, but it seems that in many cases the effect of the topography and larger-scale weather effects may have a stronger influence on the occurrence and spatial distribution of UHIs compared to the urban form character- istics. For example, a strong effect of a thermal inversion in Oslo was observed, when on days with mild or no winds the temperature increases with increasing altitude [43].

4 Discussion and Conclusions

This study has explored the availability and suitability of low-cost Netatmo station weather observations for evaluating the spatial and temporal distribution of UHIs. It proved that the very high spatial resolution of the crowdsourced weather data, coupled with appropriate data management procedures, makes it possible to provide an updated estimate of the distribution of UHIs intensity.

The increasing popularity of these personal weather stations results in high spatial coverage with data points which, even after applying rigorous data quality check schemes, are a substantial dataset very useful for further analysis. This makes it possible to evaluate and address the problem locally in order to design better mitigation strategies. Many previous case studies looking into the drivers of UHIs are available but they are very often case-specific and not easily transferable to other areas with different climatic, topographic, or spatial conditions. This work showcases the suitability of the emerging weather data collection networks for more climate-intelligent spatial planning.

The next step will be to better understand to what extent UHIs effect in Oslo can be explained by urban form characteristics and how much it is affected by the local weather patterns and topography. It will be also important to better understand the relationship between UHIs intensity and urban form characteristics as this can bring important implications for urban planning and design.

Acknowledgments This research was partly funded by the Excellence Initiative—Research Initiative programme, DEC no 11/1/2022/IDUB/II.1b/Am (beneficiary: Joanna Badach).

References

1. Yadav, N., Rajendra, K., Awasthi, A., Singh, C.: Systematic exploration of heat wave impact on mortality and urban heat island: a review from 2000 to 2022. Urban Clim. **51**, 101622 (2023)
2. Miles, V., Esau, I.: Surface urban heat islands in 57 cities across different climates in northern Fennoscandia. Urban Clim. **31**, 100575 (2020)
3. Zhang, Z., Paschalis, A., Mijic, A., Meili, N., Manoli, G., van Reeuwijk, M., Fatichi, S.: A mechanistic assessment of urban heat island intensities and drivers across climates. Urban Clim. **44**, 101215 (2022)
4. Shao, L., Liao, W., Li, P., Luo, M., Xiong, X., Liu, X.: Drivers of global surface urban heat islands: surface property, climate background, and 2D/3D urban morphologies. Build. Environ. **242**, 110581 (2023)
5. Liu, Y., Li, Q., Yang, L., Mu, K., Zhang, M., Liu, J.: Urban heat island effects of various urban morphologies under regional climate conditions. Sci. Total Environ. **743**, 140589 (2020)
6. He, B.-J., Ding, L., Prasad, D.: Relationships among local-scale urban morphology, urban ventilation, urban heat island and outdoor thermal comfort under sea breeze influence. Sustain. Cities Soc. **60**, 102289 (2020)
7. Jang, S., Bae, J., Kim, Y.J.: Street-level urban heat island mitigation: assessing the cooling effect of green infrastructure using urban IoT sensor big data. Sustain. Cities Soc. **100**, 105007 (2024)
8. Santos, L.G.R., Nevat, I., Pignatta, G., Norford, L.K.: Climate-informed decision-making for urban design: assessing the impact of urban morphology on urban heat island. Urban Clim. **36**, 100776 (2021)
9. Yuan, M., Huang, Y., Shen, H., Li, T.: Effects of urban form on haze pollution in China: spatial regression analysis based on PM2.5 remote sensing data. Appl. Geogr. **98**, 215–223 (2018)
10. Gao, Y., Zhao, J., Han, L.: Exploring the spatial heterogeneity of urban heat island effect and its relationship to block morphology with the geographically weighted regression model. Sustain. Cities Soc. **76**, 103431 (2022)
11. Gardes, T., Schoetter, R., Hidalgo, J., Long, N., Marquès, E., Masson, V.: Statistical prediction of the nocturnal urban heat island intensity based on urban morphology and geographical factors – an investigation based on numerical model results for a large ensemble of French cities. Sci. Total Environ. **737**, 139253 (2020)
12. Erdem Okumus, D., Terzi, F.: Evaluating the role of urban fabric on surface urban heat island: the case of Istanbul. Sustain. Cities Soc. **73**, 103128 (2021)
13. Oukawa, G.Y., Krecl, P., Targino, A.C.: Fine-scale modeling of the urban heat island: a comparison of multiple linear regression and random forest approaches. Sci. Total Environ. **815**, 152836 (2022)

14. Mirzaei, P.A.: Recent challenges in modeling of urban heat island. Sustain. Cities Soc. **19**, 200–206 (2015)

15. Unal Cilek, M., Cilek, A.: Analyses of land surface temperature (LST) variability among local climate zones (LCZs) comparing Landsat-8 and ENVI-met model data. Sustain. Cities Soc. **69**, 102877 (2021)

16. Aghamolaei, R., Fallahpour, M., Mirzaei, P.A.: Tempo-spatial thermal comfort analysis of urban heat island with coupling of CFD and building energy simulation. Energy Build. **251**, 111317 (2021)

17. Back, Y., Kumar, P., Bach, P.M., Rauch, W., Kleidorfer, M.: Integrating CFD-GIS modelling to refine urban heat and thermal comfort assessment. Sci. Total Environ. **858**, 159729 (2023)

18. Coney, J., Pickering, B., Dufton, D., Lukach, M., Brooks, B., Neely, R.R.: How useful are crowdsourced air temperature observations? An assessment of Netatmo stations and quality control schemes over the United Kingdom. Meteorol. Appl. **29**(3), 1–18 (2022)

19. Venter, Z.S., Krog, N.H., Barton, D.N.: Linking green infrastructure to urban heat and human health risk mitigation in Oslo, Norway. Sci. Total Environ. **709**, 136193 (2020)

20. Henriksen, A.: This is how you become your own weather forecaster [in Norwegian]. Aftenposten (2024). https://www.aftenposten.no/hytte/friluftsliv-turtips-turforslag-fra-hytta/i/dw614X/yrno-vil-ha-din-hjelp-som-vaermelder

21. Venter, Z.S., Figari, H., Krange, O., Gundersen, V.: Environmental justice in a very green city: spatial inequality in exposure to urban nature, air pollution and heat in Oslo, Norway. Sci. Total Environ. **858**(2022), 160193 (2023)

22. Kottek, M., Grieser, J., Beck, C., Rudolf, B., Rubel, F.: World map of the Köppen-Geiger climate classification updated. Meteorol. Z. **15**(3), 259–263 (2006)

23. Wai, C.Y., Muttil, N., Tariq, M.A.U.R., Paresi, P., Nnachi, R.C., Ng, A.W.M.: Investigating the relationship between human activity and the urban heat island effect in Melbourne and four other international cities impacted by COVID-19. Sustainability. **14**(1), 378 (2022)

24. Hofstad, H., Torfing, J.: Towards a climate-resilient city: collaborative innovation for a "green shift" in Oslo. In: Álvarez Fernández, R., Zubelzu, S., Martínez, R. (eds.) Carbon Footprint and the Industrial Life Cycle. Springer, Cham (2017)

25. Netetmo (2024). https://www.netatmo.com

26. Schild, P.G.: MeteoDataXL: software tool to download and process hourly meteorological data (2023). https://github.com/SchildCode/MeteoDataXL/

27. Esri Inc. 2023: ArcGIS Pro, version 3.2 (2024). https://pro.arcgis.com/

28. Meier, F., Fenner, D., Grassmann, T., Otto, M., Scherer, D.: Crowdsourcing air temperature from citizen weather stations for urban climate research. Urban Clim. **19**, 170–191 (2017)

29. Muller, C.L., et al.: Crowdsourcing for climate and atmospheric sciences: current status and future potential. Int. J. Climatol. **35**(11), 3185–3203 (2015)

30. Meier, F., Fenner, D., Grassmann, T., Jänicke, B., Otto, M., Scherer, D.: Challenges and benefits from crowdsourced atmospheric data for urban climate research using Berlin, Germany, as testbed. In: ICUC9 – 9th International Conference on Urban Climate jointly with 12th Symposium on the Urban Environment, p. 6. IAUC (2015)

31. Chapman, L., Bell, C., Bell, S.: Can the crowdsourcing data paradigm take atmospheric science to a new level? A case study of the urban heat island of London quantified using Netatmo weather stations. Int. J. Climatol. **37**(9), 3597–3605 (2017)

32. Napoly, A., Grassmann, T., Meier, F., Fenner, D.: Development and application of a statistically-based quality control for crowdsourced air temperature data. Front. Earth Sci. **6**, 1–16 (2018)

33. Feichtinger, M., et al.: Case-study of neighborhood-scale summertime urban air temperature for the City of Vienna using crowd-sourced data. Urban Clim. **32**, 1–12 (2020)

34. Fenner, D., Bechtel, B., Demuzere, M., Kittner, J., Meier, F.: CrowdQC+—a quality-control for crowdsourced air-temperature observations enabling world-wide urban climate applications. Front. Environ. Sci. **9**, 1–21 (2021)

35. Clark, M.R., Webb, J.D.C., Kirk, P.J.: Fine-scale analysis of a severe hailstorm using crowd-sourced and conventional observations. Meteorol. Appl. **25**(3), 472–492 (2018)
36. Nipen, T.N., Seierstad, I.A., Lussana, C., Kristiansen, J., Hov, Ø.: Adopting citizen observations in operational weather prediction. Bull. Am. Meteorol. Soc. **101**(1), E43–E57 (2020)
37. Zumwald, M., Knüsel, B., Bresch, D.N., Knutti, R.: Mapping urban temperature using crowd-sensing data and machine learning. Urban Clim. **35**, 100739 (2021)
38. VITO (Vlaamse Instelling voor Technologisch Onderzoek) in collaboration with partners: Demo Web-Application URBAN.1. Copernicus Climate Change Service Demo (2020)
39. Norwegian Meteorological Institute (MET Norway): Frost API (2024). https://frost.met.no/index.html
40. Norwegian Institute of Bioeconomy Research (NIBIO): AgroMetBase (2024). https://lmt.nibio.no/agrometbase/getweatherdata.php
41. Copernicus: CORINE Land Cover. Copernicus Land Monitoring Service (2018) [Online]. Available: https://land.copernicus.eu/en/products/corine-land-cover
42. Norwegian Mapping Authority: The Map Catalougue. Geonorge (2024) [Online]. Available: https://www.geonorge.no/
43. Castell, N., et al.: Localized real-time information on outdoor air quality at kindergartens in Oslo, Norway using low-cost sensor nodes. Environ. Res. **165**, 410–419 (2018)

Achieving a Net Zero Built Environment: The Need to Focus on Urban Green Footprint

Wahidul K. Biswas, Gordon D. Ingram, and Michele John

Contents

1   Introduction

Whilst urban areas occupy only 4% of land area, they contribute about 80% of greenhouse gas (GHG) emissions and consume 60% of world water resources [1]. This situation is likely to be become more severe by 2050, when around 80% of the global population is expected to be living in urban areas [2, 3] resulting in a number of sustainability pressures including increasing global warming impact, water stress, biodiversity loss and resource scarcity [1]. In addition, the urban heat island (UHI) effect, which occurs in cities due to the replacement of natural land cover with dense concentrations of pavement, buildings, and other surfaces that

W. K. Biswas (✉) · M. John
Sustainable Engineering Group, School of Civil and Mechanical Engineering, Curtin University, Perth, WA, Australia
e-mail: w.biswas@curtin.edu.au

G. D. Ingram
Western Australian School of Mines: Minerals, Energy and Chemical Engineering, Curtin University, Perth, WA, Australia

© The Author(s) 2025
M. Kioumarsi, B. Shafei (eds.), *The 1st International Conference on Net-Zero Built Environment*, Lecture Notes in Civil Engineering 237,
https://doi.org/10.1007/978-3-031-69626-8_90

absorb and retain heat, will also have a significant heating effect on our urban environment [4]. This effect increases energy costs and associated GHG emissions (e.g. for air conditioning), air pollution levels, and heat-related illness and mortality. These pressures will likely increase exponentially, unless resource consumption is reduced using innovative and efficient technologies and government policies for reducing environmental impact. Developing economies are going to experience significant rural–urban migration [3], and there will also increasingly be higher standards of urban living, which will result in increased consumption and associated GHG emissions [5]. GHG emissions resulting from fossil fuels and excessive resource consumption are directly and indirectly destroying natural resources due to sea level rise, drought, wild/bush fires, and intense flooding. Other indirect benefits are improved human health, livelihoods, water and food security resulting from reduced air pollution, particularly in urban areas [6]. To tackle climate change and resources depletion resulting from urban activities, urban metabolism (UM) approaches have been considered.

UM facilitates the analysis of the flows of materials and energy within cities, based on mass balance or energy and material flow analysis of a city [7] and identifies the ways to promote circular economy resource-efficiency measures to help in the achievement of a carbon neutral city [8]. Using this approach, Li et al. [9] found that there is a negative correlation between population size and carbon emission rates, hence large cities are usually more energy efficient than small cities due to the reduction in per capita resource consumption. In this paper, life cycle assessment (LCA), which is an environmental management assessment tool, has been combined with UM to evaluate environmental impacts and resource efficiency in modern cities [10–12]. In addition to resource efficiency, Seto et al. [13] found that urban and regional landscapes could also help achieve near net zero carbon emission targets by reducing their metabolic emissions through sequestering CO_2 from the atmosphere. This paper is unique in the way that it presents the use of LCA to help determine decarbonisation strategies for the energy-intensive central business districts (CBDs) in our cities using a UM approach. This approach will consider both the CO_2 sequestration of the vegetation coverage of the urban landscape and the integration of a sustainability assessment framework to assess the sustainability of decarbonisation pathways/strategies to assist achieving net zero emission (NZE).

2   Urban Metabolism and Decarbonisation

The UM approach considers cities as living organisms [14] and analyses the flows (inputs, stocks, and outputs) by studying production, distribution, and consumption-related processes and their corresponding outputs: growth, energy production, and waste generation [15]. Increased urban metabolic energy use is directly linked to unsustainable production and consumption, resulting in the gradual increase of carbon dioxide emissions. Urban metabolic carbon, which is directly linked to metabolic energy, has been defined as the amount of GHGs emitted from a particular urban area, produced through the activities of primary footprint processes (e.g. burning fossil fuels) and secondary footprint processes (e.g. electricity generation). LCA can be used to highlight the areas causing the most emissions so that mitigation strategies can be considered to reduce the metabolic energy and materials used so as to help decarbonise the UM.

3   Sustainable Decarbonisation Model

This model uses a life cycle assessment tool for calculating the GHG emissions from the flows of the materials and energy within cities to find out the end uses and sectors contributing the most to the overall GHG emissions, known as urban GHG hotspots. Accordingly, the decarbonisation pathways or improved technological scenarios are selected to treat these hotspots and further reduce GHG emissions. These emissions can be reduced, but cannot be reduced to zero, so vegetation coverage also needs to be considered to sequester GHG emissions. This framework will thus determine the amount of vegetation coverage needed to sequester the remaining emissions after the implementation of GHG mitigation technologies. The technological options for decarbonisation could have other environmental, social, and economic impacts. Therefore, a life cycle sustainability assessment has been considered for evaluating the sustainability performance or implications of these options. The following section discusses the procedure for decarbonising a city.

3.1   Carbon Footprint Calculation

The ISO14040-44 guideline has been followed to determine the carbon footprint of the model of a 'standardised CBD' as a representation of the NZE efforts required to decarbonise our cities. The LCA guidelines consist of four steps [16]: goal and scope definition, compilation of life cycle inventory (LCI), life cycle impact assessment, and interpretation. The first step of the guideline is to determine the goal and scope of the LCA. The goal of this LCA is to decarbonise a city. The target or the timeframe for achieving the decarbonisation will depend, however, on regional or national policies. The scope defines what inputs, processes, and activities need to be considered in the LCA analysis. The emissions from the production of capital infrastructure, such as buildings, pipe infrastructure and machinery, and transmission and distribution lines, are excluded from the system boundary of the LCA analysis due to their long lifespans [17]. Only the operational inputs used in a typical city-based CBD in terms of its urban metabolism are included, as shown in Table 1. A functional unit then needs to be defined to conduct a mass and energy balance of these inputs. The functional unit in this LCA is the CBD. Table 1 shows that energy, materials, transportation, and water are consumed by end use appliances to perform day-to-day activities in the residential, industrial, public, and service sectors of a CBD.

Table 1 Urban metabolism activities

Key inputs (*I*)	Systems (sys)	Types (*t*)
Energy	Utility—Decentralised	Renewable
	Utility—Centralised	Non-renewable
		Thermal
		Electricity
	Private transport	Conventional
	Public transport	Alternative
		Heavy vehicle
		Light vehicle
Materials	Virgin	Building
	Recycled	Ferrous
		Non-ferrous
		Composite
		Ceramic/diamond
		Nanomaterials
		Rocks/clay
Water	Centralised	Surface
	Decentralised	Desalinated
		Recycled
Sectors (*s*)	**Activities (*a*)**	**End use appliances (*e*)**
Industry	Food processing	Lamps
	Textiles	Motors
	SMEs	Machinery
	Equipment	Manufacturing technology
Residential	Apartments	Air conditioning
	Houses	Cranes
	Waste management	Treatment technology
	Construction	Cars
Service	Wholesale and retail trade	Trucks/buses
	Accommodation	Electronics
	Food services	Office equipment
	Information/telecom	Traffic system
	Finance/real estate	Harvesters
	Education/training	Elevators/conveyors
	Legal and administrative offices	Information system/server
Public sector	Health	Transmission and distribution
	Road and traffic service	Gas stations
	Waste management	Shower heads/pumps
	Construction	Heaters (water)
	Utilities	Mowers
	Parks and landscaping	Storage

The second step is to develop a life cycle inventory that estimates the amounts of the inputs used in the urban metabolism, including energy, materials, transportation, and water. The inputs consumed by the end use appliances and machinery across different sectors are quantified and then multiplied by the corresponding emission factor (EF) to determine the environmental impact, which is the third step in the ISO LCA guidelines. An emission factor (EF) is the amount of pollutants emitted to the geosphere resulting from the production of the input during all its upstream processes (e.g. mining, processing, and manufacturing of inputs such as cement, caustic soda, and urea). Emission factors are usually expressed as the weight of the pollutant produced divided by a unit weight or volume of the input or service involved (e.g. kilograms of particulate pollutant emitted per MWh of electricity generated). The units of inputs (e.g. MWh) and their corresponding emission factors (e.g. kg/MWh) should be consistent so that they can be simply multiplied together.

Different gases and effluents resulting from the production of an input cause different impacts. The carbon footprint is one of these impacts. In the case of a decarbonisation LCA, only GHG gases (global warming-causing gases; e.g. CO_2, CH_4, N_2O, etc.) are to be considered in estimating the carbon footprint. Different types of GHG are produced from the production of different inputs used in the urban metabolism process. Since global warming impact is represented in terms of CO_2 produced, all GHGs need to be converted to an equivalent amount of CO_2 (i.e. kg or tonne of CO_2 eq). Once all inputs in the urban metabolism have been converted to the equivalent amount of CO_2, they are added up to determine the total CO_2 emissions/global warming impacts resulting from the activities in the CBD.

Careful consideration needs to be made in using or selecting the emission factors as any misrepresentation could affect the reliability of the LCA data outputs. The emission factor for the same input can vary with the processes, location, and inputs used to make them. The same amount of electricity can be generated from both renewable and non-renewable sources, but their EFs are different because the electricity is generated from different types of energy sources using different energy technologies. For example, the average emissions from electricity production in Denmark were 185 g CO_2 eq/kWh in 2023, where 81% of the energy was produced from renewable sources (with 58.7% generated from wind alone) and only 9.9% was produced from coal [18]. If the same 1 kWh of electricity were produced in the USA, 296 g of CO_2 will be produced due to having a different energy mix (i.e. 41% coal, 17% natural gas, 42% petroleum, and other sources) [19]. Therefore, the same amount of a material produced in two different locations using different electricity mixes will have different emission factors. Similarly, different construction materials with varied specifications can provide the same structural strength, but their EFs will be different due to having different manufacturing processes and using different raw materials. The EF of the same input produced in the same location could vary depending on the way it was produced. For example, the equivalent of 3894 tonnes of CO_2 would be emitted from the production of 1 GL of desalinated water from a seawater desalination plant if the electricity were produced from a fossil fuel-powered electricity grid [20]. This emission can be reduced to 367 tonnes of CO_2 if 100% renewable energy were used to run the desalination plant. If recycled water is used instead of desalinated water, the emission factor will be 1300 tonnes of CO_2 eq/GL of water produced, which is 933 tonnes of CO_2 eq higher than the desalination plant at Binningup in Western Australia powered by 100% renewable electricity [21]. Therefore, the EFs should be carefully selected for the inputs used in the CBD metabolism for the LCA analysis. It is also worth mentioning that both the inputs in the inventory as well as their EFs need to be updated every 5 years to account for future technological changes [22]. The emission factor EF for an input I of type t, which is produced within a system sys is expressed as

$$EF_{sys,t} = \frac{\text{kilograms or tonnes of } CO_2 \text{ eq}}{I_{sys,t}} \quad (1)$$

For clarification, t could be the type of energy or materials or water produced in a certain system (e.g. conventional, decentralised, recycling). The emissions of the same input could potentially vary with the systems, locations, production strategies, processes, regulations, chemicals, energy mix, sources of raw materials, usage pattern, etc. Therefore, in the decarbonisation process, the inputs need to be sourced in a way that has a low emission factor.

The carbon footprint of inputs used in the CBD metabolism, which is expressed in terms of kilograms or tonnes of CO_2 eq, is obtained by converting all GHGs to an equivalent amount of CO_2 by multiplying by the equivalent emission factors (e.g. 1 kg of CH_4 = 28 kg of CO_2, 1 kg of N_2O = 265 kg of CO_2). So, EF is measured in kilograms or tonnes of CO_2 equivalent.

The carbon footprint of the CBD can be calculated as follows:

$$CBD_{CO2} = \sum_{s=1}^{S} \sum_{a_s=1}^{A} \sum_{e_a=1}^{E} I_{sys,t,e_a} \cdot EF_{sys,t} \quad (2)$$

where e_a = 1, 2, ..., E is the number of end use appliances used for an activity a, a_s = 1, 2, ..., A is the number of activities performed in a sector s, and s = 1, 2, ..., S is the number of sectors in the CBD.

Equation (2) gives the breakdown of GHG emissions to help find the sector contributing most to the emissions. Secondly, the GHG breakdown in terms of inputs for significantly contributing sectors can be presented to identify the hotspot (s). Once the hotspot(s), or carbon-intensive inputs or processes, have been identified, a range of mitigation strategies or improvement scenarios can be considered to further reduce CO_2 emissions.

3.2 Hotspot Analysis for Determining Mitigation Strategies

Hotspot identification is important to help develop a decarbonisation pathway for the city. Accordingly, measures are taken to improve the technology and processes to further reduce GHG emissions. The sector contributing the most will be investigated,

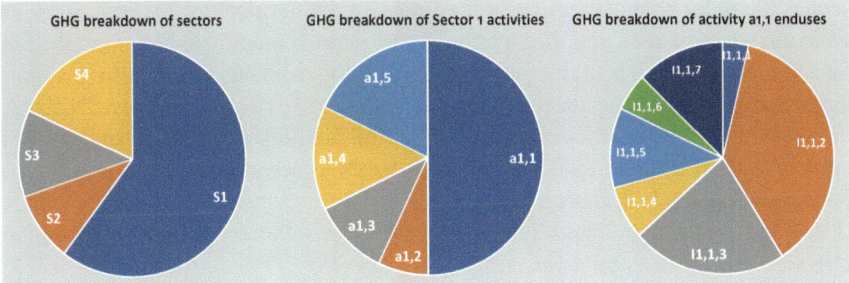

Fig. 1 Hypothetical example for hotspot analysis

Table 2 Decarbonisation strategies

Set$_1$	Set$_2$	$\cdots$	Set$_n$
O11	O21		On1
O12	O22		On2
$\vdots$	$\vdots$		$\vdots$
O1N	O2N		OnN
10 million tonnes	15 million tonnes	$\cdots$	16 million tonnes

support provided and the logistics offered to deploy appropriate mitigation measures.

Figure 1 shows the type of hotspot analysis that can be carried out on the CBD to identify the end use appliances in a sector contributing to the highest GHG emissions. This approach is aligned with Eq. (2) in the previous section. Figure 1 shows that sector S1 contributes most to the GHG emissions, then within sector S1, activity a1,1 accounts for the most significant portion of the emissions. Finally, end use appliances I1,1,2 and I1,1,3 account for the greatest fraction of the total emissions of activity a1,1 in sector S1.

A number of sets of mitigation options for reducing the emissions from these end use appliances (I1,1,2 and I1,1,3) need to be considered to determine the set(s) with the highest mitigation potential (Table 2). It is worth mentioning that it may not always be the case that only one sector or only one activity is the hotspot. In terms of mitigation options, this framework is using the CBD model as representative of urban metabolism and carbon emissions of a city as a way to facilitate reader understanding of the decarbonisation strategies needed in urban development.

The sets contributing most to the mitigation of GHG or CO_2 equivalent emissions are chosen as the decarbonising pathways for the city.

3.3 Net Zero Emission

It is not possible to reduce CBD CO_2 emissions to zero, unless we are able to utilise carbon sequestration within urban green space landscapes. It is suggested that this sequestration by green space can at least remove the balance of CO_2 emissions that were unable to be removed through the use of improved options. For example, Set_2 and Set_n have the highest mitigation potentials for reducing greenhouse gas emissions and decarbonising the city, but their use alone may not achieve NZE. As another important decarbonisation strategy, the area of urban green space that is sufficient to offset these reduced emissions needs to be estimated. The green space landscape could help achieve a balance between anthropogenic emissions by sources (UM) and removal by sinks (green spaces) of greenhouse gases. Net emissions (NE) are expressed as follows:

$$\text{Net emissions} = CO_2 \text{ sequestration} - \text{reduced emissions from end uses} \qquad (3)$$

When CO_2 sequestration = reduced GHG emissions, NE = 0 (net zero emissions).

3.4   Mapping of Urban Green Space

In order to evaluate Eq. (2) or to determine NZE, the amount of CO_2 that can be sequestered by urban green spaces needs to be estimated. Geographical Information System (GIS) software can effectively integrate, represent, and communicate carbon sequestration data to guide the planning process for urban green space (UGS) [22]. Remotely sensed data are also widely used for Land Use Cover Change (LUCC) studies. Alternatively, drones with machine learning software can fly above green spaces to provide accurate reporting of the potential carbon capture by the landscape [23].

The urban green space ecosystem could play a crucial role in carbon sequestration and providing carbon sinks. It could also provide an important role in countering the heat impacts associated with the materials we typically use in urban construction (e.g. concrete, bitumen) [4]. Although the urban green space area accounts for only a part of the urban ecosystem, it plays a very positive role in the carbon balance of urban and global ecosystems in terms of carbon sequestration and urban cooling. This research suggests that it should be increasingly noted as a carbon sequestration mechanism in urban development and, in particular, as a means of offsetting difficult-to-mitigate emissions from CBD city development like cement and construction material use, non-renewable energy use and LUCC that removes vegetation cover [24].

3.5   Sustainability of Decarbonisation Pathways

Improvement scenarios will be developed for reducing GHG emissions. These scenarios have environmental, economic, and social consequences. Therefore, a sustainability framework for decarbonisation needs to be developed to determine the decarbonisation scenarios with the best sustainability outcomes. This is based on a framework developed for palm oil and building industries in 2017 and 2020 by Biswas et al. [25, 26]. This framework was based on the 'strong' definition of sustainability as it integrates three pillars of sustainability (environmental, social, and economic) helping to support technologically focused development. A multi-criteria hierarchal analysis and bottom-up approach (i.e., commencing from lower measuring units of sustainability and focusing on micro-aspects like job losses) have been adopted. The process of the development of the life cycle sustainability assessment (LCSA) framework for this research started with a literature review of existing case studies on sustainability performance, building materials, and the service life of buildings to identify the research gaps in existing LCSA frameworks.

The steps for conducting the LCSA framework are as follows:

1. Development of triple bottom line (TBL) indicators
2. Calculation of the TBL indicators
3. Determination of the sustainability gap for further improvements

Development of Triple Bottom Line Indicators A participatory approach was carried out to select and develop the TBL sustainability indicators for cities of a region because these indicators could vary across regions due to the variation in socio-economic situation, geography, and lifestyle. Local experts were involved in the selection of these region-specific TBL indicators for improved decarbonisation scenarios. The methodology involved four main steps.

A. Preliminary Selection of TBL Sustainability KPIs A careful review of existing national and international literature for best practice guidelines on life cycle impact assessment is needed for selecting the indicators pertaining to the city's sustainability.

B. Final Selection of TBL Sustainability KPIs Once the preliminary list of indicators has been selected, a questionnaire was designed to collect data from the area experts regarding the relevancy and importance of these KPIs. The questionnaire also had the provision for participants to provide feedback and suggest additional KPIs. Potential participants need to be approached through emails to gauge their interest to participate in an online survey. The respondents agreeing to participate will be emailed a link for the online questionnaire to gather their responses. It would be preferable to gather responses from an equal number of respondents from each stakeholder category. The responses are of two types. Firstly, to uncover their agreement with the relevance of the indicator for assessing the sustainability performance of the city. If an indicator receives 50% or more votes, then this indicator is selected. Secondly is the rank of the indicator given by the experts, which is then converted to a weight. The value of the weight represents the significance of the indicator.

C. Weighting of Selected KPIs The weight of each KPI was determined based on the level of importance provided by respondents. The responses with no comments were excluded from weighting calculations and when the response for a KPI was 'irrelevant', it was excluded from further consideration. The scores for responses, including somewhat important, moderately important, important, and most important, were presented on a Likert scale and were assigned as 1, 2, 3, and 4, respectively. The weights of all KPIs under one impact category were then aggregated to the determine the weight of the respective impact category. The weights of the impact categories were aggregated to calculate the weight of the sustainability objectives.

The scores for the level of importance of each of these KPIs given by the area experts in a consensus survey were converted to corresponding weights using Eq. (4) [26].

$$w_i = \frac{\sum_{r=1}^{N} S_{ri}}{N \cdot \sum_{i=I_1}^{I_n} S_i} \tag{4}$$

where w_i = weight of KPI i; N = number of respondents $r = 1, 2, \ldots, N$; S_{ri} = score given by respondent r for indicator $i = 1, 2, \ldots, I_n$, the number of KPIs; and S_i = value of each score.

D. Threshold Values of KPIs The threshold values of the KPIs will be determined by reviewing existing case studies, sustainability guidelines and standards, national and international agreements, and government statistics and reports. The values considered to be optimal or the best value for the regional context will be selected as the threshold values and were given a maximum score of five on a five-point Likert scale.

Calculation of TBL Indicators Environmental LCA (ELCA), social life cycle assessment (SLCA), and life cycle costing (LCC) will be used to calculate the environmental, social, and economic KPIs, respectively. Both ELCA and SLCA follow the same ISO guideline. In the case of SLCA, the qualitative values are converted to numerical values. LCC uses the same LCI as ELCA for calculating the economic indicators.

Determination of the Sustainability Gap for Further Improvements The position of a TBL indicator is located on a five-point Likert scale where the gap between its position and the corresponding threshold value can then be determined.

$$P_{\text{low}} = \frac{\text{Threshold value}}{\text{Calculated value}} \times 5 \tag{5}$$

$$P_{\text{high}} = \frac{\text{Calculated value}}{\text{Threshold value}} \times 5 \tag{6}$$

Equation (5) was used to determine the value of the KPI's position on a Likert scale where less is good (e.g., carbon footprint, embodied energy, construction, and demolition waste). Equation (6) was used to determine the position of KPIs when more is good (e.g., net benefit, GDP, income). The difference between the position value of the KPI and its threshold value (i.e., five on the Likert scale) is the gap G, which was multiplied by the corresponding weight to calculate the performance score PS of the KPI. For example, if the position value of an indicator on a five-point Likert scale is 3, the gap will be $G = 3 - 5 = -2$. Suppose the weight of the indicator determined by using Eq. (4) is 0.21 then the performance score will be $PS = 0.21 \times (-2) = -0.42$. The indicators with the highest gaps need to be identified and appropriate strategies need to be considered to further reduce the gaps. This is an iterative process that keeps going until the gaps are sufficiently reduced. Policy, strategy, and technological changes are considered for reducing the sustainability gaps. That is why the LCSA framework is helpful in determining strategies and policy changes for reducing the gaps associated with the hotspot indicators to help find a sustainable decarbonisation pathway.

4   Deployment Plan

Government Departments of Planning, Land and Heritage, and Environment and Energy can work together using this decarbonisation model by collaborating with different sectors in the city, *including* utilities, suppliers, researchers, urban planners, architects, and industries to deploy this decarbonisation framework.

The biggest challenge is in gathering data from all the stakeholders in the city, including the industrial, service, residential, and government components of the CBD, to complete online questionnaires to help provide energy, water, material, transport and waste management data, which can then be converted into a carbon footprint of the CBD using the above assessment framework.

Urban planners and architects will assist in the landscape and green space design with the intention of increasing green coverage in a cost-effective manner, while designing sufficient green space for effective CO_2 sequestration and urban heat island impact reduction [4]. Industry, academia, and stakeholders should all be consulted through stakeholder workshops to help determine the technological options that can help achieve a more sustainable NZE decarbonisation pathway for a CBD environment. A range of options, such as rebates, feed-in-tariffs, and capacity building programs, can be established, enabling the various participant sectors in the CBD to switch to decarbonised options. The stakeholders also need to be aware of the economic benefits of the decarbonising operational changes, which could reduce operational costs and avoid carbon taxes.

5   Conclusions

This LCA framework was developed to review the potential for decarbonising a city through the increased usage of GHG mitigation options. A step-by-step process has been discussed, including the type of tools that will be used and the categories of stakeholders who should participate in achieving the decarbonisation pathways for important urban planning areas like major central business districts. Resource efficiency through technological improvement and the existence of a sufficient green space were found to be the key variables in the NZE target achieving pathway. This is undoubtedly a data-intensive process as it requires the collection of information on energy and materials for activities in different sectors of the CBD, and interviewing experts and stakeholders who could be directly or indirectly be affected by the decarbonisation activities. In addition, such a planning framework can assist in managing the increasing pressures of urban heat island impacts in our central business districts.

References

1. Voukkali, I., Zorpas, A.A.: Evaluation of urban metabolism assessment methods through SWOT analysis and analytical hierocracy process. Sci. Total Environ. **807**, 150700 (2021)
2. The World Bank. https://data.worldbank.org/indicator/SP.URB.TOTL.IN.ZS?view=chart. Last accessed 24 Feb 2024
3. United Nations Environment Programme and International Resource Panel: The weight of cities: resource requirements of future urbanization—summary for policymakers (2018). Available online: https://wedocs.unep.org/20.500.11822/31624. Last accessed 24 Feb 2024
4. Cheela, V.R.S., John, M., Biswas, W., Sarker, P.: Combating urban heat island effect—a review of reflective pavements and tree shading strategies. Buildings. **11**, 93 (2021)
5. Kovacic, Z., Strand, R., Völker, T.: The Circular Economy in Europe: Critical Perspectives on Policies and Imaginaries. Routledge, New York (2020)
6. Hartley, A., Turnock, S.: What are the benefits of reducing global CO_2 emissions to net-zero by 2050? Weather. **77**, 27–28 (2022) Royal Meteorological Society, https://doi.org/10.1002/wea. 4111
7. Delivandani, F., Rajabi, A., Kermani, A.N.: Process-based improvement of urban metabolism in optimizing the development cycle of the small city using MIA method. Math. Probl. Eng. **2021**, 5545307, 13 p (2021). https://doi.org/10.1155/2021/5545307
8. Fernandes, J., Ferrão, P.: Urban metabolism-based approaches for promoting circular economy in buildings refurbishment. Environments. **10**(1), 13 (2023). https://doi.org/10.3390/environments10010013
9. Li, J.S., Xia, C., Xiang, M., Cao, Y., Yang, J.: The impact of urban scale on carbon metabolism – a case study of Hangzhou, China. J. Clean. Prod. **292**, 126055 (2021) ISSN 0959-6526
10. Birkved, M., Goldstein, B.P.: Environmental sustainability assessment of urban systems applying coupled urban metabolism and life cycle assessment. In: Höfler, K., Maydl, P., Passer, A. (eds.) Proceedings of the Sustainable Buildings – Construction Products and Technologies: Collection of Full Papers, pp. 521–532. Verlag der Technischen Universität Graz (2013)
11. Maranghi, S., Parisi, M.L., Facchini, A., Rubino, A., Kordas, O., Basosi, R.: Integrating urban metabolism and life cycle assessment to analyse urban sustainability. Ecol. Indic. **112**, 106074 (2020). https://doi.org/10.1016/j.ecolind.2020.106074
12. De Toro, P., Iodice, S.: Urban metabolism evaluation methods: life cycle assessment and territorial regeneration. In: Amenta, L., Russo, M., van Timmeren, A. (eds.) Regenerative Territories. GeoJournal Library, vol. 128. Springer, Cham (2022). https://doi.org/10.1007/978-3-030-78536-9_13
13. Seto, K., Churkina, G., Hsu, A., Keller, M., Newman, P., Qin, B., Ramaswami, A.: From low-to net-zero carbon cities: the next global agenda. Annu. Rev. Environ. Resour. **46**, 377–415 (2021). https://doi.org/10.1146/annurev-environ-050120-113117

14. Zhang, Y.: Urban metabolism: a review of research methodologies. Environ. Pollut. **178**, 463–473 (2013). https://doi.org/10.1016/j.envpol.2013.03.052
15. Peña, D.O., Perrotti, D., Mohareb, E.: Advancing urban metabolism studies through GIS data: resource flows, open space networks, and vulnerable communities in Mexico City. J. Ind. Ecol. **26**, 1333–1349 (2022)
16. International Standard Organization: ISO 14040:2006: Environmental Management—Life Cycle Assessment—Principles and Framework. ISO (2006) http://www.iso.org/iso/iso_catalogue/catalogue_tc/catalogue_detail.htm?csnumber=37456
17. Frischknecht, R., Althaus, H.J., Bauer, C., Doka, G., Heck, T., Jungbluth, N., Kellenberger, D., Nemecek, T.: The environmental relevance of capital goods in life cycle assessments of products and services. Int. J. Life Cycle Assess. **12**, 7–17 (2007)
18. Nowtricity: Current emissions in Denmark 2023. https://www.nowtricity.com/country/Denmark/. Last accessed 1 Mar 2024
19. US Energy Information Administration: How much carbon dioxide is produced per kilowatthour of U.S. electricity generation? https://www.eia.gov/tools/faqs/faq.php?id=74&t=11. Last accessed 1 Mar 2024
20. Biswas, W.K.: Life cycle assessment of seawater desalinization in Western Australia. World Acad. Sci. Eng. Technol. **56**, 369–375 (2009)
21. Simms, A., Hamilton, S., Biswas, W.K.: Carbon footprint assessment of Western Australian groundwater recycling scheme. Environ. Manag. **59**, 557–570 (2017)
22. Biswas, W.K., Alhorr, Y., Lawania, K.K., Sarker, P.K., Elsarrag, E.: Life cycle assessment for environmental product declaration of concrete in the Gulf States. Sustain. Cities Soc. **35**, 36–46 (2017)
23. Lahoti, S., Kefi, M., Lahoti, A., Saito, O.: Mapping methodology of public urban green spaces using GIS: an example of Nagpur City, India. Sustainability. **11**, 2166 (2019). https://doi.org/10.3390/su11072166
24. Davar, Z.: Drone reforestation: technology for natural carbon capture (2019). https://medium.com/cleantech-rising/drone-reforestation-technology-for-natural-carbon-capture-6cb6938fefa7. Last accessed 6 Mar 2024
25. Lim, C.I., Biswas, W.K.: Development of triple bottom line indicators for sustainability assessment framework of Malaysian palm oil industry. Clean Techn. Environ. Policy. **20**, 539–560 (2018)
26. Janjua, S., Sarker, P., Biswas, W.: Development of triple bottom line indicators for life cycle sustainability assessment of residential buildings. J. Environ. Manag. **264**, 110476 (2020)

The Effect of Student Active Learning on Awareness of Students on Environmental Impacts of Material Selection for Building Application: A Case Study

Mohammad Hajmohammadian Baghban ⓘ

Contents

1 Introduction

The COVID-19 pandemic provided an unanticipated motivation for the transition towards digital pedagogy. This shift, while necessitated by the global health crisis, opened up new avenues in the realm of education. It is a testament to the resilience of educational systems in the face of unprecedented challenges and may pave the way for future advancements in digital learning environments. The utilization of various digital tools fosters a transition towards the flipped classroom model. This pedagogical approach can enhance students' comprehension of the subject matter and optimize lecturers' time by eliminating the need for repetitive instruction each time the topic is presented. This shift not only promotes effective learning but also contributes to the efficient use of educational resources.

This study investigates the possibilities and challenges of student active learning in a flipped classroom in a "concrete technology" course, a specialization course

M. H. Baghban (✉)
Department of Manufacturing and Civil Engineering, Faculty of Engineering, Norwegian University of Science and Technology (NTNU), Gjøvik, Norway
e-mail: mohammad.baghban@ntnu.no

© The Author(s) 2025
M. Kioumarsi, B. Shafei (eds.), *The 1st International Conference on Net-Zero Built Environment*, Lecture Notes in Civil Engineering 237,
https://doi.org/10.1007/978-3-031-69626-8_91

presented in the fourth semester of bachelor's study in civil engineering at Norwegian University of Science and Technology (NTNU). During the course of this study, 30 students were enrolled in the course. The curriculum primarily encompasses the understanding of concrete properties, the effects of its components, factors influencing its service life, and environmental impacts.

Remote teaching due to the pandemic has made the motivation to shift to flipped classroom method by uploading the course materials and prerecorded videos for the students and using the exercise session to guide the students to perform the expected assignments. Online course materials and affordable hardware and software have also given this opportunity to accelerate this shift [1–5]. Furthermore, measures are planned during the exercise session to keep the students active. While evaluation criteria based on performance and skills are the dominating methods in various educational domains, this evaluation method was also found to fit the given course [6].

The main objective of this study is to help students gain the learning outcome of the course in the flipped classroom teaching system in an active learning environment by improving the learning process. Among different expected learning outcomes, sustainability issues including greenhouse gas emission of the materials have been selected as the learning outcomes to be investigated in this course.

2 Method

To systematically conduct the research and achieve expected results, possible actions were selected based on the literature review, own experience, communication with colleagues, and interaction with students.

Initially, the overall learning methods that would increase student motivation, including acquisition, inquiry, practice, production, discussion, and collaboration, were considered based on the existing literature [7–10], and potentially relevant activities were listed.

Moreover, the methods were selected according to the capacity and experience of the lecturer and the nature of the course, which naturally demands being familiar with the theory and applying it in practice preferably in the laboratory as well as feedback from students and colleagues.

The students were expected to go through the course materials and prerecorded videos and be prepared for the physical classroom. To facilitate the learning environment, the students were encouraged to deliver an assignment before the physical classroom in the form of multiple-choice questions to evaluate their own competence in the theory. In the initial 3 weeks of the course, students experienced a traditional classroom setting, in order to prepare the students and the environment for the flipped classroom. During this time, quizzes and discussions in the physical classroom were solely based on the lecture delivered on the same day. However, after 3 weeks, the classroom approach was flipped, and the students were asked to be prepared and go through the multiple-choice questions before the physical classroom.

In order to ensure that the students have understood the necessary concepts in each lecture, the relevant concepts were reviewed shortly at the beginning of the physical classroom session by going through game-based individual quizzes using a digital learning platform (Kahoot!). This allowed the lecturer to observe the performance of students and the level of acquired knowledge and modify the focus area of the physical session based on the student performance and getting the opportunity to check with the students if there is any topic that needs more explanation. Furthermore, as it is expected that some of the students may not have gone through the course materials and prerecorded videos of the week, activating students' awareness by getting feedback from them as a quiz at the beginning of the physical session would keep them motivated to follow the lectures before attending the physical session. Note that the quizzes were held at the end of the physical classroom in the first 3 weeks that traditional teaching was conducted.

On the other hand, in addition to the multiple-choice questions and quizzes before and during the physical classroom, the learning outcomes were assessed by conducting discussions during the physical classrooms according to the recommendations from the literature [11–13]. Since it is expected that the students show better confidence during and after discussions in groups rather than individually discussing the topic in the classroom, the students were mainly asked to discuss in groups and then present the viewpoint of the group to the classroom. Moreover, the students were given a course project in the form of a laboratory-based group assignment, and they needed to design concrete mixes with different fresh and hardened properties.

The communication with colleagues was conducted through discussions at different stages of the course's progress. Main feedback was collected from the experienced laboratory engineer as well as an associate professor familiar with the subject. The lecturer would then try to implement the outcome of the collegial discussion and observe how it would contribute to achieving the objectives. Furthermore, the students' feedbacks were collected through meetings with the student representatives, which were a group of three volunteer students of the classroom. Figure 1 illustrates the overview of the research methodology conducted in this study.

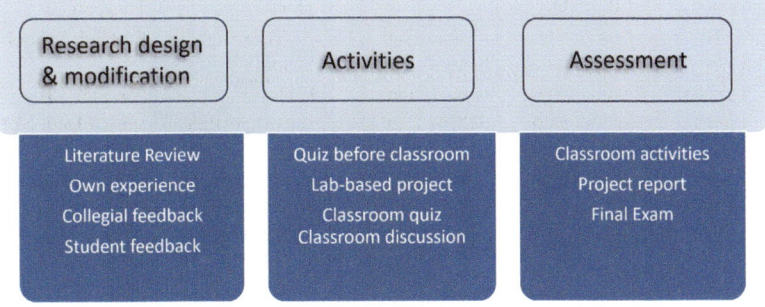

Fig. 1 Overview of the methodology

3 Results and Discussion

Different challenges were observed through performing this investigation according to the above-mentioned method. As it was expected, preparing prerecorded videos with good quality is an extremely time-consuming process that may not be feasible to be conducted during one semester during teaching the course, and it could be considered as a continuously improving process. Furthermore, to keep the quality of the teaching in the prerecorded videos, the potential questions that may arise from students need to be answered in the videos. The feedback from students showed that they were optimistic about the prerecorded videos provided that the videos have an acceptable quality. The lecturer had challenges with the time to prepare the videos and noted that students' feedback and questions could help improve the lectures. Thus, after discussion with the colleagues, it was decided to consider the Microsoft PowerPoint slides for prerecording the lectures and proceed with recording slide by slide to make it feasible to modify them in the next run.

Furthermore, the results showed that the majority of students and colleagues found it beneficial to use Kahoot! to review or discuss the lectures. The student representatives shared feedback from their peers, noting that this process helps keep them engaged and better prepared for in-person classroom sessions. Padlet and Mentimeter were also the other tools suggested by the colleagues, which were tested to a limited extent to compare with Kahoot! It's worth noting that among these digital tools, each has its unique strengths and features. Users can strike a balance in their usage by considering factors such as demand, availability, and feasibility. A noticeable improvement in student performance was observed when transitioning from traditional classroom teaching to the flipped classroom approach. Figure 2 illustrates the overall quiz performance during the initial 3 weeks of both teaching methods: traditional teaching and the flipped classroom. Notably, students performed significantly better in the flipped classroom system, likely due to their preparation before attending the physical classroom sessions.

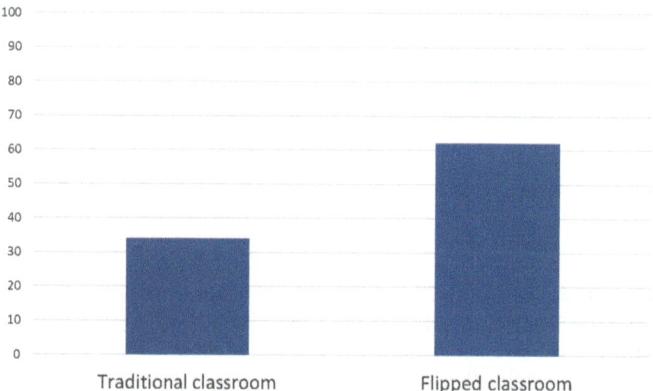

Fig. 2 Average score of the students in the quizzes (percentage) conducted in the physical classroom

On the other hand, both colleagues and the lecturer agreed that too much focus on the review of the lectures at the beginning of the physical session would take considerable time and reduce the motivation of the students to go through the uploaded lecture notes and prerecorded videos. There were different opinions concerning this issue between students and some found more focus on the review process helpful, while some others meant that there was no need for the review as they had gone through the lecture before the physical session. Thus, it was found that a more in-depth review of the lecture, which summarizes the topic, could preferably be done as a separate prerecorded video or podcast and the focus of physical sessions would be on hands-on exercises and in-depth discussions for practical application of the theory.

Furthermore, making the students to discuss practical topics and project-related tasks in groups, could facilitate a student active learning process, and the colleagues agreed with this method as well. However, some students also came with the comment that this method can be inefficient in some cases where no discussion happens in the group or if the time dedicated for the discussion is limited and hinders reaching a suitable outcome. Thus, the engagement and supervision of the lecturer in the group discussions, presenting a proper discussion plan for the students, and considering the suitable time for discussion are suggested as some measures to contribute to overcoming the challenges mentioned by the students, which is in alignment with the literature used in the method section. In addition, using some icebreakers, instructing students how to work in groups, expecting group presentations, and some competition between groups can enable communication between students in the classroom. The lecturer experienced that the automation of activities by methods such as prerecorded videos, multiple-choice questions, and group activities (peer-to-peer discussion) was beneficial for both the lecturer and the students, provided that the lecturer had enough time to design and implement these activities properly and consider a plan for improving them after each run.

Evaluation of the course project in the form of laboratory-based group assignments by the lecturer as well as the feedback received from student representatives emphasized the importance of the laboratory work to bring the theory to practice. Asking students to follow a specific template with guidelines on how to proceed with each section as the requirement to be followed in the delivery of the project report was found to be effective in enhancing the quality of learning outcomes. The requirements of the project work would also need continuous improvements over time to make it suitable for achieving the course objectives.

On the other hand, the students' performance in the quizzes and the final exam revealed that their performance in achieving the learning outcomes was enhanced when they were directly asked to discuss on different aspects of the learning outcomes during classroom discussions, as well as in the requirements of the project report. This was apparent in the selected learning outcome for this study, which was the different aspects of reducing environmental impacts of concrete. Different factors such as optimizing the mix design by reducing the matrix volume, using supplementary cementitious materials, blended cement, or the cement with a more environmentally friendly production method, selecting the right type of cement with

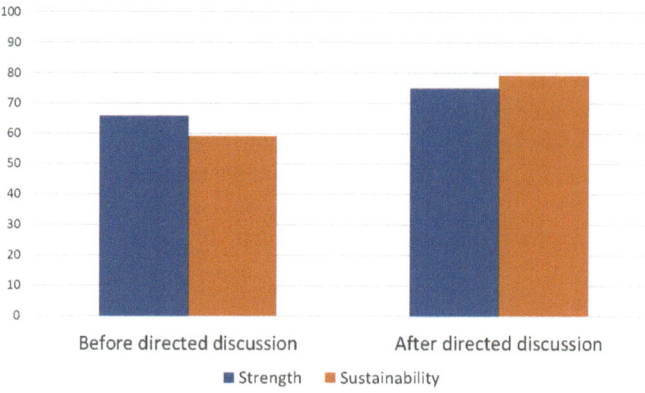

Fig. 3 Average score of the students in the quizzes (percentage) conducted in the physical classroom

minimum transportation demand, proper strength development as demanded, effective consumption of fillers and admixtures, shape and properties of the aggregates, were mentioned to be effective in greenhouse gas emissions of concrete mixes. These factors were relevant to different subjects in the course timeline; however, they were also mentioned as a collection of measures that could be taken to reduce the greenhouse gas emissions. It was found that the students could have a more in-depth reflection on the factors that were directly discussed during the physical classroom or were considered as a requirement to be discussed in the project report meaning that preparing a list of important aspects of learning outcomes and preparing some discussion arena for students to reflect on them during the course timeline can result in a considerable improvement in the students' performance in the course.

Figure 3 shows the students' performance in two distinct learning outcomes within the course: strength development and sustainability issues. Notably, when assessments were conducted after planned discussions, students demonstrated improved performance in questions related to these learning objectives. It's worth mentioning that, in addition to classroom discussions, students were specifically tasked with discussing sustainability issues in their project reports. The higher rate of performance enhancement observed for sustainability issues compared to strength development is likely due to the additional focus on sustainability in the project reports.

4 Conclusion

In this study, the main objective was to help students gain the learning outcome of the course in the flipped classroom teaching system in an active learning environment by improving the learning process. Among different expected learning outcomes, sustainability issues, including greenhouse gas emission of the materials, were selected as the learning outcome to be investigated in this course. It is expected that increasing student awareness of the environmental impacts associated with material selection for building applications can empower future experts to make informed choices that contribute to sustainability and net-zero construction. While the overall experience of the course responsible, students, and the involved colleagues were in the favor of the proposed teaching method, it was concluded that the following measures can improve the learning process of the designed teaching process:

- Since the prerecorded videos need to be continuously improved over time, recording the lectures in a format with editing feasibility such as slide-by-slide recording is found to be suitable.
- While having a short review of the lecture in the form of a quiz was found to be of students' interest at the beginning of the physical sessions, it was found that more in-depth review of the lecture, which summarizes the topic, could preferably be done as a separate prerecorded video or podcast in order to keep the focus of physical sessions on the hands-on exercises and in-depth discussions for practical application of the theory.
- The engagement and supervision of the lecturer in group discussions, presenting a proper discussion plan for the students with an emphasis on learning outcomes, and considering suitable times for discussion can enhance the quality of students' discussion in the physical classroom. Furthermore, using ice-breaker methods, instructing students on how to work in groups, expecting group presentations, and competition between groups can enable communication between students to improve the discussion performance.
- Preparing a list of important aspects of learning outcomes by the course responsible and enabling a discussion arena for students to reflect on them (such as group discussion in the class, answering to research questions in project report) could make a significant effect on the quality of students' reflections on the learning outcomes.

References

1. Komulainen, T.M., Alcocer, A., Haugen, F.: Experiences and trends in control education: a HIOA/USN perspective. In: 2016 9th EUROSIM Congress on Modelling and Simulation Oulu, Finland, 12 16 September 2016
2. Bree, R.T.: Embracing alternative formats, assessment strategies and digital technologies to revitalise practical sessions in Science & Health (2019)
3. Hite, R.L., Jones, M.G., Childers, G.M.: Classifying and modeling secondary students' active learning in a virtual learning environment through generated questions. Comput. Educ. **208**, 104940 (2024)
4. Opre, D., Şerban, C., Veşcan, A., Iucu, R.: Supporting students' active learning with a computer based tool. Act. Learn. High. Educ. **25**(1), 146978742211004 (2024)

5. Nieminen, J.H., Haataja, E., Cobb, P.J.: From active learners to knowledge contributors: authentic assessment as a catalyst for students' epistemic agency. Teach. High. Educ., 1–21 (2024). https://doi.org/10.1080/13562517.2024.2332252

6. Zughoul, O., Momani, F., Almasri, O.H., Zaidan, A.A., Zaidan, B.B., Alsalem, M.A., Albahri, O.S., Albahri, A.S., Hashim, M.: Comprehensive insights into the criteria of student performance in various educational domains. IEEE Access. **6**, 73245–73264 (2018)

7. Cakula, S.: Active learning methods for sustainable education development. In: 14th International Scientific Conference Rural Environment. Education. Personality (REEP) (2021)

8. Neto, I.R., Amaral, F.G.: Teaching occupational health and safety in engineering using active learning: a systematic review. Saf. Sci. **171**, 106391 (2024)

9. Michael, J.: Where's the evidence that active learning works? Adv. Physiol. Educ. **30**, 159 (2006)

10. Marello, C., Marchisio, M., Pulvirenti, M., Fissore, C.: Automatic assessment to enhance online dictionaries consultation skills. In: 16th International Conference on Cognition and Exploratory Learning in the Digital Age (CELDA 2019), pp. 331–338. IADIS Press (2019)

11. Bishop, J.L., Verleger, M.A.: The flipped classroom: a survey of the research. ASEE Natl. Conf. Proc., Atlanta, GA. **30**(9), 1–18 (2013)

12. Børte, K., Nesje, K., Lillejord, S.: Barriers to student active learning in higher education. Teach. High. Educ. **28**(3), 597 (2023)

13. Sivan, A., Leung, R.W., Woon, C.-c., Kember, D.: An implementation of active learning and its effect on the quality of student learning. Innov. Educ. Train. Int. **37**(4), 381 (2000)

Urban Waste Recycling: A Multi-objective Approach for Sustainable Construction in European Regions

Anastasija Komkova ⓘ, Sophie Krog Agergaard, Birgitte Holt Andersen, and Guillaume Habert ⓘ

Contents

1 Introduction

Currently, more than 40% of waste generated in Europe is landfilled [1], including residues from incineration processes of urban waste streams like municipal solid waste and sewage sludge, as well as biomass. However, these ashes can undergo valuable metal and mineral recovery, making the residues suitable for recycling as partial replacements for conventional cement, either as supplementary cementitious materials or within alternative binders, e.g., alkali-activated materials [2, 3]. This highlights the potential for urban-industrial metabolism not only in terms of energy recovery but also in waste valorization, where waste originating from the urban environment can be recycled within the construction sector, thereby reducing dependency on raw materials and minimizing GHG emissions associated with conventional cement-based construction material production. Within the European research

A. Komkova (✉)
Chair of Sustainable Construction, ETH Zurich, Zurich, Switzerland
e-mail: komkova@ibi.baug.ethz.ch

S. Krog Agergaard · B. Holt Andersen · G. Habert
Cware Aps, Copenhagen, Denmark

M. Kioumarsi, B. Shafei (eds.), *The 1st International Conference on Net-Zero Built Environment*, Lecture Notes in Civil Engineering 237,
https://doi.org/10.1007/978-3-031-69626-8_92

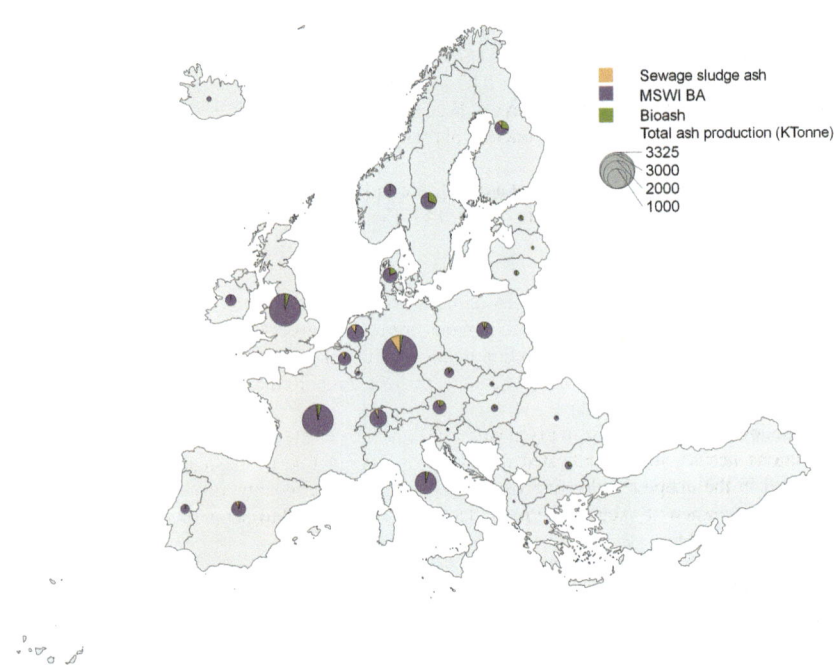

Fig. 1 Production of ashes as residues from urban waste incineration in Europe

project AshCycle [4], several ash pretreatment technologies are being tested for the recovery of valuable metals and minerals, as well as for further pretreatment of ashes for use in alternatives to cement binders, carbstone, and in clay brick production.

In Europe, there are nearly 500 waste-to-energy (WtE) plants [5] that generate almost 19 million tons per year of municipal solid waste incineration bottom ashes (MSWI BA) [6], along with 0.7 Mt/year of sewage sludge ashes (SSA) and 6 Mt/year of biomass ashes (BA). Figure 1 illustrates the amounts of ashes generated in different European countries. This figure reflects the current landscape, with a notable upward trend as more European countries transition to WtE practices, thereby incinerating larger volumes of waste [5]. This underscores the increasing importance of exploring innovative solutions for managing and repurposing these ashes in a sustainable way.

In this study, we apply material flow analysis (MFA) and multi-objective optimization methods to identify optimal supply chain network scenarios for recycling underutilized ashes in construction materials, thereby quantifying associated environmental impacts and costs.

2 Methods

To analyze the environmental and economic impacts associated with urban waste recycling within the construction sector, we focus on the urban-industrial metabolism framework illustrated in Fig. 2.

Within the examined urban-industrial metabolism framework, urban waste streams, including municipal solid waste, sewage sludge, and biomass wastes, are transported to incinerators for energy recovery. The energy generated from waste incineration processes is then used for electricity production and/or district heating [5]. Instead of being landfilled, the ashes, being the residues of incineration processes, undergo recycling where valuable substances, including phosphorus from sewage sludge and ferrous and nonferrous metals from municipal solid wastes, can be recovered. Moreover, ashes can undergo further mechanical and/or chemical pretreatment and be used in the production of alternative construction materials, which are then supplied to the urban built environment.

As a first step, we perform a material flow analysis to identify waste streams that are underutilized, for instance, landfilled or waste streams that can be upcycled. In the next step, we use a combination of geospatial analysis and life cycle assessment (LCA), followed by a multi-objective optimization model to identify optimal supply chain network designs for waste recycling while minimizing environmental impacts and costs associated with transportation, pretreatment, and production of alternative construction materials. Within this paper, we examine urban-industrial metabolism in Denmark.

2.1 Material Flow Analysis

Mass flow analysis is a methodology used to systematically track the flow of materials or substances within a system, such as a city, region, or industrial sector

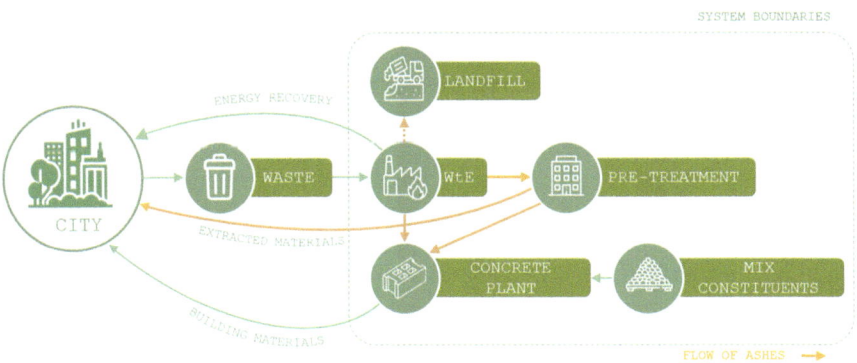

Fig. 2 Urban-industrial metabolism framework

[7]. It offers a comprehensive understanding of material flows, including their sources, destinations, and transformations throughout the system. By quantifying material flows and identifying key hotspots, MFA enables decision-makers to optimize resource management, minimize waste generation, and enhance sustainability [8].

We examine the incineration pathways of municipal solid waste [9], sewage sludge waste [10], and biomass waste [11] in Denmark. The analysis aims to establish the foundation for exploring the potential of extracting valuable materials from the ashes and repurposing the by-products in various construction materials.

2.2 Multi-objective Optimization

To identify optimal supply chain network scenarios for ash recycling, recovery of valuable substances, and the application of ashes in alternative construction materials, we apply a hybrid methodology that involves geospatial analysis and life cycle assessment, followed by a multi-objective optimization model [12, 13].

Geographical information systems (GIS) are used to identify the locations of incinerators, potential recycling facilities, and manufacturers of construction materials, as well as landfills. To determine the shortest transportation distances between the actors within the supply chains, an origin–destination matrix analysis is performed in QGIS software.

As a next step, the environmental impacts associated with transportation and waste recycling are quantified following LCA methodology, where the IPCC 2013 impact assessment method is applied to estimate global warming potential (GWP), using the Ecoinvent database v3.9 for life cycle inventory. The functional unit is defined as the total (potential) quantity of construction materials produced in Denmark using ashes, compared to conventional cement concrete with equivalent mechanical properties. In this case study we examine ash application as supplementary cementitious material and compare it to the conventional CEM III.

To assess the environmental impacts and costs associated with the proposed urban-industrial metabolism, we define a mixed-integer linear programming (MILP) model in GAMS software, building upon previous research [12, 13]. This model allows to identify the optimal locations for ash recycling and the recovery of ferrous and nonferrous metals, as well as valuable substances. Furthermore, the model allows to identify the optimal destinations for ashes to be used in construction material production. The first objective function aims to minimize environmental impacts E associated with the transportation of ashes, recovery and recycling, and concrete production (Eq. 1), and the second objective function aims to minimize corresponding costs C (Eq. 2). Weighted sum method is used to assign weights for each objective function and allows to examine trade-offs between GWP and costs.

$$\min E = \sum_i \sum_j d_{i,j} \times q_{i,j} \times f_t + \sum_i \sum_k d_{i,k} \times q_{i,k} \times f_t + \sum_k \sum_j d_{k,j} \times q_{k,j} \times f_t$$

$$+ \sum e_w \times q_w \times f_w + \sum e_s \times q_s \times f_s + \sum q_m \times f_m \qquad (1)$$

$$\min C = \sum_i \sum_j d_{i,j} \times q_{i,j} \times c_t + \sum_i \sum_k d_{i,k} \times q_{i,k} \times c_t + \sum_k \sum_j d_{k,j} \times q_{k,j} \times c_t$$

$$+ \sum e_w \times q_w \times c_w + \sum e_s \times q_s \times c_s + \sum q_m \times c_m \qquad (2)$$

$$\text{s.t.} \quad \sum_j q_{i,j} + \sum_k q_{i,k} = a_i \qquad (3)$$

$$\sum_i q_{i,j} + \sum_k q_{k,j} \le b_j \qquad (4)$$

$$\sum_i q_{i,k} \le b_k \times \partial_k \qquad (5)$$

$$\sum_j q_{k,j} \ge b_k \times \partial_k \qquad (6)$$

$$\sum_{i,k} q_{i,k} = \sum_{k,j} q_{k,j} \qquad (7)$$

where d are transportation distances between actors in supply chains and q are quantities of materials, while the total amount of ashes generated is q_w. The optimization model evaluates options of transporting waste from incinerator i directly to concrete plants j, or to centralized recycling plants k, where valuable substances s are recovered, with subsequent transportation to concrete plants. In this model, m is the vector of other construction material mix constituents, including cement and supplementary cementitious materials.

The emission factor f and cost factor c of transportation t are equal to 0.19 kg CO_2 eq. and 0.09 EUR, respectively, per ton and km traveled. The recovery of valuable substances from ashes is associated with energy consumption e_s and the treatment of ashes for use in construction materials, including milling, grinding, and sieving, is associated with electricity consumption e_w. The emission and cost factors are fixed, being f_w equal to 0.15 kg CO_2 eq. per kWh and c_w equal to 0.13 EUR per kWh, respectively. Different capacities of potential recycling plants are examined, where lower marginal costs are modeled for a bigger centralized facility.

Equations (3)–(7) define constraints to the objective functions on the capacity b of recycling and concrete plants, where ∂_k is a binary variable indicating whether the recycling plant is operating or not. Furthermore, we assume that all ashes produced at the incinerator a_i, will be fully utilized.

The optimized supply chain scenarios are then compared to the baseline scenario where ashes are landfilled and conventional mix constituents are used for the construction material production.

3   Results

In this chapter, we present the results of the material flow analysis that are used to identify waste streams that can be recycled within the construction sector in Denmark. Furthermore, we compare the environmental impacts and costs of optimized supply chain networks for ash recycling in construction materials with the baseline scenario and examine potential implications associated with urban waste stream recycling practices at the European scale.

3.1   MFA in Denmark

Denmark is among the top producers of waste per capita in Europe and has a long history with waste incineration [14]. Its waste-to-energy infrastructure enables the efficient conversion of a significant portion of nonrecyclable waste, including municipal solid waste, sewage sludge and biomass, into energy resources, providing both heat and electricity while minimizing waste volumes.

Figure 3 illustrates the mass flow diagram for Denmark. It encompasses the incineration phase, treatment, material extraction, and end-of-life stages, providing a simple overview of the lifecycle of these ashes. In total, there are approximately 3.8 Mt of MSW being incinerated, which includes 26% imported waste, which results in nearly 0.7 Mt of MSW ashes being generated yearly. While MSW bottom ashes (BA) are used mainly in road construction, fly ashes (FA) pose specific challenges, associated with high concentrations of heavy metals compared to bottom ashes, indicating the need for further improvement of pretreatment technologies [15]. As a result, a considerable portion of MSWI fly ashes are currently being landfilled.

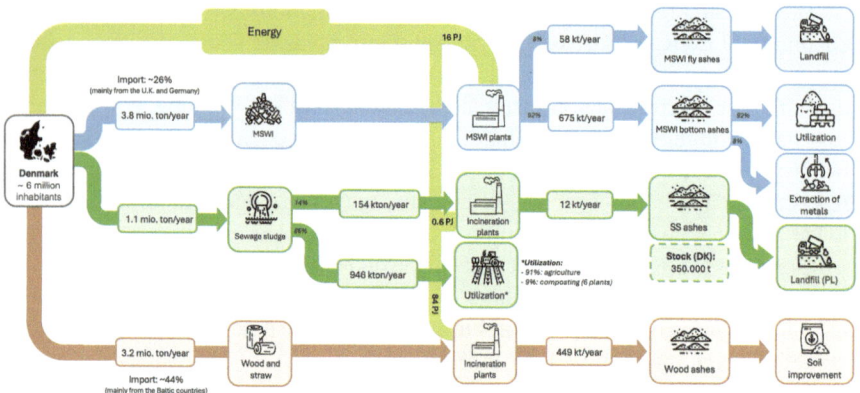

Fig. 3 MFA of incineration ashes generated in Denmark

In the case of raw sewage sludge, nearly 86% is currently utilized directly in agriculture. This approach minimizes emissions associated with pretreatment and transport, thus reducing the overall environmental footprint. However, the increasing pressure from EU directives reflects a growing concern for environmental sustainability and public health [16]. These directives aim to regulate and minimize the presence of contaminants such as microplastics and per- and poly-fluoroalkyl substances in sewage sludge, recognizing the potential risks they pose to ecosystems and human health [17]. The sewage sludge with high contamination levels is therefore incinerated, where technology advancements offer opportunities to extract valuable materials from sewage sludge ashes, particularly phosphorus [18]. If technology can effectively recover the approximately 10% of phosphorus present in sewage sludge ashes, it would represent a highly lucrative business case [19]. Phosphorus is a scarce and valuable resource crucial for agricultural fertilizers and various industrial processes. Considering that all sewage sludge ashes currently generated in Denmark (nearly 12,000 tons) are currently being landfilled, there is substantial potential for both phosphorus extraction and utilizing the ashes in construction materials. Looking ahead, with 86% of raw sewage sludge presently utilized in agriculture, there is a notable opportunity to explore alternative future scenarios, such as increased incineration, particularly if more strict regulations on the direct use of sewage sludge ashes are imminent [20].

Utilizing sewage sludge ashes in construction material presents a promising avenue for P resource recovery and waste reduction [21]. Recent studies showed that SSAs can be used as supplementary cementitious materials, replacing up to 20% of cement in mortars, without compromising mechanical properties [22, 23], and exhibiting similar durability properties to conventional cement concretes [24]. SSAs also showed the potential to partially replace clay in fired brick production (by up to 50%) [10] and to be used as co-precursors in alkali-activated materials [25]. SSAs have also proved suitable as sand replacements in mortars and concretes [26, 27]. Further environmental benefits of the application of SSAs in construction materials are associated with the immobilization of heavy metals, thus reducing leaching substantially, depending on application [28, 29]. Furthermore, there is ongoing research on optimization of phosphorous recovery technologies and increasing applications of pretreated ashes in construction materials within the AshCycle project [4].

In Denmark, nearly 0.5 Mt of biomass ashes are produced per year and have been fully incorporated into agricultural practices and are commercially available alongside various products derived from energy use based on biomass [30]. Biomass ashes are used as fertilizers, and biogas residues are used in composting, improving soil [31].

Based on the results of MFA, this study focuses on the optimization of supply chain networks for sewage sludge ashes, acknowledging their significant potential for utilization in various applications.

3.2 Optimal Supply Chain Networks

Figure 4 illustrates the main incinerators of municipal solid waste, sewage sludge, and biomass in Denmark. While the MSWI BAs are already widely used in the construction sector (e.g., as road fillers), the MSWI FAs are currently landfilled due to the presence of hazardous contaminants. Currently, there are three sewage sludge incineration plants in Denmark that generate nearly 12,000 tons of ashes per year, which are landfilled. Considering the limited capacity of landfills in Denmark, the ashes are transported to neighboring countries. For instance, MSWI fly ashes are transported to Norway, and sewage sludge ashes are exported to Poland for landfilling.

As a case study, we focus on exploring the optimal supply chain networks that can stimulate the recycling of sewage sludge ashes in alternative construction materials. We apply a multi-objective optimization model to identify the optimal locations for ash recycling, including recovery of phosphorus and pretreatment for further use at existing concrete plants.

Figure 5 illustrates the environmental impacts and costs associated with the baseline scenario and the optimized scenario. Within the baseline, all annually produced SSA are exported for landfilling, while cement concrete is produced using conventional cementitious materials, and primary fertilizers are used. Within

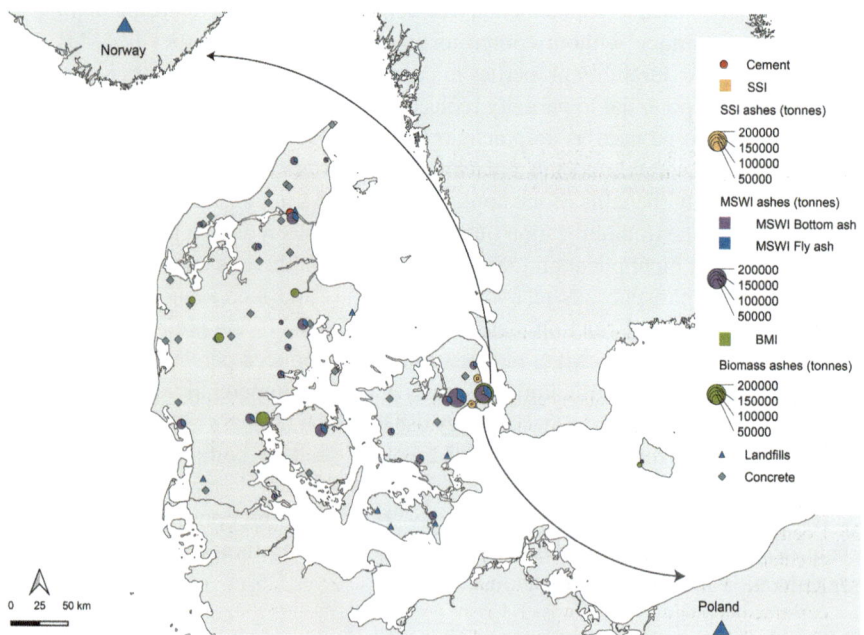

Fig. 4 Baseline scenario. MSWI fly ash and SSA are landfilled outside Denmark

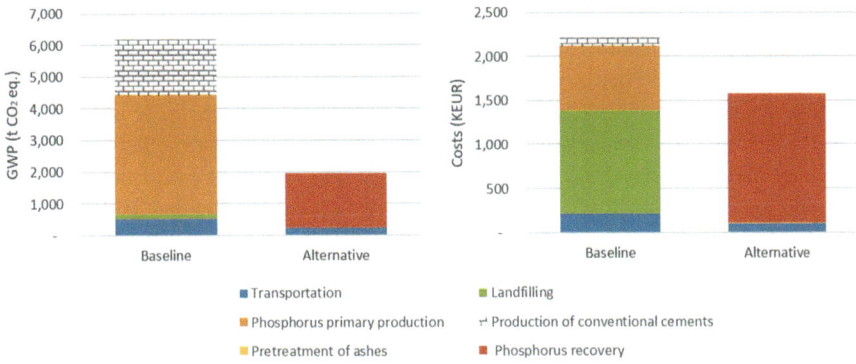

Fig. 5 GWP and costs associated with SSA utilization in Denmark

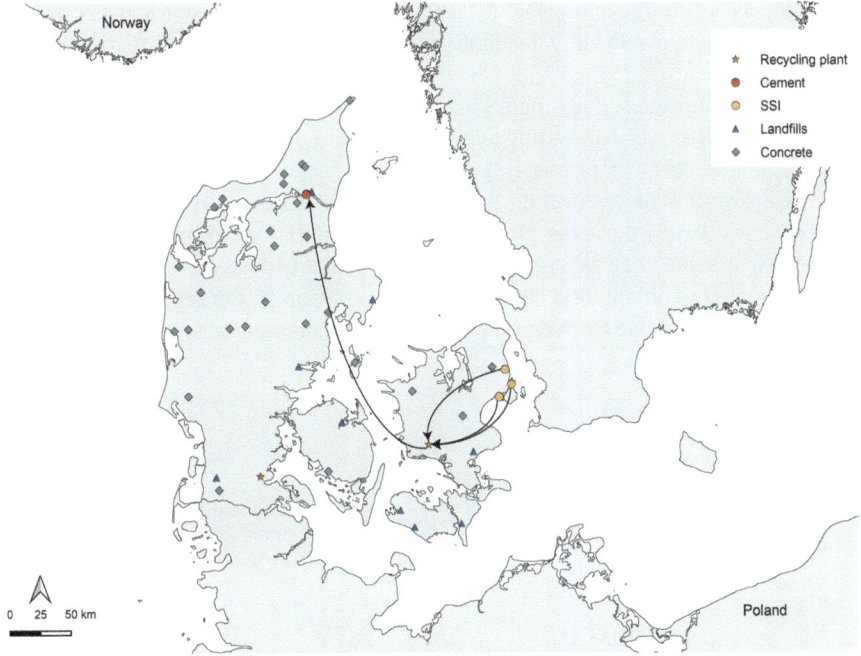

Fig. 6 Alternative scenario. Optimal supply chains for SSA recycling in Denmark

the alternative scenario, these ashes are transported to a single centralized plant (Fig. 6), where 1445 tons of P is recovered and residue ashes are pretreated for further use in concrete production.

Within the baseline scenario (Fig. 5), production of primary phosphorous-based fertilizers contributes 61% to the total GWP, production of conventional cement contributes 29%, and transportation to landfills contributes 8%, while within the economic impacts, costs of landfilling are the major contributor of 53%. Within the proposed ash recycling scenario, the main costs and environmental impacts are associated with phosphorus recovery. Pretreatment of SSAs after recovery of *P* for further use as supplementary cementitious materials has ~1% contribution to both total costs and emissions. Nonetheless, emissions associated with *P* recovery are approximately 40% lower compared to the phosphorus primary production. Overall, it can be observed that the alternative scenario, where *P* is recovered from ashes and residue ashes are used as cement substitutes, results in 69% lower GWP, and 29% lower costs, compared to the baseline.

We further compare the costs of phosphorus recovery technology estimated within the current study with the costs of two existing phosphorus recovery technologies examined in [32]. The PASCH technology involves the recovery of calcium phosphate from SSA using a wet-chemical leaching approach and is associated with the installation of a new recycling plant. The EcoPhos technology involves the recovery of *P* in the form of phosphoric acid at an existing plant: therefore, the investments in the recycling facility are omitted. The investment and operational costs of these two recycling technologies are estimated based on data provided in [32].

Figure 7 illustrates that operational costs vary significantly across examined technologies. Recovery of *P* using available technologies requires higher operational costs than the costs of primarily produced fertilizers. As observed from Fig. 7, investments in the new recycling facility would potentially constitute a substantial barrier, considering the example of existing PASCH technology. In the current study, we assume that ash recovery will be performed at an existing recycling plant. Depending on the total costs of examined scenarios, the economic incentives can be applied to nudge recovery of phosphorus and subsequent recycling of ashes in

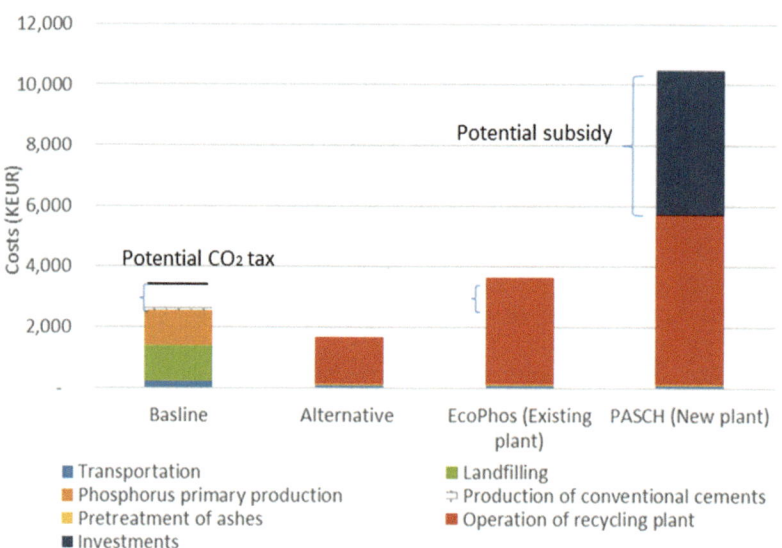

Fig. 7 Costs of SSA utilization in Denmark, considering different recycling technologies

the construction sector. For instance, the CO_2 taxes could increase the costs of the baseline scenario, while policymakers could subsidize the establishment of recycling plants.

3.3 Impacts at European Scale

Within this paper, we examined the potential case of urban-industrial symbiosis in the recycling of sewage sludge incineration ashes within construction materials, as well as quantified potential environmental and economic impacts. While transboundary waste management involving exports of waste can lead to disputes on emission allocations between countries, recycling waste streams within domestic settings can not only bring benefits in terms of recovery of phosphorus but also stimulate recycling of ashes in construction materials, thereby decreasing the demand for conventional cement-based materials and minimizing associated emissions.

In Europe, almost 1.1 Mt of phosphorus fertilizers were used in the year 2021 [33]. As global production of phosphorus takes place outside of Europe, including Russia and Belarus, European countries to a high extent depend on the imports, which not only involves long transportation distances but also due to the current geopolitical situation makes the international supply chain vulnerable and leads to higher costs [33]. The price of phosphorus used in agriculture has been fluctuating between 500 and 600 EUR per ton over the past years [34]. Assuming that the full potential for phosphorus recovery from bottom ashes will be exploited, a considerable share of EU consumption of phosphorus can be covered, and the phosphorous cycle can be partly re-established. As illustrated in Fig. 1, high amounts of bottom ashes are currently produced in central Europe. At the same time, sewage sludge is currently landfilled in multiple European regions, including South and Eastern Europe: Croatia, Slovenia, Spain, Italy, and Greece, where waste-to-energy potentials shall be examined [35]. Moreover, recovery of phosphorus can nudge domestic application of ashes in construction materials, thus minimizing dependency on primary resources or on conventional supplementary cementitious materials, such as blast furnace slag and coal fly ash. The optimization of supply chain networks is therefore an important step that can help to identify optimal locations for waste recycling and recovery of valuable elements, thus providing indication to the policymakers on nudging industrial symbiosis.

4 Conclusions

In this study, we apply a methodology that allows us to evaluate environmental and economic impacts associated with the recycling of waste incineration ashes in construction materials and the recovery of valuable minerals. We focus our analysis on ashes produced at WtE plants in Denmark, while further analysis will focus on examining patterns across European regions. The MFA enables us to identify waste streams that can be further upcycled in construction materials, including sewage sludge ashes and municipal solid waste fly ashes that are currently landfilled. An analysis of SSAs shows that the recovery of valuable substances is an important step in the urban-industrial metabolism framework that can reduce CO_2 eq. emissions associated with phosphorus production by almost 40%, and encourage further pretreatment of ashes for use in alternative construction materials, overall bringing a 69% reduction in GWP compared to business-as-usual ash landfilling and production of conventional construction materials and fertilizers. Given the high prices of primary fertilizers and the landfilling costs of ashes, recycling of ashes in alternative concrete is associated with 29% lower costs compared to the business-as-usual scenario. The optimization of supply chain networks, therefore, allows to identify optimal locations for the recycling and pretreatment of ashes, thereby minimizing environmental impacts and costs. Improvement of ash recycling technologies can further decrease associated emissions, and it is important to consider the investment and operational costs of new recycling facilities.

Acknowledgments This project has received funding from the European Union's Horizon Europe research and innovation program No. 101058162 AshCycle.

References

1. Eurostat: Treatment of waste by waste category, hazardousness and waste management operation (2023)
2. Kurda, R., Silva, R.V., de Brito, J.: Incorporation of alkali-activated municipal solid waste incinerator bottom ash in mortar and concrete: a critical review. Materials. **13**, 3428 (2020). https://doi.org/10.3390/MA13153428
3. Yin, K., Ahamed, A., Lisak, G.: Environmental perspectives of recycling various combustion ashes in cement production – a review. Waste Manag. **78**, 401–446 (2018)
4. AshCycle project. https://www.ashcycle.eu/en/project/. Last accessed 15 Mar 2024
5. Poretti, F., Stengler, E.: The climate roadmap of the European waste-to-energy sector I the path to carbon negative. SSRN Electron. J. (2022). https://doi.org/10.2139/ssrn.4284664
6. Bruno, M., Abis, M., Kuchta, K., Simon, F.G., Grönholm, R., Hoppe, M., Fiore, S.: Material flow, economic and environmental assessment of municipal solid waste incineration bottom ash recycling potential in Europe. J. Clean. Prod. **317**, 128511 (2021)
7. Graedel, T.E.: Material flow analysis from origin to evolution. Environ. Sci. Technol. **53**, 12188–12196 (2019)
8. Makarichi, L., Techato, K., Jutidamrongphan, W.: Material flow analysis as a support tool for multi-criteria analysis in solid waste management decision-making. Resour. Conserv. Recycl. **139**, 351–365 (2018)
9. OECD: OECD Environmental Performance Reviews: Denmark 2019. OECD Publishing, Paris (2019)
10. Ottosen, L.M., Bertelsen, I.M.G., Jensen, P.E., Kirkelund, G.M.: Sewage sludge ash as resource for phosphorous and material for clay brick manufacturing. Constr. Build. Mater. **249**, 118684 (2020). https://doi.org/10.1016/j.conbuildmat.2020.118684
11. Danish Energy Agency: Biomass analysis (2020)

12. Komkova, A., Habert, G.: The Urban-Industrial metabolism: contribution of waste recycling to the circular economy objectives within the construction sector. J. Phys. Conf. Ser. **2600**, 172002 (2023). https://doi.org/10.1088/1742-6596/2600/17/172002

13. Komkova, A., Habert, G.: Optimal supply chain networks for waste materials used in alkali-activated concrete fostering circular economy. Resour. Conserv. Recycl. **193**, 106949 (2023). https://doi.org/10.1016/j.resconrec.2023.106949

14. Danish EPA: Denmark without waste (2013)

15. Wang, S., Yu, L., Qiao, Z., Deng, H., Xu, L., Wu, K., Yang, Z., Tang, L.: The toxic leaching behavior of MSWI fly ash made green and non-sintered lightweight aggregates. Constr. Build. Mater. **373**, 130809 (2023)

16. Hudcová, H., Vymazal, J., Rozkošný, M.: Present restrictions of sewage sludge application in agriculture within the European Union. Soil Water Res. **14**, 104–120 (2019)

17. Pozzebon, E.A., Seifert, L.: Emerging environmental health risks associated with the land application of biosolids: a scoping review. Environ. Health. **22**, 1–15 (2023). https://doi.org/10.1186/S12940-023-01008-4/FIGURES/3

18. Havukainen, J., Nguyen, M.T., Hermann, L., Horttanainen, M., Mikkilä, M., Deviatkin, I., Linnanen, L.: Potential of phosphorus recovery from sewage sludge and manure ash by thermochemical treatment. Waste Manag. **49**, 221–229 (2016)

19. Viader, R.P., Jensen, P.E., Ottosen, L.M., Thomsen, T.P., Ahrenfeldt, J., Hauggaard-Nielsen, H.: Comparison of phosphorus recovery from incineration and gasification sewage sludge ash. Water Sci. Technol. **75**, 1251–1260 (2017). https://doi.org/10.2166/wst.2016.620

20. Dakofa: Ny oversigt over alternativer til direkte anvendelse af spildevandsslam i landbruget. https://dakofa.dk/element/ny-oversigt-over-alternativer-til-direkte-anvendelse-af-spildevandsslam-i-landbruget/. Last accessed 15 Mar 2024

21. Tominc, S., Ducman, V., Wisniewski, W., Luukkonen, T., Kirkelund, G.M., Ottosen, L.M.: Recovery of phosphorus and metals from the ash of sewage sludge, municipal solid waste, or wood biomass: a review and proposals for further use. Materials. **16**, 6948 (2023). https://doi.org/10.3390/MA16216948

22. Vouk, D., Nakic, D., Stirmer, N., Cheeseman, C.: Influence of combustion temperature on the performance of sewage sludge ash as a supplementary cementitious material. J. Mater. Cycles Waste Manag. **20**, 1458–1467 (2018)

23. Kappel, A., Viader, R.P., Kowalski, K.P., Kirkelund, G.M., Ottosen, L.M.: Utilisation of electrodialytically treated sewage sludge ash in mortar. Waste Biomass Valoriz. **9**, 2503–2515 (2018)

24. Kobayashi, K., Koyanagi, W.: 17-Year-long sewage sludge ash concrete exposure test. Infrastructures. **6**, 74 (2021)

25. Zhao, Q., Ma, C., Huang, B., Lu, X.: Development of alkali activated cementitious material from sewage sludge ash: two-part and one-part geopolymer. J. Clean. Prod. **384**, 135547 (2023). https://doi.org/10.1016/J.JCLEPRO.2022.135547

26. Zhou, X., Zhou, X., He, L., Zhang, Y.: Effect of sewage sludge ash on mechanical properties, drying shrinkage and high-temperature resistance of cement mortar. Case Stud. Constr. Mater. **20**, e03101 (2024). https://doi.org/10.1016/J.CSCM.2024.E03101

27. Lynn, C.J., Dhir, R.K., Ghataora, G.S., West, R.P.: Sewage sludge ash characteristics and potential for use in concrete. Constr. Build. Mater. **98**, 767–779 (2015)

28. Lynn, C.J., Dhir, R.K., Ghataora, G.S.: Environmental impacts of sewage sludge ash in construction: leaching assessment. Resour. Conserv. Recycl. **136**, 306–314 (2018)

29. Naamane, S., Alaoui, M.S.H., Taleb, M., Haboubi, K., Rais, Z.: Leachability of cement mortars containing sewage sludge ash. Mater. Today Proc. **58**, 1098–1103 (2022)

30. Hededanmark: Gød med aske og dæk markens behov for kalium og fosfor. https://www.hededanmark.dk/goedning/aske-til-goedning. Last accessed 15 Mar 2024

31. Hovmand, M.F., Riis-Nielsen, T., Ekelund, F., Sylvester, G., Krag, M.M.: Aske fra energitrae er naturgødning. Aktuel Naturvidenskab (2022)

32. Egle, L., Rechberger, H., Krampe, J., Zessner, M.: Phosphorus recovery from municipal wastewater: an integrated comparative technological, environmental and economic assessment of P recovery technologies. Sci. Total Environ. **571**, 522–542 (2016)
33. Eurostat: Agri-environmental indicator – mineral fertiliser consumption (2023)
34. EC: Ensuring availability and affordability of fertilisers (2024)
35. Eurostat: Sewage sludge production and disposal from urban wastewater (2023)

Sustainability and Uncertainty in Collaborative Project Delivery

Kristoffer Brattegard Narum ⓘ, Per Morten Kals, and Ola Lædre ⓘ

Contents

1 Introduction

Collaborative project delivery models (CDMs) are considered integral for building construction to contribute to reaching the United Nations sustainability goals [4]. For a project to reach its goals, it must successfully manage uncertainty [8]. The relationship between CDM elements, uncertainty and sustainability in construction projects remains insufficiently understood [1, 13]. Through a literature review and a

K. B. Narum (✉)
Norwegian University of Science and Technology (NTNU), Trondheim, Norway

Norwegian Defense Estates Agency, Oslo, Norway
e-mail: kristoffer.b.narum@ntnu.no

P. M. Kals
Oslobygg KF, Oslo, Norway

O. Lædre
Norwegian University of Science and Technology (NTNU), Trondheim, Norway

© The Author(s) 2025
M. Kioumarsi, B. Shafei (eds.), *The 1st International Conference on Net-Zero Built Environment*, Lecture Notes in Civil Engineering 237,
https://doi.org/10.1007/978-3-031-69626-8_93

comparative multiple case study of five large European collaborative projects, this chapter seeks to fill these research gaps by answering the following research questions:

1. How are the main elements of CDMs in large projects?
2. How does uncertainty impact the implementation of CDMs in large projects?
3. What are the main lessons learned on the relationship between CDM elements, the management of uncertainty and sustainability in large projects?

2   Theoretical Background

Building construction has a profound impact on the economic, social and environmental dimensions of sustainability. These dimensions encompass several sustainability indicators. Motives and competency of human resources has been identified as the most significant barrier and enabler for the advancement of sustainability in building construction. The sustainable transformation of building construction will necessarily transform the way construction projects are being delivered [3, 10, 13]. A project delivery model (PDM) is a system of elements for organizing and financing the delivery of a project in accordance with the goals of the client. It includes the organization form, the project structure, the form of specification, the procurement route and the agreement format, with the latter encompassing the contract format, conflict resolution mechanisms, risk allocation and the compensation format. PDM elements must be well adapted to one another, the project and the context in which the project is carried out [9].

Sustainable transformation of building construction delivery is accompanied by increased complexity and need for innovation [1]. Complex, innovative, large and uncertain projects are growing in size, frequency and importance and have a relatively high tendency to be unsuccessful in achieving their goals [6]. Uncertainty encompasses both opportunity and risk, which has to be managed successfully if a project is to reach its goals [8]. Uncertainty will change throughout the life cycle of a project and is particularly high in the front-end. Uncertainty emanating from external stakeholders is especially high in the front-end of large projects [15]. Collaborative project delivery models (CDMs) have been found to be particularly suitable for projects with high levels of uncertainty, complexity, risk, size, environmental challenges and number of stakeholders [16]. CDMs are characterized by the integration of inter-organizational participants towards collaboration on common goals [2]. The three most common variants of CDMs are partnering, integrated project delivery (IPD) and alliancing, with the latter being considered the most collaborative [1].

CDM elements are categorized by Engebø et al. [2] into contractual, organizational and cultural elements. Contractual elements, also called hard elements, are directly regulated by the contract or have their basis in the procurement process. Key contractual elements include early contractor involvement (ECI), value-based procurement, target costing with shared risk/reward and open book. Organizational elements are mechanisms that affect the organization and the process surrounding the project, but are not directly regulated in the contract. Key organizational elements include an integrated project organization, co-location, integrated project controls and workshops. Cultural elements, also called soft or behavioral elements, are designated towards the project culture and contribute to the relationship between the people in the project. Cultural elements frequently mentioned in literature include mutual goals, respect, equality, trust, commitment and openness [2–5, 7, 11, 12].

There is a need for more research on CDMs. No general definition or framework for CDMs has yet to become established. Engebø et al. [1] call for more research on the cause-effect relationship between CDMs, uncertainty and performance. Moradi et al. [13] request further studies on the contribution of CDMs towards the advancement of sustainability in construction.

3 Research Methods

The research design for the paper was based on the framework of Yin [20]. To gain an understanding of the relationship between CDM elements, sustainability and uncertainty in large projects, it was decided to conduct a multiple case study based on qualitative and quantitative data gathered through a triangulation of data-gathering methods.

A literature study was conducted. The literature search was carried out at Scopus, Oria and Google Scholar, with the search terms being "collaborative project delivery", "partnering", "IPD", or "Alliancing", combined with either the term "sustainability", "uncertainty" or "risk".

Five case projects described in Table 1 were selected for in-depth study due to their relevance and willingness to share experiences.

Cases 1 and 2 were studied as part of a specialization report written in collaboration with one of Norway's leading developers of public buildings, while Cases 3, 4, 5 were studied as part of a master's thesis written in collaboration with one of Norway's leading developers of public infrastructure. The projects are anonymized per request of the contact persons from the client.

Table 1 Case information

Case	Type	Status	Cost (mill. USD)	Construction time (months)
1	School	Operation	40	20
2	Swimming facility	Operation	75	29
3	Road tunnel	Operation	210	37
4	Oil and gas	Operation	7500	52
5	Railway	Construction	3100	69

Document studies were carried out for all case projects. Through collaboration with the client leadership, students and researchers gained access to relevant project documents, including contracts, project plans and lessons learned reports.

Semi-structured interviews with clients constitute the main source of data for the study. The primary interviewees were the project managers, who in all cases were engineers with decades of industrial experience. Senior managers, construction managers and assistant project managers were also interviewed, with the latter including non-engineers with only a few years of experience. The number of interviews ranged from eight interviews with seven individuals in Case 1, to three interviews with three individuals in Cases 3 and 4. Interviews for Cases 1 and 2 and Cases 3, 4 and 5 were carried out with different interview guides, which nevertheless covered similar PDM elements. Several of the interviews were carried out with the active participation of both researchers and industrial partners.

Document studies and interviews were in Cases 2 and 5 supplemented with observations. In Case 2, daily observations during the construction period were made over the course of several months, while observations in Case 5 were carried out through a physical visit by researchers and industrial partners to the construction site and a day-long workshop together with client management.

4 Results

4.1 Case 1

In Case 1, which encompassed the refurbishment of a historic school building of national significance, the contractor was involved upon startup of the pre-project. Procurement was carried out with prequalification, value-based award criteria emphasizing collaborative ability and a bidding competition. The CDM was to be carried out in two stages, with stage 1 consisting of the pre-project, and stage 2 consisting of detailed design and construction. The contractor was compensated on a reimbursable basis with open book. The contract included a target cost agreement with a shared bonus/scheme and an equal share of risk. The project had an integrated steering committee with the participation of senior management and project management from the client and contractor. The project organization was led by the project manager from the client, who was located onsite with the contractor and actively involved in all aspects of the project. The designers had been procured upon conceptual selection, and their contract was transferred to the contractor upon the start of the pre-project. Subcontractors were involved early, with the client actively participating in their selection. Project controls were strongly integrated and highly focused on hunting for opportunities, with active use of BIM for design and Last Planner for schedule management. Workshops were actively used both for both problem-solving and team building. Trust, openness and mutual goals were highlighted in the contract as key components of the project culture.

One of the main challenges during stage 1 was the task of developing a design which would both meet modern requirements and preserve the original characteristics of the school. Active dialogue between the project manager and the municipality and early involvement and good collaboration between subcontractors and the rest of the project organization through BIM contributed to the successful management of this uncertainty. Through Last Planner, the project organization developed an optimized schedule that enabled the school to open one year earlier than planned. Joint sharing of risk and opportunity, and an active hunt for opportunities, enabled the project to improve on sustainability throughout construction. Early subcontractor involvement facilitated innovative reuse of historic construction material, which was beneficial from both economic, social and environmental perspectives. Active use of BIM in collaboration with subcontractors contributed to a design which minimized the need for changes during stage 2, enabling the project to be sustainably delivered on time for the upcoming semester. Upon completion, the project was on schedule and about 20 percent below budget, with the client and contractor sharing a substantial bonus.

4.2   Case 2

The CDM of Case 2, which encompassed the construction of a swimming facility in an urban neighborhood, was based on learnings from Case 1. The contractor was involved after the completion of the pre-project, with the relatively late involvement being due to disagreements between senior management and the project organization on the choice of delivery model. Procurement was carried out with value-based award criteria emphasizing collaborative ability and a bidding competition, with a single bid being received. The CDM was to be carried out in two stages, with stage 1 consisting of detailed engineering, and stage 2 consisting of construction. The contractor was compensated on a reimbursable basis with open book. The contract included a target cost agreement with a shared bonus/scheme and a planned equal sharing of risk. The project had an integrated steering committee with the participation of senior management and project management from the client, contractor and designer. The project organization was led by the project manager from the client, who was located on-site with the contractor on a daily basis and actively involved in all aspects of the project. The designers had been procured by the client at the start of conceptual development and there was an option for them to have their contract transferred to the contractor upon completion of detailed engineering. Subcontractors were involved together with the contractor at the onset of detailed engineering. Project controls were integrated and highly focused on hunting for opportunities, with active use of BIM for design, Last Planner for schedule management and a shared digital platform for information management. Workshops were actively used for both for both problem-solving and team building, with a signed collaborative charter being developed at a startup workshop. Trust, openness and mutual goals were highlighted in the contract as key components in the project culture, with the collaborative charter also emphasizing mutual respect.

At the beginning of stage 1, the contractor and the designer disagreed on several conceptual solutions. Active client involvement was crucial to developing a mutually agreed concept and a collaborative culture. Stage 1 was extended for several months, during which the energy efficiency, life cycle cost and user-friendliness of the facility was significantly improved. Interviewees believe earlier involvement of the contractor and subcontracting of the designers to the contractor could have prevented the initial disagreements. Cost estimates for the new concept turned out to be substantially higher than the estimates for the initial concept with a reduced scope, and this cost difference made the municipal authorities hesitate to give the project permission to proceed to stage 2. Interviewees believe a more thorough conceptual review and realistic cost estimation in the front-end would have prevented the municipal hesitation. To keep the project organization together and ensure schedule performance, the client encouraged the suppliers to finalize the detailed design. As a result of concerns about cost, it was agreed to separate several elements from the target cost and make them fully reimbursable, resulting in a lower cost estimate but less equal risk sharing than planned. The design drawings were shared on the digital platform relatively late, which gave the subcontractors little time to review risk before signing their contracts. Construction start was postponed because of delayed voting in the municipal assembly, but once the cost estimates had been subjected to external review, the client secured municipal permission for an early start-up of preparatory works. During construction, changes were required both for the technical installations and the reimbursable elements separated from the target cost. Client-organized workshops were important to ensure constructive collaboration on the work for the technical installations, while the interviewees believe earlier involvement of the subcontractors and sharing of drawings on the digital platform could have reduced the need for changes. For the changes connected to the reimbursable elements, the client and the contractor disagreed on whether they should be handled through the contingency or not, but through active involvement of the steering committee it was agreed to use of the contingency. Interviewees believe the project would have required less senior management involvement if the risk allocation in the target cost had been clearer and more equal. Uncertainty emanating from global events ensured that the project received additional municipal funding and a three-month extension to stage 2. Upon completion, the project was on budget, built in accordance with the planned construction time and relatively high-performing compared to contemporary swimming facility projects. The reaching of a final settlement required renewed involvement of the steering committee, and none of the contractual parties received a bonus. Reflecting on experiences from Case 2, one interviewee interestingly stated: "A CDM should be contractually structured so that every participant emphasizes what's best for the project rather than what's best for themselves. That means that what's best for the project must be what's best for everyone. This will give you the best effect of a CDM."

4.3 Case 3

Case 3 was a road tunnel delivered through project alliancing. Contractor involvement was initiated 5 years after project startup, when the project had already received governmental and municipal planning permission. Procurement was carried out in several stages with prequalification, value-based award criteria emphasizing collaborative ability and technical and commercial negotiations, during which the collaborative ability of the bidders was tested. The CDM was to be carried out in two stages, with stage 1 consisting of project development, and stage 2 consisting of project implementation. The Alliance was compensated on a reimbursable basis with open book. The contract included a target cost agreement with a shared bonus/ scheme connected to several KPIs for economic, social and environmental sustainability, while risks were carefully allocated to either the Alliance or the clients, or shared between them. The highest decision-maker in the Alliance was an integrated steering committee chaired by the senior manager from the main client. The Alliance was led by a project manager from the contractor, who worked in close cooperation with the deputy project manager from the client. The designers were part of the Alliance, while the suppliers of technical installations were subcontracted yet tightly integrated into the organization. The project organization was co-located onsite. Project controls such as risk management, cost management and schedule management were strongly integrated, with a strong emphasis on fostering innovation. Workshops were actively used both for team building and joint problem-solving. Openness, mutual trust and respect were highlighted in the project plan as key components of the project culture.

At the startup of the development stage, a political party strongly opposed to the tunnel won local elections, putting the project at risk of not being delivered. Reimbursable payment and a collaborative culture encouraged the suppliers to contribute resources to the project despite the political uncertainty. Early involvement of the suppliers, and a joint focus on innovation, enabled an economic, social and environmental optimization of the design, which combined with joint stakeholder management helped win over enough political support for the project to proceed to the implementation stage. Key activities in the implementation stage included tunneling, construction of technical installations and commissioning. The biggest risks impacting tunneling were rock quality and groundwater conditions. Sharing of these risks between the clients and the Alliance enabled efficient management, with excavated rock being innovatively utilized for local land development, and groundwater being successfully handled in a way that minimized negative consequences for residents and a neighboring lake. Joint schedule management and good collaboration contributed to efficient tunneling. Early involvement and tight integration of suppliers of technical installations in the project organization enabled early construction of technical installations and commissioning, which was critical to ensure timely delivery. At completion, the project was 6 months ahead of schedule and on budget, with all members of the Alliance allocated a substantial bonus.

4.4 *Case 4*

Case 4 was an oil and gas project consisting of dozens of contracts, of which the largest and most complex were delivered with active use of collaborative elements. For these contracts, contractors were involved early in the pre-project as subcontractors of the main consultant. Procurement was carried out with prequalification, value-based award criteria emphasizing collaborative ability and technical and commercial negotiations, during which the collaborative ability of the bidders was tested. The collaborative contracts were to be carried out in two stages, with stage 1 consisting of detailed engineering, and stage 2 consisting of construction. They included target cost agreements with a shared bonus/scheme connected to several KPIs, including quality of collaboration. A substantial amount of risk was allocated to the client. Integrated steering committees were established for all major contracts. The project organization was led by the client, who closely involved themselves in collaboration with the suppliers. Integrated and co-located teams were established for all major contracts, with the participation of the client, the consultant, the contractors and subcontractors. Suppliers of heavy equipment were involved early as subcontractors of the main consultant. Project controls, in particular schedule and safety management, were strongly integrated, with a heavy focus on continuous improvement. Workshops were actively used, particularly for the purpose of managing interfaces and building teams. The project aspired to create a "One Team" project culture characterized by mutual goals, openness and trust.

At the startup of the pre-project, the parliamentary majority pressured the minority government and the client to adopt a conceptual solution with high cost and schedule risk, but the client was reluctant to accept this. Integrated involvement of key external stakeholders was crucial for the project at this time. Amidst a sharp market downturn, the government permitted the client to award conditional contracts to the contractors and move ahead with detailed engineering and construction, which saved several local companies and workers from respective bankruptcy and unemployment. The market downturn enabled the client to negotiate contracts on favorable terms and helped the project win over the parliamentary majority to the original concept. Familiarization processes with regular workshops for interface management and team building ensured good collaboration and smooth handover of the pre-project design for detailed engineering, while integrated schedule management enabled the development of an aggressive and optimized schedule. Integrated on-site teams and project controls enhanced quality, efficiency and worker safety during fabrication. Target costing with incentives encouraged close collaboration, which contributed to efficient marine operations and innovations in pipelaying, which in turn reduced emissions, material consumption and disturbances to marine biodiversity. Early involvement and tight integration of the project organization enabled the project to capitalize on innovative technology for the hook-up and commissioning, which significantly reduced eco-nomic, social and environmental risk. Upon completion, the project was 2 months ahead of schedule, 30 percent below budget, and one of world's highest-performing oil and gas megaprojects in terms of worker safety and CO_2 emissions.

4.5 Case 5

In Case 5, which is an ongoing railway project, the contractor was involved several years after project startup, after the project had received governmental planning permission, but before necessary municipal permissions had been secured. Procurement was carried out in several stages, with prequalification, value-based award criteria emphasizing collaborative ability and technical and commercial negotiations, during which the collaborative ability of the bidders was tested. The CDM was to be carried out in two stages, with stage 1 consisting of front-end design and stage 2 of detailed design and construction. The contractor is compensated on a reimbursable basis with open book. The contract includes a target cost agreement with incentives, while much risk was intended to be shared. The highest decision-maker in the project is an integrated management committee with representatives from the client, contractor and designer. The project organization is led by a project manager from the contractor, who works in close cooperation with client management. The designers and suppliers of many key works are subcontracted yet tightly involved. The project organization is co-located onsite. Project controls such as risk management and cost management are strongly integrated. Mutual trust is highlighted in the contract as a key component of the project culture.

Developing a design that would accommodate the needs of municipal authorities and residents was a major challenge during stage 1. Early contractor involvement, active joint stakeholder management and sharing of planning risk helped the project face the challenge. Rising cost estimates during stage 1 made the government hesitate to permit the project to proceed to stage 2, causing delays and uncertainty. Reimbursable payment encouraged the suppliers to contribute resources to the project despite the uncertainty. The assumption by the client of almost all project risk helped reduce the cost estimates to a level acceptable to the government, which gave the project permission to proceed to stage 2. Design immaturity upon procession to stage 2 caused changes during contractor-led detailed engineering, requiring active client oversight. Ground conditions, rising costs of materials and concerns of local authorities and residents were major uncertainties during construction. Much risk was borne by the client, leading to cost increases for the project, while the schedule and quality of collaboration were largely preserved. Early involvement with a co-located and integrated project organization helped build a collaborative project culture, which contributed to efficiency and innovations in land reclamation, stakeholder management and localized sustainable concrete production. Target costing with incentives helped encourage the contractor to perform sustainably, but the interviewees believe the allocation of more risk to the contractor would have incentivized them to perform even better. As the project was reaching completion, it was one year delayed and had a cost overrun of about 100%.

5 Discussion

Early contractor involvement (ECI) was characteristic of all case projects. Case 1 and Case 4 are characterized by particularly early involvement and had the highest performance on economic sustainability. This is in line with past identification of ECI as the most important CDM element and sufficiently early involvement as the most important ECI success factor [7, 17]. While the procurement method in all case projects emphasized the collaborative ability of bidders, this ability was more vigorously tested in Cases 1, 3 and 4, which had the highest performance on all sustainability indicators. The method of contractor involvement appears to be significant to succeed with ECI, which is in line with prior research [17–19]. While all the case projects were delivered through a two-stage CDM, the level of detail on the design on the transition between the stages differed between the projects, and a relationship between this level of detail and sustainability cannot be observed. Nevertheless, it can be observed from Case 2 that sufficient investment of time and effort in the first stage can contribute to a more sustainable design and collaborative culture. Reimbursable compensation with open book was characteristic of all projects and appears to be particularly important to maintain schedule and mutual commitments when collaborative projects are facing external uncertainty. While target costing with shared bonus/malus was characteristic of all case projects, Cases 1, 3 and 4 were characterized by a higher degree of risk sharing, better collaboration on the management of risk and higher performance on all sustainability indicators. This is in line with past studies that have found equitable sharing of risk as important to succeed with target costing in collaborative projects [14]. Social, political and market uncertainty was high in the front-end of most of the cases, and in Case 2 and Case 5, these uncertainties were closely connected to low front-end cost estimates and limited market capacity. These are also the two cases with the highest client risk, most challenging collaboration and lowest performance on economic sustainability. Successful engagement of social and political groups, realistic front-end estimates and the securing of sufficient market interest during procurement appear to be crucial to ensure a target cost risk allocation that encourages collaboration towards advancing sustainability.

Integrated steering committees and active client involvement were important organizational elements in all the studied cases. This was particularly the case in projects where the risk allocation was unclear or where collaboration between stakeholders constituted a major uncertainty. In such situations, a client manager supported by the steering committee can bring stakeholders together, encourage the development of joint solutions and make swift and mutually understood decisions for the best of project. The way of involvement of the designers differs among the projects, but the competence and quality of collaboration between the designers and the rest of the project organization appears to be important for performance. Early and tight involvement of subcontractors of technical installations and heavy equipment was a crucial success factor for the advancement of both economic, social and environ-mental sustainability in Cases 1, 3 and 4. In Case 2, insufficient involvement of sub-contractors resulted in changes and collaborative challenges, which required active client involvement. Integrated and continuous schedule management with early engagement of stakeholders and active focus on opportunities was a key feature of the projects with the highest performance on schedule, which was beneficial from both economic, social and environmental perspectives. Co-location was characteristic of all projects and is reported to have contributed to improved team building and higher efficiency, quality and safety of work, which is in line with theory [2, 5]. Workshops were actively used in the majority of the studied case projects, and in several cases crucial for project success. For Case 4 in particular, client-organized startup workshops enabled the development of joint understanding between stakeholders, while continuous workshops facilitated good interface management, which was critical for performance. In Case 2, active referencing to a collaborative charter from the startup workshop, and continuous client-organized workshops, helped get collaboration back on track when challenges occurred during construction.

The identification of mutual trust as a cultural element in all the studied case projects is in line with previous literature on the subject [3]. Mutual goals and openness were cultural elements of importance in four of the five cases, and the observed significance of these elements corresponds well with literature [11, 12]. While mutual respect was identified as a cultural element in just two cases, this element was considered fundamental for the maintenance of trust, which is necessary for a team to efficiently collaborate towards common goals [2]. The identification of mutual respect as a cultural element of fundamental importance in CDMs corresponds with recent findings [11, 12].

6 Conclusions and Further Research

This study demonstrates the interrelatedness between CDM elements, uncertainty and sustainability in large projects. CDM elements contribute to enhanced management of opportunities and risks in ways that can advance both economic, social and environmental sustainability. Through the contractual elements, the client can recruit suppliers with motives and competence to collaborate on the goal of advancing sustainability, and establish mechanisms that further incentivize them to achieve this goal. Social, political and market uncertainty greatly impacts the flexibility of the client to select and implement contractual elements that advance sustainability. Successful implementation of the organizational and cultural elements benefits from active client involvement in their implementation. Organizational elements contribute to the recruitment of personnel with the right motives and competence and the sustenance and enhancement of the ability of the team to collaborate towards common goals, while cultural elements help formalize the goals of the team and the way they are to collaborate in achieving them.

This chapter has documented similar experiences with CDMs in localized medium-sized construction projects and transcontinental industrial megaprojects, which supports the view that its conclusions can be generalized irrespective of size, industry and geography. In future research, it is recommended to study in detail how collaboration towards sustainability can be integrated into CDM elements and to carry out comparative quantitative studies on how projects with various forms of such integration perform economically, socially and environmentally.

References

1. Engebø, A., Lædre, O., Young, B., Larssen, P.F., Lohne, J., Klakegg, O.J.: Collaborative project delivery methods: a scoping review. J. Civ. Eng. Manag. **26**(3), 278–303 (2020)
2. Engebø, A., Klakegg, O.J., Lohne, J., Lædre, O.: A collaborative project delivery method for design of a high-performance building. Int. J. Manag. Proj. Bus. **13**(6), 1141–1165 (2020)
3. Engebø, A., Klakegg, O.J., Lohne, J., Bohne, R.A., Fyhn, H., Lædre, O.: High-performance building projects: how to build trust in the team. Archit. Eng. Des. Manag. **8**(6), 774–790 (2020)
4. Engebø, A., Rygh, J.A.M., Klakegg, O.J., Lohne, J., Lædre, O.: Experiences from implementing a collaborative project delivery method. In: Lindahl, G., Gottlieb, S.C. (eds.) SDGs in Construction Economics and Organization. CREON 2022. Springer Proceedings in Business and Economics, pp. 327–341. Springer, Cham (2023)
5. Falch, M.R., Engebø, A., Lædre, O.: Effects of partnering elements: an exploratory case study. In: Proceedings of 28th Annual Conference of the International Group for Lean Construction (IGLC), pp. 757–768 (2020)
6. Flyvbjerg, B.: What you should know about megaprojects and why: an overview. Proj. Manag. J. **45**(2), 6–19 (2014)
7. Hosseini, A., Wondimu, P., Klakegg, O., Andersen, B., Lædre, O.: Project partnering in the construction industry: theory vs. Practice. Eng. Proj. Org. J. **8**(1), 13–35 (2018)
8. Johansen, A., Olsson, N., Jergeas, G., Rolstadås, A.: Project Risk and Opportunity Management: An Owner's Perspective. Routledge (2019)
9. Klakegg, O.J.: Project delivery models—situational or fixed design? In: 2017 12th International Scientific and Technical Conference on Computer Sciences and Information Technologies (CSIT), vol. 2, pp. 202–206 (2017)
10. Lædre, O., Haavaldsen, T., Bohne, R.A., Kallaos, J., Lohne, J.: Determining sustainability impact assessment indicators. Impact Assess. Proj. Apprais. **33**(2), 98–107 (2015)
11. Moradi, S., Kähkönen, K., Sormunen, P.: Analytical and conceptual perspectives toward behavioral elements of collaborative delivery models in construction projects. Buildings. **12**(3), 1–19 (2022)
12. Moradi, S., Kähkönen, K.: Success in collaborative construction through the lens of project delivery elements. Built Environ. Proj. Asset Manag. **12**(6), 973–991 (2022)
13. Moradi, S., Kähkönen, K.: Sustainability indicators in building construction projects through the lens of project delivery elements. IOP Conf. Ser. Earth Environ. Sci. **1101**(2), 1–10 (2022)
14. Narum, K.B., Engebø, A., Lædre, O., Torp, O.: Collaborative project delivery with early contractor involvement and target cost. In: Proceedings of 30th Annual Conference of the International Group for Lean Construction (IGLC), pp. 984–995 (2022)
15. Rothengatter, W.: Megaprojects in transportation networks. Transp. Policy. **75**, A1–A15 (2019)
16. Tadayon, A., Andersen, B.: Characteristics of a project that are suitable for a relational PDM. Proc. Comput. Sci. **181**, 1089–1096 (2021)

17. Wondimu, P.A., Hailemichael, E., Hosseini, A., Lohne, J., Torp, O., Lædre, O.: Success factors for early contractor involvement (ECI) in public infrastructure projects. Energy Procedia. **96**, 845–854 (2016)
18. Wondimu, P.A., Hosseini, A., Lohne, J., Lædre, O.: Early contractor involvement approaches in public project procurement. J. Public Procure. **18**(4), 355–378 (2018)
19. Wondimu, P.A., Klakegg, O.J., Lædre, O.: Early contractor involvement (ECI): ways to do it in public projects. J. Public Procure. **20**(1), 62–87 (2020)
20. Yin, R.K.: Case Study Research and Applications: Design and Methods, 6th edn. Sage (2018)

Future-Proofing Energy Infrastructure: Power Grid Risk Assessment

Muneer Qudaisat, Dela Houssou, William Gallus, and Alice Alipour

Contents

1 Introduction

Uninterrupted power supply through robust power generation, transmission, and distribution in electric power networks (EPNs) is crucial to keep society functioning today. However, local power grids are prone to whims of weather events with undeniable impacts, such as hurricanes, strong winds, tornados, wildfires, floods, blizzards, or extreme cold. The profound impact of strong winds on power grids often leads to widespread disruptions in the electricity supply caused by breaking power lines, toppling utility poles, and damage to transmission towers [1, 2]; these threats, coupled with flying debris, pose a significant threat to the structural integrity of the grid. Extreme high or low temperatures are usually associated with higher electricity demand that can potentially overload the grid, leading to equipment failures and blackouts. For colder climates, extreme winter weather events such as ice storms and heavy wet snowfall can build up on power lines, weighing them down

M. Qudaisat · D. Houssou · W. Gallus · A. Alipour (✉)
Iowa State University, Ames, IA, USA
e-mail: muneerq@iastate.edu; alipour@iastate.edu

© The Author(s) 2025
M. Kioumarsi, B. Shafei (eds.), *The 1st International Conference on Net-Zero Built Environment*, Lecture Notes in Civil Engineering 237,
https://doi.org/10.1007/978-3-031-69626-8_94

and potentially causing them to snap. Heavy rainfall accompanying hurricanes and organized thunderstorm systems, including derechos, can lead to riverine flood events or more local flash flooding, impacting substations and underground grids and weakening the foundation of poles and towers.

As climate change accelerates, its impact exacerbates the challenges faced by power grids in the burden of extreme weather events. Rising global temperatures may contribute to the intensification of hurricanes, increased frequency, and changes in the timing and location of wildfires and tornadoes. Additionally, shifting climate patterns may impact severe weather phenomena such as destructive winds brought by derechos. The warming climate also contributes to the instability of polar vortexes, escalating the likelihood of extreme cold spells. These climate-related shifts can change the frequency and severity of weather events, underscore the urgency of power grid infrastructure adaptation to the evolving challenges posed by a changing climate, and integrate more reliable power resources in the long run through effective climate change adaptation actions [3]. In the United States, 96% of power outages in 2020 were caused by severe weather events [4]. The increased intensity and frequency of natural hazards subsequently increased the consequences of such events on electric power networks. The adverse economic, social, and environmental impacts of power supply disruptions are significant. Preparing for the effects of these weather-related events is paramount to ensuring the resilience of power grids. Pre-emptive measures, such as reinforcing power lines and implementing smart grid technologies, can be strategically employed based on the insights provided by risk assessment. Moreover, the energy sector is essential to climate change vulnerability and adaptation analyses. It is responsible for almost two-thirds of greenhouse gas emissions and most power supply disruptions, causing adverse economic, social, and environmental impacts. Figure 1 shows weather events and their associated impacts on energy infrastructures [5].

Worldwide, the burden of global climate change and increasingly frequent associated events have started to affect various infrastructures, as highlighted by the UK Institution of Civil Engineers [6]. Much of climate change vulnerability, risk, and adaptation efforts are devoted to better understanding and quantifying global climate change impacts on the regional level, given the possibility of increasing the intensity and frequency of extreme environmental events, including intense wind events [7]. The investigation of the reliability of power networks is already challenging due to the numerous involved variables and their inherited uncertainties [8]; the estimation of climate change projections adds to the complexity of the process, especially the regional variability of climate, climate change, and infrastructure properties. Climate scientists use historical data and complex models to estimate and predict future environmental conditions. This process is usually region-dependent, as generalizing or adapting climate change adaptation strategies of other locations might not be feasible or possible due to the different and complex nature of climate change impacts and the critical infrastructure performance in each region [9], such as The North American Regional Climate Change Assessment Program (NARCCAP) [10], High-resolution regional climate change projections

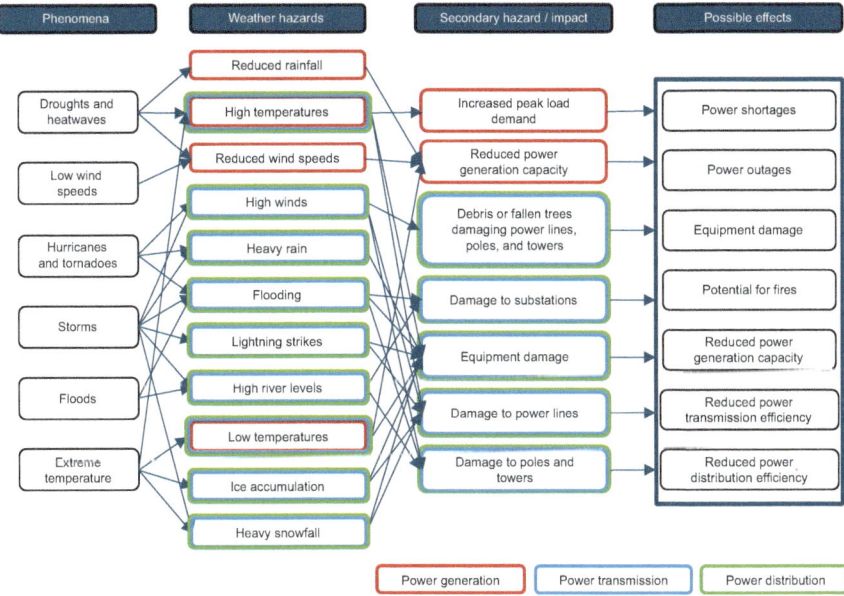

Fig. 1 Weather hazards and possible associated impacts on energy infrastructures

for the northeast USA developed using IPCC SRES emission scenarios by Hayhoe et al. [11], and for the Midwest area by Wuebbles et al. [12]. Additionally, regional projections have been developed in the UK (UK Climate Projection 2009 (UKCP09)) and KNMI climate change scenarios for the Netherlands 2006 following the fourth International Panel on Climate Change (IPCC) [13].

Numerous researchers have studied the future temperature, precipitation, and wind speed trends. Mideksa and Kallbekken investigated thermal power plants' supply sensitivity to temperature changes due to the large geographical variability [14]. Bloomfield et al. investigated the effects of the shifting climate on renewable energy resources and the supply-demand balance in future power systems [15]. Dobson et al. [16] addressed the power system blackouts and outages driven by extreme weather events and their future projections. The increasing ambient temperature associated with climate change and global warming compromises the efficiency and maximum capacity lifetime of power transformers and power lines, as they are vulnerable to high ambient air temperatures [17]. Global warming and, subsequently, ocean warming impact the complex process of wind formation, leading to changes in wind patterns and currents that can promote distribution network failure by rupturing the poles and wire lines, damaging pole-mounted equipment, or causing cascading outages [18]. Increasing participation intensity, storm surge events, and rising sea levels at some locations jeopardize the power transmission and distribution lines and substations to risk flooding [19–21].

2   Climatic Parameters

Based on the region, climate change can increase the imposed hazard on power distribution networks through increasing extreme wind speeds each year, hence making the grid assets more prone to failure, increasing temperature that expedites the decay rate of wooden poles and increases their vulnerability with time. However, annual rainfall reduction can slow the decay rate of wooden poles [22]. Therefore, assessing the impact of climate change on power distribution networks requires detailed modeling to capture the contrasting region-dependent effects of climatic changes and their differing extents. To visibly measure the effect of projected regional climate change on EPNs' performance, the baseline vulnerability status of a network needs to be established first. Hence, a case study of four cities in Iowa in the USA, namely, Muscatine, Algona, Pella, and Cedar Falls, is presented.

The WRF 3.4.1 model is a numerical weather prediction system designed to serve atmospheric research and operational forecasting needs, therefore assessing the infrastructure resilience from multiple aspects [23]. This study uses a higher resolution 4 km cell grid than is typically used operationally to give a more accurate projection for frequency and intensity of wind speed, precipitation (flooding), and freezing rain over the 13-year simulation period considering climate change. The resolution used herein is much finer compared to the 18 km resolution used in earlier models, including the UK flooding model (UKCP 18), global reanalysis (ERA5), Global-to-Regional Integrated Forecast System (GRIST), China Merged Precipitation Analysis (CMPA), and Integrated Multi-Satellite Retrievals for the Global Precipitation Measurement (IMERG) [24, 25]. The simulation period is based on a reference control period spanning from October 1, 2000, to September 30, 2013, with 6 hours and 0.7 °C and a sensitivity model considering the effects of climate change using the PWG approach ("Physics-WGNE" (Working Group on Numerical Experimentation)), with ten perturbed physical fields to account for uncertainties climate system mechanism. The future climate simulation extends from October 1, 2086, to September 30, 2099.

2.1   Maximum Wind Speed

The behavior of the wind speed shows oscillation around a mean value. Still, given the turbulent characteristic of wind, the maximum wind speed can peak away from the mean value [26]. The wind speed dataset contains maximum wind speeds 10 meters above ground level. Structural analysis showed that poles start to fail near the 15 m/s wind speed threshold, considering that the model values are averaged over the 4 × 4 km box and peak winds usually are very localized. In addition, models typically are deficient in bringing momentum to the ground. Accordingly, the projections of wind speed over 20 m/s were calculated along with their frequencies in the control and future models.

2.2   Wind Speed Simulation Results

The following Fig. 2 shows the high winds of the four study areas during the reference and projected years. There is a noticeable variety in high wind speeds across the four regions. At the same time, some areas experience higher wind speeds, indicating natural weather pattern fluctuation that can be attributed to terrains and other landscape features. For example, Muscatine generally has higher wind speeds over the years, and occasional years, such as 2090 and 2091, show significant increases in peak wind speeds. Compared to earlier years, Algona shows increased peak wind speeds in certain future years, such as 2087 and 2092, and Pella shows increased peak wind speeds in certain years, such as 2090 and 2093. Cedar Falls has less wind speed fluctuation across the study years. Hence, specific years exhibit extremely high wind speeds compared to the average, indicating the possibility of intermittent extreme weather events, and there seems to be a trend of stronger winds toward the later years of the dataset.

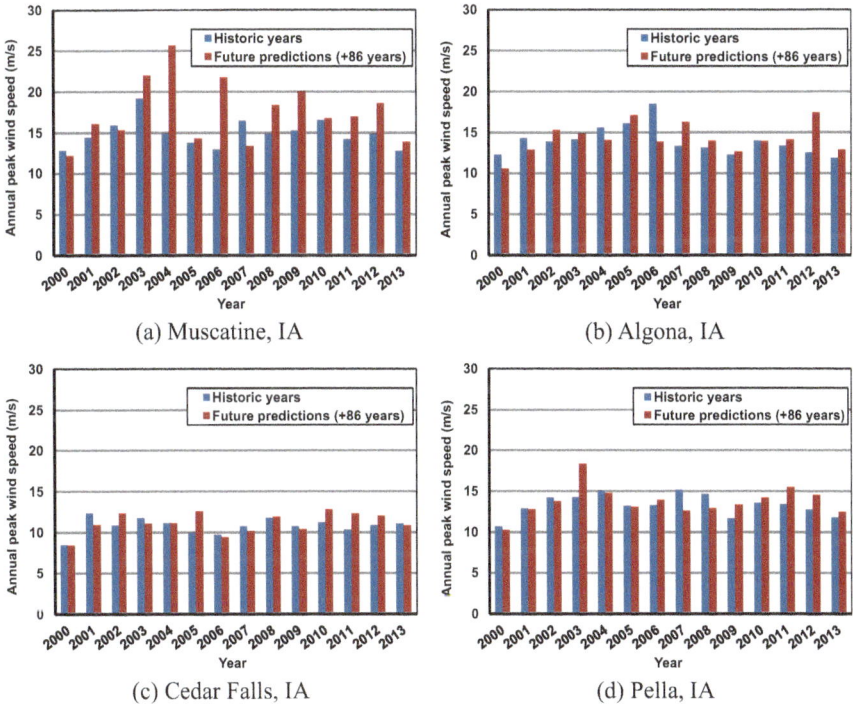

Fig. 2 Maximum wind speed in the case study areas. (**a**) Muscatine, IA. (**b**) Algona, IA. (**c**) Cedar Falls, IA. (**d**) Pella, IA

3 Maximum Wind Speeds and Electric Power Network Reliability

3.1 System Fragility Curve Development

The radial configuration of local power distribution networks leads to service outage downstream from any interruption point, i.e., network component failure. Hence, it is crucial to investigate the reliability of the network components in correspondence to various environmental event intensities. In this study, which focuses on the wooden power distribution poles, the capacity of those poles is influenced by their class, geometry, age, and the environmental conditions that expedite their deterioration. After analyzing the wind demand on the poles in the network and their capacity, it is essential to use a limit state function to describe the system's vulnerability by establishing the pole's conditional failure probability in response to increasing wind speed.

3.2 Wind Demand on EPN Components

The poles and wires of EPNs are directly exposed to wind and flying debris. Hence, they are considered highly vulnerable. To evaluate the exerted wind loads, the following relationships provided by ASCE/SEI 7-22 can be used [27]:

$$F = 0.613 \, K_z \, K_{zt} \, K_d \, K_e \, G \, Cf \, A_f \, V^2 \, (\text{N}) \tag{1}$$

G is the gust-effect factor, Cf is the force coefficient, A_f is the pole or wire area projected normal to the wind direction. K_z is the velocity pressure exposure coefficient, K_{zt} is a topographic factor, and K_d is the wind directionality. K_e is the ground elevation factor. V is the basic 3-sec gust wind speed. The distributions and coefficient of variations (CoV) of the mentioned random variables related to poles and wires are summarized by Ellingwood and Tekie [28].

3.3 EPN Components Capacity and Case Study

The vast majority of US power distribution poles are wooden due to the availability, serviceability, and lower wood cost than other pole materials [29]. The American National Standards Institute published in (ANSI-O5.1) classified wooden poles into 15 classes designed to have approximately the same load-carrying capacity regardless of their species [30]. In the four locations where the wind speed projections were analyzed, the main wood pole is the Class 3 Southern pine pole. Hence, a fragility function for a 40-year-old Class 3 pole with an average height of 13 meters and an average wire span of 100 meters is chosen as an example.

3.4   Limit State Function

The wind loads acting on the poles and wires of EPNs translate into bending moment in the pole and tensile stress in the wires; hence, the structural demand is directly proportional to wind speed and duration. On the other hand, the EPNs can be structurally analyzed to determine the structural capacity of their components in response to various weather events and external stresses. The following relationship can generally describe the limit state function, $G(x)$:

$$G(x) = C(x_c, d_c) - D(x_d, d_d) \qquad (2)$$

Where C is the structural capacity as a function of random and deterministic variables x_c and d_c, respectively. D is the structural demand on the system as a function of random and deterministic variables x_d and d_d, respectively. The system fails whenever the demand exceeds the system capacity and results in the limit state function being negative. The Latin hypercube sampling method (LHS) is used to generate ten thousand random samples for wind speeds, which are used in accordance with the provisions of ASCE/SEI 7-22 to calculate wind load on the poles and connected wires. Subsequently, the failure probability at each wind speed.

3.5   Fragility Function Development

After structurally analyzing the ten thousand realizations and determining their failure or survival based on the limit state function for three modes of failure, i.e., pole rupture, foundation failure, and wires breakage, The pole rupture turned out to be the predominant mode of failure within the range of wind speeds in this study. Afterward, the log-normal distribution was chosen to describe the relationship between wind speed and the fragility of the poles. The distribution parameters can be obtained using the maximum likelihood estimation (MLE) method [31, 32]. The following fragility function shows an increasing failure probability of the pole with increasing wind speeds. Considering the generally projected increase in wind speed in some regions in the coming years, accompanied by the degradation of wooden poles and strength decay, accounting for climate change impacts becomes necessary (Fig. 3).

3.6   The Annual Probability of Failure

The annual probability of failure is obtained by performing a mathematical convolution of the fragility and hazard curve; the last describes the instantaneous probability of failure at a given wind speed. The wind speed occurrence frequency is

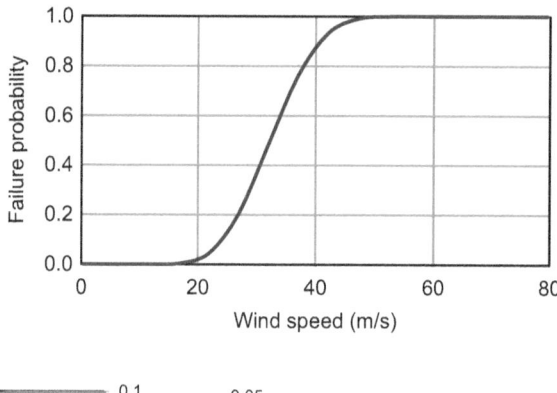

Fig. 3 Fragility function of a 13 m high Class 3 wooden pole

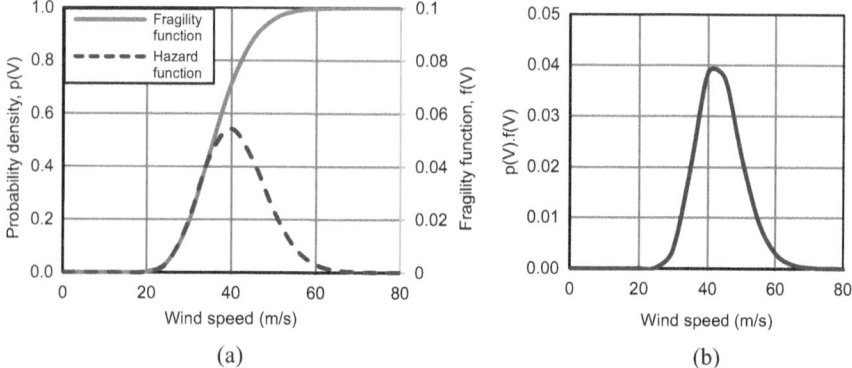

Fig. 4 (**a**) Fragility and hazard curves, (**b**) fragility and hazard functions convolution curve

presented through a probability density function (PDF) that models the available location-related wind data based on the data provided by Vickery et al. [33] in Fig. 4a. Convolution principally combines the fragility and hazard functions to show the overall likelihood of failure at different wind speeds throughout the year. The annual probability of failure, P_f, at maximum wind speed, V_{max}, can be expressed as:

$$P_f = \int_0^{V_{max}} p(V).f(V).dV \tag{3}$$

where $p(V)$ is the fragility curve, representing the probability of failure at wind speed V, and $f(V)$ represents the PDF of wind speed V, obtained from the hazard curve. Hence, the annual probability of failure, P_f, is the area under the curve in Fig. 4b up to the maximum wind speed, V_{max}.

Considering the prevailing wind conditions, the change in the annual probability of failure of the pole between the reference year and the projected year offers further insight into the changing risk of failure associated with climate change, as depicted in Fig. 5. Despite the shown fluctuations and lack of a clear linear trend between the

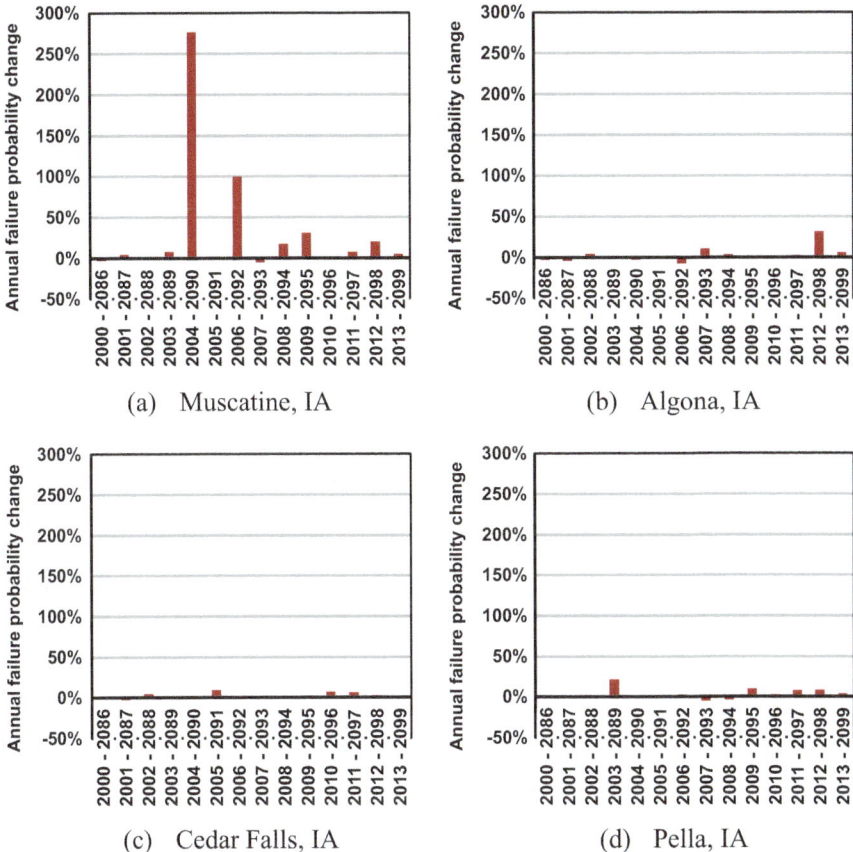

Fig. 5 Percent change of the annual failure probability in the case study areas. (**a**) Muscatine, IA. (**b**) Algona, IA. (**c**) Cedar Falls, IA. (**d**) Pella, IA

change in annual failure probability over the years, there were substantial increases, such as 275% in 2090 for Muscatine, and other decreases, such as −8% in 2092 for Algona. It can be noticed that those percentages correspond to the projected change in maximum wind speed displayed in Fig. 2 as increasing wind speed increases the wind load demand and, subsequently, the component failure probability.

4   Results and Conclusions

As the intensity and frequency of some weather events increase, it further stresses the power supply chain integrity. Distribution networks are the most vulnerable to extreme weather events as they can cause widespread power outages, overload the grid, and potentially lead to equipment failure. Fragility models are essential to estimate the likelihood of component failure and risk assessment, enabling informed decision-making for predictive maintenance and risk mitigation procedures. This research reviewed the impacts of the ongoing climate change on power distribution networks by estimating the projected maximum wind speeds, evaluating their effect on the components of the power distribution network, and establishing the vulnerability of the EPN as a function of wind speed through statistical analysis. The presented research discusses weather phenomena, their association with climate change, and their projected impacts. The numerical weather prediction model WRF 3.4.1 with a 4 km resolution cell grid gives a more accurate projection of the frequency and intensity of high winds. The percent change in the predicted annual probability of failure is associated with the percent increase or decrease of the high wind speed, therefore calling for adaptive and continuous risk assessment of the network to ensure its reliability.

References

1. Dikshit, S., Dobson, I., Alipour, A.: Cascading structural failures of towers in an electric power transmission line due to straight line winds. Reliab. Eng. Syst. Saf. **250**, 110304 (2024) https://doi.org/10.1016/j.ress.2024.110304
2. Alipour, A., Sarkar, P., Dikshit, S., Razavi, A., Jafari, M.: Analytical approach to characterize tornado-induced loads on lattice structures. J. Struct. Eng. **146**(6), (2020) https://doi.org/10.1061/(ASCE)ST.1943-541X.0002660
3. Dumas, M., Kc, B., Cunliff, C.I.: Extreme weather and climate vulnerabilities of the electric grid: a summary of environmental sensitivity quantification methods. 2019 Aug 1 [cited 2024 Mar 12]; Available from: http://www.osti.gov/servlets/purl/1558514/
4. Stone, B., Mallen, E., Rajput, M., Gronlund, C.J., Broadbent, A.M., Krayenhoff, E.S., et al.: Compound climate and infrastructure events: how electrical grid failure alters heat wave risk. Environ. Sci. Technol. [Internet]. **55**(10), 6957–6964 (2021 [cited 2024 Mar 13]) Available from: https://www.apmresearchlab.org/10x-power-climate
5. Gonçalves, A.C.R., Costoya, X., Nieto, R., Liberato, M.L.R.: Extreme weather events on energy systems: a comprehensive review on impacts, mitigation, and adaptation measures. Sustain. Energy Res. [Internet]. **11**(1), 4 (2024) Available from: https://jrenewables.springeropen.com/articles/10.1186/s40807-023-00097-6
6. Knutti, R.: Should we believe model predictions of future climate change? Philos. Trans. R. Soc. A Math. Phys. Eng. Sci. **366**(1885), 4647–4664 (2008) Available from: https://royalsocietypublishing.org/doi/10.1098/rsta.2008.0169
7. Della-Marta, P.M., Mathis, H., Frei, C., Liniger, M.A., Kleinn, J., Appenzeller, C.: The return period of wind storms over Europe. Int. J. Climatol. **29**(3), 437–459 (2009)
8. Micheli, L., Alipour, A., Laflamme, S.: Multiple-surrogate models for probabilistic performance assessment of wind-excited tall buildings under uncertainties. ASCE-ASME Journal of Risk and Uncertainty in Engineering Systems Part A: Civil Engineering 6(4), (2020) https://doi.org/10.1061/AJRUA6.0001091
9. Ryan, P.C., Stewart, M.G.: Regional variability of climate change adaptation feasibility for timber power poles. Struct. Infrast. Eng. [Internet]. **17**(4), 579–589 (2021) Available from: https://www.tandfonline.com/doi/full/10.1080/15732479.2020.1843505
10. Mearns, L.O., Gutowski, W., Jones, R., Leung, R., Mcginnis, S., Nunes, A., et al.: A regional climate change assessment program for North America. Eos (Washington DC). **90**(36), 311 (2009)

11. Hayhoe, K., Wake, C., Anderson, B., Liang, X.Z., Maurer, E., Zhu, J., et al.: Regional climate change projections for the Northeast USA. Mitig. Adapt. Strateg. Glob. Chang. [Internet]. **13**(5–6), 425–36 (2008 [cited 2024 Mar 10]) Available from: https://link.springer.com/article/10.1007/s11027-007-9133-2

12. Wuebbles, D.J., Hayhoe, K.: Climate change projections for the United States Midwest. Mitig. Adapt. Strateg. Glob. Chang. [Internet]. **9**(4), 335–63 (2004 Oct [cited 2024 Mar 10]) Available from: https://link.springer.com/article/10.1023/B:MITI.0000038843.73424.de

13. Van Den Hurk, B., Klein Tank, A., Lenderink, G., Van Ulden, A., Van Oldenborgh, J., Katsman, C., et al.: KNMI Climate Change Scenarios 2006 for the Netherlands [Internet]. 2006. Available from: www.knmi.nl

14. Mideksa, T.K., Kallbekken, S.: The impact of climate change on the electricity market: a review. Energy Policy. **38**(7), 3579–3585 (2010)

15. Bloomfield, H.C., Brayshaw, D.J., Troccoli, A., Goodess, C.M., De Felice, M., Dubus, L., et al.: Quantifying the sensitivity of european power systems to energy scenarios and climate change projections. 2020 [cited 2024 Mar 10]; Available from: http://creativecommons.org/licenses/by/4.0/

16. Dobson, I., Dong, Y., Kezunovic, M.: Impact of extreme weather on power system blackouts and forced outages: new challenges. [cited 2024 Mar 11]; Available from: https://www.researchgate.net/publication/228379681

17. Wilbanks, T.J., Fernandez, S.J.: Climate Change and infrastruCture, urban systems, and Vulnerabilities Technical Report for the US Department of Energy in Support of the National Climate Assessment national Climate assessment regional technical input report series

18. Fern, S.J., et al.: Application of hybrid geo-spatially granular fragility curves to improve power outage predictions. J. Geograph. Nat. Disast. [Internet]. **4**(2), 1–6 (2014 [cited 2024 Mar 12]) Available from: https://www.longdom.org/open-access/application-of-hybrid-geospatially-granular-fragility-curves-to-improve-power-outage-predictions-33876.html

19. Zamuda, C., Mignone, B., Bilello, D., Hallett, K.C., Lee, C., Macknick, et al.: US Department of Energy. US Energy Sector Vulnerabilities to Climate Change and Extreme Weather (2013 [cited 2024 Mar 11]) Available from: https://apps.dtic.mil/sti/citations/ADA583709

20. Zhang, N., Alipour, A.: Flood risk assessment and application of risk curves for design of mitigation strategies. Int. J. Crit. Infrastructure Prot. **36**, 100490 (2022) https://doi.org/10.1016/j.ijcip.2021.100490

21. Zhang, N., Alipour, A.: A two-level mixed-integer programming model for bridge replacement prioritization. Abstract. Comput.-Aided Civ. Infrastruct. Eng. **35**(2), 116–133 (2020) https://doi.org/10.1111/mice.v35.2; https://doi.org/10.1111/mice.12482

22. Hsiang, W.C., Wang, X.: Vulnerability of timber in ground contact to fungal decay under climate change. Clim. Change [Internet]. **115**(3–4), 777–794 (2012) Available from: http://link.springer.com/10.1007/s10584-012-0454-0

23. Zhang, G., Zhang, F., Wang, X., Zhang, X.: Fast resilience assessment of distribution systems with a non-simulation-based method. IEEE Trans. Power Deliv. [Internet]. **37**(2) (2022) Available from: https://doi.org/10.1109/TPWRD

24. Chen, T., Li, J., Zhang, Y., Chen, H., Li, P., Che, H.: Evaluation of hourly precipitation characteristics from a global reanalysis and variable-resolution global model over the Tibetan Plateau by using a satellite-gauge merged rainfall product. Remote Sens. **15**(4), 1013 (2023)

25. Liu, H., Liu, X., Liu, C., Yun, Y.: High-resolution regional climate modeling of warm-season precipitation over the Tibetan Plateau: impact of grid spacing and convective parameterization. Atmos. Res. **281**, 106498 (2023)

26. Teoh, Y.E., Alipour, A., Cancelli, A.: Probabilistic performance assessment of power distribution infrastructure under wind events. Eng. Struct. **15**, 197 (2019)

27. Structural Engineering Institute: Minimum Design Loads and Associated Criteria for Buildings and Other Structures [Internet], p. 593. American Society of Civil Engineers, Reston (2021) Available from: https://ascelibrary.org/doi/book/10.1061/9780784415788

28. Ellingwood, B.R., Tekie, P.B.: Wind load statistics for probability-based structural design. J. Struct. Eng. [Internet]. **125**(4), 453–463 (1999) Available from: https://ascelibrary.org/doi/10.1061/%28ASCE%290733-9445%281999%29125%3A4%28453%29

29. Mohammadi Darestani, Y., Shafieezadeh, A.: Multi-dimensional wind fragility functions for wood utility poles. Eng. Struct. **183**, 937–948 (2019)

30. American National Standard for Telecommunications Wood Poles-Specifications & Dimensions [Internet]. 2009 [cited 2023 Jun 7]. Available from: https://webstore.ansi.org/standards/ansi/ansio52022

31. Shinozuka, M., Feng, M.Q., Lee, J., Naganuma, T.: Statistical analysis of fragility curves. J. Eng. Mech. [Internet]. **126**(12), 1224–1231 (2000) Available from: https://ascelibrary.org/doi/10.1061/%28ASCE%290733-9399%282000%29126%3A12%281224%29

32. Dikshit, S., Alipour, A.: A moment-matching method for fragility analysis of transmission towers under straight line winds. Reliab. Eng. Syst. Saf. **236**, 109241 (2023) https://doi.org/10.1016/j.ress.2023.109241

33. Vickery, P.J., Wadhera, D., Galsworthy, J., Peterka, J.A., Irwin, P.A., Griffis, L.A.: Ultimate wind load design gust wind speeds in the United States for use in ASCE-7. J. Struct. Eng. [Internet]. **136**(5), 613–625 (2010) Available from: https://ascelibrary.org/doi/10.1061/%28ASCE%29ST.1943-541X.0000145

Energy Communities for the Decarbonization of Historical Villages: A Case Study in Italy

Silvia Brunoro ⓘ and Emanuele Piaia ⓘ

Contents

1 Introduction

In the last decade, the decline of Historical Villages (HVs) has assumed considerable dimensions, resulting in the disappearance of a secular civilization and, with it, the identity linked to both buildings and traces and elements of material and immaterial culture. This negative trend is increasing in several EU countries, and it is especially evident in Italy, which has 4.7% of the world's architectural heritage and where HVs with less than 5000 inhabitants cover a total of 60% of the territory, involves approximately 22% of its population [1].

HVs are settlements that have maintained the recognizability of their structure and the continuity of the historic building fabric, where the original typological and morphological characteristics are clear and denoted by high social, historical, architectural, and/or landscape value, establishing them as an important part of the European Cultural Heritage (CH). HVs constitute fragile environments, often located in marginal areas isolated from main urban centers, frequently abandoned

S. Brunoro · E. Piaia (✉)
University of Ferrara, Department of Architecture, Ferrara, Italy
e-mail: emanuele.piaia@unife.it

© The Author(s) 2025
M. Kioumarsi, B. Shafei (eds.), *The 1st International Conference on Net-Zero Built Environment*, Lecture Notes in Civil Engineering 237,
https://doi.org/10.1007/978-3-031-69626-8_95

due to their complex orography, their fragile economy, and high environmental risks, recently exacerbated by the effects of climate change [2].

These features led them to progressively face depopulation phenomena, unemployment, and ageing of the population, which, in turn, caused growing socioeconomic issues and the degradation of buildings and infrastructure. In addition, this significant heritage no longer answers to the new expected building energy standards required to contrast climate change [3]. This gap hinders the fulfillment of objectives advocated by the new EU Green Deal, aiming to establish Europe as the first climate-neutral continent by 2050 [4].

Unfortunately, despite the high cultural riches, the improvement and valorization of this heritage is hindered primarily by economic budget constraints.

In this context, a potential solution that has emerged in Europe, involves the establishment of participatory processes through public–private partnerships. These processes power the local inhabitants to spearhead bottom-up initiatives promoting their crucial role into the building's refurbishment, retrofit, and valorization toward decarbonization aims. The most impactful solution, successfully addressing social, cultural, and economic concerns alongside environmental considerations, is the establishment of Energy Communities (ECs) [5].

Renewable ECs[1] can be regarded as a model of organizational innovation that empowers end customers as protagonists of the energy transition. This approach enables citizens, administrations, and local businesses to collaboratively develop and manage energy projects or services. The governance and ownership structure of these communities often differs from traditional business organizations, emphasizing a more participatory and community-driven approach [6].

Fueled by an ever-evolving national and European legislative framework on the energy transition, as well as initiatives associated with the Next Generation EU plan, Italy has witnessed a notable increase in initiatives dedicated to establishing ECs in recent years. This trend is particularly pronounced in specific contexts, such as HVs [7].

Analyzing the Italian context, the paper provides an overview of the current state of the evolution of ECs, both from a legislative and practical standpoint. Through a detailed description, grounded in reference case studies, the objective is to underscore replicable and inclusive reference models that can encourage interventions such as the reuse, enhancement, and redevelopment of HVs. This includes maintenance and restoration initiatives aimed at integrating energy retrofit interventions preserving the heritage value and enhancing the historical identity of this asset.

To study the innovation potential of the Renewable Energy Communities in Italy, the following methodological steps have been followed:

---

[1] Energy Communities can be classified in two levels: the Renewable Energy Communities (RECs) can only exploit Renewable Energy Sources (RES) to fulfill different kinds of demand (e.g., electrical and thermal), whereas the Citizens Energy Communities (CEC) can manage only electricity, both from RES and fossil fuels, and can operate in an autonomous microgrid.

- The analysis of the regulation, process and factors that influences the regime and create windows of opportunity for the initiative forming and development in the Italian context.
- The identification of the main goals toward HV shifting to clean energy.
- The descriptive analysis of a best practice example, illustrating the key success factors: community engagement, funding strategies, etc.
- The potentiality of replication for revitalizing HV in Italy.

2 The Energy Communities

Due to the Europe-wide publication of the Clean Energy for all Europeans Package, energy end-users, previously the final link in the decision-making chain regarding the definition and implementation of energy policies, have now been empowered with a potential new role. This newfound role provides them with the opportunity to actively participate, for instance, in the generation of energy from renewable sources. In the short term, this shift has led to a profound transformation of the traditional model of energy system production and management [7].

The European Parliament has issued two directives that lay the foundation for the introduction of ECs: the Renewable Energy Directive 2018/2001 (known as RED II), published in December 2018, and the Directive on common rules for the internal market for electricity 2019/944 (known as the EMI Directive), published in June 2019. The two directives set the legal framework at the European level for individual and community participation, introducing specific definitions for self-consumption (including collective) schemes and ECs [8, 9].

Based on the assumptions contained in the above directives, the basic concept that defines the role of ECs emerges, namely: to promote the generation, transmission and distribution of energy produced from renewable sources, in response to energy, environmental and social needs identified in local realities.

The proliferation of initiatives geared toward establishing ECs is bolstered by specific factors: [10–12].

- Increased adoption of innovative and sustainable energy models: the wider acceptance of innovative and sustainable energy models facilitates the pursuit of objectives set by the energy transition through inclusive, community-based approaches.
- Technological advancements: the maturity of technologies for decentralized energy production on a local scale plays a pivotal role in supporting these initiatives.
- Citizen participation in the energy market: the opportunity for active involvement in the citizens' energy market enables a shift from being solely consumers to becoming prosumers, actively contributing to energy generation.
- Addressing energy poverty: ECs contribute to combating energy poverty, ensuring more equitable access to energy resources.

Finally key success factors for these initiatives lies in the inherent connection between the initiators and the targeted area. Indeed, no one is better positioned to champion the enhancement of their locality than its inhabitants, particularly when the area holds historical and cultural significance. Legislatively, members of an EC can include individuals, SMEs, territorial entities, or local authorities, including municipal governments. However, it's important to note that for private companies, participation in the renewable energy community should not constitute their primary commercial and/or industrial activity.

2.1 Challenges and Opportunities in the Italian Context

The term "Renewable Energy Community" (REC) is adopted to denote a coalition of users who, by establishing a legal entity, choose to collaborate with the objective of meeting the identified needs of the members. This collaboration involves the production, consumption, and management of energy through one or more local plants powered by renewable sources [13].

Main characteristics of the developing models of RECs, responding to different needs site specificity (stakeholder, funding, and sharing model), are summarized in Table 1.

All the assets of the REC must be connected to the same medium voltage (MV)–low voltage (LV) transformation station (i.e., sub-stations or secondary stations). The Shared Energy (SE) is equal to the minimum, at each hourly period, between the electrical energy produced and fed into the grid by RES and the electrical energy absorbed from all the customers. The economic incentives for RECs are guaranteed for 20 years. Afterward, annual tacit renewals apply. The BESS (Battery Energy Storage System) consists of an aggregation of loads connected with a Photovoltaic (PV) plant at the low voltage side of a same transformer, where the loads are

Table 1 Classification of REC models in the Italian context

REC model	Stakeholder/ proponent	Funding	Sharing model
Citizen-driven	Citizens, local private proponents, SMEs	Members' investment, feed-in premium tariffs, bank loans	Equitable redistribution according to the investment participation/amount of shared energy.
PA-driven	Public/local authority, PA, non-profit organizations	Public (national and regional) funding (in the cumulable percentage) and feed-in premium tariffs	Equitable redistribution, establishment of a social funds for disadvantaged members, community services, or energy efficiency measures
Energy/ technical operator-driven	Technical player/energy operator	Developer and members' investments, and feed-in premium tariffs	Divided between energy/technical operator and members

inflexible residential customers. Also, the PV plant could be any RES plant. Recently, the Delibera 727/2022 of ARERA proposed a new policy option to enable the contribution of BESS in standalone configuration [14].

The principles of decentralization and localization form the foundation of an EC. Through citizen involvement and the growth of local businesses and commercial activities, such a community can effectively generate, consume, and exchange energy in a model of self-consumption and collaboration [15]. After the publication of the Clean Energy for all Europeans Package and the accompanying Directives, by 2020 Italy initiated a partial transposition of the Renewable Energy Directive II (REDII)[2]. This transposition aimed to kick start the first experimental applications, albeit subject to two primary constraints: the maximum installable power and the proximity of the plants to the withdrawal points held by consumers.

In addition to national legislation, the institutionalization of ECs in the Italian context has spurred the development of numerous regionally initiated laws. These regional laws aim to locally promote the development of RECs more closely tied to the specific territorial characteristics of each region.

In Italy, the establishment of ECs encountered several challenges, with the main issues including [16]:

1. Individual Power Limits: the imposition of a maximum power limit of 200 kW for individual plants restricted the involvement of third parties. This limitation only allowed for projects that could engage a limited number of households and an even smaller number of small and medium-sized enterprises.
2. Constraints related to MV/LV Transformation Substations: the presence of MV/LV transformation substations posed a significant hurdle to the implementation of large-scale initiatives. This complication also made it more challenging to identify members to participate in these communities.
3. Diverse Regional Laws and Regulations: despite national-level efforts, ECs implementation was impacted by various regional laws and regulations. This occasionally led to uncertainty and created obstacles in standardizing practices.
4. Complexity in identifying members: the difficulty in identifying and engaging Energy Community members presented a problem. There were challenges in educating and incentivizing community participation.
5. Limited Size of Initiatives: regulatory constraints and operational challenges often restricted the size of initiatives, impeding the execution of large-scale projects.

---

[2] As of 2020, through the combined provisions of Law 8/2020 (which converts Article 42/bis DL 162/19—*Milleproroghe* Decree), the regulatory model identified by ARERA (Resolution 318/2020), and the incentives defined by the Ministry for Economic Development—MiSE (Ministerial Decree September 16, 2020), the establishment of RECs and collective self-consumption schemes became possible. Specifically, these laws predetermined the legislative framework to be followed, outlined incentive mechanisms, and established regulations for tariffs applicable to RECs and collective self-consumption schemes.

6. Limits to Third-Party Involvement: constraints on third-party participation potentially limited the breadth and diversity of actors involved in ECs.

Several limitations observed in the initial experiments have since been addressed with the provisions outlined in the drafts of the Legislative Decrees for the national transposition of the two relevant Directives in 2021. Notable among the key enhancements are:

1. Increased power limit for incentives: the power limit for plants eligible for incentives within the scheme has been raised to below 1 MW per plant. Additionally, the connection of plants and utilities under the same primary substation is now considered, providing greater flexibility.
2. Expanded perimeter: the expansion of the perimeter from MV/LV transformer substations to HV/MV substations serves a dual purpose. It not only overcomes challenges in accessing information but also allows for the construction of larger plants capable of effectively meeting the energy needs of a community.
3. Inclusion of existing plants: the option to include existing plants in the configuration, albeit with a maximum share of 30% of the power held by the community, aligns with the direction of community-based and more integrated management of the generation stock within individual communities.

Following the approval by the European Commission in November, the MASE (Ministry of Sustainable Energy) officially released the RECs decree (DM 414/2023) to actively support the emergence and growth of RECs and collective self-consumption systems. Effective from January 24, 2024, the decree is designed to encourage the proliferation of collective forms of energy production and self-consumption from renewable sources. The decree introduces two primary mechanisms to promote the development of RECs in Italy:

1. Forty percent Non-Repayable Subsidy: financed by the National Recovery and Resilience Plan[3] (PNRR), this provision offers a 40% non-repayable subsidy. The incentive is targeted at communities whose installations are established in municipalities with a population of fewer than five thousand inhabitants.
2. Incentive tariff on renewable energy: the decree also introduces an incentive tariff on renewable energy produced and shared, applicable nationwide. This aims to stimulate the generation and sharing of renewable energy throughout the country.

---

[3] The PNRR serves as the country's comprehensive recovery program in response to the crisis and socio-economic disruptions induced by the COVID-19 pandemic. Formally accepted by the European Union on June 22, 2021, and ratified through the European Council Decision on July 13, the Plan is an integral component of the Next-Generation EU program. Specifically, the PNRR allocates 2.2 billion euros for the development of Energy Communities in municipalities with a population of fewer than 5000. This substantial financial commitment underscores the strategic focus on fostering sustainable energy initiatives within smaller communities as part of the broader recovery efforts.

3 CommOn Light Energy Community: An Italian Best Practices Case

Despite the predominant administrative, bureaucratic, and economic challenges faced during the initial stages of launching ECs in Italy, there are noteworthy experimental case studies that can be considered as exemplary practices and models for potential replication in other international HVs [17].

The CommOn Light Energy Community serves as a notable exemplar of best practices within the Italian context. This pioneering initiative, situated in the Municipality of Ferla,[4] Sicily, showcases a successful model for sustainable energy development, collaboration, and community engagement.

CommOn Light emerged from a collaborative effort between the Municipality of Ferla, the University of Catania, and the interdepartmental research project TREPESL (Energy Transition and New Models of Participation and Local Development). The collaborative foundation ensured a multifaceted approach, incorporating both academic expertise and local administrative support.

Clear delineation of roles enhanced project efficiency. The Municipality of Ferla assumed responsibility for political, communication, and engagement aspects, as well as the facility's implementation. Simultaneously, the University of Catania took charge of legal-administrative facets, ensuring a seamless integration of skills and expertise (Fig. 1).

The project's commencement capitalized on funding from the Sicily FESR (Fondo Europeo per lo sviluppo regionale) Operational Plan 2014–2020, demonstrating strategic financial planning. Additionally, the integration of resources from municipal funds and the targeted utilization of PNRR funds earmarked for small municipalities underscored a creative and diversified funding strategy.

CommOn Light's primary objective is to curtail dependence on external, nonrenewable energy sources. By implementing on-site energy production and storage, the community strives for sustainability and reduced environmental impact.

An open and inclusive social structure lies at the core of CommOn Light's success. The initiative actively engages the community through cultural awareness programs, training activities, and transparent communication, fostering a sense of ownership and participation.

At the beginning, the EC involved the only installation of a 20-kW photovoltaic system on the roof of the Ferla Municipality building. This solution was made possible through discussions with the Superintendence for Cultural Heritage to identify a design solution that seamlessly integrated with the landscape (Fig. 2).

---

[4]Ferla is a small municipality situated in southern Italy within the province of Syracuse, boasting a population of just over 2000 inhabitants. It was the first Sicilian municipality to create a renewable energy community. Ferla, in conjunction with the city of Syracuse, has held the designation of a UNESCO Municipality since 2004 and has been a member of the *"Borghi più belli d'Italia"* since 2014.

Fig. 1 View of a segment of the historic village of Ferla

Fig. 2 Installation of photovoltaic panels on the roof of the Ferla city hall to support the ECs

The outcome of the initiative has sparked the creation of a "green village," wherein energy generated from renewable sources is shared through a mutual exchange. This collaborative effort involves private citizens or small and medium-sized enterprises whose users are connected to the same medium/low voltage transformation cabin.

Individuals interested in participating can join the Community by lending their photovoltaic system (thus becoming producers) or by connecting to the grid as end consumers. A portion of the incentives will be allocated to compensate the investments of fellow citizens who opt to be both consumers and energy producers. Another segment will be directed toward providing discounts on energy bills, while the remainder will be reinvested within the community itself, contributing to the expansion of energy capacity through additional systems.

Since 2015, Ferla has initiated a series of administrative and planning activities dedicated to innovative energy. These efforts include the creation of six photovoltaic systems with a total power capacity of 31 kW.

Starting from this experience, the path to establishing the first EC in Sicily begins in January 2021.

The municipal resolutions outlining the creation of the EC materialized through initial acts formulated between March, April, and May 2021. The guideline act was enacted in March 2021, followed by the expression of interest for external users keen on participating in the project.

In May 2021, the association's statute, the incorporation deed of the EC "CommOn Light," and the internal regulations, signed by the members to govern the distribution of incentives, were approved.

Authorization from the Superintendence of CH was crucial for the construction of the photovoltaic system above the Municipality building's roof, given the existing landscape restrictions.

Despite the higher costs and an anticipated lower yield, the Superintendence granted favorable approval for the system's construction, with the condition that the panels be colored in a brown shade to minimize landscape impact.

The management of this system requires the Municipality to make the photovoltaic system available to the community while retaining ownership under an indefinite loan contract. The agreement includes a final term linked to the EC's duration, emphasizing the potential for widespread replicability of this organizational model.

Citizen involvement practices were formulated by the members of the working group and were initiated during the initial meetings dating back to January 2021. The official establishment of the EC was announced to citizens on March 28, 2021. Following this announcement, invitations to participate in a call for interests were disseminated through local social media and the Municipality's official website, resulting in the first wave of participant responses.

As introduced, part of incentives will contribute to the expansion of energy potential of CommOn Light through additional systems. Based on this assumption, new initiatives have already been launched such as:

- the installation of a 30-kW photovoltaic system on the municipal *Carabinieri* barracks, further enhancing the energy potential of CommOn Light;
- additionally, there are plans for the construction of an additional 20-kW system on municipal property, transforming it into *"BorgOstello."*

Through this initiative, the Municipality of Ferla has showcased its commitment to achieving key objectives outlined in the 2030 Agenda. Notable accomplishments include:

- ensuring access for all to economic, reliable, sustainable, and modern energy systems (Objective 7);
- reducing inequalities within and between countries (Objective 10);
- cultivating inclusive, safe, resilient, and sustainable cities and human settlements (Objective 11);
- promoting models of sustainable production and consumption (Objective 12);
- taking actions at all levels to combat climate change (Objective 13);
- protecting, restoring, and promoting the sustainable use of terrestrial ecosystems (Objective 15);
- strengthening implementation methods and revitalizing the global partnership for sustainable development (Objective 17).

CommOn Light demonstrates a replicable model for other communities across Italy. The project provides a blueprint for how localities can leverage diverse funding sources for sustainable energy projects.

The CommOn Light EC exemplifies best practices in Italy by combining collaboration, innovative funding strategies, environmental consciousness, clearly defined roles, community engagement, and a scalable model for national replication. This case serves as an inspiring reference point for future endeavors in sustainable energy development and community-driven initiatives.

4   Conclusions

The paper aimed at presenting an emerging strategy for the revitalization and decarbonization of Historical Villages.

The decline of HVs in Europe, particularly in Italy, has become a pressing issue due to factors like depopulation, economic challenges, and environmental risks.

In response to this, the establishment of ECs has emerged as a potential solution. ECs empower local communities to spearhead initiatives for building refurbishment, retrofit, and valorization aimed at decarbonization [18, 19].

These ECs are part of a broader trend driven by European legislation promoting renewable energy and community participation in energy production.

Despite facing challenges such as regulatory constraints and limited involvement of third parties, ECs have seen notable developments in Italy. Recent legislative decrees, including the RECs decree (DM 414/2023), offer incentives like non-repayable subsidies and tariff incentives to promote the development of ECs.

In this context, the CommOn Light Energy Community in the Municipality of Ferla, Sicily, stands as a best practice case. The project aims to reduce dependence on nonrenewable energy sources by implementing on-site renewable energy production and storage. Collaborative efforts between the municipality, academic institutions, and research projects have facilitated the establishment of CommOn Light.

Key features contributing to CommOn Light's success include clear delineation of roles, strategic financial planning, and an inclusive social structure. The community actively engages residents through various programs and transparent communication, fostering a sense of ownership and participation.

CommOn Light's achievements align with the objectives outlined in the 2030 Agenda, demonstrating its potential for replication in other Italian communities. By leveraging diverse funding sources and emphasizing community engagement, CommOn Light serves as a model for sustainable energy development and community-driven initiatives nationwide.

Acknowledgments The chapter is the results of a common reflection of the authors, based on their expertise in the specific fields of research.

Specifically, Emanuele Piaia authored the chapter's introduction and Sects. 2 and 3, while Silvia Brunoro wrote Sect. 4. The final revision of the work and the approval of the manuscript version to be published are attributable to all the authors.

This work was possible thanks to the FIRD2023 project *"Renewable energy communities— methodological perspectives and framework models"* funded by the Department of Architecture of the University of Ferrara (local research projects).

The authors also extend their gratitude to Mauro Guerra (Coordinator of *Borghi più Belli d'Italia* for the Emilia Romagna region and the national contact person for initiatives related to Energy Communities) for generously sharing invaluable insights, particularly regarding recent Italian experiences, with a special focus on the Ferla demonstration case.

References

1. DM 555 02/12/2016, Directive of Ministry of Cultural Heritage, cultural activities, and tourism 2017 – Year of the Italian Villages. https://www.beniculturali.it/comunicato/dm-554-02-12-201 6decreto-di-rimodulazione-dei-progetti-di-valorizzazione-2016. Last accessed 3 Feb 2024
2. Sacchetti, L., Piaia, E., Frighi, V., Spasari, I.: Revitalizing small historical villages through social, economic, cultural and energy efficiency asset. Tech. Ann. **1**(2), 1–21 (2023)
3. Ascione, F., De Rossi, F., Vanoli, G.P.: Energy retrofit of historical buildings: theoretical and experimental investigations for the modeling of reliable performance scenarios. Energ. Build. **43**(8), 1925–1936 (2011)
4. Minghui, G.E., MacGill, I.: Typology of future clean energy communities: an exploratory structure, opportunities, and challenges. Energy Res. Soc. Sci. **35**, 94–107 (2018)
5. Piselli, C., Fronzetti, C.A., Segneri, L., Pisello, A.L.: Evaluating and improving social awareness of energy communities through semantic network analysis of online news. Renew. Sust. Energ. Rev. **167**, 112792 (2022)
6. De Vidovich, L., Tricarico, L., Zulianello, M.: Community Energy Map. Una ricognizione delle prime esperienze di comunità energetiche rinnovabili. Franco-Angeli, Milano (2021)

7. Mazzarella, L.: Energy retrofit of historic and existing buildings. The legislative and regulatory point of view. Energy Build. **95**, 23–31 (2015)
8. AIEE (Associazione Italiana Economisti dell'energia) e Federmanager, Il ruolo delle Comunità energetiche nel processo di transizione verso la decarbonizzazione (2021)
9. EU, Directive (EU) 2018/2001 of the European Parliament and of the Council: https://eur-lex.europa.eu/legal-content/EN/TXT/PDF/?uri=CELEX:32018L2001&from=EN. Last accessed 22 Feb 2024
10. EU, Directive (EU) 2019/944 of the European Parliament and of the Council: https://eur-lex.europa.eu/legal-content/EN/TXT/PDF/?uri=CELEX:32019L0944&from=EN. Last accessed 22 Feb 2024
11. REScoop: Community Energy: A practical guide to reclaiming power https://www.rescoop.eu/toolbox/community-energy-a-practical-guide-to-reclaiming-power. Last accessed 1 Mar 2024
12. Van der Schoor, T., Scholtens, L.J.: Power to the people: local community initiatives and the transition to sustainable energy. Renew. Sust. Energ. Rev. **43**, 666–675 (2015)
13. Barroco F., Cappellaro F., Palumbo C.: Le Comunità Energetiche in Italia. Una guida per orientare i cittadini nel nuovo mercato dell'energia. Rapporto prodotto da Climate-Kic, Agenzia per lo Sviluppo Sostenibile, ENEA e Alma Mater Studiorum Università di Bologna. ed. ENEA, Roma (2020)
14. ARERA Bilancio Energetico Nazionale: https://www.arera.it/it/dati/bilancio_en.htm (2020). Last accessed 21 Feb 2021
15. Reis, I.F.G., Gonçalves, I., Lopes, M.A.R., Henggeler Antunes, C.: Business models for energy communities: a review of key issues and trends. Renew. Sust. Energ. Rev. **144**, 11101 (2021)
16. Candelise, C., Ruggieri, G.: Community Energy in Italy: Heterogeneous Institutional Characteristics and Citizens Engagement. IEFE, Center for Research on Energy and Environmental Economics and Policy (2017)
17. Blumberga, A., Vanaga, R., Freimanis, R., Blumberga, D., Antužs, J., Krastiņš, A., Ivars, J.I., Bondars, E., Treija, S.: Transition from traditional historic urban block to positive energy block. Energy. **202**, 117485. Elsevier (2020)
18. Ceglia, F., Marrasso, E., Roselli, C., Sasso, M.: Small renewable energy community: the role of energy and environmental indicators for power grid. Sustain. For. **13**(4), 2137 (2021)
19. Dal Cin, E., Carraro, G., Volpato, G., Lazzaretto, A., Danieli, P.: A multi-criteria approach to optimize the design-operation of energy communities considering economic-environmental objectives and demand side management. Energy Convers. Manag. **63**, 115677 (2022)

Reducing Energy Consumption Through Energy Monitoring Systems: A Case Study in Norway

Emma Zheng Liang, Sergey Paramonov, and Habtamu Bayera Madessa

Contents

1 Introduction

According to a report by the United Nations Environment Programme [1], the building and construction industry is responsible for 37% of global carbon emissions, underscoring its significant role in climate change. Further, L. Pérez-Lombard et al. [2] estimate that buildings contribute to 42% of energy consumption in Western countries, highlighting their substantial impact on energy demand. While numerous studies have investigated the relationship between building materials and energy consumption regarding environmental emissions, there has been less focus on real-time energy monitoring and its potential for reducing energy use. I. Knight [3] demonstrated that energy monitoring has significantly reduced energy consumption since 1995. Despite the widespread adoption of EMS, research has predominantly

E. Z. Liang (✉) · S. Paramonov
Rambøll Norge AS, Oslo, Norway
e-mail: emma.liang@ramboll.no; sergey.paramonov@ramboll.no

H. B. Madessa
Department of Built Environment, OsloMet-Oslo Metropolitan University, Oslo, Norway
e-mail: habama@oslomet.no

M. Kioumarsi, B. Shafei (eds.), *The 1st International Conference on Net-Zero Built Environment*, Lecture Notes in Civil Engineering 237,
https://doi.org/10.1007/978-3-031-69626-8_96

focused on industrial complexes, leaving a gap in studies that assess EMS in public buildings. Aman, Simmhan, and Prasanna (2013) highlighted that real-time insights from EMS are key for stakeholders to make informed decisions and develop proactive energy management strategies [4].

This chapter investigates a pilot project that implemented EMS in 34 public buildings within the Vestfold and Telemark County municipality in Norway. The objective of the project was to reduce energy usage by closely tracking and analyzing consumption patterns on a weekly basis through EMS. The system provides real-time monitoring, in-depth data analysis, and comprehensive reporting, offering a detailed view of energy consumption trends. This capability allows for a systematic approach to identifying inefficiencies and applying targeted interventions to improve energy performance.

By examining how energy is consumed and identifying patterns through interactive diagrams, the EMS system provides a framework for optimizing energy use. This approach has the potential to inform daily practices and improve the temporal distribution of energy consumption, leading to more efficient and cost-effective energy management.

The structure of this paper is as follows: Sect. 2 provides background information on EMS in Norway. Section 3 details the case study involving 34 public buildings in the Vestfold and Telemark county municipality. Section 4 discusses the findings from the case study, including practical barriers encountered during the implementation of the EMS system, and concludes with implications for future research and policy recommendations.

2   Energy Monitoring Systems (EMS) in Norway

EMS play a critical role in monitoring, analyzing, and optimizing energy costs in mostly commercial and industrial buildings. EMS typically include sensors, meters, and communication software designed to track energy consumption patterns. The software component provides data analytics, visualization, and reporting capabilities, enabling building managers to identify operational inefficiencies and potential opportunities for improvement, leading to enhanced energy efficiency. This approach has been shown to be effective in large public buildings, as demonstrated by Zhao, Zhang, and Liang in their study of EMS in China [5].

The Norwegian government has recognized the strategic value of EMS in energy management and has implemented policies to foster their adoption. Organizations such as Enova offer subsidy programs to encourage the use of EMS, promoting energy efficiency and sustainability across the country [6]. By leveraging EMS, building managers can monitor and optimize energy consumption, contributing to Norway's national sustainability goals and helping to reduce carbon emissions.

In accordance with Norwegian Planning and Building Act of 1997, all buildings are legally required to maintain an energy and power budget, which EMS can help track. To ensure standardization and compliance, EMS systems must be designed

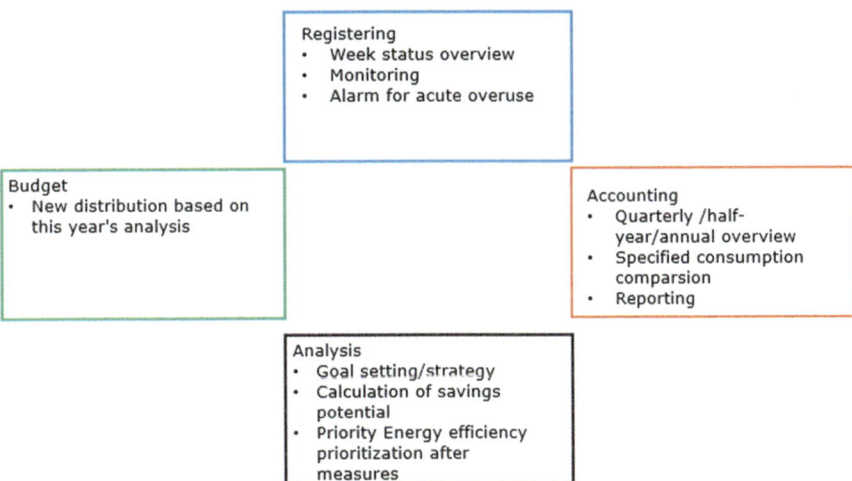

Fig. 1 Energy Monitoring Systems in Norway (Enova handbook 2004:3, page 4) translated in English by the author

according to a specific template, as outlined in "Enova Handbook 2004:3" [7]. This legal requirement also underscores the critical role of EMS in achieving energy efficiency standards in Norway. Fig. 1 suggests how a systematic energy monitoring should be conducted and organized. Experiences in Norway emphasize the importance of organizations systematic approach that is integrated into management and tailored to the routine reporting.

EMS platforms also facilitate transparent energy monitoring, which is essential for identifying consumption patterns, operational inefficiencies, and opportunities for energy optimization. The combination of real-time data collection and advanced analytics empowers building managers to implement targeted energy-saving measures, ultimately leading to cost savings and reduced environmental impact as outlined by Burgess and Nye [8, 9].

3   Case Study: A Pilot Project for Vestfold and Telemark County Municipality

3.1   The Background of a Pilot Project: The Plan for EMS

In 2021, it was decided that all energy usage in the county (Vestfold and Telemark county municipality) needs to be monitored on a regular basis. Energy monitoring was initiated at all county facilities (see the list of buildings in Table 1), as reported by Vestfold and Telemark County Council in a press release. The goal was to reduce energy consumption by 3.4%, but the actual savings amounted to 9.8% compared to normal usage, totaling 9.8 million Norwegian kroner.

Table 1 The list of the buildings for the case study

Sande high school	Holmestrand high school	Horten high school	Re high school	Kompetansesbyggern
Færder High School	Greveskogen High School	Melsom High School	Nøtterøy High School	Sandefjord High School
Sandefjord FHS	Th. Heyer-dahl High School	Haugar	Servicebygget Midgard	Gildehallen Midgard
Midgard Vikingsenter	Midgard Elfordeling	Hinderveien 10	County Hall Skien	County Hall Tønsberg
Skogmo High School	Notodden High School	Rjukan High School	Kragerø High School	Bø High School
Porsgrunn High School Dep. South	Nome High School Dep. Søve	Nome High School Dep. Lunde	Hjalmar Johansen High School	Porsgrunn High School Dep. Nord
Bamble High School Dep. Croftholmen	Skien High School	Vest-Telemark High School Dep. Seljord	Vest-Telemark High School Dep. Dalen	Bamble High School Dep. Cramyr

The Process for Applying EMS in the Case Study The process of implementing an EMS for the case study involved several key steps. To ensure comprehensive monitoring of energy consumption, energy meters were installed at facilities that lacked existing online energy delivery connections. This approach allowed for the collection of energy usage data from a broader range of sources.

Once installed, all energy meters were integrated into a common energy monitoring platform. This integration provided a unified system for real-time tracking and analysis of energy consumption across the county. The common platform enabled stakeholders to access data from all monitored facilities in a centralized manner, facilitating the identification of energy usage patterns and trends.

The EMS process, as illustrated in Fig. 2, included data collection, integration, and analysis, allowing for more effective energy management. By consolidating energy data into a single platform, the system made it easier to conduct comparative analysis, monitor performance, and implement energy-saving measures based on insights gained from the data.

3.1.1 The Energy Target of the Project

The project established energy targets for each county facility, supported by a follow-up plan. Weekly monitoring of energy consumption and quarterly meetings with facility managers were implemented to ensure adherence to these targets. Any deviations were promptly identified and addressed.

The energy target for 2022 was set at 3.4% below the energy consumption levels of 2021. These energy monitoring efforts continued into 2023, as reported by the county council, with regular tracking and management strategies to maintain and improve energy efficiency across the facilities.

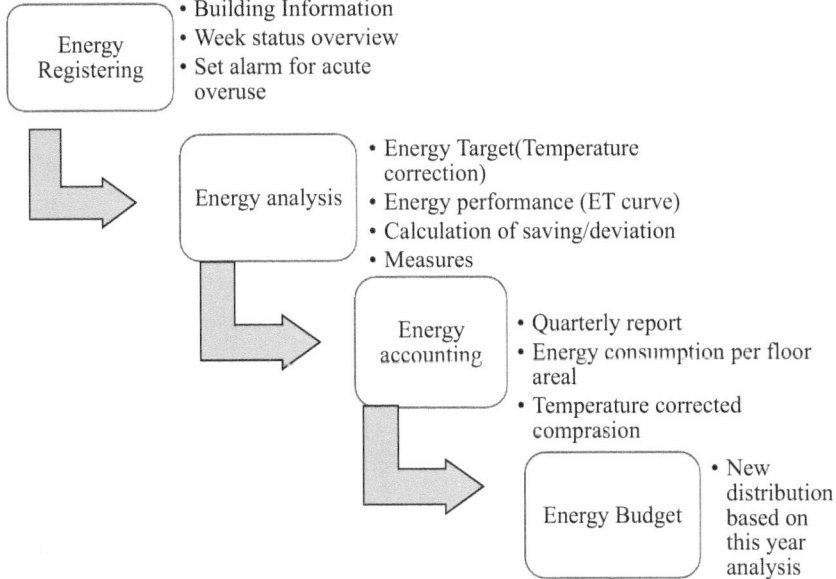

Fig. 2 The process of energy monitoring system of the case study

The assessment of additional metering requirements continued into 2023, focusing on new projects involving solar panels and heat pumps. As a result, more meters were installed throughout 2023, reflecting a proactive approach to energy monitoring. The EMS software or platform, provided by Energinet [10], displayed data from all energy sources, enhancing transparency and enabling comprehensive energy management. Certain buildings operated their own EMS in addition to Energinet, providing localized control over energy consumption.

To ensure precise energy monitoring and response, energy–temperature (ET) curves were adjusted to reduce false alarms caused by excessive energy consumption. This proactive adjustment aligns with the principles outlined in NS-EN 17267:2019, which emphasize the importance of flexible monitoring and the ability to respond to energy data in real time [11].

3.1.2 Energy Analysis: The Interactive ET Curve

The term "ET curve" refers to the energy–temperature curve. The ET curve, a practical tool in Norway, is crucial for energy monitoring, showing how a building's energy usage correlates with outdoor temperature. It is unique to each building according to its energy usage. All deviations from the ET curve exceeding control limits should be documented in terms of total energy consumption and currency. This involves multiplying the deviation in the unit of kWh/m^2 by the floor area to obtain energy usage in kWh. Fig. 3 illustrates an ET diagram with total energy

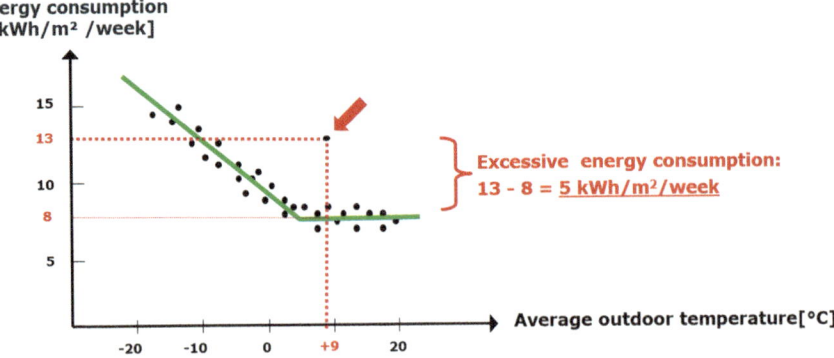

Fig. 3 ET curve example from the case study

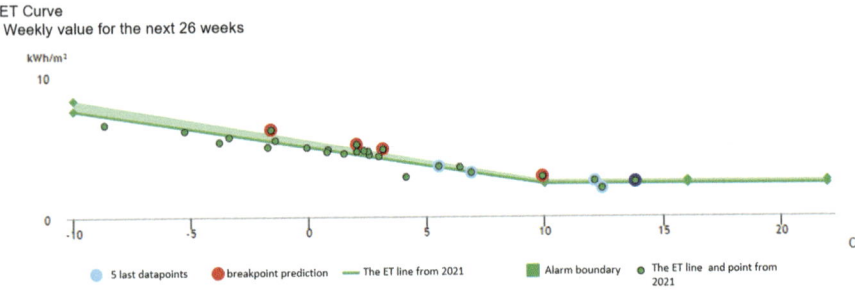

Fig. 4 The ET curve for the weekly value for the next 26 weeks

consumption per week, facilitating communications about energy usage conse-quences. For instance, a point marked with red point representing 13 kWh/m^2 usage corresponds to approximately a 5 kWh/m^2 deviation from the baseline (8 kWh/m^2).

Figure 4 illustrates how a sudden increase in energy consumption may indicate potential system failure or other operational issues. The cause of the sudden devia-tion will be identified, and thus specific measures was to be taken to restore the energy performance of the building to the predicted curve as Fig. 4 shows.

3.1.3 Comparing Energy Consumption Across Different Years: Temperature Correction

Even under identical operating conditions, energy usage will vary from one period to another. Energy usage in one period, such as January, cannot be directly compared to the same period in the previous year due to different outdoor temperatures. There-fore, energy usage must be temperature-corrected prior to intercomparison. By reflecting the outdoor temperature fluctuations in calculations, it enables the comparison of energy use for the same period across different years. This helps the stakeholders spot systematic drifts and errors such as leaks.

In the case study, the "Quarterly report" featured interactive updates on energy consumption, comparisons with previous periods, and strategies for energy saving. This facilitated direct communication with facility managers and decision-makers, promoting actionable solution toward energy efficiency.

Energy Accounting: Quarterly Report One of the buildings within County Hall, Tønsberg, with a heated floor area of 6,400 m^2, is powered by a combination of electricity and district heating. The structure was built in two phases: the first part in 1956 and the latter part in 1994.

The energy target for the first quarter of 2023 (January, February, and March), corrected for temperature variations, was set at 370,706 kWh. However, actual energy consumption during this period totaled 397,038 kWh, exceeding the target by 26,332 kWh. This deviation indicates that the building used 7% more energy than projected, suggesting negative energy performance and reduced efficiency during the first quarter.

The data presented in Table 2 further illustrate these findings:

These values demonstrate that the deviation for January and February contributed significantly to the overall increase in energy consumption for the quarter, while the reduction in March was insufficient to offset the earlier excess. The fluctuations suggest a need for targeted energy-saving strategies in the initial months of the year to meet the desired targets.

Additionally, despite this shortfall against the quarterly target, a year-over-year comparison presents a more favorable outcome. The temperature-corrected energy consumption for the first quarter of 2022 was 404,388 kWh, while the same period in 2023 saw actual energy usage reduced to 397,038 kWh, representing a reduction of 7,350 kWh. This change indicates a 2% reduction in energy consumption from 2022 to 2023, suggesting that overall energy efficiency has improved over the previous year, likely due to effective energy management practices.

Quarterly reports were compiled four times a year, offering a broader analysis of energy consumption patterns and trends. Facility manager meetings were also held on a quarterly basis, promoting collaboration and the sharing of best practices. Additionally, weekly energy follow-up reports were generated, highlighting ongoing energy usage and allowing facility managers to identify areas for improvement. Regular meetings, often held via Microsoft Teams, facilitated communication among facility managers and allowed them to discuss and propose tailored measures

Table 2 The quarterly report on the building of County Hall Tønsberg

Time (month)	Energy consumption	Target	Deviation(kWh)
January 2023	151445	125550	25895
February 2023	124359	119720	4639
March 2023	121234	125436	−4202
Quarter (tot)	**397038**	**370706**	**26332**

to reduce energy consumption. This structured approach to energy management underscores the importance of consistent monitoring and the flexibility to adapt to changing energy requirements.

In summary, the data-driven metrics and outcomes from the first quarter of 2023 indicate significant progress in energy consumption management for the building, exceeding both set targets and reducing energy use compared to the previous year. This progress not only demonstrates the building's trajectory toward greater energy efficiency but also underscores the positive impact on sustainability and reduced environmental footprints.

The achievement of these energy efficiency targets was facilitated by a combination of national support from organizations like Enova, clear energy-saving objectives, and structured, interactive EMS data monitoring through weekly tracking by Energinet. This systematic approach ensures effective communication among stakeholders, early detection of system failures, and a proactive response to optimize energy usage.

Overall, the case study exemplifies how a concerted effort in energy monitoring and management can contribute to national climate goals. The systematic approach detailed in the study serves as a model for other public and commercial buildings seeking to improve energy efficiency and reduce their carbon footprint. By adopting similar practices, buildings can not only meet but exceed energy-saving targets, supporting the broader objectives outlined in the National Climate Policy Action Plan [6].

4   Conclusion

The case study presented in this paper illustrates an effective, systematic approach to implementing an EMS to visualize and promote multidisciplinary collaboration from the national level down to the community level. By leveraging the EMS platform, various stakeholders, including facility managers, engineers, and policymakers, can identify the root causes of energy inefficiencies and develop customized measures to optimize energy use. This collaborative approach helps achieve energy-saving targets through a combination of top-down guidance and bottom-up initiatives.

The study demonstrates how the EMS system serves as an interactive platform that supports both regulatory compliance at the national level and cooperative efforts among local stakeholders. This dual approach fosters collaboration among facility managers, engineers, and local politicians, enabling them to align efforts and achieve common energy-saving goals.

However, several limitations are inherent in EMS systems. Energy consumption is influenced by a range of meteorological factors beyond just outdoor temperature, such as wind, solar radiation, and cloud cover. This variability necessitates setting control limits, typically within $\pm 10\%$, to account for expected deviations. The specific control limits vary for each building, depending on factors like age, design, location, and intended usage.

Another challenge arises when adjusting EMS data for external factors like sunlight exposure, shading, outdoor lighting, and wind conditions. While specialized monitoring systems may consider these variables, the added cost, effort, and complexity can be prohibitive. This can be particularly relevant for buildings aiming to meet rigorous standards, such as energy-efficient passive houses, where even minor fluctuations can impact overall energy performance.

Moreover, not all EMS systems provide features for calculating CO_2 emissions based on different energy sources. Since energy supply types can significantly impact carbon footprints, it is crucial for EMS platforms to address both energy efficiency and CO_2 emissions reduction. This requires sophisticated software capable of integrating energy use data with emissions metrics, allowing users to develop comprehensive strategies for reducing both energy consumption and environmental impact.

In summary, the EMS system provides a valuable framework for addressing energy efficiency through a combination of top-down and bottom-up approaches. The ability to collaborate across different levels of governance and technical expertise is key to achieving sustainable energy use. Despite its limitations, the EMS system serves as a critical tool for driving energy efficiency and reducing carbon emissions in the building sector.

References

1. Building Materials and the Climate: Constructing a new future : https://www.unep.org/resources/report/building-materials-and-climate-constructing-new-future
2. Pérez-Lombard, L., Ortiz, J., Pout, C.: A review on buildings energy consumption information. Energy Buildings. **40**(3), 394–398 (2008)
3. Knight, I.: Energy monitoring and its effect on energy consumption at the University of Wales, College of Cardiff (UWCC). Building Service Eng. **16**, 1–7 (1995)
4. Aman, S., Simmhan, Y., Prasanna, V.K.: Energy management systems: State of the art and emerging trends. 2013 IEEE Comm Surveys Tutor. **15**(3), 1447–1462 (2013)
5. Zhao, L., Zhang, J.-L, Liang, R.-b: Development of an energy monitoring system for large public buildings. Energy Buildings. **51**, 1–6 (2012)
6. Norwegian Ministry of Climate and Environment. National Climate Policy Action Plan: https://www.regjeringen.no/contentassets/a78ecf5ad2344fa5ae4a394412ef8975/en-gb/pdfs/stm202020210013000engpdfs.pdf (2019)
7. Enova. Energioppfølging i næringsbygg – en innføring. Enova håndbok 2004:3, 33 s (2004)
8. Energinet is energy management software (EMS) for energy. https://kiona.com/no/produkter/energinet
9. Burgess, J., Nye, M.: Re-materialising energy use through transparent monitoring systems☆. Energy Policy. **36**(12), 4454–4459 (2008)
10. NS-EN 17267:2019 Plan for energimåling og energiovervåking - Utforming og iverksetting - Prinsipper for innsamling av energidata
11. Local paper on the case study: https://www.reavisa.no/2023/01/25/aktiv-energiovervaking-lonner-seg-i-millionklassen/

Sustainable Service Ecosystems in Positive Energy Districts: A Conceptual Framework to Steer Long-Term Impacts

Anna Viljakainen, Katri Valkokari, Mari Hukkalainen, and Tytti Nikunen

Contents

1 Introduction

Grand societal challenges (e.g., climate change, pollution, resource depletion) are complex and multi-dimensional problems that cannot be solved with incremental improvements but require fundamental shifts in sociotechnical systems—including the energy system. These are referred to as sustainability transitions (STs) [1]. Positive Energy Districts (PEDs) have emerged to drive STs in the energy sector. Understanding how STs take place in these complex sociotechnical systems is however a challenge. This is because we lack a thorough understanding on the dynamics and interconnections between the various components of the system [1]. STs in complex systems are hindered by the institutionalization, path dependencies, and lock-ins [2], and enabled in the relational processes between public, private, and non-profit actors [3, 4].

A service ecosystems perspective is selected in this paper to address STs in the energy sector. Focusing on the service ecosystem made of overlapping and nested

A. Viljakainen (✉) · K. Valkokari · M. Hukkalainen · T. Nikunen
VTT Technical Research Centre of Finland, Espoo, Finland
e-mail: anna.viljakainen@vtt.fi

© The Author(s) 2025
M. Kioumarsi, B. Shafei (eds.), *The 1st International Conference on Net-Zero Built Environment*, Lecture Notes in Civil Engineering 237,
https://doi.org/10.1007/978-3-031-69626-8_97

ecosystems [5] allows us to tackle the complexity of STs [6]. Service ecosystems are "self-containing, self-adjusting system of resource-integrating actors connected by shared institutional arrangements and mutual value creation through service exchange" [7, p. 161]. By emphasizing the logic of value, the business model (BM) approach allows the study of value co-creation in networked environments [8]. We use the BM as unit of analysis to capture the value creation and capture arising from multiple sources [9]. BMs play an important role in the transitions toward low-carbon energy systems, given they can either hinder or drive change in sociotechnical systems [10]. Understanding their distinct role in the transformation of energy systems is therefore fundamental [11].

EU's project initiatives and the energy and climate targets have had a significant role in the emergence of PEDs as a concept. Originally, the concept was initiated in the EU's Strategic Energy Technology Plan (SET Plan) Action 3.2 in 2018 [12], including a target for establishing 100 PEDs by 2025. Several definitions in the literature and the European research field prevail. All these concepts however highlight the role of energy-efficient districts that produce renewable energy, use energy flexibility to optimize the energy use and flows within the area, and integrate to the external energy systems. In this work, the definition from JPI Urban Europe [13] is used as a starting point, where PEDs are identified as "energy-efficient and energy-flexible urban areas or groups of connected buildings, which produce net-zero greenhouse gas emissions and actively manage an annual local or regional surplus production of renewable energy. They require integration of different systems and infrastructures and interaction between buildings, the users and the regional energy, mobility, and ICT systems, while securing the energy supply and a good life for all in line with social, economic, and environmental sustainability" [13, p. 4]. It is further added that each PED is expected to "find its own optimal balance between energy efficiency, energy flexibility and local/regional energy production on its way towards climate neutrality and energy surplus" [13, p. 8]. Overall, three types of PEDs are identified [14, 15]: (1) PED autonomous (i.e., energy self-sufficient districts within geographical boundaries); (2) PED dynamic (i.e., districts within geographical boundaries that generate renewable energy over their own demands; dynamic exchanges with the hinterland to compensate momentary balance differences), and; (3) PED virtual (i.e., those that use virtual renewable energy systems and storage outside their geographical district boundaries).

Transformations taking place at a micro level may induce broader STs at the macro level [4, 16]. Therefore, we propose a conceptual framework intended to support the transition and renewal of PED service ecosystems at micro- and meso levels and demonstrate its potential by applying it to an illustrative case study on a PED located in Espoo, Finland, that focuses on dynamic exchanges of energy with power grid (PED virtual). Espoo is the second largest city in Finland (300,000 residents) with overarching sustainability objectives to reach carbon neutrality by 2030 and reduce emissions by 80%. The Espoo PED is treated as a service ecosystem made of four overlapping and nested ecosystems, namely: (i) Energy positive buildings and district; (ii) Renewable energy production, distribution, and storage; (iii) Energy management (with Virtual Power Plant); and (iv) Electric mobility.

2 Conceptual Framework

We elaborate a conceptual framework mapping the key elements of PED business models from a service ecosystems perspective. As existing business model blueprints are less suited to examine the interdependency and transformation of actors in ecosystems [17], we conducted a literature review to build a conceptualization grounded in relationships between key concepts (Fig. 1). This framework acts as a micro- and meso-level model, describing how the different actors in PED ecosystems create, deliver, and capture value. It attempts to feature both the service offerings available in the business ecosystems, as well as the ecosystem actors and their interrelatedness. The BM framework we propose serves a dual purpose [18, 19]: it acts as (i) a *static blueprint* to describe how the different actors operate and the activities they employ to generate value in the nested ecosystems (i.e., micro level); and (ii) a *transformational tool* to address changes in energy systems (i.e., meso level). The framework has been further developed into an online tool to foster its scalability on a city level and/or other PEDs.

Our research process followed an abductive approach [20], where we combined empirical and theoretical inputs with emphasis on making interpretations and comparisons to inform both theory and practice [21]. Service design was adopted as method to enable the tackling of a grand societal challenge with systemic and long-term lenses, while supporting the co-creation of knowledge and mindset change in STs [22]. Core partners from the Espoo's PED service ecosystem were interviewed. Due to a small size of data sample ($N = 4$), the interview findings are used for illustrative purpose only. The framework was externally validated with the SPARCS research project partners (funded by EU Horizon 2020 under Grant Agreement No. 864242), including the interviewees. The practice of submitting knowledge

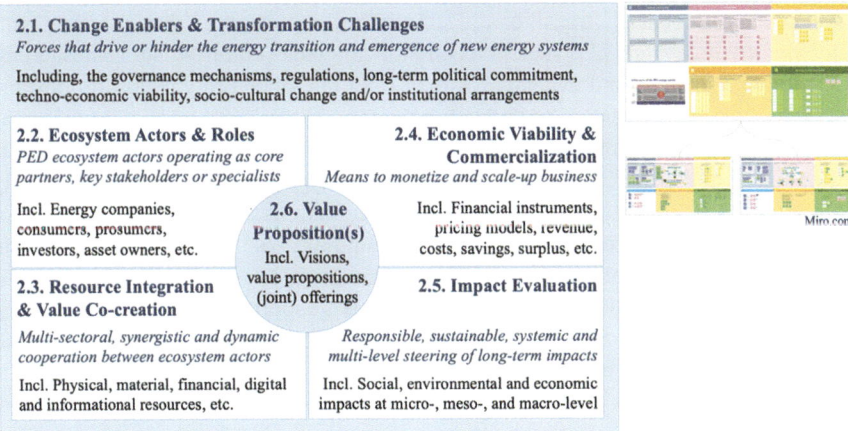

Fig. 1 Conceptual framework elements and online tool (as illustration) to assess and address changes in PED (nested) ecosystem business models

claims for discussion and falsification to members in relevant communities is referred to as communicative validation [23].

In the following sections we discuss the conceptual elements of our framework by first exploring the change enablers and transformation challenges (Sect. 2.1) supplemented with illustrative examples from our case context. Then, we identify the ecosystem actors and their roles (Sect. 2.2), and the mechanisms for resource integration and value co-creation (Sect. 2.3). Thereafter we discuss the economic viability (Sect. 2.4) of PED service ecosystems and the viewpoints for impact evaluation (Sect. 2.5). We end the chapter with discussion on value propositions (Sect. 2.6).

2.1   Change Enablers and Transformation Challenges

Energy transition is a complex challenge, which is why the enablers and challenges are widely studied in the broader context of new energy systems. Overall, they fall into four broad categories. The first category relates to the *governance mechanisms, regulations, and (geo-)political* forces, which emphasize the long-term political commitment, stable regulatory frameworks, clear vision and net-zero targets, as well as incentive mechanisms as key drivers (and challenges thereof in their absence) to enable the alignment of priorities and actor engagement in energy transitions [10, 11]. The second category discusses the *techno-economic* perspectives, for example, external financial support and funding to ensure the economic viability for new energy systems, alignment of public and private profitability and sustainability targets, adoption of emerging technologies, and the interoperability between the physical and digital infrastructures [24–26]. The third category centers around *sociocultural* aspects, with emphasis on the political and regulatory forces to foster social acceptance, skills, behaviors, confidence, and trust toward new energy systems and the energy transitions [27, 28]. The final category discusses the *institutional arrangements* [29] such as a common language, cultures, and values, cross-sectoral partnerships and evaluation targets, and interoperability of institutions and systems as either drivers or hinderers for change.

In Espoo, the regulatory and political mechanisms play a crucial role in the adoption of new solutions and infrastructure that support the city's carbon neutrality objectives. City planning guided by public regulations sets the requirements for integrated energy systems, for example, in buildings. Sanctions will follow from neglecting the directive regulations. The requirements have proliferated in the past couple of years due to increased tensions in the geopolitical environment, setting pressures to speed up the decision-making and preparation processes in organizations characterized by hierarchical and sectoral silos. In the domain of e-mobility, a great deal of effort is going into promoting sociocultural change. For example, shared mobility solutions (e.g., e-scooters) are introduced, to promote attitude shift and the penetration of e-mobility. Yet, the development of the transport system has faced great challenges due to the COVID pandemic, and the price fluctuations of fossil fuels. Espoo has attempted to solve these issues by placing public transport nodes and mobility hubs strategically to locations that will support private companies and inhabitants to boost electrification.

2.2   Ecosystem Actors in Energy Systems

Generally, the literature classifies distinct roles that actors operating in energy systems can adopt either individually, or as combination [30, 31]. *Producers* represent the energy production companies (e.g., utility companies, district heating) and local energy producers (e.g., operators of solar/wind power plants), energy *service providers* (e.g., energy management, control, demand response, energy saving companies), and *energy distributors*. *Consumers* (e.g., households) form another category. *Prosumers* are typically discussed in the literature on energy communities as the residential (i.e., small-scale households) or large-scale actors (i.e., industrial entities) operating as energy consumers, producers, and flexibility providers. *Investors* are public organizations (e.g., government subsidies, municipalities, NGOs) and private entities (e.g., local businesses, inhabitants). *Asset owners* typically include technology providers (e.g., platform operators, ICT providers). Literature further separates actors into orchestrators and followers [10]. *Orchestrators* are key players or decision-makers that ensure the overall health of the business ecosystem and/or control activities between actors, for example, by creating incentives and setting rules for knowledge sharing and interaction [32]. *Specialists* are followers or niche-players that employ their specialized capabilities in running the system.

By drawing from these disciplines, our framework categorizes three main actor groups for PEDs, namely, core partners, key stakeholders, and specialists. *Core partners* are orchestrators that ensure the overall viability and interoperability of the PED service ecosystem. They also strengthen social acceptance and action. In Espoo, these actors are energy management system providers, network operators, and the city organization at large (i.e., the sectors of traffic management and design, urban planning, etc.). *Key stakeholders* are actors that ensure the operability and functioning of the nested ecosystems (i.e., positive energy buildings, renewable energy, e-mobility, and energy management in the Espoo context). For example, energy and district heating distributors, energy infrastructure providers, energy storage operators, energy market providers, distributed energy asset owners, and land/real-estate owners, among others. The third category is *specialists* that employ specialist capabilities periodically, for example, for the purpose of research, development, and innovation activity. Actors falling in this category in our illustration are consultants, advisors, research institutes, universities, construction, and engineering companies.

2.3 Resource Integration and Value Co-creation

Literature on energy systems generally differentiates between distinct resource types [10, 11]: (1) *physical or material* resources represent grid infrastructure, electricity supply, renewable energy (e.g., groundwater, biomass), or physical equipment (e.g., energy storage, solar panels); (2) *financial* resources are, for example, capital from financial markets, federal or municipal subsidies, loans, or investments from local residents and businesses; (4) *digital* resources, for example, data or digital networks, and; (4) *information* resources, for example, specialist knowledge on local energy potentials. These resources, then, are shared and integrated through different (often technological) means—smart meters, wireless technologies, (open) data platforms, and grid infrastructure, among others. Overall, collaboration between local inhabitants, municipalities, and private companies, as well as the ability to combine different technological solutions are seen crucial to achieve efficiency, flexibility, and profitability in PEDs [33].

The framework presented in this study emphasizes that actors in service ecosystems are dynamic resources that co-create value and co-produce the service as part of their network [34]. The connectivity between ecosystem actors in smart energy systems permits the analysis of these interdependencies, roles, network positions, and power structures that drive continuous innovation and value co-creation [24]. This connectivity also nurtures a complex interaction between the different actors that is a promoter of systemic change and innovation [2]. In specific, the fostering of knowledge, skills, and innovation is effected in iterative processes where the different ecosystem actors integrate both own and stakeholder resources (i.e., effectuation; [35]). Interactions are influenced by the institutional arrangements (e.g., norms, rules, practices, beliefs) that can either drive or hinder value co-creation [7]. In other words, from a sociotechnical transition point-of-view, transformation of complex energy systems take place through value co-creation and resource integration between actors beyond system borders.

In our illustrative case, a centralized energy management system is in place to measure and control the energy use in the PED service ecosystem. By combining the loads available for demand response (e.g., shifting the time of energy demand) from different sources, the system enables the PED service ecosystem to act as Virtual Power Plant (VPP). Integrated energy systems—including distributed renewable energy production, energy storage, and district heating and cooling—provide energy for the smart buildings and e-mobility as needed. Often, however, interoperability and optimization may cause challenges in a multi-stakeholder environment, with each public and private actor influenced by their technological readiness, beliefs about the future developments, and expectations about the impacts.

2.4 Economic Viability and Commercialization

It is emphasized that actors of new energy systems should continuously explore new business opportunities and sources of finance arising from the private and public sectors to ensure their viability in the long-term (e.g., private capital investments, public subsidies, loans, crowd funding; [25, 28]). Commercial organizations are seen to hold an important role in improving the economic viability of decentralized energy systems by creating more appealing business cases, attracting outside funding, and decreasing operating expenses [30]. The lowering of costs is a key area to increase the viability of new energy systems. Here, the discussion centers around measures that increase energy efficiency and sufficiency—for example, smart energy management, self-consumption, and reduction of peak consumption [11]. The monetization mechanisms separate between revenues from energy services (e.g., mobility, heating, electricity and/or energy management services) and revenues from external services (e.g., balancing power, energy aggregation and flexibility, data valorization, and/or market trading; [31]).

Our framework puts emphasis on the bundling of the various means in the nested ecosystems to reach full operational capacity in the long-term. It looks at the roadmap for commercialization. The PED service ecosystem in Espoo interconnects and commercializes local de-centralized and centralized energy production solutions into well-functioning VPP, offering customers energy solutions as-a-service and enabling the flexibility toward the grid. Therefore, a pool of monetization mechanisms is operational (i.e., private funding, service fees, flexibility market income) due to a strong presence of private organizations, without reliance on public funding. Along with the VPP, new operating models in the production, consumption and distribution of energy are/will be introduced, to better manage energy peaks and energy saving efforts.

2.5 Impact Evaluation

Energy transition puts emphasis not only on economic, but also societal and environmental wellbeing. PEDs aim to facilitate the energy system transformation from a holistic perspective, by approaching the socioeconomic, technical, environmental, political, and institutional aspects of change [33]. This type of perceptual change highlights the role of innovation activities in the society, and the way in which we evaluate their impacts [36]. Evaluation of innovations considers their societal value that promote long-term change in STs [37]. The goal of impact evaluation is to reduce uncertainties, improve effectiveness, and support decision- and policymaking [38], while making purposeful measures for the intended outcomes of specified interventions in defined societal contexts [39]. Multiple actors are in the core of a systemic view on innovation: value is created in interactions at multiple levels [40], and guided by the norms, values, and beliefs of the actors (i.e., institutional arrangements; [41]). The traditional technological and financial evaluation approaches do not fully capture this multiplicity and systemic nature of innovations [42] that take place in energy transitions. For these reasons, our framework adopts a multi-dimensional approach by looking at the societal impacts in parallel with the more traditional techno-economic characteristics of innovations [37, 43], while considering the shared institutional arrangements among actors at the different levels of aggregation (micro-, meso-, and macro levels; [40]).

In Espoo, the goals and performance of the City Council and sectoral divisions are evaluated with key performance indicators (KPIs) derived from the UN's Sustainable Development Goals (SDGs). This indication system codifies the targets set to steer economic, social, and environmental sustainability performance within the energy system ecosystems. On a meso level, they steer the development efforts at the level of PED service ecosystems, and at micro level, align organizational targets to operate as part of nested ecosystems. For example, Espoo's target in electric mobility is to co-create an infrastructure for e-vehicles together with public and private organizations in partner networks. Here, the macro-level impact indicators detect changes in the share of electric and low-emission cars in total vehicle fleet and new-registrations, meso-level indicators enforce increases in municipal e-car infrastructure, and micro-level indicators increase in residents' satisfaction with road networks and transport routes. These targets and indicators, then, drive the research, development, and innovation activity of individual actors who wish to operate in the Espoo PED area.

2.6 Value Proposition(s)

The value proposition points the ways in which ecosystem actors contribute to the value creation of the customer, while also generating new resources and value for themselves [41]. The role of the value proposition in service ecosystems is to offer opportunities for co-creation and resource integration between the different social and economic actors, rather than act as the proposal for a service offering per se [44]. The relationships between ecosystem actors are founded on value propositions [45] suggesting that the design of new value propositions can improve existing offerings, create new offerings, and/or reconfigure the ecosystem [46]. Multi-actor interdependencies connect individual offerings under joint value propositions [47], while each actor holds distinct capacities to innovate the joint value proposition by adding novel offerings that serve different customer needs.

In Espoo, the overall vision is to reach carbon neutrality, reduce emissions, and generate surplus energy to the district energy systems, by using renewable energy sources (e.g., solar power, geothermal wells), appliances (e.g., sensors, smart meters), and by controlling the electric loads (e.g., lighting, HVAC, and elevators). The value propositions of ecosystem actors center around increasing energy efficiency, flexibility, and self-sufficiency through distributed renewable energy production, while reducing the overall energy consumption, and lowering the carbon footprint. For example, the value proposition for the VPP is the provision of affordable, sustainable, and reliable energy, while managing energy peaks and energy saving efforts by having local energy production solutions under one management system. The vision for e-mobility is to develop an e-mobility hub for large-scale EV charging systems and multiple EV types. The value proposition, then, is a cost-efficient and convenient solution that serves everyone with an e-vehicle. The joint value proposition for all nested ecosystems is however, ultimately, a good customer experience. PEDs are created for *people*. At its core is a pleasant experience of a place where people (and businesses thereof) which to reside and recreate.

3 Discussion

There are fundamental changes in how energy systems will be operated in the future to enable energy transitions and the development of PEDs. Municipalities and local authorities hold a strategic role in the transition toward renewable energy and local energy systems [30]. Transformations are taking place across siloed organizations and beyond traditional organizational boundaries [48]. BMs hold a vital—although complex—role in the ST of sociotechnical systems. Successful transition in the complex energy systems compels alignment in the different BMs, and co-evolvement of ecosystem actors in new partnerships and resource reconfigurations [10]. For example, tailor-made energy systems and individual system elements must be incorporated to function at the local levels. Transformation at the system level is difficult, requiring abilities to depict the path-dependencies and lock-ins in the present sociotechnical systems [2]. Dealings with such complex transitions necessitate increased flexibility, interconnectivity, cooperation, and decentralized governance mechanisms [42].

Our main theoretical contribution is to apply systems thinking that is fundamentally multidisciplinary [40] to the energy sector, by combining the theories on innovation and sustainability transitions. This allows us to examine the synergies between the two fields. By doing so, we acknowledge that the energy markets are shaped via activities of different actors and formed in systemic processes [2]. We apply the BM as a transformational tool to address change [18, 19] in service ecosystems around PEDs. Accordingly, we depict the managerial opportunities to influence value co-creation [8] and drive STs in PED service ecosystems. Taking this perspective is important, as business model innovation (BMI) is highlighted to contribute toward a complex change process in energy transitions [26]. They not only challenge the prevailing institutional arrangements and industry structures [49], but also the interactions between the public and private actors in PED ecosystems. A networked approach in energy transitions depicts the interconnectedness of the different actors in value co-creation. Therefore, our framework adopts a multidimensional approach [43] and analyses the interaction in PED service ecosystems at different levels of aggregation (micro- and meso levels; [40]). By doing so, we aim to make visible how the value in PED service ecosystems is co-created in complex long-term processes.

4 Conclusions and Future Research Avenues

Effective business in PEDs requires seamless cooperation between different private and public actors. We find this interoperability and optimization to be challenging in a multi-stakeholder environment, with each public and private actor influenced by their technological readiness, beliefs, and expectations. PEDs however offer a mean to combine the core elements in energy transitions. They enhance energy efficiency, minimize the energy demand, maximize the energy production locally from renewable energy sources, and balance the demand and production through energy flexibility. These core elements together necessitate the adoption of a holistic view to the provision of sustainable energy when needed. Interoperability of both energy and ICT systems is central. Implementing such complex systems requires understanding of the bigger picture and looking beyond individual business models of specific organizations.

The goal of this paper is to concretize what makes sustainable energy systems commercially successful and offer a comprehensive tool for private and public organizations—including cities—to assess and drive transformation in PED service ecosystems. A framework is created to assess the PED service ecosystem business models: (i) transformation enablers and challenges; (ii) roles of focal actors operating in PEDs; (iii) opportunities for value co-creation and resources integration; (iv) elements of viable business; (v) long-term impacts that unfold; and (vi) value propositions that increase the actors' innovation capacity. We apply a transformational approach and value-based view to the BM framework [19] in the context of STs. BMs are seen to hold a fundamental role in the transition toward low-carbon energy systems [10, 11], but existing BM blueprints are less suited to examine the interdependency and transformation of actors in ecosystems [17]. Therefore, we have built a conceptualization based on key concepts drawn from literature, to allow the analysis of (sub-)ecosystems in energy transitions. As managerial implications, the framework helps public and private organizations to evaluate how to get involved in PED service ecosystems, or alternatively, expand their ongoing operations within PEDs. The framework can also be used as a tool to scale solutions on a city level or to other PEDs. We hope that the framework will be tested and applied in other PEDs and cities to strengthen the construct validity and generalizability of the results.

Acknowledgment The research was funded by the European Union's Horizon 2020 research and innovation program LC-SC3-SCC-1-2018-2019-2020-Smart Cities and Communities under the project name SPARCS, grant number 864242. The funding body was not involved in preparing the manuscript, methods, and results.

References

1. Köhler, J., Geels, F.W., Kern, F., Markard, J., Wieczorek, A., Alkemade, F., Avelino, F., Bergek, A., Boons, F., Fünfschilling, L., Hess, D., Holtz, G., Hyysalo, S., Jenkins, K., Kivimaa, P., Martikainen, M., McMeekin, A., Mühlermeier, M.S., Nykvist, B., Wells, P.: An agenda for sustainability transitions research: state of the art and future directions. Environ. Innov. Soc. Trans. **31**, 1–32 (2019)
2. Geels, F.W.: From sectoral systems of innovation to socio-technical systems. Res. Policy. **33**(6–7), 897–920 (2004)
3. Gallan, A.S., McColl-Kennedy, J.R., Barakshina, T., Figueiredo, B., Jefferies, J.G., Gollnhofer, J., Hibbert, S., Luca, N., Roy, S., Spanjol, J., Winklhofer, H.: Transforming community wellbeing through patients' lived experience. J. Bus. Res. **100**, 376–391 (2019)
4. Ekman, P., Röndell, J., Yang, Y.: Exploring smart cities and market transformations from a service-dominant logic perspective. Sustain. Cities Soc. **51**, 1–11 (2019)
5. Vargo, S.L., Peters, L., Kjellberg, H., Koskela-Huotari, K., Nenonen, S., Polese, F., Sarno, D., Vaughan, C.: Emergence in marketing: an institutional and ecosystem framework. J. Acad. Mark. Sci. **51**, 2–22 (2023)
6. Polese, F., Frow, P., Nenonen, S., Payne, A., Sarno, D.: Emergence and phase transitions of service-ecosystems. J. Bus. Res. **127**(April), 25–34 (2021)
7. Vargo, S.L., Lusch, R.F.: Institutions and axioms: an extension and update of service-dominant logic. J. Acad. Mark. Sci. **44**, 5–23 (2016)
8. Nenonen, S., Storbacka, K.: Business model design: conceptualizing networked value co-creation. Int. J. Qual. Serv. Sci. **2**(1), 43–59 (2010)
9. Amit, R., Zott, C.: Value creation in E-business. Strateg. Manag. J. **22**(6–7), 493–520 (2001)
10. Speich, M., Ulli-Beer, S.: Applying an ecosystem lens to low-carbon energy transition: a conceptual framework. J. Clean. Prod. **398**(136429) (2023)
11. Siksnelyte-Butkiene, I., Streimikiene, D., Balezentis, T., Volkov, A.: Enablers and barriers for energy prosumption: conceptual review and an integrated analysis of business models. Sustain. Energy Technol. **57**(103163) (2023)
12. JPI Urban Europe: https://jpi-urbaneurope.eu/wp-content/uploads/2021/10/setplan_smartcities_implementationplan-2.pdf. Last accessed 2024/02/27
13. JPI Urban Europe: https://jpi-urbaneurope.eu/wp-content/uploads/2020/04/White-Paper-PED-Framework-Definition-2020323-final.pdf. Last accessed 2024/02/27
14. Wyckmans, A., Karatzoudi, K., Brigg, D., Ahlers, D.: D9.5: Report on attendance at eventsheld by other SCC-01 co-ordinators 2. +CityxChange Work Package 9, Task 9.2. (2019)
15. Lindholm, O., Rehman, H., Reda, F.: Positioning positive energy districts in European cities. Buildings. **11**(1), 19 (2021)
16. Gaziulusoy, A.I., Boyle, C.: Proposing a heuristic reflective tool for reviewing literature in transdisciplinary research for sustainability. J. Clean. Prod. **48**, 139–147 (2013)
17. Weiller, C., Neely, A.: Business Model Design in an Ecosystem Context. The Cambridge Service Alliance Publication Series (2013)
18. Demil, B., Lecocq, X.: Business model evolution: in search of dynamic consistency. Long Range Plan. **43**(2 3), 227–246 (2010)
19. Viljakainen, A., Toivonen, M., Aikala, M.: Industry Transformation Towards a Service Logic: A Business Model Approach. The Cambridge Service Alliance Publication Series (2013)
20. Van Maanen, J., Sørensen, J.B., Mitchell, T.R.: The interplay between theory and method. Acad. Manag. Rev. **32**(4), 1145–1154 (2007)
21. Dubois, A., Gadde, L.E.: Systematic combining: an abductive approach to case research. J. Bus. Res. **55**(7), 553–560 (2002)
22. Ceschin, F., Gaziulusoy, I.: Evolution of design for sustainability: from product design to design for system innovations and transitions. Des. Stud. **47**(Nov.), 118–163 (2016)
23. Kvale, S.: InterViews: An Introduction to Qualitative Research Interviewing. SAGE Publication, Thousand Oaks (1996)

24. Brea, E.: A framework for mapping actor roles and their innovation potential in digital ecosystems. Technovation. **125**(102783) (2023)

25. Brummer, V.: Community energy – benefits and barriers: a comparative literature review of Community Energy in the UK, Germany and the USA, the benefits it provides for society and the barriers it faces. Renew. Sust. Energ. Rev. **94**, 187–196 (2018)

26. de Vasconcelos Gomes, L.A., Farago, F.E., Figueiredo Facin, A.L., Flechas, X.A., Nascimento Silva, L.E.: From open business model to ecosystem business model: a process view. Technol. Forecast. Soc. Change. **194**(122668) (2023)

27. Bauwens, T., Schraven, D., Drewing, E., Radtke, J., Holstenkamp, L., Gotchev, B., Yildiz, Ö.: Conceptualizing community energy systems: a systematic review of 183 definitions. Renew. Sust. Energ. Rev. **156**(111999) (2022)

28. Berka, A.L., Creamer, E.: Taking stock of the local impacts of community owned renewable energy: a review and research agenda. Renew. Sust. Energ. Rev. **82**, 3400–3419 (2018)

29. Warbroek, B., Hoppe, T., Bressers, H., Coenen, F.: Testing the social, organizationa, and governance factors for success in local carbon energy initiatives. Energy Res. Soc. Sci. **58**(101269) (2019)

30. Gui, E.M., MacGill, I.: Typology of future clean energy communities: an exploratory structure, opportunities, and challenges. Energy Res. Soc. Sci. **35**, 94–107 (2018)

31. Kubli, M., Puranik, S.: A typology for business models for energy communities. Curr. Emerg. Des. Options. **176**(113165) (2023)

32. Autio, E.: Orchestrating ecosystems: a multi-layered framework. Innovation, 1–14 (2021)

33. Derkenbaeva, E., Halleck Vega, S., Hofstede, G.J., van Leuwen, E.: Positive Energy Districts: mainstreaming energy transition in urban areas. Renew. Sust. Energ. Rev. **153**(111782) (2022)

34. Vargo, S.L., Lusch, R.F.: From goods to service(s): divergences and convergences of logics. Ind. Mark. Manag. **37**(3), 254–259 (2008)

35. Read, S., Dew, N., Sarasvathy, S.D., Song, M., Wiltbank, R.: Marketing under uncertainty: the logic of an effectual approach. J. Mark. **73**(3), 1–18 (2009)

36. Shapira, P., Kuhlmann, S. (eds.): Learning from Science and Technology Policy Evaluation. Experiences from the United States and Europe. Edward Elgar, Cheltenham and Northampton (2003)

37. Djellal, F., Gallouj, F.: The Handbook of Innovation in Services. A Multi-disciplinary Perspective. Edward Elgar, Cheltenham (2010)

38. Guba, E.G., Lincoln, Y.S.: Fourth Generation Evaluation. Sage, Newbury Park (1989)

39. Rossi, P.H., Freeman, H.E., Lipsey, M.W.: Evaluation. A Systemic Approach, 6th edn. Sage, Thousand Oaks, London and New Delhi (1999)

40. Vargo, S.L., Wieland, H., Akaka, M.A.: Innovation through institutionalization: a service ecosystems perspective. Ind. Mark. Manag. **44**, 63–72 (2015)

41. Vargo, S.L., Lusch, R.F.: It's all B2B. . .and beyond: toward a system perspective of the market. Ind. Mark. Manag. **40**(2), 181–187 (2011)

42. Hyytinen, K.: Supporting service innovation via evaluation: a future oriented, systemic and multi-actor approach. Doctoral dissertation, Aalto University Department of Industrial Engineering and Management (2017)

43. Djellal, F., Gallouj, F.: The productivity in services: measurement and strategic perspectives. Serv. Ind. J. **33**(3–4), 282–299 (2013)

44. Frow, P., McColl-Kennedy, J.T., Hilton, T., Davidson, A., Payne, A., Brozovic, D.: Value propositions: a service ecosystems perspective. Mark. Theory. **23**, 1–25 (2014)

45. Anderson, J.C., Kumar, N., Narus, J.A.: Value Merchants: Demonstrating and Documenting Superior Value in Business Markets. Harvard Business School Press, Boston (2007)

46. Maglio, P.P., Spohrer, J.: A service science perspective on business model innovation. Ind. Mark. Manag. **42**(5), 665–670 (2013)

47. Ganco, M., Kapoor, R., Lee, G.K.: From rugged landscapes to rugged ecosystems: structure of interdependencies and firms' innovative search. Acad. Manag. Rev. **45**(3), 646–674 (2020)
48. Payne, A.F., Storbacka, K., Frow, P.: Managing the co-creation of value. J. Acad. Mark. Sci. **36**(1), 83–96 (2008)
49. Kallio, L., Heiskanen, E., Apajalahti, E.-L., Matschoss, K.: Farm power: how a new business model impacts the energy transition in Finland. Energy Res. Soc. Sci. **65**(101484) (2020)

Evaluation of Train Travel Inside Submerged Floating Tunnel Under Wave Loads

Jian Dai ⓘ and Paban Acharya

Contents

1 Introduction

Railway transport is an efficient and one of the greenest means of transportation of passengers and goods in large quantities over long distances. This mode of travel is reliable, punctual, convenient, cost-effective and sustainable, and thus has remained popular among different travel options. For places where the land is separated by wide and deep water bodies, however, the railway connections become challenging. Under such circumstances, railway lines usually have to be terminated on both sides of the water body with ferry connections. This often leads to a longer commuting time and the societal demand for more efficient ways of transportation.

Over wide and deep water bodies such as fjords and sea channels, the construction of conventional bridges can be very challenging and even inappropriate owing to the complex environmental and seabed conditions. A submerged floating tunnel (SFT), on the other hand, may prove to be an appealing option for the connection of railway lines across water bodies owing to its numerical advantages over other types

J. Dai (✉) · P. Acharya
Department of Built Environment, Oslo Metropolitan University, Oslo, Norway
e-mail: jiandai@oslomet.no

© The Author(s) 2025
M. Kioumarsi, B. Shafei (eds.), *The 1st International Conference on Net-Zero Built Environment*, Lecture Notes in Civil Engineering 237,
https://doi.org/10.1007/978-3-031-69626-8_98

of fjord- and sea-crossing options. For instance, SFTs make use of the natural buoyancy to support their weight and imposed loads in the gravitational direction, which eliminates the need for conventional pile supports and thus can cope with complex seabed conditions. Besides, they are submerged at a distance below the water surface where most of the wave energy is concentrated. Therefore, they are subjected to substantially reduced wave loads and no wind loads as compared to their surface bridge counterparts. Besides, such structures are virtually immune to climate changes and thus the water surface elevation due to global warming. The submergence also provides navigation clearance for areas with vessels passing through. Furthermore, as they do not protrude above the water surface, their visual impact on the surrounding environment is minimized. For these reasons, SFTs are perceived as novel engineering structures contributing to a net-zero society and are becoming a hot topic among researchers and engineers for the consideration of fjord- and sea-crossing solutions.

The design concept of SFTs can date back to the early twentieth century [1]. Even though no SFTs have been constructed in reality, numerous feasibility studies and research works have been conducted for various locations in the world. As SFTs are often designed to have a much longer length as compared to their cross-sectional dimensions, they are commonly considered as slender systems interacting with the fluid where the semi-empirical Morison equation is applicable to account for the hydrodynamic loads induced by the waves and current on the SFT structure [2–4]. Alternatively, the hydrodynamic actions can also be evaluated by employing the diffraction theory [5, 6]. A comparative study revealed that both approaches are able to give accurate hydrodynamic calculations under normal conditions [7].

Besides the environmental loads, the main function of SFTs is to serve as a transportation infrastructure to provide safe and comfortable passage of vehicles across water bodies. When they are designed for railway line extensions, the performance of SFT-train systems is studied by various researchers using different models. In a computational analysis, the SFT is typically modeled as Euler–Bernoulli beam segments resting on a Winkler-type foundation representing the hydrostatic stiffness. The train loads are modeled as either moving loads [8, 9] or moving sprung mass systems that account for the inertia effect of the vehicle [10, 11]. To represent more realistic scenarios, the inclusion of the railway track systems leads to a coupled vehicle-track-SFT system [12].

While a coupled SFT-train system considers the interaction between the traversing train and the sustaining SFT structure, such a complex model often hinders its applications in engineering design practice. The development of a simplified, decoupled approach is thus highly desirable. This is possible in view of the fact that the train mass is much lower than that of the SFT and that the effect of a moving train on the dynamic responses of the sustaining SFT is expected to be limited [12].

In this paper, an uncoupled approach is proposed and put forward to investigate the performance of a train travelling inside an SFT under various wave conditions. The SFT is modeled as Euler–Bernoulli beam elements subjected to wave loads modeled according to Morison's equation. The dynamic displacement of the SFT is then used as input to a train model to study the dynamic responses of the vehicle. Parametric studies are carried out to examine the effect of different wave conditions and train speeds on the vertical motions of the train.

2 Methodology

2.1 Problem Definition

Consider an SFT segment with a length of 700 m and an external diameter of 23 m, as shown in Fig. 1. Both ends of the segment are restrained by strong fixtures/towers which makes it reasonable to assume fixed boundary conditions. The SFT segment is also vertically restrained by mooring tethers equally spaced apart. The water depth is 100 m and the tunnel centerline is submerged 61.5 m below the water free surface. In view that the vertical motion of a train travelling inside an SFT is usually dominating, this study considers the vertical responses of the SFT and the train only. Table 1 lists the key parameters of the SFT. Note that these parameters are taken from the study reported in [13].

2.2 SFT Model

In this study, the SFT is modeled according to the Euler–Bernoulli beam theory, which is justifiable in view of its slenderness. The finite element method is employed by discretizing the SFT segment into a number of beam elements. The mooring

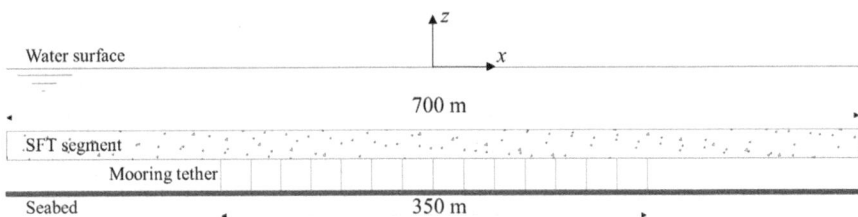

Fig. 1 Illustration of SFT segment

Table 1 Parameters of SFT

Parameter	Value
Young's modulus E	30 GPa
Sectional area A	142 m^2
Second moment of area I	7800 m^4
Mass per unit length m	3.7×10^5 kg/m
Tether stiffness	1.17×10^5 N/m

tethers are modeled as linear vertical springs. According to Newton's second law of motion, the governing equation of the SFT can be written as

$$\mathbf{M}_S\ddot{\mathbf{U}}_S + \mathbf{C}_S\dot{\mathbf{U}}_S + \mathbf{K}_S\mathbf{U}_S = \mathbf{F}_S \tag{1}$$

where $\mathbf{M}_S$, $\mathbf{C}_S$, and $\mathbf{K}_S$ are the structural mass, damping and stiffness matrices of the SFT, respectively. $\ddot{\mathbf{U}}_S$, $\dot{\mathbf{U}}_S$, and $\mathbf{U}_S$ are the acceleration, velocity and displacement vectors of the SFT, respectively. $\mathbf{F}_S$ is the external force vector including the wave loads and the train loads.

According to Morison's equation, the wave load per unit length, f_W, exerted on the SFT can be written as [14]

$$f_W = \frac{1}{4}\pi D^2 \rho_w \dot{u}(t) + \frac{1}{4}C_a\pi D^2\rho_w\left(\dot{u}(t) - \frac{\partial^2 w(x,t)}{\partial t^2}\right)$$
$$+ \frac{1}{2}C_d\rho_w D\left|u(t) - \frac{\partial w(x,t)}{\partial t}\right|\left(u(t) - \frac{\partial w(x,t)}{\partial t}\right) \tag{2}$$

where u and w denote the fluid velocity and the displacement of the SFT in the vertical direction, respectively. C_a and C_d are the added mass and drag coefficients, respectively. D is the diameter of the SFT. ρ_w refers to the seawater density.

2.3 Train Model

A quarter train model based on the multi-body system, as shown in Fig. 2, is employed to investigate the performance of the train travel inside the SFT. A quarter of the car body, half of a bogie and one wheelset are modeled as rigid bodies interconnected via spring-dashpot units representing the system systems. Again, according to Newton's second law of motion, the governing equation of the train model can be written as

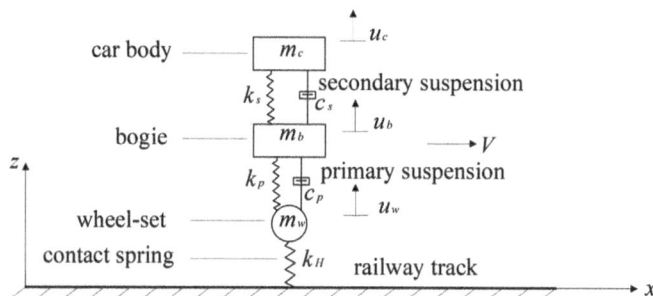

Fig. 2 Quarter train model

$$\mathbf{M_T \ddot{U}_T + C_T \dot{U}_T + K_T U_T = F_T} \tag{3}$$

where $\mathbf{M_T}$, $\mathbf{C_T}$, and $\mathbf{K_T}$ are the mass, damping and stiffness matrices of the train, respectively. $\mathbf{\ddot{U}_T}$, $\mathbf{\dot{U}_T}$, and $\mathbf{U_T}$ are the acceleration, velocity and displacement vectors of the train, respectively. $\mathbf{F_T}$ is the external force vector.

Table 2 lists the parameters of the train model [12].

2.4 Wave Model

Short-term beam sea conditions are considered in this study. The waves are assumed to be long-crested and can be described by the JONSWAP spectrum as

$$S_\zeta(\omega) = \frac{5}{16} A_\gamma H_s^2 \frac{\omega_p^4}{\omega^5} e^{-\frac{5}{4}\left(\frac{\omega}{\omega_p}\right)^{-4}} \gamma^{e^{-\frac{1}{2}\left(\frac{\omega-\omega_p}{\sigma\omega_p}\right)^2}} \tag{4}$$

where $A_\gamma = 1 - 0.287\ln(\gamma)$, H_s is the significant wave height, ω_p is the peak angular frequency which equals $2\pi/T_p$, T_p is the peak period, γ is the non-dimensional peak shape parameter, and σ is the spectrum width parameter which equals 0.07 for $\omega \leq \omega_p$ and 0.09 for $\omega > \omega_p$.

Three sea states corresponding to the design operational, rough and extreme conditions are considered. Table 3 lists the characteristic values of the three wave conditions.

Table 2 Parameters of quarter train model

Parameter	Value
Mass of car body m_c	10965.6 kg
Mass of bogie m_b	1200 kg
Mass of wheelset m_w	1850 kg
Primary suspension stiffness coefficient k_p	1.26×10^6 N/m
Primary suspension damping coefficient c_p	4×10^4 Ns/m
Secondary suspension stiffness coefficient k_s	1.41×10^5 N/m
Secondary suspension damping coefficient c_s	1.3×10^4 Ns/m

Table 3 Wave conditions

Wave condition	Significant wave height (m)	Peak period (s)
Design operational	3	5
Rough	5	10
Extreme	9	12

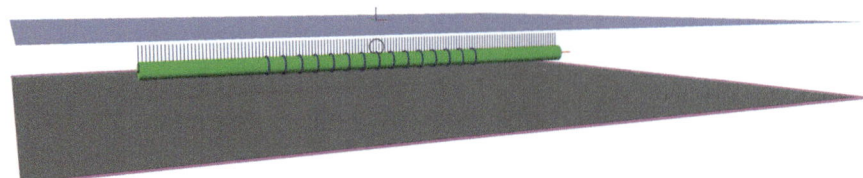

Fig. 3 SFT model

According to the Airy wave theory, the vertical fluid particle velocity associated with a harmonic wave component can be written as

$$u = \sigma a \frac{\sinh k(z + h)}{\sinh kh} \sin(\omega t + \varphi) \tag{5}$$

where $a = H/2$ represents the wave amplitude, $\omega = 2\pi/T$ is the incident wave angular frequency, and k is the wave number. h is the water depth.

2.5 Numerical Procedure

The SFT segment is constructed by using SIMA/RIFLEX [15], as shown in Fig. 3. It is discretized into 140 elements of equal length of 5 m. Rayleigh damping is employed for the structural damping with a ratio of 0.8% [16]. For each of the three wave conditions defined in Sect. 2.4, six independent one-hour wave realizations are used to reduce the statistical uncertainties. In the numerical analysis, the Newmark constant acceleration scheme is employed for the time marching integration. The analysis results of the SFT segment are then processed to remove the initial transient responses before they are used as input to the train model for the vehicle dynamic analysis.

3 Results and Discussion

3.1 Modal Analysis

Figure 4 shows the first three dry modes of the SFT segment. Note that the hydrostatic stiffness is considered in the eigenvalue analysis. However, the hydrodynamic mass is not taken into account. The natural periods of the first three modes are 8, 7.87, and 2.91 s, respectively.

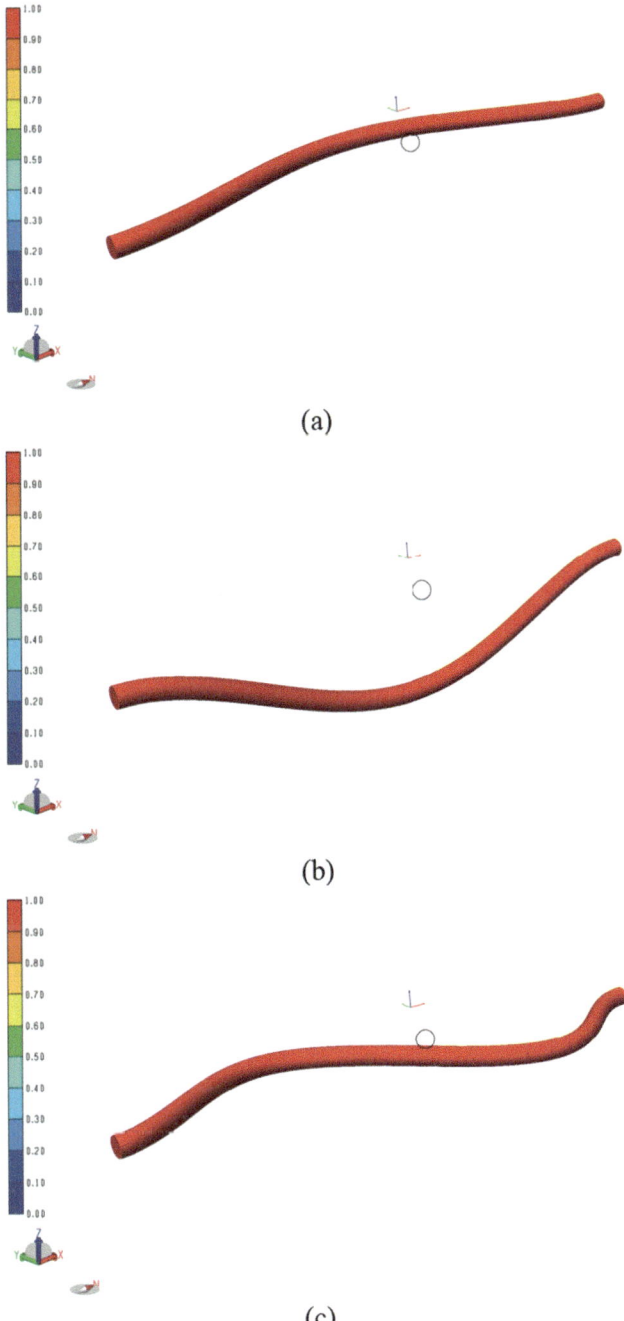

Fig. 4 Mode shapes of SFT segment: (**a**) first mode of 8 s, (**b**) second mode of 7.87 s, and (**c**) third mode of 2.91 s

3.2 SFT Responses

The structural performance of the SFT under different wave conditions is next examined. Figure 5 shows the vertical motion response statistics of the SFT segment at mid-span. Note that the statistical values are obtained by taking the average values of those from each of the six random simulations as explained earlier in Sect. 2.5. As can be seen, the SFT is almost stationary under the design operational wave condition with a maximum vertical displacement below 6 cm at its mid-span. When the wave condition becomes harsher, the vertical motions of the SFT increase substantially. For all the wave conditions considered in this study, the standard deviations of the vertical mid-span displacement, which correspond to the dynamic response components, are below 0.5 m. This can be considered as reasonably low in view of the fact that the SFT segment is 700 m long, and it is expected to vibrate slowly with a fundamental period of around 8 s.

Figure 6 shows the statistical results of the bending moment M_y of the SFT segment at its mid-span. Notably, the contribution due to the performance loads arising from the self-weight, buoyancy, and static component of the mooring tensions has a dominating role in the vertical bending moment. The wave-induced moments are found to be negligible under the design operational wave condition. For rough and extreme wave conditions, the standard deviations of M_y reach 26% and 37% of the magnitude of the mean value, respectively, showcasing a significant increase in the effect of the wave action on the SFT bending moment.

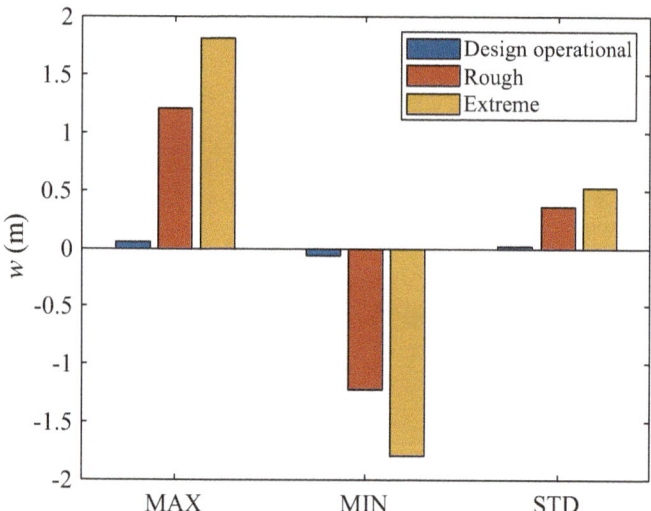

Fig. 5 Vertical displacement of SFT at mid-span

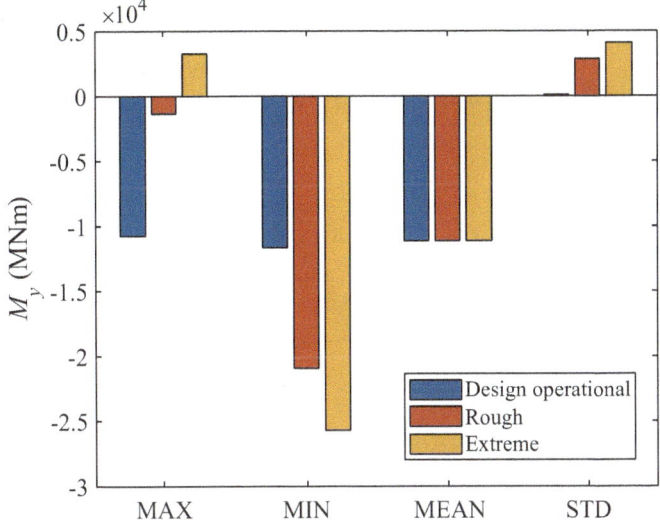

Fig. 6 Bending moment of SFT at mid-span

3.3 Train Responses

The vertical displacement of the SFT segment is next used as input in the form of a vibrating foundation to examine the train responses. In this study, the initial transient responses of the SFT are discarded. In the analysis of the train model presented in Sect. 2.3, the train is assumed to start with an unloaded condition. An initial duration of 10 s is taken into account for the train to settle to its equilibrium state before it enters the SFT segment. Again, these initial train responses are discarded in the post-processing of the results. The railway track is assumed to be perfectly smooth without any track irregularities.

Figure 7 shows the maximum vertical acceleration of the car body under different wave conditions and train speeds. Note that these accelerations are the maximum values out of the six independent random simulations. As can be seen, the vertical accelerations of the car body are rather low under the design operational condition. The train speed also has a small effect on the maximum acceleration. This indicates that train travel is generally comfortable when the track is in new or good conditions. As the wave condition becomes rough, the maximum accelerations of the car body increase significantly. Higher train speeds are also found to augment the car body accelerations. Notably, with a speed of 300 km/h, the maximum car body acceleration is close to the safe limit of 0.6 m/s^2. As only a limited number of random simulations were carried out, this limit may be exceeded in reality, therefore posing an alarming concern on train operations with high speeds under rough wave conditions. Under the extreme wave condition, the maximum car body accelerations exceed the safe limit for all train speeds considered in the study, thereby implying that the SFT segment should be closed to traffic under such harsh sea conditions.

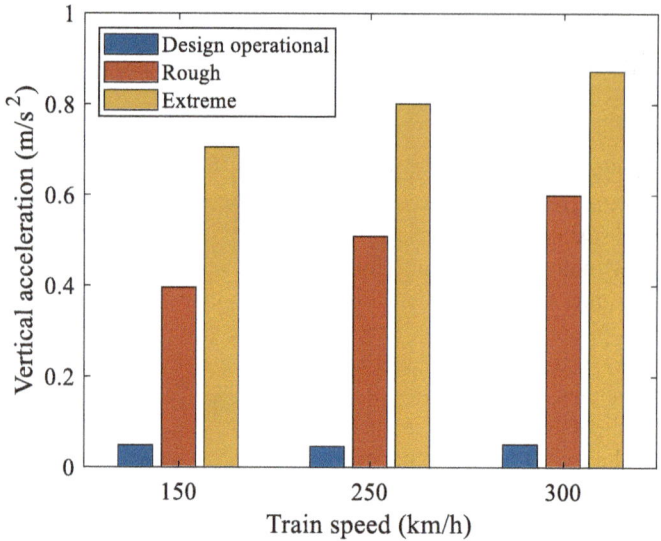

Fig. 7 Vertical acceleration of car body inside SFT

4 Conclusions

This paper is concerned with the evaluation of train travel performance inside a submerged floating tunnel (SFT) subjected to wave excitations. An SFT segment is modeled by using Euler–Bernoulli beam elements with mooring tethers represented by using linear springs. A quarter train model based on the multi-body system is employed to represent the train travelling inside the SFT. In the numerical analysis, the SFT subjected to different wave conditions is first analyzed. The vertical displacements of the SFT are then used as input in the form of a vibration foundation to the train model to examine the vertical vibrations of the train. The analysis results show that trains can travel across the SFT segment safely at a high speed under the normal design operational wave conditions. When the sea state becomes rough, the vertical train accelerations are approaching 0.6 m/s^2. Thus, reducing the train speed may be necessary in order to maintain safe travel. Under extreme or storm conditions, the SFT should be closed to train operations.

References

1. Kunisu, H., Mizuno, S., Mizuno, Y., Saeki, H.: Study on submerged floating tunnel characteristics under the wave condition. In: Proceedings of the 4th International Offshore and Polar Engineering Conference, ISOPE 1994, Osaka, Japan, 10–15 April, 1994
2. Seo, S.-I., Mun, H.-S., Lee, J.-H., Kim, J.-H.: Simplified analysis for estimation of the behavior of a submerged floating tunnel in waves and experimental verification. Marine. Structure. **44**, 142–158 (2015)

3. Luo, G., Zhang, Y., Ren, Y., Guo, Z., Pan, S.: Dynamic response analysis of submerged floating tunnel under impact-vehicle load action. Appl. Math. Model. **99**, 346–358 (2021)
4. Jin, C., Kim, M.-H.: Tunnel-mooring-train coupled dynamic analysis for submerged floating tunnel under wave excitations. Appl. Ocean Res. **94**, 102008 (2020)
5. Chakrabarti, S., Barnett, J., Kanchi, H., Mehta, A., Yim, J.: Design analysis of a truss pontoon semi-submersible concept in deep water. Ocean Eng. **34**, 621–629 (2007)
6. Paik, I.Y., Oh, C.K., Kwon, J.S., Chang, S.P.: Analysis of wave force induced dynamic response of submerged floating tunnel. KSCE J. Civ. Eng. **8**(5), 543–550 (2004)
7. Kunisu, H.: Evaluation of wave force acting on submerged floating tunnels. Procedia Eng. **4**, 99–105 (2010)
8. Tariverdilo, S., Mirzapour, J., Shahmardani, M., Shabani, R., Gheyretmand, C.: Vibration of submerged floating tunnels due to moving loads. Appl. Math. Model. **35**, 5413–5425 (2011)
9. Yang, Y., Xiang, Y., Gao, C.: Vehicle-SFT-current coupling vibration of multi-span submerged floating tunnel, part I: mode superposition and Galerkin hybrid method. Ocean Eng. **247**, 110746 (2022)
10. Xiang, Y., Lin, H., Bai, B., Chen, Z., Yang, Y.: Numerical simulation and experimental study of submerged floating tunnel subjected to moving vehicle load. Ocean Eng. **235**, 109431 (2021)
11. Yang, Y., Xiang, Y., Gao, C.: Vehicle-SFT-current coupling vibration of multi-span submerged floating tunnel, Part II: comparative analysis of finite difference method and parametric study. Ocean Eng. **249**, 110951 (2022)
12. Song, Y., Dai, J., Lau, A.: Analysis of railway vehicle-track interaction performance in submerged floating tunnel subjected to wave excitations. Ocean Eng. **303**, 117760 (2024)
13. Jin, C., Kim, M.-H.: Time-domain hydro-elastic analysis of a SFT (submerged floating tunnel) with mooring lines under extreme wave and seismic excitations. Appl. Sci. **8**(12), 2386 (2018)
14. Dai, J., Abrahamsen, B.C., Viuff, T., Leira, B.J.: Effect of wave-current interaction on a long fjord-crossing floating pontoon bridge. Eng. Struct. **226**, 114549 (2020)
15. SINTEF Ocean: RIFLEX 4.16.0 Theory Manual, Trondheim (2019)
16. Statens vegvesen: Håndbok N400 Bruprosjektering, Oslo (2015)

Exploring Multilevel Governance Networks in Deployment of Positive Energy Districts: Case of Salzburg

Caroline Cheng and Savis Gohari

Contents

1   Introduction

In today's world, urban and global challenges are increasingly being recognized as wicked problems, characterized by their complexity, interconnectivity, and urgency. The need to address these challenges has become more pressing than ever, requiring innovative approaches and collaborative efforts across various levels and sectors. The Sustainable Development Goals (SDGs) call for action by all countries, poor, rich and middle-income, to promote prosperity while protecting the planet earth and its life support system [1]. Europe is striving to become a global leader in energy

C. Cheng (✉)
Department of Architecture, Materials and Structures, SINTEF Community, Trondheim, Norway
e-mail: caroline.cheng@sintef.no

S. Gohari
Department of Civil and Environmental Engineering, Norwegian University of Science and Technology, Trondheim, Norway
e-mail: savis.gohari@ntnu.no

© The Author(s) 2025
M. Kioumarsi, B. Shafei (eds.), *The 1st International Conference on Net-Zero Built Environment*, Lecture Notes in Civil Engineering 237,
https://doi.org/10.1007/978-3-031-69626-8_99

transition and carbon footprint reduction, aligning with sustainable development goals. The European Union introduced the concept of Positive Energy Districts (PEDs) to maximize the utilization of local and renewable energy sources while positively contributing to the optimization and security of the broader electricity grid.

Many emerging studies have recognized that the main challenge of implementing PED pertains to suboptimal governance systems and silo thinking [2, 3]. Silo thinking and acting between different levels and sectors, territorial gaps between governments, and lack of interdisciplinary collaboration are among the major obstacles hindering successful outcomes for energy transition [4, 5]. Effective governance networks can be understood as enabling stakeholders to collaboratively address challenges and set priorities, ensuring commitment and ownership of planning outcomes. To achieve successful collaborative governance, a holistic systemic change and paradigm shift are necessary, along with training and education on how collaboration and knowledge exchange across levels and sectors should function [6, 7]. Breaking silo thinking and acting requires exploring governance systems empirically to understand the pitfalls, barriers, challenges, as well as the factors contributing to success and failure of collaborative governance processes [8]. This exploratory chapter aims to delve into the governance structures and functions in the deployment of a PED project.

2   Governance Processes of PED Projects

Moving beyond the building level to the neighborhood scale in PED projects not only enables the implementation and assessment of solutions such as renewable energy sources and decentralized energy production but also necessitates interaction with multiple stakeholders from different sectors [9]. Governance has stood out as crucial for achieving successful PED projects [10]. Above all, effective governance networks enable stakeholders to collaboratively address challenges and set priorities, ensuring commitment and ownership of planning outcomes. According to Potts et al. [11], there is a paucity of practical frameworks for conceptualizing the governance processes and planning processes are often perceived as idealistic, simplistic, and linear models of decision-making, in which issues of representation and the plurality of public interests are overlooked or untouched [11]. This section distills the conceptual framework that can be used to investigate the governance structures and functions in the empirical setting of a PED project.

2.1   Shift to Collaborative Network Governance

In recent years, there has been a change in how governments organize themselves to solve societal problems. Instead of using top-down approaches where the government directs actions and policies, there's a shift toward networked forms of governance [3, 12]. Unlike traditional hierarchical structures, where one entity has authority over others, network governance relies on collaboration and coordination among various organizations, as they bring together different stakeholders to address common issues. Network governance is about building sustainable relationships between these actors, who come together to solve specific problems or provide services. Thus, network governance is assumed to be better capable of handling complex societal, poorly structured (and even "wicked") problems [13, 14] than more traditional forms of governing that assume hierarchical relationships [15, 16].

In collaborative network governance, actors collectively determine the organization's structure, coordination methods, goals, and strategies to achieve them. To ensure the smooth functionality of such collaborative network, fostering win-win situations, managing actors and resources, reducing interaction costs, ensuring commitment and compliance, coordinating politically and administratively, and maintaining quality and openness of interaction are important factors [17]. This is imaginably an uphill task and actors face barriers such as conflicting political priorities, community complaints, environmental regulations, alterations in funding schemes and participant grievances [18].

To date, few studies have considered the cumulative impact of governance structure and function and their interactions in real-world, multilayered, and complex systems. Many planning theories emphasize the interrelations of structures but overlook the significant influence of functions within the system. In the existing theories, planning theories generally shape the understanding of the governance issue, primarily focusing on its structural components, while decision-making theories offer practical tools for assessing and analyzing the functional aspects of governance processes. Since our research examines the interconnection between governance structure and process, we draw on the "structural-functionalism theory" in the planning context (Potts et al. [11]) to investigate how dynamic, multilayered, and complex governance structures and processes have led to different outcomes (of planning and decision-making) over time.

2.2 Structural-Functionalism Model in Planning

The structural-functionalism model uses the rational planning approach to discuss and analyze the underlying complexities and uncertainties, in which visioning and objective-setting, strategy development, implementation and monitoring/evaluation are the recognizable stages of strategic planning. According to Potts et al. [11], structures are organized or institutionalized in a specific manner and consist of many interrelated, interdependent, but also autonomous, individuals, existing at different levels of society: national, regional and local. They can include government agencies, industry groups, nongovernment organizations, community groups and individuals. Structures run processes and are involved in different planning or policy cycles, such as goal setting, strategy development, implementation and evaluation, and produce different outputs (e.g., formal documents such as legislation, policies, strategies, plans, reports) to guide actions for the achievement of desired outcomes.

The functional dimension of the structural-functionalism approach connects the actors in a system but also represents the relationships between them. The functional components tend to be more dynamic and less robust than structures. Three cornerstone functional components can be used to investigate "governance processes," namely knowledge use, connectivity and capacity. Knowledge use indicates the importance of applying, coordinating and integrating relevant social, economic, environmental, traditional and historical knowledge, rather than one set of knowledge or one method, to solve complex problems and enable better functioning governance systems. Connectivity implies a collaborative approach of stakeholders' engagement and participation, which can develop consensus and build stronger community and institutional networks. Capacity (human, social, financial and physical) refers to the power or capability of institutions to achieve outcomes, depending on the amount and types of capital/resources that they have accrued or access.

In the structural-functionalism model, the levels of capacity, connectivity and knowledge are unequal among different actors. There are therefore different magnitude of the capacity, varying strength of connectivity, and different knowledge disciplines, both explicit and tacit, to be considered.

Structural-functionalism provides a logical, systematic and evidence-based approach to the analysis of strategic planning systems that is complementary to existing theories of planning. It recognizes the inherent complexity and iterative nature of planning systems and allows analysts to consider a plethora of interactions and other factors influencing planning processes and outcomes across scales. While structures exhibit more stability and change at a slower pace, functions are in a constant state of flux and evolution. It is essential to recognize that structures and functions influence each other reciprocally, and any study or analysis should view them not in isolation but in their interconnectedness and interaction.

2.3   Research Question

This chapter therefore poses the following research questions: What are the structural components of governance in the multiactor environment in the development of a sustainable plus energy neighborhood? How do the functional components of its governance system operate in terms of the interaction among actors for exchanging knowledge and resources, utilizing their capacity as well as building connectivity to achieve defined goals?

3   Research Design and Methodology

This section explains why the single case study design is used and outlines the data collection methods before presenting the GNICE SPEN as the empirical setting for investigating governance structures and functions.

3.1 Why Single Case Study

The research design concerns a single case study. It aims to develop in-depth descriptive-analytical perspectives toward the governance system in planning and decision-making processes of the GNICE SPEN in Salzburg. Yin [19] (p. 18) believes that the case study has a distinct advantage when a "how" or "why" question is being asked about a set of events within its real-life context, especially when the boundaries between phenomenon and context are not clearly evident [19]. When there is no pressure on the researcher to impose controls or to change circumstances [20], the best research strategy is the case study that provides the opportunity to obtain an in-depth investigation of a given phenomenon, within its context, by the strategic selection of a case [21]. Since the governance concept focuses on interconnected and interrelated structures and processes, sufficient details are required to unravel the complexities of a given situation. Thus, the case study of GNICE SPEN is specifically and strategically selected for this study, out of the many PED cases that are familiar to the authors.

This research adopts intrinsic case study [22], in which a particular case is explored to describe and analyze the background, development of the idea, interaction of governance components, and internal-external influences in the planning and decision-making processes over time. Our single case will determine whether propositions of governance in structural-functionalism theory are confirmed, challenged, or need to be extended. This single case can represent a significant contribution to knowledge and theory building and can help to refocus future investigations in the entire field.

The justification for selecting this specific case is rooted in our ongoing research and collaboration, aimed at identifying enabling and success factors across various PED cases. Through this collaborative effort, we have observed and interviewed numerous cases, leading us to recognize that the GNICE SPEN stands out for its successful governance practices and outcomes. Consequently, our aim is to delve deeper into this case to explore its success factors, glean valuable insights and clarify its deeper causes and its consequences.

3.2 Data Collection

To comprehensively understand the dynamics of governance networks, given the fluidity of actors, roles, and interactions across multiple projects and contexts, research has called for empirical consideration of contingent factors influencing governance relationships [23]. For the last 5 years, the authors have been actively engaged in governance and stakeholder analyses of neighborhood-scale projects across various contexts, including smart city, energy transition, and PEDs. Their ongoing collaboration and involvement in relevant projects and networks have kept them abreast of the latest governance theories and studies, aligning their research interests with their practical experiences. The first author's role as work package leader responsible for innovation management, exploitation and market uptake of the H2020 Innovation Action project syn.ikia project has fostered close collaboration and ongoing dialogue (since 2021) with the stakeholders involved in GNICE SPEN. The first step toward network governance is to systematically identify the key actors in the different phases of PED development [24]. Data collection has relied on site visits, in-person discussions, bilateral digital meetings and email correspondence to better understand its governance practices and outcomes. In addition to extensive desk research (project deliverables), document analysis (quality agreement document for GNICE) and literature reviews, the narrative, semistructured interviews with the project responsible of GNICE SPEN in 2022 was conducted and facilitated by the second author. Emphasizing participatory research and observation methods, the authors leveraged their collective expertise and experiences from various projects and initiatives to jointly analyze and comprehend the governance dynamics in the advancement of GNICE SPEN in Salzburg.

3.3   Sustainable Plus Energy Neighborhoods (SPENs)

Many PED projects are being implemented to deliver proofs-of-concept to demonstrate that decarbonization in the built environment is not only possible but that it also has the potential to generate value. One pathway focuses on planning and designing multistory apartment homes on a neighborhood scale than the business-as-usual conventional way of building, embarking on a novel concept called Sustainable Plus Energy Neighborhoods (SPENs), via four real-life demonstration cases in four climatic zones (Spain, Austria, Netherlands and Norway). SPENs are envisioned to be groups of interconnected multistory apartment homes that are highly energy efficient and fitted with renewable energy sources that have the potential to generate a surplus of energy [25].

3.4   GNICE SPEN in Salzburg

GNICE is a new residential development in Salzburg to be developed into a Sustainable Plus Energy Neighborhood. It is a greenfield development located on the city's outskirts in a quiet area with primarily multifamily houses. The 17 buildings on the plot consist of mostly multiresidential dwellings (251 apartments), from which half will be social housing units, and the other half will be sold at a certain fixed price. Passive design includes well insulated and airtight envelope and high-performance glazing. Active systems include heat pumps for space heating and DWH, and efficient mechanical ventilation with heat recovery. PV panels are placed on the roof and on some façade sections. Innovations include smart house technology, user engagement processes, central heat pump system based on ground and wastewater sources, photovoltaics, energy peak shaving, green roofs, and consideration of climate change adaptation. The neighborhood will have a kindergarten, a doctor, and a base zone (for low threshold neighborhood work) managed by the social aid and service organization, Caritas, and will also include a Silberstreif (for those aged 55 and older) residential group.

4 Governance Processes for the Development of GNICE SPEN

This section presents highlights of the structures and functions that enabled the governance of the multiactor environment in the development of a SPEN.

4.1 Structural Components

Governance structures shape a framework that determines how different actors relate to each other within a network and what outcomes are expected. It includes policies, rules, procedures, and organizational frameworks. By identifying the governance structure in GNICE SPEN, the goal is to identify which structural components will help to establish order, accountability, and efficiency within the project's organizations that defines roles, responsibilities, and decision-making processes. We identified five structural components to understand the governance structure for the development of GNICE SPEN: Smart Cities Network Austria, the competence center for sustainable neighborhoods, Klimaaktiv, Local Multiactor Framework and the Quality Agreement, an informal voluntary instrument.

The Positive Energy District (PED) is increasingly recognized as a vital component in the journey toward achieving climate neutrality and fostering smart cities. Derived from the principles of smart city initiatives, the PED concept serves as a complementary approach aimed at optimizing energy efficiency and sustainability within urban environments. The Austrian Smart Cities Network plays a pivotal role in advancing Austria's urban development strategies toward sustainable and resilient cities. This network consists of pioneering cities committed to the implementation of smart city concepts and proactive climate change adaptation. Among these cities, Salzburg, alongside six others, stands as a prominent member of the Smart Cities Network Austria. Supported by the Federal Ministry for Climate Action, Environment, Energy, Mobility, Innovation, and Technology (BMK) and the Climate and Energy Fund, these initiatives have empowered municipalities to prioritize urban quality of life, innovative technologies, and integrated urban planning. With a holistic vision for decarbonizing existing systems, the Smart Cities Network recognizes the importance of strategic collaboration among diverse stakeholders, including funding agencies, local authorities, decision-makers, and energy suppliers. At its core, the Smart Cities Network operates as a dynamic partnership between the BMK, cities, and the coordination office, serving as a catalyst for impactful urban transformation and innovation in Austria.

Another key structural component that operates at different levels of the governance system in Austria needs to be emphasized: The Salzburg Institute for Regional Planning and Housing or Salzburger Institut für Raumordnung und Wohnen (SIR). SIR is a competence center for sustainable neighborhoods in the Austrian National network. It is located locally in Salzburg but operates nationally and acts as an intermediary between multiple governance sectors, including the EU, national, and local levels. Through coordinated efforts facilitated by the SIR, the network fosters interdisciplinary collaborations and multilevel governance approaches.

Another significant structural component in the Salzburg case study that operates within multilevel governance systems and warrants recognition is Klimaaktiv. Serving as an initiator of climate-friendly technologies and practices in Austria, Klimaaktiv plays a pivotal role in promoting sustainability and addressing climate change challenges. Established in 2004, Klimaaktiv, under the management of the Austrian Energy Agency, has been instrumental in promoting climate-friendly initiatives in various sectors such as building and renovation, energy conservation, renewable energies, and mobility [26]. Acting as a catalyst for change, Klimaaktiv bridges the gaps between politics, business, and society through its modern governance approach. By providing advice, information, and qualification initiatives, along with transparent standards and quality assurance measures, Klimaaktiv actively engages and integrates relevant stakeholders in its efforts. Understanding Klimaaktiv as a structural component in the Salzburg case study is essential for comprehending the transformative impact it has had on the region's approach to climate protection and sustainability.

As the GNICE SPEN is not only a construction project but also an energy concept, a diversity of actors are involved. As the project progresses, a local multiactor framework was gradually identified as those most relevant for the activities backed by shared goals that may or may not derive from legal or formally prescribed responsibilities [27–30]. GNICE SPEN is being implemented by the social housing provider and project developer, Heimat Österreich. It works through and together with a Steering Committee, which is led by the City of Salzburg. This Steering committee makes the final decisions and is also the interface between the local authority and the political institutions. Energy Consulting Austria is the Energy planning company responsible for the development, planning, implementation, and monitoring of the plus energy concept. Energiewerk Baumgartner e.U. is the external energy consultant for neighborhood, and it manages the energy questions and integration with the neighbors. Raumsinn is the service provider responsible for resident engagement during the planning process as well as resident engagement in the use of the buildings. Raumsinn's sociologists have conducted important analyses of the existing neighborhood. Caritas Foundation Austria, an Austrian nonprofit charity organization, is responsible for the implementation of social concept and represents the voice of future users of the new buildings. It has provided input during the development and planning process. Wohngruppe Silberstreif (meaning silver lining) is a housing community (+55 year old community) cum residential association have been consulted upon to provide input during the development and planning process. Finally, Österreichischer Verband Gemeinnütziger Bauvereinigungen–Revisionsverband (GBV or Austrian association of nonprofit building associations), an umbrella organization of the nonprofit housing industry acting as a cooperative audit association, provides feedback as it acts as an interest group giving voice to housing associations focused on affordable housing.

Another structural component identified in our research is the Quality Agreement initiated by SIR, which is a nonlegal binding agreement or an informal voluntary instrument. In structural functionalism theory, certain legal and formal outputs such as policies, plans, and regulations are determined as safeguards for collaboration, coordination, accountability, efficiency, and legitimacy. In the case of Salzburg, the Quality Agreement is an agreement between the project developer, the city of Salzburg, the architects, and other relevant planners, that describes the goals/qualities for six topics (management, communication, urban development, buildings, energy, mobility). It consists of documents that are short, easy to understand, and serve as reference points throughout the project. At the outset of the project, various actors convene in a workshop to share their interests, goals, and strategies. If a new actor joins the network later, another workshop is conducted to ensure that the agreement is understandable for new entrants while remaining open and flexible for new ideas and insights. The Quality Agreement is the result of a communication process and expresses the goals that must be fulfilled by the end of the project to ensure that all stakeholders deem it a success. These goals include reaching the requirements of the local building code, complying with the local housing subsidy directive, adhering to the Klimaaktiv standard for the entire neighborhood, achieving the Klimaaktiv gold standard for all buildings, and meeting the Greenpass standard for the blue and green infrastructure.

4.2   Functional Dimension

In understanding the functional components of governance within the GNICE SPEN project, we recognize the significance of four key factors.

Firstly, the unique role, position, experience, functionality, and capacity of SIR within the multilevel governance system stand out prominently. SIR leads the project group, coordinates steering group and working groups to make decisions, keeps all relevant stakeholders informed, both internal and external, and ensures the defined goals are achieved. With extensive years of experience and active involvement in Austrian planning for sustainability and climate, SIR possesses relevant political, practical, and operational knowledge and skills. Familiarity with the dynamics of different actors' interests, strategies, and perspectives enables SIR to effectively function as a moderator and consultant, initiating positive strategies for communication, collaboration, participation, transparency, trust-building, and enhancing the sense of ownership, commitment, and engagement among diverse stakeholders. Additionally, SIR's role as an "external" expert in the quality assurance process, devoid of municipality or governmental biases, serves as a crucial intermediary. Collaboration among the partners of the EU project helped SIR share ideas on an international level and stay informed about international trends in the planning and construction of neighborhoods.

Secondly, the role of the quality agreement, an informal, voluntary joint agreement, emerges as another success factor for governance functionality. The agreement, which took more than a year to discuss and define during the preplanning phase preceding detailed design, fosters trust, coordination, and collaboration, effectively mediating governance challenges identified in research and practice. It is time-consuming to explain to involved parties that the document is a voluntary instrument and not legally binding. However, incoming new actors are introduced to the quality agreement to ensure that all stakeholders "have the same picture." Whenever questions arise, actors can always refer to the quality agreement.

Thirdly, the engagement of the community and their involvement in the steering group adds significant value to the governance structure. Caritas, in collaboration with the highly motivated +55-year-old community group "Wohngruppe Silberstreif," shapes the neighborhood's community, particularly addressing the needs of the increasing elderly population. The Silberstreif +55 association provides older people with the opportunity to remain independent for longer while living among like-minded individuals.

Lastly, the contribution of sociologists at Raumsinn stands out as pivotal. Their involvement in the governance network and collaboration with experts from various disciplines facilitate the integration of nontechnical and human-centric aspects into the project, establishing a culture of interdisciplinary collaboration. Raumsinn's role as a service provider hired to develop the "social concept" of the district further enhances governance functionality. Raumsinn analyzes the neighborhood infrastructure, develops common areas, organizes community activities, and engages residents in the project. The surrounding community, organized as a citizen association, is also informed and involved in the GNICE construction project and provided with information about energy advisors and consulting programs for renewable heating installations during the design phase of the project. Raumsinn has played a crucial role in gaining acceptance and commitment from the residents in the neighborhood. By involving residents in the process, their concerns were addressed, and trust began to develop. The cooperative planning process helped to minimize objections, which are common in such building projects when solutions cannot satisfy everyone. Strong relationships have been established with surrounding neighbors, future inhabitants, and companies, which are expected to play a vital role in GNICE SPEN as partners meet with external experts to discuss the next steps regarding the energy community and the involvement of neighbors. Raumsinn laid the groundwork in the early stages of the project, starting in 2019, and is expected to build upon this as tenants move in during the operational phase around 2025.

5 Discussion and Concluding Remarks

This chapter contributes to refining the structural-functionalism model for empirical investigations of governance models, particularly in the context of Positive Energy Districts (PEDs). Our findings challenge the adequacy of both total rational planning and incremental planning approaches in addressing the complexities inherent in neighborhood-scale projects like GNICE SPEN in Salzburg. The rational planning approach, characterized by its emphasis on predictability and continuity, falls short in capturing the nuanced dynamics of multilevel governance systems. In the context of Salzburg, where numerous initiatives operate beyond the formal project structure, rational planning fails to account for the temporal and informal influences that significantly impact decision-making processes. Moreover, the uncertain and interconnected nature of complex contexts renders the rigid framework of rational planning inadequate, as it struggles to accommodate dynamic power relations and ambiguous causality.

Conversely, while incremental planning offers flexibility and adaptability, our case demonstrates the necessity for some level of rational planning to provide guidance and maintain focus on overarching goals. The quality agreement established in our study exemplifies this need, serving as a flexible yet binding instrument to ensure alignment with project objectives. Thus, a balanced approach that integrates elements of both rational and incremental planning is essential to navigate the complexities of governance in projects like GNICE SPEN.

By adopting a communicative planning approach, Salzburg's case illustrates the potential for reconciling the tensions between rational and incremental planning. This approach emphasizes dialogue, negotiation, and collaboration among stakeholders, allowing for iterative decision-making processes that accommodate diverse perspectives and uncertainties. However, it also highlights the limitations of the structural-functionalism model in adequately capturing the cross-scale complexity of governance systems. The model's holistic framework fails to distinguish between intrainstitutional and interinstitutional relationships, overlooking the fluidity of power dynamics and the evolving nature of governance processes.

To conclude, our study underscores the importance of recognizing the limitations of traditional planning paradigms and advocating for more adaptive and inclusive approaches to governance. By embracing communicative planning principles and integrating elements of rational and incremental planning, stakeholders can navigate the complexities of multilevel governance systems more effectively, ultimately enhancing the success and sustainability of projects like GNICE SPEN.

Acknowledgments The research conducted in this chapter has received funding from the European Union's Horizon 2020 research and innovation program under grant agreement No 869918, SYN.IKIA Project. This chapter is partially based upon work from COST Action 19126 Positive Energy Districts European Network PED-EU-NET (see https://pedeu.net), supported by COST (European Cooperation in Science and Technology, http://www.cost.eu).

References

1. Biermann, F., Kanie, N., Kim, R.E.: Global governance by goal-setting: the novel approach of the UN sustainable development goals. Curr. Opin. Environ. Sustain. **26**, 26–31 (2017)
2. Sareen, S., et al.: Ten questions concerning positive energy districts. Build. Environ. **216**, 109017 (2022)
3. Jordan, A.: The governance of sustainable development: taking stock and looking forwards. Environ. Plann. C Gov. Policy. **26**(1), 17–33 (2008)
4. Maya-Drysdale, D., Jensen, L.K., Mathiesen, B.V.: Energy vision strategies for the EU green new deal: a case study of European cities. Energies. **13**(9), 2194 (2020)
5. Ackrill, R., Galanakis, K.: Policy Guidance for Human-Centric Urban Developments. Smart-BEEjS Consortium, Nottingham (2023)
6. Adger, W.N., Lorenzoni, I., O'Brien, K.L. (eds.): Adapting to Climate Change: Thresholds, Values, Governance. Cambridge University Press (2009)
7. O'Brien, K., Sygna, L.: Responding to climate change: the three spheres of transformation. Proceedings of transformation in a changing climate, University of Oslo, Oslo, pp. 16–23 (2013)
8. Yoo, H.K., Nguyen, M.-T., Lamonaca, L., Galanakis, K., Ackrill, R.: Socio-economic factors and citizens' practices, enabling positive energy districts. In: Challenging 'Silo Thinking' for Promoting PEDs. Smart-BEEjS Consortium, Nottingham (2020)
9. Sassenou, L.-N., Olivieri, L., Olivieri, F.: Challenges for positive energy districts deployment: a systematic review. Renew. Sust. Energ. Rev. **191**, 114152 (2024). https://doi.org/10.1016/j.rser.2023.114152
10. Krangsås, S.G., et al.: Positive energy districts: identifying challenges and interdependencies. Sustain. For. **13**(19), 10551 (2021)
11. Potts, R., Vella, K., Dale, A., Sipe, N.: Exploring the usefulness of structural–functional approaches to analyse governance of planning systems. Plan. Theory. **15**(2), 162–189 (2016). https://doi.org/10.1177/1473095214553519
12. Bellamy, R., Palumbo, A.: From Government to Governance. Ashgate, Farnham (2010)
13. Rittel, H.W., Webber, M.M.: Dilemmas in a general theory of planning. Policy. Sci. **4**, 155–169 (1973)
14. Van Bueren, E., Klijn, E., Koppenjan, J.: Dealing with wicked problems in networks: analyzing an environmental debate from a network perspective. J. Public Admin. Res. Theory. **13**, 193–212 (2003)
15. Kickert, W.J.M., Klijn, E.-H., Koppenjan, J.F.M. (eds.): Managing Complex Networks: Strategies for the Public Sector, p. 206. Sage, London, Thousand Oaks, New Delhi (1997)
16. Klijn, E.-H., Steijn, B., Edelenbos, J.: The impact of network management on outcomes in governance networks. Public Adm. **88**, 1063–1082 (2010)
17. Hoppe, T., Miedema, M.: A governance approach to regional energy transition: meaning, conceptualization and practice. Sustain. For. **12**(3), 915 (2020)
18. Lutz, L.M., Fischer, L.-B., Newig, J., Lang, D.J.: Driving factors for the regional implementation of renewable energy – a multiple case study on the German energy transition. Energy Policy. **105**, 136–147 (2017)
19. Yin, R.K.: Case Study Research: Design and Methods, 4th edn. Sage, Thousand Oaks (2009)
20. Denscombe, M.: The Good Research Guide: For Small-Scale Social Research Projects. Open University Press (1998)
21. Flyvbjerg, B.: Five misunderstandings about case-study research. Qual. Inq. **12**(2), 219–245 (2006). https://doi.org/10.1177/1077800405284363
22. Stake, R.E.: The Art of Case Study Research. Sage (1995)
23. Filatotchev, I., Wright, M.: Methodological issues in governance research: an editor's perspective. Corp. Gov. **25**(6), 454–460 (2017)

24. Cheng, C., Albert-Seifried, V., Aelenei, L., Vandevyvere, H., Seco, O., Nuria Sánchez, M., Hukkalainen, M.: A systematic approach towards mapping stakeholders in different phases of PED development—extending the PED toolbox. In: Littlewood, J.R., Howlett, R.J., Jain, L.C. (eds.) Sustainability in Energy and Buildings 2021, pp. 447–463. Springer Nature (2022). https://doi.org/10.1007/978-981-16-6269-0_38
25. Salom, J., Tamm, M., Andresen, I., Cali, D., Magyari, Á., Bukovszki, V., Balázs, R., Dorizas, P., Toth, Z., Zuhaib, S., Mafé, C., Cheng, C., Reith, A., Civiero, P., Pascual, J., Gaitani, N.: An evaluation framework for sustainable plus energy neighbourhoods: moving beyond the traditional building energy assessment. Energies. **14**, 4314 (2021). https://doi.org/10.3390/en15155646
26. Climate protection initiative "klimaaktiv" (bmk.gv.at). Accessed 10 Dec
27. Cheng, C., Kvellheim, A.K., Gaitani, N., Andresen, I.: Deliverable D6.1: a systematic approach to development, registration and reporting of innovations, syn.ikia (EU project) (2020)
28. Cheng, C., Kvellheim, A.K.: Deliverable D6.2: an exploitation strategy for syn.ikia partners and syn.ikia innovations, syn.ikia (EU project) (2022)
29. Cortes, C.: Deliverable 7.21 report on stakeholder and user engagement activities, syn.ikia (EU project) (2022)
30. Trulsrud, et al.: Deliverable 2.9 integrated energy design guidelines for sustainable plus energy neighbourhoods. syn.ikia (EU project) (2023)

Risk Management and Fault Detection Methods in Building Energy Renovation

Natalia Lastovets ⓘ, Janne Hirvonen ⓘ, Mohamed Elsayed ⓘ, and Piia Sormunen ⓘ

Contents

1 Introduction

The transition to a sustainable built environment demands a multidisciplinary approach that involves combining knowledge and techniques from different fields to tackle complex challenges. One such example is the integration of risk management and fault detection techniques in building energy renovation. This approach combines technical expertise and strategic planning to address the diverse challenges related to energy efficiency. [1].

To enhance the energy performance of existing buildings, building energy renovation practices require a nuanced understanding of risk management principles. This understanding ensures that renovation efforts meet the planned energy

N. Lastovets (✉) · J. Hirvonen · M. Elsayed
Tampere University, Tampere, Finland
e-mail: natalia.lastovets@tuni.fi

P. Sormunen
Tampere University, Tampere, Finland

Granlund Oy, Helsinki, Finland

© The Author(s) 2025
M. Kioumarsi, B. Shafei (eds.), *The 1st International Conference on Net-Zero Built Environment*, Lecture Notes in Civil Engineering 237,
https://doi.org/10.1007/978-3-031-69626-8_100

performance targets and adhere to the highest reliability, efficiency, and occupant comfort standards. The literature reveals a broad spectrum of risks associated with building energy renovation [2]. These risks range from the delivery and installation of heating, ventilation and air conditioning (HVAC) systems to the optimisation of these systems under uncertainty and the development of innovative business models for energy management in urban districts.

In addition to the risks associated with building energy renovation, there are challenges in the maintenance and operation of building energy systems. Fault detection and diagnosis are critical components of this process, as they can significantly improve operational efficiency and energy savings. The process involves identifying deviations from normal operation and diagnosing the type of problem or its location [3]. Previous studies indicate that operational faults can increase energy costs by up to 30%, highlighting a significant issue for building operators who often lack the resources for effective fault detection and resolution [4]. Fault detection can significantly improve operational efficiency and energy savings, providing a robust foundation for continuous energy management practices such as monitoring-based commissioning. However, despite the potential benefits, implementing fault detection technologies faces several hurdles, including a lack of standardised methodologies and the complexity of building energy systems [5]. The analysis of fault detection and diagnosis in building systems reveals a need for improved algorithms and consistent use of terminology and definitions [6].

The maintenance and operation of building energy systems, particularly HVAC installations, involve significant risks such as project delays and cost overruns. Research by Mosaad et al. [7] highlights the complexities and interdependencies within construction activities, while Sun et al. [8] propose an integrated framework for enhancing HVAC efficiency through uncertainty and sensitivity analysis at the buildings' level. Innovations in business models for energy management at the district level, as explored by Sepponen and Heimonen [9], and smart building energy management strategies, like the conditional value-at-risk model by Feng et al. [10], illustrate the shift towards advanced analytical methods and comprehensive risk assessment to implement energy-efficient solutions effectively.

Furthermore, the integration of smart building technology and HVAC systems in buildings offers a new perspective on managing uncertainties associated with renewable energy and fluctuating conditions, promoting innovative strategies for real-time energy optimisation. Sepponen and Heimonen's work [9] emphasises stakeholder-centric approaches to renewable energy adoption, advocating for flexible and sustainable business models to overcome barriers to energy-efficient urban development. The findings from Sepponen's study, suggest that the field of building energy renovation is moving toward an integrated approach combining technical innovation with strategic risk management and business modelling. The previous aforementioned studies collectively advocate for a shift from traditional, prescriptive methodologies towards more dynamic, analytics-driven approaches that can accommodate the complexities of modern building systems and the uncertainties inherent in their operation.

2 Methods

This study adopted a mixed-methods approach to explore risk management and fault detection methodologies in building energy renovation, focusing on identifying current practices, challenges, and suggestions for improvements. The research was conducted in two primary phases: a comprehensive literature review and targeted Finnish expert interviews.

The literature review of existing academic and industry literature to establish a foundational understanding of the current state of energy efficiency, risk management practices, and fault detection technologies within the realm of building renovations. Sources included peer-reviewed journals, industry reports, and regulatory documents, with the aim of identifying knowledge gaps, inconsistencies in terminology, and variations in methodological approaches. The findings from the review were used to formulate the interview questions by highlighting key areas of interest and concern within the field.

Following the literature review, semi-structured online interviews were conducted with a carefully selected group of professionals, including facility managers and service providers of building services and automation systems in Finland. These participants were chosen based on their extensive experience and expertise in commissioning, maintaining, and monitoring the building's energy performance. Demographic information about the interviews is shown in Fig. 1. The interviews were focused on the insights into the practical challenges of building commissioning, implementation of technologies, effective risk management strategies and fault detection techniques. Questions were open-ended to allow for the exploration of topics not covered in the literature review.

Interviews were video-recorded and subjected to thematic analysis. The responses were analysed to identify recurring themes and patterns related to risk management practices, technical challenges, and areas for improvement. The study

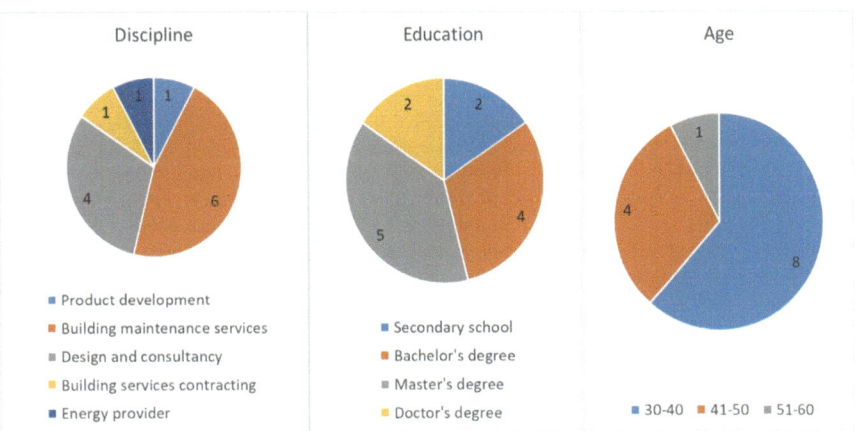

Fig. 1 Demographic information of the interviewees

was conducted in accordance with ethical guidelines for research with human participants. All interviewees provided informed consent, and measures were taken to ensure the confidentiality and anonymity of the participants and their responses.

3   Results

Table 1 organises the collected data from the interviews into three main categories: risk management practices, challenges, and suggestions for improvements. This arrangement allows for a comprehensive understanding of the current state of risk management in building energy systems and offers insights into potential areas of enhancement. The results highlight technical and operational problems that obstruct effective risk management. These issues range from technical difficulties with equipment to the complexities of managing hybrid systems, which require advanced technical knowledge to oversee modern building energy systems. Complex and

Table 1 Practices for risk management, challenges, and improvements

Challenges	Practices for risk management	Suggestions for improvements
Service Providers		
Technical issues with system equipment Navigating the complexity of hybrid systems Issues with sensor placement accuracy Integration of data from different systems and providers Lack of historical data trends or difficulty in accessing building data	Proactive management of component failures Prevention of inter-component interference Early fault detection with multiple sensors and data analysis Data collection at adequate clarity and resolution	Commitment to regular and data-driven system maintenance Enhanced staff training on fault detection and response, including hands-on experience of systems Comprehensive documentation on system operations and failure management Communication with the client on the uses and benefits of building data Sharing of good practices between rival companies
Facility Management		
Adequacy of the electric power connection in the building for use with an electric heating system Adjusting to space usage changes Energy price fluctuations High setup and maintenance costs	Ensuring stability and reliability of heating systems Maintaining a sustainable long-term temperature regime Design and control flexibility for future adjustments	Temperature, flow, and pressure measurements for benchmarking and fault detection Utilising smart building management systems for dynamic adjustments Leveraging data analytics for optimised energy use and forecasting Investment in advanced sensor technology and integration of a unified analytics platform

hybrid systems face operational difficulties, such as inadequate sensor placement and insufficient documentation, which make monitoring and managing them difficult. These challenges reveal the gaps in current practices, particularly in sensor deployment, documentation, and data analysis integration for early fault detection. They suggest a more informed and proactive approach to risk management is necessary.

To properly monitor the energy efficiency in building systems, there should be parallel metering of temperature, flow rate and pressure, which are all connected to the building automation system with appropriate alarm setup. These meters should be placed before and after heat exchangers to identify leaks, fouling, stuck valves and general suboptimal operation. Room condition monitoring is also essential to confirm system operation. Parallel metering of different variables enables the separation of sensor issues from actual system malfunctions. Data-based benchmarking can also be utilised for pre-emptive and demand-based maintenance. Integration of different systems must be done carefully to avoid conflicting operations, such as simultaneous heating and cooling from centralised and local units. It is important to collect dynamic data for long periods to establish historical trends and expected operation levels. A common problem is a too-short storage period of data, which prevents identification of the origins of observed malfunctions.

Figure 2 outlines the four phases of risk management: Risk Identification, Risk Analysis, Risk Response Planning, and Implementation and Monitoring. The figure is designed to guide stakeholders through a structured approach to managing risks in building energy renovation projects.

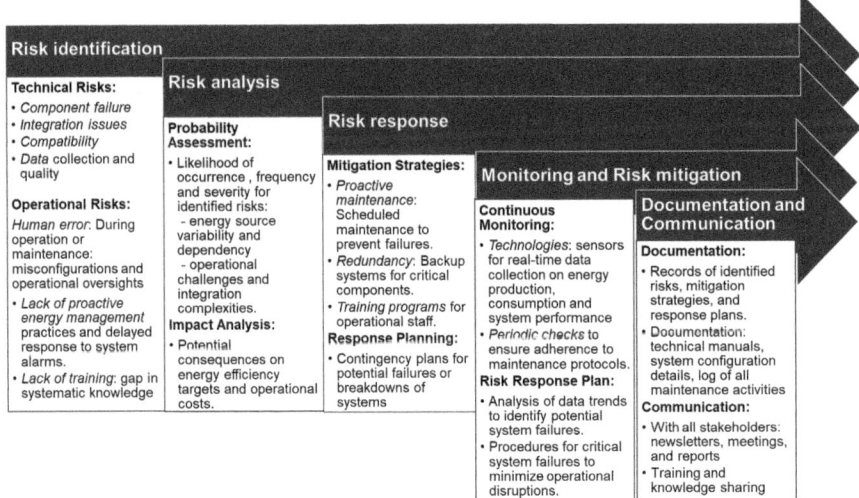

Fig. 2 Proposed process of risk management in building energy projects

The figure shows the first step of Risk Identification, where potential risks associated with building energy systems are identified. Potential issues are anticipated by analysing historical data and performing predictive analytics. A risk analysis evaluates identified risks according to their severity and probability, prioritising them for effective resource allocation and threat preparation. The Risk Response Planning process develops strategies for mitigating, transferring, or accepting prioritised risks when necessary. Implementation and Monitoring involve implementing risk response strategies, adjusting them in real time, and detecting new risks as early as possible. Continually review and adjust the risk management approach to stay aligned with objectives and external realities.

The results underscore the complexities involved in managing risks in building energy systems, highlighting significant challenges and proposing a structured approach to improvement. Table 1 reveals a gap in current practices, particularly in the areas of sensor placement, documentation, and the integration of data analysis for early fault detection. Figure 2 complements these findings by proposing a comprehensive risk management process that emphasises the need for ongoing monitoring and adaptation. Together, these elements contribute to a deeper understanding of the challenges and potential pathways for enhancing energy efficiency and operational reliability in building renovations.

4 Discussion

This research ventured into the complex terrain of risk management and fault detection in the context of building energy renovations, revealing significant insights while also encountering limitations that pave the way for future studies. The reliance on expert interviews and literature reviews, while robust, underscored the need for more empirical data to validate the proposed process of risk management. The generalisability of the findings could be enhanced through quantitative research, incorporating larger datasets from a wider array of buildings and systems.

Future research directions emanating from these limitations include the need for empirical studies that test the proposed risk management model in diverse real-world settings. The integration of quantitative data would allow for a more robust validation of the findings and could help in the development of predictive models for risk management and fault detection. Another avenue for future studies is the exploration of the impact of emerging technologies, such as artificial intelligence and machine learning, on the efficiency and effectiveness of fault detection systems. These technologies hold the potential to transform current practices but require thorough investigation to understand their practical applications and limitations.

5   Conclusions

This research focused on investigating the intricate process of managing risks and detecting faults in building energy systems. The main goal was to develop a risk management protocol to enhance energy efficiency and improve operational reliability. The research methodology involved gathering insights from experts and conducting a comprehensive review of existing literature in order to identify practical strategies for improving risk management practices in this field. As a result of the interview analysis, existing risk management practices were mapped out, prevailing challenges were identified, and a series of targeted improvements were articulated. The preliminary suggestion for the risk management model, as derived from the synthesis of expert interviews and a comprehensive literature review, emphasises an integrated, adaptive approach tailored to the unique challenges of building energy systems. This model advocates for a four-phased methodology: identification, analysis, response, and monitoring of risks, underscored by a continuous feedback loop for dynamic adaptation. These approaches are essential for addressing the intricate nature of modern building systems and the array of uncertainties they encounter. The findings revealed the necessity for a holistic strategy that combines technical innovation, strategic risk management, and flexible business modelling. Such an approach not only aims to mitigate immediate operational risks but also aligns with broader goals of sustainability and energy conservation.

References

1. Abuimara, T., Hobson, B.W., Gunay, B., O'Brien, W.: A data-driven workflow to improve energy efficient operation of commercial buildings: a review with real-world examples. Build. Serv. Eng. Res. Technol. **43**(4), 517–534 (2022)
2. Topouzi, M., Owen, A., Killip, G., Fawcett, T.: Deep retrofit approaches: managing risks to minimise the energy performance gap. In: ECEEE 2019 Summer Study Proceedings, pp. 1345–1354. European Council for an Energy Efficient Economy (2019)
3. Li, Y., O'Neill, Z.: A critical review of fault modeling of HVAC systems in buildings. In: Building Simulation, vol. 11, pp. 953–975. Springer, Berlin Heidelberg (2018)
4. Belfast, J.: Fault diagnostics tools for commercial buildings—applications, algorithms and barriers. Energy Eng. **111**(3), 57–78 (2014)
5. Granderson, J., Lin, G., Harding, A., Im, P., Chen, Y.: Building fault detection data to aid diagnostic algorithm creation and performance testing. Sci. Data. **7**(1), 65 (2020)
6. Melgaard, S.P., Andersen, K.H., Marszal-Pomianowska, A., Jensen, R.L., Heiselberg, P.K.: Fault detection and diagnosis encyclopedia for building systems: a systematic review. Energies. **15**(12), 4366 (2022)
7. Mosaad, S.A.A., Issa, U.H., Hassan, M.S.: Risks affecting the delivery of HVAC systems: identifying and analysis. J. Build. Eng. **16**, 20–30 (2018)
8. Sun, Y., Gu, L., Wu, C.J., Augenbroe, G.: Exploring HVAC system sizing under uncertainty. Energ. Buildings. **81**, 243–252 (2014)

9. Sepponen, M., Heimonen, I.: Business concepts for districts' energy hub systems with maximised share of renewable energy. Energ. Buildings. **124**, 273–280 (2016)
10. Feng, W., Wei, Z., Sun, G., Zhou, Y., Zang, H., Chen, S.: A conditional value-at-risk-based dispatch approach for the energy management of smart buildings with HVAC systems. Electr. Power Syst. Res. **188**, 106535 (2020)

Measurement and Verification Framework for Clusters of Buildings: A Comprehensive Approach to Validating and Quantifying Savings in Large-Scale Building Interventions

Georgios Siokas ⓘ, Athanasios Balomenos ⓘ, Evangelos Fekas, Georgios Triantafyllis ⓘ, and Yannis Kopsinis ⓘ

Contents

1 Introduction

By 2030, urban areas are expected to have gradually concentrated over 70% of the world's population [1], leading to a growing demand for thermal energy [2]. In the European Union (EU), buildings consume approximately 40% of EU energy for heating, cooling, and domestic hot water usage [3], of which around 35% exceed 50 years of age and demonstrate nearly 75% energy inefficiency. Simultaneously, the average annual energy renovation rate remains relatively low at approximately 1% [4].

G. Siokas (✉) · A. Balomenos · Y. Kopsinis
LIBRA AI Technologies PC, Athens, Greece
e-mail: gsiokas@libramli.ai

E. Fekas · G. Triantafyllis
National Technical University of Athens, Athens, Greece

© The Author(s) 2025
M. Kioumarsi, B. Shafei (eds.), *The 1st International Conference on Net-Zero Built Environment*, Lecture Notes in Civil Engineering 237,
https://doi.org/10.1007/978-3-031-69626-8_101

EU wants to change this situation by aiming for a more sustainable future and urging building owners and managers to minimise their energy consumption [5]. The public bodies are trying to set in motion plans to renovate buildings owned by them to meet the annual target of renovating at least 3% of their total floor area and reducing their annual energy consumption reduction by 1.9% [6]. This raised the need to invest in energy conservation measures (ECMs), energy efficiency measures (EEMs), or interventions. These measures consist of potential replacements or upgrades of existing systems that enhance efficiency in a cost-effective manner and are recommended to save or reduce utility costs, e.g. energy and gas.

When implementing an intervention in a building or a set of buildings, the actual energy savings cannot be directly quantified. Alternately, savings are calculated by comparing the energy consumption before (baseline period) and after (reporting period) the implementation of interventions while adjusting to condition changes. Based on the literature, systematic approaches for determining the true impact of an investment in energy efficiency are collectively referred to as measurement and verification (M&V) methods, supported by a well-established set of guidelines, directives, and protocols, e.g. International Performance Measurement and Verification Protocol (IPMVP) [7, 8].

M&V is the process of planning, measuring, collecting, and analysing data resulting from ECMs implementation [8]. This methodology is used as a "means of verification" for energy savings achieved by a system capable of adjusting to the user's needs and empowering building occupants by providing precise and relevant information on operational energy consumption and applied interventions [8]. At the same time, it provides a clearer estimation of the actual energy savings achieved by the interventions.

This tool can be helpful for organisations that monitor numerous buildings and are obliged to meet specific energy efficiency measures. Relying on recent advances in M&V, the paper proposes a methodology for applying M&V with large-scale aggregated data during an operational year, making suitable adjustments to the models for changes in the multiple buildings (e.g. environmental, operational changes or renovations).

The proposed methodology is applied to Danish residential buildings in the Municipality of Aalborg [9]. This way, one shared adjusted baseline model is built per cluster of buildings to predict post-intervention energy consumption in the future.

2 M&V Framework with Large-Scale Aggregated Data

The proposed methodology is divided into two phases: (1) the period before applying the interventions (pre-ECM period) and (2) the period after the implementation and activation of the interventions (post-ECM period). The pre-ECM period includes data gathering, data processing, selecting the M&V option, measurement

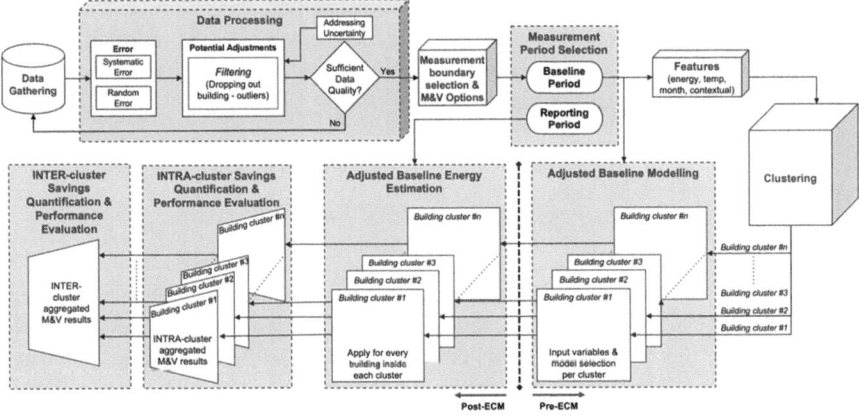

Fig. 1 Methodology for applying M&V on a dataset with multiple buildings

period selection, feature extraction, clustering, variable selection, and finally, baseline model selection for each cluster.

For the post-ECM period, the performance of each intervention is evaluated, and the energy savings are estimated per model and cluster. Multiple-building modelling focusses on a common processing method, model, and set of input variables for all buildings inside each cluster instead of simply summing individual consumptions of each building and modelling them. The proposed methodology formulated to apply M&V on a dataset with multiple buildings is presented in Fig. 1.

2.1 Step 1—Data Preparation and Project Definition

When handling data from multiple buildings, it is essential to (i) understand the functionality and typology of each building and (ii) define the necessary input data variables for adjusted baseline model development. For this step, the relevant information concerns the building's characteristics, data sources, periods with available data, energy types and consumptions, and factors impacting the performance of the intervention and monitoring process (e.g. sensor failure).

2.2 Step 2—Data Processing

The available data are validated for the minimum requirements for clustering and developing the M&V model. Regarding multiple buildings, non-routine adjustments are not easily identified (e.g. facility size, change in occupancy and occupant activities); thus, only the routine adjustments are considered input, and buildings with irregular energy consumption patterns are omitted. So, the available datapoints are processed to identify potential system failures, random errors, and corrupted data to evaluate the data quality and reduce uncertainty. The minimum datapoints recommended are the total energy, temperature, and relative humidity regarding the outdoor conditions of the buildings.

2.3 Step 3—Measurement Boundary Selection

Per the available information, the measurement boundary and type of energy savings are selected. In the case of multiple buildings, there are two measurement boundaries to be defined: (i) at a macro level, encompassing all the multiple buildings, and (ii) at a micro level, relating to each building. Adhering to the guidelines by IPMVP, option C (whole facility) is the most appropriate [10]. Additionally, several considerations impact the selection process: (i) the period of year in which the intervention occurs, influenced by weather conditions and the type of energy monitored; (ii) the selection of an appropriate statistical model; and (iii) the data frequency, as it is advised to use monthly data to avoid weekly variations [11].

2.4 Step 4—Measurement Period Selection

Two periods are needed to be defined. The period with constant and stable conditions before the ECM is characterised as the baseline, and after the intervention, it is called the reporting period. The selected baseline period should correspond to a normal and full operating cycle of the buildings concerning energy consumption without interventions, representing all operating modes (maximum energy consumption to a minimum) to characterise the savings effectiveness in normal operating conditions. Under real circumstances, such a period may not always be available; therefore, assumptions must be made. For the selection of the reporting period, there are two options: (i) the definition of a common reporting period for all the buildings in the sample or the cluster, which leads to a wider gap between the two periods or (ii) define the reporting period for each building separately.

2.5 Step 5—Explanatory Variables, Features, and Clustering

At first, this pre-processing step focusses on selecting the appropriate explanatory variables and the feature for the clustering. The explanatory variables impact the model creation and the period selection. They can either originate from the dataset to be generated, e.g. quadratic transforms of all the other explanatory variables, or index-derived variables, such as the day of the month. Usually, the most impactful variables are indicators relevant to the weather.

For the feature selection, the methodology proposes a handcrafted feature extraction scheme instead of another due to (i) dimensionality reduction of the feature space is achieved as it is computationally intensive to apply multivariate time series (MTS) clustering, (ii) interpretability of clusters with different patterns (heating vs. cooling) are grouped, and (iii) contextual information integration of buildings' characteristics.

Having the desirable features generated, all buildings are separated into clusters by applying the K-Means clustering algorithm [12, 13], based on the Elbow method [14]. The factors impacting the efficacy of the final clusters formed are [15, 16]: (i) the number of clusters (groups) (K), (ii) the initial cluster centres impact the final cluster formation, as the K-means algorithm is non-deterministic, and (iii) the outliers impact the cluster formation and thus need to be processed. Additionally, the data undergoes scaling using a Minmax scaler, followed by principal component analysis (PCA) to reduce the data dimensionality. The PCA is configured to retain 90% of the variance in the data.

Upon determining the number of clusters, a visualisation check is implemented. Using the raw multivariate time series, the variables per cluster are mapped into a 3-D histogram, with energy, weather, and month on each axis. The histogram is flattened and concatenated with additional binary-encoded contextual features, if available. This step aids in validating the selected number of clusters by revealing shared energy patterns within specific groups of buildings.

2.6   Step 6—Adjusted and Selected Baseline Modelling

Following the assignment of each building to a specific cluster, a testing procedure is conducted to allocate a baseline model and the appropriate transformation for variables within each cluster, utilising a chosen metric evaluated across a predetermined number of representatives.

During the creation of the adjusted baseline modelling, different models are tested in each cluster with different explanatory variables to identify the model with the least test errors in all the indicators [11]. The selected model for different building types or clusters can produce accurate enough adjusted baseline predictions. The baseline model for each cluster is developed by using one of the methods [11, 17, 18]:

1. Linear regression (LR).
2. Theil-Sen regressor (TSR).
3. Random sample consensus (RANSAC) regressor.
4. Huber regressor (HR).
5. Bayesian ridge regressor (BRR).
6. Least-angle regression (LARS).
7. Automatic relevance determination regression (ARD Regression).

2.7   Step 7—Performance Evaluation Per Cluster

A random selection of 10 representatives from each cluster is proposed, with each building trained in all available models and transformations. Subsequently, the model's performance is based on three dimensionless indicators $(e_{m,b}{}^x)$ specific to the comparison of different heterogeneous datasets [11, 19]: (i) the mean ratio to first (mrtf), averaging the ratio of each model error to the smallest model error obtained for each dataset, without normalising the maximum error; (ii) the mean ranking position (mrp), ranking position obtained by the model on each building; and (iii) the first ration (fr), proportion of times a model ranks first. Eq. 1, Eq. 2, and Eq. 3 calculate the metrics accordingly, with M the number of models and B the number of buildings.

$$e_m^{(\text{mrtf})} = 100 \frac{1}{B} \sum_{b=1}^{B} \frac{e_{m,b}}{\min_{m \in M(e_{m,b})}} \tag{1}$$

$$e_m^{(\text{mrp})} = \frac{1}{B} \sum_{b=1}^{B} \text{rank}_{M(e_{m,b})} \tag{2}$$

$$e_m^{(\text{fr})} = \frac{1}{B} \sum_{b=1}^{B} 1 \left(\text{rank}_{M(e_{m,b})} = 1 \right) \tag{3}$$

The cumulative errors for each cluster are computed, and the model and transformation yielding the minimum cumulative error are identified. This process facilitates the determination of the optimal model and transformation for each cluster.

Additionally, three metrics are used in conjunction to evaluate the performance and select the most accurate reference model for each cluster [8, 11, 17]: (i) the coefficient of variation of root mean square error (CV-RMSE), depending on the time scale, (ii) the mean absolute scaled error (MASE), and (iii) the normalised mean bias error (NMBE), quantifying the tendency of a model to over or underestimate across a series of values.

The model performs better when the value of CV-RMSE is the lowest, minimising the uncertainty. The metric MASE is independent of the data's scale; therefore, if the forecast is better than the naive forecast computed in-sample, the scaled error is less than one; otherwise, the scaled error is greater than one. The NMBE is more versatile and valuable for globally ranking variables and algorithms, leading to a common baseline model for several buildings. Finally, one model will be selected based on the best evaluation metrics among the algorithms and the selected baseline period.

2.8   Step 8—Intra- and Intercluster and Aggregated Saving Quantification and Performance Evaluation

At this stage, the dataset is M&V-ready and is poised for application to the selected reporting period. Based on the literature, the first results indicate the energy consumption of the buildings in each cluster. They are characterised as INTRA-cluster aggregated M&V results, and the generalised results for the total number of buildings are presented as inter-cluster aggregated M&V results [11].

Following the predictions, the differences between the predicted and actual values are calculated. The cumulative differences across all days are computed to ascertain the predicted savings. This process provides insight into the energy savings achieved after the interventions. The general Eq. (4) to calculate savings is:

$$\text{Savings} = (\text{Adjusted Baseline Energy—Post} - \text{ECM Energy}) \pm \text{Adjustments} \quad (4)$$

3 Integration, Deployment, and Results

3.1 Project Definition and Data Preparation and Processing

The proposed methodology is applied to the Aalborg Municipality, the largest city in the North Denmark region, with a population of 215.328. The dataset comprises 3127 Danish residential buildings (apartments, single-family houses, or terraced houses) equipped with smart commercial heat meters and connected to the district heating network owned and operated by Aalborg Municipality [9]. The analysis focusses on the 2459 single-family houses. Since the dataset was processed before publication, all the data was considered input. At this point, it is essential to mention that the dataset consists of information regarding the construction year, the type of building, and, if available, the energy label and the heat measure for 3021 buildings [9]. Additionally, the selected dataset has only information concerning the energy consumption before an ECM. Therefore, the methodology was applied to build one shared adjusted baseline model for several buildings belonging to a cluster.

3.2 Measurement Boundary and Period Selection

Concerning the measurement boundary, option C (whole facility) is selected for calculating the total thermal heating energy and the corresponding selected baseline period from 02/01/2018 to 31/12/2019 (2 full operational years) and the reporting period 01/01/2020 till 31/12/2020 (1 full operational year). The choice of the reporting period is made to align with the paper's requirements and facilitate the application of the methodology at each stage.

3.3 Features Selection and Cluster Definition

After filtering the available datapoints, three explanatory variables were selected: (i) outdoor temperature, (ii) enthalpy, and (iii) dew point (see Fig. 2). Along with these three variables, two more variables were created and used: (v) quadratic

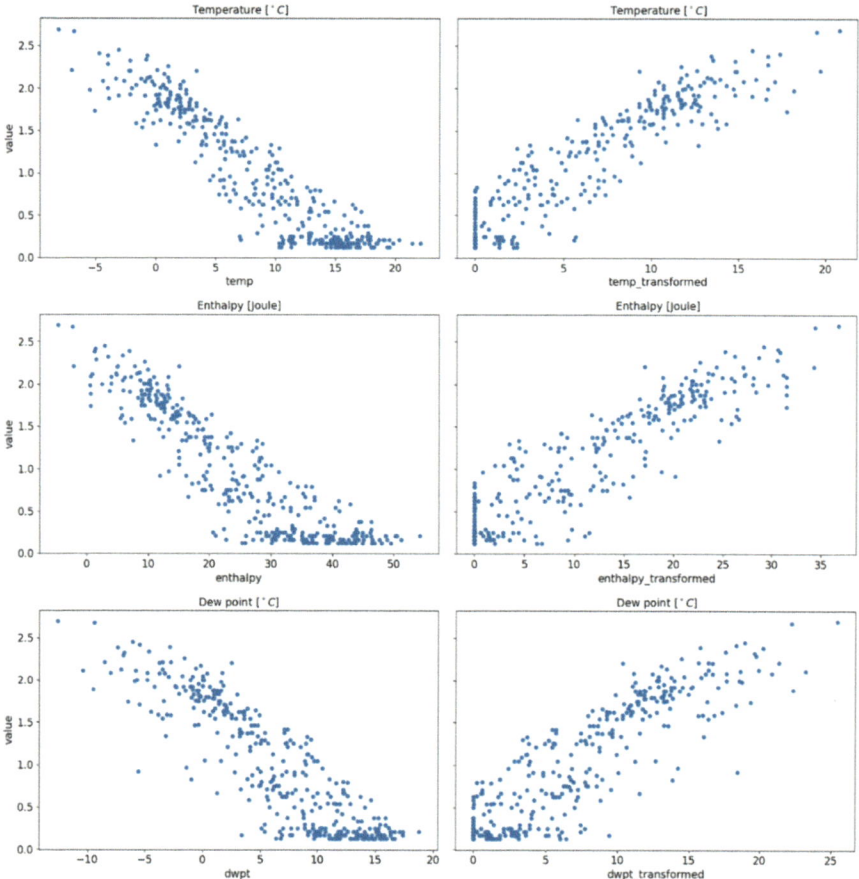

Fig. 2 Graphic representation of temperature, enthalpy, and dew points and their transformations

transforms of all above and (vi) index derived variables, such as day of week and month of the year.

According to the methodology, a series of features were selected for testing, as listed in the documentation by tsfresh Github repo [20]. A removal procedure eliminates features that do not effectively separate the data. This removal process entails discarding binary and constant data and data closely resembling an initial value, subject to a specified threshold. The final concatenated features are used as input for clustering.

For the clustering process, the selected extracted features were used by the K-Means algorithm, based on the Elbow method [21]. Several runs are recommended for sparse high-dimensional problems. Subsequently, the K-means clustering is employed with a predetermined number of clusters. The appropriate number of clusters was $k = 12$ (see Fig. 3), in which all the buildings were visually presented (see Fig. 4). The final distribution of the buildings is presented in Table 1.

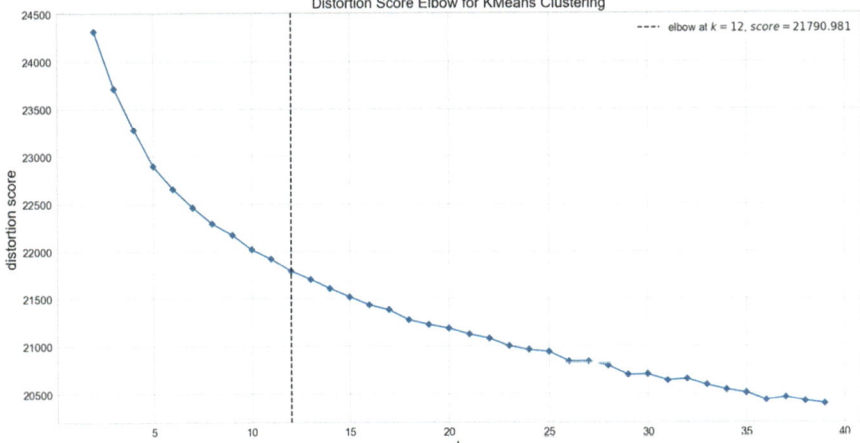

Fig. 3 Distortion score elbow for KMeans clustering

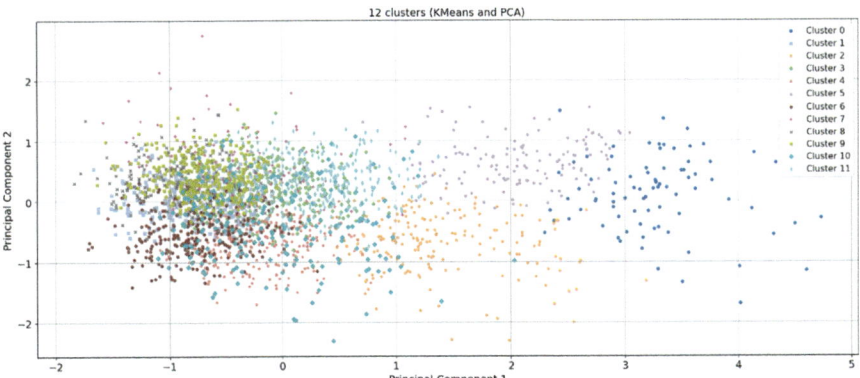

Fig. 4 12 Clusters (KMeans and PCA)

Table 1 The distribution of the single-family buildings in the clusters

Cluster number	0	1	2	3	4	5	6	7	8	9	10	11
Number of buildings	210	181	377	156	315	51	205	195	225	28	262	254

The clustering approach offers the advantage of rapid execution, attributed to the efficient computational characteristics of K-means clustering. In supplementary, the silhouette score was calculated (-0.01386).

So, a time series with measurements for one year each consists of 365 (energy, temp, month) combinations, which are assigned to a bin of the 3-D histogram. After, the 3D histogram is flattened and concatenated with additional binary-encoded contextual features given (unit type and energy label) and the construction year of the building. From the histogram, we observe the energy signatures of the different building types (clusters) for extended periods (e.g. during summer), while others have zero consumption only for a few months.

3.4 Adjusted Baseline Model Selection and Performance Evaluation

A testing procedure is conducted to allocate a baseline model and the appropriate transformation for variables within each cluster, developing a model evaluated across a predetermined number of 10 representatives.

This was followed by selecting the best combination of variables and models. The primary indicator for comparison inside the cluster was the first ration (fr) equation, and where there was a tie between models, the mrtf equation and mrp equation were applied consecutively. Additionally, three metrics were used, CVRMSE, NMBE, and MASE, to verify that the selected model for each building is appropriate. Thus, by feeding the explanatory variables of the reporting period into the selected model for each cluster, we obtain the most appropriate adjusted baseline model with the relevant transformations (see Table 2).

3.5 Results of M&V Analysis

Through the comparison of the adjusted baseline per building with the monitoring consumption in the reporting period, the final per-cluster (intra) savings are calculated. Then, the savings per cluster are combined to calculate the final (inter) savings. Based on the available dataset, the total savings are estimated to be -1.35%, indicating that the model fits the actual energy consumption closely. As no interventions were recorded, the savings should have been 0%. Consequently, the model demonstrates a well-fitted representation of the observed energy consumption.

4 Discussion

By identifying the different clusters of the buildings, a model per building was deployed based on the cluster to which it belonged. This way, each building had its baseline model, and by aggregating them, the final (inter) savings were close to 0%. This result allows us to assume that the M&V methodology was applied successfully and the adjusted baseline model is ready to be applied once the ECMs occur. In other words, one baseline model is developed per cluster, characterising them as "M&V ready", meaning that when intervention occurs and the reporting period is sufficient, the developed models can be applied to extract the savings achieved.

Table 2 The performance of the selected baseline model per cluster[a]

Cluster number	0	1	2*	3	4*	5*	6*	7*	8*	9	10	11
Baseline model (estimator)	HR	OLS	OLS	HR	RANSAC	HR	RANSAC	HR	HR	HR	HR	OLS
fr score[b]	4	5	4	5	2	4	3	4	3	5	3	5

[a]The clusters with the asterisk (*) indicate that more than one model with their transformed variable had the same fr score
[b]Number of representatives which selected the model (max: 10)

We concluded that by clustering the meters that show a similar pattern, we can choose a model and transformation for all the houses of the cluster because of their similar nature. This process may incur some inaccuracies compared to studying each building separately, but considering the time and computational power resources needed and the performance indicators, the results are satisfactory overall. The advantage is that we do not have to find each house's regression model and transformation and repeat the same process multiple times.

5   Conclusions and Future Work

The proposed methodology contributes to evaluating the energy savings of multiple buildings during an operational year, covering all the seasons, by comparing the adjusted energy consumption with the energy consumption from the previous phase of the buildings (i.e. without ECMs). This methodology can be extended to all types of energies (i.e. thermal or electrical), including gas, power, and water consumption.

When the M&V is applied in multiple buildings, the optimisation strategy of energy consumption within a set of buildings must be validated without aggregating energy data. This tool can assist public and private organisations that manage multiple buildings in better monitoring the impact and performance of energy-saving initiatives and interventions. This will help them align with the EU energy consumption and efficiency guidelines.

Summarising, the methodology helps develop a generic model per cluster based on the energy type. Therefore, it helps minimise the computational effort of calculating the energy consumption of a series of buildings separately, simultaneously achieving reliable predictions. In general, the modelling process is simplified. However, applying the methodology in a real case is essential to estimate the savings achieved in a wider area of buildings.

Acknowledgements   This work is part of the PRELUDE project. This project has received funding from the European Union's Horizon 2020 research and innovation programme under Grant Agreement N° 958345.

References

1. World Urbanization Prospects – Population Division – United Nations. https://population.un.org/wup/Download/. Last accessed 2021/05/12
2. Waite, M., Cohen, E., Torbey, H., Piccirilli, M., Tian, Y., Modi, V.: Global trends in urban electricity demands for cooling and heating. Energy. **127**, 786–802 (2017)
3. European PPP Expertise Centre – Guidance on Energy Efficiency in Public Buildings. https://www.eib.org/epec. Last accessed 2024/02/19
4. European Commission – New rules to boost energy performance of buildings. https://ec.europa.eu/commission/presscorner/detail/en/ip_23_6423. Last accessed 2024/02/28
5. European Commission – Energy Efficiency Directive. https://www.eceee.org/policy-areas/product-policy/energy-efficiency-directive/. Last accessed 2024/02/19

6. European Commission – Energy efficiency targets. https://energy.ec.europa.eu/topics/energy-efficiency/energy-efficiency-targets-directive-and-rules/energy-efficiency-targets_en. Last accessed 2024/02/28
7. multEE – Monitoring and Verification Schemes. https://multee.eu/topics/monitoring-and-verification-schemes. Last accessed 2023/11/08
8. EVO: International Performance Measurement and Verification Protocol (IPMVP) – Core Concepts. Efficiency Valuation Organization, Washington, DC (2022)
9. Schaffer, M., Tvedebrink, T., Marszal-Pomianowska, A.: Three years of hourly data from 3021 smart heat meters installed in Danish residential buildings. Scientific Data. **9**(1), 1–13 (2022)
10. EVO: International Performance Measurement and Verification Protocol. https://evo-world.org/en/products-services-mainmenu-en/protocols/ipmvp. Last accessed 2024/02/18
11. Agenis-Nevers, M., Wang, Y., Dugachard, M., Salvazet, R., Becker, G., Chenu, D.: Measurement and verification for multiple buildings: an innovative baseline model selection framework applied to real energy performance contracts. Energ. Buildings. **249**, 111–183 (2021)
12. Akhiat, Y., Asnaoui, Y., Chahhou, M., Zinedine, A.: A new graph feature selection approach. In: 2020 6th IEEE Congress on Information Science and Technology (CiSt) on Proceedings, pp. 156–161. IEEE, Agadir – Essaouira (2020)
13. Cadenas, J.M., Garrido, M.C., Martínez, R.: Feature subset selection filter-wrapper based on low quality data. Expert Syst. Appl. **40**(16), 6241–6252 (2013)
14. Shi, C., Wei, B., Wei, S., Wang, W., Liu, H., Liu, J.: A quantitative discriminant method of elbow point for the optimal number of clusters in clustering algorithm. EURASIP J. Wirel. Commun. Netw. **31**, 1–16 (2021)
15. Selecting the number of clusters with silhouette analysis on KMeans clustering – scikit-learn 1.2.1 documentation. https://scikit-learn.org/stable/auto_examples/cluster/plot_kmeans_silhouette_analysis.html#selecting-the-number-of-clusters-with-silhouette-analysis-on-kmeans-clustering. Last accessed 2023/01/30
16. Yuan, C., Yang, H.: Research on K-value selection method of K-means clustering algorithm. J MDPI. **2**(2), 226–235 (2019)
17. Gallagher, C.V., Leahy, K., O'Donovan, P., Bruton, K., O'Sullivan, D.T.J.: Development and application of a machine learning supported methodology for measurement and verification (M&V) 2.0. Energ. Buildings. **167**, 8–22 (2018)
18. Sun, Y., Haghighat, F., Fung, B.C.M.: A review of the-state-of-the-art in data-driven approaches for building energy prediction. Energ. Buildings. **221**, 110022 (2020)
19. Hastie, T., Friedman, J., Tibshirani, R.: The Elements of Statistical Learning: Data Mining, Inference, and Prediction, 2nd edn. Springer, New York (2009)
20. tsfresh: Overview on extracted features – tsfresh 0.20.1.post0.dev10+gaf039f9 documentation. https://tsfresh.readthedocs.io/en/latest/text/list_of_features.html. Last accessed 2024/02/21
21. Elbow Method: Yellowbrick v1.5 documentation. https://www.scikit-yb.org/en/latest/api/cluster/elbow.html. Last accessed 2024/02/29

An Unmanned Aerial Vehicle-Based Digital Twin Framework for Inspection and Assessment of Bridge Structures

Ibrahim Odeh and Behrouz Shafei ⓘD

Contents

1 Introduction

Predicting, evaluating, and measuring the extent of damage in different materials and structures has been a growing research interest where relying on various models such as mathematical, quantitative, material-based, and other models that can assess the damage [1–5]. Recently, implementing unmanned aerial vehicles (UAVs), artificial intelligence (AI), and machine learning (ML) to detect the damage and then incorporate that into the digital twin concept have introduced a transformative potential to various industries by enhancing predictive analytics, simulation, and real-time monitoring capabilities. A digital twin is a virtual representation of a physical object,

---

I. Odeh · B. Shafei (✉)
Department of Civil, Construction, and Environmental Engineering, Iowa State University, Ames, IA, USA
e-mail: shafei@iastate.edu

© The Author(s) 2025
M. Kioumarsi, B. Shafei (eds.), *The 1st International Conference on Net-Zero Built Environment*, Lecture Notes in Civil Engineering 237,
https://doi.org/10.1007/978-3-031-69626-8_102

system, or process that mirrors its state in the real world. Leveraging UAVs and ML/AI in digital twins for bridges significantly benefits infrastructure management, emphasizing damage detection, timely maintenance, and efficient inspection [6–10].

Building a 3D model of a bridge by using accelerometers, UAVs, and surveillance cameras was the framework proposed by [11, 12] for the inspection and maintenance of bridges. Moreover, A systematic review published by [13] consolidates the latest findings on using bridge information modeling, finite element modeling, and bridge health monitoring for creating digital twins of bridges without AI or ML capabilities. Furthermore, a 3D digital twin model for bridge structures as a part of the next generation of bridge maintenance systems was proposed by [14]. These models are crafted through 3D scanning techniques and alignment-based parametric modeling, incorporating comprehensive geometrical details. Recent studies were published that discussed and assessed the transportation networks and lifetime resilience of the bridges through proposed and developed frameworks for the entire process [15, 16].

This study aims to provide a framework and recommendations for generating a 3D bridge model by utilizing UAVs, photogrammetry software, and AI capabilities in order to have a precise and digital copy of bridges. This innovative approach aims to produce a precise digital replica of the intended bridge structure. Moreover, the study introduces a novel methodology for conducting safe, swift, accurate, and remote inspections, enabling the examination of difficult-to-access areas of the bridge. This methodology enhances the efficiency and effectiveness of bridge inspections and significantly improves decision-making, safety, and accessibility for inspectors, potentially transforming conventional practices in bridge maintenance and monitoring.

2 Digital Twin Framework

Figure 1 represents the structured framework of the present study for bridge inspection and assessment utilizing modern technology. The process begins with data collection, where drones are deployed to gather comprehensive data (2D images) of the bridge. This data is then meticulously filtered to isolate relevant information. Next, defects identified in the data are annotated, a crucial step that feeds into training a specialized AI model. Once the model is trained, it can run a detection algorithm to spot and highlight anomalies within the bridge structure. Following this, by feeding the data into the software with AI capabilities, a 3D bridge model is generated, providing a detailed and manipulable representation of the bridge for further inspection. The inspection process can be initiated using any available device (such as PCs, tablets, and even smartphones) with access to a main server, enabling inspectors to conduct thorough evaluations remotely. Finally, the gathered information and analysis aid in decision-making regarding the bridge's maintenance needs. This high-technology, systematic approach enhances bridge inspection's safety, accuracy, and efficiency.

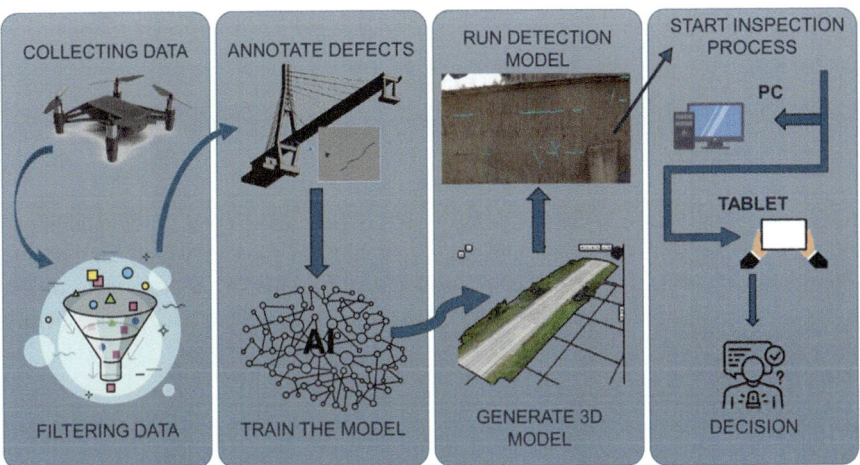

Fig. 1 The framework developed for the current study

(a)   (b)

Fig. 2 The DJI drones (**a**) Mavic 2 Pro, and (**b**) Phantom 4 RTK

3 Data Collection

Utilizing drones for image collection has stood out as a transformative technology in bridge assessment. Specifically, drones with high-resolution cameras offer a unique vantage point that facilitates unparalleled data acquisition when applied to bridge mapping. Two types of drones were used in the current study to collect the image datasets, DJI Mavic 2 Pro and DJI Phantom 4 RTK (Fig. 2).

Both drones are widely used in various professional fields. Table 1 present the specifications for Mavic 2 Pro and Phantom 4 RT, respectively.

4 Data Quality and Resolution

Data quality and resolution of images captured by drones hold an essential role in the effectiveness of bridge inspections and 3D modeling. When it comes to structural assessment and maintenance of bridges, detailed and high-resolution imagery is

Table 1 Camera properties in Mavic 2 Pro and Phantom 4 RTK drones

Camera specification	Mavic 2 Pro	Phantom 4 RTK
CMOS	1″	1″
Megapixels	20	20
Lens	f/2.8-f/11	f/2.8-f/11
FOV	77°	84°
Electronic rolling shutter	8-1/8000 s	8-1/8000 s
Resolution	5472 × 3648	5472 × 3648
ISO range	100–12,800	100–12,800
Flight time	30 min	30 min
Flight speed	44.7 mph	31 mph

Table 2 Number of images and resolution compared to the literature

Reference	Images dataset	Resolution (pixels)
[11]	458	3024 × 4032
[14]	277	4928 × 3264
[17]	15	3904 × 2196 2048 × 1152
[18]	200	4928 × 3264
[19]	250	4928 × 3264
[20]	175	6000 × 4000
[21]	435	7952 × 5304
[22]	240	4928 × 3264 5152 × 3864
[23]	602	5152 × 3864 1280 × 720
[24]	1250	4160 × 3120
[25]	150	5472 × 3648
Present study	177	5472 × 3648

critical for identifying potential issues, such as cracks, corrosion, or other structural defects. Conversely, operating the drone at a greater distance ensures safety and can provide a broader view of the bridge, which is useful for overall structural assessments. However, it may compromise the ability to capture minute details. Therefore, finding the optimal distance is key. This distance is often determined by the drone's camera specifications, the desired resolution, and the need to balance detailed inspection with operational safety. Table 2 compares the current study with the literature on both resolution and the number of images in the datasets.

The first dataset consisting of 177 images was received from the Mavic 2 Pro drone. 111 images were also received from the Phantom drone. The quality for both datasets was 5472 × 3648 pixels. Figure 3 presents an example from the Mavic 2 Pro, while Fig. 4 shows an example from the Phantom 4 RTK. The Mavic drone flew at a distance of 50 feet from the bridge, while the Phantom drone covered the area from a distance of about 180 feet.

Fig. 3 Example for an image captured by Mavic 2 Pro drone from 50 feet distance

Fig. 4 Example for an image captured by Phantom 4 RKT drone from 180 feet distance

The difference in flying distances between Mavic 2 Pro and Phantom 4 RKT from the bridge offers a valuable opportunity for comprehensive data collection. Additionally, it allows for a meaningful comparison of the quality of the 3D models generated by each drone. By examining how much detail the cameras capture at different distances, insights can be gained into the precision and level of information each drone provides. In Fig. 3, the closer proximity of the drone to the bridge allowed for a more detailed capture of the bridge's structural defects, such as cracks. This contrast appears when compared to Fig. 4, where the drone was positioned farther away (180 feet), which resulted in a reduction in capturing the bridge defects. This difference underlines the importance of drone positioning in accurately generating and assessing a 3D model, demonstrating that closer aerial flight can reveal more accurate and critical details that assist with generating an accurate model. Incorporating the optimal drone distance into the inspection strategy significantly enhances drone-based bridge inspection effectiveness and generates precise 3D models. It ensures that the images captured are of high quality and resolution, providing a reliable basis for accurate assessment and informed decision-making in bridge maintenance and safety protocols.

5 Generated 3D Models

In this study, the primary focus is on the 3D modeling of the bridge. By utilizing Bentley Software (with AI capabilities), Pix4D, and Autodesk ReCap Pro, high-quality 3D models have been generated from the images captured by drones for multiple bridges. This integration of photogrammetry software into the digital twin concept not only ensured a more detailed and accurate representation of the bridge's current condition, it also enabled a comprehensive analysis of its structural condition. The 3D models provided a multi-dimensional perspective, allowing a more thorough examination of expected damages and areas requiring maintenance that were missed or there were difficulties in reaching.

In this case, a bridge characterized by its unique construction and current condition was examined. The bridge, built in 1922, stands as a century-old structure, located 19 miles south of Ames, Iowa. It features a robust design with steel beams supported by abutments and topped with a reinforced concrete deck. Figure 5 shows a comparison among the images generated by the three software packages. The difference in the 3D model quality can be noted, where the Bentley software exhibits the highest and most accurate model among the three software packages, while all of them used the same dataset to generate the model. Bentley software has powerful measurement capabilities such as location, distance, area, and volume. Incorporating and utilizing these capabilities with high-quality 3D models offers the safest and fastest inspection and damage detection process. In order to represent the measurement tools in Bentley software, some defects were measured in the generated bridge.

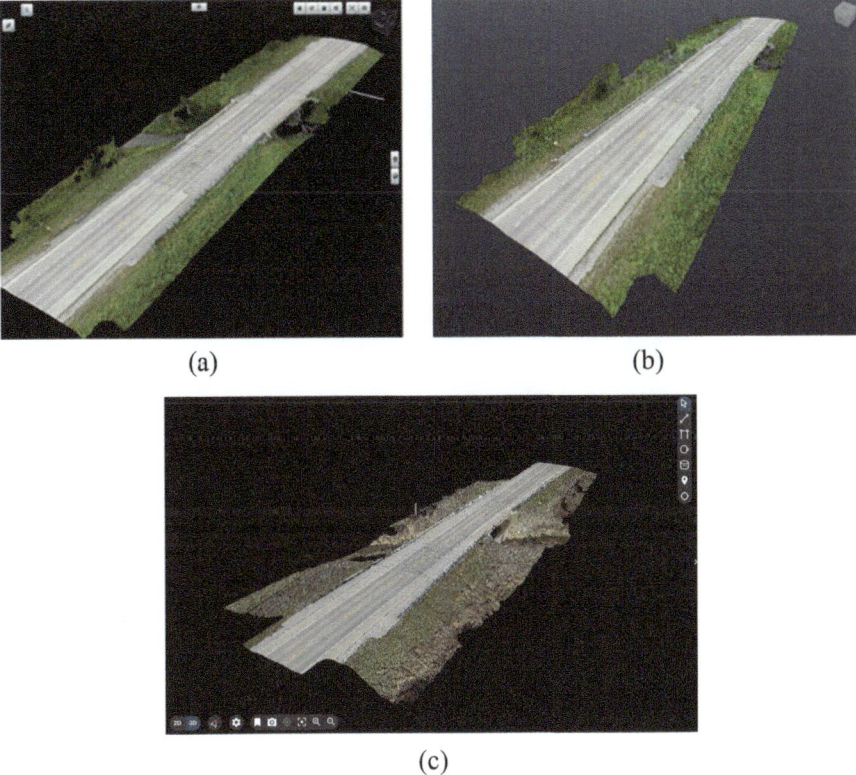

Fig. 5 A comparison among the images generated by (**a**) Bentley software, (**b**) Pix4D, and (**c**) Autodesk ReCap Pro

6 Implementing AI

The integration of AI technology represents a significant advancement in the field of infrastructure management. By preparing and training the AI algorithm, software packages like Bentley can automate and enhance various aspects of its operations. AI can be utilized for more efficient and accurate feature recognition in captured data, improving the identification of defects and anomalies during inspections. In this study, AI capabilities were applied to detect potholes and cracks in the case study bridge. Figure 6 shows the before and after applying the detection model. The software learns from the provided pre-trained weight to detect the cracks (in blue line) and the potholes (green shadow). A high accuracy was obtained because the detection method was trained to detect per pixel in the image, what is often called semantic segmentation.

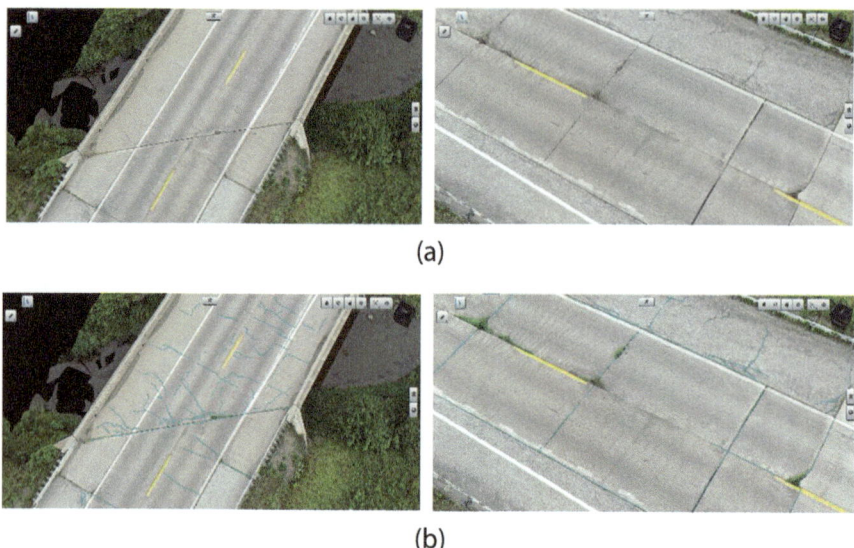

(a)

(b)

Fig. 6 The 3D model of the bridge (**a**) before and (**b**) after applying the detection algorithm

7 Conclusions

This chapter introduces creating a 3D bridge model using three distinct software packages. In addition, an AI algorithm was implemented into the model to identify defects in the bridge's 3D model. This direct approach leverages the precision of visualization capabilities alongside the analytical power of AI to ensure the integrity and safety of the bridge assessment. Implementing AI capabilities exhibits a powerful combination of measurement tools and detection algorithms. The measurement tools show varsity and accurate scales while comparing them with the actual measurements. Training the detection model on a custom dataset can capture different types of defects directly on the 3D model. This introduces tangible benefits to improve the inspection and maintenance of bridge structures.

Acknowledgments The research study, results of which reported in this chapter, was partially sponsored by the US National Science Foundation (NSF) and the Iowa Department of Transportation (Iowa DOT). The authors would like to acknowledge and thank the sponsors for their support. Opinions, findings, and conclusions expressed in this chapter are of the authors and do not necessarily reflect those of the sponsors.

References

1. Saini, D., Shafei, B.: Prediction of extent of damage to metal roof panels under hail impact. Eng. Struct. **187**, 362–371 (2019)
2. Abu-Farsakh, G.A., Odeh, I.N.: A new damage-based failure criterion for nonlinear behavior of fibrous composite materials. Int. J. Damage Mech. **32**(7), 940–961 (2023)
3. Saini, D., Shafei, B.: Damage assessment of wood frame shear walls subjected to lateral wind load and windborne debris impact. J. Wind Eng. Ind. Aerodyn. **198**, 104091 (2020)
4. Khatami, D., Shafei, B., Bektas, B.: Data-assisted prediction of deterioration of reinforced concrete bridges using physics-based models, ASCE. J. Infrastruct. Syst. **29**(2), 05023003, 1–13 (2023)
5. Khatami, D., Shafei, B.: Impact of climate conditions on deteriorating reinforced concrete bridges in the U.S. Midwest region, ASCE. J. Perform. Constr. Facil. **35**(1), 04020129, 1–11 (2021)
6. Mohammadi, M., Rashidi, M., Mousavi, V., Karami, A., Yu, Y., Samali, B.: Quality evaluation of digital twins generated based on UAV photogrammetry and TLS: bridge case study. Remote Sens. **13**(17), 3499 (2021)
7. Shim, C.S., Dang, N.S., Lon, S., Jeon, C.H.: Development of a bridge maintenance system for prestressed concrete bridges using 3D digital twin model. Struct. Infrastruct. Eng. **15**(10), 1319–1332 (2019)
8. Teng, S., Chen, X., Chen, G., Cheng, L.: Structural damage detection based on transfer learning strategy using digital twins of bridges. Mech. Syst. Signal Process. **191**, 110160 (2023)
9. Bono, A., D'Alfonso, L., Fedele, G., Filice, A., Natalizio, E.: Path planning and control of a UAV fleet in bridge management systems. Remote Sens. **14**(8), 1858 (2022)
10. Gohari, H., Berry, C., Barari, A.: A digital twin for integrated inspection system in digital manufacturing. IFAC-PapersOnLine. **52**(10), 182–187 (2019)
11. Zhou, C., Xiao, D., Hu, J., Yang, Y., Li, B., Hu, S., et al.: An example of digital twins for bridge monitoring and maintenance: preliminary results. In: Proceedings of the 1st Conference of the European Association on Quality Control of Bridges and Structures: EUROSTRUCT 2021, 1st edn, pp. 1134–1143. Springer International Publishing, Cham (2022)
12. Jiménez Rios, A., Plevris, V., Nogal, M.: Bridge management through digital twin-based anomaly detection systems: a systematic review. Front. Built Environ. **9**, 1176621 (2023)
13. Dang, N., Kang, H., Lon, S., Shim, C.: 3D digital twin models for bridge maintenance. In: Proceedings of the 10th International Conference on Short and Medium Span Bridges, Quebec City, QC, Canada, vol. 31 (July 2018)
14. Cha, Y.J., Choi, W., Büyüköztürk, O.: Deep learning-based crack damage detection using convolutional neural networks. Comput. Aided Civ. Inf. Eng. **32**(5), 361–378 (2017)
15. Alipour, A., Shafei, B.: An overarching framework to assess the life-time resilience of deteriorating transportation networks in seismic-prone regions. J. Resilient Cities Struct. **1**(2), 87–96 (2022)
16. Khatami, D., Hajilar, S., Shafei, B.: Investigation of oxygen diffusion and corrosion potential through a cellular automaton framework. J. Corros. Sci. **187**, 109496, 1–10 (2021)
17. Malek, K., Mohammadkhorasani, A., Moreu, F.: Methodology to integrate augmented reality and pattern recognition for crack detection. Comput. Aided Civ. Inf. Eng. **38**(8), 1000–1019 (2022)
18. Bao, Y., Li, J., Nagayama, T., Xu, Y., Spencer Jr., B.F., Li, H.: The 1st international project competition for structural health monitoring (IPC-SHM, 2020): a summary and benchmark problem. Struct. Health Monit. **20**(4), 2229–2239 (2021)
19. Perry, B.J., Guo, Y., Mahmoud, H.N.: Automated site-specific assessment of steel structures through integrating machine learning and fracture mechanics. Autom. Constr. **133**, 104022 (2022)

20. Zheng, Y., Gao, Y., Lu, S., Mosalam, K.M.: Multistage semisupervised active learning framework for crack identification, segmentation, and measurement of bridges. Comput. Aided Civ. Inf. Eng. **37**(9), 1089–1108 (2022)
21. Feng, C., Zhang, H., Wang, S., Li, Y., Wang, H., Yan, F.: Structural damage detection using deep convolutional neural network and transfer learning. KSCE J. Civ. Eng. **23**, 4493–4502 (2019)
22. Zhai, G.H., Narazaki, Y., Wang, S., Shajihan, S.A.V., Spencer, B.F.: Synthetic data augmentation for pixel-wise steel fatigue crack identification using fully convolutional networks. Smart Struct. Syst. **29**(1), 237–250 (2022)
23. Deng, J.H., Lu, Y., Lee, V.C.S.: Imaging-based crack detection on concrete surfaces using You Only Look Once network. Struct. Health Monit. **20**(2), 484–499 (2021)
24. Li, S., Zhao, X.: Image-based concrete crack detection using convolutional neural network and exhaustive search technique. Adv. Civ. Eng. **2019**, 6520620 (2019)
25. Liu, Y., Gao, M.: Detecting cracks in concrete structures with the baseline model of the visual characteristics of images. Comput. Aided Civ. Inf. Eng. **37**(14), 1891–1913 (2022)

Circular Bio-based Walls: Applications and Challenges

Valeria Cascione ⓘ, Barrie Dams ⓘ, Matt Roberts ⓘ, Andrew Shea ⓘ, Dan Maskell ⓘ, Stephen Allen ⓘ, Pete Walker ⓘ, and Stephen Emmitt ⓘ

Contents

1   Introduction

The construction industry is responsible of three fifths of the total UK waste [1], and it generates high resources consumption and greenhouse emissions [2]. The high environmental impacts of the construction industry are linked to the linear economy approach of the sector ('take, make, waste'). To achieve net-zero targets, it is essential for the construction sector to introduce circular economy principles [3]. Circular economy is a framework that is based on three principles: eliminate waste and pollution, reuse and recycle resource, and regenerate nature [4].

V. Cascione (✉) · D. Maskell · S. Allen · P. Walker
Department of Architecture and Civil Engineering, University of Bath, Bath, UK

Institute of Sustainability, Bath, UK
e-mail: v.cascione@bath.ac.uk

B. Dams · M. Roberts · A. Shea · S. Emmitt
Department of Architecture and Civil Engineering, University of Bath, Bath, UK

© The Author(s) 2025
M. Kioumarsi, B. Shafei (eds.), *The 1st International Conference on Net-Zero Built Environment*, Lecture Notes in Civil Engineering 237,
https://doi.org/10.1007/978-3-031-69626-8_103

Bio-based construction materials are materials that are derived from organic renewable and often fast-growing resources. Agricultural waste (flax, straw, hemp) and other recycled fibres (paper, textile), can be used to produce a variety of construction materials, such as insulation, boards, cladding, etc. [5] Because of the organic composition of bio-based materials, bio-based materials can sequester carbon from the environment, contributing to the reduction of carbon dioxide in the air [6]. Moreover, bio-based materials can be easily recycled or reintroduced in the biosphere as fertilisers and compost.

Although bio-based materials demonstrate circular economy principles [7], there is still scepticism in using them in the construction industry. Lack of confidence in their performance and overall benefits, questions on durability and costs are some of the reasons why bio-based materials have not proliferated in the construction sector [8].

To improve confidence in their performance and sustainability, this research uses bio-based materials in demountable wall panels to assess the feasibility and performances of such innovative wall technology. Three wall panel prototypes were designed to test the applicability and compatibility of different bio-based materials in walls, and to investigate different configurations to make walls demountable.

In this paper, an overview of the three panel prototypes is provided, followed by experimental results on the thermal performances of the wall panels. The wall panels were placed in a large environmental chamber (LEC) at the University of Bath, UK, to investigate the actual thermal transmittance of the walls. A life cycle assessment of the bio-based materials used in the wall panels following the EN15804(2014) is also proposed. The cradle to grave life cycle of the materials was assessed and discussed. This work highlights the benefits and challenges of using bio-based materials in buildings, while, at the same time, giving confidence in the applicability of bio-based materials in buildings.

2   Circular Bio-based Wall Panels

2.1   Design Methodology

Three wall panels were prototyped to inform the design of two living labs built in Ghent (Belgium) and in Kloetinge (Netherlands). The prototypes were designed to be adaptable and demountable. Adaptability is the ability to adapt building elements to different use and purpose, while demountablity refers to the reversibility, accessibility to all elements and reusability of the walls. The metrics for the design of the panels were based on the BS ISO 20887:2020 [9] that provides requirements and guidance for design for disassembly and adaptability in buildings. The following general principles for the wall panel design were applied to all prototypes:

- *Expandability*: design can be expanded.
- *Standardisation*: minimisation of the variations of components.

- *Access to components*: minimal work and no damage to access wall elements.
- *Safety of disassembly*: health and safety aspects considered when dismantling the panels.
- *Simplicity*: construct using the same basic structural element, so that the frame can be built in a factory environment
- *Reversible connections*: screw, gravity connections and other solutions that enable the reuse of materials and components.

2.2   Wall Panels Design

The three prototypes consist in a double timber frame panel (Fig. 1a), stainless steel frame (Fig. 1b) and cross laminated timber (CLT) panels (Fig. 1c).

2.2.1   Timber Frame Panel

The first prototype design (Fig. 1a) is based on a frame of C24 structural timber measuring a finished thickness of 42 mm. Two 1.8 m × 1.2 m frames were joined together, A 'crisscross' framing (the first frame containing horizontal timber bracing members and horizontal bracing in the second frame) was constructed to stiffen the main frame (Fig. 2a). Aluminium removable hook connectors were used to join the

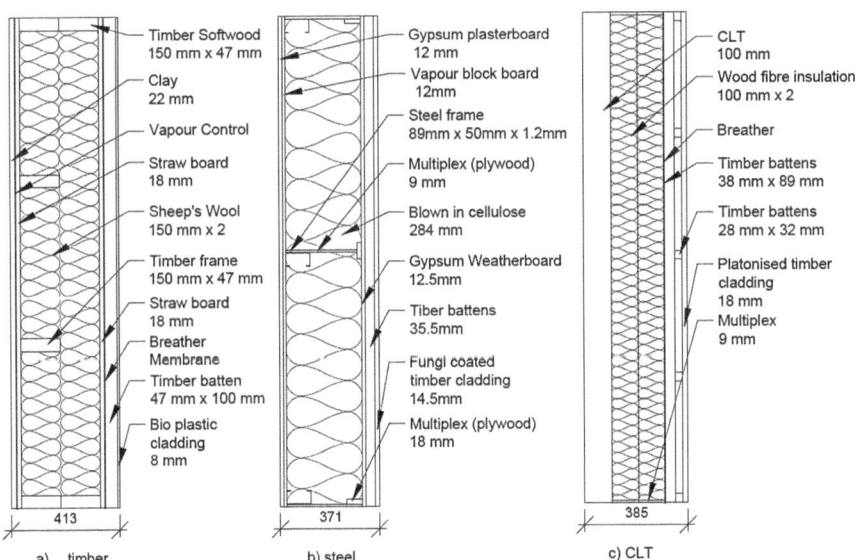

Fig. 1 Section of the three wall panels (mm): (**a**) first iteration timber frame panel, (**b**) steel frame panel, (**c**) CLT panel

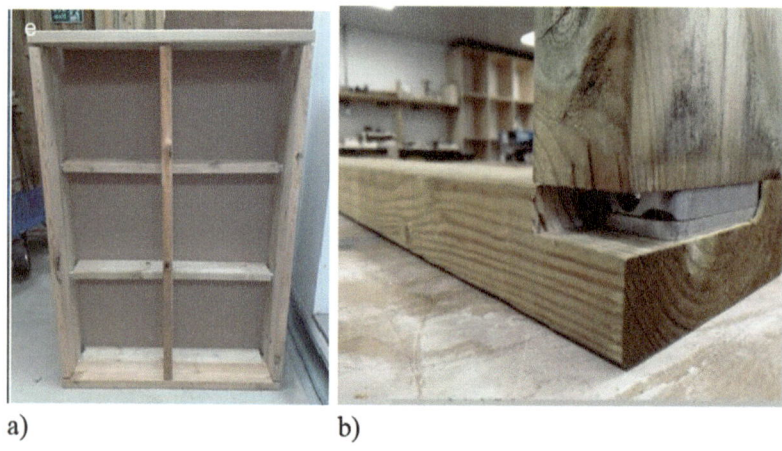

a)   b)

Fig. 2 Timber frame panel: (**a**) criss-cross design and (**b**) metal connections

perimetral frame elements together, and the bracing with the perimetral frame (Fig. 2b). These connectors are usually used to connect secondary timber beams to primary. Sheep's wool was selected as insulation material by considering cost, performances, sustainability and availability of the materials in construction-scale quantities [10, 11]. The choice of the façade material was a bio-based polymer composite material manufactured using reed, calcium carbonate precipitated from waste treatment, recovered toilet paper waste [12]. Straw based boards, produced with a soy binding agent were screwed on both sides of the timber frame. As internal finish, a clay board panel reinforced with a textile mesh and wood chips was selected. Commercially available moisture barriers (on the external side) and vapour control layers (on the internal side) were applied.

2.2.2 Stainless Steel Frame Panel

The second panel is a modular light gauge steel frame with blown-in cellulose insulation. Each steel module consists of 900 mm × 600 mm components. Two modules were used to design a 1200 mm × 900 mm wall (Fig. 3a). Internal sheathing boards were affixed to the steel frame, after the drilling of pilot holes into the frame. The internal structural sheathing board consisted of a chipboard with integrated vapour barrier, while the external one is a modified plasterboard with high resistance to fire and weathering. Plasterboard was used as internal finishing. As external façade, a fungi coated timber cladding was used. The cladding was manually coated with the mushroom coating that consists of a layer of linseed oil and a layer of a fungal cells [13].

Fig. 3 Steel frame panel: (**a**) modular light gauge steel frame, (**b**) cladding treated with fungi coating

a)   b)

2.2.3 CLT Panel

The structural element of the third panel was a Cross Laminated Timber (CLT) panel consisting of five laminations joined with a binding agent. Mycelium insulation is used for the prototype. Mycelium is the vegetative part of a fungus, consisting of a network of fine white filaments. 66 mm thick mycelium panels (Fig. 4a) were then joined using a bio-based binding agent. For the façade, a platonised timber cladding was selected (Fig. 4b). The platonisation process consists of a chemical modification of the wood surface through a drying and heating process that increase durability of timber.

3 Hygrothermal Properties

Panels were tested in a Large Environmental Chamber (LEC) at the Building Research Park of the University of Bath under controlled conditions (Fig. 5). The LEC consists of two chambers, with one simulating external environment and indoor conditions, respectively. The prototypes are placed in the central partition wall between the two rooms. The outdoor chamber was set up at 11 °C, while the indoor chamber at 25 °C. Humidity was free to vary. Heat flux, temperature and RH probes

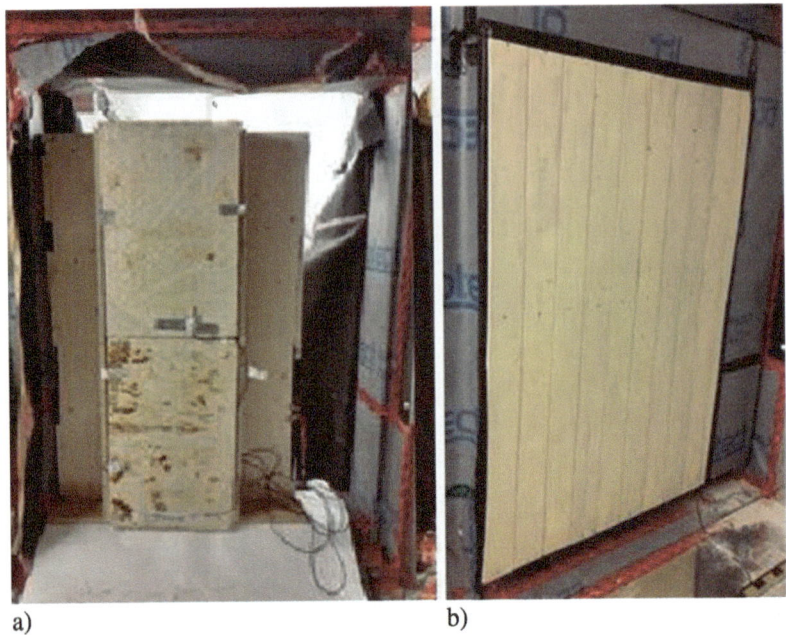

a)   b)

Fig. 4 CLT panel: (**a**) mycelium insulation and (**b**) timber cladding

Fig. 5 On the left, an external view of the LEC; on the right, a view of the partition view with the timber frame wall

were inserted in the walls in correspondence of each layer. Thermal transmittance was observed for 20 days.

From the heat flux and internal and external room temperature sensors, the average thermal transmittance for the 20 days testing was calculated. All the walls comply with the English Building Regulation (Part L) [14] that states that walls in new building should have a U-Value of 0.18 W/m^{2}K or lower. CLT and steel frame resulted in a U-Value of 0.16 and 0.17 W/m^2K, respectively. The timber frame prototype outperformed the others (0.12 W/m^2K), but this is due to mainly the significant difference in the thickness of the walls (Fig. 1). Sheep's wool insulation is 300 mm thick, compared to cellulose (284 mm) in the steel frame and mycelium (200 mm) in CLT. There are also some variations on the thermal conductivity of the insulation materials. The declared thermal conductivity for sheep wool and cellulose is 0.039 and 0.40 W/mK, respectively. The declared thermal conductivity is provided by the manufactures. Mycelium thermal conductivity was measure at the University of Bath and it resulted 0.046 W/mK.

4 Life Cycle Assessment of Bio-Based Components Materials

4.1 Life Cycle Inventory and Method

The LCA of the insulation, boards and claddings used in the three panels was performed (Table 1). The aim is to understand whether bio-based materials can help reduce overall Global Warming Potential (GWP100) of the panels. The LCA was conducted to investigate the potential environmental impacts of the circular bio-based building materials by following EN 15804, 2012 + A2, 2019 [15]. The cradle-to-grave impacts were considered by including life cycle stages within Modules A-C. Module D was not included in the study as there are many uncertainties on the loads and benefits of recycling and on the calorific energy produced by bio-based materials when incinerated. The service life of the materials was assumed to be 60 years, which is a standardised reference life span for buildings [16]. A functional unit of 0.15 W/m^2K for 1 m^2 of insulation was set up, while for the other materials a declared unit of 1 m^2 was defined, as the thickness of the materials is predefined by the manufacturers. The LCA study was conducted using an attributional framework. LCA was modelled using openLCA version 1.10.3. For background data and, when necessary, for foreground data, ecoinvent 3.6 implemented in EuGeos 15804+A2 IA v4 Unit Processes [17] was used, to enable life cycle impact assessment (LCIA) according to the EN 15804 (2012) + A2 (2019) standard.

Further information on the LCA method can be seen in [18]. The process modelling for insulation, structural, finishing, cladding, sheeting is reported in [18, 19], except for gypsum weather board, vapour block board and platonised timber, for which process modelling is described below (Tables 2, 3, and 4). For

Table 1 Materials assessed using LCA

Insulation	Finishing	Cladding	Sheeting
Mycelium	Clay board	Bioplastic	Straw board
Sheep's wool	Plasterboard	Fungi coating	Vapour block particle board
Cellulose		Platonised wood	Gypsum weather board

Table 2 Manufacturing of gypsum weather board (density 860 kg/m^3)

Flows	Quantity (kg)	Process	Source
Input			
Glass wool mat	0.0056	Market for glass wool mat—GLO	Ecoinvent
Gypsum Plasterboard	0.98	Market for gypsum plasterboard—GLO	Ecoinvent
Polyester resin	0.014	Market for polyester resin, unsaturated—RER	Ecoinvent
Output			
Gypsum weather	1		

Table 3 Manufacturing of the vapour block particle board (density 720 kg/m^3)

Flows	Quantity (kg)	Process	Source
Input			
Acrylic dispersion, without water, in 65% solution state	0.087	Market for acrylic dispersion—RER	Ecoinvent
Carbon dioxide, in air	1.52	–	Ecoinvent
Particle board, for indoor use	0.0014	Particle board production—RER	Ecoinvent
Output			
Vapour block board	1		

Table 4 Manufacturing of 1 kg of platonised wood cladding (density 460 kg/m^3)

Flows	Quantity	Process	Source
Input			
Platonisation	0.09 m^2	See Table 6	–
Carbon dioxide, in air	1.81 kg	–	Ecoinvent
Sawnwood, board, softwood, dried ($u = 10\%$), planed	0.0012 m^3	Planning, board, softwood, $u = 10\%$—RoW	Ecoinvent
Output			
Platonised cladding	1 kg		

the end-of-life of the three materials, data can be found in Tables 9 and 10. The Life Cycle Inventory (LCI) was built by collecting data from literature and manufacturers.

4.2   Life Cycle Assessment Results

The results in Table 5 show that materials like sheep wool, cellulose and platonised timber have a low environmental impact, which make the materials suitable to work towards net-zero carbon target. To achieve net-zero it is necessary to improve

Table 5 GWP_{100} of the construction materials ($kgCO_{2e}/m^2$)

Insulation		Finishing		Cladding		Sheeting	
Mycelium	51.2	Clay board	8.6	Bioplastic	46.4	Straw board	6.99
Sheep's wool	10.9	Plasterboard	2.1	Fungi coating	83.9	Vapour block board	7.35
Cellulose	15.7			Platonised	1.4	Gypsum weather board	5.70

recyclability routes for bio-based materials, reduce the amount of chemicals or find more sustainable chemical substitutes to reduce emissions. Chemicals are usually added to bio-based materials to improve durability and performances (especially fire resistance). It is, however, important to notice that not all bio-based materials (Table 5) have low carbon emissions. In this study mycelium produced 215.6 $kgCO_2e/m^2$, which is unusual for a material that is 100% composed of bio-based materials (fungi and plant-based substrate). The reason behind such high value is a combination of high uncertainty around the data collection of mycelium manufacturing and mycelium is not yet widely available on the market. Mycelium, (as well as bioplastic and fungi coating) are relatively new products that did not go through the optimisation process when production is scaled up to industrial scale. Hence, data for LCA are usually collected at laboratory scale or assumption were made [18, 19].

5 Conclusions

Integrating bio-based materials and circular economy principles into wall construction can potentially enhance indoor comfort by providing good thermal insulation and moisture regulations. This paper explored the use of bio-based materials in dismountable wall panel prototypes. The experimental testing showed that the walls have excellent thermal performances with U-Values below 0.18 W/m^2K. Further investigation on the moisture regulation performances in the long term to improve indoor comfort is recommended. The LCA, meanwhile, highlighted some challenges in accurately assessing the carbon footprint of bio-based materials, as some innovative materials are not yet well-established on the market. Some bio-based materials have not yet gone through industrial scale-up and optimisation, and/or data for LCI are not available. Good quality and primary data are necessary to assess the actual environmental impact of bio-based materials, together with life cycle costing analysis. Overall, this paper demonstrated that there is a good potential for bio-based materials to achieve net-zero targets while creating more comfortable living spaces.

Acknowledgements The Circular Bio-based Construction Industry (CBCI) project is funded by the European Union Regional Development Fund Interreg 2 Seas Mers Zeeen (2S05-036). The authors are grateful for the interchange made possible with a range of academic and industrial partners including BBRI, KU Leuven, Emergis, Vonhaut and Agrodome under the Interreg programme.

Additional Materials (Tables 6, 7, 8, 9, and 10)

Table 6 Platonisation process per m$_2$ of timber (0.025 m thick)

Flows	Quantity	Process
Input		
Accelerator	0.01 kg	See Table 7
Acrylic dispersion, without water, in 65% solution	0.14 kg	Market for acrylic dispersion, without water, in 65% solution state—RER
Wood protector	0.06 kg	See Table 8
Electricity, medium voltage	1.5 MJ	Electricity production, medium voltage, petroleum refinery operation—Europe
Heat, DISTRICT, or industrial, natural gas	49 MJ	Market for heat, DISTRICT, or industrial, natural gas—Europe
Output		
Platonisation	1 m^2	

Table 7 Accelerator production for kg of material

Flows	Quantity (kg)	Process
Input		
Methylene diphenyl diisocyanate	0.36	Market for methylene diphenyl diisocyanate—RER
Toluene diisocyanate	0.64	Market for toluene diisocyanate—RER
Output		
Accelerator	1	

Table 8 Wood protector production for kg of material

Flows	Quantity	Process
Diesel, burned in agricultural machinery	2.2 MJ	Diesel, burned in agricultural machinery—GLO
Electricity, medium voltage	4.4 MJ	Electricity production, medium voltage, petroleum refinery operation—Europe without Switzerland
Heavy fuel oil, burned in refinery furnace	3 MJ	Heavy fuel oil, burned in refinery furnace—Europe without Switzerland
Linseed	2.86 kg	Market for linseed—GLO
Nitrogen fertiliser, as N	0.05 kg	Nutrient supply from manure, liquid, cattle—GLO
Phosphoric acid, fertiliser grade, without water, in 70% solution state	0.02 kg	Market for phosphoric acid, fertiliser grade, without water, in 70% solution state—GLO
Wood protector	1 kg	
Waste organic	1.86 kg	–

Table 9 Sorting of wastes at the end of life

	Incinerated (%)	Landfill (%)	Recycling (%)
Vapour block board	96	4	0
Gypsum weather board	0	5	95
Platonised cladding	9	91	0

Table 10 End-of-life flows for incineration and landfill

Material	Percentage (%)	Waste
Vapour block board	9	Waste emulsion paint, on wood
	7	Waste polyethylene
	84	Waste wood, untreated
Gypsum weather board	98	Waste gypsum
	2	Waste polyethylene
Platonised cladding	100	Waste wood, untreated

References

1. WRAP: Construction, Demolition and Excavation Waste Arisings, Use and Disposal for England 2008, England (2010)
2. Agency., I. E.: Energy IG. CO2 Status Report 2017, England (2017)
3. Minunno, R., O'Grady, T., Morrison, G.M., Gruner, R.L., Colling, M.: Strategies for applying the circular economy to prefabricated buildings. Buildings. **8**, 125 (2018)
4. Foundation, E. M.: What Is a Circular Economy? https://www.ellenmacarthurfoundation.org/topics/circular-economy-introduction/overview
5. Yadav, M., Agarwal, M.: Biobased building materials for sustainable future: an overview. Mater. Today: Proc. **43**, 2895–2902 (2021)
6. Hawkins, W., Cooper, S., Allen, S., Roynon, J., Ibell, T.: Embodied carbon assessment using a dynamic climate model: case-study comparison of a concrete, steel and timber building structure. Structure. **33**, 90–98 (2021)
7. Dams, B., et al.: A circular construction evaluation framework to promote designing for disassembly and adaptability. J. Clean. Prod. **316**, 128122 (2021)
8. Dams, B., et al.: Upscaling circular bio-based construction: challenges and opportunities. Build. Res. Inf. (2023). https://doi.org/10.1080/09613218.2023.2204414
9. 20887:2020: B. I. Sustainability in Buildings and Civil Engineering Works. Design for Disassembly and Adaptability. Principles, Requirements and Guidance, England (2020)
10. Zach, J., Korjenic, A., Petránek, V., Hroudová, J., Bednar, T.: Performance evaluation and research of alternative thermal insulations based on sheep wool. Energ. Buildings. **49**, 246–253 (2012)
11. Parlato, M.C.M., Porto, S.M.C., Valenti, F.: Assessment of sheep wool waste as new resource for green building elements. Build. Environ. **225**, 109596 (2022)
12. Verspeek, S., Van Der Burgh, F.: Performance of bio-based facades. In: 3rd International Conference of Bio-Based Building Materials, Belfast, UK (2019)
13. Poohphajai, F., et al.: Bioinspired living coating system in service: evaluation of the wood protected with biofinish during one-year natural weathering. Coatings. **11**, 701 (2021)
14. HM Government: NBS 2010b the building regulations 2010 for England and Wales, Part L. Build. Regul. **2010**, 23–24 (2013)

15. EN. EN 15804:2012 + A2:2019: Sustainability of construction works—environmental product declarations—core rules for the product category of construction products. Int. Stand. **7**, 73 (2012)
16. Athina Papakosta, S.S.: Whole life carbon assessment for the built environment. RICS (2017)
17. Eugeos: EUGEOS' 15804 A2 IA Database: Method, England (2020)
18. Cascione, V., et al.: Comparing the carbon footprint of bio-based and conventional insulation materials. In: International Conference on Non-conventional Materials and Technologies, England (2022)
19. Cascione, V., et al.: Integration of life cycle assessments (LCA) in circular bio-based wall panel design. J. Clean. Prod. (2022). https://doi.org/10.1016/j.jclepro.2022.130938

Assessing the Global Warming Potential of a Novel Hybrid Timber-Based Façade System Through Life Cycle and Considering Future Climate Conditions

Guilherme B. A. Coelho ⓘ, Elsa Buvik, Haidar Hosamo ⓘ, and Dimitrios Kraniotis ⓘ

Contents

1 Introduction

An envelope has a key role in the behaviour of a building since it is its protective system against the outdoor climate. This statement is true for any location in which a building is to be constructed. Hence, it is important to guarantee that the envelope fulfils its role throughout its whole service life. This can be achieved by providing precisely timed building maintenance, which can be determined by having a predic-

G. B. A. Coelho (✉)
Department of Built Environment, Faculty of Technology, Art and Design, Oslo Metropolitan University, Oslo, Norway

CERIS and Departamento de Engenharia Civil, Faculdade de Ciências e Tecnologia, Universidade NOVA de Lisboa, Caparica, Portugal
e-mail: coelho@oslomet.no

E. Buvik · H. Hosamo · D. Kraniotis
Department of Built Environment, Faculty of Technology, Art and Design, Oslo Metropolitan University, Oslo, Norway

© The Author(s) 2025
M. Kioumarsi, B. Shafei (eds.), *The 1st International Conference on Net-Zero Built Environment*, Lecture Notes in Civil Engineering 237,
https://doi.org/10.1007/978-3-031-69626-8_104

tive *building maintenance system* (BMS) [1, 2]. One of the main aims of Staticus Care [3], an EEA and Norway grants financed project, is to develop this kind of tool [4].

The Staticus Care project also aims to develop a novel hybrid unitised façade (HUF) system that replaces part of its aluminium frame by timber [4]. This allows to obtain a more sustainable product [5], given the recent years focus on decarbonizing the built environment sector. It is important to tackle the emissions in this sector because it corresponds to 37% of the total emissions worldwide [6]. It is possible to decrease the carbon footprint of a product if its constituents embodied greenhouse gas (GHG) emissions are precisely quantified. This can be achieved by performing a *life cycle assessment* (LCA) of the product, which is not a novel approach, but has gain a lot of value in the recent years [7]. LCA tracks, throughout a product whole life cycle, the emissions released to nature and the extractions from nature associated to the product [7].

Buildings in the Nordic countries must fulfil even more stringent regulations in terms of heat losses through the envelope compared to other countries [8, 9]. In high-energy efficient buildings, with low energy use, the embodied impacts—production of building materials, construction, maintenance and end of life [7]—become more important than the operation impacts, for example, heating, cooling and lighting [7]. This is also connected to the main message of UN's recent publication [6], in which it is declared that the future decarbonization effort should focus on the building materials.

Buildings are normally designed to have a service life of 50 years. However, it is known that facades have to be, at least partially, retrofitted in the middle of this period, since some of its component's service life is lower than 50 years [10]. The limiting factor is, however, its sealant materials, such as gaskets. They must be replaced due to the wear caused by the environmental impacts, such UV-radiation and moisture [10]. This retrofit has, evidently, financial as well as environmental costs since it will generate waste and consume materials [7].

This study aims to quantify the carbon footprint of a novel hybrid unitized façade (HUF) system by means of developing a cradle-to-grave LCA. Real data for this system, such as material quantities, EPDs and transportation routes, have been used in this study. Ultimately, two hypotheses were tested. This will be achieved by determining the materials used to build one unit of the HUF and respective amounts. Subsequently, the product-specific *environmental product declaration* (EPD) of the used materials was also collected.

The LCA was performed using the tool One Click LCA [11]. The energy consumption was obtained for Oslo, Norway, using a generic office building with the HUF installed. This type of environmental impact studies is scarce in literature for hybrid unitized façade systems [12, 13], but of key importance, if the goals set by the Paris agreement of a more decarbonized built environment are to be met until 2030 [14].

2 Methodology

This study aims to quantify the carbon dioxide footprint of a novel hybrid unitized façade (HUF) unit by carrying out a *cradle-to-grave* LCA based on real data for this specific product. This goal is achieved by applying a multi-set procedure (Fig. 1). Firstly, the materials that compose a unit of HUF were determined as well as their respective quantities. The product-specific environmental product declarations were obtained from each product manufacturer. Secondly, a building energy simulation (BES) model of a specific case-study was run during 60 years for Oslo, Norway. Finally, the performed LCA considers only the HUF unit; no other assemblies, like ceilings or partition walls, are considered.

Climate change (CC) climate files were used, so that the model could account for the expected changes of the climate [15]. The following future RCP (*Representative Concentration Pathway*) climate scenarios were considered: (1) RCP 4.5 (mid-emission scenario) and (2) RCP 8.5 (high-emission scenario), for the period between 2025 and 2084 (60 years). These files were obtained from Meteonorm [16] software. Finally, the carbon footprint of the HUF was determined using the LCA tool, *One Click LCA* [11].

Ultimately, the study explores two strategies for long-term sustainability: (i) examining the impact of retrofitting the façade system elements in accordance with their respective service life and (ii) examining the impact of a complete retrofit of the façade system at 30 years.

The main study limitations are the following: (1) cradle-to-grave LCA is performed without accounting for stage A5, since there is still no information concerning energy spent or the GHGs emissions produced in the installation of the

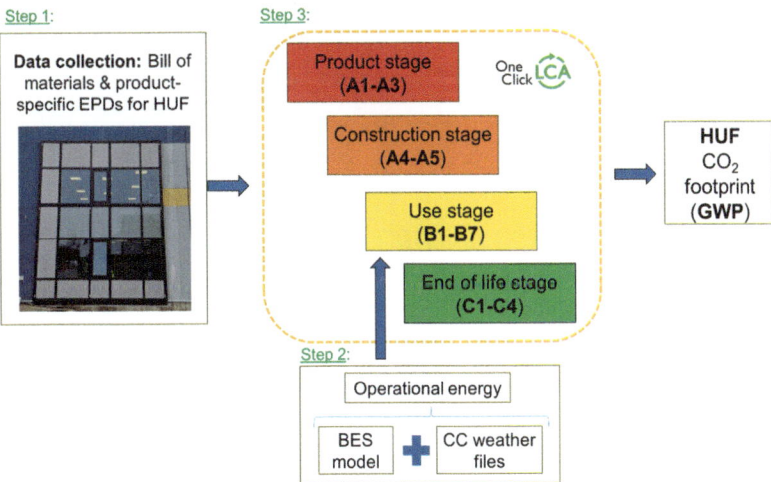

Fig. 1 Step-by-step methodology followed in this study

HUF units in situ; (2) an average energy demand for a span of 60 years is considered due to software; and (3) the influence of the material deterioration in the energy demands is not considered.

2.1 Hybrid Unitized Façade

Staticus Care has developed a HUF system [3] which replaces part of its typical aluminium frame by timber with the aim of reducing the HUF's carbon footprint. This approach is understandable given that while timber is a low-intensive energy material (4.6 MJ/kg [17]), aluminium is a highly intensive energy material (201.0 [17]) and, consequently, has a much higher CO_2 embodied value (15.1 ton CO_2/ton for aluminium and 0.35 ton CO_2/ton for timber [17]).

The HUF system has already been previously studied but more in the scope of its hygrothermal performance [18–20]. This study focuses on the quantification of the HUF carbon footprint. The HUF is composed by three different zones (Fig. 2), namely: (A) Opaque zone—external finishing, mineral wool and tin sheet (U-value of 0.129 W/(m^{2}K) [20]), (B) Frames—aluminium profiles and glue laminated timber (U-value of 0.375 W/(m^{2}K) [20]), and (C) Transparent zone—triple glazed glass unit (U_{window}-value of 0.5 W/(m^{2}K) [20]).

Fig. 2 Example of the Hybrid Unitized Façade (HUF) system

2.2 Life Cycle Assessment Database

A comprehensive *cradle-to-grave* LCA will be performed here to determine the carbon footprint of a HUF unit. The materials used to build a unit of the HUF were covered by their respective EPDs. These were obtained by contacting their respective producers. Each of the inputted materials' amounts was adopted in accordance with the respective EPD declared units.

The major contributors to the total weight of each HUF unit are the following materials: (1) windows, (2) timber, (3) steel, (4) aluminium, and (5) insulation (Table 1). Of course, these materials will have different embodied GHG emissions given their associated environmental impacts and resource use [7].

Table 1 presents the major material families that compose a unit of HUF and their global warming potential (A1–A3) per family weight. This latter index allows to ascertain the most pollutant material families while considering their weight. It is visible that the sealants family is the most polluting, while timber family is the less polluting one. A total of 16 product-specific EPDs were used to determine carbon footprint of the HUF and assess its global warming potential (GWP).

2.3 Building Energy Simulation

The building energy simulation (BES) model used in this study was previously developed by Coelho and Kraniotis [20] in WUFI®Plus. It corresponds to a room with 150 m^2 of a generic office building that has the HUF installed in its building envelope. The exterior envelope satisfies the TEK17 [8] limits imposed to attain a minimum level of energy efficiency. As it was demonstrated by Coelho and Kraniotis [20], the total net energy for office building for the three simulated Norwegian climates was well below the limit imposed by TEK17, that is, 40% for the three tested climates [20].

Table 1 Major material families and respective weights (in %) in relation to the total weight of a HUF unit and global warming potential (A1–A3) per family total weight (in kg CO_2e/kg)

Material family	Aluminium	Windows	Insulation	Sealants	Steel	Timber
Weight	10%	53%	6%	1%	11%	16%
GWP/ Weight	1.5 kg CO_2e/kg	1.4 kg CO_2e/kg	0.5 kg CO_2e/kg	7.6 kg CO_2e/kg	2.1 kg CO_2e/kg	0.3 kg CO_2e/kg

Minor materials are not represented (total < 3%)

Table 2 Legal limits (minimum requirements) in Norway for U-values and leakage number according to its regulation [8] for office buildings

Parameter	$U_{ext, wall}$ [W/(m^{2}K)]	U_{roof} [W/(m^{2}K)]	U_{floor} [W/(m^{2}K)]	$U_{windows}$ [W/(m^{2}K)]	Air leakage [h^{-1}]
Norway	0.22	0.18	0.18	1.20	1.50

Table 3 Minimum air flow rates for mechanical ventilation in office buildings in Norway and setpoint temperatures for heating and cooling according to SN-NSPEK 3031 [21] during operating hours (DOH) and outside operating hours (OOH)

Specific airflow rate (m^3/(h.m^2))	DOH		OOH	
	7		2	
Setpoint temperature (°C)	DOH		OOH	
	Min	Max	Min	Max
	21	24	21	24

In addition, it was also reviewed the legal minimum limits for U-values and leakage number (Table 2), which were considered when defining the envelopes. The opaque zone and frame zone U-values (Sect. 2.1) yield an equivalent U-value of 0.17 W/(m^{2}K), which guarantees the legal limits in Norway. Additionally, the window U-value (triple glazing) is also below the minimum value for Norway.

The model assumed the indoor gains depicted in the Norwegian specification SN-NSPEK 3031 [21] for office building by means of day profiles, which account for the occupants' use of the space. The model considers the occupants, technical equipment and lighting system influence on the heat and moisture load that the software considers (WUFI Plus). In addition, this specification also defines the ventilation during and outside operating hours and the setpoint temperatures during and outside the operating hours (Table 3). All these aspects can be a significant source of variability in LCA [22].

An efficiency of the heat recovery of the ventilation system of 80% was assumed, as it is a typical value for all systems with balanced ventilation in new buildings in Norway [8]. In addition, the same energy use as the ones depicted in SN-NSPEK 3031 for lighting, technical equipment, hot water and specific fan power was assumed. A more detailed explanation about the assumed loads and performed conversions can be found in Ref. [20].

2.4   Climate Files

Climate change, which has been largely caused by anthropogenic GHGs emissions, is expected to change considerably the outdoor meteorological conditions [23]. This change will depend on the location, but a great variability exists in future climate modelling. This is due to the long span of time that it takes into consideration and the variance of all the parameters that influence the climate models.

Hence, two climate change scenarios were selected for this study, namely, (1) RCP 4.5 corresponds to a mid GHGs emissions scenario [24] and (2) RCP 8.5 corresponds to a high GHG emissions scenario [24]. Consequently, the energy simulations were run for 60 years, which corresponds to the lifespan of the product (HUF). The climate files—current and future conditions—were downloaded from Meteonorm tool [16].

3　Results and Discussion

3.1　General Considerations

This study explores two strategies for long-term sustainability (Sect. 3.4): (i) examining the impact of retrofitting the façade system elements in accordance with their respective service life and (ii) examining the impact of a complete retrofit of the façade system at 30 years. Two future scenarios are considered—RCP 4.5 & 8.5—for one European capital—Oslo (cold climate).

The facility that produces the HUF units is in Vilnius, Lithuania. The building's location corresponds to sites in which Staticus has a lot of installed façade projects: Oslo, Norway. For this location, the HUF will be transported using articulated trucks that are partly including ferry trips. The following distance, which consist of the actual routes, was considered in stage A4 (Transport): 1261 km, including 275 km in ferry.

3.2　Future Energy Demand with the HUF System

The total energy demand (Fig. 3a), which includes heating and cooling demands obtained from the BES simulation while accounting for climate change, and the other components as described in Sect. 2.3, had to be annually average. These values include heating and cooling demands from the BES model, and hot water, technical equipment, lighting and specific fan power uses in accordance with SN-NSPEK 3031 [21]. The obtained averages were 65.8 kWh/m^2 for RCP 4.5 and 65.3 kWh/m^2 for RCP 8.5, which is atypical, but understandable if the heating and cooling demands are assessed.

It is interesting to see the substantial difference in terms of heating and cooling energy demand between the two scenarios (Fig. 3b, c), more substantial at the end of the century. While for RCP 8.5, the heating demand decreases considerable due to the increase of the global air temperature, the cooling demand also increases considerably. For RCP 4.5, these changes are not so significant. Note that this behaviour is understandable in a cold climate, but it is expected to differ for other climate typologies [25].

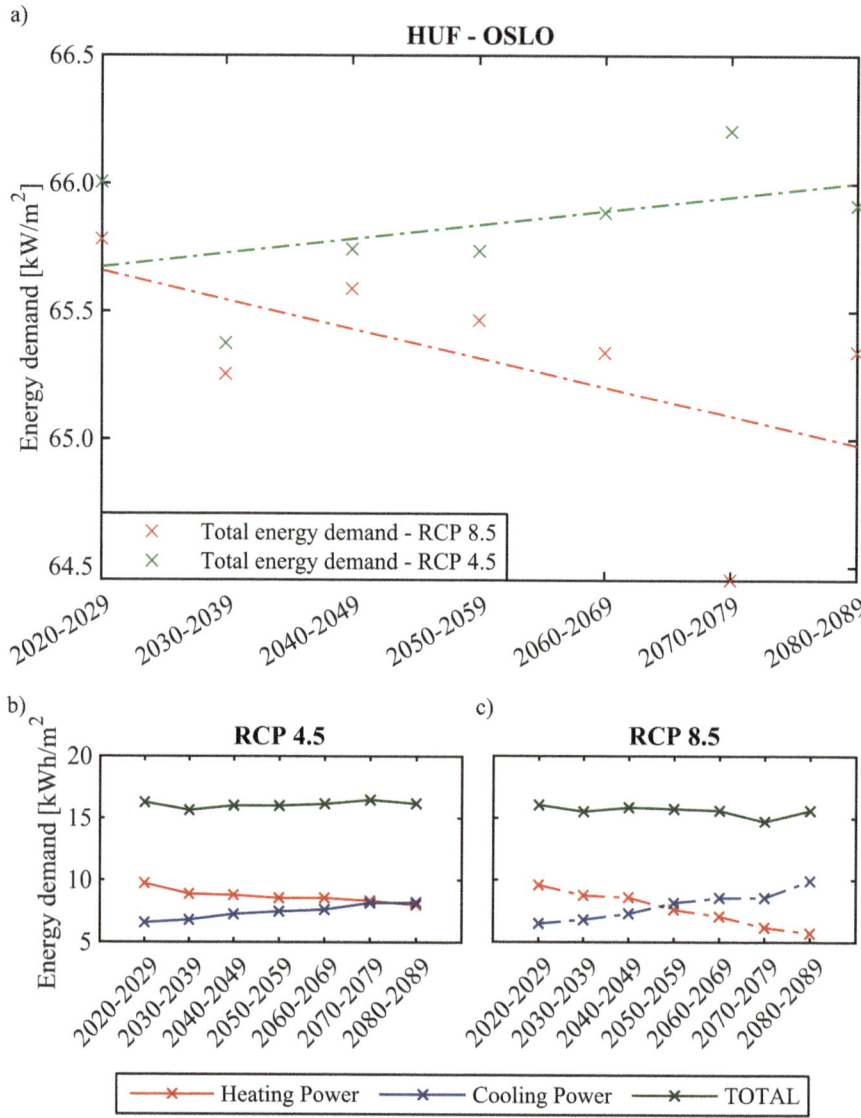

Fig. 3 Average energy demands for each 10-year period from 2020 to 2089 for RCP 4.5 & 8.5 scenarios (**a**), and heating, cooling and sum energy demand for RCP 4.5 (**b**) & 8.5 (**c**)

The limitation of having to calculate an average energy consumption occurs because the calculation methodology of One Click LCA automatically multiplies the energy demand by the admitted service life of the product. This, unfortunately, is a limitation when aiming for a dynamic LCA [22], since the energy demand has to be adopted and, therefore, the two software are not fully connected.

3.3 Carbon Footprint of the HUF System

Figure 4 shows the global warming potential (GWP) obtained for the HUF system for Oslo based on a cradle-to-grave LCA for two climate change scenarios—RCP 4.5 and RCP 8.5. It is visible that both stage A4 (*Transport to construction site*) and stage C1–C4 (*End of life*) are negligible when compared to the other stages. Stage B6 (*operational energy*) has the highest responsibility in terms GWP—72% for RCP 4.5 and 71% for RCP 8.5. This stage is then followed by A1–A3 (*Product stage*)—21% for both RCP 4.5 and RCP 8.5.

It is also visible that the differences between the two cases are slightly visible in stage B6, which is natural given that the outdoor climate differs [19]. However, it is important to state that only the HUF is accounted in stages A1–A3, neglecting the remaining envelopes' contribution to this stage and the other stages. This is a limitation of this study, but it also allows to assess the LCA results that correspond to just the HUF. Finally, in total, the HUF corresponds to a GWP of 129 kg CO_2e/m^2 for scenario RCP 4.5 and 128 kg CO_2e/m^2 for scenario RCP 8.5.

3.4 Refurbishment Measures Influence on the HUF's LCA

A LCA can be used as a tool to define a building's material replacement and refurbishment strategy by means of its CO_2 emission. This methodology was applied here by means of using two different retrofit strategies: (1) using the materials'

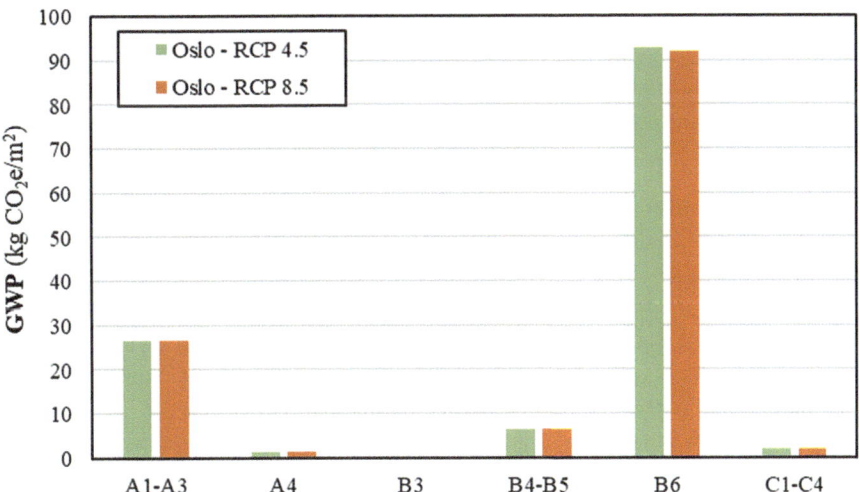

Fig. 4 Global warming potential of the HUF for Oslo (Norway) for two climate change scenarios—RCP 4.5 & RCP 8.5

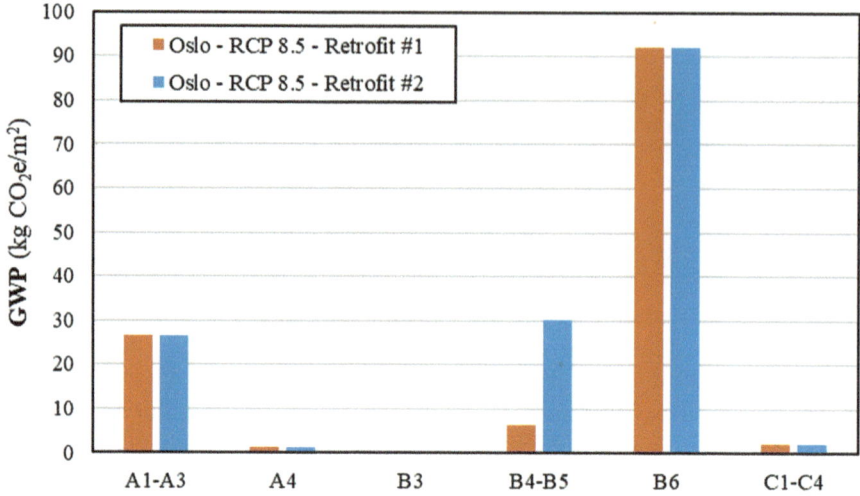

Fig. 5 Global warming potential of the HUF for Oslo (Norway) for scenario RCP 4.5 for two retrofit options

defined service life (retrofit #1) and (2) all materials are replaced at 30 years except for the glues and silicones, which must be replaced with a lower frequency (retrofit #2).

The influence of this second strategy is only visible in the B4–B5 stage (*Material replacement and refurbishment*) which goes from an individual GWP of 6 kg CO_2e/m^2 (4%) to 30 kg CO_2e/m^2 (18%) (Fig. 5). However, it is important to take into consideration that the natural wear of the building materials will cause the decrease of the building's energy performance, which means a higher consumption of energy. Hence, it is necessary to incorporate the material deterioration into stage B6 in future studies to obtain a more comprehensive value.

4   Conclusions

This study reflects on a cradle-to-grave life cycle assessment (LCA) of an innovative hybrid unitized façade (HUF) system using One Click LCA. The energy demands were determined using a representative office building model for Oslo, Norway. Two future climate scenarios were considered, namely, (1) RCP 4.5 (mid-emissions scenario) and (2) RCP 8.5 (high-emission scenario).

The global warming potential for HUF in Oslo is 129 kg CO_2e/m^2 if the world evolves according to what is depicted in scenario RCP 4.5, or 128 kg CO_2e/m^2, if it follows scenario RCP 8.5 pathway. The operational energy state is responsible for the highest emissions of the total life cycle impact—56% for RCP 4.5 and 55% for RCP 8.5, followed by the product stage—21% for both RCPs.

Two refurbishment strategies were tested: (1) using the materials' defined service life and (2) materials are all replaced at 30 years. It was shown that this second strategy increases substantially the CO_2 emissions by 24 kg CO_2e/m^2 in the material replacement and refurbishment stage (B4–B5) emissions. However, it is necessary to account in future studies for the material deterioration in the energy simulations to determine if this second strategy effectively decreases or not the total GWP of the HUF.

One of the observed limitations of the applied methodology was the fact that it is not possible to account for the energy demand for the whole period, that is, adopted service life. Consequently, it is necessary to use an average value that is representative of the whole period. This is a limitation that needs to be overcome, so that the LCA results can better reflect the influence of the stochastic nature of the outdoor climate on the energy demand. The authors are currently developing a Revit plug-in designed to execute cradle-to-cradle LCAs directly from Revit, using material quantities derived from the Building Information Modelling (BIM) and dynamically linking them to the energy needs shaped by the external climate. This integration will provide sensor data to fine-tune energy simulations within Revit, eliminating the dependency on external software by embedding this capability through a software-as-a-service (SaaS) model. Such an approach ensures the plug-in streamlines dynamic LCA studies for new buildings and embeds energy simulation tools within Revit. Furthermore, future developments will explore methodologies to quantify the impact of material/product deterioration on a building's energy demand, enhancing strategies for effective refurbishment. This tool will allow dynamic LCA studies to be more accessible, thus addressing current challenges.

Acknowledgments The authors acknowledge the EEA and Norway grants for the financial support through project LT07-1-EIM-K01-003 (*"Development of a less polluting, automated façade system integrated into building management systems"*).

References

1. Hosamo, H.H., Nielsen, H.K., Kraniotis, D., Svennevig, P.R., Svidt, K.: Digital Twin framework for automated fault source detection and prediction for comfort performance evaluation of existing non-residential Norwegian buildings. Energy Build. **281**, 112732 (2023). https://doi.org/10.1016/j.enbuild.2022.112732
2. Hosamo, H.H., Nielsen, H.K., Alnmr, A.N., Svennevig, P.R., Svidt, K.: A review of the Digital Twin technology for fault detection in buildings. Front. Built Environ. **8**, 1–23 (2022). https://doi.org/10.3389/fbuil.2022.1013196
3. StaticusCare: Project website. https://staticuscare.com/. Accessed Oct 2022 (2022)
4. Coelho, G.B.A., Ostapska, K., Kraniotis, D., Brozovsky, J., Loli, A.: Development of a hybrid timber and aluminum based unitized façade system resilient to the future weather conditions in Europe via monitoring campaigns and computational models. Procedia Struct. Integr. **55**, 39–45 (2024). https://doi.org/10.1016/j.prostr.2024.02.006

5. Gasparri, E.: Unitized Timber Envelopes: the future generation of sustainable, high-performance, industrialized facades for construction decarbonization. In: Rethinking Building Skins, pp. 231–255, Elsevier (2022). https://doi.org/10.1016/B978-0-12-822477-9.00014-0

6. United Nations Environment Programme: Building Materials and the Climate: Constructing a New Future, Nairobi (2023)

7. Simonen, K.: Life Cycle Assessment. Routledge (2014)

8. Byggteknisk forskrift (TEK17) med veiledning: Kapittel 14 Energi. https://dibk.no/regelverk/byggteknisk-forskrift-tek17/14/14-2/. Accessed Dec 2022 (n.d.)

9. Nord, N.: Building energy efficiency in cold climates. In: Encyclopedia of Sustainable Technologies, pp. 149–157. Elsevier (2017). https://doi.org/10.1016/B978-0-12-409548-9.10190-3

10. Herzog, T., Krippner, R., Lang, W.: Facade Construction Manual, 2nd edn. Birkhauser-Publishers for Architecture (2017)

11. One Click LCA: Web-based LCA tool (version 1.12.0) (2023)

12. Pastori, S., Sergio Mazzucchelli, E., Wallhagen, M.: Hybrid timber-based structures: a state of the art review. Constr. Build. Mater. **359**, 129505 (2022). https://doi.org/10.1016/j.conbuildmat.2022.129505

13. ARUP, Saint Goben: Carbon Footprint of Façades: Significance of Glass. Arup and Saint-Gobain Glass (2022)

14. Paris Agreement: https://unfccc.int/process-and-meetings/the-paris-agreement/the-paris-agreement. Accessed Jan 2023 (n.d.)

15. Coelho, G.B.A., de Freitas, V.P., Henriques, F.M.A., Silva, H.E.: Retrofitting historic buildings for future climatic conditions and consequences in terms of artifacts conservation using hygrothermal building simulation. Appl. Sci. **13**, 2382 (2023). https://doi.org/10.3390/app13042382

16. Meteonorm 8: Global Meteorological database, Version 8.1.4.25305 (2020)

17. Ahmed, A., Sturges, J.: Materials Science in Construction: an Introduction. Routledge (2015)

18. Ostapska, K., Coelho, G.B.A., Brozovsky, J., Kraniotis, D., Loli, A.: Development of climatic damage predictive tool for timber façade moisture-related damage. J. Phys. Conf. Ser. **2600**, 162002 (2023). https://doi.org/10.1088/1742-6596/2600/16/162002

19. Coelho, G.B.A., Kraniotis, D.: Numerical investigation of mould growth risk in a timber-based facade system under current and future climate scenarios. J. Phys. Conf. Ser. **2654**, 012019 (2023). https://doi.org/10.1088/1742-6596/2654/1/012019

20. Coelho, G.B.A., Kraniotis, D.: A multistep approach for the hygrothermal assessment of a hybrid timber and aluminium based facade system exposed to different sub-climates in Norway. Energy Build. **296**, 113368 (2023). https://doi.org/10.1016/j.enbuild.2023.113368

21. SN-NSPEK 3031:2023: Energy performance of buildings calculation of energy needs and energy supply (in Norwegian). Standard Norge (2023)

22. Hosamo, H., Coelho, G.B.A., Buvik, E., Drissi, S., Kraniotis, D.: Building sustainability through a novel exploration of dynamic LCA uncertainty: Overview and state of the art. Build. Environ. **264**, 111922 (2024). https://doi.org/10.1016/j.buildenv.2024.111922

23. Climate in Norway 2100: A Knowledge Base for Climate Adaptation. Norwegian Centre for Climate Services (NCCS), Report no. 1/2017 (2017)

24. Climate Change – SPM, Climate Change 2013: The Physical Science Basis, Cambridge University Press, 2014. https://doi.org/10.1017/CBO9781107415324

25. Coelho, G.B.A., Silva, H.E., Henriques, F.M.A.: Impact of climate change in cultural heritage: from energy consumption to artefacts' conservation and building rehabilitation. Energy Build., 110250 (2020). https://doi.org/10.1016/j.enbuild.2020.110250

Optimal Use of UHPC in Wall Systems to Mitigate Windborne Debris Hazard

Abhijit Kulkarni ⓘ and Behrouz Shafei ⓘ

Contents

1 Introduction

Optimizing the use of ultra-high-performance concrete (UHPC) in critical structural applications has become an important topic in the building and construction industries [1–7]. Although there are a number of studies carried out to utilize the strength and durability characteristics of UHPC [4–6], very few dedicated studies have focused on the excellent impact resistance capabilities of UHPC, which can be utilized in structures in the event of accidental impact loads [7–13]. The implementation of UHPC to resist impact loads is vital mainly where there is a high risk of windborne debris hazard, such as tornado-alley within the United States or other tornado/hurricane-prone regions within the other parts of the world. Majority of the buildings (residential and commercial) in the United States are constructed using construction friendly methods and lightweight materials. These types of buildings

A. Kulkarni · B. Shafei (✉)
Iowa State University, Ames, IA, USA
e-mail: shafei@iastate.edu

© The Author(s) 2025
M. Kioumarsi, B. Shafei (eds.), *The 1st International Conference on Net-Zero Built Environment*, Lecture Notes in Civil Engineering 237,
https://doi.org/10.1007/978-3-031-69626-8_105

have been found to be at a higher risk of damage under the debris impact [14]. Therefore, to enhance building protection and mitigate the risk of structural damage due to windborne debris impact, while also minimizing the concrete usage, UHPC wall systems present a promising solution for entire construction industry in tornado/hurricane-prone regions. With the increase in the fluctuating weather and environment, it has become imperative to assess the impacts of extreme events and also improve the resiliency of the structures.

In typical residential and commercial construction, various types of walls are employed, including wood stud walls, light-gauge metal stud walls, precast concrete walls (solid, composite, and ribbed), tilt-up panels, and cast-in-place concrete walls are used. Such conventional wall assemblies typically exhibit poor performance when subjected to impact forces from different windborne debris missiles [15, 16]. The performance of these conventional walls, however, can be improved by modifying the structural section of the wall systems or utilizing a composite section to enhance their impact resistance capability. However, such alternatives involve an increase in the material quantities, additional weight and corresponding increase in the cost of construction, which further becomes unpractical [14–16].

A review of the existing literature by Kulkarni and Shafei [12] reveals that many of the most used wall assemblies are insufficiently robust to endure both the basic and enhanced impact tests. Thus, there is ample scope to consider UHPC wall panels as a potential alternative. UHPC offers excellent compressive, tensile, and flexural strengths compared to normal-strength concrete. Although numerous studies have explored the performance of reinforced concrete (RC) walls under impact loads, there exists a gap in the literature concerning the application of UHPC walls to provide the anticipated impact protection particularly against the windborne debris. This gap served as the motivation for the present study, which aims to conduct comprehensive investigation of UHPC wall panels of varying thicknesses subjected to different debris impact scenarios. In order to address a diverse range of impact intensities, lumber debris specified in ICC (2014) [16] and a schedule 40 steel pipe outlined in DOE [17] are utilized. These objects serve as representative projectiles commonly employed for assessing windborne debris impact on both basic and enhanced protection facilities, respectively.

2 Walls Under Impact

To study the impact of windborne debris on UHPC and NSC panels, the Karagozian and Case (K&C) concrete damage model was utilized within the LS-DYNA software package. This model, developed through physics-based analyses of RC walls under blast effects, served as the foundation for the investigation [18–24]. In the simulations, eight-node solid elements with a single-point integration algorithm with a mesh size of 10 mm were used. Several validation simulations were carried out to accurately predict the values of different material parameters of UHPC material. Uniaxial compression, uniaxial tension, four-point flexure tests were simulated within the LS-Dyna package. Detailed discussion on validation at the material and full-scale levels is presented in Kulkarni and Shafei [12]. After validating the entire simulation setup, including the UHPC material model, the simulations of windborne debris impact were conducted. In these simulations, UHPC with compressive strength of 175.3 MPa was assumed. The panels were 1830 mm × 3050 mm (6 ft × 10 ft) in dimensions. The panels were assumed to have thicknesses of 80 mm (3.15 in), 100 mm (4 in), and 150 mm (6 in). The top and bottom faces of the walls were assumed to be fixed. Further, unreinforced and reinforced UHPC panels were studied.

In the case of reinforced UHPC panels, minimum reinforcement of 0.0018 times the gross cross-sectional area was assumed. The missile 2″ × 4″ (50 mm × 100 mm) 15 lb. (6.2 kg) was modeled per the specifications of ICC 500. For the enhanced protection test, pipe missile having a diameter of 6 in. (152.4 mm) with a mass of 130 kg as per the DOE standard was employed. A rigid model was used for the impacting debris in windborne debris impact simulations. The simulations were developed for a vertical surface, for which ICC 500 recommends a velocity ranging from 35.5 to 44.7 m/s and DOE further recommends using 49.1–67.0 m/s for testing of impact protective assemblies. To conduct simulations, FE models of lumber with velocities of 10, 30, 40, 50, and 70 m/s, and pipe with velocities of 22.2, 34.0, 45.1, and 55.5 m/s were developed.

3 Results of Impact on UHPC and NSC Walls

The response measures of the simulations such as deflection of the walls, type and extent of damage, and residual velocity of the projectiles are reported in Table 1 for the reinforced UHPC walls, under lumber and pipe projectiles. In the presented tables, MS refers to minor spalling, which is noted when some pieces of concrete are separated from the impacted face of the wall panel, while the panel maintains its overall integrity; PN refers to the situation where the debris begins to penetrate into the panel, forming a hole (or footprint), which is often equal to or slightly larger than the debris diameter; SB is reported if the spalling of concrete from the rear surface occurs, while the panel maintains its overall integrity; and PR is noted when the scabbing and depth of debris penetration increases to a level that a perforation is observed.

4 Findings and Discussion

From the simulation results obtained for the 76.2 mm UHPC wall (with embedded reinforcement) under the impact velocity of 70.0 m/s of the lumber missile, the projectile perforated the wall panel with a rebound velocity of −1.4 and a deflection of 6.5 mm. For the 101.6 mm UHPC wall (with embedded reinforcement), at impact

Table 1 UHPC walls with embedded reinforcement under lumber missile impact

Thickness	Missile type	Impact velocity (m/s)	Type of damage[a]	Damage (mm × mm)	Residual velocity (m/s)	Pass/fail
76.2	Lumber	10	ND	–	−2.4	Pass
		30	ND	–	−2.2	Pass
		40	ND	–	−1.4	Pass
		50	ND	–	−0.7	Pass
		70	PR	201 × 220	−1.4	Fail
	Pipe	22.2	PR	272 × 272	15.4	Fail
		34	PR	296 × 290	27.5	Fail
		45.1	PR	340 × 340	38.2	Fail
		55.5	PR	350 × 350	47.7	Fail
101.2	Lumber	10	ND	–	−3.7	Pass
		30	ND	–	−5.5	Pass
		40	ND	–	−5.8	Pass
		50	ND	–	−5.6	Pass
		70	SB	–	−4.4	Fail
	Pipe	22.2	PR	265 × 272	11.6	Fail
		34	PR	331 × 300	25.7	Fail
		45.1	PR	311 × 331	37.3	Fail
		55.5	PR	331 × 312	47	Fail
152.4	Lumber	10	ND	–	−6.6	Pass
		30	ND	–	−15.6	Pass
		40	ND	–	−19.4	Pass
		50	ND	–	−22	Pass
		70	ND	–	−27.7	Pass
	Pipe	22.2	PR	331 × 282	2	Fail
		34	PR	430 × 353	19.5	Fail
		45.1	PR	371 × 390	31.5	Fail
		55.5	PR	372 × 354	42.1	Fail
203.2	Lumber	10	ND	–	−7.7	Pass
		30	ND	–	−20.6	Pass
		40	ND	–	−25	Pass
		50	ND	–	−30.4	Pass
		70	ND	–	−39.5	Pass
	Pipe	22.2	MS	–	−2.8	Pass
		34	PR	480 × 432	10.4	Fail
		45.1	PR	451 × 441	24.1	Fail
		55.5	PR	370 × 372	39.8	Fail

[a]ND: No damage; MS: Minor spalling; PN: Penetration; SB: Scabbing; PR: Perforation.

velocity of 70.0 m/s, the 101.6 mm wall provided protection from perforation with damage to its back in the form of scabbing. In the case of 152.4 and 203.2 mm UHPC walls, the lumber missile with an impact velocity of 70.0 m/s was not able to cause any damage and rebounded. From the simulation results obtained for the

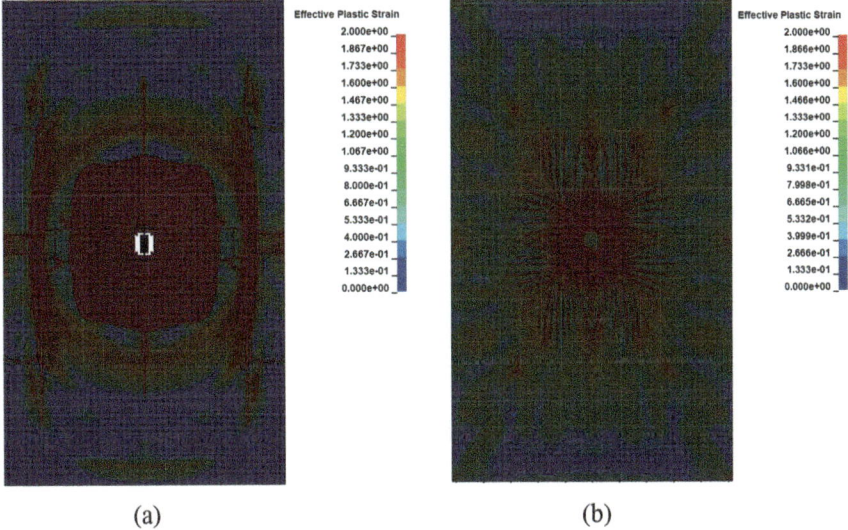

Fig. 1 Damage in wall panels under a lumber missile with an impact velocity of 70 m/s: (**a**) 76.2 mm UHPC wall, and (**b**) 101.2 mm UHPC wall

76.2 mm UHPC wall (with embedded reinforcement) under the impact velocity of 22.2 m/s of the pipe missile, the projectile perforated the wall panel, and the pipe missile was able continue with a velocity of 15.4 m/s. The size of the perforation was 272 mm × 272 mm (Fig. 1).

For all the other impact velocities (34.0, 44.1, and 55.5 m/s), the wall was perforated by the pipe missile, with the increase in perforation damage dimension. For the 101.2 mm UHPC wall (with embedded reinforcement) under the impact velocity of 22.2 m/s of the pipe missile, the projectile again perforated the wall panel, and the pipe missile was able continue with a velocity of 11.6 m/s. The size of the perforation was 265 mm × 272 mm. For all other impact velocities (34.0, 44.1, and 55.5 m/s) the wall was perforated by the pipe missile, with the increase in perforation damage dimension. For the 152.4 mm UHPC wall (with embedded reinforcement) under the impact velocity of 22.2 m/s for pipe missile, the projectile perforated the wall panel, however its residual velocity was significantly reduced to 2.0 m/s. For all other impact velocities (34.0, 44.1, and 55.5 m/s) the wall was perforated by the pipe missile, with the increase in perforation damage dimension. For the 203.2 mm UHPC wall (with embedded reinforcement) under the impact velocity of 22.2 m/s of the pipe missile, the wall panel was able to withhold the pipe missile from incurring any damage, and only minor damage was caused in the form of minor scabbing. With the increase in the impact velocities (34.0, 44.1, and 55.5 m/s) the wall was perforated by the pipe missile, with the increase in perforation damage dimension.

5 Material Saving and Embodied Carbon

The use of UHPC in applications where windborne debris hazard is significant can reduce the material usage and amount of carbon emission contribution. To calculate the amount of carbon contribution of UHPC and NSC walls, the embodied carbon of typical UHPC mix and typical NSC mix were calculated using the ICE Database. The database provides emission factors for more than 200 materials including cement, aggregates, bricks, mortars, glass, and timber. For a typical UHPC mix of 175–200 MPa compressive strength, the embodied carbon content was found to be 1031 kg/m³. For a typical concrete mix of 35 MPa compressive strength, the embodied carbon contribution was found to be 521 kg/m³. Figure 2 shows the embodied carbon prediction of typical 175 MPa UHPC and 35 MPa NSC concrete mix.

Based on only embodied carbon contribution UHPC and NSC walls, UHPC mix shows two times embodied carbon contribution than NSC concrete of 35 MPa. The required thickness of UHPC wall to resist the perforation of the lumber missile when the velocity is less than 50 m/s is 76.2 mm, whereas for NSC walls, it is 152.4 mm, which is twice the thickness of the UHPC wall. While embodied carbon contribution of UHPC and NSC walls are the same, the amount of material saved by the use of UHPC wall in lieu of NSC concrete wall is on average 0.56 kg/m³. With the advancement in the UHPC mixes, the benefit of UHPC can be maximized and its overall CO_2 contribution can be reduced further to make it viable from environmental point of view as well.

The highest contribution of CO_2 in UHPC mix comes from cement. Utilizing industrial by-products like fly ash, slag, or silica fume as partial replacements for cement, thereby reducing the carbon footprint associated with cement production can reduce the cement dependency and thereby reducing the carbon contribution. Another potential strategy in reducing the carbon contribution of UHPC is to use

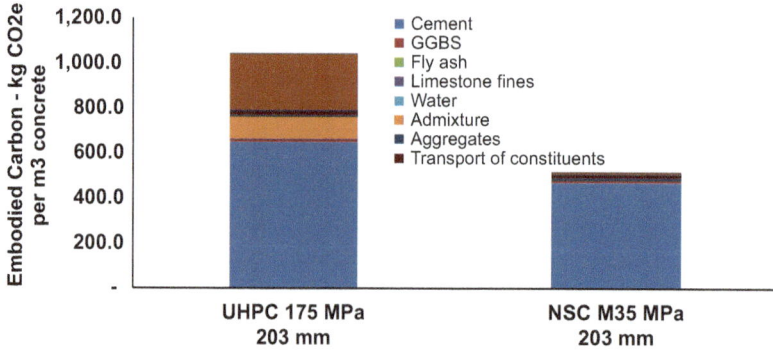

Fig. 2 Embodied carbon contribution of NSC and UHPC

locally available aggregates to reduce the emissions from transportation activities. The third potential strategy is to use optimum levels of admixtures to reduce the emissions arising from the admixture.

6   Conclusions

Small or large windborne debris travelling at a medium (20–50 m/s) to high (50 m/s and above) velocity can cause severe damage to structures within the tornado or hurricane prone regions, which can lead to partial or complete collapse of the structure. In this study, windborne debris impact assessment of UHPC and NSC walls was systematically performed considering two objects, i.e., lumber and pipe. Afterward, the embodied carbon contribution and savings in material quantities was evaluated. The results show that 76.2 mm UHPC can withstand the impact of lumber object at medium velocities (20–50 m/s). For the same range of medium velocities, NSC wall requires a thickness of 152.4 mm. At the highest impact velocity (70 m/s) of the lumber object, 101.6 mm UHPC wall can stop the object from perforating the wall barrier. For the same highest impact velocity (70 m/s), 152.4 mm NSC thick wall is required to stop the lumber object from perforating. The embodied carbon contribution of the reduced UHPC wall and NSC wall was found to be the same. However, the use of UHPC wall can offer additional advantages by reducing the required concrete volume.

References

1. Karim, R., Shafei, B.: Investigation of five synthetic fibers as potential replacements of steel fibers in ultrahigh-performance concrete. J. Mater. Civ. Eng. **34**, 04022126 (2022)
2. Saini, D., Shafei, B.: Synergistic use of ultra-high-performance fiber-reinforced concrete (UHPFRC) and carbon fiber-reinforced polymer (CFRP) for improving the impact resistance of concrete-filled steel tubes. Struct. Design Tall Spec. Build. **32**(11–12), 1541–7794 (2023)
3. Karim, R., Shafei, B.: Ultra-high performance concrete under direct tension: investigation of a hybrid of steel and synthetic fibers. Struct. Concr. **25**(1), 423–439 (2022)
4. Kulkarni, A., Shafei, B.: Impact of extreme events on transportation infrastructure in Iowa: a Bayesian network approach. Transp. Res. Rec. **2672**(48), 45–57 (2018)
5. Shafei, B., Phares, B., Sritharan, S., Najimi, M., Hosteng, T.: Laboratory and field evaluation of an alternative UHPC mix and associated UHPC bridge. IHRB Project TR-684 (2019)
6. Khaksefidi, S., Ghalehnovi, M., Brito, J.: Bond behaviour of high-strength steel rebars in normal (NSC) and ultra-high performance concrete (UHPC). J. Build. Eng. **33**, 101592 (2021)
7. Karim, R., Najimi, M., Shafei, B.: Assessment of transport properties, volume stability, and frost resistance of non-proprietary ultra-high performance concrete. Constr. Build. Mater. **227**, 117031 (2019)
8. Wang, Z., Yan, J., Lin, Y., Fan, F., Yang, Y.: Mechanical properties of steel-UHPC-steel slabs under concentrated loads considering composite action. Eng. Struct. **222**, 111095 (2020)
9. Zhu, Y., Zhang, Y., Li, X., Chen, G.: Finite element model to predict structural response of predamaged RC beams reinforced by toughness-improved UHPC under unloading status. Eng. Struct. **235**, 112019 (2021)

10. Sayyafi, E.A., Chowdhury, A.G., Mirmiran, A.: Innovative hurricane-resistant UHPC roof system. J. Archit. Eng. **24**(1), 1–11 (2018)
11. Zhai, Y.X., Wu, H., Fang, Q.: Impact resistance of armor steel/ceramic/UHPC layered composite targets against 30CrMnSiNi2A steel projectiles. Int. J. Impact Eng. **154**, 103888 (2021)
12. Kulkarni, A., Shafei, B.: Ultra-high performance concrete building wall panels engineered to resist windborne debris impact. J. Build. Eng. **42**, 103004 (2021)
13. Kulkarni, A., Shafei, B., Epackachi, S.: Structural response assessment of ultra-high performance concrete shear walls under cyclic lateral loads. Struct. Concr. **24**(2), 2699–2720 (2023)
14. Main, J., Dillard, M., Kuligowski, E., Davis, B., Dukes, J., Harrison, K., Helgeson, J., Johnson, K., Levitan, M., Mitrani-Reiser, J., Weaver, S., Yeo, D., Aponte-Bermúdez, L.D., Cline, J., Kirsch, T.: Learning from Hurricane Maria's Impacts on Puerto Rico: A Progress Report. National Institute of Standards and Technology, Gaithersburg (2021)
15. Carter, R.R.: Wind-generated missile impact on composite wall systems. M.Sc. Thesis, Texas Tech University, Lubbock, TX (1998)
16. ICC: ICC 500–2014 Standard and Commentary: ICC/NSSA Design and Construction of Storm Shelters. International Code Council, Washington, DC (2014)
17. DOE: DOE-STD-1020-2016 Natural Phenomena Hazards Analysis and Design Criteria for DOE Facilities. U.S. Department of Energy, Washington, DC (2016)
18. Crawford, J.E., Wu, Y., Choi, H.-J., Magallanes, J.M., Lan, S.: Use and validation of the release III K&C concrete material model in LS-DYNA. Technical Report TR-11-36.5, Glendale, CA (2012)
19. Markovich, N., Kochavi, E., Ben-Dor, G.: An improved calibration of the concrete damage model. Finite Elem. Anal. Des. **47**(11), 1280–1290 (2011)
20. Xu, M., Wille, K.: Calibration of K&C concrete model for UHPC in LS-DYNA. Adv. Mater. Res. **1081**, 254–259 (2014)
21. Yin, H., Shirai, K., Teo, W.: Numerical assessment of ultra-high performance concrete material. IOP Conf. Ser. Mater. Sci. Eng. **241**, 12004 (2017)
22. Wu, C., Li, J., Su, Y.: Ultra-high performance concrete columns. In: Development of Ultra-High Performance Concrete Against Blasts: From Materials to Structures, pp. 215–282. Woodhead Publishing (2018a)
23. Wu, C., Li, J., Su, Y.: Ultra-high performance concrete-filled steel tubular columns. In: Development of Ultra-High Performance Concrete Against Blasts: From Materials to Structures, pp. 283–395. Woodhead Publishing (2018b)
24. Terranova, B., Whittaker, A., Schwer, L.: Design of concrete walls and slabs for wind-borne missile loadings. Eng. Struct. **194**, 357–369 (2019)

Assessment of the Structural Potential of Nonconventional Material Alternatives in Shear Walls

Mohammad Haseeb Qureshi, Sarosh Raja Nawaz, Amirhosein Vakili, Mahdi Kioumarsi, and Behrouz Shafei ⓘ

Contents

1   Introduction

Shear walls are crucial structural members used as a part of lateral-resisting systems of buildings to resist lateral loads such as seismic and wind forces. Reinforced concrete shear walls have traditionally been constructed using normal-strength concrete (NSC); however, there has been increased research in the last decade to substitute the NSC and reinforcement with nonconventional materials of higher strength such as ultra-high-performance concrete (UHPC) and high-strength steel (HSS). The former provides greater lateral performance such as greater energy dissipation capacity, higher shear strength, and greater ductility when utilized in shear walls [1]. A transition to shear walls with nonconventional materials in the building industry would benefit sustainability by requiring less material.

M. H. Qureshi (✉) · S. R. Nawaz · M. Kioumarsi
Oslo Metropolitan University, Oslo, Norway
e-mail: s341717@oslomet.no

A. Vakili · B. Shafei
Iowa State University, Ames, IA, USA
e-mail: shafei@iastate.edu

M. Kioumarsi, B. Shafei (eds.), *The 1st International Conference on Net-Zero Built Environment*, Lecture Notes in Civil Engineering 237,
https://doi.org/10.1007/978-3-031-69626-8_106

Hung et al. [2] experimentally assessed the seismic behavior of four UHPC shear walls reinforced with HSS and tested them under displacement reversals. The study found that the specimens with steel fibers performed greater in terms of 50% better flexural strength and an increase in 70% in shear capacity, thus the utilization of UHPC in shear walls gave an advantage of better earthquake-resisting ability. Similarly, Li et al. [3] conducted cyclic tests on five squat UHPC shear walls and one high-strength concrete (HSC) counterpart. The study found that the shear capacity and ultimate deformation of the UHPC specimens significantly improved by 85% and 95%, respectively, compared to HSC.

In another comparative study, Kulkarni et al. [4] performed a holistic investigation of various UHPC shear walls exposed to cyclic loading to imitate the loading shear walls during earthquake and wind exposures. FE models were initially developed and verified with experiments. Simulations were then carried out to grasp the response of UHPC shear walls compared to NSC shear walls. The results indicated that the UHPC shear walls had on average 32% greater shear capacity than their conventional counterparts. On the other hand, the study found that a transition from normal-strength steel bars to high-strength steel-embedded bars increased the maximum shear capacity of the wall specimen as well.

The literature review reveals an increasing interest of UHPC in shear walls designed for seismic-prone zones. However, the previous investigations have been limited to only a few geometric details. Therefore, it is crucial to explore the structural performance characteristics of various shear wall sections to minimize material usage and meet design requirements. This study numerically assesses the benefits of utilizing UHPC instead of NSC by developing a set of finite elements (FE) models in the Abaqus software subjected to a nonlinear static analysis (pushover analysis). Initially, the study verifies a benchmark wall model for NSC and one for UHPC against existing research. This is followed by an examination of the load-bearing capacity and the damage progression of the walls. Furthermore, an investigation was carried out on the NSC walls, wherein their thickness was incrementally increased to evaluate if they could achieve strength levels comparable to those of UHPC.

2   Finite Element Modeling

A total of three specimens are modeled in this study using Abaqus. Table 1 provides their geometry and dimensions. The UHPC specimens are assumed to have a steel fiber content of 2%. The concrete compressive strength was 35 MPa for NSC and 140 MPa for UHPC. The steel bars, made of S500B grade steel, had a yield strength of 500 MPa and a diameter of 4.5 mm. These specimens were analyzed under a pushover load which was horizontally applied to the top section of the walls.

Table 1 List of the six different FE models

Model number	S1	S2	S3
Height	1.5 m	1.5 m	1.5 m
Thickness	0.08 m	0.08 m	0.16 m
Length	0.6 m	0.6 m	0.6 m
Aspect ratio	2.5	2.5	2.5
Concrete type	NSC	UHPC	NSC

2.1 Material Properties

The concrete damage plasticity (CDP) model is one of the possible material models that can predict the constitutive behavior of concrete in Abaqus [5]. This model assumes that the two main failure mechanisms are tensile cracking and compressive crushing of the concrete material. The plasticity parameters for NSC are adopted from the experimental investigations conducted by Krahl et al. [6]. The skeleton curve is based on the stress–strain data of the NSC material, while the damage parameters, d_c and d_t are calculated based on the "effective" tensile and compressive cohesion stresses as described in Eqs. 1 and 2 [7].

$$d_c = 1 - \frac{\sigma_c}{\bar{\sigma}_c} \tag{1}$$

$$d_t = 1 - \frac{\sigma_t}{\bar{\sigma}_t} \tag{2}$$

where σ_c and σ_t are the compressive and tensile stresses, respectively, while $\bar{\sigma}_c$ and $\bar{\sigma}_t$ are the nominal compressive and tensile stresses, respectively.

While the CDP model has proved to accurately predict the behavior of NSC, numerous studies have validated the material model for its accuracy in simulating the behavior of UHPC as well [1, 8–10]. Consequently, this study employs the CDP model based on its established effectiveness. The constitutive behavior is defined by the skeleton curves derived from Singh et al. [11]. Furthermore, the compression damage parameters are also sourced from Krahl et al. [6], given their extensive research on damage characterization. In contrast, the tensile damage parameters were calculated using Eq. 2.

For the reinforcement, the elastoplastic model was adopted. Initially, the material behaves elastically, deforming under load and returning to its original shape upon unloading. This phase continues until the stress reaches the yield point, where plastic deformation begins, and the material undergoes a permanent change in shape.

2.2 Establishment of the Model

In Abaqus, the shear wall was modeled using three distinct parts: the main wall, vertical rebars, and horizontal rebars, which were then connected using the embedded region constraint method to create an interaction between the concrete and reinforcement elements. Two reference points were placed at the top and bottom sections of the walls using coupling constraints for easier application of loads and boundary conditions. The shear wall models were simulated by eight-node reduced three-dimensional solid elements (C3D8R), and the steel bars were simulated by three-dimensional two-node truss elements (T3D2). The wall experienced a constant axial load ratio (ALR) of 6% applied at the top and was simultaneously subjected to a lateral pushover load, expressed in terms of drift ratios. The bottom section of the wall remained fixed throughout this process. The drift ratio was calculated as the ratio of the lateral displacement at the top of the specimen to the specimen height (1500 mm). For the simulation, the mesh size was established at 25 mm for both the wall and reinforcement, after a rigorous mesh sensitivity analysis.

2.3   Verification of the Normal Strength Concrete Model

The numerical model of the NSC shear wall constructed by Nguyen et al. [12] is employed for model verification. The slender shear wall was subjected to a consistent vertical compressive load of 90 kN at the top, accompanied by a progressively increasing horizontal displacement of 14.2 mm, while the bottom was fixed.

Figure 1 illustrates the damage patterns between the experimentally tested specimen and the established FE model. Initially, damage manifests along the bottom edge of the wall at a displacement of 2.02 mm. The wall experiences a bending failure which occurs at a displacement of 14.2 mm. As shown, the results from the FE model aligned well with the experimentally tested specimen. From the pushover curves that compare load and displacement for both the experimental and FE models of the NSC shear wall, it can be observed that the numerical pushover curves closely match the experimental data, with some slight differences of strength at various displacements. The experimental shear wall underwent abrupt failure at an approximate displacement of 14 mm, where it exhibited a sudden reduction in strength

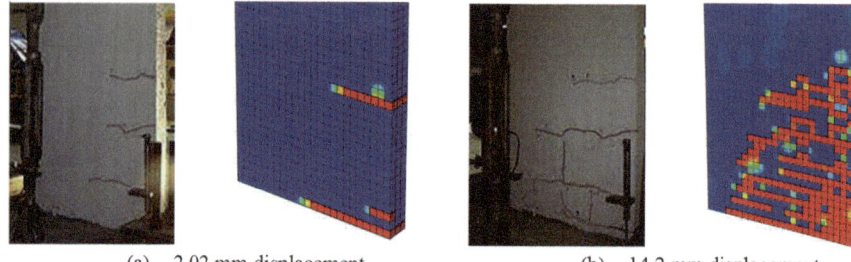

(a)   2.02 mm displacement   (b)   14.2 mm displacement

Fig. 1 Comparison of experimental test [12] and FE failure modes of NSC shear wall (**a**) 2.02 mm displacement, (**b**) 14.2 mm displacement

[12]. At this ultimate displacement, the loads recorded were 25 kN for the experimental model and 23.5 kN for the FE model, indicating a close correlation between the two specimens with a percentage error of 6%.

2.4   Verification of the UHPC Model

The shear wall model constructed with UHPC is validated through comparisons with the studies by Hu et al. [1, 13]. The bottom of the wall was fixed, and a constant vertical axial load of 1428 kN was applied at the top. Additionally, a pushover displacement of 39 mm was implemented to simulate the effect of a cyclic load.

Following the approach of Hu et al. [1], which simplified their investigation by modeling a half-scale UHPC shear wall, this study adopts the same strategy for consistency. Whereas their analysis focused on cyclic loading, this research explores the effects of pushover loads on shear walls. To ensure the methodology's reliability, this study conducts a comparison between the skeleton curves obtained from Hu et al. research and the skeleton curves generated in this investigation. These results display a notable similarity, although the verification model exhibits a lower load at various drift levels, except in the range of 0.25% to 1%, where it demonstrates higher loads. This discrepancy is likely due to simplifications in the numerical model concerning the mechanical properties of UHPC and steel rebars, in addition to considerations of geometry, loading configurations, and the material model used. The FE model and experimental outcomes exhibit maximum load capacities of 743 kN and 794 kN, respectively, resulting in a percentage error of 6.86%. This indicates that the load capacity of UHPC can be accurately predicted using the developed FE details alongside the adopted CDP model, despite the minor deviations observed.

3   Results and Discussions

This section details the results for the six models simulated in Abaqus. It begins by comparing the shear walls, focusing on the differences in damage development, and load-bearing capacities between NSC and UHPC. Subsequently, an investigation is undertaken, which involves increasing the thickness of the NSC walls, aiming to align their strength with that of UHPC walls.

3.1   Slender Shear Walls

This section examines the slender shear wall models for NSC (S1) and UHPC (S2). Figure 2 shows the damage progression of the two models and illustrates that both exhibit horizontal shear cracks at their lower sections. For the S1 model, horizontal

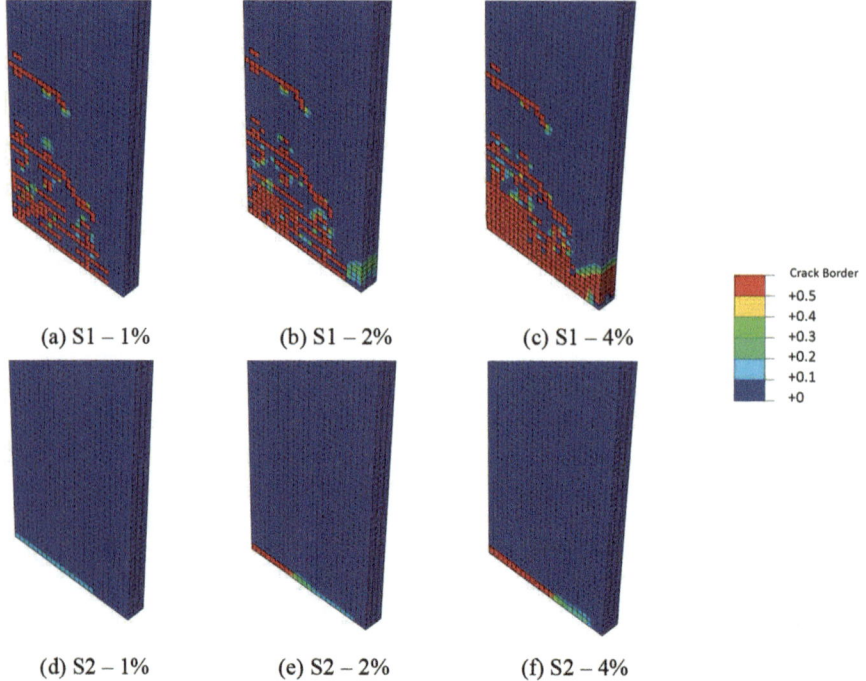

(a) S1 – 1% (b) S1 – 2% (c) S1 – 4%

Crack Border
+0.5
+0.4
+0.3
+0.2
+0.1
+0

(d) S2 – 1% (e) S2 – 2% (f) S2 – 4%

Fig. 2 Damage progression of the shear walls for models S1 (NSC) and S2 (UHPC) (**a**) S1–1% (**b**) S1–2% (**c**) S1–4% (**d**) S2–1% (**e**) S2–2% (**f**) S2–4%

cracks initiate at the bottom of the wall and propagate toward the center. Concurrently, diagonal cracks develop in the middle of the shear wall, expanding diagonally across its surface. After reaching a 2% drift, the wall started to develop cracks along the bottom edge on the opposite side which became more noticeable once the wall achieved the maximum drift ratio of 4%. In contrast to S1, cracks in S2 were solely located in the concrete elements at the base. The damage progressed linearly without significant deterioration. Based on the side-by-side comparison of the shear walls, it can be noticed that the UHPC shear wall is capable of handling much higher loads before experiencing failure compared to its NSC counterpart. The biggest factor influencing this is considered to be the addition of steel fibers, which have been noted to significantly increase the shear capacity and tensile strength of the shear walls. The UHPC shear wall showed its structural integrity with the potential to resist larger displacements than the ones investigated in this study.

Initial cracking in both models occurs at a drift ratio of 0.06%, with Model S1 experiencing a load of 10.8 kN and Model S2 a load of 14.5 kN. The enhanced material strength, along with the addition of fibers allows the material to exhibit superior crack control behavior. The peak strength achieved by the NSC shear wall was 29 kN with a 0.6% drift ratio, while the UHPC model attained a peak strength of 132 kN at a lower drift ratio of 0.95%. This indicates a substantial enhancement in

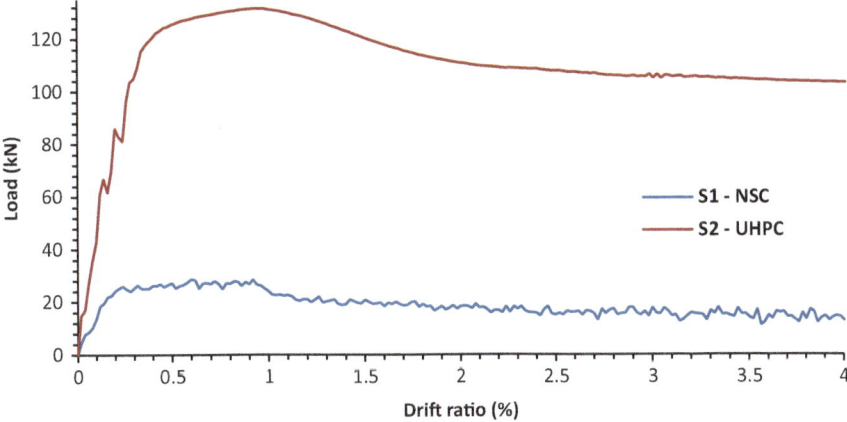

Fig. 3 Load–displacement curves comparing the FE models for S1 and S2 shear walls

peak strength for the UHPC, surpassing the NSC by 103 kN. The S1 model maintained its strength fairly consistently after reaching peak levels, exhibiting a gradual linear decrease rather than a significant drop. Upon reaching 85% of its maximum drift capacity, the load diminished to 12.4 kN, corresponding to a modest reduction of 14.8% in strength. In contrast, S2 exhibited a decline to 109 kN in strength, which was then stabilized and followed a linear decrease towards the maximum drift.

This behavior can be attributed to the inherent strength of the UHPC material. Given that the reinforcement was identical to that in the NSC model, it likely could not complement the advanced performance of UHPC, thus becoming less effective in bearing additional loads once the UHPC reached its peak strength. After reaching 85% of its maximum drift, the load dropped to 104 kN, resulting in a drop of approximately 21%. Figure 3 presents the load-displacement curves of the two models, showcasing a distinct divergence that highlights the UHPC model's substantial load-bearing advantage.

3.2   Shear Wall Thickness Evaluation

An investigation was conducted on the slender shear walls to enhance the understanding of the model's response with different thicknesses. This study focused on a single variable: increasing the thickness of the NSC shear walls to evaluate if their load-bearing capacity and damage characteristics could match those of the UHPC specimens. Model S3, designed for this scenario, features a thickness of 160 mm and is detailed in Table 1. Furthermore, when increasing the wall thickness, it is essential to evaluate two crucial parameters: the axial load ratio (ALR) and the reinforcement ratio. The ALR is assumed to be uniform across all models, calculated as per Eq. 3.

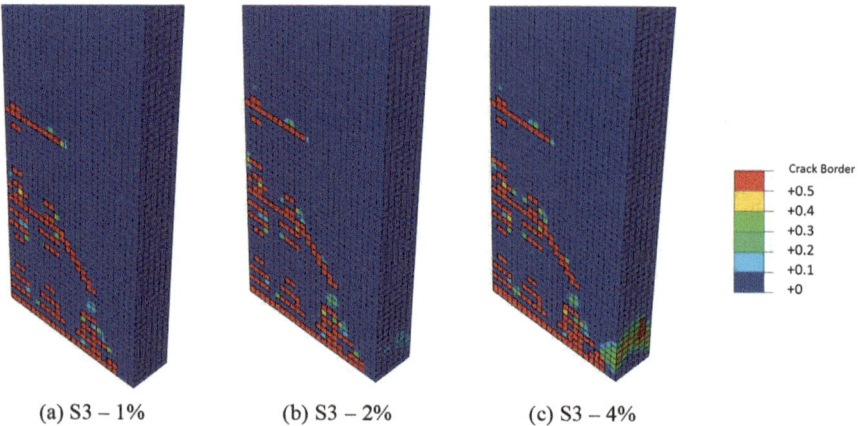

	Crack Border
	+0.5
	+0.4
	+0.3
	+0.2
	+0.1
	+0

(a) S3 – 1%   (b) S3 – 2%   (c) S3 – 4%

Fig. 4 Damage progression at different drift ratios for model S3. (a) S3–1% (b) S3–2% (c) S3–4%

$$ALR = \frac{N}{f_c A} \qquad (3)$$

where N is the constant vertical axial force, f_c is the compressive strength, and A is the cross-sectional area of the wall. This equals an axial load of 201.6 kN for model S3. The cross-section of the reinforcement is determined by calculating the reinforcement ratio which resulted in a cross-section of 31.77 mm^2.

The damage progression of S3 is shown in Fig. 4. It can be noted that the damage progression begins similarly to the NSC shear wall. However, due to the increased thickness, there was less damage progression, even at the maximum drift ratio of 4%. Notably, the extent of damage remains consistent across the three drift ratios (1%, 2%, and 4%), with the primary observable change being the emergence of minor cracks on the lower edge of the opposite side post-2% drift. The increased thickness of the wall and cross-section of the reinforcement for S3 contribute to a decelerated damage progression, attributed to enhanced robustness.

Figure 5 presents the load-displacement curves of models S1, S2, and S3. The curve for S3 indicates a marked enhancement, with a load-bearing capacity double that of S1, and a 93% increase in peak load compared to the original. This significant improvement is mainly due to the augmented wall thickness, which offers a larger cross-sectional area, facilitating a more efficient stress distribution across the wall and thus increasing its ability to resist pushover loads. Additionally, the increased thickness of the wall not only provides additional space for reinforcement but also contributes to a larger cross-sectional area for the reinforcing materials, increasing the wall's strength. While increasing the thickness of the wall enhances its material qualities compared to the standard, the UHPC in model S2 shows a distinctly superior performance, with a 138% greater peak load. This stark difference between the performance of models S2 and S3 suggests that the advanced material properties

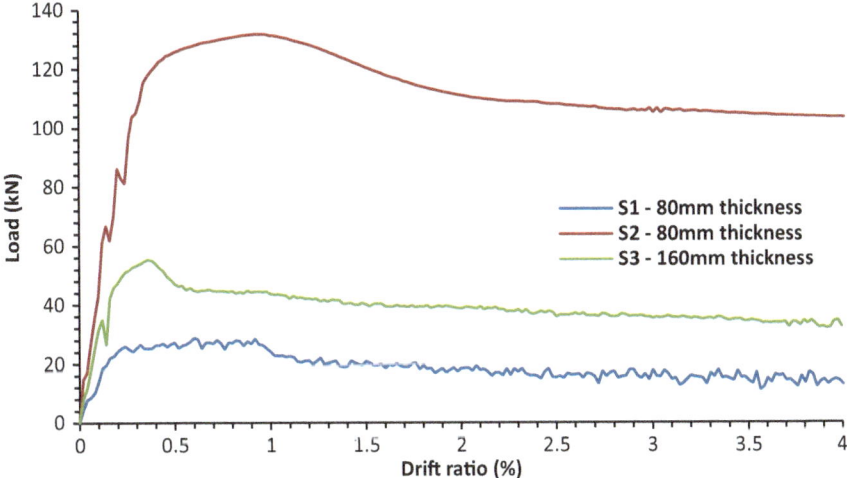

Fig. 5 Comparison of the load–displacement curves for S1, S2, and S3

of UHPC greatly enhance the structural capacity of shear walls, potentially allowing for thinner walls to be used without compromising, and even enhancing, lateral load performance.

4   Conclusions

A comparative study on the lateral load performance between UHPC and NSC shear walls was conducted. The objective of the pushover analysis conducted in Abaqus software was to evaluate the performance of UHPC relative to NSC in terms of strength and damage resistance. The findings indicated that UHPC walls exhibit superior damage resistance with damage primarily localized at the base elements and having the capacity to support increased loads due to the integration of steel fibers. Additionally, the UHPC walls demonstrated superior load-bearing capabilities, achieving a peak strength that surpassed the NSC model by 103 kN. Even when the thickness of the NSC shear wall was doubled, it failed to match the performance metrics of UHPC in terms of both damage progression and load-bearing capacity. This enhanced NSC wall exhibited damage patterns akin to the standard NSC wall, and while its peak load capacity increased by 93%, the UHPC model still outperformed it by 138%.

Acknowledgments This work is a part of the TRANSFORM project, which has received funding from the Norwegian Directorate for Higher Education and Competence (HK-dir), project number UTF-2020/10107. Additionally, we acknowledge the support from the Net-Zero Future project, an INTPART project funded by the Norwegian Directorate for Higher Education and Competence (HK-dir) and the Research Council of Norway, project number 337262.

References

1. Hu, R., Fang, Z., Benmokrane, B.: Nonlinear finite-element analysis for predicting the cyclic behavior of UHPC shear walls reinforced with FRP and steel bars. Structure. **53**, 265–278 (2023)
2. Hung, C.-C., Li, H., Chen, H.-C.: High-strength steel reinforced squat UHPFRC shear walls: cyclic behavior and design implications. Eng. Struct. **141**, 59–74 (2017)
3. Li, Y.-Y., et al.: Seismic performance of squat UHPC shear walls subjected to high-compression shear combined cyclic load. Eng. Struct. **276**, 115369 (2023)
4. Kulkarni, A., Shafei, B., Epackachi, S.: Structural response assessment of ultra-high performance concrete shear walls under cyclic lateral loads. Struct. Concr. **24**(2), 2699–2720 (2023)
5. Sümer, Y., Aktaş, M.: Defining parameters for concrete damage plasticity model. Challenge Journal of Structural Mechanics. **1**(3), 149–155 (2015)
6. Krahl, P.A., Carrazedo, R., El Debs, M.K.: Mechanical damage evolution in UHPFRC: experimental and numerical investigation. Eng. Struct. **170**, 63–77 (2018)
7. ABAQUS, I. ABAQUS Analysis User's Manual. 2006; https://classes.engineering.wustl.edu/2009/spring/mase5513/abaqus/docs/v6.6/books/usb/default.htm?startat=pt05ch18s05abm36.html
8. Singh, V., Sangle, K.: Analysis of vertically oriented coupled shear wall interconnected with coupling beams. HighTech and Innovation Journal. **3**(2), 230–242 (2022)
9. Arshadi, H., Kheyroddin, A., Nezhad, A.A.: High-strength steel effects on the behavior of special shear walls. Magazine of Civil Engineering. **111**(3), 11102 (2022)
10. Zhang, S., Li, K.: Finite Element Simulation of Seismic Behavior of Shear Wall with High Strength Steel Bar. In 2021 7th International Conference on Hydraulic and Civil Engineering & Smart Water Conservancy and Intelligent Disaster Reduction Forum (ICHCE & SWIDR). IEEE (2021)
11. Singh, M., et al.: Experimental and numerical study of the flexural behaviour of ultra-high performance fibre reinforced concrete beams. Constr. Build. Mater. **138**, 12–25 (2017)
12. Le Nguyen, K., et al.: Pushover experiment and numerical analyses on CFRP-retrofit concrete shear walls with different aspect ratios. Compos. Struct. **113**, 403–418 (2014)
13. Hu, R., Fang, Z., Xu, B.: Cyclic behavior of ultra-high-performance concrete shear walls with different axial-load ratios. ACI Struct. J. **119**, 233–246 (2022)

Non-Destructive Assessment of Reclaimed Timber Elements Using CT Scanning: Methods and Computational Modelling Framework

Martin Tamke ⓘ, Tom Svilans ⓘ, Johannes A. J. Huber ⓘ,
Wendy Wuyts ⓘ, and Mette Ramsgaard Thomsen ⓘ

Contents

1 Introduction

1.1 Reclaimed Timber

Circularity emphasizes the importance of maximizing service time for wood products through cascading use, directing them towards various destinations [1]. This approach ensures that these products contribute the highest value to society for as long as possible, primarily due to their capacity for carbon storage [2]. For wood, it is

M. Tamke (✉) · T. Svilans · M. R. Thomsen
CITA (Centre for IT and Architecture), Royal Danish Academy, Copenhagen, Denmark
e-mail: martin.tamke@kglakademi.dk

J. A. J. Huber
Wood Science and Engineering, Luleå University of Technology, Skellefteå, Sweden

W. Wuyts
Omtre AS, Hønefoss, Norway

© The Author(s) 2025
M. Kioumarsi, B. Shafei (eds.), *The 1st International Conference on Net-Zero Built
Environment*, Lecture Notes in Civil Engineering 237,
https://doi.org/10.1007/978-3-031-69626-8_107

the objective to maintain its use at the highest utility level and structural integrity before it transitions in multiple use cycles to lower-grade applications and inciner-ation. This circular concept is challenged [3, 4]. Key issues for timber reuse are concerns around the treatment, usage, and storage, and especially the quality of reclaimed wood. While virgin wood comes with certification and other data, this lacks for used timber. Here, data might have never been determined, or have been regarded as unnecessary and deleted, or lost during a building's life. This compli-cates its reintegration into the building industry [5]. Robust and automated methods for efficient non-destructive estimation of mechanical properties and quality assur-ance become crucial to bridge these gaps, ensuring reclaimed timber's reapplication in the construction, including means to ensure the longevity of data and its maxi-mized use in material passports, templates, or catalogues of secondary material suppliers.

1.2   Goal and Objectives

This study is based on an ongoing investigation with the goal of developing a non-destructive modelling pipeline for the prediction of mechanical properties of reclaimed timber based on X-ray CT scanning and numerical modelling. The objectives of this paper are to (i) provide a background of the challenges that the current practice of quality assessment of timber is confronted with regarding reclaimed material and (ii) propose and describe a data-driven modelling pipeline for the prediction of stiffness and strength of individual reclaimed timber elements based on CT scans.

2   Background

2.1   Challenges of Traditional Grading Regarding Reclaimed Timber

Timber used in the structural capacity in European buildings must adhere to the harmonized standard EN14081 and supporting standards. It sorts rectangular cross-section timber into categories based on the three grade-determining properties (GDPs) strength (MoR), stiffness (MoE), and density. These categories do not state anything about the actual properties of the individual piece of timber but solely operate on a statistical level. In particular, the lower fifth percentile of the underlying distribution defines the strength and density of a class, whereas the mean is used for stiffness. Similar approaches are used for all materials with inherent uncertainty induced, e.g. by heterogeneity as in wood (see EN 1990). Building codes consider these uncertainties and make, e.g. engineers apply structural calculations (punitive) safety factors on the mechanical properties of materials. This results—at least from a structural engineering and economic perspective—in an overengineering of timber buildings and increased consumption of resources.

2.2 Non-Destructive Assessment of Virgin Timber

The challenge of the resources' heterogeneity and uniqueness of every piece is addressed in the non-destructive assessment of each piece in the grading process. The current European system for grading allows visual and machine grading. In both cases, some measurements are used to establish a correlation to the three GDPs, which in turn must be elicited during destructive testing on a sufficiently large sample from the specific growth area [6]. Visual grading is executed by humans, who assess a range of features that affect strength, stiffness, and density, including knots, slope of grain, ring width, and reaction wood. Machine Grading assesses in contrast only one or a few properties of the timber. The measurement results of e.g. resonance or density establish indicating properties (IPs) that are then used to predict the GDPs and thus the grade. IPs typically serve as more robust indicators of quality compared to those assessable through visual grading. Machine strength grading expedites the process significantly and reduces susceptibility to human errors.

The underlying principle of both approaches is the correlation between detected features and grades. This correlation is established in large test series and is upheld through repeated testing of samples. The correlation is specific to a species and a growth area, which limits the use of the data to a specific region and even individual sawmills or other manufacturers have their own set of IPs.

2.3 Towards Non-Destructive Assessment of Reclaimed Timber

The transition of these assessment practices to reclaimed timber introduces significant challenges due to the inherent diversity and unpredictability of reclaimed materials [7]. Unlike virgin timber, reclaimed wood comes in a wide range of dimensions and conditions and contains potential contaminants such as metals in the form of nails or dust and surface modifications (e.g. brushed surfaces, impregnation, and coating). This variability, resulting from, e.g. mechanical wear, environmental exposure, and alterations from previous uses, demands flexible and robust systems for quality assurance.

The current rules for non-destructive timber assessment have been developed for a narrow set of variations in the input material with respect to, e.g. geometry and moisture content, while factors, which affect the quality of reclaimed timber, such as loading history or cracks would today in most cases disqualify the timber from use in a building context [8]. A further challenge is the classification systems focus on specific species and their place of growth to determine IPs, such as density. While this approach fits well with sawmills with a regional intake of wood, it becomes challenging when an urban area with heterogeneous buildings becomes the site for timber sourcing.

Related research and standardization efforts are going on for many years, motivated by emerging circular timber practices, or preceding this in building conservation and restoration, where the quality of old timber needs in structures needs to be assessed. Recently standards have been proposed to deal with the terminology and general rules for reclaimed timber and how contamination and treatments of reclaimed timber shall be evaluated [9]. The first standard for the visual grading of reclaimed timber shall undergo a hearing in the summer of 2024. Some existing standards, such as the Italian UNI11119, provide already means for the assessment of old timber using a combination of on-site inspections and non-destructive techniques [8], while other research shows how measurement of samples in accordance with US standards might provide means to assess the necessary mechanical properties for construction [7]. While the latter approach is cost intensive and project specific, reports about the former show that neither visual nor non-destructive testing methods, such as ultrasound or stress wave methods or local methods such as pilodyn or resistograph measurements show a sufficient correlation with mechanical properties determined in mechanical tests [8]. Bergsagel and Heisel list a series of criteria that should be considered, when characterizing reused timber: (i) the different dimensions of traditional timber elements with respect to actual sawmill products, (ii) technological characteristics, such as wane, must not exclude timber from being assessed, but simply influence the effective resistant cross-sectional area, (iii) the importance of knots on the mechanical properties and the need to take the actual size, shape, and direction of the internal knots into account and not solely their 'superficial incidence', (iv) the need to have a homogenous moisture distribution in the timber elements, and (v) the frequent use of hardwood in existing structures and the need for its improved characterization.

A transfer of the current quality assessment for timber, based on grading with its limitations regarding region and species, into a circular economy faces inhibitive challenges. It is, therefore, necessary to develop alternative assessment methods, adapted to dissimilar pieces of timber with unknown origins and with a large variety of defects and additional unknowns.

2.4   CT Scanning for Reclaimed Timber

X-ray computed tomography (CT) has originally been developed for clinical applications, but today it is also an industrially available technology which can provide information about the internal density distribution of a scanned object non-destructively. CT scanning of wood at the millimetre scale is used in the sawmill industry for optimizing the positioning of the cut [10], but research has shown that CT images at this resolution (growth ring scale) can also be used for investigations of wood physical [11–13] and wood mechanical properties, in particular, if the local variations of fibre orientations can be estimated from the image gradient. The field-based nature of the gathered data enables the creation of models based on computational mechanics, e.g. finite element (FE) models, which have been used to successfully predict the stiffness, strength, and initial crack location in individual pieces of dry virgin sawn timber [14].

CT-based modelling presents an opportunity for an accurate and detailed assessment of the mechanical properties of individual reclaimed timber elements, especially since their moisture content is typically below fibre saturation, which yields a higher contrast between the various biological features in the CT images.

CT scanning at the required scale for modelling can be performed at high feed speeds like a sawmill's processing speed, i.e. approximately 1–2 m/s. However, CT-based numerical modelling, like in [16], may depend on the model complexity and the specific hardware and software implementation, currently not be able to keep up with this pace. If individual pieces of timber can be tracked throughout a processing facility, then in-line high-speed evaluations may not be necessary. Instead, once a piece of timber has been scanned, it could be stored until its mechanical properties have been estimated and its optimal use has been determined. CT-based numerical modelling at a high level of detail may also be a precursor for more simplified data-driven modelling, e.g. based on laser scanning or discrete X-ray scanning, which both require less investment in equipment.

3 Materials and Non-Destructive Data Acquisition

The reclaimed timber for this study was supplied by the Norwegian company Omtre AS. From their storage in Hønefoss, Omtre contributed timber samples, including 56 planks with cross-sectional dimensions 49×98 mm^2 and five round barn logs, all varying in length. The materials represent two exemplary histories of reclaimed elements: one for premium barn elements and one for reclaimed construction wood.

The barn timber originated from an old barn in Brøttum, Ringsaker (Fig. 1). Constructed in 1870, the barn was a timber-framed structure with exterior cladding.

Fig. 1 Demounting the barn in 2022, provided by Omtre

This style of barn, known as an 'enhetslåve' or 'one-unit barn', became prevalent in Norway around 1850 and consolidated multiple farm functions (storage for hay, equipment, and animal shelter). The log house portion provided weather and cold protection for the animals, utilizing pine for its rot-resistant heartwood.

The original positioning of the barn logs in the building could not be reconstructed, since this information was lost during Omtre's reclaiming of the logs [17]. Nevertheless, based on knowledge concerning cultural heritage preservation, Omtre estimates that the barn logs used in this study very likely served as the supporting foundation and beams under the first floor.

The reclaimed construction timber originated from a building site of student apartments at Kringsjå (Norway) and had been temporarily used in the construction process and was subsequently regarded as waste. Due to Omtre's collaboration with contractors, this wood was collected in containers at the site and saved from incineration. Any remaining metal fasteners were removed from all timber elements prior to storage in Hønefoss, where shelter from precipitation was provided, but neither heating nor air conditioning. The timber remained in storage for approximately 2 years.

The timber was *CT scanned* without additional treatment or cleaning at the Division of Wood Science and Engineering of Luleå University of Technology (LTU) in Skellefteå, in January 2024. For the construction timber, all samples were positioned standing upright in the scanner, maintaining consistent orientation regarding the pith and the ends corresponding to the top and bottom of the tree (see Fig. 2). A consistent identification (ID) numbering system was applied and a specific positioning sequence was followed in the image plane to identify the individual beams in the scans.

The timber was scanned employing a helical scanning trajectory in an industrial CT scanner (Microtec MiTO) adapted for research, at a tube voltage of 180 kV and current of 5.55 mA. The images were reconstructed to an equally spaced voxel grid of $0.5 \times 0.5 \times 0.5$ mm^3, and the density in kg/m^3 was calculated from the CT number.

Fig. 2 56 planks and 5 barn timber logs (left), sample setup during CT scanning (middle) and a section of the scan (right)

4 Computational Modelling Framework

The workflow of the developed framework is illustrated in Fig. 3. On a conceptual level, it consists of three steps: (1) image analysis and feature detection, (2) geometry extraction and volumetric discretization, and (3) data-driven material modelling. Prior to analysis, the 3D CT images were cropped to represent only single timber elements including a small frame of 30 voxels containing the surrounding air. All image processing was performed using the free and open-source libraries SimpleITK, DIPlib, and OpenCV in Python.

For *step (1)*, the CT image data of each beam can be regarded as a volumetric density field, $J(x, y, z)$, on an equally spaced rectangular grid. First, wood needs to be separated from the background (e.g. air), which can be achieved by thresholding on J and subsequent image closing operations to remove internal holes. The segmentation results in a Boolean mask which allows a fast estimation of the element boundary and its rough shape. The total volume of the object can be derived simply by summing up the voxels contained within the mask and accounting for the voxel scale. For segmenting other regions of interest with distinctly different densities, like knots, growth rings, high-density inclusions, or contaminants, e.g. nails, thresholding may be also suitable. Some features may require a more subtle analysis of the density field and its spatial rate of change, i.e. its gradient and higher-order derivatives. These include the separation of early- and latewood, the location of the pith [18], and the estimation of the fibre tensor (longitudinal, radial, and tangential material orientations) using the gradient structure tensor [15]. For the preliminary trials of our study, only the volume mask and outer shape of the timber were extracted, and the fibre tensor was calculated.

The density field and the extracted biological regions and features—the volume mask, the knots and contaminant regions, the early- and latewood regions, and the fibre tensor—were collected in a data structure referred to as the *Body of Properties* (BoP). The BoP serves as an expandable record of relevant properties in the underlying continuum that can be 'harvested' for subsequent modelling and simulation steps. Most properties are stored as volumetric samples at various resolutions, depending on the property. For example, density is available at the full scanning resolution, while material orientation might be available at a coarser resolution

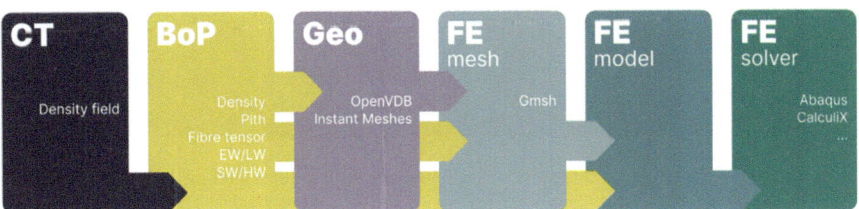

Fig. 3 Schematic overview of the prototyped framework

depending on the radius of the kernel that is used to analyze the volume gradient. Features such as knots are stored as volume regions, but they could also be represented by a lightweight parametrized shape or a computer-aided design (CAD) surface.

In *step (2)*, for usage in further analysis, simulation, and CAD environments, the volumetric data is converted to a boundary representation such as a non-uniform rational B-spline surface (NURBS) or a discrete triangle or quad mesh. The most straightforward method is to extract the isosurface of the density field at the masking threshold described above, which results in a robust and watertight mesh. For this study, the open VDB library and toolset were used to extract the surface mesh. In applications, where a surface mesh at the full scanning resolution is undesirable or too unwieldy, the volume grid may be down-sampled to a coarser resolution prior to meshing. Here, the remeshing software Instant Meshes was used to decimate the surface mesh and generate a higher-quality mesh topology to facilitate further usage. The remeshed surfaces contained only quad elements that follow the topology of the original timber element (Fig. 4). The quads were subsequently converted to NURBS patches, creating a solid NURBS model for use in traditional CAD environments, which provides high flexibility for downstream applications and modelling.

For volumetric finite element (FE) meshing of the identified geometry of the scanned object, the free and open-source meshing software Gmsh was used in Python. Its interface is particularly convenient regarding finding the locations of Gauss quadrature integration points for each element, which is required for the subsequent material modelling (Fig. 5a).

For *step (3)*, the FE mesh is incorporated into an FE model, where loads and boundary conditions were applied according to the desired analysis, and a material model was employed which was informed about the underlying properties from the BoP at each integration point. The model was written to an input file (.inp) which is interpretable by the commercial FE solver Abaqus and open-source solvers like CalculiX. The material model was implemented by user subroutines for either Abaqus or CalculiX, in which the local material orientations and the density were imported at each integration point (Fig. 5b). Since any information from the BoP

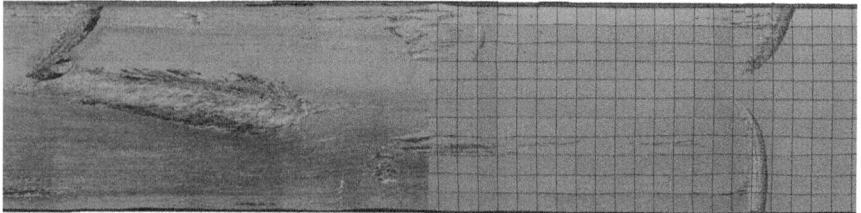

Fig. 4 The isosurface of the timber element boundary (left) is remeshed into a coarser quad-only surface mesh, suitable for further conversion into NURBS surfaces (right). The isosurfaces of knots are included for reference. The information about these is finally stored in the BoP

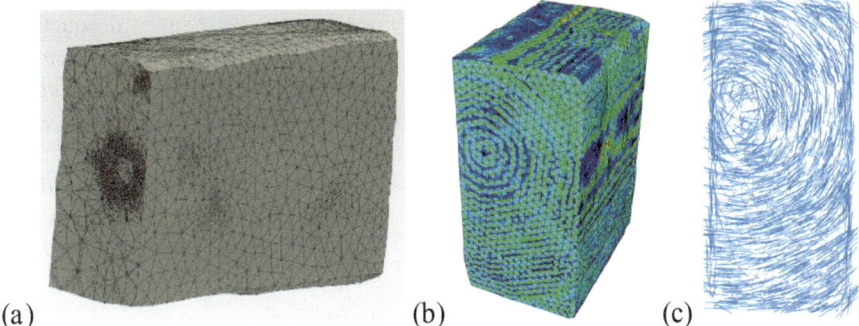

(a)   (b)   (c)

Fig. 5 Adaptive 3D meshing of a portion of the timber element using Gmsh, with a smaller element size around high-density regions, such as knots (**a**), 'harvested' density field in a FE model (**b**), and corresponding tangential material orientations (**c**)

could be accessed by the material model, arbitrarily refined approaches could be implemented, e.g. scaling the stiffness tensor by factors related to density and proximity to knots [16] or varying a hygromechanical material behaviour depending on features in the BoP [12].

For CalculiX, a custom user material was created as a stand-alone shared library which is called by the solver. Currently, it replicates the behaviour of a standard orthotropic material model; however, it can be adapted to incorporate the range of properties from the BoP, as well as different methods of loading and querying them. Typical procedures for loading external data into the solver in user material involve writing per-integration point data to one or many files and then opening these files during the model initialization. In this study, the user material accesses a block of shared memory that is populated by another process, allowing it to query the BoP directly without writing and reading files. While the primary advantage is to access speed, it also means that this procedure could be expanded to query the BoP dynamically or include other behaviours defined during runtime. The vision is to arrive at a model ecosystem where the BoP—or several BsoP—can inform and mediate between design, material analysis, and simulation models in a flexible and extensible way.

A linear elastic orthotropic material law was applied using a stiffness tensor with constant textbook values for spruce, while the material orientations were varied according to the BoP (Fig. 5c). To evaluate the mechanical properties of the modelled timber, pure bending was simulated on partial lengthwise sections along the beam and on the full beam. The mid-point deflection during bending was recorded and the local bending stiffness (MoE) was calculated from it using the approach in [16]. In addition, a frequency analysis with free boundary conditions was conducted to extract eigenfrequencies and eigenmodes.

5 Preliminary Trials and Discussion

Our investigations are part of an ongoing study, and therefore the results mainly serve demonstrative purposes. Results of the analysis of a single beam from the construction timber are presented. From the CT image analysis, an average density of 536 kg/m³, a total weight of 5.694 kg, and a length of 2.194 m were elicited. Figures 6 and 7 show the 'harvested' density field in the FE model of the beam and, respectively, for a partial section with corresponding absolute transverse stress (perpendicular to the fibre) during pure bending. All displacements are scaled, and the plotted values are qualitative for comparison (red indicates higher, and blue indicates lower values). Figure 6 additionally shows the first longitudinal mode in the frequency analysis, at a longitudinal eigenfrequency of 440 Hz. The bending stiffness was evaluated for consecutive 250 mm long sections along the beam (Fig. 8) and in the resulting profile, the bending stiffness dropped as expected in regions with large knots.

The simulation-based bending deflections, eigenfrequencies, and stiffness profile may be used to assess the stiffness of the individual pieces of reclaimed timber and serve as predictors for bending strength. The stress field during bending may be used

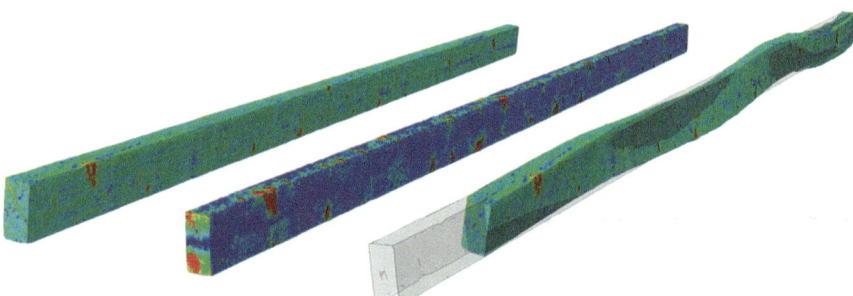

Fig. 6 Full beam FE model showing the density field (left), the absolute transverse stress field under pure bending (middle), and the first predominantly longitudinal vibration mode (right)

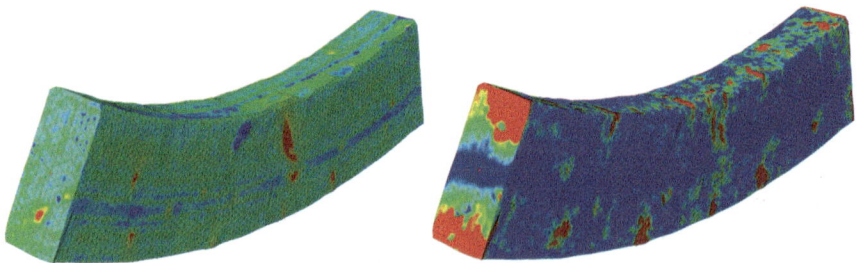

Fig. 7 FE model for a partial section of the beam, density field (left) and corresponding absolute transverse stress field (right)

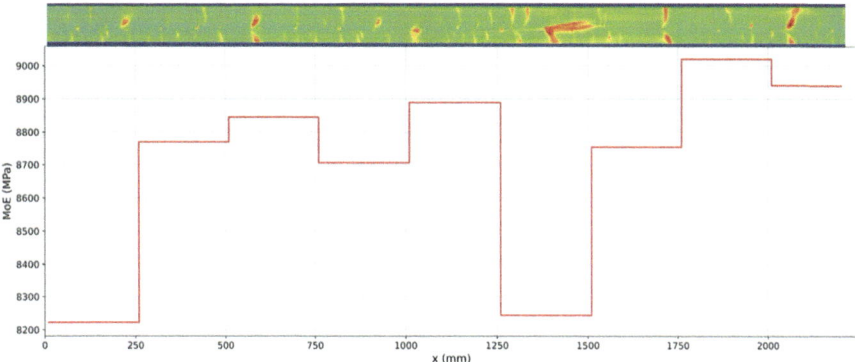

Fig. 8 Bending stiffness profiles for the evaluated sections of the studied beam. The average density along the thickness direction is plotted on top, and red colours indicate higher and blue lower values

to evaluate the susceptibility to failure under increasing load, e.g. by employing maximum stress or a Tsai-Wu anisotropic failure criterion.

The preliminary trials of our presented pipeline demonstrate the soundness of its working principle and that the simulation results already reside in a reasonable range. The used models have not yet been fine-tuned, e.g. by error checking the fibre reconstruction, in specific at boundaries, or by letting the stiffness tensor vary depending on the underlying biological tissue (knot, latewood, earlywood, etc.), which are both expected to increase the resulting bending stiffness. In addition, the models need to be adjusted for the climatic conditions monitored during storage before scanning and the resulting approximate moisture content (MC).

Once the development of the pipeline has reached a state that enables a more automated treatment of irregularities in the input geometries (diverse shapes of timber, cut or broken ends) and thus the analysis of a larger number of beams, extensive validations will be conducted. For this purpose, we are currently collecting mechanical data on all the scanned timber, i.e. longitudinal eigenfrequency by dynamic excitation, local and global bending stiffness and strength by non-destructive and destructive four-point bending, and measurements of the MC during testing. This data and the corresponding models will yield a base for validation of our framework and enable a comparison to alternative methods, like stress grading based on mechanical tests and even visual grading (using, e.g. surface visualizations from the CT images).

Sources, propagation, and impact of uncertainties need to be assessed to evaluate the framework's robustness. To achieve this, we will conduct sensitivity studies on input factors to establish uncertainty ranges for the predictions, instead of single point values. We will include factors from the data acquisition stage (e.g. scanning speed, CT artefacts, and effects of timber surface quality), the image analysis and model creation stage (e.g. image resolution, geometric assumptions, and discretization strategies), the modelling stage (e.g. material parameters and their dependency on the underlying tissue, fibre reconstruction, model boundary conditions, and load application), and the evaluation stage (e.g. stiffness evaluation and failure criterion).

We expect our framework's accuracy to be limited to the amount of information extractable from the density field of a piece of timber. In practice, also the available computational resources will limit the achievable fidelity.

6   Conclusion and Future Steps

We highlight the limitations of current timber assessment methods with reclaimed timber and the need for the development of a non-destructive assessment method that is independent of regional data and species. This paper presents a first step into the necessary groundwork by demonstrating a framework to assess the mechanical properties of individual reclaimed timber elements by models based on the extraction of properties from CT scans. The preliminary results demonstrate the modelling pipeline and outline the evaluation of mechanical properties. This reduces some uncertainty concerning the reuse of the reclaimed timber and its reuse as structural. Further research is required for validation and demonstration of the robustness of our method, specifically regarding the large variability of reclaimed timber.

The modelling framework is based on free and open-source software to promote interchangeability, extensibility, and longevity. Its design is modular, minimizing dependency on specific formats and maximizing its flexibility. It allows for varying resolutions and degrees of approximation and may in principle also be adapted to different modes of data describing the BoP, e.g. laser and X-ray scanning. Finally, this study is part of a larger research effort to compare different non-destructive methods which can be deployed to classify and sort for different destinations. Herein, the assessment and validation of reclaimed timber is a crucial component.

To this effect, we position our work within this collective effort to show how these advancements in a digital timber practice can lead to an increased utilization of reclaimed timber and a more sophisticated usage of this new resource—an integrative 'hyper-grading' of timber that spans the digital value chain between resource, analysis, design, and re-utilization.

Acknowledgements   The research by MT, TS, MR and the materials and shipment have received funding from the European Research Council (ERC) under the European Union's Horizon 2020 research and innovation program (grant agreement No 101019693). JH gratefully acknowledges the scholarship from His Majesty King Carl XVI Gustaf's 50th Anniversary Fund for Science, Technology, and Environment and the funding from the CT WOOD research program at LTU Skellefteå. The contribution by WW is co-funded by the European Commission under the DRAS-TIC project (grant agreement 101123330). The timber elements are supplied by Omtre AS. We thank Christine Jørgensen and Ivar Ragnhildstveit from Omtre AS for providing background information about the material supply.

References

1. Hudert, M., Pfeiffer, S.: Rethinking wood: future dimensions of timber assembly. Birkhäuser, Basel (2019)
2. Churkina, G., Organschi, A., Reyer, C.P.O., Ruff, A., Vinke, K., Liu, Z., Reck, B.K., Graedel, T.E., Schellnhuber, H.J.: Buildings as a global carbon sink. Nat. Sustain. **3**, 269–276 (2020)
3. Ramsgaard Thomsen, M., Nicholas, P., Rossi, G., Daugaard, A.E., Rech, A.: Extending the Circular Design Framework for Bio-Based Materials. UIACPH2023, p. 609. Springer International Publishing, Cham (2023)
4. Byers, B.S., Raghu, D., Olumo, A., De Wolf, C., Haas, C.: From research to practice: a review on technologies for addressing the information gap for building material reuse in circular construction. Sustain. Prod. Consum. **45**, 177–191 (2024)
5. Litleskare, S., Wuyts, W.: Planning reclamation, diagnosis and reuse in Norwegian timber construction with circular economy investment and operating costs for information. Sustain. For. **15**, 10225 (2023)
6. Ridley-Ellis, D., Stapel, P., Baño, V.: Strength grading of sawn timber in Europe: an explanation for engineers and researchers. Eur. J. Wood Prod. **74**, 291–306 (2016)
7. Bergsagel, D., Heisel, F.: Structural design using reclaimed wood—a case study and proposed design procedure. J. Clean. Prod. **420**, 138316 (2023)
8. Piazza, M., Riggio, M.: Visual strength-grading and NDT of timber in traditional structures. J. Build. Apprais. **3**, 267–296 (2008)
9. NS 3691-1 & NS 3691-2: https://standard.no/nyheter/nye-standarder-for-evaluering-av-returtre-pa-horing/ (2024)
10. Rais, A., Ursella, E., Vicario, E., Giudiceandrea, F.: The use of the first industrial X-ray CT scanner increases the lumber recovery value: case study on visually strength-graded Douglas-fir timber. Ann. For. Sci. **74**, 1–9 (2017)
11. Hansson, L., Couceiro, J., Fjellner, B.-A.: Estimation of shrinkage coefficients in radial and tangential directions from CT images. Wood Mater. Sci. Eng. **12**, 251–256 (2017)
12. Florisson, S., Hansson, L., Couceiro, J., Sandberg, D.: Macroscopic X-ray computed tomography aided numerical modelling of moisture flow in sawn timber. Eur. J. Wood Prod. **80**, 1351–1365 (2022)
13. Tamke, M., Gatz, S., Svilans, T., Ramsgaard Thomsen, M.: Tree-to-product: prototypical workflow connecting data from tree with fabrication of engineered wood structure – RawLam. In: Proceedings of WCTE2021, pp. 2754–2763, Santiago (2021)
14. Huber, J.A.J., Olofsson, L.: Evaluation of knots and fibre orientation by gradient analysis in X-ray computed tomography images of wood. In: 3rd ECCOMAS Thematic Conference on Computational Methods in Wood Mechanics (CompWood 2023), pp. 143–144. International Center for Numerical Methods in Engineering (CIMNE) (2023)
15. Huber, J., Abdeljaber, O., Oja, J., Olsson, A.: Evaluation of models of fibre orientation in sawn timber using synchronised computed tomography and optical scanning data. In: WCTE 2023, pp. 421–427, Oslo (2023)
16. Huber, J.A.J., Broman, O., Ekevad, M., Oja, J., Hansson, L.: A method for generating finite element models of wood boards from X-ray computed tomography scans. Comput. Struct. **260**, 106702 (2022)
17. Wuyts, W., Tomczak, A., Nore, K., Huang, L., Haavi, T.: Reuse of wood – learning about the benefits and challenges of high- and low-tech diagnostic methods through action research in Norway. In: Proceedings of the World Conference on Timber Engineering, Oslo, pp. 19–22 (2023)
18. Longuetaud, F., Leban, J.-M., Mothe, F., Kerrien, E., Berger, M.-O.: Automatic detection of pith on CT images of spruce logs. Comput. Electron. Agric. **44**, 107–119 (2004)

Sustainable Preservation and Evaluation of Burnt Brick Masonry Structures Through Condition Assessment and Retrofitting Techniques

Lakshmi Latha and Samit Ray-Chaudhuri

Contents

1 Introduction

Among the plain masonry, the structural units of burnt bricks are an age-old masonry constituent used from prehistoric times. The durability of these units is permanently found due to their continuous exposure to different climates over the years. They have gained wide popularity, and 62% of the residential units in South Asian region are already made up of burnt clay bricks [1]. It is one of the most commonly used building materials in developing countries. Despite the durability, bricks are inherently brittle materials. Mortar joints between them are also often weaker than the bricks themselves. The brick masonry in overall lacks the ductility needed to absorb and dissipate seismic energy effectively. In general, masonry constructions are highly vulnerable to lateral loads, even though they are inherently good at resisting compressive loads. Such a scenario, especially for plain masonry construction, is detrimental since they lack any reinforcing element. Even though the use of alter-

L. Latha (✉) · S. Ray-Chaudhuri
IIT Kanpur, Kanpur, India
e-mail: lakshmia@iitk.ac.in; samitrc@iitk.ac.in

© The Author(s) 2025
M. Kioumarsi, B. Shafei (eds.), *The 1st International Conference on Net-Zero Built Environment*, Lecture Notes in Civil Engineering 237,
https://doi.org/10.1007/978-3-031-69626-8_108

natives or innovative materials for new masonry construction is gaining interest in the modern era, existing masonry structures also contribute to about 70% of the building stock worldwide (whether plain, confined, or reinforced masonry).

There are different practices which are aimed at improving the performance of such existing brick masonry structure. Various conventional strengthening methods have been proposed to enhance the seismic performance of existing unreinforced brick masonry (URM) structures which includes surface treatment, stitching, grout/ epoxy injection, and re-pointing. Additionally, external reinforcement techniques have been investigated, such as reinforcing wall junctions with L-form and poly-propylene (PP) bands, rubber tire posttensioning, and employing textile-reinforced mortars. Efforts have also focused on improving performance by implementing confinement measures like external tie columns, mesh reinforcement, and similar approaches [2–4]. Some research has explored the effectiveness of techniques such as steel or fiber-reinforced mortar/concrete coating and engineered cementitious composite overlays for reinforcing masonry infill walls [4–6]. Additionally, the utilization of stainless-steel reinforcing bars and high-strength steel cords has demonstrated utility in stabilizing and enhancing the performance of masonry walls [7, 8]. In the context of load-bearing masonry structures, Indian standards (IS 4326-BIS 1993, IS 1893-BIS 2002, IS 13935-BIS 2009) recommend the incorporation of reinforced concrete (RC) elements and corner reinforcement as measures to improve seismic performance, especially in the absence of specific Indian standards for confined masonry (CM) constructions [9–11]. Confined masonry construction is all good for a new construction. Confined masonry with full confinement through the thickness of wall will be intrusive. For an existing brick masonry structure, research by authors found that partial confinement of existing masonry structure is a notable method [12]. The study clearly demonstrates that the suggested retrofitting approach for existing deficient masonry buildings is minimally invasive, easily executable, and cost-effective, while substantially improving seismic performance. This sustainable approach also minimizes material usage by focusing on nominal RC bands, reducing the demand for new resources. It also avoids generating significant construction and demolition wastes (CDW) and requires minimal energy. Moreover, it causes minimal disruption to the surrounding environment, creating little sound and dust pollution. Additionally, this approach improves the seismic performance of seismically deficient buildings, potentially avoiding the need for their demolition and subsequent environmental impacts in terms of material usage, carbon footprint due to new materials and construction, and environmental pollutions. Moreover, the utilization of partial confining elements proved highly beneficial in enhancing the overall performance of masonry walls in terms of ductility, energy dissipation capacity, integrity, and stability under slow-cyclic lateral loading tests. Further, the cyclic testing of the masonry structures resulted in the formation of cracks. The main emphasis of this study is to restore the structural integrity of the cracked masonry structures, to explore variations of the semiconfinement technique and to perform an immediate assessment of these structures. The overall idea is to set the stage for future studies of these masonry structures considering environmental sustainability and structural durability. Exploring variations of semiconfinement technique may contribute to the ongoing evolution of such retrofitting strategies for masonry structures enhancing their resilience and longevity.

For ensuring structural integrity (for safe and functional structures), the understanding of modal characteristics of structures is crucial. The insights gained from modal analysis can help in maintaining the structural integrity of buildings during earthquakes etc. The immediate effects of earthquakes as well as long-term impacts which explores the damage from past seismic events on structural as well as nonstructural elements can be well understood. Thus, modal analysis would help in understanding real-world structural responses. Experimental Modal Analysis (EMA), also referred to as frequency response function testing, is commonly considered for dynamically identifying structures. Hence, this study utilized EMA to evaluate the dynamic parameters. EMA entails stimulating the structure with known inputs, such as impulse hammers, drop weights, or electrodynamics shakers, and then measuring the resulting structural response. These techniques offer important information about modal characteristics, such as natural frequencies, damping ratios, and mode shapes, which are crucial for model analysis, validation, and ensuring overall structural safety. Obtaining modal parameters like natural frequency and mode shapes is a significant concern in designing structures to withstand dynamic loading conditions.

EMA provides the advantage of allowing precise excitation characteristics, including type, location, and amplitude. Additionally, it facilitates the selective stimulation of specific vibration modes. Genovese and Vestroni (1998) have considered low amplitude forced oscillations for investigating dynamic behavior of an existing masonry building [13]. Zonta (2000) has considered impact hammer and shaker excitations for experimental modal testing on a Roman amphitheater with structural problems due to aging and material deterioration [14]. Atamturktur (2009) considered Bayesian inference to update the parameters of a masonry vault model based on EMA data using Bayesian inference [15]. Atamturktur et al. (2010) considered modal testing of masonry vaults using operational modal analysis and experimental modal analysis [16]. Cakir et al. (2016) considered experimental modal analysis for masonry arches which was strengthened with graphene nanoplatelets-reinforced prepreg composites [17]. De Felice et al. (2022) evaluated seismic behavior of rubble masonry using shake table test, and experimental modal analysis has been used to evaluate the modal properties [18].

Here, experimental modal analysis using shaker test is considered to evaluate the effectiveness of crack sealing using grouting by considering the accelerometer data. The main idea of considering the grouting to regain the lost strength of masonry wall is lay foundation for future studies on the un-reinforced as well as retrofitted masonry wall. The repair process focuses on effective and sustainable sealing of cracks in the masonry structures by grouting. Most importantly, it emphasized on environmental considerations by minimizing material usage and needs only light application for an eco-friendly repair in order to reuse these buildings. The judicious application of it targeting needed areas may not require excessive use of materials like traditional repair methods. This may reduce the environmental footprint of the repair and will be cost-effective method. This may not introduce additional load on the structure as well which can have unintended consequences for the structural integrity of the masonry. For the reuse of such masonry structures, the method considers both structural and environmental aspects for a comprehensive and eco-friendly solution. Now, the effectiveness of the method on the performance of both buildings need to evaluated for further studies. Experimental Modal Analysis (EMA) is employed to analyze the accelerometer data, providing insights into the modal characteristics, of the masonry structures. This condition assessment helps in identifying potential improvements in the overall dynamic characteristics of the masonry structures.

2 Experimental Program

2.1 Test Buildings for Condition Assessment

The slow-cyclic lateral loading tests have been conducted on two identical one-room buildings, one without reinforcement (un-reinforced brick masonry building or URM) and the other retrofitted with semiconfining elements (semiconfined brick masonry building or SCURM) of nominal reinforced concrete bands. Tests have been conducted on both buildings to assess the in-plane behavior of their masonry walls under simulated gravity loads. The loading conditions followed FEMA 461 guidelines, aiming to replicate deformations experienced during earthquake events. The focus of the investigation was on comprehending the hysteretic behavior, strength, ductility, energy dissipation, stiffness degradation, and associated failure mechanisms. The main objective was to quantify the impact of the semiconfining bands on the in-plane behavior of masonry walls, considering a specific arrangement of vertical and horizontal semiconfining elements. Detailed experimental investigations and observation can be found in Latha et al. (2022) [12]. Both the structures had developed cracks during testing. Figures 1 and 2 show the structures and the developed cracks after cyclic lateral loading test. These cracks were further sealed using mortar grouting sustainably.

For SCURM building, the anchoring of steel bar has been considered full (for past cyclic test only partial anchoring was considered), and all the cracks were sealed. Only a minimal amount of grouting material was considered and applied to fill the cracks. Typical mortar grouting at the base of cracked-retrofitted building after cyclic lateral loading test is shown in Fig. 3, and same was followed for un-reinforced building as well. In this way, test specimens were made ready for the condition assessment. For the condition assessment, accelerometers were strategically placed at various locations of both the buildings for shaker tests and the schematic is shown in Fig. 4.

Fig. 1 Un-reinforced building (or URM) and developed cracks on the building after cyclic lateral loading test

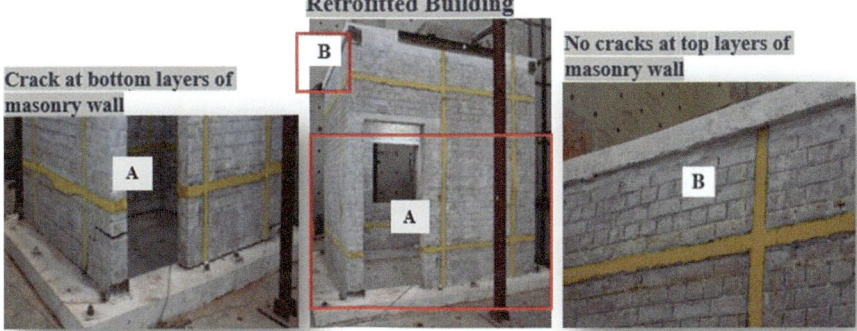

Fig. 2 Retrofitted building (or SCURM) and developed cracks on the building after cyclic lateral loading test

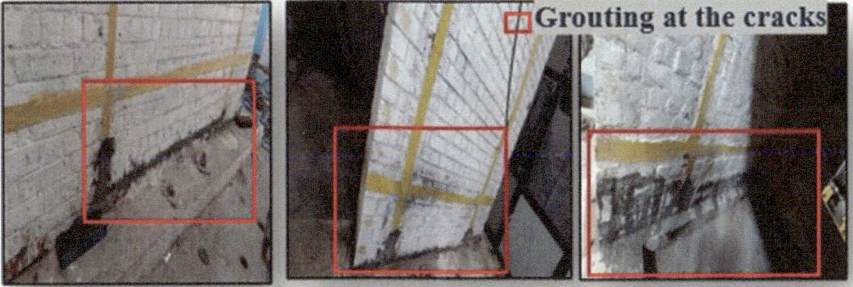

Fig. 3 Mortar grouting at the base of cracked-retrofitted building after cyclic lateral loading test

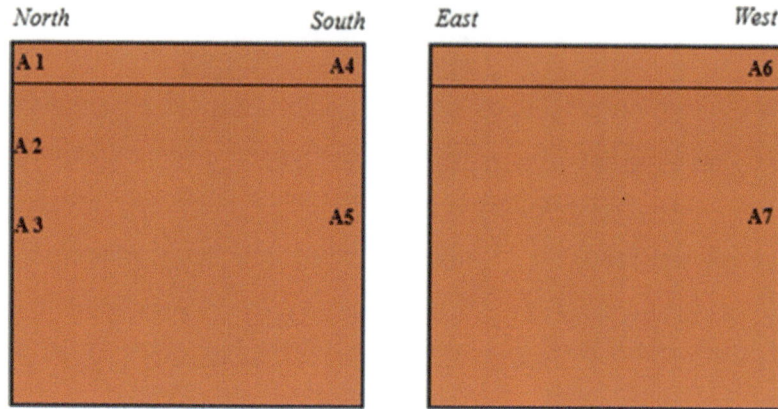

Fig. 4 Schematic for location of sensors for both the buildings (in addition to these eighth one is on the shaker as well)

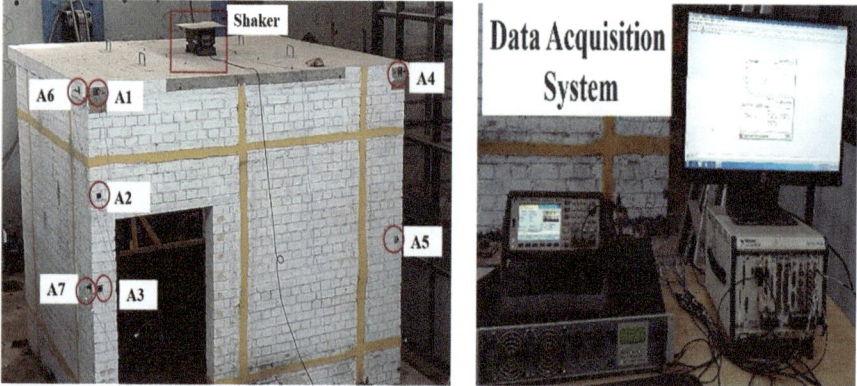

Fig. 5 Retrofitted building with shaker and accelerometers (A1 to A7 are locations of seven accelerometers, and eighth accelerometer is on the shaker) and data acquisition system with connected sensors and computer during the testing

2.2 Test Setup and Instrumentation

In order to examine the effectiveness of the repair, shaker tests have been done on both the buildings. The accelerations were recorded during the shaker test which are used to experimentally determine the modal characteristics of masonry. Shear accelerometers of model 393B04 manufactured by PCB Piezotronics, Inc. were employed, possessing a frequency range spanning from 0.02 to 1700 Hz and a sensitivity of 1000 mV/g. Figure 5 shows the retrofitted building with the shaker, accelerometers, and data acquisition system with connected sensors and computer during the testing. The same test set-up and instrumentation was considered for un-reinforced masonry building as well.

3 Results and Discussions

The acceleration data obtained from the shaker test is considered and further analyzed for the modal characteristics of masonry structures. This assessment aids in identifying areas for potential enhancements in the overall dynamic properties of these structures. The changes in the dynamic behavior of control and retrofitted buildings after the grouting are found from the comparison of test results obtained from the shaker tests. Three trials are considered for the white noise input motion for both the buildings. The transfer function plot generated from the white noise data for obtaining the modal characteristics for control building and retrofitted building are shown in Figs. 6 and 7. The fundamental frequencies of both control and retrofitted buildings obtained for three trials for both the buildings are shown in Table 1. Here, the fundamental frequency is considered as the frequency with most significant power or energy content; hence, the peaks are considered accordingly. For three trials, the percentage increase of the fundamental frequency of retrofitted building when compared with un-reinforced building are 4%, 2.8%, and 4.3%. There is a notable change in the fundamental frequency of retrofitted building when compared with un-reinforced one after grouting, which may be due to the semiconfinement by RC bands. The change in the value of percentage increase for different trials indicates that the parameters obtained from the real-time monitoring at different times are affected by some variability. The variability can be due to data acquisition or processing or may be due to environmental conditions.

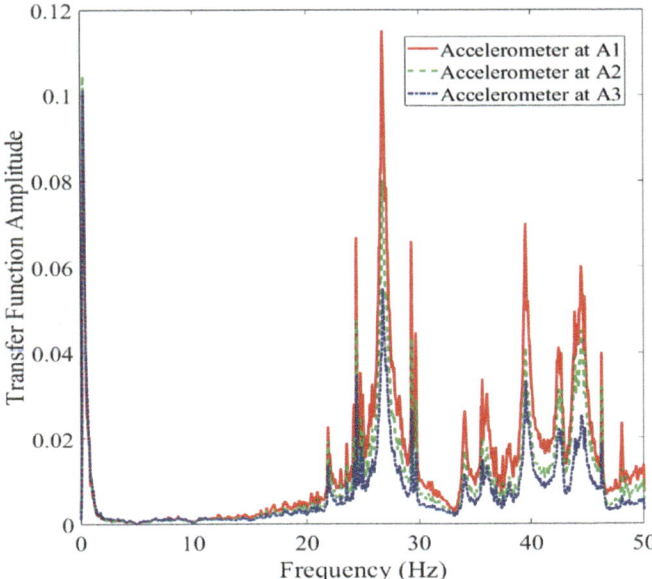

Fig. 6 Transfer function amplitude vs. frequency from white noise input for un-reinforced building

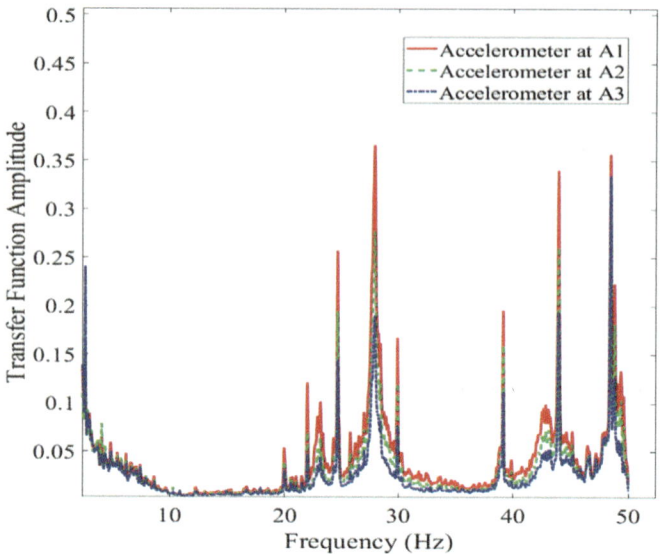

Fig. 7 Transfer function amplitude vs. frequency from white noise input for full-scale retrofitted building

Table 1 Fundamental frequencies (in Hz) obtained for control and retrofitted building at different sensor locations

Sensor locations	Trial 1	Trial 2	Trial 3
Control building at A1	26.83	26.97	26.94
Retrofitted building at A1	27.91	27.71	28.09
Control building at A2	26.83	26.96	26.94
Retrofitted building at A2	27.91	27.72	28.09
Control building at A3	26.83	26.98	26.94
Retrofitted building at A3	27.91	27.72	28.09

In order to understand the mode shape along the height of the buildings, the acceleration transfer function amplitudes obtained for sensors at various positions (A1 at 3 m, A2 at 2.25 m and A3 at 1.5 m) along the height were considered. The mode shapes of the control and retrofitted buildings are shown in Fig. 8. The fundamental mode shape of retrofitted building is observed to be pure shear type when compared with control building which may depict the lateral load resistance action due to semiconfining elements. The behavior was observed similar for control and retrofitted building during the cyclic testing as well [12].

4 Conclusions

The utilization of experimental modal analysis by shaker tests is considered in this study to assess the effectiveness of crack sealing using grouting as a repair method for already cracked unreinforced and retrofitted masonry walls during cyclic testing.

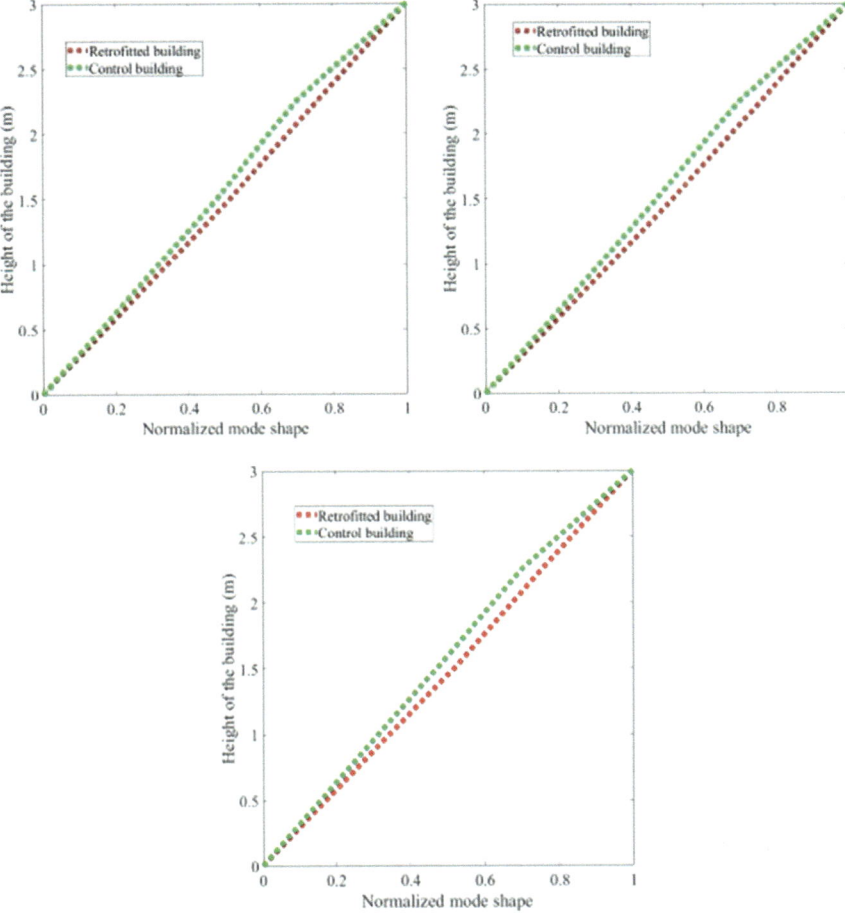

Fig. 8 Mode shapes of both control and retrofitted building obtained from: (**a**) Trial 1, (**b**) Trial 2, and (**c**) Trial 3 data

The cyclic testing was done to assess the performance of retrofitted structure when compared with un-reinforced structure. The method of retrofitting was found to be very effective in improving the overall performance of the structure with enhanced ductility, energy dissipation capacity, integrity, and stability. The findings could establish the effectiveness of the proposed retrofitting approach for existing deficient URM buildings. The approach has only minimal intrusion, possess ease of implementation, and cost-effectiveness. This sustainable retrofitting strategy reduces material consumption by concentrating on nominal RC bands, thus lessening the need for additional resources. It also sidesteps the creation of substantial construction and demolition wastes (CDW) while keeping energy requirements low. Furthermore, it minimizes disturbances to the nearby environment, resulting in limited noise and dust pollution. Moreover, it enhances the seismic resilience of buildings, possibly preventing the need to demolish them and thus reducing environmental impacts like increased material usage, carbon emissions from new materials, and pollution from construction activities. Further studies on the scheme of retrofitting are essential for its wide implementation. So, the already tested structures are further repaired. The reuse of such masonry structures is considered such that it offers both structural and environmental aspects for a comprehensive and eco-friendly solution. The emphasis of the study is on effective and sustainable way of preserving the integrity of structure which prioritizes environmental considerations (minimal material usage and eco-friendly repair method), by lowering the environmental footprint and offering a cost-effective solution compared to traditional methods.

The repair method is aimed to restore the initial behavior enabling further research on both unreinforced and retrofitted masonry walls. Modal characteristics evaluation has found the increase of global stiffness of the retrofitted building than the un-reinforced repaired building. The pure shear type mode shape of retrofitted building depicts lateral load resistance action of semiconfining elements. The behavior of the repaired buildings observed is found similar to that observed during cyclic testing. In general, it is found that the retrofitting of the masonry building using RC bands does not alter modal behavior of the buildings. However, such retrofitting significantly enhances the ductility, thereby improving the seismic performance as shown through reverse cyclic test results.

The dynamic identification of the structures is crucial for an immediate assessment of the performance by considering sustainable preservation. This may facilitate further studies on retrofitting the repaired control masonry or enhancing the scheme of retrofitting or strengthening of semiconfined masonry for improved performance and wide implementation.

References

1. Afzal, Q., Abbas, S., Abbass, W., Ahmed, A., Azam, R., Riaz, M.R.: Characterization of sustainable interlocking burnt clay brick wall panels: an alternative to conventional bricks. Constr. Build. Mater. **231**, 117190 (2020)
2. ElGawady, M., Lestuzzi, P., Badoux, M.: A review of conventional seismic retrofitting techniques for URM. In: 13th International Brick and Block Masonry Conference, vol. 4, No. 7, Amsterdam (2004, July)
3. Bhattacharya, S., Nayak, S., Dutta, S.C.: A critical review of retrofitting methods for unreinforced masonry structures. Int. J. Disast. Risk Reduct. **7**(Mar), 51–67 (2014). https://doi.org/10.1016/j.ijdrr.2013.12.004
4. Messali, F., Metelli, G., Plizzari, G.: Experimental results on the retrofitting of hollow brick masonry walls with reinforced high performance mortar coatings. Constr. Build. Mater. **141**, 619–630 (2017)
5. Facconi, L., Minelli, F.: Retrofitting RC infills by a glass fiber mesh reinforced overlay and steel dowels: experimental and numerical study. Constr. Build. Mater. **231**(Jan), 117–133 (2020)

6. Lucchini, S.S., Facconi, L., Minelli, F., Plizzari, G.: Cyclic test on a full-scale unreinforced masonry building repaired with steel fiber-reinforced mortar coating. J. Struct. Eng. **147**(6), 04021059 (2021)
7. Corradi, M., Di Schino, A., Borri, A., Rufini, R.: A review of the use of stainless steel for masonry repair and reinforcement. Constr. Build. Mater. **181**, 335–346 (2018)
8. Borri, A., Corradi, M., Castori, G., Molinari, A.: Stainless steel strip—a proposed shear reinforcement for masonry wall panels. Constr. Build. Mater. **211**(Jun), 594–604 (2019). https://doi.org/10.1016/j.conbuildmat.2019.03.197
9. BIS (Bureau of Indian Standards): Indian Standard Code of Practice for Earthquake Resistant Design and Construction of Buildings. IS 4326. BIS, New Delhi, India (1993)
10. BIS (Bureau of Indian Standards): Criteria for Earthquake Resistant Design of Structure. IS 1893. BIS, New Delhi, India (2002)
11. BIS (Bureau of Indian Standards): Seismic Evaluation, Repair and Strengthening of Masonry Buildings—Guidelines. IS 13935. BIS, New Delhi, India (2009)
12. Latha, L., Ray Chaudhuri, S., Mukhopadhyay, S., Bajpai, K.K.: Seismic performance enhancement of unreinforced brick masonry buildings by retrofitting with reinforced concrete bands: full scale experiments. J. Struct. Eng. **148**(12), 04022195 (2022)
13. Genovese, F., Vestroni, F.: Identification of dynamic characteristics of a masonry building. In: Bisch, P., Labbe, P., Pecker, A. (eds.) Proceedings of the 11th European Conference on Earthquake Engineering, Paris, France. Taylor & Francis, The Netherlands (1998)
14. Zonta, D.: Structural damage detection and localization by using vibrational measurements. PhD thesis, University of Bologna, Italy (2000)
15. Atamturktur, S.: Calibration under uncertainty for finite element models of Masonry Monuments. PhD thesis, Pennsylvania State University, PA (2009)
16. Atamturktur, S., Fanning, P., Boothby, T.E.: Traditional and operational modal testing of masonry vaults. Proc. Institut. Civil Eng. Comput. Mech. **163**(3), 213–223 (2010)
17. Cakir, F., Uysal, H., Acar, V.: Experimental modal analysis of masonry arches strengthened with graphene nanoplatelets reinforced prepreg composites. Measurement. **90**, 233–241 (2016)
18. de Felice, G., Liberatore, D., De Santis, S., Gobbin, F., Roselli, I., Sangirardi, M., Sorrentino, L.: Seismic behaviour of rubble masonry: shake table test and numerical modelling. Earthquake Eng. Struct. Dynam. **51**(5), 1245–1266 (2022)

Framework for Quality Documentation for Reusing Structural Timber Components

Hussna Satar Janjua, Amne Haura Al-Shorayer, Erik Løhre Grimsmo, Dimitrios Kraniotis, and Allen Tadayon

Contents

1 Introduction

The construction industry plays a significant role in global emissions, contributing up to 40% of total emissions and posing challenges to environmental sustainability [1]. With the increasing demand for infrastructure and urban development, the industry's carbon footprint continues to grow, highlighting the urgent need to address the impact of construction activities on climate change and ecological integrity. Greenhouse gas emissions from construction processes, material extraction, and transportation, coupled with the embodied carbon in built environments,

H. S. Janjua (✉)
Oslo Metropolitan University (OsloMet), Oslo, Norway

Norwegian University of Science and Technology (NTNU), Trondheim, Norway
e-mail: hussnasj@stud.ntnu.no

A. H. Al-Shorayer · E. L. Grimsmo · D. Kraniotis · A. Tadayon
Oslo Metropolitan University (OsloMet), Oslo, Norway

© The Author(s) 2025
M. Kioumarsi, B. Shafei (eds.), *The 1st International Conference on Net-Zero Built Environment*, Lecture Notes in Civil Engineering 237,
https://doi.org/10.1007/978-3-031-69626-8_109

underscore the complexity of mitigating the industry's environmental impact [2]. Conventional construction practices, reliant on resource-intensive methodologies and materials, present formidable barriers to sustainability efforts. Despite these challenges, opportunities for innovation and collaboration emerge, calling stakeholders to pave the way toward a more sustainable future.

The concept of circularity is one of the main promising solutions to the construction industry's environmental challenges [3]. Circularity represents a departure from the linear "take-make-dispose" model, emphasizing resource efficiency, waste reduction, and sustainable practices. Reuse, defined as using products or materials again for a similar purpose as before without significant processing [4], is a key component of circularity. It prioritizes preserving materials' value and minimizing waste generation, thus opening new avenues for innovation and resource conservation. By embracing reuse, the construction industry can reduce its reliance on virgin resources and mitigate carbon emissions.

Structural timber, including cross-laminated timber (CLT), glue-laminated timber (glulam), and solid timber (e.g. log), emerges as a highly suitable option for reuse due to its unique properties and environmental advantages. As a renewable resource, timber is sourced from sustainably managed forests, resulting in a lower environmental impact compared to non-renewable materials like concrete and steel [5–7]. Studies across Europe have shown the potential for reuse of structural timber components [8, 9]. For instance, a case study in Finland concluded that while the technical feasibility of reusing structural timber exists, the establishment of efficient and standardized assessment criteria is needed to ensure the mechanical integrity of these materials [9].

Aligned with principles of circularity and resource efficiency, timber offers versatility, flexibility, and precision in design and construction [10]. However, to ensure the effective reuse of structural timber, guidelines, and quality criteria must be established to confirm their structural integrity and compliance with technical regulations. Presently, a lack of comprehensive guidelines and standards for the reuse of structural timber poses a significant challenge. This study aims to address this gap by providing a flowchart for quality checks in the reuse of structural timber, with the following research question guiding the investigation: *"What specific quality criteria should be considered to facilitate the reuse of structural timber in construction projects?"*

This study addresses the challenge of limited guidelines for reusing structural timber by developing a flowchart for quality checks. This flowchart, informed by literature reviews, document studies, and interviews with Norwegian construction industry professionals, establishes specific quality criteria for evaluating the suitability of structural timber components for reuse in construction projects. This research contributes to the advancement of circular economy principles within the construction industry by facilitating the effective reuse of timber and promoting sustainable practices.

2 Method

This study employed a qualitative research method by conducting a series of interviews as well as literature and document study. The document study includes reports from SINTEF,[1] an independent research organization, and relevant Norwegian standards to acquire a deeper understanding of the subject matter. The aim was to have a solid foundation of knowledge about the subject before stepping into the interviews. By triangulating data from these various sources, the study aimed to enhance the validity and reliability of its findings.

2.1 Literature and Document Study

In developing the theoretical framework for this study, a literature review was conducted in accordance with the guidelines outlined by Blumberg et al. [11]. The literature review utilized the search engines Oria (Norwegian library resource) and ScienceDirect. Different keywords such as "reuse," "circularity," "timber," "CLT," "solid wood," "glulam," "regulations," and "structural" were used during the search process. This approach enabled the retrieval of relevant articles to support the topic of this study. Additionally, a document study was employed, examining reports from DiBK[2] and, as well as relevant standards such as NS-INSTA 142 [12], which focuses on the visual strength grading of timber. Moreover, unpublished standards for evaluation of reclaimed timber, such as prNS-3691-1 [13] and prNS3691–2 [14], were also analyzed. This method allowed for the extraction of valuable information aligning with thechapter study's objectives.

2.2 Interviews

In this study, a total of nine interviews with ten participants were conducted digitally, each lasting approximately 50 to 60 minutes. Following the guidelines outlined in the book "Design Methods and Practices for Research of Project Management," [15] all interviews were recorded and transcribed. The interviews were conducted as semi-structural interviews to enable a balance between flexibility and structure in gathering information. The objective was to gather information relevant to the topic of the, thus the conduct of the interview was adapted based on the dynamics between the respondent and the interviewer. Additional questions did arise based on the flow of the interview and insights gained between interviews.

---

[1] An independent research organization in Norway
[2] The Norwegian Directorate of Building Quality

Table 1 Participant information

Participants	Position	Organization
1	Chief executive officer (CEO)	Treteknisk—Norwegian institute of wood technology
2	Senior engineer	DiBK—Directorate of Building Quality
3	Research and innovative manager	OmTre—Company that works with circular economy regarding timber
4	Project manager	BundeBygg—Contractor firm
5 & 6	Construction manager, environmental adviser	Veidekke—Contractor firm
7	Production manager	Skanska—Contractor firm
8	Project manager	Statsbygg—State-owned contractor firm
9	Project developer	Betonmast—Contractor firm
10	Project manager	Seltor—Contractor firm

The selection of participants was carefully considered to ensure a diverse range of perspectives. Specific criteria were established for participant selection, including individuals with knowledge of industry standards and those familiar with relevant building regulations in Norway. Additionally, experts specializing in the circular economy regarding timber were sought after. Lastly, participants actively involved in building practices were also included. Table 1 presents the participants as well as their positions in the respective organizations.

2.3 Data Analysis

A data analysis was applied to synthesize and examine information gathered from the literature study, document study, and interviews. All interviews were transcribed to ensure the utilization of information in our findings. Subsequently, the transcribed data, the data from the literature study, and the document analysis underwent a systematic coding process to categorize content according to predefined themes relevant to the study's topic. This systematic approach allowed for the efficient utilization of data for further analysis and interpretation [16].

3 Theoretical Background

3.1 Relevant Wood Properties and Conditions

For structural timber to be reused in construction projects, it must meet specific standards for physical and mechanical properties, including strength, stiffness, and density as well as being in a certain level of condition with respect to moisture content, and durability [17]. The difference between wood and timber is that wood refers to the raw material, whereas timber refers to the processed product resulting from manufacturing [18].

Strength is a key mechanical aspect of reused structural timber. It ensures the timber can handle loads and maintain its structural integrity over time. For instance, CLT has typically strong bending and shear strength due to its layered construction and adhesive bonding. Glulam, made by bonding wood laminations, also exhibits strong tensile and bending strength, making it suitable for longer spans. Meanwhile, solid wood, that comes directly from a tree trunk retains its original characteristics such as grain pattern and knots. Understanding these differences helps in selecting timber that can withstand loads during reuse, ensuring safety and durability in construction [19].

Stiffness is also crucial for repurposed structural timber. It determines the timber's ability to resist deformation under load, crucial for stability and preventing structural failure. CLT, glulam, and solid wood have different stiffness properties based on their construction and materials. Assessing stiffness helps determine if the timber is suitable for specific reuse scenarios, ensuring good performance and structural integrity across different construction contexts [19].

Moisture content, along with the thermal performance as they are cross correlated, plays a significant role in the dimensional stability and durability of repurposed structural timber. Timber exposed to continuous sunlight may develop cracks, while during rain, it can absorb water, leading to moisture retention. Excessive moisture can cause changes in size, warping, and decay, risking structural integrity. Managing moisture is important to preserve the quality and lifespan of structural timber during reuse, preventing damage from excessive moisture [19].

Durability is essential for repurposed structural timber, defining its resistance to decay, insects, and moisture-induced problems over time. CLT, glulam, and solid wood vary in durability based on their properties and treatment methods. Understanding the durability of each timber type helps choose materials that can withstand environmental challenges and ensure long-term structural performance. By prioritizing durability in reused structural timber, stakeholders can improve the sustainability and resilience of construction practices, reducing the need for frequent replacements and promoting efficiency in resource use [19].

3.2 Regulations in Norway

To ensure the safety, sustainability, and fire resistance of buildings and structures, both old and new building materials must undergo quality documentation. In Norway, building materials must comply with regulations such as the Planning and Building Act, Documentation of Building Products (DOK), the Building Product Regulation, TEK17 (Building Technical Regulations 2017), and SAK 10 (Building Regulations) [20]. Table 2 provides an overview of the various regulations that building products must comply with in Norway.

Table 2 Regulatory framework for a building material

Regulatory framework	Purpose	Regulates
Planning and Building Act [21]	Primary law for planning and construction in Norway	Planning and executing construction and demolition work, cultural heritage, esthetics, and universal design
Documentation of building product (DOK) [22]	Ensure documentation of the quality and characteristics of the building	Technical solutions, material consumption, energy use, and environmental impact
Building Technical Regulations 2017 (TEK17) [23]	Ensure that buildings have satisfactory quality and functionality and are safe to use	For example, fire safety, structural safety, energy efficiency, health, and moisture
Building Product Regulation [24]	Establish rules for the sale and supervision of CE-marked building products	Regulates the sale and supervision of CE-marked building products throughout Europe
Building Regulations (SAK 10) [25]	How municipalities should conduct building permit processing, quality assurance, control, supervision, and approval of companies	Conducting building permit processing, quality assurance, control, supervision, and approval of companies for responsibility

From 1 July 2022, changes were made to the energy, climate, and environmental re quirements in TEK17 and corresponding changes in SAK10. These changes aim to reduce the climate and environmental footprint of buildings and increase the turnover and use of used building materials [26]. This will be achieved by having stakeholders in the building process assess whether the building material is suitable for reuse and provide documentation for relevant properties.

In general, if a building product adheres to the required documentation as per building product regulations, developers or responsible companies are exempt from conducting their own testing of the product's properties before its use in a building [27]. However, if the product lacks or contains incorrect documentation, it becomes necessary to verify that the building product possesses the required properties for the finished building to meet the stipulated requirements of the building technical regulations [27]. This verification process may entail testing or other forms of control to define the performance level. When utilizing a used product for a specific project, it is only necessary to document the properties relevant to that project, which can vary based on the product type and its intended use within the project [28]. To determine the relevant properties for documentation, one can examine the properties typically documented for similar new products, such as reviewing a performance declaration or consulting the harmonized standard for the product.

The relevant standards for the reuse of structural timber are listed in Table 3 and referenced in the flowchart shown in Fig. 1. These standards are necessary to ensure that the mechanical properties are adequate, along with other requirements needed for reusing structural timber.

Table 3 Relevant standards to consider for the reuse of structural timber

Standard	Describe
NS-INSTA-142 [12]	Visual inspection—Created for new timber
prNS 3691–1 [13]	Evaluation of reclaimed timber—General rules
prNS 3691–2 [14]	Evaluation of reclaimed timber—Impurities
NS-EN 16351 [29]	Requirements for CLT
NS-EN 14080 [30]	Requirements for glulam
NS-EN 408 [31]	Requirements for solid wood and glulam

4 Findings and Discussion

This paper aims to address the research question from Sect. 1. This is achieved by gathering and combining information from the literature review, document study, and interviews, resulting in a flowchart shown in Fig. 1. This flowchart is the main part of answering the research question. The flowchart begins with reclaimed structural timber, which is evaluated according to the standard under development known as pr-NS 3691, divided into three parts. The first part establishes terminology and general rules for evaluating reclaimed timber, while the second part addresses impurities found in reclaimed timber. The third part focuses on visual strength grading, although it is not yet finalized and is not considered in this flowchart. According to one of the participating researchers, they stated, *"For now, we use the standard NS-INSTA 142 for visual grading. However, we are working on creating a new standard that is more aligned with reclaimed timber, but that will come later."* The main difference between pr-NS 3691 and NS-INSTA 142 [12] is that the first standard is specific to reclaimed timber, while the second standard is for virgin timber.

The standard pr-NS 3691 is divided into three parallel processes, P1, P2, and P3, as shown in Fig. 1. P3 is included in the flowchart; however, its detailed content has not been yet published. Process P1 incorporates requirements from pr-NS 3691–1 [13] and recommendations from a SINTEF report [32] detailing various types of impairments that structural timber should avoid. The flowchart then assesses whether these damages affect the reclaimed timber. If yes, the timber is unsuitable for reuse; if not, it may be considered temporarily suitable for reuse.

Process P2 addresses impurities in structural timber based on pr-NS 3691–2 [14]. Impurities in timber can be hazardous waste, humus and minerals, metal fasteners, surface treatments, and impregnation. Those impurities can be categorized as clear, partially clear, or unclear based on their presence. Only clear and partially clear timber is considered temporarily suitable for reuse.

Reclaimed structural timber undergoes visual inspection according to NS-INSTA 142 (marked as P3*) and in the near future according to pr-NS 3691–3 as well. Visual strength grading is conducted by inspecting knots, annual ring width, slope of grain, and cracks. The tolerance values for these features can be found in the standard, which is currently based on virgin timber. However, one of the researchers

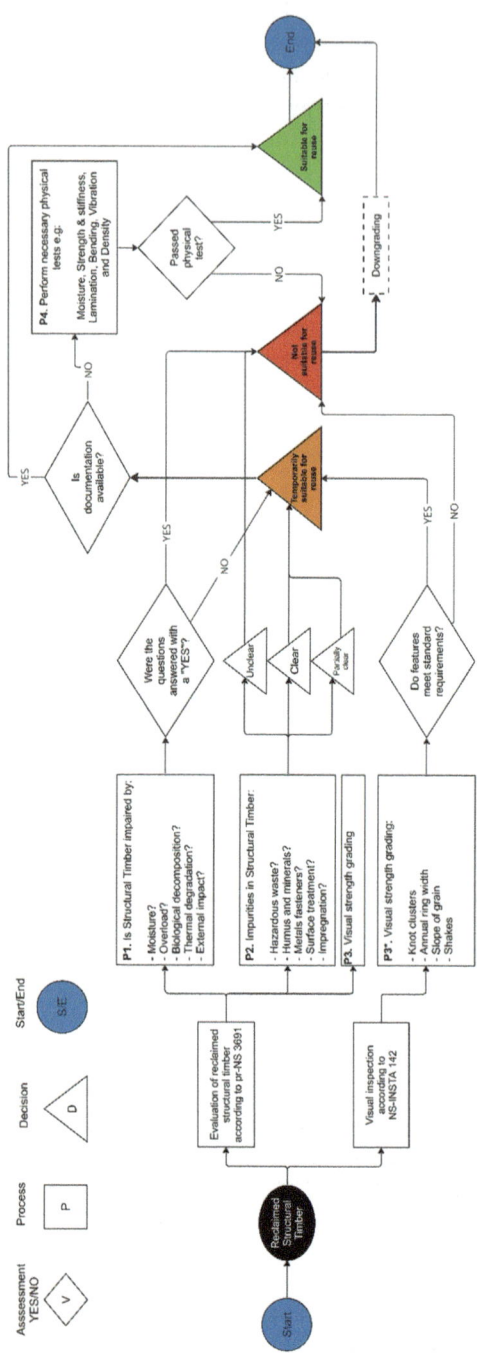

Fig. 1 Flowchart for reuse assessment for structural timber

interviewed mentioned that the standard for reclaimed timber is under process, *"It is part of the follow-up projects that come to SirkTre, which will automate and industrialize tests that correspond to today's sorting methods and are more suited to reclaimed timber."* SirkTre is a NRF[3]project that OmTre leads [33].

Timber categorized as temporarily suitable for reuse undergoes further inspection regarding documentation availability. If documentation is available, the timber is suitable for reuse; if not, it proceeds to process P4 for necessary physical tests. If the timber passes, it's suitable for reuse; if not, it's no longer suitable for reuse as structural timber.

Timber categorized as unsuitable for reuse can be downgraded for alternative uses, as highlighted during an interview with a contractor firm. *"Downgrading, meaning that if it's not sufficient as a load-bearing element for reuse, it can still be repurposed for other uses, such as decoration, playground equipment, or something else. There are many possibilities for reusing an element."* It is important to note that unclear wood from P2 is considered a waste due to potential health risks associated with impurities.

Reflecting on the findings gathered through literature review, document analysis, and interviews, the flowchart we developed emerges as a valuable tool for incorporating the reuse of structural timber in building practices. During the interviews, particularly those conducted with various contractor firms, a recurring pattern surfaced regarding their perception of whether the regulatory changes had indeed simplified material reuse. Previously, regulations posed significant barriers, making material reuse seem impractical. However, with the regulatory changes effective from July 1, 2022, material reuse became more feasible. Findings from the interviews showed that the current obstacle lies in the lack of standards for reclaimed timber, which is essential for effectively promoting the reuse of structural timber. Summarizing the changes in TEK17 as mentioned in the Sect. 3.2, new buildings must be constructed with future dismantling in mind, and materials must be identified for reuse during major renovations in existing buildings.

However, the transfer of responsibility from producers to contractors was another change in the regulations, and it may pose a challenge. The crucial question arises: who is willing to bear such a significant responsibility? Responses from interviews varied. While some expressed readiness to assume responsibility, they emphasized the importance of industry collaboration and support from clients, such as the building owner. *"I don't think going alone as a contractor, there must be industry collaboration, really"* commented a participant from a contractor firm, echoing sentiments shared by others.

Furthermore, the economic aspect of material reuse cannot be overlooked. While materials themselves may be inexpensive, the process of reusing them often incurs higher costs compared to using new materials. *"But as it stands now, in terms of economic benefits, my observation is that we haven't reached a point where the*

---

[3]The Research Council of Norway

commercial aspect of reuse is substantial enough to provide economic advantages," stated a participant from a contractor firm. The construction industry prioritizes cost-saving measures, necessitating clear directives from building owners and sufficient project funding to support material reuse initiatives.

Another significant barrier to promoting material reuse in building practices is the absence of standardized procedures for integrating reused materials into new projects. As one participant from contractor firm highlighted, "[...] *but that's also because there haven't been established standard procedures for testing and preparing it for new use.*" Establishing standardized procedures, such as those outlined in prNS-3691, could effectively address this issue. While these standards have not yet been published, they are currently under review and will be available for implementation in the near future.

Participants also highlighted a lack of information on how to effectively reuse materials. To address participants' concerns, we developed a flowchart outlining a step-by-step procedure for evaluating reused timber materials. The flowchart is designed to streamline the process of reusing structural timber. For the development of the flowchart, we used the mentioned standards listed in Table 3 as well as information gathered from the literature study.

5 Conclusion

This study investigated the challenges associated with reusing structural timber in construction projects, particularly the lack of standardized procedures for quality documentation. Key findings from this study are:

- The research question, "What specific quality criteria should be considered to facilitate the reuse of structural timber in construction projects?" was addressed through the development of a flowchart for quality checks.
- The proposed flowchart establishes a framework for evaluating the suitability of structural timber components for reuse, considering various quality criteria.
- This research contributes to the advancement of circular economy principles in construction by promoting the effective reuse of timber and reducing reliance on virgin resources.
- The flowchart currently has limitations, particularly in the area of assessing the longevity and durability of adhesives used in glued laminated timber (glulam) and cross-laminated timber (CLT).
- Additionally, future studies could explore the economic feasibility of implementing structural timber reuse on a larger scale within the construction industry.

References

1. Grønn byggallianse. https://byggalliansen.no/kunnskapssenter/publikasjoner/ infopakkeklimakjempen/. Last accessed 2023/11/03
2. Crawford, R.H.: Greenhouse gas emissions of global construction industries., IOP Conf. Ser. Mater. Sci. Eng. **1218**(1), 012047 (2022)
3. Yu, Y., Junjan, V., Yazan, D.M., Iacob, M.-E.: A systematic literature review on circular economy implementation in the construction industry: a policy-making perspective. Resour. Conserv. Recycl. **183**, 106359 (2022)
4. Miljødirektoratet. https://www.miljodirektoratet.no/ansvarsomrader/avfall/sirkular-okonomi/. Last accessed 2023/11/10
5. Abed, J., Rayburg, S., Rodwell, J., Neave, M.: A review of the performance and benefits of mass timber as an alternative to concrete and steel for improving the sustainability of structures. Sustain. For. **14**(9), Art. no. 9 (2022)
6. Svortevik, V.J., Engevik, M.B., Kraniotis, D.: Use of cross laminated timber (CLT) in industrial buildings in Nordic climate—a case study. IOP Conf. Ser. Earth Environ. Sci. **410**(1), 012082 (2020)
7. Planke, T., Nore, K., Rygh Nordhagen, V., Bockelie, A., Kraniotis, D.: Transformation of Reclaimed Materials from Barn Buildings—Design of a New Timber Building Frame. CRIStin-NTNU (2023)
8. Brol, J., Dawczyński, S., Adamczyk, K.: Possibilities of Timber Structural Members Reuse. SHATIS (2015)
9. Niu, Y., Rasi, K., Hughes, M., Halme, M., Fink, G.: Prolonging life cycles of construction materials and combating climate change by cascading: the case of reusing timber in Finland. Resour. Conserv. Recycl. **170**, 105555 (2021)
10. Schuster, S., Geier, S.: CircularWOOD—towards circularity in timber construction in the German context. IOP Conf. Ser. Earth Environ. Sci. **1078**(1), 012030 (2022). https://doi.org/ 10.1088/1755-1315/1078/1/012030
11. Blumberg, B., Cooper, D., Schindler, P.: EBOOK: Business Research Methods. McGraw Hill (2014)
12. NS-INSTA 142:2009, Nordic visual strength grading rules for timber, (2010)
13. prNS 3691–1:2023, Evaluering av returtre—Del 1: Terminologi og generelle regler
14. prNS 3691–2:2023, Evaluering av returtre—Del 2: Urenheter
15. Pasian, M.B.: Designs, Methods and Practices for Research of Project Management. Gower Publishing, Limited (2015)
16. Gioia, D.A., Corley, K.G., Hamilton, A.L.: Seeking qualitative rigor in inductive research: notes on the Gioia methodology. Organ. Res. Methods. **16**(1), 15–31 (2013)
17. Teder, M., Pilt, K., Miljan, M., Pallav, V., Miljan, J.: Investigation of the physical-mechanical properties of timber using ultrasound examination. J. Civ. Eng. Manag. **18**(6), Art. no. 6 (2012)
18. Ramage, M.H., et al.: The wood from the trees: the use of timber in construction. Renew. Sust. Energ. Rev. **68**, 333–359 (2017)
19. Ayanleye, S., Udele, K., Nasir, V., Zhang, X., Militz, H.: Durability and protection of mass timber structures: a review. J. Build. Eng. **46**, 103731 (2022)
20. Regjeringen.no. https://www.regjeringen.no/no/tema/plan-bygg-og-eiendom/ bygningsregelverket-fra-1965%2D%2D20172/forskrifter/id2590708/. Last accessed 2024/01/ 10
21. Lovdata. https://lovdata.no/dokument/NL/lov/2008-06-27-71. Last accessed 2024/01/13
22. Direktoratet for byggkvalitet. https://www.dibk.no/regelverk/dok. Last accessed 2024/01/13
23. Direktoratet for byggkvalitet. https://www.dibk.no/regelverk/byggteknisk-forskrift-tek17. Last accessed 2024/01/18
24. Direktoratet for byggkvalitet. https://www.dibk.no/regelverk/dok/byggevareforordningen/ byggevareforordningen. Last accessed 2024/01/24

25. Direktoratet for byggkvalitet. https://www.dibk.no/regelverk/sak. Last accessed 2024/02/01
26. Direktoratet for byggkvalitet. https://www.dibk.no/om-oss/Nyhetsarkiv/regelendringer-fra-1.-juli. Last accessed 2024/02/01
27. Direktoratet for byggkvalitet. https://www.dibk.no/regelverk/dok/iii/9. Last accessed 2024/02/06
28. Direktoratet for byggkvalitet. https://www.dibk.no/regelverk/dok/iii/13. Last accessed 2024/01/24
29. NS-EN 16351:2021, Timber structures — Cross laminated timber — Requirements, (2021)
30. NS-EN 14080:2013+NA:2016, Timber structures — Glued laminated timber and glued solid timber — Requirements, (2016)
31. NS-EN 408:2010+A1:2012, Timber structures — Structural timber and glued laminated timber — Determination of some physical and mechanical properties, (2012)
32. Sørnes, K., Nordby, A.S., Fjeldheim, H., Hashem, S.M.B., Mysen, M., Schlanbusch, R.D.: Anbefalinger ved ombruk av byggematerialer. SINTEF akademisk forlag (2014)
33. OMTRE. https://www.omtre.no. Last accessed 2023/03/12

Structural Response of CLT Bridge Decks to Heavy Vehicle Loads: A Serviceability Evaluation

Emil Lindersson, Mohammed Askari, Amirhosein Vakili, Justin Dahlberg, Mahdi Kioumarsi, and Behrouz Shafei ⓘ

Contents

1   Introduction

Compared to the two most commonly used construction materials; concrete and steel, cross-laminated timber (CLT) usage entails advantages such as high dimension ability, energy efficiency, high strength-to-weight ratio, and great sustainability and aesthetics [1]. The overall conclusion is that the timber product has an impressively low toll on the environment if sustainable forestry sources are utilized, which is well achievable in several parts of the world where such resources are abundant. However, despite these advantages, the application of CLT in bridge construction remains under-explored. Apart from North America and Europe, research has also proven that using CLT in other regions may be highly beneficial. A study presented by Yeh and Chiao [2], conducted in Taiwan, showed that CLT and other timber

E. Lindersson · M. Askari · M. Kioumarsi
Oslo Metropolitan University, Oslo, Norway

A. Vakili · J. Dahlberg · B. Shafei (✉)
Iowa State University, Ames, IA, USA
e-mail: shafei@iastate.edu

© The Author(s) 2025
M. Kioumarsi, B. Shafei (eds.), *The 1st International Conference on Net-Zero Built Environment*, Lecture Notes in Civil Engineering 237,
https://doi.org/10.1007/978-3-031-69626-8_110

products are significantly more environment-friendly than reinforced concrete (RC) and steel. This statement was validated using life-cycle analysis (LCA) on buildings and structures within the country. Another significant advantage is that the product does not corrode, eliminating the most common type of deterioration mechanism that a vast number of reinforced concrete (RC) bridges are prone to.

Engineers have developed various product designs to best utilize the properties and behavior of timber for particular service conditions, and the most innovative and well-known timber designs for bridge decks include Glulam, SLT, and CLT [3]. Glulam was first introduced in Europe during the 1890s and later patented in 1901. This engineered wood product consists of several laminates stacked and glued together to form a strong composite structural element. SLT is widely used for structural components such as bridge decks and was initially developed in the North American region in the late 1970s and has in the subsequent decades been utilized in Australia and various Nordic countries [4, 5]. SLT decks are composed by arranging several parallel timber beams side by side with pre-stressed steel rods applied to compress timber laminations such as structural timber or glulam beams together.

CLT is an engineered wood product and the concept of CLT originated in Austria and Germany in the early 1990s [6]. CLT is another highly prevalent timber product that is commonly utilized within the construction domain. Though it is a very popular choice and is extensively used for commercial building projects, its application, particularly in bridge structures, is relatively limited in comparison. CLT boards are commonly derived from softwood species such as spruce, pine, or fir and are fabricated by arranging multiple layers of laminates perpendicularly and bonding them using adhesives. The cross-lamination process involves gluing the boards in each layer perpendicular to the previous layer. The number of layers, the thickness of each layer, and the grade of the timber are dependent on the specific design requirements of its intended application. CLT construction elements are always prefabricated and available in the maximum dimensions allowed by transportation agencies, leading to elevated levels of quality control and efficient site practices. Commercially available thicknesses range from <100 mm to over 500 mm [7]. CLT is highly suitable for bridge applications due to its inherent strength and durability. The cross-lamination orientation of each layer provides sufficient strength in both directions, making it a versatile material for applications that require robust load-bearing capacity. Its lightweight, ease of application, and ability to have dimensions customized along with its structural integrity make CLT an excellent alternative for bridge decks.

The design process of the link slab was based on the principles of reinforced concrete materials [8, 9]. It focused on the loads that the girder end rotations exerted on the link slabs. Although different studies have researched various aspects of slabs and decks [10–12], studies on CLT bridge deck systems are limited. In response to this opportunity, this study aims to explore the characteristics of CLT bridge deck system and determine their suitability for rural bridge applications. The bridge was designed to resemble a full-scale bridge based on standardized lane dimensions, but with some reduced widths to facilitate feasible testing in a laboratory environment. It was experimentally tested while subjected to both static and cyclic loading exposure simulating American design trucks. The main objective of this research is to assess how suitable various available CLT deck systems are for rural bridge applications. The main focus is on distinguishing their advantages and limitations in terms of serviceability and structural performance. Specifically, the elastic, fatigue, and failure behaviors are to be examined while primarily focusing on vehicle-induced vertical displacements, stresses, and strains.

2 Experimental Testing

The experiments were carried out to comprehensively investigate the performance of CLT deck under varying loading conditions. To achieve this, CLT deck were carefully constructed to assess its durability, strength, and overall performance under both static and dynamic loads that simulate real-life bridge applications. The deck configuration is presented consecutively with geometry, technical information, and properties. Relevant information on the static and dynamic tests is described and finally, the results are presented with an associated discussion.

2.1 Geometry and Connections

The longitudinal panels were assembled using 36 laminates from Douglas Fir, grade select structural, with the dimensions $5\text{-}3/8'' \times 1\text{-}1/2'' \times 312''$ inch^3 ($135 \times 38 \times 7925$ mm^3). In the transverse direction, the continuous panels were assembled using 59 laminates from Douglas Fir of grade No. 2 with the dimensions $5\text{-}3/8'' \times 1\text{--}1/2'' \times 192$ inch^3 ($135 \times 38 \times 3962$ mm^3). The panels were then attached using an overlapping intersection that was bolted together using a cross-fastening method with three bolts per connection every $11''$ (280 mm) along the longitudinal direction. The two $45°$ diagonal bolts were fully threaded countersunk heads with dimensions $3/8'' \times 15\text{-}3/4''$ (10×400 mm^2) and the vertical bolt was a fully threaded cylinder head with dimensions $5/16'' \times 11\text{-}7/8''$ (8×300 mm^2). The cross-fastening bolts were made of hardened steel. The top and bottom laminates were aligned toward the longitudinal direction to support the major strength direction in which the loads were to be applied. The deck was simply supported on both ends resting on concrete abutments. Figure 1a displays the Deck 1 test setup overview. Spreader beams consisting of four layers of wooden boards with dimensions $2'' \times 8''$ (50.8×203.2 mm^2) were placed on the bottom exterior side of the deck. The beams were fastened using fully threaded zinc-plated bolts made from grade A steel (Fig. 1b). They were placed along the transverse direction with centroid distances of 1 ft (305 mm).

| (a) | (b) |

Fig. 1 Test setup: (**a**) deck and abutment, (**b**) bolted spreader beam along the transverse direction

2.2 Mechanical and Material Properties

The mechanical properties of both Douglas fir-North select Structural and No. 1/No. 2 are taken with a 12% moisture content, which is a standard reference value for well-seasoned timber with developed strength and durability characteristics. The chosen species is native to North America and extensively used for structural applications where the loading demand is high. The characteristic values vary greatly depending on the considered axes of the members. The values were either provided by or calculated using "Mechanical Properties of Wood" in the Wood Handbook [13]. The density is 480 kg/m^3, the modulus of elasticity in longitudinal, radial, and tangential axes of wood are 13,100, 891, and 655 MPa respectively.

2.3 Loading Protocols

Two loading configurations were employed to assess the behavior of CLT deck composed of the timber type Douglas Fir-North. The configurations involved static and fatigue loading, as further elaborated herein. The load was exerted on the bridge deck using a hydraulic actuator, where the loads were imposed by the actuators through two steel beams with two protrusions each. These protrusions applied 2-point vertical loads distributed over surface areas with dimensions of 8″ × 20″ (203.2 × 508 mm^2). This setup amounts to a four-point loading with 12.5-kip (55.60 kN) force per point and is conducted similarly to the AASHTO HL-93 design tandem to represent a 50-kip (222.41 kN) load from one heavy vehicle.

For the static loading configuration, the CLT bridge deck was subjected to constant quasi-static loading. As such, the vertical force was applied at a slow enough rate that inertia effects could be neglected. The static loading was performed on several locations to measure the deflection and strain on the decks when applied to a long-term constant force. Four distinct loading locations were chosen; centered on midspan (CM), offset on midspan (OM), centered on end span (CE), and offset on ends pan (OE). The static tests lasted about 20 min. The locations of CM and CE were designated to target and capture maximum midspan deflections due to bending moment, meanwhile, OM and OE were two locations anticipated to provide large horizontal elongations throughout the depth of the deck where only one panel was loaded. As for CE, the loading position is also employed to induce the greatest overall shear forces that may visually assess the system's resistance to interlaminar slippage and shear cracking.

In the second load configuration, fatigue test was performed on the CLT deck to assess properties such as durability and fatigue life, by exposing it to repeated cycling loading. The decks were subjected to 500,000 AASTHO fatigue II load cycles with a cycle duration of 2 s, the same amplitude of 50 kips, and about 277 h of continuous loading. The data acquisition system extracted 1 min of data at the start of every hour.

3   Experimental Results and Discussion

The raw data were subjected to data analysis and the results are presented in a graphical form, where the findings are accompanied by a continuous discussion. Figure 2 shows the force/displacement relationship for the two centrally positioned sensors on each side of the panel intersection. A close to perfectly linear elastic behavior is exhibited for both sensors up to maximum load. This further indicates that the system is operating safely below the yield limit, and any local defects such as drying laminar shrinkage and internal cracks have a low impact on global performance. Utilizing derivation on the graphs provides an approximate linear force/

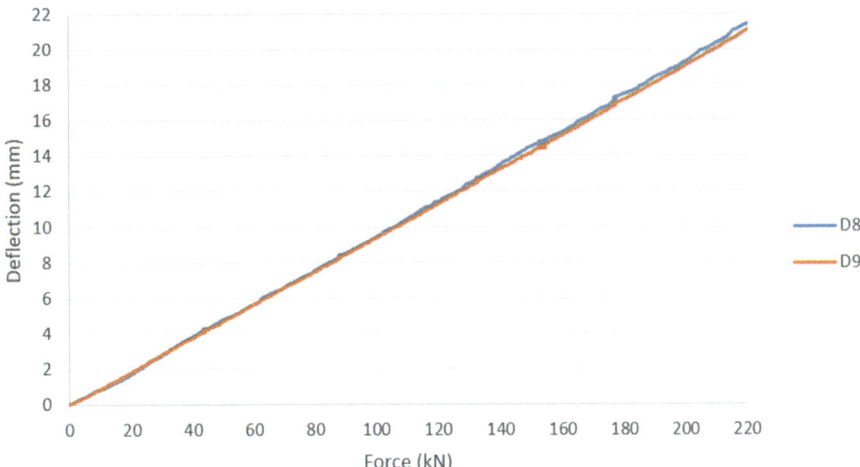

Fig. 2 Force/displacement relationship recorded by the two center gages as a function of force

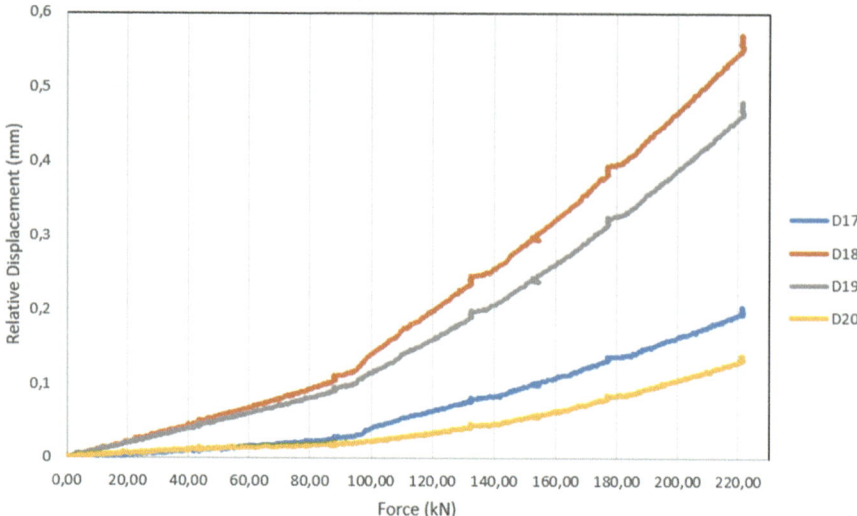

Fig. 3 Relative horizontal displacement between the panels when the load is applied at CM with a magnitude ranging from zero to maximum load

deflection relationship of 0.096 mm maximum deflection per kN of the centrally applied load. As such, in the current deck configuration, the system is evaluated to only be able to accommodate a single vehicle with a weight of 103.23 kN (less than half of the design tandem) while satisfying the AASHTO recommended displacement value of 9.91 mm with the given deck length. However, this is a simplification as the vehicle load is applied as a quasi-static force in the laboratory. In realistic applications, dynamic loading would be present, and the deformation would be influenced by the frequency of the load applied by the vehicle and the natural frequency of the system, which would impact the given values to a certain extent.

Figure 3 displays relative horizontal displacement between the panels when the load is applied at CM with a magnitude ranging from zero to maximum load. As shown in this figure, the two intermediate Sensors D18 and D19 exhibit the greatest relative horizontal displacement as functions of the applied load. The exceedingly low deflection value of 0.6 mm with a corresponding force of 222.41 kN indicates robust bonding between panels due to the cross-fastening method that is applied. Therefore, the system can be assumed to behave symmetrically and homogenously due to the marginal discrepancies between the panels and that it exhibits negligible lateral slippage.

Further to static loading, the behavior of deck under cyclic loading conditions was investigated. Figure 4 displays the increased fatigue-induced deflections in percentages as a function of transverse distance. The increases range between 4.6% and 9.9%, indicating that some stiffness reduction may have occurred over time due to the cyclic loading. A visual assessment was conducted after both the cyclic loading and the static test aiming to identify any visible damage to the structure and determine how it may have impacted the obtained results. No major cracks or

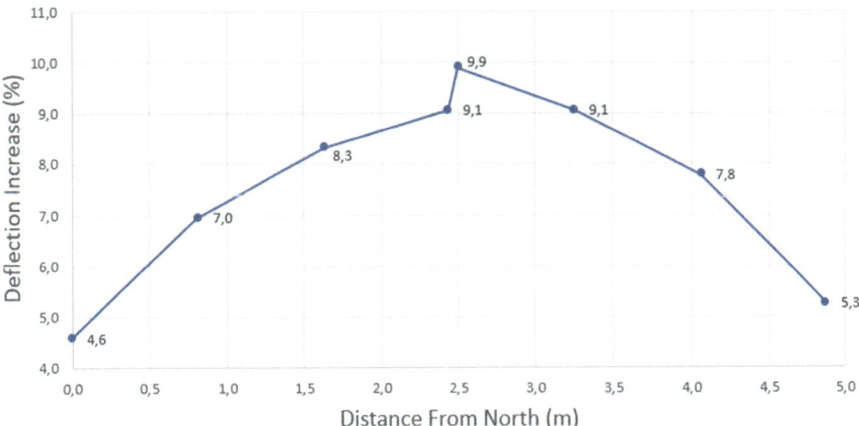

Fig. 4 Maximum displacement relative increases over time along the transverse direction

other obvious damage were detected after the loading schemes and the overall condition was concluded to be unchanged. Upon visual inspection, some splits were observed in the longitudinal direction on the bottom and top surface of deck, including splits in the annual rings of the west surface. These splits may have resulted from a 1-year storage period of the CLT deck in significantly low-humidity conditions, leading to an increased drying rate. It is plausible that the stresses and strains imposed on the deck from the cyclic loading may have resulted in the growth of these splits, although it is not immediately noticeable. Overall, the deck was found to be in good condition where only some minor delamination lines were found at a local scale.

4 Conclusions

The purpose of this study was to assess the structural performance of different CLT deck systems in terms of rural bridge applications. This included experimental testing and data processing of fabricated full-scale bridges by applying static and dynamic loads resembling design trucks in a laboratory setting. The study addressed factors such as general advantages and limitations, load-carrying capacity, durability, and long-term performance. During the experimental testing, it was found that the deck setups behaved linear elastically under the considered loading configurations. It was further subjected to 500,000 loading cycles and found that no measurable stiffness-reduction was exhibited, and no noticeable damage was shown other than some minor local cracking. It is thus concluded that CLT bridge decks may resist fatigue exceptionally well. In conclusion, this research provides valuable insights into the structural performance of CLT deck systems for rural bridge applications. The findings highlight the importance of optimizing design parameters to meet serviceability guidelines and address the challenges of fiber compression as a potential failure mode. The study addressed a research gap in the field and contributed to the understanding and future development of CLT bridges. Furthermore, the study highlighted the environmental advantages of CLT bridges. Compared to traditional materials like concrete and steel, CLT bridges have the potential to significantly reduce carbon footprint due to the carbon sequestration properties of wood. Sustainable timber resources are utilized in the construction of CLT bridges, allowing them to store carbon dioxide instead of emitting it during production. These findings emphasize the importance of using CLT technology in infrastructure projects to not only improve structural performance but also promote environmental sustainability.

Acknowledgments This work is a part of the TRANSFORM project, which has received funding from the Norwegian Directorate for Higher Education and Competence (HK-dir), project number UTF-2020/10107. Additionally, we acknowledge the support from the Net-Zero Future project, an INTPART project funded by the Norwegian Directorate for Higher Education and Competence (HK-dir) and the Research Council of Norway, project number 337262.

References

1. Zhang, L., et al.: Nondestructive assessment of cross-laminated timber using non-contact transverse vibration and ultrasonic testing. Eur. J. Wood Wood Prod. **79**, 335–347 (2021)
2. Yeh, Y.H., Chiao, C.K.: Environmental performance of timber constructions located in highly utilised area-based on realised buildings made of sawn timber or CLT. Paper presented at the 2016 World Conference on Timber Engineering, WCTE 2016, Vienna University of Technology, 2016
3. Davalos, J., Kish, D., Wolcott, M.: Bending stiffness of stress-laminated timber decks with butt joints. J. Struct. Eng. **119**(5), 1670–1676 (1993)
4. Cepelka, M.: Splicing of Large Glued Laminated Timber Elements by Use of Long Threaded Rods. NTNU (2017)
5. Massaro, F.M., Malo, K.A.: Stress-laminated timber decks in bridges: friction between lamellas, butt joints and pre-stressing system. Eng. Struct. **213**, 110592 (2020)
6. Wieruszewski, M., Mazela, B.: Cross laminated timber (CLT) as an alternative form of construction wood. Wood Ind./Drvna Ind. **68**(4), 359–367 (2017)
7. Ussher, E., Arjomandi, K., Weckendorf, J., Smith, I.: Predicting effects of design variables on modal responses of CLT floors. Structures. **11**, 40–48 (2017)
8. Shafei, B., et al.: Evaluation of the Use of Link Slabs in Bridge Projects. Iowa Department of Transportation, Ames (2023)
9. Shafei, B., Taylor, P.C., Phares, B.M., Najimi, M., Karim, R., Hajilar, S.: Material Design and Structural Configuration of Link Slabs for ABC Applications. Accelerated Bridge Construction University Transportation Center, U.S. Department of Transportation/Office of the Assistant Secretary for Research and Technology and Iowa Department of Transportation (2018)
10. Shi, W., Shafei, B., Liu, Z., Phares, B.: Longitudinal box-beam bridge joints under monotonic and cyclic loads. Eng. Struct. **220**, 110976 (2020)

11. Karim, R., Shafei, B.: Addition of partial-depth link slabs to bridge structures: role of support conditions. J. Bridg. Eng. **27**(7), 04022049 (2022)
12. Shafei, B., Phares, B., Shi, W., Azad, S.: Increase Service Life at Bridge Ends Through Improved Abutment and Approach Slab Details and Water Management Practices. Iowa State University, Ames (2023)
13. Kretschmann, D.E.: Mechanical properties of wood. Environments. **5**, 34 (2010)

Composite Facade with Timber and Concrete Connected by Bonding

Roufaida Assal, Laurent Michel, and Emanuel Ferrier

Contents

1   Introduction

The integration of hybrid facades marks a big step forward in the field of civil engineering, offering a highly efficient building envelope that enhances structural and thermal performance. Typically, they combine different materials. In the case of sandwich panels, they comprise an outer and inner layer, commonly separated by an insulating material. The connection between these layers is established either through shear transfer connection or adhesive bonding. In contemporary construction industry, a diverse range of hybrid facade types is employed, with steel connectors being the most prevalent. However, the thermal conductivity of steel connectors tends to be higher than that of the constituent materials leading to heat loss, diminishing the energy efficiency of buildings [1] and increasing heating and cooling costs [2]. Studies have shown that the use of metal fasteners spaced 16″ on center can elevate the thermal transmittance of a sandwich wall panel by over 20% [2]. To offset this effect connectors made from materials like carbon fiber and

R. Assal · L. Michel · E. Ferrier (✉)
Claud Bernard University, Villeurbanne, France
e-mail: emmanuel.ferrier@univ-lyon1.fr; lncs@springer.com

© The Author(s) 2025
M. Kioumarsi, B. Shafei (eds.), *The 1st International Conference on Net-Zero Built Environment*, Lecture Notes in Civil Engineering 237,
https://doi.org/10.1007/978-3-031-69626-8_111

fiberglass is taken into consideration, offering the potential to eliminate thermal bridges [3, 4]. However, the difficulty lies in ensuring the connector's thermal conductivity matches that of the insulation layer, a factor frequently dependent on specific suppliers, potentially leading to rising costs. Examination carried out on the studies of [5], involved testing precast concrete insulated sandwich wall panels with glass fiber-reinforced polymer (GFRP) shear connectors, revealed an increase of up to 80% in ultimate loads. Another study [6] demonstrated an 80% increase in adhesion using GRP connectors, while the research carried out by [7] by a precast concrete sandwich panel made of Ultra-high-performance concrete (UHPC) and fiberglass shear connectors showed a gradual deflection increase with rising composite action.

A challenge of this system is the difference in thermal stress caused by temperature differentials leading to the deflection of the panel. This phenomenon becomes more pronounced in panels with a higher span and composite action, where insulation limits the temperature flow, intensifying the temperature gradient [8]. In response to the thermal binding, adhesive bonding has emerged as a sophisticated method, offering advantages such as reduced weight, increased durability, and weather resistance. The use of adhesive ensures a continuous and uniform distribution of loads and stresses across the facade [9] and eliminates the need for drilling or thermally attacking materials before assembly. Research and studies on bonding applications [10] [11–13] further contribute to the diverse landscape of hybrid facades. The thermo-mechanical performances of the GFRP-glass Sandwich facade were also studied by [19], reporting improved thermal and structural performances compared to traditional systems. In related investigation, [14] aimed to characterize the thermal and structural properties, confirming linear behavior up to the ultimate load. However, environmental factors, especially humidity and heat, play a crucial role, causing alterations in adhesive materials and reducing bond strength [15]. Epoxies, commonly used in adhesive joints, lose rigidity around 45–55 °C, impacting the durability of bonded assemblies [16] [17].

In a related context, [18–20] demonstrated an improvement in the mixed behavior of the beams by using bonded connections leading to enhanced ductility.

This chapter covers an in-depth study detailing the experimental results obtained using wood–concrete composite façade panels. The main objective of this research was to analyze the thermal responses and expansions generated by the application of a specific thermal gradient to these panels. It also presents an extensive program of experimental tests, featuring two distinct facade configurations, the first incorporates a glass-fiber-reinforced concrete, while the second uses welded mesh for reinforcement.

2 Materials and Structures

The wood used in the facade design consisted of glulam elements of strength class GL24H. Table 1 shows the mechanical properties derived from EN1994 [21], bending strength $f_{m, g, k}$; tensile strength parallel to the grain $f_{t, 0, g, k}$ compressive

Table 1 Mechanical properties of the materials

Material	Parameter	Value
Concrete	$f_{ck}[MPa]$	40.8
	ε_{cu} [%]	0.38
	E_{cm} [MPa]	33,546
Timber (GL24h)	$f_{m,\,g,\,k}$ [MPa]	24
	$f_{t,\,0,\,g,\,k}$ [MPa]	16.5
	$f_{c,\,0,\,g,\,k}$ [MPa]	24
	$f_{v,\,g,\,k}$ [MPa]	2.7
	$E_{0,\,g,\,mean}$ [MPa]	11,600
	$E_{0,\,g,\,05}$ [MPa]	9400
Resin	f_c [MPa]	83 ± 4
	f_t [MPa]	32 ± 3
	ε_t [%]	1.2 ± 0.2
	E_t [MPa]	3500 ± 500

Table 2 Measurement condition

Specimens	Medium temperature (°C)	Average hygrometry (%HR)	Change in mass (%)
RC-01	23.5	51.5	0.08
RC-02	24.2	55.1	0.00
FC-01	23.5	59.9	0.15
FC-02	25.5	64.4	0.02

strength parallel to the grain $f_{c,\,0,\,g,\,k}$ shear strength $f_{v,\,g,\,k}$; mean modulus of elasticity parallel to the grain $E_{0,\,g,\,05}$; and characteristic modulus of elasticity parallel to the grain $E_{0,\,g,\,mean}$. The external skin of the concrete is strength class C40/50 concrete. The mechanical properties of the concrete are summarized below [22], where f_{ck}, ε_{cu}, E_{cm} represent the cylindrical compression strength, strain corresponding to the maximum compression stress, and Young's modulus, respectively. The adhesive used is a two-component epoxy resin (Eponal 371) whose main mechanical properties, as specified by the manufacturer.

Table 2 summarizes the test conditions for thermal conductivity on four concrete specimens each with the dimensions of 150*150 (mm). The formulations include anti-Crack fiberglass concrete dosed at 10 kg/m^3 and steel-reinforced concrete with a diameter of 6 mm, and the measuring temperature was set as 15 °C, 25 °C, and 40 °C.

2.1 Experimental Program and Test Conditions

Temperature tests were conducted on two prototypes of wood/concrete facades (Fig. 1). One of the facades was reinforced with fiberglass concrete and welded mesh, as shown in Fig. 2. These prototypes have dimensions of 2.5 m in height, 2.5 m (Fig. 2d) in width, with a concrete thickness of 7 cm and 16 cm for the GL24 h glued laminated wood elements. The purpose of these experiments was to reproduce a temperature difference between the intrados and extrados of the façade (Table 3).

(a)   (b)

Fig. 1 (**a**) Equipment survey (**b**) panel installation

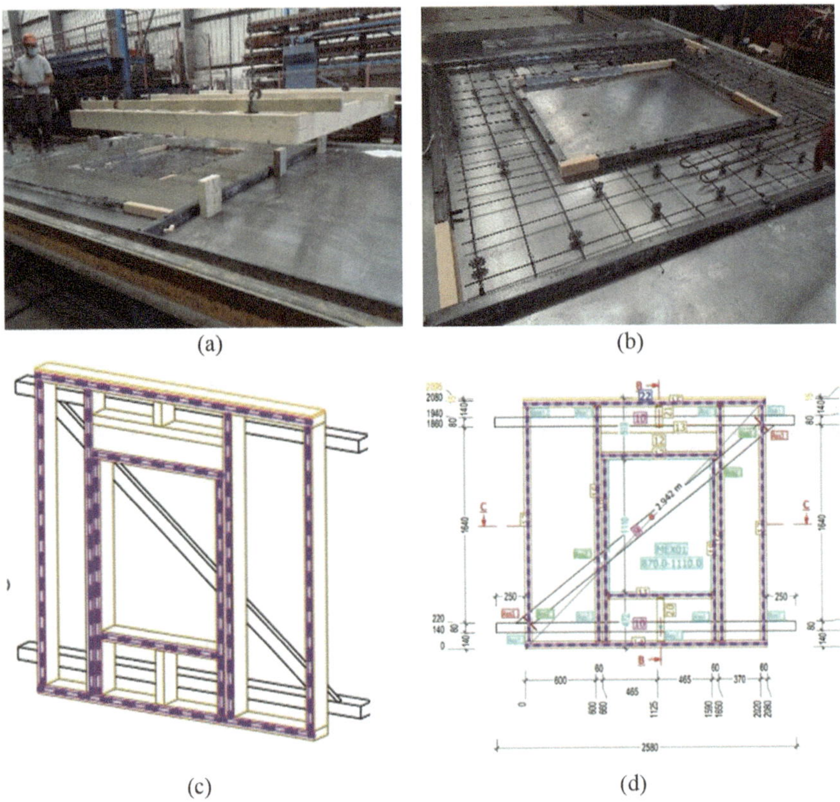

(a)   (b)

(c)   (d)

Fig. 2 Geometry of panels tested under thermal load (**a**) framework, (**b**) mesh reinforcement, (**c**) resin coating (**d**) side view of panel tested

Table 3 Test conditions in the two climatic chambers

Specimens	Pressure	Air-conditioning speed	Air speed
Hot enclosure	From 0 to 40 pa	0.5 °C/min	0.2 to 2 m/s
Cold enclosure			0.2 to 6 m/s

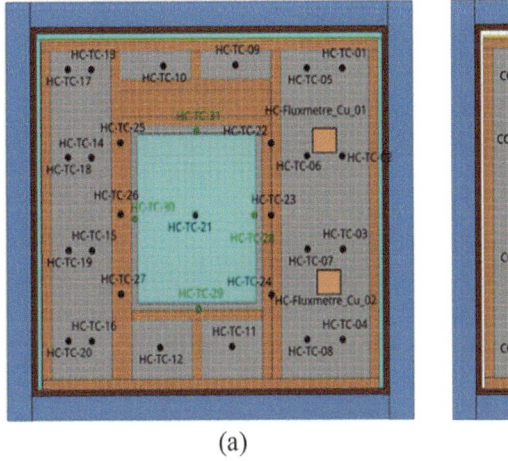 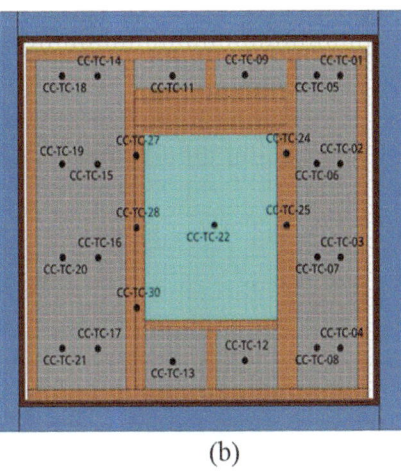

(a)   (b)

Fig. 3 Location of the sensors (**a**) Location of sensors on the hot side test tube, position of the flux meters, (**b**) Location of sensors on the cold side test tube

The initial conditions for the first two tests were considered adequate, involving temperature control in the enclosures ranging from − 10 °C to 40 °C. However, in the last test, the temperature on both sides of the facade was elevated to intensify the temperature gradient. A supporting frame was set up to secure the panels in position.

2.2 Instrumentation

In each chamber, a set of 40 T-type thermocouples and 10 Pt100 probes were positioned for temperature measurements, Pt100 probes were used to measure and transmit air parameters, as well as to validate the uniformity of boundary conditions, with reference to the hot and cold air flows generated by the two chambers. Temperature gradient measurements were conducted for the first test. Two fluxmeters were installed on the heated inner surface of the concrete to force an increase in heat flow through the wall. Of 40 thermocouples installed in each chamber, those bearing the numbers CC-TC-15, CC-TC-20, CC-TC-27, HC-TC-03, and HC-TC-06 were included in the thermal expansion study, these sensors are positioned close to the gauges.

For the evaluation of thermal expansion, 30 mm strain gauges were strategically employed. Positioned at the Mid-height of the panel and near windows, at the distance of 48 cm, covering both the cooled and heated sections (Figs. 3 and 4, Table 4).

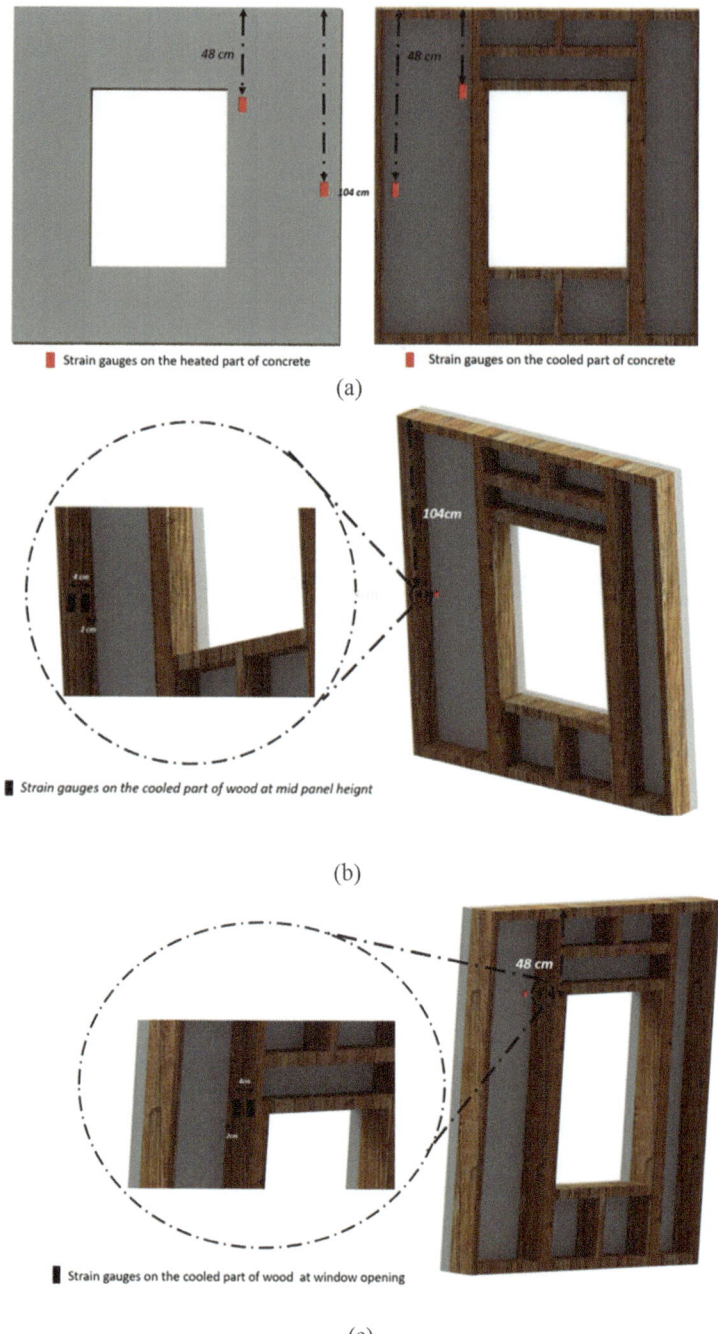

Fig. 4 Gauges positioning Gauge positioning, (**a**) heated concrete at window opening, (**b**) heated concrete at midnight of the panel, (**c**) Cooled wood at midnight, (**d**) cooled concrete near the window

Table 4 Gauges positioning summary

Specimens	Cold surface		Hot surface	
	Near window	Mid-height	Near window	Mid-height
Panel 1	G2-C	G1-C	G'7-C	G'8-C
	G5-W	G4-W		
	G4-W	G6-W		
Panel 2	G4-C	G3-C	G'7-C	G'8-C
	------	G1-W		
	G6-W	G5-W		
Panel 3	G5-C	G4-C	G'7-C	G'8-C
	G6-W	G3-W		
	G2-W	G1-W		

3   Experimental Results

3.1   Thermal Conductivity Results

Thermal conductivity values are predictably lower for fiber-reinforced concrete than for reinforced concrete. The values shown in the table below represent the average conductivity values (Table 5).

3.2   Temperature at Interface

Figure 5 illustrates the temporal evolution of the temperature gradient for the first panel. An initial rise of ΔT before stabilizing during the heating period was observed, followed by a gradual decrease during the cooling phase, reaching around 20 °C and 18 °C when the climate chambers were shut down.

A higher temperature gradient is observed at thermocouple K1, reaching up to 30 °C, positioned proximate to the window, as compared to thermocouple K2, which registers 27 °C. The recorded data manifests a discernible temperature differential of 3 °C between the initial and subsequent facades. The temperature drop is recorded. The graphs below (Fig. 6) depict the temperature rise profiles of hot and cold air and moisture. Despite pronounced similarities in the thermal profiles of each panel, panels 1 and 2 showed a temperature rise ranging from 40 °C to 45 °C on the side exposed to the heat source, compared with the side exposed to the cold, where the temperature reached -15 °C. For the last panel, the temperature conditions imposed are 40 °C on the cold side and 45 °C on the hot side, to increase the thermal gradient at the interface.

Table 5 Concrete thermal conductivity results

Specimens	ΔT (K)	Thermal conductivity (mW/mK)	Measurement uncertainty (mW/mK)
RC-01	15	1449.50	41.8
RC-02	15	1132.90	32.8
FC-01	15	935.49	27.1
FC-02	15	859.88	24.9

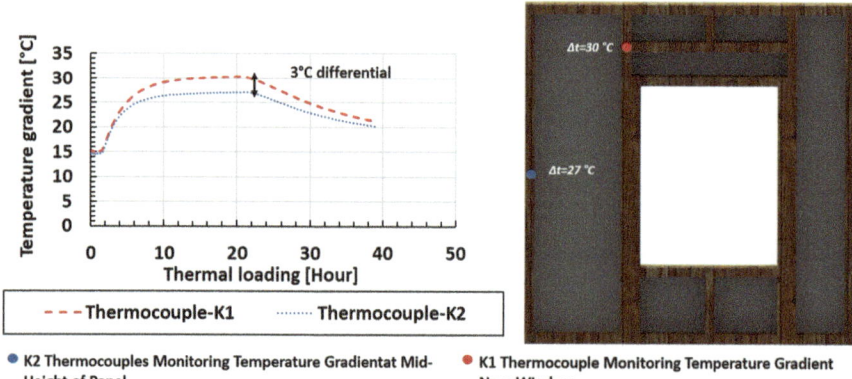

- K2 Thermocouples Monitoring Temperature Gradient at Mid-Height of Panel
- K1 Thermocouple Monitoring Temperature Gradient Near Window

Fig. 5 Evolution of the thermal gradient inside the panel

3.3   Thermomechanical Response of Panels

At a temperature variation of 30 °C (HC-TC-06), thermal strains were observed on the heated surface indicating a compression of the concrete, the thermal strains ranged from 226 με (G'8-C-P2) and 184 με (G'7-C-P2) for the fiber reinforced concrete panel (Panel1). The second panel, which had been reinforced with welded mesh, showed thermal expansions comparable to those of the first one, the thermal strains were assessed as follows to 237με (G'8-C-P3) and 185 με (G'7-C-P3), in the cooled part of the panel, the gauges show tensile strains ranging between 100 and 250 με, no improvement was reported with the addition of fiber. Regarding the last heated panel on both sides, all gauges recorded a compression (Fig. 7).

It is essential to consider the specific boundary conditions that have been set up this multidirectional stress has a significant influence on the thermomechanical behavior of materials, modifying the way wood and concrete react to temperature variations. In relation to the results obtained, no fractures were noted at the interface, and no cracking was observed at the façade.

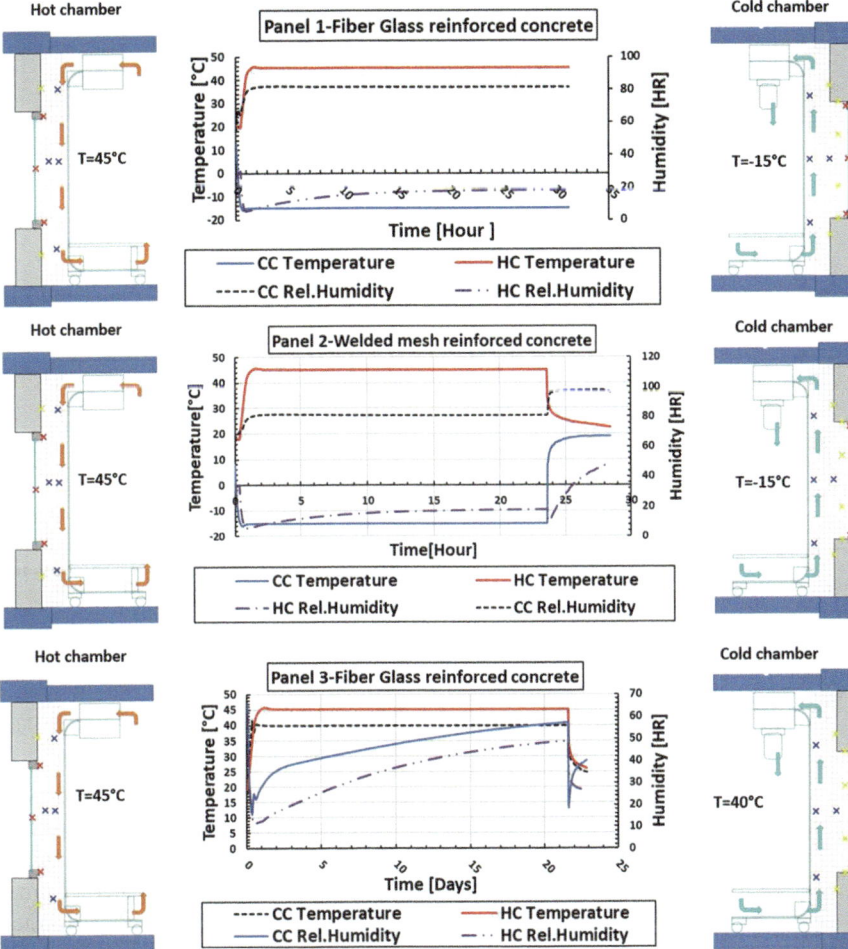

Fig. 6 Change in temperature of generated air inside the hot chamber and cold chamber and moisture for all panels

4 Conclusions

Data analysis focused on temperature variations recorded on the facade surface, as well as thermal expansion measurements taken on the wood and concrete segments.

The observed temperature variations led to diverse expansions or contractions depending on the thermal and mechanical properties of the materials involved.

The results underlined the influence of boundary conditions in influencing the distribution of the stresses between wood and concrete.

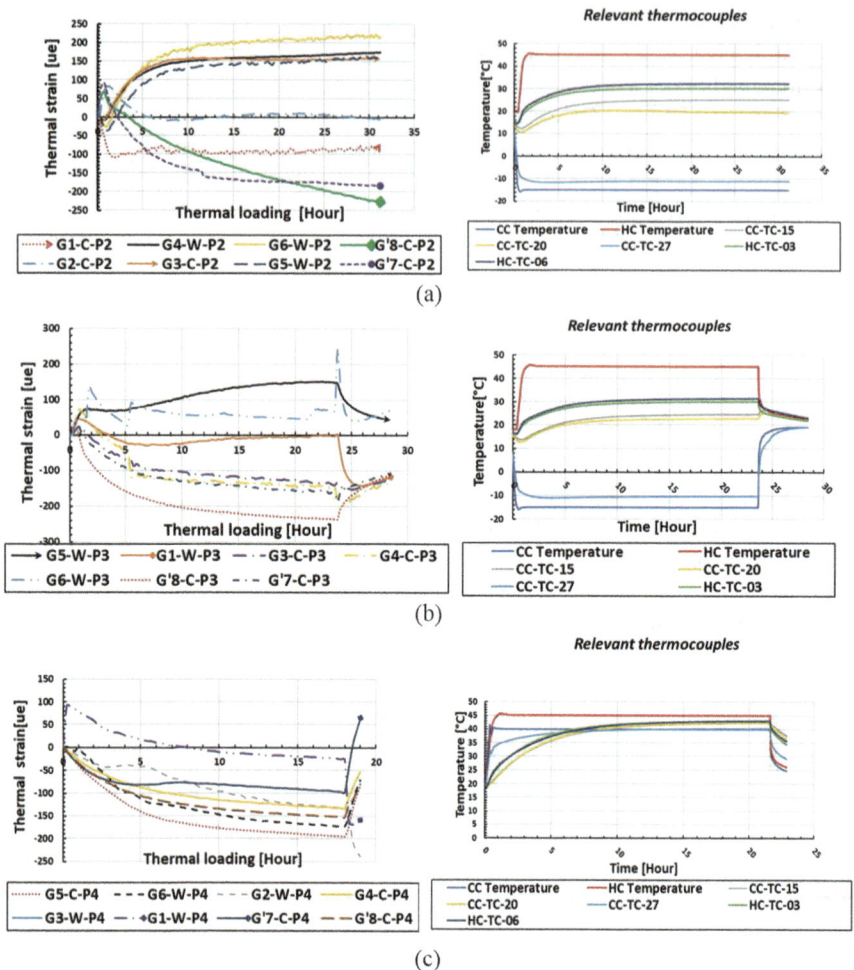

Fig. 7 Simultaneous evolution of thermal deformations and temperatures from thermocouples near strain gauges for (**a**) Panel 1, (**b**) Panel 2, (**c**) Pane 3

The correlation between the thermal expansion of wood and the contraction of heated concrete during a temperature increase raises concerns regarding the thermomechanical properties of the materials studied. Wood, with a naturally higher thermal expansion than concrete, reacts more intensely to temperature variations. Its fibrous structure and flexible mechanical properties enable it to withstand high compressive loads. In contrast, although concrete can also expand thermally, its mechanical properties, such as rigidity, restrict this expansion.

Acknowledgments The authors would like to thanks French reasearch agency ADEME, for their finalcial support of the project Hybridal 2. Authors would like to thanks the technical support of Cruard and Jousselin companies.

References

1. Olsen, J.T.: Developing a General Methodology for Evaluating Composite Action in Insulated Wall Panels. Utah State University (2017)
2. Naito, C.J., et al.: Precast/prestressed concrete experiments performance on non-load bearing sandwich wall panels. In: Air Force Research Laboratory. Materials and Manufacturing Directorate (2011)
3. Sorensen, T., Dorafshan, S., Maguire, M.: Thermal evaluation of common locations of heat loss in sandwich wall panels. In: *Congress on Technical Advancement 2017*, pp. 173–184 (2017)
4. T. J. Sorensen, R. J. Thomas, S. Dorafshan, et M. Maguire, « Thermal bridging in concrete sandwich walls », 2018
5. Tomlinson, D., Fam, A.: Experimental investigation of precast concrete insulated Sandwich panels with glass fiber-reinforced polymer shear connectors. ACI Struct. J. **111**(3) (2014)
6. Woltman, G., Tomlinson, D., Fam, A.: Investigation of various GFRP shear connectors for insulated precast concrete sandwich wall panels. J. Compos. Constr. **17**(5), 711–721 (2013)
7. Mai, Y.: Performance evaluation of sandwich panels subjected to bending compression and thermal bowing. Mater. Constr. **13**, 159–168 (1980)
8. Jawdhari, A., Fam, A.: Thermal-structural analysis and thermal bowing of double Wythe UHPC insulated walls. Energ. Buildings. **223**, 110012 (2020)
9. Augeard, E., Michel, L., Ferrier, E.: Composite wood-concrete panels-effect of cyclic loading and creep. In: International Interactive Symposium on Ultra-High Performance Concrete. Iowa State University Digital Press (2019)
10. Hoffmeister, B., Di Biase, P., Richter, C., Feldmann, M.: Innovative steel-glass components for high-performance building skins: testing of full-scale prototypes. Glass Struct. Eng. **2**, 57–78 (2017)
11. Belis, J., Inghelbrecht, B., Van Impe, R., Callewaert, D.: Cold bending of laminated glass panels. Heron. **52**(1–2), 123–146 (2007)
12. Santarsiero, M., Louter, C., Lebet, J.: The mechanical behavior of SentryGlas and TSSA laminated polymers in cured and uncured state in uniaxial tensile test. Challenging Glass. **4**, 375–384 (2014)
13. Bucak, Ö., Feldmann, M., Kasper, R., Bues, M., Illguth, M.: Das Bauprodukt "warm gebogenes Glas"–Prüfverfahren, Festigkeiten und Qualitätssicherung. Stahlbau. **78**(S1), 23–28 (2009)
14. Cueff, G., Mindeguia, J.-C., Dréan, V., Breysse, D., Auguin, G.: Experimental and numerical study of the thermomechanical behaviour of wood-based panels exposed to fire. Constr. Build. Mater. **160**, 668–678 (2018)
15. Griffiths, R., Ball, A.: An assessment of the properties and degradation behaviour of glass-fibre-reinforced polyester polymer concrete. Compos. Sci. Technol. **60**(14), 2747–2753 (2000). https://doi.org/10.1016/S0266-3538(00)00147-0
16. D. J. Barber, « Fire Resistance of Epoxied Steel Rods in Glulam Timber », 1994
17. Aicher, S., Dill-Langer, G.: Influence of moisture, temperature and load duration on performance of glued-in rods. In: *Rilem Symposium on Joints in Timber Structures*, pp. 383–392 (2001)
18. Bouazaoui, L.: Contribution à l'étude expérimentale et théorique de structures mixtes acier-béton assemblées par collage. », PhD Thesis,, Reims (2005)
19. Nordin, H., Täljsten, B.: Testing of hybrid FRP composite beams in bending. Compos. Part B. **35**(1), 27–33 (2004)
20. Jurkiewiez, B., Meaud, C., Michel, L.: Non linear behaviour of steel–concrete epoxy bonded composite beams. J. Constr. Steel Res. **67**(3), 389–397 (2011)
21. EN 1995-1-1 (2004) (English): Eurocode 5: Design of timber structures - Part 1-1: General - Common rules and rules for buildings [Authority: The European Union Per Regulation 305/2011, Directive 98/34/EC, Directive 2004/18/EC]
22. Ferrara, G., Michel, L., Ferrier, E.: Flexural behaviour of timber-concrete composite floor systems linearly supported at two edges. Eng. Struct. **281**, 115782 (2023)

Operational Modal Analysis and Finite Element Model Updating of the Naillac Tower in Rhodes, Greece

Amirhosein Shabani ⓘ and Amir Hossein Karimi ⓘ

Contents

1 Introduction

Heritage structures offer valuable insights into past architecture, construction techniques, materials used, and cultural practices [1]. These structures attract tourists and contribute to the economy, making their preservation and maintenance crucial [2]. Many historical buildings, typically made of masonry materials, lack lateral load-bearing systems to withstand forces like earthquakes. Because of their low tensile strength, they are prone to collapse under loads [3, 4]. Understanding the structural behavior of each unique historical building is essential for developing

A. Shabani (✉)
Section of Road Operation and Maintenance, Department of Road and Street, The Mobility Division, Oslo Urban Agency (Bymiljøetaten), Oslo, Norway

Department of Built Environment, Oslo Metropolitan University, Oslo, Norway
e-mail: amirhosein.shabani@bym.oslo.kommune.no

A. H. Karimi
Department of Civil Engineering, Technical and Vocational University, Tehran, Iran

© The Author(s) 2025
M. Kioumarsi, B. Shafei (eds.), *The 1st International Conference on Net-Zero Built Environment*, Lecture Notes in Civil Engineering 237,
https://doi.org/10.1007/978-3-031-69626-8_112

appropriate restoration and retrofitting plans. This knowledge helps researchers ensure the structural integrity of these buildings [5, 6].

The geometry of historical monuments is highly intricate. These structures typically consist of walls, pillars, vaults, arches, roofs, and stairs, along with numerous doors and windows [7]. Thick walls help them withstand gravity loads effectively [8]. However, the absence of lateral structural elements poses challenges for these buildings [9]. The presence of multiple doors and windows makes them susceptible to lateral loads. The connections between different structural elements, such as roofs to walls or arches and domes to walls, are complex in such structures. Laser scanning, digital cameras, or drones can be employed to obtain the precise geometry of these structures [10]. These tools allow capturing the complex geometry of buildings, which can then be translated into numerical modeling using drawing software [11].

Ambient vibration testing (AVT), a powerful tool for determining the dynamic characteristics of structures, is a noninvasive method. It provides valuable insights into structural integrity by measuring ambient vibrations, particularly for cultural heritage structures like towers, monumental buildings, and bridges, where minimal interference with normal use is crucial [12]. Operational modal analysis (OMA), a technique in structural health monitoring (SHM), extracts modal parameters from ambient vibration data [13]. OMA identifies resonant frequencies, mode shapes, and damping ratios, offering insights into a structure's dynamic properties under real-world conditions [14]. Calibrating structures using OMA results is a nondestructive method to determine the material properties of historical structures and reduce differences between the dynamic characteristics of numerical models and the actual structure [15, 16].

The main objective of this study is to introduce a comprehensive methodology for creating three-dimensional (3D) simulation-based digital twins of cultural heritage assets using various technologies such as 3D laser scanners, drones, digital cameras, total stations, and accelerometers. The methodology was applied to develop a 3D digital twin of the Naillac tower in Rhodes Island, Greece. Initially, the material properties of the stone masonry were estimated using empirical equations. However, following model calibration based on OMA results, the updated finite element (FE) model's material properties were compared with the initially assumed values. Finally, the study discusses the significance of model calibration based on OMA results and its impact on the structure's performance.

2   An Overview of the Case Study

Constructed in 1400 AD by Grand Master Philibert de Naillac, the Tower of Naillac in Rhodes was a medieval fortress and lighthouse, renowned for its architectural brilliance [17]. Serving as a strategic defense for the commercial port, the tower featured a nightly use massive chain to block the harbor entrance. With canons, vigilant knights, and defensive chains, Rhodes, under Grand Master Naillac's

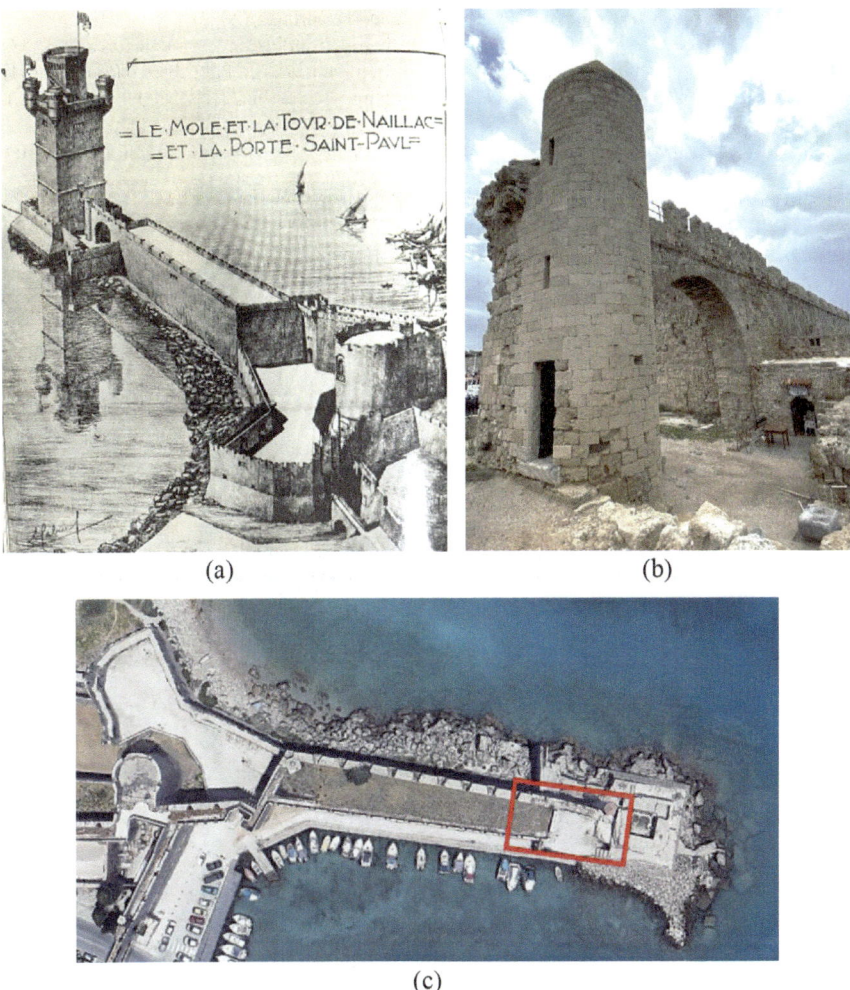

(a)   (b)

(c)

Fig. 1 (**a**) A drawing of the old Naillac tower [18]. (**b**) The remaining part of the tower and the arch (**c**) top view of the fortification taken from Google Earth

leadership, became one of the world's most fortified places [18]. Figure 1a shows the old Naillac tower and the arch bridge, which was the entrance to the tower. Unfortunately, the tower fell during a massive earthquake in 1863, leaving only its foundations visible in the sea [19]. Figure 1b shows the current corner tower, which was built during the Italian occupation to house a staircase leading to the entrance of the big rectangular tower and the arch, and Fig. 1c shows the top view of the fortification and the selected structure in the red box as the case study.

3 Numerical Modeling

3.1 3D Geometric Documentation

A holistic approach was employed for the 3D geometric documentation of the structure based on Kolokoussis et al. [20] and Shabani et al. [21]. A total of 736 ground digital images were captured, along with 1117 aerial images using drones, which had lower resolution due to difficulties in accessing certain parts of the structure. Subsequently, the digital images underwent processing in image-based modeling software, involving filtering and noise reduction, to generate a dense point cloud. Additionally, gaps in the point clouds derived from the digital images were addressed using 3D laser scanners, resulting in the final dense point cloud [20]. A total of 38 scans were captured using 2 terrestrial laser scanners. Establishing a local coordinate system involved the use of two total stations to define target points for laser scanner point clouds and ground control points for image orientation. The completion of the 3D model involved processing the triangulated irregular network representation model, converting each point into a polygon object. Following this process, a 3D dense point cloud, a 3D light model, and the cross sections were produced [20]. The height of the tower from the entrance level to the top is 11.3 m, the diameter of the tower and the thickness of its wall are 3 m and 0.2 m, respectively. The arch span is 7.65 m with a thickness of 2.7 m.

3.2 Numerical Modeling

The homogenized method was used for 3D FE modeling of the tower, neglecting the discretization of masonry units and mortar, which is commonly done for modeling full-scale structures [22, 23]. A 3D model was created using Revit software [24] based on the products from the previous step and imported into Abaqus software. The 3D FE model in Abaqus was split into two components: the tower and the fortification wall, with rigid boundary conditions assigned. The structure is built of Sfouggaria stone, a common material on the island of Rhodes in the past, with a compressive strength estimated at 9 MPa [25]. The elastic modulus of the homogenized masonry was assumed to be 1.001 GPa, considering the presence of soft mortar with a compressive strength of 1 MPa based on the empirical equations provided in Ghiassi et al. [26]. The density and Poisson's ratio of the homogenized masonry were assumed to be $2200 \frac{\text{kg}}{\text{m}^3}$ and 0.3, respectively [27]. Figure 2a shows the 3D model of the structure. The interaction between the wall of the fortification in the western part of the structure and the 3D model of the structure was simulated using axial springs. Figure 2b shows the 3D FE model and the axial springs on the side of the wall highlighted with red stripes. The axial stiffness per area of the springs was calculated to be $0.39 \frac{\text{GPa}}{\text{m}}$ considering the fortification wall with the length of 70 m based on Aghabeigi et al. [28]. After meshing, the model was then imported into FEMTools software for calibration as shown in Fig. 2c, d.

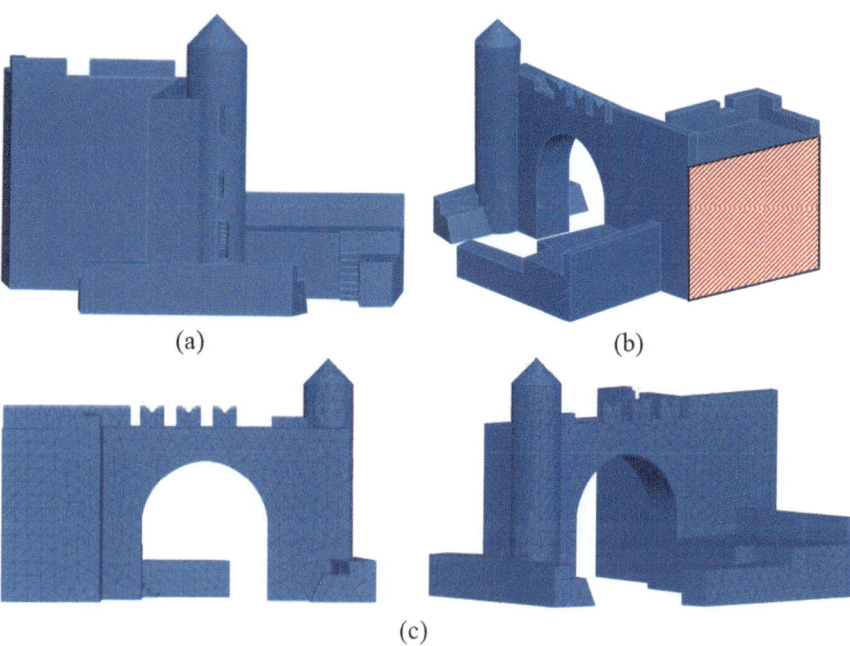

Fig. 2 (**a**) A 3D model of the structure. (**b**) A 3D model of the structure and the wall with axial springs. (**c**) A 3D mesh of the structure from different views

4 System Identification

4.1 Ambient Vibration Testing

Ambient vibration testing was carried out using 3-Axis MEMS digital Unquake accelerometers with a sampling rate of 250 Hz. Six accelerometers were placed in the tower, whereas none were placed outside due to stormy weather. The positions of the accelerometers were determined based on engineering judgment and previous studies [29, 30]. Sensors were selected based on the results of fast Fourier transform analysis, which examined the natural frequency values for each sensor measurement in each direction. After this analysis, three accelerometers were found to be functioning properly and were kept for further analysis. The locations of the three sensors were on the first opening, on the stair between the first and second opening, and on the top stair of the tower in front of the door. The accelerometers come with a Global Positioning System (GPS) antenna and a global navigation satellite system (GNSS) receiver. The recorded time from the GPS data was used to synchronize all the data collected by the accelerometers [29].

4.2 *Operational Modal Analysis*

All synchronized sensor data from the setups were brought into the ARTeMIS modal software package [31] to conduct OMA on a sensor network in order to determine the mode shapes. Three methods of frequency domain OMA were used: frequency domain decomposition (FDD), enhanced frequency domain decomposition, and curve-fitted enhanced frequency domain decomposition [32]. The results show that there was little difference in the natural frequency values obtained from these methods. Figure 3a shows the singular value decomposition (SVD) graph of the FDD method and the first two peaks [33]. The first two peaks show the first two natural frequencies of the structure, which are 5.981 Hz and 9.253 Hz. The acquired mode shapes from OMA indicate that the first and second modes correspond to the transverse and longitudinal directions of the fortification wall, respectively.

5 Finite Element Model Updating

The goal of the FE model updating is to reduce the disparities between the natural frequencies of the numerical model and the actual structure and to enhance the modal assurance criterion (MAC) values of the correlated modes by adjusting the material properties [34]. The tower is discretized into two sets: the tower itself and the remainder of the structure, which includes the arch and a small section of the fortification wall. This division is aimed at better addressing the variations in material properties across different components and facilitating the updating process by increasing the number of parameters. For model updating, a sensitivity-based

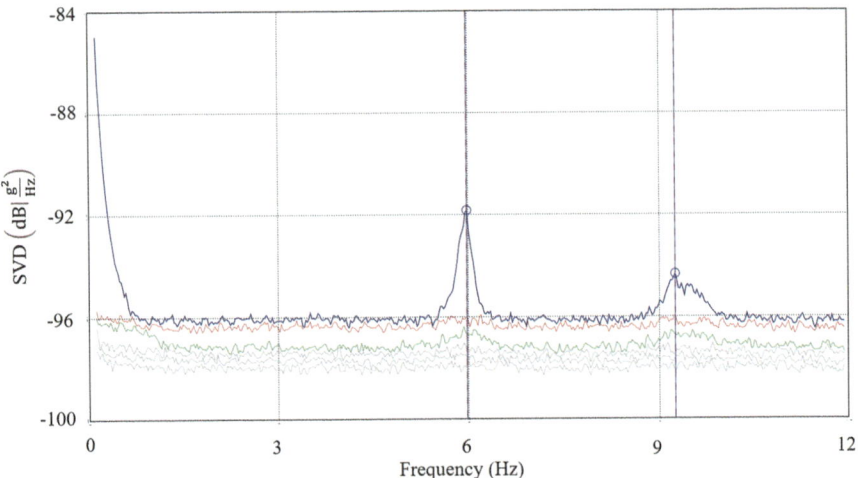

Fig. 3 An SVD graph of the FDD method with the selected two peaks

parameter estimation method was employed using the FEMTools software package [35]. Figure 4a, b shows the first and second mode shapes of the calibrated models, respectively, and Fig. 4c, d shows the OMA mode shapes correlated with the numerical model. The yellow dots in Figure (c) and (d) show the sensors' location.

Table 1 shows the absolute difference in natural frequency (ADF) for both modes. The ADF values are less than 0.03%, indicating a high degree of conformity between the frequency values obtained from the numerical model and those of the actual structure. The MAC values for the first and second modes are 87.921% and 75.568%, respectively, which shows a good correlation between the OMA and the numerical analysis results in terms of mode shapes. The MAC matrix is illustrated in Fig. 5. Higher values of the diagonal elements of the MAC matrix and lower values of the off-diagonal elements show that mode shapes from finite element analysis (FEA) follow the OMA mode shapes [29].

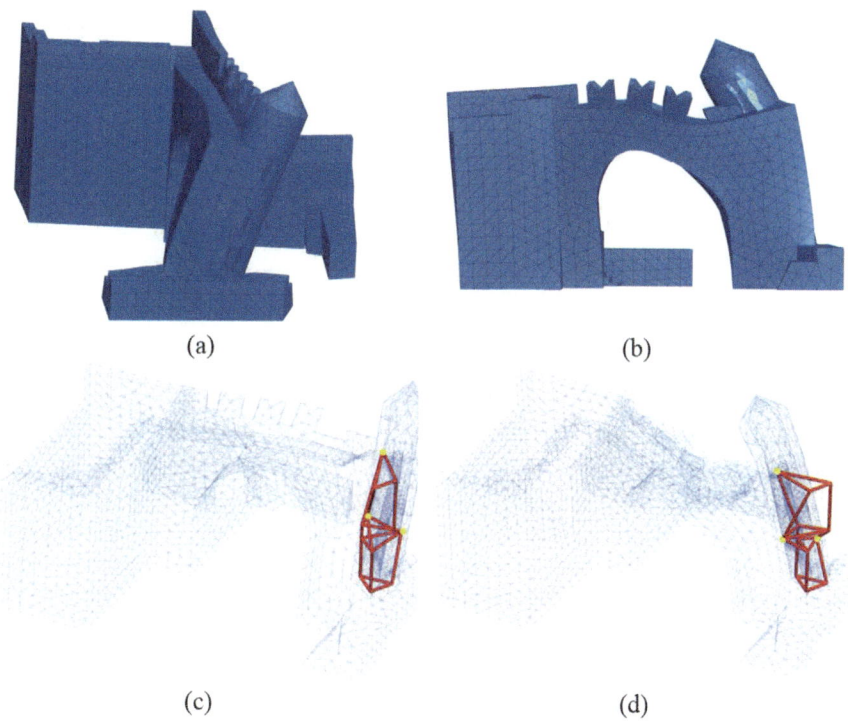

(a)   (b)

(c)   (d)

Fig. 4 (**a**) The first and (**b**) second mode shapes of the calibrated model and (**c**) the first and (**d**) second mode shapes' correlation with the test results

Table 1 Frequency values of the test and calibrated model and ADF of the first natural modes

Mode number	Test frequency (Hz)	Calibrated model frequency (Hz)	ADF (%)
1	5.981	5.982	0.023
2	9.253	9.251	0.024

Fig. 5 An MAC matrix of
the calibrated model

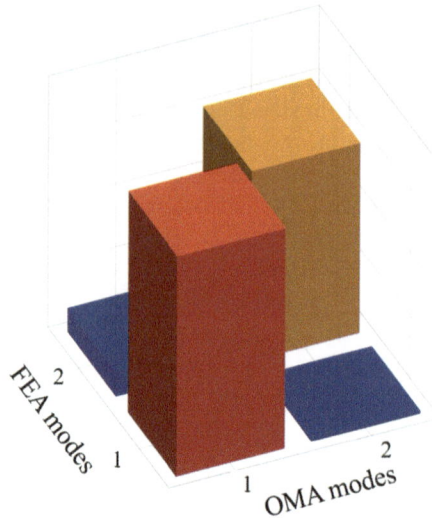

Table 2 Base and calibrated material properties and spring interface stiffness

Component	Property	Base value	Updated value
Tower	Elastic modulus (GPa)	1.001	3.03
	Density ($\frac{kg}{m^3}$)	2200	2183.76
The rest of the model	Elastic modulus (GPa)	1.001	2.08
	Density ($\frac{kg}{m^3}$)	2200	2008.09
Spring interface	Stiffness per area ($\frac{GPa}{m}$)	0.39	0.38

The updated material properties of two components of the structure and the stiffness of the assigned springs before and after calibration are presented in Table 2. The updated elastic modulus, a crucial material property affecting structural behavior, underwent significant changes. This emphasizes the importance of calibrating numerical models based on OMA. Specifically, the tower component's elastic modulus is more than three times greater than the value calculated using the empirical equations. Therefore, the structural performance is underestimated, and the structure exhibits lower initial stiffness when subjected to wind or seismic loads. Additionally, it was observed that the tower component's elastic modulus exceeded that of the rest of the structure, indicating its greater stiffness, which is against the initial assumption. Furthermore, the spring interface stiffness experienced minimal change, indicating its negligible impact on the structure's modal properties.

6 Conclusions

A holistic methodology was employed for developing a simulation-based digital twin of a cultural heritage asset. The case study is Naillac tower, which is the remnant of the older stone masonry tower built in 1400 AD that collapsed due to an earthquake in the eighteenth century. The corner tower, originally constructed to accommodate the staircase leading to the main tower, along with the arch and a portion of the fortification, was chosen for modeling. The 3D geometric documentation was conducted using digital cameras, drones, total stations, and 3D laser scanners. Subsequently, a 3D model was constructed from the gathered data. Following this, a 3D finite element (FE) model of the case study was created, applying base material properties and boundary conditions. Concurrently, operational modal analysis (OMA) was executed using accelerometer sensors to capture the first two natural frequencies and corresponding mode shapes. Lastly, the material properties of the numerical model components were adjusted based on the OMA findings. The main findings of this research study are outlined as follows:

- The absolute difference in natural frequency (ADF) values for both modes **are less than 0.03%, and** the modal assurance criterion (MAC) values for the first and second modes are more than 75%, which shows a good correlation between the OMA and the numerical analysis results.
- The updated elastic modulus, crucial for structural behavior, showed significant changes, highlighting the need for OMA-based model calibration. The elastic modulus of the tower component exceeds three times the value estimated by empirical equations. Consequently, this leads to an underestimation of structural performance, resulting in reduced initial stiffness when the structure is subjected to wind or seismic forces.
- The elastic modulus of the tower surpasses the value of the rest of the structure contrary to the initial assumptions.
- Spring interface stiffness remains minimally affected after the calibration process.

Acknowledgments This work is a part of the HYPERION project. HYPERION has received funding from the European Union's Framework Programme for Research and Innovation (Horizon 2020) under Grant Agreement No. 821054. The contents of this publication are the sole responsibility of Oslo Metropolitan University (Work Package 5, Task 2) and do not necessarily reflect the opinion of the European Union.

References

1. Miccoli, S., Gil-Martín, L., Hernández-Montes, E.: New historical records about the construction of the arch of Ctesiphon and their impact on the history of structural engineering, notes and records: the Royal Society journal of the. Hist. Sci. **77**, 113 (2021). https://doi.org/10.1098/rsnr.2021.0025
2. Gürbüz, M., Kocaman, İ.: Enhancing seismic resilience: a proposed reinforcement technique for historical minarets. Eng. Fail. Anal. **156**, 107832 (2024). https://doi.org/10.1016/j.engfailanal.2023.107832
3. Gonen, S., Pulatsu, B., Erduran, E., Pelà, L., Soyoz, S.: Dynamic characteristics of stone masonry walls before and after damage: experimental and numerical investigations. Eng. Struct. **306**, 117808 (2024). https://doi.org/10.1016/j.engstruct.2024.117808

4. Shabani, A., Kioumarsi, M., Zucconi, M.: State of the art of simplified analytical methods for seismic vulnerability assessment of unreinforced masonry buildings. Eng. Struct. **239**, 112280 (2021). https://doi.org/10.1016/j.engstruct.2021.112280

5. Yavartanoo, F., Kang, T.H.K.: Retrofitting of unreinforced masonry structures and considerations for heritage-sensitive constructions. J. Build. Eng. **49**, 103993 (2022). https://doi.org/10.1016/j.jobe.2022.103993

6. Shabani, A., Zucconi, M., Kazemian, D., Kioumarsi, M.: Seismic fragility analysis of low-rise unreinforced masonry buildings subjected to near-and far-field ground motions. Results Eng. **18**, 101221 (2023). https://doi.org/10.1016/j.rineng.2023.101221

7. Xu, D., Xie, Q., Hao, W.: Seismic damage evaluation of historical masonry towers through numerical model. Bull. Earthq. Eng. **22**(4), 2235–2266 (2024). https://doi.org/10.1007/s10518-024-01858-4

8. Tavafi, E., Mohebkhah, A., Sarhosis, V.: Seismic behavior of the cube of Zoroaster tower using the discrete element method. Int. J. Architect. Herit. **15**(8), 1097–1112 (2021). https://doi.org/10.1080/15583058.2019.1650135

9. Micelli, F., Cascardi, A.: Structural assessment and seismic analysis of a 14th century masonry tower. Eng. Fail. Anal. **107**, 104198 (2020). https://doi.org/10.1016/j.engfailanal.2019.104198

10. Korumaz, M., Betti, M., Conti, A., Tucci, G., Bartoli, G., Bonora, V., Korumaz, A.G., Fiorini, L.: An integrated terrestrial laser scanner (TLS), deviation analysis (DA) and finite element (FE) approach for health assessment of historical structures. A minaret case study. Eng. Struct. **153**, 224–238 (2017). https://doi.org/10.1016/j.engstruct.2017.10.026

11. Angjeliu, G., Coronelli, D., Cardani, G.: Development of the simulation model for digital twin applications in historical masonry buildings: the integration between numerical and experimental reality. Comput. Struct. **238**, 106282 (2020). https://doi.org/10.1016/j.compstruc.2020.106282

12. Pallarés, F.J., Betti, M., Bartoli, G., Pallarés, L.: Structural health monitoring (SHM) and nondestructive testing (NDT) of slender masonry structures: a practical review. Constr. Build. Mater. **297**, 123768 (2021). https://doi.org/10.1016/j.conbuildmat.2021.123768

13. Sarmadi, H., Entezami, A., Yuen, K.-V., Behkamal, B.: Review on smartphone sensing technology for structural health monitoring. Measurement. **223**, 113716 (2023). https://doi.org/10.1016/j.measurement.2023.113716

14. Capanna, I., Cirella, R., Aloisio, A., Alaggio, R., Di Fabio, F., Fragiacomo, M.: Operational modal analysis, model update and fragility curves estimation, through truncated incremental dynamic analysis, of a masonry belfry. Buildings. **11**(3), 120 (2021). https://doi.org/10.3390/buildings11030120

15. Altunişik, A.C., Günaydin, M., Ertürk Atmaca, E., Genç, A.F., Okur, F.Y., Sevim, B.: Experimental measurement-based FE model updating and seismic response of Santa Maria church and its guesthouse building. J. Civ. Struct. Heal. Monit., 663 (2024). https://doi.org/10.1007/s13349-023-00747-9

16. Ergün, M., Tayfur, B.: Evaluation of the seismic performance pre- and post-restoration of a masonry clock tower's FE model updated via experimental and optimization methods. Eng. Fail. Anal. **158**, 107986 (2024). https://doi.org/10.1016/j.engfailanal.2024.107986

17. Vatin, N.: Rhodes et l'ordre de Saint-Jean-de-Jérusalem. CNRS Éditions, Paris (2000)

18. Karousos, C.: Rhodos: History, Monuments, Art (Ancient Sites and Sanctuaries of Greece). Esperos Editions (1973)

19. Manoussou, E.: Le Paysage Culturel et Les Monuments Symbols Disparus de la Ville de Rhodes, vol. Europa Nostra .Bulletin Scientifique, pp. 59–74 (2012)

20. Kolokoussis, P., Skamantzari, M., Tapinaki, S., Karathanassi, V., Georgopoulos, A.: 3D and hyperspectral data integration for assessing material degradation in medieval masonry heritage buildings. Int. Arch. Photogramm. Remote. Sens. Spat. Inf. Sci. **43**, 583–590 (2021)

21. Shabani, A., Skamantzari, M., Tapinaki, S., Georgopoulos, A., Plevris, V., Kioumarsi, M.: 3D simulation models for developing digital twins of heritage structures: challenges and strategies. Proc. Struct. Integr. **37**, 314–320 (2022). https://doi.org/10.1016/j.prostr.2022.01.090

22. Valente, M.: Seismic behavior and damage assessment of two historical fortified masonry palaces with corner towers. Eng. Fail. Anal. **134**, 106003 (2022). https://doi.org/10.1016/j.engfailanal.2021.106003
23. D'Altri, A.M., Sarhosis, V., Milani, G., Rots, J., Cattari, S., Lagomarsino, S., Sacco, E., Tralli, A., Castellazzi, G., de Miranda, S.: Modeling strategies for the computational analysis of unreinforced masonry structures: review and classification. Arch. Comput. Meth. Eng. **27**(4), 1153–1185 (2020). https://doi.org/10.1007/s11831-019-09351-x
24. Revit, Autodesk, Revit BIM (Building Information Modeling), California, U.S, ed, (2024)
25. Shabani, A., Kioumarsi, M.: Seismic assessment and strengthening of a historical masonry bridge considering soil-structure interaction. Eng. Struct. **293**, 116589 (2023). https://doi.org/10.1016/j.engstruct.2023.116589
26. Ghiassi, B., Vermelfoort, A.T., Lourenço, P.B.: Chapter 7—Masonry mechanical properties. In: Ghiassi, B., Milani, G. (eds.) Numerical Modeling of Masonry and Historical Structures, pp. 239–261. Woodhead Publishing (2019)
27. Shabani, A., Kioumarsi, M., Plevris, V.: Performance-based seismic assessment of a historical masonry arch bridge: effect of pulse-like excitations, Frontiers of structural and civil. Engineering. **17**(6), 855–869 (2023). https://doi.org/10.1007/s11709-023-0972-z
28. Aghabeigi, P., Farahmand-Tabar, S.: Seismic vulnerability assessment and retrofitting of historic masonry building of Malek Timche in Tabriz Grand Bazaar. Eng. Struct. **240**, 112418 (2021). https://doi.org/10.1016/j.engstruct.2021.112418
29. Shabani, A., Feyzabadi, M., Kioumarsi, M.: Model updating of a masonry tower based on operational modal analysis: the role of soil-structure interaction, case studies. Constr. Mater. **16**, e00957 (2022). https://doi.org/10.1016/j.cscm.2022.e00957
30. Standoli, G., Giordano, E., Milani, G., Clementi, F.: Model updating of historical belfries based on Oma identification techniques. Int. J. Archit. Herit. **15**(1), 132–156 (2021). https://doi.org/10.1080/15583058.2020.1723735
31. SVIBS, ARTeMIS Modal, Structural Vibration Solution, Aalborg (2023)
32. Reynders, E.: System identification methods for (operational) modal analysis: review and comparison. Arch. Comput. Meth. Eng. **19**(1), 51–124 (2012). https://doi.org/10.1007/s11831-012-9069-x
33. Zahid, F.B., Ong, Z.C., Khoo, S.Y.: A review of operational modal analysis techniques for in-service modal identification. J. Braz. Soc. Mech. Sci. Eng. **42**(8), 398 (2020)
34. Chisari, C., Zizi, M., Rouhi, J., Lavino, A., De Matteis, G.: Ambient Vibration Testing and model updating of the bell tower of St. Michele Arcangelo Cathedral in Casertavecchia, Italy. Proc. Struct. Integr. **44**, 1100–1107 (2023). https://doi.org/10.1016/j.prostr.2023.01.142
35. FEMtools, Dynamic Design Solutions, FEMtools 4, Leuven, ed, (2023)

Structural Analysis of Glulam Frame of a Modular Timber–Aluminium Hybrid Façade System in Nordic Climate

Katarzyna Ostapska [ID], Johannes Brozovsky [ID], Domas Valiukas, and Eimantas Tinginys

Contents

1 Introduction

The ongoing effort to limit emissions in the construction industry is now focusing on the embodied energy in building materials to introduce renewable and bio-based alternatives, such as engineered wood products (EWP), to building products traditionally made with concrete, steel, or aluminium. One such substitution is a façade system bearing frame structure commonly made in aluminium. While increasing the sustainability of source material performance requirements must still be met, and the minimum 35 years of service life should be ensured to qualify as carbon removal according to the new European certification scheme [1]. While certain system elements, that is insulated panels and glass panels attached to the timber frame, can be exchanged after 20–25 years [2], the structure should withstand at least two

K. Ostapska (✉) · J. Brozovsky
SINTEF Community, Trondheim, Norway
e-mail: katarzyna.ostapska@sintef.no

D. Valiukas · E. Tinginys
Staticus UAB, Vilnius, Lithuania

© The Author(s) 2025
M. Kioumarsi, B. Shafei (eds.), *The 1st International Conference on Net-Zero Built Environment*, Lecture Notes in Civil Engineering 237,
https://doi.org/10.1007/978-3-031-69626-8_113

service lives and approach the typical building design service life of 100 years. The infill/insulation, sealing, and cladding elements that are subject to ever-increasing energy-efficiency and air and moisture tightness requirements should be installed allowing easy disassembly as specified in the standard ISO 20887:2020 [3]. Overall sustainability of the hybrid timber–aluminium façade system was shown to be superior to the aluminium-based system [2] via Life-Cycle Assessment including end-of-life scenarios of landfilling, recycling, and reuse. The benefits of the most preferable scenario, that is reuse, are subject to the uncertainties connected to the delayed realization and technical and economic feasibilities. To ensure the durability of the structural glulam façade frame with an accepted reliability level, real environmental loads should be considered and possibly adapted to include climate change scenarios. Wood exposed to moisture fluctuations over time can develop fungi, and rot and lose its functional properties. Timber can be effectively protected over many decades by providing proper façade insulation and ventilation based on good practice, as well as introducing predictive inner climate control via sensors. Especially challenging weather conditions can be experienced in Scandinavia, where high humidity and short dry seasons require climate-conscious adaptations of building solutions. Furthermore, coastal locations are exposed to high wind loads that lead to unavoidable fatigue-induced damages over longer periods. Fatigue in wood is a complex research field, where material anisotropy and typical huge variations of static mechanical properties hinder the development of universally applicable rules while accurate quantification of environmental and operational loads poses further challenges. Fatigue life is usually understood as the number of load cycles N at a given load amplitude S leading to the mechanical failure, where damage accumulates with every cycle and reaches value 1 at ultimate failure. Different fatigue lives for common wood failure modes found in literature and standards are shown in Fig. 1.

Load cycles differ in amplitude, frequency, form, direction, and distribution on the structure. Dynamically applied load, that is wind on facades is difficult to

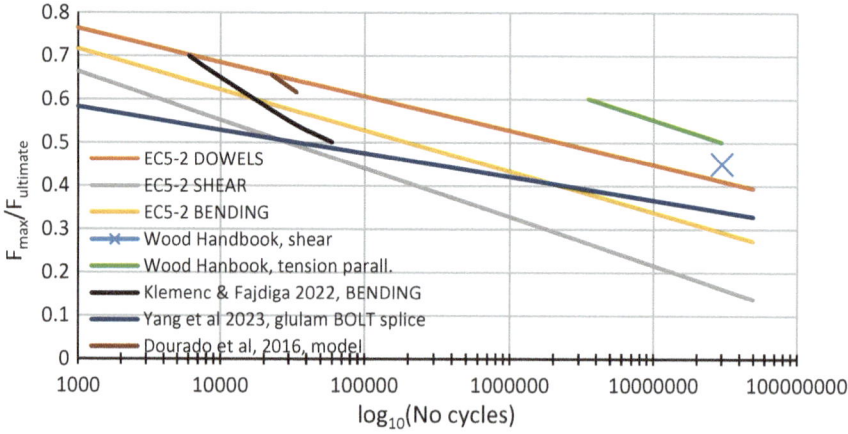

Fig. 1 Fatigue life curves for timber in literature [4–6] and EC5-2

Fig. 2 Elements of the façade system from the left: timber frame, insulation, aluminium frame, and glass pack/metal panels

quantify due to varying directions and particular aerodynamic conditions in the area. Increased exposure to high wind loads can lead to decreased service life and premature damage. Damage in wood due to various cyclic loads is underdeveloped research area. Non-linear damage accumulation models require multitude long-duration cycling experiments for a given wood species. However, linear damage accumulation model of Palmgren-Miner was found to exhibit sufficient accuracy for timber roof-to-wall connections subjected to wind load [7]. This article presents the methodology of evaluating fatigue life of façade timber frame under wind loads and its practical application for three locations in Norway (Oslo, Trondheim, Tromsø) based on 3 years of wind observations (2021, 2022, 2023). The frame and façade system build-up is depicted in Fig. 2.

2 Methodology

To ensure the successful reuse of façade elements, the capacity to withstand several environmental loads must be sufficient. Most critical loads include cycling moisture and temperature changes and wind loads. The considered hybrid timber–aluminium façade system was already assessed as safe against mould development in Nordic climates [8, 9]. This study is devoted to the mechanical performance only and assumes sufficient biological decay resistance of timber frame over the 100-year period and proper maintenance and regular inspection or monitoring.

2.1 Experimental Tests of Timber Frame Connections

Experimental static mechanical tests were performed on the internal screw connections of the frame depicted in Fig. 3. The connections are loaded vertically, in the plane of the glulam frame and represent gravity loads. Three- and four-point bending tests were performed for L-joint and T-joint, respectively. Strain gauges were used to apply the load at the speed of 0.1 kN/s. Force and displacement data were logged.

L-joint consists of a 150 × 150 × 10 mm L-bracket and nine 45-degree inclined screws of 5 mm diameter and 50 mm length. T-joint features three rows of four 5 × 140 mm screws connecting three timber members at 45-degree to the wood fibre direction.

2.2 Numerical FEM of Experimental Tests of Joints

The finite element method was used to simulate the mechanical behaviours of the connection under the conditions of the laboratory test. Continuum 3D elements with eight nodes reduced integration and a general size of 5 mm were used. Mesh sensitivity was investigated to provide sufficient accuracy within reasonable

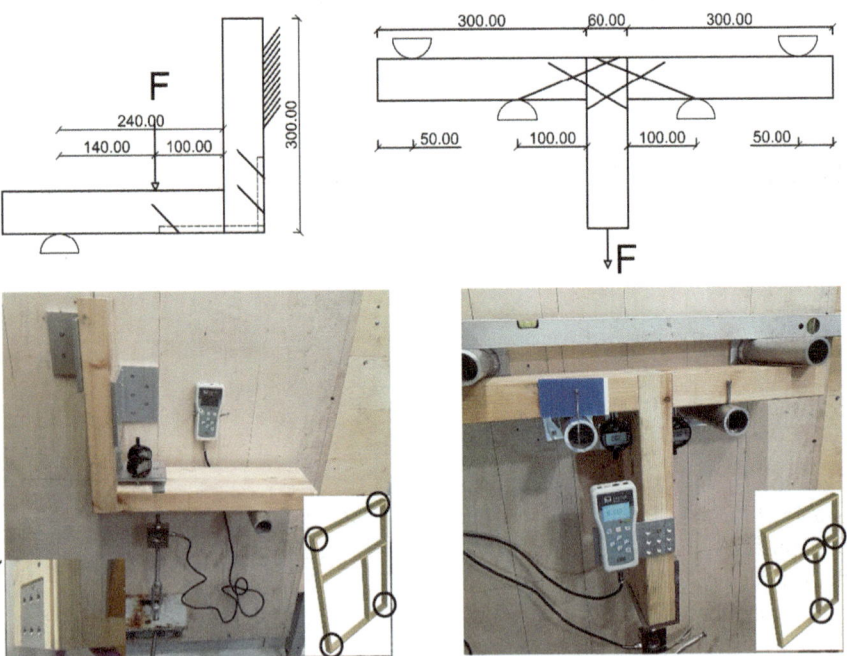

Fig. 3 Experimental setup scheme and photo for L-joint (left) and T-joint (right)

Fig. 4 Discretized numerical models of L-joint (left) and T-joint (right)

Table 1 Wood material model: elastic orthotropy constants for E modulus, Poisson ratio ν and shear modulus G, where L-longitudinal, R-radial, and T-tangential direction, f-strength

Stiffness	E_L	E_R	E_T	ν_{LR}	ν_{LR}	ν_{LR}	G_{LR}	G_{LT}	G_{RT}
[GPa]	10	0.79	0.34	0.5	0.66	0.84	0.64	0.58	0.03
Strength	$f_{L,t}$	$f_{L,c}$	$f_{R,t}$	$f_{R,c}$	$f_{T,tn}$	$f_{T,c}$	f_{LR}	f_{LT}	f_{RT}
[MPa]	63	−29	4.9	−3.6	2.8	−3.8	6.1	4.4	1.6

computation time. The discretization of the models is shown in Fig. 4. The wood material model is elastic orthotropy with nine parameters listed in Table 1.

Steel was modelled as elastic isotropic with $E = 210$ GPa, and $\nu = 0.3$. Screws are modelled without thread as a smooth shank of the screw core diameter and tied to wood.

2.3 Wind Load Model

Wind load calculations were carried out according to EN-1991-1-4 2005. The rules apply for buildings up to 200 m high. Basic wind velocity is obtained from the meteorological station (Oslo, Blindern) using API Frost as max wind velocity values in 10-min intervals at 10 m height above the ground and hourly mean velocity for reference, see Fig. 5.

Wind action on the façade is represented by the equivalent surface pressure q_p:

$$q_p(z) = \left[1 + \frac{k_r \cdot v_b \cdot k_l}{v_m(z)}\right] \cdot 0.5 \cdot \rho \cdot v_m{}^2(z), \, v_m(z) = k_r \cdot \ln\left(\frac{z}{z_0}\right) \cdot c_0(z) \cdot v_b \quad (1)$$

where in the function of height z in the coastal area with terrain category 0, where v_b is the base wind speed, see EN 1995-1 for details.

$$v_b = c_{dir} \cdot c_{season} \cdot v_{b,0} \quad (2)$$

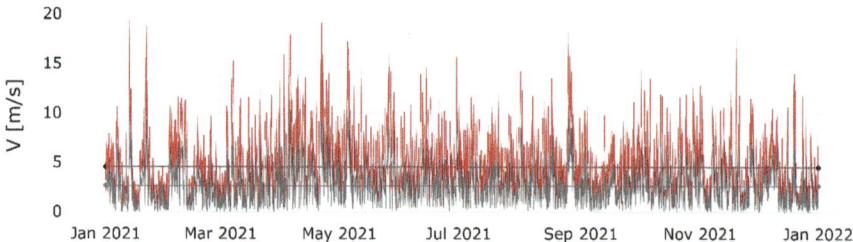

Fig. 5 Wind speed at 10-min temporal resolution (red—maximum peak, grey—mean) observed at Oslo-Blindern in 2021

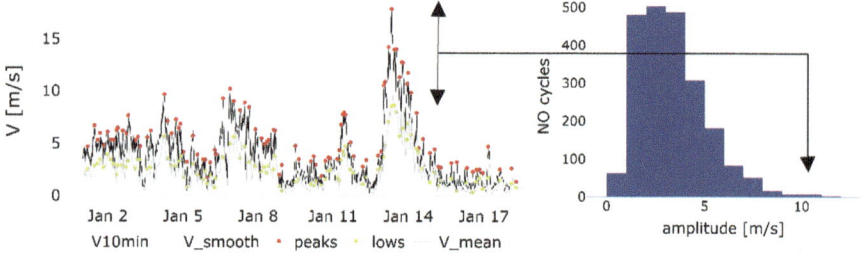

Fig. 6 Wind speed peaks (Oslo) identified from 10 min maximum values shown for 1 week (top) and amplitude histogram for all peaks during one year of data

where c_{dir}, c_{season}, are directional and seasonal wind factors, respectively. Wind load has a non-constant amplitude and requires handling damage accumulation from different cycles. Cycles are extracted from maximum wind time series using smoothing function (Savitzky-Golay filter within scipy package for Python 3) and peak (extremum) finding algorithm with peak prominence of minimum 1 m/s. The wind observations are available only for maximum and mean values at the highest 10-min resolution so the reference for amplitude estimation is chosen as the mean wind speed for each peak. Peaks and amplitude histograms for 10 amplitude ranges are depicted in Fig. 6.

2.4 Numerical Simulation of the Frame Under Static Equivalent Force

The model of the glulam frame for structural analysis with FEM was created within commercial software ABAQUS, see Fig. 7. The connection models validated with experimental tests for T-joint and L-joint were embedded in the frame model. The frame was fixed by T-plates on the top and horizontal translations are blocked at the bottom beam where bolted locks are attached. Pressure is collected from the frame

Fig. 7 Geometry, mesh, and boundary conditions of the glulam frame model for FEA

façade surface 2.45×3.4 m^2 and distributed over the front frame surface. The weight of the façade panels is not applied, and only the self-weight of timber and steel is considered.

Fatigue verification for timber structures is usually based on a simplified method from EN 1995-2-2, Annex A assuming equivalent constant amplitude fatigue loading. When several different loading amplitudes are present cumulative damage can be evaluated according to Palmgren-Miner theory of linear cumulative damage, see Eq. 3, where D denotes damage, N_{tot} is the total number of cycles, n_i is the stress level of a given load cycle i, and $N_{f,i}$ is the fatigue life corresponding to stress level n_i.

$$D = \sum_{i=1}^{N_{tot}} \frac{n_i}{N_{f,i}}, D \in\ <0, 1>$$
(3)

Information about damage induced by each load cycle is required and is evaluated by inverting EC5 formula to obtain fatigue life for a k stress level:

$$D \log(N_i) = (1 - k_i) \frac{a(b \quad R)}{1 - R}, R = \frac{\sigma_{d,\min}}{\sigma_{d,\max}}$$
(4)

Stress levels k_i is directly extracted from the wind history as wind peaks ratio to the maximum allowed wind capacity. Coefficients a and b representing the type of fatigue are chosen for dowel-type connections according to Table A.1 in EC5-2 (EN 1995-2-2). R is defined as a ratio of minimum to maximum design stress applied in the load cycle. Fatigue material factor is not accounted for.

3 Results

3.1 Experimental and Numerical Results for Timber Frame Connections

Force–displacement curves from connections experiment and simulation show good agreement in the elastic range, see Fig. 8. L-connection experiences slip, not captured by the simulation but measured at around 0.26 mm. The slip is caused by the presence of the L-shaped, $15 \times 15 \times 1$ cm steel plate connecting frame corner elements.

3.2 Numerical Simulation of Timber Frame and Fatigue Service Life Estimation

The final maximum basic wind speed for building height 12 m (third story) was 19.4 m/s and corresponds to the pressure of around 6.8 kPa. Under this statically applied pressure, the frame timber members reach 28% capacity in compression perpendicular to the grain around the T-plate bolts, see Fig. 9.

The central T-joint is the most strained connection in the frame and perpendicular to grain compression stresses are at 3 MPa and 83% capacity. L-joint has the largest value of longitudinal shear stresses with 30% capacity (see Table 2), while compression perpendicular to the grain is at 22%, see Fig. 9. Stresses in steel elements under maximum load are depicted in Fig. 10. Mises stress hypothesis was used to estimate stress level in T-brackets, L-joint, and T-joint. Most stressed steel elements are bottom bolts and the bottom part of the T-bracket that transfer most of the load of the timber frame hanging on them. However, the level of stress at just below 160 MPa was found to be below 50% of the ultimate tensile strength limit for the most typical carbon steel of grade S235, namely 360 MPa. Such a low level of stress allows us to assume that load in steel elements is below the endurance limit for fatigue defined in Eurocode 3, part 1–9 (EN 1993-1-9).

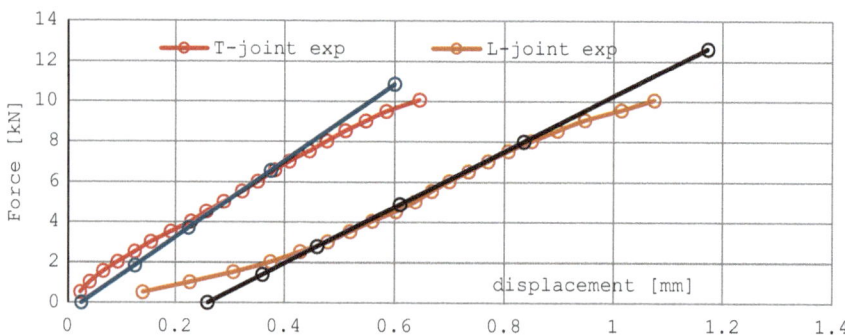

Fig. 8 Load–displacement curves for T-joint and L-joint in glulam frame from experiment (red, orange) and numerical simulation (blue, black)

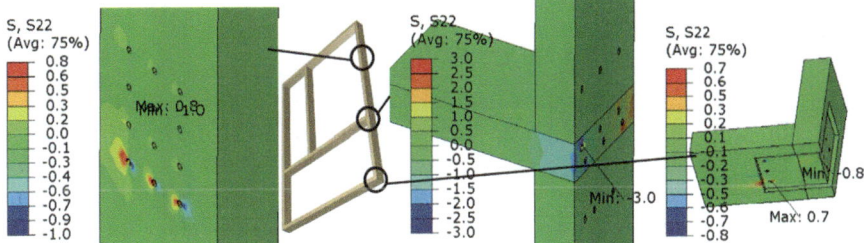

Fig. 9 Stress in timber perpendicular to grain around T-plate bolts at ca 28% capacity (left), 83% in T-joint (middle), and 22% in L-joint (right)

Table 2 Max stresses from FEA for each material direction and corresponding usage

Element		$\sigma_{L,t}$	$\sigma_{L,c}$	$\sigma_{R,t}$	$\sigma_{R,c}$	$\sigma_{T,t}$	$\sigma_{T,c}$	σ_{LR}	σ_{LT}	σ_{RT}
Beams	[MPa]	3.7	−4.9	0.8	−1	0.1	−0.2	1.3	0.4	0.1
	[−]	6%	17%	16%	28%	4%	5%	21%	9%	6%
T-joint	[MPa]	3.4	−3.4	3	−3	0.5	−0.5	1.7	0.6	0.4
	[−]	5%	12%	61%	83%	18%	13%	28%	14%	25%
L-joint	[MPa]	4.8	−4.5	0.7	−0.8	0.8	−0.9	0.7	1.3	0.4
	[−]	8%	16%	14%	22%	29%	24%	11%	30%	25%

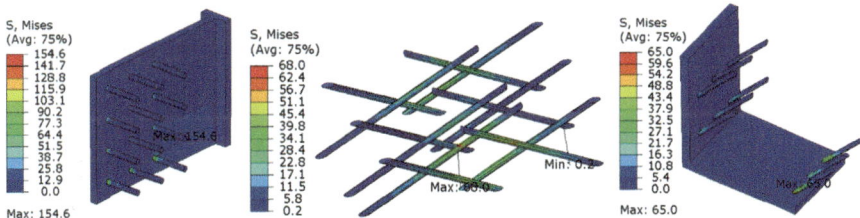

Fig. 10 Mises stress in steel T-plate bolts at ca 80% yield limit capacity (left), 36% in T-joint (middle), and 34% in L-bracket screws (right)

Maximum deformations of 4.8 mm occur in T-joint as shown in Fig. 11.

The wind speed at which the frame is at 100% of local static perpendicular to grain compressive stress capacity is set at $19.4/0.83 = 23.5$ m/s. Additionally, to account for the higher dynamic elastic modulus, the maximum wind speed was increased by 20% to 28.2 m/s. The value was based on the literature findings, where a different mean ratio between dynamic and static modulus of elasticity for wood was found, that is 1.07 [10], 1.1 [11], 1.13 [12], 1.22 [13], and 1.28 [14]. All identified load cycles are considered for total damage evaluation without assuming fatigue endurance limit. Cumulative linear damage for four consecutive years 2020–2023 in three Nordic cities is summarized in Table 3. Based on those damages, the predicted service life due to mechanical fatigue failure in the frame is estimated for each location at over 100 years for both Oslo and Trondheim and at 32 years for Tromsø.

Overview of load cycles due to wind for a given year is shown in Fig. 12.

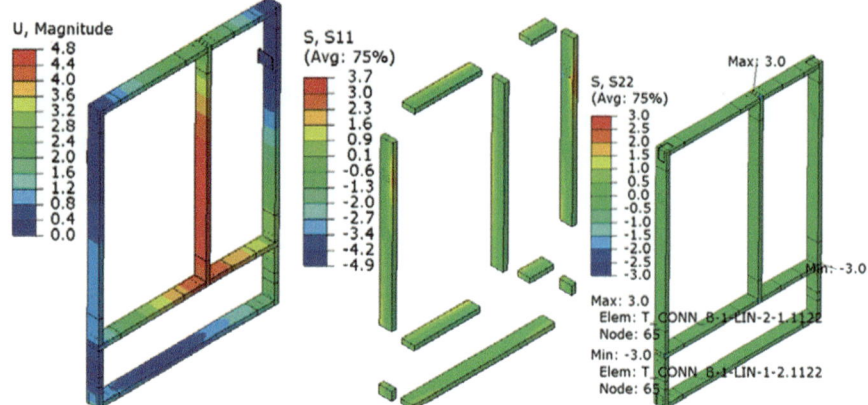

Fig. 11 Deformations [mm] of the frame under 6.8 kPa pressure (left), bending stresses in beams (middle), max perpendicular to grain stresses in joints (right) [MPa]

Table 3 Cumulative linear damage due to wind-induced fatigue

		Damage		
	Year	Oslo	Trondheim	Tromsø
	2020	7.58E–06	7.63E–05	1.05E–01
	2021	1.91E–06	4.45E–04	5.88E–03
	2022	4.33E–05	4.45E–04	1.28E–02
	2023	3.28E–07	3.63E–05	2.19E–03
Total damage		5.31E–05	1.00E–03	1.26E–01
Predicted service life		>100	>100	32

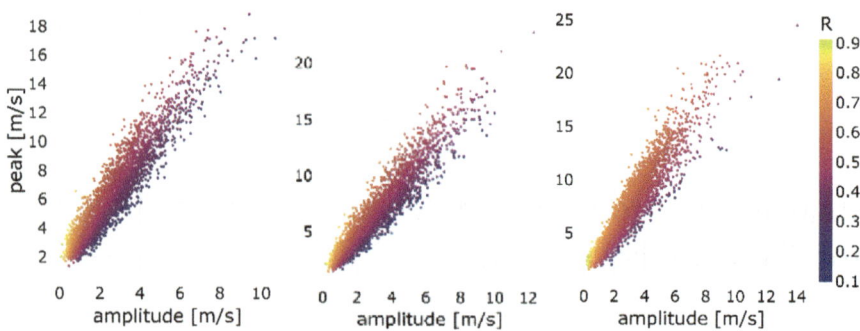

Fig. 12 Total yearly (2021) wind load cycles by amplitude and peak speed, color-coded by R value for Oslo (left), Trondheim (middle), and Tromsø (right)

4 Discussion

4.1 Discussion of Results

Service life due to damage accumulation in timber frame under wind exposure is estimated as non-critical for Oslo (Blindern) and Trondheim (Voll) observations assuming a height of 12 m above the ground. Tromsø location has a decidedly smaller service life, at 32 years. The main factor is the presence of several high peaks in wind speed compared to the estimated frame capacity of 28.5 m/s. For the more exposed locations with direct seaside locations, and stronger winds design changes are recommended, for example symmetric geometry or stiffening elements. Furthermore, experiments on the out-of-plane load response of both T-joint and L-joint would improve the estimation of capacity and service life.

4.2 Limitations of the Research

This work makes several limiting assumptions. Only static structural analysis is performed, and no dynamic effects and time-dependent material properties are accounted for in the simulation. Moisture-induced stresses, damage due to moisture cycles, and coupled moisture-mechanical stress fields were not considered. Wind loads are collected from 10 min frequency data of maximum wind speeds. Thus, actual number of cycling and real minimum wind speed for accurate amplitude estimation is not known. Wind direction was not accounted for to decrease the real pressure on the façade, which is a conservative assumption. The positive stiffening effect of façade panels was not included, and fatigue of steel T-plate connectors and screws was not studied in detail due to the assumed endurance limit at the expected stress level. Additionally, the cumulative damage was obtained based on linear theory, and the progressive increase in the damaging effect of similar cycles was not considered. The influence of the accumulated damage and duration of load can be substantial factors for fatigue life assessment. However, no sufficient data for the wood performance under different sequences of short-term loading are available in the standards and literature to allow for a non-linear damage accumulation model with a sufficient confidence level. Thus, the focus of future research within timber engineering should be damage development and accumulation under cycling loading if carbon removal goals in timber construction are to be achieved.

5 Conclusions

Structural analysis of the glulam frame supporting a novel modular façade system in the Nordic wind conditions was performed to predict fatigue service life to ensure the long-term sustainability of replacing aluminium/steel with glulam. The finite element model of the frame was calibrated using laboratory mechanical tests of the T- and L-joint with screws and brackets. Continuum models of both timber and steel were used to increase the precision of modelling complex stress states for the identification of failure modes. Linear damage accumulation calculation is shown for 4 years between 2020 and 2023 in Oslo, Trondheim, and Tromsø locations in Norway. Key conclusions from the paper are listed:

- The critical role of the major wind load events with speed over 20 m/s on the fatigue life and damage in timber frame is observed.
- Design alternation, that is introducing symmetrical geometry, re-designing, or reinforcing joint is recommended to limit the load level in the critical location.
- The most critical failure mode was identified as local perpendicular to grain compression in wood in the T-joint.
- The predicted service fatigue life for the façade system is estimated at 32 years for Tromsø and over 100 years for Oslo and Trondheim.

Acknowledgements This research was funded by the EEA Norway Grant nr LT-INNOVATION-0002: 'Developing a more environmentally friendly automated façade system that is integrated into the building's control systems'. Experimental tests were performed in 2022.

References

1. Commission welcomes political agreement on EU-wide certification scheme for carbon removals. https://ec.europa.eu/commission/presscorner/detail/en/ip_24_885, released 2024/02/20
2. Cheong, C.Y., Brambilla, A., Gasparri, E., Kuru, A., Sangiorgio, A.: Life cycle assessment of curtain wall facades: a screening study on end-of-life scenarios. J. Build. Eng. **84**, 108–600 (2024)
3. ISO 20887: Sustainability in buildings and civil engineering works—design for disassembly and adaptability—principles, requirements and guidance (2020)
4. Klemenc, J., Fajdiga, G.: Statistical modelling of the fatigue bending strength of Norway spruce wood. Materials. **15**, 536 (2022)
5. Yang, L., Chen, A., Zhou, J., He, G., Wang, H., Li, C.: Flexural fatigue behavior of glulam beams connected with steel splints and bolts. Buildings. **13**(5), 1218 (2023)
6. Dourado, N., de Moura, M.F.S.F., de Jesus, A.: Fatigue-fracture characterization of wood under mode I loading. Int. J. Fatigue. **121**, 265–271 (2019)
7. Alhawamdeh, B., Shao, X.: Fatigue performance of wood frame roof-to-wall connections with elastomeric adhesives under uplift cyclic loading. Eng. Struct. **229**, 111602 (2021)
8. Ostapska, K., et al.: Development of climatic damage predictive tool for timber façade moisture-related damage. J. Phys. Conf. Ser. **2600**(16), 162002 (2023)
9. Coelho, G.B.A., Kraniotis, D.: A multistep approach for the hygrothermal assessment of a hybrid timber and aluminium based facade system exposed to different sub-climates in Norway. Energ. Build. **296**, 113368 (2023)
10. Haines, D.W., Leban, J.M., Herbé, C.: Determination of Young's modulus for spruce, fir and isotropic materials by the resonance flexure method with comparisons to static flexure and other dynamic methods. Wood Sci. Technol. **30**, 253–263 (1996)

11. Spycher, M., Schwarze, F., Steiger, R.: Assessment of resonance wood quality by comparing its physical and histological properties. Wood Sci. Technol. **42**, 325–342 (2008)
12. Divos, F., Tanaka, T.: Relation between static and dynamic modulus of elasticity of wood. Acta Silv. Lign. Hung. **1**, 105–110 (2005)
13. Buron, I.: Modulus of elasticity of Norway spruce using different techniques. Master thesis, Lund University of Technology (1998)
14. Chauhan, S., Sethy, A.: Differences in dynamic modulus of elasticity determined by three vibration methods and their relationship with static modulus of elasticity. Maderas. Ciencia y Tecnología. **18**, 373 (2016)

Performance Evaluation of Glass Fiber-Reinforced Polymer (GFRP) Bars in Bridge Decks

Shadi Azad and Behrouz Shafei ⓘ

Contents

1 Introduction

Steel-reinforced concrete (RC) bridge decks are now widely adopted for bridges. Despite their numerous benefits, the steel reinforcement within the concrete is vulnerable to corrosion over the bridge's service life, highlighting a key factor in the bridge deterioration [1]. This issue is particularly notable in colder climates where deicing salts are employed, where the degradation of bridge decks is primarily attributed to the corrosion of steel [2]. Initiation and propagation are the two common stages of steel corrosion. In the first stage, the ions accumulate on the rebar surface due to water and aggressive agents' penetration, leading to a set of electrochemical reactions which results in the initiation of the second stage and the formation of rust [3]. The first stage was investigated by Shafei et al. [4] through developing a stochastic framework using finite element models and various sources of uncertainty. Corrosion effects on the durability of RC structures have been

S. Azad · B. Shafei (✉)
Iowa State University, Ames, IA, USA
e-mail: shafei@iastate.edu

© The Author(s) 2025
M. Kioumarsi, B. Shafei (eds.), *The 1st International Conference on Net-Zero Built Environment*, Lecture Notes in Civil Engineering 237,
https://doi.org/10.1007/978-3-031-69626-8_114

concerning for engineers and decision makers, as the cost of repair and rehabilitation can be significant [5]. Battelle-NBS, a benchmark study in 1975, showed a total of $70 billion corrosion cost (4.2% of the gross national product (GNP)) in the United States. The estimated corrosion cost increased to a total of $300 billion in 1995, which was estimated to increase even further to about twice considering the indirect costs of corrosion. Accordingly, the cost of corrosion in the infrastructure sector covered 16.4% of the total annual cost of corrosion for the industry sectors. A study conducted by Virmani [6] reported that 15% of the bridges had deficiencies majorly due to corrosion of steel, indicating 37% corrosion costs of highway bridges out of all the corrosion costs of the infrastructure sector.

Khatami and Shafei [7] investigated the deterioration of RC bridges in the Midwest due to environmental factors like temperature, humidity, and deicing salts. The authors developed a computational model to predict corrosion initiation times based on regional climate data. The study quantified climate condition effect of both crack width and the duration needed to reach durability thresholds. Shi et al. [8, 9] also studied the chloride penetration in concrete mixture with shrinkage-compensating cement, showing that increasing the dosage of the agent increased chloride penetration.

The production of carbon-intensive materials is a major contributor to the considerable greenhouse gas emissions. Cement and steel were reported to be responsible for 8% and 7% of global carbon dioxide (CO_2) emissions, respectively. Thus, a movement toward decarbonization of the construction industry can be a beneficial step in reducing greenhouse gas emissions. Several approaches have been proposed to decrease the carbon emissions of cement and steel production including using alternative fuels, carbon-capturing, energy efficiency of cement, or alternative materials [10]. The development of fiber-reinforced polymer (FRP) bars in recent years has provided the opportunity for decreasing the use of steel rebars in RC structures. FRP bars consist of extremely fine polymer bars embedded in a matrix. Discrete fibers of various types have also been used in concrete mixtures to enhance the mechanical properties [11–13]. The polymer reinforcement is not susceptible to corrosion while providing high strength-to-weight ratio, high tensile strength, and light weight compared to the conventional steel bars. Four widely-known FRP materials are glass (GFRP), carbon (CFRP), basalt (BFRP), and aramid (AFRP). The low manufacturing cost and easy availability of the glass fibers make them one of the most popular FRP materials. Research has shown that raw material supply is the main contributor to the energy use and environmental impact category results of GFRP structures, followed by the minor contribution of raw material transportation (less than 10%) and manufacturing (10% of greenhouse gases and 15–20% of the energy use). Utilizing the FRP bars in the concrete structures increase their durability and service life, decreasing the need for repair and rehabilitation, which can lead to significant carbon emission and material waste. Thus, the hidden costs of repair and rehabilitation during the service life of a structure are lower than that of a steel-reinforced structure, making FRP bars feasible alternatives despite the higher initial cost [5, 14, 15].

Several researchers studied and analyzed the use of FRP materials in RC structures in terms of sustainable metrics, environmental impacts, and costs. Daniel [16] showed that a GFRP bridge required less than half of the energy input in comparison with bridges of other traditional materials. The sustainability of a GFRP bridge deck in comparison with a composite (steel concrete) bridge was analyzed by Mara et al. [17] in terms of life cycle, maintenance, and construction process, indicating a reduction of 20% in carbon emission in case of a GFRP deck. Maxineasa et al. [18] showed that glass fibers had less negative environmental impacts on global warming, human health, and ozone depletion potential compared to carbon fibers. Barker [14] studied the feasibility of GFRP, CFRP, and AFRP as alternatives to the conventional steel in RC structures. The FRP materials were compared considering the energy requirement and greenhouse gas emissions during the construction followed by sustainable metrics within the service life and also recyclability. The study showed that GFRP minimized the energy requirement and CO_2 emissions, as it was produced at a lower temperature compared to steel. The life-cycle assessment of the design stage study of a GFRP-reinforced bridge and a conventional steel-reinforced bridge were studied by Cadenazzi et al. [19], reporting lower acidification, global warming, eutrophication, and photochemical oxidant creation for the GFRP bridge compared to the steel one. Garg and Shrivastava [5] compared the steel rebars with CFRP, BFRP, and GFRP bars in reinforced concrete beams both environmentally and economically. The results of the study showed 43% and 46% less CO_2 emission and energy consumption for GFRP-reinforced beams compared to the steel-reinforced beams, respectively, providing an environment friendly option in terms of carbon footprint (Fig. 1).

Preinstorfer et al. [20] evaluated the mass-related global warming and material cost of a broad range of concrete structures including a rail platform barrier, a retaining wall and a bridge reinforced with conventional steel and GFRP bars. The parametric design study linked the mass of longitudinal rebars to the properties of global warming. The results showed slightly more sustainable solutions for GFRP-reinforced structures for global warming. As a protective alkaline environment was not required due to non-corrosiveness of GFRP bars, an optimized concrete mix with less clinker amount could provide the same performance. The optimized mix resulted in 31% lower mass-related global warming factors compared to the steel bars. Another study by Sbanieh et al. [10] showed that using GFRP bars instead of steel reinforcement decreased CO_2 emissions by almost 23%. Ji et al. [21] compared the sustainability of GFRP, CFRP, BFRP, and steel bars for carbon emission based on both weight and volume of FRP bars. Accordingly, the weight-based factors of steel bars were slightly lower than those of GFRP bars; however, from a volumetric aspect, the carbon emission factors of the GFRP bars were lower than those of steel as the density of GFRP bars were lower.

From another perspective, the use of concrete consumes significant amount of fresh water, crushed stones, and river sand. Also, the long transportation distance of crushed stone to the marine environment increases the cost and pollution. The non-corrosive properties of the FRP bars provides the opportunity for the utilization of seawater and sea sand concrete (SWSSC) in construction, conserving the limited

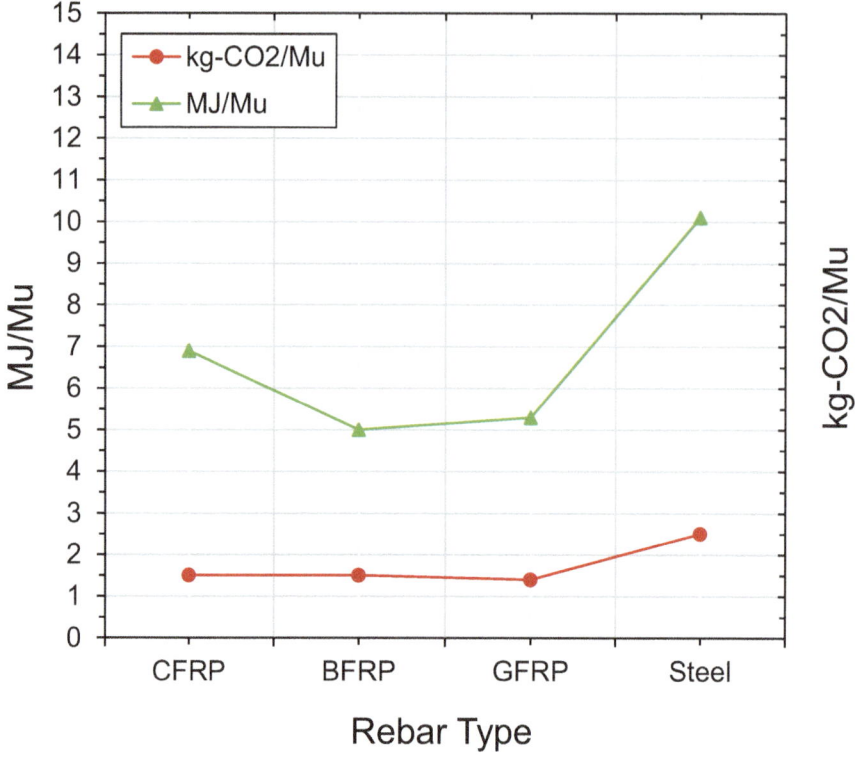

Fig. 1 Carbon emission and energy use of different rebar types [5]

source of river sands and fresh water while decreasing the transportation distance of the aggregates. Dong et al. [22] performed a life-cycle cost assessment (LCCA) of GFRP/CFRP-reinforced and also steel-reinforced concrete beams in marine environments. The LCCA included the production, transportation, construction, service life and end of life categories. The results showed that SWSSC had considerably lower environmental impacts in all the categories with reduction rates ranging from 26.3% to 48.6% compared to the conventional concrete.

Given the substantial freight activity across the US infrastructure and transportation network, road freight vehicles contribute significantly to CO_2 emissions in the transportation sector, ranking second only to passenger road vehicles (Fig. 2). Thus, options with lower freight activities provide opportunities for substantial decarbonization. Considering the lower weight of GFRP bars compared to steel bars, GFRP-reinforced structures decrease the environmental burdens in longer transportation distances as the weight of GFRP makes it possible to use fewer trucks saving several freight distances per project [22].

In this study, various aspects of the performance of a GFRP-reinforced bridge deck are compared to the counterpart aspects of a steel-reinforced bridge deck under similar environmental conditions and traffic loads. The bridges were instrumented

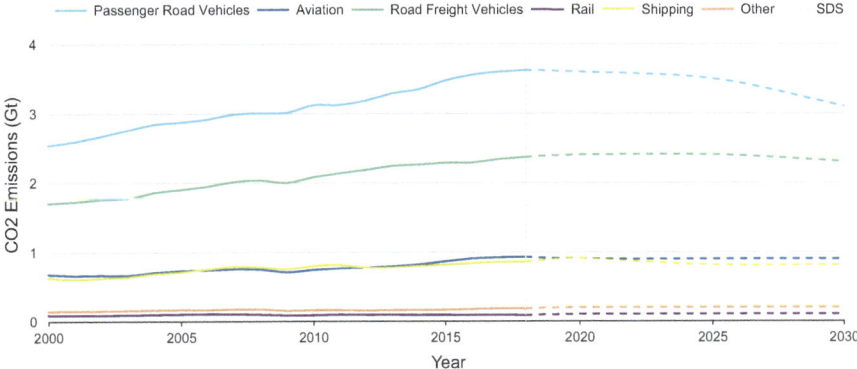

Fig. 2 Transport sector CO_2 emissions over time [23]

using various sensors and monitored over the years. The strain data was collected and analyzed to find the seasonal patterns and also anomalies in the structural response of the bridge decks. The results of the experimental tests of the GFRP bars were also examined to compare the mechanical properties of the GFRP bars with steel bars. A life-cycle cost analysis was also performed to evaluate the financial aspects of using GFRP in bridge decks instead of steel reinforcement. The results of the study sheds light on the viability of using GFRP-reinforced bridge decks instead of steel-reinforced bridge decks in terms of structural performance, long-term bridge response, and the total cost, while providing the bridge engineering community with a more durable and sustainable option that causes lower environmental impacts.

2   Long-Term Health Monitoring

Two bridges in Minnesota with similar properties were selected to study the structural performance of a GFRP-reinforced deck and a conventional steel-reinforced deck in comparison with each other. Each bridge was instrumented with vibrating wire strain gauges at several sections. The strain and temperature values at all sensor locations were collected for a period of over 48 months. The collected data was analyzed for maximum strains, seasonal variations, and abnormalities in the bridge deck's behavior. As the cracking tensile strain of concrete is considered to be 100–120 μstrain, the maximum strains extracted from the collected data was used to find the potential cracks in concrete decks.

Figure 3a shows the collected data for three temperatures of 10, 50, and 70 °F at the mid-span of the steel-reinforced bridge. The gauge was positioned on the bottom reinforcement mat of the deck close to the longitudinal axis of the bridge. The figure shows that the average compressive strain at 10 °F increased at the bottom center gauge. Under a temperature of 50 °F, the strain values generally decreased in every

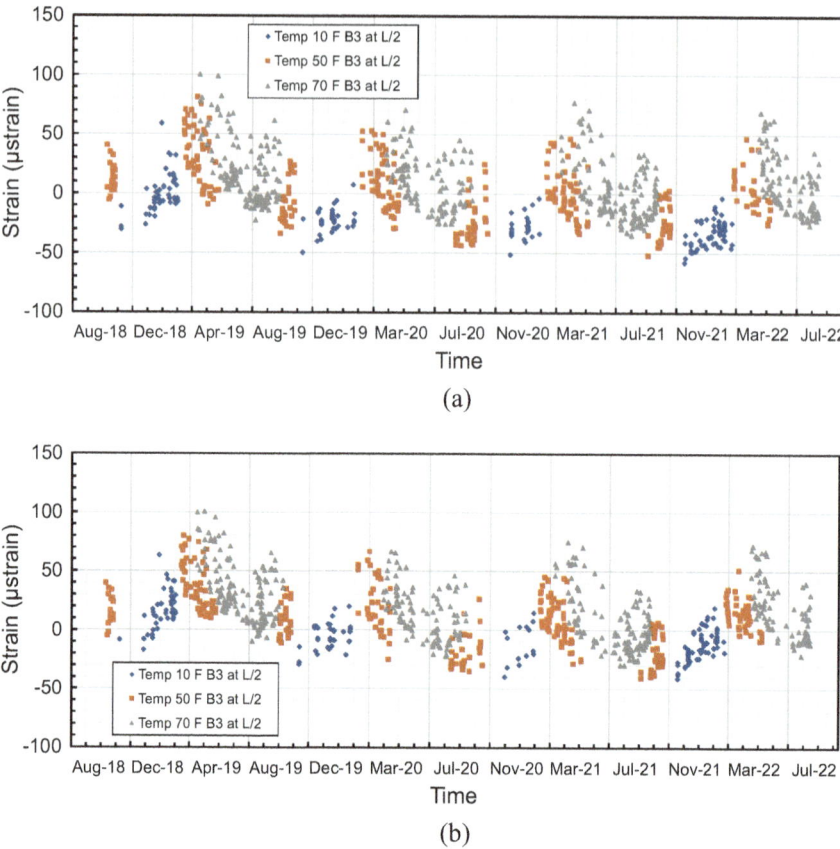

Fig. 3 Strain values of the bottom center gauge of the bridge decks at 10, 50, and 70 °F. (**a**) Steel-reinforced bridge deck. (**b**) GFRP-reinforced bridge deck

subsequent fall and spring season, trending mostly to compressive strains. However, the strains increased every subsequent fall. On the other hand, the strains decreased over the summer and generally over the years at 70 °F at mid-span. The figure shows scattered strain values with variations for different seasons with various temperatures.

The strain time history of the bottom center sensor of the GFRP-reinforced bridge deck at temperatures of 10, 50, and 70 °F is illustrated in Fig. 3b. The figure shows a decreasing trend over the years at 10 °F at mid-span. At 50 °F, the strains showed an overall decreasing pattern with an increase every spring at mid-span. At 70 °F, the strains show a decreasing trend over each spring and the subsequent summer with a peak of tensile strains close to the beginning of each spring to lower tensile or even compressive strains more through the subsequent fall.

The figures show similar behaviors of the GFRP- and steel-reinforced bridge decks with respect to low, medium, and high temperatures across the seasons. Accordingly, both bridge decks reacted to cold weather with majorly compressive strains, indicating that the reinforcing bars were mainly in compression as the concrete contracted. The strain values in both bridges covered similar ranges, with the steel-reinforced deck showing a slightly steeper decreasing slope over the years. Similar strain ranges and generally decreasing average strains were observed in different seasons at medium and high temperatures, too. The trends show that the high tensile strains occurred as the concrete expanded in warm seasons.

3 Laboratory Tests on GFRP Bars

In order to gain a better understanding of the mechanical properties and strength of the GFRP bars utilized in the construction of the GFRP-reinforced bridge deck, a series of laboratory material characterization tests was conducted. Four samples of the GFRP bars employed in the bridge construction project under study were subjected to testing. These specimens underwent tension until failure while measuring force, displacement, and strain. The collected data facilitated the development of stress–strain curves, enabling comparison of the tensile properties of the specimens with reference values obtained from prior GFRP tensile tests.

The GFRP bars utilized had a nominal diameter of 0.625 inches and an effective diameter of 0.67 inches, with each piece being cut into 28-inch segments. One longitudinal strain gauge was installed on each 28-inch specimen. Displacement rate was determined according to ASTM D7205 [24]. The specimens were loaded in tension until failure, with force, strain, and displacement being measured.

The stress–strain curves for the four GFRP test specimens are depicted in Fig. 4, demonstrating consistent mechanical properties across all tests. Stress increases linearly until reaching maximum load, followed by sudden failure marked by multiple glass fibers snapping simultaneously, necessitating a sudden decrease in load by the testing machine to maintain a constant displacement rate.

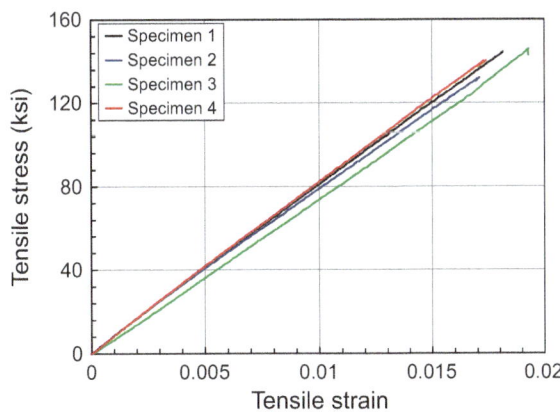

Fig. 4 Stress–strain curve of the four GFRP specimens tested in this study

4 Life-Cycle Cost Assessment

An important aspect of using GFRP reinforcement as an alternative in reinforced concrete bridge decks is the economic viability of manufacturing, construction, and service life of the material. A life-cycle cost analysis was conducted to compare the GFRP-reinforced bridge deck cost to that of the steel-reinforced bridge deck during the estimated service life. The study evaluated the cost of using GFRP and conventional steel in bridge decks in three categories including the initial cost, the cost of repair and rehabilitation, and the cost of reconstruction of the deck at the end of the service life of the bridges.

As the GFRP bars consist of fine glass fibers in a resin matrix, the manufacturing process and material composition often results in a higher initial cost compared to the conventional steel, which is widely available and commonly used. As such, the manufacturing and the production of the GFRP bars could cost 1.7 times more than those of conventional steel. Moreover, the construction costs associated with GFRP bars could increase by up to 20% when compared to those of steel bars. Thus, the initial costs of working with GFRP bars might be an unfavorable factor that prevents utilizing such bars in bridges. However, the main advantage of employing the alternative reinforcement is to protect the reinforcement against corrosion, which is a major reason for the bridge deck degradation especially in the cold regions. Therefore, the bridge deck's durability and service life increase, reducing the need for frequent repair and rehabilitation like the steel-reinforced bridge decks in corrosive environments. Also, the reconstruction cost of the GFRP deck was considered to be approximately 10% higher than that of the steel-reinforced deck. Thus, despite the higher initial manufacturing cost, total construction cost, and also deck reconstruction cost of the GFRP-reinforced deck, the LCCA showed that the advantages of utilizing the GFRP bars in decreasing the repair and rehabilitation needs and the associated costs can affect the economic viability of a GFRP-reinforced bridge deck. As such, utilizing the GFRP bars could decrease the life-cycle cost of a bridge deck by more than 20% with a similar service life period. In the meantime, using GFRP bars also increase the durability of the bridges, providing a larger life span that could decrease the life-cycle cost of the bridge deck even further by 37%. Therefore, the life-cycle cost analysis showed that the GFRP-reinforced bridge deck could be a superior choice economically. The benefits can be further realized when reducing the carbon footprint is also taken into consideration.

5 Conclusions

In this study, the GFRP bars were investigated as an environmentally friendly alternative to the conventional steel reinforcement in bridge decks. The GFRP reinforcement is non-corrosive, providing more durable concrete bridges with larger life spans. The study compared the long-term behavior, mechanical properties, and life-cycle costs of two similar bridge decks, one reinforced with GFRP and the other reinforced with conventional steel. The GFRP bars decrease the need for the production of carbon-intensive materials such as steel and cement, reducing the carbon footprint and the environmental impacts of the construction industry. Besides, the light weight of the GFRP bars compared to the steel bars decreases the required transportation distance, reducing the carbon emission even further. Increasing the service life of the bridge structures and reducing the need for reconstruction of the bridges at an early age decrease the environmental impacts of the construction process and material waste. Thus, GFRP reinforcement shows superior characteristic as a promising alternative to the steel bars in bridge decks in terms of environmental friendliness while providing satisfactory structural performance.

Acknowledgments The research study was sponsored by the Minnesota Department of Transportation. The authors would like to acknowledge and thank the sponsor for this support. Opinions, findings, and conclusions expressed in this paper are of the authors and do not necessarily reflect those of the sponsor.

References

1. Khatami, D., Shafei, B., Bektas, B.: Data-assisted prediction of deterioration of reinforced concrete bridges using physics-based models, ASCE. J. Infrastruct. Syst. **29**(2), 05023003, 1–13 (2023)
2. Shafei, B., Phares, B., Saini, D., Azad, S.: Assessment of Bridge Decks with Glass Fiber-Reinforced Polymer (GFRP) Reinforcement. Report No: MN 2023-13. Institute for Transportation, Iowa State University (2023)
3. Khatami, D., Hajilar, S., Shafei, B.: Investigation of oxygen diffusion and corrosion potential through a cellular automaton framework. J. Corros. Sci. **187**, 109496, 1–10 (2021)
4. Shafei, B., Alipour, A., Shinozuka, M.: A stochastic computational framework to investigate the initial stage of corrosion in reinforced concrete superstructures. J. Comput. Aided Civ. Infrastruct. Eng. **28**(7), 482–494 (2013)
5. Garg, N., Shrivastava, S.: Environmental and economic comparison of FRP reinforcements and steel reinforcements in concrete beams based on design strength parameter. In: Proceedings of the UKIERI Concrete Congress. Jalandhar, India (2019)
6. Virmani, Y.P.: Corrosion Cost and Preventive Strategies in the United States. Report No: FHWA-RD-01-156. CC Technologies, Inc, Federal Highway Administration (2002)
7. Khatami, D., Shafei, B.: Impact of climate conditions on deteriorating reinforced concrete bridges in the U.S. Midwest region, ASCE. J. Perform. Constr. Facil. **35**(1), 04020129, 1–11 (2021)
8. Shi, W., Najimi, M., Shafei, B.: Chloride penetration in shrinkage-compensating cement concretes. J. Cem. Concr. Compos. **113**, 103656, 1–11 (2020)
9. Shi, W., Najimi, M., Shafei, B.: Reinforcement corrosion and transport of water and chloride ions in shrinkage-compensating cement concretes. J. Cem. Concr. Res. **135**, 106121, 1–9 (2020)
10. Sbahieh, S., Serdar, M.Z., Al-Ghamdi, S.G.: Decarbonization strategies of building materials used in the construction industry. Mater. Today Proc. (2023). https://doi.org/10.1016/j.matpr.2023.08.346

11. Shafei, B., Kazemian, M., Dopko, M., Najimi, M.: State-of-the-art review of capabilities and limitations of polymer and glass fibers used in fiber-reinforced concrete. J. Mater. **14**(2), 409, 1–44 (2021)
12. Karim, R., Shafei, B.: Investigation of five synthetic fibers as potential replacements of steel fibers in ultra-high performance concrete, ASCE. J. Mater. Civ. Eng. **34**(7), 04022126, 1–14 (2022)
13. Saini, D., Shafei, B.: Synergistic use of ultra-high-performance fiber-reinforced concrete (UHPFRC) and carbon fiber-reinforced polymer (CFRP) for improving the impact resistance of concrete-filled steel tubes. J. Struct. Design Tall Spec. Build. **32**(11–12), E2036, 1–17 (2023)
14. Barker, C.: The Feasibility of Fibre Reinforced Polymers as an Alternative to Steel in Reinforced Concrete. McGill University (2016)
15. Precast/Prestressed Concrete Institute, PCI: Environmental Product Declaration for Glass-Fiber-Reinforced Concrete. ASTM International (2017)
16. Daniel, R.A.: A composite bridge is favoured by quantifying ecological impact. Struct. Eng. Int. **20**(4), 385–391 (2010)
17. Mara, V., Haghani, R., Harryson, P.: Bridge decks of fibre reinforced polymer (FRP): a sustainable solution. Constr. Build. Mater. **50**, 190–199 (2014)
18. Maxineasa, S.G., Isopescu, D.N., Entuc, I.S., Taranu, N., Lupu, L.M., Hudisteanu, I.: Environmental performances of different carbon and glass fibre reinforced polymer shear strengthening solutions of linear reinforced concrete. Bull. Transilv. Univ. Braşov. **11**(60), 107–115 (2018)
19. Cadenazzi, T., Dotelli, G., Rossini, M., Nolan, S., Nanni, A.: Life-cycle cost and life-cycle assessment analysis at the design stage of a fiber-reinforced polymer-reinforced concrete bridge in Florida. Adv. Civ. Eng. Mater. **8**(2), 128–151 (2019)
20. Preinstorfer, P., Huber, T., Reichenbach, S., Lees, J.M., Kromoser, B.: Parametric design studies of mass-related global warming potential and construction costs of FRP-reinforced concrete infrastructure. Polymers. **14**(12), 2383 (2022)
21. Ji, X.L., Chen, L.J., Liang, K., Pan, W., Su, R.K.: A review on FRP bars and supplementary cementitious materials for the next generation of sustainable and durable construction materials. Constr. Build. Mater. **383**, 131403 (2023)
22. Dong, S., Li, C., Xian, G.: Environmental impacts of glass-and carbon-fiber-reinforced polymer bar-reinforced seawater and sea sand concrete beams used in marine environments: an LCA case study. Polymers. **13**(1), 154 (2021)
23. International Energy Agency (IEA): Transport sector CO2 emissions by mode in the Sustainable Development Scenario, 2000–2030 (2019). https://www.iea.org/data-and-statistics/charts/transport-sector-co2-emissions-by-mode-in-the-sustainable-development-scenario-2000-2030
24. ASTM D7205: Standard Test Method for Tensile Properties of Fiber Reinforced Polymer Matrix Composite Bars. ASTM International, West Conshohocken (2016)

Numerical Study of a Hybrid Timber–Concrete Floor System

Themistoklis Tsalkatidis ⓘ and Mohand Morchid Alhussain

Contents

1 Introduction

During the past decade, there have been ongoing discussions of how to decrease the global CO_2 emissions output. The building sector is responsible for a significant portion of such emissions so the shift toward sustainable construction materials and systems is required [1]. These alternative sustainable solutions must also be efficient to gradually replace the traditional ones. An example of such a solution is the hybrid timber–concrete floor system. The aim to combine different materials in construction systems or elements is to create a hybrid system that maximizes the advantages and minimizes the disadvantages that each material inherently has [2]. Timber–concrete floor systems have also shown excellent prefabrication capabilities and can be classified as slender in comparison to traditional concrete floors [3]. The use of timber as construction material promotes the application of low-carbon materials, which is vital for modern and sustainable building projects [4].

T. Tsalkatidis (✉) · M. M. Alhussain
Norwegian University of Life Sciences, Ås, Akershus, Norway
e-mail: themistoklis.tsalkatidis@nmbu.no; mohand.morchid.alhussain@nmbu.no

© The Author(s) 2025
M. Kioumarsi, B. Shafei (eds.), *The 1st International Conference on Net-Zero Built Environment*, Lecture Notes in Civil Engineering 237,
https://doi.org/10.1007/978-3-031-69626-8_115

In addition, several research studies have shown the potential of the timber–concrete composite floor systems (TCC). Fragiacomo and Lukaszewska [5] underlined the importance of the connection between timber and concrete as key in the behaviour of the composite floor system as well as the option to use prefabricated elements when needed. Yeoh [6] examined the ultimate and serviceability short and long-term behaviour of timber–concrete floor systems having three different connection types and developed an analytical model that predicts the strength of potential connection types under different modes of failure. Siddika et al. [7] provided a state-of-the-art review of the available cross-laminated timber–concrete floor systems. These systems have been found to be three to five times stronger than ordinary timber–concrete systems with significant out-of-plane load-bearing capacity. Quang Mai et al. [8] discovered that the bending capacity of a hybrid CLT timber–concrete floor is three to five times higher than the corresponding capacity of a CLT floor and that the natural frequency is 25% higher too.

The scope of this study has been to investigate numerically the properties of a hybrid timber–concrete floor system. The floor system consists of an upper thin concrete floor slab part that is connected to the lower part of transverse and longitudinal glulam beams. Timber and concrete parts are connected using epoxy resin. Plywood panels have been placed between the longitudinal glulam beams, but they do not have any structural contribution. They only act as a framework for the concrete slab during the casting phase. The system has been designed based on the specifications of a test specimen from the experimental campaign by Ferrara et al. [9]. According to this research study, hybrid timber–concrete floor systems have been found to be able to maintain significant flexural capacity with small to moderate deflections when they undergo bending. The test configuration used by Ferrara et al. has been the widely used four-point bending test that has provided a good testing platform.

2   Analysis

The numerical problem has been analysed using the finite element method (FEM). The finite element method is a reliable and versatile method that has been used in complex engineering problems for decades. The philosophy of this method is to examine a complex problem as a summation of the solutions of simpler physical problems [10, 11]. Obviously, the sum of these problems forms an equivalent problem to the original one. The problems are solved using advanced mathematics until a converged solution is reached based on the convergence criteria used in the problem. The applied convergence criteria and root-finding algorithms can vary since these depend on the physical problem and the desired degree of accuracy of the solution [10, 12].

In this research study, a reference model of the hybrid floor system under investigation has been constructed and validated against the experimental results of the reference test specimen. The numerical model has considered both geometric and contact nonlinearities as well as the update of the stiffness matrix after each load iteration. The Newton–Raphson incremental-iterative method has been used to derive the numerical solution [12, 13]. After the first task of the research study had been completed, a series of parameters that influence the structural performance of the floor system were examined. These parameters are the thickness of the concrete slab, the flexural strength class of the glulam beams and the placement of the load. The effect of these parameters on the structural behaviour of the hybrid system has been quantified and important conclusions have been drawn. The presented parametric study together with the reference model, and the available test results from the literature provide an insight into the contribution of each component that composes this hybrid floor system in the overall performance of the system [9, 12, 13].

2.1 Reference Model

For the numerical simulations performed in this analysis, the ANSYS 2023 R2 Student version software package [12] has been used. As described before, the hybrid timber–concrete floor system consists of a concrete slab that is connected by means of adhesion to the underlying grid of glulam beams. The concrete slab has a thickness of 70 mm and covers an area of 2400 × 8000 mm. The external longitudinal beams have a rectangular shape with cross-section dimensions 800 × 240 mm and the rest of the beams are 100 × 240 mm. The length of the longitudinal beams is 8000 mm. The distance between the external and internal longitudinal beams is 470 mm and between the internal ones is 500 mm. The floor system is pin-supported, and the two pressing line loads are placed symmetrically, 3000 mm from each support [9]. The configuration of the model can be seen in Fig. 1.

The beams and the concrete slab have been simulated using three-dimensional structural element models; SOLID185 and SOLID65 element types have been selected from the ANSYS library, respectively. SOLID 65 is an element commonly used for the simulation of concrete since it provides both cracking and crushing options [12]. Regarding the material properties, the concrete slab is made of strength class C40/50 concrete and the flexural strength of glulam is Gl24h, as in the experiment [9, 13].

The epoxy resin has an important role since it connects the timber and concrete parts thus ensuring a hybrid system formation [9]. For modelling purposes, the resin has been replaced by a pair of contact elements at the timber–concrete interface [13]. The contact elements form a layer between the timber and concrete elements and are placed as surface-to-surface elements. CONTA174 and TARGE170 element types have been used for the contact pairs. The simulation of the connectivity conditions between concrete and timber using a contact pair has been challenging

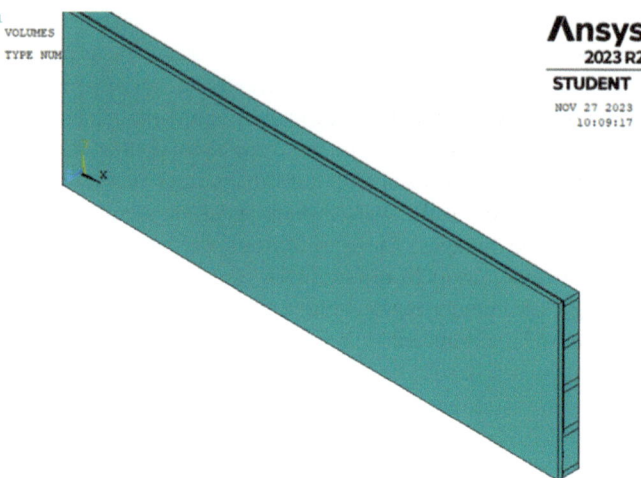

Fig. 1 The hybrid timber–concrete floor system under investigation

Table 1 Examined cases during the parametric study

Parameter	Case	Comments
Thickness of concrete slab	First parametric study-alternative 1	Decrease to 50 mm
Thickness of concrete slab	First parametric study-alternative 2	Increase to 90 mm
Glulam strength class	Second parametric study	Increase to GL30h
Load placement	Third parametric study	Loads placed at 2 m from the supports

due to necessary conversion of the experimental connectivity conditions to numerical factors. Therefore, selecting the values of several factors to some extent relied on the test and trial method, which is a limitation of the model. The Coulomb friction coefficient has been set equal to 0.3, and the 'bonded always' contact option has been used [13–15].

2.2 Parametric Investigation

After the verification of the reference numerical model by comparing its results to the experimental findings, the next step has been to perform a parametric study. The scope of this study has been to investigate, understand, and quantify the effect of certain parameters on the structural performance of the hybrid floor system. The examined parameters and the corresponding cases are presented in Table 1.

3 Results and Discussion

The numerical results from the parametric study and the reference model are presented in the following Figs. 2, 3, 4, 5, 6, 7, and 8. Emphasis has been placed on three findings, the distribution of the von Mises stresses (see Fig. 2), the

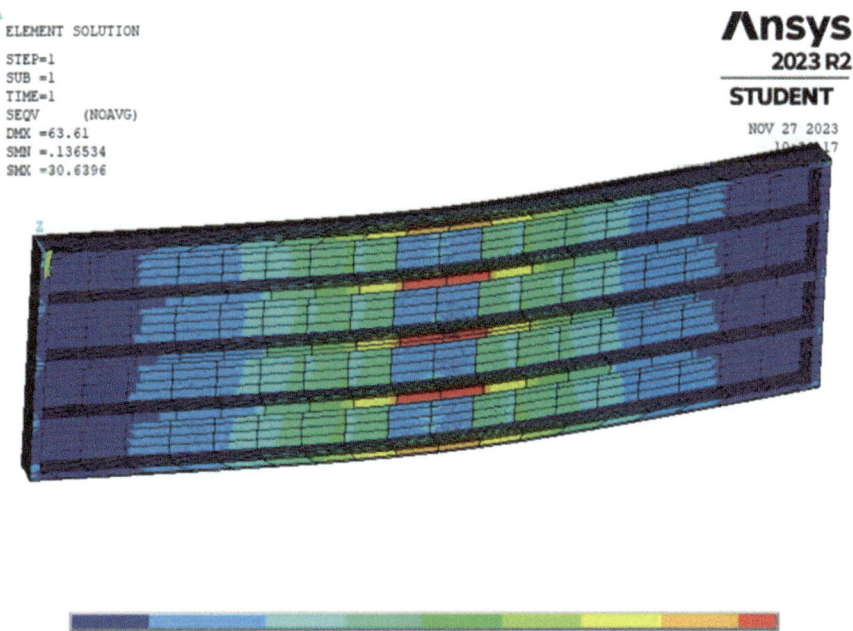

Fig. 2 Von Mises stress distribution of the reference model

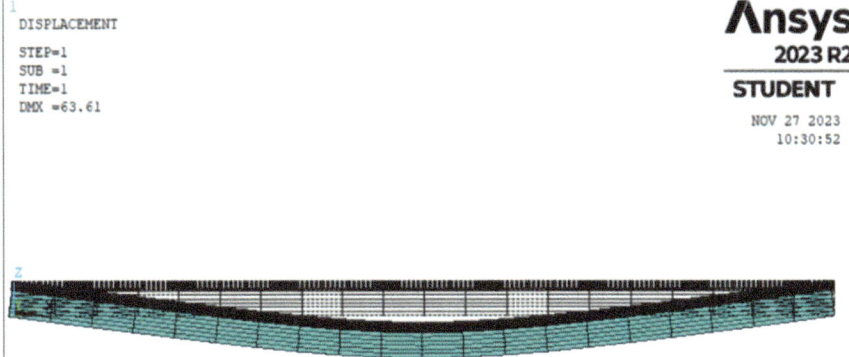

Fig. 3 Maximum deflection of the reference model

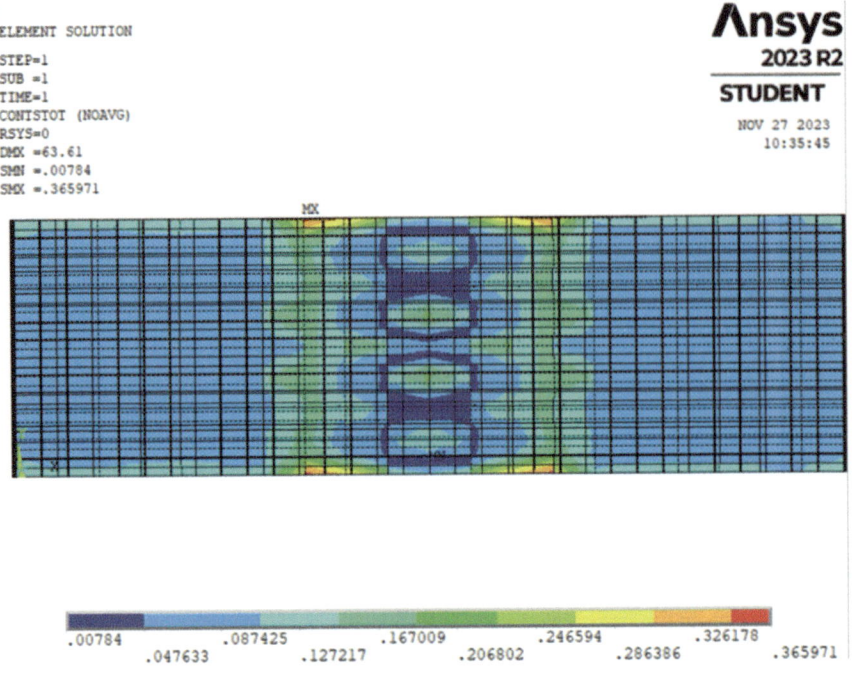

Fig. 4 Contact element stress distribution of the reference model

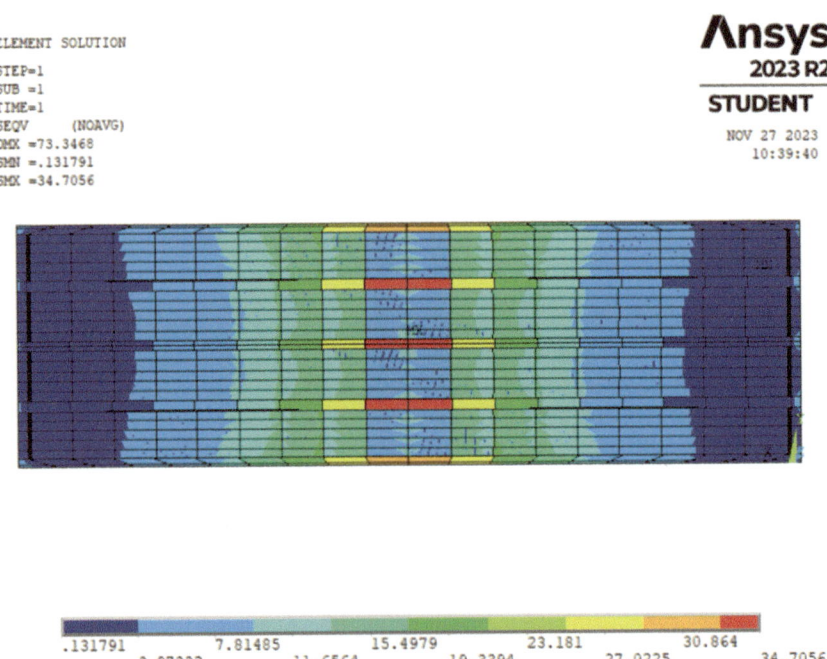

Fig. 5 Von Mises stress distribution of the first parametric study-alternative 1

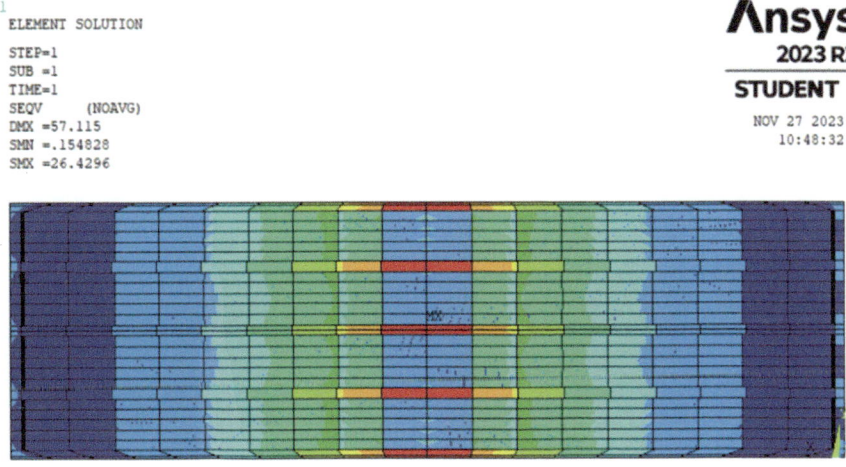

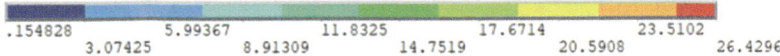

Fig. 6 Von Mises stress distribution of the first parametric study-alternative 2

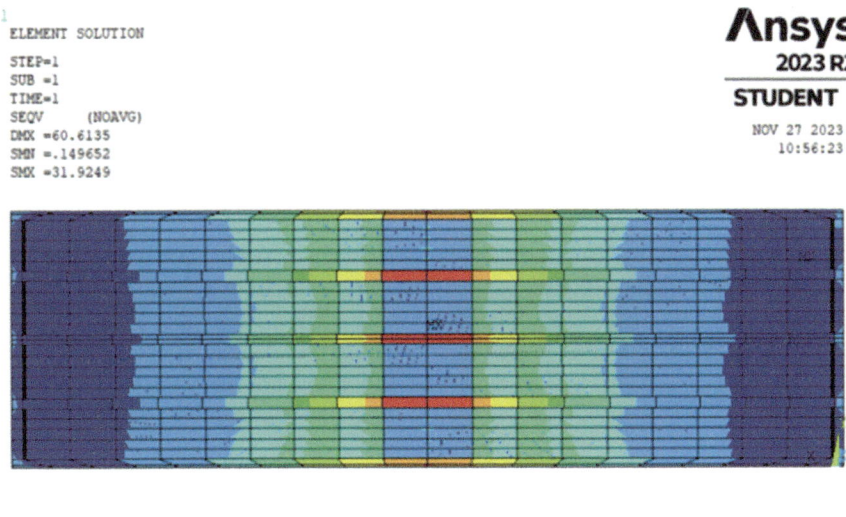

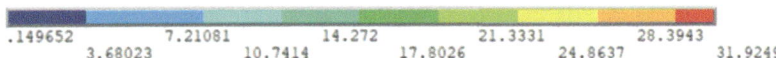

Fig. 7 Von Mises stress distribution of the second parametric study

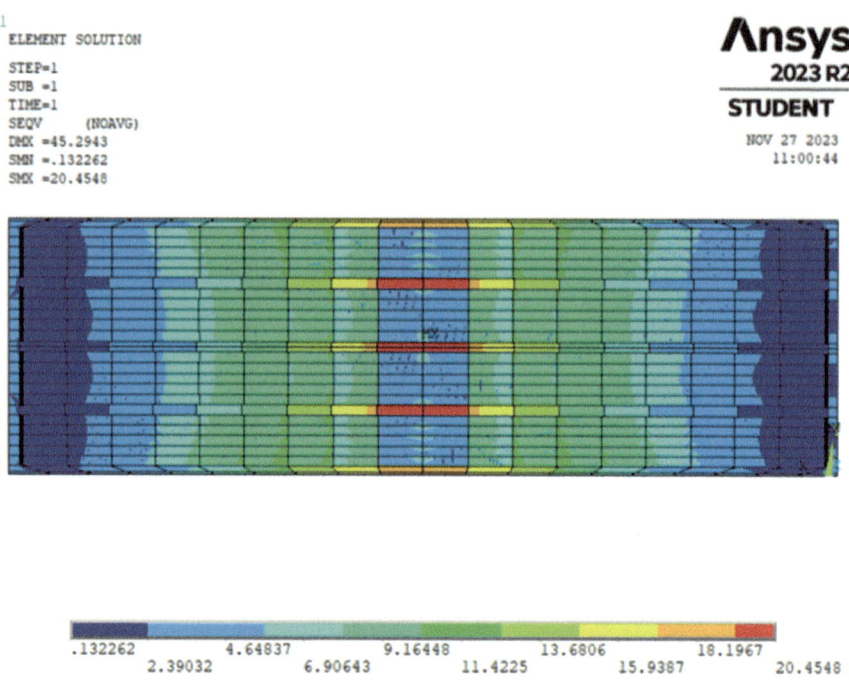

Fig. 8 Von Mises stress distribution of the third parametric study

maximum deflection at the mid-span of the floor system (see Fig. 3), and the contact stresses developed at the timber–concrete interface (see Fig. 4).

The maximum deflection of the reference model has been found to be equal to 63.61 mm, which is very close to the 63.6 mm measurement during the experiment [9]. The third parametric study has produced a lower value of the mid-span deflection of 45.29 mm, whereas the first parametric study-alternative 1 produces a higher of 73.35 mm.

Contact stresses have been detected throughout the interface of the slab, but their maximum values have been developed close to the load application areas. The third parametric study has produced a lower value of contact stresses equal to 0.25 MPa whereas in the reference model the value is 0.37 MPa. The higher value has been detected in the first parametric study-alternative 1, and it is equal to 0.41 MPa.

The shape of the von Mises stress distribution has remained the same as in the reference model (see Fig. 2) throughout the parametric study but the values vary (see Figs. 5, 6, 7, and 8), depending on the parameter under examination. The contour plot appears to have a rather symmetric shape with the peak stress values near the midpoint of the span.

The von Mises stresses have been found to have greater values in the case where the concrete slab has a decreased thickness, as shown in Fig. 6 in comparison to Fig. 5. There is a drop of about 24% in the developed maximum stress.

In Fig. 7, where a higher quality glulam grade is used the von Mises stresses appear to be quite close in value to the reference model (see Fig. 2) with a maximum value of around 31 MPa.

In Fig. 8, the placement of the load closer to the end supports has a significant effect on the reduction of the magnitude of the von Mises stresses. The maximum value in this case is 20.45 MPa.

4 Conclusions

The presented and proposed numerical simulation has managed to produce quite accurate results and provide further knowledge on the structural performance of hybrid timber–concrete floor systems. As aforementioned, the numerical results of this study have been verified against experimental ones and are in close agreement.

The parametric study has determined the effect of different parameters on the structural behaviour of the hybrid floor systems. Obviously, an increase either in the thickness of the concrete slab or in the strength class of timber has a positive contribution to the floor system. However, the effect of the thickness of the concrete slab has been found to be more significant than the effect of the strength class of the glulam beams. This must be connected to the higher stiffness of the concrete part in comparison to the timber one.

The effect of the load placement has been proven to be the most profound, both in the calculated maximum deflection and stress. This case has produced the lowest values of stress and deflection out of every case. The long span of the floor is believed to have contributed to this outcome.

The long-term and post-failure behaviour of the components used in the construction of the hybrid floor system has not been examined in this study.

Overall, hybrid timber–concrete floor systems have been proven to be a sustainable and efficient alternative to traditional floor systems. The numerical analysis of hybrid floor systems is challenging, but it can be considered as an effective tool when designing such systems.

References

1. Pye, S., Li, F.G., Price, J., Fais, B.: Achieving net-zero emissions through the reframing of UK national targets in the post-Paris Agreement era. Nat. Energy. **2**(3), 1–7 (2017)
2. Yeoh, D., Fragiacomo, M., De Franceschi, M., Heng Boon, K.: State of the art on timber-concrete composite structures: literature review. J. Struct. Eng. **137**(10), 1085–1095 (2011)
3. Lamothe, S., Sorelli, L., Blanchet, P., Galimard, P.: Lightweight and slender timber-concrete composite floors made of CLT-HPC and CLT-UHPC with ductile notch connectors. Eng. Struct. **243**, 112409 (2021)

4. Plüss, Y., Zwicky, D.: A case study on the eco-balance of a timber-concrete composite structure in comparison to other construction methods. In: Justness, H. (ed.) Concrete Innovation Conference, pp. 11–21. NB, Oslo (2014)
5. Ferrara, G., Michel, L., Ferrier, E.: Flexural behaviour of timber-concrete composite floor systems linearly supported at two edges. Eng. Struct. **281**, 115782 (2023)
6. Fragiacomo, M., Lukaszewska, E.: Development of prefabricated timber-concrete composite floor systems. Struct. Build. **164**(2), 117–129 (2011)
7. Yeoh, D.: Behaviour and design of timber-concrete composite floor system. PhD Thesis, University of Canterbury, Christchurch (2010)
8. Siddika, A., Al Mamun, M.A., Aslani, F., Zhuge, Y., Alyousef, R., Hajimohammadi, A.: Cross-laminated timber–concrete composite structural floor system: a state-of-the-art review. Eng. Fail. Anal. **130**, 105766 (2021)
9. Quang Mai, K., Park, A., Tan Nguyen, K., Lee, K.: Full-scale static and dynamic experiments of hybrid CLT–concrete composite floor. Constr. Build. Mater. **170**, 55–65 (2018)
10. Bathe, K.-J.: Finite element procedures, 2nd edn. Prentice Hall, Upper Saddle River (2006)
11. Bi, Z.: Finite element analysis applications: a systematic and practical approach, 1st edn. Academic Press, Cambridge (2017)
12. ANSYS Inc.: Ansys Manual vol. 2023, Canonsburg (2023)
13. Alhussain, M.M.: Numerical investigation of the flexural behavior of timber-concrete composite floor systems. MSc Thesis, NMBU, Ås (2023)
14. Tsalkatidis, T., Avdelas, A.: The unilateral contact problem in composite slabs. Experimental study and numerical treatment. J. Constr. Steel Res. **66**, 480–486 (2010)
15. Jaaranen, J., Fink, G.: Frictional behaviour of timber-concrete contact pairs. Constr. Build. Mater. **243**, 118273 (2020)

Influence of Openings on Seismic Failure Mechanism of URM Infilled RC Hill Buildings

Z. Naorem and P. Haldar

Contents

1 Introduction

Buildings constructed on sloping terrain are inherently susceptible to premature damage in the event of an earthquake due to the horizontal and vertical irregularities present resulting from the construction of foundations at varying levels following the hill profile. Further, extremely high seismicity of Indian Himalayan region, mostly categorized under high seismic zones in Indian standard IS 1893 (2016) [1], i.e. zones IV and V, gravely intensifies the associated seismic risk. The consequences of the combination of the inherent seismic vulnerability of such buildings and the extremely high seismicity of the Indian Himalayan region have proven to be disastrous as evident from the aftermath of past earthquakes. The extremely complex seismic response of buildings constructed on sloping terrain gets further exacerbated upon incorporating URM infills as partitions in those buildings due to the complex interaction between the frame and infill elements. However, openings in the form of doors and windows are integral parts of URM infill panels to accommodate the

Z. Naorem · P. Haldar (✉)
Department of Civil Engineering, IIT Ropar, Rupnagar, Punjab, India
e-mail: putul.haldar@iitrpr.ac.in

© The Author(s) 2025
M. Kioumarsi, B. Shafei (eds.), *The 1st International Conference on Net-Zero Built Environment*, Lecture Notes in Civil Engineering 237,
https://doi.org/10.1007/978-3-031-69626-8_116

functional requirements. To accurately assess the seismic response of such buildings, it is necessary to consider the infill-frame interaction alongside the effects of functional openings on the overall strength and stiffness of infill panels. Extensive efforts have been put forth towards the understanding of the inelastic response of such buildings under the effects of seismic excitations over the past decade [2, 3]. However, most of the scientific studies have been undertaken either on bare frames without considering the effect of infill-frame interaction or on fully infilled frames without considering the presence of openings. The aim of the present study is to acquire a deeper understanding and shed light on the effect of functional openings in infills on the dynamic properties that govern the seismic response in terms of the fundamental period of vibration and modal mass participation ratios considering the fundamental mode of vibration. Influence of various opening ratios prevalent in India for various combinations of doors and windows on the failure mechanism of the overall structure is also investigated. The present study will provide key insights into the expected structural damage and damage in infills which will be crucial for developing sustainable earthquake-resistant design strategies for buildings constructed on sloping terrain.

2   Description of Analytical Models

A generic mid-rise unreinforced masonry (URM) infilled RC building as shown in Fig. 1a with a plan dimension of 25.6 × 15 m prevalent in the Indian scenario is considered for the parametric study. Two simple vertical structural configurations prevalent in towns and cities of the Indian Himalayas are considered, viz., split-foundation (Fig. 1b) and step-back (Fig. 1c).

All the considered buildings are situated on soil type I in seismic Zone IV of IS 1893 (2016) and are designed and detailed as Special Moment Resisting Frames (SMRFs) following the guidelines of IS456 [4], IS1893 [1], and IS13920 [5]. To assess the seismic response, three-dimensional analytical models of the considered buildings are developed using the software package SAP2000 [6]. Dead and imposed loads are applied on individual structural elements as per recommendations of IS 875 [7] and IS 875 [8], respectively. Strong column-weak-beam design is considered by limiting the ratio of the moment capacity of columns to beams to a minimum of 1.4. The details of the material properties and resulting section sizes are listed in Table 1.

The presence of infill panels significantly alters the overall response of RC frames under seismic excitation [9–12] and should be considered for accurate estimation of seismic response. A number of techniques are available to incorporate the interaction of infill and frame elements [13]. In the present study, the action of infill panels under seismic excitation is modelled using single eccentrically braced diagonal struts with geometric properties as per IS 1893 (2016). A wide range of sizes of openings for doors and windows are observed throughout India. The area of opening for windows expressed as a percentage of overall size of infill panel ranges from 5.7% to 29.4%

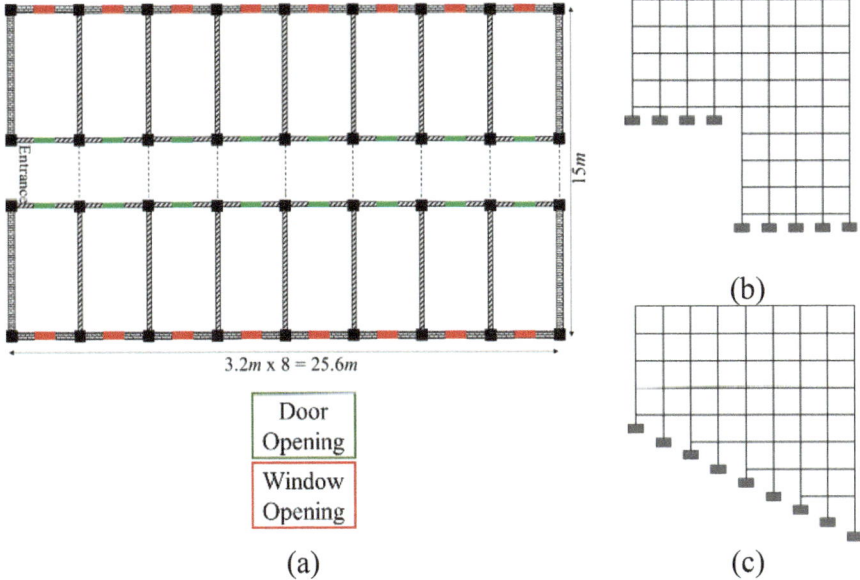

Fig. 1 (a) Plan; (b) elevation with split-foundation (SF); (c) elevation with step-back (SB) foundation

Table 1 Material properties and design details of the considered buildings

Description	Specification
Grade of concrete	M40
Grade of steel reinforcement	Fe500
Size of beam sections	300 × 400 mm
Size of column sections	350 × 350 mm
Size of section of 1.1 m short columns in split-foundation building	350 × 350 mm
Size of section of 1.1 m and 2.75 m short columns in step-back building	600 × 600 mm
Size and spacing of transverse reinforcements	8 mm @ 100 mm c/c

and in the case of doors, it ranges from 19.06% to 30.49% [14]. Considering the functional requirement of the considered buildings, the location of doors and windows are placed systematically as shown in Fig. 1a. Further, possible combinations of door type (D1, D2, and D3) and window type (W1, W2,..., and W11) are considered resulting in opening combinations ranging from W1D1 to W11D3 resulting in a total of 66 buildings with openings and percentage openings of doors and windows were varied systematically from 20% to 30% and 5% to 30%, respectively (Table 2). A number of literature is available regarding the effect of opening on the overall strength and stiffness of the infill panels [15–19]. Kurmi and Haldar [20] showed that the strength and stiffness modifiers proposed by Decanini et al. [15] had the least deviation from experimental results and was easy to use for parametric study, hence, is considered for the present study.

Table 2 Dimensions of prevalent openings and combinations of doors and windows considered

Combination of doors and windows	Width of window (m)	Height of window (m)	Width of door (m)	Height of door (m)	Door opening ratio (%)	Window opening ratio (%)
FI	0	0	0	0	0	0
W1D1	0.34	1.2	0.79	2.1	20	5
W2D1	0.52	1.2	0.79	2.1	20	7.5
W3D1	0.69	1.2	0.79	2.1	20	10
W4D1	0.86	1.2	0.79	2.1	20	12.5
W5D1	1.03	1.2	0.79	2.1	20	15
W6D1	1.21	1.2	0.79	2.1	20	17.5
W7D1	1.38	1.2	0.79	2.1	20	0.2
W8D1	1.55	1.2	0.79	2.1	20	22.5
W9D1	1.72	1.2	0.79	2.1	20	25
W10D1	1.89	1.2	0.79	2.1	20	27.5
W11D1	2.07	1.2	0.79	2.1	20	30
W1D2	0.34	1.2	0.98	2.1	25	5
W2D2	0.52	1.2	0.98	2.1	25	7.5
W3D2	0.69	1.2	0.98	2.1	25	10
W4D2	0.86	1.2	0.98	2.1	25	12.5
W5D2	1.03	1.2	0.98	2.1	25	15
W6D2	1.21	1.2	0.98	2.1	25	17.5
W7D2	1.38	1.2	0.98	2.1	25	20
W8D2	1.55	1.2	0.98	2.1	25	22.5
W9D2	1.72	1.2	0.98	2.1	25	25
W10D2	1.89	1.2	0.98	2.1	25	27.5
W11D2	2.07	1.2	0.98	2.1	25	30
W1D3	0.34	1.2	1.18	2.1	30	5
W2D3	0.52	1.2	1.18	2.1	30	7.5
W3D3	0.69	1.2	1.18	2.1	30	10
W4D3	0.86	1.2	1.18	2.1	30	12.5
W5D3	1.03	1.2	1.18	2.1	30	15
W6D3	1.21	1.2	1.18	2.1	30	17.5
W7D3	1.38	1.2	1.18	2.1	30	20
W8D3	1.55	1.2	1.18	2.1	30	22.5
W9D3	1.72	1.2	1.18	2.1	30	25
W10D3	1.89	1.2	1.18	2.1	30	27.5
W11D3	2.07	1.2	1.18	2.1	30	30

To model the inelastic deformations of beam elements, plastic moment hinges are provided at both ends of the beam as per ASCE-41 [21]. In the case of the columns, fibre-interacting P-M-M hinges are provided at each end. In addition, shear hinges, with capacities estimated as per ASCE-41 [21], are provided in each column to check the probability of shear failure of column due to shear forces developed due to strut action of infill panel under seismic excitation. All the probable failure modes of infill panels identified by Haldar [13] have also been considered. Accordingly, axial hinges are provided at the middle of each diagonal strut as per ASCE-41 [22] accounting for bed-joint failure mode of the infill panel.

3   Results and Discussion

Seismic response of buildings is characterized by a number of parameters such as fundamental period of vibration, mode shape, and modal mass participation ratio. The imparted alteration in the strength and stiffness of infill panels due to the presence of functional openings combined with the inherent structural irregularities arising from step-back and split foundation can have a significant impact on the overall seismic behaviour and consequent global failure of buildings constructed in hilly regions. The effect of variation in the opening ratio on the parameters that govern the seismic behaviour in terms of the fundamental period of vibration and modal mass participation ratio is discussed in the following sections.

3.1   Period of Vibration

The fundamental period of vibration is an important dynamic property that determines the seismic response of buildings and largely depends on the mass and stiffness properties of the considered buildings. Depending on the characteristics of the seismic excitation, buildings with varying dynamic properties will respond differently. The variation in the fundamental period of vibration of the buildings with opening ratio considering various combinations of doors and window openings are plotted in Figs. 2 and 3. In Fig. 2, the trend of variation in the fundamental period of vibration is plotted for considering specific door opening but varying the window opening ratio. The observed variations are almost completely linear with minor deviations and the trendlines are completely parallel to each other in the direction across the slope of the terrain (Fig. 2a, c). The variation in the fundamental period of vibrations among buildings with varying door opening ratio but same window opening ratio (like W1D1, W1D2 and W1D3) remains almost constant with increase in window opening ratio. An almost completely linear variation is also observed in the direction along the slope of the terrain, but the trendlines were not parallel to each other. In this direction, the variation in the fundamental period of vibrations among buildings with varying opening ratio of doors but same window opening ratio increases as the opening ratio of windows increases and is maximum between W11D1, W11D2, and W11D3 in both step-back and split-foundation buildings (Fig. 2b, d). Figure 3 shows the overall variation in the fundamental period of vibration for all considered combinations (W1D1–W11D3) including the case of fully infilled (FI) building without any functional openings in the infill panel. The

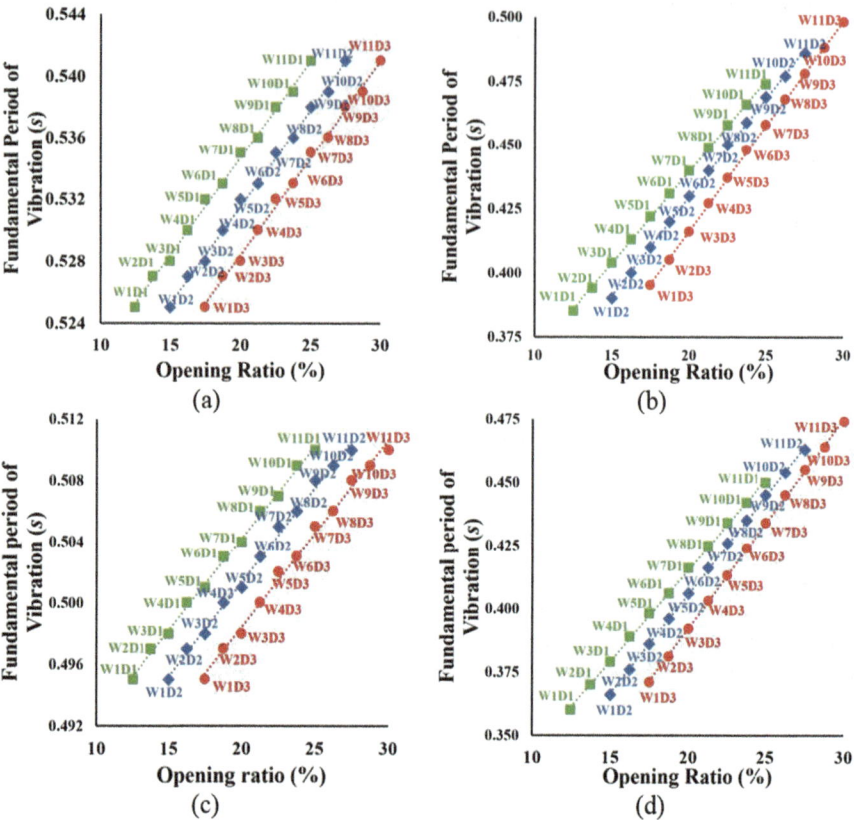

Fig. 2 Variation of period of vibration with opening ratio in the direction across the slope for (**a**) split-foundation and (**b**) step-back building; and along the slope direction for (**c**) split-foundation and (**d**) step-back building with trendlines for varying window opening but fixed door opening

variation of the fundamental period of vibration as a function of the opening ratio can be established using regression relationships using Eqs. (1) and (2) for split-foundation buildings in the direction across and along the slope, respectively. Similarly, for step-back buildings, Eqs. (3) and (4) can be used to describe the same in the direction across and along the slope, respectively.

$$T = 0.0002a^2 + 0.00004a + 0.522 \tag{1}$$

$$T = 0.0001a^2 + 0.0019a + 0.345 \tag{2}$$

$$T = 0.0002a^2 + 0.00004a + 0.492 \tag{3}$$

$$T = 0.0001a^2 + 0.0019a + 0.319 \tag{4}$$

where T is the fundamental period of vibration in s and a is the opening ratio in percentage.

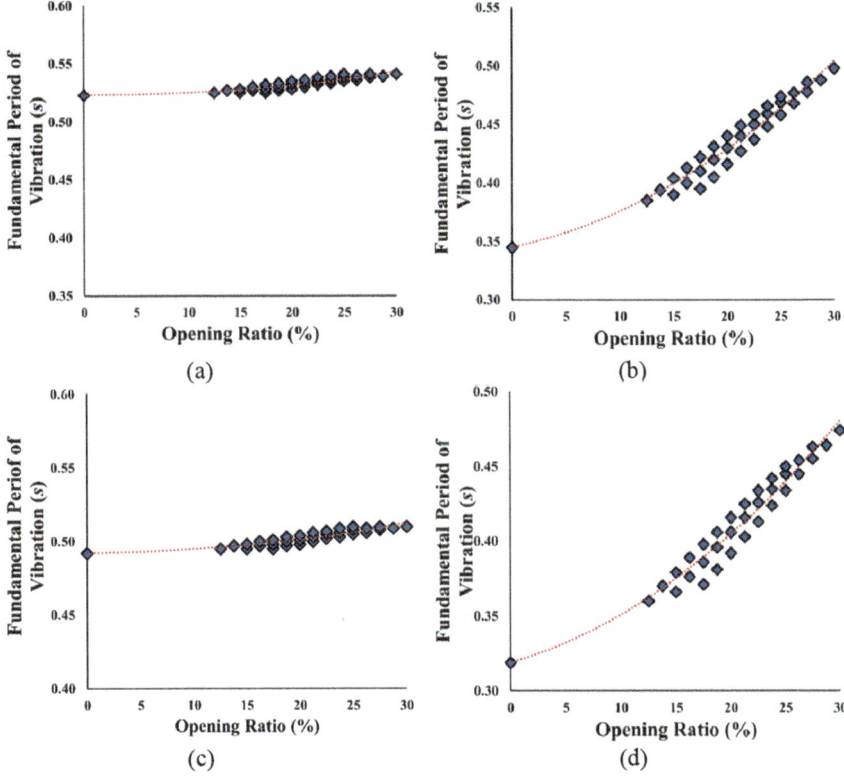

Fig. 3 Variation of period of vibration with opening ratio in the direction across the slope for (**a**) split-foundation and (**b**) step-back building; and along the slope direction for (**c**) split-foundation and (**d**) step-back building

In the direction along the slope of the terrain, a comparatively larger variation is observed in the fundamental period of vibration in both considered building typologies (i.e. step-back and split-foundation). The higher variation is a direct result of the presence of openings in the infills along the direction of slope thereby reducing the overall stiffness particularly in this direction. Conversely, the insignificant change in the fundamental period of vibration can be attributed to the lack of any functional openings in the infill panels in this direction.

3.2 Modal Mass Participation Ratio

In addition to the fundamental period of vibration, the mode shape and the modal participation ratio are another important dynamic property that determines the overall seismic response of buildings. Buildings with a very high modal mass ratio in the fundamental mode of vibration tend to vibrate very close to their fundamental mode during seismic excitation. The seismic evaluation of such buildings can be achieved using simple traditional non-linear capacity spectrum-based approaches with reasonable accuracy. However, buildings with comparatively lower modal mass participation ratio of fundamental mode shape will have a highly complex and irregular response due to the contribution of higher mode shapes on the overall seismic response. In the case of such buildings, complex and computationally expensive non-linear dynamic approaches need to be undertaken to evaluate the seismic response. The variation of the modal mass participation factor with the opening ratio is listed in Table 3. The observed modal mass participation of the fundamental mode of vibration in fully infilled buildings (both SF and SB) is lower than 85–90%, which is generally considered a minimum for the determination of the overall seismic response of buildings. This indicates a significant contribution of the higher modes of vibration (in translation and rotation, due to the torsional irregularity of hill buildings) on the seismic response of the considered buildings.

The modal mass participation ratio further reduces in both along and across the slope directions with an increase in the opening ratio ranging from 1–4% for a 1.5% opening ratio to 8–10% for a 30% opening ratio as compared to their fully infilled counterparts. It could be observed that the reduction in modal mass participation ratio is higher in the direction across the slope of the terrain resulting from the torsional irregularity of the building in this particular direction.

3.3   Assessment of Damage Under Seismic Excitation

Due to the significant contribution of higher modes of vibration on the seismic response of all considered buildings as listed in Table 3, a simple non-linear time-history analysis is considered to assess the non-linear response of the buildings under seismic excitation. The 1976 Friuli earthquake is considered for this purpose due to its moderately high ground motion parameters (6.5 Mw and 0.35 g PGA). Both components of the considered ground motion are applied simultaneously since all considered buildings have torsional irregularity resulting from the construction of foundations at varying levels.

Figure 4 shows the displacement history record at the roof of the respective buildings in the direction along the slope of the terrain. It could be observed that the decrease in the overall stiffness of the buildings due to the presence of functional openings results in an increase in the displacement response of both split-foundation and step-back buildings. Moreover, the displacement response of the respective buildings increases with an increase in the opening ratio as shown in Fig. 4. It could be observed that the displacement response of W11D3 was 62.2% higher than FI, while in the case of W1D1, it was 33.36% higher for split-foundation buildings. Similarly, for step-back buildings, the maximum displacement at the roof was 62.72% and 25.69% higher for W11D3 and W1D1, respectively, when compared to their FI counterpart. The damage observed in the split foundation buildings is

Table 3 Modal mass participation ratio of fundamental mode of vibration

Designation	Opening ratio (%)	Modal mass participation ratio (%)			
		Split-foundation building		Step-back building	
		Across the slope	Along the slope	Across slope	Along the slope
FI	0	60.58	64.64	59.87	60.67
W1D1	12.5	59.23	61.91	58.75	59.29
W2D1	13.75	58.52	61.42	58.16	59.05
W3D1	15	57.78	60.94	57.53	58.83
W4D1	16.25	57.02	60.49	56.89	58.62
W5D1	17.5	56.24	60.06	56.22	58.43
W6D1	18.75	55.45	59.66	55.55	58.25
W7D1	20	54.66	59.28	54.87	58.1
W8D1	21.25	53.61	58.92	54.2	57.92
W9D1	22.5	53.08	58.59	53.5	57.77
W10D1	23.75	52.33	58.29	52.85	57.63
W11D1	25	51.57	58	52.18	57.5
W1D2	15	59.22	61.58	58.74	59.14
W2D2	16.25	58.51	61.08	58.15	58.9
W3D2	17.5	57.77	60.59	57.52	58.67
W4D2	18.75	57.01	60.13	56.88	58.46
W5D2	20	56.23	59.69	56.21	58.26
W6D2	21.25	55.44	59.28	55.54	58.1
W7D2	22.5	54.65	58.9	54.86	57.9
W8D2	23.75	53.85	58.53	54.16	57.74
W9D2	25	53.68	58.19	53.48	57.59
W10D2	26.25	52.32	57.88	52.83	57.45
W11D2	27.5	51.56	57.58	52.16	57.31
W1D3	17.5	59.21	61.3	58.73	59.01
W2D3	18.75	58.5	60.78	58.14	58.77
W3D3	20	57.76	60.28	57.51	58.54
W4D3	21.25	57	59.82	56.87	58.32
W5D3	22.5	56.22	59.37	56.2	58.12
W6D3	23.75	55.43	58.95	55.52	57.93
W7D3	25	54.65	58.56	54.85	57.76
W8D3	26.25	53.84	58.18	54.15	57.59
W9D3	27.5	53.6	57.83	53.47	57.43
W10D3	28.75	52.31	57.52	52.81	57.29
W11D3	30	51.55	57.21	52.15	57.16

shown in Fig. 5 for varying opening ratios in infill panels. It can be observed that the damage in the structural elements (beams and columns) remains largely the same with an increase in the opening ratio. Collapse of all the considered split-foundation buildings is caused by a combination of shear and flexural failure of the short columns just above the uppermost foundation level. However, the extent of damage

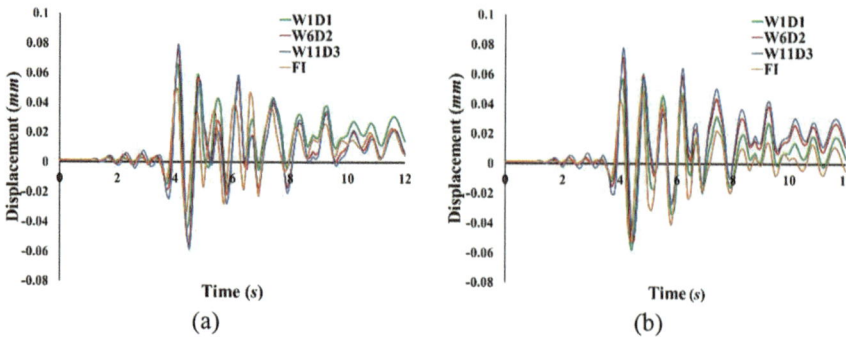

Fig. 4 Displacement response history at roof of building with (**a**) split-foundation; and (**b**) step-backs

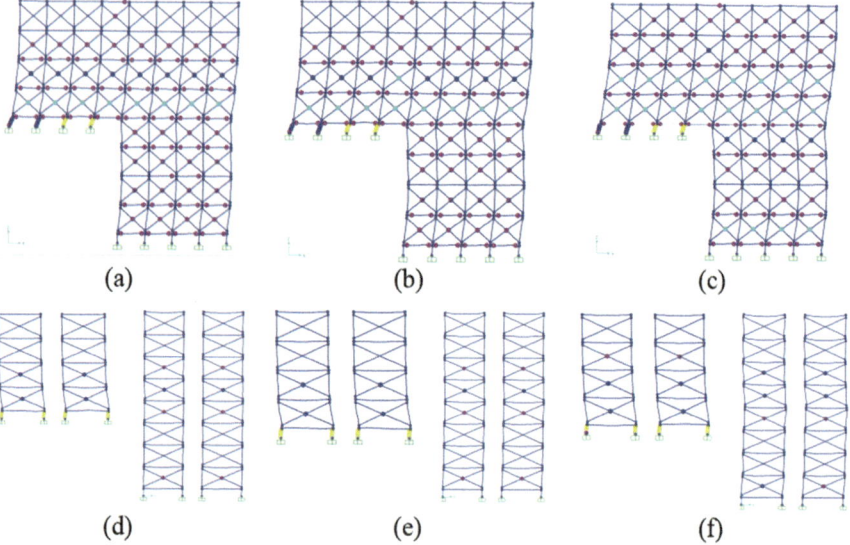

Fig. 5 Damage pattern observed in split-foundation buildings in the direction along the slope of the terrain (**a**) W1D1, (**b**) W6D2, (**c**) W11D3; and across the slope of the terrain (**d**) W1D1, (**e**) W6D2, (**f**) W11D3

in the infill panels increases significantly in both along and across the slope directions as a consequence of the reduction in strength and stiffness of the infill panels. Similarly, in the case of step-back buildings, the collapse is caused by a combination of shear and flexural failure of the 1.1 m short columns closest to the uppermost foundation level (Fig. 6). An increase in the extent of damage in infill elements can be observed in with an increase in the opening ratio. Thus, the extent of earthquake losses will increase considerably with an increase in the percentage area of openings in infill panels under moderately large seismic excitation.

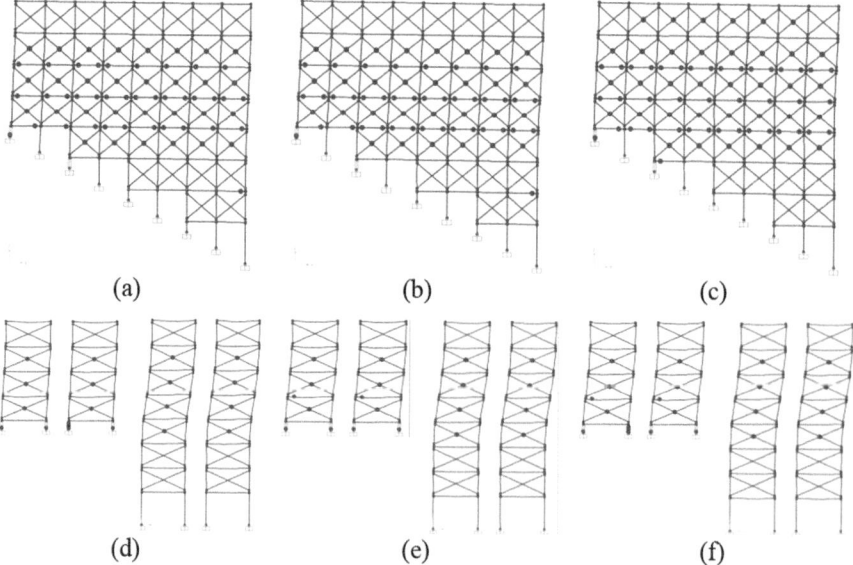

Fig. 6 Damage pattern observed in step-back buildings in the direction along the slope of the terrain (**a**) W1D1, (**b**) W6D2, (**c**) W11D3; and across the slope of the terrain (**d**) W1D1, (**e**) W6D2, (**f**) W11D3

4 Conclusion

The dynamic properties of buildings constructed on sloping terrain considering various opening ratios relevant to Indian construction practices have been compared in the present study. It was found that the fundamental period of vibration increases considerably in the direction along the slope of the terrain with an increase in the opening ratio while an insignificant variation was observed in the direction across the slope due to the fact that openings in the form of doors and windows were present in the infill panels along the direction of slope. With an increase in the opening ratio, a reduction in the modal mass participation ratio in the fundamental modes and contribution of higher modes of vibration has been observed both for split-foundation and step-back buildings. Non-linear dynamic evaluation of the 66 numbers of considered buildings with various doors and window openings combinations showed the displacement response of the considered buildings increases with an increase in the opening ratio. The maximum displacement at the roof of the buildings increased by 33.36–62.2% and 25.69–62.72%, depending on the opening ratio, for step-back and split-foundation buildings, respectively, compared to their fully infilled counterparts. As the opening ratio increases, the extent of damage in the infills also increases due to the reduction in the strength and stiffness of the infill panel. For the comparison purpose, all the 66 considered buildings are subjected to 1976 Friuli earthquake (6.5 Mw and 0.35 g PGA) due to its similarity in expected seismic hazard in the Indian Himalayan region. However, large variation exists in ground motion parameters such as amplitude of vibration, frequency content and duration of ground motion. Therefore, in order to minimize the subjectivity and to incorporate the variability in ground motion parameters, a large number of time-history analyses considering multiple earthquake records need to be performed which is out of the scope of the present study. Due to the highly non-linear and complex seismic response of URM-infilled RC buildings constructed on sloping terrain, a comprehensive seismic vulnerability assessment is required to develop proper earthquake design standards and disaster mitigation policies to reduce the expected seismic risk in the towns and cities of the Indian Himalayan region.

References

1. IS1893: Criteria for Earthquake Resistant Design of Structures. Bureau of Indian Standards, New Delhi (2016)
2. Surana, M., Singh, Y., Lang, D.H.: Seismic characterization and vulnerability of building stock in hilly regions. Nat. Hazards Rev. **19**(1), 04017024 (2018)
3. Naorem, Z., Haldar, P.: Effect of seismic design provisions of Indian standards on seismic response of URM infilled RC buildings on hill. J. Vib. Eng. Technol. **12**, 7107–7120 (2024)
4. IS456: Code of practice for plain and reinforced concrete. Bureau of Indian Standards, New Delhi (2000)
5. IS13920: Ductile design and detailing of reinforced concrete structure subjected to seismic force. In: Bureau of Indian Standard (BIS), New Delhi (2016)
6. SAP2000, Computers and Structures Inc. (CSI): Integrated Software for Structural Analysis and De-sign, SAP2000, Berkeley
7. IS875: Code of practice for design loads (other than earthquake) for buildings and structures: part 1 dead loads – unit weight of building materials and stored materials. Bureau of Indian Standards, New Delhi (1987)
8. IS875: Code of practice for design loads (other than earthquake) for buildings and structures: part 2 imposed loads. Bureau of Indian Standards, New Delhi (1987)
9. Ricci, P., Di Domenico, M., Verderame, G.M.: Effects of the in-plane/out-of-plane interaction in URM infills on the seismic performance of RC buildings designed to Eurocodes. J. Earthq. Eng. **26**(3), 1595–1629 (2022)
10. Ricci, P., Di Domenico, M., Verderame, G.M.: Out-of-plane seismic safety assessment of URM infills accounting for the in-plane/out-of-plane interaction in a nonlinear static framework. Eng. Struct. **195**, 96–112 (2019)
11. Kurmi, P.L., Haldar, P.: Effect of revised seismic design provisions on seismic performance of RC frame buildings with and without infills. Singapore, Springer (2022)
12. Haldar, P., Singh, Y.: Modelling of URM infills and their effect on seismic behaviour of RC frame buildings. Open Constr. Build. Technol. J. Bentham Science Publishers, 6(Suppl 1-M1) (Special Issue on Advances in Infilled Framed Structures: Experimental & Modelling Aspects), 35–41 (2012)
13. Haldar, P.: Seismic behavior and vulnerability of Indian RC frame buildings with URM infills, in Department of Earthquake Engineering. Indian Institute of Technology, Roorkee (2013)
14. CPWD: Manual on door and window details for residential buildings, vol. 1. Govt. of India (2006)
15. Decanini, L.D., Liberatore, L., Mollaioli, F.: Strength and stiffness reduction factors for infilled frames with openings. Earthq. Eng. Eng. Vib. **13**(3), 437–454 (2014)

16. Asteris, P.G., Giannopoulos, I.P., Chrysostomou, C.Z.: Modeling of infilled frames with openings. Open Constr. Build. Technol. J.. Bentham Science Publishers. **6**, 81–91 (2012)
17. Al-Chaar, G.: Evaluating strength and stiffness of unreinforced masonry infill structures. U.-S. Army Corps of Engineers: Report no. ERDC/CERL TR-02-1. Champaign (2002)
18. Chen, X., Liu, Y.: Numerical study of in-plane behaviour and strength of concrete masonry infills with openings. Eng. Struct. **82**, 226–235 (2015)
19. Shabani, A., Plevris, V., Kioumarsi, M.: A comparative study on the initial in-plane stiffness of masonry walls with openings. In: 17th World Conference on Earthquake Engineering (17WCEE), Sendai (2021)
20. Kurmi, P.L., Haldar, P.: Modeling of opening for realistic assessment of infilled RC frame buildings. Structure. **41**, 1700–1709 (2022)
21. ASCE-41: Seismic evaluation and retrofit of existing buildings, American Society of Civil Engineers, Virginia. American Society of Civil Engineers, Virginia (2017)
22. ASCE-41: Seismic Evaluation and Retrofit of Existing Buildings. American Society of Civil Engineers, Virginia (2013)

Seismic Collapse Risk of Stone Masonry Buildings in the Indian Himalayas

Ravi Shastri [ID], Samit Ray-Chaudhuri [ID], and Yogendra Singh [ID]

Contents

1   Introduction

Representation of seismic hazards is always challenging for structural engineers as it can be done in many ways, such as hazard curves, uniform hazard spectrum (UHS), and peak ground acceleration (PGA). Among the many options, two of the most widely used representations use UHS and hazard curves. While UHS represents the seismic hazards in terms of spectral coordinates corresponding to a fixed probability of exceedance (POE) at all spectral periods, hazard curves are constructed for each

R. Shastri (✉) · S. Ray-Chaudhuri
Indian Institute of Technology Kanpur, Kanpur, India
e-mail: rshastri22@iitk.ac.in; samitrc@iitk.ac.in

Y. Singh
Indian Institute of Technology Roorkee, Roorkee, India
e-mail: yogendra.singh@eq.iitr.ac.in

© The Author(s) 2025
M. Kioumarsi, B. Shafei (eds.), *The 1st International Conference on Net-Zero Built Environment*, Lecture Notes in Civil Engineering 237,
https://doi.org/10.1007/978-3-031-69626-8_117

spectral period, providing POEs of different amplitudes of spectral ordinates. Current code provisions use UHS as the basis for estimating design demand. However, using UHS offers good safety against the hazards corresponding to different return periods. Still, it fails to control the collapse risk for the structures subjected to any hazard in its lifetime. This question of collapse risk is answered by representing seismic hazards in terms of the risk-targeted response spectrum (RTRS), which will be discussed in the subsequent sections.

2 Risk-Targeted Response Spectrum

A structure might be exposed to different levels of seismic hazards in its lifetime, any of which can lead to its collapse. The hazard and the conditional probability of collapse in itself is not deterministic, i.e., it brings an element of uncertainty with itself. With this uncertainty associated with the hazard and the conditional collapse probability, there is always a risk associated with the collapse of structures. This risk of collapse is estimated using the hazard and the fragility curves, as discussed in [1]. A structure constructed using conventional UHS rarely yields a uniform risk of collapse of structures. However, suppose the design demand is set to result in a uniform risk of collapse in the structures with different spectral periods at a given site. In that case, there is better control over the performance of structures. A response spectrum constructed using those design demand values as spectral ordinates corresponding to the respective spectral periods is called an RTRS. The iterative adjustment of the design demand values is illustrated in [1].

RTRS depends on the targeted risk value, the type of structure for which the fragility curve is estimated, and the hazard curve. Since the hazard at a location is governed by the geological activities beneath the earth's surface, which humans cannot control. So, the targeted risk value is attained by adjusting the fragility curve of the structure, which is a function of the structure's type and capacity. In this work, masonry structures are the point of focus.

3 Literature Review

Numerous researchers have previously done seismic risk estimation for various structures. Luco et al. [2] drew a comparison between the seismic design maps available during that period and the risk-targeted seismic design maps. The study was done for the Conterminous United States (US). They adjusted the 10th percentile collapse capacity value in the structure's fragility curve to achieve the targeted risk value by keeping the other factors constant until the target collapse risk was achieved. For the first step in iteration, they assumed the 10th percentile collapse value to be equal to the hazard value corresponding to a return period of 2475 years.

A similar study was done by [3] to develop risk-targeted seismic design maps for the regions of France. They found that applying the targeted-risk methodologies for the design hazard estimation did not yield significant variation in the already existing seismic hazard maps, which is opposite to what was found for similar studies in the US. The target annual collapse probability they chose (10^{-5}) was also on the lower side when compared to the target value selected for the US.

A slightly new aspect of the study was explored by [4], where they performed disaggregation of the mean annual collapse risk. They found out that the fragility curve's lower half contributes the highest to the annual collapse risk despite having a low conditional collapse probability. This implies ground motion intensities associated with those fragility values are pivotal in increasing the collapse risk.

The conclusions of [4] also highlight the need for several ground motions for the fragility curve estimation, as they found that these values were very sensitive to slight perturbation if a small number of ground motions were used. Their proposed method suggests building the fragility curve using only two intensity levels. Thus, a greater number of ground motions can be used for the non-linear time history analyses (NLTHA) for these two intensity levels rather than using small number of ground motions for all the intensity levels.

4   Methodology of Collapse Risk Estimation

The estimation of collapse risk associated with a structure requires meticulous effort. It mainly depends on two contributors, i.e., the site hazard curve and the collapse fragility curve of the structure. The methodology is briefed here and can be studied in detail from [1].

- Obtaining the hazard curve for the selected site followed by converting the probabilities in, say, 'T' years (in general, 50 years) to the annual probabilities. Alternatively, the hazard values can only be estimated as annual probabilities at the start.
- Next, the conditional collapse probabilities of the structure, represented by the fragility curve, are estimated. Various methods of non-linear time history analyses, such as incremental dynamic analysis or multiple-stripe analysis, can be used for this purpose.
- The multiplicative integration (convolution in some texts, as in [3, 5]) is done by dividing the hazard and the fragility curves into numerous small segments to increase the accuracy of numerical integration. This yields the annual collapse risk associated with the structure.
- The collapse risk in the structure's lifetime can be obtained using the annual collapse risk and assuming its temporal distribution as the Poisson process.

5   Iterative Procedure to Achieve Targeted Risk

The collapse risk value estimated in the previous section is the initial risk value. Achieving the target collapse risk value requires the above procedure to be performed iteratively until the target is met. The median value of the fragility curve (μ) is slightly changed at the start of each new iteration, keeping all other values the same as the previous iteration. The step at which the criteria of collapse risk are met, and the value of that step is reported as the RTRS value. The following subsections describe how the hazard and the fragility values are estimated in this study for calculating collapse risk, followed by the procedure to develop RTRS.

5.1   Hazard Estimation

Two methods used for estimation of seismic hazard are deterministic seismic hazard assessment (DSHA) and probabilistic seismic hazard assessment (PSHA). DSHA quantifies the seismic hazard for a site by a single value that is obtained from the maximum of all the values due to the nearby occurring faults, thrusts, lineaments, etc. On the other hand, PSHA estimates the hazard in terms of the value and its probability of occurrence. PSHA considers uncertainty in all the three elements of the hazard estimation, namely, source model, attenuation relationships, and the local site effects. Appropriate contribution by each of the models or the attenuation relationships are provided by assigning proper weightages.

For this study, PSHA is used to obtain the hazard at a few selected locations in the northern Himalayas using the computational tool by [6]. The source models and the attenuation relationships are taken from [7], who converted the input data from [8] in the desired format executable by the computational tool OpenQuake. The logic tree for the source model is kept the same as [8]. The attenuation relationships used for the hazard estimation in this study are listed below.

- Kanno et al. [9]
- Atkinson and Boore [10]
- Zhao et al. [11]
- Atkinson and Macias [12]

5.2   Fragility Curves

A fragility curve for a structure is the plot of failure exceedance probability against the demand values for which the structure is being designed. The computation of fragility curves is a computationally extensive task which requires a lot of time. Performing these computations for different buildings at different places takes a lot of work. Instead, a lognormal assumption of the curve by taking the 10th percentile collapse capacity as the demand hazard value corresponding to a return period of 2475 years simplifies the job significantly, as discussed in [2].

The assumption of the 10th percentile collapse capacity is taken from the conclusions of the FEMA-funded ATC-63 project. After testing the structures of several structural and architectural types under maximum considered earthquake (MCE) ground motions, they found that about 10 per cent of the structures collapsed, thus assuming the 10th percentile capacity of the structure to be equal to the MCE level demand value. Although these values are average for structures constructed of different materials, the same has been assumed for the masonry structures in this chapter for simplicity.

5.3   Developing RTRS

As discussed in the preceding section, the 10th percentile collapse value forms the basis for the estimation of the median value for the fragility curve. The multiplicative integration of the hazard and the fragility curves yields the associated annual risk, which is further used to obtain the collapse risk in the structures' lifetime, assuming the temporal distribution as Poisson process. Now, the target risk is achieved by iteratively adjusting the 10th percentile value, which changes the fragility curve's median value until the target is met. Once the targeted risk value is attained, the corresponding 10th percentile value of the fragility curve in the latest step is reported as the risk-targeted value. This calculation is repeated to obtain the risk-targeted values for different spectral periods. The response spectrum constructed using these values is called RTRS for the selected target risk and structure type. For this study region, the targeted risk of 1 per cent in 50 years is chosen.

For the risk calculation while estimating the RTRS values and the collapse risk of the masonry structures in the upcoming section, the intensity measure (IM) used to obtain the hazard curve is the spectral acceleration (Sa) at the structural time period. Calculation of the RTRS requires determining the risk-targeted hazard value at each spectral period, and the hazard curves used for the calculation of those values correspond to the respective spectral period. The engineering demand parameter (s) (EDPs) such as peak inter-storey drift ratio or beam column joint rotation can be used such that it differentiates between the collapse and non-collapse cases clearly.

6   Risk Calculation for Masonry Structures

As discussed in Sect. 1, the structures designed for the code provisions yield non-uniform collapse risks across different spectral periods and locations. The results of collapse risk assessment (shown in Table 2) using the actual masonry structures' fragility curves of [13], supports the claim made above. Several stone

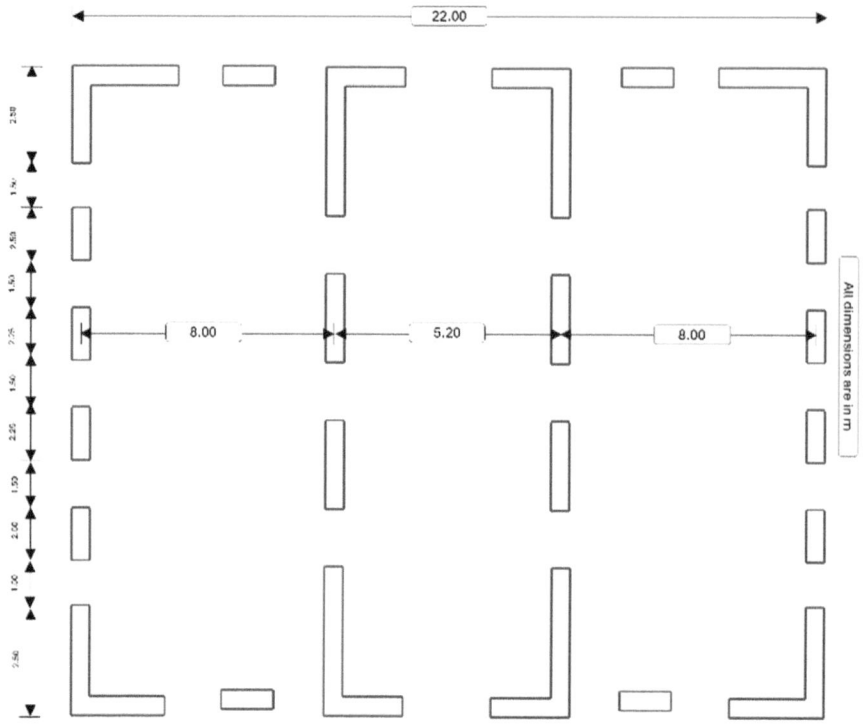

Fig. 1 Plan of the considered stone masonry building

masonry structures comprising different storey heights and aspect ratio of the openings (*h/l*) along with flexible and rigid diaphragms are compared based on associated collapse probability conditioned to MCE level hazard. The schematic plan of the masonry building is shown in Fig. 1 for reference. Also, the risk of collapse of structure due to the exceedance of any hazard level in the structure's lifetime is estimated. The collapse risk related to the exceedance of any hazard level can be obtained using the multiplicative integration of the hazard and fragility curves. However, the conditional collapse probability associated with the exceedance of MCE level hazard can be obtained using the fragility curves of [13]. For the collapse risk estimation from the annual likelihood of collapse in the structures' lifetime, the temporal distribution is assumed to be Poisson process, which is correct for seismic events.

The mean and the standard deviation values of the lognormal fragility curves are taken from Table 5.1 of the work done by [13]. The values in the tables are normalized regarding effective spectral acceleration. So, to obtain the actual median capacity of the structures, the values obtained from the table are multiplied by the effective behaviour factor given in Table 3.1 of [13], which is further multiplied by the DBE to MCE conversion factor of [14] to obtain the median capacity of the structures. A similar conversion is done with the x-axes of the fragility plots from

Fig. 2 Damage Grade 5 of the masonry structure. (Taken from [13])

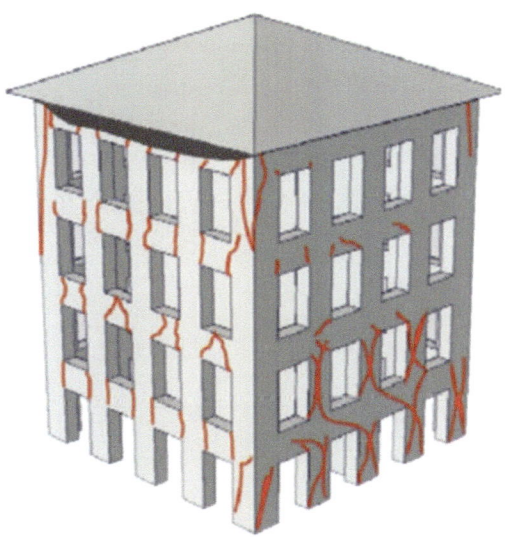

[13], and the conditional collapse probability corresponding to the MCE demand value, estimated using PSHA, is obtained. The standard deviation values are adopted as in Table 5.1 of [13]. These values are also mentioned in Table 2 of this work. The structural period of the structures is obtained using the empirical equation given in [14].

Similarly, the hazard values are obtained, as discussed in Sect. 5.1 of this chapter. The hazard curve for Mandi is considered for this comparison of collapse risks. The parameters of the fragility curves are taken for the damage grade 5 (DG5), as mentioned in [13]. The damage measure considered by them is the percentage of the damaged masonry of the entire surface area of the masonry walls under non-linear biaxial failure criterion. The structure whose 50 per cent or more area of the walls got damaged under the biaxial state of stress is classified as DG5. Figure 2 shows the damage state of the masonry structure listed under DG5. It is assumed that of all the illustrated damage states, DG5 is nearest to the collapse of the structure. The risk values thus obtained for different buildings are tabulated in Table 2 of this work. Due to the absence of certain fragility curves in [13], the conditional collapse probability values associated with the exceedance of MCE level hazard for those buildings are unavailable (N.A.) in Table 2.

7 Results

For the locations tabulated in Table 1, all in the northern Himalayan region, a comparison of the UHS and the RTRS values is made. All the listed locations lie in the Indian states of Himachal Pradesh and Uttarakhand. The upcoming

Table 1 Selected locations for the comparison of UHS and RTRS

Selected locations	Longitude	Latitude
Mandi	76.93°E	31.71°N
Shimla	77.17°E	31.10°N
Roorkee	77.88°E	29.85°N
Dehradun	78.03°E	30.32°N
Uttarkashi	78.44°E	30.73°N
Nainital	79.45°E	29.39°N
Chamoli	79.56°E	30.29°N
Almora	79.64°E	29.59°N
Bageshwar	79.77°E	29.84°N
Pithoragarh	80.22°E	29.58°N

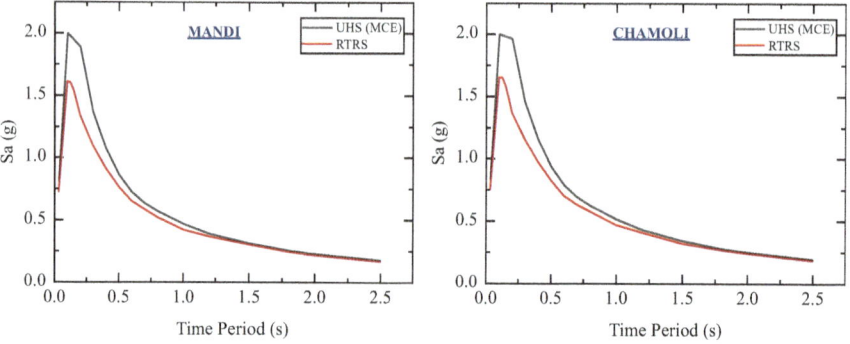

Fig. 3 UHS vs. RTRS for Mandi and Chamoli (Zone V)

subsections discuss the comparison of the UHS and the RTRS values. For the UHS plot, the seismic hazard values correspond to a hazard level with a return period of 2475 years. The RTRS values for the same region are obtained for a chosen targeted collapse risk value of 1% in 50 years and compared.

7.1   Locations of Zone V

According to the Indian code provisions of [14], Zone V is the zone with the highest seismic activity in the country. Among the selected locations for this study, the UHS and RTRS for the regions in Zone V are compared in Figs. 3 and 4. Evidently, for all the areas in Zone V, the design demand value must be decreased to achieve the targeted risk value. The difference between UHS and the RTRS values is significant for the small structural periods. For larger periods, the difference between the values is minimal.

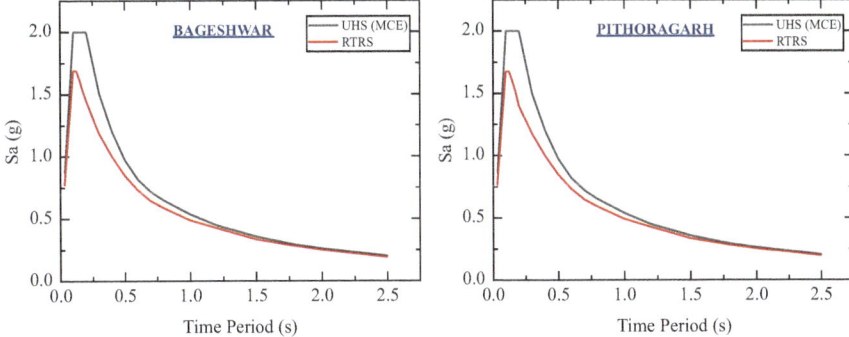

Fig. 4 UHS vs. RTRS for Bageshwar and Pithoragarh (Zone V)

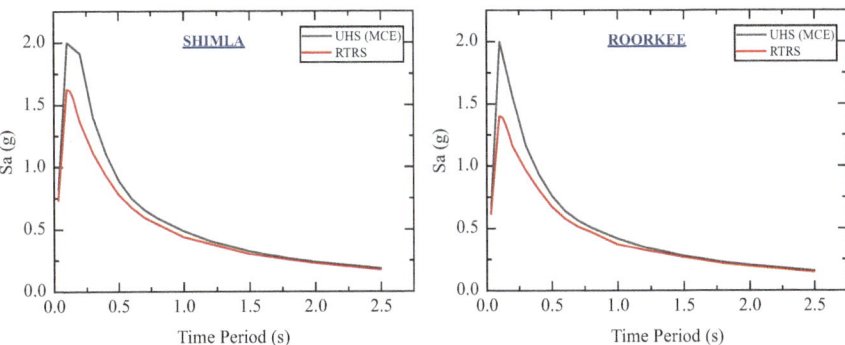

Fig. 5 UHS vs. RTRS for Shimla and Roorkee (Zone IV)

7.2 *Locations of Zone IV*

A comparison of the UHS and RTRS for the regions lying in Zone IV (less seismicity than Zone V), according to [14], is plotted in the Figs. 5, 6, and 7. Similar to the results obtained for Zone V, the RTRS values are also lower than UHS for the MCE level hazard. For the regions of Roorkee and Nainital, the reduction is larger than that of the other regions of the same seismic zone. For the larger structural periods, the trend remains the same as is the case for the locations of Zone V.

7.3 *Collapse Risk Variability*

The collapse probability conditioned to the exceedance of MCE level hazard is larger than 10% for all the cases discussed in Table 2. However, the conditional collapse probabilities suggest that structures with rigid diaphragms are safer than those with flexible diaphragms.

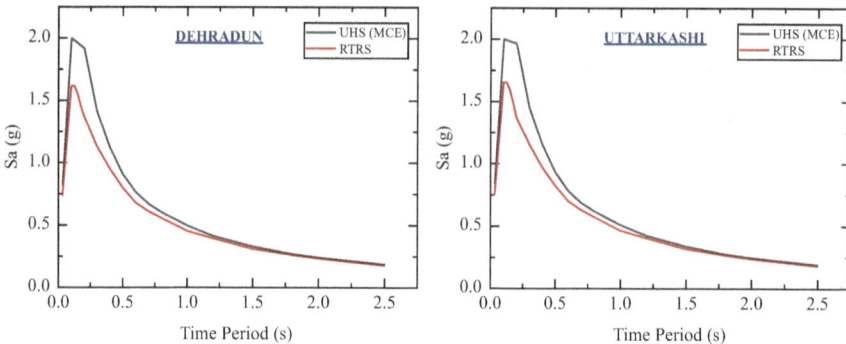

Fig. 6 UHS vs. RTRS for Dehradun and Uttarkashi (Zone IV)

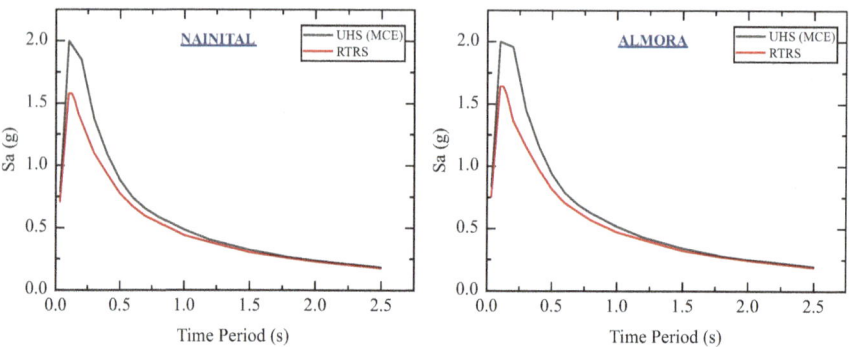

Fig. 7 UHS and RTRS for Nainital and Almora (Zone IV)

For a change in the aspect ratio of the opening (h/l), the collapse risk varies from 2.24% to 3.75% for the 2-storey buildings with flexible diaphragms. This value ranges from 2.16% to 4% for the 6-storey building. The collapse risk varies from a minimum of 2.16% to a maximum of 3.55% with number of stories in buildings with flexible diaphragms and an opening aspect ratio less than unity. These values vary significantly when the structures with different diaphragm types, i.e., flexible or rigid, are studied. For the structures with an opening aspect ratio greater than unity, the collapse risk for the 2-storey building with flexible diaphragms is 3.75%. This value is 1.53% for the structure with rigid diaphragms.

8   Conclusions

For all the locations of the study region, RTRS values are lower than UHS values. This is because the current code provisions for determining design demand are conservative, i.e., current provisions overestimate the design hazard values.

Table 2 Risk of exceedance of damage state for different masonry structures

Diaphragm	h/l	Fragility parameter	2-storey	4-storey	6-storey
Flexible	≥1.0	μ	1.86	1.26	0.90
		β	0.71	0.75	0.69
		Collapse prob. conditioned to (MCE) (%)	56	60	60
		Collapse risk (%)	3.75	4.00	4.00
	<1.0	μ	2.40	1.32	1.38
		β	0.66	0.70	0.70
		Collapse prob. conditioned to (MCE) (%)	N.A.	60	N.A.
		Collapse risk (%)	2.24	3.55	2.16
Rigid	≥1.0	μ	2.80	2.80	2.08
		β	0.63	0.63	0.65
		Collapse prob. conditioned to (MCE) (%)	20	15	20
		Collapse risk (%)	1.53	0.79	0.96
	<1.0	μ	4.48	3.44	2.24
		β	0.61	0.64	0.64
		Collapse prob. conditioned to (MCE) (%)	N.A.	20	N.A.
		Collapse risk (%)	0.40	0.48	0.81

Fragility parameters and specifications obtained from [13]

However, this practice does not ensure the level of safety against collapse. On the other hand, designing for RTRS provides a targeted safety level with optimal cost. This way, RTRS also contributes to the sustainability of the masonry structures. Knowing the collapse risk beforehand estimates the minimum performance level the structure will likely provide at the exceedance of a particular hazard level.

The conditional collapse probability (shown in Table 2) being higher than the values (10 per cent) obtained from the conclusions of the ATC-63 Project is because the masonry structures discussed in this study are not code-compliant, hence show a higher risk of collapse. Also, variability in the collapse risks obtained for the actual masonry structures highlights the need for RTRS to estimate the design demand, which ensures that the structures are designed to achieve a uniform (targeted) risk of collapse in their lifetime. Designing for RTRS will help in better planning and emergency preparedness in case of the occurrence and the exceedance of a hazard level.

9 Limitations of the Study

Although it is attempted to make this study as general as possible, due to the assumptions taken, this study has several limitations, as discussed below.

- The assumption of the 10th percentile collapse value of the fragility curve as MCE demand value for masonry structures is only sometimes valid, as the value was an average of the values of the structures of all material types.

- For simplicity, the local site conditions were assumed to be the same for the entire region, which is a possible source of error.
- The standard deviation values selected for the lognormal fragility curves are based on previous studies. Application of the values determined for the structures constructed of a particular material to the same type of structure will yield better results.
- Owing to the complex nature of the procedure, its implementation in practice requires significant simplification and adjustments to existing codes and standards.

References

1. Shastri, R., Singh, Y., Das, J.: Risk targeted seismic hazard assessment in Uttarakhand Himalayas. In: Symposium in Earthquake Engineering, pp. 247–257. Springer (2022)
2. Luco, N., Ellingwood, B.R., Hamburger, R.O., Hooper, J.D., Kimball, J.K., Kircher, C.A.: Risk-Targeted Versus Current Seismic Design Maps for the Conterminous United States. Structural Engineers Association of California (2007)
3. Douglas, J., Ulrich, T., Negulescu, C.: Risk-targeted seismic design maps for mainland France. Nat. Hazards. **65**, 1999–2013 (2013)
4. Eads, L., Miranda, E., Krawinkler, H., Lignos, D.: An efficient method for estimating the collapse risk of structures in seismic regions. Earthquake Engineering Structural Dynamics. **42**, 25–41 (2013)
5. Hirata, K., Nakajima, M., Ootori, Y.: Proposal of a simplified method for estimating evaluation of structures seismic risk of structures. Proceedings of 15th world conference on earthquake engineering, Lisboa (2012)
6. Pagani, M., Monelli, D., Weatherill, G., Danciu, L., Crowley, H., Silva, V., et al.: OpenQuake engine: an open hazard (and risk) software for the global earthquake model. Seismol. Res. Lett. **85**, 692–702 (2014)
7. Ackerley, N.: An Open Model for Probabilistic Seismic Hazard Assessment on the Indian Subcontinent. Istituto Universitario di Studi Superiori, Pavia (2016)
8. Nath, S., Thingbaijam, K.: Probabilistic seismic hazard assessment of India. Seismol. Res. Lett. **83**, 135–149 (2012)
9. Kanno, T., Narita, A., Morikawa, N., Fujiwara, H., Fukushima, Y.: A new attenuation relation for strong ground motion in Japan based on recorded data. Bull. Seismol. Soc. Am. **96**, 879–897 (2006)
10. Atkinson, G.M., Boore, D.M.: Empirical ground-motion relations for subduction-zone earthquakes and their application to Cascadia and other regions. Bull. Seismol. Soc. Am. **93**, 1703–1729 (2003)
11. Zhao, J.X., Zhang, J., Asano, A., Ohno, Y., Oouchi, T., Takahashi, T., et al.: Attenuation relations of strong ground motion in Japan using site classification based on predominant period. Bull. Seismol. Soc. Am. **96**, 898–913 (2006)
12. Atkinson, G.M., Macias, M.: Predicted ground motions for great interface earthquakes in the Cascadia subduction zone. Bull. Seismol. Soc. Am. **99**, 1552–1578 (2009)
13. Karantoni, F., Lyrantzaki, F., Tsionis, G., Fardis, M.: Seismic fragility functions of stone masonry buildings. In: Proceedings of the 15th World Conference on Earthquake Engineering, pp. 24–28 (2012)
14. IS 1893 (Part 1): 2016. Criteria for earthquake resistant design of structures. Bureau of Indian Standards.Manak Bhavan, 9 Bahadur shah zafar marg, New Delhi - 110002 (2016)

Machine Learning-Aided Prediction of Seismic Response of RC Bridge Piers Exposed to Chloride-Induced Corrosion

Pooria Poorahad Anzabi, Mahmoud R. Shiravand, and Shima Mahboubi

Contents

1 Introduction

The long-term performance of reinforced concrete (RC) bridges is significantly affected by corrosion, especially in aggressive chloride environments. Corrosion may cause serviceability issues, such as staining, cracking, and spalling of concrete, and seismic performance issues, namely strength and ductility decay or confinement degradation. These issues are amplified given that bridges play a key role in transport infrastructure networks and their functionality is vital for the operation of the whole system [1, 2]. The impact of corrosion on the seismic behavior of RC bridges is experimentally investigated through an accelerated corrosion procedure. Experimental testing is cost-intensive, and due to the time-dependent nature of corrosion, the process is accelerated in the laboratory. In recent years, numerical methods are employed in many researches to simulate the corrosion in RC bridges and take its

P. Poorahad Anzabi · M. R. Shiravand (✉) · S. Mahboubi
Faculty of Civil, Water, and Environmental Engineering, Shahid Beheshti University,
Tehran, Iran
e-mail: m_shiravand@sbu.ac.ir

© The Author(s) 2025
M. Kioumarsi, B. Shafei (eds.), *The 1st International Conference on Net-Zero Built
Environment*, Lecture Notes in Civil Engineering 237,
https://doi.org/10.1007/978-3-031-69626-8_118

time-dependent nature into account. The accuracy of the models is highly dependent on the accuracy of the material models [3].

With the development of big data techniques, machine learning (ML) has shown significant advantages in prediction and categorization in the field of engineering. ML is the specific area of artificial intelligence that allows computers to learn from data and solve a given task. Salami et al. [4] compared the performances of different ML techniques to optimize the predictive modeling of corrosion initiation time. Zhu et al. [5] used artificial neural network in conjunction with Kohonen self-organized mapping to model the chloride threshold in RC. Most recently, Xu et al. [6] estimated multilevel capacity of bridges with the aid of the ML. The trained ML models could also accurately identify the flexure and shear failure modes. In another recent work, Xi et al. [7] conducted a reliability analysis on the basis of ML considering realistic nonuniform corrosion.

Chloride contamination is the primary cause of corrosion in RC bridges. Researchers have studied the effect of corrosion on RC bridges through experimental and numerical methods. Corrosion is a time-dependent phenomenon. Thus, its process is accelerated in the laboratory to be able to take its effect into account, which is costly and accompanied by some amount of error. However, numerical methods make it possible to implement nonlinear dynamic analysis and derive seismic fragility curves of corroded RC bridge piers over time. With the evolution of computational algorithms, ML-aided numerical methods are developed to recognize or predict patterns in data. In this paper, numerical models of RC bridge piers are built in OpenSees [8] platform. The geometric, loading, and material properties of the piers are randomly chosen. Each model is nonlinearly analyzed under random ground motion records scaled to design-based earthquake (DBE) and maximum considered earthquake (MCE) spectra. Prior to the analyses, corrosion is allocated to the models by degrading the mechanical properties of steel and concrete. The percentage of the corrosion is considered as a stochastic parameter. Maximum drift ratios are stored at each analysis to build the database. Using the created database, different ML algorithms are compared to find the most accurate one. R-squared (R^2), mean absolute error (MAE), mean squared error (MSE), and root mean squared error (RMSE) metrics are considered as the criteria for tracking the efficiency of the models. Bayesian search is employed as the algorithm for tuning the hyperparameters.

2 Corrosion-Induced Degradation

Morphology of corrosion in RC structures is categorized into two types: (i) uniform and (ii) pitting. Uniform corrosion is caused by concrete carbonation while penetration of chlorides induces pitting corrosion. Figure 1a and b depict schematic illustrations of uniform and pitting corrosions. Both cases are characterized by a reduction in the mechanical properties of the steel. Yield and ultimate strengths, elastic modulus, and ultimate strain of reinforcing steels are proven to be reduced

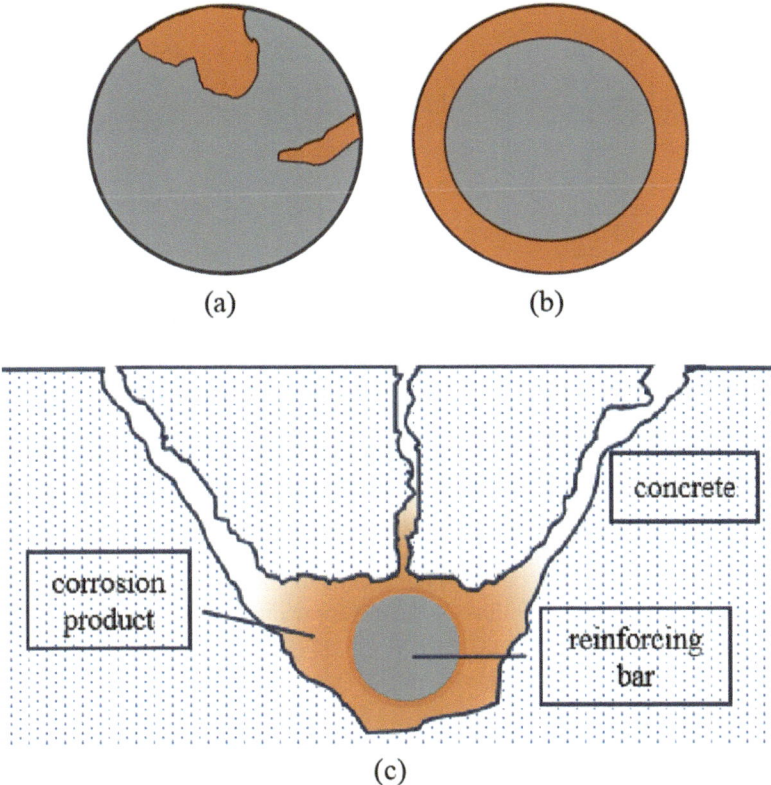

Fig. 1 Corrosion of reinforcing steel: (**a**) pitting corrosion, (**b**) uniform corrosion, (**c**) cracking, and spalling of concrete due to the steel corrosion

when corroded. For analytical and numerical purposes, decay equations have been proposed in various researches [9–11] that take the corrosion-induced degradation into account, as

$$f_y = (1.0 - \alpha_y Q_{corr}) f_{y0} \tag{1}$$

$$f_u = (1.0 - \alpha_u Q_{corr}) f_{u0} \tag{2}$$

$$E_s = (1.0 - \alpha_E Q_{corr}) E_{s0} \tag{3}$$

$$\varepsilon_u = (1.0 - \alpha_\varepsilon Q_{corr}) \varepsilon_{u0} \tag{4}$$

where f_y, f_u, E_s, and ε_u are residual yield and ultimate strengths, elastic modulus, and ultimate strain of the corroded steel, respectively, f_{y0}, f_{u0}, E_{s0}, and ε_{u0} are initial yield and residual strengths, elastic modulus, and ultimate strain of the intact steel, respectively, Q_{corr} is the corrosion percentage, and α are corresponding empirical coefficients for which different values are reported. In the present study, α_y and α_u

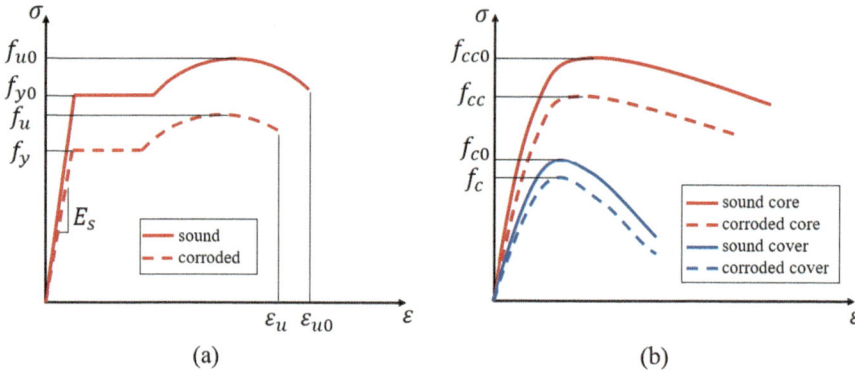

Fig. 2 Mechanical properties of corroded components: (**a**) reinforcing steel and (**b**) core and cover concrete. (Adapted from Xu et al. [6])

are determined to be 0.5 from the work of Du et al. [9], $\alpha_E = 0.75$ is adopted from the study by Lee and Cho [10], and α_ε is taken as 3.0 according to Cairns et al. [11].

During the corrosion process, rusted steel undergoes volumetric expansion leading to cracking and even spalling of the cover concrete, which in turn reduces the compressive strength of the concrete as shown in Fig. 1c. Coronelli and Gambarova [12] proposed the following equation to calculate the compressive strength of the corroded concrete:

$$f_c = \frac{f_{c0}}{1 + K \frac{\varepsilon_1}{\varepsilon_{co}}} \tag{5}$$

where f_c is the residual compressive strength of concrete after corrosion, f_{c0} is the compressive strength of sound concrete, K is the bar roughness and diameter coefficient that has been suggested to be taken as 0.1 for medium-diameter ribbed longitudinal bars [13], ε_{co} is the strain at peak compressive strain, and $\varepsilon_1 = 2\pi(v_{rs} - 1)Xn_{bars}/p_0$ is the smeared tensile strain in the cracked concrete in which $v_{rs} = 2$ is ratio of volumetric expansion of oxides with respect to sound material [14], X is the depth of the corrosion attack, n_{bars} is the number of the compressive bars, and p_0 is the perimeter of column section. The stress–strain relationships of sound and corroded reinforcing steel and concrete are illustrated in Fig. 2.

3   Numerical Method and Verification

To have an accurate simulation of an RC bridge pier, it is needed to take three main issues into account [15]; (i) correct simulation of plasticity in the pier, which matches with the actual behavior, (ii) considering buckling and low-cycle fatigue fracture in the nonlinear behavior of reinforcements, and (iii) considering bar slippage (strain

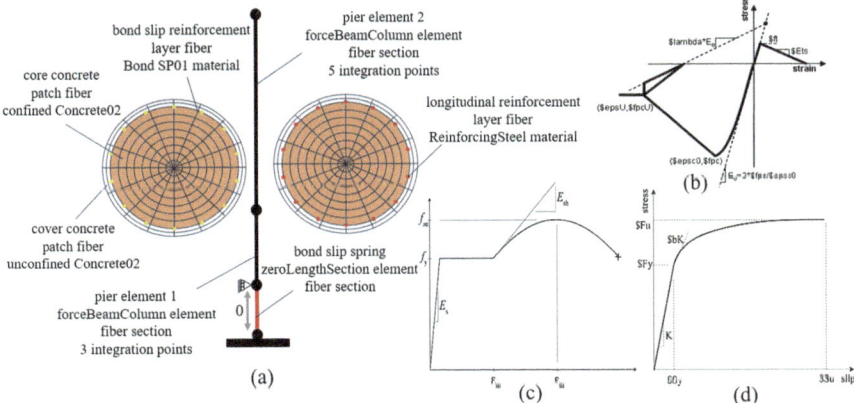

Fig. 3 Numerical modeling method: (**a**) OpenSees model, (**b**) Concrete02 material, (**c**) ReinforcingSteel material, and (**d**) Bond SP01 material

penetration). Hence, nonlinear analyses of RC bridge piers are conducted in OpenSees platform, in which it is possible to accurately consider the aforementioned behaviors.

Nonlinearities may develop anywhere along the pier during an earthquake loading Therefore, a force-based beam-column element is employed in this study, which has distributed plasticity along the element. The pier is modeled with two elements. The first element's height is $6L_b$ in which L_b is buckling length of longitudinal bar. It is modeled with three integration points to ensure that the distance to the first integration point is equal to the buckling length of longitudinal bar as shown in Fig. 3a. The remaining height of the pier is modeled with another element with 5 integration points. The fiber section allocated to both elements consists of cover concrete, core concrete, and reinforcement fibers. Concrete02 material (Fig. 3b) is adopted for cover and core concrete fibers. The confinement effect is accounted for using the method proposed by Chang and Mander [16]. Reinforcing steel material is employed for reinforcement fiber, which is capable of modeling compression buckling and low-cycle fatigue effects as shown in Fig. 3c.

At intersections such as column-to-footing connections, the bond-slip associated with strain penetration typically occurs along a portion of the anchorage length. To account for this effect, a ZeroLengthSection element with a fiber section consisting of cover and core concrete fibers and reinforcement fiber with Bond SP01 material (Fig. 3d) is placed at the intersection between the column and footing. The ZeroLength Section element used at the bottom of piers is depicted in Fig. 1a with yellow layer fibers.

In addition to the modeling issues of the RC piers, mentioned in the previous paragraphs, substructure mass effects are usually ignored during the design and analysis procedures. It means that the seismic performance of the pier is not included in the evaluation of the seismic performance of the whole pier system

[17]. Therefore, the contribution of the substructure mass is considered in the modeling of the RC piers of the present paper by allocating the pier mass using the element mass density property in the force-based beam-column elements. Furthermore, a lumped mass is applied to the topmost node of the pier to account for the deck mass. The conventional damping value of 5% is assigned to the piers through Rayleigh damping model [18].

The RC bridge pier models of the present study are adopted from antecedent research by Su et al. [19]. In the mentioned research, ten specimens with different reinforcement grades and concrete strength were cyclically loaded, half of which had circular a cross-section and the other half had rectangular a cross-section. The circular specimens are selected, numerically modeled using the described method, and quasi-static analyses are conducted according to the specified loading protocol. Prior to the lateral loading, gravity loading is applied to the models using load control to the specified values. Base shear and top node displacement values are stored and compared to the results from the mentioned experiment in Fig. 4. It is clearly seen that the modeling method is capable of simulating the nonlinear behavior of RC bridge piers with reasonable accuracy.

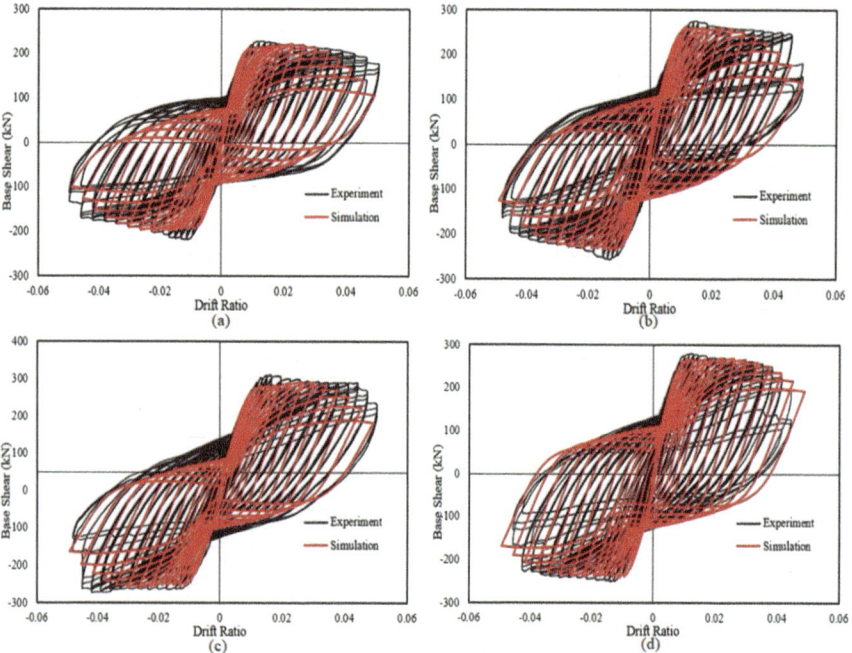

Fig. 4 Comparison of the results from the experiment and the simulation: (**a**) C-LM-C40, (**b**) C-MM-C40, (**c**) C-HM-C40, and (**d**) C-MM-C60

4 Results

The corrosion effects are applied to the circular RC pier models using the degradation equations. The parameter Q_{corr} is considered to be a stochastic parameter. Therefore, Eqs. (1)–(4) get a random nature due to their dependency on Q_{corr}. In this way, the variable nature of environmental conditions of different regions is taken into account, improving the generalization of this study. The generality of the study is further boosted by accounting for different pier designs through giving a random nature to the aspect ratio of the pier, H/B, initial diameter of the longitudinal bars, d_{b0}, and axial load applied to the pier, W. The main properties of the concrete and the steel, namely initial compressive strength of concrete (f_{c0}) and initial yield strength of steel (f_{y0}), are also randomly selected to consider the variable nature of the materials. Table 1 summarizes the stochastic parameters considered for this study. At each analysis, a random value is assigned to each stochastic parameter according to their respective distributions, and the model is analyzed under a randomly selected ground motion adopted from the records introduced in FEMA P695 [20]. All the ground motion sets including far-field and near-field records are chosen, constituting 100 ground motion records. The ground motions are scaled to the DBE and MCE spectra according to provisions of ASCE/SEI 7-16 [21].

The ML algorithms chosen for this study are in the supervised learning class. Supervised learning is used when the labeled data is available. Data collection is one of the main challenges in implementing artificial intelligence processes, due to the lack of relevant, accurate, and high-quality data. In the present research, the labeled data required for the database is created through a randomized experiment with the stochastic parameters tabulated in Table 1. This way of data creation is computationally extensive and time-consuming. However, the result is a well-structured and accurate database that leads to high-performance predictive models. To create the database necessary for the ML algorithms, 1000 analyses are conducted for each seismic level, constituting 2000 data for the database. The entire dataset is randomly divided into a training set (70%) and a testing set (30%). The workflow of the ML and the hyperparameter optimization is depicted in Fig. 5.

The codes of linear regression (LR), decision tree (DT), random forest (RF), and XGBoost algorithms are developed to build the predictive models. The training set is used for data fitting of the models, and the testing set is used for validating the

Table 1 Stochastic parameters considered in the present study

Parameter	Distribution	Mean value	Coefficient of variation
Q_{corr}	Lognormal	0.15	0.2
H/B	Uniform	8	0.3
d_{b0}	Normal	35.5 (mm)	0.05
W	Lognormal	500 (kN)	0.1
f_{c0}	Normal	43 (MPa)	0.13
f_{y0}	Normal	450 (MPa)	0.12
Ground motion	Uniform	50.5	0.57

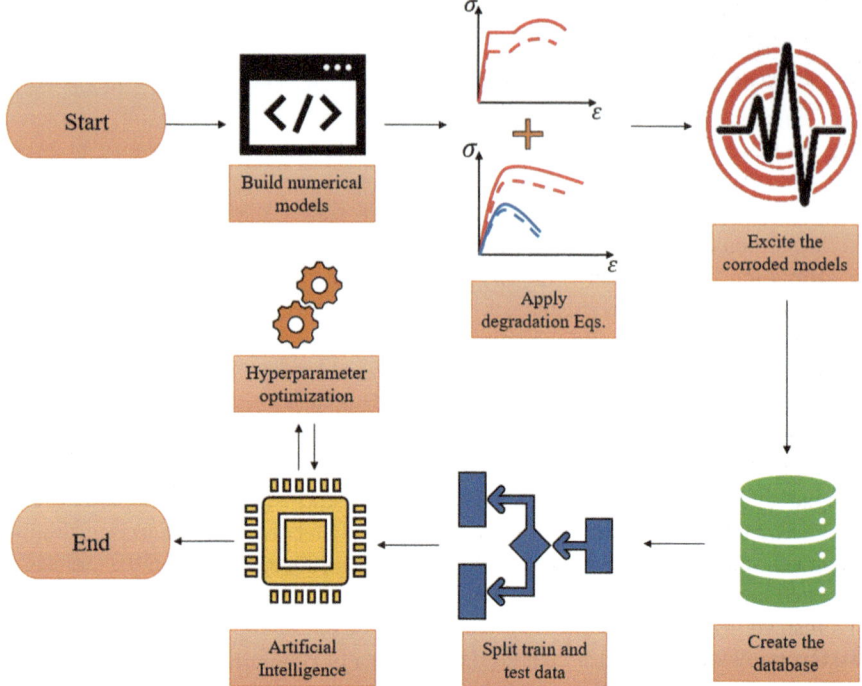

Fig. 5 Flowchart of the ML and hyperparameter optimization

models. The parameters tabulated in Table 1 are taken as feature variables, and the maximum drift experienced during the excitation, $\delta_{\max}$, is the output taken as the target variable in this study. The maximum drift is calculated as the ratio of the maximum lateral displacement to the pier height. All the dataset components are numeric values except for the seismic level which is a categorial data, consisting of two categories; (i) DBE and (ii) MCE. One-hot encoding technique is used to convert categorical variables to numeric values by creating a binary column for each category, indicating whether the category is present or not.

Every algorithm comes with additional hyperparameters that need to be config-ured to avoid problems such as overfitting, underfitting, convergence, etc. Hyperparameter optimization (HPO) algorithms are proposed to tackle these prob-lems. The goal is to find $x^* \in \text{argmin}_{x \in X} f(x)$, in which $f : X \to \mathbb{R}$ is the function that maps from the hyperparameter space $x \in X$, and x^* is the set of hyperparameters with the best performance. The most commonly used HPO algorithms are grid search, randomized search, and Bayesian search. While grid search investigates every possible combination of hyperparameters, randomized search tests a random com-bination of hyperparameters, and thus it is less time-consuming. However, these two algorithms do not consider the past results. Bayesian search improves the perfor-mance of the optimization by considering the previously selected hyperparameters

Table 2 The explored hyperparameters and their ranges for ML algorithms

Algorithm	Hyperparameters	Range
LR	N/A	N/A
DT	max_depth	[1 ... 21]
	min_samples_split	[2 ... 11]
	min_samples_leaf	[3 ... 26]
RF	n_estimator	[25 ... 100]
	max_features	[1 ... 21]
	max_depth	[2 ... 11]
	min_samples_split	[3 ... 26]
XGBoost	eta	[0.01, 0.2]
	min_child_weight	[1 ... 6]
	max_depth	[2 ... 11]
	min_samples_split	[3 ... 26]

when determining the next set. Hence, Bayesian search is utilized in this study. The explored hyperparameters and their ranges in the HPO process are summarized in Table 2.

The hyperparameters of LR algorithm are not too much in number, and those that it has cannot be tuned. Consequently, HPO is not applied to LR. In tree-based algorithms, max_depth (maximum depth) is the most important hyperparameter that indicates how big the tree can grow. Complex patterns cannot be learned provided that max_depth is set to be too low while the models with large max_depth value are prone to overfitting. Thus, HPO is needed for max_depth. The two remaining hyperparameters, min_samples_split (minimal number of samples required to divide a node) and min_samples_leaf (minimum number of samples required to form a leaf node), are also decisive factors in avoiding the overfitting issue. In RF algorithm, along with max_depth and min_samples_split, two more hyperparameters, n_estimator (the number of trees to consider) and max_features (maximum number of features provided to each tree), are explored. If the n_estimator is too low, the performance of the model will be weak. And if it is a very large number, it will increase the computational complexity of the model without any increase in the performance of the model. The same trend is true for max_features. In XGBoost, the conservativeness of the model can be adjusted by the eta (features weight) and min_child_weight (minimum sum of instance weight needed in a child).

The observed maximum drifts and the predicted ones for both training and testing sets are plotted in Fig. 6. Lines corresponding to absolute errors of ±0.1 and ±0.2 are also plotted. The scatters illustrated in Fig. 6 are accompanied by histograms of the observed and predicted values on the plot margins. It is seen that LR and DT models have a considerable number of data points with absolute errors that exceed ±0.01 and ±0.02 while the number of the same data points in RF and XGBoost models is much lesser, showing the capability of these algorithms in training accurate models.

The accuracy and performance of the models are compared in Fig. 7. The predictive model developed by the LR algorithm provides the lowest accuracy in the testing dataset with $R^2 = 0.53$, MAE = 0.0026, MSE = 1.4×10^{-5}, and

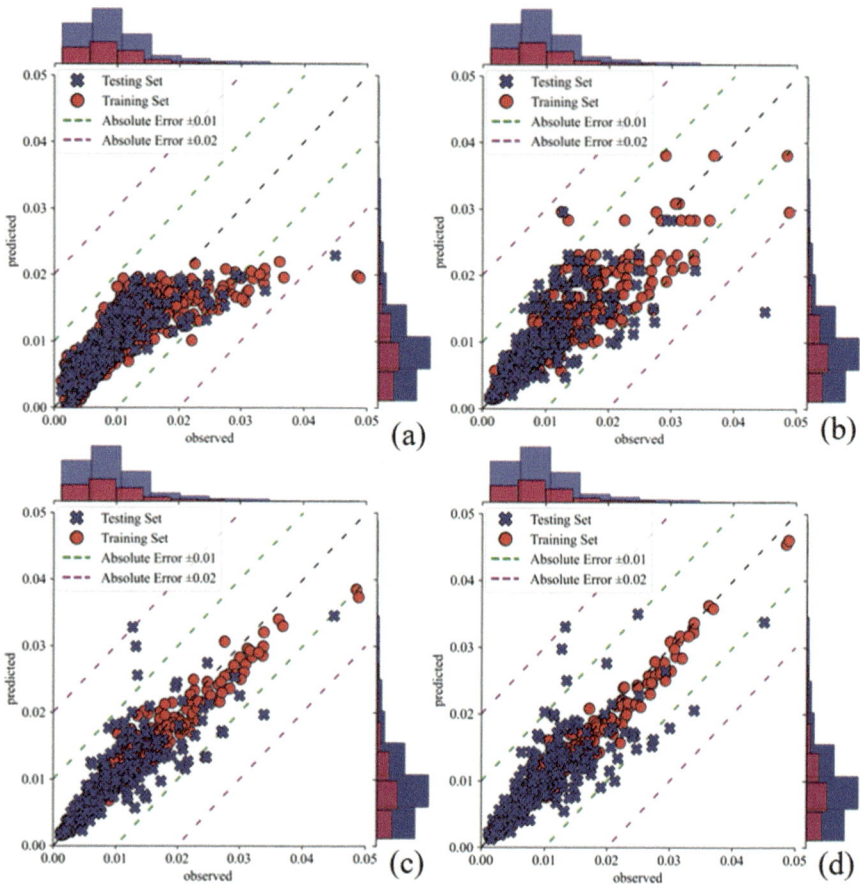

Fig. 6 Observed-predicted scatters of different ML models: (**a**) LR, (**b**) DT, (**c**) RF, and (**d**) XGBoost

RMSE $= 0.0036$. The weakness of LR in measurement of maximum drift of corroded RC bridge piers indicates the nonlinear nature of the database. Tree-based models yield more accurate predictions. Among them, DT with $R^2 = 0.7$, MAE $= 0.0024$, MSE $= 1.1 \times 10^{-5}$, and RMSE $= 0.0033$ is the least accurate one. However, according to its performance seen in Figs. 6 and 7, the HPO has helped avoid overfitting of the model. The accuracy of the predictions is further increased by using the RF algorithm. The RF model has $R^2 = 0.73$, MAE $= 0.002$, MSE $= 8 \times 10^{-6}$, and RMSE $= 0.003$. According to the metrics in Fig. 7, XGBoost algorithm with $R^2 = 0.8$, MAE $= 0.0015$, MSE $= 5 \times 10^{-6}$, and RMSE $= 0.0028$ yielded the most accurate predictions among all other models. According to Fig. 6, all the predicted data points from XGBoost algorithm have absolute errors of less than ± 0.02.

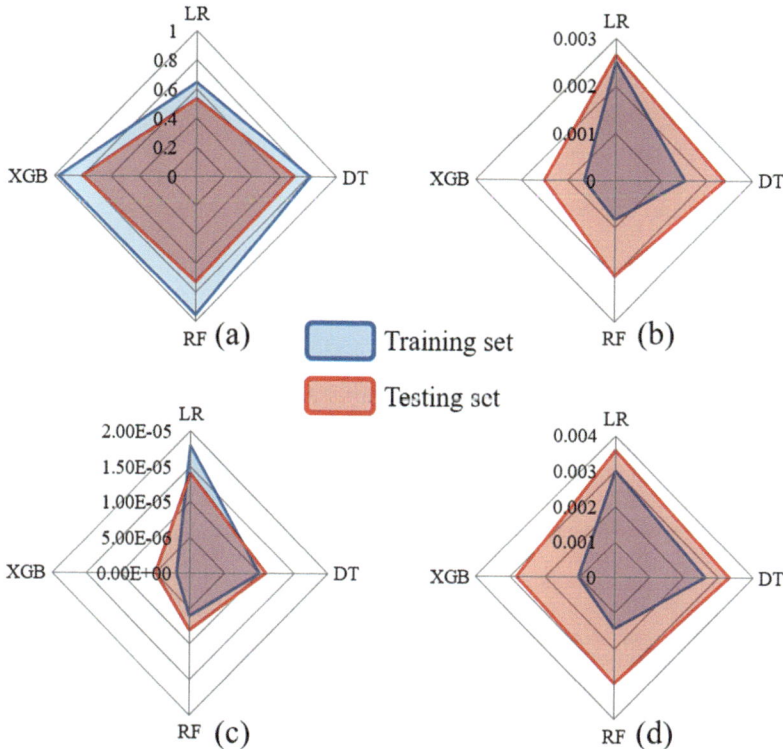

Fig. 7 Performance comparison of different ML models: (**a**) R^2, (**b**) MAE, (**c**) MSE, and (**d**) RMSE

5 Conclusion

Different ML algorithms are utilized in the present study to predict the maximum drift of corroded RC bridge piers. The database is built by developing numerical models of bridge piers whose mechanical properties have deteriorated due to corrosion. To improve the generality of the study, the key parameters affecting geometry, design, material, and environmental conditions are randomly chosen. The models are trained with 70% of the database and are tested with the remaining 30%. Bayesian search algorithm is also employed to tune the hyperparameters of the ML algorithms to obtain the best performance of them. The major findings of this study are as follows:

- Due to the nonlinear relationship between the features and the target, the LR model showed a weak performance in predicting the maximum drift of corroded RC bridge piers.
- Tree-based models, namely DT and RF, provided more accurate predictions; about 35% more accurate than the linear model. It indicates that tree-based algorithms map nonlinear relationships quite well.

- Among tree-based models, RF performed slightly better. The reason lies in the ability of the RF algorithm to avoide the overfitting of the model.
- The closeness of the performances of the DT model and the RF one indicates the significance of the HPO in avoiding the overfitting of the model.
- The comparison of the metrics used in this study exhibited that XGBoost model has the highest accuracy. The accuracy of this model is 50% and 15% higher than the linear and tree-based models, respectively.

References

1. Mahboubi, S., Shiravand, M.: Seismic evaluation of bridge bearings based on damage index. Bull. Earthq. Eng. **17**, 4269–4297 (2019)
2. Mahboubi, S., Shiravand, M., Shid, G., Kioumarsi, M.: Probabilistic assessment of RC piers considering vertical seismic excitation based on damage indices. In: International RILEM Conference on Synergising Expertise towards Sustainability and Robustness of CBMs and Concrete Structures, pp. 828–839. Springer (2023)
3. Mahboubi, S., Kioumarsi, M.: Damage assessment of RC bridges considering joint impact of corrosion and seismic loads: a systematic literature review. Constr. Build. Mater. **295**, 123662 (2021)
4. Salami, B.A., Rahman, S.M., Oyehan, T.A., Maslehuddin, M., Al Dulaijan, S.U.: Ensemble machine learning model for corrosion initiation time estimation of embedded steel reinforced self-compacting concrete. Measurement. **165**, 108141 (2020)
5. Zhu, Y., Macdonald, D.D., Qiu, J., Urquidi-Macdonald, M.: Corrosion of rebar in concrete. Part III: artificial neural network analysis of chloride threshold data. Corros. Sci. **185**, 109438 (2021)
6. Xu, B., Wang, X., Yang, C.-S.W., Li, Y.: Machine learning–aided rapid estimation of multilevel capacity of flexure-identified circular concrete bridge columns with corroded reinforcement. J. Struct. Eng. **150**(3), 04024002 (2024)
7. Xi, X., Yin, Z., Yang, S., Li, C.-Q.: Machine learning–based reliability analysis of structural concrete cracking considering realistic nonuniform corrosion development. J. Struct. Eng. **150**(1), 04023207 (2024)
8. McKenna, F.: OpenSees: a framework for earthquake engineering simulation. Comput. Sci. Eng. **13**(4), 58–66 (2011)
9. Du, Y., Clark, L., Chan, A.: Residual capacity of corroded reinforcing bars. Mag. Concr. Res. **57**(3), 135–147 (2005)
10. Lee, H.-S., Cho, Y.-S.: Evaluation of the mechanical properties of steel reinforcement embedded in concrete specimen as a function of the degree of reinforcement corrosion. Int. J. Fract. **157**, 81–88 (2009)
11. Cairns, J., Plizzari, G.A., Du, Y., Law, D.W., Franzoni, C.: Mechanical properties of corrosion-damaged reinforcement. ACI Mater. J. **102**(4), 256–264 (2005)
12. Coronelli, D., Gambarova, P.: Structural assessment of corroded reinforced concrete beams: modeling guidelines. J. Struct. Eng. **130**(8), 1214–1224 (2004)
13. Capé, M.: Residual Service-Life Assessment of Existing R/C Structures. Chalmers University of Technology/Milan University of Technology, Italy Erasmus Program, Goteborg/Milan (1999)
14. Molina, F., Alonso, C., Andrade, C.: Cover cracking as a function of rebar corrosion: Part 2—Numerical model. Mater. Struct. **26**, 532–548 (1993)
15. Rassoulpour, S., Shiravand, M., Safi, M.: Proposed seismic-resistant dual system for continuous-span concrete bridges using self-centering cores. Eng. Struct. **274**, 115181 (2023)

16. Chang, G., Mander, J.B.: Seismic Energy Based Fatigue Damage Analysis of Bridge Columns: Part I-Evaluation of Seismic Capacity. National Center for Earthquake Engineering Research, Buffalo (1994)
17. Rasouli, M., Shiravand, M.R., Ardakani, R.R.: Substructure mass participation effect on the performance-based seismic design method for isolated bridges. J. Bridg. Eng. **28**(12), 04023095 (2023)
18. Poorahad Anzabi, P., Shiravand, M.R.: Segments arrangement effect on improvement of self-centering precast post-tensioned segmental piers seismic performance. Struct. Concr. **25**(1), 185–206 (2024)
19. Su, J., Wang, J., Li, Z., Liang, X.: Effect of reinforcement grade and concrete strength on seismic performance of reinforced concrete bridge piers. Eng. Struct. **198**, 109512 (2019)
20. Applied Technology Council: Quantification of Building Seismic Performance Factors. US Department of Homeland Security, FEMA, Washington, DC (2009)
21. American Society of Civil Engineers: Minimum Design Loads and Associated Criteria for Buildings and Other Structures. American Society of Civil Engineers, Reston (2016)

Analytical and Numerical Approaches in Predicting the Flexural Behaviour of Reinforced Concrete Beams

Chinyere O. Nwankwo ⓘ and Jeffrey Mahachi ⓘ

Contents

1 Introduction

Understanding material behaviour lies at the core of structural engineering. Engineers seek to predict how different types of materials react to different kinds of load actions. The beam element is an essential structural element that can experience compression, flexure (bending), torsion, tension, or a combination of these actions. Understanding how a reinforced concrete (RC) beam resists bending loads serves as the fundamental principle for designing a plethora of structures. When a load is imposed on an RC beam, it undergoes bending, experiencing both tension and

C. O. Nwankwo (✉)
Center of Applied Research and Innovation in the Built Environment (CARINBE),
Faculty of Engineering and the Built Environment, University of Johannesburg, Johannesburg,
South Africa

J. Mahachi
Department of Civil Engineering Technology, University of Johannesburg, Johannesburg,
South Africa

© The Author(s) 2025
M. Kioumarsi, B. Shafei (eds.), *The 1st International Conference on Net-Zero Built
Environment*, Lecture Notes in Civil Engineering 237,
https://doi.org/10.1007/978-3-031-69626-8_119

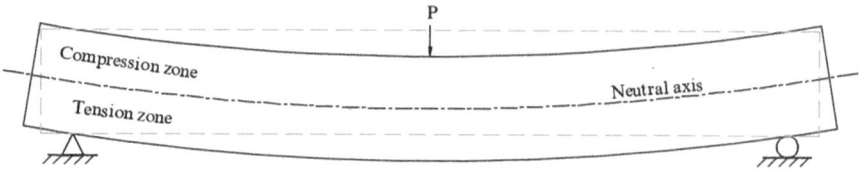

Fig. 1 RC beam in flexure

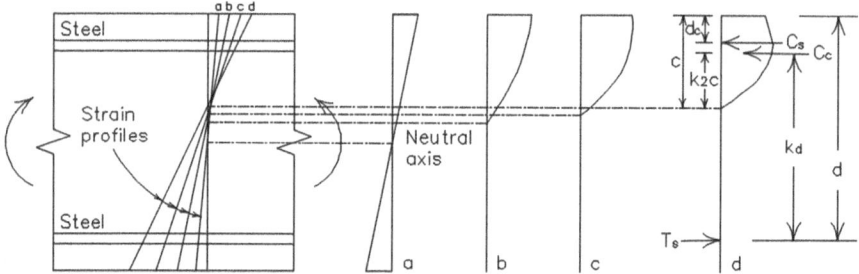

Fig. 2 Stress and strain distribution in an RC section under bending

compression. The distribution of these forces within the beam is contingent on the direction of application of the load. When a point load is applied to the top of an RC, the top of the beam undergoes compression while the bottom experiences tension, as depicted in Fig. 1.

These two zones are separated by a plane that experiences zero stress due to bending called the neutral axis. The position of this neutral axis within the beam cross-section shifts based on the distribution and magnitude of applied loads. Typically, steel reinforcement in the form of circular bars is incorporated within RC beams to better support the loads applied. As the applied load induces stress in the RC beam, corresponding stains occur. Figure 2 presents the stress profile and strain distribution for a doubly reinforced RC section under bending [1]. The compression stress blocks a, b, c and d correspond to the strain profiles a, b, c and d. When an RC beam undergoes bending, compressive stresses are borne by the concrete and reinforcement in the compression zone, while the reinforcement in the tensile zone primarily carries tensile stresses. Concrete, possessing minimal tensile capacity, resists only limited tensile forces in the tension zone before it cracks.

The flexural response of a beam describes how the beam deflects under loading. The typical moment-deflection curve of an RC beam, particularly with an under-reinforced or balanced section, is characterized by three main stages (Fig. 3). In the first stage, the beam undergoes elastic deformation from the initial loading until the concrete in tension cracks. During this stage, both the concrete and the tension reinforcement resist the tensile forces until concrete cracking occurs. Beyond this point, it is assumed that the concrete in tension ceases to contribute to the beam's flexural capacity, and the tension steel reinforcement bears all tensile stress. Some discontinuity occurs between stage one and stage two owing to the change of beam

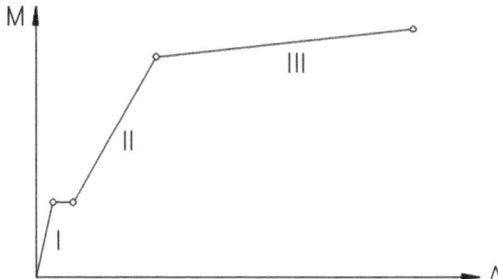

Fig. 3 Typical load–deflection curve of an RC beam

properties from the loss of concrete's tensile strength. As the beam loading continues, stage two begins after the concrete cracks and ends with the yielding of the tension reinforcement. In stage three, there is a more pronounced change in slope, and since the internal steel reinforcement used is a highly ductile material, plastic deformation occurs where the beam continues to rotate. The beam does not support significantly higher loads at this stage but experiences considerable deflections until reaching its ultimate moment capacity. Assuming the shear strength of the beam exceeds its flexural strength, the beam eventually fails beyond this point through steel rupture or concrete crushing or their combined action.

The flexural capacity of an RC beam can be assessed through experimental testing in a laboratory setting but can also be determined via analytical and numerical methods. With knowledge of concrete and steel material behaviour and how they interact in a beam system, mathematical formulations can be done to determine a beam's flexural capacity. This analytical method is well established in literature [1–4] and is based on the following assumptions: (1) Plane sections perpendicular to the bending axis before bending remain plane after bending, (2) The constitutive model (stress–strain relationship) of concrete and steel is known and used to compute the stress in concrete and steel (3) Internal forces within the beam are in equilibrium with external loads at that section (4) there is equal strain in the concrete and reinforcement at the same level. More advanced numerical techniques like finite element (FE) analysis with computer software packages like ANSYS [5], ATENA [6], Abaqus [7], DIANANFEA [8] and NX Nastran [9] can also be used to effectively model RC beams and simulate their reactions to various types of loads. Numerical modelling is usually done alongside experimental investigations to determine the structural behaviour of materials. Increasing knowledge and software capacity has made the numerical modelling of complex RC behaviour possible.

Consequently, various types of concrete beams with distinct characteristics and configurations can be modelled and analysed using these software packages, resulting in significant savings in both time and cost compared to conducting physical experiments [10]. This present study aims to predict an RC beam's flexural response using analytical and numerical methods. The two methods used are validated with results from an experimental program conducted in an existing study [2].

2   Methodology

A case study from an existing study was considered. The chosen beam was analysed using numerical and analytical approaches. The results were subsequently compared with the existing experimental results. An RC beam of 1840 × 120 × 240 mm with a 25 mm concrete cover, two 10 mm rebars at the tension side of the beam and two 8 mm rebars at the compression side was considered. Shear links of 8 mm were provided in 80 mm centres. The beam schematic with loading points shown is presented in Fig. 4. The steel reinforcement had a yield strength of 540 MPa, and the cylinder concrete compressive was 50 MPa.

2.1   Material Constitutive Models

The constitutive model for the concrete and the steel reinforcement is illustrated in Fig. 5. The model for concrete, as proposed by Hognestad [11], is characterized by two curves: an ascending parabolic branch defined from Eqs. 1 to 3 and a descending branch. The concrete elastic modulus and tensile strength are derived from Eqs. 4 and 5 [12]. For the numerical analysis, the first point of the curve was made to satisfy Hooke's law as required by ANSYS, and the second point of the curve defined the extent of the elastic region (30% of the ultimate compressive stress) [13]. The

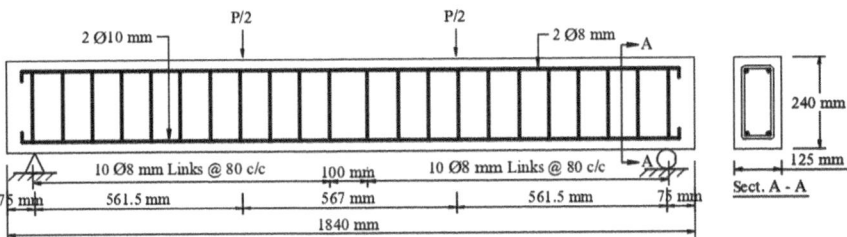

Fig. 4 Beam reinforcement details

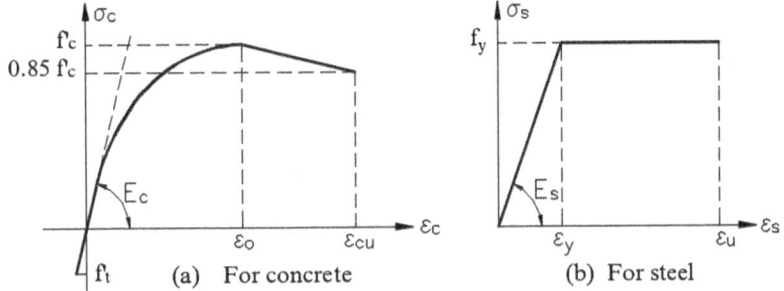

Fig. 5 Stress–strain relationship (**a**) for concrete and (**b**) for steel

following points of the curve leading to the ultimate compressive strength were derived from Eq. 2. The descending branch of the curve was ignored in the FE material model to avoid convergence issues.

$$f_c = f'_c \left[2\left(\frac{\varepsilon_c}{\varepsilon_0}\right) - \left(\frac{\varepsilon_c}{\varepsilon_0}\right)^2 \right] \qquad 0 \leq \varepsilon_c \leq \varepsilon_0 \qquad (1)$$

$$f_c = f'_c - 0.15f'_c \left[\frac{\varepsilon_c - \varepsilon_0}{\varepsilon_{cu} - \varepsilon_0}\right] \qquad \varepsilon_c > \varepsilon_0 \qquad (2)$$

$$\varepsilon_0 = \frac{2f'_c}{E_C} \qquad (3)$$

$$E_c = 4730 \times \sqrt{f'_c} \qquad (4)$$

$$f_t = 0.7 \times \sqrt{f'_c} \qquad (5)$$

where

f_c = stress at concrete strain (ε)
f'_c = Cylinder compressive stress
f_t = Splitting tensile concrete stress
ε_c = strain at concrete stress (f)
ε_0 = strain at f'_c
E_c = Concrete Modulus of elasticity

The tension and compression behaviour of the reinforcing steel is assumed to follow an elastic-plastic model, as determined by Eqs. 6 and 7. Here, f_s and ε_s are the stress and strains in the reinforcing steel, respectively, while f_y and ε_y denote the yield stress and strains in the reinforcing steel. E_s is the reinforcing steel modulus of elasticity.

$$f_s = E_s \varepsilon_s \qquad \varepsilon_s < \varepsilon_y \qquad (6)$$

$$f_s = f_y \qquad \varepsilon_s \geq \varepsilon_y \qquad (7)$$

2.2 Analytical Approach

The analytical approach to determine the behaviour of the RC beam was done by using force equilibrium and strain compatibility equations; the various points of the load–deflection curve were guided by Toutanji et al. [14]. For the start of the beam loading just before the tension steel yields, the neutral axis was calculated from the first moment of area of the transformed area. Different moments of inertia are calculated before the tension concrete cracking and the different stages after. Force

Table 1 Stages of RC beam flexural behaviour

Stage	Section—stress block–strain profile	Neutral axis depth and moment	Curvature and deflection
I		$\sum \frac{Ay}{A} = 0$ $M = \frac{f_r I}{y_b}$	$I = \sum(I_i + A_i d_i^2)$ $\varnothing = \frac{M}{E_c I}$ $\Delta = \frac{M}{24 E_c I}(3L^2 - 4a^2)$
I–II			
II		$C_c + C_s - T_s = 0$ $M = C_c k_2 c + C_s(c - d_c) + T_s(d - c)$	
III			

equilibrium is used for section analysis from the point concrete enters its plastic behaviour after the tension steel yields. Table 1 presents the parameters for the section analysis, stress block, strain profiles, moment, curvature and deflection formulas for the different stages of the RC beam loading. The sectional strain relationships are used alongside these expressions to calculate the neutral axis depths and corresponding moments.

L is the clear span of the beam, a represents the distance from the beam edge to the load point, $P/2$, y_b is the distance from the centroid of the transformed section to the bottom of the section, C_c is the concrete compressive force, C_s is the compression steel compressive force, T_s is the tension steel tensile force, c is the depth of the neutral axis, d_c is the depth of the compression steel and k_{c2} is the distance between the centroid of the compression force in concrete and the neutral axis as shown in Fig. 2.

2.3   Numerical Approach

The numerical modelling was done using the ANSYS FE analysis software. The software considers the nonlinear behaviour of structures under gradual loading with the concept of load steps. It adjusts the model's stiffness matrix after each load step to incorporate the nonlinear alterations in the structure's stiffness due to the nonlinear behaviour of the constituent materials [15]. The software updates the

Table 2 Material properties used in the ANSYS model

Material	Element type	Material properties			
Concrete	Solid65	Linear isotropic	EX—33,446 MPa	PRXY—0.2	
		Multilinear isotropic		Strain	Stress (MPa)
			Point 1	0.0001000	3.3446
			Point 2	0.0004485	15.0000
			Point 3	0.0010000	27.8528
			Point 4	0.0015000	37.5844
			Point 5	0.0020000	44.5193
			Point 6	0.0025000	48.6577
			Point 7	0.0030000	50.0000
		Concrete	Open shear transfer coefficient	0.3	
			Closed shear transfer coefficient	0.8	
			Uniaxial cracking stress	4.94 MPa	
			Uniaxial crushing stress	50 MPa	
Rebar	Link180	Linear isotropic	EX—200,000 MPa	PRXY—0.3	
		Bilinear isotropic	Yield stress— 520 MPa	Tangent modulus— 20 MPa	
Steel plates	Solid185	Linear isotropic	EX—200,000 MPa	PRXY—0.3	

stiffness matrix using Newton–Raphson equilibrium iterations. With each load step, the difference between the forces applied to the model and the element stresses is computed, known as the out-of-balance load vector. The program then checks if there is convergence. If the convergence criterion is not fulfilled, an iterative process continues, re-evaluating the out-of-balance load vector and updating the global stiffness matrix until convergence is reached [16].

Table 2 presents the various material elements used to numerically model the concrete, steel reinforcement, support and load application plates and the mechanical properties assigned to them. The FE beam was modelled with the exact dimensions and material properties as the experimental beam to represent the physical condition accurately. The concrete and steel materials were modelled with similar equations used in the analytical model: multilinear isotropic relationship (Eqs. 1–3), elastic modulus (Eq. 4), uniaxial cracking stress (Eq. 5) and uniaxial crushing stress (experimental case study). The shear transfer coefficients represent the crack face condition of concrete, and the values range from 0 to 1. 0 represents the complete loss of shear transfer (smooth crack), while 1 signifies no loss of shear transfer (rough crack) [15]. These values were chosen from existing literature [5] and adjusted to ensure convergence.

The loading and support distances and conditions of the FE model were also made to be identical to the experimental beam. Two 20 mm thick steel plates were modelled with the beam to serve as the beam support and the loading point. These plates distribute stress more evenly, preventing stress concentrations and averting

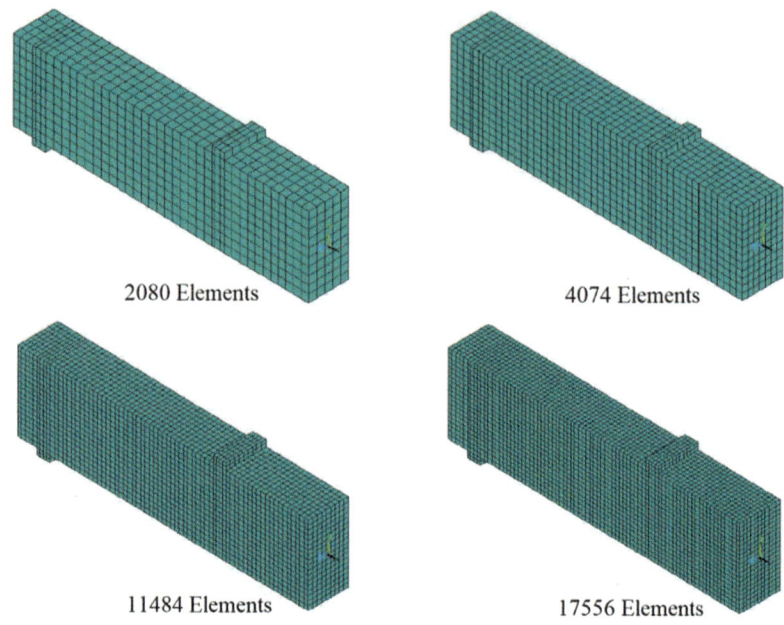

2080 Elements   4074 Elements

11484 Elements   17556 Elements

Fig. 6 Beam FE models with different mesh densities

premature beam failure. Given the symmetric nature of the beam, only half of it was modelled to reduce computation time and disk requirements. A plane of symmetry was established at the mid-span, effectively restricting its movement in the direction perpendicular to that plane.

After creating the necessary solid and line elements, the model was meshed (divided into small elements). The mesh density, the number of elements the model is divided into, is critical in obtaining an accurate solution [5]. The mesh density is determined by the mesh size, which is user-controlled in the program; reducing the element size increases the number of elements. In RC models, a coarse mesh may result in a solution that is not sufficiently accurate [15]. Though accuracy typically increases with an increased mesh density, computing time also increases. A mesh sensitivity analysis (MSA) was done to determine a suitable mesh size for the ANSYS beams, one that is not too large to compromise the analysis results but also not too small to maximize the computing process. The MSA process entails increasing the mesh density until further increase has an inconsequential impact on the beam's outcomes. Six identical half beams without internal steel reinforcements were modelled for this process. Beam models with 25 mm, 20 mm, 15 mm, 14 mm, 13 mm and 12 mm corresponded to models with 2080, 4074, 9180, 11,484, 14,030 and 17,556 number of elements. Figure 6 shows four of the six beam models with their corresponding number of elements indicated. An arbitrary 7 kN load was applied to each beam, and the resulting mid-span deflection was measured. The result of the MSA is presented in Fig. 7, which presents the mid-span deflection for

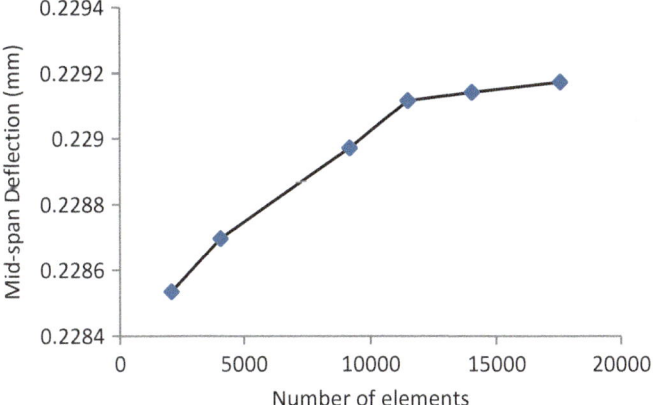

Fig. 7 Deflection at mid-span convergence study

the six beams. The results began to converge in the model with 11,484 elements, corresponding to the 14 mm element size. The 14 mm mesh size was used to model the beam. A finer mesh was also provided at the loading and support locations to prevent stress concentrations.

3 Results

Figure 8 shows the load–defection relationship derived from the analytical formulations presented in this chapter, the numerical analysis done in ANSYS and the case study experimental beam. The three stages of the load–deflection curve are evident from the three curves. The compression concrete cracking load from the experimental, numerical and analytical analysis was 21.68 kN, 22.88 kN and 23.90 kN respectively. The compression yield load and ultimate load derived from the experimental, numerical and analytical analyses are presented in Table 3. The ultimate deflection from the experimental, numerical and analytical analysis was 33.96 mm, 39.77 mm and 35.83 mm, respectively. The percentage difference of the numerical and analytical results with respect to the experimental is also presented. The numerical and analytical curves correlated with the experimental results to a reasonable degree.

4 Discussion

It can be observed from the graph that the numerical and analytical curves were stiffer than the experimental curve. This increased stiffness is because the analytical and numerical models are in perfect conditions with perfect materials. This

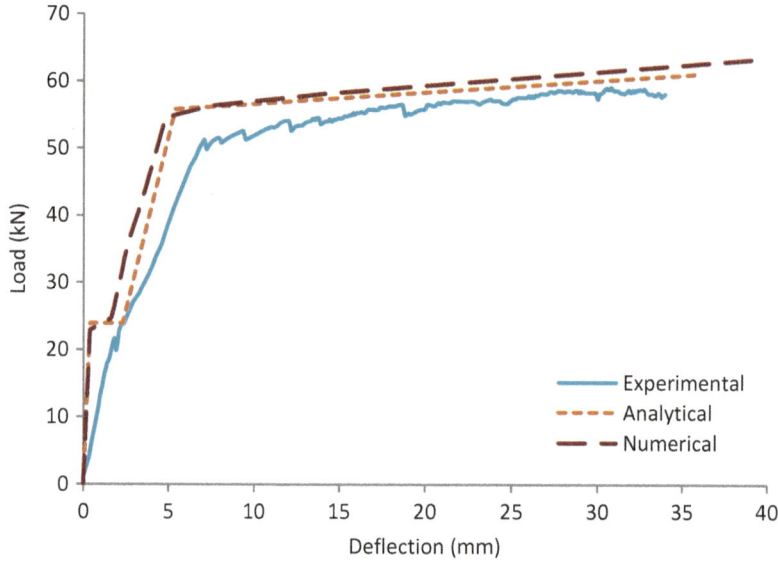

Fig. 8 Load–deflection curves

Table 3 Load–deflection response of control beam

Analysis type	Yield load, P_u (kN)	% Difference	Ultimate load, P_y (kN)	% Difference
Experimental	51.26	–	59.09	–
Numerical	54.56	6.45	63.36	7.22
Analytical	55.78	8.81	61.00	3.22

perfection does not exist in the physical condition. In the physical condition, material imperfections exist, and the bond between the concrete and steel reinforcement could also be imperfect, allowing for slippage. Though these imperfections can also exist when RC beams are used in real-life applications, there could be human error in conducting experimental tests and machine calibration errors with the testing equipment.

It can also be observed that beyond the concrete cracking point, the numerical model is slightly stiffer than the analytical model. This difference can be attributed to the concrete material model adopted in the analytical model. In the true tensile behaviour of concrete, a descending branch known as the tension softening region exists after the maximum tensile stress is reached. However, this region was omitted in the analytical model to simplify the analysis. Similarly, the steel reinforcement also has a strain hardening region when the stress goes from the yield stress to ultimate stress before a slight descent before failure. This region was overlooked in both models. It is essential to recognize that the accuracy of any analytical or numerical model hinges on the precision of the assumptions made. Material models are deliberately simplified to facilitate analysis. Therefore, when developing analytical or numerical models, it is crucial to understand the implications of the chosen material models, ensuring that they do not compromise the accuracy of the obtained results.

5 Further Considerations

In the research of RC beam flexural behaviour, it is imperative not to rely solely on any single approach in isolation. When conducting physical experiments, it is advisable to complement them with analytical or numerical methods. This helps confirm the results obtained from physical tests, considering the inherent imperfections of physical systems. Likewise, when analytical or numerical analyses are employed, validating the results against existing experimental data is essential. While analytical and numerical models provide ideal conditions often unattainable in real-world settings, their integration can significantly economize both time and costs when exploring various scenarios. Preliminary tests and parametric analyses can be conducted using analytical and numerical modelling techniques, allowing for the variation of different parameters within the system. Parametric analysis, when conducted, will help to identify optimal conditions and configurations before committing to laboratory testing.

It is also important to note that the analytical approach is restricted to specific beam systems characterized by predefined boundary conditions, loading types and beam geometries. As the complexity of the beam system increases, so does the complexity of the analytical formulations required. Conversely, the numerical approach offers an advantage in this aspect. Various beam configurations featuring diverse loading patterns can be effectively modelled and analysed using FE analysis software. Furthermore, an FE model can incorporate intricate phenomena such as bond slip. Though there are advantages to the numerical approach, there is also a cost consideration when acquiring the FE software. While certain software, like ANSYS, offers academic versions, these versions may have limitations regarding the scale of work (size of model) that can be conducted. The numerical analysis also requires computational power and memory; the extent needed depends on the size of the analysis.

6 Conclusion

Analytical and numerical approaches were used to predict the flexural behaviour of an RC beam. The analytical and numerical beam flexural behaviour depicted by a load–deflection curve compared well with the available experimental data. Effective utilization of analytical or numerical approaches necessitates a comprehensive understanding of beam mechanics and the material behaviour of concrete and reinforcing steel. The precision of predictions hinges on the accuracy of material behaviour assumptions.

Experimental, analytical and numerical approaches in determining RC beam behaviour should be used collaboratively. Analytical and numerical methods can save on cost and time but present perfect results from a perfect system. Predefined assumptions and simplified models constrain the analytical approach, while the numerical approach provides a more versatile platform for modelling more diverse beam configurations and accounts for complex phenomena.

Acknowledgements This research is funded by the Intra-Africa Mobility Scheme of the European Union in partnership with the African Union under the Africa Sustainable Infrastructure Mobility (ASIM) scheme (624204-PANAF-1-2020-1-ZA-PANAF-MOBAF). Opinions and conclusions are those of the authors and are not necessarily attributable to ASIM. The work is supported and part of collaborative research at the Centre of Applied Research and Innovation in the Built Environment (CARINBE).

References

1. Park, R., Paulay, T.: Reinforced Concrete Strutures. John Wiley & Sons, New York (1975)
2. Hawileh, R.A., Rasheed, H.A., Abdalla, J.A., Al-Tamimi, A.K.: Behavior of reinforced concrete beams strengthened with externally bonded hybrid fiber reinforced polymer systems. Mater. Des. **53**, 972–982 (2014). https://doi.org/10.1016/j.matdes.2013.07.087
3. Nerilli, F., Vairo, G.: Strengthening of reinforced concrete beams with basalt-based FRP sheets: an analytical assessment. AIP Conf. Proc. (2016). https://doi.org/10.1063/1.4952055
4. Thamrin, R.: Analytical prediction on flexural response of RC beams strengthened with steel plates. MATEC Web Conf. **103** (2017). https://doi.org/10.1051/matecconf/201710302012
5. Badiger, N.S.: Parametric study on reinforced concrete beam using ANSYS. Civ. Environ. Res. **6**(8), 88–95 (2014)
6. Tambusay, A., Suprobo, P.: Predicting the flexural response of a reinforced concrete beam using the fracture-plastic model. J. Civ. Eng. **34**(2), 61 (2019). https://doi.org/10.12962/j20861206.v34i2.6470
7. Soleimani, S.M., Roudsari, S.S.: Analytical study of reinforced concrete beams tested under quasi-static and impact loadings. Appl. Sci. **9**(14) (2019). https://doi.org/10.3390/app9142838
8. Vahida, S., Medic, S., Zlatar, M.: Experimental vs. numerical modeling of reinforced concrete beam. In: ECCOMAS MSF 2021 Thematic Conference, Croatia (2021)
9. Alekseytsev, A.V., Antonov, M.D.: Analysis of the ultimate loading on concrete beams in FEMAP NX Nastran. In: Vatin, N.I., Tamrazyan, A.G., Plotnikov, A.N., Leonovich, S.N., Pakrastins, L., Rakhmonzoda, A. (eds.) Advances in Construction and Development, pp. 13–20. Springer Nature Singapore, Singapore (2022)
10. Saifullah, I., Hossain, M.A., Uddin, S.M.K., Khan, M.R.A., Amin, M.A.: Nonlinear analysis of RC beam for different shear reinforcement patterns by finite element analysis. Int. J. Civ. Environ. Eng. **11**(01), 86–97 (2011)
11. Hognestad, E.: A study of combined bending and axial load in reinforced concrete members. Univ. Illinois Bull. **399**, 45 (1951)
12. ACI Committee 318: Building Code Requirements for Structural Plain Concrete (ACI 318-08) and Commentary (2008)
13. Bangash, M.Y.H.: Concrete and Concrete Structures: Numerical Modelling and Applications. Elsevier Science Publishers Ltd., London (1989)
14. Toutanji, H., Zhao, L., Zhang, Y.: Flexural behavior of reinforced concrete beams externally strengthened with CFRP sheets bonded with an inorganic matrix. Eng. Struct. **28**, 557–566 (2006). https://doi.org/10.1016/j.engstruct.2005.09.011

15. Kachlakev, D., Miller, T.: Finite Element Modeling of Reinforced Concrete Structures Strengthened with FRP Laminates. Oregon Department of Transportation Research Group & Federal Highway Administration, Washington, DC (2001)
16. Al-Ta'an, S.A., Mohammed, A.A., Al-Jurmaa, M.A.: Nonlinear three dimensional finite element analysis of steel fiber reinforced concrete deep beam. Iraqi J. Mech. Mater. Eng. **E**, 13–25 (2010)

Sustainable Method for Determining Shear Strength Parameters by Machine Learning

Jnanendra Vijay Kumar Chorapalli (iD) **and Soukat Kumar Das** (iD)

Contents

1 Introduction

The concept "net zero" entails the attainment of equilibrium between the quantity of greenhouse gases released into the atmosphere and those subsequently withdrawn from it [1]. A zero-carbon transition can boost both productivity and growth [2]. Renewable power could primarily replace fossil fuel-based energy consumption in order to achieve net-zero emissions [3]. Through national-level decisions, the Paris Agreement establishes an international framework with the goal of keeping the average rise in global temperature to well below 2 °C [4, 5]. The environment is significantly impacted by the building industry. Around 40% of global energy use

J. V. K. Chorapalli · S. K. Das (✉)
National Institute of Technology Rourkela, Rourkela, Odisha, India
e-mail: dassoukat@nitrkl.ac.in

© The Author(s) 2025
M. Kioumarsi, B. Shafei (eds.), *The 1st International Conference on Net-Zero Built Environment*, Lecture Notes in Civil Engineering 237,
https://doi.org/10.1007/978-3-031-69626-8_120

and greenhouse gas emissions are attributed to it [6]. Also, in terms of geotechnical investigations, it leads to a carbon footprint with the manufacturing and operation of heavy machinery on site as well as in the laboratory. Artificial intelligence (AI), machine learning, deep learning, and artificial neural networks (ANNs) are reshaping data processing and analysis [7] and can be employed as alternatives for laboratory testing.

The main aim is to suggest a technological alternative that decreases the reliance on laboratory testing equipment used for finding C and φ of soil. To this end, ANN, a machine-learning method, is being utilized. By applying this ANN and inputting the index properties of soil, including C and φ, a model is developed through training, testing, and validating its accuracy. Commonly, index properties of soil can typically be determined using basic testing equipment without the need for machinery. Thus, the model is supplied with the index properties as inputs, and it then produces precise values for C and φ. As a consequence, usage of testing machines can be avoided, leading to a reduction in net emissions.

2   Carbon Footprint of Laboratory Testing Equipment

Regarding climate change, greenhouse gas (GHG) emissions, primarily carbon dioxide (CO_2) emissions, are a global concern. Because manufacturing processes require a lot of energy by nature, electricity produced from fossil fuels contributes significantly to global warming [8]. It is widely acknowledged that the comminution of the ore accounts for roughly 70% of the energy used in conventional mining and mineral-processing circuits [9]. Currently, the industrial sector uses over half of the energy consumed worldwide [10]. Global CO_2 emissions from the manufacturing sector are summarized in Fig. 1, which also includes direct emissions for electricity, heat, and buildings. China's industrial growth after 2002, the country that entered the World Trade Organization in 2001, is the reason for the sharp increase [11].

3   Overview of ANNs

An artificial intelligence system is called ANN modeling. It has been successfully used to solve a variety of geotechnical issues [12]. The capacity of ANNs for learning allows them to uncover hidden connections. ANNs do not impose any limitations on the input variables and are capable of learning and modeling latent correlations between inputs and outputs [13]. Although there are numerous types of ANNs, the multilayer feed-forward ANN is the type that is most frequently utilized [14].

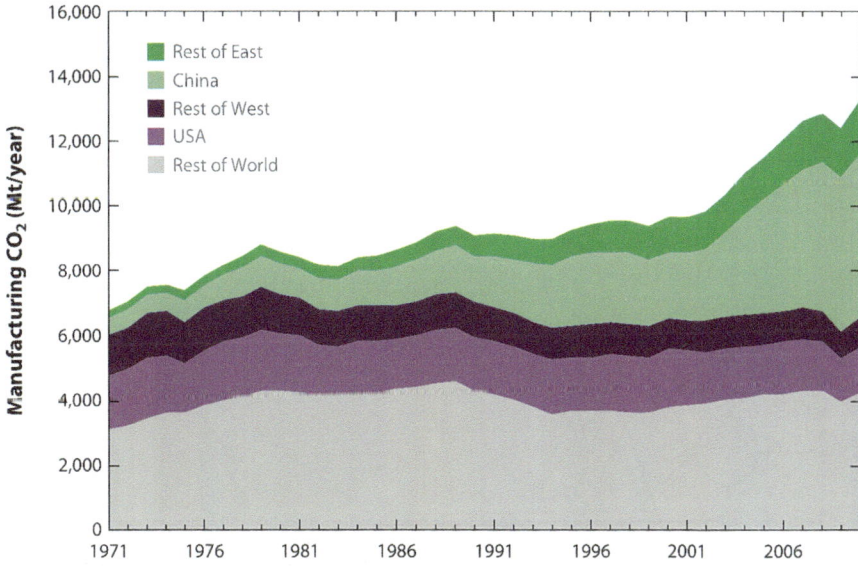

Fig. 1 Carbon dioxide emissions from various regions of the world [11]

3.1 Functioning of Neural Networks

There are three layers in Fig. 2 below: the input layer, the hidden layer, and the output layer. Neurons make up the majority of these layers. All input parameters are contained in the input layer, and goal outputs are found in the output layer. Additionally, the hidden layer is typically used to aid the network's learning process and is positioned between the input and output layers [15]. The summation function, transfer function, and connection weights of the neurons are readily built with the use of parallel computation methods [16].

3.2 Bayesian Optimization Technique

Compared with conventional back-propagation nets, Bayesian regularized artificial neural networks (BRANNs) exhibit greater resilience [18]. The objective function that is often utilized, such as mean squared error (MSE), or E_d in Eq. (1), is modified in the Bayesian approach. A term, E_w, is added to the objective function in Eq. (2) to extend it.

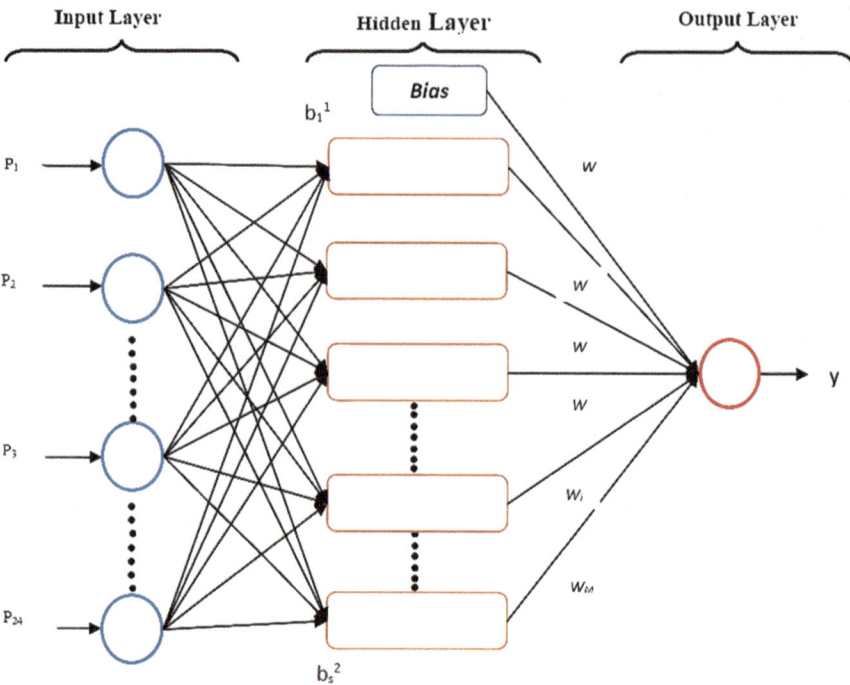

Fig. 2 Architecture of an artificial neural network [17]

$$F = E_d = \frac{1}{N} \sum_{i=1}^{N} (e_j)^2 \tag{1}$$

$$F = \beta E_d + \alpha E_w, \tag{2}$$

where e_j is the error between the target and observed values, E_d is the MSE, α and β are the objective function parameters, whereas E_w is the sum of the squares of the network weights [19].

4 Data Used for the Study

4.1 Description of the Dataset

Data were sourced and used in the model's construction from prior research endeavors [20], comprising 11 input parameters and 2 output parameters, namely, C and φ and the remaining input parameters are clearly set out in Table 1.

Table 1 Inputs utilized in the creation of the model

	FC	CC	D10	D30	D60	Cu	Cc	L	γ	ω	γ_d	C	φ
1	88	12	0.003	0.01	0.04	13.3	2.4	25	1.5	5.7	1.42	0.41	25
2	80.4	19.6	0.0001	0.004	0.033	330	6.9	30	1.8	10	1.64	1.07	31
3	95	5	0.0012	0.011	0.04	33.3	2.5	25	2.0	15	1.74	0.77	33
4	97.1	2.87	0.001	0.013	0.033	33	5.1	27	2	16	1.73	0.57	25
5	81.6	18.4	0.0011	0.011	0.047	42.7	2.3	26	2	19	1.67	0.04	23
...	...	...	...	...	...	...	...	...	...	...	...	...	...
84	78	22	0.0006	0.006	0.036	60	2.1	30	1.7	12	1.55	0.62	25
85	81.8	18.2	0.0019	0.012	0.041	21.5	1.8	29	1.7	16	1.51	0.16	25
86	94.7	5.3	0.001	0.007	0.022	22	2.2	29	2	19	1.72	0.46	15
87	85.7	14.3	0.001	0.01	0.035	35	2.8	28	1.9	21	1.57	0.33	17
88	41.4	58.6	0.004	0.028	0.48	120	0.4	25	2	23	1.66	0.32	12

FC fine grain soil content (%), *CC* coarse grain soil content (%), *D10* particle diameter corresponding to 10% finer (mm), *D30* particle diameter corresponding to 30% finer (mm), *D60* particle diameter corresponding to 60% finer (mm), *Cu* coefficient of uniformity, *Cc* coefficient of curvature, *L* liquid limit, γ soil unit weight (g/cc), ω water content (%), $γ_d$ dry unit weight of soil (g/cc), *C* cohesion, φ friction angle

Table 2 Descriptive statistics of the variables used for the model-building process

	FC	CC	D10	D30	D60	Cu	Cc	L	γ	ω	γ_d	C	φ
1	88	88	88	88	88	88	88	88	88	88	88	88	88
2	81.8	18.1	0	0.05	0.22	72.6	4.16	27.4	1.8	16.4	1.6	0.36	23
3	18.9	18.9	0.01	0.23	1.14	176	9.62	4.59	0.1	6.08	0.1	0.32	7.66
4	9.6	0.9	0	0	0.01	2.14	0.05	21	1.4	3.4	1.3	0.02	1
5	77.3	5.45	0	0.01	0.03	11.9	1.62	24.7	1.8	12.8	1.5	0.11	17.7
6	89.9	10.1	0	0.01	0.04	27.08	2.33	27	1.9	16.3	1.6	0.31	25
7	9.55	22.6	0	0.02	0.05	42.89	4.26	29	1.9	20.5	1.6	0.50	29
8	99.1	90.4	0.09	1.9	10	1150	90.2	41	2.1	30	1.8	0.75	37

1: count, 2: average, 3: std dev, 4: minimum, 5: 25%, 6: 50%, 7: 75%, 8: maximum

4.2 Characteristics of the Dataset

The statistical properties of all the parameters used for the study are illustrated in Table 2. Also, the correlation among all the input parameters were demonstrated in Fig. 3. The connection between two variables was determined using the Pearson method and depicted using a heat map. The correlation matrix revealed that dry unit weight displayed the highest correlation with cohesion, followed by liquid limit and moisture content. Conversely, the coefficient of uniformity and unit weight exhibited the weakest correlation with cohesion. Similarly, moisture content showed the strongest correlation with the angle of internal friction, whereas the coefficient of curvature demonstrated the weakest correlation.

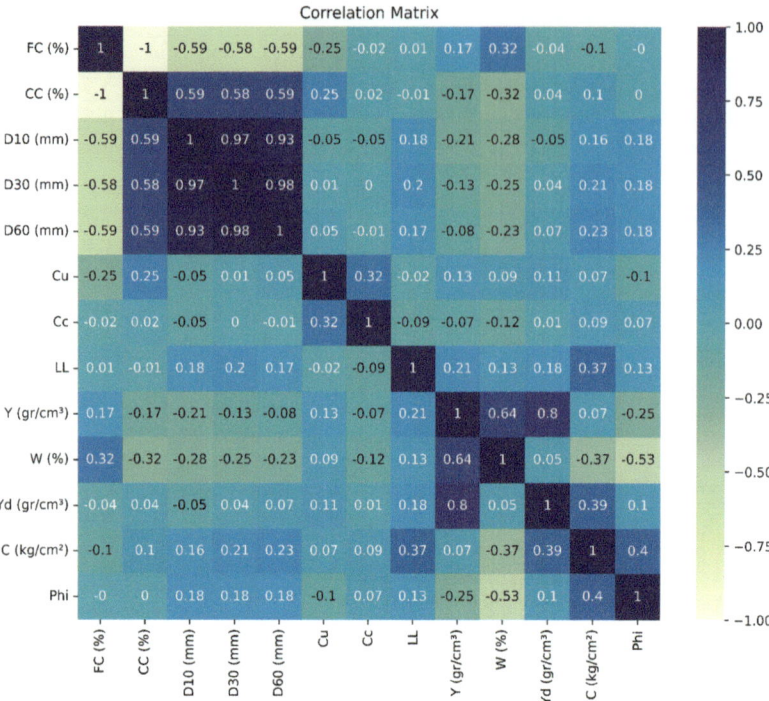

Fig. 3 Correlation among the inputs used for model creation

5 Evaluation Criteria

To assess the precision and resilience of the developed model, five evaluation metrics were employed, namely the root mean squared error (RMSE), the coefficient of determination (R^2), and the mean absolute error (MAE). These metrics are conventionally applied in regression scenarios to gauge the performance of predictive models. The equations corresponding to each metric are as follows:

$$\text{RMSE}\sqrt{\frac{1}{n}\sum_{i=1}^{n}\left(y_i - y_i^{\wedge}\right)^2} \tag{3}$$

$$R^2 = 1 - \frac{\sum_{i=1}^{n}\left(y_i - y_i^{\wedge}\right)^2}{\sum_{i=1}^{n}\left(y_i - y_i'\right)^2} \tag{4}$$

$$\text{MAE} = \frac{1}{n}\sum_{i=1}^{n}\left|y_i - y_i^{\wedge}\right|, \tag{5}$$

where n denotes the total number of observations, y_i stands for the target value, $y_i^{\wedge}$ for the predicted value, and y_i' for the mean of the target values [21, 22].

6 Results Overview and Interpretation

This section explains how successful the constructed ANNs are using the Bayesian regularization optimization algorithm for predicting the cohesion and φ of soil. In adherence to established methodologies, the training and testing subsets were partitioned in a ratio of 70% to 30% respectively, serving as foundational components for model development and subsequent validation. The database underwent segmentation into distinct training and testing subsets, comprising 62 and 26 samples respectively, to facilitate the model construction process. The assessment of the constructed models' performance ensued through meticulous evaluation, employing various parameters to ascertain their robustness and predictive capabilities.

6.1 Development of the Model

For construction of the model, MATLAB R2018a was employed as the primary computational tool. To ascertain the optimal configuration, encompassing the ideal hidden layer's number and neurons inside each layer, a meticulous trial and error procedure was carried out, guided by a systematic approach. This iterative exploration was performed to ensure the attainment of a model possessing a commendable degree of accuracy. Furthermore, the refinement of ensemble complexity was imperative to optimize the algorithmic space, warranting the inclusion of an optimal number of individuals within the suggested ensembles. As depicted in Fig. 4, the final model has iterations of 178 and the best performance of MSE is 0.026183, attained at 31st epoch. The regression plots associated with data sets for testing, training, and validation are seen in Fig. 5.

6.2 Performance of the Model

Each model's ability to learn is reflected in how well training outcomes work. The results of the testing and training phases of soil cohesion in comparison with original values are depicted in Fig. 6 for cohesion and in Fig. 7 for angle of internal friction of the soil. One way to evaluate how accurate the prediction is by looking at the correlation between the predicted and values. For cohesion, the finalized model had RMSE $= 0.293$, $R^2 = 0.890$, MAE $= 0.548$. Also for the friction angle the model had RMSE $= 0.3862$, $R^2 = 0.819$, and MAE $= 0.7739$. Thus, an ANN with the Bayesian regularization algorithm shows the precise soil shear strength parameter prediction using soil index properties.

3D surface plots of cohesion and φ calculated by RMSE values with respect to each neuron in the hidden layers were given in Figs. 8 and 9 respectively. From both graphs it was observed that the ANN models performed well for up to two hidden

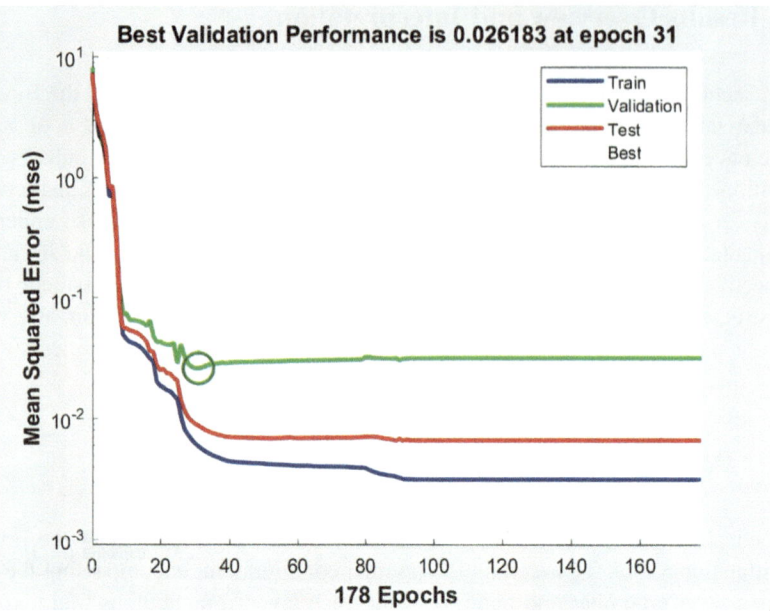

Fig. 4 Performance graph of the final model

layers corresponding to each neuron, as RMSE values close to zero indicate a good model, but from the third hidden layer onward there is an abrupt increase in RMSE values, which is undesirable and indicates low performance. In addition, for cohesion, the performance of the models is good after six neurons in the first two hidden layers, whereas in φ the performance of the models is same for all the neurons in the first two hidden layers.

6.3   Sustainability of the ANN Model in the Long Term

The accuracy and reliability of the initial model for a particular site are undoubtedly good in order to continue; expanding the database with diverse soil data from various sites should be used for testing, training, and validation. With the time-to-time monitoring of performance, data quality, and incorporating the latest optimization techniques improves the generalization and robustness of the model. Incorporating uncertainty analysis techniques enhances the prediction confidence and reveals sources of uncertainty. The traditional method of testing involves more chances of human error and requires high initial capital and time-to-time maintenance, repair, and operational costs are also added, whereas an ANN model initially requires investment in data collection, pre-processing but offers significant cost advantages over time.

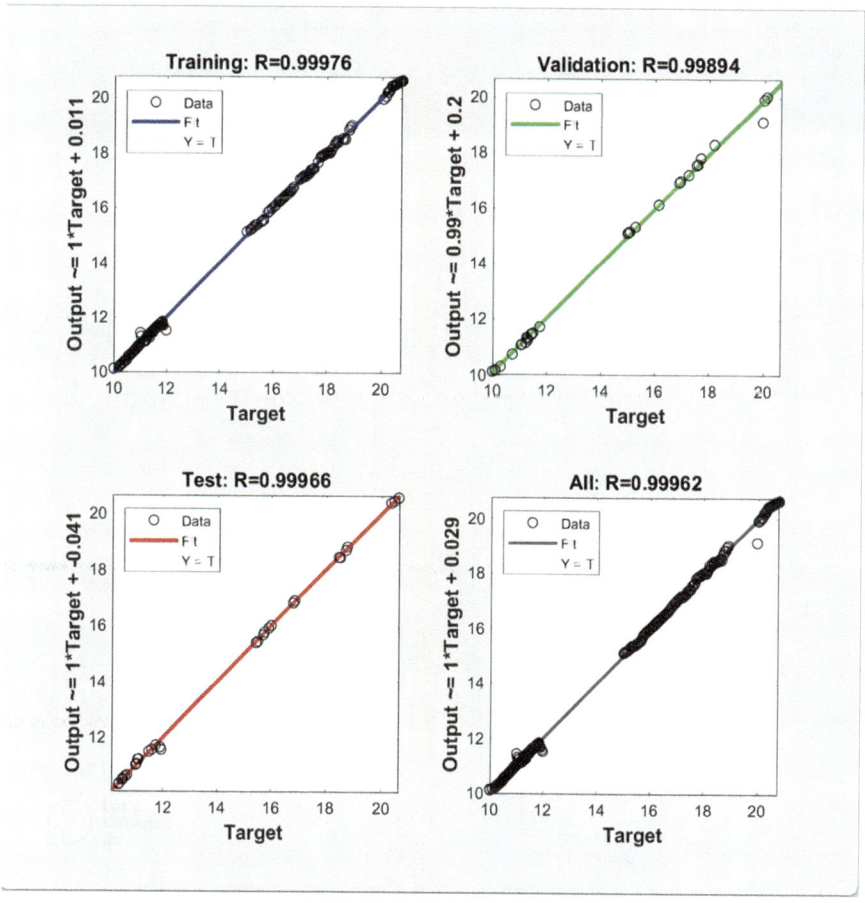

Fig. 5 Regression plots of the testing, training, and validation sets

7   Carbon Reduction Achieved by this Model

For the estimation of carbon reduction, a triaxial machine is considered at our institution. It has a power consumption of 990 W and the time required for testing one sample varies depending on the type of test, with the unconsolidated undrained (UU) test taking 5–10 min, the consolidated undrained (CU) test 26–30 h, and the consolidated drained (CD) test 7–10 days. To obtain the values of C and φ for one particular soil, at least three samples need to be tested. Energy consumption is calculated as

$$\text{Energy (kWh)} = (\text{Wattage (W)} \times \text{operation time of machine (h)})/1000. \qquad (6)$$

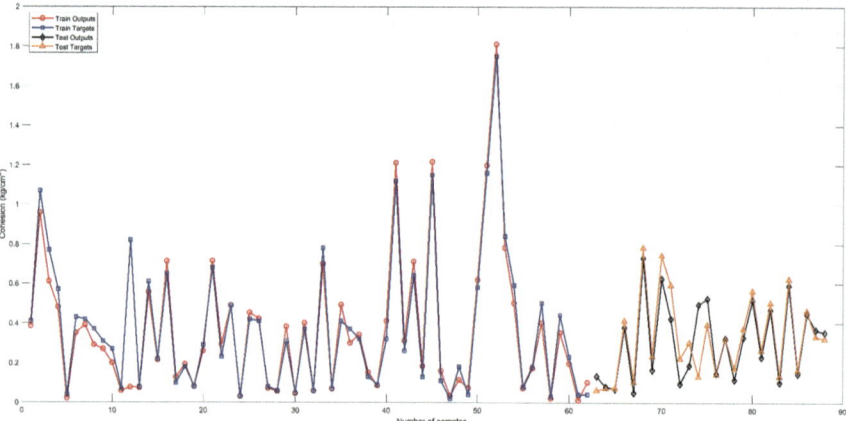

Fig. 6 Comparison of the testing and training data of actual and predicted cohesion

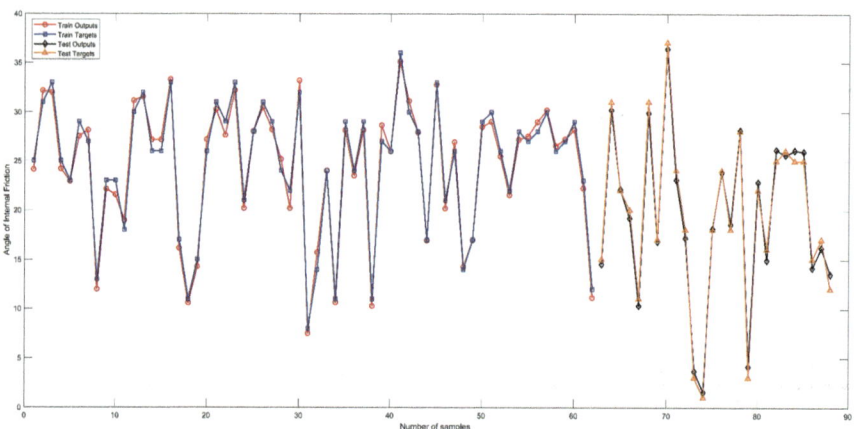

Fig. 7 Comparison of the testing and training data of actual and predicted internal friction

Using Eq. (6) energy consumption for the testing of three samples of each type of test can be calculated. Table 3 shows these values, along with GHG emission by the production of 1 kWh of energy at a thermal power plant obtained from Mittal [23]. Also, the total power consumption range and corresponding emissions range are mentioned.

The development of the model required a PC with a power consumption of 200 W, which was used for 15 days, with an average usage of 8 h, resulting in a total consumption of 24 kWh. The average time taken to use the model for a new prediction is 15 min, which accounts for 0.05 kWh of energy consumption, less than 9.9 times UU, 1782 times CU, and 14,256 times CD test energy consumption and thereby the reduction of GHG emissions by the same number of times.

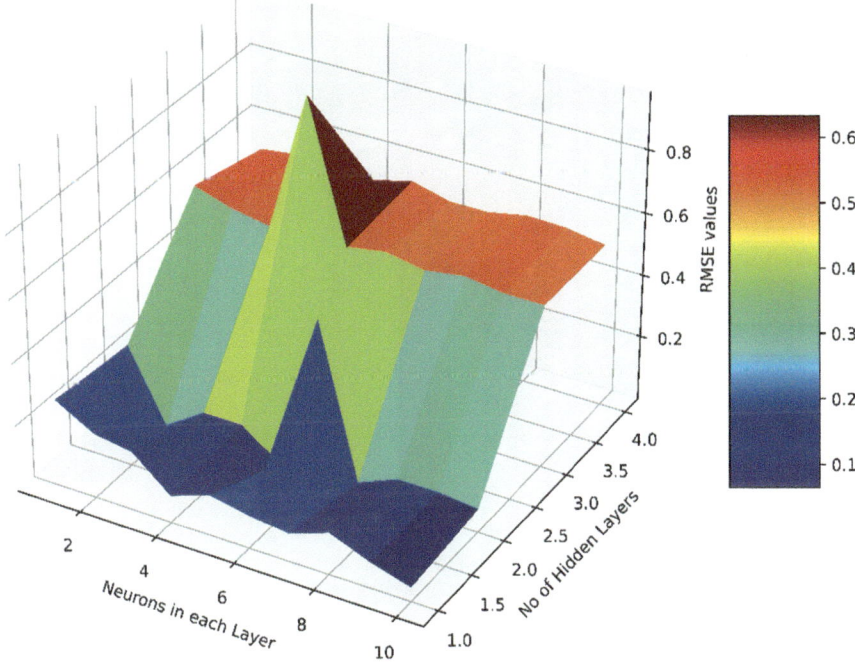

Fig. 8 Change in root mean squared error of cohesion owing to neurons in the hidden layers

7.1 Limitations

- The data utilized for the model are rather limited, and data from a single site have been incorporated.
- The testing and training data exhibit a lack of diversity in soil type and its properties.
- The accuracy of the model may diminish when applied to soils with extreme and typical properties.
- Consequently, the applicability of the model for diverse soils on a global scale may reduce its predictive capabilities.

8 Conclusion

In this paper an alternative to traditional testing for obtaining cohesion and the angle of internal friction is proposed using an ANN optimized with the Bayesian regularization algorithm. During development of the model, it was noticed that performance of the model declined significantly from the third hidden layer onward. This was primarily due to overfitting of the model; as the number of neurons and layers

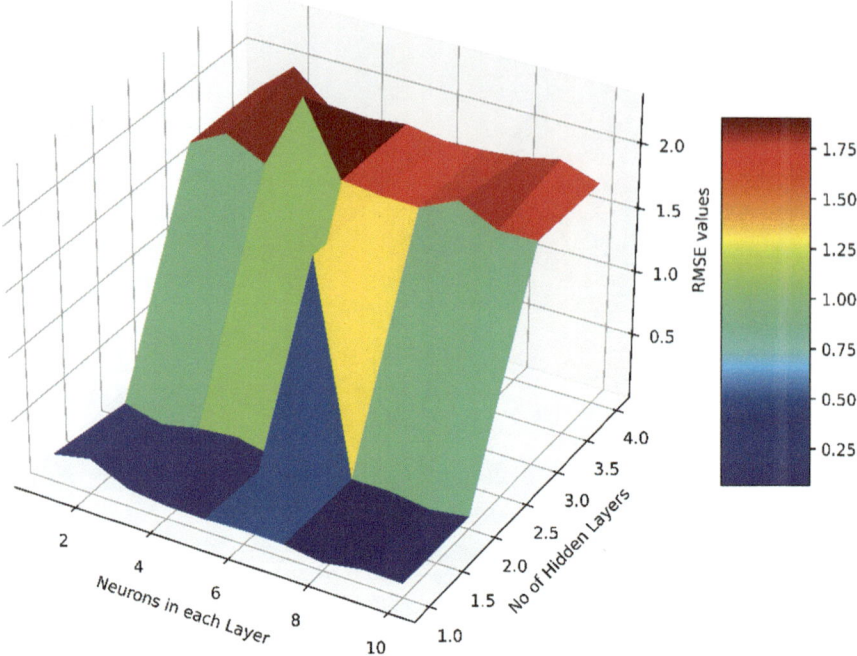

Fig. 9 Change in root mean squared error of phi owing to neurons in the hidden layers

Table 3 Energy consumption and corresponding emissions

	Ideal	UU	CU	CD
Energy (kWh)	1	0.246–0.495	77.22–89.1	499.89–712.8
CO_2 (kg/kWh)	0.95	0.234–0.470	73.36–84.64	474.89–677.16
SO_2 (g/kWh)	7.2	1.774–3.564	555.98–614.52	3599.21–5132.16
NO (g/kWh)	4.38	1.079–2.168	338.22–390.26	2189.52–3122.06

increased, it began to capture noise in the data rather than the underlying patterns. The final proposed model checked using statistical metrics such as RMSE, R^2, and MAE for cohesion values of 0.293, 0.89, and 0.548, and for angle of internal friction values of 0.386, 0.819, and 0.7739.

From comparison of the reduction in carbon footprint by the ANN model developed it is evident that it demonstrated a substantial reduction in GHG emissions. During the development phase the emissions caused were 0.021, 3.71, and 29.7 times lower than those caused by the UU, CU, and CD tests respectively. When the model was used for actual predictions, the emissions caused by the power consumption of the PC were 10.1%, 0.056%, and 0.007% of the emissions caused by the UU, CU, and CD tests respectively. These findings confirm that the utilization of the ANN model contributes to resource conservation and minimizes the environmental impact associated with laboratory testing procedures. However, the model developed here is site specific, and further research is needed to improve the model and enhance its predictive capabilities by including a wide range of soil data, thereby making it applicable across diverse geotechnical contexts worldwide.

References

1. Pye, S., et al.: Modelling net-zero emissions energy systems requires a change in approach. Clim. Pol. **21**(2), 222–231 (2021). https://doi.org/10.1080/14693062.2020.1824891
2. Stern, N., Valero, A.: Innovation, growth and the transition to net-zero emissions. Res. Policy. **50**(9), 104293 (2021). https://doi.org/10.1016/j.respol.2021.104293
3. Fasihi, M., et al.: Long-term hydrocarbon trade options for the Maghreb region and Europe. Sustainability. **9**(2), 306 (2017). https://doi.org/10.3390/su9020306
4. Pye, S., et al.: Achieving net-zero emissions through reframing of UK national targets in the post-Paris Agreement era. Nat. Energy. **2**(3), 17024 (2017). https://doi.org/10.1038/nenergy.2017.24
5. Van Soest, H.L., et al.: Net-zero emission targets for major emitting countries consistent with the Paris Agreement. Nat. Commun. **12**(1), 2140 (2021). https://doi.org/10.1038/s41467-021-22294-x
6. Ohene, E., et al.: Review of global research advances towards net-zero emissions buildings. Energ. Buildings. **266**, 112142 (2022). https://doi.org/10.1016/j.enbuild.2022.112142
7. Jobin, A., et al.: The global landscape of AI ethics guidelines. Nat Mach Intell. **1**(9), 389–399 (2019). https://doi.org/10.1038/s42256-019-0088-2
8. Jeswiet, J., Nava, P.: Applying CES to assembly and comparing carbon footprints. Int. J. Sustain. Eng. **2**(4), 232–240 (2009). https://doi.org/10.1080/19397030903311957
9. Norgate, T., Jahanshahi, S.: Reducing the greenhouse gas footprint of primary metal production. Miner. Eng. **24**(14), 1563–1570 (2011). https://doi.org/10.1016/j.mineng.2011.08.007
10. Fang, K., et al.: A new approach to scheduling in manufacturing for power consumption and carbon footprint reduction. J. Manuf. Syst. **30**(4), 234–240 (2011). https://doi.org/10.1016/j.jmsy.2011.08.004
11. Gutowski, T.G., et al.: A global assessment of manufacturing. Annu. Rev. Environ. Resour. **38**(1), 81–106 (2013). https://doi.org/10.1146/annurev-environ-041112-110510
12. Das, S., et al.: Machine learning techniques applied to prediction of residual strength of clay. Cent. Eur. J. Geosci. **3**(4), 449–461 (2011). https://doi.org/10.2478/s13533-011-0043-1
13. Kim, Y., et al.: Estimation of effective cohesion using artificial neural networks. Eng. Geol. **289**, 106163 (2021). https://doi.org/10.1016/j.enggeo.2021.106163
14. Rabbani, A., et al.: A novel hybrid model of augmented grey wolf optimizer and artificial neural network for predicting shear strength of soil. Model. Earth Syst. Environ. **9**(2), 2327–2347 (2023). https://doi.org/10.1007/s40808-022-01610-4
15. Lee, S.J., et al.: An approach to estimate unsaturated shear strength. Comput. Geotech. **30**(6), 489–503 (2003). https://doi.org/10.1016/S0266-352X(03)00058-2
16. Kayadelen, C.: Estimation of effective stress parameter of unsaturated soils by using artificial neural networks. Int. J. Numer. Anal. Methods Geomech. **32**(9), 1087–1106 (2008). https://doi.org/10.1002/nag.660
17. Kayri, M.: Predictive abilities of Bayesian regularization and Levenberg–Marquardt algorithms in artificial neural networks: a comparative empirical study on social data. Math. Comput. Appl. **21**(2), 20 (2016). https://doi.org/10.3390/mca21020020
18. Pomponi, J., et al.: Bayesian neural networks with maximum mean discrepancy regularization. Neurocomputing. **453**, 428–437 (2021). https://doi.org/10.1016/j.neucom.2021.01.090

19. Doan, C.D., Liong, S.-Y.: Generalization for multilayer neural network Bayesian regularization or early stopping. In: Proceedings of Asia Pacific association of hydrology and water resources 2nd conference (2004)
20. Mousavi, S.M., et al.: Nonlinear genetic-based simulation of soil shear strength parameters. J. Earth Syst. Sci. **120**(6), 1001–1022 (2011). https://doi.org/10.1007/s12040-011-0119-9
21. Hoque, M.J., et al.: Prediction of strength properties of soft soil considering simple soil parameters. Open J. Civ. Eng. **13**(3), 479–496 (2023). https://doi.org/10.4236/ojce.2023.133035
22. Momeni, E., et al.: Novel hybrid XGBoost model to forecast soil shear strength. Comput. Model. Eng. Sci. **136**(3), 2527–2550 (2023). https://doi.org/10.32604/cmes.2023.026531
23. Mittal, M.L.: Estimates of emissions from coal fired thermal power plants in India (2012)

Transforming Asphalt Quality Evaluation: A Digital Twin Exploration Using Microcomputer Tomography and MATLAB Analysis

Aya Kassem and Berthe Dongmo-Engeland

Contents

1 Introduction

A durable pavement is characterized by a well-constructed road surface composed of an asphalt mixture with an optimal air void content, effective binder, and well graded, high-quality aggregates [1]. Air void content is a crucial parameter that affects asphalt pavement quality and long-term performance [2] . Factors such as durability, fatigue cracking resistance, strength, and stiffness have shown to be highly influenced by air void content [1, 3].

Proper execution of the compaction process is essential to achieve good pavement performance. Regardless of perfect pavement design or high-quality asphalt mixture, improper compaction will negatively impact the pavement's overall performance [4, 5]. Over-compaction can lead to reduction in air void content, making the pavement less flexible, thus more susceptible to premature rutting under repeated traffic loads [2, 6]. Additionally, a lower air void content might also lead to

A. Kassem · B. Dongmo-Engeland (✉)
Department of Built Environment, Oslo Metropolitan University, Oslo, Norway
e-mail: bereng@oslomet.no

© The Author(s) 2025   1451
M. Kioumarsi, B. Shafei (eds.), *The 1st International Conference on Net-Zero Built Environment*, Lecture Notes in Civil Engineering 237,
https://doi.org/10.1007/978-3-031-69626-8_121

permeability loss resulting in clogging problems [2]. Neither too many nor too little air voids in the asphalt are desirable. The presence of too many air voids can increase water permeability [2], which weakens the adhesion between the binder and aggregates while also disrupting the cohesive strength within the binder itself, reducing the service life of the pavement [7].

In Norway, the Norwegian Public Roads Administration (NPRA) is responsible for road development, with a primary emphasis on guaranteeing excellent services and ensuring the safety of road users, as well as prioritizing environmental preservation. In their current practices, the NPRA relies on performing laboratory tests to compute air void content in order to assess the quality of asphalt pavements. However, research has shown that current practices can lead to inaccurate air void measurements [8–10].

With the advancement in technology in recent years, the possibility of developing digital twins has increased significantly. These technological evolvements have opened up new opportunities for digitization in the Architecture, Engineering, and Construction (AEC) sector [11]. This study aims to introduce a new method for air void calculation and observation, namely the "digital twin," created by microcomputer tomography and MATLAB analysis. This chapter aims to investigate the potential of developing a digital twin of asphalt pavements as a substitute for conventional laboratory tests in assessing asphalt pavement quality.

2   Method

To validate the accuracy of the digital twin, numerical data from laboratory investigations were gathered and compared to the data from the digital twin. Relevant road sections were selected based on asphalt composition and durability. Laboratory tests were performed for air void calculations and resistance to permanent deformations. The digital twin was generated by microcomputed tomography of asphalt samples from the road sections and processed using a MATLAB script.

2.1   Description of Chosen Road Sections

This study investigated the wearing course of two road sections, national road 4 at Gjøvik and E-road 6 at Tretten in Norway. These sections were selected based on their durability, expected lifespan, and asphalt composition. The roads are both paved with SMA 16 PMB, allowing for a better comparison between them.

National road 4 at Gjøvik (G-series) This road section is located in Gjøvik municipality in Norway. The wearing course of this section exhibits higher air void content than specified in the contractual agreement, where a shorter lifespan is expected.

E-road 6 at Tretten (H-series) The second road section that was investigated is situated in Øyer municipality in Norway. This pavement distinguishes itself from the previously mentioned pavement by demonstrating a high-performance asphalt pavement, characterized by an optimal quantity of air voids. Consequently, a longer lifespan is anticipated for this road section.

2.2 Samples

A total of 12 core samples (six from each section) were extracted from all the layers of the pavement structure. The samples were taken with a diameter of 200 mm and a height of around 300 mm. The samples were drilled in the laboratory to a height of approximately 40 mm, in order to closely examine the wearing course. Among the samples, six of them were kept with their original diameter of 200 mm, while the remaining ones were further drilled to three samples of 50 mm and three 100 mm diameter samples.

The bulk density of all 12 samples was measured using three different laboratory tests. Digital twins were generated for the 100 mm samples only, and then the maximum density was calculated as a mean value of these same samples. Given that these samples were extracted from the same core samples, it was decided to consider the maximum density of the 100 mm samples as representative for all samples. Wheel track testing was performed solely on the 200 mm samples.

2.3 Determining Air Void Content

Laboratory tests The air void content of the asphalt samples was calculated according to the laboratory procedures described in standard NS-EN 12697-8 [12]. The air void content is calculated from the maximum density and bulk density shown in the formula below:

$$V_m = \frac{\rho_m - \rho_d}{\rho_m} \cdot 100\% \tag{1}$$

where V_m is the amount of air voids (%), ρ_m is the maximum density (kg/m^3), and ρ_d is the bulk density (kg/m^3).

Maximum density The maximum density was determined using a pycnometer and water. Disparities are prone to arise between the density values obtained from the asphalt recipes and the core samples obtained from the road, due to variations in binder content and aggregate density during production. Moreover, the mathematical approach necessitates assumptions regarding binder absorption resulting in uncertainties [13]. Hence, relying on maximum density from the pycnometer ensures that an accurate value of the actual density is acquired, preventing potential inaccuracies from the mathematical approach.

The bulk density was determined by three different methods: measurement of dimensions, hydrostatic surface-dry method, and sealing using CoreLok. It is worth noting that the first two methods are standardized methods by the NPRA, whereas CoreLok is a method that had been under testing by the NPRA.

Measurement of dimensions This method as its name implies involves the calculation of the volume of the asphalt sample based on measuring its dimensions using a caliper. This method is suitable for very porous samples with air voids over 10% or for samples abundant with connected pores. The bulk density calculated by measurement of dimensions is given by the equation below:

$$\rho_d = \frac{m_1}{\left(\frac{d}{2}\right)^2 \cdot \pi \cdot h} \cdot 10^3 \tag{2}$$

where m_1 is the mass of dry sample (g), d is the average diameter of the cylindrical sample (mm), and h is the average height of the cylindrical sample (mm).

Saturated surface-dry method The saturated surface-dry method relies on determining the volume of sample based on its buoyancy in water. The bulk density determined by the hydrostatic surface-dry method is calculated using the following equation:

$$\rho_d = \frac{m_1 \cdot \rho_w}{m_3 - m_2} \tag{3}$$

where ρ_w is the density of water at 25 °C (Mg/m^3), m_1 is the mass of dry sample (g), m_2 is the mass of sample immersed in water (g), and m_3 is the mass of wet sample (g).

CoreLok vacuum-sealing method CoreLok is a device designed to seal asphalt samples in specialized polymer bags [14]. To inquire more about this method, the reader is directed to this article [9].

To calculate the bulk density using CoreLok, the following formula is employed:

$$\rho_d = \frac{m_1}{\frac{(m_2 - m_3)}{\rho_w} - \frac{(m_2 - m_1)}{\rho_{sm}}} \tag{4}$$

where ρ_{sm} is the density of the plastic polymer bag at 25 °C (Mg/m^3), m_1 is the mass of unsealed sample in air (g)], m_2 is the mass of sealed sample in air (g)], and m_3 is the mass of sealed sample in water at $t = 1$ min (g).

Microcomputed tomography Microcomputed tomography (micro-CT) is a nondestructive imaging method that utilizes X-ray radiation to generate three-dimensional images of objects [15]. It makes it possible to visualize the internal microstructure of objects, making it particularly suitable for examining the internal structure of asphalt samples. The approach is based on distinguishing different components of the asphalt structure based on differences in X-ray absorption rates leading to density changes [16].

Equipment The scanning equipment employed in this study is a Nikon Metrology XT H 225 LC industrial scanner located at the Norwegian Geotechnical Institute (NGI) in Oslo. The scanner is capable of delivering a maximum voltage of 225 kV and electrical current of 1 mA [17].

Scanner setup and parameter selections The scanning settings were consistent for all samples, where each scan lasted for roughly 43 minutes per sample. The scans were conducted at a voltage of 225 kV and 240 µA. To filter out the X-ray radiation, a 2.5 mm copper filter was employed. A total radiation of 360° was selected with 2500 projections and an angular step about 0.14°. The exposure time was 1000 ms.

Limitations in sample dimensions The used equipment imposes restrictions on allowable dimensions for samples due to available space inside the scanner, weight restrictions of the object turntable, and image quality.

The maximum allowed object dimensions for the Nikon Metrology XT H 225 LC industrial scanner from [17] are presented in Table 1.

The choice of sample size and distance from the X-ray source must be carefully considered to achieve adequate 3D image resolution. Image resolution relies on voxel size, where a smaller voxel size corresponds to a higher image resolution [18]. Scanning the entire sample width in a single scan is desirable; thus, using small samples is preferable. Optimal placement of the test sample involves positioning it as close as possible to the X-ray source [17]. This results in a larger magnification of the sample on the detector, thereby achieving a better image resolution.

Scanned samples A total of six asphalt samples (1G, 2G, 3G, 1H, 2H, and 3H) were scanned in which three were from each road segment. The obtained samples possessed dimensions of 100 mm in diameter and an approximate height of

Table 1 Allowed sample dimensions for the Nikon Metrology XT H 225 LC scanner

Sample dimension	Requirement
Maximum diameter in a single scan[a]	Ca. 30 cm
Weight limit of object turntable	Up to 70 kg
Maximum height[b]	1 m

[a]Maximum diameter during a single scan to achieve an adequate image resolution is ca. 30 cm which corresponds to a resolution of 150 µm
[b]Objects with a maximum height of 1 m can be entirely scanned by merging multiple scans with a height of up to 135 cm or with 165 cm height if additional restrictions on width and resolution are applied

40 mm. An attempt to scan 200 mm and 50 mm samples was also done. Samples of 200 mm were found to be inappropriate for micro-CT due to volumetric constraints and high density. These constraints hindered the effectiveness of penetration of X-rays, which consequently resulted in poor image quality.

Image processing using MATLAB The image processing of the micro-CT data was done in a MATLAB code created by [19]. This code generates 3D digital representations of air void distribution in asphalt samples with the help of binarization from manual histogram-based thresholding. Additionally, the code is able to calculate mean values for air void percentage present in a sample while also presenting graphical representations of distribution of porosity throughout the height of the sample. The initial cross-sectional images obtained from the scanning process usually appear unclear, often due to the inclusion of the object turntable that needs to be excluded. The inclusion of all cross-sectional images in the image processing may hinder the code's ability to correctly differentiate between voids, aggregate, and binder during binarization, thereby leading to inaccurate results. Thus, the image selection begins with the first high-quality image and progresses to the final image.

Wheel track test Wheel tracking is a method used to evaluate the resistance against permanent deformations of an asphalt mixture and is performed under specific load, speed, and temperature to simulate field conditions [20, 21].

Wheel track testing was performed on six samples with a diameter of 200 mm (4G, 5G, 6G, 4H, 5H, and 6H) at a temperature of 50 °C according to procedure B in NS-EN12697-22 [22]. The equipment used in this study was WTEN2 produced by Cooper Research Technology, located in the NPRA's laboratory in Trondheim. The test computes the rut depth (RD), proportional rut depth (PRD), and wheel tracking slope (WTS). The maximum allowed PRD for the wearing course of a stone mastic asphalt with AADT between 5001 and 10,000 is 7% and 5% for AADT larger than 10,000 [23].

3   Results and Discussion

3.1   Wheel Track Results

The results of PRD for the 200 mm samples are shown in Table 2.

The H-series resulted in an average PRD value of 1.8%, while the G-series had a slightly higher PRD value of 3.7%. However, both road sections have a PRD within the 5% PRD limit value for stone mastic asphalt. Surprisingly, the G-series demonstrated favorable wheel track results, despite the high air void content observed in the experimental results. This is most likely due to the robust aggregates present in the SMA that contribute to the section's resistance to permanent deformations. Consequently, it can be interpreted that wheel track results alone, provide an incomplete

Table 2 PRD values for H-series

Sample	PRD (%)
H-series	
H4 200 mm	1.8
H5 200 mm	1.9
H6 200 mm	1.7
Average	**1.8**
G-series	
G4 200 mm	3.4
G5 200 mm	4.1
G6 200 mm	3.7
Average	**3.7**

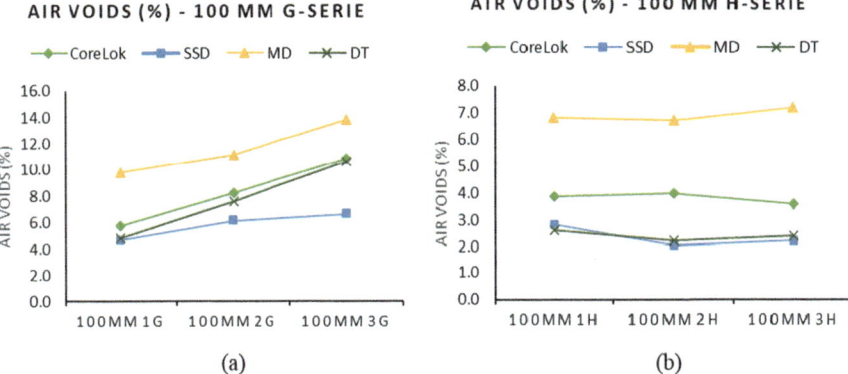

Fig. 1 Air void measurements of 100 mm diameter samples from (**a**) G-series and (**b**) H-series

evaluation of the quality of asphalt, particularly regarding durability. Therefore, wheel track results should be complemented by additional computations of air void percentages for a complete assessment.

3.2 Comparison of Air Void Content from Laboratory Tests and the Digital Twin

The following graphs represent a comparison between the air void content from the different measurements (Figs. 1 and 2).

A noticeable trend in the results from testing both different sample sizes and two road sections is evident, that is, measurements of dimensions are always giving the highest air void percentage, CoreLok is giving a middle value, and surface-dry method is giving the lowest percentages.

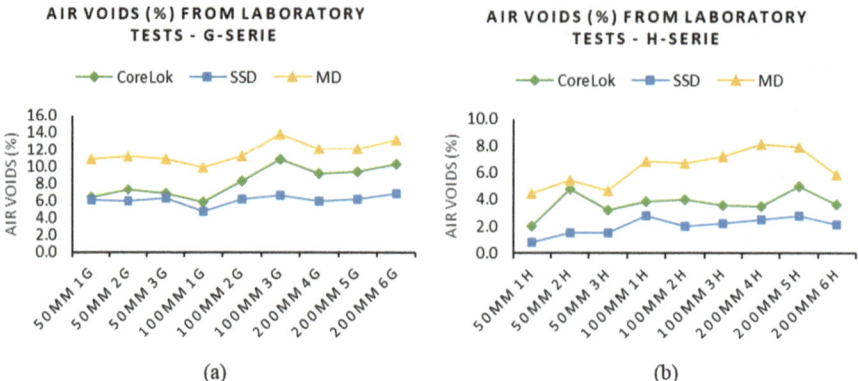

Fig. 2 Air void measurements from laboratory tests for all samples in (**a**) the G-series and (**b**) the H-series

Measurement of dimensions Both sections demonstrate high void content in the dimensional method. For the G-series, the air void content is above 10%; thus, according to the regulations stated in [24], this method is the one that should be applied for this section. Current practices by the NPRA are that asphalt pavements with over 10% air void content will lead to a 100% contractual deduction. Despite this, the G-series presents good resistance against permanent deformations from the wheel track testing. This can be assumed to be related to the high durability and structural stability properties of stone mastic asphalt [25]. It should be kept in mind that the coarse surface structure is included in this calculation method, thus leading to too high air void percentages. Therefore, it could be said that this method is too strict in this case. Hence, a decision regarding how much of the coarse surface structure that should be included in the computations should be made.

Saturated surface-dry method As stated in the methodology section, the saturated surface-dry method is applicable for samples with air voids up to 7%. For the G-series, the average void percentage from the samples is 6.1%; therefore, the saturated surface-dry method is applicable. However, water leakage from the sample on the scale was observed during experimentation. This means that the results from this method are inapplicable for the G-series. On the other hand, the H-series had an average void percentage of 2%, making this method applicable for it. This method accounts only for the isolated voids within the sample, since water flows through connected pores, whereas isolated voids contribute to buoyancy when immersed in water. Given that the H-series exhibits fewer isolated voids, a lower void percentage is shown in this method.

CoreLok Issues were encountered during the execution of CoreLok, specifically in the G-series. Problems such as water penetration into the polymer bag sealed samples or mistakes when calibrating the weight are assumed to be the reasons,

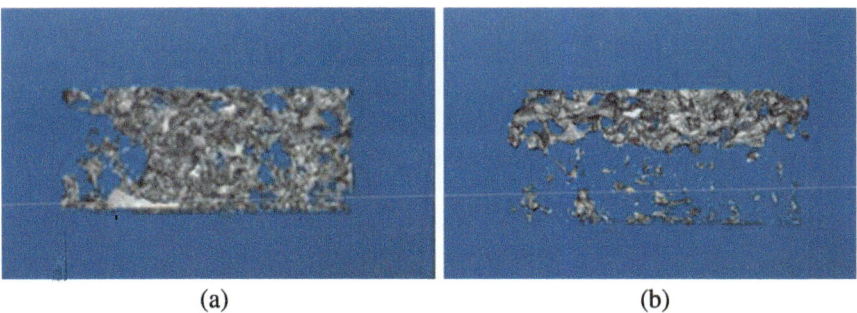

                    (a)   (b)

Fig. 3 3D views of 100 mm samples: (**a**) 1G and (**b**) 1H from MATLAB

and these are difficult to discover during execution. Nevertheless, these errors can be identified in the results. The experimentation for test sample 100 mm 3G was repeated several times, until the result was acceptable.

Digital twin Figures from the three-dimensional views of the void distribution from the digital twin for two samples from each road section are presented below. The first observation is the great difference between the void distribution of the two road sections. It is evident that the G-series has connected voids divided evenly through-out the height of the sample, indicating possible mistakes during compaction of the road. On the other hand, the voids in the H-series are mostly concentrated at the top constituting the rough surface layer, which is normal for SMA (Fig. 3).

Validation of digital twin results with laboratory tests Examining the results of the digital twin in the H-series, its results are aligned with the results from the saturated surface-dry method, which is the applicable method for this section. This indicates that the H-series is well compacted with few isolated air voids which is also proven by the digital twin.

In the case of the G-series, the results from the digital twin show a high degree of similarity with those from CoreLok. This could be explained by the exclusion of some of the rough surface layer of the SMA during vacuuming of the sample during the CoreLok procedure. This was also done during the creation of the digital twin, where some of the cross-sectional images were excluded to find the actual air void percentage of the samples. As discussed earlier, although measurement of dimen-sions is the applicable method for the G-series according to current regulations, it can be considered too strict which is proven by the digital twin.

4 Conclusion

In this chapter, an in-depth comparison of two road sections was examined for air void content and resistance against permanent deformations. The study also presents the digital twin technology obtained by microcomputer tomography as a new method for evaluating air void content in asphalt samples. The main conclusions from this study can be drawn as follows:

1. Based on the qualitative assessment, the digital twin demonstrated highly accurate air void measurements compared with those from conventional laboratory tests. However, more simulations might need to be performed with more samples to establish a statistical quantitative assessment.
2. The digital twin provided unique insights into the internal microstructure of asphalt samples, that is, new information that is unachievable/unavailable from conventional methods. This can be used to control the quality of the asphalt compaction process.
3. The digital twin proved to be a more flexible method in terms of accommodating a wider range of air void percentage in samples, eliminating constraints of traditional laboratory procedures.
4. Reduced physical demands, less repeatability, and less labor-intensive computations by utilizing the digital twin, rather than conventional methods.

5 Future Work

Digital twins from microcomputer tomography can still not fully replace laboratory tests due to sample size constraints of the employed equipment; however, its multiple benefits present promising potential for future applications. A suggestion for future work is to examine alternative scanning methods for creating digital twins to overcome the sample size constraints.

Acknowledgments The authors wish to thank the Norwegian Public Roads Administration for providing asphalt samples and access to their laboratory for conducting tests. Furthermore, we want to thank the Norwegian Geotechnical Institute for allowing us to use their micro-CT scanner.

References

1. Bonaquist, R.: Impact of mix design on Asphalt pavement durability. In: Enhancing the Durability of Asphalt Pavements, pp. 1–17. Transportation Research Board, Washington, D.C (2013)
2. Zaltuom, A.M.: A review study of the effect of air voids on asphalt pavement life. In: First Conference for Engineering Sciences and Technology, pp. 618–625. CEST-2018, Libya (2018)

3. Pellinen, T.K., Song, J., Xiao, S.: Characterization of hot mix asphalt with varying air voids content using triaxial shear strength test. In: Proceedings of the 8th Conference on Asphalt Pavements for Southern Africa (CAPSA'04). Sun City, South Africa (2004)
4. Yiqiu, T., Haipeng, W., Shaojun, M., Huining, X.: Quality control of asphalt pavement compaction using fibre Bragg grating sensing technology. Constr. Build. Mater. **54**, 53–59 (2014)
5. Khan, Z.A., Al-Abdul Wahab, H.I., Asi, I., Ramadhan, R.: Comparative study of asphalt concrete laboratory compaction methods to simulate field compaction. Constr. Build. Mater. **12**(6), 373–384 (1998)
6. Brown, E.R., Cross, S.A.: Comparison of laboratory and field density of asphalt mixtures. In: Annual Meeting of the Transportation Research Board. National Center for Asphalt Technology, Washington, DC (1991)
7. Kassem, E., Masad, E., Lytton, R., Chowdhury, A.: Influence of air voids on mechanical properties of Asphalt mixtures. Road Mater. Pavement Des. **12**(3), 493–524 (2011)
8. Bhattacharjee, S., Mallick, R.B.: An alternative approach for the determination of bulk specific gravity and permeability of Hot Mix Asphalt (HMA). Int. J. Pavement Eng. **3**(3), 143–152 (2010)
9. Buchanan, M.S., White, T.D.: Hot Mix Asphalt mix design evaluation using the CoreLok vacuum-sealing device. J. Mater. Civ. Eng. **17**(2), 137–142 (2005)
10. Hall, K.D., Griffith, F.T., Williams, S.G.: Examination of operator variability for selected methods for measuring bulk specific gravity of Hot-Mix Asphalt concrete. Transp. Res. Rec. **1761**(1), 81–85 (2001)
11. Pregnolato, M., Gunner, S., Voyagaki, E., De Risi, R., Carhart, N., Gavriel, G., Tully, P., Tryfonas, T., Macdonald, J., Taylor, C.: Towards civil engineering 4.0: concept, workflow and application of digital twins for existing infrastructure. Autom. Constr. **141**, 104421 (2022)
12. Bituminous mixtures — Test methods — Part 8: Determination of void characteristics of bituminous specimens, NS-EN 12697-8, 2018
13. Vegdirektoratet: Laboratorieundersøkelser. Statens vegvesen, Handbook R210 (2016)
14. InstroTek Inc.: CoreLok®., https://www.instrotek.com/products/corelok, last accessed 2024/02/20
15. Vicente, M.A., Mínguez, J., González, D.C.: The use of computed tomography to explore the microstructure of materials in civil engineering: from rocks to concrete. In: Computed Tomography - Advanced Applications. IntechOpen, Rijeka (2017)
16. Taheri-Shakib, J., Al-Mayah, A.: A review of microstructure characterization of asphalt mixtures using computed tomography imaging: prospects for properties and phase determination. Constr. Build. Mater. **385**, 131419 (2023)
17. Soldal, M., Johnsen, Ø.: The NGI Micro-Focus CT Scanner. Norwegian Geotechnical Institute, Tech Rep
18. Guo, X.L., Li, G., Zheng, J.Q., Ma, R.H., Liu, F.C., Yuan, F.S., Lyu, P.J., Guo, Y.J., Yin, S.: Accuracy of detecting vertical root fractures in non-root filled teeth using cone beam computed tomography: effect of voxel size and fracture width. Int. Endod. J. **52**(6), 887–898 (2019)
19. Allen, R., Dongmo-Engeland, B., Al-Batat, S.: Development of a MATLAB-based code for quantification of effective void space in porous pavement. In: 64th International Conference of Scandinavian Simulation Society, SIMS 2023, pp. 386–392. LiU E-press, Västerås, Sweden (2023)
20. Fontes, L.P.T.L., Trichês, G., Pais, J.C., Pereira, P.A.A.: Evaluating permanent deformation in asphalt rubber mixtures. Constr. Build. Mater. **24**(7), 1193–1200 (2010)
21. Mashaan, N.S.M., Karim, R.: Evaluation of permanent deformation of CRM-reinforced SMA and its correlation with dynamic stiffness and dynamic creep. Sci. World J. **2013**, 1–7 (2013)
22. Bituminous mixtures — Test methods — Part 22: Wheel tracking, NS-EN 12697-22: 2020, 2020

23. Statens vegvesen N200 Vegbygging, https://store.vegnorm.vegvesen.no/n200_2022, last accessed 2024/02/20
24. Myre, J., Evensen, R.: Reseptorienterte asfaltkontrakter - Kontroll og dokumentasjon av utførelse. Statens vegvesen, Tech Rep, 2505 (2008)
25. Nejad, F.M., Aflaki, E., Mohammadi, M.A.: Fatigue behavior of SMA and HMA mixtures. Constr. Build. Mater. **24**(7), 1158–1165 (2010)

Sustainable Vibration Screening with Dual Trenches Infilled with Geofoam and Aggregate: Full-Scale Field Experiments

Pradipta Chakrabortty (ID)**, Nitish Jauhari** (ID)**, Raja Kumar, and Diego Maria Barbieri** (ID)

Contents

1 Introduction

Ground-borne vibrations derived from several sources, such as machine foundations [1], road and rail traffic [2], and construction operations [3, 4], can cause harmful effects on the built environment. Therefore, it is necessary to adopt some isolation techniques to tackle the issue. Generally, the vibrations are screened out either at the origin or along the propagation path. Several researchers have recommended ground improvement techniques wherein sandbags, geocell-reinforced soil beds, and rubber

P. Chakrabortty (✉) · N. Jauhari · R. Kumar
Department of Civil and Environmental Engineering, Indian Institute of Technology Patna, Patna, Bihar, India
e-mail: pradipt@iitp.ac.in; nitish_2021ce15@iitp.ac.in; raja_2211ce41@iitp.ac.in

D. M. Barbieri
Department of Built Environment, Oslo Metropolitan University, Oslo, Norway
e-mail: diego.barbieri@oslomet.no

© The Author(s) 2025
M. Kioumarsi, B. Shafei (eds.), *The 1st International Conference on Net-Zero Built Environment*, Lecture Notes in Civil Engineering 237,
https://doi.org/10.1007/978-3-031-69626-8_122

sheets were placed under the excitation source to provide a dampening effect [5–7]. These approaches are generally labour-intensive due to the significant material modification required to improve the dynamic properties of soil. Furthermore, these processes generate a substantial amount of carbon footprint [8]. Therefore, a simple, cost-effective, and environmentally friendly solution represented by wave barriers which can be constructed open or infilled has attained significant attention in recent decades.

Several field studies have assessed the isolation potential of single open and infilled trenches [9–12]. The former ones were observed as the best mitigation solution as they create a discontinuity in the medium that stops the wave propagation. Several studies have shown that a small discontinuity in the soil medium is effective since the surface motions require a solid body to advance [13, 14]. However, this solution may lead to some environmental challenges, e.g., the trenches may get filled up with water and may facilitate the growth of vermin and insects [10]. In addition, side empty walls may collapse if not properly designed. The opportunity to fill the empty trenches using different sustainable materials is currently being explored by several researchers [9, 10, 15].

Çelebi et al. [9] conducted field tests to explore the suitability of softer (bentonite) and stiffer (cement concrete) materials as infilling agents in a single empty trench. The results showed that the isolation performance of the bentonite was superior to that of the cement concrete in isolating the vibrations. Ulgen and Toygar [12] conducted field tests to evaluate the performance of open, water-filled and geofoam-filled trenches. The isolation potential pertaining to water was observed to be inferior to open and geofoam-filled solutions. Mahdavisefat et al. [10] conducted a series of field experiments to examine the performance of sand-rubber mixture (SRM) under a wide range of frequencies (10–600 Hz). The margin of isolation efficiencies (IE) between the single open and SRM-filled trenches was only 10%, with the former being superior. The researchers recommended a depth of $1.50L_R$ (L_R = Rayleigh wavelength) for the efficient mitigation of vibrations. Furthermore, the optimum rubber content was suggested to be 30%. Alzawi and El Naggar [13] conducted outdoor field tests to assess the performance of open and geofoam-infilled barriers and observed that the mitigation improved significantly for depths greater than $0.6L_R$. Naghizadehrokni et al. [14] conducted field experiments using thin foam and suggested a required depth of $1.0L_R$.

Current solutions to vibration mitigation using single trenches have recommended a depth ranging from $0.6L_R$ to $1.5L_R$. However, the suggested depth is found unfeasible in many practical scenarios due to poor subsoil conditions and stability issues. Jauhari et al. [15] conducted full-scale field tests to assess the isolation efficacy of dual open and geofoam-infilled trenches. The depth requested for the former ones was 55% less than the single open trench for an IE of 50%. Furthermore, the IE for dual open and geofoam-infilled trenches for a depth of $0.3L_R$ was noted as 76% and 71%, respectively.

It is apparent that the performance of dual trenches is significantly higher compared to that of single ones. However, there is a significant lack of thorough research dwelling on dual trenches and finding the most suitable sustainable infill material. Therefore, an attempt has been made in this study to appraise the performance of a combination of softer and stiffer materials placed inside the trenches compared to the natural soil. Geofoam and aggregate were used as the soft and stiff materials, respectively. A harmonically-varying dynamic load was applied using a Lazan-type mechanical oscillator within a frequency range of 20–45 Hz. Furthermore, the infill materials were interchanged in the dual trenches to assess the most suitable location for each.

2 Material Properties

Full-scale field vibration tests were conducted on a dedicated site in the campus of the Indian Institute of Technology Patna (Bihta, India). The soil stratigraphy of Patna city is closely related to the meandering and migrating nature of two rivers, Sone and Ganges. Therefore, the study area lies in the Indo-Gangetic alluvial plain consisting mostly sand and silt particles. A soil sample was collected from a depth of 0.5 meters from the experimental site and was subjected to different laboratory tests to assess the physical properties. The bulk density (ρ_S) of the soil was calculated as 1.79 g/cc. The soil was classified as *SM* as per the Unified Soil Classification System. Multichannel Analyses of Surface Wave (*MASW*) tests were performed to investigate the dynamic properties [16]. The waves with an average shear wave velocity (V_s) of 157.84 m/s propagated near the surface and down to a depth of 3 m. The acoustic impedance ($Z_S = \rho_S \times V_S$) of soil was calculated as 2.83×10^5 kgm^{-2}s^{-1}.

The expanded polystyrene geofoam, supplied by Gupta polymers private limited, Patna, Bihar, India, was used as an infilling material. Its physical properties were calculated using ASTM standards. The density (ρ_g) was 0.015 g/cc as per ASTM C303 [17]. The primary (V_{pg}) and secondary (V_{sg}) wave velocities were 710 m/s and 476.6 m/s, respectively as per ASTM C597 [18]. The acoustic impedance ($Z_g = \rho_g \times V_{sg}$) was 7.15×10^3 kgm^{-2}s^{-1}.

Natural aggregates, procured from a quarry near Gaya, Bihar, India, were also used as an infill material. The particle size ranges between 31 and 50 mm as per IS 2386-1 [19]. The mean particle size (D_{50}) was 42 mm. The bulk specific gravity was 2.79 as per IS 2386-3 [20].

3 Test Procedure

The Lazan-type mechanical oscillator, rated to operate up to a speed of 3000 revolutions per minute, was used to generate the dynamic forces. The device was placed on a concrete block of dimensions $0.6 \times 0.6 \times 0.5$ m^3 and exerted frequencies ranging from 20 to 45 Hz. A unit (1 kN) vertical dynamic force was maintained throughout the study.

Initially, the amplitudes of vertical vibrations were noted for the condition "without trench" (*WT*). The recorded data was used to compare the results in the presence of the two mitigation systems. The two trenches, at distances of 1.5 m and

3 m from the source, were excavated to a depth of 1 m. The width of the trenches was 0.5 m. In the first trench, a material softer than in-situ soil, i.e., geofoam (G), was placed, whereas a stiffer material, i.e., aggregate (A), was allocated in the second trench. Therefore, the incoming waves were first subjected to interact with the soft material and then the stiffer one. This mitigation system has been abbreviated as $G + A$ in the following discussion. Subsequently, the materials in the trenches were interchanged, with the former being filled with the aggregate and the latter with the geofoam. In this case, the waves firstly encountered the stiff material and afterward the softer one ($A + G$). Figure 1 schematically shows the dimensions and location of the dual trenches and the point of interest.

In order to record the amplitude of vibrations, a piezoelectric constant current line drive accelerometer operating in the frequency range of 0.1–6000 Hz and a sensitivity of 1 mV/ms^{-2} was placed at a distance of 1 m from the edge of the second trench. The position was maintained at the same place for all the experiments. The vibration data were recorded for a duration of 15 seconds with a sampling rate of 65,536 Hz. The $A + G$ and $G + A$ setups are shown in Fig. 2.

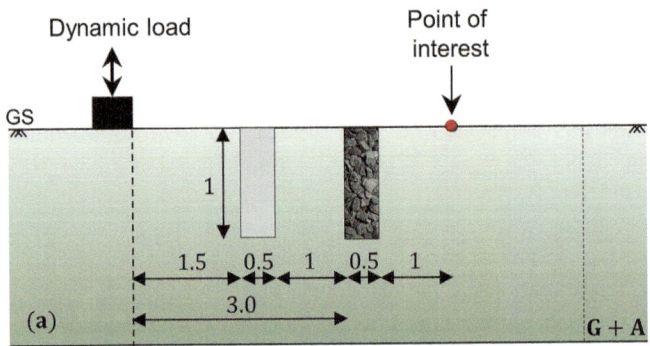

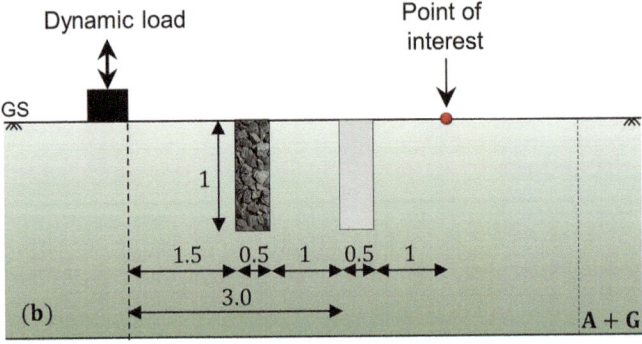

Fig. 1 Schematics of barrier systems: (**a**) geofoam and aggregate ($G + A$); (**b**) aggregate and geofoam ($A + G$). All dimensions are in meters

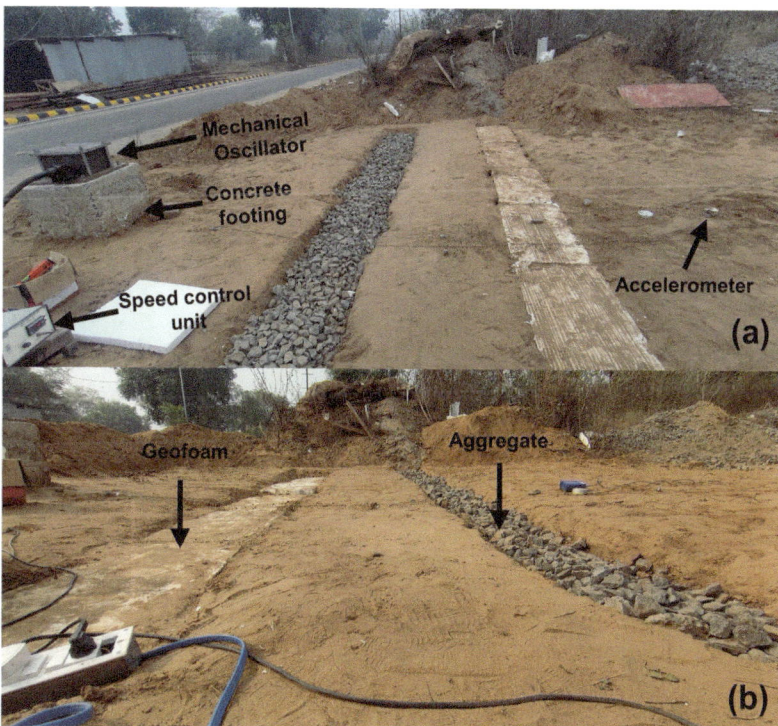

Fig. 2 Field photographs of the vibration mitigation systems: (**a**) $A + G$; (**b**) $G + A$

Fig. 3 Interrelations
between actual and
normalized parameters of
the dual trench system

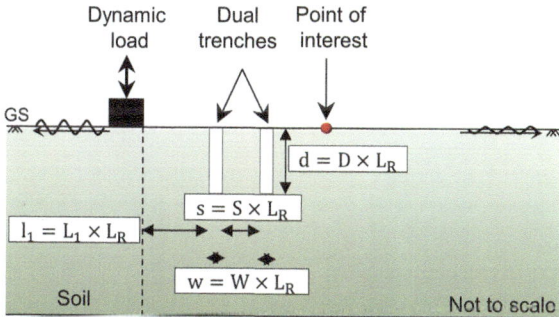

The geometric parameters, namely, location (l), width (w), depth (d), and spacing (s) associated with the mitigation systems, were normalized by means of the Rayleigh wavelength (L_R) to get the respective dimensionless parameters, namely normalized location (L), normalized width (W), normalized depth (D), and normalized spacing (S). Such a normalization process was implemented to remove the dependency of the results on the excitation frequencies. Figure 3 presents the correlations between the actual and dimensionless parameters of the dual trench systems.

4   Results and Discussion

4.1   Insertion Loss

Insertion loss (IL) can be defined as the reduction in the vibration levels due to the introduction of the mitigation systems. The IL in decibels can be calculated using the equation suggested by Hoorickx et al. [21] as

$$IL\ (dB) = 20\ \log_{10}\left(\frac{A_{without}}{A_{with}}\right) \tag{1}$$

where A_{with} and $A_{without}$ symbolize the peak acceleration amplitudes of the vibrations in the presence and absence of a barrier system, respectively.

Figure 4 shows the insertion loss variation with the excitation frequency at the point of interest for two mitigation systems. A higher insertion loss was observed for the vibrations generated due to higher excitation frequencies. An insertion loss of 12 dB and 16 dB was observed for the $G + A$ and $A + G$ setups for an excitation frequency of 45 Hz.

4.2   Time History and Fast Fourier Transform

Figure 5 depicts the variation of time history and frequency domain data at the point of interest. The plots were developed for an excitation frequency of 45 Hz. The time history plots, as depicted in Fig. 5a, show that the presence of mitigation systems weakens the vibration amplitudes significantly at the measuring point. The amplitudes of the vertical vibrations were noted as 1.00 m/s^2, 0.25 m/s^2, and 0.16 m/s^2 for WT, $G + A$, and $A + G$ mitigation systems, respectively.

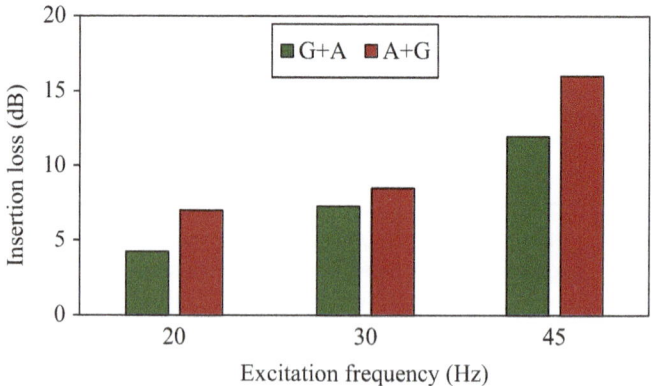

Fig. 4 Variation of insertion loss with the excitation frequency

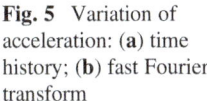

Fig. 5 Variation of acceleration: (**a**) time history; (**b**) fast Fourier transform

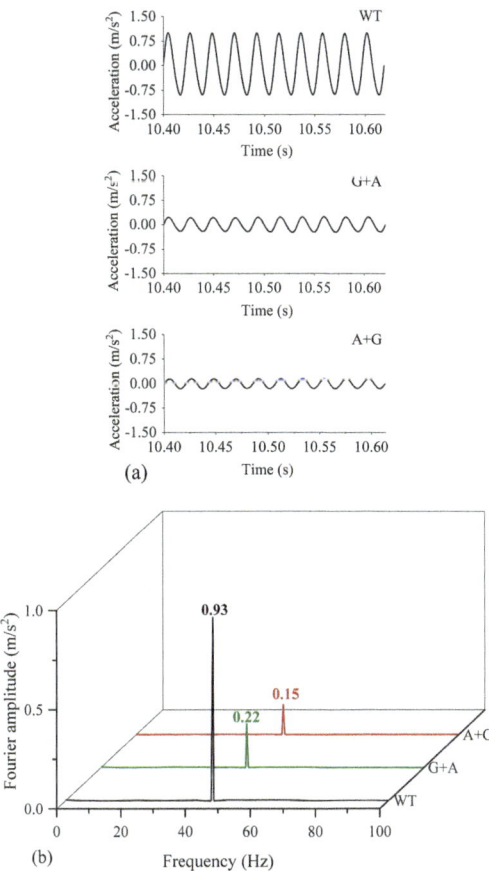

Furthermore, Fig. 5b shows the frequency domain plots representing the distribution of vibration energy over a range of frequencies. The result shows that the ground response matches the same frequency as that of the applied dynamic load. The peak Fourier amplitude was 0.93 m/s^2, 0.22 m/s^2, and 0.15 m/s^2 for WT, $G + A$, and $A + G$ mitigation systems, respectively.

4.3 Effect of the Normalized Depth of Trenches

The normalized depth of the trenches was varied between 0.14 and 0.31 during the tests. The influence of the increase in D on the isolation performance of the mitigation systems was observed in terms of the Amplitude Reduction Ratio (ARR), which was calculated at the point of interest. This parameter is the ratio

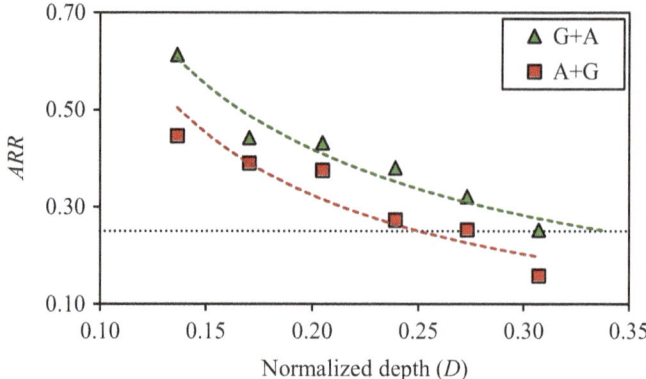

Fig. 6 Variation of *ARR* with normalized depth

between the amplitude of the ground vibrations in the presence of a mitigation system and the corresponding one in its absence as follows:

$$ARR = \frac{\text{Amplitude of ground motion with a mitigation system}}{\text{Amplitude of ground motion without a mitigation system}} \qquad (2)$$

Its variation with the normalized depth for the two mitigation systems is shown in Fig. 6. It is possible to note a negative correlation between *ARR* and the normalized depth.

The performance of the dual trenches filled successively with aggregate and geofoam ($A + G$) was superior to that of system filled with geofoam and aggregate ($G + A$). Several researchers have suggested that a barrier system would be efficient if it attains an *ARR* value equal to or less than 0.25 [22, 23]. The required depth for the $A + G$ and $G + A$ setups to reach this limiting criterion was noted as $0.25L_R$ and $0.34L_R$, respectively.

In the case of the $A + G$ mitigation system, the incoming waves from the source initially confront the trench filled with aggregate, which exhibits a random arrangement of air pockets, thus weakening the advancement of the wave energy. Furthermore, a significant amount of vibration becomes absorbed, diffracted and scattered at the interfaces of particles. Thereafter, the transferred waves encounter the geofoam-filled trench, undermining the propagation toward the point of interest due to its low acoustic impedance value.

5   Conclusions

The current study discussed the use of wave barriers as mitigation systems in isolating ground vibrations generated due to machine foundations. Full-scale field experiments have been conducted to examine the isolation efficiency of dual barriers infilled with geofoam and aggregate. These two filling materials were also interchanged to assess the most suitable location for each.

- The amplitudes of the vertical vibrations were 1.00 m/s^2, 0.25 m/s^2, and 0.16 m/s^2 for WT, $G + A$, and $A + G$ mitigation systems, respectively. The Fourier amplitude was reduced by 76% and 84% for the $G + A$ and $A + G$ setups compared to the free field condition.
- Maximum insertion losses of 12 dB and 16 dB were recorded at the point of interest under an operating frequency of 45 Hz for $G + A$ and $A + G$ setups, respectively.
- The optimal depth required to reach an ARR less than or equal to 0.25 was $0.25L_R$ and $0.34L_R$ for $A + G$ and $G + A$ mitigation systems, respectively.
- The results suggest that infilling the first trench with aggregate and the second one with geofoam achieves the best performance.

Eventually, it may be worth stressing that the findings are only valid for the soil condition and stratigraphy considered in this research, which aimed at mitigating continuous harmonic vibrations. Further experimental studies can be performed to assess the efficacy of dual trenches in the isolation of ground vibrations due to other types of loading, e.g., impact.

References

1. Jauhari, N., Hegde, A., Chakrabortty, P.: Numerical investigation of vibration screening using single and dual open trenches in layered soil media. In: 10th European Conference on Numerical Methods in Geotechnical Engineering. Imperial College, London, UK (2023)
2. Majumder, M., Ghosh, P.: Screening of train-induced vibration with open trench—a numerical study. In: Prashant, A., Sachan, A., Desai, C.S. (eds.) Advances in Computer Methods and Geomechanics. Springer Singapore, Singapore (2020)
3. Das, A., Chakrabortty, P.: Artificial neural network and regression models for prediction of free-field ground vibration parameters induced from vibroflotation. Soil Dyn. Earthq. Eng. **148**, 106823 (2021)
4. Jauhari, N., Raj, A., Hegde, A.: Efficacy of dual barrier systems in mitigating ground-borne vibrations induced by impact loading. Geo-Congress. **2024**, 202–211 (2024)
5. Liu, S.-H., Gao, J.-J., Wang, Y.-Q., Weng, L.-P.: Experimental study on vibration reduction by using soilbags. Geotext. Geomembr. **42**(1), 52–62 (2014)
6. Tafreshi, S.N.M., Zakeri, R., Dawson, A.R., Hegde, A.: Compressive wave isolation of lightweight equipment very close to a vibration source using a rubber sheet. Int. J. Geomech. **22**(1), 04021263 (2022)
7. Venkateswarlu, H., Hegde, A.: Isolation prospects of geosynthetics reinforced soil beds subjected to vibration loading: experimental and analytical studies. Geotech. Geol. Eng. **38**(6), 6447–6465 (2020)
8. Chester, M.V., Horvath, A.: Environmental assessment of passenger transportation should include infrastructure and supply chains. Environ. Res. Lett. **4**(2), 024008 (2009)
9. Çelebi, E., Firat, S., Beyhan, G., Çankaya, İ., Vural, İ., Kirtel, O.: Field experiments on wave propagation and vibration isolation by using wave barriers. Soil Dyn. Earthq. Eng. **29**(5), 824–833 (2009)

10. Mahdavisefat, E., Salehzadeh, H., Heshmati, A.A.: Full-scale experimental study on screening effectiveness of SRM-filled trench barriers. Geotechnique. **68**, 869 (2018)
11. Nappa, V., Bilotta, E., Flora, A.: Experimental and numerical investigation on the effectiveness of polymeric barriers to mitigate vibrations. Geotech. Geol. Eng. **37**(6), 4687–4705 (2019)
12. Ulgen, D., Toygar, O.: Screening effectiveness of open and in-filled wave barriers: a full-scale experimental study. Constr. Build. Mater. **86**, 12–20 (2015)
13. Alzawi, A., El Naggar, M.H.: Full scale experimental study on vibration scattering using open and in-filled (GeoFoam) wave barriers. Soil Dyn. Earthq. Eng. **31**(3), 306–317 (2011)
14. Naghizadehrokni, M., Ziegler, M., Sprengel, J.: A full experimental and numerical modelling of the practicability of thin foam barrier as vibration reduction measure. Soil Dyn. Earthq. Eng. **139**, 106416 (2020)
15. Jauhari, N., Hegde, A., Chakrabortty, P.: Full scale field studies for assessing the vibration isolation performance of single and dual trenches. Transp. Geotech. **39**(2023), 1–15 (2023)
16. Nilay, N., Chakrabortty, P., Popescu, R.: Liquefaction hazard mapping using various types of field test data. Indian Geotech. J. **52**(2), 280–300 (2022)
17. ASTM C303: Standard Test Method for Dimensions and Density of Preformed Block and Board–Type Thermal Insulation. American Society of Testing and Materials, West Conshohocken, Pennsylvania, USA (2021)
18. ASTM C597: Standard Test Method for Pulse Velocity through Concrete. American Society of Testing and Materials, West Conshohocken, Pennsylvania, USA (2009)
19. IS 2386-1: Methods of Test for Aggregates for Concrete: Particle Size and Shape. Bureau of Indian Standards, New Delhi (1963)
20. IS 2386-3: Methods of Test for Aggregates for Concrete: Specific Gravity, Density, Voids, Absorption and Bulking. Bureau of Indian Standards, New Delhi (1963)
21. Hoorickx, V.C., Schevenels, M., Lombaert, G.: Double wall barriers for the reduction of ground vibration transmission. Soil Dyn. Earthq. Eng. **97**, 1–13 (2017)
22. Woods, R.D.: Screening of surface wave in soils. J Soil Mech. Found. Div. **94**, 951 (1968)
23. Jauhari, N., Hegde, A., Chakrabortty, P.: Vibration mitigation using dual-open and infilled trenches in layered soil media: Field tests and numerical simulations. Comput. Geotech. **170**, 1–16 (2024)

Exploring the Predictive Performance of Simple Regression Models and ANN in 2D Truss Analysis

Vagelis Plevris ⓘ, Alejandro Jiménez Rios ⓘ, and Usama A. Ebead ⓘ

Contents

1 Objective and Literature Review

Artificial intelligence (AI) has garnered considerable interest in various scientific domains in recent years, spanning from managing large datasets to aiding in medical diagnostics. Its integration into everyday life is evident through personalized advertisements, virtual assistants, autonomous vehicles, and more. Unsurprisingly, AI techniques have permeated engineering disciplines, including civil and structural engineering [1, 2], showcasing remarkable achievements [3, 4].

Málaga-Chuquitaype [5] authored a thought-provoking review article on the application of machine learning (ML) in structural design. The article raises a compelling query regarding the future necessity of human engineers in the field of

V. Plevris (✉) · U. A. Ebead
Qatar University, Doha, Qatar
e-mail: vplevris@qu.edu.qa; uebead@qu.edu.qa

A. J. Rios
Oslo Metropolitan University, Oslo, Norway
e-mail: alejand@oslomet.no

© The Author(s) 2025
M. Kioumarsi, B. Shafei (eds.), *The 1st International Conference on Net-Zero Built Environment*, Lecture Notes in Civil Engineering 237,
https://doi.org/10.1007/978-3-031-69626-8_123

structural design. Notably, the paper refrains from providing a direct answer to this question but rather operates on the assumption that the role of human engineers in traditional design practices is steadily diminishing. In another work, Nguyen and Vu [6] undertook a comparative analysis of ML algorithms aimed at predicting the behavior of truss structures. They conducted this comparison through the examination of two numerical examples, evaluating the performance of multiple ML algorithms using three error metrics. The outcomes of their investigation demonstrated that AdaBoost outperformed the other six algorithms that were considered in the study.

Mai et al. [7] introduced a deep unsupervised learning framework for optimizing truss structures under various constraints. They explored several illustrative examples to showcase the effectiveness of their proposed framework, highlighting its ability to deliver high-quality optimal solutions while significantly reducing computational costs compared to traditional methods. Kang and Yoon [8] focused on the configuration and training of a two-layer ANN tailored for truss design applications, highlighting its potential roles in structural design problem-solving. Mai et al. [9] introduced a straightforward and resilient unsupervised ANN framework designed for conducting geometrically nonlinear analyses of inelastic truss structures. The fundamental concept involved utilizing the ANN to directly predict nonlinear structural responses without resorting to time-consuming incremental-iterative algorithms typical in standard finite element (FE) methods.

Khodadadi et al. [10] developed an innovative enhanced ANN model tailored for optimizing the design of truss structures which featured two distinct characteristics. Firstly, an improved initialization mechanism was introduced, leveraging opposite-based learning. Secondly, the algorithm incorporated a small set of tunable parameters to enhance its ability for exploration and exploitation. The efficacy of the method was evaluated in engineering design scenarios.

The primary objective of this study is to assess and compare the performance of various regression models in the context of predicting critical structural parameters within a plane truss model. Specifically, we aim to examine the effectiveness of linear, polynomial (both second and third degrees), and artificial neural network (ANN) regression models in estimating the maximum displacement, maximum (tensile) stress, and minimum (compressive) stress exhibited by the truss structure. By subjecting these diverse regression techniques to a rigorous evaluation, we seek to identify which model excels in capturing the intricate and nonlinear relationships that govern the behavior of the truss under varying conditions. This research objective is crucial for providing valuable insights into the selection of the most suitable regression model for structural analysis and design, with potential applications spanning fields such as civil engineering, architecture, and materials science.

Furthermore, this study also endeavors to investigate the role of model complexity in enhancing predictive performance. By systematically transitioning from simpler linear models to more intricate polynomial and ANN models, we aim to discern how adding complexity to the regression models affects their accuracy and generalization capabilities. Understanding the interplay between model complexity and predictive accuracy is essential, as it can guide practitioners and researchers in selecting the most appropriate modeling approach for diverse structural scenarios.

The remainder of this paper is structured as follows: Section 2 provides an in-depth overview of the truss analysis problem being examined, Sect. 2.2 delves into the framework for data generation, accompanied by a discussion of its primary attributes. Section 3 outlines the various regression models that have been used and assessed, whereas Sect. 4 presents and deliberates upon the findings obtained. Lastly, Sect. 5 encapsulates the conclusions drawn from this study.

2 Truss Structure and Dataset

The truss considered is a standard 10-bar plane truss, depicted in Fig. 1. The model has six nodes and ten elements in total, numbered from 1 to 10, as shown in the figure. The members have a modulus of elasticity (E) equal to 10,000 ksi (≈ 68.95 GPa), while the length L is equal to 360 in (9.144 m). Each of the applied nodal loads is equal to 100 kip (≈ 444.82 kN). This truss structure has been extensively used as a benchmark problem in structural optimization, with the objective being to minimize the weight of the structure under constraints on stresses and displacements [11–13].

The dataset comprises 2000 data points, each featuring five input variables and three output variables, resulting in a tabular format with dimensions of 2000 rows and 8 columns. The outputs of the dataset are computed using an FE analysis code written in MATLAB. The whole dataset is stored as a text file with 253 kB file size. A portion of the data, specifically 15% (or 300 data points), is reserved for testing purposes and remains excluded from the training of any of the models. For comprehensive information about the input and output variables, please refer to the subsequent subsections.

Fig. 1 The 10-bar truss structure considered

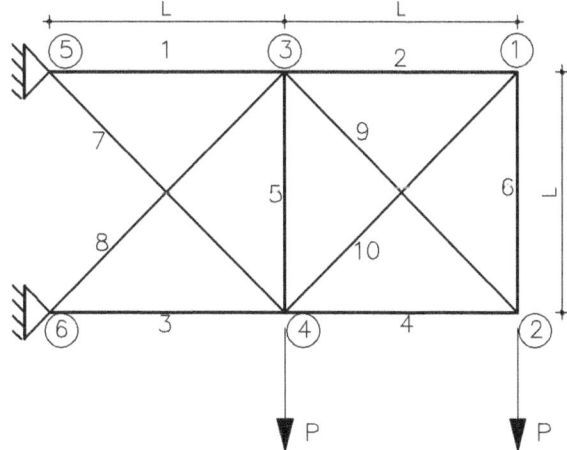

2.1   Input and Output Variables

There are five input variables related to the cross-sectional area of individual members. Given that there are a total of ten members, it becomes necessary to organize them into specific groups. Table 1 provides an overview of both the input variables and the grouping arrangement for these elements. For example, input variable I-4 is the section area of members 7 and 9 of the truss (diagonals in the NW-SE direction).

There are three output variables: (1) Maximum absolute vertical displacement (in the y direction), in inches (in), (2) Maximum stress (ksi) of all members (maximum tensile stress), (3) Minimum stress (ksi) of all members, in absolute terms (i.e., maximum compressive stress).

2.2   Generation of the Dataset

The five input variables correspond to section areas, each falling within the specified range of [0.1, 35]. These inputs are created as random values within this range using a uniform distribution and are then randomly combined with each other. Consequently, the first five columns of the dataset consist of random numbers selected from this specified range (2000 × 5 = 10,000 random numbers in total for the first five columns). The output variables, on the other hand, are determined through FE analysis of the truss structure. This analysis is carried out 2000 times, resulting in the generation of 2000 distinct output values.

In Fig. 2, box plots depict the input and output variables, highlighting distinct characteristics. It is evident that the input variables exhibit a uniform distribution

Table 1 Input variables and grouping of the truss elements	Input variable	Elements	Range	Units
	I-1 (A_1, A_2)	1, 2 (top, horizontal)	[0.1, 35]	in^2
	I-2 (A_3, A_4)	3, 4 (bottom, horizontal)	[0.1, 35]	in^2
	I-3 (A_5, A_6)	5, 6 (vertical)	[0.1, 35]	in^2
	I-4 (A_7, A_9)	7, 9 (diagonal NW-SE)	[0.1, 35]	in^2
	I-5 (A_8, A_{10})	8, 10 (diagonal SW-NE)	[0.1, 35]	in^2

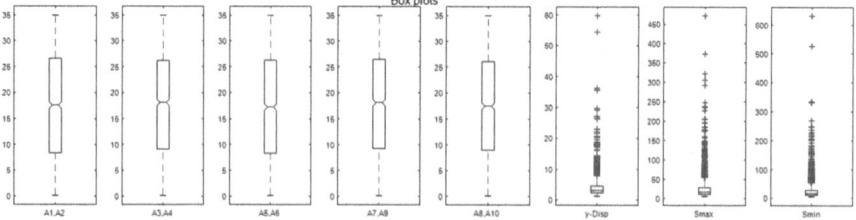

Fig. 2   Box plots of input and output variables

within the [0.1, 35] range. In contrast, the distribution of the output variables shows considerable non-uniformity, marked by the presence of numerous outliers.

3 Regression Models

3.1 Linear (First Order) and Polynomial Regression (Second and Third Order)

Linear regression [14] (LR) is a fundamental statistical method used to model the relationship between a dependent variable and one or more independent variables. This method assumes that the relationship between the variables is linear, i.e., can be represented by a straight line. In its simplest form, known as simple linear regression, the model predicts the dependent variable based on a single independent variable (one input). It calculates the best-fitting line through the data points, where "best-fitting" is defined as minimizing the sum of the squares of the differences between the observed and predicted values. This line is represented by the equation "$y = mx + c$," where y is the dependent variable, x is the independent variable, m is the slope of the line, and c is the y-intercept. Linear regression is widely used in various fields for predictive analysis and inferential statistics, owing to its simplicity and interpretability. In our specific analysis, we deal with a more intricate scenario involving five input variables and three output variables. Consequently, the number of independent variables amounts to five.

Polynomial regression [14] is a versatile statistical method employed to capture nonlinear relationships between a dependent variable and one or more independent variables. Unlike linear regression, which assumes a linear relationship, polynomial regression allows for the modeling of more intricate patterns by introducing polynomial terms of the independent variables into the equation.

In our analysis, we employ two distinct polynomial regression models to better capture complex relationships within our dataset. The first model is a second-degree polynomial regression model, which introduces second-degree terms and interactions between the independent variables. For instance, in a second-degree model featuring two independent variables, x and y, the terms incorporated would include the constant, x, x^2, y, xy, and y^2. This model extends beyond the simplicity of the linear (first degree) model, allowing us to account for nonlinearities and more intricate patterns.

Moreover, we also utilize a third-degree polynomial regression model, which incorporates third-degree terms and further enriches the model by including additional interactions between the independent variables. This third-degree model surpasses the complexity of the second-degree model, enabling us to comprehensively explore and represent the nonlinear relationships and higher-order interactions among our five independent variables, thereby enhancing our understanding of the underlying data dynamics.

To accommodate the three outputs, we run the linear and polynomial regression models three times each, generating three distinct sub-models for each model, each tailored to predict one of the three output variables. This is not the case with the Artificial Neural Network model, which can handle all output variables simultaneously.

3.2   Artificial Neural Network

Artificial Neural Network (ANN) regression is a robust ML technique used to model complex relationships between input variables and predict output values, which has several applications in structural engineering problems [15–19]. In our study, we implement a back-propagation ANN regression model with specific architectural characteristics to suit our research needs. Our ANN model consists of two hidden layers, each containing 25 neurons, making it a deep neural network capable of capturing intricate patterns and nonlinear dependencies within the data. The input layer comprises five neurons, corresponding to our five independent variables, while the output layer consists of three neurons, aligning with the three output variables we aim to predict.

For the training process, we employ the Levenberg–Marquardt training algorithm [20], a widely used optimization method for fine-tuning neural network weights and biases. To ensure the model's robustness and generalization capability, we divide our dataset into three distinct subsets: 15% of the data is used for testing, the same as with the previous models, 70% of the data is allocated for training itself, while another 15% is used (during training) for validation purposes. The validation procedure monitors the performance of the network on the validation data during training to prevent overfitting. This comprehensive approach ensures that our ANN regression model not only learns from the available data but also generalizes well to make accurate predictions on new and unseen data points, ultimately enhancing the reliability of the research findings. It has to be noted that the testing set is exactly the same for all regression models. In all cases, the testing is done after the model has been fully developed and the testing data remains unseen to the model until then.

3.3   Performance Metrics

To assess the performance of each model, we utilize three separate metrics: (i) the Root Mean Squared Error (RMSE), (ii) the Pearson correlation coefficient (R), and (iii) the Mean Normalized Gross Error (MNGE). The formulas for these three metrics are outlined in Eqs. (2)–(4):

$$\bar{p} = \frac{1}{N} \sum_{i=1}^{N} p_i, \quad \bar{r} = \frac{1}{N} \sum_{i=1}^{N} r_i \tag{1}$$

$$\text{RMSE} = \sqrt{\frac{1}{N} \sum_{i=1}^{N} (p_i - r_i)^2} \tag{2}$$

$$R = \frac{\sum_{i=1}^{N} (p_i - \bar{p})(r_i - \bar{r})}{\sqrt{\sum_{i=1}^{N} (p_i - \bar{p})^2} \cdot \sqrt{\sum_{i=1}^{N} (r_i - \bar{r})^2}} \tag{3}$$

$$\text{MNGE} = \frac{1}{N} \sum_{i=1}^{N} \frac{|p_i - r_i|}{r_i} \tag{4}$$

In these equations, N represents the total count of data points in the entire dataset, or within a specific subset such as the training or testing subset, depending on the subset chosen for evaluating the regression model. In addition, r_i represents the real (target) value, p_i denotes the predicted value, and $\bar{r}, \bar{p}$ denote the mean values of the real and predicted values, respectively. RMSE is expressed in the units of the corresponding output, while R and MNGE are unitless quantities. More detailed information about these metrics is available in reference [21].

Additionally, we employ the Taylor diagram, which integrates three statistical parameters—namely, (i) the Centered Root Mean Square Difference (CRMSD), (ii) R, and (iii) the Standard Deviation σ—into a single, easily interpretable diagram. The Taylor diagram proves valuable for summarizing and comparing the relative strengths of various models, as discussed in Ref. [21]. The formula for the calculation of CRMSD is given in Eq. (5).

$$\text{CRMSD} = \sqrt{\frac{1}{N} \sum_{i=1}^{N} [(p_i - \bar{p}) - (r_i - \bar{r})]^2} \tag{5}$$

4   Numerical Results

Figure 3 displays plots depicting the predicted values contrasted with the actual (target) values for each model and output variable. Additionally, the graph features a reference line, $y = x$, representing an ideal match, accompanied by two error boundary lines signifying a 10% prediction error tolerance.

Table 2 presents the error metrics values for each model and each output. The target values, denoting a perfect match, are also shown in the table. Figure 4

y-Displacement

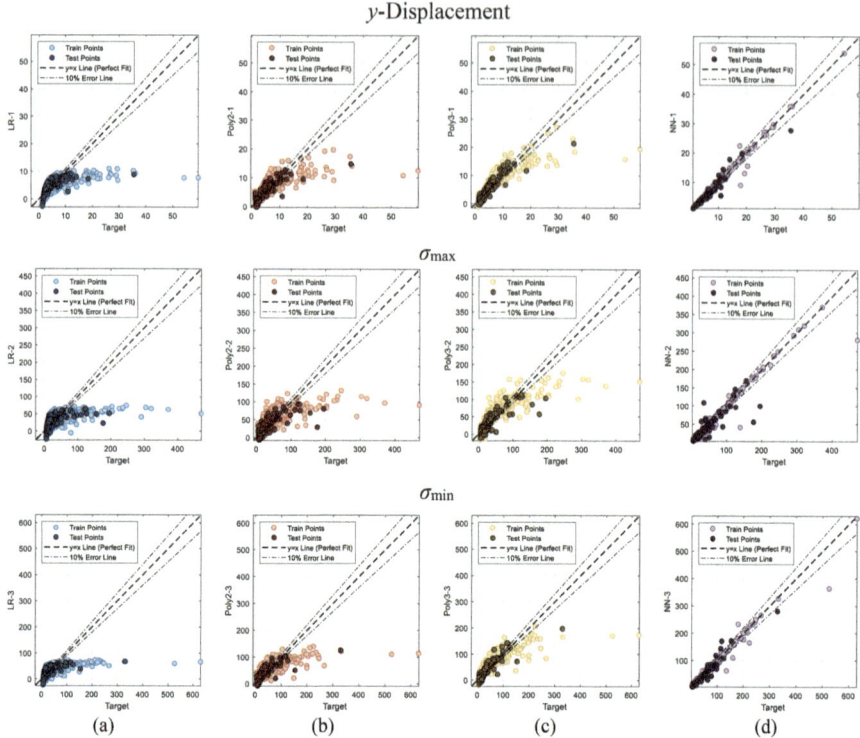

Fig. 3 Predicted vs. actual (target) values for all four models, for all three outputs: (**a**) LR Model, (**b**) Poly2 Model, (**c**) Poly3 Model, (**d**) NN Model

Table 2 Error metrics values

		All data (train and test)			Test data		
	Model	RMSE	R	MNGE	RMSE	R	MNGE
Output variable	Target	0	1	0	0	1	0
1 (y-displacement)	LR	2.96	0.603	0.403	2.32	0.638	0.408
	Poly2	2.41	0.760	0.307	1.81	0.798	0.311
	Poly3	1.96	0.849	0.263	1.39	0.888	0.264
	NN	0.62	0.987	0.030	0.71	0.971	0.043
2 (σ_{max})	LR	25.54	0.571	0.701	20.19	0.570	0.677
	Poly2	21.08	0.735	0.551	16.46	0.740	0.538
	Poly3	17.03	0.837	0.459	13.86	0.826	0.470
	NN	6.80	0.976	0.072	11.39	0.884	0.106
3 (σ_{min})	LR	28.27	0.559	0.737	22.25	0.625	0.693
	Poly2	23.58	0.722	0.572	17.97	0.782	0.580
	Poly3	19.58	0.819	0.484	13.93	0.880	0.489
	NN	5.62	0.987	0.075	6.72	0.972	0.092

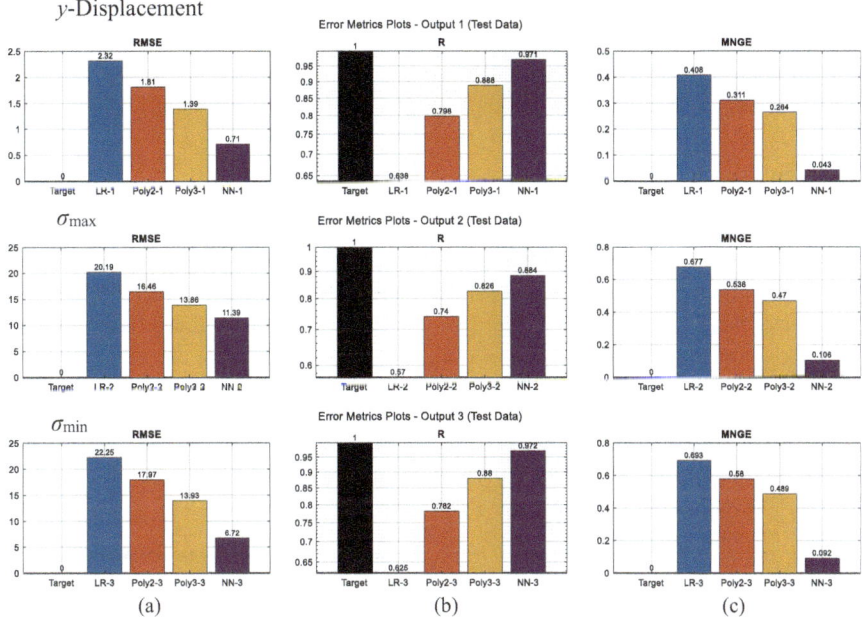

Fig. 4 Error metrics for all models and all outputs (Test data only): (**a**) RMSE, (**b**) R, (**c**) MNGE

illustrates the error metrics for all four models across each output variable, exclusively for the test dataset. To facilitate comparison, we have included an initial column that serves as a reference point denoting the "target" value for each metric. In other words, it represents the value associated with a perfect fit, which is zero for RMSE and MNGE and 1 for R. Importantly, these diagrams exclusively pertain to test predictions, omitting the training data employed by the models. This deliberate exclusion ensures a fair assessment of the models' generalization capabilities, highlighting their performance on unseen data points. While it is evident that all models perform optimally on the training set, our focus here centers on the test set, where models typically exhibit comparatively lower performance.

Notably, the ANN model outperforms the other models across all considered metrics. Additionally, the introduction of complexity to simpler models proves beneficial, with the second-degree polynomial model demonstrating superior performance compared to the linear model, and the third-degree polynomial model surpassing the second-degree model in terms of predictive accuracy.

The MNGE metric provides valuable insights into the model's real-world performance. Specifically, when analyzing the ANN model's MNGE value of 0.043 for the first output, it signifies an average prediction error of 4.3% across all test data set predictions for the maximum y-displacement of the nodes.

Digging deeper into the output data, we find that the median error is even lower than the MNGE, at 2.9%. It is important to note that in this context, MNGE for a single observation represents the bias error (the difference between the predicted value and the real value) over the real value, in absolute terms. In stark contrast, the MNGE values for the other models consistently exceed 25%, rendering them impractical for real-world applications.

4.1 Taylor Diagrams

In Fig. 5, we display three Taylor diagrams, each corresponding to an individual output variable, and exclusively based on the data of the test set. Within these diagrams, every model is symbolized as a distinct point, with the reference target point positioned at the diagram's bottom. In a Taylor's diagram, the proximity of a model point to this target point serves as an indicator of the model's predictive accuracy—closer points signify better predictions. As before, it remains evident that the ANN model consistently outperforms the other models. Furthermore, the trends

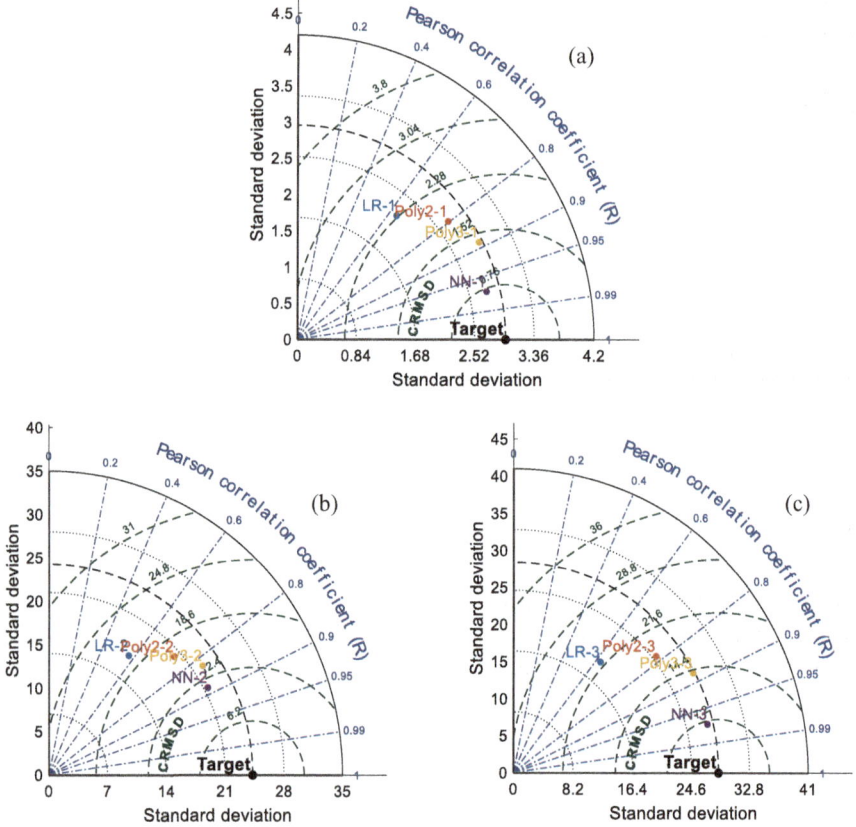

Fig. 5 Taylor diagrams for all models, for all three outputs (test data only): (**a**) output 1 (y-displacement), (**b**) output 2 (σ_{max}), (**c**) output 3 (σ_{min})

observed in the previous paragraph, regarding error metrics, are reaffirmed here, underscoring the ANN model's superior performance across these visual representations as well.

5   Conclusions

Our study has undertaken a comprehensive evaluation of various regression models, encompassing linear, second- and third-degree polynomial, and ANN models, to predict crucial parameters within a plane truss model. Our findings unequivocally establish the supremacy of the ANN model, demonstrating its exceptional aptitude for capturing intricate nonlinear relationships within the dataset. Notably, our exploration has shed light on the pivotal role of model complexity. The augmentation of complexity, observed in the transition from linear to polynomial models, has yielded tangible improvements in predictive performance. The superior performance exhibited by the ANN model holds promise for a wide array of practical engineering applications.

While our study represents a significant stride in comprehending the performance of regression models, promising avenues for further investigation persist. Future research might delve deeper into hyperparameter tuning, refined feature selection techniques, and the exploration of model architectures, with the aim of further optimizing the ANN model's capacity. Moreover, the realm of regression modeling extends far beyond the models assessed in this study. The incorporation of additional models, such as Support Vector Machine (SVM), K-nearest neighbors (KNN), and Decision Tree, among others, could offer additional insights into predictive accuracy.

The ANN's predictions of structural behavior serve a dual purpose. They serve as valuable surrogate modeling tools, reducing the computational demands for the analysis of intricate structural models with minimal compromise in accuracy, but also open doors to various practical applications. For instance, they can be integrated into the context of structural optimization, furnishing predictions of structural responses and obviating the need for FE analysis in every optimization iteration. This capability can prove highly beneficial, particularly in the initial phases of optimization.

However, it is imperative to approach such predictions with discernment. While the overall performance of advanced regression models remains commendable, certain cases may yield predictions significantly divergent from actual output values. This is substantiated by the presence of outliers observed in the "Predictions vs. Targets" plots, even in the case of the best-performing model. Hence, while regression models offer remarkable predictive capabilities, their outputs should be interpreted cautiously, with consideration for potential outliers and/or anomalies.

References

1. Lagaros, N.D., Plevris, V.: Artificial intelligence (AI) applied in civil engineering. Appl. Sci. **12**(15), 7595 (2022). https://doi.org/10.3390/app12157595
2. Lagaros, N.D., Plevris, V. (eds.): Artificial Intelligence (AI) Applied in Civil Engineering, p. 698. MDPI, Basel (2022). https://doi.org/10.3390/books978-3-0365-5084-8
3. Lu, X., Plevris, V., Tsiatas, G., De Domenico, D.: Editorial: artificial intelligence-powered methodologies and applications in earthquake and structural engineering. Front. Built Environ. **8**, 876077 (2022). https://doi.org/10.3389/fbuil.2022.876077
4. Lagaros, N.D., Tsompanakis, Y. (eds.): Intelligent Computational Paradigms in Earthquake Engineering. Idea Group Publishing (2006)
5. Málaga-Chuquitaype, C.: Machine learning in structural design: an opinionated review. Front. Built Environ. **8**, 815717 (2022). https://doi.org/10.3389/fbuil.2022.815717
6. Nguyen, T.-H., Vu, A.-T.: A Comparative Study of Machine Learning Algorithms in Predicting the Behavior of Truss Structures. Springer Singapore, Singapore (2021)
7. Mai, H.T., Lieu, Q.X., Kang, J., Lee, J.: A novel deep unsupervised learning-based framework for optimization of truss structures. Eng. Comput. **39**(4), 2585–2608 (2023). https://doi.org/10.1007/s00366-022-01636-3
8. Kang, H.-T., Yoon, C.J.: Neural network approaches to aid simple truss design problems. Comput. Aided Civ. Inf. Eng. **9**(3), 211–218 (1994). https://doi.org/10.1111/j.1467-8667.1994.tb00374.x
9. Mai, H.T., Lieu, Q.X., Kang, J., Lee, J.: A robust unsupervised neural network framework for geometrically nonlinear analysis of inelastic truss structures. Appl. Math. Model. **107**, 332–352 (2022). https://doi.org/10.1016/j.apm.2022.02.036
10. Khodadadi, N., Talatahari, S., Gandomi, A.H.: ANNA: advanced neural network algorithm for optimisation of structures. Proc. Inst. Civ. Eng. Struct. Build. **177**, 1–23 (2023). https://doi.org/10.1680/jstbu.22.00083
11. Ghasemi, M.R., Hinton, E., Wood, R.D.: Optimization of trusses using genetic algorithms for discrete and continuous variables. Eng. Comput. **16**(3), 272–303 (1997). https://doi.org/10.1108/02644409910266403
12. El-Sayed, M.E.M., Jang, T.S.: Structural optimization using unconstrained nonlinear goal programming algorithm. Comput. Struct. **52**(4), 723–727 (1994). https://doi.org/10.1016/0045-7949(94)90353-0
13. Plevris, V.: Innovative Computational Techniques for the Optimum Structural Design Considering Uncertainties, p. 312. National Technical University of Athens, Athens (2009). https://doi.org/10.12681/eadd/17936
14. Montgomery, D.C., Peck, E.A., Vining, G.G.: Introduction to Linear Regression Analysis, 6th edn. Wiley, Hoboken (2021)
15. Solorzano, G., Plevris, V.: An open-source framework for modeling RC shear walls using deep neural networks. Adv. Civ. Eng. **2023**, 7953869 (2023). https://doi.org/10.1155/2023/7953869
16. Solorzano, G., Plevris, V.: DNN-MLVEM: a data-driven macromodel for RC shear walls based on deep neural networks. Mathematics. **11**(10), 2347 (2023). https://doi.org/10.3390/math11102347
17. Solorzano, G., Plevris, V.: ANN-based surrogate model for predicting the lateral load capacity of RC shear walls. In: 8th European Congress on Computational Methods in Applied Sciences and Engineering (ECCOMAS 2022), 5–9 June 2022, Oslo, Norway (2022). https://doi.org/10.23967/eccomas.2022.050
18. Plevris, V., Solorzano, G.: Prediction of the eigenperiods of MDOF shear buildings using neural networks. In: 8th ECCOMAS Thematic Conference on Computational Methods in Structural Dynamics and Earthquake Engineering (COMPDYN 2021), 28–30 June 2021, ECCOMAS: Athens, Greece, pp. 3894–3911. Eccomas Proceedia. https://doi.org/10.7712/120121.8755.20415

19. Sharib, S., Ahmad, N., Plevris, V., Ahmad, A.: Prediction models for load carrying capacity of RC wall through neural network. In: 14th ECCOMAS Thematic Conference on Evolutionary and Deterministic Methods for Design, Optimization and Control (EUROGEN 2021), 28–30 June 2021, ECCOMAS: Streamed from Athens, Greece, pp. 132–142. Eccomas Proceedia. https://doi.org/10.7712/140121.7956.18529
20. Hagan, M.T., Menhaj, M.B.: Training feedforward networks with the Marquardt algorithm. IEEE Trans. Neural Netw. **5**(6), 989–993 (1994)
21. Plevris, V., Solorzano, G., Bakas, N.P., Ben Seghier, M.E.A.: Investigation of performance metrics in regression analysis and machine learning-based prediction models. In: 8th European Congress on Computational Methods in Applied Sciences and Engineering (ECCOMAS 2022), 5–9 June 2022, Oslo, Norway (2022). https://doi.org/10.23967/eccomas.2022.155

Resource-Efficient and Climate-Friendly Design of Concrete Structures Through Advanced Structural Safety Concepts

Ramon Hingorani ⓘ and Jochen Köhler ⓘ

Contents

1   Introduction

Many current and future challenges are related to the efficient management of limited financial and natural resources or the appropriate mitigation of consequences due to climate change. Engineering structures play a fundamental role in this regard since they consume large amounts of raw materials and contribute significantly to greenhouse gas emissions worldwide, e.g., [1–4]. As the challenges will become bigger, the demand for innovative and sustainable solutions well beyond traditional structural engineering practice will increase.

Structural design codes, which regulate the safety of our built environment and the corresponding use of structural materials, are crucial in this context. They feature safety concepts that are highly generalized to make them simple and applicable to a large variety of structures at the same time. This leads to structures that are

R. Hingorani (✉)
SINTEF Community, Trondheim, Norway
e-mail: ramon.hingorani@sintef.no

J. Köhler
Norwegian University of Science and Technology (NTNU), Trondheim, Norway

© The Author(s) 2025
M. Kioumarsi, B. Shafei (eds.), *The 1st International Conference on Net-Zero Built Environment*, Lecture Notes in Civil Engineering 237,
https://doi.org/10.1007/978-3-031-69626-8_124

sufficiently safe and functional, but where structural material is not utilized in an optimal manner. For a sustainable built environment, this is not good enough. We need to unlock the potential offered by risk-based decision approaches in order to utilize our resources in structural design procedures as efficient as possible. Intensively discussed in the scientific literature over the past decades, e.g., [5–8], and by now supported by standardized guidance [9], these approaches facilitate optimized solutions such that a balance is achieved between a specific structural safety level and the corresponding resources invested.

In the past, the objective of such optimization procedures was mainly of economic nature, i.e., with focus on minimizing costs. While this approach may contribute likewise to mitigation of material consumption and greenhouse gas emissions, it was recently suggested to explore the potential benefits of addressing such environmental objectives explicitly, since the corresponding optimum might significantly deviate from the solution based on an economic decision framework [10, 11].

On the grounds of an emission-based objective function (Sect. 2), the paper assesses the material and greenhouse gas emission-saving potential of a risk-informed design of floor structures with reinforced concrete members. The environmental benefits compared to the application of the standardized design rules (Partial Factor Method) as defined in the Eurocodes [12] are demonstrated and discussed (Sect. 3).

2 Optimization Framework

2.1 Objective Function

Taking the widely accepted economic optimization framework proposed by Rackwitz [5] as a basis, an objective function for a risk-informed design based on carbon emissions (CE) is being defined, see Eq. (1). In this formulation, the benefits have been considered independent of the decision parameter p (e.g., a cross-section dimension), wherefore the optimization problem simplifies to the determination of $p = p_{opt}$ which minimizes the expected total emissions CE_{tot} associated with a specific decision, as illustrated in Fig. 1. The total emissions CE_{tot} are constituted by the sum of the emissions related to investments into safety (CE_s) and those expected in consequence of a failure event (CE_f). The former are defined as the sum of the emissions associated with the production stage, CE_c (cradle-to-gate emissions), in the following referred to as *construction-related emissions*, and the expected emissions due to structural obsolescence, CE_o.

$$p_{opt} = \arg \min_{p} (CE_c + CE_o + CE_f) \tag{1}$$

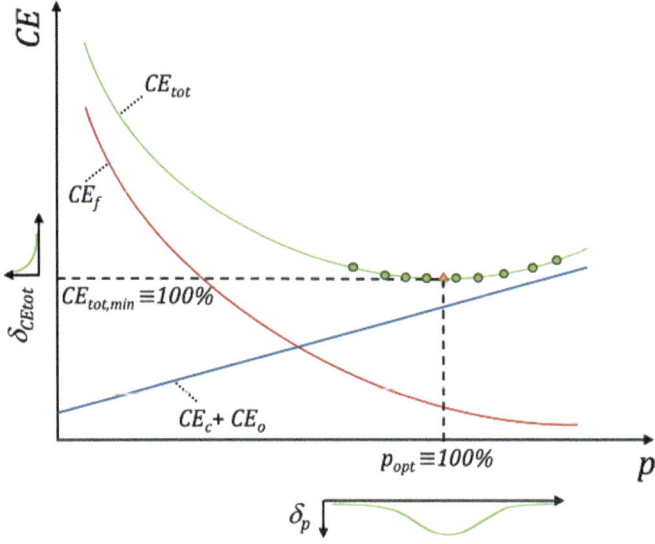

Fig. 1 Illustration of the CE-based optimization principle and deviations $\delta_{CE,tot}$ and δ_p from the minimum total emissions ($CE_{tot,min}$) and optimal decision parameter (p_{opt}), respectively

In Eq. (2), the construction-related emissions CE_c are defined as the sum of term $CE_1 p$, which is proportional to the decision parameter p, and a p-independent constant CE_0. The expected emissions due to failure (CE_f) and obsolescence (CE_o) defined by, respectively, Eqs. (3) and (4), are estimated based on the assumption that structures are continuously renewed either after failure or after becoming obsolete respectively. A discussion and justification of this assumption can be found in [5]. Emissions in consequence of reconstruction are multiplied by the yearly obsolescence-rate ω resulting in the expected yearly obsolescence emissions (CE_o). The expected yearly failure emissions (CE_f) are defined as the sum of reconstruction-related emissions CE_c and indirect failure emissions H, independent of p, multiplied by the yearly failure rate $\lambda P_f(p)$. Like costs, future emissions are to be discounted, see, e.g., [13]. The net present value of the sum of annual expectations with the annual interest rate i is computed by utilizing the asymptotic solution expressed by Eq. (5).

$$CE_c = CE_0 + CE_1 p \tag{2}$$

$$CE_o = CE_c \frac{\omega}{i} \tag{3}$$

$$CE_f = (CE_c + H) \frac{\lambda P_f(p)}{i} \tag{4}$$

$$\sum_{\tau=1}^{\infty} \frac{1}{(1+i)^\tau} = \frac{1}{i} \tag{5}$$

2.2 Life Safety Constraints

While the objective function given by Eq. (1) pursues an optimum design from an environmental perspective (in this case related to embodied greenhouse gas emissions), it does not guarantee that this optimum (p_{opt}) is consistent with the societal preferences in regard to life safety investments. As illustrated in Fig. 2, this can be considered by imposing a corresponding acceptance criterion to the failure probability, $P_{f,opt} < P_{f,acc}$, e.g., based on the marginal life-saving costs principle, see [7, 9] for further guidance.

2.3 Saving Potential

The resource- and emission-saving potential of the described optimization framework in relation to the sub-optimal decisions based on the standardized design rules defined in structural codes and standards, such as the Eurocodes, is illustrated in Fig. 1. Generally, these sub-optimal decisions, represented by the dots, entail a certain deviation, in terms of the expected total emissions CE_{tot} from their respective optimum, $\delta_{CE,tot}$. Further, it is of interest, how the sub-optimal solutions deviate from the optimum in terms of the decision parameter p, which is representative of the material consumption and of the associated embodied greenhouse gas emissions at the structural design stage, i.e., when the decision is made. In contrast to $\delta_{CE,tot}$, the deviation δ_p can theoretically occur on both sides of the optimum p_{opt}, see Fig. 1.

Fig. 2 Illustration of life-safety acceptance (acc) criteria as constraint to the optimization (opt)

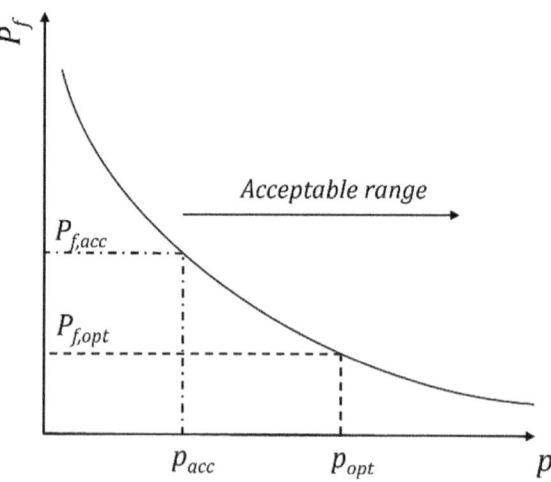

3 Case Study

3.1 Structural System and Loads

The case study assumes two different types of floor systems in office buildings (Fig. 3): (A) RC floor beams which provide support to one-way RC slabs spanning in the direction perpendicular to the beam orientation and (B) a floor system supported by T-beams. In both systems, the beams of span L are simply supported and spaced at a distance D.

Permanent loads g acting on the beams include the structural self-weight (g_{sw}) of both the beams, $g_{sw,b}$ (limited to the web in case of the T-beam system) and the slabs, $g_{sw,sl}$, as well as the weight of non-structural permanent loads, g_{nspl}, including the flooring and roofing system attached to the slabs. In addition, the beams sustain a uniformly distributed imposed load q (Fig. 4).

3.2 Limit State Function

Bending failure at mid-span of the beams is considered. The limit state function (LSF) represents the state of equilibrium between acting and resisting bending moments at mid-span of the beams, respectively M_E and M_R. These are specified in Eqs. (6) and (7), where Θ_E and Θ_R denote the corresponding model uncertainties. Variable Θ_q in Eq. (6) represents the uncertainties associated with the imposed load

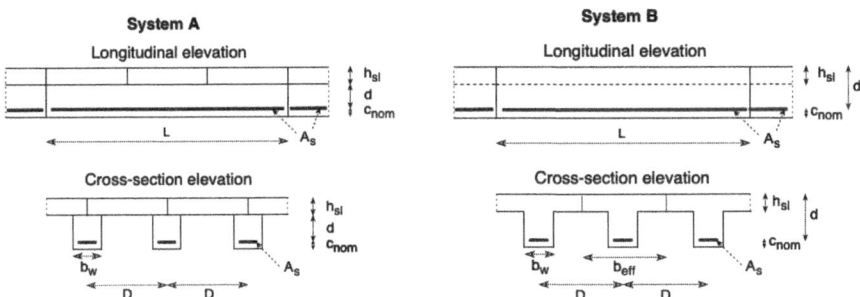

Fig. 3 Sketch of the floor systems: (**a**) RC beams supporting one-way RC slabs; (**b**) RC T-beam slab

Fig. 4 Structural system and loads

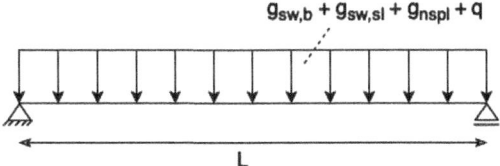

model. The bending resistance M_R is specified in a generic way such that it represents both systems analyzed, A and B (Fig. 3). In the latter, the beam flange contributes to resistance of the concrete compression stresses over an effective width $b = b_{eff}$ according to EN1992-1-1 [14], in the former $b = b_w$, see Eq. (8). The reinforcement ratio ρ_s is defined accordingly in Eq. (9). Variables f_y and f_c represent the yield strength of the reinforcing steel and the compressive strength of concrete, respectively.

$$M_E = \Theta_E \frac{(g_{sw} + g_{nspl} + \Theta_q q)DL^2}{8} \qquad (6)$$

$$M_R = \Theta_R b \rho_s d^2 f_y \left(1 - 0.5 \rho_s \frac{f_y}{f_c}\right) \qquad (7)$$

$$b = \begin{cases} b_w & \text{(A)} \\ b_{eff} & \text{(B)} \end{cases} \qquad (8)$$

$$\rho_s = \begin{cases} \rho_{s,w} = \dfrac{A_s}{b_w d} & \text{(A)} \\ \rho_{s,eff} = \dfrac{A_s}{b_{eff} d} & \text{(B)} \end{cases} \qquad (9)$$

3.3   Characteristic Values

With the aim to define a representative set of design situations, the parameters most relevant to the design of the concrete floor beams are varied within reasonable ranges.

Three spans L are distinguished (6, 12, and 18 m) in addition to four different spacings D between members ($L/1$, $L/2$, $L/3$, $L/4$) (Fig. 3). The beams effective cross-section depth d is kept constant equal to $L/15$. A nominal concrete cover to the tensile reinforcement of 30 mm is adopted. Reinforcing steel grade B500 ($f_{yk} = 500$ N/mm^2) is assumed in combination with concrete of type C20/25 ($f_{ck} = 20$ N/mm^2).

In line with EN1991-1-1 [15], a characteristic value of the imposed load of $q_k = 3$ kN/m^2 is adopted. A wide range of ratios a_q of variable (q_k) to total ($g_k + q_k$) characteristic loads between 0.05 and 0.95 is considered, with an interval size $\Delta a_q = 0.05$. In addition, three ratios a_g of structural self-weight $g_{sw,k}$ to total permanent loads ($g_{sw,k} + g_{nspl,k}$) ranging between 0.6 and 1.0 are adopted ($\Delta a_g = 0.2$). Based on these assumptions, $g_{sw,k}$ and $g_{nspl,k}$ are computed.

An array for beam tensile reinforcement ratios $\rho_{s,w}$ varying from $\rho_{s,w} = 0.002$ to 0.03 ($\Delta a_q = 0.004$) is defined, within the limits of the prescribed $\rho_{s,w,min}$ and $\rho_{s,w,max}$ thresholds according to EN 1992-1-1 [14]. Note that for system B, this results in a corresponding $\rho_{s,eff}$ array ($\rho_{s,eff} = \rho_{s,w} b_w/b_{eff}$).

3.4 Eurocode Design

Given the characteristic values described under Sect. 3.3, a strict Eurocode design ($M_{Ed} = M_{Rd}$) of the beam's mid-span cross-section is performed based on the Partial Factor Method (PFM) according to EN1990 [12], with partial factors for f_c and f_y taken from EN1992-1-1 [14]. This results in the strictly required cross-section width b, see Eq. (8)—note that in case of system B, the design procedure is iterative. The geometry of the beam is thereby established, allowing for a distinction of the structural self-weight of beam and slab, respectively. The depth of the latter, h_{sl}, can then be inferred under the assumption of typical densities for RC members ($\rho_{sl} = 2500$ kg/m^3).

To confine the scope of the study to practically relevant design situations, maximum and minimum slab depth to span ratios (h_{sl}/D) are imposed, as specified in Table 1. In addition, a minimum slab depth h_{sl} of 100 mm is required. Furthermore, boundaries for the ratio of effective beam cross-section depth (d) to width (b_w) are specified (Table 1). Finally, it is ensured that only ductile cross-sections complying with a $x/d < 0.45$ are considered, where x is the neutral axis depth of the mid-span cross-section.

For system A, the latter of the described boundary conditions is the most stringent and only 24 relevant design situations are identified. For system B, where the contribution of the flange to the compression forces in the concrete leads generally to relatively small neutral axis depths x (generally $x < h_{fl}$), a total of 145 design situations comply with the imposed boundary conditions.

3.5 Risk-Informed Design

3.5.1 Decision Parameter

In the risk-based optimization of the RC beams, the dimensions of the, respectively, 24 (system A) and 145 (system B), concrete cross-sections obtained in the Eurocode design (Sect. 3.4) are adopted. The decision parameter (p) corresponds to the optimal amount of tensile reinforcement (A_s) according to the framework established by Eqs. (1)–(5). The corresponding assumptions are described below.

Table 1 Boundary conditions for selection of design situations included in the study

Variable	Min	Max
h_{sl}/D	20	30
h_{sl} [mm]	100	–
d/b_w	1	5

3.5.2   Construction-Related Emissions

Construction-related carbon emissions CE_c according to Eq. (2) are specified by Eq. (10) as the sum of the contribution of concrete (c) and reinforcing steel (s), respectively. These respective contributions are computed as the product of the material mass M ($= V\rho$, where $\rho_c = 2400$ kg/m^3 and $\rho_s = 7850$ kg/m^3) and the corresponding carbon emission intensities (CEI_c and CEI_s). Since the optimization procedure is limited to the beam, V_c excludes the volume of the RC slabs in system A, i.e., $V_c = V_{c,b}$ (concrete volume of the beam), while the emissions embodied in the RC slabs are accounted for in the indirect failure costs term, see Sect. 3.5.3. In system B, however, the slab (i.e., the flange of the T-beam) contributes to the resistance mechanism of the beam and is hence included in CE_c, i.e., $V_c = V_{c,b+sl}$. Note that the term $V_{c,b+sl}\rho_c CEI_c$ in Eq. (10) corresponds to the p-independent component C_0 in Eq. (2).

The emission intensities for concrete, CEI_c and reinforcing steel, CEI_s, reflect cradle-to-gate emissions and have been adopted or inferred from Environmental Product Declarations (EPD) according to EN 15804, issued by the German Institute for Construction and Environment (IBU) [16]. The values are given in Table 2.

$$CE_c = \begin{cases} V_{c,b}\rho_c CEI_c + V_s\rho_s CEI_s & \text{(A)} \\ V_{c,b+sl}\rho_c CEI_c + V_s\rho_s CEI_s & \text{(B)} \end{cases} \qquad (10)$$

3.5.3   Failure Emissions

The study considers carbon emissions in consequence of beam failure, CE_f, see Eq. (4). In addition to those embodied in the re-constructed system (after failure), considered in Eq. (10), indirect failure emissions H need to be accounted for. These are estimated according to Eqs. (11) and (12) as the sum of the emissions embodied in: the collapsed RC slabs (only for system A following the explanation given in Sect. 3.5.2), the non-structural permanent loads g_{nspl}, i.e., the flooring and roofing systems attached to the slabs, and the non-permanent building equipment, such as furniture or movable partitions. The latter is approximated as a function of the 5 year mean value of the sustained load contribution in office buildings, $q_s = 0.5$ kN/m^2 [17].

$$H = c_e A_{col} \qquad (11)$$

$$c_e = \begin{cases} g_{sw,sl}CEI_{sl} + g_{nspl}CEI_{nspl} + q_s CEI_{q,s} & \text{(A)} \\ g_{nspl}CEI_{nspl} + q_s CEI_{q,s} & \text{(B)} \end{cases} \qquad (12)$$

Table 2 Carbon emission intensities [kgCO$_{2,eq}$/kg]

CEI_c	CEI_s	CEI_{sl}	CEI_{nspl}	$CEI_{q,s}$
0.074	1.23	0.12	1.13	1.65

In Eq. (11), A_{col} denotes the area affected by the collapse of a beam, which, for both systems (A and B) is approximated as $A_{col} = 4LD$ (floor area supported by and below the collapsing beam). The CEI involved in Eq. (12) are given in Table 2. The value for the RC slabs (system A), CEI_{sl}, represents a weighted average of CEI_c and CEI_s, whereas the value for non-structural permanent loads, CEI_{nspl}, corresponds to an average out of 32 different configurations of such loads, comprising four different flooring (ceramic tiles [18], laminate [19], PEC floor [20], in all cases including screed [21], and raised floor [22]) and roofing systems (suspended ceiling [23], plaster boards [24], mineral boards [25], and metal ceiling [26], in all cases including metal ducts for ventilation or installations [27]), respectively. Each of the system components is represented by two different EPDs. This procedure leads to a weighted average value of $CEI_{nspl} = 1.13$ $kgCO_{2,eq}/kg$ (weighted with regard to the contribution of each component to the total weight of flooring and roofing system).

The carbon emission intensity corresponding to non-permanent building equipment, $CEI_{q,s} = 1.65$ $kgCO_{2,eq}/kg$, has been estimated as an average value of the CEI reflected in a variety of EPDs for different furniture types (e.g., office chair, desks, shelves, meeting tables), compiled in [28].

3.5.4 Obsolescence, Interest Rate, and Probabilistic Models

The calculation of expected emissions due to structural obsolescence CE_o based on Eq. (3) assumes an annual obsolescence rate $\omega = 1/50$. The considered annual discount rate i is 3%. The probabilistic models for the loads (g_{sw}, g_{nspl}, q), material strengths (f_y, f_c) and model uncertainties (Θ_E, Θ_R, Θ_q) are adopted from [17].

3.6 Results

The difference δ_{CEtot} (see Fig. 1) between the expected value of the total emissions corresponding to, respectively, the optimized (OPT) and the Eurocode (EC) solution is depicted in Fig. 5 depending on the ratio of variable to total loads a_q, which is found to range between 0.2 and 0.45 (System A, Fig. 5a) and 0.15 and 0.5 (System B, Fig. 5b), respectively—design situations with smaller or larger a_q led to non-compliance with the imposed boundary conditions described in Sect. 3.4). In both systems, A and B, the obtained δ_{CEtot} are of the same order. For system A, a mean value δ_{CEtot} of around 2% is found, the maximum reaches nearly 4%. The corresponding mean and max values for system B are <1% and approximately 2%.

With respect to the EC solution, the optimization entails *significant material (reinforcement) and emission reductions at the design stage* of the floor systems. This can be observed in Fig. 5c, d, which show the difference δ_p (Fig. 1) between the realizations of the decision parameter p according to EC and optimized design, respectively. In case of system A, δ_p attains on average about 10%, with a maximum

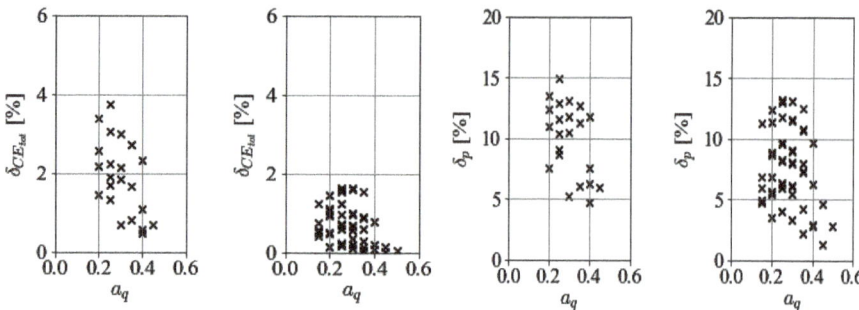

Fig. 5 Savings δ_{CEtot} in terms of expected total emissions CE_{tot} for (**a**) System A; (**b**) System B; Savings δ_p in terms of reinforcement consumption and associated emissions at the design stage for (**c**) System A; (**d**) System B

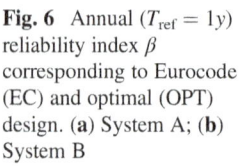

Fig. 6 Annual ($T_{ref} = 1y$) reliability index β corresponding to Eurocode (EC) and optimal (OPT) design. (**a**) System A; (**b**) System B

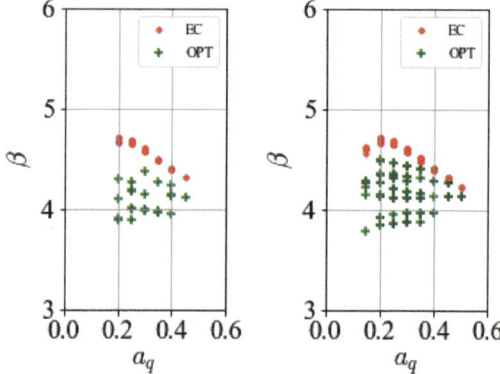

of up to 15%. Similarly, for the T-beam system (B), mean and max δ_p are around 7% and 13%, respectively. In all cases, $\delta_p > 0$, i.e., $p_{EC} > p_{OPT}$ (Fig. 1). Accordingly, the optimal (annual) reliability indices β_{OPT} are in all cases below the corresponding β_{EC} values. As Fig. 6 shows, β_{OPT} vary between around 3.9 and 4.4 for system A, and between 3.8 and 4.5 for system B. For larger a_q ratios the difference between β_{OPT} and β_{EC} diminishes and so does the saving potential δ_p.

3.7 Discussion

3.7.1 Emission-Saving Potential

The results presented in Fig. 5a, b point to a relatively low saving potential of the risk-based optimization (with respect to the EC solution) in terms of the total expected carbon emissions, CE_{tot}. The reason for this is that emission savings at the design stage, due to lower material consumption, are partially offset by increased failure-emissions, due to lower structural safety levels, i.e., those emissions which arise in case of a structural failure event and the corresponding need for reconstruction of the floor system and the replacement of both permanent and non-permanent building equipment affected by the collapse.

While this appears to be at first sight a sobering observation, the promising result is that the potential emission savings at the design stage, δ_p, are found to be significant. Given the need for urgent measures to reduce embodied greenhouse gas emissions in constructions, the opportunity for potential "immediate" emission reductions of the order of $\delta_p = 7$–10% (average of results in Fig. 5c, d), achieved through application of advanced and rational structural safety concepts in structural design procedures, should not be ignored. In any case, it should be ensured that the safety level exceeds the human safety requirements (Sect. 2.2). According to the results shown in Fig. 6, it might be possible that this is not fulfilled in all analyzed situations. In this case, the corresponding δ_p are expected to decrease *slightly* when imposing human safety constraints to the failure probability (Fig. 2). However, given the large number of structures erected every day all over the globe, the resource- and emission-saving potential with respect to a standardized Eurocode design is, even then, to be judged as large, at least when assessed in absolute terms.

3.7.2 Limitations and Outlook

The presented case study is limited to a specific set of design situations comprising reinforce concrete members. A similar study on the flexural design of steel beams supporting prestressed hollow-core slabs in buildings [11] revealed similar results as found herein. Future studies should assess to which extend these findings can be extrapolated to design situations characterized by other failure modes and loading scenarios, including situations governed by environmental loads, such as snow and wind.

In addition, it is imperative for future endeavors to address the inherent complexity that impedes the seamless integration of risk-based approaches into daily structural design procedures. Beyond advancing understanding through education, there are two primary ways to mitigate this challenge. One approach involves leveraging higher-level safety concepts to calibrate simplified decision rules such as the Partial Factor method [10]. However, it is contended that a significantly greater degree of optimization could be attained by directly applying risk-based decision analysis to individual structural design projects. The emergence of digital tools presents a transformative opportunity to efficiently implement this approach on a wide scale. Departing from the current analogue approach to structural decision-making, reliant on codified rules and coarse information, the adoption of user-friendly and operationally intuitive IT tools would make the identification of optimal solutions by means of flexible risk-based design accessible for routine applications in daily engineering practice. For a more detailed exploration of strategies to surmount the complexities inherent in risk-based design approaches, refer to [10].

4 Conclusions

Within the context of limited resources and a strong need for mitigation of greenhouse gas emissions, the development of a sustainable built environment in general, and of its loadbearing systems in particular, is of paramount importance. Modifications to the safety concepts and rules for design of such systems bear the potential for significant material savings. A case study addressing risk-based design of RC floor systems has illustrated this. Future studies should increase the scope of the study to other materials, failure modes, and loading scenarios, providing the basis for a consistent risk-based calibration of the standardized design rules and hence for a broad implementation of sustainability principles in structural engineering decision-making.

Acknowledgments This study has received funding from the European Union's Horizon 2020 research and innovation program under the Marie Sklodowska-Curie grant agreement No. 893527.

References

1. Joint Committee on the Globe Consensus (Globe): Decarbonizing Global Construction – Draft. http://globe-consensus.com (2022)
2. Gibbons, O.P., Orr, J.J.: How to Calculate Embodied Carbon. The Institution of Structural Engineers. https://www.istructe.org (2022)
3. De Wolf, C., Yang, F., Cox, D., Charlson, A., Hattan, A.S., Ochsendorf, J.: Material quantities and embodied carbon dioxide in structures. Proc. Inst. Civ. Eng. Eng. Sustain. **169**(4), 150–161 (2016)
4. Collings, D.: The carbon footprint of bridges. Struct. Eng. Int. **32**(4), 501–506 (2022)
5. Rackwitz, R.: Optimization — the basis of code-making and reliability verification. Struct. Saf. **22**(1), 27–60 (2000)
6. Rackwitz, R., Lentz, A., Faber, M.: Socio-economically sustainable civil engineering infrastructures by optimization. Struct. Saf. **27**(3), 187–229 (2005)
7. Fischer, K., Viljoen, C., Köhler, J., Faber, M.H.: Optimal and acceptable reliabilities for structural design. Struct. Saf. **76**, 149–161 (2019)
8. Köhler, J., Baravalle, M.: Risk-based decision making and the calibration of structural design codes – prospects and challenges. Civ. Eng. Environ. Syst. **36**(1), 55–72 (2019)
9. ISO 2394: General Principles on Reliability for Structures, 4th edn. International Organization for Standardization (ISO) (2015)
10. Hingorani, R., Köhler, J.: Towards optimized decisions for resource and carbon-efficient structural design. Civ. Eng. Environ. Syst. **40**(1–2), 1–31 (2023)
11. Hingorani, R., Köhler, J.: Sustainability potential of risk-informed decisions in structural design. In: 14th International Conference on Application of Statistics and Probability in Civil Engineering (ICASP), Dublin (2023)
12. EN 1990: Eurocode – Basis of Structural Design. European Committee for Standardization, Brussels (2002)
13. Adhikari, P., Mahmoud, H., Xie, A., Simonen, K., Ellingwood, B.: Life-cycle cost and carbon footprint analysis for light-framed residential buildings subjected to tornado hazard. J. Build. Eng. **32**, 101657 (2020)

14. EN 1992-1-1: Eurocode 2: Design of concrete structures – Part 1–1: General Rules and Rules for Buildings. European Committee for Standardization, Brussels (2004)
15. EN 1991-1-1: Eurocode. Actions on structures. General Actions – Densities, Selfweight, Imposed Loads for Buildings. European Committee for Standardization, Brussels (2002)
16. EPD-IZB-20180097-IBG1-DE; EPD-ARM-20160051-IBD3-EN. www.ibu-epd.com
17. CEN TC 250 SC10 N 5553: JRC Report – Reliability Background of the Eurocodes – Draft 10-2023. Rotterdam (2023)
18. EPD-BKF-20220184-ICG1-DE; EPD-COI-20220297-ICG1-EN. www.ibu-epd.com
19. EPD-PAR-20180193-IBC1-DE; EPD-EGG-20180194-IBC1-DE. www.ibu-epd.com
20. EPD-ERF-20180183-CCII-EN; EPD-JHA-20180056-IBA1-EN. www.ibu-epd.com
21. EPD-JAM-20220072-CBD1-DE; EPD-IWM-20190151-IBG1-DE. www.ibu-epd.com
22. EPD-LIN-20210022-IBA1-DE; EPD-LIN-20170194-IBD1-DE. www.ibu-epd.com
23. EPD-RWI-20200018-CBD6-EN; EPD-SPH-20210147-IBA1-DE. www.ibu-epd.com
24. EPD-BVG-20220090-IAG1-DE; EPD-JAM-20220071-CBD1-DE. www.ibu-epd.com
25. EPD-WTC-20170171-IAF1-DE; EPD-AKC-20200124-ICA1-EN. www.ibu-epd.com
26. EPD-TAI-20180162-IBG2-DE; EPD-TAI-20180164-IBG2-EN. www.ibu-epd.com
27. NEPD-2145-971-NO; NEPD-2989-1669-EN. www.epd-norge.no
28. Lauvland, H.J.: The Carbon Footprint of Furniture. Master thesis. Norwegian University of Science and Technology (NTNU) (2021)

Optimizing Corbel Reinforcement Through Nonlinear Analysis: Determining Superiority of Steel Bars or CFRP

Ali Kheyroddin, Shakiba Raygan, and Mahdi Kioumarsi

Contents

1   Introduction

Reinforced concrete corbels and brackets play a crucial role in the structural integrity of beam-column connections, commonly appearing as projecting cantilevers from walls or columns. These structural members are engineered to resist horizontal and vertical forces arising from diverse expected conditions. Traditional design approaches for corbels encompass techniques like the strut-and-tie model (STM) and the utilization of shear friction [1, 2]. Following the seismic events of the 1985 Mexico City earthquake, a necessity emerged to fortify the resilience of beam-column connections [3]. The durability of these connections holds significant importance in reinforced concrete frame structural systems, given their critical function in

A. Kheyroddin · S. Raygan
Department of Civil Engineering, Semnan University, Semnan, Iran

M. Kioumarsi (✉)
Department of Built Environment, OsloMet – Oslo Metropolitan University, Oslo, Norway
e-mail: mahdi.kioumarsi@oslomet.no

© The Author(s) 2025

M. Kioumarsi, B. Shafei (eds.), *The 1st International Conference on Net-Zero Built Environment*, Lecture Notes in Civil Engineering 237,
https://doi.org/10.1007/978-3-031-69626-8_125

the distribution and transmission of seismic forces [4, 5]. However, the dense arrangement of reinforcement hoops, bars, and ties in these junctions presents obstacles in the placement and consolidation processes [6]. Additionally, the design complexity of abbreviated corbels is compounded by the presence of floor slabs, transverse beams, and restricted entry to connection sites [3, 7].

Fiber-reinforced polymer (FRP) sheets are considered optimal for enhancing the strength of corbels due to their improved resistance to inter-story drift, stiffness, and strength. FRP sheets exhibit considerable potential to improve the structural efficacy of RC corbels, notably by augmenting their shear capacity [8, 9]. As per existing literature [10, 11] and shear friction theory [2], strengthening RC corbels with diagonal FRP boosts ultimate capacity by over 30% against sheets/plates alone. Empirical tests confirm the efficiency of pairing inclined and horizontal FRP [10–12].

To prevent brittle failure in construction, adhering to specific ranges for reinforcing bar percentages of primary and secondary reinforcements, including steel ties and stirrups, is crucial [9]. As per ACI 318-19 [2], the minimum area for horizontal or vertical stirrups should exceed half of that of steel ties ($A_{sh} > 0.5A_s$). Secondary bars are predominantly utilized to control the width of cracks [13], which can increase the first cracking load and ultimate load in normal and high-strength self-compacting concrete corbels (NSCC and HSCC) [14] or high-strength fiber-reinforced concrete corbels [15].

Due to the considerable expenses entailed in experimental inquiry, Finite Element (FE) analysis stands out as one of the foremost efficient approaches in projecting the behavior of reinforced concrete (RC) corbels. The findings from the numerical investigations by El-Maddawy and Sherif [11] demonstrated that increasing the steel reinforcement reduced the involvement of CFRP and that the use of horizontal stirrups influenced the development of cracks, causing the shear cracks to be more curved toward the outer surface of the corbel [11].

In a study on secondary steel reinforcement options for enhancing RC corbels with CFRPs, it was observed that the presence of such reinforcement not only prevented the development of cracks but also enhanced the ultimate load capacity by as much as 24%, primarily through the action of stirrups bridging the primary cracks [12]. However, a dearth of research exists on comparing the impacts of secondary steel reinforcements with different CFRP configurations, as well as the effects of increasing the geometrical transverse steel ratios (ρ_{sh}) of horizontal closed stirrups.

While prior work has studied the effects of secondary steel bars or discrete CFRP configurations independently, the direct assessment and contrast of the structural performance and load capacity improvements provided by each technique individually or in combination is lacking. Identifying the optimal strengthening scheme through a comparative evaluation of horizontal closed stirrups and CFRP configurations using nonlinear FEA modeling with validation from experimental data can help inform reinforcement design practices and material selection decisions to enhance structural integrity and service life. Additionally, apart from the detailed experimental case, two different ρ_{sh} values for the same CFRP sheet configuration were investigated, and the optimal ρ_{sh} was confirmed. The research compares the influence of the optimal CFRP sheet configuration with equal volume and ρ_{sh} on the ultimate load capacity of strengthened RC corbels.

2 Optimized Reinforcement Strategies of Strengthened Corbels

To ascertain the optimized horizontal closed stirrups for corbels and compare them with the optimal CFRP configuration, two additional ρ_{sh} values were proposed, modeled, and analyzed using ABAQUS software. In alignment with the research by Kheyroddin et al. [16], experimental data presented in the work of Assih et al. [17] was employed to validate the numerical model, and four distinct CFRP sheet configurations were proposed and analyzed. The CFRP sheet configuration suggested as optimal in the study by Kheyroddin et al. [16] was selected to compare the effects of increasing ρ_{sh} on the ultimate load capacity.

3 Numerical Modeling

The work of Assih et al. [17] was utilized to validate the numerical model, which is made of concrete columns supporting reinforced concrete cantilevers. The dimensions and reinforcement details of the test specimens are shown in Figs. 1 and 2. The chosen specimen was strengthened with three layers of CFRP sheets using the EBR technique (CB3u) with the mechanical properties of Table 1. In the experiments, The specimens underwent three-point flexural loading at a rate of 0.2 kN/s, during which strain gauges were affixed to the primary reinforcing bar of the corbel (designated as point G1) to quantify displacements. The steel rebar diameters of 6, 10, and 14 mm were used with a yield strength of 500 MPa.

The concrete components were modeled using 8-node brick finite elements featuring linear geometry approximation and reduced integration. Reinforcing steel elements, including primary corbel bars, secondary bars, longitudinal reinforcement, and column ties, were represented by 2-node truss elements to account for their load-carrying behavior in the axial direction only. FRP sheets were simulated using 4-node thin shell elements. The bond between the concrete and steel reinforcement was enforced through embedded constraints. An ideal adherence was defined at the concrete-FRP interface using tying in ABAQUS. Refinement of the discretization scheme was found to impact load-deformation response as cracking initiation and propagation depends sensitively on stress distributions which are governed by element size and shape. Ultimate capacity predictions also varied with mesh characteristics due to altered failure mechanics within the composite structure, which, according to the study of Kheyroddin et al. [16], the optimal mesh size was

Fig. 1 Test specimen
(CB3u) [17]

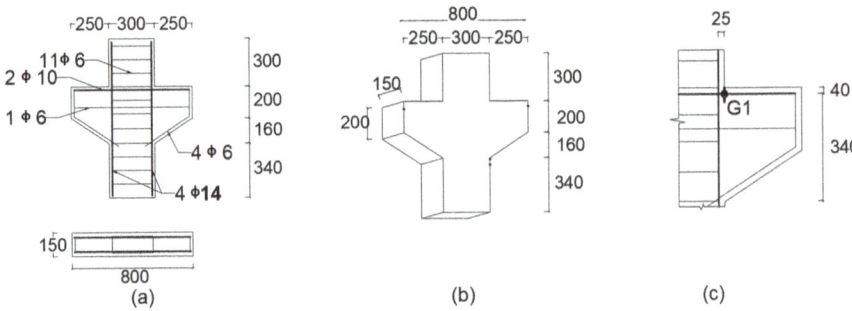

Fig. 2 (**a**) Details of steel reinforcement, (**b**) the geometric delineation, and (**c**) the location of strain gauge, after [16]. All dimensions are in mm

Table 1 The mechanical characteristics of CB3u [16]

Material	Young modulus (GPa)	Ultimate strength (MPa)	Poisson ratio
CFRP	86	972	0.45
Concrete	30	33.2	0.25
Steel bar	210	610	0.3

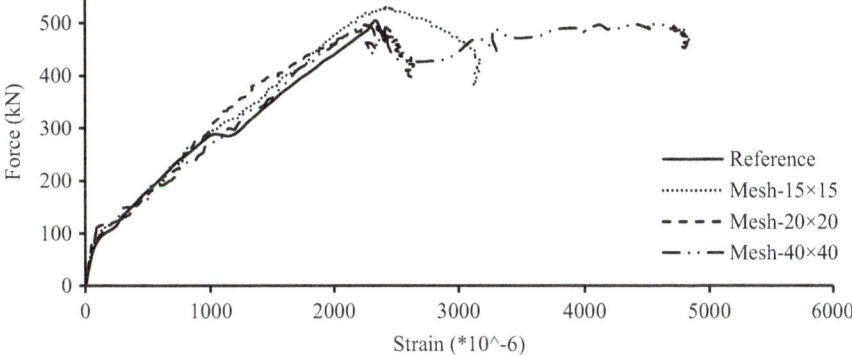

Fig. 3 Verification of the reference sample

Fig. 4 Boundary
conditions of the model

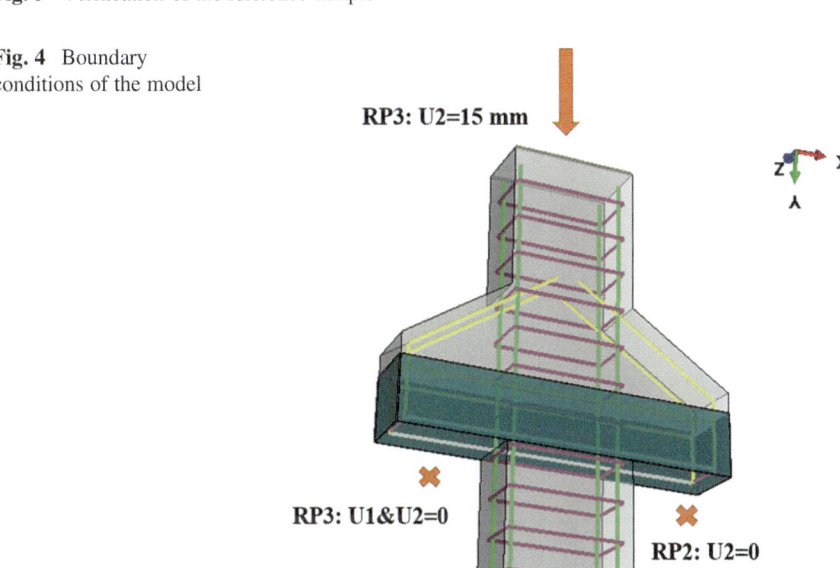

20×20 mm, see Fig. 3. In addition, the displacement and rotational constraints at the base of the corbel were individually imposed on distinct reference points, RP1 and RP2, facilitating controlled analysis of the structure's behavior, see Fig. 4.

3.1 Material Properties

The investigation involved numerical simulations performed on the experimental sample, utilizing the documented uniaxial compressive strength of the concrete ($f'_C = 33$ MPa) as documented in the experimental research publication [17]. In alignment with guidelines outlined in ACI 318-19 [2], the concrete's modulus of

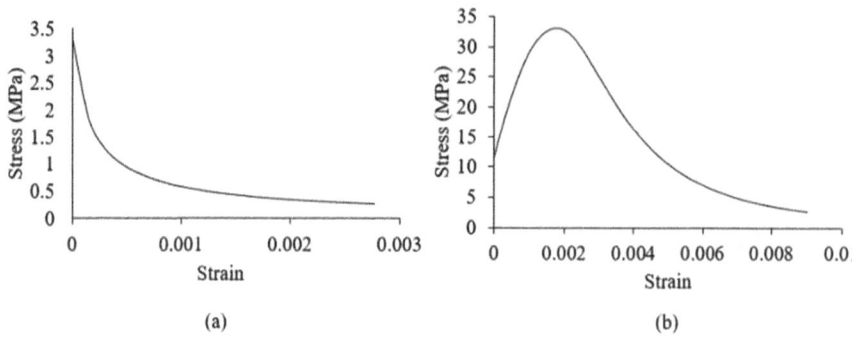

Fig. 5 (**a**) Response of concrete to uniaxial loading in tension (only plastic strains) and (**b**) response of concrete to uniaxial loading in compression (only plastic strains)

Table 2 CDP input parameters

Parameter	$\frac{f_{bu}}{f_{c0}}$	Viscosity	Dilation angle (Ψ)	K_c	Eccentricity ($\in$)	Poisson ratio (ν)
Value	1.16	0.01	37	0.667	0.1	0.25

Table 3 Mechanical characteristics of S500 steel reinforcement

Parameter	Ultimate stress (f_u) MPa	Yield stress (f_y) MPa	Poisson ratio (ν)	Elasticity modulus (E) GPa
Value	610	520	0.3	210

Table 4 Mechanical specifications of CFRP sheets

Parameter	Thickness mm	Poisson ratio (ν_{12})	$G_{13} = G_{23}$ MPa	G_{12} MPa	Density g/mm^2	E_1 GPa	E_2 GPa	Elongation at rupture
Value	0.2	0.3	2038	1019	1600	165	86	1.8%

elasticity (E_C) was determined to be 26.9 GPa. The concrete damaged plasticity (CDP) model (Fig. 5 and Table 2) was utilized to generate more accurate representations of concrete cracking behavior, accounting for rate effects. The CDP model simulates degradation in stiffness upon damage initiation, enabling the prediction of strength and fracturing as loads increase. The steel rebar diameters were 6, 8, and 10 mm, and their mechanical properties are presented in Table 3.

Given the orthotropic nature of FRP composites, thin shell elements with an orthotropic "LAMINA" material formulation were employed in the simulation [18, 19]. While FRPs generally exhibit anisotropic behavior, the unidirectional fiber architecture of the carbon FRP used permitted simplification to an orthotropic model. The mechanical properties in Table 4 [19, 20] characterized the linear elastic and pre-failure response, assuming the specified moduli for the carbon FRP reinforcement. This modeling approach captured the directionally-dependent stiffness of the FRP material.

4 Proposed Strengthening Method

Two steel reinforcement ratios (ρ_{sh}) were introduced to discern the most effective method for strengthening RC corbels and to assess how the variations in CFRP configurations and geometric secondary steel reinforcement ratios (ρ_{sh}) impact the ultimate load capacity and behavior of strengthened RC corbels. In a study by Kheyroddin et al. [16], the optimal CFRP configuration for RC corbels, among four proposed configurations with the same volume, was the 3R-45-50+4Triangle specimen. This specimen featured six inclined strips of CFRPs at a 45-degree angle and with a width of 50 millimeters, and it was reinforced at the corner of the RC corbel by four triangle-shaped CFRP sheets, as illustrated in Fig. 6.

In this study, two models were presented, both sharing the same CFRP configuration as the modeled experimental case (comprising three unidirectional CFRP layers wrapped horizontally), but with the diameters of the closed stirrups modified to 8 and 10 mm, named 1R-0-200+ST8 and 1R-0-200+ST10, respectively. The steel reinforcement ratios (ρ_{sh}) for the proposed specimens were 0.33% and 0.52%, respectively. Initially, this study explored the impact of augmenting the steel reinforcement ratio (ρ_{sh}) on the load capacity of RC corbels, and subsequently, the outcomes were compared with those of the study by Kheyroddin et al. [16] to delineate the optimal reinforcement means for strengthening RC corbels.

Fig. 6 The optimal CRP configuration (3R-45-50+4Triangle)

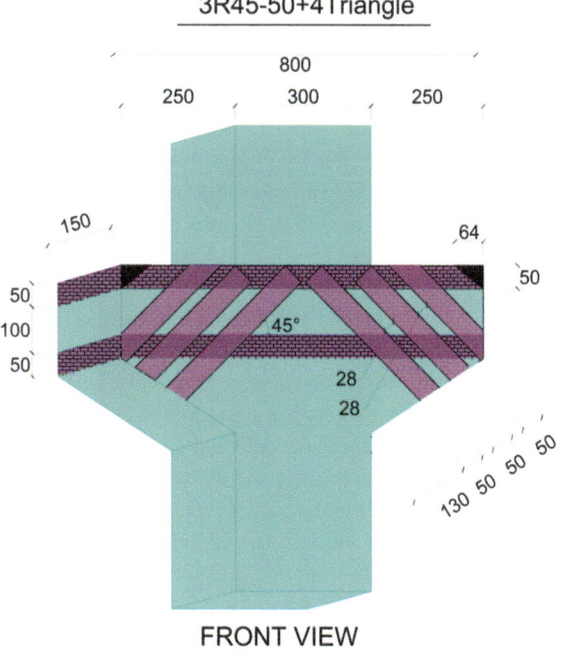

5   Analysis and Interpretation of Finite Element Results

5.1   *Enhanced Corbel Load Capacity Analysis*

By increasing the diameter of the closed stirrups from 6 to 8 mm and 10 mm in the 1R-0-200+ST8 and 1R-0-200+ST10 specimens, a respective 5.52% and 11.55% enhancement in the ultimate load capacity was achieved through a higher steel reinforcement ratio (ρ_{sh}). On average, for specimens featuring the same CFRP configuration, the results demonstrated a 5.6% increase in capacity for every 0.17% rise in the steel reinforcement ratio. Nonetheless, in a study by Kheyroddin et al. [16], the median increase in load capacity for four proposed CFRP arrangements was 28%, with the 3R-45-50+4Triangle configuration showing a substantial 33% capacity boost.

To ascertain the relationship between the CFRP configuration and the increase in steel reinforcement ratio (ρ_{sh}), the specimen 3R-45-50+4Triangle+ST10, with a 10 mm diameter of closed stirrups (resulting in a ρ_{sh} increase up to 0.52%), was analyzed, revealing a 4.39% surge in capacity.

5.2   *Behavioral Analysis: Load–Displacement*

The load–displacement curves of the five investigated samples are presented in Fig. 7. The research findings reveal that, in comparison, the most effective strengthening method for increasing the ultimate load entailed an optimal CFRP configuration in conjunction with an increase in the geometrical secondary steel reinforcement ratio (ρ_{sh}). Overall, specimens with higher (ρ_{sh}) but the same CFRP configurations demonstrated greater stiffness. The extent of stiffness enhancement varied, particularly across different CFRP configurations. For instance, in the case of the 1R-0-200+ST10 specimen (initial configuration), the percentage increase in load capacity was more than double that of the same (ρ_{sh}) in the optimized CFRP configuration (11.55% versus 4.39%).

The process of loading experienced by the simulated and investigated samples can be delineated into five stages, all uniformly undergone by each specimen. These stages can be identified as initial flexural/shear cracking, splitting cracking, concrete failure on the flange-to-column joint, yielding of the primary steel reinforcement, and cracking of the vertical-sided area of the corbel flange, as illustrated in Fig. 8.

The notable distinction between the two prime reinforcement approaches, CFRP configurations and the secondary steel reinforcement ratio, pertains to the Mises stress contour. The findings suggest that alterations in the secondary steel reinforcement ratio do not impact the Mises stress contour, while changes in the CFRP configuration do, as depicted in Fig. 9.

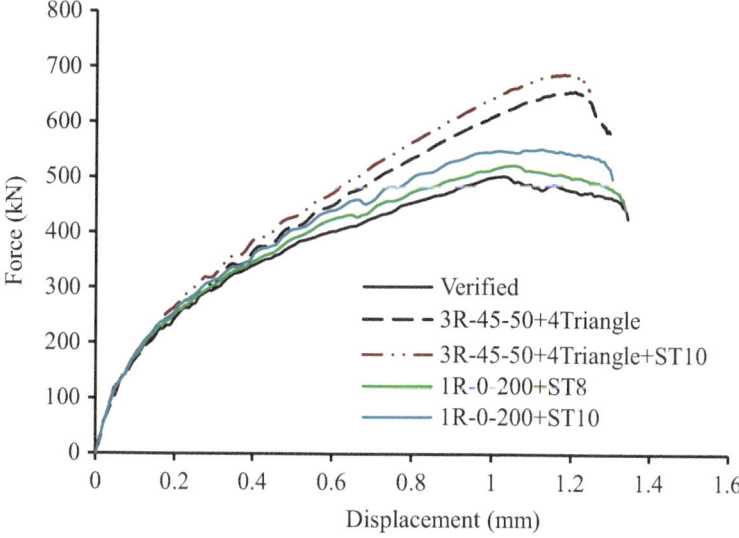

Fig. 7 Evaluation of load–displacement behavior of samples

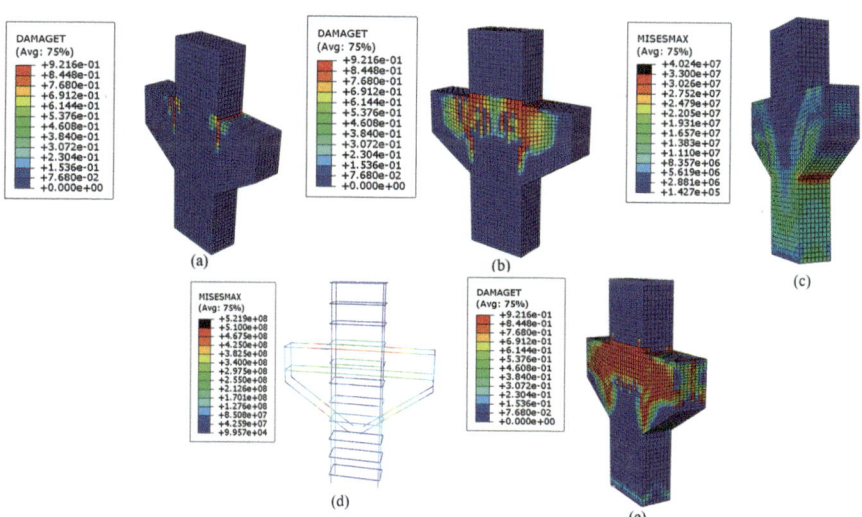

Fig. 8 1R-0-200+ST10 sample: (**a**) the flexural/shear crack, (**b**) the splitting cracks, (**c**) the failure of concrete on the flange-to-column joint, (**d**) yielding of the primary steel reinforcement, and (**e**) the cracked portion of the corbel flange slope

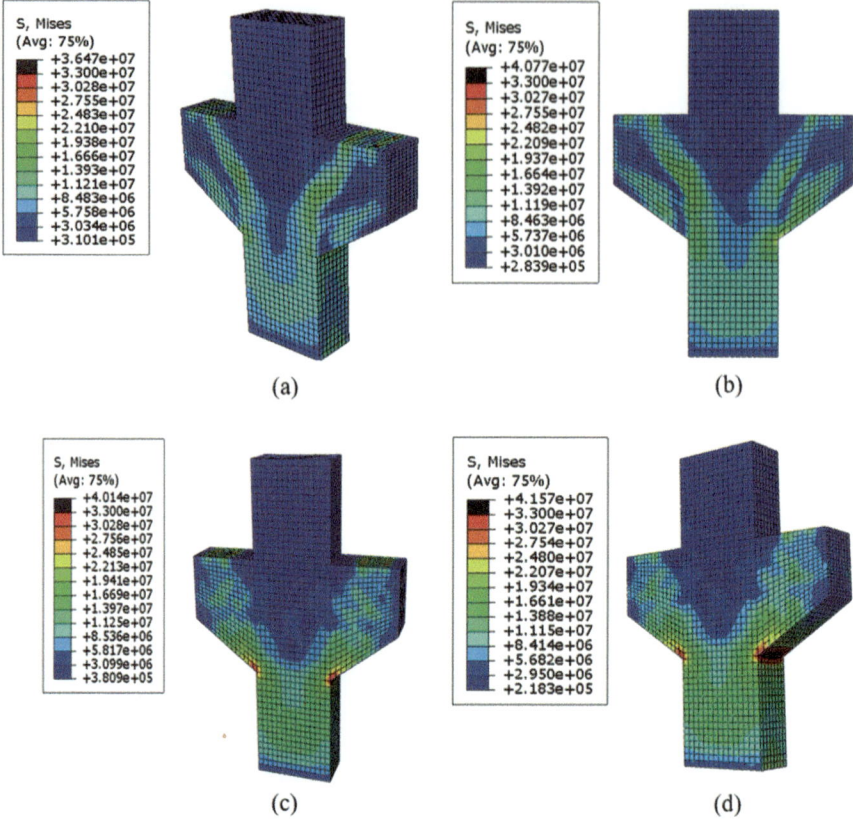

Fig. 9 Mises stress distribution in samples: (**a**) experimentally validated sample, (**b**) 1R-0-200+ST10, (**c**) 3R-45-50+4Triangle, and (**d**) 3R-45-50+4Triangle+ST10

5.3 Evaluation of Primary Reinforcement Load–Strain

Per the guidelines outlined in the CEP-FIP code [21], it is possible to convert the stress–strain characteristics of reinforcement bars confined by concrete into a standard load–strain curve. Similar to the impact of the CFRP configuration, an increase in the ρ_{sh} parameter is observed in post-yield zone D. However, it is noted that specimens with an increased ρ_{sh} exhibit a reduced load–strain slope compared to those with an altered CFRP configuration (3R-45-50+4Triangle). This observation underscores the significant influence of the CFRP configuration despite the utilization of the same amount of material, as depicted in Fig. 10.

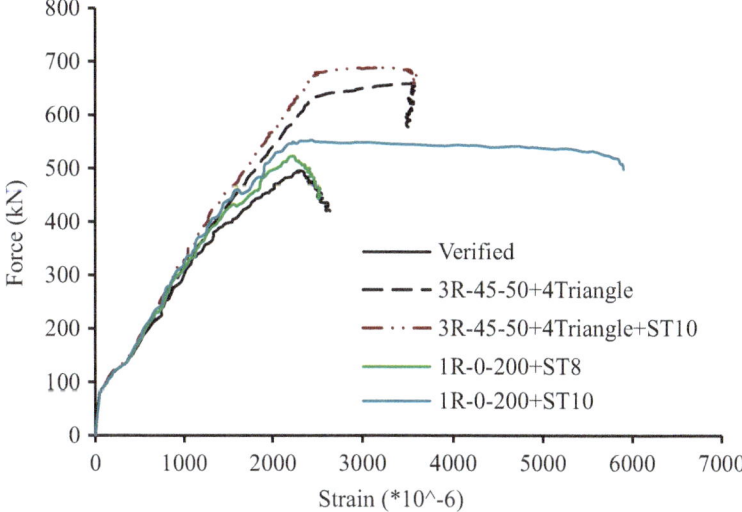

Fig. 10 Load–strain curves of proposed strengthened specimens

6 Conclusion

This study introduced two steel reinforcement ratios (ρ_{sh}) for a particular CFRP configuration. The optimal ρ_{sh} was determined through an analysis of the proposed reinforcement models utilizing nonlinear finite element (FE) techniques. Subsequently, the optimized CFRP configurations were modeled with the selected optimal ρ_{sh}. Their stress–strain behavior, stiffness, maximum load-bearing capability, Mises stress contours, and displacement were compared. The outcomes of the research substantiate the following conclusions:

- The CFRP configurations can impact the amount of influence of horizontal closed stirrups.
- Despite utilizing the same amount of material in different CFRP configurations, the general Mises stress contour of specimens alters. However, the mentioned factor is the same with changing the (ρ_{sh}).

In comparison to employing various CFRP configurations, the effect of modifying the (ρ_{sh}) during the crack-formation phase of the specimens is somewhat more pronounced, albeit not statistically significant. Nonetheless, the substantial impact of CFRP configurations becomes noticeable in the post-yield phase and the bearing capacity. It is advisable for future research to explore the influence of closed stirrups arrangements like vertical or the combination of horizontal and vertical, and in which CFRP configurations (horizontal, vertical, and inclined) the effect of the horizontal closed stirrups are the most.

Notation

A_s = Area of longitudinal steel bars
A_{sh} = Area of horizontal or vertical stirrups
ρ_{sh} = The geometrical transverse steel ratios

Acknowledgments This work was part of the HYPERION project. HYPERION has received funding from the European Union's Framework Programme for Research and Innovation (Horizon 2020) under grant agreement No. 821054. The contents of this publication are the sole responsibility and do not necessarily reflect the opinion of the European Union.

References

1. Mattock, A.H., Chen, K.C., Soongswang, K.: The behavior of reinforced concrete corbels. PCI J. **21**(2), 52–77 (1976). https://doi.org/10.15554/pcij.03011976.52.77
2. 318-19: Building Code Requirements for Structural Concrete and Commentary. American Concrete Institute (2019). https://doi.org/10.14359/51716937
3. Engindeniz, M., Kahn, L.F., Zureick, A.-H.: Repair and strengthening of reinforced concrete beam-column joints: state of the art. ACI Struct. J. **102**(2), 1–14 (2005)
4. Barbhuiya, S., Choudhury, A.M.: A study on the size effect of RC beam-column connections under cyclic loading. Eng. Struct. **95**, 1–7 (2015). https://doi.org/10.1016/j.engstruct.2015.03.052
5. Rezaee Azariani, H., Shariatmadar, H., Reza Esfahani, M.: Exterior concrete beam-column connection reinforced with glass fiber reinforced polymers (GFRP) bars under cyclic loading. AUT J. Civ. Eng. **2**(2), 161–176 (2018). https://doi.org/10.22060/ajce.2018.14611.5487
6. Kang, T.H.-K., Ha, S.-S., Choi, D.-U.: Bar pullout tests and seismic tests of small-headed bars. ACI Struct. J. **107**(1), 32–42 (2010). https://doi.org/10.14359/51663386
7. Romanichen, R.M., Souza, R.A.: Reinforced concrete corbels strengthened with external prestressing. Rev. IBRACON Estrut. Mater. **12**(4), 812–831 (2019). https://doi.org/10.1590/s1983-41952019000400006
8. Mortezaei, A., Ronagh, H.R., Kheyroddin, A.: Seismic evaluation of FRP strengthened RC buildings subjected to near-fault ground motions having fling step. Compos. Struct. **92**(5), 1200–1211 (2010). https://doi.org/10.1016/j.compstruct.2009.10.017
9. Campione, G., Cannella, F.: Analytical model for flexural response of reinforced concrete corbels externally strengthened with fiber- reinforced polymer. ACI Struct. J. **117**(4), 91–102 (2020). https://doi.org/10.14359/51721374
10. Sayhood, E.K., Abdul Majeed Hassan, Q., Gh Yassin, L.A.: Enhancement in the load-carrying capacity of reinforced concrete corbels strengthened with CFRP strips under monotonic or repeated loads. Eng. Technol. J. **34**(14) (2016)
11. El-Maaddawy, T.A., Sherif, E.-S.I.: Response of concrete corbels reinforced with internal steel rebars and external composite sheets: experimental testing and finite element modeling. J. Compos. Constr. **18**(1), 04013020 (2014). https://doi.org/10.1061/(asce)cc.1943-5614.0000403
12. Al-Kamaki, Y.S.S., Hassan, G.B., Alsofi, G.: Experimental study of the behaviour of RC corbels strengthened with CFRP sheets. Case Stud. Constr. Mater. **9** (2018). https://doi.org/10.1016/j.cscm.2018.e00181
13. Khosravikia, F., et al.: Experimental and numerical assessment of corbels designed based on strut-and-tie provisions. J. Struct. Eng. **144**(9), 04018138 (2018). https://doi.org/10.1061/(asce)st.1943-541x.0002137

14. Abass, J.M.A.: Behavior and strength of self-compacting fiber reinforced concrete corbels. PhD thesis. Al-Mustansiriya University, Baghdad (2014)
15. Muhammad, A.H.: Behavior and strength of high-strength fiber reinforced concrete corbels subjected to monotonic or cyclic (repeated) loading. Doctoral dissertation. University of Technology, Baghdad (1998)
16. Kheyroddin, A., Raygan, S., Kioumarsi, M.: Strut and tie model for CFRP strengthened reinforced concrete corbels. Eng. Struct. **304**, 117609 (2024). https://doi.org/10.1016/j.engstruct.2024.117609
17. Assih, J., Ivanova, I., Dontchev, D., Li, A.: Concrete damaged analysis in strengthened corbel by external bonded carbon fibre fabrics. Appl. Adhes. Sci. **3**(1) (2015). https://doi.org/10.1186/s40563-015-0045-1
18. Vahidpour, M., Kheyroddin, A., Kioumarsi, M.: Experimental investigation on flexural capacity of reinforced concrete beams strengthened with 3D-fiberglass, CFRP and GFRP. Int. J. Concr. Struct. Mater. **16**(1) (2022). https://doi.org/10.1186/s40069-022-00508-w
19. Ivanova, I., Assih, J., Stankov, V., Dontchev, D.: Numerical simulation of shear strength in a short reinforced concrete corbel strengthened with composite material compared with experimental results. Comput. Modell. Concr. Struct., 585–594 (2018). https://doi.org/10.1201/9781315182964-71
20. C. SIKA: S-Pultruded Carbon Fibre Plates for Structural Strengthening as Part of the Sika® CarboDur® System: Product Data Sheet (2018)
21. Comité Euro-International du Béton: CEB-FIP Model Code 1990. Thomas Telford Publishing (1993)

Dynamic Behavior of Imperfect FGM Beams with Various Porosity Distribution Rates: Analysis and Modeling

Lazreg Hadji ⓘ, Vagelis Plevris ⓘ, and Royal Madan ⓘ

Contents

1 Introduction

Functionally graded materials (FGMs) signify a significant advancement in engineering and scientific domains, offering solutions to intricate challenges encountered across diverse industries, notably aerospace and biomedical applications [1–3]. It is crucial to recognize that porosities may arise within FGMs during the sintering phase of fabrication, mainly because of the differences in solidification temperatures between the different materials [4, 5]. In the design of FGM structures exposed to dynamic loads, accounting for the influence of porosity is paramount [6, 7]. The

L. Hadji
University of Tiaret, Tiaret, Algeria

Laboratory of Geomatics and Sustainable Development, University of Tiaret, Tiaret, Algeria
e-mail: lazreg.hadji@univ-tiaret.dz

V. Plevris (✉)
Qatar University, Doha, Qatar
e-mail: vplevris@qu.edu.qa

R. Madan
Graphic Era (Deemed to be University), Dehradun, India

© The Author(s) 2025
M. Kioumarsi, B. Shafei (eds.), *The 1st International Conference on Net-Zero Built Environment*, Lecture Notes in Civil Engineering 237,
https://doi.org/10.1007/978-3-031-69626-8_126

disparity in solidification temperatures between metals and ceramics gives rise to the formation of metal phase grains, while ceramics persist as interspersed particles. Furthermore, the varied sizes and configurations of the reinforcement (ceramics) powders can engender pore formation in proximity to the reinforced particles, leading to divergent levels of porosity within both phases [6].

This investigation examines the repercussions of distinct porosity types within both ceramic and metal constituents. Each discussed porosity type manifests differing percentages of porosity in ceramic and metal phases. The analysis encompasses how the stiffness of functionally graded beams is affected by the power-law index, furnishing elucidations into their dynamic response. Furthermore, scrutiny of the length-to-thickness ratio yields significant insights into the ramifications of geometric proportions. Collectively, this study offers novel insights into the behavior of porous functionally graded beams, considering a spectrum of parameters and underscoring their relevance in real-world applications.

2 Problem Formulation

2.1 Constitutive Relations of FG Beams Made of Metal and Ceramic

We examine an imperfect FGM characterized by the porosity volume fraction, a (where $a \ll 1$), evenly distributed between the two constituents. The rule of mixture (modified), as proposed by [8], is

$$P = P_m \left(V_m - \frac{\alpha}{2} \right) + P_c \left(V_c - \frac{\alpha}{2} \right) \tag{1}$$

$$V_c = \left(\frac{z}{h} + \frac{1}{2} \right)^k \tag{2}$$

$$V_c + V_m = 1 \Rightarrow V_c = 1 - V_m \tag{3}$$

The power law of volume fraction is described in detail in Table 1. The properties of the imperfect FGM can be formulated as

$$P = (P_c - P_m) \left(\frac{z}{h} + \frac{1}{2} \right)^k + P_m - (P_c + P_m) \frac{\alpha}{2} \tag{4}$$

The parameter k, a non-negative real number ($0 \le k \le \infty$), represents the volume fraction or power-law index, while z denotes the distance from the mid-plane of the beam. The FG beam transitions to a fully ceramic one as k approaches zero, and to a fully metallic one as k becomes large. The equations for Elastic Modulus (E) and the density of the material (ρ) of the imperfect FGM beam are detailed in [9]. Table 1 shows the equations used for E for the various porosity distributions present in the

Table 1 Porosity distribution in the FGM's (ceramic/metal): different types

	Porosity rate distribution		
Types	Ceramic	Metal	Elastic modulus, $E(z) =$
T-1	Perfect FG beam, without porosity $(a = 0)$		$(E_c - E_m)\left(\frac{z}{h} + \frac{1}{2}\right)^k + E_m$
T-2	50%	50%	$E_m\left(V_m - \frac{\alpha}{2}\right) + E_c\left(V_c - \frac{\alpha}{2}\right)$
			$(E_c - E_m)\left(\frac{z}{h} + \frac{1}{2}\right)^k + E_m - (E_c + E_m)\frac{\alpha}{2}$
T-3	60%	40%	$E_m\left(V_m - \frac{2\alpha}{5}\right) + E_c\left(V_c - \frac{3\alpha}{5}\right)$
			$(E_c - E_m)\left(\frac{z}{h} + \frac{1}{2}\right)^k + E_m - (3E_c + 2E_m)\frac{\alpha}{5}$
T-4	40%	60%	$E_m\left(V_m - \frac{3\alpha}{5}\right) + E_c\left(V_c - \frac{2\alpha}{5}\right)$
			$(E_c - E_m)\left(\frac{z}{h} + \frac{1}{2}\right)^k + E_m - (2E_c + 3E_m)\frac{\alpha}{5}$
T-5	75%	25%	$E_m\left(V_m - \frac{\alpha}{4}\right) + E_c\left(V_c - \frac{3\alpha}{4}\right)$
			$(E_c - E_m)\left(\frac{z}{h} + \frac{1}{2}\right)^k + E_m - (3E_c + E_m)\frac{\alpha}{4}$
T-6	25%	75%	$E_m\left(V_m - \frac{3\alpha}{4}\right) + E_c\left(V_c - \frac{\alpha}{4}\right)$
			$(E_c - E_m)\left(\frac{z}{h} + \frac{1}{2}\right)^k + E_m - (E_c + 3E_m)\frac{\alpha}{4}$

FGMs. The Poisson's ratio (ν) is assumed to remain constant. In the special case where $a = 0$, we obtain material properties corresponding to a perfect FG beam.

2.2 Theoretical Formulation

2.2.1 Assumptions

The theory operates under the following assumptions:

- Displacements are significantly smaller in magnitude compared to the height of the beam, thus resulting in infinitesimal strains.
- The displacement in the transverse direction, w, comprises two components for bending (w_b) and shear (w_s), which are both solely functions of the coordinates x and t:

$$w(x, z, t) = w_b(x, t) + w_s(x, t) \tag{5}$$

- Normal stress σ_z (in the transverse direction) is considerably smaller in magnitude compared to the in-plane stresses σ_x.
- Axial displacement u (in the x-direction), comprises extension, bending, and shear components.

$$u = u_0 + u_b + u_s \qquad (6)$$

- The component of bending, u_b, is assumed to closely resemble the displacements predicted by the beam theory. Hence, u_b can be expressed as

$$u_b = -z \frac{\partial w_b}{\partial x} \qquad (7)$$

- The combination of the shear component u_s with w_s results in a hyperbolic variation of shear strain γ_{xz}, causing shear stress τ_{xz} to distribute across the thickness of the beam. This distribution ensures that shear stress τ_{xz} is zero at both the top and bottom surfaces of the beam. Thus, the expression for u_s can be stated as follows:

$$u_s = -f(z) \frac{\partial w_s}{\partial x} \qquad (8)$$

$$f(z) = -\frac{z}{4} + \frac{5z^3}{3h^2} \qquad (9)$$

2.2.2 Constitutive Equations and Kinematics

Utilizing the formulations outlined in the previous section, the field of displacements can be derived from Eqs. (5), (6), (7), (8), and (9) as

$$u(x, z, t) = u_0(x, t) - z \frac{\partial w_b}{\partial x} - f(z) \frac{\partial w_s}{\partial x} \qquad (10)$$

$$w(x, z, t) = w_b(x, t) + w_s(x, t) \qquad (11)$$

The strains corresponding to the displacements in Eqs. (10) and (11) are as follows:

$$\varepsilon_x = \varepsilon_x^0 + z \, k_x^b + f(z) \, k_x^s \qquad (12)$$

$$\gamma_{xz} = g(z) \, \gamma_{xz}^s \qquad (13)$$

where

$$\varepsilon_x^0 = \frac{\partial u_0}{\partial x}, \quad k_x^b = -\frac{\partial^2 w_b}{\partial x^2}, \quad k_x^s = -\frac{\partial^2 w_s}{\partial x^2}, \quad \gamma_{xz}^s = \frac{\partial w_s}{\partial x} \tag{14}$$

$$g(z) = 1 - f'(z), \quad f'(z) = \frac{df(z)}{dz} \tag{15}$$

Assuming adherence to Hooke's law for the material of the FG beam, the stresses within the beam can be determined:

$$\sigma_x = Q_{11}(z)\,\varepsilon_x \tag{16}$$

$$\tau_{xz} = Q_{55}(z)\,\gamma_{xz} \tag{17}$$

$$Q_{11}(z) = E(z) \tag{18}$$

$$Q_{55}(z) = E(z)/[2(1+\nu)] \tag{19}$$

2.2.3 Motion Equations

In this context, Hamilton's principle is utilized to end up to the equations of motion, as follows [10]:

$$\delta \int_{t_1}^{t_2} (U - T)dt = 0 \tag{20}$$

Here, t represents the time, t_1 is the initial, t_2 denotes the final time, δU signifies the virtual variation of the strain energy, and δT represents the virtual variation of the kinetic energy. The variation in strain energy of the beam is

$$\delta U = \int_0^L \int_{-h/2}^{h/2} \left(\sigma_x \delta \varepsilon_x + \tau_{xz} \delta \gamma_{xz}\right) dz dx$$

$$= \int_0^L \left(N_x \delta \varepsilon_x^0 - M_x^b \delta k_x^b - M_x^s \delta k_x^s + Q_{xz} \delta \gamma_{xz}^s\right) dx \tag{21}$$

where the four stress resultants can be defined as

$$\left(N_x, M_x^b, M_x^s\right) = \int_{-h/2}^{h/2} (1, z, f(z))\,\sigma_x dz \tag{22}$$

$$Q_{xz} = \int_{-h/2}^{h/2} \tau_{xz} g(z) dz \tag{23}$$

The kinetic energy variation is given by

$$\delta T = \int_0^L \int_{-h/2}^{h/2} \rho(z) [\dot{u}\delta \dot{u} + \dot{w}\delta \dot{w}] \, dz_{ns} dx$$

$$= \int_0^L \left\{ I_0[\dot{u}_0 \delta \dot{u}_0 + (\dot{w}_b + \dot{w}_s)(\delta \dot{w}_b + \delta \dot{w}_s)] - I_1 \left(\dot{u}_0 \frac{d\delta \dot{w}_b}{dx} + \frac{d\dot{w}_b}{dx} \delta \dot{u}_0 \right) \right.$$

$$+ I_2 \left(\frac{d\dot{w}_b}{dx} \frac{d\delta \dot{w}_b}{dx} \right) - J_1 \left(\dot{u}_0 \frac{d\delta \dot{w}_s}{dx} + \frac{d\dot{w}_s}{dx} \delta \dot{u}_0 \right) + K_2 \left(\frac{d\dot{w}_s}{dx} \frac{d\delta \dot{w}_s}{dx} \right)$$

$$\left. + J_2 \left(\frac{d\dot{w}_b}{dx} \frac{d\delta \dot{w}_s}{dx} + \frac{d\dot{w}_s}{dx} \frac{d\delta \dot{w}_b}{dx} \right) \right\} dx$$

$$\tag{24}$$

In the provided context, the dot-superscript notation means differentiation with respect to the time variable t. $\rho(z)$ represents the mass density, and the mass inertias are defined as

$$(I_0, I_1, J_1, I_2, J_2, K_2) = \int_{-h/2}^{h/2} \left(1, z, f, z^2, zf, f^2\right) \rho(z) dz \tag{25}$$

By using Eqs. (21) and (24) into Eq. (20), we have

$$\delta u_0 : \frac{dN_x}{dx} = I_0 \ddot{u}_0 - I_1 \frac{d\ddot{w}_b}{dx} - J_1 \frac{d\ddot{w}_s}{dx} \tag{26}$$

$$\delta w_b : \frac{d^2 M_b}{dx^2} = I_0(\ddot{w}_b + \ddot{w}_s) + I_1 \frac{d\ddot{u}_0}{dx} - I_2 \frac{d^2 \ddot{w}_b}{dx^2} - J_2 \frac{d^2 \ddot{w}_s}{dx^2} \tag{27}$$

$$\delta w_s : \frac{d^2 M_s}{dx^2} + \frac{dQ_{xz}}{dx} = I_0(\ddot{w}_b + \ddot{w}_s) + J_1 \frac{d\ddot{u}_0}{dx} - J_2 \frac{d^2 \ddot{w}_b}{dx^2} - K_2 \frac{d^2 \ddot{w}_s}{dx^2} \tag{28}$$

Introducing Eqs. (22) and (23) into Eqs. (26), (27), and (28), the motion equations can be expressed as follows, in terms of u_0, w_b, w_s:

$$A_{11} \frac{\partial^2 u_0}{\partial x^2} - B_{11} \frac{\partial^3 w_b}{\partial x^3} - B_{11}^s \frac{\partial^3 w_s}{\partial x^3} = I_0 \ddot{u}_0 - I_1 \frac{d\ddot{w}_b}{dx} - J_1 \frac{d\ddot{w}_s}{dx} \tag{29}$$

$$B_{11} \frac{\partial^3 u_0}{\partial x^3} - D_{11} \frac{\partial^4 w_b}{\partial x^4} - D_{11}^s \frac{\partial^4 w_s}{\partial x^4} = I_0(\ddot{w}_b + \ddot{w}_s) + I_1 \frac{d\ddot{u}_0}{dx} - I_2 \frac{d^2 \ddot{w}_b}{dx^2} - J_2 \frac{d^2 \ddot{w}_s}{dx^2}$$

$$(30)$$

$$B_{11}^s \frac{\partial^3 u_0}{\partial x^3} - D_{11}^s \frac{\partial^4 w_b}{\partial x^4} - H_{11}^s \frac{\partial^4 w_s}{\partial x^4} + A_{55}^s \frac{\partial^2 w_s}{\partial x^2} = I_0(\ddot{w}_b + \ddot{w}_s) + J_1 \frac{d\ddot{u}_0}{dx}$$

$$- J_2 \frac{d^2 \ddot{w}_b}{dx^2} - K_2 \frac{d^2 \ddot{w}_s}{dx^2}$$

$$(31)$$

Where the beam stiffnesses are defined by

$$\left(A_{ij}, A_{ij}^s, B_{ij}, D_{ij}, B_{ij}^s, D_{ij}^s, H_{ij}^s \right) = \int\limits_{-h/2}^{h/2} Q_{ij}\left(1, g^2(z), z, z^2, f(z), zf(z), f^2(z) \right) dz \quad (32)$$

2.2.4   Analytical Solution

The analytical solutions in the Navier-type format are derived for the free vibration analysis of FG beams. Following this approach, the variables (unknown displacements) are expanded into a Fourier series as follows:

$$\begin{Bmatrix} u_0 \\ w_b \\ w_s \end{Bmatrix} = \sum_{m=1}^{\infty} \begin{Bmatrix} U_m \cos(\lambda x)\, e^{i\omega t} \\ W_{bm} \sin(\lambda x)\, e^{i\omega t} \\ W_{sm} \sin(\lambda x)\, e^{i\omega t} \end{Bmatrix} \quad (33)$$

In the above, U_m, W_{bm}, and W_{sm} denote arbitrary parameters that are to be calculated, ω is the frequency associated with the m-th eigenmode, and $\lambda = m{\cdot}\pi/L$. By substituting Eq. (33) into Eqs. (29), (30), and (31), the analytical solution can be derived through the eigenvalue equations below, for any given value of the eigenmode m.

$$\left([K] - \omega^2 [M] \right) \{\Delta\} = \{0\} \quad (34)$$

$$[K] = \begin{bmatrix} a_{11} & a_{12} & a_{13} \\ a_{12} & a_{22} & a_{23} \\ a_{13} & a_{23} & a_{33} \end{bmatrix}, \quad (35)$$

$$[M] = \begin{bmatrix} m_{11} & m_{12} & m_{13} \\ m_{12} & m_{22} & m_{23} \\ m_{13} & m_{23} & m_{33} \end{bmatrix}, \quad (36)$$

$$\{\Delta\} = \begin{Bmatrix} U_m \\ W_{bm} \\ W_{sm} \end{Bmatrix}, \tag{37}$$

$$a_{11} = A_{11}\lambda^2, \quad a_{12} = -B_{11}\lambda^3, \quad a_{13} = -B_{11}^s\lambda^3, \tag{38}$$

$$a_{22} = D_{11}\lambda^4, \quad a_{23} = D_{11}^s\lambda^4, \quad a_{33} = H_{11}^s\lambda^4 + A_{55}^s\lambda^2 \tag{39}$$

$$m_{11} = I_0, \quad m_{12} = -I_1\alpha, \quad m_{13} = -J_1\alpha, \tag{40}$$

$$m_{22} = I_0 + I_2\alpha^2, \quad m_{23} = I_0 + J_2\alpha^2, \quad m_{33} = I_0 + K_2\alpha^2 \tag{41}$$

3   Results

We present the results for the numerical frequencies of imperfect FGM beams with various rates of porosity distribution, with the aim of validating the accuracy of the present formulation. The properties of the beam's material are outlined below:

- Ceramic part: Al_2O_3 (Alumina) with $\nu = 0.3$, $E_c = 380$ GPa, $\rho = 3960$ kg/m^3.
- Metal part: Al (Aluminum) with $\nu = 0.3$, $E_m = 70$ GPa, $\rho = 2702$ kg/m^3.

The following non-dimensional parameter has been used for simplicity:

$$\bar{\omega} = \frac{\omega \cdot L^2}{h}\sqrt{\frac{\rho_m}{E_m}} \tag{42}$$

The natural frequencies of both imperfect and perfect beams were examined for $L/h = 5$ and $L/h = 20$ across various power-law indices (k), with results summarized in Table 2. The present theory's outcomes are corroborated and demonstrate excellent alignment with previously published findings. Additionally, the analysis extends to different porosity types categorized as T-1 to T-6, as detailed in Table 1. The results reveal that the natural frequency attains its maximum for T-6, followed by the cases T-4, T-2, T-3, T-5, and T-1.

When the beam is composed entirely of ceramic ($k = 0$), an escalation in grading indices results in a greater proportion of metal within the beam, consequently diminishing its stiffness and subsequently reducing its natural frequency. This trend mirrors the behavior observed in the non-dimensional flexural natural frequencies of porous functionally graded beams, as depicted in Table 3 (for $L/h = 5$) and Table 4 (for $L/h = 20$). Across all instances, the natural frequency is higher for porous beams in comparison to non-porous beams.

Figure 1 shows the evolution of the frequency of imperfect FG beams across a range of power-law indices (k). The analysis reveals a decline in frequency with the augmentation of the porosity fraction k. Specifically, when k is less than 5, a sharp

Table 2 Non-dimensional frequencies of simply supported porous FG beams ($a = 0.1$)

L/h	Theory	$k = 0$	$k = 1$	$k = 2$	$k = 5$	$k = 10$
5	Bernoulli–Euler (1744) [11]	5.3953	4.1484	3.7793	3.5949	3.4921
	Timoshenko (1921) [12]	5.1524	3.9902	3.6343	3.4311	3.3134
	Simsek (2010) [13]	5.1527	3.9904	3.6261	3.4012	3.2816
	Reddy (1984) [14]	5.1527	3.9904	3.6264	3.4012	3.2816
	Sayyad et al. (2018) [15]	5.1453	3.9826	3.6184	3.3917	3.2727
	T-1	5.1527	3.9904	3.6264	3.4012	3.2816
	T-2	5.2223	3.9070	3.4418	3.1479	3.0292
	T-3	5.2087	3.8712	3.3889	3.0813	2.9627
	T-4	5.2359	3.9419	3.4928	3.2113	3.0924
	T-5	5.1879	3.8158	3.3058	2.9745	2.8561
	T-6	5.2559	3.9929	3.5659	3.3011	3.1819
20	Bernoulli–Euler (1744) [11]	5.4777	4.2163	3.8472	3.6628	3.5547
	Timoshenko (1921) [12]	5.4603	4.2050	3.8367	3.6508	3.5415
	Simsek (2010) [13]	5.4603	4.2050	3.8361	3.6485	3.5389
	Reddy (1984) [14]	5.4603	4.2050	3.8361	3.6485	3.5389
	Sayyad et al. (2018) [15]	5.4603	4.2050	3.8361	3.6485	3.5389
	T-1	5.4603	4.2051	3.8361	3.6485	3.5389
	T-2	5.5341	4.1117	3.6335	3.3776	3.2809
	T-3	5.5196	4.0732	3.5764	3.3059	3.2113
	T-4	5.5484	4.1494	3.6885	3.4458	3.3470
	T-5	5.4976	4.0137	3.4866	3.1907	3.0993
	T-6	5.5696	4.2042	3.7674	3.5421	3.4405

Table 3 Flexural natural frequencies (non-dimensional) of porous FG beams ($a = 0.1$, $L/h = 5$)

Mode	Theory	$k = 0$	$k = 1$	$k = 2$	$k = 5$	$k = 10$
1	Sayyad et al. (2018) [15]	5.1453	3.9826	3.6184	3.3917	3.2727
	T-1	5.1527	3.9904	3.6264	3.4012	3.2816
	T-2	5.2223	3.9070	3.4418	3.1479	3.0292
	T-3	5.2087	3.8712	3.3889	3.0813	2.9627
	T-4	5.2359	3.9419	3.4928	3.2113	3.0924
	T-5	5.1879	3.8158	3.3058	2.9745	2.8561
	T-6	5.2559	3.9929	3.5659	3.3011	3.1819
2	Sayyad et al. (2018) [15]	17.589	13.754	12.388	11.260	10.748
	T-1	17.881	14.009	12.641	11.543	11.024
	T-2	18.123	13.755	12.049	10.685	10.103
	T-3	18.075	13.635	11.873	10.462	9.870
	T-4	18.169	13.873	12.219	10.898	10.326
	T-5	18.003	13.449	11.596	10.106	9.497
	T-6	18.239	14.044	12.463	11.202	10.642
3	Sayyad et al. (2018) [15]	32.324	25.538	22.812	20.117	19.003
	T-1	34.209	27.098	24.315	21.716	20.556
	T-2	34.672	26.675	23.276	20.124	18.748
	T-3	34.581	26.453	22.953	19.711	18.301
	T-4	34.761	26.892	23.587	20.519	19.175
	T-5	34.443	26.109	22.446	19.056	17.589
	T-6	34.895	27.209	24.035	21.082	19.783

Table 4 Non-dimensional flexural natural frequencies of porous FG beams ($a = 0.1$, $L/h = 20$)

Mode	Theory	$k = 0$	$k = 1$	$k = 2$	$k = 5$	$k = 10$
1	Sayyad et al. (2018) [15]	5.4603	4.2050	3.8361	3.6484	3.5389
	T-1	5.4603	4.2051	3.8361	3.6485	3.5389
	T-2	5.5341	4.1117	3.6335	3.3776	3.2809
	T-3	5.5196	4.0732	3.5764	3.3059	3.2113
	T-4	5.5484	4.1494	3.6885	3.4458	3.3470
	T-5	5.4976	4.0137	3.4866	3.1907	3.0993
	T-6	5.5696	4.2042	3.7674	3.5421	3.4405
2	Sayyad et al. (2018) [15]	21.571	16.631	15.158	14.370	13.922
	T-1	21.573	16.634	15.162	14.375	13.926
	T-2	21.865	16.270	14.367	13.306	12.898
	T-3	21.807	16.118	14.143	13.024	12.622
	T-4	21.921	16.418	14.584	13.575	13.159
	T-5	21.721	15.884	13.789	12.571	12.178
	T-6	22.005	16.634	14.895	13.954	13.530
3	Sayyad et al. (2018) [15]	47.569	36.740	33.440	31.543	30.505
	T-1	47.593	36.768	33.469	31.578	30.537
	T-2	48.236	35.979	31.737	29.228	28.239
	T-3	48.109	35.646	31.245	28.609	27.628
	T-4	48.361	36.304	32.211	29.818	28.819
	T-5	47.918	35.131	30.470	27.615	26.646
	T-6	48.546	36.778	32.893	30.652	29.642

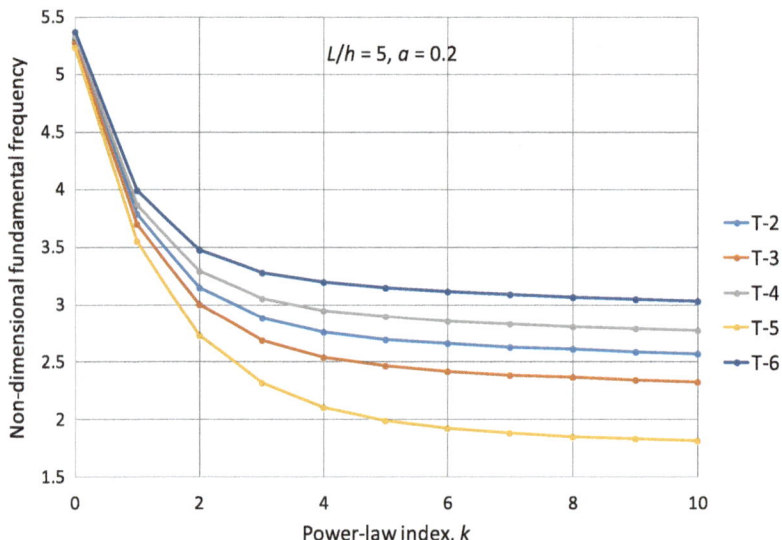

Fig. 1 Fundamental frequency $\overline{\omega}$ of imperfect beams versus power-law index, k ($L/h = 5$, $a = 0.2$)

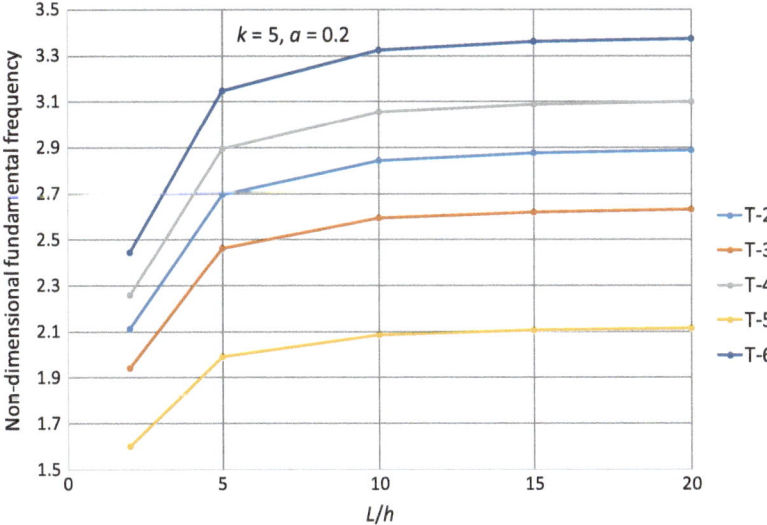

Fig. 2 Fundamental frequency $\overline{\omega}$ of imperfect beams versus L/h ratio ($k = 5$, $a = 0.2$)

decrease in frequency occurs, whereas beyond k greater than 5, a consistent decrease in natural frequency is observed. Furthermore, the investigation into natural frequency encompassed various L/h ratios and different porosity types, demonstrating an increase in frequency with higher L/h ratios (Fig. 2).

4 Conclusions

This study conducted a comprehensive analysis of the free vibration behavior of FG beams with porosities, considering various types of porosity. The investigation focused on evaluating the impact of differing levels of porosity within ceramic and metal components. Specifically, the analysis examined how the power-law index, length-to-thickness ratio, and porosity distribution types influenced the natural frequency.

Validation of the obtained results was performed by comparing them with existing literature for cases where the beam exhibited no porosity. The findings revealed that an increase in the power-law indices resulted in a decrease in the stiffness of functionally graded beams, leading to a corresponding reduction in natural frequency. Additionally, it was observed that a higher length-to-thickness ratio (L/h ratio) was associated with an increase in natural frequency.

Furthermore, the study demonstrated a significant decrease in the natural frequency as the percentage of porosity increased, regardless of the specific porosity types examined. These findings provide valuable insights for industries engaged in the manufacturing of porous beams, aiding in the decision-making process for selecting the most suitable porosity type to achieve optimal performance objectives. Overall, this analysis contributes to advancing our understanding of the complex interplay between porosity, material properties, and geometric parameters in functionally graded structures, with implications for various engineering applications.

References

1. Abdelall, E.S., Hayajneh, M., Almomani, M.: Fabrication of functionally graded injection molds using friction stir molding process of AA5083/brass-laminated composite. Prog. Addit. Manuf. **8**(2), 169–177 (2023). https://doi.org/10.1007/s40964-022-00320-8
2. Adachi, T., Higuchi, M.: Fabrication of bulk functionally-graded syntactic foams for impact energy absorption. Mater. Sci. Forum. **706–709**, 711–716 (2012). https://doi.org/10.4028/www.scientific.net/MSF.706-709.711
3. Boss, J.N., Ganesh, V.K.: Fabrication and properties of graded composite rods for biomedical applications. Compos. Struct. **74**(3), 289–293 (2006). https://doi.org/10.1016/j.compstruct.2005.04.030
4. Al Jahwari, F., Anwer, A.A.W., Naguib, H.E.: Fabrication and microstructural characterization of functionally graded porous acrylonitrile butadiene styrene and the effect of cellular morphology on creep behavior. J. Polym. Sci. B Polym. Phys. **53**(11), 795–803 (2015). https://doi.org/10.1002/polb.23698
5. Arefi, M., Firouzeh, S., Mohammad-Rezaei Bidgoli, E., Civalek, Ö.: Analysis of porous microplates reinforced with FG-GNPs based on Reddy plate theory. Compos. Struct. **247**, 112391 (2020). https://doi.org/10.1016/j.compstruct.2020.112391
6. Madan, R., Bhowmick, S.: Fabrication and microstructural characterization of Al-SiC based functionally graded disk. Aircr. Eng. Aerosp. Technol. **95**(2), 292–301 (2023). https://doi.org/10.1108/AEAT-03-2022-0096
7. Hadji, L., Plevris, V., Madan, R.: A static and free vibration analysis of porous functionally graded beams. In: 2nd International Conference on Civil Infrastructure and Construction (CIC 2023). QU Press, Doha, Qatar (2023). https://doi.org/10.29117/cic.2023.0059
8. Wattanasakulpong, N., Ungbhakorn, V.: Linear and nonlinear vibration analysis of elastically restrained ends FGM beams with porosities. Aerosp. Sci. Technol. **32**(1), 111–120 (2014). https://doi.org/10.1016/j.ast.2013.12.002
9. Zouatnia, N., Hadji, L., Kassou, A.: An analytical solution for bending and vibration responses of functionally graded beams with porosities. Wind Struct. **25**(4), 329–342 (2017). https://doi.org/10.12989/was.2017.25.4.329
10. Thai, H.-T., Vo, T.P.: Bending and free vibration of functionally graded beams using various higher-order shear deformation beam theories. Int. J. Mech. Sci. **62**(1), 57–66 (2012). https://doi.org/10.1016/j.ijmecsci.2012.05.014
11. Euler, L.: Methodus inveniendi lineas curvas maximi minimive proprietate gaudentes, sive solutio problematis isoperimetrici lattissimo sensu accepti, vol. 24. 1744, pp. 1–308. apud Marcum-Michaelem Bousquet & Socios, Lausannae & Genevae
12. Timoshenko, S.P.: LXVI. On the correction for shear of the differential equation for transverse vibrations of prismatic bars. Lond. Edinb. Dublin Philos. Mag. J. Sci. **41**(245), 744–746 (1921). https://doi.org/10.1080/14786442108636264
13. Şimşek, M.: Fundamental frequency analysis of functionally graded beams by using different higher-order beam theories. Nucl. Eng. Des. **240**(4), 697–705 (2010). https://doi.org/10.1016/j.nucengdes.2009.12.013

14. Reddy, J.N.: A simple higher-order theory for laminated composite plates. J. Appl. Mech. **51**(4), 745–752 (1984). https://doi.org/10.1115/1.3167719

15. Sayyad, A.S., Ghugal, Y.M.: An inverse hyperbolic theory for FG beams resting on Winkler-Pasternak elastic foundation. Adv. Aircraft Spacecraft Sci. **5**(6), 671–689 (2018). https://doi.org/10.12989/aas.2018.5.6.671

Design Optimization to Minimize Material Usage in Steel Buildings Subjected to Lateral Loads

Bahareh Dokhaei, Dikshant Saini, Behrouz Shafei ⓘ, and Alice Alipour

Contents

1 Introduction

As the global construction industry steers toward sustainable practices, optimizing structural designs becomes paramount in achieving net-zero buildings while aligning with sustainable development goals. With the increasing complexity of modern structures and the necessity to meet stringent performance criteria, optimization tools play a pivotal role in assisting engineers to make informed design decisions efficiently. This chapter presents a comprehensive approach to optimize steel usage, leading to a lightweight structural system that minimizes carbon footprints in the built environment. This optimization design works with the integration of a nonlinear programming solver with ETABS, a structural analysis and design software [1]. This integration streamlines the optimization process, enabling engineers to systematically identify optimal design solutions while considering structural capacities and serviceability constraints.

B. Dokhaei (✉) · D. Saini · B. Shafei · A. Alipour
Iowa State University, Ames, IA, USA
e-mail: dokhaei@iastate.edu; alipour@iastate.edu

© The Author(s) 2025
M. Kioumarsi, B. Shafei (eds.), *The 1st International Conference on Net-Zero Built Environment*, Lecture Notes in Civil Engineering 237,
https://doi.org/10.1007/978-3-031-69626-8_127

This study focuses on identifying optimum cross-sectional dimensions of structural sections to withstand lateral wind loads. The optimization objective function defined as the volume/weight of structural steel to ensure sustainability, drives the search for configurations that minimize material usage, while two constraints of the demand to capacity index (DCI) of structural members as a strength condition and inter-story drift ratio as a serviceability constraint are taken into account to meet performance criteria, safety, and cost-effectiveness designing process. The results show that this optimized design can effectively reduce the volume/weight of the structural steel usage to achieve sustainable buildings. Section 2 of this chapter outlines the optimization problem and its components, while Sect. 3 provides a case study on design iterations for a 15-story building. Section 4 explains the results obtained from the case study, and Sect. 5 summarizes the main conclusions.

2 Optimization Formulation

In this study, a nonlinear programming solver in MATLAB was utilized to optimize the thickness of structural members. This solver offers a powerful tool designed to find the minimum of a constrained nonlinear multivariable function. Specifically, a nonlinear sequential quadratic programming (SQP) method was used. This method has proven to be more efficient, accurate, and successful than other existing methods. This was confirmed through numerous test problems [2, 3]. Previous studies showed that this method closely resembles Newton's method of constrained optimization, which is similar to unconstrained optimization. This is an iterative process that involves approximating the Hessian of the Lagrangian function using the quasi-Newton updating method. Consequently, a quadratic programming subproblem was generated [4]. The optimization problem can be defined as follows:

Minimize steel volume/weight
Subjected to : $DCI \leq 1.0$ (strength condition)

Inter-story drift ratio $\leq 1/400$ (serviceability condition)

Critical factors such as structural strength, represented by the DCI, and serviceability constraints like inter-story drift ratio were defined in the first step. The DCIs were calculated using Eqs. (1) and (2) from AISC 360-22 [5]:

$$DCI = \frac{P_r}{\varnothing_p P_n} + \frac{8}{9} \left(\frac{M_{rx}}{\varnothing_m M_{nx}} + \frac{M_{ry}}{\varnothing_m M_{ny}} \right) \quad \frac{P_r}{\varnothing_p P_n} \geq 0.2 \tag{1}$$

$$DCI = \frac{P_r}{2 \times \varnothing_p P_n} + \left(\frac{M_{rx}}{\varnothing_m M_{nx}} + \frac{M_{ry}}{\varnothing_m M_{ny}} \right) \quad \frac{P_r}{\varnothing_p P_n} < 0.2 \tag{2}$$

where P_r and P_n are the demand and nominal axial strength; M_{rx} and M_{nx} are the demand and nominal flexural strength about the major axis; M_{ry} and M_{ry} are the

Table 1 Acceptance criteria for various performance levels [7]

Performance levels	Story drift response	Residual story drift
Occupant comfort	–	–
Operational	$H/400$	–
Continuous occupancy	$H/200$	$H/1000$

demand and nominal flexural strength about the minor axis; and $\varnothing_p$ and $\varnothing_m$ are the resistance factors.

The inter-story drift ratio is a crucial metric in structural engineering that measures the relative translational displacement observed between two consecutive floors in a building, divided by the height of each story. This parameter significantly evaluates structural performance, especially when exposed to seismic or wind forces [6]. Saini et al. [7] conducted a study on the performance of a steel building based on three criteria: occupant comfort, operational efficiency, and continuous occupancy. The referenced study [8–12] calculated different performance measures, such as inter-story drift ratio, as one of the most critical criteria. The calculated inter-story drift ratio was compared to the acceptance criteria of the desired performance objectives, as listed in Table 1. The performance assessment revealed that the building met the requirements for strength design, as well as performance limitations for continuous occupancy and limited interruption in Risk Category II. However, the building failed to meet the serviceability requirement for the performance level of operation, which has a limited story drift of $H/400$, where H is the story height. As a result, the building needed to be redesigned to meet this performance level. In the current study, to optimize the building design for serviceability requirements, a limit for the story drift equal to $H/400$ was considered.

Initially, within the ETABS software environment, frame elements and floor decks were created, along with their respective constraints and boundary conditions. Subsequently, preliminary design parameters were defined. The numbers of property types for beams, columns, and braces were chosen, and initial section dimensions were assigned to members. Then, all the crucial data such as node and frame coordinates, frame labels, and section properties were extracted from ETABS and stored within MATLAB for further analysis.

Load patterns and combinations were then defined, and lateral loads were applied to model joints. The maximum values of DCIs and inter-story drift ratios were determined by considering the wind loads acting laterally and the gravity loads, such as dead and live loads. The load combination specified in ASCE 7-22 [13] was used (Eq. 3), involving the consideration of dead load (DL), live load (LL), and wind load (WL).

$$\text{Load combination} = 1.2\,\text{DL} + 1.0\,\text{LL} + 1.0\,\text{WL} \tag{3}$$

At the next step, the optimization phase commenced to determine the optimal thickness for each section property. Using the solver in MATLAB, member thickness was optimized to minimize steel volume/weight while limiting the predefined criteria, such as the DCI or DCI and drift ratio within specified thresholds. Finally, after finding the thicknesses, the volume/weight of steel and DCI for the optimized frame members were recalculated and presented, signifying the conclusion of the optimization process, and providing insights into an enhanced structural design.

3   Steel Building Structure

In this section, the optimization process is studied for a 15-story moment resisting frame building. This building had a floor height of 4.0 m and bay width of 7.62 m. All the beams of the building were assumed to have one HSS section with outer dimensions of 300 mm × 300 mm, while the column sections were assumed to be HSS section with outer dimensions of 400 mm × 400 mm for the first five stories (stories 1–5), 300 mm × 3000 mm for the next five stories (stories 6–10), and 200 mm × 200 mm for the last five stories (stories 11–15), respectively. Floor diaphragms were modeled. In addition to the dead loads, a live load of 2.4 kN/m^2 is applied on floor slabs of each floor, which are typical live load expected for the office building. On one face of the building, a constant point load of 50 kN was applied on all nodes in the horizontal direction.

Before optimization, the first step was to create a model of the beam and column elements using ETABS software. This model includes specific section properties and the geometry of the model, which can be seen in Fig. 1. Once the model was complete, the optimization code was initiated. It began by extracting all section

Fig. 1 A 15-story building model developed in ETABS with necessary beam, column, and connection details

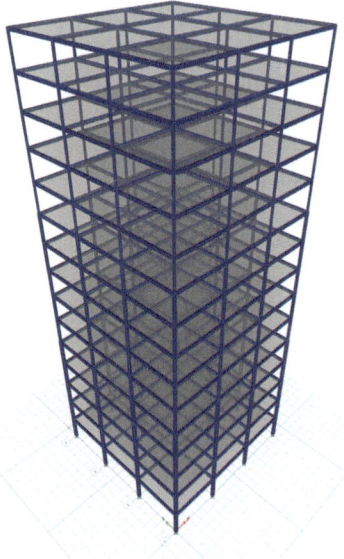

properties from the ETABS model and then iterated through each beam and column type, retrieving important parameters such as area, torsional properties, moments of inertia, and other relevant data.

The optimization code then sets out to determine the optimal thickness for both beam and column sections including T_{C1}, T_{C2}, and T_{C3}, which refer to the thickness of columns at first five stories (story 1 to story 5), second five stories (story 6 to story 10), and third five stories (story 11 to story 15), respectively, and the beam thickness which refers to as T_b. Determining the optimal thickness was achieved by formulating an objective function based on the volume/weight of the steel sections, while also applying constraints on the DCI indices and story drift ratios. This constraint framework is necessary to ensure the performance of the building design. Specifically, the constraints of this study ensured that the DCI indices for structural members did not exceed one and that the story drift remained within acceptable limits (H/400).

4   Results and Findings

In this section the results of design optimization are discussed. Section 4.1 evaluates the optimization results for strength constraint, and Sect. 4.2 examines the results for both strength and serviceability constraints.

4.1   Optimization Results with Strength Constraint

The design of the building was investigated for achieving optimal section properties while adhering to strength constraints, ensuring that the DCI indices for all the structural members do not surpass one. The lower and upper bounds for the flange/web thickness of beam and column sections were set at 12.7 mm and 100 mm, respectively. These bounds allowed the optimization algorithm to explore a range of thickness options while maintaining structural integrity within specified limits. Initially, the beam and column sections are configured with flange and web thicknesses set to 12.7 mm. After the optimization code was executed, the results indicated a shift in the thickness values of both column and beam sections from their initial starting points to the end points. The results of the optimization are summarized in Table 2. In the referenced table, the *Feasibility* column indicates the maximum constraint violation, with lower values reflecting closer adherence to

Table 2 Optimization results for member thicknesses with strength constraint

Iter.	Feasibility	T_b (mm)	T_{C1} (mm)	T_{C2} (mm)	T_{C3} (mm)	Weight (ton)
0	6.60E-01	12.70	12.70	12.70	12.70	424.21
7	1.92E-13	15.60	23.21	19.63	21.14	563.40

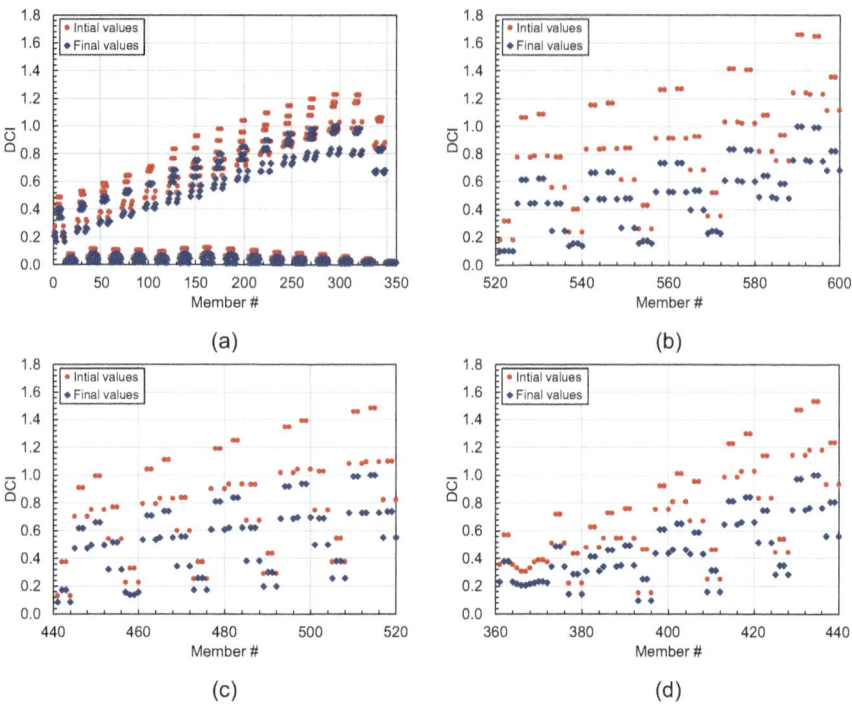

Fig. 2 Optimization results of DCI indices for building with strength constraint: (**a**) beams, (**b**) columns at stories 1–5, (**c**) columns at stories 6–10, and (**d**) columns at stories 11–15

constraints. A decrease in *Feasibility* indicates a closer match and convergence in the results. The optimization process was initiated with an initial thickness of 12.7 mm for each component. Through iterative adjustments, the thickness values evolved until convergence or meeting the specified stopping criteria; and shifted to a stable solution, which was between the defined lower and upper bound, meeting all the constraints. Column thicknesses were shifted to 23.21, 19.63, and 21.14 mm, while beam thickness became 15.60 mm after seven iterations. The objective function of this case, which was the total steel volume/weight, had an initial value of 424.21 ton and a final value of 563.40 ton.

Figure 2 illustrates the comparison between the initial and final DCI indices. Initially, the DCI indices for beam and column members surpassed 1.0, implying member failure. After optimization, the maximum DCI for these members was reduced to 1.0. For instance, the maximum DCI for stories 1–5 columns decreased from 1.65 to 1.00, as depicted in Fig. 2b. This shows that the structural elements now offer adequate capacity to withstand applied loads without risking failure, while having the optimum member size with the minimum steel usage.

4.2 Optimization Results with Both Strength and Serviceability Constraints

The optimization results for the same 15-story building were further evaluated by considering both serviceability and strength constraints. To verify the serviceability constraint, larger member sizes were found to be needed to meet the requirements compared to Sect. 4.1. Therefore, for the beams, HSS sections were employed with outer dimensions of 400 mm × 400 mm. As for the column sections, HSS sections were assumed with outer dimensions of 500 mm × 500 mm for the stories at floors 1–5, 450 mm × 450 mm for the next five stories at floors 5–10, and 400 mm × 400 mm for the last five stories at floors 11–15. The lower and upper bounds for the flange/web thickness of beam and column sections were set at 12 mm and 100 mm, respectively. These bounds allowed the optimization algorithm to explore a range of thickness options while maintaining structural integrity within specified limits. Initially, the beam and column sections were configured with flange and web thicknesses set to 16 mm. After the optimization code was executed, the results indicated a shift in the thickness values of both column and beam sections from their initial starting points to the end points. The results of the optimization for iteration 0 (start point) and final iteration are summarized in Table 3.

The optimization process commenced with an initial uniform thickness setting of 16 mm for all columns and beams. The optimization algorithm continuously improved the solution by adjusting the thickness of the members to minimize the steel volume/weight, while meeting strength and serviceability criteria. In Table 3, *Feasibility* shows how much a solution violates the constraints. It quantifies the degree of constraint violation, with lower values indicating closer adherence to constraints. A decrease in *Feasibility* indicates a better convergence in results.

After 29 iterations, column thicknesses evolved to 64.59, 29.81, and 12.00 mm for T_{C1}, T_{C2}, and T_{C3}, respectively, while beam flange and web thicknesses became 39.17 mm. The initial objective function was 738.53 ton, which ultimately reached a final value of 1672.68 ton steel. Table 3 also shows the initial and final inter story drift ratios, confirming that serviceability constraints have been satisfied. The final story drift ratio matched the specified limit of 0.0025H, meeting the requirements set forth by the serviceability constraint. This outcome ensured that the structure maintains adequate performance under operational conditions, safeguarding against excessive drifts that could compromise its usability and safety.

Figure 3 shows the DCI values for structural members. All elements exhibited final DCI values within acceptable limits, confirming compliance with strength constraints and ensuring an economically efficient structural design. The final DCI

Table 3 Optimization results for member thicknesses with both strength and serviceability constraints

Iter.	Feasibility	T_b (mm)	T_{C1} (mm)	T_{C2} (mm)	T_{C3} (mm)	Weight (ton)	Max IDR
0	3.20E-03	16.00	16.00	16.00	16.00	738.53	0.0057H
29	1.65E-12	39.17	64.59	29.81	12.00	1672.68	0.0025H

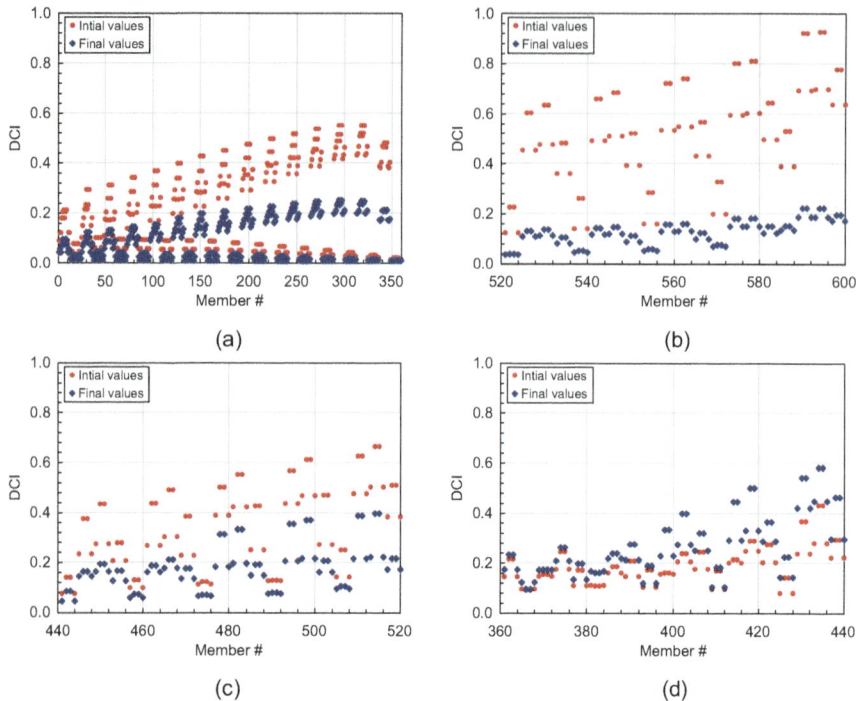

Fig. 3 Optimization results of DCI indices for building with strength and serviceability constraints: (**a**) beams, (**b**) columns at stories 1–5, (**c**) columns at stories 6–10, and (**d**) columns at stories 11–15

indices were in the range of 0–0.6, which is less than 1.0. It shows that inter story drift ratio governed the member size in a way that maximum inter story drift ratio was less than $H/400$.

In summary, the optimization method effectively generates cost-effective designs for structural members of a 15-story building under different scenarios, including both strength and serviceability constraints. These results highlight the versatility of the approach in achieving sustainable outcomes. Sustainable design practices, including minimizing material usage and optimizing energy efficiency, are essential for achieving net-zero buildings. By reducing steel usage while maintaining structural integrity, engineers contribute to environmental sustainability and align with broader goals of sustainable development.

5   Conclusions

This study evaluated the design optimization of steel buildings to optimize steel usage for achieving net-zero buildings, aligning with sustainable development goals. The outcome presented a robust framework for engineers to efficiently identify the optimal cross-sectional dimensions of structural sections through automating the iterative design refinement process, while ensuring compliance with performance criteria, and minimizing carbon footprints in the built environment. The results obtained from applying the optimization tool to a 15-story building demonstrate its effectiveness in meeting both structural yield requirements and operational performance criteria, including a maximum story drift limited to $H/400$. This indicates that the presented approach offers a viable solution for designing buildings that not only fulfill structural demands but also uphold stringent operational standards, contributing to sustainable construction practices.

Acknowledgments This research study was partially sponsored by the U.S. National Science Foundation (NSF) and the National Institute of Standards and Technology (NIST). The authors would like to acknowledge and thank the sponsors for their support. Opinions, findings, and conclusions expressed in this chapter are of the authors and do not necessarily reflect those of the sponsors.

References

1. ETABS: CSI Application Programming Interface (API) ETABS v1. Computers and Structures, Inc. (2023)
2. Schittkowski, K.: More Test Examples for Nonlinear Programming Codes, vol. 282. Springer Science & Business Media (2012)
3. Schittkowski, K.: NLPQL: a FORTRAN subroutine solving constrained non-linear programming problems. Ann. Oper. Res. **5**, 485–500 (1986)
4. MATLAB R2023b: Optimization ToolboxTM User's Guide. The MathWorks, Inc., Natick (2023)
5. American Institute of Steel Construction: ANSI/AISC 360-22, Specification for Structural Steel Buildings. American Institute of Steel Construction (2022)
6. American Society of Civil Engineers: Prestandard for Performance-Based Wind De-sign. American Society of Civil Engineers (2019)
7. Saini, D., Dokhaei, B., Shafei, B., Alipour, A.: Performance evaluation of high-rise buildings using database-assisted design approach. Struct. Saf. **109**, 102447 (2024)
8. Preetha Hareendran, S., Alipour, A., Shafei, B., Sarkar, P.: Performance-based wind design of tall buildings considering the nonlinearity in building response. J. Struct. Eng. **148**(9), 04022119 (2022)
9. Saini, D., Shafei, B.: Damage assessment of wood frame shear walls subjected to lateral wind load and windborne debris impact. J. Wind Eng. Ind. Aerodyn. **198**, 104091 (2020)
10. Kulkarni, A., Shafei, B., Epackachi, S.: Structural response assessment of ultra-high performance concrete shear walls under cyclic lateral loads. Struct. Concr. **24**(2), 2699–2720 (2023)
11. Hareendran, S.P., Alipour, A., Shafei, B., Sarkar, P.: Characterizing wind-structure interaction for performance-based wind design of tall buildings. Eng. Struct. **289**, 115812 (2023)
12. Safaei, S., Shamlu, M., Vakili, A.: Modeling methods and constitutive laws for nonlinear analysis of steel moment-resisting frames. J. Constr. Steel Res. **199**, 107583 (2022)
13. American Society of Civil Engineers: Minimum Design Loads and Associated Criteria for Buildings and Other Structures (ASCE/SEI 7-22). American Society of Civil Engineers (2022)

Intensity Measures for Flood Hazards in Fragility Assessments of Bridges

Tahereh Torabi, Behrouz Shafei ⓘ, and Alice Alipour

Contents

1 Introduction

To design a bridge structure, the following permanent and transient loads shall be often considered: horizontal earth pressure load, earth surcharge load, vertical pressure from dead load of earth fill, blast load, vehicular force (collision), ice load, water load and stream pressure, wave load, scour, and wind load [1] In the event of the occurrence of an extreme event, the specified load(s) will be allocated or augmented to exacerbate the pre-existing loads. This phenomenon underscores the importance of understanding and mitigating the impact of hazards on structural systems. Fragility functions are dependent on hazard intensity measures (IMs), which are used to create a fragility surface representing the probability of bridge failure. Khatami and Shafei [2] proposed a risk-based life-cycle analysis framework to enhance current bridge management systems by incorporating the consequences of sudden extreme events, such as environmental stressors and seismic hazards, into condition state predictions, thus improving performance and cost predictions, particularly in scenarios where the impact of extreme events cannot be overlooked.

T. Torabi · B. Shafei · A. Alipour (✉)
Iowa State University, Ames, IA, USA
e-mail: alipour@iastate.edu

© The Author(s) 2025
M. Kioumarsi, B. Shafei (eds.), *The 1st International Conference on Net-Zero Built Environment*, Lecture Notes in Civil Engineering 237,
https://doi.org/10.1007/978-3-031-69626-8_128

Khatami and Shafei [3] investigated how climate conditions influence the deterioration of reinforced concrete bridges specifically in the US Midwest region. The referenced study aimed to understand the relationship between climate factors and the degradation of bridges in service to provide insights into improving maintenance practices and structural design considerations. A comprehensive study that involved collecting data on climate conditions, such as temperature, precipitation, humidity, and freeze-thaw cycles, in the US Midwest region was completed. Also, data on the condition of reinforced concrete bridges, including factors such as cracking, spalling, and corrosion, was gathered through field inspections and existing databases to correlate climate data with the observed deterioration of bridges. The study found significant correlations between certain climate conditions and the deterioration of reinforced concrete bridges. The results also highlighted specific vulnerabilities of these bridges to certain climate stressors, providing valuable insights for infrastructure management and maintenance planning. The current study aimed to explore the implications of load allocation and intensification in the context of hazards, specifically flood, and whether frequency has been considered to investigate the results of that.

Flood hazards can be defined in terms of the chance that a certain magnitude (area and depth) of flooding is exceeded in any given year (probabilistic event) or as a specific, deterministic event, such as a historic peak flow or recent event captured with high water marks. Depth, duration, and velocity of water in the floodplain are the primary factors contributing to flood losses. Other hazards associated with flooding that contribute to flood losses include channel erosion and migration, sediment deposition, bridge scour, and the impact of flood-borne debris. Flood magnitude is usually measured as a discharge value, flood elevation, or depth. Using the flood frequency convention, flood hazard is defined by a relation between the depth of flooding and the annual likelihood of inundation greater than that depth [4]. Flood hazards emphasize the challenges posed by scour, hydrodynamic loading, inundation, and debris accumulation.

2   Intensity Measures of Flood Hazard

2.1   Without Considering Climate Change

Lee et al. [5] investigated the vulnerability of bridges to floods, emphasizing factors such as water pressure, scour, corrosion, and debris as common causes of bridge destruction. The water pressure was identified as a critical water load, with AASHTO [6] defining it as the pressure of flowing water acting in the longitudinal direction of bridge substructures. The debris accumulation leads to increased water pressure on the bridge, affecting piers and girders. Water velocity was introduced as an intensity measure, and AASHTO [7] provides an equation to calculate water pressure based on flow velocity and drag coefficients for different pier types. The water pressure P (MPa) as Eq. (1), and the force exerted on the bridge was determined by multiplying the calculated hydrodynamic pressure by the projected area subjected to the pressure. Also, Liao et al. [8] considered loads, including the vertical load, wind load, and hydrodynamic pressure.

The hydrodynamic pressure was calculated according to Eq. (1).

$$P = 5.14 \times 10^{-4} \, C_D V^2 \tag{1}$$

where C_D denotes the drag coefficient for piers, depending on the geometric shape of the bridge and the accumulated debris or absence thereof, and V denotes the flow velocity (m/s) during a flood. Kim et al. [9] calculated scour depths around each pier by using an empirical equation, and the stiffness values of springs below the calculated scour depth remained unchanged. The study experimentally observed scour depth ratios for multiple piles to account for variations in scour depths based on pile arrangement and geometric shapes. Kim et al. [9] considered the increase in water velocity due to debris accumulation around bridge piers. The water pressure was estimated using equations recommended by AASHTO [7] and the Korean Highway Bridge Design Specification [10], incorporating the drag coefficient that depends on the pier type. The calculated water pressure as Eq. (1) was applied to bridge piers as an external load. An accurate consideration of scour is crucial, as it directly impacts bridge stability calculations. The scour depth for a single pier can be calculated using an empirical formula, e.g., Eq. (2) [9]:

$$S = 1.564 \, X^{0.405} \left(\frac{V}{\sqrt{gd}} \right)^{0.413} \tag{2}$$

where S is the scour depth, X is the relative approach flow depth, V is the water velocity, g is the gravitational acceleration, and d is the depth of the approach flow. Oppong et al. [11] investigated the vulnerability of bridges over navigable waterways to the combined effects of scour and barge collision. Scour scenarios were developed for 100- and 500-year flood events (determined from the annual peak discharge data reported by the United States Geological Survey (USGS)), following the procedure in the Hydraulic Engineering Circular No. 18 [12]. Alipour et al. [13] developed a multi-hazard reliability-based framework to assess the structural response of reinforced concrete (RC) bridges under combined pier scour and earthquake events, aiming to calibrate scour load-modification factors for bridges in high seismic areas. A deterministic approach utilizes established equations to calculate scour depth, while a probabilistic approach introduces uncertainties inherent in model parameters. Field data suggested discrepancies between observed and calculated scour depths, addressed by introducing a model factor with a normal distribution. Uncertainties in influential model parameters such as streambed condition coefficient and Manning roughness coefficient were considered with assumed probability distributions. Moreover, variability in flow discharge rate, a key factor in scour depth determination, was accounted for using peak annual discharge rate data from USGS, represented by rivers with low-flow and high-flow discharge rates, respectively.

Tubaldi et al. [14] developed a three-dimensional mesoscale representation of masonry arch bridges. Water flow imposed various actions on arch bridges during floods, including hydrodynamic pressure on submerged surfaces, reduced foundation capacities, buoyant forces affecting load-carrying capacity, impact from debris, and scour at bridge foundations, the latter being a common cause of collapse for arch bridges due to their vulnerability to foundation settlements. Arch bridges' low clearance makes them susceptible to debris accumulation, exacerbating scour and hydrodynamic forces. Khandel and Soliman [15] calculated hydrodynamic flood-induced load (F_{dyn}) based on FEMA [4] recommendations as Eq. (3).

$$F_{dyn} = \frac{1}{2} C_D \rho V^2 A \tag{3}$$

where C_D drag coefficient; ρ mass density of fluid (9.81 kN/m^3 for freshwater); V water velocity; and A surface area of obstruction normal to flow. Ahamed et al. [16] utilized HEC-RAS (River Analysis System) for one-dimensional surface profile analysis, which incorporated Manning's equation for discharge estimation and the energy equation for water surface profile computation. Hydraulic models, such as HEC-RAS [17] developed by the US Army Corps of Engineers, simulate river hydraulics by considering factors like discharge, velocity, and flow depth. HEC-RAS generates flow hydraulic properties, including velocity, flow depth, and discharge at the bridge location, and calculates local, and contraction scours around bridge piers using established bridge scour relations from [12]. The discharge was estimated using Manning's equation as Eq. (4).

$$Q = \left(\frac{1}{n}\right) A R^{2/3} \sqrt{S} \tag{4}$$

where n is Manning's roughness coefficient, A is the area of the flow, R is the hydraulic radius, S is the slope of energy line.

Anisha et al. [18] reported that the stochastic responses of a bridge are influenced by random geometric and material parameters and flood loading parameters. Parameters like pile diameter, pier depth, pier width, deck width, span, pile height, steel yield strength, modulus of elasticity, concrete mass density, and compressive strength are considered random. Various flood-related parameters such as hydrodynamic loads, scouring, water table depth, buoyancy, debris accumulation, turbulence, and frequency of flood events, were treated as highly uncertain. The water pressure was determined by a formula incorporating flow velocity and a random parameter, Q.

$$F(h) = Q.F_0(h) \tag{5}$$

$$F_0(h) = 52KV(h)^2 \tag{6}$$

where $V(h)^2$ in m/s represents the flow velocity at a height h, and K is a constant that depends on the shape of the pier. The randomness in water pressure $F_0(h)$ is incorporated by the parameter Q, which was assumed to have a normal distribution with a unit mean and a 10% standard deviation.

2.2 With Considering Climate Change

Kulkarni and Shafei [19] investigated Iowa's transportation infrastructure, particularly focusing on its susceptibility to extreme weather events such as flooding, utilized Bayesian belief networks to assess the impact and predict the probability of damage to roads and bridges, considering climate projections. Future climate conditions in Iowa were projected to include increased precipitation, more extreme temperatures, and more frequent heat waves, leading to higher risks of flooding, landslides, and infrastructure deterioration due to thermal effects, emphasizing the need for long-term design considerations and the assessment of existing infrastructure's resilience to projected environmental stressors. Loli et al. [20] used the Digital Elevation Model (DEM), specifically the FABDEM (Forest and Buildings removed Copernicus Digital Elevation Model) with improved accuracy, which serves as a base layer for mapping elevation, ground slope, and hydrography. Satellite imagery from Copernicus Sentinel-2 aids in land use/cover mapping. The USGS (United States Geological Survey) provides the shear wave velocity map, while soil permeability data are sourced from the global hydrogeology map GLHYMPS 2.0. Using the ERA5-Land dataset, a map of climatological maximum daily accumulated precipitation during Ianos was generated, illustrating the exceptionally intense rainfall in the study area.

Comparison with precipitation records since 1950 underscores the unprecedented intensity, with the Ianos event producing over twice the annual maximum daily precipitation observed from 1950 to 2022 (Fig. 1). The estimated precipitation peak closely aligns with the measured rain gauge value reported in the study. Loli et al.

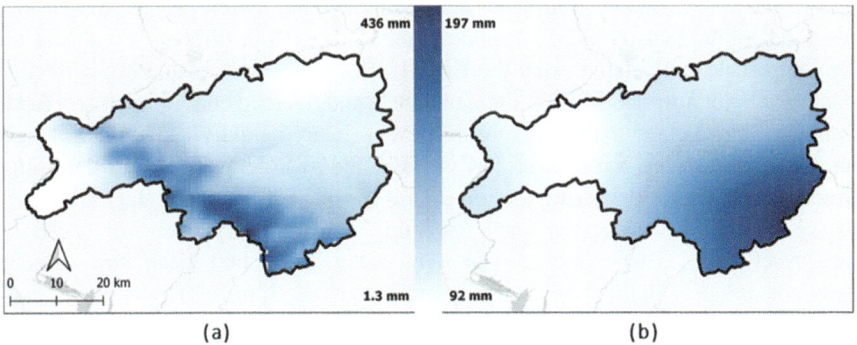

Fig. 1 Rainfall: (**a**) accumulated daily precipitation on the day that Medicane Ianos stroke (September 18, 2020) compared to (**b**) the annual maximum daily precipitation between 1950 and 2022 (Note: the two contour plots have different scales) [20]

[20] considered the hydraulic actions on bridges due to floods with two primary components: Component 1, consisting of hydrodynamic loads (drag and lift forces), and Component 2, involving scour. Hydrodynamic loads act in the direction of and perpendicular to the flow determined by Eq. (3). Scour (d_s) weakens the foundation's bearing capacity due to general scour, contraction scour, and local scour. The scour depth depended on various parameters, and these elements collectively contribute to the formulation of a new index of flood hazard intensity (IFHI). IFHI integrates indicators related to flow intensity, general scour, contraction scour, and local scour, offering a comprehensive assessment of flood hazard on bridges. This index considered parameters, such as flow velocity (U), flow depth (h_w), width of flow obstruction (B), median grain size (D_{50}), and critical flow velocity (U_c). The Index of Flow Intensity (IFI) is a significant component of IFHI, accounting for hydrodynamic loads based on multi-criteria decision-making analysis. Scour indicators (A_1, A_2, A_3) factored into IFHI are estimated following the approach proposed by the AROSA guidelines [21] but modified appropriately to incorporate U in the scoring of scour susceptibility, in agreement with the relevant theoretical framework described by Eq. (7).

$$d_s = f(U, h_w, D_{50}, B, U_c) \tag{7}$$

$$\text{IFHI} = \text{IFI}(A_1 + A_2 + A_3) \tag{8}$$

Kosic et al. [22] utilized numerical simulations to gather hydraulic data for vulnerability assessment, focusing on extreme flooding events in 2014 and 2015. They depicted complex hydraulic conditions at bridge locations through relationships between water height, water discharge, and flow velocities, considering unsteady flow conditions due to data dispersion and limited observations. The assessment incorporated 95% prediction intervals from steady-flow Manning's equation regression to explore severe flooding scenarios. Loads on the bridge structure included gravity, hydrodynamic forces (drag and lift), buoyancy forces, and wood debris accumulation, computed according to the Australian standard [23]. Buoyancy force varied with deck submergence, while debris dimensions were determined by the debris Froude number. The depth of local scour near the bridge pier was calculated using the HEC-18 [12] equation, considering correction factors for pier nose shape, flow angle of attack, and riverbed conditions. The Froude number was defined based on mean flow velocity, gravity, and water height upstream of the pier. Kosic et al. [24] sourced the hydrological data for the bridge from the Karlovac measuring station 20 m upstream, the stream stage-discharge curve exhibits hysteretic behavior, reflecting non-stationary flow influenced by nearby effluents. The relationship between water height and mean flow velocity was not uniquely defined, and an estimation derived from numerical simulations conducted for historical flood events (2005, 2014, 2015) was considered. The depth of local pier scour for each flood scenario was determined using HEC-18 [12] guidelines, incorporating correction factors for pier nose shape, flow angle, and riverbed conditions.

Also, Khatami et al. [25] developed a data-assisted framework that incorporates physics-based models for predicting the deterioration of reinforced concrete bridges, and to improve the condition state predictions made by bridge management systems, which relied heavily on inspection data and expert opinions to develop deterioration models based on Markov processes. The system had limitations due to the unavailability of long-term inspection data, low quality of available data, and potential errors in expert opinions. The physics-based method models the two stages of chloride-induced corrosion in reinforced concrete bridges: (1) The corrosion initiation stage uses Fick's second law to estimate chloride ingress and corrosion initiation time based on diffusion coefficient, surface chloride, and threshold, determining hazard rates for Condition States; (2) The corrosion propagation stage employs Butler-Volmer kinetics and oxygen diffusion to estimate steel mass loss, crack widths from a thick-walled cylinder model mapped to higher condition states, with hazard rates based on Tafel slopes and oxygen concentrations. As mentioned, most existing studies on transportation network resilience did not account for the effects of aging and deterioration of bridges, leading to inaccurate predictions. Alipour and Shafei [26] Developed an integrated simulation-based framework for assessing resilience of deteriorating transportation networks subjected to combined effects of environmental stressors like corrosion, temperature variations, etc. and seismic events.

Despite efforts to enhance bridge resiliency, challenges such as a lack of knowledge, gaps between research and practical approaches, budget constraints, and insufficient risk assessments, urge joint efforts among academia, decision-makers, and technical specialists.

3   Conclusions

The comprehensive exploration of load allocation and intensification within the context of hazards, particularly flood events, underscores the critical importance of understanding and mitigating the impacts of natural disasters on bridge infrastructure. The studies reviewed factors such as hydrodynamic forces, scouring and debris accumulation. Despite the advancements in hydraulic modeling and structural analysis techniques, challenges persist, including uncertainties in hazard prediction, gaps between research and practical applications, and limitations in current risk assessment methodologies. Notably, the absence of consideration for climate change in past studies points to a critical area for future research and intervention. Given the projected increase in the intensity and frequency of extreme weather events due to climate change, incorporating climate variability and long-term trends into hazard assessments and infrastructure design is imperative for ensuring the resilience and longevity of bridge structures. Moving forward, collaborative efforts are essential to address the complex challenges posed by flood hazards and climate change on bridge infrastructure. By integrating advancements in modeling techniques, data analytics, and risk assessment methodologies, alongside proactive adaptation strategies, the resilience of bridges can be enhanced. This will help mitigate the potential impacts of future flood events, ultimately safeguarding critical transportation networks and promoting sustainable infrastructure development.

Acknowledgments This research study was partially sponsored by the US Federal Highway Administration (FHWA). The authors would like to acknowledge and thank the sponsor for their support. Opinions, findings, and conclusions expressed in this chapter are of the authors and do not necessarily reflect those of the sponsor.

References

1. AASHTO: LRFD Bridge Design Specifications. AASHTO (2020)
2. Khatami, D., Shafei, B., Smadi, O.: Management of bridges under aging mechanisms and extreme events: risk-based approach. Transp. Res. Rec. **2550**(1), 89–95 (2016)
3. Khatami, D., Shafei, B.: Impact of climate conditions on deteriorating reinforced concrete bridges in the US Midwest region. J. Perform. Constr. Facil. **35**(1), 04020129 (2021)
4. FEMA: Coastal Construction Manual: Principles and Practices of Planning, Siting, Designing, Construction and Maintaining Residential Building in Coastal Areas. FEMA, Washington, DC (2011)
5. Lee, J., Lee, Y.-J., Kim, H., Sim, S., Kim, J.: A new methodology development for flood fragility curve derivation considering structural deterioration for bridges. Smart Struct. Syst. **17**(1), 149–165 (2016)
6. AASHTO: LRFD Bridge Design Specifications. AASHTO (2017)
7. AASHTO: Guide Specifications for Bridges Vulnerable to Coastal Storms. AASHTO (2008)
8. Liao, K.-W., Muto, Y., Chen, W.-L., Wu, B.-H.: A probabilistic bridge safety evaluation against floods. SpringerPlus. **5**, 1–19 (2016)
9. Kim, H., Sim, S.-H., Lee, J., Lee, Y.-J., Kim, J.-M.: Flood fragility analysis for bridges with multiple failure modes. Adv. Mech. Eng. **9**(3), 1–11 (2017)
10. Korea Road & Transportation Association (KRTA): Korean Highway Bridge Design Specification. KRTA (2010)
11. Oppong, K., Saini, D., Shafei, B.: Multihazard performance assessment of scoured bridges subjected to barge collision. J. Bridg. Eng. **27**(7), 04022048 (2022)
12. Arneson, L.: Evaluating Scour at Bridges–Fifth Edition, Federal Highway Administration Hydraulic Engineering Circular No. 18 (Tech. Rep. FHWA HIF-12-003). Federal Highway Administration (2012)
13. Alipour, A., Shafei, B., Shinozuka, M.: Reliability-based calibration of load and resistance factors for design of RC bridges under multiple extreme events: scour and earthquake. J. Bridg. Eng. **18**(5), 362–371 (2013)
14. Tubaldi, E., Macorini, L., Izzuddin, B.: Flood risk assessment of masonry arch bridges. In: 2nd International Conference on Uncertainty Quantification in Computational Sciences and Engineering, UNCECOMP, 2017, pp. 15–17. Eccomas Proceedia (2017)
15. Khandel, O., Soliman, M.: Integrated framework for assessment of time-variant flood fragility of bridges using deep learning neural networks. J. Infrastruct. Syst. **27**(1), 04020045 (2021)
16. Ahamed, T., Duan, J.G., Jo, H.: Flood-fragility analysis of instream bridges–consideration of flow hydraulics, geotechnical uncertainties, and variable scour depth. Struct. Infrastruct. Eng. **17**(11), 1494–1507 (2021)
17. U.S. Army Corps of Engineers–Hydrologic Engineering Center: HEC-RAS River Analysis System: Hydraulic Reference Manual, Version 5.0, vol. 547. U.S. Army Corps of Engineers–Hydrologic Engineering Center, Davis (2016)

18. Anisha, A., Sahu, D.K., Sarkar, P., Mangalathu, S., Davis, R.: High dimensional model representation for flood fragility analysis of highway bridge. Eng. Struct. **281**, 115817 (2023)
19. Kulkarni, A.R., Shafei, B.: Impact of extreme events on transportation infrastructure in Iowa: a Bayesian network approach. Transp. Res. Rec. **2672**(48), 45–57 (2018)
20. Loli, M., Kefalas, G., Dafis, S., Mitoulis, S.A., Schmidt, F.: Bridge-specific flood risk assessment of transport networks using GIS and remotely sensed data. Sci. Total Environ. **850**, 157976 (2022)
21. Durand, E., Davi, D., Delgado, J.: AROSA: a new French guideline for scour at bridges risk-based analysis. In: Scour and Erosion IX, p. 671. Taylor & Francis (2018)
22. Kosic, M., Anzlin, A., Water, V.B.: Flood vulnerability study of a roadway bridge subjected to hydrodynamic actions, local scour and wood debris accumulation. Waters. **15**(1), 129 (2023)
23. Australian Standard: Bridge Design Part 2: Design Loads (AS5100. 2). Standards Australia International Ltd, Sydney (2004)
24. Kosic, M., Prendergast, L., Anzlin, A.: Analysis of the response of a roadway bridge under extreme flooding-related events: scour and debris-loading. Eng. Struct. **279**, 115607 (2023)
25. Khatami, D., Shafei, B., Bektas, B.: Data-assisted prediction of deterioration of reinforced concrete bridges using physics-based models. J. Infrastruct. Syst. **29**(2), 05023003 (2023)
26. Alipour, A., Shafei, B.: An overarching framework to assess the life-time resilience of deteriorating transportation networks in seismic-prone regions. Resilient Cities Struct. **1**(2), 87–96 (2022)

Life Cycle Carbon Assessment Methods for Structural Materials in Bridges

Alessandro Sterpellone (iD)**, Randi Christensen** (iD)**,
Arne Frederiksen Bæksted, and Julie Rønholt Lange**

Contents

1   Introduction

Emissions from the production of traditional building materials, and especially structural materials like reinforced concrete, play an important role in the climate crisis that is facing us today. Cement production alone contributes to almost 7% of global greenhouse gas emissions, and its demand is expected to increase by 12–23% by 2050, whereas the steel and iron industry is responsible to 7–9% of emissions [1]. Because of this, the construction sector is looking for alternative materials for structural applications, and wood is a natural candidate.

A. Sterpellone (✉) · R. Christensen · A. F. Bæksted · J. R. Lange
COWI, Kongens Lyngby, Denmark
e-mail: aose@cowi.com

© The Author(s) 2025
M. Kioumarsi, B. Shafei (eds.), *The 1st International Conference on Net-Zero Built Environment*, Lecture Notes in Civil Engineering 237,
https://doi.org/10.1007/978-3-031-69626-8_129

1.1 Wood as a Sustainable Building Material

Wood has been used for centuries in construction as an essential, versatile, and truly renewable building material. This is particularly true for the Nordics, whose abundant forest resources produce around 20% of lumber in Europe [2]. In addition to their material availability and long-standing woodworking tradition, the Nordics have climate commitments to carbon-neutral building solutions, which have been propelling the use of timber in buildings.

Furthermore, advancements in engineered wood products and innovative construction techniques have expanded the potential applications of wood in construction projects, making it a viable option for structural elements in large infrastructure with more severe strength and durability requirements, such as bridges.

A few important considerations must be made regarding the sustainability of wood, which will fall outside the scope of the current analysis. Following the projected growth in demand in the coming years, and since its production capabilities are limited, there will be a need to prioritize the use of wood in the coming years, especially in the case of projects that will require a very intensive use of this material. It is also crucial to note that wood has significant impacts across several environmental impact categories during its life cycle. As an example, only recently is the building sector starting to develop tools to assess the biodiversity impacts of wood, as intensive management of forests, land use change, and the subsequent habitat destruction are all causes of biodiversity loss.

The current analysis sets out to focus only on impacts on the climate, so the Global Warming Potential (GWP) is the only indicator considered—from now on referred to as carbon, as it is expressed in kg CO_2-equivalent.

1.2 Climate Impacts of Timber in Bridges

There are currently no standards or guidelines on how to best assess the sustainability performance of timber bridges. It is left to sustainability professionals to decide on what methodology and approach to take, to get around the intricacies of this task.

Only a very limited number of LCA studies have been conducted to assess the sustainability of bridges with timber as a primary construction material, and many intricacies regarding their carbon profile emerged from a review study from 2018 [3]. The authors state in their findings that "the consideration of carbon storage of wood material needs to be carefully treated" and that depending on the main purpose of the investigation, one must consider both options of assuming:

- No carbon storage, thus excluding biogenic carbon from the analysis and treating wood as carbon-neutral
- Carbon storage, thus including biogenic carbon flows in the analysis

All reviewed studies in the chapter except for one fall in the first group, using an approach that arguably oversimplifies the carbon flows from the life cycle of wood products. Our study sets out to include these carbon flows and explore different ways of doing so.

2 Materials and Methods

This section will include a brief description of the subject bridge, one subsection outlining some important concepts in LCA, and the other following the setup of the three comparative LCA studies performed.

2.1 Bridge Design

The original project considered in this analysis is the Remmevej highway overpass, a glued laminated timber bridge opened in 2001 in Jutland, Denmark. As shown in Fig. 1, the bridge has an overall length of 54 m, divided into a main span of 26 m and two side spans of each 14 m.

In the period preceding its construction, the Danish Road Directorate had taken interest in investigating advantages of other types of construction materials used in infrastructures. COWI was thus involved in all stages of the Remmevej bridge project, carrying out conceptual design, preliminary studies, detailed design, and supervising construction operations of this timber bridge. Parameters such as construction and operation costs, service life, aesthetics, climate and ecosystem impacts, and scarcity of local resources (e.g., aggregates) were discussed during the design and construction of the bridge to demonstrate the performance of the structure over its 100 years' service life.

For the purpose of this study, the Remmevej design was compared to a typical pre-stressed concrete bridge with the same use and dimensions, assumed to be built in Denmark in the same year, and for which the bill of quantities was known by

LÆNGDESNIT A-A, 1: 200

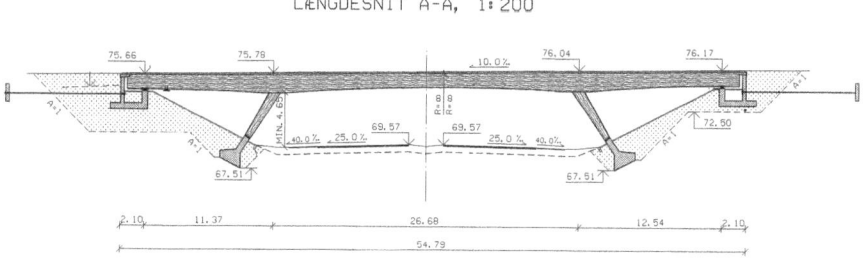

Fig. 1 Drawing of the Remmevej bridge (measures in meters)

COWI. Both bridges were designed to carry normal vehicular loads per codified standards comparable to EN 1991-2 and with a typical geometry to serve such purpose.

2.2 Service Life

The classical definition for service life of concrete bridges in Denmark is 100 years for the primary load carrying structure (concrete, reinforcement, post-tensioning) when operated and maintained at normal and foreseeable efforts and measures. This is confirmed by the Danish Road Directorate's experience with designing, building, and maintaining similar concrete structures.

There is no such proven track record for road carrying bridges using glue laminated timber as the primary construction material. In the case of the Remmevej bridge, Danbro (Danish Road Directorate Bridge Management System, BMS) shows no reported maintenance issues, and the next forecasted primary maintenance effort will be replacement of waterproofing and pavement which is due to happen in some years. This is not different from comparable concrete bridges.

Reviewing the experience from other countries, the service life of timber bridges has a wide range, which can be summarized by the Norwegian requirements: a default service life 100 years, which may be reduced to 50 years if replacement is planned. For this reason, it was decided to assign a service life of 100 years to the concrete bridge, and two alternative service lives of, respectively, 100 and 50 years to the wooden bridge, to cover different potential scenarios. For the shorter service life, it is assumed that the bridge will see a 1 to 1 replacement of structural elements after 50 years, doubling its service life to reach the study's time horizon.

2.3 LCA Concepts Used in the Study

Attributional and Consequential Approaches to LCA
There are two accepted approaches to performing an LCALife cycle assessment (LCA), which aim at answering two different questions. An attributional spiepr IndexRangeStart ID="ITerm48"spiepr IndexRangeStart ID="ITerm13"LCALife cycle assessment (LCA) aims to describe the environmentally relevant physical flows to and from a life cycle and its subsystems, answering the question: what share of the global environmental burdens belongs to a product, process, or system? A consequential LCALife cycle assessment (LCA) aims to describe how environmentally relevant flows will change in response to possible decisions, answering the question: how are the global environmental burdens affected by the production and use of the product? [4, 5].

Following the attributional principle, by compiling the LCA of every activity in the world, the sum of all these results would amount to the actual total anthropogenic

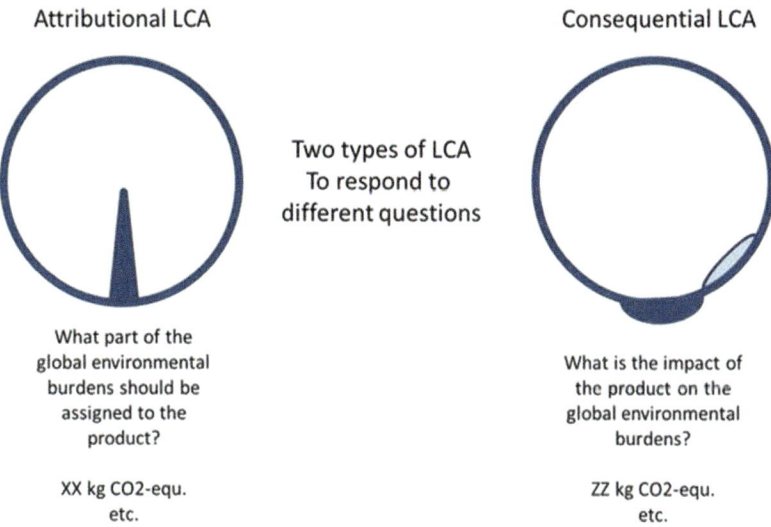

Fig. 2 Illustration of attributional and consequential approaches to LCA. The large circles symbolize the total environmental burdens of the world [5]

environmental impacts at that given time. The consequential principle is instead interested in changes to global production, and it aims at including in its analysis the ripple effects in other product systems reflecting market dynamics and technological change. This is illustrated in Fig. 2.

It must be noticed that there are still some unresolved issues in giving a clear definition to either system. One of the reference documents used in Europe for LCA, the ILCD Handbook [6], contains a distinction between the applicability of each model that has received strong criticism from consequential academia [7]. It is thus important to be extremely clear in formulating the goal, scope, and assumption of each LCA study.

Discounting Emissions
When building with wood, the biogenic carbon absorbed by the tree during its life is stored in the wood for the whole lifetime of the asset, and possibly beyond. To account for this carbon storage, the European Commission's ILCD Handbook [6] and the British Publicly Available Specification (PAS) 2050 [8] allow to discount emissions over time. This approach is based on the IPCC models of the residence time of a CO_2 pulse in the atmosphere following the 100-year criterion. With this approach, delayed emissions are accounted for with a linear discounting rate of 1% per year during a maximum of 100 years. At this point of time there is a cutoff, and all emissions occurring after this will have zero effects on the time interval [9].

It must be noted that the discounting of emissions is a practice that raises some methodological, scientific, and ethical controversies in the world of LCA, and for a more detailed discussion on this, please refer to [9, 10]. This optional method is however tested in this study to explore its implications on the result.

Logging Effect

This is a more conservative approach for accounting for the embodied carbon in wood, which was developed for this analysis based on a study that COWI conducted for the Norwegian Water Resources and Energy Directorate (NVE) in 2015 [11]. This approach includes the following assumptions:

- The decision to cut down a tree does not affect any carbon emissions up to that point; hence, no credits (negative emissions) are assigned for uptake during growth.
- The counterfactual scenario — had the tree not been cut down, how much CO_2 would it have taken up from this point on? This missed carbon sink will be deducted.
- The future uptake from a new tree growing where the old tree was cut will be included.
- Only ¼ of trunks end up as solid wood material, while the rest is planed off or discarded for quality reasons, and instead used for other purposes.
- The emissions from rotting branches left on the forest floor.
- The albedo effect — a clear-cut, snow-covered plot of land reflects more sunlight than an evergreen canopy.

All the above has been modeled in the COWI-NVE study and included in the emission factors in the third LCA presented below, together with the discounting approach.

2.4   Methodology of the LCA Study

Goal

The study sets out to provide a use case for integrating carbon considerations in the choice between wood and reinforced concrete in the construction of road bridges, and a methodological starting point for future LCA investigations on large timber infrastructure. It compares two real bridge designs, one constructed with traditional reinforced concrete and the other utilizing wood as the principal material, with a double goal:

1. Assessing their respective climate performance
2. Assessing the changes in such results when repeating the study through different approaches and assumptions:

   (a) Attributional approach
   (b) Consequential approach
   (c) Consequential approach, logging effect, and discounting of the emissions

   In all three approaches, two options were considered for the timber bridge: one assuming a structure with a life span of 100 years, another assuming a life span of 50 years, with complete replacement of timber elements at 50 years to complete

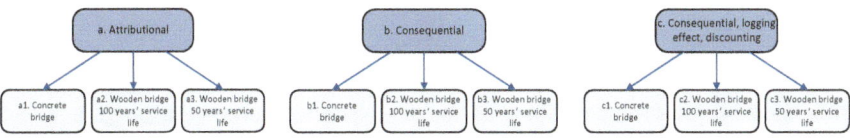

Fig. 3 Structure of the study

the 100 years' service life of the bridge. The structure of the study is illustrated in Fig. 3.

The functional unit is one bridge for road use.

Scope

The LCA is at screening-level and follows the general LCA structure of ISO 14040 and ISO 14044. The life cycle stages used in the study are:

- Production (A1–A3) includes raw materials extraction (including uptake of greenhouse gasses by the trees in the forests), processing of raw materials, transportation, and production of materials and products.
- Replacement (B4) includes replacement of construction elements where needed. The phase includes the production of the element and the end-of-life processes of the removed element.
- End of life (C3–C4) includes waste processing of the materials, such as recycling of steel, downcycling of concrete to produce gravel for roads, or incineration for wood.
- Next product system (D) includes environmental burdens and benefits beyond the defined system boundary, in this case after the bridge's service life, and captures the potential for recycling or utilization of waste materials, e.g., the energy recovered from wood incineration.

In the case of "c," the logging effect was included in the system definition and considered in the analysis, and the emissions for the time series were modeled following the principle of linear discounting over the study's time horizon. In all cases, emissions from traffic on the bridge are excluded from the use stage.

The time horizon of the study is also 100 years, and the geographical scope is Denmark, with lumber sourced from Norway or other Nordic countries. The results are not intended to be directly applied to future construction projects, as only one impact category is considered and the design for the wooden bridge is over 20 years old.

LCI and LCIA

For all three studies, the LCI and LCIA are integrated and sometimes implicit in the EPDs and databases used to extract relevant GWP values for the different materials and processes in each life cycle stage. The method and sources for each of the three studies is illustrated here:

(a) In the attributional study, the inventory of emissions for wood and concrete products was compiled using EPDs for Danish products, while for the rest of materials and processes, as well as for the missing life cycle stages, the Norwegian VejLCA and German Ökobaudat databases were used.

(b) Study "b" uses a consequential version of the Ecoinvent database.

(c) Study "c" uses same database as "b," but also incorporates the "logging effect" model to calculate the emission factors of wood products. In this model, the discounting of carbon follows the factors taken from the CIRAIG Dynamic Carbon Footprinter tool.

3   Results and Discussion

(a) *Attributional Approach*

The results of the LCIA for the first approach, shown in Fig. 4, show that the concrete bridge has the worst performance in terms of GWP impacts.

In the emissions for production and replacement, all wood, including formwork, contributes with a negative sign. The EPDs used for the inventories of the wooden products include the carbon storage of the wood, and for all of them the carbon sink during the growth of the wood more than offsets the emissions caused during logging, manufacturing, and transport of the wooden materials. In the case of the wooden bridge, the negative emissions from the wood more than offset all other construction materials as well, giving an overall negative sign to production.

The production stage accounts for 89% of the total emissions of the concrete bridge, and 89% of the production emissions is attributed to concrete and steel. This is due to their large material quantities and high carbon intensity.

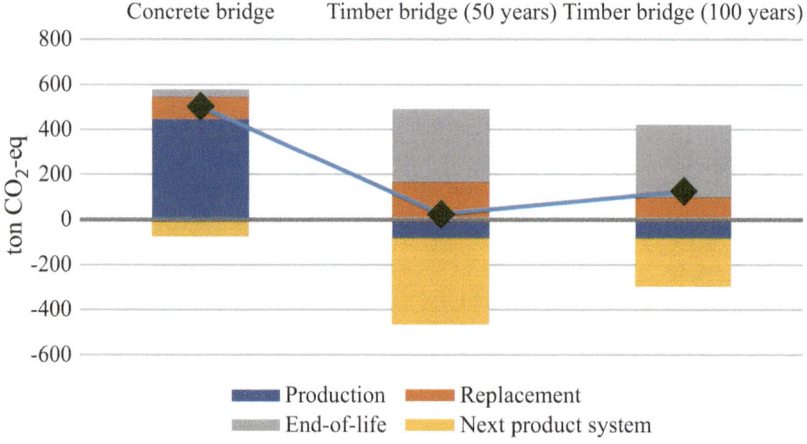

Fig. 4 Attributional approach results

Replacement captures those elements that need to be substituted during the life cycle of the bridge, including their end of life. In the case of the 50-year wooden bridge, this quantity has substantially higher emissions as it takes into account the carbon emitted to the atmosphere by the incineration of the old wood.

The end of life of steel and concrete has a much lower carbon intensity than that of wood, and it generates negative emissions (credits) in the next product system because of recycling. This factor is very low in the case of concrete and significant in the case of steel. The energy produced from wood combustion is credited to the scenario in module D. As the electricity and heat production in Denmark is projected to become greener over the coming years, the credit decreases and becomes very low after year 2050. This means that the benefit of using timber in construction becomes smaller and smaller as the benefit of energy production is being reduced.

It can be concluded that, at a project-level, a wooden bridge has a better carbon profile than a concrete one. It is important to note that this result is not appropriate for considering larger, systemic changes, for example a decision at the national level.

(b) *Consequential Approach*

The overall picture looks similar when using the consequential approach, as the concrete bridge is still the worst-performing in terms of carbon, as shown in Fig. 5.

Its savings in the next product system, however, are larger in all cases. This is because the consequential approach considers the ripple effects of waste recycling and utilization, such as changes in market dynamics, resource allocation, and waste management practices. This holistic approach allows the model to capture indirect impacts that may not be fully accounted for in attributional LCA.

Steel is a particularly good example, as it is observed that in its next product system, the scrap steel's recycling, and subsequent displacement of average steel on the market offsets more than half of the life cycle emissions from steel.

It is interesting to observe that, as in the previous case, the more wood is used, the lower the environmental burdens. An extreme case of this is provided by the 50 year

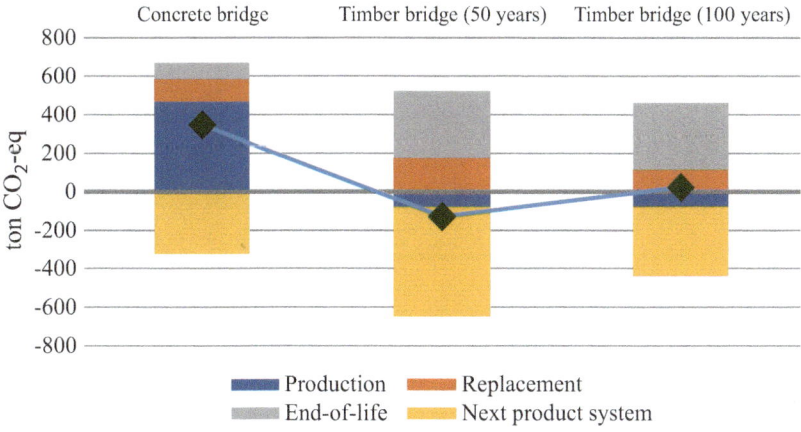

Fig. 5 Consequential approach results

bridge, where the total emissions appear to be negative, suggesting that constructing a shorter-lived wooden bridge would actually result in a net carbon sink of -127 t CO_2-eq, as opposed to 347 t CO_2-eq that the concrete bridge would contribute to.

(c) Consequential Approach, Logging Effect and Discounting

The radically different set of assumptions used in this third approach is reflected in the results, which show an entirely new picture compared to the previous two cases. In this case, it appears that the short-lived timber bridge is by far the highest emitter, as it is responsible for the emission of 1961 t CO_2-eq. This can be observed in Fig. 6.

This approach takes into consideration the opportunity cost of not extracting timber, but it also uses some specific assumptions, setting the cut-off at 100 years and carbon discounting until that point. This allows to fully discount all end-of-life and next product system emissions, as they will in fact be responsible for no radiative forcings during the time horizon of the study. The only exception would be the end-of-life of old material after replacement, which is in replacement, and their beyond life cycle impacts found in next product system, in this case only structural wood, which is however discounted by 59% following the carbon discounting rule set up in this study.

This allows to focus on the emissions from the production of all materials, which is similar to the previous studies for the concrete bridge, and entirely different for the wooden ones. This is in turn a direct consequence of the assumptions expressed in the logging effect, which try to represent the opportunity cost of logging as opposed to not extracting wood and letting the forest grow.

From this result, it can be concluded that when considering the opportunity cost of logging and for how long emitted CO_2 will actually cause radiative forcing within the 100-year study period, a timber bridge will cause more global warming than a concrete bridge.

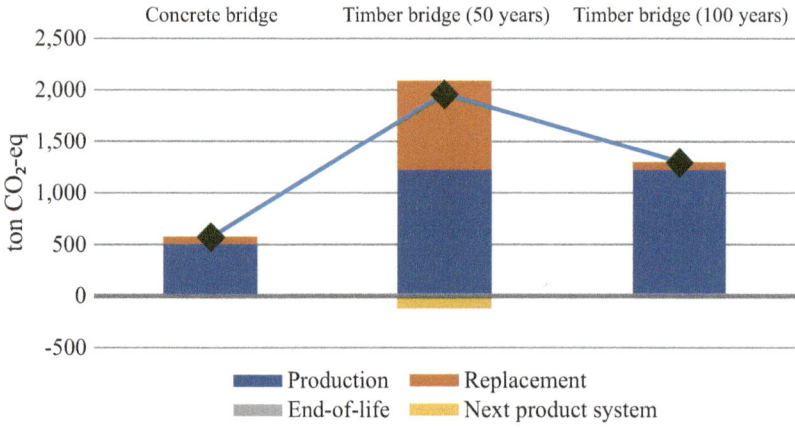

Fig. 6 Consequential approach, logging effect, discounting results

Table 1 Comparison of the results from the three LCAs

	Concrete bridge [ton CO₂-eq]	Timber bridge (50 years) [ton CO₂-eq]	Timber bridge (100 years) [ton CO₂-eq]
(a) Attributional approach	502	25	126
(b) Consequential approach	347	−127	23
(c) Consequential approach, logging effect, discounting	384	2457	1323

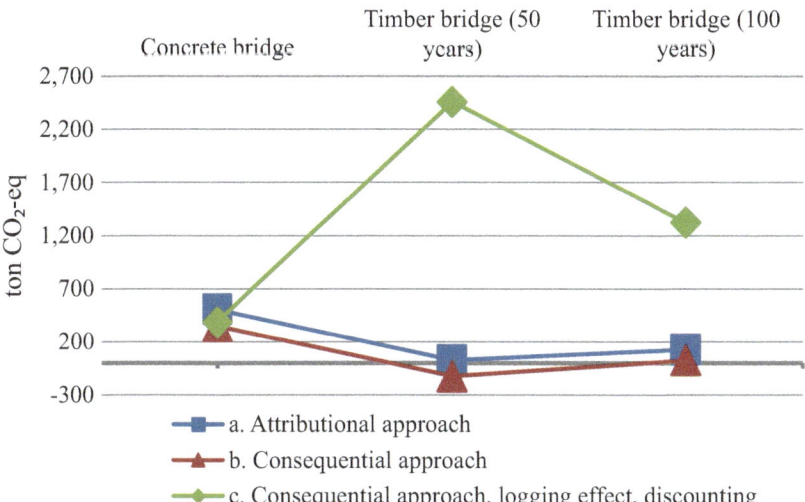

Fig. 7 Comparison of the results from the three LCAs

Comparison

Whereas the embodied carbon of the concrete bridge is comparable across the studies, the 100 years wooden bridge varies by 2 orders of magnitude and the 50 years bridge even changes the sign of its contribution—from carbon sink to huge climate offender. More detail on the results can be observed in Table 1 and Fig. 7.

4   Conclusions

The study successfully conducted all three life cycle assessments. Comparing the first two studies showed that, notwithstanding the deep methodological and conceptual differences between the attributional and consequential approaches, their application to this specific bridge project yielded some overall similar results. In all three cases, the consequential study tends to provide an overall lower global warming potential, which is due to the consequential approach's ability to include more indirect impacts in its analysis, with a particular emphasis on recycling and waste utilization.

Looking at the results of the study, approach "c" is clearly an outlier. Its importance is to provide an example of how a different framing of the initial research question can lead to wildly different results. The considerations at the base of this approach can help us acknowledge that current carbon accounting practices are varying widely, especially when analyzing biogenic carbon, and that by changing the initial question, the answer might be just opposite.

When looking at a single project like Remmevej, it is common to adopt a framing that is closer to the attributional question: how much carbon is this project responsible for? But when addressing the larger question: what role will wood play in the future of bridges? This question must stem for a deeper reflection on how a substantial increase in the demand of wood will affect its supply and change land use. For this type of analysis, it might be more appropriate to use the consequential approach and to include more impact categories in the analysis to address all relevant impacts on ecosystems, human health, and resource availability.

References

1. UNECE: COP27: UN Report Shows Pathways to Carbon-Neutrality in "Energy Intensive" Steel, Chemicals and Cement Industries. https://unece.org/media/press/372890 (2022)
2. Lundmark, T., Hannerz, M.: Den nordiska skogens klimatnytta. Nordiska ministerrådet (2017)
3. Niu, Y., Fink, G.: Life cycle assessment on modern timber bridges. Wood Mater. Sci. Eng. (2018). https://doi.org/10.1080/17480272.2018.1501421
4. Finnveden, G., Hauschild, M.Z., Ekvall, T., Guinée, J., Heijungs, R., Hellweg, S., et al.: Recent developments in life cycle assessment. J. Environ. Manag. **91**(1), 1–21 (2009). https://doi.org/10.1016/j.jenvman.2009.06.018
5. Ekvall, T.: Attributional and Consequential Life Cycle Assessment. IntechOpen (2020). https://doi.org/10.5772/intechopen.89202
6. Chomkhamsri, K., Wolf, M.A., Pant, R.: International reference life cycle data system (ILCD) handbook: review schemes for life cycle assessment. In: Finkbeiner, M. (ed.) Towards Life Cycle Sustainability Management. Springer, Dordrecht (2011). https://doi.org/10.1007/978-94-007-1899-9_11
7. Schmidt, J., Weidema, B.P.: Response to the Public Consultation on a Set of Guidance Documents of the International Reference Life Cycle Data System (ILCD) Handbook. 2.-0 LCA Consultants. https://lca-net.com/p/184 (2009)
8. PAS 2050: Specification for the Assessment of the Life Cycle Greenhouse Gas Emissions of Goods and Services. BSI System (2011)
9. Vogtländer, J.G., van der Velden, N.M., van der Lugt, P.: Carbon sequestration in LCA, a proposal for a new approach based on the global carbon cycle; cases on wood and on bamboo. Int. J. Life Cycle Assess. **19**, 13–23 (2014). https://doi.org/10.1007/s11367-013-0629-6
10. Lueddeckens, S., Saling, P., Guenther, E.: Discounting and life cycle assessment: a distorting measure in assessments, a reasonable instrument for decisions. Int. J. Environ. Sci. Technol. **19**, 2961–2972 (2022). https://doi.org/10.1007/s13762-021-03426-8
11. COWI: Analyse av klimagassutslipp fra utnyttelse av skog til energiformål: Litteraturgjennomgang og livsløpsvurderinger. Norges vassdrags- og energidirektorat (NVE) (2015)

Data-Driven Decision-Making for Road Maintenance in Norway

Henri Giudici

Contents

1   Introduction

The United Nations envision in the Agenda 2030 a roadmap toward a peaceful and prosperous world for people, a planet for today and future generations [1]. The transport sector is an important component for the achievement of this vision. A well-functioning transport sector promotes economic advancement, social inclusivity, and well-being of nations and regions [2]. The transport sector does so by providing services and infrastructures capable to enhance mobility of goods and people in a safe, resilient, accessible, efficient, and affordable way while minimizing the related environmental impact [3]. Countries are aligning with this vision in their respective national transport plans.

Over the life cycle of a road infrastructure, the operational stage is the longest one and requires maintenance and rehabilitations activities to keep the road at a

H. Giudici (✉)
University of South-Eastern Norway (USN), Department of Science and Industry Systems, Kongsberg, Norway
e-mail: Henri.Giudici@usn.no

© The Author(s) 2025
M. Kioumarsi, B. Shafei (eds.), *The 1st International Conference on Net-Zero Built Environment*, Lecture Notes in Civil Engineering 237,
https://doi.org/10.1007/978-3-031-69626-8_130

satisfactory level of service. Significant national resources are spent every year to keep the infrastructure at the expected levels [4–6]. To maintain and/or rehabilitate the infrastructure, there is the need to monitor and assess the condition of the road network and finally evaluate potential need for maintenance and/or rehabilitation activities. Road monitoring systems can be deployed to inspect the status of the road network. Each monitoring system produces big data that are collected, stored, and analyzed by appropriate Pavement Management Systems (PMS). These PMS can support road decision-makers while prioritizing the road network sections to be maintained and/or rehabilitated. The decision-making process is not trivial as it is a balance among different aspects to be considered simultaneously [32].

The present chapter presents an overview of the decision-making process of road maintenance using as a case study the Norwegian case. At first, this work presents a background on road monitoring systems and pavement management systems, and then it focuses on the decision-making process of road maintenance in Norway. This chapter aims to bring the attention to the opportunities and challenges of maintenance decision-making process in line with the Norwegian national transport plan.

2 Background

2.1 Road Surface Characteristics

There are multiple road characteristics that may be inspected. These characteristics include road friction, texture, smoothness, unevenness, and others. These characteristics are represented by parameters such as road friction, International Roughness Index (IRI) [7], Mean Profile Depth (MPD) [8], rut depth [9], and road defects (e.g., cracks and potholes). By measuring these parameters, it is possible to assess and evaluate the condition of the road network.

2.2 Road Monitoring Systems

Monitoring activities of a road network is an essential task to detect poor road conditions. Different approaches have been developed during the years to monitor roads, from static measurements to continuous measurements [8, 10]. Continuous measurements are typically sensor systems mounted on vehicle platforms (e.g., cars) which may drive at normal driving speed and collect continuous measurements. The main components of these systems comprehend a measurement sensor (laser-based, mechanical, camera-based, etc.), geo-localization sensor (GPS/GNSS), and orientation sensor (IMU) [10, 11]. Vehicle platforms include Unmanned Aerial Vehicles (UAVs) [13, 14]. In recent years, the literature also observes the use of mobile phones to inspect roads [12, 15, 16]. These monitoring systems are adopted from operators to inspect the roads continuously. Each inspection activity collects a high amount of road information which may be collected multiple times over a period of time (1 or more years). This big data is handled by PMS.

2.3 Pavement Management Systems

The management of the pavement infrastructure is defined as a process which assesses and evaluates with a given priority the maintenance and rehabilitation tasks of a road network, optimizing the relative resources [17, 18]. Through the use of heterogeneous big data (e.g., road conditions, annual traffic, budgets, and policics), PMSs assess the quality of the road network pavement, with the use of engineered indexes [e.g., Pavement Condition Indexes (PCIs)] and designed models. These indexes and models predict the quality and performance of the road network and assign prioritization levels of segments in the road network [19–21]. There are different PMSs that can be adopted from road authorities [20, 22, 23]. The authors [24] investigate the accuracy of the predictive models with their relative consequences in case the actual and predicted values of the investigated models vary, quantifying the budgetary implications of the relative impact. With a focus on highway pavement management, the authors [25] describe the potentialities of the Geographic Information System (GIS) for PMS purpose. In their work, the authors [26] studied and evaluated the potentialities of data mining and knowledge discovery (DMKD) combined with GIS for pavement management purposes. Their approach may provide consistent decisions and optimize resources (time and costs related) by speeding up the decision-making process. However, this approach requires verification in post-processing phases. While [27] describes how the GIS-PMS approach can be a valuable tool for evaluating the quality of pavement condition of road network in proximity to the city of Lisbon, the authors in [28] propose a database management system which might assist small municipalities facing budget constraints. The work [29] presents a model of GIS in PMS integrated with the Life Cycle Analysis (LCA) to predict pavement distresses and support decision-making processes.

3 Case Study: The Norwegian Road Transport Sector

The Norwegian road network comprehended 10,500 km of national roads and 45,000 km of county roads. Prior 2020, Norway comprised 19 counties. The Norwegian Public Road Administration (NPRA) was responsible for the whole Norwegian road network, on behalf of government and counties. In 2020 there was a reorganization of the road transport sector and the NPRA became responsible for the national road network (10,500 km), while the counties (downsized from 19 to 11) became responsible for the county road network (45,000 km) [30].

The budget for the road maintenance is based on the national funding (amount per year designated to the maintenance). Based on the budget, the section of roads within the road network is classified with a given priority. Based on their priority, the budget is allocated to maintain the roads with the highest priority. Technical requirements of the road network, and related level of service, can be found in [31].

3.1 Power Relations in Road Infrastructure Maintenance

Figure 1 shows the power relations between operators, decision-makers, and executors. The below conceptual model contains a color categorization (color boxes) to depict the correlations between the actors and their functions.

Based on the road network they are responsible for, operators (NPRA, counties or consultancies) perform road data collections (or inspections) and store the collected data. Then, decision-makers (e.g., national road authority and/or counties) adopt the stored data to assess and evaluate potential actions to improve the quality of the road network. After it is decided which road sections require maintenance and rehabilitations, executors (e.g., road contractors) require to perform actions oriented to improve the quality of the road sections.

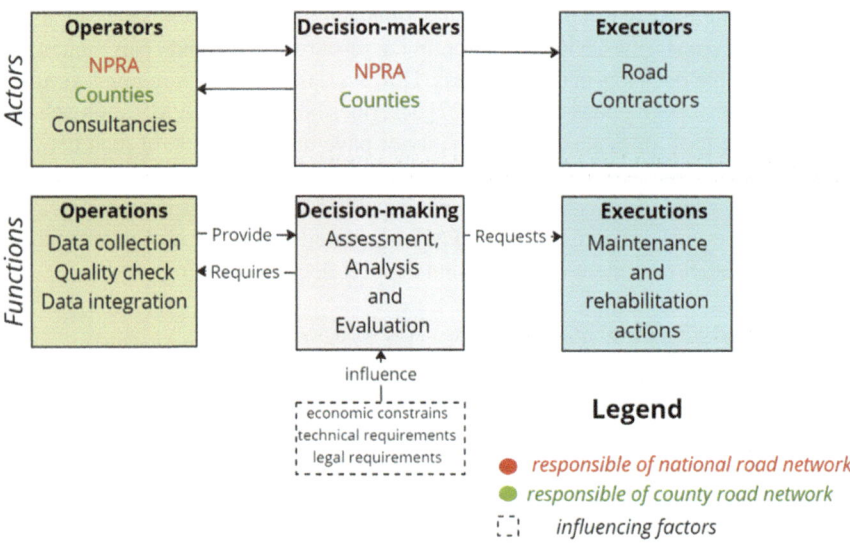

Fig. 1 Power roles, actors, and functions in road maintenance decision-making, inspired from [32]

3.2 Maintenance Decision-Making Process

The process begins with the selection of a desired road characteristic(s) to be assessed over the interested road network. After that, the road network is inspected, and road information is retrieved. After road inspections, reports containing geo-localized road data are generated. The operator generates and stores the reports in the national database, Rosita. Rosita is a data bank system that stores any road characteristics data retrieved from the Norwegian network according to international standard criteria. Rosita includes national road and counties road network information.

Further step is the analysis of road deterioration over the years. The historical reports related to the interested road characteristic(s) are retrieved in Rosita, and the progressive deterioration is analyzed. The PMS analyze, assess, and evaluate the condition of the road network by merging Rosita's data with the annual traffic conditions, technical/legal requirements [31] for the interested road network, and economic yearly budgets. The PMS categorizes each section of the road network based on the severity of the road condition and its potential impact on road traffic. In case the quality level is critical (road characteristic(s) value(s) is near or below requirement(s) legal/technical threshold limits), road maintenance/rehabilitation actions are mandatory. In contrary cases, the prioritization of the road maintenance follows the PMS's categories of priority. NVDB stores road network's information such as geometry, topology, road's constituents (equipment, drainage, etc.), infrastructure layers' properties, traffic volumes, and accidents. NVDB's information is also included in the PMS analysis and evaluations. Figure 2 shows the system constituents to perform a data-driven decision-making, and Fig. 3 shows the roam maintenance workflow.

4 Discussions

The Norwegian transport plan envisions transport infrastructures that are safe, effective, valuable for end user, innovative, and environmentally friendly [35]. This applies also to road infrastructure. Maintenance activities and related decision-making are of primary importance to keep road infrastructure at acceptable

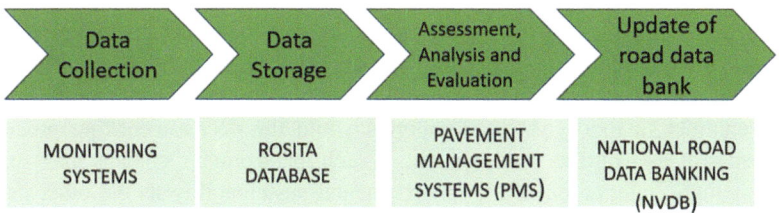

Fig. 2 Systems constituents of road maintenance decision-making process

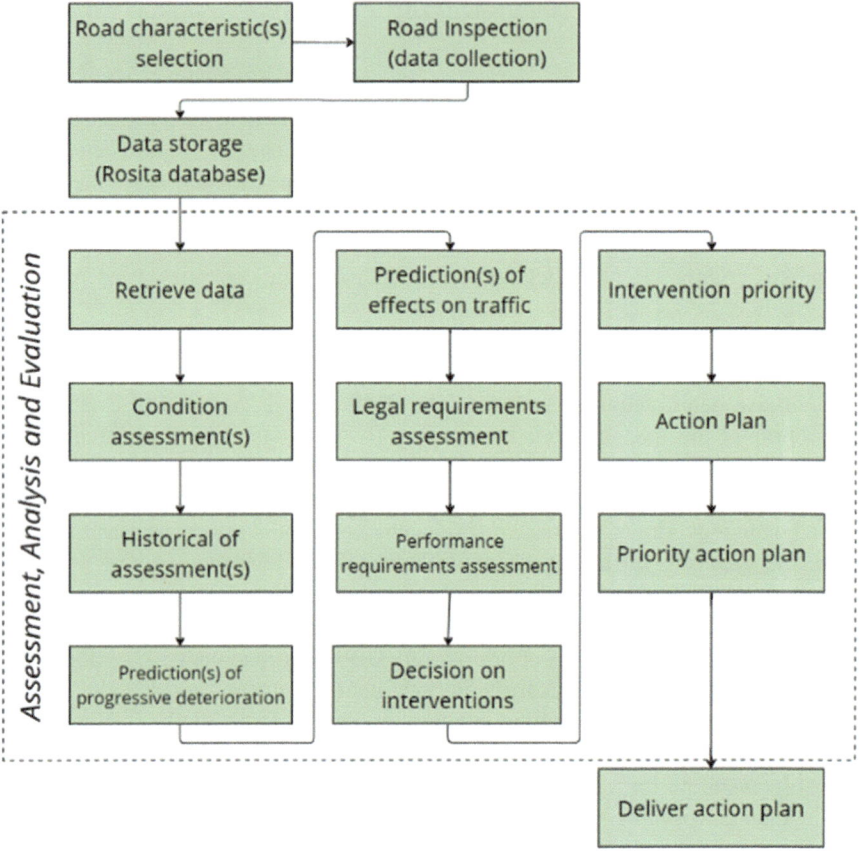

Fig. 3 Road maintenance workflow

level of services enhancing thus the safety, effectiveness, and value for the road end user. However, to align with the national transport vision, it is required to satisfy at the same time its multiple qualities. The environmental sustainability is one of these qualities which needs to be considered in the maintenance decision-making process. Currently, during the decision-making process, the environmental sustainability is indirectly addressed by the executor's operations. To obtain permission for operations, executors need to document their practices and related environmental impact. This is done, for instance, by providing documentation related to their machinery (e.g., electric or fuel-based vehicles) and Environmental Production Declaration (EPD). However, to integrate the quality of environmental sustainability and the other qualities, a more systematic approach into the decision-making process is required [32, 34]. Further development of such management systems may support the systematic integration of the environmental sustainability as the other desired transport system's qualities. This may potentially include considerations related to the environmental impact of winter maintenance practices.

Innovation, technologies with relative (big) data and advanced materials may contribute to the development of a (smart) mobility and (smart) transportation with the mentioned qualities [33]. Implementing innovative technologies in the decision-making process, such as using vehicle-to-vehicle and/or vehicle-to-infrastructure for monitoring purposes, may require an alteration of the data-driven process of the systems constituents shown in Fig. 2.

The reorganization of the transport infrastructure poses an additional challenge to be mentioned. After the reorganization, there was a decentralization of responsibilities from national level to county level. Counties became responsible for keeping the quality of their relative road network at acceptable quality levels, without putting too much stress on their budgets. Yet these counties require competence in road engineering which may be replaced by not-road engineering personnel. The latter needs to make decisions based on highly engineering road technologies and PMS often counterintuitive for those not expert in the field [28]. Systems thinking approaches may support the comprehension of the holistic complexity focusing on the intervention that fit the desired purpose [34].

5 Conclusion

This chapter presents an overview of road maintenance decision-making. The presented Norwegian case describes the maintenance and operations processes for the national road network. The chapter discussed the power relations, the decision-making constituents, and processes in light of the reorganization of the Norwegian transport sector. The chapter observed the need to better integrate the environmental sustainability, as well as other quality, in the road maintenance decision-making process. The alignment of these qualities is a valuable aspect to align the decision-making of maintenance and/or rehabilitation toward the vision of the Norwegian transport plan.

References

1. United Nations: Transforming our world: the 2030 Agenda for Sustainable Development. A/RES/70/1. A/RES/70/1 Transforming our world: the 2030 Agenda for Sustainable Development (un.org) (2015)
2. United Nations: Sustainable transport, sustainable development. Interagency report for second Global Sustainable Transport Conference (2021)
3. High-level Advisory Group on Sustainable Transport: Mobilizing Sustainable Transport for Development. Report (sustainabledevelopment.un.org) (2016)
4. Bull, A., NU. CEPAL, N: Traffic Congestion: the Problem and how to Deal with it. Economic Commission for Latin America and the Caribbean. ECLAC (2003) ISBN: 92-1-121432-7
5. Burningham, S., Stankevich, N.: Why Road Maintenance Is Important and How to Get it Done. Transport Notes Series; No. TRN 4. World Bank, Washington, DC (2005) © World Bank

6. OECD and European Commission: Cities in the World (2020) https://doi.org/10.1787/d0efcbda-en
7. Pawar, P.R., Mathew, A.T., Saraf, M.R.: IRI (International Roughness Index): an indicator of vehicle response. Mater. Today Proc. **5**(5), 11738–11750 (2018)
8. Sandberg, U.: Influence of Road Surface Texture on Traffic Characteristics Related to Environment, Economy, and Safety: a State-of-the-Art Study Regarding Measures and Measuring Methods, VTI Report 53A-1997. Swedish National Road Administration, Borlange, Sweden (1998)
9. Gillespie, T.D., Sayers, M.W., Hagan, M.R.: Methodology for road roughness profiling and rut depth measurement (1987)
10. Sayers, M.W., Karamihas, S.M.: The Little Book of Profiling: Basic Information about Measuring and Interpreting Road Profiles. University of Michigan, Ann Arbor, Transportation Research Institute (1998)
11. Giudici, H., Mocialov, B., Myklatun, A.: Towards sustainable smart cities: the use of the ViaPPS as road monitoring system. In: Sustainable Smart Cities: Theoretical Foundations and Practical Considerations, pp. 135–153. Springer International Publishing, Cham (2022)
12. Hanson, T., Cameron, C., Hildebrand, E.: Evaluation of low-cost consumer-level mobile phone technology for measuring international roughness index (IRI) values. Can. J. Civ. Eng. **41**(9), 819–827 (2014)
13. Prosser-Contreras, M., Atencio, E., Muñoz La Rivera, F., Herrera, R.F.: Use of unmanned aerial vehicles (UAVs) and photogrammetry to obtain the international roughness index (IRI) on roads. Appl. Sci. **10**(24), 8788 (2020)
14. Nappo, N., Mavrouli, O., Nex, F., van Westen, C., Gambillara, R., Michetti, A.M.: Use of UAV-based photogrammetry products for semi-automatic detection and classification of asphalt road damage in landslide-affected areas. Eng. Geol. **294**, 106363 (2021)
15. Zeng, J., Zhang, J., Cao, Q., Guo, W.: Research on vibration index of IRI detection based on smart phone. In: International Conference on Wireless Communications, Networking and Applications, pp. 1067–1076. Springer, Singapore (2022)
16. Varadharajan, S., Jose, S., Sharma, K., Wander, L., Mertz, C.: Vision for road inspection. In: IEEE Winter Conference on Applications of Computer Vision, pp. 115–122. IEEE (2014, March)
17. Pavement Management - A Manual for Communities, U. S. Department of Transportation, Metropolitan Area Planning Council, Boston (1986)
18. Wolters, A., Zimmerman, K., Schattler, K., Rietgraf, A.: Implementing pavement management systems for local agencies (2011)
19. Kulkarni, R.B.: Dynamic decision model for a pavement management system. Transp. Res. Rec. **997**, 11–18 (1984)
20. Ismail, N., Ismail, A., Atiq, R.: An overview of expert systems in pavement management. Eur. J. Sci. Res. **30**(1), 99–111 (2009)
21. Kulkarni, R.B., Miller, R.W.: Pavement management systems: Past, present, and future. Transp. Res. Rec. **1853**(1), 65–71 (2003)
22. Wang, Z., Pyle, T.: Implementing a pavement management system: the Caltrans experience. Int. J. Transport. Sci. Technol. **8**(3), 251–262 (2019)
23. Sarsam, S.I.: Pavement maintenance management system: a review. Trends Transp. Eng. Appl. **3**(2), 19–30 (2016) (check)
24. Hosseini, S.A., Smadi, O.: How prediction accuracy can affect the decision-making process in pavement management system. Inf. Dent. **6**(2), 28 (2021)
25. Parida, M., Aggarwal, S., Jain, S.S.: Enhancing pavement management systems using GIS. In: Proceedings of the Institution of Civil Engineers-Transport, vol. 158, No. 2, pp. 107–113. Thomas Telford Ltd. (2005, May)
26. Zhou, G., Wang, L., Wang, D., Reichle, S.: Integration of GIS and data mining technology to enhance the pavement management decision making. J. Transp. Eng. **136**(4), 332–341 (2010)

27. Picado-Santos, L., Ferreira, A., Antunes, A., Carvalheira, C., Santos, B., Bicho, M., Quadrado, I., Silvestre, S.: Pavement management system for Lisbon. In: Proceedings of the Institution of Civil Engineers-Municipal Engineer, vol. 157, No. 3, pp. 157–165. Thomas Telford Ltd (2004, Sept)
28. Zagvozda, M., Dimter, S., Moser, V., Barišić, I.: Application of GIS technology in pavement management systems. Građevinar. **71**(04), 297–304 (2019)
29. Al-Mansoori, T., Abdalkadhum, A., Al-Husainy, A.S.: A GIS-enhanced pavement management system: a case study in Iraq. J. Eng. Sci. Technol. **15**(4), 2639–2648 (2020)
30. Gryteselv, D.: Norwegian road (pavement) maintenance, an overview. European Road Profiling User's Group (ERPUG) Conference proceedings. PowerPoint-presentation (erpug.org) (2022)
31. NPRA (Norwegian Public Roads Administration): Standard for drift og vedlikehold av riksveger. NPRA, Oslo (2014)
32. Muller, G., Giudici, H.: Social systems of systems thinking to improve decision-making processes towards the sustainable transition. In: CSER 2024 Conference Proceedings. Springer Nature (2024). pp 341–353.
33. Giudici, H., Pérez-Fortes, A.P.: How recent developments in smart road technologies and construction materials can contribute to the sustainability of road infrastructure. J. Infrastruct. Syst. **28**(4), 02522002 (2022)
34. Giudici, H., Falk, K., Muller, G., Helle, D.E., Drilen, E.: A systems thinking perspective on the obstacles faced by industrial organizations towards sustainability. Highlights of Sustainability. **3**(2), 240–254.
35. Det Kongelege Samferdeselsdepartement: Nasjonal transportplan (2021). https://www.regjeringen.no/no/dokumenter /meld.-st.-20-20202021/id2839503/

Bridging the Shift from Linear to Circular Economy in Road Infrastructure: The Norwegian Stakeholders' Perspectives on Challenges and Opportunities

Alexander Grødum Vetnes and Reyn O'Born (iD)

Contents

1 Introduction

The global construction industry is responsible for consuming over 30% of virgin resources and is a significant source of greenhouse gas (GHG) emissions and waste generation, contributing 40% and 25%, respectively [1, 2]. Modern road infrastructure (RI) is particularly resource-intensive, requiring substantial amounts of land, steel, asphalt, concrete, aggregates, and energy in construction, thereby contributing significantly to the construction industry's GHG emissions [3]. The industry faces increasing pressure to adopt sustainable practices, spurred by international agreements like the Paris Agreement, the United Nations Sustainable Development Goals, and the European Union's European Green Deal. The shift from the current linear economy, which is marked by a "take-make-dispose" approach, to a circular economy (CE) where resources are utilized to their maximum potential and kept in use for

A. G. Vetnes (✉) · R. O'Born
University of Agder, Grimstad, Norway
e-mail: Alexander.G.Vetnes@Uia.no; Reyn.oborn@uia.no

M. Kioumarsi, B. Shafei (eds.), *The 1st International Conference on Net-Zero Built Environment*, Lecture Notes in Civil Engineering 237,
https://doi.org/10.1007/978-3-031-69626-8_131

as long as possible, is globally recognized as vital for achieving and sustaining future environmental sustainability [4]. Given the escalating focus on reducing emissions and limiting resource use, reevaluating how RI is built and maintained is imperative.

The global road network is currently estimated at approximately 16.3 million kilometers, with an expected increase of 25 million kilometers by 2050 [5]. The total emissions from the transport sector since the industrial age account for about 15% of all GHG emissions [6]. In light of these global challenges, this paper focuses on Norway, a country with its own set of challenges and opportunities, in transitioning to CE in RI. Highlighting the massive resource consumption of the current practices in Norway, an average of 50 tons of stone aggregates are used per meter of a new two-lane road [7]. To transition from the traditional linear economy to a more circular one, a fundamental shift in the design, construction, use, maintenance, and eventual decommissioning of RI is required [8].

Although various barriers and drivers for circular infrastructure have been identified across political, economic, technological, social, knowledge-based, and attitudinal categories [9], the most substantial barriers seem to be political and economic. However, these barriers could potentially transform into significant drivers through revised regulations and standards [10]. Facilitating sustainable choices in projects via incentives, emphasizing CE in public procurement, and incorporating CE criteria in tender competitions are some key strategies proposed for this transition [11].

While extensive research exists on isolated components of RI construction or materials (e.g., novel asphalt mixtures and aggregates, electrification of machinery), a systematic overview integrating CE principles into the entire RI is scarce. Research that fully integrates these principles is necessary, particularly in adopting a holistic approach that seeks to reduce not just GHG emissions but also the consumption of virgin resources and land use [12]. This study aims to explore the broader challenges hindering Norwegian actors in transitioning toward more sustainable and circular RI and to identify potential opportunities for circular growth.

2 Methods

This study employs a mixed-methods approach. Initially, a preliminary literature search on the barriers and drivers for CE in the broader construction industry was conducted. The results from the literature were used as a basis to develop a web-based survey targeting specific groups of professionals in the Norwegian road industry. This survey was distributed to 117 actors in the sector, yielding a response rate of 48.7%, with 57 respondents completing the survey. An additional 13.7% of recipients (16 individuals) partially completed it. The survey was designed to gather knowledge from their experiences and was supplemented with six semi-structured interviews to gain further insight.

2.1 Survey

A web-based survey was designed using the program SurveyXact and distributed to various actors working with RI. The surveys were intended to get a clearer understanding of which focus areas the different Norwegian actors themselves perceived as promoting and inhibiting the development of sustainable and CE-oriented RI. The respondents targeted were from various groups with experiences with Norwegian RI, such as developers, contractors, and material suppliers. These groups were selected due to their potential to provide valuable insights and firsthand knowledge, which are essential to reveal how the sector itself perceives sustainability, CE, and the factors that may facilitate or impede the transition toward these themes. Several of the respondents were contacted via an ongoing research project related to sustainable RI.

The survey was divided into three themes: (1) background information on the respondents' roles and the region of their workplace, (2) general views on sustainability and its importance for the construction industry, and (3) perceptions on CE, including barriers and drivers for transitioning to CE in Norwegian RI. A total of 13 questions were asked, structured as multiple-choice questions without requiring free-text responses. However, to capture aspects that might fall outside the provided alternatives, several questions included the option 'other,' where respondents had the opportunity to further explain in their own words.

2.2 Qualitative Interviews

Six semi-structured interviews were conducted, featuring two participants each from developers, contractors, and material suppliers. These interviewees were selected from a group of 11 survey respondents who had indicated their willingness to participate in further interviews. In an effort to maintain equal representation among the different stakeholders, invitations were sent to nine of the respondents to verify their continued availability for an interview. However, due to non-responses from three individuals, the final tally of interview subjects was reduced to six.

The questions used in the qualitative interview were designed to follow up on the initial survey results and to allow the interviewees to provide more depth of understanding than the questions responded to in the initial survey. The survey results were analyzed to identify areas requiring deeper exploration, guiding the formulation of the interview questions. This approach aimed to delve into aspects not fully captured by the survey, such as nuanced experiences and detailed perspectives, thereby enriching the study's findings with a more comprehensive understanding of the participants' views and experiences.

The interview guide consisted of 12 open-ended questions, and the 45-to-60-min interviews were conducted online due to geographical distribution, using web-based platforms and audio recordings. To ensure confidentiality and data protection, personal identifiers were anonymized. The interviews were transcribed and analyzed using NVivo, assisting in organizing and identifying patterns in qualitative data and ensuring a thorough and accurate representation of the interviewees' insights.

3   Results and Discussion

The combined results from the mixed-methods approach were intended to provide a series of perspectives on the transition toward CE in Norwegian RI. These diverse sources of information reveal consistent themes and challenges, as well as potential pathways for the successful implementation of CE practices.

The survey, which targeted 117 individuals as previously detailed in the Methods section, elicited responses from 57 respondents and partial responses from 16, with their position within the firm presented in Table 1. Due to the varying levels of survey completion, there were some variations in the total number of respondents for specific questions, as indicated in the accompanying figure texts. The respondents were evenly distributed geographically across Norway. Among the "Other" category, eight were researchers, and the remaining four aligned with the "Leader" category (subject manager, COO, and two in administration).

3.1   Motivations

This section will investigate the motivations of the actors to transition toward CE. Figure 1 presents the survey results for the question *"What benefits do you consider circular economy can have in your field of work?"*. The perceived benefits identified were predominantly reduced environmental impact, improved resource efficiency, and social responsibility (Fig. 1).

When asked, *"What measures have you implemented to make your projects more sustainable?"* the results highlighted a focus on reducing GHG emissions (72%) in their current projects. Other significant circular economy (CE) measures included

Table 1 Position within the firm of respondents

Position	Number of respondents
Leader	42
Advisor/consultant	7
Project lead	5
Marketing	1
Engineer	5
Finance/accountant	1
Architect	0
Other	12

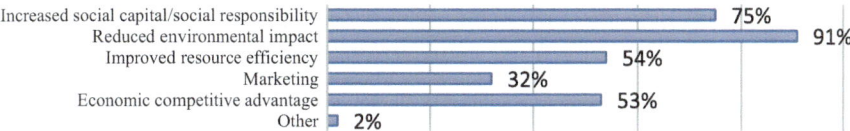

Fig. 1 What benefits do you consider the circular economy to have in your field of work? (multiple choices possible) (Respondents: 57)

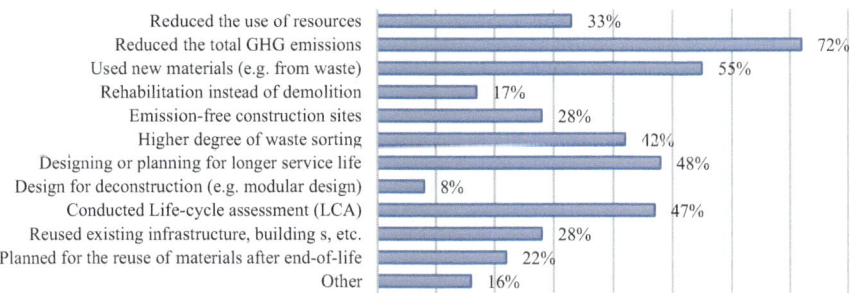

Fig. 2 What measures have you implemented to make your projects more sustainable? (multiple answers possible) (Respondents: 64)

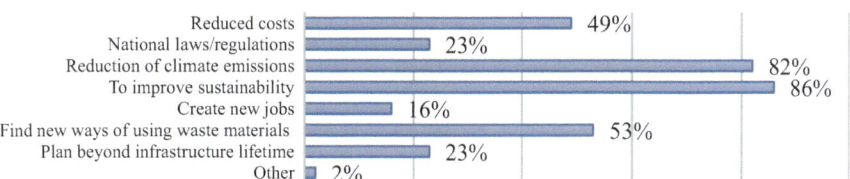

Fig. 3 What is your main objective in adopting a circular economy in your field of work? (multiple choices possible) (Respondents: 57)

using new material types (55%), increasing recycling rates (42%), designing for extended lifespans (48%), and conducting life-cycle analysis (47%) (Fig. 2).

When the respondents were asked to identify what they believe is the sole most effective measure to reduce waste generation from RI, there was an even distribution between planning for the longest possible lifespan (42%) and reusing materials (38%) as the standout choices. The remaining options, "to plan for future maintenance" (9%), "designing for future disassembly" (6%), and "more stringent requirements for the degree of sorting of generated waste" (5%), were not considered to be as effective.

The question "*What is your main objective in adopting a circular economy in your field of work?*" showed that the motivations for implementing CE principles are seen as a way to both improve sustainability (86%) and reduce GHG emissions (82%). Fewer see the possibilities of reducing costs (49%) and closing the resource loops by using waste materials (53%) as a motivation for change (Fig. 3). The contractors interviewed did not only perceive sustainability as a way to win contracts but as a responsibility to do better.

The transition toward CE in RI is motivated by a combination of environmental, economic, and sustainability concerns [12]. Survey responses reflect a strong awareness of the importance of this transition. However, the prioritization for a transition to CE varied, with an almost linear distribution from no (4%) to high (40%) priority.

Personal interviews reinforce these findings, revealing a general consensus among RI professionals about the importance of resource efficiency and the need to keep used resources in circulation, thereby emphasizing the sustainability aspect of CE.

3.2 Drivers

To reveal what the actors themselves experienced that would drive the transition to CE, the question *"What do you think will be the biggest driver for a transition to a circular economy in your field of work?"* was asked. The results reveal that the most significant driver for CE adoption is identified as regulatory demands (77%), followed by emphasis on CE in tendering processes (61%), and improved profitability (54%) (Fig. 4). The survey results show that regulatory demands and tendering processes can have a significant impact on implementing CE. Some of the interviewees reinforced this result by suggesting that project owners can significantly influence CE adoption through contract criteria.

Drivers for the adoption of CE in RI are complex. Governmental policies and regulations emerge as potential drivers, as indicated by both the literature and interviews, with regulatory demands and policy initiatives identified as key factors influencing CE adoption. The survey data further substantiates this, indicating that regulatory requirements significantly impact the prioritization of CE practices. Economic incentives, such as favorable tendering conditions and financial support mechanisms, are highlighted as crucial in encouraging the adoption of CE practices. Transitioning to CE also has significant economic implications. The survey and literature suggest potential cost savings and long-term financial benefits. For instance, efficient resource use can lead to reduced material costs, while innovative recycling technologies can create new revenue streams.

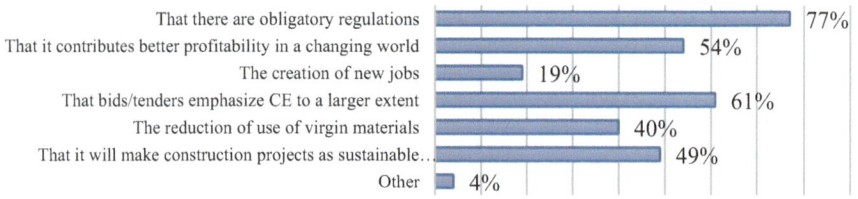

Fig. 4 What do you think will be the biggest driver for a transition to a circular economy in your field of work? (multiple choices possible) (Respondents: 57)

3.3 Barriers

In line with the literature, both the survey and interviews showed that Norwegian RI actors regard political and regulatory frameworks as prominent barriers to CE (Fig. 5). Furthermore, both the survey and interviews highlight economic factors as critical barriers, with a lack of incentive schemes being particularly prominent. However, the economic barriers also hold potential as drivers, should regulatory changes and incentives align favorably (see Fig. 4 and Table 2).

The literature, survey, and interviews all point to attitudinal resistance and a lack of comprehensive knowledge as obstacles [9, 10]. Showing that there is a need for greater awareness and understanding of CE's broader benefits beyond environmental impact. Several of the interviews pointed to encountering attitudinal difficulties when trying to make use of a novel approach or material, which ultimately led to the traditional alternative being chosen.

The survey revealed that CE and sustainability are currently regarded as costs and not as sources of possible cost savings. This is in line with the literature, where the economy is regarded as one of the main barriers to transitioning to CE [9]. A cause of this may be that it has been difficult to illuminate the cost-saving possibilities that a fully integrated CE could bring. A more effective use of available resources can lead to cost savings by not ordering more goods than is necessary. This will mean that the generation of waste that can occur on the construction site is minimized. Once goods and materials have become waste, they no longer have the same value as the costs associated with purchasing them. In other words, there are a lot of materials that today are only transported to the construction site to be transformed into worthless waste. This will mainly result in additional costs being incurred in the projects.

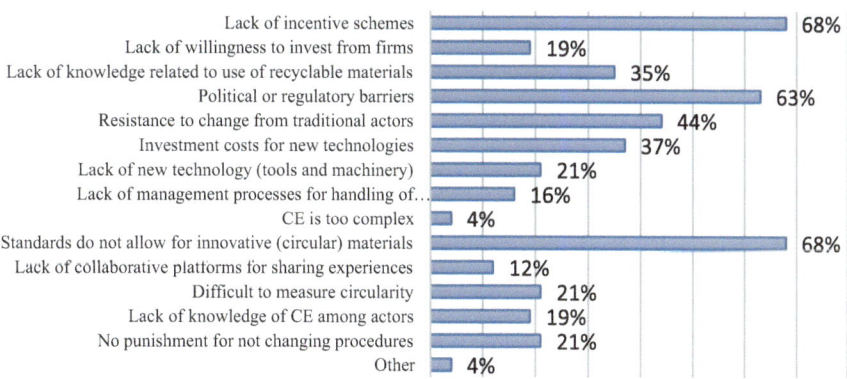

Fig. 5 What will be the biggest obstacles/barriers that could limit a transition to a circular economy? (multiple choices possible) (Respondents: 57)

Table 2 Summary of interview findings

Theme	Findings
Understanding of CE	There was unanimous recognition of CE's sustainability aspect, emphasizing resource efficiency and the necessity of recycling and reusing resources rather than extracting virgin resources. However, the economic and social dimensions of CE were notably absent from the discussions.
Perceived barriers and drivers	One key barrier identified is the rigid regulatory framework and lack of incentives, hindering the adoption of more sustainable alternatives. Conversely, the presence of strong will and desire among industry players to implement greener alternatives was noted as a significant driver, although hindered by current limitations.
Role of competition, and contractual requirements	The importance of the role of project owners in setting contractual requirements for sustainability was highlighted. Some respondents noted a shift toward sustainability when such criteria were included in contracts.
Regulatory framework, manuals, and standards	The existing regulatory framework and standards are seen as the most significant barriers to making sustainable choices. These standards are perceived as far too rigid to accommodate new materials or approaches, making sustainable choices challenging to implement.
Economic considerations	From a private sector perspective, economic viability is a primary consideration. The potential for economic incentives or penalties from project owners is seen as a critical factor in driving sustainable choices.
Technology and innovation	The current state of technology, particularly in electric construction machinery, is seen as not yet mature enough for widespread adoption. Hydrogen technology is considered a more promising alternative for future development.
Attitudinal barriers	Attitudes within the industry are seen as both a barrier and a driver. Conservative approaches often favor traditional methods over newer, potentially more sustainable options.
Political factors	The complexity of navigating various support schemes and the need for more targeted governmental actions and policies to facilitate the transition toward sustainability were emphasized.

3.4 Summary of Interview Findings

This section presents the key findings from the semi-structured interviews, as detailed in Table 2. These interviews yielded deeper insights into the initial survey results. Through analysis in NVivo, themes and patterns were identified.

As shown in Table 2, while there is a clear understanding and willingness within the industry to move toward a more sustainable CE, significant barriers such as rigid regulations, economic considerations, technological limitations, and conservative attitudes need to be addressed. These findings suggest a need for more flexible regulations, economic incentives, technological advancements, and a shift in industry attitudes to fully realize the potential of CE.

3.5 Summary and Discussion of Findings

The results of the survey showed that a large majority of respondents were "Leaders." This may have led to the results being shifted toward the more economic considerations that a change to CE would bring about. Management is often responsible for ensuring that the profitability of the company is considered, and they often make targeted choices so that the allocated budget is kept. This possible bias could have given greater weight to the financial aspects when asked to weigh advantages and disadvantages against each other in a survey.

Owing to the diverse dependencies and fragmentation amongst stakeholders in the broader construction industry, it's challenging to identify a singular action that would optimally facilitate the increased adoption of CE. Researchers advocate for various categories to classify the barriers and drivers, yet these categories are interrelated, making it impractical to focus solely on one as the definitive solution. Barriers may become drivers with altered premises, and inaction could turn current drivers into future barriers. A holistic approach is needed to overcome these barriers and ensure the activation of necessary drivers.

3.6 Further Work

Further research is required to address the broader scope of CE in RI comprehensively. Defining what constitutes a "circular road." This involves establishing a clear, comprehensive understanding of design, materials, and life-cycle impacts. Equally important is the development of methods to measure the circularity of RI projects. Effective metrics and indicators to assess circularity across various life-cycle stages will be instrumental in evaluating progress and shaping policy and practice in RI construction in the CE framework. Additionally, comparing materials and practices aligned with CE principles against traditional methods will yield insights into efficiency, sustainability, and cost-effectiveness.

Moreover, there is a need for increased knowledge and awareness about the full scope of CE, which extends beyond environmental benefits. This highlights the importance of education and information dissemination as drivers for CE adoption.

4 Conclusions

This study underscores the need for a paradigm shift in the RI industry, transitioning from its current linear model, which is heavily resource-intensive and a significant contributor to GHG emissions, to a more sustainable circular approach. Focusing on the Norwegian context, this research highlights that the transitions to CE in RI promise significant environmental benefits, including a minimized carbon footprint, reduced resource extraction, and waste generation. Economically, it can lead to cost savings through collaboration, efficient resource use, and potentially open new market opportunities. Despite these benefits, several challenges hinder the transition. High initial costs, the need for technological innovation, resistance to change, and a lack of clear guidelines and standards are prominent barriers. Furthermore, the current economic framework often does not incentivize sustainable practices, necessitating policy and regulatory changes. Policy and regulation play a pivotal role in facilitating this transition. Policies encouraging the use of recycled materials, promoting sustainable practices, and standardizing CE approaches are recommended.

In conclusion, the shift toward CE in RI presents a promising pathway toward sustainability. While there are challenges to be addressed, the benefits in terms of environmental conservation, economic savings, and resource efficiency are substantial. Continued research, innovation, and collaboration among stakeholders are essential to realizing the full potential of this transition. This study underscores the urgent need for policy reforms and collaborative efforts in Norway to harness the full potential of a circular economy in road infrastructure.

Acknowledgments This study is financed by the research project "Green Platform—Sustainable Value Chain and Materials in Road Construction," as financed by the Norwegian Research Council grant project number 340901.

References

1. Benachio, G.L.F., Freitas, M. do C.D, Tavares, S.F.: Circular economy in the construction industry: a systematic literature review. J. Clean. Prod. **260** (2020). https://www.sciencedirect.com/science/article/pii/S0959652620310933?via%3Dihub
2. Huang, L., Krigsvoll, G., Johansen, F., Liu, Y., Zhang, X.: Carbon emission of global construction sector. Renew. Sust. Energ. Rev. **81**, 1906–1916 (2018)
3. Liu, N., Wang, Y., Bai, Q., Liu, Y., Wang, P., (Slade), Xue, S., Yu, Q., Li, Q.: Road life-cycle carbon dioxide emissions and emission reduction technologies: a review. J. Traffic Transp. Eng. (English Edition). **9**, 532–555 (2022)
4. Ghufran, M., Khan, K.I.A., Ullah, F., Nasir, A.R., Al Alahmadi, A.A., Alzaed, A.N., Alwetaishi, M.: Circular economy in the construction industry: a step towards sustainable development. Buildings. **12** (2022). https://www.sciencedirect.com/science/article/pii/S0959652620310933?via%3Dihub
5. Barbieri, D.M., Lou, B., Wang, F., Hoff, I., Wu, S., Li, J., Vignisdottir, H.R., Bohne, R.A., Anastasio, S., Kristensen, T.: Assessment of carbon dioxide emissions during production, construction and use stages of asphalt pavements. Transp. Res. Interdiscip. Perspect. **11** (2021). https://www.sciencedirect.com/science/article/pii/S0959652620310933?via%3Dihub
6. Hasan, U., Whyte, A., Al Jassmi, H.: Critical review and methodological issues in integrated life-cycle analysis on road networks. J. Clean. Prod. **206**, 541–558 (2019)
7. Rise, T., Alnæs, L., Rambæk, I.: Kortreist stein - Oppnådde resultater (2016–2019). SINTEF akademisk forlag, Trondheim (2019)
8. Mhatre, P., Gedam, V., Unnikrishnan, S., Verma, S.: Circular economy in built environment – literature review and theory development. J. Build. Eng. **35** (2021). https://www.sciencedirect.com/science/article/pii/S0959652620310933?via%3Dihub

9. Munaro, M.R., Tavares, S.F.: A review on barriers, drivers, and stakeholders towards the circular economy: the construction sector perspective. Clean. Responsib. Consum. **8** (2023). https://www.sciencedirect.com/science/article/pii/S0959652620310933?via%3Dihub

10. Ababio, B.K., Lu, W.: Barriers and enablers of circular economy in construction: a multi-system perspective towards the development of a practical framework. Constr. Manag. Econ. **41**, 3–21 (2023)

11. Adabre, M.A., Chan, A.P.C., Darko, A., Hosseini, M.R.: Facilitating a transition to a circular economy in construction projects: intermediate theoretical models based on the theory of planned behaviour. Build. Res. Inf. **51**, 85–104 (2023)

12. Mantalovas, K., Di Mino, G., Del Barco Carrion, A.J., Keijzer, E., Kalman, B., Parry, T., Presti, D.L.: European national road authorities and circular economy: an insight into their approaches. Sustainability. **12** (2020). https://www.sciencedirect.com/science/article/pii/S0959652620310 933?via%3Dihub

A Transition to Sustainable Built Environment: A Framework for Modular Building Construction Designed for Disassembly

Bilkisu Ali-Gombe ⓘ, Serik Tokbolat ⓘ, and Jon Mckechnie ⓘ

Contents

1 Introduction

By substituting the traditional linear economic system of raw material usage, a circular economy provides good alternative to the current linear system with a focus on efficient use of resources at all stages of products life cycle where consumption is kept within a circular and technological cycle [1]. Circular economy is a system with a regenerative approach to resource consumption, allowing extended life span and multiple reuse of products [2]. The circular strategies are typically considered in the R frameworks derived from the 3R's strategies, the 9R's as presented by [3] and extended 10R's [4]. The strategies are prioritised upon on the extent in which the value of product is maintained, this implies that keeping a product in its original form results in better resource efficiency than downcycling material for recycling [5], which leads to loss of greater value coupled with required energy and resource for the recycling process. Therefore, the main goal of circularity

B. Ali-Gombe (✉) · S. Tokbolat · J. Mckechnie
Faculty of Engineering, University of Nottingham, Nottingham, UK
e-mail: bilkisu.ali-gombe@nottingham.ac.uk

© The Author(s) 2025
M. Kioumarsi, B. Shafei (eds.), *The 1st International Conference on Net-Zero Built Environment*, Lecture Notes in Civil Engineering 237,
https://doi.org/10.1007/978-3-031-69626-8_132

in products is to preserve for as long as possible material and components with highest value, with subsequent extension of life of product achieved through repair, refurbishment and remanufacturing [6]. Circular strategies with focus on retaining maximum resource value are focal point in the EU environmental policies [7].

The built environment consumes almost a half of material ever used in human history [8], thereby making buildings at the end of life a likely source of material stock to serve future demand [9]. Previous sustainability practices have fallen short when it comes to effectively handling the end-of-life phase of resources. Hence, most of buildings materials are disposed to the landfills at the end of life because most of the materials used do not have any reuse potential [10]. Therefore, the end of life of buildings becomes the stage with most environmental impact with construction and demolition waste constituting over 25% of land fill waste [11]. The construction industry does not have a clear approach to disassembly of buildings, with primary emphasis on recycling rather than embracing a comprehensive approach that could yield more substantial environmental benefits [12]. The reuse of building components from existing built assets has been recognised as an effective approach from a multiple life cycle perspective to improve waste elimination and resource efficiency within the concept of circular economy in the construction industry [13, 14].

Modular design is an integral method or approach for product designed with multiple lifecycle [15], to enhance recovery and reuse of material and products [15]. Modular design and construction have been adopted in the construction industry, but most are not aimed at multiple life cycle. This is largely attributed to the complex nature of construction products as not all materials and components are suitable for multiple life cycle design [16]. Products that are targeted for multiple life cycles requires material with high inherent value capable of easy collection at the end of life, standardised components, disassembly and reassembly potentials, and modularity function, among others [17]. Design for disassembly also referred to design for deconstruction in the construction industry supports a sustainable reuse and remanufacturing of components at the end of life through effective disassembly or deconstruction [2]. An important element in buildings designed for disassembly or deconstruction to enable reuse at the end of life is the use of prefabricated modular assemblies with reversible connections rather than composite systems that will necessitate demolishing at the end of life [18]. Designing modular buildings for disassembly and reuse can potentially reduce waste and material consumption alongside other sustainability impacts of construction projects in line with the circular economy framework [19]. In recent times, there is an increase in modular approaches in building projects, largely influenced by government policy [20], rather than a systematic performance evaluation of environmental, economic and social impacts. Therefore, to aid this transition, a comprehensive framework to support a holistic evaluation of impacts with the three-dimensional sustainability (environmental, economic and social) approach is required for an effective decision-making support.

2 Methodology

In this study, a systematic review was conducted to analyse the existing literature on modular construction and design for disassembly. The methodology shown in Fig. 1 consists of the following phases: (i) defining the research question, (ii) determining the search process, (iii) defining the article selection criteria, (iv) result classification and analysis, and (v) developing the proposed framework.

2.1   Search Process and Selection

The study was conducted to identify the current trends in studies and application of modular buildings designed for disassembly and identify the sustainability goals as well as various indicators applicable to the studies. The evaluation of various types of material inputs for these strategies was conducted to identify the extent to which it

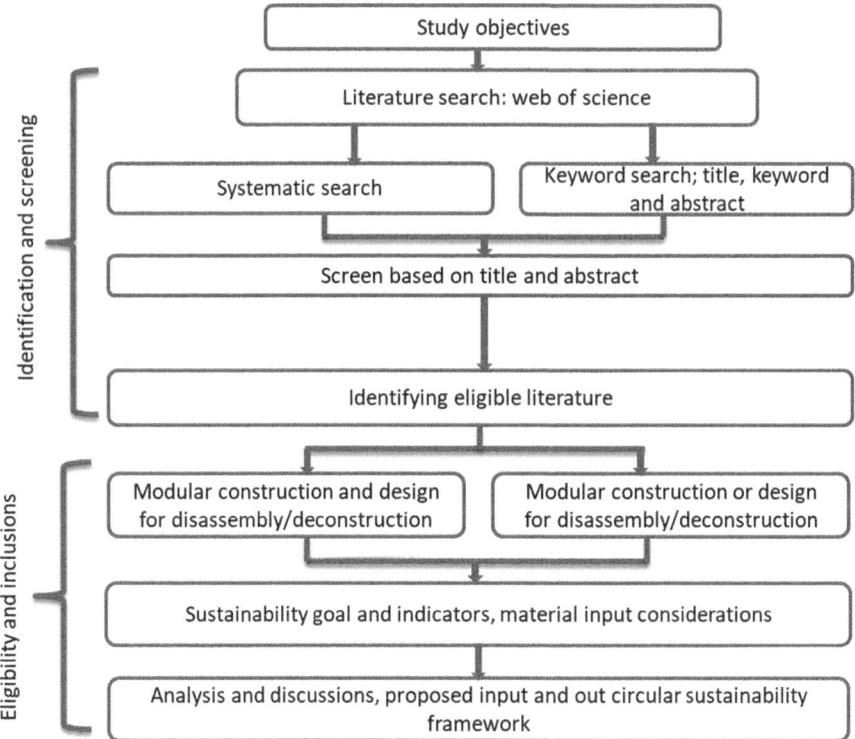

Fig. 1 Adopted methodology for the study

affects the sustainability performance. The study further evaluates the extent to which the existing literature addresses these circular approaches from the three-dimensional sustainability perspective.

Relevant key words for the literature search were developed based on preliminary search in the subject area from the web of science data base in the field of construction building technology base on title, keyword and abstract. The theme based specific key words are circular*, modular* build* or construct*, 'design for disassembly' or 'design for deconstruction'. The search yielded a lot of studies in general fields, but 35 research articles were considered in the field of construction building technologies. The selection of relevant article was based on certain criteria; the study most have a case study application/analysis, framework, or experimental test for at least design for disassembly/deconstruction or modular design/construction processes or both.

3 Result and Analysis

3.1 Temporal and Geospatial Trends

The study reveals a rapid increase in studies and applications of modular buildings and design for disassembly from 2003 to 2023 (see Fig. 2). This can be largely attributed to the rapid increase on the need to develop effective circular solutions to the resource-intensive construction industry, as the industry has a great potential for implementation of these solutions to solve the problems of resource efficiency, waste management, energy and emissions problems.

From the review of literature highlighted in Fig. 3, more studies and applications are found in Europe with 51%, followed by North America with 16%, 14% in Asia and 8%, 5%, 3% and 3% in Australia, Africa, South America and Middle East, respectively. Although there is a rapid increase in the studies in the area of circular modular buildings designed for disassembly, this study reveals significant regional

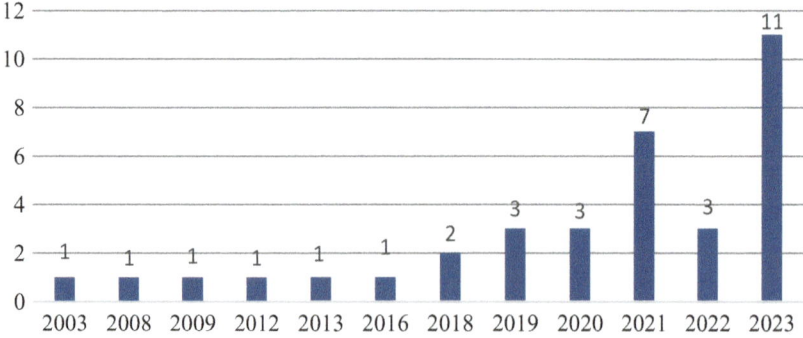

Fig. 2 Temporal trend

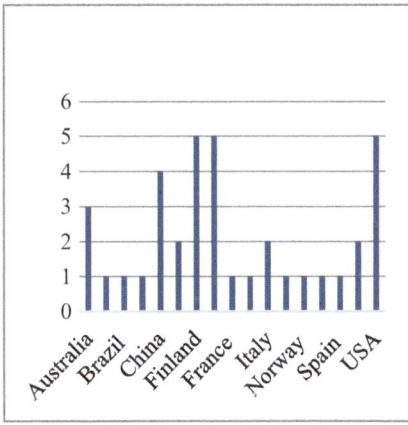
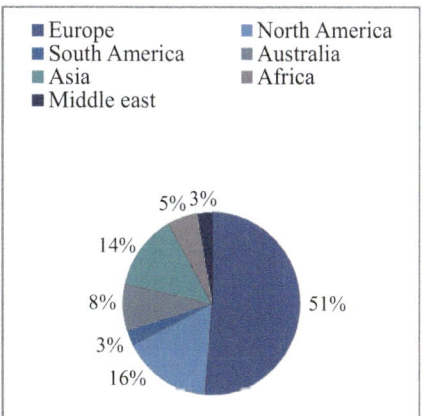

Fig. 3 Geospatial trend

imbalance with fewer research and applications in developing countries (Africa, Asia, Middle East) where infrastructure deficits and construction demand are high. The European countries have already established national circular specific legislations such as the EU-Action plan which supports a top-down implementation of these strategies. Whereas in the USA and other parts of the developing world are yet to develop such a top-down approach. Hence, the bottom-up efforts from stakeholders is the main driving force towards this progress [21].

3.2 Sustainability Goals and Indicators

The analysis of literature shows that studies in modular construction and design for disassembly/deconstruction mostly focus on environmental sustainability (ENV) as the main goal, highlighting indicators such emissions and energy reduction [18, 22, 23], raw material and waste reduction and material circularity [24–26]. Some studies featured a combination of environmental and economic (ECO) goals with cost performance being the main economic indicator [20, 27, 28]. Few studies embraces the triple bottom line sustainability approach [20] having environmental, economic and social (SOC) aspects of energy and waste reduction, cost, as well as safety and welfare. In summary, 71% of the reviewed studies focused on environmental aspects as the only sustainability goal, 26% featured environmental and economic goals, and only 3% addresses all the three aspects of environmental, economic and social impacts as shown in Fig. 4. Sustainability goals are strongly interconnected, and hence, the achievement of sustainable development in the built environment requires a holistic three dimensional approach impact evaluation and determination of appropriate solution.

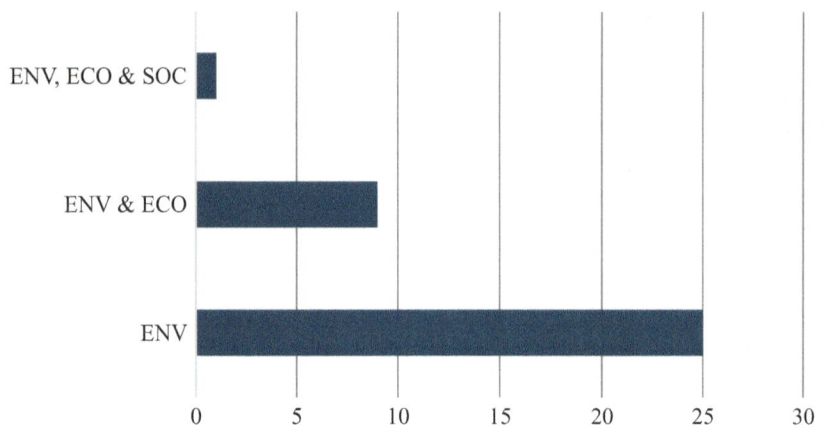

Fig. 4 Sustainability goals

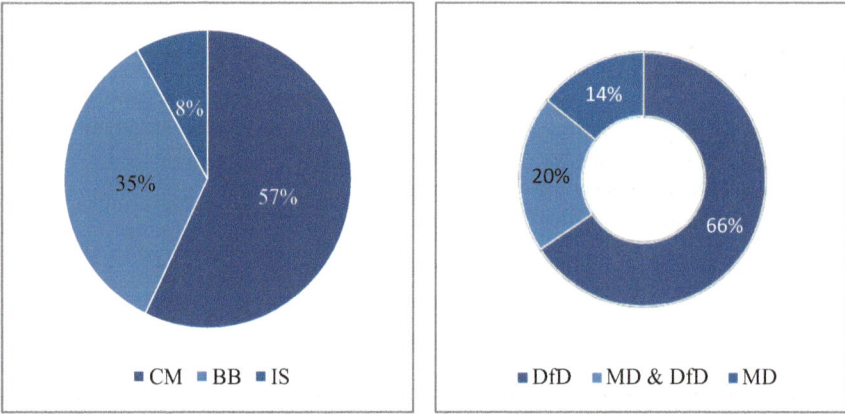

Fig. 5 Material inputs (left), design and construction strategies (right)

3.3 Material Inputs for Modular Buildings and Design for Disassembly

The analysis of the review reveals that the concepts of modular building construction and design for disassembly/deconstruction are often considered separately in most studies although the application of the two concepts presents more benefits from circularity and sustainability perspectives. Design for disassembly/deconstruction application remains the most popular from multiple lifecycle consideration representing 66% of the studies. Modular design/construction application present 14% followed by the combination of both presenting in 20% of the studies.

The result of this review showcases several types of material inputs (see Fig. 5), some of which highlighted benefits from reuse and circular perspective [25, 26, 29, 30], others from pollution, waste elimination and health safety indicators [20, 31–33].

3.4 Conventional Materials (CM)

The use of conventional building material such as concrete and steel used for modular prefabricated construction provided as reinforced concrete modules or steel framed structures, offers great resource efficiency, durability and reuse/recyclability potentials of buildings [20, 34]. Concrete and steel composite building elements base on experimental tests also provides effective deconstructability potentials [21], as well as material and component recoverability during disassembly at the end of building lifecycle [25]. Brick as a building component has a great reusability and flexibility potential with effective design and construction targeted at multiple life cycles [35]. High durability characteristic of bricks as a building material gives a good environmental payback over a period of life cycles with high potential of upcycling for multiple life cycles [35].

3.5 Biobased Material (BB)

The use of wood materials to promote reuse and increase lifetime of elements in panelised timber frames with reversible connections in prefabricated modular building assemblies enables effective reuse and adaptability of panels, with emission and raw material reduction and enhanced circularity [26, 36]. The use of wood/timber concrete composite elements designed for disassembly supports circularity from reuse, remanufacturing and recycling perspective [37, 38]. Biobased material used as design inputs at the early design stages provides a great control over the environmental impact of the building material as well as the possible end of life scenarios [39]. The use of other forms of natural materials such as stones, sand, clay, lime and straw has mostly been proven to have high environmental and economic benefits. Stone and gabions used as structural element designed for deconstruction with simple construction methods also provided high environmental and economic benefits [33].

3.6 Industrial Symbiosis Material (IS)

Material from industrial symbiosis, i.e. materials from other industrial waste/by products as building material presented in studies can lead to improved circularity across all industries. The use of recycled aggregate designed from reuse and recycling perspectives in structural elements [22, 30] offers sustainable solutions to concrete waste generation and embodied carbon emission over lifecycle. The use of organic agricultural waste like rice husk as building materials in modular green wall enhances circularity goals as well as reduced environmental impacts and offer economic benefits [40]. Also sheet metal waste from automobile bodies production adopted for modular living walls system results in positive effect with reduced greenhouse gases emission and energy usage from traditional metal recycling processes [41]. Other industrial symbiosis application includes partition walls from recycled plastic bottles, recycled carton pipe from waste stream [39] and many more offers long term circular and sustainable solutions.

4   The Proposed Framework

In the quest to improve sustainability performance of buildings, a holistic evaluation of impacts in all phases of lifecycle from all the three sustainability dimensions (environmental, economic and social) becomes necessary for effective decision-making. Previous sustainability studies in the built environment have fallen short of providing a comprehensive evaluation from all the three sustainability aspects. The proposed input–output circular sustainability framework shown in Fig. 6 supports holistic evaluation from the triple bottom line sustainability point of view as well as circularity assessment to ensure effective plan is made for end of life the building. The proposed framework put forward material selection as the main input driver at early design stage. Typically, material consideration follows design stage thereby making it more complex to implement the circular strategies.

The framework gives preference to low carbon building materials to achieve reduced output of greenhouse gas emission and waste, considering biobased material and reusable material from the building industry and cross other industries. The framework presents circular design and construction strategies (design for disassembly and modular construction) to ensure adaptability for extended lifecycle and suitable end-of-life scenario to achieve highest level of value through recovery and reuse of building components and modules.

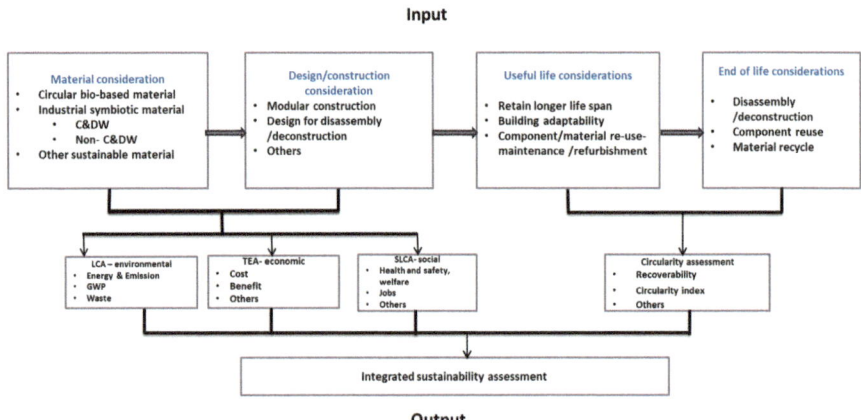

Fig. 6 An input–output circular sustainability framework

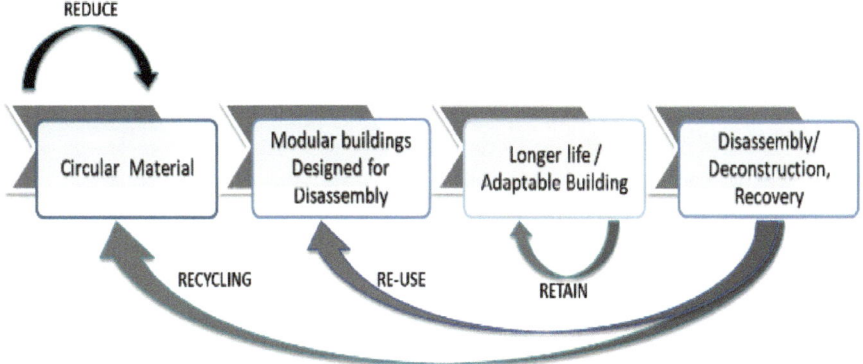

Fig. 7 An input–output circular sustainability framework within R- strategies

From the proposed framework, the adoption of circular low carbon reused/upcycled or recycled material will potentially reduce raw material usage and reduce embodied carbon. The design and construction strategies can support reuse at the end of life and retain longer building life through adaptability as shown in Fig. 7.

5 Conclusion

This study presented a systematic literature review on modular construction and design for disassembly in relation to material inputs and sustainability targets (environmental, economic, social). The outcome of the review shows that studies within this scope mainly focused on addressing environmental sustainability issues then environmental and economic. Few studies address all the three sustainability dimensions, and hence, the social sustainability aspect largely suffers neglect in research. Sustainable building construction presents numerous social sustainability benefits including up to 50% and 80% risk reduction in concrete and steel modular building construction respectively [20]. Material consideration is an integral aspect in achieving improved ecological impacts of building from reduce, reuse and recycle perspectives [39]. Modular construction and design for disassembly applications mostly adopts conventional material inputs rather than more sustainable circular material. Designing building with circular material reduces raw material inputs and emission output and hence, supports effective circularity and successful transition to sustainable built environment and net-zero. Modular building and design for disassembly are promising circular concepts and they are becoming popular in research and application, but mostly in Europe and North America, therefore this present significant regional imbalance. Despite the potential benefits of adopting modular buildings design for disassembly, the concepts are mostly applied separately. Hence the study presents a comprehensive circular sustainability evaluation framework, with triple bottom line sustainability approach. Sustainability performance is site specific, therefore critical evaluation of environmental alongside socio economic impacts of circular solutions will be beneficial especially in developing countries with high demand of infrastructure projects. Hence, further studies can be conducted to implement the proposed framework at early design stage for effective decision-making support.

References

1. Kayaçetin, N.C., Verdoodt, S., Lefevre, L., Versele, A.: Integrated decision support for embodied impact assessment of circular and bio-based building components. J. Build. Eng. **63**, 105427 (2023)
2. Cruz Rios, F., Grau, D., Chong, W.K.: Reusing exterior wall framing systems: a cradle-to-cradle comparative life cycle assessment. Waste Manag. **94**, 120–135 (2019)
3. Potting, J., Hekkert, M., Worrell, E., Hanemaaijer, A.: Circular Economy: Measuring Innovation in the Product Chain. PBL Netherlands Assessment Agency (2017)
4. Reike, D., Vermeulen, W.J.V., Witjes, S.: The circular economy: new or refurbished as CE 3.0? — exploring controversies in the conceptualization of the circular economy through a focus on history and resource value retention options. Resour. Conserv. Recycl. **135**, 246–264 (2018)
5. Dokter, G.: Circular Design in Practice Towards a Co-created Circular Economy Through Design. Chalmers University of Technology, Gothenburg (2021)
6. Ghoreishi, M., Happonen, A.: New promises AI brings into circular economy accelerated product design: a review on supporting literature. E3S Web Conf. **158**, 06002 (2020)
7. Circular Economy Action Plan (2020) Environment. https://environment.ec.europa.eu/strategy/circular-economy-action-plan_en. Accessed 29 Apr 2024
8. Sanchez, B., Haas, C.: A novel selective disassembly sequence planning method for adaptive reuse of buildings. J. Clean. Prod. **183**, 998–1010 (2018)
9. Eberhardt, L.C.M., Birgisdóttir, H., Birkved, M.: Life cycle assessment of a Danish office building designed for disassembly. Build. Res. Inf. **47**(6), 666–680 (2019)
10. Akanbi, L.A., et al.: Salvaging building materials in a circular economy: a BIM-based whole-life performance estimator. Resour. Conserv. Recycl. **129**, 175–186 (2018)
11. Kabirifar, K., Mojtahedi, M., Wang, C., Tam, V.W.Y.: Construction and demolition waste management contributing factors coupled with reduce, reuse, and recycle strategies for effective waste management: a review. J. Clean. Prod. **263**, 121265 (2020). Elsevier Ltd
12. Minunno, R., O'Grady, T., Morrison, G.M., Gruner, R.L.: Exploring environmental benefits of reuse and recycle practices: a circular economy case study of a modular building. Resour. Conserv. Recycl. **160**, 104855 (2020)
13. Cruz Rios, F., Grau, D., Bilec, M.: Barriers and enablers to circular building design in the US: an empirical study. J. Constr. Eng. Manag. **147**(10), 04021117 (2021)
14. Witjes, S., Lozano, R.: Towards a more Circular Economy: proposing a framework linking sustainable public procurement and sustainable business models. Resour. Conserv. Recycl. **112**, 37–44 (2016)
15. Asif, F.M.A., Roci, M., Lieder, M., Rashid, A., Mihelič, A., Kotnik, S.: A methodological approach to design products for multiple lifecycles in the context of circular manufacturing systems. J. Clean. Prod. **296**, 126534 (2021)
16. Yang, Y., Guan, J., Nwaogu, J.M., Chan, A.P.C., Lin Chi, H., Luk, C.W.H.: Attaining higher levels of circularity in construction: scientometric review and cross-industry exploration. J. Clean. Prod. **375**, 133934. Elsevier Ltd (2022)
17. Go, T.F., Wahab, D.A., Hishamuddin, H.: Multiple generation life-cycles for product sustainability: the way forward. J. Clean. Prod. **95**, 16–29 (2015)

18. Eckelman, M.J., Brown, C., Troup, L.N., Wang, L., Webster, M.D., Hajjar, J.F.: Life cycle energy and environmental benefits of novel design-for-deconstruction structural systems in steel buildings. Build. Environ. **143**, 421–430 (2018)
19. O'Grady, T.M., Minunno, R., Chong, H.Y., Morrison, G.M.: Interconnections: an analysis of disassemblable building connection systems towards a circular economy. Buildings. **11**(11), 535 (2021)
20. Pan, W., Zhang, Z.: Benchmarking the sustainability of concrete and steel modular construction for buildings in urban development. Sustain. Cities Soc. **90**, 104400 (2023)
21. Wang, L., Webster, M.D., Hajjar, J.F.: Design for deconstruction using sustainable composite beams with precast concrete planks and clamping connectors. J. Struct. Eng. **146**(8), 04020158 (2020)
22. Xiao, J., Chen, Z., Ding, T., Xia, B.: Effect of recycled aggregate concrete on the seismic behavior of DfD beam-column joints under cyclic loading. Adv. Struct. Eng. **24**(8), 1709–1723 (2021)
23. Joensuu, T., Leino, R., Heinonen, J., Saari, A.: Developing buildings' life cycle assessment in circular economy-comparing methods for assessing carbon footprint of reusable components. Sustain. Cities Soc. **77**, 103499 (2022)
24. Atta, I., Bakhoum, E.S., Marzouk, M.M.: Digitizing material passport for sustainable construction projects using BIM. J. Build. Eng. **43**, 103233 (2021)
25. Denis, F., Vandervaeren, C., De Temmerman, N.: Using network analysis and BIM to quantify the impact of Design for Disassembly. Buildings. **8**(8), 113 (2018)
26. Incelli, F., Cardellicchio, L., Rossetti, M.: Circularity indicators as a design tool for design and construction strategies in architecture. Buildings. **13**(7), 1706 (2023)
27. Antwi-Afari, P., Ng, S.T., Chen, J., Oluleye, B.I., Antwi-Afari, M.F., Ababio, B.K.: Enhancing life cycle assessment for circular economy measurement of different case scenarios of modular steel slab. Build. Environ. **239**, 110411 (2023)
28. Pereiro, X., Cabaleiro, M., Conde, B., Riveiro, B.: BIM methodology for cost analysis, sustainability, and management of steel structures with reconfigurable joints for industrial structures. J. Build. Eng. **77**, 107443 (2023)
29. Laasonen, S., Pajunen, S.: Assessment of load-bearing timber elements for the design for disassembly. Buildings. **13**(7), 1878 (2023)
30. Ding, T., Xiao, J., Chen, E., Khan, A.-R.: Experimental study of the seismic performance of concrete beam-column frame joints with DfD connections. J. Struct. Eng. **146**(4), 04020036 (2020)
31. Lehmann, S.: Low carbon construction systems using prefabricated engineered solid wood panels for urban infill to significantly reduce greenhouse gas emissions. Sustain. Cities Soc. **6**(1), 57–67 (2013)
32. Hendriks, C.F., Janssen, G.M.T.: Use of recycled materials in constructions. Mater. Struct./ Matdriaux et Constr. **36**, 604–608 (2003)
33. Conti, L., Barbari, M., Monti, M.: Design of sustainable agricultural buildings: a case study of a wine cellar in Tuscany, Italy. Buildings. **6**(2), 17 (2016)
34. Olipitz, M.: Hochleistungsfertigteile für den Hochbau: Ein Beitrag zum disruptiven Wandel bei mineralischen Bauteilen. Beton- Stahlbetonbau. **118**(3), 201–213 (2023)
35. Nordby, A.S., Berge, B., Hakonsen, F., Hestnes, A.G.: Criteria for salvageability: the reuse of bricks. Build. Res. Inf. **37**(1), 55–67 (2009)
36. Yan, Z., Ottenhaus, L.M., Leardini, P., Jockwer, R.: Performance of reversible timber connections in Australian light timber framed panelised construction. J. Build. Eng. **61**, 105244 (2022)
37. Derikvand, M., Fink, G.: Design for deconstruction: benefits, challenges, and outlook for timber–concrete composite floors. Buildings. **13**(7), 1754 (2023)
38. Eslami, H., Jayasinghe, L.B., Waldmann, D.: Experimental and numerical investigation of a novel demountable timber–concrete composite floor. Buildings. **13**(7), 1763 (2023)

39. Dahy, H.: 'Materials as a design tool' design philosophy applied in three innovative research pavilions out of sustainable building materials with controlled end-of-life scenarios. Buildings. **9**(3), 64 (2019)
40. De Lucia, M., Treves, A., Comino, E.: Rice husk and thermal comfort: design and evaluation of indoor modular green walls. Dev. Built Environ. **6**, 100043 (2021)
41. Kio, P., Ali, A.K.: In situ experimental evaluation of a novel modular living wall system for industrial symbiosis. Energy Build. **252**, 111405 (2021)

Towards Net-Zero Construction Projects by Applying BIM-Enabled Circular Economy

Ana Julie Foseid Bjerke and Omar Amoudi

Contents

1 Background

The built environment is currently in a dire moment of change to achieve further sustainable solutions and to reduce the carbon footprint [1–3]. The Linear economic model has been predominantly adopted in the construction industry, in addition to the introduction of recycling of construction waste. In alignment with the findings in UNEP [4] where the built environment was found to account for 39% of the global carbon emissions. Giorgi et al. [5] stated that linear economic models and recycling are no longer a solution due to limited resources available. The concept of the CE is relatively new and generating further recognition across many respective fields, including the construction industry [6]. Some existing challenges interluding the further adoption of CE principles within the AEC industry include but not limited to;

A. J. F. Bjerke (✉) · O. Amoudi
Oxford Brookes University, Oxford, UK
e-mail: 19222475@brookes.ac.uk; oamoudi@brookes.ac.uk

© The Author(s) 2025
M. Kioumarsi, B. Shafei (eds.), *The 1st International Conference on Net-Zero Built Environment*, Lecture Notes in Civil Engineering 237,
https://doi.org/10.1007/978-3-031-69626-8_133

the traditional construction and design strategies, construction and demolition waste management, poor supply chain management, Ignorance of End-of-Life (EOL) principles, lack of regulations and strategies to encourage the adoption of the CE principles, lack of collaboration, and lack of information exchange of CE practices (Al [7, 8]).

Building Information Modelling has been a vital element in the construction industry allowing for improved communication, collaboration, and interoperability in construction projects. Digitalisation, especially BIM as a promising tool, facilitates the collaboration between project teams to incorporate and manage information of building processes that may lead towards more sustainable construction projects, and reinvents contemporary design [9]. This paper aims to assess the potential utilisation of BIM as an enabler to achieve further implementation of CE principles within the built environment. BIM's role, when aiming to achieve CE-related goals within construction projects, is to be clarified, showing the emerging challenges and barriers, recommendations, and key findings.

2   Concept of Circular Economy

Ellen MacArthur Foundation [10] founded the 3Rs principles that describe circular economy processes to facilitate the transition towards more circularity: reduce, recycle, and reuse. Stated in the works of Ellen MacArthur Foundation [11], CE is a regenerative economy where the target is to retain the highest values for products and materials in a permanent circular system for their optimal reuse, recycling, remanufacturing, and refurbishment. In general, CE illustrates the impact on the economy that natural resources sustain, and give further insight into waste, production, economic systems, and waste within the construction industry [12]. Xue et al. [13] proposed the alignment of CE principles and the technical requirements within the construction industry in terms of resource mass consumption. Similarly, Suárez-Eiroa et al. [14] found supporting evidence that the adoption of CE concept maintains the economic growth, creates further job opportunities, and reduces CO_2-emissions. Conclusively, CE maintains and retains added high value in materials, products, and components in a restorative and regenerative design system until their EOL [15, 16].

2.1   Digitalisation and Enabling CE Adoption

Digitalisation is considered as one of the key enablers of CE as it offers improved intelligence, information management, design and construction process visibility, and condition of assets and products. This in turn assists in managing resources and the facility information and reduces the uses of resources [17, 18]. Another key enabler is the 3DR index coined by O'Grady et al. [19]: Disassembly,

Deconstruction, Resilience. Enabling the principles of 3DR generates increased knowledge on implementation of CE in the context of transforming waste into new suitable resources. Furthermore, O'Grady et al. [19] found emerging benefits in all stages within a project's lifecycle when 3DR is applied. Digitalisation could hold a significant role in empowering the 3DRs, because these principles require strong information management which could be missed or mis-coordinated if the traditional tools/practices used for managing the information is required. Depending on the traditional tool of storing and managing information could lead to more errors in the 3DRs processes, that results in extra cost and more waste.

2.2 Challenges Facing CE

The biggest challenges facing the adoption of CE in the construction industry, is the lack of knowledge and awareness among project parties, lack of clear design incentives for reuse of assets and materials' EOL, and lack of the concept of design for disassembly [20–22]. Purchase et al. [23] found other challenges prohibiting the further adoption of CE principles that include poor governance, high cost, lack of awareness and poor information management, and lack of clear specifications on recycled assets and materials.

The existing structure and characteristics of the construction industry were mentioned by Adams et al. [20] as leading to the fragmentation within the supply chain, which is again one of the key barriers facing the CE implementation [24]. Fulford and Standing [25] found that the efficiency and productivity are severely impacted by the supply chain fragmentation within the construction industry and the lack of standardisation. To obtain a comprehensive understanding of standardisation and apply it properly into projects, BIM emerges as a tool to meet the continuous improvement that standardisation acquires. Bradley et al. [26] found BIM use is commonly applied for any standards in projects where the BIM model is utilised for design purposes, rather than using 4D and 5D BIM model for cost and programme management. It is a requirement in the construction industry for standardisation regarding BIM, which is found within the exchange information between used software applications [27]. BIM's flexibility could provide the tools to monitor and manage the required sustainability measures in terms of waste reduction during the building's entire lifespan [28].

2.3 Synergies Between BIM and CE

The current use of BIM during an asset's life cycle could enhance the possibility of recovering the asset's history in the events of partial or whole destruction in an asset refurbishment, which is one of the current existing synergies between BIM and CE [29]. It is argued by Göswein et al. [30] that dependent factors already in use within the built environment, such as facility and design management, resource management, and waste management, altogether sustain the synergy currently between BIM and CE. BIM utilisation for waste management is already enabling some of these factors such as quantity take-off, site utilisation planning, 3D coordination and planning, digital fabrication, phase planning, and design review [31]. Performing a BIM-Deconstructability Assessment Score allows for the further enabling of BIM as a tool for Circularity to prolong the EOL of materials, additionally with the use of as-built BIM models for building operations and maintenance using laser scanners [32, 33]. Previous research could identify the as-built BIM models as tools for cost analysis for demolitions plans in selected waste systems including design for disassembly, assets, and waste management [34, 35].

2.4 Barriers and Opportunities of Current BIM-Enabled CE Principles

To help the Built Environment shift from the linear economic model to the circular economic model, Hai et al. [36] suggested assessing benefits such as increased CE awareness and improved recycling rates to aid this shift. Enhancing the building's performance by encouraging further usage of BIM to match EOL assessments of projects, will reduce waste and help preserve the embodied energy in construction projects [37]. Xue et al. [13] found in general that the widespread use of BIM and LCA within the construction industry to not being fully applied. Existing concerns regarding BIM-enabled CE revolve around the whether industry professionals obtain sufficient BIM knowledge, BIM competency, and BIM software training, and the BIM documentation and models in cloud-share collaboration [6]. However, Göswein et al. [30] purport that the main challenges facing BIM-enabled CE are lack of data, poor procedures, and interoperability issues. It is clear that several studies referred to the potential relations between BIM features and the principles of the CE. This will be further explored in this paper.

3 Research Methodology

The research methodology adopted in this study was a simple mixed methodology consisting of a quantitative and qualitative approach to obtain the primary data through two stages: (1) A questionnaire survey and (2) An interview. A pilot study was carried out to validate both the questionnaire survey questions and interview questions. The questionnaire survey was employed to obtain professionals opinion on the current use of BIM in the industry, its use with link to sustainability practices, and its potential use to enhance CE principles. Thirty-seven questionnaire responses were received where the 3 biggest positions at 20% each were identified as working as BIM Modeller or Coordinator, Architects, and Engineers as shown in Fig. 1.

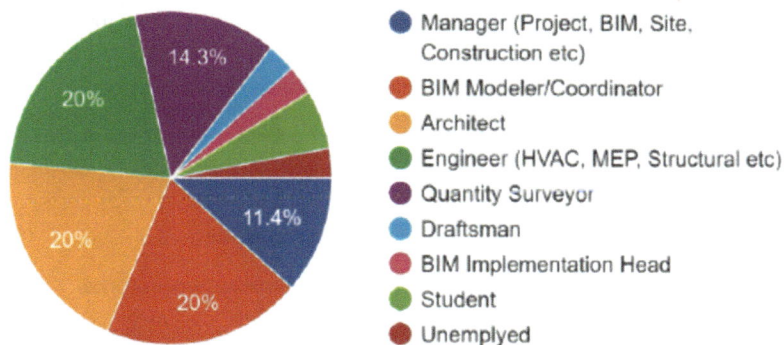

Fig. 1 Participants' current role within their organisations

Then, five interviews were carried out with experts (i.e. 2 BIM Managers, 2 BIM Coordinators, and 1 Asset Manager) to explore in depth the potential utilisation of BIM in facilitating the adoption of CE in the AEC industry. The collected data from the questionnaire survey was statistically analysed to demonstrate the significance of various variables in each theme. Then, thematic analysis was employed to analyse the data collected from interviews. Then, a conceptual framework is produced to show the potentials of BIM in facilitating the CE in the AEC industry.

4 Data Analysis and Findings

4.1 Current BIM Practices

Invaluable BIM insights were provided by the participants as a total of 62% of the participants were revealed to be directly involved in the daily utilisation of BIM, and they could reveal the main utilisation of BIM is for design purposes as shown in Fig. 2. However, the use of BIM for waste recycling, reuse of building components, and deconstruction receives less importance and understanding on how to use BIM for these practices.

4.2 Awareness of CE

In general, the participants' perception of the CE was for recycling and reuse of construction elements and materials. It can be seen from Fig. 3 that there is significant awareness of the participants with a score of 77% on the utilisation of the 3Rs (recycle, repurpose, reuse) which is already applied in construction projects, with the other common CE principles such as design for lessened energy consumption and prefabrication of elements.

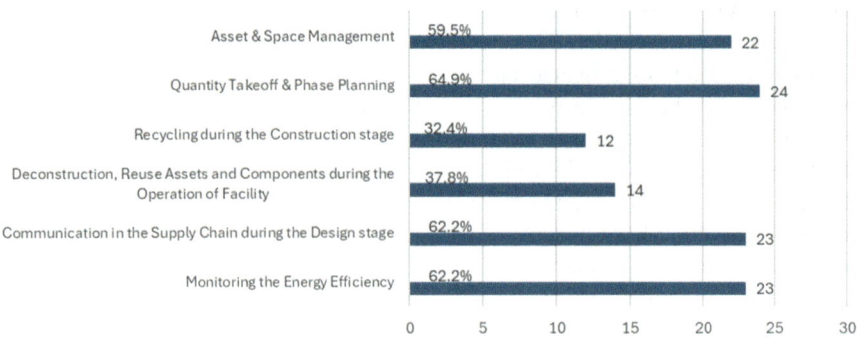

Fig. 2 Current practices of BIM in your organisation

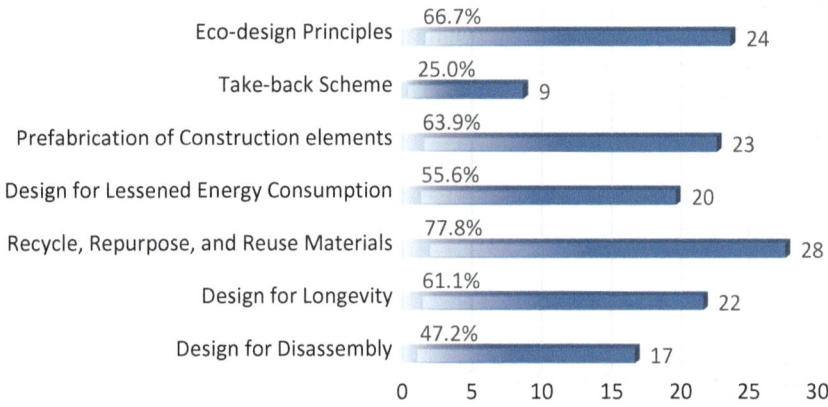

Fig. 3 Participant's awareness regarding CE principles

4.3 Perceived Benefits from Utilising BIM as a Tool for Circularity

Figure 4 demonstrates the participants' perceived benefits for construction projects adopting BIM for facilitating the CE principles include; improved interoperability and communication within the supply chain, improved life cycle cost estimation of assets, and standardisation using BIM in projects. The general recommendation is the need for a general shift in the paradigm in addition to complete EOL review when BIM is used. This is aligned with 3 out of 5 interviewees who stated that BIM has high potential in facilitation circularity as quoted by one professional: 'We've had success applying principles such as Design for Disassembly, mainly due to the

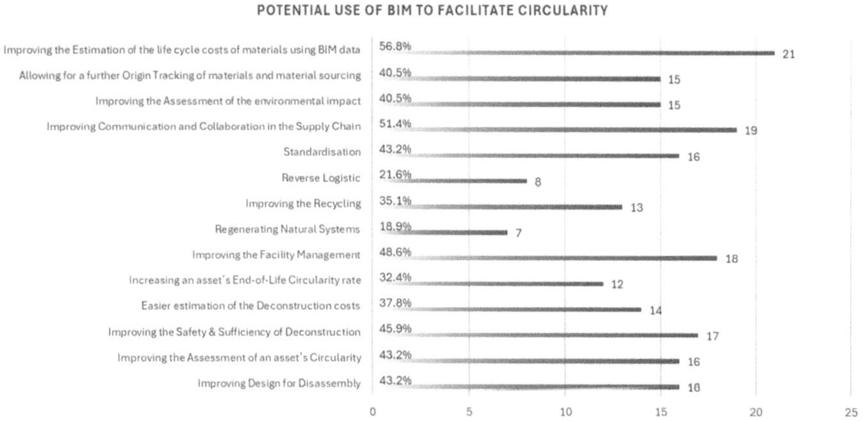

POTENTIAL USE OF BIM TO FACILITATE CIRCULARITY

Improving the Estimation of the life cycle costs of materials using BIM data	56.8%	21
Allowing for a further Origin Tracking of materials and material sourcing	40.5%	15
Improving the Assessment of the environmental impact	40.5%	15
Improving Communication and Collaboration in the Supply Chain	51.4%	19
Standardisation	43.2%	16
Reverse Logistic	21.6%	8
Improving the Recycling	35.1%	13
Regenerating Natural Systems	18.9%	7
Improving the Facility Management	48.6%	18
Increasing an asset's End-of-Life Circularity rate	32.4%	12
Easier estimation of the Deconstruction costs	37.8%	14
Improving the Safety & Sufficiency of Deconstruction	45.9%	17
Improving the Assessment of an asset's Circularity	43.2%	16
Improving Design for Disassembly	43.2%	16

Fig. 4 Potential use of BIM to facilitate circularity

central origin of data and information BIM provides'. And another quoted: *'We used BIM in a recent project to develop digital twin to optimise and stimulate the building energy performance. Due to the creation of that digital twin. It allowed us to alter the design swiftly and assess the sustainability metrics. This resulted in the achievement of a LEED Platinum certification for our organisation'*.

The results reveal there is a reasonable awareness on the potential use of BIM to facilitate circularity within the AEC industry. However, there are some barriers such as the absence of demands to apply the CE principles from clients, professional bodies, and the government. This is aligned with Charef's [6] finding. Additionally, the client was identified as holding the main responsibility for the encouragement and development for adopting CE within built environment projects. It was suggested that stakeholders might not comprehend the desired awareness and knowledge of the CE principles to enable it.

5 Discussion

BIM utilisation was identified by all interviewees to foster stakeholders' communication and collaboration that yields a sustainable work environment for improved, sustainable construction projects, which aligns with Kuzina's [38] suggestion of enabling BIM for life cycle management. 4 out of 5 of the interviewees reported on BIM generated benefits includes material and energy efficiency, energy simulation, risk analysis, cost savings, improved design and decision making. Sustainability efforts emerging from their organisations BIM use was energy analysis, data integration, resource management, and design optimisation.

In terms of stakeholder engagement and perception of BIM's sustainability potential, 80% of the interviewees could acknowledge that the perception among stakeholders varies due to differences in knowledge, willingness, and adjustment to adopt BIM completely for sustainable procedures. Among the identified barriers were legal concerns, change resistance, high cost, and complete knowledge of BIM. Designers, engineers, and architects demonstrate a high interest in enabling the complete sustainable potentials that BIM offers. One of the interviewees suggested that to combat stakeholders' reservations, BIM specialists could display the benefits to gain further support from stakeholders. One of the interviewees suggests the identified barriers could be solved by organisations introducing proper data integration and BIM training.

Sixty per cent of the interviewees agreed to the existing alignment between the CE principles and the BIM features in the construction industry that can be found in procedures such as resource efficiency, and material consumption, selection and tracking. BIM was particularly identified as an enabler for practices of Design for Disassembly (DFD), with an interviewee describing the emerging success within a project with DFD was solely due to the central origin of project information data provided by BIM. Although, another interviewee stated that BIM in 4D which is very useful for resource utilisation in various locations in construction projects, is a level of BIM maturity yet to be seen. Eighty per cent of the participants stated that increased building performance and waste reduction are the two commonly witnessed BIM enabling the CE principles. Interviewee 4 particularly stated that their use of digital twin allowed for quick design alterations and assessment of sustainability metrics yielding their organisation's a LEED Platinum certification.

A total of 60% of the interviewees identified security, management, quality, and data interoperability to be the most common areas to causes challenges when BIM was promoted for sustainability efforts and CE procedures. One interviewee stated the main issue was still resistance and sceptics among stakeholders. Interviewee 4 provided valuable insight of the further need of complete demonstration of BIM's advantages to combat these challenges, that aligned with Charef's [6] common user concern including the lack of BIM expertise and knowledge within the project team.

Applicable demonstration of BIM's benefits will gain further awareness and decrease the occurrence of challenges. This concern was discovered by Xue et al. [13] on how BIM in coordination with a LCA is not utilised to its desired outcome. Applying standardisation and BIM training was suggested by 20% of the interviewees to prevent these challenges, where standardisation was found promising in Poljanšek [27] for exchange information in the application of different software.

Emerging examples of BIM-enabled CE principles, also found by Kevin van Langen et al. [39], are reported by all interviewees, including economic growth, reduced carbon footprint, lessened energy consumption, lessened environmental impact, and reduced costs on a material, operational, and constructional level. This theme is in accordance with what was purported by Manzoor et al. [40], that BIM offers a vast selection of tools to enhance that sustainability in construction projects. Based on the above findings, in the following section a proposed conceptual framework is produced to demonstrate the potential relationship between BIM features and circularity principles within the built environment.

6 Conceptual Framework

A proposed framework presented in Fig. 5 is developed in combination with the analysis and results emerging from the above existing literature and data analysis, in order to display how the competencies of BIM can aid the further adoption of the CE principles in the BE. The various BIM competencies are highlighted in colours on the left, and the CE principles are located on the right hand side in non-coloured segments. Highlighted texts in red within the links between BIM competencies CE principles show the faciliated outcome that in the link could achieve sustainable aspects within the CE principles.

The conceptual framework in Fig. 5 sheds light on the existing gap between the competencies of BIM and the CE principles, in exhibiting an interconnection that could aid further sustainable approaches, resilient projects, and resource efficiency within the BE. BIM's data driven capacities aligns with procedures to achieve CE, and the framework can be used as a practical tool by industry professionals and stakeholders to analyse and navigate for more sustainable, net-zero carbon projects. For example, 'data and information management' as one of the key BIM

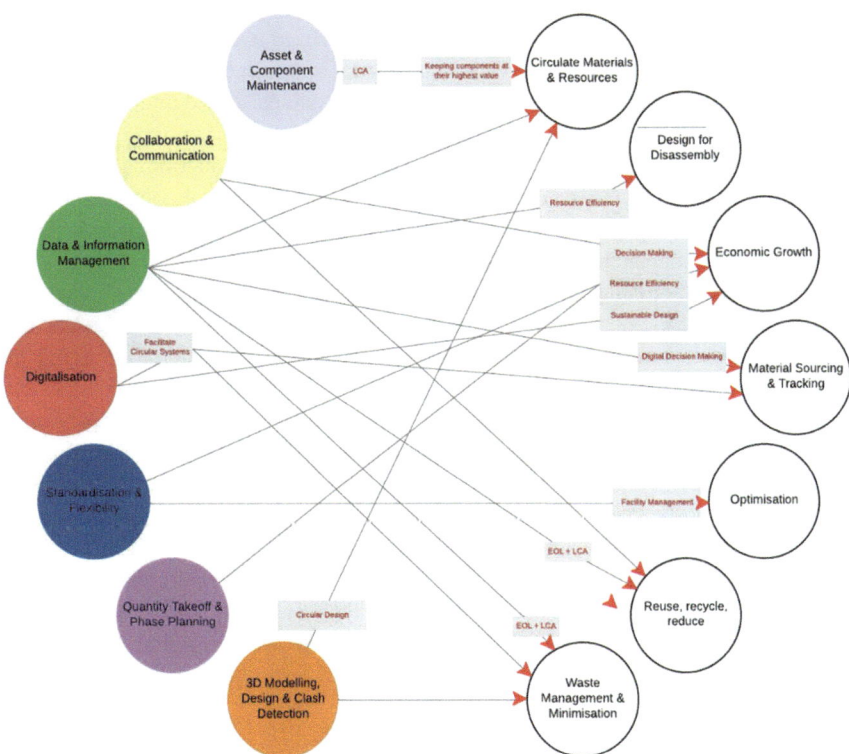

Fig. 5 Links between BIM Competencies and the CE principles

competences plays a crucial role in enabling 'Design for Disassembly', 'tracking materials and provide information on life span and circularity', and updating the data on 'reuse, recycle, reduce principles' which in turn enable waste management and minimisation. No doubt DfD and Standardisation goes hand in hand, which are essentials for circularity, reuse, repurpose and waste minimisation. The potentials of using BIM to enable circularity principles are many and difficult to cover them here.

Vital synergy, patterns, and themes between the BIM competencies and the CE principles can be identified when applying the framework which facilitates through knowledge of current synergies and challenges within the AEC sector. Critical insight emerging from the framework: circular strategies, collaboration & communication, design for adaptability, decision-making, efficiency gains, interdependence of competencies and principles, and lifecycle perspectives.

7   Conclusion

The most vital role as an enabler can be found in stakeholders and their resilience to apply BIM requirements that facilitate and foster sustainable procedures, and the shared BIM model's data allows for sufficient interoperability between stakeholders. To achieve BIM-enabled CE principles within the BE, it is crucial that stakeholders, and mainly the client, encourage various construction project participants to increase their training of BIM's use and explore its wide potentials. These findings align with both relevant literature and the suggested growing interest of BIM-enabled CE principles for preservation of embodied energy and waste reduction, and reports of increased recycling rates that occur when shifting to a Circular Economic Model.

Some of the frequently used BIM competencies emerging from the data include asset & space management, communication & collaboration in the supply chain, energy efficiency management, phase planning, and quantity take-off. These competencies enable the possibility of design and resource optimisation, deconstruction, energy analysis for operational costs and waste reduction, material sourcing and tracking, and displays the interconnection between BIM and CE principles. As seen in the conceptual framework, these links can facilitate the proper adoption of circularity in construction projects. Emerging from the qualitative data, most frequent principles of CE currently utilised within the BE include the 3Rs, design for eco-principles, design for lessened energy consumption, design for longevity, and prefabrication of construction elements. Furthermore, the findings of this research and the proposed conceptual framework could be used as a basis for further research and investigation, where some case studies could be used to demonstrate the most significant contributions of BIM towards enhancing circularity and sustainability within the built environment.

References

1. Fridrich, J., Kubečka, K.: Bim – the process of modern civil engineering in higher education. Procedia. Soc. Behav. Sci. **141**, 763–767 (2014). Available at: https://doi.org/10.1016/j.sbspro. 2014.05.134

2. Mukherjee, A., Muga, H.: An integrative framework for studying sustainable practices and its adoption in the AEC industry. a case study. J. Eng. Technol. Manag. **27**(3–4), 197–214 (2010). Available at: https://doi.org/10.1016/j.jengtecman.2010.06.006

3. Opoku, D.-G.J., Agyekum, K., Ayarkwa, J.: Drivers of environmental sustainability of construction projects: a thematic analysis of verbatim comments from built environment consultants. Int. J. Constr. Manag. **22**(6), 1033–1041 (2019). https://doi.org/10.1080/15623599.2019. 1678865

4. UNEP: Guidelines on Education Policy for Sustainable Built Environments – United Nations Environment Programme, Sustainable Buildings and Climate Initiative. [Preprint] (2017). Available at: https://wedocs.unep.org/bitstream/handle/20.500.11822/7997/-Guidelines%20on %20Education%20Policy%20for%20Sustainable%20Built%20Environments-2010993.pdf

5. Giorgi, S., Lavagna, M., Wang, K., Osmani, M., Liu, G., Campioli, A.: Drivers and barriers towards circular economy in the building sector: stakeholder interviews and analysis of five European countries policies and practices. J. Clean. Prod. **336**, 130395 (2022). Available at: https://doi.org/10.1016/j.jclepro.2022.130395

6. Charef, R.: The use of building information modelling in the circular economy context: several models and a new dimension of BIM (8D). Clean. Eng. Technol. **7**, 100414 (2022). Available at: https://doi.org/10.1016/j.clet.2022.100414

7. Hossain, M.U., Ng, S.T., Antwi-Afari, P., Amor, B.: Circular economy and the construction industry: existing trends, challenges and prospective framework for sustainable construction. Renew. Sust. Energ. Rev. **130**, 109948 (2020). Available at: https://doi.org/10.1016/j.rser.2020. 109948

8. Yu, Y., Yazan, D.M., Junjan, V., Iacob, M.E.: Circular economy in the construction industry: a review of decision support tools based on Information & Communication Technologies. J. Clean. Prod. **349**, 131335 (2022). Available at: https://doi.org/10.1016/j. jclepro.2022.131335

9. Ghaffarianhoseini, A., Tookey, J., Ghaffarianhoseini, A., Naismith, N., Azhar, S., Efimova, O., Raahemifar, K.: Building information modelling (BIM) uptake: clear benefits, understanding its implementation, risks and challenges. Renew. Sust. Energ. Rev. **75**, 1046–1053 (2017). Available at: https://doi.org/10.1016/j.rser.2016.11.083

10. Ellen MacArthur Foundation: How the Circular Economy Tackles Climate change (2019)

11. Ellen MacArthur Foundation: Growth Within: A Circular Economy Vision for a competitive Europe (2015)

12. Geissdoerfer, M., Savaget, P., Bocken, N.M., Hultink, E.J.: The circular economy – a new sustainability paradigm? J. Clean. Prod. **143**, 757–768 (2017). Available at: https://doi.org/10. 1016/j.jclepro.2016.12.048

13. Xue, K., Hossain, M.U., Liu, M., Ma, M., Zhang, Y., Hu, M., et al.: Bim integrated LCA for promoting circular economy towards sustainable construction: an analytical review. Sustain. For. **13**(3), 1310 (2021). Available at: https://doi.org/10.3390/su13031310

14. Suárez-Eiroa, B., Fernández, E., Méndez-Martínez, G., Soto-Oñate, D.: Operational principles of circular economy for sustainable development: linking theory and practice. J. Clean. Prod. **214**, 952–961 (2019). Available at: https://doi.org/10.1016/j.jclepro.2018.12.271

15. Akhimien, N.G., Latif, E., Hou, S.S.: Application of circular economy principles in buildings: a systematic review. J. Build. Eng. **38**, 102041 (2021). Available at: https://doi.org/10.1016/j. jobe.2020.102041

16. Velenturf, A.P.M., Purnell, P.: Principles for a sustainable circular economy. Sustain. Prod. Consum. **27**, 1437–1457 (2021). Available at: https://doi.org/10.1016/j.spc.2021.02.018

17. Antikainen, M., Uusitalo, T., Kivikytö-Reponen, P.: Digitalisation as an enabler of circular economy. Proc. CIRP. **73**, 45–49 (2018). Available at: https://doi.org/10.1016/j.procir.2018. 04.027

18. Glöser-Chahoud, S., Huster, S., Rosenberg, S., Baazouzi, S., Kiemel, S., Singh, S., et al.: Industrial disassembling as a key enabler of circular economy solutions for obsolete electric vehicle battery systems. Resour. Conserv. Recycl. **174**, 105735 (2021). Available at: https://doi. org/10.1016/j.resconrec.2021.105735

19. O'Grady, T., et al.: Design for disassembly, deconstruction and resilience: a circular economy index for the built environment. Resour. Conserv. Recycl. **175**, 105847 (2021). Available at: https://doi.org/10.1016/j.resconrec.2021.105847

20. Adams, K.T., Osmani, M., Thorpe, T., Thornback, J.: Circular economy in construction: current awareness, challenges and enablers. Proc. Inst. Civ. Eng. Waste Resour. Manage. **170**(1), 15–24 (2017). Available at: https://doi.org/10.1680/jwarm.16.00011

21. Al Hosni, I.S., Amoudi, O., Callaghan, N.: An exploratory on challenges of circular economy in the built environment in Oman. Proc. Inst. Civ. Eng. Manag. Procure. Law. **173**(3), 104–113 (2020)

22. Thomas, D., Amoudi, O.: Barriers and drivers associated with the adoption of circular economy in Kuwait's construction industry. The International Conference on Advancing Sustainable Future (ICASF 2023), Organised by Abu Dhabi University in the UAE from 5–6 December (2023)

23. Purchase, C.K., Al Zulayq, D.M., O'Brien, B.T., Kowalewski, M.J., Berenjian, A., Tarighaleslami, A.H., Seifan, M.: Circular economy of construction and demolition waste: a literature review on lessons, challenges, and benefits'. Materials. **15**(1), 76 (2021). https://doi. org/10.3390/ma15010076. Radaelli, C.M., Pasquier, R.: (2008) 'Conceptual issues', Europeanization, pp. 35–45. Available at: https://doi.org/10.1057/9780230584525_3

24. Shooshtarian, S. Hosseini, M., Kocaturk, T.: The circular economy in the australian built environment; the state of play and a research agenda. Deakin University (2021)

25. Fulford, R., Standing, C.: Construction industry productivity and the potential for collaborative practice. Int. J. Proj. Manag. **32**(2), 315–326 (2014). Available at: https://doi.org/10.1016/j. ijproman.2013.05.007

26. Bradley, A., Li, H., Lark, R., Dunn, S.: BIM for infrastructure: an overall review and constructor perspective. Autom. Constr. **71**, 139–152 (2016). Available at: https://doi.org/10.1016/j.autcon. 2016.08.019

27. Poljanšek, M.: Building Information Modelling (BIM) Standardization, JRC Technical Reports – European Commission [Preprint] (2017)

28. Chen, G., Chen, J., Tang, Y., Ning, Y., Li, Q.: Collaboration strategy selection in BIM-enabled construction projects: a perspective through typical collaboration profiles. Eng. Constr. Archit. Manag. **29**(7), 2689–2713 (2021). Available at: https://doi.org/10.1108/ecam-01-2021-0004

29. Charef, R., Emmitt, S.: Uses of building information modelling for overcoming barriers to a circular economy. J. Clean. Prod. **285**, 124854 (2021). Available at: https://doi.org/10.1016/j. jclepro.2020.124854

30. Göswein, V., Carvalho, S., Lorena, A., Fernandes, J., Ferrão, P.: Bridging the gap – a database tool for BIM-based circularity assessment. IOP Conf. Ser. Earth Environ. Sci. **1078**(1), 012099 (2022). Available at: https://doi.org/10.1088/1755-1315/1078/1/012099

31. Handayani, T.N., Putri, K.N., Istiqomah, N.A., Likhitruangsilp, V.: The Building Information Modeling (BIM)-based system framework to implement circular economy in construction waste management. J. Civ. Eng. Forum.. Universitas Gadjah Mada. (2021). Available at: https://doaj. org/article/38ba0e152a1c4c209c3c069f3f36b408. Accessed 1 Mar 2023

32. Jung, J. et al.: Productive modeling for development of AS-built BIM of existing indoor structures. Autom. Constr. **42**, 68–77 (2014). Available at: https://doi.org/10.1016/j.autcon. 2014.02.021

33. Akinade, O.O., Oyedele, L.O., Bilal, M., Ajayi, S.O., Owolabi, H.A., Alaka, H.A., Bello, S.A.: Waste minimisation through deconstruction: a BIM based deconstructability assessment score (BIM-DAS). Resour. Conserv. Recycl. **105**, 167–176 (2015). Available at: https://doi.org/10.1016/j.resconrec.2015.10.018

34. Ge, X.J., Livesey, P., Wang, J., Huang, S., He, X., Zhang, C.: Deconstruction waste management through 3D reconstruction and BIM: a case study. Vis. Eng. **5**(1) (2017). Available at: https://doi.org/10.1186/s40327-017-0050-5

35. Hamidi, B., Bulbul, T., Pearce, A., Thabet, W.: Potential application of BIM in cost-benefit analysis of demolition waste management. In: Construction Research Congress 2014 [Preprint] (2014). Available at: https://doi.org/10.1061/9780784413517.029

36. Hai, Y., Yang, L., Sepasgozar, S.: Bim applications in waste and demolition management in circular economy concept. The 3rd Built Environment Research Forum [Preprint] (2022). Available at: https://doi.org/10.3390/environsciproc2021012013

37. Akanbi, L.A., Oyedele, L.O., Akinade, O.O., Ajayi, A.O., Delgado, M.D., Bilal, M., Bello, S.A.: Salvaging building materials in a circular economy: a BIM based whole life performance estimator. Resour. Conserv. Recycl. **129**, 175–186 (2018). Available at: https://doi.org/10.1016/j.resconrec.2017.10.026

38. Kuzina, O.: Information technology application in the construction project life cycle. IOP Conf. Ser. Earth Environ. Sci. **869**(6), 062044 (2020). Available at: https://doi.org/10.1088/1757-899x/869/6/062044

39. Van Langen, S.K., Vassillo, C., Ghisellini, P., Restaino, D., Passaro, R., Ulgiati, S.: Promoting circular economy transition: a study about perceptions and awareness by different stakeholders groups. J. Clean. Prod. **316**, 128166 (2021). Available at: https://doi.org/10.1016/j.jclepro.2021.128166

40. Manzoor, B., Othman, I., Gardezi, S.S.S., Harirchian, E.: Strategies for adopting building information modeling (BIM) in sustainable building projects—a case of Malaysia. Buildings. **11**(6), 249 (2021). Available at: https://doi.org/10.3390/buildings11060249

Advancing Sustainability Through Structural Optimization: Innovations in Material Efficiency and Environmental Impact Reduction

Vagelis Plevris ⓘ, Abdulaziz Almutairi ⓘ, and Alejandro Jiménez Rios ⓘ

Contents

1 Introduction

Concrete stands as not only the most utilized building material globally but also ranks second in overall material usage, following only water—an unsurprising status given its abundance, affordability, and widespread availability. Its versatility allows for a myriad of applications, rendering it indispensable in various construction projects. Cement serves as a binding agent utilized in concrete and other construction materials. The total volume of cement production worldwide amounted to an estimated 4.1 billion metric tons in 2022 [1]. At present, China stands as the foremost producer of cement globally, with an output of approximately 2.1 billion

V. Plevris (✉) · A. Almutairi
Qatar University, Doha, Qatar
e-mail: vplevris@qu.edu.qa; aa1903166@qu.edu.qa

A. J. Rios
Oslo Metropolitan University, Oslo, Norway
e-mail: alejand@oslomet.no

© The Author(s) 2025
M. Kioumarsi, B. Shafei (eds.), *The 1st International Conference on Net-Zero Built Environment*, Lecture Notes in Civil Engineering 237,
https://doi.org/10.1007/978-3-031-69626-8_134

tons in 2022 (more than half of the global production) [2]. Following behind, India emerges as the second-largest producer of cement with its production reaching 370 million tons. Vietnam secures the third position, with 120 million tons in 2022. Among the remaining cement-producing countries, none surpasses the threshold of 100 million tons.

Steel, as a construction material, offers advantages in sustainability when compared to concrete. While both materials are widely used in construction, steel's recyclability stands out as a key factor contributing to its sustainability. Unlike concrete, which relies heavily on cement production—a process associated with significant carbon emissions—steel can be recycled infinitely without loss of quality, reducing the need for raw material extraction and minimizing environmental impact. On the other hand, the production of crude steel also has a significant cost to the environment. One of the primary contributors to environmental degradation is the extraction and processing of raw materials, particularly iron ore and coal, which are essential inputs in steelmaking. Moreover, the manufacturing process itself involves high energy consumption and emissions of greenhouse gases, such as CO_2 and CH_4. Overall, the environmental cost of steel production underscores the urgent need for sustainable practices and technological innovations to mitigate its adverse impacts [3]. In 2022, global crude steel production reached nearly 1.89 billion tons [4]. China emerged as the largest steel-producing country, contributing approximately 1 billion tons, accounting for approximately 54% of the world's production.

Figure 1 illustrates the annual cement and steel production worldwide from 1995 to 2022. Comparatively, in 1995, global cement output was a modest 1.39 billion tons, or 34% of the 2022 production. Similarly, crude steel production in 1995 totaled 753 million tons, representing 40% of the output witnessed in 2022. These statistics underscore the recent expansion experienced in the construction sector globally.

Despite the finite nature of the planet's resources, both concrete and steel production continue to escalate annually. This trend underscores the urgent need for responsible resource management and sustainable construction practices. In the face of burgeoning demand, it becomes imperative to utilize these materials efficiently and judiciously, avoiding wastage wherever possible. Furthermore, it is

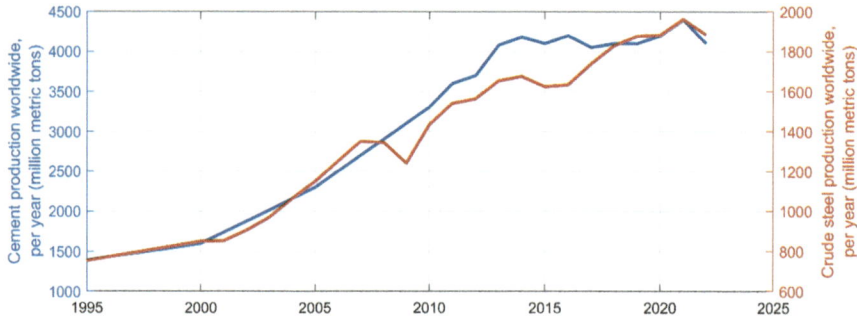

Fig. 1 Cement and steel production worldwide, per year (1995–2022) [1, 4]

essential to acknowledge that the building sector in general consumes approximately 40% of the world's energy, 25% of its water, and 40% of its resources, while also serving as the leading contributor to global greenhouse gas (GHG) emissions, accounting for 30%. Additionally, the construction industry is a significant source of waste generation [5].

2 Structural Optimization and Sustainability

Structural optimization emerges as a vital tool in the quest for sustainability in our constructions, offering a pathway toward constructing structures that maximize efficiency while meeting stringent safety standards. Structural optimization involves applying optimization techniques to the design of load-bearing engineering structures such as buildings, bridges, and others. Before the adoption of computer-assisted optimization procedures, structural elements like beams and plates were designed optimally using manual trial-and-error methods, which were time-consuming. Today, by leveraging advanced optimization techniques, engineers can design buildings and infrastructures that require minimal material inputs without compromising structural integrity or performance. In doing so, structural optimization paves the way for a more sustainable future, where our constructions are not only resource-efficient but also resilient and environmentally conscious.

2.1 Literature Review

Structural optimization has emerged as an innovative technology in recent decades, as demonstrated by the wealth of research articles dedicated to the subject in the literature. A search on Scopus conducted on March 16, 2024, using the query "TITLE-ABS-KEY ("structural optimization") "AND PUBYEAR>1994 AND PUBYEAR<2024" yielded a total of 52,404 published documents from 1995 to 2023. By limiting the search to engineering only, using the additional instruction "AND (LIMIT-TO (SUBJAREA, "ENGI"))" we similarly get 33,670 published documents. Figure 2 depicts the trend of published papers on structural optimization according to Scopus for the period 1995–2023, both overall and within the engineering field. Within the engineering domain, the figure reveals a consistent increase in the number of papers published each year, from 117 papers in 1995 to 1365 papers in 2009 and 3439 papers in 2023.

Lagaros [6] sought to demonstrate the profound environmental benefits and economic advancements achievable through the adoption of optimization-based design procedures within the construction industry. Russo and Rizzi [7] introduced a computer-aided methodology, which integrates Structural Optimization and Life Cycle Assessment (LCA) tools. They used optimization strategies that convert environmental objectives and constraints into structural and geometrical parameters, enabling the generation of alternative green scenarios based on shape, material, and

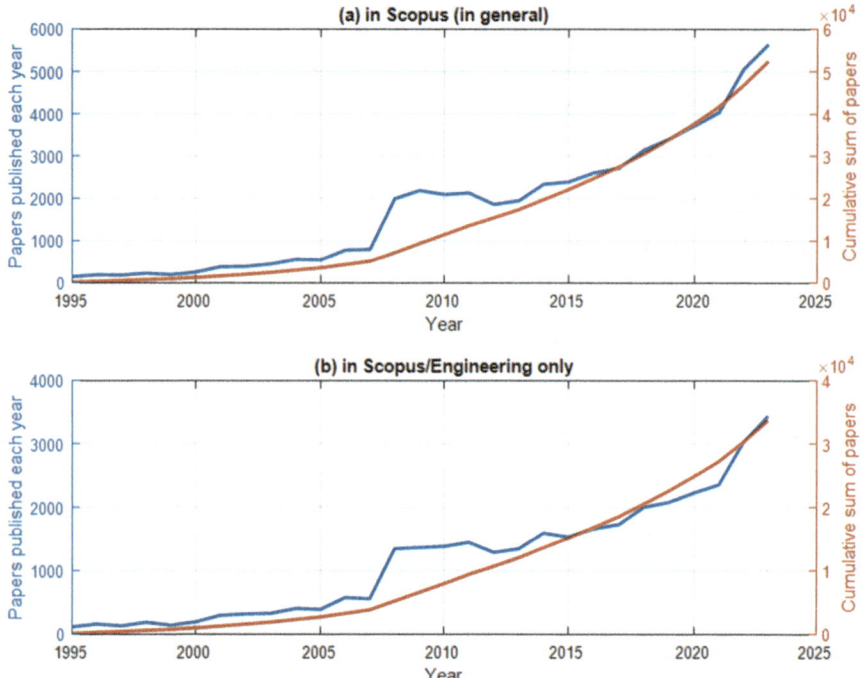

Fig. 2 Published papers in "Structural Optimization," according to Scopus (1995–2023)

production. Rempling et al. [8] attempted automatic structural design by combining set-based design, parametric design, finite element analysis (FEA), and multi-criteria decision analysis. The method was tested on three existing bridges. Tien and Van Tung [9] used building information modeling (BIM) for the multidisciplinary design optimization of sustainable structures, employing the Non-dominated Sorting Genetic Algorithm II. Islam et al. [10] described an optimization approach for balancing life cycle cost and environmental impacts for typical Australian houses, employing single- and multi-objective optimization techniques. Afzal et al. [11] investigated the potential of BIM and optimization algorithms to optimize structural systems and improve design outcomes, following the PRISMA systematic review methodology.

3 Optimization Problems, Algorithms, and Objectives

3.1 Types of Structural Optimization Problems

The basic types of structural optimization problems encompass topology, shape, and sizing optimization, as shown in Fig. 3. Topology optimization is a computational method that aims to determine the optimal distribution of material within a given

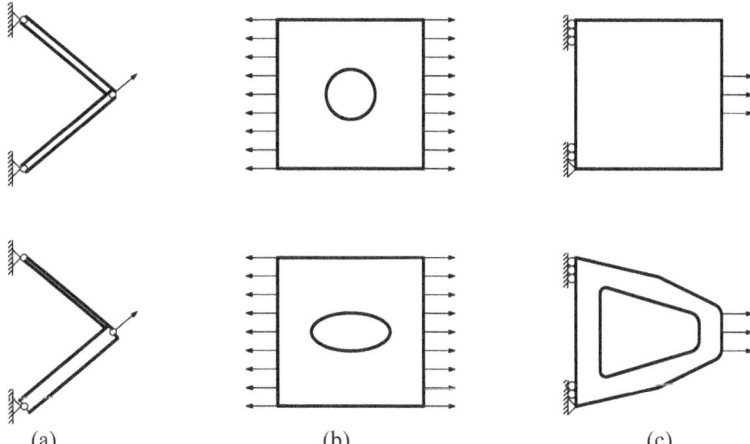

Fig. 3 Basic types of structural optimization problems [16] (Top: Original structure, Bottom: Optimized structure): (**a**) Sizing, (**b**) Shape, and (**c**) Topology optimization

design space to achieve predefined performance objectives while minimizing material usage [12]. By iteratively removing unnecessary material and redistributing load paths, topology optimization enables the creation of structurally efficient designs with optimized strength-to-weight ratios. This approach allows engineers to explore unconventional complex geometries, leading to the development of lightweight and resource-efficient structures that can meet stringent sustainability criteria.

Shape optimization focuses on refining the geometry of structural components to improve their performance characteristics, such as stiffness, strength, or aerodynamic efficiency [13]. By adjusting the shape of individual elements or entire structures, engineers can achieve desired functional requirements while minimizing material usage and environmental impact. Shape optimization techniques often involve parametric modeling and FEA to explore a vast design space and identify the most efficient geometric configurations. This approach facilitates the creation of streamlined and aerodynamic structures that enhance sustainability by reducing energy consumption and material waste.

Sizing optimization involves determining the optimal dimensions of structural members, such as beams, columns, or trusses [14, 15], to maximize performance while minimizing material usage and cost. By systematically adjusting the cross-sectional properties of components based on loading conditions and design constraints, engineers can achieve optimal structural efficiency. Sizing optimization techniques consider factors such as strength, stiffness, and stability to ensure that the resulting structures meet safety standards and functional requirements. This approach allows for the creation of lightweight and resource-efficient designs that contribute to overall sustainability by reducing material consumption and carbon emissions.

3.2   Optimization Algorithms

Optimization algorithms can be broadly categorized into mathematical and metaheuristic ones. *Mathematical algorithms*, also known as deterministic algorithms, are systematic procedures that rely on mathematical principles to find optimal solutions. One of the most widely used mathematical optimization techniques in structural optimization is the gradient-based optimization method. Gradient-based algorithms, such as the method of steepest descent and Newton's method, iteratively update the design variables in the direction of the gradient of the objective function to converge toward the optimal solution. These algorithms are efficient for convex optimization problems with smooth, continuous objective functions.

On the other hand, *metaheuristic algorithms* [17] are stochastic search techniques inspired by natural phenomena or human behavior [18]. These algorithms explore the solution space using randomized search strategies, making them well-suited for non-convex, multimodal optimization problems commonly encountered in structural optimization. They can be broadly classified into the categories shown in Table 1 which also includes some example algorithms for each class. The list is not exhaustive.

In structural optimization, mathematical algorithms are often preferred for problems with well-defined objectives and constraints which can be often expressed analytically. Gradient-based methods are particularly effective for problems with smooth, continuous objective functions and explicit constraints. However, for more complex, non-convex optimization problems with discontinuous or non-smooth objective functions, metaheuristic algorithms offer a more robust and versatile solution approach due to their ability to effectively handle non-convex, multimodal

Table 1   Categories of metaheuristic optimization algorithms

Category	Examples and related works
Evolution-based algorithms	Genetic algorithms (GA) [19, 20]
	Differential evolution (DE) [21, 22]
Swarm intelligence-based algorithms	Particle swarm optimization (PSO) [23]
	Firefly algorithm (FA) [24]
	Grasshopper optimization algorithm (GOA) [25]
	Cuckoo search (CS) [26]
	Whale optimization algorithm (WOA) [27]
Physics-based algorithms	Simulated annealing (SA) [28]
	Lightning search algorithm (LSA) [29]
	Gravitational search algorithm (GSA) [30]
	Electromagnetic field optimization (EFO) [31]
Human-related algorithms	Teaching-based learning optimization (TBLO) [32]
	Sharing knowledge-based algorithm (GSKA) [33]

objective functions and complex design spaces. Hybrid methods can also be employed, aiming to harness the strengths of both categories. Typically, these methods utilize a metaheuristic algorithm for global exploration, followed by a mathematical optimizer for fine-tuning through localized search around the global optimum [23].

3.3   Optimization Objectives

Optimization criteria, also known as objective functions, play a critical role in structural optimization by defining the goals and performance metrics that guide the design process. In optimum structural design, these criteria typically encompass various factors such as structural efficiency, safety, cost-effectiveness, and environmental sustainability. Structural engineers often aim to maximize structural performance while minimizing material usage, weight, or construction costs. Additionally, considerations may include constraints related to permissible stress levels, deflection limits, and geometric configurations. By carefully selecting and formulating optimization criteria, engineers can effectively balance competing objectives and tailor designs to meet specific project requirements, ultimately yielding more sustainable structures.

In *single-objective optimization* (SOO) within structural engineering, the focus is on optimizing a single performance metric or objective function, such as minimizing material usage or cost, maximizing structural strength, or minimizing deflection under load. Engineers typically formulate the optimization problem as a mathematical function, seeking the optimal solution that satisfies predefined constraints while optimizing the chosen objective. SOO techniques, such as gradient-based methods or evolutionary algorithms, facilitate the exploration of design alternatives to achieve the desired performance targets. While SOO provides a straightforward approach for addressing specific design goals, it may overlook trade-offs between conflicting objectives and fail to capture the full spectrum of design possibilities.

In contrast, *multi-objective optimization* (MOO) considers multiple conflicting objectives simultaneously, aiming to identify a set of solutions that represent trade-offs between competing criteria [34]. These objectives often encompass diverse aspects such as structural performance, cost, sustainability, and aesthetic considerations. MOO techniques enable engineers to explore the trade-off space and generate a range of design alternatives known as the Pareto front [35]. By evaluating trade-offs between different objectives, MOO facilitates informed decision-making and helps identify design solutions that offer superior overall performance across multiple criteria. This approach is particularly valuable in complex design scenarios where conflicting objectives must be balanced to achieve optimal outcomes.

4 Benefits of Optimization in Structural Design

Optimization in structural design offers numerous benefits that contribute to the development of efficient, cost-effective, and sustainable structures. By leveraging advanced computational techniques and mathematical algorithms, engineers can systematically refine designs to enhance performance, reduce material usage, and minimize environmental impact. Optimization enables the exploration of a vast design space, leading to innovative solutions that prioritize structural integrity, safety, and functionality. It facilitates informed decision-making by quantifying trade-offs between conflicting objectives, allowing engineers to achieve optimal outcomes tailored to specific requirements. Ultimately, the adoption of optimization methodologies empowers engineers to push the boundaries of structural design, resulting in more resilient and resource-efficient built environments.

Efficiency Optimization improves the efficiency of structural systems by maximizing performance while minimizing resource consumption and waste. Through iterative refinement of designs, engineers can achieve higher structural efficiency, leading to lighter, more streamlined structures that require fewer materials and resources to construct. This enhanced efficiency translates to reduced construction time, lower energy consumption, and improved overall sustainability.

Cost Savings Optimization helps to reduce construction costs by optimizing material usage, minimizing the need for expensive materials, and streamlining construction processes. By identifying cost-effective design alternatives and eliminating unnecessary elements, engineers can realize significant savings in material procurement, labor costs, and project overheads. Furthermore, optimized designs often require less maintenance over their lifespan, further reducing life-cycle costs and enhancing long-term affordability.

Sustainability Optimization plays a crucial role in promoting sustainability in structural design by minimizing environmental impact and resource depletion. By optimizing material usage, structural configurations, and construction methodologies, engineers can reduce the carbon footprint of buildings and infrastructure projects. Additionally, optimization enables the integration of sustainable materials, renewable energy systems, and passive design strategies, further enhancing environmental performance and resilience to climate change.

Innovative Designs Optimization fosters creativity and innovation in structural design by pushing the boundaries of conventional practices and exploring new possibilities. By leveraging advanced computational tools and generative design techniques, engineers can generate novel structural forms and geometries that maximize performance and visual appeal. Optimization encourages experimentation with unconventional materials, fabrication methods, and construction techniques, leading to the realization of iconic and landmark structures that inspire future generations.

Safety Optimization can enhance structural safety by optimizing designs to withstand a wide range of loading conditions and environmental hazards. By conducting rigorous analysis and optimization iterations, engineers can identify potential weaknesses, improve structural robustness, and mitigate risks of failure. Optimization also allows for the incorporation of safety factors and design redundancies, ensuring that structures meet or exceed regulatory standards and withstand unforeseen challenges throughout their lifespan.

5 Challenges and Limitations

Optimization in structural design, while offering significant benefits, also presents various challenges and limitations that engineers must navigate to achieve successful outcomes. Below we examine and analyze some of these obstacles in detail.

Computational Intensity One of the primary challenges in optimization-driven structural design is the computational intensity required to solve complex optimization problems. The iterative nature of optimization algorithms and the need for detailed FEA in each step contribute to significant computational demands, requiring substantial computing resources and time. As a result, engineers must carefully manage computational resources, employ efficient algorithms, and leverage parallel computing techniques to overcome this challenge and expedite the optimization process without compromising accuracy. For instance, if a single FEA iteration of a building requires approximately 5 s to finish, and an optimization algorithm demands 10,000 iterations to reach convergence, then the optimization process could potentially consume up to 14 h to complete—a substantial duration of time.

Multiple Objectives Balancing multiple conflicting objectives and constraints poses a significant challenge in structural optimization, particularly in real-world construction problems. Engineers must navigate trade-offs between competing design criteria, such as performance, cost, sustainability, and aesthetics, to identify Pareto-optimal solutions that represent acceptable compromises. Achieving consensus among stakeholders and reconciling conflicting preferences can further complicate the optimization process, requiring robust decision-making frameworks and stakeholder engagement strategies to address diverse perspectives and priorities effectively.

Inherent Uncertainties The existence of uncertainties in material properties, loading conditions, and modeling assumptions introduces challenges to the accuracy and reliability of optimization results [36]. Sensitivity analysis techniques are essential for assessing the sensitivity of optimized designs to input parameters and identifying sources of uncertainty that may affect performance and safety. Engineers must employ robust probabilistic methods, such as Monte Carlo simulation [37] or stochastic optimization, to account for uncertainties and optimize designs that exhibit resilience and robustness [38] in the face of varying operating conditions and environmental factors. Consequently, this introduces further computational complexities to the optimization process [39].

Constraint Handling Handling constraints presents a significant challenge in optimization-driven structural design, as balancing optimization objectives with real-world constraints can be complex and difficult to fully automate. Structural engineers must navigate a diverse array of constraints, including fabrication limitations, regulatory requirements, material availability, and construction logistics, which can significantly impact the feasibility and practicality of designs. While optimization algorithms excel at finding solutions that optimize specified performance metrics, integrating constraints into the optimization process requires careful consideration and often involves trade-offs. Engineers must strike a delicate balance between achieving optimal performance and adhering to practical constraints, such as budgetary limitations, site-specific conditions, and safety standards. Furthermore, constraints may vary in nature and complexity across different projects, necessitating customized approaches and expert judgment while optimizing structural designs for sustainability, efficiency, and safety. Despite these challenges, advancements in constraint handling techniques, continue to enhance the capabilities of optimization algorithms to accommodate diverse constraints and facilitate the development of innovative and practical structural solutions.

6 Conclusions

In conclusion, structural optimization emerges as a pivotal asset in contemporary engineering, offering a pathway toward the creation of stronger, more efficient, and sustainable structures. By harnessing the power of optimization algorithms and computational tools, engineers can unlock innovative design solutions that maximize performance while minimizing environmental impact. Optimization techniques enable the strategic allocation of materials and resources, resulting in structures that exhibit superior resilience, structural integrity, and resource efficiency. Moreover, optimization has the potential to revolutionize the construction industry by streamlining design processes, reducing material waste, and enhancing project cost-effectiveness.

With advancements in technology, hardware, and software, coupled with the increasing availability of computational resources, the field of structural optimization is poised for significant expansion. The integration of cutting-edge computational methods and predictive modeling techniques promises to propel structural design into a new era of unprecedented creativity and efficiency. By leveraging these technological advancements, engineers can explore complex design spaces, optimize structural configurations, and push the boundaries of conventional construction practices. As a result, the forthcoming era of structural design holds the promise of seamlessly blending artistic ingenuity with scientific precision, yielding structures that not only meet functional requirements but also embody ecological mindfulness and cost-efficiency.

In essence, structural optimization represents a transformative paradigm shift in the built environment, offering a harmonious convergence of engineering excellence and sustainable design principles. By embracing optimization methodologies, engineers can catalyze positive change in the construction industry, fostering the development of resilient, adaptable, and environmentally conscious structures. As we embark on this journey toward a more sustainable future, structural optimization stands as a beacon of innovation, driving the evolution of architectural design toward a greener, more efficient, and harmonious built environment.

References

1. Ige, O.E., Von Kallon, D.V., Desai, D.: Carbon emissions mitigation methods for cement industry using a systems dynamics model. Clean Techn. Environ. Policy. (2024). https://doi.org/10.1007/s10098-023-02683-0
2. worldpopulationreview.com. Cement Production by Country 2024. 2024. Available from: https://worldpopulationreview.com/country-rankings/cement-production-by-country. Accessed 16 Mar 2024
3. Conejo, A.N., Birat, J.-P., Dutta, A.: A review of the current environmental challenges of the steel industry and its value chain. J. Environ. Manag. **259**, 109782 (2020). https://doi.org/10.1016/j.jenvman.2019.109782
4. World Steel Association. World Steel in Figures 2023. 2023. Available from: https://worldsteel.org/steel-topics/statistics/world-steel-in-figures-2023/. Accessed 16 Mar 2024
5. Hussain, C.M., Paulraj, M.S., Nuzhat, S.: Chapter 5 – Source reduction and waste minimization in construction industry. In: Hussain, C.M., Paulraj, M.S., Nuzhat, S. (eds.) Source Reduction and Waste Minimization, pp. 111–126. Elsevier (2022). https://doi.org/10.1016/B978-0-12-824320-6.00005-8
6. Lagaros, N.D.: The environmental and economic impact of structural optimization. Struct. Multidiscip. Optim. **58**(4), 1751–1768 (2018). https://doi.org/10.1007/s00158-018-1998-z
7. Russo, D., Rizzi, C.: Structural optimization strategies to design green products. Comput. Ind. **65**(3), 470–479 (2014). https://doi.org/10.1016/j.compind.2013.12.009
8. Rempling, R., Mathern, A., Tarazona Ramos, D., Luis Fernández, S.: Automatic structural design by a set-based parametric design method. Autom. Constr. **108**, 102936 (2019). https://doi.org/10.1016/j.autcon.2019.102936
9. Tien, L.H., Van Tung, N.: Multidisciplinary design optimization for sustainable design using building information modeling. IOP Conf. Ser. Mater. Sci. Eng. **1109**(1), 012013 (2021). https://doi.org/10.1088/1757-899X/1109/1/012013
10. Islam, H., Jollands, M., Setunge, S., Bhuiyan, M.A.: Optimization approach of balancing life cycle cost and environmental impacts on residential building design. Energ. Buildings. **87**, 282–292 (2015). https://doi.org/10.1016/j.enbuild.2014.11.048
11. Afzal, M., Li, R.Y.M., Ayyub, M.F., Shoaib, M., Bilal, M.: Towards BIM-based sustainable structural design optimization: a systematic review and industry perspective. Sustain. For. **15**(20), 15117 (2023). https://doi.org/10.3390/su152015117
12. Wang, M.Y., Wang, X., Guo, D.: A level set method for structural topology optimization. Comput. Methods Appl. Mech. Eng. **192**(1), 227–246 (2003). https://doi.org/10.1016/S0045-7825(02)00559-5

13. Bendsøe, M.P.: Optimal shape design as a material distribution problem. Struct. Optim. **1**(4), 193–202 (1989). https://doi.org/10.1007/BF01650949

14. Bekdaş, G., Nigdeli, S.M., Yang, X.-S.: Sizing optimization of truss structures using flower pollination algorithm. Appl. Soft Comput. **37**, 322–331 (2015). https://doi.org/10.1016/j.asoc.2015.08.037

15. Degertekin, S.O.: Improved harmony search algorithms for sizing optimization of truss structures. Comput. Struct. **92-93**, 229–241 (2012). https://doi.org/10.1016/j.compstruc.2011.10.022

16. Plevris, V.: Innovative Computational Techniques for the Optimum Structural Design Considering Uncertainties, p. 312. National Technical University of Athens, Athens (2009). https://doi.org/10.12681/eadd/17936

17. Kaveh, A.: Advances in Metaheuristic Algorithms for Optimal Design of Structures, 3rd edn. Springer, Cham (2021). https://doi.org/10.1007/978-3-030-59392-6

18. Lagaros, N.D., Plevris, V., Kallioras, N.A.: The mosaic of metaheuristic algorithms in structural optimization. Arch. Comput. Methods Eng. **29**, 5457–5492 (2022). https://doi.org/10.1007/s11831-022-09773-0

19. Solorzano, G., Plevris, V.: Optimum design of RC footings with genetic algorithms according to ACI 318-19. Buildings. **10**(6), 1–17 (2020). https://doi.org/10.3390/buildings10060110

20. Papazafeiropoulos, G., Plevris, V., Papadrakakis, M.: Optimum design of cantilever walls retaining linear elastic backfill by use of Genetic Algorithm. In: Computational Methods in Structural Dynamics and Earthquake Engineering 2013 (COMPDYN 2013), pp. 2731–2750, Kos Island (2013). https://doi.org/10.7712/120113.4700.C1746

21. Georgioudakis, M., Plevris, V.: A comparative study of differential evolution variants in constrained structural optimization. Front. Built Environ. **6**(102), 1–14 (2020). https://doi.org/10.3389/fbuil.2020.00102

22. Georgioudakis, M., Plevris, V.: On the performance of differential evolution variants in constrained structural optimization. Proc. Manuf. **44**, 371–378 (2020). https://doi.org/10.1016/j.promfg.2020.02.281

23. Plevris, V., Papadrakakis, M.: A hybrid particle swarm – gradient algorithm for global structural optimization. Comput. Aided Civ. Inf. Eng. **26**(1), 48–68 (2011). https://doi.org/10.1111/j.1467-8667.2010.00664.x

24. Gandomi, A.H., Yang, X.-S., Alavi, A.H.: Mixed variable structural optimization using firefly algorithm. Comput. Struct. **89**(23), 2325–2336 (2011). https://doi.org/10.1016/j.compstruc.2011.08.002

25. Saremi, S., Mirjalili, S., Lewis, A.: Grasshopper optimisation algorithm: theory and application. Adv. Eng. Softw. **105**, 30–47 (2017). https://doi.org/10.1016/j.advengsoft.2017.01.004

26. Gandomi, A.H., Yang, X.-S., Alavi, A.H.: Cuckoo search algorithm: a metaheuristic approach to solve structural optimization problems. Eng. Comput. **29**(1), 17–35 (2013). https://doi.org/10.1007/s00366-011-0241-y

27. Kaveh, A., Ghazaan, M.I.: Enhanced whale optimization algorithm for sizing optimization of skeletal structures. Mech. Based Des. Struct. Mach. **45**(3), 345–362 (2017). https://doi.org/10.1080/15397734.2016.1213639

28. Leite, J.P.B., Topping, B.H.V.: Parallel simulated annealing for structural optimization. Comput. Struct. **73**(1), 545–564 (1999). https://doi.org/10.1016/S0045-7949(98)00255-7

29. Panigrahy, D., Samal, P.: Modified lightning search algorithm for optimization. Eng. Appl. Artif. Intell. **105**, 104419 (2021). https://doi.org/10.1016/j.engappai.2021.104419

30. Guo, H., Wei, J.: Modified gravitational search algorithm and its application to structural damage detection. J. Phys. Conf. Ser. **1635**(1), 012018 (2020). https://doi.org/10.1088/1742-6596/1635/1/012018

31. Abedinpourshotorban, H., Mariyam Shamsuddin, S., Beheshti, Z., Jawawi, D.N.A.: Electromagnetic field optimization: a physics-inspired metaheuristic optimization algorithm. Swarm Evolut. Comput. **26**, 8–22 (2016). https://doi.org/10.1016/j.swevo.2015.07.002

32. Degertekin, S.O., Hayalioglu, M.S.: Sizing truss structures using teaching-learning-based optimization. Comput. Struct. **119**, 177–188 (2013). https://doi.org/10.1016/j.compstruc.2012.12.011

33. Chalabi, N.E., Attia, A., Alnowibet, K.A., Zawbaa, H.M., Masri, H., Mohamed, A.W.: A multi–objective gaining–sharing knowledge-based optimization algorithm for solving engineering problems. Mathematics. **11**(14), 3092 (2023). https://doi.org/10.3390/math11143092

34. Papadrakakis, M., Lagaros, N.D., Plevris, V.: Multi-objective optimization of skeletal structures under static and seismic loading conditions. Eng. Optim. **34**(6), 645–669 (2002). https://doi.org/10.1080/03052150215716

35. Lagaros, N.D., Plevris, V., Papadrakakis, M.: Multi-objective design optimization using Cascade evolutionary computations. Comput. Methods Appl. Mech. Eng. **194**(30–33), 3496–3515 (2005). https://doi.org/10.1016/j.cma.2004.12.029

36. Papadrakakis, M., Lagaros, N.D., Plevris, V.: Design optimization of steel structures considering uncertainties. Eng. Struct. **27**(9), 1408–1418 (2005). https://doi.org/10.1016/j.engstruct.2005.04.002

37. Papadrakakis, M., Lagaros, N.D., Plevris, V.: Structural optimization considering the probabilistic system response. Theor. Appl. Mech. **31**(3–4), 361–394 (2004). https://doi.org/10.2298/TAM0404361P

38. Lagaros, N.D., Plevris, V., Papadrakakis, M.: Reliability based robust design optimization of steel structures. Int. J. Simul. Multidiscip. Des. Optim. **1**(1), 19–29 (2007). https://doi.org/10.1051/ijsmdo:2007003

39. Lagaros, N.D., Tsompanakis, Y., Fragiadakis, M., Plevris, V., Papadrakakis, M.: Metamodel-based computational techniques for solving structural optimization problems considering uncertainties. In: Tsompanakis, Y., Lagaros, N.D., Papadrakakis, M. (eds.) Structural Design Optimization Considering Uncertainties, pp. 567–597. Taylor and Francis (2008)

Unleashing the Potential of AI in Sustainable Urban Planning and Design

Arefeh Mortzavi Rad, Elsa Haagensen Karlsen, and Mohammed Nazar

Contents

1   Introduction

AI's ability to analyze vast data sets, predict trends, and automate tasks presents a promising avenue for smarter, more sustainable urban planning [1].

Moreover, AI's potential in discovering environmentally friendly materials and methods is still being explored [2].

A significant overlooked opportunity is the use of AI to identify innovative collaborators for building sustainable cities [3]. This article aims to illuminate how AI can revolutionize urban planning and design, exploring strategies to overcome these barriers and maximize AI's benefits.

A. M. Rad (✉)
Monash University, Clayton, VIC, Australia
e-mail: amor0085@student.monash.edu

E. H. Karlsen · M. Nazar
Western Norway University of Applied Sciences, Bergen, Norway
e-mail: ehk@hvl.no; mnaz@hvl.no

© The Author(s) 2025
M. Kioumarsi, B. Shafei (eds.), *The 1st International Conference on Net-Zero Built Environment*, Lecture Notes in Civil Engineering 237,
https://doi.org/10.1007/978-3-031-69626-8_135

By simplifying complex concepts and focusing on practical applications, we aim to make the discussion about AI and sustainable cities more engaging and comprehensible.

2 Theoretical Framework

Artificial Intelligence (AI) has significantly transformed our interaction with the environment, evolving from basic computational tools to advanced systems capable of tasks like language and image recognition. Originating from Claude Shannon's work in the 1950s, AI has progressed through remarkable milestones, now operating predominantly in the realm of Artificial Narrow Intelligence (ANI), excelling in tasks like powering voice assistants and chatbots. The ambition extends to Artificial General Intelligence (AGI) and Artificial Superintelligence (ASI), aiming to match or exceed human cognitive abilities, thus potentially reshaping societal norms [4]. AI's applications extend across healthcare, finance, and transportation, driven by advancements in Natural Language Processing (NLP), Computer Vision, Machine Learning (ML), and Deep Learning (DL), enhancing our problem-solving capacity and fostering innovation [5].

In sustainable urban planning, AI plays a pivotal role, leading the development of smart cities that integrate environmental sustainability with technological advancements [6].

AI enhances urban efficiency by optimizing energy use, improving waste management, and refining public transportation through detailed analyses, with cities like Stockholm, Copenhagen, and Amsterdam exemplifying AI's ability to support sustainable, resilient urban areas [7]. Nonetheless, challenges persist, including misconceptions about AI's capabilities, data privacy concerns, financial costs, and potential biases in AI algorithms. Overcoming these obstacles requires targeted educational efforts, transparent data governance, cost mitigation collaborations, and a dedication to ethical AI practices [8].

AI's impact on sustainable urban planning is profound, especially in introducing innovative materials and methods, such as machine learning for land use optimization and energy-efficient building designs. Supported by sensor and satellite data, this global research and development collaboration underscores AI's vital role in fostering sustainable, livable cities [9].

In procurement within urban planning, selecting the right suppliers is crucial. Decision-making models like Multi-Criteria Decision-Making (MCDM) and techniques such as ELECTRE and PROMETHEE help evaluate suppliers based on various criteria, including environmental impact. Generative AI further revolutionizes procurement by enabling the rapid production of high-quality content for supplier evaluation, marking a new procurement era and enhancing selection strategies [10].

2.1   Case Study: AI in Amsterdam's Urban Planning

Amsterdam's AI application in urban planning, particularly in flood prediction and green infrastructure planning, showcases AI's potential to address specific challenges. AI algorithms allow for accurate flood pattern forecasting and the design of urban initiatives aligned with natural water systems, highlighting a sophisticated approach to integrating technology with environmental sustainability. This not only demonstrates AI's problem-solving capabilities but also contributes significantly to the city's climate resilience [11].

3   Methodology

This investigation explores Artificial Intelligence's (AI) role in evaluating and selecting innovative contractors and suppliers for sustainable urban planning. Using a scoping literature review, based on Arksey and O'Malley's framework [12], it examines AI's integration in supplier selection and urban design supply chain management. This methodological approach is ideal for the evolving field of AI, which spans diverse methodologies and outcomes. Our methodology commenced with formulating a research question to explore AI's potential in sustainable urban design. A rigorous literature search followed, employing databases like Semantic Scholar, Google Scholar, and ABI/INFORM, and using Boolean operators for precision (Table 1).

This strategic search, favoring Semantic Scholar for its relevance to our objectives, aimed to cover the last 25 years, capturing AI's advancements and seminal contributions to the field. We prioritized peer-reviewed articles relevant to AI in supplier selection, deliberately excluding unrelated studies. The Elicit tool was instrumental in filtering over 126 million papers, identifying 32 articles crucial for our study. These selections were based on their methodological rigor and innovative insights into AI's procurement role within urban planning. This process allowed for a focused review, ensuring comprehensive coverage and relevance.

Our refined strategy involved specific inclusion and exclusion criteria, streamlining the review process. The choice of individual search terms over a combined string enhanced the breadth of our exploration, enriching the study's depth. This meticulous approach facilitated the identification of 32 articles that

Table 1 Result of our Boolean operators from Semantic Scholar

Boolean operators	Number of articles
AI and Urban Design	Approximately 49,400 articles
AI and Supplier Selection	Approximately 10,200 articles
AI and Order Allocation	Approximately 29,500 articles
AI and Supply Chain Management	Approximately 4240 articles

were pivotal due to their focus on AI in procurement, showcasing methodological sophistication and global applicability.

Data charting adhered to Arksey and O'Malley's [12] framework, summarizing findings to offer a structured overview of AI's role in sustainable urban planning. The final stage collated and summarized results, presenting a synthesized view of the research on AI's transformative potential in sustainable urban development. This methodology, detailed for clarity and depth, and leveraging a Large Language Model (LLM) for textual refinement, maintained scholarly integrity while enhancing readability.

By outlining the use of Boolean operators and strategic database choices, the revision underscores the selection process's rigor. Discussion on methodology reflects transparently on the applied criteria, offering a systematic review that integrates references [13], ensuring the academic integrity of our concise exploration (Fig. 1 and Tables 2, 3, 4 and 5).

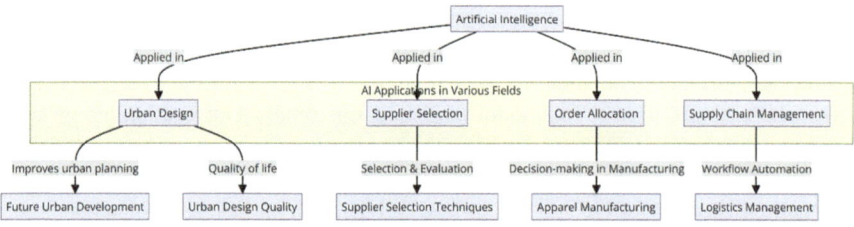

Fig. 1 Relationships between the topics of artificial intelligence (AI) applications in various fields

Table 2 Articles on artificial intelligence and urban design

Title	Abstract summary
AI for Future Urban Development	Highlights AI and machine learning's role in enhancing urban planning through advanced data analysis and inference [14]
AI's Impact on Urban Design	Discusses AI's contributions to urban planning and design, with an emphasis on improving city life quality [15]
Urban AI: Interdisciplinary Research Agenda	Calls for a dedicated research community at the AI and urban design intersection [16]
AI in Urban Design: The Egypt Case	Investigates the potential and challenges of employing AI in urban planning, focusing on Egypt [17]
AI in Australian Urban Planning	Analyzes the adoption and effects of AI technologies in urban planning and development in Australia [18]
Review of Urban Planning and AI	Reviews methods of applying AI to tackle urban dynamics [19]
Hybrid AI Systems in Urban Planning	Describes the development of a hybrid AI system for automating urban planning decisions [20]
Urban-GAN: AI for Urban Design	Introduces a system that uses AI to create distinctive urban designs, showcased in cities such as Manhattan, Portland, and Shanghai [21]

Table 3 Articles on artificial intelligence and supplier selection

Title	Abstract summary
AI and Optimization in Supplier Selection	Develops a system that uses AI and particle swarm optimization for precise supplier selection based on both qualitative and quantitative factors [22]
Neural Networks in Supplier Selection	Explains how the integration of neural networks with the analytic hierarchy process can reduce the amount of training data needed for supplier selection [23]
Intelligent Management at Honeywell	Showcases how Honeywell Consumer Product (Hong Kong) Limited utilized an intelligent system to reduce outsource cycle times [24]
Predictive Model for Supplier Negotiations	Introduces an ANN-based model to improve the supplier selection negotiation process [25]
Green Supplier Selection in a Fuzzy Environment	Combines data envelopment analysis and genetic programming for evaluating supplier efficiency in uncertain conditions [26]
Supplier Management for Product Development	Highlights the effectiveness of Honeywell's intelligent system in reducing outsource cycle times for new product development [27]

Table 4 Articles on artificial intelligence and order allocation

Title	Abstract summary
Cloud-Based System for Apparel Manufacturing	Developed to streamline order tracking and allocation across multiple apparel manufacturing sites [28]
Intelligent Order Allocation	Utilizes fuzzy numbers to enhance decision-making for order fulfillment amid changing customer demands [29]
Symbiotic AI for Order Picking	Introduces an egocentric Symbiotic AI that uses ambient sensing for agent training, improving order picking efficiency without direct supervision [30]
AI in System Order Discrimination	Employs learning machines to identify features for discriminating the order of multivariable systems [31]
Intelligent Warehouse Management Algorithms	Optimizes warehouse order picking by reducing picker travel distance through intelligent algorithms [32]
Carbon Tax-Conscious Order Allocation	Evaluates the economic and environmental impacts of order allocation within online freight platforms under carbon tax constraints [33]
Autonomous Warehouse Robots	Investigates the use of robots for uninterrupted order collection in large warehouse environments [34]
Order Placement and Resource Allocation	Studies the influence of different order placement strategies on resource allocation and processing times in multi-user warehouses [35]

The study highlights the significant role AI plays in supplier management and urban planning, with particular emphasis on innovative systems and models, like Urban-GAN, and focuses on geographical or unique research areas. Notable mentions include the methodologically rigorous and innovative applications of AI, especially in Honeywell's operations in Hong Kong, demonstrating AI's potential

Table 5 Articles on artificial intelligence and supply chain management

Title	Abstract summary
AI in Logistics and SCM	AI enhances automation in logistics and supply chain workflows [36]
AI's Role in Logistics and SCM	Highlights AI's potential in improving decision-making and efficiency in logistics and SCM [37]
AI Solutions for SCM	Discusses AI's emerging portfolio of solutions for supply chain management [38]
AI in SCM: An Exploratory Case Study	Examines the importance of AI-driven research and applications in SCM [39]
AI, SCM, and FinTech	Explores AI integration into analytics and CRM for better customer service in SCM [40]
AI and Blockchain in Digital SCM	Advocates a strategic approach to adopting AI and blockchain for SCM benefits [41]
AI for SCM Execution	Applies AI approaches to improve supply chain execution [42]
Systematic Review of AI in SCM	Reviews current and potential AI techniques for enhancing SCM studies and practices [43]

to enhance efficiency and decision-making in supply chain management. Through a combination of automated and manual screening, the research navigates a vast literature landscape effectively. By employing Arksey and O'Malley's framework, the research develops comprehensive tables summarizing the collected articles, addressing how AI aids in evaluating and selecting contractors and suppliers for sustainable urban development. This synthesized overview provides a concise, yet informative look at AI's current state in supplier selection and order allocation, including the sustainability criteria addressed by AI tools.

Additionally, this research integrates the use of a Large Language Model (LLM) for the refinement of linguistic expression within our materials, enhancing the clarity and accuracy of the content. This careful enhancement ensures that the dissemination of findings remains both comprehensible and adherent to scholarly standards. Such detailed attention to linguistic quality not only elevates the coherence of the presented research but also enriches the discourse on the influence of AI in advancing sustainable urban planning initiatives.

4 Findings and Conclusion

Our study provides a synthesized overview of the pivotal role of artificial intelligence (AI) in the evaluation and selection of innovative contractors and suppliers within the sphere of sustainable urban development. This summary distills key insights into the methodologies, applications, and overall impact of AI technologies on sustainable urban planning and supply chain management, highlighting its essential role in propelling toward a more sustainable future (Table 6).

Table 6 Synthesized overview of findings

Field	Capabilities	Methods/usage	Advantages/impact	Innovations
AI in Urban Design	Enhances urban planning through data analysis and promotes interdisciplinary research	N/A	Automates decision-making and fosters innovative designs across diverse areas	N/A
AI in Supplier Selection	Streamlines the selection process with sophisticated methods	Employs artificial neural networks, genetic algorithms, and fuzzy neural networks	Enhances supplier evaluation and management; supports green procurement decisions	N/A
AI in Order Allocation	Optimizes order allocation and fulfillment in logistics and manufacturing	AI-driven systems improve operational processes	Enhances efficiency, sustainability, and resource allocation	N/A
AI in Supply Chain Management	Automates workflows and enhances decision-making	Ranges from logistics to integrating financial technology	Improves operational efficiency	Utilizes analytics and blockchain to revolutionize supply chain practices

These findings highlight AI's transformative impact across various domains, advancing theoretical understanding and showcasing practical applications, especially in optimizing supplier selection and management processes, affirming AI's contribution to sustainable urban development. The integration of AI in urban design and supply chain management marks a significant transformation toward enhanced efficiency, sustainability, and resilience. However, leveraging AI's full potential requires the development of scalable, ethically responsible solutions. The need for comprehensive strategies to enhance supply chain resilience, efficiency, and sustainability is underscored by the absence of systematic reviews addressing the intricacies of supplier selection and order allocation processes [6]. This gap represents an opportunity for research that enriches our understanding and addresses the challenges of urban logistics and sustainability [44].

Future research should focus on developing specific AI algorithms and conducting empirical studies to assess their real-world applications and challenges. Investigating the ethical, social, and regulatory dimensions of AI deployment is essential for ensuring its widespread acceptance [45]. Additionally, understanding geographic and cultural differences is crucial for creating adaptable AI solutions for diverse urban settings [46].

In conclusion, this analysis presents a strategic roadmap for stakeholders committed to fostering more sustainable, efficient, and resilient urban and supply chain systems through AI. This pursuit enriches both the theoretical and practical understanding of these areas, preparing them to thrive amidst the future's dynamic changes.

References

1. Gupta, S., Degbelo, A.: An empirical analysis of AI contributions to sustainable cities (SDG 11). In: Mazzi, F., Floridi, L. (eds.) The Ethics of Artificial Intelligence for the Sustainable Development Goals. Philosophical Studies Series, vol. 152. Springer, Cham (2023). https://doi.org/10.1007/978-3-031-21147-8_25

2. Kim, J.: Google DeepMind's new AI tool helped create more than 700 new materials. (2023). https://www.technologyreview.com/2023/11/29/1084061/deepmind-ai-tool-for-new-materials-discovery/. Accessed 13.04.2024

3. Ufberg, M.: The 10 most innovative companies in artificial intelligence of 2023. (2023). https://www.fastcompany.com/90846670/most-innovative-companies-artificial-intelligence-2023. Accessed 13.04.2024

4. HISTORY OF AI: History of AI, https://historyof.ai/. Accessed 15.02.2024

5. Ramezani, M., Takian, A., Bakhtiari, A., et al.: The application of artificial intelligence in health financing: a scoping review. Cost Eff. Resour. Alloc. **21**, 83 (2023). https://doi.org/10.1186/s12962-023-00492-2

6. Bibri, S.E., Alexandre, A., Sharifi, A., et al.: Environmentally sustainable smart cities and their converging AI, IoT, and big data technologies and solutions: an integrated approach to an extensive literature review. Energy Inform. **6**, 9 (2023). https://doi.org/10.1186/s42162-023-00259-2

7. Jevinger, Å., Zhao, C., Persson, J.A., et al.: Artificial intelligence for improving public transport: a mapping study. Public Transp. **16**, 99–158 (2024). https://doi.org/10.1007/s12469-023-00334-7

8. ArchDaily: Artificial Intelligence and Urban Planning: Technology as a Tool for City Design. ArchDaily. https://doi.org/10.1007/s43681-021-00043-6. Accessed 15.02.2024

9. Van Wynsberghe, A.: Sustainable AI: AI for sustainability and the sustainability of AI. AI Ethics. **1**, 213–218 (2021). https://doi.org/10.1007/s43681-021-00043-6

10. Roy, B.: The outranking approach and the foundations of electre methods. Theor. Decis. **31**, 49–73 (1991). https://doi.org/10.1007/BF00134132

11. Ye, X., Wang, S., Lu, Z., et al.: Towards an AI-driven framework for multi-scale urban flood resilience planning and design. Comput. Urban Sci. **1**, 11 (2021). https://doi.org/10.1007/s43762-021-00011-0

12. Arksey, H., O'Malley, L.: Scoping studies: towards a methodological frame-work. Int. J. Soc. Res. Methodol. **8**(1), 19–32 (2005). https://doi.org/10.1080/1364557032000119616

13. Naqvi, M.A., Amin, S.H.: Supplier selection and order allocation: a literature review. J. Data Inf. Manag. **3**, 125–139 (2021). https://doi.org/10.1007/s42488-021-00049-z

14. Dewi, A.C., Zagloel, T.Y.M.: Sustainable supplier selection and order allocation: a systematic literature review. AIP Conf. Proc. **2693**, 030025 (2023) https://pubs.aip.org/aip/acp/article-abstract/2693/1/030025/2920071/Sustainable-supplier-selection-and-order?redirectedFrom=fulltext. Accessed 15.02.2024

15. Haldorai, A., Ramu, A., Murugan, S.: Artificial Intelligence and Machine Learning for Future Urban Development. Urban Computing (2019)

16. Nematollahi, M.A., Shahbazi, S., Nabian, N.: Application of AI in Urban Design. Computer Vision and Audition in Urban Analysis Using the Remorph Framework (2018)

17. Luusua, A., Ylipulli, J.: Urban AI: formulating an agenda for the interdisciplinary research of artificial intelligence in cities. In: Companion Publication of the 2020 ACM Designing Interactive Systems Conference (2020)

18. Ismaeel, W.S.: Challenges for Applying Artificial Intelligence in the Field of Urban Design and Planning; Case of Egypt, vol. 2022, pp. 1–4. Engineering and Technology for Sustainable Architectural and Interior Design Environments (ETSAIDE) (2022)

19. Regona, M., Ruiz Maldonado, A., Rowan, B., Ryu, A.J., Desouza, K.C., Corchado, J.M., Mehmood, R., Li, R.Y.: Artificial intelligence technologies and related urban planning and development concepts: how are they perceived and utilized in Australia? J. Open Innov. Technol. Market Complex. (2020)

20. Jain, D.K. (2011). A Review Study on Urban Planning & Artificial Intelligence
21. Feng, S., Xu, L.D.: Hybrid artificial intelligence approach to urban planning. Expert. Syst. **16** (1999)
22. Quan, S.: Urban-GAN: an artificial intelligence-aided computation system for plural urban design. Environ. Plan. B Urban Anal. City Sci. **49**, 2500–2515 (2022)
23. Asthana, N., Gupta, M.: Supplier selection using artificial neural network and genetic algorithm. Int. J. Indian Cult. Bus. Manag. **11**, 457–472 (2015)
24. Ahmad, I., Liu, Y., Javeed, D., Shamshad, N., Sarwr, D., Ahmad, S.: A review of artificial intelligence techniques for selection & evaluation. IOP Conf. Ser. Mater. Sci. Eng. **853** (2020)
25. Kuo, R.J., Hong, S.Y., Huang, Y.C.: Integration of particle swarm optimization-based fuzzy neural network and artificial neural network for supplier selection. Appl. Math. Model. **34**, 3976–3990 (2010)
26. Kar, A.K.: Using artificial neural networks and analytic hierarchy process for the supplier selection problem. In: 2013 IEEE International Conference on Signal Processing, Computing and Control (ISPCC), pp. 1–6 (2013)
27. Choy, K.L., Lee, W.B., Lo, V.H.: An intelligent supplier relationship management system for selecting and benchmarking suppliers. Int. J. Technol. Manag. **26**, 717–742 (2003)
28. Xu, L., Qian, F.: Supplier selection using ANN-based predictive model. In: 2010 Sixth International Conference on Natural Computation, vol. 4, pp. 1873–1876 (2010)
29. Fallahpour, A., Olugu, E.U., Musa, S.N., Khezrimotlagh, D., Wong, K.Y.: An integrated model for green supplier selection under fuzzy environment: application of data envelopment analysis and genetic programming approach. Neural Comput. & Applic. **27**, 707–725 (2015)
30. Choy, K.L., Lee, W.B., Lau, H.C., Lu, D., Lo, V.H.: Design of an intelligent supplier relationship management system for new product development. Int. J. Comput. Integr. Manuf. **17**, 692–715 (2004)
31. Guo, Z.X., Wong, W.K., Guo, C.: A cloud-based intelligent decision-making system for order tracking and allocation in apparel manufacturing. Int. J. Prod. Res. **52**, 1100–1115 (2014)
32. Leung, K.H., Choy, K.L., Lam, H.: An Intelligent Order Allocation System for Effective Order Fulfilment Under Changing Customer Demand. MATEC Web of Conferences (2019)
33. Chng, Z.M., Tang, C., Yang, D.K., Chopra, S., Womack, J., Starner, T.: Symbiotic artificial intelligence: order picking and ambient sensing. In: 2023 IEEE International Conference on Acoustics, Speech, and Signal Processing Workshops (ICASSPW), pp. 1–5 (2023)
34. Thiga, R.S., Gough, N.E.: Artificial intelligence applied to the discrimination of the order of multivariable linear systems. Int. J. Control. **20**, 961–969 (1974)
35. Bottani, E., Montanari, R., Rinaldi, M., Vignali, G.: Intelligent Algorithms for Warehouse Management. Intelligent Techniques in Engineering Management (2015)
36. Jiang, C., Xu, J., Li, S., Zhang, X., Wu, Y.: The order allocation problem and the algorithm of network freight platform under the constraint of carbon tax policy. Int. J. Environ. Res. Public Health. **19** (2022)
37. Lengare, G.D.: Order collecting robot with automatic allocation. Int. J. Res. Appl. Sci. Eng. Techn. (2019)
38. Elbert, R., Knigge, J.: How order placement influences resource allocation and order processing times inside a multi-user warehouse. Winter Simulation Conference (WSC). **2018**, 2921–2932 (2018)
39. Boute, R.N., Udenio, M.: AI in Logistics and Supply Chain Management. PROD: Empirical (Supply) (Topic) (2021)
40. Niranjan, D., Narayana, D.K., Rao, M.M.: Role of Artifical intelligence in logistics and supply chain. In: International Conference on Computer Communication and Informatics (ICCCI), vol. 2021, pp. 1–3 (2021)
41. Torres-Franco, M.: Artificial intelligence and supply chain management application, development, and forecast. Adv. Market. Custom. Relationsh. Manag. E-Services. (2021)
42. Helo, P.T., Hao, Y.: Artificial intelligence in operations management and supply chain management: an exploratory case study. Prod. Plan. Control. **33**, 1573–1590 (2021)

43. Soleimani, S.: A Perfect Triangle with: Artificial Intelligence, Supply Chain Management, and Financial Technology, vol. 6. Arch. Bus. Res. (2018)
44. Adah, W.A., Ikumapayi, N.A., Muhammed, H.B.: The ethical implications of advanced artificial general intelligence: ensuring responsible AI development and deployment. SSRN. (2023). https://doi.org/10.2139/ssrn.4457301. Accessed 15.02.2024
45. Negri, M., Cagno, E., Colicchia, C., Sarkis, J.: Integrating sustainability and resilience in the supply chain: a systematic literature review and a research agenda. Bus. Strateg. Environ. (2021). https://doi.org/10.1002/bse.2776. https://onlinelibrary.wiley.com/doi/epdf/10.1002/bse.2776. Accessed 15.02.2024
46. Sontake, A., Jain, N., Singh, A.R.: Sustainable supplier selection and order allocation considering discount schemes and disruptions in supply chain. In: Sachdeva, A., Kumar, P., Yadav, O., Garg, R., Gupta, A. (eds.) Operations Management and Systems Engineering Lecture Notes on Multidisciplinary Industrial Engineering. Springer, Singapore (2021). https://doi.org/10.1007/978-981-15-6017-0_5

Tactical Urbanism: A Means of Enacting Mobility Transition? A Literature Review of International Practice

Jarvis Suslowicz (iD) and **Helge Hillnhütter** (iD)

Contents

1   Introduction

In recent years, active travel has seen increased interest in research, policy and practice, presented as a vehicle for improving health, liveability, safety and sustainability in cities by altering the way we move within them [1]. These changes are typically approached through alterations to the built environment, not least through street space reallocation projects to create desirable spaces for alternative mobilities from previously unclassed road space [2]. However, in attempting to create structural change in mobility norms, practisers encounter significant barriers execution of plans [3]. A proposed solution lies in institutional mobilisation of tactical urbanism and associated concepts, which utilise designed temporariness to test and softly introduce new layouts for urban space for alternative uses [4].

J. Suslowicz (✉) · H. Hillnhütter
Norwegian University of Science and Technology (NTNU), Trondheim, Norway
e-mail: jarvis.suslowicz@ntnu.no

© The Author(s) 2025
M. Kioumarsi, B. Shafei (eds.), *The 1st International Conference on Net-Zero Built Environment*, Lecture Notes in Civil Engineering 237,
https://doi.org/10.1007/978-3-031-69626-8_136

Institutional mobilisation of temporary and tactical interventions in the urban realm increased in relevance during the COVID-19 pandemic in response to changes in the demand for intra-urban movement [5, 6]. The conditions of the pandemic provided a unique opportunity for expanding provision for walking, wheeling and cycling, where lower demands for street space for mobility permitted the introduction of new temporary measures under the legitimation of the acute emergency [7, 8]. The increased interest in tactical urbanism in practice has been accompanied by a similar, co-constitutive trend in research [9–11].

However, the transformative potential of tactical urbanism remains unclear. Rather, instead of acting as precursors to permanent or systemic change, interventions are often removed with the previous street design and mobility norm reinstated. This suggests that in practice, tactical urbanism is not yet able to achieve results that proponents describe. This chapter examines this problem across case studies from international practice in order to determine where difficulties remain and what changes may then lead to more effective utilisation. To structure findings, cases are weighed against five assessment criteria derived from Bertolini [12]: characteristics which must be present in street-based experiments that possess transitionary potential.

The following research questions are approached:

1. What are the main drivers behind short-termism in tactical urbanism planning?
2. To what degree can planning be generalised or is otherwise dependent on local structural conditions?
3. (How) can planning process be adapted to account for these blockages?

The chapter confirms earlier suggestions of a barriers in practice to radical change, where pressure is not placed on the existing regime while supporting the intended norms, and instead, decision-makers still display risk aversion found in permanent planning [13]. Rather, in practice, it is more typical for interventions to follow a path of least resistance in efforts to avoid conflict, finding 'spare space' where demand is lower [14]. Under the COVID-19 pandemic, demand for mobility initially dropped, providing room for experimentation without significant risk to decision-makers, an opportunity exploited by city authorities, particularly in North America and Europe [8, 15, 16], but increasingly in South America and Oceania [17–19]. The critical point then comes during recovery, when demand for space by other modes returns and temporary changes must find additional justification for remaining.

The findings indicate long-term goals have not universally been incorporated into the development of temporary projects. Temporary interventions, while more feasible to implement, remain vulnerable to removal in response to public and political will [20].

Finally, a note on language. Success and failure of projects in this context are measured against their ability to expand and continue along a trajectory which supports uptake of active travel modes, but this is not a definition used across cases. Many cases address short-term issues via short-term means, with little attention to upscaling, achieving their goals without further development. Success measures across cases are more diverse than the one implemented here and is, as such, not a blanket judgement of subjective quality.

2   Theoretical Framework

This review is grounded in insight from transition studies, including the concept of transition management (TM), which provides basis for determining necessary inputs and dynamics for temporary public space interventions to set into process long-term and potentially permanent changes [21].

Tactical urbanism itself has evolved greatly from its conceptual origins as citizen-initiated alterations in the local environment, based upon collective problematisation and dissatisfaction with official channels of orchestrating change, as described by Lydon & Garcia [22]. Rather, the term has been arguably co-opted for use by the very structures the concept aimed to find alternatives to. However, the usefulness of temporary and inexpensive design in supporting change towards active travel is increasingly recognised, particularly during a period of widespread austerity putting spending under scrutiny [19, 23]. Temporariness as used by official authorities allows for testing of uncertain concepts before committing to a redesign, supporting a strategic transition going beyond the local area. While doing so, socio-political change can be initiated to challenge and disrupt existing norms. Therefore, critics argue only urban experimentation at the institutional scale can drive a long-term transition [24].

Much of the understanding behind the role of tactical urbanism as enabling transition derives from the multi-level perspective (MLP) framework, used for analysing (potential) system change [21, 25]. Previously, a common public space catering to diverse uses, automobility has dominated the mobility system as the primary mode and use for space: remaking the public in its own image, designed to serve a singular purpose in moving vehicles as efficiently as possible. Alternative modes of different speeds, sizes and directionality come into conflict with this goal [26].

Transition studies propose that change is possible via breaks in the third level of the MLP, the landscape. The landscape is external and encompassing, but periodical breaks and crises provide space to undermine the hold of the regime. In this context, the key landscape opportunity is the pandemic. The sudden and widespread challenge of the COVID-19 pandemic had such a radical effect on patterns of urban mobility that it both forced the hand of policymakers to experiment and provided space to do so under exceptional conditions.

With experimentation in mind, Roorda et al. [27] developed a guidance manual for transition management derived from case studies in Western European cities. The characteristics used here to assess transitional capacity in practice are derived from this and assembled by Bertolini [12]: radical; challenge driven; feasible; strategic; communicative/mobilising. A degree of reinterpretation of certain characteristics in light of lessons from the pandemic is applied during discussion of results.

3   Methodology

The review was conducted semi-systematically using thematic analysis based on the pre-established research questions. Search terms were established through a precursive review. Strings using Boolean operators (Table 1) were used to ensure a focus on (active) mobility and to include articles identifiable only through temporariness and types or focus of intervention. Literature was collected and assembled in NVIVO, with further exclusions made during the initial read-through. Ninety-two works were included.

Search terms were chosen to provide as broad an interpretation as possible in order to include results not under the Euro-American vocabulary used to describe temporary active travel projects. Geographic bias is present in the included literature but is reflective of the field. In a recent commentary, von Schönfeld [28] highlights this bias as one connected to expectations about development in planning. In the Global North view mobility experiments as undoing perceived mistakes of a modernism which was never universal. In further development of the field, inclusion of diverse forms of southern urban informality can offer a more holistic view of utilising temporariness.

4   Results

As noted by Stevens, Awepuga & Dovey [19], temporary and tactical urbanisms, when referred to as such, are centred in Europe and North America within the English-language academic literature. Case studies featured in the collected literature reflect this, with the highest number of case studies based on Paris (6), London (6), Berlin (5), Milan (5) and San Francisco (5). A geographical shift can be noted, however, as the focus interest in tactical urbanism moved from bottom-up instances,

Table 1 Boolean operators used in literature search

Main term	Operator	Cycling terms	Operator	Walking terms	Operator	General terms
'tactical urbanism'	AND	'cycling' 'bicycle' 'bike'	OR	'pedestrian' 'walking'	OR	'active travel' 'mobility'
'street experiments'	AND	'cycling' 'bicycle' 'bike'	OR	'pedestrian' 'walking'	OR	'active travel'
'temporary'	AND	'bike lane' 'bike path' 'cycle path'	OR	'walkway' 'pavement' 'sidewalk'	OR	'active travel'
'pop-up'	AND	'bike lane' 'bike path' 'cycle path'	OR	'walkway' 'pavement' 'sidewalk'	OR	'active travel'

of which San Fransisco was a hotspot as the origin of the *parklet* concept, towards municipalities actively working to expand active travel networks and reorient space via tactical means, particularly during the pandemic [22, 29].

A key finding of the review is that municipalities and planners have, in practice, still limited capacity to position temporary projects towards supporting long-term mobility transitions locally. Public support is vital in justifying political decisions. Authorities employ varying participatory processes, from none at all under the 'emergency' mandate of the pandemic [7, 30], to co-creative strategies intending to provide citizen ownership over the projects throughout the entire development process [18].

4.1 Radical

Radical character is essential to socio-technical transitions of all kinds, as it represents a clear break with the existing status quo or dominant regime. Radical interventions in mobile space not only provide facilitation for active travel but also destabilise the automobility's dominance over space.

Using tactical urbanism to initiate changes in street space can permit more radical changes to occur, compared to their permanent counterparts, which can quickly become politically contentious [31, 32]. The aim of the temporary intervention in such cases, then, is to demonstrate the feasibility of the change both to the public and decision-makers, creating a base for further development.

Often, radical aspects of projects are lost during the process of experimentation, or in the transition to a permanent incantation following evaluation of the temporary version. Ohlund et al. [33] describe alterations made to a temporary bike lane to better accommodate vehicle traffic following neighbourhood 'bikelash' in Zapopan, Mexico. Landgrave-Serrano & Stoker [34] find that efforts to make a Slow Streets programme more widely palatable in the transition to a permanent version led to a drop in use, as restrictions for automobile traffic were removed in favour of traffic-calming measures. A temporary bike lane in London reviewed by Nikitas et al. [17] experienced a similar fate following poor levels of acceptance. Several articles identify major reasons for public backlash against interventions which aim to support active modes, namely concerns about limitations on parking and perceived increases in traffic as a result of spatial reallocation of road space [18, 33–36].

According to proponents, more radical changes can be tested when done temporarily as there is room for failure if initial versions do not function as well as expected [37, 38]. However, under a transitionary lens, this requires willingness to adapt measures based on evaluation, a contingency which must be planned into timelines and budgets.

The notion of 'pop-up' infrastructures exemplifies this risk aversion type of temporary urbanisms. A planner interviewed by Stevens, Awepuga & Dovey [19] connects this with short-termism, a version used by policymakers unprepared for long-term change. This was a key point of contention for many authors writing under

COVID-19 [39]. While some pop-up infrastructures were explicitly used as a precursive measure in a long-term strategy towards permanent change, in others it performed the role of a stopgap, addressing time-sensitive demands [40, 41].

Radical character, then, is not a universally applicable category, instead dependent upon how developed active travel provision is in the local context and citizens' perceptions of and willingness to accept change. Rather, it necessitates a fundamental break from dominant practices [12]. For cities with no or very weak cycling provision, testing using tactical urbanism may be relatively limited, but should focus on providing routes where most valuable [6, 42]. On the other hand, in cities with existing networks, tactical urbanism is used radically to strengthen weak points, creating safe routes for longer journeys [6, 11, 38]. It was these cities which achieved the most promising results while exploiting pandemic conditions, adapting existing plans to the context with clear lines for future development.

4.2   Challenge Driven

Bertolini defines the characteristic of 'challenge driven' as answerable to the question 'Is the experiment a step toward a potential long-term change pathway to address a societal challenge?' [12: 744].

While pandemic-era mobility interventions centre around the short- to mid-term challenges to urban movement presented by virus transmission, for many case studies the issues presented a landscape opportunity to catalyse pre-existing plans for a sustainable mobility transition. Buck [5] notes that the pandemic drew attention to existing health-related arguments for supporting active travel as a means of increasing physical activity, alongside environmental changes which support safer journeys.

This characteristic is essential when considering public and political support for street alterations, linking with communication. Projects must be justifiable, with intentions for further development clearly articulated.

The challenge of improving safety functions as a strong justification for temporary alterations, particularly when the foci of improvements are oriented around children's needs in residential areas and around schools. Positive results for children's recreation have served as mandate for the continuation of temporary projects while challenging motor normativity as the dominant use for street space [4, 37, 43]. Programmes such as School Streets are then often accompanied by complimentary policies on surrounding streets such as lower speed limits [38, 44, 45], recognising that discrete street changes are not adequate to bring about change of the scale intended.

Sets of complimentary measures arranged around the same challenge may more effectively support spatial changes, including campaigns to support lasting competencies for trying new modes like cycling, alongside wider measures to make driving less attractive [45].

4.3 Feasible

Feasibility can come under two interpretations. The first, as put by Bertolini [12], deals with the feasibility of the temporary intervention itself—it must be feasible to take place. The other interpretation looks longer term, where a project must be feasible to upscale and continue a productive trajectory, designed to be reproducible.

Temporariness is the key factor in what allows such interventions to become feasible in comparison to their permanent counterparts, as they are easily removable without the loss of a great investment, making them popular for creating space for cycling under the 'pop-up' bike lane concept [37, 45]. Temporariness can be material, or even discursive, using permanent materials but assuring that the space can and will be returned to its earlier state if it's deemed a failure. The Sarphatistraat Zuid project in Amsterdam is an example of this, laying a new bicycle street in red asphalt, removing traffic lights and lowering the speed limit—but still 'vowed to change the street back to its original state if unsuccessful' [13: 6]. It's unclear whether this assertion enabled feasibility amongst the public, as there was no formal consultation; a choice which in turn, did improve feasibility.

Public consultation generally improves feasibility in the short term [12, 46], but risks alienating the public when the project is considered for subsequent stages [35]. Where approval is high, however, it can be used to justify continued expansion of the scheme [27] or to create a base level of trust between the community and decision-makers useful in further iterations [19]. Some projects begin with participation as a key element but it is lost over time as the plan develops and authorities seek to lower the risks of public scrutiny [47]. In sum, the connection between feasibility and communicative/mobilising in practice is complex and variable, with the nature of this relationship discussed in more detail later.

Aesthetic purpose and alternative uses for street space 'beyond mobility' come into consideration of materials for reorganising space [12]. Plant pots, flowers and street furniture are popular for both creating new pathways for active travel from road space, but also for blocking through traffic from accessing whole stretches in the case of low traffic neighbourhoods and slow/active/school streets [31, 48, 49]. Aesthetics play a role in overall environmental quality, which can influence public perception and future feasibility of projects which may otherwise be oriented towards a specific user base who already have interest in walking or cycling [18, 48, 50].

4.4 Strategic

Strategic links with feasibility as defined as enabling a more forward-looking approach. According to Bertolini [12], it must be present in urban experiments as that which generates lessons for future practice, indicating that an intervention is even temporary in physical lifespan and openness in eventual directionality towards goals.

The role of strategy is highlighted under the question of goal formulation and framing. As noted by Gregg et al. [39], cities with most long-term success such as Paris, Milan and Bogota framed their street space alterations as transportation projects, while those in North America were largely positioned as responses to the sudden requirement for social distancing in the public realm. The emergency justification is certainly not new, but the COVID-19 crisis was unique as an opportunity for utilising the drop in demand for mobility itself to create entirely new habits of moving.

Varying practices of governing and funding projects are visible across the reviewed case studies. In Liverpool, funding from the National active travel fund supported pop-up cycling measures which, alongside lower initial traffic levels, allowed for the safe development of new mobility habits for cycling [5]. However, supportive measures were not continuously developed to account for the return of traffic following the first lockdowns, and as such, ground made in behavioural adaptation was lost.

Similarly, momentum can be lost during the initial project when temporary interventions are neglected. Implementation of a Thessaloniki pop-up bike lane failed to include supporting measures like enforcement of parking, drainage maintenance and connectivity at junctions, leading to very low overall satisfaction [41, 51]. Maintenance of interventions with temporary materials is necessary to maintain usability.

Paris can be considered a success case for COVID-era mobility intervention because of its precursive strategy to expand its city-wide cycling network, using the pandemic conditions as mandate to fill in gaps and improve connectivity. A higher share of protected, bi-directional lanes was implemented, with 58% connecting with four other bike lanes, compared to 28% of those built between 2005 and 2020 [20]. Improvements are reflected in an increase in cycling levels: '60% greater in 2020 and 2021 than in 2019' [37: 20]. Continual investment and support of measures, including making temporary infrastructure permanent, form part of Paris' strategy for maintaining the trajectory of the trend once the risk of transmission has subsided.

Communication between actors arises as a key challenge in maintaining momentum and longevity. During the process, actor networks do not remain consistent, being necessary to continuously expand as new competencies become required for subsequent stages, particularly in the eventual transition to permanence [52].

4.5   Communicative/Mobilising

In the literature, attention to communication within planning of tactical urbanism interventions varies greatly. Strategic avoidance of participatory processes was enabled during the COVID-19 pandemic as a crisis reaction, permitting suspension of normal requirements [31, 53, 54].

Some evidence exists to support participation throughout the entire process, where citizens may accept subsequent stages of development and change, even when initial efforts are not entirely successful [10, 13]. Others find 'more positive outcomes' and acceptance of change when communities were involved from the beginning, again being more willing to work with failures when there is some sense of ownership [18]. In these cases, higher levels of communication with the public lower the vulnerability.

In the same vein, poor communication can be a death sentence for projects-in-progress, halting them from reaching any subsequent stage of development. Brovarone, Staricco and Verlinghieri [32] highlight a lack of public knowledge of projects' justification as a key factor in project failure. Kyriakidis [55] cites the lack of a clear message about the positive impacts that road reallocation would have on the image of Athens as reason for low approval of a pilot version. In essence, the proposed benefits of changing street space must be clearly justifiable.

Beyond public participation, communication within scales of governance and external interest groups are shown by several case studies to greatly impact the post-intervention viability of networked projects [10, 56, 57]. Municipalities without direct decision-making power over their streets must negotiate with higher-level governmental authorities—provincial or national—for permission or funding to initiate the change [57]. Where goal formulations are inconsistent across different levels of responsible parties, it may result in continuation of projects being dependent on conditions impossible to meet [56]. Rather, new connections and often funding sources must be sought to maintain established momentum [5]. Power for the continuation or renewal of temporary projects ultimately lies within networks which are constantly evolving, leaving experiments vulnerable even when broadly supported [12, 28].

An ideal sense of communication may then be dependent upon what is to be achieved and the context of implementation. For practitioners, keeping internal actors informed of progress and goals at each point in the process is necessary for further development of short-term projects into significant changes.

5 Discussion

The aims of this review are to gain an overview of whether city authorities have been able to utilise tactical urbanism practices to set into motion longer-term plans for transitioning to active travel-based mobility norms. In cases where that has not been achieved, the paper aims to identify key blockages for practice, leading to pathways for alleviating these issues with attention to success cases.

Good practice for tactical urbanism in many ways mirrors standards for mobility planning more generally. Projects which achieved the greatest impact on transforming mobility locally were those which acknowledged and accounted for the characteristics of mobility, considering needs for measures beyond the intervention and beyond the period of initial implementation. Paris effectively utilised landscape opportunity of the breakdown in mobility habits to provide grounds for building new ones in line with municipal goals. The city government implemented a strategy from established plans and adapted to the pandemic period, creating viable networks for cycling where demand was highest [11, 20], rather than along paths of least resistance [14].

Consideration of strategy for implementation is, however, dependent upon existing institutional factors: the nature of funding schemes; local powers and political willingness to enact change; stage of local progression towards that change and existing public support. Communication is emphasised here as critical but variable in suitable application. Certain conflicts can be anticipated and managed during the intervention stage, recognising that they challenge fundamental power relationships in cities.

6   Conclusions

The main objective of this literature study is to provide a wide overview of mobility-oriented tactical urbanism practice by institutional authorities, evaluated against assessment criteria which weigh transitional capacity of experimental urban interventions. Temporary projects can benefit from borrowing processual characteristics from their permanent counterparts, better utilising a strategy suited to the complexities of planning for mobility, recognising long-term change and multiscalar impacts.

Future research could build upon the points raised in this chapter in the creation of guidelines for practice, testing whether these issues can be eased. Rather, it may be that in many contexts, structural issues associated with funding, decision-making and implementation power must be resolved before full potential can be reached.

The field could benefit from follow-up study of eventual outcomes of projects, with attention to the evaluation and decision-making of subsequent iterations.

References

1. Macmillen, J., Givoni, M., Banister, D.: Evaluating active travel: decision-making for the sustainable city. Built Environ. **36**(4, 5), 519–536 (2010)
2. Aldred, R.: Built environment interventions to increase active travel: a critical review and discussion. Curr. Environ. Health Rep. **6**, 309–315 (2019)
3. Hickman, R., Sallo, K.: The political economy of streetspace reallocation projects: Aldgate Square and Bank Junction, London. J. Urban Des. **27**(4), 397–420 (2022)
4. Barata, A.F., Fontes, A.S.: Tactical urbanism and sustainability: tactical experience in the promotion of active transportation. Int. J. Urban Civ. Eng. **11**(6), 734–739 (2017)
5. Buck, M.: Disruption, an opportunity to facilitate a long-term modal shift to cycling? Stories, lessons and reflections from the COVID-19 pandemic. Active Travel Stud. **3**(2), 1–20 (2023)
6. Fenu, N.: Bicycle and urban design. A lesson from COVID-19. TeMA J. Land Use Mobil. Environ. **14**(1), 69–92 (2021)

7. Aquilué, I., Caicedo, A., Moreno, J., Estrada, M., Pagès, L.: A methodology for assessing the impact of living labs on urban design: the case of the furnish project. Sustainability. **13**, 4562 (2021)
8. Dunning, R., Nurse, A.: The surprising availability of cycling and walking infrastructure through COVID-19. Town Plan. Rev. **92**(2), 149–155 (2020)
9. Becker, S., von Schneidemesser, D., Caseiro, A., Götting, K., Schmitz, S., von Schneidemesser, E.: Pop-up cycling infrastructure as a niche innovation for sustainable transportation in European cities: an inter- and transdisciplinary case study of Berlin. Sustain. Cities Soc. **87**, 104168 (2022)
10. Blewden, M., MacArthur-Beadle, S., Haines, G., Raja, A., Nord, A., Hawley, G.: Streets for tomorrow. . .today. J. Road Saf. **33**(4), 21–31 (2022)
11. Creutzig, F., Lohrey, S., Franza, M.V.: Shifting urban mobility patterns due to COVID-19: comparative analysis of implemented urban policies and travel behaviour changes with an assessment of overall GHG emissions implications. Environ. Res.: Infrastruct. Sustain. **2**, 041003 (2022)
12. Bertolini, L.: From "streets for traffic" to "streets for people": can street experiments transform urban mobility? Transp. Rev. **40**(6), 734–753 (2020)
13. VanHoose, K., de Gante, A.R., Bertolini, L., Kinigadner, J., Büttner, B.: From temporary arrangements to permanent change: assessing the transitional capacity of city street experiments. J. Urban Mobil. **2**, 100015 (2022)
14. Lovelace, R., Talbot, J., Morgan, M., Lucas-Smith, M.: Methods to prioritise pop-up active transport infrastructure. Findings. (2020). https://findingspress.org/article/33765-treating-covid-with-bike-lanes-design-spatial-and-network-analysis-of-pop-up-bike-lanes-in-paris (Online-only journals)
15. Fields, B., Cradock, A.L., Barrett, J.L., Hull, T., Melly, S.J.: Active transportation pilot program evaluation: a longitudinal assessment of bicycle facility density changes on use in Minneapolis. Transp. Res. Interdiscip. Perspect. **14**, 100604 (2022)
16. Glaser, M., Krizek, K.J.: Can street-focused emergency response measures trigger a transition to new transport systems? Exploring evidence and lessons from 55 US cities. Transp. Policy. **103**, 146–155 (2021)
17. Nikitas, A., Tsigdinos, S., Karolemeas, C., Kourmpa, E., Bakogiannis, E.: Cycling in the era of COVID-19: lessons learnt and best practice policy recommendations for a more bike-centric future. Sustainability. **13**(9), 4620 (2021)
18. Sargisson, R.J., Brown, G.S., Hanna, C., Charlton, S.G., Kurian, P., Barrett, P., Milfont, T.L.: Citizen responses to tactical urbanism initiatives in Aotearoa New Zealand. SSRN Electron. J. (2022). https://findingspress.org/article/33765-treating-covid-with-bike-lanes-design-spatial-and-network-analysis-of-pop-up-bike-lanes-in-paris (Online-only journals)
19. Stevens, Q., Awepuga, F., Dovey, K.: Temporary and tactical urbanism in Australia: perspectives from practice. Urban Policy Res. **39**(3), 262–275 (2021)
20. Moran, M.E.: Treating COVID with bike lanes: design, spatial, and network analysis of 'pop-up' bike lanes in Paris. Findings. (2022). https://doi.org/10.32866/001c.33765
21. Geels, F.W.: A socio-technical analysis of low-carbon transitions: introducing the multi-level perspective into transport studies. J. Transp. Geogr. **24**, 471–482 (2012)
22. Lydon, M., Garcia, A.: Tactical Urbanism: Short-Term Action for Long-Term Change. Island Press, Washington, DC (2015)
23. Angelidou, M.: Tactical urbanism: reclaiming the right to use public spaces in Thessaloniki, Greece. In: Proceedings of 4th Conference on Sustainable Urban Mobility (CSUM2018), pp. 241–248. Springer, Cham (2019)
24. Savini, F., Bertolini, L.: Urban experimentation as a politics of niches. Environ. Plann. A: Econ. Space. **51**(4), 831–848 (2019)
25. Whitmarsh, L.: How useful is the multi-level perspective for transport and sustainability research? J. Transp. Geogr. **24**, 483–487 (2012)
26. Urry, J.: The 'system' of automobility. Theory Cult. Soc. **21**(4–5), 25–39 (2004)

27. Roorda, C., Wittmayer, J., Henneman, P., van Steenbergen, F., Frantzeskaki, N., Loorbach, D.: Transition Management in the Urban Context: Guidance Manual. DRIFT, Erasmus University Rotterdam, Rotterdam (2014)
28. Von Schönfeld, K.C.: On the 'Impertinence of impermanence' and three other critiques: reflections on the relationship between experimentation and lasting – or significant? – change. J. Urban Mobil. **5**, 100070 (2024)
29. Douglas, G.C.C.: Privilege and participation: on the democratic implications and social contradictions of bottom-up urbanisms. In: Arefi, M., Kickert, C. (eds.) The Palgrave Handbook of Bottom-Up Urbanism, pp. 305–321. Palgrave Macmillan, Cham (2018)
30. Flynn, A., Thorpe, A.: Pandemic pop-ups and the performance of legality. In: Doucet, B., van Melik, R., Filion, P. (eds.) Global Reflections on COVID-19 and Urban Inequalities: Volume 1: Community and Society, pp. 25–35. Bristol University Press, Bristol (2021)
31. Stevens, Q.: Temporality. In: Dovey, K., Stevens, Q. (eds.) Temporary and Tactical Urbanism: (Re)assembling Urban Space, pp. 129–148. Routledge, Abingdon (2022)
32. Brovarone, E.V., Staricco, L., Verlinghieri, E.: Whose is this street? Actors and conflicts in the governance of pedestrianisation processes. J. Transp. Geogr. **107**, 103528 (2023)
33. Ohlund, H., El-Samra, S., Amezola, D., Morfin, J.C.S., Zaragoza, C.Z., Gónzalez, S.A.: Building emergent cycling infrastructure during the COVID-19 pandemic: the case of Zapopan, Méxic. Front. Sustain. Cities. **4** (2022). https://doi.org/10.3389/frsc.2022.805125
34. Landgrave-Serrano, M., Stoker, P.: Increasing physical activity and active transportation in an Arid City: slow streets and the COVID-19 pandemic. J. Urban Des. **28**(2), 155–173 (2023)
35. Rérat, P., Haldiman, L., Widmer, H.: Cycling in the era of COVID-19: the effects of the pandemic and pop-up cycle lanes on cycling practices. Transp. Res. Interdiscip. Perspect. **15**, 100677 (2022)
36. VanHoose, K., Bertolini, L.: The role of municipalities and their impact on the transitional capacity of city street experiments: lessons from Ghent. Cities. **140**, 104402 (2022)
37. Buehler, R., Pucher, J.: Cycling through the COVID-19 pandemic to a more sustainable transport future: evidence from case studies of 14 large bicycle-friendly cities in Europe and North America. Sustainability. **14**(12), 7293 (2022)
38. Harris, M., McCue, P.: Pop-up cycleways. J. Am. Plan. Assoc. **89**(2), 240–252 (2023)
39. Gregg, K., Hess, P., Brody, J., James, A.: North American street design for the coronavirus pandemic: a typology of emerging interventions. J. Urban.: Int. Res. Placemaking Urban Sustain. (2022). https://doi.org/10.1080/17549175.2022.2071970
40. Hassen, N.: Leveraging built environment interventions to equitably promote health during and after COVID-19 in Toronto, Canada. Health Promot. Int. **37**(2), 1–15 (2022)
41. Kinis, D., Palantzas, G., Nalmpantis, D.: Evaluation of the functionality, safety, and environmental impact of the new pop-up bicycle lane on Konstantinos Karamanlis Avenue in Thessaloniki, Greece. IOP Conf. Series: Earth Environ. Sci. **1123**(1), 012054 (2022)
42. Francke, A.: Cycling during and after the COVID-19 pandemic. Adv. Transp. Policy Plann. **10**, 265–290 (2022)
43. Cruz, S.S., Paulino, S.R.: Experiences of innovation in public services for sustainable urban mobility. J. Urban Manag. **11**, 108–122 (2022)
44. Fontes, A.S.: We protect schools: tactical urbanism actions in the school surroundings of Barcelona, Spain. J. Environ. Manag. Sustain. **10**(1), 1–21 (2021)
45. Kim, J.: COVID-19's impact on local planning and urban design practice: focusing on tactical urbanism and the public realm with respect to low income communities. J. Urban.: Int. Res. Placemaking Urban Sustain. (2022). https://doi.org/10.1080/17549175.2022.2146155
46. Zieff, S.G., Hipp, J.A., Eyler, A.A., Kin, M.: Ciclovía initiatives: engaging communities, partners, and policy makers along the route to success. J. Public Health Manag. Pract. **19**(3), 74–82 (2013)
47. Schaller, S., Guinand, S.: Pop-up landscapes: a new trigger to push up land value? Urban Geogr. **39**(1), 54–74 (2018)

48. Marcheschi, E., Vogel, N., Larsson, A., Perander, S., Koglin, T.: Residents' acceptance towards car-free street experiments: focus on perceived quality of life and neighborhood attachment. Transp. Res. Interdiscip. Perspect. **14**, 100585 (2022)
49. Thorpe, A.: Prefigurative infrastructure: mobility, citizenship, and the agency of objects. Int. J. Urban Reg. Res. **47**, 183–199 (2023)
50. Fyhri, A., Karlsen, K., Pokorny, P.: Protected Bike Lanes and Pop-Up Cycling Infrastructure, TØI Report 1837/2021 (2021)
51. Katsavounidou, G., Papagiannakis, A., Christakidis, I., Mavros, O.: Emergent Bicycle Infrastructure During the COVID-19 Pandemic: The Karamanli Avenue Pop-Up Cycle Lane in Thessaloniki, Greece. CSUM 2022: Smart Energy for Smart Transport, pp. 714–727. Springer, Cham, Switzerland (2022)
52. Cariello, A., Ferorelli, R., Rotondo, F.: Tactical urbanism in Italy: from grassroots to institutional tool—assessing value of public space experiments. Sustainability. **13**, 11482 (2021)
53. Oluyede, L., Combs, T.S., Pardo, C.F.: The why and how of COVID streets: a city-level review of research into motivations and approaches during a crisis. Transp. Rev. **44**(2), 345–367 (2024)
54. Madanipour, A.: Cities in Time: Temporary Urbanism and the Future of the City. Bloomsbury, London (2017)
55. Kyriakidis, C., Chatziioannou, I., Iliadis, F., Nikitas, A., Bakogiannis, E.: Evaluating the public acceptance of sustainable mobility interventions responding to COVID-19: the case of the Great Walk of Athens and the importance of citizen engagement. Cities. **132**, 103966 (2013)
56. Dudley, G., Bannister, D., Schwanen, T.: Low traffic neighbourhoods and the paradox of UK government control of the active travel agenda. Polit. Q. **93**(4), 585–593 (2022)
57. Sagaris, L.: Governance, human agency and other blindspots in active transport practice and research. Active Travel Stud. **1**(1), 1–12 (2021)

Emergy Framework, from Building Stock Model to Retrofit Model: A Study in Mexico's City Public Office Buildings

Ivett Flores

Contents

1 Introduction

Transforming the existing building stock to enhance energy efficiency is a crucial strategy for climate change mitigation and adaptation. This transformation largely involves the reduction of operational emissions through energy-efficient building operations and the integration of on-site renewable energy generation, making

---

Emergy (short for embodied energy) is a concept that represents the total amount of energy that is consumed directly or indirectly to produce a product or service, expressed in terms of one type of energy, often solar energy. It was developed by systems ecologist Howard T. Odum in the 1980s as part of his work on ecological economics and systems theory.

I. Flores (✉)
Technical University of Braunschweig, Braunschweig, Germany
e-mail: i.flores-nunez@tu-braunschweig.de

© The Author(s) 2025
M. Kioumarsi, B. Shafei (eds.), *The 1st International Conference on Net-Zero Built Environment*, Lecture Notes in Civil Engineering 237,
https://doi.org/10.1007/978-3-031-69626-8_137

Building Energy Retrofit (BER) strategies vital for substantial emission reductions across environmental, energy, and societal dimensions. Notably, the quantity of embodied carbon dioxide equivalent ($CO_{2\text{-eq}}$), which arises throughout the lifecycle of building materials and components, often surpasses operational emissions, accounting for 69% in residential, 76% in warehouses, and 67% in commercial buildings [1]. Global initiatives aiming for a net-zero carbon building stock by 2050 require a 50% reduction in direct building CO_2 emissions and a 60% reduction in indirect emissions from power generation by 2030, correlating to a consistent annual decrease of about 6% from 2020 to 2030 [2].

Despite the focus on new constructions, retrofitting existing buildings is essential to meet the 2050 energy reduction targets, especially since 90% of current buildings will remain by that year [3]. The potential of retrofit strategies is emphasized by their proven capability to significantly enhance environmental performance, with documented energy savings of 66%, 50%, and 75% in various studies [4–6]. However, current tools for building evaluation often fail to fully recognize the socio-economic and environmental impacts, collectively known as Non-Energy Benefits, which include improved Indoor Air Quality (IAQ), resilience, health, and productivity [7]. These aspects are increasingly relevant in the evolving climate change economy and require greater emphasis in adaptation scenarios and decision-making.

In Mexico, a significant energy consumer and CO_2 emitter, ranking 11th globally in greenhouse gas emissions and heavily reliant on fossil fuels—with oil and natural gas comprising 45% and 38% of its energy supply, respectively [8]—the importance of retrofit strategies is accentuated. These strategies are critical not just for reducing fossil fuel dependency and enhancing energy security but also for stimulating economic benefits, improving indoor comfort, realizing long-term cost savings, and fostering urban planning and renewal efforts. Retrofitting in Mexico could thus catalyze a transformative shift toward a sustainable and resilient built environment.

The overarching goal of this study is to contribute to the body of knowledge by formulating a framework and evaluation method system for retrofit models tailored to temperate climates akin to the conditions prevalent in Mexico City throughout the year. This research endeavors to bridge the existing gaps in understanding, ultimately guiding the formulation of cost-benefit efficient retrofit strategies that align with the city's unique environmental context (Fig. 1).

2   Research Statement and Objectives

The landscape of Building Retrofit Models, while robust in many respects, reveals significant potential for enhancing the way Energy Efficiency improvements and decarbonization are approached. Existing models have yet to fully integrate a cohesive framework that effectively balances Energy Efficiency, embodied carbon, and the broader architectural context, often resulting in fragmented retrofit strategies. This is compounded by the absence of established methodologies to perform the Emergy analysis concept at an urban scale, limiting the ability to thoroughly assess environmental impacts of retrofit interventions. Moreover, the challenges in

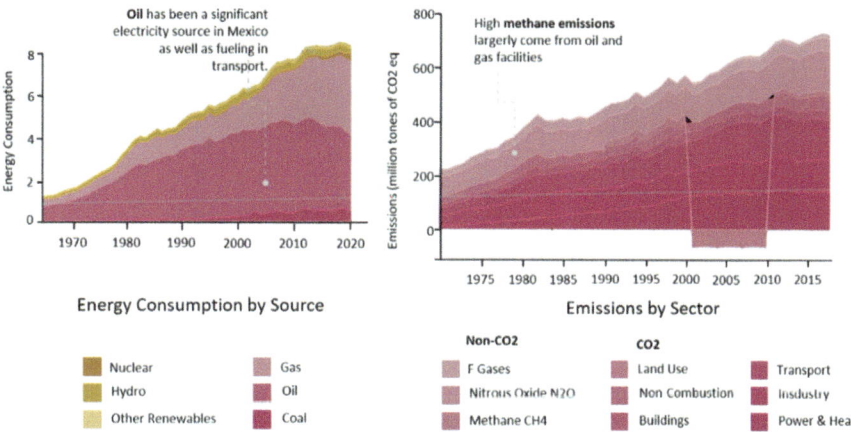

Fig. 1 Energy consumption by source and emissions by sector in Mexico (Carbon Brief Profile 2023)

establishing reliable baselines for evaluating retrofit impacts due to inconsistent and inaccessible data further exacerbate the difficulty in measuring true retrofit efficacy. Disparities in accuracy between urban-scale and individual-scale retrofit analysis undermine effective decision-making and hinder the adoption of impactful strategies. Additionally, current models do not sufficiently account for non-energy benefits—such as improved indoor air quality, occupant health, and resilience—which are critical for a holistic assessment of retrofit interventions. Addressing these aspects within a value-based framework designed to enhance retrofit strategies offers a significant opportunity to improve the efficacy and comprehensive nature of building retrofits, driving forward the agenda for sustainable urban development.

The research objectives of this study are designed to address the gaps and challenges identified within the realm of Building Retrofit Models, with a clear focus on enhancing energy efficiency and decarbonization strategies.

Developing a Robust Building Stock Model Framework One of the primary objectives of this research is to establish a comprehensive and robust Building Stock Model framework. This framework will be intricately tied to a Baseline Approach, enabling accurate comparisons and assessments of retrofit interventions. By creating a systematic and standardized model, this objective aims to provide a reliable foundation for evaluating and comparing building efficiency, from the baseline to its performance after the retrofit strategies

Emergy analysis of Retrofit Proposals This study recognizes the crucial significance of considering the entire lifecycle of buildings, including their embodied carbon and operational emissions. To address this, an essential research objective is to formulate a method for the Emergy calculation of the retrofit proposals. This involves conducting thorough assessments encompassing various stages of a building's lifecycle, providing a holistic perspective on its environmental impact.

Fig. 2 Selected buildings mapping (Cuauhtemoc District, Mexico City)

Developing a Retrofit Model Framework as a Library of Solutions The development of an effective Retrofit Model framework is another core objective of this research. This framework will function as a dynamic "Library of Solutions," [1] offering a diverse range of retrofit strategies tailored to specific contexts. Through meticulous research and analysis, this objective aims to compile a comprehensive repository of technically sound and environmentally impactful retrofit options (Fig. 2).

3 State of the Art

The review of the current state of the art in the field of Building Retrofit Models reveals key concepts and research works that have significantly influenced and provided valuable insights in the development of this study:

Retrofit Strategies Implementation for Existing Buildings A noteworthy observation from the current body of literature is the prevalence of tools and methods designed to facilitate energy-efficient features primarily during the design and predesign phases of existing buildings. However, the literature considerable reduce when it is related to develop retrofit strategies for existing buildings, especially in an urban scale. In this context, the research of Claudio Nägeli et al. holds substantial importance, particularly, their works titled "Towards agent-based building stock modeling: Bottom-up modeling of long-term stock dynamics affecting the energy and climate impact of building stocks, Energy and Buildings" [1] and "Synthetic building stocks as a way to assess the energy demand and greenhouse gas emissions of national building stocks. Energy and Buildings" [2] have offered critical insights into retrofit strategies through their case studies, such as the retrofit study conducted in Switzerland. Furthermore, Claudio Nägeli et al.'s article "Methodologies for Synthetic Spatial, Building Stock Modelling: Data-Availability-Adapted Approaches for the Spatial Analysis of Building Stock Energy Demand" forms the foundational basis for my forthcoming synthetic urban-scale analysis.

Data-Driven Analysis Tools for Building Retrofit Models The development of effective Building Retrofit Models is significantly influenced by the availability of instruments and tools. In shaping my own model as a Data-Driven Analysis Tool, the work of C. Szum et al. has provided valuable guidance. Their research, especially the texts "Data-Driven Analysis Tool Plays Critical Role in Climate Neutral Buildings Advances in Applied Energy [9] and "Targeting Building Energy Efficiency Opportunities—An Open-source Analytical and Benchmarking Tool," [8] offers insights into open-source analytical tools that play a pivotal role in advancing climate-neutral building practices.

Building Stock Models and Retrofit Focus in Latin America While the research landscape is rich with studies focusing on retrofit strategies in the residential and non-residential sectors across European and North American countries, the perspective is notably limited in the context of Latin America, particularly Mexico. In this regard, the works of researchers such as Luis Eduardo Medrano-Gómez, Azucena Escobedo Izquierdo [10], Miguel Flores et al. [3], Itzell Torres et al. [4], and Rodrigo Mercado Fernández et al. [5] have emerged as pivotal resources. Their research, including non-domestic Passivhaus retrofit proposals, addresses the unique challenges and opportunities of retrofitting in the Latin American context.

4 Theoretical Framework

The theoretical foundations of this research are rooted in the dynamics and hierarchies of energy systems that are fundamental to achieving significant environmental impacts in architecture and urban planning. Recognizing buildings as open thermodynamic systems, this study leverages the concept of emergy—short for "energy memory" [6]—as a practical tool to assess the cumulative energy quality and historical energy use within these systems. Emergy provides a critical lens through which the magnitudes and hierarchies of energy systems can be comprehensively understood, highlighting their capacity to influence architectural and urban outcomes.

Central to this framework are the laws of thermodynamics, which are pivotal in the design of energy systems. The first law, emphasizing energy conservation, and the second law, which introduces the inevitability of increasing entropy in isolated systems, together guide the development of sustainable non-isolated energy systems in built environments. This research extends beyond traditional Life Cycle Assessment by focusing on how energy system analysis and design can transform the overall power and efficiency of urban architectural systems.

For existing buildings, the theoretical approach supports the visualization of retrofit strategies through two primary avenues to improve a building's Emergy Sustainability Index (ESI):

Maximizing Site Renewable Source This involves harnessing the renewable energy potential of the site to optimally reduce operational energy use.

Evaluating Carbon Impacts in Usage Stage and Maintenance Phase This focuses on assessing and mitigating the carbon footprint associated with potential retrofit strategies over the building's lifecycle.

5   Method

This section outlines the methodology employed to address the complexities inherent in assessing building energy demand and greenhouse gas emissions at an urban scale, particularly in the face of limited information availability. The method is structured into three primary steps, collectively aimed at creating an "Archetype Building Information" framework. These steps are conducted iteratively, guided by the pursuit of uncovering an accurate representation of the urban context.

Step 1—Synthetic Stock Generation The initial step entails the transition from the architectural to the urban scale, necessitating a paradigm shift to cope with information scarcity. Buildings are treated as fundamental model entities, forming the basis for further analysis. This step is directly related to the 1st research goal, "Developing a robust building stock model framework." It is made up of three parts:

Characterization of the Buildings. This involves defining the key attributes that distinguish one building unit from another. These attributes encompass physical features, usage patterns, and potential retrofit possibilities.

Building Agents: Incorporating the human element, building owners, and users into the model adds a layer of realism. Their behavior, preferences, and decisions play a pivotal role in shaping retrofit choices and energy-related strategies.

Archetype classification. This subset recognizes the need for practicality. It involves categorizing the urban building stock based on critical characteristics and finding archetype buildings.

Step 2—Building Stock Dynamics The second step is related to the second goal of the research, "Developing an emergy analysis of retrofit proposals." The intricacies of this step are informed by economic, environmental, and policy considerations. Three distinct facets constitute this step:

Identification of Retrofit Strategy. Each building unit is subjected to analysis to identify the most suitable retrofit strategy as a function of its typology.

Selection of Retrofit Strategy. The benefit-to-cost ratio guides the selection process, underlining the significance of economic feasibility.

Development of Retrofit Strategy. This aspect considers technological feasibility, policy restrictions, and the availability of market options.

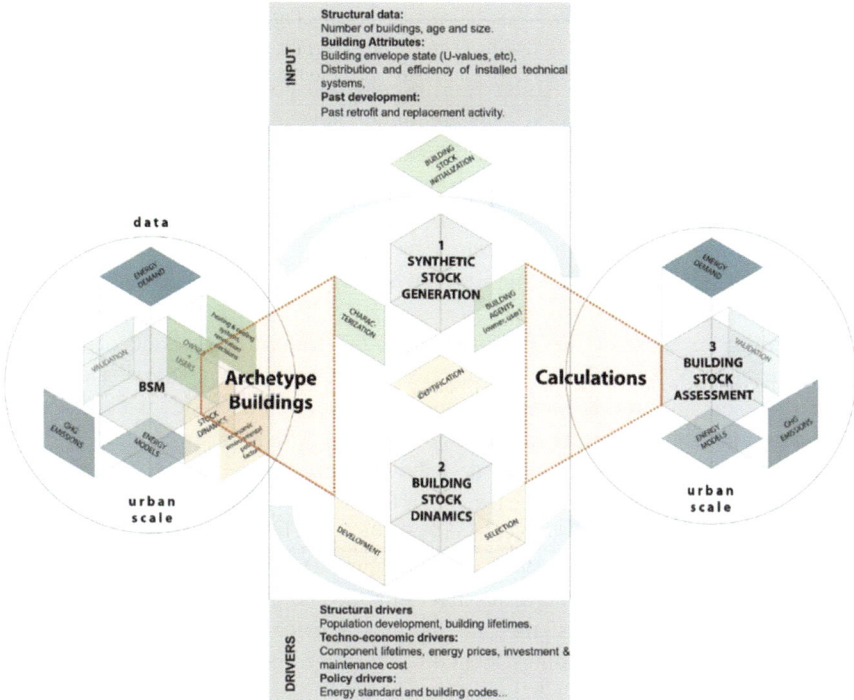

Fig. 3 Method overview

Step 3—Building Stock Assessment It involves validating the synthesized data and refining the Archetype Building Information framework. This step ensures the alignment of the framework with real-world data and its accuracy in reflecting urban building stock dynamics. The iterative nature of this research process is essential, as it embodies a continuous quest for the most accurate representation of the urban context (Fig. 3).

6 Hypothesis

Within the context of urban sustainability and building stock retrofitting, this study posits that by employing a value-based framework capable of amalgamating building stock performance data with the intricate interplay of Building Stock Dynamics (BSD), municipal decision-making processes can be significantly streamlined. Consequently, the urban environment will be empowered with an efficient approach to assess and compute retrofit scenarios on a citywide scale. The anticipated outcome of this approach is not solely limited to the enhancement of building performance, but also extends to encompass Non-Energy Benefits (NEBs) which will be quantitatively appraised, positioning them as indispensable as the evaluation of CO_2 emissions.

7   Preliminary Results

Regarding the lack of information from private and governmental buildings, the definition of Archetypes has a relevant importance in the development of the analysis, because they will represent the stock. As explained in the methodology, the retrofit strategies must be a function of each typology. Archetypes were defined under three main categories: Location, building characterization and renewable energy. Each of them will be measured by indicators, those related to the NEB (Non-Energy Benefits) concept and Emergy Sustainability Index (ESI) that includes cost/benefit analysis.

According to Mexico City's climatic data, the comfortable period of the year corresponds to 67%, 33% for hot days, and 0% for too cold days. This is an important fact, since too hot means temperatures from 25 to 35 °C, which exceeds the Environmental comfort temperature according to ASHRAE, from 20 to 25.5 °C. Mexico's temperature projection models show that the temperature will rise yearly, besides the correspondent increase use of Air Conditioning (Fig. 4). According to the CBE Thermal Comfort Tool Psychrometric chart, Air Conditioning mechanical system is the last source to reach comfort (Fig. 5). Sixty-seven percent of the building stock uses this mechanism, increasing the energy consumption, especially considering that HVAC systems and lighting, are the most energy consuming systems in Mexico City buildings, with 48 and 24% respectively.

Regarding the Building Characterization, four main periods were identified: Barroco Colonial (1521–1821), nineteenth century, early twentieth (1821–1920), Modernism and International Movement (1920–1985), and the present times after Mexico City 1985 Earthquake. The information regarding to constructive systems, materials and Thermal Resistance U-Values are described in Fig. 5. Each of the 65 buildings from the stock were classified according to its construction year. It is quite clear the relation between building's characterization and its year of construction since Architecture is an expression of its time.

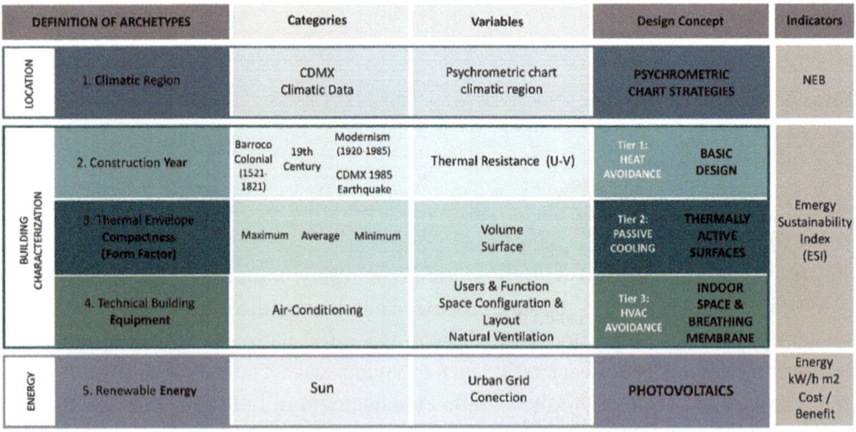

DEFINITION OF ARCHETYPES		Categories	Variables	Design Concept		Indicators
LOCATION	1. Climatic Region	CDMX Climatic Data	Psychrometric chart climatic region	PSYCHROMETRIC CHART STRATEGIES		NEB
BUILDING CHARACTERIZATION	2. Construction Year	Barroco Colonial (1521-1821) 19th Century Modernism (1920-1985) CDMX 1985 Earthquake	Thermal Resistance (U-V)	Tier 1: HEAT AVOIDANCE	BASIC DESIGN	Emergy Sustainability Index (ESI)
	3. Thermal Envelope Compactness (Form Factor)	Maximum Average Minimum	Volume Surface	Tier 2: PASSIVE COOLING	THERMALLY ACTIVE SURFACES	
	4. Technical Building Equipment	Air-Conditioning	Users & Function Space Configuration & Layout Natural Ventilation	Tier 3: HVAC AVOIDANCE	INDOOR SPACE & BREATHING MEMBRANE	
ENERGY	5. Renewable Energy	Sun	Urban Grid Conection	PHOTOVOLTAICS		Energy kW/h m2 Cost / Benefit

Fig. 4 Direct relation between the Archetypes definition and the design concepts for retrofit strategies

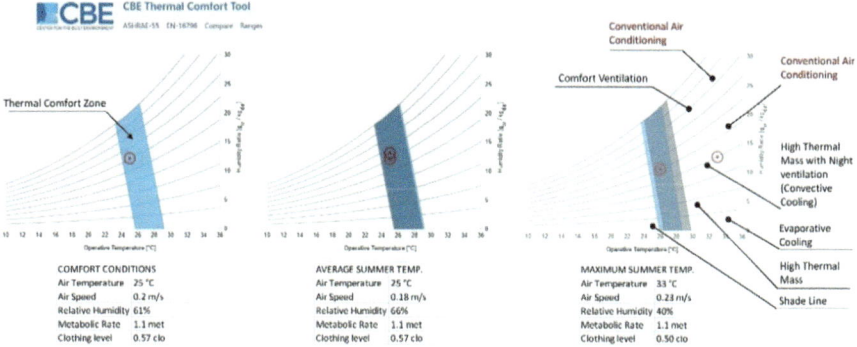

Fig. 5 Psychrometric chart strategies for Mexico City Summer Temperature. CBE Thermal Comfort Tool. CBE Thermal Comfort Tool for ASHRAE-55 (berkeley.edu)

As part of the building characterization, de degree of Compactness of building forms, must be considered as a function of their building shape, regardless their size. According to the square-cube lay, this means that compactness is a scale independent metric to measure the degree of building forms. Three classification of compactness were chosen: Maximum (1), average (1.18), and minimum (3.20).

The fact that a building uses or no Air conditioning will be taken into account in the global energy performance. The challenge is avoiding this system with low-tech strategies and bring buildings thermal comfort.

As a preliminary result of this stage in the research, were identified six archetype buildings, which will be analyzed in their baseline energy performance and in the development of retrofit strategies, according to the proposed design concepts:

Tier 1: Heat avoidance—Basic Design
Tier 2: Passive Cooling—Thermally Active Surfaces
Tier 3: HVAC Avoidance—Indoor Space and Breathing membrane.

The use of photovoltaics as a form to capture renewable energy will be explored since Mexico has a solar radiation of 5.5 kWh/m^2, five times the solar radiation of Germany; however, solar energy generated in the European country is 44.2 times higher.

8 Conclusions

Finding the correct and more representative Archetype buildings represents an inflection point in the research. The results will be proportionally assigned to the corresponding building among the stock. Not only the energy performance as a baseline but also their variations and feasibility after retrofit strategies.

The natural correspondence between the year or period of construction and its building characterization, as well as the proportional relation between the degree of compactness and optimality of building forms, makes possible the replicability of this model in any context. However, the renewable energy source must be reconsidered according to the location (Fig. 6).

ARCHITECTURAL STAGES IN CDMX & CHARACTERISTICS							
PERIOD		**CONSTRUCTIVE MATERIALS**	**CONSTRUCTIVE SYSTEMS & ELEMENTS**	**ELEMENT**	**U-VALUE**	**HIGHLIGHTS**	**EXAMPLES**
1	**Barroco Colonial (1521-1821)**	Stone — Volcanic Tezontle Cantera — Used in fachades and structural elements; Brick — Often used for internal structures and decorative; Wood — Commonly used for roofing frameworks, floors, and ornate doors and windows; Stucco and Plaster — Applied to walls for smooth finishes; Gold Leaf and Paint — Used for decorative purposes, especially in church interiors.	Vaulted Ceilings — vaults and groin vaults, these were common in churches and important; Dome Construction — Often constructed with a drum or on pendentives, decorated with Ornamental; Façade Treatments — complexity, including carved stonework, statuary, and portal framing; Mural Painting — Murals depicting religious and historic themes in walls and ceilings in many	WALL	1.8-2.2 W/(m²K)	Made of thick stone and brick, the U-value would be relatively low, suggesting good insulation properties.	Mexico City Metropolitan Cathedral; Church of Santo Domingo; Palacio de Iturbide; The Church of San Francisco
				ROOF	2.5-3.5 W/(m²K)	Colonial buildings often used tile roofs over wood structures, which might have a higher U-value, depending on construction.	
				SLAB	1.8-2.2 W/(m²K)	Floor slabs made of stone or heavy brick would align with the wall values, also featuring good thermal mass and similar U-values.	
2	**19th Century, Beginnings of the 20th (1821-1920)** — NEOCLASIC	Stone; Brick; Plaster and Stuco — Internal structures like cornices and capitals; Wood — Structural elements in floors and roofs; Iron and Steel — Structural frameworks and ornamentation	Masonry Walls — Thick walls constructed from; Symmetrical Layouts — Central axis and symmetrical design; Classical Columns and Pilasters — Often featuring Ionic, Doric, or Corinthian capitals; Pediments and Porticos — Grand and formal appearance; Domes and Vaulted Ceilings — Less ornate than Baroque. More geometrically precise.	WALL	1.5-1.9 W/(m²K)	Given the thickness and dense materials used, such as stone and brick, the U-value would be relatively low.	Palacio de Minería; Palacio Nacional; Alameda Central and Surrounding Buildings; Old Portal de Mercaderes
				ROOF	2.0-3.0 W/(m²K)	Roofs constructed with timber and covered with tiles or metal would have higher U-values, depending on the materials and construction quality.	
				SLAB	1.5-1.9 W/(m²K)	Floor slabs made of heavy masonry would similarly have low U-values, comparable to the walls, due to their significant thermal mass.	
	ART NOUVEAU	Iron and Steel — frameworks and ornamentation; Glass — Light-filled interiors and decorative stained glass features; Stone — Structural and decorative elements; Ceramic Tiles — Interior and exterior decoration; Plaster — Ornamental interior details and smooth	Steel Frameworks — Skeletal structures; Reinforced Concrete — Organic forms and shapes; Masonry Walls; Decorative Facades — Ceramics and glass	WALL	1.5-2.0 W/(m²K)	Typically constructed from a combination of masonry, plaster, and decorative tiles, providing moderate insulation.	Edificio La Nacional; Museo Nacional de Arte (MUNAL); Casa de los Azulejos; Palacio de Bellas Artes
				ROOF	2.5-3.5 W/(m²K)	Roofs in Art Nouveau buildings often featured decorative elements and used materials such as tiles or metal.	
				SLAB	1.5-2.0 W/(m²K)	Floor slabs were usually of reinforced concrete or masonry, offering similar U-values to the walls, benefiting from good	
	ECLECTIC	Stone — Volcanic Cantera; Brick — Structural support and façade elements; Stucco — Applied over brick or stone for a smooth; Iron and Steel — Ornate balconies, railings, and other; Glass — Large windows, often stained or etched	Masonry Construction — Thick walls made of brick or stone; Decorative Facades — Elements from Gothic, Baroque; Ironwork — Decorative iron flat and pitched; Complex Roofs — sections, with decorative elements such as cupolas and	WALL	1.5-2.0 W/(m²K)	Constructed from thick masonry, reflecting good insulation properties due to the mass of the materials.	Postal Palace of Mexico City; Palacio de Bellas Artes; Gran Hotel Ciudad de Mexico; Edificio La Nacional
				ROOF	2.5-3.5 W/(m²K)	Often featured tiles or metal, which can have higher U-values.	
				SLAB	1.5-2.0 W/(m²K)	Floor slabs made of reinforced concrete or heavy masonry would likely have U-values similar to the walls, benefiting	
3	**Modernism & International Movement (1920-1985)**	Glass — To create clean, light-filled spaces. Often employed in large panels and curtain walls; Steel — Allowed for the creation of slim profiles and wide spans; Concrete; Aluminum — Window frames	Frame Structures — Steel and reinforced concrete frames allowed for open floor plans, free of load-bearing walls; Curtain Walls — Non-structural outer coverings of buildings which are hung from the structure; Flat Roofs — Enabled the creation of clean and horizontal roof lines, often used as; Open Floor Plans — Lack of interior walls often used allowed adaptable interior spaces.	WALL	2.5-5.5 W/(m²K)	Curtain walls primarily made of glass and steel have higher U-values, indicating poorer insulation compared to traditional materials.	Torre Latino; Museo Nacional de Antropología
				ROOF	0.5-1.0 W/(m²K)	Modernist flat roofs often constructed with reinforced concrete, covered with a waterproof membrane, depending on insulation.	Ciudad Universitaria (UNAM); Centro Cultural Universitario Tlatelolco
				SLAB	1.0-2.0 W/(m²K)	Concrete slabs, often used in floors and ceilings, benefiting from the thermal mass of concrete.	
4	**After Mexico City 1985 Earthquake to the Present (1985-Present)**	Reinforced Concrete; Steel; Glass; High-performance fibers and polymers used for their strength, lightweight, and corrosion resistance; Composite Materials	Seismic Isolation Systems — Base isolators and dampers that absorb seismic energy; Moment-resisting Frames — Steel or reinforced concrete frames designed to resist bending during earthquakes; Shear Walls — Reinforced concrete walls that add rigidity and strength to structures; High-tech Facades — Double-skin facades	WALL	0.3-1.0 W/(m²K)	Modern materials and technologies have lowered the U-value, depending on the material and insulation techniques used.	Torre Reforma (2016); Torre Mayor (2003)
				ROOF	0.2-0.5 W/(m²K)	Innovations in green roofing and insulation have brought U-values, significantly improving thermal performance.	BBVA Bancomer Tower (2015)
				SLAB	0.2-0.5 W/(m²K)	Concrete slabs with modern insulation can achieve low U-values, which help in maintaining thermal comfort and reducing energy costs.	

Fig. 6 Architectural stages in Mexico City and buildings characterization

Acknowledgments This research would not have been possible without the support of the National Council of Science and Technology of Mexico (CONACYT) through the scholarship corresponding to the Ph.D. Program in Sciences and Humanities Abroad 2022.

References

1. Nägeli, C., et al.: Towards agent-based building stock modeling: bottom-up modeling of long-term stock dynamics affecting the energy and climate impact of building stocks. Energy Build. **211**, 109763 (2020)
2. Nägeli, C., et al.: Synthetic building stocks as a way to assess the energy demand and greenhouse gas emissions of national building stocks. Energy Build. **173**, 443–460 (2018)
3. Flores, M., Flores, S., Rodríguez López, L.: Estimating a National Energy Security Index in Mexico: a quantitative approach and public policy implications. SSRN Electron. J. (2022). https://doi.org/10.2139/ssrn.4185658
4. Torres, I., Niewöhner, J.: Whose energy sovereignty? Competing imaginaries of Mexico's energy future. Energy Res. Soc. Sci. **96**, 102919 (2023)
5. Mercado Fernandez, R., Baker, E.: The sustainability of decarbonizing the grid: a multi-model decision analysis applied to Mexico. Renew. Sustain. Energy Trans. **2**, 100020 (2022)
6. Moe, S.R.K.: The Hierarchy of Energy in Architecture. Emergy analysis. PocketArchitecture: Technical Design Series (2015). Routledge. 711 Third Avenue, New York, NY 10017
7. Cagno, E., et al.: Only non-energy benefits from the adoption of energy efficiency measures? A novel framework. J. Clean. Prod. **212**, 1319–1333 https://doi.org/10.1016/j.jclepro.2018.12.049 (2019)
8. Li, H., et al.: Targeting Building Energy Efficiency Opportunities: An Open- source Analytical & Benchmarking Tool 2019. 2019 ASHRAE Winter Conference. Georgia World Congress Center. Atlanta, Georgia. USA.
9. Ding, C., et al.: Data-driven analysis tool plays critical role in climate neutral buildings. Adv. Appl. Energy. **2**, 100014 (2021)
10. Medrano Gomez, L., Izquierdo, A.: Social housing retrofit: improving energy efficiency and thermal comfort for the housing stock recovery in Mexico. Energy Procedia. **121**, 41–48 (2017)

Exploring Urban Mobility Trends Using Cellular Network Data

Oluwaleke Yusuf ⓘ, Adil Rasheed ⓘ, and Frank Lindseth ⓘ

Contents

1 Introduction

The expansion of urban centers, such as Trondheim, Norway's third-largest city with a 2023 population of approximately 206,000 [2], underscores the pressing need for infrastructure modernization to address challenges such as traffic congestion and pollution driven by rising populations and mobility demands. Trondheim's growth—spurred by the presence of educational and research institutions like the Norwegian University of Science and Technology, SINTEF, and St. Olav's University Hospital—places pressure on its transportation infrastructure and mobility

O. Yusuf (✉) · A. Rasheed
Department of Engineering Cybernetics, Norwegian University of Science and Technology,
Trondheim, Norway
e-mail: oluwaleke.u.yusuf@ntnu.no; adil.rasheed@ntnu.no

F. Lindseth
Department of Computer Science, Norwegian University of Science and Technology,
Trondheim, Norway
e-mail: frankl@ntnu.no

© The Author(s) 2025
M. Kioumarsi, B. Shafei (eds.), *The 1st International Conference on Net-Zero Built
Environment*, Lecture Notes in Civil Engineering 237,
https://doi.org/10.1007/978-3-031-69626-8_138

system. This development has led to a shift toward sustainable mobility solutions, with initiatives like Miljøpakken and MobilitetsLab Stor-Trondheim (MoST) emphasizing public transit and active transportation over private car usage. Achieving such goals necessitates a data-driven approach toward understanding and improving the spatiotemporal dynamics of urban mobility systems for enhanced efficiency, sustainability, and human well-being.

One significant hurdle faced by such mobility initiatives is the difficulty in precisely evaluating the effectiveness and necessity of projects and interventions in alignment with their mobility goals. Such evaluation demands the collection of detailed mobility data across the entire transportation and mobility system. Yet, traditional data collection methods such as traffic surveys, sensor networks, and traffic cameras not only incur significant costs when deployed at scale but also often provide insufficient coverage. These methods typically yield either highly detailed temporal data for narrow segments (e.g., specific areas or intersections) or overly generalized, aggregated data for the whole system. To enable truly effective, data-driven decision-making, there is a critical need for cost-effective methods that can deliver detailed temporal insights spanning the full geographical extent of the mobility system.

2   Cellular Network Data

Network data—derived from the cellular network activity of millions of subscribers across extensive areas—can serve as a source of rich mobility information. This data captures the movements and behaviors of large groups, offering insights into the spatial and temporal aspects of population flows within and across regions. The cellular network data explored in this research was obtained from Telia's Crowd Insights platform.

2.1   *Methodology*

To transform cellular signals into mobility data, a network operator collects temporal data generated during routine mobility subscriber activities (such as calls, texts, and movements within network coverage), anonymizing and aggregating this data to respect privacy and meet GDPR standards. Spatial data from network cell coverage is used to estimate geospatial positions, facilitating movement pattern analysis without traditional triangulation. Such a dataset typically excludes sensitive, roaming, and inactive subscriptions, focusing on active mobile subscriptions and extrapolating this information to reflect the broader population in an area. Advanced algorithms distinguish between stationary and moving signals, classifying them as dwells and transits, thus providing a foundation for understanding wider mobility trends via reports such as [3]:

- *Activity Reports*, which shed light on the locations, times, and origins or destinations of groups
- *Trip Reports*, which offer an origin–destination matrix to understand movement volumes between pairs of locations
- *Routing Reports*, which provide insights into the most likely travel routes taken, encompassing various modes of transportation

2.2 Advantages and Challenges

As discussed, a serious challenge in analyzing mobility trends is the acquisition of high-resolution temporal data that also spans extensive geographical areas. Cellular network data, collected during routine telecom operations, provides a rich source of mobility information which offers valuable insights into both real-time and historical patterns of population movement. This information is invaluable for researchers, policymakers, and planners in understanding traffic flows, thus facilitating informed, data-driven decisions and policies in urban transportation and mobility planning.

However, leveraging cellular data for mobility analysis comes with some drawbacks and challenges, including variable spatial resolution from differing network coverages and the potential loss of granularity due to anonymization and aggregation. Data processing assumptions, such as stationary signals and proximity to the nearest cell, may not always reflect the complexities of urban mobility. Furthermore, policies excluding data from smaller groups to protect privacy can lead to an underrepresentation of certain areas or times. These limitations necessitate a careful evaluation of the data's utility against the backdrop of privacy considerations and the inherent characteristics of cellular networks.

3   Routing Reports for Trondheim Municipality

Routing Reports analyze *peopleFlow*—journey patterns of groups of people—inferring the most probable travel routes to gain insights into travel behaviors, highlighting preferences in routes and transportation modes over various regions and times. This analysis incorporates multiple transportation modes—such as road, rail, ferry, and pedestrian pathways—leveraging OpenStreetMap for mapping probable trip paths over 1 km. The routing reports employ an open trip planner to ascertain the quickest route between two points, considering the operator's network coverage and penalizing unnecessary mode switches.

3.1 Data Description

The routing report dataset used in this study spans from January 17, 2019, to November 30, 2023, covering 1744 dates, with 35 dates omitted due to data gaps or poor signal quality. The raw dataset is aggregated into hourly and daily intervals, resulting in 82,798,311 and 3,792,160 data points, respectively. As shown in Fig. 1, the dataset geographically covers the entire Trondheim municipality and extends into parts of the neighboring Malvik municipality, due to the location of Trondheim Airport just beyond the municipal boundary. The dataset contains six (or seven) attributes for the daily (or hourly) temporal aggregations, as follows:

- *Date and Hour*: These refer to the date and hour the mobility data was recorded. The *Hour* attribute is only present in the hourly temporal aggregation.
- *wayID*: This is a unique OpenStreetMap identifier assigned to each road segment. A stretch of road, rail, or ferry route is thus composed of several connected *ways* (segments) each encoded as a `LineString` object consisting of the longitude and latitude coordinates of its nodes. There are 2210 and 2212 unique *wayID* values in the hourly and daily aggregations, respectively.
- *tagKey*: This refers to the transportation mode, one of three options: road, rail, or ferry.
- *tagValue*: This is a breakdown of each *tagKey* into its subcategories as defined by OpenStreetMap, such as primary/trunk/secondary for roads and expressboat/ cruise for ferry, among others.

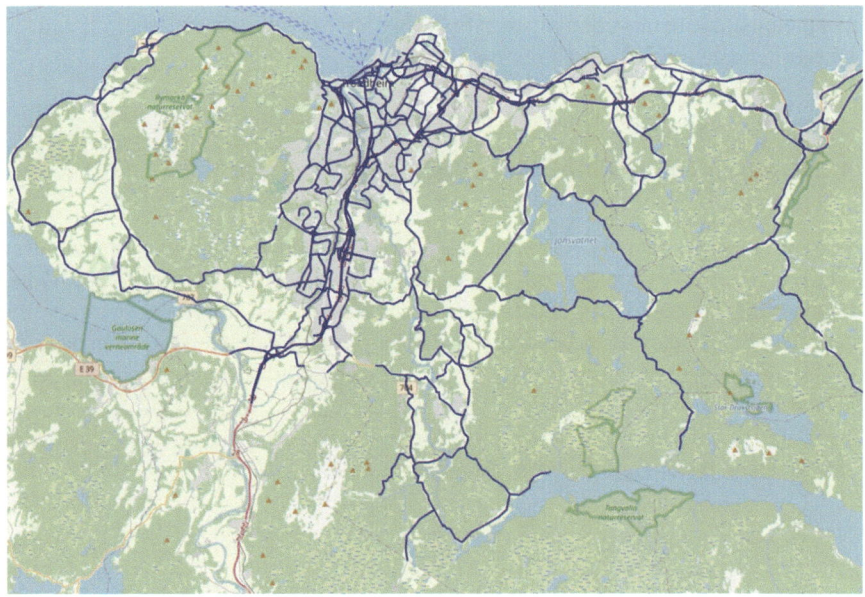

Fig. 1 Satellite map of ways covered by the routing reports

- *Municipality*: This refers to the municipality the *wayID* falls under based on the geographical coordinates of its nodes, either Trondheim or Malvik.
- *peopleFlow*: This is the estimated number of people passing through a specific *wayID* at the daily or hourly aggregation.

3.2 Data Preprocessing, Feature Engineering, and Graph Augmentation

The raw dataset was preprocessed to rectify missing *tagValue* entries in the hourly and daily aggregations, supplementing them with data from OpenStreetMap. To facilitate direct comparisons, the daily dataset was constrained to *wayIDs* consistent with those in the hourly data. The excluded *wayIDs* from the hourly dataset typically had such low *peopleFlow* volumes in the daily data that the hourly numbers would not meet the minimum threshold of five individuals. In addition, some discrepancies were identified in the daily data, which were corrected to align with the total hourly *peopleFlow* values for the respective days.

Subsequently, the dataset was enriched with temporal attributes by extracting metadata from the *Date* feature—including *Day*, *Month*, *WeekNumber*, *Year*, and *HolidayName*—for subsequent analysis. Historical weather data [4] was incorporated to assess the impact of weather conditions on mobility patterns. Additionally, population statistics for Trondheim and Malvik municipalities were sourced from Statistisk Sentralbryå [2] to examine the relationship between mobility patterns and demographic trends over time.

Routing reports enable the analysis of *peopleFlow* trends by selecting origin and destination coordinates and calculating routes within a mobility network represented as a MultiDiGraph—a graph that allows multiple directed edges between the same pair of nodes. However, the geospatial graph associated with the routing reports has discontinuities, preventing some nodes from being reachable. To address such gaps, a fully connected graph from OpenStreetMap, covering road, rail, and ferry modes was obtained and adapted to prioritize ways in the routing reports while ensuring all nodes are accessible. Subsequent trend analyses were then confined to ways (and *wayIDs*) documented in the routing reports and associated *peopleFlow* data.

4 Analyses and Discussion

This section details our investigation into the spatiotemporal dynamics of Trondheim's mobility system, utilizing historical *peopleFlow* volumes obtained from the routing reports. Our analysis encompasses geospatial trends, temporal patterns, and the impact of external factors. In addition, we explore the dynamics of specific routes and areas of interest in the city.

4.1 Geospatial Trends

Figure 2 presents the average (mean) *peopleFlow* volumes derived from routing reports between January 7, 2019, and November 30, 2023, juxtaposed with annual population data. This reveals a consistent increase in *peopleFlow* volumes, mirroring the average yearly population growth of approximately 3000 residents. The COVID-19 pandemic's effect is pronounced, showing a dip and subsequent recovery in *peopleFlow* volumes from March 2020 to March 2021. Excluding this period, seasonal declines in *peopleFlow* are evident during Easter, Summer, Christmas, and New Year holidays. The rationale for using mean *peopleFlow* rather than total sums is due to the overlapping volumes across different routes. For instance, a single individual traversing a road segmented into three ways would be counted in the *peopleFlow* for each segment, thereby overestimating the total mobility if sums were used.

Figure 3 breaks down the annual *peopleFlow* volumes by municipality, offering insight into the mobility patterns to and from Trondheim. Notably, much of the data from Malvik municipality pertains to road and rail traffic on the E6 trunk highway east of Trondheim which extends to Trondheim Airport. This analysis reveals that Malvik's traffic volumes generally remain lower than Trondheim's, except during significant holiday seasons when there is increased travel in and out of Trondheim. Furthermore, analyzing the yearly volumes by transportation mode (*tagKey*) in Fig. 4 highlights periods of increased volume in one mode compared to others. Such fluctuations may signal shifts in the transportation preferences of the city's inhabitants, prompting further investigations.

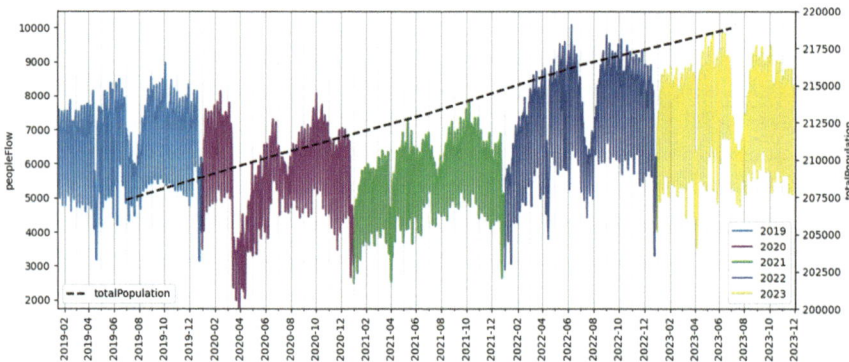

Fig. 2 Yearly breakdown of *peopleFlow* volumes from January 2019 to November 2023

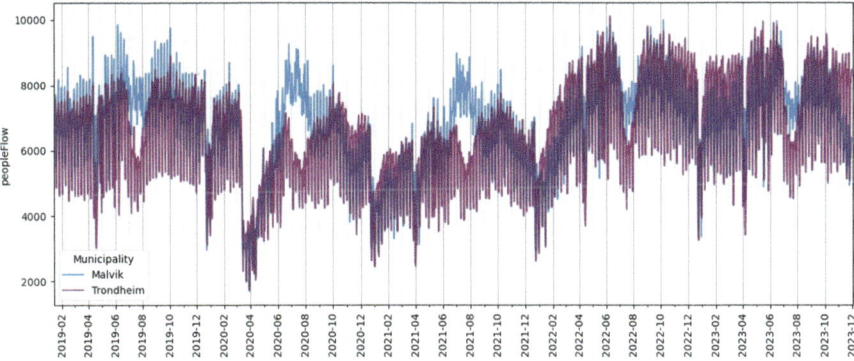

Fig. 3 Detailed overview of *peopleFlow* volumes across municipalities

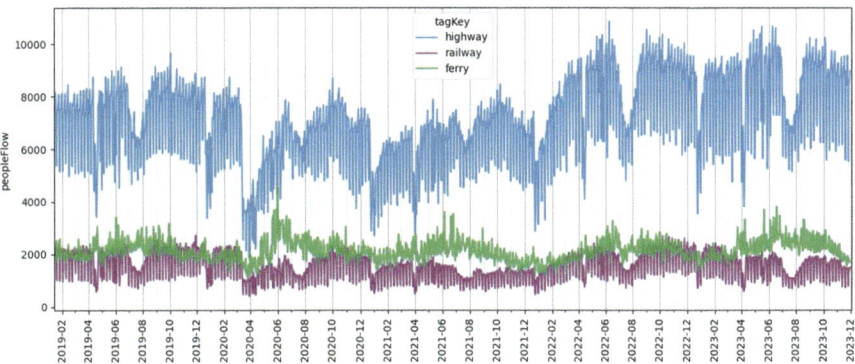

Fig. 4 Detailed overview of *peopleFlow* volumes across transportation modes

4.2 Temporal (Normalized) Patterns

The temporal analysis of the routing reports encompasses hourly, daily, and weekly levels across different transportation modes, focusing on data from March 1, 2022, to November 30, 2023, with the COVID-19 period analyzed separately in Sect. 4.3. Due to significant volume disparities among the transportation modes, normalization within the range [0,1] was applied to each mode's data for a fair comparison of flow trends. Figure 5 shows that the hourly patterns for rail and road traffic are similar, while ferry traffic tends to increase during road and rail's off-peak hours.

Similarly, Fig. 6 reveals that ferry traffic peaks on Sundays, contrasting with the weekend decline in road and rail flows. Meanwhile, Fig. 7 indicates a less pronounced correlation among the three modes, though road and rail still exhibit parallel trends. Notably, road and rail traffic decrease during the summer holidays, but ferry traffic remains consistent due to tourism around the city. The monthly trends, not shown here, offer a broader view which smoothens out the weekly fluctuations.

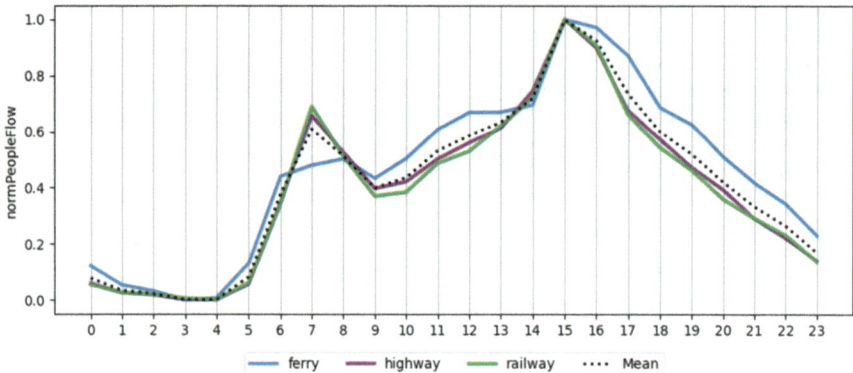

Fig. 5 Normalized hourly variation of *peopleFlow* volumes across transportation modes

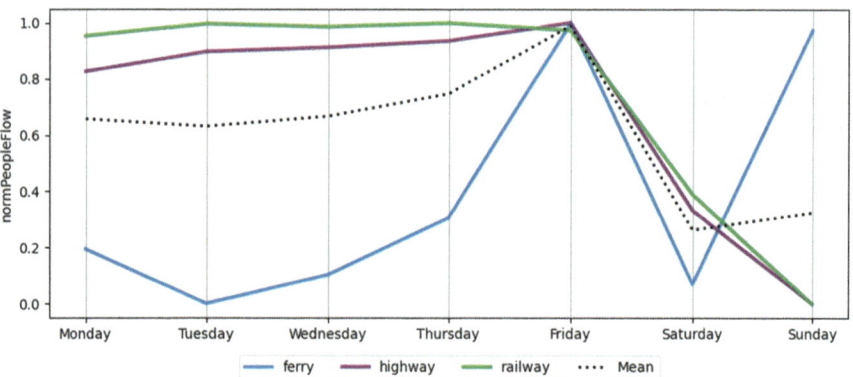

Fig. 6 Normalized daily variation of *peopleFlow* volumes across transportation modes

4.3 External Factors

This section delves into how external factors like the COVID-19 pandemic, weather conditions, and road attributes (speed limits and lane counts) affect *peopleFlow* volumes across Trondheim.

COVID-19 Pandemic The pandemic's impact on *peopleFlow* volumes is discernible across Figs. 2, 3, and 4. The correlation between the Norwegian government's COVID-19 measures (from March 13, 2020, to March 1, 2022) and *peopleFlow* variations is visible in Fig. 8, which presents a 7-day rolling average of the *peopleFlow* volumes. The clear correlation between policy changes and mobility patterns justifies the exclusion of this timeframe from earlier analyses.

Weather Conditions An analysis incorporating historical weather data shows a slight impact of weather on *peopleFlow* trends. For precipitation types, Fig. 9

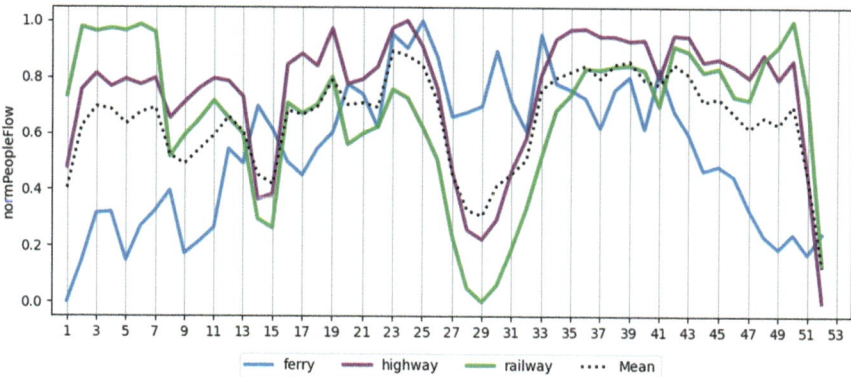

Fig. 7 Normalized weekly variation of *peopleFlow* volumes across transportation modes

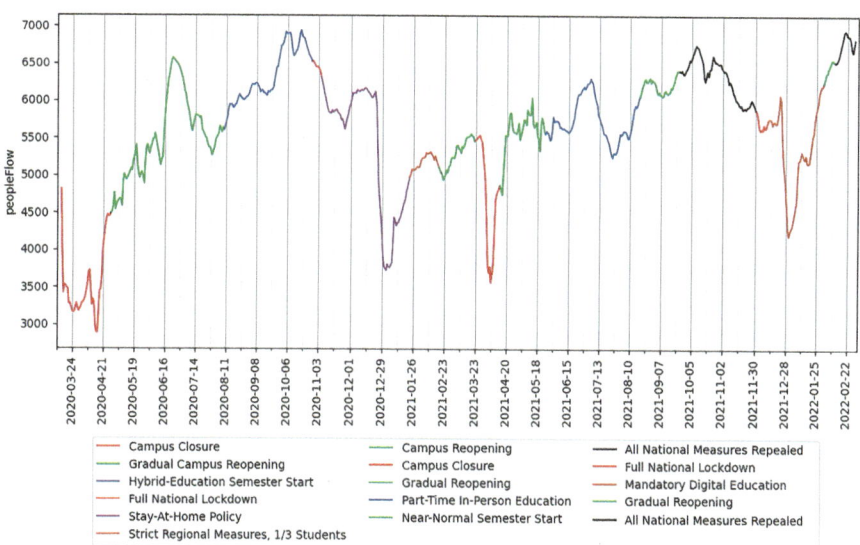

Fig. 8 Effect of COVID-19 measures on *peopleFlow* volumes

indicates that flow volumes are highest in clear weather, followed by rain, snow, and mixed rain/snow conditions. On the other hand, the lowest flows are associated with rainy, snowy, or cloudy weather conditions.

Speed Limits and Lane Counts OpenStreetMap data, linked via *wayIDs* from the routing reports, provided information on speed limits and lane counts for each way. The analysis, depicted in Fig. 10, reveals a positive correlation between *peopleFlow* volumes and both higher speed limits and greater lane counts. This trend might be influenced by the routing algorithm's preference for quicker routes. Notably, the effect of speed limits on *peopleFlow* plateaus beyond 80 km/h due to the reduced number of routes supporting such speeds.

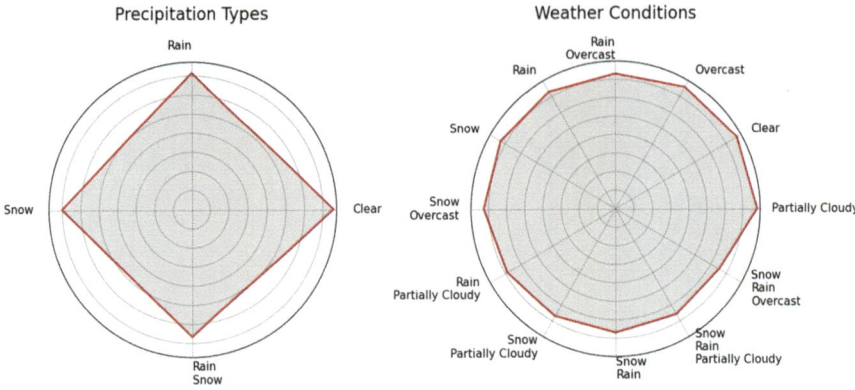

Fig. 9 Variation of *peopleFlow* volumes with weather conditions

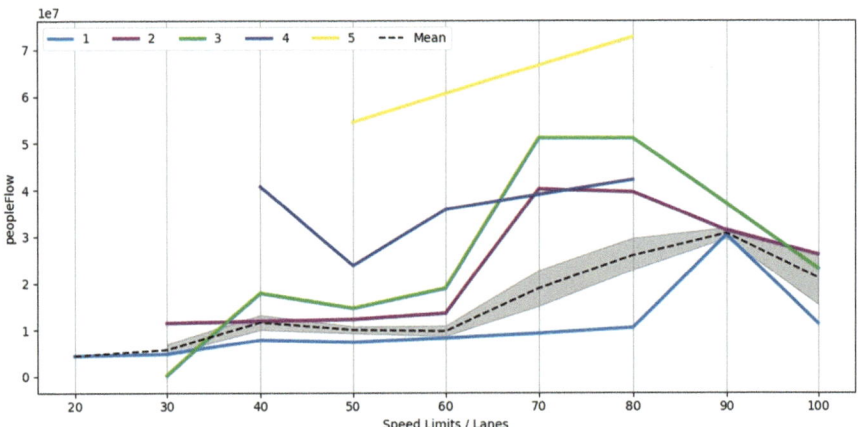

Fig. 10 Combined effect of speed limits and lane count on *peopleFlow* volumes

4.4 Specific Routes and Areas

This section provides a temporal analysis of *peopleFlow* trends along specific routes and areas identified as relevant to the Miljøpakken and MoST initiatives.

Comparison with Public Transit Leveraging Automated Passenger Counting (APC) data from AtB, Trondheim's public transport authority, enabled a comparison between public transit and overall *peopleFlow* volumes along identical routes.

The AtB APC data—covering the period from May 1, 2020, to November 30, 2023—included 1,112,221 unique bus trips over 1295 days, spanning 6 lines and 204 stops. For the analysis in Fig. 11, both AtB and routing report volumes were

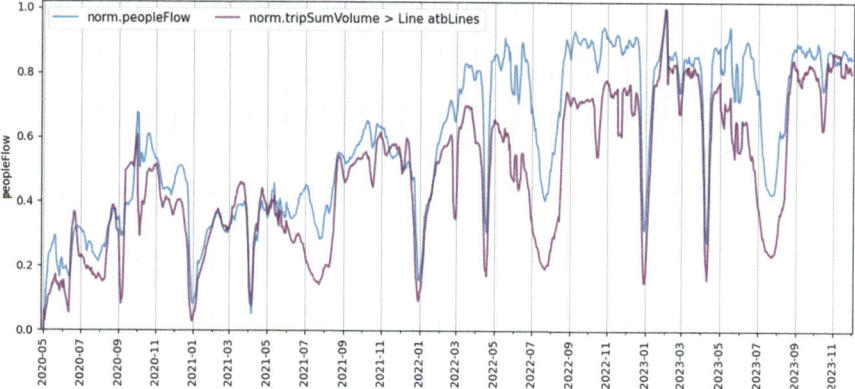

Fig. 11 Comparison of public transit and *peopleFlow* volumes along selected bus routes

smoothed using a 7-day rolling average to eliminate the weekday/weekend variations and subsequently normalized to address the considerable scale differences
between them. While a strong correlation exists between AtB and Telia volumes,
the instances where AtB volumes surpass Telia volumes stand out and require
further investigation, suggesting public transit might be experiencing traffic
increases not reflected in the broader mobility system. It is also instructive that the
majority of these occurrences took place during the COVID-19 period, analyzed
separately earlier.

Miljøpakken Bromstadruta Project Miljøpakken is developing a 3.2 km cycle
path and sidewalk project in Trondheim named Bromstadruta [1], depicted in Fig. 12
(left) with the cycle path outlined in red and the corresponding routing report ways in
blue.

Such infrastructure projects stand to benefit from a thorough analysis of mobility
patterns along the proposed route early in the planning phase. For example, Fig. 13
shows normalized *peopleFlow* trends along Bromstadruta versus the city at large,
uncovering unique dynamics that could be crucial for planning and achieving project
goals.

Trondheim City Center A similar analysis for specific urban areas can be
conducted, in this case focusing on the city center as shown in Fig. 12 (right) with
the target area highlighted in gray. Figure 14 reveals the city center's pronounced
daily (unnormalized) *peopleFlow* volumes compared to other areas. An in-depth
examination of these central urban mobility patterns could provide valuable insights
for optimizing traffic flow, managing congestion, and allocating public transit
resources more effectively.

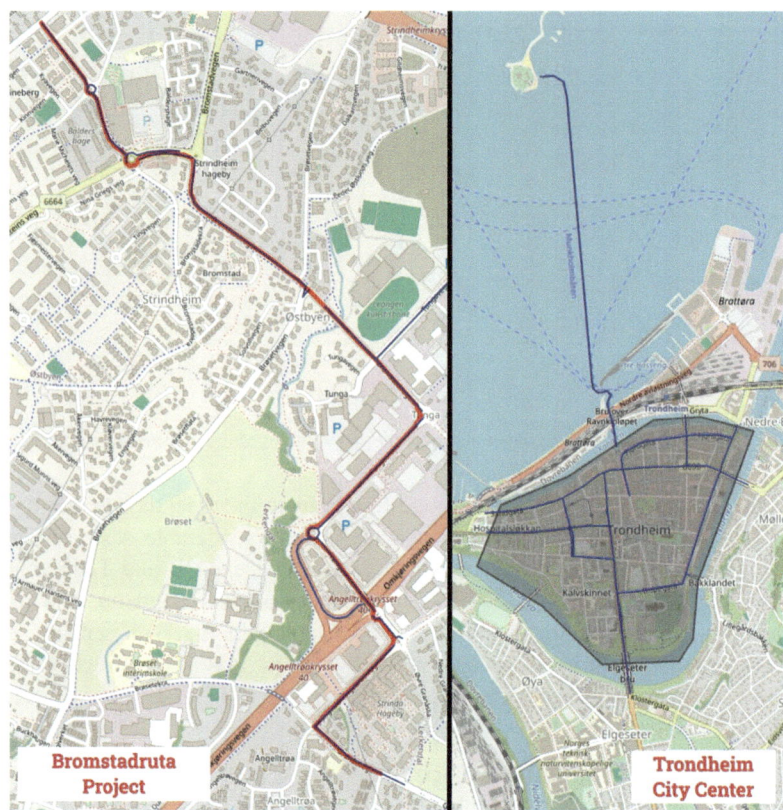

Fig. 12 Bromstadruta cycle path (left) and Trondheim city center (right)

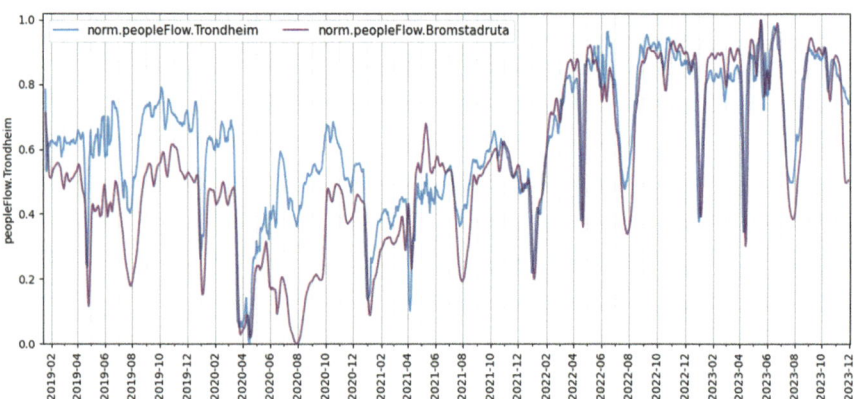

Fig. 13 Analysis of normalized *peopleFlow* trends along the planned Bromstadruta cycle path in Trondheim

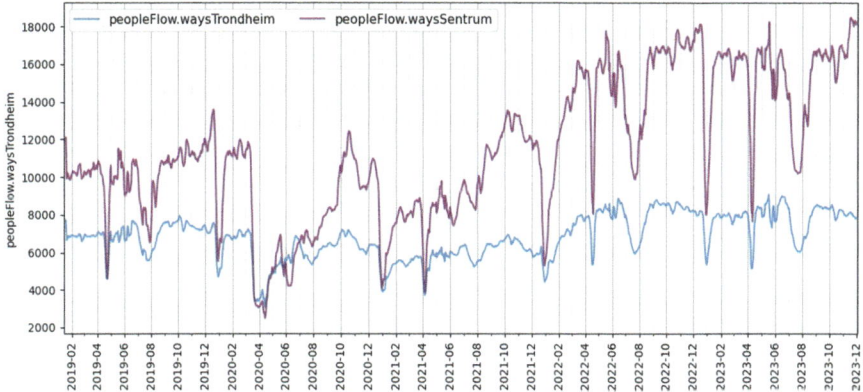

Fig. 14 Analysis of *peopleFlow* trends within Trondheim's city center

5 Conclusion and Future Work

This chapter explores the utility of cellular network data to support efficient and sustainable mobility initiatives like Miljøpakken and MobilitetsLab StorTrondheim, offering a cost-effective, detailed, and wide-ranging source of mobility information. Through the analysis of Trondheim, Norway's routing reports, the potential of such data for gaining insights into mobility flows within and across regions becomes evident, highlighting:

1. Geospatial trends of the mobility dynamics of specific areas and routes within and across municipalities
2. Temporal patterns across different transportation modes from across hourly and daily scales
3. The impact of external factors like pandemics, weather, public transit, and transportation infrastructure on mobility volumes

These insights are crucial for data-driven decision-making in urban mobility planning and policy formulation. However, the inherent nature of cellular network signals and practical realities such as privacy concerns bake some assumptions into the mobility data which require careful consideration in its application. Furthermore, data preprocessing and other enhancements are required to fully capitalize on its potential.

This study is part of a larger initiative to develop a digital twin of Trondheim's transportation infrastructure and mobility system, enriched with comprehensive data-driven historical insights and predictive modelling capabilities. Future research will focus on a thorough spatiotemporal analysis of the mobility network—extending to other modes of transport such as cycling and walking—to pinpoint key ways critical to the *peopleFlow* dynamics across the network.

Acknowledgements This research received funding from the PERSEUS project, a European Union's Horizon 2020 research and innovation program under the Marie Skłodowska-Curie grant agreement No. 101034240. The authors also acknowledge MobilitetsLab Stor-Trondheim (MoST) for their financial contribution and AtB for providing mobility data, which has been instrumental in our research.

References

1. Miljøpakken: Bromstadruta for Sykkel. https://miljopakken.no/prosjekter/ bromstadruta-for-sykkel (2024)
2. Statistisk Sentralbryå: 04861: Area and Population of Urban Settlements (M) 2000–2023. https://www.ssb.no/en/statbank/table/04861/ (2024)
3. Telia: Crowd Insights Methodology. https://coda.io/@data-insights/ telia-webinars-and-training/crowd-insights-methodology-training-27 (2021)
4. Visual Crossing Corporation: Visual Crossing Weather (2019–2024). Data Service. https://www.visualcrossing.com/ (2024)

Evaluating Digital Citizen Participation in Smart Cities

Aashish Adhikari, Mahgol Afshari, Dave Collins, Alenka Temeljotov Salaj, and Agnar Johansen

Contents

1 Introduction

Cities of the future are envisioned to be smart cities, where digital technology and data-driven societies are at the central role, and the sustainability of cities is attributed to their smartness [6]. This explains future cities to be sustainable through their intelligence fundamentally by using digital technologies and their interconnections to optimize infrastructure, services, and resident's quality of life [15]. Beyond technological advancement, the essence of smart cities lies in smarter citizen participation at the core [4]. Adopting a participatory planning approach becomes imperative to integrate the diverse needs, aspirations, and values of people within the city. The use of new and innovative digital technologies to include individuals in planning and decision-making related to their city is one of the fundamental requirements

A. Adhikari (✉) · D. Collins · A. T. Salaj · A. Johansen
Norwegian University of Science and Technology (NTNU), Trondheim, Norway
e-mail: aashish.adhikari@ntnu.no

M. Afshari
Trøndelag County Municipality, Trondheim, Norway

M. Kioumarsi, B. Shafei (eds.), *The 1st International Conference on Net-Zero Built Environment*, Lecture Notes in Civil Engineering 237,
https://doi.org/10.1007/978-3-031-69626-8_139

of the future of smart cities [4]. This approach to building and shaping future smart cities with a data-driven participatory framework empowers communities to shape their cities and neighbourhoods through a collaborative, inclusive, and transparent process. As such, data-driven smart cities with participatory urban planning and development at the core are imperative to developing and envisioning the cities of tomorrow [4].

On the other hand, digital technologies have also evolved significantly, becoming smarter in recent years, presenting opportunities to involve more individuals and communities in the participatory process. This evolution has been particularly more pronounced in recent years, with advancements in technology enabling greater connectivity, accessibility, and functionality, leading to more diverse and inclusive data-driven planning and participation [16]. Acknowledging the fact that, the COVID-19 pandemic accelerated the adoption and utilization of digital technologies for individuals and cities; the participation in the urban planning process as the traditional mechanisms of participation like public events, workshops, and seminars were restricted due to social distancing and restrictions on physical gathering. Cities around the globe started using this digital shift, leveraging it to engage citizens in tasks and meetings that would otherwise be held in town halls and city centres [14]. As such, studies of participatory processes during the pandemic have highlighted that the use of digital technologies increased the number of individuals in such processes [14]. This increase was attributed to the implementation of different measures, such as effective communication using social media by governments [7] and the utilization of digital citizen participation tools like discussion forums, reporting tools for urban issues, satisfaction questionnaires, and more [18].

However, while there is a strong emphasis on leveraging digital tools and technologies to make cities smarter and more efficient, the major challenge lies in the risk of relying solely on data-driven insights without considering the nuanced needs, aspirations, and values of the individuals and community. The ultimate risk of solely relying on attaining more data-driven solutions has the potential for expert-driven decision-making, top-down processes that prioritize technocratic decision-making [17]. In such a scenario, data may be used to justify and defend the participatory processes and demonstrate accountability while nudging actual citizen participation [10, 17]. Additionally, this framework can lead to one-way communication, and decisions leading to disconnections of city development and aspirations of the individuals or losing the essence of citizen control which is the notion of participatory smart cities of the future while requirements are pushing the threshold higher at each step through learning from previous engagement [22].

The following question has guided the research presented in this paper:

How can digital citizen participation be evaluated and what lessons can be drawn from the participatory process?

This study addresses the research question by performing an interpretative case study of a development area called Torskeholmen in Grimstad, a coastal municipality in the southern part of Norway. To translate "Torskeholmen" from Norwegian literally into English would be "cod-islet"; highlighting that the case area studied is essentially an urban redevelopment project of a historical and cultural fishing area and fishing market. The islet lies in the middle of the town of Grimstad and is owned by the municipality themselves. The municipal goal for developing the area is to revitalize the seaside area to make the area an appealing part of the city for future generations to live, work, and study in [8].

Primarily, the public participation process for the discussion of alternatives and designs was initially planned as a public meeting but due to the COVID-19 pandemic, the participation scheme was changed to a digital survey accompanied by a live digital meeting [2]. The survey was performed by a consultant company Asplan-Viak. The digital survey was able to generate a total of 3474 responses, nearly 15% of the total population (23544) of Grimstad in 2020 [11]. As stated in the survey results, one of the many positive effects of the digital survey was the inclusion of the younger generation, who were generally not present in the town hall and public meetings in a proportionate number [2] (Figs. 1 and 2).

As such, to perform the case study, firstly a theoretical background for evaluating digital citizen participation is presented in Sect. 2 of this study. The theoretical notions presented were used to evaluate the participation scheme of the case study area. The methodological approach to this study is further described in Sect. 3. Remaining within the theoretical framework, an analysis of the participatory scheme in the case area is done and presented in Sect. 4 of this paper. A brief discussion and conclusion are summarized in the final section with implications and recommendations for further studies.

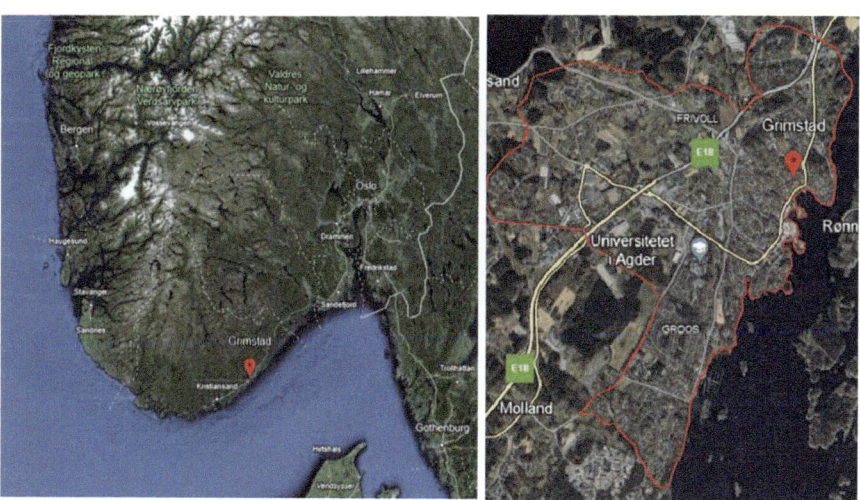

Fig. 1 Location map of Grimstad Municipality in Norway [9]

Fig. 2 The islet of Torskeholmen (Left), Development alternative for future (Right) [2, 11]

2 Theoretical Background

This section highlights the overview of theoretical notions that guided this study. They are different frameworks extracted from the literature and presented, which was done remaining within the context of the relevance of the theoretical models in the setting of participation, use of digital technologies and the Norwegian context.

2.1 Degree of Citizen Participation

The perspective of the community about the digital participation process being able to represent nuanced needs, aspirations, and values of the individuals and community can be discussed through one of the most famous and relevant frameworks of the rungs of the ladder by Arnstein [1]. In participatory planning processes, citizen engagement is facilitated through various mechanisms like town hall meetings, digital tools, workshops, and so on. The different mechanisms used in such processes can be analysed through this perspective of eight levels of participation. The rungs or steps in the ladder here represent the extent of power citizens have in the decision-making process [1]. As described in the concept, participation extends from non-participation to tokenism, and up to the highest level of citizen control and citizen power.

Contextually, this framework was adopted in the Norwegian setting, and the ladder of citizen participation has been discussed as five rungs consisting of Information, Consultation, Dialogue, Agenda Setting, and Co-management [20]. Information dissemination advertises planning processes, informing, and engaging citizens from the start. The consultation makes plan proposals publicly available for feedback, ensuring inclusivity. Dialogue fosters communication between citizens and decision-makers, encouraging discussions on issues. Agenda setting allows citizens to influence planning strategies by suggesting themes. Co-management involves citizen initiatives in initiating or revising planning processes [20].

2.2 Evaluating Digital Participation

The evaluation perspective of digital participation for this study is to be able to understand the underlying notions beyond data and recognize engagement effectiveness to capture the cultural notion of participation. One of the very first well-recognized frameworks to evaluate participation highlights that before starting the evaluation procedure of any participatory scheme, it is essential to identify the goal of the evaluation process [19]. If the evaluation is being done for management, auditing, or learning objectives, it must be made clear. For this study, the main purpose of the evaluation was to draw lessons and learnings from the evaluation of digital participation.

As discussed, in data-driven planning, a more dominant approach to evaluating the success of the participatory process would be to look at the numbers or counting the heads of people involved in such processes. Although this is an important metric through which the impact of the use of digital technologies is evident and is quantitative and measurable, the overall quality of participation should also be assessed in terms of metrics like representativeness, inclusiveness, transparency, and facilitation in the process of participation itself [13].

From the quantitative side, the framework from Sæbø et al. [21], for the case of "e-Participation" can be used as a solid standpoint for this research. The amount of participation is determined by tallying the total number of users and contributions made online. In addition, participant demographics can be used to examine and compare offline involvement. Age, gender, educational level, and other demographic variables were deemed to be important for the analysis [21]. This has been represented similarly, through the standpoint of representativeness and inclusiveness in the planning process quality [13]. Representativeness entails ensuring that participants reflect the demographic diversity of the affected population, while inclusiveness involves providing equal opportunities for all participants to contribute, considering factors like gender, age, and education.

Additionally, the other aspect introduces the depth of participation to assess the manner and tone of online interactions, seeking out characteristics like openness, respect for differing viewpoints, and coherent reasoning [21]. As such, understanding and analysing these aspects enables the customization of participation initiatives, ultimately contributing to the enhancement of engagement in the digital age. Further parameters like clear and accessible information and objectives, facilitation to foster constructive exchanges and understanding among participants, digitally mediated settings, and promoting consensus-building or respectful disagreement can be discussed as criteria collectively aiming to ensure the legitimacy, effectiveness, and fairness of the digital participation process [13, 21]. The qualitative parameters of questions like knowledge, feeling, opinion and behaviour-question-based evaluation can also be used to frame the evaluation study [19].

3   Research Design and Methods

The methodological approach for this study consists of an interpretive case study. A single case design was selected, and a theoretical background based on literature was used to understand the participation using digital technology. A qualitative interpretation of the phenomenon is done to highlight the meaning embedded in the analysis of the case. This interpretation is based on the focus of strategy adopted during the information flow and interpretation guided by questionnaires and analysis of digital technology use. The study draws specific implications or lessons from the generalizable results of the participatory scheme in the case area based on digital participation.

The case was chosen as a success story due to its interesting approach to recording a huge number of responses as data from citizens and the utilization of interesting and innovative methodologies to gain more data-driven insights from the community [2]. As such, the process was also framed by the municipality as an approach to incorporate value-based planning for the redevelopment of the area [8]. This study interprets the participatory scheme of the redevelopment project using interpretation from document studies as a method.

For this study, documents from the digital survey were accessed through the municipal website [12] which were then analysed through the lens of the theoretical background presented in this study. The overall design of the survey is analysed, and lessons are drawn from this case for future recommendations. To sort out, data triangulation newspaper narratives were also drawn from the local newspaper Grimstad Adressetidende (www.gat.no). The keyword "Torskeholmen" was used with dates defined including September and October, the whole period of the digital survey undertaken. This study is also based on the 47 results obtained from the process.

4   Digital Participation in Torskeholmen Case

The theoretical frameworks developed were used to analyse the case study from the citizen participation process in the project. This is done in parts, Firstly, this study analyses how information flow or dissemination of information about the survey was done to reach individuals. Then the participatory tools used are analysed using the degree of participation framework, afterwards, the depth of the survey and analysis performed in the survey report are discussed in this section.

Firstly, the dissemination of information regarding the survey was done using live meetings, SMS, and social media handles of the municipality. A general strategy using low-priced advertisements on social media is used along with links to in-depth websites and detailed information [5]. This was exactly done in the Torskeholmen project where SMS was sent out to around 16,000 numbers registered in the address of Grimstad, with a link to the digital survey and municipal website. A huge response from the individuals was seen with a drastic increase in the daily visits to the municipal website [2]. Strong messages like "What will Torskeholmen become in the future?" [3] as a leading question with incentives like a set of Air pods and 10 pizzas were the efforts used to generate more responses to the survey, which can be regarded as to encouragement factor for youths under the age of 34 to be represented quite well in this process. From the perspective of the theory of degree of citizen participation, although this may lie in the lower rungs of the ladder, good information flow in the early stages is the first step to a good participatory process [20].

Another part of the process was the survey itself, which consisted of 13 questions, 11 of which were quantitative, closed-ended questionnaires and the last two qualitative free-text questions were used [2]. These included questions from the demographics to choosing different alternatives or models of development for the city. Evaluating through the framework of OECD [19], behaviour-based questions like "Living situation" with close-ended answers "alone, with a partner and with children, student, single parent" were asked to know the individuals better from the perspective of what types of housing requirements would individuals have and what solutions would be required in the development. Likewise, "means of livelihood", visits to the centre of Grimstad were asked to understand the behaviour patterns of the individuals of the city. Feeling-based questions were used like "Relationship with the Grimstad centre", with closed options "I live here", "I do my job here", study, work and tourist were presented as options to know the respondents better. Although this may be argued to be opinion-based questions the evaluation can be done based on the feelings of the individuals. Opinion-based questions which were majorly important were "What do you like best about Grimstad city centre today?", "What has the greatest potential for improvement in the centre of Grimstad?" where individuals opined for "location by the sea" as the best thing about the city and opted for improvement requirements for shops and outdoor areas. These opinions help shape the city better through value-based interpretation of the individuals rooted in the culture of the community. The use of mixed questions like "What is most important for you in the development of Torskeholmen" can be argued to lie in a mix of opinion, knowledge and feeling-based questions.

Similarly, The questionnaires described were followed by an opinion-based digital survey questionnaire where respondents had to choose the development type they wanted from the city. This was done after different "development alternatives" as options for different development models were presented and alternatives from low-rise to high-rise developments had to be chosen by the individuals filling the survey. Visual cues were developed in 3D and 2D, and "Development alternatives", and interactive development design were presented in the survey. Following the development alternative question a deeper insight, feeling-based questionnaire had to be provided with an open-ended typology, in which individuals could comment on why they had chosen the particular development alternative and their thoughts and comments on the development alternative. The interactive maps and tools for visualization can make engagement interesting, but it is a one-way communication or a tool with asynchronous communication and lies in the consultation in rungs of the ladder typology, although the presence of open-ended questions or narration element may be argued to be a dialogue to foster communication and encourage further discussions on the issues [20].

Hence, the depth and value of the project from the residents' perspectives were laid in the cultural dimensions of the planning area. The area had been a place for fishing for generations and was also an important part of the landscape of the town. Hence, citizen participation and their thoughts on the development of the project were an imperative part of the project planning.

The survey questions were also centred on this fact, and they revolved around these dimensions. Closed questions like "What do you like best about Grimstad city centre today?", "What is most important to you in the development of Torskeholmen?" was used to know the tangible things people expected from the development of Torskeholmen [2]. These questions were used as prompts to attach values, desires, and aspirations individuals and the community had from the development of the area, which is shown in Fig. 3.

The analysis of the survey results was presented concerning the demographics broken down by age and gender. The development alternatives chosen were discussed from the perspective of how and what forms of development people with ages below 50 and after 50 wanted and these were discussed from the narratives of the open-ended questions. This was done to understand better the underlying meanings associated with the area by different individuals [3].

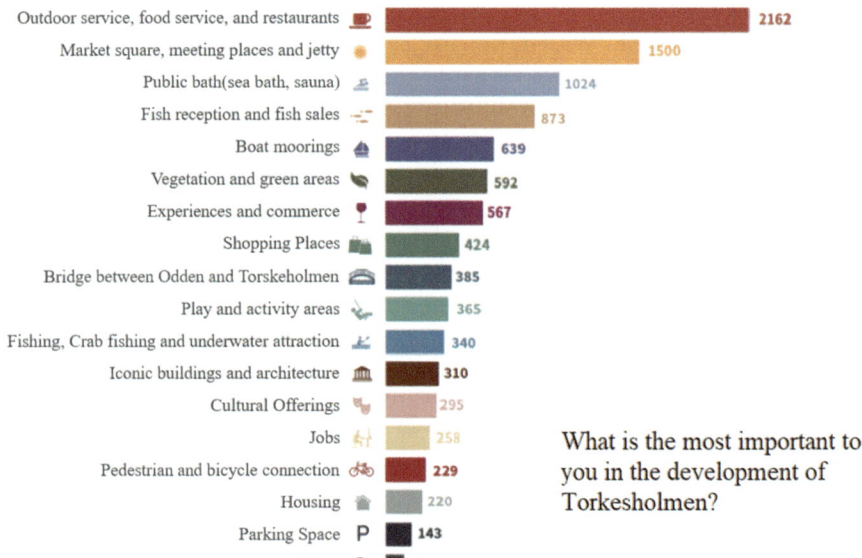

Fig. 3 Results from the questionnaire. (Source: [2])

5 Discussion and Conclusion

This study posed a question regarding the evaluation of digital participation, which was sorted through the theoretical framework presented and drawing lessons through the evaluation of the case area. This study highlights how digital technologies may effectively promote citizen participation with innovative methods that are not only able to represent a wide number of participants but also deeper lying notions through good practices. The evaluation of participation through the degree of participation framework has been to demonstrate meaningful participation oriented towards empowering citizens in the decision-making process. The analysis beyond mere numbers and parameters like representativeness, and inclusiveness is critical to ensuring legitimacy and effectiveness of the process. Hence, the study was able to develop a framework for a value-based approach to enhance participation. From the case, digital participation in the Torskeholmen project was in a similar line as described and overall, Grimstad had a successful participatory process with a good response rate and engagement. The framing of the questions was designed to include the cultural aspect of the fishing town and the findings were further used as a starting point for discursive urban planning and the use of digital tools was performed in a way that was able to represent the notion of need of involving citizens. The findings of the survey were used to enhance further public meetings and for the development of a planning proposal that incorporated the survey findings in line with the development alternative which was chosen by most people as their preferred choice [8]. This process was further worked together with fishermen through physical meetings in the area to locate fish reception and finally ended with two alternatives for consideration in the municipal council in the last quarter of 2022 [8].

Moreover, the major highlight can be seen as the use of digital tools, which can be used to engage young citizens as demonstrated in the case, which had a huge presence in this planning process, despite the presence in physical meetings being rather low for the group of young participants. The correlation of data from different demographics and their choices provided valuable insight into their aspirations from the project, presenting how different groups of people in the area had associated the values of their environment. The use of interactive maps and communications can be a major lesson for future projects in a similar manner. The questionnaires and message set up by the team of surveyors.

Nonetheless, it is important to acknowledge the fact that the Torskeholmen project is not distant from controversies surrounding participation in the project. The project has been criticized and several newspaper articles have been published with reports regarding the failure of the project to capture needs, aspirations, and values into the planning process.

Hence, the evolution towards smarter cities necessitates a careful balance between technological innovations and value-based urban planning principles, which are reliant on the cultural, needs, desires, and aspirations of the citizens. For the cities to become smarter as claimed and to be sustainable through the smartness. Digital participation can be used to encapsulate the visions of a large group of citizens all at once at a fair and cheap price. But it must be highlighted that digital technologies are the future and citizen empowerment in future cities lies in this participation being able to present the needs, suggestions and demands of citizens. This study argues that the need for a participatory framework in planning requires tools and measures that truly create a genuine collaboration between citizens and cities. Relying on digital participation that is not designed properly runs the risk the participation becoming one way, there is a risk that it may be headed towards tokenism if not carefully implemented, using data from citizens as a form of consultation [17]. To truly empower citizens and achieve meaningful engagement, efforts should aim to foster genuine partnerships and enable citizen control through digital platforms. Simply generating more responses on digital platforms does not necessarily translate to genuine engagement, and although lean digital tools can generate a larger response, seemingly from a larger audience, this may not capture meaningful participation [14]. Furthermore, these findings may be used to analyse the nature of citizen participation and improve the participatory process in smart cities. Special attention must be provided to include citizens for meaningful partic- ipation, thus leading to smarter cities through the use of smarter technologies and a net zero sustainable future.

Acknowledgement This study would like to thank the research project "CaPs- Citizens as Pilots of Smart Cities", funded by Nordforsk (project number 95576) and Grimstad Municipality along with its residents and others associated with the Torskeholmen project.

References

1. Arnstein, S.R.: A ladder of citizen participation. J. Am. Inst. Plann. **35**(4), 216–224 (1969). https://doi.org/10.1080/01944366908977225
2. Asplan-Viak: Spørreundersøkelse oppsummering_2020-11-05 [Survey Summary] (2020a)
3. Asplan-Viak: Fremtidens Torskeholmen (2020b). https://storymaps.arcgis.com/stories/4063 aa998ef24300ab81e0bca35f9bc9
4. Batty, M., Axhausen, K.W., Giannotti, F., Pozdnoukhov, A., Bazzani, A., Wachowicz, M., Ouzounis, G., Portugali, Y.: Smart cities of the future. Eur. Phys. J. Spec. Topics. **214**(1), 481–518 (2012). https://doi.org/10.1140/epjst/e2012-01703-3
5. Bonsón, E., Perea, D., Bednárová, M.: Twitter as a tool for citizen engagement: an empirical study of the Andalusian municipalities. Gov. Inf. Q. **36**(3), 480–489 (2019). https://doi.org/10. 1016/j.giq.2019.03.001
6. Bouzguenda, I., Alalouch, C., Fava, N.: Towards smart sustainable cities: a review of the role digital citizen participation could play in advancing social sustainability. Sustain. Cities Soc. **50**, 101627 (2019). https://doi.org/10.1016/j.scs.2019.101627
7. Chen, Q., Min, C., Zhang, W., Wang, G., Ma, X., Evans, R.: Unpacking the black box: how to promote citizen engagement through government social media during the COVID-19 crisis. Comput. Hum. Behav. **110**, 106380 (2020). https://doi.org/10.1016/j.chb.2020.106380
8. GEU: Presentasjonfor kommuneplanutvalget - 03.02.2022 [Presentation of the plan for the Grimstad Regeneration Project] (2022)
9. Google: Google Earth. https://earth.google.com/web/search/Grimstad/ (Date Accessed 2024-04-21) (2024)

10. Granier, B., Kudo, H.: How are citizens involved in smart cities? Analysing citizen participation in Japanese "smart communities". Inf. Polity. **21**(1), 61–76 (2016). https://doi.org/10.3233/IP-150367
11. Grimstad municipality: Årsmelding (Annual report) 30-03-2020 (Vol. 6(12)) (2021). https://doi.org/10.3934/math.2021814
12. Grimstad municipality: Torskeholmen. 20.06.2023 (2023). https://www.grimstad.kommune.no/tjenester/plan-bygg-brann-og-eiendom/reguleringsplaner-planarbeid-og-kommuneplan/reguleringsplaner-horinger-og-vedtak/torskeholmen.26140.aspx
13. Hofmann, M., Münster, S., Noennig, J.R.: A theoretical framework for the evaluation of massive digital participation systems in urban planning. J. Geovis. Spat. Anal. **4**(1), 3 (2019). https://doi.org/10.1007/s41651-019-0040-3
14. Hofstra, R., Michels, A., Meijer, A.: Online democratic participation during COVID-19. Inf. Polity. **28**(3), 395–410 (2022). https://doi.org/10.3233/IP-211540
15. Höjer, M., Wangel, J.: Smart sustainable cities: definition and challenges. In: Hilty, L.M., Aebischer, B. (eds.) ICT Innovations for Sustainability, pp. 333–349. Springer International Publishing (2015)
16. Javed, A.R., Shahzad, F., ur Rehman, S., Zikria, Y.B., Razzak, I., Jalil, Z., Xu, G.: Future smart cities: requirements, emerging technologies, applications, challenges, and future aspects. Cities. **129**, 103794 (2022). https://doi.org/10.1016/j.cities.2022.103794
17. Kitchin, R.: The real-time city? Big data and smart urbanism. GeoJournal. **79**(1), 1–14 (2014). https://doi.org/10.1007/s10708-013-9516-8
18. Kopackova, H., Komarkova, J., Horak, O.: Enhancing the diffusion of e-participation tools in smart cities. Cities. **125**, 103640 (2022). https://doi.org/10.1016/j.cities.2022.103640
19. OECD: Evaluating Public Participation in Policy Making. OECD Publishing (2005). https://doi.org/10.1787/9789264008960-en
20. Ringholm, T., Nyseth, T., Hanssen, G.S.: Participation according to the law. Eur. J. Spat. Dev. **67**, 1–20 (2018). https://doi.org/10.30689/EJSD2018
21. Sæbø, Ø., Rose, J., Skiftenes Flak, L.: The shape of eParticipation: characterizing an emerging research area. Gov. Inf. Q. **25**(3), 400–428 (2008). https://doi.org/10.1016/j.giq.2007.04.007
22. Senior, C., Jowkar, M., Temeljotov-Salaj, A., Johansen, A.: Empowering citizens in a smart city project one step at a time: a Norwegian case study. In: 2021 IEEE European Technology and Engineering Management Summit (E-TEMS), pp. 10–15 (2021). https://doi.org/10.1109/E-TEMS51171.2021.9524892

Emission-Based Relocation Strategies for Mobile Prefabrication Factories

Jianxiang Ma ⓘ, Andrea Revolti ⓘ, Lorenzo Benedetti ⓘ,
Edwin Zea Escamilla ⓘ, and Guillaume Habert ⓘ

Contents

1 Introduction

Linear infrastructure construction projects, such as railways, roads, and tunnels, often encounter challenges such as budget overruns, delays, high environmental impact, and subsequent public resistance [1]. Conventionally, prefabricated elements are manufactured in a stationary factory and then transported to the construction site. However, in linear infrastructure projects, the construction site shifts as the project advances. This dynamic characteristic can result in extensive shipping distances, leading to negative impacts on the economic and environmental aspects of the project [2]. Mobile factories are well suitable for situations in the construction

J. Ma (✉) · E. Z. Escamilla · G. Habert
ETH Zürich, Zürich, Switzerland
e-mail: ma@ibi.baug.ethz.ch

A. Revolti
Freie Universität Bozen, Bolzano, Italy

L. Benedetti
EuroTube Foundation, Dübendorf, Switzerland

© The Author(s) 2025
M. Kioumarsi, B. Shafei (eds.), *The 1st International Conference on Net-Zero Built Environment*, Lecture Notes in Civil Engineering 237,
https://doi.org/10.1007/978-3-031-69626-8_140

industry with long distances and high logistics costs [3]. However, factory relocation introduces significant variability in the distances for transporting materials to the factory and for delivering the prefabricated elements from the factory to the project site. This distance variability further introduces variability of transportation-associated carbon emissions.

In this paper, we present a two-stage integer linear programming (ILP) model to optimize the overall transportation carbon emissions. We aim to estimate the potential reduction in the carbon emissions related to the potential implementation of mobile prefabrication factories to supply large-scale linear infrastructure projects. Additionally, we compare the carbon emissions reductions associated with various relocation frequencies to identify feasible mobile factory relocation strategies.

2   Methodology

Figure 1 presents the conceptual framework used in this paper. The first step involves gathering project information data including details on the bill of quantities of the prefabricated elements, the locations of material suppliers, the potential factory sites, and the construction site locations for the prefabricated elements. In the second step, we calculate the distances of materials transportation and the prefabricated elements supply transportation using geographic information system (GIS) tools. We estimate the carbon emissions associated with the reconfiguration processes using life cycle assessment (LCA) following the methodology described in the ISO14040 [4]. Subsequently, this data is fed into the two-stage ILP model to determine the optimal relocation strategy of the mobile factory for different relocation numbers, which was developed in Pyomo [5] and solved using the COIN-OR

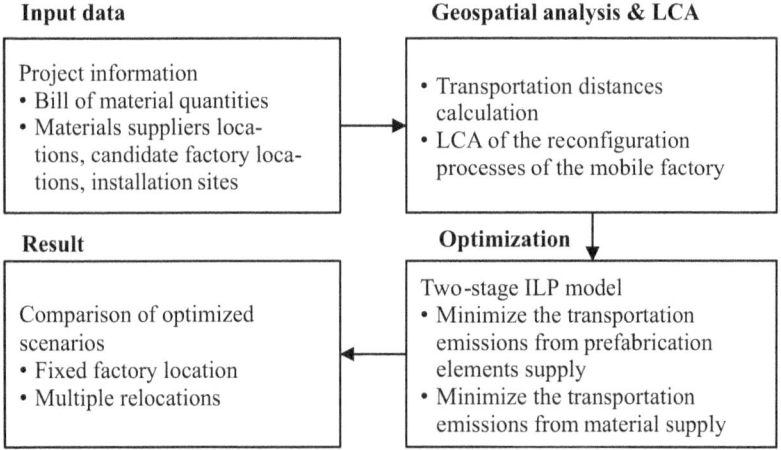

Fig. 1   Conceptual framework

Branch-and-Cut (CBC) [6] solver. In the final stage, we compare the relocation strategies to find the transportation scheme with the lowest carbon footprint. This process accounts for multiple suppliers for various materials at each potential factory deployment location.

2.1 Mathematical Model

The relocation of the mobile factory impacts both the material supply distance and the distance for supplying prefabricated elements, subsequently resulting in variability in the carbon emissions from the transportation of materials and prefabricated elements. Simultaneously, the reconfiguration processes of the mobile factory, along with the transportation of the factory, could also contribute to the overall carbon emissions. To address this computational complexity, we split this optimization problem into two stages. In the first stage, we construct an ILP model to determine the deployment locations for the mobile factory aiming at minimizing the carbon emissions from the transportation of the prefabricated elements, as well as the transportation and the reconfiguration processes of the mobile factory. In the second stage of the ILP model, material suppliers for the factory deployment locations are selected to minimize the carbon emissions from the transportation of materials.

First Stage Model To determine the deployment locations for the mobile factory, we need to find the supply pattern of the prefabricated elements with the lowest carbon emissions. This pattern involves three types of activities: transporting prefabricated elements from the mobile factory to the construction sites, the reconfiguration processes of the mobile factory, and transporting the mobile factory itself. We introduce three decision variables, X_{fi}, Y_f, and Z_{fk}, to quantify the carbon emissions from these three activities, respectively. X_{fi} is a binary variable that equals one if only the prefabricated elements required at the construction site i is supplied by the factory location f. Y_f is a binary variable that equals one if only the factory is deployed at the location f. Z_{fk} is a binary variable that equals one if only the mobile factory moves from the factory location f to the factory location k. With these three decision variables, we can formulate this optimization problem as follows:

$$\text{Min} \sum_{f \in F} \sum_{i \in I} et_{fi} m_i D_{fi} X_{fi} + \sum_{f \in F} e_f Y_f + \sum_{f \in F} \sum_{k \in F} et_{fk} m_f D_{fk} Z_{fk} \quad (1)$$

subject to:

$$\sum_{f \in F} X_{fi} = 1, \forall i \in I \quad (2)$$

$$Y_f \leq \sum_{i \in I} X_{fi}, \forall f \in F \quad (3)$$

$$Y_f \geq X_{fi}, \quad \forall f \in F, \forall i \in I \quad (4)$$

$$\sum_{f \in F} Y_f = f_{se} \tag{5}$$

$$Z_{fk} \le Y_f, \forall f, k \in F \tag{6}$$

$$Z_{fk} \le Y_k, \quad \forall f, k \in F \tag{7}$$

$$Z_{fk} = 0, \quad \forall f \ge k, \forall f, k \in F \tag{8}$$

$$\sum_{f < k} Z_{fk} = Y_f, \quad f = 1, \ldots, n-1, k = 2, \ldots, n \tag{9}$$

$$\sum_{f < k} Z_{fk} = Y_k, \quad f = 1, \ldots, n-1, k = 2, \ldots, n \tag{10}$$

The first term in Formula (1) calculates the total carbon emissions from the transportation of prefabricated elements from the mobile factory to all the construction sites. Here, D_{fi} denotes the transportation distance from the factory location f to the construction site i, et_{fi} represents the carbon emissions indicator corresponding to the transportation mode for D_{fi}, and m_i denotes the mass of required prefabricated elements at the construction site i. F and I are the sets of the candidate factory locations and the set of the construction sites, respectively. Both sets are indexed in an ascending order along the linear infrastructure. The second term quantifies the total carbon emissions associated with reconfiguring the mobile factory for mobilization, including site preparation, factory assembly and disassembly, and site restoration. Here, e_f represents the carbon emissions from the factory reconfiguration processes at the location f. The third term calculates carbon emissions from transporting the factory throughout the construction project. Similar to the first term, D_{fk} denotes the transportation distance from factory locations f to k, et_{fk} corresponds to the carbon emissions indicator for D_{fk}, and m_f denotes the mass of the mobile factory.

Constraint (2) ensures that every construction site is supplied by exactly one factory location. Constraints (3) and (4) state that the factory should only be deployed to the locations that supply at least one construction site. Constraint (5) controls the relocation times of the factory, where f_{se} is the desired number of deployment locations for the factory. Constraints (6)–(10) regulate the factory moves through all the selected deployment locations in ascending order of the factory location indexes, where n is the number of candidate factory locations.

Second Stage Model The second stage optimization model locates the nearest materials suppliers for the selected factory deployment locations determined in the first stage. Similar to the first term in Formula (1), we can formulate this optimization problem as follows:

$$\text{Min} \sum_{j \in J} \sum_{f' \in F'} \sum_{s_j \in S_j} et_{s_j f'} \cdot m_{jf'} \cdot D_{s_j f'} \cdot X_{s_j f'} \tag{11}$$

subject to:

$$\sum_{s_j \in S_j} X_{s_j f'} = 1, \forall f' \in F', \forall j \in J \tag{12}$$

In Formula (11), $X_{s_jf'}$ is a binary decision variable that equals one only if the factory deployment location f' receives material j from supplier s_j, $D_{s_jf'}$ denotes the corresponding transportation distance, and $et_{s_jf'}$ represents the carbon emissions indicator for $D_{s_jf'}$, $m_{jf'}$ denotes the mass of the material j required at the deployment location f'. Here, $F' = \{f : X_{fi} = 1\}$ represents the set of the selected factory deployment locations. J denotes the set of materials required for manufacturing the prefabricated elements and S_j represents the set of suppliers for material j. We assume that, at each selected factory deployment location, the mobile factory receives each material from only one supplier. Therefore, we can use Constraint (12) to guarantee that the factory receives sufficient materials at each selected deployment location.

3 Case Study

In this paper, we developed the case study from applying the mobile factory to supply a hyperloop infrastructure project. The hyperloop concept consists of a passenger pod traveling through a tube under a light vacuum while being propelled and levitated by a combination of permanent and electro-magnets [7]. EuroTube Foundation (ETF) [8], a Swiss non-profit research organization, is dedicated to the development of hyperloop technology. Their hyperloop infrastructure solution utilizes concrete pipes to enable lower production costs and facilitate easier maintenance and management throughout the infrastructure's life cycle [1]. In this case study, the prefabricated elements are the concrete pipes produced in the mobile factory. The candidate factory locations were selected using Multi-Criteria Decision Making (MCDM) in collaboration with the factory design team from the Free University of Bozen-Bolzano (unibz) and ETF [8]. The main criteria include proximity to the infrastructure line, sufficient space for the deployment of the mobile factory, and good accessibility to main road networks.

Figure 2 depicts the geographic information near the candidate factory locations 17–19 in this case study. The numbers below the red cycles are the indexes of the candidate factory locations. The studied infrastructure line (dashed line) from Geneva to Zurich was defined in collaboration with ETF [8], with a total length of 248 km. Considering that each concrete pipe has a length of 0.02 km, the model would require 12,400 construction sites, posing a heavy computation burden. To address this, we divided the infrastructure into 248 segments, each with a length of 1 km. We assume that for each segment, the construction site for every concrete pipe is represented by the segment's midpoint (grey square). The locations of materials suppliers were obtained from Google Maps [9]. The red line demonstrates the transportation routes for concrete pipes from the candidate factory locations to the construction sites. The transportation distance in this case study is the direct distance between two locations. The distance calculation was performed using the GeoPandas package [10], with the projected coordinate system for distance calculation being EPSG 3035 [11].

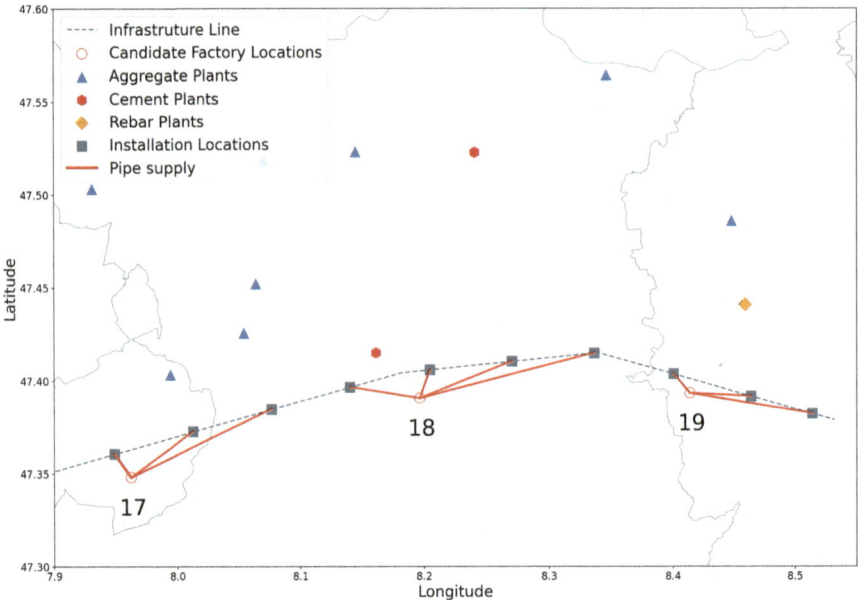

Fig. 2 Case study of hyperloop construction near candidate factory locations 17–19

The bill of quantities for the concrete pipes was provided by ETF [8]. ρ_p, ρ_a, ρ_c, and ρ_r are the required mass for constructing 1 km of the infrastructure for concrete pipes, aggregate, cement, and rebar, respectively. The concrete pipes mass m_i required by the infrastructure segment i with a length of l_i can be calculated with $m_i = \rho_p l_i$, and the mass of the material j required by the factory at the deployment location f' can be calculated with Eq. (13):

$$m_{jf'} = \rho_j \sum_{i \in I} l_i X_{f'i} \tag{13}$$

Here, ρ_j represents the required mass of material j for constructing 1 km of the infrastructure, $X_{f'i}$ is the decision variable resulting from the first stage model at the selected factory deployment location f'.

The transportation carbon emissions indicators were obtained from the ecoinvent 3.9.1 database [12]. Due to the lack of detailed information, we assume the transportation modes for both materials and concrete pipes remain unchanged throughout the whole project. The transportation modes for material supply and factory transportation are set to use "lorries of all sizes" with a carbon emission indicator (et_a, et_c, et_r, and et_f) of 0.1478 kg CO_2-eq/tkm. Given that the mass of one piece of concrete pipe is approximately 125 tons, the transportation mode for pipes is set to be "lorries larger than 32 tons" with a carbon emission indicator et_p of 0.1002 kg CO_2-eq/tkm. Details of the mobile factory were provided by the factory design team from unibz, with further information available in [1, 13]. The mass of the mobile factory m_f is

Table 1 Key input parameters

et_a, et_c, et_r	et_f	et_p	e_f	ρ_a	ρ_c	ρ_r	ρ_p	m_f
0.1478	0.1478	0.1002	3822	4741.4	1208	219.9	6269.3	575

575 tons. The carbon emissions from the factory reconfiguration processes e_f were estimated to be 3822 kg CO_2-eq per relocation for the mobile factory that uses a pneumatic tent solution. e_f is assumed to remain the same for all the candidate factory locations. The LCA calculation was conducted using the IPCC 2021 GWP100 method [14] based on the ecoinvent database 3.9.1 [12] with the Brightway version bw25 package [15]. Table 1 summarizes the key input parameters for the two-stage ILP model. The model was built with Pyomo package 6.7.1 [5] and solved by the COIN-OR Branch-and-Cut (CBC) [6] solver 2.10.10.

4 Result and Discussion

Figure 3 illustrates the optimal total carbon emissions with respect to different factory relocation numbers. The overall carbon emissions demonstrate a significant decrease of 58% as the factory relocation number increases from zero to six. However, beyond six factory relocations, the reduction of the overall carbon emissions becomes insignificant. Specifically, as the number of relocations increases from zero to six, the carbon emissions from the transportation for the concrete pipes decrease by 84%, while the carbon emissions of the transportation for the materials fluctuate around 4000 tons CO_2-eq. These fluctuations are insignificant compared to the reduction of carbon emissions from pipe transportation. This is due to the relatively uniform geographic distribution of material suppliers around the candidate factory locations. Except for zero relocation, the carbon emissions from the transportation of the mobile factory remain around 20 tons CO_2-eq for all relocation numbers. This consistency arises because the factory is consistently deployed at the first and the last candidate factory locations, resulting in only slight changes in the total transportation distance of the mobile factory. Considering the relocation of the mobile factory can be time consuming, four to six relocations are efficient for reducing the overall carbon emissions from transportation in this case study.

Figure 4 shows the contributions of the six carbon emissions categories to the overall carbon emissions with respect to different relocation numbers. The carbon emissions from pipe transportation are the predominant contributor until four relocations at which point the carbon emissions from aggregate transportation become the largest contributor. The contribution of the carbon emissions from factory transportation is insignificant for all the relocation numbers.

Figure 5 illustrates the transportation distances for materials from suppliers to the factory, as well as the total transportation distance for concrete pipes from the factory to the construction sites in four, six, and eight relocations. The total transportation distance for concrete pipes is calculated as the sum of the distances from each

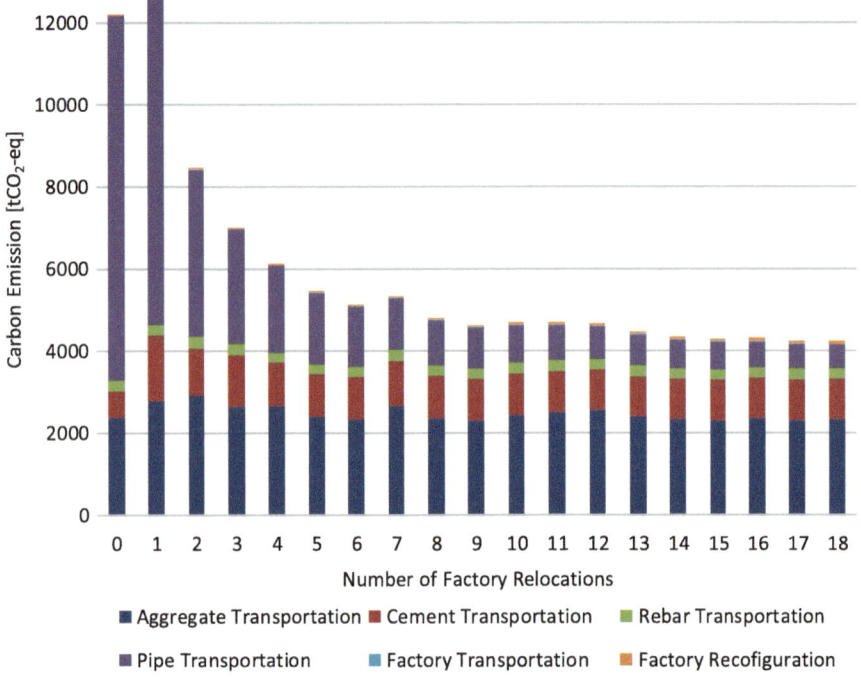

Fig. 3 Optimized total carbon emissions for different factory relocation numbers

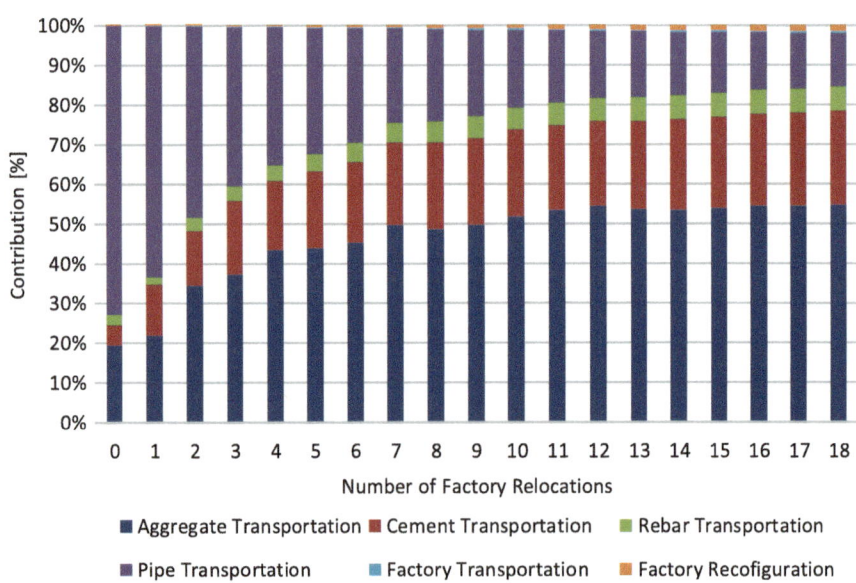

Fig. 4 Contribution analysis for various factory relocation numbers

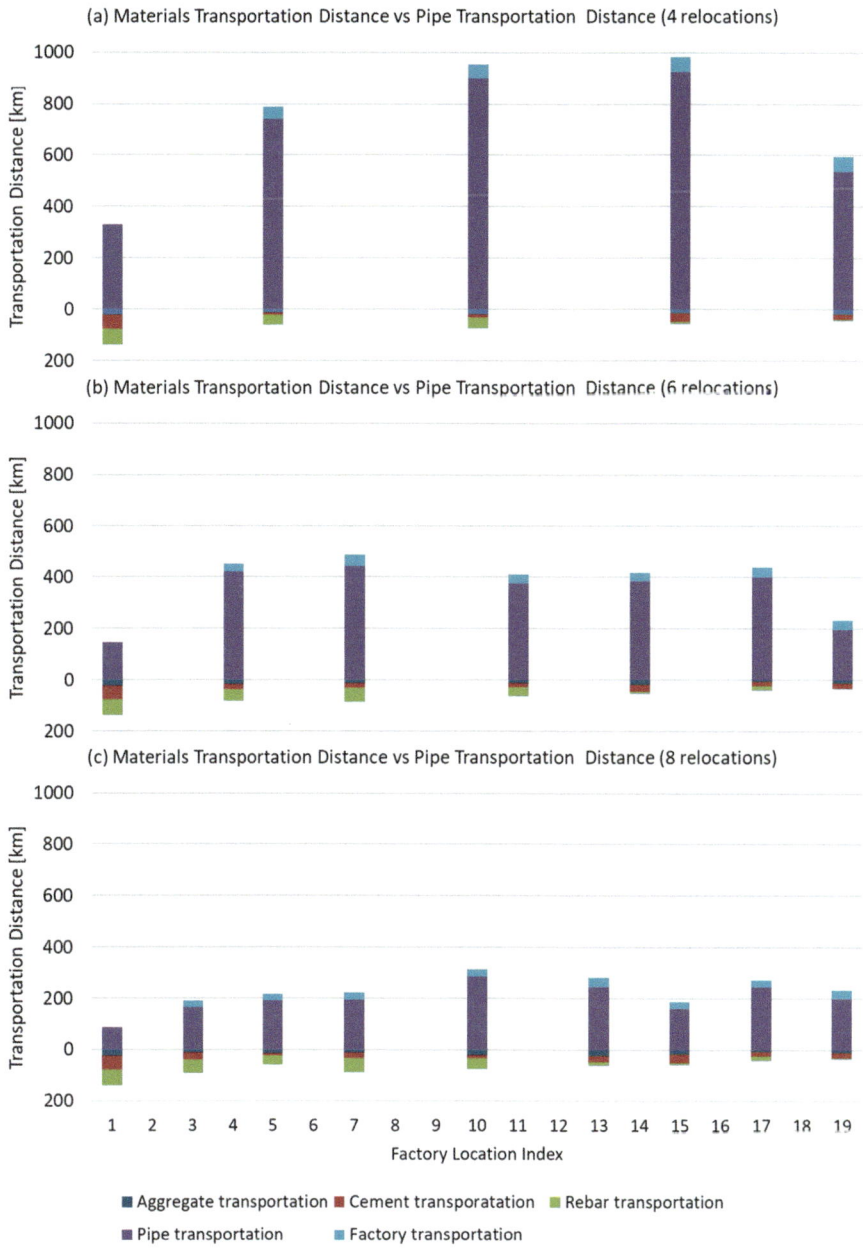

Fig. 5 Transportation distances for selected factory locations in (**a**) four-relocation scenario, in (**b**) six-relocation scenario, and in (**c**) eight-relocation scenario

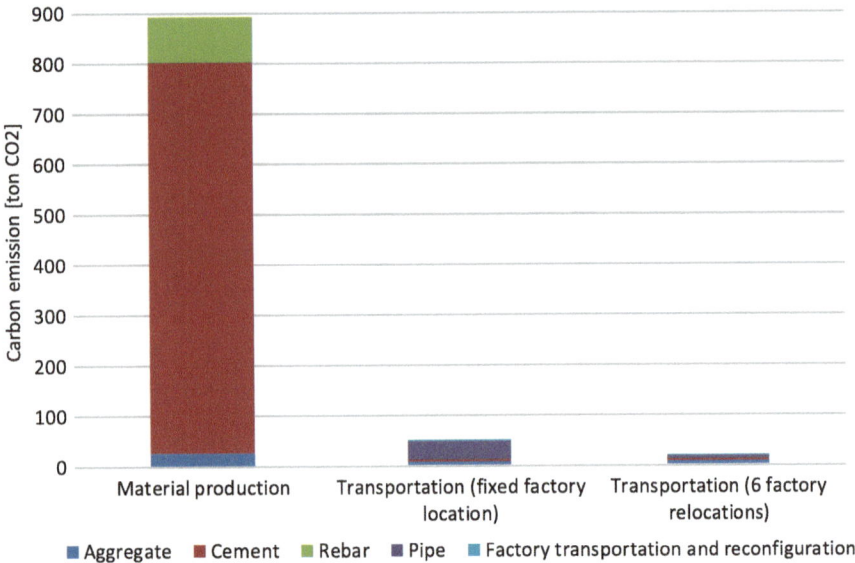

Fig. 6 Carbon emissions from materials production and transportation (averaged on 1 km construction of the infrastructure)

construction site to its corresponding factory location. Notably, the total transportation distance for pipes is significantly greater than that for materials at each selected factory deployment location. As the number of relocations increases, there is a significant decrease in the total transportation distance for pipes at each selected factory deployment location. However, the transportation distances for materials at each factory deployment location only show slight variations. Together with Fig. 3, it is evident that the reduction of carbon emissions by increasing the relocation number is primarily due to the reduction of the transportation distance for pipes from the factory to the construction sites.

Figure 6 illustrates the carbon emissions from both material production and transportation. Notably, carbon emissions from transportation, including material transportation, pipe transportation, factory transportation, and factory reconfigurations, are significantly lower compared to emissions from materials production. This suggests that while the use of a mobile factory for constructing the hyperloop infrastructure can reduce its carbon footprint, most emissions are still associated with materials production.

The main research limitation of this study is that the two-stage ILP model cannot guarantee a globally optimal solution for minimizing the total transportation carbon emissions, as the minimization of carbon emissions from materials supply relies on the predetermined factory locations from the first-stage model. Integrating the selection of materials suppliers into the decision variables could address this limitation in future studies. Additionally, the assumption that, at each selected factory location, the mobile factory receives each material from only one supplier may underestimate carbon emissions if the selected supplier cannot provide sufficient material. However, this underestimation would not affect the recommended relocation strategy significantly, as the primary carbon reduction comes from reducing transportation distance for delivering prefabricated elements. Another limitation is the use of segment midpoints to represent construction sites, which simplifies computation but may underestimate pipe transportation distance. Despite this, this impact is minor given the small carbon emissions indicator for pipe transportation. To improve accuracy, future studies can run the model with exact construction site locations for each pipe. Finally, assuming unchanged transportation modes for materials and pipes in the case study may not accurately reflect reality, particularly in instances where roads are not well constructed. In the future study, we will conduct a sensitivity analysis to estimate its impact on the mobile factory relocation strategy.

5 Conclusion

In this paper, we developed a methodology that combines GIS and an optimization model to analyze the potential reduction in transportation carbon emissions achievable using the mobile prefabrication factory in large-scale linear infrastructure construction projects. A two-stage ILP model was established to identify the optimal relocation strategy for minimizing transportation carbon emissions. A case study of constructing a hyperloop from Geneva to Zurich was utilized to evaluate this model. The results indicate that, in this case study, the implementation of a four-relocation strategy can potentially reduce the total transportation carbon emissions by 50%. This reduction in carbon emissions is primarily due to the optimized transportation distance of the concrete pipes from the mobile factory to the construction sites.

Acknowledgments The research presented in this article was carried out within the research project "Smart Mobile Factory for Infrastructure Projects (SMF4INFRA)," which has received funding from the Swiss National Science Foundation under grant agreement no.204852 and Autonomous Province of Bolzano/Bozen—South Tyrol. The authors are grateful to EuroTube Foundation for the support and insights on this project.

References

1. Dallasega, P., Revolti, A., Schulze, F., Benedetti, L., de Morsier, D.: Requirement analysis and concept design of a smart mobile factory for infrastructure projects. In: Alfnes, E., Romsdal, A., Strandhagen, J.O., von Cieminski, G., Romero, D. (eds.) Advances in Production Management Systems. Production Management Systems for Responsible Manufacturing, Service, and Logistics Futures, pp. 19–33. Springer Nature, Switzerland, Cham (2023). https://doi.org/10.1007/978-3-031-43670-3_2

2. Alix, T., Benama, Y., Perry, N.: A framework for the design of a reconfigurable and mobile manufacturing system. Proc. Manuf. **35**, 304–309 (2019). https://doi.org/10.1016/j.promfg.2019.05.044

3. Rauch, E., Matt, D.T., Dallasega, P.: Mobile factory network (MFN)—network of flexible and agile manufacturing systems in the construction industry. Appl. Mech. Mater. **752–753**, 1368–1373 (2015). https://doi.org/10.4028/www.scientific.net/AMM.752-753.1368

4. ISO 14040:2006(en): Environmental Management—Life Cycle Assessment—Principles and Framework. https://www.iso.org/standard/37456.html

5. Pyomo: http://www.pyomo.org

6. COIN-OR: Computational Infrastructure for Operations Research. https://www.coin-or.org/

7. Decker, K., Chin, J., Peng, A., Summers, C., Nguyen, G., Oberlander, A., Sakib, G., Sharifrazi, N., Heath, C., Gray, J.S., Falck, R.D.: Conceptual sizing and feasibility study for a magnetic plane concept. In: 55th AIAA Aerospace Sciences Meeting. American Institute of Aeronautics and Astronautics (2017). https://doi.org/10.2514/6.2017-0221

8. EuroTube – Hyperloop Foundation: https://eurotube.org/. Last accessed 26 Feb 2024

9. Google Maps: https://www.google.com/maps/. Last accessed 28 Feb 2024

10. GeoPandas: https://geopandas.org/en/stable/

11. ETRS89-extended / LAEA Europe - EPSG:3035, https://epsg.io. Last accessed 26 Feb 2024

12. ecoinvent – Data with purpose. https://ecoinvent.org/. Last accessed 26 Feb 2024.

13. Bataleblu, A.A., Rauch, E., Revolti, A., Dallasega, P., Puik, E., Cochran, D.S., Foley, J.T., Foith-Förster, P.: Proceedings of the 15th International Conference on Axiomatic Design 2023 Smart Mobile Factory Design Decomposition Using Model-Based Systems Engineering. Springer Nature Switzerland Cham 15–25 (2024)

14. Intergovernmental Panel on Climate Change: Climate Change 2021 – The Physical Science Basis: Working Group I Contribution to the Sixth Assessment Report of the Intergovernmental Panel on Climate Change. Cambridge University Press (2023). https://doi.org/10.1017/9781009157896

15. Brightway LCA Software Framework: https://docs.brightway.dev/en/latest/

Sustainability Potentials of the Precast Industry in Kenya

Joseph Mwiti Marangu ⓘ, Andrew Onderi Nyabuto ⓘ, Thomas Pfeiffer ⓘ, Sabine Kruschwitz ⓘ, Christoph Völker ⓘ, and Wolfram Schmidt ⓘ

Contents

1 Introduction

1.1 Background

Kenya is a developing country located to the east of the African continent, along the Indian Ocean coastline. It is home to the four major cities Nairobi, Mombasa, Kisumu, and the latest conceptualized Nakuru city. In the East Africa block, Nairobi

J. M. Marangu (✉) · A. O. Nyabuto
Meru University of Science & Technology, Meru, Kenya
e-mail: jmarangu@must.ac.ke

T. Pfeiffer · C. Völker
SRH University of Applied Sciences Berlin, Berlin, Germany

S. Kruschwitz · W. Schmidt
Bundesanstalt für Materialforschung und –prüfung, Berlin, Germany

© The Author(s) 2025   1699
M. Kioumarsi, B. Shafei (eds.), *The 1st International Conference on Net-Zero Built Environment*, Lecture Notes in Civil Engineering 237,
https://doi.org/10.1007/978-3-031-69626-8_141

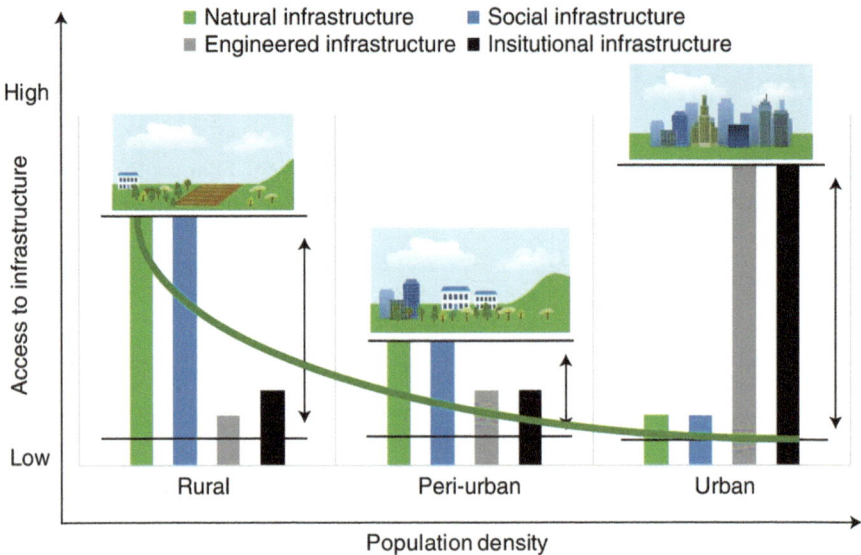

Fig. 1 Access to services varies across individuals within each area (arrows), and nature may act as a safety net in many areas across the Global South (green line) [4]

is the second most populated after Dar-es-salaam. The four cities are relatively straining from an increase in demand for infrastructure and decent housing [1] due to massive urbanization. According to Hope (2012), urbanization is due to social pressures for better and improved living standards (as illustrated in Fig. 1) but many people end up in undesirable living conditions [2, 3]. Urbanization comes along with many construction activities that cannot compare with global conditions such as in developed continents, i.e., Europe and North America. For example, in Kenya, cement manufacturers are located in the cities, with Nairobi accounting for 6 out of 14 registered plants as per the global cement report in 2023 as shown in Fig. 2.

The mushrooming construction activities around cities and towns within Kenya use Portland cement as the main binder. Cement production in Kenya is done according to the KS EAS-18-1-2017 [5] standard adopted from European Norm EN 197 for cement standards. The construction industry in Kenya uses Ordinary Portland Cement (OPC) CEM I, Portland Pozzolana Cement (PPC) CEM II/B-P of 21–35% pozzolana, CEM IV/A containing 11–35% pozzolana and Portland Limestone Cement (PLC) CEM II/A-LL having 6–20% limestone [6]. Different companies introduce pozzolans into cement blends to reduce the amount of clinker to save on marginal production costs. There is limited production of OPC (CEM I) unless on special orders and use [6].

However, the cement factories are trying to go green by introducing more pozzolana and finer grinding while keen not to compromise strength, consistency, and durability. Some of the pozzolana introduced include lime, volcanic ash, limestone, and clays. It is because Portland cement contributes 80–90% of the

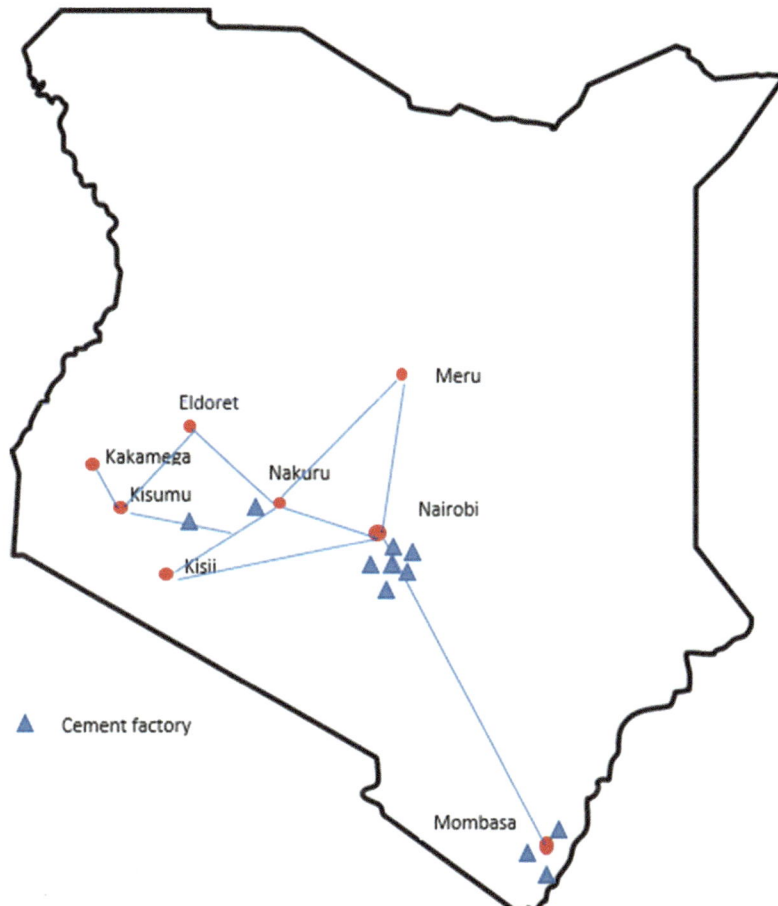

Fig. 2 A map of Kenya showing the distribution of cement factories

estimated 10% of anthropogenic gas emissions responsible for climate change [7]. There is an urgent need to decarbonize cement production despite increased demand by the government agenda to address the housing shortage. In addition, decarbonizing cement is difficult as 77% of the CO_2 emissions occur during chemical reactions in production. The other percentage of emissions are tied to operating equipment and transportation of aggregates [8]. Transportation of concrete elements such as aggregates and equipment has been linked to contribute about 7% of the 8–9% anthropogenic gas emissions [8]. Workable solutions must deploy low-carbon clinker, alternative SCMs, circular technology and AI. The cement industry has introduced targets to realize net-zero CO_2 emissions by 2050.

The distribution of cement plants in cities reflects on two major considerations that include materials and supply chain. The situation draws large construction activities to concentrate in major cities [9] including the precast products industries.

The status quo is not going to hold long enough as the population growth is increasing sharing the demand for both housing and infrastructure in the devolved government units such as counties and small towns like Eldoret, Meru, Nyeri, Kisii, and Kakamega shown in Fig. 2. Increasing demand for concrete from rural regions implies an increase in carbon emissions from the transport and storage of cement during distribution. The demand for materials such as limestone and clinker influence the locating of cement plants in the city. The transportation of raw materials to the factory and cement to customers is a reason for high cement prices as the cost of production is cascaded to the consumers. The location of a few plants outside the cities is linked to the main transport corridor between the cities due to the ease of the distribution matrix. Therefore, using supplementary cementitious materials (SCMs) would ease the pressure of having to set up plants in cities within proximity of major supply streams of raw materials. Innovatively introducing biomass from agricultural waste and natural resources such as abundant clay deposits are milestones toward affordable and ecological cement as traditional SCMs such as fly ash and slag are deemed unavailable and unaffordable. Clay has been proven to naturally occur in many parts of Kenya for manufacturing limestone calcined clay cement (LC3) [10]. Elsewhere, rice farmers are potential suppliers of rice husk that has proven to be rich in silica from the ash as a pozzolana [11]. Cassava peels are also converted to ash and can be substituted for about 75% due to the high contents of silicates and aluminates [12]. All the above solutions are presented but the country is stuck in the status quo from a poor supply chain due to a lack of infrastructure, low technology and lack of diverse national codes and standards in support.

In developed continents such as Europe, solutions are even traditionally feasible as the cement manufacturing plants are evenly distributed enhancing a good supply chain for industry waste such as blast furnace slag (GBBS) and fly ash (FA). They have developed codes of practice and standards for adopting the use of low-carbon types of cement with the potential of reducing the use of clinker by up to 50% as per EN 197-5 like LC3 [13]. The precast concrete (PC) technology involves designing structural components, manufacturing in a factory, transporting, delivering, and erecting on site [14].

Precast technology has been classified to be among the top ten innovations in the construction sector as it follows a modular approach [15]. The precast industry (PI) has been in practice for over a century with numerous advantages. Some of the benefits are due to its versatility in many applications such as road bridges, channels, towers, and buildings. The precast technology is efficient in terms of quality finishes as elements and connections are determined in a controlled environment. Also, the cost of construction of the overall structure is saved from reduced onsite labor, and saving on disarray scheduling. The expenses expected on formwork and smooth detailed finishes such as plastering is eliminated. The other secondary benefits include low repair and maintenance costs as the quality of components is guaranteed. Safety and health concerns are addressed through minimal onsite interactions with unskilled labor, thus reducing construction-related accidents. PI also is environmentally friendly due to flexibility in allowing the recycling and reuse of aggregates [16] and demolished components. In some aspects, the damping of aggregate waste from demolitions is also minimized. In general, the PI sector is more beneficial and can save on the time and cost of completing a construction project and impart sustainability.

1.2 The Precast Industry in Kenya

Concrete has long been identified as the most used construction material [17, 18] across the globe due to its versatility in shape and form. The flexibility in the construction industry allowed the inception of precast products which means no more onsite ready mixes. Structural elements of defined shape, form, and size are manufactured in the plant, transported to the field, and fixed. The precast industry has gained prominence in Kenya for the past few decades. However, it is not well documented. The available sources imply that precast products have been largely on fixtures and fitting and not major structural elements. However, some companies are developing precast panels for walls, beams, and slabs. The sustainability of the precast industry is attributed to reduced construction time, low labor costs and overall benefits to the clean environment due to a controlled mix by avoiding heavy onsite construction machinery [7].

The global view of precast technology is well documented but lowly used in developing regions. PI needs to align its activities with the societal needs to improve uptake. Developing economies perceive concrete works as high-risk, labor-intensive, and technology-intensive [19]. The slow integration of technology in business advancement affects the uptake of precast technology—skilled labor in designing and managing PI components. The main objective of this study was to identify the potential of using AI in increasing the uptake of precast technology and sustainability in Kenya.

2 Methodology

The study focuses on identifying and analyzing the challenges and/or opportunities available for the adoption of precast products through available literature and artificial intelligence (AI). However, a qualitative research approach was used in obtaining data through questionnaires, observations, and cost comparison on a case study. In the case of questionnaires, four key research questions were posed and twelve corresponding companies' samples for their work on precast. One case study was picked for precise size and available quantitative information for a cost comparison. The research framework was adopted as in Fig. 3. The objectives of the study included the following:

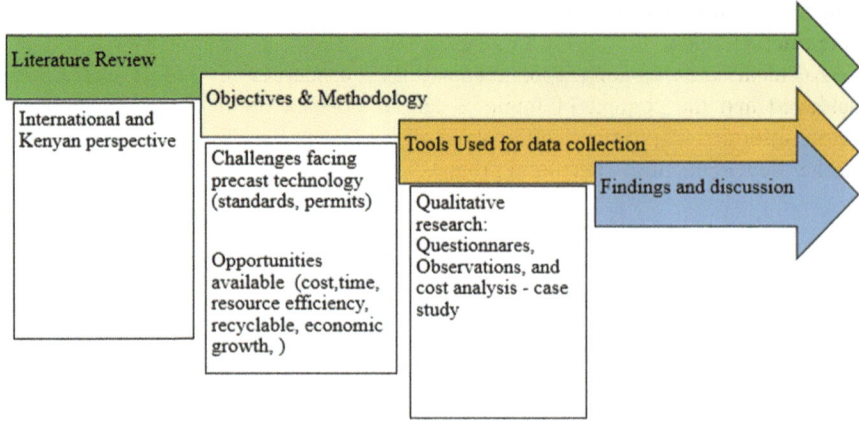

Fig. 3 Research framework

- To analyze the size and coverage of precast technology in Kenya
- To determine the challenges faced by the precast industry in applying the technology in Kenya
- To establish the potential application of AI in increasing the adoption of the precast technology

The study questionnaire involved four fundamental questions directed to the precast components manufacturing companies registered in Kenya.

1. What types of precast products does your company manufacture?
2. What are primary factors influencing your decision to adopt precast technology?
3. What are main challenges hindering the wider adoption of precast technology?
4. How would you rate the government's support for the precast industry in Kenya?

The cost comparison between the conventional concrete building methods and precast structure was done on the same area structure. The cost of the precast sampled building was as per the precast manufacturing company and the Kenyan government cost handbook for constructing the same structure. A 64 square meters 2-bedroom house was opted for cost analysis. The pictorial representation for conventional concrete walls and precast walls is shown in Fig. 4.

3 Discussion of Findings

3.1 Precast Components, Primary Factors, Main Challenges, and Government Support

While only 50 questionnaires were emailed to registered companies and contractors that advertised themselves to be handing precast products on the active websites, 24 responses were received. The responses were received through direct dialed calls

Fig. 4 Tuff stone conventional house vs. precast wall house

Table 1 Types of components prepared by different companies

Precast components	Wall panels	Beams and columns	Slab	Hollow blocks	Channels	Others
Number of companies	4	4	5	24	12	21

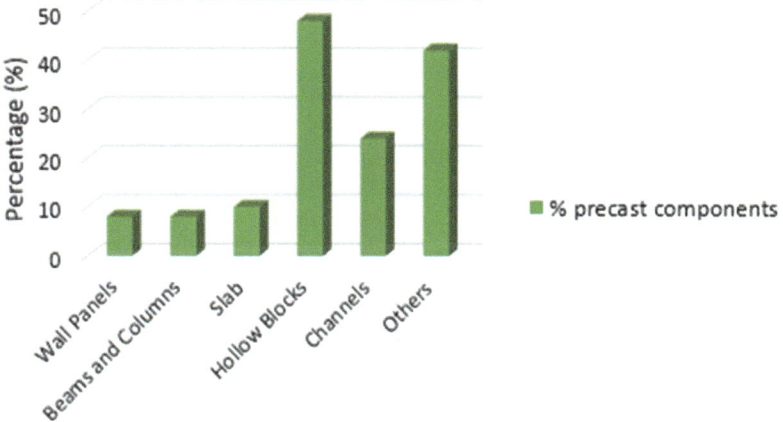

Fig. 5 Precast components

as many follow-ups preferred instant responses. The results of the data collected are in Table 1 and Fig. 5, respectively.

As per Fig. 5, 8% of the companies sampled were found to make wall panels, 8% of beams, 10% slab (suspended and foundation), 48% hollow blocks, 24% channels and 42% others. The results imply that only five companies can construct houses. At least all companies that responded are producing precast elements on a small scale. Many companies seem to concentrate on replacing tuff stones preferred in wall construction. The structural uptake can account for 8% on a large scale.

The respondents to the questionnaires agreed to have a unanimous choice to adopt precast to save on construction cost, period, and durability. The 8% of major players expressed optimism about saving on material waste, environmental degradation and

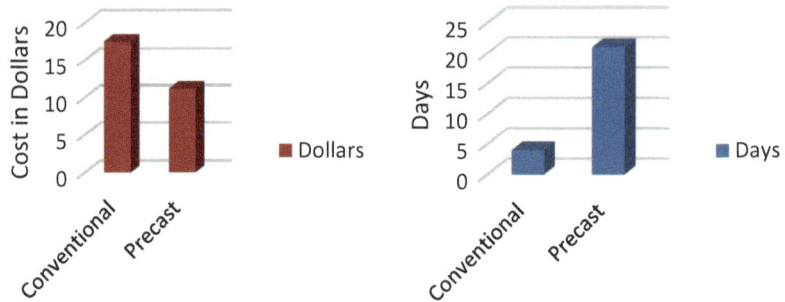

Fig. 6 Cost and time comparison for conventional and precast wall panel

reducing repair and rehabilitation costs. The cost component comparison was conducted on wall panels for a 64-meter square house. The comparison involved a wall panel component. The wall panel was selected as it involved a comparison of the time of construction and cost per square meter of a conventional wall as shown in Fig. 6.

The two projects of the same plinth area were compared against cost per square area and construction time according to Kenya's building code. It was noted that conventional concrete construction is 34% more expensive than precast concrete. The duration of erecting a precast wall was found to be 17 days slower for the same plinth area. The results are in line with Nyanam et al. (2017) findings on precast over conventional construction methods on overall projects [7]. The opportunities, challenges, pros and cons, and potential use of AI on PI in Kenya are further discussed in the next section.

4   The Opportunities Available for the PI in Kenya

PI products have been identified by many manufacturers, researchers, and consumers to offer versatility in applications. There are many opportunities present for Kenya PI in building products and construction elements. The opportunities include the following:

- Precast technology is versatile and flexible in recycling and reusing aggregates hence minimizing demolition and onsite mix waste. Precast components can be reused in case of demolition saving up to 50% embodied energy [20].
- The precast plant requires low initial capital investment compared to starting a cement plant as it accommodates the low supply streams of SCMs for the quality of precast components. Kenya is rich in natural pozzolana such as clay and limestone that enhance concrete durability.
- The precast industry offers opportunities for recycling and reuse of agricultural waste such as concrete fibers, admixtures, and SCMs. Hence PI can be built in any part of the country by utilizing local materials such as sisal fibre reeds from Kilifi, Kenya.

- The PC is a resource-efficient venture through saving on time of construction, minimizing material waste, and being less laborious for low-income earning households. The materials are easy to transport and install fast and safely at the site.
- The PI is a revenue stream for the local communities through selling aggregates and collecting agricultural and regenerative waste such as rice husks, cassava peels and water hyacinths from Lake Victoria, Kenya respectively. This supports the circular economy. The growth in precast components can be imported and enhance the growth of the country's gross domestic production (GDP).
- An opportunity exists for government support through creating national annex codes, standards, and consumer incentives, especially in the ongoing Boma Yangu Affordable housing projects, that are devolved to counties.

5 Pros and Cons of Precast Concrete

In both ways, blending cement as a binder from factory and precast production has both pros and cons. The precast industry (PI) will provide more advantages to the small towns as are reliable and durable, enhancing the quality and timely smooth finishes. Secondly, the precast concrete will enhance a clean built environment by minimizing waste and recyclable structural elements. Incorporating SCMs into OPC enhances reliability in building strong structures with reduced embedded CO_2 footprint from conventional concrete.

Blending types of cement onsite bear a risk of unwanted reactions that can retard [13] or enhance flash settings that lead to early concrete cracks resulting in many active phases. Therefore, blends produced during concrete production and molding elements enhance a homogeneous and ideal mixture. This is possible by controlling OPC and SCM quantities in the concrete production process. The main constraint in concrete production is the low supply streams of SCMs in manufacturing cement processes. This is the reap for precast concrete as locally available materials can substitute large amounts of cement for structural components.

6 Challenges Facing the Precast Industry in Kenya

The precast industry is on the verge of penetrating small towns in Kenya as time goes on. Like any business model and technology, the PI faces several challenges hindering its fast growth. Despite having a good percentage of Kenyans connected to the internet, 26% don't have access or even use it [21]. A few have embraced the internet and integrated AI to improve businesses like precast technology. Some of the challenges facing PI in Kenya include the following:

- Limited knowledge of contractors and vast unskilled labor in adopting precast technology.

- The engineers and technicians show technical fears from the lack of codes and standards in supporting PI components.
- Setting up a precast plant is expensive among many low-income households as access to credit is difficult due to lack of collateral.
- Access to grants is limited to an informed population such as research institutions that lack a clear framework and government initiatives on enhancing business startups.
- There are no outlined government initiatives and incentives for the precast industry in place for either producers or consumers in Kenya.

7   The Future and Outlook of the Precast Industry in Kenya

The construction industry in Kenya contributes significantly to the gross domestic product (GDP). The PI as a sector in construction has witnessed notable growth over the past few decades. The country is driven by increasing population and demand for decent and affordable housing. The demand is increasing for overall improved infrastructure that requires fast, affordable, and durable construction methods. The current precast concrete in Kenya focuses on structural elements such as road furniture, beams, slabs, wall and slab panels, and columns. Opportunities for growing PI arise mainly from government projects. The projects that have contributed to the growth so far are roads and affordable houses under the government's bottom-up transformative agenda (BETA). It is a result of quality control, speed construction and durability on government projects that require longevity in performance. The rise of computer-aided design and AI-enhancing graphic representation will eventually enhance the two-fold growth of PI in Kenya.

8   Application Potential of Artificial Intelligence to Precast Concrete

Artificial intelligence (AI) is the ability to make machines mimic and perform human tasks. Engineering playing a major role in design and construction is inseparable from technology like the use of AI. Technological applications are multifaceted from mechanical applications in cars, computers, 3D printing and medicine. Technology has been integrated well into building information management (BIM) for the construction industry in applications for design, schedule, and management construction [22]. All these emerging technologies include the current rapid growth of the internet with an integrated algorithm of improving many operation ways faster than human brains.

Artificial intelligence (AI) has been made available by the high access to the internet. Kenya has already rolled out 5G internet connection which has boosted the connection and expanded the bandwidth for faster upload and download speed

[21]. AI aims at picking human traits and applies to tasks way faster than humans. It implies that AI can learn, plan, and perceive environmental language to execute a task. AI is getting into the construction industry to improve productivity, quality, and safety. Currently, AI can use cloud-based applications that are accessed remotely through the internet to capture key aspects like progress and resource allocation and management. The precast industry would not be different as machine learning (ML) models can be fitted into mixing and casting machines. The machines will be able to maintain quality by intelligently integrating algorithms that would execute leveraged mix designs and achieve prismatic shapes and sizes.

With logistics as key to implementing precast technology, AI can alter business models. Customer response and uptake of technology combinations will only be possible through automation systems. The overall implications will improve quality, minimize waste, reduce injuries, increase efficiency and reduce environmental impact to enhance sustainability. Generally, large projects will be realized within the timeframe.

Eventually, PI is the futuristic approach to the construction industry in Kenya. Been around for over a decade. Many companies have ventured into fabricating road drainage channels, road kerbs, fencing poles, slab waffles, fencing panels, bridge beams, and wall panels. PI in Kenya is about 8% successive as companies can form most elements and construct. Despite many challenges, PI has the potential to grow with an increase in skills, machinery, and alternative materials. The documentation of PI in market share is not well documented in terms of numbers. However, the market is growing gradually as many players are trying to modernize and localize materials used for specific uses. In addition, the precast components are cost-saving, eco-friendly, and time-saving for prefabricated ways of construction.

References

1. Shitote, S.M., et al.: A pre-cast concrete Technology for Affordable Housing in Kenya. In: Proceedings from the International Conference on Advances in Engineering and Technology, pp. 680–695. Elsevier (2006). https://doi.org/10.1016/B978-008045312-5/50073-X
2. Hope, K.R.: Urbanisation in Kenya. Afr. J. Econ. Sustain. Dev. 1(1), 4 (2012). https://doi.org/10.1504/AJESD.2012.045751
3. Nyaura, J.E.: Urbanization process in Kenya: The effects and consequences in the 21st century. Int. J. Novel Res. Humanit. Soc. Sci. 1(2), 33–42. [Online]. Available: www.noveltyjournals.com (2013)
4. Hutchings, P., et al.: Understanding rural–urban transitions in the Global South through peri-urban turbulence. Nat. Sustain. 5(11), 924–930 (2022). https://doi.org/10.1038/s41893-022-00920-w
5. KS EAS_148_1: Cement — Test methods — Part 1: Determination of strength (2017)
6. Okumu, V.A.: Suitability of the Kenyan blended Portland cements for structural concrete production. J. Sustain. Res. Eng. 4(2), 55–68. [Online]. Available: http://sri.jkuat.ac.ke/ojs/index.php/sri (2018)
7. Nanyam, V.P.S.N., Basu, R., Sawhney, A., Vikram, H., Lodha, G.: Implementation of precast technology in India–opportunities and challenges. Procedia Eng. 196, 144–151 (2017). https://doi.org/10.1016/j.proeng.2017.07.184

8. Griffiths, S., et al.: Decarbonizing the cement and concrete industry: a systematic review of socio-technical systems, technological innovations, and policy options. Renew. Sust. Energ. Rev. **180**, 113291 (2023). https://doi.org/10.1016/j.rser.2023.113291

9. Mbaka, J.G.: Impact of cement industry on water quality in the Athi River, Machakos County, Kenya. East Afr. J. Environ. Nat. Resour. **6**(1), 232–242 (2023). https://doi.org/10.37284/eajenr.6.1.1322

10. Marangu, J.M., Riding, K., Alaibani, A., Zayed, A., Thiong'o, J., Muthengia, W.: Potential for selected Kenyan clay in production of limestone calcined clay cement, October 2019

11. Okoya, B.O., Abuodha, S.O., Mumenya, S.W., Dulo, S.O.: Characteristics of Kenyan rice husk ash produced under controlled burning. Int. J. Eng. Res. Technol. IJERT. **10**(11) (2021)

12. Familusi, A., Adewumi, B., Olusami, J., Mujedu, K., Ogundare, D.: Converting waste to wealth: cassava peel ash as potential replacement for cement in concrete, November 2019

13. Schmidt, W., et al.: Sustainability potentials of the precast industry in West Africa. CPI—Concr. Plant Int. **Industry Review**(Part 1). [Online]. Available: www.cpi-worldwide.com (2023)

14. Murari, S.S., Joshi, A.M.: Precast construction methodology in construction industry. SSRN Electron. J. (2017). https://doi.org/10.2139/ssrn.3496019

15. Gautam, B.P.: Precast and prestressed concrete for the future construction of Sudurpaschim. Far West. Rev. **1**(2), 84–100 (2023). https://doi.org/10.3126/fwr.v1i2.62131

16. Hahn, L., Cherif, C.: New potentials for the precast industry through the use of innovative carbon reinforcement, June 2022

17. Smarzewski, P., Stolarski, A.: Properties and performance of concrete materials and structures. Crystals. **12**(9), 1193 (2022). https://doi.org/10.3390/cryst12091193

18. Attri, G.K., Gupta, R.C., Shrivastava, S.: Sustainable precast concrete blocks incorporating recycled concrete aggregate, stone crusher, and silica dust. J. Clean. Prod. **362**, 132354 (2022). https://doi.org/10.1016/j.jclepro.2022.132354

19. Polat, G.: Precast concrete systems in developing vs. industrialized countries. J. Civ. Eng. Manag. **16**(1), 85–94 (2010). https://doi.org/10.3846/jcem.2010.08

20. Akduman, S., Aktepe, R., Aldemir, A., Ozcelikci, E., Alam, B., Sahmaran, M.: Opportunities and challenges in constructing a demountable precast building using C&D waste-based geopolymer concrete: a case study in Türkiye. J. Clean. Prod. **434**, 139976 (2024). https://doi.org/10.1016/j.jclepro.2023.139976

21. Mureithi, M.: The internet journey for Kenya: the interplay of disruptive innovation and entrepreneurship in fueling rapid growth. In: Ndemo, B., Weiss, T. (eds.) Digital Kenya: An Entrepreneurial Revolution in the Making, pp. 27–53. Palgrave Macmillan UK, London (2017). https://doi.org/10.1057/978-1-137-57878-5_2

22. Mahajan, A.G.: Applications of artificial intelligence in construction management. Int. J. Res. Eng. **32**, 32–1541 (2019)

Prefabricated Vertical Drain-Supported Railway Embankment on Thick Marshy Deposit: A Case Study of Udaipur Railway Station Project, Tripura, India

Samrat Ghose and Arindam Dey

Contents

1   Introduction

In the current era of infrastructure development, the construction of geotechnical engineering systems (roads and railway embankments) is growing rapidly over the lands comprising soft soils that act as a week foundation medium. Such saturated soft cohesive deposits have poor geotechnical characteristics (such as low bearing capacity, high compressibility, and low permeability) and are generally found near the marine coast and in marshy lowlands around lakes and river beds. Hence, reclamation of lands is carried out by implementing suitable ground improvement techniques by the practicing community [1]. Application of prefabricated vertical drains (PVDs) with surcharge preloading is one of the efficient ways to enhance the ground conditions of saturated soft cohesive deposits [2, 3]. Introduction of PVDs into the low permeable soft soils creates an artificial drainage path that facilitates

S. Ghose · A. Dey (✉)
Indian Institute of Technology, Guwahati, Assam, India
e-mail: arindam.dey@iitg.ac.in

© The Author(s) 2025
M. Kioumarsi, B. Shafei (eds.), *The 1st International Conference on Net-Zero Built Environment*, Lecture Notes in Civil Engineering 237,
https://doi.org/10.1007/978-3-031-69626-8_142

quicker dissipation of pore water and excess pore pressure due to a shortened drainage length, thereby resulting in accelerated consolidation settlement [4]. Application of PVDs in managing the settlement of railway embankments passing through soft soils has gained profound popularity [5].

In most of the conventional studies involving PVD treatment, subsurface stratification is considered horizontal, barring very few literatures that consider the realistic undulations in the substratum [6, 7]. In this chapter, a case study of the Udaipur railway embankment in Tripura, India, raised over a marshy lowland, is presented. A finite element (FE) analysis is conducted to predict the consolidation of PVD-improved layered subsoils under the embankment. As per the data obtained from multiple borehole conducted at the site, the subsurface was found to comprise undulated stratification. The Mohr–Coulomb (MC) constitutive model is used to represent the cohesionless deposit and the embankment fill, while the Soft Soil (SS) material model is used to represent the soft cohesive subsoil layers. Assessment of PVD-improved subsoils is carried out to reflect upon the influence of realistic undulated soil profile over the conventional consideration of horizontally layered subsoil layers toward the estimation of consolidation settlement and excess pore-water pressure (EPWP). Furthermore, it is observed that an approach of implementing PVDs to varying depths, depending on site-specific scenario, should be adopted to attain a more sustainable solution in improving the consolidation behavior of the soft compressible soils.

2 Rehabilitation of Railway Embankment: Udaipur, India

2.1 History of the Problem and Proposed Ground Improvement

A new broad gauge (BG) line was proposed to be built as a part of the project for connecting Agartala and Sabroom (110 km) in the state of Tripura, India by Northeast Frontier Railways (NFR), Government of India (Fig. 1) [8]. At the chainage of 41.85–42.8 km from Agartala, Udaipur railway station yard and station building were to be constructed. The site for construction was proposed over a low-lying marshy land, referred to as Sukhsagar lake. The earthwork for constructing the railway embankment commenced without any pretreatment to the existing ground. While the embankment was being raised up to 5–6 m, a massive settlement exceeding 2 m was triggered at the site [9]. As a result, the embankment exhibited an unsustainable subsidence. Finally, after thorough site reconnaissance and based on geotechnical investigation, ground improvement through PVD was suggested to improve the properties of the underlying subsoil for reconstruction of the embankment [9].

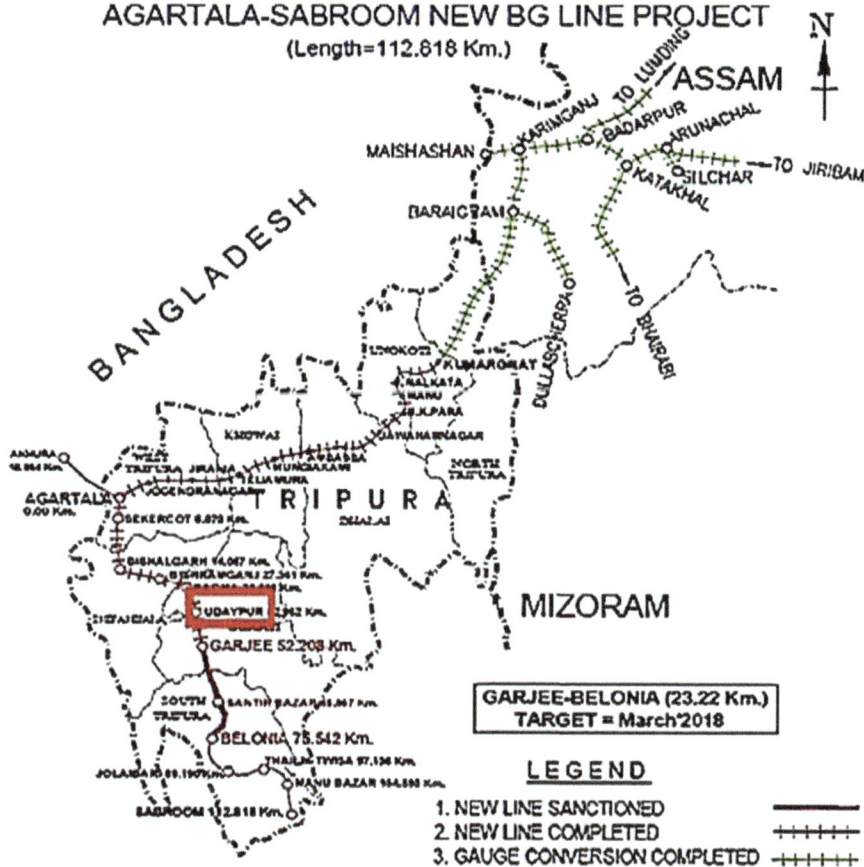

Fig. 1 Map of Agartala–Sabroom new BG line project [8]

2.2 Subsoil Conditions

The typical cross-section of the borehole soil profiles reflects the presence of embankment fill material over the soft cohesive deposits. Due to settlements of the underlying soft compressible layered medium below the embankment prior to the ground improvement, the embankment material has subsided and deposited over the existing subsoil strata [9]. As can be observed from the borehole data below the embankment centerline, the underlying subsoil consists of moderately to highly compressible clayey silt layers with traces of decomposed wood and organic matter overlain by sandy silt fill layer. The average SPT-N values range from 2 to 6 until 19 m below the existing ground level (EGL) [10]. Multiple boreholes along the embankment cross-section show that the soil profile is undulating with varying thickness of each soil layer. Figure 2 shows the soil profile at chainage location of 42.2 km. Table 1 shows the subsoil properties [10].

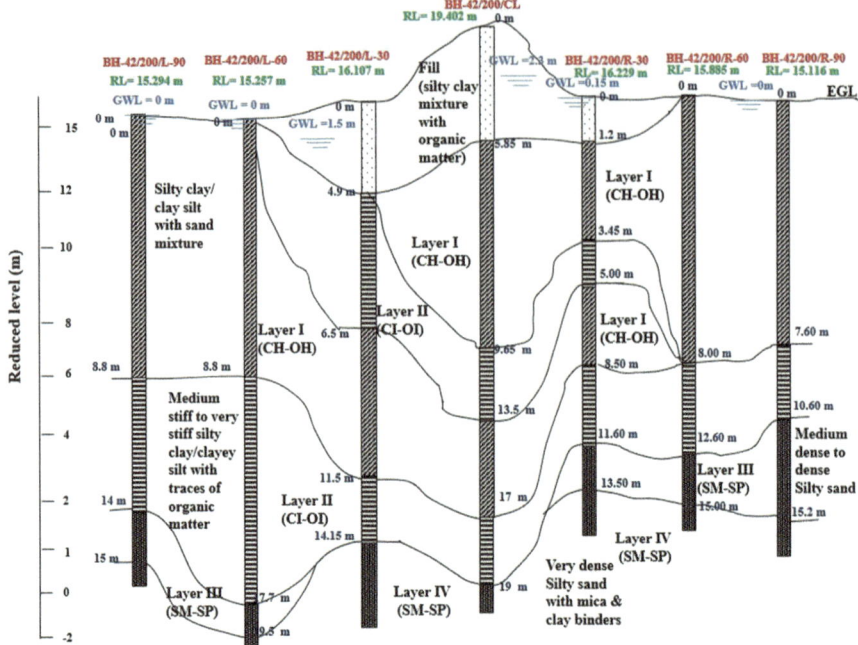

Fig. 2 Borehole stratigraphy showing actual soil profile at 42.2 km chainage of the Udaipur embankment site [10]

Table 1 Subsoil properties at Udaipur site for the considered borehole location [10]

	Fill layer	Layer I	Layer II	Layer III	Layer IV
Bulk density (kN/m³)	17.3	16.7	18	18.78.7	–
Dry density (kN/m³)	10.6	10.1	12.912.9	15.615.6	–
Void ratio	0.606	1.204	1.215	–	–
Specific gravity	2.62	2.59	–	2.65	2.64
Liquid limit (%)	41	51	45	–	–
Plastic limit (%)	23	28	25	–	–

2.3 PVD Installation Beneath the Embankment Constructed by Staged Loading

PVDs were proposed to be installed up to a depth of 13 m from the EGL, owing to the presence of underlying compressible soft cohesive layers [11]. The PVDs are typically rectangular in shape with cross-section of 100 × 4 mm in size installed to the desired depth at a spacing of 1 m c/c in a triangular pattern. A 300 mm compacted sand mat, acting as a horizontal drainage, is constructed at the ground surface over the PVDs for draining out the pore water collected and transported by the PVDs from underlying layers [11].

Table 2 Stage wise embankment construction plan and assessed degree of consolidation [11]

Stage	Height of embankment (m)	Time for staged construction (days)	Cumulative time (days)	Degree of consolidation (%)
1	0.5	20	20	45
2	1	20	40	70
3	1.5	20	60	84
4	2	20	80	91
5	2.5	20	100	95

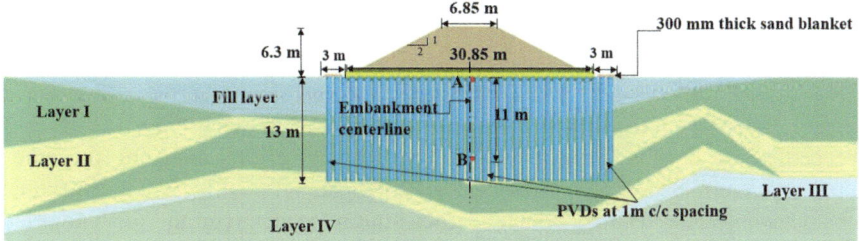

Fig. 3 2D FE model of Udaipur railway embankment supported over PVD-improved subsoil

The embankment was proposed to be built in stages, considering 20 days of time interval during execution of each stage of 0.5 m height. The degree of consolidation during the staged construction is computed through conventional analytical assessments, considering the coefficient of vertical consolidation for the compressible soft soil layer as $C_v = 3.8 \times 10^{-3}$ m^2/day [11]. Table 2 shows the stage-wise embankment construction plan with degree of consolidation achieved after completion of each stage.

3 Finite Element Analysis of Udaipur Railway Embankment Supported upon PVD-Improved Layered Subsoil with Realistic Undulated Stratification

Two-dimensional plane strain finite element (FE) analysis of the Udaipur railway embankment (located at 42.2 km from Agartala) is carried out using PLAXIS 2D v.2018.Typical FE-based model of the PVD-improved subsoil layers and embankment load is presented in Fig. 3. In the present analysis, a medium coarse mesh is selected with an average element size of 0.06 m. For modeling the soil, 15-noded triangular elements are employed, while PVDs are modeled with 1D drain elements inbuilt in PLAXIS 2D. The drain elements act as ideal drains, allowing infinite discharge of the pore water generated in its influence zone of the surrounding soil. The PVDs are assumed to be arranged in triangular pattern with 1 m c/c spacing, implying an equivalent diameter of the influencing surrounding soil around each PVD to be $D_e = 1.05S$ ($S =$ c/c spacing between the PVDs) [4]. The soil layers are allowed to deform freely at the top of the ground surface when subjected to compressive load from the embankment, while the vertical boundaries are allowed to deform vertically and the bottom boundary is fully restrained from any deformation. During radial consolidation of the subsoil layers takes, the pore-water dissipation from the subsoil takes place through the PVDs and is directed to the top horizontal drainage layer under the embankment (i.e., through the sand blanket).

As mentioned earlier, a suitable constitutive model for soil layers is adopted in the FE analysis; the corresponding model parameters are obtained from bore log reports (Fig. 2) and the corresponding laboratory tests conducted on the collected undisturbed soil samples [10]. Due to unavailability of relevant laboratory/field permeability test data, the vertical permeability coefficients have been adopted based on the type of soil and particle size distribution of the soil layers [12]. Table 3 shows the material models and parameters for the embankment material and each of the subsoil layers considered for the present study. The material properties are obtained from an exhaustive laboratory tests including the index property determination and assessment of shear strength, stiffness, and hydraulic characteristics [10]. The horizontal coefficient of permeability of the soil, subjected to radial consolidation through PVDs, is computed based on Hird's equivalent plane strain permeability compatibility approach [13], as given by the following expression (i.e., Eq. 1):

$$\frac{k_{hp}}{k_h} = \frac{2}{3\left[\ln(n) - \frac{3}{4}\right]} \tag{1}$$

where k_{hp} is the equivalent plane strain permeability neglecting the smear effect, k_h is the axisymmetric horizontal permeability coefficient of the soil surrounding each PVD at the given cross-section neglecting smear effect, and n is the drain spacing ratio (i.e., the ratio of equivalent circular diameter of the influencing surrounding soil to the equivalent circular diameter of band-shaped PVD, i.e., $n = D_e/d_w$).

The installation of PVDs results in smearing of the soft cohesive soil around the PVDs. In the smeared zone, the permeability reduces to magnitudes lesser than the surrounding undisturbed soil. For computing the permeability coefficient in the smear zone (neglecting the well resistance), an equivalent plane strain smear permeability coefficient (k_{sp}) is evaluated from Eq. 2 proposed by Indraratna and Redana [14] The smear zone permeability is considered to be 1/3rd of the horizontal coefficient of permeability of surrounding undisturbed soil (i.e., $k_s = k_h/3$), and the diameter of the smeared zone (d_s) is assumed twice that of the equivalent drain diameter (d_w) (i.e., $s = d_s/d_w = 2$) [4].

$$k_{sp} = \frac{\beta \times k_{hp}}{\frac{k_{hp}}{k_h}\left[\ln(n) + \frac{k_h}{k_s}\ln(s) - 0.75\right] - \alpha} \tag{2}$$

where $\beta = \frac{2(s-1)}{n^2(n-1)}\left[n(n-s-1) + \frac{1}{3}(s^2+s+1)\right]$ and $\alpha = \frac{2}{3}\frac{(n-s)^3}{n^2(n-1)}$.

Table 3 Material properties of embankment and soil layers

Parameters	Embankment	Fill layer	Layer II	Layer II	Layer III	Layer IV
Material model	MC[a]	MC	SS[b]	SS	MC	MC
Drainage type	Drained	Undrained	Undrained	Undrained	Undrained	Undrained
E' (kPa)	30×10^3	3450	–	–	0.265×10^5	43×10^3
ν'	0.3	0.3	–	–	0.3	0.3
c' (kPa)	30	52.95	21.57	27.45	25	10
φ' (°)	35	10	10	3	30	35
λ^*	–	–	0.072	0.06	–	–
κ^*	–	–	0.013	0.010	–	–
k_h (m/day)	0.01673	1.296×10^{-3}	1.296×10^{-4}	1.296×10^{-4}	0.1296	0.864
k_v (m/day)	0.01673	8.64×10^{-4}	8.64×10^{-5}	8.64×10^{-5}	0.0864	0.864
c_k	10^{15}	10^{15}	0.6	0.3	10^{15}	10^{15}

[a]MC Mohr–Coulomb model
[b]SS Soft soil model

4   Results and Discussion

4.1   Effect of Installing Uniform Length PVD in Layered Subsoil Comprising Undulated Stratification

In line with the conventional practice of FE analysis of embankment resting on PVD-treated soft soil, the underlying subsurface is assumed comprising horizontally stratified layers. Such stratification is either generally based on the information from a single borehole data or based on an equivalent horizontally layered soil profile from multiple borehole data. Consequently, in the process, the influence of the realistic soil profile, which is mostly undulated, is lost, and the accuracy of consolidation prediction is often compromised by ignoring the spatial variability of the soil layers and their properties. In this regard, a benchmark analysis is carried out in PLAXIS 2D considering horizontal subsurface stratification conforming only to the borehole data pertaining to the centerline of the embankment (Borehole 42/200/CL as in Fig. 2). As in the case in Fig. 3, the FE analysis for horizontally stratified substrata is carried out by considering the PVDs installed up to 13 m depth below EGL (Fig. 4). For symmetricity, one-half of the PVD-treated subsoil from the centerline of the embankment is considered in the present analysis.

Through the present FE-based study, a comparative assessment is carried out between the vertical settlement and EPWP evolution during the consolidation of the underlying stratified soil deposits with and without PVDs. As indicated in Figs. 3 and 4, the settlement and EPWP magnitudes for the underlying subsurface soil layers are considered at point A (just below the embankment, at its centerline) and point B

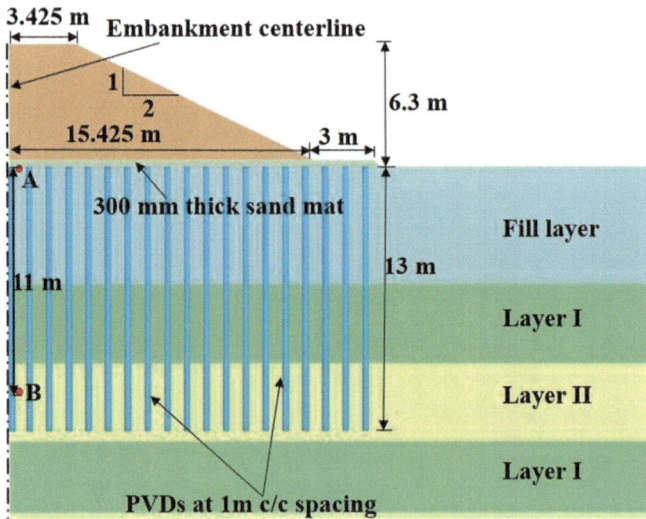

Fig. 4   2D FE model of the Udaipur railway embankment supported over PVD-treated horizontally stratified soil layers

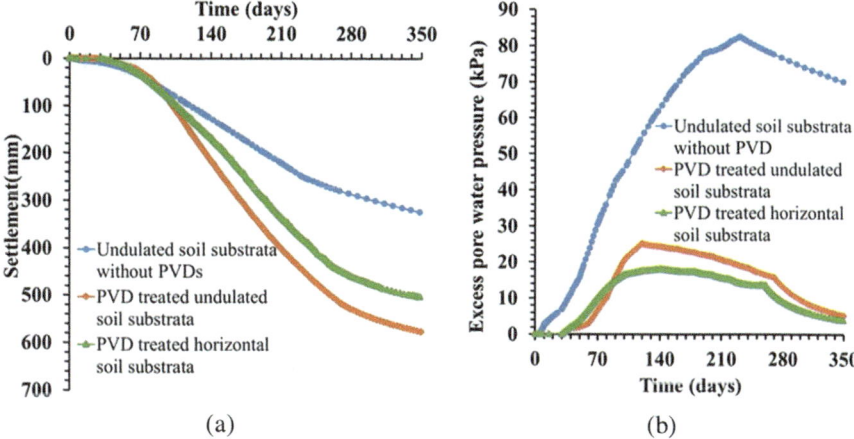

Fig. 5 Effect of PVD installation on consolidation characteristics of soft soil layers with and without considering the presence of undulations in the subsoil profile

(at 11 m depth from the EGL along the embankment centerline), respectively. The rate of settlement is noticeably rapid in case of embankment with PVDs due to creation of an artificial drainage path, leading to the faster dissipation of generated EPWP in the relatively low permeable soil layer (Fig. 5). It can be further noticed that for the case with actual undulating soil profile, a settlement of 577 mm is attained at the end of 350 days (i.e., 250 days after the completion of embankment construction), as compared to 300 mm for the case of unimproved soil (without PVDs). Similarly, for the case with actual undulating soil profile, the developed EPWP for the low permeable soil layers without the PVDs is much higher (peak EPWP of 83 kPa) as compared to that in case with PVD treatment (peak EPWP of 25 kPa). Moreover, the time for attaining peak EPWP is much quicker (i.e., 28% faster) in case of subsoil layers with PVDs compared to the unimproved subsoil case. On the other hand, the vertical settlement and EPWP attained at the end of 350 days is underestimated approximately by 25% when simplified horizontal subsurface layering is considered in contrary to the undulated subsurface profile.

4.2 Sustainable Practice of Ground Improvement Using Nonuniform Length of PVDs for Consolidating Soft Soil Under Udaipur Railway Embankment

The most conventional practice of PVD-treated ground improvement is carried out by installing PVDs of uniform length within the base width of the embankment at a given cross-section. However, the efficacy of PVD-based ground improvement also lies with optimized usage of PVD installation below the embankment (i.e., extending the intermediate PVDs to underlying deeper low permeable strata), governed by the

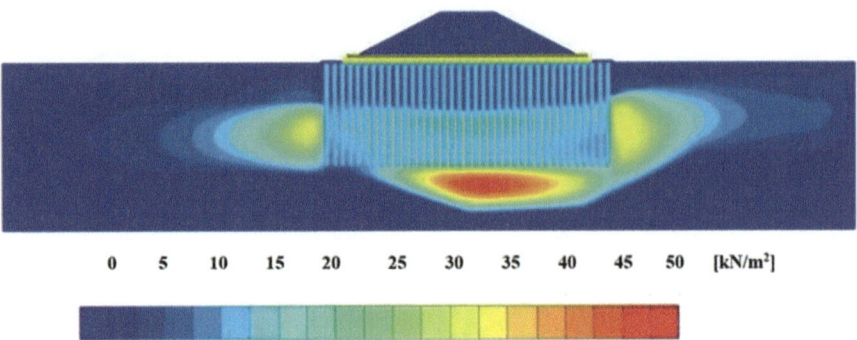

0   5   10   15   20   25   30   35   40   45   50   [kN/m²]

Fig. 6 EPWP accumulation below 13 m depth of PVDs with actual undulated stratified subsoil at the completion of embankment construction in Udaipur railway embankment site

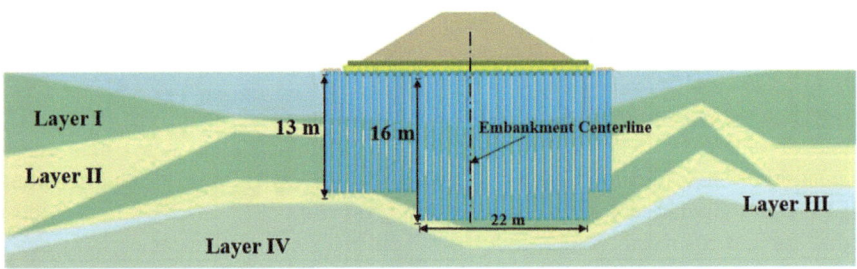

Fig. 7 FE-based model of Udaipur railway embankment considering nonuniform length of PVDs (i.e., a combination of 13 m and 16 m) below the EGL

spatial distribution of the soil layers and nature of subsurface stratification. Figure 6 shows the EPWP distribution at the end of embankment construction. It is evident that substantial magnitude of EPWP (approximately 50 kPa) is accumulated beyond the depth of installation of PVDs (i.e., below 13 m from EGL) This observation is attributed to the basin effect developed due to the undulating Stratum-I below the PVD-treated substrata (as shown in Fig. 3) that acts as an underground reservoir to store the pore water. In this case, it would not dissipate easily as it remains beyond the reach of the PVDs. As a matter of fact, it is noticed from FE analysis that even after allowing a consolidation time up to 350 days, the decrement in EPWP is merely 32%.

Therefore, in order to arrive at a more sustainable solution of ground improvement through PVD treatment, it would be prudent to extend the length of the PVDs deeper into the undulating Layer-I below the embankment. Accordingly, an FE analysis is conducted similar to earlier scenario but extending the PVDs up to 16 m depth within the zone of accumulated EPWP (for a width of 22 m, around the centerline of embankment), as shown in Fig. 7. The length of PVDs in the remaining locations is maintained to 13 m as earlier. Figure 8 shows the comparison of time rate of settlement and evolution of EPWP for the underlying consolidating

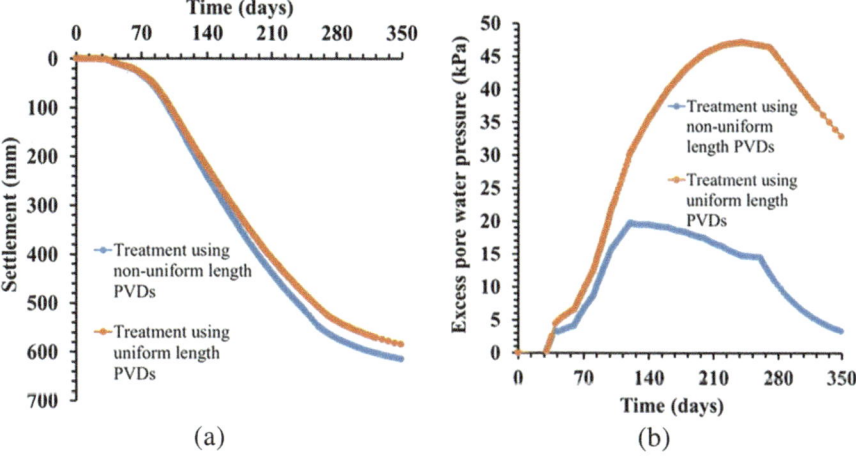

Fig. 8 (**a**) Time rate of settlement and (**b**) EPWP evolution for PVD implemented up to 16 m depth below the embankment near the embankment centerline with that at 13 m depth of PVD

soil deposits (pertaining to the locations A and B, respectively, as indicated in Fig. 3) when nonuniform length of PVDs is used (a combination of 13 m and 16 m lengths) with respect to the uniform length PVDs. It can be observed that although the difference in settlement magnitudes at the end of 350 days is less than 0.05% (Fig. 8a), a substantial difference in peak EPWP generated in both the cases is noted (Fig. 8b). The maximum EPWP generated for the case of treatment with nonuniform PVD lengths is approximately 60% lesser than that observed from uniform length PVDs. It can be observed that much lower EPWP is attained in much quicker time, which is substantiated by a higher time rate of consolidation for the case with nonuniform length PVDs, as reflected in Fig. 9. In comparison to the uniform length PVDs, the rate of consolidation at each stage of embankment construction is observed to be faster, in the range of 34–49%. Figure 10 shows the EPWP distribution for the case with nonuniform length of PVDs considering the undulated soil profile at the end of embankment construction. As can be noticed that, the EPWP dissipation takes place at the designated depth of PVD insertion, with remaining EPWP magnitude within the soil surrounding the PVDs to be lesser than 20 kPa.

5   Conclusion

The present study aims toward a more sustainable solution of ground improvement through PVDs by considering the consolidation behavior of fully saturated and low permeable deposits by incorporating the actual undulated substrata profile and nonuniform length of installed PVDs. Based on the present study, the following inferences can be drawn out of present study:

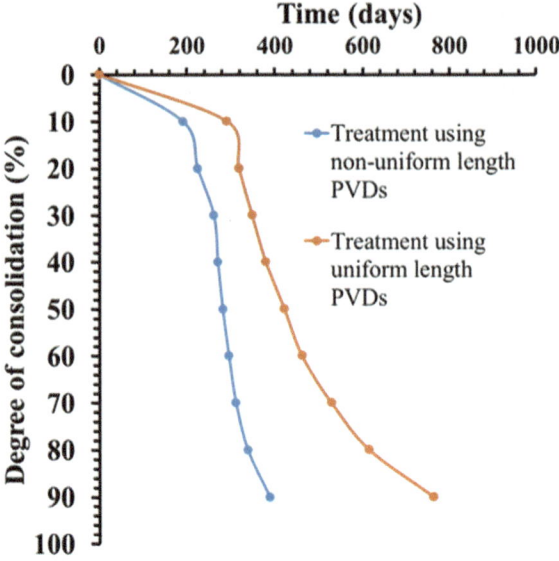

Fig. 9 Time rate of consolidation for PVDs of uniform length (extending to 13 m depth) and nonuniform length (a combination of 13 m and 16 m depth) below the embankment

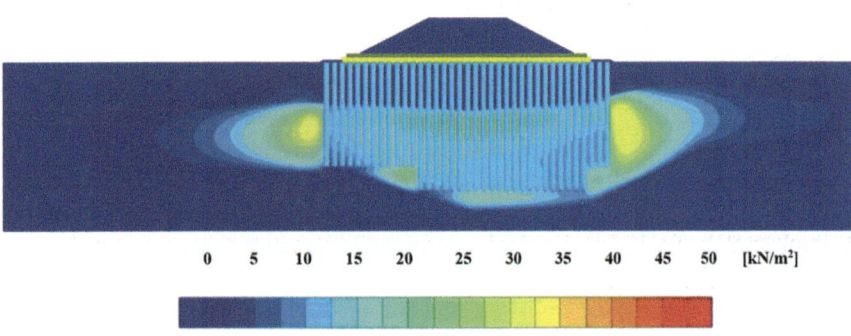

Fig. 10 EPWP distribution at the completion of embankment construction for undulated soil profile treated with nonuniform length PVDs (up to 13 m and 16 m depth below EGL)

1. The rate of settlement of the underlying fill layer below the embankment center is much quicker for PVD-treated soft subsoil layers as compared without PVD treatment. In this study, with PVD treatment, the settlement at the end of 350 days is observed to be 48% higher in comparison to untreated substratum.

2. Application of PVD leads to significant and quick dissipation of pore-water pressure from the deep-seated low-permeable substrata. For the present study, reduction in peak magnitude of EPWP is observed to be more than 50% for the PVD-treated subsoil. Besides, the time of evolution of the peak EPWP is 28% quicker than subsoil without PVDs.

3. It is very important to develop detailed knowledge about the spatially varying soft substrata beneath the embankment as it significantly affects the time rate of consolidation and EPWP evolution. In the present case, the settlement and EPWP is observed to be underestimated by 25% for conventionally adopted horizontally layered subsoil profile in comparison to the undulated substrata profile.

4. Uniform length PVDs might not always be sufficient to reduce the EPWP from deeper strata due to the basin effect of specifically undulated layers as was evident from the present study in which significant EPWP is noticed below PVDs of 13 m depth.

5. In comparison to uniform length, PVDs with nonuniform length (with a hybridization of 13 and 16 m long PVDs) below the embankment leads to a substantial improvement in the time rate of consolidation, nearly 34–49% for the present study.

6. Consideration of the actual and realistic undulating subsurface profile comprising soft soils is immensely important in developing sustainable solutions to long-term ground improvement measures. Conventional practice of considering uniform length PVDs for accelerated consolidation of thick soft undulating deposits might not be effective in specific scenarios. In such cases, extending the PVDs to larger depths would be required, thereby leading to providing nonuniform length of PVDs along the cross-section of the embankment. Such provision can yield more efficient and sustainable practice in improving the consolidation behavior of the soft cohesive soils.

References

1. Han, J.: Principles and Practice of Ground Improvement. Wiley, Hoboken (2015)
2. Bergado, D.T., Balasubramanium, A.S., Fanin, R.J., Holtz, R.D.: Prefabricated vertical drains (PVDs) in soft Bangkok clay: a case study of the new Bangkok international airport project. Can. Geotech. J. **39**(2), 304–315 (2002)
3. Wang, J., Fang, Z., Cai, Y., Chai, J.C., Wang, P., Geng, X.: Preloading using fill surcharge and prefabricated vertical drains for an airport. Geotext. Geomembr. **46**(5), 575–585 (2018)
4. Hansbo, S.: Consolidation of fine-grained soils by prefabricated vertical drains. In: Proceedings of 10th International Conference on Soil Mechanics, Stockholm, vol. 3, pp. 677–682 (1981)
5. Mridakh, A.H., Ejjaaouani, H., Nguyen, B.P., Lahlou, F., Labied, H.: Soft soil behavior under high-speed railway embankment loading using numerical modelling. Geotech. Geol. Eng. **40**(5), 2751–2767 (2022)
6. Indraratna, B., Rujikiatkamjorn, C., Wijeyakulasuriya, V.: Soft clay stabilization using prefabricated vertical drains and the role of viscous creep at the site of sunshine motorway, Queensland. Paper No. 0157, pp. 1–6 (2007)
7. Oh, Y.N.E.: Geotechnical and ground improvement aspects of motorway embankments in soft clay, Southeast Queensland. PhD Thesis, Griffiths University (2006)
8. Northeast Frontier Railway, Official account. nfr.indianrailways.gov.in
9. Murali Krishna, A., Dey, A., Sreedeep, S.: A Report on Geotechnical Assessment, Design and Remedial Measures Related to the Construction of Railway Embankment in Udaipur Station Yard. India Institute of Technology, Guwahati (2015) (Unpublished)

10. CE Testing Company Pvt. Limited: A report on Geotechnical investigation work at Udaipur station yard area at km 42.00 in connection with Agartala-Sabroom new BG line project, Tripura (2014) (Unpublished)
11. TECHFAB India: A case history on Rehabilitation of railway track at Udaipur station yard in Agartala-Udaipur-Sabroom new B.G. line at Udaipur, Tripura (2020) (Unpublished)
12. Das, B.M.: Advanced Soil Mechanics, 3rd edn. Taylor and Francis, New York (2008)
13. Hird, C.C., Pyrah, I.C., Russell, D., Cinicioglu, F.: Modelling the effect of vertical drain in two-dimensional finite element analysis of embankments on soft ground. Can. Geotech. J. **32**, 795–807 (1995)
14. Indraratna, B., Redana, I.W.: Plane strain modelling of smear effects associated with vertical drains. J. Geotech. Geoenviron. **123**(5), 474–478 (1997)

Citizen Engagement and Co-creation in a Net-Zero Built Environment Transitions: Challenges and Best Practices

Christian Wolfgang Kunze, Alemu Moges Belay, Ahmed Samir Hedar, and Aaditya Dandwate

Contents

1   Introduction

Co-creation and stakeholder engagement are crucial for shaping corporate compet-itive strategies and for sustainable transitions. Unlike the market-based view (MBV), the resource-based view (RBV) research reinforces the relevance of co-creation for corporate success [1–3]. Originally, industrial co-creation processes evolved from co-production, where consumer participation was integrated in the supply chain with the aim of cost minimization (e.g., IKEA) [4] in the 1990s. This evolution marked a significant milestone as IKEA launched "Co-Create IKEA" in 2018, establishing a digital platform for customer-driven product development. This initiative was further expanded with the opening of the first Co-create Hub at the IKEA Home of Tomorrow in Szczecin, Poland, in 2020, giving everyone a space to co-create better living futures with IKEA [5]. The effectiveness of co-creation in creating customer-friendly solutions is well supported by empirical evidence, especially within corpo-rate environments. Therefore, empirical evidence of successful co-creation activities

C. W. Kunze · A. M. Belay (✉) · A. S. Hedar · A. Dandwate
Smart Innovation Norway, Halden, Norway
e-mail: alemu.belay@smartinnovationnorway.com

© The Author(s) 2025
M. Kioumarsi, B. Shafei (eds.), *The 1st International Conference on Net-Zero Built Environment*, Lecture Notes in Civil Engineering 237,
https://doi.org/10.1007/978-3-031-69626-8_143

within various EU-funded projects focuses on applied co-creation, i.e., the different steps in a co-creation process [4].

Before delving deeper into co-creation, it is important to highlight that the success of these initiatives lies not only in the mechanisms of collaboration but also in who is involved. The focus shifts from the broad perspective of stakeholder engagement to an important subset, the citizens. In this context, citizen engagement emerged as a separate study in the late 1990s and early 2000s and is in line with the RBV interpretation of the source of competitive advantage. It is considered as a long-term sustainability strategy. Engagement strategies have progressively shifted towards more interactive methodologies, ranging from two-way dialogues to active stakeholder participation, encompassing strategic, normative, moral, or pragmatic approaches [6, 7].

Among these, the moral approach emphasizes the empowerment of less visible stakeholders, advocating for the democratization of science and projects to ensure inclusive participation. Methods for engaging citizens in this manner include representation by public officials, direct involvement of individuals and community members, and advocacy by NGOs, all aimed at fostering a deeper understanding of citizen motivations and values for sustained participation [8]. The significance of citizen engagement becomes more pronounced in the energy transition context, where legitimacy and social acceptance are essential. This narrative is supported by the energy democracy movement and the concept of "energy citizenship," advocating for a more participatory role for citizens beyond mere consumption. Emphasizing the social dimensions of energy systems encourages citizens to become producers, investors, and activists, thereby enriching the dialogue on energy transition and sustainability [9].

1.1 Definitions of Co-creation

Co-creation can refer to diverse practices with the shared goal of engaging the stakeholders from the design or creation stage. Therefore, it involves stakeholder engagement for the creation of a product or value. Co-creation is often defined through two primary lenses: The business literature refers to "value co-creation" as creating a shared value between the business and the customers leading to mutual benefits and the "co-creation in design" as a research and innovation tool involving multiple stakeholders [10]. Similar definitions have been proposed in various contexts and applications. In urban planning and public services, co-creation enhances democratic processes and addresses societal complexities. Successful initiatives stress the need for a clear mission, emphasizing genuine collaboration over treating co-creation as a rhetorical tool to achieve meaningful outcomes [10].

At its core, co-creation involves high levels of citizen participation, in which citizens play a major role in identifying problems, generating solutions, and/or determining policies [11, 12], aligning with the concepts of stakeholders' empowerment. Co-creation may take various forms of participation: "Co-initiation" when citizens take the initiative, "Co-design" when they participate in the design phase, and "Co-implementation" when they are involved in the implementation phase [13]. J. Eckhardt et al. [14] model this stakeholder involvement in co-creation as an ecosystem of transdisciplinary cooperation that can have organizational and cultural effects across various sectors.

The authors define co-creation, in the context of this chapter, as an approach that tailors the design of products or service concepts to the needs and expectations of customers. By actively involving future users and key stakeholders in interactive workshops, focus group meetings, or other events, co-creation aims to unleash creativity, share ideas, and enhance product development and concepts in an organized process. In summary, regardless of the co-creation application across various disciplines and domains, active citizen engagement emerges as a common thread. Whether in business, urban planning, or public services, the process is dependent upon involving diverse participants from various sectors.

1.2 Research Questions (RQ) and Objectives

To achieve the overarching objectives of speeding up the green transition in a net-zero built environment, the two key research questions will be addressed.

- *RQ1*. What are the experiences and practices in the co-creation and citizen engagement activities, challenges, enablers, and tools applied in selected EU-funded projects and how do they support green transitions?
- *RQ2*. How can identified effective co-creation and engagement processes be structured and organized to expand/replicate from electricity to a sustainable and net-zero built environment?

2 Methodology (Case Studies and Review)

In this research, a qualitative approach is used by combining case studies and a scoping review. In this regard, the chapter employed Intrinsic and Extrinsic Case Studies complemented by a topical literature review focusing on the co-creation, stakeholders, and citizen engagement in the transition towards a Net-Zero Built Environment. In this connection, a tailored EU-funded project (SENDER) is considered for the intrinsic case analysis to understand deeply the evolvement of the co-creation and citizen engagement concepts. The rationale for selecting SENDER is that it focuses on sustainable consumer engagement and demand response.

For extrinsic cases, the chapter considers relevant projects in the green neighbourhood sharing common net-zero objectives. Moreover, co-creation-focused literature review articles on the electricity and net-zero built environment area are considered and discussed to incorporate a broader perspective of the topic and understand various contexts before generalization and reaching a conclusion. In this regard, the chapter used Google Scholar with a publish or perish database and search review articles published between 2003 and 2023 using keywords: Co-creation, Citizen Engagement, Demand Response, Net-Zero Transition, and Net-Zero Built Environment. Based on the authors' content analysis from 121 articles found in the database, selected papers are used for the result and discussion and to drive a conclusion.

2.1 Overview from the EU-Funded SENDER Project

The project is considered for the intrinsic analysis of co-creation and citizen engagement strategies. SENDER (Sustainable Consumer Engagement and Demand Response) [15] aims to develop and test novel solutions and tools that engage consumers and prosumers in demand–response mechanisms and energy services. SENDER puts household consumers at the heart of the energy market, using behaviour change measures, addressing specific loads, maximizing the use of next-generation energy services, creating a digital "twin" of the consumer, and developing new business models.

SENDER Social Innovation Meta-Level Approach
The SENDER project, along with several previous projects coordinated by Smart Innovation Norway, follows a meta-level multiple helix approach. This represents a way of broadening and opening former relatively closed policy-making processes and innovation systems, thereby making them more democratic. Early innovation systems, such as the "triple helix" based on collaboration between universities, government, and industry [16], have evolved. Later development to this model added a "fourth helix"—civil society—to form a quadruple helix model [17]. Further expansions of the quadruple helix model point out that definitions of democracy must also include sustainable development, not only energy but also other broader perspectives [18]. This has led to the creation of a quintuple helix, including also environmental settings of a specific region. In the context of the SENDER project, the quintuple helix acts as the meta-model for the selection of stakeholders (e.g., Academia, Civil Society, Government, etc.) that provide relevant contributions to the project as shown in Fig. 1.

SENDER's engagement strategy, focusing on consumers, technology providers, and electricity market actors, employs a three-phased approach (see Fig. 2). The initial phase centres on recruitment, disseminating materials, hosting events and workshops, and conducting individual energy counselling and surveys. The second phase focuses on social pressure and gamification to underscore participation benefits and engage citizens in rollout and installation. The third phase targets persistence, establishing an online forum, enrolling beta users, arranging exhibitions and seminars, and showcasing the real impact of the demand response (DR) actions of the citizens [19].

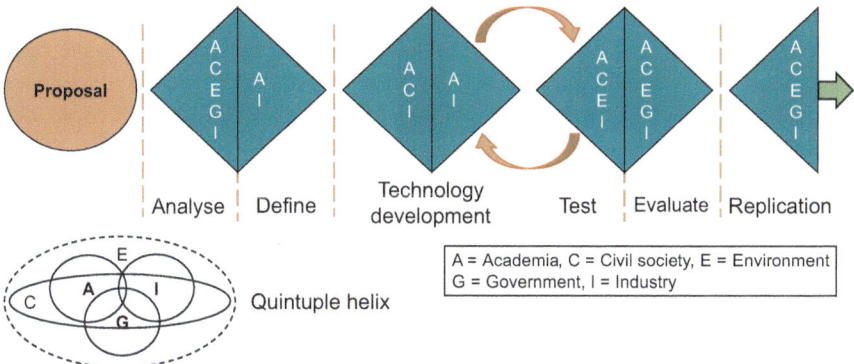

Fig. 1 The quintuple helix innovation model used in the SENDER project

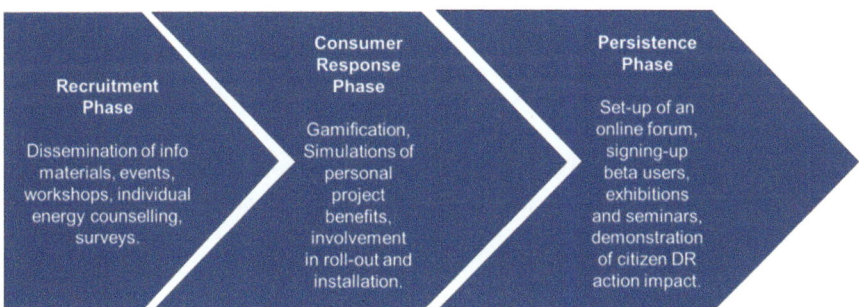

Fig. 2 SENDER consumer engagement phases

2.2 Overview from the EU-Funded HESTIA Project

HESTIA [20] has created an inclusive and participatory strategy for engaging households in the design and development of the HESTIA platform and the proposed DR solutions. Co-creation, as the ongoing involvement of all relevant stakeholders, not just users/householders in the development of the platform, has proven to be a fundamental and challenging task, e.g., to create DR projects with a bottom-up approach. Challenges, like the pandemic restrictions that took place during the first months of the project, brought about the need to rethink the engagement approach (e.g., face-to-face interactions were partly replaced with virtual home tours and online interviews to keep a continuous line of communication with users).

The key lesson learnt is related to the need for keeping project participants in continuous communication with the project partners and engaging them—making them aware of the process and the requirements of their involvement at each stage. Community and collective engagement have proven to be particularly important to participants, rather than just individual engagement of single households. Some of the pilot sites were already established energy communities, such as a Dutch pilot, whereas others, such as an Italian pilot, were in the process of being established.

The HESTIA platform interface, which is co-designed with the users, therefore expresses the input of the community, as households can monitor not only their own energy but also watch the contribution or consumption of their neighbours. Furthermore, the co-creation activities have helped the project partners uncover issues such as the gendering of energy technologies and the unequal distribution of household labour at home between operators of energy technologies and those performing household tasks. Such challenges relating to socio-cultural background, gender, and digital literacy have informed recommendations for the design of DR solutions and community engagement for different household typologies. Examples of these recommendations include the flexibility potential for different households, through the empirical insights of the HESTIA pilots, according to the time householders spend at home, and the recommendations for the coordination of everyday housekeeping with digital housekeeping, which involves the interactions with the energy systems.

2.3   Overview from the EU-Funded ReDREAM Project

ReDREAM's [21] engagement strategy is based on three pillars: first, to provide benefits over time to maintain user engagement throughout the project; second, to offer added value so that benefits obtained are greater than the costs involved; and third, to build a trustful ecosystem. In addition to these pillars, a complementary set of principles guides the design of engagement strategies and the ecosystem: personalization, feasibility, transparency, simplicity, discoverability, and automatization.

Regarding recruitment strategy, ReDREAM defined a threefold approach based on target, messages, and channels. Two strategic recruitment approaches were applied. Initially, a "motivation-to-eligibility" strategy targets users with allegedly mid-to-high levels of motivation and then screens eligible users with electrical assets that can be interrupted for short periods. To recruit these motivated users, messages emphasized the main benefits unearthed in the initial research: potential monetary savings, reduced environmental emissions, community self-sufficiency and resilience, or the anticipated pride in being an early adopter of these services. This approach eventually evolved into an "eligibility-to-motivation strategy," targeting users with the necessary equipment and using messages to demonstrate the value they will glean from participating in the project. To identify these eligible households, ReDREAM proposed five strategies that could be used by demo managers to identify and locate households with the appropriate equipment.

2.4   Overview from the EU-Funded ACCEPT Project

The ACCEPT project [22] is focusing its engagement strategy on awareness-raising campaigns, recruitment—primarily via pilot partners acting as "gatekeepers" to their communities—and planning for engagement activities including workshops, one-to-one meetings, and presentations. Furthermore, the elaboration of communication material plays a crucial role in ACCEPT's engagement strategy, as it enables the process of building trust between the pilot leaders and the local communities. The recruitment strategy is still being implemented.

2.5 Overview from the EU-Funded iFLEX Project

iFLEX's [23] engagement strategy focuses on the recruitment of end users (residential energy consumers and prosumers) from the three pilot sites that are being deployed in Finland, Slovenia, and Greece, by using personal and direct communications and concise surveys during registration to gather information on devices, energy-awareness, and incentives; offering free equipment installation, prizes, and chances to join iFLEX's environmental and innovation; and creating communication campaigns on the pilot websites and social media.

2.6 Overview from the EU-Funded PROBONO Project

The EU-Funded Horizon 2020 PROBONO project [24] caters to the integration of green buildings in a delimited area or district level to develop energy-positive and zero-carbon Green Building Neighbourhoods (GBNs). The concept of GBNs focuses on several domains related to stakeholder engagement, investments in green energy, green mobility, sustainable built environment, etc. The project also demonstrates co-creation benefits through its living labs serving as pilot sites and first users of the PROBONO's project results. The project focuses on introducing innovations to address technological challenges, enhance energy savings, reduce emissions, and develop sustainable infrastructure meanwhile including citizens as active participants and crucial stakeholders in the co-creation of the sustainable GBNs. The PROBONO project aims to deliver five GBN Transition Acceleration Enablers of which Enabler 2: Social Engagement and Innovation Clusters [24] facilitates the engagement of citizens and stakeholders in the GBN initiatives utilizing different tools like workshops, interviews, and site visits.

2.7 Lessons Learned from EU-Funded Projects

In conclusion and answering RQ1, the engagement strategies of the European projects reflect a commitment to inclusivity and participatory methods in end user recruitment. External challenges, such as the pandemic, prompted innovative adaptations, showcasing the projects' resilience. The emphasis on community engagement, as seen in HESTIA, underscores the importance of a bottom-up approach.

SENDER's phased strategy, ReDREAM's pillars, and ACCEPT's focus on aware-ness and trust-building highlight the multifaceted nature of engagement. iFLEX's personalized communication and incentives demonstrate a tailored approach to end user recruitment. Collectively, these projects offer valuable insights into effective approaches for engaging end users, emphasizing adaptability, personalization, and community collaboration as key components of successful recruitment strategies that support a bottom-up approach towards a green transition.

Drawing from the lessons learned and to provide valuable insights to projects currently implementing or considering engagement strategies, the following recom-mendations [19] are derived:

1. *Diverse engagement strategies*: Employ a range of strategies for tailored and impactful user engagement.
2. *Phased implementation*: Organize engagement strategies into distinct phases for enhanced efficacy and adaptability, aligning with the timeline of engagement.
3. *Keyword utilization*: Incorporate targeted keywords
   and focus on specific daily life aspects to facilitate personalization and adaptability
4. *Adaptability in crisis*: Be prepared to adapt strategies during challenges, maintaining flexibility for continued effectiveness, particularly in the face of unforeseen circumstances like the COVID-19 pandemic.
5. *Motivation and eligibility*: Craft engaging recruitment materials recognizing the vital role of motivation and eligibility criteria.
6. *Continuous assessment*: Navigate diverse contexts through continuous assess-ment, ensuring alignment with community needs and infrastructural support.
7. *Sustained collaboration*: Foster collaboration and knowledge sharing among project partners, leveraging shared on-field experiences for enhanced effectiveness.
8. *Overcoming barriers*: Proactively address encountered barriers and provide strat-egies for effective user involvement in DR initiatives.

These recommendations serve as a guide for project developers and decision-makers, offering valuable insights to enhance citizen engagement in the evolving landscape of energy transition projects. The following discussion serves to analyse the transferability of the recommendations from the energy domain to the net-zero built environment.

3   Discussion of Co-creation in Net-Zero Built Environment

While the idea of co-creation has been explored in various fields, its systematic consideration within the built environment is still in its infancy [25]. The limited number of publications on co-creation projects in this domain outlines applications with a small number of involved stakeholder groups. Co-creation is presented as a tool for collaboration between researchers and industry players during the innovation development process [26], as an approach to creating built environment knowledge for sustainable development [27], and as an approach to service concept design and architectural practice [28].

The noted underrepresentation of co-creation in the construction and buildings industry, as well as the limited number of involved stakeholder groups discovered in research publications, indicates a potential area for further exploration and implementation. Therefore, the following paragraphs serve to outline how the transfer of experiences from various EU-funded, energy-related projects can serve to develop a more holistic co-creation approach for projects in the building industry.

Net-zero buildings are structures that offset their consumed energy, creating a self-sustaining system. They embody a holistic approach to sustainability, integrating high-efficiency appliances, HVAC systems, and building insulation, along with renewable energy systems [29]. These initiatives aim to minimize the carbon footprint of the building, both during use and the embodied emissions of building materials.

To achieve a net-zero built environment, strategies can be categorized into efficiency (optimizing energy demand and using highly efficient appliances), consistency (utilizing renewable energy and sustainable materials), and sufficiency (reducing living space and heating/cooling needs) [30]. This categorization reveals an overlap in the objectives and methods of energy-related projects and net-zero efforts.

Both energy-related projects and net-zero efforts rely on stakeholder engagement, particularly with citizens as the main end users. Many net-zero initiatives in Europe focus on retrofitting existing buildings. Like DR projects, these initiatives aim to incentivize citizens to optimize their consumption, use smart and efficient appliances, and indirectly enhance the grid's capacity to accommodate more renewable energy.

3.1 Proposed Framework for Replicable Co-creation

With respect to RQ2, it is relevant to outline how identified effective co-creation and engagement processes from selected EU-funded projects can be structured and organized to expand/replicate from electricity to sustainable and net-zero built environments. From an overarching point of view, the proposed phased implementation of co-creation activities serves also in net-zero environments to structure engagement strategies and to enhance their efficacy and adaptability.

Recruitment Phase Due to the very high stakeholder variety in net-zero initiatives, the number of involved stakeholders is much higher compared to EU-funded DR projects. The utilization of a City Portrait Canvas is proposed to foster big-picture thinking on how a particular city strategy may impact the world—both socially and ecologically, locally and globally, and how to shape approaches to reducing both sector-based and consumption-based emissions, as well as to climate change adaptation [31]. Therefore, the requirement to diverse engagement strategies applies to a much higher degree in net-zero initiatives. The employment of a range of strategies for tailored and impactful user engagement requires the crafting of engaging recruitment materials, recognizing the central role of motivation and eligibility criteria, which must be based on a highly detailed stakeholder analysis.

Consumer Response Phase In line with the definition of stakeholder-specific motivation and eligibility criteria, targeted keywords that focus on specific daily life aspects must also be used in net-zero ecosystems to facilitate personalization and adaptability. Furthermore, a sustained collaboration and knowledge-sharing culture among project partners will serve to leverage on-field experiences for enhanced effectiveness. This counts especially for realized barriers during the consumer response phase in net-zero implementations. The project duration of EU-funded electricity projects is usually limited to 3–4 years and an exploitation phase of up to 5 years. The impact of net-zero building implementations will last for generations and, therefore, it is of utmost importance to identify barriers and challenges experienced by consumers during the response phase and to immediately apply corrective actions. This is also fostered by the fact that in contrast to EU DR projects, where implementations strive for a technological readiness level (TRL) of 7–8 out of a scale of 10, net-zero building implementations rely on market-ready technology solutions. Therefore, it will be difficult to re-assess and adapt technologies once they are implemented.

Persistence Phase Finally, also co-creation activities in net-zero built environment approaches must provide fall-back solutions to provide adaptability in crisis, maintaining flexibility for continued effectiveness, particularly in the face of unforeseen circumstances like the COVID-19 pandemic, supply shortages for tech components and appliances and primary energy resources [32]. Especially, the latter aspect is of utmost importance as regulatory and political impacts on net-zero built environments must not be under-evaluated. Due to the long-term characteristics of net-zero building implementations, it is inevitable to model and forecast mid- and long-term regulatory and political objectives, e.g., in terms of national or European energy supply strategies. This serves to create future proof solutions that do not require major adjustments over time. Also, in the context of the SENDER and other EU DR projects, this should be based on continuous assessments during the persistence phase to simulate and navigate diverse contexts, ensuring net-zero approaches alignment with community needs and infrastructural support.

A co-creation framework that includes the three identified phases, quintuple helix, and guiding recommendations towards the net-zero built environment transition is presented in Fig. 3.

Based on the guiding recommendations, the co-creation steering committee manages a co-creation space that consists of various co-creation clusters. These clusters' work is supported by social science and ecological and environmental analysis, partly provided by external partners. The processing of internal and external knowledge is organized according to the three-phased co-creation approach

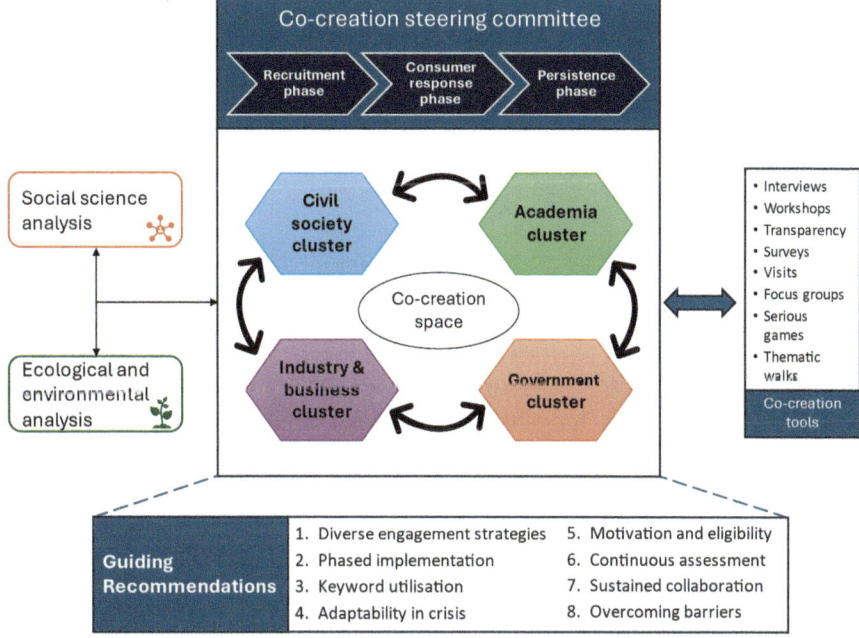

Fig. 3 Proposed framework for co-creation in net-zero built environment initiatives

and will provide jointly agreed solutions, reflecting the democracy aspect of co-creation.

The limitation of the study is the lack of concrete indicators and metrices as most of the social and behavioural science studies involve people and are based on human factors. There are efforts in some of the projects, such as PROBONO, on identifying some measurable indicators. However, they need to be validated and tested in more similar cases, especially within the net-zero built environment.

4 Conclusion

In conclusion, this chapter outlined, according to RQ1, experiences and practices in co-creation and citizen engagement activities, challenges, enablers, and tools applied in selected EU-funded projects. Green transition is supported by active citizen participation and motivation for green transitions.

In addition, referring to RQ2, the co-creation guidelines developed within EU DR projects can be expanded to and replicated in a net-zero building environment in a three-phased approach. Nevertheless, compared to EU DR projects, net-zero building implementations are characterized by a much longer impact period, an overly complex stakeholder variety, and the risk of political and regulatory changes over time.

Therefore, the authors propose to adjust the structured three-phase approach for consumer engagement and co-creation to the net-zero building ecosystems. Their complexity requires the integration of highly detailed expertise from different knowledge domains. An approach defining required knowledge clusters and applying appropriate co-creation techniques is suggested. Co-created recommendations of the various clusters should then be bundled and provided to a co-creation steering committee comprising strategic net-zero building initiative decision-makers. This approach will serve to facilitate democratic decisions based on integrating expertise from all relevant knowledge domains of the net-zero building ecosystem.

Future research should implement, document, and analyse co-creation approaches in the net-zero build environment. This field research approach might create insights for a building industry-specific adjustment of the proposed, three-phased approach.

References

1. Czepiel, J.A.: Managing relationships with customers: a differentiation philosophy of marketing. In: Bowen, D.E., Chase, R.B., Cummings, T.G. (eds.) Service Management Effectiveness, pp. 299–323. Jossey-Bass, San Francisco (1990)
2. Prahalad, C.K., Ramaswamy, V.: Co-creating unique value with customers. Strategy Leadersh. **32**, 3–9 (2004)
3. Prahalad, C.K., Ramaswamy, V.: Co-opting customer competence. Harv. Bus. Rev. **78**(1), 79–90 (2000)
4. De Koning, J., Crul, M., Wever, R.: Models of co-creation, Conference Proceedings, ServDes, Fifth Service Design and Innovation Conference 2016, pp. 266 (2016)
5. The not-so-secret ingredient to accelerating sustainability (& why IKEA is the king of co-creation). https://www.limelights.com/blog-post/the-not-so-secret-ingredient-to-accelerating-sustainability-why-ikea-is-the-king-of-co-creation, 2024/02/16
6. Kujala, J., Sachs, S., Leinonen, H., Heikkinen, A., Laude, D.: Stakeholder engagement: past, present, and future. Bus. Soc. **61**(5), 1136–1196 (2022)
7. Reed, M.S.: Stakeholder participation for environmental management: a literature review. Biol. Conserv. **141**(10), 2417–2431 (2008)
8. Huttunen, S., Ojanen, M., Ott, A., Saarikoski, H.: What about citizens? A literature review of citizen engagement in sustainability transitions research. Energy Res. Soc. Sci. **91**(102714), 2214–6296 (2022)
9. Araújo, K.M.: Participation in the energy system. In: Routledge Handbook of Energy Transitions, 1st edn. Routledge (2022)
10. Jones, P.: Contexts of co-creation: designing with system stakeholders. In: Systemic Design: Theory, Methods, and Practice, pp. 3–52. Springer, Tokyo (2018)
11. Leino, H., Puumala, E.: What can co-creation do for the citizens? Applying co-creation for the promotion of participation in cities. Politics Space. **39**(4), 781–799 (2021)
12. De Jong, M.D., Neulen, S., Jansma, S.R.: Citizens' intentions to participate in governmental co-creation initiatives: comparing three co-creation configurations. Gov. Inf. Q. **36**(3), 490–500 (2019)
13. Brandsen, T., Marlies, H.: Definitions of co-production and co-creation. In: Co-Production and Co-Creation, pp. 9–17. Routledge, New York (2018)

14. Eckhardt, J., Kaletka, C., Krüger, D., Maldonado-Mariscal, K., Schulz, A.C.: Ecosystems of co-creation. Front. Sociol. **6**, 642289 (2021)
15. The SENDER project. https://www.sender-h2020.eu/. Last accessed 27 Feb 2024
16. Leydesdorff, L., Etzkowitz, H.: Emergence of a triple helix of university–industry–government relations. Sci. Public Policy. **23**(5), 279–286 (1996)
17. Etzkowitz, H., Leydesdorff, L.: The dynamics of innovation: from nation-al systems and "mode 2" to a triple helix of university–industry–government relations. Res. Policy. **29**(2), 109–123 (2000)
18. Carayannis, E.G., Rakhmatullin, R.: The quadruple/quintuple innovation helixes and smart specialisation strategies for sustainable and inclusive growth in Europe and beyond. J. Knowl. Econ. **5**(2), 212–239 (2014)
19. Ospina, Y., Aprà, F.M., Aggeli, A.: Lessons learnt during the workshop "empowerment of citizens: fostering user engagement for innovative demand response for effective flexibility" [version 2; peer review: 3 approved with reservations]. Open Res. Eur. **3**, 1 (2024)
20. The HESTIA project. https://hestia-eu.com/. Last accessed 27 Feb 2024
21. The ReDREAM project. https://www.redream.energy. Last accessed 29 Feb 2024. The ReDREAM project. https://www.redream.energy/. Last accessed 29 Feb 2024
22. The ACCEPT project. https://www.accept-project.eu. Last accessed 29 Feb 2024. The ACCEPT project. https://www.accept-project.eu/. Last accessed 29 Feb 2024
23. The iFLEX project. https://www.iflex-project.eu. Last accessed 29 Feb 2024
24. The PROBONO project. https://www.probonoh2020.eu. Last accessed 27 Feb 2024
25. Liu, A., Fellows, R., Chan, I.: Fostering value co-creation in construction: a case study of an airport project in India. Int. J. Archit. Eng. Constr. **2**(3), 120–130 (2014)
26. Suija-Markova, I.: A methodological framework for co-creation of government-research-industry innovation. Scientific Conference on Economics and Entrepreneurship Proceedings, pp. 100–109 (2022)
27. Prompen, C., Arunotai, N., Nuangchalerm, P.: Co-creating built environment knowledge for sustainable development: case study report from Thailand. J. Educ. Issues. **8**(2), 698 (2022)
28. Malakhatka, E., Sopjani, L., Lundqvist, P.: Co-creating service concepts for the built environment based on the end-user's daily activities analysis, kth live-in-lab explorative case study. Sustain. For. **13**(4), 1942 (2021)
29. Net-Zero Buildings: Guiding the Way Ahead for Sustainable Architecture. https://www.novatr.com/blog/what-are-net-zero-buildings. Last accessed 14 Feb 2024
30. Scherz, M., Passer, A., Kreiner, H.: Challenges in the achievement of a net zero carbon built environment – a systemic approach to support the decision-aiding process in the design stage of buildings. IOP Conf. Ser. Earth Environ. Sci. **588**(3), 032034 (2020)
31. The City Portrait Canvas Guide. https://www.c40knowledgehub.org/s/article/The-City-Portrait-Canvas-Guide. Last accessed 29 Feb 2024
32. How to Handle the Construction Material Shortage. https://www.dustyrobotics.com/articles/how-to-handle-the-construction-material-shortage. Last accessed 29 Feb 2024

Life Cycle Analysis of Floating Offshore Wind Turbine Concepts

Ramin Shakori and Arnab Chaudhuri

Contents

1 Introduction

The International Energy Agency (IEA) Net Zero Emissions by 2030, which is a normative scenario showing a pathway for the world's energy industry to achieve net-zero CO_2 emissions by the year of 2050, meets the United Nation's Sustainable Development Goals (SDGs) [1]. According to the IEA, by 2021, of the total 830 GW wind capacity installed in the world, only 7% were offshore wind farms, while 93% were located onshore [2]. While onshore wind is a developed technology with a location in 115 countries, offshore wind is at a relatively early expansion stage with 19 countries currently having such systems. Offshore presence is however expected to increase in the upcoming years as more and more countries are in the developing stages of their first offshore wind farms. Windmills, used for grinding grain and pumping water, have a long history dating back to seventh-century Persia (Iran) [3]. However, generating electricity with wind is a much newer development. Large-scale wind turbines for electricity generation only emerged in the 1930s, primarily in

R. Shakori · A. Chaudhuri (✉)
Oslo Metropolitan University, Department of Built Environment, Oslo, Norway
e-mail: arnab.chaudhuri@oslomet.no

© The Author(s) 2025
M. Kioumarsi, B. Shafei (eds.), *The 1st International Conference on Net-Zero Built Environment*, Lecture Notes in Civil Engineering 237,
https://doi.org/10.1007/978-3-031-69626-8_144

Europe and the USA. There are several benefits of offshore wind power compared to onshore wind power. Firstly, offshore wind turbines can generate more electricity due to higher wind speeds and less turbulence over water. Secondly, offshore wind farms can be located closer to population centers, reducing transmission costs, and improving energy security. Thirdly, offshore wind turbines have less visual impact on the landscape and can be placed in deeper waters where they are less likely to interfere with shipping lanes or fishing grounds. The FOWT farms can create jobs in coastal communities and act as a reliable source of clean energy while minimizing environmental impacts.

Life Cycle Assessment (LCA) is a quintessential tool to assess the environmental impacts associated with all stages of a product's life cycle and helps to improve sustainability (see Curran [4]). Yildiz et al. [5] performed LCA of a Barge-Type Floating Wind Turbine and Comparison with Other Types of Wind Turbines. The authors made an environmental comparison of three scenarios of a 2 MW wind energy system—onshore, offshore floating, and jacket offshore wind (OW) turbines. The study revealed that the greatest environmental impact corresponds to the manufacturing stage of the system components steel and concrete emissions, and total GHG emissions stated in GWP for the offshore floating concept was 18.6 g CO_2-eq./kWh, the onshore concept had 7.09 g CO_2-eq./kWh, and the jacket offshore concept had 9.49 g CO_2-eq./kWh. Authors from the leading French OW development firm Ideol SA (now BW Ideol, Japanese) studied an offshore unit involving square ring floater. They have recommended that the construction could either be a steel-concrete combination or could be made purely of steel [6]. Calculated results were 3376 and 1468 tons (tot/8250 T = 0.409 t CO_2/T and 0.177 t CO_2/T) of CO_2 emissions for the concrete concept, and 7018 and 2420 tons (3.19 T and 1.1 T CO_2/T) of CO_2 emission for the steel concept—with new and recycled materials, respectively. The study found that concrete hulls are heavier but offer better stability, while steel hulls are lighter but require more maintenance, and that a steel hull's carbon content is twice as large as a concrete hull. The comparative study of concrete and steel substructures for FOWT was reported in DNV's project study [7] to investigate the possibility of establishing an industrial plant to produce concrete structures for the offshore wind industry. The objective of the study was the comparison of two offshore floating concepts made in steel or concrete, and results indicate the concrete concept has both lower financial costs and carbon emissions compared to steel. The study concludes that Norwegian firms are capable to deliver concrete concepts target of 67 units per year, while the steel industry is not capable of the same. Calculated results show that the carbon emission of concrete is 2.5–3 times lower than those of steel; even with even a 70% recycled steel used in the Chinese plants, concrete remains the more favorable. For the Norwegian National Framework for Wind power, the NVE performed a study on wind power emissions [8]. Their report provides an insight into LCA of wind turbines mounted on Norwegian land and compared wind power to other sources of energy. The report concluded with the fact that the leading environmental impact of these turbine systems came from the production of concrete for the foundation and transport of wind farm components. GWP values stated in the framework were that wind power has a GWP of 3–46 g

CO_2/kWh compared to gas- and coal energy sources with 500 and 1000 g CO_2/kWh, respectively. The life-cycle energy and environmental emissions of a typical offshore wind farm in China is reported by Yang et al. [9] where the authors have presented the emission from China's first offshore wind power farm. They performed an LCA study in which Life Cycle Inventory (LCI) assessment is reported to list the system components based on GHG emissions and reported a GWP emissions of 25.5 g CO_2-eq/kWh over the lifetime.

On March 29th, 2023, the Office of the Prime Minister, and OED—the Ministry of Petroleum, announced the first public tender for offshore wind projects in two areas of the North Sea—Utsira Nord and Sørlige Nordsjø II. The aim is to increase renewable energy production and reduce GHG emissions, a big step toward Norway's aim of 30 GW offshore wind by 2040. The government will propose a proposition to Parliament with a proposal for a cost framework and commitment authorization. The projects that receive support are awarded support for 500 MW each but can build up to 750 MW, while the project that does not receive state funding retains the right to the area and may use the public support system and participate in future competitions.

In the OW industry, both steel and concrete are critical, and while the sector has a low environmental impact compared to traditional fossil fuel energy sources [10], there are still carbon emissions to consider. The manufacturing process of the OW turbine structure requires significant amounts of energy and resources, including steel, concrete, surface treatment chemicals and the transportation to site. The extraction and processing of building materials and chemicals can result in GHG emissions. Transportation of the structures to the installation site also requires large vessels with emissions. The objective of this work is therefore to assess the carbon emission effects on the environment, comparing a hull made of steel vs. concrete, with the aim of filling the knowledge gap in LCA studies and raising carbon footprint awareness.

An industrial case scenario was used to conduct an LCA study, project layout projected (see Fig. 1), of the following: Case study scenario 1: steel hull, installed in Singapore, then shipped to Hanøytangen, Norway—Chinese steel origin (see Fig. 2a, b). Partial goal: determine EN-15804- and NORSOK-compliant surface coating system. Case study scenario 2: concrete hull, produced, installed, in periphery of Hanøytangen, Norway—Norwegian concrete origin (see Fig. 2c). Before gaining interest in this research study, a relevancy analysis was performed to use an industry specific case scenario applicable to the LCA study (see Fig. 1). An approach was made to map engineering consultancy firms in the offshore structural field, and a finalized reach out was done to SEVAN SSP—with whom a correspondence resulted in the study task, and the following project data and system description (see Table 1).

The article is organized as follows. A brief methodology is described in Sect. 2, with the subsection of steel surface coating, followed by Sect. 3 of results and discussions, with the subsection of a result sensitivity analysis. Finally, conclusions are drawn in Sect. 4.

Fig. 1 SEVAN SSP SWACH™ wind

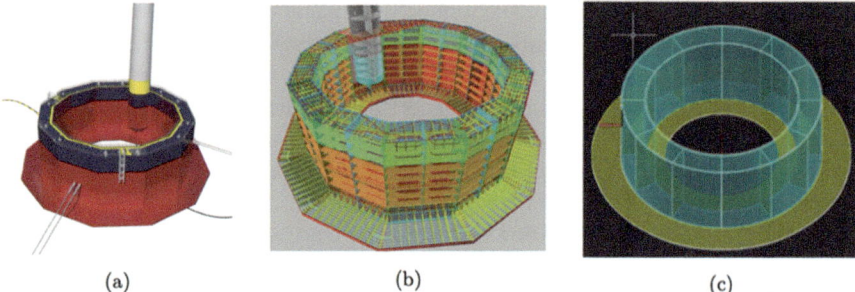

(a)   (b)   (c)

Fig. 2 SEVAN SSP-engineered FOWT concepts - a, b are steel, and c is concrete

Table 1 Project system description

Functional unit	Value	Description
Lifetime	50 years	The average technical lifetime of whole wind farm
No. of platform units	30	Study consists of 30 in steel versus 30 in concrete
Freight distance, SIN-OSL	10,000 nm	10,000 nautical miles equals 18,520 km
Freight distance, HAN-UTS	50 nm	50 nautical miles from Hanøytangen to Utsira

2   Methodology

In this section, we present the methodology for calculating CO_{2e}/T for steel vs. concrete FOWT units. It outlines the LCA analysis using chosen software and breaks down the steps of the study to find CO_{2e}/T, a material data assessment for

Fig. 3 Study-specific process flow chart

30 units steel vs. concrete, and GWP calculation in CO_{2e}/T. A flow chart (sec Fig. 3) outlines the process. Iterations ensured compatibility with the SEVAN SSP case study.

As sustainability, climate change, and circular economy practices become more prevalent in the Engineering, Procurement, Construction and Installation (EPCI) industry, the demand for a time and cost-efficient method continues to grow. LCAs can be complex calculations, but there are digital tools and databases available to make the process easier. Ecoinvent is an online database that provides comprehensive information on the environmental impact of various products and services and contains data on the entire life cycle of a product, from raw material extraction to disposal. For this study the database search tool used was Ecoinvent's section for EN15804-compatible allocation and cut-off—since the system gives LCI describing waste and resource categories, excluding excessive by-products.

With so many LCA-software options on the market, it is important to determine a suitable one for our assessment. Tools considered were Ecochain Mobius, Ecochain Helix, GaBI, OneClickLCA, openLCA, and SimaPro, based on a selection criterion of having free trials, BIM compatibility, being web-based, while having a user-friendly Interface (UI). The appropriate software for this study was OneClickLCA, justified by the fact it has free online courses, help center with common FAQ, open-source learning material (YouTube, Vimeo), BIM-compatibility (IFC-file readability), and discounted academic and trial licenses.

OneClickLCA sources its data from a selection of sources, and they can be generic or product-specific if a producer has uploaded an Environmental Product Declaration (EPD) available for each country's EPD Program Operator and publisher. To shortly describe the software tool use process, a Life Cycle Inventory Assessment (LCIA) was done based on case study project specifications. Concrete concept hull design weighs 20,500 tons, of which 2000 tons are steel reinforcement bars—rebar/concrete ratio subject to confidential research. The 18,500 tons of concrete (high-strength C60/C70-grade NORSOK-compliant with 10900PSI compressive strength) was given A2 mode transport distance from producer to storage of 70 kilometers onboard a 40-ton trucks with EURO-6 engine with 100% capacity. Subsequently, A4 shows 26 kilometers transport distance from storage to project site.

2.1 Coating System

A partial objective is to determine an EN15804- and NORSOK M-501-system1-compliant surface treatment system for the steel hull. Method is to manually

Fig. 4 Coating system selection process chart

determine the case study needs (see Fig. 4), of a specific product and its feasible availability, GWP given in CO_{2e} per Ton, and total GWP for the 30-unit steel case scenario.

The NORSOK M-501 system 1 [11] is the requirement for the selection of a compliant coating system given by communication with engineers from SEVAN SSP—selected from the range of choice spanning from system no. 1 to system no. 9.

Selection criteria determined coating system no. 1, and to continue the determination of a product-specific EPD compliant with the PCR rules of EN15804, GWP stated in CO_{2e}, and a factory proximity to East-Asian Sea ports, a survey of available coating systems on the EPD Norway library was done. This resulted in the elimination of brands and factory locations from Norway, Germany, India and resulted in Jotun Jotamastic 70, Zhangjiagang, China, with EPD number: 1599-624-EN. The basis of the choice of the given factory location roots in the fact that there is a seaport proximity, for the easiness of shipping to the Singapore installation site.

Results found in the EPD [11] with a EN15804-product stage system boundary A1-A3, *cradle-to-gate*, consist of A1: raw materials extraction, production, packaging, then A2: transport of raw materials and packaging, and finally A3: product stage: preassembly, pigments mixing dispersion, quality control, filling, and labeling. The calculations necessary to find GWP is based on; the EPD data of 3.19 kg CO_{2e} per kg paint, film thickness range 7.7–3.1 m^2/l, density (ρ) of 1.6 kg/l, with a project specific steel hull surface Area $= 30,430$ m^2, resulting in total masses of 6323.12 kg and 15705.81 kg paint, simply using $m = \rho/V$ for both film thicknesses.

3   Results and Discussions

Steel concept LCA study resulted in having the environmental impact parameter GWP of 13,821.257 T CO_{2e}/T, for the 5000-ton steel concept which is equivalent to the study selected unit of 2.76 T CO_{2e}/T (see Fig. 5a).

As for the NORSOK-compliant Jotun Jotamastic 70, the results are calculated to: Total GWP for spread rate 1 with 7.7 $m^2/l = 30 \times 6323.12$ kg $\times 3.19$ kg $CO_{2e}/$ kg $= 6051$ T CO_{2e} Total GWP for spread rate 2 with 3.1 $m^2/l = 30 \times 15,705.81$ kg $\times$ 3.19 kg $CO_{2e}/kg = 1503.1$ T CO_{2e}. The steel concept surface coating system necessary for 30 units has GWP ranging from 605.1 T to 1503.1 T CO_{2e}/T.

The concrete concept gives a GWP of 8298.743 T CO_{2e}, for the 20,500-ton concrete structure of which 2000 tons are reinforcement bars which is equivalent to the study selected unit of 0.404 T CO_{2e}/T (see Fig. 5b).

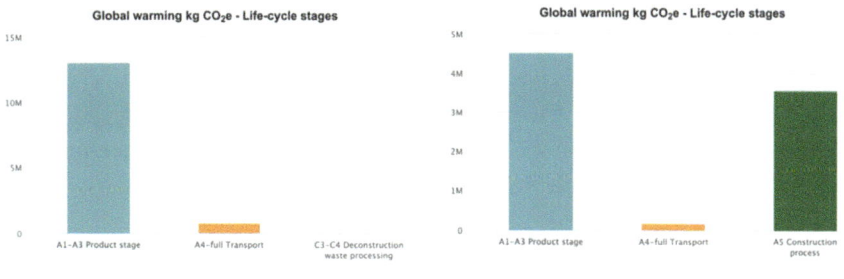

(a) LCA Results Steel units.   (b) LCA Results Concrete units.

Fig. 5 (**a, b**) LCA results

Note that this is per system boundary of A1-A3. Installation, operation, mainte-nance stage is not defined in this case. The steel concept GWP per ton is 6–7 times higher, but that is only per weight. The project concept states 30 units, of which each concrete hull has 20,500 T while the steel concept only has 5000 T—equaling a 1/4 steel vs. concrete hull weight ratio which makes the steel concept only 1.77-times higher than concrete—equaling 70% worse than the concrete in total project mass.

Few LCA studies are performed on FOWT structures involving steel or concrete—results from this study are therefore relevant as supplementary scientific material. In relation to the aforementioned literature studies, the GWP calculation of this study with 0.404 T CO_{2e}/T for concrete and 2.76 T CO_{2e}/T for the steel concept falls within Choisnet et al.'s published article where the impacts amounted in 1.1–3.19 T CO_{2e}/T for steel and 0.177–0.41 for concrete. Simultaneously the NVE National Framework concluded with the fact that concrete production has a signif-icant environmental impact, which confirms the graph from this study's result section where stage A1-A3 is shown in the results (see Fig. 5a, b).

Regarding the selection of the 10,900-PSI C60/C70 grade high-density concrete selected in OneClickLCA for the study, an estimated 10% recycled additive per-centage was a composition of the cement used in the concrete mix—typical additives being hydraulic binders, such as fly ash, fume, and slag—all by-products from industrial processes. In case of a change in percentage of the reuse, a potential carbon emission reduction can occur. Typically, these additives also have their own GWP, not stated transparently although they are sources of CO_2 emissions.

Regarding the surface coating, read on page 8 of the source [11], the electricity mix in China is stated with a GHG of 1194.03 g CO_{2ekv}/kWh, by Ecoinvent. In China the GHG emissions from electricity is presumably higher than electricity mixes in Norway where 98% of energy sources are from renewable sources, while we in China have traditional fossil sources such as coal, heavy fuel, etc. An alternative solution is to treat the surface at dock in Hanøytangen. Note that this alternative must consider labor cost changes and shipping distance recalculations.

3.1 Sensitivity Analysis of the LCA Results

As the steel reinforcing bars of the concrete hull are a matter of early-stage research, even stated by DNV as a 2023-Q1-starting industrial collaborative research project, an awareness disclaimer must be made to the reader that the 9% steel content was used for calculation simplicity while DNV uses 2% [12]. As 8% of total emission in the steel concept LCA calculations represents the steel reinforcement bars, this is a source of emission which could possibly affect the total GWP if a higher recycling rate is used in its production.

As steel is the main material used to manufacture the 30-unit hulls, the upstream process is highly determinant of the total emission stated in the results. Even a low change in the steel recycling grade percentage during production in China could be significant to total emissions.

4 Conclusions

In today's world, amidst growing environmental concerns and resource scarcity, sustainability in engineering and energy efficiency has become more paramount than ever before. With the increasing demand for energy and the depletion of natural resources, it is crucial that we find ways to reduce our carbon footprint and preserve the environment for future generations.

Wind power is one of the most promising sources of renewable energy, and it has the potential to provide clean, reliable power and a significant portion of our global energy needs. Wind power is also a promising step toward reaching the UN SDGs, including SDG 7: Affordable and Clean Energy, SDG 9: Industry, Innovation and Infrastructure, and SDG 12: Responsible Consumption and Production. By investing in sustainable engineering practices and embracing energy efficiency, we can create a brighter future for ourselves and our planet. We have the power to make a difference, and by working together toward a common goal, we can create a world that is cleaner, healthier, and more sustainable for all.

An LCA study has been a great hands-on approach to achieving and comprehending the actual carbon footprint when choosing between concept 1 (steel) and concept 2 (concrete). This study has allowed us to compare the environmental impacts of the two materials over their life cycle, from the extraction of raw materials to the manufacturing, transportation, and installation, of the FOWT structures.

An environmental approach to industrial decision-making can make a difference and help the energy sector reach the targets of the UN SDGs. By considering the environmental impacts of our choices, we can select materials and processes that minimize our impact on the planet.

This work aims to share the knowledge gained during this research, hoping it proves beneficial for readers from both academic and industrial backgrounds. The aim was to make it comprehensible for the reader, by formulating it apprehensible without the reader being required to possess significant knowledge on LCA as a scientific field. As a conclusion, LCA studies can be a valuable tool for identifying opportunities to reduce the carbon footprint of FOWT structures.

4.1   Limitations and Future Work

As a disclaimer, it must be noted that challenges of the study were mainly limitations of material database availability and data variations on the A1-A3 analysis specter.

- Lack of sufficient EPD's from steel and concrete plant material databases.
- Limited data availability from the actual steel plants in China other than the single source used in this project in OneClickLCA.
- Average travel speed variations: shipping vessel carrying the two steel structures.
- Emission reduction of the steel industry will further reduce the GWP of FOWT farms. In the future, more data needs to be collected to improve LCA reliability.
- According to the NVE, in southern parts of China, the raw material stage's mining processes has affected loss of ecosystems and biodiversity [8].
- Jotun Jotamastic 70: the product service life is highly dependent on the use conditions; the GWP was calculated for a one-time application, and it depends on whether it needs reapplication or maintenance within the 50-year lifespan.

4.2   Declaration of Interests

As transparency in research is important, the signatory author has performed a review process of a number of questions on matter of interest conflicts— from a financial, professional, and personal point of view. Ethical approval is not applicable for this theoretical study, and no monetary funding or endowment was involved. The author hereby declares that there are no known interests or personal relationships that have occurred to in any way influence the angle, methodology, or results of this study.

Acknowledgments This work is an extraction of the bachelor thesis of the first author, and we would hereby like to thank SEVAN SSP—whom we must acknowledge for entrusting us with the case-specific project-specifications and industry-relevant knowledge.

References

1. IEA: Wind Power Capacity in the Net Zero Scenario 2010–2030. www.iea.org/data-and-statistics/charts/wind-power-capacity-in-the-net-zero-scenario-2010-2030. Last accessed 10 Feb 2023
2. IEA: Wind Electricity Technology Deep Dive. www.iea.org/reports/wind-electricity. Last accessed 1 Mar 2023
3. US Energy Information Administration: Wind Explained – History of Wind Power. www.eia.gov/energyexplained/wind/history-of-wind-power. Last accessed 15 Jan 2023
4. Curran, M.A.: Life Cycle Assessment Handbook, 1st edn. Scrivener, Beverly (2012)
5. Yildiz, N., Hemida, H., Baniotopoulos, C.: Life cycle assessment of a barge-type floating wind turbine and comparison with other types of wind turbines. Energies. **14**(18), 5656 (2021). In: Blaabjerg F. MDPI, Basel
6. Choisnet, T., Geschier, B., Vetrano G.: Initial comparison of concrete and steel hulls in the case of ideol's square ring floating substructure. 15th WWEC, Tokyo, Japan, 1 Nov 2016
7. DNV: Comparative Study of Concrete and Steel Substructures for FOWT. www.windworks-jelsa.no/app/uploads/2022/01/Comparative-study-of-concrete-and-steel-substructure-for-FOWT_final-for-distribusjon. Last accessed 10 Feb 2023.
8. NVE: National Framework for Wind Power. www.publikasjoner.nve.no/rapport/2019/rapport2019_17, 2019. Last accessed 1 May 2023.
9. Yang, J., Chang, Y., Zhang, L.: The life-cycle energy and environmental emissions of a typical offshore wind farm in China. J. Clean. Prod. **180**, 316–324. Elsevier, Amsterdam (2018)
10. DoE: Advantages and Challenges of Wind Energy. US DoE Homepage. www.energy.gov/eere/wind/advantages-and-challenges-wind-energy. Last accessed 13 Feb 2023
11. JOTUN: Norsok Technical Info 2023. www.jotun.com/ww-en/industries/technical-information-and-services/information. Last accessed 1 May 2023
12. DNV: Concrete for Floating Offshore Wind (FLOW). www.dnv.com/article/concrete-for-floating-offshore-wind-flow-227327. Last accessed 4 May 2023

Life Cycle Management and Probabilistic Levelized Cost of Energy Analysis of Floating Offshore Wind Farms

Hadi Amlashi and Omid Lotfizadeh

Contents

1 Introduction

Offshore wind farms initially had higher costs and lower reliability than did onshore wind farms, making the offshore wind industry less attractive with limited industrial support. Due to these issues, it was expected that offshore wind farms would not be financially viable until 2020. However, by the early 2000s, events such as the European Union (EU)'s acceptance of the Kyoto Protocol, energy security concerns, and economic crises had shifted public attitudes. This led to an increase in political and societal support for European offshore wind farms [1].

While offshore wind power is known for being environmentally friendly during operation, it actually produces emissions throughout its life cycle. This includes manufacturing, installation, transportation, maintenance, and decommissioning [2]. The study of the life cycle can be approached from various perspectives, such as economic and environmental considerations [3]. The environmental impact and energy efficiency of offshore wind technology can be analyzed using a variety of

H. Amlashi (✉) · O. Lotfizadeh
University of South-Eastern Norway (USN), Porsgrunn, Norway
e-mail: hadi.amlashi@usn.no

© The Author(s) 2025
M. Kioumarsi, B. Shafei (eds.), *The 1st International Conference on Net-Zero Built Environment*, Lecture Notes in Civil Engineering 237,
https://doi.org/10.1007/978-3-031-69626-8_145

1749

accessible wind turbine models, with life cycle assessment (LCA) being the most common assessment approach [2]. The information gained from these analyses can be used for a more accurate economic viability assessment, such as the levelized cost of energy (LCOE).

2 The Life Cycle Assessment Concept in Offshore Wind Energy

2.1 General

Life cycle assessment (LCA) has versatile applications and can be helpful in a variety of ways, including [4]:

- Identifying strategies to enhance the environmental performance of products throughout their entire life cycle
- Informing decision-makers in the industry, government, or other sectors
- Choosing appropriate environmental performance indicators and assessment methods
- Supporting marketing efforts

A product's life cycle consists of five phases: extraction, manufacturing, distribution, usage, and end of life. The assessment approach is typically applied from cradle to grave but can also be utilized from cradle to gate, gate to gate, cradle to customer, gate to grave, or cradle to cradle. The cradle-to-grave approach is defined as a full LCA, starting from resource extraction (cradle), through the use phase, and ultimately to the disposal phase (grave) [4].

2.2 The LCA Framework

According to International Organization for Standardization (ISO) 14040 and ISO 40444, the LCA framework comprises four phases [5]:

1. Goal and scope definition
2. The life cycle inventory (LCI) analysis phase
3. The life cycle impact Assessment (LCIA) phase
4. Interpretation

Goal and Scope Definition The first stage of an LCA is defining the goals and scope, which is considered the most important phase because it establishes the research context, sets requirements for the modeling that will be performed, and plans the project [6].

The goals of this study are defined as follows:

- Filling the gap in the literature as there are few comprehensive LCAs of offshore wind farms. This can help developers and scholars track their progress toward sustainability goals and identify areas for improvement.
- Comparison with other renewable energies
- Public education
- Life cycle optimization
- Data collection and methodology improvement for further research

Moreover, design practitioners must identify four key activities throughout the goal and scope definition stage: (1) defining a functional unit, (2) establishing a clear system boundary, (3) selecting the environmental impact types, and (4) determining the level of complexity and data required for the study. The functional unit acts as a point of reference for the inputs and outputs of the product under investigation [4].

For an LCA, a logical functional unit could be the amount of energy produced. In this study, the functional unit is 1 MWh of electricity produced and sent to the grid through the wind farm's substation. In this step, the system boundary is determined. The system boundaries specify what is included and omitted from the evaluation. Small amounts of substances may be excluded from the analysis due to their minimal impact on the overall footprint. Figure 1 illustrates the defined system boundaries in this study.

Life Cycle Inventory Analysis The second stage of LCA is Life cycle inventory analysis, as stated in ISO 14040 [5]:

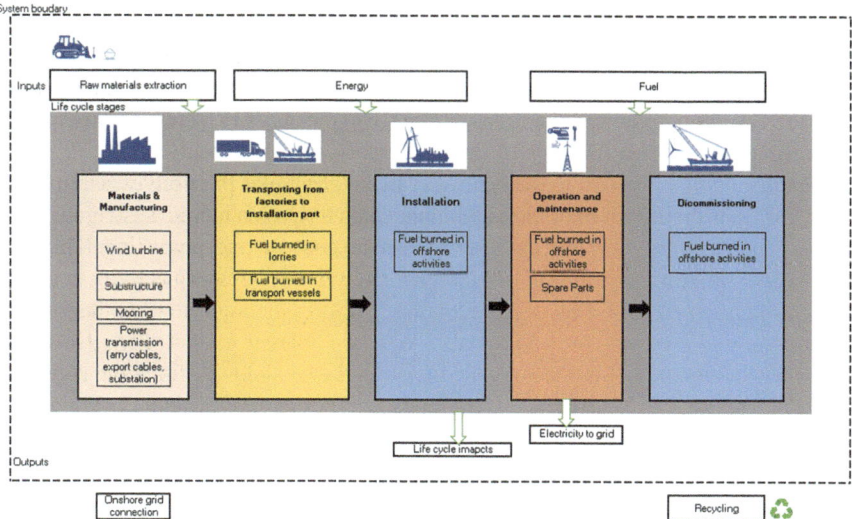

Fig. 1 System boundaries from Garcia-Teruel et al. [7] with some changes

Phase of life cycle assessment involving the compilation and quantification of inputs and outputs for a product throughout its entire life cycle.

The output of this stage is an inventory of basic flows, which serves as the foundation of the life cycle impact assessment. Data were gathered for each unit process under evaluation. These data include energy inputs, raw materials, emissions, and waste.

Life Cycle Impact Assessment (LCIA) The third phase of an LCA study is LCIA where the data from the life cycle inventory regarding basic flows are transformed into environmental impact scores. Typically, the life cycle impact assessment (LCIA) stage is fully automated, with the expert selecting an LCIA method and a few additional parameters using menus and buttons in LCA software.

The global warming potential (GWP) is the most well-known impact category. Several life cycle impact assessment (LCIA) models have been developed and used to assess the environmental impacts of products and activities, including ReCiPe (Resource Use, Emissions, and Health Impacts), IMPACT (Integrated Methodology for Impact Assessment of Chemicals, CML (Centre for Environmental Studies), Eco-indicator, EPS (Environmental Priority Strategies), USEtox (Unified System for the Evaluation of Toxicity), and TRACI (Tool for the Reduction and Assessment of Chemical and Other Environmental Impacts) [4]. The chosen methodology for this study is ReCiPe 2016 v1.03, midpoint (H) using the ecoinvent database by OpenLCA software [6].

Interpretation The interpretation stage is the final phase of an LCA, where the results of the previous phases are combined and evaluated in consideration of the uncertainties present in the data and the documented assumptions made during the study.

By applying the LCA methodology, the GWP of the wind farm is calculated to be 36.5 kg CO_2/MWh. Table 1 shows the contribution of each life cycle stage to the total emissions. Each of these stages is categorized into capital expenditure (CAPEX), operational expenditure (OPEX), or decommissioning expenditure (DECEX).

The results are in good agreement with those of similar previous studies. However, the contribution of operation and maintenance (O&M) is much lower in many studies due to their simplifying assumptions (except for the study conducted by

Table 1 Contribution of each life cycle stage to the total emissions

Stage	Category	Contribution (%)	Contribution (kg CO_2/MWh)
Manufacturing	CAPEX	50.64	18.48
Transportation			
Installation			
O&M vessel	OPEX	40.06	14.62
O&M spare parts			
Decommissioning	DECEX	9.3	3.4

Garcia-Teruel et al. [7], which has concluded a similar contribution of O&M to the total emissions). This study has shed light on this important stage of the wind farm life cycle.

Some previous studies on LCOE have concluded that O&M costs are in the range of 30% of the total cost [8]. The assumption of CAPEX of approximately 4 MEUR/MW (Million Euros per Megawatt) can be justified by utilizing historical patterns in adjusted and normalized offshore wind capital costs, whereas DECEX can be in the order of 0.5 MEUR/MW [9].

3 Levelized Cost of Energy

3.1 General

The levelized cost of energy (LCOE) is an economic metric often presented as a monetary unit per kilowatt-hour (kWh) and is used for comparing different competing energy resources [10, 11]. Parameters that may influence LCOE of a wind farm include the type of renewable energy resource, the technologies used and their efficiencies, the type of project, the countries in which they will be built and operated, and capital and operating expenditures [12]. LCOE is defined as the ratio of the net present value of lifetime costs to the net present value of lifetime generated energy.

3.2 Probabilistic LCOE Methodology

Methodology Based on the definition provided above, the levelized cost of energy (LCOE) can be probabilistically calculated as shown in Eq. (1)

$$\text{LCOE} = \frac{\frac{1}{\gamma_a}\widehat{X}_c.\text{CAPEX} + \frac{1}{\gamma_a}\widehat{X}_d.\text{DECEX} + \widehat{X}_p.\text{OPEX}}{X_{eff}.X_w.8760.N_{\text{WT}}.P_{\text{WT,nom}}.\gamma_c} \quad (1)$$

in which, $\widehat{X}_c$ is the random variable representing uncertainty in the capital expenditure, $\widehat{X}_d$ represents uncertainty in the decommissioning expenditure, $\widehat{X}_p$ represents uncertainty in the operational and maintenance expenditure, $\widehat{X}_{eff}$ is the random variable representing the uncertainty in predicting the average energy production in a wind farm, $\widehat{X}_w$ is the random variable accounting for the decrease in energy production due to the wake effect in subsequent wind turbines, N_{WT} is the number of wind turbines in a wind farm, and $P_{\text{WT, nom}}$ is the nominal power capacity of the wind turbine.

γ_a is the annuity factor defined as in Eq. (2):

$$\gamma_a = \frac{1}{d}\left[1 - (1+d)^{-N_{\mathrm{LT}}}\right] \tag{2}$$

d is the discount rate and N_{LT} is the lifetime of the offshore wind farm. Here, a simple discounted cash flow approach is used taking into consideration the time value of money. It is assumed that the total investment cost can be uniformly distributed over the lifetime of the offshore wind farm and that annual operating expenditure (OPEX) will remain the same for all years during the lifetime. The variable nature of power generation and energy mix in the energy market and grid management is not addressed in this analysis. The decision criterion for LCOE (EURc/KWh) can be presented as explained below in Eq. (3):

$$\mathrm{LCOE} \leq \gamma_p P_c \tag{3}$$

where γ_p is defined as the correction factor for LCOE due to the uncertainties involved in the above-stated random variables and P_c is defined as the characteristic LCOE. The characteristic value is usually specified with a specific confidence level. Due to a lack of proper statistical data, a typical characteristic value of 5–15 EURc/KWh is assumed for LCOE when compared to other energy sources. This assumption may be somewhat arbitrary, but it is utilized in this chapter to evaluate the impact of this definition on the probabilistic analysis. It should be noted that no calibration was performed in this study to account for the single safety factor of γ_p. Therefore, a factor of 1.0 is considered here.

The Monte Carlo simulation (MCS) technique is applied in this study to accurately model the LCOE distribution, particularly its tail behavior. The advantage of the MCS method is that it converges on the exact results when enough simulations are carried out. However, the method can be time-consuming, especially when estimating small failure probabilities. In this study, a sufficient number of simulations is considered. For an overview of the computational methods for probabilistic analysis, refer to Melchers et al. [13] and Faber [14]. The Strurel program is used for probabilistic analysis, whereas Strurel 3.14 is used for sample analysis and parameter estimations [15].

Modeling Random Variables

Uncertainties. The following section discusses and quantifies model uncertainties in expenditures, prediction of average energy production, wake effects, and the capacity factor. Due to a lack of sufficient statistical data, a normal distribution (with a bias of 1.0) is assumed for all expenditure random variables. This assumption is made to illustrate the effect of randomness on the results. For the expenditure random variables, it is assumed that OPEX can range roughly from 0.5 to 2 EURc/KWh and DECOM from 4 to 5 EURc/KWh, whereas LCOE can range from 50 to 80 EURc/KWh [16]. Energy losses due to the wake effect may vary between 2% and 7% depending on the wind farm configuration and the correlation between wind turbine availability and wind power production [17]. Both capacity factor and annual energy estimation introduce uncertainty in the energy output of a wind farm. Different sources provide different capacity factors ranging from 0.2 to 0.6 [18, 19]. A modern

Table 2 Summary of random variables and design parameters used in the probabilistic formulation

Variable	Description	Distribution type	Mean value (μ)	Coefficient of variation (CoV)
$\widehat{X}_c$	Capital expenditure estimation error	Normal	1.0	0.2
$\widehat{X}_d$	Decommissioning expenditure estimation error	Normal	1.0	0.2
$\widehat{X}_p$	Operational and maintenance expenditure estimation error	Normal	1.0	0.2
$\widehat{X}_{eff}$	Predicting the average energy production in a wind farm	Normal	0.4	0.5
$\widehat{X}_w$	Wake effects	Normal	0.85	0.5
CAPEX	Typical CAPEX estimate	Fixed value	4 MEUR/ MW	–
OPEX	Typical OPEX estimate	Fixed value	0.1 MEUR/ MW/year	–
DECEX	Typical DECEX estimate	Fixed value	0.4 MEUR/ MW	–
N_{WT}	Number of wind turbines in the farm	Fixed value	250	–
$P_{\mathrm{WT, nom}}$	Nominal power of the wind turbine	Fixed value	5 MW	–
γ_c	Capacity factor	Normal	0.6	0.5
N_{LT}	Design lifetime for the wind farm	Fixed value	25 yrs	–
γ_a	Annuity factor	Fixed value	10%	–
P_c	Characteristic LCOE	Fixed value	5, 10, 15 EURc/KWh	–

wind turbine typically has a capacity factor of around 0.6 [19]. It is important to acknowledge that the available data for statistical analysis are limited. When estimating expenditure errors, a coefficient of variation (CoV) of 0.2 is used, whereas a CoV of 0.5 is assumed for prediction errors in average energy production and wake effects. For the capacity factor, we assume a mean value of 0.6 and a CoV of 0.5. It is crucial to highlight that the selection of random uncertainty (CoV) and design parameters in this example is arbitrary. However, as a general rule, more uncertainties are assigned to energy output variables than to expenditure variables.

Random Variables. Based on the data discussed above, Table 2 outlines the key parameters for a hypothetical design case for a wind farm consisting of 250 5-MW floating wind turbines.

3.3 Probabilistic Analysis

Results and Discussion Monte Carlo simulations (MCSs) are used for probabilistic analysis. A fitted distribution of LCOE was created by fitting the parameters of a library distribution to sampled data. In this chapter, the simulation results are fitted to

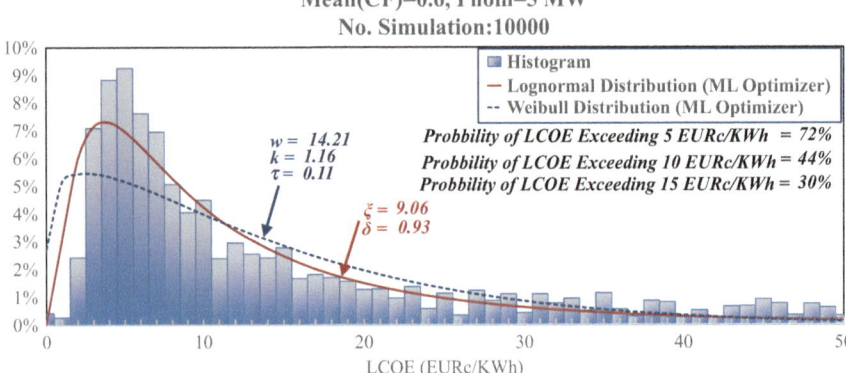

Fig. 2 Log-normal and Weibull fits to the simulated LCOE distribution histogram; wind farm with 250 wind turbines with a nominal capacity of 5 MW; capacity factor $\gamma_c = N(0.6, 0.2)$; Monte Carlo simulation (1.0E + 4); characteristic LCOE (P_c) assumed to be 5, 10, or 15 EURc/KWh

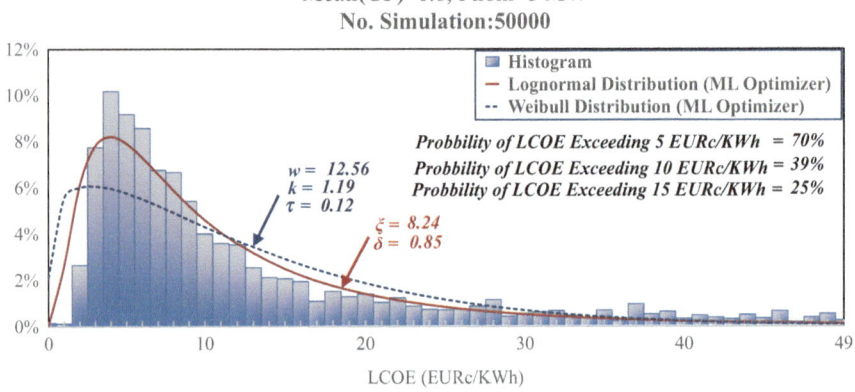

Fig. 3 Log-normal and Weibull fits to the simulated LCOE distribution histogram; wind farm with 250 wind turbines with a nominal capacity of 5 MW; capacity factor $\gamma_c = N(0.6, 0.2)$; Monte Carlo simulation (5.0E + 4); characteristic LCOE (P_c) assumed to be 5, 10, or 15 EURc/KWh

Weibull (minimum) and log-normal distribution by means of a maximum likelihood (ML) optimizer.

Figure 2 shows the results of MCS with 10^4 simulations using six random variables as outlined in Table 2. It is evident that the log-normal distribution provides a more precise fit compared to a Weibull distribution. The probability of surpassing the characteristic design value is also depicted in Fig. 2. To confirm the adequacy of the number of simulations for this analysis, another analysis with 5×10^4 simulations was performed. The result is shown in Fig. 3. The findings indicate that 10^4 simulations are adequate for comparing cases, as the percentage of exceeding the characteristic value only marginally decreased.

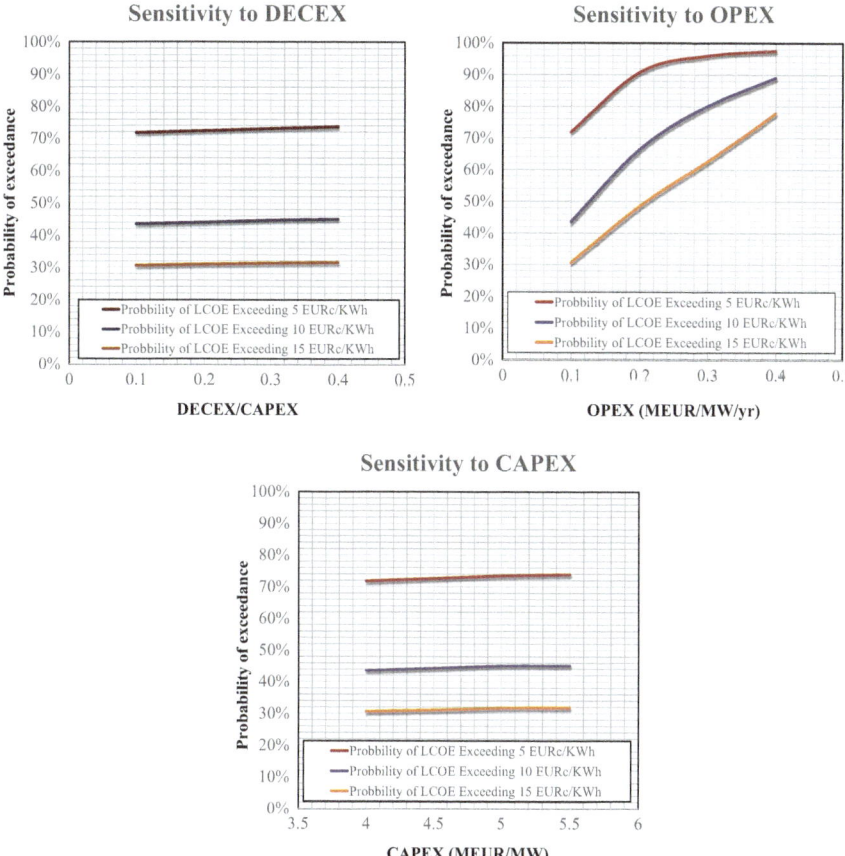

Fig. 4 Sensitivity analysis of variations in expenditures

Sensitivity Studies Sensitivity analysis was conducted to assess the impact of expenditure assumption on LCOE. Sensitivity involves variation in DECEX/CAPEX ranging from 0.1 to 0.4, OPEX ranging from 0.1 to 0.4 MEUR/MW/yr, and CAPEX ranging from 4.0 to 5.5 MEUR/MW. The results are shown in Fig. 4. It is evident that the probability of LCOE exceeding the characteristic value is not significantly affected by variations in the DECEX/CAPEX and CAPEX values. However, for OPEX, it is observed that LCOE is sensitive to an increase in annual OPEX. However, the probability of LCOE exceeding 5 EURc/KWh becomes less sensitive when OPEX is increased beyond 0.3 MEUR/MW/yr.

4 Conclusions and Recommendations

A life cycle assessment methodology was proposed for a typical wind farm. By applying this methodology, the global warming potential (GWP) of the wind farm was calculated to be 36.5 kg CO_2/MWh. In general, a good agreement was observed with the literature regarding CAPEX, DECEX, and OPEX. The results highlight the significance of different types of expenditures. A new methodology was introduced for probabilistically estimating the levelized cost of energy (LCOE) of the wind farm. Monte Carlo simulations were performed to predict the distribution of LCOE. The assessments were based on simplified assumptions regarding uncertainties in the random variables. Sensitivity studies were also conducted to evaluate the impact of assumptions in the types of expenditures on LCOE.

Based on the assumptions made in this analysis, the calculations indicate that LCOE will not be significantly affected by varying the level of DECEX/CAPEX or by increasing the amount of CAPEX. For OPEX, the probability of LCOE exceeding a defined characteristic LCOE (P_c) increases by increasing the level of OPEX from 0.1 to 0.4 MEUR/MWh/yr. Other wind farms may yield different results. It is important to note that the results are dependent on the assumptions made regarding the uncertainty in the capacity factor, expenditure estimation, error in predicting wind farm efficiency, and wake losses, as actual data for these parameters are limited.

The methodology presented in this chapter will be useful for determining the feasibility of wind farms during preliminary studies. More effort should be made to gather and organize investment data and address the uncertainties associated with them. By conducting a more accurate evaluation of the LCOE distribution, a more informed decision can be made to determine the viability of offshore wind farms.

Acknowledgment The authors acknowledge the financial support received from the University of South-Eastern Norway.

References

1. Poudineh, R., Brown, C., Foley, B.: Economics of Offshore Wind Power. Springer International Publishing, Cham (2017). https://doi.org/10.1007/978-3-319-66420-0
2. Bhandari, R., Kumar, B., Mayer, F.: Life cycle greenhouse gas emission from wind farms in reference to turbine sizes and capacity factors. J. Clean. Prod. **277**, 123385 (2020). https://doi.org/10.1016/j.jclepro.2020.123385
3. Alsubal, S., Alaloul, W.S., Musarat, M.A., Shawn, E.L., Liew, M.S., Palaniappan, P.: Life cycle cost assessment of offshore wind farm: Kudat Malaysia case. Sustainability (Switzerland). **13**(14) (2021). https://doi.org/10.3390/su13147943
4. Hesan, M.: Life cycle assessment of an NPK fertilizer production with the focus on principal harmful substances. Master thesis, University of South-Eastern Norway. [Online]. Available: https://openarchive.usn.no/usn-xmlui/handle/11250/3077225 (2023). Accessed 05 Mar 2024
5. ISO 14044:2006: ISO. [Online]. Available: https://www.iso.org/standard/38498.html. Accessed 6 Mar 2024
6. Bang, J.-I., Ma, C., Tarantino, E., Vela, A., Yamane, D.: Life Cycle Assessment of Greenhouse Gas Emissions for Floating Offshore Wind Energy in California. University of California Santa Barbara (2019)
7. Garcia-Teruel, A., Rinaldi, G., Thies, P.R., Johanning, L., Jeffrey, H.: Life cycle assessment of floating offshore wind farms: an evaluation of operation and maintenance. Appl. Energy. **307**, 118067 (2022). https://doi.org/10.1016/j.apenergy.2021.118067

8. Chitteth Ramachandran, R., Desmond, C., Judge, F., Serraris, J.-J., Murphy, J.: Floating wind turbines: marine operations challenges and opportunities. Wind Energy Sci. **7**(2), 903–924 (2022). https://doi.org/10.5194/wes-7-903-2022
9. Kaiser, M.J., Snyder, B.: Offshore Wind Energy Installation and Decommissioning Cost Estimation in the U.S. Outer Continental Shelf. Energy Research Group, LLC/Minerals Management Service, Baton Rouge/Herndon, PB2011103554 (2010). [Online]. Accessed 18 Mar 2024
10. Wind Energy Engineering: A Handbook for Onshore and Offshore Wind Turbines, 2nd edn. Editor: Trevor Letcher, Language: English, Hardback ISBN: 9780323993531, eBook ISBN: 9780323958301 (2023)
11. IRENA Insights on Renewables: Installed capacity trends: Available online: https://www.irena.org/Data (2022)
12. Pantusa, D., Francone, A., Tomasicchio, G.R.: Floating offshore renewable energy farms. A life-cycle cost analysis at Brindisi, Italy. Energies. **13**, 6150 (2020)
13. Melchers, R.E., Beck, A.T.: Structural Reliability Analysis and Prediction, 3rd edn, p. 528. ISBN: 978-1-119-26599-3 (2018)
14. Faber, M.H.: Statistics and Probability Theory Topics in Safety, Risk, Reliability and Quality, 1st edn, p. 192. Springer. ISBN: 978-94-017-8530-3 (2012). https://doi.org/10.1007/978-94-017-4056-3
15. Structural reliability analysis programs: STRUREL. http://www.strurel.de/index.html
16. Yildiz, N., Hemida, H., Baniotopoulos, C.: Operation, maintenance, and decommissioning cost in life-cycle cost analysis of floating wind turbines. Energies. **17**, 1332 (2024). https://doi.org/10.3390/en17061332
17. Ali, M., Matevosyan, J., Milanović, J.V.: Probabilistic assessment of wind farm annual energy production. Electr. Pow. Syst. Res. **89**, 70–79, ISSN 0378-7796 (2012). https://doi.org/10.1016/j.epsr.2012.01.019
18. Rubert, T., McMillan, D., Niewczas, P.: A decision support tool to assist with lifetime extension of wind turbines. Renew. Energy. **120**, 423–433, ISSN 0960-1481 (2018). https://doi.org/10.1016/j.renene.2017.12.064
19. Miller, L., Carriveau, R., Harper, S., Singh, S.: Evaluating the link between LCOE and PPA elements and structure for wind energy. Energy Strat. Rev. **16**, 33–42 (2017). https://doi.org/10.1016/j.esr.2017.02.006

Sustainable and Carbon Neutral Built Environment Through ECBC Compliance

Anil Kumar and Rohit Thakur

Contents

Nomenclature

A_G	Area of windows
A_R	Area of the roof
A_W	Area of the wall
c_{1Fenes}	Fenestration U-factor coefficient
c_{Roof}	Coefficient value for the roof
c_{wall}	Coefficient value for the wall
EPF_{Fenest}	Envelope performance factor for fenestrations
EPF_{Roof}	Envelope performance factor for roofs
EPF_{wall}	Envelope performance factor walls

A. Kumar (✉)
Department of Mechanical Engineering, Delhi Technological University, Delhi, India

Division-Clean Energy: Nodal Centre of Excellence in Energy Transition (NCEET), Delhi Technological University, Delhi, India
e-mail: anilkumar76@dtu.ac.in

R. Thakur
Department of Mechanical Engineering, Delhi Technological University, Delhi, India

© The Author(s) 2025
M. Kioumarsi, B. Shafei (eds.), *The 1st International Conference on Net-Zero Built Environment*, Lecture Notes in Civil Engineering 237,
https://doi.org/10.1007/978-3-031-69626-8_146

SEF_G	A multiplier for the window SHGC that depends on the projection factor of an overhang or side-fin
$SHGC_G$	The solar heat gain coefficient for windows
U_G	The U-factor for the window
U_R	The U-factor for the roof
U_W	The U-factor for the wall

1 Introduction

The rapid urbanization of modern society has prompted a global concern over energy consumption. Since the first energy crisis in the 1970s, there has been a substantial international emphasis on researching building energy efficiency [1]. Building operations account for 30% of global final energy consumption and 26% of energy-related emissions. Of these, 8% are direct emissions in buildings, and 18% are indirect emissions from the generation of electricity and heat that is utilized in buildings. While severe temperatures drove increased heating-related emissions in specific places, direct emissions from the building sector dropped in 2022 compared to the previous year. This was the case, notwithstanding climate change. The energy consumption of the building industry grew by around 1% in 2022 [2]. With a commitment to reducing emissions intensity, India aims to achieve a 33–45% decrease in national gross domestic product (GDP) emissions by 2030 compared to 2005 levels. Despite the impending construction of 70% of the required structures by 2030, this sector continues to contribute significantly to greenhouse gas emissions [3]. Buildings in India currently consume around 34% of energy, with commercial buildings accounting for 8.29% of total consumption [4]. By 2030, we expect the total built-up area of commercial buildings to reach 1.9 billion square meters, a threefold increase from the 2015 figure of 847 million square meters [5]. India's pioneering energy conservation building code (ECBC) aims to achieve energy neutrality in commercial buildings within the Indian context [6]. The building envelope plays a crucial role in heat transfer into a building, impacting the energy load in conditioned buildings. To address heat transmission challenges and facilitate ECBC adoption in the nation, suitable construction solutions for opaque exterior walls are imperative.

Thermal transmittances of walls are vital thermophysical properties significantly influencing a building's energy efficiency and play a critical role in determining a building's energy needs and understanding heat loss [7, 8]. They regulate heat exchange with the external environment, impacting the energy demands of heating, ventilation, and air conditioning (HVAC) systems [9–11]. Despite insufficient ECBC implementation and slow code acceptance in India, the long-term viability of these regulations depends on market sustainability. The deployment of ECBC and other energy efficiency measures has the potential to save a large amount of energy in buildings that are located in India [12, 13]. Drywall, also known as plasterboard, wallboard, gypsum board, or gyprock, is a composite panel made of gypsum plaster sandwiched between two robust paper sheets. With nearly half the mixture consisting of paraffin capsules, the new drywall incorporates minuscule paraffin beads to collect and emit heat, representing a significant advancement in sustainable construction technology. This versatile construction method, weighing eight to ten times less than traditional systems, offers faster construction while providing a smooth finish and crack-free surfaces [14, 15]. Additionally, up to 20% of the gypsum used in drywall production can be repurposed from waste generated at manufacturing facilities or construction sites. This study employs an evolutionary optimization model to evaluate the suitability of various materials for assessing thermal transmittance in commercial building opaque wall assemblies. The goal is to ensure ECBC compliance in different Indian climatic zones, promoting wider acceptance within the building industry. This information is valuable for practitioners conducting energy audits and scholars seeking contemporary insights. ECBC for commercial buildings in 2017 specifies U-factor or thermal transmittance requirements in a parametric framework for different climatic zones in India. Compliance levels include ECBC, ECBC+, and ECBC Super, each offering varying energy efficiency improvements. This research provides an overview of compliance requirements and energy-saving uses of plasterboard, emphasizing its sustainable growth and administrative advantages on building sites.

2 Methodology

Every new building greater than a connected load of 100 kW is required to adhere to the energy conservation building code (ECBC) compliance procedure, encompassing a set of obligatory regulations related to energy utilization in the building. Following consultation with experts on ECBC and referencing the ECBC user guide, diverse methods for achieving ECBC compliance have been delineated. To allow for a degree of flexibility in design and construction, the code compliance requirements can be satisfied through the following approaches as deemed appropriate (see Fig. 1) [16].

2.1 Prescriptive Approach

The prescribed technique establishes the required minimum energy efficiency features for different components and systems within the envisioned buildings. Specific standards are outlined for the building envelope, HVAC systems, service hot water and pumps, lighting systems, and electrical power. Compliance can be achieved by

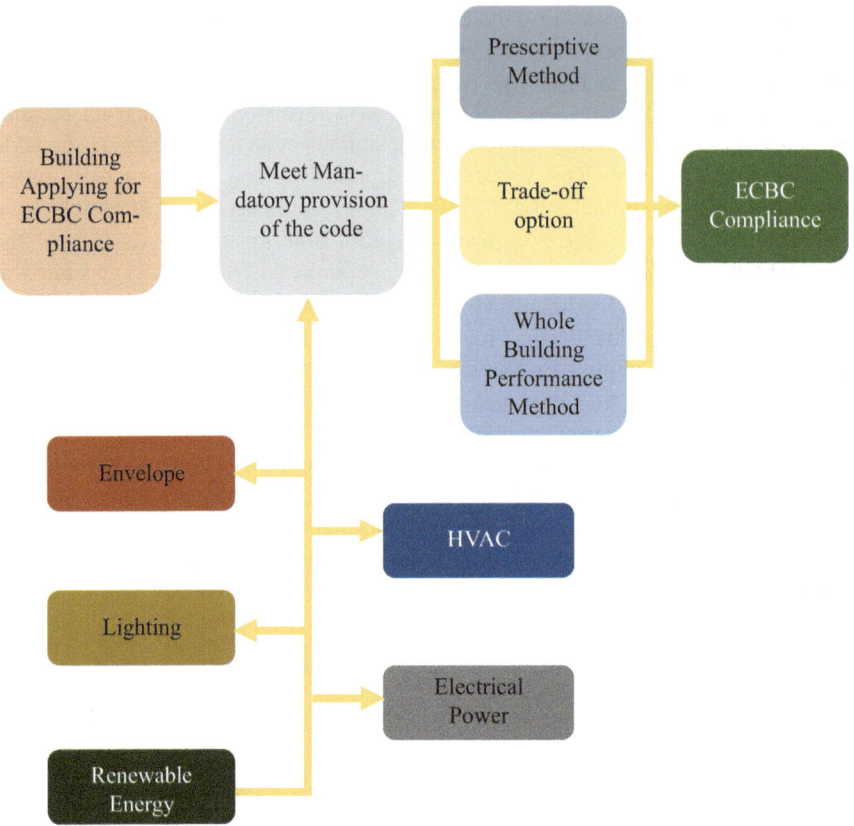

Fig. 1 Methods for ECBC code compliance

meeting the fundamental requirements outlined in the code. However, the code also allows for surpassing these standards, leading to enhanced energy efficiency. The energy conservation building code (ECBC) introduces the trade-off option for the building envelope, enabling a trade-off between the efficacy of one envelope component and another without compromising the overall efficiency level stipulated by the code.

2.2   Building Envelope Trade-Off Approach

This approach utilizes a systems-based method, enabling a decrease in the thermal performance standards of specific exterior components while enhancing the efficiency of other building elements. For example, choosing higher wall insulation could potentially ease the U-factor requirement for windows, and vice versa. Trade-offs typically apply to significant building components such as roofs, walls, fenestration, overhangs, etc. Compared to adhering to recommended values for each element, this strategy provides more flexibility in building design. However, trade-offs are limited to building envelope components, and adjustments to lighting or HVAC systems, for instance, are not feasible through this approach. Consequently, opting for the trade-off choice necessitates more computations than selecting the prescriptive approach. The surface area of each external and semi-exterior surface must be individually calculated, with computations carried out separately for each direction.

The building envelope is considered compliant with the code when the envelope performance factor (EPF) of the proposed building is lower than that of the standard building, which strictly adheres to the prescriptive requirements for the building envelope. This approach does not apply to buildings with a window-to-wall ratio (WWR) exceeding 40%. The trade-off method does not include provisions for skylights. To ensure compliance with ECBC using the trade-off method, the calculation of the envelope performance factor should be conducted in accordance with Eqs. (1) through (4) [6]. EPF of the envelope is the summation of the EPF of the roof, wall and fenestration. The EPF of both the wall and roof is assessed by taking into account the individual area and thermal transmittance values of each component. Additionally, a multiplication coefficient is applied to adjust for the influence of the local climate. Similarly, to evaluate the EPF of fenestration components, coefficients are employed for each direction, and other parameters such as thermal transmittance, area, and solar heat gain coefficient (SHGC) considered.

$$EPF_{Total} = EPF_{Roof} + EPF_{Wall} + EPF_{Fenest} \tag{1}$$

$$EPF_{Roof} = C_{Roof} \sum_{s=1}^{n} U_R A_R \tag{2}$$

$$EPF_{Wall} = C_{Wall} \sum_{s=1}^{n} U_W A_W \tag{3}$$

$$
\begin{aligned}
EPF_{Fenest} = {}& C_{1Fenset,North} \sum_{w=1}^{n} U_G A_G + C_{2Fenset,North} \sum_{w=1}^{n} \frac{SHGC_G}{SEF_G} A_G \\
& + C_{1Fenset,South} \sum_{w=1}^{n} U_G A_G + C_{2Fenset,South} \sum_{w=1}^{n} \frac{SHGC_G}{SEF_G} A_G \\
& + C_{1Fenset,East} \sum_{w=1}^{n} U_G A_G + C_{2Fenset,East} \sum_{w=1}^{n} \frac{SHGC_G}{SEF_G} A_G \\
& + C_{1Fenset,West} \sum_{w=1}^{n} U_G A_G + C_{2Fenset,West} \sum_{w=1}^{n} \frac{SHGC_G}{SEF_G} A_G
\end{aligned}
\tag{4}
$$

2.3   Whole Building Performance Approach

An alternative method for ECBC compliance is the Whole Building Performance (WBP). Under this approach, a computer-based hourly energy simulation model of the proposed design is created, and its energy usage is compared with that of a typical design to ensure compliance with ECBC, as illustrated in Fig. 2. The simulation encompasses thermal, visual, ventilation, and other energy-consuming aspects of the building, factoring in its orientation, construction materials, façade, climatic conditions, and internal and external environments. Following all the necessary prescriptive criteria of the code, if the energy consumption in the standard design aligns with the maximum allowed for that specific building, compliance is achieved. With the WBP technique, a "Proposed Design" needs to be modelled, its annual energy usage calculated, and then compared to the "Standard Design" using their Energy Performance Index (EPI). Additionally, the simulations must adhere to the ECBC's standard energy modelling technique.

2.4   Building Representation

This study examined the energy consumption of a hospitality building using eQUEST software. The utilization of plasterboard technology has seen substantial growth in the Western world over the past decades. However, its adoption in India remains limited due to a lack of awareness. Considering that nearly 70% of the building stock expected by 2030 is yet to be constructed in India, plasterboard

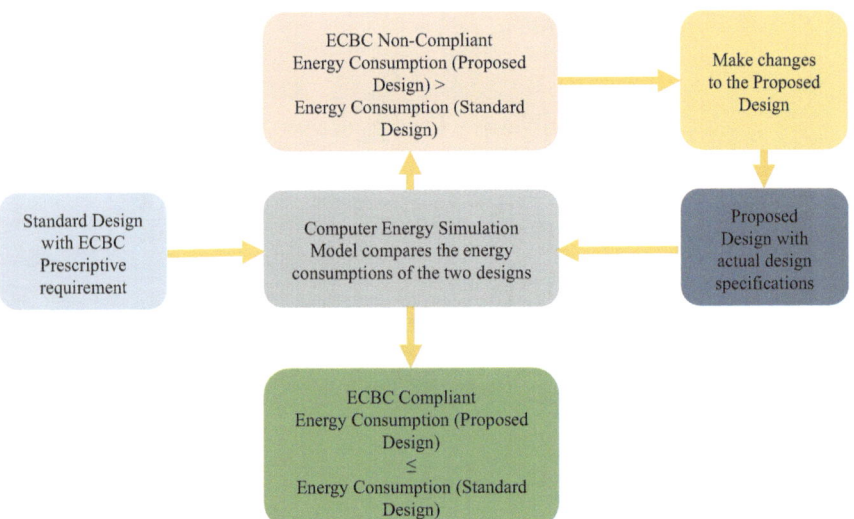

Fig. 2 Whole building performance approach

technology emerges as a viable option for sustainable development. This technology can replace traditional systems, reducing the reliance on fossil fuels, electricity, and emissions associated with building energy use. Additionally, these systems are introduced as environmentally friendly materials, with a focus on recycling the majority of used materials, thereby diminishing environmental pollutants after the building's demolition. A zoning plan was developed for every floor and integrated into the simulation model as illustrated in Fig. 3. Each zone received a specific set of characteristics, encompassing lighting power density, equipment power density, occupant density, infiltration rate, outside air requirement, and an occupancy schedule. Additionally, each zone was attributed physical properties, such as floor-to-floor height, material density, conductivity, and fenestration area. The occupancy schedule for these zones has been referenced from ECBC [6]. The overall built-up area considered for the analyses is 26,315 m². Table 1 elucidates the diverse composition of materials under consideration, while Table 2 presents the thermal properties of these compositions and other input parameters employed in the analysis.

Fig. 3 3D-view showing model graphic rendering

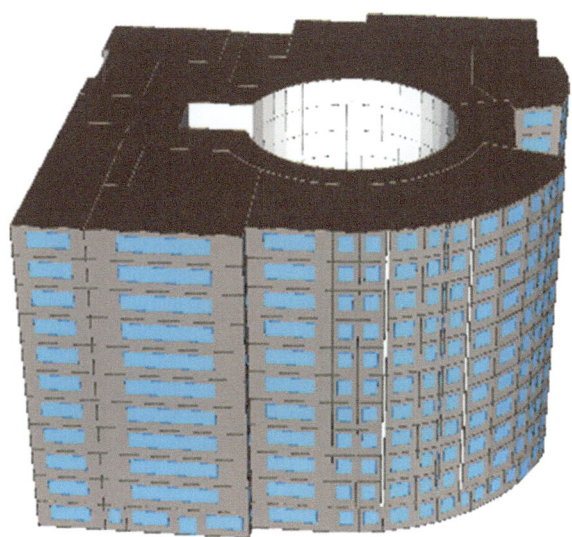

Table 1 Different external and internal wall composition

Case no.	External wall	Internal wall
Case 1	Fired clay brick (FCB)	Fired clay brick (FCB)
Case 2	Autoclaved aerated concrete (AAC)	Autoclaved aerated concrete (AAC)
Case 3	Autoclaved aerated concrete (AAC)	Plaster board (PB)
Case 4	Plaster board (PB)	Plaster board (PB)
Case 5	Fired clay Brick (FCB)	Plaster board (PB)
Case 6	Plaster board (PB)	Autoclaved aerated concrete (AAC)
Case 7	Plaster board (PB)	Fired clay brick (FCB)

Table 2 Input parameters for simulation

	Case						
	1	2	3	4	5	6	7
Exterior loads							
Exterior walls U-value (W/m²·K)	2.11	2.11	0.37	0.37	0.67	0.67	0.37
Roof U-value (W/m²·K)	0.33	0.33	0.33	0.33	0.33	0.33	0.33
Ground floor U-value (W/m²·K)	1.33	1.33	1.33	1.33	1.33	1.33	1.33
First floor U-value (W/m²·K)	2.05	2.05	2.05	2.05	2.05	2.05	2.05
Partition wall U-value (W/m²·K)	2.24	0.61	2.24	0.61	1.05	0.61	1.05
Glazing							
Window wall ratio	31.6%						
Glass—U-value (W/m²·K)	3						
Glass—shading coefficient (SC)	0.27						
Glass—visual light transmittance (VLT)	27%						
Internal loads							
Lighting power density (LPD) (W/m²)	11						
Equipment power density (EPD) (W/m²)	10						
HVAC system							
HVAC system type	Fan coil unit (FCU) & variable air volume (VAV)						
	COP—5.2 water cooled chiller						

3 Results and Discussion

The study involved creating a model with consistent dimensions but different combinations of external and internal wall materials, followed by performing energy simulations specific to New Delhi. The subsequent sections present the analyzed data from diverse material compositions. Figure 4 illustrates the monthly energy consumption for different scenarios, considering traditional materials such as fired clay brick (FCB), plasterboard (PB), and autoclaved aerated concrete (AAC). Based on the findings of the investigation, it is evident that plaster boards and AAC materials use less energy when compared to fire clay bricks.

As shown in Fig. 5, the building that was created using PB on both the exterior and internal walls was able to save 9% of the total amount of energy demand. This is in comparison to the structure that was made with conventional construction materials, which included FCB on both the exterior and interior walls. Additionally, in comparison to the conventional type of building, the building that was constructed with PB on the exterior and FCB on the interior walls system was able to save 8.03% of the desired amount of energy. When compared to a structure that was created using conventional construction materials, it was discovered that the energy demand reduction in the case of FCB on external walls and PB on inner walls was 1.17%. When compared to the structure that was created using conventional construction materials, such as FCB on both the exterior and interior walls, the energy reduction was determined to be 8.69% when the building was constructed using PB on the outside and AAC block on the inner walls. Buildings that are constructed with AAC

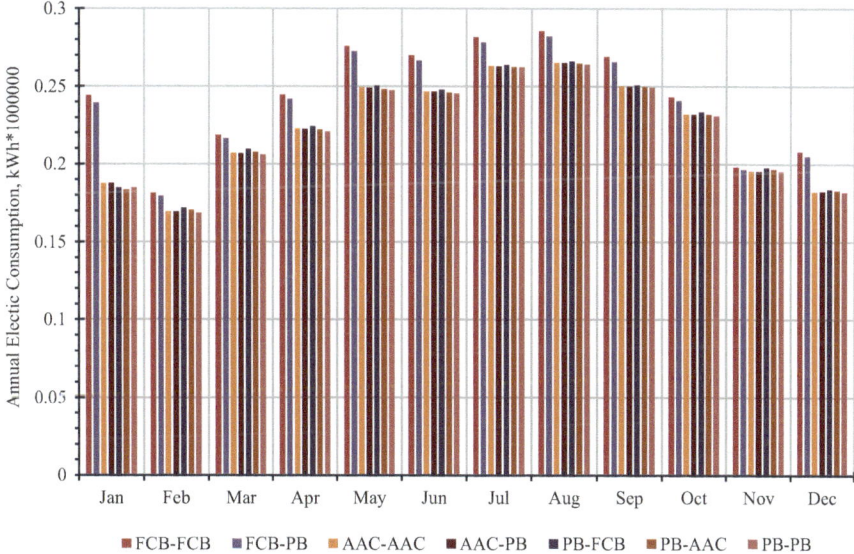

Fig. 4 Monthly energy comparison of different exterior and interior wall configurations

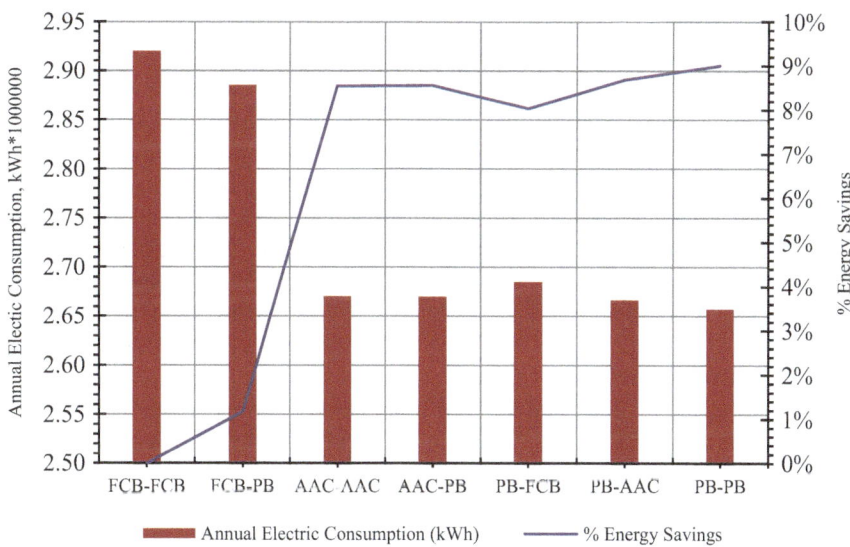

Fig. 5 Annual energy comparison of different exterior and interior wall configurations

blocks on both the exterior and interior as well as the walls save 8.54% of the amount of energy demand when compared to buildings that are constructed with traditional building materials. On the other hand, buildings that are constructed with AAC blocks on the exterior and PB blocks on the interior and walls save 8.56% of the amount of energy demand when compared to buildings that are constructed with traditional building materials.

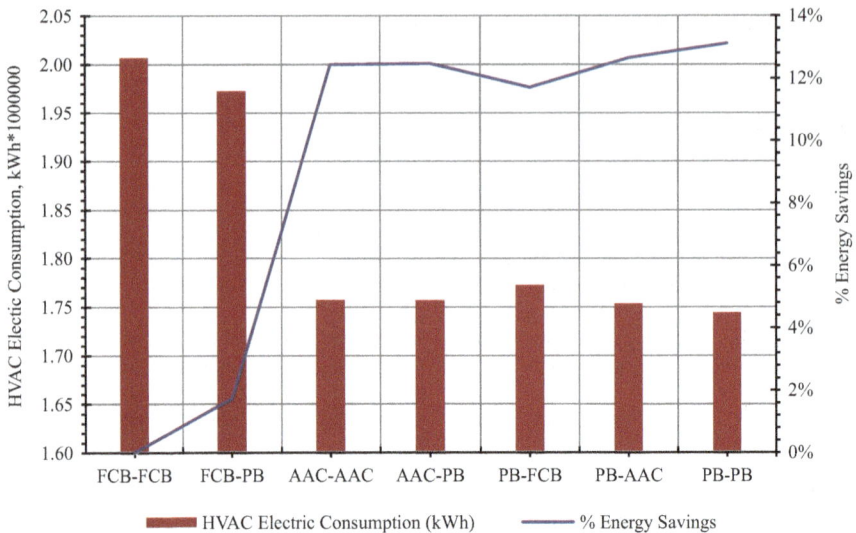

Fig. 6 Annual HVAC energy comparison of different exterior and interior wall configurations

As can be seen in Fig. 6, the structure that was built with PB for both the exterior and internal walls demonstrated a 13.10% reduction in the energy demand for heating, ventilation, and air conditioning (HVAC) in comparison to the structure that was built with conventional materials such as FCB on both sides. Additionally, in comparison to the traditional construction, the building that used PB on the outer walls and FCB on the interior walls was able to achieve an 11.69% reduction in the amount of energy that was required for HVAC.

When the energy demand for HVAC systems was compared to the building that was constructed using standard construction materials, the combination of FCB on the outside walls and PB on the interior walls resulted in a decrease of 1.71%. Furthermore, in comparison to the standard FCB construction, the use of PB on the outer walls and AAC block on the internal walls led to a reduction of 12.64% in the amount of energy required for HVAC.

When compared to conventional construction materials, a structure that was made completely out of AAC block, both on the outside and the inside, demonstrated a 12.43% reduction in the amount of energy used for HVAC. When compared to the traditional FCB construction, the building that had AAC block on the outside and PB on the inside walls was able to save 12.45% of the energy consumption for heating, ventilation, and air conditioning.

4 Conclusion

The utilization of plasterboard technology has witnessed a substantial surge in the Western world over recent decades. However, in India, its adoption remains limited primarily due to a lack of awareness. This innovative technology has the potential to replace traditional building systems, thereby mitigating the consumption of fossil fuels, electricity, and emissions associated with energy use in buildings. Moreover, these systems are introduced as environmentally friendly materials, emphasizing the majority of the used materials can be recycled, reducing environmental pollutants following building demolition.

The findings of this study underscore a significant reduction in electricity consumption achieved by implementing a plasterboard system instead of traditional brick materials in the exterior walls of hotel buildings across different climatic zones in India.

Beyond reducing energy consumption, the use of plasterboards also facilitates the effective implementation of Energy Conservation Building Codes. Traditional materials, such as fired clay brick with a total thickness of 280 mm (25 mm plaster + 230 mm FCB + 25 mm plaster), exhibit a thermal transmittance of 2.11 $W/m^2 \cdot K$, surpassing the values stipulated in the codes. In contrast, the thermal transmittance of AAC blocks with similar thickness and layer composition is 0.67 $W/m^2 \cdot K$. Furthermore, plasterboards with a total thickness of 157 mm demonstrate a thermal transmittance of 0.37 $W/m^2 \cdot K$, falling below the prescribed value of 0.40 $W/m^2 \cdot K$ specified in the codes. This highlights the potential of plasterboard technology not only in enhancing energy efficiency but also in aligning with and exceeding regulatory standards for sustainable construction practices. This study demonstrated a noteworthy 9% decrease in electricity consumption when employing a plasterboard system instead of traditional brick materials for the exterior walls of hotel buildings. Furthermore, there was a 13% reduction in cooling load by replacing traditional bricks with plasterboards.

Acknowledgement The authors express sincere gratitude to the Department of Mechanical Engineering and the Centre for Energy and Environment at Delhi Technological University, Delhi, India, for providing the necessary facilities to compile this work.

References

1. International Energy Agency: Global energy crisis. https://www.iea.org/topics/global-energy-crisis. Last accessed 24 Feb 2024
2. International Energy Agency: Buildings. https://www.iea.org/energy-system/buildings. Last accessed 24 Feb 2024
3. Government of India: India's Updated First Nationally Determined Contribution Under Paris Agreement (2021–2030). Submission to UNFCCC (August 2022)
4. Government of India: Energy Statistics India 2023. Ministry of Statistics and Programme Implementation National Statistical Office (2023)
5. TERI: Green Growth and Buildings Sector in India, 39 pp. The Energy and Resources Institute, New Delhi (2015)
6. Bureau of Energy Efficiency: Energy Conservation Building Code 2017 (With Amendments Upto 2020). Bureau of Energy Efficiency, Ministry of Power, Government of India (2017)
7. Yilmaz, Z.: The limit U values for building envelope related to building form in temperate and cold climatic zones. Build. Environ. **37**, 1173–1180 (2002)

8. Prada, A., Cappelletti, F., Baggio, P., Gasparella, A.: On the effect of material un-certainties in envelope heat transfer simulations. Energy Build. **71**, 53–60 (2014)

9. International Organization for Standardization: ISO 7345:1987: Thermal Insulation – Physical Quantities and Definitions. ISO, Geneva (1987)

10. Park, K., Kim, M.: Energy demand reduction in the residential building sector: a case study of Korea. Energies. **10**, 1–11 (2017)

11. Battista, G., Evangelisti, L., Guattari, C., Basilicata, C., de Lieto Vollaro, R.: Buildings energy efficiency: interventions analysis under a smart cities approach. Sustainability. **6**, 4694–4705 (2014)

12. Chedwal, R., Mathur, J., Agarwal, G.D., Dhaka, S.: Energy saving potential through Energy Conservation Building Code and advance energy efficiency measures in hotel buildings of Jaipur City, India. Energy Build. **92**, 282–295 (2015) ISSN 0378-7788

13. Tulsyan, A., Dhaka, S., Mathur, J., Yadav, J.V.: Potential of energy savings through implementation of Energy Conservation Building Code in Jaipur city, India. Energy Build. **58**, 123–130 (2013) ISSN 0378-7788

14. Hansen, S., Sadeghian, P.: Recycled gypsum powder from waste drywalls combined with fly ash for partial cement replacement in concrete. J. Clean. Prod. **274**, 122785 (2020) ISSN 0959-6526

15. Manzello, S.L., Park, S.-H., Mizukami, T., Bentz, D.P.: Measurement of thermal properties of gypsum board at elevated temperatures. In: Proceedings of the Fifth International Conference on Structures in Fire (SiF'08)

16. Bureau of Energy Efficiency: Energy Conservation Building Code, 2017 Users' Manual. Bureau of Energy Efficiency, Ministry of Power, Government of India (2017)

Nyhavna: A Harbour Area on Its Way to Climate Neutrality: Empirical Insights and Learnings for Different Stakeholders

Marianne Skaar, Lars Arne Bø, and Judith Thomsen

Contents

1   Introduction

As per reports from the IEA, OECD, and the UN Environment Programme, cities account for two-thirds of global energy consumption, 50% of global waste, and over 70% of worldwide energy-related CO_2 emissions. Key contributors to these figures are the transportation and building sectors. Projections indicate that by 2050, over 70% of the global population will reside in cities, leading to an increased demand for spaces such as housing, retail, and industry, accompanied by the requisite urban mobility and energy infrastructure [1]. Given this scenario, cities, including their buildings and neighbourhoods, present a significant opportunity for substantial carbon emissions reduction if planned with that goal in mind. A key aspect of this effort involves the reuse and transformation of existing urban areas, aiming to curtail

M. Skaar (✉) · L. A. Bø · J. Thomsen
SINTEF Community, Trondheim, Norway
e-mail: marianne.skaar@sintef.no

© The Author(s) 2025
M. Kioumarsi, B. Shafei (eds.), *The 1st International Conference on Net-Zero Built Environment*, Lecture Notes in Civil Engineering 237,
https://doi.org/10.1007/978-3-031-69626-8_147

Fig. 1 Nyhavna [6]

the necessity for new construction on undeveloped land and thus contribute to the preservation of nature and farmland [2].

This chapter investigates the transformation process of an urban harbour area, Nyhavna (New Harbour), in Trondheim, Norway, with a focus on converting it into a neighbourhood reducing its carbon emissions towards zero as defined by the Research Centre on Zero Emission Neighbourhoods in Smart Cities (ZEN) [3]. Urban transformations are complex, involving numerous stakeholders with diverse interests and ambitions. Little existing research explores how to organize and lead such a process at the neighbourhood level to achieve ambitious environmental goals like ZEN or other related ambitions such as Positive Energy Districts (PEDs) [4].

The research questions guiding this study are: How does the chosen pathway for the planning process of Nyhavna impact the harbour area's transformation into an attractive, zero emission neighbourhood? What challenges arise, and which strategies are essential for maintaining high ambitions over time? First, literature on transformation of harbour areas is presented, before giving the context of Nyhavna. Then the case study research approach is explained [5], involving a document analysis and qualitative interviews with experts engaged in the pre-early-planning process. The analysis focuses on (1) the early planning phase, (2) the different perceptions of the future of Nyhavna and (3) the early efforts to bring the plans and visions to life. The chapter contributes to understanding factors influencing climate-neutral neighbourhood development and concludes with practical recommendations for development of urban transformation projects based on the case of Nyhavna (Fig. 1).

2   Transformation of Harbour Areas: Commodifying the Waterfront

The urban transformation of harbour areas into housing and retail areas is a phenomenon that has gained momentum over the last few decades. This process involves repurposing formerly industrial or underutilized harbour zones into mixed-use spaces that combine residential, commercial, and recreational elements. The factors that have contributed to the emergence of this trend are multiple. Many harbour areas were historically industrial zones, hosting shipping ports, warehouses, and factories. As cities have shifted away from heavy industry, these waterfront areas became obsolete, prompting the need for alternative land use. The harbour areas are often centrally located and increasing populations in urban areas demand living space. Waterfront areas became attractive, and usually as well economically profitable and often expensive living spaces [7]. Over the last 50 years, urban regeneration has evolved from a focus on physical transformation to a more comprehensive vision that emphasizes improving the quality of life for inhabitants as well as an environmental focus, highlighting the process as placemaking an existing urban context alongside sustainable effects [8].

In the Norwegian context, Aker Brygge and the subsequent waterfront developments in Oslo, have been criticized for becoming neighbourhoods for the affluent. This is also something that is feared at Nyhavna [9]. Despite the initial aspirations, the Fjordbyplan ("Fjord-city-plan", approved 2008) has not realized its goal of social diversity in the Oslo harbour area [10]. Oslo, like many other cities, has witnessed a shift towards market-driven urban planning strategies, prioritizing economic development, private investments, and the market forces in shaping the cityscape. Røe [10] suggests that the implemented strategies may not have effectively addressed the socio-economic and cultural diversity desired for the district. This discrepancy can be attributed to various factors, such as market dynamics, policy implementation challenges, or unforeseen consequences. One significant aspect highlighted by Røe [10] is that the Fjordbyplan was not legally binding. This lack of legal obligation might have allowed for deviations from the initial vision during the implementation phase, allowing developers and investors to prioritize profit-driven projects over the inclusivity and diversity goals outlined in the plan. This suggests that the planning decisions were rather influenced by economic considerations than by ideas of social equity and diversity.

Andersen and Skrede [11] demonstrate how private interests are 'de facto' city makers in the Norwegian planning regime, while the municipalities' planning authority mostly have a regulatory role. It is a fact that planning proposals are mainly submitted by private corporations, pursuing own interests, and executing the plans. Andersen and Skrede [11] describe this planning regime as adhering to a neoliberal logic, where the strong position of private interests is shaping the city development. In their analysis of the municipal master plan in Oslo, they state (in line with Røe [10]) that the municipality has failed to realize its ideals of a socially just municipality, due to the strong positions of private interests in city development. This is despite Norway being an affluent and equality-oriented country [11]. One example Andersen and Skrede [11] describe, is the development at Furuset, located East in Oslo. Here, the municipality planned for an ambitious environmental concept, but experienced at the time a lack of interest from the private housing developers. The developers' perceptions were that the potential buyers of this area, with a lower socio-economic status, would not have the ability to pay the accompanying extra costs. These kinds of concepts for buildings were regarded by them as more suitable for a waterfront development, where they estimated a higher profit due to more affluent customers.

Stakeholders may have competing interests and different conceptions of what an 'inclusive and attractive', 'zero-emission', and 'sustainable neighbourhood' is. The complexity of competing conceptions is explored by Guy and Farmer [12] showing how different logics of ecological placemaking is competing in representations of sustainable architecture. Through a study of buildings and a literature review of books, articles, and reports, Guy and Farmer [12] identify six distinct and competing conceptions of ecological placemaking. They treat these competing views as different environmental discourses, not as static concepts. Of these, three are relevant to this chapter: the eco-technical, the eco-centric, and the eco-social. The eco-technical logic suggests that the environmental problems can be addressed through science and technological- and structural solutions. In contrast, both eco-centric and eco-social logics, emphasize the need for a fundamental change of values to solve environmental issues. The eco-centric view is more individual and lifestyle oriented, with a perception that the built environment in general has a negative environmental impact. The eco-social advocates for a community model that serves common needs and goals.

Guy and Farmer studied the social construction of sustainable architecture at building level, but the concepts can also be a fruitful lens on sustainable urban planning [13] and representations of sustainable neighbourhoods [14]. Kittang [13] illustrates the contradictions between the municipality's eco-technical discourse conflicting with consultants and architects more eco-centric logics in the development of a low-emission neighbourhood in Trondheim.

Urban transformations are complex, affecting the social, institutional, cultural, political, economic, technological, and ecological context of an area [8]. Multiple actors such as local governments, companies, entrepreneurs, households, as well as regional or national governments represent different roles, interests, and loyalties, that shape an urban development [8]. Against this backdrop, it's evident becomes why there is no simple picture of an urban transformation process that all would agree on.

3   The Case of Nyhavna

In Trondheim, Nyhavna's transformation is the city's third urban waterfront development, following Nedre Elvehavn (1990–2016) and the ongoing Brattøra port. Nyhavna was established in the end of the nineteenth century and is a diverse port- and industrial area near the waterfront, just a 15-minutes' walk from the historical city centre. The peculiarity of the area are the submarine bunkers from WWII, Dora 1 and Dora 2. Around 2000 people have their workplace at Nyhavna today [15]. Warehouses and large industry buildings are in use by both traditional- and small-scale industries, maritime businesses, and various creative industries. Many of the companies present at Nyhavna have long term lease agreements, the longest lasts until 2120.

For understanding the planning process, it is relevant to know that there are three major landowners at Nyhavna: Trondheim Municipality, Trondheim Port Authority and Bane NOR Eiendom AS. The port authority is a self-financed intermunicipal company. Bane NOR Eiendom AS is a state-owned company responsible for the national railway infrastructure. Changes in the Norwegian port legislation ('Havne-og farvannsloven') in 2020 allowed for the redeveloped of port areas. This led to the establishment of Nyhavna Utvikling AS in January 2021, a real estate company with two shareholders: Trondheim municipality (67%) and Trondheim Port Authority (33%). Their properties at Nyhavna are demerged into Nyhavna Utvikling AS. Essentially, most of the plots (ca 85%) within the municipal sub-plan are publicly owned, but the ownerships are organized as AS/companies.

The Planning and Building Act determines how the country's land should be used and regulated. Spatial planning is important to ensure that land use is purposeful, efficient, and rational. A typical municipal planning hierarchy starts with the municipal master plan, which includes both a societal part and a land use part. The land use part should show where development can take place in the municipality and which areas should be used for what purposes. The next step is a municipal sub-plan or an area regulation, depending on whether it is a private or public developer. These are used if the area to be regulated encompasses multiple properties or larger areas. The final step in spatial planning is detailed regulation for each individual sub-area. For all types of plans under the Planning and Building Act, it applies that only mapped information, along with any accompanying regulations, can have legal effect. Like most major port areas, Nyhavna follows this structure, and in addition, there is developed an ambitious Quality program (QP, in the following) [16] to guide the development of Nyhavna. A Quality program for urban development is a tool used to plan and manage the development of urban areas with a focus on ensuring high quality in design and functionality, often aiming to integrate various aspects of urban development, including architectural design, infrastructure, sustainability, social inclusion, and economic viability. The application of such programs can vary depending on local conditions, legal frameworks, and political priorities. The QP at Nyhavna is advisory, not legal, and forms the basis for further work on plans and initiatives in Nyhavna [16].

4   Methodology

This study of the case of Nyhavna started off by unravelling the threads of the process that lead to description of the future vision for Nyhavna in the QP. The main stakeholders involved in the pre-planning phase of the area were identified. These were relevant for explaining the process until this milestone and the way envisioned ahead for the early planning phase.

Qualitative interviews were conducted with seven respondents in six interviews with main players influencing this stage of the process represented—the Municipality (planners), the plot owners and a politician. Qualitative interviews are a useful method when there is not only one answer to a question and when the intention is to inquire on people's personal experiences and motivations [17]. The interview data were transcribed and then manually coded and analysed with the approach of a thematic analysis [18]. The data of the interview is supplemented through information from documents and text analysis of public available information, such as newspaper articles, planning documents, and public hearing documents, validating information from the interviews from a different angle [5].

5 Planning for an Ambitious Nyhavna

5.1 Planning Process: The Municipality's High Ambitions for Nyhavna

The planning process for the transformation of Nyhavna began as early as 2009, initiated by the Building Council. The municipal sector plan has been developed by the Municipality of Trondheim in collaboration with Trondheim Port and was adopted in 2016. The purpose of the plan 'is to ensure overarching, long-term strategies and frameworks for the transformation of Nyhavna from an industrial port to urban purposes, while also safeguarding long-term needs for port activities' [19]. The plan description also clearly states that Nyhavna should be developed into a densely built urban area, including housing.

Nyhavna is divided into 10 different sub-areas, and a comprehensive regulatory plan will be developed for each sub-area. There are specific requirements for the preparation of regulatory plans. Regulatory provision § 3.2, basis for all regulatory plans, states that before the first regulatory plan, a Quality program for public spaces and an environmental monitoring program must be approved by the municipality and serve as the basis for all plan proposals. Additionally, § 3.3 states that a cultural heritage program must be approved before the approval of the first regulatory plan within the preservation of cultural heritage consideration zone [19]. A municipal sector plan is standard for such areas at that time, aiming to address the structure at an overarching level while details will be addressed at the regulatory plan level for each sub-area.

In 2019, a three-year project leader position under the Municipal Director's professional staff was established to facilitate participation and communication between the different actors at Nyhavna. One of the tasks was developing a strategy- and action plan for different initiatives, and implement these, contributing to make Nyhavna a more attractive district both temporary and in the long run.

The parallel assignment 'Public Spaces in Nyhavna' [20] was commissioned by the Municipality of Trondheim and Trondheim Port Authority as a follow-up of the Municipality sub-plan. Four chosen teams delivered a wealth of ideas for the design

Table 1 Simplified process timeline

Planning process started		Project City Life Nyhavna		Nyhavna Development AS established		Comprehensive plan for Nyhavna (August)
2009	2016	2019-22	2020	2021		2022
	Municipal Sector plan for Nyhavna		Parallel assignment "Public Spaces in Nyhavna"		Quality Program for Nyhavna (May)	Architectural competition "Transittkaia"

of urban spaces, both as principles and as concrete solutions. The parallel assignment, together with other studies on cultural heritage, cultural industries, marine industries, environment/energy, technical infrastructure, sea level rise, and urban life form the basis for the QP for Nyhavna. The approval of the QP in May 2022 [16] (Table 1).

5.2 The Battle of the Narrative for Nyhavna

With the fabulous, holistic ambitions for the area, there come along many stakeholders and possible conflicting interests and perceptions of the future of Nyhavna. The first public hearing of the QP initiated a more extensive public debate on the plans for Nyhavna, but also city planning in Trondheim in general. A general agreement on the overall strategic aims is identified, but the interpretation of values and practical implications varies.

From the perspective of the urban planning office, the interviewees stated that the primary purpose of the QP is to communicate their overall visions and ambitions, particularly to landowners, politicians, and other stakeholders in Nyhavna, as well as to the public. With ten strategic aims in the QP, the representatives at the urban planning office wanted to 'formulate intentions that will stand the test of time, are easy to communicate and are setting a mood' for the future of Nyhavna. They aimed at being both concrete and open to interpretations at the same time. One interviewee says: '[The Municipality] plays a crucial part in creating the narrative that one is involved in something greater than just individual [organizations] goals'. According to Sandercock [21], storytelling as a tool in public planning process can work as a catalyst for change by being an inspirational example and shaping a new imagination of alternatives.

The QP visualize future scenarios of Nyhavna and targets a broad audience with a reader-friendly layout accompanied by pictures, maps and architectural illustrations. Additionally, it incorporates examples from inspirational international city development projects. People are prominently featured in most of the pictures and illustrations. A vibrant city district is illustrated, and the document can be viewed as a representation of an idealized future city life [10] in Nyhavna.

Developers at Nyhavna raised concerns in the public hearing of the QP [22] about illustrations used in the QP, not being in line with the legal guidelines in the Municipal sub-plan, giving the impression of unrealistic future solutions. One stakeholder highlights the lack of a plan illustrating how the prominence of private cars in the neighbourhood can be substantially reduced. An envisioned mobility hub is seen as part of the infrastructural development prioritizing less private car use, but this analysis is not yet carried out. Another stakeholder points out a lack of clear area designation in the QP. This stakeholder states they are looking for a perspective that is rooted in a pragmatic approach to urban development, emphasizing the importance of realistic, achievable plans. Concerns are also raised about the reliance on repurposing areas without adequately addressing the challenges and timelines associated with major infrastructural changes. This critique underscores the need for a unified approach that ensures all planning efforts are synchronized, thereby avoiding conflicting directions that could hinder the area's development in the long run.

Actors within music, arts and various creative industries, and the alternative neighbourhood of the adjacent urban ecological experimental area Svartlamoen express concerns about Nyhavna becoming a place like other harbour development areas, with only the upper class able to reside, or rent businesses [23]. The QP emphasizes the importance of preserving diversity and culture through its strategic aim, number 7, 'letting arts and culture characterize Nyhavna'. It suggests investigating various business models for culture-led regeneration. However, if wanting to preserve and support the current renting (sub-)cultural actors at Nyhavna, these proposals would necessitate taking concrete supporting actions, a fact that commentators claim they haven't observed thus far [23].

The discussion surrounding Nyhavna has evolved from a culture-driven regeneration strategy to one that also includes perspectives from the industrial sector [24]. This represents a shift in the urban discourse concerning Nyhavna. There is broad political agreement that the heavy industry is conflicting with urban city development of the area and will be relocated, but the future of lighter and small-scale industry has gained political attention to remain at Nyhavna. During the initial public hearing of the QP, actors within lighter industries and maritime businesses expressed their dissatisfaction with having been grouped together with arts-related entities in the Program for Creative Industries [25]. They advocated for a separate program tailored to the needs of the future of ocean space related actors and small-scale industry at Nyhavna, emphasizing the importance of access to test facilities near water. This feedback led to the development of the Industry program, which is on public hearing until March 2024. These actors also argue that their presence will contribute with an environment for innovations and green technology, aligning with the ambitions outlined in the QP. Nyhavna is pictured as the future of 'Silicon Harbour of Norway' and as a 'Norwegian ocean space technology capital' [26].

Each of these perspectives contributes to the complex tapestry of Nyhavna's future. The discussions around the QP highlight both the challenges and opportunities inherent in balancing various interests and visions for a district on the verge of significant transformation.

5.3 Bringing Nyhavna's Visions to Life

The QP states that '*many actors will have to contribute and stretch far if we are to achieve an ambitious urban development at Nyhavna where environmental, social and economic sustainability aspects are taken care of*' [16, p. 6]. The representatives from the urban planning office also emphasize that many of the qualities they aim to achieve cannot be realized without the cooperation of landowners.

Once the QP, with its innovative ideas, was unleashed, the municipality's role underwent a significant transformation. They transitioned from being the inspirational actor to becoming the majority owner of Nyhavna Utvikling AS, which operates as a commercial developer with the expectation of financial returns. Consequently, their influence, now exercised through this ownership position, shifted from a frontstage to a backstage role. While municipal influence persists, the dynamics of the arena have changed. The planning authority is subsequently tasked with approving the plan proposals submitted through Nyhavna Utvkling AS. This dual responsibility poses challenges, acknowledged by Municipal employees and other involved parties, as highlighted in the interviews. From the developers' side, one interviewee stated that the lack of understanding of other parties' roles and the accompanying responsibilities, is one of the major problems they experienced in the early planning phase. Once the roles change, or are not clearly defined, this can be a breeding ground for conflicts.

The transparency and detailed nature of the QP are considered assets, yet they also raise concerns among landowners about managing the multitude of goals. Demanding terms like 'must' and 'should' are often employed, indicating equal importance for all described actions. One developer highlights a concern regarding the QP being taken too literally, with an expectation for the fulfilment of every ambition, due to the lack of a clear hierarchy of goals. The QP is politically adopted as a *vision* for the Municipality's ambition for the district [16]. Meanwhile, representatives of the municipality refer to it as 'an operating document'. In the interviews, the developers report difficulties in navigating between various documents guiding Nyhavna's development, including the Municipal sector plan and the QP. There is ambiguity regarding which documents hold authority and significance. The lack of juridical status of the QP, and misalignment with the municipal sector plan are highlighted as concerning in the interviews.

The expected financial return for Nyhavna Utvikling AS by developing the area is not quantified, but according to one interviewee, it should be comparable to private development companies, where 15% margin is common. The developers operate within market principles. At the same time, there are expectations that Nyhavna should provide affordable housing accessible to the general public, including socially disadvantaged groups. However, there is little information on how this will be achieved beyond municipal rental housing. The study on housing-related social measures, planned in spring 2024, may provide potential solutions. Affordable housing can be a challenge once return margin should be achieved, and goals such as high energy efficiency and zero-emission goals can increase the costs due to new technologies [27].

The topics addressed by the ZEN-related goal number 6 have attracted numerous stakeholders interested in the environmental dimension. Statkraft stated in the (restricted) public hearing comments [22] that they wish to be involved in further planning of solutions for thermal energy for Nyhavna. Consequently, the support and collaboration of actors striving for the future decarbonization of the energy system for the area are promising. The active application of the ZEN centre's definition and indicators serve as a guideline in this case [3]. This goal appears therefore to be among those that are concrete enough for implementation. Based on the interview data, this seems not the case for goals missing a practical approach or operationalization.

Perceived challenges in balancing ambitious goals, discussing returns, and dealing with financial uncertainties complicate the ongoing processes. Comments show that costs associated with implementation appear to be unclear and the hearing input to the QP from two of the developers highlight the challenges associated with unclear and extensive costs related to implementation. Control over land-use and land development is key to the success of international harbour transformation projects [7]. According to one interviewee (former employee, Municipality), a less fragmented ownership structure could have been beneficial for sharing targets and economic risks. Even though there are few actors and a good starting point, a common ownership structure could, according to the interviewee, have a) led to the need to agree on a common picture of goals, and b) to financially greater strength. This would in turn as well have given the muscles to take culture actors seriously, instead of displacing the economically weak cultural segment.

The engagement and interest among the energy- and technology actors related to the technological development of the ZEN ambition, in combination with a possible willingness to invest, may indicate that the more lifestyle- and value-oriented eco-centric and ecosocial logics [12] will be less prominent than the eco-technical in the further development of Nyhavna.

6   Conclusions

Experience from the work of the ZEN centre illustrates that there is no one-fits-all recipe to transform a neighbourhood to an ambitious, sustainable area. Drawing on the experience from ZEN there are however recurring tendencies when working with the implementation of high ambitions. Not only must the technical systems be in place, but how the holistic planning process is structured to achieve common goals and ambitions is equally important. Ambitious projects at the district level deal with process related complexities such as diverse stakeholder interests, lack of procurement practices, and uncertainty in technical and management processes [27]. The early phase is the most relevant time for anchoring ambitions and to forge a common vision and objective [27, 28].

It has been put down a considerable effort in definition of ambitions and visions for Nyhavna. The complexity seems however, not be able to be processed through the tool chosen—the QP. The subsequent hands-on and practical implementation document appears to lack, leaving many doors open for guessing on how to move forward. A relatively open municipal district plan, followed by an ambitious QP, has proven to be quite challenging to manage, according to several of those who have been interviewed. Whether this would have been easier with a different sequence, where the QP had resulted in a municipal district plan, can be considered, as prioritization would have been necessary and there could have been less discussion regarding legal status. This point was underscored by Røe [10] who pointed to the Fjordbyplan in Oslo lacking legal binding. This absence of legal obligation may have permitted departures from the initial vision during the implementation phase.

In the context of Nyhavna, the ZEN ambition appears currently to be among the more tangible goal of the QP, due to its defined categories and indicators, while several other goals remain diffused. Even though, it is still demanding to apply these throughout the steps of the planning and decision-making processes in the respective projects. It is important to point out that urban development in Nyhavna will occur gradually and unfold over several decades. How the goals are implemented in future zoning plans and further realized through construction remains uncertain.

In summary, the main conclusions are:

- The hierarchy of goals must be clearly defined.
- The ownership structure of a neighbourhood is crucial determinant.
- Continuous stakeholder involvement must be defined clearly.
- The planning hierarchy can impose barriers to ambitious goals.

Acknowledgments   The authors gratefully acknowledge the support from partners in the Research Centre on Zero Emission Neighbourhoods in Smart Cities (FME ZEN).

References

1. IEA. Empowering Cities for a Net Zero Future. Available from: https://www.iea.org/reports/empowering-cities-for-a-net-zero-future. Licence: CC BY 4.0 (2021)
2. United Nations Convention on Biological Diversity 2022. Available from: https://www.cbd.int/doc/decisions/cop-15/cop-15-dec-04-en.pdf
3. Wiik, M.R.K., Fjellheim, K., Vandervaeren, C., Lien, S.K., Meland, S., Nordström, T., Cheng, C.Y., Brattebø, H., Thiis, T.K.: Zero Emission Neighbourhoods in Smart Cities. Definition, Key Performance Indicators and Assessment Criteria: Version 4.0 SINTEF Report No 45. SINTEF, Trondheim (2022) Available from: https://sintef.brage.unit.no/sintef-xmlui/bitstream/handle/11250/3043357/ZEN%2bReport%2bno%2b45.pdf?sequence=1&isAllowed=y

4. Vandevyvere, H., Ahlers, D., Wyckmans, A.: The sense and non-sense of PEDs – feeding back practical experiences of positive energy district demonstrators into the European PED framework definition development process. Energies. **15**(12), 4491 (2022)
5. Yin, R.K.: Case study research. Design and methods. Sage, London (2003)
6. Tømmervold, J.A.H. (2023). Nyhavna og DORA 1. Trondheim municipality: https://foto.trondheim.kommune.no/p/galleri/album/4052/28460047
7. Smith, H., Garcia Ferrari, M.: Lessons from shared experiences in sustainable waterfront regeneration around the North Sea. In: Smith, H., Garcia Ferrari, M. (eds.) Waterfront Regeneration. Experiences in City-Building, p. 203. Routledge, New York (2012)
8. Hölscher, K., Frantzeskaki, N.: Perspectives on urban transformation research: transformations in, of, and by cities. Urban Transform. **3**(2) (2021)
9. Lundemo, T.: Bydel for folk flest, eller bare for de rike? Adresseavisen. 2021. Available from: https://www.adressa.no/midtnorskdebatt/i/qWAlv0/her-kan-vi-fa-enda-en-haug-med-brune-blokker-der-bare-rikfolk-kan-bo
10. Røe, P.G.: Iscenesettelser av den kompakte byen – som visuell representasjon, arkitektur og salgsobjekt. In: Sandskjær Hanssen, G., Hofstad, H., Saglie, I.-L. (eds.) Kompakt byutvikling—muligheter og utfordringer, pp. 48–57. Universitetsforlaget, Oslo (2015)
11. Andersen, B., Skrede, J.: Planning for a sustainable Oslo: the challenge of turning urban theory into practice. Local Environ. **22**(5), 581–594 (2017)
12. Guy, S., Farmer, G.: Reinterpreting sustainable architecture: the place of technology. J. Archit. Educ. **54**(3), 140–148 (2001)
13. Kittang, D.: How does the carbon-neutral settlement of Brøset contribute to a new paradigm in planning the urban fabric? In: Kristjánsdóttir, S. (ed.) Nordic Experiences of Sustainable Planning. Policy and Practice. Routledge, London (2017)
14. Henriksen, H.M.: From science to sales: changing representations of zero emission housing. Build. Cities. **4**(1), 594–611 (2023)
15. Arentz, R., Vorkinnslien, R.: Kultur og kjeledresser på Nyhavna. Adresseavisen. 2022 April 28
16. Trondheim Municipality: Kvalitetsprogram for Nyhavna. Trondheim kommune, Trondheim (2022) Available from: https://www.trondheim.kommune.no/globalassets/10-bilder-og-filer/10-byutvikling/byplankontoret/1d_kunngj-annet-plan/2022/kvalitetsprogram-for-nyhavna/kvalitetsprogram-for-nyhavna-vedtatt-19.05.22.pdf
17. Kvale, S.: InterViews. An Introduction to Qualitative Research Interviewing. Sage, London (1996)
18. Braun, V., Clarke, V.: Using thematic analysis in psychology. Qual. Res. Psychol. **3**(2), 77–101 (2006)
19. Trondheim Municipality: Nyhavna, Kommunedelplan. KDP- k20110005, 28.04.2016
20. Norske Arkitekters landsforbund: Parallelloppdrag "Offentlige rom på Nyhavna", p. 523. NAL. Norske arkitektkonkurranser, Oslo (2020) Available from: https://arkitektforbundet.no/media/t5vj0dpw/nak-523-nyhavna-trondheim-po-03-02-20-kl1045.pdf
21. Sandercock, L.: Out of the closet. The importance of stories and storytelling in planning practice. In: Stiftel, B., Watson, V. (eds.) Dialogues in Urban & Regional Planning, pp. 299–321. Routledge, London/New York (2005)
22. Trondheim Municipality: Høringsinnspill og kommunedirektørens vurdering. Trondheim kommune, Trondheim (2021) Available from: https://drive.google.com/file/d/1Y9dEJTTY6n6MsTkZ-MZ55isMh25vFjEh/view
23. Kristensen, K.: Hvis alt skal være posh og nytt, dør Trondheim som by. Adresseavisen. 2022 April 26
24. Borgen, A., Rødde, G.B.: Blåskjorter og blådresser midt i byen. Adresseavisen. 2022 March 8
25. Trondheim Municipality: Kunnskapsgrunnlag for et kulturnæringsprogram for Nyhavna. Trondheim kommune, Trondheim (2021) https://drive.google.com/file/d/12MTVXl-8C13J5dqIXr1BP8hpdbPdpCL6/view
26. Fremtidens Industri. Nyhavna kan bli Norges «Silicon Harbour». Fremtidens Industri. 2023 Nov 29. Available from: https://fi-nor.no/nyhavna-kan-bli-norges-silicon-harbor/

27. Vergerio, G., Knotten, V.: Ensuring Ambitious Goals: Barrier and Good Practices in the Planning and Building Process SINTEF Report No 56. SINTEF, Trondheim (2024) Available from: https://sintef.brage.unit.no/sintef-xmlui/handle/11250/3120507

28. Baer, D., Ekambaram, A.: Integrating user needs in sustainable neighbourhood transition of the smart city – expanding knowledge and insight among professional stakeholders. In: Schrenk, M., Popovich, V.V., Zeile, P., Elisei, P., Beyer, C., Ryser, J., Stöglehner, G. (eds.) CITIES 20.50. Creating Habitats for the 3rd Millennium. Smart – Sustainable – Climate Neutral. Proceedings of 26th International Conference on Urban Planning, Regional Development and Information Society, pp. 423–432. REAL CORP (2021)

Leadership and Orchestration of PED Projects: An Organizational Perspective

Micol Pezzotta and Anders Riel Müller

Contents

1 Introduction

With 50% of the global population living in urban areas, cities consume about 75% of the world's energy production and account for 70% of the annual global carbon emissions [1]. Cities encompass a huge potential to advance the reduction of energy consumption and support the clean energy transition, by, for example, improving energy efficiency [1]. Therefore, energy planning in cities plays a pivotal role and needs to be aligned with long-term urban strategies [2] and better integrated with urban planning [3].

Positive Energy Districts/Neighborhoods (PEDs/PENs) is one of the concepts put forward in 2018 by the European Union's Strategic Energy Technology Plan

M. Pezzotta (✉)
Department of Energy and Petroleum Technology, University of Stavanger, Stavanger, Norway
e-mail: micol.pezzotta@uis.no

A. R. Müller
Department of Safety, Economics and Planning, University of Stavanger, Stavanger, Norway

© The Author(s) 2025
M. Kioumarsi, B. Shafei (eds.), *The 1st International Conference on Net-Zero Built Environment*, Lecture Notes in Civil Engineering 237,
https://doi.org/10.1007/978-3-031-69626-8_148

(SET-Plan) to help realize the objective of a net-zero carbon future in our cities [4–6]. A reference framework definition for PEDs/PENs was proposed in the White Paper on "PED Reference Framework for Positive Energy Districts and Neighbourhoods" prepared by Joint Programming Initiative (JPI) Urban Europe and SET-Plan 3.2 Programme on Positive Energy Districts [3]:

"Positive Energy Districts are energy-efficient and energy-flexible urban areas or groups of connected buildings which produce net zero greenhouse gas emissions and actively manage an annual local or regional surplus production of renewable energy. They require integration of different systems and infrastructures and interaction between buildings, the users and the regional energy, mobility, and ICT systems, while securing the energy supply and a good life for all in line with social, economic, and environmental sustainability." [3]

The PED's reference framework is quite wide to allow for local variations, still aiming at creating a common vision, and does not include quantitative criteria, which may make categorizing an urban development project as a PED, difficult. Aside from the relatively broad definition in the SET-Plan, a clearly agreed definition of PED does not yet exist (see, e.g., [5, 7–9]).

By February 2020, 29 PEDs and 32 sites with aspects relevant to PEDs, were implemented in Europe, as registered by the Joint Programming Initiative (JPI) Urban Europe [7]. The implementation of PEDs calls for new solutions not just in the form of innovative and integrated technologies but also within "social, economic, financial, environmental and legal/regulatory" areas [2]. Thus, PEDs provide a broad set of challenges, including new regulatory frameworks, new planning practices, financing, and organizational forms.

The challenges in the implementation of PEDs have been treated in the existing literature [4, 6, 8], and the following elements emerge as crucial to enable the deployment of PEDs: integration of technical and non-technical solutions, holistic planning, regulatory protocols and business models, stakeholder involvement at an early stage, and collaboration [6].

Sassenou et al. [4] review the existing literature and point out that there is an urgent need for clarification of terminology with regard to energy concepts to support PED deployment. According to a study by Krangsås et al. [8], challenges in the field of governance score the highest, also confirming the findings of Bossi et al. [10]. The development of PEDs demands new forms of collaborative governance and a holistic approach. The plurality of the stakeholders involved and their interests and goals, as well as the complexity of the decision-making processes, the need to find the right business model(s) and to design an appropriate market, demand for coordination [8].

This paper contributes to the discussion around collaborative governance, multi-stakeholder orchestration, and leadership in PED projects by exploring what forms of leadership have been adopted and/or can be identified in the existing literature on PED projects and how leadership is exercised. The structure is the following: in Sect. 2, we lay out our analytical framework by presenting in more detail the topics of stakeholder collaboration, co-creation, and leadership in sustainability projects in general and in PEDs in particular and we suggest introducing theories from strategic management to better understand leadership roles in PED projects. In Sect. 3, we present the methodology adopted for searching existing published research on leadership and co-creation of PEDs; and in Sect. 4, we summarize the findings. In Sect. 5, we discuss the findings and suggest how the proposed framework of leadership roles may aid in the future to study formal and informal leadership roles in PEDs and we draw the conclusions in Sect. 6.

2 Co-Creation and Leadership of Positive Energy Districts: An Analytical Framework

2.1 Stakeholder Collaboration and Orchestration in Sustainability Projects

Finding innovative solutions to solve complex problems requires the involvement of several actors from both the public and private sectors [11]. However, the role of the stakeholders in PED development projects is not strictly defined and may shift during the initiation, design, planning, and implementation stages of the process [12, 13]. The stakeholders participate with resources, competencies, and knowledge, and hold at the same time their own power, interests, goals, and definitions of value [8]; therefore, it can be complex to coordinate them [4]. Factors that foster collaboration, such as stakeholders' alignment and knowledge integration, are reviewed by Hamdan et al. [13]. Collaboration among stakeholders is recognized as a vital enabler for the planning of PEDs, but, from a project management perspective, the variety of stakeholders contributes to increasing the complexity of the project [13]. It is also suggested that the dynamics and the outcomes of interactions among stakeholders are less understood [13].

The core characteristics of co-creation in green transition-related projects are analyzed in Hofstad et al. [11], where co-creation is understood as a form of collaborative innovation [11] and heterogeneous actors are working together to generate innovative solutions. Spanning from this analysis, a list of core tasks for co-creation leaders was proposed, which includes, among other functions: establishing trust; building collaborative platforms and arenas; managing conflicts; exerting boundary spanning; stimulating mutual learning; motivating and unifying stakeholders around creating common public value; and managing risk [11]. It is also stated that leadership is usually distributed (i.e., different involved actors take on specific responsibilities), and public actors have a coordinating role. Furthermore, Hofstad et al. [11] stress that co-creation leadership is challenging, and common barriers include the absence of formal authority, complexity of processes, poor or weak alignment among actors, and goals and solutions that may change in time [11].

The aspect of collaboration among stakeholders from different organizations was studied by Gray [14] with a domain- and process-oriented approach. The domain approach shifts the focus from the single organization to the interdependencies among stakeholders' perspectives, toward the challenges at hand. Conditions facilitating inter-organizational collaboration are suggested for each stage of the collaboration process (problem-setting, direction-setting, and structuring). Gray suggests that analyzing the negotiation processes among stakeholders belonging to "non-formally established networks", could be beneficial for the conductor of the collaboration for understanding how to promote collaboration, and it is concluded that comparative studies are needed to derive general principles [14].

2.2 Public Sector Leadership and Co-Creation in Sustainability Projects

The role of the public sector has evolved in the last decades from a bureaucratic and top-down fashion to a more inclusive and collaborative one. The management paradigms the public sector has been through are known in the literature as Traditional Public Administration (TPA), New Public Management (NPM), and New Public Governance (NPG) as presented, for example by Tortzen [15]. Briefly summarized, as compared to TPA, NPM emphasizes the efficiency and quality of services, while the focus of NPG is co-creation and collaboration in networks and partnerships. Torfing et al. [16] suggest that the role of municipalities has drifted toward a form of governance similar to the system that existed before current welfare systems were established and link this change to the rise of "co-creation" as a way to solve problems by gathering competencies also from outside the municipalities and which is characterized by an active participation of diverse actors. The tendency for the co-creation mode of working seems to be the result of an increase in expectations from citizens, in combination with a decrease in available resources in municipalities [16], and of the increased complexity of problems to be solved [17, 18]. These are, among other factors, also seen as drivers for public innovation [19].

The concept of co-creation was originally developed in the private sector, and it was concerned with the involvement of the customer in creating a product that would satisfy its needs. Torfing et al. [17] point out the relevance of co-creation for the public sector with regard to the aspect of creating services for the satisfaction of the user; further, they highlight a distinction between co-creation and co-production, as in the latter the actors are the providers and the users of the service, and the focus is toward a specific service/product. Co-creation, instead, refers to the process between public and private actors aimed at collectively solving shared problems and creating public value. Furthermore, Torfing et al. [17] draw attention to the difference between the concepts of social innovation and collaborative governance. Social innovation comprises initiatives by civil society that arise to respond to social needs not supplied by the public sector, while co-creation involves interaction between, and concerted efforts by, private and public actors. On the other hand, collaborative governance does not aim to create innovative solutions, but multi-stakeholder collaboration is rather a way of governing. It "aims to make or implement public policy or manage public programs or assets" [20].

Co-creation has been extensively studied [15–17, 21, 22]. Despite this, Sillak et al. [22] claim that a more structured approach to co-creation is needed and proposed a framework for the assessment of co-creation processes for energy transitions based on three aspects, that is, involved actors, tools to enhance collaboration, and outcomes. Torfing et al. [16] also argue that attention should be directed toward the effects of co-creation and the leadership of co-creation processes.

2.3   Leadership Roles in PED Projects and Theory

Collaboration among stakeholders that belong to "non-formally established networks" is necessary when complex problems arise [14], often also called "wicked problems" in the literature (see, e.g., [23, 24]). A diverse set of stakeholders must be involved to solve "wicked problems" (see, e.g., [14]): engagement will also contribute to create a sense of ownership [11] and increased acceptance [13]. Nonetheless, collaboration among stakeholders needs to be coordinated and supported.

Leadership by the public sector is considered important to initiate and support PED initiatives [2]. However, the formal leader of PED projects is not always the public sector in practice and the leadership needed for a PED project to be successful may not rest only with the formal leaders and managers. PED projects can be initiated and led by either public, private, or non-profit actors, including associations, public institutions, or private companies, and the heterogeneity of actors involved in PEDs makes it hard to point out an ideal lead organization a priori. Nevertheless, we argue, that understanding leadership roles and functions is of a more generic character and will be relevant no matter who is the formal initiator of a PED project. The different leadership roles needed to implement a PED project are thus an important area for research. Who does what, when, and how? And are these leadership roles definable as part of a formal organizational structure, or do they emerge more informally and ad hoc?

Torfing et al. [17] highlight aspects of public leadership of co-creation processes and suggest established leadership theories (e.g., distributed leadership and horizontal leadership, among others) that emphasize that aspect. Many aspects of leadership of co-creation highlighted in the study by Torfing et al. [17] are relevant also for the leadership of PED projects, for example, delegation of tasks among stakeholders, framing stakeholders' contributions, and leading without a strong traditional authority.

One perspective for researching effective leadership, as presented by Jacobsen and Thorsvik [25], is to study what the leader typically does, and it is referred to as a leader's role. One such approach to defining leadership roles is Mintzberg's theory on managerial roles [26], developed from the observation of corporate managers' and leaders' practices together with an analysis of existing literature on the topic. Mintzberg suggests leadership roles can be categorized into three main groups, that is, interpersonal roles, informational roles, and decisional roles [25, 26].

Table 1 Leadership roles according to [26]

Main categories	Domain	Sub-categories
Interpersonal roles	Making contacts and creating a network	Figurehead role Leader role Liaison role
Informational roles	Processing information	Monitor role Dissemination role Spokesperson role
Decisional roles	Making decisions and strategy	Entrepreneur role Disturbance handler role Resources allocator role Negotiator role

Interpersonal roles are related to activities for establishing or maintaining contact with external actors, nurturing relations with the core work team, and fostering potential new collaborations. Informational roles regard the "processing of information" [26], where the flow of information is going in both directions from and to the organization, while decisional roles revolve around the process of decision-making, which in turn is related to identifying possibilities, discarding non-productive issues, or solving problems that may affect the efficiency of processes, choosing how to use resources, and negotiating collaborations [25, 26]. Table 1 illustrates the three main categories and the sub-categories; more details can be found in [25].

Earlier, we discussed the need to further understand leadership roles in PED projects. The existing literature emphasizes the need for multi-stakeholder collaboration and co-creation as a dominant way to orchestrate stakeholder contributions. In the previous section, it was shown how a strand of literature focuses on inter-institutional/-organizational collaboration and co-creation and sheds light on the importance, as well as complexity, of co-creation, the leadership of co-creation processes, and the coordination of stakeholders' resources. To enhance our understanding of leadership roles, we introduced the theory by Mintzberg, taken from strategic management literature, which focuses on leadership roles. However, we do not necessarily argue that all these leadership roles are needed in every project, and we also do not assume that the leadership roles above are necessarily attached to a formal position, that is, project manager or communication manager. Some leadership roles are only needed for short periods and may be filled informally by a person or group that "steps" up. How leadership roles are enacted and filled in practice is thus important to be understood to advance our knowledge of PED co-creation processes. We suggest that combining elements from the collaborative governance literature with approaches from the management literature could contribute to the existing PED pilots and cases in Europe and common practices of leadership therein. This will build a starting point to draw guidelines for effective leadership of PED processes and stakeholder coordination.

3 Methodology

Literature reviews (broad and narrow) were chosen as the method to investigate: the importance of the leadership and co-creation topics for PED development projects and the existence of research on the topics, with a special focus on the coordination of people and processes (as discussed in the previous section). A broad search was carried out with Google Scholar (as the search is done everywhere in the publication) and a narrower one with Scopus. The keywords "Positive Energy Districts" and "leadership" or "co-creation" were searched for. As expected, the search with Google Scholar gave a larger number of entries than Scopus, but as the relevance was decreasing, we confirmed by reading the abstract that many of the entries did not contain research or results on the topics we were searching for.

A literature search was carried out with the search engine Google Scholar using the keywords "Positive Energy Districts" and "leadership" and the exact phrase "Positive Energy Districts", in the time range 2018–2024, as 2018 is the year when the concept of PED appeared in the SET-Plan (see introduction). The findings amounted to 391, sorted by relevance and in pages of 10 entries each. The decrease in relevance was confirmed by reading the titles and, if unclear, the abstracts of the papers; the first three pages of entries were analyzed, and it was decided to analyze four entries in depth. These papers were selected as they discuss challenges and the state-of-the-art of PEDs; we deemed them very relevant, especially to confirm that leadership and collaboration among stakeholders are topics that affect the deployment of PEDs. One paper discussing stakeholders' interaction was also included. Papers with an applied approach or focusing more on technical planning were not prioritized in the analysis. We found that sometimes the word leadership was used in a more generic way, for example, to describe the role of cities and PEDs in the energy transition. The keyword Positive Energy District was found somehow to be limiting; some relevant results were found referring to zero-emission neighborhoods or sustainability projects instead of PEDs, and these were included in the analysis.

Similarly, a literature search was carried out using the keywords "Positive Energy Districts" and "co-creation" and the exact phrase "Positive Energy Districts," in the time range 2018 to 2024. The search returned 258 results sorted by relevance. As above, the decrease in relevance was confirmed by reading the titles and, if unclear, the abstracts of the papers. Thus, the first three pages of entries, where each page had 10 entries, were examined and it was decided to analyze 9 papers in depth. Three of these overlap with the previous search. As above, recent reviews or papers reviewing challenges in PEDs were included to confirm the importance of the topic. The other 6 selected papers were included in the in-depth study because they discuss governance, frameworks for collaboration, or tools or methods used to map or coordinate stakeholders. It is worth noting that, despite the large number of entries found containing the keywords, the word "co-creation" referred in several publications to a way of establishing specifically citizen engagement, which is not as broad as stakeholder collaboration, and for this reason, after reading the abstract, they did not seem relevant to our study and were excluded from the in-depth study.

Similar searches were carried out in Scopus. The searches of the same combination of keywords are limited in Scopus to the abstract, title, and keywords and only among peer-reviewed publications. The searches returned one entry for leadership and seven for co-creation. Three out of the seven entries about co-creation overlapped with the results from the Google Scholar searches. Only the paper on leadership was added to the in-depth study because it discusses leadership approaches (top-down vs. bottom-up) and factors that help or hinder stakeholder collaboration according to practitioners involved in PED pilot cases.

In total, based on the two searches, 11 papers were analyzed in depth.

4   Findings

The selected literature confirms the importance of (collaborative) governance and stakeholders' involvement in PEDs [6, 8] and underscores a need for further research focusing on the topic. In a recent review paper by Sassenou et al. [4], the same is attested also when the search is extended to nearly and net-zero districts literature. Furthermore, finding appropriate governance models and promoting stakeholder collaboration will have a high impact on most of the other challenges involved with PED development projects, such as market design, planning and decision-making processes, and balance of energy demand and supply [8]. However, methods and tools for motivating stakeholders and establishing "a common vision and shared values" are needed according to Sareen et al. [6]. Stakeholder collaboration is sometimes interpreted as a dimension of social innovation [27] or a factor that can contribute to reaching the social objectives of PEDs [28].

The examined literature supports the statement that co-creation and stakeholder engagement are crucial for developing PEDs (e.g., [4]). Most of the papers focus on conditions to enable co-creation or ways to promote it. As an example, an open innovation approach, a quadruple helix participation model of collaboration, early involvement of stakeholders (already in the idea/project development stage), and making results available are indicated in [29]. In [30], stakeholders' analysis and "business models for sustainability" are presented as tools to map interests and align goals, and therefore coordinate and support co-creation processes.

Mattsson et al. [31] investigated what kind of organization appears to be in a better position to lead PEDs and the type of PED initiative (top-down or bottom-up) that turns out to be more successful by interviewing practitioners involved in PED pilots in Sweden. The study found that communication and transparency support PED development projects and that PED pilots contribute to improved collaboration and knowledge sharing. Nonetheless, different goals among stakeholders may hinder collaboration and it is suggested that "creating a common picture" could help counteract this [31].

A better definition of energy concepts (meanings and methodologies for calculations) in relation to PEDs, for example, energy demand and consumption, energy production and supply, energy balance, and energy performance, as well as a common terminology, is also seen as crucial for the deployment of PEDs and communication among stakeholders [4]. This will also ease the transferability of results and experiences among pilots. As an example, efforts on collecting and comparing results are ongoing, for example, database development [9], but the categorization is complicated by the lack of agreed terminology. Different terminologies (low, net/nearly zero energy/emission) also make it difficult to compare results and for non-expert, readers to follow the development in the field [7]. Mattsson et al. [32] organized a workshop with practitioners and researchers working on PEDs in Sweden, and from the discussion on drivers and challenges to setup PEDs, they concluded that collaboration and leadership were "mentioned as important drivers" [32], and their lack is seen as a challenge.

In sum, we can conclude that the literature has so far given more attention to formal structures and framework conditions in facilitating processes and negotiations among stakeholders or actors, than to leadership roles or dynamics among stakeholders. Little was found that explicitly discussed how leadership is exercised so we could not identify, in the analyzed material, forms of leadership or leadership roles.

5 Discussion

Complexity and a lack of formalized regulatory and institutional settings strengthen the need for co-creation and collaboration as important framework conditions for PED development. However, this literature review shows that there seems to be an absence of literature on PEDs that focuses on the orchestration of processes and leadership roles, although the literature on multi-stakeholder collaboration and co-creation suggests that leadership is important. One of the possible weaknesses of the existing literature is the focus on collaborative governance structures rather than collaborative processes, that is, there is a stronger focus on what conditions need to be in place and less on the actual practices. For this reason, this study introduced theories of leadership roles from strategic management literature that have a stronger focus on roles, functions, and processes.

The leadership roles' framework was proposed by Mintzberg in 1973 [26] after observing the practice of several managers/leaders. These leaders had a formal leadership position in a corporate setting where responsibilities were formally distributed and an organizational hierarchy or structure was defined. The stakeholders involved in the creation of a PED participate in the development process and often make decisions under the orchestration of a formal leader, such as a public sector actor. The collaboration framework in most PED projects is, however, much less well defined than in a formal organization. The formal leader in a PED project cannot always dictate what other participants should do, and the roles or responsibilities, unless agreed by all the stakeholders, are much more diffuse. Thus, while the leadership roles are clearly assigned in the organization of a corporation, PED projects may entail more uncertainty on how these roles are enacted or assigned, and how they may be distributed among stakeholders.

Research is thus needed on leadership roles and functions by looking at practices so far in PEDs to understand what leadership roles are usually necessary, who assigns them or who takes them on, and to answer the question if more formal structures would enable more efficient PED development processes or not. Such insights into the organizational processes of leadership may help the acceleration of PED implementation, and other collaborative sustainability initiatives, in the future.

Based on the literature review and theories of leadership roles, we propose further research that is more focused on organizational processes rather than framework conditions for PED deployment, posing the following questions for further research:

- What leadership roles are needed and when they are needed?
- How are leadership roles "filled" both formally and informally, and when?
- Do certain organizations or actors take on some leadership roles more naturally than others, and why?

As PED pilots and PED-relevant cases around Europe are building experience in the implementation processes around the development of PEDs, the investigation of these processes, including the involved actors, responsibilities, and decision-making criteria, in existing PED pilots seen from a leadership roles perspective could help improve our understanding of success factors and thus support and ease transferability of knowledge and experience.

The main limitation of this study is represented by the fact that the literature search was quite narrow as well as the in-depth analysis focused on the keywords Positive Energy Districts and Leadership provided few results. As discussed above, more results could have been found by extending the search to a wider area (zero neighborhood). Results from Smart Cities that can be transferred to the goals of PEDs could have also been relevant. However, this study was meant as an exploratory exercise and wanted to highlight needs and suggest the potential of using existing strategic tools to support stakeholder collaboration in PED development.

6 Conclusions

The role of the public sector has been well-defined in traditional urban development projects, where the public sector has, for example, supported developers following procedures and filling requirements, and actors involved in such processes have had clear roles with assigned responsibilities. However, the increased complexity of urban projects, additional requirements called for by climate change and restricted budgets, have forced a change in the role of the public sector and in the type of collaboration with external actors. PED initiatives are still in an early phase and involve uncertainties, partly due to a still vague definition of the PED concept, which makes assessment of the results difficult. The roles and responsibilities of the involved actors are often vague or not well-defined and may change during the design and development process. Collaboration is voluntary and is driven by interests of different kinds. Leadership is not bound to any established structure of governance, and the way decisions are taken, is not clear.

Tools for promoting co-creation and stakeholders (especially citizens) engagement are developed and tested in several research and innovation projects, and research on the role of cities and new tasks for public managers was found. Nonetheless, the way leadership roles are exercised needs to receive more attention to ease the transfer of learnings from pilots and tests and create best practices and guidelines for practitioners. While there is an emphasis on the importance of the engagement of multiple stakeholders in designing PED projects and a co-creation process in making decisions, our literature search showed that little attention is paid to the various leadership roles and the involved processes in these projects. A multi-stakeholder perspective is present no matter who leads the initiative (public, or private; for-profit or non-profit organization); however, leadership roles and tasks, and who takes them on, seem to be only superficially discussed in the existing literature on PEDs. It is therefore suggested to conduct further research on the leadership aspects of PED planning and implementation. The leadership roles framework proposed by Mintzberg [26] provides an existing analytical framework to analyze processes and roles in PED development projects from a strategic management perspective. Studies examining leadership roles, as they appear in different PED projects—paying attention to relations between actors, flow of information and decision-making arenas, criteria, and steps could provide valuable insights into what leadership roles are needed for successful PED implementation.

References

1. International Energy Agency: Empowering Cities for a Net Zero Future: Unlocking Resilient, Smart, Sustainable Urban Energy Systems. OECD Publishing, Paris (2021)
2. SET-Plan Action 3.2: Implementation Plan: Europe to become a global role model in integrated, innovative solutions for the planning, deployment, and replication of Positive Energy Districts (2018)
3. JPI Urban Europe and SET-Plan 3.2 Programme on Positive Energy Districts: White Paper on Reference Framework for Positive Energy Districts and Neighbourhoods (2020)
4. Sassenou, L.N., Olivieri, L., Olivieri, F.: Challenges for positive energy districts deployment: a systematic review. Renew. Sust. Energ. Rev. **191**, 114152 (2024)
5. Albert-Seifried, V., Murauskaite, L., Massa, G., Aelenei, L., Baer, D., Krangsås, S.G., Alpagut, B., Mutule, A., Pokorny, N., Vandevyvere, H.: Definitions of positive energy districts: a review of the status quo and challenges. In: Sustainability in Energy and Buildings 2021, pp. 493–506. Springer Nature Singapore (2022)
6. Sareen, S., Albert-Seifried, V., Aelenei, L., Reda, F., Etminan, G., Andreucci, M.-B., Kuzmic, M., Maas, N., Seco, O., Civiero, P., Gohari, S., Hukkalainen, M., Neumann, H.-M.: Ten questions concerning positive energy districts. Build. Environ. **216**, 109017 (2022)
7. Brozovsky, J., Gustavsen, A., Gaitani, N.: Zero emission neighbourhoods and positive energy districts—a state-of-the-art review. Sustain. Cities Soc. **72**, 103013 (2021)
8. Krangsås, S.G., Steemers, K., Konstantinou, T., Soutullo, S., Liu, M., Giancola, E., Prebreza, B., Ashrafian, T., Murauskaitė, L., Maas, N.: Positive energy districts: identifying challenges and interdependencies. Sustainability. **13**, 10551 (2021)
9. Turci, G., Alpagut, B., Civiero, P., Kuzmic, M., Pagliula, S., Massa, G., Albert-Seifried, V., Seco, O., Soutullo, S.: A comprehensive PED-database for mapping and comparing positive energy districts experiences at European level. Sustainability. **14**, 427 (2022)

10. Bossi, S., Gollner, C., Theierling, S.: Towards 100 positive energy districts in Europe: preliminary data analysis of 61 European cases. Energies. **13**, 6083 (2020)
11. Hofstad, H., Sørensen, E., Torfing, J., Vedeld, T.: Leading co-creation for the green shift. Public Money Manag. **43**, 357–366 (2023)
12. Cheng, C., Albert-Seifried, V., Aelenei, L., Vandevyvere, H., Seco, O., Nuria Sánchez, M., Hukkalainen, M.: A systematic approach towards mapping stakeholders in different phases of PED development—extending the PED toolbox. In: Sustainability in Energy and Buildings 2021, pp. 447–463. Springer (2022)
13. Hamdan, H., Andersen, P.H., De Boer, L.: Stakeholder collaboration in sustainable neighborhood projects—a review and research agenda. Sustain. Cities Soc. **68**, 102776 (2021)
14. Gray, B.: Conditions facilitating interorganizational collaboration. Hum. Relat. **38**, 911–936 (1985)
15. Tortzen, A.: Samskabelse i kommunale rammer: hvordan kan ledelse understøtte samskabelse? Roskilde Universitet, Roskilde (2016)
16. Torfing, J., Sørensen, E., Røiseland, A.: Samskapelse er bedre og billigere. Stat & Styring. **30**, 31–35 (2020)
17. Torfing, J., Sørensen, E., Røiseland, A.: Transforming the public sector into an arena for co-creation: barriers, drivers, benefits, and ways forward. Adm. Soc. **51**, 795–825 (2019)
18. Klijn, E.-H.: Governance and governance networks in Europe. Public Manag. Rev. **10**, 505–525 (2008)
19. Sørensen, E., Torfing, J.: Metagoverning collaborative innovation in governance networks. Am. Rev. Public Adm. **47**, 826–839 (2017)
20. Ansell, C., Gash, A.: Collaborative governance in theory and practice. J. Public Adm. Res. Theory. **18**, 543–571 (2007)
21. Hofstad, H., Vedeld, T.: Exploring city climate leadership in theory and practice: responding to the polycentric challenge. J. Environ. Policy Plan. **23**, 496–509 (2021)
22. Sillak, S., Borch, K., Sperling, K.: Assessing co-creation in strategic planning for urban energy transitions. Energy Res. Soc. Sci. **74**, 101952 (2021)
23. Termeer, C.J.A.M., Dewulf, A., Biesbroek, R.: A critical assessment of the wicked problem concept: relevance and usefulness for policy science and practice. Polic. Soc. **38**, 167–179 (2019)
24. Lönngren, J., van Poeck, K.: Wicked problems: a mapping review of the literature. Int. J. Sustain. Dev. World Ecol. **28**, 481–502 (2021)
25. Jacobsen, D.I., Thorsvik, J.: Hvordan organisasjoner fungerer. Fagbokforl., Bergen (2013)
26. Mintzberg, H.: A new look at the chief executive's job. Organ. Dyn. **1**, 21–30 (1973)
27. Baer, D., Loewen, B., Cheng, C., Thomsen, J., Wyckmans, A., Temeljotov-Salaj, A., Ahlers, D.: Approaches to social innovation in positive energy districts (PEDs)—a comparison of Norwegian projects. Sustainability. **13**, 7362 (2021)
28. Derkenbaeva, E., Halleck Vega, S., Hofstede, G.J., van Leeuwen, E.: Positive energy districts: mainstreaming energy transition in urban areas. Renew. Sust. Energ. Rev. **153**, 111782 (2022)
29. Ahlers, D., Driscoll, P., Wibe, H., Wyckmans, A.: Co-creation of positive energy blocks. IOP Conf. Ser. Earth Environ, Sci. **352**, 012060 (2019)
30. Mihailova, D., Schubert, I., Burger, P., Fritz, M.M.C.: Exploring modes of sustainable value co-creation in renewable energy communities. J. Clean. Prod. **330**, 129917 (2022)
31. Mattsson, M., Olofsson, T., Lundberg, L., Korda, O., Nair, G.: An exploratory study on Swedish stakeholders' experiences with positive energy districts. Energies. **16**, 4790 (2023)
32. Mattsson, M., Lundberg, L., Olofsson, T., Kordas, O., Nair, G.: Challenges and drivers for Positive Energy Districts in a Swedish context. In: ECEEE Summer Study Proceedings, pp. 633–639 (2022)

Navigating Sustainability: Engineers' Views Within the Norwegian Construction Industry

Izma Ahmad, Kristine Lilleløkken, Zdravka Savcheva, Makarena Saavedra, and Allen Tadayon ⓘ

Contents

1 Introduction

The term "sustainability" was first introduced in the Brundtland Commission re-port "Our Common Future" in 1987 [1]. Here, sustainability was defined as "meeting the needs of the present without compromising the ability of future generations to meet their own needs." The report divided sustainability into three parts, environmental, economic, and social aspects. Later, during the conference "World Summit on Social Development" in 1995, the understanding was further developed to emphasize that "economic development, social development, and environmental protection interact and reinforce each other mutually" [2].

I. Ahmad · K. Lilleløkken · Z. Savcheva · M. Saavedra · A. Tadayon (✉)
Oslo Metropolitan University, Oslo, Norway
e-mail: allentad@oslomet.no

© The Author(s) 2025
M. Kioumarsi, B. Shafei (eds.), *The 1st International Conference on Net-Zero Built Environment*, Lecture Notes in Civil Engineering 237,
https://doi.org/10.1007/978-3-031-69626-8_149

Norway as a nation has a considerable amount of history in terms of sustainable development, going all the way back to 2002 when the first national strategy was drafted [3]. The strategy plan was developed with United Nations sustainability goals in mind. Currently, Norway ranks 7th with a score of 82 out of 100 among all other UN member states [4]. The score is a representation of the total progress made for UN's sustainable development goals. With this in mind we set out to find what sustainability means for engineers in Norway on a personal and work-related basis.

The goal of this study is to analyze the position of sustainability in the construction industry in Norway by mapping engineers' perspectives on the matter in this field. The study aims to encompass the significance of sustainability in individuals' personal lives, in the professional sphere, and generally within the industry. The following research question is formulated to steer toward this goal: *What is the perspective on sustainability in the construction industry today, and how can it be improved further?*

To assess the role of sustainability in the construction industry in Norway, data was collected by conducting interviews. Ten interviews have been conducted with actively working engineers, varying in levels of experience and holding different roles in projects.

2 Background and Theory

2.1 Origin of the Term Sustainability

Sustainable development was first introduced by the conference "World Commission on Environment and Development (WCED)" in 1987 [1]. The report "Our common future" which was issued during the conference defined sustainable development as "... development that meets the needs of the present without compromising the ability of future generations to meet their own needs." The key aspects that were recognized as a part of sustainable development were environmental, economic and social aspects. They were used to establish the foundation upon which further development could be made. The purpose of the report was to shed light on the negative impact that human activity has on the planet and highlight the importance of altering our practices. Today the conference is largely seen as an important milestone that sets the tone for future progress regarding sustainability [2].

During the conference "World Summit on Social Development" in 1995, the understanding of sustainability was updated [2]. The aim was to put a greater focus on the social aspects of sustainability. As a result of the conference, the three aspects of sustainability were considered equally important and interdependent on each other. All three had to be included and well-balanced in order to achieve sustainability.

2.2   Three Aspects Within Sustainability

The social element of sustainability is deeply intertwined with both the environmental and economic aspects. Achieving social equity is an important pillar for establishing a well-functioning society [5]. Areas highlighted as crucial are poverty alleviation, health care, quality education and protection of human rights. By adhering to policies that protect human capital, a well-adjusted society can be created. Policies that ensure that the working conditions are humane, and that people are compensated in a fair manner are the cornerstones of achieving social sustainability.

Environmental sustainability often takes center stage when sustainability is mentioned. One reason for this can be the growing concern about climate change. There is a strong link between the environmental aspects and climate change. It is heavily publicized that the recent changes in weather patterns are the result of human activity. These changes are often intricately linked to increasing greenhouse emissions resulting from polluting industrial processes and improper waste management [6].

Economic sustainability is to ensure economic growth without compromising on the environmental and social aspects. Some aspects of it are balanced economic growth and efficient resource use. A balanced economic growth must benefit all segments of society. It ought to ensure growth in the whole world, not just individual societies. A challenge while pursuing a sustainable economy is to cover human needs while staying within the limits of the earth's tolerance [7].

2.3   Previous Studies

Abdulmaksoud et al. [8] carried out a study on perceptions of sustainability in the construction sector [8]. The report stated that "Sustainability has a solid understanding on a broad level; however, in construction, there is no consensus about what it encompasses." The study indicates that it is necessary to increase the common understanding of the word and raise more consciousness around the topic. The study also shows that the social and economic aspects do not get the same attention and end up as less important as the environmental aspects. They have pointed out that it is essential to develop guidelines and common definitions to make it easier to interact with the building sector.

Fufa et al. [9] presenting user perspectives on the reuse of construction products in Norway [9]. The study indicates that there is a struggle in the Norwegian market between containing costs and at the same time front sustainability in projects. The findings show that customers seem to be more concerned about the economy than the environmental impact [10]. Another factor is a lack of understanding of how to reuse, and the need for a clear and balanced definition. The report cites that "*Laws and regulations, testing, documentation and certification, and economic subsidies are mentioned as the top three measures to address the current barriers.*"

Moreover, Knoth [11] has studied perspectives on the reuse of construction products in Norway. They have done interviews, and "lack of information" is mentioned as the biggest challenge related to reuse. It is also suggested that *"reuse in Norway could be greatly advanced by more communication and cooperation between different actors in the value chain."* Additionally, the government offers little to no incentives or regulations. They need to adjust the policies in favor of reuse.

Cerin et al. [12] have studied business incentives as an approach to sustainability [12]. The report states that *"Public and private demands for sustainable development put pressure on firms to develop strategies that include environmental concerns."* Furthermore, companies often possess the competence to optimize the total life cycle environmental performance of their products. The lack of requirements and incentives from the authorities may be a factor that contributes to sustainability being opted out.

3   Method

This study utilizes a qualitative method based on a series of semi-structure interviews. The semi-structured method gives the possibility to follow a set of predefined questions, and at the same time gives flexibility to make changes during the interviews. The goal is to get the participants to feel safe and to answer freely and express honestly about the subject [13].

For the study, 10 interviews have been conducted with active engineers in the Norwegian construction sector. Seventy percent of the participants work in big-size companies, while 30% work in small to medium-sized companies with less than 100 employees. All the respondents have an engineering title and at least 4 years of relevant experience working as an engineer.

An interview guide was developed with three parts; the introduction, the main part and the cool down [13]. The order of the questions was changed depending on the answers that were given. This way the participants could answer more freely, and it contributed to a natural flow in the conversation. Toward the end of the interviews, some lighter questions were included to ease the conversation, facilitating a natural conclusion. Participants were given the opportunity to ask questions or add comments by the end of the session.

Before collecting the main data, the interview guide was tested with a pilot interview to assess the effectiveness of the questions. After the pilot interview, the questions were reviewed and adjusted slightly to improve the flow of the interviews. Ahead of the interviews, all respondents received a consent form to sign, together with the interview guide and some information about the study by email. The interviews were conducted one-on-one and recorded via Teams, lasting between 45 and 60 min.

3.1 Review and Analysis of Collected Data

Following data collection, the transcript program "Autotekst" from the University of Oslo was utilized to transcribe the interviews automatically. Subsequently, each interview transcript was thoroughly analyzed and summarized. Given the semi-structured interview method, participants' opinions were closely monitored throughout the text, employing inductive reasoning to draw general conclusions when necessary.

After closely examining the individual interviews, they were compared to each other to look for similarities and differences in the answers, checking for patterns and thematic areas.

4 Results

The informants' opinions are reflected and presented in five categories throughout the chapter, where relevant quotations will be presented. They are as follows: Sects. 4.1, 4.2, 4.3, 4.4, and 4.5.

4.1 Individual Understanding of Sustainability

In order to assess the participants' understanding of sustainability, they were asked how they understand the word sustainability and what sustainability and environmental awareness mean to them personally and in everyday life.

When the participants were inquired about their understanding of sustainability, 8 out of 10 primarily emphasized the environmental dimensions of sustainability. The majority talked about energy use, reduction in CO_2 emissions, the future of the planet and natural resources availability. Some mentioned the importance of taking care of the planet, not just for the present generation, but also for future generations to come. The participants also talked about the importance of finding sustainable solutions, and responsible use of natural resources.

Two participants brought up social aspects of sustainability. They talked about the importance of people having their basic needs covered. They mentioned that it is important to take into account the working conditions and how those can affect the health of the workers. One participant said that sustainability is also to "... make sure that you have sustainable production with the people who stands and works with the products, tests the products, supervising, and making sure that there are good working conditions and that whole bit too." He added that humans are resources and if people have their basic needs covered, they will be better members of society.

Some of the participants said that they gave importance to sustainability and the environment in their everyday life, while others believed that it was not a thought in theirs. One interviewee said that he tried to contribute by focusing on everyday choices that are environment friendly. For example, he said "... I don't have a car. I usually cycle or walk if I'm going somewhere, or I just use public transportation. I'm frugal with the use of energy and things like that. Sometimes I'm wondering if I'm a bit stingy...." Another participant spoke about this by saying "... I'm probably not one of the most environmentally conscious, but I find it more rewarding when I know something is environmentally friendly than doing things that are not environmentally friendly."

While some participants actively made choices that they considered environment friendly, others talked about being affected by economic factors. A participant shared his thoughts on electric cars and heat pumps, which are generally considered more environment friendly than the alternatives as a car on fossil fuel, or not having a heat pump. He stated: "I drive an electric car. Whether the motivation behind it is to be environmentally conscious, or simply because of all of the incentives that are available ..." and "I haven't installed a heat pump at home, but if the electricity gets expensive enough, then it can be considered...." He later added that he thinks electric cars and heat pumps are better for the environment, but his choices are affected by the prices. There were several people who brought up electric cars as an example, and they mostly reasoned their choices with the prices or subsidies.

One participant additionally linked his understanding of sustainability to the upbringing of his children, by expressing "... I am concerned that my children should have a safe and good place to grow up. And sustainability is important for that to happen." The social aspect was brought up by a few, and the importance of sustainable food production was also referred to.

4.2   General Thoughts About Sustainability

The participants were asked general questions about sustainability in projects. These questions were about sustainable projects, the implementation of sustainability in projects, and who is responsible for ensuring implementation of sustainability in projects.

In response to what the participants considered a sustainable project, there was a strong focus on the environmental aspect. One participant mentioned, "One should try to use materials with low carbon footprint. Attempt to use recyclable materials. Recycled materials. And then, it's about the energy use in the building. It should be energy neutral." Another participant spoke about "... ensuring that there is not a significant impact on society or resources" and being mindful of ecosystems that are affected. Additionally, attention was drawn to "CO_2, transportation to and from places, the entire life cycle, life cycle costs, and emissions." The choice of energy source was also highlighted as crucial in defining a sustainable project, as "the building is supposed to stand there and live for 60–100 years" and therefore "it's important to look at the entire period, not just the construction period."

Another participant, also focusing on the environmental aspect, stated, "A lot revolves around the choice of products. This means minimizing the impact on the environment when producing products and making the right choice when selecting the components these products are built from. It is important to consider the product's life cycle and that it is possible to recycle the product when we are done with it. Transport is very important. Transport is a significant thief when it comes to energy consumption and materials." However, he concluded by bringing in social aspects, saying, "Another important factor is the people who work in the production of these products; they must work under good working conditions. This is also sustainability."

A couple of participants also mentioned economic aspects associated with the concept. It was stated that a sustainable project is "...one that achieves economic gain, so the company doesn't go under. If you look at it from an environmental perspective, a sustainable project is a project with high environmental ambitions." The participant brings in the environmental aspect, similar to the majority of those interviewed.

In response to the question of who is responsible for promoting sustainable solutions in construction projects, most participants agreed that the project owner bears responsibility for their project and, consequently, has a duty to set requirements for other parties involved in the construction process. However, it was acknowledged that no single party carries the responsibility for everything.

Another participant, who also believes that implementing sustainability in projects is a shared responsibility among all actors, says, "It's probably not just one of them. It's probably everyone. The municipality and the state have a responsibility to force the industry over through regulations. The project owner has a responsibility to facilitate the implementation of sustainable construction projects. The consultant is crucial in the planning phase... The contractor has a responsibility to make sustainable purchases."

The government's responsibility was emphasized, "They have the authority to impose on the industry through mandates and regulations." Furthermore, it is mentioned that the government can demand more from the construction industry, but "Many politicians may not fully understand sustainability and often use it as a trendy term without much knowledge." Several agree that it needs to be more requirements, but also incentives and support to implement measures for more sustainability, because "Where funds are limited, sustainability is deprioritized."

It is also noted that, in addition to requirements, flexibility should be shown in cases where the requirements are unreasonable. A highlighted example is that the construction of zero-emission buildings can be costly and not an absolute necessity.

4.3 Attitudes on Sustainability in the Construction Industry

The industry's stance toward sustainability has been examined and the main features of thoughts and clear extremes are presented. It has also been explored if competence development is encouraged, whether the industry is willing to choose less tested solutions, and what hinders and promotes sustainable development.

One participant noted, "We are evolving. We have started with emission-free construction sites and waste management. Something is happening. I believe the industry's stance is that it should change. But it changes very slowly, because there are few who take the lead and responsibility." Several participants expressed that the industry has a visible wish for progress in a sustainable direction. They mention that an increasing number of companies talk about working on sustainability in their projects. Within companies, participants experience that there are more demands for actions that contribute to reducing the companies' carbon footprint but adds that he thinks it's due to the opportunity to promote his own company.

Some participants emphasized that companies talk about being sustainable but have few projects to showcase as examples of their sustainability efforts. One participant says, "...there are some projects where we earn extremely little money, and we know it before we enter the project, but it's a landmark project ...." She is talking about how companies sometimes enter projects they know will yield very little profit, but they do so to use the project to send specific signals externally. These few projects are often the ones repeatedly highlighted in the companies' marketing.

Many believe that companies focus on measures that provide them with visual documentation of the effort they have put in. Several certifications have been developed, which can be achieved by meeting requirements in specific categories. Certifications are often considered indicators of how environmentally and sustainable a company is. However, it may contribute to a focus on visible measures rather than the entire value chain. One participant notes that "It seems like they are more concerned about the products or solutions they buy to save energy. It's not so much about what has happened along the way before the product reaches them."

There can also be ethical dilemmas associated with proposing sustainable solutions, such as reuse, as it may lead to fewer hours and less work for the company. One participant has considered the thought that "The best thing for my company is to tear everything down and start over because then I get a lot of work. But that's not what's best for society." She mentions that it's a thought she has had, although she hasn't changed her approach due to it. However, she acknowledges that it can be a challenging issue, especially during economic downturns.

Another issue raised is the lack of a defined responsibility allocation for who would be held accountable if a solution ends up being unsuccessful. One participant says, "If you choose a solution that turns out to be poor, expensive, or doesn't work, the responsibility falls on the advisor." She continues by stating, "There's probably a need to look at a distribution of responsibility. If you're going to be willing to push forward less proven solutions, the project owner probably has to expect to share some of that responsibility."

Another challenging issue mentioned is reuse and ensuring the quality of reused products and materials. While reuse is perceived as a sustainable alternative, it requires that products maintain a certain quality in accordance with the country's requirements. "There is much more that can be done regarding things related to the building, such as floors, roofs, and windows. When you look at an electrical installation in a building, there is limited room for reuse. Because there is very little that can be reused, and there is not much flexibility in the choice of materials. Much is determined through regulations and laws." He believes that some disciplines are not well suited for reuse, and in those cases, there is limited scope for employees working in those areas.

Difficulties can also arise when attempting to incorporate reused products in projects, as it requires extensive collaboration between parties. "we don't have any processes for reuse ... Where should we store products before assembling them again?" The participant mentions that effective and clear logistics solutions are required for storing reused products throughout the process. She further asks, "Who will take responsibility for complaints?" When there isn't a well-established system for reuse, including storage, transportation, and warranty possibilities, it becomes more challenging to choose it as an option.

4.4 Sustainability in Own Work

Engineers' individual initiatives to choose sustainable solutions differ. So does the training they receive in the workplace on this topic. Some prioritize sustainability in their work, while others give it minimal consideration. Some participants said that they have received offers to take courses centered around sustainability. Others mentioned that their company focused on the internal transfer of knowledge. A few said that they hadn't heard about any offers to take courses to extend their knowledge of sustainability related to their work.

The respondents placed a pronounced importance on the individuals as well. It was made clear that engagement with sustainability-related courses can vary considerably. Overall, it was noted that organizational commitment to sustainability is typically manifested through initiatives, such as lunchtime lectures. Such measures were intended to serve as forums for deeper exploration and understanding of sustainability.

Several participants expressed that they didn't have an influence over decisions since the project's framework was already established. When asked if they felt heard when proposing an idea, some respondents answered, "No, if it has economic disadvantages, it can probably be stopped" and "Not always. In my role as a consultant, I have no influence. But if we have a specific standard that we follow, that's something we can point to." Often, economics was pointed out as a recurring factor influencing decisions in a project. However, several participants also felt that there was room to suggest sustainable solutions, but the feasibility was not always straightforward.

The theme of recycling and reuse was repeatedly brought up as a topic that many participants considered sustainable but was not well-executed, or hard to implement in projects. One participant mentioned, "Especially reuse is talked about a lot. It's easy for me to talk about, but it's very difficult for those I collaborate with to relate to" and another stated, "We need a registration system to assess capacity for reused material so that we can incorporate it into the project. When I suggested this, I got no response. That left me a bit frustrated." They expressed a need for regulations before they can use it actively in their work.

4.5 Perspective of Future Outlook on Sustainability

The participants were asked about their thoughts on how sustainability within the construction industry will develop over the next 10 years. There are varying views on the anticipated progress. Two participants who believe there will be an increasing focus on sustainability commented: "... a lot because of the climate changes we are now witnessing. It's no longer just something Greta Thunberg talks about, but something we actually see and experience, causing massive destruction for countries undergoing it." By being drivers and active in sustainable development, companies can use this to promote their own businesses and gain economic benefits.

Another participant believes that if not Norway, the EU might introduce policies that force the state to impose stricter requirements on the construction industry. It seems that there is not much confidence in Norwegian authorities significantly tightening requirements but rather that demands and initiatives must come from elsewhere.

One participant also linked the expected development in sustainability to the economic situation. He discussed how an uncertain economy could deprioritize efforts to achieve long-term goals, instead focusing on short-term concerns. He believed that climate goals might be challenging to attain and could risk being sidelined when the economy declines. Without clear and stringent requirements from authorities, he thought that sustainability in the construction industry might be deprioritized.

The majority predicted significant changes related to requirements and standards. The rehabilitation of buildings and the reuse of components were also mentioned as possible focal points in the industry. Furthermore, a better market for reused components and a significant improvement in legislated rights were suggested.

But for sustainable development it is expressed that we need a clear understanding of what it is. This is proposed through a comment that "In 10 years, we may have progressed further in specifying what it is, and the public understanding may have significantly increased."

5 Discussion

When asked about the participants' actions related to sustainability in private life, profitable alternatives were prioritized, over alternatives that are usually seen as more environment friendly. An example is electric cars which seem to be popular due to government subsidies. Similarly, energy use was deemed important, but the participants did not show any desire to reduce their energy consumption with the reasoning that it can have a positive effect on the environment. The same applies to heat pumps, which were only seen as an option if it was economically profitable to install. In other words, sustainable choices tend to be prioritized, when they are also considered to be favorable for the individuals, not because they are understood to be sustainable. This is supported by previous research, where findings suggest that people do think about their energy usage, but do not sacrifice their comfort to save energy.

The study has been conducted in Norway, where a high quality of life may impact the perspectives on sustainability [14]. The participants overwhelmingly focused on the environmental aspect of sustainability, with less attention given to social and economic aspects. It indicates that the term is defined based on the situation of the individual or the society. Therefore, a need to concretize what the term "sustainability" actually means, may arise. This will contribute to an increased awareness and better understanding of what the term entails.

The government can regulate the construction industry by reinforcing laws and regulations. They should take part in the responsibility and make it viable for companies to try out new solutions in an effort to help companies transition to more sustainable practices. This can be achieved through subsidies or ensuring that banks loan out money with more favorable terms. By interpreting the answers from the participants, it was clear that the industry requires regulations from the state. It is important to note that such incentives should not have a negative impact on the market while they are developed.

Several companies seem to have the desire to appear sustainably conscious. Internally it is not always reflected. Some of the participants talked about frequent courses and skill development related to sustainable practice, while others were unfamiliar with such courses being available to them through their workplace. The participants who got the chance for further development through their work seemed more positively inclined toward sustainability and were visibly more motivated about the impact of their work. The employee's point of view often tends to influence whether there is a focus on implementing sustainable solutions in projects or not. Therefore, companies should take responsibility for developing their employees.

The different parties involved in a project have individual responsibilities for implementing sustainability, but this can lead to conflicts over who is responsible if something goes wrong. One solution is to reconsider the distribution of responsibility, so all parties would share some of the blame. Under the current distribution, a participant experienced that consultants may hold back less proven solutions in fear of being left to bear the responsibility all alone. To encourage parties to propose and implement innovative solutions, one can explore solutions where responsibility is shared among all of the involved parties. It must be rewarding for everyone to implement sustainability in projects. Tenants can also have more strict and clear requirements for the building owners. Then the builder has to build based on market demand.

There are no effective systems related to reuse, with slow development in the area. It also appears that the industry does not perceive the existing regulations as adequate. It was made clear that the requirements did not address the needs of the industry. This indicates a need for more guidelines and requirements to facilitate a smooth project flow and make material reuse more straightforward.

6 Conclusion

The goal of the study has been to analyze the perspectives on sustainability in the construction industry. This has been done by conducting interviews with ten engineers with varying levels of experience. The interviews have included questions related to the engineers' individual understanding of sustainability in their private and working life, attitudes toward sustainability in the industry, sustainability in their own work and their perspective of future outlook on sustainability.

Even though most of the participants did not perceive themselves as environmentally conscious, there were indications that they considered sustainable and environmentally friendly alternatives to be more positive and rewarding. This also reflects the building industry, where companies have an increasing activity in regard to sustainability. A dominant factor mentioned during the interviews is the marketing of one's own company as sustainable to secure a larger market share. However, this contributes to a willingness for development in the field. To force the construction industry the government needs to impose stricter requirements and regulations. Subsidies and better loan terms can be used as encouragement and a light push of the industry in the right direction. The industry should also take responsibility into its own hands and start implementing measures that results in providing the necessary opportunities for the development of new and innovative solutions. It is proposed to do it through an even distribution of responsibility.

To answer the research question, the industry is showing an increasing interest in ensuring sustainable societal development but still faces many obstacles that need to be overcome. In this progression, it is essential not to wait for others, but for companies and individuals to take the lead themselves. The responsibility lies with everyone, and the work must be done collaboratively.

References

1. Imperatives, S.: Report of the World Commission on Environment and Development: Our common future. Accessed Feb 1987. **10**(42,427)
2. United Nations: World Summit for Social Development, pp. 1–129 (1995)
3. Ministry of Foreign Affairs, National Strategy for Sustainable Development: The Royal Norwegian Ministry of Foreign Affairs (2002)
4. Sachs, J.L., Lafortune, G., Fuller, G., Drumm, E.: Implementing the SDG Stimulus: Sustainable Development Report 2023. Dublin University Press, Paris (2023)
5. Eizenberg, E.J., Jabareen, Y.: Social Sustainability: A New Conceptual Framework. Sustain. For. **9**(1), 68 (2017)
6. Office of Legacy Management: Environmental Sustainability. [cited 2 Dec 2023]. Available from: https://www.energy.gov/lm/listings/environmental-sustainability
7. FN-SAMBANDET: Bærekraftig utvikling. Available from: https://fn.no/tema/baerekraftig-utvikling-fattigdom-og-befolkning/baerekraftig-utvikling
8. Abdulmaksoud, S.B., Beheiry, S.: Perceptions governing sustainability in the UAE construction sector. Buildings. **13**(3), 683 (2023)

9. Fufa, S.M.B., Kristine, M., Hauge, Å.L., Johnsen, S.Å., Fjellheim, K.: User perspectives on reuse of construction products in Norway: results of a national survey. J. Clean. Prod. **408**, 137067 (2023)
10. Sørensen, K.H.L., Anette, V., Hojem, T.S.M.: Articulations of mundane transition work among consulting engineers. Environ. Innov. Soc. Trans. **28**, 70–78 (2018)
11. Knoth, K.F., Mamo, S., Seilskjær, E.: Barriers, success factors, and perspectives for the reuse of construction products in Norway. J. Clean. Prod. **337**, 130494 (2022)
12. Cerin, P.K., Karlson, L.: Business incentives for sustainability: a property rights approach. Ecol. Econ. **40**(1), 13–22 (2002)
13. Pasian, M.B.: Designs, Methods and Practices for Research of Project Management. Gower Publishing, Ltd (2015)
14. OECD Better Life Index: Learn more about Norway. [cited 2 Dec 2023]. Available from: https://www.oecdbetterlifeindex.org/countries/norway/

Post-Project Evaluation: A Perspective on Effective and Sustainable Healthcare Design

Laura Sacchetti ⓘ **and Roberto Di Giulio** ⓘ

Contents

1   Introduction

In recent decades, European countries have been facing shared healthcare challenges arising from several fronts that require significant improvements in the efficiency and effectiveness of national healthcare service delivery.

The latest pandemic emergency has highlighted the need to improve the emergency preparedness of national healthcare services, thus requiring more flexible and resilient systems. This awareness is coupled with a growing demand for healthcare services—due to ageing populations, the prevalence of chronic diseases, medical advancements, etc.—which, in turn, clashes with economic constraints and workforce shortages. In light of these challenges, it has become increasingly evident that primary care plays a crucial role in improving the efficiency, effectiveness and adaptability in response to emergencies and evolving health needs [1–3]. Consequently, many European countries have been reorienting their healthcare policies towards enhancing local care services, strengthening existing or developing new

L. Sacchetti (✉) · R. Di Giulio
Department of Architecture, University of Ferrara, Ferrara, Italy
e-mail: laura.sacchetti@unife.it; roberto.digiulio@unife.it

© The Author(s) 2025
M. Kioumarsi, B. Shafei (eds.), *The 1st International Conference on Net-Zero Built Environment*, Lecture Notes in Civil Engineering 237,
https://doi.org/10.1007/978-3-031-69626-8_150

facilities based on primary care. This model aims to bring care closer to where people live, thereby offering a more convenient, accessible, reliable and familiar point of contact for healthcare services, ultimately leading to reduced unnecessary hospitalisations and increased patient satisfaction.

Second, there is an urgent need for healthcare buildings to become more energy-efficient, aligning with broader global efforts to mitigate climate change and reduce carbon emissions. It is widely acknowledged that healthcare facilities are among the most energy-intensive buildings, mainly due to their continuous operation [4–7]. At the same time, in the building sector, there is a noticeable trend towards considering sustainability and comfort levels as interconnected priorities because embracing environmental and sustainable building practices also entails designing high-quality spaces that promote better health conditions for building users [8].

This is tied to another observation: in healthcare facilities, the built environment plays a significant role with a direct impact on both the end users' well-being and clinical outcomes [9–11]. Certainly, the design of physical environments can facilitate healthcare and non-healthcare activities to be performed. Additionally, the scientific literature in this field demonstrates that the healthcare-built environment can have both physical and psychological impacts on end users, including patients, staff, visitors and the local community, with consequences on their experience and satisfaction [12, 13]. As a result, the perceived comfort level for users is increasingly being regarded as a key metric of the effectiveness of healthcare buildings [8].

1.1 The Role of Building Performance Evaluation

In this evolving context, in which efficiency and effectiveness—partly derived from economic constraints—are the driving forces behind a transition towards new organisational models, building performance evaluation has become a critical aspect of the construction and operation of buildings.

In the past decades, the assessment of built environments has evolved from traditional methods like post-occupancy evaluation (POE) to a more comprehensive approach known as building performance evaluation (BPE) [14]. POE focuses on gathering feedback once the building has been used for a certain period of time, to assess whether design requirements and users' needs are met by the design. In contrast, BPE involves assessments that may occur throughout the building process, from the initial brief to the post-occupancy phases, to ensure that built environments fulfil users' needs and achieve clinical outcomes, economic feasibility and long-term sustainability.

In the present context, in which new organisational models have few prior examples to reference, uncertainty may arise regarding the most effective design and construction choices. Particularly in healthcare facilities, design entails a delicate balance between various needs and requirements. For example, functional layouts may not always align perfectly with the most effective energy-efficient solutions or environmental requirements can sometimes pose challenges to flexible design solutions. Then, it becomes significant to determine relevant priorities to achieve optimal building performance.

This underscores the need for BPE tools that may provide a rational basis to guide designers and local public authorities to optimise care facility design choices according to the needs and objectives to be prioritised. By adopting these tools, evaluations can generally result in the early detection and resolution of potential design issues and provide assurance that the facilities not only adhere to regulatory standards but also provide a high-quality environment that serves their intended purpose.

1.2   Research Objectives

At the moment, in the European context, existing evaluation tools primarily focus on hospitals or healthcare facilities in general, while there is a lack of studies centred on the evaluation of primary care environments. Although tools developed for hospitals or healthcare premises, as well as those originating from distant contexts, may offer relevant methodologies and outcomes, it is important to recognise that the primary care facilities in Europe have unique objectives and characteristics that require specific considerations when evaluating these healthcare settings (e.g. they cater to community health needs, they place emphasis on prevention and health promotion, etc.).

This chapter is part of a broader research that, in light of previous considerations, aims to streamline the development of evaluation tools for healthcare-built environments that are suitable for the current transition phase underway in European countries.

The goal is to establish a consistent and systematic approach to evaluating the design quality of primary care facilities throughout the design process, as well as post-construction and occupancy, to support effective implementation of high-quality and sustainable interventions, and to facilitate the exchange of information that can inform future design and projects.

Consequently, the aim of this chapter is to analyse validated evaluation tools for healthcare-built environments to provide a comprehensive understanding of their main characteristics, thereby developing a guideline framework that prompts reflection on the potential to either adapt existing tools or develop new ones tailored to assessing local care facilities.

2   Methods

This research comprised a literature review conducted through keyword searches across three main repositories (i.e. Scopus, Science Direct and Google Scholar), aimed at gathering relevant information regarding the existing tools and methodologies for evaluating healthcare-built environments. The literature search yielded the

Table 1 Healthcare built environment evaluation tools selected for analysis

	Tool	Year	Developed by	Country	Sources[a]
T1	A Staff and Patient Environment Calibration Toolkit (ASPECT)	2004	DH Estates and Facilities	UK	[12, 15, 16]
T2	Achieving Excellence Design Evaluation Toolkit (AEDET)	2004	University of Sheffield; NHS Estate	UK	[12, 16, 17]
T3	Clinic Design Post-Occupancy Evaluation Toolkit	2011	The Center for Health Design	USA	[18, 19]
T4	Evaluation of Older people's Living Environments (EVOLVE)	2010	University of Sheffield; others	UK	[20]
T5	Patient Room Design Checklist and Evaluation Tool	2015	The Center for Health Design	USA	[21]
T6	Perceived Hospital Environment Quality Indicators (PHEQI)	2006	University of Cagliari; Lisbon University Institute	IT	[22–24]
T7	Community Health Center Facility Evaluation Tool (CHD-CHC)	2017	The Center for Health Design, Kresge Foundation	USA	[25]
T8	Design Quality Indicator for Healthcare (DQI Healthcare)	2014	Construction Industry Council	UK	[26–29]
T9	Sustainable High Quality Healthcare Environments version 2 (SustHealth v2)	2021	Politecnico di Milano	IT	[30, 31]

[a]For a more comprehensive understanding of the analysed tools, this column directly references the tools' websites as well as the reports and research papers outlining the tools' development process and the their features

identification of several tools, among which nine were selected for further analysis, based on their international relevance and reliability as well as availability of information (Table 1).

Moreover, additional grey literature was retrieved by specifically searching for each of the selected tools; this included the main websites, reports and documents provided by the developers and users of the tools as well as research papers presenting or adopting the selected tools. The available information was then screened, read and analysed.

Additionally, thematic analysis was applied using a comparison grid aimed at collecting and comparing the attributes of the selected tools, as listed in Table 2.

Each attribute was analysed using comparative tables in Excel spreadsheets. The features of each tool corresponding to the attributes were listed and summarised to derive the aggregated results for each attribute.

3   Results and Discussion

This research reviewed and analysed the existing healthcare built environment evaluation tools to discuss the relevance of building performance evaluation and post-occupancy evaluation in contributing to the future development of more

Table 2 Attributes included in the comparison grid

	Attribute		Attribute		Attribute
A1	Name	A7	Target audience	A13	Scoring system
A2	Year and last update	A8	Application phase	A14	Weighting system
A3	Developed by	A9	Application scale	A15	Customisation
A4	Country	A10	Data collection methods	A16	Data analysis
A5	Target facilities	A11	Users involved	A17	Visualisation format
A6	Focus/aspects investigated	A12	Tool format	A18	Expected outcomes and impacts

effective and sustainable design of local healthcare facilities. This review allows summarising the key characteristics of the existing tools based on their identified attributes, extracted from information sourced from websites, reports and research papers.

The following paragraphs illustrate the main attributes in relation to the aim of this chapter. In the following paragraphs, the numbers in brackets indicate the quantity of tools associated with the specific attribute or feature.

3.1 Target Audience

The user groups explicitly mentioned as targets by the different selected tools are presented in Table 3. They mainly refer to five different categories: (i) healthcare stakeholders, (ii) design and construction stakeholders, (iii) academia, (iv) public authorities and (v) the general public.

Healthcare stakeholders (e.g. clients, facility owners and managers, planners, etc.) are cited by eight tools, with the majority focusing on healthcare organisations and providers (6). Architects, designers and contractors are one of the groups targeted by seven tools, primarily focusing on design firms and teams. Mention of governmental bodies is limited to one tool, whereas another underscores the applicability of the tools for healthcare trusts and local authorities. Researchers are also a category of interest, with few tools (2) including them within their scope. Lastly, healthcare facility end users, encompassing patients, staff and visitors and the general public, are referenced by four tools.

3.2 Focus and Aspects Investigated

The focus of the evaluation tools has been categorised according to the four aspects investigated: (i) design quality, (ii) user perceptions, (iii) health outcomes and (iv) sustainability. Almost all of these tools mention focusing on design quality (8), whereas few focus on user perceptions (2), health outcomes (2) and

Table 3 Target audience addressed by the selected tools

Target audience/tool number	T1	T2	T3	T4	T5	T6	T7	T8	T9	TOT
Clients, healthcare organisations or providers	x	x	x	x			x	x		6
Healthcare (HC) facilities managers and administrators	x	x				x		x	x	5
Healthcare trusts	x									1
Healthcare planners							x			1
Healthcare facilities owners							x			1
Developers	x	x					x	x		4
Financiers								x		1
Design teams or architecture firms	x	x		x		x	x	x		6
Project managers	x	x								2
Design champions	x	x								2
Interior designers					x					1
Contractors								x		1
Researchers				x		x				2
Governmental boards or bodies		x								1
Users or occupants, patients, carers, visitors	x	x						x		3
General public	x	x		x						3

sustainability (1). Only one tool assesses building performance based solely on user perceptions.

Most tools explicitly affirm to focus on one of these aspects, with only a few reviewing multiple aspects (3), particularly: (i) design quality, user perceptions and health outcomes; (ii) design quality and health outcomes or (iii) design quality and sustainability. The more comprehensive tools are typically designed to be adopted across various stages of the design process, beginning with an expert evaluation of project quality during the initial design phase and extending to gathering feedback on the building performance during post-occupancy phases, when they may also resort to user perceptions and assessments of health outcomes.

It must be noted that, although "sustainability" is directly mentioned by only one tool, when looking at the tools' evaluation domains and criteria, environmental performance and passive or active sustainable design solutions to reduce the environmental impact and enhance energy efficiency are often included in the assessment (6).

3.3 Application Phases

The tools are designed to be used in single or multiple phases of the design and construction process, in particular: (i) planning, (ii) brief, (iii) design,

(iv) construction and (v) post-occupancy. The majority of them are relevant to the briefing (5) and design (8) phases, whereas only few of them are expected to be adopted during the planning (1) and construction (2) stages. Additionally, all tools are suitable for post-occupancy studies, whereas only two tools are exclusively devoted to POE. This stands in contrast to most tools, which—in line with the progressive shift towards BPE—offer at least two or three potential application stages, usually including the briefing, design and operation phases.

3.4 Data Collection Methods

The data collection methods proposed and employed by the selected tools for the assessment of healthcare built environments (Table 4) can be classified as "objective", based on objective physical measures (i.e. taken by means of instruments) or expert judgements as the result of analytic processes of knowledge, or as "subjective", based on users' observations and perceptions shaped by the interpretation and experience of places by those who use them [23].

Among these, checklists are generally used to collect information about project quality (7), based on scoring systems that allow translating qualitative data and observations into quantitative metrics and results. On the other hand, design quality in existing buildings is often evaluated through questionnaires administered to patients, staff or even caregivers or visitors assisting patients within the facility (3). Sometimes, the scorings resulting from checklists or from the sum of individual questionnaire responses are discussed through workshops (3) or focus groups (1).

Table 4 Data collection methods adopted by the selected tools

Data collection methods/tool number	T1	T2	T3	T4	T5	T6	T7	T8	T9	TOT
Document review			x							1
Walkthrough			x	x			x			3
Standalone checklist	x	x	x	x	x		x		x	7
Interviews			x							1
Workshops	x	x						x		3
Focus groups							x			1
Questionnaires:			x			x		x		3
Patient questionnaire			x			x				2
Staff questionnaire			x			x				2
Visitor questionnaire						x				1
Measurements			x							1
Benchmarking		x								1

3.5 *Expected Outcomes and Impacts*

The tools discussed in this research propose and articulate several outcomes and impacts they seek to achieve, which can be categorised as both direct and indirect. Direct outcomes are relevant to the facility under assessment, whereas indirect impacts extend to broader implications for the design and construction practice beyond the specific project.

In this research, the outcomes and impacts explicitly mentioned in the tools' presentations and reports have been categorised into five main areas of influence: (i) appraise, (ii) compare, (iii) support, (iv) improve and (v) contribute.

Appraise Evaluating the quality of healthcare spaces, albeit from different perspectives, is the primary goal of all the discussed tools. Under this category, the illustrated outcomes are:

- Evaluate the design quality (7): The majority of tools resolve to evaluate the "quality" of the design. The concept of quality is indeed multifaceted and challenging to define [32]. Therefore, it is not surprising that this concept is interpreted differently across various tools, either in terms of building performance and effectiveness, as well as environmental quality, as the performance of design spaces against key healthcare goals, or as a synthetic overview of different "layers" of quality (environmental, organisational, social, etc.).
- Ensure that the design requirements are met (4): Several evaluation tools are aimed at verifying that the project or facility is effective in meeting the preset design and performance goals, whereas others tend to assess the design against specific evidence-based design (EBD) goals.
- Evaluate design against user needs (5): This category includes the evaluation of users' perceptions or satisfaction levels, the assessment of how well the design contributes to the physical support and well-being of end users or to population health or the extent to which spatial–physical features are successfully based on user-centred design.
- Identify strengths and weaknesses (6): Most tools list the identification of issues and weaknesses among their intents and may also pinpoint design strengths. This is oriented to the identification of areas for improvement either during the initial design stages or when opportunities arise for facility renovation or improvement.

Compare The comparison of healthcare facility performance is presented as an opportunity to assess design alternatives or for benchmarking purposes. More specifically:

- Evaluating design alternatives (5): The evaluation of building performance can be beneficial when faced with several design proposals or competing schemes, by facilitating direct comparisons, by informing design choices that align more closely with the project's vision and requirements and by understanding the trade-offs between choices; moreover, adopting assessment tools may provide a rational basis for the evaluation and selection of alternative proposals in competitive procurement processes. Overall, this practice may also provide a rationale for long-term planning, prioritising investments in new projects or existing building stocks.

- Benchmarking (3): Some tools propose their utilisation for benchmarking, either of cost and quality, against the results from other users of the tool, or benchmarking against facilities of other healthcare providers.

Support The use of evaluation tools can be advantageous in guiding and supporting stakeholders along the implementation of different phases of the design process. More specifically, as often argued by the tool developers:

- Inform decision-making (5): Building performance evaluation offers stakeholders a structured approach to inform, support, facilitate and optimise decision-making processes. Evaluation tools serve as supportive instruments to manage existing healthcare facilities or invest in new projects, especially for existing facilities to be renovated. They also inform client and user choices during the project briefing stage and facilitate decision-making or justification regarding future actions and expenditures.
- Facilitate stakeholder engagement and discussion (4): Another advantage of adopting standardised tools is that they provide a means to engage stakeholders with a structured discussion agenda. They establish a framework to guide stakeholders in effectively communicating, discussing, setting and tracking design quality goals for the project; they may also aid in monitoring design aspirations and managing expectations throughout the process. For instance, a checklist can be used as a basis for collaborating with the facility owner to determine the appropriate standards, thus facilitating comprehensive discussions about "quality" and requirements. In addition, this approach ensures clarity of understanding and a shared vision, allowing for the comparison and integration of views among different participants during the design, procurement and construction processes.

Improve The term "improve" encompasses efforts directed towards enhancing the quality of the project, with a focus on achieving optimal value and integrating evidence into the design process.

- Improve quality (8): The majority of tools claim their effectiveness in driving improvement and enhancing the quality and value of built environments in healthcare settings: first of all, they aid in identifying and resolving issues or identifying key areas for improvement; they provide consistent and useful feedback to the design, construction and maintenance processes for effective building production and use and also inform ongoing building adjustments to accommodate changing organisation needs; they allow fine-tuning the building in response to user needs and feedback and also support constant monitoring of quality improvement processes and, lastly, in the long term, these tools foster the upgrading of existing estates.
- Aim for the best value (5): According to developers and users, the utilisation of several tools can result in projects deriving increased value from building design, ultimately resulting in the best value for investment and achieving excellent

design outcomes. This is primarily attributed to their ability to collect regular and usable feedback that can support in better planning the investments based on different priorities.

- Apply evidence-based design (2): Incorporating research and evidence derived from best practices can direct design choices towards the achievement of high-quality built environments. The use of tools can guide design teams in consciously focusing on lessons learnt derived from evidence-based design, thus underpinning the design and commissioning processes with a firm evidence base. This ensures that essential evidence-based design elements are integrated into the design.

Contribute This term refers to broader impacts, beyond individual projects, which comprise generating knowledge and evidence to support EBD implementation in the healthcare design practice and leading to a potential improvement of future designs.

- Support evidence-based design (2): Obtaining feedback from existing projects and facilities contributes to the ongoing cycle of evidence-based design and construction, by providing insights into the correlation between the physical built environment and its effects on health outcomes, health organisations, user satisfaction and occupant well-being.
- Improve future designs (4): Establishing an association between a particular design solution and its effectiveness can present valuable insights into how to adjust repetitive design solution that can be used on a recurring basis; it can also inform future interventions, by capitalising on user feedback and preferences about the environment. Eventually, this knowledge can inform future projects and contribute to the improvement of the overall design quality in similar types of facilities.

4   Conclusions

Based on the presented analysis of tool attributes, a framework has been conceived with the aim of guiding the future development of evaluation tools for local care facilities.

This framework aids researchers in designing new tools by drawing on past experiences, offering an "orientation map" detailing the relationships between the following attributes: (i) target audience, (ii) aspects investigated, (iii) application phase, (iv) data collection methods and (v) expected outcomes and impacts (Fig. 1).

The development of these tools should be approached with the awareness that the presented methods may facilitate the assessment of different domains, ranging from the functional to the social and environmental aspects of building performance. However, not all domains may be equally significant in the specific context of local care facilities. For instance, functional requirements such as flexibility may

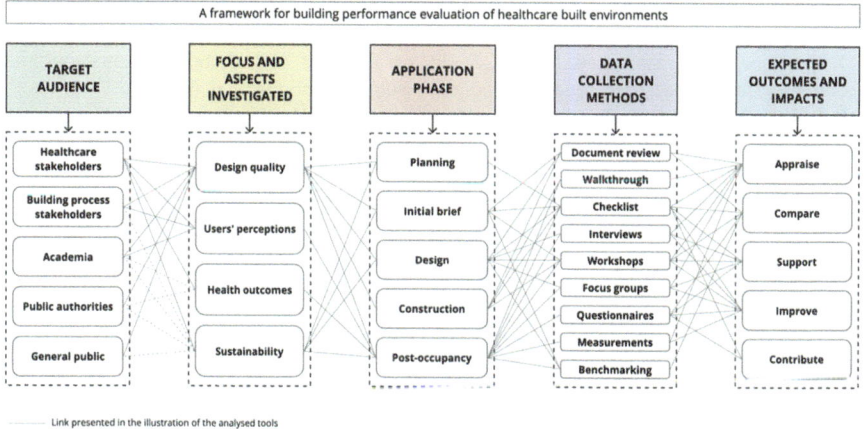

Fig. 1 A framework to support the development of healthcare building performance evaluation tools based on existing experiences

conflict with the optimisation of building systems; in the case of community facilities, resilience and adaptability present great opportunities for the reorganisation of the layouts according to contextual needs. This may enhance their overall utilisation, effectiveness and cost-efficiency, albeit at the expense of energy efficiency.

Consequently, evaluation tools can offer feedback on several evaluation "areas", pinpointing those that are the most critical for the specific typology of care facilities being addressed, therefore establishing a "weight" for each domain. As design poses challenges related to the right balance between functional, social, energy performance, this information can be then used to develop and refine decision support tools to be adopted in the briefing and design phase, to support design decisions by streamlining the prioritisation of design objectives.

References

1. Starfield, B., Shi, L., Macinko, J.: Contribution of primary care to health systems and health. Milbank Q. **83**(3), 457–502 (2005)
2. WHO: Primary Health Care Now More Than Ever. https://apps.who.int/iris/handle/10665/44349 (2008)
3. Kringos, D.S., Boerma, W.G.W., Hutchinson, A., Saltman, R.B. (eds.): Building Primary Care in a Changing Europe. European Observatory on Health Systems and Policies, Copenhagen (2015)
4. Santamouris, M., Dascalaki, E., Balaras, C., Argiriou, A., Gaglia, A.: Energy performance and energy conservation in health care buildings in hellas. Energy Convers. Manag. **35**(4), 293–305 (1994)
5. Castro, M.D.F., Mateus, R., Bragança, L.: A critical analysis of building sustainability assessment methods for healthcare buildings. Environ. Dev. Sustain. **17**(6), 1381–1412 (2015)

6. Eckelman, M.J., Sherman, J.: Environmental impacts of the U.S. health care system and effects on public health. PLoS One. **11**(6), e0157014 (2016)

7. González, A., García-Sanz-Calcedo, J., Rodríguez Salgado, D.: Evaluation of energy consumption in German hospitals: benchmarking in the public sector. Energies. **11**(9), 2279 (2018)

8. Buffoli, M., Bellini, E., Bellagarda, A.: Listening to people to cure people the LpCp – tool, an instrument to evaluate hospital humanization. Ann. Ig. Med. Prev. Comunità. **5**, 447–455 (2014). https://doi.org/10.7416/ai.2014.2004

9. Hamilton, D.K.: The four levels of evidence-based practice. Healthc. Des. **11**(3), 19–26 (2003)

10. Ulrich, R.S., Berry, L.L., Quan, X., Parish, J.T.: A conceptual framework for the domain of evidence-based design. HERD. **4**(1), 95–114 (2010)

11. Mahmood, F.J.: The role of evidence-based design in informing health-care architects. J. Facil. Manag. **19**(2), 249–262 (2021)

12. Ruddock, S., Aouad, G.: Creating impact in healthcare design: assessment through design evaluation. In: Amaratunga, R. (ed.) Proceedings of the 6th International Postgraduate Research Conference (IPRC-6), International Built and Human Environment Research Week. Technische Universiteit Delft, Delft (2006)

13. Zhang, Y., Tzortzopoulos, P., Kagioglou, M.: Healing built-environment effects on health outcomes: environment–occupant–health framework. Build. Res. Inf. **47**(6), 747–766 (2019)

14. Preiser, W.F.E., Hardy, A.E., Schramm, U. (eds.): Building Performance Evaluation: From Delivery Process to Life Cycle Phases. Springer International Publishing, Cham (2018)

15. NHS DH Estates and Facilities: A Staff and Patient Environment Calibration Toolkit (ASPECT). Instructions, Scoring and Guidance. (2004)

16. Abbas, M.Y., Ghazali, R.: Healing environment: paediatric wards – status and design trend. Procedia. Soc. Behav. Sci. **49**, 28–38 (2012)

17. NHS Scotland: Achieving Excellence Design Evaluation Toolkit. Instructions, Scoring and Guidance. (2017)

18. The Center for Health Design: A Guide to Clinic Design Post-Occupancy Evaluation Toolkit. (2015)

19. The Center for Health Design: Clinic Design Post-Occupancy Evaluation Toolkit. https://www.healthdesign.org/insights-solutions/clinic-design-post-occupancy-evaluation-toolkit-pdf-version. Last accessed 24 Feb 2024

20. Housing LIN: EVOLVE Tool – Evaluation of Older People's Living Environments. https://www.housinglin.org.uk/Topics/type/EVOLVE-Tool-Evaluation-of-Older-Peoples-Living-Environments/. Last accessed 25 Feb 2024

21. The Center for Health Design: Patient Room Design Checklist and Evaluation Tool. https://www.healthdesign.org/insights-solutions/patient-room-design-checklist-and-evaluation-tool. Last accessed 26 Feb 2024

22. Andrade, C., Lima, M.L., Fornara, F., Bonaiuto, M.: Users' views of hospital environmental quality: validation of the perceived hospital environment quality indicators (PHEQIs). J. Environ. Psychol. **32**(2), 97–111 (2012)

23. Fornara, F., Bonaiuto, M., Bonnes, M.: Perceived hospital environment quality indicators: a study of orthopaedic units. J. Environ. Psychol. **26**(4), 321–334 (2006)

24. Fornara, F., Bonnes, M., Bonaiuto, M.: Perceived hospital environment quality indicators: a comparison among general surgery units. Psicol. della Salute. **1**, 39–60 (2012)

25. The Center for Health Design: Community Health Center Facility Evaluation Tool. https://www.healthdesign.org/insights-solutions/population-health-clinic-evaluation-tool-pdf-version. Last accessed 26 Feb 2024

26. Construction Industry Council: Design Quality Indicator. (2003)

27. Gann, D., Salter, A., Whyte, J.: Design quality indicator as a tool for thinking. Build. Res. Inf. **31**(5), 318–333 (2003)

28. Construction Industry Council: Design Quality Indicator for Health 2. https://www.cic.org.uk/services/the-design-quality-indicator/design-quality-indicator-for-health-2. Last accessed 26 Feb 2024

29. Construction Industry Council: DQI for Health 2- What It Is and Why Design Appraisal Is Essential to Your Project. (2020)
30. Brambilla, A., Lindahl, G., Dell'Ovo, M., Capolongo, S.: Validation of a multiple criteria tool for healthcare facilities quality evaluation. Facilities. **39**(5/6), 434–447 (2020)
31. Brambilla, A., Apel, J.M., Schmidt-Ross, I., Buffoli, M., Capolongo, S.: Testing of a multiple criteria assessment tool for healthcare facilities quality and sustainability: the case of German hospitals. Sustainability. **14**(24), 16742 (2022)
32. Anåker, A., Heylighen, A., Nordin, S., Elf, M.: Design quality in the context of healthcare environments: a scoping review. HERD. **10**(4), 136–150 (2017)

Framework for Combined Life Cycle Environmental, Economic, and Social Assessment of Reclaimed Construction Products

Camille Vandervaeren ⓘ, Selamawit Mamo Fufa ⓘ, and Nilay Elginoz ⓘ

Contents

1 Introduction

A relevant approach to improve the sustainability of new construction and renovation projects is to limit the consumption of virgin materials by using reclaimed construction products and elements. A reclaimed construction product is one that has been previously used in a building or infrastructure project and that has been dismantled and prepared for being used again in a building or infrastructure project. In this study, we consider that a reclaimed construction product can be used in part or in whole, for the same or a different function, and on the same site or a different site, it then becomes a reused product. Reuse differs from recycling, which implies that energy is used to reprocess products into new products, materials, or substances

---

C. Vandervaeren (✉) · S. M. Fufa
SINTEF, Oslo, Norway
e-mail: camille.vandervaeren@sintef.no

N. Elginoz
IVL – Swedish Environmental Research Institute, Stockholm, Sweden

M. Kioumarsi, B. Shafei (eds.), *The 1st International Conference on Net-Zero Built Environment*, Lecture Notes in Civil Engineering 237,
https://doi.org/10.1007/978-3-031-69626-8_151

[1]. The chosen reuse definition differs from the definition in the EU Waste Frame-work Directive (WFD), which restricts reuse of products for their original intended purpose [1]. However, following the WFD definition may hinder the rate of reuse, as current practices demonstrate the challenges and limitations associated with using products again for their original purpose. An early example of the extensive use of reclaimed materials in contemporary architecture is the Villa Welpeloo built in Enschede, the Netherlands, in 2005 (architect: 2012Architecten) [2].

Reusing construction products avoids the consumption of virgin products (with all resource extraction, manufacturing, and transport it usually requires) and delays the waste treatment of that product, but also often involves some preparation activities and transport. Whether reuse will lead to environmental, economic, and social benefits depends on various parameters, including the transport to their new use, the interventions needed to put them back into the value chain, such as testing, cleaning, and repair, their remaining service life, and functional performance [3].

Currently, the sustainability of using reclaimed products in construction works is evaluated following various methods and assumptions. Additionally, there is a need for a harmonized sustainability assessment framework to compare the environmen-tal, economic, and social impact of reclaimed and new construction products over their whole life cycle [4, 5].

2   Objective and Scope

The objective of the research was (1) to evaluate the maturity of life cycle environ-mental assessment (LCA), life cycle costing (LCC), and social LCA (S-LCA) and their compatibility in the context of reclaimed and reused construction products and (2) to develop a harmonized framework for evaluating and comparing side-by-side the environmental, economic, and social impacts of reclaimed and new construction products. This chapter presents the results of the literature review, the developed life cycle sustainability assessment (LCSA) framework and an example of how the LCSA framework can support experts in comparing the sustainability of a pavement in Oslo, Norway, in reclaimed versus new paving stones.

3   Literature Review

3.1   Environmental LCA

LCA is an acknowledged and standardized method to assess the environmental impact of products and services over their whole life cycle [6]. LCA for construction products is covered by ISO 21930 [7] and EN 15804+A2 [8] standards, which are complement by several standards at building and framework levels. There exist several product-specific and generic databases of life cycle inventory and impact

Table 1 Original and adapted nomenclature of the life cycle stages

Life cycle stage	Life cycle stage name [8]	Adapted name for reused parts
A1	Raw material supply	Dismantling of the product
A2	Transport (to manufacturing)	Transport to reconditioning and testing facilities
A3	Manufacturing	Reconditioning, incl. testing
A4	Transport (to construction site)	Transport from seller to building site
A5	Construction/installation	Construction/installation
B1–B7	Operational stage	Operational stage
C1–C4	End-of-life stage	End-of-life stage
D	Potential benefits and loads outside the system boundary	Potential benefits and loads outside the system boundary

assessments of construction materials and products, such as Ecoinvent [9] and Environmental Product Declarations (EPDs). The EN 15804+A2 standard for LCA of construction products provides a nomenclature for the different product life cycle stages of construction products, ranging from A1 to C4, and D (Table 1), but this nomenclature is usually not used in LCC and S-LCA of construction products.

Many studies have shown the environmental benefits of using reclaimed construction products through LCA [3, 10]. A methodological aspect that can greatly affect LCA results is the allocation approach [11]. The EN 15804+A2 [8] standard requires that loads and benefits related to the reuse of secondary materials and products to be allocated following a cutoff (or recycled content) approach, meaning that the environmental impacts from the primary production are allocated to the primary products, while the secondary products are considered burden-free, except for any impacts resulting from the reuse process. Impacts related to future reusability are not part of the product system under study but are reported in a separate module (Module D).

3.2　Life Cycle Costing

Some authors consider that there are three different approaches of LCC [4, 12, 13]. Conventional LCC considers the costs borne by one actor (usually the producer or the consumer) and does not cover all life cycle stages [4, 13]. It is the LCC approach covered in ISO 15686-5 [14] and NS 3454 [15], which is commonly used in Norway. The scope of a conventional LCC is considered less comprehensive than those of an LCA [13]. Environmental LCC is meant to have scope (system boundary and functional unit) in line with the scope of an LCA and includes direct costs for the different actors over the value chain, hence internalizing the costs of externalities (such as for a manufacturer, the cost linked to repairs, and final disposal) for different actors [13]. Societal LCC considers direct costs and costs of externalities for the whole society and aims to covers environmental and social effects; hence, it is not complemented by an LCA [13]. EN 16627 [16] specifies the calculation methods to assess the economic performance of a building. It is based on the principles of LCC defined in ISO 15686-5, adapted to the European context [16].

3.3   Social LCA

Social LCA (S-LCA) follows the ISO 14040 framework for LCA [6] and includes four phases: goal and scope definition, life cycle inventory, impact assessment, and interpretation [17]. Social-LCI is developed for indicators (e.g., regular and documented payment of workers, number of employees earning wages below poverty line) linked to impact subcategories (e.g., fair salary) that can be linked to impact categories (e.g., working conditions) [17, 18]. The impact categories and subcategories assessed in S-LCA are linked to stakeholder categories (workers, local community, society, consumers, value chain actors, and children), which can be directly affected positively or negatively during the life cycle of a product [17].

UNEP Social LCA guidelines published in 2009 and revised in 2020 [17] is a widely used and up-to-date guidance document, which is accompanied by Methodological Sheets for Subcategories in Social Life Cycle Assessment (S-LCA) published in 2021 [18]. The ISO standards "Principles and framework for social life cycle assessment" is under development. Two of the "pilot projects on guidelines for S-LCA of products and organizations," published in 2022 by life cycle initiative and social life cycle alliance are aluminum profile and steel production cases. These pilot projects can be used for guidance for construction product S-LCAs. Although S-LCA has been used for evaluation of building materials [19], building construction [20], and recycled construction materials [21], its application on reused construction materials is not common.

S-LCA can be applied in two levels, either together or separately: (1) hot spot analysis using available databases like Social Hot Spots Database (SHDB) and Product Social Impact Life Cycle Assessment (PSILCA), which provide social impacts of industries in different countries, and (2) site-specific S-LCA by collecting inventory data for the investigated facility. Compared to LCA databases, the current social databases are less detailed, with social impacts reported per sector and country. For example, it is not possible to distinguish between different types of mineral construction products. However, social databases allow quick comparison of risks related to different producer countries for the same sector or risk mapping.

3.4   Combined LCA-LCC-SLCA Assessments

The European Standard EN 15643:2021 [22] provides a framework for the sustainability assessment of buildings, including their environmental, economic, and social performance; it does not apply to construction products. Some methodological choices that assessors can make within the framework defined by the standard, such as the definition of the system boundary, affect the way the benefits of reuse are calculated [11, 23]. These benefits also depend on assumptions on uncertain parameters, such as the technical performance of the reclaimed product or its remaining service life.

4 Methods

4.1 LCSA Framework Development

The LCSA framework has been developed as part of the REBUS research project [24], which identified user needs, developed guidelines, and recommendations to increase the reuse of building products in Norway. The project lasted from 2020 to 2023 and focused on five types of products: paving stones [10], interior glass partitions, windows and doors, ventilation components, and sanitary equipment. The scope of the project has influenced the development of the LCSA framework (especially, as it builds on the CEN standards), its applicability to other countries, and product types has not been assessed. The LCSA framework was developed based on findings from the review of the literature (summarized in Sect. 3) and on expert knowledge among the REBUS project partners, including discussions with a consultant specialized in the reclaim of construction products. Methodologically, the underlying principle is that the framework should not prevent the assessor from testing different methods (e.g., different allocation, system boundary, and life cycle costing approach), but that the assessment should be conducted following at least one specific approach: a cutoff (also called "100:0" and recycled content) allocation in the LCA and S-LCA and a conventional LCC approach.

4.2 Pavement Case Study

The case study consists of the comparison of reclaimed and new paving stones for their environmental, economic, and social performance. The functional unit is defined as 1 m^2 of pavement, built in Oslo in 2021 with granite stones, sand, and gravel. The options compared are (1) a pavement with 100% new granite stones (small version) produced in Sweden and (2) a pavement with 100% used granite stones (small version) reclaimed from a street in Oslo. In both cases, the layer of paving stones (10 cm) lies on a layer of new sand (5 cm) and gravel (40 cm).

For the LCA and LCC part, the inventory and assessment data were extracted from a previous report by Fufa et al. [10] where different options and scenarios are evaluated. In the report, the LCA and LCC analyses are conducted in Microsoft Excel and following the standards EN 15804+A2 and NS 3454. The environmental indicator they assess is Global Warming Potential (in CO_{2e}). Due to lack of EPD for granite paving stones produced in Sweden, Fufa et al. [10] use EPD data for granite paving stones produced in a neighboring country (Finland). The present study uses the now available EPD (number S-P-04621) for granite paving stones produced in Sweden [25], in order to be consistent with the results for the LCC and S-LCA. We follow here the same assessment method as Fufa et al. [10]. Since both options are expected to have the same impact during use and end-of-life, the system boundary for the LCSA covers the production, transport, and installation of the pavement (stages A1 to A5). For the LCA system boundary, Fufa et al. [10] follow a cutoff (recycled content) allocation approach and consider that, once deconstructed and stacked, the reclaimed stones are functionally equivalent to new ones. Therefore, the system boundary for the pavement in reclaimed stones only includes storing the stones in an open-air local storage, transporting some of them to a laboratory and testing them, transporting the rest of them to the re-installation site in Oslo and manually installing them. Fufa et al. [10] consider that, in the reclaimed option, only transport to laboratory and construction site are emitting GHG, the rest being mostly manual activities with neglectable energy consumption, material loss and waste treatment. The system boundary for the pavement in new stones produced in Sweden included material extraction, transport, and manufacturing of the stones, transport to site and manual installation. The system boundary for both options also included the production (in Norway), transport, and installation of the new sand and gravel.

For the LCC, A1–A3 covers the purchasing cost of stones, gravel, and sand. A4 includes the cost of transporting stones, sand, and gravel to the construction site, while A5 includs the cost of installing the pavement with new or reclaimed stones. The cost of testing new stones was assumed to be included in the purchased price. However, for reused stones, testing costs were not included in the purchased price. Even if the cost analysis from testing of the reused stone samples were conducted, it was excluded from the LCC analysis due to uncertainties in the cost scenarios. Data collection method for LCA and LCC can be found in the report [10].

For the S-LCA part, OpenLCA [26] was used with social impact data from SHDB 2019 with "Category Method with Weights" [27]. The social impact categories covered by this database and method are listed and described by Pelletier et. al [28] as: "Labour Rights and Decent Work (with indicator data for child labor; forced labor; excessive working time; wage assessment; poverty; migrant labor; freedom of association, right to strike, and collective bargaining rights); Health and Safety (with indicator data for injuries and fatalities; and toxics and hazards); Human Rights (with indicator data for indigenous rights; gender equity; and high conflicts); Governance (with indicator data for legal system and corruption); and Community Infrastructure (with indicator data for access to hospital beds; improved drinking water; and improved sanitation)." The cost data used as an input in S-LCA was the collected cost data [10], converted to USD_{2021} and without taxes. The reclaimed stones were considered to have zero social impact. The social impacts were accounted for the production of sand and gravel, transport of sand and gravel, and construction of the pavement.

5 Results

5.1 LCSA Framework

Based on the current state of the literature, we decided that the LCSA framework should support consultants and researchers with expertise in LCA, LCC, and/or S-LCA in conducting and reporting the assessment. Therefore, we developed (1) a set of methodological choices, (2) a data collection spreadsheet for general information about the product to assess and its inventory of processes and material flows, and (3) a reporting spreadsheet. The data collection and reporting spreadsheets are combined in a Microsoft Excel file, complemented by a glossary and user guide.

The spreadsheet is only meant for collecting and structuring the product, product inventory, and impact assessment data, and not to generate this data. The impact assessment data should be imported into the spreadsheet from an LCA, LCC, or S-LCA calculation tool. We developed the structure of the Excel spreadsheet based on the four-step approach in LCA of goal and scope definition, inventory, impact assessment, and interpretation [8].

Although Environmental LCC aligns best with LCA on a methodological level [13], we chose a conventional LCC approach (from the perspective of one actor) to simplify the assessment and avoid a formal distinction between internal costs and externalities. This way the costs reported by an actor can be directly reported in the Excel spreadsheet. Regarding the LCA allocation approach between use cycles, our framework requires to apply at least the cutoff approach, which is in line with the "polluter pay" principle in EN 15804+A2. The assessors may complement the assessment with other allocation approaches if it seems relevant for the case. In the inventory data collection sheet for the reclaimed and new products, the different life cycle stages are listed vertically. We also have adapted the name of the life cycle stages for the reclaimed/reused construction product (Table 1). Under each stage, the assessor can list the material flows and processes occurring within the stage. Each material flow and process can be described with the following categories: material flow or process unit (e.g., kg, m^2), amount per unit, cost per unit (both in NOK, including taxes and in USD, excluding taxes), country (where the material flow or the process takes place), and year (when the material flow or the process takes place). Under each life cycle stage, we have provided some examples of material flows and processes that are typically occurring in the different life cycle stages. Some of them might not be relevant for the product case to be assessed. To simplify the data collection, the assessor can answer a questionnaire which may eliminate irrelevant life cycle stages, material flows and processes, and copy values from the other product data collection sheet (Fig. 1).

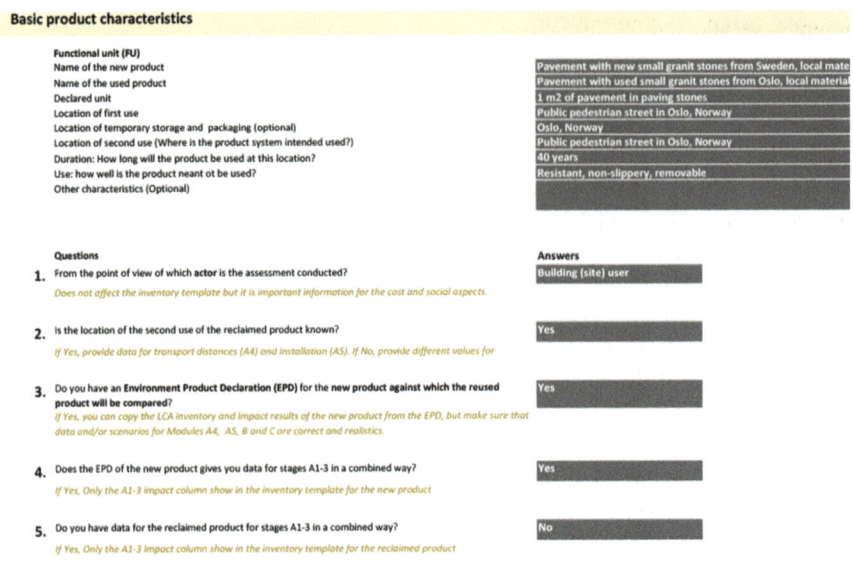

Fig. 1 Screenshot from the LCSA framework for the product characteristics

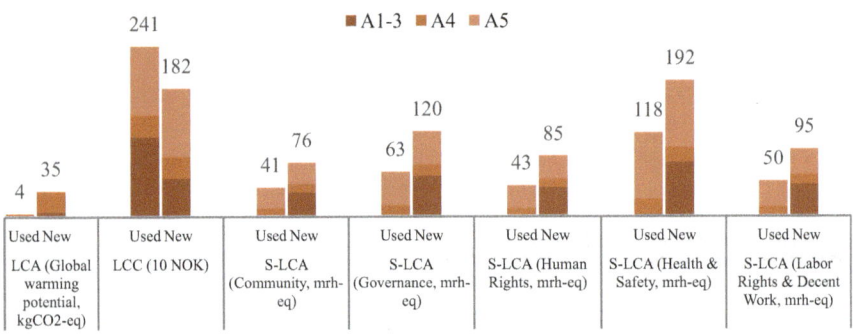

Fig. 2 LCSA assessment case study results for a pavement in used and new granite stones

5.2   Pavement Assessment Results

The results of the LCA show that using reclaimed instead of new paving stones in pavements can result in a 90% reduction in GHG emissions (Fig. 2, LCA). These LCA results are similar to those obtained by Fufa et al. [10], which indicate a 93% reduction in GHG emissions by using reclaimed stones. The main contributor to GHG emissions in pavement with new paving stones is the transport of the paving stones (80%). The A1-A5 GHG emissions from pavement in reused stones mainly come from gravel production (37%) and transport (34%). There are no GHG emissions from the manual installation (A5) of the stones. By choosing new stones from Sweden instead of Finland, the A1–A3 GHG emissions for the pavement increase by 40%, while A4 GHG emissions decrease by 38%. The production, transport, and construction of the pavement with new stones from Sweden potentially emits 35.4 $kgCO_2e$, while the one with stones from Finland emits 50.3 $kgCO_2e$.

The LCC results indicate that the reuse of paving stones increases the total costs (A1–A5) by more than 33% compared to using new stones (Fig. 2, LCC). The total costs of the pavement in reused stones originate primarily from the purchase of the pavement elements (47%) then from the installation onsite (40%). In contrast, for a pavement with new stones, it is first the installation (54%) then the purchase of the materials (30%). While reuse can significantly reduce the GHG emissions of the pavement, it makes it more costly, as in the KA13 building case in Norway [29].

Since the social impacts significantly differ from country to country for the same sector, S-LCA results depend on the country where the paving stone is produced in the new stones case. Here it is assumed that the paving stones are produced in Sweden in the new stone case, and in both cases, gravel is produced in Norway. The results show that the social impacts in all categories are 62% to 97% higher in the new stones case compared to reuse case (Fig. 2, S-LCA). In the new stone produced in Sweden case, 40% to 50% of the impacts come from production of the new stones and gravel, 38% to 49% come from construction, and 11% to 14% from transportation. To see the country differences in social impacts, two more options were evaluated; new stones produced in China and Portugal. Compared to a pavement in reused stones, a pavement in stones produced in Portugal scores 54% to 78% higher and a pavement in stones produced in China scores 105% to 340% higher.

6   Conclusion

The REBUS LCSA framework was developed to support assessors when comparing and reporting the sustainability of reused and new construction products. The framework shows it is possible to harmonize the various assessment methods and expand the scope of product assessment from purely GHG emissions to overall sustainability.

The case study demonstrates how the LCSA framework can be utilized to compare GHG emissions, costs, and social impacts using both existing and new LCI data and LCIA results. It also shows that the additional data needed to expand from GHG assessment to sustainability assessment in the case study is relatively limited. The results of the LCSA assessment of the pavement in reclaimed and new stones show that reuse can bring lower environmental and social impacts, while being more costly. Here it should be noted that, this study assumed a 100% reuse rate for the reclaimed paving stones and slightly higher GHG emissions can be expected when mixing new and reclaimed paving stones [10].

Among the limitations of the framework, the scopes of the LCC, the LCA, and S-LCA do not align perfectly. The LCC is conducted from the perspective of one actor in the value chain, while the LCA takes a global perspective. The S-LCA scope lies between these two and takes the perspective of a set of stakeholders within the value chain. To better understand the economic and social impacts, it would be necessary to take a multistakeholder perspective. Also, the framework has only been tested on one product (the paving stone) and has not yet been tested outside the REBUS research project. In the future, the framework should be tested by experts in one or more life cycle approaches for different actors in the reuse value chain (e.g., product resellers, building clients, architects, contractors, and waste handlers). It should also be noted that the framework does not set any rule for setting of the system boundary between the primary product and the reclaimed product (i.e., end of waste state) because this boundary is dependent on the type of product and the local market. The assessor must still choose the system boundary and ideally analyze the sensitivity of the results to different choices of system boundary.

Besides inherent limitations, this case study is, to our knowledge, the first comparative LCA, LCC, and S-LCA of a reclaimed construction product, and this paper contributes to the so far very limited exploration of LCC and S-LCA of reclaimed construction products. Looking at multiple aspects over the life cycle of the product reduces the risks of shifting the problem to another stage of the life cycle (e.g., from production to end-of-life) or to another sustainability indicator (e.g., economically beneficial but with higher social impacts). The LCSA framework can contribute to promoting a more holistic approach to sustainability that encompasses different environmental, economic, and social issues.

Acknowledgments This research was supported by The Research Council of Norway through the project "REBUS—Reuse of building materials—a user perspective," under grant no. 302754.

References

1. European Commission: Waste Framework Directive. https://environment.ec.europa.eu/topics/waste-and-recycling/waste-framework-directive_en. Last accessed 2023/01/04
2. van Andel, F.: Villa Welpeloo Enschede: 2012Architecten. DASH Delft Archit. Stud. Hous., 148–155 (2012)
3. Brütting, J., Vandervaeren, C., Senatore, G., De Temmerman, N., Fivet, C.: Environmental impact minimization of reticular structures made of reused and new elements through Life Cycle Assessment and Mixed-Integer Linear Programming. Energy Build. **215**, 109827 (2020). https://doi.org/10.1016/j.enbuild.2020.109827
4. Wouterszoon Jansen, B., van Stijn, A., Gruis, V., van Bortel, G.: A circular economy life cycle costing model (CE-LCC) for building components. Resour. Conserv. Recycl. **161**, 104857 (2020). https://doi.org/10.1016/j.resconrec.2020.104857
5. Larsen, V.G., Tollin, N., Sattrup, P.A., Birkved, M., Holmboe, T.: What are the challenges in assessing circular economy for the built environment? A literature review on integrating LCA, LCC and S-LCA in life cycle sustainability assessment, LCSA. J. Build. Eng. **50**, 104203 (2022). https://doi.org/10.1016/j.jobe.2022.104203

6. ISO 14040+A1: ISO 14040:2006/Amd 1:2020(en) Environmental management – Life cycle assessment – Principles and framework – AMENDMENT 1. International Organization for Standardization (ISO), Geneva. (2020)
7. ISO 21930:2017: Sustainability in buildings and civil engineering works Core rules for environmental product declarations of construction products and services. (2017)
8. EN 15804+A2:2019: Sustainability of construction works – Environmental product declarations – Core rules for the product category of construction products. European Committee for Standardisation, Brussels. (2019)
9. Ecoinvent: Ecoinvent Database v3.8. https://ecoinvent.org/the-ecoinvent-database/. (2021)
10. Fufa, S.M., Plesser, T.S.W., Grytli, T.: Ombruk av gatestein. Kartlegging, prøving, LCA og kostnadsanalyser (2021)
11. Malabi Eberhardt, L.C., van Stijn, A., Nygaard Rasmussen, F., Birkved, M., Birgisdottir, H.: Development of a life cycle assessment allocation approach for circular economy in the built environment. Sustainability. **12**, 9579 (2020). https://doi.org/10.3390/su12229579
12. Swarr, T.E., Hunkeler, D., Klöpffer, W., Pesonen, H.-L., Ciroth, A., Brent, A.C., Pagan, R.: Environmental life-cycle costing: a code of practice. Int. J. Life Cycle Assess. **16**, 389–391 (2011). https://doi.org/10.1007/s11367-011-0287-5
13. Hunkeler, D., Lichtenvort, K., Rebitzer, G.: Environmental Life Cycle Costing. (2008)
14. ISO 15686-5:2017: Building and construction assets – service life planning. Part 5: Life-cycle costing. (2017)
15. NS 3454: 2013: Livssykluskostnader for byggverk – Prinsipper og klassifikasjon/Life cycle costs for construction works – Principles and classification. (2013)
16. NS-EN 16627:2015: NS-EN 16627 Sustainability of construction works – Assessment of economic performance of buildings – Calculation methods. (2015)
17. UNEP: Guidelines for Social Life Cycle Assessment of Products and Organisations 2020 – Life Cycle Initiative. UNEP. (2020)
18. UNEP: Methodological Sheets for Subcategories in Social life cycle assessment (S-LCA) 2021. United Nations Environment Programme (UNEP). (2021)
19. Hosseinijou, S.A., Mansour, S., Shirazi, M.A.: Social life cycle assessment for material selection: a case study of building materials. Int. J. Life Cycle Assess. **19**, 620–645 (2014). https://doi.org/10.1007/s11367-013-0658-1
20. Dong, Y.H., Ng, S.T.: A social life cycle assessment model for building construction in Hong Kong. Int. J. Life Cycle Assess. **20**, 1166–1180 (2015). https://doi.org/10.1007/s11367-015-0908-5
21. Hossain, M.U., Poon, C.S., Dong, Y.H., Lo, I.M.C., Cheng, J.C.P.: Development of social sustainability assessment method and a comparative case study on assessing recycled construction materials. Int. J. Life Cycle Assess. **23**, 1654–1674 (2018). https://doi.org/10.1007/s11367-017-1373-0
22. EN 15643: EN 15643:2021 – Sustainability of construction works – Framework for assessment of buildings and civil engineering works. (2021)
23. Hoogmartens, R., Van Passel, S., Van Acker, K., Dubois, M.: Bridging the gap between LCA, LCC and CBA as sustainability assessment tools. Environ. Impact Assess. Rev. **48**, 27–33 (2014). https://doi.org/10.1016/j.eiar.2014.05.001
24. REBUS: REBUS – Reuse of Building Materials – a USer Perspective. https://www.sintef.no/projectweb/rebus/. Last accessed 2024/02/26
25. Naturstenskompaniet Sverige, A.B.: Naturstenskompaniet Sverige AB Natural Stone Products of Granite and Limestone. The International EPD® System (2021)
26. GreenDelta: openLCA. https://www.openlca.org/
27. SHDB: Social Hotspot Database. http://www.socialhotspot.org/. Last accessed 2024/02/26
28. Pelletier, N., Ustaoglu, E., Norris, C., Norris, G., Rosenbaum, E., Vasta, A., Sala, S.: Social sustainability in trade and development policy. Int. J. Life Cycle Assess. **23** (2018). https://doi.org/10.1007/s11367-016-1059-z
29. Entra: Erfaringsrapport ombruk Kristian August gate 13. (2021)

The Renaissance of Reuse in Norway: The Future Is Back

James Kallaos [ID] and Camille Vandervaeren [ID]

Contents

1 Introduction

Society has historically valued both human labor and natural resources, including the intermediate and end products produced. Discarding these artifacts was seen as squandering valuable effort and resources, and waste was historically an uncommon and avoided activity. This tradition was rapidly lost in many western countries, replaced by a more rapid linear approach whereby materials quickly went from cradle to grave in a linear system or economy. There has been a recent revival of the idea of moving back from this modern linear model, however, with increasing pressures from environmental degradation, climate change, material scarcity, and population growth.

Much of the current work surrounding circularity in the construction industry appears to track with the recommendations made with the last wave of research into

J. Kallaos (✉) · C. Vandervaeren
SINTEF Community, Oslo, Norway
e-mail: james.kallaos@sintef.no

© The Author(s) 2025
M. Kioumarsi, B. Shafei (eds.), *The 1st International Conference on Net-Zero Built
Environment*, Lecture Notes in Civil Engineering 237,
https://doi.org/10.1007/978-3-031-69626-8_152

reuse—from the mid-1990s to the mid-2000s, several research projects were conducted, and a large body of research was published. This chapter draws heavily on that period of research, as it helps highlight that many of the reuse discoveries being made today have been well known for decades.

2   Reclamation Audits: *Ombrukskartlegginger*

The recent revival of Norwegian circularity has resulted in a flurry of activity, with one of the most notable being the new requirement for reclamation audits, requiring renovation and demolition activities in apartment blocks or commercial buildings meeting certain size or mass criteria (100 m^2 or 10 tons) to conduct an assessment of the reuse potential of the materials and elements scheduled for disposal [1, 2]. The few audits which have been published in Norway (available via interactive map [3]) appear to have started from scratch, without building on the decades of research which could inspire or inform a modern approach. The results of the few publicly available reports are qualitative, text-based observations and assessments, which don't necessarily follow European [4, 5] or Norwegian recommendations [2, 6], with regard to consistency of formatting and the provision of standardized or quantitative values.

Nearly 25 years ago, researchers at the University of Florida presented their "Building Deconstruction Assessment Tool" developed for residential buildings based on experiences derived from a short project, which showed that deconstruction was cost-competitive with demolition if the resale of recovered materials was included [7]. The assessment tool was made up of three sections, with increasing detail, including a preliminary evaluation, a regulatory cost estimate, and finally a detailed (component level) assessment and inventory [7]. The output was a building deconstruction feasibility score which yielded different recommendations (demolish, deconstruct, or move) based on the resulting score ranges [7].

The current approach toward reclamation audits in Norway is admittedly a first step, with no actual requirements beyond conducting the audit itself. The process is expected to grow and adapt in both content and application area, with more requirements expected to be introduced including greenhouse gas accounting [8].

Several minor revisions to the process could be expected to simplify and yield both better data and better results. The development and use of a common software tool for reclamation audits in Norway could provide more consistent formats and comparable results with systematic outputs. At a minimum, this consistency and comparability could be expanded with the use of a standardized, mandated spreadsheet for reporting (e.g., [4]). Along with a follow-up post-reclamation audit, the data could be used for both internal and external comparisons—such as comparing the total amounts indicated in the table with the quantities sent for reuse or attributed to different waste fractions, and developing a database of expectations based on different building attributes and archetypes.

3   Reuse and Availability

The new Norwegian requirement for reclamation audits highlights a known problem with timing and quality assurance: the items are generally documented in situ where they exist in a sort of Schrodinger's dilemma. The item and its quality are documented only at the time and place of the actual audit, which should identify and document reusable components [6], but the documented attributes and quality may change when the component is deconstructed and delivered to the site. The timing is also difficult without an adequate storage area—the timing of the deconstruction and the new construction would need to line up perfectly. With new materials purchased from logistics companies or middlemen, the deliveries can be scheduled to arrive exactly as they are needed. The old proverb "a bird in the hand is worth two in the bush" translates into Norwegian as "…worth ten on the roof" but rings equally true—uncertainty increases risk and decreases perceived value. Construction companies and contractors are restricted by deadlines and space limitations, and probable reusable materials do not fit well with rapid, choreographed modern construction practices—highlighting the importance of having public or private "reuse hubs" dealing in controlled or controllable reclaimed construction materials.

Increasing the risk and uncertainty from reclamation audits is the legacy of the past 75 years of modern methods of construction (MMC), which have involved rapid technological developments, including the introduction of new materials and methods which do not promote deconstruction, disassembly, and reuse [9]. For decades, researchers have been pointing out the flaws in MMC and calling for more methodical and conscientious approaches to construction—with an eye toward the future instead of the immediate present (see e.g., [10–13]).

A set of strategies (Fig. 1) developed decades ago shows an early version of the circular economy and EU waste hierarchy [14] applied to buildings [15, 16].

The four cascading strategies for material reuse apply at different processing levels and feed back into the relevant stage of the building life cycle:

1. Relocation (building level).
2. Reuse (component level—assembly stage).
3. Reprocessing (material level—manufacturing stage).
4. Recycling (material level—processing stage) [19].

The waste hierarchy itself has been around for nearly 50 years and has gone through multiple iterations. The Ladder of Lansink presented a seven-level top-down list: Prevention, Element reuse, Material reuse, Useful application, Incineration with energy recovery, Incineration, and Landfill [17, 18]. The Delft Ladder followed with ten stages, and the current EU waste framework directive (Fig. 2) trimmed back down to present a five-step "waste hierarchy" [14, 18]. The common denominator in all approaches to the waste hierarchy is that prevention is generally the preferred option, and disposal in a landfill is generally last. The current approach to reuse is well intentioned and a much needed (re)development, but it is attempting to put the cart before the horse. The top approach "prevention" should have been

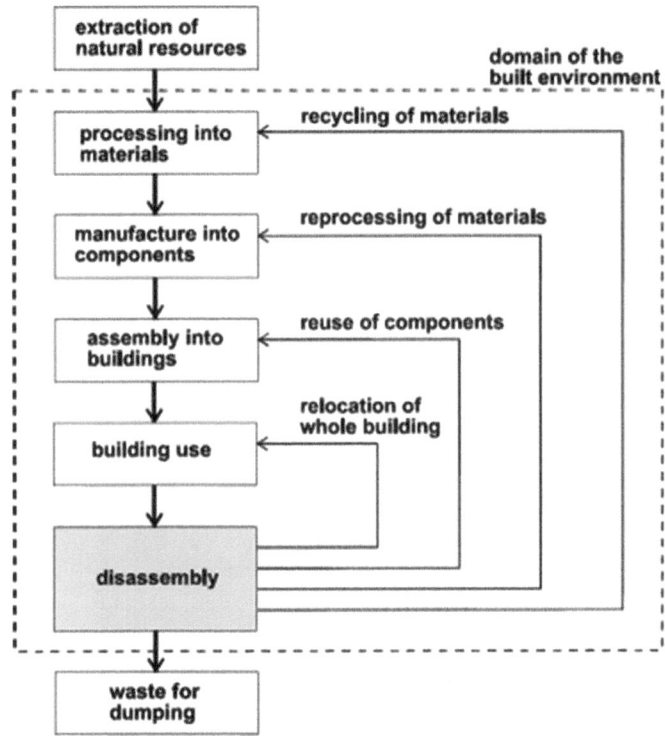

Fig. 1 Crowther's "Scenarios for Reuse in the Life Cycle of the Built Environment" [20]—diagram reused with author permission

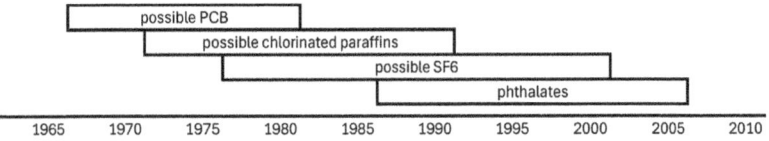

Fig. 2 Timeline of selected hazardous materials in insulated windows [41–46]

introduced at the design stage: "Prevention tries to prevent the production of waste. This step must be taken before a building is demolished, in the design and building stage" [18, p. 96]. As Crowther noted in 1999, with regrettable premonition, there are plenty of "...forgotten technological strategies that can enhance the possibilities of separating materials, components and whole buildings for reuse and recycling" [19, p. 35]. Most buildings being demolished today were not built with prevention in mind. Those that were designed with prevention in mind were built in a different era, before the rapid rise of industrialization, and are often old enough to be given protected status.

The traditions of building with reversible connections can be considered the original form of Design for Disassembly (DfD), whether ancient (e.g., stacking, layering, interlocking) or more modern (e.g., screws, bolts). These assembly techniques allowed for straightforward disassembly or deconstruction, whether the design decision was intentional [20] or not [21]. Ancient forms of transportable housing (e.g., tipi, yurt, and lavvu) and their more modern equivalents, like the Dymaxion House in 1927, the BLPS House in 1937, the Geodesic dome in 1947, or the Nakagin Capsule Tower in 1970, also integrated reversibility and modularity into the designs [22–24].

Design for Disassembly reemerged in the 1970s as a response to environmental pressures, initially in the realm of product design—when extended to buildings this is also called Design for Deconstruction, both abbreviated DfD. Depending on construction and material type, reuse will likely be a losing effort without some form of DfD. Stick frame wood buildings, as are common in single family residential housing in much of the United States, may be the easiest of modern construction techniques to disassemble (deconstruct), though generally built with nails, and demolition remains the norm. The buildings are relatively small and low to the ground. Considering these conventional wood-framed houses in a study in the United States, Guy and McLendon found that their deconstruction is generally more expensive than demolition, but the extra expense may be offset by the sale of salvaged materials [25]. In their review of deconstruction projects for 12 wood-frame buildings built between 1900 and 1950, Chini and Nguyen found a wide range of recovery rates (28–82%) owing to a variety of factors [26]. They considered the effects of project factors on the diversion/recovery rates: building age, condition, and size, as well as lot size and site conditions, along with the presence of hazardous materials [26]. Lot size and access conditions are positively correlated, with larger lots and easier access leading to higher recovery rates [26]. The presence of hazardous materials entails extra time and costs, and lower material recovery rates [26], and complicates any attempt at generalization, as demolition with some materials (e.g., asbestos) may cause the entire waste load to be categorized as hazardous waste [27].

Considering that most everyone agrees that prevention is the most important part of the waste hierarchy, it is disappointing that, outside of academics endlessly pleading, nearly nothing has been done to promote it. This may be due to the principal-agent problem, whereby the one who pays is not the one who obviously benefits. "... the potential carbon benefits of reusing construction materials are not gained immediately but at the point of future retrieval" [28, p. 6]. Adding extra effort to design and construct buildings which will only benefit an unknown future owner and an unknown future society is a difficult proposition for private actors. The tragedy of the commons has been described since Aristotle [29], and yet we seem no closer to resolving it.

Back in 1995, Spence & Mulligan proposed tandem regulatory and industry approaches to help resolve some of these issues, including:

- Regulatory:
  - Carbon tax.
  - Resource extraction tax (minerals, trees, etc.)

- Landfill waste charges.
- Fines/charges for pollution.

- Industry (to reduce overall materials use).

  - Consider embodied energy (use low-energy materials and structural systems).
  - Low rise instead of high-rise buildings.
  - Use waste or recycled materials.
  - Reduce site and recycling waste.
  - Design for recycling, long life, and adaptability [30].

In the 30 years since these recommendations were proposed, some have been enacted, and others have been ignored. Oslo, for example, has been at the forefront of promoting the use of recycled and reused material, while at the same promoting a new policy toward more tall buildings downtown [31].

Laws and regulations are another approach to influencing (or imposing) change. The EU ecodesign regulation proposal is a promising step forward toward harmonizing European approaches to reducing waste and increasing circularity. It includes an important statement regarding the restriction of "substances that hinder circularity" [32, p. 26]. While the ecodesign proposal is an important step for sustainability and circularity in the EU and Norway, it includes text deferring to the Construction Products Regulation (CPR) [33], where except in "exceptional cases" ... "...this regulation should not set requirements on final construction products when requirements on environmental sustainability for such products have already been developed under the [CPR]" [32, p. 96]. It will be interesting to see if the goals of the ecodesign proposal, which is written clearly and directly, will be properly filled through adherence to the CPR, which is a much more opaque and legally technical document, and one which "does not set product requirements" but defers responsibility to the Member States [33, p. 1].

The Norwegian Government Agency for Financial Management (DFØ) that directs public procurement has released a criteria set and web-based tool to guide sustainable procurement [34]. The criteria set includes a section on Circular economy, with a subsection on building for adaptability and reuse. The specific minimum criteria to be addressed for adaptability and reuse are "Adaptable buildings, Correct lifetime of component, Flexible joints, Labelling of materials and components for reuse, Sources of hazardous substances that restrict future reuse, Homogeneous materials, & Sensible layering" [34].

4   Risky Reuse

Many construction products currently in use in buildings contain materials or substances which are no longer acceptable or legal for introduction into buildings. While these products were accepted products at the time of installation, they are no longer approved and generally cannot be reused in Norway. It is likely that this trend will continue, with products which are currently considered legal and safe being determined unsafe and/or having their approval revoked over time. Even well-intentioned changes may result in "regrettable substitution"—as we have seen with replacement plastics for water bottles [35–38]. Reuse professionals (and the circular economy "industry") will need to deal with the existing and future threat to circularity of currently inbuilt materials. REACH legislation has reduced the future risk, but the fields of health and science are in an endless state of discovery. Reversible connections and DfD could help reduce this unknown future risk by simplifying remediation, removal, and replacement. Windows and concrete slabs are used here as examples, but the list of risky materials and components is not short [39, 40].

4.1 Windows

The reuse utility of insulated windows is diminished owing to both the degradation of insulating properties over time and increasing insulation requirements in new construction. Insulated windows removed from their planned use (between interior and exterior) can find themselves reused in spaces where there is no formal requirement for insulating properties (e.g., partition walls), yielding their technical qualities mostly irrelevant. The possible reuse of these windows then becomes a question of whether it is safe or allowed to reuse them. With the investigation and testing of each product likely prohibitively time consuming or expensive, the simpler approach is to find and check the manufacturing date to see whether currently prohibited substances were used at the time of manufacture. With insulated windows, the construction dates where reuse should be avoided are overlapping and cover most of the past 60 years (Fig. 2), with lead, asbestos, and PCBs also occurring in paint, sealants, and putty [41–46].

4.2 Hollow-Core Concrete Slabs

In a reuse pilot, 21 hollow core slabs were removed from a government building being demolished in Oslo, cleaned and tested, and then used a few blocks away, in the ambitious project known as KA13 [47]. Lessons learned from the slab reuse in KA13 helped inform a new Norwegian standard on reuse of prefabricated hollow-core concrete slabs [48]. The slab reuse involved direct costs estimated to be 5–6 times the cost of new elements, not including extra time [49]. The extra costs were associated with disassembly and removal, extra processes due to bracing, supporting, drilling, cutting, and hoisting, testing, transportation, cutting, processing, cleaning, preparation, and assembly, extra labor for design and administration [49]. This was a novel project, and future costs will likely decrease due to experience, but much of the extra expense was due to the lack of planned reversibility and DfD.

4.3   Connections

Connections that are not intended to be reversible, such as those with glues, adhesives, mortars, or one-way mechanical fasteners, are problematic for multiple reasons. Not only do these connections make disassembly difficult or impossible, limiting reusability, but larger components (which could otherwise be reusable) may be rendered unusable due to attachment to a small but unusable fraction. One example of this problematic issue is with several decades worth of structural blocks which are built into Norwegian homes. Leca Isoblokk (expanded clay structural block with polyurethane foam core) from the 1980s to the 2000s contained CFCs and HCFCs in the foam insulation. The entire assembly is hazardous waste due to the irreversible bond between the hazardous and nonhazardous components. Removed blocks cannot be reused and must be destroyed in a specific manner. At least 17 different types of integrated hazardous materials have been identified in or attached to heavy building materials, spanning different usage periods—concentrated in the decades after World War 2, but ranging from years to centuries, up to the present day [50].

5   The Return

The recent flurry of activity toward circularity and reuse in the Norwegian building sector represents a rapid and ambitious return to the principles and design traditions common in Norway in the past [51]. The process is hindered, however, by a building industry that moves fast in some regard (such as integrating materials and methods that quickly become obsolete or contain future banned substances) and very slow in others (e.g., integrating digitalization, building typologies with a very long stock time). The implementation of reuse incentives needs to be matched with incentives for changing the construction process—to include a return to "innovations" such as buildings designed for ease of disassembly and adaptability [52]. This (return to logical design of long-life products) will require/entail a process change in the industry now, with benefits deferring to the future.

The current approach to circularity in the Norwegian construction industry appears to be an attempt to change one part of an interconnected system, without corresponding effort to change the rest. The actors appear to be adapting quickly, however, so there is reason to be optimistic. For example, one of the businesses providing reused products and reuse services showed an extremely complicated and convoluted pathway for the reuse of building materials—12 steps for the buyer, 8 steps for the seller, and 9 steps for the material [53]. This process has been addressed to some extent through both policy and business innovation in just 2 years. If everything moved this quickly the problem could likely be quickly solved—but the system is not generally so fast acting. One issue is that buildings generally have very long service lives, especially compared to most consumer products, which means that it can take several decades before changes in building codes significantly affect the building stock. Another issue is meeting market demands, where the Norwegian government building commissioner, manager, and developer (Statsbygg) found that adaptive reuse would be chosen in only 1 of 6 scenarios, with decisions based on technical and functional factors, as well as cost, competition, and ability to "convince our client" [54, p. 107].

5.1 Incentives

Reuse is clearly societally preferable to waste, but if the economic system does not provide immediate, recognizable benefits, that preference is difficult to translate into action for actors driven or constrained by economic decisions, and other systems should be introduced to incentivize reuse. One approach to overcome the split incentive (or principal-agent problem) and narrow the barrier to action between those who endure costs (or take voluntary action) now to ensure that others in the future can benefit is to find a way to transfer some of those future benefits to the present. Two recent approaches in Norway have introduced methods for reducing this barrier. The FutureBuilt Circularity Index aims to "motivate circular principles in rehabilitation, demolition, and new construction, and set a standard for what should be the ambition level for a circular building" [55, p. 2]. An earlier version of the Circularity Index is integrated into the BREEAM-NOR v6.0 (Mat 06) certification for material efficiency and reuse [56]. The result is documented circularity potential certified by FutureBuilt and/or BREEAM.

6 Discussion/Conclusions

Many recent reports have been published regarding reuse and addressing the barriers to reuse in Norway [42, 57–60]. These reports provide an excellent and specific description of the issues facing attempts to increase reuse in Norway. Some of the main issues have been addressed through legislation since some of the reports were written. A recent change in Norwegian regulations has simplified reuse slightly by reducing the legal requirements for reusing construction materials [8]. But one of the fundamental barriers that remains is that buildings were not built with the intention of reusing the materials or components. They may be glued or nailed or mortared or otherwise irreversibly attached to other materials. As knowledge increases, we find that some of the technological marvels we were quick to integrate into our buildings turn out to have drawbacks. Many of these are now classified as hazardous waste. When these disparate materials are connected, they lose their salvageability and their reusability [61].

The strides being made toward reuse in Norway, and especially in Oslo, include incredible advancement in desire, intent, and ambition toward reuse. But these ambitions are hobbled by the past 50 years of architectural "progress"—where buildings were built quickly and efficiently, but with little to no thought for future generations, for material scarcity, or for a planet in peril. Economic decisions with life-cycle costing, high discount rates, and no regard for uncertainty have led to a situation where the existing building stock, with embedded materials and building methods, is a hindrance to the present. While the current generation tries to do better, the decisions of those in the past slow or prevent them from success.

6.1 Looking Forward

Legislation like the REACH directive, the EU ecodesign proposal, and the EU taxonomy and Level(s) may help with the retransition to a more circular economy. But what will likely be needed to have a larger effect is a more profound shift in material valuation. One of the pioneers of the circular economy, Walter Stahel, stated that this shift could be accelerated with "….one simple shift in public policy—adapting the tax system to the principles of sustainability by not taxing renewable resources including work" [62, p. 1]. Omitted but implied is that taxation of nonrenewable resources should make up the lost tax income. In their article on rebound, Ottelin et al. are much more specific about the economic shift that is needed: "Our findings highlight the limitations of circular consumption in today's economic systems, and the need for stronger policy incentives, such as shifting taxation from renewable resources and labour to non-renewable resources" [63, p. 1]. This would change the entire valuation structure of building materials and elements going forward and put us on the pathway toward the sustainable future of construction—a future we have known how to achieve for decades but have drifted away from. Unfortunately, it can do little to help with the oft-polluted outflows from the buildings being removed from stock today.

References

1. DiBK: § 9-7. Kartlegging av farlig avfall, bygningsfraksjoner som må fjernes og materialer som er egnet for ombruk. Krav til rapportering. In: Byggteknisk forskrift (TEK17) med veiledning. Direktoratet for byggkvalitet (DiBK) (2024)
2. SINTEF Community: Ombrukskartlegging av bygninger. SINTEF Community, Oslo (2023)
3. Rygh, E.: Deling av ombrukskartleggingsrapporter. https://padlet.com/emil498/ deling-av-ombrukskartleggingsrapporter-usnyma3faogxqqpz. Last accessed 2024/02/20
4. Smeyers, T., Deweerdt, M., Mertens, M.: Reuse Toolkit: The Reclamation Audit. Interreg FCRBE – Facilitating the Circulation of Reclaimed Building Elements (2022)
5. Wahlström, M., Hradil, P., Zu Castell-Rudenhausen, M., Bergmans, J., van Cauwenberghe, L., Van Belle, Y., Siáková, A., Struková, Z., Li, J.: Pre-demolition Audit – Overall Guidance Document. VTT Technical Research Centre of Finland – EIT RawMaterials Project – PARADE, Helsinki (2019)
6. NGBC, Statsbygg: Bestilling av ombrukskartlegging – slik gjør du det. Norwegian Green Building Council (NGBC) and Statsbygg, Oslo (2023)

7. Guy, B.: Paper 9. Building deconstruction assessment tool. In: Chini, A.R. (ed.) Deconstruction and Materials Reuse: Technology, Economic, and Policy, pp. 125–137. International Council for Research and Innovation in Building Construction (CIB), Wellington (2001)
8. Hauge, Å., Brown, M., Rønning, M., Flagstad, I., Plesser, T.: Ombruk av byggevarer – innspill til statlige føringer. SINTEF Community, Oslo (2023)
9. Lowe, T.: MMC makes it harder to reuse materials, say demolition firms. https://www.building.co.uk/news/mmc-makes-it-harder-to-reuse-materials-say-demolition-firms/5118786.article. (2022)
10. Chini, A. (ed.): Deconstruction and Materials Reuse: Technology, Economic, and Policy. CIB, Wellington (2001)
11. EPA: Deconstruction – Building Disassembly and Material Salvage: The Riverdale Case Study. National Association of Home Builders Research Center (NAHBRC) for US Environmental Protection Agency (EPA), Washington, DC (1997)
12. Nakajima, S., Russell, M.: Barriers for Deconstruction and Reuse/Recycling of Construction Materials. Working Commission 115 of the CIB, Rotterdam (2014)
13. Kibert, C., Chini, A. (eds.): Overview of Deconstruction in Selected Countries. CIB, Task Group 39 on Deconstruction (2000)
14. European Commission (EC): Waste Framework Directive. https://environment.ec.europa.eu/topics/waste-and-recycling/waste-framework-directive_en. Last accessed 2023/07/12
15. Crowther, P.: Developing guidelines for designing for deconstruction. In: Deconstruction – Closing the Loop, Watford (2000)
16. Crowther, P.: Building disassembly and the lessons of industrial ecology. In: Shaping the Sustainable Millennium: Collaborative Approaches Conference Proceedings. Queensland University of Technology Publications, Brisbane (2000)
17. Lansink, A.: Motie Van Het Lid Lansink Cs. Voorgesteld 1 November 1979. (1979)
18. van Dijk, K., Boedianto, P., te Dorsthorst, B., Kowalczyk, A.: Chapter 6. State of the art deconstruction in The Netherlands. In: [13]
19. Crowther, P.: Historic trends in building disassembly. In: Technology in Transition: Proceeding from the ACSA/CIB Technology Conference, Montreal (1999)
20. SINTEF Byggforsk: Flytting av trehus ved demontering. SINTEF Byggforsk, Oslo (2017)
21. Fang, D., Berglund-Brown, J., Iwakuni, D., Mueller, C.: Carbon and craft: lessons from the deconstruction, relocation, and reuse of a traditional Japanese house's timber structure. J. Phys. Conf. Ser. **2600**, 192002 (2023)
22. BFI: R. Buckminster Fuller, 1895–1983. https://www.bfi.org/about-fuller/biography/. Last accessed 2024/03/14
23. Jean Prouvé: Jean Prouvé l 1937. https://www.jeanprouve.com/en/fiche/1937-13. Last accessed 2024/03/14
24. McCurry, J.: Legacy of Japan's Nakagin capsule tower lives on in restored pods, The Guardian, 2023/01/06. (2023)
25. Guy, B., McLendon, S.: How cost effective is deconstruction? Biocycle. **42**, 75–82 (2001)
26. Chini, A., Nguyen, H.: Optimizing deconstruction of lightwood framed construction. In: Chini, A.R. (ed.) Deconstruction and Materials Reuse. CIB, Gainesville (2003)
27. Guy, B., McLendon, S.: Building Deconstruction: Reuse and Recycling of Building Materials. University of Florida, Center for Construction and Environment (CCE) for the Florida Department of Environmental Protection (DEP), Gainesville (1999)
28. Gallego-Schmid, A., Chen, H., Sharmina, M., Mendoza, J.: Links between circular economy and climate change mitigation in the built environment. J. Clean. Prod. **260**, 121115 (2020)
29. Aflaki, S.: The effect of environmental uncertainty on the tragedy of the commons. Games Econ. Behav. **82**, 240–253 (2013)
30. Spence, R., Mulligan, H.: Sustainable development and the construction industry. Habitat Int. **19**, 279–292 (1995)
31. Oslo PBE: Høyhus i Oslo. Strategi for bærekraftige høyhus. Oslo kommune, Plan- og bygningsetaten (Oslo PBE), Oslo (2023)

32. EC: Proposal for a Regulation of the European Parliament and of the Council Establishing a Framework for Setting Ecodesign Requirements for Sustainable Products and Repealing Directive 2009/125/EC. EC, Brussels (2023)

33. EC: Proposal for a Regulation of the European Parliament and of the Council Laying Down Harmonised Conditions for the Marketing of Construction Products, Amending Regulation (EU) 2019/1020 and repealing Regulation (EU) 305/2011. EC, Brussels (2022)

34. DFØ: Kriterieveiviseren. https://kriterieveiviseren.anskaffelser.no/valg/nybygg-og-rehabilitering. Last accessed 2023/06/21

35. Trasande, L.: Exploring regrettable substitution: replacements for bisphenol A. Lancet Planet. Health. 1, e88–e89 (2017)

36. Pang, Q., Li, Y., Meng, L., Li, G., Luo, Z., Fan, R.: Neurotoxicity of BPA, BPS, and BPB for the hippocampal cell line (HT-22): an implication for the replacement of BPA in plastics. Chemosphere. 226, 545–552 (2019)

37. Ji, Z., Liu, J., Sakkiah, S., Guo, W., Hong, H.: BPA replacement compounds: current status and perspectives. ACS Sustain. Chem. Eng. 9, 2433–2446 (2021)

38. Bittner, G., Yang, C., Stoner, M.: Estrogenic chemicals often leach from BPA-free plastic products that are replacements for BPA-containing polycarbonate products. Environ. Health. 13, 41 (2014)

39. MD: Farlig avfall fra bygg og anlegg. Miljødirektoratet (MD). (2013)

40. MD: Kartlegging av farlig avfall og miljøsaneringsbeskrivelse. https://www.miljodirektoratet.no/ansvarsomrader/avfall/for-naringsliv/massehandtering/betong-og-tegl-fra-riveprosjekter/. Last accessed 2024/04/16

41. byggemiljo: Farlig avfall – vinduer/Hazardous waste – windows. https://www.byggemiljo.no/farlig-avfall-vinduer/. Last accessed 2022/05/19

42. Kilvær, L., Sunde, O., Eid, M., Rydningen, O., Fjeldheim, H.: Forsvarlig ombruk av byggevarer. Team Resirqel for Direktoratet for byggkvalitet (DiBK) (2019)

43. Kron, M., Plesser, T., Risholt, B., Stråby, K., Thunshelle, K.: Ombruk av byggematerialer: Veileder for dokumentasjon av ytelser. SINTEF Community, Oslo (2022)

44. NHP: Handlingsplan 2013–2016. Nasjonal handlingsplan for bygg- og anleggsavfall (NHP3). Vedlegg område 1 – Farlig avfall og miljøgifter. NHP-nettverket. (2014)

45. Risholt, B.: Ombruk av vinduer og dører. REBUS webinar, Oslo (2022)

46. Syed, S.: Kartlegging av utvalgte typer farlig avfall. Byggavfallskonferansen (BAK) 2019, Oslo (2019)

47. FutureBuilt: Kristian August gate 13. https://www.futurebuilt.no/Forbildeprosjekter#!/Forbildeprosjekter/Kristian-August-gate-13. Last accessed 2024/03/14

48. NS 3682:2022: Hulldekker av betong til ombruk. Standard Norge, Lysaker (2022)

49. entra: Erfaringsrapport Ombruk: Kristian Augusts gate 13. FutureBuilt collaboration (utarbeidet av entra), Oslo (2021)

50. Wærner, E., Mejlgaard Ulla, K.: Betongveilederen: Prøvetakingsstrategi, regelverk, tolking av analyseresultater, miljøkartlegging, og søknad om nyttiggjøring av betongavfall. Versjon 3.0. Forum for miljøkartlegging og -sanering, Oslo (2021)

51. Kallaos, J.: The renaissance of reuse in Norway – what's old is new again. In: RILEM International Conference on Sustainable Materials & Structures, Toulouse (2024)

52. Jahren, S., Nørstebø, V., Simas, M., Wiebe, K.: Studie av potensialet for lavere klimagassutslipp og omstilling til et lavutslippssamfunn gjennom sirkulærøkonomiske strategier. SINTEF Industry for Enova (2020)

53. Rehub: Rehub. https://www.rehub.no/. Last accessed 2022/05/18

54. Bingh, L.: Towards circularity in Statsbygg. Stakeholder workshop – Measuring the application of circular approaches in construction industry ecosystem. (2023)

55. Nordby, A., Stoknes, S., Aasen Vadseth, R., Seilskjær, E., Holand Hay, N.: FutureBuilt Circular – criteria for circular buildings. FutureBuilt collaboration (2023)

56. NGBC: BREEAM-NOR v6.0 New Construction. NGBC, Oslo (2022)

57. Asplan Viak, N.H.P.: Utredning av barrierer og muligheter for ombruk av byggematerialer og tekniske installasjoner i bygg. Asplan Viak for NHP-nettverket, Sandvika (2018)
58. Knoth, K., Fufa, S., Seilskjær, E.: Barriers, success factors, and perspectives for the reuse of construction products in Norway. J. Clean. Prod. **337**, 10 (2022)
59. Nørstebø, V., Wiebe, K., Andersen, T., Grytli, T., Johansen, U., Rocha Aponte, F., Perez-Valdes, G., Jahren, S.: Studie av potensialet for verdiskaping og sysselsetting av sirkulærøkonomiske tiltak. Utvalgte tiltak og case. SINTEF (2020)
60. Sandberg, E., Kvellheim, A.: Ombruk av byggematerialer – marked, drivere og barrierer. SINTEF Community (2021)
61. Nordby, A.: Salvageability of Building Materials, PhD Thesis. Norwegian University of Science and Technology (NTNU) (2009)
62. Stahel, W.: The business angle of a circular economy – higher competitiveness, higher resource security and material efficiency. In: A New Dynamic: Effective Business in a Circular Economy, pp. 11–32. Ellen MacArthur Foundation (EMF), Cowes, Isle of Wight (2012)
63. Ottelin, J., Cetinay, H., Behrens, P.: Rebound effects may jeopardize the resource savings of circular consumption: evidence from household material footprints. Environ. Res. Lett. **15**, 104044 (2020)

Applying Value Engineering Function Analysis to the Process for Building Disassembly and the Recovery of Wood and Timber for Construction

Raymond L. Sheen

Contents

1   Introduction

Recovery and recycling of wood and timber products is an application of circular economy principles to the construction industry [1]. However, the construction industry has not yet widely adopted the principles. According to a recent European Union (EU) Commission report on the construction industry [2], only 29% of companies engage in any form of reuse or recycling of construction and demolition waste (CDW). Within CDW, wood is one of the best materials with respect to circularity [3]. Wood and timbers are renewable resources that can be recycled multiple times in multiple formulations for multiple building lifecycles. At its end of life, wood can be converted to energy through incineration.

R. L. Sheen (✉)
Produktif AS, Raufoss, Norway

Worcester Polytechnic Institute, Worcester, MA, USA
e-mail: ray@produktif.com

© The Author(s) 2025
M. Kioumarsi, B. Shafei (eds.), *The 1st International Conference on Net-Zero Built Environment*, Lecture Notes in Civil Engineering 237,
https://doi.org/10.1007/978-3-031-69626-8_153

The specific circularity principles vary from industry to industry. Within construction, the circular economy is often described using ten "R" principles. These are described in the United Nations (UN) report [3, p. 4]. They are often listed in a hierarchy with recovery being the most basic form of circularity and each additional "R" level increasing circular value. Mrad and Frölén Ribeiro provide a definition of the ten "R" principles [4, p. 4].

- Recover: Material is incinerated to recover energy.
- Recycle: Material is processed to obtain new material.
- Repurpose: Using the discarded product or its parts for a different purpose.
- Remanufacture: Using abandoned parts for a new product with the same function.
- Refurbish: Updating or restoring an existing product.
- Repair: Repairing a product to continue its function.
- Reuse: Reusing a product discarded but in good condition and fulfilling its original function by another user.
- Reduce: Increasing the efficiency in product manufacture or use by consuming fewer natural resources and materials to make a product.
- Rethink: Enhancing product usage.
- Refuse: Removing functions to make products redundant.

These "R" principles form the basis for analyzing the value embedded in processes that are associated with the recovery and reuse of wood and timber from CDW.

2   Background

2.1   System Elements

The value engineering discipline seeks to optimize the perceived customer value as compared to the cost of a component or system. Value analysis is a core element of value engineering. When doing value engineering, the foundation of value analysis is the function analysis of a product or process [5]. Within the construction industry, value analysis is often used as a form of cost containment [6]. Through value analysis, unnecessary building components can be identified and eliminated to reduce costs. Value analysis has been used with various forms of construction. Heiza et al. [7] used value analysis with high-rise buildings as a problem-solving tool that contained cost and increased quality. The area of the world where value analysis is most commonly applied to construction projects is Asia [8]. In their study, Chen et al. found that although the value analysis application is highly variable, when used, it delivers excellent results. Gimenez et al. [9] highlight the benefit of value analysis at the time of the building design. Once the design is completed, the cost of changes often negates any benefit that could be realized from adopting value analysis recommendations.

2.2 Element Interactions

The recovery and recycling of wood from CDW is a focus of the advocates for circular economy adoption. "The greatest opportunity for improved circularity of wood in existing buildings is in the recovery, reuse and/or recycling of building demolition waste" [3, p. xi]. However, recycling wood from CDW is a complex business decision. There are many options for what to do with the wood [10]. The direct costs, logistics costs, and energy costs for each option frequently change. In addition, some of the options create potential revenue, and some of the options have permitting or other regulatory costs. A key CDW recycling subprocess is sorting and segregating different construction materials to determine what form of reuse or recycling is appropriate. This sorting can be done either manually or with automation [11]. However, there is a cost and purity tradeoff with the automation equipment. Wood waste that cannot be clearly identified and segregated can be used for filler in composite materials. However, the CDW wood is usually contaminated, and the mechanical properties of the composite material with that type of filler are inferior [12]. Another key for successful wood recycling is to develop an intentional process. Oliveira Neto and Correia found that an intentional reverse logistics plan contributed to the benefit from recycling [13].

3 Methodology

3.1 Process Flow Diagrams

Seven of the ten "Re" principles involve additional processing of CDW. Those seven are recover, recycle, repurpose, remanufacture, refurbish, repair, and reuse [4]. Each of these is the result of a process that starts with the wood elements of CDW and ends with that wood being in some state that creates new value. In addition, there is an eighth process that does not result in a new value. That process sends the wood products to a landfill for eventual decomposition. These eight processes can be represented with process flow diagrams. The process flow diagram starts with the generation of CDW and shows the steps required to complete each "R" process. The steps are shown in sequential order and are connected with arrows. Process flow diagrams create a visual image of a process that may take a long time and be spread across many locations.

Jahan et al. [14] created a high-level flow for the lifecycle of construction products that had five steps in sequence. These steps are (1) raw material extraction, (2) preconstruction, (3) construction, (4) renovation/demolition, and (5) end of life. The first two steps are associated with the initial fabrication of the construction products, and waste in those steps would be considered industrial waste, not CDW. The remaining three steps can and often do create CDW. For the purposes of this study, the starting point of the process flow will be the generation of the CDW, regardless of which step in the construction lifecycle the waste was created.

3.2   Value Engineering

The value engineering approach used in this study is based on the methodology recommended by the leading professional organization that provides certification in value engineering techniques, SAVE International. They recommend a value engineering process that starts with a function analysis using a Function Analysis System Technique (FAST) diagram. This identifies unneeded functions, missing functions, and prioritizes the existing functions. The identified functions become one side of a Function Cost Matrix. The other side of the matrix is the product architecture for a product analysis or the process flow chart elements for a process analysis. The matrix is used to create a functional value for each element.

Function Analysis System Technique (FAST)   The FAST analysis is a method for identifying and organizing the functions of a product or process. Since this study is focused on the recovery and recycling of wood from CDW, the process perspective is used. The FAST technique starts by setting the boundaries or scope of the process being analyzed. Outside the process boundary on the left side is a higher-order function that normally reflects the ultimate purpose of the process. Outside the boundary on the right side are any causative functions or events that would initiate the process.

Within the boundaries are the functions or activities that occur within the process. On the left side are the primary function(s). These function(s) normally represent a final step in a process path. They are the outputs of the process. The primary function delivers the result of the process, secondary functions are enablers or enhancers of the primary function. One goal of a FAST analysis is to determine if there are more efficient ways to accomplish the secondary functions. If the primary function can be achieved without some of the secondary functions, those secondary functions should be eliminated since they are contributing little if any value.

When creating a FAST analysis, it is often helpful to navigate the diagram from left to right and then from right to left to identify any unneeded or missing functions. When going from left to right, the question "How?" is asked with each function to identify the preceding functions. When going from right to left, the question "Why?" is asked to understand the relationship with any succeeding function. Also, with each function, the question "When?" is asked to determine if there are other secondary functions that should be connected with each other. An illustration is found in Fig. 1.

Function Cost Matrix   The functional cost matrix (FCM) combines the functions identified with the FAST analysis and the steps associated with processing CDW to identify the inherent value of each step. The format of the Function Cost Matrix is similar to that used with Quality Function Deployment (QFD) [15]. In the FCM, functions are listed across the top of the matrix and the process steps are listed down the side. In one variation of an FCM, actual cost values to accomplish each process step are allocated across the functions. A second variation of the FCM uses the QFD approach for analysis. In this case, the functions are prioritized using a scale of 1 to 5 with "5" being the highest priority. The process steps are then correlated to the

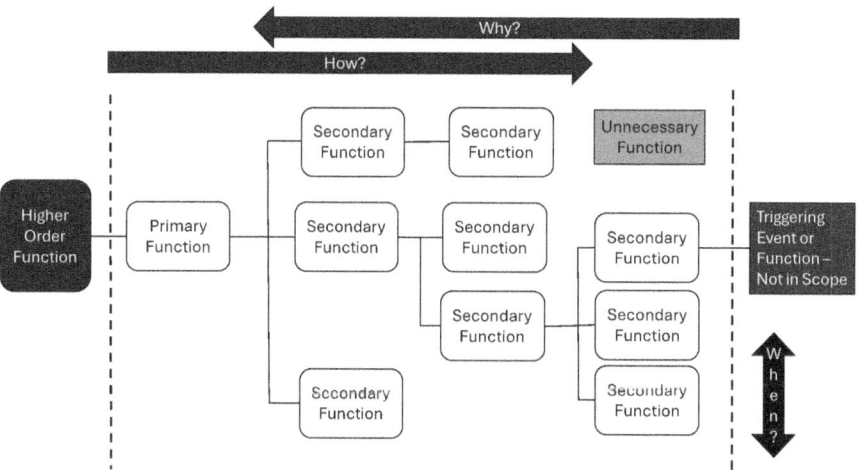

Fig. 1 FAST analysis template

functions based on how significantly they directly contribute to the functional performance. The correlations are rated as high, medium, low, or no correlation. Using the standard QFD approach, the high correlation is given a multiplier of "9," the medium has a multiplier of "3," the low has a multiplier of "1," and no correlation has a multiplier of zero. These multipliers are used with the priority of the functions to determine a correlation score for that step/function relationship. The total score of a step reflects that step's contributed value to the overall process. When using the QFD approach, the scores reflect a relative value between seps.

The perceived customer value or score of each step is then compared to the total cost for that step. If the cost is significantly higher than the value, consider redesigning the process to minimize the cost at that step. If the value is significantly higher than the cost, assess whether the method of accomplishing that step is fully meeting the customer quality expectations. If not, consider improving the performance of that step, even if it means spending more money to do that step well. Improving the performance of that step will increase the value of the entire process in the eyes of the customer. If the cost and value are approximately the same, the effort associated with accomplishing that step is appropriate.

4 Analysis

The analysis in this study is based on using information from other studies and reports created by the EU or UN. This is an analysis of a generalized model of the industry, not a specific company or case study. Therefore, the analysis provides relative values for the process steps, not absolute values. Absolute values will depend on the specific organizational capabilities, geographical location of sites, and the current regulatory environment for the region being studied.

4.1   Value Engineering

The process flow diagrams show an idealized view of the processes that result in the
eight potential "R" dispositions of wood that are found in CDW. The steps are at a
relatively high level. In many cases, when a company decides to adopt one of these
processes, it will need to establish a set of detailed steps to accomplish the task
described in this high-level step. The particular steps often depend on the construction site and the organization's infrastructure and capabilities. For instance,
depending on site locations, shipping waste to an incinerator may be a short haul
by truck or a long haul involving rail shipments and multiple trucks. For this reason,
it is impossible to set exact detailed steps that apply in every case. The process map is
found in Fig. 2.

The process starts with the creation of CDW during construction or deconstruction. A decision is made to recover the wood and use it again in the construction of
another building or to allow it to be transformed into another product, material, or
physical state. If it will not be used for production, it must be shipped to the
appropriate facility and either dumped in a landfill or processed to be transformed
into energy, a different composite material, or a different fiber-based product that is
not used in construction. If it is to be reused in construction, then a decision is made
concerning how it is to be used. If no modification is required, once it is inspected, it
is sent to an appropriate distribution channel for resale. If modifications are required,
then it must be sent to the correct facility. Often minor repairs can be made at the
point where the sorting occurs. An example would be to trim a beam to a shorter
length and remove an end that was compromised during demolition. That beam can
then be inspected and returned to the supply chain. In some cases, the modification
may be more extensive, so the wood is sent to a manufacturing facility to be
refurbished into construction materials. Again, once inspected, it can be returned
to the supply chain.

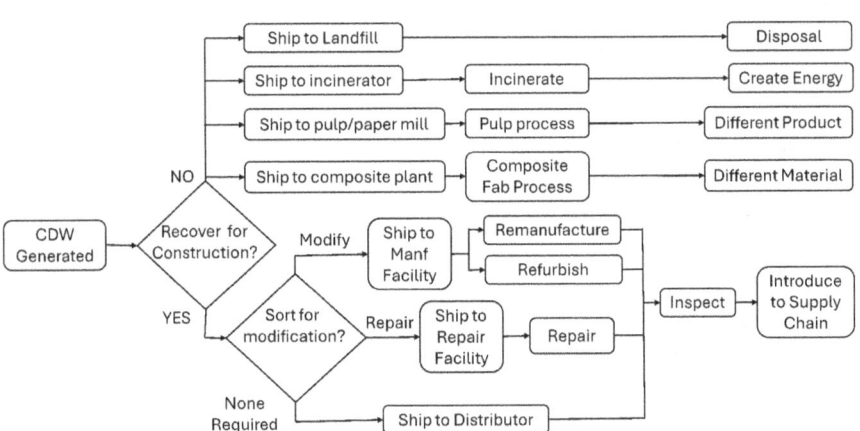

Fig. 2 Process map for CDW waste recovery and disposition

4.2 FAST Analysis

The FAST analysis of wood recovered from CDW is somewhat complex. There are two higher-order functions that are mutually exclusive. One is the reintroduction of the wood items into the construction supply chain. The other is the transformation of the wood into another form. A specific wood item could go through the first higher-order function many times throughout its lifecycle. However, once it has been transformed, it is no longer possible for it to be a wood element in construction so it will not be CDW. In fact, depending on the transformation, it may not exist at all.

Since there are two higher order functions, there are two FAST analysis diagrams. Figure 3 is the diagram for wood components reentering the construction supply chain. This occurs either as a component in its original family or as a wooden item that has been remanufactured into a new wooden construction component. The component may be undamaged and ready for immediate reuse. It could require minor repairs before being returned to the supply chain. In some cases, it requires restoration or refurbishment before reentering the supply chain. In all cases, the documentation for that item should be updated to show the history of the component. Once the CDW has been selected for disposition back into the construction supply chain, a decision is required to determine which approach is appropriate. This FAST analysis diagram is the ideal state, so no unnecessary functions are shown. If an organization does their own internal FAST analysis, they may discover other unneeded functions embedded in their process.

The second higher-order function is to transform the wooden elements within the CDW into a new state. In this case, the FAST analysis diagram has four primary functions that reflect the four different types of transformation. Each of these primary functions has a secondary function that represents the transformation process and a function that represents the selection process. In addition, there is again the function

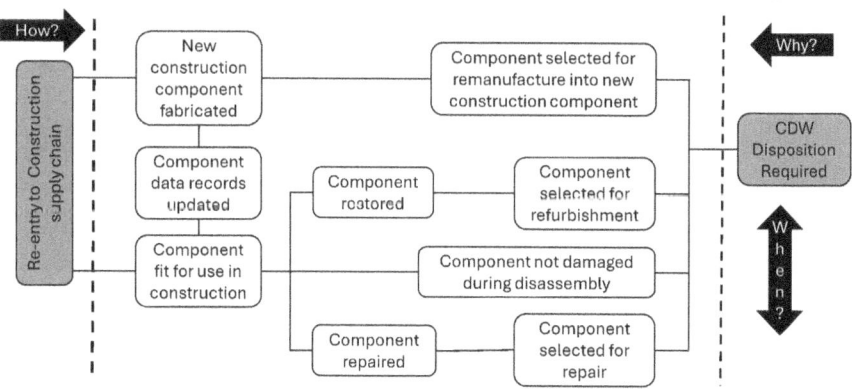

Fig. 3 FAST analysis when wood component reenters the supply chain

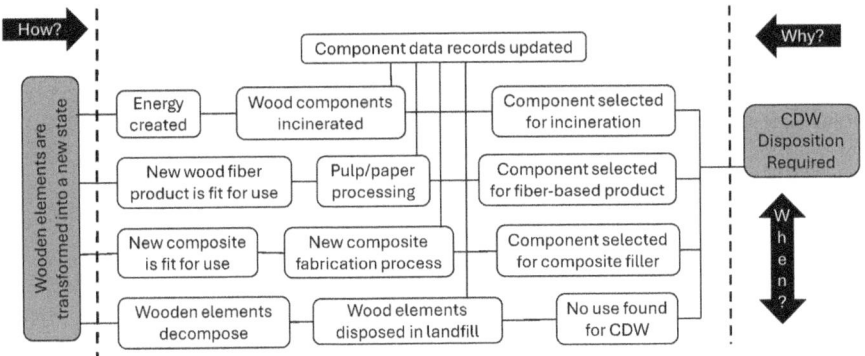

Fig. 4 FAST analysis when wood components will not reenter the supply chain

of updating the documentation based on the disposition of the CDW wood elements. This analysis is represented in Fig. 4. As with the previous FAST analysis, an organization may find unneeded functions when they do an internal FAST analysis.

4.3   Function Cost Matrix

The FCM combines the steps from the process flow with the functions from the FAST analysis. In this case, there are 21 functions and 22 process steps. The FCM functions are listed across the top of the matrix. Following the QFD model, the functions are prioritized using the 1 to 5 scale. The process steps are listed down the side of the matrix. The steps are correlated to the functions. The low/medium/high correlations are based on assessing the degree to which completing the step in an excellent manner will directly affect the performance of the function.

When prioritizing the functions, the UN report's assessment of the circular economy "R" principles was used [3, p. 25]. The perceived value is consistent with industry analysis conducted by López Ruiz et al. [16]. They compiled data from 15 studies on circular economy and construction. These studies covered the preceding 15 years and spanned the globe. The UN study placed the highest priority on the complete reuse of the building, but in that case, there would be no CDW. The next highest priority was component reuse, and this study assigned functions that operate at that level the priority rating of "5." These functions were associated with the immediate reuse of components as soon as the component is removed from the construction or deconstruction site. These items are intact and retain their full technical qualities. The other function to receive the highest priority of "5" was the update to the data records. With the widespread adoption of Environmental Performance Declarations (EPDs) in the construction industry, it is essential that correct records of the origin and use of wood construction components are maintained.

The next highest priority from the UN report was material reprocessing [3]. The functions associated with this were given a priority of "4." These were functions directly associated with ensuring the wood components were returned to the construction industry after minor modifications. These would be functions associated with the repair and refurbishment of the recovered item. A rating of "4" was also given to the function, Inspection. This function is needed as support for the repair and refurbishment of items to rate the items correctly and reintroduce them to the supply chain. Finally, the rating of "4" was used with all the functions that selected wood elements in the CDW for some level of circularity, regardless of the process selected. The only selection function not included was the decision to not select a wood item for any type of circular processing and instead just send it to a landfill for decomposition.

The third priority level from the UN report is material recycling [3]. Priority level "3" was used with the functions that transform the wood items recovered from the CDW into another material. This included remanufacturing the items for the construction supply chain and remanufacturing into another type of product that is normally not associated with the construction industry. This latter approach usually includes reprocessing the recovered items into another fiber-based product. This is primarily done by a pulp and paper mill factory. The recovered material is used as stock in those mills. Another function is the use of wood components in the manufacture of new composite materials. Composites rely on a matrix or filler material to provide the overall composite material properties. Generally, CDW wood is less practical than new wood for this application because of the contaminants that are often in the recovered wood items [12].

The lowest two levels in the UN report were energy recovery and disposal. Therefore, energy recovery functions received a "2" rating [3]. With wood items from the CDW, the energy recovery is normally accomplished through incineration. The functions associated with landfills have the priority of "1." Unfortunately, CDW accounts for 36% of all solid waste in Europe, which is an indication of why these functions are so low in priority [16].

The FCM is found in Table 1. The first step in the process, the generation of CDW, is considered a low relationship with all the functions. It does not directly affect any function, but the functions would not be needed if there was no CDW. The next two process steps determine the circularity function to be applied to the CDW. These steps have a high impact on which the "R" function is used. In addition, since they select the function, they also have a low impact on the final result of each function. The "shipping" steps have a low impact on the functions to which they ship the CDW. They must ship the items to the right location, but they do not directly affect the operation of the "R" function. Not surprisingly, the "R" step has a high correlation with its respective "R" function. Finally, the relationship to the documentation function for the "R" steps that return the product to the construction supply chain is high because this function must complete the EPD. The correlation with the other "R" function is medium since all that is documented is the end of the

Table 1 Function cost matrix

Functions / Process Steps	Component fit for Use	New Component Fabricated	Energy Created	New Wood-fiber product	Composite material	Decompose	Data records updated	Component restored	Component remanufactured	Component repaired	Incineration	Pulp/paper processing	Composite fabrication	Dispose in landfill	Component undamaged	Slected for repair	Selected for refurbishment	Selected forremanufacture	Selected for incineration	Slected for fpaper product	Selected for composite	No use found	Total
Priority	5	4	2	3	3	1	5	4	3	4	2	3	3	1	5	4	4	4	4	4	4	1	
CDW Generated	L	L	L	L	L	L	L	L	L	L	L	L	L	L	L	L	L	L	L	L	L		73
Recover for Construction?	L	L	L	L	L	L	H								H	H	H	H	H	H	H	H	333
Sort for Modification?	L	L					H	H	H	H					H	H	H	H					306
Ship to Distributor	L	L					L																14
Ship to Manufacturer							L	L	L														12
Shipt to Repair							L			L													9
Remanufacture		H					H		H														108
Refurbish	H						H	H															126
Repair	H						H			H													126
Inspect	H	H					M	M	M	M													129
Introduce to Supply Chain	H	H		L	L		H																132
Ship to Comoposite Plant							L						L										8
Ship to Pulp/Paper Mill							L					L											8
Ship to Incinerator							L				L												7
Ship to Landfill							L							L									6
Incienrate							M				H												33
Pulp Processing							M					H											42
Composite Fabrication							M						H										42
Dispsoal							M							H									24
Create Energy							L				H												23
Create different Product							L					H											32
Create Composite Material							L						H										32

construction life of the wood item. The correlation of the documentation function with the shipping process steps is low since there should not be any modification to the item while being shipped.

The process steps scores are relative scores, not actual cost, or value amounts. They represent the relative value of these steps with respect to each other. The higher the score, the more important that step is to achieve full functional value from the process.

5　Discussion

This analysis is intended to provide a methodology and framework for understanding where value is to be created in the process of recovering and recycling wood from CDW. This framework can be used by those in the market or those intending to enter the market. The result of the analysis is a tool that prioritizes the process steps.

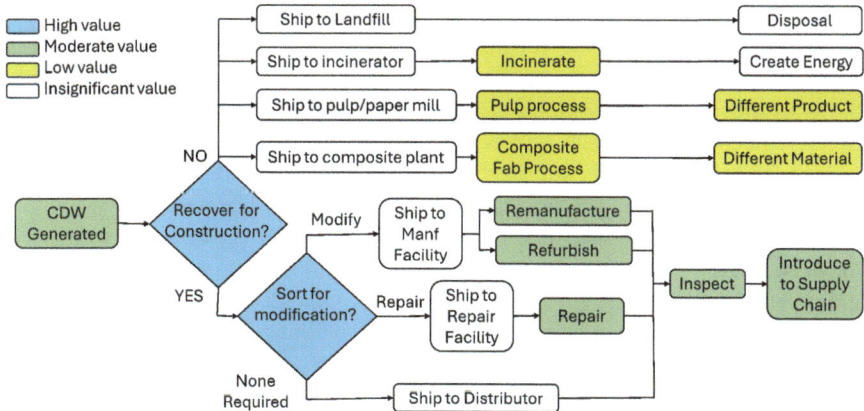

Fig. 5 Prioritized process map for waste recovery and disposition

Organizations can use this tool to identify areas for improvement or new business opportunities. Figure 5 shows the prioritized steps in the process of recovering wood from CDW.

The Function Cost Matrix does not contain any surprises, but it places clear emphasis on several process steps. In this case, the steps that decide what wood items are to be recovered and what "R" process will be used are the steps with the highest scores. This is consistent with the recommendations by López Ruiz et al. on the importance of creating a Site Waste Management Plan and developing collection and segregation techniques [16]. To create maximum value in the context of the circular economy, the two decision steps that determine which wood items can be recovered and identify the best path for circularity must be done in an excellent and efficient manner. That may mean investing in techniques to improve the performance of those steps. Since those steps are early in the process, they are also one of the first points of resistance noted in the UN report [3]. This is understandable. Without clear guidelines and methods for making these decisions, these steps could create delays, added costs, and risk if the wrong decision is made. These two steps should be the focus of further research and development.

The next set of steps with similar moderately high values are the "R" processing steps that create wood products that return to the construction supply chain following inspection. These include the steps of repair, refurbishment, and remanufacture. Equally important to these steps are the inspection capabilities to qualify and accept these items into the supply chain and the corresponding updates to the EPD. Finally, the actual process of introducing and marketing these recovered items is of moderate importance. Each of these process steps is important and adds value to the construction industry. The relative value of each of these steps is very similar. However, if investment in one of these "R" processes results in that process becoming more efficient with lower costs than others, that process will have a distinct value advantage over the other processes.

The low-value steps are the other "R" processes that do not result in the CDW wood item reentering the construction supply chain. Except for disposal in a landfill, these items represent an element of circularity. However, it is not circularity within the construction industry, which is why their score left them with a low rating. Improvements in these areas will be beneficial for a circular economy, but those benefits will not be directly attributable to the construction industry.

Many of the steps in the process are related to shipping. These steps had very low scores because they did not create value within the circular economy. Nevertheless, they can create a great deal of cost. The costs in the shipping process steps should be minimized or, if possible, eliminated. To do this, one or more of the "R" processes would need to be done at or near the site of the building deconstruction. Another possible option to reduce shipping costs would be to create a recovery site where CDW is shipped. That site could do the decision processes of determining which "R" approach to use and have the capability to perform some or all the "R" functions at that facility. This would likely lower the total shipping costs and speed up the process of returning the recovered wood item back into the construction supply chain.

6 Conclusion

The creation and disposition of CDW is currently an issue for the construction industry as it attempts to adopt the principles of a circular economy. Based on value analysis, the most important step in the process of disposing and recovering CDW is the early decision of which "R" path each wood item in the CDW should follow. The highest value paths are those in which the wood item is returned to the construction supply chain. There are four options to achieve this objective, reuse, repair, refurbish, and remanufacture. Investment in the tools or techniques used to improve disposition decision-making or these "R" processing would likely improve the value of the construction wood recovery process. To create maximum value from a circular economy perspective, options for disposing of wood items—such as repurposing, recycling, recovering, or disposing of them in a landfill—should be carefully considered; however, these options may not align with the construction industry's value priorities. Furthermore, any measures to reduce or streamline the shipping costs associated with the disposition and recovery of CDW should be implemented, as these costs do not inherently add value.

This study was conducted as a qualitative analysis of the functional value of each step in a generic process for recovering and disposing of wood items found in CDW. Further research should be done in specific local regions using the actual costs of functions in that region. The access and capability of different "R" function capabilities could change the scores on the process steps.

References

1. Yu, Y., Junjan, V., Yazan, D.M., Iacob, M.: A systematic literature review on Circular Economy implementation in the construction industry: a policy-making perspective. Resour. Conserv. Recycl. **183**, 106359 (2022). https://doi.org/10.1016/j.resconrec.2022.106359
2. European Commission, European Innovation Council and SMEs Executive Agency: In: Brincat, C., Graaf, I., León Vargas, C. (eds.) Study on Measuring the Application of Circular Approaches in the Construction Industry Ecosystem: Final Study. Publications Office of the European Union (2023) https://data.europa.eu/doi/10.2826/488711
3. Bowyer, J., Fernholz, K., Kacprzak, A.: Circularity Concepts in Wood Construction. United Nations Economic Commission for Europe and Food and Agricultural Organization (2023) https://unece.org/sites/default/files/2023-05/ECE_TIM_DP95E_web.pdf
4. Mrad, C., Frölén Ribeiro, L.: A review of Europe's circular economy in the building sector. Sustainability (Basel, Switzerland). **14**(21), 14211 (2022). https://doi.org/10.3390/su142114211
5. Spaulding, W.M., Bridge, A., Skitmore, M.: The use of function analysis as the basis of value management in the Australian construction industry. Constr. Manag. Econ. **23**(7), 723–731 (2005). https://doi.org/10.1080/01446190500040679
6. Altaf, M., Alaloul, W.S., Khan, S., Liew, M.S., Musarat, M.A., Mohsen, A.A.: Value analysis in construction projects with BIM implementation: a systematic review. Paper presented at the 2021 international conference on decision aid sciences and application (DASA), pp. 51–56 (2021). https://doi.org/10.1109/DASA53625.2021.9682253
7. Heiza, K.M., Abo Elenen, N.E., Mahdi, I.M.: State of the art review on application of value engineering, value analysis and value management on construction projects: high rise buildings. In: The International Conference on Civil and Architecture Engineering (Vol. 11, No. 11PthP International Conference on Civil and Architecture Engineering, pp. 1–15). Military Technical College (2016, April)
8. Chen, W.T., Merrett, H.C., Liu, S.S., Fauzia, N., Liem, F.N.: A decade of value engineering in construction projects. Adv. Civ. Eng. **2022**, 2324277 (2022)
9. Gimenez, Z., Mourgues, C., Alarcon, L.F., Mesa, H., Pellicer, E.: Value analysis model to support the building design process. Sustainability. **12**(10), 4224 (2020)
10. Morris, J.: Recycle, bury, or burn wood waste biomass?: LCA answer depends on carbon accounting, emissions controls, displaced fuels, and impact costs. J. Ind. Ecol. **21**(4), 844–856 (2017). https://doi.org/10.1111/jiec.12469
11. Hyvärinen, M., Ronkanen, M., Kärki, T.: Sorting efficiency in mechanical sorting of construction and demolition waste. Waste Manag. Res. **38**(7), 812–816 (2020). https://doi.org/10.1177/0734242X20914750
12. Sormunen, P., Kärki, T.: Recycled construction and demolition waste as a possible source of materials for composite manufacturing. J. Build. Eng. **24**, 100742 (2019). https://doi.org/10.1016/j.jobe.2019.100742
13. Oliveira Neto, G.C., Correia, J.M.: Environmental and economic advantages of adopting reverse logistics for recycling construction and demolition waste: a case study of Brazilian construction and recycling companies. Waste Manag. Res. **37**(2), 176–185 (2019). https://doi.org/10.1177/0734242X18816790
14. Jahan, I., Zhang, G., Bhuiyan, M., Navaratnam, S.: Circular economy of construction and demolition wood waste—a theoretical framework approach. Sustainability (Basel, Switzerland). **14**(17), 10478 (2022). https://doi.org/10.3390/su141710478
15. Elhegazy, H., Ebid, A., Mahdi, I., Haggag, S., Abdul-Rashied, I.: Implementing QFD in decision making for selecting the optimal structural system for buildings. Constr. Innov. **21**(2), 345–360 (2021). https://doi.org/10.1108/CI-12-2019-0149
16. López Ruiz, L.A., Roca Ramón, X., Gassó Domingo, S.: The circular economy in the construction and demolition waste sector—a review and an integrative model approach. J. Clean. Prod. **248**, 119238 (2020). https://doi.org/10.1016/j.jclepro.2019.119238

Carbon Trading for the Construction Industry: A Systems Theory Approach

Augustine Senanu Komla Kukah, Xiaohua Jin, Robert Osei-Kyei, and Srinath Perera

Contents

1 Introduction

The nature of construction activities makes the sector resource intensive [1]. According to Kukah et al. [2], construction and its activities are responsible for a large proportion of carbon emissions. Globally, the construction industry consumes 25% of timber, 16% of water, and 40% of sand, stones, and gravel [3]. Carbon is also released by the construction process due to usage of fuel and electricity [4]. Furthermore, cement is the key component contributing to CO_2 emissions during concrete production [5]. As the world strives towards attaining carbon neutrality and net zero, it is imperative to identify ways in which the construction industry can reduce or totally eliminate its emissions [6, 7].

Schwartz et al. [8] explain lifecycle carbon cycle (LCC) emissions to mean the carbon dioxide that is emitted through the life cycle of buildings via all the processes

A. S. K. Kukah (✉) · X. Jin · R. Osei-Kyei · S. Perera
Centre for Smart Modern Construction, School of Engineering Design & Built Environment, Western Sydney University, Penrith, NSW, Australia
e-mail: a.kukah@westernsydney.edu.au; xiaohua.jin@westernsydney.edu.au; r.osei-kyei@westernsydney.edu.au; srinath.perera@westernsydney.edu.au

© The Author(s) 2025
M. Kioumarsi, B. Shafei (eds.), *The 1st International Conference on Net-Zero Built Environment*, Lecture Notes in Civil Engineering 237,
https://doi.org/10.1007/978-3-031-69626-8_154

the building goes through. LCC has also been explained by Rodrigo et al. [9] and De Wolf et al. [10] to include operational carbon (OC) for ventilation, heating, lighting, cooling as well as embodied carbon (EC) for production, transport, material, construction, and disassembling. Schwartz et al. [8] in their study classified LCC under three groups: (i) EC emissions during extraction of raw material, transporting, construction, maintaining, and refurbishing, (ii) OC emissions cooling, heating, lighting, and (iii) carbon emissions from demolition of building and transporting of the waste generated.

Carbon trading is a market mechanism that seeks to facilitate the reduction of emissions of global greenhouse gases such as carbon dioxide [11]. The principle underlying carbon emissions trading is that entities must not exceed or go beyond the set carbon emission quotas that have been distributed by the government [12]. Firms whose emissions exceed cap would have to purchase emission quotas/allowances in order to emit [13].

According to Ng and Luk [14], the high quantity of emissions arising from the construction industry provides a justification for usage of carbon trading schemes in increasing emission reduction possibilities. The successful implementation of carbon trading in the construction industry is however dependent on the presence of transparent and rigorous standards plus the ability in setting realistic baselines or emission caps [15]. It is, therefore, expedient to have the engagement of all relevant stakeholders involved before carbon trading for the construction industry is implemented [16, 17].

Even though carbon trading has been identified as a panacea for mitigation of greenhouse gases and subsequent decrease in climate change menace, yet current carbon trading systems were originally not developed for the construction industry and carbon trading is a new concept to the construction industry [15, 18–20]. Even though the European Union Emission Trading System (EU ETS) is the largest multinational implemented carbon trading scheme in the world, it covers a number of economic sectors but little on construction [21]. Other typical emission trading schemes also do not have much focus on the construction industry. The aim of this study is therefore to develop a carbon trading emissions system for the construction industry based on systems theory.

2   Literature Review

2.1   Carbon Trading

The concept of carbon trading originates from "emissions trading" proposed by Dales who was an American economist [22]. Emissions trading is a government environmental regulatory policy and market instrument that seems to reduce environmental pollution. In 2002, the Netherlands was the first to launch carbon trading market and was followed by United States in 2005 [22].

Countries around the world have adopted a variety of market mechanisms to internalize environmental externalities, such as the European Union (EU) emissions trading system, the United Kingdom (UK) emissions trading system, the California carbon market in the United States, and the New Zealand carbon emissions trading system [23].

The underlying concept behind carbon emissions trading is to provide incentives for companies/nations to cut down on their carbon emissions so as to have leftover permits they can sell [24]. The wealthier and bigger companies/nations efficiently subsidize the efforts of the poorer and high-polluting companies/nations by purchasing their credits [25]. With time, these wealthier companies/nations are incentivized to also reduce their emissions, so they do not have to purchase as many carbon credits in the market.

2.2 Systems Theory

System refers to the collection of components that are inter-related and work together in attaining a common goal [26]. Systems may be complex and can consist of other smaller systems called sub-systems. Systems do not operate completely in isolation [27]. Systems are contained within environment which contains other systems and external agents. A boundary defines the scope of a system. The constituents outside the boundary are part of the system's environment and everything contained within the boundary also parts of the system itself [28]. System boundary also serves as the interface between the environment and the system [29]. Interface represents the exchanges that take place between the system and other systems or with the environment.

When component of a system is changed, it affects other components within the system or even the whole system [30]. It is possible to predict changes in behavior patterns in the system. For systems that engage in learning and adaptation, the level of growth and degree of adaptation is influenced by how effectively the system engages with the environment and other contexts that influence the organization [31]. Some systems support other systems and maintain the other systems in order to prevent failure. The key aims of systems theory are to model a system's dynamics, conditions, constraints, and relations [32]. Systems theory also makes use of principles such as tools, methods, and measures and applies these principles to other systems [33].

Input of a system encompasses the raw materials that are processed to deliver a particular output [34]. Inputs come in varying forms and are not always physical in nature. Examples of inputs comprise knowledge, data, machinery, personnel, and physical location among others [35]. Process involves the transformation of inputs into outputs [36]. Output of a system is the final outcome or finished product that the system creates. Outputs can take different forms. They can be products, services, information, etc. [37].

3 Carbon Trading System for the Construction Industry

The carbon trading system for the construction industry is modeled based on systems theory. Figure 1 illustrates the high-level model of the carbon trading system. The system comprises of an input, process, and an output. Feedback is from the output to input. Feedback is the element of control in the system. Feedback details out information on performance of the system and helps in adjusting behaviors in the system.

3.1 Inputs of the Carbon Trading System

Input consists of variables that are put in a system. Relating this to carbon trading system for the construction industry, the inputs refer to all the components and constituents that need to be in place for carbon trading to take place. The inputs range from marketplace variables, policies on carbon trading, stakeholders for the trading among others. Since the input for carbon trading is plenty and diverse, a categorization method was deemed expedient to place these various variables under. This study adopts the PROMISE framework for categorizing the system's inputs.

The PROMISE framework is a system thinking approach to sustainability strategy developed by Massachusetts Institute of Technology (MIT) Sloan School of Management researchers John D. Sterman, Jayson Jay, and Roberto Rigobon [38]. It is an acronym representing Personal, Relational, Organizational, Market, Institutional, Social, and Environmental [39]. In this study, it was adopted for categorizing the input variables for the carbon trading system because PROMISE framework is based on systems thinking which belongs to the family of systems theory.

The framework explains how sustainability projects work. Projects that are sustainability-oriented start with people (personal) and the interactions among stakeholders (relational). Firms are involved in these projects (organizational), and they are influenced by market conditions (market). Governments and other stakeholders are involved in these sustainability projects (institutional). A broader group of

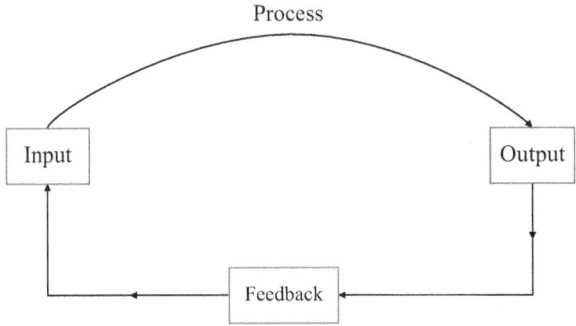

Fig. 1 High-level model of carbon trading system. (Source: Authors' construct (2024))

stakeholders involve the society that is affected by these projects (social). Natural resources further have a role to play in sustainability-related projects (environment). These six factors have an influence on each other [38, 39].

"Personal/relational" category input variables in carbon trading involves both personal and relational aspects of carbon trading in construction. Personal carbon trading underscores the rights and roles of consumers through household use of energy [40]. It operates on the idea that individuals are allocated carbon quota. The quota can be used entirely or partially sold to others [41]. Relational explains the relationship between stakeholders in the carbon trading scheme [42]. Stakeholders for carbon trading in the construction industry include central government, local government, construction firms, individuals, environmental protection departments, non-governmental organizations, and financial institutions [43].

"Organizational" category in carbon trading describes elements that influence the way organizations/firms work and behave. According to Wade et al. [44], organizational factors in carbon trading include corporate culture, environmental attitudes of organization, historical engagement, leadership abilities, innovation adoption, research and development, and technology use.

"Market" category input variables explain where buying and selling takes place. Market category variables encompass economic trends that affect demand and supply. Market category explains the elements in place whereby parties engage in exchange. In carbon trading, market mainly relates to the production and buying and selling of carbon credits [24].

"Institutional" category input variables include the irreplaceable role of government in carbon emissions reduction through carbon trading schemes [45]. Institutional factors include executive orders, laws, penalties, rewards, incentives, and compensations related to carbon trading for construction [46]. Carbon emission targets and policy tools based on these emission targets for the construction industry are encompassed in the input category called institutional [47].

"Social" category input variables are the elements of carbon trading that affect people within a society. Variables in carbon trading under social category include population, urbanization, and social cost. [48]. "Environmental" categorization variables in carbon trading relate to the environmental considerations of carbon trading including emissions of carbon dioxide.

Incorporating these six components in sustainability projects leads to better and successful projects. At each level of the framework, variables can be optimized [39]. The PROMISE framework, however, demonstrates at each level the qualities of a system: interdependencies, uncertainty, and feedback loops [49]. Some recent studies on sustainability have adopted the PROMISE framework for classifying their variables in sustainability-related projects. A notable example is the study by Dumitriu [50]. The study conducted research on sustainability adoption in universities. Figure 2 illustrates the grouping of input variables for the system based on PROMISE framework.

Fig. 2 Classification of grouping for input variables based on PROMISE framework. (Source: [38, 39])

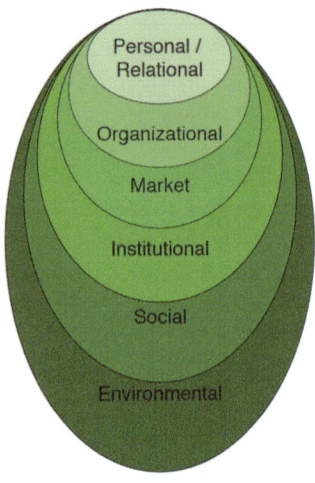

3.2 Process of the Carbon Trading System

Process in systems theory refers to what happens in the system that transforms the inputs into the final output. In carbon trading system, there are multiple processes that transform the input. In this study, the process was based on the interaction and influence among the categorization of input variables using the PROMISE framework. A foremost process in carbon trading is setting of cap. The cap implies the uppermost limit of greenhouse gas that is allowed in the trading scheme. It also represents the total number of allowances that covered entities can have access to. The setting of cap corresponds with the baseline against which emissions must be reduced [51]. In setting the cap, historical emissions are set against a base year or projected future emissions. Governments set cap on the maximum level of permits or allowances under the cap. Emitting firms must therefore obtain and surrender permit for every unit they emit. Permits can be obtained from government or through trading with other firms in the carbon trading market [19]. The allowances from government are either free or auctioned. In Fig. 3, the variables under setting of cap fall under the categories: market, institutional, organizational, and environmental. These variables have interactions among other variables in their categories and other categories. Some of the variables include emission cap, carbon allowance price cap, free quota, carbon intensity reduction rate, CO_2 emissions, auction allocation, and level of free allocation of permits [52].

   Purchasing of economic carbon credits or carbon offsets is another major process of carbon trading through which buildings in the construction industry can minimize their net environmental impact. These carbon credits come in the form of tradable certificates awarded to firms to lower the content of carbon released in the atmosphere. This can come in the form of carbon capture and sequestration. From the PROMISE framework categorization, majority of the variables for purchasing carbon credits fall under market categorization. However, they have influence on

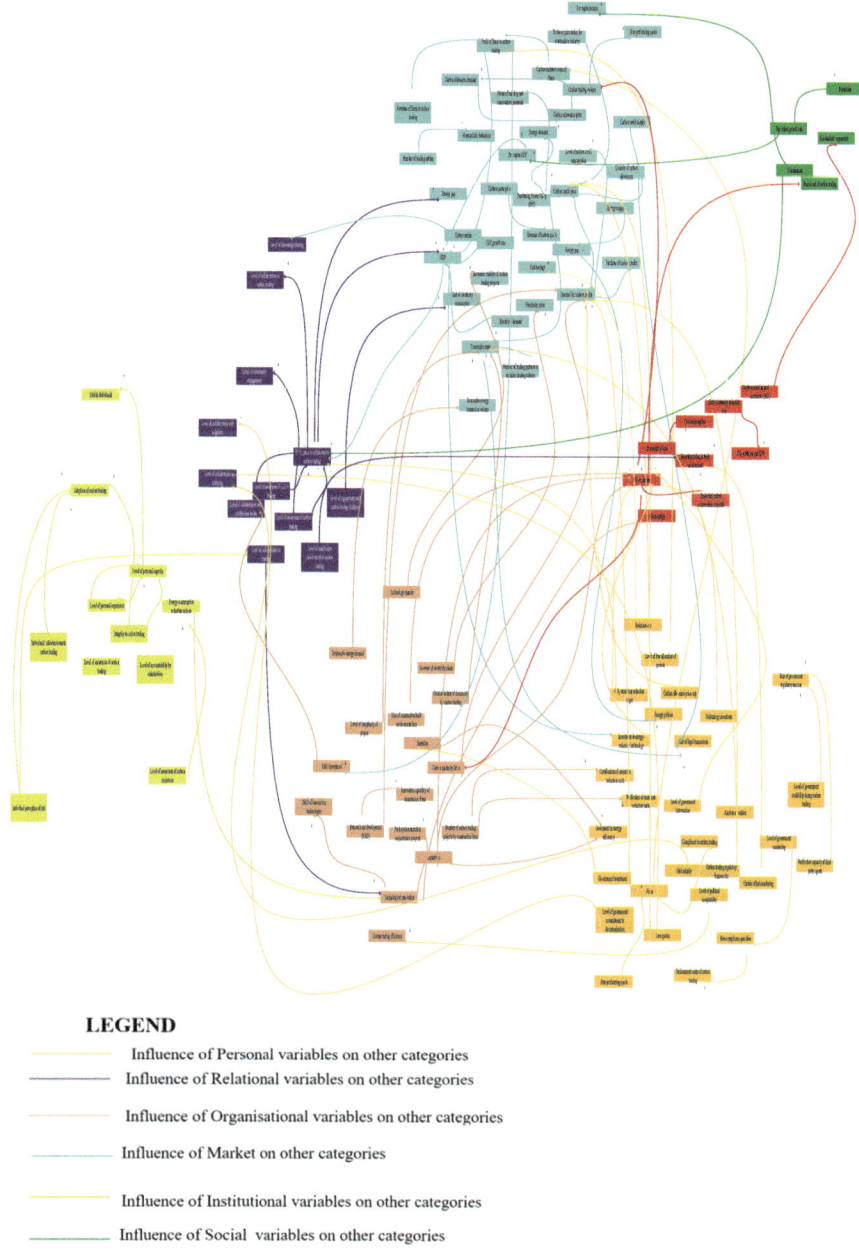

LEGEND

Influence of Personal variables on other categories

Influence of Relational variables on other categories

Influence of Organisational variables on other categories

Influence of Market on other categories

Influence of Institutional variables on other categories

Influence of Social variables on other categories

Influence of Environmental variables on other categories

Fig. 3 Process in carbon trading system. (Source: Authors' construct (2024))

other factors in the market categorization and variables in the other categories. Some of these variables include carbon credit price, demand for carbon credits, level of carbon credit consumption, quantity of carbon allowances, profit of firms in carbon trading, carbon trading volume, and transaction costs [21, 22].

A further major process in carbon trading system for the construction industry is Monitoring, Reporting, and Verification (MRV). MRV consists of multiple steps involved in measuring emissions removed by polluting firms and certifying that the removals are truly real, additional, permanent, and verifiable. This is then reported to a third-party agent that is accredited to certify that the removal was completed according to the right MRV protocols. The verification of carbon credits is a rigorous process, and the end goal is to ensure legitimacy of the credits. This verification process starts with the project developers and firms that implement activities that reduce carbon emissions. The developers/firms have to provide evidence of reduction in carbon through the monitoring of data, project reports and other required documents. In Fig. 3, MRV and its associated variables are found in the Institutional category. Some of the variables in the system include verification capacity of third-party agents, verification of emissions reduction costs, certification of emissions reduction costs, and participation in carbon sequestration projects [53].

Figure 3 below shows the process for carbon trading in the construction industry based on the variables in the input categorization using PROMISE.

3.3 Output of the Carbon Trading System

Output in systems theory is the final outcome obtained after running an entire process or a part of the process. In the trading system, the output will be measured by two indicators. These are the efficiency of the trading system and its ability to contribute to mitigation of greenhouse gas emissions. According to Kukah et al. [19], when carbon trading system for the construction industry is operational and running, it is appropriate for the efficiency of the system to be measured in order to ascertain if it is attaining its set goals. Xia et al. [54] also posit that the end goal of a carbon trading system is to contribute to the mitigation of greenhouse gas emissions. The built environment sector contributes 39% of global carbon emissions comprising of both embodied carbon and operational carbon [55]. It is expedient, therefore, to ensure the carbon trading system developed for the construction industry is able to contribute to reduction in these emissions. Figure 4 below shows the output for carbon trading in the construction industry.

Fig. 4 Output in carbon trading system. (Source: Authors' construct (2024))

4 Conclusions

Emissions trading has been identified to be a reasonable panacea for curbing future levels of greenhouse gas emissions. This paper presented a conceptual model of carbon trading system for the construction industry based on systems theory. The system comprises input, transformed through process and having an output. There is feedback from the output to the input in the system. This study has useful practical, empirical, theoretical, and wider implications. Overall, this research makes a unique contribution to climate change mitigation agenda by modeling carbon trading system for the construction industry. This study is relevant to policymakers, industry, and academic researchers since the outcome of the study enhances knowledge and gives sound basis for future works on reduction in greenhouse gas emissions through the application of carbon trading.

References

1. Agyekum, K., Kukah, A.S., Amudjie, J.: The impact of COVID-19 on the construction industry in Ghana: the case of some selected firms. J. Eng. Des. Technol. **20**(1), 222–244 (2021)
2. Kukah, A.S.K., Blay, A.V.K.J., Opoku, A.: Strategies to reduce the impact of resource consumption in the Ghanaian construction industry. Int. J. Real Estate Stud. **16**(1), 51–59 (2022)
3. Ding, G., Thuy, N.: Carbon management. Adv. Carbon Manag.Technol. Carbon Remov. Renew. Nucl. Energy. **1**, 40 (2020)
4. Sizirici, B., Fseha, Y., Cho, C.-S., Yildiz, I., Byon, Y.-J.: A review of carbon footprint reduction in construction industry, from design to operation. Materials. **14**(20), 6094 (2021)
5. Karadumpa, C.S., Pancharathi, R.K.: Study on energy use and carbon emission from manufacturing of OPC and blended cements in India. Environ. Sci. Pollut. Res. **31**(4), 5364–5383 (2024)
6. Blay Jnr, A.V.K., Kukah, A.S.K., Opoku, A., Asiedu, R.: Impact of competitive strategies on achieving the sustainable development goals: context of Ghanaian construction firms. Int. J. Constr. Manag. **23**(13), 2209–2220 (2023)
7. Kukah, A.S.K., Jin, X., Osei-Kyei, R., Perera, S.: A conceptual framework for carbon trading in the construction industry. Paper Presented at 45th Australasian Universities Building Education Association (AUBEA) Conference. Global Challenges in a Disrupted World: Smart, Sustainable and Resilient Approaches in the Built Environment, Sydney, Australia (2022)
8. Schwartz, Y., Raslan, R., Mumovic, D.: The life cycle carbon footprint of refurbished and new buildings–a systematic review of case studies. Renew. Sust. Energ. Rev. **81**, 231–241 (2018)
9. Rodrigo, M., Perera, S., Senaratne, S., Jin, X.: Embodied carbon mitigation strategies in the construction industry. CIB World Build. Congress. **2019**, 1–10 (2019)
10. De Wolf, C., Pomponi, F., Moncaster, A.: Measuring embodied carbon dioxide equivalent of buildings: a review and critique of current industry practice. Energy Build. **140**, 68–80 (2017)
11. Pan, Y., Zhang, X., Wang, Y., Yan, J., Zhou, S., Li, G., Bao, J.: Application of blockchain in carbon trading. Energy Procedia. **158**, 4286–4291 (2019)
12. Huang, W., Wang, Q., Li, H., Fan, H., Qian, Y., Klemeš, J.J.: Review of recent progress of emission trading policy in China. J. Clean. Prod. **349**, 131480 (2022)

13. Fang, G., Tian, L., Liu, M., Fu, M., Sun, M.: How to optimize the development of carbon trading in China—enlightenment from evolution rules of the EU carbon price. Appl. Energy. **211**, 1039–1049 (2018)
14. Ng, S.T., Luk, W.S.: Applicability of existing carbon trading schemes to the construction industry. In: CESB 2013 PRAGUE-Central Europe Towards Sustainable Building 2013: Sustainable Building and Refurbishment for Next Generations, pp. 753–756. Czech Technical University in Prague (2013)
15. Oke, A.E., Aigbavboa, C.O., Dlamini, S.A.: Carbon emission trading in South African construction industry. Energy Procedia. **142**, 2371–2376 (2017)
16. Falana, J., Osei-Kyei, R., Tam, V.W.: Towards achieving a net zero carbon building: a review of key stakeholders and their roles in net zero carbon building whole life cycle. J. Build. Eng. **82**, 108223 (2023)
17. Oke, A.E., Oyediran, A.O., Koriko, G., Tang, L.M.: Carbon trading practices adoption for sustainable construction: a study of the barriers in a developing country. Sustain. Dev. **32**, 1120 (2024)
18. Koriko, O.: Appraisal of Carbon Trading Practices in the Construction Industry in Lagos State, Nigeria. Federal University of Technology, Akure (2021)
19. Kukah, A.S.K., Jin, X., Osei-Kyei, R., Perera, S.: Towards developing a carbon trading system for the construction industry: identification of major components. In: EC3 Conference 2023, European Council on Computing in Construction, pp. 0–0. (2023)
20. Yang, Z.H., Li, S.J., Zhou, Y.P., Wang, Q., Liu, Q.Z.: Clean development mechanism and carbon trading market construction in China. Adv. Mater. Res. **616**, 1500–1504 (2013)
21. Song, X., Shen, L., Yam, M.C., Zhao, Z.: SNA based identification of key factors affecting the implementation of emission trading system (ETS) in building sector: a study in the context of China. In: Proceedings of the 20th International Symposium on Advancement of Construction Management and Real Estate, pp. 595–606. Springer (2017)
22. Liu, B., Ding, C.J., Hu, J., Su, Y., Qin, C.: Carbon trading and regional carbon productivity. J. Clean. Prod. **420**, 138395 (2023)
23. Zhang, Y.J., Liang, T., Jin, Y.L., Shen, B.: The impact of carbon trading on economic output and carbon emissions reduction in China's industrial sectors. Appl. Energy. **260**, 114290 (2020)
24. Chi, Y.-Y., Zhao, H., Hu, Y., Yuan, Y.-K., Pang, Y.-X.: The impact of allocation methods on carbon emission trading under electricity marketization reform in China: a system dynamics analysis. Energy. **259**, 125034 (2022)
25. Dibie, R.: Environmental Sustainability and Solutions. In: Comparative Perspectives on Environmental Policies and Issues, pp. 450–484. Routledge (2014)
26. Becvar, R.J., Becvar, D.S., Reif, L.V.: Systems Theory and Family Therapy: a Primer. Rowman & Littlefield (2023)
27. Kuntsevich, V., Gubarev, V., Kondratenko, Y.: Control Systems: Theory and Applications. CRC Press (2022)
28. Skyttner, L.: General Systems Theory: Ideas & Applications. World Scientific (2001)
29. Cabrera, D., Colosi, L., Lobdell, C.: Systems thinking. Eval. Program Plann. **31**(3), 299–310 (2008)
30. Kramer, N.J., De Smit, J.: Systems Thinking: Concepts and Notions. Springer Science & Business Media (2012)
31. Ramos, G., Aguiar, A.P., Pequito, S.: An overview of structural systems theory. Automatica. **140**, 110229 (2022)
32. Renn, O., Laubichler, M., Lucas, K., Kröger, W., Schanze, J., Scholz, R.W., Schweizer, P.J.: Systemic risks from different perspectives. Risk Anal. **42**(9), 1902–1920 (2022)
33. Giachetti, R.: Design of Enterprise Systems: Theory, Architecture, and Methods. CRC Press (2016)
34. Midgley, G.: Systems Thinking. Sage London (2003)
35. Jackson, M.C.: Systems Thinking: Creative Holism for Managers. Wiley. (2016)

36. Shaked, H., Schechter, C., Shaked, H., Schechter, C.: Definitions and development of systems thinking. In: Systems Thinking for School Leaders: Holistic Leadership for Excellence in Education, pp. 9–22. Springer, Cham (2017)
37. Zadeh, L., Desoer, C.: Linear System Theory: the State Space Approach. Courier Dover Publications (2008)
38. Jay, J.: PROMISE: a systems approach to sustainability strategy. In: Sustainability Initiative. MIT Sloan School of Management (2012)
39. Fischhoff, M., Jay, J., Sterman, J.: How to Handle Complexity. Network for Business Sustainability (2018)
40. Dally, D., Rohayati, Y., Kazemian, S.: Personal carbon trading, carbon-knowledge management and their influence on environmental sustainability in Thailand. Int J Energy Econ Policy. **10**(6), 609–616 (2020)
41. Jiang, J., Xie, D., Ye, B., Shen, B., Chen, Z.: Research on China's cap-and-trade carbon emission trading scheme: overview and outlook. Appl. Energy. **178**, 902–917 (2016)
42. Yang, Z., Ju, M., Zhou, Y., Wang, Q., Ma, N.: An analysis of greenhouse gas emission trading system from the perspective of stakeholders. Procedia Environ. Sci. **2**, 82–91 (2010)
43. Nidhin, B.K.S.N., Domingo, N., Bui, T.T.P., Wilkinson, S.: Construction stakeholders' knowledge on zero carbon initiatives in New Zealand. Int. J. Build. Pathol. Adapt. (2023). https://doi.org/10.1108/IJBPA-08-2022-0119
44. Wade, B., Dargusch, P., Griffiths, A.: Defining best practice carbon management in an Australian context. Aust. J. Environ. Manag. **21**(1), 52–64 (2014)
45. Tang, D., Gong, X., Liu, M.: The impact of government behaviors on the transition towards carbon neutrality in the construction industry: a perspective of the whole life cycle of buildings. Front. Environ. Sci., 1041 (2022)
46. Saka, N., Olanipekun, A.O., Omotayo, T.: Reward and compensation incentives for enhancing green building construction. Environ. Sustain. Indic. **11**, 100138 (2021)
47. Liu, G., Li, X., Tan, Y., Zhang, G.: Building green retrofit in China: policies, barriers and recommendations. Energy Policy. **139**, 111356 (2020)
48. Tebourbi, I., Thi Truc Nguyen, A., Yuan, S.-F., Huang, C.-Y.: How do social and economic factors affect carbon emissions? New evidence from five ASEAN developing countries. Econ. Res.-Ekonomska istraživanja. **36**(1) (2023)
49. MIT, Jay, J.: PROMISE: A systems approach to sustainability strategy (Webinar powerpoint slides by Jason Jay). Available at https://s3.amazonaws.com/sbweb/slideshow/SB+webinar+SSB+exec+ed+preview.pdf (2015)
50. Dumitriu, C.: Sustainable development at universities as viewed through the lens of the PROMISE framework for sustainability In: Handbook of Theory and Practice of Sustainable Development in Higher Education, vol. 2(1), pp. 35–48. (2017)
51. Chen, D., Yin, J., Xu, F., Huang, C., Li, Z.: A market-based framework for CO_2 emissions reduction in China's civil aviation industry. Transp. Policy. **143**, 150–158 (2023)
52. Shi, B., Li, N., Gao, Q., Li, G.: Market incentives, carbon quota allocation and carbon emission reduction: evidence from China's carbon trading pilot policy. J. Environ. Manag. **319**, 115650 (2022)
53. Bellassen, V., Stephan, N., Afriat, M., Alberola, E., Barker, A., Chang, J.-P., Chiquet, C., Cochran, I., Deheza, M., Dimopoulos, C.: Monitoring, reporting and verifying emissions in the climate economy. Nat. Clim. Change. **5**(4), 319–328 (2015)
54. Xia, Q., Li, L., Dong, J., Zhang, B.: Reduction effect and mechanism analysis of carbon trading policy on carbon emissions from land use. Sustainability. **13**(17), 9558 (2021)
55. Sun, Z., Ma, Z., Ma, M., Cai, W., Xiang, X., Zhang, S., Chen, M., Chen, L.: Carbon peak and carbon neutrality in the building sector: a bibliometric review. Buildings. **12**(2), 128 (2022)

SirkTRE's Evolution or Circulution?: Diverse Pathways of Circular Systemic Solutions for a Net-Zero Timber-Built Environment

Wendy Wuyts ⓘ, Nhat Strøm-Andersen ⓘ, Shumaila Khatri ⓘ,
Arild Eriksen, Per F. Jørgensen, Arild Øvergaard, Emil Rygh,
Angelica Kveen, Alexander Mertens, Jannicke Stadaas, Inger Gamme,
Veronique Vasseur ⓘ, Anders Q. Nyrud ⓘ, and Kristine Nore ⓘ

Contents

W. Wuyts (✉) · K. Nore
Omtre AS, Hønefoss, Norway
e-mail: wendy@omtre.no

N. Strøm-Andersen
Norwegian Institute of Bioeconomy Research NIBIO, Ås, Norway

S. Khatri
Inland Norway University of Applied Sciences, Koppang, Norway

A. Eriksen
Fragment AS, Oslo, Norway

P. F. Jørgensen
Vill Energi AS, Oslo, Norway

A. Øvergaard
Norsk Massivtre AS, Bekkestua, Norway

E. Rygh
Sirkulær Resurssentral, Oslo, Norway

A. Kveen · A. Mertens · J. Stadaas
Grape Architects AS, Oslo, Norway

© The Author(s) 2025   1881
M. Kioumarsi, B. Shafei (eds.), *The 1st International Conference on Net-Zero Built
Environment*, Lecture Notes in Civil Engineering 237,
https://doi.org/10.1007/978-3-031-69626-8_155

1   Introduction

As the planet faces escalating environmental challenges, particularly in the realm of the built environment, the urgency to transition toward zero waste and zero emission strategies is more critical than ever. These strategies are essential to mitigate the profound ecological footprint of human activities, aiming to alleviate the environmental burdens we impose on our planet. Within this pressing context, the circular economy (CE) model emerges as a model of transformation, advocating for the establishment of sustainable, self-sufficient systems that significantly reduce environmental impacts while simultaneously fostering economic resilience. This model, fundamentally, is about creating an economic system dedicated to the elimination of waste and enabling the continual use of resources. It encompasses an extensive array of practices designed to extend the lifecycle of materials, promote efficient recycling and reuse, and decisively reduce the extraction and consumption of resources [1].

Introducing the 'Circulation' concept in this discourse underscores the application of CE principles specifically within the timber and construction sectors. This novel term encapsulates the necessity for both evolutionary and revolutionary approaches in the quest to foster a Net-Zero built environment. 'Circulation' embodies the dual pathways of innovation within the construction sector concerning timber use: evolution, which indicates a gradual, continuous development of sustainable practices and technologies; and resolution, which signals a significant, transformative shift toward groundbreaking innovations. This dual approach underscores the necessity for comprehensive and multifaceted strategies to accelerate the transition to a Net-Zero built environment, blending incremental improvements with radical, systemic changes.

The transition toward a CE necessitates profound shifts not only in our production and consumption patterns but also across the entire value chain. It demands the adaptation of market strategies, the innovation of business models, and a thorough reevaluation of management strategies to align with the diverse visions of stakeholders involved in the circular built environment. However, the concept of circularity is rife with contestation, characterized by diverse interpretations and expectations among stakeholders [2]. This contested nature of circularity highlights the importance of stakeholder expectations and consumer perceptions as pivotal elements that can significantly accelerate the transition process to a robust circular economy but can also hinder it if there are no shared visions and values

I. Gamme
Hunton, Gjøvik, Norway

V. Vasseur
Maastricht University, Maastricht, The Netherlands

A. Q. Nyrud
Norwegian University of Life Sciences, Ås, Norway

[3]. Transitioning to new business models involves redefining value propositions for customers and reconfiguring value chains with partners to create and capture new value, indicating a strategic overhaul of traditional business practices [4, 5]. Achieving a true 'circulation' within the timber building industry necessitates the integration of governance structures, innovative business models, and active policy interaction. It highlights the need for collaboration among a multitude of actors across different sectors and the adoption of collective action. A CE organization cannot operate in isolation across the product or service lifecycle; previous research highlighted that it must engage with its stakeholders and comply with existing regulations, embodying the essence of collective action [6–9]. Observing a CE organization unfolds as an exploration of a collective project, involving regulations, practices, tools, devices, and the amalgamation of diverse values held by stakeholders. This collective endeavor prompts a deeper inquiry into the fundamental reasons behind the existence and emergence of enterprises and, more broadly, into the dynamics of collective action in the circular economy.

The Norwegian SirkTRE project (2022–2025) is at the forefront of pioneering a CE within the timber construction sector, with a clear mission to transform the way timber is utilized, reused, and recycled in the building industry [10]. The SirkTRE is co-funded by the Norwegian Green Platform initiative, which provides funding for enterprises and research institutes engaged in green growth and restructuring driven by research and innovation. SirkTRE was funded to address critical environmental challenges in the construction sector by promoting circular economy principles, particularly focusing on timber, to mitigate the ecological footprint of human activity. There are five focus areas, which are divided into 24 different work packages (Fig. 1). The five focus areas are wood-based plates and connections (SirkRESSURS), solid wood (SirkHELTRE), demonstration projects (SirkREALISERING), digital technology (SirkTEK), and standards (SirkINN). SirkINN envelopes also required coordination, communication, dissemination, and

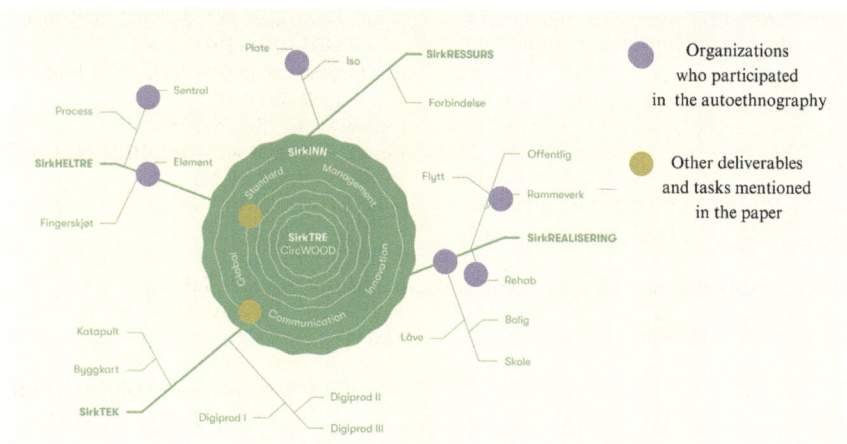

Fig. 1 The structure of SirkTRE, original design by the SirkTRE consortium, adapted by the authors

exploitation activities. The consortium counts almost 30 Norwegian industry partners from across the whole value chain (forestry, wood waste logistics, storage, prefab construction, product manufacturers, and architects), including a cluster organization, which represents more companies. Mostly only one to two partners are allocated to a work package. This means that SirkTRE has overall project objectives, but also envelopes dozens of sub-objectives (linked with the 24 different work packages). Most focus areas are managed by the project manager, Omtre AS, a startup that was founded to address some of the missing roles and responsibilities identified in the proposal writing stage in 2021.

However, this study was inspired by previous research [6–9] highlighting the necessity of stakeholder orchestration and compliance to achieve circularity in the construction sector. Additionally, it was driven by the recognition of the need for a midterm evaluation to assess hypotheses regarding collective learning and coordination, particularly in business model development and upscaling, given the initial structural disconnection of the project. One hypothesis posits that learning by doing addresses these less favorable initial conditions, while another suggests that the absence of companies providing digital enablers may hinder progress. Originally, the project aimed to have another Green Platform project complementing the focus on wood technology and architectural solutions in SirkTRE. Thus, the study serves as a reflective assessment for partners to evaluate necessary actions in the project's final phase, both internally and externally.

Hence, the main objective of this study is to analyze the evolution and resolution pathways within the project, focusing on its endeavor to establish a circular timber value chain for the construction sector and to identify policy recommendations, but also for the ongoing project. This objective will be addressed through the following research questions: How do the evolution pathways pursued by SirkTRE contribute to the transition towards a Net-Zero built environment? What practical strategies and interventions are employed by the SirkTRE consortium to implement circular systemic solutions within the construction sector, particularly concerning timber use? What are the main barriers, success factors, and key learnings derived from the SirkTRE project's implementation so far (in the first two years), and how do they inform future initiatives aiming to scale up circular systemic solutions within the construction industry?

2   Methodology

2.1   Analytical Framework: Evolutionary Lens, Termed 'Circulution'

As aforementioned, we propel our own evolutionary framework, 'circulution'. We build further on other existing evolutionary or economic evolutionary geography frameworks that have been used to analyze the governance of resources in the transition toward a circular economy in various regions (e.g [12]). Marjanović and Williams [12] explored the uptake of circularity in a Dutch and Finnish region: Evolutionary Governance Theory provides a dynamic lens to understand the evolutionary trajectory of circular systemic solutions, capturing both gradual evolution and transformative shifts akin to the concepts of evolution and resolution highlighted in the title. By emphasizing the continuous interaction among actors, institutions, and discourses, Evolutionary Governance Theory offers insights into how governance structures and strategies evolve over time. However, the lens that [12] uses is based on observations of other phenomena, not on circular economy transitions per se. This conference paper takes a more grounded approach and, through autoethnographic investigation of a specific case (see Sects. 2.2. and 2.3), focuses on economic, political, technological, and strategic factors influencing the scalability and implementation of circular systemic solutions. In this way, we develop our own evolutionary framework, which we term 'circulation', to explore how these varied resources are orchestrated to facilitate the adoption of circular wood solutions in construction.

2.2   Methodological Choice: Collaborative Autoethnography

We opted for a collaborative autoethnographic approach to delve into the multifaceted dynamics of the SirkTRE project. This methodological choice allowed us to blend personal experiences, insights from interviews, and reflections on collective actions within the project context. By engaging with various stakeholders and social scientists involved in the project, we aimed to co-create a rich narrative that captures the complexities, challenges, and successes of SirkTRE's journey towards circular systemic solutions. Through this collaborative endeavor, we sought to provide a holistic understanding of the project's evolution and its implications for sustainable transitions in the timber-built environment. Reflecting on the collaborative autoethnographic approach employed in this study, several strengths and limitations come to light. One notable strength is the rich and nuanced understanding of the research subject derived from the integration of multiple perspectives and experiences. Additionally, the collaborative nature of the approach fosters a sense of ownership and empowerment among participants, enhancing the credibility and relevance of the research findings. However, the approach also presents certain limitations, including the potential for bias or subjectivity inherent in self-reflection and interpretation. One mitigation measure was the omission of solutions developed by Organization G, which was the organization of the first author, which means less representation of, for example, SirkTEK. The focus of this chapter was on external challenges, but there were also hints of internal challenges, for example, lack of entrepreneurship/intrapreneurship in the involved organization and lack of internal resources (financial, competence, capacity, etc.) to scale up inventions to innovations. The investigation of internal challenges might be rather done by researchers in outsider roles. Moreover, coordinating multiple voices and viewpoints in the analysis process can be complex and time-consuming, requiring careful navigation of interpersonal dynamics and power relations. Due to its autoethnographic approach, his study focused on one consortium but did not compare with other consortia focusing on other products, services and materials, which omits insights about other reasons hindering innovation.

2.3   Data Collection and Validation

This study utilized a multiple-case study approach [11], focusing on interviews with representatives from six partner organizations within the SirkTRE consortium. They are labelled as Organizations A, B, C, D, E and F (see also Fig. 1 for their position in the SirkTRE consortium). The six interviewees are mostly engineers or general managers of architectural firms, manufacturing companies, storage places and environmental consultancy. The first author contacted many contact persons of the SirkTRE consortium for a study on mid-term evaluation of the barriers and key factors for upscaling their solutions., but only these representatives replied to the call. The first author, coming from Organization G, conducted these interviews, transcribed them, and then shared the transcripts with the respective partners for validation and consensus to ensure accuracy and alignment with their perspectives. Based on the insights gained from these validated interviews, the first author crafted an initial draft, capturing the developmental history, and personal reflections on the successes and challenges faced by SirkTRE and its participants. Following the creation of this initial draft, a series of feedback rounds were initiated, involving the interviewees from the partner organizations as well as social scientists who have been closely observing the SirkTRE project. This collaborative review process allowed for a richer, more multifaceted exploration of the themes identified in the interviews, ensuring that the manuscript accurately reflects the collective experiences and insights of those deeply involved in the project. Through this iterative process of drafting, feedback, and revision, the study aims to provide a comprehensive and nuanced understanding of the SirkTRE project's endeavors to foster circular systemic solutions within the construction industry.

3   Results: Diverse Pathways of Circular Systemic Solutions for a Net-Zero Timber-Built Environment for 2021 to 2023

The exploration of circular systemic solutions within the SirkTRE project unveils a rich menu of innovative approaches aimed at achieving a net-zero timber-built environment. Pathways in sustainability and innovation contexts are strategic approaches designed to achieve long-term goals by addressing complex challenges

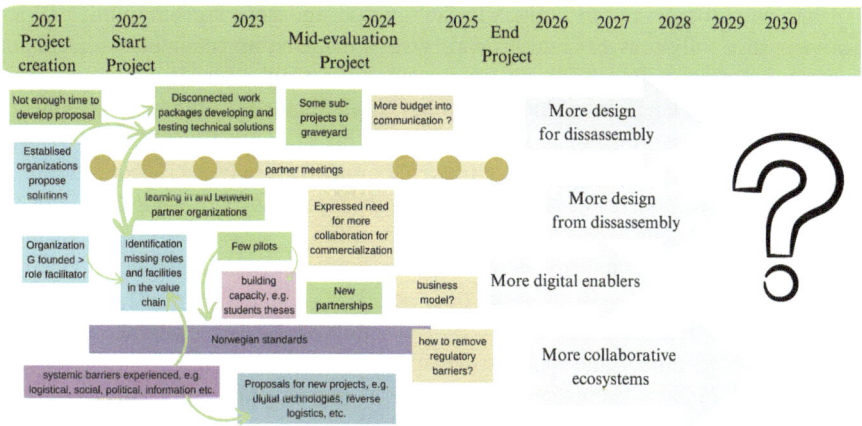

Fig. 2 Timeline of observations in SirkTRE and the possible diverse pathways unfolding for a circular future with net-zero timber built environment

such as climate change or resource scarcity. They are multidimensional, incorporating technological, economic, social, and environmental aspects, and are dynamic, evolving over time to adapt to new challenges and insights. These pathways are systems-oriented, focusing on the interconnectedness of various factors within a system, and are developed through the engagement of diverse stakeholders to ensure inclusivity and broad support. Specifically, in projects aiming for a net-zero timber-built environment, pathways would integrate circular economy principles, innovative construction techniques, and digital tools for material management, emphasizing collaboration across the construction value chain to achieve sustainability objectives efficiently. This section presents an overview of these pathways for SirkTRE (Fig. 2).

3.1   Design for Disassembly and Reuse

The development of modular and reusable interior wall systems represents a significant innovation within the SirkTRE project, particularly highlighted in the work of organizations like D and E. These systems, designed for ease of disassembly and reassembly, stand at the forefront of extending material lifespans and embodying adaptability and flexibility in architectural design. Organization E's approach to this challenge includes the conceptualization and creation of a building model that supports resident participation and transformation over time. Their system is not just about constructing a space but enabling it to evolve according to the needs of its occupants, thereby reducing the need for new materials and minimizing waste. This model is inspired by broader, systemic solutions that consider the entire lifecycle of a building, emphasizing the importance of user interaction and the potential for spaces to adapt and change. Organization D delved into the technical and practical aspects of implementing circular solutions in the building industry, focusing on the development of a fully reusable interior wall system made from reclaimed timber. Their prototype demonstrates the feasibility of using reclaimed materials in new constructions, offering a tangible solution to the industry's challenge of interior fit-outs that typically have a short lifespan. By designing for disassembly, Organization D's wall system can be easily relocated, reconfigured, or updated, significantly contributing to waste reduction and resource efficiency. Both organizations, through their work, highlight the crucial role of collaboration across the value chain, from material sourcing to construction and beyond. Their projects underscore the potential of modular and reusable systems to not only reduce waste but also to foster innovation in building design and construction. By pushing the boundaries of what is possible with reclaimed timber and modular construction techniques, these initiatives pave the way for a more sustainable, adaptable, and circular building industry.

3.2 Design from Disassembly: Building Models Utilizing Reclaimed Timber

Another pathway is the comprehensive building model that integrates reclaimed timber into its core design. This model not only champions the use of recycled materials but also emphasizes the importance of building systems that can adapt to future needs without requiring extensive modifications. The commitment to using reclaimed timber not only reduces demand for virgin resources but also highlights a valuable shift towards valuing existing materials in the construction sector.

A notable omission within SirkTRE's initial stages was the lack of focused task forces on developing viable business models for these circular solutions. None of the interviewed organizations had any input on business model development for reclaimed timber as it was out of scope for them. This omission was visible after conducting all interviews. This gap has hindered the broader adoption and commercialization of the innovations. However, insights from 2024 suggest a promising shift, with SirkTRE increasingly directing its attention towards overcoming commercial barriers and fostering market readiness for circular construction practices. This evolution marks a crucial step in moving from concept to commercial viability, underscoring the project's role in leading the construction industry towards a more sustainable and circular future.

3.3 Digital Enablers Like Product Passports, Tracing and Tracking Technologies

In the quest for circularity within the SirkTRE project, Organizations A and B have highlighted the pivotal role of digital innovations, particularly emphasizing the potential of digital product passports. Organization A sees digital tools as essential for tracking the lifecycle of wood products, from production through reuse. Their project focuses on understanding how products can be effectively returned to the manufacturing process or repurposed, with digital tracking systems offering a way to monitor product history and facilitate this circular process. In the journey toward circularity, Organization A encountered significant challenges related to information and data management, which have implications for their product development processes. One of the primary hurdles was the lack of detailed, accessible data on the lifecycle and usage history of wood products. This gap made it difficult for Organization A to accurately assess the potential for reuse and remanufacturing of their existing products and to develop products that could easily be integrated into circular systems. Additionally, the absence of standardized data formats and platforms for sharing information across the construction industry ecosystem further compounded these challenges. As a result, these information and data challenges not only slowed down Organization A's product development but also highlighted the critical need for comprehensive digital solutions, such as the digital product passports discussed, to support the transition to a circular economy in the timber construction sector. Organization B, on the other hand, brings a practical perspective to the implementation of digital passports. They underscore the challenges faced in the current regulatory environment, which often does not favor the reuse of materials due to stringent classification standards. However, they also see digital passports as a solution to these challenges, by providing a transparent and accessible record of material quality, safety, and reuse potential. This could, in turn, influence policy changes and encourage greater acceptance of reused materials in the construction industry. Both organizations underline the necessity of robust, user-friendly digital platforms that can integrate seamlessly with existing industry workflows. The envisioned digital passports would not only store critical material data but also include information on dismantling and reassembly instructions, further supporting the circular economy model. This approach aims to bridge the gap between the potential for material reuse and the current practices dominated by linear consumption. Despite the enthusiasm for digital enablers, there is a gap in SirkTRE's initial focus on these technologies. The early stages of the project did not prioritize the development of digital tools as a central component of circular solutions. However, the evolving insights from Organization A and Organization B suggest a growing recognition of the importance of digitalization in achieving the project's circular goals. This shift points towards a more integrated approach, where digital enablers are seen as critical to the success of circular construction practices, potentially transforming the way materials are managed across their lifecycle.

3.4 Collaborative Ecosystems and Value Chains

Expanding the collaborative ecosystems and innovative value chains within the timber construction industry, SirkTRE exemplifies the transformative potential of cross-sectoral collaboration for advancing circularity. The initiative notably includes contributions from policy entities like Standards Norway [13] and spans to hands-on industry practitioners, establishing a broad, collaborative network. However, insights from consortium members like Organizations C and D underscore a missed opportunity for more integrated collaborative tasks or work packages that could have fostered deeper synergies across the value chain. They highlight a longing for structured collaborations where partners representing diverse segments of the value chain are intentionally brought together to leverage their unique contributions towards circular solutions. Organization F's experiences further illuminate the challenges of consortium formation, noting the time-intensive nature of assembling a diverse group and the difficulty in identifying and exploiting linkages between partners unfamiliar with each other or who are nascent in their circularity journey. This feedback underscores a critical gap in the consortium's operation, suggesting a need for mechanisms that foster closer collaboration and knowledge exchange among partners. It points to the potential benefits of incorporating collaborative tasks explicitly designed to bridge different areas of expertise, encouraging innovation through shared knowledge and co-creation. The initiative's focus on establishing supportive standards and advocating for policy adjustments further showcases its dedication to overcoming regulatory hurdles and fostering an industry-wide shift towards circular solutions. While the journey reveals areas for improvement, particularly in enhancing partner collaboration and integration, SirkTRE's comprehensive approach continues to drive the timber construction industry closer to achieving a net-zero, circularly built environment, demonstrating the critical role of partnerships in achieving systemic change.

4   Major Challenges and Key Success Factors So Far

The SirkTRE project, with its diverse consortium and ambitious goals, has fostered the development of multiple circular solutions within the construction sector, addressing the urgent need for sustainable practices. A more grounded lens, with the aim to propose a circulation framework, provides a comprehensive understanding of both the evolution and revolution of the outcomes, challenges, and the path forward for these solutions. The different pathways within SirkTRE illustrate not only the successes but also the challenges, including how economic, political, technological, and strategic decisions can impact the scaling up/out or ending.

4.1   Technical–Economic Dimension

A recurring theme across the experiences of the SirkTRE partners is the balancing act between technical feasibility and economic viability. Innovations like modular and reusable interior wall systems and comprehensive building models utilizing reclaimed timber exemplify technical successes with significant environmental benefits. Technologically, the SirkTRE project showcases an array of innovations aimed at enhancing circularity in construction. The progression from conceptual stages to real-world prototypes in some projects illustrates a successful translation of innovative ideas into practical applications. However, technological limitations are also apparent, particularly in scaling up these solutions for broader market adoption. The economic acceptability of these solutions often hinges on overcoming market inertia, regulatory barriers, and the upfront costs associated with adopting new technologies. One of the key challenges, remarked by several interviewees, facing the reuse of reclaimed timber is the cost competitiveness relative to virgin materials. The collaboration with stakeholders such as Standards Norway indicates a movement towards establishing supportive policies and standards, yet the pathway to economic viability remains a challenge, underlined by concerns over the scalability of solutions within current market structures. Challenges related to the integration of digital enablers, such as digital product passports, and the need for more advanced technological solutions to streamline the reuse of materials are underscored. Moreover, the disparity in Technology Readiness Levels among different solutions indicates varying degrees of maturity and market readiness, which could impact the project's overall success.

4.2   Ethical–Political Dimension

The ethical–political dimension highlights the project's alignment with broader societal, economic, and environmental goals. SirkTRE partners reveal a collective ambition to contribute significantly to Norway's CO_2 reduction targets and to foster a more sustainable construction industry. A strong emphasis on environmental sustainability as a primary motivation behind the circular solutions is evident. Projects ranging from reusable interior wall systems to comprehensive building models utilizing reclaimed timber exemplify a commitment to reducing waste and carbon emissions. However, environmental challenges such as the classification and standardization of reclaimed materials, as well as the need for broader life cycle assessments, are highlighted. The difficulty in measuring the direct environmental impact of the SirkTRE project on the broader value chain and sector points to a gap in comprehensive environmental reporting and assessment methodologies within circular construction practices. In addition, the realization of these goals is contingent upon navigating political landscapes, securing governmental support, and influencing policy changes to favor circular practices. The CO_2 accounting system remains a work in progress with significant political dimensions, as major corporations lobby intensively to shape a system that prioritizes their interests. As [14] illustrated, the current policy landscape in Norway does not view construction material circularity as a policy objective at any governance level and lacks economic, capacity building and regulatory tools to advance wood circularity in construction under-utilizing the influence of top-down mechanisms. The engagement with policymakers and standard-setting bodies marks steps towards these ends, but the pace of political and regulatory support is a noted barrier.

4.3   Developmental Dimension

SirkTRE's impact is also measured through its developmental growth, encompassing geographical expansion, stakeholder engagement, and performance metrics. While the project has successfully fostered a collaborative ecosystem, expanding its influence beyond initial boundaries remains challenging. The diversity of stakeholders from various sectors enhances the project's richness but also complicates coordination efforts. These observations in SirkTRE are aligned with other observations in other ecosystems and projects and previous research highlighting stakeholder coordination as an enabler of CE in the construction sector [6–9]. Growth in terms of legal status changes, sales evolution, or geographical presence is indirectly influenced by the project's ability to navigate and adapt to the evolving landscape of the timber construction industry. Concerning the geographical dimension, the Norwegian ecosystem has path dependencies (e.g. forestry infrastructures, and incineration plants competing for wood waste resources) that can hinder the uptake. In addition, spatial capacity building and planning for a circular built environment are not present in Norwegian circular economy discourse and actualization, while these are required policy instruments [15, 16]. Litleskare and Wuyts [16] addressed findings from qualitative interviews and focus groups with architects, stakeholders in the wood industry and inhabitants of urban timber buildings. They found that the choice of using wood efficiently and adapting circular construction goals in the wood industry is partially guided by values, ideas and convictions of implementing practices that benefit society. The governance aspect within SirkTRE reveals a complex interplay of regulatory frameworks, industry standards, and collaborative dynamics. The project's goal to create new value chains and influence policy through the development of new standards and practices is commendable. Yet, the interviews disclose a notable barrier in the form of existing regulations that do not fully accommodate or incentivize circular solutions, thereby hindering wider adoption. Furthermore, the interviews reflect a need for enhanced governance mechanisms within the SirkTRE consortium itself. A more integrated approach, facilitating better cooperation and knowledge exchange among partners, is essential for overcoming silos and leveraging the collective expertise toward achieving the project's ambitious goals.

4.4   Inventiveness and Regulationist Dimension

The project's capacity for innovation and the establishment of new rules and coordination mechanisms is critical. The development of digital enablers like product passports and the exploration of novel construction models demonstrate a high degree of inventiveness. Yet, the formation of collective action and the implementation of new practices are impeded by existing regulatory frameworks and industry standards. Efforts to negotiate and coordinate within the consortium and with external entities reveal the complexities of aligning diverse interests and visions for a circular future.

5 Conclusion, Implications and Recommendations

The findings offer valuable insights into the governance of circular systemic solutions in the timber-built environment. SirkTRE's journey can be aptly described as a 'Circulation', a term that encapsulates its evolutionary path towards fostering circular systemic solutions within the construction industry. This progression is characterized by the project's ambitious aim to integrate circular economy principles into the timber construction sector, navigating through a complex landscape of technological, economic, ethical–political, and developmental dimensions. SirkTRE has successfully sparked innovation by developing modular and reusable interior wall systems and comprehensive building models utilizing reclaimed timber and leveraging digital enablers like product passports. However, the solutions are still at a low technology readiness level (TRL), and some solutions score lower at the market readiness level (MRL).

These inventions not only showcase the project's technical and economic achievements but also its commitment to ethical and political values, emphasizing sustainability and resource efficiency. Participants got access to knowledge on diverse topics from wood technology to circular business models, training programs, educational initiatives, and networking opportunities in promoting awareness and understanding of circular economy principles and practices. However, the circulation of SirkTRE is not without its challenges. The project has encountered obstacles in fully realizing its goals, stemming from a need for more robust collaborative ecosystems and value chains. Various partners shared a desire for more integrated efforts and knowledge sharing among partners, highlighting the importance of collaboration in overcoming barriers to circularity. Moreover, there's a recognized need for addressing the scale-up challenges, particularly in terms of business models and regulatory support, which were less emphasized in the project's early stages. The SirkTRE partners underscore the importance of supportive policy frameworks in facilitating the adoption of circular practices in timber construction. They highlighted the need for clearer regulations on material reuse, standardized guidelines for sustainable timber sourcing, and streamlined permitting processes to incentivize circularity. To advance the adoption of circular systemic solutions in the construction sector, policymakers and the science-policy interface are advised to refine regulations and standards to better support circular practices, foster public-private partnerships for co-creating circular economy models and introduce financial incentives for circular solutions. Additionally, bridging the gap between science and policy involves disseminating SirkTRE's innovations and sustainable construction practices widely, establishing collaborative platforms for stakeholder engagement across the construction value chain, and promoting education and training focused on circular economy principles and sustainable building techniques. Several key implications emerge, shedding light on the challenges and opportunities for advancing circularity in the sector. The recommendations for the final phase of SirkTRE focus on enhancing these collaborative efforts, suggesting the formation of task forces or work packages that foster inter-partner collaboration, thus ensuring a more cohesive approach to circularity. Additionally, the project is encouraged to deepen its engagement with policymakers and the science-policy interface to facilitate the broader adoption of circular systemic solutions. In conclusion, SirkTRE's successes and challenges alike offer valuable lessons for future endeavors in this field, underscoring the need for continued innovation, policy support, and, most critically, collaborative action. This study demonstrates a way for practitioners to evaluate their project-based collective learning and innovation and the diverse pathways to reach their bigger and smaller goals. As SirkTRE enters its final phase, the focus on overcoming existing barriers and leveraging the project's achievements can further propel the construction industry toward a more sustainable and circular future. Strategic emphasis on cross-project collaboration, stakeholder engagement, scalability, and digital innovation is essential. It is recommended that the consortium intensify efforts to foster cross-work package synergies through regular interdisciplinary workshops, complemented by a unified communication strategy for disseminating project achievements. Active engagement with policymakers, industry partners, and end-users is crucial to demonstrate the practical applicability and benefits of circular solutions, thereby facilitating policy support and market adoption. Developing sustainable business models and conducting market analyses will be key to assessing commercial viability and ensuring the long-term impact of innovations. Additionally, documenting lessons learned and investing in training and educational initiatives can enhance knowledge transfer within the construction sector. Leveraging digital tools, such as product passports and material databases, will further support the optimization of resource use throughout the construction lifecycle. Focused efforts in these areas can significantly contribute to achieving SirkTRE's vision for a sustainable, circular timber-built environment, setting a precedent for future construction practices.

Acknowledgements This conference paper is funded by the Norwegian Research Council, Innovation Norway and Siva under the two sister projects CircWood (project No. 328698) and SirkTRE (project No. 328731) and by the European Union under the DRASTIC project (grant agreement No 101123330). We want to thank Ola Rostad and Marjan Marjanović for providing comments on earlier drafts.

References

1. Potting, J., Hekkert, M.P., Worrell, E., Hanemaaijer, A.: Circular Economy: Measuring Innovation in the Product Chain, vol. 2544. Planbureau voor de Leefomgeving (2017)
2. Korhonen, J., Nuur, C., Feldmann, A., Birkie, S.E.: Circular economy as an essentially contested concept. J. Clean. Prod. **175**, 544–552 (2018)

3. Verbiest, E., Marin, J., De Meulder, B., Vande, M.A.: Untangling stakeholder dynamics in circularity of the built environment. Spool. **10**(1), 57–70 (2023)
4. Jørgensen, S., Tynes Pedersen, L.J.: RESTART Sustainable Business Model Innovation. Springer Nature (2018)
5. Fet, A.M.: Business Transitions: a Path to Sustainability: the Capsem Model. Springer Nature (2023)
6. Giesekam, J., Barrett, J.R., Taylor, P.: Construction sector views on low carbon building materials. Build. Res. Inform. **44**(4), 423–444 (2016)
7. Knoth, K., Fufa, S.M., Seilskjær, E.: Barriers, success factors, and perspectives for the reuse of construction products in Norway. J. Clean. Prod. **337**, 130494 (2022)
8. Giorgi, S., Lavagna, M., Wang, K., Osmani, M., Liu, G., Campioli, A.: Drivers and barriers towards circular economy in the building sector: stakeholder interviews and analysis of five European countries policies and practices. J. Clean. Prod. **336**, 130395 (2022)
9. Anastasiades, K., Dockx, J., van den Berg, M., Rinke, M., Blom, J., Audenaert, A.: Stakeholder perceptions on implementing design for disassembly and standardisation for heterogeneous construction components. Waste Manag. Res. **41**(8), 1372–1381 (2023)
10. SirkTRE Homepage, https://www.sirktre.no/en. Last accessed 2024/03/15
11. Yin, R.K.: Case Study Research: Design and Methods, vol. 5. Sage (2009)
12. Marjanović, M., Williams, J.: Mapping the emergence of the circular economy within the governance paths of shrinking cities and regions: a comparative study of Parkstad Limburg (NL) and Satakunta (FI). Camb. J. Reg. Econ. Soc. **XX**, 1 (2024)
13. Standard Norge: Nye standarder for evaluering av returtre på høring. https://standard.no/nyheter/nye-standarder-for-evaluering-av-returtre-pa-horing/. Last accessed 2024/01/29
14. Khatri, S., Nyrud, A.Q., Sjølie, H.K.: Policy analysis of wood circularity in construction: Current policy objectives and instruments in Norway. (Under review)
15. Heltorp, K., Groba, U., Nyrud, A.Q.: Circular wood construction – expectations, experiences and acceptance among users, architects and industry representatives. In: Nyrud, A.Q., Malo, K.A., Nore, K. (eds.) World Conference on Timber Engineering 2023, Oslo, Norway (2023)
16. Litleskare, S., Wuyts, W.: Planning reclamation, diagnosis and reuse in Norwegian timber construction with circular economy investment and operating costs for information. Sustain. For. **15**(13), 10225 (2023)

Design Science Research for Digital Information Management System (DIMS) in Construction and Demolition Waste Management

Ali Nader Saad ⓘ, Jason Underwood ⓘ, and Juan Ferriz-Papi ⓘ

Contents

1 Introduction

The construction industry is intricately connected to economic growth, employing approximately 7% of the global workforce and contributing to 13% of the gross domestic product (GDP) [1]. This sector consumes vast amounts of raw materials annually. Worldwide statistics show that construction produces 30–40% of all solid waste, with an average of 35% ending up in landfills. Consequently, this situation presents a substantial environmental challenge for the construction sector worldwide [1]. Therefore, the industry must move from its linear economy (Take-Make-Waste) to the circular economy. The circular economy encompasses concepts like industrial ecosystems and symbioses, the 3R principle (reduce, reuse, recycle), circular material flows, zero emissions, and more [2]. In the realm of construction, the impact of cutting-edge digital tools and methodologies is becoming increasingly evident [3].

A. N. Saad (✉) · J. Underwood · J. Ferriz-Papi
School of the Built Environment, University of Salford, Manchester, UK
e-mail: a.saad3@edu.salford.ac.uk

© The Author(s) 2025
M. Kioumarsi, B. Shafei (eds.), *The 1st International Conference on Net-Zero Built Environment*, Lecture Notes in Civil Engineering 237,
https://doi.org/10.1007/978-3-031-69626-8_156

Industry 4.0 technologies play a pivotal role in the journey toward Circular Economy (CE) [4]. By redefining waste management practices, the convergence of Industry 4.0 and CE holds immense promise for environmental, health, and societal gains. The European Commission [5] has pinpointed three key areas where Industry 4.0 decision support systems can bolster CE efforts: (1) facilitating circular production through resource flow tracking, (2) enhancing the resilience and adaptability of built assets in alignment with CE principles via robust material information management, and (3) driving innovation in data spaces and establishing the architecture and governance framework for intelligent applications. Simultaneously, as a new technological era unfolds, digital advancements are revolutionising national and global economies by expediting decision-making and enhancing core business processes [6]. The recent surge in novel technologies propels the built environment into a contemporary data-driven landscape characterised by five distinct phases: data acquisition, mobile data highways, data security, data analysis, and data realisation [7–9].

This chapter focuses on the design science research (DSR) strategy to develop a conceptual framework for a digital information management system (DIMS) to facilitate construction and demolition waste management.

2 Research Philosophy for Information Systems Through Design Science Research

Research and its conclusions may be deceptive or meaningless if philosophy is not given enough thought or comprehension [10]. This is supported by Dawood & Underwood [11], who state that the failure of a great deal of research arises from the researcher needing first to understand their own philosophical assumptions. On the other hand, understanding philosophical questions can help Information Systems (IS) researchers ensure that their work is insightful and comprehensive. It can also improve the quality of the real work [12]. Ontology, epistemology, and axiology are the philosophical groundings of design science in research [13, 14]. Ontology pertains to underlying beliefs concerning the fundamental nature of reality. Ontological assumptions aid the researcher in moulding the perspective through which research objects are seen and examined [15]. After gaining an understanding of the fundamental concept of reality (ontology), the researcher can then delve into exploring its nature (i.e. epistemology) [16]. Epistemology, as a foundational component of research philosophy, delves into the essence of knowledge. It is defined as 'the theory of knowledge' and places its emphasis on the procedure involved in constructing, understanding, and advancing knowledge [15]. According to Mardiana [17], there are three paradigms within the philosophy of IS research: positivism, interpretivism, and pragmatism.

Positivism and DSR in IS share a focus on research rigour and relevance [18, 19]. While positivist research emphasises hypothesis testing and quantitative methods, DSR incorporates interpretive and critical perspectives, allowing for a comprehensive understanding [19]. Therefore, interpretivism emerged in response to positivism's limitations, emphasising contextual variables and human experiences [20]. In DSR within IS, interpretivism complements positivism by emphasising the subjective experiences of individuals, acknowledging the complexity of human sense-making and the importance of context in studying IS [21]. However, pragmatism is also recognised as a paradigm for DSR in IS, emphasising practical solutions to real-world problems [22, 23]. Pragmatism supports the creation and evaluation of I.T. artefacts, aligning with DSR's goal to address organisational challenges and contribute to both theoretical knowledge and practical value [21, 24]. Pragmatism emerged as the preferred research philosophy for this study. This decision was based on its ability to blend elements of various philosophies, including qualitative and quantitative approaches, and its strong alignment with applied research and practical problem-solving. Given the goal of developing a DIMS that serves the RECONMATIC consortium and contributes to knowledge, pragmatism offers the flexibility to use a mix of methods, consider multiple stakeholder perspectives, and prioritise practical outcomes.

3   Design Science Research for Information Systems

DSR within the field of IS represents a research paradigm focused on developing and evaluating innovative I.T. artefacts designed to address practical, real-world issues. The foundational principle is that knowledge and comprehension of a problem domain and its solution are acquired through the construction and application of the designed artefact. [22].

Several authors, including Hevner and Chatterjee [22], Vaishnavi and William Jr Kuechler [13], and March and Smith [25], have collaborated to outline common approaches in DSR within IS. Their collective work focuses on the step-by-step process of DSR, covering problem identification, artefact design, artefact evaluation, and knowledge contribution Fig. 1. Moreover, they unanimously agree on the key outcomes of DSR. They emphasise the creation of practical elements like constructs, models, methods, and real-world applications, along with the continuous improvement of overarching theories. This shared perspective underscores a widespread consensus in academia regarding the importance of tangible contributions through constructs, models, methods, real-world applications, and refined theoretical frameworks within the realm of DSR.

Holmstrom et al. [26] assert that the initial stage in design science involves pinpointing the problem through 'diagnosing the primary research problem'. Johannesson [27] suggests that the foremost consideration for a design science researcher should be the realisation that 'something is not quite right with the world, and it has to be changed'. The output of this phase is a proposal, formal or informal, for a new research effort [13]. Furthermore, the second step, according to Voordijk [28] and Hevner et al. [22], is to create the 'technological rule' (artefact) that will solve the real-world issue. Designing and creating this artefact, according to

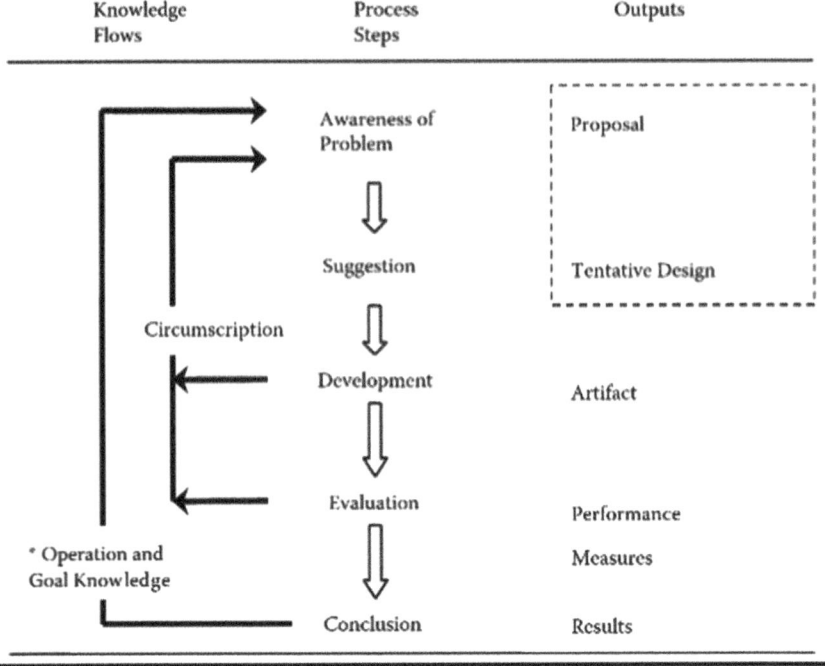

Fig. 1 The general methodology of design science research [13]

Hevner et al. [22], is the process of creating a solution concept (method or system) for a specific goal. This stage comes right after the proposal and is closely connected to it through the tentative design that is expanded upon and put into practice during the development stage [13]. Kehily and Underwood [29] state that the process of creating the artefact needs to be transparent for the solution to be meaningful from an academic perspective. This calls for an explanation of the choices made during the artefact's evolution and the development process. According to Hevner et al. [22], a design artefact's utility, quality and efficacy must be thoroughly demonstrated through well evaluation techniques. This is supported by Peffers et al. [30], who mentioned that these techniques are such as observation, analysis, experimentation, or testing, whereas Arthur [31] divided the evaluation into two components based on his belief that design science researchers have a greater chance of producing high-quality research outcomes if they employ a high-quality research process. The two components are (1) evaluating the quality of the process used and (2) evaluating the quality of the outcomes Fig. 2.

The evaluation results should provide evidence of the artefact's utility, quality, and efficacy. DSR in IS should also contribute to theoretical and practical knowledge of the problem and solution domains [30].

Finally, the last phase of DSR is the conclusion. As stated by Vaishnavi and William Jr Kuechler [13], this phase marks the culmination of a distinct research endeavour, typically resulting from satisficing, wherein discrepancies in the

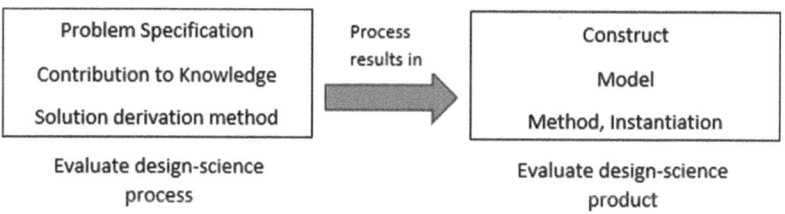

Fig. 2 Evaluating quality of DSR [31]

artefact's behaviour from the (repeatedly) revised hypothetical predictions are considered 'satisfactory'. Furthermore, Hevner et al. [22] emphasise the importance of presenting DSR effectively to both technology-oriented and management-oriented audiences.

4 DSR to Develop DIMS for CDWM

To address the challenges in CDWM, a comprehensive DIMS is being developed using the DSR method. Following DSR Fig. 3., the first phase is awareness of problem, the journey began with a critical review of the existing literature and the latest studies in CDWM, exploring the barriers and drivers in CDWM, the application of Industry 4.0 in CDWM, and the role of IS in the field to understand the problem. Through this critical examination, a noticeable gap emerged, concerning the integration of Industry 4.0 technologies to support the waste management, waste quantification, auditing and waste diversion in the C&D sector [31–33]. Specifically, this gap revolves around the absence of a comprehensive digital information management system that encompasses the entire project lifecycle, extending from inception to demolition or deconstruction phases. Furthermore, there is a notable lack of clarity regarding the roles and engagement of various stakeholders within these systems. Most of the information management systems reviewed in the existing literature, while promising in their potential, have exhibited shortcomings. One significant limitation is the absence of a secure and robust database infrastructure for effectively handling the vast datasets inherent in the integration of technologies like BIM and IoT or real-time data collection in BIM and Blockchain integration. Moreover, many systems have faltered in achieving real-time data collection, a critical feature to ensure the timely and accurate tracking of waste and materials in construction and demolition processes which will help analysing the material flow. In response to this gap, a proposal was formulated, suggesting the integration of BIM, IoT, and BC to enhance material flow analysis throughout the lifecycle of the project.

Looking towards the future, implementing the proposed solution will undergo a rigorous evaluation as part of the DSR process outlined by Arthur [34]. This process begins with a preliminary study to verify the identified problem's existence and

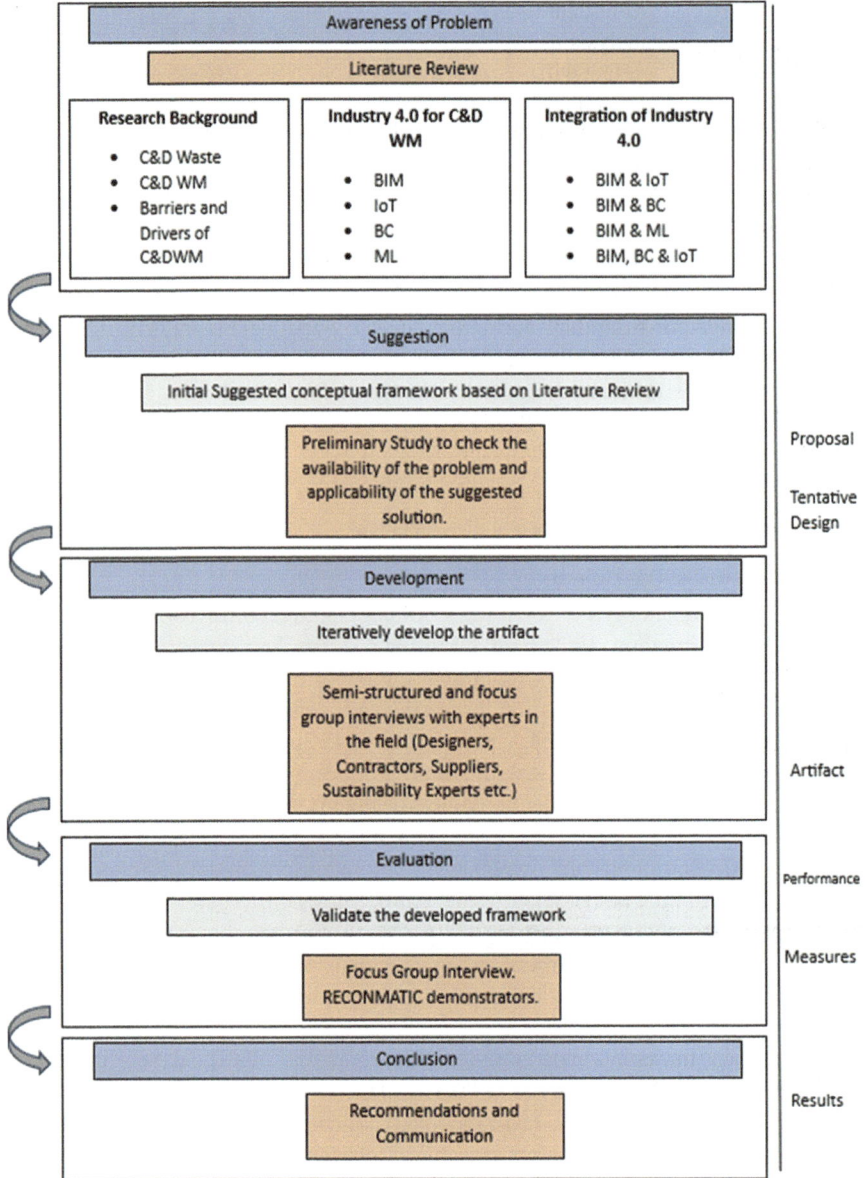

Fig. 3 Design science research process flow

assess the proposed solution's applicability. Two distinct groups will be targeted for this study through a questionnaire survey. RECONMATIC partners will be engaged to examine the problem's existence, the solution's applicability, and its support for RECONMATIC DEMONSTRATORS [35], and the wider industry's perspective will also be considered to check the problem's existence and the solution's applicability. Following data collection, the results from the preliminary study will undergo statistical analysis using SPSS to analyse the data and assess its reliability. Subsequently, the artefact will develop as a co-creation between the industry experts, involving semi-structured interviews with RECONMATIC partners and a focus group in a workshop with the wider industry. The framework will undergo iterative refinement by thematically analysing the qualitative data collected and then will be evaluated in the RECONMATIC annual meeting workshop, incorporating insights from RECONMATIC case studies. This will be done by following the Framework for Evaluation in DSR developed by Venable et al. (2016), by defining the explicit goals, choosing evaluation strategies determining properties that will be evaluated and designing the individual evaluation episodes [36]. Following this, the framework will progress to implementation, culminating in the development of the system architecture. This will be done by following the Togaf standard [37]. The outcomes and results of this endeavour will be disseminated through various channels, including journal papers, conferences, and the RECONMATIC website, solidifying its contribution to the field of CDWM, CE and Digital Technologies.

5 Conclusion

In conclusion, the aim of this research is to develop a comprehensive conceptual framework for a DIMS that incorporates Industry 4.0 technologies to enhance CDWM, encompassing the entire project lifecycle. This approach aligns seamlessly with the principles of DSR, where the central focus is on designing and exploring the use of an artifact to address a specific problem within a given context [38]. In DSR, researchers propose and develop an artifact to engage with a problem context, ultimately aiming to enhance elements within that context [38, 39]. This chapter has delved into the steps followed within the DSR strategy to develop this system as part of the Horizon Europe Project RECONMATIC.

Moreover, developing the DIMS presents challenges that blend academic standards with human interactions. Engaging stakeholders through interviews requires balancing diverse needs and opinions, often requiring negotiation to reach consensus. Whereas the development of the framework will be a co-creation between RECONMATIC Partners and the wider industry, this could also be a challenge since the RECONMATIC Partners would give their insights based on RECONMATIC project only. This potentially limits the scope of perspectives and experiences considered in the development process, posing a challenge in ensuring the inclusivity and comprehensiveness of the DIMS framework.

Thus far, the researcher has completed the problem detection phase, identifying key issues, including the lack of integration of Industry 4.0 covering the entire project lifecycle, a lack of understanding of stakeholder roles within these systems, and the absence of systems supporting material flow analysis. The researcher has proposed a framework that integrates BIM, IoT, and BC specifically to strategically manage information across the whole lifecycle of the project, supporting material flow analysis. This proposed solution will be followed by an evaluation of the identified problem and the suggested framework with RECONMATIC partners and the wider industry. Subsequently, the research will move on to the development of the artifact and, finally, its evaluation.

Funding This research was funded by Horizon Europe (grant agreement number 101058580, 2022) and by Innovate UK as part of the UK Guarantee programme for UK Horizon Europe participation (project number 10038579). Views and opinions expressed are, however, those of the author(s) only and do not necessarily reflect those of the HORIZON-RIA. Neither the European Union, The UK nor the granting authority can be held responsible for them.

References

1. Miller, N.: The industry creating a third of the world's waste available online: https://www.bbc.com/future/article/20211215-the-buildings-made-from-rubbish
2. Díaz-López, C., Bonoli, A., Martín-Morales, M., Zamorano, M.: Analysis of the scientific evolution of the circular economy applied to construction and demolition waste. Sustainability. **13**, 9416 (2021). https://doi.org/10.3390/su13169416
3. Maskuriy, R., Selamat, A., Ali, K.N., Maresova, P., Krejcar, O.: Industry 4.0 for the construction industry—how ready is the industry? Appl Sci. **9**, 2819 (2019). https://doi.org/10.3390/app9142819
4. Singh, S., Babbitt, C., Gaustad, G., Eckelman, M.J., Gregory, J., Ryen, E., Mathur, N., Stevens, M.C., Parvatker, A., Buch, R., et al.: Thematic exploration of sectoral and cross-cutting challenges to circular economy implementation. Clean Tech Envir Policy. **23**, 915–936 (2021). https://doi.org/10.1007/s10098-020-02016-5
5. European Commission Circular Economy Action Plan Available online: https://environment.ec.europa.eu/strategy/circular-economy-action-plan_en
6. Aleksandrova, E., Vinogradova, V., Tokunova, G.: Integration of digital technologies in the field of construction in the Russian federation. Eng Manag Prod Serv. **11**, 38–47 (2019). https://doi.org/10.2478/emj-2019-0019
7. Woo, J., Shin, S., Asutosh, A.T., Li, J., Kibert, C.J.: An overview of state-of-the-art technologies for data-driven construction. Lect Notes Civil Eng, 1323–1334 (2020). https://doi.org/10.1007/978-3-030-51295-8_94
8. Belle, I.: The architecture, engineering and construction industry and blockchain technology available online: http://www.iris-belle.com/architecture-engineering-construction-industry-blockchain-technology/. Accessed 10 Aug 2023
9. Wang, C.-X., Haider, F., Gao, X., You, X.-H., Yang, Y., Yuan, D., Aggoune, H., Haas, H., Fletcher, S., Hepsaydir, E.: Cellular architecture and key technologies for 5G wireless communication networks. IEEE Comm Mag. **52**, 122–130 (2014). https://doi.org/10.1109/mcom.2014.6736752
10. Hassan, N.R., Mingers, J., Stahl, B.: Philosophy and information systems: where are we and where should we go? Eur J Infor Syst. **27**, 263–277 (2018). https://doi.org/10.1080/0960085x.2018.1470776
11. Dawood, I., Underwood, J.: W04 research methodology explained; (2010), p. 177
12. Lee, A.: Thinking about social theory and philosophy for information systems; (2004)
13. Vaishnavi, V.K.: William Jr Kuechler design science research methods and patterns innovating information and communication technology. Boca Raton Crc Press, (2015). ISBN 9781498715256
14. Denicolo, P., Becker, L.M.: Developing research proposals. Los Angeles, Sage (2012) ISBN 9780857028662.

15. Grix, J.: The foundations of research. Macmillan International Higher Education, Red Globe Press, London (2004) ISBN 9781352002003
16. Saunders, M.N.K., Lewis, P., Thornhill, A.: Research methods for business students, 8th edn, pp. 128–280. Pearson, New York (2019) ISBN 9781292208787.
17. Mardiana, S.: Modifying research onion for information systems research. Solid State Tech. **63**, 5304–5313 (2020)
18. Panagiotopoulos, P., Al-Debei, M.M., Fitzgerald, G., Elliman, T.: A business model perspective for ICTs in public engagement. Government Information Quarterly. **29**, 192–202 (2012). https://doi.org/10.1016/j.giq.2011.09.011
19. Gregor, S., Jones, D.: The anatomy of a design theory. J Assoc Infor Syst. **8** (2007). https://doi.org/10.17705/1jais.00129
20. Alharahsheh, H., Pius, A.: A review of key paradigms: positivism vs interpretivism. Global Acad J Human Soc Sci. **2**, 39–43 (2020). https://doi.org/10.36348/gajhss.2020.v02i03.001
21. Goldkuhl, G.: Pragmatism vs interpretivism in qualitative information systems research. Eur J Infor Syst. **21**, 135–146 (2012)
22. Hevner, A.R., Chatterjee, S.: Springerlink (Online service design research in information systems: theory and practice). Springer, New York, London (2010) ISBN 9781441956538.
23. Arnott, D., Pervan, G.: Design science in decision support systems research: an assessment using the Hevner, March, Park, and Ram Guidelines. J Assoc Infor Syst. **13**, 923–949 (2012). https://doi.org/10.17705/1jais.00315
24. Miah, S.J., McGrath, M., Kerr, D.: Design science research for decision support systems development: recent publication trends in the premier IS journals. Aus J Infor Syst. **20** (2016). https://doi.org/10.3127/ajis.v20i0.1482
25. March, S.T., Smith, G.F.: Design and natural science research on information technology. Decision Supp Syst. **15**, 251–266 (1995). https://doi.org/10.1016/0167-9236(94)00041-2
26. Holmström, J., Ketokivi, M., Hameri, A.-P.: Bridging practice and theory: a design science approach. Decision Sci. **40**, 65–87 (2009). https://doi.org/10.1111/j.1540-5915.2008.00221.x
27. Johannesson, P.: Introduction to design science. Springer International Publisher (2016) ISBN 9783319361109.
28. Voordijk, H.: Construction management and economics: the epistemology of a multidisciplinary design science. Const Manag Eco. **27**, 713–720 (2009). https://doi.org/10.1080/01446190903117777
29. Kehily, D., Underwood, J.: Design science: choosing an appropriate methodology for research in BIM. Conference papers; (2015). https://doi.org/10.21427/fde9-tj97
30. Peffers, K., Tuunanen, T., Rothenberger, M.A., Chatterjee, S.: A design science research methodology for information systems research. J Manag Infor Syst. **24**, 45–77 (2007). https://doi.org/10.2753/mis0742-1222240302
31. Yu, Y., Yazan, D.M., Junjan, V., Iacob, M.-E.: Circular economy in the construction industry: a review of decision support tools based on information & communication technologies. J Cleaner Prod. **349**, 131335 (2022). https://doi.org/10.1016/j.jclepro.2022.131335
32. Li, C.Z., Zhao, Y., Xiao, B., Yu, B., Tam, V.W.Y., Chen, Z., Ya, Y.: Research trend of the application of information technologies in construction and demolition waste management. J Cleaner Prod. **263**, 121458 (2020). https://doi.org/10.1016/j.jclepro.2020.121458
33. Ratnasabapathy, S., Perera, S., Alashwal, A.: A review of smart technology usage in construction and demolition waste management. dl.lib.uom.lk. (2019). https://doi.org/10.31705/WCS.2019.5
34. Arthur, R.: Design-science research. Elsevier eBooks, 267–288 (2018). https://doi.org/10.1016/b978-0-08-102220-7.00011-x
35. RECONMATIC Demonstrators Available online: https://www.reconmatic.eu/demonstrators
36. Venable, J., Pries-Heje, J., Baskerville, R.: FEDS: a framework for evaluation in design science research. Eur J Infor Syst. **25**, 77–89 (2016). https://doi.org/10.1057/ejis.2014.36

37. The Open Group TOGAF | the Open Group Available online: https://www.opengroup.org/togaf
38. Wieringa, R.: Design science methodology for information systems and software engineering, pp. 3–11. Springer, Heidelberg (2014) ISBN 9783662438398.
39. Alta, Gerber, A., Smuts, H.: Guidelines for conducting design science research in information systems. Comm Comp Infor Sci, 163–178 (2019). https://doi.org/10.1007/978-3-030-35629-3_11

Construction Industry and the Circular Economy: A Systems Analysis

Raymond L. Sheen

Contents

1 Introduction

The concept of a circular economy strives for an efficient use of materials and a reduction of waste [1]. A difficulty in making progress toward the goal of circularity is that there has often been confusion concerning key principles that should be followed [2]. In fact, the definition and measurement of a circular economy and its impact often lack scientific and research rigor [3]. Within this context, the construction industry is a major contributor to waste and pollution [4]. In their study, the EU Commission expressed concern that the construction industry has not fully embraced the principles of circularity. The study included a model of the ecosystem and identified areas of concern, in particular, the slow uptake of circularity in construction.

R. L. Sheen (✉)
Produktif AS, Raufoss, Norway

Worcester Polytechnic Institute, Worcester, MA, USA
e-mail: ray@produktif.com

© The Author(s) 2025
M. Kioumarsi, B. Shafei (eds.), *The 1st International Conference on Net-Zero Built Environment*, Lecture Notes in Civil Engineering 237,
https://doi.org/10.1007/978-3-031-69626-8_157

The report by the European Commission found the construction industry is "currently responsible for over one-third of total waste generation in the EU" [4: 3]. This study noted that the industry has been relying on voluntary targets and measures to reduce waste and pollution, but these have been ineffective. The construction industry in Europe represents 9.6% of the total economic activity and employs 24.9 million people [4: 3]. However, the industry is highly fragmented and distributed with thousands of contractors and construction material providers.

To transform, the industry needs to change from linear thinking to systems thinking [4]. The traditional mindset of those in the construction industry has not been consistent with a circular economy [5]. Systems thinking approaches the industry as a network of many elements, where the actions of one element influence the actions of others [6]. These interactions can be modeled with a causal loop diagram (CLD). This leads to this study's research objectives. The first is to create a model of the construction industry when applying circularity. The second is to analyze the model to identify control loops and potential dysfunctional behavior. The third is to assess the likely impact of adopting the recommendations identified by the European Commission [4].

2   Background

The concept of a circular economy is not new. Civilizations such as ancient Egypt, India, Peru, and the Byzantine Empire had programs for recycling and reuse [7]. At its core, the modern circular economy mindset is concerned with sustainability [1]. Despite the interest in the circular economy, progress is still needed to develop viable business models that achieve high performance and are grounded in the circular economy [8].

Galle et al. note that the classic chicken and egg dilemma exists in the construction industry [5]. There is little demand for circular construction solutions, so there are few product offerings. Since there are few products, customer awareness is low, resulting in little demand. The technical and social challenges facing the construction industry as it transforms into a circular economy are well documented [3]. These challenges are throughout the industry and point to the need for a holistic change [9]. One of the challenges is the fragmentation throughout the industry. Most builders or suppliers do not recognize the impacts their actions can have on the entire industry ecosystem [10]. Fragmentation also exists on the regulatory front. There are many policymakers with many policies and standards being developed [11]. For instance, there is little harmonization of standards and processes with regard to construction and demolition waste disposal [12]. In addition, past decisions made by architects and builders when a building was built are the cause of high levels of current construction waste. However, there is no current accountability for future demolition problems [4, 13]. This is to be expected since most construction design decisions are made based on short-term cost/benefit tradeoffs [5]. Buyers and financial institutions seldom include circular economy criteria in their decision-making. Osei-Tutu et al. [14] addressed this in their scientometric analysis of available literature concerning the barriers to the construction industry's adoption of circularity. The six categories of barriers they identified were social, cultural, economic, technical, environmental, and technological.

Several studies have approached the construction industry from a systems perspective. Ghufran et al. analyzed the construction industry supply chain using systems analysis [13]. This study found 12 enablers of circularity for the construction industry. The study found that the reinforcing action of system loops was slow-acting. In addition, a strong fast-acting balancing loop keeps the industry in a steady-state condition. Ding et al. considered the forward and reverse logistics cycles as a system and identified the similarities and differences between the two flows [15]. The process steps identified by Ding et al. closely mirror the nine process steps identified in the EU report [4].

Many circular economy business models are built on linear thinking [16]. System thinking incorporates feedback loops and multiple interaction effects. A system model consists of elements and their interactions. System elements are events, processes, states, or objects [17]. The critical characteristic of system elements is the interaction between elements [6: 14]. Systems thinking often uses visualizations to provide stakeholders with insight into how to transition an industry to a circular economy [18]. In a CLD, the interactions are shown with arrows indicating the interaction between elements. These arrows are depicted with a plus sign to show a positive correlation or a minus sign indicating a negative correlation. An element can be connected to many elements, causing all of them to be impacted simultaneously. Similarly, one element may be impacted by multiple elements and the actual effect will be a balance between those elements.

Feedback loops occur in a CLD when two or more elements are connected in a chain that doubles back on itself. When all the arrows in the chain have the same sign, the loop is a reinforcing loop. Interactions in this loop will continue until they exceed a system constraint or until an element is consumed. When there are both positive and negative signs, the loop is likely to be a balancing loop. A balancing loop tends to hold the elements at a specified level or steady-state condition. Meadows [6] identifies several system performance challenges and issues. One of these, layers of limits, occurs when the performance of an element is based on multiple interactions that must work in concert. Another challenge is ubiquitous delays, where the impact of a relationship within a system is slow-acting, resulting in feedback loops that may be ineffective at influencing the behavior of the system. Another performance issue is the tragedy of the commons. It is associated with an element of the system that is consumed by other elements. Success to the successful problem occurs when a reinforcing feedback loop grows one aspect of the system to a dominant position and changes the structure of the system. One additional problem is seeking the wrong goal. It is associated with establishing goals for system elements that are not aligned with the overall system goal.

3   Methodology

3.1   System Elements

As noted earlier, system elements are events, processes, states, or objects. An example of this is the implementation of a new industry regulation. An example of this process is the demolition of a building. A state example is the end-of-life condition of construction materials. Object examples are recovered construction materials. The CLD uses elements from the EU report [4]. That report identified nine phases in a construction lifecycle [4; 12–13]. Each of those phases is a process element in the CLD.

- Concept—Initial ideas, goals, and constraints are identified.
- Procurement—Acquisition of goods and services for construction is conducted.
- Design—Detailed plans and schematics of the construction (and deconstruction, when applicable) are created.
- Manufacture—Fabrication of construction materials, including the refurbishment of previously used materials.
- Demolition—Dismantling an existing construction asset to make room for new construction. This is the first step in a renovation, rather than a deconstruction.
- Construction—Assembly and erection of the designed structure.
- Handover for Use—User/occupant takes possession of the constructed asset and uses it for their purpose.
- Refurbishment, reuse, renovation, repair—Modification and replacement of elements in the constructed asset.
- Deconstruction and end-of-life—Dismantle and dispose of materials from the constructed asset.

A critical element of circularity in the construction industry is the disassembly and reuse of construction materials. This area required more detail in the CLD. The study conducted by López Ruiz et al. identified five stages related to the deconstruction and end-of-life of buildings [12]. These stages are added as process elements to the CLD to enhance the overall process of demolition, as listed in the EU report [4].

- Preconstruction—Designing to minimize waste and simplify disassembly and deconstruction. Within the CLD, this is synonymous with the design element.
- Construction and Renovation—Creating and implementing site waste management plans. Within the CLD, this is considered to be the construction element.
- Collection and Distribution—Techniques and controls for collecting and segregating all demolition materials and optimizing the transportation of those materials.
- End-of-Life—Techniques and procedures to determine whether materials are at end-of-life or if they can be reused, refurbished, repaired, or recycled.
- Material Recovery and Production—Establishing processes for the reuse and recycling of demolition materials and the restoration of the site.

The CLD also required more detail in the supply chain. The literature review identified a study by Ghufran et al., that included a CLD for the construction industry supply chain [13]. That diagram used 12 elements that included processes and a number of organizational or cultural states. These were merged into the CLD being developed using causal relationships identified by Ghufran et al.

- Agility—The ability to quickly pivot to new products or processes.
- Adaptability—The level of difficulty in changing systems and processes.
- Collaboration—The willingness and capacity to work with other organizations.
- Compatibility—The ability of physical products, digital systems, and personnel resources to integrate with other organizations.
- Information Sharing—The willingness and capability to release product and project information to other interfacing organizations.
- Just in Time—A management and scheduling approach for materials and components that ensures the materials are available when needed, not early and not late.
- Strategic Risk Planning—Proactive risk management to identify organizational threats and opportunities within the design, construction, and renovation processes.
- Leadership—The ability to influence the actions of others to achieve a goal.
- Top Management Support—The level of interest and engagement by senior managers with respect to an initiative or goal.
- Corporate Social Responsibility—The degree to which corporate decision-making includes social and environmental issues as major considerations.
- Flexible Structure—The ability of the organization to reallocate and deploy resources based on current needs and issues.
- Visibility—The extent to which the current status of a project or process is made known to various levels of decision-makers in an organization.

Additional elements used to enhance the CLD were based on the inherent industry barriers identified by Osei-Tutu et al. [14]. Their research identified six barrier categories that create challenges for the construction industry's transition to circularity.

Cultural Barriers—The cultural barriers included the perception that reclaimed/refurbished materials were of poor quality and an unwillingness to collaborate within the value chain. This latter barrier is a long-standing characteristic of the industry.

- Economic Barriers—The economic barriers addressed profitability. First, reclaimed materials are often more expensive than new materials. Second, developing new circular designs adds costs for existing builders with old designs. Finally, in many areas, the landfill costs for construction materials are low.
- Technical Barriers—These barriers were related to the lack of design standards and codes. This creates uncertainty and delays for both architects and builders.
- Social Barriers—These barriers are due to the lack of demand for construction projects that embrace the principles of a circular economy. The study found a lack of knowledge about the construction impacts on the environment. In addition,

builders are unaware of construction practices that align with circular economy principles.

- Environmental Barriers—The environmental barriers were also two-fold. There is widespread ignorance of environmental impacts with respect to construction waste. In addition, there are few incentives to encourage the adoption of construction practices that minimize negative environmental impacts.
- Technological Barriers—These barriers are associated with recovery, reuse, recycling, and refurbishment technologies. There is limited availability of reliable technologies and little incentive for producers to adopt or improve these technologies.

3.2 Element Interactions

In a CLD, every interaction is represented with an arrow that indicates the nature of the correlation. The ecosystem relationships represented in this diagram are based on the relationships identified in the literature. The goal of this system analysis is the adoption of circular economy principles in construction. Therefore, the positive relationships are those that promote circularity, and the negative ones inhibit circularity.

Many of the high-level lifecycle process steps have multiple predecessor arrows, some positive and some negative. When the influence of the positive causes is greater than the negative causes, that element aligns with circular economy principles. For instance, at the design element, when the positive causal predecessors are stronger than the negative predecessors, the building designer(s) will likely incorporate circular economy design features. If the negative predecessors predominate, the designs will continue with conventional construction designs and processes.

3.3 Causal Loop Diagram

The construction industry CLD contains 30 elements representing the lifecycle flow of a construction project (see Fig. 1). This CLD includes causal effects that encourage the development of buildings relying on circularity concepts and barriers or challenges that limit the adoption of circularity principles.

A quick assessment of the CLD shows that the design element is the most connected with six successor relationships and four predecessor relationships. Next is visibility with nine relationships, followed closely by construction with eight and both collection and deconstruction with seven. Only six elements follow the classic linear thinking approach of a single predecessor and successor. None of those are part of the construction industry lifecycle. The complexity of this CLD indicates the construction ecosystem cannot be successfully managed with linear thinking; it must be managed as a system.

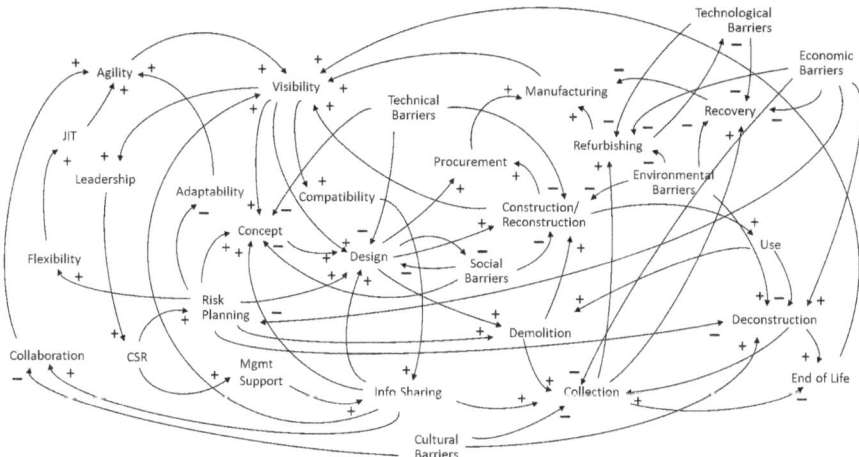

Fig. 1 Construction Industry Causal Loop Diagram

4 Analysis

4.1 Loops

One method of analyzing a CLD is to identify the balancing and reinforcing loops. Based on the level of complexity of this diagram, there are many loops. For the purposes of this study, the focus will be on loops that include the primary elements of the construction lifecycle. Ten reinforcing loops were identified.

- Concept to Design to Construction to Visibility to Concept—This loop illustrates that publicizing buildings that incorporate the circular economy principle should lead to more buildings that follow that approach. However, the technical barriers due to standards are suppressing activity in concept, design, and construction. In addition, social barriers limit the demand for circular concepts, designs, and construction.
- Concept to Design to Demolition to Construction to Visibility to Concept—This variation on the previous loop includes demolition. As buildings are renovated or retrofitted with circular designs, that trend is amplified. However, the technical barriers due to standards and social barriers that limit demand are suppressing activity in concept, design, and construction.
- Concept to Design to Procurement to Manufacturing to Visibility to Concept—This loop is a variation on the first loop. The difference is a shift from the contractors doing the building to suppliers who provide materials that are fabricated using circular economy principles. However, the technical barriers due to standards are suppressing activity in concept and design. In addition, social barriers are limiting the demand for concepts and designs that embrace circular economy principles.

- Concept to Design to Demolition to Collecting to Refurbishing to Manufacturing to Visibility to Concept—This loop is based on taking reclaimed construction materials from the demolition process and remanufacturing or refurbishing them to be reintroduced into the supply chain. However, technical and social barriers are again limiting the demand for concepts and designs. Collecting is constrained by economic and cultural barriers. Refurbishing is constrained by technological and economic barriers.
- Concept to Design to Construction to Visibility to Leadership to CSR to Risk Planning to Concept—This loop steps away from the building lifecycle after a building is constructed to use the visibility to spur on leadership to set new goals and objectives. However, technical barriers and social barriers are still limiting the demand for concepts, designs, and construction that embrace circular economy principles.
- Concept to Design to Procurement to Manufacturing to Visibility to Leadership to CSR to Risk Planning to Concept—This loop is like the previous loop only the emphasis now is on the suppliers and manufacturers of building components instead of the builder/contractors. Once again, the technical and social barriers are limiting the demand for concepts and designs that embrace circular economy principles.
- Design to Social Barriers to Design—This is a short loop with negative signs on both arrows. This means the more circular designs are completed, the lower the social barriers, in particular the lack of customer demand, and when social barriers are lowered and demand increases, more circular designs will be created.
- Construction to Use to Demolition to Construction—This loop is focused on the renovation aspect of construction. The building user removes (demolishes) items in the existing building and replaces them with items that are. However, the technical barriers due to standards are suppressing circular economy construction. In addition, social barriers limit the demand for construction that embraces circular economy principles.
- Collection to Refurbishing to Manufacturing to Visibility to Compatibility to Info Sharing with Collection—This loop is focused on the growth of refurbished or remanufactured recovered materials. As the availability of remanufactured materials increases, their visibility leads to greater demand for those materials, identifying even more candidates for refurbishment. However, the collecting function is suppressed by economic and cultural barriers that create difficulties in sorting the construction materials.
- Refurbishing to Technological Barriers to Refurbishing—This is another short reinforcing loop with negative relationships. The more refurbishing, the lower the technological barriers for refurbishing, which will further increase the amount of refurbishing.

The element visibility was found in seven of the ten reinforcing loops. The greater the visibility of actual building projects and actual refurbishment of construction materials, the more momentum there is for the entire circular economy movement. The design element was also found in seven reinforcing loops. This highlights the need for design tools, techniques, and templates to encourage architects and designers to use circular economy approaches. In contrast, eight of the loops are impacted by social barriers, which are typically constraining the demand for design, concepts, and construction. In addition, the technical barriers impacted seven loops. The technical barriers create friction in the system, making it difficult for architects, engineers, and contractors to shift to circular economy construction approaches.

Four balancing loops were identified. These tend to hold elements of the system at a steady-state level. Balancing loops are advantageous if the loop in the system is meeting its desired goal or objective. They are detrimental if they prevent desired change. Again, only loops that affect primary processes in the construction lifecycle are included.

- Concept to Design to Demolition to Collection to Recovery to Manufacturing to Visibility to Concept—The negative correlation is between recovery and manufacturing. The more materials that are recovered, the less demand for new or refurbished materials. This loop assumes a constant overall demand for construction materials. While this loop dampens manufacturing, it does not reduce the shift to circular economy construction since the reuse of recovered material is often better from an environmental perspective. However, social barriers limit the demand for this concept, regardless of material origin. The key to this loop is to set the goal for the desired amount of recovered material, not manufactured material. The technical barriers constrain the concept and design elements of this loop.
- Construction to Use to Deconstruction to End-of-Life to Visibility to Concept to Design to Construction—The negative correlation relationship in this loop is between building use and deconstruction. Buildings that stay in use do not require deconstruction. The steady-state stabilization of this loop is based on the need for buildings. This loop becomes interesting when a building's use changes. If a building were designed for disassembly, instead of the balancing loop that goes through deconstruction and end-of-life, the reinforcing loop that goes through demolition and renovation would be activated. While the system goal is in the building use element, the place to break or suppress this loop is in the design element. The technical barriers limit the efficacy of the concept, design, and construction elements. In addition, social barriers constrain the demand for this type of construction. Finally, the cultural and economic barriers impact the deconstruction element.
- Demolition to Construction to Use to Deconstruction to End-of-Life to Visibility to Leadership to CSR to Risk Planning to Demolition—The negative correlation relationship in this loop is between building use and deconstruction. The steady-state stabilization of this loop again happens in that relationship. However, this loop does not go through design; it goes through leadership, CSR, and risk planning. This loop is susceptible to economic barriers that impact risk planning and social and technical barriers that constrain construction. The stabilization goals that are the focus of this loop will occur at the use and deconstruction elements.

- Deconstruction to End-of-Life to Visibility to Leadership to CSR to Risk Planning to Deconstruction—The negative relationship is between risk planning and deconstruction. When circular economy risk planning is low deconstruction will be high. However, risk planning often leads to the consideration of alternatives to deconstruction. The steady-state level of deconstruction is based on cultural and economic barriers that tend to increase deconstruction. In addition, economic barriers suppress risk planning.

4.2 System Performance and Archetypes

The system is subject to performance problems associated with its design and behaviors. One of these problems is layers of limits on many elements. For instance, the design element is based on interactions between six elements. At any given time, one of these will be the primary limit on the creation of circular designs. Improvement in that interaction just shifts the focus to the next one. Changes must be made in all areas simultaneously to see significant changes in the nature of building designs.

Ubiquitous delays are inherent in any manual system. In this system, the delays are magnified because many of the steps in the construction lifecycle take months or years. This effect is further amplified by the fragmented nature of the industry. Even if one builder or architect adopts circularity, there are still thousands of others who do not.

The archetype known as the tragedy of the commons occurs when a key system resource is consumed by multiple system actors with no one actor taking responsibility. This archetype is at the core of circularity. Civilization is consuming non-renewable resources with no way to replenish them. The reuse and refurbishment of construction materials reduce this effect. Education and communication are needed to overcome this archetype. This highlights the importance of the visibility element of the system.

The archetype of success refers to elements of the system that grow and squeeze out other elements. When the growth is with elements that adopt circularity, this is not a problem. However, as mentioned earlier, this industry has thousands of actors and those who do not adopt circular principles could find themselves shut out of a portion of the industry, reducing their employment. This will likely increase the cultural or social barriers, which would depress some of the reinforcing loops.

The final archetype is seeking the wrong goal. The EU report identifies several goals for the construction industry as it adopts circular economy principles [4]. These include a significant reduction in construction waste, a reduction in CO_2 emissions, economic growth through job creation, and targets for renovation. Each of these goals is beneficial. Therefore, they must be kept in balance. If any one goal rises to ascendancy, it will likely cause the system to be unable to achieve the other goals.

4.3 Loops

The EU report recommended the introduction of metrics that show the level of adoption of circularity [4]. Publicly tracking these metrics would strengthen the visibility element, which is prominent in many of the feedback loops. The EU report listed the recommended measures in Table 8 [4: 58 64]. These recommendations have been summarized in Table 1, which also identifies the system elements and loops impacted. The nature of the leverage point for improvement that the metrics address is also identified.

Nine of the 11 recommendations impact the design elements of the system. In fact, the most significant impact of the EU recommendations from a systems perspective is the enhancement of the design element. The next most significant system impact from these recommendations will be the establishment of a system goal. The danger is that if there are too many goals, one or more of the reinforcing loops will lock onto one goal and drive the system to a state of suboptimization.

5 Conclusion

The construction industry's transitional adoption of circular economy principles can be modeled using systems thinking. The CLD shows a complex system with many interactions that invalidate a linear approach to this ecosystem. The reinforcing loops in the system are susceptible to ubiquitous delays and are relatively weak because of barriers. The balancing loops are also weak. However, they could assist in moving the industry forward with visible measurable goals, particularly with respect to waste and reuse. The EU is recommending the adoption of many new measures to stimulate the transition. These will increase visibility, which will likely improve system performance, but there is a danger of seeking the wrong goal if the set of measurements does not stay balanced.

The most significant leverage point from a systems perspective is to enhance the design element in the system. It is the most connected element, and the EU metrics prominently focus on design measures. The barriers constraining that element are technical barriers, primarily standards, and social barriers, primarily customer demand.

This study was conducted as a qualitative review of literature that was then used to create the CLD. Further research should include a quantitative analysis to establish the strength of the interaction relationships. In addition, this study represents a broad industry overview. Local region analyses should be done to identify regional differences.

Table 1 EU Recommendations and Systems Impact

Recommendation	Element affected	System loop impact	Leverage point
Product as service, new business models	Concept, Risk Planning, Flexibility, Economic Barriers	Concept impacts 7 loops, risk planning impacts 4 loops, and Economic Barriers impact 4 loops	Restructure the system with a new business model
Designing for future disassembly and reuse	Design	Design impacts 8 loops	Improve the most strategic element
Designing for flexibility and adaptability	Design, Refurbishing	Design impacts 8 loops Refurbishing impact 3 loops	Improve the most strategic element
Improving material characteristics of materials used	Design	Design impacts 8 loops	Improve the most strategic element
Improving durability, lifespan, and repairability of construction works	Concept, Design, Construction, Demolition, Use, Refurbishing	Concept impacts 7 loops, Design impacts 8 loops, Construction impacts 6 loops, Demolition impacts 4 loops, Use impacts 3 loops, Refurbishing impacts 3 loops	Improve the reinforcing loops that emphasize renovation and refurbishing
Increasing recycled and secondary content of construction materials	Design, End-of-Life, Collection, Recovery	Design impacts 8 loops, End-of-Life impacts 3 loops, Collection impacts 2 loops, Recovery impacts 1 loop	Expand recovery and improve those reinforcing loops
Increasing direct reuse of products and materials	Concept, Design, Demolition, Collection, Recovery, Refurbishing	Concept impacts 7 loops, Design impacts 8 loops, Demolition impacts 4 loops, Collection impacts 2 loops, Recovery impacts 1 loop, Refurbishing impacts 3 loops	Improve reinforcing loops that emphasize renovation and refurbishment over deconstruction
Increasing reuse/recycling of waste from construction works	Design, Construction, Recovery, Refurbishing, Technical Barriers	Design impact 8 loops, Construction impacts 6 loops, Recovery impacts 1 loop, Refurbishing impacts 3 loops, Technical Barriers impact 9 loops	The focus is on the construction process, not the demolition
Increasing reuse/recycling of waste from demolition works	Design, Demolition, Collection, Recovery, Refurbishing, End-of-Life	Design impacts 8 loops, Demolition impacts 4 loops, Collection impacts 2 loops, Recovery impacts 1loop, Refurbishing impacts 3 loops	The focus is on demolition, not construction

(continued)

Table 1 (continued)

Recommendation	Element affected	System loop impact	Leverage point
Reducing waste/ wastage rates/ waste generation from construction activities	Concept, Construction, Demolition, Deconstruction, End-of-Life	Concept impact 7 loops, Construction impacts 6 loops, Demolition impacts 4 loops, Deconstruction impacts 3 loops, End-of-Life impacts 3 loops	Use of measures to set waste goals at critical steps/elements
Lifetime extension e.g., through retaining and refurbishing	Concept, Design, Construction, Demolition, Recovery, Refurbishment, End-of-Life	Concept impacts 7 loops, Design impacts 8 loops, Construction impacts 6 loops, Demolition impacts 4 loops, Recovery impacts 1 loop, Refurbishing impacts 3 loops, End-of-Life impacts 3 loops	A new set of lifetime measurements for materials and wastes will change design goals

References

1. Ghisellini, P., Cialani, C., Ulgiati, S.: A review on circular economy: the expected transition to a balanced interplay of environmental and economic systems. J. Clean. Prod. **114**, 11–32 (2016). https://doi.org/10.1016/j.jclepro.2015.09.007
2. Reike, D., Vermeulen, W.J.V., Witjes, S.: The circular economy: new or refurbished as CE 3.0? — exploring controversies in the conceptualization of the circular economy through a focus on history and resource value retention options. Resour. Conserv. Recycl. **135**, 246–264 (2018). https://doi.org/10.1016/j.resconrec.2017.08.027
3. Korhonen, J., Honkasalo, A., Seppälä, J.: Circular economy: the concept and its limitations. Ecol. Econ. **143**, 37–46 (2018). https://doi.org/10.1016/j.ecolecon.2017.06.041
4. European Commission, European Innovation Council and SMEs Executive Agency, Brin-Cat, C., Graaf, I., León Vargas, C.: Study on Measuring the Application of Circular Approaches in the Construction Industry Ecosystem: Final Study. Publications Office of the European Union (2023) https://data.europa.eu/doi/10.2826/488711
5. Galle, W., Debacker, W., Weerdt, Y.D.: Co-creating systemic changes for a circular economy in/under construction, preliminary lessons from the Flemish living lab on circular construction. IOP Conf. Ser. Earth Environ. Sci. **855**(1), 12012 (2021). https://doi.org/10.1088/1755-1315/855/1/012012
6. Meadows, D.H.: Thinking in Systems: a Primer. Chelsea Green Publishing (2008)
7. Hajoary, P.K., Ramani, V., Nuur, C.: New for some, old for others: circular economy practices in ancient time. Circ. Econ. Sustain. (2023). https://doi.org/10.1007/s43615-023-00323-9
8. Lewandowski, M.: Designing the business models for circular economy—towards the conceptual framework. Sustainability (Basel, Switzerland). **8**(1), 43 (2016). https://doi.org/10.3390/su8010043
9. Charef, R., Ganjian, E., Emmitt, S.: Socio-economic and environmental barriers for a holistic asset lifecycle approach to achieve circular economy: a pattern-matching method. Technol. Forecast. Soc. Chang. **170**, 120798 (2021). https://doi.org/10.1016/j.techfore.2021.120798
10. Ababio, B.K., Lu, W.: Barriers and enablers of circular economy in construction: a multi-system perspective towards the development of a practical framework. Constr. Manag. Econ. **41**(1), 3–21 (2023). https://doi.org/10.1080/01446193.2022.2135750

11. Yu, Y., Junjan, V., Yazan, D.M., Iacob, M.: A systematic literature review on circular economy implementation in the construction industry: a policy-making perspective. Resour. Conserv. Recycl. **183**, 106359 (2022). https://doi.org/10.1016/j.resconrec.2022.106359

12. López Ruiz, L.A., Roca Ramón, X., Gassó Domingo, S.: The circular economy in the construction and demolition waste sector—a review and an integrative model approach. J. Clean. Prod. **248**, 119238 (2020). https://doi.org/10.1016/j.jclepro.2019.119238

13. Ghufran, M., Khan, K.I.A., Ullah, F., Alaloul, W.S., Musarat, M.A.: Key enablers of resilient and sustainable construction supply chains: a systems thinking approach. Sustain. For. **14**(19) (2022). https://doi.org/10.3390/su141911815

14. Osei-Tutu, S., Ayarkwa, J., Osei-Asibey, D., Nani, G., Afful, A.E.: Barriers impeding circular economy (CE) uptake in the construction industry. Smart Sustain. Built Environ. **12**(4), 892–918 (2023). https://doi.org/10.1108/SASBE-03-2022-0049

15. Ding, L., Wang, T., Chan, P.W.: Forward and reverse logistics for circular econ-omy in construction: a systematic literature review. J. Clean. Prod. **388**, 135981 (2023). https://doi.org/10.1016/j.jclepro.2023.135981

16. De Angelis, R.: Circular economy business models as resilient complex adaptive systems. Bus. Strateg. Environ. **31**(5), 2245–2255 (2022). https://doi.org/10.1002/bse.3019

17. Schlüter, L., Kørnøv, L., Mortensen, L., Løkke, S., Storrs, K., Lyhne, I., Nors, B.: Sustainable business model innovation: design guidelines for integrating systems thinking principles in tools for early-stage sustainability assessment. J. Clean. Prod. **387**, 135776 (2023). https://doi.org/10.1016/j.jclepro.2022.135776

18. AlMashaqbeh, S., Munive-Hernandez, J.E.: Risk analysis under a circular econ-omy context using a systems thinking approach. Sustainability (Basel, Switzerland). **15**(5), 4141 (2023). https://doi.org/10.3390/su15054141

Assessing the Environmental Benefit of Circular Economy Decisions for Managing Waste in Road Construction: The Norwegian GoGreen Case Study

Reyn O'Born (iD) **and Alexander Vetnes**

Contents

1 Introduction

The materials used in construction are responsible for approximately 10% of total global carbon emissions per year and are the industry with the largest consumption of raw materials and the highest amount of waste generated [1, 2]. In Norway, the construction industry is only 2.5% circular and consumes upwards of 43 million tons of materials each year as construction material recycling and reuse schemes are nascent and underdeveloped [3, 4]. The overall material footprint of the construction industry urgently needs to be reduced to reduce the overall ecological footprint, which is estimated to be over one million tons of CO_2 annually from road construction in Norway alone [5]. Simultaneously, the existing built environment presents an

R. O'Born (✉) · A. Vetnes
University of Agder, Grimstad, Norway
e-mail: reyn.oborn@uia.no

© The Author(s) 2025
M. Kioumarsi, B. Shafei (eds.), *The 1st International Conference on Net-Zero Built Environment*, Lecture Notes in Civil Engineering 237,
https://doi.org/10.1007/978-3-031-69626-8_158

opportunity for the construction industry as a large reservoir of materials within the built environment is available for future use. The quantities are potentially enormous, as the total estimated material stock of roads in Norway is more than 420 gigatons [6].

The Norwegian national statistics organization reports that more than 2.1 million tons of construction waste are generated each year, with less than 954,000 tons being sent to recycling, 447,000 tons sent to landfills, and 537,000 tons sent to incineration [7]. Indeed, there is a large potential benefit for reusing and recycling construction materials for use again, and collecting and using these materials represents a first step in a circular construction industry [8, 9]. The challenge for construction actors is both planning for circularity within smarter design and developing systems for collecting, sorting, and recovering materials from existing building stocks on construction (or de-construction) sites.

The pilot project, known as GoGreen, is an attempt to further develop material recovery from existing building stocks slated for demolition and to allow for further reuse and recycling of these materials. The GoGreen pilot is part of a larger road construction project where several existing structures were demolished and the waste was collected, sorted, and sent for recycling and reuse instead of being sent directly to landfill or incineration. The purpose of this chapter is to understand the environmental benefits of implementing thorough material recovery and recycling processes as part of a greater push towards circularity in the construction industry.

2 Methods

This study is an LCA study that is based on the GoGreen tent activities. LCA methods are used in combination with data from the Betna-Hestnes site.

2.1 LCA Methodology

This study follows LCA methods in ISO14040 (2006), which have four main phases: goal and scope, inventory analysis, life cycle impact assessment, and interpretation [10] (Fig. 1).

The goal of this study is to evaluate the environmental impacts of the GoGreen tent, which was in operation from November 2022 to December 2023, and sort materials collected within this time period. Thus, the functional unit of the study is 13 months of operation of the GoGreen tent, or alternatively, the entire waste collection and sorting of demolition materials in the Betna-Hestnes road project. The scope of this study is based on a case study of the GoGreen pilot located in Betna, Norway. The system boundaries are shown in Fig. 2.

Fig. 1 Four steps within an LCA study

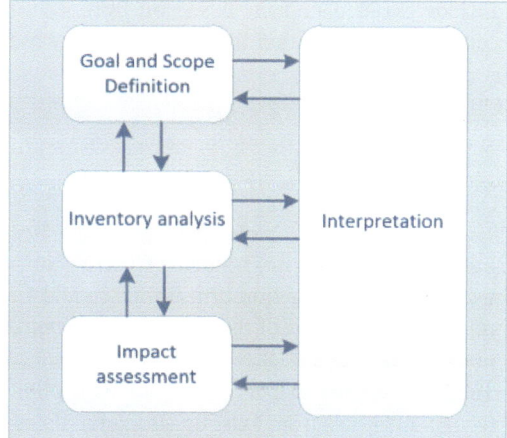

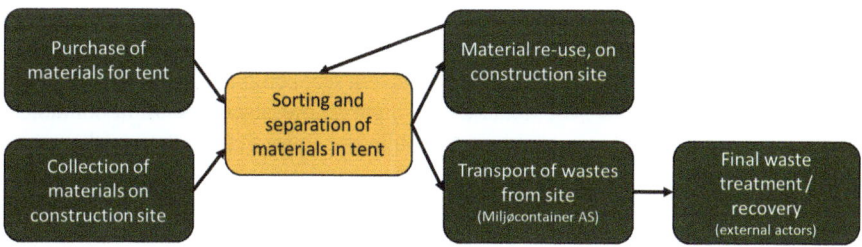

Fig. 2 System boundaries of the GoGreen LCA study

The purchase of materials pertains to the materials used for waste sorting, primarily plastic bags, used on-site. The collection of materials from demolished structures happens on the construction site with machinery, and the materials are transported further for sorting and separation in the tent. The sorting process is done manually and is not considered to have an environmental impact. Some of the materials on site are reused on the construction site, primarily wood used in formwork, before being taken back to the tent and resorted. The transport of the waste to the waste treatment facility is done by the company Miljøcontainer AS, which collects and delivers the waste to additional sorting and processing approximately 150 km away in the city of Trondheim. This waste is then further transported for recycling, while some are also incinerated according to local laws on processing contaminated materials.

The study was carried out using SimaPro version 9.5.0.1 with EcoInvent database version 3.9.1. The global impact assessment method ReCiPe 2016 Midpoint (Hierarchical) was used for this study. This study is intended to be a consequential LCA study, where avoided impacts from production are explicitly included in the model where material recycling and reuse are included. Several assumptions were made to include avoided impacts based on European and Norwegian market average processes contained within the EcoInvent databases.

2.2 Case Study

The GoGreen pilot is part of a larger Betna-Hestnes road construction project on the Europavei 39 (E39) in Trøndelag County, Norway. This 12.8 km highway project is managed by construction firm Bertelsen and Garpestad (B&G) and began construction in 2022. As part of reducing the environmental impacts of the overall construction process, several circular and "green" innovations were implemented by B&G. The focus of this paper is the GoGreen pilot, which was developed by B&G to recover materials from existing structures slated for demolition for the new road. The pilot project is designed to be easily implemented in other projects, flexible, and effective at recovering materials, with a stated goal of recovering 100% of materials from the demolished structures to be sent for further recycling and reuse.

The GoGreen pilot consists of a 300 m^2 portable tent structure, which acts as a facility for dismantling, sorting, and warehousing recovered materials. Two workers at a time sort and separate the materials by hand before placing them in containers, which are sent for further processing. Although not a part of the scope of this paper, the workers are hired in collaboration with the Norwegian social services, as part of an integration process for individuals who have struggled to find work. The GoGreen pilot also works with these employees to offer on-the-job training for further advancement in B&G as part of the social sustainability measures. The workers use hand tools and electric tools to aid in the separation of materials and sorting materials based on material type. In 2022 and 2023, the workers sorted through 354 tons of material collected from three demolished houses, one shed, one garage, one small barn, one cabin, and several other smaller constructions. Figure 3 shows the GoGreen tent in use, including the sorted materials, material collection bins, and overall structure and design. The GoGreen system is designed to be simple to implement for B&G in future projects, and they would also like to push the construction industry towards greater material recovery and sorting.

The GoGreen pilot intends to test the efficacy of the material collection and sorting process, which is intended to be further implemented in future construction projects. This paper is an evaluation of the environmental impacts and potential benefits of the GoGreen system seen in the context of the larger Betna-Hestnes construction project.

Fig. 3 Visual overview of the GoGreen tent activities

2.3 Data Collection

The data for this study was collected from the operators of the GoGreen tent, the Betna-Hestnes project manager, and the waste collection company Miljø Container AS. The waste quantities are based on data from Miljø Container AS received from the Betna-Hestnes site. Miljø Container AS weighed containers of sorted waste to form the basis of the life cycle inventory. The waste collected is transported 154 km by truck to the waste processing facility in the city of Trondheim. Additional information on how the waste is treated comes from the manager of Miljø Container

AS based on his communication with the waste processing facility in Trondheim where the waste is delivered.

3 Results

3.1 Life Cycle Inventory Results

Table 1 shows an overview of the waste materials collected on the Betna-Hestnes site as delivered to the waste collection process as well as their assumed final product

Table 1 Life cycle inventory of materials collected at the GoGreen tent

Waste type	kg	Treatment process	Final product	Notes
Mixed timber wastes	245,030	Material recycling	Wood chips	Assumed 5% loss
Mixed metals	20,960	Material recycling	Reinforcing steel	Assumed 5% loss
Metals (unclassified)	18,340	Material recycling	Reinforcing steel	Assumed 5% loss
Asbestos	14,760	Landfilling	Inert waste	
Insulation/mineral wool	11,270	Material recycling	New insulation	Assumed 20% in new insulation
CCA-impregnated timber	10,400	Energy recovery	District heat	Assumed net 14,4 MJ per kg
Hard plastic	7880	Material recycling	Plastic bags	
Cardboard and paperboard	7560	Material recycling	Recycled paper	Assumed 5% loss
Tractor and truck tires	7300	Material recycling	Rubber granulate	Assumed 75% loss of quality
Other tires	3200	Material recycling	Rubber granulate	Assumed 75% loss of quality
Mixed hard plastics/plastic waste	2200	Material recycling	Plastic bags	
Cables and wires	1500	Unknown	Unknown	Transport included
Plastic wrapping	1300	Material recycling	Plastic bags	
Window glass, not laminated	540	Material recycling	Glass cullet	
Floor sealer	359	Energy recovery	District heat	
Mixed electronic wastes	350	Unknown	Unknown	Transport included
Remaining materials	1581	Unknown	Unknown	Transport included
In total, all materials	**354,530**	–	–	All transport included

once treated. The largest quantity of materials was mixed timber wastes from structural elements of the dismantled buildings (walls, roofs, etc.). This material would typically be incinerated, but the waste treatment company sends these materials to be chipped and made into new oriented strand boards and particle boards. Mixed metals and unclassified metals can be a variety of metal materials but are mostly steel, which is then melted down and reused as reinforcing steel. Hard plastics and mixed plastic wastes are turned into plastic bags, while truck and tractor tires used on site are recycled into rubber granulate, which can be used for flooring and on artificial sports fields. Some materials are not able to be recycled, primarily asbestos, which is banned for use and is landfilled. Other materials that are not suitable for recycling are incinerated with heat being produced. CCA-impregnated wood is bound by regulations to be incinerated. The treatment of some wastes was not known at the time of the study, so it is assumed they have been landfilled, with transport included in these processes. Assumptions have also been made in the model regarding system efficiency, quality loss, and energy content in combustion.

3.2   Impact Assessment Results

The life cycle impact assessment (LCIA) results show impacts from the three main GoGreen processes that are included in this model (Fig. 4). The "*material reuse*" process, shown in green, shows the impact of the materials being reused on site, which is the reuse of concrete formwork for bridge construction. The assumption was that formwork was reused five times for each of the identical bridge designs on the Betna-Hestnes project. The orange represents the impacts of "*purchased*

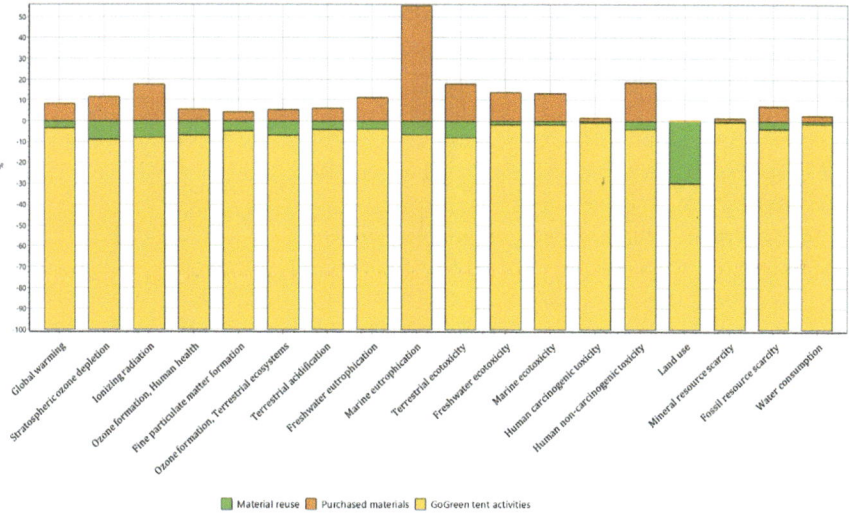

Fig. 4 Normalized LCIA results, by the GoGreen process

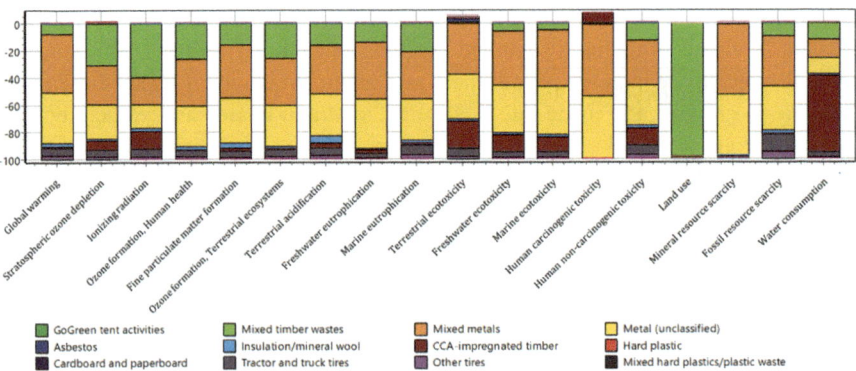

Fig. 5 Normalized LCIA results, by material type

materials" required for the GoGreen tent, which are primarily plastic bags and energy for heating and lighting of the tent. The yellow refers to the "*GoGreen tent activities,*" which are the sorting processes and the environmental benefit of avoided impacts. The overall results show a mix of both environmental impacts and environmental benefits based on the consequential modelling of avoided impacts. The results in Fig. 4 show that in all impact categories, there was an environmental benefit to recycling and reusing materials, respectively, while there was an environmental impact from the use of virgin materials and energy. Using avoided impacts in the assessment showed that the overall impact of implementing the GoGreen pilot led to a net environmental benefit (i.e. negative emissions vs the standard use of new materials). While the model makes some estimates on the efficiency of the recycling process and the loss of product quality, the overall results show that the GoGreen pilot has environmental benefits versus using newly produced raw materials.

Figure 5 shows the normalized LCIA results based on each material type. The scale of the figure shows that the environmental impacts are negative (i.e. environmental benefits) for most materials within most impact categories. The only processes with notable environmental impacts are the CCA-impregnated timber, which is due to emissions from combustion. Most other materials have impacts that are negative.

The materials with the largest environmental benefit are mixed metals and metals (unclassified), where the production of new materials is the greatest, with more than 39 tons of production for reinforcing steel assumed to be avoided. The second most environmentally beneficial product is from mixed timber wastes, which had the highest total mass of materials that were collected and sorted in the GoGreen tent.

Figure 6 shows the overall global warming potential (GWP) for all GoGreen processes on the Betna-Hestnes site, while Fig. 7 shows the GWP for all the materials processed in the GoGreen tent. The largest environmental benefit comes from the avoided impact of new material production due to material recycling. Reuse of formwork also has a small but relevant environmental benefit, while the purchase of materials (plastic bags and heating and lighting of the tent) and transport of

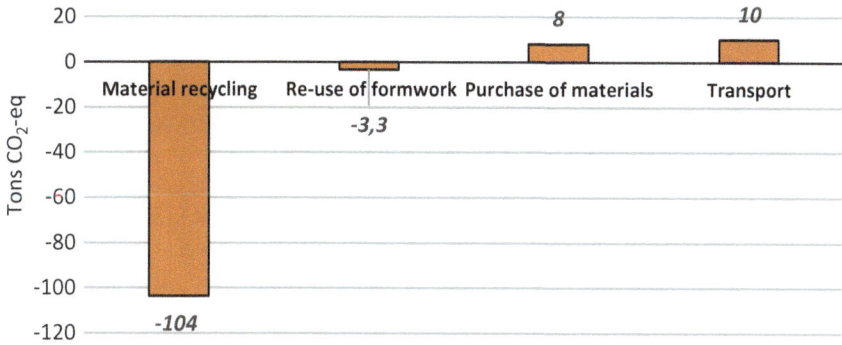

Fig. 6 GWP in tons CO_2-eq, by GoGreen process

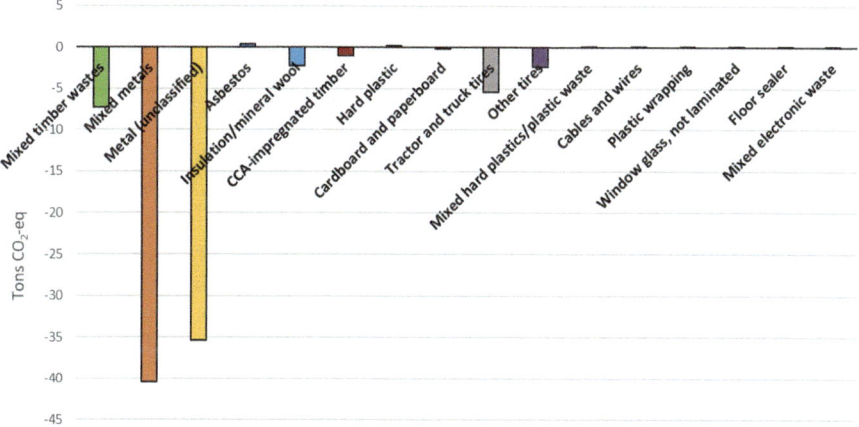

Fig. 7 GWP in tons CO_2-eq, by material type

products to the waste treatment facility had an environmental impact. The net overall global warming potential benefit for the GoGreen pilot, including material recycling, reuse of concrete formwork, purchase of materials, and transport, was -88.9 tons of CO_2-equivalents based on a sorting and final waste treatment (landfilling, incineration, and recycling) of 354 tons of materials collected in the GoGreen tent.

Figure 7 shows the GWP for the material from the collection, sorting, and recycling with avoided impacts of new material production included. This figure shows the dominance of the avoided impacts from metals in the overall result. The mixed timber wastes and tires also have a noticeable environmental benefit.

4 Discussion

The overall result of this study shows that there is an environmental benefit to collecting and sorting materials on the construction site as long as these materials are sent for further processing and recycling and can avoid the impacts of new material production. The GoGreen pilot also facilitated the reuse of materials on site, such as concrete formwork, which would not have been possible without a sorting facility. The GoGreen pilot reported a sorting of 100% of all waste materials with a small fraction of materials lost, sent to landfills, or incinerated. The materials that were lost for recycling were lost primarily due to regulations on waste for the treatment of toxic materials (i.e. asbestos and CCA-impregnated timber). This GoGreen pilot project shows that it is possible to implement effective sorting and collection of materials in the demolition phase of infrastructure and drastically reduce the amount of construction waste that is sent to landfills.

The environmental benefits and impacts of collecting and sorting waste are relatively simple to determine when the fate of the waste is known and if there is a protocol for documenting waste collected. This study has made several assumptions on how the waste is treated, which materials avoid impacts, and on system efficiency and loss in material quality. These assumptions have not accounted for any market expansion due to lower prices leading to increased material demand (i.e. price effect), which may not actually lead to reduced consumption of new materials due to positive feedback loops [11, 12]. This study is intended to only evaluate the GoGreen pilot, but an LCA study of upscaling of this method should include market dynamics as an additional model parameter.

The choice of using avoided impacts implies a value judgement that the entirety of the environmental benefits of the GoGreen pilot is allocated to the construction firm B&G. This allocation method may be unfair, as B&G does not actually carry out the recycling process [9, 13]. B&G were chosen as the beneficiaries of this assessment as they are responsible for recovering the materials and choose not to send them to landfills or incineration. The use of the EcoInvent database for modelling generic processes for avoided impacts also adds to the uncertainty of the results. EcoInvent is only a representation of the waste treatment and avoided production processes, not necessarily the most accurate measurement of real-world activities. The generic processes chosen for avoided impacts were region-specific, and processes were modified to account for Norwegian conditions (primarily changing energy mixes and transport distances) when information was available. Future work should involve carrying out additional studies on the waste treatment of collected materials and a better understanding of the final fate of the collected materials.

The scale of the GoGreen pilot project is not particularly large, as this project is primarily a road construction project in which existing structures that were dismantled had a relatively small building stock (354 tons total vs 2.1 million tons of construction waste generated in Norway annually). This sorting method, while useful on the GoGreen project, should be further expanded to be implemented on buildings with larger stocks that are slated for demolition and have more complex structures, like concrete and steel buildings. Thus, the environmental benefit of this single pilot study is positive; the overall effects of a more thorough dismantling and sorting system will be more effective when scaled up for the entire construction industry.

The environmental savings from the GoGreen pilot are also relatively small compared to the overall estimated environmental impact of the entire Betna-Hestnes project, which was estimated to produce more than 10,500 tons of CO_2 (in 2022 and 2023 combined) from diesel fuel consumption in construction machinery alone. This implies that B&G still has more to do to reduce its environmental impact further. B&G has implemented additional innovative methods for reducing environmental impacts from construction on the Betna-Hestnes site, such as reusing construction masses, reducing the area footprint of the construction site, using electric machines when possible, and avoiding construction on boglands, which are outside of the scope of this study but can potentially have large environmental benefits within a circular economy framework. The GoGreen pilot also contributes to creating new jobs within the circular industry and training the future workforce to work towards sustainability. Additional metrics for measuring circularity in road construction projects that go beyond the traditional life cycle assessment framework are needed to fully understand the impact and benefits of circular activities [4, 14].

5   Conclusion

This study used life cycle assessment to quantify the environmental impacts and benefits of implementing dismantling, collection, and sorting procedures on a road construction project in Norway. The results show that there is an environmental benefit to implementing better sorting procedures so that more materials may be sent to recycling. The scale of the GoGreen pilot is relatively small but can have potentially large benefits for reducing material consumption and emissions if implemented on a larger scale. Further work should focus on evaluating emissions from additional circular economy strategies in road construction and defining better metrics for measuring circularity in road construction projects.

References

1. United Nations Environment Programme: 2022 global status report for buildings and construction: Towards a zero-emission, efficient and resilient buildings and construction sector, Nairobi (2022)
2. Circle Economy: The circularity gap report 2024, Amsterdam (2024) [Online]. Available: https://drive.google.com/file/d/15droT_mBFK6Kkd1aO5kPzYFUqLdul2qM/view
3. Circle Economy and Circular Norway: The circularity gap report of Norway, Oslo (2020) [Online]. Available: https://www.circularnorway.no/gap-report

4. Abadi, M., Moore, D.R., Sammuneh, M.A.: A framework of indicators to measure project circularity in construction circular economy. Proc. Inst. Civ. Eng. Manag. Procure. Law. **175**(2), 54–66 (2021). https://doi.org/10.1680/jmapl.21.00020

5. Barbieri, D.M., et al.: Assessment of carbon dioxide emissions during production, construction and use stages of asphalt pavements. Transp. Res. Interdiscip. Perspect. **11**(15), 1469–1481 (2021). https://doi.org/10.1016/j.jclepro.2006.03.005

6. Ebrahimi, B., Rosado, L., Wallbaum, H.: Machine learning-based stocks and flows modeling of road infrastructure. J. Ind. Ecol. **26**(1), 44–57 (2022). https://doi.org/10.1111/jiec.13232

7. Statistics Norway: Table 09781: Treatment of waste from construction, rehabilitation and demolition of buildings (tonnes), by material, contents, year and treatment, Statistics Norway, Oslo (2024) [Online]. Available: https://www.ssb.no/statbank/table/09781/

8. Andersen, C.E., Kanafani, K., Zimmermann, R.K., Rasmussen, F.N., Birgisdóttir, H.: Comparison of GHG emissions from circular and conventional building components. Build. Cities. **1**(1), 379–392 (2020). https://doi.org/10.5334/bc.55

9. Malabi Eberhardt, L.C., Van Stijn, A., Rasmussen, F.N., Birkved, M., Birgisdottir, H.: Towards circular life cycle assessment for the built environment: a comparison of allocation approaches. IOP Conf. Ser. Earth Environ. Sci. **588**(3) (2020). https://doi.org/10.1088/1755-1315/588/3/032026

10. ISO: ISO14040: Environmental Management-Life Cycle Assessment Principles and Frameworks. British Standards Institution, London (2006)

11. Rajagopal, D.: A step towards a general framework for consequential life cycle assessment. J. Ind. Ecol. **21**(2), 261–271 (2017). https://doi.org/10.1111/jiec.12433

12. Sandén, B.A., Karlström, M.: Positive and negative feedback in consequential life-cycle assessment. J. Clean. Prod. **15**(15), 1469–1481 (2007). https://doi.org/10.1016/j.jclepro.2006.03.005

13. Eberhardt, L.C.M., van Stijn, A., Rasmussen, F.N., Birkved, M., Birgisdottir, H.: Development of a life cycle assessment allocation approach for circular economy in the built environment. Sustain. For. **12**(22), 1–16 (2020). https://doi.org/10.3390/su12229579

14. Heisel, F., Rau-Oberhuber, S.: Calculation and evaluation of circularity indicators for the built environment using the case studies of UMAR and Madaster. J. Clean. Prod. **243**, 118482 (2020). https://doi.org/10.1016/j.jclepro.2019.118482

Index